Advanced
Engineering
Mathematics

$ 30 - 20
¢

ADVANCED

ENGINEERING

MATHEMATICS

THIRD EDITION

Peter V. O'Neil

University of Alabama at Birmingham

Wadsworth Publishing Company
Belmont, California

A division of Wadsworth, Inc.

Mathematics Editor: Anne Scanlan-Rohrer

Assistant Editor: Tamiko Verkler

Editorial Assistant: Leslie With

Production: Greg Hubit Bookworks

Print Buyer: Martha Branch

Copy Editor: Patricia Harris

Cover Design: Stuart Paterson

Cover Photo Research: Judy Mason

Cover Photos: Stuart Cohen/COMSTOCK (front top and lower right; back top and lower right; spine); Thomas Wear/COMSTOCK (front lower left); Hartman-DeWitt/COMSTOCK (back lower left); COMSTOCK (back lower center); Patrice Astier/Pei Cobb Freed & Partners (front lower center)

Cover: THE LOUVRE PYRAMID

Printed in the United States of America

1 2 3 4 5 6 7 8 9 10—95 94 93 92 91

Library of Congress Cataloging in Publication Data

O' Neil, Peter V.
  Advanced engineering mathematics / Peter V. O'Neil. — 3rd ed.
    p. cm.
  Includes bibliographical references and index.
  ISBN 0-534-13584-6
  1. Engineering mathematics. I. Title.
TA330.053 1991   90-24178
515—dc20   CIP

Preface

This book is intended to introduce students of engineering, science, and applied mathematics to some of the mathematics and mathematical modeling employed by these disciplines, and to do so in a way that displays the mathematics as being useful and interesting while maintaining its integrity.

The prerequisite is a standard course in calculus, including partial derivatives, multiple integrals, improper integrals (for transforms and Fourier integrals), and infinite series (for series solutions, Fourier series, and complex analysis). Power series are reviewed in Section 5.1, and multiple integrals and changes of variables in multiple integrals are reviewed in Appendices C, D, and E following Chapter 16.

Many sections from the second edition have been rewritten, with new problem sets, and some new topics have been added. Some of the changes made in this edition are as follows:

- New chapters on calculus of variations, difference equations, and solution of systems of differential equations by matrix methods.

- A treatment of Fourier transforms and complex Fourier series, with applications to electrical engineering and to boundary value problems.

- Sections on the fast Fourier transform and the discrete Fourier transform.

- Expanded discussion of nonlinear differential equations and stability, including a section on limit cycles.

- Numerical solution of initial value problems, with programs for their execution.

- A more detailed discussion of expansions in series of orthogonal functions, and additional properties of Legendre polynomials and Bessel functions.

- A residue formula for the inverse Laplace transform.

- An emphasis on constructing mathematical models of physical phenomena, as well as solving the resulting equations.

- A discussion of the rationale behind series solutions in the method of Frobenius.

Some sections have been reorganized to provide a better logical structure or a clearer exposition. For example, reduction of order now precedes the solution of constant coefficient second order differential equations so that it is available to derive second solutions when the characteristic equation has repeated roots. In the same vein, the notation in the treatment of the Heaviside step function has been improved to make it easier to use in representing piecewise-continuous functions.

The book is set in six parts according to main themes, to make course construction and inclusion or omission of optional material easy for the instructor. The intent is not only to provide flexibility for the instructor in using the book for a variety of courses, but also to provide a reference for the student who may later need a source for additional concepts or methods in a particular area.

Care has been taken to maintain a balance between intuition and rigor. Although one objective is to introduce ideas intuitively and with appeal to the student's scientific and/or engineering experience, it is also an objective to state mathematical ideas correctly. A distinction is made between a mathematical model and a statement of a theorem or definition. The former involves not only mathematics but physical background, intuition, and approximation, while the latter should retain the precision expected of mathematics.

The author believes that some results can be understood and correctly used only if the student has the opportunity at least to read a proof. Thus, many proofs are given in detail; the instructor can make judgments in deciding which proofs to include and which to omit in a course. For some deeper theorems, such as the Riemann mapping theorem, no proof is attempted.

The six parts of the book are organized as follows.

Part I is an introductory course in differential equations, including series solutions, the Laplace transform, and numerical methods and programs for their execution.

Part II deals with additional ideas from ordinary differential equations and constitutes a set of chapters which can be covered in any order following the material of Part I. Included are special functions and Sturm-Liouville theory, orthogonal expansions, nonlinear differential equations and stability, difference equations, and calculus of variations.

Part III introduces vectors and vector spaces, matrices and their use in solving systems of linear equations, and the solution of systems of differential equations by matrix methods, including the matrix exponential function.

Part IV discusses vector differential and integral calculus, culminating in the theorems of Gauss and Stokes and a variety of their uses.

Part V is devoted to Fourier analysis, including Fourier series and multiple Fourier series, Fourier integrals, finite and discrete Fourier transforms, Fourier sine and cosine transforms, the Fourier transform, and the fast Fourier transform. As might be expected, Fourier methods are used to solve boundary value problems, but they are also applied to problems of interest to electrical engineers, such as analysis of the amplitude spectrum of a signal, construction of filters, and reconstruction of band-limited signals.

The material of Part V is written to be self-contained. However, the student who has studied Sturm-Liouville theory in Part II will have an added perspective in recognizing Fourier series as part of a general theory of eigenfunction expansions.

Part VI is an introduction to complex analysis, including series expansions, residues, complex integration, and conformal mappings, with applications to the

Dirichlet problem, the Laplace transform, summation of series, evaluation of improper integrals, and the analysis of fluid flows.

All parts of the book have been class tested, and all problems were solved before their inclusion in the final version. There is a large number of problems, both of a routine/drill type and thought-provoking and applications-oriented problems. Some of these extend ideas in the text or point out unusual or surprising relationships. For example, Problems 27 and 28 of Section 7.5 discuss an intriguing relationship between the distribution of charged beads on a wire and zeros of Legendre and Laguerre polynomials. And Problem 30 of Section 17.10 pursues a surprising relationship between Bessel functions and Fourier transforms of Legendre polynomials. Specifically, if $f_n(t)$ is set equal to $P_n(t)$ (the nth Legendre polynomial for $|t| < 1$ and to zero for $|t| > 1$), the Fourier transform of f_n is $(-i)^n[\sqrt{2\pi/\omega}]J_{n+1/2}(\omega)$.

Answers to odd-numbered problems are included at the back of the book, and a solutions manual provides solutions of even-numbered problems.

I want to acknowledge the assistance of Dr. Thomas O'Neil of California Polytechnic State University in preparing this edition. Tom has offered opinions and suggestions on many parts of the book, worked with the problem sets and proofreading, and helped write certain sections. The staff at Wadsworth has also been helpful.

I would like to thank the many users of the second edition of the text who responded to a survey and provided many helpful comments. I would also like to acknowledge the reviewers of the revised manuscript: Tuncay Aktosun, Duke University; Harvey Charlton, North Carolina State University at Raleigh; Chris Frenzen, Naval Postgraduate School; Vuryl Klassen, California State University, Fullerton; Lawrence Levine, Stevens Institute of Technology; Mauro Pierucci, San Diego State University; Tom Schulte, California State University, Sacramento; Jacob Weinberg, University of Lowell.

I would also like to acknowledge the following individuals for checking the accuracy of the text and exercises: Lianjun An, Duke University; Bob Bass, University of Tennessee, Chattanooga; Bob Broschat, South Dakota State University; Stan Byrd, University of Tennessee, Chattanooga; Peter Colwell, Iowa State University; John Conly, San Diego State University; Chris Frenzen, United States Naval Postgraduate School; Feiyue He, University of Cincinnati; Oivind Heggli; William Hull, University of Cincinnati; Eleanor Killam, University of Massachusetts, Amherst; Chul Kim, North Carolina State University; Istvan Kovacs, University of Southern Alabama; Ravinber Kumar, South Dakota State University; Jianao Lian, Texas A&M University; Jill Macari, Richard Millspaugh, University of North Dakota; Tom Schulte, California State University, Sacramento; Paul Shawcroft, Brigham Young University; Robert Vanderheyden, Virginia Polytechnic Institute and State University; William Yslas Velez, University of Arizona; Terry Walters, University of Tennessee, Chattanooga; Jacob Weinburg, University of Lowell; Yun-Gang Ye, Duke University.

Finally, I have benefited from numerous comments by users of the first two editions. Comments on this edition will be most welcome.

Peter V. O'Neil
Department of Mathematics
University of Alabama at Birmingham
Birmingham, AL 35216

Contents

Chapter 3   Higher Order Differential Equations   180

Chapter 4   The Laplace Transform   209

Chapter 5   Series Solutions of Differential Equations   291

Advanced
Engineering
Mathematics

PART I

Introduction to Ordinary Differential Equations

Differential Equations and _____ Mathematical Modeling _____

In attempting to describe, understand, and predict behavior of a physical process or system, we often construct a mathematical model. The reasoning involved in doing this may be outlined as follows.

First, we attempt to identify the main components of the system being studied. For an electrical circuit, these would typically be resistance, inductance, capacitance, current, electromotive force, and configuration of the circuit. We observe the behavior of the system and attempt to relate the variables by means of one or more equations. Often these equations involve derivatives, because we observe systems in motion, and derivatives measure rates of change. Such equations are called *differential equations*.

The differential equations, together with any other equations needed to specify information about the system, are said to constitute a *mathematical model* of the system. We attempt to solve the equations of the model in order to understand how the various components of the system behave and interact and in the hope of predicting how the system will behave at later times.

Parts I and II of this book are devoted to methods for solving ordinary differential equations (involving total, not partial, derivatives) and to models containing such equations. We will also consider methods for approximating their solutions or obtaining information about them when it is impossible to obtain exact solutions in terms of elementary functions we can work with comfortably.

For the remainder of this introductory chapter, we will construct some fairly simple but important mathematical models. As we acquire more experience and mastery of techniques, we will construct models of more complicated phenomena.

EXAMPLE 0.1   An Electrical Circuit Model

Suppose we are interested in the flow of current through a series circuit, such as that of Figure 0.1. Usually we are given the voltage $E(t)$ at time t and the values of the

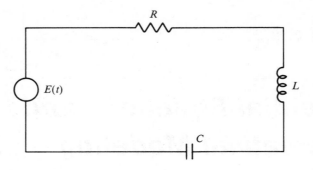

Figure 0.1. RLC circuit.

resistance, capacitance, and inductance and want to find the current $i(t)$ at any time t. We therefore set out to derive a relationship between $i(t)$ and these other quantities.

Assume that the resistance R, the capacitance C, and the inductance L are constant and that at time t the capacitor has a charge $q(t)$. If current is measured in amperes, then R is in ohms, C is in farads, L is in henrys, and q is in coulombs. For the circuit under consideration, experiment tells us that

> the voltage drop across the resistance equals iR;

> the voltage drop across the capacitor equals $\dfrac{q}{C}$;

and

> the voltage drop across the inductance equals $L\dfrac{di}{dt}$.

Further, Kirchhoff's second law for circuits states that the impressed voltage $E(t)$ is the sum of the voltage drops in the circuit. Therefore,

$$E(t) = L\frac{di}{dt} + Ri + \frac{1}{C}\,q. \tag{0.1}$$

This is a differential equation because it involves the derivative of $i(t)$. Equation (0.1) also involves another function, the charge $q(t)$, which is also unknown. We can reduce equation (0.1) to an equation involving just one unknown function if we use the fact that the current is the rate of change of the charge with respect to time:

$$i = \frac{dq}{dt}.$$

Substitute this result into equation (0.1) to obtain the differential equation

$$L\frac{d^2q}{dt^2} + R\frac{dq}{dt} + \frac{1}{C}\,q = E(t). \tag{0.2}$$

This equation involves only one unknown function, $q(t)$, and its derivatives. We would attempt to solve equation (0.2) for $q(t)$, then obtain $i(t)$ as dq/dt.

Equation (0.2) by itself does not uniquely specify the charge $q(t)$. In addition to equation (0.2), we must also know the values of $q(t)$ and $q'(t)$ at some particular time t_0. The mathematical model for the circuit consists of the differential equation together with two *initial conditions*

$$L\frac{d^2q}{dt^2} + R\frac{dq}{dt} + \frac{1}{C}q = E(t); \qquad q(t_0) = A, \qquad q'(t_0) = B,$$

in which t_0 is a given time and A and B are known.

We will solve a simplified version of this model in Section 1.8 and analyze the model in much more detail in Section 2.10 and throughout Chapter 4. ■

EXAMPLE 0.2   A Model of Free Oscillations of a Mass-Spring System

Suppose we suspend a ball from a spring, pull down on the ball, and then release it to oscillate in a vertical plane. We would like to describe the resulting motion.

Imagine the spring as shown in Figure 0.2(a). In its unstretched state, it has length L. If we hang a ball of mass m from the spring and leave it until the system comes to rest, the spring will be stretched an amount d. In this equilibrium position, the spring has length $L + d$ [Figure 0.2(b)]. We will measure vertical displacement $y(t)$ at time t from this equilibrium position. Thus, in Figure 0.2(c), the origin of the y-axis is at the length $L + d$. We choose downward as the positive direction on this axis and upward as negative (simply a convention). At spring length $L + d$ ($y = 0$), we say that the spring is in *static equilibrium*.

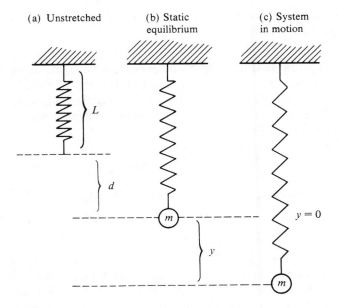

Figure 0.2

Now look at the forces acting on the ball.

1. Gravity pulls the ball downward. The magnitude of the attraction due to gravity is mg, where g is the constant magnitude of acceleration due to gravity and is about 9.8 meters/second^2.

2. There is also the restoring force due to the spring. Hooke's law states that the magnitude of this force is proportional to the distance the spring is stretched. The constant of proportionality k is called the *spring modulus* and varies from one spring to another. The stiffer the spring, the larger the value of k.

At static equilibrium, the force of the spring is $-kd$ (minus because the spring tends to pull the ball upward, in the negative direction). If the ball is pulled downward a distance y from the static equilibrium position, an additional force $-ky$ is exerted on the ball. Thus, the total force on the ball due to the spring is

$$-kd - ky.$$

The total force on the ball due to gravity and the spring is

$$mg - kd - ky.$$

When the ball is at the static equilibrium position, the forces balance because the ball is not moving. Therefore, when $y = 0$, we must have

$$mg - kd = 0,$$

and therefore

$$mg = kd.$$

The total force acting on the ball is $mg - mg - ky$, or just $-ky$.

Now consider two cases.

Case 1: The System is Undamped In this case, we imagine that damping effects (such as air resistance) are negligible and omit them from the model. Use Newton's second law of motion (force equals mass times acceleration), with $t =$ time. The acceleration of the ball is d^2y/dt^2, since y measures the displacement of the ball from the static equilibrium position. Therefore,

$$m \frac{d^2y}{dt^2} = -ky.$$

The differential equation for the displacement function is

$$\frac{d^2y}{dt^2} + \frac{k}{m} y = 0. \tag{0.3}$$

The position of the ball at time t will also be influenced by the position of the ball when it was released (initial position) and its velocity at the time of release (initial velocity). We must therefore specify $y(t_0)$ and $y'(t_0)$, with t_0 the time of release, in order to determine the displacement function uniquely.

A model for the motion of the ball consists of the differential equation (0.3) together with information about the initial position and velocity:

$$\frac{d^2y}{dt^2} + \frac{k}{m}y = 0; \qquad y(t_0) = A, \qquad y'(t_0) = B.$$

We will see in Chapter 2 that all solutions of equation (0.3) are of the form

$$y(t) = c_1\cos\left(\sqrt{\frac{k}{m}}\,t\right) + c_2\sin\left(\sqrt{\frac{k}{m}}\,t\right), \tag{0.4}$$

in which c_1 and c_2 can be any constants. The initial conditions $y(t_0) = A$ and $y'(t_0) = B$ are used to solve for c_1 and c_2 to obtain the appropriate displacement function for the given initial position and velocity.

As a specific example, let $t_0 = 0$, and suppose that the ball is initially stretched downward 4 centimeters and released from rest, with no push in either direction. The model for this case is

$$\frac{d^2y}{dt^2} + \frac{k}{m}y = 0; \qquad y(0) = 4, \qquad y'(0) = 0. \tag{0.5}$$

The general form of the solution of this differential equation was given in equation (0.4). We must choose the constants c_1 and c_2 to satisfy the initial conditions. First, we must have

$$y(0) = c_1\cos(0) + c_2\sin(0) = 4,$$

hence $c_1 = 4$. Thus far, $y(t) = 4\cos(\sqrt{k/m}\,t) + c_2\sin(\sqrt{k/m}\,t)$.

Now calculate

$$y'(t) = -4\sqrt{\frac{k}{m}}\,\sin\left(\sqrt{\frac{k}{m}}\,t\right) + c_2\sqrt{\frac{k}{m}}\,\cos\left(\sqrt{\frac{k}{m}}\,t\right).$$

Then

$$y'(0) = c_2\sqrt{\frac{k}{m}} = 0,$$

and we must choose $c_2 = 0$.

Therefore, at time t, an undamped ball of mass m, released from rest after being stretched 4 centimeters downward from static equilibrium, is at the position given by

$$y(t) = 4\cos\left(\sqrt{\frac{k}{m}}\,t\right).$$

The ball exhibits a periodic motion called a *harmonic oscillation*; it oscillates up and down through the static equilibrium position, reaching a maximum of 4 centimeters above this position, then returning to 4 centimeters below this position, and repeating this motion indefinitely.

This is an accurate description of what would happen in this ideal case in which there are no damping forces. Of course, this case does not actually occur in nature.

Case 2: There Are Damping Forces   Suppose that the ball is connected to a dashpot, as in Figure 0.3. Now a new force comes into play, tending to damp out the motion. Experiment shows that the damping force is proportional in magnitude to the velocity dy/dt of the ball. Now the total force acting on the ball is

$$-ky - c\frac{dy}{dt}$$

for some constant c called the *damping constant*. The equation for y, from Newton's second law, is

$$m\frac{d^2y}{dt^2} = -ky - c\frac{dy}{dt},$$

or, as it is more customarily written,

$$\frac{d^2y}{dt^2} + \frac{c}{m}\frac{dy}{dt} + \frac{k}{m}y = 0. \tag{0.6}$$

The mathematical model for this damped motion would consist of the differential equation (0.6) together with initial conditions specifying $y(t_0)$ and $y'(t_0)$. As we might expect, the motion in this damped case can be quite complicated. We will discuss a complete mathematical analysis of this motion in Section 2.11.

Figure 0.3.
Damped
spring system.

EXAMPLE 0.3   Radioactive Decay

Experiments show that radioactive elements, such as radium and plutonium, decay at a rate proportional to the mass present. Let $M(t)$ be the mass at time t. If the mass decreases at a rate proportional to itself, then for some positive constant k, which depends on the element, we must have

$$\frac{dM}{dt} = -kM, \tag{0.7}$$

with the minus sign indicating that the mass decreases with time.

For a given element, we must determine k by experiment. Assuming that k is known, the differential equation (0.7) enables us to write a formula for $M(t)$ if we know the mass at any given time t_0. Thus, a model for the mass of the decaying element is

$$\frac{dM}{dt} = -kM; \qquad M(t_0) = m,$$

in which k, m, and t_0 are known. This model is quite easy to solve, and we will use it to discuss carbon dating of artifacts in Section 1.2.   ■

EXAMPLE 0.4   The Simple Pendulum

Suppose we suspend a ball of mass m at the end of a straight rod of length L and set it in motion swinging back and forth. We would like to write an equation describing the motion of the ball.

The pendulum is shown in Figure 0.4; it is called *simple* to distinguish it from a compound pendulum, such as that of Figure 0.5. In Figure 0.4, θ is the angle of

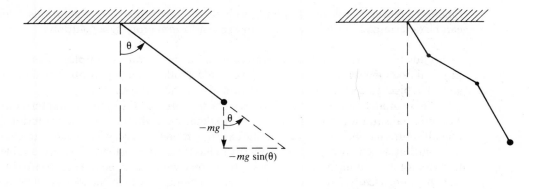

Figure 0.4. Simple pendulum.   *Figure 0.5.* Compound pendulum.

displacement of the rod from the vertical position. This angle varies with time and is measured in radians.

To derive an equation for θ, we analyze the forces acting on the ball and use Newton's second law of motion. The tangential component of the force of gravity acting on the ball is $-mg \sin(\theta)$, minus because this force tends to restore the pendulum to equilibrium position, in which the ball hangs straight down. We will neglect other forces, such as air resistance.

Now write the acceleration in terms of θ. If s is the displacement of the pendulum from the equilibrium position, then $s = L\theta$. The acceleration of the ball is d^2s/dt^2, which can be written in terms of θ as $L\,(d^2\theta/dt^2)$, assuming that L is constant. By Newton's second law of motion,

$$mL\,\frac{d^2\theta}{dt^2} = -mg\,\sin(\theta),$$

or

$$\frac{d^2\theta}{dt^2} + \frac{g}{L}\,\sin(\theta) = 0. \tag{0.8}$$

This is a differential equation for the displacement angle $\theta(t)$. Notice that m does not appear in this equation, implying that the motion is independent of the mass of the ball. This surprising result was observed by Galileo in the early seventeenth century. (Galileo may also have been the first to observe that falling objects accelerate at a constant rate independent of the mass. It is said that he made this observation by dropping objects of different mass from the leaning tower of Pisa.)

Clearly, the position of the pendulum at time t will also depend on its position at its time of release as well as whether it was released from rest or pushed. One can show that $\theta(t)$ is uniquely determined by the differential equation (0.8) and the conditions $\theta(t_0) = A$, $\theta'(t_0) = B$ specifying its position and velocity at some time t_0. The equation (0.8) and these initial conditions constitute a mathematical model for the simple pendulum.

The fairly simple looking differential equation (0.8) is extremely difficult to solve. In fact, it is impossible to write a solution as a finite sum of multiples of elementary functions. Nevertheless, in Chapter 8, we will see how properties of solutions of (0.8) can be determined in the absence of an explicit formula for these solutions. ∎

Thus far, we have looked at four simple but important examples of models of physical phenomena. In each model, we derived a differential equation, which must be solved subject to certain other information about the phenomenon.

These models suggest several issues we will address in the coming chapters. Given a differential equation, we want to be able to determine whether there are solutions; to find all solutions which exist; and, in the case of a model of a physical system, to choose the solution which correctly describes the behavior of the system. If we are unable to find the solution in terms of elementary functions, we want to be able to approximate its value numerically. Further, we will develop techniques which enable us to determine properties of solutions (for example, whether solutions are periodic) even when we do not have an expression for the solution. We will begin in Chapter 1 with differential equations involving only first derivatives.

PROBLEMS FOR CHAPTER 0

In each of the following problems, a physical process is described. In some instances, observations which have been verified experimentally are provided, together with simplifying assumptions. Use this information to write a differential equation governing the process.

1. *Newton's law of cooling*   Newton verified experimentally that the surface temperature of an object changes at a rate proportional to the difference between the temperature of the object and that of the surrounding medium (such as air). Assuming that the medium temperature is constant, write a differential equation for the temperature $T(t)$ of an object cooling in air according to Newton's law.

2. *A population model*   Assume that, at any time t, the population of bacteria in a petri dish changes at a rate proportional to the population at that time. Write a differential equation for the population $P(t)$ at time t.

3. *Another population model*   Assume that, at any time t, both the birth rate and the death rate in the United States are constant multiples of the population at that time (though not necessarily the same constants). Write a differential equation for the population $P(t)$ at any time t, assuming that no members of the population are gained or lost through emigration or immigration.

4. *Boyle-Mariotte gas law*   At constant temperature and low pressure p, the rate of change of the volume V of a gas with respect to pressure has been observed to be proportional to $-V/p$. Write a differential equation for V as a function of p.

5. A ball is thrown vertically upward from the surface of the earth. The forces acting on the ball are (1) the constant acceleration due to gravity and (2) air resistance, which has been found to be proportional to the velocity. Write a differential equation for the motion of the ball in terms of the position function $y(t)$ at time t, with $y(t)$ measured upward from the surface of the earth.

6. A light beam directed downward into the ocean is partially absorbed as it passes through the water. Its intensity decreases at a rate proportional to the intensity at any given depth. Write a differential equation for the intensity as a function of the depth.

7. The thickness of ice on a frozen lake increases at a rate proportional to the square root of the number of degrees Fahrenheit the temperature is below freezing. Write a differential equation for the thickness of the ice at any temperature t.

8. A bullet with mass 0.4 gram is shot vertically upward from the earth's surface. The muzzle velocity is 1500 meters/second. The forces acting on the bullet are air resistance, which is assumed to be $v(t)^2/1000$ [with $v(t)$ the velocity at time t] and the gravitational pull of the earth. Neglecting other factors, such as the wind, write a

differential equation for the velocity $v(t)$ at time t. Also, write an initial condition for $v(t)$, that is, the value of $v(t)$ at a particular time t_0.

9. *Compound interest*   A man invests \$4000 at 6% interest, compounded continuously. This means that the rate of change of the amount of money present at time t, with respect to t, is proportional to the amount itself. Write a differential equation for the amount he has at any later time t, assuming no withdrawals.

10. *Motion on an inclined plane*   A 50-pound block is released from rest at the top of an inclined plane making an angle of 30 degrees with the horizontal. The force due to air resistance has magnitude $v/2$, where v is the velocity, and the coefficient of friction is 0.34. The force of friction on the object is of magnitude $0.34N$, where N is the force exerted by the surface on the block perpendicular to the inclined plane. Assuming that gravity, air resistance, and friction are the only forces acting on the block, write a differential equation for its velocity at time t.

11. *Mixing of salt in water*   At an initial time $t = 0$, S_0 pounds of salt are dissolved in 500 gallons of water stored in a tank. At $t = 0$, a tap is opened, and a solution containing 0.2 pound/gallon of salt is poured into the tank at a rate of 6 gallons/minute. The solution is continuously stirred to dissolve the salt. At the same time, a tap is opened in the bottom of the tank, and brine solution is drawn off at the rate of 6 gallons/minute. Write a differential equation for the amount of salt in the tank at time t. *Hint:* Let $S(t)$ be the amount of salt in the tank at time t. Then dS/dt can be calculated as the rate at which salt is introduced into the tank minus the rate at which salt is allowed to pour out of the tank. That is, dS/dt equals rate in minus rate out.

12. A block of wood is falling through an oil mixture. The forces acting on the block are gravity and a resistant force due to the oil. This force is proportional to v, where v is the velocity at time t. Derive a differential equation for v.

13. A spherical body of radius R and mass m is falling in a dense grease. The forces acting on the sphere are (1) gravity, (2) a buoyant force equal to the weight of the grease displaced by the sphere, and (3) a resistive force of magnitude $6\pi\mu Rv$, in which v is the velocity of the sphere and μ is a constant called the *viscosity coefficient* of the grease. Write a differential equation for v as a function of time t.

14. Write a differential equation for the motion of a rectangular block of wood floating in water, assuming that it bobs up and down so that its top and bottom remain horizontal and its sides remain vertical. *Hint:* Archimedes' principle states that an object submerged in a fluid is buoyed up by a force equal to the weight of the fluid displaced. (This principle was used by Archimedes to determine the proportions of gold and silver in the crown of King Hiero.)

15. Imagine that we have a simple pendulum, as in Example 0.4, but that the length of the rod is $L(t)$, a function of time, and increases at a steady rate of V inches-minute. Let $\theta(t)$ be the angle of displacement from the vertical. Derive the differential equation

$$\frac{d^2\theta}{dx^2} + \frac{2}{x}\frac{d\theta}{dx} + \frac{g}{V}\frac{1}{x}\theta = 0,$$

in which

$$x = \frac{1}{V}(L_0 + Vt)$$

and L_0 is the initial length of the rod.

First Order Differential Equations

1.0 Introduction

A *first order ordinary differential equation* is an equation of the form

$$F(x, y, y') = 0,$$

involving one dependent and one independent variable and a first derivative. Examples are

$$y' = 4xy - x^2,$$
$$yy' = -x,$$

and

$$(y')^2 = y.$$

The term "ordinary" simply means that no partial derivatives are present. Often we refer to a first order ordinary differential equation as simply a *first order equation* or a *first order differential equation*.

A first order differential equation $F(x, y, y') = 0$ must contain a first derivative y', but it need not explicitly contain either x or y. For example, $y' = 4$ is a first order equation in which x and y do not explicitly appear. Similarly,

$$y' = -y^2$$

does not explicitly contain x, although x is implicit in y, which is thought of as a function of x.

Of course, other symbols may be used for the variables. The radioactive decay equation $M'(t) = -kM$ (Example 0.3) is a first order differential equation with the mass M as a function of t.

A *solution* of a differential equation is any function satisfying the differential equation. A solution may be defined on the entire real line or on only part of the line, often on an interval. We say that $f(x)$ is a solution of $F(x, y, y') = 0$ on an interval I if $F(x, f(x), f'(x)) = 0$ for all x in I. For example, $y = e^{2x}$ is a solution of $y' - 2y = 0$ for all x, and $y = \ln(x)$ is a solution of $y' = 1/x$ for $x > 0$. These statements are easily verified by substituting the function into the differential equation.

Sometimes a solution is defined implicitly by an equation involving the independent and dependent variables. For example, the equation

$$x^2 + y^2 = 4 \tag{1.1}$$

implicitly defines y as a function of x, and this function is a solution of the first order differential equation

$$yy' = -x. \tag{1.2}$$

To verify this assertion, differentiate equation (1.1) implicitly with respect to x:

$$2x + 2yy' = 0.$$

Then $yy' = -x$, which is exactly the differential equation (1.2). In this example, equation (1.1) actually defines *two* solutions of equation (1.2), namely,

$$y = \varphi_1(x) = \sqrt{4 - x^2} \quad \text{and} \quad y = \varphi_2(x) = -\sqrt{4 - x^2},$$

both defined for $-2 \le x \le 2$. Again, the fact that both of these functions are solutions of the first order equation (1.2) can be verified by substituting them into equation (1.2).

Since a first order differential equation involves a first derivative, we expect intuitively that any process of producing a solution must at some stage involve an integration and hence give rise to an arbitrary constant of integration. A solution of $F(x, y, y') = 0$ which contains one arbitrary constant is called the *general solution* of this differential equation. For example, $y = Ce^{2x}$ is a solution of $y' - 2y = 0$ for any choice of the constant C. Thus, $y = Ce^{2x}$ is the general solution of this differential equation.

Similarly, we can check by substitution that $y = 4 + Ke^{-x}$ is a solution of $y' + y = 4$ for any choice of the constant K. Thus, $y = 4 + Ke^{-x}$ is the general solution of this differential equation.

A solution obtained by making a specific choice of the constant in the general solution is called a *particular solution*. For example, $y = 4 - e^{-x}$ is a particular solution of $y' + y = 4$, obtained by letting $K = -1$ in the general solution $y = 4 + Ke^{-x}$. Similarly, $y = -5e^{2x}$, $y = \sqrt{2}e^{2x}$, and $y = 16\pi e^{2x}$ are particular solutions of $y' - 2y = 0$.

In some models of physical systems, we want to solve a first order differential equation subject to the condition that the solution pass through a given point (x_0, y_0). That is, we want to solve the problem

$$F(x, y, y') = 0; \qquad y(x_0) = y_0, \tag{1.3}$$

consisting of a first order differential equation and a value the solution must assume at a given point. Here, x_0 and y_0 are given numbers.

A condition $y(x_0) = y_0$ is called an *initial condition*, and the problem (1.3) consisting of a differential equation and an initial condition is called an *initial value problem*. Often we solve an initial value problem by finding the general solution of the

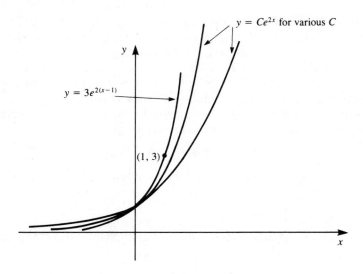

Figure 1.1

differential equation and then choosing the constant to find a particular solution satisfying the initial condition. As a purely mathematical illustration of this, suppose we want to solve the initial value problem

$$y' - 2y = 0; \qquad y(1) = 3. \tag{1.4}$$

We have seen previously that the general solution of $y' - 2y = 0$ is $y = Ce^{2x}$. To satisfy the initial condition $y(1) = 3$, we must choose C so that

$$y(1) = Ce^2 = 3.$$

Thus, $C = 3e^{-2}$. A solution of the initial value problem (1.4) is

$$y = 3e^{-2}e^{2x} = 3e^{2(x-1)}.$$

Figure 1.1 shows the graphs of functions $y = Ce^{2x}$ for various choices of C together with the graph of the particular solution $y = 3e^{2(x-1)}$. Geometrically, graphs of $y = Ce^{2x}$ form a family of curves in the plane, called *solution curves*, or *integral curves*, of the differential equation. By choosing C equal to $3e^{-2}$, we specify the particular integral curve passing through the point $(1, 3)$.

In this example, there is only one integral curve of the differential equation passing through each point in the plane. This need not always be the case, as we will see in Section 1.10 when we consider whether an initial value problem must have a solution and, if so, how many it may have.

Even with first order differential equations, unless the equation has a fairly simple form, finding a solution may be difficult or even impossible in terms of elementary functions. In Sections 1.1 through 1.7, we will discuss methods for solving certain types of first order differential equations, studying some applications in Section 1.2. The remainder of the chapter will be devoted to presenting additional applications of first order differential equations and providing some theoretical background.

PROBLEMS FOR SECTION 1.0

In each of Problems 1 through 10, determine whether the function φ is a solution of the differential equation. C denotes an arbitrary constant.

1. $2yy' = 1$; $\varphi(x) = \sqrt{x - 1}$ for $x > 1$

2. $y' + y = 0$; $\varphi(x) = Ce^{-x}$

3. $y' = -\dfrac{2y + e^x}{2x}$ for $x > 0$; $\varphi(x) = \dfrac{C - e^x}{2x}$ for $x > 0$

4. $y' = \dfrac{2xy}{2 - x^2}$ for $x \neq \pm\sqrt{2}$; $\varphi(x) = \dfrac{C}{x^2 - 2}$ for $x \neq \pm\sqrt{2}$

5. $xy' = x - y$: $\varphi(x) = \dfrac{x^2 - 3}{2x}$ for $x \neq 0$

6. $y' + y = 1$; $\varphi(x) = 1 + Ce^{-x}$

7. $x^2yy' = -1 - xy^2$; $\varphi(x) = \dfrac{4 - x^2}{2x}$ for $x \neq 0$

8. $y' + 2y = 0$; $\varphi(x) = \sin(3x) - 4$

9. $\sinh(x)y' + y\cosh(x) = 0$; $\varphi(x) = \dfrac{-1}{\sinh(x)}$

10. $y' - 3y = x$; $\varphi(x) = -\frac{1}{3}x + \frac{1}{3} + Ce^{3x}$

In each of Problems 11 through 15, verify by implicit differentiation that the given equation implicitly defines a solution of the differential equation.

11. $y^2 + xy - 2x^2 - 3x - 2y = C$; $(y - 4x - 3) + (x + 2y - 2)y' = 0$

12. $xy^3 - y = C$; $y^3 + (3xy^2 - 1)y' = 0$

13. $y^2 - 4x^2 + e^{xy} = C$; $(8x - ye^{xy}) - (2y + xe^{xy})y' = 0$

14. $8\ln|x - 2y + 4| - 2x + 6y = C$; $y' = \dfrac{x - 2y}{3x - 6y + 4}$

15. $\tan^{-1}\left(\dfrac{y}{x}\right) + x^2 = C$; $\dfrac{2x^3 + 2xy^2 - y}{x^2 + y^2} + \dfrac{x}{x^2 + y^2}y' = 0$

In each of Problems 16 through 19, find the general solution of the differential equation by direct integration. Using the same set of axes, sketch graphs of several integral curves obtained by making choices for the arbitrary constant in the general solution. Then find the particular solution satisfying the given initial condition and sketch the graph of this particular solution as well.

16. $y' = 2x$; $y(2) = 1$

17. $y' = e^{-x}$; $y(0) = 2$

18. $y' = 2(x + 1)$; $y(-1) = 1$

19. $y' = 4\cos(x)\sin(x)$; $y\left(\dfrac{\pi}{2}\right) = 0$

1.1 Separable Equations _____

A differential equation $F(x, y, y') = 0$ is called *separable* if it can be written in the form

$$B(y)y' = A(x). \tag{1.5}$$

In this event, we can integrate both sides of equation (1.5) to get

$$\int B(y)y'(x)\, dx = \int A(x)\, dx. \tag{1.6}$$

Assuming that we can carry out these integrations, we obtain an equation which implicitly or explicitly defines the general solution of the differential equation.

It is often convenient to carry out this process in differential notation. To do this, first rewrite equation (1.5) using Leibniz notation for the derivative:

$$B(y)\frac{dy}{dx} = A(x).$$

This equation suggests the differential form of equation (1.5).

$$B(y)\, dy = A(x)\, dx. \tag{1.7}$$

Since the left side of equation (1.7) involves only y and the right side involves only x, we say at this point that the variables have been *separated*. Now we can rewrite equation (1.6) as

$$\int B(y)\, dy = \int A(x)\, dx.$$

We are, in effect, integrating the left side of (1.7) with respect to y and the right side with respect to x. This results in an equation which explicitly or implicitly defines the general solution of the differential equation.

EXAMPLE 1.1

The first order differential equation

$$\frac{dy}{dx} = 3x^2 + 1$$

is separable. In differential form, it is

$$dy = (3x^2 + 1)\, dx,$$

in which x and y are isolated on opposite sides of the equation. Now integrate:

$$\int dy = \int (3x^2 + 1)\, dx,$$

or

$$y = x^3 + x + C,$$

an explicit expression for the general solution.

Sometimes we want a particular solution satisfying an initial condition. For example, if we want a solution such that $y(1) = 4$, we must choose C so that

$$y(1) = 1^3 + 1 + C = 2 + C = 4,$$

hence $C = 2$. The particular solution satisfying $y(1) = 4$ is

$$y = x^3 + x + 2. \quad \blacksquare$$

EXAMPLE 1.2

Consider the differential equation

$$\frac{dy}{dx} = 8x^3 y^2.$$

In differential form, we have

$$\frac{1}{y^2}\,dy = 8x^3\,dx$$

if $y \neq 0$. The variables have been separated, and the original differential equation is therefore separable. Now integrate:

$$\int \frac{1}{y^2}\,dy = \int 8x^3\,dx,$$

or

$$-\frac{1}{y} = 2x^4 + C.$$

This equation can be solved explicitly for y:

$$y = \frac{-1}{2x^4 + C},$$

the general solution of the differential equation. ∎

Continuing the example, suppose now we want to solve the initial value problem

$$\frac{dy}{dx} = 8x^3 y^2; \qquad y(2) = 3.$$

We must choose C in the general solution so that $y(2) = 3$. We need

$$y(2) = \frac{-1}{32 + C} = 3;$$

hence, we must choose $C = -\frac{97}{3}$. The solution of the initial value problem is

$$y = \frac{-1}{2x^4 - \frac{97}{3}}.$$

We may also think of this as a particular solution of the differential equation whose graph passes through the point $(2, 3)$.

EXAMPLE 1.3

The first order equation

$$\sin(xy)y' + y^2 - x = 0$$

is not separable. There is no way of manipulating this equation into the form of equation (1.7). ∎

EXAMPLE 1.4

The differential equation

$$\frac{dy}{dx} = \frac{\cos(x) + 2}{\sin(y) + y} \tag{1.8}$$

is separable because it can be written in the form

$$[\sin(y) + y]\, dy = [\cos(x) + 2]\, dx.$$

Then

$$\int [\sin(y) + y]\, dy = \int [\cos(x) + 2]\, dx,$$

yielding

$$-\cos(y) + \tfrac{1}{2}y^2 = \sin(x) + 2x + C. \tag{1.9}$$

We cannot solve this equation for y explicitly as a function of x. Equation (1.9) implicitly defines the general solution of the differential equation (1.8). The point is that, even with separable differential equations, we cannot always expect an explicitly defined solution.

We may still, however, be able to solve for C to obtain particular solutions satisfying initial conditions. For example, suppose we want a solution of (1.8) satisfying $y(0) = 3$. Put $x = 0$ and $y = 3$ into equation (1.9) to get

$$-\cos(3) + \tfrac{9}{2} = C.$$

A solution of equation (1.8) satisfying the initial condition $y(1) = 3$ is defined implicitly by the equation

$$-\cos(y) + \tfrac{1}{2}y^2 = \sin(x) + 2x - \cos(3) + \tfrac{9}{2}. \quad \blacksquare$$

One difficulty we can encounter with separable differential equations is that the algebra involved in separating the variables may impose one or more conditions not contained in the original equation. The following example demonstrates how this can occur.

EXAMPLE 1.5

Consider again the differential equation

$$\frac{dy}{dx} = 8x^3 y^2.$$

In Example 1.2, we found the general solution

$$y = \frac{-1}{2x^4 + C}.$$

Now review the mathematics involved in finding this solution. When we separated the variables, we wrote

$$\frac{1}{y^2}\, dy = 8x^3\, dx.$$

Division of the differential equation by y^2 requires the assumption that $y \neq 0$, an assumption not contained in the original differential equation. Thus, the algebra of separating the variables has imposed a new condition.

Now observe that in fact $y = 0$ *is a solution* of the differential equation (if $y = 0$, both sides of $y' = 8x^3y^2$ are zero). The solution $y = 0$ is called a *singular solution* because we cannot obtain it from the general solution for any choice of the constant C.

∎

The point to this example is that, whenever we separate variables in a differential equation, we must be alert to new conditions imposed in carrying out the separation. Any such conditions must be examined to determine whether they lead to singular solutions, that is, additional solutions not contained in the general solution.

There is no simple test to determine easily whether a first order equation is separable. In practice, we usually try to separate the variables until we either succeed or become convinced that the equation is not separable.

PROBLEMS FOR SECTION 1.1

In each of Problems 1 through 16, determine whether the differential equation is separable. If it is, find the general solution (explicitly or implicitly defined). If it is not separable, do not attempt to solve it at this time.

In view of Example 1.5, note any conditions imposed in carrying out the algebra of separating the variables, and find any singular solutions these conditions give rise to.

1. $\dfrac{1}{x} y^2 \dfrac{dy}{dx} = 1 + x^2$

2. $y + x\dfrac{dy}{dx} = 0$

3. $\cos(y)\dfrac{dy}{dx} = \sin(x + y)$

4. $(2y + xy)\dfrac{dy}{dx} = 1 + \dfrac{4}{x} + \dfrac{4}{x^2}$

5. $3\dfrac{dy}{dx} = \dfrac{4x}{y^2}$

6. $\dfrac{dy}{dx} = \dfrac{(x + 1)^2 - 2y}{y}$

7. $\dfrac{dy}{dx} + y = y[\sin(x) + 1]$

8. $\dfrac{dy}{dx} = \dfrac{4x + xy^2}{2y + x^2y}$

9. $\dfrac{dy}{dx} = y\dfrac{x^2 - 2x + 1}{y + 3}$

10. $\ln(y^x)\dfrac{dy}{dx} = 3x^2y$

11. $e^{x+y}\dfrac{dy}{dx} = 2x$

12. $(x + y)^2\dfrac{dy}{dx} = x^2 + y^2$

13. $x\sin(y)\dfrac{dy}{dx} = \sec(y)$

14. $3x^2y - 6x^3 - y^2 + 2xy + (2x^2 - xy)\dfrac{dy}{dx} = 0$

15. $\dfrac{1}{y} x\dfrac{dy}{dx} = \dfrac{2y^2 + 1}{x + 1}$

16. $[\cos(x + y) + \sin(x - y)]\dfrac{dy}{dx} = \cos(2x)$

In each of Problems 17 through 26, find the general solution (perhaps implicitly defined) of the separable differential equation. Then find a particular solution (perhaps implicitly defined) satisfying the initial condition.

17. $\dfrac{dy}{dx} = 3x^2(y + 2);\quad y(4) = 8$

18. $x\dfrac{dy}{dx} = y^2;\quad y(3) = 5$

19. $\dfrac{dy}{dx} = \dfrac{x^2 + 2}{y};\quad y(1) = 7$

20. $y\dfrac{dy}{dx} = \dfrac{x^2}{y + 4};\quad y(3) = 2$

21. $\dfrac{dy}{dx} = \dfrac{x - 1}{y + 2};\quad y(-1) = 6$

22. $x^2\dfrac{dy}{dx} = \dfrac{1}{y};\quad y(4) = 9$

23. $\dfrac{dy}{dx} = \dfrac{1}{x} e^y;\quad y(1) = 4$

24. $\dfrac{dy}{dx} = \dfrac{-3x}{y + 4};\quad y(2) = 7$

25. $\dfrac{dy}{dx} = \dfrac{-2\sin(1 + x)}{y^2};\quad y(\pi) = 4$

26. $2y\dfrac{dy}{dx} = e^{x-y^2};\quad y(4) = -2$

27. Find all real-valued functions g which are continuous and positive on $[0, \infty)$ and have the property that, for all $x > 0$, the y-coordinate of the centroid of the set R_x is the same as the average value of $g(x)$ on $[0, x]$, where R_x consists of all points (s, t) such that $0 \le s \le x$ and $0 \le t \le g(s)$. (This problem appeared on the 1984 William Lowell Putnam Mathematical Competition.) *Hint:* Express the centroid and the average as integrals. Once an equation has been established containing these integrals, with no fractions, use the fundamental theorem of calculus, then multiply by x and make a substitution from the original equation to obtain a quadratic equation in $\int_0^x g(t)\,dt$. Solve this equation, then use the fundamental theorem again to obtain a separable differential equation for g.

28. Evaluate $\int_0^\infty e^{-[t^2 + (9/t^2)]}\,dt$. *Hint:* Let $I(x) = \int_0^\infty e^{-[t^2 + (x/t)^2]}\,dt$. Find $I'(x)$ by differentiating under the integral sign with respect to x, then let $u = x/t$. Show that $I'(x) = 2I(x)$, and solve this separable differential equation for $I(x)$. Assume as known the fact that $\int_0^\infty e^{-x^2}\,dx = \sqrt{\pi}/2$ to obtain the initial condition $I(0) = \sqrt{\pi}/2$, and find the solution of the differential equation satisfying this condition. Finally, note that the integral requested is $I(3)$.

1.2 Separable Differential Equations in Electrical Circuits, Melting, and Carbon Dating

In this section, we will discuss and solve three models of physical phenomena involving separable differential equations.

EXAMPLE 1.6　An Electrical Circuit Equation

Differential equations modeling electrical circuits generally are not separable (see Example 0.1). For the simple circuit in Figure 1.2, however, the differential equation for the current i is

$$L\frac{di}{dt} = E(t),$$

which can be written in differential form with the variables separated:

$$di = \frac{1}{L}E(t)\,dt.$$

Here, the inductance L is assumed to be constant.

A commonly seen impressed voltage is $E(t) = A\cos(\omega t)$, in which case we have

$$di = \frac{A}{L}\cos(\omega t)\,dt$$

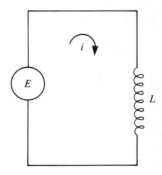

Figure 1.2

Then

$$\int di = \frac{A}{L}\int \cos(\omega t)\,dt,$$

or

$$i(t) = \frac{A}{\omega L}\sin(\omega t) + C.$$

We can determine C if we know the current at any particular time. For example, if $i(0) = 0$, then $C = 0$, and the current is a sinusoidal wave:

$$i(t) = \frac{A}{\omega L} \sin(\omega t). \quad \blacksquare$$

EXAMPLE 1.7   Melting

Assume that a sphere of ice melts at a rate proportional to its surface area. We want an expression for the volume of the sphere at any time t.

Let $V(t)$ be the volume at time t. We interpret melting as the rate of change of volume with respect to time, hence as dV/dt. If $r(t)$ is the radius of the ball of ice at time t, then for some constant of proportionality k we have

$$\frac{dV}{dt} = k \cdot (\text{surface area}) = 4k\pi r^2. \tag{1.10}$$

This equation involves two unknowns, V and r, which are both thought of as functions of t. To eliminate one unknown, use the fact that

$$V(t) = \tfrac{4}{3}\pi r^3(t).$$

(This assumes symmetrical melting so that the ice retains a spherical shape.) Solve this equation for r in terms of V to get

$$r = \left(\frac{3V}{4\pi}\right)^{1/3}. \tag{1.11}$$

Substitute (1.11) into (1.10) to get

$$\frac{dV}{dt} = 4k\pi \left(\frac{3V}{4\pi}\right)^{2/3} = (4\pi)^{1/3}3^{2/3}kV^{2/3}.$$

This is a separable differential equation, since we can write it in differential form as

$$\frac{1}{V^{2/3}}\,dV = (4\pi)^{1/3}3^{2/3}k\,dt.$$

Then

$$\int V^{-2/3}\,dV = (4\pi)^{1/3}3^{2/3}k \int dt,$$

or

$$3V^{1/3} = (4\pi)^{1/3}3^{2/3}kt + C.$$

Then

$$V^{1/3} = \left(\frac{4}{3}\pi\right)^{1/3}kt + \frac{C}{3}.$$

We can solve this equation explicitly for V:

$$V(t) = \left[\left(\frac{4}{3}\pi\right)^{1/3}kt + C^*\right]^3$$

in which $C^* = C/3$ is again an arbitrary constant. This is the general solution of the differential equation. To determine C^*, we need to be given the volume at some particular time. ∎

EXAMPLE 1.8   Carbon Dating

Separable differential equations are applicable to the problem of dating artifacts. We will discuss the idea behind carbon dating before carrying out a mathematical analysis.

Radioactive elements decay into new elements, or isotopes of the same element, at a rate proportional to their mass. The rate at which decay takes place is measured by detecting the number of alpha and beta particles emitted as part of the decay process.

The *half-life* of an isotope is the time it takes for one-half of the atoms in a sample of the isotope to decay. If m grams are present at year t_0 and the half-life is H years, there will be $\frac{1}{2}m$ grams present at $t_0 + H$ years.

Strictly speaking, the process of radioactive decay is not a continuous one because the number of nuclei is finite, and atoms decay one at a time. However, the number of atoms in any sample of material is so large that the decaying process can be thought of as a continuous one and modeled by a differential equation as follows.

Suppose that $M(t)$ is the mass of radioactive material present at time t. Since the rate of change of M with respect to time is proportional to M,

$$\frac{dM}{dt} = kM(t)$$

for some constant k. This differential equation is separable, since it can be written

$$\frac{1}{M} \frac{dM}{dt} = k,$$

or, in differential form,

$$\frac{1}{M} dM = k\, dt.$$

Upon integrating, we get

$$\ln|M| = kt + c.$$

Since $M > 0$, $|M| = M$ and we have $\ln(M) = kt + c$, or

$$M(t) = e^{kt+c} = e^{kt}e^c = Ae^{kt},$$

in which $A = e^c$ is a constant to be determined.

Sometimes we know the mass at some time, which we often designate as time zero. Then $M(0) = m$, a known quantity, and we have

$$M(0) = A = m.$$

This determines the constant A in terms of a measured quantity and yields the relationship

$$M(t) = me^{kt}.$$

To find k, we must know the mass at some other time. For our purposes, it is convenient to express k in terms of the half-life of the substance (the time it takes for half of the mass to be converted to energy). Suppose we have measured the half-life and found it to be H. Since $M(0) = m$, $M(H) = \frac{1}{2}m$, and we have

$$M(H) = me^{kH} = \tfrac{1}{2}m.$$

Then

$$e^{kH} = \tfrac{1}{2}.$$

Take the natural logarithm of both sides of this equation to get

$$kH = \ln(\tfrac{1}{2}) = -\ln(2).$$

Then

$$k = -\frac{1}{H}\ln(2).$$

Recalling that H is assumed known (by measurement), we now have the mass at any time $t \geq 0$:

$$M(t) = me^{-\ln(2)t/H}.$$

If we write $-\ln(2)t/H = \ln(2^{-t/H})$, we can write $M(t)$ as

$$M(t) = m2^{-t/H}. \tag{1.12}$$

Equation (1.12), combined with some facts about radioactive carbon and its presence in organic materials, forms the basis for the carbon-dating technique for estimating the age of certain prehistoric objects. We will discuss this technique, then look at a specific example.

High-energy cosmic rays, which consist of high-velocity atomic nuclei from outer space, bombard the upper atmosphere of the earth, producing large numbers of neutrons. These neutrons collide with nitrogen in the air, changing some of it into carbon 14, or C-14. This substance is radioactive, with a half-life of approximately 5570 years.

C-14 combines with oxygen in the upper atmosphere to form carbon dioxide, which drifts to earth and is absorbed by plants. The radioactive C-14 becomes fixed in the plants' protoplasm and is then ingested by animals eating the plants. Once a plant or animal dies, it ceases its intake of C-14. By comparing the radioactivity of carbon from present-day plants and animals with that of ancient samples, scientists can estimate the age of these samples. This process is outlined in Figure 1.3.

For example, if the number of beta particles emitted from a gram of carbon from an ancient campfire is one-fourth that of carbon from a living plant, then two half-life periods have passed since the wood used in the fire was living, so the campfire would have burned about $2 \cdot 5570$, or 11,140, years ago.

We will now make a specific calculation. Suppose measurements show that a piece of wood is emitting beta particles at a rate of 5 per minute. It is known that C-14 emits about 15 beta particles per minute per gram. Then the amount of C-14 now in the wood is about $\frac{1}{3}$ gram. Suppose we have also determined that, if the same piece of wood were alive now, it would have 2 grams of C-14. When did the tree die?

Put $m = 2$ and $H = 5570$ into equation (1.12) to get

$$M(t) = 2 \cdot 2^{-t/5570}.$$

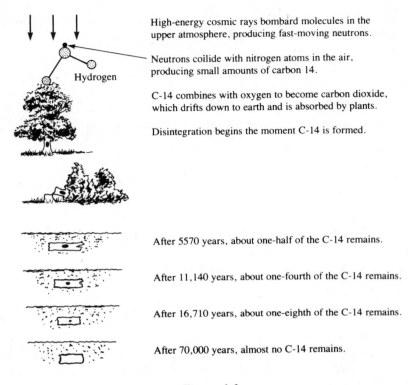

High-energy cosmic rays bombard molecules in the upper atmosphere, producing fast-moving neutrons.

Neutrons collide with nitrogen atoms in the air, producing small amounts of carbon 14.

C-14 combines with oxygen to become carbon dioxide, which drifts down to earth and is absorbed by plants.

Disintegration begins the moment C-14 is formed.

After 5570 years, about one-half of the C-14 remains.

After 11,140 years, about one-fourth of the C-14 remains.

After 16,710 years, about one-eighth of the C-14 remains.

After 70,000 years, almost no C-14 remains.

Figure 1.3

Let T denote the present time (T years after the time designated time zero, when the tree died). We know that there is now $\frac{1}{3}$ gram of C-14 present, so

$$M(T) = \tfrac{1}{3} = 2 \cdot 2^{-T/5570}.$$

Solve this equation for T. We get

$$T = \frac{5570 \ln(6)}{\ln(2)}.$$

Thus, the tree died about 14,000 years ago.

As a check of the reasonableness of this figure, make another estimate of the time as follows. Only one-sixth of the original C-14 is present now; this is between one-fourth and one-eighth, so the wood has been dead between two and three half-lives of C-14. This puts the time of death of the tree somewhere between $2 \cdot 5570$ and $3 \cdot 5570$ years ago, or between 11,140 and 16,710 years ago. The number we calculated is in this range. ∎

PROBLEMS FOR SECTION 1.2

1. Solve for the current in the circuit of Example 1.6 if $i(0) = 0$ and

$$E(t) = \alpha \cos(\omega_1 t) + \beta \sin(\omega_2 t),$$

in which α, β, ω_1 and ω_2 are given constants.

2. Solve for the current in the circuit of Example 1.6 if $i(0) = 4$ amperes and

$$E(t) = 2 \cos(\omega_1 t) + 4 \sin(\omega_2 t).$$

3. Suppose that the population P of bacteria in a culture at time t changes at a rate proportional to $P^2 - P$. Suppose that $P^2 - P > 0$.

(a) Let the constant of proportionality be k. Write a differential equation for $P(t)$ and find the general solution. This general solution will contain k, together with an arbitrary constant.

(b) Assume that there are 1000 bacteria at time $t = 0$ hours. Use this information to solve for the arbitrary constant in the general solution of (a).

(c) Suppose that, at $t = 5$ hours, there are 100 bacteria. Use this information to solve for k.

(d) Determine $\lim_{t \to \infty} P(t)$.

4. *Pressure in a perfect gas*   For an adiabatic process involving a perfect gas, pressure p is related to volume V by the equation

$$\frac{dp}{dV} = -\left(\frac{c_p}{c_V}\right)\frac{p}{V},$$

in which c_p is the specific heat of the gas at constant pressure and c_V is the specific heat at constant volume. Solve for pressure as a function of volume, assuming that the pressure is 4 pounds per cubic inch when the volume is 1 cubic inch. (This solution will contain c_p and c_V but no other arbitrary constants.) *Note:* An adiabatic process is one involving a change of matter without heat transfer. The specific heat of a substance is the amount of heat energy required to impart a unit increase in temperature to one unit of mass of the substance. A perfect gas is one in which the molecules experience no force except that due to collisions with other molecules of the gas or with sides of the container holding the gas. Such a gas is a theoretical construction and cannot exist in this form in the real world.

5. *Blackbody radiation*   Experiment has shown that if u is the energy density of a blackbody and T is its absolute temperature,

$$\frac{du}{dT} = 4\frac{u}{T}.$$

Find a general expression for the dependence of u on T. (The Stefan-Boltzmann law states that the rate at which thermal energy is radiated by a blackbody of surface area A and Stefan-Boltzmann constant σ is $\sigma A T^4$. How is this result related to the general solution for u?) *Note:* A blackbody is an object which absorbs all radiation falling upon its surface, reflecting none.

6. Find a general expression for the volume V of a gas as a function of the pressure p if the rate of change of the volume with respect to the pressure is proportional to $-V/p^2$.

7. A spherical raindrop falling through a cloud accumulates moisture. Assume that its volume increases with respect to distance fallen at a rate proportional to its cross-sectional area at that distance. Derive an expression for the volume of the drop at any distance h.

8. A spherical hailstone melts at a rate proportional to its surface area. Suppose it was originally $\frac{1}{8}$ inch in radius and 40 minutes later its radius is $\frac{1}{24}$ inch.

(a) Find an expression for the radius of the hailstone at any time t.

(b) Find expressions for the surface area and volume of the hailstone at any time t.

(c) When is the hailstone reduced to $\frac{1}{100}$ inch radius?

(d) At what time is the surface area reduced to half of its original value?

(e) At what time is the volume reduced to half of its original value?

9. If food and living space are unlimited, some populations grow at a rate proportional to the population. It is estimated that the world population in 1900 was 1600 million people and that by 1950 it had risen to 2510 million people. What will be the world population in the year 2000, assuming unlimited food and living space?

10. A bacterial culture has a population density of 100,000 per square inch. A culture which covered an area of 1 square inch at 10:00 A.M. on Tuesday was found to have grown to 3 square inches by noon the following Thursday. How many bacteria will be present at 3:00 P.M. the following Sunday, assuming that the population density changes at a rate proportional to itself? How many will be present on Monday at 4:00 P.M.?

11. Newton's law of cooling states that the rate at which a body cools is proportional to the difference in temperature between the body and the surrounding medium. An object of temperature 90 degrees Fahrenheit is placed in a 60-degree environment. Ten minutes later, the object has cooled to 80 degrees Fahrenheit. What will the body's temperature be after it has been subjected to this environment for a period of 20 minutes? At what time will the body's temperature be 65 degrees Fahrenheit?

12. A thermometer is taken outside from a house in which the ambient temperature is 70 degrees Fahrenheit. At the end of 5 minutes, the thermometer reads 60 degrees Fahrenheit, and 5 minutes later, it reads 54 degrees Fahrenheit. What is the outside temperature?

13. The police discover the body of a differential equations instructor. Critical to solving the crime is the determination of when the murder was committed. The coroner arrives at noon and immediately finds that the body temperature is 94.6 degrees Fahrenheit. She waits 1 hour and finds the temperature of the body to be 93.4 degrees Fahrenheit. She also notes that the temperature of the room is a constant 70 degrees Fahrenheit. Assuming that the victim was normal (at least in temperature) up to the time of his demise, determine the time when the murder was committed.

14. The radioactive element O'Neillium has a half-life of ln(2) weeks. If e^3 tons of the element are present at a given time, how much will be left 3 weeks later?

15. The half-life of uranium 238 is approximately 4.5×10^9 years. How much of a 10-kilogram block of U-238 will be present 1 billion years from now?

16. Given that 12 grams of U-228 decays to 9.1 grams in a time span of only 4 minutes, compute the half-life of this relatively unstable isotope of uranium.

17. Air pressure decreases with altitude at a rate proportional to the pressure and is half as much at 18,000 feet above sea level as it is at sea level. If an automobile engine has a compression ratio of 7.9:1, determine the reading that should be obtained on a compression tester applied to this engine in Laramie, Wyoming, which is at an altitude of 7200 feet above sea level. Assume a standard pressure of 14.7 pounds per square inch at sea level.

1.3 *Homogeneous and Nearly Homogeneous Differential Equations*

"Most" first order differential equations are not separable. In some instances, a change of variables can be used to transform a differential equation which is not separable into one which is separable. In this section, we will study a class of differential equations to which this idea applies.

A first order differential equation of the form

$$\frac{dy}{dx} = f\left(\frac{y}{x}\right) \tag{1.13}$$

is called *homogeneous*. Here, dy/dx is isolated on one side of the equation, while the other side is some expression in which y and x always appear in the combination y/x. For example,

$$\frac{dy}{dx} = \left(\frac{y}{x}\right)^2 + 1$$

is homogeneous, as is

$$\frac{dy}{dx} = e^{y/x} - 2\frac{y}{x}.$$

As these examples show, a homogeneous equation need not be separable. However, a homogeneous equation is always transformed into a separable equation by the change of variables

$$u = \frac{y}{x}. \tag{1.14}$$

To see this, write equation (1.14) as $y = ux$. Then

$$\frac{dy}{dx} = u + x\frac{du}{dx}. \tag{1.15}$$

Substitute these expressions for y/x and dy/dx into equation (1.13) to get

$$u + x\frac{du}{dx} = f(u). \tag{1.16}$$

In differential form, equation (1.16) can be written

$$\frac{1}{f(u) - u}\,du = \frac{1}{x}\,dx.$$

Therefore, equation (1.16) is separable. The strategy is to attempt to solve the differential equation (1.16). The solution will be in terms of u and x. Upon letting $u = y/x$, we obtain a solution of the original equation (1.13).

EXAMPLE 1.9

Solve

$$x\frac{dy}{dx} = \frac{y^2}{x} + y \qquad (x \neq 0).$$

This differential equation is not separable. However, write it as

$$\frac{dy}{dx} = \frac{y^2}{x^2} + \frac{y}{x},$$

which we recognize as being homogeneous. Let $u = y/x$ to get

$$u + x\frac{du}{dx} = u^2 + u,$$

or

$$x\frac{du}{dx} = u^2.$$

This separable equation in terms of x and u has differential form

$$\frac{1}{u^2}\,du = \frac{1}{x}\,dx.$$

Integrate:

$$\int \frac{1}{u^2}\, du = \int \frac{1}{x}\, dx.$$

Then

$$-\frac{1}{u} = \ln|x| + C,$$

so

$$u = \frac{-1}{\ln|x| + C},$$

the general solution for u in terms of x. Since $u = y/x$, the general solution of the original differential equation in terms of x and y is

$$y = \frac{-x}{\ln|x| + C}. \quad \blacksquare$$

EXAMPLE 1.10

Solve

$$x^3 \frac{dy}{dx} = x^2 y - 2y^3.$$

Upon dividing this equation by x^3 (assuming that $x \neq 0$), we have

$$\frac{dy}{dx} = \frac{y}{x} - 2\frac{y^3}{x^3}.$$

Let $u = y/x$ to get

$$u + x\frac{du}{dx} = u - 2u^3.$$

In differential form, this equation can be written

$$-\frac{1}{2}\frac{1}{u^3}\, du = \frac{1}{x}\, dx.$$

Upon integrating, we get

$$\frac{1}{4}\frac{1}{u^2} = \ln|x| + C.$$

Let $u = y/x$ to obtain

$$\frac{x^2}{4y^2} = \ln|x| + C.$$

This equation implicitly defines the general solution of the original differential equation. If we wish, we can in this example solve for y to get the general solution

$$y = \frac{|x|}{2\sqrt{\ln|x| + C}}. \tag{1.17}$$

As further illustration, we can now solve the initial value problem

$$x^3 \frac{dy}{dx} = x^2 y - 2y^3; \qquad y(1) = 6. \tag{1.18}$$

Let $x = 1$ and $y = 6$ in the general solution (1.17) to obtain

$$y(1) = 6 = \frac{1}{2\sqrt{C}}.$$

Therefore, choose $C = (\frac{1}{12})^2 = \frac{1}{144}$. The solution of the initial value problem (1.18) is

$$y = \frac{|x|}{2\sqrt{\ln|x| + \frac{1}{144}}}. \quad \blacksquare$$

As the next example shows, the integrations need not always be as simple as in the two examples we have just seen.

EXAMPLE 1.11

Solve

$$\frac{dy}{dx} = \left(2 + \frac{y}{x}\right)^2.$$

This is a homogeneous differential equation. Let $u = y/x$ to get

$$u + x \frac{du}{dx} = (2 + u)^2 = u^2 + 4u + 4.$$

Then

$$x \frac{du}{dx} = u^2 + 3u + 4,$$

so

$$\frac{1}{u^2 + 3u + 4} du = \frac{1}{x} dx.$$

Then

$$\int \frac{1}{u^2 + 3u + 4} du = \int \frac{1}{x} dx. \tag{1.19}$$

The integral on the left is of a type that we encounter frequently when dealing with homogeneous differential equations. For reference, we find from a table of integrals that

$$\int \frac{1}{au^2 + bu + c} du = \begin{cases} \dfrac{2}{\sqrt{q}} \tan^{-1}\left(\dfrac{2au + b}{\sqrt{q}}\right) & \text{if} \quad q > 0 \\[3mm] \dfrac{-2}{\sqrt{-q}} \tanh^{-1}\left(\dfrac{2au + b}{\sqrt{-q}}\right) & \text{if} \quad q < 0, \end{cases}$$

in which $q = 4ac - b^2$.

To apply this formula to the integral on the left side of equation (1.19), let $a = 1$, $b = 3$, and $c = 4$. Then $q = 16 - 9 = 7$, and from equation (1.19) we have

$$\frac{2}{\sqrt{7}} \tan^{-1}\left(\frac{2u + 3}{\sqrt{7}}\right) = \ln|x| + C.$$

Since $u = y/x$, the general solution of the original differential equation is defined implicitly by the equation

$$\frac{2}{\sqrt{7}} \tan^{-1}\left(\frac{2y + 3x}{\sqrt{7}x}\right) = \ln|x| + C.$$

This equation can be solved explicitly for y in terms of x:

$$y(x) = \frac{1}{2}\left[\sqrt{7}x \tan\left(\frac{\sqrt{7}}{2}(\ln|x| + C)\right) - 3x\right]. \quad \blacksquare$$

NEARLY HOMOGENEOUS DIFFERENTIAL EQUATIONS

Consider the differential equation

$$\frac{dy}{dx} = f\left(\frac{ax + by + c}{dx + ey + h}\right), \tag{1.20}$$

in which a, b, c, d, e, and h are constants. If $x \neq 0$, we can write this equation as

$$\frac{dy}{dx} = f\left(\frac{a + b\left(\dfrac{y}{x}\right) + \left(\dfrac{c}{x}\right)}{d + e\left(\dfrac{y}{x}\right) + \left(\dfrac{h}{x}\right)}\right).$$

It is apparent that when equation (1.20) is written in this form it is homogeneous exactly when $c = h = 0$; if $c \neq 0$ or $h \neq 0$, equation (1.20) is not homogeneous. However, as we shall now show, equation (1.20) is "near enough" to being homogeneous that a simple change of variables will transform it into a homogeneous equation. The idea is to let

$$X = x - \alpha \quad \text{and} \quad Y = y - \beta \tag{1.21}$$

and attempt to choose the constants α and β so that equation (1.20) is homogeneous in terms of X and Y.

Suppose, then, that at least one of c and/or h is nonzero, so that equation (1.20) is not homogeneous. Substitute $X = x - \alpha$ and $Y = y - \beta$ into equation (1.20) to get

$$\frac{dY}{dX} = f\left(\frac{a(X + \alpha) + b(Y + \beta) + c}{d(X + \alpha) + e(Y + \beta) + h}\right)$$

$$= f\left(\frac{aX + bY + (a\alpha + b\beta + c)}{dX + eY + (d\alpha + e\beta + h)}\right).$$

This equation is homogeneous exactly when

$$\begin{aligned} a\alpha + b\beta + c &= 0 \\ d\alpha + e\beta + h &= 0. \end{aligned} \tag{1.22}$$

The issue now is whether we can choose α and β satisfying equations (1.22). The system (1.22) has a solution for α and β exactly when $ae - bd \neq 0$. In this event, we use the solutions for α and β in the change of variables (1.21), resulting in a homogeneous differential equation in terms of X and Y.

Before looking at the case in which $ae - bd$ is zero, we will consider an example of the procedure up to this point.

EXAMPLE 1.12

Solve

$$\frac{dy}{dx} = \left(\frac{2x + y - 1}{x - 2}\right)^2. \tag{1.23}$$

This equation is neither separable nor homogeneous but is of the form of equation (1.20), with $a = 2$, $b = 1$, $c = -1$, $d = 1$, $e = 0$, $h = -2$, and $f(t) = t^2$. We have

$$ae - bd = -1 \neq 0.$$

Let $x = X + \alpha$ and $y = Y + \beta$ in equation (1.23) to obtain

$$\begin{aligned}
\frac{dY}{dX} &= \left(\frac{2(X + \alpha) + Y + \beta - 1}{X + \alpha - 2}\right)^2 \\
&= \left(\frac{2X + Y + 2\alpha + \beta - 1}{X + \alpha - 2}\right)^2.
\end{aligned} \tag{1.24}$$

Now choose α and β so that

$$2\alpha + \beta - 1 = 0$$
$$\alpha - 2 = 0.$$

Then $\alpha = 2$ and $\beta = -3$. Thus, let

$$x = X + 2 \quad \text{and} \quad y = Y - 3.$$

Equation (1.24) becomes

$$\frac{dY}{dX} = \left(\frac{2X + Y}{X}\right)^2 = \left(2 + \frac{Y}{X}\right)^2, \tag{1.25}$$

a homogeneous differential equation in terms of Y and X. From Example 1.11, the general solution of this equation is

$$Y(X) = \frac{1}{2}\left[\sqrt{7}X \tan\left(\frac{\sqrt{7}}{2}(\ln|X| + C)\right) - 3X\right].$$

Since $X = x - 2$ and $Y = y + 3$, the general solution of the original differential equation (1.23) is

$$y(x) = \frac{1}{2}\left[\sqrt{7}(x - 2)\tan\left(\frac{\sqrt{7}}{2}(\ln|x - 2| + C)\right) - 3(x - 2)\right] - 3. \ \blacksquare$$

If we encounter a differential equation of the form (1.20) in which $ae - bd = 0$, we cannot proceed as we have just done because we cannot solve for α and β. In this case, however, there is another change of variables we can use to transform equation (1.20) into a separable differential equation.

Suppose, then, that $ae - bd = 0$ in equation (1.20). Now let

$$v = \frac{ax + by}{a}, \tag{1.26}$$

assuming that $a \neq 0$. Since $ae = bd$, $b/a = e/d$; hence,

$$v = \frac{dx + ey}{d}.$$

Calculate

$$\frac{dv}{dx} = 1 + \frac{b}{a}\frac{dy}{dx}$$

so that

$$\frac{dy}{dx} = \frac{a}{b}\left(\frac{dv}{dx} - 1\right). \tag{1.27}$$

Substitute (1.26) and (1.27) into equation (1.20) to get

$$\frac{a}{b}\left(\frac{dv}{dx} - 1\right) = f\left(\frac{av + c}{dv + h}\right),$$

or

$$\frac{dv}{dx} = 1 + \frac{b}{a}\, f\left(\frac{av + c}{dv + h}\right).$$

This equation is separable; in differential form, it can be written

$$\frac{1}{1 + \dfrac{b}{a}\, f\left(\dfrac{av + c}{dv + h}\right)}\, dv = dx.$$

We attempt to solve this equation, then write the general solution in terms of y and x.

EXAMPLE 1.13

Solve

$$\frac{dy}{dx} = \frac{2x + y - 1}{4x + 2y - 4}.$$

This has the form of equation (1.20), with $a = 2$, $b = 1$, $c = -1$, $d = 4$, $e = 2$, $h = -4$, and $f(t) = t$. Note that $ae - bd = 0$. Thus, let

$$v = \frac{ax + by}{a} = \frac{2x + y}{2}.$$

Then

$$\frac{dv}{dx} = 1 + \frac{1}{2}\frac{dy}{dx},$$

and we obtain

$$\frac{dy}{dx} = 2\left(\frac{dv}{dx} - 1\right).$$

With these changes of variables, the differential equation becomes

$$\frac{dv}{dx} = 1 + \frac{b}{a}\, f\left(\frac{av + c}{dv + h}\right) = 1 + \frac{1}{2}\left(\frac{2v - 1}{4v - 4}\right).$$

In differential form, we have

$$\frac{8v - 8}{10v - 9}\, dv = dx.$$

Upon integrating both sides of this equation, we get

$$\tfrac{4}{5}v - \tfrac{2}{25}\ln|10v - 9| = x + C.$$

Since $v = (2x + y)/2$, this equation yields

$$\frac{4}{5}\frac{2x + y}{2} - \frac{2}{25}\ln\left|10\left(\frac{2x + y}{2}\right) - 9\right| = x + C,$$

or

$$\tfrac{2}{5}(2x + y) - \tfrac{2}{25}\ln|10x + 5y - 9| = x + C.$$

This equation implicitly defines the general solution of the original differential equation.

If we have the initial value problem

$$\frac{dy}{dx} = \frac{2x + y - 1}{4x + 2y - 4}; \quad y(1) = 0,$$

we obtain a solution by putting $x = 1$ and $y = 0$ into the general solution. This yields

$$\tfrac{4}{5} - \tfrac{2}{25}\ln|10 - 9| = 1 + C.$$

Since $\ln(1) = 0$, we have $C = -\tfrac{1}{5}$, and a solution of the initial value problem is defined implicitly by the equation

$$\tfrac{2}{5}(2x + y) - \tfrac{2}{25}\ln|10x + 5y - 9| = x - \tfrac{1}{5}. \quad \blacksquare$$

PROBLEMS FOR SECTION 1.3

In each of Problems 1 through 10, determine whether the differential equation is homogeneous. If it is, find an expression containing the general solution. If it is not homogeneous, do not attempt to solve it at this time.

1. $\dfrac{dy}{dx} = \dfrac{x}{y} + \dfrac{y}{x}$

2. $\dfrac{dy}{dx} = \dfrac{y}{x + y}$

3. $x\dfrac{dy}{dx} = y^2$

4. $\dfrac{dy}{dx} = \dfrac{x + y}{x}$

5. $(x + y)\dfrac{dy}{dx} = y$

6. $(x - 2y)\dfrac{dy}{dx} = 2x - y$

7. $x^3 \dfrac{dy}{dx} = x^2 y - y^3$ **8.** $x \dfrac{dy}{dx} = x \cos\left(\dfrac{y}{x}\right) + y$ **9.** $x^2 \dfrac{dy}{dx} = x^2 + y^2$

10. $x^2 \dfrac{dy}{dx} = y^2$

In each of Problems 11 through 21, find an expression containing the general solution of the nearly homogeneous differential equation.

11. $\dfrac{dy}{dx} = \dfrac{y - 3}{x + y - 1}$ **12.** $\dfrac{dy}{dx} = \dfrac{x - 3y - 7}{x - 4}$ **13.** $\dfrac{dy}{dx} = \dfrac{x - y + 2}{x - y + 3}$

14. $\dfrac{dy}{dx} = \dfrac{x + 2y + 7}{-2x + y - 9}$ **15.** $\dfrac{dy}{dx} = \dfrac{3x + y - 1}{6x + 2y - 3}$ **16.** $\dfrac{dy}{dx} = \dfrac{x - 2y}{3x - 6y + 4}$

17. $\dfrac{dy}{dx} = \dfrac{3x - y - 9}{x + y + 1}$ **18.** $\dfrac{dy}{dx} = \dfrac{x - y + 6}{3x - 3y + 4}$ **19.** $\dfrac{dy}{dx} = \dfrac{2x - 5y - 9}{-4x + y + 9}$

20. $\dfrac{dy}{dx} = \left[\dfrac{x - y + 1}{x + 1}\right]^2$ **21.** $\dfrac{dy}{dx} = \left[\dfrac{x - y + 1}{2x - 2y}\right]^2$

In each of Problems 22 through 30, find the solution (perhaps implicitly defined) of the initial value problem.

22. $x \dfrac{dy}{dx} = \sqrt{x^2 + y^2} + y; \quad y(3) = 0$ **23.** $x \dfrac{dy}{dx} = y + 4\sqrt{xy}; \quad y(e) = 9e$

24. $x \dfrac{dy}{dx} = y + \sqrt{x^2 - y^2}; \quad y(1) = \dfrac{1}{2}$ **25.** $y^2 \dfrac{dy}{dx} = x^2 + \dfrac{1}{x} y^3; \quad y(e) = 2e$

26. $x \dfrac{dy}{dx} = 3y + \dfrac{1}{x} y^2; \quad y(1) = 4$ **27.** $(x - 2y)\dfrac{dy}{dx} = x - y; \quad y(1) = 4$

28. $x \dfrac{dy}{dx} - y = \dfrac{y}{\ln(y) - \ln(x)}; \quad y(1) = e$ **29.** $\dfrac{dy}{dx} = \dfrac{x^2 - y^2}{x^2}; \quad y(2) = 8$

30. $(x + 2y)\dfrac{dy}{dx} = y - 3x; \quad y(1) = 6$

1.4 Exact Differential Equations _____

Consider a first order differential equation of the form

$$\frac{dy}{dx} = f(x, y). \tag{1.28}$$

In this section, we will develop a method of solution which is sometimes effective even when previous methods do not apply. The rationale depends on writing the differential equation in a certain way.

We can always write $dy/dx = f(x, y)$ in the form

$$\frac{dy}{dx} = -\frac{M(x, y)}{N(x, y)} \tag{1.29}$$

by choosing $M(x, y) = -f(x, y)$ and $N(x, y) = 1$. In terms of differentials, we have

$$M(x, y) \, dx + N(x, y) \, dy = 0. \tag{1.30}$$

If we can find a function $F(x, y)$ whose total differential dF is equal to $M(x, y)\, dx + N(x, y)\, dy$, equation (1.30) becomes simply

$$dF = 0,$$

implying that

$$F(x, y) = C. \tag{1.31}$$

Equation (1.31) implicitly defines the general solution of the differential equation (1.28) [or (1.29)].

This is an effective strategy *if* we can find a function F such that $dF = M(x, y)\, dx + N(x, y)\, dy$. If such a function exists, we call the differential equation $dy/dx = -M(x, y)/N(x, y)$ *exact*, and we call F a *potential function* for this differential equation. As a convenience, we will also refer to equation (1.30) as exact and to F as a potential function for this equation, even though $M(x, y)\, dx + N(x, y)\, dy = 0$ is not actually a differential equation (it contains differentials but no derivative).

EXAMPLE 1.14

Consider the differential equation

$$\frac{dy}{dx} = \frac{-2xy^3 - 2}{3x^2y^2 + e^y}. \tag{1.32}$$

In differential form, we have

$$(2xy^3 + 2)\, dx + (3x^2y^2 + e^y)\, dy = 0.$$

Now consider the function $F(x, y) = x^2y^3 + 2x + e^y$, and observe that

$$dF = (2xy^3 + 2)\, dx + (3x^2y^2 + e^y)\, dy.$$

(We will see shortly how to find this function F.) Then $F(x, y)$ is a potential function for the differential equation (1.32), and the general solution of (1.32) is defined implicitly by the equation

$$x^2y^3 + 2x + e^y = C. \tag{1.33}$$

We will verify this assertion. Differentiate equation (1.33) implicitly with respect to x, thinking of y as a function of x. We get

$$2xy^3 + 3x^2y^2\frac{dy}{dx} + 2 + e^y\frac{dy}{dx} = 0.$$

Solve for dy/dx to get

$$\frac{dy}{dx} = \frac{-2xy^3 - 2}{3x^2y^2 + e^y},$$

which is exactly the differential equation (1.32). ∎

How do we find a potential function if one exists? Consider again the differential equation (1.29), written in differential form (1.30):

$$M(x, y)\, dx + N(x, y)\, dy = 0.$$

If $F(x, y)$ is a potential function, then

$$dF = \frac{\partial F}{\partial x}\, dx + \frac{\partial F}{\partial y}\, dy = M(x, y)\, dy + N(x, y)\, dy.$$

Therefore,

$$\frac{\partial F}{\partial x} = M(x, y) \quad \text{and} \quad \frac{\partial F}{\partial y} = N(x, y). \tag{1.34}$$

Since we know $M(x, y)$ and $N(x, y)$, we attempt to integrate the equations (1.34) to determine F.

EXAMPLE 1.15

Consider again the differential equation of Example 1.14. In differential form,

$$(2xy^3 + 2)\, dx + (3x^2y^2 + e^y)\, dy = 0.$$

A potential function F must satisfy

$$\frac{\partial F}{\partial x} = M(x, y) = 2xy^3 + 2 \quad \text{and} \quad \frac{\partial F}{\partial y} = N(x, y) = 3x^2y^2 + e^y.$$

Choose one of these equations, say,

$$\frac{\partial F}{\partial x} = 2xy^3 + 2.$$

Integrate with respect to x, thinking of y as constant (because y is held constant in computing the partial derivative $\partial F / \partial x$). We get

$$F(x, y) = \int (2xy^3 + 2)\, dx = x^2y^3 + 2x + k(y).$$

The constant of integration is denoted $k(y)$ because it may involve y, which was held constant in the integration.

Thus far, $F(x, y)$ has the form

$$F(x, y) = x^2y^3 + 2x + k(y).$$

We also require that

$$\frac{\partial F}{\partial y} = 3x^2y^2 + k'(y) = N(x, y) = 3x^2y^2 + e^y.$$

This can hold only if

$$k'(y) = e^y,$$

and we may choose $k(y) = e^y$. A potential function for this differential equation is

$$F(x, y) = x^2y^3 + 2x + e^y,$$

as we stated without explanation in Example 1.14. ∎

EXAMPLE 1.16

Solve

$$\frac{dy}{dx} = \frac{-\cos(xy) + xy\,\sin(xy)}{-x^2\sin(xy) + 2y}.$$

In differential form,

$$[\cos(xy) - xy\,\sin(xy)]\,dx + [-x^2\sin(xy) + 2y]\,dy = 0.$$

A potential function F must satisfy

$$\frac{\partial F}{\partial x} = \cos(xy) - xy\,\sin(xy) \quad \text{and} \quad \frac{\partial F}{\partial y} = -x^2\sin(xy) + 2y.$$

We choose one of these to integrate. If we choose the second, we integrate with respect to y, thinking of x as fixed:

$$F(x, y) = \int [-x^2\sin(xy) + 2y]\,dy$$

$$= x\,\cos(xy) + y^2 + k(x).$$

Here, the "constant" of integration may involve x because x is held constant in the integration with respect to y.

We will have F if we can find $k(x)$. We must have

$$\frac{\partial F}{\partial x} = \frac{\partial}{\partial x}[x\,\cos(xy) + y^2 + k(x)]$$

$$= \cos(xy) - xy\,\sin(xy) + k'(x) = \cos(xy) - xy\,\sin(xy).$$

This equation is true if $k'(x) = 0$, and we may choose $k(x) = 0$.

A potential function for this differential equation is

$$F(x, y) = x\,\cos(xy) + y^2.$$

The general solution of the differential equation is defined implicitly by

$$x\,\cos(xy) + y^2 = C.$$

To continue the example, suppose we have the initial value problem

$$\frac{dy}{dx} = \frac{-\cos(xy) + xy\,\sin(xy)}{-x^2\sin(xy) + 2y}; \qquad y(2) = 6.$$

Put $x = 2$ and $y = 6$ into the general solution we have just found, yielding

$$2\,\cos(12) + 36 = C.$$

The solution of the initial value problem is implicitly defined by the equation

$$x\,\cos(xy) + y^2 = 2\,\cos(12) + 36. \quad \blacksquare$$

In the last two examples, we carried out integrations and let the constant of integration be zero. The reason we could do this is that, if $F(x, y)$ is a potential function for a differential equation, so is $F(x, y) + K$ for any constant K (a proof is requested in

Problem 31 at the end of this section). Since the general solution of the differential equation is defined by $F(x, y) = C$, replacing $F(x, y)$ by $F(x, y) + K$ does not contribute any new information.

EXAMPLE 1.17

Solve

$$\frac{dy}{dx} = \frac{-2xy^3 - 2xy}{3x^2y^2 + e^y},$$

a variation on the differential equation of Examples 1.14 and 1.15. In differential form,

$$(2xy^3 + 2xy)\, dx + (3x^2y^2 + e^y)\, dy = 0.$$

We will attempt to find a potential function. We want $F(x, y)$ so that

$$\frac{\partial F}{\partial x} = 2xy^3 + 2xy \quad \text{and} \quad \frac{\partial F}{\partial y} = 3x^2y^2 + e^y.$$

Integrate the first equation with respect to x:

$$F(x, y) = \int (2xy^3 + 2xy)\, dx = x^2y^3 + x^2y + k(y).$$

Now we must have

$$\frac{\partial F}{\partial y} = 3x^2y^2 + x^2 + k'(y) = N(x, y) = 3x^2y^2 + e^y.$$

This equation requires that

$$k'(y) = e^y - x^2,$$

which is impossible if k is a function of y alone. We conclude that the differential equation of this example is not exact and that there is no potential function. ∎

This example suggests one way of determining whether a differential equation is exact. Attempt to find a potential function, and if a contradiction arises, the equation is not exact. There is, however, a test which is easier to apply.

THEOREM 1.1   Test for Exactness

Suppose that M, N, $\partial M/\partial y$, and $\partial N/\partial x$ are continuous over a rectangle R. Then the differential equation $dy/dx = -M(x, y)/N(x, y)$ is exact for (x, y) in R if and only if

$$\frac{\partial M}{\partial y} = \frac{\partial N}{\partial x}$$

for (x, y) in R.

In practice, we often encounter differential equations in which M and N and their first partial derivatives are continuous over the entire plane, in which case they are continuous over any rectangle in the plane. ∎

A proof of this theorem will be given when we discuss potential theory in the plane in Section 16.3.

EXAMPLE 1.18

In Example 1.17, we showed by direct calculation that the differential equation

$$\frac{dy}{dx} = \frac{-2xy^3 - 2xy}{3x^2y^2 + e^y}$$

is not exact. We will illustrate the use of the theorem to show this.

We have $M(x, y) = 2xy^3 + 2xy$ and $N(x, y) = 3x^2y^2 + e^y$. Compute

$$\frac{\partial M}{\partial y} = 6xy^2 + 2x \quad \text{and} \quad \frac{\partial N}{\partial x} = 6xy^2 + e^y.$$

These partial derivatives are equal at the single point $(\frac{1}{2}, 0)$ but are not equal throughout any rectangle, as required by the theorem. Therefore, the differential equation is not exact over any rectangle in the plane. ∎

You can check that $\partial M/\partial y = \partial N/\partial x$ in those examples in which the differential equation was exact.

In concluding this section, we will reiterate that, if $F(x, y)$ is a potential function for $dy/dx = -M(x, y)/N(x, y)$, the general solution is defined implicitly by the equation $F(x, y) = C$. The solution is *not* $y = F(x, y)$, a commonly seen mistake.

In the next section, we will consider a way of approaching the differential equation $dy/dx = -M(x, y)/N(x, y)$ when this equation is not exact.

PROBLEMS FOR SECTION 1.4

In each of Problems 1 through 13, test the differential equation (which may be written in differential form) for exactness, using Theorem 1.1. If it is exact, find a potential function and an equation which defines the general solution. Check your solution, using implicit differentiation if necessary. If the differential equation is not exact, do not attempt a solution at this time.

1. $(2y^2 + ye^{xy})\, dx + (4xy + xe^{xy} + 2y)\, dy = 0$
2. $[2\cos(x + y) - 2x\sin(x + y)]\, dx - 2x\sin(x + y)\, dy = 0$
3. $y + e^x + x\dfrac{dy}{dx} = 0$
4. $4xy + 2x + (2x^2 + 3y^2)\dfrac{dy}{dx} = 0$
5. $(4xy + 2x^2y)\, dx + (2x^2 + 3y^2)\, dy = 0$
6. $\cos(y)e^{x\cos(y)} - x\sin(y)e^{x\cos(y)}\dfrac{dy}{dx} = 0$
7. $\dfrac{x}{x^2 + y^2} + \dfrac{y}{x^2 + y^2}\dfrac{dy}{dx} = 0$
8. $(3x^2 + 3y)\, dx + (2y + 3x)\, dy = 0$
9. $\sinh(x)\sinh(y) + \cosh(x)\cosh(y)\dfrac{dy}{dx} = 0$
10. $e^x\sin(y^2) + xe^x\sin(y^2) + [2xye^x\cos(y^2) + e^y]\dfrac{dy}{dx} = 0$
11. $\dfrac{1}{x} + y + (3y^2 + x)\dfrac{dy}{dx} = 0$
12. $6x - ye^{xy} + (e^y - xe^{xy})\dfrac{dy}{dx} = 0$
13. $\cos(4y^2)\, dx - 8xy\sin(4y^2)\, dy = 0$

In Problems 14 through 20, continue as with the preceding problems; the only difference is that different letters are used for the variables.

14. $[16\theta^3 r - 3\cos(\theta)]\, d\theta + [4\theta^4 + 3\cos(r)]\, dr = 0$

15. $\dfrac{2s^3 + 2st^2 - t}{s^2 + t^2}\, ds + \dfrac{s}{s^2 + t^2}\, dt = 0$

16. $(2x^3 u - 3x)\, du - (3u + 3x^2 u^2)\, dx = 0$

17. $(2xt - e^t)\, dx + (x^2 - t^2)\, dt = 0$

18 $[2u\sin(2uv) + 2u^2 v\cos(2uv)]\, du + 2u^3\cos(2uv)\, dv = 0$

19. $\dfrac{t}{z^2}\, dt - \dfrac{t^2}{z^3}\, dz = 0$

20. $\left[1 - \dfrac{y}{t^2}\sec^2\left(\dfrac{y}{t}\right)\right] dt + \dfrac{1}{t}\sec^2\left(\dfrac{y}{t}\right) dy = 0$

In each of Problems 21 through 30, show that the differential equation is exact, find the general solution (perhaps implicitly defined), and use the general solution to find a solution of the initial value problem.

21. $3y^4 - 1 + 12xy^3 \dfrac{dy}{dx} = 0; \quad y(1) = 2$

22. $x\cos(2y - x) - \sin(2y - x) - 2x\cos(2y - x)\dfrac{dy}{dx} = 0; \quad y\!\left(\dfrac{\pi}{12}\right) = \dfrac{\pi}{8}$

23. $e^y + (xe^y - 1)\dfrac{dy}{dx} = 0; \quad y(5) = 0$

24. $y + (x - 3y^2)\dfrac{dy}{dx} = 0; \quad y(3) = -2$

25. $2x - y\sin(xy) + [3y^2 - x\sin(xy)]\dfrac{dy}{dx} = 0; \quad y(0) = 2$

26. $1 + (2e^y - 3y^2)\dfrac{dy}{dx} = 0; \quad y(1) = 4$

27. $ye^{xy} - 8x + (2y + xe^{xy})\dfrac{dy}{dx} = 0; \quad y(2) = -6$

28. $\cosh(x - y) + x\sinh(x - y) - x\sinh(x - y)\dfrac{dy}{dx} = 0; \quad y(4) = 4$

29. $2y - y^2\sec^2(xy^2) + [2x - 2xy\sec^2(xy^2)]\dfrac{dy}{dx} = 0; \quad y(1) = 2$

30. $1 + e^{y/x} - \dfrac{y}{x}e^{y/x} + e^{y/x}\dfrac{dy}{dx} = 0; \quad y(1) = -5$

31. Suppose that $F(x, y)$ is a potential function for $M(x, y)\, dx + N(x, y)\, dy = 0$. Show that $F(x, y) + C$ is a potential function for any constant C.

32. Suppose that $M(x, y) + N(x, y)\, dy/dx = 0$ is both homogeneous and exact and that $xM(x, y) + yN(x, y)$ is not constant. Show that the general solution of the differential equation is defined implicitly by the equation

$$xM(x, y) + yN(x, y) = C.$$

1.5 Integrating Factors and the Bernoulli Differential Equation _____

Suppose we want to solve the first order differential equation

$$\frac{dy}{dx} = -\frac{M(x, y)}{N(x, y)}, \tag{1.35}$$

or, in differential form,

$$M(x, y)\, dx + N(x, y)\, dy = 0. \tag{1.36}$$

If equation (1.36) is not exact, we can attempt to find a nonzero function $\mu(x, y)$ such that multiplying equation (1.36) by μ results in an exact equation. That is, we want $\mu\ (\neq 0)$ such that

$$\mu(x, y)M(x, y)\, dx + \mu(x, y)N(x, y)\, dy = 0 \tag{1.37}$$

is exact. Such a function μ is called an *integrating factor*.

What will this achieve? If equation (1.37) is exact, it has a potential function F. The equation $F(x, y) = C$ defines the general solution of $dy/dx = -\mu M(x, y)/\mu N(x, y) = -M(x, y)/N(x, y)$. Therefore, $F(x, y) = C$ also defines the general solution of the original differential equation (1.35)!

In summary, if $M(x, y)\, dx + N(x, y)\, dy = 0$ is not exact, we seek a nonzero function $\mu(x, y)$ such that $(\mu M)\, dx + (\mu N)\, dy = 0$ is exact. If F is a potential function for this exact equation, the equation $F(x, y) = C$ implicitly defines the general solution of the original differential equation.

EXAMPLE 1.19

Consider

$$(y^2 - 6xy)\, dx + (3xy - 6x^2)\, dy = 0. \tag{1.38}$$

Here,

$$M(x, y) = y^2 - 6xy \quad \text{and} \quad N(x, y) = 3xy - 6x^2.$$

Notice that $\partial M/\partial y = 2y - 6x$ and $\partial N/\partial x = 3y - 12x$, and these are not equal throughout any rectangle. Thus, equation (1.38) is not exact on any rectangle.

Multiply equation (1.38) by $\mu(x, y) = y$ to get

$$(y^3 - 6xy^2)\, dx + (3xy^2 - 6x^2y)\, dy = 0. \tag{1.39}$$

Equation (1.39) is exact because

$$\frac{\partial}{\partial y}[y^3 - 6xy^2] = 3y^2 - 12xy = \frac{\partial}{\partial x}[3xy^2 - 6x^2y].$$

Thus, $\mu(x, y) = y$ is an integrating factor for equation (1.38) over any rectangle in which $y \neq 0$. We will discuss later how we found this integrating factor; for now, observe how it helps us solve the original differential equation. We find that a potential function for the exact equation (1.39) is

$$F(x, y) = xy^3 - 3x^2y^2.$$

The general solution of equation (1.39) is defined implicitly by the equation

$$xy^3 - 3x^2y^2 = C. \tag{1.40}$$

This equation also defines the general solution of the original equation (1.38). To demonstrate this, differentiate equation (1.40) implicitly with respect to x:

$$y^3 + 3xy^2\frac{dy}{dx} - 6xy^2 - 6x^2y\frac{dy}{dx} = 0.$$

In differential form,

$$(y^3 - 6xy^2)\,dx + (3xy^2 - 6x^2y)\,dy = 0,$$

or

$$y[(y^2 - 6xy)\,dx + (3xy - 6x^2)\,dy] = 0.$$

If $y \neq 0$, we can divide this equation by y, obtaining the original equation (1.38). ∎

How do we find an integrating factor? If μ is an integrating factor for $M\,dx + N\,dy = 0$, then $(\mu M)\,dx + (\mu N)\,dy = 0$ is exact. Therefore, we must have

$$\frac{\partial}{\partial y}[\mu M] = \frac{\partial}{\partial x}[\mu N].$$

Then

$$M\frac{\partial \mu}{\partial y} + \mu\frac{\partial M}{\partial y} = N\frac{\partial \mu}{\partial x} + \mu\frac{\partial N}{\partial x}. \tag{1.41}$$

This is a partial differential equation for μ, since M and N are known. If M and N are reasonably well behaved, equation (1.41) will have a solution, at least in theory. In practice, solving this equation for μ can be at least as difficult as solving the original differential equation. This means that as a practical matter we can find integrating factors only in some cases. (This is why many differential equations are difficult to solve.)

Here are two special cases to look for.

1. Sometimes we can find an integrating factor which is a function of just x or just y [as in Example 1.19, in which $\mu(y) = y$ was an integrating factor]. In such a case, equation (1.41) is simplified.

   For example, if there is an integrating factor $\mu = \mu(x)$, then $\partial\mu/\partial y = 0$ in equation (1.41), which becomes

   $$\mu\frac{\partial M}{\partial y} = \mu\frac{\partial N}{\partial x} + N\frac{d\mu}{dx}.$$

   This equation can be written

   $$\frac{1}{\mu}\frac{d\mu}{dx} = \frac{1}{N}\left(\frac{\partial M}{\partial y} - \frac{\partial N}{\partial x}\right). \tag{1.42}$$

   If the right side of this equation is independent of y, equation (1.42) is a separable differential equation for an integrating factor $\mu(x)$. Similarly, if there is an integrating factor $\mu = \mu(y)$, independent of x, equation (1.41) becomes

   $$\frac{1}{\mu}\frac{d\mu}{dy} = \frac{1}{M}\left(\frac{\partial N}{\partial x} - \frac{\partial M}{\partial y}\right). \tag{1.43}$$

   If the right side of (1.43) is independent of x, we have a separable differential equation to solve for an integrating factor $\mu(y)$.

2. It may be that we cannot find an integrating factor depending on only one variable.

Then we try simple combinations of familiar functions, for example,

$$\mu(x, y) = x^a y^b.$$

We multiply the differential equation by this function and attempt to find a and b so that the resulting equation is exact. If this fails, we must try other possibilities, such as e^{ax+by} or $x^a e^{by}$ or $e^{ax}y^b$, and so on. We will not, however, always succeed in finding an integrating factor in terms of elementary functions.

EXAMPLE 1.20

Consider the differential equation

$$\frac{dy}{dx} = \frac{6xy - y^2}{3xy - 6x^2}.$$

In differential form,

$$(y^2 - 6xy)\,dx + (3xy - 6x^2)\,dy = 0.$$

Notice that

$$\frac{\partial M}{\partial y} - \frac{\partial N}{\partial x} = (2y - 6x) - (3y - 12x) = -y + 6x,$$

which is not zero throughout any rectangle. Thus, the differential equation is not exact. However, observe that

$$\frac{1}{M}\left(\frac{\partial N}{\partial x} - \frac{\partial M}{\partial y}\right) = \frac{y - 6x}{y^2 - 6xy} = \frac{1}{y}$$

is independent of x, suggesting [from equation (1.43)] that there is an integrating factor depending on only y. Equation (1.43) for such an integrating factor is

$$\frac{1}{\mu}\frac{d\mu}{dy} = \frac{1}{y}.$$

One solution of this equation is $\mu(y) = y$, the integrating factor we used in Example 1.19. ∎

EXAMPLE 1.21

Consider the equation

$$(3x^2 y + 6xy + \tfrac{1}{2}y^2)\,dx + (3x^2 + y)\,dy = 0.$$

We have

$$\frac{\partial M}{\partial y} - \frac{\partial N}{\partial x} = (3x^2 + 6x + y) - (6x) = 3x^2 + y,$$

which is not zero throughout any rectangle. Thus, the differential equation is not exact. However, notice that

$$\frac{1}{N}\left(\frac{\partial M}{\partial y} - \frac{\partial N}{\partial x}\right) = 1,$$

and this is independent of y. Thus, there is an integrating factor $\mu(x)$ which is independent of y, and equation (1.42) for μ is

$$\frac{1}{\mu}\frac{d\mu}{dx} = 1.$$

This has solution $\mu(x) = e^x$. Multiply the original equation by e^x to get

$$(3x^2y + 6xy + \tfrac{1}{2}y^2)e^x\,dx + (3x^2 + y)e^x\,dy = 0.$$

It is routine to check that this equation is exact and that a potential function is

$$F(x, y) = 3x^2ye^x + \tfrac{1}{2}y^2e^x.$$

The general solution of the original differential equation is defined implicitly by

$$3x^2ye^x + \tfrac{1}{2}y^2e^x = C. \quad\blacksquare$$

EXAMPLE 1.22

Solve

$$(2y^2 - 9xy)\,dx + (3xy - 6x^2)\,dy = 0.$$

Here,

$$\frac{\partial M}{\partial y} - \frac{\partial N}{\partial x} = (4y - 9x) - (3y - 12x) = y + 3x,$$

which is not zero throughout any rectangle. First, check to see if equations (1.42) or (1.43) will yield an equation for an integrating factor. We have

$$\frac{1}{N}\left(\frac{\partial M}{\partial y} - \frac{\partial N}{\partial x}\right) = \frac{y + 3x}{3xy - 6x^2},$$

and this is not independent of y. Thus, equation (1.42) cannot be solved for an integrating factor $\mu(x)$. Further,

$$\frac{1}{M}\left(\frac{\partial N}{\partial x} - \frac{\partial M}{\partial y}\right) = \frac{-y - 3x}{2y^2 - 9xy}.$$

This is not independent of x, so equation (1.43) cannot be solved for an integrating factor $\mu(y)$.

We will try a simple function of both x and y:

$$\mu(x, y) = x^a y^b.$$

Substitute μ, M, and N into equation (1.41) to obtain

$$4x^a y^{b+1} - 9x^{a+1}y^b + 2bx^a y^{b+1} - 9bx^{a+1}y^b = 3x^a y^{b+1} - 12x^{a+1}y^b$$
$$+ 3ax^a y^{b+1} - 6ax^{a+1}y^b.$$

Divide this equation by $x^a y^b$ to get

$$4y - 9x + 2by - 9bx = 3y - 12x + 3ay - 6ax.$$

This equation can be written

$$(1 + 2b - 3a)y + (3 - 9b + 6a)x = 0.$$

Since x and y are independent variables, their coefficients must vanish:

$$1 + 2b - 3a = 0$$
$$3 - 9b + 6a = 0.$$

Solve these equations to get $a = b = 1$; hence, $\mu(x, y) = xy$ is an integrating factor if $x \neq 0$ and $y \neq 0$. Multiply the original equation by xy to get

$$(2xy^3 - 9x^2y^2)\, dx + (3x^2y^2 - 6x^3y)\, dy = 0.$$

This equation is exact, with potential function

$$F(x, y) = x^2y^3 - 3x^3y^2.$$

The general solution of the differential equation is defined implicitly by

$$x^2y^3 - 3x^3y^2 = C. \quad \blacksquare$$

THE BERNOULLI DIFFERENTIAL EQUATION

Occasionally, we can identify an important type of differential equation for which an integrating factor can be found. One such type is the *Bernoulli equation*

$$P(x)\frac{dy}{dx} + Q(x)y = R(x)y^\alpha,$$

in which α is a constant (not necessarily an integer). In differential form, the Bernoulli equation is

$$[Q(x)y - R(x)y^\alpha]\, dx + P(x)\, dy = 0.$$

In general, this equation is not exact. We can, however, find an integrating factor as follows. Equation (1.41) for an integrating factor gives us

$$\mu \frac{\partial}{\partial y}[Q(x)y - R(x)y^\alpha] + [Q(x)y - R(x)y^\alpha]\frac{\partial \mu}{\partial y} = \mu \frac{\partial}{\partial x} P(x) + P(x)\frac{\partial \mu}{\partial x}.$$

This can be written

$$\mu[Q(x) - \alpha y^{\alpha-1}R(x)] + [Q(x)y - R(x)y^\alpha]\frac{\partial \mu}{\partial y} = \mu P'(x) + P(x)\frac{\partial \mu}{\partial x}. \qquad (1.44)$$

It is easy to check that we cannot usually find μ as a function of just x or just y and that $x^a y^b$ will not work for every α. We will try something more general but still simple enough to be tractable:

$$\mu(x, y) = f(x)y^b.$$

Substitute $\mu(x, y)$ into equation (1.44) to get

$$f(x)y^b[Q(x) - \alpha y^{\alpha-1}R(x)] + [Q(x)y - R(x)y^\alpha]f(x)by^{b-1} = f(x)y^bP'(x) + P(x)f'(x)y^b.$$

Then

$$f'(x)P(x)y^b + f(x)P'(x)y^b = f(x)Q(x)(b+1)y^b - R(x)f(x)(b+\alpha)y^{b+\alpha-1}.$$

Now observe that by our choosing $b = -\alpha$, the last term on the right becomes zero, giving us

$$f'(x)P(x)y^{-\alpha} + f(x)P'(x)y^{-\alpha} = f(x)Q(x)(1 - \alpha)y^{-\alpha}.$$

Further, each term has the common factor $y^{-\alpha}$. Divide by $f(x)P(x)y^{-\alpha}$ to get

$$\frac{f'(x)}{f(x)} = -\frac{P'(x)}{P(x)} + (1 - \alpha)\frac{Q(x)}{P(x)}.$$

Integrate this equation to get

$$\ln|f(x)| = -\ln|P(x)| + (1 - \alpha)\int \frac{Q(x)}{P(x)}\,dx,$$

or

$$|f(x)| = \frac{1}{|P(x)|}\,e^{(1 - \alpha)\int Q(x)/P(x)\,dx}.$$

We choose $f(x)$ to satisfy this equation; then, $\mu(x, y) = f(x)y^{-\alpha}$ is an integrating factor for the Bernoulli equation.

EXAMPLE 1.23

Solve

$$\frac{dy}{dx} + y = y^4.$$

This is a Bernoulli equation with $P(x) = Q(x) = R(x) = 1$ and $\alpha = 4$. Calculate

$$\mu(x, y) = \frac{1}{y^4}\,e^{-3\int dx} = y^{-4}e^{-3x}.$$

Write the differential equation as

$$(y - y^4)\,dx + dy = 0$$

and multiply by the integrating factor $y^{-4}e^{-3x}$ to get

$$(y^{-3} - 1)e^{-3x}\,dx + y^{-4}e^{-3x}\,dy = 0.$$

This is an exact equation, and we find that a potential function is

$$F(x, y) = -\tfrac{1}{3}e^{-3x}y^{-3} + \tfrac{1}{3}e^{-3x}.$$

The general solution of the differential equation is defined implicitly by

$$-\tfrac{1}{3}e^{-3x}y^{-3} + \tfrac{1}{3}e^{-3x} \doteq C,$$

or, upon letting $K = -3C$ (so K is also an arbitrary constant),

$$y^{-3} - 1 = Ke^{3x}. \quad \blacksquare$$

EXAMPLE 1.24

It is not necessary for α to be an integer in the Bernoulli equation. Consider

$$x^2 \frac{dy}{dx} + xy = -y^{-3/2},$$

or

$$(xy + y^{-3/2})\,dx + x^2\,dy = 0. \tag{1.45}$$

An integrating factor is

$$\mu(x, y) = \frac{1}{y^{-3/2}x^2}\,e^{5/2\int 1/x\,dx} = y^{3/2}x^{1/2}.$$

Multiply equation (1.45) by $y^{3/2}x^{1/2}$ to get

$$(x^{3/2}y^{5/2} + x^{1/2})\,dx + x^{5/2}y^{3/2}\,dy = 0.$$

This exact equation has potential function

$$F(x, y) = \tfrac{2}{5}x^{5/2}y^{5/2} + \tfrac{2}{3}x^{3/2},$$

so the general solution of equation (1.45) is implicitly defined by

$$\tfrac{2}{5}x^{5/2}y^{5/2} + \tfrac{2}{3}x^{3/2} = C. \quad \blacksquare$$

Here is an example in which x and y are interchanged as independent and dependent variables. In fact, the differential form $M(x, y)\,dx + N(x, y)\,dy = 0$ does not contain any intrinsic preference of x or y as the independent variable.

EXAMPLE 1.25

Solve

$$y^2 \frac{dx}{dy} + 2yx = x^4. \tag{1.46}$$

Recognize that this is a Bernoulli equation of the form

$$P(y)\frac{dx}{dy} + Q(y)x = R(y)x^4,$$

in which x is dependent upon y. With $P(y) = y^2$, $Q(y) = 2y$, and $R(y) = 1$, an integrating factor is

$$\mu(x, y) = \frac{1}{x^4y^2}\,e^{-3\int 2/y\,dy} = \frac{1}{x^4y^8}.$$

Multiply equation (1.46) by $1/x^4y^8$ to get

$$\frac{1}{x^4y^6}\,dx + \left(\frac{2}{x^3y^7} - \frac{1}{y^8}\right)dy = 0,$$

an exact equation with potential function

$$F(x, y) = \frac{-1}{3x^3 y^6} + \frac{1}{7y^7}.$$

The general solution of the differential equation is defined implicitly by

$$\frac{-1}{3x^3 y^6} + \frac{1}{7y^7} = C. \quad \blacksquare$$

PROBLEMS FOR SECTION 1.5

In each of Problems 1 through 11, (a) show that the given equation is not exact; (b) find an integrating factor; (c) find an expression for the general solution of the original differential equation; and (d) check the solution by differentiation.

1. $x + y + \dfrac{dy}{dx} = 0$

2. $x \dfrac{dy}{dx} - 3y = 2x^3$

3. $(2y^2 - 9xy)\, dx + (3xy - 6x^2)\, dy = 0$

4. $(6x^2 y + 12xy + y^2)\, dx + 2(3x^2 + y)\, dy = 0$

5. $1 + (3x - e^{-2y})\dfrac{dy}{dx} = 0$

6. $(4xy + 6y^2)\, dx + (2x^2 + 6xy)\, dy = 0$

7. $(2xy^2 + 2xy)\, dx + (x^2 y + x^2)\, dy = 0$

8. $y^2 + y - x \dfrac{dy}{dx} = 0$

9. $(6xy + 2y + 8)\, dx + x\, dy = 0$

10. $(3y^2 + 6y + 10x^2)\, dx + x(y + 1)\, dy = 0$

11. $2x - 2y - x^2 + 2xy + (2x^2 - 4xy - 2x)\dfrac{dy}{dx} = 0$ *Hint:* Try $\mu(x, y) = e^{ax}e^{by}$.

In each of Problems 12 through 21, determine an integrating factor for the differential equation, find an expression for the general solution of the differential equation, and then find an expression for a solution of the initial value problem.

12. $1 + x \dfrac{dy}{dx} = 0;\quad y(e^4) = 3$

13. $3y + 4x \dfrac{dy}{dx} = 0;\quad y(1) = 6$

14. $2(y^3 - 2) + 3xy^2 \dfrac{dy}{dx} = 0;\quad y(3) = 1$

15. $y(1 + x) + 2x \dfrac{dy}{dx} = 0;\quad y(4) = 6$
  Hint: Try $\mu = y^a e^{bx}$.

16. $2xy + 3 \dfrac{dy}{dx} = 0;\quad y(0) = 4$ *Hint:* Try $\mu = y^a e^{bx^2}$

17. $2y(1 + x^2) + x \dfrac{dy}{dx} = 0;\quad y(2) = 3$ *Hint:*
  Try $\mu = x^a e^{bx^2}$.

18. $\sin(x - y) + \cos(x - y) - \cos(x - y)\dfrac{dy}{dx} = 0;\quad y(0) = \dfrac{7\pi}{6}$

19. $4xy + 6y^2 + (x^2 + 6xy)\dfrac{dy}{dx} = 0;\quad y(-1) = 2$

20. $3x^2 y + y^3 + 2xy^2 \dfrac{dy}{dx} = 0;\quad y(2) = 1$

21. $3x^2 y + y^3 + 2xy^2 \dfrac{dy}{dx} = 0;\quad y(2) = 0$

22. Find conditions on $M(x, y)$ and $N(x, y)$ sufficient to ensure that $M + N(dy/dx) = 0$ has an integrating factor of the form $\mu(x, y) = x^a y^b$ for some numbers a and b.

23. Consider the differential equation $y - x(dy/dx) = 0$.
  (a) Show that this differential equation has an integrating factor which is a function of x alone.
  (b) Show that this differential equation also has an integrating factor which is a function of y alone.
  (c) Show that this differential equation also has integrating factors of the form $\mu(x, y) = x^a y^b$, with a and b both nonzero, and find all such integrating factors.

24. Suppose that μ is an integrating factor for $M + N\,(dy/dx) = 0$. Show that $c\mu$ is an integrating factor for any nonzero constant c.

25. Suppose that μ is an integrating factor for $M + N\,(dy/dx) = 0$ and that the general solution of this differential equation is defined implicitly by an equation $\varphi(x, y) = C$. Show that $\mu(x, y)g(\varphi(x, y))$ is also an integrating factor for the differential equation for any differentiable function g of one variable.

26. Suppose the differential equation $M + N\,(dy/dx) = 0$ has a solution. Show that this differential equation must have an integrating factor.

27. Suppose that μ and v are integrating factors for $M + N\,(dy/dx) = 0$ and that μ is not a constant multiple of v. Show that the equation

$$\frac{\mu(\dot{x}, y)}{v(x, y)} = C$$

implicitly defines the general solution of the differential equation. Explain why this statement fails to be true in general if μ is a constant multiple of v.

28. For the differential equation $y - x\,(dy/dx) = 0$, find two integrating factors μ and v which are not constant multiples of each other. Show that the equation

$$\frac{\mu(x, y)}{v(x, y)} = C$$

implicitly defines the general solution of the differential equation.

In each of Problems 29 through 38, solve the Bernoulli equation.

29. $x^3 \dfrac{dy}{dx} + x^2 y = 2y^{-4/3}$

30. $\dfrac{dy}{dx} + xy = xy^2$

31. $2x^2 \dfrac{dy}{dx} - xy = y^{3/2}$

32. $x^3 \dfrac{dy}{dx} + 2x^2 y = y^{-3}$

33. $x^2 \dfrac{dy}{dx} - xy = y^3$

34. $x^2 \dfrac{dy}{dx} - xy = e^x y^3$

35. $x^4 \dfrac{dy}{dx} + x^3 y = y^{-3/4}$

36. $x^2 \dfrac{dy}{dx} - 3xy = -2y^{5/3}$

37. $2x^3 \dfrac{dy}{dx} + 4x^2 y = -5y^4$

38. $x \dfrac{dy}{dx} + 2y = 3y^2$

1.6 Linear Equations _____

A first order differential equation of the form

$$y' + p(x)y = q(x) \tag{1.47}$$

is called a *linear* differential equation, or just a linear equation.

A linear equation is a Bernoulli equation with $\alpha = 0$ and $P(x) = 1$. Linear equations, however, are sufficiently important to be considered in their own right. We can derive an explicit formula for the solution of any linear equation as follows.

Multiply equation (1.47) by $e^{\int p(x)\,dx}$. (This is exactly the integrating factor we get from previous work on Bernoulli equations.) We obtain

$$y'e^{\int p(x)\,dx} + p(x)ye^{\int p(x)\,dx} = q(x)e^{\int p(x)\,dx}.$$

Now notice that the left side of this equation is the derivative of the product $ye^{\int p(x)\,dx}$.

Therefore,

$$\frac{d}{dx}\left[ye^{\int p(x)\,dx}\right] = qe^{\int p(x)\,dx}.$$

Integrate to get

$$ye^{\int p(x)\,dx} = \int qe^{\int p(x)\,dx}\,dx + C.$$

Solve this equation for y to get

$$y(x) = e^{-\int p(x)\,dx}\int qe^{\int p(x)\,dx}\,dx + Ce^{-\int p(x)\,dx}.$$

This is the general solution of the linear equation (1.47). You can memorize this formula, but it is probably easier to remember to multiply the differential equation by $e^{\int p(x)\,dx}$, write one side of the resulting equation as the derivative of a product, and integrate.

EXAMPLE 1.26

Solve

$$y' + y = \sin(x).$$

This is a linear equation with $p(x) = 1$ and $q(x) = \sin(x)$. Compute

$$\int p(x)\,dx = \int dx = x$$

and multiply the differential equation by e^x to get

$$y'e^x + ye^x = e^x\sin(x).$$

Then

$$[ye^x]' = e^x\sin(x).$$

Integrate to obtain

$$ye^x = \int e^x\sin(x)\,dx = \tfrac{1}{2}[\sin(x) - \cos(x)]e^x + C.$$

Finally, solve for y to write the general solution

$$y(x) = \tfrac{1}{2}[\sin(x) - \cos(x)] + Ce^{-x}. \quad\blacksquare$$

EXAMPLE 1.27

Solve

$$y' + \frac{1}{x}y = 3x^2 \qquad (x \neq 0).$$

Calculate $e^{\int 1/x\,dx} = e^{\ln(x)} = x$. Multiply the differential equation by x to get

$$xy' + y = 3x^3,$$

or

$$[xy]' = 3x^3.$$

Integrate to get

$$xy = \tfrac{3}{4}x^4 + C$$

and solve for y to obtain the general solution

$$y = \frac{3}{4}x^3 + \frac{C}{x}. \quad \blacksquare$$

In solving a linear equation, we may encounter an integral we cannot evaluate in terms of elementary functions.

EXAMPLE 1.28

Solve

$$y' + xy = \cos(x).$$

Multiply this equation by $e^{\int x\,dx} = e^{x^2/2}$ to get

$$y'e^{x^2/2} + xye^{x^2/2} = e^{x^2/2}\cos(x),$$

or

$$[e^{x^2/2}y]' = e^{x^2/2}\cos(x).$$

Then

$$e^{x^2/2}y = \int e^{x^2/2}\cos(x) + C.$$

The general solution is

$$y(x) = e^{-x^2/2}\int e^{x^2/2}\cos(x)\,dx + Ce^{-x^2/2},$$

although we cannot carry out the integration as a finite sum of constants times elementary functions. $\blacksquare$

PROBLEMS FOR SECTION 1.6

In each of Problems 1 through 12, find the general solution of the differential equation.

1. $y' = \dfrac{1}{x}y = x^2 + 2$ **2.** $y' - \dfrac{3}{x}y = 2x^2$ **3.** $2y' + 3y = e^{2x}$

4. $y' - y = \sinh(x)$ **5.** $y' + 2y = x$ **6.** $\sin(2x)y' + 2\sin^2(x)y = 2\sin(x)$

7. $y' + \dfrac{1}{x}y = 3e^{-x^2}$ **8.** $y' - 2y = -8x^2$ **9.** $(x^2 - x - 2)y' + 3xy = x^2 - 4x + 4$

10. $y' - 6y = x - \cosh(x)$ **11.** $y' - xy = 3x$ **12.** $xy' + \dfrac{2}{x}y = 4$

In each of Problems 13 through 20, find a solution of the initial value problem.

13. $y' - y = 2e^{4x}; \quad y(0) = -3$ **14.** $y' + 3y = 5e^{2x} - 6; \quad y(0) = 2$

15. $y' + \dfrac{5y}{9x} = 3x^3 + x;$ $y(-1) = 4$

16. $y' + \dfrac{3y}{2x} = x^3 - 1;$ $y(1) = 2$

17. $y' + \dfrac{4}{x}y = 2;$ $y(1) = -4$

18. $y' + \dfrac{2}{x+1}y = 3;$ $y(0) = 5$

19. $y' + \dfrac{1}{x-2}y = 3x;$ $y(3) = 4$

20. $y' - 4y = x + \sin(x);$ $y(0) = 3$

21. Find the general solution of

$$(1 - 2xe^{2y})\frac{dy}{dx} - e^{2y} = 0.$$

Hint: Reverse roles of x and y as independent and dependent variables, using the fact that $dy/dx = 1/(dx/dy)$.

22. Use the idea of Problem 21 to find the general solution of

$$(4y^3 - x)\frac{dy}{dx} = y.$$

23. Show that the Bernoulli equation

$$\frac{dy}{dx} + P(x)y = Q(x)y^\alpha$$

is transformed into a linear differential equation in terms of u and x by the change of variables $u = y^{1-\alpha}$. Assume that $\alpha \neq 0$ and $\alpha \neq 1$. (If $\alpha = 0$, this Bernoulli equation is already linear; if $\alpha = 1$, it is separable.)

In each of Problems 24 through 29, use the idea of Problem 23 to find the general solution of the Bernoulli equation (in terms of x and y).

24. $\dfrac{dy}{dx} + \dfrac{1}{x}y = 3x^2y^3$

25. $\dfrac{dy}{dx} - 2y = 4xy^2$

26. $\dfrac{dy}{dx} + \dfrac{3}{x}y = -2xy^{5/2}$

27. $\dfrac{dy}{dx} + 8y = 2x^3y^{3/4}$

28. $\dfrac{dy}{dx} + \dfrac{4}{x}y = xy^4$

29. $x^3\dfrac{dy}{dx} + x^2y = 2y^{-4/3}$

30. Find all functions f whose domains are intervals having zero as left endpoint and whose average values on $[0, x]$ are the geometric means of $f(0)$ and $f(x)$. (This problem appeared on the William Lowell Putnam Mathematical Competition for 1962.) Recall that the geometric mean of a and b is $\sqrt{ab}$.

31. Suppose that ψ is a solution of $y' + p(x)y = q(x)$ and that φ is a solution of $y' + p(x)y = 0$. Show that ξ is a solution of $y' + p(x)y = q(x)$ if and only if, for some constant C,

$$\xi = \psi + C\varphi.$$

32. Consider the linear differential equation

$$y' + p(x)y = 0.$$

Assume that p is continuous on an open interval I.

(a) Show that $y = 0$ is a solution of the differential equation. This is called the *trivial solution*.

(b) Suppose that φ is any solution satisfying $\varphi(x_0) = 0$ for some x_0 in I. Show that φ must be the trivial solution.

(c) Suppose that φ and ψ are solutions of the differential equation and that $\varphi(x_0) = \psi(x_0)$ for some x_0 in I. Show that $\varphi(x) = \psi(x)$ for all x in I. (That is, if two solutions agree at some point in I, they are identical on I.)

(d) Suppose that φ is any solution of the differential equation on I. Show that, for any constant C, $C\varphi$ is also a solution.

(e) Suppose that φ is a solution of the differential equation on I but is not the trivial solution. Suppose that ψ is any solution of the differential equation on I. Show that, for some constant C, $\psi(x) = C\varphi(x)$ for all x in I.

33. Suppose that φ and ψ are solutions of $y' + p(x)y = q(x)$ on an open interval I and that x_0 is in I.

(a) Show that

$$\varphi(x) - \psi(x) = [\varphi(x_0) - \psi(x_0)]e^{\int_{x_0}^{x} p(t)\,dt}.$$

(b) If $\varphi(x_0) = \psi(x_0)$ for some x_0 in I, how are φ and ψ related on I?

34. Find all functions with the property that the y-intercept of the tangent line to the graph at (x, y) is $2x^2$.

35. *Moment of inertia of a crankshaft*   Suppose that the crankshaft in a diesel engine revolves at R radians/second. The torque applied to the crankshaft by the gases is called the throttle torque and is of the form kx, where x is the coordinate of the throttle position and k is constant. If I is the moment of inertia of the crankshaft, the inertial torque acting against the throttle torque is $I(dR/dx)$. Also opposing the throttle torque is a damping torque of the form cR. Assume Newton's law in torsion form:

(moment of inertia) · (angular acceleration) = torque.

We get

$$I\frac{dR}{dx} + cR = kx.$$

Assuming that $c \neq 0$ and $I = $ constant, find the general solution of this linear differential equation for R. Then find the particular solution satisfying the initial condition $R(0) = 0$.

36. *Rotation of a propellor*   Consider rotation of a propellor on a diesel-driven ship. If M is the moment of inertia of the rotating shaft of the engine, N is the number of revolutions per minute, and x is the throttle coordinate, then N can be shown to satisfy a differential equation of the form

$$M\frac{dN}{dx} + AN^{\alpha} = f(x),$$

in which A and α are constant. Assuming that M is constant, $A = 10$, $\alpha = 1$, and $f(x) = x^2$, find the general solution for $N(x)$.

37. *Permanent magnet DC motor*   Imagine a permanent magnet direct-current motor with an armature moving between the poles of the magnet and a coil wound around the armature. Suppose that the coil has n turns and that the armature has $2n$ conductors of length s parallel to the axis of the armature and at a distance r from the axis. The total torque on the armature is $2nrBsi$, where B is the constant average flux density of the magnetic field and i is the current. It has been found that if N is the speed of the rotor,

$$M\frac{dN}{dt} + kN = 2nrBsi,$$

where k is a constant of proportionality and M is the constant moment of inertia of the armature. The term kN accounts for friction. Let E be the voltage, L be the inductance, and R be the resistance of the coil. Then i satisfies

$$L\frac{di}{dt} + Ri = E - E_c,$$

where E_c is the counter electromotive force generated in the coil by the conductors moving in the magnetic field. Assuming that $E - E_c$, L, and R are constant, solve the linear differential equation for i. Substitute this solution into the differential equation for N and solve the resulting linear differential equation.

1.7 *The Riccati Equation*

A differential equation of the form

$$y' = P(x)y^2 + Q(x)y + R(x) \tag{1.48}$$

is called a *Riccati equation*. A Riccati equation is linear if $P(x) = 0$ but is nonlinear otherwise. We will now see how to produce the general solution of a Riccati equation *if* we are somehow able (by observation, guessing, trial and error, or some other means) to produce one specific solution $S(x)$.

Given one solution $S(x)$, change variables in equation (1.48) by letting

$$y = S(x) + \frac{1}{z}. \tag{1.49}$$

Then

$$y' = S'(x) - \frac{1}{z^2} z',$$

where $z' = dz/dx$. Substitute these quantities for y and y' into equation (1.48) to get

$$S'(x) - \frac{1}{z^2} z' = P(x)\left(S(x) + \frac{1}{z}\right)^2 + Q(x)\left(S(x) + \frac{1}{z}\right) + R(x).$$

This equation can be written

$$S'(x) - \frac{1}{z^2} z' = [P(x)S(x)^2 + Q(x)S(x) + R(x)]$$

$$+ \left[P(x)\frac{1}{z^2} + 2P(x)S(x)\frac{1}{z} + Q(x)\frac{1}{z}\right]. \tag{1.50}$$

Since $S(x)$ is a solution of equation (1.48), we must have

$$S'(x) = P(s)S(x)^2 + Q(x)S(x) + R(x).$$

Therefore, the first term on the left side of equation (1.50) cancels the terms in the first brackets on the right side, leaving us with

$$-\frac{1}{z^2} z' = P(x)\frac{1}{z^2} + 2P(x)S(x)\frac{1}{z} + Q(x)\frac{1}{z}.$$

Multiply this equation by $-z^2$ to obtain

$$z' + (2PS + Q)z = -P. \tag{1.51}$$

This is a linear differential equation for z, since P, S, and Q are known. We attempt to find the general solution of this equation for z, then substitute this general solution into equation (1.49) to find the general solution for y.

There are two difficulties we can encounter. First, we have to find one solution, $S(x)$, of the original Riccati equation (1.48). Second, we must be able to carry out the integrations needed to find the general solution of equation (1.51).

EXAMPLE 1.29

Solve the Riccati equation

$$y' = \frac{1}{x} y^2 + \frac{1}{x} y - \frac{2}{x}.$$

By observation, we see that $S(x) = 1$ is a solution of this equation. Let

$$y = 1 + \frac{1}{z}.$$

Then $y' = -(1/z^2)z'$, and the differential equation becomes

$$-\frac{1}{z^2} z' = \frac{1}{x}\left(1 + \frac{1}{z}\right)^2 + \frac{1}{x}\left(1 + \frac{1}{z}\right) - \frac{2}{x}.$$

After some algebra, this equation can be written

$$z' + \frac{3}{x} z = -\frac{1}{x},$$

a linear equation for z in terms of x. Calculate $e^{\int 3/x\,dx} = x^3$, and multiply the linear equation for z by x^3 to get

$$x^3 z' + 3x^2 z = -x^2,$$

or

$$[x^3 z]' = -x^2.$$

Then

$$x^3 z = -\tfrac{1}{3}x^3 + C,$$

so

$$z = -\frac{1}{3} + \frac{C}{x^3}.$$

The general solution of the Riccati equation is

$$y = 1 + \frac{1}{z} = 1 + \frac{1}{-\dfrac{1}{3} + \dfrac{C}{x^3}},$$

in which C is an arbitrary constant. This solution may also be written

$$y(x) = \frac{2x^3 + K}{K - x^3},$$

in which $K = 3C$ is still an arbitrary constant. ∎

EXAMPLE 1.30

Solve

$$y' = y^2 - 2xy + x^2 + 1.$$

By inspection, we see that one solution of this Riccati equation is

$$S(x) = x.$$

Let

$$y = x + \frac{1}{z}.$$

Then

$$y' = 1 - \frac{1}{z^2} z'.$$

Substitute these expressions into the differential equation to get

$$1 - \frac{1}{z^2} z' = \left(x + \frac{1}{z}\right)^2 - 2x\left(x + \frac{1}{z}\right) + x^2 + 1.$$

After some manipulation, this equation becomes

$$-\frac{1}{z^2} z' = \frac{1}{z^2},$$

or just

$$z' = -1.$$

Then $z = -x + C$, and the general solution of the Riccati equation is

$$y = x + \frac{1}{C - x} = \frac{Cx - x^2 + 1}{C - x}. \quad \blacksquare$$

In the next two sections, we will discuss some applications of first order differential equations to a variety of problems.

PROBLEMS FOR SECTION 1.7

In each of Problems 1 through 10, find the general solution of the Riccati equation and use this general solution to find a solution of the initial value problem.

1. $y' = \frac{1}{x^2} y^2 - \frac{1}{x} y + 1;$ $y(1) = 3$ *Hint:* One solution is $S(x) = x.$

2. $y' = \frac{1}{2x} y^2 - \frac{1}{x} y - \frac{4}{x};$ $y(2) = -6$ *Hint:* One solution is $S(x) = 4.$

3. $y' = -\frac{1}{x} y^2 + \frac{2}{x} y;$ $y(1) = 4$ **4.** $y' = y^2 + \left(\frac{1}{x} - 2\right) y - \frac{1}{x} + 1;$ $y(1) = 2$

5. $y' = -e^{-x} y^2 + y + e^x;$ $y(0) = 6$ *Hint:* Try $S(x) = -e^x.$

6. $y' = e^{2x} y^2 - 2y - 9e^{-2x};$ $y(0) = 4$ *Hint:* Try $S(x)$ of the form $ae^{bx}.$

7. $y' = \frac{1}{16x^2} y^2 - y + 4x(x + 4);$ $y(1) = -3$ *Hint:* Try $S(x)$ of the form $ax^b.$

8. $y' = y^2 - \frac{6}{x^2} y - \frac{6}{x^3} + \frac{9}{x^4};$ $y(2) = 4$ **9.** $y' = xy^2 + \left(-8x^2 + \frac{1}{x}\right) y + 16x^3;$ $y(2) = 6$

10. $y' = \dfrac{1}{x} e^{-3x} y^2 - \dfrac{1}{x} y + 3e^{3x}; \quad y(3) = -4$

11. Show that the Riccati equation has an integrating factor

$$u(x, y) = \frac{1}{[y - S(x)]^2} e^{\int [2P(x)S(x) + Q(x)] dx}$$

if S is one solution.

12. Let $S_1(x)$ and $S_2(x)$ be particular solutions of the Riccati equation, and suppose that S_1 is not a constant multiple of S_2. Let

$$u_1(x, y) = \frac{1}{[y - S_1(x)]^2} e^{\int [2P(x)S_1(x) + Q(x)] dx}$$

and

$$u_2(x, y) = \frac{1}{[y - S_2(x)]^2} e^{\int [2P(x)S_2(x) + Q(x)] dx}.$$

Show that the general solution of the Riccati equation is defined implicitly by the equation

$$\frac{u_1(x, y)}{u_2(x, y)} = C.$$

Hint: Use the result given in Problem 27 of Section 1.5.

13. Use the result of Problem 12 to show that the general solution of a Riccati equation always has the form

$$y = \frac{F(x) + CG(x)}{H(x) + CJ(x)}$$

for some functions F, G, H, and J and C an arbitrary constant. Review Problems 1 through 5 and show that the general solution has this form in each case.

14. Consider the special Riccati equation

$$y' = ay^2 + bx^\alpha,$$

in which a and b are constants. Show that the general solution can be obtained in closed form when α is of the form $-4n/(2n + 1)$ or $-4n/(2n - 1)$ for some nonnegative integer n. (Closed form means that all integrations can be explicitly performed in terms of finite sums of elementary functions.) It is a fact that the general solution of this Riccati equation cannot be written in closed form when α is not in one of these two forms, but a proof of this is more difficult.

15. Riccati equations are related to second order differential equations, which we will study in the next chapter. Suppose that F is one solution of

$$y'' - \left(Q(x) + \frac{P'(x)}{P(x)} \right) y' + P(x)R(x)y = 0.$$

Show that

$$y = -\frac{F'(x)}{P(x)F(x)}$$

is a solution of the Riccati equation $y' = P(x)y^2 + Q(x)y + R(x)$.

1.8 *First Order Differential Equations in Mechanics and Electrical Circuit Theory*

APPLICATION TO MECHANICS

In order to apply first order differential equations to problems in mechanics, we will review some facts from physics.

Newton's second law of motion states that

> the rate of change of momentum (mass times velocity) of a body is proportional to the resultant external force acting on the body.

Although force and velocity are vectors, for now we will consider motion along a straight line trajectory and will therefore consider force and velocity as real-valued functions of time, acting along the line of motion. If F denotes the force, m the mass, and v the velocity, then Newton's law may be written

$$F = k \frac{d}{dt}[mv],$$

with k some constant of proportionality. If any of the units of Table 1.1 are employed, $k = 1$. We will assume that this is the case.

TABLE 1.1 Units of Measurement

	English System	*mks System*	*gcs System*
Force	Pounds	Newtons	Dynes
Length	Feet	Meters	Centimeters
Mass	Slugs	Kilograms	Grams
Time	Seconds	Seconds	Seconds

The mass of a moving object need not be constant. For example, a rocket or airplane consumes fuel as it moves, decreasing its total mass. If the mass is constant, however, Newton's law is

$$F = m \frac{dv}{dt}.$$

Newton also stated the law of gravitational attraction between objects. If two bodies have masses m_1 and m_2, and r is the distance between their centers of mass, the force of gravitational attraction between them has magnitude

$$F = G \frac{m_1 m_2}{r^2}.$$

G is a constant of proportionality which has the same value for all pairs of objects and is called the *universal gravitational constant*. If one of the bodies is the earth,

$$F = G \frac{mM}{(R + x)^2},$$

where M is the mass of the earth, R is its radius (about 3960 miles), m is the mass of the second body, and x is its distance from the surface of the earth. This equation assumes that the earth is spherical and that its center of mass is at the center of this sphere.

If x is small compared with R, then $R + x \approx R$, and we often use the approximation

$$F = \frac{GM}{R^2} m = mg,$$

where $GM/R^2 = g$, the traditional symbol for the constant magnitude of acceleration due to gravity on the earth ($g \approx 32$ feet/second^2, or 9.8 meters/second^2).

This discussion applies to free fall, in which the only force on the object of mass m is the gravitational attraction between it and the earth. Sometimes we see the last equation written as

$$F = -mg,$$

depending on the sign given the direction of motion.

We are now prepared to look at some examples, beginning with a simple one often seen in calculus.

EXAMPLE 1.31

An object of mass m is thrown straight downward from the top of a building 400 feet high with an initial velocity of 12 feet/second. Assuming that the only force acting on the object is gravity, determine the location of the object at any time t, and also determine when it strikes the ground.

If we denote the downward direction as positive, then

$$F = mg$$

from Newton's law of gravitational attraction.

But we also have, from Newton's second law of motion,

$$F = m\frac{dv}{dt}.$$

Therefore,

$$m\frac{dv}{dt} = mg,$$

or

$$\frac{dv}{dt} = g.$$

Integrate both sides of this equation with respect to t to get

$$v(t) = gt + C.$$

If $x(t)$ is the position of the object at time t on its line of descent as measured from the top of the building, with downward the positive direction, as indicated in Figure 1.4, then $v(t) = dx/dt$, so

$$\frac{dx}{dt} = gt + C.$$

Integrate again with respect to t to get

$$x(t) = \tfrac{1}{2}gt^2 + Ct + K.$$

Now we must determine the integration constants C and K. We are told that the initial velocity of the object is 12 feet/second. In these units, we will use the value $g = 32$ feet/second^2. Further, for convenience, we will start measuring time at the instant when the object is thrown, so that this is time $t = 0$. Then

$$v(0) = g \cdot 0 + C = 12$$

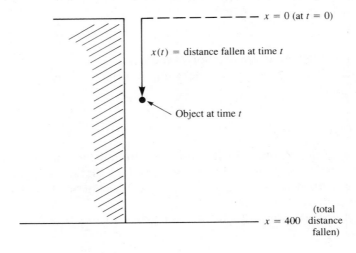

Figure 1.4

and $C = 12$. Then

$$x(t) = 16t^2 + 12t + K.$$

To evaluate K, note that $x(0) = 0$, so $K = 0$ with the coordinate system chosen. Therefore,

$$v(t) = 32t + 12$$
$$x(t) = 16t^2 + 12t.$$

Now, the building is 400 feet high. Thus, the object strikes the ground after it has traveled 400 feet. When does this happen? Solve for t in the equation

$$x(t) = 400 = 16t^2 + 12t.$$

This is the quadratic equation $16t^2 + 12t - 400 = 0$, with solutions

$$t = \frac{-12 \pm \sqrt{144 + 4(16)(400)}}{32},$$

approximately -5.3890 or 4.6390 seconds. Since we agreed that the object was thrown at time zero, we reject the negative root and keep the positive one, which is exactly

$$\frac{-12 + \sqrt{25744}}{32}$$

seconds, or about 4.64 seconds.

The object strikes the ground at the time we have just found. For $0 \le t \le (-12 + \sqrt{25744})/32$, the object at time t is at $x(t) = 16t^2 + 12t$ feet below the release point. After this time, the object is no longer in motion (assuming no recoil from the ground). ∎

This very simple problem illustrates the use of Newton's laws but involves only differential equations we can solve by direct integration. Here is a more sophisticated example in which both the question and the analysis required to answer it are more interesting.

EXAMPLE 1.32   Descent of a Parachutist

Imagine a parachutist falling toward the earth with velocity $v(t)$. We would like an explicit expression for v, assuming that the parachute opens at the beginning of the fall.

The main forces acting on the parachutist are the pull of gravity (acting downward) and the drag caused by the parachute (acting upward). The gravitational pull has magnitude mg (with downward chosen as the positive direction). Experiments tell us that, for most parachutists, the magnitude of the drag force equals kv^2 for some constant k. Neglecting other factors, the total force acting on the parachutist is

$$mg - kv^2.$$

By Newton's second law of motion,

$$m\frac{dv}{dt} = mg - kv^2.$$

We can write this differential equation as

$$\frac{m}{mg - kv^2}\frac{dv}{dt} = 1$$

and recognize it as a separable equation. To help with the integration, let $a^2 = mg/k$ and write the last equation in differential form as

$$\frac{1}{v^2 - a^2}\,dv = -\frac{k}{m}\,dt.$$

Factor $v^2 - a^2 = (v - a)(v + a)$ and use partial fractions to write

$$\frac{1}{v^2 - a^2} = \frac{1}{2a}\left[\frac{1}{v - a} - \frac{1}{v + a}\right].$$

Then

$$\frac{1}{2a}\left[\frac{1}{v - a} - \frac{1}{v + a}\right]dv = -\frac{k}{m}\,dt.$$

Integrate both sides of this equation to get

$$\frac{1}{2a}\left[\ln|v - a| - \ln|v + a|\right] = -\frac{k}{m}t + C,$$

or

$$\ln\left|\frac{v - a}{v + a}\right| = -\frac{2ak}{m}t + 2aC.$$

Take the natural exponential of both sides of this equation to get

$$\left|\frac{v - a}{v + a}\right| = e^{-2akt/m}e^{2aC}.$$

Then

$$\frac{v - a}{v + a} = \pm e^{2aC}e^{-2akt/m}.$$

Let $b = \pm e^{2aC}$ for convenience, and write

$$\frac{v - a}{v + a} = be^{-2akt/m}. \tag{1.52}$$

The sign of b is determined by the sign of $(v - a)/(v + a)$. Finally, solve equation (1.52) for v to obtain

$$v(t) = \frac{a(1 + be^{-2akt/m})}{1 - be^{-2akt/m}}. \tag{1.53}$$

Notice that, as $t \to \infty$, $e^{-2akt/m} \to 0$; hence, $v(t) \to a = \sqrt{mg/k}$. This means that, as time increases, the velocity tends toward a constant value which is independent of the initial velocity. This limiting velocity, or *terminal velocity* does depend on m, however, and so will vary from one person to another.

We will now assign values to the constants in equation (1.53). We know that g is approximately 9.8 meters/second^2. For the earth's atmosphere and a standard parachute, measurements show that k is about 30 kilograms/meter. Suppose that the mass of the person plus parachute is 80 kilograms, just to be specific.

To determine b, we must have more information, for example, the initial velocity of the parachutist. Suppose that the velocity at time zero, when the parachute opens, is 15 meters/second. Then

$$v(0) = 15 = a\,\frac{1+b}{a-b},$$

from which we obtain $b = (15 - a)/(15 + a)$. With the values we have stated for m, g, and k, we obtain $a \approx 5.112$ ($\approx$ means "approximately equal"), so $b \approx 0.492$. Then

$$v(t) \approx (5.112)\left[\frac{1 + 0.492e^{-3.83t}}{1 - 0.492e^{-3.83t}}\right].$$

This is only an approximation because we have used approximate values for the constants in the exact solution (1.53). Using this approximate solution for $v(t)$, we can answer various questions about the parachutist's descent. For example, we can determine the distance fallen at any time t after the drop from the plane by integrating $v(t)$ and using an initial displacement of zero to evaluate the resulting constant. ∎

EXAMPLE 1.33   Block Sliding on a Ramp

A block weighing 96 pounds is released from rest at the top of an inclined plane of slope 30 degrees and slope length 50 feet. Assume a coefficient of friction of $\mu = \sqrt{\sqrt 3}/4$. Assume also that air resistance acts to retard the block's descent down the plane with a force equal to one-half the velocity. We want to determine the velocity of the block at any time t and the time required for the block to reach the bottom of the ramp.

Begin by analyzing the forces acting on the block, referring to Figure 1.5.

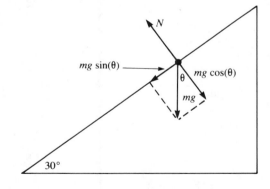

Figure 1.5

1. The force due to gravity acts downward along the plane of the diagram and has magnitude

$$mg\,\sin(\theta),$$

which is $96\sin(30°)$, or 48, pounds.

2. The drag due to friction is

$$-\mu N = -\mu mg\,\cos(\theta) = -\frac{\sqrt 3}{4}(96)\frac{\sqrt 3}{2} = -36$$

pounds. The minus sign indicates that this force resists the motion.

3. The drag due to air resistance has magnitude $v/2$. Since it impedes the motion, we denote this force as $-v/2$.

Since the block weighs 96 pounds, it has a mass of $\frac{96}{32}$, or 3, slugs. By Newton's second law of motion,

$$m \frac{dv}{dt} = 3 \frac{dv}{dt} = F = mg \sin(\theta) - \mu mg \cos(\theta) - \frac{1}{2} v$$

$$= 48 - 36 - \frac{1}{2} v = 12 - \frac{1}{2} v.$$

The differential equation

$$3 \frac{dv}{dt} = 12 - \frac{1}{2} v$$

is separable, and we could use this fact to solve it. But it is probably easier to write the differential equation in linear form:

$$\frac{dv}{dt} + \frac{1}{6} v = 4.$$

Multiply this equation by $e^{\int 1/6 \, dt} = e^{t/6}$ to get

$$e^{t/6} \frac{dv}{dt} + \frac{1}{6} e^{t/6} v = 4 e^{t/6},$$

or

$$\frac{d}{dt} [e^{t/6} v] = 4 e^{t/6}.$$

Integrate to get

$$e^{t/6} v(t) = 24 e^{t/6} + C.$$

Then

$$v(t) = 24 + C e^{-t/6}.$$

Since the block is assumed to begin sliding from rest, the initial velocity is zero, so we have the initial condition

$$v(0) = 0.$$

Then

$$v(0) = 24 + C = 0,$$

so $C = -24$, and the solution for the velocity function is

$$v(t) = 24(1 - e^{-t/6}).$$

We can now determine the displacement $x(t)$ by integrating the velocity function:

$$x(t) = \int v(t) \, dt = \int [24 - 24 e^{-t/6}] \, dt = 24t + 144 e^{-t/6} + k.$$

If we measure the displacement from the top of the ramp, where the block began its descent, then $x(0) = 0$ and

$$0 = 144 + k,$$

so $k = -144$ and the solution for the displacement function is

$$x(t) = 24t + 144(e^{-t/6} - 1).$$

To calculate the time of descent, let $x = 50$ (the length of the ramp), and solve for t in the equation

$$x(t) = 50 = 24t + 144(e^{-t/6} - 1).$$

We cannot solve this equation algebraically for t and must resort to a numerical approximation. Using Newton's method for approximating the zeros of a function, we find that the time required for the block to slide the 50 feet to the bottom of the ramp is about 5.8 seconds.

This analysis can be used to determine how fast the block is moving when it reaches the bottom of the ramp, how far the block has moved after t seconds, how variations in the ramp angle θ and the coefficient of friction μ affect the velocity, and various other facts about the motion of the block.   ■

APPLICATION TO ELECTRICAL CIRCUITS

Before looking at electrical circuits, we will review some facts about resistors, capacitors, and inductors.

A *resistor* is a device, often made of carbon, which resists the flow of electrons in a way analogous to the way a narrow pipe resists the flow of water. The voltage drop across a resistor is proportional to the current flow, and the constant of proportionality is called the *resistance*, which is measured in ohms.

A *capacitor* is a storage device consisting of two plates of conducting material isolated from each other by an insulating material called the *dielectric*. Electrons cannot move between the plates through the dielectric (unless the capacitor leaks). Nevertheless, they can be transferred from one plate to the other via external circuitry by applying an electromotive force to the circuit. In this way, electrons can be accumulated on one of the plates. Which plate accumulates electrons, and the amount of accumulation, are determined by the sign and magnitude of the applied potential.

Each electron carries with it a charge of 1.6×10^{-10} coulomb. The charge on a capacitor (in coulombs) is essentially a count of how many more electrons there are on one plate than on the other, multiplied by the value of the charge of one electron. For instance, if a capacitor has a charge of 10^{-5} coulomb, there are approximately 6.2×10^{13} more electrons on one plate than on the other.

The charge on a capacitor is proportional to the applied electromotive force, with the constant of proportionality called the *capacitance* of the capacitor. The capacitance depends on the area of the plates as well as the distance between them and the composition of the dielectric separating them. Capacitance is typically an extremely small number, often stipulated as so many microfarads (10^{-6} farad) or picofarads (10^{-12} farad). To simplify calculations, we will sometimes diverge from reality and use much larger values of capacitance in our examples.

Since the difference in the number of electrons between capacitor plates cannot be changed instantaneously, the charge on a capacitor is considered as a continuous function, a fact we will use in determining initial currents in circuits containing capacitors.

An *inductor* provides the inertia in a circuit. It is constructed by winding a conductor such as wire around a core of magnetic material such as iron. When a current is conducted in the wire, a magnetic field is created in the core and around the inductor. The strength of the magnetic field is a function of the amount of current, the number of turns of wire in the inductor, and the properties of the magnetic core, primarily its length, cross-sectional area, and permeability.

The strength of the magnetic field will remain the same as long as the current flow remains constant. If the current increases, the field strength increases, and if the current is reduced, the field collapses by a proportional amount. The voltage drop across an inductor is proportional to the change in current flow, and the constant of proportionality is called the *inductance* of the inductor. Inductance is measured in henrys.

Since the strength of the magnetic field about an inductor cannot be changed instantly, the current through an inductor is continuous. If the current is not changing, the only voltage drop across the inductor is due to the resistance of the wire in the coil. If in a particular inductor this resistance is too large to ignore, it can be accounted for in the circuit as a resistor in series with the inductor.

Current is measured in amperes, with one ampere equivalent to a rate of electron flow of one coulomb per second. Charge q and current i are related by

$$i = \frac{dq}{dt}$$

Kirchhoff's current law states that the algebraic sum of the currents at any junction of a circuit is zero. This means that the total current entering the junction equals the total current leaving the junction.

Kirchhoff's voltage law states that the algebraic sum of the potential rises and drops around any closed loop in the circuit is zero.

We will apply these laws to the circuit of Figure 1.6. Starting at a point A, we travel clockwise around the circuit, first crossing the battery, where we see an increase in

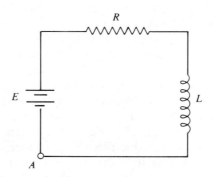

Figure 1.6

potential of E volts. Next, there will be a decrease in potential of iR volts as we cross the resistor. Then there is a decrease in potential of $L(di/dt)$ volts in the inductor. This brings us back to point A and completes the loop. Kirchhoff's voltage law gives us the equation

$$E - iR - L\frac{di}{dt} = 0,$$

or

$$\frac{di}{dt} + \frac{R}{L}i = \frac{1}{L}E.$$

This is a linear first order differential equation for the current $i(t)$.

An easier way to apply Kirchhoff's laws to derive a mathematical model of an electrical circuit is to designate one of the components in a loop as a source, then set the voltage drop across that component equal to the sum of the voltage drops across the remaining components. This technique becomes especially efficient when dealing with circuits involving several loops.

Returning to Figure 1.6, we designate one of the components as the source. The natural one to choose is the battery. Now we set the voltage drop of the source equal to the sum of the voltage drops in the remainder of the loop. We get

$$E = iR + L\frac{di}{dt},$$

which is equivalent to equation (1.54). But, with this derivation, we did not have to consider whether the potential was rising or falling as the current traversed the circuit.

To appreciate this method, we will construct a mathematical model of the circuit of Figure 1.7. Label the current in the first loop i_1 and that flowing in the second loop i_2. Now use Kirchhoff's current law. The current entering point A through the resistor R_1 is i_1, and the current leaving A toward the resistor R_2 is i_2. Then the current entering this junction from the inductor must have a value of $-(i_1 - i_2)$. Equivalently, the current going downward (in the diagram) through the inductor must be $i_1 - i_2$. In the first loop, we have

$$E(t) = R_1 i_1 + L\frac{d}{dt}[i_1 - i_2].$$

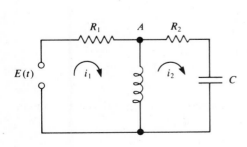

Figure 1.7

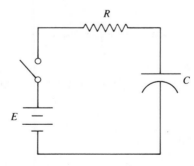

Figure 1.8

In the second loop, we designate the inductor as the loop source. Then

$$L \frac{d}{dt}[i_1 - i_2] = R_2 i_2 + \frac{1}{C} q_2.$$

Our mathematical model for this circuit is a system of two differential equations, which we are not prepared to solve at this time. The point to the derivation is to illustrate the process of modeling a circuit. Normally we would also accompany these differential equations with initial conditions giving the status of the circuit at some initial time.

Here is an example involving a circuit containing a capacitor (see Figure 1.8). Suppose that the switch is initially open so that no current flows, and further assume that the charge on the capacitor is zero. Close the switch at time zero. We want to determine the resultant charge on the capacitor at any time $t > 0$.

Notice that we have to close the switch before there is a loop. Using the battery for a source, we have

$$iR + \frac{q}{C} = E. \tag{1.55}$$

Since $i = dq/dt$, this equation becomes

$$R \frac{dq}{dt} + \frac{1}{C} q = E,$$

or, in linear form,

$$\frac{dq}{dt} + \frac{1}{RC} q = \frac{E}{R}. \tag{1.56}$$

Since the charge on the capacitor was zero prior to closing the switch, and the charge on a capacitor is a continuous function of time, $q(0) = 0$. This is the initial condition.

An integrating factor for the differential equation (1.56) is

$$e^{\int 1/RC \, dt},$$

or $e^{t/RC}$. Multiply equation (1.56) by $e^{t/RC}$ to obtain

$$\frac{d}{dt}[q e^{t/RC}] = \frac{1}{R} E e^{t/RC}.$$

Integrate this equation to get

$$q e^{t/RC} = ECe^{t/RC} + K.$$

The general solution of equation (1.56) is

$$q(t) = EC + K e^{-t/RC}.$$

Since $q(0) = 0 = EC + K$, $K = -EC$, and the solution of the initial value problem for q is

$$q(t) = EC(1 - e^{-t/RC}).$$

From this solution, we can determine a good deal of information about the circuit. For example, at any time $t > 0$, the voltage on the capacitor is $q(t)/C$, or $E(1 - e^{-t/RC})$, volts. Notice that this quantity approaches E as $t \to \infty$, and E is the battery potential. We conclude that the difference between battery and capacitor voltages becomes negligible as time increases, indicating a very small voltage drop across the resistor.

Now suppose that we want the amount of current flowing in the circuit at time t. One way of obtaining $i(t)$ if the charge has already been determined is to use equation (1.55), substitute the value of q just obtained, and solve for i. This works well enough if the circuit is simple, as this one is. Another method is to differentiate the charge, which we have obtained, and find i directly:

$$i(t) = \frac{dq}{dt} = \frac{E}{R} e^{-t/RC}.$$

Here is another approach which is effective whenever the source potential is a differentiable function. Differentiate equation (1.56) with respect to t to obtain

$$\frac{d^2q}{dt^2} + \frac{1}{RC} \frac{dq}{dt} = 0.$$

Since $i = dq/dt$, we can write this differential equation in terms of i as

$$\frac{di}{dt} + \frac{1}{RC} i = 0,$$

assuming that E is constant. This differential equation is linear (also separable), and we easily find the general solution

$$i(t) = ke^{-t/RC}. \tag{1.57}$$

To determine k, we need an initial condition on i. Recall that the current was initially zero. However, if we let $i(0) = 0$, we find that $k = 0$ in equation (1.57) and obtain $i(t) = 0$ for $t > 0$, contrary to the result obtained earlier by differentiating q (and also contrary to reality). The difficulty is that the current was zero only up to the time the switch was closed. Equation (1.56) is valid only *after* the switch is closed, completing the circuit.

We can use equation (1.57), and the fact that the charge on the capacitor, which was zero prior to closure, is continuous, to determine the current in the circuit at time $t = 0+$ (that is, immediately after the circuit is completed). Put $t = 0+$ into equation (1.55) to get

$$Ri(0+) + \frac{1}{C} q(0+) = E.$$

Here, $i(0+) = \lim_{t \to 0+} i(t)$ and $q(0+) = \lim_{t \to 0+} q(t)$. Since $q(0+) = q(0-) = 0$, by continuity of q, we have $Ri(0+) = E$; hence,

$$i(0+) = \frac{E}{R}.$$

Use this value to determine k in equation (1.57). We get

$$i(0+) = k = \frac{E}{R}.$$

This gives us the solution

$$i(t) = \frac{E}{R} e^{-t/RC}$$

for $t > 0$, consistent with the result derived previously using equation (1.55) and the solution for q.

Actually, for any time t, we have

$$i(t) = \begin{cases} 0 & \text{if} \quad t < 0 \\ \dfrac{E}{R} e^{-t/RC} & \text{if} \quad t > 0. \end{cases}$$

Notice that, in this expression, the current approaches E/R as t approaches zero from the right, consistent with the fact that $i(0+) = E/R$. Further, as $t \to \infty$, $i(t) \to 0$. Since the voltage drop across the resistor is $i(t)R$, it too will diminish to zero as t increases, in agreement with the conclusions we drew earlier when discussing the capacitor charge.

Often, in dealing with electrical circuits we encounter discontinuous currents, as in this case, and nondifferentiable or even discontinuous potential sources. In Chapter 4 we will develop Laplace transform techniques which are particularly effective in treating such problems. Until then, we must analyze circuits using methods such as those we have just illustrated. Table 1.2 summarizes some symbols and terminology commonly used in dealing with electrical circuits.

TABLE 1.2 Electrical Symbols and Terminology

Device	Symbol	Units	Voltage Drop
Resistor R	—⋀⋀⋀—	Ohms (Ω)	$E_R = iR$
Capacitor C	—⊣⊢—	Farads (f)	$E_C = \dfrac{q}{C}$
Inductor L	—⟲⟲⟲⟲⟲—	Henrys (hy)	$E_L = L\dfrac{di}{dt}$

PROBLEMS FOR SECTION 1.8

1. A 10-pound ballast bag is dropped from a hot air balloon which is at an altitude of 342 feet and is ascending at a rate of 4 feet/second. Assuming that there is no air resistance, determine the maximum height attained by the bag, the length of time it remains aloft, and the speed with which it strikes the ground.

2. A 6-ounce softball is thrown upward from a height of 7 feet with an initial velocity of 84 feet/second. If the ball is subjected to an air resistance equal (in pounds) to $\frac{3}{128}$ the speed of the ball (in feet/second), how high will it rise before falling back to earth?

3. A ship weighing 86,400 tons starts from rest under a constant propellor thrust of 66,000 pounds. If the resistance due to the water is $4500v$ pounds, where v is the velocity in feet/second, find the velocity of the ship as a function of time, its terminal velocity in miles per hour, and the distance it will have traveled by the time it has achieved 85% of its terminal velocity.

4. A girl and her sled weigh a total of 96 pounds. They begin at rest at the top of a long slope which makes an angle of 30 degrees with the horizontal. The coefficient of friction between the sled runners and the snow is $\sqrt{3}/24$, and there is a drag due to air resistance equal to the velocity. Nine seconds after she starts at the top of the slope, she glides out onto a level surface. How far from the base of the hill will she travel before her sled comes to rest?

5. A 96-pound box is placed on a long inclined plane making an angle of 30 degrees with the horizontal and is then given an initial push of 2 feet/second down the plane. If the coefficient of friction between the box and the plane is $\sqrt{3}/12$ and the drag due to air resistance is twice the velocity, determine the velocity of the box and how far it has traveled for any time $t > 0$.

6. A 48-pound box is given an initial push of 16 feet/second down an inclined plane which has a gradient of $\frac{7}{24}$. If there is a coefficient of friction of $\frac{1}{3}$ between the box and the plane and an air resistance equal to $\frac{3}{2}$ the velocity of the box, determine how far the box will travel before coming to rest.

7. A parachutist and her equipment together weigh 192 pounds. Before her chute is opened, there is an air drag equal to six times her velocity. Four seconds after stepping from the plane, she opens her chute, producing a drag equal to three times the square of the velocity. Determine the velocity of the parachutist and how far she has fallen for any time $t > 0$. What is her terminal velocity?

8. *Archimedes' principle of buoyancy* states that an object submerged in a fluid is buoyed up by a force equal to the weight of the fluid displaced. A rectangular box 1 foot by 2 feet by 3 feet and weighing 384 pounds is dropped into a 100-foot-deep freshwater lake (water has a density of 62.5 pounds/foot^3). The box begins to sink with a drag due to the water equal to one-half its velocity. Calculate the terminal velocity of the box. Will the box have attained a velocity of 10 feet/second by the time it reaches the bottom of the lake?

9. Suppose that the box in Problem 8 cracks open upon hitting the bottom of the lake and 32 pounds of the contents fall out. Approximate the velocity with which the box surfaces.

10. A horizontal spring of negligible mass has one of its ends attached to a wall and a weight of mass m attached to its free end. Assume that the spring exerts a restoring force of kx pounds when stretched an amount x and that there is no friction between the weight and the surface upon which it glides. Find the relationship between the resulting velocity and position of the mass at any time $t > 0$ if the spring is stretched x_0 feet and then released from rest. *Hint:* Since t does not explicitly appear in the equation of motion, write $dv/dt = (dv/dx)(dx/dt) = v\,(dv/dx)$ in terms of Newton's law.

11. How much would an object weighing 100 pounds at sea level weigh at an altitude of 990 feet? How much would it weigh at 3690 feet?

12. Determine the escape velocity from the earth. *Hint:* Assume that a rocket is fired vertically upward from the earth's surface with an initial velocity v_0. If we neglect the drag due to air resistance, we have the equation of motion

$$m\frac{dv}{dt} = -\frac{GMm}{(R + x)^2},$$

in which $x(t)$ is the altitude of the rocket at time t (measured from the surface of the earth) and R is the radius of the earth. Replace dv/dt with v (dv/dx) (as in Problem 10), and solve the resulting separable differential equation. Now choose v_0 just large enough to ensure that v remains positive for all time.

13. Determine the escape velocity from the earth's moon. The moon's mass is approximately 0.0123 that of the earth's, and it has a radius of about 0.2725 that of the earth.

14. The acceleration due to gravity inside the earth is proportional to the distance from the center of the earth. An object is dropped from the surface of the earth into a hole extending through the center of the earth. Calculate the speed it will achieve by the time it reaches the center of the earth.

15. A particle starts from rest at the highest point of a vertical circle and slides under only the influence of gravity along a chord to any other point on the circle. Show that the time taken is independent of the choice of the terminal point. What is this common time?

16. An oil tanker of mass M is sailing in a straight line. At time zero it shuts off its engines and coasts. Assume that the water tends to slow the tanker with a force proportional to v^α, in which v is the velocity at time t and α is a constant.
    (a) Derive a differential equation for v as a function of time.
    (b) Show that the tanker moves in a straight line and eventually comes to a full stop if $0 < \alpha < 1$. What happens if $\alpha > 1$?

17. A ball of mass m is thrown upward from the surface of the earth. The initial velocity is v_0, and the forces acting on the ball are air resistance, which is proportional to the square of the velocity, and gravity.
    (a) Write and solve a differential equation for the height of the ball at time t.
    (b) Find the maximum height attained by the ball and the time it takes to reach this height.
    (c) Is it true that the time it takes the ball to reach its maximum height is less than the time it takes for the ball to fall back to earth?

The following problems deal with electrical circuits.

18. Determine each of the currents in the circuit shown in the diagram.

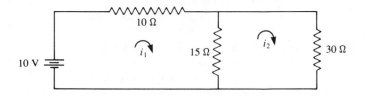

19. In the circuit shown in the diagram, the capacitor is initially discharged. How long after the switch is closed will it take for the capacitor voltage to reach 76 volts? Determine the current in the resistor at that time. (Here, kΩ denotes 1000 ohms and μF denotes 10^{-6} farad.)

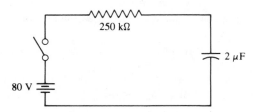

20. Suppose that, in Problem 19, the capacitor had a potential of 50 volts at the time the switch was closed. How long would it take for the capacitor voltage to reach 76 volts?

21. Referring to the circuit shown in the diagram, find all three current values immediately after the switch is closed, assuming that all of these currents, as well as the charges on the capacitors, are zero just prior to the closing of the switch.

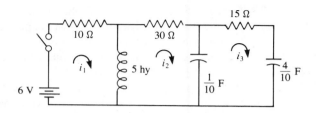

22. In a constant electromotive force RL circuit, we find that the current is given by

$$i(t) = \frac{E}{R}(1 - e^{-Rt/L}) + i(0)e^{-Rt/L}.$$

Assume, for convenience, that $i(0) = 0$.

(a) Show that $i(t)$ is increasing with time.

(b) Find t_0 such that $i(t_0)$ is 63% of E/R, the limiting value of $i(t)$ as $t \to \infty$. This value of t_0 is called the *inductive time constant* of the circuit.

(c) Would the inductive time constant change if $i(0)$ were nonzero? If so, in what way?

23. Recall that the charge $q(t)$ in an RC circuit satisfies the differential equation

$$\frac{dq}{dt} + \frac{1}{RC}q(t) = \frac{1}{R}E(t).$$

(a) Solve this differential equation for q in the case in which E is constant. Evaluate the constant of integration by assuming that $q(0) = q_0$.

(b) Show that $q(t)$ has a limiting value which is independent of q_0 as $t \to \infty$. Find this limiting value.

(c) Graph q as a function of time. For what value of t does $q(t)$ have its maximum value? What is this maximum value?

(d) Determine the time at which q is within 1% of its steady-state value. Does this value depend on q_0? If so, in what way? (The steady-state value is the limit as $t \to \infty$.)

24. Using the differential equation of Problem 23, determine the charge in an RC circuit with electromotive force $E(t) = A \cos(\omega t)$, with A and ω positive constants. Evaluate the constant of integration by assuming that $q(0) = q_0$.

25. Solve for the current $i(t)$ in an RL circuit in which $R = 2$ ohms, $L = 25$ henrys, and the electromotive force is $E(t) = Ae^{-t}$, with A a positive constant. Graph i as a function of time.

26. Solve for the current in an RL circuit in which the electromotive force is $E(t) = A \cos(\omega t) + Be^{-t}$, with A and B positive constants. Treat the general case, in which R is any resistance and L is any inductance. Graph i as a function of time and describe its behavior at t increases.

27. Find the current in an RL circuit having electromotive force $E(t) = A \sin(\omega_1 t) + B \cos(\omega_2 t)$, with A, B, ω_1, and ω_2 positive constants. How is the solution affected by the choices of ω_1 and ω_2? Assume that $i(0) = 0$.

28. Solve for the current in an RC circuit having a resistance of R ohms, a capacitance of C farads, and electromotive force of $E(t) = 1 - \cos(2t)$. Assume that $q(0) = 0$.

29. Find the current $i(t)$ in an RL circuit if $i(0) = 0$ and $E(t) = 1 + e^{-t}$.

POSTSCRIPT

It is interesting to note that current flow is not a flow of electrons, as is commonly thought, but rather a flow of "holes" left as a result of the electrons moving in the opposite direction.

Electrons have negative charge and flow from a negative point, such as the negative side of a battery, toward a point with greater positive potential, such as the positive side of a battery. By contrast, current flows from a positive source toward a stronger negative source. This convention stems from an incorrect guess by Benjamin Franklin as to the direction of current flow in a circuit.

We will comply with tradition and common practice and denote the direction of the flow of current with an arrow pointing from positive to negative.

1.9 *Mixing Problems and Orthogonal Trajectories*

Mixing Problems

Sometimes we want to know how much of a given material or substance is present in a container in which various items are being added, mixed, and removed. Such problems are called *mixing problems*. Here is a typical example.

EXAMPLE 1.34

A tank contains 200 gallons of brine in which are dissolved 100 pounds of salt. A mixture consisting of $\frac{1}{8}$ pound of salt per gallon is flowing into the tank at a rate of 3 gallons/minute, and the brine mixture in the tank is continuously stirred. Meanwhile, brine is allowed to empty out of the tank at the same rate of 3 gallons/minute. This process is illustrated in Figure 1.9. What is the amount of salt in the tank at time t?

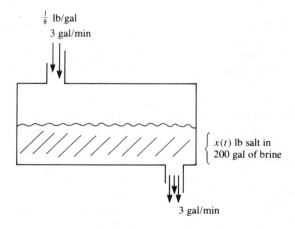

Figure 1.9

We begin with some observations. First, the initial ratio of salt to brine in the tank is 100 pounds per 200 gallons, or $\frac{1}{2}$ pound/gallon. Since the incoming mixture is in a constant ratio of $\frac{1}{8}$ pound/gallon, we expect the brine mixture to dilute toward the incoming ratio, with a "terminal" amount of salt in the tank of $\frac{1}{8}$ pound/gallon times 200 gallons, or 25 pounds. We can use this analysis to check our eventual solution for the amount of salt at time t.

Now let $Q(t)$ be the amount of salt in the tank at time t. Then $Q(0) = 100$ pounds. Also, dQ/dt is the rate of change, with respect to time, of the amount of salt in the tank and therefore equals the difference between the rate at which salt is pumped into the tank and the rate at which salt runs out of the tank. Therefore,

$$
\begin{aligned}
\frac{dQ}{dt} &= \text{rate of change with respect to time} \\
&\quad\; \text{of the amount of salt in the tank} \\
&= [\text{rate in}] - [\text{rate out}] \\
&= \left(\frac{1}{8}\ \text{lb/gal}\right)(3\ \text{gal/min}) - \left(\frac{Q(t)}{200}\ \text{lb/gal}\right)(3\ \text{gal/min}) \\
&= \frac{3}{8} - \frac{3}{200}\ Q(t)\ \text{lb/min}.
\end{aligned}
$$

Omitting the dimensions, we have the differential equation

$$
\frac{dQ}{dt} = \frac{3}{8} - \frac{3}{200}\ Q(t).
$$

This can be written as the linear differential equation

$$
Q' + \frac{3}{200}\ Q = \frac{3}{8}.
$$

We want a particular solution satisfying the initial condition

$$
Q(0) = 100.
$$

An integrating factor for the differential equation is $e^{\int 3/200\, dt}$, or $e^{3t/200}$. Multiply the differential equation by $e^{3t/200}$ to obtain

$$
\frac{d}{dt}\left[e^{3t/200}Q(t)\right] = \frac{3}{8}\ e^{3t/200}.
$$

Integrate both sides of this equation to get

$$
Q(t)e^{3t/200} = \frac{3}{8}\ \frac{200}{3}\ e^{3t/200} + C,
$$

or

$$
Q(t) = 25 + Ce^{-3t/200}.
$$

To find C, we have

$$
Q(0) = 25 + C = 100,
$$

so $C = 75$. Then

$$
Q(t) = 25 + 75e^{-3t/200}.
$$

Notice that $\lim_{t \to \infty} Q(t) = 25$, as we had observed should be the case.

An examination of the derivation of the differential equation for Q, and its solution, reveals that this limiting value is independent of the initial amount of salt in the tank. It does depend on the input rate of salt flow. The term 25 in the solution for $Q(t)$ is called the *steady-state* part of the solution, and the term $75e^{-3t/200}$ is the *transient part* (since it decreases to zero with time). ■

ORTHOGONAL TRAJECTORIES

Sometimes we encounter two families of curves in which each curve of one family is orthogonal (perpendicular) to each curve of the other family. Examples are parallels and meridians on a globe. Each meridian is orthogonal to each parallel wherever they intersect (that is, the tangents are perpendicular there). Similarly, equipotential lines and electric lines of force are orthogonal families.

We will now see how to construct a family of curves which is orthogonal to a given family. A family of curves in the plane consists of the graphs of an equation

$$F(x, y, K) = 0,$$

in which K is a parameter (constant). For each choice of K, we obtain an equation $F(x, y, K) = 0$, and the graph of this equation is a member of the family. We have seen a family of curves associated with an exact differential equation. The integral curves (graphs of solutions) of a first order differential equation form a family of curves in the plane.

As a specific example of a family of curves, consider the function

$$F(x, y, K) = x^2 - y^2 - K^2.$$

The family $F(x, y, K) = 0$ consists of the hyperbolas $x^2 - y^2 = K^2$ if $K \neq 0$ and the straight lines $y = \pm x$ if $K = 0$. Some members of this family are sketched in Figure 1.10.

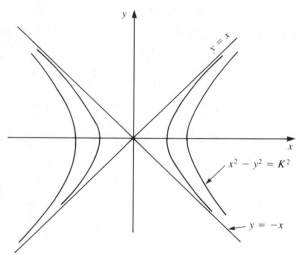

Figure 1.10. Some members of the family $x^2 - y^2 - K = 0$.

In this illustration, we have simply a family of curves, without reference to a differential equation. However, given a family of curves we can often determine a differential equation whose integral curves are exactly the curves of the family. This differential equation is quite naturally called the *differential equation of the family.*

We can find the differential equation of a family $F(x, y, K) = 0$ by differentiating the equation $F(x, y, K) = 0$ implicitly with respect to x and eliminating the parameter K. This may involve solving for K from the original equation of the family, as the following example illustrates.

EXAMPLE 1.35

Consider the family defined by $F(x, y, K) = x^2 + Ky^3 = 0$.

Differentiate this equation implicitly with respect to x to get

$$2x + 3Ky^2y' = 0. \tag{1.58}$$

Now solve for K in terms of x and y from the equation of the family:

$$K = -\frac{x^2}{y^3}.$$

Use this expression for K in equation (1.58) to obtain the differential equation

$$2x + 3\left(-\frac{x^2}{y^3}\right)y^2y' = 0,$$

or

$$2x - \frac{3x^2}{y}\, y' = 0.$$

This is the differential equation of the family $x^2 + Ky^3 = 0$. It is a differential equation whose integral curves are exactly the curves in the family we began with. ∎

With this as background, we now pose the following problem. Given a family $F(x, y, K) = 0$ of curves, find a second family of curves each of which is orthogonal (at every point of intersection) to every curve of the given family. These curves form the *family of orthogonal trajectories* of the family $F(x, y, K) = 0$.

To determine the family of orthogonal trajectories of $F(x, y, K) = 0$, first find the differential equation $y' = f(x, y)$ of the family $F(x, y, K) = 0$. The curves of $F(x, y, K) = 0$ are integral curves of $y' = f(x, y)$. At each point (x_0, y_0), $f(x_0, y_0)$ is the slope of the curve in the family $F(x, y, K) = 0$ passing through (x_0, y_0). A curve through (x_0, y_0) orthogonal to this curve must therefore have slope

$$\frac{-1}{f(x_0, y_0)}$$

wherever $f(x_0, y_0) \neq 0$. (Remember that two lines are orthogonal if their slopes are negative reciprocals.)

We conclude that the family of orthogonal trajectories of $F(x, y, K) = 0$ has differential equation

$$y' = \frac{-1}{f(x, y)}.$$

The integral curves of $y' = -1/f(x, y)$ therefore form the family of orthogonal trajectories of the family $F(x, y, K) = 0$.

EXAMPLE 1.36

Find the family of orthogonal trajectories of the family

$$F(x, y, K) = x^2 + y^2 - K^2 = 0.$$

This is a family of circles of radius $|K|$ about the origin (for $K \neq 0$).

First, we need the differential equation of this family. Differentiate $x^2 + y^2 - K^2 = 0$ implicitly with respect to x:

$$2x + 2yy' = 0.$$

Then

$$y' = -\frac{x}{y}.$$

This is the differential equation of the given family. It happened that here we did not need to use the equation of the original family to eliminate K, as in Example 1.35.

The family of orthogonal trajectories has differential equation

$$y' = \frac{y}{x}. \tag{1.59}$$

This differential equation is separable (and also linear) and has general solution $y = Cx$, a family of straight lines through the origin. Some curves in both families are shown in Figure 1.11. Each straight line is orthogonal to each circle wherever they intersect. ∎

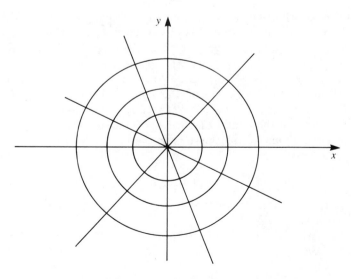

Figure 1.11. Orthogonal families: circles and lines.

EXAMPLE 1.37

Find the orthogonal trajectories of the family

$$F(x, y, K) = y - Kx^2 = 0.$$

First, find the differential equation of the family. Differentiate $y - Kx^2 = 0$ with respect to x to get

$$y' - 2Kx = 0,$$

or

$$y' = 2Kx.$$

From $y = Kx^2$, the equation of the family, we get $K = y/x^2$. Substitute this for K in $y' = 2Kx$ to get

$$y' = 2\left(\frac{y}{x^2}\right)x = 2\frac{y}{x}$$

for $x \neq 0$. This is the differential equation of the given family. The differential equation of the family of orthogonal trajectories is

$$y' = -\frac{1}{2}\frac{x}{y}.$$

Write this separable differential equation in differential form as

$$2y\,dy = -x\,dx$$

and integrate both sides to get

$$y^2 = -\tfrac{1}{2}x^2 + C,$$

or

$$\tfrac{1}{2}x^2 + y^2 = C.$$

This family of ellipses is orthogonal to the given family of parabolas. Some curves from these families are sketched in Figure 1.12. ■

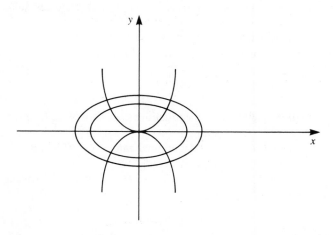

Figure 1.12. Orthogonal families: parabolas and ellipses.

PROBLEMS FOR SECTION 1.9

1. A 500-gallon tank initially contains 100 gallons of brine in which 5 pounds of salt have been dissolved. Brine containing 2 pounds/gallon is being added at the rate of 5 gallons/minute, and the mixture is pouring out of the tank at the rate of 3 gallons/minute. Determine how much salt is in the tank at the moment it overflows.

2. A 400-gallon tank is filled with brine containing 45 pounds of salt. At a certain time, the brine begins draining from an open valve at the bottom at a rate of 5 gallons/minute. Simultaneously, a brine mixture containing $\frac{1}{8}$ pound/gallon is added to the tank at the rate of 3 gallons/minute. Three hours later, a freshwater valve is turned on, supplying 2 gallons/minute to the tank in addition to the brine mixture already being added. Calculate the amount of salt in the tank for any time $t \geq 0$. What is the steady-state amount of salt in the tank?

3. Two tanks are cascaded as shown in the accompanying diagram. Tank 1 initially contains 20 pounds of salt dissolved in 100 gallons of brine, and tank 2 initially contains 150 gallons of brine in which 90 pounds of salt are dissolved. At time zero, a brine solution containing $\frac{1}{2}$ pound of salt per gallon is added to tank 1 at the rate of 5 gallons/minute. Tank 1 has an output which discharges brine into tank 2 at the rate of 5 gallons/minute, and tank 2 also has an output of 5 gallons/minute. Determine the amount of salt in each tank for any time $t \geq 0$. Also determine when the concentration of salt in tank 2 is minimum and how much salt is in the tank at that time. *Hint:* Solve for the amount of salt in tank 1 first, then use this to determine the amount of salt in tank 2.

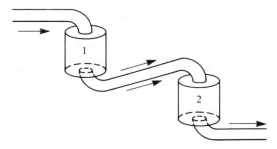

4. Let H and K be chemicals which react together. At time t, let $x(t)$ and $y(t)$ be the concentrations (in moles per liter) of H and K, respectively. Assume that one molecule of H combines with one molecule of K to form a new product, and let $z(t)$ be the amount by which $x(t)$ and $y(t)$ have decreased in time t. *Note:* A *mole* is one gram-molecule of any substance, that is, the molecular weight expressed in grams. One mole of sulfuric acid weighs about 98.08 grams.

   (a) Explain why $x(t) = x(0) - z(t)$ and $y(t) = y(0) - z(t)$.

   (b) Under constant temperature, the rate of reaction dz/dt satisfies the equation

   $$\frac{dz}{dt} = kxy$$

   for some constant k. Use this equation, together with the equations from (a), to solve for $z(t)$, then for $x(t)$ and $y(t)$.

   (c) Find expressions for the times T_x and T_y it takes for the original amounts of H and K, respectively, to be halved.

5. Suppose that in Problem 4 two molecules of K combine with one of H to form a new product.

   (a) Explain why $x(t) = x(0) - z(t)$ and $y(t) = y(0) - 2z(t)$.

   (b) Experiments show that

   $$\frac{dz}{dt} = kxy^2$$

for some constant k. Use this equation and information from (a) to solve for z, hence also for x and y at any time t.

(c) Find an expression for the time it takes for the original amount of H to be halved and for the original amount of K to be halved.

6. Suppose that in Problem 4 α molecules of H combine with β molecules of K to form a new product. Now, we find by experiment that

$$\frac{dz}{dt} = kx^\alpha y^\beta.$$

Write a differential equation involving just z, noting that $x(t) = x(0) - \alpha z(t)$ and $y(t) = y(0) - \beta z(t)$. Why is this differential equation more difficult to solve than those in the special cases of Problems 4 and 5?

In each of Problems 7 through 24, find the family of orthogonal trajectories of the given family of curves.

7. $x + 2y = K$
8. $2x^2 - 3y = K$
9. $x^2 + 2y^2 = K$
10. $x + Ky = 1$

11. $x^2 - 3y^2 = K$
12. $y = 2Kx^2 + 1$
13. $x + 2y^2 = K$
14. $x^2 - Ky^2 = 1$

15. $y = e^{Kx}$
16. $y = Ke^x$
17. $y = \dfrac{1 + Kx}{1 - Kx}$
18. $x - 2y = K$

19. $y = (x - K)^2$
20. $y^2 - x^2 = K$
21. $y^2 = Kx^3$
22. $2x - 3(y + K) = 4$

23. $x^2 - Ky = 1$
24. $y = Kx^n$ (n any positive integer)

25. Given a family $F(x, y, K) = 0$, a curve $\mathscr{C}$ is called an *oblique trajectory* of this family if the angle between $\mathscr{C}$ and any curve of the family it intersects is α, with $0 < \alpha < \pi/2$. The totality of all such curves constitutes the *family of oblique trajectories* of the given family.

Assume that $y' = f(x, y)$ is the differential equation of the family $F(x, y, K) = 0$. Show that the family of oblique trajectories of $F(x, y, K) = 0$ is the family of integral curves of the differential equation

$$y' = \frac{f(x, y) + \tan(\alpha)}{1 - f(x, y)\tan(\alpha)}.$$

Hint: Let C be a curve in the family $F(x, y, K) = 0$ passing through (x, y). The slope of C at (x, y) is $f(x, y)$. The tangent to C at (x, y) makes an angle $\theta = \tan^{-1}[f(x, y)]$ with the x-axis. An oblique trajectory curve C^* intersecting C at an angle α at (x, y) has tangent making an angle α with the tangent to C at (x, y). The slope of C^* at (x, y) is therefore $\tan(\theta + \alpha)$, or $\tan\{\tan^{-1}[f(x, y) + \alpha]\}$. Now use a formula for the tangent of a sum of two angles.

In each of Problems 26 through 30, find the family of oblique trajectories making the given angle with the given family.

26. $x^2 + y^2 = K;$ $\alpha = \dfrac{\pi}{3}$
27. $y = 2x + K;$ $\alpha = \dfrac{\pi}{4}$
28. $y = Ke^x;$ $\alpha = \dfrac{\pi}{4}$

29. $x^2 + y^2 = Ky;$ $\alpha = \dfrac{\pi}{3}$
30. $x = Ke^y;$ $\alpha = \dfrac{\pi}{6}$

31. Suppose we are given a family of curves $F(\theta, r, K) = 0$ in which θ and r are polar coordinates. Let the differential equation of this family be $f(\theta, r, r') = 0$, in which $r' = dr/d\theta$. Show that the family of orthogonal trajectories has differential equation

$$f\left(\theta, r, -\frac{r^2}{r'}\right) = 0.$$

Hint: Recall that, if C is the graph of a curve in polar coordinates, the angle ψ between the tangent to C at P and the line from the origin to P satisfies $\tan(\psi) = r/r'$, assuming that $r \neq 0$.

In each of Problems 32 through 36, find the family of orthogonal trajectories of the given family.

32. $r = K\theta$ **33.** $r = Ke^{-\theta}$ **34.** $r = K[1 - \cos(\theta)]$

35. $r = K\theta^2$ **36.** $r = K \ln(\theta)$

1.10 *Existence and Uniqueness of Solutions of Initial Value Problems*

In this section, we will examine questions involving existence and uniqueness of solutions of initial value problems.

Consider the initial value problem

$$y' = f(x, y); \qquad y(x_0) = y_0, \tag{1.60}$$

in which x_0 and y_0 are given numbers. The general solution of the differential equation $y' = f(x, y)$ defines a family of integral curves in the plane. Any integral curve passing through the point (x_0, y_0) is the graph of a solution of the initial value problem (1.60).

Until this time, we have solved an initial value problem by finding the general solution of the differential equation, then selecting the arbitrary constant to obtain a solution satisfying the initial condition. Here is another example.

EXAMPLE 1.38

Consider the initial value problem

$$y' + y = 1; \qquad y(0) = 6.$$

The general solution of the linear differential equation $y' + y = 0$ is

$$y = 1 + Ke^{-x}.$$

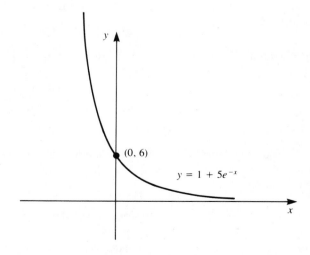

Figure 1.13

To solve the initial value problem, choose K so that

$$y(0) = 1 + Ke^{-0} = 1 + K = 6.$$

Therefore, $K = 5$, and the solution of the initial value problem is

$$y = 1 + 5e^{-x}.$$

The graph of $y = 1 + 5e^{-x}$ represents a particular solution of the differential equation passing through the point (0, 6) and is shown in Figure 1.13.

In this example, it seems plausible that $y = 1 + 5e^{-x}$ is the only solution the initial value problem can have, although we have not proved this. It is possible, however, for an initial value problem to have more than one solution, as the following example shows.

EXAMPLE 1.39

Consider the initial value problem

$$y' = 2\sqrt{y}; \qquad y(2) = 0.$$

Obviously, $y = 0$ is a solution of this initial value problem. But

$$y = \begin{cases} 0 & \text{if} \quad x \le 2 \\ (x - 2)^2 & \text{if} \quad x \ge 2 \end{cases}$$

also defines a solution of this initial value problem, which therefore has at least two solutions.

This example shows that an initial value problem may not have a unique solution. It is also possible to write an initial value problem which has no solution at all. ∎

EXAMPLE 1.40

Consider the initial value problem

$$y' = 2\sqrt{y}; \qquad y(0) = -1.$$

This problem has no solution. We cannot have $y(0) = -1$, because the differential equation is defined only if $y \ge 0$. ∎

Examples 1.39 and 1.40 demonstrate the need to address the questions of existence and uniqueness of the solution of an initial value problem. We will now state a theorem which does this, giving sufficient conditions for an initial value problem to have exactly one solution. In the statement of the theorem, a closed rectangle in the plane consists of all points on or inside a rectangle in the plane.

THEOREM 1.2   Existence and Uniqueness for an Initial Value Problem

Suppose that f and $\partial f / \partial y$ are continuous in a closed rectangle R centered at (x_0, y_0) and that R has its sides parallel to the coordinate axes. Then there exists a positive number h such that the initial value problem

$$y' = f(x, y); \qquad y(x_0) = y_0$$

has a unique solution in the interval $(x_0 - h, x_0 + h)$. ∎

We will not give a proof of this theorem. Geometrically, existence of a solution of the initial value problem is equivalent to the statement that there is an integral curve passing through (x_0, y_0). Uniqueness of the solution is equivalent to the assertion that there is only one integral curve passing through (x_0, y_0).

EXAMPLE 1.41

Armed with this theorem, consider again the initial value problem

$$y' = 2\sqrt{y}; \qquad y(x_0) = y_0.$$

From Examples 1.39 and 1.40, this problem has no solution if $y_0 < 0$, and it has at least two solutions if $y_0 = 0$. However, if $y_0 > 0$, this problem has a unique solution.

To relate these results to the theorem, write $f(x, y) = 2\sqrt{y}$. Then $\partial f/\partial y = 1/\sqrt{y}$. If $y_0 > 0$, we can find a closed rectangle about (x_0, y_0) in which both f and $\partial f/\partial y$ are continuous (see Figure 1.14). The theorem guarantees that there exists exactly one solution in this case.

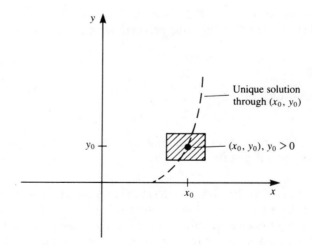

Figure 1.14

If, however, $y_0 \leq 0$, there is no closed rectangle throughout which f and $\partial f/\partial y$ are continuous because such a rectangle must contain points with negative y-coordinates (see Figure 1.15). The theorem does not guarantee either existence or uniqueness if $y_0 \leq 0$. We have seen that uniqueness fails if $y_0 = 0$ and existence fails if $y_0 < 0$.

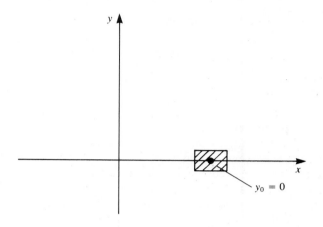

Figure 1.15   ■

Now consider the number h in the conclusion of Theorem 1.2. The theorem gives sufficient conditions for there to be an interval $(x_0 - h, x_0 + h)$ in which the initial value problem has a unique solution. However, *we have no a priori control over the size of h*. In some problems, h may be quite large; in others, it may be very small. Thus, the theorem gives a *local* result, valid in some (possibly very small) interval about x_0. The following example dramatizes this point.

EXAMPLE 1.42

Consider the initial value problem

$$y' = y^2; \qquad y(0) = 1.$$

Both $f(x, y) = y^2$ and $\partial f/\partial y = 2y$ are continuous everywhere, hence certainly in any closed rectangle about $(0, 1)$. By Theorem 1.2, this initial value problem has a unique solution in *some* interval about zero. In fact, we have no difficulty in deriving the solution

$$y = \frac{1}{1 - x}$$

of this initial value problem. We know from the theorem that this is the unique solution in *some* interval $(-h, h)$. We do not know h, but we can certainly say that $h \leq 1$, because this solution is not defined at 1 and so $(-h, h)$ cannot contain 1.

Now change the initial condition and consider the different initial value problem

$$y' = y^2; \qquad y(0) = 2.$$

This problem has the unique solution

$$y = \frac{2}{1 - 2x}$$

in some interval $(-h, h)$. As before, we do not know a numerical value for h. But now certainly we must have $h \leq \frac{1}{2}$, because this solution is undefined at $x = \frac{1}{2}$.

In general, if n is any positive integer, the initial value problem

$$y' = y^2; \qquad y(0) = n \tag{1.61}$$

has the solution

$$y = \frac{n}{1 - nx}.$$

This solution is unique in some interval $(-h, h)$, and we must have $h \leq \frac{1}{n}$.

The point is that the initial value problem (1.61) has a unique solution on *some* interval $(-h, h)$, by Theorem 1.2. However, at least in this example, the interval in which this solution is defined must become smaller as n is chosen larger. This fact is readily apparent from the solution $n/(1 - nx)$, which we can find explicitly in this example but which would not be at all obvious if we had only the initial value problem itself, without the solution.   ∎

PROBLEMS FOR SECTION 1.10

In each of Problems 1 through 5, show that the conditions of the existence/uniqueness theorem are satisfied by the initial value problem. Do not attempt to solve these problems explicitly. You may assume familiar facts from calculus about continuity of functions of two variables.

1. $y' = 2y^2 + 3xe^y \sin(xy); \quad y(2) = 4$

2. $y' = 4xy + \cosh(x); \quad y(1) = -1$

3. $y' = (xy)^3 - \sin(y); \quad y(2) = 2$

4. $y' = x^5 - y^5 + 2xe^y; \quad y(3) = \pi$

5. $y' = x^2 y e^{-2x} + y^2; \quad y(3) = 8$

6. Consider the initial value problem

$$|y'| = 2y; \qquad y(x_0) = y_0.$$

(a) Find two solutions of this initial value problem, assuming that $y_0 > 0$.

(b) Explain why this example does not contradict the conclusion of Theorem 1.2.

7. Show that the initial value problem

$$y' = y^{1/3}; \qquad y(0) = 0$$

has two solutions. Why does this not contradict Theorem 1.2?

The existence/uniqueness theorem stated in this section is usually proved by using Picard iterates. We will now define Picard iterates and give some problems related to them. Given the initial value problem

$$y' = f(x, y); \qquad y(x_0) = y_0,$$

define a sequence of functions as follows. Let

$$y_1(x) = y_0 + \int_{x_0}^{x} f(t, y_0) \, dt,$$

$$y_2(x) = y_0 + \int_{x_0}^{x} f(t, y_1(t)) \, dt,$$

$$y_3(x) = y_0 + \int_{x_0}^{x} f(t, y_2(t)) \, dt,$$

and, for any positive integer n,

$$y_n(x) = y_0 + \int_{x_0}^{x} f(t, y_{n-1}(t))\, dt.$$

Thus, we determine y_1 from y_0 and f, y_2 from y_1 and f, and so on, until y_n is generated from y_{n-1} and f. One way of proving Theorem 1.2 is to show that this sequence of functions $y_n(x)$ converges and that $\lim_{n \to \infty} y_n(x)$ is the solution of the initial value problem. The functions $y_1(x), y_2(x), \ldots, y_n(x), \ldots$ are called *Picard iterates* of the initial value problem.

We will not carry out the details of this proof. However, it is possible to get some feeling for the convergence of the Picard iterates to the solution of the initial value problem by working through some examples.

8. Consider the initial value problem

$$y' + y = 2; \qquad y(0) = 1.$$

(a) Use the theorem to show that this problem has exactly one solution.
(b) Determine the solution of the initial value problem explicitly.
(c) Determine the first five Picard iterates $y_0(x), \ldots, y_4(x)$ by explicitly carrying out the integrations. Note that in this problem $x_0 = 0$.
(d) From the first five Picard iterates, guess what $y_n(x)$ will look like.
(e) Write the solution determined in (b) as an infinite series by expanding the exponential function in its Maclaurin series.
(f) Relate the series expansion of the solution found in (e) to the Picard iterate $y_n(x)$ found in (d). You should find that the Picard iterates are exactly the partial sums of the Maclaurin expansion of the solution and, hence, that the Picard iterates converge to the solution.

Carry out the instructions of Problem 8 for each of the following initial value problems.

9. $y' = xy; \quad y(0) = 1$ **10.** $y' = ye^x; \quad y(1) = 1$ **11.** $y' = 2 - x; \quad y(2) = -2$
12. $y' = x^2 y; \quad y(0) = 1$ **13.** $y' = 2x^2; \quad y(1) = 3$

1.11 Direction Fields

We will now discuss the method of direction fields, which is used to get a qualitative idea of the behavior of solutions of a differential equation $y' = f(x, y)$, even when we are unable to produce a solution explicitly or find an equation implicitly defining a solution.

The idea is that we can sometimes picture the graph of a curve if we have a sketch showing segments of the tangent to the curve at various points. This idea is illustrated in Figure 1.16, in which the short tangent segments suggest the shape of the curve.

Now suppose that the initial value problem

$$y' = f(x, y); \qquad y(x_0) = y_0$$

has a solution at each point of a set R of points in the plane. At each point (x, y) in R, draw a short line segment, called a *lineal element*, having slope $f(x, y)$. Since $y' = f(x, y)$, the lineal element through (x, y) has the same slope as the integral curve passing through (x, y). Each lineal element is therefore a segment of the tangent to the integral curve through that point. If we draw enough lineal elements

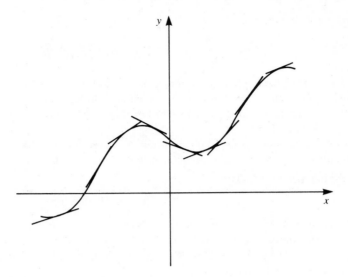

Figure 1.16

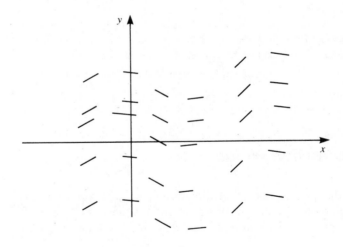

Figure 1.17

(tangent segments), we can obtain "dashed-line" sketches of integral curves, as shown in Figure 1.17.

The totality of the lineal elements is called the *direction field* of the differential equation. Of course, in practice we can draw only some of the lineal elements. We try to draw enough of them to visualize the flow pattern of the integral curves.

EXAMPLE 1.43

Consider the differential equation

$$y' = y.$$

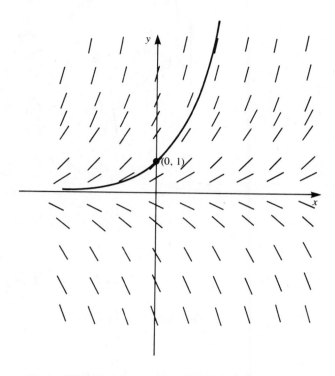

Figure 1.18. Direction field of $y' = y$.

The general solution of this differential equation is

$$y = Ce^x.$$

For illustration, we will sketch some of the lineal elements of the direction field for $y' = y$. At any point (x, y), draw a short line segment of slope y. Some of these lineal elements are shown in Figure 1.18. The particular integral curve passing through $(1, 0)$ is shown, as is the graph of the unique solution of the initial value problem $y' = y$; $y(0) = 1$. This solution is, of course, $y = e^x$. ■

EXAMPLE 1.44

Consider the differential equation

$$y' = y^2.$$

Again, we easily solve this equation, obtaining the general solution

$$y = \frac{-1}{x + C}.$$

With $f(x, y) = y^2$, a lineal element through (x, y) is a short line segment of slope y^2. Some of these lineal elements are shown in Figure 1.19, in which the particular integral curve through $(2, 1)$ is highlighted. This curve is the graph of the unique solution $y = 1/(3 - x)$ of the initial value problem $y' = y^2$; $y(2) = 1$.

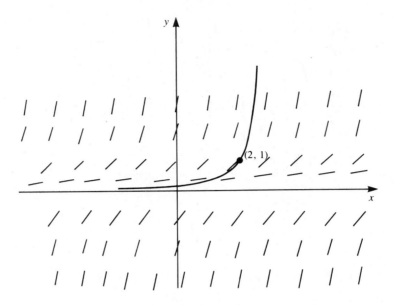

Figure 1.19. Direction field of $y' = y^2$.   ∎

Sometimes a direction field gives us information not immediately apparent from the differential equation itself. Here is an illustration of this assertion.

EXAMPLE 1.45

Consider the differential equation

$$y' = 1 - y^2.$$

Notice that values of y divide the plane into regions, where y' has constant sign. Specifically,

$$y' < 0 \quad \text{if} \quad |y| > 1;$$
$$y' = 0 \quad \text{if} \quad y = \pm 1;$$
$$y' > 0 \quad \text{if} \quad |y| < 1.$$

Suppose we want information about the solution φ of the initial value problem

$$y' = 1 - y^2; \qquad y(x_0) = y_0$$

as x becomes very large. Notice that, if $y_0 = 1$, then $\varphi(x) = 1$ for all x. By the same token, if $y_0 = -1$, then $\varphi(x) = -1$ for all x.

From the direction field in Figure 1.20 we see that, if $|y_0| < 1$, then

$$\lim_{x \to \infty} \varphi(x) = 1.$$

It also appears from the direction field that $\lim_{x \to \infty} \varphi(x) = 1$ if $y_0 > 1$. Finally, if $y_0 < -1$, there will be a finite value of x, say, x_1, at which $\lim_{x \to x_1-} \varphi(x) = -\infty$.

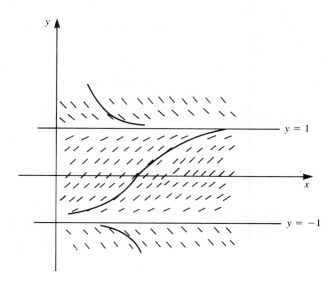

Figure 1.20. Direction field of $y' = 1 - y^2$.

All of this information has been obtained without solving the differential equation! In fact, the general solution of the separable differential equation $y' = 1 - y^2$ is

$$y = \frac{1 + Ce^{-2x}}{1 - Ce^{-2x}}. \tag{1.62}$$

To obtain the unique solution of the initial value problem

$$y' = 1 - y^2; \qquad y(x_0) = y_0,$$

we must choose

$$C = \frac{y_0 - 1}{y_0 + 1} e^{2x_0}, \tag{1.63}$$

provided that $y_0 \neq -1$. We will use this explicit solution to check the conclusions we have drawn from the direction field.

First, it is clear from the general solution (1.62) that $\lim_{x \to \infty} y(x) = 1$, since $e^{-2x} \to 0$ as $x \to \infty$. This conclusion holds as long as $y_0 \neq -1$. The solution $y = -1$ of the differential equation is a singular solution, not obtainable from the general solution (1.62) for any choice of C.

Next, suppose that $y_0 < -1$. Then, from equation (1.63), we have $C > 0$. Look at the denominator $1 - Ce^{-2x}$ in the general solution (1.62). If $C > 0$, the equation

$$1 - Ce^{-2x} = 0$$

has a solution, namely, $x = \ln(C)/2$. At this value of x, the numerator in the general solution is $1 + Ce^{-2x}$, or 2. Further, for $x < \ln(C)/2$, $1 - 2^{-2x} < 0$. Thus, as $x \to \ln(C)/2$ from the left, we get

$$\lim_{x \to \ln(C)/2\,-} y(x) = -\infty.$$

This establishes the existence of a number x_1 at which $\lim_{x \to x_1 -} \varphi(x) = 0$, as predicted from the direction field. ∎

As this example demonstrates, we can sometimes use a direction field to make observations about the behavior of solutions of a differential equation without actually solving the equation. The study of techniques of this type falls under the heading of qualitative behavior of solutions of differential equations, some aspects of which are discussed in Chapter 8.

PROBLEMS FOR SECTION 1.11

In each of Problems 1 through 10, (a) draw enough lineal elements to sketch the integral curve passing through the given point, (b) solve the differential equation and determine the solution whose graph passes through the given point, and (c) sketch the graph of the solution found in (b) and compare this sketch with the curve drawn using lineal elements in (a).

1. $y' = x + y$; $(2, 2)$

2. $y' + \dfrac{2}{x} y = e^x$; $(1, 1)$

3. $y' = e^{2x} + 2y$; $(0, 4)$

4. $y' = e^x - y$; $(1, 1)$

5. $y' = xy$; $(0, 4)$

6. $y' = y(1 + x)$; $(2, 3)$

7. $y' + \dfrac{1}{x} y = 2x$; $(3, 3)$

8. $y' - xy = 2x$; $(3, 1)$

9. $y' - \dfrac{3}{x} y = x^4 + 2$; $(3, 2)$

10. $y' + 2xy = 4x$; $(3, 4)$

In each of Problems 11 through 16, use the direction field method to sketch the solution of the initial value problem. Do not solve these problems exactly.

11. $y' = xy^2 - 2$; $y(1) = 3$

12. $y' = 8x^2 y + 4e^{-x}$; $y(0) = 1$

13. $y' = x \cos(2x) - y$; $y(1) = 1$

14. $y' = 2x + e^x y$; $y(-3) = 1$

15. $y' = 3x^2 y - 2y$; $y(0) = 1$

16. $y' = y \sin(x) - 3x^2$; $y(0) = 1$

17. Show that the direction field of the linear differential equation $y' + p(x)y = q(x)$ has the following property. The lineal elements on any vertical line $x = x_0$ [with $p(x_0) \neq 0$] all pass through a single point (ξ, η) in which

$$\xi = x_0 + \frac{1}{p(x_0)} \quad \text{and} \quad \eta = \frac{q(x_0)}{p(x_0)}.$$

Summary _____

This chapter is devoted to various aspects of first order differential equations, including initial value problems, an existence and uniqueness theorem, and techniques for explicitly or implicitly producing solutions for certain types of first order equations.

Probably the easiest type of first order equation to solve is the separable equation. In this case, the process of solution reduces to an exercise in integration. A homogeneous equation $y' = f(y/x)$ can be transformed into a separable one by the change of variables $u = y/x$. A nearly homogeneous equation $y' = f[(ax + by + c)/(dx + ey + h)]$ can be transformed into a homogeneous equation (if $ae - bd \neq 0$) or to a separable equation (if $ae - bd = 0$).

If $y' = f(x, y)$ is exact, we attempt to find a potential function F. The general solution is implicitly defined by the equation $F(x, y) = C$. If a differential equation is not exact, we can sometimes find an integrating factor. Often we attempt an integrating factor which is a function of just x or just y; failing this, we attempt "simple" functions of x and y. For the Bernoulli equation, we can write a formula for an integrating factor.

The general solution of a first order differential equation contains an arbitrary constant and defines a family of curves in the plane called integral curves. Each curve in this family is the graph of a particular solution obtained by choosing a value for the arbitrary constant. A first order equation may also have singular solutions, which are solutions not obtainable from the general solution by making a specific choice of the constant.

A solution of the initial value problem $y' = f(x, y)$; $y(x_0) = y_0$ is a solution of the differential equation whose graph passes through the point (x_0, y_0). Under the conditions of Theorem 1.2, this problem has a unique solution in some interval about x_0.

If $F(x, y, K) = 0$ is a family of curves having differential equation $y' = f(x, y)$, the family of orthogonal trajectories has differential equation $y' = -1/f(x, y)$. The family of oblique trajectories making an angle α with curves of the family $F(x, y, K) = 0$ has differential equation $y' = [f(x, y) + \tan(\alpha)]/[1 - f(x, y)\tan(\alpha)]$.

Here is an outline of some of the methods of this chapter.

Type of Differential Equation	*Method of Solution*
Separable: $$A(x)\, dx = B(y)\, dy$$	Integrate.
Homogeneous: $$\frac{dy}{dx} = f\left(\frac{y}{x}\right)$$	Let $u = y/x$, $dy/dx = u + x\,(du/dx)$: $$\frac{1}{f(u) - u}\, du = \frac{1}{x}\, dx,$$ a separable equation.
Nearly homogeneous: $$\frac{dy}{dx} = f\left(\frac{ax + by + c}{dx + ey + h}\right)$$	If $ae - bd \neq 0$, let $$X = x - \alpha, \quad Y = y - \beta$$ to transform into a homogeneous equation. If $ae - bd = 0$, let $$v = \frac{1}{a}(ax + by)$$ to get a separable equation.
Exact equations and integrating factors: $$\frac{dy}{dx} = -\frac{M(x, y)}{N(x, y)}$$	Exact if $\partial M/\partial y = \partial N/\partial x$ throughout a rectangle; $F(x, y) = C$ defines the general solution; if the equation is not exact, seek an integrating factor.

(Continued)

Type of Differential Equation	Method of Solution
Bernoulli equation:	An integrating factor is
$P(x)y' + Q(x)y = R(x)y^\alpha$	$\mu = \dfrac{1}{y^\alpha P(x)} e^{\int (1-\alpha)Q(x)/P(x)\,dx}.$
Linear equation:	An integrating factor is $e^{\int p(x)\,dx}$.
$y' + p(x)y = q(x)$	
Riccati equation:	Let $y = S(x) + 1/z$, with $S(x)$ one solution.
$y' = P(x)y^2 + Q(x)y + R(x)$	

ADDITIONAL PROBLEMS

In each of Problems 1 through 80, find an expression which implicitly or explicitly defines the general solution of the differential equation.

The differential equations in Problems 1 through 80 include all types encountered in this chapter, so identifying the method to use is the first step. In doing this, it is sometimes useful to write the differential equation in various ways.

1. $(x^4 - 6xe^y)y' = 6e^y - 4x^3 y$

2. $(2x + y)y' = \dfrac{4}{y}x^2 + y + 4x$

3. $y' = 8x^3 - 3y$

4. $2(y - x)y' = 2y - 7x$

5. $yy' = x^2 - y^2$

6. $(x + 2)y' = y + x^2 + 4x$

7. $(4y - 2x + 3)y' = 2y - x - 4$

8. $x^3 + xy = x^2 y'$

9. $y^2 + 2xy - x^2 y' = 0$

10. $(x^2 - 4)y' = y + 3$

11. $5x - y + 4 + (x - 5y - 4)y' = 0$

12. $e^{-y}\cos(x) + xe^{-y}\sin(x) - xe^{-y}\cos(x)y' = 0$

13. $x^2 y' + x^2 y + 2xy - 1 = 0$

14. $xy' + y = 2y^{3/2}$

15. $12x^2 e^y - \sin(x - y) + [4x^3 e^y + \sin(x - y)]y' = 0$

16. $y' = \dfrac{4x - 2y + 1}{2x - y + 6}$

17. $6x - 2yy' = 0$

18. $y' + y^2 = x^{-2} - \dfrac{y}{x}$

19. $(3y^5 + 2x)y' = y$

20. $y' = \dfrac{x + 2y - 4}{2x + y + 6}$

21. $(x + 2)y' = y + 2(x + 2)^2(x + 3)$

22. $y^3 + (4xy^2 - 2)y' = 0$

23. $y' = xy^2 - 2x^2 y + x^3 + 1$

24. $x - 3y = (2x - 6y - 1)y' - 2$

25. $2x - 2y + 4 = (4x - 4y + 6)y'$

26. $\dfrac{y}{x^2 + y^2} = \dfrac{x}{x^2 + y^2}y'$

27. $(x^2 y - 2y + 3x + xy + 3x^2 - 6)y' = 1$

28. $y' = \dfrac{4y}{4x - y}$

29. $(2xy + e^{y^2})y' = 1$

30. $y' = \dfrac{1}{x}e^{-3x}y^2 - \dfrac{1}{x}y + 3e^{3x}$

31. $3y - x + xy' = 0$

32. $3x^2 y - 6x^3 - y^2 + 2xy + (2x^2 - xy)y' = 0$

33. $x^2 y' = y^2 + xy$

34. $y\cosh(xy) + [1 + x\cosh(xy)]y' = 0$

35. $y' = xy^2 + (1 - 2x)y + x - 1$

36. $y = xy' + \dfrac{y}{1 + y}$

37. $x^2 y' = xy + e^x y^3$

38. $y' = -\dfrac{1}{x} y + e^{-x}$

39. $xy' - y - x \cos\left(\dfrac{y}{x}\right) = 0$

40. $y' = \dfrac{3x + 2y + 1}{x + 2y + 5}$

41. $y' = \dfrac{-8x - 9x^2 y^2}{6x^3 y}$

42. $y' = \dfrac{1 - x}{y}$

43. $xy' + 4y = \dfrac{\cos(x)}{x^2}$

44. $(1 + y^2)\tan^{-1}(y) = xy'$

45. $24x^2 y + e^x \sin(y) + [8x^3 + e^x \cos(y)]y' = 0$

46. $2x^2 y + y^3 + (xy^2 - 2x^3)y' = 0$

47. $y' = 8y - e^{-x}$

48. $xy' = 4y - 3x$

49. $(8x - x^2)y + 4x^2 y' = 0$

50. $(x + 2y^2)y' = 1$

51 $xy' = \dfrac{e^y}{xy}$

52. $2xyy' = x + 8y^2$

53. $xyy' + x^2 + y^2 = 0$

54. $y' - 2xy = e^{x^2}$

55. $2 + y \cos(xy) + x \cos(xy)y' = 0$

56. $\sin(y) + x \cos(y)y' = 0$

57. $xy' + y = x^3 y^{4/3}$

58. $y = (y^4 + 3x)y'$

59. $y' = y - 3x^2$

60. $(x + 2y - 4)y' = 3x - 2y + 1$

61. $y' = y + 2x^2$

62. $\dfrac{1}{3x} y' = \dfrac{(y + 5)x}{y + 1} + xy$

63. $y' = \dfrac{e^{x-y}}{e^{x-y} - 1}$

64. $x^3 y^3 y' + 2x^2 y^4 = 1$

65. $y' - 5y = e^{5x}\sin(x)$

66. $\cos(x)y' - y = 5$

67. $y' = \dfrac{(x + y)^2}{x^2}$

68. $xy' - y = \dfrac{y}{\ln(y) - \ln(x)}$

69. $y' = 2 + \dfrac{y^2}{x^2}$

70. $y' = \dfrac{3x + 2y + 4}{x - y + 6}$

71. $xy' + 3y = x^2 \sin(x)$

72. $2x^3 y' - 4x^2 y + 5y^3 = 0$

73. $y = (y^2 - x)y'$

74. $y' = \dfrac{2x + y}{x - y}$

75. $y' = \dfrac{2y + y \cos(x)}{2x + \sin(x)}$

76. $y' = \dfrac{6e^y - 4x^3 y}{x^4 - 6xe^y}$

77. $y' = \dfrac{x - 2y + 8}{x + y + 1}$

78. $[2y \cos(x) + 2]y' = y^2 \sin(x)$

79. $y' = (x + y)^2$

80. $(x - 2)y' = x - y$

81. Determine values of C and K so that the following equation is exact:

$$[Cx^2 ye^y + 2 \cos(y)] \, dx + [x^3 e^y y + x^3 e^y + Kx \sin(y)] \, dy = 0.$$

82. The differential equation

$$y = xy' + f(y')$$

is called *Clairaut's equation.*

(a) Show that, for any constant k such that $f(k)$ is defined, $y = kx + f(k)$ is a solution of Clairaut's equation.

(b) Use the result of (a) to find the general solution of $y = xy' + (y')^2$.

(c) Graph several members of the family of solutions in (b).

(d) Show that $y = -\frac{1}{4}x^2$ is a solution of the equation in (b) but is not obtainable by any choice of k. How does the graph of this singular solution relate to the graphs of the other solutions found in (c)? The graph of $y = -\frac{1}{4}x^2$ is called the *envelope* of the family of curves found in (c).

(e) Solve the Clairaut equation

$$y = xy' + 2\sqrt{1 + (y')^2}.$$

(f) Follow the instructions of (c) and (d) with the function defined implicitly by the equation $x^2 + y^2 = 4$, $y \geq 0$.

83. A radioactive substance decays at a rate proportional to the mass present at time t. At time zero, there were 2 grams of the substance. In 80 years, there will be 1.7 grams.

(a) How many grams will there be in 200 years?

(b) In how many years will there be exactly 1 gram left?

84. A 6-pound ball falls from rest at the top of a tall building. As it falls, it is pulled downward by gravity, but the fall is also impeded by air resistance, which has a force of magnitude 0.6 times the velocity (measured in feet per second). Find a general formula for the velocity as a function of time.

CHAPTER 2

Linear Second Order Differential Equations

2.0 Introduction

A *second order differential equation* is an equation of the form

$$F(x, y, y', y'') = 0$$

involving a second, but no higher, derivative. Some examples are

$$y'' + 2xy' = e^x\cos(y)$$

and

$$y'' - y = x.$$

We have seen the second order differential equations

$$L\frac{d^2q}{dt^2} + R\frac{dq}{dt} + \frac{1}{C}q = E(t)$$

for charge in an *RLC* circuit (Example 0.1),

$$\frac{d^2y}{dt^2} + \frac{c}{m}\frac{dy}{dt} + \frac{k}{m}y = 0$$

for displacement in a mass-spring system (Example 0.2), and

$$\frac{d^2\theta}{dt^2} + \frac{g}{L}\sin(\theta) = 0$$

for the angle of displacement of a simple pendulum (Example 0.4).

A second order differential equation must explicitly contain a second derivative term, but it can have no higher derivatives. The quantities x, y, and y' need not appear explicitly. For example, $y'' = 10$ is a second order differential equation, even though x, y, and y' do not explicitly appear.

A *solution* of $F(x, y, y', y'') = 0$ is a function satisfying the equation. For example, $\cos(2x)$ is a solution of $y'' + 4y = 0$ for all x, and $e^x - x$ is a solution of $y'' - y = x$ for all x. These facts can be verified by substituting the function into the differential equation.

Sometimes we must restrict the independent variable to an interval I. We say that φ is a *solution of* $F(x, y, y', y'') = 0$ *on an interval* I if $F(x, \varphi(x), \varphi'(x), \varphi''(x)) = 0$ for all x in I. For example, it is routine to show by substitution into the differential equation that

$$\varphi(x) = \sqrt{\frac{2}{\pi x}} \, \sin(x)$$

is a solution of

$$y'' + \frac{1}{x} y' + \left(1 - \frac{1}{4x^2}\right) y = 0$$

on $(0, \infty)$.

As suggested by Examples 0.1, 0.2, and 0.4, second order differential equations can be used to model a variety of interesting and important physical phenomena. Recall that, in the first order case, we were able to solve only equations having certain forms. Similarly, we are not able to treat the second order differential equation $F(x, y, y', y'') = 0$ in all its generality but must restrict ourselves to special cases. For much of this chapter, we shall consider linear second order differential equations, which we will begin to develop in the next section.

PROBLEMS FOR SECTION 2.0

In each of Problems 1 through 10, show that the given function φ is a solution of the differential equation for any choices of the constants c_1 and c_2.

1. $y'' - 5y' + 4y = 0$; $\varphi(x) = c_1 e^x + c_2 e^{4x}$

2. $y'' - y' - 2y = 0$; $\varphi(x) = c_1 e^{-x} + c_2 e^{2x}$

3. $y'' + 8y' + 16y = 0$; $\varphi(x) = e^{-4x}(c_1 + c_2 x)$

4. $y'' + 16y = 0$; $\varphi(x) = c_1 \cos(4x) + c_2 \sin(4x)$

5. $y'' - 16y = 0$; $\varphi(x) = c_1 e^{4x} + c_2 e^{-4x}$

6. $x^2 y'' + 2xy' - y = 0$; $\varphi(x) = \dfrac{1}{\sqrt{x}}(c_1 x^{\sqrt{5}/2} + c_2 x^{-\sqrt{5}/2})$ for $x > 0$

7. $y'' + 4y = 2x$; $\varphi(x) = c_1 \cos(2x) + c_2 \sin(2x) + \frac{1}{2}x$

8. $y'' - 9y = 3$, $\varphi(x) = c_1 e^{3x} + c_2 e^{-3x} - \frac{1}{3}$

9. $y'' + \dfrac{2}{x} y' - \dfrac{3}{x^2} y = 0$; $\varphi(x) = \dfrac{1}{\sqrt{x}}(c_1 x^{\sqrt{13}/2} + c_2 x^{-\sqrt{13}/2})$ for $x > 0$

10. $y'' + y' - 12y = x^2 - 1$; $\varphi(x) = c_1 e^{-4x} + c_2 e^{3x} - \frac{1}{12}x^2 - \frac{1}{72}x + \frac{59}{864}$

2.1 *Linear Second Order Differential Equations and Initial Value Problems*_____

A second order differential equation is *linear* if it has the form

$$y'' + p(x)y' + q(x)y = f(x). \qquad (2.1)$$

The functions p, q, and f are called *coefficient functions*, or *coefficients*, of the differential equation, and f is also called a *forcing function*. Usually these are assumed to be continuous (on an interval, or perhaps on the entire real line).

For example, the equations

$$y'' - 2xy' + 3y = 4e^{-2x}$$

and

$$y'' - 2y = 0$$

are linear and are defined for all x because the coefficient functions are defined for all x. The equation

$$y'' + \frac{1}{x}y = \frac{1}{x^2 - 4}$$

is linear on any intervals not containing 0, 2, or -2, where the coefficient functions $1/x$ and $1/(x^2 - 4)$ are not defined.

The second order differential equation

$$y'' - yy' + 6xy = 0$$

is not linear because of the yy' term (the coefficient of y' must be a function of x alone, or perhaps a constant). The equation

$$(y'')^2 = e^x$$

is not linear because y'' is squared. We refer to such equations as *nonlinear*.

Since a second order differential equation contains a second derivative, we intuitively expect that any process of finding a solution must involve two integrations and hence must give rise to two arbitrary and independent constants. We will require two additional pieces of information to assign unique values to these constants. The two pieces of information we usually specify are the solution value at a point, $y(x_0)$, and its derivative at the same point, $y'(x_0)$. Analogous to the first order case, this leads us to define a *linear second order initial value problem* as one of the form

$$y'' + p(x)y' + q(x)y = f(x); \qquad y(x_0) = A, \qquad y'(x_0) = B \qquad (2.2)$$

in which x_0, A and B are given numbers. The conditions $y(x_0) = A$ and $y'(x_0) = B$ are called *initial conditions*. Geometrically, they are equivalent to specifying that the graph of the solution of the differential equation must pass through the point (x_0, A) with slope B.

We can consider more general initial value problems in which the differential equation may be nonlinear. For the most part, however, we shall be dealing with the linear case.

EXAMPLE 2.1

Consider the initial value problem

$$y'' + 4y = 4x; \qquad y(0) = 1, \qquad y'(0) = 7.$$

We can check by substitution into the differential equation that

$$y = c_1\cos(2x) + c_2\sin(2x) + x$$

is a solution of the differential equation for any numbers c_1 and c_2. We must choose the constants to satisfy the initial conditions. First, to have $y(0) = 1$, we need

$$y(0) = c_1 = 1.$$

Thus far,

$$y = \cos(2x) + c_2\sin(2x) + x.$$

Now calculate

$$y' = -2 \sin(2x) + 2c_2\cos(2x) + 1.$$

Then

$$y'(0) = 2c_2 + 1 = 7,$$

and we must choose $c_2 = 3$. One solution of this initial value problem is

$$y = \cos(2x) + 3 \sin(2x) + x.$$

The following theorem will enable us to assert that this is the only solution.   ∎

THEOREM 2.1   Existence and Uniqueness for the Second Order Linear Initial Value Problem

Suppose that p, q, and f are continuous on an open interval I, and let x_0 be any point of I. Let A and B be any real numbers. Then the initial value problem

$$y'' + p(x)y' + q(x)y = f(x); \qquad y(x_0) = A, \qquad y'(x_0) = B$$

has a unique solution, and it is defined on the entire interval I.   ∎

The first order analogue of this theorem, given in Section 1.10, asserted existence and uniqueness only on some interval $(x_0 - h, x_0 + h)$, while this theorem asserts existence and uniqueness on the entire interval I over which the coefficient functions are continuous. The reason for this stronger conclusion is that we have restricted our attention here to the case in which the differential equation is linear. The theorem in Section 1.10 did not assume linearity. We will not prove this theorem, but we will make use of it in deriving properties of solutions in subsequent sections.

In the next section, we will obtain some facts about solutions of the linear differential equation $y'' + p(x)y' + q(x)y = f(x)$ and lay the foundations for finding solutions, at least in some important special cases.

PROBLEMS FOR SECTION 2.1

In each of Problems 1 through 10, an initial value problem is given. In the same-numbered problem of Section 2.0, a solution of the differential equation is given, involving two arbitrary constants. Using this solution, find the constants to produce the unique solution of the initial value problem.

1. $y'' - 5y' + 4y = 0$; $y(0) = 1$, $y'(0) = -4$

2. $y'' - y' - 2y = 0$; $y(0) = 3$, $y'(0) = 2$

3. $y'' + 8y' + 16y = 0$; $y(0) = y'(0) = 3$

4. $y'' + 16y = 0$; $y(\pi) = y'(\pi) = 1$

5. $y'' - 16y = 0$; $y(0) = 1$, $y'(0) = -2$

6. $x^2 y'' + 2xy' - y = 0$; $y(1) = y'(1) = 3$

7. $y'' + 4y = 2x$; $y\left(\dfrac{\pi}{4}\right) = 0$, $y'\left(\dfrac{\pi}{4}\right) = 1$

8. $y'' - 9y = 3$; $y(0) = 1$, $y'(0) = 4$

9. $y'' + \dfrac{2}{x} y' - \dfrac{3}{x^2} y = 0$; $y(1) = y'(1) = 0$

10. $y'' + y' - 12y = x^2 - 1$; $y(0) = 1$, $y'(0) = -3$

2.2 *Theory of Linear Homogeneous Second Order Differential Equations*

A linear second order differential equation

$$y'' + p(x)y' + q(x)y = f(x) \tag{2.3}$$

is *homogeneous* on an interval I if $f(x) = 0$ for all x in I. Thus, a homogeneous linear equation has the form

$$y'' + p(x)y' + q(x)y = 0. \tag{2.4}$$

For example,

$$y'' + xy = 0$$

is homogeneous on the whole real line, and

$$y'' + \frac{1}{x} y' - \ln(x)y = 0$$

is homogeneous on the interval $(0, \infty)$.

The word "homogeneous" is not to be confused with its use in connection with homogeneous first order differential equations. Here, the word simply means that $f(x)$ is identically zero in equation (2.3). It should be clear from context whether we are talking about equation (2.3) or about a homogeneous first order equation $y' = f(y/x)$.

A linear second order differential equation which is not homogeneous is called *nonhomogeneous*. For example,

$$y'' - y' + 2y = x - 1$$

is nonhomogeneous. Even though $x - 1$ is zero at $x = 1$, $x - 1$ is not identically zero throughout any interval over which we might be considering this differential equation.

We will now look at properties of solutions of the homogeneous linear equation (2.4). These properties will tell us what to look for in attempting to solve differential equations of this form.

THEOREM 2.2

Let y_1 and y_2 be solutions of the homogeneous linear equation

$$y'' + p(x)y' + q(x) = 0$$

on an interval I. Then

1. $y_1 + y_2$ is a solution on I.
2. For any constant c, cy_1 is a solution on I.

Proof   To prove (1), substitute $y_1 + y_2$ into the differential equation to get

$$(y_1 + y_2)'' + p(x)(y_1 + y_2)' + q(x)(y_1 + y_2)$$
$$= [y_1'' + p(x)y_1' + q(x)y_1] + [y_2'' + p(x)y_2' + q(x)y_2] = 0 + 0 = 0,$$

since y_1 and y_2 are both solutions.

For conclusion (2), substitute cy_1 into the differential equation to get

$$(cy_1)'' + p(x)(cy_1)' + q(x)(cy_1) = c[y_1'' + p(x)y_1' + q(x)y_1] = c \cdot 0 = 0,$$

proving that cy_1 is also a solution. ∎

We can combine the two conclusions of Theorem 2.2 into a single statement:

If y_1 and y_2 are solutions of $y'' + p(x)y' + q(x)y = 0$, then

$$c_1 y_1 + c_2 y_2$$

is a solution for any constants c_1 and c_2.

Any function of the form $c_1 y_1 + c_2 y_2$, with c_1 and c_2 constants, is called a *linear combination* of y_1 and y_2. Thus, Theorem 2.2 may be rephrased as follows: Any linear combination of two solutions of the homogeneous linear equation (2.4) is also a solution of (2.4).

EXAMPLE 2.2

It is routine to check that e^{2x} and e^{-2x} are solutions of

$$y'' - 4y = 0 .$$

for all x. Thus, $c_1 e^{2x} + c_2 e^{-2x}$ is a solution for any constants c_1 and c_2, as can also be verified by direct substitution into the differential equation. For example, if we choose

$c_1 = c_2 = \frac{1}{2}$, we have the solution $\frac{1}{2}(e^{2x} + e^{-2x})$, or $\cosh(2x)$. If we choose $c_1 = \frac{1}{2}$ and $c_2 = -\frac{1}{2}$, we get the solution $\frac{1}{2}(e^{2x} - e^{-2x})$, or $\sinh(2x)$.   ∎

The conclusion of Theorem 2.2 *fails to hold* if the differential equation is non-homogeneous. For example,

$$y_1(x) = e^x + x$$

and

$$y_2(x) = -e^x + x$$

are solutions of the nonhomogeneous linear equation

$$y'' - y = -x.$$

However,

$$y_1(x) + y_2(x) = 2x$$

is *not* a solution of $y'' - y = -x$, since

$$(2x)'' - (2x) = -2x$$

and, in general, $-2x \neq -x$.

Theorem 2.2 enables us to manufacture new solutions $c_1 y_1 + c_2 y_2$ of equation (2.4) from two solutions y_1 and y_2 on an interval I. However, observe that the linear combination $c_1 y_1 + c_2 y_2$ is again just a multiple of y_1, if y_2 is itself a constant multiple of y_1. If, say,

$$y_2(x) = k y_1(x)$$

for all x in I, then

$$c_1 y_1(x) + c_2 y_2(x) = c_1 y_1(x) + c_2 k y_1(x) = (c_1 + k c_2) y_1(x),$$

again just a multiple of y_1. Thus, a linear combination $c_1 y_1 + c_2 y_2$ does not provide us any information we did not already have from y_1 if y_2 is itself a constant multiple of y_1. This leads us to distinguish the case in which y_1 and y_2 are multiples of each other and give it a special name.

DEFINITION

Two functions y_1 and y_2 are *linearly dependent* on an interval I if one of these functions is a constant multiple of the other for each x in I. Two functions which are not linearly dependent are called *linearly independent*.   ∎

Linear independence of y_1 and y_2 on I means that one function cannot be written as a constant multiple of the other for all x in the relevant interval.

EXAMPLE 2.3

We can check that

$$y_1(x) = e^{2x} + e^{-2x}$$

and

$$y_2(x) = 4\cosh(2x)$$

are solutions of $y'' - 4y = 0$ for all x. Now, y_1 and y_2 are linearly dependent, because $y_2(x) = 2y_1(x)$, or, equivalently, $y_1(x) = \frac{1}{2}y_2(x)$ for all x. This means that we already know one solution, given the other one. If we found the solution y_1, we would have the second solution y_2 automatically from conclusion (2) of Theorem 2.2 because $2y_1$ is also a solution. ■

On the other hand, e^{2x} and e^{-2x} are also solutions of $y'' - 4y = 0$, but these solutions are linearly independent because e^{2x} and e^{-2x} are not constant multiples of each other for all x. The linear combination $c_1e^{2x} + c_2e^{-2x}$ is a solution for any choice of the constants c_1 and c_2 and yields solutions different from both e^{2x} and e^{-2x} if we choose c_1 and c_2 nonzero. In the case in which we have two linearly independent solutions, the linear combination $c_1y_1 + c_2y_2$ yields new solutions, in contrast to what occurred when we had two linearly dependent solutions.

Two linearly independent solutions provide much more information about other solutions of the linear equation (2.4) than do two linearly dependent solutions. Often we can quite easily tell whether two functions are linearly independent or dependent. In the case of very complicated functions (such as functions defined by infinite series), however, linear independence or dependence is sometimes unclear, and the following test is important.

THEOREM 2.3   Wronskian Test for Linear Independence

Let y_1 and y_2 be solutions of $y'' + p(x)y' + q(x)y = 0$ on an interval $[a, b]$. Let

$$W[y_1, y_2](x) = y_1(x)y_2'(x) - y_2(x)y_1'(x)$$

for x in $[a, b]$. Then

1. Either $W[y_1, y_2](x) = 0$ for all x in $[a, b]$ or $W[y_1, y_2](x) \neq 0$ for all x in $[a, b]$. (That is, $W[y_1, y_2](x)$ either vanishes identically on $[a, b]$ or is never zero on $[a, b]$.)
2. y_1 and y_2 are linearly independent on $[a, b]$ if and only if $W[y_1, y_2](x) \neq 0$ for some x in $[a, b]$. ■

The function $W[y_1, y_2]$ is called the *Wronskian* of y_1 and y_2 and can be written as the determinant

$$W[y_1, y_2](x) = \begin{vmatrix} y_1(x) & y_2(x) \\ y_1'(x) & y_2'(x) \end{vmatrix}.$$

This determinant has the functions along the first row and their derivatives along the second.

The value of conclusion (1) of Theorem 2.3 is that we need only determine whether the Wronskian is zero or nonzero at *any* point of $[a, b]$; the Wronskian is then either zero or nonzero at all points of this interval. In working with functions defined by

Taylor series, often the Wronskian is easy to evaluate numerically at the center of the series but difficult to evaluate at any other point. This is the real value of conclusion (1).

Conclusion (2) tells us that the solutions y_1 and y_2 are linearly independent exactly when the Wronskian is nonzero. If the Wronskian vanishes at some point of the interval, the solutions are linearly dependent.

We will prove Theorem 2.3 at the end of this section. Here are two illustrations of the theorem.

EXAMPLE 2.4

The functions

$$y_1(x) = e^{2x} \quad \text{and} \quad y_2(x) = e^{-2x}$$

are solutions of $y'' - 4y = 0$ for all x. The Wronskian is

$$W[y_1, y_2](x) = \begin{vmatrix} e^{2x} & e^{-2x} \\ 2e^{2x} & -2e^{-2x} \end{vmatrix} = -4,$$

which is nonzero for all x. Thus, y_1 and y_2 are linearly independent over the entire real line. ∎

In this example, the conclusion is obvious without the Wronskian because e^{2x} and e^{-2x} are not constant multiples of each other. Here is an example which shows the real power of the Wronskian test.

EXAMPLE 2.5

Consider the linear differential equation

$$y'' + xy = 0.$$

By differentiating term by term and substituting the series into the differential equation, we can verify that the following two functions are solutions:

$$y_1(x) = 1 - \frac{1}{6}x^3 + \frac{1}{180}x^6 - \frac{1}{12960}x^9 + \cdots$$

and

$$y_2(x) = x - \frac{1}{12}x^4 + \frac{1}{504}x^7 - \frac{1}{45360}x^{10} + \cdots$$

Both of these series converge for all x.

Since $y_1(x)$ and $y_2(x)$ are defined by infinite series, linear independence may not be as obvious as it was in the preceding example. However, these series are easy to evaluate numerically at the origin:

$$y_1(0) = 1 \quad \text{and} \quad y_2(0) = 0.$$

Further, their derivatives are easy to evaluate at the origin:

$$y_1'(0) = 0 \quad \text{and} \quad y_2'(0) = 1.$$

Then

$$W[y_1, y_2](0) = \begin{vmatrix} 1 & 0 \\ 0 & 1 \end{vmatrix} = 1 \neq 0;$$

hence, y_1 and y_2 are linearly independent. We need not evaluate the Wronskian at any other point to establish this conclusion. ∎

The next theorem provides the rationale for this discussion of linear independence. It tells us that we know *all* solutions of $y'' + p(x)y' + q(x)y = 0$, on a given interval, if we know just two linearly independent solutions.

THEOREM 2.4

Let y_1 and y_2 be linearly independent solutions of $y'' + p(x)y' + q(x)y = 0$ on an interval I. Then *every* solution of this equation on I is a linear combination of y_1 and y_2. That is, every solution is of the form

$$c_1 y_1 + c_2 y_2$$

for constants c_1 and c_2.

Proof   Let φ be any solution of $y'' + p(x)y' + q(x)y = 0$. Choose any x_0 in I, and let

$$\varphi(x_0) = A \quad \text{and} \quad \varphi'(x_0) = B.$$

Now consider the system of algebraic equations

$$y_1(x_0)X + y_2(x_0)Y = A$$
$$y_1'(x_0)X + y_2'(x_0)Y = B,$$

with X and Y as unknowns. This system of two equations in two unknowns has a solution exactly when the determinant of the coefficients is nonzero. That is, there is a solution for X and Y if and only if

$$\begin{vmatrix} y_1(x_0) & y_2(x_0) \\ y_1'(x_0) & y_2'(x_0) \end{vmatrix} \neq 0.$$

But this determinant is the Wronskian of y_1 and y_2, evaluated at x_0. Since y_1 and y_2 are assumed to be linearly independent, this Wronskian is nonzero. Therefore, the system (2.5) has a solution, say, $X = c_1$ and $Y = c_2$.

We will complete the proof by showing that $\varphi = c_1 y_1 + c_2 y_2$. First, observe from Theorem 2.2 that $c_1 y_1 + c_2 y_2$ is a solution of the differential equation. Further, by the way we chose the constants c_1 and c_2, the function $c_1 y_1 + c_2 y_2$ satisfies

$$c_1 y_1(x_0) + c_2 y_2(x_0) = A$$
$$c_1 y_1'(x_0) + c_2 y_2'(x_0) = B.$$

Therefore, $c_1 y_1 + c_2 y_2$ is a solution of the initial value problem

$$y'' + p(x)y' + q(x)y = 0; \qquad y(x_0) = A, \qquad y'(x_0) = B.$$

But φ is also a solution of this initial value problem, since we defined A and B by letting $\varphi(x_0) = A$ and $\varphi'(x_0) = B$. By Theorem 2.1, this initial value problem has only one solution. Therefore, $\varphi = c_1 y_1 + c_2 y_2$, and the proof is complete. ∎

Because of the conclusion of Theorem 2.4, we call the linear combination

$$c_1 y_1 + c_2 y_2$$

the *general solution* of $y'' + p(x)y' + q(x)y = 0$ whenever y_1 and y_2 are linearly independent solutions. Theorem 2.4 tells us that this general solution contains *all possible solutions* of the differential equation. The linearly independent solutions y_1 and y_2 are said to constitute a *fundamental set of solutions* for this differential equation.

We now have a strategy for finding the general solution (all solutions) of the linear homogeneous equation $y'' + p(x)y' + q(x)y = 0$. We must find *two* linearly independent solutions y_1 and y_2. These form a fundamental set of solutions, and the linear combination $c_1 y_1 + c_2 y_2$ contains all possible solutions, as c_1 and c_2 are allowed to take on any numerical values.

This strategy also suggests one way of approaching the linear initial value problem

$$y'' + p(x)y' + q(x)y = 0; \qquad y(x_0) = A, \qquad y'(x_0) = B.$$

We attempt to find the general solution $c_1 y_1 + c_2 y_2$ of the differential equation and then choose c_1 and c_2 to satisfy the initial conditions.

EXAMPLE 2.6

Solve the initial value problem

$$y'' + 4y = 0; \qquad y(\pi) = 3, \qquad y'(\pi) = -2.$$

We know from previous examples that two solutions of this differential equation are $y_1(x) = \cos(2x)$ and $y_2(x) = \sin(2x)$. These functions are linearly independent. [Clearly, neither is a constant multiple of the other; we can also apply the Wronskian test, obtaining

$$W[y_1, y_2](x) = \begin{vmatrix} \cos(2x) & \sin(2x) \\ -2\sin(2x) & 2\cos(2x) \end{vmatrix}$$
$$= 2\cos^2(2x) + 2\sin^2(2x) = 2 \neq 0.]$$

The general solution of $y'' + 4y = 0$ is

$$y(x) = c_1 \cos(2x) + c_2 \sin(2x).$$

Every solution of $y'' + 4y = 0$ is contained in this expression, for appropriate choices of c_1 and c_2. Now we must choose c_1 and c_2 to satisfy the initial conditions. Since $y(\pi) = 3$, we have

$$y(\pi) = c_1 \cos(2\pi) + c_2 \sin(2\pi) = c_1 = 3.$$

Therefore,

$$y(x) = 3\cos(2x) + c_2 \sin(2x).$$

To solve for c_2, calculate

$$y'(x) = -6\sin(2x) + 2c_2 \cos(2x).$$

Then

$$y'(\pi) = 2c_2 \cos(2\pi) = 2c_2 = -2;$$

hence, $c_2 = -1$. The unique solution of the initial value problem is

$$y(x) = 3\cos(2x) - \sin(2x). \quad \blacksquare$$

The theory we have just developed is fine *if* we can produce two linearly independent solutions of the linear differential equation $y'' + p(x)y' + q(x)y = 0$. In the next section, we will develop three techniques called reduction of order. Two of these apply to nonlinear as well as linear equations. The third will help us produce fundamental sets of solutions for certain linear differential equations.

We conclude this section with a proof of Theorem 2.3, the Wronskian test.

Proof of Theorem 2.3   For part (1), write $W[y_1, y_2] = W$ for simplicity, and calculate

$$W' = y_1'y_2' + y_1y_2'' - y_1''y_2 - y_1'y_2' = y_1y_2'' - y_1''y_2.$$

Now recall that y_1 and y_2 are both solutions of $y'' + p(x)y' + q(x)y = 0$. Then

$$y_1'' + p(x)y_1' + q(x)y_1 = 0$$

and

$$y_2'' + p(x)y_2' + q(x)y_2 = 0.$$

Multiply the first of these equations by $-y_2$ and the second by y_1 to get

$$-y_2y_1'' - p(x)y_2y_1' - q(x)y_2y_1 = 0$$

and

$$y_1y_2'' + p(x)y_1y_2' + q(x)y_1y_2 = 0.$$

Add these two equations to get

$$y_1y_2'' - y_2y_1'' + p(x)[-y_2y_1' + y_1y_2'] = 0.$$

In terms of the Wronskian, this is exactly

$$W' + p(x)W = 0.$$

Thus, the Wronskian satisfies a linear first order differential equation. An integrating factor for this differential equation is $e^{\int p(x)\,dx}$, and we find that the general solution is

$$W = Ce^{-\int p(x)\,dx},$$

in which C is any constant. Since the exponential factor is never zero, W must be either never zero (if $C \neq 0$) or always zero (if $C = 0$), proving conclusion (1) of the theorem.

To prove (2), suppose first that y_1 and y_2 are linearly dependent on $[a, b]$. Then $y_2(x) = cy_1(x)$ for some constant c, and $a \leq x \leq b$. Then

$$W = y_1y_2' - y_1'y_2 = cy_1y_1' - cy_1y_1' = 0$$

for all x in $[a, b]$. Conversely, suppose that $W(x) = 0$ for all x in $[a, b]$. We want to show that y_1 and y_2 are linearly dependent. Certainly, if $y_2(x) = 0$ for $a \leq x \leq b$, then $y_2(x) = 0 \cdot y_1(x)$, and y_1 and y_2 are linearly dependent. Thus, we may assume that $y_2(x_0) \neq 0$ for some x_0 in $[a, b]$. By continuity of y_2, there is some subinterval $[c, d]$ in $[a, b]$ containing x_0 such that $y_2(x) \neq 0$ for $c \leq x \leq d$.

For x in $[c, d]$, we can divide $W(x)$ by $y_2^2(x)$ to get

$$\frac{W(x)}{y_2^2(x)} = \frac{y_1(x)y_2'(x) - y_1'(x)y_2(x)}{y_2^2(x)}. \tag{2.6}$$

This quotient is identically zero on $[c, d]$ because $W(x)$ is assumed to be zero for all x in $[a, b]$. Now recognize that the right side of equation (2.6) is the derivative of the quotient y_1/y_2. We conclude that

$$\frac{d}{dx}\left[\frac{y_1(x)}{y_2(x)}\right] = 0$$

for x in $[c, d]$. Therefore, this quotient must be constant on $[c, d]$. For some constant K,

$$\frac{y_1(x)}{y_2(x)} = K$$

for $c \leq x \leq d$. Then $y_1(x) = Ky_2(x)$ for $c \leq y \leq d$.

We will now show that this equality holds for all x in the perhaps larger interval $[a, b]$. Observe that x_0 is in $[c, d]$ and hence is also in $[a, b]$. Further, $y_1(x)$ and $Ky_2(x)$ are both solutions of the initial value problem

$$y'' + p(x)y' + q(x)y = 0; \qquad y(x_0) = y_1(x_0), \qquad y'(x_0) = y_1'(x_0),$$

which is defined on $[a, b]$. By Theorem 2.1, this problem has a unique solution on $[a, b]$; hence, $y_1(x) = Ky_2(x)$ for $a \leq x \leq b$. Therefore, y_1 and y_2 are linearly dependent on $[a, b]$, as was to be proved. ∎

PROBLEMS FOR SECTION 2.2

In each of Problems 1 through 8, (a) verify that y_1 and y_2 are solutions of the differential equation; (b) use the Wronskian to show that y_1 and y_2 are linearly independent; (c) write the general solution of the differential equation; and (d) find the unique solution of the initial value problem consisting of the differential equation and the initial conditions.

1. $y'' - k^2y = 0$; $\quad y_1 = \cosh(kx)$, $\quad y_2 = \sinh(kx)$, $\quad$ with k, any positive constant; $\quad y(0) = 1$, $\quad y'(0) = 0$
2. $y'' + k^2y = 0$; $\quad y_1 = \sin(kx)$, $\quad y_2 = \cos(kx)$, $\quad$ with k any positive constant; $\quad y(\pi) = 0$, $\quad y'(\pi) = 1$
3. $y'' + 4y' - 12y = 0$; $\quad y_1 = e^{2x}$, $\quad y_2 = e^{-6x}$; $\quad y(0) = 1$, $\quad y'(0) = -1$
4. $y'' + 11y' + 24y = 0$; $\quad y_1 = e^{-8x}$, $\quad y_2 = e^{-3x}$; $\quad y(0) = 2$, $\quad y'(0) = -4$
5. $y'' - y' - 6y = 0$; $\quad y_1 = e^{-2x}$, $\quad y_2 = e^{3x}$; $\quad y(-1) = 3$, $\quad y'(-1) = 6$
6. $y'' + 11y' - 42y = 0$; $\quad y_1 = e^{3x}$, $\quad y_2 = e^{-14x}$; $\quad y(0) = -5$, $\quad y'(0) = 0$
7. $y'' - \dfrac{7}{x}y' + \dfrac{16}{x^2}y = 0$; $\quad y_1 = x^4$, $\quad y_2 = x^4\ln(x)$ for $x > 0$; $\quad y(1) = 2$, $\quad y'(1) = 4$
8. $y'' + \dfrac{1}{x}y' + \left(1 - \dfrac{1}{4x^2}\right)y = 0$; $\quad y_1 = \sqrt{\dfrac{2}{\pi x}}\sin(x)$, $\quad y_2 = \sqrt{\dfrac{2}{\pi x}}\cos(x)$ $\quad$ for $x > 0$; $\quad y(\pi) = -5$, $\quad y'(\pi) = 8$
9. Let $y_1 = x^2$ and $y_2 = x^3$ for all x. Show that $W[y_1, y_2](x) = x^4$. Why does it not contradict Theorem 2.3 that the Wronskian is zero at $x = 0$ but nonzero if $x \neq 0$?
10. Verify that $y_1 = x$ and $y_2 = x^2$ are linearly independent solutions of $x^2y'' - 2xy' + 2y = 0$ on $[-1, 1]$ but that the Wronskian is zero at $x = 0$. Why does this not contradict Theorem 2.3?
11. Give an example to show that the product of two solutions of a linear equation $y'' + p(x)y' + q(x)y = 0$ is not necessarily a solution.

12. Verify that both $y_1 = 3e^{2x} - 1$ and $y_2 = e^{-x} + 2$ are solutions of the nonlinear differential equation $yy'' + 2y' - (y')^2 = 0$. However, show that neither $2y_1$ nor $y_1 + y_2$ is a solution.

13. Verify that $y_1 = e^{2x} - 2x$ and $y_2 = 3e^{-2x} - 2x$ are solutions of the nonhomogeneous linear differential equation $y'' - 4y = 8x$ but that a linear combination $c_1 y_1 + c_2 y_2$ is not a solution if $c_1 + c_2 \neq 1$.

14. Let $y_1 = x^2$ and $y_2 = x|x|$. Show that y_1 and y_2 are linearly dependent on $[a, b]$ if $a > 0$ but that these functions are linearly independent if $a < 0$ and $b > 0$. Explain why Theorem 2.3, the Wronskian test, does not apply to these functions.

15. Let φ be any continuous function defined on an interval I. Prove that the zero function and φ are linearly dependent on I.

16. Let y_1 be a solution of the initial value problem

$$y'' + p(x)y' + q(x)y = 0; \qquad y(x_0) = A, \qquad y'(x_0) = 0.$$

Let y_2 be a solution of the initial value problem

$$y'' + p(x)y' + q(x)y = 0; \qquad y(x_0) = 0, \qquad y'(x_0) = B.$$

Prove that $y_1 + y_2$ is a solution of the initial value problem

$$y'' + p(x)y' + q(x)y = 0; \qquad y(x_0) = A, \qquad y'(x_0) = B.$$

Also, evaluate the Wronskian of y_1 and y_2.

17. Suppose that y_1 and y_2 are solutions of $y'' + p(x)y' + q(x)y = 0$ on $[a, b]$ and that y_1 and y_2 both have a relative extremum at x_0 in (a, b). Prove that y_1 and y_2 do not form a fundamental set of solutions of this differential equation on $[a, b]$.

18. Suppose that y_1 is a solution of $y'' + p(x)y' + q(x)y = 0$ on $[a, b]$ and that y_1 is not identically zero, but $y_1(x_0) = 0$ for some x_0 in (a, b). Prove that $y_1'(x_0) \neq 0$.

19. Suppose that y_1 and y_2 are distinct solutions of $y'' + p(x)y' + q(x)y = 0$ on $[a, b]$ and that, for some x_0 in (a, b), $y_1(x_0) = y_2(x_0) = 0$. Prove that y_1 and y_2 are linearly dependent on $[a, b]$.

20. Suppose that y_1 and y_2 are solutions of $y'' + p(x)y' + q(x)y = 0$ on (a, b) and that x_0 is in (a, b). Prove that

$$W[y_1, y_2](x) = e^{-\int_{x_0}^{x} p(t)\,dt} W[y_1, y_2](x_0).$$

Hint: Use part of the proof of Theorem 2.3.

21. Use the result of Problem 20 to show that, if y_1 and y_2 are solutions of $y'' + p(x)y' + q(x)y = 0$ on (a, b), the Wronskian of y_1 and y_2 is constant on (a, b) if and only if $p(x) = 0$ for all x in (a, b).

2.3 Reduction of Order

In this section we will discuss a set of techniques which are all grouped under the heading of *reduction of order*. The idea is to reduce the problem of solving a second order differential equation to one of solving one or more first order differential equations. We will consider three circumstances in which this method of approach is effective.

ABSENT DEPENDENT VARIABLE

Consider the second order differential equation

$$F(x, y, y', y'') = 0,$$

in which x is the independent variable and y is dependent on x. If y does not appear explicitly in this differential equation (although y' may, and y'' must), we say that the dependent variable is absent. In this event, the differential equation has the form

$$f(x, y', y'') = 0.$$

Let

$$u = y' \quad \text{and} \quad u' = y''$$

to transform the second order equation $f(x, y', y'') = 0$ into the first order differential equation

$$f(x, u, u') = 0.$$

We attempt to solve this equation for u, then obtain $y(x) = \int u(x)\, dx$.

EXAMPLE 2.7

Solve

$$xy'' + 2y' = 4x^3.$$

Since y does not appear explicitly, let $u = y'$ and $u' = y''$ to get

$$xu' + 2u = 4x^3.$$

If $x \neq 0$, we can write this as a linear first order differential equation

$$u' + \frac{2}{x}u = 4x^2.$$

An integrating factor is

$$e^{\int (2/x)\,dx} = e^{2\ln(x)} = e^{\ln(x^2)} = x^2.$$

Multiply the differential equation by x^2 to get

$$x^2 u' + 2xu = 4x^4,$$

or

$$[x^2 u]' = 4x^4.$$

Integrate to get

$$x^2 u = \tfrac{4}{5}x^5 + C,$$

or

$$u(x) = \frac{4}{5}x^3 + \frac{C}{x^2},$$

for $x \neq 0$. On any interval not containing zero, the original differential equation has general solution

$$y(x) = \int u(x)\, dx = \frac{1}{5}x^4 - \frac{C}{x} + K,$$

with C and K arbitrary constants. ■

ABSENT INDEPENDENT VARIABLE

If the independent variable x does not appear explicitly in the second order differential equation $F(x, y, y', y'') = 0$, the equation has the form $f(y, y', y'') = 0$. Now proceed as follows.

Let $u = y'$, as when the independent variable is absent. Unlike that case, however, we now put the original differential equation in terms of y, u, and u', with y as the new *independent* variable. To do this, we have $u = y'$ by choice and, by the chain rule,

$$y'' = \frac{d}{dx}[y'] = \frac{du}{dx} = \frac{du}{dy}\frac{dy}{dx} = u\frac{du}{dy}.$$

Then

$$f(y, y', y'') = 0 \quad \text{becomes} \quad f\left(y, u, u\frac{du}{dy}\right) = 0.$$

Solve the last differential equation for u in terms of y, then use $u = y'$ to solve for y in terms of x.

EXAMPLE 2.8

Solve

$$y'' - 2yy' = 0.$$

Since x does not appear explicitly, replace y' with u and y'' with $u\,(du/dy)$ to get

$$u\frac{du}{dy} - 2yu = 0.$$

Assuming that $u \neq 0$, we get

$$\frac{du}{dy} = 2y,$$

which we integrate directly with respect to y to get

$$u = y^2 + C.$$

Since $u = dy/dx$,

$$\frac{dy}{dx} = y^2 + C,$$

another separable differential equation, which we can write as

$$\frac{1}{C + y^2}\frac{dy}{dx} = 1.$$

Integrate both sides of this equation with respect to x. The result of the integration depends on the sign of the arbitrary constant C. If $C > 0$, we get

$$\frac{1}{\sqrt{C}}\tan^{-1}\left(\frac{y}{\sqrt{C}}\right) = x + K.$$

If $C = 0$, we get

$$-\frac{1}{y} = x + K.$$

And, if $C < 0$, we get

$$\frac{1}{\sqrt{-C}} \tanh^{-1}\left(\frac{y}{\sqrt{-C}}\right) = x + K.$$

In each case, we can solve for y as a function of x, obtaining

$$y(x) = \begin{cases} \sqrt{C} \tan[\sqrt{C}(x + K)] & \text{if} \quad C > 0 \\ \dfrac{-1}{x + K} & \text{if} \quad C = 0 \\ \sqrt{-C} \tanh[\sqrt{-C}(x + K)] & \text{if} \quad C < 0. \end{cases}$$

These expressions give us the general solution of the original second order differential equation.

To complete the example, we must consider what happens in the case $u = 0$, which we excluded in the above analysis. If $u = dy/dx = 0$, then $y = $ constant. It is easy to check that $y = K$ is also a solution of $y'' - 2yy' = 0$. ∎

USING ONE SOLUTION TO FIND A SECOND SOLUTION

In the preceding discussions, the differential equation did not have to be either linear or homogeneous. Now suppose we have a linear homogeneous second order equation

$$y'' + p(x)y' + q(x)y = 0.$$

Suppose also that we know one solution y_1. We will see how to use y_1 to find a second, linearly independent solution y_2. The idea is to look for a second solution of the form

$$y_2(x) = u(x)y_1(x),$$

in which u is a nonconstant function of x. Compute

$$y_2' = u'y_1 + uy_1' \quad \text{and} \quad y_2'' = u''y_1 + 2u'y_1' + uy_1''.$$

Substitute these quantities into the differential equation to get

$$u''y_1 + 2u'y_1' + uy_1'' + pu'y_1 + puy_1' + quy_1 = 0.$$

Write this equation as

$$y_1u'' + (2y_1' + py_1)u' + [y_1'' + py_1' + qy_1]u = 0.$$

Since y_1 is assumed to be a solution, the quantity in square brackets is zero, and we are left with

$$y_1u'' + (2y_1' + py_1)u' = 0.$$

Remembering that y_1, y_1', and p are known, this is a differential equation for u. Further, the dependent variable u is not explicitly present. Following the first method of this section, let $v = u'$ and $v' = u''$ to get

$$y_1 v' + (2y_1' + py_1)v = 0.$$

On any interval I in which $y_1(x) \neq 0$, write this equation as

$$v' + \left(2\frac{y_1'}{y_1} + p\right)v = 0, \tag{2.7}$$

a linear first order differential equation for v. An integrating factor for this equation is

$$e^{\int[2(y_1'/y_1) + p]\,dx} = e^{2\ln(y_1) + \int p(x)\,dx} = e^{\ln(y_1^2)}e^{\int p(x)\,dx} = y_1^2 e^{\int p(x)\,dx}.$$

Multiply equation (2.7) by this integrating factor to get

$$v'y_1^2 e^{\int p(x)\,dx} + \left(2\frac{y_1'}{y_1} + p\right)vy_1^2 e^{\int p(x)\,dx} = 0,$$

or

$$v'y_1^2 e^{\int p(x)\,dx} + (2y_1 y_1' + y_1^2 p)ve^{\int p(x)\,dx} = 0.$$

This equation can be written

$$\frac{d}{dx}\, y_1^2 v e^{\int p(x)\,dx} = 0.$$

Then

$$y_1^2 v e^{\int p(x)\,dx} = C;$$

hence,

$$v(x) = \frac{C}{y_1(x)^2}\, e^{-\int p(x)\,dx}.$$

We retrieve u as the integral of v:

$$u(x) = \int v(x)\,dx.$$

Once we know u, $y_2 = uy_1$ is a second solution of $y'' + p(x)y' + q(x)y = 0$ on I. Further, y_1 and uy_1 are linearly independent because u is not a constant if $c \neq 0$. Thus, y_1 and uy_1 form a fundamental set of solutions of $y'' + p(x)y' + q(x)y = 0$ on I, and the general solution is $y = c_1 y_1 + c_2 uy_1$, or $y = (c_1 + c_2 u)y_1$.

Rather than memorizing formulas for v and u in this derivation, we recommend substituting uy_1 into the differential equation and solving the resulting differential equation for u, noting that some terms cancel because y_1 is a solution of the original differential equation.

EXAMPLE 2.9

One solution of $y'' + 6y' + 9y = 0$ is $y_1 = e^{-3x}$. We will use the method just discussed to find a second, linearly independent solution.

Let $y_2(x) = u(x)e^{-3x}$. Compute

$$y_2'(x) = u'(x)e^{-3x} - 3u(x)e^{-3x}$$

and

$$y_2''(x) = u''(x)e^{-3x} - 6u'(x)e^{-3x} + 9u(x)e^{-3x}.$$

Substitute these quantities into the differential equation to get

$$u''(x)e^{-3x} - 6u'(x)e^{-3x} + 9u(x)e^{-3x} + 6[u'(x)e^{-3x} - 3u(x)e^{-3x}] + 9u(x)e^{-3x} = 0.$$

Divide this equation by e^{-3x} and cancel terms where possible to obtain

$$u''(x) = 0.$$

Then $u(x) = cx + d$ for any constants c and d. Since we need only one noncon-stant function u satisfying $u''(x) = 0$, choose $c = 1$ and $d = 0$ to get $u(x) = x$. Then $y_2(x) = xe^{-3x}$ is a second solution of $y'' + 6y' + 9y = 0$. The general solution is $y(x) = (c_1 + c_2 x)e^{-3x}$. ∎

PROBLEMS FOR SECTION 2.3

In each of Problems 1 through 27, solve the differential equation, using reduction of order. All of these problems have absent dependent or independent variable.

1. $xy'' = 2 + y'$ 　　　　　　**2.** $2y'' + e^x = 3y'$ 　　　　**3.** $xy'' - 2y' = 1$

4. $1 - y' = 4y''$ 　　　　　　**5.** $-3y'' - 2y' = 8x + 2$ 　**6.** $-3y'' + 2y' = 4x - 5$

7. $xy'' + 2y' = x$ 　　　　　　**8.** $1 - y'' = 2y' + x^2$ 　　**9.** $2y'' = 1 + y$

10. $y'' = 3y - 2$ 　　　　　　**11.** $yy'' + (y')^2 = 0$ 　　　**12.** $yy'' - (y')^2 = 0$

13. $y'' - 4y = 0$ 　　　　　　**14.** $xy'' - y' = 0$ 　　　　　**15.** $yy'' = y^2y' + (y')^2$

16. $y'' + (y')^2 = 0$ 　　　　　**17.** $y'' - (y')^2 = 0$ 　　　　**18.** $y'' - k^2y = 0$

19. $y'' + k^2y = 0$ 　　　　　**20.** $y'' - 2yy' = 0$ 　　　　**21.** $y'' = 1 + (y')^2$

22. $yy'' + 3(y')^2 = 0$ 　　　　**23.** $y'' + e^{2y}(y')^3 = 0$ 　**24.** $yy'' + (y + 1)(y')^2 = 0$

25. $2yy'' = (y')^2$ 　　　　　　**26.** $y'' = -y$ 　　　　　　**27.** $yy'' = (y')^3$

In each of Problems 28 through 37, find a second solution of $y'' + p(x)y' + q(x)y = 0$ which is linearly independent from the given solution on the interval. Then write the general solution of the differential equation.

28. $y'' - \dfrac{3}{x}y' + \dfrac{4}{x^2}y = 0;$ 　$y_1(x) = x^2$ 　for $x > 0$

29. $y'' - \dfrac{1}{x}y' - \dfrac{8}{x^2}y = 0;$ 　$y_1(x) = x^4$ 　for $x > 0$

30. $y'' - \dfrac{2x}{1 + x^2}y' + \dfrac{2}{1 + x^2}y = 0;$ 　$y_1(x) = x$ 　for all x

31. $y'' + \dfrac{2x}{1 - x^2}y' - \dfrac{2}{1 - x^2}y = 0;$ 　$y_1(x) = x$ 　for $-1 < x < 1$

32. $y'' - \dfrac{4x}{2x^2 + 1}y' + \dfrac{4}{2x^2 + 1}y = 0;$ 　$y_1(x) = x$ 　for all x

33. $y'' + \left(1 - \dfrac{2}{x}\right)y' - \left(\dfrac{1}{x} - \dfrac{2}{x^2}\right)y = 0;$ 　$y_1(x) = x$ 　for all $x > 0$

34. $y'' + \dfrac{1}{x}y' + \left(1 - \dfrac{1}{4x^2}\right)y = 0;$ 　$y_1(x) = \dfrac{1}{\sqrt{x}}\cos(x)$ 　for $x > 0$

35. $y'' - \left(2\tan(x) + \dfrac{2}{x}\right)y' + \left(\dfrac{2\tan(x)}{x} + \dfrac{2}{x^2}\right)y = 0;$ 　$y_1(x) = x$ 　for $0 < x < \dfrac{\pi}{2}$

36. $y'' - \dfrac{1}{x}(x + 2)y' + \left(\dfrac{1}{x} + \dfrac{2}{x^2}\right)y = 0;\quad y_1(x) = x$　for $x > 0$

37. $y'' + \dfrac{2x}{2x^2 + 3x + 1}\, y' - \dfrac{2}{2x^2 + 3x + 1}\, y = 0;\quad y_1(x) = x$　for x on any interval not containing -1 or $\dfrac{-1}{2}$.

38. In using reduction of order to find a second solution of $y'' + p(x)y' + q(x)y = 0$, using a first solution, we let constants of integration be zero each time we integrated. Suppose we retained the constants of integration throughout the process. Show that the second solution obtained in this way is actually the general solution of the original differential equation.

39. Verify that $y_1(x) = e^{-ax}$ is a solution of

$$y'' + 2ay' + a^2 y = 0.$$

Use reduction of order to find a second, linearly independent solution, assuming that a is a positive constant.

40. Determine a two-parameter family of curves having constant curvature.

41. Find an equation for the curve in the xy-plane which passes through the point $(0, 2)$ with slope 0 and whose curvature at any point (x, y) is $\cos(x)$.

42. The second order differential equation

$$A(x)y'' + B(x)y' + C(x)y = 0$$

is said to be *exact* if it can be written in the form

$$[A(x)y']' + [F(x)y]' = 0$$

for some function F.

(a) Show that if $A(x)y'' + B(x)y' + C(x)y = 0$ is exact, then

$$A''(x) - B'(x) + C(x) = 0.$$

(b) The point to the concept of exactness for second order differential equations is that

$$[A(x)y']' + [F(x)y]' = 0$$

can be integrated once, reducing the problem to one of solving a first order differential equation. Write a formula for the general solution of $A(x)y'' + B(x)y' + C(x)y = 0$, assuming that it is exact.

In each of Problems 43 through 47, show that the differential equation is exact and use the results of Problem 42 to find the general solution.

43. $2x^2 y'' + 5xy' + y = 0$　　　**44.** $x^2 y'' + 3xy' + y = 0$　　　**45.** $2x^2 y'' + 8xy' + 4y = 0$

46. $xy'' + (3x + 1)y' + 3y = 0$　　　**47.** $4xy'' + x^2 y' + 2xy = 0$

In each of Problems 48 and 49, show that the differential equation is not exact. Multiply the differential equation by the given integrating factor $\mu(x)$, and use the results of Problem 42 to find the general solution of the original differential equation.

48. $y'' + 4y' + 4y = 0;\quad \mu(x) = e^{2x}$　　　　　**49.** $y'' + y' - 6y = 0;\quad \mu(x) = e^{3x}$

50. Show that the change of variables $u = y'/y$ transforms

$$y'' + p(x)y' + q(x)y = 0$$

into a Riccati equation

$$\dfrac{du}{dx} = -q(x) - p(x)u - u^2.$$

2.4 *The General Solution of $y'' + Ay' + By = 0$ When $A^2 - 4B \geq 0$*

At this point, the linear differential equation $y'' + p(x)y' + q(x)y = 0$ is too difficult for us to solve without some assumptions about the coefficient functions p and q. Until Section 2.7, we will consider the special case in which p and q are constant. Thus, consider the equation

$$y'' + Ay' + By = 0, \qquad (2.8)$$

with A and B any real numbers. Equation (2.8) *always* has exponential solutions. To see why, attempt a solution

$$y = e^{rx}. \qquad (2.9)$$

The idea is to substitute this function into equation (2.8) and attempt to solve for r to produce a solution. Substituting (2.9) into (2.8) gives us

$$r^2 e^{rx} + Are^{rx} + Be^{rx} = 0,$$

or

$$[r^2 + Ar + B]e^{rx} = 0.$$

Since $e^{rx} \neq 0$ for all x, no matter what r is, we obtain

$$r^2 + Ar + B = 0. \qquad (2.10)$$

This is a quadratic equation for r called the *characteristic equation* of the differential equation (2.8). Its roots yield values of r for which e^{rx} is a solution of equation (2.8). These roots are

$$r = \frac{-A \pm \sqrt{A^2 - 4B}}{2}. \qquad (2.11)$$

The characteristic equation (2.10) has

1. Two real, distinct roots if $A^2 - 4B > 0$.
2. One real, repeated root if $A^2 - 4B = 0$.
3. Complex conjugate roots if $A^2 - 4B < 0$.

Case (3) is more complicated than the other two, and we will deal with it in the next two sections. We will treat cases (1) and (2) now.

In case (1), $A^2 - 4B > 0$. Now, equation (2.11) gives two real, distinct values for r,

$$r_1 = \frac{-A + \sqrt{A^2 - 4B}}{2} \quad \text{and} \quad r_2 = \frac{-A - \sqrt{A^2 - 4B}}{2}.$$

Then $y_1 = e^{r_1 x}$ and $y_2 = e^{r_2 x}$ are solutions of the differential equation (2.8). Further, these solutions are linearly independent on any interval because

$$W[y_1, y_2](x) = \begin{vmatrix} e^{r_1 x} & e^{r_2 x} \\ r_1 e^{r_1 x} & r_2 e^{r_2 x} \end{vmatrix} = (r_2 - r_1)e^{(r_1 + r_2)x},$$

and this is nonzero because $r_1 \neq r_2$. Thus, $e^{r_1 x}$ and $e^{r_2 x}$ form a fundamental set of solutions, and the general solution of (2.8) in this case is

$$y = c_1 e^{r_1 x} + c_2 e^{r_2 x}.$$

EXAMPLE 2.10

Solve $y'' + 4y' - 2y = 0$.
    The characteristic equation is

$$r^2 + 4r - 2 = 0,$$

with roots $-2 + \sqrt{6}$ and $-2 - \sqrt{6}$. The general solution of the differential equation is

$$y = c_1 e^{(-2+\sqrt{6})x} + c_2 e^{(-2-\sqrt{6})x}. \quad \blacksquare$$

In the case in which $A^2 - 4B = 0$, the characteristic equation for r has only one root, and it is $r = -A/2$. One solution of the differential equation in this case is

$$y_1 = e^{-Ax/2}.$$

To find a second, linearly independent solution, we will use reduction of order (the third case treated in Section 2.3). Let

$$y_2(x) = u(x)y_1(x).$$

Calculate

$$y_2' = u'(x)e^{-Ax/2} - \frac{A}{2} u(x)e^{-Ax/2}$$

and

$$y_2'' = u''(x)e^{-Ax/2} - Au'(x)e^{-Ax/2} + \tfrac{1}{4}A^2 u(x)e^{-Ax/2}.$$

Substitute these into $y'' + Ay' + By = 0$ to get

$$u''(x)e^{-Ax/2} - Au'(x)e^{-Ax/2} + \frac{1}{4} A^2 u(x)e^{-Ax/2}$$

$$+ A\left[u'(x)e^{-Ax/2} - \frac{A}{2} u(x)^{-Ax/2} \right] + Bu(x)e^{-Ax/2} = 0.$$

Since we are assuming that $A^2 - 4B = 0$, we can replace B with $\frac{1}{4}A^2$ to get

$$u''(x)e^{-Ax/2} - Au'(x)e^{-Ax/2} + \frac{1}{4} A^2 u(x)e^{-Ax/2}$$

$$+ A\left[u'(x)e^{-Ax/2} - \frac{A}{2} u(x)e^{-Ax/2} \right] + \frac{1}{4} A^2 u(x)e^{-Ax/2} = 0.$$

Some terms in this equation cancel, and we are left with just

$$u''(x)e^{-Ax/2} = 0,$$

or

$$u''(x) = 0.$$

We may therefore choose $u(x) = x$ and obtain $y_2(x) = xe^{-Ax/2}$ as a second, linearly independent solution. The general solution in this case is

$$y = c_1 e^{-Ax/2} + c_2 xe^{-Ax/2},$$

or

$$y = (c_1 + c_2 x)e^{-Ax/2}.$$

It is not necessary to repeat this derivation each time $A^2 - 4B = 0$. Just remember that, in this case, one solution is e^{rx} and a second is xe^{rx}, where r is the repeated root of the characteristic equation.

EXAMPLE 2.11

Solve

$$y'' + 4y' + 4y = 0.$$

The characteristic equation is

$$r^2 + 4r + 4 = 0,$$

with just one root, $r = -2$. The general solution is

$$y = (c_1 + c_2 x)e^{-2x}. \quad \blacksquare$$

EXAMPLE 2.12

Solve the initial value problem

$$y'' + 2y' - 3y = 0; \qquad y(0) = 1, \qquad y'(0) = 4.$$

The characteristic equation of the differential equation is

$$r^2 + 2r - 3 = 0,$$

with roots $r = 1$ and $r = -3$. The general solution of the differential equation is

$$y = c_1 e^x + c_2 e^{-3x}.$$

From the initial conditions, we have

$$y(0) = c_1 + c_2 = 1$$

and

$$y'(0) = c_1 - 3c_2 = 4.$$

Solve these equations to obtain $c_1 = \frac{7}{4}$ and $c_2 = -\frac{3}{4}$. The initial value problem has the unique solution

$$y = \tfrac{1}{4}(7e^x - 3e^{-3x}). \quad \blacksquare$$

EXAMPLE 2.13

Solve the initial value problem

$$y'' - 8y' + 16y = 0; \qquad y(1) = 3, \qquad y'(1) = -2.$$

The characteristic equation of the differential equation is

$$r^2 - 8r + 16 = 0,$$

with one root, $r = 4$. The general solution of the differential equation is

$$y = (c_1 + c_2 x)e^{4x}.$$

We have

$$y(1) = (c_1 + c_2)e^4 = 3.$$

Now calculate $y'(x) = 4(c_1 + c_2 x)e^{4x} + c_2 e^{4x}$, so

$$y'(1) = 4(c_1 + c_2)e^4 + c_2 e^4 = -2.$$

Write these equations for c_1 and c_2 as

$$c_1 + c_2 = 3e^{-4}$$

and

$$4c_1 + 5c_2 = -2e^{-4}.$$

Solve these equations to obtain $c_1 = 17e^{-4}$ and $c_2 = -14e^{-4}$. The unique solution of the initial value problem is

$$y = 17e^{-4}e^{4x} - 14e^{-4}xe^{4x}$$
$$= [17 - 14x]e^{4(x-1)}. \quad \blacksquare$$

Returning to the general discussion of equation (2.8), we have yet to produce a fundamental set of solutions in the case in which $A^2 - 4B < 0$. The roots of the characteristic equation in this case are complex conjugate numbers $r_1 = p + iq$ and $r_2 = p - iq$.

Formally, we would expect that $y_1 = e^{(p+iq)x}$ and $y_2 = e^{(p-iq)x}$ make up a fundamental set of solutions. In fact, they do. However, these solutions require that we be familiar with the complex exponential function. We will discuss this function in the next section and then see how to replace this fundamental set of solutions with another involving only real-valued functions. Such solutions are sometimes preferable to work with (for example, in graphing solutions).

PROBLEMS FOR SECTION 2.4

In each of Problems 1 through 20, find the general solution of the differential equation.

1. $y'' - y' - 6y = 0$

2. $y'' + 6y' - 40y = 0$

3. $y'' - 16y' + 64y = 0$

4. $y'' + 6y' + 9y = 0$

5. $y'' - 3y' = 0$

6. $y'' + 3y' - 18y = 0$

7. $y'' - 9y' + 20y = 0$

8. $y'' + 16y' + 64y = 0$

9. $y'' + 2y' - 16y = 0$

10. $y'' - 6y' + 3y = 0$

11. $y'' - 14y' + 49y = 0$

12. $y'' - 12y' + y = 0$

13. $y'' + 12y' + 36y = 0$

14. $y'' + 2y' + y = 0$

15. $y'' + 7y' - 5y = 0$

16. $y'' + 11y' - 11y = 0$ **17.** $y'' - 4y' + y = 0$ **18.** $y'' + y' - 3y = 0$
19. $y'' - 10y' + 25y = 0$ **20.** $y'' - 6y' + 7y = 0$

In each of Problems 21 through 30, solve the initial value problem.

21. $y'' + 3y' = 0;$ $y(0) = 3,$ $y'(0) = 6$ **22.** $y'' + 2y' - 3y = 0;$ $y(0) = y'(0) = 1$
23. $y'' - 4y' + 4y = 0;$ $y(0) = 3,$ $y'(0) = 5$ **24.** $y'' - 2y' + y = 0;$ $y(1) = 1,$ $y'(1) = -3$
25. $y'' - 2y' + y = 0;$ $y(1) = y'(1) = 0$ **26.** $y'' + 12y' + 36y = 0;$ $y(0) = -2,$ $y'(0) = -3$
27. $y'' - 2y' - 5y = 0;$ $y(0) = 0,$ $y'(0) = 3$ **28.** $y'' + 3y' - 2y = 0;$ $y(0) = 2,$ $y'(0) = -3$
29. $y'' - 7y' + 2y = 0;$ $y(2) = 1,$ $y'(2) = 0$ **30.** $y'' - 6y' + 9y = 0;$ $y(-1) = 1,$ $y'(-1) = 7$

31. Write the general solution of $y'' - k^2y = 0$ in terms of hyperbolic functions. Here, k is any positive constant.
32. Show that the general solution of

$$y'' + Ay' + By = 0$$

can always be written in the form

$$y(x) = [c_1\cosh(\beta x) + c_2\sinh(\beta x)]e^{\alpha x}$$

for appropriate choices of α and β, assuming that $A^2 - 4B > 0$. *Hint:* α and β will be functions of A and B.
33. A ball of mass m is thrown upward from the earth with initial velocity v_0. Acceleration due to gravity g is constant and air resistance is proportional to the velocity v.
     (a) Write an initial value problem for the height $x(t)$ of the ball (from the earth's surface) at time t. *Hint:* Remember Newton's second law of motion, which states for constant mass that force equals mass times acceleration. The velocity is $v = dx/dt$, and the acceleration is d^2x/dt^2, with x measured along the line of motion of the ball.
     (b) Solve the initial value problem for $x(t)$.
     (c) Find the maximum height achieved by the ball (as a function of v_0).
     (d) Show that the time required for the ball to reach the maximum height is less than the time required for the ball to fall back to earth from this height.

2.5 *The Complex Exponential Function* _____

In this section, we will define what it means to raise e to a complex power. We encounter this problem in solving $y'' + Ay' + By = 0$ when $A^2 - 4B < 0$.

For a complex number $a + ib$, with a and b real, define

$$e^{a+ib} = e^a[\cos(b) + i\sin(b)].$$

This is called *Euler's formula*, and a justification for it is given in Problem 15 at the end of this section. It enables us to write the complex exponential e^{a+ib} in terms of the real-valued quantities e^a, $\cos(b)$ and $\sin(b)$. For example,

$$e^{2+i\pi} = e^2[\cos(\pi) + i\sin(\pi)] = -e^2 \quad \text{(a real number!)},$$

$$e^{2+i\pi/2} = e^2\left[\cos\left(\frac{\pi}{2}\right) + i\sin\left(\frac{\pi}{2}\right)\right] = ie^2 \quad \text{(pure imaginary)},$$

and

$$e^{2+3i} = e^2[\cos(3) + i\sin(3)].$$

If $b = 0$, then $a + ib = a$, and Euler's formula yields

$$e^a[\cos(0) + i \sin(0)],$$

or just e^a. Thus, Euler's formula gives us the real exponential function when $a + ib$ is the real number a.

We want to define the complex-valued exponential function $e^{(a+ib)x}$, in which a and b are given real constants and x is a real variable. Using Euler's formula, write

$$e^{(a+ib)x} = e^{ax+ibx}$$
$$= e^{ax}[\cos(bx) + i \sin(bx)]$$
$$= e^{ax}\cos(bx) + ie^{ax}\sin(bx).$$

Further,

$$\frac{d}{dx} e^{(a+ib)x} = \frac{d}{dx}[e^{ax}\cos(bx)] + i\frac{d}{dx}[e^{ax}\sin(bx)]$$
$$= ae^{ax}\cos(bx) - be^{ax}\sin(bx) + i[ae^{ax}\sin(bx) + be^{ax}\cos(bx)]$$
$$= e^{ax}[a\cos(bx) - b\sin(bx) + i[a\sin(bx) + b\cos(bx)]]$$
$$= (a + ib)e^{ax}[\cos(bx) + i\sin(bx)]$$
$$= (a + ib)e^{(a+ib)x}.$$

Therefore, the complex-valued function $e^{(a+ib)x}$ obeys the same differentiation rule that the familiar real-valued exponential e^{ax} does:

$$\frac{d}{dx} e^{(a+ib)x} = (a + ib)e^{(a+ib)x}.$$

We can also check that the familiar formula

$$e^{\alpha x}e^{\beta x} = e^{(\alpha+\beta)x}$$

remains true when α and β are complex numbers (see Problem 14).

In the next section, we will apply these ideas to finding the general solution of $y'' + Ay' + By = 0$ when $A^2 - 4B < 0$.

PROBLEMS FOR SECTION 2.5

In each of Problems 1 through 10, use Euler's formula to write the given complex exponential in the form $\alpha + i\beta$, with α and β real numbers.

1. $e^{1-i\pi}$ **2.** $e^{\pi+2i\pi}$ **3.** $e^{4i\pi}$ **4.** $e^{1-i\sqrt{2}}$

5. $e^{\pi i}$ **6.** e^{8-2i} **7.** e^{1+4i} **8.** e^{2-i}

9. e^i **10.** $e^{1-i(\pi/2)}$

11. Use Euler's formula to show that

$$e^{a-ib} = e^a[\cos(b) - i \sin(b)].$$

12. Use the result of Problem 11, together with Euler's formula, to show that

$$\cos(\theta) = \frac{1}{2}(e^{i\theta} + e^{-i\theta}) \quad \text{and} \quad \sin(\theta) = -\frac{i}{2}(e^{i\theta} - e^{-i\theta}).$$

Hint: Let $a = 0$ and $b = \theta$ in Euler's formula; then let $a = 0$ and $b = \theta$ in the result of Problem 11. Solve the resulting equations for $\cos(\theta)$ and $\sin(\theta)$.

13. Use the result of Problem 12 to derive the results

$$\cos(\theta) = \cosh(i\theta) \quad \text{and} \quad i\sin(\theta) = \sinh(i\theta).$$

14. Use Euler's formula and trigonometric identities to show that

$$e^{a+ib}e^{c+id} = e^{(a+c)+i(b+d)}.$$

Hint: Write

$$e^{(a+c)+i(b+d)} = e^{a+c}[\cos(b+d) + i\sin(b+d)]$$

and use trigonometric identities to write $\cos(b+d)$ and $\sin(b+d)$ in terms of $\cos(b)$, $\cos(d)$, $\sin(b)$, and $\sin(d)$.

15. The following sequence of steps provides a motivation for Euler's formula.
   (a) Write the Maclaurin series for $\sin(b)$, $\cos(b)$, and e^b.
   (b) In the Maclaurin series for e^b, replace b with ib.
   (c) Using the fact that $i^2 = -1$, write the Maclaurin series for e^{ib} as a sum of two power series, one containing only even powers of b and the other containing i times odd powers of b.
   (d) Use the result of (a), recognize the sum of power series in (c) as a sum of the form $\cos(b) + i\sin(b)$. Conclude that $e^{ib} = \cos(b) + i\sin(b)$ if b is real.
   (e) Using the result of (d), and assuming that the rule $e^{\alpha}e^{\beta} = e^{\alpha+\beta}$ holds true if α and β are complex, derive Euler's formula.

2.6 The General Solution of $y'' + Ay' + By = 0$ When $A^2 - 4B < 0$ _____

Continuing from Section 2.4, we want to solve the differential equation

$$y'' + Ay' + By = 0 \tag{2.12}$$

when the characteristic equation has complex roots. This occurs when $A^2 - 4B < 0$ and the complex roots are

$$r_1 = \frac{-A + i\sqrt{4B - A^2}}{2} \quad \text{and} \quad r_1 = \frac{-A - i\sqrt{4B - A^2}}{2}.$$

For convenience, write

$$p = -\frac{A}{2} \quad \text{and} \quad q = \frac{\sqrt{4B - A^2}}{2}.$$

Then

$$r_1 = p + iq \quad \text{and} \quad r_2 = p - iq.$$

The functions $e^{(p+iq)x}$ and $e^{(p-iq)x}$ form a fundamental set of solutions of equation (2.12), whose general solution may be written

$$y = c_1 e^{(p+iq)x} + c_2 e^{(p-iq)x}. \tag{2.13}$$

It is sometimes convenient (for example, in graphing solutions) to write the general solution in terms of real-valued functions. This can be done as follows. Use Euler's formula to write the general solution (2.13) as

$$y = c_1 e^{px}[\cos(qx) + i\sin(qx)] + c_2 e^{px}[\cos(qx) - i\sin(qx)].$$

If we choose $c_1 = c_2 = \frac{1}{2}$ in this equation, we obtain the particular solution

$$y_1 = e^{px}\cos(qx).$$

If we choose $c_1 = -i/2$ and $c_2 = i/2$, we obtain another particular solution,

$$y_2 = e^{px}\sin(qx).$$

The Wronskian of these two particular solutions is

$$W[y_1, y_2](x) = \begin{vmatrix} e^{px}\cos(qx) & e^{px}\sin(qx) \\ pe^{px}\cos(qx) - qe^{px}\sin(qx) & pe^{px}\sin(qx) + qe^{px}\cos(qx) \end{vmatrix} = qe^{2px}.$$

This is nonzero for all x because $q = A^2 - 4B < 0$ in this case. Therefore, $e^{px}\cos(qx)$ and $e^{px}\sin(qx)$ also form a fundamental set of solutions of equation (2.12) for all x, and the general solution may be written in the form

$$y = c_1 e^{px}\cos(qx) + c_2 e^{px}\sin(qx), \tag{2.14}$$

with c_1 and c_2 arbitrary constants. Often we use this general solution, instead of the (equally valid) general solution (2.13), in the case in which $A^2 - 4B < 0$.

EXAMPLE 2.14

Solve $y'' + 2y' + 6y = 0$.
   The characteristic equation is

$$r^2 + 2r + 6 = 0,$$

with roots

$$r_1 = -1 + i\sqrt{5} \quad \text{and} \quad r_2 = -1 - i\sqrt{5}.$$

The general solution of the differential equation is

$$y = c_1 e^{-x}\cos(\sqrt{5}x) + c_2 e^{-x}\sin(\sqrt{5}x). \quad \blacksquare$$

EXAMPLE 2.15

Solve the initial value problem

$$y'' + 2y' + 3y = 0; \qquad y(0) = 2, \qquad y'(0) = -3.$$

   The characteristic equation is

$$r^2 + 2r + 3 = 0,$$

with roots $-1 + i\sqrt{2}$ and $-1 - i\sqrt{2}$. The general solution of the differential equation is

$$y = c_1 e^{-x}\cos(\sqrt{2}x) + c_2 e^{-x}\sin(\sqrt{2}x).$$

Now, we want

$$y(0) = c_1 = 2.$$

This gives us

$$y = 2e^{-x}\cos(\sqrt{2}x) + c_2 e^{-x}\sin(\sqrt{2}x).$$

Now calculate

$$y' = -2e^{-x}\cos(\sqrt{2}x) - 2\sqrt{2}e^{-x}\sin(\sqrt{2}x) - c_2 e^{-x}\sin(\sqrt{2}x) + \sqrt{2}c_2 e^{-x}\cos(\sqrt{2}x).$$

The second initial condition requires that

$$y'(0) = -2 + \sqrt{2}c_2 = -3;$$

hence,

$$c_2 = -\frac{1}{\sqrt{2}}.$$

The unique solution of the initial value problem is

$$y = 2e^{-x}\cos(\sqrt{2}x) - \frac{1}{\sqrt{2}} e^{-x}\sin(\sqrt{2}x). \quad \blacksquare$$

From Sections 2.4 through 2.6, it is now routine to write the general solution of any constant coefficient linear equation $y'' + Ay' + By = 0$. The procedure is as follows.

$$\boxed{y'' + Ay' + By = 0}$$

$$\downarrow$$

$$\boxed{r^2 + Ar + B = 0}$$

$$\downarrow$$

$$\boxed{\text{solve for } r} \quad \begin{cases} \text{Real roots } r_1 \neq r_2: & y = c_1 e^{r_1 x} + c_2 e^{r_2 x} \\ \text{One real root } r: & y = e^{rx}(c_1 + c_2 x) \\ \text{Roots } p \pm iq: & y = e^{px}[c_1\cos(qx) + c_2\sin(qx)] \end{cases}$$

EXAMPLE 2.16

Solve $y'' + 6y' - 3y = 0$.

The characteristic equation is $r^2 + 6r - 3 = 0$, with roots $-3 \pm 2\sqrt{3}$. The general solution is

$$y = c_1 e^{(-3+2\sqrt{3})x} + c_2 e^{(-3-2\sqrt{3})x}$$
$$= e^{-3x}(c_1 e^{2\sqrt{3}x} + c_2 e^{-2\sqrt{3}x}). \quad \blacksquare$$

EXAMPLE 2.17

Solve $y'' - 3y' + 8y = 0$.

The characteristic equation is $r^2 - 3r + 8 = 0$, with roots $(3 \pm i\sqrt{23})/2$. The general solution is

$$y = c_1 e^{3x/2}\cos(\tfrac{1}{2}\sqrt{23}\,x) + c_2 e^{3x/2}\sin(\tfrac{1}{2}\sqrt{23}\,x). \quad \blacksquare$$

EXAMPLE 2.18

Solve $y'' - 12y' + 36y = 0$.

The characteristic equation is $r^2 - 12r + 36 = 0$, with the repeated root $r = 6$. The general solution is

$$y = e^{6x}(c_1 + c_2 x). \quad \blacksquare$$

In the next section, we will solve a particular kind of linear second order differential equation in which the coefficients are not constant.

PROBLEMS FOR SECTION 2.6

In each of Problems 1 through 16, find the general solution of the differential equation. (These equations include all cases presented in Sections 2.4 and 2.6.)

1. $y'' - 4y' + 8y = 0$ **2.** $y'' - y = 0$ **3.** $y'' + 22y' + 121y = 0$

4. $y'' + 9y = 0$ **5.** $y'' + 10y' + 26y = 0$ **6.** $y'' - 2y' + 10y = 0$

7. $y'' + 10y' + 29y = 0$ **8.** $y'' + 6y' + 9y = 0$ **9.** $y'' - 4y' = 0$

10. $y'' + 2y' + 3y = 0$ **11.** $y'' + y' + y = 0$ **12.** $y'' + 3y' = 0$

13. $y'' - 4y' + 2y = 0$ **14.** $y'' + 3y' + 5y = 0$ **15.** $y'' + 10y' - y = 0$

16. $y'' + 20y' + 100y = 0$

In each of Problems 17 through 26, solve the initial value problem.

17. $y'' + 2y' - 3y = 0$; $y(0) = 0$, $y'(0) = -2$ **18.** $y'' - 2y' + 3y = 0$; $y(0) = 0$, $y'(0) = 3$

19. $y'' + 4y = 0$; $y(0) = -1$, $y'(0) = -8$ **20.** $y'' - y' - 2y = 0$; $y(0) = 5$, $y'(0) = 7$

21. $y'' - 4y' + 4y = 0$; $y(0) = 3$, $y'(0) = 2$ **22.** $y'' - 4y' + 5y = 0$; $y(0) = 2$, $y'(0) = 1$

23. $y'' + 2y' + 4y = 0$; $y(0) = 1$, $y'(0) = 0$ **24.** $y'' + y' + y = 0$; $y(0) = 2$, $y'(0) = 0$

25. $y'' - y' - 6y = 0$; $y(1) = 4$, $y'(1) = 7$ **26.** $y'' - 6y' + 9y = 0$; $y(3) = 3$, $y'(3) = 2$

27. The differential equation for a simple pendulum is

$$\theta''(t) + \frac{g}{L}\sin(\theta) = 0$$

(see Example 0.4). If θ is small, then $\sin(\theta) \approx \theta$, and we can approximate this differential equation by the constant coefficient linear differential equation $\theta'' + (g/L)\theta = 0$. Find the general solution of this differential equation, and show that the motion of the pendulum (for small θ) is periodic with period $2\pi\sqrt{L/g}$.

28. Suppose that φ is a solution of the initial value problem

$$y'' + Ay' + By = 0; \qquad y(x_0) = a, \qquad y'(x_0) = b.$$

Suppose that $A > 0$. Show that

$$\lim_{x \to \infty} \varphi(x) = 0.$$

29. Suppose that $A^2 - 4B \neq 0$. Show that the general solution of

$$y'' + Ay' + By = 0$$

can always be written in the form

$$y = e^{-Ax/2}[c_1\cosh(\beta x) + c_2\sinh(\beta x)],$$

in which β depends on A and B.

30. Use trigonometric identities to determine C and δ, with $0 \leq \delta \leq 2\pi$, such that

$$c_1\cos(\omega x) + c_2\sin(\omega x) = C \cos(\omega x + \delta).$$

The right side of this equation is called the *phase angle*, or *harmonic*, form of the left side.

In each of Problems 31 through 34, use the result of Problem 30 to write the given function in phase angle form, possibly multiplied by an exponential function.

31. $-4\sqrt{3} \cos(6x) + 12 \sin(6x)$ **32.** $e^{-5x}[-3 \cos(\pi x) - \sqrt{3} \sin(\pi x)]$

33. $8 \cos(2x) - 8 \sin(2x)$ **34.** $5e^{-6x}\sin(4x)$

In each of Problems 35, 36, and 37, construct a second order linear differential equation having the given function as general solution.

35. $c_1e^{-2x} + c_2e^{3x}$ **36.** $e^{-3x}[c_1\cos(2x) + c_2\sin(2x)]$ **37.** $c_1e^{-4x} + c_2xe^{-4x}$

2.7 The Cauchy-Euler Differential Equation _____

We will now consider another type of linear second order equation which we can solve.

The *Cauchy-Euler* differential equation has the form

$$y'' + \frac{1}{x} Ay' + \frac{1}{x^2} By = 0 \tag{2.15}$$

for $x > 0$ and A and B any constants. Often we write this differential equation as

$$x^2y'' + Axy' + By = 0 \tag{2.16}$$

for $x > 0$. We also refer to the differential equation (2.16) as the Cauchy-Euler differential equation.

There are two approaches commonly used to solve the Cauchy-Euler equation. We will pursue the details of a transformation method and mention the other method at the end of this section.

The transformation method consists of a change of variables which converts a Cauchy-Euler equation into a constant coefficient second order linear differential equation, which we know how to solve. Let

$$t = \ln(x)$$

for $x > 0$. Equivalently, $x = e^t$. If we replace x by e^t in $y(x)$, we obtain a new function Y of the independent variable t,

$$Y(t) = y(e^t) = y(x).$$

To write the Cauchy-Euler equation in terms of Y and t, we must compute some chain rule derivatives. We have

$$y'(x) = \frac{dy}{dx} = \frac{dY}{dx} = \frac{dY}{dt}\frac{dt}{dx} = Y'(t)\frac{1}{x}$$

and

$$y''(x) = \frac{d^2y}{dx^2} = \frac{d}{dx}\left[\frac{dY}{dt}\frac{dt}{dx}\right] = \frac{d^2Y}{dt^2}\left[\frac{dt}{dx}\right]^2 + \frac{dY}{dt}\frac{d^2t}{dx^2}$$

$$= Y''(t)\frac{1}{x^2} + Y'(t)\frac{-1}{x^2} = \frac{1}{x^2}[Y''(t) - Y'(t)].$$

As a result of these two differentiations, we conclude that

$$xy'(x) = Y'(t)$$

and

$$x^2y''(x) = Y''(t) - Y'(t).$$

Substitute these relationships into the Cauchy-Euler equation (2.16) to get

$$Y''(t) - Y'(t) + AY'(t) + BY(t) = 0,$$

or

$$Y'' + (A - 1)Y' + BY = 0. \tag{2.17}$$

This is a linear second order constant coefficient differential equation, whose general solution $Y(t)$ we can always find. The general solution of the original Cauchy-Euler equation is then obtained as

$$y(x) = Y[\ln(x)]$$

for $x > 0$.

In practice, we do not repeat the chain rule computations of $Y'(t)$ and $Y''(t)$ each time we meet a Cauchy-Euler equation. Just remember the following process:

$$x^2y'' + Axy' + By = 0$$
$$\downarrow \quad t = \ln(x)$$
$$Y''(t) + (A - 1)Y'(t) + BY(t) = 0$$
$$\downarrow$$
$$\text{general solution } Y(t)$$
$$\downarrow$$
$$\text{general solution } y(x) = Y[\ln(x)]$$
$$\text{of the Cauchy-Euler equation.}$$

Note that the coefficient of Y' in the transformed equation (2.17) is $A - 1$, where A is the coefficient of xy' in (2.16); the coefficient of Y is B, the same as the coefficient of y in (2.16). Thus, we can read the coefficients in the transformed equation (2.17) directly from the coefficients in the original Cauchy-Euler equation (2.16).

EXAMPLE 2.19

Find the general solution of $x^2y'' + 2xy' - 6y = 0$ for $x > 0$.

With $A = 2$ and $B = -6$, the transformed equation is

$$Y'' + (2 - 1)Y' - 6Y = 0,$$

or

$$Y'' + Y' - 6Y = 0.$$

This differential equation has characteristic equation $r^2 + r - 6 = 0$, with roots 2 and -3. The general solution for Y is

$$Y(t) = c_1 e^{2t} + c_2 e^{-3t}.$$

Let $t = \ln(x)$ to obtain the general solution for y;

$$y(x) = c_1 e^{2\ln(x)} + c_2 e^{-3\ln(x)} = c_1 x^2 + c_2 x^{-3}$$

for $x > 0$. ∎

In this example, we used the fact that

$$e^{\alpha \ln(x)} = e^{\ln(x^\alpha)} = x^\alpha$$

for any real number α and for $x > 0$. This relationship is used often when dealing with Cauchy-Euler equations.

EXAMPLE 2.20

Solve $x^2 y'' - xy' + 10y = 0$ for $x > 0$.
Put $t = \ln(x)$ for $x > 0$. The transformed equation is

$$Y'' - 2Y' + 10Y = 0.$$

This differential equation has characteristic equation $r^2 - 2r + 10 = 0$, with roots $1 \pm 3i$. The general solution for Y is

$$Y(t) = e^t [c_1 \cos(3t) + c_2 \sin(3t)].$$

Let $t = \ln(x)$ to get the general solution for y,

$$\begin{aligned}
y(x) &= e^{\ln(x)} \{ c_1 \cos[3\ln(x)] + c_2 \sin[3\ln(x)] \} \\
&= x \{ c_1 \cos[3\ln(x)] + c_2 \sin[3\ln(x)] \}
\end{aligned}$$

for $x > 0$. ∎

As usual, we can solve an initial value problem by finding the general solution of the differential equation, then using the initial values to solve for the constants.

EXAMPLE 2.21

Solve the initial value problem

$$x^2 y'' - 3xy' + 4y = 0; \qquad y(1) = 4, \qquad y'(1) = 5.$$

Letting $t = \ln(x)$, the differential equation transforms to

$$Y'' - 4Y' + 4Y = 0.$$

This equation has general solution

$$Y(t) = e^{2t}(c_1 + c_2 t).$$

Let $t = \ln(x)$ to get the general solution for y;

$$y(x) = e^{2\ln(x)}[c_1 + c_2\ln(x)] = x^2[c_1 + c_2\ln(x)].$$

Now we need

$$y(1) = c_1 = 4.$$

Thus far,

$$y(x) = x^2[4 + c_2\ln(x)].$$

Compute

$$y'(x) = 2x[4 + c_2\ln(x)] + x^2 c_2 \frac{1}{x};$$

hence,

$$y'(1) = 8 + c_2 = 5,$$

and we have $c_2 = -3$. The solution of the initial value problem is

$$y(x) = x^2[4 - 3\ln(x)].$$

By Theorem 2.1, this solution is unique in the interval $(0, \infty)$. This interval cannot be extended beyond zero because the solution is undefined at zero. ■

In this example, we used the transformation $t = \ln(x)$ to solve the Cauchy-Euler equation but wrote the general solution in terms of x before using the initial data. Sometimes we choose instead to transform the *entire* initial value problem (differential equation *and* initial conditions) in terms of t. We will pursue this idea in Problems 31 through 35.

We can solve the Cauchy-Euler differential equation for $x < 0$ by using the transformation $t = \ln(-x)$. The details are similar to those for $x > 0$, and we will not pursue them.

Another approach to finding the general solution of the Cauchy-Euler equation is to attempt a solution of the form $y = x^r$. Substitute $y = x^r$ into equation (2.16). Upon division by x^r, we obtain an algebraic equation (characteristic equation) for r yielding values of r for which x^r is a solution. We will pursue the details of this approach in Problem 48.

PROBLEMS FOR SECTION 2.7

In each of Problems 1 through 20, find the general solution of the Cauchy-Euler equation, assuming that $x > 0$.

1. $x^2y'' + 2xy' - 6y = 0$ **2.** $x^2y'' - 6xy' + 12y = 0$ **3.** $x^2y'' + 3xy' + y = 0$

4. $x^2y'' - 3xy' + 4y = 0$ **5.** $x^2y'' + xy' + 4y = 0$ **6.** $x^2y'' - 3xy' + 5y = 0$

7. $x^2y'' + xy' - 4y = 0$ **8.** $x^2y'' + 7xy' + 13y = 0$ **9.** $x^2y'' + 7xy' + 9y = 0$

10. $x^2y'' - xy' - 3y = 0$ **11.** $x^2y'' + 3xy' + 10y = 0$ **12.** $x^2y'' + 15xy' + 49y = 0$

13. $x^2y'' + 6xy' + 6y = 0$ **14.** $x^2y'' + xy' - 16y = 0$ **15.** $x^2y'' - 5xy' + 58y = 0$

16. $x^2y'' - 9xy' + 24y = 0$ **17.** $x^2y'' + 25y' + 144y = 0$ **18.** $x^2y'' + 7xy' + 24y = 0$

19. $x^2y'' - 11xy' + 35y = 0$ **20.** $x^2y'' + xy' - 6y = 0$

In each of Problems 21 through 30, solve the initial value problem.

21. $x^2y'' + 3xy' + 2y = 0$; $y(1) = 3$, $y'(1) = 3$ **22.** $x^2y'' + 5xy' + 20y = 0$; $y(1) = 0$, $y'(1) = 2$

23. $x^2y'' + 5xy' - 21y = 0$; $y(2) = 1$, $y'(2) = 0$ **24.** $x^2y'' + 25xy' + 144y = 0$; $y(1) = -3$, $y'(1) = 0$

25. $x^2y'' + xy' - y = 0$; $y(2) = 1$, $y'(2) = -3$ **26.** $x^2y'' - 9xy' + 24y = 0$; $y(1) = 1$, $y'(1) = 10$

27. $x^2y'' - 3xy' + 4y = 0$; $y(1) = 4$, $y'(1) = 5$ **28.** $x^2y'' + xy' - 4y = 0$; $y(1) = 3$, $y'(1) = 2$

29. $x^2y'' + 7xy' + 13y = 0$; $y(1) = 1$, $y'(1) = 3$ **30.** $x^2y'' - 3xy' - 5y = 0$; $y(1) = 8$, $y'(1) = 4$

31. Show that the change of variables $t = \ln(x)$ transforms the initial value problem

$$x^2y'' + Axy' + By = 0; \qquad y(x_0) = a, \qquad y'(x_0) = b,$$

with $x_0 > 0$, into the initial value problem

$$Y'' + (A - 1)Y' + BY = 0; \qquad Y[\ln(x_0)] = a, \qquad Y'[\ln(x_0)] = x_0 b.$$

In each of Problems 32 through 35, use the result of Problem 31 to transform the initial value problem into a constant coefficient second order initial value problem, then solve this problem and transform the solution to a solution of the original initial value problem.

32. $x^2y'' - 5xy' + 10y = 0$; $y(1) = 5$, $y'(1) = 17$ **33.** $x^2y'' - 3xy' + 4y = 0$; $y(1) = 5$, $y'(1) = 7$

34. $x^2y'' - 3xy' + (4 + \pi^2)y = 0$; $y(e) = e^2$, $y'(e) = 2e$

35. $x^2y'' - 3xy' + 3y = 0$; $y(2) = 2$, $y'(2) = 9$

In Problems 36 and 37, devise a transformation which converts the differential equation into a constant coefficient differential equation, then solve the initial value problem.

36. $(x + 2)^2y'' - 4(x + 2)y' + 6y = 0$; $y(0) = -4$, $y'(0) = 8$

37. $(3x - 4)^2y'' + 3(3x - 4)y' + 36y = 0$; $y\left(\dfrac{5}{3}\right) = 3$, $y'\left(\dfrac{5}{3}\right) = 12$

38. Determine conditions on $p(x)$ and $q(x)$ so that it is possible to transform the linear differential equation $y'' + p(x)y' + q(x)y = 0$ into a constant coefficient differential equation. *Hint:* Suppose that $t = f(x)$ is a twice-differentiable transformation that achieves this end and that f has an inverse function f^{-1}, so $x = f^{-1}(t)$. Let $Y(t) = y[f^{-1}(t)] = y(x)$. Imitate the discussion of the transformation of the Cauchy-Euler equation to show that

$$y'(x) = Y'(t)f'(x)$$

and

$$y''(x) = Y''(t)[f'(x)]^2 + Y'(t)f''(x).$$

Now use the original differential equation to show that

$$Y''(t) + \frac{f''(x) + p(x)f'(x)}{[f'(x)]^2} Y'(t) + \frac{q(x)}{[f'(x)]^2} Y(t) = 0.$$

This will be a constant coefficient differential equation if and only if

$$\frac{f''(x) + p(x)f'(x)}{[f'(x)]^2} = c_1 \quad \text{and} \quad \frac{q(x)}{[f'(x)]^2} = c_2$$

for some constants c_1 and c_2. Differentiate $q(x) = c_2[f'(x)]^2$ to produce $f''(x)$, and also use this equation to solve for $f'(x)$ in terms of known functions. Use these expressions for $f''(x)$ and $f'(x)$ to replace all $f'(x)$ and $f''(x)$

terms in the equation involving $p(x)$ to derive the requirement that

$$\frac{q'(x) + 2p(x)q(x)}{[q(x)]^{3/2}} = \frac{2c_1}{\sqrt{c_2}} = \text{constant.}$$

Verify that this requirement on $p(x)$ and $q(x)$ is met in the case of the Cauchy-Euler equation.

Finally, use the equation $|q(x)| = [f'(x)]^2$ to derive a formula for the transformation function f. If any transformation can be found, any constant multiple of it will also work, so it is not necessary to know the value of c_2 before finding f. Show that, in the case of the Cauchy-Euler equation, the transformation obtained in this way is $t = \ln(x)$ (or some constant multiple thereof), exactly the transformation we used.

In each of Problems 39 through 44, use the result of Problem 38 to find a transformation that converts the differential equation into one with constant coefficients. Use this transformation to solve the differential equation

39. $x(1 - x^2)^2 y'' - (1 - x^2)^2 y' + x^3 y = 0$

40. $y'' + (e^x - 1)y' + e^{2x} y = 0$

41. $4y'' + 4(e^x - 1)y' + e^{2x} y = 0$

42. $xy'' + (x^2 - 1)y' + x^3 y = 0$

43. $y'' + \tan(x)y' + \cos^2(x)y = 0$

44. $xy'' - y' + 4x^3 y = 0$

In each of Problems 45, 46, and 47, find a Cauchy-Euler differential equation having the given function as general solution.

45. $c_1 x^2 + c_2 x^{-3}$

46. $x^{-2}(c_1 \cos[3 \ln(x)] + c_2 \sin[3 \ln(x)])$

47. $c_1 x^4 + c_2 x^4 \ln(x)$

48. Substitute $y = x^r$ into the Cauchy-Euler differential equation (2.16). Show that x^r is a solution if and only if r satisfies the algebraic equation

$$r^2 + (A - 1)r + B = 0. \tag{2.18}$$

Now consider cases.

(a) Suppose that equation (2.18) has two distinct roots r_1 and r_2. Write the general solution for (2.16) in this case.

(b) Suppose that (2.18) has a repeated root r. Conclude that $y_1(x) = x^r$ is one solution. Use reduction of order to find a second, linearly independent solution $y_2(x) = \ln(x)x^r$.

(c) Suppose that equation (2.18) has complex roots $p \pm iq$. Show that x^{p+iq} and x^{p-iq} are linearly independent, and write the general solution of (2.16) in this case.

(d) Continuing from (c), use the fact that $x^p = e^{p \ln(x)}$ for $x > 0$, together with Euler's formula, to define

$$x^{p+iq} = x^p x^{iq} = x^p e^{iq \ln(x)} = x^p \{\cos[q \ln(x)] + i \sin[q \ln(x)]\}.$$

By replacing q with $-q$, show that

$$x^{p-iq} = x^p \{\cos[q \ln(x)] - i \sin[q \ln(x)]\}.$$

By taking linear combinations of x^{p+iq} and x^{p-iq}, show that $x^p \cos[q \ln(x)]$ and $x^p \sin[q \ln(x)]$ are solutions of (2.16) for $x > 0$, in this case that equation (2.18) has complex conjugate roots. Show that these two solutions are linearly independent on $(0, \infty)$, and write the general solution of the Cauchy-Euler equation in terms of these solutions.

2.8 Second Order Differential Equations and Mechanical Systems

Suppose that a helical spring of natural (unstretched) length L is suspended vertically from a support and a mass m is attached to the lower end. This mass stretches the spring an additional s units after it is allowed to come to rest (Figure 2.1). The spring is then

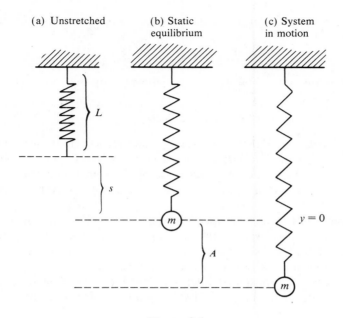

(a) Unstretched   (b) Static equilibrium   (c) System in motion

Figure 2.1

pulled down by an amount A and released with a given initial velocity B. We want to describe the ensuing motion of the mass.

ANALYSIS OF UNDAMPED SPRING MOTION

Everyday experience suggests that the mass will oscillate up and down, with speed and amount of displacement dependent upon the mass, the initial displacement A, the stiffness of the spring, and whether the weight is released from rest or struck in some manner. In reality, the motion will die out eventually because of damping forces which are always present. We will assume initially that there are no damping forces and determine what the motion would be in this circumstance. Then we will add damping and other forces and determine how they influence the motion.

In Example 0.2 we derived the differential equation for the displacement function $y(t)$ for this undamped case. It is

$$y'' + \frac{k}{m}\, y = 0,$$

in which k is the spring constant. The initial conditions take the form

$$y(0) = A, \qquad y'(0) = B,$$

in which we have let $t = 0$ be the time at which the system is set in motion and A and B are given numbers. The general solution of $y'' + (k/m)y = 0$ is

$$y = c_1\cos(\omega_0 t) + c_2\sin(\omega_0 t),$$

in which $\omega_0 = \sqrt{k/m}$.

As a simple first case, suppose that the mass is released from rest. Now the initial conditions are

$$y(0) = A, \qquad y'(0) = 0.$$

Then $y(0) = c_1 = A$ and $y'(0) = c_2\omega_0 = 0$, and we obtain the unique solution for the displacement function,

$$y(t) = A \cos(\omega_0 t).$$

This tells us the position of the weight at any time t and says that the weight oscillates up and down with period $T = 2\pi/\omega_0$. The frequency is $1/T = \omega_0/2\pi$. Note that the stiffer the spring, the larger k is, resulting in a larger frequency and smaller period. This coincides with our intuition about how the stiffness of the spring will influence the motion.

The amplitude of the displacements is $|y(t)| = A$. In real life, we would not expect the spring to continue to oscillate the same amount as this initial displacement, but in this model it does because we have deliberately omitted damping factors. In fact, damping factors such as air resistance are always present.

If we consider the same model but now suppose that the initial displacement is A, and the weight is given an initial velocity B, the initial value problem is

$$y'' + \frac{k}{m} y = 0; \qquad y(0) = A, \qquad y'(0) = B.$$

We find that this problem has unique solution

$$y(t) = A \cos(\omega_0 t) + \frac{B}{\omega_0} \sin(\omega_0 t).$$

This solution can be written in the equivalent form

$$y(t) = \sqrt{A^2 + \omega_0^{-2} B^2} \cos(\omega_0 t + \delta)$$

for some constant δ (see Problem 22 at the end of this section). This is called the *phase angle*, or *harmonic*, form of the solution, and it shows that, in the absence of damping, the motion is always oscillatory, regardless of initial velocity and displacement. In this case, we say that the system experiences *undamped free motion*. Here are two examples involving this type of motion.

EXAMPLE 2.22

A 5-pound object is attached to the lower end of a helical spring, stretching it $\frac{3}{2}$ inches from its natural length. After the object has come to rest, it is pulled down an additional 6 inches and released. Find the position of the object at any time t, and determine the period, frequency, and amplitude of the resulting oscillations.

Since 5 pounds of force stretch the spring $\frac{3}{2}$ inches (or $\frac{1}{8}$ foot), then by Hooke's law,

$$5 = \tfrac{1}{8}k,$$

in which k is the spring constant. Thus, $k = 40$ pounds/foot. The weight of the object is the product of its mass and the acceleration due to gravity (32 feet/second^2), so the mass of the object is $\frac{5}{32}$ slug. Substitute these numbers for k and m into the

differential equation

$$y'' + \frac{k}{m} y = 0$$

to get

$$y'' + 256y = 0.$$

The initial conditions are

$$y(0) = \tfrac{1}{2} \quad \text{and} \quad y'(0) = 0,$$

the latter due to the assumption that the object was released from rest. The solution of this initial value problem is

$$y = \tfrac{1}{2} \cos(16t),$$

giving the position on the vertical y-axis at any time $t \geq 0$. The period of oscillation is $2\pi/16$, or $\pi/8$, seconds/cycle. The frequency is $8/\pi$ cycles/second, or $8/\pi$ hertz. The amplitude of the motion is $\tfrac{1}{2}$ foot. ■

EXAMPLE 2.23

Keep everything as in Example 2.22 except that, after the object has been pulled down $\tfrac{1}{2}$ foot, it is struck upward with a force giving it an initial vertical velocity of 6 feet/second instead of being released from rest. How does the system react?

Now the initial value problem is

$$y'' + 256y = 0; \qquad y(0) = \tfrac{1}{2}, \qquad y'(0) = -6$$

(minus because downward was arbitrarily chosen as positive and the weight is struck upward). This initial value problem is routine to solve, and we obtain

$$y(t) = \tfrac{1}{2} \cos(16t) - \tfrac{3}{8} \sin(16t).$$

The frequency of the motion is the same as in the preceding example, but now the amplitude of the oscillations is $\sqrt{(\tfrac{1}{2})^2 + (-\tfrac{3}{8})^2}$, or $\tfrac{5}{8}$ foot.

It is left for the student to derive the amplitude of the oscillations if the weight is struck a downward, instead of an upward, blow. ■

DAMPED SPRING MOTION

We will now extend our analysis to include damping, leading to much more interesting and realistic results. Damping can be caused in many ways. Even the twisting of the wire in the spring itself causes some damping. Damping can also be caused by use of a dashpot (shock absorber), as in Figure 2.2, or by pulling the object through water, grease, or some other fluid. Experiments have shown that the damping forces of these types are proportional to the velocity of the weight and act opposite to the direction of motion. If this constant of proportionality is a positive number c (called the *damping constant*), the damping force has magnitude cy', and the equation of motion is

$$my'' = -ky - cy'.$$

Figure 2.2

This equation differs from that of the undamped case by the inclusion of the damping term cy'. Write this differential equation as

$$y'' + \frac{c}{m} y' + \frac{k}{m} y = 0.$$

The characteristic equation is

$$r^2 + \frac{c}{m} r + \frac{k}{m} = 0,$$

with roots

$$r = -\frac{c}{2m} \pm \frac{1}{2m} \sqrt{c^2 - 4km}.$$

The general solution for y, and therefore the resulting motion, will differ greatly depending on the nature of these roots, which in turn depends on the relative sizes of the mass, spring constant, and damping constant. Dependence of the motion on these three factors should not be surprising.

Consider three cases.

Case 1: $c^2 - 4km > 0$ Now the characteristic equation has two real, distinct roots

$$r_1 = -\frac{c}{2m} + \frac{1}{2m} \sqrt{c^2 - 4km} \quad \text{and} \quad r_2 = -\frac{c}{2m} - \frac{1}{2m} \sqrt{c^2 - 4km}.$$

The equation $y'' + (c/m)y' + (k/m)y = 0$ has general solution

$$y(t) = c_1 e^{r_1 t} + c_2 e^{r_2 t}.$$

Since m, c, and k are positive constants, $c^2 - 4km < c^2$; hence,

$$\sqrt{c^2 - 4km} < c.$$

Therefore, $r_1 < 0$. Clearly, $r_2 < 0$ also. This means that

$$\lim_{t \to \infty} y(t) = 0$$

and the motion dies out with time.

This case in which the characteristic equation has two real roots is called *overdamping*. As a specific example of overdamping, suppose that $c = 6$, $k = 5$, and $m = 1$. Then $c^2 - 4km = 16$, and we get $r_1 = -1$ and $r_2 = -5$. The general solution is

$$y(t) = c_1 e^{-t} + c_2 e^{-5t}.$$

To be even more specific, suppose at time zero the mass was drawn four units upward from the equilibrium position and released downward with a speed of two units per second. This gives us the initial conditions

$$y(0) = -4 \quad \text{and} \quad y'(0) = 2.$$

From these initial conditions, we obtain $c_1 = -\frac{9}{2}$ and $c_2 = \frac{1}{2}$, so the initial value problem has the unique solution

$$y(t) = -\tfrac{9}{2} e^{-t} + \tfrac{1}{2} e^{-5t} = \tfrac{1}{2} e^{-t}[-9 + e^{-4t}].$$

This function is graphed in Figure 2.3 for $t \geq 0$.

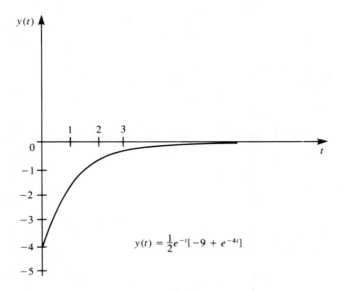

$$y(t) = \frac{1}{2}e^{-t}[-9 + e^{-4t}]$$

Figure 2.3. Overdamping (free motion).

What does this solution tell us about the motion of the mass? Since $-9 + e^{-4t} < 0$ for $t \geq 0$, the mass always remains above the equilibrium point. Since $y(t) \to 0$ as $t \to \infty$, the mass begins moving downward from its starting point four units above the equilibrium point and continues moving downward. The graph of $y(t)$ is concave downward, so the velocity $y'(t)$ is decreasing. We conclude that the object moves downward at a decreasing rate of speed, never quite coming to rest under the conditions of the model and never reaching the equilibrium point. ∎

Case 2: $c^2 - 4km = 0$ Now the characteristic equation has repeated root $r = -c/2m$, and the general solution of $y'' + (c/m)y' + (k/m)y = 0$ is

$$y(t) = (c_1 + c_2 t)e^{-ct/2m}.$$

This case is called *critical damping*. Again, $y(t) \to 0$ as $t \to \infty$, so the motion dies out with time. As a specific example of critical damping, suppose that $c = 2$ and $k = m = 1$. Suppose also that the mass is initially pulled up to a position four units above the equilibrium position and then pushed downward with a speed of five units per second. With these values of c, k, and m, the general solution of $y'' + (c/m)y' + (k/m)y = 0$ is

$$y(t) = (c_1 + c_2 t)e^{-t}.$$

The initial conditions are

$$y(0) = -4, \qquad y'(0) = 5.$$

We find that the unique solution of the initial value problem is

$$y(t) = (-4 + t)e^{-t}.$$

Note that $y(4) = 0$, which means that the mass passes through the equilibrium point $(y = 0)$ 4 seconds after being released. In fact, $y(t)$ reaches its maximum value at

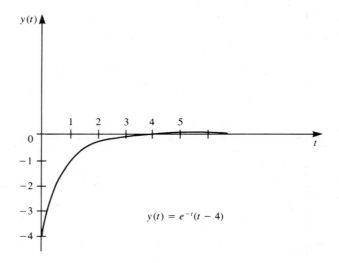

Figure 2.4. Critical damping (free motion).

$t = 5$ seconds, and this maximum value is $y(5) = e^{-5} \approx 0.007$. After this, y decreases to zero as t increases toward infinity. Figure 2.4 shows a graph of $y(t)$.

   With these initial conditions and values of c, k, and m, the object begins four units above the equilibrium point, moves downward through this point at the 4-second mark, and continues downward at a decreasing speed until, at the 5-second mark, it reaches 0.007 unit below equilibrium. The mass then begins moving back up toward the equilibrium point with continually decreasing speed, never reaching this point again. ∎

   It is left for the student to show that, in the critical damping case, the object either passes through the equilibrium point exactly once (as in this example) or never reaches it at all.

Case 3: $c^2 - 4km < 0$ This case is called *underdamping*. The characteristic equation has complex roots

$$-\frac{c}{2m} \pm \frac{1}{2m}\sqrt{4km - c^2}\,i,$$

and the general solution of $y'' + (c/m)y' + (k/m)y = 0$ is

$$y(t) = e^{-ct/2m}[c_1\cos(\beta t) + c_2\sin(\beta t)],$$

where $\beta = \sqrt{4km - c^2}/2m$. Because c and m are positive and $t \geq 0$, $y(t) \to 0$ as $t \to \infty$. Thus, in all cases in which there is damping, the motion dies out eventually. In this case of underdamping, however, the motion is oscillatory, because of the sine and cosine terms. Because of the exponential factor, the motion is *not* periodic.

   As a specific example of underdamping, suppose that $c = k = 2$ and $m = 1$. The general solution is

$$y(t) = e^{-t}[c_1\cos(t) + c_2\sin(t)].$$

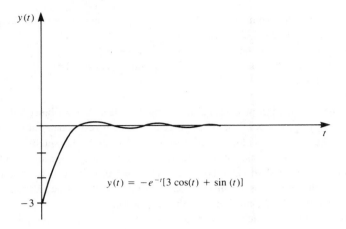

Figure 2.5. Underdamping (free motion).

Suppose that the object is driven downward from a point three units above the equilibrium point with an initial speed of two units per second. These initial conditions are

$$y(0) = -3, \qquad y'(0) = 2.$$

The unique solution of this initial value problem is found to be

$$y(t) = -e^{-t}[3 \cos(t) + \sin(t)]$$
$$= -\sqrt{10}e^{-t}\cos[t - \tan^{-1}(\tfrac{1}{3})].$$

This function is graphed in Figure 2.5. Although the motion decays with time, it also oscillates, but with decreasing amplitude. Further, the mass will pass back and forth through the equilibrium point. In fact, the mass passes through equilibrium at exactly those times at which $y(t) = 0$, which in turn occurs when $\tan(t) = -3$. This equation has infinitely many positive solutions, some of which are approximately 1.89 seconds, 5.03 seconds, 8.17 seconds, and 11.32 seconds. ■

Although this analysis of spring motion has yielded some interesting information, it has omitted the case in which the motion is driven by some forcing function. We will consider a mathematical model for this possibility after we learn to solve certain nonhomogeneous linear differential equations. At that point, we will also be able to apply differential equations to the analysis of electrical circuits.

PROBLEMS FOR SECTION 2.8

1. In Example 1, determine the time at which the mass first passes through the equilibrium position, moving downward. Determine the speed of the mass at that time.
2. A 2-pound weight is attached to the lower end of a spring, stretching it $\frac{8}{3}$ inches. The weight is then pulled down an additional 4 inches and released from rest.
   (a) Set up and solve the differential equation for the displacement function of the weight.
   (b) How long after the weight has been released does it first pass through the equilibrium point?
   (c) What is the speed of the weight when it first passes through the equilibrium point?

3. An object weighing 8 pounds is suspended from a spring, thus stretching it 6 inches. Once the weight has come to rest at the equilibrium position, it is pushed downward at 4 feet/second.
   (a) Solve the initial value problem describing the displacement of the weight.
   (b) What is the velocity of the weight when it is 3 inches above equilibrium and falling?
   (c) What is the period of the oscillations?
   (d) Express the solution of the initial value problem in phase angle form.

4. An object weighing 4π pounds is suspended from a spring, stretching it $1/\pi$ feet. A new object is then attached to the spring, replacing the old one, pulled down 5 inches, and released from rest. The resulting frequency of oscillation is 4 hertz. Assuming that there is no damping in the system, determine the weight of the new object.

5. When a 3-pound weight is suspended from a spring, it stretches the spring 1 inch. A 32-pound weight is attached to the spring and the entire system is lowered into an oil bath, causing a damping force of $12v$ pounds, where v is the velocity of the weight. The weight is then raised 6 inches from its equilibrium position and released from rest. Find and solve the differential equation of the motion. Graph the solution.

6. An object having a mass of 1 gram is attached to the lower end of a spring having spring modulus 29 dynes/centimeter. The weight is in turn adhered to a dashpot which imposes a damping force of $10v$ dynes. Determine the motion of the weight if it is pulled down 3 centimeters from equilibrium and then struck upward with a blow sufficient to impart a velocity of 1 centimeter/second. Graph the solution of this problem.

7. An object of mass 1 kilogram is suspended from a spring which has spring constant 24 newtons/meter. Attached to the object is a shock absorber which induces a drag of $11v$ newtons. The system is set in motion by lowering the weight 25/3 centimeters and then striking upward hard enough to impart a velocity of 5 meters/second. Solve for and graph the displacement function.

8. When an 8-pound weight is suspended from a spring, it stretches the spring 2 inches. Determine the equation of motion when an object with a mass of 7 kilograms is suspended from this spring and the system is set in motion by striking the object an upward blow imparting a velocity of 4 meters/second.

9. Graph the general solution of $my'' + cy' + ky = 0$ in the case of overdamping.

10. How many times can the weight pass through the equilibrium point in the case of overdamping? What condition can you place on the initial displacement $y(0)$ to guarantee that the weight never passes through the equilibrium position?

11. Graph the general solution of $my'' + cy' + ky = 0$ in the case of critical damping.

12. How many times can the weight pass through the equilibrium position in the case of critical damping? What condition can you place on $y(0)$ to guarantee that the weight never passes through the equilibrium position? How does the initial velocity $y'(0)$ influence whether the weight passes through the equilibrium position?

13. Let A be a positive constant. Find solutions of the initial value problem

$$my'' + cy' + ky = 0; \qquad y(0) = A, \qquad y'(0) = 0$$

in all four cases: undamped $(c = 0)$, overdamped, critically damped, and underdamped. Graph all four solutions on the same set of axes to compare the effects of the the the initial displacement on the solution. Be careful to mark each graph clearly according to the case it represents.

14. Let B be a positive constant. Find solutions of the initial value problem

$$my'' + cy' + ky = 0; \qquad y(0) = 0, \qquad y'(0) = B$$

for the following cases: undamped $(c = 0)$, overdamped, critically damped, and underdamped motion. Graph all four solutions on the same set of axes, clearly marking each curve, to compare effects of the initial velocity on the displacement function.

15. Let A be a positive constant. In order to gauge comparative effects of initial position and velocity on the resulting motion, find and graph solutions of $my'' + cy' + ky = 0$ in the overdamped case
    (a) when $y(0) = A$ and $y'(0) = 0$;

(b) when $y(0) = 0$ and $y'(0) = A$.

How does the motion differ in these two cases?

16. Find the critically damped solutions of $my'' + cy' + ky = 0$ when
    (a) $y(0) = A,$ $y'(0) = 0$;
    (b) $y(0) = 0,$ $y'(0) = A$,

    with A any positive constant. Graph both solutions on the same set of axes. What conclusions do you draw about the influence of initial displacement and initial velocity on the displacement function?

17. Redo Problem 16 with underdamping in place of critical damping.

18. Redo Problem 16 with undamped replacing critical damping.

19. In underdamped motion, what effect does the damping constant c have on the frequency of the motion?

20. Suppose that $y(0) = y'(0) \neq 0$. Determine the maximum displacement of the weight in the critically damped solution, and show that the time at which this maximum occurs is independent of the actual value of $y(0)$.

21. As m increases in the underdamped case, the solution appears to die out more quickly with time. Intuitively, however, it would seem that a heavier mass would stretch the spring more and cause greater oscillations. Which is correct?

22. In the treatment of the undamped equation $y'' + (k/m)y = 0$ at the beginning of this section, we obtained the general solution

$$y(t) = c_1\cos(\omega_0 t) + c_2\sin(\omega_0 t),$$

in which $\omega_0 = \sqrt{k/m}$. Show in detail, by producing expressions for d and δ, that this solution can be written as $y(t) = d \cos(\omega_0 t + \delta)$.

23. In the case of underdamping, we obtained the general solution

$$y(t) = e^{-ct/2m}\left[c_1\cos\left(\sqrt{4km - c^2}\,\frac{t}{2m}\right) + c_2\sin\left(\sqrt{4km - c^2}\,\frac{t}{2m}\right)\right].$$

Show in detail how this solution can be written in the form

$$y(t) = de^{-ct/2m}\cos(\bar{\omega}t + \delta)$$

for appropriate choices of $\bar{\omega}$ and δ.

24. Consider underdamped motion, with the general solution written as

$$y(t) = de^{-ct/2m}\cos(\bar{\omega}t + \delta).$$

(a) Graph this solution, recalling that d, δ, and $\bar{\omega}$ are constants.

(b) The *natural period* of an undamped system, whose differential equation is $y'' + (k/m)y = 0$, is defined to be $T = 2\pi\sqrt{m/k}$. In the underdamped case, the motion does not exhibit this periodicity. Define the *quasi period* as the time between successive maxima and minima of $y(t)$. Show that the quasi period T_q is given by

$$T_q = \frac{4m\pi}{\sqrt{4km - c^2}}.$$

(c) Show that

$$T_q = \frac{T}{\sqrt{1 - \dfrac{c^2}{4mk}}}.$$

Hence, conclude that when $c^2/4mk$ is very small, effects of damping are negligible in computing the quasi period, and in this case T_q is approximately equal to T.

25. Prove the principle of conservation of energy for the undamped harmonic oscillator. This principle states that

the total energy of the system remains constant throughout the motion. *Hint:* The energy is a sum of the kinetic energy $\frac{1}{2}mv^2$ (v is the velocity) and the potential energy $\frac{1}{2}ky^2$. Show that the differential equation $my'' + ky = 0$ can be written as $mv(dv/dy) = -ky$, and integrate.

26. Suppose that the acceleration of the weight on the spring at a distance d from the equilibrium position is a. Prove that the period of the motion is

$$2\pi\sqrt{-\frac{d}{a}}$$

in the case of undamped motion.

27. A mass m_1 is attached to a spring and allowed to vibrate with undamped motion having period p. Suppose that at some time a second mass m_2 is instantaneously glued onto m_1. Prove that the total mass $m_1 + m_2$ now exhibits simple harmonic motion with period

$$p\sqrt{1 + \frac{m_2}{m_1}}.$$

In Problems 28 through 33, use a calculator to find the needed constants to four decimal places.

28. Write the solution for the undamped case in the form $y = d\cos(\omega_0 t + \delta)$ if $k = 1.3$, $m = 2$, $y(0) = 1.72$, and $y'(0) = 2.4$.

29. Write the solution for the displacement function if $m = 3$, $c = 2.4$, $k = \sqrt{3}$, $y(0) = 3.72$, and $y'(0) = 1.6$.

30. Write the solution for the displacement function if $m = 1$, $c = 3.45$, $k = 2.4$, $y(0) = 1.2$, and $y'(0) = 2.74$.

31. Suppose that $k = 1.8$, $m = 2.2996$, $c = 4.069$, $y(0) = 3.24$, and $y'(0) = 1.52$. We find that $c^2 - 4km$ is approximately 0.0004, which we take to be approximately zero. With this assumption, find a critical damped solution for $y(t)$, with $y(0) = 3.24$ and $y'(0) = 1.52$.

32. Find a solution for $y(t)$ satisfying $y(0) = 1.8$ and $y'(0) = 0.32$ in the case in which $c = \sqrt{5}$, $k = 2.46$, and $m = 1.92$.

33. Find a solution for $y(t)$ satisfying $y(0) = 2.36$ and $y'(0) = 1.4$ if $m = 2.31$, $c = \sqrt{7}$, and $k = 3.02$.

2.9 *Theory of Nonhomogeneous Second Order Equations*

We will now consider the nonhomogeneous linear differential equation

$$y'' + p(x)y' + q(x)y = f(x), \tag{2.19}$$

in which f is continuous and not identically zero on some interval I. Much of our work will be based on the following theorem, which suggests how we might approach finding the general solution of equation (2.19).

THEOREM 2.5

Suppose that y_1 and y_2 form a fundamental set of solutions of the homogeneous equation $y'' + p(x)y' + q(x)y = 0$ on I. Let y_p be any solution of equation (2.19) on I. Then every solution of equation (2.19) on I is of the form

$$y = c_1 y_1(x) + c_2 y_2(x) + y_p(x)$$

for some choices of the constants c_1 and c_2. ∎

Before proving the theorem, we will consider its ramifications. The theorem states that we know every solution of equation (2.19) on I if we can find a fundamental set of solutions of the *homogeneous* equation $y'' + p(x)y' + q(x)y = 0$, together with *just one* (any) particular solution of the *nonhomogeneous* equation $y'' + p(x)y' + q(x)y = f(x)$. We call the equation

$$y'' + p(x)y' + q(x)y = 0$$

the *associated homogeneous equation* of equation (2.19), and $c_1 y_1 + c_2 y_2$ is the *associated homogeneous solution* of equation (2.19). For this reason, we often denote $c_1 y_1 + c_2 y_2$ as $y_h(x)$. Once we have found y_h, we need *only one* solution y_p of equation (2.19). The sum $y_h + y_p$ is called the *general solution* of $y'' + p(x)y' + q(x)y = f(x)$ and contains all solutions of this equation.

Proof of Theorem 2.5   Let ξ be any solution of equation (2.19) on I. Then

$$\xi'' + p(x)\xi' + q(x)\xi = f(x).$$

Since y_p is also a solution of equation (2.19),

$$y_p'' + p(x)y_p' + q(x)y_p = f(x).$$

Subtract these two equations to get

$$(\xi - y_p)'' + p(x)(\xi - y_p)' + q(x)(\xi - y_p) = f(x) - f(x) = 0.$$

Therefore, $\xi - y_p$ is a solution of the associated homogeneous equation $y'' + p(x)y' + q(x)y = 0$ on I. But y_1 and y_2 form a fundamental set of solutions of this equation. Therefore, there are constants c_1 and c_2 such that

$$\xi - y_p = c_1 y_1 + c_2 y_2,$$

so $\xi = c_1 y_1 + c_2 y_2 + y_p$, as we wanted to prove.   ∎

With a little more work, we could show that, given a solution ξ of (2.19), there are *unique* constants c_1 and c_2 such that $\xi = y_p + c_1 y_1 + c_2 y_2$.

EXAMPLE 2.24

Solve the differential equation $y'' - 4y = 8x$.

The associated homogeneous equation is

$$y'' - 4y = 0,$$

with general solution

$$y_h(x) = c_1 e^{2x} + c_2 e^{-2x}.$$

This is the associated homogeneous solution of $y'' - 4y = 8x$. One particular solution of $y'' - 4y = 8x$ is $y_p(x) = -2x$ (we will see in the next section how to obtain this). The general solution of $y'' - 4y = 8x$ is

$$y = y_h + y_p = c_1 e^{2x} + c_2 e^{-2x} - 2x. \quad ∎$$

EXAMPLE 2.25

Solve the initial value problem

$$x^2 y'' - 4xy' + 6y = 36 \ln(x); \qquad y(1) = 2, \qquad y'(1) = 3.$$

The associated homogeneous equation is

$$x^2 y'' - 4xy' + 6y = 0,$$

a Cauchy-Euler equation which has general solution

$$y_h(x) = c_1 x^3 + c_2 x^2.$$

This is the associated homogeneous solution of $x^2 y'' - 4xy' + 6y = 36 \ln(x)$. A particular solution of the nonhomogeneous differential equation is

$$y_p(x) = 6 \ln(x) + 3x^2 + 5$$

for $x > 0$. (Again, we will say later how to obtain y_p.) The general solution of the nonhomogeneous Cauchy-Euler equation is $y = y_h + y_p$:

$$y = c_1 x^3 + c_2 x^2 + 6 \ln(x) + 3x^2 + 5$$

for $x > 0$.

Now solve for c_1 and c_2 to satisfy the initial conditions. We have

$$y(1) = c_1 + c_2 + 3 + 5 = 2,$$

or

$$c_1 + c_2 = -6.$$

Since $y'(x) = 3c_1 x^2 + 2c_2 x + 6/x + 6x$, we get

$$y'(1) = 3c_1 + 2c_2 + 12 = 3,$$

or

$$3c_1 + 2c_2 = -9.$$

These equations for c_1 and c_2 have solution $c_1 = 3$ and $c_2 = -9$. The solution of the initial value problem is

$$y(x) = 3x^3 - 9x^2 + 6 \ln(x) + 3x^2 + 5$$
$$= 3x^3 - 6x^2 + 6 \ln(x) + 5.$$

This solution is unique on $(0, \infty)$. ∎

Note of caution: The values of c_1 and c_2 *cannot* be determined from the associated homogeneous solution y_h alone. The initial values apply only to the solution of the given (in this case, nonhomogeneous) differential equation.

EXAMPLE 2.26

Find the general solution of $y'' + 4y = e^x$.

The associated homogeneous equation is $y'' + 4y = 0$, with general solution

$$y_h = c_1\cos(2x) + c_2\sin(2x).$$

In the next section, we will see how to find the particular solution

$$y_p = \tfrac{1}{5}e^x$$

of the nonhomogeneous equation $y'' + 4y = e^x$. The general solution of $y'' + 4y = e^x$ is

$$y = y_h + y_p = c_1\cos(2x) + c_2\sin(2x) + \tfrac{1}{5}e^x. \quad \blacksquare$$

THE PRINCIPLE OF SUPERPOSITION

Sometimes it is difficult to find a particular solution of equation (2.19). It may be, however, that f is a finite sum of functions

$$f = f_1 + f_2 + \cdots + f_n,$$

and we may be able to find a particular solution y_j of each problem

$$y'' + p(x)y' + q(x)y = f_j(x).$$

In this case, the sum of these particular solutions is a particular solution of $y'' + p(x)y' + q(x)y = f(x)$, due to the linearity of the differential equation. This idea is called the *principle of superposition*, and it is extremely useful in decomposing a problem into "smaller" problems which individually might be easier to solve than the original problem. Here is a more careful statement of this principle.

THEOREM 2.6   (Principle of Superposition)

Suppose that $f = f_1 + f_2 + \cdots + f_n$ on I and that y_j is a solution of

$$y'' + p(x)y' + q(x)y = f_j(x)$$

on I for $j = 1, 2, \ldots, n$. Then

$$y_p = y_1 + y_2 + \cdots + y_n$$

is a solution of $y'' + p(x)y' + q(x)y = f(x)$ on I.

Proof   Upon substitution of y_p into the differential equation, we get

$$\begin{aligned}
[y_1 + y_2 + \cdots &+ y_n]'' + p(x)[y_1 + y_2 + \cdots + y_n]' + q(x)[y_1 + y_2 + \cdots + y_n] \\
&= [y_1'' + p(x)y_1' + q(x)y_1] + [y_2'' + p(x)y_2' + q(x)y_2] + \cdots \\
&\quad + [y_n'' + p(x)y_n' + q(x)y_n] \\
&= f_1(x) + f_2(x) + \cdots + f_n(x) = f(x),
\end{aligned}$$

proving the theorem. $\quad \blacksquare$

In summary, here is a schematic diagram of our results on solving the nonhomogeneous linear equation $y'' + p(x)y' + q(x)y = f(x)$.

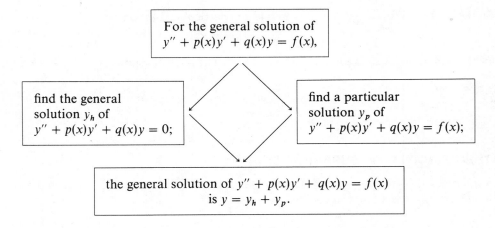

We now have a procedure for finding the general solution of $y'' + p(x)y' + q(x)y = f(x)$: find y_h and a particular solution y_p. We already know how to produce y_h in some cases. In the next two sections, we will discuss methods for finding y_p.

PROBLEMS FOR SECTION 2.9

In each of Problems 1 through 25, (a) verify that y_p is a solution of the differential equation; (b) find the general solution of the differential equation; and (c) use this general solution to find a solution of the initial value problem.

1. $y'' - y' - 6y = 26 - 24x$; $y_p(x) = 4x - 5$; $y(0) = 0$, $y'(0) = 4$
2. $y'' - 4y' + 4y = 6 \sin(x) - 8 \cos(x)$; $y_p(x) = 2 \sin(x)$; $y(0) = 3$, $y'(0) = 4$
3. $y'' + 4y = 2x^2 - 12x + 9$; $y_p(x) = \frac{1}{2}x^2 - 3x + 2$; $y(0) = 5$, $y'(0) = 5$
4. $y'' + 2y' + 10y = 72e^{2x}$; $y_p(x) = 4e^{2x}$; $y(0) = 9$, $y'(0) = -3$
5. $x^2 y'' - 4xy' + 6y = 6 \ln(x) + 7$; $y_p(x) = \ln(x) + 2$; $y(1) = 5$, $y'(1) = 8$
6. $x^2 y'' - 2y = 8x^3 - 4x + 6$; $y_p(x) = 2x^3 + 2x - 3$; $y(1) = 0$, $y'(1) = 8$
7. $x^2 y'' - xy' + y = 3x^2 + 6x$; $y_p(x) = 3x^2 + 3x[\ln(x)]^2$; $y(1) = 2$, $y'(1) = 19$
8. $x^2 y'' - 2xy' + 2y = 10 \sin[\ln(x)]$; $y_p(x) = 3 \cos[\ln(x)] + \sin[\ln(x)]$; $y(1) = 6$, $y'(1) = 3$
9. $y'' + 4y = 6 \cos(x)$; $y_p(x) = 2 \cos(x)$; $y(\pi) = 2$, $y'(\pi) = 6$
10. $y'' + y = x^3$; $y_p(x) = x^3 - 6x$; $y(0) = -2$, $y'(0) = 3$
11. $y'' + 2y' + y = 4e^x$; $y_p(x) = e^x$; $y(0) = y'(0) = 0$
12. $y'' + 3y' - 4y = 8e^{2x} + 1$; $y_p(x) = \frac{4}{3}e^{2x} - \frac{1}{4}$; $y(0) = -4$, $y'(0) = 0$
13. $y'' - 3y' + 2y = x$; $y_p(x) = \frac{1}{2}x + \frac{3}{4}$; $y(0) = -5$, $y'(0) = 6$
14. $y'' - 4y = 8e^{2x}$; $y_p(x) = 2xe^{2x}$; $y(0) = 4$, $y'(0) = 6$
15. $y'' + 9y = 5 \cos(2x) - 10 \sin(2x)$; $y_p(x) = \cos(2x) - 2 \sin(2x)$; $y\left(\frac{\pi}{2}\right) = 2$, $y'\left(\frac{\pi}{2}\right) = 6$
16. $y'' + 2y' + 8y = 3e^{4x} + 1$; $y_p(x) = \frac{3}{32}e^{4x} + \frac{1}{8}$; $y(0) = y'(0) = 0$
17. $y'' - 4y' + 2y = 3x + 2$; $y_p(x) = \frac{3}{2}x + 4$; $y(0) = 1$, $y'(0) = 0$
18. $y'' - y' + 10y = 4x^2 + 2x$; $y_p(x) = \frac{2}{5}x^2 + \frac{7}{25}x - \frac{13}{250}$; $y(3) = 1$, $y'(3) = 0$
19. $y'' + 2y' - 5y = 20$; $y_p(x) = -4$; $y(0) = -4$, $y'(0) = 0$

20. $y'' + 2y' = \tan(2x);$ $y_p(x) = -\frac{1}{4}\cos(2x)\ln|\sec(2x) + \tan(2x)|;$ $y(0) = 6,$ $y'(0) = \frac{3}{2}$

21. $y'' + y = \tan(x);$ $y_p(x) = -\cos(x)\ln|\sec(x) + \tan(x)|;$ $y(0) = 2,$ $y'(0) = 0$

22. $x^2 y'' + 3xy' + y = 4x;$ $y_p(x) = x;$ $y(1) = 6,$ $y'(1) = -7$

23. $x^2 y'' + 2xy' - 6y = x^4 - 2x^2;$ $y_p(x) = \frac{1}{14}x^4 + \frac{2}{25}x^2 - \frac{2}{5}x^2\ln(x);$ $y(1) = 5,$ $y'(1) = 2$

24. $y'' + 2y' + y = \dfrac{1}{x}e^{-x};$ $y_p(x) = xe^{-x}\ln(x);$ $y(1) = y'(1) = 0$

25. $y'' - 4y' + 3y = -3\sin(x + 2);$ $y_p(x) = -\frac{3}{5}\cos(x + 2) - \frac{3}{10}\sin(x + 2);$ $y(-2) = 2,$ $y'(-2) = 2$

26. Show that $y_1(x) = x^3 - 6x$ is a solution of $y'' + y = x^3$ and that $y_2(x) = -\cos(x)\ln|\sec(x) + \tan(x)|$ is a solution of $y'' + y = \tan(x)$. Use the principle of superposition to help find the general solution of $y'' + y = 5x^3 + 2\tan(x)$.

27. Suppose that α is *not* a root of the characteristic equation of $y'' + Ay' + By = 0$. Determine a constant b so that $be^{\alpha x}$ is a solution of $y'' + Ay' + By = ae^{\alpha x}$. *Hint: b* will be in terms of a, A, and B.

Use the idea of Problem 27 to find the general solution of each of the following.

28. $y'' - y' - 6y = 8e^{2x}$

29. $y'' - 4y' = 2e^{3x}$

30. $y'' - 4y' + 13y = 3e^{2x} - 5e^{3x}$

31. $y'' - y' - 2y = 3e^x + 5e^{-4x}$

2.10 The Method of Undetermined Coefficients

In this section, we will consider the nonhomogeneous linear differential equation

$$y'' + Ay' + By = f(x), \tag{2.20}$$

in which A and B are constant and f is continuous on an interval I. Since we can always find the general solution y_h of $y'' + Ay' + By = 0$, we need produce only *one* solution y_p of equation (2.20) to find its general solution $y_h + y_p$. The *method of undetermined coefficients*, which we will now discuss, sometimes enables us find such a particular solution.

The idea is that the form of $f(x)$ may suggest what $y_p(x)$ should look like in general appearance, usually involving one or more unknown constants (coefficients), which we try to choose so that $y_p'' + Ay_p' + By_p = f(x)$. We will look at some examples to illustrate the idea and then examine the method more critically.

EXAMPLE 2.27

Find a solution of $y'' - 4y = 8x^2 - 2x$.

Observe that $f(x) = 8x^2 - 2x$ is a polynomial. Since derivatives of polynomials are polynomials, we conjecture that some polynomial can be inserted into $y'' - 4y$ to yield $8x^2 - 2x$. Further, the polynomial we try should not contain powers x^3 or higher because then $y'' - 4y$ would have terms of degree 3 or higher, and we know that it does not (because $y'' - 4y = 8x^2 - 2x$, which is of degree 2). We therefore attempt a particular solution which is a polynomial of degree 2:

$$y_p(x) = ax^2 + bx + c.$$

Although $8x^2 - 2x$ has a zero constant term, we cannot assume that y_p will also have a zero constant term. At this point, we simply do not know. Calculate

$$y_p'(x) = 2ax + b \quad \text{and} \quad y_p''(x) = 2a.$$

Now let $y_p'' - 4y_p = 8x^2 - 2x$ to get

$$2a - 4[ax^2 + bx + c] = 8x^2 - 2x.$$

Collect terms to write this equation as

$$(-4a - 8)x^2 + (-4b + 2)x + (2a - 4c) = 0.$$

This must hold for all x. But the only way a second degree polynomial can be zero for all x is for each coefficient to be zero. Therefore,

$$-4a - 8 = 0$$
$$-4b + 2 = 0$$
$$2a - 4c = 0.$$

Solve these equations to get

$$a = -2, \quad b = \tfrac{1}{2}, \quad c = -1.$$

We conclude that

$$y_p(x) = -2x^2 + \tfrac{1}{2}x - 1$$

is a solution of $y'' - 4y = 8x^2 - 2x$. This can be checked by substitution.

It is routine to find that the general solution of $y'' - 4y = 0$ is $y_h(x) = c_1 e^{2x} + c_2 e^{-2x}$. By Theorem 2.5, the general solution of $y'' - 4y = 8x^2 - 2x$ is

$$y = c_1 e^{2x} + c_2 e^{-2x} - 2x^2 + \tfrac{1}{2}x - 1. \quad \blacksquare$$

EXAMPLE 2.28

Solve $y'' + 2y' - 3y = 4e^{2x}$.

Since derivatives of e^{2x} are always constants times e^{2x}, we conjecture that some constant multiple of e^{2x} may be a solution. This leads us to try

$$y_p(x) = ke^{2x}.$$

Substitute $y = ke^{2x}$ into the differential equation to get

$$4ke^{2x} + 2(2ke^{2x}) - 3ke^{2x} = 4e^{2x}.$$

Then

$$5ke^{2x} = 4e^{2x},$$

which holds if $k = \tfrac{4}{5}$. Therefore, $y_p(x) = \tfrac{4}{5}e^{2x}$ is a solution. The general solution of $y'' + 2y' - 3y = 0$ is $y_h(x) = c_1 e^{x} + c_2 e^{-3x}$. Therefore, the general solution of $y'' + 2y' - 3y = 4e^{2x}$ is

$$y = c_1 e^{x} + c_2 e^{-3x} + \tfrac{4}{5}e^{2x}. \quad \blacksquare$$

EXAMPLE 2.29

Solve $y'' - 5y' + 6y = -3 \sin(2x)$.

Here we must be careful. Derivatives of $\sin(2x)$ may be constant multiples of $\sin(2x)$ or of $\cos(2x)$, depending on how many times we differentiate. Thus, instead of trying just a multiple of $\sin(2x)$, we will try a linear combination of both $\sin(2x)$ and $\cos(2x)$:

$$y_p(x) = c \sin(2x) + k \cos(2x).$$

Compute

$$y_p'(x) = 2c \cos(2x) - 2k \sin(2x)$$

and

$$y_p''(x) = -4c \sin(2x) - 4k \cos(2x).$$

Substitute y_p into the differential equation to get

$$-4c \sin(2x) - 4k \cos(2x) - 5[2c \cos(2x) - 2k \sin(2x)]$$
$$+ 6[c \sin(2x) + k \cos(2x)] = -3 \sin(2x).$$

Collect terms to write

$$[2c + 10k + 3]\sin(2x) + [2k - 10c]\cos(2x) = 0.$$

But $\sin(2x)$ and $\cos(2x)$ are linearly independent; therefore,

$$2c + 10k + 3 = 0 \quad \text{and} \quad 2k - 10c = 0.$$

Then $c = -\frac{3}{52}$ and $k = -\frac{15}{52}$, and a particular solution of $y'' - 5y' + 6y = -3 \sin(2x)$ is

$$y_p(x) = \tfrac{-3}{52} \sin(2x) - \tfrac{15}{52} \cos(2x).$$

Note that y_p contains both a $\sin(2x)$ and a $\cos(2x)$ term, even though $f(x)$ in the differential equation contained only a $\sin(2x)$ term. The general solution of $y'' - 5y' + 6y = 0$ is $y_h(x) = c_1 e^{2x} + c_2 e^{3x}$. Therefore, the general solution of $y'' - 5y' + 6y = -3 \sin(2x)$ is

$$y = c_1 e^{2x} + c_2 e^{3x} - \tfrac{3}{52} \sin(2x) - \tfrac{15}{52} \cos(2x). \quad \blacksquare$$

We will now examine the method more carefully. It would appear that the method is effective if $f(x)$ has a certain form, suggesting an initial guess for y_p. We encounter a difficulty, however, if this leads to a guess for y_p which also satisfies the homogeneous equation $y'' + Ay' + By = 0$. Example 2.30 illustrates what happens in this case.

EXAMPLE 2.30

Consider

$$y'' + 2y' - 3y = 8e^x. \tag{2.21}$$

From our success in Example 2.28, we are led to attempt a solution

$$y_p(x) = ke^x.$$

However, ke^x is a solution of $y'' + 2y' - 3y = 0$. If we substitute ke^x into the equation (2.21), we obtain the absurd "equation" $0 = 8e^x$. This leads us to conclude that we cannot find a particular solution of (2.28) of the form $y_p(x) = ke^x$. ■

In this example, we reached an impasse because the "natural" choice for y_p turned out to satisfy the associated homogeneous equation. We claim that, when this happens, we can still succeed in finding a particular solution by multiplying the "natural" choice of y_p by x, or, if the resulting function is also a solution of the associated homogeneous equation, by x^2. Observe how this strategy works in Example 2.30.

CONTINUATION OF EXAMPLE 2.30

We failed to find a particular solution of the form ke^x because this is a solution of $y'' + 2y' - 3y = 0$ for any k. Multiply the choice ke^x by x and try a solution $y_p(x) = kxe^x$. Compute

$$y_p' = ke^x + kxe^x \quad \text{and} \quad y_p'' = 2ke^x + kxe^x.$$

Substitute y_p into equation (2.21) to get

$$2ke^x + kxe^x + 2[ke^x + kxe^x] - 3kxe^x = 8e^x,$$

or

$$4ke^x = 8e^x.$$

Then $k = 2$, and $y_p = 2xe^x$ is a particular solution of equation (2.21). ■

To understand why this strategy has worked, reason as follows. We know that e^x is a solution of $y'' + 2y' - 3y = 0$. Recalling the reduction of order strategy, attempt a particular solution of (2.21) of the form $y_p = u(x)e^x$. Compute

$$y_p' = u'(x)e^x + u(x)e^x \quad \text{and} \quad y_p'' = u''(x)e^x + 2u'(x)e^x + u(x)e^x.$$

Substitute y_p into equation (2.21) to get

$$u''e^x + 2u'e^x + ue^x + 2[u'e^x + ue^x] - 3ue^x = 8e^x.$$

Divide this equation by e^x and collect terms to get

$$u'' + 4u' = 8.$$

Let $v = u'$ to get $v' + 4v = 8$, a linear first order equation for v. Write this equation as $d/dx\,[ve^{4x}] = 8e^{4x}$, hence $ve^{4x} = 2e^{4x} + C$. Then $v = u' = 2 + Ce^{-4x}$. Then $u(x) = 2x - (C/4)e^{-4x} + K$, in which C and K can be any constants. Since we need only one nonconstant solution for u, choose $C = K = 0$ and use just $u(x) = 2x$. This gives us $y_p(x) = 2xe^x$ as a particular solution of equation (2.21), exactly as we obtained in the continuation of Example 2.30. Incidentally, the general solution of equation (2.21) is now easily found to be

$$y = c_1e^x + c_2e^{-3x} + 2xe^x.$$

Here is an example in which we must multiply the natural choice of y_p by x^2.

EXAMPLE 2.31

Solve

$$y'' - 6y' + 9y = 8e^{3x}. \tag{2.22}$$

We are tempted to try $y_p(x) = ke^{3x}$. However, e^{3x} satisfies the associated homogeneous equation $y'' - 6y' + 9y = 0$. The next thing to try is kxe^{3x}. However, xe^{3x} is also a solution of the associated homogeneous equation. Therefore, try

$$y_p(x) = kx^2 e^{3x}.$$

Calculate

$$y_p'(x) = 2kxe^{3x} + 3kx^2 e^{3x}$$

and

$$y_p''(x) = 2ke^{3x} + 12kxe^{3x} + 9kx^2 e^{3x}.$$

Upon substitution of y_p into equation (2.22), we get

$$2ke^{3x} + 12kxe^{3x} + 9kx^2 e^{3x} - 6[2kxe^{3x} + 3kx^2 e^{3x}] + 9kx^2 e^{3x} = 8e^{3x}.$$

Terms involving kxe^{3x} and $kx^2 e^{3x}$ cancel, and we are left with

$$2ke^{3x} = 8e^{3x};$$

hence, $k = 4$. A particular solution of $y'' - 6y' + 9y = 8e^{3x}$ is

$$y_p(x) = 4x^2 e^{3x}.$$

The general solution of $y'' - 6y' + 9y = 0$ is $(c_1 + c_2 x)e^{3x}$, and we can write the general solution of $y'' - 6y' + 9y = 8e^{3x}$ as

$$y = (c_1 + c_2 x)e^{3x} + 4x^2 e^{3x} = (c_1 + c_2 x + 4x^2)e^{3x}. \quad \blacksquare$$

In view of Examples 2.30 and 2.31 and the rule stated immediately following Example 2.30, *it is advisable to solve* $y'' + Ay' + By = 0$ *before guessing* y_p *from the form of* $f(x)$. Knowing a fundamental set of solutions of $y'' + Ay' + By = 0$ lets us know whether to multiply the "natural" choice of $y_p(x)$ by x or x^2 before proceeding to substitute y_p into the nonhomogeneous differential equation and solve for the coefficients.

Here is a short table of functions to try for $y_p(x)$, corresponding to the form of $f(x)$. The table is followed by a summary of steps to use in applying the method. In the table, $p(x)$, $q(x)$, and $r(x)$ are general nth degree polynomials

$$p(x) = c_0 + c_1 x + \cdots + c_n x^n,$$
$$q(x) = a_0 + a_1 x + \cdots + a_n x^n,$$
and
$$r(x) = b_0 + b_1 x + \cdots + b_n x^n.$$

In a specific problem, the numbers $c_0, c_1, \ldots, c_n$ are given and we must find $a_0, a_1, \ldots, a_n$ and/or $b_0, b_1, \ldots, b_n$.

$f(x)$	$y_p(x)$
$p(x)$	$q(x)$
$ce^{\alpha x}$	$ke^{\alpha x}$
$p(x)\sin(\beta x)$ or $p(x)\cos(\beta x)$	$q(x)\sin(\beta x) + r(x)\cos(\beta x)$
$\left.\begin{array}{c} p(x)e^{\alpha x}\sin(\beta x) \\ \text{or} \\ p(x)e^{\alpha x}\cos(\beta x) \end{array}\right\}$	$q(x)e^{\alpha x}\sin(\beta x) + r(x)e^{\alpha x}\cos(\beta x)$

Steps to Follow in the Method of Undetermined Coefficients

1. Check that the differential equation is *linear* with *constant coefficients*.
2. Determine whether the form of $f(x)$ is in the left column of the table and, if so, note the corresponding entry in the right column.
3. Solve $y'' + Ay' + By = 0$ to determine whether the entry in the right column should be multiplied by x or x^2.
4. Having decided upon a form for y_p, let $y_p'' + Ay_p' + By_p = f(x)$ and solve for the coefficients.
5. If $f = f_1 + f_2 + \cdots + f_n$, and each f_j appears in the left column of the table, apply the method separately to each of the problems $y'' + Ay' + By = f_j$ and apply the principle of superposition, adding the solutions of each of these problems to determine a solution of $y'' + Ay' + By = f$.

Here is an example of this procedure.

EXAMPLE 2.32

Find the general solution of

$$y'' - 8y' + 16y = 8\sin(2x) + 3e^{4x}. \tag{2.23}$$

First, the general solution of $y'' - 8y' + 16y = 0$ is $y_h = (c_1 + c_2 x)e^{4x}$. We need one particular solution of equation (2.23). We will split the problem of finding a particular solution into two problems and apply the principle of superposition.

$$\text{Problem 1:} \quad y'' - 8y' + 16y = 8\sin(2x).$$
$$\text{Problem 2:} \quad y'' - 8y' + 16y = 3e^{4x}.$$

A sum of particular solutions of these problems is a particular solution of (2.23).

Solution of Problem 1   Try $\psi_p = A\sin(2x) + B\cos(2x)$. Compute

$$\psi_p' = 2A\cos(2x) - 2B\sin(2x) \quad \text{and} \quad \psi_p'' = -4A\sin(2x) - 4B\cos(2x).$$

Substitute ψ_p into the differential equation of problem 1 to get

$$-4A\sin(2x) - 4B\cos(2x) - 16A\cos(2x) + 16B\sin(2x)$$
$$+ 16A\sin(2x) + 16B\cos(2x) = 8\sin(2x).$$

Collect terms to write

$$[12A + 16B - 8]\sin(2x) + [-16A + 12B]\cos(2x) = 0$$

for all x. But $\sin(2x)$ and $\cos(2x)$ are linearly independent; hence,

$$12A + 16B - 8 = 0$$
$$-16A + 12B = 0.$$

Then $A = \frac{6}{25}$ and $B = \frac{8}{25}$, so a particular solution of problem 1 is

$$\psi_p(x) = \frac{6}{25}\sin(2x) + \frac{8}{25}\cos(2x).$$

Solution of Problem 2 A natural choice for a particular solution is ke^{4x}. However, e^{4x} is a solution of $y'' - 8y' + 16y = 0$, as is xe^{4x}. Thus, attempt a particular solution to problem 2 of the form $\xi_p = kx^2e^{4x}$. Compute

$$\xi_p' = 2kxe^{4x} + 4kx^2e^{4x} \quad \text{and} \quad \xi_p'' = 16kxe^{4x} + 2ke^{4x} + 16kx^2e^{4x}.$$

Substitute ξ_p into the differential equation of problem 2 to get

$$16kxe^{4x} + 2ke^{4x} + 16kx^2e^{4x} - 16kxe^{4x} - 32kx^2e^{4x} + 16kx^2e^{4x} = 3e^{4x}.$$

Then $2ke^{4x} = 3e^{4x}$; hence, $k = \frac{3}{2}$. A particular solution of problem 2 is

$$\xi_p(x) = \frac{3}{2}x^2e^{4x}.$$

By the principle of superposition, a particular solution of equation (2.23) is

$$\frac{6}{25}\sin(2x) + \frac{8}{25}\cos(2x) + \frac{3}{2}x^2e^{4x}.$$

The general solution of equation (2.23) is

$$y = (c_1 + c_2x)e^{4x} + \frac{6}{25}\sin(2x) + \frac{8}{25}\cos(2x) + \frac{3}{2}x^2e^{4x}. \quad \blacksquare$$

The method of undetermined coefficients applies only if the differential equation has constant coefficients, and it sometimes involves a good deal of calculation. Nevertheless, whenever the method applies, *it works*. This is a virtue not to be underestimated. Although the method applies only to constant coefficient differential equations, sometimes a change of variables will transform a differential equation with nonconstant coefficients to one with constant coefficients (we have seen this with the Cauchy-Euler equation). In such a case, we may use the method on the constant coefficient equation, then transform the solution back into a solution of the original equation. Here is an illustration of this idea.

EXAMPLE 2.33

Solve

$$x^2y'' - 5xy' + 8y = 2\ln(x) \qquad (x > 0).$$

The differential equation $x^2y'' - 5xy' + 8y = 0$ is a Cauchy-Euler equation, suggesting that we try the transformation $t = \ln(x)$. Using results from Section 2.7, we obtain the transformed differential equation

$$Y''(t) - 6Y'(t) + 8Y(t) = 2\ln(e^t),$$

or

$$Y'' - 6Y' + 8Y = 2t,$$

in which $Y(t) = y[e^t]$. The method of undetermined coefficients *does not apply* to the Cauchy-Euler equation for y; it *does* apply to the *transformed* differential equation for Y. The associated homogeneous differential equation for Y is $Y'' - 6Y' + 8Y = 0$, with general solution

$$Y_h(t) = c_1 e^{4t} + c_2 e^{2t}.$$

For a particular solution of $Y'' - 6Y' + 8Y = 2t$, assume a solution $Y_p(t) = ct + d$. Then

$$Y_p'(t) = c \quad \text{and} \quad Y_p''(t) = 0,$$

and letting $Y_p'' - 6Y_p' + 8Y_p = 2t$ results in the equation

$$-6c + 8ct + 8d = 2t.$$

Then

$$(8c - 2)t + (8d - 6c) = 0.$$

For this to be true for all t, we need

$$8c - 2 = 0$$
$$8d - 6c = 0.$$

Therefore, $c = \frac{1}{4}$ and $d = \frac{3}{16}$, and

$$Y_p(x) = \frac{1}{4}t + \frac{3}{16}.$$

The general solution of the transformed differential equation is

$$Y(t) = c_1 e^{2t} + c_2 e^{4t} + \frac{1}{4}t + \frac{3}{16}.$$

Put $t = \ln(x)$ to obtain the general solution of the original differential equation:

$$y(x) = c_1 x^2 + c_2 x^4 + \frac{1}{4}\ln(x) + \frac{3}{16}. \quad \blacksquare$$

We emphasize again that the method of undetermined coefficients applies only to differential equations with constant coefficients. In the preceding example, the technique was applied not to the Cauchy-Euler equation but to the transformed, constant coefficient equation.

In the next section, we will develop a method which can sometimes be used to solve a linear second order differential equation in which the coefficients are not constant.

PROBLEMS FOR SECTION 2.10

In each of Problems 1 through 25, find the general solution of the differential equation by using the method of undetermined coefficients.

1. $y'' - y' - 2y = 2x^2 + 5$

2. $y'' - y = x^3 - 4x^2 - 11x + 9$

3. $y'' + 4y = 8x^3 - 20x^2 + 16x - 18$

4. $y'' - y' - 6y = 8e^{2x}$

5. $y'' - 2y' + 10y = 20x^2 + 2x - 8$

6. $y'' - 4y' + 5y = 21e^{2x}$

7. $y'' - 6y' + 8y = 3e^x$

8. $y'' + 6y' + 9y = 9\cos(3x)$

9. $y'' - 3y' + 2y = 10\sin(x)$

10. $y'' - 4y' = 8x^2 + 2e^{3x}$

11. $y'' - 4y' + 13y = 3e^{2x} - 5e^{3x}$

12. $y'' - 2y' + y = 3x + 25\sin(3x)$

13. $y'' - 4y' = 36x^2 - 2x + 24$

14. $y'' + 4y' = x^2e^{-x}\sin(3x)$

15. $y'' + 4y' = -3\cos(3x) + \sin(2x)$

16. $y'' - y' - 6y = 12xe^x$

17. $y'' + 2y' + y = -3e^{-x} + 8xe^{-x} + 1$

18. $y'' + y' = 5 + 10\sin(2x)$

19. $y'' - 4y = x^2e^x\cos(x)$

20. $y'' + 4y' - 12y = x + \cos(3x)$

21. $y'' + 2y = 4\cos^2(3x)$

22. $y'' - 3y' + 2y = 60e^{2x}\cos(3x)$

23. $y'' - 4y = 5\sinh(2x)$

24. $y'' - 2y' - 3y = 8e^{3x}$

25. $y'' + y = e^{-x}\sin(x) - e^{3x}\cos(5x)$

In each of Problems 26 through 33, solve the initial value problem.

26. $y'' - 4y = -7e^{2x} + x$; $y(0) = 1$, $y'(0) = 3$

27. $y'' + 4y' = 34\cos(x) + 8$; $y(0) = 3$, $y'(0) = 2$

28. $y'' + 8y' + 12y = e^{-x} + 7$; $y(0) = 1$, $y'(0) = 0$

29. $y'' + y = 5\sin(2x) - 4$; $y(0) = 0$, $y'(0) = -3$

30. $y'' + 10y' + 24y = 1$; $y(2) = -3$, $y'(2) = 5$

31. $y'' - 3y' = 2e^{2x}\sin(x)$; $y(0) = 1$, $y'(0) = 2$

32. $y'' - 2y' - 8y = 10e^{-x} + 8e^{2x}$; $y(0) = 1$, $y'(0) = 4$

33. $y'' - 6y' + 9y = 4e^{3x}$; $y(0) = 1$, $y'(0) = 2$

34. Use the principle of superposition to find a particular solution of

$$y'' - 3y' + 2y = 4\sin(x) - 2\sin(2x) + 3\sin(3x).$$

35. Assuming that the principle of superposition extends to infinite sums, find a particular solution of

$$y'' - 4y = \sum_{n=1}^{\infty} \frac{1}{n}\sin(nx).$$

In Problems 36, 37, and 38, we have a differential equation $y'' + Ay' + By = f$, but f does not fall into any of the categories in the table of this section. However, in each case it is possible to find a number c such that on $(-\infty, c)$, $f(x)$ is in the table, and on (c, ∞), $f(x)$ is a (possibly different) item in the table. In each of these problems, solve the differential equation on $(-\infty, c)$ and then separately on (c, ∞), using the method of undetermined coefficients. Use these solutions to define a solution valid for all x.

36. $y'' - 3y' + 2y = f$, where

$$f(x) = \begin{cases} x^2 & \text{for} \quad x \le 3 \\ 2x + 3 & \text{for} \quad x \ge 3. \end{cases}$$

37. $y'' + 4y' = f$, where

$$f(x) = \begin{cases} 3e^x & \text{for} \quad x \le 0 \\ x^2 - 2x + 3 & \text{for} \quad x \ge 0. \end{cases}$$

38. $y'' - 4y = f$, where

$$f(x) = \begin{cases} 2x & \text{for} \quad x \le 2 \\ 4 & \text{for} \quad x \ge 2. \end{cases}$$

39. Use the result of Problem 38 to solve the initial value problem

$$y'' - 4y = f; \qquad y(0) = 2, \qquad y'(0) = 0,$$

in which f is the function defined in Problem 38.

In each of Problems 40 through 46, use the method of Example 2.33 to find the general solution of the differential equation for $x > 0$.

40. $x^2 y'' - 2xy' + 2y = \ln(x) + 1$

41. $x^2 y'' - 12y = 4x$

42. $x^2 y'' + 6xy' + 4y = x - 3$

43. $x^2 y'' - 4xy' + 6y = x^2 - x$

44. $x^2 y'' - xy' - 8y = x^4 - 3 \ln(x)$

45. $x^2 + 10xy' + 20y = 4 \ln(x) - x$

46. $x^2 y'' - 3xy' + 4y = x + 3$

47. Suppose that φ is a solution of

$$y'' - 2y' + y = 2e^x.$$

(a) If $\varphi(x) > 0$ for all x, is it true that $\varphi'(x) > 0$ for all x?

(b) If $\varphi'(x) > 0$ for all x, is it true that $\varphi(x) > 0$ for all x?

(This problem appeared on the William Lowell Putnam Mathematical Competition for 1987.)

2.11 *Variation of Parameters*

Consider the equation

$$y'' + p(x)y' + q(x)y = f(x), \tag{2.24}$$

with p, q, and f continuous on some interval I. Assuming that we can find two linearly independent solutions y_1 and y_2 of $y'' + p(x)y' + q(x)y = 0$, the *variation of parameters* method attempts to find a particular solution of equation (2.24) of the form

$$y_p(x) = u(x)y_1(x) + v(x)y_2(x),$$

in which u and v are twice-differentiable functions.

We will need two equations to determine the unknown functions u and v. In fact, we will obtain equations we can solve for u' and v'. We must then obtain u and v by integration. In order to substitute $u(x)y_1(x) + v(x)y_2(x)$ into equation (2.24), first compute

$$y_p'(x) = u(x)y_1'(x) + v(x)y_2'(x) + u'(x)y_1(x) + v'(x)y_2(x).$$

To simplify both y_p' and $y_p''(x)$, we will require that u and v must satisfy

$$u'(x)y_1(x) + v'(x)y_2(x) = 0,$$

or, written more simply,

$$u'y_1 + v'y_2 = 0. \tag{2.25}$$

This gives us one equation for u' and v'. Now,

$$y_p''(x) = u'(x)y_1'(x) + v'(x)y_2'(x) + u(x)y_1''(x) + v(x)y_2''(x).$$

The equation $y_p'' + p(x)y_p' + q(x)y_p = f(x)$ is

$$u'y_1' + v'y_2' + uy_1'' + vy_2'' + p[uy_1' + vy_2'] + q[uy_1 + vy_2] = f.$$

Collect terms to rewrite this equation as

$$u[y_1'' + py_1' + qy_1] + v[y_2'' + py_2' + qy_2] + u'y_1' + v'y_2' = f.$$

Since y_1 and y_2 are assumed to be solutions of the associated homogeneous equation $y'' + p(x)y' + q(x)y = 0$, the terms in square brackets are zero, and we have

$$u'y_1' + v'y_2' = f, \tag{2.26}$$

a second equation for u' and v'.

Equations (2.25) and (2.26) give us two equations for u' and v'. For ease of reference, we will rewrite these equations together:

$$\begin{aligned} y_1u' + y_2v' &= 0 \\ y_1'u' + y_2'v' &= f. \end{aligned} \tag{2.27}$$

This system of equations has a unique solution if and only if the determinant of the coefficients is nonzero. This determinant is

$$\begin{vmatrix} y_1 & y_2 \\ y_1' & y_2' \end{vmatrix},$$

which is exactly the Wronskian W of y_1 and y_2. Since y_1 and y_2 are assumed linearly independent on I, $W(x) \neq 0$ for all x in I. The solutions of equations (2.27) for u' and v' are

$$u' = \frac{\begin{vmatrix} 0 & y_2 \\ f & y_2' \end{vmatrix}}{\begin{vmatrix} y_1 & y_2 \\ y_1' & y_2' \end{vmatrix}} \quad \text{and} \quad v' = \frac{\begin{vmatrix} y_1 & 0 \\ y_1' & f \end{vmatrix}}{\begin{vmatrix} y_1 & y_2 \\ y_1' & y_2' \end{vmatrix}}.$$

Since the determinant in the denominator is the Wronskian (which we will denote as just W), these solutions for u' and v' can be written

$$u' = \frac{\begin{vmatrix} 0 & y_2 \\ f & y_2' \end{vmatrix}}{W} = -\frac{y_2 f}{W} \quad \text{and} \quad v' = \frac{\begin{vmatrix} y_1 & 0 \\ y_1' & f \end{vmatrix}}{W} = \frac{y_1 f}{W}. \tag{2.28}$$

If we can carry out the necessary integrations, we have functions u and v such that

$$y_p = uy_1 + vy_2$$

is a particular solution of $y'' + p(x)y' + q(x)y = f(x)$. Usually we let the constants of integration be zero in integrating to obtain u and v because we do not need more than one u or v having the required properties.

EXAMPLE 2.34

Find a solution of $y'' + 4y = \tan(2x)$ on the interval $(-\pi/4, \pi/4)$.

The associated homogeneous equation $y'' + 4y = 0$ has $y_1(x) = \cos(2x)$ and $y_2(x) = \sin(2x)$ as a fundamental set of solutions. Their Wronskian is

$$\begin{aligned} W(x) &= \begin{vmatrix} \cos(2x) & \sin(2x) \\ -2\sin(2x) & 2\cos(2x) \end{vmatrix} \\ &= 2\cos^2(x) + 2\sin^2(x) = 2. \end{aligned}$$

With $f(x) = \tan(2x)$, equations (2.28) give us

$$u'(x) = -\frac{y_2(x)f(x)}{W(x)} = -\frac{1}{2}\sin(2x)\tan(2x)$$

and

$$v'(x) = \frac{y_1(x)f(x)}{W(x)} = \frac{1}{2}\cos(2x)\tan(2x).$$

Then

$$u(x) = \int -\frac{1}{2}\sin(2x)\tan(2x)\,dx = -\frac{1}{2}\int \frac{\sin^2(2x)}{\cos(2x)}\,dx$$

$$= -\frac{1}{2}\int \frac{1 - \cos^2(2x)}{\cos(2x)}\,dx = -\frac{1}{2}\int \sec(2x)\,dx + \frac{1}{2}\int \cos(2x)\,dx$$

$$= -\frac{1}{4}\ln|\sec(2x) + \tan(2x)| + \frac{1}{4}\sin(2x)$$

and

$$v(x) = \frac{1}{2}\int \cos(2x)\tan(2x)\,dx = \frac{1}{2}\int \sin(2x)\,dx = -\frac{1}{4}\cos(2x).$$

A particular solution of $y'' + 4y = \tan(2x)$ is

$$\begin{aligned}
y_p(x) &= u(x)y_1(x) + v(x)y_2(x) \\
&= [-\tfrac{1}{4}\ln|\sec(2x) + \tan(2x)| + \tfrac{1}{4}\sin(2x)]\cos(2x) - \tfrac{1}{4}\cos(2x)\sin(2x) \\
&= -\tfrac{1}{4}\ln|\sec(2x) + \tan(2x)|\cos(2x).
\end{aligned}$$

If we wish, we can write the general solution of $y'' + 4y = \tan(2x)$ on $(-\pi/4, \pi/4)$ as

$$y = c_1\cos(2x) + c_2\sin(2x) - \tfrac{1}{4}\ln|\sec(2x) + \tan(2x)|\cos(2x). \quad \blacksquare$$

Although the differential equation in Example 2.34 had constant coefficients, this is not a requirement for the method of variation of parameters, as it was for the method of undetermined coefficients.

EXAMPLE 2.35

Find a solution of $y'' - (4/x)y' + (4/x^2)y = x^2 + 1$ for $x > 0$.

The associated homogeneous equation $y'' - (4/x)y' + (4/x^2)y = 0$ is a Cauchy-Euler equation, and we find the fundamental set of solutions $y_1(x) = x$ and $y_2(x) = x^4$ for $x > 0$. The Wronskian of y_1 and y_2 is

$$W = \begin{vmatrix} x & x^4 \\ 1 & 4x^3 \end{vmatrix} = 4x^4 - x^4 = 3x^4,$$

and this is nonzero for $x > 0$. With $f(x) = x^2 + 1$, equations (2.28) give us

$$u'(x) = -\frac{x^4(x^2 + 1)}{3x^4} = -\frac{x^2 + 1}{3}$$

and

$$v'(x) = \frac{x(x^2 + 1)}{3x^4} = \frac{1}{3}\frac{1}{x} + \frac{1}{3x^3}.$$

Then

$$u(x) = \int -\frac{x^2 + 1}{3}\,dx = -\frac{1}{9}x^3 - \frac{1}{3}x$$

and

$$v(x) = \int \left[\frac{1}{3}\frac{1}{x} + \frac{1}{3x^3}\right]dx = \frac{1}{3}\ln(x) - \frac{1}{6x^2}.$$

A particular solution of $y'' - (4/x)y' + (4/x^2)y = x^2 + 1$ is

$$y_p(x) = \left(-\frac{1}{9}x^3 - \frac{1}{3}x\right)x + \left(\frac{1}{3}\ln(x) - \frac{1}{6x^2}\right)x^4$$

$$= -\frac{1}{9}x^4 - \frac{1}{2}x^2 + \frac{1}{3}x^4\ln(x).$$

The general solution of $y'' - (4/x)y' + (4/x^2)y = x^2 + 1$ is

$$y = c_1 x + c_2 x^4 - \tfrac{1}{9}x^4 - \tfrac{1}{2}x^2 + \tfrac{1}{3}x^4\ln(x). \quad \blacksquare$$

We have considered the method of variation of parameters for the differential equation

$$y'' + p(x)y' + q(x)y = f(x).$$

Although p and q need not be constant to use the method, we have assumed that the coefficient of y'' is 1. If we encounter a differential equation in which the coefficient of y'' is a nonconstant function $r(x)$, we must divide the equation by $r(x)$ [in an interval in which $r(x) \neq 0$] before we can apply the method to the resulting equation.

PROBLEMS FOR SECTION 2.11

In each of Problems 1 through 20, use the method of variation of parameters (even if another method applies) to find a particular solution. Then write the general solution of the differential equation.

1. $y'' - y' - 2y = e^{2x}$ **2.** $y'' + y = \tan(x)$ **3.** $y'' + y' - 6y = x$

4. $x^2 y'' + 5xy' - 12y = \ln(x)$ **5.** $x^2 y'' - 5xy' + 8y = 3x$ **6.** $y'' - y' - 12y = 2\sinh^2(x)$

7. $y'' - 4y' + 3y = 2\cos(x + 3)$ **8.** $x^2 y'' + 3xy' + y = 4/x$ **9.** $x^2 y'' + xy' + 4y = \sin[2\ln(x)]$

10. $x^2 y'' + 2xy' - 6y = x^2 - 2$ **11.** $y'' - 4y' + 4y = 6e^{2x}$ **12.** $y'' - y' - 2y = 3e^{2x}$

13. $y'' + 9y = 3\sec(3x)$ **14.** $y'' + 2y' + y = \dfrac{1}{x}e^{-x}$ **15.** $x^2 y'' - xy' + y = 6x$

16. $y'' - 2y' - 3y = 2\sin^2(x)$ **17.** $y'' - 3y' + 2y = \cos(e^{-x})$ **18.** $y'' - 5y' + 6y = 8\sin^2(4x)$

19. $x^2 y'' + 3xy' + y = 9x^2 + 8x + 5$ **20.** $x^2 y'' - 3xy' + 3y = 6x^4 e^{-3x}$

In each of Problems 21 through 26, solve the initial value problem.

21. $y'' - y = 5\sin^2(x);$ $y(0) = 2,$ $y'(0) = -4$ **22.** $y'' + y = \tan^2(x);$ $y(0) = 4,$ $y'(0) = 3$

23. $x^2 y'' - 6y = 8x^2;$ $y(1) = 1,$ $y'(1) = 0$

24. $x^2 y'' + 7xy' + 9y = 27 \ln(x)$; $y(1) = 1$, $y'(1) = -4$

25. $x^2 y'' - 2xy' + 2y = 10 \sin[\ln(x)]$; $y(1) = 3$, $y'(1) = 0$

26. $x^2 y'' - 4xy' + 6y = x^4 e^x$; $y(1) = 2$, $y'(1) = 4$

27. Show that $y_1(x) = x^2$ and $y_2(x) = x - 1$ are solutions of

$$(x^2 - 2x)y'' + 2(1 - x)y' + 2y = 0.$$

Use this fact to find the general solution of

$$(x^2 - 2x)y'' + 2(1 - x)y' + 2y = 6(x^2 - 2x)^2.$$

28. Show that $y_1(x) = x$ and $y_2(x) = x^2 - 1$ are solutions of

$$(x^2 + 1)y'' - 2xy' + 2y = 0.$$

Use this fact to find the general solution of

$$(x^2 + 1)y'' - 2xy' + 2y = 6(x^2 + 1)^2.$$

29. Evaluate

$$\int_1^x \int_1^t \frac{\ln(x)}{z^2}\, dz\, dt$$

as follows. Set this integral equal to $y(x)$. Determine $y''(x)$ by using the fundamental theorem of calculus, and determine a nonhomogeneous differential equation satisfied by y. Now evaluate $y(1)$ and $y'(1)$ and solve the initial value problem consisting of the differential equation and these initial conditions.

2.12 *Forced Oscillations, Resonance, Beats, and Electrical Circuits*

In Section 2.8, we considered free motion of a mass on a spring. By "free" we mean there was no external driving force acting on the mass.

We will now extend our analysis to allow for a driving force, concentrating on the commonly encountered case in which the mass is acted on by a periodic driving force

$$f(t) = A \cos(\omega t),$$

with A and ω positive constants. The differential equation for the displacement $y(t)$ from equilibrium at time t is

$$y'' + \frac{c}{m} y' + \frac{k}{m} y = \frac{A}{m} \cos(\omega t), \tag{2.29}$$

in which k is the spring constant and c is the damping constant. The general solution of equation (2.29) is of the form $y = y_h + y_p$, in which y_h is the general solution of $y'' + (c/m)y' + (k/m)y = 0$ and y_p is any solution of $y'' + (c/m)y' + (k/m)y = A/m \cos(\omega t)$. We know y_h in all possible cases from the work of Section 2.8. We will now determine a particular solution y_p.

In order to apply the method of undetermined coefficients, let

$$y_p(t) = a \cos(\omega t) + b \sin(\omega t).$$

Substitute y_p into equation (2.29) and rearrange terms to obtain

$$\left[-a\omega^2 + \frac{b\omega c}{m} + a\frac{k}{m} - \frac{A}{m}\right]\cos(\omega t) - \left[b\omega^2 + \frac{a\omega c}{m} - b\frac{k}{m}\right]\sin(\omega t) = 0.$$

Since $\cos(\omega t)$ and $\sin(\omega t)$ are linearly independent on $[0, \infty)$, this equation can hold for $t \geq 0$ only if the coefficients of $\cos(\omega t)$ and $\sin(\omega t)$ are zero. Therefore,

$$-a\omega^2 + \frac{b\omega c}{m} + a\frac{k}{m} - \frac{A}{m} = 0$$

and

$$b\omega^2 + \frac{a\omega c}{m} - b\frac{k}{m} = 0.$$

Collect terms and write these equations as

$$(k - m\omega^2)a + \omega cb = A$$
$$-\omega ca + (k - m\omega^2)b = 0.$$

Solve these equations for a and b to obtain

$$a = \frac{A(k - m\omega^2)}{(k - m\omega^2)^2 + \omega^2 c^2} \quad \text{and} \quad b = \frac{A\omega c}{(k - m\omega^2)^2 + \omega^2 c^2}.$$

As before, let $\omega_0 = \sqrt{k/m}$. A particular solution of

$$y'' + (c/m)y' + (k/m)y = (A/m)\cos(\omega t)$$

may be written

$$y_p(t) = \frac{mA(\omega_0^2 - \omega^2)}{m^2(\omega_0^2 - \omega^2)^2 + \omega^2 c^2}\cos(\omega t) + \frac{A\omega c}{m^2(\omega_0^2 - \omega^2)^2 + \omega^2 c^2}\sin(\omega t),$$

assuming that $m^2(\omega_0^2 - \omega^2) + \omega^2 c^2 \neq 0$, which is the case if $c \neq 0$ or if $\omega_0 \neq \omega$.

We can now look at the general solution $y_h + y_p$ in the cases of overdamping, critical damping, and underdamping. This can be done for arbitrary $k, m, c,$ and A, but for clarity we will restrict our attention to specific, but representative, cases.

AN EXAMPLE OF OVERDAMPED, FORCED MOTION

Let $c = 6$, $k = 5$, and $m = 1$, as we had with overdamping in Section 2.8. There we found that $y_h(t) = c_1 e^{-t} + c_2 e^{-5t}$. Suppose also that $A/m = 6\sqrt{5}$ and $\omega = \sqrt{5}$. The driving force is $6\sqrt{5}\cos(\sqrt{5}t)$. Further, $k - m\omega^2 = 0$, and the general solution is

$$y(t) = c_1 e^{-t} + c_2 e^{-5t} + \sin(\sqrt{5}t).$$

If the mass is released from rest at the equilibrium position, the initial conditions are

$$y(0) = y'(0) = 0.$$

From these we obtain $c_1 = -\sqrt{5}/4$ and $c_2 = \sqrt{5}/4$, so the displacement function is

$$y(t) = \frac{\sqrt{5}}{4}(-e^{-t} + e^{-5t}) + \sin(\sqrt{5}t).$$

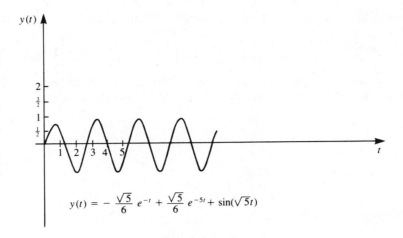

$$y(t) = -\frac{\sqrt{5}}{6}e^{-t} + \frac{\sqrt{5}}{6}e^{-5t} + \sin(\sqrt{5}t)$$

Figure 2.6. Overdamping (forced motion).

A graph of this function is shown in Figure 2.6. Notice that the exponential terms decrease very quickly, exerting less influence on $y(t)$ as t increases. After some time, the solution tends to behave more like $\sin(\sqrt{5}t)$, with the mass moving up and down with approximate period $2\pi/\sqrt{5}$.

AN EXAMPLE OF CRITICALLY DAMPED, FORCED MOTION

Next, suppose $c = 2$ and $m = k = 1$. Now $c^2 - 4km = 0$, and we have critical damping. The general solution for $y(t)$ is

$$y_h(t) = (c_1 + c_2 t)e^{-t},$$

using results from Section 2.8. We will choose $\omega = 1$ and $A/m = 2$ to get the general solution

$$y(t) = (c_1 + c_2 t)e^{-t} + \sin(t).$$

If we again suppose that $y(0) = y'(0) = 0$, the solution is found to be

$$y(t) = -te^{-t} + \sin(t).$$

A graph of $y(t)$ is shown in Figure 2.7. Notice that the exponential term exerts a significant influence at first, but that, as time increases, the graph more closely resembles that of $\sin(t)$, indicating an almost periodic motion as time passes.

AN EXAMPLE OF UNDERDAMPED, FORCED MOTION

Finally, suppose that $c = k = 2$ and $m = 1$. Choose $\omega = \sqrt{2}$ and $A/m = 2\sqrt{2}$. Again using results from Section 2.8, we have the general solution

$$y(t) = e^{-t}[c_1\cos(t) + c_2\sin(t)] + \sin(\sqrt{2}t).$$

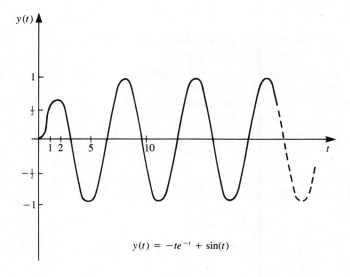

$$y(t) = -te^{-t} + \sin(t)$$

Figure 2.7. Critical damping (forced motion).

If we assume again that $y(0) = y'(0) = 0$, the solution is found to be

$$y = -\sqrt{2}e^{-t}\sin(t) + \sin(\sqrt{2}t).$$

Figure 2.8 shows a graph of this solution. As time passes, $y(t)$ more closely resembles a sine function.

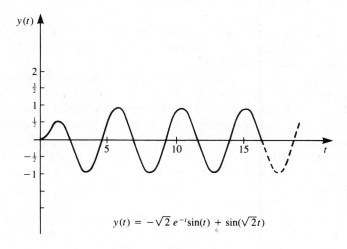

$$y(t) = -\sqrt{2}\,e^{-t}\sin(t) + \sin(\sqrt{2}t)$$

Figure 2.8. Underdamping (forced motion).

RESONANCE

The physical system we have been studying will exhibit a variety of interesting properties under different assumptions on m, c, k, and the driving force. In the absence of damping, the system can experience a phenomenon called *resonance*, which has considerable practical importance. We will now examine this concept.

To begin, suppose that $c = 0$ and that the driving force still has the form $A \cos(\omega t)$, so that the differential equation for the displacement function is

$$y'' + \frac{k}{m} y = \frac{A}{m} \cos(\omega t).$$

The assumption that $c = 0$ means that the system is not damped, and the general homogeneous solution of the differential equation is

$$y_h(t) = c_1 \cos(\omega_0 t) + c_2 \sin(\omega_0 t),$$

in which $\omega_0 = \sqrt{k/m}$, as usual. Assuming that $\omega \neq \omega_0$, the method of undetermined coefficients leads to the particular solution

$$y_p(t) = \frac{A}{m(\omega_0^2 - \omega^2)} \cos(\omega t).$$

The general solution of the differential equation is

$$y(t) = c_1 \cos(\omega_0 t) + c_2 \sin(\omega_0 t) + \frac{A}{m(\omega_0^2 - \omega^2)} \cos(\omega t).$$

Certain conclusions about the motion become more apparent if we let $d = \sqrt{c_1^2 + c_2^2}$ and $\delta = \tan^{-1}(-c_2/c_1)$ to write this general solution in the phase angle form

$$y(t) = d \cos(\omega_0 t + \delta) + \frac{A}{m(\omega_0^2 - \omega^2)} \cos(\omega t).$$

In this form, we see that the motion is a sum of two harmonic oscillations, the first with frequency $\omega_0/2\pi$ (called the *natural frequency* of the system) and the second with frequency $\omega/2\pi$ (called the *input frequency*).

In deriving this solution, we have assumed that the natural and input frequencies of the system are different. Notice that, if these frequencies are chosen closer together, the amplitude of $y_p(t)$ becomes larger. Thus, a system with no damping (or even with c very close to zero) will experience larger and larger oscillations as natural and input frequencies become more nearly equal.

Although we can choose ω as close to ω_0 as we like in the above analysis, the solution we have just derived for $y(t)$ is not valid in the case in which $\omega = \omega_0$. In this case, we must return to the differential equation, which becomes

$$y'' + \omega_0^2 y = \frac{A}{m} \cos(\omega_0 t).$$

The general solution of $y'' + \omega_0^2 y = 0$ is still

$$y_h(t) = c_1 \cos(\omega_0 t) + c_2 \sin(\omega_0 t),$$

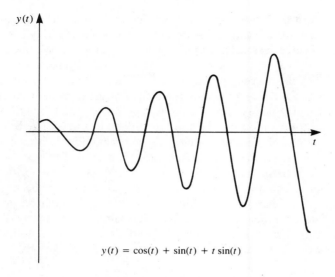

$$y(t) = \cos(t) + \sin(t) + t\sin(t)$$

Figure 2.9. Resonance.

as in the case in which $\omega \neq \omega_0$. However, now $\cos(\omega_0 t)$ is a solution of the associated homogeneous equation $y'' + \omega_0^2 y = 0$, so in applying the method of undetermined coefficients to find a particular solution, we must try

$$y_p(t) = ct\cos(\omega_0 t) + dt\sin(\omega_0 t).$$

We find after a routine calculation that

$$y_p(t) = \frac{A}{2m\omega_0}\, t\sin(\omega_0 t).$$

Thus, the general solution of $y'' + \omega_0^2 y = A/m\cos(\omega_0 t)$ is

$$y(t) = c_1\cos(\omega_0 t) + c_2\sin(\omega_0 t) + \frac{A}{2m\omega_0}\, t\sin(\omega_0 t).$$

In this case in which $\omega = \omega_0$, the factor of t in the particular solution term causes the amplitude to increase without bound as t increases. This phenomenon is known as *resonance.* As a specific illustration, Figure 2.9 shows the graph of $y(t)$, with constants chosen for convenience so that

$$c_1 = c_2 = \omega_0 = \frac{A}{2m\omega_0} = 1.$$

In the real world, of course, there is always some damping. Nevertheless, if c is close to zero compared with other significant factors, such as the mass, and if the natural and input frequencies are almost equal, then oscillations can build up to a large enough magnitude to damage the system. This phenomenon is also called resonance.

A famous example of this occurred in 1940 in Tacoma, Washington. A newly constructed bridge across the Tacoma Narrows experienced vibrations of increasing violence until it finally collapsed, only 4 months after it was opened to traffic.

The energy for the vibrations in this case was supplied by the wind, and the bridge began to experience small oscillations soon after it was opened. Engineers assured the public that the bridge was safe, and the gentle undulations of the roadbed were regarded as so curious that people came considerable distances to drive across it. Finally, a support near the center of the bridge buckled, and the oscillations began to increase in amplitude. Toward the end, the bridge thrashed violently back and forth. The last person on the bridge was a newspaper reporter who had driven halfway across, only to be forced to abandon his car and crawl back to safety. The bridge collapsed shortly thereafter, the only casualty being a dog who went down with the structure. The entire sequence of events was well documented, and many physics and engineering departments have films showing the dramatic last moments of the bridge.

Another classic example of resonance occurred near Manchester, England, in 1831. A column of soldiers marching in step across the Broughton bridge set up a periodic force whose frequency happened to approximate very closely the normal frequency of the bridge, causing its collapse. Nowadays, a column of soldiers crossing a bridge is usually ordered to fall out of step to avoid such oscillations.

BEATS

A second phenomenon we will discuss is that of *beats*. Assume that there is no damping ($c = 0$), and consider the differential equation

$$y'' + \omega_0^2 y = \frac{A}{m}\cos(\omega t),$$

with $\omega \neq \omega_0$. We have seen that the general solution can be written

$$y(t) = c_1\cos(\omega_0 t) + c_2\sin(\omega_0 t) + \frac{A}{m(\omega_0^2 - \omega^2)}\cos(\omega t).$$

Suppose that the mass is initially at rest and that its initial velocity is zero. Then $y(0) = y'(0) = 0$, and the initial value problem has the unique solution

$$y(t) = \frac{A}{m(\omega_0^2 - \omega^2)}\left[\cos(\omega t) - \cos(\omega_0 t)\right].$$

This is a constant times a sum of cosine terms having the same amplitudes but different frequencies.

The phenomenon of beats is more readily apparent from this solution if we rewrite it in a different form. Recall the trigonometric identity

$$\sin(\alpha)\sin(\beta) = \tfrac{1}{2}\left[\cos(\alpha - \beta) - \cos(\alpha + \beta)\right].$$

Choose α and β so that $\alpha + \beta = \omega_0 t$ and $\alpha - \beta = \omega t$. That is, let

$$\alpha = \tfrac{1}{2}(\omega_0 + \omega)t \quad \text{and} \quad \beta = -\tfrac{1}{2}(\omega_0 - \omega)t.$$

Then

$$y(t) = \frac{2A}{m(\omega_0^2 - \omega^2)}\sin\left(\frac{1}{2}(\omega_0 + \omega)t\right)\sin\left(-\frac{1}{2}(\omega_0 - \omega)t\right)$$

$$= \frac{2A}{m(\omega^2 - \omega_0^2)}\sin\left(\frac{1}{2}(\omega_0 + \omega)t\right)\sin\left(\frac{1}{2}(\omega_0 - \omega)t\right).$$

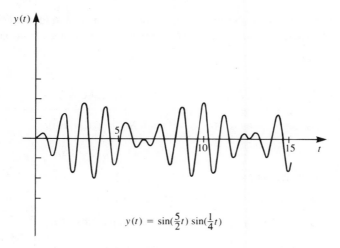

$$y(t) = \sin(\tfrac{5}{2}t)\sin(\tfrac{1}{4}t)$$

Figure 2.10. Beats.

This solution graphs as a wave having a periodic variation of amplitude depending on the relative sizes of $\omega_0 + \omega$ and $\omega_0 - \omega$. The resulting motion is called a *beat*, and the variation of amplitude with time is called *amplitude modulation*.

To see a beat in a specific case, refer to Figure 2.10, which shows the graph of $y(t)$ with $\omega_0 + \omega = 5$ and $\omega_0 - \omega = 0.5$, with the other constants chosen so that $2A/[m(\omega^2 - \omega_0^2)] = 1$.

APPLICATION TO ELECTRICAL CIRCUITS

A typical *RLC* circuit is shown in Figure 2.11. The differential equations for the charge $q(t)$ and the current $i(t)$ are

$$Lq'' + Rq' + \frac{1}{C}q = E(t) \qquad (2.30)$$

and

$$Li'' + Ri' + \frac{1}{C}i = E'(t).$$

Recall also that $i(t) = q'(t)$.

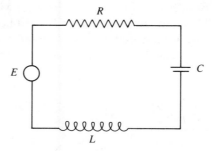

Figure 2.11

In these equations, L is the inductance, R is the resistance, and C is the capacitance, all assumed constant, and $E(t)$ is the electromotive force, which may vary with t. We can solve these differential equations with various choices of $E(t)$, but it is useful to notice an analogy between the differential equations of an electrical circuit and the spring equation

$$my'' + cy' + ky = f(t)$$

we have just analyzed. The differential equations are identical in form, differing only in the names of the quantities involved. We can make the following comparisons.

Spring System	Electrical Circuit
$my'' + cy' + ky = f(t)$	$Lq'' + Rq' + q/C = E(t)$ or $Li'' + Ri' + i/C = E'(t)$
Displacement $y(t)$	Charge $q(t)$
Velocity $y'(t)$	Current $i(t)$
Driving force $f(t)$	Electromotive force $E(t)$
Mass m	Inductance L
Damping constant c	Resistance R
Spring modulus k	Reciprocal $1/C$ of capacitance

These comparisons mean that general solutions of one type of problem can be readily adapted to the other type. There are practical ramifications of these comparisons as well. In the laboratory, it is usually easier and less expensive to do experiments with circuits than with spring systems. In addition, one can usually make more accurate measurements of an electrical system than of a mechanical system. Here is an example involving an RLC circuit.

EXAMPLE 2.36   Forced Oscillations in an Electrical Circuit

Consider the circuit of Figure 2.12, with resistance $R = 120$ ohms, capacitance $C = 10^{-3}$ farad, and inductance $L = 10$ henrys, driven by the potential $E(t) = 17 \sin(2t)$ volts. Assume that at time $t = 0$ the current is zero and the charge on the capacitor is $\frac{1}{2000}$ coulomb. That is, at time $t = 0$, the voltage drop across the capacitor is 0.5 volt.

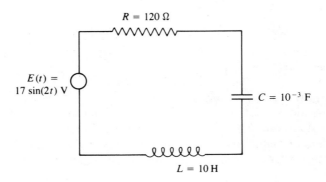

Figure 2.12

If we want to know the charge $q(t)$ on the capacitor for $t > 0$, we must solve the initial value problem

$$10i' + 120i + 1000q = 17 \sin(2t); \qquad q(0) = \tfrac{1}{2000}, \qquad i(0) = 0.$$

Since $q' = i$, we have the following initial value problem in q alone:

$$10q'' + 120q' + 1000q = 17 \sin(2t); \qquad q(0) = \tfrac{1}{2000}, \qquad q'(0) = 0.$$

If we want to know the current $i(t)$ for $t > 0$, we have two options. First, we can solve for $q(t)$ in the initial value problem we have just posed and then obtain $i(t) = q'(t)$. Second, we can use the fact that, in this circuit, the driving potential is differentiable. We can therefore establish an initial value problem for $i(t)$ and solve for the current directly. This initial value problem for i is

$$10i'' + 120i' + 1000i = E'(t) = 34 \cos(2t); \qquad i(0) = 0, \qquad i'(0) = K,$$

in which K is a number that we must determine. The given initial data provide only $i(0)$ and $q(0)$. However, notice that the differential equation (2.30) for q contains an i' term, since $i' = q''$. Solve this equation for i' to get

$$i'(t) = \frac{1}{L}\left[E(t) - Ri(t) - \frac{1}{C}\, q(t) \right].$$

Here we have that

$$i'(t) = \tfrac{17}{10} \sin(2t) - 12i(t) - 100q(t)$$

for $t > 0$. Since we know $i(0)$, $q(0)$, and $E(0)$, we can calculate $i'(0+)$, which is $\lim_{t \to 0+} i(t)$. We get

$$i'(0) = i'_0 = -100q(0) = -100\tfrac{1}{2000} = -\tfrac{1}{20}.$$

The initial value problem for the current function is therefore

$$i'' + 12i' + 100i = \tfrac{17}{5} \cos(2t); \qquad i(0+) = 0, \qquad i'(0+) = -\tfrac{1}{20}.$$

The associated homogeneous equation $i'' + 12i' + 100i = 0$ has general solution

$$i_h(t) = e^{-6t}[a \cos(8t) + b \sin(8y)].$$

For a particular solution of the differential equation for i, we will use the method of undetermined coefficients, attempting

$$i_p(t) = A \cos(2t) + B \sin(2t).$$

After a routine calculation, which we omit, we find that

$$i_p(t) = \tfrac{1}{30} \cos(2t) + \tfrac{1}{120} \sin(2t).$$

The general solution of the differential equation for i is

$$i(t) = e^{-6t}[a \cos(8t) + b \sin(8t)] + \tfrac{1}{30} \cos(2t) + \tfrac{1}{120} \sin(2t).$$

Notice that, as $t \to \infty$, $i_h(t) \to 0$, and this general solution behaves more like the particular solution. For this reason, $i_h(t)$ is called the *transient* part of the solution and $i_p(t)$ is called the *steady-state* part. Figure 2.13 shows a graph of i_h, and Figure 2.14 shows one of i_p.

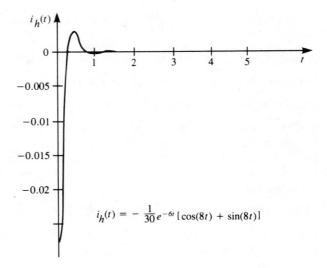

$$i_h(t) = -\frac{1}{30}e^{-6t}[\cos(8t) + \sin(8t)]$$

Figure 2.13

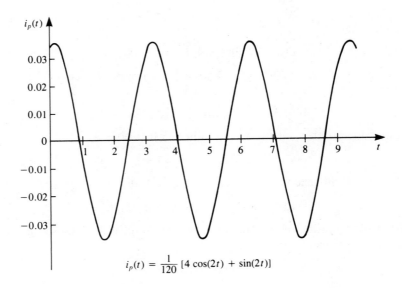

$$i_p(t) = \frac{1}{120}[4\cos(2t) + \sin(2t)]$$

Figure 2.14

From the initial condition, we have

$$i(0+) = 0 = a + \tfrac{1}{30}$$

and

$$i'(0) = -\tfrac{1}{20} = -6a + 8b + \tfrac{1}{60}.$$

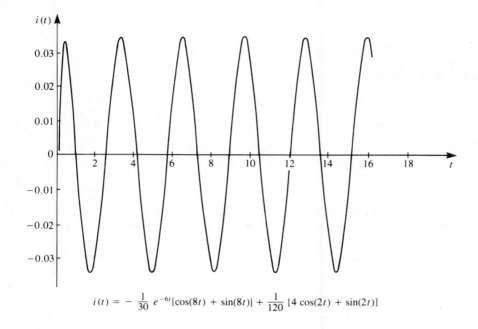

$$i(t) = -\frac{1}{30} e^{-6t}[\cos(8t) + \sin(8t)] + \frac{1}{120}[4\cos(2t) + \sin(2t)]$$

Figure 2.15

Solve these equations to get $a = b = -\frac{1}{30}$. The solution for the current is

$$i(t) = -\frac{1}{30}e^{-6t}[\cos(8t) + \sin(8t)] + \frac{1}{120}[4\cos(2t) + \sin(2t)].$$

A graph of this solution is shown in Figure 2.15. It is a sum of the graphs in Figures 2.13 and 2.14, the transient and steady-state solutions. ∎

EXAMPLE 2.37

Consider the circuit of Figure 2.16. The switch is closed at time $t = 0$, and prior to that the capacitors have zero charge and no current is flowing. We will find the charge and current in each loop for $t > 0$.

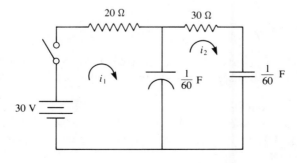

Figure 2.16

We will use Kirchhoff's voltage and current laws. In the first loop,

$$30 = 20i_1 + 60(q_1 - q_2) = 20q_1' + 60q_1 - 60q_2.$$

In the second loop,

$$60q_1 - 60q_2 = 30i_2 + 60q_2 = 30q_2' + 60q_2.$$

Because the circuit has two loops, we have a system of two (first order) differential equations:

$$2q_1' + 6q_1 - 6q_2 = 3 \qquad\qquad (2.31)$$
$$2q_1 - 4q_2 = q_2'. \qquad\qquad (2.32)$$

We need the initial values for q_1 and q_2. Recall that these differential equations are valid only after the switch is closed, at which time Kirchhoff's laws apply. The charge on the capacitor in the second loop prior to the time of switch closure is zero, so $q_2(0-) = 0$. Also, the charge on the capacitor in the first loop is zero before the switch is closed. Since this charge is $q_1(0-) - q_2(0-)$, we have $q_1(0-) = q_2(0-) = 0$. Since $q(t)$ is assumed continuous, $q_1(0+) = q_1(0-) = 0$ and $q_2(0+) = q_2(0-) = 0$. We conclude that $q_1(0) = q_2(0) = 0$.

A word of caution is needed here. It is not always the case that the initial values to be used in the initial value problem are those stated as the conditions prior to the time the switch is closed. As an example, suppose that we were interested in the currents in this circuit rather than the charges. Then we would change the differential equation to one involving current. We would then need values for the initial currents. We know them to be zero prior to the time the switch is closed, but what values should be used in the initial value problem, which is valid only after the switch is closed?

Return to the equations (2.31) and (2.32). Equation (2.31) states that, for all $t > 0$,

$$2i_1(t) + 6q_1(t) - 6q_2(t) = 3.$$

Then

$$i_1(0+) = \tfrac{1}{2}[3 - 6q_1(0+) + 6q_2(0+)] = \tfrac{3}{2} \text{ amperes.}$$

Using equation (2.32), we get

$$2q_1(t) - 4q_2(t) = i_2(t)$$

for $t > 0$, so

$$i_2(0+) = 2q_1(0+) - 4q_2(0+) = 0.$$

In fact, these values of initial current are not needed to solve the problem, but they are easy to obtain and may be used as verification of the derived solution for the charges in the circuit.

Now return to equations (2.31) and (2.32). In Chapter 14 we will learn to deal with systems of first order differential equations. For now we will incorporate equations (2.31) and (2.32) into a single second order differential equation. Differentiate equation (2.31) to get

$$2q_1'' + 6q_1' - 6q_2' = 0.$$

Replace q_2' with $2q_2 - 4q_2$, from (2.32), to get

$$2q_1'' + 6q_1' - 6(2q_1 - 4q_2) = 0,$$

or

$$q_1'' + 3q_1' - 6q_1 + 12q_2 = 0. \qquad (2.33)$$

But, from equation (2.31), $6q_2 = 2q_1' + 6q_1 - 3$, so

$$12q_2 = 4q_1' + 12q_1 - 6.$$

Use this to eliminate q_2 from equation (2.33), giving us

$$q_1'' + 3q_1' - 6q_1 + 4q_1' + 12q_1 - 6 = 0,$$

or

$$q_1'' + 7q_1' + 6q_1 = 6.$$

The general solution of this second order differential equation for q_1 is

$$q_1(t) = c_1 e^{-t} + c_2 e^{-6t} + 1.$$

Now use this solution to obtain q_1', and substitute q_1' and q_1 into equation (2.31) to get

$$q_2(t) = \tfrac{2}{3} c_1 e^{-t} - c_2 e^{-6t} + \tfrac{1}{2}.$$

Finally, use the initial conditions $q_1(0) = q_2(0) = 0$ to obtain the equations

$$c_1 + c_2 + 1 = 0$$
$$\tfrac{2}{3} c_1 - c_2 + \tfrac{1}{2} = 0,$$

from which we find that $c_1 = -\tfrac{9}{10}$ and $c_2 = -\tfrac{1}{10}$. Therefore, for $t \geq 0$, we have

$$q_1(t) = -\tfrac{9}{10} e^{-t} - \tfrac{1}{10} e^{-6t} + 1$$
$$q_2(t) = -\tfrac{3}{5} e^{-t} + \tfrac{1}{10} e^{-6t} + \tfrac{1}{2}.$$

Since $i = q'$, we easily find the solution for the current in each loop:

$$i_1(t) = q_1'(t) = \tfrac{9}{10} e^{-t} + \tfrac{3}{5} e^{-6t};$$
$$i_2(t) = q_2'(t) = \tfrac{3}{5} e^{-t} - \tfrac{3}{5} e^{-6t}.$$

From this explicit expression for $i_1(t)$, valid for $t > 0$, we obtain

$$i_1(0+) = \lim_{t \to 0+} i_1(t) = \tfrac{9}{10} + \tfrac{3}{5} = \tfrac{3}{2}$$

and

$$i_2(0+) = \lim_{t \to 0+} i_2(t) = \tfrac{3}{5} - \tfrac{3}{5} = 0,$$

in agreement with the result we obtained previously. ∎

PROBLEMS FOR SECTION 2.12

1. Recall that $y(t) = (A/[m(\omega_0^2 - \omega^2)])[\cos(\omega t) - \cos(\omega_0 t)]$ is the solution of the initial value problem

$$y'' + \omega_0^2 y = \frac{A}{m} \cos(\omega t); \qquad y(0) = 0, \qquad y'(0) = 0,$$

in the case in which $\omega \neq \omega_0$. Determine $\lim_{\omega \to \omega_0} y(t)$. How does this limit compare with the solution of the initial value problem $y'' + \omega_0^2 y = (A/m)\cos(\omega_0 t)$, $y(0) = y'(0) = 0$?

2. A 16-pound weight is suspended from a spring, stretching it 6 inches. The weight is pulled down an additional 3 inches below this equilibrium position and released. At this instant, an external force equal to $\frac{1}{4}\cos(6t)$ is applied to the system. Compute and sketch a graph of the displacement function, assuming no damping.

3. Calculate the motion of the weight in Problem 2 if there is a damping force of $8v$ pounds, where v is the velocity of the weight.

4. A 16-pound weight is suspended from a spring, stretching it $\frac{8}{11}$ foot. Then the weight is submerged in a liquid which imposes a drag of $2v$ pounds. The entire system is subjected to an external force $4\cos(\omega t)$. Determine the value of ω which maximizes the amplitude of the steady-state solution. What is this maximum amplitude?

In each of Problems 5 through 8, find the transient and steady-state solutions for $y(t)$, given m, c, k, and the driving force and initial conditions.

5. $m = k = 1$, $c = 2$, $F(t) = 2\sin(t)$, $y(0) = 2$, $y'(0) = 1$
6. $m = k = 2$, $c = 5$, $F(t) = \sin(4t)$, $y(0) = y'(0) = A$
7. $m = 5$, $c = 7$, $k = 2$, $F(t) = t + e^{-t}$, $y(0) = y'(0) = A$
8. $m = 1$, $c = 2$, $k = 4$, $F(t) = 3\sin(4t)$, $y(0) = 2$, $y'(0) = 0$

In each of Problems 9 through 12, find the current $i(t)$ in the RLC circuit shown below, with the given values of R, L, and C and assuming zero initial current and capacitor charge.

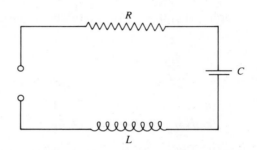

9. $R = 400$ ohms, $L = 0.12$ henry, $C = 0.004$ farad, $E(t) = 120\sin(20t)$ volts
10. $R = 200$ ohms, $L = 0.10$ henry, $C = 0.006$ farad, $E(t) = te^{-t}$ volt
11. $R = 450$ ohms, $L = 0.95$ henry, $C = 0.007$ farad, $E(t) = e^{-t}$ volts
12. $R = 150$ ohms, $L = 0.2$ henry, $C = 0.050$ farad, $E(t) = 1 - e^{-t}$ volt
13. Determine the current in each loop of the circuit shown below, assuming that the currents and capacitor charge are zero until the switch is closed.

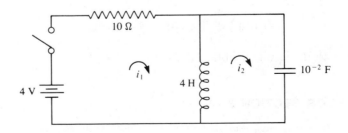

14. Calculate the steady-state currents in the circuit shown following Problem 13 if the battery is replaced with a sine wave generator which forces the circuit with the signal $4\cos(5t)$.

15. Compute the current in each loop of the circuit shown below, assuming that both loop currents and the capacitor charge are zero prior to the time the switch is closed.

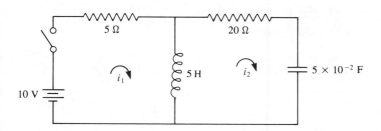

16. Suppose we have overdamped motion forced by the function $F(t) = A \cos(\omega t)$.
  (a) Find a solution satisfying $y(0) = B$, $y'(0) = 0$.
  (b) Find a solution satisfying $y(0) = 0$, $y'(0) = B$.
  (c) Sketch these solutions on the same set of axes to compare the relative influence of initial displacement and initial velocity on the resulting displacement function. What conclusions do you draw from these graphs?
  (d) On the same set of axes, sketch the graphs of the solution, the transient solution, and the steady-state solution in (a).
  (e) On the same set of axes, sketch the graphs of the solution, the transient solution, and the steady-state solution in (b).
  (f) From (d) and (e), how do the initial displacement and initial velocity differ in their influence on the transient and steady-state solutions?

17. Carry out the instructions of Problem 16 for the case of critical damping.
18. Carry out the instructions of Problem 16 for the case of underdamping.
19. Completely analyze the motion of a body of mass m suspended from a spring with spring constant k if the damping constant of the system is c and the external driving force is $F(t) = e^{-t}$. Consider all cases that arise and discuss what the mathematical solution for the displacement function tells you about the motion of the object.
20. Carry out the instructions of Problem 19 if the forcing function is $F(t) = t$.
21. Carry out the instructions of Problem 19 if the forcing function is $F(t) = Ae^{-t}\sin(\omega t)$.
22. Show that the distance between successive maxima of $y(t)$ in the case of beats is

$$\frac{4\pi}{\omega_0 + \omega}.$$

Sketch a graph of the solution for $y(t)$ obtained in the discussion of beats.

23. Show that damped, forced motion of a mass on a spring, with forcing function $A \cos(\omega t)$, is always bounded.
24. Consider damped, forced motion governed by the differential equation

$$my'' + cy' + ky = A \cos(\omega t).$$

Show that the maximum amplitude of the steady-state solution is achieved if ω is chosen so that

$$\omega^2 = \frac{k}{m} - \frac{c^2}{2m^2}.$$

This result has practical uses. In building a seismic detector, we would try to choose k, c, and m so that ω^2 is as near to this value as possible, in order to maximize the response.

Summary

Here is a schematic summary of some of the methods of this chapter.

Constant coefficient linear equation:

$$\boxed{y'' + Ay' + By = 0} \xrightarrow[\text{try } y = e^{rx}]{} \boxed{r^2 + Ar + B = 0}$$

$$\longrightarrow \begin{cases} r_1, r_2 \text{ (real, distinct):} \quad y = c_1 e^{r_1 x} + c_2 e^{r_2 x} \\ \text{one root } r: \quad y = (c_1 + c_2 x) e^{rx} \\ \text{complex roots } p \pm iq: \quad y = e^{px}[c_1 \cos(qx) + c_2 \sin(qx)]. \end{cases}$$

Cauchy-Euler equation:

$$\boxed{x^2 y'' + Axy' + By = 0} \xrightarrow[Y(t) = y(e^t)]{t = \ln(x)} Y'' + (A - 1) Y' + BY = 0$$

$$\longrightarrow \text{ solution } Y(t) \longrightarrow \text{ solution } y(x) = Y[\ln(x)].$$

Reduction of order:

$$F(x, y', y'') = 0 \xrightarrow{u = y'} f(x, u, u') = 0$$

$$F(y, y', y'') = 0 \xrightarrow[y'' = u\frac{du}{dy}]{u = y'} f\left[y, u, u\left(\frac{du}{dy}\right)\right] = 0.$$

If y_1 is one solution of $y'' + p(x)y' + q(x)y = 0$, let $y_2 = u(x)y_1(x)$ and solve for nonconstant $u(x)$.

General solution of $y'' + p(x)y' + q(x)y = f(x)$:

$$\boxed{y'' + p(x)y' + q(x)y = f(y)} \longrightarrow y = y_h + y_p$$

$$y_p \begin{cases} \text{undetermined coefficients (requires} \\ \text{constant coefficients)} \\ \text{variation of parameters } [y_p(x) = u(x)y_1(x) + v(x)y_2(x), \\ \text{with } y_1 \text{ and } y_2 \text{ a fundamental set for } y'' + py' + qy = 0]. \end{cases}$$

ADDITIONAL PROBLEMS

In each of Problems 1 through 60, find the general solution of the differential equation. These problems will require techniques from throughout the entire chapter.

1. $y'' - 3y' + 8y = 0$

2. $y'' + 3y' - 4y = 0$

3. $x^2 y'' - 9xy' + 25y = 0$

4. $x^2 y'' + 7xy' + 6y = 0$

5. $yy'' = (y')^2$

6. $yy'' - 2(y')^2 = 0$

7. $y'' - 14y' + 49y = 0$

8. $y'' + y' + 4y = 0$

9. $x^2y'' - 3xy' + 2y = 0$

10. $x^2y'' - 7xy' + 16y = 0$

11. $yy'' + 2y' = (y')^2$

12. $xy'' - 4y' = 14x^6$

13. $y'' - y' - 2y = 5e^{4x} + 6x$

14. $x^2y'' - 9xy' + 25y = 0$

15. $y'' + 5y' - 14y = 0$

16. $y'' - 3y' + 2y = 5e^{2x} + 4e^{-2x}$

17. $x^2y'' + xy' - y = x^2e^{-x}$

18. $x^2y'' + xy' + 4y = 0$

19. $2y' - y'' = 12x$

20. $y'' + 4y = 4\sec^2(2x)$

21. $x^2y'' - 2xy' + 10y = 0$

22. $yy'' + (y')^2 = 2yy'$

23. $y'' - 6y' + 9y = 0$

24. $x^2y'' - 6xy' + 12y = 3x^5\sin(x)$

25. $y'' + y = \tan^2(x)$

26. $y'' - y' - 8y = 3x^2 + 5$

27. $y'' - 3y' = 18x^2 + 6x - 3$

28. $y'' + 25y = 0$

29. $x^2y'' - xy' - 8y = 0$

30. $y'' - y = xe^x$

31. $x^2y'' + 11xy' + 41y = 0$

32. $y'' + 2y' - 3y = 13\cos(2x)$

33. $y'' - 4y' + 13y = 0$

34. $y'' + 4y = 8\sin(2x)$

35. $y'' - 2y' - y = 0$

36. $x^2y'' + 10xy' - 10y = 0$

37. $y'' + 4y' + 9y = 0$

38. $x^2y'' + 2xy' - 2y = 4x^2 + 6$

39. $y'' + y' - 2y = 8x^2 + 2x - 1$

40. $y'' - 3y' + 2y = 3e^x + 4x$

41. $y'' + y' = -3x^2$

42. $y'' - 6y' = 8e^{2x}$

43. $y'' - 8y' + 16y = \cosh(4x)$

44. $y'' + 9y = \sin(3x)$

45. $2y' - y'' = x^2$

46. $x^2y'' - 9xy' + 2y = x^2$

47. $x^2y'' - 3xy' + 2y = 0$

48. $x^2y'' - 2xy' + 2y = 2x^3e^{-3x}$

49. $x^2y'' - 5xy' + 8y = \dfrac{1}{x}$

50. $y'' + y = \sec^3(x)$

51. $y'' + 2y' + y = \dfrac{1}{x}e^{-x}$

52. $y'' + 8y' + 16y = x\cos(2x)$

53. $y'' + 2y' - 4y = \cos(3x) - e^x$

54. $y'' + 18y = \cos(2x)$

55. $y'' - 5y' + 9y = x^2 - \sin(x)$

56. $y'' - 18y' + 81y = 3x^3 - 5$

57. $x^2y'' - 2xy' + 14y = 0$

58. $y'' + 10y' - 4y = e^x - 5$

59. $y'' + 3y' + 2y = \dfrac{1}{1 + e^x}$

60. $y'' - 3y' + 2y = \sin(e^{-x})$

In each of Problems 61 through 80, solve the initial value problem.

61. $y'' + 22y' + 121y = 28$; $\quad y(0) = -5$, $\quad y'(0) = 2$

62. $x^2y'' + 13xy' + 45y = 0$; $\quad y(1) = 2$, $\quad y'(1) = 2$

63. $y'' + 4y' - 96y = 3e^{8x} - \cos(3x)$; $\quad y(0) = 0$, $\quad y'(0) = 0$

64. $y'' - y' - 20y = x^2 - 1$; $\quad y(0) = 0$, $\quad y'(0) = 0$

65. $y'' - 2y' + y = 2\sin(3x)$; $\quad y(0) = 2$, $\quad y'(0) = 1$

66. $y'' - 4y' + 2y = x^2 + 1$; $\quad y(0) = 4$, $\quad y'(0) = 0$

67. $y'' + y = \tan(x)$; $\quad y(0) = 0$, $\quad y'(0) = 0$

68. $y'' + 2y' + y = \dfrac{1}{x}e^{-x}$; $\quad y(1) = 0$, $\quad y'(1) = 0$

69. $x^2y'' + 3xy' + y = 4x$; $\quad y(1) = 6$, $\quad y'(1) = -7$

70. $x^2y'' + 2xy' - 6y = x^4 - 2x^2$; $\quad y(1) = 5$, $\quad y'(1) = 2$

71. $y'' - 4y' + 3y = -3\sin(x + 2)$; $\quad y(-2) = 2$, $\quad y'(-2) = 2$

72. $y'' + 2y' - 8y = 10e^x - 32x$; $\quad y(0) = 3$, $\quad y'(0) = 16$

73. $x^2y'' - xy' - 2y = x^3 + 4\ln(x)$; $\quad y(1) = 9$, $\quad y'(1) = 7$

74. $y'' + 9y = \tan(3x);$ $y(0) = 5,$ $y'(0) = 4$

75. $yy'' - 2(y')^2 = 0;$ $y(0) = 6,$ $y'(0) = 2$

76. $y'' - 2y' + y = 2e^x;$ $y(0) = 4,$ $y'(0) = 6$

77. $y'' - 2y' + 7y = e^x;$ $y(0) = -2,$ $y'(0) = -2$

78. $y'' + 8y' - 2y = x + 5;$ $y(-1) = -3,$ $y'(-1) = 2$

79. $y'' + y = 2\sec^3(x);$ $y(0) = 4,$ $y'(0) = 2$

80. $y'' - 2y' + y = 4e^x\ln(x);$ $y(1) = 2e,$ $y'(1) = 0$

81. A particle of mass m moves along a straight line and has coordinate $x(t)$ at time t. The total force acting on the particle is $F(x)$.

(a) Use Newton's second law of motion to write a differential equation for x.

(b) Show that the energy of the particle is a constant of the motion (conservation of energy). *Hint:* The kinetic energy is $\frac{1}{2}m(dx/dt)^2$ and the potential energy is $-\int_0^x F(t)\,dt$.

82. A particle of mass m inside the earth at a distance r from the center experiences a gravitational force of $F = -mgr/R$, where R is the radius of the earth (assuming that the earth is a sphere). Show that a particle in an evacuated tube through the earth's center will execute simple harmonic motion, and determine the period of this motion.

83. A ball of mass m is thrown vertically downward from a stationary dirigible hovering h feet above the ground. The initial velocity of the ball is v_0. Neglecting air resistance, show that the ball will impact on the ground at time

$$\frac{1}{g}(\sqrt{v_0^2 + 2gh} - v_0).$$

84. A 4-pound brick is dropped from the top of a building h feet tall. After it has fallen k feet ($k < h$), a 6-pound brick is dropped from the same point.

(a) Neglecting air resistance, show that, when the first brick strikes the ground, the second still has $2\sqrt{kh} - k$ feet to fall.

(b) Of what importance is the weight of each brick in reaching the conclusion asked for in (a)?

85. For a simple pendulum, we obtained in Example 0.4 that

$$\frac{d^2\theta}{dt^2} + \frac{g}{L}\sin(\theta) = 0.$$

Assume that the pendulum bob is released from rest at time $t = 0$ from the position $\theta = -\alpha$, with $0 < \alpha < \pi/2$. (That is, the bob is drawn back α radians to the left and released from rest.) Show that, on the first half-swing of the pendulum,

$$t = \sqrt{\frac{L}{2g}} \int_{-\alpha}^{\theta} \frac{1}{\sqrt{\cos(\varphi) - \cos(\alpha)}}\,d\varphi.$$

86. Prove the conservation of energy law for the simple pendulum. *Hint:* The kinetic energy is $\frac{1}{2}m(ds/dt)^2$, where $s = L\theta$, and the potential energy is $mgL[1 - \cos(\theta)]$.

87. If θ is small, then $\sin(\theta) \approx \theta$, and the pendulum equation is approximated by the linear differential equation

$$\frac{d^2\theta}{dt^2} + \frac{g}{L}\theta = 0.$$

This equation has sine and cosine solutions, with period $2\pi\sqrt{L/g}$. If we do not assume that $\sin(\theta) \approx \theta$, the pendulum equation cannot be solved exactly as a linear combination of elementary functions. However, we can carry out the following analysis.

(a) Let $u = d\theta/dt$. Show that

$$u \frac{du}{d\theta} + \frac{g}{L} \sin(\theta) = 0.$$

(b) Integrate the equation of (a) to conclude that

$$\frac{1}{2} u^2 - \frac{g}{L} \cos(\theta) = C,$$

in which C is constant.

(c) Suppose that $\theta = \theta_0$ at $t = 0$ and that $\theta'(0) = 0$. Show that, as the pendulum moves from its initial position θ_0 to $\theta = 0$ (the vertical position),

$$\frac{d\theta}{dt} = -\sqrt{\frac{2g}{L}} \sqrt{\cos(\theta) - \cos(\theta_0)}.$$

(d) Suppose that as θ moves from θ_0 to 0, the time consumed is one-fourth the period p of the pendulum. Show that

$$p = 4\sqrt{\frac{L}{2g}} \int_0^{\theta_0} \frac{1}{\sqrt{\cos(\theta) - \cos(\theta_0)}} \, d\theta.$$

(e) Use trigonometric identities to show that

$$p = 2\sqrt{\frac{L}{g}} \int_0^{\theta_0} \frac{1}{\sqrt{\sin^2\left(\frac{\theta_0}{2}\right) - \sin^2\left(\frac{\theta}{2}\right)}} \, d\theta.$$

(f) Let $\sin(\theta/2) = \sin(\theta_0/2)\sin(\alpha)$. Show that

$$p = 4\sqrt{\frac{L}{g}} \int_0^{\pi/2} \sqrt{1 - k^2\sin^2(\alpha)} \, d\alpha,$$

where $k = \sin(\theta_0/2)$. The integral on the right is an *elliptic integral of the second kind* and cannot be evaluated as a finite linear combination of elementary functions. There are tables giving values of this integral for various values of k.

(g) If $\theta \approx 0$, show that $k \approx 0$ also and that the result of (f) again gives the value $2\pi\sqrt{L/g}$ for the period of the pendulum.

88. Use the result of Problem 87(f) to show that the period p of a simple pendulum is given by

$$p = 2\pi\sqrt{\frac{L}{g}} \left[1 + \frac{1^2}{2^2} k^2 + \frac{1^2 3^2}{2^2 4^2} k^4 + \frac{1^2 3^2 5^2}{2^2 4^2 6^2} k^2 + \cdots \right],$$

in which $k = \sin(\theta_0/2)$ and $\theta_0 = \theta(0)$, as in Problem 87. *Hint:* Recall from the binomial theorem that

$$(1 + \alpha)^r = 1 + r\alpha + \frac{r(r-1)}{2!} \alpha^2 + \frac{r(r-1)(r-2)}{3!} \alpha^3 + \cdots$$

if $|\alpha| < 1$. Use this result to expand the integrand in Problem 87(f), and integrate the resulting series term by term, using the fact that

$$\int_0^{\pi/2} \sin^{2n}(\alpha) \, d\alpha = \frac{(1)(3)(5)\cdots(2n-1)}{(2)(4)(6)\cdots(2n)} \frac{\pi}{2}.$$

CHAPTER 3

Higher Order Differential Equations

3.0 Introduction

Thus far, we have seen some practical and some theoretical results involving first and second order differential equations. We will now look at higher order differential equations. This is a mildly ambiguous phrase (one can ask, higher than what?). By a "higher order" differential equation, we will mean one in which derivatives of order at least 3 occur.

For convenience, denote the kth derivative of y as $y^{(k)}$. An nth order differential equation is an equation

$$F(x, y, y', y'', \ldots, y^{(n)}) = 0$$

in which an nth derivative $y^{(n)}$ must explicitly appear but $x, y, y', \ldots, y^{(n-1)}$ may or may not be explicitly present. For example,

$$y^{(4)} - 3y' + 2y = 0$$

is a fourth order differential equation. In this example, y, y', and (necessarily) $y^{(4)}$ appear explicitly, while x, y'', and $y^{(3)}$ do not.

Higher order differential equations appear in many contexts. For example, problems involving deflections of loaded beams may lead to fourth order differential equations. Mechanical spring systems having several springs connected in tandem, or series electrical circuits containing several loops, can also be modeled using higher order differential equations.

The solution of higher order differential equations is often approached by converting the differential equation into a system of first order differential equations, which are then solved using matrix methods. We will pursue this approach in Chapter 14. However, it is also useful to be able to solve certain kinds of higher order equations without converting to a system of first order equations. Often we use methods which are direct generalizations of those we have seen for second order equations. This chapter develops some of these ideas and methods.

3.1 *Theoretical Considerations*

The general nth order differential equation

$$F(x, y, y', \ldots, y^{(n)}) = 0$$

is beyond our means at this point. As we did with first and second order differential equations, we will look for special cases which are important and which we have some hope of solving. In carrying out this program, we will imitate to a large extent our results and experience with linear and Cauchy-Euler equations in Chapter 2.

An nth order differential equation is said to be *linear* if it is of the form

$$y^{(n)} + p_{n-1}(x)y^{(n-1)} + \cdots + p_1(x)y' + p_0(x)y = f(x). \tag{3.1}$$

We call $p_{n-1}, \ldots, p_0$ and f the *coefficient functions*, or *coefficients*, of this linear equation. Often these functions are defined only on an interval I, in which case the discussion is restricted to this interval. A *solution* of this linear equation is a function satisfying the equation, perhaps on an interval I. Thus, φ is a solution of the linear equation on I if

$$\varphi^{(n)}(x) + p_{n-1}(x)\varphi^{(n-1)}(x) + \cdots + p_1(x)\varphi'(x) + p_0(x)\varphi(x) = f(x)$$

for all x in I.

Since an nth order equation involves an nth (but no higher) order derivative, we might expect that a solution process will involve n integrations and hence give rise to n arbitrary integration constants. To determine these constants, and in turn to determine a unique solution, we must specify n pieces of information. These are usually given in terms of values of the solution and its first $n - 1$ derivatives at a particular point. In view of this, and our experience with the second order case, we define the nth order *linear initial value problem* to be the problem

$$y^{(n)} + p_{n-1}(x)y^{(n-1)} + \cdots + p_1(x)y' + p_0(x)y = f(x);$$
$$y(x_0) = A_0, \qquad y'(x_0) = A_1, \qquad \ldots, \qquad y^{(n-1)}(x_0) = A_{n-1},$$

in which $x_0, A_0, A_1, \ldots, A_{n-1}$ are given constants. The conditions

$$y(x_0) = A_0, \qquad y'(x_0) = A_1, \qquad \ldots, \qquad y^{(n-1)}(x_0) = A_{n-1}$$

are called *initial conditions*.

The following theorem is analogous to the existence/uniqueness theorem for second order linear initial value problems.

THEOREM 3.1   Existence and Uniqueness for the Linear Initial Value Problem

Let n be an integer, with $n \geq 2$. Let $p_0, p_1, \ldots, p_{n-1}$, and f be continuous on an open interval I, and let x_0 be in I. Let $A_0, A_1, \ldots, A_{n-1}$ be any numbers. Then the initial value problem

$$y^{(n)} + p_{n-1}(x)y^{(n-1)} + \cdots + p_1(x)y' + p_0(x)y = f(x);$$
$$y(x_0) = A_0, \qquad y'(x_0) = A_1, \qquad \ldots, \qquad y^{(n-1)}(x_0) = A_{n-1}$$

has a unique solution defined over the entire interval I. ∎

As with second order equations, we will now separate our study of the linear nth order differential equation into two cases: homogeneous and nonhomogeneous. We call the linear differential equation (3.1) *homogeneous* on I if $f(x) = 0$ for all x in I; otherwise, the equation is called *nonhomogeneous*.

THE HOMOGENEOUS LINEAR nTH ORDER EQUATION

Consider the homogeneous linear nth order differential equation

$$y^{(n)} + p_{n-1}(x)y^{(n-1)} + \cdots + p_1(x)y' + p_0(x)y = 0. \tag{3.2}$$

Because this equation is both linear and homogeneous, sums of solutions are solutions, and constants times solutions are solutions.

THEOREM 3.2

Let $y_1, \ldots, y_k$ be solutions of equation (3.2). Then

1. $y_1 + y_2 + \cdots + y_k$ is a solution.

2. αy_1 is a solution for any number α. ∎

We can prove conclusion (1) by direct substitution of $y_1 + \cdots + y_k$ into the differential equation, as we did in the case $n = 2$. A similar tactic can be used to prove conclusion (2). The conclusions of Theorem 3.2 may fail if the differential equation is not homogeneous or not linear.

A more efficient way of stating the conclusion of Theorem 3.1 is that any function

$$c_1 y_1 + \cdots + c_k y_k,$$

in which $y_1, \ldots, y_k$ are solutions and $c_1, \ldots, c_k$ are constants, is also a solution. We call an expression $c_1 y_1 + \cdots + c_k y_k$ a *linear combination* of the functions $y_1, \ldots, y_k$. Thus, any linear combination of solutions of equation (3.2) is again a solution.

We say that functions $y_1, \ldots, y_k$ are *linearly dependent* on an interval I if there is a linear combination $c_1 y_1 + \cdots + c_k y_k$ which is identically zero for all x in I but has at least one coefficient c_j nonzero. If the only way $c_1 y_1(x) + \cdots + c_k y_k(x)$ can be zero for all x in I is for $c_1 = c_2 = \cdots = c_k = 0$, then we call the functions $y_1, \ldots, y_k$ *linearly independent* on I.

If $y_1, \ldots, y_k$ are linearly dependent on I, then for some constants $c_1, \ldots, c_k$ not all zero, we have

$$c_1 y_1(x) + \cdots + c_k y_k(x) = 0 \qquad \text{for all } x \text{ in } I.$$

Suppose, to illustrate a point, that $c_1 \neq 0$. In this event, we can solve for $y_1(x)$ and write

$$y_1(x) = -\frac{c_2}{c_1} y_2(x) - \frac{c_3}{c_1} y_3(x) - \cdots - \frac{c_k}{c_1} y_k(x)$$

for x in I. That is, on I we have

$$y_1 = -\frac{c_2}{c_1} y_2 - \frac{c_3}{c_1} y_3 - \cdots - \frac{c_k}{c_1} y_k,$$

and one of the functions (in this case, y_1) can be written as a linear combination of the others. This means that, in a sense, y_1 becomes irrelevant; we already know it as a sum of constants times $y_2, \ldots, y_k$. By contrast, if $y_1, \ldots, y_k$ are linearly independent, no one of the functions can be written as a sum of constants times the others.

EXAMPLE 3.1

The functions e^x, $\cos(x)$, and $\sin(x)$ are linearly independent on the entire real line. This may be obvious, because we cannot write one of these functions as a sum of constants times the other two. However, as an exercise, we will apply the definition.

Suppose that there are constants c_1, c_2, and c_3 such that

$$c_1 e^x + c_2 \sin(x) + c_3 \cos(x) = 0 \tag{3.3}$$

for all x. We will show that necessarily $c_1 = c_2 = c_3 = 0$. First, putting $x = 0$ in equation (3.3), we get

$$c_1 + c_3 = 0;$$

hence, $c_3 = -c_1$. Next, put $x = \pi$ into equation (3.3) to get

$$c_1 e^\pi - c_3 = 0;$$

hence, $c_3 = c_1 e^\pi$. We now have $c_3 = -c_1 = c_1 e^\pi$. Since $e^\pi \neq -1$, the equation $-c_1 = c_1 e^\pi$ requires that $c_1 = 0$ and hence $c_3 = 0$ also. But now equation (3.3) states that $c_2 \sin(x) = 0$ for all x, and this can be true only if $c_2 = 0$ also. Since equation (3.3) forces $c_1 = c_2 = c_3 = 0$, e^x, $\sin(x)$, and $\cos(x)$ are linearly independent. ∎

EXAMPLE 3.2

The functions

$$5\cos(2x), \qquad 3\cos^2(x), \quad \text{and} \quad -4\sin^2(x)$$

are linearly dependent because, for all x,

$$\tfrac{1}{5}[5\cos(2x)] - \tfrac{1}{3}[3\cos^2(x)] - \tfrac{1}{4}[-4\sin^2(x)] = 0.$$

This equation is true because of the identity

$$\cos(2x) = \cos^2(x) - \sin^2(x). \quad ∎$$

We will now state an nth order version of the Wronskian test. The Wronskian of n functions $y_1, y_2, \ldots, y_n$ is defined to be the $n \times n$ determinant

$$W[y_1, \ldots, y_n] = \begin{vmatrix} y_1 & y_2 & \cdots & y_n \\ y_1' & y_2' & \cdots & y_n' \\ \vdots & \vdots & & \vdots \\ y_1^{(n-1)} & y_2^{(n-1)} & \cdots & y_n^{(n-1)} \end{vmatrix}.$$

Usually we will denote this determinant as just W. The first row of W consists of the functions, the second row consists of their derivatives, the third contains the second derivatives, and so on, until the nth row, which contains the derivatives of order $(n-1)$.

THEOREM 3.3 Wronskian Test

Let $y_1, \ldots, y_n$ be solutions of

$$y^{(n)} + p_{n-1}(x)y^{(n-1)} + \cdots + p_1(x)y' + p_0(x)y = 0$$

on an open interval I. Then

1. Either $W(x) = 0$ for all x in I or $W(x) \neq 0$ for all x in I.

2. $y_1, \ldots, y_n$ are linearly independent on I if and only if $W(x_0) \neq 0$ for some x_0 in I. ∎

Thus, we need check the value of the Wronskian at only one point (any point) in I. If the Wronskian is zero at that point, the solutions are linearly dependent; if it is nonzero, the solutions are linearly independent.

As an illustration, look again at the functions in Example 3.1. These functions are solutions of the third order linear differential equation

$$y^{(3)} - y'' + y' - y = 0,$$

as can be verified by substitution. The Wronskian of these solutions is

$$W = \begin{vmatrix} e^x & \sin(x) & \cos(x) \\ e^x & \cos(x) & -\sin(x) \\ e^x & -\sin(x) & -\cos(x) \end{vmatrix},$$

which equals $-2e^x$. Since this is not zero, these solutions are linearly independent over the entire real line.

Note that the Wronskian test applies only to n solutions of an nth order linear homogeneous differential equation; it cannot be used to show that just any n functions are linearly dependent or independent.

The next theorem tells us how to use n linearly independent solutions to write an expression containing all possible solutions of an nth order linear homogeneous differential equation.

THEOREM 3.4

Let $y_1, \ldots, y_n$ be linearly independent solutions of

$$y^{(n)} + p_{n-1}(x)y^{(n-1)} + \cdots + p_1(x)y' + p_0(x)y = 0$$

on an open interval I. Then every solution on I is a linear combination of $y_1, \ldots, y_n$. ∎

A proof of this theorem can be modeled after that of Theorem 2.4 in Section 2.2 and is left to the student.

In view of Theorem 3.4, we call n linearly independent solutions of the homogeneous linear nth order equation a *fundamental set of solutions*. The linear combination $c_1y_1 + \cdots + c_ny_n$, in which $c_1, \ldots, c_n$ are arbitrary constants, is called the *general solution* of this equation. We obtain all solutions by making different choices of the constants $c_1, \ldots, c_n$.

EXAMPLE 3.3

We know from preceding remarks that e^x, $\cos(x)$, and $\sin(x)$ are three linearly independent solutions of $y^{(3)} - y'' + y' - y = 0$. These functions form a fundamental set of solutions, and the general solution is

$$y = c_1e^x + c_2\cos(x) + c_3\sin(x). \quad \blacksquare$$

THE NONHOMOGENEOUS LINEAR nTH ORDER EQUATION

Now consider the nonhomogeneous linear equation

$$y^{(n)} + p_{n-1}(x)y^{(n-1)} + \cdots + p_1(x)y' + p_0(x)y = f(x).$$

We will see that, as in the case $n = 2$, the key to finding the general solution lies in finding *one* particular solution, together with the general solution of the associated homogeneous equation

$$y^{(n)} + p_{n-1}(x)y^{(n-1)} + \cdots + p_1(x)y' + p_0(x)y = 0.$$

THEOREM 3.5

Let y_p be any solution of

$$y^{(n)} + p_{n-1}(x)y^{(n-1)} + \cdots + p_1(x)y' + p_0(x)y = f(x) \qquad (3.4)$$

on an open interval I. Let $y_1, \ldots, y_n$ be linearly independent solutions of

$$y^{(n)} + p_{n-1}(x)y^{(n-1)} + \cdots + p_1(x)y' + p_0(x)y = 0.$$

Then any solution of the nonhomogeneous equation can be written in the form

$$c_1y_1 + c_2y_2 + \cdots + c_ny_n + y_p$$

for some constants $c_1, \ldots, c_n$. $\quad \blacksquare$

Following our experience with $n = 2$, we call the linear combination $c_1y_1 + c_2y_2 + \cdots + c_ny_n$ the *associated homogeneous solution* of the nonhomogeneous equation (3.4) and denote it y_h. The *general solution* of equation (3.4) is

$$y = y_h + y_p,$$

a sum of the associated homogeneous solution and any particular solution of (3.4).

EXAMPLE 3.4

It is routine to check that $y_p = x$ is one solution of

$$y^{(3)} - 2y'' - y' + 2y = 2x - 1.$$

Further, e^{2x}, e^x, and e^{-x} are linearly independent solutions of the associated homogeneous equation $y^{(3)} - 2y'' - y' + 2y = 0$. Thus, let

$$y_h = c_1 e^{2x} + c_2 e^x + c_3 e^{-x}.$$

The general solution of the nonhomogeneous equation is

$$y = y_h + y_p = c_1 e^{2x} + c_2 e^x + c_3 e^{-x} + x. \quad \blacksquare$$

Theorems 3.4 and 3.5 tell us what to look for to obtain the general solution of a homogeneous or nonhomogeneous nth order linear differential equation. For the remainder of this chapter, we will look at special types of linear equations which we may be able to solve explicitly.

Finally, we approach an nth order linear initial value problem much as we did the second order linear initial value problem. We attempt to find the general solution of the differential equation, then use the initial data to solve for the constants to produce the unique solution of the initial value problem.

EXAMPLE 3.5

Solve the initial value problem

$$y^{(3)} - 2y'' - y' + 2y = 2x - 1; \quad y(0) = 1, \quad y'(0) = -3, \quad y''(0) = 4.$$

We know from Example 3.4 that the general solution of this differential equation is

$$y = c_1 e^{2x} + c_2 e^x + c_3 e^{-x} + x.$$

From the initial data, we require that

$$y(0) = c_1 + c_2 + c_3 = 1,$$
$$y'(0) = 2c_1 + c_2 - c_3 + 1 = -3,$$

and

$$y''(0) = 4c_1 + c_2 + c_3 = 4.$$

Solve these equations to obtain $c_1 = 1$, $c_2 = -3$, and $c_3 = 3$. The unique solution of the initial value problem is

$$y = e^{2x} - 3e^x + 3e^{-x} + x. \quad \blacksquare$$

PROBLEMS FOR SECTION 3.1

1. Show that $e^{\alpha x}$, $e^{\beta x}$, and $e^{\gamma x}$ are linearly independent over the entire real line if α, β, and γ are distinct real numbers. *Hint:* This can be done directly from the definition of linear independence, using an argument similar to that of Example 3.1. To use the Wronskian test, we are required first to produce a linear third order differential equation having these functions as solutions.

2. Show that $e^{\alpha x}$, $\sin(\beta x)$, and $\cos(\beta x)$ are linearly independent over the entire real line if $\beta \neq 0$.

In each of Problems 3 through 9, (a) verify that the functions form a fundamental set of solutions of the differential equation and (b) write the general solution of the differential equation.

3. $y^{(3)} - 5y'' + 2y' + 8y = 0$; e^{4x}, e^{2x}, e^{-x} 4. $y^{(3)} - 2y'' - 7y' - 4y = 0$; e^{4x}, e^{-x}, xe^{-x}

5. $y^{(4)} - y^{(3)} - 9y'' - 11y' - 4y = 0$; e^{4x}, e^{-x}, xe^{-x}, $x^2 e^{-x}$

6. $y^{(3)} + y' = 0$; 1, $\cos(x)$, $\sin(x)$

7. $y^{(3)} + y'' - 7y' - 15y = 0$; e^{3x}, $e^{-2x}\cos(x)$, $e^{-2x}\sin(x)$

8. $x^3 y^{(3)} + 7x^2 y'' + 2xy' - 18y = 0$ $(x > 0)$; x^{-3}, $x^{-3}\ln(x)$, x^2

9. $x^2 y^{(3)} - x(x+2)y'' + (x+2)y' = 0$ $(x > 0)$; 1, x^2, $(x-1)e^x$

In each of Problems 10 through 16, verify that y_p is a solution of the nonhomogeneous differential equation.

10. $y^{(3)} - 5y'' + 2y' + 8y = x^2$; $y_p = \frac{1}{64}(8x^2 - 4x + 11)$

11. $y^{(3)} - 2y'' - 7y' - 4y = 3\cos(2x)$; $y_p = \frac{1}{250}[6\cos(2x) - 33\sin(2x)]$

12. $y^{(4)} - y^{(3)} - 9y'' - 11y' - 4y = 2 - x^3$; $y_p = \frac{1}{4}x^3 - \frac{33}{16}x^2 + \frac{255}{32}x - \frac{1729}{128}$

13. $y^{(3)} + y' = -4e^{2x}$; $y_p = -\frac{2}{5}e^{2x}$

14. $y^{(3)} + y'' - 7y' - 15y = 8\sin(3x)$; $y_p = \frac{2}{15}\cos(3x) - \frac{1}{15}\sin(3x)$

15. $x^3 y^{(3)} + 7x^2 y'' + 2xy' - 18y = 2\ln(x) - 5$ $(x > 0)$; $y_p = -\frac{1}{9}\ln(x) + \frac{8}{27}$

16. $x^2 y^{(3)} - x(x+2)y'' + (x+2)y' = 6x^3$ $(x > 0)$; $y_p = -2x^3 - 6$

In each of Problems 17 through 23, two previous problems are cited. Use results from these problems to write the general solution of the differential equation. Then use the initial data to produce the unique solution of the initial value problem.

17. $y^{(3)} - 5y'' + 2y' + 8y = x^2$; $y(0) = 0$, $y'(0) = 1$, $y''(0) = 0$ (Problems 3 and 10)

18. $y^{(3)} - 2y'' - 7y' - 4y = 3\cos(2x)$; $y(0) = y'(0) = 0$, $y''(0) = 1$ (Problems 4 and 11)

19. $y^{(4)} - y^{(3)} - 9y'' - 11y' - 4y = 2 - x^3$; $y(0) = 1$, $y'(0) = y''(0) = 0$, $y^{(3)}(0) = 2$ (Problems 5 and 12)

20. $y^{(3)} + y' = -4e^{2x}$; $y(0) = y'(0) = 1$, $y''(0) = -2$ (Problems 6 and 13)

21. $y^{(3)} + y'' - 7y' - 15y = 8\sin(3x)$; $y(0) = 1$, $y'(0) = -1$, $y''(0) = -4$ (Problems 7 and 14)

22. $x^3 y^{(3)} + 7x^2 y'' + 2xy' - 18y = 2\ln(x) - 5$ $(x > 0)$; $y(1) = 0$, $y'(1) = -1$, $y''(1) = \frac{2}{3}$ (Problems 8 and 15)

23. $x^2 y^{(3)} - x(x+2)y'' + (x+2)y' = 6x^3$ $(x > 0)$; $y(1) = y'(1) = y''(1) = 0$ (Problems 9 and 16)

3.2 The Constant Coefficient Homogeneous Equation

Assume that the coefficient functions are constant, and consider the homogeneous linear nth order differential equation

$$y^{(n)} + a_{n-1}y^{(n-1)} + \cdots + a_2 y'' + a_1 y' + a_0 y = 0. \tag{3.5}$$

As in the case $n = 2$, we will attempt exponential solutions $y = e^{rx}$. Note that the kth derivative of e^{rx} is $r^k e^{rx}$. Assuming that e^{rx} is a solution of equation (3.5), we obtain upon substitution that

$$r^n e^{rx} + a_{n-1}r^{n-1}e^{rx} + \cdots + r^2 a_2 e^{rx} + ra_1 e^{rx} + a_0 e^{rx} = 0,$$

or

$$[r^n + a_{n-1}r^{n-1} + \cdots + a_2 r^2 + a_1 r + a_0]e^{rx} = 0.$$

Since $e^{rx} \neq 0$, this yields the *characteristic equation* of equation (3.5),

$$r^n + a_{n-1}r^{n-1} + \cdots + a_2 r^2 + a_1 r + a_0 = 0.$$

This nth degree polynomial equation has n roots (some may be repeated). For each root r of this equation, e^{rx} is a solution of equation (3.5). This idea is similar to the case $n = 2$. There are, however, practical difficulties in carrying out the method when $n > 2$ because for $n \geq 3$ it becomes more difficult to find roots of the nth degree characteristic equation.

EXAMPLE 3.6

Solve $y^{(4)} - 2y^{(3)} - y'' + 2y' = 0$.

The characteristic equation is

$$r^4 - 2r^3 - r^2 + 2r = 0,$$

with roots 0, 1, -1, and 2. Four solutions are

$$e^{0x} = 1, \qquad e^x, \qquad e^{-x}, \quad \text{and} \quad e^{2x}.$$

These solutions are linearly independent (their Wronskian is $12e^{2x}$) over the entire real line. The general solution of the differential equation is

$$y = c_1 + c_2 e^x + c_3 e^{-x} + c_4 e^{2x}$$

for all x. ∎

If we have an initial value problem, we obtain the unique solution by using the initial data to solve for the constants appearing in the general solution.

EXAMPLE 3.7

Solve the initial value problem

$$y^{(4)} - 2y^{(3)} - y'' + 2y' = 0;$$
$$y(0) = 0, \qquad y'(0) = 1, \qquad y''(0) = -1, \qquad y^{(3)}(0) = 7.$$

From Example 3.6, the general solution of the differential equation is

$$y = c_1 + c_2 e^x + c_3 e^{-x} + c_4 e^{2x}.$$

From the initial data, we obtain four equations for the constants c_1 through c_4:

$$y(0) = c_1 + c_2 + c_3 + c_4 = 0$$
$$y'(0) = c_2 - c_3 + 2c_4 = 1$$
$$y''(0) = c_2 + c_3 + 4c_4 = -1$$
$$y^{(3)}(0) = c_2 - c_3 + 8c_4 = 7.$$

Solve these equations to get

$$c_1 = 4, \qquad c_2 = -3, \qquad c_3 = -2, \qquad c_4 = 1.$$

The unique solution of the initial value problem is

$$y(x) = 4 - 3e^x - 2e^{-x} + e^{2x}. ∎$$

EXAMPLE 3.8

Solve the initial value problem

$$y^{(4)} - 3y^{(3)} - y'' + 13y' - 10y = 0;$$
$$y(0) = 5, \qquad y'(0) = 12, \qquad y''(0) = -5, \qquad y^{(3)}(0) = -6.$$

The characteristic equation of the differential equation is

$$r^4 - 3r^3 - r^2 + 13r - 10 = 0,$$

with roots $1, -2, 2 + i$, and $2 - i$. Four solutions of the differential equation are

$$e^x, \quad e^{-2x}, \quad e^{(2+i)x}, \quad \text{and} \quad e^{(2-i)x}.$$

These solutions form a fundamental set. However, usually we prefer to replace the complex exponential solutions with real-valued solutions. As in the case $n = 2$, we obtain from the complex conjugate roots the two solutions $e^{2x}\cos(x)$ and $e^{2x}\sin(x)$. Thus, another set of four solutions is

$$e^x, \quad e^{-2x}, \quad e^{2x}\cos(x), \quad \text{and} \quad e^{2x}\sin(x).$$

These solutions are also linearly independent. Therefore, the general solution of the differential equation can be written

$$y = c_1 e^x + c_2 e^{-2x} + c_3 e^{2x}\cos(x) + c_4 e^{2x}\sin(x).$$

From the initial data, we find that

$$y(0) = c_1 + c_2 + c_3 = 5$$
$$y'(0) = c_1 - 2c_2 + 2c_3 + c_4 = 12$$
$$y''(0) = c_1 + 4c_2 + 3c_3 + 4c_4 = -5$$
$$y^{(3)}(0) = c_1 - 8c_2 + 2c_3 + 11c_4 = -6.$$

Solve these equations to obtain

$$c_1 = 3, \quad c_2 = -2, \quad c_3 = 4, \quad c_4 = -3.$$

The unique solution of the initial value problem is

$$y = 3e^x - 2e^{-2x} + 4e^{2x}\cos(x) - 3e^{2x}\sin(x). \quad \blacksquare$$

In the case $n = 2$, if the characteristic equation had a repeated real root $r = \alpha$, we found by reduction of order that two linearly independent solutions were $e^{\alpha x}$ and $xe^{\alpha x}$. A similar analysis holds if $n > 2$. If α is a real root of the characteristic equation, repeated k times, then $e^{\alpha x}, xe^{\alpha x}, \ldots, x^{k-1}e^{\alpha x}$ are linearly independent solutions corresponding to the root α. The part of a general solution corresponding to these solutions will be $c_1 e^{\alpha x} + c_2 x e^{\alpha x} + \cdots + c_k x^{k-1}e^{\alpha x}$, or

$$(c_1 + c_2 x + \cdots + c_k x^{k-1})e^{\alpha x}.$$

Of course, if $k < n$, the general solution will also contain terms corresponding to the other roots of the characteristic equation.

EXAMPLE 3.9

Solve $y^{(4)} - 4y^{(3)} + 6y'' - 4y' + y = 0$.

The characteristic equation is

$$r^4 - 4r^3 + 6r^2 - 4r + 1 = 0,$$

or

$$(r - 1)^4 = 0.$$

The roots of the characteristic equation are 1, 1, 1, 1. Four linearly independent solutions are

$$e^x, \qquad xe^x, \qquad x^2e^x, \quad \text{and} \quad x^3e^x.$$

Since 4 is the order of the differential equation, these solutions constitute a fundamental set, and the general solution is

$$y = (c_1 + c_2x + c_3x^2 + c_4x^3)e^x. \quad \blacksquare$$

EXAMPLE 3.10

Solve $y^{(5)} + 2y^{(4)} - 3y^{(3)} - 4y'' + 4y' = 0$.

The characteristic equation is

$$r^5 + 2r^4 - 3r^3 - 4r^2 + 4r = 0,$$

which has roots 1, 1, -2, -2, 0. The functions

$$e^x, \qquad xe^x, \qquad e^{-2x}, \qquad xe^{-2x}, \quad \text{and} \quad e^0 = 1$$

are linearly independent and form a fundamental set of solutions. The general solution is

$$y = (c_1 + c_2x)e^x + (c_3 + c_4x)e^{-2x} + c_5. \quad \blacksquare$$

The only real obstacle to solving a constant coefficient equation

$$y^{(n)} + a_{n-1}y^{(n-1)} + \cdots + a_1y' + a_0y = 0$$

is in solving the characteristic equation. The problems and examples in this book have been designed so that roots of polynomials can be found by inspection or a little trial and error. For the most part, roots of characteristic equations encountered here are integers or complex numbers $\alpha + i\beta$ with α and β integers. In general, it is useful to remember that, if $p(r)$ is a polynomial of degree n, and a is a root [that is, $p(a) = 0$], then $r - a$ is a factor of $p(r)$, and we can write

$$p(r) = (r - a)q(r),$$

where $q(r)$ has degree $n - 1$. We attempt to find other roots of $p(r) = 0$ by solving $q(r) = 0$, which may be a simpler problem.

For example, consider the equation

$$p(r) = r^3 + 9r^2 + 10r - 24 = 0.$$

By inspection, or trial and error, we find that $r = -3$ is a root. Factor $r + 3$ from $p(r)$ to get

$$p(r) = (r + 3)(r^2 + 6r - 8).$$

Now use the quadratic formula to solve $r^2 + 6r - 8 = 0$, getting $-3 \pm \sqrt{17}$. The roots of $p(r) = 0$ are therefore -3, $-3 + \sqrt{17}$, and $-3 - \sqrt{17}$.

If a polynomial has integer coefficients, there is an aid to finding any rational roots it may have. Suppose that $p(r) = c_nr^n + c_{n-1}r^{n-1} + \cdots + c_1r + c_0$. In order for a rational number $r = a/b$ (in lowest terms) to be a root, b must be a factor of c_n and

a a factor of c_0. For example, consider

$$p(r) = 6r^3 - 11r^2 + 58r + 10 = 0.$$

Here, $n = 3$, $c_3 = 6$, and $c_0 = 10$. For a/b to be a root, b must be a factor of 6 and a a factor of 10. Thus, b must be one of $\pm 1, \pm 2, \pm 3$, or ± 6 and a must be one of $\pm 1, \pm 2, \pm 5$, or ± 10. This limits the possibilities for a/b, and we find by trial and error that $-\frac{1}{6}$ is a root. Now factor

$$p(r) = (r + \tfrac{1}{6})(6r^2 - 12r + 60) = 6(r + \tfrac{1}{6})(r^2 - 2r + 10),$$

and use the quadratic formula to find the other two roots, $1 \pm 3i$.

Problems arising from the real world often generate differential equations whose characteristic polynomials have irrational roots. In this case, it is sometimes necessary to resort to a numerical method (such as Newton's method) to approximate the roots. Of course, once one root is found, we can use the above strategy of factoring to obtain a polynomial of lower degree and attempt to find more roots.

As a reminder of Newton's method, suppose we have a function f and want to find a number z such that $f(z) = 0$. Choose a number x_0 (if possible, use calculus to choose x_0 close to a value of z). Define the recursive sequence by

$$x_{n+1} = x_n - \frac{f(x_n)}{f'(x_n)}$$

for $n = 1, 2, 3, \ldots$. Many calculus texts discuss conditions on f and x_0 for this sequence to converge to a number z such that $f(z) = 0$. Here is an example of Newton's method in solving a differential equation.

EXAMPLE 3.11

Find the general solution of

$$y^{(3)} - 2y'' + y' + 3y = 0.$$

The characteristic equation is

$$p(r) = r^3 - 2r^2 + r + 3 = 0.$$

A sketch of the graph of p is shown in Figure 3.1. There appears to be a root near -1, so we will choose $x_0 = -1$ and define the sequence $\{x_n\}$ recursively by

$$x_{n+1} = x_n - \frac{x_n^3 - 2x_n^2 + x_n + 3}{3x_n^2 - 4x_n + 1}.$$

After calculating some x_n's, we find that -0.8637 is approximately a root of p. Now use long division to factor p as

$$p(r) \approx (r + 0.8637)(r^2 - 2.8637r + 3.4734),$$

where $\approx$ means that this is an approximation (because -0.8637 is approximately a root). By the quadratic formula, two roots of $r^2 - 2.8637r + 3.4734 = 0$ are approximately $1.4319 \pm 1.1930i$. With these approximate roots of the characteristic polynomial, we obtain the approximate general solution of the differential equation:

$$y \approx c_1 e^{-0.8637x} + e^{1.4319x}[c_1\cos(1.1930x) + c_2\sin(1.1930x)]. \quad \blacksquare$$

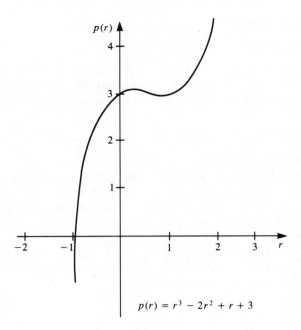

$$p(r) = r^3 - 2r^2 + r + 3$$

Figure 3.1

Finally, it is useful to be able to write the nth root of a number. For example, suppose we want the fourth roots of -16. Use Euler's formula to write

$$-16 = 16 \cos(\pi) + 16i \sin(\pi)$$
$$= 16 \cos(\pi + 2k\pi) + 16i \sin(\pi + 2k\pi) = 16e^{i(\pi + 2k\pi)},$$

in which k is any integer. The fourth roots of -16 are

$$(-16)^{1/4} = [16e^{i(\pi + 2k\pi)}]^{1/4} = 2e^{i(\pi + 2k\pi)/4}.$$

Letting $k = 0, 1, 2, 3$, we get the fourth roots

$$2e^{\pi i/4} = 2\left(\frac{\sqrt{2}}{2} + i\frac{\sqrt{2}}{2}\right) = \sqrt{2} + i\sqrt{2},$$

$$2e^{3\pi i/4} = -\sqrt{2} + i\sqrt{2}, \qquad 2e^{5\pi i/4} = -\sqrt{2} - i\sqrt{2},$$

and

$$2e^{7\pi i/4} = \sqrt{2} - i\sqrt{2}.$$

Other values of k repeat these four roots. Thus, for example, the differential equation $y^{(4)} + 16y = 0$ has characteristic equation $r^4 + 16 = 0$, with roots equal to the fourth roots of -16, which we have just found. The general solution of $y^{(4)} + 16y = 0$ is

$$y = e^{\sqrt{2}x}[c_1\cos(\sqrt{2}x) + c_2\sin(\sqrt{2}x)] + e^{-\sqrt{2}x}[c_3\cos(\sqrt{2}x) + c_4\sin(\sqrt{2}x)].$$

PROBLEMS FOR SECTION 3.2

In each of Problems 1 through 15, find the general solution of the differential equation. (The roots of these characteristic polynomials should be fairly easy to find if you are observant and try "small" integers.)

1. $y^{(5)} - 10y^{(4)} + 40y^{(3)} - 80y'' + 80y' - 32y = 0$
2. $y^{(3)} - 3y' + 2y = 0$
3. $y^{(3)} - 4y'' - y' + 4y = 0$
4. $8y^{(3)} + 6y'' - 29y' + 15y = 0$
5. $y^{(4)} + 5y'' + 4y = 0$
6. $y^{(3)} - 8y'' + 5y' + 50y = 0$
7. $y^{(3)} - 2y' + 4y = 0$
8. $6y^{(4)} + y^{(3)} + 22y'' + 4y' - 8y = 0$
9. $y^{(3)} + 4y'' - 11y' + 6y = 0$
10. $y^{(5)} - 8y^{(3)} + 16y' = 0$
11. $y^{(4)} + 3y^{(3)} - 18y'' + 4y' + 24y = 0$
12. $y^{(8)} - 256y = 0$
13. $y^{(4)} - 5y'' + 4y = 0$
14. $16y^{(4)} - 32y^{(3)} + 88y'' - 72y' + 117y = 0$
15. $y^{(4)} + 2y'' + y = 0$

In each of Problems 16 through 25, solve the initial value problem.

16. $y^{(3)} - 8y = 0$; $y(0) = 1$, $y'(0) = y''(0) = 0$
17. $y^{(3)} - y'' + 2y' - 2y = 0$; $y(0) = -1$, $y'(0) = 2$, $y''(0) = 5$
18. $y^{(3)} + y'' + 4y' + 4y = 0$; $y(0) = 0$, $y'(0) = y''(0) = 6$
19. $y^{(3)} + 2y'' + 29y' + 148y = 0$; $y(\pi) = y'(\pi) = 0$, $y''(\pi) = 8$
20. $y^{(4)} - 256y = 0$; $y(1) = 0$, $y'(1) = 2$, $y''(1) = 0$, $y^{(3)}(1) = -1$
21. $y^{(3)} - 11y'' + 7y' + 147y = 0$; $y(0) = 4$, $y'(0) = 0$, $y''(0) = 6$
22. $y^{(3)} - 14y'' + 69y' - 90y = 0$; $y(0) = y'(0) = 0$, $y''(0) = -4$
23. $y^{(3)} - 8y'' + y' + 42y = 0$; $y(0) = -2$, $y'(0) = y''(0) = 3$
24. $y^{(3)} - 8y'' + 5y' + 50y = 0$; $y(-2) = 3$, $y'(-2) = 1$, $y''(-2) = -6$
25. $y^{(3)} + 17y'' + 40y' - 300y = 0$; $y(0) = y'(0) = y''(0) = -1$

In each of Problems 26 through 29, determine approximate roots of the characteristic polynomial and use these to find an approximate general solution of the differential equation.

26. $y^{(3)} - 3y'' - y' + 12y = 0$
27. $y^{(3)} - 4y'' + 3y' + 3y = 0$
28. $y^{(4)} - 4y^{(3)} + 2y'' + 12y' - 13y = 0$
29. $y^{(4)} + y'' - 36y' + 46y = 0$

In each of Problems 30 through 34, construct a homogeneous constant coefficient linear differential equation having the given function as general solution.

30. $y(x) = c_1 e^{2x} + c_2 x e^{2x} + c_3 e^{-x} + c_4 e^{3x}\cos(2x) + c_5 e^{3x}\sin(2x)$
31. $y(x) = c_1 e^{-4x} + c_2 x e^{-4x} + c_3 x^2 e^{-4x} + c_4 x^3 e^{-4x} + c_5 e^{-x}\cos(\sqrt{2}x) + c_6 e^{-x}\sin(\sqrt{2}x)$
32. $y(x) = c_1 + c_2\cos(5x) + c_3\sin(5x)$
33. $y(x) = c_1\cos(2x) + c_2\sin(2x) + c_3\cos(3x) + c_4\sin(3x)$
34. $y(x) = c_1 + c_2 x + c_3 e^x + c_4 x e^x$

3.3 nTH ORDER CAUCHY-EULER EQUATIONS _____

The nth order Cauchy-Euler equation has the form

$$y^{(n)} + \frac{a_{n-1}}{x} y^{(n-1)} + \cdots + \frac{a_1}{x^{n-1}} + \frac{a_0}{x^n} y = 0,$$

with $a_0, a_1, \ldots, a_{n-1}$ all constant. Notice that the coefficient functions are continuous in any interval which does not contain 0. Usually we will assume that $x > 0$ and look for solutions on $(0, \infty)$.

The nth order Cauchy-Euler equation can be written as

$$x^n y^{(n)} + a_{n-1} x^{n-1} y^{(n-1)} + \cdots + a_2 x^2 y'' + a_1 x y' + a_0 y = 0.$$

As in the case $n = 2$, we will use the transformation $t = \ln(x)$ for $x > 0$. Let

$$y(x) = y(e^t) = Y(t). \tag{3.6}$$

We found in Section 2.6 that

$$xy'(x) = Y'(t) \quad \text{and} \quad x^2 y''(x) = Y''(t) - Y'(t). \tag{3.7}$$

In the nth order case, we must continue this computation to find $x^3 y^{(3)}$ in terms of derivatives of Y, and so on, up to and including $x^n y^{(n)}$.

A general expression for $y^{(k)}(x)$ in terms of derivatives of $Y(t)$ is quite complicated, and we will not attempt to write it for each k. For $k = n$, it can be shown that

$$x^n y^{(n)}(x) = Y^{(n)}(t) + b_{n-1} Y^{(n-1)}(t) + \cdots + b_1 Y'(t),$$

where b_k is the coefficient of r^k in the product

$$r(r - 1)(r - 2) \cdots (r - n + 1).$$

For reference, we will write the results of this calculation for $n = 3$ and $n = 4$. For $n = 3$, we get

$$x^3 y^{(3)}(x) = Y^{(3)}(t) - 3Y''(t) + 2Y'(t), \tag{3.8}$$

and for $n = 4$,

$$x^4 y^{(4)}(x) = Y^{(4)}(t) - 6Y^{(3)}(t) + 11Y''(t) - 6Y'(t). \tag{3.9}$$

EXAMPLE 3.12

Solve $x^3 y^{(3)} + x^2 y'' - 2xy' + 2y = 0$ for $x > 0$.

Put $t = \ln(x)$ and $y(e^t) = Y(t)$. From equations (3.7) and (3.8), the original differential equation in terms of y transforms to

$$[Y^{(3)} - 3Y'' + 2Y'] + [Y'' - Y'] - 2[Y'] + 2Y = 0,$$
$$x^3 y^{(3)} \qquad + \quad x^2 y'' \quad - \quad 2xy' \; + 2y = 0$$

or

$$Y^{(3)} - 2Y'' - Y' + 2Y = 0.$$

This linear constant coefficient differential equation has characteristic equation

$$r^3 - 2r^2 - r + 2 = 0,$$

with roots -1, 1, and 2. The general solution for $Y(t)$ is

$$Y(t) = c_1 e^{-t} + c_2 e^t + c_3 e^{2t}.$$

The general solution of the Cauchy-Euler equation for $x > 0$ is

$$y(x) = c_1 e^{-\ln(x)} + c_2 e^{\ln(x)} + c_3 e^{2\ln(x)}$$

$$= c_1 \frac{1}{x} + c_2 x + c_3 x^2. \quad \blacksquare$$

EXAMPLE 3.13

Solve $x^3y^{(3)} + 9x^2y'' + 19xy' + 8y = 0$ for $x > 0$.

Let $t = \ln(x)$ to use equations (3.7) and (3.8) to obtain

$$[Y^{(3)} - 3Y'' + 2Y'] + 9[Y'' - Y'] + 19[Y'] + 8Y = 0.$$
$$x^3y^{(3)} \qquad + \qquad 9x^2y'' \quad + \quad 19xy' \; + \; 8y = 0$$

Collect terms and write this differential equation for Y as

$$Y^{(3)} + 6Y'' + 12Y' + 8Y = 0.$$

The characteristic equation is

$$r^3 + 6r^2 + 12r + 8 = 0,$$

with root -2 (repeated three times). The general solution for Y is

$$Y(t) = (c_1 + c_2 t + c_3 t^2)e^{-2t}.$$

Since $t = \ln(x)$, the Cauchy-Euler equation has the general solution

$$y(x) = \{c_1 + c_2\ln(x) + c_3[\ln(x)]^2\}e^{-2\ln(x)}$$
$$= \frac{1}{x^2}\{c_1 + c_2\ln(x) + c_3[\ln(x)]^2\}$$

for $x > 0$. ∎

EXAMPLE 3.14

Solve $x^3y^{(3)} - 5x^2y'' + 18xy' - 26y = 0$ for $x > 0$.

Use equations (3.7) and (3.8) to get

$$[Y^{(3)} - 3Y'' + 2Y'] - 5[Y'' - Y'] + 18Y' - 26Y = 0,$$

or

$$Y^{(3)} - 8Y'' + 25Y' - 26Y = 0.$$

The characteristic equation is

$$r^3 - 8r^2 + 25r - 26 = 0,$$

with roots 2, $3 + 2i$, and $3 - 2i$. The general solution for Y is

$$Y(t) = c_1e^{2t} + c_2e^{3t}\cos(2t) + c_3e^{3t}\sin(2t).$$

The general solution for y is

$$y(x) = c_1e^{2\ln(x)} + c_2e^{3\ln(x)}\cos[2\ln(x)] + c_3e^{3\ln(x)}\sin[2\ln(x)]$$
$$= c_1x^2 + c_2x^3\cos[2\ln(x)] + c_3x^3\sin[2\ln(x)]$$

for $x > 0$. ∎

In this discussion, we have assumed that $x > 0$. An *n*th order Cauchy-Euler equation may also be solved on the interval $(-\infty, 0)$ by using the transformation $t = \ln(-x)$.

If the Cauchy-Euler equation is of order 4, we must use equation (3.9) to write $x^4 y^{(4)}$ in terms of $Y^{(4)}$, $Y^{(3)}$, Y'', and Y'. For $n > 4$, we must determine the appropriate substitutions for terms $x^5 y^{(5)}, \ldots, x^n y^{(n)}$, and this will involve considerable computation if n is large. In addition, in solving for Y, we encounter an nth degree characteristic equation, and the roots may be difficult to determine.

PROBLEMS FOR SECTION 3.3

In each of Problems 1 through 15, find the general solution of the Cauchy-Euler equation.

1. $x^3 y^{(3)} - 3xy' + 3y = 0$

2. $x^3 y^{(3)} + 2x^2 y'' - xy' + y = 0$

3. $x^3 y^{(3)} - 7x^2 y'' + 27xy' - 40y = 0$

4. $x^3 y^{(3)} - 9x^2 y'' + 37xy' - 64y = 0$

5. $x^3 y^{(3)} + 4x^2 y'' - 6xy' - 12y = 0$

6. $x^3 y^{(3)} + 7x^2 y'' + 4xy' - 4y = 0$

7. $x^3 y^{(3)} - 2x^2 y'' - 8xy' + 60y = 0$

8. $x^3 y^{(3)} + \frac{1}{2}x^2 y'' - \frac{7}{2}xy' + 6y = 0$

9. $x^3 y^{(3)} + 12x^2 y'' + 37xy' + 27y = 0$

10. $x^3 y^{(3)} + x^2 y'' + 9xy' = 0$

11. $x^3 y^{(3)} + 2x^2 y'' + 17xy' + 87y = 0$

12. $x^4 y^{(4)} - 2x^3 y^{(3)} + 7x^2 y'' - 15xy' + 16y = 0$

13. $x^3 y^{(3)} - 8x^2 y'' + 31xy' - 51y = 0$

14. $x^3 y^{(3)} + 8x^2 y'' - 2xy' - 48y = 0$

15. $x^3 y^{(3)} - 7x^2 y'' + 19xy' + 104y = 0$

In each of Problems 16 through 25, solve the initial value problem.

16. $x^3 y^{(3)} + 2x^2 y'' - xy' + y = 0$; $y(2) = y'(2) = 0$, $y''(2) = 4$

17. $x^3 y^{(3)} - 3x^2 y'' + 44xy' + 130y = 0$; $y(1) = y'(1) = -3$, $y''(1) = 0$

18. $x^3 y^{(3)} + 4x^2 y'' - 2xy' + 6y = 0$; $y(1) = -2$, $y'(1) = 3$, $y''(1) = 0$

19. $x^3 y^{(3)} - 3x^2 y'' + 6xy' - 12y = 0$; $y(3) = 0$, $y'(3) = 1$, $y''(3) = 0$

20. $x^3 y^{(3)} + 5x^2 y'' + 2xy' - 14y = 0$; $y(1) = y'(1) = 4$, $y''(1) = -6$

21. $x^3 y^{(3)} + x^2 y'' - 20xy' - 30y = 0$; $y(1) = y'(1) = y''(1) = -3$

22. $x^3 y^{(3)} + 4x^2 y'' - 8xy' + 8y = 0$; $y(2) = 1$, $y'(2) = -2$, $y''(2) = 2$

23. $x^4 y^{(4)} + 14x^3 y^{(3)} + 55x^2 y'' + 65xy' + 16y = 0$; $y(1) = y'(1) = 2$, $y''(1) = y^{(3)}(1) = -2$

24. $x^3 y^{(3)} - 9x^2 y'' - 11xy' + 256y = 0$; $y(1) = y'(1) = 0$; $y''(1) = -5$

25. $x^3 y^{(3)} + 5x^2 y'' - 14xy' + 42y = 0$; $y(1) = y'(1) = 0$; $y''(1) = 7$

In each of Problems 26, 27, and 28, construct a Cauchy-Euler differential equation having the given function as the general solution.

26. $y(x) = c_1 x^2 + c_2 x^{-3} + c_3 x^{-3}\ln(x)$

27. $y(x) = c_1 x + c_2 x \ln(x) + c_3 x \ln^2(x) + c_4 \cos[3 \ln(x)] + c_5 \sin[3 \ln(x)]$

28. $y(x) = c_1 x^4 \cos[\sqrt{3} \ln(x)] + c_2 x^4 \sin[\sqrt{3} \ln(x)] + c_3 x^{-5} + c_4 x^{-2}$

29. Use chain rule differentiation to derive equation (3.8). *Hint:* See the derivation of equation (2.17) in Section 2.7.

30. Use chain rule differentiation to derive equation (3.9).

3.4 The Methods of Undetermined Coefficients and Variation of Parameters

In this section, we will consider two methods for producing a particular solution of the nonhomogeneous linear differential equation

$$y^{(n)} + p_{n-1}(x)y^{(n-1)} + \cdots + p_1(x)y' + p_0(x)y = f(x). \tag{3.10}$$

We will assume that the coefficient functions $p_0, p_1, \ldots, p_{n-1}$, and f are continuous on an interval I and that $f(x) \neq 0$ for some point in I. The two methods we will consider are direct generalizations of the methods of undetermined coefficients and variation of parameters.

THE METHOD OF UNDETERMINED COEFFICIENTS

As in the case $n = 2$, *assume that the coefficient functions $p_{n-1}(x), \ldots, p_0(x)$ are constant*, and write the differential equation as

$$y^{(n)} + a_{n-1}y^{(n-1)} + \cdots + a_1 y' + a_0 y = f.$$

The table given in Section 2.10 also provides the key to the method when $n > 2$ as well. If $f(x)$ is of a form given in this table (a polynomial, sine or cosine, exponential, or product of such functions), then the table suggests an appropriate form to attempt for a particular solution y_p. Substitute the conjectured form for y_p into the differential equation and solve for the coefficients to obtain a solution.

In carrying out this strategy, there are two important points to keep in mind.

1. The principle of superposition applies to nth order linear equations as it does to linear equations of order 2. Here is a statement of this principle.

PRINCIPLE OF SUPERPOSITION

Suppose that $f = f_1 + f_2 + \cdots + f_N$ and that y_j is a solution of

$$y^{(n)} + p_{n-1}(x)y^{(n-1)} + \cdots + p_1(x)y' + p_0(x)y = f_j(x)$$

for $j = 1, 2, \ldots, N$. Then $y_1 + y_2 + \cdots + y_N$ is a solution of

$$y^{(n)} + p_{n-1}(x)y^{(n-1)} + \cdots + p_1(x)y' + p_0(x)y = f(x).$$

As we did with the case $n = 2$, this principle can sometimes be used to convert a difficult problem into a sum of individually easier ones. Example 3.17 illustrates this. (The coefficient functions need not be constant in using the principle of superposition. However, in using the method of undetermined coefficients, the functions $p_{n-1}(x), \ldots, p_0(x)$ must be constant.)

2. It may happen that the form of $f(x)$ suggests that we try a particular solution y_p but that one or more terms in y_p are solutions of the homogeneous equation

$$y^{(n)} + a_{n-1}y^{(n-1)} + \cdots + a_1 y' + a_0 y = 0.$$

When this happened with $n = 2$, we attempted solutions of the form xy_p or $x^2 y_p$ for the nonhomogeneous equation. If this occurs with $n > 2$, we try $x^s y_p$, where s is the smallest integer such that $x^s y_p(x)$ is not a solution of the associated homogeneous equation. (See Examples 3.15 and 3.18.) Because of this, we suggest finding the general solution of the associated homogeneous equation before attempting a particular solution y_p of the nonhomogeneous equation, in order to see if the initial guess for y_p must be multiplied by an appropriate factor x^s.

EXAMPLE 3.15

Find the general solution of $y^{(3)} - y'' - 8y' + 12y = 7e^{2x}$.

First, solve the associated homogeneous equation

$$y^{(3)} - y'' - 8y' + 12y = 0.$$

The characteristic equation is

$$r^3 - r^2 - 8r + 12 = 0,$$

with roots 2, 2, and -3. The general homogeneous solution is

$$y_h(x) = c_1 e^{2x} + c_2 x e^{2x} + c_3 e^{-3x}.$$

Since $f(x) = 7e^{2x}$, we are tempted to try $y_p(x) = ke^{2x}$. However, e^{2x} and xe^{2x} are both solutions of the homogeneous equation. The smallest power s of x such that $x^s e^{2x}$ is not a solution of the homogeneous equation is 2. Thus, try a particular solution

$$y_p(x) = kx^2 e^{2x}.$$

Calculate

$$y_p'(x) = 2kxe^{2x} + 2kx^2 e^{2x}, \qquad y_p''(x) = 2ke^{2x} + 8kxe^{2x} + 4kx^2 e^{2x},$$

and

$$y_p^{(3)}(x) = 12ke^{2x} + 24kxe^{2x} + 8kx^2 e^{2x}.$$

Substitute these quantities into the original, nonhomogeneous differential equation. We get

$$12ke^{2x} + 24kxe^{2x} + 8kx^2 e^{2x} - 2ke^{2x} - 8kxe^{2x} - 4kx^2 e^{2x}$$
$$- 16kxe^{2x} - 16kx^2 e^{2x} + 12kx^2 e^{2x} = 7e^{2x}.$$

Terms involving xe^{2x} and $x^2 e^{2x}$ cancel, leaving us with

$$10ke^{2x} = 7e^{2x}.$$

Therefore, $k = \frac{7}{10}$, and a particular solution is

$$y_p(x) = \frac{7}{10}x^2 e^{2x}.$$

The general solution of $y^{(3)} - y'' - 8y' + 12y = 7e^{2x}$ is

$$y = y_h + y_p = c_1 e^{2x} + c_2 x e^{2x} + c_3 e^{-3x} + \frac{7}{10}x^2 e^{2x}$$
$$= [c_1 + c_2 x + \frac{7}{10}x^2]e^{2x} + c_3 e^{-3x}. \quad \blacksquare$$

EXAMPLE 3.16

Find the general solution of $y^{(3)} - 4y'' + y' + 6y = x^3 - 4x + 2$.

The general solution of the associated homogeneous equation is

$$y_h(x) = c_1 e^{2x} + c_2 e^{-x} + c_3 e^{3x}.$$

Since $x^3 - 4x + 2$ is a polynomial of degree 3, attempt a particular solution

$$y_p(x) = ax^3 + bx^2 + cx + d.$$

Compute

$$y_p'(x) = 3ax^2 + 2bx + c, \qquad y_p''(x) = 6ax + 2b, \qquad y_p^{(3)}(x) = 6a.$$

Substitute these quantities into the differential equation to get

$$6a - 24ax - 8b + 3ax^2 + 2bx + c + 6ax^3 + 6bx^2 + 6cx + 6d = x^3 - 4x + 2.$$

This equation can be written

$$(6a - 1)x^3 + (3a + 6b)x^2 + (-24a + 2b + 6c + 4)x$$
$$+ (6a - 8b + c + 6d - 2) = 0.$$

Since different powers of x are linearly independent, all the coefficients must be zero. Therefore,

$$6a - 1 = 0$$
$$3a + 6b = 0$$
$$-24a + 2b + 6c + 4 = 0$$
$$6a - 8b + c + 6d - 2 = 0.$$

Solve these four equations to obtain

$$a = \tfrac{1}{6}, \qquad b = -\tfrac{1}{12}, \qquad c = \tfrac{1}{36}, \quad \text{and} \quad d = \tfrac{11}{216}.$$

A particular solution of $y^{(3)} - 4y'' + y' + 6y = x^3 - 4x + 2$ is

$$y_p(x) = \tfrac{1}{6}[x^3 - \tfrac{1}{2}x^2 + \tfrac{1}{6}x + \tfrac{11}{36}].$$

The general solution is

$$y = c_1 e^{2x} + c_2 e^{-x} + c_3 e^{3x} + \tfrac{1}{6}[x^3 - \tfrac{1}{2}x^2 + \tfrac{1}{6}x + \tfrac{11}{36}]. \quad \blacksquare$$

EXAMPLE 3.17

Find the general solution of $y^{(3)} + 2y'' - y' = 4e^x - 3\cos(2x)$.
    We find that the general solution of $y^{(3)} + 2y'' - y' = 0$ is

$$y_h(x) = c_1 + c_2 e^{(-1+\sqrt{2})x} + c_3 e^{(-1-\sqrt{2})x}.$$

We will use the principle of superposition to find a particular solution of $y^{(3)} + 2y'' - y' = 4e^x - 3\cos(2x)$. Consider two problems.

$$\text{Problem 1:} \quad y^{(3)} + 2y'' - y' = 4e^x.$$
$$\text{Problem 2:} \quad y^{(3)} + 2y'' - y' = -3\cos(2x).$$

Solution of Problem 1   Attempt a particular solution $\psi_p(x) = ke^x$ of problem 1. Substitute this into the differential equation $y^{(3)} + 2y'' - y' = 4e^x$ to get

$$ke^x + 2ke^x - ke^x = 4e^x.$$

Then $2ke^x = 4e^x$, so $k = 2$, and a particular solution is $\psi_p(x) = 2e^x$.

Solution of Problem 2   Attempt a particular solution

$$\varphi_p(x) = a\cos(2x) + b\sin(2x).$$

Substitute φ_p into $y^{(3)} + 2y'' - y' = -3 \cos(2x)$ to get

$$8a \sin(2x) - 8b \cos(2x) - 8a \cos(2x) - 8b \sin(2x)$$
$$+ 2a \sin(2x) - 2b \cos(2x) = -3 \cos(2x),$$

or

$$[8a - 8b + 2a]\sin(2x) + [-8b - 8a - 2b + 3]\cos(2x) = 0.$$

Since $\sin(2x)$ and $\cos(2x)$ are linearly independent for all x, the coefficients of $\sin(2x)$ and $\cos(2x)$ must be zero, and

$$10a - 8b = 0$$
$$-8a - 10b = -3.$$

Solve these equations to get

$$a = \tfrac{6}{41} \quad \text{and} \quad b = \tfrac{15}{82}.$$

A particular solution of problem 2 is

$$\varphi_p(x) = \tfrac{1}{82}[12 \cos(2x) + 15 \sin(2x)].$$

By the principle of superposition, a particular solution of $y^{(3)} + 2y'' - y' = 4e^x - 3 \cos(2x)$ is

$$y_p(x) = \psi_p(x) + \varphi_p(x) = 2e^x + \tfrac{1}{82}[12 \cos(2x) + 15 \sin(2x)].$$

The general solution of $y^{(3)} + 2y'' - y' = 4e^x - 3 \cos(2x)$ is

$$y = c_1 + c_2 e^{(-1+\sqrt{2})x} + c_3 e^{(-1-\sqrt{2})x} + 2e^x$$
$$+ \tfrac{1}{82}[12 \cos(2x) + 15 \sin(2x)]. \quad \blacksquare$$

EXAMPLE 3.18

Find the general solution of

$$y^{(3)} + 2y'' + 9y' + 18y = -7 \cos(3x).$$

First, we find that the general solution of $y^{(3)} + 2y'' + 9y' + 18y = 0$ is

$$y_h(x) = c_1 e^{-2x} + c_2 \cos(3x) + c_3 \sin(3x).$$

Because $\cos(3x)$ is a solution of this homogeneous equation, we will attempt a particular solution of the nonhomogeneous equation of the form

$$y_p(x) = x[a \cos(3x) + b \sin(3x)].$$

That is, we multiply the "natural" choice, $a \cos(3x) + b \sin(3x)$, by x. Calculate

$$y_p'(x) = a \cos(3x) + b \sin(3x) - 3ax \sin(3x) + 3bx \cos(3x),$$
$$y_p''(x) = -6a \sin(3x) + 6b \cos(3x) - 9ax \cos(3x) - 9bx \sin(3x),$$

and

$$y_p^{(3)}(x) = -27a \cos(3x) - 27b \sin(3x) + 27ax \sin(3x) - 27bx \cos(3x).$$

Substitute these quantities into the differential equation to obtain

$$-27a\cos(3x) - 27b\sin(3x) + 27ax\sin(3x) - 27bx\cos(3x)$$
$$-12a\sin(3x) + 12b\cos(3x) - 18ax\cos(3x) - 18bx\sin(3x)$$
$$+9a\cos(3x) + 9b\sin(3x) - 27ax\sin(3x) + 27bx\cos(3x)$$
$$+18ax\cos(3x) + 18bx\sin(3x) = -7\cos(3x).$$

All terms involving $x\cos(3x)$ and $x\sin(3x)$ cancel, and we can collect the remaining terms to write

$$[-27a + 12b + 9a + 7]\cos(3x) + [-27b - 12a + 9b]\sin(3x) = 0.$$

The coefficients of $\cos(3x)$ and $\sin(3x)$ must be zero; hence,

$$-18a + 12b = -7$$
$$-18b - 12a = 0.$$

Then $a = \frac{7}{26}$ and $b = -\frac{7}{39}$, and we have a particular solution

$$y_p(x) = \frac{7x}{13}\left[\frac{1}{2}\cos(3x) - \frac{1}{3}\sin(3x)\right].$$

The general solution of $y^{(3)} + 2y'' + 9y' + 18y = -7\cos(3x)$ is

$$y = c_1 e^{-2x} + c_2\cos(3x) + c_3\sin(3x) + \frac{7x}{13}\left[\frac{1}{2}\cos(3x) - \frac{1}{3}\sin(3x)\right]. \quad \blacksquare$$

THE METHOD OF VARIATION OF PARAMETERS

We will now generalize the method of variation of parameters (see Section 2.11). Consider a linear nonhomogeneous differential equation

$$y^{(n)} + p_{n-1}(x)y^{(n-1)} + \cdots + p_1(x)y' + p_0(x)y = f(x)$$

in which the coefficient functions p_j and f are continuous on the relevant interval I but *need not be constant*.

Assume that we have found a fundamental set $y_1, y_2, \ldots, y_n$ of solutions of the associated homogeneous equation. We attempt to produce nonconstant functions $u_1, u_2, \ldots, u_n$ such that

$$y_p = u_1 y_1 + u_2 y_2 + \cdots + u_n y_n$$

is a solution of $y^{(n)} + p_{n-1}(x)y^{(n-1)} + \cdots + p_1(x)y' + p_0(x)y = f(x)$. We will actually produce n equations for $u_1', \ldots, u_n'$. These must be solved for $u_1', \ldots, u_n'$, and then $u_1, \ldots, u_n$ must be found by n integrations. The condition that

$$y_p^{(n)} + p_{n-1}(x)y_p^{(n-1)} + \cdots + p_1(x)y_p' + p_0(x)y_p = f(x)$$

will give us one equation. We will obtain an additional $n - 1$ equations by imposing conditions which simplify successive derivatives of y_p.

To begin, calculate

$$y_p' = u_1' y_1 + \cdots + u_n' y_n + u_1 y_1' + \cdots + u_n y_n'.$$

Let

$$u_1' y_1 + \cdots + u_n' y_n = 0.$$

Then

$$y_p' = u_1 y_1' + \cdots + u_n y_n', \tag{3.11}$$

hence,

$$y_p'' = u_1' y_1' + \cdots + u_n' y_n' + u_1 y_1'' + \cdots + u_n y_n''.$$

To simplify this derivative, let

$$u_1' y_1' + \cdots + u_n' y_n' = 0.$$

Then

$$y_p'' = u_1 y_1'' + \cdots + u_n y_n''. \tag{3.12}$$

Differentiate again:

$$y_p^{(3)} = u_1' y_1'' + \cdots + u_n' y_n'' + u_1 y_1^{(3)} + \cdots + u_n y_n^{(3)}.$$

Let

$$u_1' y_1'' + \cdots + u_n' y_n'' = 0.$$

Then

$$y_p^{(3)} = u_1 y_1^{(3)} + \cdots + u_n y_n^{(3)}. \tag{3.13}$$

Continue in this way. After differentiating $y_p^{(j-1)}$ to obtain $y_p^{(j)}$, let the sum of all terms involving $u_1', \ldots, u_n'$ in $y_p^{(j)}$ equal zero. This step is performed for $j = 1, 2, \ldots, n-1$. This yields $n-1$ equations involving the unknown functions $u_1', \ldots, u_n'$, with derivatives of the known functions $y_1, \ldots, y_n$ as coefficients. The nth equation for $u_1', \ldots, u_n'$ is obtained by letting $y_p^{(n)} + p_{n-1}(x) y_p^{(n-1)} + \cdots + p_1(x) y_p' + p_0(x) y_p = f(x)$. In view of equations (3.11) through (3.13), this equation is

$$y_p^{(n)} + p_{n-1}(x) y_p^{(n-1)} + p_{n-2}(x) y_p^{(n-2)} + \cdots + p_1(x) y_p' + p_0(x) y_p$$
$$= [u_1' y_1^{(n-1)} + u_2' y_2^{(n-1)} + \cdots + u_n' y_n^{(n-1)}]$$
$$\quad + [u_1 y_1^{(n)} + u_2 y_2^{(n)} + \cdots + u_n y_n^{(n)}]$$
$$\quad + p_{n-1}(x) [u_1 y_1^{(n-1)} + u_2 y_2^{(n-1)} + \cdots + u_n y_n^{(n-1)}]$$
$$\quad + \cdots + p_1(x) [u_1 y_1' + u_2 y_2' + \cdots + u_n y_n']$$
$$\quad + p_0(x) [u_1 y_1 + u_2 y_2 + \cdots + u_n y_n]$$
$$= u_1' y_1^{(n-1)} + u_2' y_2^{(n-1)} + \cdots + u_n' y_n^{(n-1)}$$
$$\quad + u_1 [y_1^{(n)} + p_{n-1}(x) y_1^{(n-1)} + \cdots + p_1(x) y_1' + p_0(x) y_1]$$
$$\quad + u_2 [y_2^{(n)} + p_{n-1}(x) y_2^{(n-1)} + \cdots + p_1(x) y_2' + p_0(x) y_2]$$
$$\quad + \cdots + u_n [y_n^{(n)} + p_{n-1}(x) y_n^{(n-1)} + \cdots + p_1(x) y_n' + p_0(x) y_n] = f(x).$$

Because each of $y_1, \ldots, y_n$ is a solution of the homogeneous equation $y^{(n)} + p_{n-1}(x) y^{(n-1)} + \cdots + p_1(x) y' + p_0(x) y = 0$, all the terms in square brackets are equal to zero, and the last equation becomes simply

$$u_1' y_1^{(n-1)} + u_2' y_2^{(n-1)} + \cdots + u_n' y_n^{(n-1)} = f,$$

which is the nth equation for $u_1', \ldots, u_n'$.

In summary, the n equations for $u'_1, \ldots, u'_n$ are

$$y_1 u'_1 + y_2 u'_2 + \cdots + y_n u'_n = 0$$
$$y'_1 u'_1 + y'_2 u'_2 + \cdots + y'_n u'_n = 0$$
$$y''_1 u'_1 + y''_2 u'_2 + \cdots + y''_n u'_n = 0$$
$$\cdots\cdots\cdots\cdots\cdots\cdots\cdots\cdots\cdots\cdots\cdots$$
$$y_1^{(n-2)} u'_1 + y_2^{(n-2)} u'_2 + \cdots + y_n^{(n-2)} u'_n = 0$$
$$y_1^{(n-1)} u'_1 + y_2^{(n-1)} u'_2 + \cdots + y_n^{(n-1)} u'_n = f.$$

Notice that the $n \times n$ determinant of the coefficients in this system of equations is exactly the Wronskian $W[y_1, \ldots, y_n]$, which is nonzero because $y_1, \ldots, y_n$ are assumed to be linearly independent. Therefore, this system of equations has a solution for $u'_1, \ldots, u'_n$. In terms of determinants, with $W[y_1, \ldots, y_n]$ written as just W, we have

$$u'_1 = \frac{1}{W} \begin{vmatrix} 0 & y_2 & \cdots & y_n \\ 0 & y'_2 & \cdots & y'_n \\ \vdots & \vdots & & \vdots \\ f & y_2^{(n-1)} & \cdots & y_n^{(n-1)} \end{vmatrix},$$

$$u'_2 = \frac{1}{W} \begin{vmatrix} y_1 & 0 & \cdots & y_n \\ y'_1 & 0 & \cdots & y'_n \\ \vdots & \vdots & & \vdots \\ y_1^{(n-1)} & f & \cdots & y_n^{(n-1)} \end{vmatrix},$$

and

$$u'_n = \frac{1}{W} \begin{vmatrix} y_1 & y_2 & \cdots & 0 \\ y'_1 & y'_2 & \cdots & 0 \\ \vdots & \vdots & & \vdots \\ y_1^{(n-1)} & y_2^{(n-1)} & \cdots & f \end{vmatrix}.$$

These expressions are easy to remember by observing the pattern. The denominator in each expression is the Wronskian W. The numerator in the expression for u'_k is the $n \times n$ determinant obtained by replacing column k of the Wronskian with the column having all zeros except in row n, where the entry is f. Assuming that we can expand these $n \times n$ determinants and obtain expressions for $u'_1, \ldots, u'_n$, we integrate to obtain $u_1, \ldots, u_n$. A particular solution is

$$y_p = u_1 y_1 + u_2 y_2 + \cdots + u_n y_n.$$

EXAMPLE 3.19

Find the general solution of

$$x^3 y^{(3)} - 4x^2 y'' + 8xy' - 8y = 6x^3 (x^2 + 1)^{-3/2}$$

for $x > 0$.

Notice that the method of undetermined coefficients does not apply here because the coefficients are not all constant. First, divide the differential equation by x^3 to get

$$y^{(3)} - \frac{4}{x} y'' + \frac{8}{x^2} y' - \frac{8}{x^3} y = 6(x^2 + 1)^{-3/2}.$$

This step is necessary because we derived the method of variation of parameters under the assumption that the coefficient of $y^{(n)}$ is 1. From this equation, we have $f(x) = 6(x^2 + 1)^{-3/2}$ in the preceding discussion. Now solve the associated homogeneous differential equation

$$y^{(3)} - \frac{4}{x} y'' + \frac{8}{x^2} y' - \frac{8}{x^3} y = 0.$$

This is a third order Cauchy-Euler equation

$$x^3 y^{(3)} - 4x^2 y'' + 8xy' - 8y = 0,$$

with fundamental set of solutions $y_1(x) = x$, $y_2(x) = x^2$, and $y_3(x) = x^4$. The Wronskian of these functions is

$$W(x) = \begin{vmatrix} x & x^2 & x^4 \\ 1 & 2x & 4x^3 \\ 0 & 2 & 12x^2 \end{vmatrix} = 6x^4,$$

which is nonzero on $(0, \infty)$. We now have

$$u_1' = \frac{1}{6x^4} \begin{vmatrix} 0 & x^2 & x^4 \\ 0 & 2x & 4x^3 \\ 6(x^2 + 1)^{-3/2} & 2 & 12x^2 \end{vmatrix} = 2x(x^2 + 1)^{-3/2},$$

$$u_2' = \frac{1}{6x^4} \begin{vmatrix} x & 0 & x^4 \\ 1 & 0 & 4x^3 \\ 0 & 6(x^2 + 1)^{-3/2} & 12x^2 \end{vmatrix} = -3(x^2 + 1)^{-3/2},$$

and

$$u_3' = \frac{1}{6x^4} \begin{vmatrix} x & x^2 & 0 \\ 1 & 2x & 0 \\ 0 & 2 & 6(x^2 + 1)^{-3/2} \end{vmatrix} = x^{-2}(x^2 + 1)^{-3/2}.$$

Integrate to get

$$u_1(x) = \int 2x(x^2 + 1)^{-3/2} \, dx = \frac{-2}{(x^2 + 1)^{1/2}},$$

$$u_2(x) = \int -3(x^2 + 1)^{-3/2} \, dx = \frac{-3x}{(x^2 + 1)^{1/2}},$$

and

$$u_3(x) = \int x^{-2}(x^2 + 1)^{-3/2} \, dx = -\frac{2x^2 + 1}{x(x^2 + 1)^{1/2}}.$$

A particular solution of the differential equation is

$$y_p(x) = -\frac{2x}{(x^2 + 1)^{1/2}} - \frac{3x^3}{(x^2 + 1)^{1/2}} - \frac{x^3(2x^2 + 1)}{(x^2 + 1)^{1/2}}$$

$$= \frac{-2x(x^4 + 2x^2 + 1)}{(x^2 + 1)^{1/2}} = -2x(x^2 + 1)^{3/2}.$$

The general solution of the differential equation is

$$y = c_1 x + c_2 x^2 + c_3 x^4 - 2x(x^2 + 1)^{3/2}. \quad \blacksquare$$

PROBLEMS FOR SECTION 3.4

In each of Problems 1 through 20, find the general solution of the differential equation.

1. $y^{(4)} - y = -4\cosh(2x)$

2. $y^{(4)} - 10y^{(3)} + 25y'' = -4$

3. $y^{(3)} - 9y'' + 16y' - 4y = x - 1$

4. $y^{(3)} - 5y'' - 13y' - 7y = 8x^2 - 4$

5. $y^{(3)} - 4y'' + 20y' = x^2 + 4x - 10$

6. $y^{(3)} - 4y'' + y' + 6y = \cos(3x) - 4e^{-2x} + 6e^{2x}$

7. $y^{(3)} - 2y'' + y' - 2y = x^2 - 2x + 4 - 3\cos(x)$

8. $y^{(3)} - 12y' - 16y = 3x^2 - \cos(4x)$

9. $y^{(3)} + 6y'' + 12y' + 8y = 8e^{-2x}$

10. $y^{(4)} - 2y^{(3)} - 3y'' + 4y' + 4y = 3x^4 - 2x^3 + x - 5$

11. $y^{(3)} + y'' - 14y' - 24y = x^3 - 2\cos(x) + 7e^{4x}$

12. $y^{(3)} + 5y'' + 19y' - 25y = x - \cosh(3x)$

13. $y^{(3)} + 12y'' + 36y' = \sin(2x) - \cos(2x)$

14. $y^{(3)} - 8y'' + 25y' - 26y = -7e^{3x}\cos(2x)$

15. $y^{(3)} - 7y' - 6y = \sin(2x) - 5$

16. $x^3 y^{(3)} - 2x^2 y'' + 9xy' - 5y = 4\cos[\ln(x)] - 6\sin[\ln(x)]$ $(x > 0)$

17. $x^3 y^{(3)} + 2x^2 y'' + 4xy' - 4y = 3x^{-2}$ $(x > 0)$

18. $x^4 y^{(4)} + 4x^3 y^{(3)} + x^2 y'' + xy' - y = 3\ln(x)$ $(x > 0)$

19. $x^3 y^{(3)} + 5x^2 y'' + xy' - 4y = 6x\ln(x)$ $(x > 0)$

20. $x^4 y^{(4)} + 7x^3 y^{(3)} + 8x^2 y'' = 4x^{-3}$ $(x > 0)$

In each of Problems 21 through 34, solve the initial value problem.

21. $y^{(3)} - 4y'' - 3y' + 18y = x - e^{2x}$; $y(0) = -1$, $y'(0) = 1$, $y''(0) = 0$

22. $y^{(4)} - 16y = 0$; $y(0) = -2$; $y'(0) = y''(0) = 0$, $y^{(3)}(0) = 3$

23. $y^{(3)} + 9y'' + 15y' - 25y = x^2 + 2$; $y(0) = y'(0) = -3$, $y''(0) = 1$

24. $y^{(3)} + 6y'' + y' + 6y = -3\sin(x) + 4$; $y(\pi) = y'(\pi) = 0$, $y''(\pi) = 6$

25. $y^{(3)} - 6y'' + 25y' = -3e^{-2x}$; $y(0) = 1$, $y'(0) = 0$, $y''(0) = 0$

26. $y^{(3)} - 5y'' + y' + 7y = -\cos(3x)$; $y(0) = -2$, $y'(0) = 1$, $y''(0) = 0$

27. $y^{(4)} + 4y^{(3)} + 6y'' + 4y' + y = 3e^{-x}$; $y(0) = y'(0) = y''(0) = 0$, $y^{(3)}(0) = 1$

28. $y^{(3)} + 2y'' - 4y' - 8y = x^3 - 3x$; $y(2) = 1$, $y'(2) = -3$, $y''(2) = 0$

29. $x^3 y^{(3)} - 2x^2 y'' + 5xy' - 5y = 5$; $y(1) = y'(1) = y''(1) = 0$

30. $x^3 y^{(3)} - x^2 y'' - 6xy' + 18y = \ln(x) - x^3$; $y(1) = -1$, $y'(1) = 1$, $y''(1) = -1$

31. $x^3 y^{(3)} + 4x^2 y'' - 3xy' + 3y = 2x^{-3} + x$; $y(1) = -2$, $y'(1) = 1$, $y''(1) = 2$

32. $x^2 y^{(3)} - x(x + 2)y'' + (x + 2)y' = 6x^3$; $y(1) = y'(1) = y''(1) = 0$

Hint: Notice that there is no y-term; let $u = y'$. Next, observe that $u = x$ is a solution of the reduced homogeneous equation. Use this fact to reduce the order further. Once a set of three linearly independent solutions of the original homogeneous equation has been found, use variation of parameters to find a particular solution.

33. $y^{(3)} - 5y'' + 9y' - 5y = 5$; $y(0) = y'(0) = y''(0) = 0$

34. $y^{(3)} + y'' - 5y' + 3y = 2e^{-3x} + e^x$; $y(0) = -2$, $y'(0) = y''(0) = 1$

Summary

Here is a schematic summary of some of the methods of this chapter.

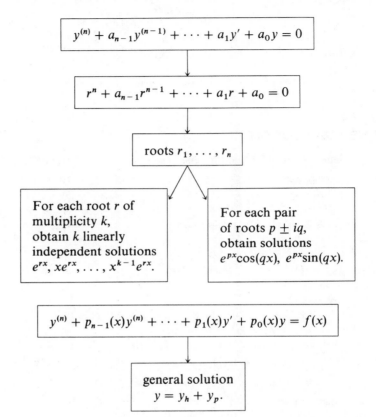

$$y^{(n)} + a_{n-1}y^{(n-1)} + \cdots + a_1 y' + a_0 y = 0$$

$$r^n + a_{n-1}r^{n-1} + \cdots + a_1 r + a_0 = 0$$

roots $r_1, \ldots, r_n$

For each root r of multiplicity k, obtain k linearly independent solutions $e^{rx}, xe^{rx}, \ldots, x^{k-1}e^{rx}$.

For each pair of roots $p \pm iq$, obtain solutions $e^{px}\cos(qx), e^{px}\sin(qx)$.

$$y^{(n)} + p_{n-1}(x)y^{(n)} + \cdots + p_1(x)y' + p_0(x)y = f(x)$$

general solution
$$y = y_h + y_p.$$

For y_p: $\begin{cases} \text{undetermined coefficients} \\ \text{(in constant coefficient case only)} \\ \text{variation of parameters} \end{cases}$

For variation of parameters, $y_p = u_1 y_1 + \cdots + u_n y_n$, where

$$y_1 u'_1 + y_2 u'_2 + \cdots + y_n u'_n = 0$$
$$y'_1 u'_1 + y'_2 u'_2 + \cdots + y'_n u'_n = 0$$

$$\cdots\cdots\cdots\cdots\cdots$$

$$y_1^{(n-2)}u'_1 + y_2^{(n-2)}u'_2 + \cdots + y_n^{(n-2)}u'_n = 0$$
$$y_1^{(n-1)}u'_1 + y_2^{(n-1)}u'_2 + \cdots + y_n^{(n-1)}u'_n = f.$$

For the Cauchy-Euler equation

$$x^n y^{(n)} + a_{n-1}x^{n-1}y^{(n-1)} + \cdots + a_1 xy' + a_0 y = 0,$$

let $t = \ln(x)$ for $x > 0$ and let $Y(t) = y(e^t)$. Terms $x^k y^{(k)}(x)$ must be related to sums of derivatives of Y with respect to t, through chain rule differentiation.

ADDITIONAL PROBLEMS

In each of Problems 1 through 60, find the general solution of the differential equation. If initial values are given, solve the initial value problem.

1. $y^{(3)} + 18y'' + 108y' + 216y = x - 4$; $y(0) = y'(0) = 1$, $y''(0) = -2$
2. $x^3 y^{(3)} + 21x^2 y'' + 127xy' + 216y = 0$; $y(1) = y'(1) = 0$, $y''(1) = -3$
3. $y^{(3)} + 5y'' + 3y' - 9y = 4e^{-3x} - e^x$; $y(0) = 2$, $y'(0) = y''(0) = -1$
4. $y^{(4)} - 3y'' = 12$; $y(0) = 2$, $y'(0) = -3$, $y''(0) = 2$, $y^{(3)}(0) = 0$
5. $y^{(3)} - 8y'' + 85y' - 146y = 146$; $y(0) = y'(0) = -3$, $y''(0) = 4$
6. $x^3 y^{(3)} - 6x^2 y'' + 65xy' - 65y = 0$; $y(1) = -1$, $y'(1) = 2$, $y''(1) = -5$
7. $x^3 y^{(3)} - 3x^2 y'' + 6xy' - 6y = \ln(x)$; $y(1) = 4$, $y'(1) = y''(1) = 1$
8. $x^3 y^{(3)} - 2x^2 y'' + 13xy' - 13y = 17x^{-3}$
9. $x^3 y^{(3)} + 7x^2 y'' + 4xy' - 4y = 2x \ln(x)$
10. $x^3 y^{(3)} - 8x^2 y'' + 23xy' - 35y = 0$; $y(1) = y'(1) = y''(1) = -3$

11. $y^{(3)} - 6y'' + 45y' - 148y = 0$; $y\left(\dfrac{\pi}{6}\right) = -3$, $y'\left(\dfrac{\pi}{6}\right) = y''\left(\dfrac{\pi}{6}\right) = 0$

12. $y^{(3)} - 3y' + 2y = 4e^x - 3$; $y(-1) = 1$, $y'(-1) = 4$, $y''(-1) = -3$
13. $x^3 y^{(3)} - 3x^2 y'' + 7xy' - 8y = 0$; $y(1) = -5$, $y'(1) = y''(1) = 0$
14. $y^{(3)} - y'' + y' - y = 4x$; $y(\pi) = 1$, $y'(\pi) = y''(\pi) = 0$
15. $x^4 y^{(4)} + 2x^3 y^{(3)} + 10x^2 y'' - 10xy' + 10y = 0$; $y(1) = -4$, $y'(1) = 2$, $y''(1) = 7$, $y^{(3)}(1) = 0$
16. $y^{(3)} + 6y'' + 15y' + 14y = 13 \sin(x)$
17. $y^{(3)} - 6y'' - 11y' + 116y = 3e^{-4x}$; $y(0) = y'(0) = y''(0) = 0$
18. $x^3 y^{(3)} + 3x^2 y'' - 6xy' - 6y = 10x^{-2}$; $y(1) = 0$, $y'(1) = 2$, $y''(1) = 0$
19. $y^{(4)} - 2y^{(3)} + 4y'' + 2y' - 5y = 10$; $y(0) = y'(0) = y''(0) = y^{(3)}(0) = 0$
20. $x^3 y^{(3)} - 3x^2 y'' + 6xy' - 6y = 16x^5 \sin(2x)$

21. $y^{(4)} + 2y^{(3)} - 59y'' - 60y' + 900y = 0$
22. $y^{(3)} - 6y'' + 18y' - 40y = 2 \sinh(4x) - x^3 + 2$
23. $x^3 y^{(3)} + xy' - y = 0$
24. $y^{(3)} + 4y'' + 11y' + 14y = 0$
25. $y^{(3)} - 4y'' + 13y' + 50y = -4 \cos(2x)$
26. $y^{(3)} - y'' + 2y' - 2y = 0$
27. $y^{(3)} + y'' - 4y' + 6y = 7e^{3x} - 2x + 4$
28. $x^3 y^{(3)} + 7x^2 y'' - 24xy' + 24y = 0$
29. $y^{(3)} - 4y'' - 19y' - 14y = 0$
30. $y^{(3)} - y'' - 40y' + 112y = x^2 + 8e^{-3x}$
31. $y^{(3)} + 6y'' - 69y' - 154y = 0$
32. $y^{(3)} + 2y'' - 20y' + 24y = 4 \cos(3x)$
33. $x^3 y^{(3)} - 3x^2 y'' - 20xy' - 8y = 0$
34. $y^{(3)} + 49y' = 0$
35. $y^{(3)} - 2y'' - y' + 2y = \sin^2(x)$
36. $y^{(3)} + 5y'' - 31y' + 21y = 0$
37. $x^3 y^{(3)} - 2x^2 y'' - 20y = 0$; $y(1) = 1$, $y'(1) = -3$, $y''(1) = 4$
38. $x^3 y^{(3)} + 4x^2 y'' + 13xy' + 26y = 0$
39. $y^{(4)} - 2y^{(3)} - 12y'' - 14y' - 5y = 0$
40. $x^3 y^{(3)} - 5x^2 y'' - 41xy' + 300y = 0$
41. $y^{(3)} + 6y'' + 6y' + 36y = 0$
42. $y^{(3)} + 5y'' - 8y' + 42y = 3 \sin^2(x)$
43. $y^{(3)} - 13y' + 12y = \cosh(2x)$
44. $y^{(3)} - 2y'' - y' + 14y = 0$
45. $y^{(3)} + 9y'' + 26y' + 24y = x^3$
46. $x^3 y^{(3)} - 8x^2 y'' + 55xy' - 123y = 0$; $y(1) = -3$, $y'(1) = y''(1) = 2$
47. $y^{(3)} - 2y'' - y' + 2y = e^{-2x}$
48. $x^4 y^{(4)} + 6x^3 y^{(3)} + 20x^2 y'' + 14xy' + 36y = 8x^{-2}$
49. $x^3 y^{(3)} + 4x^2 y'' - 39xy' - 105y = 0$; $y(1) = 0$, $y'(1) = -1$, $y''(1) = 4$

50. $y^{(3)} - 3y'' + 3y' - y = \dfrac{1}{x} e^x$
51. $x^3 y^{(3)} - 3x^2 y'' + 6xy' - 6y = 12x^4 \sin^2\left(\dfrac{x}{2}\right)$
52. $x^4 y^{(4)} - x^3 y^{(3)} + 3x^2 y'' - 6xy' + 6y = 8x^3$
53. $y^{(3)} - 6y'' - 59y' + 424y = 0$

The remaining problems require numerical techniques (such as Newton's method) or clever factoring to determine the roots of the characteristic equation. A calculator or microcomputer may be needed to do the calculations. Write answers to four decimal places.

54. $x^3 y^{(3)} + 6x^2 y'' + 2xy' - 3y = \ln(x^9)$

55. $4y^{(3)} - 3y'' - 18y' + 20y = 0$

56. $2y^{(3)} + 3y'' - 12y' + 8y = 0$

57. $9y^{(4)} - 42y^{(3)} + 22y'' + 126y' - 147y = 0;$ $y(0) = 3,$ $y'(0) = 45,$ $y''(0) = -9,$ $y^{(3)}(0) = 49$

58. $y^{(4)} + 7y^{(3)} + 16y'' + 15y' + 3y = 0$

59. $x^4 y^{(4)} + 12x^3 y^{(3)} + 148x^2 y'' + 216xy' - 216y = 0$

60. $x^4 y^{(4)} + 4x^3 y^{(3)} + 8x^2 y'' + 2xy' + 10y = 0;$ $y(1) = -5,$ $y'(1) = 2,$ $y''(1) = 6,$ $y^{(3)}(1) = 2$

CHAPTER 4

The Laplace Transform

4.0 Introduction

In mathematics, as well as in engineering, the term *transform* usually refers to some device, either mechanical, electrical, or theoretical, which changes an object from one form into another.

We have already seen transforms applied to differential equations. For example, putting $t = \ln(x)$ transforms a Cauchy-Euler differential equation into a constant coefficient linear differential equation, and letting $u = y^{1-\alpha}$ transforms a Bernoulli equation into a linear differential equation. In each case, the objective of the transformation is to replace one problem with another which is easier to solve or, perhaps, which has already been solved.

This chapter is devoted to the Laplace transform, which is defined in terms of an improper integral and can be used to transform certain initial value problems into algebra problems. The idea is to solve the algebra problem, then work back to a solution of the initial value problem through the inverse Laplace transform. We may diagram this process as follows.

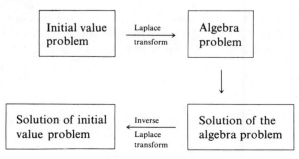

There are several advantages to this method of solving an initial value problem.

1. Sometimes algebra problems are easier to solve than initial value problems.
2. Thus far, in dealing with equations such as

$$y^{(n)} + a_{n-1} y^{(n-1)} + \cdots + a_1 y' + a_0 y = f,$$

we have always assumed that the forcing function f is continuous. With the Laplace transform, we will be able to solve certain problems involving discontinuous forcing functions. This is important in many applications. For example, if f represents an electromotive force which is switched on at a given instant t_0, then f may be discontinuous at t_0.
3. There are extensive tables of Laplace transforms and inverse Laplace transforms, facilitating their use in solving problems.

Finally, there are many types of integral transforms, of which the Laplace transform is just one. The Laplace transform therefore serves as an introduction to an important technique which is useful throughout mathematics, the sciences, and engineering. In particular, in Part V we will see various kinds of Fourier transforms and some of their uses.

4.1 Definition and Theory of the Laplace Transform _____

Suppose that $f(t)$ is defined for $t \geq 0$. The *Laplace transform* $\mathcal{L}[f]$ of f is a function defined by

$$\mathcal{L}[f](s) = \int_0^\infty e^{-st} f(t)\, dt$$

for all s such that this improper integral converges.

We emphasize that $\mathcal{L}[f]$ is a function. Thus, the Laplace transform takes a function f and produces a new function, which we denote $\mathcal{L}[f]$.

We will adopt the convention of using t (for time) as the independent variable of a function, denoted by a lowercase letter such as $f, g,$ or h. The Laplace transform of such a function will be denoted by the same letter but in uppercase:

$$\mathcal{L}[f] = F, \qquad \mathcal{L}[g] = G, \qquad \mathcal{L}[h] = H,$$

and so on.

Further, we will usually let s denote the variable of the transformed function obtained by applying $\mathcal{L}$ to f. Then $\mathcal{L}[f]$ is a function whose independent variable is denoted s, and $\mathcal{L}[f](s)$ is this function evaluated at s. For example, if $\mathcal{L}[f](s) = 1/s$, then $\mathcal{L}[f](1) = 1$ and $\mathcal{L}[f](\pi) = 1/\pi$. If we write $\mathcal{L}[f]$ as simply F, letting the capital letter denote the Laplace transform of the lowercase letter, then $F(s) = 1/s$ in this example, so $F(1) = 1$ and $F(\pi) = 1/\pi$. In general, we will have

$$\mathcal{L}[f](s) = F(s), \qquad \mathcal{L}[g](s) = G(s), \qquad \mathcal{L}[h](s) = H(s),$$

and so on.

EXAMPLE 4.1

$\mathscr{L}[1](s) = 1/s$ for $s > 0$.

That is, $1/s$, for $s > 0$, is the Laplace transform of the constant function 1. We can obtain this result by direct integration:

$$\mathscr{L}[1](s) = \int_0^\infty e^{-st}(1)\, dt = \lim_{k \to \infty} \int_0^k e^{-st}\, dt$$

$$= \lim_{k \to \infty} \left[-\frac{1}{s} e^{-st} \right]_0^k = \lim_{k \to \infty} \frac{1}{s}(1 - e^{-sk}) = \frac{1}{s},$$

provided that $s > 0$. If we denote $f(t) = 1$, then the Laplace transform is $F(s) = 1/s$. ∎

EXAMPLE 4.2

$\mathscr{L}[t](s) = 1/s^2$ if $s > 0$.

That is, the Laplace transform of $g(t) = t$ is $G(s) = 1/s^2$ for $s > 0$. Again, we can calculate this Laplace transform by direct integration:

$$\mathscr{L}[t](s) = \int_0^\infty e^{-st} t\, dt = \lim_{k \to \infty} \int_0^k t e^{-st}\, dt$$

$$= \lim_{k \to \infty} \left[\frac{1}{s^2} e^{-st}(-st - 1) \right]_0^k$$

$$= \lim_{k \to \infty} \left[\frac{1}{s^2} e^{-ks}(-sk - 1) + \frac{1}{s^2} \right] = \frac{1}{s^2}$$

if $s > 0$. ∎

Continuing in this way, we can show by repeated integration that

$$\mathscr{L}[t^n] = \frac{n!}{s^{n+1}}.$$

for any positive integer n and for $s > 0$. Here, $n!$ (read "n factorial") is the product of the integers from 1 to n, inclusive: ⟶ otherwise first term will not be

$$n! = 1 \cdot 2 \cdot \cdots \cdot n.$$ zero.

EXAMPLE 4.3

Let a be any real number. Then $\mathscr{L}[e^{at}](s) = 1/(s - a)$ for $s > a$.

To derive this result, calculate

$$\mathscr{L}[e^{at}](s) = \int_0^\infty e^{-st} e^{at}\, dt = \int_0^\infty e^{-(s-a)t}\, dt$$

$$= \lim_{k \to \infty} \int_0^k e^{-(s-a)t}\, dt = \lim_{k \to \infty} \left[\frac{-1}{s - a} e^{-(s-a)t} \right]_0^k$$

$$= \lim_{k \to \infty} \frac{1}{s - a} [-e^{-(s-a)k} + 1] = \frac{1}{s - a}$$

if $s > a$ (so that $s - a > 0$). ∎

EXAMPLE 4.4

$\mathscr{L}[\cos(t)](s) = s/(1 + s^2)$ for $s > 0$.

To derive this result from the definition, compute

$$\mathscr{L}[\cos(t)](s) = \int_0^\infty e^{-st}\cos(t)\, dt = \lim_{k \to \infty} \int_0^k e^{-st}\cos(t)\, dt$$

$$= \lim_{k \to \infty} \left[\frac{e^{-st}[-s\cos(t) + \sin(t)]}{1 + s^2} \right]_0^k$$

$$= \lim_{k \to \infty} \frac{e^{-sk}[-s\cos(k) + \sin(k)] + s}{1 + s^2} = \frac{s}{1 + s^2}. \quad \blacksquare$$

Here are two properties of the Laplace transform.

THEOREM 4.1

Suppose that $\mathscr{L}[f](s) = F(s)$ and $\mathscr{L}[g](s) = G(s)$ for $s > a$. Then

$$\mathscr{L}[f + g](s) = F(s) + G(s) \quad \text{for} \quad s > a. \tag{4.1}$$

For any number c,

$$\mathscr{L}[cf] = c\mathscr{L}[f]. \quad \blacksquare \tag{4.2}$$

That is, the Laplace transform of a sum is the sum of the Laplace transforms, and constants factor through the Laplace transform.

We can combine conclusions (4.1) and (4.2) into the single statement

$$\mathscr{L}[af + bg] = a\mathscr{L}[f] + b\mathscr{L}[g] \tag{4.3}$$

for any constants a and b. This property is called *linearity* of the Laplace transform, and $\mathscr{L}$ is called a *linear operator*. This terminology comes from algebra, where functions having properties (4.1) and (4.2) are called linear functions. For example, the Riemann integral is linear.

Proof of Theorem 4.1   Both conclusions follow easily from properties of the integral. For (4.1), note that by hypothesis, $\int_0^\infty e^{-st}f(t)\, dt$ and $\int_0^\infty e^{-st}g(t)\, dt$ converge if $s > a$. Then $\int_0^\infty e^{-st}[f(t) + g(t)]\, dt$ also converges if $s > a$, and

$$\mathscr{L}[f + g](s) = \int_0^\infty e^{-st}[f(t) + g(t)]\, dt$$

$$= \int_0^\infty e^{-st}f(t)\, dt + \int_0^\infty e^{-st}g(t)\, dt = \mathscr{L}[f](s) + \mathscr{L}[g](s).$$

For (4.2), we have $\mathscr{L}[cf](s) = \int_0^\infty e^{-st}cf(t)\, dt = c \int_0^\infty e^{-st}f(t)\, dt = c\mathscr{L}[f](s).$ $\quad \blacksquare$

We can extend equation (4.3) to any finite sum:

$$\mathscr{L}[c_1f_1 + c_2f_2 + \cdots + c_nf_n](s) = c_1\mathscr{L}[f_1](s) + c_2\mathscr{L}[f_2](s) + \cdots + c_n\mathscr{L}[f_n](s)$$

for $s > a$, assuming that each $\mathscr{L}[f_j]$ is defined for $s > a$ and that $c_1, \ldots, c_n$ are constants.

EXAMPLE 4.5

By Example 4.3 and Theorem 4.1,

$$\mathscr{L}[e^{-4t} + e^{6t}](s) = \frac{1}{s + 4} + \frac{1}{s - 6} \quad \text{if} \quad s > 6.$$

This follows from the fact that $\mathscr{L}[e^{-4t}](s) = 1/(s + 4)$ if $s > -4$ and $\mathscr{L}[e^{6t}](s) = 1/(s - 6)$ if $s > 6$. Both transforms are defined for $s > 6$. ∎

When we can, we usually obtain the Laplace transform of a function by using a table, or one of several methods to be studied soon, rather than by direct integration. The results of Examples 4.1, 4.2, 4.3, and 4.4 can all be read from Table 4.1. Following the table is an example of its use.

Table 4.1   Laplace Transforms

	$f(t)$	$F(s) = \mathscr{L}(f(t))$
(1)	1	$\dfrac{1}{s}$
(2)	t	$\dfrac{1}{s^2}$
(3)	t^n	$\dfrac{n!}{s^{n+1}}$,
(4)	$\dfrac{1}{\sqrt{t}}$	$\sqrt{\dfrac{\pi}{s}}$
(5)	e^{at}	$\dfrac{1}{s - a}$
(6)	te^{at}	$\dfrac{1}{(s - a)^2}$
(7)	$t^n e^{at}$	$\dfrac{n!}{(s - a)^{n+1}}$
(8)	$\dfrac{1}{a - b}(e^{at} - e^{bt})$	$\dfrac{1}{(s - a)(s - b)}$
(9)	$\dfrac{1}{a - b}(ae^{at} - be^{bt})$	$\dfrac{s}{(s - a)(s - b)}$
(10)	$\dfrac{(c - b)e^{at} + (a - c)e^{bt}}{(a - b)(b - c)(c - a)}$ $+ \dfrac{(b - a)e^{ct}}{(a - b)(b - c)(c - a)}$	$\dfrac{1}{(s - a)(s - b)(s - c)}$

(continued)

Table 4.1 Laplace Transforms (*Continued*)

$f(t)$	$F(s) = \mathscr{L}(f(t)]$
(11) $\sin(at)$	$\dfrac{a}{s^2 + a^2}$
(12) $\cos(at)$	$\dfrac{s}{s^2 + a^2}$
(13) $1 - \cos(at)$	$\dfrac{a^2}{s(s^2 + a^2)}$
(14) $at - \sin(at)$	$\dfrac{a^3}{s^2(s^2 + a^2)^2}$
(15) $\sin(at) - at\cos(at)$	$\dfrac{2a^3}{(s^2 + a^2)^2}$
(16) $t\sin(at)$	$\dfrac{2as}{(s^2 + a^2)^2}$
(17) $t\cos(at)$	$\dfrac{(s - a)(s + a)}{(s^2 + a^2)^2}$
(18) $\dfrac{\cos(at) - \cos(bt)}{(b - a)(b + a)}$	$\dfrac{s}{(s^2 + a^2)(s^2 + b^2)}$
(19) $e^{at}\sin(bt)$	$\dfrac{b}{(s - a)^2 + b^2}$
(20) $e^{at}\cos(bt)$	$\dfrac{s - a}{(s - a)^2 + b^2}$
(21) $\sinh(at)$	$\dfrac{a}{s^2 - a^2}$
(22) $\cosh(at)$	$\dfrac{s}{s^2 - a^2}$
(23) $\sin(at)\cosh(at) - \cos(at)\sinh(at)$	$\dfrac{4a^3}{s^4 + 4a^4}$
(24) $\sin(at)\sinh(at)$	$\dfrac{2a^2 s}{s^4 + 4a^4}$
(25) $\sinh(at) - \sin(at)$	$\dfrac{2a^3}{s^4 - a^4}$
(26) $\cosh(at) - \cos(at)$	$\dfrac{2a^2 s}{s^4 - a^4}$

(*continued*)

Table 4.1   Laplace Transforms (*Continued*)

$f(t)$	$F(s) = \mathscr{L}(f(t))$
(27) $\dfrac{e^{at}(1 + 2at)}{\sqrt{\pi t}}$	$\dfrac{s}{(s - a)^{3/2}}$
(28) $J_0(at)$	$\dfrac{1}{\sqrt{s^2 + a^2}}$
(29) $a^n J_n(at)$	$\dfrac{(\sqrt{s^2 + a^2} - s)^n}{\sqrt{s^2 + a^2}}$
(30) $J_0(2\sqrt{at})$	$\dfrac{1}{s} e^{-a/s}$
(31) $\dfrac{1}{t}\sin(at)$	$\tan^{-1}\left(\dfrac{a}{s}\right)$
(32) $\dfrac{2}{t}[1 - \cos(at)]$	$\ln\left[\dfrac{s^2 + a^2}{s^2}\right]$
(33) $\dfrac{2}{t}[1 - \cosh(at)]$	$\ln\left[\dfrac{s^2 - a^2}{s^2}\right]$
(34) $\dfrac{1}{\sqrt{\pi t}} - ae^{a^2 t}\text{erfc}(a\sqrt{t})$	$\dfrac{1}{\sqrt{s} + a}$
(35) $\dfrac{1}{\sqrt{\pi t}} + ae^{a^2 t}\text{erf}(a\sqrt{t})$	$\dfrac{\sqrt{s}}{s - a^2}$
(36) $e^{a^2 t}\text{erf}(a\sqrt{t})$	$\dfrac{a}{\sqrt{s}(s - a^2)}$
(37) $e^{a^2 t}\text{erfc}(a\sqrt{t})$	$\dfrac{1}{\sqrt{s}(\sqrt{s} + a)}$
(38) $\text{erfc}\left(\dfrac{a}{2\sqrt{t}}\right)$	$\dfrac{1}{s} e^{-a\sqrt{s}}$
(39) $\dfrac{1}{\sqrt{\pi t}} e^{-a^2/4t}$	$\dfrac{1}{\sqrt{s}} e^{-a\sqrt{s}}$
(40) $\dfrac{1}{\sqrt{\pi(t + a)}}$	$\dfrac{1}{\sqrt{s}} e^{as}\text{erfc}(\sqrt{as})$
(41) $\dfrac{1}{\pi t}\sin(2a\sqrt{t})$	$\text{erf}\left(\dfrac{a}{\sqrt{s}}\right)$
(42) $f\left(\dfrac{t}{a}\right)$	$aF(as)$

(*continued*)

Table 4.1   Laplace Transform (*Continued*)

$f(t)$	$F(s) = \mathscr{L}(f(t)]$
(43) $e^{bt/a}f\left(\dfrac{t}{a}\right)$	$aF(as - b)$
(44) $f^{(n)}(t)$	$s^n F(s) - s^{n-1}f(0) - s^{n-2}f'(0) - \cdots - sf^{(n-2)}(0)$ $- f^{(n-1)}(0)$
(45) $\delta(t - a)$	e^{-as}
(46) $\delta(t)$	1

The following functions are defined by displaying their graphs.

(47) (Triangular wave) $\dfrac{1}{as^2}\left[\dfrac{1 - e^{-as}}{1 + e^{-as}}\right]$ or $\dfrac{1}{as^2}\tanh\left(\dfrac{as}{2}\right)$

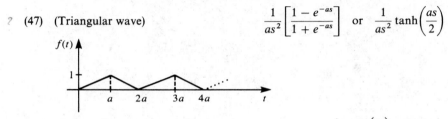

(48) (Square wave) $\dfrac{1}{s}\tanh\left(\dfrac{as}{2}\right)$

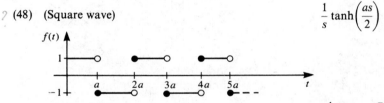

(49) (Sawtooth wave) $\dfrac{1}{as^2} - \dfrac{e^{-as}}{s(1 - e^{-as})}$

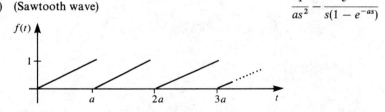

In entries (28), (29), and (30), $J_n(x)$ is the Bessel function of the first kind of order n; in entries (34) through (38), (40), and (41), $\text{erf}(x)$ is the error function and $\text{erfc}(x)$ is the complementary error function, defined by

$$\text{erf}(x) = \frac{2}{\sqrt{\pi}}\int_0^x e^{-r^2}\,dr, \qquad \text{erfc}(x) = 1 - \text{erf}(x).$$

EXAMPLE 4.6

Let $f(t) = -9\cos(t) - 4t^2 + 1$. Using Theorem 4.1 and entries (12), (3), and (1), respectively, of Table 4.1, we have

$$\mathscr{L}[f](s) = -9\mathscr{L}[\cos(t)](s) - 4\mathscr{L}[t^2](s) + \mathscr{L}[1](s)$$

$$= \frac{-9s}{s^2 + 1} - \frac{8}{s^3} + \frac{1}{s}. \quad \blacksquare$$

The existence of the Laplace transform of a function depends on convergence of an improper integral. We will examine sufficient conditions for the Laplace transform of a function to exist. Recall that

$$\mathscr{L}[f](s) = \int_0^\infty e^{-st}f(t)\,dt = \lim_{k \to \infty} \int_0^k e^{-st}f(t)\,dt.$$

In order for this limit to exist and be finite for a given s, we must at least have existence of $\int_0^k e^{-st}f(t)\,dt$ for each $k > 0$. For this to be the case, it is sufficient for f to be continuous on $[0, \infty)$, except possibly at finitely many points $c_1, \ldots, c_n$, at each of which f has a finite left and right limit. Such a function is said to be *piecewise continuous* on $[0, \infty)$.

For example, let

$$f(t) = \begin{cases} t & \text{if } 0 \le t < 4 \\ 1 & \text{if } 4 \le t \le 8 \\ \sin(t) & \text{if } t > 8 \end{cases}$$

A graph of f is shown in Figure 4.1. This function is continuous on $[0, \infty)$ except at 4 and 8. At both of these points, both one-sided limits exist and are finite:

$$\lim_{t \to 4-} f(t) = 4, \qquad \lim_{t \to 4+} f(t) = 1,$$

$$\lim_{t \to 8-} f(t) = 1, \qquad \lim_{t \to 8+} f(t) = \sin(8).$$

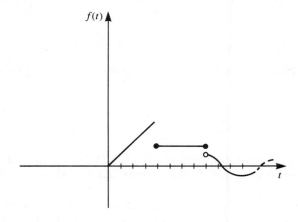

Figure 4.1

Therefore, f is piecewise continuous on $[0, \infty)$.

Even if $\int_0^k e^{-st}f(t)\,dt$ exists for each $k > 0$, it is still possible for $\int_0^\infty e^{-st}f(t)\,dt$ to diverge. To derive a condition on f for this integral to converge, for a given s, recall the comparison test for improper integrals.

$$\text{if } |h(t)| \leq g(t) \text{ for } t \geq a, \text{ and if } \int_a^\infty g(t)\,dt \text{ converges,}$$

$$\text{then } \int_a^\infty h(t)\,dt \text{ converges also.}$$

Since we are interested in convergence of $\int_0^\infty e^{-st}f(t)\,dt$, we want a function g and a number a such that

$$|e^{-st}f(t)| \leq g(t) \quad \text{for} \quad t \geq a \tag{4.4}$$

and such that $\int_0^\infty g(t)\,dt$ converges. For such a function g, the inequality (4.4) can be written

$$|f(t)| \leq e^{st}g(t).$$

One way to find such a function g is to look for a bound on f of the form

$$|f(t)| \leq Me^{bt} \tag{4.5}$$

for $t \geq a$ and for some constants M and b. If this inequality holds, by multiplying both sides of inequality (4.5) by e^{-st} we obtain

$$|e^{-st}f(t)| \leq Me^{-st}e^{bt} = Me^{(b-s)t} = g(t). \tag{4.6}$$

Certainly $\int_0^\infty Me^{(b-s)t}\,dt$ converges if $b - s < 0$ or $s > b$. Thus, an inequality of the form (4.6) implies convergence of $\int_0^\infty e^{-st}f(t)\,dt$ for $s > b$ also.

A bound of the form $|f(t)| \leq Me^{bt}$ is easily found for many functions commonly encountered. For example, if $f(t) = t^2$, we can choose $M = b = 1$, since $t^2 \leq e^t$ for t sufficiently large (recall that $\lim_{t\to\infty} t^2/e^t = 0$). By contrast, e^{t^2} does not have a bound of the form (4.6) because e^{t^2} increases with t faster than Me^{bt} for any choices of the constants M and b.

This discussion suggests the following conditions sufficient for the existence of the Laplace transform of a function.

THEOREM 4.2   Existence of the Laplace Transform

Suppose that f is piecewise continuous on $[0, \infty)$ and that there are constants M, b, and t_0 such that

$$|f(t)| \leq Me^{bt} \quad \text{for} \quad t \geq t_0.$$

Then $\mathscr{L}[f](s)$ exists for $s > t_0$. ∎

Theorem 4.2 guarantees the existence of a large class of functions having Laplace transforms. It should be noted, however, that the hypotheses of the theorem are sufficient, but not necessary, for existence. For example, let $f(t) = t^{-1/2}$ for $t > 0$. Then f is not piecewise continuous on any interval $[0, k]$ because the limit of $t^{-1/2}$ as t approaches zero from the right is not finite. Even so, the improper integral $\int_0^k e^{-st}t^{-1/2}\,dt$ converges for any $k > 0$. In fact, by making the change of variables

$x = t^{1/2}$, we obtain

$$\mathcal{L}[f](s) = \int_0^\infty e^{-st} t^{-1/2}\, dt = 2 \int_0^\infty e^{-sx^2}\, dx$$

$$= \frac{2}{\sqrt{s}} \int_0^\infty e^{-z^2}\, dz \qquad \text{(let } z = x\sqrt{s})$$

$$= \sqrt{\frac{\pi}{s}}$$

for $s > 0$. Here we have used the fact that $\int_0^\infty e^{-z^2}\, dz = \frac{1}{2}\sqrt{\pi}$.

In treatments of the Laplace transform, it is customary to say that a function f is *of exponential order as t approaches infinity* if there exist constants M, b, and t_0 such that

$$|f(t)| \le Me^{bt} \quad \text{for} \quad t \ge t_0.$$

We write $f(t) = O(e^{bt})$, read "oh of e^{bt}," if this inequality is satisfied for some M, b, and t_0. For example, we have seen that $t^2 = O(e^t)$. In this terminology, Theorem 4.2 may be phrased as follows: If f is piecewise continuous on $[0, \infty)$ and $f(t) = O(e^{bt})$, then the Laplace transform of f exists.

We will now consider the inverse Laplace transform. The *inverse Laplace transform* of F is a function f whose Laplace transform is F. If we denote the inverse Laplace transform as $\mathcal{L}^{-1}$, then

$$\mathcal{L}^{-1}[F] = f \quad \text{if} \quad \mathcal{L}[f] = F.$$

In practice, we operate with the symbol $\mathcal{L}^{-1}$ very much as we do in the algebraic statement "$ax = b$ implies that $x = a^{-1}b$."

Notice that every formula $\mathcal{L}[f](s) = F(s)$ automatically gives us a formula $\mathcal{L}^{-1}[F](t) = f(t)$. To illustrate, from Examples 4.1, 4.2, 4.3, and 4.4, we have

$$\mathcal{L}^{-1}\left[\frac{1}{s}\right](t) = 1$$

$$\mathcal{L}^{-1}\left[\frac{1}{s^2}\right](t) = t$$

$$\mathcal{L}^{-1}\left[\frac{1}{s - a}\right](t) = e^{at}$$

$$\mathcal{L}^{-1}\left[\frac{s}{1 + s^2}\right](t) = \cos(t).$$

The following theorem states that $\mathcal{L}^{-1}$ is also linear.

THEOREM 4.3

$$\mathcal{L}^{-1}[aF + bG] = a\mathcal{L}^{-1}[F] + b\mathcal{L}^{-1}[G]$$

whenever all three inverse Laplace transforms are defined. ∎

EXAMPLE 4.7

Let $F(s) = 1/(s^2 - 9)$ for $s > 3$ and $G(s) = 1/s^4$ for $s > 0$. Find $\mathcal{L}^{-1}[4F - G]$.

From Table 4.1 [entries (21) and (3)], we have

$$\mathcal{L}^{-1}[F](t) = \frac{1}{3}\sinh(3t) \quad \text{and} \quad \mathcal{L}^{-1}[G](t) = \frac{1}{3!}t^3 = \frac{1}{6}t^3.$$

Therefore,

$$\mathcal{L}^{-1}[4F - G](t) = 4\mathcal{L}^{-1}[F](t) - \mathcal{L}^{-1}[G](t) = \tfrac{4}{3}\sinh(3t) - \tfrac{1}{6}t^3. \quad \blacksquare$$

In the next section, we will see how the Laplace transform and its inverse are used to solve certain initial value problems. This will raise a subtle point. How many inverse Laplace transforms can a function have? That is, is it possible for two different functions to have the same Laplace transform?

In fact, two different functions *can* have the same Laplace transform. If we change the values of $f(t)$ at finitely many points, we produce a new function g. Figure 4.2 shows typical graphs of f and g. However, $\int_0^\infty e^{-st}f(t)\,dt = \int_0^\infty e^{-st}g(t)\,dt$, so $\mathcal{L}[f] = \mathcal{L}[g]$ even though $f \neq g$.

In this example, it is significant that g is discontinuous. The following result implies that two *continuous* functions having the same Laplace transform must be identical for $t \geq 0$.

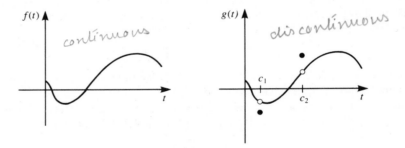

Figure 4.2

THEOREM 4.4   Lerch's Theorem

If $\mathcal{L}[f] = \mathcal{L}[g]$, and f and g are continuous on $[0, \infty)$, then $f(t) = g(t)$ for $t \geq 0$.

$\blacksquare$

For a proof of Lerch's theorem, refer to R. V. Churchill, *Operational Mathematics*, 3rd ed. (New York. McGraw-Hill, 1972).

As a consequence of Lerch's theorem, given F, there can be only one *continuous* function f such that $\mathcal{L}[f] = F$, or, equivalently, $f = \mathcal{L}^{-1}[F]$. In using the Laplace transform to solve an initial value problem, when we ask for $\mathcal{L}^{-1}[F]$, we want the continuous function f whose Laplace transform is F.

We will conclude this section with a theorem we will use in applying the Laplace transform to the solution of initial value problems.

THEOREM 4.5

Suppose that f is piecewise continuous on $[0, k]$ for every $k > 0$ and that $f(t) = O(e^{bt})$.

Let $\mathscr{L}[f](s) = F(s)$. Then

$$\lim_{s \to \infty} F(s) = 0. \quad \blacksquare$$

That is, under the given conditions, the Laplace transform of f tends to zero as s approaches infinity. A proof of Theorem 4.5 is outlined in Problem 39 at the end of this section.

NOTATIONAL CONVENTION

When working with the Laplace transform, it is often convenient to be informal with notation. For example, we know that $\mathscr{L}[1](s) = 1/s$. We will often write this as just $\mathscr{L}[1] = 1/s$. Similarly, we can write $\mathscr{L}[t] = 1/s^2$ instead of the more careful $\mathscr{L}[t](s) = 1/s^2$. As long as it is clear what is meant by the notation, this convention will facilitate our writing equations and carrying out calculations with the Laplace transform.

PROBLEMS FOR SECTION 4.1

In each of Problems 1 through 14, use Table 4.1, Theorem 4.1, and, where necessary, integration, to find $\mathscr{L}[f](s)$. In each of Problems 12, 13, and 14, the function is described by its graph.

1. $2\sinh(t) - 4$

2. $\cos(t) - \sin(t)$

3. $4t\sin(2t)$

4. $t^2 - 3t + 5$

5. $t - \cos(5t)$

6. $2t^2 e^{-3t} - 4t + 1$

7. $(t + 4)^2$

8. $3e^{-t} + \sin(6t)$

9. $t^3 - 3t + \cos(4t)$

10. $-3\cos(2t) + 5\sin(4t)$

11. $4\cos^2(3t)$

12.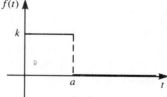

13.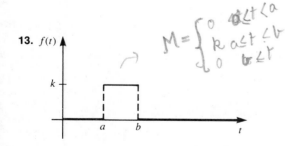

14.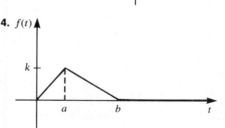

In each of Problems 15 through 24, use Theorem 4.2 and Table 4.1 to find the (continuous) inverse Laplace transform of $F(s)$.

15. $\dfrac{-2}{s + 16}$

16. $\dfrac{4s}{s^2 - 14}$

17. $\dfrac{2s - 5}{s^2 + 16}$

18. $\dfrac{3s + 17}{s^2 - 7}$

19. $\dfrac{3}{s - 7} + \dfrac{1}{s^2}$

20. $\dfrac{5}{(s + 7)^2}$

21. $\dfrac{1}{s - 4} - \dfrac{6}{(s - 4)^2}$

22. $\dfrac{2}{s^4}\left[\dfrac{1}{s} + \dfrac{3}{s^2} + \dfrac{4}{s^6}\right]$

23. $\dfrac{s - 4}{(s^2 + 5)^2} + \dfrac{s}{s^2 + 2}$

24. $\dfrac{1}{s} + \dfrac{s}{s^2 + 8}$

In each of Problems 25 through 28, derive the requested entry from Table 4.1; a denotes a constant.

25. Use the definition of the Laplace transform and integration by parts to obtain $\mathscr{L}[\sin(at)]$, which is entry (11) of the table.

26. Use the definition of the Laplace transform and integration by parts to obtain $\mathscr{L}[\cos(at)]$, which is entry (12) of the table.

27. Use the fact that $\sinh(x) = \frac{1}{2}(e^x - e^{-x})$, and the result of Example 4.3, to obtain $\mathscr{L}[\sinh(at)]$, which is entry (21) of the table.

28. Use the fact that $\cosh(x) = \frac{1}{2}(e^x + e^{-x})$, and the result of Example 4.3, to obtain $\mathscr{L}[\cosh(at)]$, which is entry (22) of the table.

29. Let $F(s) = \mathscr{L}[f](s)$. Prove that, for any positive constant a,

$$\mathscr{L}[f(at)](s) = \frac{1}{a} F\left(\frac{s}{a}\right).$$

30. Assume that the Laplace transform is defined for complex values of t. Use Euler's formula to write $\cos(at) = \frac{1}{2}(e^{iat} + e^{-iat})$, and use the Laplace transform of e^{at} to provide another derivation of a formula for $\mathscr{L}[\cos(at)]$.

31. Use the method of Problem 30 to give another derivation of the formula for $\mathscr{L}[\sin(at)]$.

32. Assume that the Laplace transform of a Taylor series is the sum of the Laplace transform of each term of the series (where the series converges). Use the Maclaurin series for e^{at} to give another derivation of the formula for $\mathscr{L}[e^{at}]$ valid for $s > a$.

In each of Problems 33 through 37, produce numbers M, b, and t_0 such that $f(t) = O(e^{bt})$.

after Improper Integral

33. $\cos(at)$, a any positive number
34. t^3 **35.** e^{5t} **36.** $\sinh(t)$ **37.** t^4

38. For each of the functions in Problems 33 through 37, determine $F(s) = \mathscr{L}[f](s)$, using Table 4.1 where possible, and verify the conclusion of Theorem 4.5 by showing that $\lim_{s \to \infty} F(s) = 0$.

39. Prove Theorem 4.5. *Hint:* Write

$$|\mathscr{L}[f](s)| = \left| \int_0^{\infty} e^{-st} f(t)\, dt \right|$$

$$= \left| \int_0^{t_0} e^{-st} f(t)\, dt + \int_{t_0}^{\infty} e^{-st} f(t)\, dt \right|$$

$$\leq \left| \int_0^{t_0} e^{-st} f(t)\, dt \right| + \left| \int_{t_0}^{\infty} e^{-st} f(t)\, dt \right|.$$

Explain why, for some number K, $|f(t)| \leq K$ for $0 \leq t \leq t_0$, and also $|f(t)| \leq Me^{bt}$ for $t \geq t_0$. Use these inequalities in the last two integrals, integrate, and take the limit as $s \to \infty$.

4.2 Solving Initial Value Problems Using the Laplace Transform

We will now see how the Laplace transform converts certain initial value problems into algebra problems. First, we must relate the Laplace transform of f to that of f'.

THEOREM 4.6

Suppose that f is continuous on $[0, \infty)$ and that f' is piecewise continuous on $[0, k]$ for every $k > 0$. Suppose also that $f(t) = O(e^{bt})$. Then

$$\mathcal{L}[f'](s) = s\mathcal{L}[f](s) - f(0). \quad \blacksquare \tag{4.7}$$

If we write $\mathcal{L}[f] = F(s)$, this conclusion may be written

$$\mathcal{L}[f'](s) = sF(s) - f(0).$$

Proof of Theorem 4.6   We will prove the theorem for the case in which f' is continuous on $[0, \infty)$. If f' is piecewise continuous, but not continuous, the argument is more involved; we will discuss this case at the end of this section.

If f' is continuous on $[0, \infty)$, we can integrate by parts as follows:

$$\mathcal{L}[f'](s) = \int_0^\infty \underbrace{e^{-st}}_{u}\underbrace{f'(t)\,dt}_{dv}$$

$$= \left[e^{-st}f(t)\right]_0^\infty - \int_0^\infty -se^{-st}f(t)\,dt$$

$$= \lim_{t\to\infty} e^{-st}f(t) - \lim_{t\to 0+} e^{-st}f(t) + s\int_0^\infty e^{-st}f(t)\,dt.$$

Now, $|f(t)| \le Me^{bt}$ for t sufficiently large. Then $|f(t)e^{-st}| \le Me^{(b-s)t}$ for t sufficiently large. Since $b - s < 0$ if $s > b$, $Me^{(b-s)t} \to 0$ as $t \to \infty$. Hence, $e^{-st}f(t) \to 0$ as $t \to \infty$. Further, since f is continuous at 0, $\lim_{t\to 0+} e^{-st}f(t) = f(0)$. Therefore,

$$\mathcal{L}[f'](s) = -f(0) + s\mathcal{L}[f](s),$$

or

$$\mathcal{L}[f'](s) = sF(s) - f(0),$$

as we wanted to show. $\blacksquare$

We often encounter the case in which f has a jump discontinuity at zero. That is, $\lim_{t\to 0+} f(t)$ exists but does not equal $f(0)$. This occurs, for example, if f is an electromotive force which is zero up until time zero and is then switched on. When f has a jump discontinuity at zero, the statement in the preceding argument that $\lim_{t\to 0} e^{-st}f(t) = f(0)$ must be replaced by

$$\lim_{t\to 0+} e^{-st}f(t) = f(0+),$$

where

$$f(0+) = \lim_{t\to 0+} f(t).$$

In this case, the conclusion of the theorem reads

$$\mathcal{L}[f'] = s\mathcal{L}[f] - f(0+),$$

or

$$\mathcal{L}[f'] = sF(s) - f(0+).$$

Since many initial value problems involve differential equations of order 2 or higher, we will also relate the Laplace transform of higher derivatives of f to the Laplace transform of f, together with values of f and its derivatives at zero.

THEOREM 4.7

[handwritten: because $f^{(n-1)}$ is not cont. $f^{(n-1)}$ is not differentiable ∴ $f^{(n)}$ does not exist]

Suppose that $f, f', f'', \ldots, f^{(n-1)}$ are continuous for $t \geq 0$ and that $f^{(n)}$ is piecewise continuous on $[0, k]$ for every $k > 0$. Suppose that each of $f, f', \ldots, f^{(n-1)}$ is $O(e^{bt})$. Then $\mathcal{L}[f^{(n)}]$ exists for $s > b$, and

$$\mathcal{L}[f^{(n)}] = s^n \mathcal{L}[f](s) - s^{n-1}f(0) - s^{n-2}f'(0) - \cdots - sf^{(n-2)}(0) - f^{(n-1)}(0). \quad \blacksquare$$

If we write $\mathcal{L}[f] = F$, this conclusion can be written

$$\mathcal{L}[f^{(n)}] = s^n F(s) - s^{n-1}f(0) - s^{n-2}f'(0) - \cdots - sf^{(n-2)}(0) - f^{(n-1)}(0).$$

Theorem 4.7 can be proved by using mathematical induction and Theorem 4.6.

The formula of Theorem 4.7 for the case $n = 2$ is used so frequently that we will state it separately:

$$\mathcal{L}[f''] = s^2 F(s) - sf(0) - f'(0), \tag{4.8}$$

under the conditions given in the theorem.

Theorems 4.6 and 4.7 are the key to using the Laplace transform to solve initial value problems in which the differential equation has constant coefficients. Here are two typical examples.

EXAMPLE 4.8

Solve the initial value problem

$$y'' + y = t; \qquad y(0) = 1, \qquad y'(0) = 0.$$

We already have ways of solving this problem, but our purpose here is to illustrate the Laplace transform method. Apply the Laplace transform to the differential equation to get

$$\mathcal{L}[y'' + y] = \mathcal{L}[t].$$

By linearity of the Laplace transform, we can write this equation as

$$\mathcal{L}[y''] + \mathcal{L}[y] = \mathcal{L}[t].$$

Now, $\mathcal{L}[t] = 1/s^2$. Write $\mathcal{L}[y] = Y(s)$, and apply equation (4.8) to get

$$s^2 Y(s) - sy(0) - y'(0) + Y(s) = \frac{1}{s^2}. \tag{4.9}$$

Notice that this equation contains $y(0)$ and $y'(0)$, which are given as initial data of the problem. Since $y(0) = 1$ and $y'(0) = 0$, equation (4.9) becomes

$$s^2 Y(s) - s + Y(s) = \frac{1}{s^2}.$$

Solve this equation for $Y(s)$ to get

$$Y(s) = \frac{1}{s^2(s^2 + 1)} + \frac{s}{s^2 + 1}.$$

We now have an explicit expression for $Y(s)$, the Laplace transform of the function y that we want. The equation $\mathscr{L}[y] = Y$ can be solved for y by applying the inverse Laplace transform:

$$y = \mathscr{L}^{-1}[Y].$$

The solution of the initial value problem is

$$y(t) = \mathscr{L}^{-1}\left[\frac{1}{s^2(s^2 + 1)}\right] + \mathscr{L}^{-1}\left[\frac{s}{s^2 + 1}\right].$$

We can read these inverse Laplace transforms from Table 4.1 if we first do some algebraic manipulation. By the method of partial fractions (which we will review in Section 4.6), we can write

$$\frac{1}{s^2(s^2 + 1)} = \frac{a}{s} + \frac{b}{s^2} + \frac{cs + d}{s^2 + 1}.$$

If the fractions on the right are added, we get

$$\frac{1}{s^2(s^2 + 1)} = \frac{as(s^2 + 1) + b(s^2 + 1) + (cs + d)s^2}{s^2(s^2 + 1)}.$$

The numerators of these fractions must be equal. Therefore,

$$as(s^2 + 1) + b(s^2 + 1) + (cs + d)s^2 = 1.$$

Group terms to write this equation as

$$(a + c)s^3 + (b + d)s^2 + as + b = 1.$$

By matching coefficients of like powers of s on both sides, we conclude that

$$a + c = 0$$
$$b + d = 0$$
$$a = 0$$
$$b = 1;$$

hence, $a = c = 0$, $b = 1$, and $d = -1$. Then

$$\frac{1}{s^2(s^2 + 1)} = \frac{1}{s^2} - \frac{1}{s^2 + 1}.$$

We can now write

$$Y(s) = \frac{1}{s^2} - \frac{1}{s^2 + 1} + \frac{s}{s^2 + 1}.$$

We finally have

$$y(t) = \mathscr{L}^{-1}\left[\frac{1}{s^2}\right] - \mathscr{L}^{-1}\left[\frac{1}{s^2 + 1}\right] + \mathscr{L}^{-1}\left[\frac{s}{s^2 + 1}\right].$$

Now use Table 4.1 [entries (2), (11), and (12), respectively] to write the solution of the initial value problem:

$$y(t) = t - \sin(t) + \cos(t). \quad \blacksquare$$

By Lerch's theorem (Section 4.1), this is the unique continuous inverse Laplace transform of Y, consistent with the fact that this initial value problem has a unique solution.

EXAMPLE 4.9

Solve the initial value problem

$$y'' + 4y' + 3y = e^t; \qquad y(0) = 0, \qquad y'(0) = 2.$$

Apply the Laplace transform to the differential equation to get

$$\mathscr{L}[y''] + 4\mathscr{L}[y'] + 3\mathscr{L}[y] = \mathscr{L}[e^t].$$

Apply equation (4.8) to the $\mathscr{L}[y'']$ term and Theorem 4.6 to the $\mathscr{L}[y']$ term to get

$$s^2 Y(s) - sy(0) - y'(0) + 4[s Y(s) - y(0)] + 3 Y(s) = \mathscr{L}[e^t]. \qquad (4.10)$$

From Table 4.1 [entry (5)], we read that $\mathscr{L}[e^t] = 1/(s - 1)$. Substitute this result, together with the initial data, into equation (4.10) to get

$$s^2 Y(s) - 2 + 4s Y(s) + 3 Y(s) = \frac{1}{s - 1},$$

or

$$(s^2 + 4s + 3)Y(s) = \frac{1}{s - 1} + 2.$$

Solve this equation for $Y(s)$ to get

$$Y(s) = \frac{2s - 1}{(s - 1)(s^2 + 4s + 3)}.$$

The solution of the initial value problem is the inverse Laplace transform of Y. We must do some manipulation to write $Y(s)$ in a form allowing us to read the inverse Laplace transform from the table. Write

$$\frac{2s - 1}{(s - 1)(s^2 + 4s + 3)} = \frac{2s - 1}{(s - 1)(s + 3)(s + 1)}$$

$$= \frac{a}{s - 1} + \frac{b}{s + 3} + \frac{c}{s + 1}$$

$$= \frac{a(s + 3)(s + 1) + b(s - 1)(s + 1) + c(s - 1)(s + 3)}{(s - 1)(s + 3)(s + 1)}.$$

The numerators of these fractions must be equal; therefore,

$$a(s + 3)(s + 1) + b(s - 1)(s + 1) + c(s - 1)(s + 3) = 2s - 1.$$

We can solve for a, b, and c in several ways. If we let $s = 1$, then

$$8a = 1,$$

or $a = \frac{1}{8}$. Letting $s = -3$, we get

$$8b = -7,$$

or $b = -\frac{7}{8}$. Finally, letting $s = -1$, we get

$$-4c = -3,$$

or $c = \frac{3}{4}$. This calculation enables us to write

$$Y(s) = \frac{1}{8}\frac{1}{s-1} - \frac{7}{8}\frac{1}{s+3} + \frac{3}{4}\frac{1}{s+1}.$$

The solution of the initial value problem is

$$y(t) = \mathscr{L}^{-1}[Y(s)]$$

$$= \frac{1}{8}\mathscr{L}^{-1}\left[\frac{1}{s-1}\right] - \frac{7}{8}\mathscr{L}^{-1}\left[\frac{1}{s+3}\right] + \frac{3}{4}\mathscr{L}^{-1}\left[\frac{1}{s+1}\right].$$

Finally, from entry (5) of Table 4.1, we have

$$y(t) = \tfrac{1}{8}e^t - \tfrac{7}{8}e^{-3t} + \tfrac{3}{4}e^{-t}. \quad \blacksquare$$

Notice that, in the Laplace transform method, we did not find the general solution of the differential equation, then use the initial data to determine the constants. Instead, the initial conditions were immediately incorporated into the equation for $Y(s)$, through the use of Theorems 4.6 and 4.7.

The Laplace transform method of solving an initial value problem has its limitations. For example, it is best suited to problems in which the differential equation has constant coefficients. But it also has its advantages. In particular, when the differential equation contains a discontinuous forcing function f, as is often the case, the Laplace transform may provide a means to a solution when previously discussed methods do not apply or become very unwieldy.

Sometimes we want to take the Laplace transform of a function defined by an integral, say, $f(t) = \int_0^t g(r)\, dr$. Theorem 4.6 enables us to do this. If g is continuous for $t \geq 0$, then $f'(t) = g(t)$ by the fundamental theorem of calculus. By theorem 4.6,

$$\mathscr{L}[f'] = s\mathscr{L}[f] - f(0).$$

Since $f(0) = 0$ and $f'(t) = g(t)$,

$$\mathscr{L}[f'] = \mathscr{L}[g] = s\mathscr{L}\left[\int_0^t g(r)\, dr\right].$$

We can write this equation as

$$\mathscr{L}\left[\int_0^t g(r)\, dr\right] = \frac{1}{s}\mathscr{L}[g].$$

We will record this result as a theorem.

THEOREM 4.8

Suppose that g is continuous for $t \geq 0$ and that $g(t) = O(e^{bt})$. Let $\mathscr{L}[g] = G(s)$ for $s > b$. Then

$$|g(t)| \leq Me^{bt}$$

$$\mathscr{L}\left[\int_0^t g(r)\, dr\right] = \frac{1}{s} G(s) \qquad \text{for} \qquad s > b. \quad \blacksquare$$

This section has provided a brief glimpse of how we use the Laplace transform to solve initial value problems. The method requires some facility in working with the Laplace transform and in computing inverse Laplace transforms. We will begin to acquire more experience in these areas in the next section.

We conclude this section with a proof of Theorem 4.6 for the case in which f' is piecewise continuous, but not necessarily continuous, on $[0, k]$ for every $k > 0$.

Proof of Theorem 4.6   Write

$$\mathscr{L}[f'] = \int_0^\infty e^{-st}f'(t)\, dt = \lim_{k \to \infty} \int_0^k e^{-st}f'(t)\, dt.$$

Let $k > 0$, and let the points in $[0, k]$ at which f' has jump discontinuities be $\beta_1, \ldots, \beta_n$, with $0 < \beta_1 < \cdots < \beta_n < k$. Now write

$$\int_0^k e^{-st}f'(t)\, dt = \int_0^{\beta_1} e^{-st}f'(t)\, dt + \int_{\beta_1}^{\beta_2} e^{-st}f'(t)\, dt + \cdots + \int_{\beta_n}^k e^{-st}f'(t)\, dt.$$

Integrate each term on the right by parts [with $u = e^{-st}$ and $dv = f'(t)\, dt$] to get

$$\int_0^k e^{-st}f'(t)\, dt = e^{-st}f(t)\Big]_0^{\beta_1} + e^{-st}f(t)\Big]_{\beta_1}^{\beta_2} + \cdots + e^{-st}f(t)\Big]_{\beta_n}^k$$

$$+ s\int_0^{\beta_1} e^{-st}f(t)\, dt + s\int_{\beta_1}^{\beta_2} e^{-st}f(t)\, dt + \cdots + s\int_{\beta_n}^k e^{-st}f(t)\, dt$$

$$= [e^{-s\beta_1}f(\beta_1) - f(0)] + [e^{-s\beta_2}f(\beta_2) - e^{-s\beta_1}f(\beta_1)] + \cdots$$

$$+ [e^{-sk}f(k) - e^{-s\beta_n}f(\beta_n)] + s\int_0^k e^{-st}f(t)\, dt.$$

The sum of terms in square brackets is telescoping, with the second term in each bracket canceling the first term in the preceding bracket (except for the first and last brackets). We get

$$\int_0^k e^{-st}f'(t)\, dt = e^{-sk}f(k) - f(0) + s\int_0^k e^{-st}f(t)\, dt. \qquad (4.11)$$

Now, $f(t) = O(e^{bt})$, so for some constants M and t_0,

$$|f(t)| \leq Me^{bt} \quad \text{for} \quad t \geq t_0.$$

Then

$$|e^{-st}f(t)| \leq Me^{-st}e^{bt} = Me^{-(s-b)t} \quad \text{for} \quad t \geq t_0.$$

Then

$$|e^{-sk}f(k)| \leq Me^{-(s-b)k} \quad \text{for} \quad t \geq t_0.$$

If $s > b$, then $-(s - b) < 0$, so $e^{-(s-b)k} \to 0$ as $k \to \infty$. Then we also have $e^{-sk}f(k) \to 0$ as $k \to \infty$. Upon taking the limit as $k \to \infty$ in equation (4.11), we have

$$\mathcal{L}[f'] = \lim_{k \to \infty} \left[e^{-sk}f(k) - f(0) + s \int_0^k e^{-st}f(t)\, dt \right]$$
$$= -f(0) + s\mathcal{L}[f],$$

as we wanted to show. If f has a jump discontinuity at zero, $f(0)$ is replaced by $f(0+)$. ∎

PROBLEMS FOR SECTION 4.2

In each of Problems 1 through 10, use the Laplace transform and Table 4.1 to solve the initial value problem.

1. $y' + 4y = 1$; $y(0) = -3$

2. $y' - 9y = t$; $y(0) = 5$

3. $y' + 4y = \cos(t)$; $y(0) = 0$

4. $y' + 2y = e^{-t}$; $y(0) = 1$

5. $y' - 2y = 1 - t$; $y(0) = 1$

6. $y'' + y = 1$; $y(0) = 6$, $y'(0) = 0$

7. $y'' - 4y' + 4y = 1$; $y(0) = 1$, $y'(0) = 4$

8. $y'' + 9y = t$; $y(0) = y'(0) = 0$

9. $y'' + 16y = 1 + t$; $y(0) = y'(0) = 0$

10. $y'' - 5y' + 6y = 0$; $y(0) = 1$, $y'(0) = 0$

11. Suppose that f satisfies the hypotheses of Theorem 4.6 except that f has a finite jump discontinuity at zero. Prove the assertion made in the text that $\mathcal{L}[f'] = sF(s) - f(0+)$.

12. Suppose that f satisfies the hypotheses of Theorem 4.6 except that f has a finite jump discontinuity at t_0, with $t_0 > 0$. Prove that

$$\mathcal{L}[f'] = sF(s) - f(0) - e^{-st_0}[f(t_0+) - f(t_0-)],$$

in which $f(t_0+) = \lim_{t \to t_0+} f(t)$ and $f(t_0-) = \lim_{t \to t_0-} f(t)$.

13. Show that

$$\mathcal{L}\left[\int_a^t f(r)\, dr \right] = \frac{1}{s} F(s) - \frac{1}{s} \int_0^a f(r)\, dr.$$

14. Determine $\mathcal{L}\left[\int_0^t \int_0^\tau \cos(3x)\, dx\, d\tau \right]$.

15. Determine

$$\mathcal{L}^{-1}\left[\frac{2}{s(s^2 + 4)} \right].$$

Suppose that f is defined on $[0, \infty)$. We say that f is *periodic with period T* if, for $t \geq 0$, $f(t + T) = f(t)$. An example is $\sin(t)$, which has period 2π.

In Problems 16 through 19, suppose that f is periodic with period T on $[0, \infty)$.

16. Show that

$$\mathcal{L}[f](s) = \sum_{n=0}^{\infty} \int_{nT}^{(n+1)T} e^{-st}f(t)\, dt.$$

17. Show that

$$\int_{nT}^{(n+1)T} e^{-st}f(t)\, dt = e^{-nsT} \int_0^T e^{-st}f(t)\, dt.$$

18. Use the results of Problems 16 and 17 to show that

$$\mathscr{L}[f](s) = \left[\sum_{n=0}^{\infty} e^{-nsT}\right] \int_0^T e^{-st} f(t)\, dt.$$

19. Use the formula for the sum of a geometric series $\left[\sum_{n=0}^{\infty} r^n = 1/(1-r) \text{ if } |r| < 1\right]$ and the result of Problem 18 to show that

$$\mathscr{L}[f](s) = \frac{1}{1 - e^{-sT}} \int_0^T e^{-st} f(t)\, dt.$$

In Problems 20 through 27, a periodic function is given (with the period T varying from one problem to another) on $[0, \infty)$. Sometimes the function is defined by a graph. In each problem, use the result of Problem 19 to determine the Laplace transform of the function.

20. $f(t) = \begin{cases} 5 & \text{if } 0 < t \le 3 \\ 0 & \text{if } 3 < t \le 6, \end{cases}$

with $f(t + 6) = f(t)$ for $t \ge 0$.

21. $f(t) = |E \sin(\omega t)|$, with E and ω positive constants. *Hint:* f is periodic of period π/ω.

22.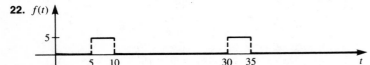

23.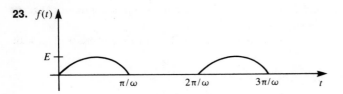

Half-wave rectification of $E \sin(\omega t)$

24.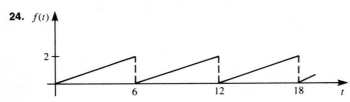

25.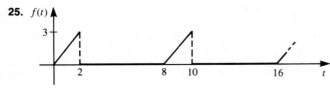

26.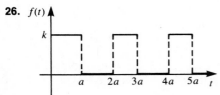

27.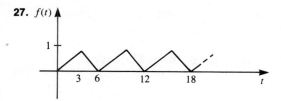

4.3 *The First Shifting Theorem*

In calculating a Laplace transform, we use tables whenever we can. Tables, however, are necessarily limited, and there are rules for manipulating Laplace transforms which must be mastered in order to use the transform effectively. In this and the next section we will develop two such rules, called *shifting theorems*.

The first theorem is called *shifting in the s-variable*, or the *first shifting theorem*.

THEOREM 4.9 Shifting in the *s*-Variable

If $F(s) = \mathscr{L}[f]$ exists for $s > b \geq 0$, and a is any constant, then

$$\mathscr{L}[e^{at}f] = F(s - a) \quad \text{for} \quad s > a + b.$$

That is, the Laplace transform of $e^{at}f$ is obtained by replacing s by $s - a$ in the Laplace transform of f. Put another way, the effect of multiplying f by e^{at} is to shift the Laplace transform of f a units to the right.

Proof We have

$$\mathscr{L}[e^{at}f] = \int_0^\infty e^{-st}e^{at}f(t)\,dt = \int_0^\infty e^{-(s-a)t}f(t)\,dt = F(s - a)$$

for $s - a > b$ or $s > a + b$. ∎

EXAMPLE 4.10

We know from entry (12) of Table 4.1 that $\mathscr{L}[\cos(bt)] = s/(s^2 + b^2)$ for $s > 0$. Then

$$\mathscr{L}[e^{at}\cos(bt)] = F(s - a) = \frac{s - a}{(s - a)^2 + b^2} \quad \text{for} \quad s > a.$$

That is, for the Laplace transform of $e^{at}\cos(bt)$, simply replace s with $s - a$ in the Laplace transform of $\cos(bt)$. ∎

EXAMPLE 4.11

We know that $\mathscr{L}[t^3] = 6/s^4$. Then

$$\mathscr{L}[e^{7t}t^3] = \frac{6}{(s - 7)^4} \quad \text{for} \quad s > 7.$$

(Replace s with $s - 7$ in the Laplace transform of t^3.) ∎

The shifting formula of Theorem 4.9 suggests a corresponding formula for the inverse Laplace transform.

THEOREM 4.10   Inverse Version of the First Shifting Theorem

If $\mathscr{L}^{-1}[F(s)] = f(t)$, then

$$\mathscr{L}^{-1}[F(s - a)] = e^{at}f(t). \quad \blacksquare$$

Stated another way,

$$\mathscr{L}^{-1}[F(s - a)] = e^{at}\mathscr{L}^{-1}[F(s)].$$

Thus, if we replace s by $s - a$ in $F(s)$, the inverse Laplace transform of the resulting function $F(s - a)$ is e^{at} times the inverse Laplace transform of F. If we know that $\mathscr{L}^{-1}[F] = f$, we can immediately find the inverse transform of $F(s - a)$ by multiplying $f(t)$ by e^{at}.

EXAMPLE 4.12

Find

$$\mathscr{L}^{-1}\left[\frac{4}{s^2 + 4s + 20}\right].$$

Since $4/(s^2 + 4s + 20)$ does not appear in the table, we will do some algebraic manipulation. Complete the square in the denominator to write

$$\frac{4}{s^2 + 4s + 20} = \frac{4}{(s + 2)^2 + 16}.$$

Think of the quotient on the right as a function of $s + 2$,

$$F(s + 2) = \frac{4}{(s + 2)^2 + 16},$$

with

$$F(s) = \frac{4}{s^2 + 16}.$$

We see from Table 4.1 [entry (11)] that $F(s)$ is the Laplace transform of $\sin(4t)$. Write

$$F(s) = \frac{4}{s^2 + 16} = \mathscr{L}[\sin(4t)],$$

or

$$\mathscr{L}^{-1}[F(s)] = \sin(4t).$$

Then

$$\mathscr{L}^{-1}\left[\frac{4}{s^2 + 4s + 20}\right] = \mathscr{L}^{-1}[F(s + 2)] = e^{-2t}\mathscr{L}^{-1}[F(s)] = e^{-2t}\sin(4t). \quad \blacksquare$$

EXAMPLE 4.13

Find

$$\mathscr{L}^{-1}\left[\frac{3s - 2}{s^2 + 4s + 20}\right].$$

Complete the square in the denominator and write

$$\frac{3s - 2}{s^2 + 4s + 20} = \frac{3s - 2}{(s + 2)^2 + 16}.$$

To apply the shifting theorem, we will write the fraction on the right as a function of $s + 2$. To do this, write

$$\frac{3s - 2}{(s + 2)^2 + 16} = \frac{3(s + 2) - 8}{(s + 2)^2 + 16}$$

$$= \frac{3(s + 2)}{(s + 2)^2 + 16} - \frac{8}{(s + 2)^2 + 16}.$$

We will take the inverse Laplace transform of each of these two terms and add the results.

Step 1   Calculating

$$\mathscr{L}^{-1}\left[\frac{3(s + 2)}{(s + 2)^2 + 16}\right].$$

Write

$$F(s + 2) = \frac{3(s + 2)}{(s + 2)^2 + 16},$$

so

$$F(s) = \frac{3s}{s^2 + 16} = \mathscr{L}[3 \cos(4t)].$$

Therefore,

$$\mathscr{L}^{-1}\left[\frac{3(s + 2)}{(s + 2)^2 + 16}\right] = 3e^{-2t}\cos(4t).$$

Step 2   Calculating

$$\mathscr{L}^{-1}\left[\frac{-8}{(s + 2)^2 + 16}\right].$$

Write

$$G(s + 2) = \frac{-8}{(s + 2)^2 + 16},$$

so

$$G(s) = \frac{-8}{s^2 + 16} = -2\frac{4}{s^2 + 16} = \mathscr{L}[-2\sin(4t)].$$

Then,

$$\mathscr{L}^{-1}\left[\frac{-8}{(s + 2)^2 + 16}\right] = -2e^{-2t}\sin(4t).$$

From these two results, we conclude that

$$\mathscr{L}^{-1}\left[\frac{3s - 2}{s^2 + 4s + 16}\right] = 3e^{-2t}\cos(4t) - 2e^{-2t}\sin(4t). \quad \blacksquare$$

EXAMPLE 4.14

Solve the initial value problem

$$y'' + 4y' + 13y = 26e^{-4t}; \qquad y(0) = 5, \qquad y'(0) = -29.$$

Apply the Laplace transform to the differential equation to get

$$\mathscr{L}[y''] + 4\mathscr{L}[y'] + 13\mathscr{L}[y] = 26\,\mathscr{L}[e^{-4t}].$$

Denote $\mathscr{L}[y] = Y$ and apply Theorems 4.6 and 4.7 of the preceding section to get

$$s^2Y(s) - sy(0) - y'(0) + 4[sY(s) - y(0)] + 13Y(s) = \frac{26}{s + 4}.$$

Substitute the initial data into this equation to get

$$s^2Y(s) - 5s + 29 + 4[sY(s) - 5] + 13Y(s) = \frac{26}{s + 4},$$

or

$$[s^2 + 4s + 13]Y(s) - 5s + 9 = \frac{26}{s + 4}.$$

Solve this equation for $Y(s)$ to get

$$Y(s) = \frac{1}{s^2 + 4s + 13}\left[5s - 9 + \frac{26}{s + 4}\right],$$

or

$$Y(s) = \frac{5s^2 + 11s - 10}{(s + 4)(s^2 + 4s + 13)}.$$

Write this fraction in the form

$$\frac{as + b}{s^2 + 4s + 13} + \frac{c}{s + 4}$$

and solve for a, b, and c to obtain

$$Y(s) = \frac{3s - 9}{s^2 + 4s + 13} + \frac{2}{s + 4}.$$

The solution is $y = \mathscr{L}^{-1}[Y]$. Immediately $\mathscr{L}^{-1}[2/(s+4)] = 2e^{-4t}$, but the inverse transform of the other term in $Y(s)$ is not obvious. Complete the square in the denominator to write

$$\frac{3s - 9}{s^2 + 4s + 13} = \frac{3s - 9}{(s + 2)^2 + 9} = 3\frac{(s + 2) - 5}{(s + 2)^2 + 9}$$

$$= 3\frac{s + 2}{(s + 2)^2 + 9} - 5\frac{3}{(s + 2)^2 + 9} = 3F(s + 2) - 5G(s + 2),$$

where

$$F(s) = \frac{s}{s^2 + 9} = \mathscr{L}[\cos(3t)]$$

and

$$G(s) = \frac{3}{s^2 + 9} = \mathscr{L}[\sin(3t)].$$

By the first shifting theorem,

$$y(t) = \mathscr{L}^{-1}[Y] = 2e^{-4t} + 3e^{-2t}\cos(3t) - 5e^{-2t}\sin(3t). \quad \blacksquare$$

The next section is devoted to the second shifting theorem, which will enable us to solve certain initial value problems with discontinuous forcing functions.

PROBLEMS FOR SECTION 4.3

In each of Problems 1 through 6, use the table in conjunction with the first shifting theorem to find the Laplace transform of the function.

1. te^{-2t}

2. $e^{-3t}(t - 2)$

3. $e^{4t}[t - \cos(t)]$

4. $e^{-t}[1 - t^2 + \sin(t)]$

5. $e^{t}[1 - \cosh(t)]$

6. $e^{-5t}[t^4 + 2t - 1]$

In each of Problems 7 through 14, use the inverse version of the first shifting theorem to find the inverse Laplace transform of the function.

7. $\dfrac{1}{s^2 + 4s + 12}$

8. $\dfrac{1}{s^2 - 4s + 5}$

9. $\dfrac{1}{s^2 + 6s + 7}$

10. $\dfrac{s - 4}{s^2 - 8s + 10}$

11. $\dfrac{s + 2}{s^2 + 6s + 1}$

12. $\dfrac{s - 3}{s^2 + 10s + 9}$

13. $\dfrac{2s + 4}{s^2 - 4s + 4}$

14. $\dfrac{s}{s^2 - 14s + 1}$

In each of Problems 15 through 24, use the Laplace transform to solve the initial value problem.

15. $y'' + 6y' + 2y = 1;$ $y(0) = 1,$ $y'(0) = -6$

16. $y'' - 4y' + 3y = 0;$ $y(0) = 2,$ $y'(0) = 4$

17. $y'' + 8y' + 15y = 2;$ $y(0) = 1,$ $y'(0) = -4$

18. $y'' - 2y' + 2y = 0;$ $y(0) = 6,$ $y'(0) = 6$

19. $y'' - 10y' + 26y = 4;$ $y(0) = 3,$ $y'(0) = 15$

20. $y'' + 4y' + 5y = 8;$ $y(0) = -6,$ $y'(0) = 6$

21. $y'' - 6y' + 8y = e^{t};$ $y(0) = 3,$ $y'(0) = 9$

22. $y'' + 6y' + 8y = 4;$ $y(0) = -4,$ $y'(0) = 12$

23. $y'' + 4y' + 3y = t;$ $y(0) = 9,$ $y'(0) = -18$

24. $y'' + 4y = e^{-t}\sin(t);$ $y(0) = 1,$ $y'(0) = 4$

25. Determine $\mathscr{L}\left[e^{-2t}\displaystyle\int_0^t e^{2\tau}\cos(3\tau)\,d\tau\right]$. *Hint:* Use the first shifting theorem and Theorem 4.8 in Section 4.2.

4.4 *The Heaviside Function and the Second Shifting Theorem*

We will now develop machinery particularly suited to solving problems involving functions which have a finite number of jump discontinuities.

Recall that a function has a jump discontinuity at *a* if $\lim_{t \to a+} f(t)$ and $\lim_{t \to a-} f(t)$ both exist and are finite but have different values. The graph of such a function has a "jump" at *a*. Figure 4.3 shows a typical jump discontinuity. The magnitude of the jump discontinuity is the "width of the gap" between the ends of the graph at *a*, or, more carefully,

$$\left| \lim_{t \to a+} f(t) - \lim_{t \to a-} f(t) \right|.$$

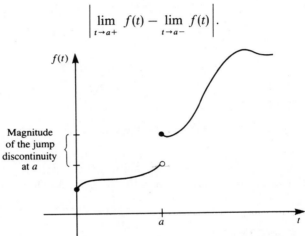

Figure 4.3

In treating functions with jump discontinuities, the following function plays an important role.

DEFINITION

The *Heaviside function*, or *unit step function*, *H*, is defined by

$$H(t) = \begin{cases} 0 & \text{if} \quad t < 0 \\ 1 & \text{if} \quad t \geq 0. \end{cases} \blacksquare$$

Oliver Heaviside was an electrical engineer who did much to introduce Laplace transform methods into electrical engineering; hence the name Heaviside function and its designation *H*. A graph of *H* is shown in Figure 4.4; it has a jump discontinuity of magnitude 1 at the origin.

Because of its on-off nature, the unit step function may be thought of as a switching function. As we will see later, any function having finitely many jump discontinuities can be written as a sum of multiples of *H*. This will be quite useful in applying the Laplace transform to a variety of problems. First, however, we will use the Heaviside function to treat the idea of a shifted function.

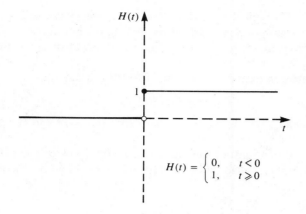

$$H(t) = \begin{cases} 0, & t < 0 \\ 1, & t \geq 0 \end{cases}$$

Figure 4.4

Let a be any number. From the definition of H, we have

$$H(t - a) = \begin{cases} 0 & \text{if} \quad t < a \\ 1 & \text{if} \quad t \geq a. \end{cases}$$

This *shifted Heaviside function* is shown in Figure 4.5. We may think of $H(t - a)$ as representing a signal of magnitude 1 which is initially off and is then switched on and left on at time $t = a$.

The shifted Heaviside function can be used conveniently to express the idea of a shifted (or translated) function in general. Suppose that f is defined for all $t \geq 0$, and let a be a positive number. Consider the new function g defined by

$$g(t) = \begin{cases} 0 & \text{if} \quad 0 \leq t < a \\ f(t - a) & \text{if} \quad\quad t \geq a. \end{cases}$$

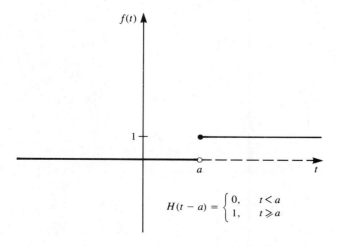

$$H(t - a) = \begin{cases} 0, & t < a \\ 1, & t \geq a \end{cases}$$

Figure 4.5

The graphs of a typical f and the resulting g are shown in Figures 4.6 and 4.7, respectively. The graph of g is obtained by shifting the graph of f a units to the right and assigning zero values to $g(t)$ for $0 \le t < a$. Another way of saying this is that g is the function f delayed a units.

We can write g in terms of f and the Heaviside function as

$$g(t) = f(t - a)H(t - a).$$

This function equals $f(t - a)$ if $t \ge a$ [where $H(t - a) = 1$] and is zero for $0 \le t < a$ [where $H(t - a) = 0$].

The following theorem tells us how to take the Laplace transform of a shifted function. The formula is called "shifting in the t-variable," and the theorem is called the *second shifting theorem*.

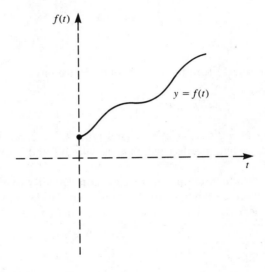

Figure 4.6

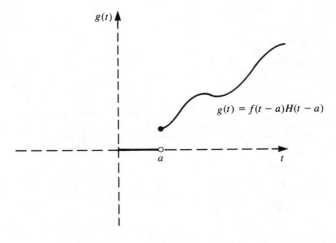

Figure 4.7

THEOREM 4.11   Shifting in the *t*-Variable

Suppose that $\mathcal{L}[f] = F(s)$ for $s > b$. Then

$$\mathcal{L}[f(t-a)H(t-a)] = e^{-as}F(s) \quad \text{for} \quad s > b.$$

That is, the Laplace transform of f shifted a units to the right is the Laplace transform of f multiplied by e^{-as}. We therefore know the Laplace transform of the shifted function if we know the Laplace transform of f.

Proof   Work directly from the definition of the Laplace transform to write

$$\mathcal{L}[f(t-a)H(t-a)] = \int_0^\infty e^{-st}f(t-a)H(t-a)\,dt$$

$$= \int_0^a e^{-st}f(t-a)H(t-a)\,dt + \int_a^\infty e^{-st}f(t-a)H(t-a)\,dt$$

$$= \int_a^\infty e^{-st}f(t-a)\,dt,$$

since $H(t-a) = 0$ if $t < a$, and $H(t-a) = 1$ if $t \geq a$. In the last integral, let $\tau = t - a$ to obtain

$$\mathcal{L}[f(t-a)H(t-a)] = \int_0^\infty e^{-s(\tau+a)}f(\tau)\,d\tau$$

$$= e^{-as}\int_0^\infty e^{-s\tau}f(\tau)\,d\tau = e^{-as}\mathcal{L}[f](s) = e^{-as}F(s). \quad \blacksquare$$

EXAMPLE 4.15

Determine the Laplace transform of the function g defined by

$$g(t) = \begin{cases} 0 & \text{if} \quad 0 \leq t < 6 \\ (t-6)^2 & \text{if} \quad\quad t > 6. \end{cases}$$

Observe that g is just the function $f(t) = t^2$ $(t \geq 0)$ shifted six units to the right. We may write

$$g(t) = f(t-6)H(t-6) = (t-6)^2 H(t-6).$$

By Theorem 4.11,

$$\mathcal{L}[g] = e^{-6s}\mathcal{L}[f](s).$$

Now, $\mathcal{L}[t^2] = 2/s^3$, from Table 4.1 [entry (3)]. Therefore,

$$\mathcal{L}[g] = \frac{2e^{-6s}}{s^3}. \quad \blacksquare$$

EXAMPLE 4.16

Determine $\mathcal{L}[g]$ if

$$g(t) = \begin{cases} 0 & \text{if} \quad 0 \leq t < 2 \\ t^2 + 1 & \text{if} \quad\quad t \geq 2. \end{cases}$$

Write

$$g(t) = (t^2 + 1)H(t - 2).$$

To apply the second shifting theorem, we need something of the form $f(t - 2)H(t - 2)$. We will do some algebraic manipulation to write $t^2 + 1$ in terms of $t - 2$. Write

$$t^2 + 1 = (t - 2 + 2)^2 + 1$$
$$= (t - 2)^2 + 4(t - 2) + 5.$$

The form of this function of $t - 2$ suggests that we let

$$f(t) = t^2 + 4t + 5.$$

Then

$$f(t - 2) = (t - 2)^2 + 4(t - 2) + 5 = t^2 + 1$$

and

$$g(t) = f(t - 2)H(t - 2).$$

Now write

$$\mathscr{L}[g] = e^{-2s}\mathscr{L}[f].$$

But, from the table,

$$\mathscr{L}[f] = \mathscr{L}[t^2 + 4t + 5] = \frac{2}{s^3} + \frac{4}{s^2} + \frac{5}{s}.$$

Therefore,

$$\mathscr{L}[g] = e^{-2s}\left[\frac{2}{s^3} + \frac{4}{s^2} + \frac{5}{s}\right]. \quad \blacksquare$$

Note that, in particular,

$$\mathscr{L}[H(t - a)] = \frac{1}{s}e^{-as},$$

a result we will use often. To derive this result, we have $H(t - a) = f(t - a)H(t - a)$, with $f(t) = 1$. Since $\mathscr{L}[1] = 1/s$, Theorem 4.11 gives us immediately that

$$\mathscr{L}[H(t - a)] = e^{-as}\mathscr{L}[1] = \frac{1}{s}e^{-as}.$$

As usual, a formula for a Laplace transform also generates a formula for an inverse Laplace transform. Recall that, by Theorem 4.11, we have

$$\mathscr{L}[f(t - a)H(t - a)] = e^{-as}F(s).$$

If we apply the inverse Laplace transform $\mathscr{L}^{-1}$ to both sides of this equation, we get

$$f(t - a)H(t - a) = \mathscr{L}^{-1}[e^{-as}F(s)].$$

That is, the inverse Laplace transform of $e^{-as}F(s)$ is the inverse Laplace transform of F shifted a units to the right.

THEOREM 4.12   Inverse Form of the Second Shifting Theorem

If $a > 0$, and $\mathscr{L}^{-1}[F](t) = f(t)$, then

$$\mathscr{L}^{-1}[e^{-as}F(s)](t) = f(t - a)H(t - a). \quad \blacksquare$$

EXAMPLE 4.17

Determine

$$\mathscr{L}^{-1}\left[\frac{se^{-3s}}{s^2 + 4}\right].$$

The presence of the factor e^{-3s} suggests use of the inverse version of the second shifting theorem. We know that the inverse Laplace transform of $s/(s^2 + 4)$ is $\cos(2t)$. Therefore, the inverse Laplace transform of $[s/(s^2 + 4)]e^{-3s}$ is $\cos(2t)$, shifted three units to the right:

$$\mathscr{L}^{-1}\left[\frac{se^{-3s}}{s^2 + 4}\right] = \cos[2(t - 3)]H(t - 3). \quad \blacksquare$$

Once this process is understood, we can proceed very quickly to the inverse Laplace transform of certain functions when the factor e^{-as} is present. Here is a diagram of the process.

$$\overbrace{\mathscr{L}^{-1}[e^{-as}F(s)]}$$

$$e^{-as}F(s) \longrightarrow \mathscr{L}^{-1}[F] = f \longrightarrow f(t - a)H(t - a)$$

With this technique at our disposal, we can solve certain initial value problems involving discontinuous forcing functions.

EXAMPLE 4.18

Solve the initial value problem

$$y'' + 4y = f(t); \qquad y(0) = y'(0) = 0$$

in which

$$f(t) = \begin{cases} 0 & \text{if} \quad 0 \le t < 3 \\ t & \text{if} \qquad t \ge 3. \end{cases}$$

Note that f has a jump discontinuity at 3. First, write f in a form needed to apply the second shifting theorem:

$$f(t) = tH(t - 3) = (t - 3)H(t - 3) + 3H(t - 3).$$

Now apply $\mathscr{L}$ to the differential equation:

$$\mathscr{L}[y''] + 4\mathscr{L}[y] = \mathscr{L}[(t - 3)H(t - 3)] + 3\mathscr{L}[H(t - 3)].$$

Let $\mathscr{L}[y] = Y$ and apply Theorems 4.6 and 4.7 of Section 4.2 to get

$$s^2Y(s) - sy(0) - y'(0) + 4Y(s) = \frac{1}{s^2}e^{-3s} + \frac{3}{s}e^{-3s}.$$

Since $y(0) = y'(0) = 0$, we have

$$(s^2 + 4)Y(s) = \frac{3s + 1}{s^2} e^{-3s}.$$

Then

$$Y(s) = \frac{3s + 1}{s^2(s^2 + 4)} e^{-3s}.$$

The initial value problem has the unique solution

$$y(t) = \mathscr{L}^{-1}\left[\frac{3s + 1}{s^2(s^2 + 4)} e^{-3s}\right].$$

To compute this inverse Laplace transform, notice first the factor e^{-3s}. By Theorem 4.7, we can compute the inverse Laplace transform g of just $(3s + 1)/[s^2(s^2 + 4)]$, and then $y(t) = g(t - 3)H(t - 3)$. Thus, concentrate on finding

$$\mathscr{L}^{-1}\left[\frac{3s + 1}{s^2(s^2 + 4)}\right].$$

Write

$$\frac{3s + 1}{s^2(s^2 + 4)} = \frac{A}{s} + \frac{B}{s^2} + \frac{Cs + D}{s^2 + 4}$$

and solve for A, B, C, and D to obtain

$$\frac{3s + 1}{s^2(s^2 + 4)} = \frac{3}{4}\frac{1}{s} + \frac{1}{4}\frac{1}{s^2} - \frac{3}{4}\frac{s}{s^2 + 4} - \frac{1}{4}\frac{1}{s^2 + 4}.$$

Now read directly from Table 4.1 that

$$g(t) = \tfrac{3}{4}(1) + \tfrac{1}{4}t - \tfrac{3}{4}\cos(2t) - \tfrac{1}{4}\tfrac{1}{2}\sin(2t).$$

The solution of the initial value problem is

$$\begin{aligned}
y(t) &= g(t - 3)H(t - 3) \\
&= \{\tfrac{3}{4} + \tfrac{1}{4}(t - 3) - \tfrac{3}{4}\cos[2(t - 3)] - \tfrac{1}{8}\sin[2(t - 3)]\}H(t - 3) \\
&= \tfrac{1}{8}\{2t - 6\cos[2(t - 3)] - \sin[2(t - 3)]\}H(t - 3).
\end{aligned}$$

Since

$$H(t - 3) = \begin{cases} 0 & \text{if} \quad t < 3 \\ 1 & \text{if} \quad t \geq 3, \end{cases}$$

this solution can be written

$$y(t) = \begin{cases} 0 & \text{if} \quad 0 \leq t < 3 \\ \tfrac{1}{8}\{2t - 6\cos[2(t - 3)] - \sin[2(t - 3)]\} & \text{if} \quad t \geq 3. \end{cases}$$

Figure 4.8 shows a graph of this solution. ∎

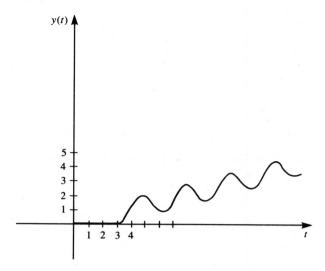

Figure 4.8

It is interesting to notice that, although f is discontinuous at 3 in this example, the solution y is not only continuous everywhere but is also twice differentiable everywhere except at 3, where only the first derivative exists. This type of behavior is typical of initial value problems with a forcing function f having jump discontinuities. If the differential equation has order n, and φ is the solution, then $\varphi, \varphi', \ldots, \varphi^{(n-1)}$ will be continuous, while $\varphi^{(n)}$ will have jump discontinuities where f does, and these jump discontinuities will agree in direction and magnitude with those of f.

In some problems, a forcing function has an arbitrary (but finite) number of jump discontinuities. We claim that any such function can be written as a sum of multiples of Heaviside functions. We will start with some simple examples and work up to more complicated functions.

EXAMPLE 4.19

Let a and b be real numbers with $a < b$. Suppose we want to describe a signal of magnitude 1 that is switched on at $t = a$ and then switched off at $t = b$. We can think of such a signal as a function which is zero up to time a, then has value 1, minus the function which is zero up to time b, then has value 1. This function is called a *pulse function* and can be written

$$p(t) = H(t - a) - H(t - b).$$

If $t < a$, then $p(t) = 0$ because $H(t - a) = H(t - b) = 0$. If $a \le t < b$, then $H(t - a) = 1$ and $H(t - b) = 0$, so $p(t) = 1$. And if $t \ge b$, then $H(t - a) = H(t - b) = 1$, so $p(t) = 0$. That is,

$$p(t) = \begin{cases} 0 & \text{if} \quad t < a \quad \text{or} \quad t \ge b \\ 1 & \text{if} \qquad\qquad a \le t < b. \end{cases}$$

This pulse function is graphed in Figure 4.9.   ■

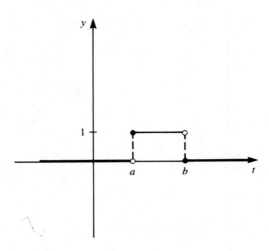

Figure 4.9

EXAMPLE 4.20

Suppose we have a function f defined for all t, and we multiply $f(t)$ by $H(t - a)$, where a is any real number. This forms a new function

$$g(t) = H(t - a)f(t).$$

If $t < a$, then $H(t - a) = 0$, so $g(t) = 0$. If $t \geq a$, then $H(t - a) = 1$, so $g(t) = f(t)$. Thus,

$$y(t) = H(t - a)f(t) = \begin{cases} 0 & \text{if } t < a \\ f(t) & \text{if } t \geq a. \end{cases}$$

(handwritten: (means function is cut to 0 for t < a))

(handwritten left margin:
but
$y(t) = H(t-a) f(t-a)$
$= \begin{cases} 0 & t < a \\ f(t-a) & t \geq a \end{cases}$
means that the function shifted to position t=0
$y(t)$)

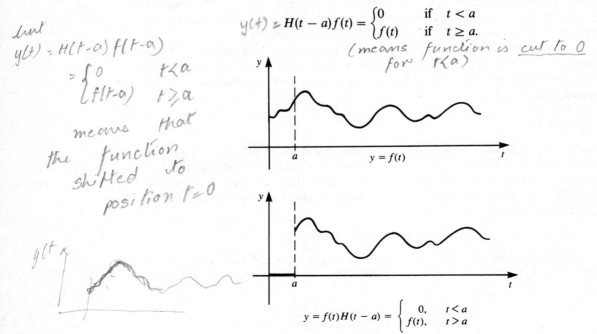

$$y = f(t)H(t - a) = \begin{cases} 0, & t < a \\ f(t), & t > a \end{cases}$$

Figure 4.10

Multiplying $f(t)$ by $H(t - a)$ has the effect of turning $f(t)$ "off" for times $t < a$, then turning $f(t)$ "on" at $t = a$ and leaving it on. The graph of a typical function f, together with that of $f(t)H(t - a)$, is shown in Figure 4.10. ■

EXAMPLE 4.21

Now consider the effect of multiplying a function f by the pulse function of Example 4.19. We have

$$f(t)[H(t - a) - H(t - b)] = \begin{cases} 0 & \text{if} & t < a \\ f(t) & \text{if} & a \leq t < b \\ 0 & \text{if} & t > b. \end{cases}$$

This product remains at the value zero until $t = a$, at which time the function f is "switched on" and left on until time $t = b$, when f is "switched off." A typical f and the product of f with the pulse function are graphed in Figure 4.11. ■

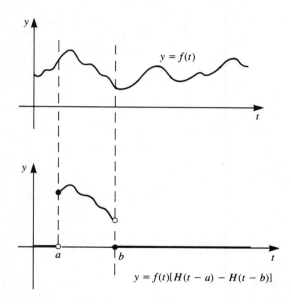

$$y = f(t)[H(t - a) - H(t - b)]$$

Figure 4.11

Here is an example of a function with two jump discontinuities. We will write it as a sum of multiples of Heaviside functions.

EXAMPLE 4.22

Let

$$f(t) = \begin{cases} 0 & \text{if} & t < 2 \\ t - 1 & \text{if} & 2 \leq t < 3 \\ -4 & \text{if} & t \geq 3. \end{cases}$$

A graph of f is shown in Figure 4.12. There are jump discontinuities at 2 (of magnitude 1) and at 3 (of magnitude 6). To write f as a sum of multiples of Heaviside functions, think of f as consisting of two nonzero parts: the part that is $t - 1$ on $[2, 3)$ and the part that is -4 on $[3, \infty)$. Before $t = 2$, $f(t) = 0$. Define two functions, each having just one of these "parts" as its only nonzero values:

$$f_1(t) = \begin{cases} 0 & \text{if} & t < 2 \\ t - 1 & \text{if} & 2 \le t < 3 \\ 0 & \text{if} & t \ge 3 \end{cases}$$

and

$$f_2(t) = \begin{cases} 0 & \text{if} & t < 3 \\ -4 & \text{if} & t \ge 3. \end{cases}$$

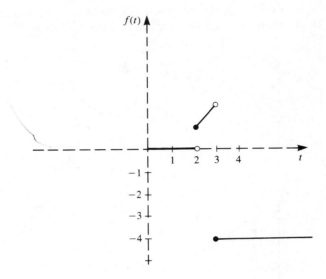

Figure 4.12

The graphs of f_1 and f_2 are shown in Figure 4.13. If the graph of f_1 is placed over the graph of f_2, the sum, or superposition, yields exactly the graph of f:

$$f(t) = f_1(t) + f_2(t).$$

Now, we can write f_1 as $(t - 1)$ times a pulse function having the value 1 on $[2, 3)$ and zero everywhere else:

$$f_1(t) = (t - 1)[H(t - 2) - H(t - 3)].$$

And f_2 is -4 times the Heaviside function shifted to be nonzero for $t \ge 3$:

$$f_2(t) = -4H(t - 3).$$

Therefore,

$$\begin{aligned} f(t) &= f_1(t) + f_2(t) \\ &= (t - 1)[H(t - 2) - H(t - 3)] - 4H(t - 3). \end{aligned} \tag{4.12}$$

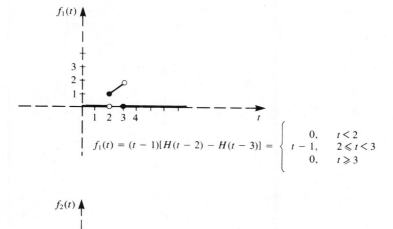

$$f_1(t) = (t - 1)[H(t - 2) - H(t - 3)] = \begin{cases} 0, & t < 2 \\ t - 1, & 2 \leqslant t < 3 \\ 0, & t \geqslant 3 \end{cases}$$

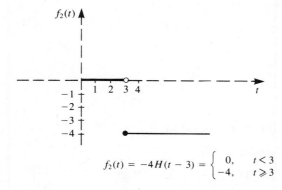

$$f_2(t) = -4H(t - 3) = \begin{cases} 0, & t < 3 \\ -4, & t \geqslant 3 \end{cases}$$

Figure 4.13

The analysis we have carried out indicates steps we can use to write a function with jump discontinuities as a sum of multiples of Heaviside functions. Having arrived at the formulation (4.12), we can now combine terms and write

$$f(t) = (t - 1)H(t - 2) + (-t + 1 - 4)H(t - 3)$$
$$= (t - 1)H(t - 2) - (t + 3)H(t - 3). \quad \blacksquare$$

We will discuss how these ideas apply to the analysis of electrical circuits.

EXAMPLE 4.23

Consider the circuit shown in Figure 4.14. Assume that the capacitor initially has no charge and that there is zero initial current. At time $t = 2$ seconds, the switch is thrown from position B to A, held there for 1 second, and then switched back to position B. We want to determine the output voltage E_{out} (that is, the voltage on the capacitor).

The forcing function E_{in} is

$$E_{\text{in}} = E(t) = \begin{cases} 0 & \text{if } 0 \leq t < 2 \\ 10 & \text{if } 2 \leq t < 3 \\ 0 & \text{if } t \geq 3. \end{cases}$$

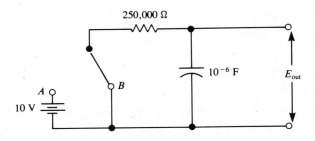

Figure 4.14

E_{in} is a pulse function. In terms of the Heaviside function,

$$E(t) = 10H(t - 2) - 10H(t - 3).$$

Now use Kirchhoff's voltage law to write

$$i(t)R + \frac{q(t)}{C} = E(t),$$

or

$$250,000i(t) + 10^6q(t) = E(t).$$

Since $i = q'$, this equation is equivalent to

$$250,000q'(t) + 10^6q(t) = E(t).$$

We want to solve this differential equation for q, subject to the initial condition

$$q(0) = 0.$$

This is a first order initial value problem for q. Take the Laplace transform of the differential equation to get

$$2.5 \times 10^5[sQ(s) - q(0)] + 10^6Q(s) = \mathscr{L}[E](s),$$

where $Q = \mathscr{L}[q]$. Now,

$$\mathscr{L}[E](s) = 10\mathscr{L}[H(t - 2)](s) - 10\mathscr{L}[H(t - 3)](s)$$

$$= \frac{10}{s}e^{-2s} - \frac{10}{s}e^{-3s}.$$

Since $q(0) = 0$, we have the following equation for $Q(s)$:

$$2.5 \times 10^5sQ(s) + 10^6Q(s) = \frac{10}{s}e^{-2s} - \frac{10}{s}e^{-3s}.$$

Then

$$Q(s) = \frac{4 \times 10^{-5}}{s(s + 4)}e^{-2s} - \frac{4 \times 10^{-5}}{s(s + 4)}e^{-3s}.$$

In order to take the inverse Laplace transform and solve for q, write

$$\frac{1}{s(s + 4)} = \frac{a}{s} + \frac{b}{s + 4}$$

and solve for a and b to obtain

$$\frac{1}{s(s+4)} = \frac{1}{4}\frac{1}{s} - \frac{1}{4}\frac{1}{s+4}.$$

Then

$$Q(s) = 10^{-5}\left[\frac{1}{s}e^{-2s} - \frac{1}{s+4}e^{-2s} - \frac{1}{s}e^{-3s} + \frac{1}{s+4}e^{-3s}\right].$$

Notice that each term on the right is multiplied by e^{-2s} or e^{-3s}, suggesting the use of Theorem 4.12. Since $\mathscr{L}^{-1}[1/s] = 1$,

$$\mathscr{L}^{-1}\left[\frac{e^{-2s}}{s}\right] = 1 \cdot H(t-2) \quad \text{and} \quad \mathscr{L}^{-1}\left[\frac{e^{-3s}}{s}\right] = 1 \cdot H(t-3).$$

Since $\mathscr{L}^{-1}[1/(s+4)] = e^{-4t}$,

$$\mathscr{L}^{-1}\left[\frac{e^{-2s}}{s+4}\right] = e^{-4(t-2)}H(t-2)$$

and

$$\mathscr{L}^{-1}\left[\frac{e^{-3s}}{s+4}\right] = e^{-4(t-3)}H(t-3).$$

Therefore,

$$\begin{aligned}
q(t) &= \mathscr{L}^{-1}[Q] \\
&= 10^{-5}[H(t-2) - e^{-4(t-2)}H(t-2) - H(t-3) + e^{-4(t-3)}H(t-3)] \\
&= 10^{-5}\{[1 - e^{-4(t-2)}]H(t-2) - [1 - e^{-4(t-3)}]H(t-3)\}.
\end{aligned}$$

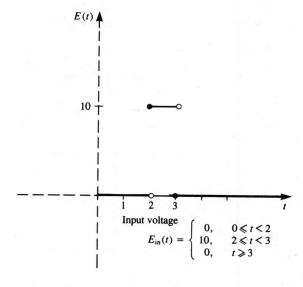

$$E_{in}(t) = \begin{cases} 0, & 0 \leqslant t < 2 \\ 10, & 2 \leqslant t < 3 \\ 0, & t \geqslant 3 \end{cases}$$

Figure 4.15

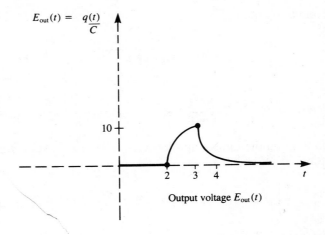

Figure 4.16

Since the output voltage is $E_{out} = 10^6 q$,

$$E_{out}(t) = 10\{[1 - e^{-4(t-2)}]H(t-2) - [1 - e^{-4(t-3)}]H(t-3)\}.$$

The input and output voltages are graphed in Figures 4.15 and 4.16. ∎

EXAMPLE 4.24

Consider the circuit of Figure 4.17, in which we have interchanged the roles of resistor and capacitor and kept the same input configuration as in Example 4.23. We want to know the output voltage $i(t)R$ at any time t.

We have the same differential equation as in Example 4.23, but now we are interested in $i(t)$. Thus, consider the equation

$$2.5 \times 10^5 i(t) + 10^6 q(t) = E(t); \quad i(0) = q(0) = 0.$$

Since we want the solution for $i(t)$, write

$$q(t) = \int_0^t i(\tau)\, d\tau + q(0) = \int_0^t i(\tau)\, d\tau.$$

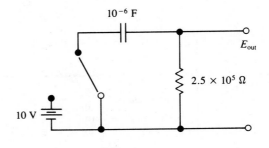

Figure 4.17

The problem to solve for the current is

$$2.5 \times 10^5 i(t) + 10^6 \int_0^t i(\tau)\, d\tau = E(t); \qquad i(0) = 0. \tag{4.13}$$

Recall that, in earlier problems, if we wanted to know the current in a circuit, we could differentiate the differential equation for q and obtain a differential equation for i. This approach will not work here because the electromotive force is not differentiable; hence, we cannot compute $E'(t)$.

Alternatively, we could try to solve the differential equation for q, then differentiate q to obtain i. This approach will not work here either, because the solution of the first order differential equation for q will not be differentiable at points where E has a jump discontinuity.

The Laplace transform and Theorem 4.8 provide us a way around these difficulties. From that theorem, we know that

$$\mathscr{L}\left[\int_0^t i(\tau)\, d\tau\right] = \frac{1}{s}\,\mathscr{L}[i].$$

Using this result, we can take the Laplace transform of equation (4.13) to get

$$2.5 \times 10^5 I(s) + 10^6\,\frac{1}{s}\,I(s) = 10\,\frac{1}{s}\,e^{-2s} - 10\,\frac{1}{s}\,e^{-3s},$$

in which $I = \mathscr{L}[i]$. Solve this equation for $I(s)$ to get

$$I(s) = \frac{4 \times 10^{-5} e^{-2s}}{s+4} - \frac{4 \times 10^{-5} e^{-3s}}{s+4}.$$

Apply the inverse Laplace transform to this expression to obtain the current:

$$i(t) = 4 \times 10^{-5} e^{-4(t-2)} H(t-2) - 4 \times 10^{-5} e^{-4(t-3)} H(t-3).$$

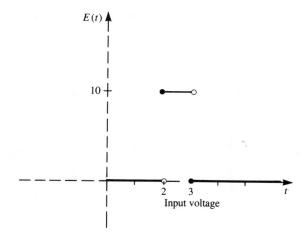

Figure 4.18

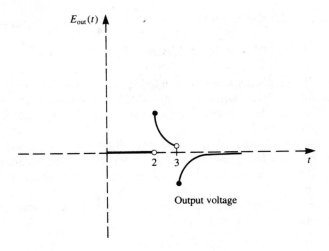

Figure 4.19

Then

$$E_{\text{out}}(t) = i(t)R = 10[e^{-4(t-2)}H(t-2) - e^{-4(t-3)}H(t-3)].$$

The input and output voltages are graphed in Figures 4.18 and 4.19. ■

Finally, observe that in these two examples we had the same equation; hence, we actually could have determined the resistor voltage drop by subtracting the capacitor voltage drop from the input voltage.

PROBLEMS FOR SECTION 4.4

In each of Problems 1 through 14, write the function in terms of the Heaviside function.

1. $f(t) = \begin{cases} 0 & \text{if} \quad 0 \le t < 3 \\ 6 & \text{if} \quad\quad t \ge 3 \end{cases}$

2. $f(t) = \begin{cases} 0 & \text{if} \quad 0 \le t < 2 \\ t & \text{if} \quad\quad t \ge 2 \end{cases}$

3. $f(t) = \begin{cases} 0 & \text{if} \quad 0 \le t < 5 \\ 1 - t & \text{if} \quad\quad t \ge 5 \end{cases}$

4. $f(t) = \begin{cases} 0 & \text{if} \quad 0 \le t < 2 \\ t^2 + t & \text{if} \quad\quad t \ge 2 \end{cases}$

5. $f(t) = \begin{cases} 0 & \text{if} \quad 0 \le t < 6 \\ e^t & \text{if} \quad\quad t \ge 6 \end{cases}$

6. $f(t) = \begin{cases} 0 & \text{if} \quad 0 \le t < 4 \\ t^3 & \text{if} \quad\quad t \ge 4 \end{cases}$

7. $f(t) = \begin{cases} 0 & \text{if} \quad 0 \le t < 1 \\ t & \text{if} \quad 1 \le t < 5 \\ 2 & \text{if} \quad\quad t \ge 5 \end{cases}$

8. $f(t) = \begin{cases} 0 & \text{if} \quad 0 \le t < 4 \\ 1 - t & \text{if} \quad 4 \le t < 8 \\ 1 + t & \text{if} \quad\quad t \ge 8 \end{cases}$

9. $f(t) = \begin{cases} 0 & \text{if} \quad 0 \le t < 3 \\ t + t^2 & \text{if} \quad 3 \le t < 4 \\ 2 + t & \text{if} \quad\quad t \ge 4 \end{cases}$

10. $f(t) = \begin{cases} 0 & \text{if} \quad 0 \le t < 2 \\ 1 - t^2 & \text{if} \quad 2 \le t < \pi \\ \cos(t) & \text{if} \quad\quad t \ge \pi \end{cases}$

11. $f(t) = \begin{cases} 4 & \text{if} \quad 0 \le t < 6 \\ 0 & \text{if} \quad\quad t \ge 6 \end{cases}$

12. $f(t) = \begin{cases} t & \text{if} \quad 0 \le t < 3 \\ 0 & \text{if} \quad\quad t \ge 3 \end{cases}$

13. $f(t) = \begin{cases} e^{-t} & \text{if } 0 \le t < 4 \\ 0 & \text{if } t \ge 4 \end{cases}$

14. $f(t) = \begin{cases} t^2 + 2t - 1 & \text{if } 0 \le t < 3 \\ 0 & \text{if } t \ge 3 \end{cases}$

In each of Problems 15 through 24, determine $\mathscr{L}[f]$.

15. $f(t) = \begin{cases} 0 & \text{if } 0 \le t < 5 \\ 1 + t^2 & \text{if } t \ge 5 \end{cases}$

16. $f(t) = \begin{cases} 0 & \text{if } 0 \le t < 4 \\ -3e^{-2t} & \text{if } t \ge 4 \end{cases}$

17. $f(t) = \begin{cases} 0 & \text{if } 0 \le t < 1 \\ 1 + 5t & \text{if } t \ge 1 \end{cases}$

18. $f(t) = \begin{cases} 0 & \text{if } 0 \le t < 7 \\ t^2 & \text{if } t \ge 7 \end{cases}$

19. $f(t) = \begin{cases} 0 & \text{if } 0 \le t < 3\pi \\ \sin(t) & \text{if } t \ge 3\pi \end{cases}$

20. $f(t) = \begin{cases} 0 & \text{if } 0 \le t < 2 \\ 1 & \text{if } 2 \le t < 5 \\ -1 & \text{if } t \ge 5 \end{cases}$

21. $f(t) = \begin{cases} 0 & \text{if } 0 \le t < 4 \\ t^2 & \text{if } 4 \le t < 5 \\ 2t & \text{if } t \ge 5 \end{cases}$

$f(t-4) = t^2$
and $f(t-5) = 2t \cdot t^2$

22. $f(t) = \begin{cases} 0 & \text{if } 0 \le t < 3 \\ e^{-3t} & \text{if } 3 \le t < 7 \\ 1 + t^2 & \text{if } t \ge 7 \end{cases}$

23. $f(t) = \begin{cases} 0 & \text{if } 0 \le t < 12 \\ 3 & \text{if } 12 \le t < 18 \\ -2 & \text{if } t \ge 18 \end{cases}$

24. $f(t) = \begin{cases} 0 & \text{if } 0 \le t < 4 \\ e^{-3t} & \text{if } 4 \le t < 6 \\ 1 + t & \text{if } t \ge 6 \end{cases}$

In each of Problems 25 through 30, find $\mathscr{L}^{-1}[F]$.

25. $\dfrac{1}{s^3} e^{-5s}$

26. $\dfrac{se^{-2s}}{s^2 + 9}$

27. $\dfrac{3}{s + 2} e^{-5s}$

28. $\dfrac{1}{(s - 5)^3} e^{-s}$

29. $\dfrac{1}{s(s^2 + 16)} e^{-4s}$

30. $\dfrac{se^{-10s}}{(s^2 + 4)^2}$

In each of Problems 31 through 40, use the Laplace transform to solve the initial value problem.

31. $y'' + 4y = f(t);$ $y(0) = 1,$ $y'(0) = 0,$ $f(t) = \begin{cases} 0 & \text{if } 0 \le t < 4 \\ 3 & \text{if } t \ge 4 \end{cases}$

32. $y'' - 2y' - 3y = f(t);$ $y(0) = 1,$ $y'(0) = 0,$ $f(t) = \begin{cases} 0 & \text{if } 0 \le t < 4 \\ 12 & \text{if } t \ge 4 \end{cases}$

33. $y^{(3)} - 8y = f(t);$ $y(0) = y'(0) = y''(0) = 0,$ $f(t) = \begin{cases} 0 & \text{if } 0 \le t < 4 \\ 2 & \text{if } t \ge 4 \end{cases}$

34. $y'' + 5y' + 6y = f(t);$ $y(0) = y'(0) = 0,$ $f(t) = \begin{cases} -2 & \text{if } 0 \le t < 3 \\ 0 & \text{if } t \ge 3 \end{cases}$

35. $y^{(3)} - y'' + 4y' - 4y = f(t);$ $y(0) = y'(0) = 0,$ $y''(0) = 1,$ $f(t) = \begin{cases} 1 & \text{if } 0 \le t < 5 \\ 2 & \text{if } t \ge 5 \end{cases}$

36. $y'' - 4y' + 4y = f(t);$ $y(0) = -2,$ $y'(0) = 1,$ $f(t) = \begin{cases} t & \text{if } 0 \le t < 3 \\ t + 2 & \text{if } t \ge 3 \end{cases}$

37. $y'' + 2y' - 7y = f(t);$ $y(0) = -2,$ $y'(0) = 0,$ $f(t) = \begin{cases} 0 & \text{if } 0 \le t < 5 \\ 2 & \text{if } t \ge 5 \end{cases}$

38. $y'' + 9y = f(t);$ $y(0) = y'(0) = 1,$ $f(t) = \begin{cases} 0 & \text{if } 0 \le t < \pi \\ \cos(t) & \text{if } t \ge \pi \end{cases}$

39. $y'' + 4y' + 4y = f(t);$ $y(0) = 1,$ $y'(0) = 2,$ $f(t) = \begin{cases} 1 & \text{if } 0 \le t < 2 \\ 0 & \text{if } t \ge 2 \end{cases}$

40. $y'' + 5y' + 6y = f(t);$ $y(0) = 0,$ $y'(0) = -4,$ $f(t) = \begin{cases} t^2 & \text{if} \quad 0 \le t < 3 \\ 0 & \text{if} \qquad t \ge 3 \end{cases}$

41. Solve the initial value problem

$$y'' - 3y' + 2y = g(t);\quad y(0) = y'(0) = 0,$$

in which

$$g(t) = \begin{cases} 0 & \text{if} \quad 0 \le t < 3 \\ 2 & \text{if} \qquad t \ge 3. \end{cases}$$

Let φ be the solution. Show that φ and φ' are continuous for $t \ge 0$ but that φ'' does not exist at 3. In fact, use l'Hôpital's rule to show that

$$\lim_{t \to 3+} \frac{\varphi'(t) - \varphi'(3)}{t - 3} = 2 \quad \text{and} \quad \lim_{t \to 3-} \frac{\varphi'(t) - \varphi'(3)}{t - 3} = 0.$$

Thus, conclude that the second derivative of φ has a jump discontinuity at 3 equal in direction and magnitude to the jump discontinuity experienced by g at 3.

42. Solve the initial value problem

$$y' - 3y = g(t);\qquad y(0) = 2,$$

with

$$g(t) = \begin{cases} 0 & \text{if} \quad 0 \le t < 4 \\ 3 & \text{if} \qquad t \ge 4, \end{cases}$$

without using the Laplace transform, as follows. First, find a solution φ of the homogeneous equation defined for $0 \le t < 4$, and use the initial condition to determine the constant of integration. Then find a solution ψ of the nonhomogeneous equation defined for $t \ge 4$. Use the limiting value, as $t \to 4-$, of φ to choose the constant of integration in ψ so that a solution ξ of the initial value problem is given by

$$\xi(t) = \begin{cases} \varphi(t) & \text{if} \quad 0 \le t \le 4 \\ \psi(t) & \text{if} \qquad t \ge 4. \end{cases}$$

43. Calculate and graph the output voltage in the circuit of Example 4.23, under the assumption that at time zero the capacitor is charged to a potential of 5 volts.

44. Calculate and graph the output voltage in the circuit of Example 4.24, under the assumption that at time zero the capacitor is charged to a potential of 5 volts.

45. Solve for the current in the RL circuit shown in the diagram if the current is initially zero and

$$E(t) = \begin{cases} 0 & \text{if} \quad 0 \le t < 5 \\ 2 & \text{if} \qquad t \ge 5. \end{cases}$$

46. Solve for the current in the RL circuit of Problem 45, if the current is initially zero and

$$E(t) = \begin{cases} k & \text{if} \quad 0 \le t < 5 \\ 0 & \text{if} \qquad t \ge 5. \end{cases}$$

47. Solve for the current in the RL circuit of Problem 45 if the initial current is zero and

$$E(t) = \begin{cases} 0 & \text{if} \quad 0 \le t < 4 \\ Ae^{-t} & \text{if} \qquad t \ge 4. \end{cases}$$

In Problems 48, 49, and 50, express the function whose graph is shown, in terms of the Heaviside function, and then find its Laplace transform.

48.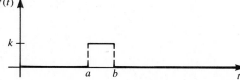

49.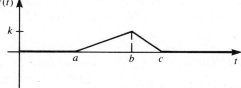

50.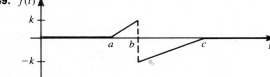

51. Determine $\mathscr{L}[\sin(t)H(t - \pi/6)]$.

52. Solve for the current in the RL circuit shown in the diagram if the initial current is zero,

$$E(t) = \begin{cases} 10 & \text{if} \quad 0 \le t < 2 \\ 0 & \text{if} \quad 2 \le t < 4, \end{cases}$$

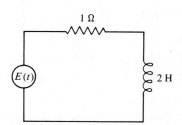

and $E(t + 4) = E(t)$, so that E is periodic of period 4. *Hint:* After setting up the differential equation for the circuit current, use the formula for the Laplace transform of a periodic function (see Problem 18, Section 4.2). Some algebra should yield

$$I(s) = F(s) \frac{1}{1 + e^{-2s}}$$

for an appropriate function F. Use a geometric series to write

$$\frac{1}{1 + e^{-4s}} = 1 - e^{-2s} + e^{-4s} - \cdots.$$

Then

$$i(t) = \sum_{n=0}^{\infty} i_n(t),$$

where

$$i_n(t) = \mathscr{L}^{-1}[(-1)^n e^{-2ns} F(s)] = (-1)^n \mathscr{L}^{-1}[F(s)] \cdot H(t)\Big|_{t \to t - 2n}$$

Graph the current for $0 \le t < 8$.

4.5 *Partial Fractions and Heaviside's Formulas for Inverse Laplace Transforms* ⎯⎯⎯⎯⎯⎯

It is often helpful to write a quotient of polynomials as a sum of "simpler" quotients (note Examples 4.8, 4.9, and 4.18). Although this idea is used in calculus as a technique of integration, we will review it here and then discuss examples of its use in finding inverse Laplace transforms.

Following this, we will discuss Heaviside's formulas, which provide an alternative method for finding inverse Laplace transforms of quotients of polynomials.

THE METHOD OF PARTIAL FRACTIONS

Suppose we have a rational function (quotient of polynomials)

$$F(s) = \frac{P(s)}{Q(s)} = \frac{a_0 + a_1 s + \cdots + a_p s^p}{b_0 + b_1 s + \cdots + b_r s^r}.$$

Assume that

1. All coefficients are real.
2. P and Q have no common factors (divide out any that occur).
3. $r > p$. That is, Q has higher degree than P.

It is a fact of algebra that (at least in theory) we can always write $Q(s)$ as a product in which each factor has the form

$$(s - a)^m$$

(power of a linear factor), with a real, or

$$(s^2 + \alpha s + \beta)^n$$

(power of a quadratic factor), with α and β real.

A power of a quadratic factor is included only if the quadratic cannot be further factored into a product of linear factors *with real coefficients*. Such a quadratic factor is called *irreducible*. For example,

$$s^3 - 3s^2 + 2s - 6 = (s - 3)(s^2 + 2)$$

is a product of a linear factor $s - 3$ and an irreducible quadratic factor $s^2 + 2$, while

$$s^3 - 4s^2 - 3s + 18 = (s - 3)^2(s + 2)$$

is a product of linear factors.

Using these ideas, we can write $P(s)/Q(s)$ as a sum of simpler quotients as follows:

1. To each linear factor $(s - a)^m$ of $Q(s)$, assign a sum of terms

$$\frac{A_1}{s - a} + \frac{A_2}{(s - a)^2} + \cdots + \frac{A_m}{(s - a)^m}.$$

2. To each irreducible quadratic factor $(s^2 + \alpha s + \beta)^n$, assign a sum of terms

$$\frac{B_1 s + C_1}{s^2 + \alpha s + \beta} + \frac{B_2 s + C_2}{(s^2 + \alpha s + \beta)^2} + \cdots + \frac{B_n s + C_n}{(s^2 + \alpha s + \beta)^n}.$$

3. Solve for the constants $A_1, \ldots, B_1, \ldots, C_1, \ldots$.

The entire sum of terms thus obtained is called the *partial fractions decomposition*, or *partial fractions expansion*, of $P(s)/Q(s)$.

EXAMPLE 4.25

Find

$$\mathscr{L}^{-1}\left[\frac{s^2 + 2}{s^4 - 6s^3 + 32s}\right].$$

Write

$$\frac{s^2 + 2}{s^4 - 6s^3 + 32s} = \frac{s^2 + 2}{s(s + 2)(s - 4)^2}$$

$$= \frac{A}{s} + \frac{B}{s + 2} + \frac{C}{s - 4} + \frac{D}{(s - 4)^2}. \tag{4.14}$$

All the factors of $s^4 - 6s^3 + 32s$ are linear (although one is squared). If the fractions on the right of equation (4.14) are added, the numerator must equal $s^2 + 2$, the numerator on the left side of (4.14). Therefore,

$$A(s + 2)(s - 4)^2 + Bs(s - 4)^2 + Cs(s + 2)(s - 4) + Ds(s + 2) = s^2 + 2.$$

Collect terms to write this equation as

$$(A + B + C)s^3 + (-6A - 8B - 2C + D)s^2 + (16B - 8C + 2D)s + 32A = s^2 + 2.$$

Match coefficients of like powers of s on both sides of this equation to get

$$\begin{aligned}
A + B + C &= 0 & \text{(coefficient of } s^3) \\
-6A - 8B - 2C + D &= 1 & \text{(coefficient of } s^2) \\
16B - 8C + 2D &= 0 & \text{(coefficient of } s) \\
32A &= 2 & \text{(constant term).}
\end{aligned}$$

Solve these equations to get

$$A = \tfrac{1}{16}, \quad B = -\tfrac{1}{12}, \quad C = \tfrac{1}{48}, \quad \text{and} \quad D = \tfrac{3}{4}.$$

Therefore,

$$\frac{s^2 + 2}{s^4 - 6s^3 + 32s} = \frac{1}{16}\frac{1}{s} - \frac{1}{12}\frac{1}{s+2} + \frac{1}{48}\frac{1}{s-4} + \frac{3}{4}\frac{1}{(s-4)^2}.$$

The inverse Laplace transform of each term on the right is easily read from Table 4.1, giving us

$$\mathscr{L}^{-1}\left[\frac{s^2 + 2}{s^4 - 6s^3 + 32s}\right] = \frac{1}{16} - \frac{1}{12}e^{-2t} + \frac{1}{48}e^{4t} + \frac{3}{4}te^{4t}. \quad \blacksquare$$

EXAMPLE 4.26

Determine

$$\mathscr{L}^{-1}\left[\frac{s + 10}{s^3 - 3s^2 + 4s - 12}\right].$$

First, write

$$\frac{s + 10}{s^3 - 3s^2 + 4s - 3} = \frac{s + 10}{(s - 3)(s^2 + 4)}$$
$$= \frac{A}{s - 3} + \frac{Bs + C}{s^2 + 4}. \tag{4.15}$$

Add the quotients on the right side of (4.15) and equate the numerator to $s + 10$:

$$A(s^2 + 4) + (s - 3)(Bs + C) = s + 10.$$

Then

$$(A + B)s^2 + (C - 3B)s + (4A - 3C) = s + 10.$$

Then

$$A + B = 0$$
$$C - 3B = 1$$
$$4A - 3C = 10.$$

Solve these equations to get $A = 1$, $B = -1$, $C = -2$. We may now write

$$\frac{s + 10}{s^3 - 3s^2 + 4s - 12} = \frac{1}{s - 3} - \frac{s}{s^2 + 4} - \frac{2}{s^2 + 4}.$$

Again, the inverse Laplace transform of each term on the right is easy to find, and we have

$$\mathscr{L}^{-1}\left[\frac{s + 10}{s^3 - 3s^2 + 4s - 12}\right] = e^{3t} - \cos(2t) - \sin(2t). \quad \blacksquare$$

One limitation of the partial fractions method is that we must factor $Q(s)$, exactly the problem we encountered in solving an nth order constant coefficient linear homogeneous differential equation in Section 3.2. The easily stated problem of finding the roots of a polynomial is a difficult one which pervades all of mathematics!

THE HEAVISIDE FORMULAS

The Heaviside formulas provide an alternative method for calculating the inverse Laplace transform of a rational function. The formulas enable us to write $\mathscr{L}^{-1}[P(s)/Q(s)]$ as a sum of terms, each obtained from a factor of $Q(s)$ by a formula. Thus, to apply the method, we must again be able to factor $Q(s)$.

Here are the formulas for three types of factors commonly encountered.

THEOREM 4.13   Heaviside's Formulas

Let P and Q be polynomials, and let $f = \mathscr{L}^{-1}[P/Q]$.

1. If $Q(s)$ has a factor $s - a$, but not $(s - a)^2$, then $f(t)$ contains the corresponding term $Z(a)e^{at}$, where $Z(s) = P(s)/Q'(s)$.

2. If $k \geq 2$, and $Q(s)$ has a factor $(s - a)^k$, but not $(s - a)^{k+1}$, then $f(t)$ contains the corresponding term

$$\left[\frac{Z^{(k-1)}(a)}{(k-1)!} + \frac{Z^{(k-2)}(a)}{(k-2)!}t + \frac{Z^{(k-3)}(a)}{(k-3)!}\frac{t^2}{2!} + \cdots \right.$$
$$\left. + \frac{Z'(a)}{1!}\frac{t^{k-2}}{(k-2)!} + Z(a)\frac{t^{k-1}}{(k-1)!}\right]e^{at},$$

in which

$$Z(s) = \frac{P(s)(s-a)^k}{Q(s)}.$$

3. If $Q(s)$ contains the irreducible quadratic factor $(s - a)^2 + b^2$, but not the square of this factor, then $f(t)$ contains the term

$$\frac{1}{b}e^{at}[\beta \cos(bt) + \alpha \sin(bt)],$$

in which α and β are defined by the equations

$$Z(s) = \frac{P(s)[(s-a)^2 + b^2]}{Q(s)}$$

and

$$Z(a + ib) = \alpha + i\beta. \quad \blacksquare$$

There are also Heaviside formulas in which $Q(s)$ contains higher powers of irreducible quadratic factors, but these become increasingly unwieldy as the power increases; we will not write these formulas.

We will discover how these formulas are derived after looking at an example.

EXAMPLE 4.27

Find

$$\mathscr{L}^{-1}\left[\frac{3s}{(s+1)(s^2 - 2s + 5)}\right].$$

In the notation we have been using, $P(s) = 3s$, and $Q(s) = (s + 1)(s^2 - 2s + 5)$. $Q(s)$ is a product of a linear factor to the first power and an irreducible quadratic factor, also to the first power.

The term in $\mathscr{L}^{-1}[P(s)/Q(s)]$ corresponding to the linear factor $s + 1$ is obtained from conclusion (1) of the theorem, with $a = -1$. This term is $Z(-1)e^{-t}$, where

$$Z(-1) = \frac{P(-1)}{Q'(-1)} = \frac{3(-1)}{(-1)^2 - 2(-1) + 5} = -\frac{3}{8}.$$

Thus, one term of $\mathscr{L}^{-1}[P(s)/Q(s)]$ is $-\frac{3}{8}e^{-t}$.

For the term in $\mathscr{L}^{-1}[P(s)/Q(s)]$ corresponding to the quadratic factor, let

$$Z(s) = \frac{P(s)(s^2 - 2s + 5)}{Q(s)} = \frac{3s}{s + 1}.$$

Now write

$$s^2 - 2s + 5 = (s - 1)^2 + 4 = (s - a)^2 + b^2.$$

This is in the form used in conclusion (2) of the theorem if we let $a = 1$ and $b = 2$. Now compute

$$Z(a + ib) = Z(1 + 2i) = \frac{3(1 + 2i)}{1 + 2i + 1} = \frac{3 + 6i}{2 + 2i}$$

$$= \frac{3 + 6i}{2 + 2i}\frac{2 - 2i}{2 - 2i} = \frac{18 + 6i}{8} = \frac{9}{4} + \frac{3}{4}i.$$

From (3) of the theorem, the term in $\mathscr{L}^{-1}[P(s)/Q(s)]$ corresponding to the quadratic term in $Q(s)$ is

$$\tfrac{1}{2}e^t[\tfrac{3}{4}\cos(2t) + \tfrac{9}{4}\sin(2t)].$$

We have now accounted for the terms in $\mathscr{L}^{-1}[P(s)/Q(s)]$ corresponding to each factor of $Q(s)$. Therefore,

$$f(t) = -\tfrac{3}{8}e^{-t} + \tfrac{1}{2}e^t[\tfrac{3}{4}\cos(2t) + \tfrac{9}{4}\sin(2t)]. \quad \blacksquare$$

We will conclude this section by deriving conclusion (1) of Theorem 4.13. We will leave (2) and (3) to Problems 41 and 42.

Proof of (1)   Since $s - a$ is a factor of $Q(s)$, but $(s - a)^2$ is not, we can write

$$\frac{P(s)}{Q(s)} = \frac{A}{s - a} + G(s),$$

where A is constant and $G(s)$ has no factor of $s - a$ in either numerator or denominator. Then

$$f(t) = \mathscr{L}^{-1}\left[\frac{P}{Q}\right] = A\mathscr{L}^{-1}\left[\frac{1}{s - a}\right] + \mathscr{L}^{-1}[G]$$

$$= Ae^{at} + \mathscr{L}^{-1}[G].$$

To determine A, recall that $Q(a) = 0$ and write

$$\frac{P(s)(s - a)}{Q(s) - Q(a)} = A + (s - a)G(s). \qquad (4.16)$$

Take the limit of both sides of equation (4.16) as $s \to a$. Since $s - a$ is not a factor of either the numerator or the denominator of $G(s)$,

$$\lim_{s \to a}[A + (s - a)G(s)] = A = \lim_{s \to a} \frac{P(s)(s - a)}{Q(s) - Q(a)}$$

$$= \lim_{s \to a} P(s)\left[\frac{1}{\dfrac{Q(s) - Q(a)}{s - a}}\right] = \frac{P(a)}{Q'(a)}.$$

Therefore,

$$f(t) = \frac{P(a)}{Q'(a)} e^{at} + \mathcal{L}^{-1}[G](t),$$

and $f(t)$ contains the term $[P(a)/Q'(a)] e^{at}$, as stated in conclusion (1) of the theorem. ∎

PROBLEMS FOR SECTION 4.5

In each of Problems 1 through 20, use a partial fractions decomposition and, if necessary, Table 4.1, to find the inverse Laplace transform of the function.

1. $\dfrac{3s + 2}{s^2 + 6s + 8}$

2. $\dfrac{-2s^2 - 12s + 16}{s^3 + 4s^2 + 16s + 64}$

3. $\dfrac{2s^2 + 12s - 46}{(s - 1)^2(s - 5)^2}$

4. $\dfrac{3s^2 - 24s + 27}{(s^2 + 9)(s - 3)^2}$

5. $\dfrac{3s + 1}{s^2 + 4s}$

6. $\dfrac{s^2 - 2s + 3}{s(s^2 - 3s + 2)}$

7. $\dfrac{s^2 + 1}{(s - 1)(s^2 + 2)}$

8. $\dfrac{3s^3 - 15s^2 + s - 81}{(s + 1)(s^2 + 4)^2}$

9. $\dfrac{4s - 5}{s^3 - s^2 - 5s - 3}$

10. $\dfrac{s^2 + 6}{s^3 + 6s^2 + 9s}$

11. $\dfrac{2s - 3}{s^2 + s - 2}$

12. $\dfrac{2s - 4}{s^2 + 3s - 4}$

13. $\dfrac{3s^2 + 2s - 1}{(s^2 + 4)(s^2 - 2s + 1)}$

14. $\dfrac{2s^2 + 3s - 4}{(s - 3)(s^2 + 4)^2}$

15. $\dfrac{s^2 - 2s + 4}{s^4 + 4s^3 - 2s^2 - 12s + 9}$

16. $\dfrac{8s^2 - 3s + 2}{s^3 - 3s^2 - 10s + 24}$

17. $\dfrac{2s + 3}{(s - 5)(s^2 + 4)^2}$

18. $\dfrac{1}{(s^2 + 4)(s + 12)}$

19. $\dfrac{8s^3 - 3s + 2}{s^4 - 3s^3 - 20s^2 + 84s - 80}$

20. $\dfrac{8s^3 - s}{(s^2 + 2)^2(s^2 - 4s + 4)}$

In each of Problems 21 through 30, use the Heaviside formulas to find the inverse Laplace transform of the function.

21. $\dfrac{s - 3}{(s^2 + 4)(s + 7)}$

22. $\dfrac{s^2}{(s + 2)^2(s + 3)}$

23. $\dfrac{3s - 4}{(s - 1)(s + 2)^2}$

24. $\dfrac{s^2 + 1}{(s + 2)^3(s - 1)}$

25. $\dfrac{-3s - 2}{(s + 4)^2}$

26. $\dfrac{8}{(s - 1)^2(s + 3)}$

27. $\dfrac{-s}{(s - 4)^2(s - 5)}$

28. $\dfrac{s^2 + 4s + 1}{(s - 2)^2(s + 3)}$

29. $\dfrac{s^3}{(s + 3)^2(s + 2)^2}$

30. $\dfrac{s + 2}{(s + 4)^2(s - 2)}$

In each of Problems 31 through 40, solve the initial value problem, using the Laplace transform.

31. $y'' + 3y' - 4y = e^{-t}$; $y(0) = y'(0) = 0$

32. $y^{(3)} - 4y'' - 2y' + 8y = 1$; $y(0) = y'(0) = y''(0) = 0$

33. $y'' + 2y' - 3y = e^{-3t}$; $y(0) = y'(0) = 0$

34. $y'' - 4y' + 5y = -2\cos(3t)$; $y(0) = y'(0) = 0$

35. $y'' + 6y' + 8y = 0$; $y(0) = 1$, $y'(0) = 0$

36. $y^{(3)} - 3y'' + 3y' - y = 1 - e^{-t}$; $y(0) = y'(0) = y''(0) = 0$

37. $y^{(3)} + 2y'' - 11y' - 12y = 4$; $y(0) = y'(0) = y''(0) = 0$

38. $y'' + 2y' - 8y = \sin(t)$; $y(0) = y'(0) = 0$

39. $y'' - y' - 6y = \cos(2t)$; $y(0) = y'(0) = 0$

40. $y'' + 5y' + 6y = \cosh(3t)$; $y(0) = 1$, $y'(0) = 0$

41. Derive conclusion (2) of Theorem 4.13. *Hint:* At least in theory, we can write

$$\frac{P(s)}{Q(s)} = \frac{A_1}{s - a} + \frac{A_2}{(s - a)^2} + \cdots + \frac{A_k}{(s - a)^k} + G(s),$$

in which $s - a$ is not a factor of either the numerator or the denominator of $G(s)$. Show that

$$Z(s) = A_1(s - a)^{k-1} + A_2(s - a)^{k-2} + \cdots$$
$$+ A_{k-1}(s - a) + A_k + (s - a)^k G(s).$$

Thus, conclude that $A_k = \lim_{s \to a} Z(s)$.

Now compute $Z'(s)$ and show that $\lim_{s \to a} Z'(s) = A_{k-1}$ and, in general, that

$$A_{k-j} = \frac{1}{j!} \lim_{s \to a} Z^{(j)}(s)$$

for $j = 1, 2, \ldots, k - 1$.

At this point, write $\lim_{s \to a} Z^{(j)}(s) = Z^{(j)}(a)$, assuming continuity of $Z^{(j)}$. Now write $P(s)/Q(s)$ as a sum of terms involving $G(s)$ and constants times powers of $1/(s - a)$, and take the inverse Laplace transform of this sum.

42. Derive conclusion (3) of Theorem 4.13. *Hint:* Write

$$\frac{P(s)}{Q(s)} = \frac{As + B}{(s - a)^2 + b^2} + G(s),$$

with $(s - a)^2 + b^2$ not a factor of either numerator or denominator of $G(s)$. Write

$$Z(s) = As + B + [(s - a)^2 + b^2]G(s).$$

Next, look at $Z(a + ib) = \alpha + i\beta$, and solve for A and B in terms of α, β, a, and b. Finally, substitute these values of A and B into $P(s)/Q(s)$ and take the inverse Laplace transform.

4.6 *The Convolution Theorem*

Usually the Laplace transform of a product is not the product of the Laplace transforms of the individual factors. There is, however, a kind of "product" of f and g, denoted $f * g$, such that $\mathscr{L}[f * g]$ is the product of $\mathscr{L}[f]$ with $\mathscr{L}[g]$.

If f and g are defined on $[0, \infty)$, and if $\int_0^t f(t - \tau)g(\tau)\,d\tau$ is defined for $t \geq 0$, then the convolution of f and g is the function $f * g$ defined by

$$(f * g)(t) = \int_0^t f(t - \tau)g(\tau)\,d\tau$$

for $t \geq 0$. The convolution theorem states that the Laplace transform of the convolution of two functions is the product of the Laplace transforms of the functions.

THEOREM 4.14   The Convolution Theorem

Let $\mathscr{L}[f] = F$ and $\mathscr{L}[g] = G$. Then

$$\mathscr{L}[f * g] = FG.$$

Proof   By definition of $\mathscr{L}$,

$$F(s) = \mathscr{L}[f] = \int_0^\infty e^{-st} f(t)\, dt \quad \text{and} \quad G(s) = \mathscr{L}[g] = \int_0^\infty e^{-st} g(t)\, dt.$$

Then

$$F(s)G(s) = F(s) \int_0^\infty e^{-st} g(t)\, dt = \int_0^\infty e^{-st} F(s) g(t)\, dt.$$

Replace the integration variable t with τ in the last integral to get

$$F(s)G(s) = \int_0^\infty e^{-s\tau} F(s) g(\tau)\, d\tau. \tag{4.17}$$

Now recall that

$$e^{-s\tau} F(s) = \mathscr{L}[f(t - \tau) H(t - \tau)]. \tag{4.18}$$

Substitute the expression on the right of equation (4.18) in place of $e^{-s\tau} F(s)$ in the integral in equation (4.17) to get

$$F(s)G(s) = \int_0^\infty \mathscr{L}[f(t - \tau) H(t - \tau)](s) g(\tau)\, d\tau.$$

In this integral, substitute the integral definition of $\mathscr{L}[f(t - \tau) H(t - \tau)]$ to get

$$F(s)G(s) = \int_0^\infty \left[\int_0^\infty e^{-st} f(t - \tau) H(t - \tau)\, dt \right] g(\tau)\, d\tau$$

$$= \int_0^\infty \int_0^\infty e^{-st} g(\tau) H(t - \tau) f(t - \tau)\, dt\, d\tau.$$

But

$$H(t - \tau) = \begin{cases} 0 & \text{if} \quad 0 \leq t < \tau \\ 1 & \text{if} \qquad t \geq \tau. \end{cases}$$

Therefore,

$$F(s)G(s) = \int_0^\infty \int_\tau^\infty e^{-st} g(\tau) f(t - \tau)\, dt\, d\tau. \tag{4.19}$$

Figure 4.20 shows the $t\tau$-plane. The last integration is over the shaded region, which consists of points (t, τ) with $0 < \tau \leq t < \infty$. Reverse the order of integration in

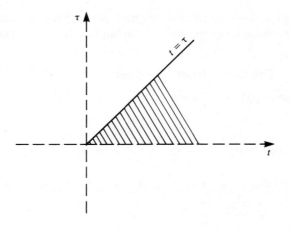

Figure 4.20

equation (4.19) to get

$$F(s)G(s) = \int_0^\infty \int_0^t e^{-st} g(\tau) f(t - \tau)\, d\tau\, dt$$

$$= \int_0^\infty e^{-st} \left[\int_0^t g(\tau) f(t - \tau)\, d\tau \right] dt$$

$$= \int_0^\infty e^{-st} (f * g)(t)\, dt$$

$$= \mathscr{L}[f * g].$$

Thus, $FG = \mathscr{L}[f * g]$, and the proof is complete. ∎

Here is the inverse transform version of the convolution theorem.

THEOREM 4.15   Inverse Version of the Convolution Theorem

Suppose that $\mathscr{L}^{-1}[F] = f$ and $\mathscr{L}^{-1}[G] = g$. Then

$$\mathscr{L}^{-1}[FG] = f * g. \quad ∎$$

This conclusion follows immediately from the convolution theorem.
    Convolution is commutative; that is, $f * g = g * f$.

THEOREM 4.16   Commutativity of the Convolution Operation

If the convolution of f and g is defined, then

$$f * g = g * f.$$

Proof Let $z = t - \tau$ in the integral defining the convolution to get

$$(f * g)(t) = \int_0^t f(t - \tau)g(\tau)\, d\tau = \int_t^0 f(z)g(t - z)(-1)\, dz$$

$$= \int_0^t g(t - z)f(z)\, dz = (g * f)(t). \quad \blacksquare$$

Convolutions are sometimes used to calculate inverse Laplace transforms and to solve initial value problems.

EXAMPLE 4.28

Determine

$$\mathscr{L}^{-1}\left[\frac{1}{s(s - 4)^2}\right].$$

We could use partial fractions or Heaviside's formulas. But we can also use the convolution theorem. Think of the problem as one of finding the inverse Laplace transform of a product. Choose

$$F(s) = \frac{1}{s} \quad \text{and} \quad G(s) = \frac{1}{(s - 4)^2}.$$

We know the inverse Laplace transform of each of these functions:

$$f(t) = \mathscr{L}^{-1}[F] = 1 \quad \text{and} \quad g(t) = \mathscr{L}^{-1}[G] = te^{4t}.$$

Therefore,

$$\mathscr{L}^{-1}[FG](t) = 1 * te^{4t} = \int_0^t \tau e^{4\tau}\, d\tau = \frac{1}{4}e^{4t}\left[t - \frac{1}{4}\right] + \frac{1}{16}. \quad \blacksquare$$

To appreciate the commutativity of convolution, try performing this calculation in the reverse order: $\mathscr{L}^{-1}[FG](t) = te^{4t} * 1$. It can be done, and the answer is the same, but it involves more work.

EXAMPLE 4.29

Consider the mass-spring system of Figure 4.21, with a periodic driving force $F(t) = A\sin(\omega t)$ and no damping. Assuming that the mass is initially at rest in the static equilibrium position, the motion is governed by the initial value problem

$$my'' + ky = A\sin(\omega t); \qquad y(0) = y'(0) = 0.$$

Write the differential equation as

$$y'' + \omega_0^2 y = B\sin(\omega t),$$

where $\omega_0^2 = k/m$ and $B = A/m$. Take the Laplace transform of this differential equation to get

$$s^2 Y(s) + \omega_0^2 Y(s) = \frac{B\omega}{s^2 + \omega^2}.$$

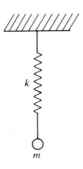

Figure 4.21

Solve this equation for $Y(s)$, obtaining

$$Y(s) = B\omega \left[\frac{1}{s^2 + \omega^2} \right] \left[\frac{1}{s^2 + \omega_0^2} \right].$$

Since $\mathscr{L}^{-1}[1/(s^2 + \omega^2)] = (1/\omega)\sin(\omega t)$ and $\mathscr{L}^{-1}[1/(s^2 + \omega_0^2)] = (1/\omega_0)\sin(\omega_0 t)$,

$$y(t) = B\omega \mathscr{L}^{-1} \left[\frac{1}{s^2 + \omega^2} \frac{1}{s^2 + \omega_0^2} \right]$$

$$= B\omega \left[\frac{1}{\omega} \sin(\omega t) \right] * \left[\frac{1}{\omega_0} \sin(\omega_0 t) \right]$$

$$= \frac{B}{\omega_0} \int_0^t \sin(\omega \tau) \sin[\omega_0(t - \tau)] \, d\tau.$$

To perform this integration, we must consider two cases.

Case 1   If $\omega \neq \omega_0$, then

$$y(t) = \frac{B}{2\omega_0} \left[\frac{-\sin[(\omega - \omega_0)\tau + \omega_0 t]}{\omega - \omega_0} + \frac{\sin[(\omega + \omega_0)\tau - \omega_0 t]}{\omega + \omega_0} \right]_0^t.$$

$$= \frac{B}{2\omega_0} \left[\frac{\sin(\omega_0 t) - \sin(\omega t)}{\omega - \omega_0} + \frac{\sin(\omega_0 t) + \sin(\omega t)}{\omega + \omega_0} \right].$$

This is the solution in the nonresonance case. With no damping, the motion is simply a periodic oscillation.

Case 2   If $\omega = \omega_0$, then

$$y(t) = \frac{B}{\omega_0} \int_0^t \sin(\omega_0 \tau) \sin[\omega_0(t - \tau)] \, d\tau$$

$$= \frac{B}{\omega_0} \left[\frac{1}{4\omega_0} \sin(\omega_0 t) - \frac{1}{2} t \cos(\omega_0 t) \right].$$

This is the resonance case. As t grows larger, so does the amplitude of the oscillation. ■

Sometimes the convolution theorem can be used to solve very generally stated initial value problems.

EXAMPLE 4.30

Solve the initial value problem

$$y'' - 2y' - 8y = f(t); \qquad y(0) = 1, \qquad y'(0) = 0.$$

Take the Laplace transform of the differential equation to get

$$s^2 Y(s) - s - 2[Y(s) - 1] - 8Y(s) = \mathcal{L}[f](s) = F(s).$$

Then

$$Y(s) = \frac{F(s)}{s^2 - 2s - 8} + \frac{s - 2}{s^2 - 2s - 8}.$$

Factor $s^2 - 2s - 8 = (s + 2)(s - 4)$ and use partial fractions to write

$$Y(s) = \frac{1}{6} F(s) \frac{1}{s - 4} - \frac{1}{6} F(s) \frac{1}{s + 2} + \frac{1}{3} \frac{1}{s - 4} + \frac{2}{3} \frac{1}{s + 2}.$$

Now take the inverse Laplace transform of this equation to get

$$y(t) = \tfrac{1}{6} f(t) * e^{4t} - \tfrac{1}{6} f(t) * e^{-2t} + \tfrac{1}{3} e^{4t} + \tfrac{2}{3} e^{-2t}.$$

This expression is a formula for the solution, assuming that $f(t) * e^{4t}$ and $f(t) * e^{-2t}$ both exist. ■

Convolutions have other uses than those we have discussed thus far. One of these is in the solution of certain types of integral equations, in which the function to be determined occurs in an equation involving an integral. We will explore this type of equation in Problems 26 through 35.

PROBLEMS FOR SECTION 4.6

In each of Problems 1 through 15, use Table 4.1 and the convolution theorem to find the inverse Laplace transform of the function. An integral table can be used to help evaluate the convolution integral. Wherever they occur, a and b are positive constants.

1. $\dfrac{1}{(s^2 + 4)(s^2 - 4)}$

2. $\dfrac{e^{-2s}}{s^2 + 4}$

3. $\dfrac{s}{(s^2 + a^2)(s^2 + b^2)}$

4. $\dfrac{1}{(s^2 + a^2)(s^2 + b^2)}$

5. $\dfrac{s^2}{(s - 3)(s^2 + 5)}$

6. $\dfrac{s}{(s^2 + a^2)(s^2 - b^2)}$

7. $\dfrac{s^2}{(s^2 + a^2)(s^2 - b^2)}$

8. $\dfrac{s^2}{(s^2 - a^2)(s^2 - b^2)}$

9. $\dfrac{1}{s(s^2 + a^2)^2}$

10. $\dfrac{1}{s^2(s^2 + 2)}$

11. $\dfrac{1}{(s + 2)(s^2 - 9)}$

12. $\dfrac{1}{s^4(s - 5)}$

13. $\dfrac{-2}{(s^2 - 5)(s - 9)^2}$

14. $\dfrac{e^{-4s}}{s(s + 2)}$

15. $\dfrac{2}{s^3(s^2 + 5)}$

In each of Problems 16 through 25, use the convolution theorem to write a formula for the solution of the initial value problem, as in Example 4.30. Assume that f is piecewise continuous on $[0, \infty)$.

16. $y'' - 5y' + 6y = f(t);$ $y(0) = y'(0) = 0$
17. $y'' + 10y' + 24y = f(t);$ $y(0) = 1,$ $y'(0) = 0$
18. $y^{(3)} - y'' - 4y' + 4y = f(t);$ $y(0) = y'(0) = 1,$ $y''(0) = 0$
19. $y^{(4)} - 11y'' + 18y = f(t);$ $y(0) = y'(0) = y''(0) = y^{(3)}(0) = 0$
20. $y'' - k^2 y = f(t);$ $y(0) = 2,$ $y'(0) = -4$
21. $y'' - 8y' + 12y = f(t);$ $y(0) = -3,$ $y'(0) = 2$
22. $y^{(3)} + 2y'' - 40y' + 64y = f(t);$ $y(0) = 1,$ $y'(0) = 0,$ $y''(0) = -5$
23. $y'' + 10y' + 24y = f(t);$ $y(0) = 0,$ $y'(0) = -2$
24. $y'' - 4y' - 5y = f(t);$ $y(0) = 2,$ $y'(0) = 1$
25. $y^{(3)} - 3y'' + 6y' - 18y = f(t);$ $y(0) = y'(0) = y''(0) = 0$

In each of Problems 26 through 35, solve for f by applying the Laplace transform to both sides of the equation and using the convolution theorem on the term involving the integral.

26. $f(t) = -1 + \int_0^t f(t - \alpha)e^{-3\alpha} \, d\alpha$

27. $f(t) = -t + \int_0^t f(t - \alpha)\sin(\alpha) \, d\alpha$

28. $f(t) = e^{-t} + \int_0^t f(t - \alpha) \, d\alpha$

29. $f(t) = 2t^2 + \int_0^t f(t - \alpha)e^{-\alpha} \, d\alpha$

30. $f(t) = \cos(t) + \int_0^t f(t - \alpha)e^{-2\alpha} \, d\alpha$

31. $f(t) = -1 + t - 2\int_0^t f(t - \alpha)\sin(\alpha) \, d\alpha$

32. $f(t) = e^{-2t} - 3\int_0^t f(t - \alpha)e^{-3\alpha} \, d\alpha$

33. $f(t) = 2t + 1 + \int_0^t f(t - \alpha)e^{-\alpha} \, d\alpha$

34. $f(t) = \sin(2t) - 4\int_0^t (t - \alpha)^2 f(\alpha) \, d\alpha$

35. $f(t) = 3 + \int_0^t \cos[2(t - \alpha)]f(\alpha) \, d\alpha$

36. Theorem 4.8 of Section 4.2 stated that

$$\mathscr{L}\left[\int_0^t f(\tau) \, d\tau\right](s) = \frac{1}{s} F(s),$$

under certain conditions on f. Use the convolution theorem to prove this result.

37. Prove that the convolution operation is distributive:

$$f * (g + h) = f * g + f * h.$$

38. Prove that the convolution operation is associative:

$$f * (g * h) = (f * g) * h.$$

39. Show that $f * 0 = 0 * f = 0$, where 0 denotes the function that is identically zero on $[0, \infty)$.
40. Show by an example that in general $f * 1 \neq f$, where 1 denotes the function whose value is 1 for $t \geq 0$. *Hint:* Try $f(t) = \cos(t)$.
41. Show by an example that $f * f$ may not be nonnegative. *Hint:* Let $f(t) = \sin(t)$ for $\pi/2 \leq t \leq \pi$, and let $f(t) = 0$ for $0 \leq t < \pi/2$ and $\pi < t$.
42. Use the convolution theorem to determine

$$\mathscr{L}\left[e^{-2t}\int_0^t e^{2\tau}\cos(3\tau) \, d\tau\right].$$

43. Use the convolution theorem to show that

$$\mathscr{L}^{-1}\left[\frac{1}{s^2}F(s)\right] = \int_0^t \int_0^\tau f(\alpha)\, d\alpha\, d\tau,$$

in which $f = \mathscr{L}^{-1}[F]$.

44. Derive a formula for $\mathscr{L}^{-1}[F(s)/(s^2 + a^2)]$ in terms of f, where $f = \mathscr{L}^{-1}[F]$.

45. Derive a formula for $\mathscr{L}^{-1}[F(s)/(s^2 - a^2)]$ in terms of f, where $f = \mathscr{L}^{-1}[F]$.

46. Suppose that u is the solution of the initial value problem

$$au'' + bu' + cu = 0; \qquad u(0) = 0, \qquad u'(0) = \frac{1}{a}.$$

Show that $y = f * u$ is the solution of the initial value problem

$$ay'' + by' + cy = f(t); \qquad y(0) = y'(0) = 0.$$

47. Use the result of Problem 46 to solve each of the following initial value problems.
   (a) $2y'' + 4y' + y = 1 + t^2;$ $y(0) = y'(0) = 0$
   (b) $y'' - 6y' + 2y = 4;$ $y(0) = y'(0) = 0$
   (c) $y'' - 3y' + y = e^{-t};$ $y(0) = y'(0) = 0$

4.7 The Dirac Delta Function and Differential Equations with Polynomial Coefficients _____

In this section, we will discuss two additional topics involving the Laplace transform.

UNIT IMPULSES AND THE DIRAC DELTA FUNCTION

Many problems in physics and engineering involve the concept of an impulse. We will express this concept mathematically as follows. First, for any positive number ϵ, define the function δ_ϵ by letting

$$\delta_\epsilon(t) = \begin{cases} \dfrac{1}{\epsilon} & \text{if } 0 \le t < \epsilon \\[2mm] 0 & \text{if } t < 0 \quad \text{or} \quad t \ge \epsilon. \end{cases}$$

Figure 4.22 shows the graph of the function $\delta_\epsilon(t - a)$, which is δ_ϵ shifted a units to the right. We have

$$\delta_\epsilon(t - a) = \begin{cases} \dfrac{1}{\epsilon} & \text{if } 0 \le t - a < \epsilon \quad (\text{or } a \le t < a + \epsilon) \\[2mm] 0 & \text{if } t < a \quad \text{or} \quad t \ge a + \epsilon. \end{cases}$$

In terms of the Heaviside function,

$$\delta_\epsilon(t) = \frac{1}{\epsilon}[H(t) - H(t - \epsilon)].$$

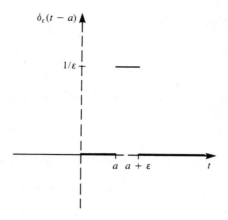

Figure 4.22

Thus,

$$\delta_\epsilon(t - a) = \frac{1}{\epsilon}[H(t - a) - H(t - a - \epsilon)].$$

Immediately, then,

$$\mathscr{L}[\delta_\epsilon(t - a)] = \frac{1}{\epsilon}\left[\frac{1}{s}e^{-as} - \frac{1}{s}e^{-(a+\epsilon)s}\right]$$

$$= \frac{e^{-as}(1 - e^{-\epsilon s})}{\epsilon s}. \tag{4.20}$$

Now define

$$\delta(t) = \lim_{\epsilon \to 0+} \delta_\epsilon(t).$$

For historical reasons, δ is called the *Dirac delta function*. Strictly speaking, δ is not a function but is an object called a *distribution*. However, it is traditional to continue to use the word function in this context. We may think of $\delta(t - a)$ as being zero for $t \neq a$ and infinity at $t = a$. Envision the graph of Figure 4.22 as ϵ is chosen smaller through positive values. The "height" of the graph, $1/\epsilon$, goes to infinity, while the interval $[a, a + \epsilon)$ over which the function is nonzero shrinks to the single point a. The result is a graph which is identically zero except at a, where it has an "infinite" spike or impulse. Such a "spike" function is useful in representing an instantaneous impulse at time $t = a$.

The Dirac delta function does not satisfy the hypotheses we have used in stating theorems about the Laplace transform. However, we can define $\mathscr{L}[\delta]$ as follows. Since $\delta(t - a) = \lim_{\epsilon \to 0+} \delta_\epsilon(t - a)$, it is natural to define

$$\mathscr{L}[\delta(t - a)] = \lim_{\epsilon \to 0+} \mathscr{L}[\delta_\epsilon(t - a)].$$

Use l'Hôpital's rule in equation (4.20) to calculate this limit. Notice that the limit is as ϵ approaches zero; hence, the derivatives are taken with respect to ϵ. We get

$$\lim_{\epsilon \to 0+} \mathcal{L}[\delta_\epsilon(t-a)] = \lim_{\epsilon \to 0+} \frac{e^{-as}(1-e^{-\epsilon s})}{\epsilon s}$$

$$= \lim_{\epsilon \to 0+} \frac{e^{-as}[se^{-\epsilon s}]}{s} = e^{-as}.$$

This leads us to define the Laplace transform of the shifted Dirac delta function $\delta(t-a)$ as

$$\mathcal{L}[\delta(t-a)] = e^{-as}.$$

In particular, by choosing $a = 0$, we get the Laplace transform of the Dirac delta function

$$\mathcal{L}[\delta(t)] = 1.$$

The following result is known as the *filtering property* of the Dirac delta function. It states that, if we hit a signal (function) with an impulse at time a by multiplying it by $\delta(t-a)$ and "sum" the resulting signal over $t \geq 0$ by integrating from zero to infinity, we obtain exactly $f(a)$, the value of the signal at a.

THEOREM 4.17 Filtering Property of the Dirac Function

Let $a > 0$. Suppose that f is integrable on $[0, \infty)$ and continuous at a. Then

$$\int_0^\infty f(t)\, \delta(t-a)\, dt = f(a).$$

Proof   Begin by calculating

$$\int_0^\infty f(t)\, \delta_\epsilon(t-a)\, dt = \frac{1}{\epsilon} \int_a^{a+\epsilon} f(t)\, dt.$$

By the mean value theorem for integrals, there is some t_0 between a and $a + \epsilon$ such that

$$\int_a^{a+\epsilon} f(t)\, dt = \epsilon f(t_0).$$

Therefore,

$$\int_0^\infty f(t)\, \delta_\epsilon(t-a)\, dt = f(t_0).$$

Now take the limit of both sides of this equation as $\epsilon \to 0+$. Since t_0 is between a and $a + \epsilon$, $t_0 \to a$. Since f is continuous at a, $f(t_0) \to f(a)$. Further, $\delta_\epsilon(t-a) \to \delta(t-a)$, and we have

$$\int_0^\infty f(t)\, \delta(t-a)\, dt = f(a),$$

as we wanted to show.   ∎

If we apply the filtering property to $f(t) = e^{-st}$, we get

$$\int_0^\infty e^{-st}\,\delta(t - a)\,dt = e^{-as},$$

consistent with the definition of the Laplace transform of $\delta(t - a)$ as e^{-as}.

Notice also that, if we change notation in the filtering property and write

$$\int_0^\infty f(\tau)\,\delta(\tau - t)\,d\tau = f(t),$$

we can recognize the convolution formula

$$f * \delta = f.$$

Thus, the Dirac delta function is an "identity" for the convolution operation in the sense that the convolution of f with δ yields f again.

Here is an example of an initial value problem involving the Dirac delta function.

EXAMPLE 4.31

Solve the initial value problem

$$y'' + 2y' + 2y = \delta(t - 3); \qquad y(0) = y'(0) = 0.$$

Upon applying the Laplace transform to the differential equation, we get

$$(s^2 + 2s + 2)Y(s) = e^{-3s}.$$

Then

$$Y(s) = \frac{e^{-3s}}{s^2 + 2s + 2} = \frac{1}{(s + 1)^2 + 1}\, e^{-3s}.$$

To determine $\mathscr{L}^{-1}[Y]$, we will use both shifting theorems. First, recall that

$$\mathscr{L}^{-1}\left[\frac{1}{s^2 + 1}\right] = \sin(t).$$

Therefore,

$$\mathscr{L}^{-1}\left[\frac{1}{(s + 1)^2 + 1}\right] = e^{-t}\sin(t)$$

by shifting in the s-variable. Then

$$\mathscr{L}^{-1}\left[\frac{1}{(s + 1)^2 + 1}\, e^{-3s}\right] = H(t - 3)e^{-(t - 3)}\sin(t - 3)$$

by shifting in the t-variable. The solution of the initial value problem is

$$y(t) = H(t - 3)e^{-(t - 3)}\sin(t - 3).$$

This may be written

$$y(t) = \begin{cases} 0 & \text{if} \quad 0 \le t < 3 \\ e^{-(t - 3)}\sin(t - 3) & \text{if} \qquad t \ge 3. \end{cases}$$

Notice that this solution is continuous and differentiable for $t \geq 0$ but that y' has a jump discontinuity of magnitude 1 at $t = 3$. The magnitude of the jump is the coefficient of $\delta(t - 3)$ in the statement of the problem.   ■

The Dirac delta function is often used to study the behavior of circuits which are subjected to transients. Transients are generated during switching, and the high input voltages which are often associated with them can create excessive current in the components, damaging the circuit.

There is another way that transients can damage a circuit. Recall the resonance phenomenon we saw when a circuit or mechanical system is exposed to forced oscillations whose frequency is near the natural frequency of the system. If the driving force is sufficiently large, or if the system contains insufficient damping, the resultant response may include oscillations large enough to damage the system. Transients contain a wide spectrum of frequencies, and the introduction of a transient into a circuit is equivalent to forcing the circuit with many different frequencies, one of which may be the natural frequency of the circuit.

Before a circuit is constructed, engineers sometimes use the Dirac delta function to model a transient and study its effect on the system.

EXAMPLE 4.32

Consider the circuit of Figure 4.23. Assume that the circuit current and capacitor charge are both zero at time zero. We want to determine the output voltage response to a transient of unit magnitude at the input.

The output voltage is $q(t)/C$, so we will try to determine q. Using Kirchhoff's voltage law, we have

$$Li' + Ri + \frac{q}{C} = i' + 10i + 100q = q'' + 10q' + 100q = \delta(t).$$

We have assumed that $i(0-) = q(0-) = 0$. Since the current through an inductor is continuous, we have $i(0+) = i(0-) = 0$. Since the charge on a capacitor is also continuous, we have $q(0+) = q(0-) = 0$. Therefore, the initial value problem for the charge is

$$q'' + 10q' + 100q = \delta(t); \qquad q(0+) = q'(0+) = 0.$$

Apply the Laplace transform to the differential equation to get

$$s^2 Q(s) - sq(0+) - q'(0+) + 10sQ(s) - 10q(0+) + 100Q(s) = \mathscr{L}[\delta(t)].$$

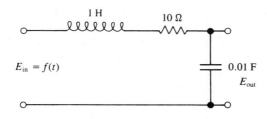

Figure 4.23

Recall that $\mathscr{L}[\delta(t)] = 1$ and insert the initial data into the equation for $Q(s)$ to get

$$(s^2 + 10s + 100)Q(s) = 1.$$

Then

$$Q(s) = \frac{1}{s^2 + 10s + 100} = \frac{1}{(s + 5)^2 + 75}.$$

Since

$$\mathscr{L}^{-1}\left[\frac{1}{s^2 + 75}\right] = \frac{1}{5\sqrt{3}}\sin(5\sqrt{3}\,t),$$

we have

$$q(t) = \mathscr{L}^{-1}\left[\frac{1}{(s + 5)^2 + 75}\right] = \frac{1}{5\sqrt{3}}e^{-5t}\sin(5\sqrt{3}\,t).$$

The output voltage is therefore

$$\frac{1}{C}q(t) = 100q(t) = \frac{20}{\sqrt{3}}e^{-5t}\sin(5\sqrt{3}\,t).$$

We can now obtain the current $i(t)$ directly from the fact that $i = q'$. A graph of this output is given in Figure 4.24. The circuit output displays damped oscillations at its natural frequency, even though it was not forced by oscillations of this frequency.

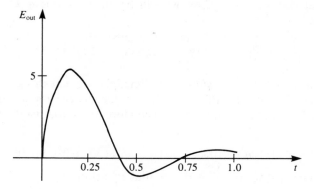

Figure 4.24   ∎

Notice that, in Example 4.32, the current changed instantly from 0 to 1 ampere. What we are actually observing here is the mathematical result of driving the circuit with a Dirac delta function. This is a purely mathematical concept which is useful in modeling but does not reflect reality with complete accuracy. In reality, we will never see an instantaneous change in the inductor current because we will never see a potential equivalent to a Dirac delta function. Nevertheless, the idea of modeling transients with this function has some merit. If a very large pulse of short duration should hit the circuit, the circuit current and output voltage would be very near those calculated in this example.

INITIAL VALUE PROBLEMS WITH POLYNOMIAL COEFFICIENTS

Sometimes the Laplace transform is effective in solving initial value problems in which the differential equation has polynomial coefficients. The key lies in the following theorem.

THEOREM 4.18

If $\mathscr{L}[f] = F$ is defined for $s > b$, then

$$\mathscr{L}[tf(t)](s) = -F'(s) \quad \text{for} \quad s > b.$$

That is, the Laplace transform of $tf(t)$ is the negative of the derivative of the Laplace transform of f.

Proof   Recall that $F(s) = \int_0^\infty e^{-st}f(t)\, dt$. Use Leibniz's rule to differentiate under the integral sign and obtain

$$F'(s) = \int_0^\infty \frac{\partial}{\partial s}[e^{-st}f(t)]\, dt = \int_0^\infty -e^{-st}[tf(t)]\, dt = -\mathscr{L}[tf(t)];$$

this is equivalent to what we want to show.   ∎

By successively applying Theorem 4.18, we obtain the following result.

COROLLARY

If $\mathscr{L}[f] = F$ for $s > b$, and n is any positive integer, then

$$\mathscr{L}[t^n f(t)] = (-1)^n F^{(n)}(s) \quad \text{for} \quad s > b. \quad ∎$$

EXAMPLE 4.33

Solve the initial value problem

$$y'' + 2ty' - 4y = 1; \qquad y(0) = y'(0) = 0.$$

Apply the Laplace transform to the differential equation to get

$$s^2 Y(s) - sy(0) - y'(0) + 2\mathscr{L}[ty'(t)](s) - 4Y(s) = \mathscr{L}[1](s) = \frac{1}{s}.$$

Substitute the values of $y(0)$ and $y'(0)$ to get

$$s^2 Y(s) + 2\mathscr{L}[ty'(t)](s) - 4Y(s) = \frac{1}{s}.$$

Use Theorem 4.18 to compute

$$\mathscr{L}[ty'(t)](s) = -\frac{d}{ds}\mathscr{L}[y'(t)](s) = -\frac{d}{ds}[sY(s) - y(0)]$$

$$= -Y(s) - sY'(s).$$

The Laplace transform of the differential equation is therefore

$$s^2 Y(s) - 2Y(s) - 2s Y'(s) - 4Y(s) = \frac{1}{s}.$$

Then

$$(s^2 - 6)Y - 2s Y' = \frac{1}{s}.$$

    Applying $\mathscr{L}$ to the differential equation having polynomial coefficients has led to a differential equation for Y instead of the algebraic equation we encounter in the constant coefficient case. Divide this differential equation for Y by $-2s$ to get

$$Y' + \left(\frac{3}{s} - \frac{s}{2}\right) Y = \frac{-1}{2s^2}. \tag{4.21}$$

This is a linear first order differential equation for Y. An integrating factor is

$$e^{\int (3/s - s/2)\, ds} = e^{3 \ln(s) - s^2/4} = s^3 e^{-s^2/4}.$$

Multiply equation (4.21) by $s^3 e^{-s^2/4}$ to get

$$s^3 e^{-s^2/4} Y' + \left(\frac{3}{s} - \frac{s}{2}\right) s^3 e^{-s^2/4} Y = -\frac{1}{2} s e^{-s^2/4},$$

or

$$s^3 e^{-s^2/4} Y' + (3s^2 - \tfrac{1}{2}s^4) e^{-s^2/4} Y = -\tfrac{1}{2} s e^{-s^2/4}.$$

Write this equation as

$$[s^3 e^{-s^2/4} Y]' = -\tfrac{1}{2} s e^{-s^2/4}.$$

Integrate both sides of this equation to get

$$s^3 e^{-s^2/4} Y = \int -\frac{1}{2} s e^{-s^2/4}\, ds = e^{-s^2/4} + C.$$

The general solution for Y is

$$Y(s) = \frac{1}{s^3} + \frac{C}{s^3} e^{s^2/4}.$$

But we must have $\lim_{s \to \infty} Y(s) = 0$ by Theorem 4.5 of Section 4.1. For this to occur, we must have $C = 0$. Therefore,

$$Y(s) = \frac{1}{s^3}$$

and

$$y(t) = \mathscr{L}^{-1}\left[\frac{1}{s^3}\right] = \frac{1}{2} t^2.$$

This solution can be checked by substitution into the initial value problem. ∎

EXAMPLE 4.34

Solve the problem

$$ty'' + (4t - 2)y' - 4y = 0; \qquad y(0) = 1.$$

Our existence/uniqueness theorem does not apply to this initial value problem because the coefficient t of y'' is zero at the point where the initial data is given. Nevertheless, we will attempt to find a solution.

Apply the Laplace transform to the differential equation to get

$$\mathcal{L}[ty''] + 4\mathcal{L}[ty'] - 2\mathcal{L}[y'] - 4\mathcal{L}[y] = 0.$$

Calculate these terms individually:

$$\mathcal{L}[ty''] = -\frac{d}{ds}\,\mathcal{L}[y''] = -\frac{d}{ds}[s^2Y - sy(0) - y'(0)]$$

$$= -2sY - s^2Y' + 1 \qquad [\text{since } y(0) = 1];$$

$$\mathcal{L}[ty'] = -\frac{d}{ds}\,\mathcal{L}[y'] = -\frac{d}{ds}[sY - y(0)] = -sY' - Y;$$

and

$$\mathcal{L}[y'] = sY - y(0) = sY - 1.$$

The Laplace transform of the differential equation is

$$-2sY - s^2Y' + 1 - 4sY' - 4Y - 2sY + 2 - 4Y = 0.$$

Collect terms to get

$$(s^2 + 4s)Y' + (4s + 8)Y = 3,$$

or

$$Y' + \frac{4s + 8}{s(s + 4)}\,Y = \frac{3}{s(s + 4)}.$$

This is a linear first order differential equation for Y. Omitting the routine but tedious details, we obtain the general solution

$$Y(s) = \frac{s}{(s + 4)^2} + \frac{6}{(s + 4)^2} + \frac{C}{s^2(s + 4)^2}$$

$$= \frac{1}{s + 4} + 2\frac{1}{(s + 4)^2} + C\left[-\frac{1}{32}\frac{1}{s} + \frac{1}{16}\frac{1}{s^2} + \frac{1}{32}\frac{1}{s + 4} + \frac{1}{16}\frac{1}{(s + 4)^2}\right].$$

Upon applying the inverse Laplace transform to Y, we obtain the solution

$$y(t) = e^{-4t} + 2te^{-4t} + K[-1 + 2t + e^{-4t} + 2te^{-4t}],$$

in which $K = C/32$. It can be verified by substitution that this is a solution of the initial value problem for any value of K. The solution is not unique; nor did we have any right to expect it to be.   ∎

We will conclude with a result similar to Theorem 4.18 except that $f(t)$ is divided by t instead of multiplied by t.

THEOREM 4.19

Let $\mathscr{L}[f] = F$ for $s > b$, and suppose that $\lim_{t \to 0+}(1/t)f(t)$ exists and is finite. Then

$$\mathscr{L}\left[\frac{1}{t}f(t)\right] = \int_s^\infty F(\sigma)\, d\sigma \quad \text{for} \quad s > b. \quad \blacksquare$$

A proof of this theorem is requested in Problem 18.

PROBLEMS FOR SECTION 4.7

In each of Problems 1 through 10, solve the problem using the Laplace transform.

1. $t^2 y'' - 2y = 2$

2. $y'' + 4ty' - 4y = 0; \quad y(0) = 0, \quad y'(0) = -7$

3. $y'' + 2ty' - 4y = 6; \quad y(0) = y'(0) = 0$

4. $y'' - 16ty' + 32y = 14; \quad y(0) = y'(0) = 0$

5. $y'' - 8ty' + 16y = 3; \quad y(0) = 0, \quad y'(0) = 0$

6. $y'' + 8ty' = 0; \quad y(0) = 4, \quad y'(0) = 0$

7. $y'' - 4ty' + 4y = 0; \quad y(0) = 0, \quad y'(0) = 10$

8. $y'' + 8ty' - 8y = 0; \quad y(0) = 0, \quad y'(0) = -4$

9. $ty'' + (t - 1)y' + y = 0; \quad y(0) = 0$

10. $t(1 - t)y'' + 2y' + 2y = 6t; \quad y(0) = 0, \quad y(2) = 0$

11. Use the Laplace transform to solve *Laguerre's equation*,

$$ty'' + (1 - t)y' + ny = 0,$$

in which n is any positive integer.

12. Use the Laplace transform to solve Bessel's equation of order zero,

$$ty'' + y' + ty = 0,$$

subject to the condition that $y(0) = 1$. *Hint:* Obtain a differential equation for Y which has general solution $Y(s) = C/\sqrt{1 + s^2}$. Write this as

$$Y(s) = \frac{C}{s\sqrt{1 + \dfrac{1}{s^2}}}.$$

Assume that $s > 1$ and use a binomial expansion to express $Y(s)$ as an infinite series, then take the inverse Laplace transform term by term to obtain an infinite series solution. This results in a Bessel function of the first kind of order zero.

13. Use the Laplace transform to solve Bessel's equation of order n,

$$t^2 y'' + ty' + (t^2 - n^2)y = 0,$$

in which n is any positive integer. *Hint:* Remember the condition that

$$\lim_{s \to \infty} Y(s) = 0.$$

The solution will be an infinite series (a Bessel function of order n).

14. Use the Laplace transform to solve

$$ty'' + (4t - 2)y' - 4y = 0; \qquad y(0) = 1.$$

15. Use the Laplace transform to solve the Cauchy-Euler equation

$$x^2 y'' + Axy' + By = 0.$$

16. Use the Laplace transform to solve the problem

$$y^{(4)}(x) = \frac{M}{EI} \delta(x - a); \qquad y(0) = 0, \qquad y'(0) = B_0, \qquad y^{(3)}(0) = F_0, \qquad y(L) = 0.$$

This set of equations models the deflection of a beam of length L, horizontally restrained at both ends, with a load M at $x = a$ and with the weight of the beam neglected. E is Young's modulus, and I is the moment of inertia of the cross-section at x with respect to a horizontal line through the centroid of the beam.

17. Use the Laplace transform to solve

$$y^{(4)}(x) = \frac{M}{EI} \delta(x - a); \qquad y(0) = y''(0) = y(L) = 0; \qquad y^{(3)}(0) = F_0.$$

F_0 measures the shearing force of the beam at the left end, $x = 0$.

18. Prove Theorem 4.19. *Hint:* Define $g(t) = (1/t)f(t)$ and use Theorem 4.18 to show that $G'(s) = -F(s)$ for $s > b$. Thus, $G(s) = -\int_a^s F(\sigma) \, d\sigma + C$ for some $a \geq b$. Now use the fact that $\lim_{s \to \infty} G(s) = 0$ to determine C.

19. Determine $\mathscr{L}[(1/t)\sin(t)]$.

20. Let $F = \mathscr{L}[f]$. Prove that $\mathscr{L}[t^n f(t)] = (-1)^n F^{(n)}(s)$ for any positive integer n.

21. Determine $\mathscr{L}[te^{-3t}\cos(4t)]$.

22. Determine $\mathscr{L}[t^3 e^{2t}]$ by using the result of Problem 20. Also find this Laplace transform by using the first shifting theorem, and compare the results.

In each of Problems 23 through 26, solve the initial value problem.

23. $y'' + 5y' + 6y = 3\delta(t - 2) - 4\delta(t - 5); \quad y(0) = y'(0) = 0$

24. $y'' - 4y' + 13y = 4\delta(t - 3); \quad y(0) = y'(0) = 0$

25. $y^{(3)} + 4y'' + 5y' + 2y = 6\delta(t); \quad y(0) = y'(0) = y''(0) = 0$

26. $y'' + 16y = 12\delta(t - 5\pi/8); \quad y(0) = 3, \quad y'(0) = 0.$ In this problem, simplify the solution as much as possible.

27. Suppose that f is not continuous at a but that $\lim_{t \to a^+} f(t)$ exists finite. Denote this limit as $f(a+)$. Prove that

$$\int_0^\infty f(t) \, \delta(t - a) \, dt = f(a+).$$

28. Evaluate $\int_0^\infty [\sin(t)/t] \, \delta(t - \pi/6) \, dt$.

29. Evaluate $\int_0^2 t^2 \, \delta(t - 3) \, dt$.

30. Evaluate $\int_0^\infty f(t) \, \delta(t - s) \, dt$, where

$$f(t) = \begin{cases} t & \text{if} \quad 0 \leq t < 2 \\ 5 & \text{if} \qquad t = 2 \\ t^2 & \text{if} \qquad t > 2. \end{cases}$$

31. It is sometimes convenient to consider $\delta(t)$ as the derivative of the nondifferentiable function $H(t)$. Use the definition of the derivative and of H, and the definition of δ as the limit of δ_ϵ, to give a heuristic justification for this practice. Assume that all the limits involved exist.

32. Use the idea that $H'(t) = \delta(t)$ (from Problem 31) to determine the output voltage of the circuit of Example 4.24 in Section 4.4 by differentiating the relevant equation to obtain an equation in i rather than writing the charge as an integral.

33. If $H'(t) = \delta(t)$, then $\mathscr{L}[H'(t)] = \mathscr{L}[\delta(t)] = 1$. Show that not all of the operational rules for the Laplace transform are compatible with this equation. *Hint:* Consider $\mathscr{L}[H'(t)] = s\mathscr{L}[H(t)](s) - H(0+)$.

34. Evaluate $\delta(t - a) * f(t)$.

35. Suppose an object of mass m is attached to the lower end of a spring of modulus k. Assume that there is no damping. Solve the resulting equation of motion for the position of the object at any time $t \geq 0$ if, at time zero, the object is pushed downward from the equilibrium position with an initial velocity of v_0. With what momentum does the object leave equilibrium?

36. Suppose an object of mass m is attached to the lower end of a spring of modulus k. Assume that there is no damping present. Solve the resulting equation of motion for the position of the weight for any time $t \geq 0$ if, at time zero, the weight is struck a downward blow of magnitude mv_0. Compare the position of the object in Problem 35 with that of the object in this problem at any time $t > 0$.

37. A 2-pound weight is attached to the lower end of a spring, stretching it $\frac{8}{3}$ inches. The weight is allowed to come to rest in the equilibrium position. At some time, which we designate as time zero, the weight is then struck a downward blow of magnitude $\frac{1}{4}$ pound (an impulse). Assume that there is no damping in the system. Determine the velocity with which the weight leaves the equilibrium position, as well as the frequency and magnitude of the resulting oscillations.

38. An object weighing 8 pounds is suspended from a spring, thus stretching it 6 inches. The weight, which is at rest in the equilibrium position, is struck an upward blow of magnitude 1 pound. Assuming that there is no damping in the system, determine the velocity with which the weight leaves the equilibrium position, as well as the frequency and magnitude of the resulting oscillations.

39. An object of mass 1 kilogram is suspended from a spring which has spring constant 24 newtons/meter. Attached to the object is a shock absorber which induces a drag of $11v$ newtons, where v is the velocity of the object. The system is set in motion from the equilibrium position by striking the weight with a downward blow of 4 newtons. Determine the velocity with which the weight leaves the equilibrium position, the time at which it obtains its maximum displacement from the equilibrium position, and the displacement at that time.

4.8 *Solution of Systems by the Laplace Transform*

We will illustrate how the Laplace transform can be used to solve certain systems of differential equations. The idea is straightforward and consists of simply applying the Laplace transform to each equation of the system and incorporating the initial data.

EXAMPLE 4.35

Solve the system

$$x'' - 2x' + 3y' + 2y = 4$$
$$2y' - x' + 3y = 0$$
$$x(0) = x'(0) = y(0) = 0.$$

Apply the Laplace transform to each of the differential equations, making use of the initial data, to get

$$s^2 X - 2sX + 3sY + 2Y = \frac{4}{s}$$
$$2sY - sX + 3Y = 0.$$

Write these equations in the form

$$(s^2 - 2s)X + (3s + 2)Y = \frac{4}{s}$$
$$-sX + (2s + 3)Y = 0.$$

Solve these equations for X and Y to obtain

$$X(s) = \frac{4s + 6}{s^2(s + 2)(s - 1)} \quad \text{and} \quad Y(s) = \frac{2}{s(s + 2)(s - 1)}.$$

It is routine to find the partial fractions decompositions

$$X(s) = -\frac{7}{2}\frac{1}{s} - 3\frac{1}{s^2} + \frac{1}{6}\frac{1}{s + 2} + \frac{10}{3}\frac{1}{s - 1}$$

and

$$Y(s) = -\frac{1}{s} + \frac{1}{3}\frac{1}{s + 2} + \frac{2}{3}\frac{1}{s - 1}.$$

Apply the inverse Laplace transform to each of these functions to get

$$x(t) = -\tfrac{7}{2} - 3t + \tfrac{1}{6}e^{-2t} + \tfrac{10}{3}e^t$$
$$y(t) = -1 + \tfrac{1}{3}e^{-2t} + \tfrac{2}{3}e^t. \quad \blacksquare$$

EXAMPLE 4.36

Consider the mass-spring system of Figure 4.25. Establish a coordinate system for each weight, with $x_1 = 0$ and $x_2 = 0$ denoting the equilibrium positions of the respective weights. Choose the direction to the right as positive.

When the weights are located at the points x_1 and x_2, two applications of Hooke's law give us a restoring force on the first weight equal to

$$-k_1 x_1 + k_2(x_2 - x_1).$$

Similarly, the restoring force on the second weight is

$$-k_2(x_2 - x_1) - k_3 x_2.$$

Now use Newton's second law to obtain the system

$$m_1 x_1'' = -(k_1 + k_2)x_1 + k_2 x_2 + f_1(t)$$
$$m_2 x_2'' = k_2 x_1 - (k_2 + k_3)x_2 + f_2(t).$$

In these equations, we have assumed no damping but have allowed for external forces of magnitudes $f_1(t)$ and $f_2(t)$ acting on m_1 and m_2, respectively.

We will analyze a specific example. Suppose that both masses are one unit; that the spring constants are $k_1 = 4$, $k_2 = \tfrac{5}{2}$, and $k_3 = 4$; and that $f_1(t) = 2[1 - H(t - 3)]$ and

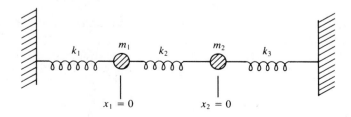

Figure 4.25

$f_2(t) = 0$. The system of differential equations for this problem is

$$x_1'' = -\frac{13}{2}x_1 + \frac{5}{2}x_2 + 2[1 - H(t - 3)]$$
$$x_2'' = \frac{5}{2}x_1 - \frac{13}{2}x_2.$$

Assume that the system starts from rest in the equilibrium position, so that

$$x_1(0) = x_2(0) = x_1'(0) = x_2'(0) = 0.$$

Apply the Laplace transform to get

$$s^2 X_1 = -\frac{13}{2} X_1 + \frac{5}{2} X_2 + \frac{2(1 - e^{-3s})}{s}$$

$$s^2 X_2 = \frac{5}{2} X_1 - \frac{13}{2} X_2.$$

Write this system of equations in the form

$$\left[s^2 + \frac{13}{2} \right] X_1 - \frac{5}{2} X_2 = \frac{2(1 - e^{-3s})}{s}$$

$$-\frac{5}{2} X_1 + \left[s^2 + \frac{13}{2} \right] X_2 = 0.$$

Solve this system of equations for X_1 and X_2 to get

$$X_1(s) = \frac{(2s^2 + 13)(1 - e^{-3s})}{s(s^2 + 9)(s^2 + 4)}$$

and

$$X_2(s) = \frac{5(1 - e^{-3s})}{s(s^2 + 9)(s^2 + 4)}.$$

Use a partial fractions decomposition to write

$$X_1(s) = (1 - e^{-3s}) \left[\frac{13}{36} \frac{1}{s} - \frac{1}{4} \frac{s}{s^2 + 4} - \frac{1}{9} \frac{s}{s^2 + 9} \right]$$

and

$$X_2(s) = (1 - e^{-3s}) \left[\frac{5}{36} \frac{1}{s} - \frac{1}{4} \frac{s}{s^2 + 4} + \frac{1}{9} \frac{s}{s^2 + 9} \right].$$

These are easier to invert than they may first appear. Carry out the indicated multiplications to write

$$X_1(s) = \frac{13}{36} \frac{1}{s} - \frac{1}{4} \frac{s}{s^2 + 4} - \frac{1}{9} \frac{s}{s^2 + 9} - \frac{13}{36} \frac{1}{s} e^{-3s}$$

$$+ \frac{1}{4} \frac{s}{s^2 + 4} e^{-3s} + \frac{1}{9} \frac{s}{s^2 + 9} e^{-3s}$$

and

$$X_2(s) = \frac{5}{36}\frac{1}{s} - \frac{1}{4}\frac{s}{s^2+4} + \frac{1}{9}\frac{s}{s^2+9} - \frac{5}{36}\frac{1}{s}e^{-3s}$$
$$+ \frac{1}{4}\frac{s}{s^2+4}e^{-3s} - \frac{1}{9}\frac{s}{s^2+9}e^{-3s}.$$

We can now apply the inverse transform, using a shifting theorem on the terms involving e^{-3s}. We get

$$x_1(t) = \frac{13}{36} - \frac{1}{4}\cos(2t) - \frac{1}{9}\cos(3t)$$
$$+ H(t-3)\left[-\frac{13}{36} + \frac{1}{4}\cos[2(t-3)] - \frac{1}{9}\cos[3(t-3)] \right]$$

and

$$x_2(t) = \frac{5}{36} - \frac{1}{4}\cos(2t) + \frac{1}{9}\cos(3t)$$
$$+ H(t-3)\left[-\frac{5}{36} + \frac{1}{4}\cos[2(t-3)] - \frac{1}{9}\cos[3(t-3)] \right].$$

These functions determine the positions of the masses at times $t \geq 0$. ∎

EXAMPLE 4.37

Consider the circuit shown in Figure 4.26. The switch is closed at time zero, and we want to know the value of the current in each loop for $t \geq 0$. Assume that both loop currents and the charges on both of the capacitors are initially zero. Apply Kirchhoff's voltage and current laws to each loop to arrive at the system of differential equations

$$40i_1 + 120(q_1 - q_2) = 10$$
$$60i_2 + 120q_2 = 120(q_1 - q_2). \tag{4.22}$$

Since the charge on the capacitor in the second loop is assumed to be zero prior to switch closure, and the charge on a capacitor is continuous, we have $q_2(0+) = 0$. Using the same reasoning, since the charge on the middle capacitor is $q_1(t) - q_2(t)$, we have

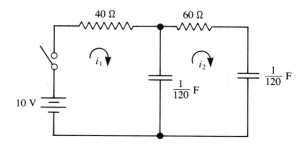

Figure 4.26

$q_1(0+) = q_2(0+) = 0$. This information can be used in the first equation of the system to get $i_1(0+) = \frac{1}{4}$ and in the second equation to get $i_2(0+) = 0$. The calculation of the initial currents here is not necessary to solve the problem we have posed, but this information can be used to check our results later.

Now recall that $i = q'$, so we can write $q(t) = \int_0^t i(\tau)\,d\tau + q(0)$. We may therefore write the system (4.22) in the form

$$40i_1 + 120 \int_0^t [i_1(\tau) - i_2(\tau)]\,d\tau + 120[q_1(0) - q_2(0)] = 10$$

$$60i_2 + 120 \int_0^t i_2(\tau)\,d\tau + 120q_2(0) = 120 \int_0^t [i_1(\tau) - i_2(\tau)]\,d\tau + 120[q_1(0) - q_2(0)].$$

Take the Laplace transform of each of these equations, using the fact that $q_1(0) = q_2(0) = 0$, to get

$$40I_1 + \frac{120}{s} I_1 - \frac{120}{s} I_2 = \frac{10}{s}$$

$$60I_2 + \frac{120}{s} I_2 = \frac{120}{s} I_1 - \frac{120}{s} I_2.$$

These equations can be written more neatly as

$$(s+3)I_1 - 3I_2 = \tfrac{1}{4}$$
$$2I_1 - (s+4)I_2 = 0.$$

Solve these equations for I_1 and I_2 to get

$$I_1(s) = \frac{s+4}{4(s+1)(s+6)} = \frac{3}{20}\frac{1}{s+1} + \frac{1}{10}\frac{1}{s+6}$$

and

$$I_2(s) = \frac{1}{2(s+1)(s+6)} = \frac{1}{10}\frac{1}{s+1} - \frac{1}{10}\frac{1}{s+6}.$$

Apply the inverse Laplace transform to these functions to get

$$i_1(t) = \tfrac{3}{20}e^{-t} + \tfrac{1}{10}e^{-6t}$$
$$i_2(t) = \tfrac{1}{10}e^{-t} - \tfrac{1}{10}e^{-6t}.$$

Notice that indeed $i_1(0+) = \frac{1}{4}$ and $i_2(0+) = 0$, in agreement with our previous observation. ∎

PROBLEMS FOR SECTION 4.8

In each of Problems 1 through 15, use the Laplace transform to solve the system.

1. $x' - 2y' = 1$
$x' + y - x = 0$
$x(0) = y(0) = 0$

2. $2x' - 3y + y' = 0$
$x' + y' = t$
$x(0) = y(0) = 0$

3. $x' + 2y' - y = 1$
$2x' + y = 0$
$x(0) = y(0) = 0$

4. $x' + y' - x = \cos(2t)$
$x' + 2y' = 0$
$x(0) = y(0) = 0$

5. $x' + 3x - y = 1$
$x' + y' + 3x = 0$
$x(0) = 2, \quad y(0) = 0$

6. $x' - x - y = 4$
$2x' - y' = 1$
$x(0) = y(0) = 0$

7. $3x' - y = 2t$
   $x' + y' - y = 0$
   $x(0) = y(0) = 0$

8. $x' + 4y' - y = 0$
   $x' + 2y = e^{-t}$
   $x(0) = y(0) = 0$

9. $x' + 2x - y' = 0$
   $x' + y + x = t^2$
   $x(0) = y(0) = 0$

10. $x' + 4x - y = 0$
    $x' + y' = t$
    $x(0) = y(0) = 0$

11. $x' + 2y' - y = t$
    $x' + 2y = 0$
    $x(0) = y(0) = 0$

12. $x' + y' - x = 0$
    $x' - y' + 2y = 3$
    $x(0) = y(0) = 0$

13. $x' + y' + x = 0$
    $x' + 2y' = 1$
    $x(0) = y(0) = 0$

14. $x' + y' + x - y = 0$
    $x' + 2y' + x = 1$
    $x(0) = y(0) = 0$

15. $x' + 2y' - x = 0$
    $4x' + 3y' + y = -6$
    $x(0) = -1, \quad y(0) = 1$

16. Use the Laplace transform to solve the initial value problem

$$y_1' - 2y_2' + 3y_3 = 0,$$
$$y_1 - 4y_2' + 3y_3' = t,$$
$$y_1 - 2y_2' + 3y_3' = -1;$$
$$y_1(0) = y_2(0) = y_3(0) = 0$$

17. Solve for the currents in the circuit shown in the diagram, assuming that the currents and charges are initially zero and that $E(t) = 2H(t - 4) - H(t - 5)$.

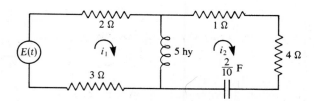

18. Solve for the currents in the circuit of Problem 17, assuming that the currents and charges are initially zero, if $E(t) = 1 - H(t - 4)\sin[2(t - 4)]$.

19. Solve for the currents in the circuit shown, assuming that the currents are initially zero and that $E(t) = 2H(t - 4)$.

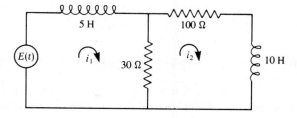

20. Solve for the currents in the circuit shown below, assuming that the currents and charge are initially zero and that $E(t) = 3 \sin(t)$.

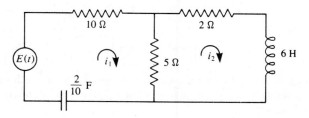

21. Solve for the displacement functions of the masses in the mass-spring system shown below. Assume no damping, external forces $f_1(t) = 2$ and $f_2(t) = 0$, and zero initial displacements and velocities.

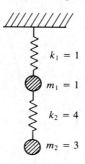

$$k_1 = 1$$
$$m_1 = 1$$
$$k_2 = 4$$
$$m_2 = 3$$

22. Solve for the displacement functions of the masses in the system of Problem 21 if $f_1(t) = 1 - H(t - 2)$ and $f_2(t) = 0$.

23. Imagine a double mass-spring system as shown. The spring attached to M has spring constant k_1 and damping constant c_1; that attached to m has spring constant k_2 and is undamped. Let M be subjected to a periodic driving force $f(t) = A \sin(\omega t)$. The masses are initially at rest in the equilibrium position.

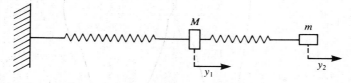

(a) Derive the initial value problem describing the motion of the masses.
(b) Solve this initial value problem for the displacement functions.
(c) Show that, if m and k_2 are chosen so that $\omega = \sqrt{k_2/m}$, the mass m cancels out the forced vibrations of the mass M. In this case, we call m a *vibration absorber*.

24. Imagine a single mass m suspended from a spring having spring constant k. Assume no damping, and suppose there is a driving force $f(t)$. Solve the equation of motion to get

$$y(t) = \frac{1}{\sqrt{mk}} \sin(\sqrt{k/m}\ t) * f(t),$$

assuming that $y(0) = y'(0) = 0$. Now determine the resulting motion if the driving force is in the form of an impulse of magnitude P at time t_0.

25. Two objects of masses m_1 and m_2 are attached to opposite ends of a spring having spring constant k, as shown. The entire apparatus is placed on a highly varnished table. Show that, if stretched and released from rest, the masses oscillate with respect to each other with period

$$2\pi \sqrt{\frac{m_1 m_2}{k(m_1 + m_2)}}.$$

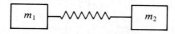

Each of Problems 26 through 30 deals with the mass-spring system shown. In each problem, derive and solve the differential equation for the displacement functions of the weights, under the assumption that there is no damping.

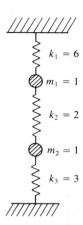

26. The weights have zero initial displacement and velocity, with no external driving force.

27. Each weight is pulled down one unit and then released from rest with no external driving force.

28. The upper weight is pulled down one unit and the lower one is raised one unit; then both are released from rest, with no external driving force.

29. The weights have zero initial displacement and velocity. The lower weight is subjected to an external force $f(t) = 1 - u_1(t)$, and the upper weight has no external driving force applied to it.

30. The weights have zero initial displacement and velocity. The lower weight is subjected to an external force $f(t) = 4\sin(t)$, and the upper weight has no external driving force applied to it.

31. Solve for the currents in the circuit shown if $E(t) = 5H(t - 2)H(t - k)$.

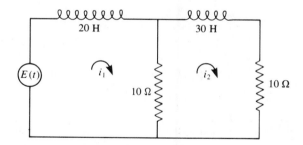

32. Solve for the currents in the circuit shown in Problem 31 if $E(t) = 5\delta(t - 1)$. Assume zero initial currents.

33. Solve for the currents in the circuit shown if the initial currents and charges are zero and $E(t) = k$.

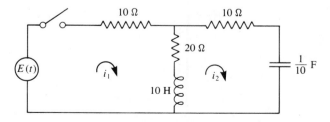

34. Solve for the currents in the circuit of Problem 33 if the initial currents are zero and $E(t) = 10\sin(t)$.

35. Solve for the currents in the circuit of Problem 33 if $E(t) = AH(t - k)$.

36. Solve for the currents in the circuit shown, assuming that the currents are initially zero, if the electromotive force is $E(t) = k$.

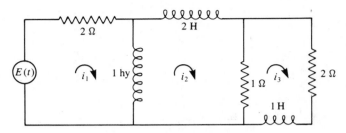

37. Solve for the currents in the circuit of Problem 36 if $E(t) = 2[1 - H(t - 3)]$.

38. Solve for the currents in the circuit of Problem 36 if $E(t) = \sin(t)$.

39. Set up the system of differential equations modeling the mass-spring system shown below. Assume no damping, zero driving forces, and zero initial displacement and velocity, except for the top weight, which is pulled down one unit and released from rest. There are n objects of equal mass m attached by $n + 1$ springs all having spring constant k. As usual, let y_j be the displacement from static equilibrium of the jth mass (counting from the top), with the positive direction chosen as downward.

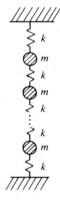

40. Develop an electrical network which models the mass-spring system of Problem 39 in the sense that the governing differential equations are the same if we make the correct analogies.

41. Referring to the diagram shown, derive the system of differential equations and initial conditions for the displacement of the weights at any time $t \geq 0$. Assume that the system is set in motion by raising the bottom

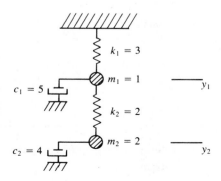

weight one unit and then striking it a downward blow sufficient to give it a velocity of four units per second. The top weight is not initially displaced but is driven by an external force $f_1(t) = 2e^{-4t}$. Solve this system for the displacement functions.

Hint: Obtain

$$Y_1(s) = \frac{18}{(s + 4)(s^4 + 7s^3 + 16s^2 + 15s + 3)}.$$

Use Newton's method to find two real roots r_1 and r_2 of the fourth degree polynomial factor in the denominator, then divide this polynomial by $(s - r_1)(s - r_2)$ and use the quadratic formula to find the remaining roots. Use partial fractions to find $y_1(t)$; then use one of the differential equations of the system to find $y_2(t)$.

Summary

Listed for reference below are some of the formulas developed in this chapter. Most hypotheses are omitted in this summary.

$\mathscr{L}[f'] = sF(s) - f(0)$ (f continuous at 0)

$\mathscr{L}[f^{(n)}] = s^n F(s) - s^{n-1}f(0) - s^{n-2}f'(0) - \cdots - f^{(n-1)}(0)$

If f has period ω, then

$$\mathscr{L}[f] = \frac{1}{1 - e^{-\omega s}} \int_0^\infty e^{-st}f(t)\, dt.$$

(s-shift) $\mathscr{L}[e^{at}f(t)] = F(s - a)$

(t-shift) $\mathscr{L}[f(t - a)H(t - a)] = e^{-as}F(s)$

$$\mathscr{L}\left[\int_0^t f(r)\, dr\right] = \frac{1}{s} F(s)$$

Convolution $\mathscr{L}^{-1}[F(s)G(s)] = f * g = \int_0^t f(t - \alpha)g(\alpha)\, d\alpha$

Polynomial coefficient $\mathscr{L}[t^n f(t)] = (-1)^n F(s)$

Heaviside formulas   To compute the term in $\mathscr{L}^{-1}[P(s)/Q(s)]$ corresponding to certain factors in $Q(s)$:

1. For a linear factor $s - a$ of $Q(s)$, the term is $Z(a)e^{at}$, where

$$Z(s) = \frac{P(s)}{Q'(s)}.$$

2. For a factor $(s - a)^k$ of $Q(s)$, the term is

$$\left[\frac{Z^{(k-1)}(a)}{(k-1)!} + \frac{Z^{(k-2)}(a)}{(k-2)!}\frac{t}{1} + \cdots + Z'(a)\frac{t^{k-2}}{(k-2)!} + Z(a)\frac{t^{k-1}}{(k-1)!}\right]e^{at},$$

in which

$$Z(s) = \frac{P(s)(s - a)^k}{Q(s)}.$$

3. For an unrepeated irreducible quadratic factor $(s - a)^2 + b^2$ of $Q(s)$, the term is

$$\frac{1}{b}[\alpha_i \cos(bt) + \alpha_r \sin(bt)]e^{at},$$

where

$$Z(s) = \frac{P(s)[(s - a)^2 + b^2]}{Q(s)}$$

and

$$\alpha_i = \text{imaginary part of } Z(a + ib), \qquad \alpha_r = \text{real part of } Z(a + ib).$$

CHAPTER 5

Series Solutions of Differential Equations

5.0 Introduction

Consider the second order linear differential equation

$$P(x)y'' + Q(x)y' + R(x)y = S(x),$$

in which P, Q, R, and S are continuous on some interval I. We have been successful in solving this equation only in very special cases [constant coefficient or Cauchy-Euler type, if $S(x)$ is "not too complicated"]. In this chapter, we will consider how to proceed with fewer restrictive assumptions on the coefficient functions. Here is a brief outline of what we will do.

First, we will usually look for solutions in some interval about a point x_0 at which initial data are specified. In this interval, the coefficient functions will generally be required to have certain properties, which we will explicitly list as we proceed. Thus, solutions will usually be local ones, valid only in some interval about x_0. We have seen this kind of local result before, with the existence/uniqueness theorem for first order initial value problems in Chapter 1.

In many instances, we will not be able to obtain solutions as finite combinations of familiar functions. In such cases, we will look for solutions in the form of power series expansions about x_0. When $P(x_0) \neq 0$, such solutions are often possible. We will see that, when $P(x_0) = 0$, power series solutions may not be possible. In this event, we will use a generalization of power series called the Frobenius series.

Although the necessity of obtaining series solutions may seem forbidding at first, remember that we have already encountered equations we could not explicitly solve in terms of familiar functions. For example, many solutions of first order differential equations are defined implicitly by equations too complicated to solve explicitly. In such cases, we are often hard pressed actually to evaluate the solution at any point. By

291

contrast, a power series solution can often be used to approximate a solution value at a point, to any desired degree of accuracy. At least in this respect, a series solution is to be preferred to some types of solutions we have already seen.

5.1 *Review of Power Series*

We will assume general familiarity with power series. This section is a review of some facts we will need.

A *power series* about x_0 is a series of the form

$$\sum_{n=0}^{\infty} a_n(x - x_0)^n.$$

The number x_0 is the *center* of the series, and the numbers $a_0, a_1, \ldots$ are the *coefficients*.

CONVERGENCE OF POWER SERIES

THEOREM 5.1   Convergence of Power Series

Suppose that $\sum_{n=0}^{\infty} a_n(x - x_0)^n$ converges at some $x_1 \neq x_0$. Then $\sum_{n=0}^{\infty} a_n(x - x_0)^n$ converges absolutely for all x such that

$$|x - x_0| < |x_1 - x_0|. \quad \blacksquare$$

That is, if the power series converges at some x_1 different from x_0, it converges absolutely at all x closer to x_0 than x_1.

From Theorem 5.1, we can show that there is a number R such that the power series converges for all x with $|x - x_0| < R$ and diverges if $|x - x_0| > R$. It may happen that $R = \infty$, in which case the series converges for all x. If R is finite, convergence must be tested at $x_0 + R$ and $x_0 - R$ individually for each power series.

R is the *radius of convergence* of the power series. The interval about x_0 consisting of all x such that the power series converges is called the *interval of convergence*. This may be the entire real line or an open, half-open, or closed interval. In the event that the series converges only for $x = x_0$, we do not speak of an interval of convergence. The radius of convergence in this case is zero.

Often we apply a form of the ratio test to determine the radius of convergence of a power series. Here is a statement of this test.

RATIO TEST

Suppose that $b_n \neq 0$ for $n \geq N$ and that

$$\lim_{n \to \infty} \left| \frac{b_{n+1}}{b_n} \right| = L,$$

in which L may be ∞. Then the series $\sum_{n=0}^{\infty} b_n$ converges absolutely if $L < 1$ and diverges if $L > 1$. The test is inconclusive if $L = 1$. $\quad \blacksquare$

Here is an illustration of how this test can be applied to a power series.

EXAMPLE 5.1

Find the radius and interval of convergence of

$$\sum_{n=0}^{\infty} \frac{(-1)^n}{(n+1)9^n} (x-2)^{2n}.$$

With

$$b_n = \frac{(-1)^n}{(n+1)9^n} (x-2)^{2n},$$

we have

$$\left| \frac{b_{n+1}}{b_n} \right| = \left| \frac{(-1)^{n+1}}{(n+2)9^{n+1}} (x-2)^{2n+2} \frac{(n+1)9^n}{(-1)^n} \frac{1}{(x-2)^{2n}} \right|$$

$$= \frac{n+1}{n+2} \frac{1}{9} |x-2|^2.$$

Then

$$\lim_{n\to\infty} \left| \frac{b_{n+1}}{b_n} \right| = \lim_{n\to\infty} \frac{n+1}{n+2} \frac{1}{9} |x-2|^2$$

$$= \frac{1}{9} |x-2|^2 \lim_{n\to\infty} \frac{n+1}{n+2} = \frac{1}{9} |x-2|^2.$$

By the ratio test, the power series converges absolutely if this limit is less than 1, hence if $\frac{1}{9}|x-2|^2 < 1$. This occurs if $|x-2| < 3$ or if $-1 < x < 5$. The power series diverges if the limit is greater than 1, which occurs if $|x-2| > 3$. This is the case if $x > 5$ or if $x < -1$. The radius of convergence is 3.

This leaves convergence still in question at -1 and at 5. At $x = 5$, we have the series

$$\sum_{n=0}^{\infty} \frac{(-1)^n 3^{2n}}{(n+1)9^n} = \sum_{n=0}^{\infty} (-1)^n \frac{1}{n+1},$$

a convergent alternating series. At $x = -1$, we obtain the same series. Thus, the interval of convergence for this power series is $[-1, 5]$. ∎

Depending on the coefficients of a power series, the limit needed in applying the ratio test may not exist. The power series still has a radius of convergence, but we must determine it by some other means.

ALGEBRA AND CALCULUS OF POWER SERIES

At points common to both intervals of convergence about x_0, two power series can be added and subtracted term by term, and the resulting series converges in the common interval of convergence. In particular, if

$$f(x) = \sum_{n=0}^{\infty} a_n(x-x_0)^n \quad \text{and} \quad g(x) = \sum_{n=0}^{\infty} b_n(x-x_0)^n,$$

then at points at which both power series converge we have

$$f(x) + g(x) = \sum_{n=0}^{\infty} (a_n + b_n)(x - x_0)^n$$

and

$$f(x) - g(x) = \sum_{n=0}^{\infty} (a_n - b_n)(x - x_0)^n.$$

We can also multiply a power series term by term by a constant:

$$kf(x) = \sum_{n=0}^{\infty} ka_n(x - x_0)^n,$$

and, if $k \neq 0$, this series has the same interval of convergence as the series for f.

Power series can be multiplied according to the definition

$$(fg)(x) = \sum_{n=0}^{\infty} \sum_{k=0}^{n} a_k b_{n-k}(x - x_0)^n$$
$$= a_0 b_0 + (a_1 b_0 + a_0 b_1)(x - x_0)$$
$$+ (a_2 b_0 + a_1 b_1 + a_0 b_2)(x - x_0)^2 + \cdots$$

The coefficients are exactly what we get by multiplying

$$[a_0 + a_1(x - x_0) + a_2(x - x_0)^2 + \cdots] \times [b_0 + b_1(x - x_0) + b_2(x - x_0)^2 + \cdots]$$

and collecting the constant term and the coefficients of $x - x_0, (x - x_0)^2$, and so on. The coefficient of $(x - x_0)^n$ in this product series is

$$\sum_{j=0}^{n} a_j b_{n-j},$$

or

$$a_0 b_n + a_1 b_{n-1} + \cdots + a_{n-1} b_n + a_n b_0,$$

a sum of all products $a_\alpha b_\beta$ in which α and β are nonnegative integers with $\alpha + \beta = n$. This product series converges at points common to the interiors of the intervals of convergence of both f and g.

In any open interval about x_0 in which the series for f converges, we may differentiate the power series term by term, and the new series converges to $f'(x)$ in this interval:

$$f'(x) = \sum_{n=1}^{\infty} na_n(x - x_0)^{n-1}$$

The series for $f'(x)$ begins at $n = 1$ because the $n = 0$ term is zero (corresponding to the derivative of the constant term a_0 of the series for f).

Continuing in this way, we can compute

$$f''(x) = \sum_{n=2}^{\infty} n(n - 1)a_n(x - x_0)^{n-2},$$

and, in general, the kth derivative,

$$f^{(k)}(x) = \sum_{n=k}^{\infty} n(n - 1)(n - 2) \cdots (n - k + 1)a_n(x - x_0)^{n-k}.$$

All of these series have the same radius of convergence as the series for f.

Within an open interval about x_0 in which the series converges, we may also integrate a power series term by term:

$$\int f(x)\,dx = \sum_{n=0}^{\infty} a_n \int (x-x_0)^n\,dx = \sum_{n=0}^{\infty} \frac{1}{n+1} a_n (x-x_0)^{n+1} + C.$$

TAYLOR SERIES

Suppose that $f(x) = \sum_{n=0}^{\infty} a_n (x-x_0)^n$ in some interval about x_0. Let $x = x_0$ in the formula for $f^{(k)}(x)$ to obtain

$$f^{(k)}(x_0) = k(k-1)(k-2)\cdots(2)(1)a_k = k!a_k,$$

where $k!$ denotes k factorial, the product of the integers from 1 through k, inclusive. Then

$$a_k = \frac{1}{k!} f^{(k)}(x_0). \tag{5.1}$$

We can let $k = 0$ in this formula by using the notational convention that $0! = 1$ and agreeing that $f^{(0)}(x_0) = f(x_0)$. That is, the "zero" derivative of f at x_0 is just $f(x_0)$.

Formula (5.1) is *Taylor's formula* for the kth coefficient of the power series expansion of f about x_0. If $f(x) = \sum_{n=0}^{\infty} a_n (x-x_0)^n$, then necessarily the a_n's are given by this formula, and the power series is the *Taylor series* for f about x_0. Put another way, the Taylor series is the only power series expansion a function can have about a point x_0.

A Taylor series about zero (that is, $x_0 = 0$) is called a *Maclaurin series*. If in some interval about x_0 we have

$$\sum_{n=0}^{\infty} a_n (x-x_0)^n = \sum_{n=0}^{\infty} b_n (x-x_0)^n,$$

then necessarily $a_n = b_n$ for $n = 0, 1, 2, 3, \ldots$ This fact is crucial in the application of power series to the solution of differential equations, and it follows from the fact that both a_n and b_n must be the nth Taylor coefficient of the function represented by these series.

If we have a function g which is infinitely differentiable in some open interval about x_0, we can write the Taylor series

$$\sum_{n=0}^{\infty} \frac{1}{n!} g^{(n)}(x_0)(x-x_0)^n$$

for g about x_0. Under appropriate conditions, this series will converge to $g(x)$ in some open interval about x_0. Care must be exercised, however, because outside of this interval the series and the function may not agree. Here is an illustration.

EXAMPLE 5.2

Let $g(x) = \ln(1+x)$ for $x > -1$.

By differentiating repeatedly, we find that

$$g^{(n)}(x) = \frac{(-1)^{n+1}(n-1)!}{(x+1)^n}.$$

In particular,

$$g^{(n)}(1) = \frac{(-1)^{n+1}(n-1)!}{2^n}$$

for $n = 1, 2, 3, \ldots$. The coefficients in the Taylor expansion of g about 1 are

$$\frac{1}{n!}\, g^{(n)}(1) = \frac{(-1)^{n+1}}{n2^n}$$

for $n = 1, 2, 3, \ldots$, and the constant term is $g(1) = \ln(2)$.

The Taylor series for g about 1 is

$$\ln(2) + \sum_{n=1}^{\infty} \frac{(-1)^{n+1}}{n2^n}\,(x-1)^n.$$

This series has radius of convergence 2 (apply the ratio test) and can be shown to equal $\ln(1 + x)$ for $-1 < x < 3$. For $x > 3$, the series diverges and therefore does not represent $\ln(1 + x)$, even though $\ln(1 + x)$ is defined for $x > 3$. ∎

A function g is said to be *analytic* at x_0 if g has a Taylor series expansion which converges to g in some interval about x_0. For example, $\ln(1 + x)$ is analytic at 1 because it has a Taylor series representation in the interval $(-1, 3)$, which contains 1.

In some cases, the radius of convergence of the Taylor series representation of a function about a point can be determined directly from the function without having to actually write the series. It can be shown that *the radius of convergence of the Taylor expansion of g about x_0 is the distance between x_0 and the nearest real or complex point at which the function is not analytic.*

To illustrate, in Example 5.2 we found that 2 is the radius of convergence of the Taylor expansion of $\ln(1 + x)$ about $x_0 = 1$. This is the distance from the center of the series (which is 1) to -1, the point nearest to 1 at which the function is not analytic [in fact, $\ln(1 + x)$ is not defined at -1]. Note that $\ln(1 + x)$ is actually analytic on $(-1, \infty)$; however, its Taylor series about 1 converges only on $(-1, 3)$. The interval of convergence of the Taylor series about 1 cannot extend beyond -1; hence, the radius of convergence cannot be greater than 2 (it turns out to be equal to 2).

This idea also explains why the Maclaurin expansion of $1/(1 + x^2)$, which is $\sum_{n=0}^{\infty}(-1)^n x^n$, has radius of convergence 1 even though $1/(1 + x^2)$ is defined (and differentiable) for all x. The reason is that the complex function $1/(1 + z^2)$ is not defined at $\pm i$, and the distance from zero to i (or to $-i$) is 1. Even though $1/(1 + x^2)$ is defined for all x, its Taylor series about zero only has radius of convergence 1.

By the same token, the radius of convergence of the Taylor expansion of $1/(1 + x^2)$ about 3 is $\sqrt{10}$. This may seem mysterious at first, but it is explained again by the fact that $1/(1 + z^2)$ is analytic everywhere except where $1 + z^2 = 0$, namely, at $z = \pm i$. The distance from 3 to i (or to $-i$) is the distance in the plane from the point $(3, 0)$ to $(0, 1)$ [or to $(0, -1)$], and this distance is $\sqrt{(3 - 0)^2 + (0 - 1)^2}$, or $\sqrt{10}$.

SHIFTING INDICES

In using series to solve differential equations, it is often convenient to rewrite certain terms so that common powers of $x - x_0$ can be factored and coefficients combined. Here is an example showing some of the manipulations involved in doing this.

EXAMPLE 5.3

Add the series $\sum_{n=0}^{\infty} a_n x^{n+2}$ and $\sum_{n=0}^{\infty} b_n x^n$.

We want to write the sum so that we can factor out like powers of x and combine coefficients. To do this, write the first series so that the power of x is n, as occurs in the second series. This involves changing the starting point of the series and also the index on each a_n.

One way to do this is to recognize that

$$\sum_{n=0}^{\infty} a_n x^{n+2} = a_0 x^2 + a_1 x^3 + a_2 x^4 + \cdots = \sum_{n=2}^{\infty} a_{n-2} x^n,$$

shifting the index back by two and starting at $n = 2$ instead of $n = 0$. If this result is not clear intuitively, it can be done more systematically as follows. Let $k = n + 2$, so that $k = 2$ when $n = 0$, and write

$$\sum_{n=0}^{\infty} a_n x^{n+2} = \sum_{k=2}^{\infty} a_{k-2} x^k.$$

Now remember that k and n are just dummy summation indices and can be replaced by any reasonable letter. Thus, we can write

$$\sum_{k=2}^{\infty} a_{k-2} x^k = \sum_{n=2}^{\infty} a_{n-2} x^n.$$

Now we can add the two series as follows:

$$\sum_{n=0}^{\infty} a_n x^{n+2} + \sum_{n=0}^{\infty} b_n x^n = \sum_{n=2}^{\infty} a_{n-2} x^n + \sum_{n=0}^{\infty} b_n x^n$$

$$= \sum_{n=2}^{\infty} a_{n-2} x^n + b_0 + b_1 x + \sum_{n=2}^{\infty} b_n x^n$$

$$= b_0 + b_1 x + \sum_{n=2}^{\infty} (a_{n-2} + b_n) x^n.$$

It is important to notice that in shifting indices in the first summation, the new series began at $n = 2$, not $n = 0$. When adding $\sum_{n=2}^{\infty} a_{n-2} x^n$ to $\sum_{n=0}^{\infty} b_n x^n$, we have to start both series at $n = 2$ and must therefore include separately the $n = 0$ and $n = 1$ terms from the second series (or else lose them without justification, incurring an error). Care must be taken not to lose existing terms, or invent new ones, when combining series after shifting indices. ∎

PROBLEMS FOR SECTION 5.1

In each of Problems 1 through 16, find the radius of convergence and the interval of convergence of the power series.

1. $\sum_{n=0}^{\infty} \frac{(-1)^n}{n+1} (x - 4)^n$

2. $\sum_{n=0}^{\infty} \frac{2^n}{n!} x^n$

3. $\sum_{n=0}^{\infty} \frac{1}{n+2} (x + 1)^n$

4. $\sum_{n=0}^{\infty} n^2 x^n$

5. $\sum_{n=0}^{\infty} \frac{2n+1}{2n-1} x^n$

6. $\sum_{n=0}^{\infty} n^n x^n$

7. $\sum_{n=0}^{\infty} (-\tfrac{3}{2})^n (x - \tfrac{5}{2})^n$

8. $\sum_{n=0}^{\infty} \left(\frac{n^2 - 3n}{n^2 + 4} \right) x^n$

9. $\sum_{n=1}^{\infty} \left(\frac{n+1}{n} \right)^n x^n$

10. $\sum_{n=0}^{\infty} \frac{3^n}{(2n)!} x^{2n}$

11. $\sum_{n=2}^{\infty} \frac{\ln(n)}{n} x^n$

12. $\sum_{n=0}^{\infty} \frac{e^n}{n!} x^{n+2}$

13. $\sum_{n=0}^{\infty} \frac{(-1)^n}{(n+2)4^n} (x+4)^{2n+1}$

14. $\sum_{n=5}^{\infty} \frac{(-1)^n}{8^n(n-4)} (x-3)^{3n}$

15. $\sum_{n=1}^{\infty} \frac{(-1)^n}{n^2 3^n} (x-2)^n$

16. $\sum_{n=1}^{\infty} \frac{n!}{n^n} x^n$ *Hint:* Recall that $\lim_{n \to \infty} \left(1 + \frac{1}{n}\right)^n = e$.

In each of Problems 17 through 28, find the terms up to and including the $(x - x_0)^5$ term in the Taylor expansion of the function about the given point x_0.

17. $\ln(x + 2)$;   0

18. xe^x;   0

19. $e^x\cos(x)$;   0

20. $(x + 1)^4$;   0

21. $\tan^{-1}(x)$;   0

22. $\cos^2(2x)$;   0

23. $x - \cos(2x)$; $\dfrac{\pi}{2}$

24. $\tan(2x)$; $\dfrac{\pi}{6}$

25. $x \sin(x)$; $\dfrac{\pi}{2}$

26. $x^2 e^x$; -1

27. $(x + 1)^2$; $\frac{1}{2}$

28. $\dfrac{1}{1 - 2x}$;   1

In each of Problems 29 through 34, shift indices so that all powers of x under the summation are x^n.

29. $\sum_{n=1}^{\infty} \frac{(-1)^{n+1}}{2n + 4} x^{n+1}$

30. $\sum_{n=2}^{\infty} \frac{(n+1)^n}{2^n} x^{n+3}$

31. $\sum_{n=2}^{\infty} \frac{2n + 3}{n} x^{n-1}$

32. $\sum_{n=4}^{\infty} \frac{(-1)^{n+1}}{2 + n^2} x^{n-3}$

33. $\sum_{n=1}^{\infty} \frac{2^n}{n^3 - 3} x^{n+1}$

34. $\sum_{n=1}^{\infty} \frac{2n}{n!} x^{n+1}$

In each of Problems 35 through 44, add the series and combine as many terms as possible under one summation, factoring out a common power of x, as in Example 5.3. There are various ways of writing correct answers to these problems.

35. $\sum_{n=1}^{\infty} 2^n x^{n+1} + \sum_{n=0}^{\infty} (n+1)x^n$

36. $\sum_{n=0}^{\infty} \frac{n!}{2^n} x^{n+3} + \sum_{n=1}^{\infty} \frac{1}{n+1} x^{n-1}$

37. $\sum_{n=1}^{\infty} \frac{n!}{n^2} x^{n-1} + \sum_{n=2}^{\infty} 2^n x^n$

38. $\sum_{n=2}^{\infty} \frac{1}{n} x^n + \sum_{n=1}^{\infty} (-1)^{n+1} x^{n+2}$

39. $\sum_{n=1}^{\infty} \frac{1}{2n} x^n - \sum_{n=3}^{\infty} n^n x^{n-3}$

40. $\sum_{n=1}^{\infty} (2n - 1)x^{n+3} + \sum_{n=0}^{\infty} \frac{1}{n+1} x^n$

41. $\sum_{n=2}^{\infty} x^{n+3} + \sum_{n=3}^{\infty} n! x^{n+2}$

42. $\sum_{n=1}^{\infty} (-1)^n x^{n+1} + \sum_{n=2}^{\infty} (\tfrac{3}{2})^n x^{n+2}$

43. $\sum_{n=0}^{\infty} \frac{1}{n+2} x^n + \sum_{n=1}^{\infty} (-2)^n x^{n+1}$

44. $\sum_{n=0}^{\infty} x^{n+3} + \sum_{n=1}^{\infty} (-1)^n \frac{1}{n+1} x^{n+2}$

45. Determine the first four nonzero terms in the Maclaurin expansion of $\tan(x)$ by using long division to divide the Maclaurin series for $\cos(x)$ into the Maclaurin series for $\sin(x)$.

46. Write the Taylor series of $1/(1 + x)$ about 2. *Hint:* Write

$$\frac{1}{1 + x} = \frac{1}{3 + (x - 2)} = \frac{1}{3} \frac{1}{1 + \dfrac{x - 2}{3}}$$

and use the geometric series $\sum_{n=0}^{\infty} (-1)^n r^n = 1/(1 + r)$ if $|r| < 1$.

47. Find the radius of convergence of the Taylor expansion of $\ln(1 + x)$ about 6. *Hint:* What is the distance from 6 to the nearest point at which $\ln(1 + x)$ is not analytic?

48. What is the radius of convergence of the Maclaurin expansion of $1/(9 + x^2)$? *Hint:* Look at $1/(9 + z^2)$ and determine the distance from zero to the nearest point at which the denominator vanishes.

49. What is the radius of convergence of the Taylor expansion of $1/(1 + x^2)$ about 5?

50. Let $f(x) = 1/(4 + x^2)$. Then f is analytic at x_0 for each real number x_0.
   (a) Expand f in a Maclaurin series and determine the radius of convergence of this series.
   (b) Note that $f(x)$ is defined for all real x. However, let $f(z) = 1/(4 + z^2)$, in which z can be complex. Find all the points at which $f(z)$ is not defined, considered as a function of the complex variable z. Such points are called *singularities*. Determine the distance from zero to the nearest singularity of $f(z)$, and compare this with your answer to (a).

51. (a) Determine the radius of convergence of the Maclaurin series for

$$f(x) = (x^2 - 4x + 13)^{-1}$$

without actually finding this series.
   (b) What is the radius of convergence of the series expansion of f about 2?

52. Define a function g by

$$g(x) = \begin{cases} e^{-1/x^2} & \text{if } x \neq 0 \\ 0 & \text{if } x = 0. \end{cases}$$

Use the definition of the derivative to show that $g'(0) = g''(0) = 0$. In fact, it can be shown that $g^{(n)}(0) = 0$ for every nonnegative integer n. Conclude that the Maclaurin series for g is identically zero and hence does not represent g except at the origin. This is an example of a function which is not analytic at zero, even though it has derivatives of all orders there.

5.2 Power Series Solutions of Differential Equations _____

The *power series method* of solving a differential equation consists of substituting a power series

$$\sum_{n=0}^{\infty} a_n(x - x_0)^n$$

into the differential equation and attempting to solve for the coefficients $a_0, a_1, a_2, \ldots$ so that this series is a solution. Not every differential equation can be solved in this way. However, the following two theorems give sufficient conditions for certain differential equations to have power series solutions.

THEOREM 5.2

If p and q are analytic at x_0, then every solution of the first order linear differential equation

$$y' + p(x)y = q(x)$$

is also analytic at x_0 and hence may be represented as a power series $\sum_{n=0}^{\infty} a_n(x - x_0)^n$.

   Further, the radius of convergence of this series solution is at least as large as the smaller of the radii of convergence of the Taylor series expansions of p and q about x_0. ∎

There is a similar result for second order linear differential equations.

THEOREM 5.3

Let p, q, and f be analytic at x_0. Then every solution of the second order linear differential equation

$$y'' + p(x)y' + q(x)y = f(x)$$

is also analytic at x_0 and hence has a power series representation $\sum_{n=0}^{\infty} a_n(x - x_0)^n$. This series has radius of convergence at least as large as the smaller of the radii of the Taylor expansions of p, q, and f about x_0. ∎

There are similar theorems for higher order linear differential equations. The point to these theorems is that we can expect power series solutions about x_0 at least when the differential equation is linear, its coefficients are analytic at x_0, and the coefficient of the highest derivative term is 1.

We will discuss a proof of Theorem 5.3 at the end of this section. Before that, we will look at examples of the power series method.

EXAMPLE 5.4

Solve $y' + ky = 0$, in which k is any nonzero constant.

Of course, we know that the general solution is $y = Ce^{-kx}$. Nevertheless, we will apply the power series method as an illustration. Choose $x_0 = 0$ for convenience. We know by Theorem 5.2 that solutions of $y' + ky = 0$ are analytic at zero. Assume a solution of the form

$$y = \sum_{n=0}^{\infty} a_n x^n.$$

Then

$$y' = \sum_{n=1}^{\infty} n a_n x^{n-1}$$

(this summation starts at $n = 1$). Substitute the series for y and y' into the differential equation to get

$$\sum_{n=1}^{\infty} n a_n x^{n-1} + \sum_{n=0}^{\infty} k a_n x^n = 0. \tag{5.2}$$

Shift indices to write

$$\sum_{n=1}^{\infty} n a_n x^{n-1} = \sum_{n=0}^{\infty} (n + 1) a_{n+1} x^n.$$

Now equation (5.2) can be written

$$\sum_{n=0}^{\infty} (n + 1) a_{n+1} x^n + \sum_{n=0}^{\infty} k a_n x^n = 0,$$

or

$$\sum_{n=0}^{\infty} [(n + 1) a_{n+1} + k a_n] x^{n+1} = 0.$$

Since this series must equal zero for all x in some interval about zero, every coefficient in the series must vanish. Therefore,

$$(n + 1)a_{n+1} + ka_n = 0 \qquad \text{for} \qquad n = 0, 1, 2, 3, \ldots,$$

so

$$a_{n+1} = -\frac{k}{n + 1}\, a_n \qquad \text{for} \qquad n = 0, 1, 2, 3, \ldots.$$

This is called a *recurrence relation*. It gives each coefficient in terms of (in this case) the preceding one. We have

$$a_1 = -ka_0,$$

$$a_2 = -k\left[\frac{1}{2} a_1\right] = -k\,\frac{1}{2}(-ka_0) = \frac{1}{2} k^2 a_0,$$

$$a_3 = -k\left[\frac{1}{3} a_2\right] = -k\,\frac{1}{3}\left[\frac{1}{2} k^2 a_0\right] = -\frac{1}{2 \cdot 3} k^3 a_0,$$

$$a_4 = -k\left[\frac{1}{4} a_3\right] = -k\,\frac{1}{4}\left[-\frac{1}{2 \cdot 3} k^3 a_0\right] = \frac{1}{2 \cdot 3 \cdot 4} k^4 a_0,$$

and so on. The pattern emerging is that

$$a_n = (-1)^n \frac{k^n}{n!}\, a_0$$

for $n = 1, 2, 3, \ldots$. The solution may therefore be written as

$$y = \sum_{n=0}^{\infty} (-1)^n \frac{k^n}{n!}\, a_0 x^n$$

$$= a_0 \sum_{n=0}^{\infty} \frac{(-1)^n}{n!} (kx)^n. \tag{5.3}$$

Now recall that the Taylor series for e^x about zero is

$$e^x = \sum_{n=0}^{\infty} \frac{1}{n!}\, x^n.$$

Replacing x by $-x$ yields

$$e^{-x} = \sum_{n=0}^{\infty} \frac{(-1)^n}{n!}\, x^n. \tag{5.4}$$

Now compare (5.3) with (5.4). We recognize that the solution may be written as

$$y = a_0 e^{-kx}.$$

In this example, the power series method has produced a series form of the solution we obtained by previous methods, with a_0 as the arbitrary constant in place of C. ∎

EXAMPLE 5.5

Solve $y'' + k^2 y = 0$, in which k is any positive constant.

We know that the general solution of this differential equation is

$$y = c_1\cos(kx) + c_2\sin(kx). \tag{5.5}$$

For illustration, we will seek power series solutions, again about zero. Write

$$y = \sum_{n=0}^{\infty} a_n x^n.$$

Then

$$y' = \sum_{n=1}^{\infty} na_n x^{n-1} \quad \text{and} \quad y'' = \sum_{n=2}^{\infty} n(n-1)a_n x^{n-2}.$$

Substitute these series into the differential equation to get

$$\sum_{n=2}^{\infty} n(n-1)a_n x^{n-2} + \sum_{n=0}^{\infty} k^2 a_n x^n = 0. \tag{5.6}$$

Shift indices in the first summation to write it as

$$\sum_{n=2}^{\infty} n(n-1)a_n x^{n-2} = \sum_{n=0}^{\infty} (n+2)(n+1)a_{n+2} x^n.$$

Now equation (5.6) becomes

$$\sum_{n=0}^{\infty} (n+2)(n+1)a_{n+2} x^n + \sum_{n=0}^{\infty} k^2 a_n x^n = 0,$$

or

$$\sum_{n=0}^{\infty} [(n+2)(n+1)a_{n+2} + k^2 a_n]x^n = 0.$$

For this series to converge to zero in some interval about zero, the coefficient of each power of x must be zero; hence,

$$(n+2)(n+1)a_{n+2} + k^2 a_n = 0$$

for $n = 0, 1, 2, 3, \ldots$. Write this equation as

$$a_{n+2} = \frac{-k^2}{(n+2)(n+1)} a_n \tag{5.7}$$

for $n = 0, 1, 2, 3, \ldots$.

Equation (5.7) is the recurrence relation. It gives each coefficient a_{n+2} in terms of a_n. Calculate some of these coefficients to try to discern a pattern:

$$a_2 = \frac{-k^2}{2 \cdot 1} a_0, \qquad\qquad a_3 = \frac{-k^2}{3 \cdot 2} a_1,$$

$$a_4 = \frac{-k^2}{4 \cdot 3} a_2 = \frac{+k^4}{4 \cdot 3 \cdot 2 \cdot 1} a_0, \qquad a_5 = \frac{-k^2}{5 \cdot 4} = \frac{+k^4}{5 \cdot 4 \cdot 3 \cdot 2 \cdot 1} a_1,$$

$$a_6 = \frac{-k^2}{6 \cdot 5} a_4 = \frac{-k^6}{6 \cdot 5 \cdot 4 \cdot 3 \cdot 2 \cdot 1} a_0, \quad a_7 = \frac{-k^2}{7 \cdot 6} a_5 = \frac{-k^6}{7 \cdot 6 \cdot 5 \cdot 4 \cdot 3 \cdot 2 \cdot 1} a_1,$$

and so on. The emerging pattern is that all even-indexed coefficients are multiples of a_0 and all odd-indexed coefficients are multiples of a_1. Specifically,

$$a_{2n} = \frac{(-1)^n k^{2n}}{(2n)!} a_0 \quad \text{and} \quad a_{2n+1} = \frac{(-1)^n k^{2n}}{(2n+1)!} a_1$$

for $n = 0, 1, 2, 3, \ldots$.

The power series solution may be written

$$y = \sum_{n=0}^{\infty} a_n x^n = \sum_{n=0}^{\infty} a_{2n} x^{2n} + \sum_{n=0}^{\infty} a_{2n+1} x^{2n+1}$$

$$= \sum_{n=0}^{\infty} \frac{(-1)^n k^{2n}}{(2n)!} a_0 x^{2n} + \sum_{n=0}^{\infty} \frac{(-1)^n k^{2n}}{(2n+1)!} a_1 x^{2n+1}.$$

In order to recognize these series as familiar functions, multiply and divide the series on the right by k and rewrite:

$$y = a_0 \sum_{n=0}^{\infty} \frac{(-1)^n}{(2n)!} (kx)^{2n} + \frac{1}{k} a_1 \sum_{n=0}^{\infty} \frac{(-1)^n}{(2n+1)!} (kx)^{2n+1}.$$

Now recall the Maclaurin expansions of $\sin(x)$ and $\cos(x)$. They are

$$\sin(x) = \sum_{n=0}^{\infty} \frac{(-1)^n}{(2n+1)!} x^{2n+1} \quad \text{and} \quad \cos(x) = \sum_{n=0}^{\infty} \frac{(-1)^n}{(2n)!} x^{2n}.$$

The solution may therefore be written

$$y = a_0 \cos(kx) + \frac{1}{k} a_1 \sin(kx). \tag{5.8}$$

Again, the power series method has produced the general solution of the differential equation. Note that (5.8) is identical with (5.5) if we write $c_1 = a_0$ and $c_2 = (1/k)a_1$. Here, a_0 and a_1 are arbitrary constants. ∎

We will now employ the power series method with more difficult differential equations rather than ones we can easily solve by other means.

EXAMPLE 5.6

Solve $y'' + xy' - y = 0$.

The coefficient functions x, -1, and 0 are analytic everywhere, and we can look for a power series solution about any point. We will seek solutions of the form $y = \sum_{n=0}^{\infty} a_n x^n$. Since

$$y' = \sum_{n=1}^{\infty} n a_n x^{n-1} \quad \text{and} \quad y'' = \sum_{n=2}^{\infty} n(n-1) a_n x^{n-2},$$

substitution into the differential equation yields

$$\sum_{n=2}^{\infty} n(n-1) a_n x^{n-2} + \sum_{n=1}^{\infty} n a_n x^n - \sum_{n=0}^{\infty} a_n x^n = 0.$$

Shift indices in the left-most series to write this equation as

$$\sum_{n=0}^{\infty} (n+2)(n+1) a_{n+2} x^n + \sum_{n=1}^{\infty} n a_n x^n - \sum_{n=0}^{\infty} a_n x^n = 0.$$

Write the $n = 0$ terms in the first and last series separately, and combine the series from $n = 1$ to infinity under one summation to get

$$2a_2 x^0 - a_0 x^0 + \sum_{n=1}^{\infty} [(n+2)(n+1) a_{n+2} + n a_n - a_n] x^n = 0.$$

For this equation to hold for all x in some interval about zero, the coefficient of each power of x on the left must be zero. Then

$$2a_2 - a_0 = 0 \quad \text{and} \quad (n+2)(n+1)a_{n+2} + na_n - a_n = 0 \qquad (5.9)$$

for $n = 1, 2, 3, \ldots$. From the equation $2a_2 = a_0$, we conclude that

$$a_2 = \tfrac{1}{2}a_0.$$

From the second equation in (5.9), we find that

$$a_{n+2} = \frac{1-n}{(n+2)(n+1)}\, a_n \quad \text{for} \quad n = 1, 2, 3, \ldots.$$

This is the recurrence relation for this problem. We use it to solve for a_3 in terms of a_1 (let $n = 1$); then for a_4 in terms of a_2 ($n = 2$); for a_5 in terms of a_3 ($n = 3$); and so on. Specifically, we get

$$a_3 = 0 \qquad\qquad\qquad\qquad\qquad\qquad (\text{let } n = 1),$$

$$a_4 = -\frac{1}{3 \cdot 4}\, a_2 = -\frac{1}{2 \cdot 3 \cdot 4}\, a_0 \qquad (\text{let } n = 2),$$

$$a_5 = -\frac{2}{4 \cdot 5}\, a_3 = 0 \qquad\qquad\qquad (\text{let } n = 3),$$

$$a_6 = -\frac{3}{5 \cdot 6}\, a_4 = \frac{-3}{5 \cdot 6}\frac{-1}{2 \cdot 3 \cdot 4}\, a_0 = \frac{3}{6!}\, a_0 \qquad (\text{let } n = 4),$$

and so on.

Observe the patterns that are forming. We will get

$$a_7 = a_9 = a_{11} = \cdots = a_{\text{odd}} = 0$$

because each odd-indexed coefficient beyond a_1 is a multiple of a_3, which is zero.

Next, let $n = 6$ to get

$$a_8 = \frac{-5}{7 \cdot 8}\, a_6 = -\frac{3 \cdot 5}{8 \cdot 7}\frac{1}{6!}\, a_0 = -\frac{3 \cdot 5}{8!}\, a_0.$$

Looking back at a_2, a_4, a_6, and a_8, the pattern suggests that

$$a_{2n} = \frac{(-1)^{n+1}(3)(5) \cdots (2n-3)}{(2n)!}\, a_0 \qquad (5.10)$$

for $n = 2, 3, \ldots$. This can be verified by mathematical induction.

Since there are no equations which a_0 or a_1 must satisfy, these are both arbitrary constants and can be assigned any values we like. Once a_0 is assigned a value, $a_2, a_4, a_6, \ldots$, and all even indexed coefficients are determined by equation (5.10). Recalling that $a_3 = a_5 = \cdots = 0$, we have the power series solution

$$y = \sum_{n=0}^{\infty} a_n x^n = a_0 + a_1 x + a_2 x^2 + a_4 x^4 + a_6 x^6 + \cdots$$

$$= a_0 + a_1 x + a_2 x^2 + \sum_{n=2}^{\infty} a_{2n} x^{2n}$$

$$= a_0 + a_1 x + \tfrac{1}{2}a_0 x^2 + a_0 \sum_{n=2}^{\infty} \frac{(-1)^{n+1}(3)(5) \cdots (2n-3)}{(2n)!}\, x^{2n}.$$

We can write this solution as

$$y = a_1 x + a_0 \left[1 + \tfrac{1}{2}x^2 + \sum_{n=2}^{\infty} \frac{(-1)^{n+1}(3)(5) \cdots (2n-3)}{(2n)!} x^{2n} \right]$$
$$= a_1 y_1(x) + a_0 y_2(x),$$

in which $y_1(x) = x$ and $y_2(x)$ is the function defined by the power series in square brackets. By the ratio test, this power series converges for all x.

It is easy to check that $y_2(0) = 1$ and $y_2'(0) = 0$; therefore,

$$W[y_1, y_2](0) = \begin{vmatrix} y_1(0) & y_2(0) \\ y_1'(0) & y_2'(0) \end{vmatrix} = \begin{vmatrix} 0 & 1 \\ 1 & 0 \end{vmatrix} = -1 \neq 0.$$

Hence, y_1 and y_2 are linearly independent (for all x) and the expression $a_1 y_1 + a_0 y_2$ is the general solution of the differential equation (recall that a_0 and a_1 are arbitrary constants). ∎

The Wronskian test is important in Example 5.6. Although y_1 is a simple function, y_2 is not, and the fact that we need to evaluate the Wronskian at only one point enables us show quite easily that y_1 and y_2 are linearly independent.

EXAMPLE 5.7

Solve the differential equation $y'' + y' - 2xy = 0$.

The coefficients 1, $-2x$, and 0 are analytic at zero. Substitute $y = \sum_{n=0}^{\infty} a_n x^n$ into the differential equation to get

$$\sum_{n=2}^{\infty} n(n-1)a_n x^{n-2} + \sum_{n=1}^{\infty} na_n x^{n-1} + \sum_{n=0}^{\infty} -2a_n x^{n+1} = 0.$$

Shift indices in each of these series to write

$$\sum_{n=0}^{\infty} (n+2)(n+1)a_{n+2} x^n + \sum_{n=0}^{\infty} (n+1)a_{n+1} x^n - 2a_0 x + \sum_{n=2}^{\infty} -2a_{n-1} x^n = 0.$$

Combine these series (from the $n=2$ term), being careful to retain the $n=0$ and $n=1$ terms in the first two series, to get

$$-2a_0 x + 2a_2 x^0 + a_1 x^0 + 3 \cdot 2a_3 x + 2a_2 x$$

$$+ \sum_{n=2}^{\infty} [(n+2)(n+1)a_{n+2} + (n+1)a_{n+1} - 2a_{n-1}]x^n = 0.$$

The coefficient of each power of x on the left side must be zero; hence,

$$2a_2 + a_1 = 0,$$
$$6a_3 + 2a_2 - 2a_0 = 0,$$

and

$$(n+2)(n+1)a_{n+2} + (n+1)a_{n+1} - 2a_{n-1} = 0$$

for $n = 2, 3, 4, \ldots$.

The last equation is the recurrence relation. For this differential equation, it gives a_{n+2} in terms of a_{n+1} and a_{n-1}. From the coefficient of the x^0 term, we get

$2a_2 + a_1 = 0$; hence,

$$a_2 = -\tfrac{1}{2}a_1.$$

From the coefficient of the x-term, we get $6a_3 + 2a_2 - 2a_0 = 0$; hence,

$$a_3 = \tfrac{1}{3}[a_0 - a_2] = \tfrac{1}{3}[a_0 + \tfrac{1}{2}a_1].$$

Now write the recurrence relation in the form

$$a_{n+2} = \frac{2a_{n-1} - (n+1)a_{n+1}}{(n+2)(n+1)} \quad \text{for} \quad n = 2, 3, 4, \dots.$$

From this equation, we can compute as many coefficients in the solution series as we like. Some of these coefficients are

$$a_4 = \frac{2a_1 - 3a_3}{4 \cdot 3} = \frac{2a_1 - a_0 - \tfrac{1}{2}a_1}{4 \cdot 3} = \frac{\tfrac{3}{2}a_1 - a_0}{12},$$

$$a_5 = \frac{2a_2 - 4a_4}{5 \cdot 4} = \frac{-a_1 - \tfrac{1}{2}a_1 + \tfrac{1}{3}a_0}{20} = \frac{-3}{40}a_1 + \frac{1}{60}a_0,$$

and

$$a_6 = \frac{2a_3 - 5a_5}{6 \cdot 5} = \frac{2(\tfrac{1}{3}a_0 + \tfrac{1}{6}a_1) - 5(-\tfrac{3}{40}a_1 + \tfrac{1}{60}a_0)}{30}$$

$$= \frac{7}{360}a_0 + \frac{17}{720}a_1.$$

Unlike Example 5.5, there is no simple pattern emerging, and we see no simple formula for a_n as a function of n. For example, to compute a_{32}, we must actually have a_{29} and a_{31} in hand. This difficulty is typical of solutions by the power series method. In fact, if the coefficients have a form we can easily recognize and write as a simple expression in terms of n, often there are solutions in terms of elementary functions such as sines, cosines, or exponentials, and we do not have to resort to the power series method.

Notice that a_0 and a_1 are arbitrary constants, since there is no equation restricting their values. We can obtain two specific solutions y_1 and y_2 of the differential equation by choosing specific values for a_0 and a_1. For example, if we choose $a_0 = 2$ and $a_1 = 0$, we obtain the solution

$$y_1(x) = 3 + x^3 - \tfrac{1}{4}x^4 + \tfrac{1}{20}x^5 + \tfrac{7}{120}x^6 + \cdots.$$

And, by letting $a_0 = 0$ and $a_1 = 2$, we get a second solution

$$y_2(x) = 2x - x^2 + \tfrac{1}{3}x^3 + \tfrac{1}{4}x^4 - \tfrac{3}{20}x^5 + \tfrac{17}{360}x^6 + \cdots.$$

Although we have computed only a few terms of these series solutions, we can make some general statements about y_1 and y_2. The coefficient functions in the differential equation $y'' + y' - 2xy = 0$ are 1, $-2x$, and 0, each of which is analytic at zero. The Maclaurin series of each of these coefficient functions converges for all x. Therefore, the solutions y_1 and y_2 both converge for all x. Further, we can easily compute

$$W[y_1, y_2](0) = \begin{vmatrix} y_1(0) & y_2(0) \\ y_1'(0) & y_2'(0) \end{vmatrix} = \begin{vmatrix} 3 & 0 \\ 0 & 2 \end{vmatrix} = 6.$$

Therefore, y_1 and y_2 are linearly independent and form a fundamental set of solutions of the differential equation. ∎

A differential equation need not be homogeneous for the power series method to apply.

EXAMPLE 5.8

Solve $y'' + xy' - y = 1 + x^2$.

The coefficient functions are analytic at zero, and we will seek solutions $y = \sum_{n=0}^{\infty} a_n x^n$. Substitute this power series into the differential equation and shift indices. Borrowing some work from Example 5.6, we obtain

$$\sum_{n=1}^{\infty} [(n + 2)(n + 1)a_{n+2} + (n - 1)a_n]x^n + 2a_2 x^0 - a_0 x^0 = 1 + x^2.$$

On the right side, we have $1 + x^2$, which is a power series about zero. In order for the power series on both sides of the equation to be equal, the coefficients of each power of x must be equal (because they are the Taylor coefficients of $1 + x^2$). Equate coefficients of equal powers of x to get

$$2a_2 - a_0 = 1 \qquad \text{(from the constant term)},$$
$$3(2)a_3 + (1 - 1)a_1 = 0 \qquad \text{(from the coefficient of } x\text{)},$$
$$4(3)a_4 + (2 - 1)a_2 = 1 \qquad \text{(from the coefficient of } x^2\text{)},$$

and, for $n = 3, 4, 5, \ldots,$

$$(n + 2)(n + 1)a_{n+2} + (n - 1)a_n = 0.$$

The recurrence relation comes from the coefficients of x^n on both sides of the equation, for $n = 3, 4, 5, \ldots.$

From $2a_2 - a_0 = 1$, we get

$$a_2 = \tfrac{1}{2}(1 + a_0).$$

From the equation $6a_3 + 0 \cdot a_1 = 0$, we conclude that

$$a_3 = 0.$$

Write the recurrence relation in the form

$$a_{n+2} = \frac{1 - n}{(n + 2)(n + 1)} a_n \quad \text{for} \quad n = 3, 4, 5, \ldots.$$

From this recurrence relation, and the fact that $a_3 = 0$, we get $a_5 = 0$; then $a_7 = 0$ and, in general, $a_k = 0$ if k is odd and $k \geq 3$.

From the coefficient of x^2, we get $12a_4 + a_2 = 1$, or

$$a_4 = \tfrac{1}{12}(1 - a_2) = \tfrac{1}{12}[1 - \tfrac{1}{2} - \tfrac{1}{2}a_0] = \tfrac{1}{24}(1 - a_0).$$

Now we obtain, by successive application of the recurrence relation,

$$a_6 = \frac{-3}{6 \cdot 5} a_4 = \frac{-3}{6 \cdot 5 \cdot 24}(1 - a_0),$$

$$a_8 = \frac{-5}{8 \cdot 7} a_6 = \frac{3 \cdot 5}{8 \cdot 7 \cdot 6 \cdot 5 \cdot 24}(1 - a_0),$$

and

$$a_{10} = \frac{-7}{10 \cdot 9} a_8 = \frac{-3 \cdot 5 \cdot 7}{10 \cdot 9 \cdot 8 \cdot 7 \cdot 6 \cdot 5 \cdot 24}(1 - a_0).$$

The pattern is

$$a_{2n} = \frac{(-1)^n (3)(5) \cdots (2n - 3)}{(2n)!}(1 - a_0)$$

for $n = 3, 4, 5, \ldots$. The power series solution we have found is

$$y = a_0 + a_1 x + \frac{1}{2}(1 + a_0)x^2 + \frac{1}{4!}(1 - a_0)x^4$$

$$- \frac{3}{6!}(1 - a_0)x^6 + \frac{3 \cdot 5}{8!}(1 - a_0)x^8 - \frac{3 \cdot 5 \cdot 7}{10!}(1 - a_0)x^{10} + \cdots,$$

with a_0 and a_1 arbitrary constants. If we choose $a_0 = a_1 = 1$, we get the specific solution

$$y_1 = 1 + x + x^2,$$

which can be verified by substitution into the differential equation. If we let $a_0 = 1$ and $a_1 = 0$, we get the solution

$$y_2 = 1 + x^2.$$

In this instance, we have found particular solutions of the nonhomogeneous problem which involve only finitely many terms (they are, in fact, power series with only a finite number of nonzero coefficients). This is not to be expected in general. In fact, we obtain series solutions with infinitely many nonzero coefficients in this example if we choose $a_0 \neq 1$. ∎

For some differential equations, it may be necessary to expand one or more of the coefficient functions in a power series in order to apply the power series method.

EXAMPLE 5.9

Solve $y'' + xy' - y = e^{2x}$.

Again, we have borrowed the left side of this differential equation from Example 5.6 to save some computation. We will attempt a solution $y = \sum_{n=0}^{\infty} a_n x^n$, since the coefficient functions x, -1, and e^{2x} are analytic at zero. Substitute this power series into the differential equation to get (from Example 5.6)

$$2a_2 x^0 - a_0 x^0 + \sum_{n=1}^{\infty} [(n + 2)(n + 1)a_{n+2} + (n - 1)a_n]x^n = e^{2x}.$$

Since the right side is not a power series, we have no way of comparing coefficients of powers of x on both sides of the equation. Expand e^{2x} in a Maclaurin series

$$e^{2x} = \sum_{n=0}^{\infty} \frac{2^n}{n!} x^n.$$

We now have

$$2a_2 x^0 - a_0 x^0 + \sum_{n=1}^{\infty} [(n+2)(n+1)a_{n+2} + (n-1)a_n]x^n = \sum_{n=0}^{\infty} \frac{2^n}{n!} x^n. \quad (5.11)$$

Write a few terms of the series on both sides of this equation:

$$(2a_2 - a_0)x^0 + 6a_3 x + (12a_4 + a_2)x^2 + (20a_5 + 2a_3)x^3$$
$$+ (30a_6 + 3a_4)x^4 + (42a_7 + 4a_5)x^5 + \cdots$$
$$= 1 + 2x + 2x^2 + \tfrac{4}{3}x^3 + \tfrac{2}{3}x^4 + \tfrac{4}{15}x^5 + \cdots.$$

By matching coefficients of x^n on both sides of this equation, we get

$$2a_2 - a_0 = 1$$
$$6a_3 = 2$$
$$12a_4 + a_2 = 2$$
$$20a_5 + 2a_3 = \tfrac{4}{3}$$
$$30a_6 + 3a_4 = \tfrac{2}{3}$$
$$42a_7 + 4a_5 = \tfrac{4}{15}.$$

Then

$$a_2 = \frac{1}{2}(1 + a_0)$$

$$a_3 = \frac{1}{3}$$

$$a_4 = \frac{1}{24}(3 - a_0)$$

$$a_5 = \frac{1}{30}$$

$$a_6 = \frac{1}{240}\left(\frac{7}{3} + a_0\right)$$

$$a_7 = \frac{1}{315}.$$

By writing more terms on both sides of equation (5.11), we can generate more equations which can be solved for more of the coefficients.

We now have a power series solution

$$y = a_0 + a_1 x + \tfrac{1}{2}(1 + a_0)x^2 + \tfrac{1}{3}x^3 + \tfrac{1}{24}(3 - a_0)x^4$$
$$+ \tfrac{1}{30}x^5 + \tfrac{1}{240}(\tfrac{7}{3} + a_0)x^6 + \tfrac{1}{315}x^7 + \cdots.$$

Since the Maclaurin series for each coefficient function has infinite radius of convergence, this solution also converges for all x by Theorem 5.3. By making different choices of a_0 and a_1, we obtain different particular solutions. ∎

Sometimes application of the power series method will require that we multiply two power series. Here is an example involving this.

EXAMPLE 5.10

Find a solution of $y'' + e^x y = 1$.

Both e^x and 1 are analytic at zero, so there are solutions

$$y = \sum_{n=0}^{\infty} a_n x^n.$$

But e^x is not a polynomial, so we must insert the Maclaurin series for e^x into the differential equation in order to collect coefficients of like powers of x. We get

$$\sum_{n=2}^{\infty} n(n-1)a_n x^{n-2} + \left(\sum_{n=0}^{\infty} \frac{1}{n!} x^n \right)\left(\sum_{n=0}^{\infty} a_n x^n \right) = 1.$$

This equation may be written

$$2a_2 + 6a_3 x + 12a_4 x^2 + 20a_5 x^3 + \cdots$$
$$+ (1 + x + \tfrac{1}{2}x^2 + \tfrac{1}{6}x^3 + \cdots)(a_0 + a_1 x + a_2 x^2 + a_3 x^3 + \cdots) = 1.$$

By calculating a few terms of this product of series, we get

$$2a_2 + 6a_3 x + 12a_4 x^2 + 20a_5 x^3 + \cdots + a_0 + (a_0 + a_1)x + \left(\frac{a_0}{2} + a_1 + a_2 \right)x^2$$
$$+ \left(\frac{a_0}{6} + \frac{a_1}{2} + a_2 + a_3 \right)x^3 + \cdots = 1.$$

Collect coefficients of like powers of x to write

$$(2a_2 + a_0) + (6a_3 + a_0 + a_1)x + \left(12a_4 + \frac{a_0}{2} + a_1 + a_2 \right)x^2$$
$$+ \left(20a_5 + \frac{a_0}{6} + \frac{a_1}{2} + a_2 + a_3 \right)x^3 + \cdots = 1.$$

Equate coefficients of like powers of x on both sides of the equation (noting that we have 1, not zero, on the right side). We get

$$2a_2 + a_0 = 1,$$
$$6a_3 + a_0 + a_1 = 0,$$
$$12a_4 + \frac{a_0}{2} + a_1 + a_2 = 0,$$
$$20a_5 + \frac{a_0}{6} + \frac{a_1}{2} + a_2 + a_3 = 0,$$

and so on. Solve these equations to get

$$a_2 = \frac{1}{2}(1 - a_0)$$

$$a_3 = -\frac{1}{6}(a_0 + a_1)$$

$$a_4 = -\frac{1}{12}\left(\frac{a_0}{2} + a_1 + a_2\right) = -\frac{1}{24}(1 + 2a_1)$$

$$a_5 = -\frac{1}{20}\left(\frac{a_0}{6} + \frac{a_1}{2} + a_2 + a_3\right) = -\frac{1}{120}(3a_0 + 2a_1 + 3).$$

Continuing in this way, we can write a_n in terms of a_0 and a_1 for n as large as we want; a_0 and a_1 are arbitrary constants. We obtain specific solutions by assigning values to a_0 and a_1. If we let $a_0 = 1$ and $a_1 = 0$, we get the solution

$$y_1(x) = 1 - \tfrac{1}{6}x^3 - \tfrac{1}{24}x^4 + \cdots.$$

If we let $a_0 = 0$ and $a_1 = 1$, we get the second solution

$$y_2(x) = x + \tfrac{1}{2}x^2 - \tfrac{1}{6}x^3 - \tfrac{1}{8}x^4 - \tfrac{1}{24}x^5 + \cdots.$$

The Wronskian test can be used to show that these solutions are linearly independent. ■

In these examples, we have always sought a solution about zero, primarily because the notation was easier in describing the power series method. However, if we have an initial value problem in which data are given at a point $x_0 \neq 0$, we attempt power series solutions about x_0. Alternatively, we can introduce a change of variables $t = x - x_0$ and translate the problem so that we still deal with expansions about zero.

EXAMPLE 5.11

Consider the initial value problem

$$xy'' - y' + y = 0; \qquad y(1) = 2, \qquad y'(1) = -4.$$

The coefficient functions are x, -1, 1, and 0. Since the initial data are prescribed at $x_0 = 1$ and the coefficient functions are analytic at 1, there are power series solutions about 1:

$$y = \sum_{n=0}^{\infty} a_n(x - 1)^n.$$

We can substitute this series into $xy'' - y' + y = 0$ to solve for the a_n's.

Alternatively, we can change variables in the initial value problem by letting $t = x - 1$ and $Y(t) = y(1 + t)$. Compute

$$Y'(t) = \frac{dy}{dx}\frac{dx}{dt} = y'(x) \quad \text{and} \quad Y''(t) = y''(x).$$

Further, $Y(0) = y(1) = 2$ and $Y'(0) = y'(1) = -4$.

The initial value problem for Y in terms of t is

$$(t + 1)Y'' - Y' + Y = 0; \qquad Y(0) = 2, \qquad Y'(0) = -4,$$

or

$$Y'' - \frac{1}{t + 1} Y' + \frac{1}{t + 1} Y = 0; \qquad Y(0) = 2, \qquad Y'(0) = -4.$$

By Theorem 5.3, this differential equation has solutions analytic at zero:

$$Y(t) = \sum_{n=0}^{\infty} a_n t^n.$$

Substitute this power series into the differential equation for Y to get

$$\sum_{n=2}^{\infty} n(n - 1)a_n t^{n-1} + \sum_{n=2}^{\infty} n(n - 1)a_n t^{n-2} - \sum_{n=1}^{\infty} na_n t^{n-1} + \sum_{n=0}^{\infty} a_n t^n = 0.$$

Shift indices in the first three summations to get

$$\sum_{n=1}^{\infty} (n + 1)na_{n+1} t^n + \sum_{n=0}^{\infty} (n + 2)(n + 1)a_{n+2} t^n - \sum_{n=0}^{\infty} (n + 1)a_{n+1} t^n + \sum_{n=0}^{\infty} a_n t^n = 0.$$

Write this equation as

$$2a_2 - a_1 + a_0 + \sum_{n=1}^{\infty} [(n + 1)na_{n+1} + (n + 2)(n + 1)a_{n+2} - (n + 1)a_{n+1} + a_n] t^n = 0.$$

The coefficient of each power of t^n must be zero. Then

$$2a_2 - a_1 + a_0 = 0$$

and

$$a_{n+2} = \frac{-a_n + (1 - n^2)a_{n+1}}{(n + 2)(n + 1)} \quad \text{for} \quad n = 1, 2, 3, \cdots.$$

Unlike the previous examples, we have specific values for a_0 and a_1 from the initial conditions

$$Y(0) = a_0 = 2 \quad \text{and} \quad Y'(0) = a_1 = -4.$$

Then

$$a_2 = \frac{1}{2}(a_1 - a_0) = -3,$$

$$a_3 = \frac{4 + 0}{3 \cdot 2} = \frac{2}{3},$$

$$a_4 = \frac{3 + (-3)(\frac{2}{3})}{4 \cdot 3} = \frac{1}{12},$$

$$a_5 = \frac{-\frac{2}{3} + (-8)(\frac{1}{12})}{5 \cdot 4} = -\frac{1}{15},$$

and so on. The initial value problem for Y has the unique solution

$$Y(t) = 2 - 4t - 3t^2 + \tfrac{2}{3}t^3 + \tfrac{1}{12}t^4 - \tfrac{1}{15}t^5 + \cdots,$$

and we can compute more coefficients explicitly if we wish.

The original initial value problem has the unique solution

$$\begin{aligned} y(x) &= Y(x-1) \\ &= 2 - 4(x-1) - 3(x-1)^2 + \tfrac{2}{3}(x-1)^3 + \tfrac{1}{12}(x-1)^4 + \cdots. \quad\blacksquare \end{aligned}$$

We will conclude this section with a sketch of a proof of Theorem 5.3. The proof actually contains a useful method for generating terms in the series solution of an initial value problem.

Sketch of a Proof of Theorem 5.3   Choose any two numbers A and B and consider the initial value problem

$$y'' + p(x)y' + q(x)y = f(x); \qquad y(x_0) = A, \qquad y'(x_0) = B.$$

Assume that p, q, and f have power series representations about x_0 valid in an interval $(x_0 - R, x_0 + R)$. We will show that the unique solution of the initial value problem has a power series expansion $y(x) = \sum_{n=0}^{\infty} a_n(x - x_0)^n$ in this interval. The idea is to compute the a_n's in terms of A, B, and known values of p, q, and f.

From the initial data, we know that

$$y(x_0) = a_0 = A \quad \text{and} \quad y'(x_0) = a_1 = B.$$

From the differential equation, we have

$$y'' = f - py' - qy.$$

Therefore,

$$\begin{aligned} y''(x_0) &= f(x_0) - p(x_0)y'(x_0) - q(x_0)y(x_0) \\ &= f(x_0) - p(x_0)B - q(x_0)A. \end{aligned}$$

Thus, we have computed $y''(x_0)$ in terms of known quantities. Now $a_2 = (1/2!)y''(x_0)$ is known.

Next, differentiate the differential equation to get

$$y^{(3)} + p'y' + py'' + q'y + qy' = f';$$

hence,

$$y^{(3)}(x_0) = f'(x_0) - p'(x_0)B - p(x_0)y''(x_0) - q'(x_0)A - q(x_0)B,$$

which gives $y^{(3)}(x_0)$ in terms of known quantities. Therefore, $a_3 = (1/3!)y^{(3)}(x_0)$ is known.

Since p, q, and f are analytic at x_0, we can continue to differentiate the differential equation and solve for $y^{(n)}(x_0)$ in terms of quantities given or previously computed. Therefore, each Taylor coefficient $a_n = (1/n!)y^{(n)}(x_0)$ can be computed in terms of known quantities, yielding the Taylor series for y about x_0. Finally, it can be shown that this series converges to $y(x)$ in $(x_0 - R, x_0 + R)$. $\blacksquare$

Here is an illustration of this method of generating a series solution of an initial value problem.

EXAMPLE 5.12

Find the first five nonzero terms of the series solution about zero of

$$y'' - xy' + e^x y = 4; \qquad y(0) = 1, \qquad y'(0) = 4.$$

The Maclaurin series of the solution is

$$y = \sum_{n=0}^{\infty} \frac{1}{n!} y^{(n)}(0) x^n.$$

We must compute the derivatives $y^{(n)}(0)$. We already know that $y(0) = 1$ and $y'(0) = 4$. Put these values into the differential equation and solve for $y''(0)$ to get

$$y''(0) = 4 + 0 \cdot y'(0) - e^0 y(0) = 4 - 1 = 3.$$

Next, differentiate the differential equation to get

$$y^{(3)}(x) - y'(x) - xy''(x) + e^x y(x) + e^x y'(x) = 0. \tag{5.12}$$

Let $x = 0$ and solve for $y^{(3)}(0)$ to get

$$y^{(3)}(0) = y'(0) + 0 \cdot y''(0) - e^0 y(0) - e^0 y'(0) = -1.$$

Differentiate equation (5.12):

$$y^{(4)}(x) - y''(x) - xy^{(3)}(x) - y''(x)$$
$$+ e^x y(x) + e^x y'(x) + e^x y'(x) + e^x y''(x) = 0. \tag{5.13}$$

Substitute $x = 0$ into equation (5.13) and solve for $y^{(4)}(0)$ to get

$$y^{(4)}(0) = 2y''(0) - 2e^0 y'(0) - e^0 y''(0) - e^0 y(0)$$
$$= 2(3) - 2(4) - 3 - 1 = -6.$$

The Maclaurin series of the solution is

$$y = 1 + 4x + \frac{3}{2!} x^2 - \frac{1}{3!} x^3 - \frac{6}{4!} x^4 + \cdots,$$

in which the dots represent terms we can compute by continuing the above process. ∎

The power series method is valid if the coefficient functions in the differential equation are analytic at a point. When this condition fails, we may still be able to obtain series solutions. We will consider this possibility in the next section.

PROBLEMS FOR SECTION 5.2

In each of Problems 1 through 14, consider a power series solution about zero. Find a recurrence relation for the coefficients and write the first five nonzero terms of each of two linearly independent power series solutions expanded about zero. Check linear independence using the Wronskian.

1. $y'' + xy = 0$

2. $y'' + xy' + 2y = 0$

3. $y'' - 2y' + xy = 0$

4. $y'' - y' + x^2 y = 0$

5. $y'' - x^3 y = 0$

6. $y'' - x^2 y' + 2y = 0$

7. $y'' + (1 - x)y' + 2xy = 0$

8. $y'' + y' + (x - 4)y = 0$

9. $y'' + xy' + 2xy = 0$

10. $y'' + x^2 y' + 2y = 0$

11. $y'' + y' - x^2 y = 0$

12. $y'' + x^2 y' + (x^2 - 1)y = 0$

13. $y'' - 8xy = 1 + 2x$

14. $y'' - 4y' + xy = -4 + 6x$

15. Apply the power series method to find the general solution of the first order differential equation $y' - 2xy = 0$.

In each of Problems 16 through 25, find the first five nonzero terms of the Maclaurin series of the general solution of the differential equation. Also find the recurrence relation for the coefficients in the series solution.

16. $y'' + 2x^2 y' - 3x^2 y = 0$

17. $2y'' - 4xy' + 8x^2 y = 0$

18. $y'' + 8x^2 y' - xy = 0$

19. $y'' + 12y' + x^2 y = 0$

20. $y'' + 2y' - 4x^2 y = 0$

21. $y'' + 2\cos(x)y' = x$

22. $y'' - e^x y' + 2y = 1$

23. $y'' - 2\tan(x)y' + y = 0$

24. $y'' - y' + \sin(2x)y = 1 + x$

25. $y'' - e^{-3x} y = 2x^2$

In each of Problems 26 through 40, find the first five nonzero terms of the solution of the initial value problem, using the method of Example 5.12.

26. $y'' - xy' + y = 0$; $y(0) = 1$, $y'(0) = 0$

27. $y'' + y' - xy = 0$; $y(0) = -2$, $y'(0) = 0$

28. $y'' + x^2 y' - y = 0$; $y(0) = 0$, $y'(0) = 1$

29. $y'' + 2xy' + (x - 1)y = 0$; $y(0) = 1$, $y'(0) = 2$

30. $y'' + 2y' - x^2 y = 0$; $y(0) = 4$, $y'(0) = 2$

31. $y'' - xy = 2x$; $y(1) = 3$, $y'(1) = 0$

32. $y'' + x^2 y' - 4y = 1$; $y(1) = 0$, $y'(1) = -1$

33. $y'' + xy' = -1 + x$; $y(2) = 1$, $y'(2) = -4$

34. $y'' - x^2 y' + (x + 2)y = x$; $y(0) = 2$, $y'(0) = 1$

35. $y'' - \dfrac{1}{x^2} y' + \dfrac{1}{x} y = 0$; $y(1) = 7$, $y'(1) = 3$ *Hint:* Multiply the differential equation by x^2 before proceeding.

36. $y'' + xy' - \sin(x)y = 1$; $y(0) = 1$, $y'(0) = -1$

37. $y'' + x^2 y = e^x$; $y(0) = -2$, $y'(0) = 7$

38. $y'' + xy' - xy = 1 - 4x$; $y(1) = 1$, $y'(1) = 0$

39. $y'' - e^x y' + 2y = 1$; $y(0) = -3$, $y'(0) = 1$

40. $y'' + y' - x^4 y = \sin(2x)$; $y(0) = 0$, $y'(0) = -2$

In each of Problems 41 through 47, use Theorem 5.3 to determine an interval about x_0 in which the initial value problem has a solution which is analytic at x_0. You need not attempt to produce this solution.

41. $y'' + \dfrac{1}{x + 2} y' - xy = 0$; $y(0) = y'(0) = 1$

42. $y'' - y' + \dfrac{1}{x} y = 1$; $y(4) = 0$, $y'(4) = 2$

43. $y'' - \ln(x)y' = -1 + x$; $y(3) = 1$, $y'(3) = \pi$

44. $y'' + x^2 y' - xy = e^x$; $y(-1) = 14$, $y'(-1) = 0$

45. $y'' + \dfrac{1}{x - 1} y' + \dfrac{1}{x + 2} y = 2$; $y(0) = y'(0) = 3$

46. $(x - 2)y'' + xy' - y = 0$; $y(0) = -3$, $y'(0) = 5$

47. $x^2 y'' - y' + xy = e^x$; $y(-2) = 1$, $y'(-2) = 6$

48. Solve Airy's equation: $y'' = xy$.

Problems 49 through 56 indicate how properties of $\sin(x)$ and $\cos(x)$ can be obtained directly from the fact that these functions are solutions of certain initial value problems. Imagine for the moment that $\sin(x)$ and $\cos(x)$ are unfamiliar. Let $S(x)$ be the unique solution of the initial value problem

$$y'' + y = 0; \qquad y(0) = 0, \qquad y'(0) = 1. \tag{5.14}$$

Let $C(x)$ be the unique solution of

$$y'' + y = 0; \qquad y(0) = 1, \qquad y'(0) = 0. \tag{5.15}$$

49. Show that $S'(x) = C(x)$. *Hint:* Show that S' is a solution of (5.15).

50. Show that $C'(x) = -S(x)$.

51. Show that $S(x)^2 + C(x)^2 = 1$ for all x. *Hint:* Consider the derivative of $\varphi(x) = S^2(x) + C^2(x)$.

52. Show that C and S are linearly independent solutions of $y'' + y = 0$.

53. Show that $S(a + b) = S(a)C(b) + C(a)S(b)$ for every pair of real numbers a and b. *Hint:* Consider the initial value problem

$$y'' + y = 0; \qquad y(0) = S(a), \qquad y'(0) = C(a).$$

Show that $\varphi(x) = c_1 C(x) + c_2 S(x)$ is the general solution of the differential equation, then determine c_1 and c_2 from the initial data. Next, show that $\psi(x) = S(x + a)$ is also a solution of this initial value problem.

54. Show that $C(a + b) = C(a)C(b) - S(a)S(b)$ for every real a and b.

55. Show that $S(2x) = 2S(x)C(x)$.

56. Show that $C(2x) = C(x)^2 - S(x)^2$.

57. Derive the binomial series expansion of $(1 + x)^\alpha$ about zero as follows. Show that $(1 + x)^\alpha$ is a solution of the differential equation

$$(1 + x)y' - \alpha y = 0,$$

and derive a series solution for this equation.

5.3 *Singular Points and the Method of Frobenius*

For the remainder of this chapter, we will deal with the homogeneous second order differential equation

$$P(x)y'' + Q(x)y' + R(x)y = 0. \tag{5.16}$$

We call x_0 an *ordinary point* of equation (5.16) if $P(x_0) \neq 0$ and P, Q, and R are analytic at x_0. If x_0 is an ordinary point, we can divide (5.16) by $P(x)$ in some interval about x_0 to obtain $y'' + p(x)y' + q(x)y = 0$ and obtain power series solutions about x_0.

If x_0 is not an ordinary point of equation (5.16), we call it a *singular point*. The study of the behavior of solutions in an interval about a singular point has led to the following definition, which delineates two kinds of singular points.

DEFINITION

x_0 is a *regular singular point* of the differential equation

$$P(x)y'' + Q(x)y' + R(x)y = 0$$

if x_0 is a singular point and if both $(x - x_0)[Q(x)/P(x)]$ and $(x - x_0)^2[R(x)/P(x)]$ are analytic at x_0.

x_0 is an *irregular singular point* of this differential equation if x_0 is a singular point but not a regular singular point. ∎

EXAMPLE 5.13

Legendre's differential equation is

$$(1 - x^2)y'' - 2xy' + \alpha(\alpha + 1)y = 0,$$

in which α is any constant. This is of the form of equation (5.16) with

$$P(x) = 1 - x^2, \qquad Q(x) = -2x, \quad \text{and} \quad R(x) = \alpha(\alpha + 1).$$

We have

$$\frac{Q(x)}{P(x)} = \frac{-2x}{1 - x^2} \quad \text{and} \quad \frac{R(x)}{P(x)} = \frac{\alpha(\alpha + 1)}{1 - x^2}.$$

At every point except 1 and -1, Q/P and R/P are analytic, so such points are ordinary points of Legendre's equation. Both 1 and -1 are singular points.

To determine whether 1 is a regular or irregular singular point, look at

$$(x - 1)\frac{Q(x)}{P(x)} = \frac{2x}{1 + x}$$

and

$$(x - 1)^2 \frac{R(x)}{P(x)} = -\frac{\alpha(\alpha + 1)(x - 1)}{1 + x}.$$

Both $(x - 1)[Q(x)/P(x)]$ and $(x - 1)^2[R(x)/P(x)]$ are analytic at 1. Thus, 1 is a regular singular point of Legendre's equation. Similarly, -1 is a regular singular point because

$$(x + 1)\frac{Q(x)}{P(x)} = \frac{-2x}{1 - x}$$

and

$$(x + 1)^2 \frac{R(x)}{P(x)} = \frac{\alpha(\alpha + 1)(x + 1)}{1 - x}$$

are both analytic at -1. ∎

EXAMPLE 5.14

Consider the differential equation

$$x^3(x - 2)^2 y'' + 5(x + 2)(x - 2)y' + 3x^2 y = 0.$$

This is of the form of equation (5.16) with

$$P(x) = x^3(x - 2)^2, \qquad Q(x) = 5(x + 2)(x - 2), \quad \text{and} \quad R(x) = 3x^2.$$

Now,

$$\frac{Q(x)}{P(x)} = \frac{5(x + 2)(x - 2)}{x^3(x - 2)^2} = \frac{5(x + 2)}{x^3(x - 2)}$$

and

$$\frac{R(x)}{P(x)} = \frac{3}{x(x - 2)^2}.$$

Every point except zero and 2 is an ordinary point. The differential equation has singular points at zero and 2. To classify the singular point at 2, examine

$$(x - 2)\frac{Q(x)}{P(x)} = \frac{5(x + 2)}{x^3}$$

and

$$(x - 2)^2 \frac{R(x)}{P(x)} = \frac{3}{x}.$$

Both of these functions are analytic at 2; thus, 2 is a regular singular point.

To classify the singular point at zero, look at

$$x\frac{Q(x)}{P(x)} = \frac{5(x + 2)}{x^2(x - 2)}.$$

This function is not analytic at zero; therefore, zero is an irregular singular point of the differential equation. ■

If x_0 is an irregular singular point of equation (5.16), the various aspects of finding a solution about x_0 are too delicate for us to treat at this time, and we will leave this case for a more advanced course in differential equations.

If x_0 is a regular singular point, we attempt a *Frobenius series* solution, which is a series of the form

$$y = \sum_{n=0}^{\infty} c_n(x - x_0)^{n+r}. \tag{5.17}$$

This is the power series $\sum_{n=0}^{\infty} c_n(x - x_0)^n$ multiplied by $(x - x_0)^r$. Since r need not be a nonnegative integer, a Frobenius series need not be a power series.

The idea is to attempt to choose the c_n's *and the number* r so that the Frobenius series defines a solution in some interval, which may be of the form $(x_0, x_0 + R)$ or $(x_0 - R, x_0)$. This procedure is called the *method of Frobenius*.

Some of the mechanics of finding the c_n's and r are similar to the power series method, but there are also essential differences between the power series method and the method of Frobenius. One significant difference is that a Frobenius series has $c_0(x - x_0)^r$ as its first term, and this is not constant if $r \neq 0$. Thus, the derivative of a Frobenius series is

$$y' = \sum_{n=0}^{\infty} (n + r)c_n(x - x_0)^{n+r-1},$$

with the series beginning at $n = 0$, *not* $n = 1$.

Another difference is that the method may not yield two linearly independent solutions, in which case we must find some other way of producing a second solution. We will deal with the problem of finding a second solution in the next section.

Here are two examples of the method.

EXAMPLE 5.15

Find a solution of $x^2y'' + 5xy' + (x + 4)y = 0$.

It is routine to check that zero is a regular singular point of this differential equation. We will attempt a Frobenius solution about zero:

$$y = \sum_{n=0}^{\infty} c_n x^{n+r}.$$

We have

$$y' = \sum_{n=0}^{\infty} (n+r)c_n x^{n+r-1} \quad \text{and} \quad y'' = \sum_{n=0}^{\infty} (n+r)(n+r-1)c_n x^{n+r-2}.$$

Substitute these series into the differential equation to get

$$\sum_{n=0}^{\infty} (n+r)(n+r-1)c_n x^{n+r} + \sum_{n=0}^{\infty} 5(n+r)c_n x^{n+r}$$

$$+ \sum_{n=0}^{\infty} c_n x^{n+r+1} + \sum_{n=0}^{\infty} 4c_n x^{n+r} = 0.$$

Shift indices in the third summation to write this equation as

$$\sum_{n=0}^{\infty} (n+r)(n+r-1)c_n x^{n+r} + \sum_{n=0}^{\infty} 5(n+r)c_n x^{n+r}$$

$$+ \sum_{n=1}^{\infty} c_{n-1} x^{n+r} + \sum_{n=0}^{\infty} 4c_n x^{n+r} = 0.$$

Combine these series under one summation, from $n = 1$, and write the terms corresponding to $n = 0$ separately, to get

$$r(r-1)c_0 x^r + 5rc_0 x^r + 4c_0 x^r$$

$$+ \sum_{n=1}^{\infty} [(n+r)(n+r-1)c_n + 5(n+r)c_n + c_{n-1} + 4c_n]x^{n+r} = 0.$$

Let the coefficient of each power of x be zero to get

$$[r(r-1) + 5r + 4]c_0 = 0 \tag{5.18}$$

and, for $n = 1, 2, 3, \ldots,$

$$(n+r)(n+r-1)c_n + 5(n+r)c_n + c_{n-1} + 4c_n = 0. \tag{5.19}$$

Assume that $c_0 \neq 0$. Then equation (5.18) implies that

$$r(r-1) + 5r + 4 = 0,$$

or

$$r^2 + 4r + 4 = 0. \tag{5.20}$$

Equation (5.20) is called the *indicial equation*; it is this equation we solve for r.

Equation (5.19) is a recurrence relation. Combine terms in (5.19) to get

$$[(r+n)^2 + 4(r+n) + 4]c_n + c_{n-1} = 0. \tag{5.21}$$

Now observe a relationship between the indicial equation (5.20) and the recurrence relation (5.21). If we define

$$F(r) = r^2 + 4r + 4,$$

the indicial equation is just $F(r) = 0$ and the recurrence relation is

$$F(r + n)c_n + c_{n-1} = 0$$

for $n = 1, 2, 3, \ldots$. Solve the indicial equation for r to get $r = -2$ [one repeated root of $F(r) = 0$]. Put $r = -2$ into the recurrence relation to get

$$F(-2 + n)c_n + c_{n-1} = 0,$$

or

$$c_n = -\frac{1}{F(-2 + n)} c_{n-1}$$

for $n = 1, 2, 3, \ldots$. Calculate

$$F(-2 + n) = (-2 + n)^2 + 4(-2 + n) + 4 = n^2.$$

The recurrence relation is therefore

$$c_n = -\frac{1}{n^2} c_{n-1} \qquad (n \geq 1).$$

Then

$$c_1 = -c_0,$$

$$c_2 = \frac{-c_1}{4} = \frac{+c_0}{4},$$

$$c_3 = \frac{-c_2}{9} = \frac{-c_0}{4 \cdot 9},$$

$$c_4 = \frac{-c_3}{16} = \frac{+c_0}{4 \cdot 9 \cdot 16},$$

$$c_5 = \frac{-c_4}{25} = \frac{-c_0}{4 \cdot 9 \cdot 16 \cdot 25},$$

and so on. The pattern is that

$$c_n = \frac{(-1)^n}{(n!)^2} c_0, \qquad n = 1, 2, 3, \ldots .$$

We therefore have one Frobenius solution,

$$y = \sum_{n=0}^{\infty} c_n x^{n+r} = \sum_{n=0}^{\infty} c_n x^{n-2} = c_0 x^{-2} + c_0 \sum_{n=1}^{\infty} \frac{(-1)^n}{(n!)^2} x^{n-2}.$$

Since $0! = 1$ by convention, we can think of $c_0 x^{-2}$ as the $n = 0$ term and write this solution under one summation:

$$y = c_0 \sum_{n=0}^{\infty} \frac{(-1)^n}{(n!)^2} x^{n-2}$$

$$= c_0 \left[\frac{1}{x^2} - \frac{1}{x} + \frac{1}{4} - \frac{1}{36} x + \frac{1}{576} x^2 + \cdots \right]. \quad \blacksquare$$

Here are some observations about the method.

1. Assuming that $c_0 \neq 0$, we obtained the indicial equation

$$F(r) = 0$$

for r. This was a second degree polynomial equation because the differential equation was of second order.

2. The recurrence relation involved $F(r + n)$.

3. In this example, the indicial equation had only one root, so the method, as discussed thus far, produced only one solution. This is important because ultimately we want to find two linearly independent solutions. We will address this issue later.

EXAMPLE 5.16

Solve $x^2 y'' + x(\frac{1}{2} + 2x)y' + (x - \frac{1}{2})y = 0$.

Note that zero is a regular singular point. Attempt a Frobenius series solution about zero:

$$y = \sum_{n=0}^{\infty} c_n x^{n+r}.$$

Substitute this series into the differential equation to get

$$\sum_{n=0}^{\infty} (n+r)(n+r-1)c_n x^{n+r} + \sum_{n=0}^{\infty} \tfrac{1}{2}(n+r)c_n x^{n+r}$$

$$+ \sum_{n=0}^{\infty} 2(n+r)c_n x^{n+r+1} + \sum_{n=0}^{\infty} c_n x^{n+r+1} + \sum_{n=0}^{\infty} -\tfrac{1}{2}c_n x^{n+r} = 0.$$

Shift indices in the third and fourth summations to obtain

$$[r(r-1)c_0 + \tfrac{1}{2}c_0 r - \tfrac{1}{2}c_0]x^r + \sum_{n=1}^{\infty} [(n+r)(n+r-1)c_n + \tfrac{1}{2}(n+r)c_n$$

$$+ 2(n+r-1)c_{n-1} + c_{n-1} - \tfrac{1}{2}c_n]x^{n+r} = 0.$$

Let the coefficient of each power of x be zero. If we assume that $c_0 \neq 0$, we read the indicial equation from the coefficient of x^r:

$$F(r) = r(r-1) + \tfrac{1}{2}r - \tfrac{1}{2} = 0.$$

From the coefficient of x^{n+r}, we get the recurrence relation

$$[(n+r)(n+r-1) + \tfrac{1}{2}(n+r) - \tfrac{1}{2}]c_n + [2(n+r-1) + 1]c_{n-1} = 0$$

for $n = 1, 2, 3, \ldots$. In terms of F, this recurrence relation is

$$F(r+n)c_n + (2n + 2r - 1)c_{n-1} = 0$$

for $n = 1, 2, 3, \ldots$. Solve this equation for c_n to write

$$c_n = -\frac{2(r+n) - 1}{F(r+n)}c_{n-1} \qquad (n = 1, 2, 3, \ldots).$$

Now solve the indicial equation $F(r) = 0$ to get two roots:

$$r_1 = 1 \quad \text{and} \quad r_2 = -\tfrac{1}{2}.$$

Substitute $r_1 = 1$ into the recurrence relation to get

$$c_n = -\frac{2(n+1) - 1}{F(n+1)} c_{n-1}.$$

Some of the coefficients are

$$c_1 = -\frac{3}{\frac{5}{2}} c_0 = -\frac{6}{5} c_0,$$

$$c_2 = -\frac{5}{7} c_1 = \frac{6}{7} c_0,$$

$$c_3 = - -\frac{7}{\frac{27}{2}} c_2 = -\frac{14}{27} c_2 = -\frac{4}{9} c_0,$$

and so on. Therefore, one Frobenius solution is

$$y_1 = c_0[x - \tfrac{6}{5}x^2 + \tfrac{6}{7}x^3 - \tfrac{4}{9}x^4 + \cdots].$$

In this example, we have a second root of the indicial equation $r_2 = -\frac{1}{2}$. Further, $F(r_2 + n) \neq 0$ for $n = 1, 2, 3, \ldots$, so the recurrence relation remains valid with this value of r. To avoid confusion with the first solution, write the second solution as

$$y_2 = \sum_{n=0}^{\infty} c_n^* x^{n+r_2} = \sum_{n=0}^{\infty} c_n^* x^{n-1/2}.$$

The c_n^*'s satisfy the recurrence relation with $r_2 = -\frac{1}{2}$ in place of r:

$$c_n^* = -\frac{2(-\frac{1}{2} + n) - 1}{F(-\frac{1}{2} + n)} c_{n-1}^* = \frac{-2(n-1)}{F(-\frac{1}{2} + n)} c_{n-1}^*.$$

Notice that $c_1^* = 0$; therefore, $c_n^* = 0$ for $n = 1, 2, 3, \ldots$, and a second solution of the differential equation is simply

$$y_2 = c_0^* x^{-1/2},$$

with c_0^* any nonzero constant. ■

Thus far, we have simply demonstrated a method with no theoretical justification. We will now state a theorem which provides part of the basis for the method of Frobenius.

THEOREM 5.4   The Method of Frobenius

Suppose that x_0 is a regular singular point of

$$P(x)y'' + Q(x)y' + R(x)y = 0.$$

Then there exists at least one Frobenius solution

$$y = \sum_{n=0}^{\infty} c_n(x - x_0)^{n+r},$$

with $c_0 \neq 0$.

Further, if the Taylor series for $(x - x_0)[Q(x)/P(x)]$ and $(x - x_0)^2[R(x)/P(x)]$ converge in $(x_0 - R, x_0 + R)$, the Frobenius series solution also converges in this interval, except possibly at x_0 itself.

Sketch of a Proof   Examples 5.15 and 5.16 suggest why the theorem says that there is "at least one" Frobenius solution. The method as stated thus far may produce only one solution. The general discussion we will now examine will explain why this happens as well as exactly when it will occur.

To simplify the notation, assume that $x_0 = 0$. This is not a loss of generality because we can always translate axes by letting $t = x - x_0$ if $x_0 \neq 0$. For convenience in carrying out some calculations with series, write the differential equation as

$$x^2 y'' + x^2 \frac{Q(x)}{P(x)} y' + x^2 \frac{R(x)}{P(x)} y = 0.$$

Let $Q(x)/P(x) = f(x)$ and $R(x)/P(x) = g(x)$ to obtain

$$x^2 y'' + x^2 f(x) y' + x^2 g(x) y = 0.$$

Write this equation as

$$x^2 y'' + x[x f(x)] y' + [x^2 g(x)] y = 0. \tag{5.22}$$

Since zero is a regular singular point of the differential equation, $x f(x)$ and $x^2 g(x)$ are analytic at zero, and we can expand these functions in Maclaurin series:

$$x f(x) = \sum_{n=0}^{\infty} a_n x^n = a_0 + a_1 x + a_2 x^2 + \cdots \quad \text{for} \quad |x| < R_1$$

and

$$x^2 g(x) = \sum_{n=0}^{\infty} b_n x^n = b_0 + b_1 x + b_2 x^2 + \cdots \quad \text{for} \quad |x| < R_2.$$

Substitute the proposed Frobenius series solution, and the Maclaurin series for $x f(x)$ and $x^2 g(x)$, into equation (5.22) to get

$$\sum_{n=0}^{\infty} (n + r)(n + r - 1) c_n x^{n+r} + \sum_{n=0}^{\infty} a_n x^n \sum_{n=0}^{\infty} (n + r) c_n x^{n+r}$$

$$+ \sum_{n=0}^{\infty} b_n x^n \sum_{n=0}^{\infty} c_n x^{n+r} = 0.$$

It is easier to follow the details if we write out a few terms of the series in this equation:

$$r(r - 1) c_0 x^r + (r + 1) r c_1 x^{r+1} + (r + 2)(r + 1) c_2 x^{r+2} + \cdots$$
$$+ [a_0 + a_1 x + a_2 x^2 + \cdots][r c_0 x^r + (r + 1) c_1 x^{r+1} + (r + 2) c_2 x^{r+2} + \cdots]$$
$$+ [b_0 + b_1 x + b_2 x^2 + \cdots][c_0 x^r + c_1 x^{r+1} + c_2 x^{r+2} + \cdots] = 0.$$

Carry out some of the multiplications and collect coefficients of like powers of x to get

$$[r(r - 1) + a_0 r + b_0] c_0 x^r + [r(r + 1) c_1 + a_1 c_0 r + a_0 (r + 1) c_1 + b_1 c_0 + b_0 c_1] x^{r+1}$$
$$+ [(r + 2)(r + 1) c_2 + a_2 c_0 r + a_1 c_1 (r + 1) \tag{5.23}$$
$$+ a_0 c_2 (r + 2) + b_0 c_2 + b_1 c_1 + b_2 c_0] x^{r+2} + \cdots = 0.$$

We can write this equation more compactly if we define

$$F(r) = r(r - 1) + a_0 r + b_0.$$

Equation (5.23) becomes

$$c_0 F(r)x^r + \sum_{n=1}^{\infty} \left\{ F(r+n)c_n + \sum_{j=0}^{n-1} [(r+j)a_{n-j} + b_{n-j}]c_j \right\} x^{r+n} = 0.$$

Now let the coefficient of each power of x be zero. *Assuming that $c_0 \neq 0$, we get the* indicial equation

$$F(r) = 0$$

together with the recurrence relation

$$F(r+n)c_n + \sum_{j=0}^{n-1} [(r+j)a_{n-j} + b_{n-j}]c_j = 0$$

for $n = 1, 2, 3, \ldots$.

The indicial equation $F(r) = 0$ is a second degree polynomial equation, which has two real, distinct roots, or one real, repeated root, or complex conjugate roots. If the roots are real, label them r_1 and r_2, with $r_1 \geq r_2$.

Now, $r_1 + n$ is not a root of the indicial equation for any positive integer n. Therefore, $F(r_1 + n) \neq 0$ for $n = 1, 2, 3, \ldots$, and we can substitute r_1 into the recurrence relation and solve (at least in theory) for the coefficients c_n, obtaining one Frobenius solution. Thus, we always obtain one Frobenius solution.

If $r_1 = r_2$, we do not obtain a second solution using the method as described so far. If $r_1 \neq r_2$, but $r_1 - r_2 = N$, a positive integer, $F(r_2 + N) = F(r_1) = 0$. In this event, the coefficient of c_N in the recurrence relation is zero and we cannot use the recurrence relation to solve for c_N or for c_{N+1} or any subsequent coefficients. In this case, as with the case $r_1 = r_2$, the method as described thus far fails to produce a second solution. ∎

We will not prove the final part of the theorem about the interval of convergence of the Frobenius solution.

We now know how to produce one Frobenius solution about a regular singular point as well as why we do not always obtain two linearly independent Frobenius solutions (this happens when the roots of the indicial equation differ by a nonnegative integer). In the next section, we will deal with the problem of producing a second, linearly independent solution about a regular singular point.

PROBLEMS FOR SECTION 5.3

In each of Problems 1 through 6, find all of the singular points of the differential equation and classify each singular point as either regular or irregular.

1. $x^2(x-3)^2 y'' + 4x(x^2 - x - 6)y' + (x^2 - x - 2)y = 0$
2. $(x^3 - 2x^2 - 7x - 4)y'' - 2(x^2 + 1)y' + (5x^2 - 2x)y = 0$
3. $x^2(x-2)y'' + (5x - 7)y' + 2(3 + 5x^2)y = 0$
4. $[(9 - x^2)y']' + (2 + x^2)y = 0$
5. $[(x-2)^{-1}y']' + x^{-5/2}y = 0$
6. $x^2 \sin^2(x - \pi)y'' + \tan(x - \pi)\tan(x)y' + (7x - 2)\cos(x)y = 0; \quad |x| < \dfrac{\pi}{2}$

In each of Problems 7 through 27, (a) show that zero is a regular singular point of the differential equation; (b) find and solve the indicial equation; (c) determine the recurrence relation; (d) use the roots of the indicial equation and the recurrence relation to find the first six nonzero terms of two linearly independent solutions of the differential equation valid in some interval about zero, except possibly at zero. (In these problems, the indicial equation will have two distinct roots which do not differ by a nonnegative integer, so the method of Frobenius will produce two linearly independent solutions.)

7. $4x^2y'' + 2xy' - xy = 0$

8. $16x^2y'' - 4x^2y' + 3y = 0$

9. $9x^2y'' + 2(2x + 1)y = 0$

10. $12x^2y'' + 5xy' + (1 - 2x^2)y = 0$

11. $2xy'' + (2x + 1)y' + 2y = 0$

12. $2x^2y'' - xy' + (1 - x^2)y = 0$

13. $2x^2y'' + x(2x + 1)y' - (2x^2 + 1)y = 0$

14. $3x^2y'' + 4xy' - (3x + 2)y = 0$

15. $9x^2y'' + 9xy' + (9x^2 - 4)y = 0$

16. $3x^2y'' + x(1 - 2x^2)y' - 4x^2y = 0$

17. $2xy'' + y' + 6xy = 0$

18. $12x^2y'' + x(12x^3 - 5)y' + 6y = 0$

19. $2x^2y'' - 3xy' - (2x^2 + 3)y = 0$

20. $6x^2y'' + x(5 - 3x)y' + x(3 - 2x)y = 0$

21. $4x^2y'' - x(x + 3)y' + (3x - 2)y = 0$

22. $2x^2y'' - x(5 + 3x^2)y' + (5 + 9x)y = 0$

23. $x^2(3x - 2)y'' - 7xy' + 3y = 0$

24. $x^2y'' + xy' - (3x + 2)y = 0$

25. $x^2y'' + xy' + (x^2 - 6)y = 0$

26. $x^2y'' + xy' + (1 - x)y = 0$

27. $x^2y'' + x(1 - 2x)y' + y = 0$

28. The differential equation

$$x(1 - x)y'' + [c - (1 + a + b)x]y' - aby = 0$$

is called the *hypergeometric equation*; a, b, and c are constants.

(a) Show that zero is a regular singular point of the hypergeometric equation.

(b) Assuming that c is not an integer, use the method of Frobenius to obtain the first six nonzero terms of each of two linearly independent solutions about zero.

29. We say that a differential equation $P(x)y'' + Q(x)y' + R(x)y = 0$ has a *regular singular point at infinity* if the transformation $u = 1/x$ yields a differential equation in terms of u and y having a regular singular point at zero.

  Show that Legendre's differential equation (Example 5.13) has a regular singular point at infinity. Use the method of Frobenius to find a Frobenius solution in terms of u and y of the transformed Legendre equation, then let $u = 1/x$ to find a solution in terms of x and y. Show that the resulting series solution converges for $|x| > 1$.

30. Show that the Cauchy-Euler differential equation

$$x^2y'' + axy' + by = 0$$

has a regular singular point at zero.

5.4 *Second Solutions and Logarithm Terms* _____

We will continue from the last section, where we were considering the differential equation

$$P(x)y'' + Q(x)y' + R(x)y = 0.$$

Assume that zero is a regular singular point. As we saw, if the indicial equation $F(r) = 0$ has only one real root, or if the roots differ by a nonnegative integer, we may obtain only one Frobenius series solution.

  In each of these cases, we can find a second, linearly independent solution, but we must know what form it will take. The following theorem gives us this information. For completeness, the theorem includes the results of the preceding section.

THEOREM 5.5 Two Linearly Independent Solutions by the Method of Frobenius

Suppose that zero is a regular singular point of the differential equation

$$P(x)y'' + Q(x)y' + R(x)y = 0.$$

Let r_1 and r_2 be roots of the indicial equation. In the case in which these are real, assume that $r_1 \geq r_2$.

Case 1 If $r_1 - r_2$ is not an integer, there are two linearly independent Frobenius series solutions obtained by using first r_1, then r_2, in the recurrence relation to generate the coefficients. These Frobenius solutions have the form

$$y_1 = \sum_{n=0}^{\infty} c_n x^{n+r_1} \quad \text{and} \quad y_2 = \sum_{n=0}^{\infty} c_n^* x^{n+r_2},$$

with $c_0 \neq 0$ and $c_0^* \neq 0$. These series converge in some interval $(0, R)$ or $(-R, 0)$.

Case 2 If $r_1 = r_2$, there is a Frobenius series solution

$$y_1(x) = \sum_{n=0}^{\infty} c_n x^{n+r_1} \qquad (c_0 \neq 0)$$

in which we obtain the c_n's by letting $r = r_1$ in the recurrence relation.
    There is also a second, linearly independent solution of the form

$$y_2(x) = y_1(x)\ln(x) + \sum_{n=1}^{\infty} c_n^* x^{n+r_1}$$

convergent in some interval $(0, R)$. We can solve for the coefficients c_n^* by substituting this expression for $y_2(x)$ into the differential equation.

Case 3 If $r_1 - r_2$ is a positive integer, there is a Frobenius series solution

$$y_1(x) = \sum_{n=0}^{\infty} c_n x^{n+r_1} \qquad (c_0 \neq 0)$$

obtained by letting $r = r_1$ in the recurrence relation and solving for the coefficients c_n.
    There is also a second, linearly independent solution of the form

$$y_2(x) = ky_1(x)\ln(x) + \sum_{n=0}^{\infty} c_n^* x^{n+r_2}$$

in some interval $(0, R)$. We can obtain k and the coefficients c_n^* by substituting this expression into the differential equation. ∎

Before looking at some examples, note the following.

1. In case 2, the series in $y_2(x)$ starts at $n = 1$, *not* $n = 0$. Further, the second solution necessarily contains a logarithm term when the roots of the indicial equation are repeated.

2. In case 3, the number k may or may not be zero. Thus, if the roots of the indicial equation differ by a positive integer, the second solution *may* have a logarithm term, or it may not.

We will sketch a proof of cases 2 and 3 at the end of this section. First, we will look at examples illustrating case 2 of the theorem and both cases of 3 (one with $k = 0$, the other with $k \neq 0$).

EXAMPLE 5.17   Equal Roots of the Indicial Equation

Consider the differential equation

$$x^2 y'' + 5xy' + (x + 4)y = 0.$$

In Example 5.15 of the preceding section, we found the indicial equation

$$F(r) = r^2 + 4r + 4 = 0$$

and the recurrence relation

$$F(r + n)c_n + c_{n-1} = 0 \qquad (n = 1, 2, 3, \ldots).$$

The roots of the indicial equation are $r_1 = r_2 = -2$. We found one solution as the Frobenius series

$$y_1(x) = c_0 \sum_{n=0}^{\infty} \frac{(-1)^n}{(n!)^2} x^{n-2} = c_0 \left[\frac{1}{x^2} - \frac{1}{x} + \frac{1}{4} - \frac{1}{36} x + \frac{1}{576} x^2 + \cdots \right],$$

valid in some interval $(0, R)$. By case 2 of Theorem 5.5, there is a second solution,

$$y_2(x) = y_1(x)\ln(x) + \sum_{n=1}^{\infty} c_n^* x^{n-2}.$$

(The summation begins at $n = 1$, not $n = 0$.) Substitute this expression for $y_2(x)$ into the differential equation to get

$$4y_1 + 2xy_1' + \sum_{n=1}^{\infty} (n - 2)(n - 3)c_n^* x^{n-2} + \sum_{n=1}^{\infty} 5(n - 2)c_n^* x^{n-2}$$

$$+ \sum_{n=1}^{\infty} c_n^* x^{n-1} + \sum_{n=1}^{\infty} 4c_n^* x^{n-2} + \ln(x)[x^2 y_1'' + 5xy_1' + (x + 4)y_1] = 0.$$

The bracketed coefficient of $\ln(x)$ is zero because y_1 is a solution of the differential equation. Now write

$$\sum_{n=1}^{\infty} c_n^* x^{n-1} = \sum_{n=2}^{\infty} c_{n-1}^* x^{n-2}.$$

Let $c_0 = 1$ for convenience, and substitute the Frobenius series for y_1 previously obtained. We get

$$-2x^{-1} + c_1^* x^{-1} + \sum_{n=2}^{\infty} \left[\frac{4(-1)^n}{(n!)^2} + \frac{2(-1)^n}{(n!)^2} (n - 2) + (n - 2)(n - 3)c_n^* \right.$$

$$\left. + 5(n - 2)c_n^* + c_{n-1}^* + 4c_n^* \right] x^{n-2} = 0.$$

Set the coefficient of each power of x equal to zero. From the coefficient of x^{-1}, we get

$$c_1^* = 2.$$

From the coefficient of x^{n-2} in the summation, we get

$$\frac{2(-1)^n}{(n!)^2} n + n^2 c_n^* + c_{n-1}^* = 0 \quad \text{for} \quad n = 2, 3, 4, \ldots.$$

The recurrence relation is

$$c_n^* = -\frac{1}{n^2} c_{n-1}^* - \frac{2(-1)^n}{n(n!)^2}$$

for $n = 2, 3, 4, \ldots$. From this relationship, we can calculate as many of the coefficients c_n^* as we wish. Here is the solution y_2 with a few terms calculated:

$$y_2(x) = y_1(x)\ln(x) + \frac{2}{x} - \frac{3}{4} + \frac{11}{108} x - \frac{25}{3456} x^2 + \frac{137}{432000} x^3 + \cdots.$$

This solution converges on some interval $(0, R)$. It is linearly independent from y_1 because of the logarithm term. The general solution of the differential equation is

$$y(x) = [a + b \ln(x)] \sum_{n=0}^{\infty} \frac{(-1)^n}{(n!)^2} x^{n-2}$$

$$+ b \left[\frac{2}{x} - \frac{3}{4} + \frac{11}{108} x - \frac{25}{3456} x^2 + \frac{137}{432000} x^3 + \cdots \right]. \quad \blacksquare$$

EXAMPLE 5.18   Case 3 of Theorem 5.5, with $k = 0$

Consider the differential equation $x^2 y'' + x^2 y' - 2y = 0$, which has zero as a regular singular point. Substitute $y = \sum c_n x^{n+r}$ into the series and shift indices to obtain

$$[r(r - 1) - 2]c_0 x^r + \sum_{n=1}^{\infty} [(n + r)(n + r - 1)c_n + (n + r - 1)c_{n-1} - 2c_n]x^{n+r} = 0.$$

Assuming that $c_0 \neq 0$, we obtain the indicial equation

$$F(r) = r(r - 1) - 2 = 0,$$

with roots $r_1 = 2$ and $r_2 = -1$. Notice that $r_1 - r_2 = 3$, a positive integer. From the coefficient of x^{n+r} in the summation, we obtain

$$(n + r)(n + r - 1)c_n + (n + r - 1)c_{n-1} - 2c_n = 0.$$

In terms of F, this recurrence relation is

$$F(r + n)c_n + (n + r - 1)c_{n-1} = 0.$$

Put $r_1 = 2$ into the recurrence relation to get

$$F(2 + n)c_n + (n + 1)c_{n-1} = 0,$$

or

$$c_n = -\frac{n + 1}{F(2 + n)} c_{n-1} = -\frac{n + 1}{n(n + 3)} c_{n-1}$$

for $n = 1, 2, 3, \ldots$. From this equation, we can calculate each c_n in terms of c_0. One solution is

$$y_1 = c_0 x^2 [1 - \tfrac{1}{2}x + \tfrac{3}{20}x^2 - \tfrac{1}{30}x^3 + \tfrac{1}{168}x^4 - \tfrac{1}{1120}x^5 + \tfrac{1}{8640}x^6 - \tfrac{1}{75600}x^7 + \cdots].$$

Since $r_2 = -1$, case 3 of the theorem tells us that there is a second, linearly independent solution of the form

$$y_2(x) = ky_1(x)\ln(x) + \sum_{n=0}^{\infty} c_n^* x^{n-1}.$$

After substituting this expression into the differential equation and shifting indices, we obtain

$$kx^2 y_1'' \ln(x) + 2kxy_1' - ky_1 + \sum_{n=0}^{\infty} (n-1)(n-2)c_n^* x^{n-1}$$

$$+ kx^2 y_1' \ln(x) + kxy_1 + \sum_{n=1}^{\infty} (n-2)c_{n-1}^* x^{n-1} \tag{5.24}$$

$$- 2\sum_{n=0}^{\infty} c_n^* x^{n-1} - 2ky_1 \ln(x) = 0.$$

The terms involving $k \ln(x)$ are

$$k \ln(x)[x^2 y_1'' + x^2 y_1' - 2y_1],$$

and this is zero because y_1 is a solution of the differential equation.

Now substitute the series for $y_1(x)$ into equation (5.24), letting $c_0 = 1$ for convenience (y_1 can be *any* nontrivial solution). After some computation and rearranging of terms, we get

$$(2c_0^* - 2c_0^*)x^{-1} + (-c_0^* - 2c_1^*)x^0 - 2c_2^* x$$

$$+ (4k - k + 2c_3^* + c_2^* - 2c_3^*)x^2$$

$$+ \left(-3k - \frac{1}{2}k + 6c_4^* + k + 2c_3^* - 2c_4^*\right)x^3$$

$$+ \left(\frac{6}{5}k + \frac{3k}{20} + 12c_5^* - \frac{k}{2} + 3c_4^* - 2c_5^*\right)x^4$$

$$+ \left(-\frac{k}{3} - \frac{k}{20} + 20c_6^* + \frac{3k}{20} + 4c_5^* - 2c_6^*\right)x^5$$

$$+ \left(\frac{k}{14} + \frac{k}{168} + 30c_7^* - \frac{k}{30} + 5c_6^* - 2c_7^*\right)x^6 + \cdots = 0.$$

Set the coefficient of each power of x equal to zero. The resulting equations can be solved to get

$$k = c_2^* = 0,$$

$$c_1^* = -\tfrac{1}{2}c_0^*,$$

$$c_4^* = -\tfrac{1}{2}c_3^*,$$

$$c_5^* = -\tfrac{3}{10}c_4^* = \tfrac{3}{20}c_3^*,$$

$$c_6^* = -\tfrac{4}{18}c_5^* = -\tfrac{1}{30}c_3^*,$$

$$c_7^* = -\tfrac{5}{28}c_6^* = \tfrac{1}{168}c_3^*,$$

and so on. We can recognize the coefficients from y_1 appearing in this solution. A second solution is

$$y_2(x) = c_0^* x^{-1} - \frac{1}{2} c_0^*$$

$$+ c_3^* \left[x^2 - \frac{1}{2} x^3 + \frac{3}{20} x^4 - \frac{1}{30} x^5 + \frac{1}{168} x^6 + \cdots \right]$$

$$= c_0^* \frac{1}{x} - \frac{1}{2} c_0^* + c_3^* y_1(x).$$

In this expression, c_0^* and c_3^* are arbitrary constants (but $c_0^* \neq 0$). If we wish, we can let $c_3^* = 0$ and obtain the simple second solution

$$y_2(x) = c_0^* \left(\frac{1}{x} - \frac{1}{2} \right).$$

This solution is valid in $(0, \infty)$. ∎

EXAMPLE 5.19   Case 3 of Theorem 5.5, with $k \neq 0$

The differential equation $xy'' - y = 0$ has a regular singular point at zero.
     Substitute $y = \sum_{n=0}^{\infty} c_n x^{n+r}$ into the differential equation to get

$$\sum_{n=0}^{\infty} (n + r)(n + r - 1) c_n x^{n+r-1} - \sum_{n=0}^{\infty} c_n x^{n+r} = 0.$$

Shift indices in the second summation to obtain

$$(r^2 - r) c_0 x^{r-1} + \sum_{n=1}^{\infty} [(n + r)(n + r - 1) c_n - c_{n-1}] x^{n+r-1} = 0.$$

The indicial equation is

$$F(r) = r^2 - r = r(r - 1) = 0,$$

with roots $r_1 = 1$ and $r_2 = 0$. Since $r_1 - r_2 = 1$, we are in case 3 of the theorem. The recurrence relation is

$$F(r + n) c_n - c_{n-1} = 0 \quad \text{for} \quad n = 1, 2, 3, \ldots .$$

Let $r = 1$ in the recurrence relation to obtain

$$c_n = \frac{1}{F(1 + n)} c_{n-1} = \frac{1}{n(n + 1)} c_{n-1}$$

for $n = 1, 2, 3, \ldots$. By using this to calculate some of the c_n's, we find that in general

$$c_n = \frac{1}{n!(n + 1)!} c_0.$$

One solution is the Frobenius series

$$y_1(x) = c_0 \sum_{n=0}^{\infty} \frac{1}{n!(n + 1)!} x^{n+1}$$

$$= c_0 \left[x + \frac{1}{2} x^2 + \frac{1}{12} x^3 + \frac{1}{144} x^4 + \cdots \right].$$

There is a second solution,

$$y_2(x) = k y_1(x) \ln(x) + \sum_{n=0}^{\infty} c_n^* x^n.$$

Substitute this expression into the differential equation, and use the fact that y_1 is a solution, to obtain

$$2k y_1' - \frac{k}{x} y_1' + \sum_{n=2}^{\infty} n(n-1) c_n^* x^{n-1} - \sum_{n=0}^{\infty} c_n^* x^n = 0.$$

Substitute the series for y_1, with c_0 chosen as 1, to obtain

$$(k - c_0^*) + \sum_{n=1}^{\infty} \left[\frac{k(2n+1)}{n!(n+1)!} + n(n+1) c_{n+1}^* - c_n^* \right] x^n = 0.$$

Set the coefficient of each power of x equal to zero. Then

$$k = c_0^*$$

and, for $n = 1, 2, 3, \ldots,$

$$c_{n+1}^* = \frac{1}{n(n+1)} \left[c_n^* - \frac{k(2n+1)}{n!(n+1)!} \right].$$

From this recurrence relation, we get

$$c_2^* = \tfrac{1}{2} c_1^* - \tfrac{3}{4} c_0^*,$$
$$c_3^* = \tfrac{1}{12} c_1^* - \tfrac{7}{36} c_0^*,$$
$$c_4^* = \tfrac{1}{144} c_1^* - \tfrac{35}{1728} c_0^*,$$

and so on. A second solution is

$$
\begin{aligned}
y_2(x) &= c_0^* y_1(x) \ln(x) + c_0^* + c_1^* x - \tfrac{3}{4} c_0^* x^2 + \tfrac{1}{2} c_1^* x - \tfrac{7}{36} c_0^* x^3 \\
&\quad + \tfrac{1}{12} c_1^* x^3 - \tfrac{35}{1728} c_0^* x^4 + \tfrac{1}{144} c_1^* x^4 + \cdots \\
&= c_0^* y_1(x) \ln(x) + c_0^* [1 - \tfrac{3}{4} x^2 - \tfrac{7}{36} x^3 - \tfrac{35}{1728} x^4 - \cdots] \\
&\quad + c_1^* [x + \tfrac{1}{2} x^2 + \tfrac{1}{12} x^3 + \tfrac{1}{144} x^4 + \cdots].
\end{aligned}
$$

Notice that the coefficient of c_1^* in this solution is just $y_1(x)$. We therefore lose nothing by choosing $c_1^* = 0$. We can also let $c_0^* = 1$, obtaining a specific second solution

$$y_2 = y_1(x) \ln(x) + 1 - \tfrac{3}{4} x^2 - \tfrac{7}{36} x^3 - \tfrac{35}{1728} x^4 - \cdots.$$

This solution contains a logarithm term and is linearly independent from y_1 on some interval $(0, R)$, with R chosen so that the series converges. ∎

We will conclude this section with an outline of an argument explaining cases 2 and 3 of Theorem 5.5. First, we will argue for case 2, with some routine details omitted. We will use the notation of the argument presented at the end of Section 5.3. There we wrote the differential equation as

$$x^2 y'' + x[x f(x)] y' + [x^2 g(x)] y = 0,$$

and, assuming a Frobenius series solution, we found the indicial equation

$$F(r) = r^2 + (a_0 - 1) r + b_0 = 0.$$

This equation has a repeated root if and only if

$$(a_0 - 1)^2 - 4b_0 = 0.$$

In this event, the repeated root is

$$r = \tfrac{1}{2}(1 - a_0).$$

The Frobenius series corresponding to r gives us one solution,

$$y_1(x) = \sum_{n=0}^{\infty} c_n x^{n+r}.$$

We will use reduction of order to find a second solution, y_2. Let

$$y_2(x) = u(x)y_1(x) = u(x) \sum_{n=0}^{\infty} c_n x^{n+r}.$$

Substitute y_2 into the differential equation. From the work of Section 2.3, u must satisfy the differential equation

$$u'' + \left[\frac{2y_1'}{y_1} + f(x) \right] u' = 0.$$

Since $xf(x) = \sum_{n=0}^{\infty} a_n x^n$, we get

$$u'' + \left[\frac{2y_1'}{y_1} + a_0 \frac{1}{x} + a_1 + a_2 x + a_3 x^3 + \cdots \right] u' = 0.$$

Using the series for y_1 and y_1', we can carry out the division y_1'/y_1 indicated in the last equation to get

$$\frac{2y_1'}{y_1} = \frac{2x^{r-1}[rc_0 + (r+1)c_1 x + (r+2)c_2 x^2 + \cdots]}{x^r[c_0 + c_1 x + c_2 x^2 + \cdots]}$$

$$= 2r \frac{1}{x} + \frac{2c_1}{c_0} + \frac{2(c_2 c_0 - c_1^2)}{c_0^2} x + \cdots.$$

From the last two steps, we obtain the differential equation

$$u'' + \left[\left(\frac{2r + a_0}{x} \right) + \left(\frac{2c_1}{c_0} + a_1 \right) + \left(\frac{2(c_2 c_0 - c_1^2)}{c_0^2} + a_2 \right) x + \cdots \right] u' = 0. \quad (5.25)$$

It is here that we first use the assumption that the indicial equation has a repeated root r. As noted previously (in slightly different form), $2r + a_0 = 1$. Using this, we can rearrange terms in equation (5.25) to write it as

$$u'' + \left(\frac{1}{x} + k_1 + k_2 x + \cdots \right) u' = 0, \quad (5.26)$$

in which $k_1, k_2, \ldots$ are constants whose specific values (in terms of a_0 and the coefficients c_k) are not important here. Write equation (5.26) as

$$\frac{u''}{u'} = -\frac{1}{x} - (k_1 - k_2 x - k_3 x^2 - \cdots)$$

and integrate both sides to obtain

$$\ln|u'| = -\ln|x| - (k_1 x + \tfrac{1}{2}k_2 x^2 + \tfrac{1}{3}k_3 x^3 + \cdots), \tag{5.27}$$

in which we let the constant of integration be zero.

Assuming that $x > 0$, take the exponential of both sides of equation (5.27) to get

$$u' = \frac{1}{x} \exp\left[-k_1 x - \frac{1}{2} k_2 x^2 - \frac{1}{3} k_3 x^3 - \cdots \right], \tag{5.28}$$

in which $\exp[A] = e^A$. Use the Maclaurin series for e^A, with A the series in brackets in equation (5.28). We get an equation of the form

$$u' = \frac{1}{x} - k_1 + \frac{k_1 - 2k_2}{2} x - \cdots.$$

Integrate this equation to get

$$u(x) = \ln(x) - k_1 x + \tfrac{1}{4}(k_1 - 2k_2)x^2 - \cdots$$

for x in some interval $(0, R)$. From this equation, we conclude that $u(x)$ has the form

$$u(x) = \ln(x) + \sum_{n=1}^{\infty} d_n x^n.$$

Therefore, a second solution has the form

$$y_2(x) = y_1(x)\ln(x) + \sum_{n=1}^{\infty} c_n^* x^{n+r},$$

as stated in case 2 of the theorem.

We will now show why the second solution in case 3 of Theorem 5.5 has the form that it does. Suppose that the indicial equation has real roots r_1 and r_2, with $r_1 \geq r_2$, and that $r_1 - r_2 = N$, a positive integer. As before, attempt a reduction of order, with y_1 one Frobenius solution and $y_2(x) = u(x)y_1(x)$. The analysis proceeds exactly as before until we reach equation (5.25), which we repeat here for ease of reference:

$$u'' + \left[\left(\frac{2r + a_0}{x} \right) + \left(\frac{2c_1}{c_0} + a_1 \right) + \left(\frac{2(c_2 c_0 - c_1^2)}{c_0^2} + a_2 \right)x + \cdots \right]u' = 0.$$

This differential equation has the form

$$u'' + \left[\frac{2r + a_0}{x} + k_1 + k_2 x + \cdots \right]u' = 0. \tag{5.29}$$

But now, instead of $2r + a_0 = 1$, as in the case of repeated roots, we have

$$2r + a_0 = N + 1. \tag{5.30}$$

(Recall that the sum of the roots of a polynomial equation $r^2 + Ar + B = 0$ is $-A$.) From equations (5.29) and (5.30), we obtain

$$\frac{u''}{u'} = -\frac{N + 1}{x} - (k_1 + k_2 x + \cdots).$$

Integrate this equation to get

$$\ln|u'| = -(N + 1)\ln|x| - (k_1 x + \tfrac{1}{2}k_2 x^2 + \cdots).$$

Assuming that $x > 0$, we obtain from this equation that

$$u' = \frac{1}{x^{N+1}} \exp[-k_1 - \tfrac{1}{2}k_2 x^2 - \cdots].$$

Expand the exponential factor in a Maclaurin series to get a series of the form

$$u' = \frac{1}{x^{N+1}} - k_1 \frac{1}{x^N} + \frac{1}{2}(k_1 - 2k_2)\frac{1}{x^{N-1}} + \cdots$$
$$+ d_{-1}x^{-1} + d_0 + d_1 x + d_2 x^2 + \cdots,$$

where the d_j's are constants whose specific values are in terms of $k_1, k_2, \ldots$. Integrate this equation to get

$$u(x) = -\frac{1}{N} x^{-N} - k_1 \frac{1}{1-N} x^{1-N} + \cdots + d_{-1}\ln(x)$$
$$+ d_0 x + \frac{1}{2} d_1 x^2 + \frac{1}{3} d_2 x^3 + \cdots.$$

From this equation, we conclude that $y_2(x)$ has the form

$$y_2(x) = k y_1(x)\ln(x) + \sum_{n=0}^{\infty} c_n^* x^{n+r_2},$$

as stated in the theorem. The logarithm term will appear in y_2 if $d_{-1} \neq 0$ and will not appear if $d_{-1} = 0$.

One final note: Whenever we can write $y_1(x)$ as a "simple" combination of elementary functions, it is usually easier to find $y_2(x)$ by reduction of order rather than use the form for $y_2(x)$ suggested by the theorem.

PROBLEMS FOR SECTION 5.4

In each of Problems 1 through 6, find the roots of the indicial equation and use Theorem 5.5 to determine the *form* of each of two linearly independent solutions of the differential equation. Do not attempt to solve for the coefficients in the solutions.

1. $25x(1 - x^2)y'' - 20(5x - 2)y' + \left[25x - \dfrac{4}{x}\right]y = 0$

2. $6(3x - 4)(5x + 8)y'' + (2x - 21)\left[3 + \dfrac{16}{x}\right]y' + \left[4 - \dfrac{27}{x^2}\right]y = 0$

3. $12x(4 + 3x)y'' - 2(5x + 7)(7x - 2)y' + 24\left[5 - \dfrac{1}{3x}\right]y = 0$

4. $3x(2x + 3)y'' + 2(6 - 5x)y' + 7\left[2x - \dfrac{8}{x}\right]y = 0$

5. $x[\sqrt{2} - 3x]y'' + 2(2 + 5x)y' + \left[\dfrac{9\sqrt{2} - 8}{4x}\right]y = 0$

6. $(3x^3 + x^2)y'' - x(10x + 1)y' + (x^2 + 2)y = 0$

In each of Problems 7 through 28, (a) show that zero is a regular singular point of the differential equation; (b) find and solve the indicial equation; and (c) find two linearly independent solutions of the differential equation for x in some interval $(0, R)$. These problems include all three cases of the theorem. In (c), write at least four nonzero terms of each series solution.

7. $xy'' + (1 - x)y' + y = 0$

8. $y'' - 2xy' + 2y = 0$

9. $x(x - 1)y'' + 3y' - 2y = 0$

10. $4x^2y'' + 4xy' + (4x^2 - 9)y = 0$

11. $4xy'' + 2y' + y = 0$

12. $4x^2y'' + 4xy' - y = 0$

13. $x^2y'' - 2xy' - (x^2 - 2)y = 0$

14. $xy'' - y' + 2y = 0$

15. $x(2 - x)y'' - 2(x - 1)y' + 2y = 0$

16. $x^2y'' + x(x^3 + 1)y' - y = 0$

17. $x^2y'' + x(x - 2)y' + (x^2 + 2)y = 0$

18. $3x^2y'' + (6x^2 - 7x)y' + 3(1 + x^3)y = 0$

19. $x^2y'' + (x^2 - 3x)y' + (x - 4)y = 0$

20. $x^2y'' - 3xy' + 4(1 + x)y = 0$

21. $xy'' + y' + xy = 0$

22. $x^2y'' + x(x^2 + 1)y' - y = 0$

23. $x^2y'' + xy' - (x + 6)y = 0$

24. $x^2y'' - 2xy' + (x^2 + 2)y = 0$

25. $x^2y'' + x(1 - x)y' + 2xy = 0$

26. $4x^2y'' - 2x(x + 2)y' + (x + 3)y = 0$

27. $16x^2y'' + 16xy' + (16x^2 - 1)y = 0$

28. $x^2y'' + xy' + (x^2 - 4)y = 0$

Summary

An nth order equation $y^{(n)} + p_{n-1}(x)y^{(n-1)} + \cdots + p_1(x)y' + p_0(x)y = f(x)$ has power series solutions about a point x_0 if each of the coefficient functions is analytic at x_0. This means that each of the coefficient functions is represented by a Taylor series expansion about x_0, in some interval about x_0.

If a linear differential equation has a coefficient of the highest order derivative which vanishes at x_0, the equation may not have power series solutions about x_0. In particular, consider the second order differential equation

$$P(x)y'' + Q(x)y' + R(x) = 0. \tag{5.31}$$

If $P(x_0) \neq 0$, and P, Q, and R are analytic at x_0, then x_0 is an ordinary point of equation (5.31). We call x_0 a singular point if it is not an ordinary point.

In turn, singular points are classified as regular or irregular. The singular point x_0 is regular if both $(x - x_0)[Q(x)/P(x)]$ and $(x - x_0)^2[R(x)/P(x)]$ are analytic at x_0. The singular point x_0 is irregular if it is not regular.

Because of technicalities which we did not elaborate upon, we did not attempt to treat solutions of equation (5.31) in an interval about an irregular singular point. If x_0 is a regular singular point, we attempt Frobenius solutions

$$y(x) = \sum_{n=0}^{\infty} c_n(x - x_0)^{n+r},$$

in which $c_0 \neq 0$. Upon substituting this series into the differential equation, we obtain an indicial equation $F(r) = 0$ together with a recurrence relation for the coefficients c_n.

The indicial equation for equation (5.31) is a quadratic equation in r and hence has two distinct, real roots, or a repeated root, or complex roots.

If the indicial equation has two roots which do not differ by an integer, we obtain two linearly independent Frobenius solutions by using in turn each root in the recurrence relation.

If the indicial equation has a repeated root r (necessarily real), we obtain one Frobenius solution y_1 by using this root in the recurrence relation. A second, linearly independent solution has the form

$$y_2 = y_1 \ln(x) + \sum_{n=1}^{\infty} c_n^* x^{n+r},$$

and we solve for the coefficients c_n^* by substituting y_2 into the differential equation. We can simplify the computations in this process by choosing $c_0 = 1$ in y_1 (*any* Frobenius solution y_1 can be used in this formulation of y_2).

If the roots of the indicial equation are r_1 and r_2 and differ by a positive integer, we obtain one Frobenius solution y_1 using the larger root r_1, then obtain a second, linearly independent solution of the form

$$y_2 = k y_1 \ln(x) + \sum_{n=0}^{\infty} c_n^* x^{n+r_2}.$$

Here, we must solve for the constant k also, and in individual cases k may or may not be zero.

Numerical Solution of
Initial Value Problems

6.0 Introduction _____

In many instances, we want to produce either a graph or numerical values of the solution of an initial value problem. One approach is to try to find an expression which implicitly or explicitly determines the solution in a form from which we can compute numerical values. This approach is not always successful.

Another approach, which is the object of this chapter, is to develop schemes for approximating numerical values of solutions to prescribed degrees of accuracy.

Although the idea of a numerical approximation is not new, the development of high-speed computers has made this approach the success that it is today. Many problems considered inaccessible 10 or 20 years ago are now considered "solved" from a practical point of view. Using supercomputers and numerical approximations, we now have more accurate models for weather prediction, analysis of national and global economies, design of nuclear reactors, ecology studies, descriptions of fluid flow around airplane wings and ship hulls, surface properties of materials, and many other phenomena of interest and importance.

A method for approximating numerical values of the solutions of an initial value problem will typically involve many considerations, including the following.

1. A numerical method will usually begin with a partition of an interval $[a, b]$ into subintervals $[a, x_1], [x_1, x_2], \ldots, [x_{n-1}, b]$, with initial data given at a. One approach is to use the initial data at a to approximate the value of the solution at x_1, then use this information to approximate the value at x_2, and so on, proceeding by steps through the interval. Accuracy of the method will depend in part on the distance between successive partition points and on the method used to approximate the value of the solution at the next point from values obtained at preceding points.

2. Each method should include an estimate of or bound on the error incurred in the approximation in order to estimate the accuracy of the method or to choose certain parameters (such as distance between successive partition points) to improve the accuracy to within desired tolerances.

3. The method must be implemented on a computer. Only simple examples, devised purely for illustrative purposes, can be done satisfactorily by hand. Although new technology has produced computers with increasing speed and accuracy, a computer is a finite machine capable of carrying out only finitely many arithmetic operations and of carrying only finitely many decimal places. It is therefore important to analyze overflow and round-off errors incurred in the computations used to implement a numerical method. These errors will depend not only on the numerical method being used but also on the hardware and software chosen to implement the method.

In one chapter, we cannot discuss all of these ideas in depth. The approach here is to discuss several methods for approximating solutions, with enough detail to give some feeling for what is involved in using a numerical method and in making error estimates. An appendix at the end of this chapter contains computer programs for implementing these methods on small computers in common use.

6.1 Euler's Method

In this section, we will discuss Euler's method, including an analysis of the error incurred in using it to approximate the solution of an initial value problem.

Euler's method is not very accurate, and there are much better methods which are to be preferred. However, the method has two features which cause us to discuss it first. One is that the method can be motivated geometrically in terms of tangent lines, making it conceptually simple. The other is that it is relatively easy (compared with other approximation methods) to derive bounds for the error resulting from using the method.

Thus, we will use Euler's method as a prototype to demonstrate how an approximation method works and what is involved in implementing it. We will develop more accurate methods in the sections that follow.

Suppose we want to approximate the solution of the initial value problem

$$y' = f(x, y); \qquad y(a) = y_0.$$

We will assume that we want to approximate the solution at points on an interval $[a, b]$. The key observation is that, if we know the solution value $y(x)$ for some x, we can compute $f(x, y(x))$, and therefore we know the slope $y'(x)$ of the tangent to the graph of the solution at that point.

This suggests that we proceed as follows. The discussion is illustrated in Figure 6.1, where the solid graph represents the exact solution $y = y(x)$ as an aid in understanding the idea behind the approximation scheme.

First, subdivide $[a, b]$ into n subintervals of equal length. Let $h = (b - a)/n$ and denote

$$x_0 = a, \qquad x_1 = a + h, \qquad x_2 = a + 2h, \qquad \ldots, \qquad x_n = a + nh = b.$$

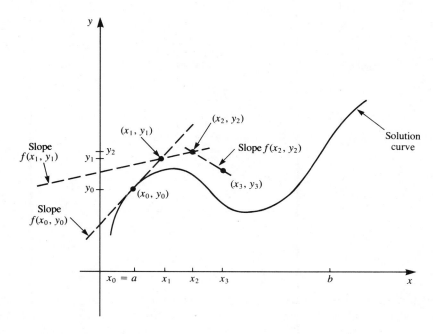

Figure 6.1

In general,

$$x_k = a + kh$$

for $k = 0, 1, 2, \ldots, n$. The number h is called the *step size* of the approximation. The points $x_1, \ldots, x_n$ are the numbers at which we will approximate the solution values.

We know $y(a) = y_0$, given as initial data. Compute $f(a, y_0)$ and draw the line having this slope through (a, y_0). This is the tangent to the solution at a. Move along this tangent line to the point (x_1, y_1). We will use y_1 as an approximation of the solution value $y(x_1)$ at x_1.

Now compute $f(x_1, y_1)$. This is approximately the slope of the tangent line to the graph of the solution which passes through (x_1, y_1). Draw the line through (x_1, y_1) with this slope. This line is "almost" parallel to the tangent at x_1 of the solution of the initial value problem. The discrepancy between the slopes of the tangent lines at $(x_1, y(x_1))$ and (x_1, y_1) depends on how sensitive f is to changes in y. Assuming that f does not change "much" with respect to y and that y_1 is a "close" approximation of $y(x_1)$, we can move along the tangent line through (x_1, y_1) to (x_2, y_2) and use y_2 as an approximation to $y(x_2)$.

Next, compute $f(x_2, y_2)$. This is approximately the slope of the tangent to the solution curve at x_2. Draw the line through (x_2, y_2) with this slope. Notice that this line is almost parallel to the tangent line to the solution at x_2. Move along this line to (x_3, y_3), thus obtaining y_3 as an approximation of $y(x_3)$.

Continue in this way. At each step, use the approximate value y_k at x_k to compute $f(x_k, y_k)$, and draw the line with this slope through (x_k, y_k). This line is approximately parallel to the tangent line to the solution at x_k. Move along this line to the point (x_{k+1}, y_{k+1}) and use y_{k+1} as an approximation of the solution value at x_{k+1}.

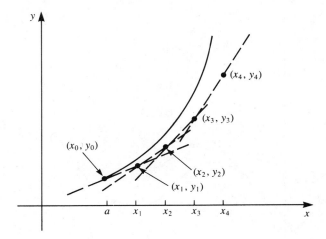

Figure 6.2

Before deriving a formula for the successive approximations y_k, we can see immediately one source of error in this method. In Figure 6.2, notice that the successively drawn line segments move *away from* the solution as x increases from a, accumulating error from one step to the next. This results in y_k being a *less* accurate approximation to $y(x_k)$ as k increases. Following segments of lines is conceptually simple and appealing but not sophisticated enough to be very accurate in general.

We will now derive an analytic expression for the approximate solution values y_k at the points x_k. From Figure 6.1, we see that

$$y_1 = y_0 + f(x_0, y_0)(x_1 - x_0).$$

Similarly,

$$y_2 = y_1 + f(x_1, y_1)(x_2 - x_1).$$

In general, after we have obtained the approximate value y_k for $y(x_k)$, we obtain

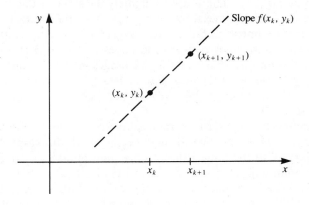

Figure 6.3

(see Figure 6.3)

$$y_{k+1} = y_k + f(x_k, y_k)(x_{k+1} - x_k).$$

Since each $x_{k+1} - x_k = h$, we can summarize Euler's method in the formula

$$y_{k+1} = y_k + f(x_k, y_k)h$$

for $k = 0, \ldots, n - 1$; y_0 given.

EXAMPLE 6.1

Use Euler's method to approximate the solution of

$$y' = \sqrt{y}\,x; \qquad y(2) = 4$$

on the interval $[2, 3]$.

This problem is easy to solve exactly. We get

$$y(x) = \left(1 + \frac{x^2}{4}\right)^2.$$

We will use this solution to check the accuracy of the Euler approximations.

First, we must decide how many points x_k to insert in $[2, 3]$. To be specific, choose $n = 6$, so that $h = (3 - 2)/6 = \frac{1}{6}$. We have

$$x_0 = 2, \qquad x_1 = 2 + \tfrac{1}{6} = \tfrac{13}{6}, \qquad x_2 = 2 + \tfrac{2}{6} = \tfrac{7}{3}, \qquad x_3 = \tfrac{15}{6} = \tfrac{5}{2},$$
$$x_4 = \tfrac{16}{6} = \tfrac{8}{3}, \qquad x_5 = \tfrac{17}{6}, \qquad x_6 = \tfrac{18}{6} = 3.$$

Calculate

$$y_{k+1} = y_k + \tfrac{1}{6}\sqrt{y_k}\,x_k.$$

Table 6.1 shows the Euler approximations. Using x_0 and y_0, calculate y_1; then use y_1 and x_1 to calculate y_2; use y_2 and x_2 to calculate y_3; and so on. In general, use x_k and y_k to compute y_{k+1} and repeat this process until y_n is approximated.

For comparison, we also include (column 4) the actual solution values $y(x_k)$ at the x_k's and the percentage error between the exact values and the approximated values

Table 6.1

k	x_k	y_k	$y(x_k)$	Percentage Error
0	2	4.00000000	4.00000000	0
1	$\frac{13}{6}$	4.66666667	4.72458527	1.2
2	$\frac{7}{3}$	5.44675583	5.57484568	2.3
3	$\frac{5}{2}$	6.35435583	6.56640625	3.2
4	$\frac{8}{3}$	7.40468282	7.71604938	4.0
5	$\frac{17}{6}$	8.61408490	9.04171489	4.7
6	3	10.00004585	10.56250000	5.3

(column 5). All calculations have been carried to eight decimal places, so the "exact" values are actually values computed from the closed form of the solution but rounded to eight places.

   The accuracy in this computation is not very good, as can be seen from the percentage errors in the last column. One reason for this is that h was chosen "fairly large." As we will soon see, we can expect better accuracy by choosing h smaller. This is intuitively clear from the geometric idea behind the method. Figure 6.4 shows a graph of the solution together with a smooth curve drawn through the approximate solution values obtained from the Euler approximations in Table 6.1.

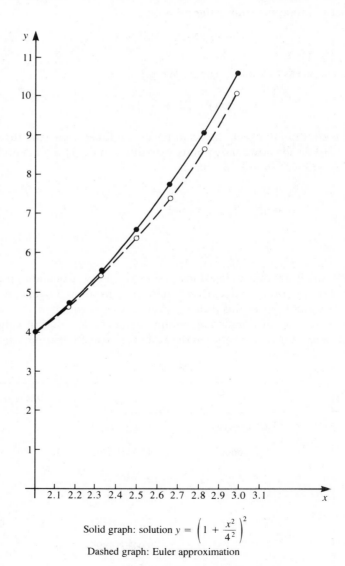

Solid graph: solution $y = \left(1 + \dfrac{x^2}{4^2}\right)^2$

Dashed graph: Euler approximation

Figure 6.4

Table 6.2

k	x_k	y_k	$y(x_k)$	Percentage Error
0	2.0	4.00000000	4.00000000	0
1	2.1	4.40000000	4.45050625	0.46
2	2.2	4.84049972	4.88410000	0.89
3	2.3	5.32452470	5.39400625	1.29
4	2.4	5.85524813	5.95360000	1.65
5	2.5	6.43599101	6.56640625	1.99
6	2.6	7.07022239	7.23610000	2.29
7	2.7	7.76155952	7.96650625	2.57
8	2.8	8.51376806	8.76160000	2.83
9	2.9	9.33076220	9.62550625	3.06
10	3.0	10.21660479	10.56250000	3.59

We said that the method should show improved accuracy if we choose h smaller (and, correspondingly, n larger). For comparison, examine the results in Tables 6.2 and 6.3. In Table 6.2 we have used $h = 0.1$, and in Table 6.3, $h = 0.05$. In addition to improving the accuracy, choosing h smaller provides approximate values of the solution at more points. Of course, this exacts a price in computation.

Even with h chosen smaller, however, the accuracy of the approximations decreases as we move from 2 toward 3. The tangent line approximations accumulate error from one step to the next and move away from the graph of the actual solution. This accumulation of error is worse for some problems than others, depending upon the curvature of the solution graph. ∎

ERROR ANALYSIS OF EULER'S METHOD

For the remainder of this section, we will analyze the error in Euler's method. The analysis we will carry out typifies one approach to estimating the error in a numerical scheme, although the details are more involved for more sophisticated approximation methods.

The idea is to obtain, as a function of step size h, a bound on the error in Euler's method. Intuitively this is a reasonable goal to try to achieve because we expect the accuracy to improve as h is chosen smaller; hence, the error should be a function of h.

In order to state results precisely, we will introduce the O notation. We say that $g(h)$ is $O(h^p)$ if $g(h)/h^p$ is bounded as $h \to 0$ but $g(h)/h^r$ is unbounded as $h \to 0$ for $r > p$. For example, $\sin(h)$ is $O(h)$. We know that $\sin(h)/h$ is bounded as $h \to 0$ [in fact, $\sin(h)/h$ has limit 1]. However, if $r > 1$, then $\sin(h)/h^r = [\sin(h)/h] \, (1/h^{r-1})$ is unbounded as $h \to 0$.

Table 6.3

k	x_k	y_k	$y(x_k)$	Percentage Error
0	2.00	4.00000000	4.00000000	0
1	2.05	4.20000000	4.20506289	0.12
2	2.10	4.41006249	4.42050625	0.24
3	2.15	4.63056405	4.64671914	0.34
4	2.20	4.86189057	4.88410000	0.45
5	2.25	5.10443721	5.13305664	0.56
6	2.30	5.35860848	5.39400625	0.66
7	2.35	5.62481817	5.66737539	0.75
8	2.40	5.90348938	5.95360000	0.84
9	2.45	6.19505455	6.25312539	0.93
10	2.50	6.49995542	6.56640625	1.01
11	2.55	6.81864304	6.89390664	1.09
12	2.60	7.15157782	7.23610000	1.16
13	2.65	7.49922947	7.59346914	1.24
14	2.70	7.86207702	7.96650625	1.31
15	2.75	8.24060886	8.35571289	1.38
16	2.80	8.63532269	8.76160000	1.44
17	2.85	9.04672557	9.18468789	1.50
18	2.90	9.47533386	9.62550625	1.56
19	2.95	9.92167330	10.08459414	1.62
20	3.00	10.38627895	10.56250000	1.67

If a quantity is $O(h)$, we expect it to decrease to zero at the same rate that h does. Similarly, if something is $O(h^3)$, it will decrease to zero at the same rate as h^3 as h is chosen smaller.

When we must use an approximation method, it is usually to our advantage to select a method having an error $O(h^p)$ for as large a p as possible because then the error decreases to zero at the same rate that h^p does, and this will be faster for larger p.

We will now show that the error in Euler's method is $O(h)$. This means, for example, if we halve h, we should expect to decrease the error by a factor of 2. That this actually occurs can be observed in the results of Tables 6.2 and 6.3. In Table 6.2, with $h = 0.1$, the percentage error at the right endpoint is about 3.59%; in Table 6.3, with $h = 0.05$ (half of 0.1), the percentage error at the right endpoint is about 1.67%, which is roughly half of 3.59%.

To show that the error is $O(h)$, we must introduce some terminology. Let

$$E(h) = \max_{1 \leq k \leq n} |y_k - y(x_k)|,$$

where y_k is the Euler approximation at x_k and $y(x_k)$ is the solution evaluated at x_k. $E(h)$ is called the *global discretization error* and is the maximum difference between the approximated and exact values of the solution on $[a, b]$. It is $E(h)$ that we want to bound.

The *local discretization error at* x is defined to be

$$L(x, h) = \frac{1}{h}[y(x + h) - y(x)] - f(x, y(x)).$$

Because $y'(x) = f(x, y(x))$, $L(x, h)$ is the difference between the difference quotient for y due to the increment h and the derivative $y'(x)$.

Notice that, if $y_k = y(x_k)$, so that y_k is the exact solution value at x_k, then the difference between $y(x_{k+1})$, the exact solution value at x_{k+1}, and y_{k+1}, the Euler approximation at x_{k+1}, is

$$
\begin{aligned}
y(x_{k+1}) - y_{k+1} &= y(x_{k+1}) - [y_k + hf(x_k, y_k)] \\
&= y(x_k + h) - y(x_k) - hf(x_k, y_k) \\
&= h\left(\frac{1}{h}[y(x_k + h) - y(x_k)] - f(x_k, y_k)\right) = hL(x_k, h).
\end{aligned}
$$

Thus, if we begin at some x_k where we have the exact value of the solution, $hL(x_k, h)$ is the error incurred in approximating the value of the solution at the next partition point x_{k+1}.

Using a Taylor expansion, we can write

$$
\begin{aligned}
L(x, h) &= \frac{1}{h}[y(x + h) - y(x)] - f(x, y(x)) \\
&= \frac{1}{h}\left\{\left[y(x) + hy'(x) + \frac{1}{2}h^2 y''(\xi)\right] - y(x)\right\} - f(x, y(x))
\end{aligned}
$$

for some ξ in (a, b). Then

$$
\begin{aligned}
L(x, h) &= \frac{1}{h}\left[hy'(x) + \frac{1}{2}h^2 y''(\xi)\right] - f(x, y(x)) \\
&= \frac{1}{2}hy''(\xi),
\end{aligned}
$$

since $y'(x) - f(x, y(x)) = 0$.

We will now make the assumption that, for some number M,

$$|y''(x)| \leq M \quad \text{for} \quad a \leq x \leq b.$$

Then

$$|L(x, h)| \leq \tfrac{1}{2}hM$$

for $a \leq x \leq b$. Now define the *local discretization error for Euler's method* to be

$$L(h) = \max_{a \leq x \leq b} |L(x, h)|.$$

For a given h, this measures the magnitude of the maximum local discretization error at x as x varies over $[a, b]$. Since $|L(x, h)| \leq \frac{1}{2}hM$ for $a \leq x \leq b$,

$$L(h) \leq \tfrac{1}{2}hM.$$

We will use this result to obtain a bound for $E(h)$ and hence on the error resulting from the use of the Euler approximation. Write

exact solution value at x_{k+1} minus the Euler approximate value

$$= y(x_{k+1}) - y_{k+1}$$
$$= y(x_k + h) - y_{k+1}.$$

Solve for $y(x_k + h)$ in $L(x_k, h)$ and substitute the result into the last line to get

$$
\begin{aligned}
y(x_{k+1}) - y_{k+1} &= hL(x_k, h) + hf(x_k, y(x_k)) + y(x_k) - y_{k+1} \\
&= hL(x_k, h) + hf(x_k, y(x_k)) + y(x_k) \\
&\quad - [y_k + hf(x_k, y_k)] \qquad \text{(from Euler's method)} \\
&= y(x_k) - y_k + hL(x_k, h) + h[f(x_k, y(x_k)) - f(x_k, y_k)].
\end{aligned}
\tag{6.1}
$$

Now assume that f is continuous and that $|(\partial f/\partial y)(x, y)| \leq R$ for some number R. By the mean value theorem, there is a number θ, with $0 < \theta < 1$, such that

$$
|f(x_k, y(x_k)) - f(x_k, y_k)| = \left| \frac{\partial f}{\partial y}(x_k, \theta y(x_k) + (1 - \theta)y_k)[y(x_k) - y_k] \right|
\tag{6.2}
$$
$$
\leq R|y(x_k) - y_k|.
$$

From results (6.1) and (6.2), we have

$$
\begin{aligned}
|y(x_{k+1}) - y_{k+1}| &\leq |y(x_k) - y_k| + h|L(x_k, h)| + h|f(x_k, y(x_k)) - f(x_k, y_k)| \\
&\leq |y(x_k) - y_k| + h|L(x_k, h)| + hR|y(x_k) - y_k| \\
&= (1 + hR)|y(x_k) - y_k| + h|L(x_k, h)|.
\end{aligned}
$$

For convenience, let $C = 1 + hR$, and denote $e_k = y(x_k) - y_k$. Then the last inequality can be written

$$|e_{k+1}| \leq C|e_k| + h|L(x_k, h)|.$$

Finally, since $|L(x_k, h)| \leq L(h)$ for $a \leq x \leq b$,

$$|e_{k+1}| \leq C|e_k| + hL(h). \tag{6.3}$$

This inequality relates the error in the approximation of the solution at x_{k+1} to the error in the approximation at x_k and the local discretization error $L(h)$.

Now use inequality (6.3) repeatedly to write

$$
\begin{aligned}
|e_{k+1}| &\leq C|e_k| + hL(h) \\
&\leq C[C|e_{k-1}| + hL(h)] + hL(h) \\
&= C^2|e_{k-1}| + ChL(h) + hL(h) \\
&\leq C^2[C|e_{k-2}| + hL(h)] + ChL(h) + hL(h) \\
&= C^3|e_{k-2}| + C^2hL(h) + ChL(h) + hL(h).
\end{aligned}
$$

Continuing in this manner, we finally obtain

$$|e_{k+1}| \le C^k|e_1| + [C^{k-1} + C^{k-2} + \cdots + C + 1]hL(h).$$

Now,

$$e_1 = y(x_0) - y_0 = 0$$

because $y(x_0) = y_0$ is the initial data given in the problem. Thus,

$$|e_{k+1}| \le [C^{k-1} + C^{k-2} + \cdots + C + 1]hL(h).$$

Since $L(h) \le hM/2$, we have

$$|e_{k+1}| \le [C^{k-1} + C^{k-2} + \cdots + C + 1]h\frac{hM}{2}$$

$$= [C^{k-1} + C^{k-2} + \cdots + C + 1]\frac{1}{2}Mh^2.$$

Finally, choose $k = n - 1$. Then

$$|e_n| \le [C^{n-2} + C^{n-3} + \cdots + C + 1]\tfrac{1}{2}Mh^2.$$

Let P be the largest of the numbers $1, C, C^2, \ldots, C^{n-2}$. Then

$$|e_n| \le nP\tfrac{1}{2}Mh^2.$$

Since $n = (b - a)/h$,

$$|e_n| \le P\frac{1}{2}M\left(\frac{b-a}{h}\right)h^2 = P\frac{M}{2}(b-a)h.$$

Finally, it is not difficult to show that $C^n = (1 + hR)^n$ remains bounded as $h \to 0$ to conclude that $E(h) = O(h)$, the result we wanted to demonstrate.

Because the error decreases at a linear rate with h, we say that Euler's method is a *first order* approximation method. If we have a method whose error decreases with h^2 as h is chosen smaller, we say that this method is *second order*. In the next section, we will develop methods which are second order or better.

PROBLEMS FOR SECTION 6.1

In each of Problems 1 through 10, use Euler's method to approximate a solution on $[a, b]$, using $n = 10$. Carry out all calculations to seven significant digits.

1. $y' = y\sin(x)$; $y(0) = 1$, on $[0, \pi]$

2. $y' = x + y$; $y(1) = -3$, on $[1, 2]$

3. $y' = x - y^2$; $y(0) = 4$, on $[0, 2]$

4. $y' = x^2\ln(y)$; $y(1) = 3$, on $[1, 3]$

5. $y' = x + \sinh(y)$; $y(0) = 2$, on $[0, \frac{1}{4}]$

6. $y' = 2xy$; $y(0) = 5$, on $[0, 2]$

7. $y' = \sqrt{x + y}$; $y(4) = 2$, on $[2, 4]$ *Hint:* Use a negative value for h.

8. $y' = x^2 + y^2$; $y(2) = 1$, on $[1, 2]$

9. $y' = 2 - x$; $y(0) = 1$, on $[0, 1]$

10. $y' = y - \cos(x)$; $y(1) = -3$, on $[1, 3]$

11. Find an analytic expression for the solution of the initial value problem of Problem 1. Make a table (like Table 6.2) containing the approximate solution values determined in Problem 1 together with the "exact" values calculated using the solution, and include a column listing percentage errors.

12. Carry out the instructions of Problem 11, using Problem 2 in place of Problem 1.

13. Carry out the instructions of Problem 11, using Problem 6 in place of Problem 1.

14. Carry out the instructions of Problem 11, using Problem 9 in place of Problem 1.

15. Carry out the instructions of Problem 11, using Problem 10 in place of Problem 1.

16. Consider the initial value problem

$$y' = y - 2; \qquad y(0) = 3.$$

(a) Find the solution of this problem.

(b) Use Euler's method to make a table of approximate solution values at the points 0.5, 1, 1.5, and 2, using $h = 0.5$. Using the solution found in (a), compute percentage errors.

(c) Repeat the program of (b), using $h = 0.25$.

(d) Repeat the program of (b), using $h = 0.125$.

(e) Repeat the program of (b), using $h = 0.0625$.

17. Approximate e as follows. Use Euler's method, with $n = 100$, to approximate $y(1)$, where y is the solution of the initial value problem $y' = y; y(0) = 1$. Sketch a graph of the solution before applying Euler's method and guess whether the approximation of e in this way will be too high or too low.

18. Approximate $\ln(2)$ by applying the idea of Problem 17 but using the initial value problem $y' = 1/x; y(1) = 0$. Will this approximation be too high or too low?

19. Approximate π by applying the idea of Problem 17 to the initial value problem $y' = 4/(1 + x^2); y(0) = 0$. Will this approximation be too high or too low?

20. Approximate π another way by applying the idea of Problem 17 to the initial value problem $y' = 6/\sqrt{1 - x^2}$; $y(0) = 0$. Will this approximation be too high or too low?

6.2 *One-Step Methods*

Euler's method is called a *one-step method* because the approximation at x_{k+1} is a function of the value approximated at x_k, one step back. In this section, we will develop other one-step methods which are in general more accurate than Euler's method.

As before, consider the initial value problem

$$y' = f(x, y); \qquad y(a) = y_0.$$

We want to approximate the solution at selected, equally spaced points on an interval $[a, b]$.

THE SECOND ORDER TAYLOR METHOD

As with Euler's method, partition $[a, b]$ into n subintervals of equal length h, letting $h = (b - a)/n$, $x_0 = a$, $x_n = b$, and $x_{k+1} = x_k + kh$ for $k = 0, 1, 2, \ldots, n - 1$. On the interval $[x_k, x_{k+1}]$, use Taylor's theorem to write

$$y(x_{k+1}) = y(x_k + h) = y(x_k) + hy'(x_k) + \frac{1}{2!} h^2 y''(x_k) + \cdots$$

$$+ \frac{1}{n!} h^n y^{(n)}(x_k) + \frac{1}{(n + 1)!} h^{n+1} y^{(n+1)}(\xi_k)$$

for some ξ_k in (x_k, x_{k+1}). If the $n + 1$ derivative of y is assumed to be bounded, we can make the last term $[1/(n + 1)!]h^{n+1}y^{(n+1)}(\xi_k)$, as small in magnitude as we like by choosing h sufficiently small. We therefore form the approximation

$$y(x_{k+1}) \approx y(x_k) + hy'(x_k) + \frac{1}{2!}h^2 y''(x_k) + \cdots + \frac{1}{n!}h^n y^{(n)}(x_k).$$

Notice that, with $n = 1$, we have Euler's method. With $n = 2$, we have

$$y(x_{k+1}) \approx y(x_k) + hy'(x_k) + \tfrac{1}{2}h^2 y''(x_k). \tag{6.4}$$

Now, we know that $y'(x) = f(x, y(x))$. This suggests that, in equation (6.4), we consider $f(x_k, y_k)$ as an approximation of $y'(x_k)$ if y_k is an approximation of $y(x_k)$:

$$y'(x_k) \approx f(x_k, y_k).$$

To approximate $y''(x_k)$ in equation (6.4), differentiate the equation $y' = f(x, y(x))$ with respect to x, using the chain rule, to obtain

$$y''(x) = \frac{\partial f}{\partial x} + \frac{\partial f}{\partial y} y'(x).$$

Evaluate the right side at x_k and the approximate value y_k, and use this as an approximation of $y''(x_k)$, with $y'(x_k)$ approximated by $f(x_k, y_k)$:

$$y''(x_k) \approx \frac{\partial f}{\partial x}(x_k, y_k) + \frac{\partial f}{\partial y}(x_k, y_k)f(x_k, y_k).$$

Put these approximations of $y'(x_k)$ and $y''(x_k)$ into the approximation (6.4) to get

$$y_{k+1} \approx y_k + hf(x_k, y_k) + \frac{1}{2}h^2\left[\frac{\partial f}{\partial x}(x_k, y_k) + \frac{\partial f}{\partial y}(x_k, y_k)f(x_k, y_k)\right]. \tag{6.5}$$

The expression on the right side of (6.5), which approximates y_{k+1} from x_k, y_k, and the given function and its partial derivatives, is called the *second order Taylor method*. It is a one-step method because all the information at x_{k+1} is inferred from information one step back, at x_k. We start the method at $k = 0$, using the fact that $y(x_0)$ is given as data in the problem; hence, we can compute $f(x_0, y(x_0))$. We then successively let $k = 1, 2, \ldots, n - 1$, obtaining the approximate solution y_{k+1} at y_k from equation (6.5).

We can simplify the notation in two ways. First, denote

$$f_k = f(x_k, y_k).$$

Second, use subscript notation for partial derivatives:

$$\frac{\partial f}{\partial x} = f_x \quad \text{and} \quad \frac{\partial f}{\partial y} = f_y.$$

Either of these partial derivatives, written with a k subscript, means the partial derivative evaluated at (x_k, y_k). Thus, $(f_x)_k = (\partial f/\partial x)(x_k, y_k)$ and $(f_y)_k = (\partial f/\partial y)(x_k, y_k)$. In this notation, the second order Taylor formula (6.5) is

$$y_{k+1} = y_k + hf_k + \tfrac{1}{2}h^2[f_x + f_k f_y]_k.$$

EXAMPLE 6.2

Consider the initial value problem

$$y' = y^2\cos(x); \qquad y(0) = \tfrac{1}{5}.$$

This initial value problem has the unique solution

$$y(x) = \frac{-1}{\sin(x) - 5}.$$

We will use the second order Taylor method to approximate the solution on $[0, 1]$, then compare the results with the exact solution. If we choose $n = 10$, then $h = (1 - 0)/10 = 0.1$. Since $f(x, y) = y^2\cos(x)$,

$$f_x = -y^2\sin(x) \quad \text{and} \quad f_y = 2y\cos(x).$$

The second order Taylor formula for this problem, with $h = 0.1$, is

$$\begin{aligned}
y_{k+1} &= y_k + hy_k^2\cos(x_k) + h^2y_k^3\cos^2(x_k) - \tfrac{1}{2}h^2y_k^2\sin(x_k) \\
&= y_k + 0.1y_k^2\cos(x_k) + 0.01y_k^3\cos^2(x_k) - 0.005y_k^2\sin(x_k)
\end{aligned}$$

for $k = 0, 1, 2, \ldots, 9$. Further, $y_0 = y(0) = \tfrac{1}{5} = 0.2$.

Table 6.4 lists approximate solution values y_k and "exact" solution values $y(x_k)$, computed to eight decimal places. It also includes Euler's method approximations with $h = 0.1$ for purposes of comparison. Euler's formula for this problem is

$$y_{k+1} = y_k + 0.1y_k^2\cos(x_k).$$

Table 6.4

k	x_k	Taylor	$y(x_k)$	Euler's
0	0	0.20000000	0.20000000	0.20000000
1	0.1	0.20408000	0.20407469	0.20400000
2	0.2	0.20828742	0.20827559	0.20814081
3	0.3	0.21258301	0.21256335	0.21238761
4	0.4	0.21692122	0.21689237	0.21669606
5	0.5	0.22125023	0.22121082	0.22102110
6	0.6	0.22551221	0.22546097	0.22530812
7	0.7	0.22964406	0.22957988	0.22949783
8	0.8	0.23357854	0.23350062	0.23352620
9	0.9	0.23724587	0.23715379	0.23732565
10	1.0	0.24057578	0.24046965	0.24082677

The second order Taylor method is in general more accurate than Euler's method. It can be shown that the error in the second order Taylor method is $O(h^2)$, which means that the error decreases as h^2 when we choose h smaller. Recall that the error in Euler's method is $O(h)$. One drawback of the Taylor method is the need to compute the derivatives (other methods just use the function itself).

THE MODIFIED EULER METHOD

Near the end of the nineteenth century, the German mathematician Carl David Runge noticed a similarity between part of the formula for the second order Taylor method and another Taylor polynomial approximation. Write the second order Taylor formula as

$$y_{k+1} = y_k + h\{f_k + \tfrac{1}{2}h[f_x(x_k, y_k) + f_k f_y(x_k, y_k)]\}. \tag{6.6}$$

Runge observed that the term in braces on the right resembles the Taylor approximation

$$f(x_k + \alpha h, y_k + \beta h) \approx f_k + \alpha h f_x(x_k, y_k) + \beta h f_k f_y(x_k, y_k).$$

In fact, the term in braces in equation (6.6) is exactly the right side of this approximation if we choose $\alpha = \beta = \tfrac{1}{2}$. This suggests the approximation scheme

$$y_{k+1} = y_k + hf\left(x_k + \frac{h}{2}, y_k + \frac{hf_k}{2}\right).$$

This formula for generating y_{k+1} from x_k, y_k, and f is called the *modified Euler method*. It is similar in form to Euler's method except that $f(x, y)$ is evaluated at $x = x_k + h/2$ and $y = y_k + hf_k/2$ instead of at $x = x_k$ and $y = y_k$. Notice that $x_k + h/2$ is midway between x_k and x_{k+1}.

EXAMPLE 6.3

Consider the initial value problem

$$y' - \frac{1}{x}y = 2x^2; \qquad y(1) = 4.$$

We will approximate solution values on the interval $[1, 3]$.

Write the differential equation as

$$y' = \frac{1}{x}y + 2x^2$$

so that $f(x, y) = (1/x)y + 2x^2$ in the notation we have been using. This initial value problem has exact solution

$$y = x^3 + 3x.$$

For purposes of illustration, choose $n = 10$ and $h = (3 - 1)/10 = 0.2$. Table 6.5 compares approximations using Euler's method, the modified Euler method, and the second order Taylor method. The approximation formulas for this problem and choice of h are as follows:

$$\text{Euler's:}\quad y_{k+1} = y_k + 0.2\left[\frac{1}{x_k}\,y_k + 2(x_k)^2\right];$$

$$\text{modified Euler:}\quad y_{k+1} = y_k + 0.2\left[\frac{y_k + 0.1\left(\dfrac{1}{x_k}\,y_k + 2(x_k)^2\right)}{x_k + 0.1} + 2(x_k + 0.1)^2\right];$$

$$\text{second order Taylor:}\quad y_{k+1} = y_k + 0.2\left[\frac{y_k}{x_k} + 2(x_k)^2\right]$$

$$+\, 0.02\left[\frac{-y_k}{x_k} + 4x_k + \left(\frac{y_k}{x_k} + 2(x_k)^2\,\frac{1}{x_k}\right)\right].$$

Table 6.5

k	x_k	Modified Euler	Taylor	Euler's	Exact
0	1.0	4.00000000	4.00000000	4.00000000	4.000000
1	1.2	5.32036364	5.32000000	5.20000000	5.328000
2	1.4	6.92739860	6.92666667	6.64266667	6.944000
3	1.6	8.86929364	8.86819048	8.37561905	8.896000
4	1.8	11.19419064	11.19271429	10.44657143	11.232000
5	2.0	13.95020013	13.94834921	12.90330159	14.000000
6	2.2	17.18541062	17.18318413	15.79363175	17.248000
7	2.4	20.94789459	20.94529178	19.16541645	21.024000
8	2.6	25.28571247	25.28273276	23.06653449	25.376000
9	2.8	30.24691542	30.24355836	27.54488330	30.352000
10	3.0	35.87944731	35.87581253	35.64837496	36.000000

■

RUNGE-KUTTA METHODS

The modified Euler method may be thought of as an Euler's method in which the "modified" slope $f(x_k + h/2, y_k + hf_k/2)$ is used in place of the approximate slope $f(x_k, y_k)$.

A number of other approximation methods have been developed using the idea of modifying the slope approximation in Euler's method, usually basing the modification on a Taylor approximation. In particular, there is a class of methods called *Runge-*

Kutta methods which use the weighted approximation

$$af_k + bf(x_k + \alpha h, y_k + \beta h f_k)$$

in place of the term in braces in formula (6.6). The idea is to attempt to choose the constants a, b, α, and β to obtain an approximation with as favorable an error bound as possible.

The RK4 (for fourth order Runge-Kutta) method has proved both accurate and computationally efficient and is obtained by a clever choice of these constants in approximating slopes at various points. Without derivation, we will give the formula:

$$y_{k+1} = y_k + \frac{1}{6}h\left[W_1 + 2W_2 + 2W_3 + W_4\right],$$

where

$$W_1 = f_k$$

$$W_2 = f\left(x_k + \frac{h}{2}, y_k + \frac{hW_1}{2}\right)$$

$$W_3 = f\left(x_k + \frac{h}{2}, y_k + \frac{hW_2}{2}\right)$$

$$W_4 = f(x_k + h, y_k + hW_3).$$

EXAMPLE 6.4

Consider the initial value problem

$$y' = \frac{\cos(y)}{2 + x}; \qquad y(3) = 0.$$

We will use RK4 to approximate the solution at points in $[3, 4]$. Choose $n = 10$ and $h = 0.1$. With $f(x, y) = \cos(y)/(2 + x)$, we obtain

$$W_1 = \frac{\cos(y_k)}{2 + x_k}, \qquad W_2 = \frac{\cos(y_k + 0.05W_1)}{2.05 + x_k},$$

$$W_3 = \frac{\cos(y_k + 0.05W_2)}{2.05 + x_k}, \quad \text{and} \quad W_4 = \frac{\cos(y_k + 0.1W_3)}{2.1 + x_k}.$$

For comparison purposes, we will also compute values using the modified Euler method, which for this problem is

$$y_{k+1} = y_k + 0.1\left[\frac{\cos\left(y_k + \dfrac{0.05\cos(y_k)}{2 + x_k}\right)}{2.05 + x_k}\right].$$

The solution of this initial value problem is implicitly defined by the equation

$$\sec(y) + \tan(y) = \tfrac{1}{5}(x + 2).$$

We can check the numerical solutions in Table 6.6 by substituting the x- and y-values into this equation, computing both sides, and comparing the results.

Table 6.6

k	x_k	RK4	Modified Euler
0	3.0	0.00000000	0.00000000
1	3.1	0.01980133	0.01980099
2	3.2	0.03921066	0.03920996
3	3.3	0.05823596	0.05823488
4	3.4	0.07688518	0.07688372
5	3.5	0.09516621	0.09516435
6	3.6	0.11308687	0.11308461
7	3.7	0.13065494	0.13065226
8	3.8	0.14787808	0.14787499
9	3.9	0.16476386	0.16476036
10	4.0	0.18131977	0.18131586

It has been mentioned that the Taylor method is $O(h^2)$. This means that the error decreases proportionally to h^2 as h is chosen smaller. It can be shown that the modified Euler method is also $O(h^2)$ and that the RK4 method is $O(h^4)$. When using RK4, then, the error decreases proportionally to h^4 as h is chosen smaller. All of these methods improve in accuracy faster than Euler's method as h is chosen smaller.

There are also higher order Runge-Kutta methods in which the error is $O(h^p)$ for larger values of p. Such methods offer improved accuracy but of course cost more in computing time and in complexity.

All of the methods discussed in this section have been one-step methods. In its most general form, a one-step method has the appearance

$$y_{k+1} = y_k + \varphi(x_k, y_k),$$

in which y_{k+1} is calculated by a formula involving x_k and y_k. We proceed one step at a time, to y_1 in terms of x_0 and the initial data y_0; then to y_2 in terms of x_1 and the just-calculated y_1; to y_3 in terms of x_2 and y_2, and so on. The second order Taylor, modified Euler, and RK4 methods are all of this form for various choices of φ.

In the next section, we will discuss multistep methods, in which the calculation of y_{k+1} depends on values of x and y for two or more steps behind x_{k+1}.

PROBLEMS FOR SECTION 6.2

In each of Problems 1 through 10, use the modified Euler, Taylor, and RK4 approximation schemes to approximate the solution of the initial value problem on the interval. Use $n = 10$ and carry out calculations to seven significant digits.

1. $y' = \sinh(x + y)$; $y(0) = 2$, on $[0, 0.2]$

2. $y' = y - x^2$; $y(1) = -4$, on $[1, 3]$

3. $y' = \cos(y) + e^{-x};$ $y(0) = 1,$ on $[0, 1]$

4. $y' = y^3 - 2xy;$ $y(3) = 2,$ on $[1, 3]$

5. $y' = -y + e^{-x};$ $y(0) = 4,$ on $[0, 1]$

6. $y' = \sec(y) - xy^2;$ $y\left(\dfrac{\pi}{4}\right) = 1,$ on $\left[\dfrac{\pi}{4}, \dfrac{\pi}{3}\right]$

7. $y' = e^y - x - y;$ $y(1) = 2,$ on $[1, 1.1]$

8. $y' = x^3 + 4xy - x^2;$ $y(-1) = 3,$ on $[-1, 0]$

9. $y' = e^{x-y} + y;$ $y(0) = 1,$ on $[0, 2]$

10. $y' = y^2 - \cosh(x + y);$ $y(0) = 4,$ on $[0, 1]$

11. In Problem 2, obtain an analytic expression for the exact solution. Use this to calculate each $y(x_k)$ to seven significant digits, and calculate the percentage errors in the results from the modified Euler, Taylor, and RK4 approximations.

12. In Problem 5, obtain an analytic expression for the solution. Use this to calculate the numbers $y(x_k)$ to seven significant digits, then calculate the percentage errors in the results from the modified Euler, Taylor, and RK4 approximations.

13. Derive the *improved Euler method*, also known as the *Heun method*, as follows. Begin with Euler's method and replace f_k with $\frac{1}{2}[f_k + f_{k+1}]$. Next, replace y_{k+1} in f_{k+1} with $y_{k+1} = y_k + hf_k$ from Euler's method. This leads to the approximation scheme

$$y_{k+1} = y_k + \tfrac{1}{2}h[f_k + f(x_{k+1}, y_k + hf_k)].$$

In each of Problems 14 through 18, use the Euler, modified Euler, and improved Euler methods to approximate the solution of the initial value problem on the interval, using $n = 10$ and carrying out calculations to seven significant digits. Also determine an analytic expression for the solution and use it to calculate "exact" values of the solution to compare with the approximated values.

14. $y' = 1 - y;$ $y(0) = 2,$ on $[0, 1]$

15. $y' = -\dfrac{1}{x}y + x;$ $y(1) = 1,$ on $[1, 2]$

16. $y' = y - e^x;$ $y(-1) = 4,$ on $[-1, 2]$

17. $y' = \dfrac{y^2}{x^2} - \dfrac{y}{x};$ $y(1) = -4,$ on $[1, 2]$

18. $y' = 2xe^{-x-y};$ $y(0) = 0,$ on $[0, 1]$

19. Suppose that f is a linear function in both x and y [that is, f has the form $f(x, y) = \alpha x + \beta y$)]. Show that in this case the improved Euler and modified Euler methods are identical.

20. In this section, we derived a method called a second order Taylor method by using Taylor's theorem to expand $y(x_k + h)$ in a finite series ending with a second derivative term. Derive a third order Taylor method by imitating this argument but using Taylor's theorem to expand $y(x_k + h)$ in a finite series ending with a third derivative term.

21. Approximate e as follows. Use the RK4 method with $n = 20$ to approximate $y(1)$, with y the solution of the initial value problem $y' = y;$ $y(0) = 1$. Compare the accuracy of this approximation with that obtained using Euler's method with $n = 100$ in Problem 17 of Section 6.1.

22. Approximate $\ln(2)$ as follows. Repeat the process outlined in Problem 21, except use the initial value problem $y' = 1/x;$ $y(1) = 0$. Compare this result with the approximation obtained in Problem 18 of Section 6.1.

23. Approximate π as follows. Follow the method of Problem 21, but use the initial value problem $y' = 4/(1 + x^2);$ $y(0) = 0$. Compare this result with that of Problem 19 of Section 6.1.

24. Approximate π by following the process of Problem 21 but using the initial value problem $y' = 6/\sqrt{1 - x^2};$ $y(0) = 0$. Compare this result with that obtained in Problem 20 of Section 6.1.

6.3 *Multistep and Predictor-Corrector Methods* _____

In this section, we will sketch the ideas behind two additional types of approximation methods: multistep methods and predictor-corrector methods.

MULTISTEP METHODS

As before, consider the initial value problem

$$y' = f(x, y); \qquad y(a) = y_0.$$

We want to approximate the value of the solution at points in an interval $[a, b]$. Subdivide $[a, b]$ into n subintervals of equal length $h = (b - a)/n$, and consider a typical subinterval $[x_k, x_{k+1}]$. The basis for many multistep approximation methods is the simple observation that, if $p(x)$ is a polynomial which approximates $f(x, y(x))$ on $[x_k, x_{k+1}]$, then the integral of $p(x)$ from x_k to x_{k+1} approximates the integral of $f(x, y(x))$ from x_k to x_{k+1}.

Now write

$$y(x_{k+1}) - y(x_k) = \int_{x_k}^{x_{k+1}} y'(x)\, dx$$

$$= \int_{x_k}^{x_{k+1}} f(x, y(x))\, dx \approx \int_{x_k}^{x_{k+1}} p(x)\, dx.$$

Therefore,

$$y(x_{k+1}) \approx y(x_k) + \int_{x_k}^{x_{k+1}} p(x)\, dx. \tag{6.7}$$

Of course, this is quite vague, since we have not made any attempt to say how good an approximation $p(x)$ is to $f(x, y(x))$ and therefore cannot say how closely (6.7) provides an approximation of $y(x_{k+1})$. Nevertheless, (6.7) contains the germ of a powerful method, which we will now state more precisely.

First, how should we choose a suitable polynomial p? Suppose that somehow we have arrived at satisfactory approximations of the exact values of the solution at $x_k, x_{k-1}, \ldots, x_{k-r}$. That is, we have numbers $y_k, y_{k-1}, \ldots, y_{k-r}$ such that

$$y_k \approx y(x_k), \qquad y_{k-1} \approx y(x_{k-1}), \qquad \ldots, \qquad y_{k-r} \approx y(x_{k-r}).$$

Then

$$f_k = f(x_k, y_k) \approx f(x_k, y(x_k)), \qquad f_{k-1} = f(x_{k-1}, y_{k-1}) \approx f(x_{k-1}, y(x_{k-1})), \ldots,$$

$$f_{k-r} = f(x_{k-r}, y_{k-r}) \approx f(x_{k-r}, y(x_{k-r})).$$

Choose p to be the polynomial of degree r passing through the points

$$(x_k, f_k), \qquad (x_{k-1}, f_{k-1}), \qquad \ldots, \qquad (x_{k-r}, f_{k-r}).$$

We call p an *interpolating polynomial* for these points. By the way p is selected,

$$p(x_k) = f_k, \qquad p(x_{k-1}) = f_{k-1}, \qquad \ldots, \qquad p(x_{k-r}) = f_{k-r}.$$

When this polynomial is inserted into the integral in (6.7), we obtain a multistep approximation method in which we let

$$y_{k+1} = y_k + \int_{x_k}^{x_{k+1}} p(x)\, dx. \tag{6.8}$$

We obtain different methods for different choices of r. Here are some specific examples of this process.

r = 0   Now p is a zero degree polynomial, or constant, and $p(x) = f_k$ for all x. Then equation (6.8) becomes

$$y_{k+1} = y_k + \int_{x_k}^{x_{k+1}} f_k \, dx = y_k + f_k[x_{k+1} - x_k] = y_k + hf_k.$$

For this very simple case in which $r = 0$, we obtain Euler's method!

r = 1   If we let $r = 1$, then p is the polynomial of degree 1 passing through the points (x_k, f_k) and (x_{k-1}, f_{k-1}). The straight line through these points has equation

$$p(x) = -\frac{1}{h}(x - x_k)f_{k-1} + \frac{1}{h}(x - x_{k-1})f_k.$$

Substitute this polynomial into equation (6.8) to obtain the approximation formula

$$y_{k+1} = y_k + \int_{x_k}^{x_{k+1}} \left[-\frac{1}{h}(x - x_k)f_{k-1} + \frac{1}{h}(x - x_{k-1})f_k \right] dx.$$

A routine integration yields

$$y_{k+1} = y_k + \frac{h}{2}[3f_k - f_{k-1}].$$

In this method, y_{k+1} depends not only on information at x_k but also on information at x_{k-1}. This is a *two-step method*.

For larger r, the idea is the same, but the calculation of $p(x)$ and of the integral in equation (6.8) are more complicated. For $r = 2$ and $r = 3$, we obtain the following results.

r = 2

$$y_{k+1} = y_k + \frac{h}{12}[23f_k - 16f_{k-1} + 5f_{k-2}]$$

r = 3

$$y_{k+1} = y_k + \frac{h}{24}[55f_k - 59f_{k-1} + 37f_{k-2} - 9f_{k-3}] \qquad (6.9)$$

These are three-step and four-step methods, respectively. As r increases, we increase the number of data points involved in the determination of y_{k+1}.

Intuitively, we might expect multistep methods to be more accurate than single-step methods because they involve use of information at more points. This often turns out to be the case. We saw before that the error in Euler's method is $O(h)$. One can show that the multistep method we have discussed, using an interpolating polynomial of degree r on each subinterval, has error of order $O(h^{r+1})$. In particular, the schemes for $r = 1, 2$, and 3 are called *Adams-Bashforth multistep methods*.

One drawback to a multistep method is that we must use some other method to start it. For example, look at the Adams-Bashforth method (6.9). Since the formula involves f_{k-3}, we must have $k \geq 3$; hence, the first value we can use it to approximate is y_4. This means that we must use some other method (such as RK4) to approximate y_1, y_2, and y_3.

Another class of multistep methods, called *Adams-Moulton methods*, is obtained by using different data points to determine the interpolating polynomial p to use in equation (6.8). For the case $r = 2$, choose p as the second degree polynomial passing through

$$(y_{k+1}, f_{k+1}), \qquad (y_k, f_k), \quad \text{and} \quad (y_{k-1}, f_{k-1}).$$

The integration formula (6.8) now yields the approximation scheme

$$y_{k+1} = y_k + \frac{h}{24}[9f_{k+1} + 19f_k - 5f_{k-1} + f_{k-2}]. \tag{6.10}$$

The Adams-Moulton method given by equation (6.10) is a four-step method, and it can be shown that the error is $O(h^4)$.

There is a significant difference between the Adams-Bashforth method (6.9) and the Adams-Moulton method (6.10). The method given by (6.9) determines y_{k+1} in terms of previously calculated quantities and is therefore *explicit*. The method (6.10) contains y_{k+1} on both sides of the equation [because $f_{k+1} = f(x_{k+1}, y_{k+1})$] and therefore defines y_{k+1} *implicitly*. Thus, equation (6.10) provides an equation for y_{k+1} which must be solved for y_{k+1}.

PREDICTOR-CORRECTOR METHODS

Using equations (6.9) and (6.10), we can develop an interesting and effective strategy called a *predictor-corrector method*. The idea is to begin with the Adams-Bashforth method (6.9) to make a first approximation of y_{k+1}. This first approximation is called a *predicted value* of y_{k+1}; to emphasize this, we shall denote it as $y_{k+1}^{[1]}$.

Next, use $y_{k+1}^{[1]}$ to compute $f(x_{k+1}, y_{k+1}^{[1]})$, the value of f at x_{k+1}, and the predicted value $y_{k+1}^{[1]}$. Denote this value of f as $f_{k+1}^{[1]}$:

$$f_{k+1}^{[1]} = f(x_{k+1}, y_{k+1}^{[1]}).$$

Finally, the predicted values $y_{k+1}^{[1]}$ and $f_{k+1}^{[1]}$ are *corrected* by inserting them into the Adams-Moulton formula (6.10).

Thus far, we have, for $k \geq 3$,

$$y_{k+1}^{[1]} = y_k + \frac{h}{24}[55f_k - 59f_{k-1} + 37f_{k-2} - 9f_{k-3}], \tag{6.11}$$

$$f_{k+1}^{[1]} = f(x_{k+1}, y_{k+1}^{[1]}), \tag{6.12}$$

$$y_{k+1}^{[2]} = y_k + \frac{h}{24}[9f_{k+1}^{[1]} + 19f_k - 5f_{k-1} + f_{k-2}]. \tag{6.13}$$

The strategy is now to correct the predicted value even further by cycling this value back into equation (6.12), producing a new value

$$f_{k+1}^{[2]} = f(x_{k+1}, y_{k+1}^{[2]}),$$

which in turn is placed into equation (6.13) to get a new $y_{k+1}^{[3]}$. This procedure is repeated until the quantity

$$\left| \frac{y_{k+1}^{[i+1]} - y_{k+1}^{[i]}}{y_{k+1}^{[i+1]}} \right|$$

is smaller than some prescribed value. At this time, y_{k+1} is assigned the value $y_{k+1}^{[i+1]}$ and the process is repeated for the next value of k.

One problem with this procedure is that we have no guarantee that the sequence $y_{k+1}^{[i]}$ will converge, that is, improve in accuracy. There is a technique called *variable step sizing* that can be used in conjunction with predictor-corrector methods (as well as some other methods) to improve their efficiency. Recall that, in general, a smaller step size h provides an increase in accuracy but at the expense of additional computing time. The corrector equation provides a value which can be used to determine whether the step size is sufficiently small to furnish the required accuracy. The user determines a minimum local discretization error E_1 deemed acceptable and a maximum local error E_2 which can be tolerated. Then the corrector formula is used once (not repeatedly) to provide a value $y_{k+1}^{[2]}$ which is used to approximate the local error as

$$|y(x_{k+1}) - y_{k+1}^{[2]}| \approx K|y_{k+1}^{[2]} - y_{k+1}^{[1]}| = D_{k+1},$$

where the value of K depends upon the predictor-corrector method being employed. In the case of the Adams-Bashforth and Adams-Moulton combination, $K = \frac{1}{14}$.

We can now accept $y_{k+1} = y_{k+1}^{[2]}$ as close enough if

$$E_1 \le D_{k+1} \le E_2.$$

To accomplish this, the step size is adjusted.

Here is a summary of the steps in this technique.

1. Use (6.11) to obtain $y_{k+1}^{[1]}$.
2. Use (6.12) to calculate $f_{k+1}^{[1]}$.
3. Use (6.13) to determine $y_{k+1}^{[2]}$.
4. Evaluate D_{k+1}.

If $E_1 \le D_{k+1} \le E_2$, assign $y_{k+1} = y_{k+1}^{[2]}$ and proceed to the next iteration.

If $D_{k+1} > E_2$, the step size is too large; replace it with $h/2$ and recompute the previous four values of y_k, which are used to make a new prediction of y_{k+1}.

If $D_{k+1} < E_1$, the step size is smaller than necessary for the desired degree of accuracy, and computing time can be saved by doubling the step size from h to $2h$. Again, the previous four values of y_k must be computed using this new step size.

To implement this predictor-corrector variable step size procedure successfully, it is necessary to have a scheme which can be used to determine new values of the previous four y_k in the event the step size is either doubled or halved. There are approximation methods available to accomplish this, but they necessarily complicate the procedure. We will be satisfied with a fixed-step procedure involving only one pass through the correction process, reserving a deeper study of these processes for a course in numerical analysis.

We will call the procedure which derives y_{k+1} by applying equations (6.11), (6.12), and (6.13) each one time the *Adams-Bashforth-Moulton predictor-corrector method*. Observe that this method is entirely explicit because in the last step we use $f_{k+1}^{[2]}$, which we have computed explicitly from the preceding two steps. Note also that k varies from 3 to $n - 1$ because the formulas involve f_{k-3}, requiring that $k \ge 3$. This means that we must approximate y_1, y_2, and y_3 by some other method before beginning this predictor-corrector scheme. Often this is done using RK4.

EXAMPLE 6.5

Approximate the solution of the initial value problem

$$y' + y = 2\cos(x); \qquad y(0) = 4$$

on $[0, 1]$

We can solve this problem exactly, obtaining the solution

$$y = \cos(x) + \sin(x) + 3e^{-x}.$$

We will use this solution to test how good the predictor-corrector approximation scheme is. Choose $n = 10$ and $h = (3 - 2)/10 = 0.1$. We will use RK4 to approximate y_1, y_2, and y_3 and then use the Adams-Bashforth-Moulton predictor-corrector method to approximate $y_4, \ldots, y_{10}$. Table 6.7 also compares these approximate values with the "exact" values (to eight decimal places) obtained from the solution.

Table 6.7

k	x_k	y_k		Exact	Exact-approximation
0	0	4.00000000	(initial data)	4.00000000	
1	0.1	3.80934996⎤		3.80934984	−0.00000012
2	0.2	3.63492838⎬	RK4	3.63492817	−0.00000021
3	0.3	3.47331163⎦		3.47331136	−0.00000027
4	0.4	3.32143908		3.32143947	0.00000039
5	0.5	3.17659913		3.17660008	0.00000095
6	0.6	3.03641157		3.03641300	0.00000143
7	0.7	2.89881395		2.89881579	0.00000184
8	0.8	2.76204749		2.76204969	0.00000220
9	0.9	2.62464333		2.62464586	0.00000253
10	1.0	2.48540879		2.48541162	0.00000283

■

PROBLEMS FOR SECTION 6.3

In each of Problems 1 through 10, use the Adams-Bashforth-Moulton predictor-corrector method, with $n = 10$, to obtain approximate solution values at points in the interval, carrying calculations to seven significant digits.

1. $y' = y - x^3$; $\quad y(-2) = -4$, $\quad$ on $[-2, -1]$

2. $y' = 2xy - y^3$; $\quad y(0) = 2$, $\quad$ on $[0, 1]$

3. $y' = \ln(x) + x^2 y$; $\quad y(2) = 1$, $\quad$ on $[2, 3]$

4. $y' = x^2 y^2 - x - y$; $\quad y(0) = 2$, $\quad$ on $[0, 2]$

5. $y' = \sin^2(y) - x$; $\quad y\left(\dfrac{\pi}{2}\right) = 1$, $\quad$ on $\left[\dfrac{\pi}{2}, \pi\right]$

6. $y' = x^2 e^x - \cos(y)$; $\quad y(1) = -2$, $\quad$ on $[1, 3]$

7. $y' = y^2 - x$; $\quad y(-1) = 0$, $\quad$ on $[-1, 0]$

8. $y' = (x - y)^2 - \cos(x)$; $\quad y(\pi) = 1$, $\quad$ on $\left[\pi, \dfrac{3\pi}{2}\right]$

9. $y' = 1 - xy + \cosh(x);$ $y(0) = 2,$ on $[0, 2]$

10. $y' = \dfrac{1}{x + \cosh(y)};$ $y(3) = 1,$ on $[3, 5]$

In each of Problems 11 through 15, make a table listing approximations of the solution using the Taylor, modified Euler, RK4, and predictor-corrector methods. Use $n = 10$, and carry out calculations to seven places.

11. $y' = 4y^2 - x;$ $y(3) = 0,$ on $[3, 4]$

12. $y' = x \sin(y) - x^2;$ $y(1) = -3,$ on $[1, 3]$

13. $y' = x^2 + 4y;$ $y(0) = -2,$ on $[0, 1]$

14. $y' = 1 - \cos(x - y) + x^2;$ $y(3) = 6,$ on $[3, 5]$

15. $y' = 4x^3 - xy + \cos(y);$ $y(0) = 4,$ on $[0, 1]$

16. Obtain an analytic expression for the solution of the initial value problem in Problem 1, and use it to calculate the solution values at the partition points to seven significant digits. Compare these values with those obtained from the predictor-corrector method, listing percentage errors.

17. For the case in which $r = 2$, obtain the interpolating polynomial p, then use equation (6.8) to derive the three-step approximation formula

$$y_{k+1} = y_k + \frac{h}{12}[23f_k - 16f_{k-1} + 5f_{k-2}],$$

which is stated in the text.

18. For the case in which $r = 3$, obtain the interpolating polynomial p, then use equation (6.8) to derive the four-step approximation formula

$$y_{k+1} = y_k + \frac{h}{24}[55f_k - 59f_{k-1} + 37f_{k-2} - 9f_{k-3}].$$

19. Every one-step and multistep method we have considered in this and the preceding section is a special case of the general formula

$$y_{k+1} = \sum_{j=1}^{m} \alpha_j y_{k+1-j} + h\varphi(x_{k+1-m}, \ldots, x_k, x_{k+1}, y_{k+1-m}, \ldots, y_k, y_{k+1}).$$

By making appropriate choices of m, the constants α_j, and the function φ, show how this formula yields each of the following methods:

(a) Euler's method;

(b) the modified Euler method;

(c) the Taylor method;

(d) RK4;

(e) Adams-Bashforth.

Appendix: Computer Programs _____

This appendix contains nine computer programs for implementing and comparing the numerical methods of this chapter. The programs are all written in "standard" BASIC (which may differ slightly from one system to another) and have been run on a variety of IBM-compatibles. The first line of the program indicates the relevant method.

Each of these programs is designed to solve a specific initial value problem involving the differential equation $y' = y/x + y$. A printout of the numerical results is included; this can be used as a check of the program on other hardware. To run the program for other initial value problems, make the necessary changes in the function $f(x, y)$ in the differential equation $y' = f(x, y)$ and also in the initial value for $y(x_0)$ and the initial point x_0.

PROGRAM 1

```
10   REM:EULER METHOD
20   REM:DEFINE F WHERE Y' = F(X,Y)
30   DEF FNF(X,Y) = Y/X + X
40   REM:SET THE NUMBER OF SUBINTERVALS N AND STEP SIZE
     H = (B-A)/N
50   N = 10
60   H = .4
70   REM:SET THE INITIAL CONDITIONS X = X0 AND Y = Y(X0)
80   X = 1
90   Y = 0
100  LPRINT "K"," XK"," YK"
110  LPRINT "0",X,Y
120  REM:LOOP FROM 1 TO N
130  FOR I = 1 TO N
140    Y = Y + H*FNF(X,Y)
150    X = X + H
160    LPRINT I,X,Y
170  NEXT I
180  END
```

K	XK	YK
0	1	0
1	1.4	.4
2	1.8	1.074286
3	2.2	2.033016
4	2.6	3.282655
5	3.0	4.827679
6	3.4	6.67137
7	3.8	8.816236
8	4.2	11.26426
9	4.6	14.01705
10	5.0	17.07592

PROGRAM 2

```
10   REM:TAYLOR METHOD
20   REM:DEFINE F WHERE Y' = F(X,Y)
30   DEF FNF(X,Y) = Y/X + X
40   REM:DEFINE THE PARTIAL OF F WITH RESPECT TO X
50   DEF FNFX(X,Y) = -Y/(X*X) + 1
60   REM:DEFINE THE PARTIAL OF F WITH RESPECT TO Y
70   DEF FNFY(X,Y) = 1/X
80   REM:SET THE NUMBER OF SUBINTERVALS N AND STEP SIZE H =
     (B-A)/N
90   N = 10
100  H = .4
110  REM:SET THE INITIAL CONDITIONS X = X0 AND Y = Y(X0)
120  X = 1
130  Y = 0
140  LPRINT "K"," XK"," YK"
150  LPRINT "0",X,Y
160  REM:LOOP FROM 1 TO N
```

```
170 FOR I = 1 TO N
180    Y = Y + H*FNF(X,Y) + .5*H*H*(FNFX(X,Y)+FNF(X,Y)*FNFY(X,Y))
190    X = X + H
200    LPRINT I,X,Y
210 NEXT I
220 END
```

K	XK	YK
0	1	0
1	1.4	.56
2	1.8	1.44
3	2.2	2.64
4	2.6	4.16
5	3.0	6
6	3.4	8.16
7	3.8	10.64
8	4.2	13.44
9	4.6	16.56
10	5.0	20

PROGRAM 3

```
10   REM:MODIFIED EULER METHOD
20   REM:DEFINE F WHERE Y' = F(X,Y)
30   DEF FNF(X,Y) = Y/X + X
40   REM:SET THE NUMBER OF SUBINTERVALS N AND STEP SIZE H =
     (B-A)/N
50   N = 10
60   H = .4
70   REM:SET THE INITIAL CONDITIONS X = XO AND Y = Y(XO)
80   X = 1
90   Y = 0
100 LPRINT "K"," XK"," YK"
110 LPRINT "0",X,Y
120 REM:LOOP FROM 1 TO N
130 FOR I = 1 TO N
140    Y = Y + H*FNF(X+H/2,Y+H*FNF(X,Y)/2)
150    X = X + H
160    LPRINT I,X,Y
170 NEXT I
180 END
```

K	XK	YK
0	1	0
1	1.4	.5466667
2	1.8	1.412857
3	2.2	2.598826
4	2.6	4.104673
5	3.0	5.930446
6	3.4	8.076172
7	3.8	10.54187
8	4.2	13.32754
9	4.6	16.43319
10	5.0	19.85883

PROGRAM 4

```
10   REM:RK4 METHOD
20   REM:DEFINE F WHERE Y' = F(X,Y)
30   DEF FNF(X,Y) = Y/X + X
40   REM:SET THE NUMBER OF SUBINTERVALS N AND STEP SIZE H =
     (B-A)/N
50   N = 10
60   H = .4
70   REM:SET THE INITIAL CONDITIONS X = XO AND Y = Y(XO)
80   X = 1
90   Y = 0
100 LPRINT "K"," XK"," YK"
110 LPRINT "0",X,Y
120 REM:LOOP FROM 1 TO N
130 FOR I = 1 TO N
140    W1 = FNF(X,Y)
150    W2 = FNF(X+H/2,Y+H*W1/2)
160    W3 = FNF(X+H/2,Y+H*W2/2)
170    W4 = FNF(X+H,Y+H*W3)
180    Y = Y + H*(W1+2*W2+2*W3+W4)/6
190    X = X + H
200    LPRINT I,X,Y
210 NEXT I
220 END
```

K	XK	YK
0	1	0
1	1.4	.5597884
2	1.8	1.439635
3	2.2	2.639506
4	2.6	4.159388
5	3.0	5.999275
6	3.4	8.159166
7	3.8	10.63906
8	4.2	13.43895
9	4.6	16.55885
10	5.0	19.99875

PROGRAM 5

```
10   REM:IMPROVED EULER METHOD
20   REM:DEFINE F WHERE Y' = F(X,Y)
30   DEF FNF(X,Y) = Y/X + X
40   REM:SET THE NUMBER OF SUBINTERVALS N AND STEP SIZE H =
     (B-A)/N
50   N = 10
60   H = .4
70   REM:SET THE INITIAL CONDITIONS X = XO AND Y = Y(XO)
80   X = 1
90   Y = 0
100 LPRINT "K"," XK"," YK"
```

```
110 LPRINT "0",X,Y
120 REM:LOOP FROM 1 TO N
130 FOR I = 1 TO N
140    Y = Y + H*(FNF(X,Y)+FNF(X+H,Y+H*FNF(X,Y)))/2
150    X = X + H
160    LPRINT I,X,Y
170 NEXT I
180 END
```

K	XK	YK
0	1	0
1	1.4	.5371429
2	1.8	1.392834
3	2.2	2.567808
4	2.6	4.062374
5	3.0	5.876688
6	3.4	8.010835
7	3.8	10.46487
8	4.2	13.23881
9	4.6	16.33269
10	5.0	19.74653

PROGRAM 6

```
10    REM:MODIFIED EULER METHOD WITH COMPARISON TO EXACT
      SOLUTION
20    REM:DEFINE F WHERE Y' = F(X,Y)
30    DEF FNF(X,Y) = Y/X + X
40    REM:DEFINE THE EXACT SOLUTION
50    DEF FNFE(X) = X*X - X
60    REM:SET THE NUMBER OF SUBINTERVALS N AND STEP SIZE H =
      (B-A)/N
70    N = 10
80    H = .4
90    REM:SET THE INITIAL CONDITIONS X = X0 AND Y = Y(X0)
100   X = 1
110   Y = 0
120   LPRINT "K"," XK"," YK","Y(XK)"," %ERROR"
130   LPRINT "0",X,Y,Y,"0"
140   REM:LOOP FROM 1 TO N
150   FOR I = 1 TO N
160      Y = Y + H*FNF(X+H/2,Y+H*FNF(X,Y)/2)
170      X = X + H
180      Z = FNFE(X)
190      REM:FIND PERCENT ABSOLUTE ERROR
200      IF Z = 0 GOTO 240
210      E = ABS((Z-Y)*100/Z)
220      LPRINT I,X,Y,Z,E
230      GOTO 250
240      LPRINT I,X,Y,Z
250      NEXT I
260      END
```

K	XK	YK	Y(XK)	% ERROR
0	1	0	0	0
1	1.4	.5466667	.56	2.38094
2	1.8	1.412857	1.44	1.884898
3	2.2	2.598826	2.64	1.559646
4	2.6	4.104673	4.16	1.330009
5	3.0	5.930446	6	1.159262
6	3.4	8.076172	8.16	1.027327
7	3.8	10.54187	10.64	.9223479
8	4.2	13.32754	13.44	.8368065
9	4.6	16.43319	16.56	.7657842
10	5.0	19.85883	20	.7058715

PROGRAM 7

```
10   REM:SIMULTANEOUS EULER-IMPROVED EULER-MODIFIED EULER
     METHODS
20   REM:DEFINE F WHERE Y' = F(X,Y)
30   DEF FNF(X,Y) = Y/X + X
40   REM:SET THE NUMBER OF SUBINTERVALS N AND STEP SIZE H =
     (B-A)/N
50   N = 10
60   H = .4
70   REM:SET THE INITIAL CONDITIONS X = XO AND Y = Y(XO)
80   X = 1
90   Y = 0
100  Z = Y
110  W = Y
120  REM: Y-EULER Z-IMPROVED EULER W-MODIFIED EULER
130  LPRINT "","",", EULER","IMP EULER","MOD EULER"
140  LPRINT "K"," XK"," YK"," YK"
150  LPRINT "0",X,Y,Z,W
160  LOOP FROM 1 TO N
170  FOR I = 1 TO N
180    Y = Y + H*FNF(X,Y)
190    Z = Z + H*(FNF(X,Z)+FNF(X+H,Z+H*FNF(X,Z)))/2
200    W = W + H*FNF(X+H/2,W+H*FNF(X,W)/2)
210    X = X + H
220    LPRINT I,X,Y,Z,W
230  NEXT I
240  END
```

K	XK	EULER YK	IMP EULER YK	MOD EULER YK
0	1	0	0	0
1	1.4	.4	.5371429	.5466667
2	1.8	1.074286	1.392834	.5466667
3	2.2	2.033016	2.567808	2.598826
4	2.6	3.282655	4.062374	4.104673
5	3.0	4.827679	5.876688	5.930446
6	3.4	6.67137	8.010835	8.076172
7	3.8	8.816236	10.46487	10.54187
8	4.2	11.26426	13.23881	13.32754
9	4.6	14.01705	16.33269	16.43319
10	5.0	17.07592	19.74653	19.85883

PROGRAM 8

```
10   REM:SIMULTANEOUS TAYLOR-MODIFIED EULER-RK4 METHODS
20   REM:DEFINE F WHERE Y' = F(X,Y)
30   DEF FNF(X,Y) = Y/X + X
40   REM:DEFINE THE PARTIAL OF F WITH RESPECT TO X
50   DEF FNFX(X,Y) = -Y/(X*X) + 1
60   REM:DEFINE THE PARTIAL OF F WITH RESPECT TO Y
70   DEF FNFY(X,Y)=1/X
80   REM:SET THE NUMBER OF SUBINTERVALS N AND STEP SIZE H =
     (B-A)/N
90   N = 10
100  H = .4
110  REM:SET THE INITIAL CONDITIONS X = X0 AND Y = Y(X0)
120  X = 1
130  Y = 0
140  Z = Y
150  W = Y
160  REM: Y-TAYLOR Z-MODIFIED EULER W-RK4
170  LPRINT "","",," TAYLOR","MOD EULER"," RK4"
180  LPRINT "K"," XK"," YK"," YK"," YK"
190  LPRINT "0",X,Y,Z,W
200  REM:LOOP FROM 1 TO N
210  FOR I = 1 TO N
220     Y = Y + H*FNF(X,Y)+.5*H*H*(FNFX(X,Y)+FNF(X,Y)*FNFY(X,Y))
230     Z = Z + H*FNF(X+H/2,Z+H*FNF(X,Z)/2)
240     W1 = FNF(X,W)
250     W2 = FNF(X+H/2,W+H*W1/2)
260     W3 = FNF(X+H/2,W+H*W2/2)
270     W4 = FNF(X+H,W+H*W3)
280     W = W + H*(W1+2*W2+2*W3+W4)/6
290     X = X + H
300     LPRINT I,X,Y,Z,W
310  NEXT I
320  END
```

K	XK	TAYLOR YK	MOD EULER YK	RK4 YK
0	1	0	0	0
1	1.4	.56	.5466667	.5597884
2	1.8	1.44	1.412857	1.439635
3	2.2	2.64	2.598826	2.639506
4	2.6	4.16	4.104673	4.159388
5	3.0	6	5.930446	5.999275
6	3.4	8.16	8.076172	8.159166
7	3.8	10.64	10.54187	10.63906
8	4.2	13.44	13.32754	13.43895
9	4.6	16.56	16.43319	16.55885
10	5.0	20	19.85883	19.99875

PROGRAM 9

```
10   REM:ADAMS-BASHFORTH-MOULTON PREDICTOR-CORRECTOR
     METHOD WITH RK4
20   REM:DEFINE F WHERE Y' = F(X,Y)
```

```
30    DEF FNF(X,Y) = Y/X + X
40    REM:SET THE NUMBER OF SUBINTERVALS N AND STEP SIZE H =
      (B-A)/N
50    N = 10
60    H = .4
70    REM:SET THE INITIAL CONDITIONS X = X0 AND Y = Y(X0)
80    X = 1
90    Y = 0
100   LPRINT "K"," XK"," YK"
110   LPRINT "0",X,Y
120   Y(0) = Y
130   REM:LOOP FROM 1 TO 3 WITH RK4
140   FOR I = 1 TO 3
150      W1 = FNF(X,Y)
160      W2 = FNF(X+H/2,Y+H*W1/2)
170      W3 = FNF(X+H/2,Y+H*W2/2)
180      W4 = FNF(X+H,Y+H*W3)
190      Y = Y + H*(W1+2*W2+2*W3+W4)
200      X = X + H
210      LPRINT I,X,Y
220      Y(I) = Y
230   NEXT I
240   REM:LOOP FROM 3 TO N-1
250   FOR K = 3 TO N-1
260      YP=Y(K)+H*(55*FNF(X,Y(K))-59*FNF(X-H,Y(K-1))
                    +37*FNF(X-2*H,Y(K-2))-9*FNF(X-3*H,Y(K-3)))/24
270      Y(K+1) = Y(K)+H*(9*FNF(X+H,YP)+19*FNF(X,Y(K))
                    -5*FNF(X-H,Y(K-1))+FNF(X-2*H,Y(K-2)))/24
280      X = X + H
290      LPRINT K+1,X,Y(K+1)
300   NEXT K
310   END
```

K	XK	YK
0	1	0
1	1.4	.5597884
2	1.8	1.439635
3	2.2	2.639506
4	2.6	4.159415
5	3.0	5.999326
6	3.4	8.159235
7	3.8	10.63915
8	4.2	13.43906
9	4.6	16.55897
10	5.0	19.99888

Special Functions, Sturm-Liouville Theory, and Eigenfunction Expansions

7.0 Introduction

When a function turns up frequently in important applications or discussions, we sometimes find it useful to compile the properties of the function and to prepare tables of its values rather than constantly rederive this information. Such a function is called a *special function.*

The familiar trigonometric functions comprise one class of special functions. In addition, they are examples of *transcendental functions*—that is, $\sin(x)$ and $\cos(x)$ cannot be defined by elementary algebraic operations on x, as are polynomials and rational functions. This is typical of special functions. For this reason, properties and numerical values of special functions often require considerable ingenuity to derive.

Many (though certainly not all) special functions arise as solutions of important differential equations or initial value problems. Indeed, $\sin(x)$ may be defined as the unique solution of the initial value problem.

$$y'' + y = 0; \qquad y(0) = 0, \qquad y'(0) = 1,$$

while $\cos(x)$ is defined as the unique solution of

$$y'' + y = 0; \qquad y(0) = 1, \qquad y'(0) = 0.$$

All of the properties of $\sin(x)$ and $\cos(x)$ can be derived from these definitions (see Problems 49 through 56 of Section 5.2).

In this section, we will consider other special functions derived as solutions of differential equations or initial value problems which have been found to be particularly important in modeling physical phenomena. Solutions of these differential equations will be special functions. For example, Bessel's differential equation will have Bessel functions as solutions, and Legendre's differential equation will have Legendre polynomial solutions. Once we derive expressions (often infinite series) for solutions of these differential equations, we will see how to obtain properties of these solutions and relationships between them. These properties and relationships form the basis for the use of these special functions in solving many problems in physics and engineering.

We will begin with Bessel functions and some of their applications, then discuss Legendre polynomials and other kinds of orthogonal polynomials. These treatments are independent of each other. Following them, however, we will develop a general concept of orthogonal functions, Sturm-Liouville theory, and eigenfunction expansions. These ideas will reveal a common thread running through much of the theory of special functions. They also provide a vantage point from which to develop Fourier methods in the solution of boundary value problems in partial differential equations.

7.1 *Bessel's Equation and Bessel Functions of the First Kind* _____

The differential equation

$$x^2 y'' + xy' + (x^2 - v^2)y = 0$$

is called *Bessel's equation of order v*, with $v \geq 0$. Bessel's equation is a second order differential equation, but it is also traditional to refer to v as the *order* of the equation. Thus, for example, if we speak of Bessel's equation of order 6, we mean the second order differential equation.

$$x^2 y'' + xy' + (x^2 - 36)y = 0.$$

Bessel's equation occurs in studies of radiation of energy and in other contexts, particularly those in which the mathematical model is most naturally posed in cylindrical coordinates (see, for example, Section 18.5).

It is routine to check that zero is a regular singular point of Bessel's equation, so we will attempt a Frobenius solution $y(x) = \sum_{n=0}^{\infty} c_n x^{n+r}$. Substitute this series into Bessel's equation and rearrange terms to get

$$[r(r-1) + r - v^2]c_0 x^r + [r(r+1) + (r+1) - v^2]c_1 x^{r+1}$$

$$+ \sum_{n=2}^{\infty} \{[(n+r)(n+r-1) + (n+r) - v^2]c_n + c_{n-2}\} x^{n+r} = 0.$$

Let the coefficient of each power of x be zero. Assuming that $c_0 \neq 0$, we find from the coefficient of x^r that the indicial equation is

$$F(r) = r^2 - v^2 = 0.$$

This equation has roots $r = v$ and $r = -v$. Let $r = v$ in the coefficient of x^{r+1} to get

$$(2v + 1)c_1 = 0.$$

Since $v \geq 0$, this equation implies that

$$c_1 = 0.$$

From the coefficient of x^{n+r}, we obtain the recurrence relation

$$F(r + n)c_n + c_{n-2} = 0 \quad \text{for} \quad n = 2, 3, 4, \ldots.$$

Put $r = v$ into the recurrence relation to get

$$F(v + n)c_n + c_{n-2} = 0,$$

or

$$c_n = -\frac{1}{F(v + n)} c_{n-2}$$

for $n = 2, 3, 4, \ldots$. Since $F(v + n) = (v + n)^2 - v^2 = n^2 + 2nv = n(n + 2v)$,

$$c_n = -\frac{1}{n(n + 2v)} c_{n-2}$$

for $n = 2, 3, 4, \ldots$.

Since $c_1 = 0$, the recurrence relation (with $n = 3$) gives us $c_3 = 0$, then $c_5 = 0$, and so on. In general,

$$c_{\text{odd}} = 0.$$

For an even-indexed term, we have

$$c_{2n} = -\frac{1}{2n(2n + 2v)} c_{2n-2} = -\frac{1}{2^2 n(n + v)} c_{2n-2}$$

$$= -\frac{1}{2^2 n(n + v)} \frac{-1}{2(n - 1)[2(n - 1) + 2v]} c_{2n-4}$$

$$= \frac{1}{2^4 n(n - 1)(n + v)(n + v - 1)} c_{2n-4}$$

$$= \cdots = \frac{(-1)^n}{(2^{2n})(1)(2) \cdots n(1 + v)(2 + v) \cdots (n + v)} c_0.$$

This gives us one solution of Bessel's equation of order v,

$$y_1(x) = c_0 \sum_{n=0}^{\infty} \frac{(-1)^n}{2^{2n} n!(1 + v) \cdots (n + v)} x^{2n+v},$$

valid in some interval $(0, R)$. Before pursuing a second solution, we will discuss a function which is useful in writing solutions of Bessel's equation.

A BRIEF EXCURSION—THE GAMMA FUNCTION

This and other solutions of Bessel's equation are often written in terms of the gamma function, which we will now discuss. For $x > 0$, define

$$\Gamma(x) = \int_0^{\infty} t^{x-1} e^{-t} \, dt.$$

Γ is the capital Greek letter gamma, and $\Gamma(x)$ is read "gamma of x." The integral defining $\Gamma(x)$ is improper at zero and at infinity, and it can be shown that this integral converges for all $x > 0$.

The most important property of $\Gamma(x)$ for our purposes is the *factorial property*:

$$\text{if} \quad x > 0, \quad \text{then} \quad \Gamma(x + 1) = x\Gamma(x).$$

This can be derived by integration by parts. With $u = t^x$ and $dv = e^{-t}\,dt$, we have $du = xt^{x-1}\,dt$ and $v = -e^{-t}$, so

$$\int_a^b t^x e^{-t}\,dt = -t^x e^{-t}\Big]_a^b - \int_a^b -e^{-t}xt^{x-1}\,dt$$

$$= -b^x e^{-b} + a^x e^{-a} + x\int_a^b t^{x-1}e^{-t}\,dt.$$

Now take the limit in the last equation as $a \to 0$ and $b \to \infty$. Then $b^x e^{-b} \to 0$ and $a^x e^{-a} \to 0$, while $x\int_a^b t^{x-1}e^{-t}\,dt \to x\int_0^\infty t^{x-1}e^{-t}\,dt = x\Gamma(x)$. Using the factorial property, we have, for any positive integer n,

$$\Gamma(n+1) = n\Gamma(n) = n(n-1)\Gamma(n-1) = \cdots$$
$$= n(n-1)(n-2)\cdots(1)\Gamma(1) = n!\Gamma(1).$$

But

$$\Gamma(1) = \int_0^\infty e^{-t}\,dt = 1;$$

hence,

$$\Gamma(n+1) = n!.$$

If $v \geq 0$, but v is not necessarily an integer, a similar property holds:

$$\Gamma(n+v+1) = (n+v)\Gamma(n+v)$$
$$= (n+v)(n+v-1)\Gamma(n+v-1)$$
$$= \cdots = (n+v)(n+v-1)\cdots(1+v)\Gamma(1+v).$$

That is,

$$\Gamma(n+v+1) = (n+v)(n+v-1)\cdots(1+v)\Gamma(1+v). \tag{7.1}$$

Equation (7.1) is also called the *factorial property of the gamma function.*

Although the improper integral defining $\Gamma(x)$ converges only if $x > 0$, it is possible to define $\Gamma(x)$ if x is negative (but not an integer) by utilizing the factorial property. From the fact that $\Gamma(x+1) = x\Gamma(x)$, we can write

$$\Gamma(x) = \frac{1}{x}\Gamma(x+1). \tag{7.2}$$

This holds for all $x > 0$. Now, if $-1 < x < 0$, then $0 < x + 1 < 1$, and $\Gamma(x+1)$ is defined. We can therefore *define* $\Gamma(x)$, for $-1 < x < 0$, by equation (7.2). For example,

$$\Gamma\left(-\frac{1}{2}\right) = \frac{1}{-\frac{1}{2}}\Gamma\left(-\frac{1}{2}+1\right) = -2\Gamma\left(\frac{1}{2}\right).$$

Having defined $\Gamma(x)$ on $(-1, 0)$, suppose now that $-2 < x < -1$. Then $-1 < x + 1 < 0$, and we can again use equation (7.2) to define $\Gamma(x)$. For example,

$$\Gamma\left(-\frac{3}{2}\right) = \frac{1}{-\frac{3}{2}}\Gamma\left(-\frac{3}{2}+1\right) = -\frac{2}{3}\Gamma\left(-\frac{1}{2}\right) = +\frac{4}{3}\Gamma\left(\frac{1}{2}\right).$$

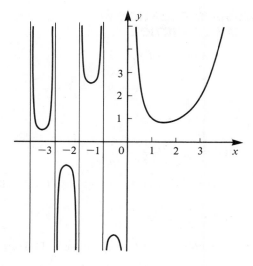

Figure 7.1. The gamma function $y = \Gamma(x)$.

Clearly, we can continue this process, moving to the left over the real line and defining $\Gamma(x)$ on every interval $(k - 1, k)$ once it has been defined on the interval $(k, k + 1)$ to the right of it, for any negative integer k.

Figure 7.1 shows a graph of this extended gamma function, defined for all $x > 0$ and for negative noninteger values of x. Some values of $\Gamma(x)$ are given in Table 7.1.

Table 7.1   Selected Values of $\Gamma(x)$ (To Five Decimal Places)

x	$\Gamma(x)$	x	$\Gamma(x)$	x	$\Gamma(x)$
1.00	1.00000	1.35	0.89115	1.70	0.90864
1.05	0.97350	1.40	0.88726	1.75	0.91906
1.10	0.95135	1.45	0.88566	1.80	0.93138
1.15	0.93304	1.50	0.88623	1.85	0.94561
1.20	0.91817	1.55	0.88887	1.90	0.96177
1.25	0.90640	1.60	0.89352	1.95	0.97988
1.30	0.89747	1.65	0.90012	2.00	1.00000

For $\Gamma(x)$ at other values of x, use the factorial property. For example, to compute $\Gamma(2.5)$, write $\Gamma(2.5) = \Gamma(1.5 + 1) = 1.5\Gamma(1.5) = 1.5(0.88623) = 1.32935$.

These values were taken from *CRC Standard Mathematical Tables*, 25th ed. (Boca Raton, FL: CRC Press, 1981), p. 414.

SOLUTIONS OF BESSEL'S EQUATION IN TERMS OF THE GAMMA FUNCTION

Notice that part of the denominator of $y_1(x)$ is

$$(1 + v)(2 + v) \cdots (n + v) = \frac{\Gamma(n + v + 1)}{\Gamma(1 + v)}.$$

We may therefore write the solution y_1 as

$$y_1(x) = c_0 \sum_{n=0}^{\infty} \frac{(-1)^n \Gamma(v + 1)}{2^{2n} n! \Gamma(n + v + 1)} x^{2n+v}.$$

It is customary to choose $c_0 = 1/[2^v \Gamma(1 + v)]$ and obtain the solution denoted $J_v(x)$:

$$J_v(x) = \sum_{n=0}^{\infty} \frac{(-1)^n}{2^{2n+v} n! \Gamma(n + v + 1)} x^{2n+v}.$$

J_v is called a *Bessel function of the first kind of order* v. This series converges for all positive x.

In this first solution, v can be any nonnegative number. We will now consider the problem of finding a second, linearly independent solution of Bessel's equation. Recall that the indicial equation is $F(r) = r^2 - v^2 = 0$, with roots $\pm v$. In the notation of Section 5.4, we have $r_1 = v$ and $r_2 = -v$, so that $r_1 - r_2 = 2v$. In seeking a second solution, Theorem 5.5 of Section 5.4 leads to the following two cases.

Case 1: $2v$ Is Not an Integer   Now $r_1 - r_2 = 2v$ is not an integer. We can repeat the work done in deriving J_v, with v replaced with $-v$, to derive the second, linearly independent solution J_{-v}. We will state this result as a theorem.

THEOREM 7.1   Solutions of Bessel's Equation of Order v When $2v$ Is Not an Integer

If $2v$ is not an integer, then two linearly independent solutions of Bessel's equation of order v are J_v and J_{-v}, where

$$J_v(x) = \sum_{n=0}^{\infty} \frac{(-1)^n}{2^{2n+v} n! \Gamma(n + v + 1)} x^{2n+v}.$$

Thus, the general solution of Bessel's equation of order v, when $2v$ is not an integer, is

$$y(x) = c_1 J_v(x) + c_2 J_{-v}(x). \quad \blacksquare$$

Case 2: $2v$ Is an Integer   Because of a special property of $J_v(x)$, this case splits into two subcases:

Case 2(a): $v = n + \frac{1}{2}$ for some nonnegative integer n;

Case 2(b): $2v$ is an integer, but v is not of the form $n + \frac{1}{2}$.

Case 2(a) If $v = n + \frac{1}{2}$ for a nonnegative integer n, it can be shown (see Problems 22 and 23 at the end of this section) that J_v and J_{-v} are linearly independent.

THEOREM 7.2

Suppose that $v = n + \frac{1}{2}$ for some nonnegative integer n. Then J_v and J_{-v} are linearly independent, and the general solution of Bessel's equation of order $n + \frac{1}{2}$ is

$$y(x) = c_1 J_{n+1/2}(x) + c_2 J_{-n-1/2}(x). \quad \blacksquare$$

In fact, if $v = n + \frac{1}{2}$, then J_v and J_{-v} can be expressed in closed form as a sum of terms involving square roots, sines, and cosines. For example, from the series for J_v, we can derive the formulas

$$J_{1/2}(x) = \sqrt{\frac{2}{\pi x}} \sin(x), \qquad J_{-1/2}(x) = \sqrt{\frac{2}{\pi x}} \cos(x),$$

$$J_{3/2}(x) = \sqrt{\frac{2}{\pi x}} \left[\frac{\sin(x)}{x} - \cos(x) \right],$$

and

$$J_{-3/2}(x) = \sqrt{\frac{2}{\pi x}} \left[-\sin(x) - \frac{\cos(x)}{x} \right].$$

These relationships can be verified by inserting the Maclaurin series for $\cos(x)$ and $\sin(x)$ and manipulating the resulting series to obtain $J_v(x)$.

Case 2(b)   If $2v$ is an integer, but $v \neq n + \frac{1}{2}$, then $2v$ is an even integer; hence, v itself is a nonnegative integer. In this case, J_v and J_{-v} are solutions of Bessel's equation of order v, but they are not linearly independent. [In fact, in this case, one can check from the series that $J_{-v} = (-1)^v J_v$.]

We will complete case 2(b), in which v is an integer, in the next section, where we will derive second solutions called Bessel functions of the second kind.

PROBLEMS FOR SECTION 7.1

1. Use the fact that J_v is a solution of Bessel's equation of order v to show that $x^a J_v(bx^c)$ is a solution of the differential equation

$$y'' - \left(\frac{2a-1}{x} \right) y' + \left(b^2 c^2 x^{2c-2} + \frac{a^2 - v^2 c^2}{x^2} \right) y = 0.$$

In each of Problems 2 through 11, determine $a, b, c,$ and v so that the differential equation has the form of that in Problem 1. Then use the result of Problem 1 to find the general solution of the differential equation in terms of powers of x times Bessel functions of the first kind.

2. $y'' + \dfrac{1}{3x} y' + \left(1 + \dfrac{7}{144x^2} \right) y = 0$

3. $y'' + \dfrac{1}{x} y' + \left(4x^2 - \dfrac{4}{9x^2} \right) y = 0$

4. $y'' - \dfrac{5}{x} y' + \left(64x^6 + \dfrac{5}{x^2} \right) y = 0$

5. $y'' + \dfrac{3}{x} y' + \left(16x^2 - \dfrac{5}{4x^2} \right) y = 0$

6. $y'' - \dfrac{3}{x} y' + 9x^4 y = 0$

7. $y'' - \dfrac{7}{x} y' + \left(36x^4 + \dfrac{175}{16x^2} \right) y = 0$

8. $y'' + \dfrac{1}{x} y' - \dfrac{1}{16x^2} y = 0$

9. $y'' + \dfrac{5}{x} y' + \left(81x^4 + \dfrac{7}{4x^2} \right) y = 0$

10. $y'' - \dfrac{5}{x} y' + \left(32x^2 + \dfrac{8}{x^2} \right) y = 0$

11. $y'' + \dfrac{1}{x} y' + \left(36x^4 - \dfrac{81}{25x^2} \right) y = 0$

12. Use the change of variables $y = (1/u)(1/b)(du/dx)$ to transform the differential equation

$$\frac{dy}{dx} + by^2 = cx^m$$

into the differential equation

$$\frac{d^2u}{dx^2} - bcx^m u = 0.$$

Now use the result of Problem 1 to find the general solution of this differential equation in terms of Bessel functions, and obtain a solution of the original differential equation. Assume that b is a positive constant.

13. Show that, for any real number v,

$$[x^{-v}J_v(x)]' = -x^{-v}J_{v+1}(x)$$

and

$$[x^v J_v(x)]' = x^v J_{v-1}(x).$$

14. Show that, for any real number v,

$$xJ'_v(x) = -vJ_v(x) + xJ_{v-1}(x)$$

and

$$xJ'_v(x) = vJ_v(x) - xJ_{v+1}(x).$$

Hint: Use the result of Problem 13.

15. Show that, for any real number v,

$$x^2 J''_v(x) = (v^2 - v - x^2)J_v(x) + xJ_{v+1}(x).$$

16. Show that, for any positive number v,

$$\frac{2}{x} vJ_v(x) = J_{v-1}(x) + J_{v+1}(x)$$

and

$$2J'_v(x) = J_{v-1}(x) - J_{v+1}(x).$$

17. Use the result of Problem 13 to prove that, for any positive integer n,

$$\int x^{n+1}J_n(x)\, dx = x^{n+1}J_{n+1}(x) + C.$$

18. Show that, for any nonnegative integer n,

$$\int x[J_n(x)]^2\, dx = \tfrac{1}{2}x^2\{[J_n(x)]^2 - J_{n-1}(x)J_{n+1}(x)\} + C.$$

19. Show that

$$J_{5/2}(x) = \sqrt{\frac{2}{\pi x}} \left[\left(\frac{3}{x^2} - 1 \right) \sin(x) - \frac{3}{x} \cos(x) \right].$$

Hint: Use the expression for $J_{3/2}(x)$ in terms of $\cos(x)$ and $\sin(x)$ given in the text, together with the result of Problem 13.

20. Show that

$$J_{-5/2}(x) = \sqrt{\frac{2}{\pi x}} \left[\left(\frac{3}{x^2} - 1 \right) \cos(x) + \frac{3}{x} \sin(x) \right].$$

21. Suppose that $J_0(\alpha) = 0$. Show that

$$\int_0^1 J_1(\alpha x)\, dx = \frac{1}{\alpha}.$$

22. Show that

$$W[J_\nu, J_{-\nu}](x) = \frac{C}{x}$$

for any positive number ν and $x > 0$. *Hint:* Write the Bessel equation of order ν in the form

$$y'' + \frac{1}{x} y' + \left(\frac{x^2 - \nu^2}{x^2} \right) y = 0$$

and assume Abel's identity, which states that, if φ and ψ are solutions of $y'' + p(x)y' + q(x)y = 0$ on an open interval I, then

$$W[\varphi, \psi](x) = Ce^{-\int p(x)\,dx}$$

for some constant C and all x in I.

23. Let ν be a positive number which is not an integer. Write the series representations of J_ν, $J_{-\nu}$, J'_ν, and $J'_{-\nu}$ and use these to find a series representation of $xW[J_\nu, J_{-\nu}](x)$. Evaluate this expression at $x = 0$ to show that $C = 2/\Gamma(-\nu)\Gamma(\nu + 1)$ in Problem 22.

24. Let $u = J_0(\alpha x)$ and $v = J_0(\beta x)$.

(a) Show that

$$xu'' + u' + \alpha^2 xu = 0$$

and

$$xv'' + v' + \beta^2 xv = 0.$$

(b) Multiply the first of these equations by v and the second by u and subtract the resulting equations to show that

$$[x(u'v - v'u)]' = (\beta^2 - \alpha^2)xuv.$$

(c) Show from (b) that

$$(\beta^2 - \alpha^2) \int xJ_0(\alpha x)J_0(\beta x)\, dx = x[\alpha J'_0(\alpha x)J_0(\beta x) - \beta J'_0(\beta x)J_0(\alpha x)].$$

(This is one of a set of formulas called *Lommel's integrals*.)

25. Recall the Maclaurin expansion

$$e^z = \sum_{k=0}^{\infty} \frac{1}{k!} z^k$$

for all z. Show that, for each nonnegative integer n, $J_n(x)$ is the coefficient of t^n in the Maclaurin expansion of

$$e^{x[t - (1/t)]/2}.$$

Hint: Write this exponential function as a product $e^{xt/2}e^{-x/2t}$, expand both factors in Maclaurin series, and collect the terms forming the coefficient of t^n.

26. Derive the integral formula

$$J_0(x) = \frac{2}{\pi} \int_0^{\pi/2} \cos[x \sin(\theta)] \, d\theta.$$

Hint: Begin with $e^{ix\sin(\theta)} = \sum_{n=0}^{\infty}(1/n!)[ix\sin(\theta)]^n$. Integrate both sides from 0 to 2π, and use the formula

$$\int_0^{2\pi} \sin^n(\theta) \, d\theta = \begin{cases} 0 & \text{if } n \text{ is odd} \\ \dfrac{(n-1)(n-3)\cdots(3)(1)(2\pi)}{n(n-2)\cdots(4)(2)} & \text{if } n \text{ is even} \end{cases}$$

to conclude that

$$J_0(x) = \frac{1}{2\pi} \int_0^{2\pi} e^{ix\sin(\theta)} \, d\theta.$$

Now separate $e^{ix\sin(\theta)}$ into real and imaginary parts, using Euler's formula, to derive the requested result.

27. This problem consists of a sequence of steps leading to a conclusion about the positive roots of the equation $J_n(x) = 0$, with n any nonnegative integer.

(a) Show that, if $k > 1$ and a is sufficiently large, there is at least one x in $(a, a + \pi)$ such that $J_n(kx) = 0$. *Hint:* First show that $u = \sqrt{kx}J_n(kx)$ satisfies

$$u'' = -\left(k^2 - \frac{n^2 - \frac{1}{4}}{x^2}\right)u$$

(see Problem 1). Next, show that $v = \sin(x - a)$ satisfies $v'' = -v$. Show from these two differential equations that

$$[uv' - vu']' = k^2 - 1 - \left(\frac{n^2 - \frac{1}{4}}{x^2}\right)uv.$$

From this, show that

$$-u(a + \pi) - u(a) = \int_a^{a+\pi} \left(k^2 - 1 - \frac{n^2 - \frac{1}{4}}{x^2}\right)u(x)v(x) \, dx.$$

Now use the mean value theorem for integrals to conclude that, for some α in $(a, a + \pi)$,

$$-u(a + \pi) - u(a) = u(\alpha) \int_a^{a+\pi} \left(k^2 - 1 - \frac{n^2 - \frac{1}{4}}{x^2}\right)v(x) \, dx.$$

Finally, show that, for sufficiently large a, this integral is positive; hence, conclude that $u(\alpha)$, $u(a)$, and $u(a + \pi)$ cannot all have the same sign.

(b) Show that the equation $J_n(x) = 0$ has infinitely many positive roots.

(c) Show that the equations $J_n(x) = 0$ and $J_n'(x) = 0$ cannot have a common positive root. *Hint:* Use Bessel's equation to show that, if $J_n(\alpha) = J_n'(\alpha) = 0$, then $J_n''(\alpha) = 0$ also. Differentiate Bessel's equation to show that $J_n^{(3)}(\alpha) = 0$. In this way, conclude that $J_n^{(k)}(\alpha) = 0$ for every positive integer k. Now derive a contradiction.

(d) Show that the equation $J_n(x) = 0$ has no positive solution in common with the equations $J_{n-1}(x) = 0$ and $J_{n+1}(x) = 0$.

(e) Show that the equation $J_{n+1}(x) = 0$ has at least one solution between each pair of positive solutions of $J_n(x) = 0$. *Hint:* Apply Rolle's theorem to the function

$$f(x) = J_n(x)x^{-n}$$

on $[a, b]$, where a and b are consecutive positive solutions of $J_n(x) = 0$.

(f) Show that, between any pair of positive solutions of $J_n(x) = 0$, there is a solution of $J_{n+1}(x) = 0$.

In view of results (e) and (f), the positive zeros of $J_n(x)$ and $J_{n+1}(x)$ are said to *interlace*. This interlacing can be seen by examining the zeros given in Table 7.3 at the end of Section 7.2. This table gives the first nine zeros (in increasing order) of $J_0(x)$ through $J_4(x)$. In fact, it can be shown, from the argument in (a), that there exists a function p such that $\lim_{x \to \infty} p(x) = 0$ and, for $x > 0$,

$$J_0(x) = \sqrt{\frac{2}{\pi x}} \left[\cos\left(x - \frac{\pi}{4} \right) + p(x) \right].$$

As x is chosen larger, $p(x)$ becomes smaller, and the zeros of $J_0(x)$ are more closely approximated by the zeros of $\cos(x - \pi/4)$, which occur at the numbers $(4n + 3)\pi/4$ for integer values of n.

7.2 *Bessel Functions of the Second Kind* _____

In this section, we will derive a second solution of Bessel's equation of order v,

$$x^2 y'' + xy' + (x^2 - v^2)y = 0,$$

in which $v \geq 0$. From the work of the preceding section, we have the general solution

$$y(x) = c_1 J_v(x) + c_2 J_{-v}(x)$$

if $2v$ is not an integer and also if $v = n + \frac{1}{2}$ for some nonnegative integer n. In the case in which v is a nonnegative integer, say $v = n$, we have one solution J_n of Bessel's equation but have not yet derived a second, linearly independent solution. We will do this now.

A SECOND SOLUTION OF BESSEL'S EQUATION IN THE CASE $n = 0$

We will carry out the details for the case in which $n = 0$. This will convey the idea without involving us in the more complicated details of the more general case in which n is any nonnegative integer.

Recall from the preceding section that the roots of the indicial equation of Bessel's equation of order n are n and $-n$, with difference $2n$. In this case in which $n = 0$, the indicial equation has only a single, repeated root. From the method of Frobenius, we know that we should look for a second solution of the form

$$y_2(x) = y_1(x)\ln(x) + \sum_{k=1}^{\infty} c_k^* x^k,$$

in which

$$y_1(x) = J_0(x) = \sum_{k=0}^{\infty} \frac{(-1)^k}{2^{2k}[k!]^2} x^{2k}.$$

This series is obtained by putting $v = 0$ into the series for $J_v(x)$ in the preceding section. Compute

$$y_2'(x) = J_0'(x)\ln(x) + \frac{1}{x} J_0(x) + \sum_{k=1}^{\infty} k c_k^* x^{k-1}$$

and

$$y_2''(x) = J_0''(x)\ln(x) + \frac{2}{x} J_0'(x) - \frac{1}{x^2} J_0(x) + \sum_{k=2}^{\infty} k(k-1)c_k^* x^{k-2}.$$

Substitute $y_2(x) = J_0(x)\ln(x) + \sum_{k=1}^{\infty} c_k^* x^k$ into Bessel's equation of order zero, which is

$$x^2 y'' + xy' + x^2 y = 0,$$

or, upon dividing by x,

$$xy'' + y' + xy = 0.$$

We get

$$xJ_0''(x)\ln(x) + 2J_0'(x) - \frac{1}{x} J_0(x) + \sum_{k=2}^{\infty} k(k-1)c_k^* x^{k-1}$$

$$+ J_0'(x)\ln(x) + \frac{1}{x} J_0(x) + \sum_{k=1}^{\infty} kc_k^* x^{k-1}$$

$$+ xJ_0(x)\ln(x) + \sum_{k=1}^{\infty} c_k^* x^{k+1} = 0.$$

The terms involving $\ln(x)$ are

$$\ln(x)[xJ_0''(x) + J_0'(x) + xJ_0(x)],$$

and the expression in brackets equals zero because J_0 is a solution of Bessel's equation of order zero. Further, the $(1/x)J_0(x)$ terms cancel. We are left with

$$2J_0'(x) + \sum_{k=2}^{\infty} k(k-1)c_k^* x^{k-1} + \sum_{k=1}^{\infty} kc_k^* x^{k-1} + \sum_{k=1}^{\infty} c_k^* x^{k+1} = 0.$$

Since $k(k-1) = k^2 - k$, part of the first summation cancels all terms except the $k = 1$ term of the second summation, and we obtain

$$2J_0'(x) + \sum_{k=2}^{\infty} k^2 c_k^* x^{k-1} + c_1^* + \sum_{k=1}^{\infty} c_k^* x^{k+1} = 0.$$

Substitute

$$J_0'(x) = \sum_{k=1}^{\infty} \frac{(-1)^k}{2^{2k-1}k!(k-1)!} x^{2k-1}$$

into this equation to get

$$2 \sum_{k=1}^{\infty} \frac{(-1)^k}{2^{2k-1}k!(k-1)!} x^{2k-1} + \sum_{k=2}^{\infty} k^2 c_k^* x^{k-1} + c_1^* + \sum_{k=1}^{\infty} c_k^* x^{k+1} = 0.$$

Shift indices in the last series to write this equation as

$$\sum_{k=1}^{\infty} \frac{(-1)^k}{2^{2k-2}k!(k-1)!} x^{2k-1} + c_1^* + 4c_2^* x + \sum_{k=3}^{\infty} (k^2 c_k^* + c_{k-2}^*)x^{k-1} = 0. \quad (7.3)$$

On the left side of this equation, the only constant term is c_1^*. This must equal the constant term on the other side of the equation, which is zero:

$$c_1^* = 0.$$

Next, the only even powers of x appearing on the left side of the equation occur in the last series, when k is odd. The coefficients of these powers of x must be zero and must satisfy

$$k^2 c_k^* + c_{k-2}^* = 0, \qquad k = 3, 5, 7, \ldots.$$

We conclude that all odd-subscripted coefficients depend on c_1^* and hence are also zero:

$$c_{2k+1}^* = 0 \quad \text{for} \quad k = 0, 1, 2, 3, \ldots.$$

We will now determine the even-indexed coefficients. Replace k by $2j$ in the second summation of equation (7.3), and k with j in the first, to get

$$\sum_{j=1}^{\infty} \frac{(-1)^j}{2^{2j-2} j! (j-1)!} x^{2j-1} + 4c_2^* x + \sum_{j=2}^{\infty} [4j^2 c_{2j}^* + c_{2j-2}^*] x^{2j-1} = 0.$$

Write this equation as

$$(4c_2^* - 1)x + \sum_{j=2}^{\infty} \left[\frac{(-1)^j}{2^{2j-2} j! (j-1)!} + 4j^2 c_{2j}^* + c_{2j-2}^* \right] x^{2j-1} = 0.$$

Equate the coefficient of each power of x to zero. We get

$$c_2^* = \tfrac{1}{4}$$

and the recurrence relation

$$c_{2j}^* = \frac{(-1)^{j+1}}{2^{2j}[j!]^2 j} - \frac{1}{4j^2} c_{2j-2}^*$$

for $j = 2, 3, 4, \ldots.$ We can see a pattern if we write a few of these coefficients:

$$\text{with} \quad j = 2, \quad c_4^* = -\frac{3}{128} = \frac{-1}{2^2 4^2} \left[1 + \frac{1}{2} \right],$$

$$\text{with} \quad j = 3, \quad c_6^* = \frac{1}{2^2 4^2 6^2} \left[1 + \frac{1}{2} + \frac{1}{3} \right],$$

and, in general, we find that

$$c_{2j}^* = \frac{(-1)^{j+1}}{(2^2)(4^2)(6^2) \cdots (2j)^2} \left[1 + \frac{1}{2} + \frac{1}{3} + \cdots + \frac{1}{j} \right].$$

It is customary to denote

$$\varphi(j) = 1 + \frac{1}{2} + \frac{1}{3} + \cdots + \frac{1}{j}.$$

This enables us to write c_{2j} more compactly:

$$c_{2j}^* = \frac{(-1)^{j+1}}{2^{2j}[j!]^2} \varphi(j)$$

for $j = 1, 2, 3, \ldots.$

A second solution of Bessel's equation of order zero may be written

$$y_2(x) = J_0(x)\ln(x) + \sum_{k=1}^{\infty} \frac{(-1)^{k+1}}{2^{2k}[k!]^2} \varphi(k)x^{2k}$$

for $x > 0$. Because of the logarithm term, this solution is linearly independent from $J_0(x)$. Instead of $y_2(x)$ for a second solution, it is customary to use a particular linear combination of $J_0(x)$ and $y_2(x)$, denoted $Y_0(x)$ and defined by

$$Y_0(x) = \frac{2}{\pi}\{y_2(x) + [\gamma - \ln(2)]J_0(x)\},$$

in which γ is Euler's constant,

$$\gamma = \lim_{k \to \infty} [\varphi(k) - \ln(k)] \approx 0.577215664901533 \ldots.$$

Since $Y_0(x)$ is a sum of constants times solutions of Bessel's equation of order zero, it is also a solution. Further, $Y_0(x)$ is linearly independent from $J_0(x)$. Thus, the general solution of Bessel's equation of order zero is

$$y(x) = c_1 J_0(x) + c_2 Y_0(x).$$

In terms of the series derived above for $y_2(x)$,

$$Y_0(x) = \frac{2}{\pi}\left\{J_0(x)\ln(x) + \sum_{k=1}^{\infty} \frac{(-1)^{k+1}}{2^{2k}[k!]^2} \varphi(k)x^{2k} + [\gamma - \ln(2)]J_0(x)\right\}$$

$$= \frac{2}{\pi}\left\{J_0(x)\left[\ln\left(\frac{x}{2}\right) + \gamma\right] + \sum_{k=1}^{\infty} \frac{(-1)^{k+1}}{2^{2k}[k!]^2} \varphi(k)x^{2k}\right\}.$$

Y_0 is a *Bessel function of the second kind of order zero*; with this choice of constants, Y_0 is also called *Neumann's function of order zero*.

A SECOND SOLUTION OF BESSEL'S EQUATION OF ORDER v IF v IS A POSITIVE INTEGER

If v is a positive integer, $v = n$, then a similar, but more involved, calculation leads us to the following second solution of Bessel's equation of order n:

$$Y_n(x) = \frac{2}{\pi}\left\{J_n(x)\left[\ln\left(\frac{x}{2}\right) + \gamma\right] + \sum_{k=0}^{\infty} \frac{(-1)^{k+1}[\varphi(k) + \varphi(n+k)]}{2^{2k+n+1}k!(k+n)!}x^{2k+n}\right.$$

$$\left. - \sum_{k=0}^{n-1} \frac{(n-k-1)!}{2^{2k-n+1}k!}x^{2k-n}\right\}.$$

Y_n and J_n are linearly independent for $x > 0$, and the general solution of Bessel's equation of order n is

$$y(x) = c_1 J_n(x) + c_2 Y_n(x).$$

Although J_n is simply J_v in the case $v = n$, our derivation of Y_n does not suggest how Y_v might be defined if v is not a nonnegative integer. However, it is possible to define Y_v, if v is not an integer, by letting

$$Y_v(x) = \frac{1}{\sin(v\pi)} [J_v(x)\cos(v\pi) - J_{-v}(x)].$$

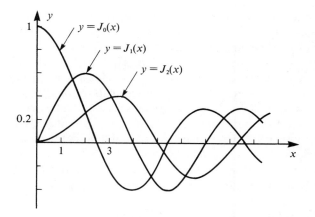

Figure 7.2. Bessel functions of the first kind.

This is a linear combination of J_v and J_{-v}, two solutions of Bessel's equation of order v, and hence is also a solution of Bessel's equation of order v. It can be shown that one can obtain Y_n, for n a nonnegative integer, from this definition by taking a limit:

$$Y_n(x) = \lim_{v \to n} Y_v(x).$$

Y_v is called *Neumann's Bessel function of order v*. It is linearly independent from J_v and has the unifying property of allowing us to write the general solution of Bessel's equation of order v as

$$y(x) = c_1 J_v(x) + c_2 Y_v(x),$$

which holds whether or not v is an integer.

Graphs of some of the Bessel functions of both kinds are shown in Figures 7.2 and 7.3. There are also extensive tables of values of many of these functions which

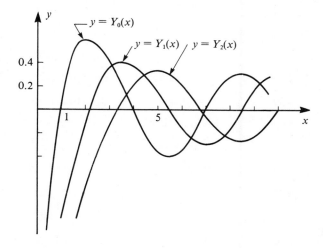

Figure 7.3. Bessel functions of the second kind.

have been compiled because of their use in a variety of applications. Table 7.2 gives values of $J_0(x)$ and $J_1(x)$ for selected values of x, and Table 7.3 gives the first nine positive solutions of the equation $J_n(x) = 0$ for $n = 0, 1, 2, 3,$ and 4. A value of x at which $J_n(x) = 0$ is called a *zero* of J_n. Problem 27 of Section 7.1 sketched a proof of the fact that $J_n(x)$ has infinitely many positive zeros for any nonnegative integer n.

In the next section, we will discuss three applications of Bessel functions. We will see then that, in some models of physical systems, a zero of a Bessel function can reveal important information about the system. We will also exploit zeros of Bessel functions when we solve boundary value problems in partial differential equations using eigenfunction expansions.

It is interesting to observe that our work on solutions of Bessel's equation illustrates all of the possibilities in the conclusions of Theorem 5.5 in Section 5.4

Table 7.2 Selected Values of $J_0(x)$ and $J_1(x)$

x	$J_0(x)$	$J_1(x)$
0.00	1.000000	0.000000
0.20	0.990025	0.099501
0.40	0.960398	0.196027
0.60	0.912005	0.286701
0.80	0.846287	0.368842
1.00	0.765197	0.440051
1.50	0.511828	0.557937
2.00	0.223891	0.576725
2.50	−0.048384	0.497094
3.00	−0.260052	0.339059
3.50	−0.380127	0.137378
4.00	−0.397149	−0.066043
4.50	−0.320543	−0.231060
5.00	−0.177597	−0.327579
5.50	−0.006844	−0.341438
6.00	0.150645	−0.276684
6.50	0.260095	−0.153841
7.00	0.300079	−0.004683
7.50	0.266339	0.135248
8.00	0.171651	0.234636

Table 7.2 (Continued)

x	$J_0(x)$	$J_1(x)$
8.50	0.041939	0.273121
9.00	−0.090334	0.245312
9.50	−0.193929	0.161264
10.0	−0.245936	0.043473
10.5	−0.236648	−0.078850
11.0	−0.171190	−0.176785
11.5	−0.067654	−0.228379
12.0	0.047689	−0.223447

Values in this table were taken from Andrew Gray and G. B. Matthews, *A Treatise on Bessel Functions and Their Applications to Physics*, 2nd ed., prepared by Andrew Gray and T. M. MacRobert (New York: Dover Publications, 1966).

Table 7.3   Positive Zeros of Bessel Functions

j	$J_0(x)$	$J_1(x)$	$J_2(x)$	$J_3(x)$	$J_4(x)$
1	2.405	3.832	5.135	6.379	7.586
2	5.520	7.016	8.417	9.760	11.064
3	8.654	10.173	11.620	13.017	14.373
4	11.792	13.323	14.796	16.224	17.616
5	14.931	16.470	17.960	19.410	20.827
6	18.071	19.616	21.117	22.583	24.018
7	21.212	22.760	24.270	25.749	27.200
8	24.353	25.903	27.421	28.909	31.813
9	27.494	29.047	30.571	32.050	33.512

In this table, row j gives (to three decimal places) the jth positive root of the listed Bessel function. For example, 10.173 is the third positive root (in increasing order from zero) of $J_1(x)$.

These values were taken from Andrew Gray and G. B. Matthews, *A Treatise on Bessel Functions and Their Applications to Physics*, 2nd ed., prepared by Andrew Gray and T. M. MacRobert (New York: Dover Publications, 1966).

(linearly independent solutions about a regular singular point). Specifically, we get case 1 of the theorem if $2v$ is not an integer; case 2 if $2v = 0$; case 3 with no logarithm term in the solution if $v = n + \frac{1}{2}$ for some nonnegative integer n; and case 3 with a logarithm term in the solution if v is a positive integer.

We will conclude this section by noting that we sometimes encounter *modified Bessel functions*, which are obtained as follows. First, notice that

$$y = c_1 J_0(kx) + c_2 Y_0(kx)$$

is the general solution of the differential equation

$$y'' + \frac{1}{x} y' + k^2 y = 0.$$

Let $k = i$, where $i^2 = -1$. Then

$$y = c_1 J_0(ix) + c_2 Y_0(ix)$$

is the general solution of

$$y'' + \frac{1}{x} y' - y = 0.$$

This differential equation is called a *modified Bessel equation of order zero*, and $J_0(ix)$ is a *modified Bessel function of the first kind of order zero*. Usually we denote

$$I_0(x) = J_0(ix).$$

Since $i^2 = -1$, substitution of ix for x in the series for J_0 yields

$$I_0(x) = 1 + \frac{1}{2^2} x^2 + \frac{1}{2^2 4^2} x^4 + \frac{1}{2^2 4^2 6^2} x^6 + \cdots.$$

Usually $Y_0(ix)$ is not used; instead, we use the function

$$K_0(x) = [\ln(2) - \gamma] I_0(x) - I_0(x) \ln(x) + \tfrac{1}{4} x^2 + \cdots.$$

K_0 is called a *modified Bessel function of the second kind of order zero*; the quantity γ appearing in the definition is Euler's constant, which we defined previously. Figure 7.4 shows graphs of I_0 and K_0.

We can now write the general solution of

$$y'' + \frac{1}{x} y' - y = 0$$

as

$$y = c_1 I_0(x) + c_2 K_0(x).$$

It is routine to check by chain rule differentiation that the general solution of

$$y'' + \frac{1}{x} y' - b^2 y = 0$$

is

$$y = c_1 I_0(bx) + c_2 K_0(bx).$$

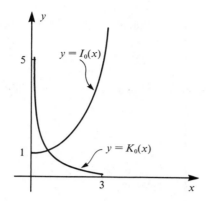

Figure 7.4. Modified Bessel functions.

We will encounter modified Bessel functions when we look at a mathematical analysis of the skin effect in the next section.

PROBLEMS FOR SECTION 7.2

In each of Problems 1 through 8, use the given change of variables to transform the differential equation into a Bessel equation of appropriate order, write the general solution of the Bessel equation, and then write the general solution of the original equation.

1. $4x^2y'' + 4xy' + (x - 9)y = 0$; $z = \sqrt{x}$

2. $4x^2y'' + 4xy' + (9x^3 - 36)y = 0$; $z = x^{3/2}$

3. $9x^2y'' + 9xy' + (4x^{2/3} - 16)y = 0$; $z = 2x^{1/3}$

4. $9x^2y'' - 27xy' + (9x^2 + 35)y = 0$; $u = \dfrac{y}{x^2}$

5. $36x^2y'' - 12xy' + (36x^2 + 7)y = 0$; $u = x^{-2/3}y$

6. $4x^2y'' + 8xy' + (4x^2 - 35)y = 0$; $u = y\sqrt{x}$

7. $4x^2y'' + 20xy' + (9x + 7)y = 0$; let $z = 3\sqrt{x}$; then $u = z^4Y$, where $Y(z) = y\left(\dfrac{z^2}{9}\right)$

8. $9x^{4/3}y'' + 33x^{1/3}y' + y = 0$; let $z = x^{1/3}$; then $u = z^4Y$, where $Y(z) = y(z^3)$

9. Show that $y(x) = x^{1/2}J_{1/3}\left(\dfrac{2kx^{3/2}}{3}\right)$ is a solution of $y'' + k^2xy = 0$.

10. Show that $y_1(x) = x^aJ_n(bx^c)$ and $y_2(x) = x^aY_n(bx^c)$ are solutions of

$$y'' - \left(\frac{2a - 1}{x}\right)y' + \left(b^2c^2x^{2c-2} + \frac{a^2 - n^2c^2}{x^2}\right)y = 0$$

for any nonnegative integer n and constants a, b, and c.

In each of Problems 11 through 20, use the result of Problem 10 to write the general solution of the differential equation.

11. $y'' - \dfrac{1}{x}y' + \left(1 - \dfrac{3}{x^2}\right)y = 0$

12. $y'' - \dfrac{3}{x}y' + \left(4 - \dfrac{5}{x^2}\right)y = 0$

13. $y'' - \dfrac{3}{x}y' + \left(\dfrac{1}{4x} + \dfrac{3}{x^2}\right)y = 0$

14. $y'' - \dfrac{5}{x}y' + \left(1 - \dfrac{7}{x^2}\right)y = 0$

15. $y'' - \dfrac{7}{x}y' + \left(1 + \dfrac{15}{x^2}\right)y = 0$

16. $y'' - \dfrac{1}{x}y' + \left(16x^2 - \dfrac{15}{x^2}\right)y = 0$

17. $y'' + \dfrac{3}{x} y' + \dfrac{1}{16x} y = 0$

18. $y'' - \dfrac{3}{x} y' + \left(1 + \dfrac{4}{x^2}\right) y = 0$

19. $y'' - \dfrac{3}{x} y' + \left(4x^2 - \dfrac{60}{x^2}\right) y = 0$

20. $y'' - \dfrac{7}{x} y' + \left(36x^4 - \dfrac{20}{x^2}\right) y = 0$

21. Write the Maclaurin series for $I_0(x)$ and show that I_0 is a solution of the differential equation
$y'' + (1/x)y' - y = 0$.

22. Show that $[xI_0'(x)]' = xI_0(x)$.

23. Use the result of Problem 22 to show that $\int xI_0(\alpha x)\, dx = (x/\alpha)I_0'(\alpha x)$ for any positive constant α.

24. Show that, for any positive numbers α and β,

$$(\beta^2 - \alpha^2)\int xI_0(\alpha x)I_0(\beta x)\, dx = x[\beta I_0'(\beta x)I_0(\alpha x) - \alpha I_0'(\alpha x)I_0(\beta x)].$$

Hint: Use the reasoning suggested in Problem 24 of Section 7.1.

7.3 *Three Applications of Bessel Functions*

In this section, we will use Bessel functions to model three physical phenomena.

DISPLACEMENT OF A SUSPENDED CHAIN

Suppose we have a uniform heavy flexible chain, as shown in Figure 7.5. The chain is fixed at the upper end and free at the bottom. We want to describe the oscillations caused by a small displacement in a vertical plane from the stable equilibrium position. We will assume that each particle of the chain oscillates in a horizontal straight line. Let m be the mass of the chain per unit length, L be the length of the chain, and $y(x, t)$ be the horizontal displacement at time t of the particle of chain whose distance from the point of suspension is x.

First, we will derive a differential equation for y. Consider an element of chain of length Δx. If the forces acting on the ends of this element are T and $T + \Delta T$, the

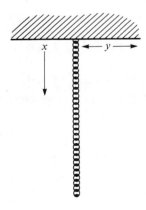

Figure 7.5. A suspended chain.

horizontal component in Newton's second law of motion (force equals the rate of change of momentum with respect to time) is

$$m(\Delta x)\frac{\partial^2 y}{\partial t^2} = \frac{\partial}{\partial x}\left(T\frac{\partial y}{\partial x}\right)\Delta x.$$

Upon dividing both sides of this equation by Δx, we have

$$m\frac{\partial^2 y}{\partial t^2} = \frac{\partial}{\partial x}\left(T\frac{\partial y}{\partial x}\right). \tag{7.4}$$

The weight of chain below x, where T acts, is $T = mg(L - x)$. Upon substituting this expression for T into equation (7.4), we get

$$\frac{\partial^2 y}{\partial t^2} = -g\frac{\partial y}{\partial x} + g(L - x)\frac{\partial^2 y}{\partial x^2}. \tag{7.5}$$

This is a partial differential equation, since it involves partial derivatives. However, we can reduce the problem of solving it to a problem involving an ordinary differential equation as follows. Change variables by letting

$$z = L - x \quad \text{and} \quad u(z, t) = y(L - z, t).$$

Then

$$\frac{\partial^2 y}{\partial t^2} = \frac{\partial^2 u}{\partial t^2}, \qquad \frac{\partial y}{\partial z} = \frac{\partial u}{\partial z}\frac{\partial z}{\partial x} = -\frac{\partial u}{\partial z}; \quad \text{and} \quad \frac{\partial^2 y}{\partial x^2} = \frac{\partial^2 u}{\partial z^2}.$$

With these substitutions, equation (7.5) becomes

$$\frac{\partial^2 u}{\partial t^2} = g\frac{\partial u}{\partial z} + gz\frac{\partial^2 u}{\partial z^2}. \tag{7.6}$$

We now anticipate a method to be developed when we study partial differential equations in Chapter 18. Since we expect the oscillations to be periodic in t, we will attempt a solution of equation (7.6) of the form

$$u(z, t) = f(z)\cos(\omega t - \delta).$$

Substitute this expression into equation (7.6) to get

$$-\omega^2 f(z)\cos(\omega t - \delta) = gf'(z)\cos(\omega t - \delta) + gzf''(z)\cos(\omega t - \delta).$$

Upon dividing by $\cos(\omega t - \delta)$, this equation becomes

$$-\omega^2 f(z) = gf'(z) + gzf''(z),$$

which we can write as

$$f''(z) + \frac{1}{z}f'(z) + \frac{\omega^2}{gz}f(z) = 0. \tag{7.7}$$

We can solve this equation using the method of Problem 1, Section 7.1. In the differential equation of that problem, let $2a - 1 = -1$, $b^2c^2 = \omega^2/g$, $2c - 2 = -1$, and $a^2 - v^2c^2 = 0$. We therefore choose $a = v = 0$, $c = \frac{1}{2}$, and $b = 2\omega/\sqrt{g}$. The general solution of (7.7) is in terms of Bessel functions of order zero:

$$f(z) = c_1 J_0\left(2\omega\sqrt{\frac{z}{g}}\right) + c_2 Y_0\left(2\omega\sqrt{\frac{z}{g}}\right).$$

Now, we know that $Y_0(2\omega\sqrt{z/g}) \to -\infty$ as $z \to 0+$ (that is, as $x \to L$). We must therefore choose $c_2 = 0$ in order to have a bounded solution, as we expect from the physical setting of the problem. This leaves us with

$$f(z) = c_1 J_0\left(2\omega\sqrt{\frac{z}{g}}\right).$$

Then

$$u(z, t) = c_1 J_0\left(2\omega\sqrt{\frac{z}{g}}\right)\cos(\omega t - \delta);$$

hence,

$$y(x, t) = c_1 J_0\left(2\omega\sqrt{\frac{L-x}{g}}\right)\cos(\omega t - \delta).$$

The frequencies of the normal oscillations of the chain are determined by using this general form of the solution for $y(x, t)$ together with the condition that the upper end of the chain is fixed and therefore does not move. For all t, we must have

$$y(0, t) = 0.$$

Assuming that $c_1 \neq 0$, this requires that we choose ω so that

$$J_0\left(2\omega\sqrt{\frac{L}{g}}\right) = 0. \tag{7.8}$$

Equation (7.8) determines those values of ω which can be frequencies of the oscillations. To find these admissible values of ω, we must consult a table of zeros of J_0. From Table 7.3 at the end of Section 7.2, the first five zeros of J_0 are approximately 2.405, 5.520, 8.654, 11.792, and 14.931. Using the first zero, 2.405, choose $\omega = \omega_1$, where ω_1 satisfies

$$2\omega_1\sqrt{\frac{L}{g}} = 2.405.$$

Therefore,

$$\omega_1 = 1.203\sqrt{\frac{g}{L}}.$$

Similarly, using the other zeros in turn in equation (7.8), we get

$$\omega_2 = 2.76\sqrt{\frac{g}{L}}, \quad \omega_3 = 4.327\sqrt{\frac{g}{L}}, \quad \omega_4 = 5.896\sqrt{\frac{g}{L}}, \quad \text{and} \quad \omega_5 = 7.466\sqrt{\frac{g}{L}}.$$

These are all admissible values of ω [that is, consistent with equation (7.8)], and they represent approximate frequencies of the normal modes of vibration. The approximate period associated with ω_j is $2\pi/\omega_j$.

Two features of this example are typical of many applications of Bessel functions. First, a change of variables was used to write solutions in terms of Bessel functions. Further, much of the information about the motion of the system was obtained from the zeros of a Bessel function.

THE CRITICAL LENGTH OF A VERTICAL ROD

Suppose we have a thin elastic rod of uniform density. The rod is clamped in a vertical position. Intuitively, if the rod is "too long" and the upper end is displaced slightly and held fast until the rod is at rest, the rod will remain in the displaced position after being released. On the other hand, if the rod is "short enough," it will spring back to the vertical position after being released. We would like to know where the transition occurs between being too long or short enough. That is, we want the minimum length at which the rod remains displaced after being released. This length is called the *critical length* of the rod and of course will depend on the material of the rod.

To derive a mathematical model from which we can solve for this critical length, let L be the length of the rod and a be the radius of its circular cross-section. Let w be the weight per unit length and E be the Young's modulus for the rod. E depends on the material of the rod. We should expect that this will influence the critical length. Finally, $I = \frac{1}{4}\pi a^4$ for the circular cross-section.

Now assume that the rod is in equilibrium and is displaced slightly from the vertical, as in Figure 7.6. The origin is as shown, and the x-axis is vertical, with downward as positive. Let $P(x, y)$ be a point on the rod, and let $Q(X, Y)$ be a point slightly above P, as shown. The moment about P of the weight of an element $w\Delta X$ at Q is $w\Delta X(Y - y)$. Assume from the theory of elasticity the fact that the moment of the elastic forces about P is $EI(d^2y/dx^2)$. Since the part of the rod above P is in equilibrium,

$$EI \frac{d^2y}{dx^2} = \int_0^x w(Y - y) \, dX.$$

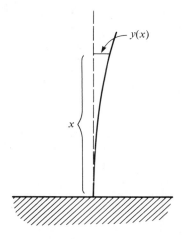

Figure 7.6. Displaced vertical rod.

Differentiate this equation with respect to x to get

$$EI \frac{d^3 y}{dx^3} = w(Y - y)\Big|_{X=x} - \int_0^x w \frac{dy}{dx} \, dX = -wx \frac{dy}{dx}.$$

This yields the third order differential equation

$$EI \frac{d^3 y}{dx^3} + wx \frac{dy}{dx} = 0,$$

or

$$\frac{d^3 y}{dx^3} + \frac{w}{EI} x \frac{dy}{dx} = 0.$$

Let $u = y'$ to obtain the second order equation

$$\frac{d^2 u}{dx^2} + \frac{w}{EI} xu = 0. \tag{7.9}$$

Compare this with the differential equation in Problem 1 of Section 7.1. We want to have $2a - 1 = 0$, $a^2 - v^2 c^2 = 0$, $2c - 2 = 1$, and $b^2 c^2 = w/EI$. Thus, choose $a = \frac{1}{2}$, $c = \frac{3}{2}$, $v = \frac{1}{3}$, and $b = \frac{2}{3}\sqrt{w/EI}$. The general solution of equation (7.9) is

$$u = \frac{dy}{dx} = c_1 x^{1/2} J_{1/3}\left(\frac{2}{3}\sqrt{\frac{w}{EI}} \, x^{3/2}\right) + c_2 x^{1/2} J_{-1/3}\left(\frac{2}{3}\sqrt{\frac{w}{EI}} \, x^{3/2}\right).$$

There is no bending moment at the upper end of the rod, so we must have

$$\frac{d^2 y}{dx^2}\Big|_{x=0} = 0.$$

This condition requires that we choose $c_1 = 0$ (details are requested as Problem 1 at the end of this section). Therefore,

$$\frac{dy}{dx} = c_2 x^{1/2} J_{-1/3}\left(\frac{2}{3}\sqrt{\frac{w}{EI}} \, x^{3/2}\right).$$

Further, the lower end of the rod is clamped and so does not move. Then

$$\frac{dy}{dx}\Big|_{x=L} = 0.$$

In order to satisfy this condition with $c_2 \neq 0$, we must have

$$J_{-1/3}\left(\frac{2}{3}\sqrt{\frac{w}{EI}} \, L^{3/2}\right) = 0. \tag{7.10}$$

The critical length is the smallest positive number L which satisfies equation (7.10). We find from a table that the smallest positive zero of the Bessel function of first kind of order $-\frac{1}{3}$ is approximately 1.8663. Thus, the critical length L is determined approximately by the equation

$$\frac{2}{3}\sqrt{\frac{w}{EI}} \, L^{3/2} = 1.8663.$$

The critical length is approximately

$$\left(1.8663\tfrac{3}{2}\sqrt{\frac{EI}{w}}\right)^{2/3},$$

or about $1.9863(EI/w)^{1/3}$.

ALTERNATING CURRENT IN A CIRCULAR WIRE: THE SKIN EFFECT

Consider an alternating current of period $2\pi/\omega$, given by $D\cos(\omega t)$. Let R be the radius of the wire, ρ be its specific resistance, μ be the permeability, $x(r, t)$ be the current density at radius r and time t, and $H(r, t)$ be the magnetic intensity at radius r from the center at time t. To derive an equation for x, we will need two laws of electromagnetic field theory. Ampère's law states that the line integral of the electric force around a closed path equals 4π times the integral of the electric current through the path. Faraday's law states that the line integral of the electric force around a closed path equals the negative of the partial derivative with respect to time of the magnetic induction through the path.

Now consider a circular path of radius r within the wire, centered about the midpoint of the wire. By Ampère's law,

$$2\pi rH = 4\pi \int_0^r 2\pi rx \, dr.$$

Differentiate both sides of this equation with respect to r to get

$$\frac{\partial}{\partial r}[rH] = 4\pi rx. \tag{7.11}$$

Consider a closed rectangular path in the wire, with two sides along the axis of the cylinder and of length L, and the other sides of length r. By Faraday's law,

$$\rho L[x(0, t) - x(r, t)] = -\frac{\partial}{\partial t}\int_0^r \mu LH \, dr.$$

Differentiate this equation with respect to r to get

$$\rho L \frac{\partial x}{\partial r} = \mu L \frac{\partial H}{\partial t}. \tag{7.12}$$

We want to eliminate H from these equations to obtain an equation for x alone. Multiply equation (7.12) by r and differentiate the resulting equation with respect to r to get

$$\rho \frac{\partial}{\partial r}\left(r\frac{\partial x}{\partial r}\right) = \mu \frac{\partial}{\partial r}\left(r\frac{\partial H}{\partial t}\right). \tag{7.13}$$

In this equation, we can write $r(\partial H/\partial t) = \partial/\partial t[rH]$ because $\partial r/\partial t = 0$. Equation (7.13) becomes

$$\rho \frac{\partial}{\partial r}\left(r\frac{\partial x}{\partial r}\right) = \mu \frac{\partial}{\partial r}\left(\frac{\partial}{\partial t}[rH]\right).$$

Assuming that we can reverse the order of the differentiations on the right side of this equation, we obtain

$$\rho \frac{\partial}{\partial r}\left(r \frac{\partial x}{\partial r}\right) = \mu \frac{\partial}{\partial t}\left(\frac{\partial}{\partial r}[rH]\right). \tag{7.14}$$

Use equation (7.11) to write equation (7.14) as

$$\rho \frac{\partial}{\partial r}\left(r \frac{\partial x}{\partial r}\right) = \mu \frac{\partial}{\partial t}[4\pi rx].$$

We now have a differential equation involving x but not H. This differential equation is usually written in the form

$$\frac{1}{r}\frac{\partial}{\partial r}\left(r \frac{\partial x}{\partial r}\right) = \frac{4\pi\mu}{\rho}\frac{\partial x}{\partial t}. \tag{7.15}$$

To solve this equation, we will employ a device which is quite standard in mathematical treatments of electricity and magnetism. Let $z(r, t) = x(r, t) + iy(r, t)$, where $i^2 = -1$, and consider x as the real part of z. (Sometimes electrical engineers use j instead of i, leaving i to denote current.) Replace x with z in equation (7.15) to get

$$\frac{1}{r}\frac{\partial}{\partial r}\left(r \frac{\partial z}{\partial r}\right) = \frac{4\pi\mu}{\rho}\frac{\partial z}{\partial t}. \tag{7.16}$$

Now recall Euler's formula. We can write

$$\cos(\omega t) + i\sin(\omega t) = e^{i\omega t}.$$

Anticipating periodic dependence of the current density on time, we will attempt solutions of (7.16) of the form

$$z(r, t) = f(r)e^{i\omega t}.$$

Substitute this expression for z into equation (7.16) to get

$$\frac{1}{r}\frac{\partial}{\partial r}[rf'(r)e^{i\omega t}] = \frac{4\pi\mu}{\rho}i\omega f(r)e^{i\omega t}.$$

Divide this equation by $e^{i\omega t}$ to get

$$\frac{1}{r}\frac{d}{dr}[rf'(r)] = \frac{4\pi\mu i\omega}{\rho}f(r),$$

or

$$f''(r) + \frac{1}{r}f'(r) - \frac{4\pi\mu i\omega}{\rho}f(r) = 0. \tag{7.17}$$

Now let

$$k = \frac{1+i}{\sqrt{2}}\sqrt{\frac{4\pi\mu\omega}{\rho}}.$$

It is routine to check that $k^2 = 4\pi\mu i\omega/\rho$ and that equation (7.17) becomes

$$f''(r) + \frac{1}{r}f'(r) - k^2 f(r) = 0.$$

This is a modified Bessel equation (see Section 7.2) with general solution

$$f(r) = c_1 I_0(kr) + c_2 K_0(kr).$$

In order for $f(r)$ to remain finite as $r \to 0+$, we must have $c_2 = 0$. Thus, $f(r)$ is of the form $c_1 I_0(kr)$, with I_0 the modified Bessel function of first kind of order zero. Then $z(r, t)$ has the form

$$z(r, t) = c_1 I_0(kr)e^{i\omega t}.$$

The current density $x(r, t)$ is the real part of this expression.

We have not yet used the assumption that the alternating current in the wire is given by $D\cos(\omega t)$, except for using ω in the exponential term $e^{i\omega t}$. We will now use this assumption. Think of $D\cos(\omega t)$ as the real part of $D\cos(\omega t) + iD\sin(\omega t)$, or $De^{i\omega t}$. Since $De^{i\omega t}$ represents the total current,

$$De^{i\omega t} = \int_0^R 2\pi r z\, dr = 2\pi c_1 \int_0^R r I_0(kr)e^{i\omega t}\, dr.$$

Upon dividing $e^{i\omega t}$ from this equation, we get

$$D = 2\pi c_1 \int_0^R r I_0(kr)\, dr.$$

Recall that D is known. Thus, we obtain c_1:

$$c_1 = \frac{D}{2\pi \displaystyle\int_0^R r I_0(kr)\, dr}. \tag{7.18}$$

We will look more closely at the integral in the denominator of equation (7.18). First, write the series for $I_0(kr)$:

$$I_0(kr) = 1 + \frac{k^2 r^2}{2^2} + \frac{k^4 r^4}{2^2 4^2} + \frac{k^6 r^6}{2^2 4^2 6^2} + \cdots. \tag{7.19}$$

Multiply this series by r and integrate the resulting series term by term:

$$\int_0^R r I_0(kr)\, dr = \int_0^R r\, dr + \int_0^R \frac{k^2 r^3}{2^2}\, dr + \int_0^R \frac{k^4 r^5}{2^2 4^2}\, dr + \int_0^R \frac{k^6 r^7}{2^2 4^2 6^2}\, dr + \cdots$$

$$= \frac{1}{2}R^2 + \frac{k^2}{2^2}\frac{1}{4}R^4 + \frac{k^4}{2^2 4^4}\frac{1}{6}R^6 + \frac{k^6}{2^2 4^2 6^2}\frac{1}{8}R^8 + \cdots. \tag{7.20}$$

Now differentiate the series (7.19) term by term with respect to r to get

$$I_0'(kr) = \frac{1}{2}k^2 r + \frac{k^4}{2^2 4}r^3 + \frac{k^6}{2^2 4^2 6}r^5 + \cdots.$$

Therefore,

$$\frac{R}{k^2} I_0'(kR) = \frac{1}{2} R^2 + \frac{k^2}{2^2 4} R^4 + \frac{k^4}{2^2 4^2 6} R^6 + \cdots. \tag{7.21}$$

Upon comparing the results of equations (7.20) and (7.21), we see that

$$\int_0^R r I_0(kr)\, dr = \frac{R}{k^2} I_0'(kR). \tag{7.22}$$

Substitute this result into equation (7.18) to get

$$c_1 = \frac{Dk^2}{2\pi R I_0'(kR)}.$$

Thus,

$$z(r, t) = \frac{Dk^2}{2\pi R I_0'(kR)} I_0(kr) e^{i\omega t}.$$

The current density $x(r, t)$ is the real part of this expression.

We will now apply this analysis to a mathematical derivation of the skin effect. It has been observed that, for sufficiently high frequencies, the current flowing through a circular wire at radius r is small compared with the total current, even for r nearly equal to R. This means that "most" of the current in a cylindrical wire flows through a thin layer at the outer surface, at the "skin" of the wire.

To derive this effect from the model, begin with the solution for z. The total current through a coaxial cylinder of radius r is

$$\int_0^r 2\pi z(r, t)\, dr = \int_0^r \frac{Dk^2}{2\pi R I_0'(kR)} (2\pi r) I_0(kr) e^{i\omega t}\, dr$$

$$= \frac{Dk^2}{R I_0'(kR)} e^{i\omega t} \int_0^r r I_0(kr)\, dr.$$

By equation (7.22), with r in place of R, we have

$$\int_0^r r I_0(kr)\, dr = \frac{r}{k^2} I_0'(kr).$$

Thus, the total current through a coaxial cylinder of radius r is

$$D e^{i\omega t} \frac{r I_0'(kr)}{R I_0'(kR)}.$$

But the total current in the wire is $D e^{i\omega t}$. Thus, the ratio of the current in the coaxial cylinder of radius r to the total current is

$$\frac{r I_0'(kr)}{R I_0'(kR)}.$$

We want to know how this quotient behaves for large k. In Problem 4, the student is asked to show that, for large x, $I_0(x)$ is approximated by

$$\frac{A e^x}{\sqrt{x}} \left(1 + \frac{1}{8x} + \frac{3^2}{2!(8x)^2} + \frac{3^2 5^2}{3!(8x)^3} + \cdots \right),$$

in which A is a positive constant whose value does not matter for present purposes. For large x, then,

$$I_0(x) \approx \frac{Ae^x}{\sqrt{x}}.$$

Then, for large x,

$$I_0'(x) \approx \frac{Ae^x}{\sqrt{x}}\left(1 - \frac{1}{2x}\right) \approx \frac{Ae^x}{\sqrt{x}}.$$

Thus, if ω is large, k is large in magnitude, and

$$\frac{rI_0'(kr)}{RI_0'(kR)} \approx \frac{r}{R}\frac{e^{kr}}{\sqrt{r}}\frac{\sqrt{R}}{e^{kR}} = \sqrt{\frac{r}{R}}\,e^{-k(R-r)}.$$

Given any $r < R$, $e^{-k(R-r)}$ can be made as small in magnitude as we like by choosing ω larger. Thus, for ω large, the ratio of the current in the coaxial cylinder of radius r to the total current is near zero, and this is the skin effect. In fact, we can choose r as close as we like to R, and this conclusion continues to hold, for sufficiently large frequencies.

PROBLEMS FOR SECTION 7.3

1. Show that we must choose $c_1 = 0$ in the solution of equation (7.9) if the solution is to satisfy the condition

$$\left.\frac{d^2y}{dx^2}\right|_{x=0} = 0.$$

2. Verify that $k^2 = 4\pi\mu i\omega/\rho$ if

$$k = \frac{1+i}{\sqrt{2}}\sqrt{\frac{4\pi\mu\omega}{\rho}}.$$

3. In treating alternating current in the coaxial cable, the current density $x(r, t)$ was shown to be the real part of $[Dk^2/2\pi RI_0'(kR)]I_0(kr)e^{i\omega t}$. Find an expression for $x(r, t)$.

4. Show that, for large x and some positive constant A, $I_0(x)$ is approximated by the series

$$\frac{1}{\sqrt{x}}Ae^x\left(1 + \frac{1}{8x} + \frac{3^2}{2!(8x)^2} + \frac{3^2 5^2}{3!(8x)^3} + \cdots\right).$$

Hint: Let $u(x) = \sqrt{x}I_0(x)$, and show that $u'' = (1 - 1/4x^2)u$. Note that, for large x, $u(x) \approx Ae^x$ for some constant A. Next, let $u(x) = v(x)e^x$, and show that

$$v'' + 2v' + \frac{1}{4x^2}v = 0.$$

Attempt a solution of this differential equation of the form

$$v(x) = 1 + a_1\frac{1}{x} + a_2\frac{1}{x^2} + a_3\frac{1}{x^3} + \cdots.$$

Substitute this proposed solution into the differential equation for v, and solve for the coefficients.

7.4 *Legendre's Equation and Legendre Polynomials*

The differential equation

$$(1 - x^2)y'' - 2xy' + \alpha(\alpha + 1)y = 0,$$

in which α is constant, is called *Legendre's equation*. It occurs in a variety of problems involving quantum mechanics, astronomy (see Problem 10 at the end of this section), and analysis of heat conduction, and is often seen in settings in which it is natural to use spherical coordinates (see, for example, Section 18.6).

Write Legendre's equation as

$$y'' - \frac{2x}{1 - x^2} y' + \frac{1}{1 - x^2} \alpha(\alpha + 1)y = 0.$$

The coefficient functions $-2x/(1 - x^2)$ and $[1/(1 - x^2)] \alpha(\alpha + 1)$ are analytic at every point except 1 and -1. In particular, both functions have Maclaurin series expansions in $(-1, 1)$. Since zero is an ordinary point of Legendre's equation, there are two linearly independent solutions which are analytic in $(-1, 1)$ and which can be found by the power series method. Let

$$y = \sum_{n=0}^{\infty} a_n x^n.$$

Substitute this series into Legendre's equation, written in the form

$$(1 - x^2)y'' - 2xy' + \alpha(\alpha + 1)y = 0,$$

to obtain

$$\sum_{n=2}^{\infty} n(n - 1)a_n x^{n-2} - \sum_{n=2}^{\infty} n(n - 1)a_n x^n - \sum_{n=1}^{\infty} 2na_n x^n + \sum_{n=0}^{\infty} \alpha(\alpha + 1)a_n x^n = 0.$$

Shift indices in the first summation to write this equation as

$$\sum_{n=0}^{\infty} (n + 2)(n + 1)a_{n+2} x^n - \sum_{n=2}^{\infty} n(n - 1)a_n x^n - \sum_{n=1}^{\infty} 2na_n x^n + \sum_{n=0}^{\infty} \alpha(\alpha + 1)a_n x^n = 0.$$

Combine summations from $n = 2$ on, writing terms for $n = 0$ and $n = 1$ separately, to obtain

$$2a_2 + 6a_3 x - 2a_1 x + \alpha(\alpha + 1)a_0 + \alpha(\alpha + 1)a_1 x$$
$$+ \sum_{n=2}^{\infty} [(n + 2)(n + 1)a_{n+2} - [n^2 + n - \alpha(\alpha + 1)]a_n]x^n = 0.$$

The coefficient of each power of x on the left side of this equation must be zero:

$$2a_2 + \alpha(\alpha + 1)a_0 = 0,$$
$$6a_3 - 2a_1 + \alpha(\alpha + 1)a_1 = 0,$$

and, for $n = 2, 3, 4, \ldots,$

$$(n + 2)(n + 1)a_{n+2} - [n^2 + n - \alpha(\alpha + 1)]a_n = 0.$$

The last equation is the recurrence relation. Write

$$n^2 + n - \alpha(\alpha + 1) = -(n + \alpha + 1)(\alpha - n),$$

so that the recurrence relation is

$$a_{n+2} = -\frac{(n + \alpha + 1)(\alpha - n)}{(n + 2)(n + 1)} a_n \quad \text{for} \quad n = 2, 3, 4, \ldots.$$

The recurrence relation expresses a_{n+2} as a multiple of a_n. Thus, a_2 is a multiple of a_0; a_4 is a multiple of a_2 (hence of a_0); and so on, with every even-indexed coefficient a multiple of a_0. Similarly, every odd-indexed coefficient is a multiple of a_1. For the even-indexed coefficients, we have

$$a_2 = -\frac{(\alpha + 1)(\alpha)}{1 \cdot 2} a_0,$$

$$a_4 = -\frac{(\alpha + 4)(\alpha - 2)}{3 \cdot 4} a_2 = \frac{(\alpha + 3)(\alpha + 1)(\alpha)(\alpha - 2)}{4!} a_0,$$

and, in general,

$$a_{2n} = (-1)^n \frac{(\alpha + 2n - 1)(\alpha + 2n - 3) \cdots (\alpha + 1)(\alpha)(\alpha - 2) \cdots (\alpha - 2n + 4)(\alpha - 2n + 2)}{(2n)!} a_0.$$

By examining some of the odd-indexed coefficients, we find the pattern

$$a_{2n+1} = (-1)^n \frac{(\alpha + 2n) \cdots (\alpha + 4)(\alpha + 2)(\alpha - 1)(\alpha - 3) \cdots (\alpha - 2n + 1)}{(2n + 1)!} a_1.$$

We can obtain two linearly independent solutions of Legendre's equation by making choices for a_0 and a_1. If we choose $a_0 = 1$ and $a_1 = 0$, we get one solution,

$$y_1(x) = \sum_{n=0}^{\infty} (-1)^n \frac{(\alpha + 2n - 1)(\alpha + 2n - 3) \cdots (\alpha + 1)(\alpha)(\alpha - 2) \cdots (\alpha - 2n + 4)(\alpha - 2n + 2)}{(2n)!} x^{2n}$$

$$= 1 - \frac{(\alpha + 1)\alpha}{2} x^2 + \frac{(\alpha + 3)(\alpha + 2)(\alpha + 1)\alpha(\alpha - 2)}{24} x^4$$

$$- \frac{(\alpha + 5)(\alpha + 3)(\alpha + 1)\alpha(\alpha - 2)(\alpha - 4)}{720} x^6 + \cdots.$$

By letting $a_0 = 0$ and $a_1 = 1$, we obtain a second solution,

$$y_2(x) = \sum_{n=0}^{\infty} (-1)^n \frac{(\alpha + 2n) \cdots (\alpha + 4)(\alpha + 2)(\alpha - 1)(\alpha - 3) \cdots (\alpha - 2n + 1)}{(2n + 1)!} x^{2n+1}$$

$$= x - \frac{(\alpha + 2)(\alpha - 1)}{6} x^3 + \frac{(\alpha + 4)(\alpha + 2)(\alpha - 1)(\alpha - 3)}{120} x^5 + \cdots.$$

These power series converge for all x in $(-1, 1)$.

By appropriately choosing α, one or the other of these series solutions is a polynomial. For example, if $\alpha = 2$, then $a_4 = 0$; hence, $a_6 = a_8 = a_{10} = \cdots = 0$, and y_1 is just a second degree polynomial

$$y_1(x) = 1 - 3x^2.$$

If $\alpha = 3$, then $a_5 = 0$, so $a_7 = a_9 = \cdots = 0$, and now y_2 is a polynomial

$$y_2(x) = x - \tfrac{5}{3}x^3.$$

In fact, whenever α is a nonnegative integer, the power series for either y_1 (if α is even) or y_2 (if α is odd) reduces to a finite series, and we obtain a polynomial solution of Legendre's equation.

Such polynomial solutions are useful in many applications, including methods for approximating solutions of equations $f(x) = 0$. In such applications, it is helpful to standardize specific polynomial solutions so their values can be tabulated. The convention is to multiply y_1 or y_2 for each n by a constant which makes the value of the polynomial 1 at $x = 1$. The resulting polynomials are called *Legendre polynomials* and are denoted $P_n(x)$. The first few Legendre polynomials are

$$
\begin{aligned}
P_0(x) &= 1, \\
P_1(x) &= x, \\
P_2(x) &= \tfrac{1}{2}(3x^2 - 1), \\
P_3(x) &= \tfrac{1}{2}(5x^3 - 3x), \\
P_4(x) &= \tfrac{1}{8}(35x^4 - 30x^2 + 3), \\
P_5(x) &= \tfrac{1}{8}(63x^5 - 70x^3 + 15x).
\end{aligned}
\tag{7.23}
$$

and

Figure 7.7 shows graphs of these functions for $-1 \le x \le 1$.

Although these polynomials are defined for all x, they are solutions of Legendre's equation only for $-1 < x < 1$. It can be shown that α must be chosen as a nonnegative integer in order to obtain nontrivial solutions of Legendre's equation which are bounded on $[-1, 1]$. This is particularly important in models of phenomena in physics and engineering, where boundedness of the solution is a natural expectation.

For some applications, the most important property of Legendre polynomials is given by the following theorem.

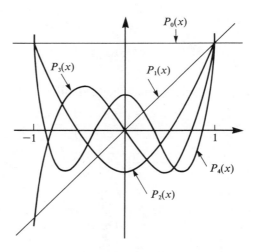

Figure 7.7 Legendre polynomials $P_0(x)$ through $P_4(x)$, $-1 \le x \le 1$.

THEOREM 7.3

If m and n are distinct nonnegative integers, then

$$\int_{-1}^{1} P_n(x)P_m(x)\, dx = 0.$$

Proof   We know that P_n and P_m are solutions of Legendre's differential equation. Therefore,

$$(1 - x^2)P_n'' - 2xP_n' + n(n + 1)P_n = 0$$

and

$$(1 - x^2)P_m'' - 2xP_m' + m(m + 1)P_m = 0.$$

Multiply the first equation by P_m and the second by P_n to get

$$(1 - x^2)P_n''P_m - 2xP_n'P_m + n(n + 1)P_nP_m = 0$$

and

$$(1 - x^2)P_m''P_n - 2xP_m'P_n + m(m + 1)P_mP_n = 0.$$

Subtract these two equations to get

$$(1 - x^2)P_n''P_m - (1 - x^2)P_m''P_n - 2xP_n'P_m + 2xP_m'P_n$$
$$+ [n(n + 1) - m(m + 1)]P_nP_m = 0.$$

This equation can be written

$$[(1 - x^2)P_n']'P_m - [(1 - x^2)P_m']'P_n + [n(n + 1) - m(m + 1)]P_nP_m = 0.$$

Integrate this equation from -1 to 1:

$$\int_{-1}^{1} [(1 - x^2)P_n'(x)]'P_m(x)\, dx - \int_{-1}^{1} [(1 - x^2)P_m'(x)]'P_n(x)\, dx$$
$$+ [n(n + 1) - m(m + 1)] \int_{-1}^{1} P_n(x)P_m(x)\, dx = 0. \tag{7.24}$$

Integrate by parts, with $u = P_m(x)$ and $dv = [(1 - x^2)P_n'(x)]'\, dx$, to get

$$\int_{-1}^{1} [(1 - x^2)P_n'(x)]'P_m(x)\, dx = P_m(x)[(1 - x^2)P_n'(x)]\Big|_{-1}^{1}$$
$$- \int_{-1}^{1} (1 - x^2)P_n'(x)P_m'(x)\, dx$$
$$= - \int_{-1}^{1} (1 - x^2)P_n'(x)P_m'(x)\, dx.$$

Similarly,

$$\int_{-1}^{1} [(1 - x^2)P_m'(x)]'P_n(x)\, dx = - \int_{-1}^{1} (1 - x^2)P_m'(x)P_n'(x)\, dx.$$

Thus, equation (7.24) gives us the equation

$$-\int_{-1}^{1} (1 - x^2)P_n'(x)P_m'(x) \, dx + \int_{-1}^{1} (1 - x^2)P_m'(x)P_n'(x) \, dx$$

$$+ [n(n + 1) - m(m + 1)]\int_{-1}^{1} P_n(x)P_m(x) \, dx = 0.$$

The first two terms in this equation cancel each other, and we are left with

$$[n(n + 1) - m(m + 1)]\int_{-1}^{1} P_n(x)P_m(x) \, dx = 0.$$

Since $n \neq m$, $\int_{-1}^{1} P_n(x)P_m(x) \, dx = 0$, as we wanted to show. ∎

Some additional properties of Legendre polynomials are developed in the following problems and in the next section.

PROBLEMS FOR SECTION 7.4

1. Verify for $n = 0, 1, 2, 3, 4,$ and 5 that $P_n(x)$, as given by formulas (7.23), is a solution of Legendre's differential equation with $\alpha = n$.
2. It can be shown that

$$P_n(x) = \sum_{k=0}^{[n/2]} \frac{(-1)^k(2n - 2k)!}{2^n k!(n - k)!(n - 2k)!} x^{n - 2k},$$

in which $[a]$ = greatest integer $\leq a$. For example, $[4] = 4$ and $[\frac{3}{2}] = 1$. Use this formula to compute $P_0(x)$ through $P_5(x)$, and compare the results with the expressions given in the text for these Legendre polynomials [equations (7.23)].
3. Show that Legendre's differential equation can be written

$$[(1 - x^2)y']' + \alpha(\alpha + 1)y = 0.$$

4. *Rodrigues's formula* states that

$$P_n(x) = \frac{1}{2^n n!} \frac{d^n}{dx^n}[(x^2 - 1)^n]$$

for $n = 1, 2, 3, \ldots$. Verify this for $P_1(x)$ through $P_5(x)$. Show that the formula also holds true for $n = 0$, with the convention that the d^0/dx^0 means simply to multiply the expression that follows by 1.
5. Prove Rodrigues's formula (Problem 4) for n any positive integer, assuming the formula for $P_n(x)$ from Problem 2. *Hint:* Using the result of Problem 2, reason that

$$P_n(x) = \frac{1}{2^n n!} \sum_{k=0}^{[n/2]} \frac{(-1)^k}{k!} \frac{n!}{(n - k)!} \frac{(2n - 2k)!}{(n - 2k)!} x^{n - 2k}$$

$$= \frac{1}{2^n n!} \sum_{k=0}^{[n/2]} \frac{(-1)^k}{k!} \frac{n!}{(n - k)!} \frac{d^n}{dx^n}[x^{2n - 2k}].$$

Explain why $[n/2]$ can be replaced with n in this summation, and use the fact that

$$(x^2 - 1)^n = \sum_{k=0}^{n} \frac{(-1)^k}{k!} \frac{n!}{(n - k)!}(x^2)^{n-k}.$$

6. Put $(x^2 - 1)^n = (x - 1)^n(x + 1)^n$ into Rodrigues's formula (Problem 4) to show that

$$P_n(x) = \frac{1}{2^n n!} \sum_{k=0}^{\infty} \frac{n!}{k!(n-k)!} \frac{d^k}{dx^k} [(x+1)^n] \frac{d^{n-k}}{dx^{n-k}} [(x-1)^n].$$

Hint: Use Leibniz's rule for differentiating a product n times:

$$\frac{d^n}{dx^n} [f(x)g(x)] = \sum_{k=0}^{n} \frac{n!}{k!(n-k)!} \frac{d^k}{dx^k} [f(x)] \frac{d^{n-k}}{dx^{n-k}} [g(x)].$$

7. Legendre polynomials satisfy the recurrence relation

$$(n+1)P_{n+1}(x) + nP_{n-1}(x) = (2n+1)xP_n(x)$$

for $n = 1, 2, 3, \ldots$. Verify this for $n = 1, 2, 3,$ and 4, using the formulas (7.23). (A proof of this recurrence relation is given in the next section.)

8. Let n be a nonnegative integer. Use reduction of order and the fact that P_n is one solution of Legendre's equation with $\alpha = n$ to obtain a second, linearly independent solution

$$Q_n(x) = P_n(x) \int \frac{1}{P_n(x)^2(1-x^2)} \, dx.$$

9. Use the formula of Problem 8 to derive the results

$$Q_0(x) = -\frac{1}{2} \ln\left(\frac{1+x}{1-x}\right), \qquad Q_1(x) = 1 - \frac{x}{2} \ln\left(\frac{1+x}{1-x}\right),$$

and

$$Q_2(x) = \frac{1}{4}(3x^2 - 1)\ln\left(\frac{1+x}{1-x}\right) - \frac{3}{2}x.$$

10. The gravitational potential at a point $P: (x, y, z)$ due to a unit mass at (x_0, y_0, z_0) is

$$\varphi(x, y, z) = \frac{1}{\sqrt{(x - x_0)^2 + (y - y_0)^2 + (z - z_0)^2}}.$$

For some problems (for example, in astronomy), it is convenient to expand φ in powers of r or $1/r$, where $r = \sqrt{x^2 + y^2 + z^2}$ is the distance from P to the origin of the coordinate system. To do this, introduce an angle θ as shown in the diagram below. Let

$$d = \sqrt{x_0^2 + y_0^2 + z_0^2}$$

and

$$R = \sqrt{(x - x_0)^2 + (y - y_0)^2 + (z - z_0)^2}.$$

(a) Use the law of cosines to write

$$\varphi(x, y, z) = \frac{1}{d\sqrt{1 - 2\dfrac{r}{d}\cos(\theta) + \left(\dfrac{r}{d}\right)^2}}.$$

(b) Use a binomial expansion (see Problem 57 of Section 5.2) to expand

$$\frac{1}{\sqrt{1 - 2at + t^2}}$$

in a series in t about zero, valid for $|t| < 1$. Observe that the coefficient of t^n is $P_n(a)$.

(c) If $r < d$, let $a = \cos(\theta)$ and $t = r/d$ in the series expansion in (b) to get

$$\varphi(t) = \sum_{n=0}^{\infty} \frac{1}{d^{n+1}} P_n[\cos(\theta)] r^n$$

for $r < d$.

(d) If $r > d$, show that

$$\varphi(r) = \frac{1}{r} \sum_{n=0}^{\infty} d^n P_n[\cos(\theta)] r^{-n}.$$

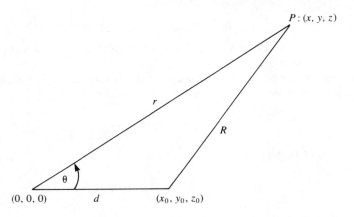

7.5 Additional Properties of Legendre Polynomials

In this section, we will derive some important properties of Legendre polynomials. One motivation is that this is useful information to have, but there is another motivation as well. The program we will follow models one way of proceeding with a class of special functions called *orthogonal polynomials*, of which the Legendre polynomials are just one example. Thus, another objective of this section is to illustrate the types of properties that are important and how we can go about establishing them.

There are several approaches to Legendre polynomials. We can use the fact that Legendre polynomials are solutions of Legendre's differential equation to derive many of their properties (this is how we proved Theorem 7.3). Another approach, which we will take, is to use a generating function. Consider the function

$$H(x, r) = (1 - 2xr + r^2)^{-1/2}. \tag{7.25}$$

Recall the binomial expansion

$$(1 - z)^{-1/2} = 1 + \frac{1}{2} z + \frac{1}{2!} \frac{1}{2} \frac{3}{2} z^2 + \frac{1}{3!} \frac{1}{2} \frac{3}{2} \frac{5}{2} z^3 + \cdots.$$

If we let $z = 2xr - r^2$ in this expansion, we get

$$H(x, r) = 1 + \frac{1}{2}(2xr - r^2) + \frac{3}{8}(2xr - r^2)^2 + \frac{5}{16}(2xr - r^2)^3 + \cdots.$$

Rewrite the series on the right in ascending powers of r. This means that we must collect all coefficients of each power of r. Each such coefficient will be a function of x. After a little work, we get

$$H(x, r) = 1 + xr + \left(-1 + \frac{3}{2}x^2\right)r^2 + \left(-\frac{3}{2}x + \frac{5}{2}x^3\right)r^3 + \cdots.$$

Now observe that

$$H(x, r) = P_0(x) + P_1(x)r + P_2(x)r^2 + P_3(x)r^3 + \cdots.$$

It can be shown that the coefficient of r^n is exactly $P_n(x)$, so

$$H(x, r) = \sum_{n=0}^{\infty} P_n(x)r^n. \tag{7.26}$$

This is the reason for calling $H(x, r)$ a *generating function* for the Legendre polynomials. By using this generating function, it is possible to derive many properties of these polynomials. We will begin with a recurrence relation which was mentioned in Problem 7 of the preceding section.

THEOREM 7.4 Recurrence Relation for Legendre Polynomials

For each positive integer n,

$$(n + 1)P_{n+1}(x) - (2n + 1)xP_n(x) + nP_{n-1}(x) = 0 \tag{7.27}$$

for $-1 \leq x \leq 1$.

Proof Differentiate equation (7.25) with respect to r to get

$$\frac{\partial H}{\partial r} = -\frac{1}{2}(1 - 2xr + r^2)^{-3/2}(-2x + 2r) = (1 - 2xr + r^2)^{-3/2}(x - r).$$

Now observe from equation (7.25) that

$$(1 - 2xr + r^2)\frac{\partial H}{\partial r} - (x - r)H = 0. \tag{7.28}$$

Substitute the series (7.26) for $H(x, r)$ into this equation to get

$$\sum_{n=1}^{\infty} (1 - 2xr + r^2)nP_n(x)r^{n-1} - \sum_{n=0}^{\infty} (x - r)P_n(x)r^n = 0.$$

Carry out the indicated multiplications to obtain

$$\sum_{n=1}^{\infty} nP_n(x)r^{n-1} + \sum_{n=1}^{\infty} -2xnP_n(x)r^n + \sum_{n=1}^{\infty} nP_n(x)r^{n+1}$$

$$- \sum_{n=0}^{\infty} xP_n(x)r^n + \sum_{n=0}^{\infty} P_n(x)r^{n+1} = 0.$$

Rearrange the first, third, and fifth series and rewrite this equation as

$$\sum_{n=0}^{\infty} (n + 1)P_{n+1}(x)r^n + \sum_{n=1}^{\infty} -2nxP_n(x)r^n + \sum_{n=2}^{\infty} (n - 1)P_{n-1}(x)r^n$$

$$- \sum_{n=0}^{\infty} xP_n(x)r^n + \sum_{n=1}^{\infty} P_{n-1}(x)r^n = 0.$$

Collect terms to write

$$\sum_{n=2}^{\infty} [(n+1)P_{n+1}(x) - 2nxP_n(x) + (n-1)P_{n-1}(x) - xP_n(x) + P_{n-1}(x)]r^n$$
$$+ P_1(x) + 2P_2(x)r - 2xP_1(x)r - xP_0(x) - xP_1(x)r + P_0(x)r = 0.$$

The coefficient of each power of r must be zero in this expansion. Hence,

$$P_1(x) - xP_0(x) = 0$$

and

$$2P_2(x) - 2xP_1(x) - xP_1(x) + P_0(x) = 0$$

This is equation (7.27) for $n = 1$. For $n = 2, 3, \ldots$, the coefficient of r^n yields

$$(n+1)P_{n+1}(x) - (2n+1)xP_n(x) + nP_{n-1}(x) = 0.$$

This completes the proof of the recurrence relation. ∎

Using the generating function and the recurrence relation (7.27), we can obtain other information about Legendre polynomials. First, we will derive a formula for the coefficient of x^n in $P_n(x)$, a number which we will use in a later calculation.

THEOREM 7.5

The coefficient of x^n in $P_n(x)$ is $[(1)(3)(5) \cdots (2n-1)]/n!$.

Proof Let c_n be the coefficient of x^n in $P_n(x)$ and consider the recurrence relation (7.27). The x^{n+1} term in $(n+1)P_{n+1}(x)$ has coefficient $(n+1)c_{n+1}$, and the x^{n+1} term in $-(2n+1)xP_n(x)$ has coefficient $-(2n+1)c_n$. There are no other x^{n+1} terms in equation (7.27). Thus, the coefficient of x^{n+1} in equation (7.27) is

$$(n+1)c_{n+1} - (2n+1)c_n,$$

and this must be zero because the coefficient of each power of x in equation (7.27) must vanish. Therefore,

$$c_{n+1} = \frac{2n+1}{n+1} c_n.$$

Working back, we have

$$c_n = \frac{2n-1}{n} c_{n-1} = \frac{2n-1}{n} \frac{2n-3}{n-1} c_{n-2}$$
$$= \frac{2n-1}{n} \frac{2n-3}{n-1} \frac{2n-5}{n-2} c_{n-3} = \cdots = \frac{(2n-1) \cdots (5)(3)(1)}{n!} c_0.$$

But c_0 is the coefficient of x^0 in $P_0(x) = 1$, so $c_0 = 1$, and the theorem is proved. ∎

The next theorem relates $P_n(x)$, $P_n'(x)$, and $P_{n-1}'(x)$.

THEOREM 7.6

For each positive integer n, $nP_n(x) - xP_n'(x) + P_{n-1}'(x) = 0$.

Proof   Differentiate equation (7.25) with respect to x to get

$$(1 - 2xr + r^2)\frac{\partial H}{\partial x} - rH = 0.$$

Compare this equation with equation (7.28) to find that

$$r\frac{\partial H}{\partial r} - (x - r)\frac{\partial H}{\partial x} = 0.$$

Now follow the method used to prove Theorem 7.4, substituting the series (7.26) into this equation and equating the coefficient of r^n to zero to prove the theorem. The details are left to the student. ∎

THEOREM 7.7

For each positive integer n,

$$nP_{n-1}(x) - P'_n(x) + xP'_{n-1}(x) = 0. \quad ∎ \qquad (7.29)$$

This result can be proved by substituting the series (7.26) into the relationship

$$r\frac{\partial}{\partial r}[rH] - (1 - rx)\frac{\partial H}{\partial x} = 0.$$

Again, the details are left to the student.

Theorem 7.3 in the preceding section stated that

$$\int_{-1}^{1} P_n(x)P_m(x)\,dx = 0 \qquad (7.30)$$

if m and n are distinct nonnegative integers. In view of this integral, the Legendre polynomials are said to be *orthogonal* to each other on $[-1, 1]$, or *mutually orthogonal* on $[-1, 1]$. We also say that the Legendre polynomials form a set of *orthogonal polynomials* on the interval $[-1, 1]$.

This orthogonality property can be used to write many functions as series of Legendre polynomials. This will be important in solving certain boundary value problems in partial differential equations. For now, we will see how to write any polynomial as such a series. Let q be a polynomial of degree m. We will see how to choose numbers $\alpha_0, \ldots, \alpha_m$ such that

$$q(x) = \sum_{n=0}^{m} \alpha_n P_n(x) \qquad (7.31)$$

for $-1 \le x \le 1$. Multiply equation (7.31) by $P_j(x)$, where j is any integer from zero to m, inclusive:

$$q(x)P_j(x) = \alpha_0 P_0(x)P_j(x) + \alpha_1 P_1(x)P_j(x) + \alpha_2 P_2(x)P_j(x) + \cdots + \alpha_m P_m(x)P_j(x).$$

Integrate both sides of this equation from -1 to 1:

$$\int_{-1}^{1} q(x)P_j(x)\,dx = \alpha_0 \int_{-1}^{1} P_0(x)P_j(x)\,dx + \alpha_1 \int_{-1}^{1} P_1(x)P_j(x)\,dx$$

$$+ \alpha_2 \int_{-1}^{1} P_2(x)P_j(x)\,dx + \cdots + \alpha_m \int_{-1}^{1} P_m(x)P_j(x)\,dx.$$

Because of equation (7.30), each integral $\int_{-1}^{1} P_k(x)P_j(x)\,dx$ on the right side is zero, except for that one integral in which $k = j$. This leaves us with

$$\int_{-1}^{1} q(x)P_j(x)\,dx = \alpha_j \int_{-1}^{1} [P_j(x)]^2\,dx;$$

hence,

$$\alpha_j = \frac{\displaystyle\int_{-1}^{1} q(x)P_j(x)\,dx}{\displaystyle\int_{-1}^{1} [P_j(x)]^2\,dx} \quad \text{for} \quad j = 0, 1, 2, 3, \ldots. \tag{7.32}$$

These numbers can be computed because we know each $P_j(x)$ and we are given $q(x)$. Using these numbers, we can write $q(x)$ as a series (7.31) of Legendre polynomials. As specific examples, straightforward calculations show that

$$x^2 = \tfrac{2}{3}P_2(x) + \tfrac{1}{3}P_0(x)$$

and

$$x^3 = \tfrac{2}{5}P_3(x) + \tfrac{3}{5}P_1(x).$$

Using these two expansions, we can write, for example,

$$1 - 4x^2 + 2x^3 = P_0(x) - \tfrac{8}{3}P_2(x) - \tfrac{4}{3}P_0(x) + \tfrac{4}{5}P_3(x) + \tfrac{6}{5}P_1(x)$$
$$= -\tfrac{1}{3}P_0(x) + \tfrac{6}{5}P_1(x) - \tfrac{8}{3}P_2(x) + \tfrac{4}{5}P_3(x).$$

The orthogonality relationship (7.30), together with the fact that any polynomial can be written as a finite series of Legendre polynomials, leads to the following fact, which may seem quite remarkable at first.

THEOREM 7.8

Let m and n be nonnegative integers, with $m < n$. Let $q(x)$ be any polynomial of degree m. Then $\int_{-1}^{1} q(x)P_n(x)\,dx = 0$. That is, the integral, from -1 to 1, of a Legendre polynomial, multiplied by any polynomial of lower degree, is zero.

Proof   Equation (7.32) shows that there are constants $\alpha_0, \alpha_1, \ldots, \alpha_m$ such that

$$q(x) = \alpha_0 P_0(x) + \alpha_1 P_1(x) + \cdots + \alpha_m P_m(x).$$

Then

$$\int_{-1}^{1} q(x)P_n(x)\,dx = \alpha_0 \int_{-1}^{1} P_0(x)P_n(x)\,dx$$
$$+ \alpha_1 \int_{-1}^{1} P_1(x)P_n(x)\,dx + \cdots + \alpha_m \int_{-1}^{1} P_m(x)P_n(x)\,dx,$$

and each of the integrals on the right is zero by equation (7.30) because $m < n$. ∎

Using these facts, we can derive a simple expression for $\int_{-1}^{1} [P_n(x)]^2\,dx$, which appears in the denominator of equation (7.32) and which we will use in later work.

THEOREM 7.9

$$\int_{-1}^{1} [P_n(x)]^2 \, dx = \frac{2}{2n+1}$$

for $n = 0, 1, 2, \ldots$.

Proof Let c_n be the coefficient of x^n in $P_n(x)$. Then the coefficient of x^{n-1} in $P_{n-1}(x)$ is c_{n-1}. Let

$$q(x) = P_n(x) - \frac{c_n}{c_{n-1}} x P_{n-1}(x).$$

The x^n term in $P_n(x)$ is canceled by the x^n term in $-(c_n/c_{n-1})xP_{n-1}(x)$; thus, $q(x)$ has degree $n - 1$ or lower. We therefore have

$$P_n(x) = \frac{c_n}{c_{n-1}} x P_{n-1}(x) + q(x),$$

in which $q(x)$ has degree $\leq n - 1$. Then

$$[P_n(x)]^2 = P_n(x) \left[\frac{c_n}{c_{n-1}} x P_{n-1}(x) + q(x) \right]$$

$$= \frac{c_n}{c_{n-1}} x P_{n-1}(x) P_n(x) + q(x) P_n(x).$$

Integrate this equation from -1 to 1:

$$\int_{-1}^{1} [P_n(x)]^2 \, dx = \frac{c_n}{c_{n-1}} \int_{-1}^{1} x P_{n-1}(x) P_n(x) \, dx + \int_{-1}^{1} q(x) P_n(x) \, dx.$$

The second integral on the right is zero by Theorem 7.8; hence

$$\int_{-1}^{1} [P_n(x)]^2 \, dx = \frac{c_n}{c_{n-1}} \int_{-1}^{1} x P_{n-1}(x) P_n(x) \, dx. \tag{7.33}$$

Now use the recurrence relation stated in Theorem 7.4 to write

$$x P_n(x) = \frac{n+1}{2n+1} P_{n+1}(x) + \frac{n}{2n+1} P_{n-1}(x).$$

Substitute the right side of this identity for $xP_n(x)$ into the integral on the right side of equation (7.33). We get

$$\int_{-1}^{1} [P_n(x)]^2 \, dx = \frac{c_n}{c_{n-1}} \int_{-1}^{1} P_{n-1}(x) \left[\frac{n+1}{2n+1} P_{n+1}(x) + \frac{n}{2n+1} P_{n-1}(x) \right] dx$$

$$= \frac{c_n}{c_{n-1}} \int_{-1}^{1} \frac{n+1}{2n+1} P_{n-1}(x) P_{n+1}(x) + \frac{c_n}{c_{n-1}} \frac{n}{2n+1} \int_{-1}^{1} [P_{n-1}(x)]^2 \, dx.$$

The first integral on the right is zero because $\int_{-1}^{1} P_{n-1}(x) P_{n+1}(x) \, dx = 0$. We therefore have

$$\int_{-1}^{1} [P_n(x)]^2 \, dx = \frac{c_n}{c_{n-1}} \frac{n}{2n+1} \int_{-1}^{1} [P_{n-1}(x)]^2 \, dx.$$

Now recall that we derived an expression for c_n in Theorem 7.5:

$$c_n = \frac{(1)(3)(5) \cdots (2n - 1)}{n!}.$$

Then

$$\frac{c_n}{c_{n-1}} = \frac{(1)(3)(5) \cdots (2n - 1)}{n!} \frac{(n - 1)!}{(1)(3)(5) \cdots (2n - 3)} = \frac{2n - 1}{n}.$$

Therefore,

$$\int_{-1}^{1} [P_n(x)]^2 \, dx = \frac{2n - 1}{n} \frac{n}{2n + 1} \int_{-1}^{1} [P_{n-1}(x)]^2 \, dx$$

$$= \frac{2n - 1}{2n + 1} \int_{-1}^{1} [P_{n-1}(x)]^2 \, dx.$$

Now we can work back, as we have done before. We get

$$\int_{-1}^{1} [P_n(x)]^2 \, dx = \frac{2n - 1}{2n + 1} \int_{-1}^{1} [P_{n-1}(x)]^2 \, dx$$

$$= \frac{2n - 1}{2n + 1} \frac{2n - 3}{2n - 1} \int_{-1}^{1} [P_{n-2}(x)]^2 \, dx$$

$$= \frac{2n - 1}{2n + 1} \frac{2n - 3}{2n - 1} \frac{2n - 5}{2n - 3} \int_{-1}^{1} [P_{n-3}(x)]^2 \, dx \cdots.$$

In view of the cancellations, we finally obtain

$$\int_{-1}^{1} [P_n(x)]^2 \, dx = \frac{1}{2n + 1} \int_{-1}^{1} [P_0(x)]^2 \, dx.$$

Since $P_0(x) = 1$, the integral on the right equals 2, and we have

$$\int_{-1}^{1} [P_n(x)]^2 \, dx = \frac{2}{2n + 1},$$

as we wanted to show. ∎

We will make use of many of the facts we have derived here when we discuss Sturm-Liouville theory and eigenfunction expansions in the next three sections. In the problems that follow, we will pursue some additional properties of Legendre polynomials. Starting with Problem 18, we will also develop some facts about two other sets of orthogonal polynomials, the Hermite and Laguerre polynomials.

PROBLEMS FOR SECTION 7.5

1. Use the recurrence relation of Theorem 7.4, together with equations (7.23), to determine $P_6(x)$ through $P_8(x)$.
2. Use the recurrence relation of Theorem 7.4 to prove that $P_n(1) = 1$ for $n = 0, 1, 2, 3, \ldots$. *Hint:* Use mathematical induction.
3. Use the recurrence relation of Theorem 7.4 to show that $P_n(-1) = (-1)^n$ for $n = 0, 1, 2, 3, \ldots$.
4. Prove that $P_n(x)$ has no repeated roots. *Hint:* Suppose that $P_n(x)$ has a repeated root at x_0. Then $P_n(x_0) =$

$P_n'(x_0) = 0$. Thus, P_n is a solution of the initial value problem

$$(1 - x^2)y'' - 2xy' + n(n + 1)y = 0; \qquad y(x_0) = y'(x_0) = 0.$$

Now use the uniqueness theorem for solutions of linear initial value problems to conclude that $P_n(x) = 0$ for all x, a contradiction.

5. Prove that $P_n(x)$ has n distinct real roots, all lying on the interval $(-1, 1)$, for $n = 1, 2, 3, \ldots$. *Hint:* First observe that $\int_{-1}^{1} P_n(x)P_0(x)\, dx = 0$ and conclude that $P_n(x)$ must change sign at least once in $(-1, 1)$. Conclude (by the intermediate value theorem) that $P_n(x) = 0$ for at least one value of x in $(-1, 1)$. This establishes the existence of at least one root of $P_n(x)$ in $(-1, 1)$.

Now suppose that the roots of $P_n(x)$ in $(-1, 1)$ are $x_1, \ldots, x_m$. Let

$$q(x) = (x - x_1) \cdots (x - x_m).$$

Use the fact that P_n has no repeated root (see Problem 4), and the fact that $P_n(x)$ and $q(x)$ change sign at the same points in $(-1, 1)$, to conclude that $P_n(x)q(x)$ must be nonnegative, or nonpositive, for all x in $(-1, 1)$. Conclude that $\int_{-1}^{1} P_n(x)q(x)\, dx$ must be either positive or negative. But, if $m < n$, show that $\int_{-1}^{1} P_n(x)q(x)\, dx = 0$, a contradiction. Therefore, $m = n$, and $x_1, \ldots, x_n$ must be all the zeros of $P_n(x)$.

6. Prove that $P_n(-x) = (-1)^n P_n(x)$ for all x in $[-1, 1]$.

7. Prove Theorem 7.6.

8. Prove Theorem 7.7.

9. Prove that $\int_{-1}^{1} P_n(x)\, dx = 0$ for $n = 1, 2, \ldots$. *Hint:* $P_n(x) = P_n(x)P_0(x)$.

10. Prove that $(1 - x^2)P_n'(x) = nP_{n-1}(x) - nxP_n(x)$.

11. Prove that $\int_{-1}^{1} xP_n(x)P_{n-1}(x)\, dx = 2n/(4n^2 - 1)$ for $n = 1, 2, 3, \ldots$.

12. Prove that $\int_{-1}^{1} xP_n'(x)P_n(x)\, dx = 2n/(2n + 1)$ for $n = 0, 1, 2, 3, \ldots$.

13. Prove that $P_{2n+1}(0) = 0$ and $P_{2n}(0) = (-1)^n(2n)!/2^{2n}(n!)^2$ for $n = 0, 1, 2, \ldots$.

14. Expand $1 + 2x - x^2$ in a series of Legendre polynomials.

15. Write $2x + x^2 - 5x^3$ as a series of Legendre polynomials.

16. Write $4x^4 - x^2 + 2$ as a series of Legendre polynomials.

17. Carry out the following sequence of steps to prove that

$$P_n(x) = \frac{1}{\pi} \int_0^\pi [x + \sqrt{x^2 - 1}\, \cos(\varphi)]^n\, d\varphi$$

for $n = 0, 1, 2, 3, \ldots$ and $-1 \le x \le 1$.

(a) Use Euler's formula to show that

$$\int_{-\pi}^{\pi} e^{ik\varphi}\, d\varphi = \begin{cases} 0 & \text{if} \quad k = 1, 2, 3, \ldots \\ 2\pi & \text{if} \quad k = 0 \end{cases}$$

(b) Use the binomial series to expand $(a + be^{i\varphi})^k$ in a finite series, and integrate term by term to show that

$$\frac{1}{2\pi} \int_{-\pi}^{\pi} (a + be^{i\varphi})^k\, d\varphi = a^k$$

for $k = 0, 1, 2, \ldots$.

(c) Let $q(x)$ be a polynomial of degree n. Use the result of (b) to show that

$$q(x) = \frac{1}{2\pi} \int_{-\pi}^{\pi} q(x + be^{i\varphi})\, d\varphi.$$

(d) Now use Rodrigues's formula (Problem 4 of Section 7.4) to write

$$P_n(x) = \frac{1}{2^n n!} \frac{d^n}{dx^n}[(x^2 - 1)^n]$$

for $n = 1, 2, 3, \ldots$. For convenience, let $p(x) = (1/2^n n!)(x^2 - 1)^n$ so that Rodrigues's formula is written more simply as $P_n(x) = (d^n/dx^n)p(x)$. Using this fact, and the result of (c), show that

$$P_n(x) = \frac{1}{2\pi} \int_{-\pi}^{\pi} \frac{d^n}{dx^n} [p(x + be^{i\varphi})] \, d\varphi.$$

(e) Integrate by parts and use the result of (d) to show that

$$\int_{-\pi}^{\pi} e^{-ik\varphi} \frac{d^{n-k}}{dx^{n-k}} [p(x + be^{i\varphi})] \, d\varphi = \frac{k+1}{b} \int_{-\pi}^{\pi} e^{-i(k+1)\varphi} \frac{d^{n-k-1}}{dx^{n-k-1}} [p(x + be^{i\varphi})] \, d\varphi,$$

in which k is any integer with $0 \le k \le n - 1$.

(f) Use the result of (e) to write

$$P_n(x) = \frac{n!}{2\pi b^n} \int_{-\pi}^{\pi} e^{-in\varphi} p(x + be^{i\varphi}) \, d\varphi.$$

(g) Assume that $-1 < x < 1$ and write $b = \sqrt{x^2 - 1}$. Show that

$$\frac{n!}{b^n} e^{-in\varphi} p(x + be^{i\varphi}) = [x + \sqrt{x^2 - 1} \cos(\varphi)]^n.$$

(h) Use the results of (f) and (g) to write

$$P_n(x) = \frac{1}{2\pi} \int_{-\pi}^{\pi} [x + \sqrt{x^2 - 1} \cos(\varphi)]^n \, d\varphi.$$

(i) Use the result of (h) to establish the formula to be proved, for $-1 < x < 1$. Finally, show that this formula also holds for $x = 1$ and for $x = -1$, completing the proof.

The following problems deal with two other kinds of orthogonal polynomials.

Hermite Polynomials

The second order differential equation

$$y'' - xy' + \lambda y = 0$$

is called *Hermite's equation*. It occurs, for example, in treating the wave mechanics of the harmonic oscillator.

18. Obtain series solutions of Hermite's equation, expanded about zero. Show that, if λ is a nonnegative integer n, there is a polynomial solution $H_n(x)$ of degree n, having only even powers of x if n is even and only odd powers of x if n is odd. Further, show that we can choose H_n so that the coefficient of x^n is 1. We call $H_n(x)$ the nth *Hermite polynomial*.

19. Determine $H_0(x)$ through $H_5(x)$.

20. Let $h_n(x) = (-1)^n e^{x^2/2} (d^n/dx^n)[e^{-x^2/2}]$. Show that $h_n(x) = H_n(x)$ for $n = 0, 1, 2, 3, 4, 5$. In fact, $h_n(x)$ is the nth Hermite polynomial for each nonnegative integer n. (This is a Rodrigues-type formula for Hermite polynomials; see Problem 4 of Section 7.4.)

21. Use the formula of Problem 20, and integration by parts, to show that

$$\int_{-\infty}^{\infty} e^{-x^2/2} [H_n(x)]^2 \, dx = n! \sqrt{2\pi}.$$

Hint: Assume as known the integral $\int_{-\infty}^{\infty} e^{-x^2/2} \, dx = \sqrt{2\pi}$.

22. Show that $\int_{-\infty}^{\infty} e^{-x^2/2} H_n(x) H_m(x) \, dx = 0$ if $n \ne m$. [We say that H_n and H_m are *orthogonal* on $(-\infty, \infty)$ with respect to the *weight function* $e^{-x^2/2}$.]

Laguerre Polynomials

23. *Laguerre's differential equation is*

$$xy'' + (1 - x)y' + \lambda y = 0.$$

Show that, when $\lambda = n$, a nonnegative integer, there is a polynomial solution $L_n(x)$ of degree n with coefficient of x^n equal to 1. These polynomials are called *Laguerre polynomials*. Determine the Laguerre polynomials $L_0(x)$ through $L_5(x)$.

24. Let $\ell_n(x) = (-1)^n e^x (d^n/dx^n)[x^n e^{-x}]$. Determine $\ell_0(x)$ through $\ell_5(x)$, and show that these are Laguerre polynomials. It is true that $\ell_n(x) = L_n(x)$ for each nonnegative integer n.

25. Show that $\int_0^\infty e^{-x}[L_n(x)]^2 \, dx = n!$.

26. Show that $\int_0^\infty e^{-x} L_n(x) L_m(x) \, dx = 0$ if $n \neq m$. Thus, the Laguerre polynomials are *orthogonal on* $[0, \infty)$ *with respect to the weight function* e^{-x}.

27. Imagine N charged beads strung on a wire which is stretched between -1 and $+1$ on the real line. The beads are free to move along the wire. Each end of the wire has a planar charge of $+1$, and each bead has a charge of $+2$. Show that the equilibrium positions of the beads coincide with the zeros of the Nth Legendre polynomial P_N. *Hint:* Particles of planar charge q_1 and q_2 located r units apart exert a force of magnitude $K(q_1 q_2/r)$ on one another. Now carry out the following steps.

(a) If $N = 1$, there is only one bead, and it will come to rest midway between the charges at -1 and $+1$, hence at $x = 0$. This is the zero of P_1. Therefore, the proposition to be proved holds if $N = 1$. Prove that two beads come to rest at $\pm 1/\sqrt{3}$, the zeros of P_2.

(b) Now assume that N beads are located at $x_1, \ldots, x_N$, with $x_k < x_{k+1}$. Show that, for $k = 1, 2, \ldots, N$,

$$\sum_{\substack{i=1 \\ i \neq k}}^{N} \frac{4k}{x_k - x_i} + \frac{2k}{x_k + 1} + \frac{2k}{x_k - 1} = 0.$$

Using this result, write

$$\frac{2x_k}{1 - x_k^2} = \sum_{\substack{i=1 \\ i \neq k}}^{N} \frac{2}{x_k - x_i}.$$

(c) Use the fact that P_N satisfies Legendre's differential equation to write

$$(1 - x^2)P_N''(x) - 2xP_N'(x) + N(N+1)P_N(x) = 0.$$

With $|x| < 1$, divide this equation by $1 - x^2$ to get

$$P_N''(x) - \frac{2x}{1 - x^2} P_N'(x) + \frac{N(N+1)}{1 - x^2} P_N(x) = 0.$$

Now suppose that x_k is a zero of P_N. Show that $P_N'(x_k) \neq 0$. [Use Legendre's differential equation to show that, if $P_N'(x_k) = 0$, then $P_N''(x_k) = 0$ also. Differentiate Legendre's differential equation to show that $P_N^{(3)}(x_k) = 0$, and proceed in this way to show that $P_N^{(n)}(x_k) = 0$ for every positive integer n, deriving a contradiction.] Now solve the last equation for $P_N''(x_k)/P_N'(x_k)$ and compare the result with that of (b). Think about what is left to prove at this point.

(d) Suppose that a polynomial $p(x)$ of degree N has N distinct zeros $x_1, \ldots, x_N$ and that the coefficient of x^N in $p(x)$ is 1. It is a fact of algebra that $p(x)$ can be factored

$$p(x) = (x - x_1)(x - x_2) \cdots (x - x_N).$$

Show that

$$p'(x) = \sum_{i=1}^{N} \prod_{\substack{m=1 \\ m \neq i}}^{N} (x - x_m)$$

and

$$p''(x) = \sum_{j=1}^{N} \sum_{\substack{i=1 \\ i \neq j}}^{N} \prod_{\substack{m=1 \\ m \neq i \\ m \neq j}}^{N} (x - x_m),$$

in which $\prod_{\substack{m=1 \\ m \neq i}}^{N}$ denotes a product as m takes on the values $1, 2, \ldots, N$ but $m \neq i$.

(e) Proceeding from (d), show that

$$p'(x_k) = \prod_{\substack{m=1 \\ m \neq k}}^{N} (x_k - x_m) \neq 0$$

and

$$p''(x_k) = 2 \sum_{\substack{i=1 \\ i \neq k}}^{N} \prod_{\substack{m=1 \\ m \neq i \\ m \neq k}}^{N} (x_k - x_m).$$

Use these two facts to evaluate $p''(x_k)/p'(x_k)$.

(f) Recall that P_N has N distinct roots and that P_N is just one of two linearly independent solutions of Legendre's equation. From this, and the results of (b), (c), (d), and (e), complete the proof.

28. Suppose that N charged beads are strung on a wire which coincides with the positive x-axis. Each bead has planar charge $+2$, and the origin has charge $+1$. If the beads are allowed to slide freely on the wire and come to rest in an equilibrium position, their locations will coincide with the zeros of the Nth Laguerre polynomial L_N, which is given by

$$L_N(x) = N! \sum_{m=1}^{N} \binom{N}{N-m} (-1)^{N-m} \frac{1}{m!} x^m.$$

(a) Verify the assertion made about the distribution of the beads in the case in which there is just one bead ($N = 1$).

(b) Verify the assertion about the distribution of the beads if N is an arbitrary positive integer.

7.6 Series Expansions and Orthogonal Sets of Functions _____

We will see later that the solution of certain problems requires that we be able to expand a function g in a series of given functions $f_0, f_1, f_2, \ldots, f_n, \ldots$. Such a series has the form

$$g(x) = \sum_{n=0}^{\infty} a_n f_n(x) = a_0 f_0(x) + a_1 f_1(x) + \cdots. \tag{7.34}$$

We saw a simple example of this type of expansion when we showed that any polynomial can be written as a series of Legendre polynomials.

Sturm-Liouville theory, which we will develop in the next section, enables us to understand the kinds of differential equations we might approach in this way and how the functions $f_0, f_1, \ldots,$ are generated. In this section, we will explore some background for this theory by developing a general concept of orthogonality as it applies to functions and see how it enables us to determine the constants a_n in the expansion (7.34). The process is very similar to what we have just seen for Legendre polynomials.

The property of Legendre polynomials that made a series of the form (7.34) possible, at least for polynomials, was that $\int_{-1}^{1} P_n(x)P_m(x)\, dx = 0$ if $m \neq n$. We described this relationship by saying that the Legendre polynomials are *orthogonal* on $[-1, 1]$. We will generalize this concept of orthogonality as follows.

DEFINITION

Suppose that p is continuous on $[a, b]$ and that $p(x) \geq 0$ for $a \leq x \leq b$. Suppose that f and h are defined on $[a, b]$. Then f and h are *orthogonal on $[a, b]$ with weight function p* if

$$\int_a^b p(x)f(x)h(x)\, dx = 0.$$

We say that $f_0, f_1, f_2, \ldots$ form an *orthogonal set of functions on $[a, b]$ with weight function p* if f_n is orthogonal to f_m with weight function p for $n \neq m$. This means that

$$\int_a^b p(x)f_n(x)f_m(x)\, dx = 0 \quad \text{if} \quad n \neq m. \quad \blacksquare$$

EXAMPLE 7.1

The Legendre polynomials are orthogonal on the interval $[-1, 1]$ with weight function $p(x) = 1$ because

$$\int_{-1}^{1} P_m(x)P_n(x)\, dx = 0$$

if $m \neq n$. $\blacksquare$

EXAMPLE 7.2

The Hermite polynomials $H_n(x)$ (see Problems 18 through 22 of Section 7.5) are orthogonal on $(-\infty, \infty)$ with weight function $p(x) = e^{-x^2/2}$ because

$$\int_{-\infty}^{\infty} e^{-x^2/2}H_n(x)H_m(x)\, dx = 0 \quad \text{if} \quad n \neq m. \quad \blacksquare$$

EXAMPLE 7.3

The Laguerre polynomials $L_n(x)$ (see Problems 23 through 26 of Section 7.5) are orthogonal on $[0, \infty)$ with weight function $p(x) = e^{-x}$ because

$$\int_0^{\infty} e^{-x}L_n(x)L_m(x)\, dx = 0 \quad \text{if} \quad n \neq m. \quad \blacksquare$$

EXAMPLE 7.4

The functions $f_n(x) = \cos(nx)$, for $n = 0, 1, 2, 3, \ldots$, are orthogonal on $[-\pi, \pi]$ with weight function $p(x) = 1$. It is a routine integration to check that

$$\int_{-\pi}^{\pi} \cos(nx)\cos(mx)\, dx = 0$$

if $n \neq m$. ∎

We will now use the concept of orthogonality to write a formula for the coefficients in the series expansion (7.34), which is reproduced here for ease of reference:

$$g(x) = \sum_{n=0}^{\infty} a_n f_n(x). \tag{7.34}$$

Assume that the functions $f_0, f_1, f_2, \ldots$ are orthogonal on $[a, b]$ with weight function p. Proceed as we did in Section 7.5 with Legendre polynomials. Choose some particular f_k, and multiply both sides of this proposed expansion by $p(x)f_k(x)$ to get

$$p(x)g(x)f_k(x) = \sum_{n=0}^{\infty} a_n p(x) f_n(x) f_k(x).$$

Integrate both sides from a to b, interchanging the summation and the integral, a step which is valid under certain conditions:

$$\int_a^b p(x)g(x)f_k(x)\, dx = \sum_{n=0}^{\infty} a_n \int_a^b p(x) f_n(x) f_k(x)\, dx. \tag{7.35}$$

By the orthogonality of the functions f_n, we have

$$\int_a^b p(x) f_n(x) f_k(x)\, dx \quad \text{if} \quad k \neq n.$$

Therefore, in the summation on the right in equation (7.35), all terms are zero except the $n = k$ term, and we are left with

$$\int_a^b p(x)g(x)f_k(x)\, dx = a_k \int_a^b p(x)[f_k(x)]^2\, dx. \tag{7.36}$$

Assuming that each $\int_a^b p(x)f_k(x)^2\, dx \neq 0$ (which occurs, for example, if p and f_k are continuous on $[a, b]$ and both nonzero at some x_0 in this interval), we can solve equation (7.36) for a_k:

$$a_k = \frac{\displaystyle\int_a^b p(x)g(x)f_k(x)\, dx}{\displaystyle\int_a^b p(x)[f_k(x)]^2\, dx}. \tag{7.37}$$

Notice that this argument has mirrored almost exactly the argument we used in Section 7.5 to find the coefficients in writing a polynomial $q(x)$ as a series $\alpha_0 P_0(x) + \cdots + \alpha_m P_m(x)$ of Legendre polynomials. Indeed, the formula we found for the coefficients α_k is exactly equation (7.37) applied to the Legendre polynomials and the polynomial $q(x)$.

Thus far, we have dealt with the mechanics of expanding a function g in a series $\sum_{n=0}^{\infty} a_n f_n$. However, there are important questions we have not considered. Among these are the following.

1. Why do we want to expand a given function g in a series of functions $f_0, f_1, \ldots$?
2. How do we choose the functions f_n and the weight function p?
3. Under what conditions will the series $\sum_{n=0}^{\infty} a_n f_n(x)$, with the a_n's chosen according to equation (7.37), converge to $g(x)$ for $a \leq x \leq b$?

These questions are considered under the heading of *Sturm-Liouville theory*, which we will discuss in the next section. We will see there that certain types of problems involving differential equations give rise to solutions $f_0, f_1, \ldots$, which are called *eigenfunctions* of the problem. These eigenfunctions form an orthogonal set with respect to a weight function which occurs as one of the coefficients in the differential equation. The solution we seek, satisfying specified conditions at the ends of the interval, is obtained as an expansion of the form (7.34) in a series of the eigenfunctions. The series (7.34) is called an *eigenfunction expansion* of g. This technique is particularly important in solving many partial differential equations arising in physics and engineering, and it is the basis for Fourier series.

PROBLEMS FOR SECTION 7.6

1. Let L be any positive number. Show that the functions $f_n(x) = \cos(n\pi x/L)$, $n = 0, 1, 2, 3, \ldots$, form an orthogonal set on $[0, L]$ with weight function $p(x) = 1$.
2. Let L be any positive number. Show that the functions $f_n(x) = \sin(n\pi x/L)$, $n = 1, 2, 3, \ldots$, form an orthogonal set on $[0, L]$ with weight function $p(x) = 1$.
3. Let L be any positive number. Show that the functions 1, $\cos(n\pi x/L)$, and $\sin(n\pi x/L)$, for $n = 1, 2, 3, \ldots$, form an orthogonal set on $[-L, L]$ with weight function $p(x) = 1$.
4. Let $f(x) = x$ for $0 \leq x \leq L$. How should the numbers a_n be chosen in a proposed expansion

$$f(x) = \sum_{n=0}^{\infty} a_n \cos\left(\frac{n\pi x}{L}\right)$$

for $0 < x < L$? *Hint:* Use equation (7.37) and the conclusion of Problem 1.
5. Let $h(x) = e^x$. How should the numbers b_n be chosen in an expansion

$$h(x) = \sum_{n=1}^{\infty} b_n \sin\left(\frac{n\pi x}{L}\right)$$

for $0 < x < L$? *Hint:* Use the conclusion of Problem 2.
6. Let $k(x) = 4$. How should the numbers $a_0, a_1, \ldots, b_1, b_2, \ldots$ be chosen in an expansion

$$k(x) = a_0 + \sum_{n=1}^{\infty} \left[a_n \cos\left(\frac{n\pi x}{L}\right) + b_n \sin\left(\frac{n\pi x}{L}\right) \right]$$

for $-L < x < L$? *Hint:* Use the conclusion of Problem 3.
7. Show that 1 and x^2 are orthogonal on $[-1, 1]$ with respect to the weight function $p(x) = x$.
8. How should the numbers A and B be chosen so that $\varphi(x) = 1 + Ax + Bx^2$ is orthogonal to both 1 and x on $[-1, 1]$ with weight function $p(x) = x$?
9. *Gram-Schmidt orthogonalization process*   Let $p_0(x), p_1(x), \ldots, p_m(x)$ be functions which are continuous and

linearly independent on an interval $[a, b]$. We will produce from these functions a new set of $m + 1$ functions $q_0(x), \ldots, q_m(x)$ which are orthogonal to each other on $[a, b]$ with weight function $p(x) = 1$ and which also have the property that each $\int_a^b [q_j(x)]^2 \, dx = 1$. First, let $k_0 = \int_a^b [p_0(x)]^2 \, dx$ and define $q_0(x) = (1/\sqrt{k_0})p_0(x)$. Now let

$$Q_1(x) = p_1(x) - \alpha q_0(x).$$

Determine how to choose α so that $Q_1(x)$ is orthogonal to $q_0(x)$. With this choice of α, let

$$q_1(x) = \frac{1}{\int_a^b [Q_1(x)]^2 \, dx} Q_1(x).$$

Now let $Q_2(x) = p_2(x) - \beta q_1(x) - \gamma q_0(x)$, and determine β and γ so that $Q_2(x)$ is orthogonal to both $q_0(x)$ and $q_1(x)$. Then let

$$q_2(x) = \frac{1}{\int_a^b [Q_2(x)]^2 \, dx} Q_2(x).$$

Continuing in this way, produce a general formula for determining $q_k(x)$, assuming that $q_0, q_1, \ldots, q_{k-1}$ have already been found. This method of producing an orthogonal set of functions from a given linearly independent set is called the *Gram-Schmidt orthogonalization process*.

10. Use the Gram-Schmidt process to produce a set of orthogonal polynomials (with weight function 1) from the polynomials 1, x, and x^2 on the interval $[0, 1]$.
11. Use the Gram-Schmidt process to produce a set of orthogonal polynomials (with weight function 1) from the polynomials 1, x, x^2, and x^3 on the interval $[-1, 1]$.
12. Use the Gram-Schmidt process to produce a set of orthogonal polynomials (with weight function 1) from the polynomials 2, $1 - x$, and $x + x^2$ on the interval $[0, 1]$.

7.7 Sturm-Liouville Theory and Boundary Value Problems

In this section, we will study a class of problems called Sturm-Liouville problems. A typical Sturm-Liouville problem involves a differential equation defined on an interval together with conditions the solution and/or its derivative is to satisfy at the endpoints of the interval. As we will see, the differential equation will have a function p as one of its coefficients, and functions $f_0, f_1, \ldots$, orthogonal with weight function p, will arise as solutions of the differential equation.

To begin, consider the differential equation

$$y'' + R(x)y' + [Q(x) + \lambda P(x)]y = 0. \tag{7.38}$$

We seek values of the constant λ for which there exist nontrivial solutions on a given interval $[a, b]$. We will also want these solutions and their first derivatives to satisfy certain conditions, called *boundary conditions*, at a and b. The differential equation (7.38) and the boundary conditions constitute a *boundary value problem*.

Specifying conditions on a solution and its derivative at the ends of an interval (boundary value problem) is quite different from specifying the value of a solution and its derivative at a given point (initial value problem). Boundary value problems usually do not have unique solutions, and it is exactly this lack of uniqueness which makes certain boundary value problems important in solving partial differential equations of physics and engineering.

A discussion of equation (7.38) is more easily carried out if we write it in a different form. Multiply equation (7.38) by $e^{\int R(x)\,dx}$ to get

$$y'' e^{\int R(x)\,dx} + R(x)e^{\int R(x)\,dx} y' + [Q(x) + \lambda P(x)]y e^{\int R(x)\,dx} = 0. \tag{7.39}$$

Since $e^{\int R(x)\,dx} > 0$, this differential equation has the same solutions as equation (7.38). Now recognize that

$$y'' e^{\int R(x)\,dx} + R(x)e^{\int R(x)\,dx} y' = [y' e^{\int R(x)\,dx}]'.$$

Let

$$r(x) = e^{\int R(x)\,dx}, \qquad q(x) = Q(x)e^{\int R(x)\,dx}, \quad \text{and} \quad p(x) = P(x)e^{\int R(x)\,dx}.$$

Now equation (7.39) can be written

$$[ry']' + (q + \lambda p)y = 0.$$

This differential equation is called the *Sturm-Liouville differential equation* and is also the *Sturm-Liouville form* of equation (7.38).

EXAMPLE 7.5

Consider the differential equation

$$y'' + \frac{2}{x}\,y' + (x^2 - \lambda x)y = 0$$

on any interval not containing zero. This is in the form of equation (7.38) with $R(x) = 2/x$, $Q(x) = x^2$, and $P(x) = -x$.

To obtain the Sturm-Liouville form of this equation, multiply it by $e^{\int(2/x)\,dx}$. Compute

$$e^{\int(2/x)\,dx} = e^{2\ln(x)} = e^{\ln(x^2)} = x^2.$$

We get

$$x^2 y'' + 2xy' + (x^4 - \lambda x^3)y = 0.$$

Write this as

$$[x^2 y']' + (x^4 - \lambda x^3)y = 0,$$

the Sturm-Liouville form of the differential equation we began with. This has the same solutions as the original equation on any interval not containing zero. ∎

We will consider three kinds of problems involving the Sturm-Liouville differential equation. Assume that p and q are continuous on $[a, b]$.

Case 1: The Regular Sturm-Liouville Problem on [a, b]   Assume that $r(x) > 0$ and $p(x) > 0$ for $a \le x \le b$. We seek numbers λ and nontrivial solutions of

$$[ry']' + (q + \lambda p)y = 0$$

satisfying boundary conditions of the form

$$A_1 y(a) + A_2 y'(a) = 0, \qquad B_1 y(b) + B_2 y'(b) = 0,$$

in which A_1 and A_2 are given and not both zero and B_1 and B_2 are given and not both zero.

Case 2: The Periodic Sturm-Liouville Problem on [a, b] Suppose that $r(x) > 0$ and $p(x) > 0$ for $a \leq x \leq b$. We seek numbers λ and nontrivial solutions of

$$[ry']' + (q + \lambda p)y = 0$$

satisfying boundary conditions

$$y(a) = y(b), \qquad y'(a) = y'(b).$$

Case 3: The Singular Sturm-Liouville Problem on [a, b] Suppose that $r(x) > 0$ and $p(x) > 0$ for $a < x < b$. We seek numbers λ and nontrivial solutions of

$$[ry']' + (q + \lambda p)y = 0$$

satisfying *one* of the following three types of boundary conditions:

a. If $r(a) = 0$, then

$$B_1 y(b) + B_2 y'(b) = 0,$$

with B_1 and B_2 given and not both zero; solutions must be bounded at a.

b. If $r(b) = 0$, then

$$A_1 y(a) + A_2 y'(a) = 0,$$

with A_1 and A_2 given and not both zero; solutions must be bounded at b.

c. If $r(a) = r(b) = 0$, we have no boundary conditions specified at a or b but require that solutions be bounded on $[a, b]$.

In each of these problems, a number λ for which a nontrivial solution exists is called an *eigenvalue* of the problem; the corresponding nontrivial solution is called an *eigenfunction associated with the eigenvalue λ*. By definition, the zero function can *never* be an eigenfunction (although zero can be an eigenvalue).

In many Sturm-Liouville problems arising in physics and engineering, the eigenvalues have a physical significance related to the problem. For example, in dealing with wave propagation, eigenvalues may represent fundamental modes of vibration.

EXAMPLE 7.6 A Regular Sturm-Liouville Problem

Consider the boundary value problem

$$y'' + \lambda y = 0; \qquad y(0) = y\left(\frac{\pi}{2}\right) = 0.$$

The interval is $[0, \pi/2]$ and, in the previous notation,

$$r(x) = p(x) = 1 \quad \text{and} \quad q(x) = 0.$$

This is a regular Sturm-Liouville problem on $[0, \pi/2]$, with $A_1 = B_1 = 1$ and $A_2 = B_2 = 0$ in the definition. We will find the eigenvalues and eigenfunctions for this problem. The characteristic equation of $y'' + \lambda y = 0$ is

$$r^2 + \lambda = 0,$$

with roots $r = \pm\sqrt{-\lambda}$. This leads us to consider three cases.

Case 1: $\lambda = 0$ Now the differential equation is $y'' = 0$, with general solution

$$y = ax + b.$$

Then

$$y(0) = b = 0$$

and

$$y\left(\frac{\pi}{2}\right) = \frac{a\pi}{2} = 0 \quad \text{implies} \quad a = 0.$$

The only solution in the case in which $\lambda = 0$ is the trivial solution. Thus, zero is not an eigenvalue of this problem (because there are no eigenfunctions, which must be nontrivial solutions, associated with $\lambda = 0$).

Case 2: $\lambda > 0$ Write $\lambda = k^2$, with $k > 0$. The general solution of $y'' + k^2 y = 0$ is

$$y(x) = a \cos(kx) + b \sin(kx).$$

Now

$$y(0) = a = 0,$$

so $y(x) = b \sin(kx)$.

Next, we require that

$$y\left(\frac{\pi}{2}\right) = b \sin\left(\frac{k\pi}{2}\right) = 0.$$

If we let $b = 0$, we have the trivial solution. If $b \neq 0$, we can satisfy $y(\pi/2) = 0$ by choosing k so that $\sin(k\pi/2) = 0$. Thus, $k\pi/2$ must be a positive integer multiple of π, or

$$k = 2n \quad \text{for} \quad n = 1, 2, 3, \ldots.$$

The numbers

$$\lambda = k^2 = 4n^2$$

are eigenvalues of this problem for $n = 1, 2, 3, \ldots$. Corresponding to the eigenvalue $2n$ is an eigenfunction $\varphi_n(x) = b_n \sin(2nx)$, with b_n any *nonzero* constant.

Case 3: $\lambda < 0$ Write $\lambda = -k^2$, with $k > 0$. The general solution of $y'' - k^2 y = 0$ is

$$y(x) = ae^{kx} + be^{-kx}.$$

Now

$$y(0) = 0 = a + b,$$

so $b = -a$, and $y(x) = a(e^{kx} - e^{-kx}) = 2a \sinh(kx)$. But then

$$y\left(\frac{\pi}{2}\right) = 2a \sinh\left(\frac{k\pi}{2}\right) = 0$$

requires that $a = 0$ because $k > 0$ implies that $\sinh(k\pi/2) > 0$. Thus, the case $\lambda < 0$ leads to only the trivial solution, and this problem has no negative eigenvalues.

In summary, the regular Sturm-Liouville problem

$$y'' + \lambda y = 0; \qquad y(0) = y\left(\frac{\pi}{2}\right) = 0$$

has eigenvalues $\lambda = 4n^2$, $n = 1, 2, 3, \ldots$, and, corresponding to the eigenvalue $4n^2$, eigenfunctions $\varphi_n(x) = b_n \sin(2nx)$, with b_n nonzero but otherwise arbitrary. ∎

EXAMPLE 7.7 A Periodic Sturm-Liouville Problem

Consider the periodic Sturm-Liouville problem on $[-\pi, \pi]$

$$y'' + \lambda y = 0; \qquad y(-\pi) = y(\pi), \qquad y'(-\pi) = y'(\pi).$$

As in Example 7.6, we have the following cases on λ.

Case 1: $\lambda = 0$ Then $y = ax + b$. Now,

$$y(-\pi) = -a\pi + b = y(\pi) = a\pi + b$$

implies that $a = 0$, so $y = b$. This constant function satisfies both boundary conditions, since

$$y'(-\pi) = y'(\pi) = 0$$

in this case. Thus, zero is an eigenvalue of this problem, with corresponding eigenfunctions $y = b$, for any nonzero constant.

Case 2: $\lambda > 0$ Write $\lambda = k^2$, with $k > 0$, to obtain the general solution

$$y = a \cos(kx) + b \sin(kx)$$

of the differential equation. Now, $y(-\pi) = y(\pi)$ gives us

$$a \cos(k\pi) - b \sin(k\pi) = a \cos(k\pi) + b \sin(k\pi).$$

Then

$$2b \sin(k\pi) = 0. \tag{7.40}$$

The condition $y'(-\pi) = y'(\pi)$ gives us

$$ak \sin(k\pi) + bk \cos(k\pi) = -ak \sin(k\pi) + bk \cos(k\pi),$$

or

$$2ak \sin(k\pi) = 0. \tag{7.41}$$

Since $k > 0$, equations (7.40) and (7.41) tell us that

$$b \sin(k\pi) = a \sin(k\pi) = 0.$$

If $\sin(k\pi) \neq 0$, then $a = b = 0$, and we have the trivial solution. If $\sin(k\pi) = 0$, then k must be a positive integer, and a and b can be nonzero. Therefore, $\lambda = n^2$ is an eigenvalue for $n = 1, 2, 3, \ldots$, and corresponding eigenfunctions are

$$\varphi_n(x) = a_n\cos(nx) + b_n\sin(nx),$$

in which a_n and b_n are constants which cannot both be zero but are otherwise arbitrary.

Case 3: $\lambda < 0$ Write $\lambda = -k^2$, with $k > 0$, to get the general solution of the differential equation

$$y(x) = ae^{kx} + be^{-kx}.$$

Now, $y(-\pi) = y(\pi)$ implies that

$$ae^{-k\pi} + be^{k\pi} = ae^{k\pi} + be^{-k\pi}. \tag{7.42}$$

And $y'(-\pi) = y'(\pi)$ gives us

$$kae^{-k\pi} - kbe^{k\pi} = kae^{k\pi} - kbe^{-k\pi}. \tag{7.43}$$

We can write equation (7.42) as

$$a(e^{k\pi} - e^{-k\pi}) = b(e^{k\pi} - e^{-k\pi})$$

and conclude that $a = b$. From this, equation (7.43) becomes

$$2kae^{-k\pi} = 2kae^{k\pi}.$$

Since $k > 0$, this implies that $ae^{-k\pi} = ae^{k\pi}$, which can hold only if $a = 0$. But then $b = 0$ also, and we have only the trivial solution. Therefore, this periodic Sturm-Liouville problem has no negative eigenvalues.

   In summary, the periodic Sturm-Liouville problem

$$y'' + \lambda y = 0; \qquad y(-\pi) = y(\pi), \qquad y'(-\pi) = y'(\pi)$$

has eigenvalue $\lambda = 0$ with eigenfunctions $\varphi_0(x) = b \neq 0$ and, for $n = 1, 2, 3, \ldots$, eigenvalues $\lambda = n^2$ and eigenfunctions $\varphi_n(x) = a_n\cos(nx) + b_n\sin(nx)$, with not both a_n and b_n zero.   ■

EXAMPLE 7.8   A Singular Sturm-Liouville Problem

Consider Legendre's differential equation, which can be written in Sturm-Liouville form as

$$[(1 - x^2)y']' + \lambda y = 0$$

for $-1 \leq x \leq 1$. Now, $r(x) = 1 - x^2$, and $r(-1) = r(1) = 0$. This puts us in case (c) of the singular Sturm-Liouville problem.

   In Section 7.4, we showed that, if $\lambda = n(n + 1)$, with $n = 0, 1, 2, 3, \ldots$, there is a Legendre polynomial solution $P_n(x)$. This problem therefore has eigenvalues $n(n + 1)$ for n any nonnegative integer and corresponding eigenfunctions $P_n(x)$, or any nonzero constant multiples of these polynomials. It can be shown that these are the only solutions of Legendre's equation which are bounded on $[-1, 1]$ and that the only eigenvalues of this problem are the numbers $\lambda = n(n + 1)$ for $n = 0, 1, 2, \ldots$.   ■

Thus far, we have discussed the type of problem we want to solve and looked at some examples. We will now examine the fundamental theorem on Sturm-Liouville problems.

THEOREM 7.10   Sturm-Liouville Theorem

1. For the regular and periodic Sturm-Liouville problems, there exists an infinite number of eigenvalues. Further, the eigenvalues can be labeled $\lambda_1, \lambda_2, \ldots$ so that

$$\lambda_n < \lambda_m \quad \text{if} \quad n < m$$

and

$$\lim_{n \to \infty} \lambda_n = \infty.$$

2. If λ_n and λ_m are distinct eigenvalues of any of the three types of Sturm-Liouville problems, with corresponding eigenfunctions φ_n and φ_m, then φ_n and φ_m are orthogonal on $[a, b]$ with weight function p. That is,

$$\int_a^b p(x)\varphi_n(x)\varphi_m(x)\, dx = 0 \quad \text{if} \quad n \neq m.$$

3. For all three Sturm-Liouville problems, all eigenvalues are real.

4. For a regular Sturm-Liouville problem, any two eigenfunctions corresponding to a given eigenvalue are linearly dependent.   ∎

Conclusion (1) of the theorem assures us of the existence of nontrivial solutions of regular and periodic Sturm-Liouville problems, since with each eigenvalue there is associated a nontrivial eigenfunction. Conclusion (1) also implies that the eigenvalues are "spread out" and do not accumulate about a finite point (as, for example, the numbers $1/n$ accumulate about zero). The weight function p occurring in conclusion (2) is the coefficient of λ in the Sturm-Liouville differential equation.

Notice how conclusions (1) and (2) are borne out in the examples we have seen. In Example 7.6, the eigenvalues were $\lambda = 4n^2$ for $n = 1, 2, 3, \ldots$, with eigenfunctions $\varphi_n(x) = b_n \sin(2nx)$. If we let $\lambda_n = 4n^2$, certainly $\lim_{n \to \infty} \lambda_n = \infty$. Further, the differential equation in that example was $y'' + \lambda y = 0$, so $p(x) = 1$, and we have

$$\int_0^{\pi/2} p(x)\varphi_n(x)\varphi_m(x)\, dx = \int_0^{\pi/2} \sin(2nx)\sin(2mx)\, dx$$

$$= \frac{\sin[(2n - 2m)x]}{2(n - m)} - \frac{\sin[(2n + 3m)x]}{2(n + m)} \Bigg]_0^{\pi/2} = 0$$

if $n \neq m$.

As another example, the Legendre polynomials are orthogonal on $[-1, 1]$ with weight function 1. These polynomials are eigenfunctions of a singular Sturm-Liouville problem.

Conclusion (4) states that any two eigenfunctions corresponding to the same eigenvalue are constant multiples of each other, in the case of a regular Sturm-Liouville problem. This is not necessarily true for nonregular Sturm-Liouville problems. In Example 7.7, $\cos(x)$ and $\sin(x)$ are linearly independent eigenfunctions associated with the same eigenvalue, 1.

In the examples thus far, we have always been able to find the eigenvalues explicitly. Often the best we can do is write an equation for the eigenvalues and approximate their numerical values. The following example illustrates this fact.

EXAMPLE 7.9

Consider the regular Sturm-Liouville problem on $[0, 1]$

$$y'' + \lambda y = 0; \qquad y(0) = 0, \qquad 3y(1) + y'(1) = 0.$$

Take cases on λ.

Case 1: $\lambda = 0$ Then $y = ax + b$. Since $y(0) = 0$, $b = 0$. Then

$$3y(1) + y'(1) = 3a + a = 4a = 0.$$

Then $a = 0$, so we obtain only the trivial solution in this case. Therefore, zero is not an eigenvalue.

Case 2: $\lambda < 0$ Write $\lambda = -\alpha^2$, with $\alpha > 0$. The general solution of the differential equation is

$$y(x) = ae^{\alpha x} + be^{-\alpha x}.$$

Since $y(0) = a + b = 0$, $b = -a$, and

$$y(x) = 2a \sinh(\alpha x).$$

Next,

$$3y(1) + y'(1) = 6a \sinh(\alpha) + 2a\alpha \cosh(\alpha) = 0.$$

If $a \neq 0$, this equation implies that

$$\tanh(\alpha) = -\tfrac{1}{3}\alpha.$$

This is impossible because, for $\alpha > 0$, $\tanh(\alpha) > 0$ and $-\tfrac{1}{3}\alpha < 0$. Therefore, there are no nontrivial solutions with $\lambda < 0$, and this problem has no negative eigenvalues.

Case 3: $\lambda > 0$ Write $\lambda = \alpha^2$, with $\alpha > 0$. The general solution of $y'' + \alpha^2 y = 0$ is

$$y(x) = a \cos(\alpha x) + b \sin(\alpha x).$$

Since $y(0) = a = 0$, $y(x) = b \sin(\alpha x)$. Further,

$$3y(1) + y'(1) = 3b \sin(\alpha) + \alpha b \cos(\alpha) = 0.$$

Assuming that $b \neq 0$, this implies that

$$\tan(\alpha) = -\tfrac{1}{3}\alpha.$$

This equation cannot be solved algebraically. However, Figure 7.8 suggests that the graphs of $y = \tan(\alpha)$ and $y = -\alpha/3$ have infinitely many points of intersection in the half-plane $\alpha > 0$. The first coordinates of these points of intersection are eigenvalues of this problem. If we use Newton's method, we find that the first five positive solutions of $\tan(\alpha) = -\alpha/3$ are approximately 2.4556, 5.2239, 8.2045, 11.2560, and 14.3434. The first five eigenvalues are approximately the squares of these numbers, giving us

$$\lambda_1 \approx 6.0300, \qquad \lambda_2 \approx 27.3822, \qquad \lambda_3 \approx 67.3138,$$
$$\lambda_4 \approx 126.6975, \quad \text{and} \quad \lambda_5 \approx 205.7331.$$

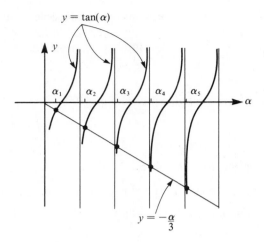

Figure 7.8

For each positive α satisfying $\tan(\alpha) = -\alpha/3$, we get an eigenvalue $\lambda = \alpha^2$ and a corresponding eigenfunction $\varphi(x) = b\,\sin(\alpha x)$, in which b is any nonzero constant. From conclusion (2) of Theorem 7.10, if λ_1 and λ_2 are distinct eigenvalues,

$$\int_0^1 \sin(\sqrt{\lambda_1}\,x)\sin(\sqrt{\lambda_2}\,x)\,dx = 0.$$

This fact can be demonstrated by integration, but is not entirely straightforward because of the equation determining the eigenvalues. ∎

We will conclude this section with proofs of portions of the Sturm-Liouville theorem. Conclusion (1) is an existence result which we will not attempt to prove.

Proof of Conclusion (2) of Theorem 7.10   From the differential equation, we have

$$[r\varphi_n']' + (q + \lambda_n p)\varphi_n = 0$$
$$[r\varphi_m']' + (q + \lambda_m p)\varphi_m = 0.$$

Multiply the first equation by φ_m and the second by φ_n and subtract to get

$$[r\varphi_n']'\varphi_m - [r\varphi_m']'\varphi_n + (\lambda_n - \lambda_m)p\varphi_n\varphi_m = 0.$$

This equation can be written

$$[r(\varphi_m\varphi_n' - \varphi_n\varphi_m')]' = (\lambda_m - \lambda_n)p\varphi_n\varphi_m.$$

Integrate both sides of this equation from a to b to get

$$\left[r(\varphi_m\varphi_n' - \varphi_n\varphi_m')\right]_a^b = (\lambda_m - \lambda_n)\int_a^b p(x)\varphi_n(x)\varphi_m(x)\,dx.$$

Suppose that $\lambda_n \neq \lambda_m$. The last equation will imply conclusion (2) if we can show that the left side is zero. To do this, we must make individual use of each type of boundary condition, giving rise to three cases.

In the regular Sturm-Liouville problem, the boundary conditions are

$$A_1 y(a) + A_2 y'(a) = 0$$
$$B_1 y(b) + B_2 y'(b) = 0,$$

in which A_1 and A_2 cannot both be zero and B_1 and B_2 are not both zero. These boundary conditions must be satisfied by both φ_n and φ_m. First, apply just the condition $A_1 y(a) + A_2 y'(a) = 0$ to both eigenfunctions. We get

$$A_1 \varphi_n(a) + A_2 \varphi'_n(a) = 0$$
$$A_1 \varphi_m(a) + A_2 \varphi'_m(a) = 0.$$

Since A_1 and A_2 are not both zero, the system of equations

$$\varphi_n(a)X + \varphi'_n(a)Y = 0$$
$$\varphi_m(a)X + \varphi'_m(a)Y = 0$$

has a nontrivial solution (namely, $X = A_1$, $Y = A_2$). Therefore,

$$\begin{vmatrix} \varphi_n(a) & \varphi'_n(a) \\ \varphi_m(a) & \varphi'_m(a) \end{vmatrix} = \varphi_n(a)\varphi'_m(a) - \varphi_m(a)\varphi'_n(a) = 0. \qquad (7.44)$$

Now apply the same reasoning, using the condition $B_1 y(b) + B_2 y'(b) = 0$, to get

$$\varphi_n(b)\varphi'_m(b) - \varphi_m(b)\varphi'_n(b) = 0. \qquad (7.45)$$

Equations (7.44) and (7.45) give us

$$r(x)[\varphi_m(x)\varphi'_n(x) - \varphi_m(x)\varphi'_n(x)]\Big]_a^b = 0,$$

and this is what we need to conclude that $\int_a^b p(x)\varphi_n(x)\varphi_m(x)\, dx = 0$.

For periodic and singular Sturm-Liouville problems, the reasoning is similar, utilizing the boundary conditions to show that

$$r(x)[\varphi_m(x)\varphi'_n(x) - \varphi_m(x)\varphi'_n(x)]\Big]_a^b = 0. \quad \blacksquare$$

Proof of Conclusion (3) of Theorem 7.10 Suppose that a Sturm-Liouville problem has a complex eigenvalue

$$\lambda = \alpha + i\beta.$$

Let φ be a corresponding eigenfunction, and write

$$\varphi(x) = u(x) + iv(x).$$

By substitution into the Sturm-Liouville problem, we find that $\bar{\lambda} = \alpha - i\beta$ is also an eigenvalue, with corresponding eigenfunction $\bar{\varphi}(x) = u(x) - iv(x)$. If $\beta \neq 0$, then λ and $\bar{\lambda}$ are distinct eigenvalues, and, by conclusion (2),

$$\int_a^b p(x)\varphi(x)\bar{\varphi}(x)\, dx = 0.$$

But

$$\varphi(x)\bar{\varphi}(x) = [u(x) + iv(x)][u(x) - iv(x)] = u(x)^2 + v(x)^2.$$

Therefore,

$$\int_a^b p(x)[u(x)^2 + v(x)^2]\, dx = 0.$$

Now, $p(x) > 0$ on (a, b), by assumption. Further, since φ is nontrivial, $u(x)^2 + v(x)^2$ cannot be identically zero on $[a, b]$. Then $p(x)[u(x)^2 + v(x)^2]$ is positive on some subinterval of $[a, b]$; hence, $\int_a^b p(x)[u(x)^2 + v(x)^2]\, dx > 0$, a contradiction. We conclude that $\beta = 0$ and that each eigenvalue must be real. ∎

Proof of Conclusion (4) of Theorem 7.10   Suppose that λ is an eigenvalue of the regular Sturm-Liouville problem

$$[ry']' + (q + \lambda p)y = 0; \qquad A_1 y(a) + A_2 y'(a) = 0, \qquad B_1 y(b) + B_2 y'(b) = 0$$

on $[a, b]$. Let φ and ψ be eigenfunctions associated with λ. We want to show that φ and ψ are linearly dependent. From the boundary condition at a, we known that

$$A_1 \varphi(a) + A_2 \varphi'(a) = 0$$
$$A_1 \psi(a) + A_2 \psi'(a) = 0,$$

with A_1 and A_2 not both zero. Therefore, the system of equations

$$\varphi(a)X + \varphi'(a)Y = 0$$
$$\psi(a)X + \psi'(a)Y = 0$$

has a nontrivial solution. Therefore,

$$\begin{vmatrix} \varphi(a) & \varphi'(a) \\ \psi(a) & \psi'(a) \end{vmatrix} = 0.$$

But this is $W[\varphi, \psi](a)$. Since φ and ψ satisfy a linear homogeneous second order differential equation, φ and ψ are linearly dependent on $[a, b]$. ∎

In the next section, we will relate Sturm-Liouville theory to properties of Legendre polynomials and Bessel functions.

PROBLEMS FOR SECTION 7.7

In each of Problems 1 through 17, classify the Sturm-Liouville problem as regular, periodic, or singular, specify the relevant interval, and find the eigenvalues (or an equation satisfied by the eigenvalues). Corresponding to each eigenvalue, determine an eigenfunction. If eigenvalues cannot be determined explicitly, give a graphical argument for their existence, as in Example 7.9.

1. $y'' + \lambda y = 0;$ $y(0) = y(L) = 0$ (L is any positive number)
2. $y'' + \lambda y = 0;$ $y(0) = y'(\pi) = 0$
3. $y'' + \lambda y = 0;$ $y'(0) = y(4) = 0$
4. $y'' + \lambda y = 0;$ $y(0) = y(\pi),$ $y'(0) = y'(\pi)$
5. $y'' + \lambda y = 0;$ $y(-3\pi) = y(3\pi),$ $y'(-3\pi) = y'(3\pi)$
6. $y'' + \lambda y = 0;$ $y(0) = 0,$ $y(\pi) + 2y'(\pi) = 0$
7. $y'' + \lambda y = 0;$ $y(0) - 2y'(0) = 0,$ $y'(1) = 0$
8. $y'' + 2y' + (\lambda + 1)y = 0;$ $y(0) = y(1) = 0$
9. $y'' - 12y' + 4(\lambda + 7)y = 0;$ $y(0) = y(5) = 0$
10. $[xy']' + \lambda x^{-1}y = 0;$ $y(1) = y(e) = 0$
11. $[xy']' + \lambda x^{-1}y = 0;$ $y(1) = y'(e) = 0$
12. $[x^{-3}y']' + (\lambda + 4)x^{-5}y = 0;$ $y(1) = y(e^2) = 0$
13. $[e^{2x}y']' + \lambda e^{2x}y = 0;$ $y(0) = y(\pi) = 0$
14. $[e^{-6x}y']' + (\lambda + 1)e^{-6x}y = 0;$ $y(0) = y(8) = 0$
15. $[x^3y']' + \lambda xy = 0;$ $y(1) = y(e^3) = 0$

16. $[4xy']' + \lambda xy = 0$; $y(1) = 0$, and the solutions are to be bounded on $(0, 1]$. *Hint:* Perform the indicated differentiation, multiply by x, and transform the resulting equation into a differential equation we have seen before by letting $t = \sqrt{\lambda}\,x/2$. A table may be useful in finding the first few eigenvalues.

17. $[x^{-1}y']' + (\lambda + 4)x^{-3}y = 0$; $y(1) = y(e^4) = 0$

18. Write Hermite's differential equation (see Problem 18 of Section 7.5) in Sturm-Liouville form. Use the Sturm-Liouville theorem to give another proof that the Hermite polynomials are orthogonal on $(-\infty, \infty)$ with weight function $e^{-x^2/2}$ (note Problem 22 of Section 7.5).

19. Write Laguerre's differential equation in Sturm-Liouville form (see Problem 23 of Section 7.5). The relevant interval is $[0, \infty)$. Use the Sturm-Liouville theorem to prove that the Laguerre polynomials are orthogonal on $[0, \infty)$ with weight function e^{-x}.

7.8 *Legendre Polynomials and Bessel Functions in Sturm-Liouville Theory*

In this section, we will discuss how Legendre polynomials and Bessel functions fit into the context of Sturm-Liouville theory.

LEGENDRE POLYNOMIALS AND STURM-LIOUVILLE THEORY

Legendre's differential equation is

$$(1 - x^2)y'' - 2xy' + \lambda y = 0.$$

The Sturm-Liouville form of Legendre's equation is

$$[(1 - x^2)y']' + \lambda y = 0.$$

In the notation of the preceding section, we have $r(x) = 1 - x^2$, $q(x) = 0$, and $p(x) = 1$, and the interval is $[-1, 1]$. Since $r(-1) = r(1) = 0$, we are in case (3) of the boundary conditions for a singular Sturm-Liouville problem. From results of Section 7.4, if n is any nonnegative integer, $\lambda = n(n + 1)$ is an eigenvalue of this problem, with associated eigenfunction $P_n(x)$, the nth Legendre polynomial. It can be shown that

$$P_n(x) = \sum_{k=0}^{[n/2]} \frac{(-1)^k (2n - 2k)!}{2^n k!(n - k)!} x^{n - 2k},$$

in which $[n/2]$ is the greatest integer less than or equal to $n/2$.

We have not shown that the numbers $n(n + 1)$, for $n = 0, 1, 2, 3, \ldots$, are the *only* eigenvalues of this Sturm-Liouville problem. This can be shown to be true if we require that the eigenfunctions be bounded on $[-1, 1]$. This is a natural requirement in many applications of Legendre's differential equation.

Since $p(x) = 1$, the Sturm-Liouville theorem tells us that the Legendre polynomials are orthogonal on $[-1, 1]$ with weight function 1. That is,

$$\int_{-1}^{1} P_n(x)P_m(x)\,dx = 0 \quad \text{if} \quad n \neq m,$$

a fact we have already established in Section 7.4.

Now we will return to the idea of series expansions we discussed in Section 7.6. We have seen that any polynomial can be written as a finite series in terms of Legendre polynomials. More generally, suppose g is any continuous function defined on $[-1, 1]$ and we want to expand g in a series of Legendre polynomials:

$$g(x) = \sum_{n=0}^{\infty} a_n P_n(x). \tag{7.46}$$

Such a series is called a *Fourier-Legendre series* for g (or expansion of g). From equation (7.37) in Section 7.6, we should choose

$$a_n = \frac{\displaystyle\int_{-1}^{1} g(x) P_n(x)\, dx}{\displaystyle\int_{-1}^{1} [P_n(x)]^2\, dx}. \tag{7.47}$$

By Theorem 7.9 in Section 7.5,

$$\int_{-1}^{1} [P_n(x)]^2\, dx = \frac{2}{2n+1} \quad \text{for} \quad n = 0, 1, 2, 3, \ldots.$$

Therefore, the formula (7.47) can be written more simply as

$$a_n = \frac{2n+1}{2} \int_{-1}^{1} g(x) P_n(x)\, dx. \tag{7.48}$$

Here is an example of a Fourier-Legendre expansion of a specific function in which we evaluate some of the coefficients by direct integration, using equation (7.48).

EXAMPLE 7.10

Find the Fourier-Legendre series for $\cos(\pi x/2)$ on $[-1, 1]$.

The series will have the form

$$\sum_{n=0}^{\infty} a_n P_n(x)\, dx,$$

where, for $n = 0, 1, 2, \ldots$,

$$a_n = \frac{2n+1}{2} \int_{-1}^{1} \cos\left(\frac{n\pi x}{2}\right) P_n(x)\, dx.$$

We will compute some of these coefficients. First, let $n = 0$. Since $P_0(x) = 1$,

$$a_0 = \frac{1}{2} \int_{-1}^{1} \cos\left(\frac{\pi x}{2}\right) dx = \frac{1}{\pi} \sin\left(\frac{\pi x}{2}\right)\Big]_{-1}^{1} = \frac{2}{\pi}.$$

Now let $n = 1$. Since $P_1(x) = x$, we have

$$a_1 = \frac{3}{2} \int_{-1}^{1} x \cos\left(\frac{\pi x}{2}\right) dx = 0.$$

Next, $P_2(x) = \frac{1}{2}(3x^2 - 1)$, so

$$a_2 = \frac{5}{2} \int_{-1}^{1} \cos\left(\frac{\pi x}{2}\right) \frac{1}{2}(3x^2 - 1)\, dx = \frac{10\pi^2 - 120}{\pi^3}.$$

In Problem 1, we outline an argument showing that $a_n = 0$ whenever n is odd. Thus far, we have

$$\sum_{n=0}^{\infty} a_n P_n(x) = a_0 P_0(x) + a_1 P_1(x) + a_2 P_2(x) + \cdots$$

$$= \frac{2}{\pi} + \left[\frac{10\pi^2 - 120}{\pi^3} \right] \frac{1}{2}(3x^2 - 1) + \cdots.$$

With some effort, we can compute $a_4, a_6, \ldots, a_{2k}$ for k as large as we like. It is interesting to note, however, that although we have explicitly computed only two nonzero terms of the Fourier-Legendre series for $\cos(\pi x/2)$, the sum of just these two terms is quite a good approximation of $\cos(\pi x/2)$. In fact, if we approximate

$$\frac{2}{\pi} \approx 0.6336 \quad \text{and} \quad \frac{10\pi^2 - 120}{\pi^3} \approx -0.6871,$$

the sum of the first four terms of the Fourier-Legendre series is

$$a_0 P_0(x) + a_1 P_1(x) + a_2 P_2(x) + a_3 P_3(x) \approx 0.9802 - 1.0306x^2.$$

Figure 7.9 shows the graphs of $0.9802 - 1.0306x^2$ and $\cos(\pi x/2)$. Notice that the graphs are almost indistinguishable for $-1 \le x \le 1$ (the interval on which the Legendre polynomials are solutions of the Sturm-Liouville problem). Of course, both this second degree polynomial and $\cos(\pi x/2)$ are defined for all x, but their values diverge as x is chosen further outside $[-1, 1]$. This can also be seen in Figure 7.9. In this example, the sum of just the first two nonzero terms of the Fourier-Legendre expansion of $\cos(\pi x/2)$ is a good approximation of $\cos(\pi x/2)$ on $[-1, 1]$. ∎

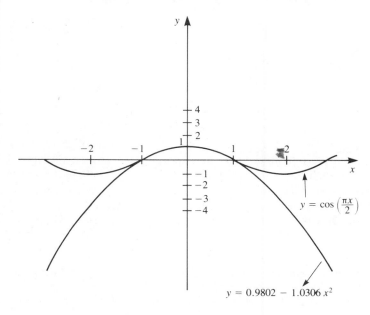

Figure 7.9

In general, given g, an explicit calculation of the coefficients using equation (7.48) can be difficult, and often we must employ recurrence relationships and various other facts about Legendre polynomials rather than a direct integration.

The following theorem gives sufficient conditions for the Fourier-Legendre series of a function g to converge to $g(x)$ for $-1 < x < 1$.

THEOREM 7.11   Convergence of Fourier-Legendre Series

If g is differentiable on $(-1, 1)$, then the Fourier-Legendre series for g converges to $g(x)$ for $-1 < x < 1$. In this event, we can write the equality

$$g(x) = \sum_{n=0}^{\infty} a_n P_n(x) \quad \text{for} \quad -1 < x < 1,$$

with

$$a_n = \frac{2}{2n + 1} \int_{-1}^{1} g(x) P_n(x) \, dx \quad \text{for} \quad n = 0, 1, 2, \dots . \quad \blacksquare$$

BESSEL FUNCTIONS AND STURM-LIOUVILLE THEORY

We will now look at Bessel functions from the perspective of Sturm-Liouville theory. Many applications (for example, a mathematical model of heat radiation from a cylindrical tank) involve the differential equation

$$x^2 y'' + x y' + (\lambda x^2 - n^2) y = 0. \tag{7.49}$$

In this equation, n is often a nonnegative integer.

The Sturm-Liouville form of this differential equation is

$$[x y']' + \left[\frac{-n^2}{x} + \lambda x \right] y = 0.$$

Typically, we want to solve this equation on an interval $(0, R]$ subject to a boundary condition

$$y(R) = 0. \tag{7.50}$$

(In the study of heat radiation from a cylindrical tank, R would be the radius of the tank and x would measure distance outward from the axis of the cylinder.) We also require that solutions remain bounded as x approaches zero from the right. (That is, the function describing heat conduction must remain bounded along the axis of the tank.)

The Sturm-Liouville differential equation (7.49) together with the boundary condition (7.50) constitute a singular Sturm-Liouville problem, since $r(x) = x$ for this equation and $r(0) = 0$. Suppose that $\lambda > 0$, and apply the transformation

$$t = \sqrt{\lambda} x$$

to the differential equation (7.49). We obtain

$$t^2 T'' + t T' + (t^2 - n^2) T = 0,$$

in which $T(t) = y(t/\sqrt{\lambda}) = y(x)$. This differential equation for T is a Bessel equation of order n. The general solution is

$$T(t) = c_1 J_n(t) + c_2 Y_n(t),$$

in which J_n and Y_n are Bessel functions of order n of the first and second kinds. Therefore,

$$y(x) = T(\sqrt{\lambda} x) = c_1 J_n(\sqrt{\lambda} x) + c_2 Y_n(\sqrt{\lambda} x).$$

Now recall that $Y_n(\sqrt{\lambda} x)$ contains a logarithm term and hence is unbounded as x approaches zero from the right, while $J_n(\sqrt{\lambda} x)$ remains bounded. In order to have a bounded solution, we must choose $c_2 = 0$, giving us

$$y(x) = c_1 J_n(\sqrt{\lambda} x).$$

Now we must satisfy the boundary condition

$$y(R) = 0.$$

This requires that

$$c_1 J_n(\sqrt{\lambda} R) = 0.$$

Further, we must have $c_1 \neq 0$, or else the solution is trivial. Therefore, we must choose the numbers λ (the eigenvalues of the Sturm-Liouville problem) so that

$$J_n(\sqrt{\lambda} R) = 0.$$

By the result of Problem 27 of Section 7.1, the equation $J_n(x) = 0$ has infinitely many positive solutions. If these numbers are denoted $z_1, z_2, \ldots$, with $z_k < z_{k+1}$, then for each positive integer k we must choose λ_k so that

$$\sqrt{\lambda_k} R = z_k,$$

or

$$\lambda_k = \frac{z_k^2}{R^2}.$$

With each eigenvalue λ_k, we have a corresponding eigenfunction

$$\varphi_k(x) = J_n(\sqrt{\lambda_k} x).$$

It is important to realize here that n is fixed in the differential equation, while k is the index, taking on all positive integer values.

Because of their wide use in applications, zeros of Bessel functions have been extensively tabulated. Table 7.3 in Section 7.2 lists the first nine positive zeros of J_0, J_1, J_2, J_3, and J_4. For example, the first five solutions of the equation $J_1(x) = 0$ are approximately 3.832, 7.016, 10.173, 13.323, and 16.470. For $n = 1$, then, the first five eigenvalues of this Sturm-Liouville problem are

$$\lambda_1 \approx \frac{14.684}{R^2}, \qquad \lambda_2 \approx \frac{49.224}{R^2}, \qquad \lambda_3 \approx \frac{103.490}{R^2},$$

$$\lambda_4 \approx \frac{177.502}{R^2}, \qquad \lambda_5 \approx \frac{271.261}{R^2}.$$

Now look again at the Sturm-Liouville differential equation (7.49), which is repeated for ease of reference

$$[xy']' + \left[\frac{-n^2}{x} + \lambda x\right]y = 0.$$

This is in the standard Sturm-Liouville form $[ry']' + (q + \lambda p)y = 0$, with $p(x) = x$. The Sturm-Liouville theorem gives us the following orthogonality relationship, with weight function $p(x) = x$, for the eigenfunctions of this problem:

$$\int_0^R xJ_n(\sqrt{\lambda_k}x)J_n(\sqrt{\lambda_j}x)\,dx = 0 \quad \text{if} \quad k \neq j.$$

(Again, remember that n is fixed; k and j are the indices.)

We can now look at the possibility of expanding a function g, defined on $(0, R)$, in a series

$$\sum_{k=1}^{\infty} a_k J_n(\sqrt{\lambda_k}x).$$

This series begins at $k = 1$ simply because we labeled the first eigenvalue λ_1 instead of λ_0. From equation (7.37) in Section 7.6, we should choose the coefficients as

$$a_k = \frac{\displaystyle\int_0^R xg(x)J_n(\sqrt{\lambda_k}x)\,dx}{\displaystyle\int_0^R x[J_n(\sqrt{\lambda_k}x)]^2\,dx}. \tag{7.51}$$

These numbers are called the *Fourier-Bessel coefficients* of g, and the series is called the *Fourier-Bessel* series of order n for g, or expansion of g. Wherever this series converges to $g(x)$, we can write

$$g(x) = \sum_{k=1}^{\infty} a_k J_n(\sqrt{\lambda_k}x).$$

This type of expansion is used in studying elasticity, heat radiation, and a variety of other phenomena.

We will state sufficient conditions for the Fourier-Bessel expansion of g to converge to $g(x)$ on $(0, R)$.

THEOREM 7.12　Convergence of Fourier-Bessel Series

If g is differentiable on $[0, R]$, then the Fourier-Bessel series for g converges to $g(x)$ on $(0, R)$. ∎

The integral $\int_0^R xJ_n(\sqrt{\lambda_k}x)^2\,dx$ appearing in the denominator of equation (7.51) can be evaluated in terms of a Bessel function of the first kind of order $n + 1$. We will state and prove this result and then discuss an example of a Fourier-Bessel expansion.

THEOREM 7.13

Let z_k be the kth positive solution of the equation of $J_n(x) = 0$, and let $\lambda_k = z_k^2/R^2$. Then

$$\int_0^R xJ_n(\sqrt{\lambda_k}x)^2\,dx = \tfrac{1}{2}R^2[J_{n+1}(\sqrt{\lambda_k}R)]^2.$$

Proof   When we proved a similar result for Legendre polynomials (Theorem 7.9 in Section 7.5), we employed a recurrence relationship. We will use a similar approach here. The two relationships we will use are

$$J_{n+1}(x) = \frac{2}{x} nJ_n(x) - J_{n-1}(x) \tag{7.52}$$

and

$$\int x[J_n(x)]^2 \, dx = \tfrac{1}{2}x^2\{[J_n(x)]^2 - J_{n-1}(x)J_{n+1}(x)\} + C. \tag{7.53}$$

(See the first formula of Problem 16 and Problem 18 of Section 7.1, respectively.)

To begin, use the change of variables $t = \sqrt{\lambda_k}x$, followed by equation (7.53) to get

$$\int_0^R xJ_n(\sqrt{\lambda_k}\,x)^2 \, dx = \frac{1}{\lambda_k} \int_0^{\sqrt{\lambda_k}R} t[J_n(t)]^2 \, dt$$

$$= \frac{t^2}{2\lambda_k} \{[J_n(t)]^2 - J_{n-1}(t)J_{n+1}(t)\}\big|_0^{\sqrt{\lambda_k}R}$$

$$= \frac{\lambda_k R^2}{2\lambda_k} \{[J_n(\sqrt{\lambda_k}R)]^2 - J_{n-1}(\sqrt{\lambda_k}R)J_{n+1}(\sqrt{\lambda_k}R)\}. \tag{7.54}$$

In the last line, $J_n(\sqrt{\lambda_k}R) = 0$ by choice of the numbers λ_k. Further, from equation (7.52), we have

$$J_{n+1}(\sqrt{\lambda_k}R) = \frac{2n}{\sqrt{\lambda_k}R} J_n(\sqrt{\lambda_k}R) - J_{n-1}(\sqrt{\lambda_k}R) = -J_{n-1}(\sqrt{\lambda_k}R).$$

Put these results into equation (7.54) to get

$$\int_0^R xJ_n(\sqrt{\lambda_k}\,x)^2 \, dx = \tfrac{1}{2}R^2[J_{n+1}(\sqrt{\lambda_k}R)]^2,$$

as we wanted to prove.   ■

We may therefore write the coefficient a_k in equation (7.51) for the Fourier-Bessel expansion of g as

$$a_k = \frac{2}{R^2} \frac{\displaystyle\int_0^R xg(x)J_n(\sqrt{\lambda_k}x) \, dx}{[J_{n+1}(\sqrt{\lambda_k}R)]^2}. \tag{7.55}$$

Fourier-Bessel expansions are difficult to compute for specific functions because of the difficulty in calculating $\int_0^R xg(x)J_n(\sqrt{\lambda_k}R) \, dx$. Generally, we attempt to employ relationships between various Bessel functions and their derivatives and tables of values to approximate the coefficients. Here is an example of this process.

EXAMPLE 7.11

Expand $g(x) = 4x - x^3$ in a Fourier-Bessel series of order 1:

$$4x - x^3 = \sum_{k=1}^{\infty} a_k J_1(\omega_k x)$$

for $0 < x < 2$. Here, $R = 2$. For convenience, we have written $\omega_k = \frac{1}{2}z_k$, with z_k the kth positive zero of $J_1(x)$. We will use the following relationships to calculate the coefficients:

$$\int_0^R x[J_n(\omega_k x)]^2 \, dx = \frac{1}{2}R^2[J_{n+1}(\omega_k R)]^2 \qquad (7.56)$$

$$\int t^{n+1}J_n(t) \, dt = t^{n+1}J_{n+1}(t) + C \qquad (7.57)$$

$$\frac{2n}{x} J_n(x) = J_{n+1}(x) + J_{n-1}(x). \qquad (7.58)$$

We obtain equation (7.57) by letting $v = n + 1$ in the second formula of Problem 13, Section 7.1; we get equation (7.58) from the first formula in Problem 16, Section 7.1; and equation (7.56) follows from Theorem 7.13. We need to compute

$$a_k = \frac{1}{2} \frac{\displaystyle\int_0^2 x(4x - x^3)J_1(\omega_k x) \, dx}{[J_2(2\omega_k)]^2}$$

$$= \frac{1}{2} \frac{4\displaystyle\int_0^2 x^2 J_1(\omega_k x) \, dx - \displaystyle\int_0^2 x^4 J_1(\omega_k x) \, dx}{[J_2(2\omega_k)]^2}.$$

We will deal individually with each of the integrals in the numerator. For the first integral, write

$$4\int_0^2 x^2 J_1(\omega_k x) \, dx = \frac{4}{\omega_k^3} \int_0^{2\omega_k} t^2 J_1(t) \, dt \qquad (\text{let } t = \omega_k x)$$

$$= \frac{4}{\omega_k^3} t^2 J_2(t) \Big]_0^{2\omega_k} = \frac{16}{\omega_k} J_2(2\omega_k) \qquad [\text{by equation (7.57)}].$$

For the second integral, first let $t = \omega_k x$ to get

$$\int_0^2 x^4 J_1(\omega_k x) \, dx = \frac{1}{\omega_k^5} \int_0^{2\omega_k} t^4 J_1(t) \, dt.$$

Now use integration by parts on the right side, with $u = t^2$, $dv = t^2 J_1(t)$, $du = 2t \, dt$, and $v = t^2 J_2(t)$ [by equation (7.57)], to get

$$\int_0^2 x^4 J_1(\omega_k x) \, dx = \frac{t^4 J_2(t)}{\omega_k^5} \Big]_0^{2\omega_k} - \frac{2}{\omega_k^5} \int_0^{2\omega_k} t^3 J_2(t) \, dt$$

$$= \frac{16 J_2(2\omega_k)}{\omega_k} - \frac{2t^3 J_3(t)}{\omega_k^5} \Big]_0^{2\omega_k} \qquad [\text{equation (7.57) again}]$$

$$= \frac{16 J_2(2\omega_k)}{\omega_k} - \frac{16 J_3(2\omega_k)}{\omega_k^2}.$$

Now use equation (7.58) with $n = 2$ and $x = 2\omega_k$ to get

$$\frac{4}{2\omega_k} J_2(2\omega_k) = J_3(2\omega_k) + J_1(2\omega_k) = J_3(2\omega_k),$$

since

$$J_1(2\omega_k) = J_1\left(\frac{2z_k}{2}\right) = J_1(z_k) = 0,$$

by choice of z_k. Putting these results together, we have

$$a_k = \frac{\dfrac{16}{\omega_k} J_2(2\omega_k) - \dfrac{16}{\omega_k} J_2(2\omega_k) + \dfrac{32}{\omega_k^3} J_2(2\omega_k)}{2[J_2(2\omega_k)]^2}$$

$$= \frac{16}{\omega_k^3 J_2(2\omega_k)} = \frac{128}{z_k^3 J_2(z_k)}. \tag{7.59}$$

There are tables giving $J_0(x)$ evaluated at zeros of J_1, so we have access to numerical values for $J_0(z_k)$. Therefore, our next objective in computing a_k is to express $J_2(z_k)$ in terms of $J_0(z_k)$. We will do this by using equation (7.58) once more. Let $n = 1$ and $x = z_k$ in equation (7.59) to get

$$J_2(z_k) + J_0(z_k) = \frac{2}{z_k} J_1(z_k) = 0.$$

Therefore,

$$J_2(z_k) = -J_0(z_k).$$

Put this result into equation (7.58) to obtain the much-simplified expression

$$a_k = \frac{-128}{z_k^3 J_0(z_k)}.$$

The first five zeros of J_1 and numerical values of $J_0(z_k)$ are tabulated as follows.

k	z_k	$J_0(z_k)$
1	3.8317	−0.4028
2	7.0156	0.3001
3	10.1735	−0.2497
4	13.3237	0.2184
5	16.4706	−0.1965

Using these values, we obtain

$$4x - x^3 = \sum_{k=1}^{\infty} a_k J_1(\omega_k x) = -128 \sum_{k=1}^{\infty} \frac{J_1\left(\dfrac{z_k x}{2}\right)}{z_k^3 J_0(z_k)}$$

$$\approx 5.6492 J_1(1.9159x) - 1.2352 J_1(3.5078x) + 0.4868 J_1(5.0867x)$$

$$- 0.2478 J_1(6.6618x) + 0.1458 J_1(8.2353x) - \cdots.$$

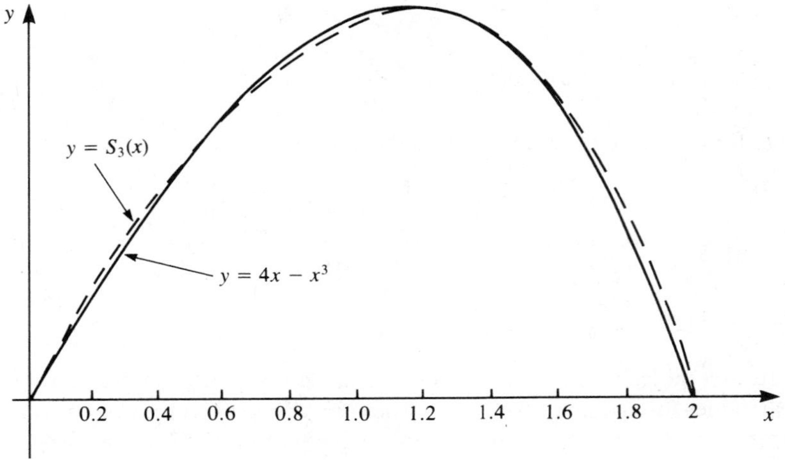

Figure 7.10

Figure 7.10 shows the graph of $g(x)$ compared with the graph of $S_3(x)$, the partial sum of the first three terms of this series. By choosing n larger, we can get $S_n(x)$ to approximate $g(x)$ on $(0, 2)$ better.

This example is fairly typical of Bessel function manipulations. We use recurrence relationships and other properties of Bessel functions to write the coefficients in terms of Bessel functions of order zero or 1, for which tables of zeros and values are available.

PROBLEMS FOR SECTION 7.8

1. Show that $\int_{-1}^{1} \cos(\pi x/2) P_n(x)\, dx = 0$ if n is an odd positive integer. *Hint:* $\cos(-\pi x/2) P_n(-x) = -\cos(\pi x/2) P_n(x)$, because $P_n(x)$ contains only odd powers of x if n is odd. Now write

$$\int_{-1}^{1} \cos\left(\frac{\pi x}{2}\right) P_n(x)\, dx = \int_{-1}^{0} \cos\left(\frac{\pi x}{2}\right) P_n(x)\, dx + \int_{0}^{1} \cos\left(\frac{\pi x}{2}\right) P_n(x)\, dx,$$

and make the change of variables $t = -x$ in the first integral on the right.

2. Let

$$g(x) = \begin{cases} 1 & \text{for} \quad -1 \le x < 0 \\ -1 & \text{for} \quad 0 \le x \le 1. \end{cases}$$

Calculate the first two nonzero terms in the Fourier-Legendre expansion of g on $[-1, 1]$. Graph this sum and the function on the same set of axes.

3. Let $h(x) = 1$ for $-1 \le x \le 1$. Find the Fourier-Legendre series for h on $[-1, 1]$.

4. Let $f(x) = |x|$ for $-1 \le x \le 1$. Find the first three nonzero terms of the Fourier-Legendre series for f on $[-1, 1]$. Graph this sum of terms and f on the same set of axes for $-1 \le x \le 1$.

5. Let $g(x) = x$ for $-1 \le x \le 1$. Find the Fourier-Legendre series for g on $[-1, 1]$.

6. Find the Fourier-Legendre series for $g(x) = 1 + x^2$ on $[-1, 1]$.

7. Let $g(x) = 4x - x^3$ for $0 \le x \le 5$. Find the first four terms of the Fourier-Bessel series of order 1 for g on $[0, 5]$.

8. Let $f(x) = 2x$ for $0 \le x \le 3$. Find the first four terms of the Fourier-Bessel series of order 1 for f on $[0, 3]$.

9. Find the first four nonzero terms in the expansion of x^2 in a series of eigenfunctions of the Sturm-Liouville problem

$$y'' + \lambda y = 0; \qquad y(0) = 0, \qquad 3y(1) + y'(1) = 0.$$

10. Find the first four nonzero terms in the expansion of e^x in a series of Hermite polynomials on $(-\infty, \infty)$. (See Problem 18 of Section 7.5.)

11. Expand $2x^3 - 5x^2 + 3x - 4$ in a series of Hermite polynomials on $(-\infty, \infty)$.

12. Expand $x^2 - x$ in a series of Laguerre polynomials on $[0, \infty)$. *Hint:* See Problem 23 of Section 7.5.

13. Find the first four nonzero terms in the expansion of e^{-2x} in a series of Laguerre polynomials on $[0, \infty)$.

14. The second order differential equation

$$(1 - x^2)y'' - xy' + \lambda y = 0 \qquad (-1 \le x \le 1)$$

is called *Tchebycheff's equation.*

(a) Write Tchebycheff's equation in Sturm-Liouville form.

(b) Obtain two linearly independent solutions expanded about the origin, using the method of Frobenius.

(c) Show that, when λ is a nonnegative integer, Tchebycheff's equation has polynomial solutions.

(d) Let

$$T_n(x) = \sum_{k=0}^{[n/2]} (-1)^k \frac{n!}{(2k)!(n - 2k)!} (1 - x^2)^k x^{n-2k},$$

in which $[n/2] = $ greatest integer $\le n/2$. These functions are called *Tchebycheff polynomials.* Calculate $T_0(x), \ldots, T_5(x)$.

(e) Show that $T_n(x)$ is a solution of Tchebycheff's equation when $\lambda = n^2$.

(f) Show that

$$\int_{-1}^{1} (1 - x^2)^{-1/2} T_n(x) T_m(x) \, dx = \begin{cases} 0 & \text{if} & n \ne m \\ \pi & \text{if} & n = m = 0 \\ \frac{1}{2}\pi & \text{if} & n = m > 0. \end{cases}$$

Hint: For $n \ne m$, use (a) and the Sturm-Liouville theorem.

Nonlinear Differential _____
Equations and Stability _____

8.1 An Informal Introduction to the Concepts of Stability, Critical Points, and Qualitative Behavior _____

Up to this point, we have concentrated on methods for determining solutions (implicitly or explicitly defined) of differential equations or approximating them numerically. We saw another aspect to the study of differential equations in Chapter 1, when we briefly discussed direction fields. These provided a means of obtaining qualitative information about solutions from the differential equation itself, without actually producing the solution or a numerical approximation of a solution.

In this chapter, we will develop this theme much further, and along different lines. Our objective is to develop techniques for determining qualitative properties of solutions of differential equations without explicitly or implicitly obtaining the solutions themselves.

There are several approaches that can be taken. In one approach, the form of the differential equation itself can be exploited to yield information about solutions. Typically, we want information about solutions near points at which we have determined that something particularly interesting is happening. If the independent variable is t (usually for time), we might also want to investigate the behavior of solutions as time increases ($t \to \infty$) or decreases ($t \to -\infty$). This type of behavior is called *asymptotic behavior* of solutions.

Another approach is to *linearize* problems. The differential equations modeling many interesting phenomena (the weather, nuclear reactions, and so on) are nonlinear. Sometimes it is possible to replace a nonlinear problem by a suitably chosen linear problem in such a way that solutions of the linear problem reveal properties of solutions of the nonlinear problem. This process is called *linearization*.

In the next section, we will begin developing the vocabulary and concepts we will use to study systems of differential equations from a qualitative point of view. Before doing this, we will look at two examples intended to provide some intuition to serve as background and help relate the mathematics we will develop to phenomena we have some practical experience with.

CRITICAL POINTS OF A DAMPED PENDULUM

Consider a damped pendulum. If the bob is initially positioned as in Figure 8.1 and released, it will move downward, eventually coming to rest with the bob at the bottom. If, however, the pendulum is placed vertically, with the bob either at the bottom (Figure 8.2) or the top (Figure 8.3), the pendulum will remain in these positions unless disturbed. Obviously, there is something distinctive about these two positions. These positions are called *critical points*, or *equilibrium points*, for the pendulum.

In Figure 8.2, if the bob is perturbed slightly (moved to one side) and then released, it will eventually return to the critical position with the pendulum vertical and the bob at the bottom. It may oscillate through the vertical position several times, or, if the damping coefficient is large enough, it may return to this equilibrium position without passing through the vertical. Because the bob eventually returns to its original position after a slight perturbation, this critical point is said to be *stable*.

Contrast this with the behavior at the other critical point. Suppose, as in Figure 8.3, the pendulum is initially vertical, with the bob at the top. If the bob is moved slightly to one side and released, it will fall away from this critical position and not return to it. For this reason, the critical point of Figure 8.3 is said to be *unstable*.

Thus, the term "stable" is used to describe a position or state of a system having the property that the system returns to this position or state after a small perturbation from it. The term "unstable" conveys the idea that the system does not return to its initial state, even after a very small perturbation from this state.

As another example from familiar experience, take a football and try to balance it on one "end" on a flat table. This can lead to some frustration, but with a little patience and luck it is usually possible to get the ball to hold a vertical position. Now perturb the ball slightly (just touch it somewhere). The football will roll around on the table and never resume its original position (vertical state) without outside

Figure 8.1   Figure 8.2   Figure 8.3

influence. The position of the ball balanced on one end is unstable. The same experiment can be performed with an egg, although it is more difficult to get an egg to balance on one "end."

CRITICAL POINTS OF A POPULATION MODEL

We will now leave the pendulum illustration for a moment and discuss a different phenomenon in which we can also recognize certain points as being distinctive, or critical. Further, one of the critical points will be stable, the other unstable.

Suppose that the population of a colony living in an unrestricted environment is $X(t)$ at time t. Assuming that the population changes at a rate proportional to itself, $X'(t)$ is proportional to $X(t)$, and X satisfies the first order linear differential equation $X' = kX$ for some constant k. The constant of proportionality k is the net growth rate, determined as the difference between birth and death rates. Further, if the population at some time t_0 is X_0, X is a solution of the linear initial value problem

$$X' = kX; \qquad X(t_0) = X_0.$$

Of course, we can solve this problem easily. However, observe that we can learn some aspects of the behavior of the solution just by studying the differential equation itself. First, if X is zero for any time T, the rate of change in population is also zero at T and remains zero for all $t > T$. Further, if $k < 0$ and $X_0 > 0$, the solution must decay, and the population (being necessarily integer valued) will die out with time. If $k > 0$ and $X_0 > 0$, the population will grow without bound.

The differential equation $X' = kX$ may adequately model population growth when the population is much smaller than the environment can support. For example, from census figures for the United States in the early nineteenth century, we can observe that the population fits an exponential solution of $X' = kX$ quite well (assuming that we use the data to choose k appropriately). Growths of cultures in petri dishes also follow an exponential rate of growth for a limited period of time.

In most instances, however, the model $X' = kX$ is too simplistic, failing to account for environmental and other factors which influence the growth of a population. A more realistic model can be developed as follows.

Assume that the carrying capacity of the environment is K. This means that the population should grow in an unrestricted manner as long as X is much smaller than K, with a decrease in growth rate as X nears K. Further, the population should decline (exhibit a negative rate of growth) if X exceeds K. These requirements are met by the differential equation

$$X' = kX(K - X),$$

which is called the *logistic model*, or *logistic equation*, of population growth. Write the logistic equation as

$$X' = aX - bX^2 = X(a - bX).$$

Now notice that, if $X = 0$ or $X = a/b$ at any time T, the change in population is zero at T, and the population will remain at these respective values at all later times. For this reason we call zero and a/b *critical points of the logistic equation*. Behavior of X at these values is analogous to the behavior of the damped pendulum at the vertical positions. Something distinctive is happening at these values.

If $0 < X < a/b$ for any t, $X'(t)$ is positive and X increases. X cannot exceed a/b, however. In order for the graph of X to cross the horizontal line $X = a/b$ at some time t_0, its slope would have to be positive at t_0, and this cannot happen because, from the differential equation, $X'(t_0) = 0$ if $X(t_0) = a/b$.

By the same token, if $X > a/b$ for some value of t, the population will have to decrease because $X' < 0$. In this event, however, the population cannot decrease below a/b. In order to decrease below a/b, its slope would have to be negative at that time, and the differential equation tells us that the slope then is zero.

Suppose that the population at some time we designate as $t = 0$ is a known value X_0. The initial value problem for X is

$$X' = aX - bX^2; \qquad X(0) = X_0.$$

Although we have discussed certain conclusions we can draw about X just from the differential equation itself, this initial value problem is easily solved to yield an explicit formula for $X(t)$. The solution is

$$X(t) = \begin{cases} 0 & \text{if} \quad X_0 = 0 \\ \dfrac{a}{b} & \text{if} \quad X_0 = \dfrac{a}{b} \\ \dfrac{a}{b + \dfrac{a - bX_0}{X_0} e^{-at}} & \text{if} \quad X_0 \neq 0 \quad \text{and} \quad X_0 \neq \dfrac{a}{b}. \end{cases} \qquad (8.1)$$

The two values zero and a/b are critical points for the logistic differential equation. As with the pendulum, the terminology is motivated by the fact that, if the solution has either of these values at some time, it will remain at this value (since its rate of change is zero there).

If X represents the population of some species, then $a > 0$, and e^{-at} will decrease with time. Therefore, any solution having an initial value different from either zero or a/b will approach the value a/b, with solutions less than this value increasing toward a/b and solutions greater than this value decreasing toward a/b.

Notice that the two critical points of the logistic equation have quite different properties. The point a/b seems to *attract* solutions (in the sense that solutions approach it as t increases). This is similar to what we saw with the stable critical point of the pendulum, with the bob at the bottom position.

The critical point zero has exactly the opposite property: solutions near zero tend to move *away from* this value. This behavior is like the unstable critical point for the pendulum. In that case, the pendulum moved away from the unstable critical point once perturbed from it.

A LINEARIZATION OF THE LOGISTIC EQUATION

We will use the logistic equation to illustrate some of the ramifications of linearizing a nonlinear equation. First, observe that, if we linearize the differential equation

$X' = aX - bX^2$ by simply discarding the nonlinear term $-bX^2$, we are actually re-turning to the model $X' = aX$ for an unrestricted environment, in which we found no stable critical points. This shows that significant features of a differential equation may be lost if we linearize it and indicates that we must carry out linearizations with careful attention to how the behavior of solutions of the linearized problem relates to behavior of solutions of the original nonlinear problem.

We will make one final observation about the effects on solutions of changing certain parameters within a differential equation. Notice what happens if we allow a and b to be negative in the logistic equation (ignore the interpretation of the logistic equation as a population model requiring that both a and b be positive). Thus, consider the differential equation

$$X' = aX - bX^2,$$

in which a and b are both negative. We now have $a < 0$ and $b < 0$ in the solution (8.1) given previously. Then zero and a/b are still critical points, but now zero is a stable critical point, while a/b is unstable. Further, in this case, the linearized problem $X' = aX$ does have solutions with properties similar to those of the nonlinear equation, provided that $|X_0| < a/b$. Since $a < 0$, $e^{-at} \to \infty$ as $t \to \infty$, and the solution approaches zero for any initial value X_0 satisfying $|X_0| < a/b$.

The purpose of this discussion has been to make some observations and suggest lines of thought which might be fruitful to pursue. In the next section, we will define what we mean by a critical point of a system of differential equations, preparing the way for a classification of different types of critical points and a discussion of the concept of stability.

PROBLEMS FOR SECTION 8.1

In each of Problems 1 through 4, solve the differential equation and determine all of the critical points as we did for the logistic equation. Classify each critical point as either stable or unstable in the intuitive sense we have discussed.

1. $X' = 1 - X^2$ **2.** $X' = X^2 - 4$ **3.** $X' = 2X - X^2$ **4.** $X' = X^3 - X^2 - 2X$

5. Solve the initial value problem

$$X' = X - 3; \qquad X(0) = 3.$$

Notice that the solution is a bounded function. Now solve the initial value problems

$$X' = X - 3; \qquad X(0) = 3.01$$

and

$$X' = X - 3; \qquad X(0) = 2.99.$$

The solutions should illustrate what can happen in an initial value problem if the initial data are perturbed slightly.

6. Suppose that φ is a solution of the logistic equation with $a > 0$ and $b > 0$, satisfying $\varphi(0) = X_0$, with $0 < X_0 < a/b$. Determine the time at which the population is increasing most rapidly.

7. Graph several solutions of the logistic equation for the case in which $a < 0$ and $b < 0$.

8.2 *Autonomous Systems of Differential Equations*

Consider the interaction of two species, one of which has an unlimited supply of food and in turn serves as food for the second species. This is the classical predator-prey setting, as typified by a population of foxes and rabbits.

To be specific, suppose that there are $x(t)$ rabbits and $y(t)$ foxes present at time t. Assume that there is no intervention from outside sources (such as hunters) and that the rabbits have what is for practical purposes an unlimited food supply (vegetation). In the absence of foxes, the rabbit population satisfies the equation

$$x' = ax$$

for some positive number a. Their population simply grows exponentially. If the foxes depend on the rabbits as their sole source of food, then in the absence of rabbits, the fox population will die out according to an exponential rule, and

$$y' = -ky$$

for some $k > 0$.

We have a more interesting situation if both foxes and rabbits are present. Now we must account for the interaction between them in the differential equations modeling their populations. The rate of increase in foxes should be proportional to their food supply. We will assume that each encounter between a fox and a rabbit results in the fox eating the rabbit. Further, assume that the number of such encounters at time t is proportional to $x(t)y(t)$, the product of the rabbit and fox populations at time t. This means that, if the fox population increases at a rate proportional to the number of rabbits eaten, we should introduce into the equation for the fox population a term cxy, in which c is a positive constant. Similarly, we will assume that the rate at which the rabbit population decreases is proportional to the number killed. To account for this decrease, we will introduce into the equation for the rabbit population a term $-bxy$, with $b > 0$. This leads to the system of differential equations

$$\begin{aligned} x' &= ax - bxy \\ y' &= cxy - ky, \end{aligned} \tag{8.2}$$

in which a, b, c, and k are positive constants. Given an initial population of each species, we want to know the population of rabbits and foxes at any later time.

First order nonlinear systems such as this predator-prey model (8.2) arise frequently. For example, the motion of an undamped pendulum (Figure 8.4) is modeled by the second order differential equation

$$\theta''(t) + \frac{g}{L} \sin(\theta) = 0. \tag{8.3}$$

If we let $x = \theta$ and $y = \theta' = x'$, we have the nonlinear system

$$\begin{aligned} x' &= y \\ y' &= -\frac{g}{L} \sin(x). \end{aligned} \tag{8.4}$$

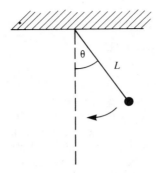

Figure 8.4

The system (8.4) is equivalent to the differential equation (8.3) in the sense that solutions of (8.3) can be recovered from solutions of (8.4).

The predator-prey system (8.2) and the system (8.4) for the pendulum both have the property that t does not appear explicitly in the equations. Such a system is called *autonomous*. We will restrict ourselves in this chapter to autonomous systems of the form

$$\frac{dx}{dt} = F(x, y)$$

$$\frac{dy}{dt} = G(x, y). \tag{8.5}$$

Before analyzing the predator-prey and pendulum models in detail, we will make some general observations about systems of the form (8.5). Any solution $x = \varphi(t)$, $y = \psi(t)$ of (8.5) is called a *trajectory* of the system. In the case in which $\varphi(t) = $ constant and $\psi(t) = $ constant, the trajectory has a single point as its graph.

We may also think of the functions $x = \varphi(t)$, $y = \psi(t)$ as parametric equations of a curve in the plane. In discussions, it is convenient to identify a trajectory with this graph and speak of the trajectory and the curve, or graph, interchangeably. We also often refer to "a trajectory through a point," by which we mean a trajectory whose graph passes through the point. Here is an illustration of this terminology.

EXAMPLE 8.1

Consider the system

$$x' = x + y$$
$$y' = 2x.$$

This is an autonomous system with $F(x, y) = x + y$ and $G(x, y) = 2x$.

Although we will later have matrix methods for solving linear systems, we can solve this simple system in a straightforward manner. First, differentiate $x' = x + y$ to get

$$x'' = x' + y'.$$

Substitute $y' = 2x$ from the second equation of the system to get

$$x'' = x' + 2x,$$

or

$$x'' - x' - 2x = 0.$$

This differential equation has general solution

$$x = \varphi(t) = c_1 e^{-t} + c_2 e^{2t}.$$

Since $y' = 2x$,

$$y' = 2c_1 e^{-t} + 2c_2 e^{2t},$$

and we can integrate to get

$$y = \psi(t) = -2c_1 e^{-t} + c_2 e^{2t} + C.$$

Finally, substitute these solutions into the first equation of the system to get

$$x' = -c_1 e^{-t} + 2c_1 e^{2t} = x + y = -c_1 e^{-t} + 2c_2 e^{2t} + C,$$

implying that $C = 0$. We now have the general solution of the system:

$$x = \varphi(t) = c_1 e^{-t} + c_2 e^{2t}$$
$$y = \psi(t) = -2c_1 e^{-t} + c_2 e^{2t}.$$

For each choice of the constants c_1 and c_2, the functions $x = \varphi(t)$ and $y = \psi(t)$ are parametric equations of a curve in the plane. These curves are graphs of the trajectories of this system of differential equations and are also referred to as trajectories. For example, if $c_1 = 0$ and $c_2 = 1$, we get

$$x = e^{2t} = y,$$

whose graph is shown in Figure 8.5. This trajectory is a half-line from the origin, because $e^{2t} > 0$ for all t. Further, the origin is not actually on this trajectory, although there are points on the trajectory as close to $(0, 0)$ as we like because $e^{2t} \to 0$ as $t \to -\infty$. Similarly, if we choose $c_1 = 1$ and $c_2 = 0$, we get

$$y = -2e^{-t} = -2x.$$

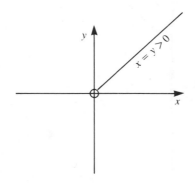

Figure 8.5

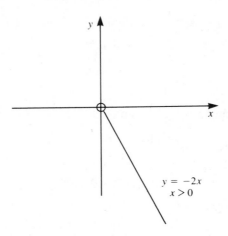

Figure 8.6

This trajectory is shown in Figure 8.6; it is a half-line from the origin having slope -2.

Given any point (x_0, y_0) and any t_0, we can find a solution $x = \varphi(t)$, $y = \psi(t)$ of the system satisfying the initial conditions $x(t_0) = x_0$, $y(t_0) = y_0$. Thus, there is a trajectory passing through each point (x_0, y_0) in the plane. For example, let $t_0 = 0$, $x_0 = 1$, and $y_0 = 0$. We find that $c_1 = \frac{1}{3}$ and $c_2 = \frac{2}{3}$, so the trajectory satisfying $x(0) = 1$ and $y(0) = 0$ is given parametrically by

$$x = \tfrac{1}{3}[e^{-t} + 2e^{2t}]$$
$$y = \tfrac{2}{3}[-e^{-t} + e^{2t}].$$

This trajectory passes through $(1, 0)$ at time $t = 0$ and is shown in Figure 8.7. If we let

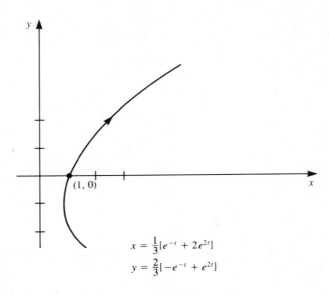

$$x = \tfrac{1}{3}[e^{-t} + 2e^{2t}]$$
$$y = \tfrac{2}{3}[-e^{-t} + e^{2t}]$$

Figure 8.7

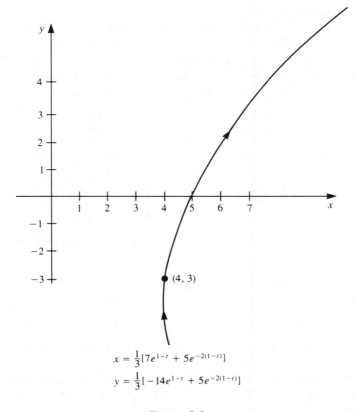

$$x = \tfrac{1}{3}[7e^{1-t} + 5e^{-2(1-t)}]$$

$$y = \tfrac{1}{3}[-14e^{1-t} + 5e^{-2(1-t)}]$$

Figure 8.8

$t_0 = 1$, $x_0 = 4$, and $y_0 = -3$, we find that $c_1 = \tfrac{7}{3}e$ and $c_2 = \tfrac{5}{3}e^{-2}$, and

$$x = \tfrac{1}{3}[7e \cdot e^{-t} + 5e^{-2}e^{2t}] = \tfrac{1}{3}[7e^{1-t} + 5e^{-2(1-t)}]$$
$$y = \tfrac{1}{3}[-14e \cdot e^{-t} + 5e^{-2}e^{2t}] = \tfrac{1}{3}[-14e^{1-t} + 5e^{-2(1-t)}].$$

This trajectory passes through $(4, -3)$ at time $t = 1$ and is shown in Figure 8.8.   ∎

Here are some facts about trajectories. We will assume that F and G in system (8.5) are continuous for all (x, y), with continuous first partial derivatives.

1. *Given any point P_0 in the plane, there is a trajectory through P_0.* To prove this, let P_0 have coordinates (x_0, y_0). The initial value problem

$$x' = F(x, y)$$
$$y' = G(x, y)$$
$$x(0) = x_0, \qquad y(0) = y_0$$

has a solution $x = \varphi(t)$, $y = \psi(t)$, as we will see when we treat systems in Chapter 14. These are parametric equations of a curve which passes through (x_0, y_0).

2. *A translation of a trajectory is a trajectory.* Here is what this means. Suppose that $x = \varphi(t)$, $y = \psi(t)$ is a solution of the system (8.5). If c is any constant, we can

check by substitution that $x = \varphi(t + c)$ and $y = \psi(t + c)$ also satisfy the system and hence also determine a trajectory. We call the solutions $x = \varphi(t + c)$, $y = \psi(t + c)$ a *translation* of $x = \varphi(t)$, $y = \varphi(t)$ by a constant c.

3. *If two trajectories pass through the same point* (x_0, y_0), *one trajectory is a translation of the other.* To prove this, suppose we have two trajectories passing through (x_0, y_0), one at time t_0 and the other at time t_1. Write these trajectories:

$$
\begin{aligned}
x &= \varphi(t) & x &= \alpha(t) \\
y &= \psi(t) & y &= \beta(t) \\
\varphi(t_0) &= x_0 & \alpha(t_1) &= x_0 \\
\psi(t_0) &= y_0 & \beta(t_1) &= y_0.
\end{aligned}
$$

We will show that one trajectory is a translation of the other. Let $c = t_0 - t_1$, and let

$$X(t) = \varphi(t + c), \qquad Y(t) = \psi(t + c).$$

Since this is a translation of a solution, $x = X(t)$, $y = Y(t)$ is also a solution of the system (8.5). Further,

$$X(t_1) = \varphi(t_0) = x_0 \quad \text{and} \quad Y(t_1) = \psi(t_0) = y_0.$$

By the uniqueness of solutions of an initial value problem, we must have

$$\alpha(t) = X(t) = \varphi(t + c) \quad \text{and} \quad \beta(t) = Y(t) = \psi(t + c).$$

The second solution is therefore a translation of the first. By the same token, the first solution is a translation of the second.

We may combine results (1) and (3) by saying that there is a trajectory through each point of the plane, and all trajectories through a given point are translations of each other.

4. *Trajectories are directed, in the following sense.* Suppose the parametric equations are $x = \varphi(t)$, $y = \psi(t)$. As t increases, we can envision the point $(\varphi(t), \psi(t))$ moving along the trajectory in a certain direction. This imposes a sense of direction along the trajectory, usually indicated by putting an arrow on the graph, as in Figure 8.9.

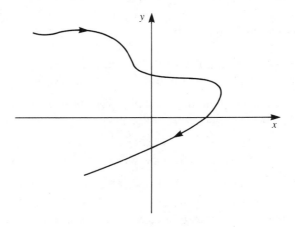

Figure 8.9

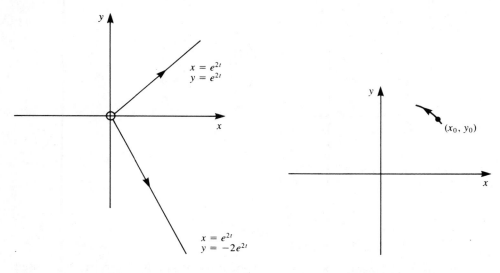

Figure 8.10   Figure 8.11

Figure 8.10 shows the trajectories of Figures 8.5 and 8.6, drawn with the arrows indicating direction as t increases.

We can actually get a good idea of direction along a trajectory just from the system (8.5) itself. For example, if $F(x_0, y_0) < 0$ and $G(x_0, y_0) > 0$, then $x' < 0$ (so x is moving to the left) and $y' > 0$ (so y is moving upward) at (x_0, y_0). Under these circumstances, the trajectory through (x_0, y_0) has the direction indicated by the arrow in Figure 8.11.

5. *A trajectory may not cross itself.* Intuitively, this means that the curve shown in Figure 8.12 cannot be the graph of a trajectory. Trajectories may, however, be

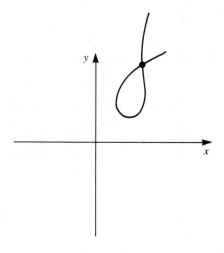

Figure 8.12

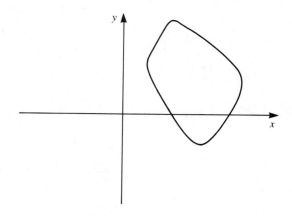

Figure 8.13

closed curves. The graph of Figure 8.13 could be the graph of a trajectory. Closed trajectories represent periodic solutions of the system (8.5).

In the case of the autonomous system (8.5), we refer to the xy-plane as the *phase plane* of the system. A plot or sketch of trajectories in the phase plane is called the *phase portrait* of the system.

The following two examples illustrate some of the terminology we have developed, as well as the properties of trajectories we have just discussed. For purposes of illustration, we will use systems which are easy to solve explicitly, so we can look at the solutions and examine their behavior. Of course, the theory we are developing is designed to give us information about systems we cannot solve explicitly.

EXAMPLE 8.2

Consider the system

$$x' = x + y$$
$$y' = -4x + y. \tag{8.6}$$

To solve this system, first subtract the equations to get

$$x' - y' = 5x.$$

From the first equation of the system, $y = x' - x$, so $y' = x'' - x'$, and

$$x' - y' = 5x = x' - (x'' - x').$$

This gives us the second order differential equation in x alone

$$x'' - 2x' + 5x = 0$$

with general solution

$$x = \varphi(t) = c_1 e^t \cos(2t) + c_2 e^t \sin(2t).$$

From the first equation of the system, we find that

$$y = x' - x = -2c_1 e^t \sin(2t) + 2c_2 e^t \cos(2t).$$

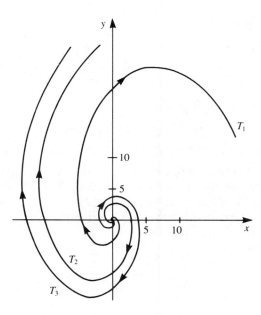

Figure 8.14

The general solution of the system is

$$x = \varphi(t) = e^t[c_1\cos(2t) + c_2\sin(2t)]$$
$$y = \psi(t) = 2e^t[-c_1\sin(2t) + c_2\cos(2t)].$$

The phase portrait of this system consists of drawings of trajectories in the xy-plane obtained by choosing different values of c_1 and c_2.

Figure 8.14 shows three trajectories T_1, T_2, and T_3. T_1 is the graph of

$$x = e^t\cos(2t)$$
$$y = -2e^t\sin(2t),$$

obtained by letting $c_1 = 1$ and $c_2 = 0$. T_2 is the graph of

$$x = e^t\sin(2t)$$
$$y = 2e^t\cos(2t),$$

obtained by letting $c_1 = 0$ and $c_2 = 1$. Finally, T_3 is the graph of

$$x = 2e^t[\cos(2t) - 2\sin(2t)]$$
$$y = 4e^t[-\sin(2t) - 2\cos(2t)],$$

obtained by choosing $c_1 = 2$ and $c_2 = -4$.

The arrows on the trajectories indicate direction of motion along the trajectory as t increases. Although near the origin the trajectories are "very close" to each other, two distinct trajectories cannot cross each other. The trajectories spiral about the origin. In the terminology to be developed in the next section, the origin is a *spiral point* of the system (8.6). ■

EXAMPLE 8.3

Consider the system

$$x' = 3x + y$$
$$y' = -13x - 3y.$$

The general solution of this system is

$$x = \varphi(t) = c_1\cos(2t) + c_2\sin(2t)$$
$$y = \psi(t) = (2c_2 - 3c_1)\cos(2t) + (-2c_1 - 3c_2)\sin(2t).$$

Figure 8.15 shows some trajectories of this system. The trajectories are closed curves, consistent with the fact that the solutions are periodic (having period π). The fact that all of the trajectories enclose the origin leads us to call the origin a *center* of this system. Again, we will define this term in the next section.

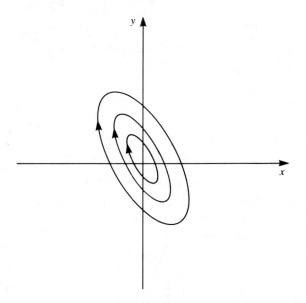

Figure 8.15 ∎

The distinction between an autonomous system and a nonautonomous system is that the former does not explicitly contain t. As the following example shows, trajectories of nonautonomous systems may not exhibit the properties we have seen for trajectories of autonomous systems.

EXAMPLE 8.4

The system

$$x' = \frac{x}{t}$$

$$y' = -\frac{y}{t} + x$$

(8.7)

is nonautonomous. We can solve this system by first solving the separable equation $x' = x/t$ for x to get $x = c_1 t$. Substitute this for x in the second equation of (8.7) to get

$$y' = -\frac{y}{t} + c_1 t.$$

We solve this linear first order differential equation for y to get

$$y = c_1 \frac{1}{3} t^2 + c_2 \frac{1}{t}.$$

The general solution of the system (8.7) is

$$x = \varphi(t) = c_1 t$$
$$y = \psi(t) = \frac{1}{3} c_1 t^2 + c_2 \frac{1}{t}.$$

Now observe that there may be many different trajectories through a single point. For example, consider the point $(1, 0)$. Let t_0 be any number different from zero, and look for a solution of the system passing through $(1, 0)$ at time t_0. We find, upon letting $\varphi(t_0) = 1$ and $\psi(t_0) = 0$, that the solution through $(1, 0)$ at t_0 is

$$x(t) = \frac{1}{t_0} t$$
$$y(t) = \frac{1}{3} \frac{1}{t_0} t^2 - \frac{1}{3} t_0^2 \frac{1}{t}.$$

Unlike what we saw for autonomous systems, this solution depends on t_0. We obtain different trajectories which are not translations of one another, passing through $(1, 0)$, by choosing different values of $t_0 \neq 0$. For example, the trajectory passing through $(1, 0)$ at $t_0 = 1$ is

$$x = t$$
$$y = \frac{1}{3}\left(t^2 - \frac{1}{t}\right),$$

shown as the trajectory T_1 in Figure 8.16. At $t_0 = 2$, the trajectory T_2 through $(1, 0)$ shown in Figure 8.16 has equations

$$x = \frac{1}{2} t$$
$$y = \frac{1}{6} t^2 - \frac{4}{3} \frac{1}{t}.$$

And, at $t_0 = 3$, the trajectory T_3 through $(1, 0)$ in Figure 8.16 is given by

$$x = \frac{1}{3} t$$
$$y = \frac{1}{9} t^2 - \frac{3}{t}.$$

These distinct trajectories all pass through $(1, 0)$ but do so at different times.

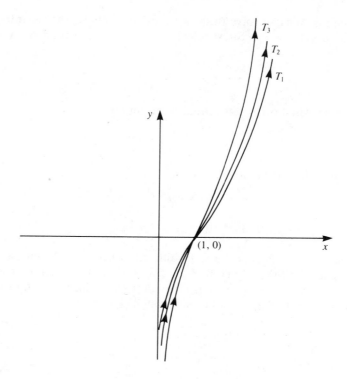

Figure 8.16

Another way of looking at these trajectories through $(1, 0)$ is that a particle starting from $(1, 0)$ at different times t_0 follows different paths if constrained to move according to the system (8.7). This cannot happen with an autonomous system. ∎

We call a point (x_0, y_0) a *critical point* (or an *equilibrium point*) of the autonomous system (8.5) if $F(x_0, y_0) = G(x_0, y_0) = 0$. A critical point is said to be *isolated* if there is a disk of positive radius about (x_0, y_0) containing no other critical point of the system. We will deal only with systems whose critical points are isolated and will therefore use the term *critical point* to mean *isolated critical point*. We will often assume that F and G are continuous, with continuous first partial derivatives, throughout the entire plane.

As the examples of the pendulum and the logistic population model suggest, the behavior of solutions about a critical point may be of particular interest and may contain a good deal of information about the behavior of solutions of the system. After we classify critical points in the next section, we will include in Sections 8.4 and 8.5 a detailed analysis of the critical points of the simple pendulum and predator-prey models.

Note that, if (x_0, y_0) is a critical point of (8.5), the solution of the initial value problem consisting of the system (8.5) and the conditions $x(t_0) = x_0$, $y(t_0) = y_0$, has a one-point trajectory. Since $x'(t_0) = F(x(t_0), y(t_0)) = F(x_0, y_0) = 0$ and, similarly,

$y'(t_0) = G(x_0, y_0) = 0$, the solution cannot leave the point (x_0, y_0) at any time t. The trajectory through a critical point is therefore the single point (x_0, y_0).

Since different trajectories of an autonomous system cannot cross each other, no other trajectory can pass through a critical point. Thus, if a trajectory begins at a noncritical point, it can never get to a critical point. It may, however, approach arbitrarily close to a critical point in a variety of ways.

In the next section, we will classify critical points according to the behavior of trajectories which begin near them or approach them. These classifications will in turn enable us to describe different types of behavior of solutions.

PROBLEMS FOR SECTION 8.2

In each of Problems 1 through 6, solve the system of differential equations and then use different values for the constants in the general solution to draw some of the graphs making up the phase portrait of the system.

1. $x' = 4x + y$
   $y' = -17x - 4y$

2. $x' = 2x$
   $y' = 8x + 2y$

3. $x' = 4x - 7y$
   $y' = 2x - 5y$

4. $x' = 3x - 2y$
   $y' = 10x - 5y$

5. $x' = 5x - 2y$
   $y' = 4y$

6. $x' = -4x - 6y$
   $y' = 2x - 11y$

In each of Problems 7 through 12, use the fact that $dy/dx = (dy/dt)/(dx/dt)$ to convert the system of differential equations into a first order differential equation of the form $dy/dx = f(x, y)$. Solve this differential equation and graph several solution curves, thus obtaining some curves in the phase portrait of the system. Use the sign of dx/dt and dy/dt at various points (x, y) to determine the direction along the trajectories in the phase portrait, and indicate the direction on each trajectory drawn by inserting an arrow on the graph.

7. $x' = 3y$
   $y' = 4x$

8. $x' = 2xy$
   $y' = y^2 - x^2$

9. $x' = y + 2$
   $y' = x - 1$

10. $x' = \csc(x)$
   $y' = y$

11. $x' = x$
   $y' = x + y$

12. $x' = x^2$
   $y' = y$

Frequently, when using the technique suggested for Problems 7 through 12, the general solution can be recognized as the equation of a conic. As an aid to plotting the phase portrait of such a system, recall that a conic whose equation is

$$Ax^2 + Bxy + Cy^2 = K, \tag{8.8}$$

with $B \neq 0$, can be rewritten in the form

$$A^*(x^*)^2 + C^*(y^*)^2 = K, \tag{8.9}$$

with no cross product x^*y^*-term, if we rotate axes counterclockwise through an angle α given by $\cot(2\alpha) = (A - C)/B$. If this rotation is carried out, we find that

$$A^* = A \cos^2(\alpha) + B \sin(\alpha)\cos(\alpha) + C \sin^2(\alpha)$$

and

$$C^* = A \sin^2(\alpha) - B \sin(\alpha)\cos(\alpha) + C \cos^2(\alpha).$$

The quantity $B^2 - 4AC$ is called the *discriminant* of the conic (8.8). This quantity is invariant under rotation of axes, as we can check that

$$B^2 - 4AC = -4A^*C^*.$$

The discriminant helps us classify a conic from its equation. The conic (8.8) [and its rotated version (8.9)] is

$$\text{a parabola if } B^2 - 4AC = 0;$$
$$\text{a hyperbola if } B^2 - 4AC > 0;$$
$$\text{an ellipse if } B^2 - 4AC < 0.$$

This method can be used to read directly from equation (8.8) the type of conic represented by the equation. The rotation of axes then puts (8.8) into the more standard form (8.9) to facilitate sketching the curve.

In each of Problems 13 through 18, find the general solution of the system in terms of an equation involving x and y, using the method given for Problems 7 through 12. Then use the ideas from analytic geometry reviewed above to sketch the phase portrait of the system.

13. $x' = 36x - 52y$
$\ y' = 73x - 36y$

14. $x' = x - 2y$
$\ y' = 2x - y$

15. $x' = 3\sqrt{3}x - 2y$
$\ y' = 8x - 3\sqrt{3}y$

16. $x' = 12x - 2y$
$\ y' = 9x - 12y$

17. $x' = 3\sqrt{3}x - 7y$
$\ y' = 13x - 3\sqrt{3}y$

18. $x' = 12x - 9y$
$\ y' = 16x - 12y$

19. Compare the phase portraits of the following two systems of differential equations.
(a) $x' = F(x, y)$ (b) $x' = -F(x, y)$
    $y' = G(x, y)$ $y' = -G(x, y).$

20. In Example 8.1, the statement is made that, given any number t_0 and any point (x_0, y_0) in the plane, there exists a solution $x = \varphi(t), y = \psi(t)$ of the system such that $\varphi(t_0) = x_0$ and $\psi(t_0) = y_0$. Prove this assertion by explicitly finding appropriate values of c_1 and c_2 in the general solution of the system.

21. Derive an autonomous system of differential equations which models a predator-prey relationship between two species in which the prey lives in a restricted environment. *Hint:* Incorporate the logistic model discussed in Section 8.1.

22. Derive a system of differential equations which models a predator-prey relationship between two species with indiscriminate harvesting by an outside agent. Assume that each species is removed continuously at a rate proportional to its population.

8.3 Stability and the Classification of Critical Points

Recall that the *phase portrait* of the autonomous system

$$x' = F(x, y)$$
$$y' = G(x, y) \tag{8.10}$$

consists of drawings of trajectories in the phase plane (xy-plane). In such a drawing, we usually attach an arrow to each trajectory to indicate the direction of motion of the point $(x(t), y(t))$ moving along the trajectory as t increases. Figure 8.17 shows a typical phase portrait.

We will now look at how various configurations of trajectories, such as those shown in Figure 8.17, might arise. Information about the phase portrait gives us information about solutions of the system. For example, trajectories that are closed curves correspond to periodic solutions of the system. Two important questions we will

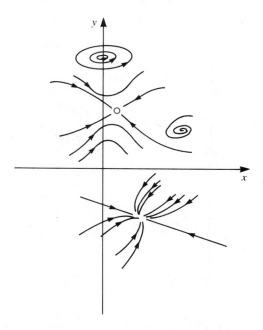

Figure 8.17. Phase portrait.

address are the following:

1. If two trajectories are "close" at some time, will they remain close for later times, or will they move away from each other? This is the question of *stability*. At a stable point of the system, trajectories will remain close in a sense we will specify soon. At an unstable point, trajectories will diverge from each other.

2. How do trajectories behave as $t \to \infty$ or as $t \to -\infty$? This is the question of *asymptotic behavior* and is considered under the heading of *asymptotic stability*.

   We will now formulate definitions which will enable us to pose these questions more accurately and to obtain useful results. Assume throughout the rest of this section that (x_0, y_0) is an isolated critical point of the autonomous system (8.10). We will let $x(t)$ and $y(t)$ denote solutions of the system (8.10) and consider the trajectory $(x(t), y(t))$.

DEFINITION

A trajectory $(x(t), y(t))$ is said to *approach* (x_0, y_0) if

$$\lim_{t \to \infty} x(t) = x_0 \quad \text{and} \quad \lim_{t \to \infty} y(t) = y_0$$

or

$$\lim_{t \to -\infty} x(t) = x_0 \quad \text{and} \quad \lim_{t \to -\infty} y(t) = y_0. \quad \blacksquare$$

This definition is illustrated in Figure 8.18. If we think of the trajectory as the path of a

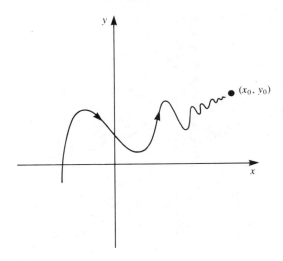

Figure 8.18 Trajectory approaching (x_0, y_0).

particle, the trajectory approaches a point if the particle can be located in an arbitrarily small circle about the point by choosing the time sufficiently large (or letting t approach minus infinity, that is, going back far enough in the particle's history).

One way of thinking of this concept is that the point $(x(t), y(t))$ approaches (x_0, y_0) if it eventually (that is, for sufficiently large t) moves within an arbitrarily small disk about (x_0, y_0) or emanates from (as t increases from $-\infty$) an arbitrarily small disk about (x_0, y_0). The trajectory cannot cross the critical point and so never actually reaches (x_0, y_0).

The next definition distinguishes a particular way a trajectory may approach a critical point.

DEFINITION

A trajectory $(x(t), y(t))$ *enters* (x_0, y_0) if it approaches (x_0, y_0) and if

$$\frac{y(t) - y_0}{x(t) - x_0}$$

has a finite limit as $t \to \infty$ or as $t \to -\infty$. ∎

This definition has a geometric interpretation. Since $(y - y_0)/(x - x_0)$ is the slope of the straight line through (x, y) and (x_0, y_0), the definition requires that the line $\overline{PP_0}$ approach a definite direction as t increases (or as $t \to -\infty$). This is illustrated in Figure 8.19. Note the difference between Figures 8.18 and 8.19. In each, the trajectory approaches (x_0, y_0). In Figure 8.18, however, the trajectory does not approach from a specific direction, as it does in Figure 8.19.

We will now distinguish four kinds of critical points, according to the behavior of trajectories near the point.

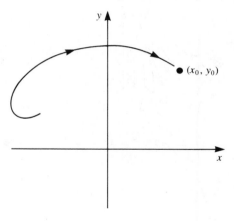

Figure 8.19. Trajectory entering (x_0, y_0).

DEFINITION   Center

A critical point (x_0, y_0) is a *center* if it is enclosed by an infinite family of closed trajectories and there are closed trajectories of the family lying arbitrarily close to (x_0, y_0) but no trajectory if the family approaches (x_0, y_0). ∎

A typical center is shown in the phase portrait of Figure 8.20. The origin is a center of the system in Example 8.3 of Section 8.2.

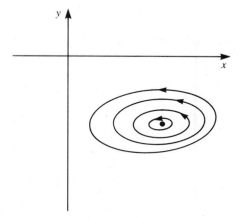

Figure 8.20. Center.

DEFINITION   Saddle Point

A critical point (x_0, y_0) is a *saddle point* if the trajectories are as shown in Figure 8.21. There are two half-line trajectories $\overline{AP_0}$ and $\overline{BP_0}$ approaching (x_0, y_0) as $t \to \infty$ and two half-line trajectories $\overline{CP_0}$ and $\overline{DP_0}$ approaching (x_0, y_0) as $t \to -\infty$; the other trajectories have these half-lines as asymptotes and are similar in shape to hyperbolas.

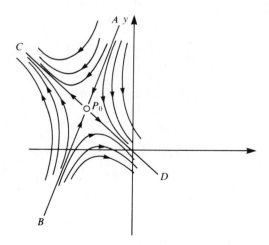

Figure 8.21. Saddle point. ∎

Note the directions along the half-lines: *toward* (x_0, y_0) as $t \to \infty$ and *away from* (x_0, y_0) as $t \to -\infty$.

DEFINITION   Spiral Point

A critical point (x_0, y_0) is a *spiral point* if there is a circle C about (x_0, y_0) such that every trajectory within C spirals about (x_0, y_0) infinitely many times and also approaches (x_0, y_0). ∎

Figure 8.22 shows a typical spiral point. The origin is a spiral point in Example 8.2 of Section 8.2.

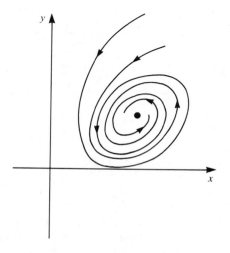

Figure 8.22. Spiral point.

DEFINITION   Node

A critical point (x_0, y_0) is a *node* if it is entered by an infinite family of trajectories.   ■

Figure 8.23 shows a typical node.

We will now define what it means for a critical point to be stable. Recall that, intuitively, a critical point is stable if trajectories which are close to the point at some time remain in the vicinity of the point for all later times. (As a model for the definition, keep in mind that small perturbations of the damped pendulum with the bob in the vertical, lowest position eventually result in the bob returning to this position. This position is therefore stable.)

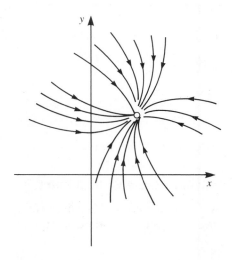

Figure 8.23. Node.

DEFINITION   Stable Critical Point

A critical point (x_0, y_0) is *stable* if, for each $R > 0$, there exists some r, with $0 \leq r \leq R$, such that every trajectory entering the circle of radius r about (x_0, y_0) at some time t_0 remains inside the circle of radius R about (x_0, y_0) for all times $t > t_0$.   ■

Note that $r \leq R$ in this definition. If (x_0, y_0) is stable, any trajectory coming within r units of (x_0, y_0) at some time t_0 remains forever after within R units of (x_0, y_0) (but not necessarily within r units). This concept is illustrated in Figure 8.24.

A critical point that is not stable is called *unstable*. With the pendulum, the vertical position, with the bob at the top, is a critical point which is unstable, because a small perturbation from this position results in the system moving away from this position and not returning to it.

Asymptotic stability refers to a stable point having the additional property that all trajectories which come sufficiently close to (x_0, y_0) must actually approach (x_0, y_0).

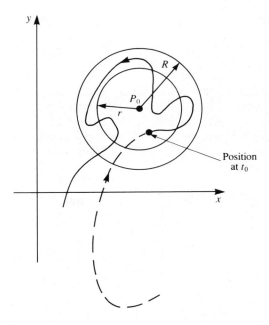

Figure 8.24. Stable point.

DEFINITION   Asymptotically Stable Critical Point

A critical point (x_0, y_0) is *asymptotically stable* if it is stable and there is a circle C about (x_0, y_0) such that every trajectory inside C at some t_0 approaches (x_0, y_0) as $t \to \infty$. ∎

This concept is illustrated in Figure 8.25.

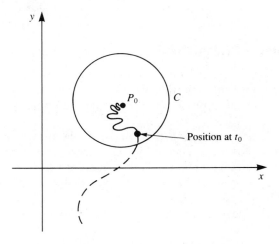

Figure 8.25. Asymptotic stability.

Our intention in formulating these concepts is to study nonlinear systems. However, for linear systems, it is possible to state a definitive result which will be used later when we linearize problems and which also provides us with a wealth of examples of the terminology we have developed.

THEOREM 8.1   Classification of Critical Points of a Linear System

Let a, b, c, and d be real numbers such that $ad - bc \neq 0$. Then the origin is the only critical point of the linear system

$$\begin{aligned} x' &= ax + by \\ y' &= cx + dy. \end{aligned} \tag{8.11}$$

Further, let λ_1 and λ_2 be the roots of the auxiliary equation

$$\lambda^2 - (a + d)\lambda + (ad - bc) = 0.$$

Then the following conclusions can be drawn.

1. $(0, 0)$ is a node if λ_1 and λ_2 are real, distinct, and of the same sign. In this event, $(0, 0)$ is asymptotically stable if λ_1 and λ_2 are both negative and unstable if λ_1 and λ_2 are both positive.

2. $(0, 0)$ is a saddle point if λ_1 and λ_2 are real, distinct, and of opposite sign. In this event, $(0, 0)$ is unstable.

3. $(0, 0)$ is a spiral point if λ_1 and λ_2 are complex conjugates with nonzero real parts. Further, $(0, 0)$ is asymptotically stable if λ_1 has a negative real part and unstable if the real part is positive.

4. $(0, 0)$ is a node if λ_1 and λ_2 are equal. In this event, $(0, 0)$ is asymptotically stable if $\lambda_1 < 0$ and unstable if $\lambda_1 > 0$.

5. $(0, 0)$ is a center if λ_1 and λ_2 are pure imaginary. Further, $(0, 0)$ is stable but not asymptotically stable. ∎

Since we have a system of linear equations, we can prove this theorem by explicitly writing the general solution and examining its behavior in the various cases.

Proof of Conclusion (1)   Suppose that λ_1 and λ_2 are real, distinct, and of the same sign. The general solution in the case $b \neq 0$ is

$$\begin{aligned} x(t) &= c_1 e^{\lambda_1 t} + c_2 e^{\lambda_2 t} \\ y(t) &= c_1\left(\frac{\lambda_1 - a}{b}\right)e^{\lambda_1 t} + c_2\left(\frac{\lambda_2 - a}{b}\right)e^{\lambda_2 t}. \end{aligned}$$

We will show that $(0, 0)$ is a node by producing an infinite family of trajectories entering the origin. Suppose, to be specific, that λ_1 and λ_2 are both negative. Since $\lambda_1 \neq \lambda_2$ by hypothesis, we may suppose that $\lambda_1 < \lambda_2 < 0$.

First, choose $c_2 = 0$ to get

$$\begin{aligned} x(t) &= c_1 e^{\lambda_1 t} \\ y(t) &= c_1\left(\frac{\lambda_1 - a}{b}\right)e^{\lambda_1 t}. \end{aligned}$$

Then

$$y = \frac{\lambda_1 - a}{b} x.$$

If $c_1 > 0$, then $x > 0$ and we get a half-line as shown in Figure 8.26. As $t \to \infty$, $x(t) \to 0$ and $y(t) \to 0$, because λ_1 is negative. This trajectory therefore enters $(0, 0)$ with slope $(\lambda_1 - a)/b$.

If $c_1 < 0$, then $x < 0$, but we still get a half-line trajectory entering $(0, 0)$ with slope $(\lambda_1 - a)/b$ as $t \to \infty$.

If $c_1 = 0$ but $c_2 \neq 0$, we get a similar conclusion, with two half-line trajectories of slope $(\lambda_2 - a)/b$, both entering $(0, 0)$ as $t \to \infty$.

If c_1 and c_2 are both nonzero, we still have $x(t) \to 0$ and $y(t) \to 0$ as $t \to \infty$, because λ_1 and λ_2 are both negative. Further, we can write

$$\frac{y}{x} = \frac{c_1\left(\dfrac{\lambda_1 - a}{b}\right)e^{\lambda_1 t} + c_2\left(\dfrac{\lambda_2 - a}{b}\right)e^{\lambda_2 t}}{c_1 e^{\lambda_1 t} + c_2 e^{\lambda_2 t}}$$

$$= \frac{c_1\left(\dfrac{\lambda_1 - a}{b}\right)e^{(\lambda_1 - \lambda_2)t} + c_2\left(\dfrac{\lambda_2 - a}{b}\right)}{c_1 e^{(\lambda_1 - \lambda_2)t} + c_2}.$$

Since $\lambda_1 < \lambda_2 < 0$, $\lambda_1 - \lambda_2 < 0$, and as $t \to \infty$, we get

$$\frac{y}{x} \to \frac{\lambda_2 - a}{b}.$$

All of these trajectories enter the origin with limiting slope $(\lambda_2 - a)/b$, and we conclude that the origin is an asymptotically stable node.

Returning to the case $b = 0$, we now find the general solution

$$x(t) = c_1 e^{at}$$

$$y(t) = c_1\left[\frac{c}{a - d}\right]e^{at} + c_2 e^{dt}.$$

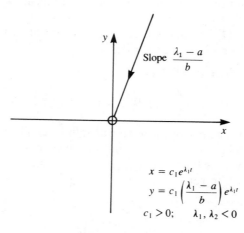

Figure 8.26. Stable node.

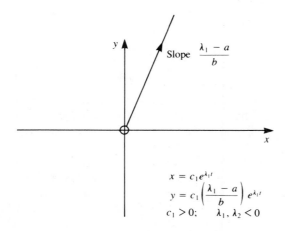

Figure 8.27. Unstable node.

If λ_1 and λ_2 are both positive, we find the same types of expressions for the solutions, but now we conclude that the trajectories enter the origin as $t \to -\infty$. Thus, the origin is a node in this case as well. However, now the trajectories are unbounded as $t \to \infty$, so the origin is an unstable node (see Figure 8.27).

Proof of Conclusion (3)   Suppose that λ_1 and λ_2 are complex conjugates with nonzero real parts. Write $\lambda_1 = p + iq$. We find that the general solution of the system can now be written

$$x(t) = e^{pt}\{c_1[\alpha_1\cos(qt) - \alpha_2\sin(qt)] + c_2[\alpha_1\sin(qt) + \alpha_2\cos(qt)]\}$$
$$y(t) = e^{pt}\{c_1[\beta_1\cos(qt) - \beta_2\sin(qt)] + c_2[\beta_1\sin(qt) + \beta_2\cos(qt)]\},$$

in which c_1 and c_2 are arbitrary constants and α_1, α_2, β_1, and β_2 are constants whose specific values are not needed for this proof.

Suppose first that $p < 0$. Then $x(t) \to 0$ and $y(t) \to 0$ as $t \to \infty$; hence, all trajectories approach $(0, 0)$. They *do not enter* $(0, 0)$, however. To see this, introduce the polar coordinate angle θ. Since $\theta = \tan^{-1}(y/x)$,

$$\theta'(t) = \frac{xy' - yx'}{x^2 + y^2}.$$

Since $x' = ax + by$ and $y' = cx + dy$, we get, upon substitution and rearrangement of terms, that

$$\theta' = \frac{cx^2 + (d - a)xy - by^2}{x^2 + y^2}.$$

Now, b and c must have opposite signs. Indeed, if b and c have the same sign, it is routine to check that the auxiliary equation has real roots. Suppose, to be specific, that $c > 0$ and $b < 0$. We claim that $\theta'(t) \neq 0$ for all t. For, if $y = 0$, the above equation for θ' gives us $\theta' = c > 0$. If $y \neq 0$, and $\theta'(t) = 0$ for some t,

$$cx^2 + (d - a)xy - by^2 = 0;$$

hence,

$$c\left(\frac{x}{y}\right)^2 + (d - a)\left(\frac{x}{y}\right) - b = 0.$$

Therefore, this equation has a real solution for x/y, contradicting the fact that this is the auxiliary equation (with x/y in place of λ), and we know that the roots of the auxiliary equation are complex. Thus, we conclude that $\theta'(t) \neq 0$ for all t. In fact, $\theta'(t) > 0$, since $\theta'(t) = c > 0$ when $y = 0$, and $\theta'(t)$ cannot change sign without being equal to zero for some t.

By a similar argument, $\theta'(t) < 0$ for all t if $c < 0$. But, from our explicit solution for x and y, we see that $x(t)$ and $y(t)$ change sign infinitely often as $t \to \infty$. We conclude that all trajectories spiral in toward the origin, counterclockwise if $c > 0$ and clockwise if $c < 0$. This is illustrated for $c > 0$ in Figure 8.28. Therefore, the origin is an asymptotically stable critical point. If $p > 0$, a similar analysis shows that the trajectories spiral toward the origin as $t \to -\infty$ but not as $t \to \infty$, and we conclude that in this event the origin is unstable.

A similar analysis can be used to prove conclusions (2), (4), and (5).

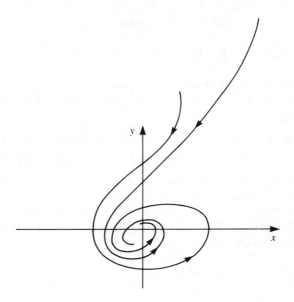

Figure 8.28 ∎

EXAMPLE 8.5 An Asymptotically Stable Node

Consider the system

$$x' = -3x + y$$
$$y' = x - 3y.$$

The auxiliary equation is

$$\lambda^2 + 6\lambda + 8 = 0,$$

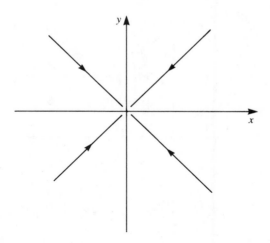

Figure 8.29. Some trajectories of

$$x' = -3x + y$$
$$y' = x - 3y.$$

with roots $\lambda_1 = -4$, $\lambda_2 = -2$. The general solution of the system is

$$x(t) = c_1 e^{-4t} + c_2 e^{-2t}$$
$$y(t) = -c_1 e^{-4t} - c_2 e^{-2t}.$$

If $c_1 \neq 0$ and $c_2 = 0$, we get half-line trajectories $y = -x$, x having the same sign as c_1.

If $c_2 \neq 0$ and $c_1 = 0$, we get half-line trajectories $y = -x$, x having the same sign as c_2. These trajectories enter the origin.

If $c_1 \neq 0$ and $c_2 \neq 0$, we obtain, after some manipulation,

$$\frac{y}{x} = \frac{-c_1 e^{-2t} + c_2}{c_1 e^{-2t} + c_2} \to 1 \quad \text{as} \quad t \to \infty.$$

The trajectories formed when c_1 and c_2 are both nonzero therefore enter the origin with slope 1, and the origin is an asymptotically stable node. Some of these trajectories are shown in Figure 8.29. ■

EXAMPLE 8.6   An Unstable Saddle Point

Consider the system

$$x' = -x + 3y$$
$$y' = 2x - 2y. \tag{8.12}$$

The auxiliary equation is $\lambda^2 + 3\lambda - 4 = 0$, with roots $\lambda_1 = 1$ and $\lambda_2 = -4$. The general solution is

$$x(t) = c_1 e^t + c_2 e^{-4t}$$
$$y(t) = \tfrac{2}{3} c_1 e^t - c_2 e^{-4t}.$$

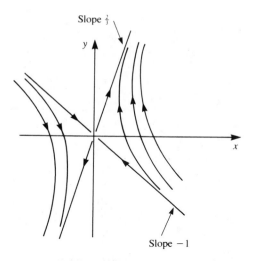

Figure 8.30. Trajectories of

$$x' = -x + 3y$$
$$y' = 2x - 2y.$$

If $c_1 = 0$ and $c_2 \neq 0$, then $y = -x = -c_2 e^{-4t}$. These half-line trajectories enter the origin with slope -1 as $t \to \infty$.

If $c_1 \neq 0$ and $c_2 = 0$, then $y = \frac{2}{3}x = \frac{2}{3}c_1 e^t$. These half-line trajectories enter the origin with slope $\frac{2}{3}$ as $t \to -\infty$ (not as $t \to \infty$).

If c_1 and c_2 are both nonzero, then x and y do not go to zero as $t \to \infty$ or as $t \to -\infty$. We get the trajectories shown in Figure 8.30.

The system (8.12) has a saddle point at the origin. Further, the origin is unstable, because any trajectories (other than the two half-lines discussed above) entering a circle about the origin must leave at some later time. ∎

EXAMPLE 8.7   A Stable, But Not Asymptotically Stable, Center

Consider the system

$$x' = 3x + y$$
$$y' = -13x - 3y.$$

The general solution of this system is

$$x(t) = c_1 \cos(2t) + c_2 \sin(2t)$$
$$y(t) = (2c_2 - 3c_1)\cos(2t) + (-2c_1 - 3c_2)\sin(2t).$$

The trajectories are closed paths about the origin, as shown in Figure 8.31. Neither x nor y approaches zero as $t \to \infty$ or as $t \to -\infty$. For this system, $(0, 0)$ is a stable, but not asymptotically stable, center.

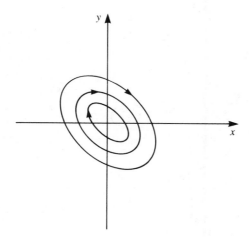

Figure 8.31. Trajectories of

$$x' = 3x + y$$
$$y' = -13x - 3y.$$ ∎

It is possible to summarize the conclusions of Theorem 8.1 succinctly in a diagram. Recall that the auxiliary equation of the linear system 8.11 is

$$\lambda^2 - (a + d)\lambda + (ad - bc) = 0.$$

Write this equation in the form

$$\lambda^2 + p\lambda + q = 0,$$

in which $p = -(a + d)$ and $q = ad - bc$. Note that $q \neq 0$, by assumption. The roots of the auxiliary equation are

$$\lambda_1 = \frac{-p + \sqrt{p^2 - 4q}}{2} \quad \text{and} \quad \lambda_2 = \frac{-p - \sqrt{p^2 - 4q}}{2}.$$

These roots are real or complex, depending upon whether $p^2 - 4q$ is positive or negative. The "boundary" of these two cases is the parabola $p^2 - 4q = 0$ in the pq-plane. Thus, in the pq-plane, the parabola $p^2 = 4q$ plays a key role.

Above this parabola ($p^2 < 4q$), the auxiliary equation has complex conjugate roots [case (3) of the theorem]; on the parabola ($p^2 = 4q$), it has real, equal roots [case (4)]; on the q-axis ($p = 0$), it has pure imaginary roots [case (5)]; below the p-axis, the roots are real and distinct, with opposite signs [case (2)]; and between the p-axis and the parabola, the roots are real and distinct, with the same sign [case (1)]. This information is summarized in Figure 8.32.

Although we have developed certain concepts with a view to studying nonlinear systems, Theorem 8.1 deals only with linear systems, which we can solve anyway. In the next section, we will turn our attention to nonlinear systems. In this discussion,

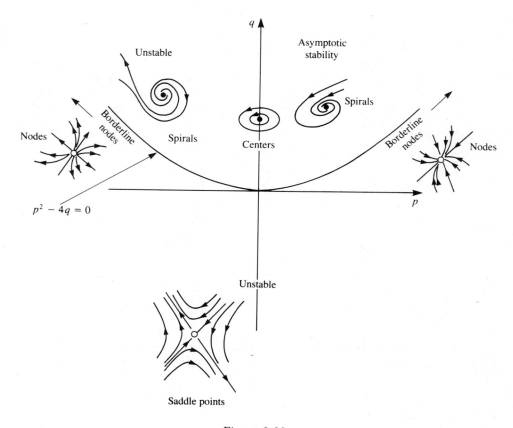

Figure 8.32

Theorem 8.1 will play an important role when we discuss the ramifications of replacing a nonlinear system with a linear one and using solutions of the linear system to draw conclusions about solutions of the nonlinear system.

PROBLEMS FOR SECTION 8.3

For each of the linear systems in Problems 1 through 16, classify the critical point at the origin and determine whether the origin is stable, asymptotically stable, unstable, or stable but not asymptotically stable.

1. $x' = 3x - 5y$
   $y' = 5x - 7y$

2. $x' = x + 4y$
   $y' = 3x$

3. $x' = x - 5y$
   $y' = x - y$

4. $x' = 9x - 7y$
   $y' = 6x - 4y$

5. $x' = 7x - 17y$
   $y' = 2x + y$

6. $x' = 2x - 7y$
   $y' = 5x - 10y$

7. $x' = 4x - y$
   $y' = x + 2y$

8. $x' = 3x - 5y$
   $y' = 8x - 3y$

9. $x' = -2x - y$
   $y' = 3x - 2y$

10. $x' = -6x - 7y$
   $y' = 7x - 20y$

11. $x' = 2x + y$
   $y' = x - 2y$

12. $x' = 3x - y$
   $y' = 5x + 3y$

13. $x' = x - 3y$
   $y' = 3x - 7y$

14. $x' = 10x - y$
   $y' = x + 12y$

15. $x' = 6x - y$
   $y' = 13x - 2y$

16. $x' = 3x - 2y$
   $y' = 11x - 3y$

For Problems 17 and 18, recall that the equation which describes the motion of the weight in a mass-spring problem is $my'' + cy' + ky = 0$. To use current notation, rewrite this differential equation as $z'' + (c/m)z' + (k/m)z = 0$ and construct an autonomous system of first order differential equations by putting $x = z$ and $y = z'$.

17. Classify the singularity at the origin and determine the type of stability of the equilibrium position of an undamped mass-spring system $(c = 0)$.

18. Classify the singularity at the origin and determine the type of stability of the equilibrium position of the damped mass-spring system. Consider three cases (overdamped, underdamped, and critically damped motion).

19. Prove conclusion (1) of Theorem 8.1 for the case $b = 0$.

20. Prove conclusion (2) of Theorem 8.1.

21. Prove conclusion (4) of Theorem 8.1.

22. Prove conclusion (5) of Theorem 8.1.

8.4 *Almost Linear Systems* _____

We will now return to the general autonomous system in x and y, which depend upon t. Since t does not appear explicitly in the system, we have

$$x' = F(x, y)$$
$$y' = G(x, y). \tag{8.13}$$

If the system has a critical point at (x_0, y_0), we can translate axes by letting $X = x - x_0$, $Y = y - y_0$. This results in a new autonomous system

$$X' = F(X + x_0, Y + y_0)$$
$$Y' = G(X + x_0, Y + y_0),$$

which has a critical point at $(0, 0)$. Further, solutions of the new system have the same behavior at $(0, 0)$ that solutions of the original system have at (x_0, y_0). Thus, we lose no generality by supposing that the system (8.13) has an isolated critical point at the origin. (There may, however, be other critical points as well. We will see later that the model of the damped pendulum has infinitely many.)

The main result of this section enables us to draw conclusions about the behavior of solutions of the system (8.13) when it is "not too different" from a linear system. To define what we mean by "not too different," consider the system

$$x' = ax + by + P(x, y)$$
$$y' = cx + dy + Q(x, y), \tag{8.14}$$

in which a, b, c, and d are constants.

Intuitively, if $P(x, y)$ and $Q(x, y)$ approach zero "fast enough" as (x, y) approaches the origin, the system (8.14) may be approximated by the linear system

$$x' = ax + by$$
$$y' = cx + dy, \tag{8.15}$$

and solutions of the two systems will exhibit similar behavior at the origin.

The condition we will require is that both $P(x, y)$ and $Q(x, y)$ approach zero faster than the distance between the origin and (x, y). This condition may be written

$$\lim_{(x, y) \to (0, 0)} \frac{P(x, y)}{\sqrt{x^2 + y^2}} = \lim_{(x, y) \to (0, 0)} \frac{Q(x, y)}{\sqrt{x^2 + y^2}} = 0.$$

When this condition holds, we call the system (8.14) *almost linear*. In order for these limits to be zero, it is necessary, but not sufficient, that $P(0, 0) = Q(0, 0) = 0$. This is exactly what is required for the origin to be a critical point of the system (8.14). We will now state our main result on almost-linear systems.

THEOREM 8.2

Suppose that the system (8.14) is almost linear and that $ad - bc \neq 0$. Assume that P and Q have first partial derivatives that are continuous for all x and y and that $(0, 0)$ is an isolated critical point of the system.

Let λ_1 and λ_2 be the roots of the auxiliary equation

$$\lambda^2 - (a + d)\lambda + ad - bc = 0.$$

Then solutions of the systems (8.14) and (8.15) have the same behavior at $(0, 0)$, in the following sense.

1. If λ_1 and λ_2 are real, distinct, and of the same sign, the origin is a node of both systems.
2. If λ_1 and λ_2 are real, distinct, and of opposite sign, the origin is a saddle point of both systems.
3. If $\lambda_1 = \lambda_2$, the origin is a node of both systems, unless $a = d \neq 0$ and $b = c = 0$. In this event, the origin is a node of the linear system but may be either a node or a spiral point of the nonlinear system.
4. If λ_1 and λ_2 are complex, with nonzero real parts, the origin is a spiral point of both systems.
5. If λ_1 and λ_2 are pure imaginary, the origin is a center of the linear system, while the origin is either a center or a spiral point of the nonlinear system. ∎

We will not prove this theorem. However, we will discuss two mathematical illustrations and later apply the theorem to an electrical circuit and to an analysis of the simple, damped pendulum.

EXAMPLE 8.8

Consider the nonlinear system

$$x' = 4x - 2y - 4xy$$
$$y' = x + 6y - 8x^2y.$$

We may think of this as the linear system

$$x' = 4x - 2y$$
$$y' = x + 6y,$$

perturbed by the addition of the nonlinear terms $P(x, y) = -4xy$ and $Q(x, y) = -8x^2y$. P and Q are continuous with continuous first partial derivatives over the entire plane. Further, $P(0, 0) = Q(0, 0) = 0$, and $(0, 0)$ is an isolated critical point of both the linear and nonlinear systems. Finally,

$$\lim_{(x, y) \to (0, 0)} \frac{-4xy}{\sqrt{x^2 + y^2}} = \lim_{(x, y) \to (0, 0)} \frac{-8x^2y}{\sqrt{x^2 + y^2}} = 0.$$

We can therefore use Theorem 8.2 to analyze the nature of the critical point. The auxiliary equation of the linear system is

$$\lambda^2 - 10\lambda + 26 = 0,$$

with roots $5 \pm i$. From conclusion (4) of the theorem, $(0, 0)$ is a spiral point of the nonlinear system. ∎

EXAMPLE 8.9

Consider the nonlinear system

$$x' = x - 3y + 4xy$$
$$y' = x + 7y - xy^{5/3}.$$

Note that $4xy$ and $-xy^{5/3}$ are continuous, with continuous first partial derivatives, for all x and y. Further,

$$\lim_{(x, y) \to (0, 0)} \frac{4xy}{\sqrt{x^2 + y^2}} = \lim_{(x, y) \to (0, 0)} \frac{-xy^{5/3}}{\sqrt{x^2 + y^2}} = 0.$$

Thus, consider the linear system

$$x' = x - 3y$$
$$y' = x + 7y.$$

This has auxiliary equation

$$\lambda^2 - 8\lambda + 10 = 0,$$

with roots $4 \pm \sqrt{6}$. By conclusion (1) of Theorem 8.2, the origin is a node of the nonlinear system. ∎

The following result of Liapunov enables us to draw conclusions about stability of the origin, for almost linear systems satisfying the hypotheses of Theorem 8.2.

THEOREM 8.3 Liapunov's Theorem

Under the hypotheses of Theorem 8.2, we may draw the following conclusions.

1. If λ_1 and λ_2 are either real and both negative, or complex with negative real parts, the origin is asymptotically stable for both systems (8.13) and (8.14).
2. If λ_1 and λ_2 are both positive, or both complex with positive real parts, the origin is an unstable critical point of both systems (8.13) and (8.14). ∎

For example, using Liapunov's theorem, we conclude that the origin is an unstable spiral point in Example 8.8 and an unstable node in Example 8.9.

EXAMPLE 8.10

Consider the system

$$x' = -x + y + x^3 y$$
$$y' = -2x - 3y - x^2 y^2.$$

Observe that $x^3 y$ and $-x^2 y^2$ satisfy the hypotheses required to use Theorems 8.2 and 8.3. The associated linear system is

$$x' = -x + y$$
$$y' = -2x - 3y.$$

This system has auxiliary equation

$$\lambda^2 + 4\lambda + 5 = 0,$$

with roots $-2 \pm i$. Since the roots are complex with negative real parts, the origin is an asymptotically stable spiral point of both systems. ∎

These examples illustrate the theorems but have no other significance. To provide some physical insight into trajectories and stability for nonlinear systems, we will consider an electrical circuit and then do a detailed analysis of the simple, damped pendulum.

EXAMPLE 8.11

Consider the electrical circuit of Figure 8.33. The differential equation for the current i is

$$Li' + Ri + \frac{1}{C} q = g(q, i),$$

where $g(q, i)$ accounts for any nonlinear terms (arising, for example, from an applied electromotive force). Let $x = q$ and $y = i = q'$ to obtain the system

$$x' = y$$
$$y' = -\frac{1}{LC} x - \frac{R}{L} y + \frac{1}{L} g(x, y).$$

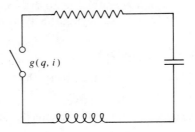

Figure 8.33

Assume that $g(0, 0) = 0$. Then the system has a critical point at the origin. The origin corresponds to the state $q = 0$, $i = 0$, in which the system has zero current and zero charge.

Also assume that

$$\lim_{(x, y) \to (0, 0)} \frac{g(x, y)}{\sqrt{x^2 + y^2}} = 0.$$

The associated linear system is

$$x' = y$$

$$y' = -\frac{1}{LC} x - \frac{R}{L} y.$$

The auxiliary equation of this system is

$$\lambda^2 + \frac{R}{L} \lambda + \frac{1}{LC} = 0,$$

with roots

$$\lambda = \frac{-\frac{R}{L} \pm \sqrt{\frac{R^2}{L^2} - \frac{4}{LC}}}{2}.$$

If $R^2/L^2 < 4/LC$, these roots are complex with negative real parts, and the origin is an asymptotically stable spiral point.

If $R^2/L^2 = 4/LC$, there is a repeated, negative root, and the origin is an asymptotically stable node or spiral point.

If $R^2/L^2 > 4/LC$, there are two distinct, negative real roots, and the origin is an asymptotically stable node.

It should not be surprising that the origin is asymptotically stable for this circuit. We would expect the charge to decay to zero with time if the input goes to zero, due to the consumption of energy in the resistor. ∎

EXAMPLE 8.12 The Simple, Damped Pendulum

Consider a simple, damped pendulum as shown in Figure 8.34. We will assume that the magnitude of the damping force is equal to $cL|\theta'(t)|$, in which c is the positive damping constant. We will derive a differential equation for θ.

The rate of change of the angular momentum about any point equals the moment of the resultant force about the point. The angular momentum about the origin is $mL^2\theta''(t)$. Therefore,

$$mL^2\theta'' = -cL\theta' - mgL \sin(\theta).$$

This equation is usually written

$$\theta''(t) + \frac{c}{mL} \theta'(t) + \frac{g}{L} \sin[\theta(t)] = 0,$$

a second order differential equation which is nonlinear due to the $\sin(\theta)$ term.

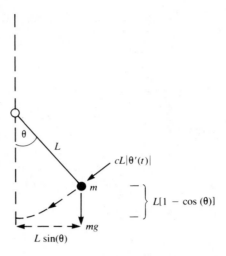

Figure 8.34

To convert this equation to a system of first order equations, let $x(t) = \theta(t)$ and $y(t) = \theta'(t)$. Then

$$x' = y$$
$$y' = -\frac{c}{mL} y - \frac{g}{L} \sin(x).$$

To make use of Theorem 8.2, add and subtract $(g/L)x$ from the equation for y' to obtain the system

$$x' = y$$
$$y' = -\frac{g}{L} x - \frac{c}{mL} y + \frac{g}{L}[x - \sin(x)].$$

We want to apply Theorem 8.2 with $P(x, y) = 0$ and $Q(x, y) = (g/L)[x - \sin(x)]$. We have

$$0 \le \left| \frac{x - \sin(x)}{\sqrt{x^2 + y^2}} \right| \le \left| \frac{x - \sin(x)}{x} \right| = \left| 1 - \frac{\sin(x)}{x} \right| \to 0$$

as $(x, y) \to (0, 0)$. Therefore,

$$\lim_{(x, y) \to (0, 0)} \frac{P(x, y)}{\sqrt{x^2 + y^2}} = \lim_{(x, y) \to (0, 0)} \frac{Q(x, y)}{\sqrt{x^2 + y^2}} = 0.$$

The associated linear system is

$$x' = y$$
$$y' = -\frac{g}{L} x - \frac{c}{mL} y.$$

This linear system has auxiliary equation

$$\lambda^2 + \frac{c}{mL}\,\lambda + \frac{g}{L} = 0,$$

with roots

$$\frac{-\dfrac{c}{mL} \pm \sqrt{\dfrac{c^2}{m^2L^2} - \dfrac{4g}{L}}}{2}.$$

Now look at the various cases, assuming that $c > 0$.

1. If $c^2/m^2L^2 - 4(g/L) > 0$, the roots of the auxiliary equation are real and distinct. Further, both roots are negative. Therefore, the origin is an asymptotically stable node.

2. If $c^2/m^2L^2 - 4(g/L) = 0$, the auxiliary equation has two equal, negative roots, and $(0, 0)$ is an asymptotically stable node of the linear system but may be an asymptotically stable node or asymptotically stable spiral point of the almost linear system.

3. If $c^2/m^2L^2 - 4(g/L) < 0$, the auxiliary equation has complex conjugate roots with negative real parts. Then $(0, 0)$ is an asymptotically stable spiral point of both the linear and almost linear systems.

What do these conclusions mean physically? Since $x = \theta$ and $y = \theta'$, the origin in the phase plane corresponds to the pendulum in the vertical position with the bob at the bottom ($\theta = 0$) and at rest ($\theta' = 0$).

The fact that the origin is asymptotically stable in all three cases means that, with a small displacement from this position, the pendulum will eventually return to this vertical position. Depending upon how large c is, it may do this in any of several ways. The bob may move to the vertical position with decreasing velocity, eventually coming to rest without passing through the vertical position (asymptotically stable node); or it may oscillate back and forth a number of times, finally coming to rest at the vertical position (asymptotically stable spiral point).

We will analyze the case in which the damping coefficient is "small" in more detail. Suppose that $c^2/m^2L^2 - 4(g/L) < 0$, or, equivalently, $c < 2m\sqrt{gL}$. We have noted that the nonlinear system has a critical point at the origin. It also has critical points $(n\pi, 0)$ for $n = 0, \pm 1, \pm 2, \ldots$.

On physical grounds, we expect these critical points to split into two categories. Each critical point $(2n\pi, 0)$, $n = 0, \pm 1, \pm 2, \ldots$, should be an asymptotically stable spiral point, as is the origin. These points all correspond to the pendulum in the vertical position with bob at the bottom, and the damped pendulum should return to this state if perturbed slightly. By contrast, each critical point $((2n + 1)\pi, 0)$, $n = 0, \pm 1, \pm 2, \ldots$, corresponds to the pendulum in a vertical position with bob at the top. After a perturbation, the pendulum will not return to this position, and we expect such points to be unstable saddle points. We will verify this explicitly for the point $(\pi, 0)$.

One way to analyze the critical point $(\pi, 0)$ is to translate the nonlinear system so that this point is moved to the origin. Let

$$X = x - \pi, \qquad Y = y.$$

The translated system is

$$X' = Y$$

$$Y' = -\frac{g}{L}(X + \pi) - \frac{c}{mL}Y + \frac{g}{L}[(X + \pi) - \sin(X + \pi)].$$

Since

$$\sin(X + \pi) = \sin(X)\cos(\pi) + \cos(X)\sin(\pi) = -\sin(X),$$

the system for X and Y can be written

$$X' = Y$$

$$Y' = +\frac{g}{L}X - \frac{c}{mL}Y - \frac{g}{L}[X - \sin(X)].$$

The associated linear system is

$$X' = Y$$

$$Y' = \frac{g}{L}X - \frac{c}{mL}Y.$$

This is identical to the associated linear system for x and y, except that $-(g/L)x$ has been replaced by $+(g/L)X$. We can therefore get the roots of the auxiliary equation for this new system by replacing $-g/L$ with g/L in the roots previously derived. We get roots

$$\frac{-\dfrac{c}{mL} \pm \sqrt{\left(\dfrac{c}{mL}\right)^2 + \dfrac{4g}{L}}}{2}.$$

The effect of the sign change of g/L is that these roots are real and distinct, with opposite signs. Therefore, $(0, 0)$ is an *unstable* saddle point of both the linear and nonlinear systems for X and Y. We conclude that $(\pi, 0)$ is an unstable saddle point for the damped pendulum. A similar analysis shows that each critical point $((2n + 1)\pi, 0)$ is an unstable saddle point.

Thus far, we have the critical points

$(2n\pi, 0)$—asymptotically stable spiral points,

$((2n + 1)\pi, 0)$—unstable saddle points

for $n = 0, \pm 1, \pm 2, \ldots$. We will examine the trajectories about these points in more detail. For the spiral point at $(0, 0)$, we can determine the direction of motion of the spirals from the system

$$x' = y$$

$$y' = -\frac{g}{L}\sin(x) - \frac{c}{mL}y.$$

Suppose the spiral intersects the y-axis at $(0, y)$, with $y > 0$, as in Figure 8.35. Then $x' > 0$ and $y' < 0$, so the spiral must be moving clockwise.

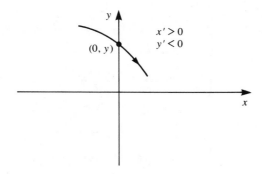

Figure 8.35

Similar reasoning holds for the trajectories about each spiral point $(2n\pi, 0)$. We can translate the system by letting

$$X = x - 2n\pi$$
$$Y = y$$

to obtain the new system

$$X' = Y$$

$$Y' = -\frac{g}{mL}\sin(X + 2n\pi) - \frac{c}{mL}Y$$

$$= -\frac{g}{L}\sin(X) - \frac{c}{mL}Y.$$

This system has the same form as the original one and hence has a spiral point at $(0, 0)$, with trajectories spiraling clockwise toward $(0, 0)$ about $(2n\pi, 0)$, as in Figure 8.36.

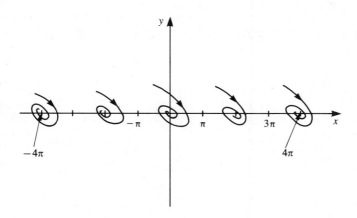

Figure 8.36. Spiral trajectories about asymptotically stable spiral points $(2n\pi, 0)$.

Now look at trajectories about the saddle points $((2n + 1)\pi, 0)$. To be specific, consider the point $(\pi, 0)$. When we translated axes previously, we obtained the system

$$X' = Y$$

$$Y' = \frac{g}{L} X - \frac{c}{mL} Y.$$

This system can be solved explicitly to get

$$X(t) = c_1 e^{\lambda_1 t} + c_2 e^{\lambda_2 t}$$
$$Y(t) = c_1 \lambda_1 e^{\lambda_1 t} + c_2 \lambda_2 e^{\lambda_2 t},$$

in which

$$\lambda_1 = \frac{-\dfrac{c}{mL} + \sqrt{\left(\dfrac{c}{mL}\right)^2 + \dfrac{4g}{L}}}{2} \quad \text{and} \quad \lambda_2 = \frac{-\dfrac{c}{mL} - \sqrt{\left(\dfrac{c}{mL}\right)^2 + \dfrac{4g}{L}}}{2}.$$

If we choose $c_1 = 0$ and $c_2 \neq 0$, we obtain the solution

$$X(t) = c_2 e^{\lambda_2 t}$$
$$Y(t) = c_2 \lambda_2 e^{\lambda_2 t},$$

with the property that $X(t) \to 0$ and $Y(t) \to 0$ as $t \to \infty$, because $\lambda_2 < 0$. Further,

$$\frac{Y}{X} = \lambda_2,$$

so the slopes of these trajectories are negative. These trajectories are pictured in Figure 8.37 for the cases $c_2 > 0$ and $c_2 < 0$.

If we choose $c_2 = 0$ and $c_1 \neq 0$, we get trajectories

$$X(t) = c_1 e^{\lambda_1 t}$$
$$Y(t) = c_1 \lambda_1 e^{\lambda_1 t}.$$

These exit from $(0, 0)$. Because $\lambda_1 > 0$, we have $X(t) \to 0$ and $Y(t) \to 0$ as $t \to -\infty$. Because $Y/X = \lambda_1$, these trajectories exit the origin with slope λ_1. They have the appearance shown in Figure 8.38, depending upon whether c_1 is positive or negative.

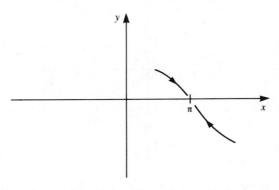

Figure 8.37

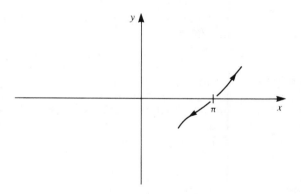

Figure 8.38

These conclusions have, of course, been drawn at the origin for the system for X and Y; they hold at $(\pi, 0)$ for the original system in x and y. A similar analysis can be made at each of the other unstable saddle points $((2n + 1)\pi, 0)$.

We can now draw the phase portrait (Figure 8.39) for the simple, damped pendulum (with small damping coefficient) by assimilating the drawings we have just made of trajectories about the stable and unstable critical points.

If we specify the system of differential equations for the damped pendulum, along with initial conditions $\theta(t_0) = \alpha$, $\theta'(t_0) = \beta$, we are in fact specifying a point (α, β) in the phase plane. The unique solution of this initial value problem is represented

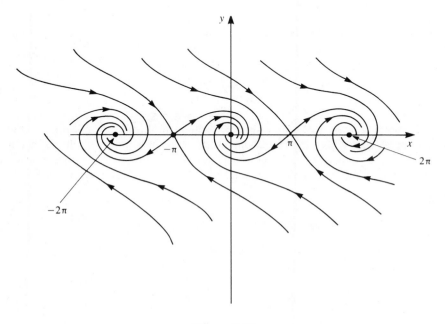

Figure 8.39

by the trajectory in the phase plane passing through this point. This trajectory contains the information about the motion of the system with those initial conditions, with one important exception.

Notice the trajectories entering the saddle points $((2n + 1)\pi. 0)$. Mathematically, we may choose (α, β) as a point on such a trajectory. The mathematical analysis tells us that eventually the pendulum will assume a vertical position, with the bob in the top position, which we know to be unstable. This is physically impossible. This mathematical model of the pendulum yields interesting and realistic results, except in this case, which cannot occur physically. ∎

PROBLEMS FOR SECTION 8.4

In each of Problems 1 through 8, apply Theorem 8.2 to determine the nature of the critical point $(0, 0)$, or explain why the theorem yields no conclusion. Then use Liapunov's theorem to determine the type of stability at the origin.

1. $x' = x + 3y + xy^2$
$y' = x + 5y + x^2$

2. $x' = 3x + 2y + 3xy$
$y' = 2x + y + 4x^2y^3$

3. $x' = -3x - y$
$y' = 4x - 3y + 2xy^2$

4. $x' = 3x - y + x\sin(y)$
$y' = -4x + 3y + x^2 - y^2$

5. $x' = 2x - 4y + 3x^3y^2$
$y' = x - 2y + 8x^4y^4$

6. $x' = 8x - 2y + x^3y$
$y' = 4x + 6y - x^2 + y^2$

7. $x' = -3x + 2y - 3x^2$
$y' = x - y - xy$

8. $x' = 2x - 6y - x^3$
$y' = 4x + 2y + \sqrt{xy}$

In each of Problems 9 through 12, determine the critical point of the system, translate the system so that the origin is the critical point of the translated system, and then determine the nature of the critical point of the original system as well as the type of stability it possesses.

9. $x' = x + y + 1$
$y' = 3x - y - 9$

10. $x' = x + y + 4$
$y' = 2x + y$

11. $x' = 3 - y$
$y' = x + 4y - 4$

12. $x' = 3x - 2y - 17$
$y' = 5x + y - 11$

In each of Problems 13 through 16, find all of the critical points of the system, then determine the nature of each critical point and the type of stability it possesses.

13. $x' = 40 - 5x - 8y + xy$
$y' = 5x + 2y$

14. $x' = x - 2y + xy - 2$
$y' = x - 2y$

15. $x' = 2x - y$
$y' = xy - 2y - 3x + 6$

16. $x' = xy - 3y + 3x - 9$
$y' = xy - 2y - x + 2$

Taylor series are a useful tool for linearizing almost linear systems, providing a means of expressing the functions in a system of equations in the form $ax + by + k(x, y)$, where k contains the terms of higher order than 1 in x and y. Suppose that F is a real-valued function of two real variables which has continuous partial derivatives of all orders. Then, with certain convergence hypotheses which we will not state here, the Taylor series representation of $F(x, y)$ about the point (x_0, y_0) is

$$F(x, y) = F(x_0, y_0) + \frac{\partial F}{\partial x}(x_0, y_0)(x - x_0) + \frac{\partial F}{\partial y}(x_0, y_0)(y - y_0)$$

$$+ \frac{\partial^2 F}{\partial x^2}(x_0, y_0)(x - x_0)^2 + 2\frac{\partial^2 F}{\partial x\,\partial y}(x_0, y_0)(x - x_0)(y - y_0)$$

$$+ \frac{\partial^2 F}{\partial y^2}(x_0, y_0)(y - y_0)^2 + \cdots.$$

If (x_0, y_0) is a critical point of the system (8.13), $F(x_0, y_0) = 0$. If the system is translated so that the critical point is

(0, 0), the series representation of the translated function $F(X + x_0, Y + y_0)$ about the new critical point has the form

$$\frac{dX}{dt} = \frac{dx}{dt} = F(x, y) = F(X + x_0, Y + y_0) = \frac{\partial F}{\partial x}(x_0, y_0)X + \frac{\partial F}{\partial y}(x_0, y_0)Y + k(X, Y),$$

where $k(X, Y)$ contains only powers of X and Y higher than the first power. We can similarly represent $G(x, y)$ in a Taylor series expansion about (x_0, y_0).

In each of Problems 17 through 20, use Taylor series expansions [of both $F(x, y)$ and $G(x, y)$] to linearize the system, then use Theorem 8.2 to determine the nature of the critical point and Theorem 8.3 to determine the type of stability.

17. $x' = 2x + x^2 y$
   $y' = e^{4x + 3y} - \cos(x)$

18. $x' = \sin(y - 3x)$
   $y' = y + 13\sin(x)$

19. $x' = \sinh(y - 4x)$
   $y' = -xe^{3y}$

20. $x' = 5x + 2y + e^{-x} - 1$
   $y' = \cos(x) - \sin(3y) - 1$

21. Theorems 8.2 and 8.3 are inconclusive in the case in which the critical point of an almost linear system is a center of the associated linear system. The systems (8.16) and (8.17) that follow can be used to show that no conclusion is possible in this case. Show this by carrying out the argument outlined in (a) through (e).

$$x' = y - x\sqrt{x^2 + y^2}$$
$$y' = -x - y\sqrt{x^2 + y^2} \tag{8.16}$$
$$x' = y + x\sqrt{x^2 + y^2}$$
$$y' = -x + y\sqrt{x^2 + y^2} \tag{8.17}$$

(a) Show that (0, 0) is a center for the associated linear system of both (8.16) and (8.17).

(b) Show that each system is almost linear.

(c) Introduce polar coordinates with $x = r\cos(\theta)$, $y = r\sin(\theta)$, and use the chain rule to get

$$x' = \frac{dx}{dr}\frac{dr}{dt} = \cos(\theta)r'(t) \quad \text{and} \quad y' = \frac{dy}{dr}\frac{dr}{dt} = \sin(\theta)r'(t).$$

Now use these derivatives to evaluate the expression $x(dx/dt) + y(dy/dt)$ in terms of r and r', where $r' = dr/dt$.

(d) Evaluate $x(dx/dt) + y(dy/dt)$ in system (8.16) and write the result in terms of r. Combine this result with the result of (c) to obtain a separable differential equation in terms of r and t. Conclude from this that $r'(t) < 0$ for all t. Now solve the differential equation for r, and show that $r \to 0$ as $t \to \infty$. Thus, conclude that, in system (8.16), the origin is asymptotically stable.

(e) Apply the procedure of (d) to system (8.17). Conclude that $r'(t) > 0$ for all t. Solve the separable differential equation for r in terms of t, with the initial condition $r(t_0) = r_0$. Conclude that $r \to \infty$ as $t \to -\infty$. Therefore, conclude that the origin is unstable for system (8.17).

22. Convert the equation $\theta'' + (g/L)\sin(\theta) = 0$ into a nonlinear first order system and find all of the critical points.

(a) Use Taylor series about the point $(2n\pi, 0)$, with n any integer, to linearize the system. Show that each critical point $(2n\pi, 0)$ is a stable center of the linear system and hence that no information pertaining to stability can be obtained from theorems of this section. What does your intuition suggest is happening at the points $(2n\pi, 0)$?

(b) Use Taylor series to linearize the almost linear system about a point $((2n + 1)\pi, 0)$, with n any integer. Classify these critical points. Does this information agree with your intuition about what is happening at these points?

23. A magnet of mass m is suspended from the lower end of a spring with spring modulus k. The magnetic weight comes to rest in an equilibrium position L units above a small piece of iron. The weight is then pulled down and released from rest. Assume that the attractive force between the magnet and the iron is K/r^2 when they are separated by r units and determine the smallest value of K which will prevent oscillations of the magnet. Do not forget to include the force due to the magnetic attraction when establishing the equilibrium position (critical point).

24. An iron weight of mass m is suspended from the lower end of a spring with modulus k and allowed to come to rest in an equilibrium position. A magnetic ring of radius r is then placed in a horizontal position with the weight at its center. The attractive force between the weight and the ring is K/z^2 when they are separated by a distance z. The weight also has a dashpot attached to it which has a damping constant c. The weight is pulled down y_0 units and released from rest.
    (a) Write the equation which describes the motion of the weight.
    (b) Convert the equation found in (a) to a system of first order differential equations.
    (c) Use a binomial expansion (see Problem 57 of Section 5.2) to linearize the equations.
    (d) Classify the critical point for various values of K.

25. Suppose the weight in Problem 24 is replaced with a magnet of the same mass, and polarity opposite that of the magnetic ring, so that the force between the ring and weight is repellent.
    (a) Write a differential equation which describes the motion of the weight.
    (b) Convert the differential equation of (a) to a system of first order differential equations.
    (c) Classify the critical point for various values of K.

8.5 *The Predator-Prey and Competing Species Models*

In this section, we will apply the methods of the preceding section to an analysis of a predator-prey model and also consider behavior of a population in which two species compete for the same resources.

A PREDATOR-PREY MODEL

Suppose we have an environment in which there are, at time t, $x(t)$ rabbits and $y(t)$ foxes. Assume that the environment has, for all practical purposes, an unlimited supply of vegetation to serve as food for the rabbits and that the rabbits in turn serve as food for the foxes. The foxes are the predators and the rabbits are the prey. Also assume that there are no external factors (such as hunters) serving to decrease or increase these populations and that each encounter of a rabbit with a fox results in the fox eating the rabbit.

In Section 8.2, we argued that x and y should satisfy a nonlinear system of differential equations of the form

$$x' = ax - bxy$$
$$y' = cxy - ky, \tag{8.18}$$

in which a, b, c, and k are positive constants. The nonlinear xy-terms account for the interaction between the two species. Meetings of foxes with rabbits tend to decrease the rabbit population and, by providing food, increase the fox population, facts accounted for by the signs in the xy-terms of the model (8.18).

Make the natural assumption that the populations of both rabbits and foxes must be nonnegative for all time. Thus, the entire phase portrait for this system lies in the first quadrant of the phase plane. The system (8.18) has two critical points, $(0, 0)$ and $(k/c, a/b)$. These are one-point trajectories in the phase portrait.

If $x(0) = x_0 > 0$ and $y(0) = 0$ (x_0 rabbits, no foxes), the solution of the system (8.18) is

$$x = \varphi(t) = x_0 e^{at}$$
$$y = \psi(t) = 0.$$

Thus, the positive x-axis, with direction to the right, is the trajectory associated with the case in which $y_0 = 0$. In this event, the initial (time zero) population of foxes is zero, and the rabbit population, lacking a restrictive predator, simply grows exponentially.

If the initial populations are $x_0 = 0$ and $y_0 > 0$ (y_0 foxes, no rabbits), the solution is

$$x = \varphi(t) = 0$$
$$y = \psi(t) = y_0 e^{-kt}.$$

This trajectory is the positive y-axis, directed downward. In the absence of its source of food (the rabbits), the fox population eventually dies out.

Any initial values $x(0) = x_0 > 0$, $y(0) = y_0 > 0$ are represented by a point (x_0, y_0) in the first quadrant. A trajectory $x = \varphi(t)$, $y = \psi(t)$ passing through (x_0, y_0) cannot cross either axis (because the positive x- and y-axes are trajectories) and must therefore remain in the first quadrant. Thus, $\varphi(t) > 0$ and $\psi(t) > 0$ for all t.

We will analyze the critical point $(k/c, a/b)$. Translate the system by letting

$$X = x - \frac{k}{c} \quad \text{and} \quad Y = y - \frac{a}{b}.$$

Because $x > 0$ and $y > 0$, we must have

$$X > -\frac{k}{c} \quad \text{and} \quad Y > -\frac{a}{b}.$$

We obtain the new nonlinear system

$$X' = -\frac{bk}{c} Y - bXY$$

$$Y' = \frac{ac}{b} X + cXY, \tag{8.19}$$

with a critical point at the origin corresponding to the critical point $(k/c, a/b)$ of the original system (8.18). The linear system associated with (8.18) is

$$X' = -\frac{bk}{c} Y$$

$$Y' = \frac{ac}{b} X. \tag{8.20}$$

This system has auxiliary equation

$$\lambda^2 + ak = 0,$$

with imaginary roots $\pm\sqrt{ak}i$, suggesting periodic solutions with $(0,0)$ as a center for the translated, linearized system (8.20). Unfortunately, this is a case in which solutions of the nonlinear and associated linear systems may not have the same behavior. We will therefore study the linearized system further, then show that we can implicitly define a solution of the nonlinear system for this particular model.

Eliminate t in the translated linearized system (8.20) by writing

$$\frac{\dfrac{dX}{dt}}{\dfrac{dY}{dt}} = \frac{dX}{dY} = \frac{-\dfrac{bk}{c}Y}{\dfrac{ac}{b}X} = -\frac{b^2 kY}{ac^2 X}.$$

This is a separable differential equation, which can be written

$$ac^2 X + b^2 kY\frac{dY}{dX} = 0.$$

This equation is easily integrated (with respect to X) to yield

$$ac^2 X^2 + b^2 kY^2 = K,$$

in which K is a positive, but otherwise arbitrary, constant. This is the equation of an ellipse in the XY-plane. We conclude that the phase portrait of the linearized system (8.19) for x and y has elliptical trajectories centered about the critical point $(k/c, a/b)$, shown in Figure 8.40.

We will now return to the original autonomous system (8.18) and derive an implicitly defined solution. Eliminate t in the system (8.18) by dividing the second

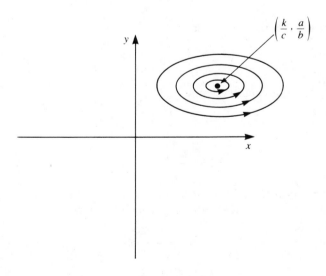

Figure 8.40

equation by the first to get

$$\frac{\frac{dy}{dt}}{\frac{dx}{dt}} = \frac{cxy - ky}{ax - bxy},$$

or

$$\frac{dy}{dx} = \frac{cxy - ky}{ax - bxy} = \frac{y(cx - k)}{x(a - by)}.$$

This is a separable differential equation, which can be written

$$\frac{a - by}{y}\frac{dy}{dx} = \frac{cx - k}{x}.$$

Integrate both sides of this equation with respect to x to get

$$a\ln(y) - by = cx - k\ln(x) + C,$$

or

$$\ln(y^a x^k) = cx + by + C.$$

This can be written

$$y^a x^k = Ke^{cx + by},$$

in which $K = e^C$ is an arbitrary positive constant.

If $x(0) = x_0$ and $y(0) = y_0$, then

$$(y_0)^a(x_0)^k = Ke^{cx_0 + by_0};$$

hence,

$$K = \frac{(y_0)^a(x_0)^k}{e^{cx_0 + by_0}}.$$

The solution of (8.18), subject to the initial condition $x(0) = x_0$, $y(0) = y_0$, is implicitly defined by the equation

$$\frac{y^a x^k}{e^{cx + by}} = \frac{(y_0)^a(x_0)^k}{e^{cx_0 + by_0}}.$$

We cannot explicitly solve this equation for y as a function of x, or vice versa. However, in Problem 1 at the end of this section we will show that this equation determines closed (but not elliptical) curves about the equilibrium point $(k/c, a/b)$.

We conclude that $(k/c, a/b)$ is a center of the system (8.18) and that the rabbit and fox populations are cyclic. The amount of variation in the populations will depend on their initial values. An initial population pair near the critical point will produce a system in which the populations are almost constant with time. A phase portrait of this system is shown in Figure 8.41, with $a = 2$, $b = 1$, $c = 1$, $k = 3$, and various initial values. Notice that the curve near the critical point is almost elliptical, in agreement

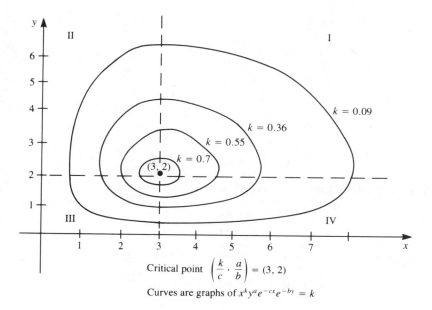

Critical point $\left(\dfrac{k}{c}, \dfrac{a}{b}\right) = (3, 2)$

Curves are graphs of $x^k y^a e^{-cx} e^{-by} = k$

Figure 8.41. Predator-prey phase portrait ($a = 2, b = c = 1,$ $k = 3$).

with the solution of the linearized system. These closed trajectories are graphs of periodic solutions of the system.

The horizontal and vertical lines through the critical point (3, 2) in Figure 8.41 separate the first quadrant into four regions. A trajectory will pass through all four regions as it makes one complete cycle. If a population pair is in region I ($x > k/c$, $y > a/b$), the foxes and rabbits are both in abundance. This will produce more encounters, which brings with it more rabbit kills. In this region, $x'(t) < 0$ and $y'(t) > 0$, so there is a decrease in rabbits and an increase in foxes.

Once the rabbit population reaches a value of k/c, the foxes find insufficient food and their numbers begin to decline from the maximum value obtained at this time. This brings us into region II, where both populations are on the decline.

When the fox population reaches the value a/b, their numbers are small enough to allow the rabbits to multiply faster than they are being consumed, and (x, y) is in region III. Here the foxes are declining and the rabbits are increasing in number.

When the fox population obtains its minimum value, the rabbit population is increasing at its fastest rate. That is, when $y'(t) = 0$ (when $x = k/c$), $x'(t)$ is a maximum. At this time, the population pairs move into region IV, where the fox population begins to increase due to the increased availability of rabbits.

This process continues to repeat cyclically, with the fox population increasing anytime the rabbit population exceeds the value k/c and decreasing with a lack of food. By the same token, the rabbit population increases whenever the fox population is below a/b.

Much of this information about the behavior of the populations can be obtained directly from the original system (8.18). For example, when $y > a/b$, $x'(t) =$

$ax - bxy = bx[a/b - y] < 0$, so the rabbit population decreases, in agreement with our observations.

It is interesting that there exists a predator-prey situation for which there are good records to test the model and its conclusions. The Hudson Bay Company of Canada has kept records of pelts traded at its stations beginning near the middle of the nineteenth century. Assuming that the actual populations of lynx and snowshoe hare are proportional to the number of pelts obtained by hunters, records over the period 1845–1935 indicate clear periodic variations in the lynx and hare populations, having about a 10-year cycle. As predicted by the model, the two populations are out of phase, with the lynx reaching a peak as the hares decline, followed by an increase in the hare population and a decline in the lynx population.

Currently, predator-prey models, usually analyzed using computer simulation, are used by biologists to model populations in an effort to predict ecological ramifications of such phenomena as construction, mining projects, and migration of human populations.

A MODEL FOR COMPETING SPECIES

In the predator-prey model, the predator species preys on the other species, which in turn has a source of food independent of the predator. In a competing species model, neither species preys on the other but both compete for a common type of food.

For an unrestricted environment, a model for competing species is the system

$$
\begin{aligned}
x' &= ax - bxy \\
y' &= ky - cxy,
\end{aligned}
\tag{8.21}
$$

in which a, b, c, and k are positive constants. Now an xy-term, accounting for a reduction in population due to competition, is *subtracted* in both equations.

As with the predator-prey model, this system has two critical points, $(0, 0)$ and $(k/c, a/b)$. Further, we have $x(t)$ and $y(t)$ nonnegative for all t, and the positive x and y axes are trajectories (with the x-axis directed outward, as before, but now the y-axis also directed outward, opposite that of the predator-prey case).

A significant difference between the models is seen with the behavior of trajectories near the critical point $(k/c, a/b)$. Translate the system (8.21) by letting $X = x - k/c$ and $Y = y - a/b$ to obtain

$$
\begin{aligned}
X' &= -\frac{bk}{c} Y - bXY \\
Y' &= -\frac{ac}{b} X - cXY.
\end{aligned}
\tag{8.22}
$$

The associated linear system is

$$
\begin{aligned}
X' &= -\frac{bk}{c} Y \\
Y' &= -\frac{ac}{b} X.
\end{aligned}
\tag{8.23}
$$

The roots of the auxiliary equation of system (8.23) are $\pm\sqrt{ak}$, which are real and of opposite sign. (Recall that, in the predator-prey model, we obtained pure imaginary roots.) This result does provide information about the behavior of the trajectories near the critical point $(0, 0)$: the origin is a saddle point of both systems (8.22) and (8.23). We conclude that $(k/c, a/b)$ is a saddle point of the system (8.21).

We can obtain implicitly defined solutions of the translated, linearized system (8.23) as follows. Divide X' by Y' to get

$$\frac{dX}{dY} = \frac{\left(\dfrac{bk}{c}\right)Y}{\left(\dfrac{ac}{b}\right)X}.$$

Solve this separable differential equation to get

$$\frac{X^2}{\dfrac{b}{ac}} - \frac{Y^2}{\dfrac{c}{bk}} = K.$$

This is a family of hyperbolas with center $(0, 0)$ in the XY-plane. Therefore, trajectories of the almost linear system (8.21) will look very much like these near $(k/c, a/b)$ in the xy-plane. Trajectories approach $(k/c, a/b)$ and then veer away, becoming unbounded with time. This behavior is shown in Figure 8.42.

Those trajectories entering or leaving $(k/c, a/b)$ are of special interest. It is possible to study these trajectories in detail because we can solve the original system (8.21).

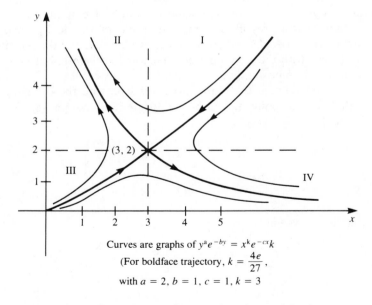

Curves are graphs of $y^a e^{-by} = x^k e^{-cx} k$
(For boldface trajectory, $k = \dfrac{4e}{27}$,
with $a = 2, b = 1, c = 1, k = 3$

Figure 8.42. Competing species phase portrait.

Divide the equation for y by that for x to get

$$\frac{dy}{dx} = \frac{y(k - cx)}{x(a - by)}.$$

Write this equation as

$$\frac{y}{a - by}\frac{dy}{dx} = \frac{k - cx}{x}.$$

Integrate both sides with respect to x and rearrange terms to get

$$y^a e^{-by} = K x^k e^{-cx},$$

in which we must choose

$$K = \frac{(y_0)^a e^{cx_0}}{(x_0)^k e^{by_0}}$$

to satisfy initial conditions $x(0) = x_0$ and $y(0) = y_0$.

If we choose $x_0 = k/c$ and $y_0 = a/b$, we obtain an equation whose graph is shown as the heavier curves in Figure 8.42 for the case in which $a = 2$, $b = c = 1$, and $k = 3$. This graph divides the first quadrant into four regions, I through IV. As with the predator-prey model, we can draw interpretations about the physical problem from the graph (although we can obtain much of this information from the system of differential equations itself, without having the luxury of a solution).

In regions I and II, both competitors begin with large populations which decrease with time due to the large number of animals having to share resources. One species will dominate the other and, unlike the predator-prey problem, one group will die out (though theoretically not in finite time). The initial populations determine which will dominate. Recall that the coefficients a, b, c, and k play a strong role in determining the regions, so it takes more than just a large population to win.

In regions III and IV, the initial populations are small, and in an environment with ample food supplies both species grow rapidly, but at some point one will dominate.

In the problems that follow, we will pursue the behavior of both the predator-prey and competing population models in which the resources are assumed to be limited in a specified way.

PROBLEMS FOR SECTION 8.5

1. Show that the trajectories in the predator-prey problem are closed paths, as follows.
   (a) Graph the function

$$f(x) = x^k e^{-cx},$$

   with $k = 3$ and $c = 1$ and $0 < x < 10$. Determine the maximum value of f and where this maximum value occurs, for arbitrary positive k and c, in terms of k and c.
   (b) Write the solution of the predator-prey problem as

$$f(x) = x^k e^{-cx} = \frac{K e^{by}}{y^a}.$$

Using the graph in (a), explain why, for each value of y such that

$$0 < \frac{Ke^{by}}{y^a} < \left(\frac{k}{c}\right)^k e^{-k},$$

there are exactly two distinct values of x satisfying the equation, one smaller than k/c and the other larger than k/c. Remember that k/c is the x-coordinate of the nontrivial critical point of the system.

(c) Repeat (a) and (b) with the function

$$g(y) = y^a e^{-by},$$

using values $a = 2$ and $b = 1$ for the graph.

2. In the predator-prey example, the roots of the auxiliary equation were found to be complex. Use the material developed in Section 8.4 to express the trajectories (ellipses) explicitly in terms of sines and cosines. Remember that this system is a translation of the original predator-prey problem, so it is necessary to translate back to have a solution of the linearized original problem. Show that every trajectory has the same period even though the length of the closed path will vary with different choices of the constants in the general solution.

3. Find the average value of the fox population in the predator-prey problem. *Hint:* Denote the time for a cycle as T. Then the average value of the fox population is

$$\bar{y} = \frac{1}{T} \int_0^T y(t) \, dt.$$

To evaluate this integral, write the equation for the rabbit population as

$$\frac{1}{x} \frac{dx}{dt} = a - by.$$

Integrate this equation with respect to t to get

$$\int_{t=0}^{t=T} \frac{1}{x} \frac{dx}{dt} \, dt = \int_0^T (a - by) \, dt.$$

What is the difference in the two population values $x(T)$ and $x(0)$? Now evaluate the second integral and solve for $\bar{y}$.

4. Duplicate the calculation of Problem 3 to compute the average rabbit population.

5. Derive the system of equations which models a predator-prey relationship, with the prey living in a restricted environment. You may wish to incorporate the logistic equation discussed in Section 8.1. What is the minimum carrying capacity of the prey to sustain life in the system? Express the predator's average population as a function of the carrying capacity of the prey and the average population of the prey. Under what circumstances is the average population of the predator greater when the prey is restricted than when it is not restricted?

6. Classify the critical point of the linearized predator-prey problem with the prey in the restricted environment of Problem 5. What can be said about the critical point of the almost linear system?

7. Derive the system of differential equations which models the predator-prey relationship in an unrestricted environment with indiscriminate harvesting. That is, there is some outside agent that removes members of both species from the system at a rate proportional to the populations, with the same constant of proportionality for both species.

8. Find the critical points of the system of Problem 7. Compare the location of the nontrivial critical point to that of the one in the predator-prey system in the absence of harvesting. In which direction has the prey's coordinate moved? In light of the results of Problems 3 and 4, who benefits more from indiscriminate harvesting, the predator or the prey? How does this result apply to the application of pesticides?

9. Study the behavior of the populations of two competing species in which both groups live in a restricted environment. Do not assume that the carrying capacity of the environment is the same for both species.

8.6 *Limit Cycles*

In this section, we will consider another important aspect of the qualitative study of differential equations.

Consider the following experiment, which is not difficult to carry out. Measure the distance L between the points where a bicycle's wheels touch the ground. Let R be a positive number greater than L, and draw a circle of radius R on the ground. Now push the bicycle so that its front wheel always remains on this circle of radius R. What path does the back wheel follow?

There are no critical points here (unless $R = L$). However, as we will see, the back wheel does have a definite goal, albeit neither a point nor infinity: it will approach the circle C of radius $\sqrt{R^2 - L^2}$ concentric with the one on which the front wheel is moved. If the back wheel begins *outside C*, it will spiral inward toward C; if it begins *inside C*, it will work its way outward toward C. Finally, if the rear wheel begins on C, it will remain on C.

This inner circle C has many of the properties possessed by a stable critical point. Trajectories which start near C move toward it, and, if a trajectory starts on C, it stays there. This is "almost" like what we saw with perturbations of a damped pendulum from the vertical position with the bob at the bottom. The major difference between C and a critical point is that C is not a point but instead is a closed curve. C is an object called a *limit cycle*.

To develop this idea a little further, we will analyze a mathematical example. Consider the autonomous almost linear system

$$x' = x + y - x\sqrt{x^2 + y^2}$$
$$y' = -x + y - y\sqrt{x^2 + y^2}. \tag{8.24}$$

To determine the critical points of the system (8.24), we must solve the equations

$$x + y - x\sqrt{x^2 + y^2} = 0$$
$$-x + y - y\sqrt{x^2 + y^2} = 0.$$

Multiply the first equation by y and the second by x to get

$$xy + y^2 - xy\sqrt{x^2 + y^2} = 0$$
$$-x^2 + xy - xy\sqrt{x^2 + y^2} = 0.$$

Subtract the second equation from the first to get

$$x^2 + y^2 = 0,$$

hence $x = y = 0$. The origin is the only critical point of this system.

The associated linear system of (8.24) is

$$x' = x + y$$
$$y' = -x + y. \tag{8.25}$$

The system (8.25) has auxiliary equation $\lambda^2 - 2\lambda + 2 = 0$, with roots $1 \pm i$. Therefore, $(0, 0)$ is an unstable critical point of both the linear system (8.25) and the nonlinear system (8.24), by Liapunov's theorem. Further, the trajectories spiral outward.

In every example we have seen thus far, trajectories which spiraled outward have grown without bound. We will show that this is *not* the case for the system (8.25). To do this, convert the system (8.24) to polar coordinates by letting $x = r \cos(\theta)$ and $y = r \sin(\theta)$. Assume that $r \geq 0$. Since $x^2 + y^2 = r^2$, we have, upon differentiating with respect to t, that

$$x \frac{dx}{dt} + y \frac{dy}{dt} = r \frac{dr}{dt},$$

or

$$xx' + yy' = rr'.$$

In the system (8.24), multiply the first equation by x and the second by y, then add the resulting equations to get

$$xx' + yy' = x^2 + xy - x^2\sqrt{x^2 + y^2} - xy + y^2 - y^2\sqrt{x^2 + y^2}$$
$$= x^2 + y^2 - (x^2 + y^2)\sqrt{x^2 + y^2}$$
$$= r^2 - r^3.$$

Since $xx' + yy' = rr'$, we conclude that

$$rr' = r^2 - r^3.$$

Assuming that $r \neq 0$, this yields a differential equation involving r,

$$r' = r(1 - r).$$

If $r > 1$, then $r' < 0$; hence, trajectories lying outside the unit circle tend to decrease in magnitude with increasing t. If $0 < r < 1$, then $r' > 0$, so trajectories lying inside the unit circle tend to grow with time. The precise path followed by such trajectories is not clear from the analysis done so far.

To find a differential equation involving θ, differentiate the equations $x = r \cos(\theta)$ and $y = r \sin(\theta)$ with respect to t to get

$$x' = r'\cos(\theta) - r \sin(\theta)\theta', \qquad y' = r'\sin(\theta) + r \cos(\theta)\theta'.$$

Now observe that

$$yx' - xy' = r \sin(\theta)[r'\cos(\theta) - r \sin(\theta)\theta'] - r \cos(\theta)[r'\sin(\theta) + r \cos(\theta)\theta']$$
$$= -r^2[\sin^2(\theta) + \cos^2(\theta)]\theta' = -r^2\theta'.$$

But, from the system (8.24), we also have

$$yx' - xy' = xy + y^2 - xy\sqrt{x^2 + y^2} + x^2 - xy + xy\sqrt{x^2 + y^2}$$
$$\cdot \quad = x^2 + y^2 = r^2.$$

From the last two equations, we conclude that

$$-r^2\theta' = r^2.$$

Again assuming that $r \neq 0$, we obtain

$$\theta' = -1.$$

The differential equations we have found for r and θ (written together below) are uncoupled (that is, one equation involves only θ and the other, only r):

$$\theta' = -1$$
$$r' = r(1 - r). \tag{8.26}$$

This system is equivalent to the original system (8.24), in the sense that we can obtain solutions of one system from solutions of the other, using the equations relating rectangular and polar coordinates.

Solve each equation of the uncoupled system (8.26) to get

$$\theta = -t + c_1$$
$$\frac{r-1}{r} = c_2 e^{-t},$$

in which c_1 and c_2 are arbitrary constants. Then

$$\theta = -t + c_1$$
$$r = \frac{1}{1 - c_2 e^{-t}}.$$

From the second equation, it is obvious that $r \to 1$ as $t \to \infty$. Given a point with polar coordinates (r_0, θ_0) in the plane, with $r_0 > 0$, the solution of (8.26) passing through this point at time $t = 0$ is

$$\theta = \theta_0 - t$$
$$r = \frac{1}{1 - \dfrac{r_0 - 1}{r_0} e^{-t}}.$$

If $r_0 = 1$, so that the initial point is on the unit circle about the origin, the solution is $\theta = \theta_0 - t$, $r = 1$ and remains on the unit circle, traveling counterclockwise as t increases.

If $r_0 < 1$, the denominator in r is larger than 1 for all t and decreases toward 1. Thus, $r < 1$, and the trajectory is inside the unit circle and approaches it from within (Figure 8.43).

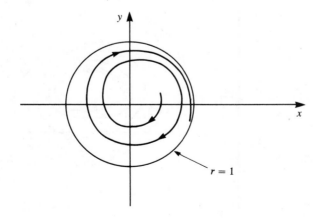

$r = 1$

Figure 8.43

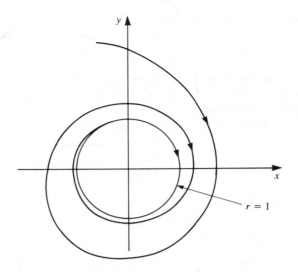

Figure 8.44

If $r_0 > 1$, the denominator in r is less than 1 for all t and approaches 1, so the solution spirals inward toward the unit circle from outside this circle (Figure 8.44).

A closed trajectory, such as the unit circle in this example, which has nonclosed trajectories spiraling toward it from either inside or outside as t increases, is called a *limit cycle*. Typical limit cycles are shown in Figures 8.45 and 8.46.

We say that a limit cycle is *stable* if all trajectories which start within a certain distance from the limit cycle spiral toward it. This is illustrated in Figure 8.47. If all

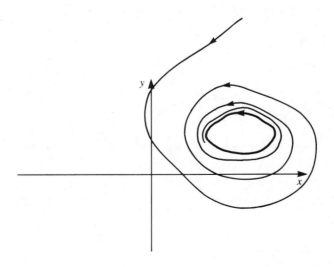

Figure 8.45. Limit cycle.

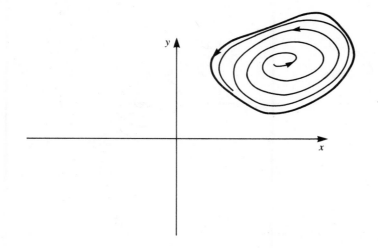

Figure 8.46. Limit cycle.

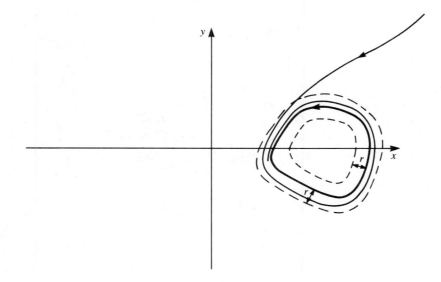

Figure 8.47. Stable limit cycle.

trajectories on either side of the limit cycle spiral away from it as t increases, we call the limit cycle *unstable*. Figure 8.48 shows a typical unstable limit cycle.

If all the trajectories sufficiently close to the limit cycle and on one side of it spiral toward it as t increases, and those on the other side of the limit cycle spiral away from it as t increases, we say that the limit cycle is *semistable*. Figure 8.49 shows a typical semistable limit cycle.

Finally, if a closed trajectory of a system is neither approached nor receded away from by other trajectories as t increases, we call the closed trajectory *neutrally stable*. Figure 8.50 shows neutrally stable closed trajectories.

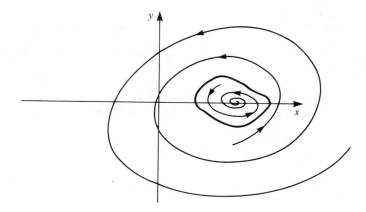

Figure 8.48. Unstable limit cycle.

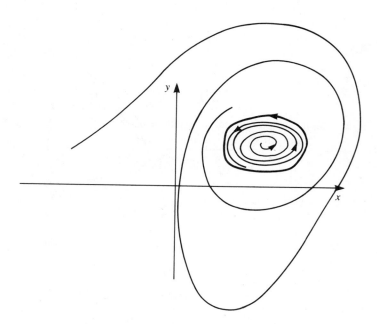

Figure 8.49. Semistable limit cycle.

The example we have just discussed was atypical because we were able to determine the trajectories explicitly. There are theorems which help us determine the presence or absence of limit cycles in a nonlinear autonomous system in certain instances when we cannot produce a solution explicitly or implicitly. The first two theorems help us identify where an autonomous system *cannot* have a closed trajectory.

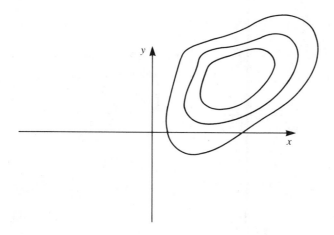

Figure 8.50. Neutrally stable closed trajectories.

THEOREM 8.4

Suppose that C is a closed trajectory of the autonomous system

$$x' = F(x, y)$$
$$y' = G(x, y),$$

in which F, G, and their first partial derivatives are continuous for all (x, y). Then there is at least one critical point of the system enclosed by C. ∎

This theorem can be used to determine that certain regions of the plane cannot contain closed trajectories of the system. For example, if the origin is the only critical point, there can be no closed trajectory lying completely in one quadrant of the phase plane, because such a closed trajectory cannot enclose a critical point.

THEOREM 8.5   Bendixson's Theorem

Let F and G and their first partial derivatives be continuous over the entire phase plane. Suppose that

$$\frac{\partial F}{\partial x} + \frac{\partial G}{\partial y}$$

is either strictly positive or strictly negative for all (x, y) in a certain region R of the plane. Then the autonomous system

$$x' = F(x, y)$$
$$y' = G(x, y)$$

has no closed trajectory in R.

Proof   Suppose that R contains a closed trajectory C: $x = \varphi(t), y = \psi(t)$ for $a \leq t \leq b$.

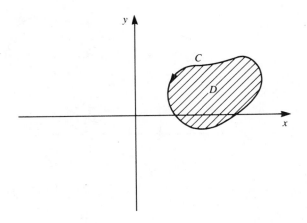

Figure 8.51

Suppose that C bounds a region D of the plane (Figure 8.51). By Green's theorem,

$$\int_C -G\,dx + F\,dy = \iint_D \left(\frac{\partial F}{\partial x} + \frac{\partial G}{\partial y}\right) dx\,dy.$$

Since $\partial F/\partial x + \partial G/\partial y$ has the same sign throughout D, the double integral on the right cannot be zero.

Now look at the line integral. From the system of differential equations,

$$dx = F(x, y)\,dt \quad \text{and} \quad dy = G(x, y)\,dt.$$

Then

$$\int_C -G\,dx + F\,dy = \int_a^b [-G(\varphi(t), \psi(t))F(\varphi(t), \psi(t))$$
$$+ F(\varphi(t), \psi(t))G(\varphi(t), \psi(t))]\,dt = 0.$$

This contradiction proves the theorem. ∎

Here is an illustration of the use of Bendixson's theorem.

EXAMPLE 8.13

Consider the system

$$x' = 3x + 4y + x^3$$
$$y' = 5x - 2y + y^3.$$

With $F(x, y) = 3x + 4y + x^3$ and $G(x, y) = 5x - 2y + y^3$, F and G and their first partial derivatives are continuous for all (x, y). Further,

$$\frac{\partial F}{\partial x} + \frac{\partial G}{\partial y} = 3 + 3x^2 - 2 + 3y^2 = 1 + 3(x^2 + y^2) > 0$$

for all (x, y). Therefore, this system has no closed trajectory in the plane. ∎

Theorems 8.4 and 8.5 provide negative criteria; they help us determine when an autonomous system has no closed trajectory in a certain region of the plane. The next theorem gives sufficient conditions for a closed trajectory to exist.

THEOREM 8.6   Poincaré-Bendixson Theorem

Let R be a bounded region of the phase plane, containing no critical point of the autonomous system

$$x' = F(x, y)$$
$$y' = G(x, y).$$

Assume that F and G and their first partial derivatives are continuous for all (x, y). Let C be a trajectory that is in R at all times $t > t_0$. Then either C is itself a closed trajectory or C spirals toward a closed trajectory as $t \to \infty$.   ∎

The system (8.24) considered previously provides an illustration of this powerful result. Recall that the system is

$$x' = x + y - x\sqrt{x^2 + y^2}$$
$$y' = -x + y - y\sqrt{x^2 + y^2}.$$

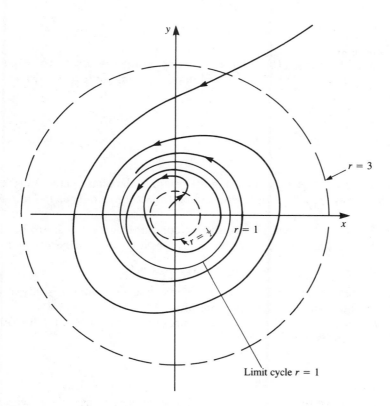

Figure 8.52

Upon introducing polar coordinates, we found that $r' = r(1 - r)$. Let R be the region between the concentric circles $r = \frac{1}{2}$ and $r = 3$ shown in Figure 8.52. The only critical point of this system is $(0, 0)$, which is not in R.

Note that $r' > 0$ on the inner circle and $r' < 0$ on the outer circle. A trajectory entering R across either circle will remain in R and spiral toward the unit circle $r = 1$. This is the closed trajectory (actually a limit cycle) whose existence is guaranteed by the Poincaré-Bendixson theorem.

We will conclude this section with Lienard's theorem, which deals with solutions of the differential equation

$$x'' + p(x)x' + q(x) = 0.$$

To place this differential equation into the present context, we will convert it to a system of first order differential equations. Let $y = x'$ to get

$$x' = y$$
$$y' = -p(x)y - q(x).$$

THEOREM 8.7   Lienard's Theorem

Let p and q be continuous and have continuous derivatives for all x. Suppose also that

1. $q(x) = -q(-x)$ for all x.
2. $q(x) > 0$ for $x > 0$.
3. $p(x) = p(-x)$ for all x.
4. The function $F(x) = \int_0^x p(\xi)\, d\xi$ has exactly one positive zero, say at $x = a$, $F(x) < 0$ for $0 < x < a$, and $F(x)$ is positive and nondecreasing for $x > a$. Further, assume that $F(x) \to \infty$ as $x \to \infty$.

   Then the system

$$x' = y$$
$$y' = -p(x)y - q(x)$$

has a unique closed trajectory enclosing the origin in the phase plane. Further, every other trajectory approaches this unique closed trajectory spirally as $t \to \infty$. ∎

Thus, Lienard's theorem gives sufficient conditions for a system to have a limit cycle. As an illustration of Lienard's theorem, consider the *van der Pol equation*

$$x'' + \alpha(x^2 - 1)x' + x = 0,$$

derived by Balthasar van der Pol in the 1920s in connection with the study of vacuum tubes. In this equation, α is a positive constant.

With $p(x) = \alpha(x^2 - 1)$ and $q(x) = x$, it is routine to verify that the hypotheses of Lienard's theorem hold. For example,

$$F(x) = \int_0^x \alpha(\xi^2 - 1)\, d\xi = \alpha\left(\frac{1}{3}x^3 - x\right),$$

with a single positive zero at $x = \sqrt{3}$. The first order system associated with van der

Pol's equation is

$$x' = y$$
$$y' = -\alpha(x^2 - 1)y - x.$$

By Lienard's theorem, van der Pol's equation has a unique closed trajectory about the critical point (0, 0). This closed trajectory represents a periodic solution. Existence of a periodic solution of van der Pol's equation is not otherwise obvious.

For the case $\alpha = 1$, Figure 8.53 shows the limit cycle of van der Pol's equation, together with some of the trajectories approaching the limit cycle from both within and without.

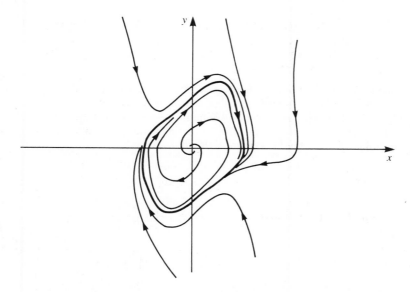

Figure 8.53. Limit cycle of van der Pol's equation for $\alpha = 1$.

PROBLEMS FOR SECTION 8.6

In each of Problems 1 through 4, use Bendixson's theorem to prove that the system has no closed trajectory.

1. $x' = -2x - y + x^3$
   $y' = 10x + 5y + x^2y - 2y \sin(x)$

2. $x' = -x - 3y + e^{2x}$
   $y' = x + 2y + \cos(y)$

3. $x' = 2x - 7y + \sinh(x)$
   $y' = x - 3y + 5e^{3y}$

4. $x' = y$
   $y' = -x + y(9 - x^2 - 3y^2)$ for (x, y) in the elliptical region R given by $x^2/9 + y^2 \le 1$.

5. With the transformation from rectangular to polar coordinates, we obtain

$$x\frac{dx}{dt} + y\frac{dy}{dt} = r\frac{dr}{dt} \quad \text{and} \quad y\frac{dx}{dt} - x\frac{dy}{dt} = -r^2\frac{d\theta}{dt}. \tag{8.27}$$

Use these equations to show that the system

$$x' = y + \frac{x}{\sqrt{x^2 + y^2}} f(\sqrt{x^2 + y^2})$$

$$y' = -x + \frac{y}{\sqrt{x^2 + y^2}} f(\sqrt{x^2 + y^2})$$

has limit cycles associated with zeros of the function f (assume that f is continuous for all $t \geq 0$). What is the direction of these cycles?

In each of Problems 6 through 9, use equations (8.27) of Problem 5 to find all of the closed trajectories of the system, determine the type of stability each possesses, and sketch the phase portrait of the system.

6. $x' = y + x \sin(\sqrt{x^2 + y^2})$ **7.** $x' = y(1 - x^2 - y^2)$ **8.** $x' = x(1 - x^2 - y^2)$

     $y' = -x + y \sin(\sqrt{x^2 + y^2})$ $y' = -x(1 - x^2 - y^2)$ $y' = y(1 - x^2 - y^2)$

9. $x' = y + x(1 - x^2 - y^2)(4 - x^2 - y^2)(9 - x^2 - y^2)$

     $y' = -x + y(1 - x^2 - y^2)(4 - x^2 - y^2)(9 - x^2 - y^2)$

In each of Problems 10 through 13, use the Poincaré-Bendixson theorem to establish the existence of a closed trajectory of the system. In each problem, find an annular region R, consisting of points (x, y) with $a \leq x^2 + y^2 \leq b$ for some positive numbers a and b, such that solutions within R remain within R (to do this, check the sign of $xx' + yy'$ on the circles bounding the annulus). By using polar coordinates, show that R contains a closed trajectory.

10. $x' = x - y - x\sqrt{x^2 + y^2}$ **11.** $x' = 4x - 4y - x(x^2 + 9y^2)$

      $y' = x + y - y\sqrt{x^2 + y^2}$ $y' = 4x + 4y - y(x^2 + 9y^2)$

12. $x' = y$

      $y' = -x + y - y(x^2 + 2y^2)$ *Hint:* Consider the region $\frac{1}{2} \leq x^2 + y^2 \leq 1$.

13. $x' = 4x - 2y - x(4x^2 + y^2)$

      $y' = 2x + 4y - y(4x^2 + y^2)$

In each of Problems 14 through 22, determine whether the system has a closed trajectory.

14. $x' = 3x + 4xy + xy^2$ **15.** $x' = -y + x + x(x^2 + y^2)$ **16.** $x' = -y^3$

      $y' = -2y^2 + x^4y$ $y' = x + y + y(x^2 + y^2)$ $y' = 3x + 2x^3$

17. $x' = y$ **18.** $x' = y$ **19.** $x' = x - 5y + x^3$

      $y' = x^2 + e^{\sin(x)}$ $y' = -x + y - x^2y$ $y' = x - y + y^3 + 7y^5$

20. $x' = y$ **21.** $x' = y$ **22.** $x' = -9x - 5y - x(x^2 + 9y^2)$

      $y' = -x + ye^{-y}$ $y' = -x^3$ $y' = 5x + 9y - y(x^2 + 9y^2)$

23. (a) Suppose that a solution of the autonomous system $x' = F(x, y)$, $y' = G(x, y)$ has a closed trajectory. Prove that this trajectory is a periodic solution.

      (b) Let C be the graph of a solution $x = \varphi(t)$, $y = \psi(t)$ of the autonomous system, and suppose that C is a closed curve. Suppose that $x(t_0) = x_0$ and $y(t_0) = y_0$. Show that there exists a positive number T such that $\varphi(t_0 + T) = x_0$ and $\psi(y_0 + T) = y_0$.

A differential equation with x as a function of t has a periodic solution if there is a solution $x = \varphi(t)$ and a positive number T such that $\varphi(t + T) = \varphi(t)$ for all t. In each of Problems 24 through 30, prove that the differential equation has a periodic solution.

24. $x'' + (5x^4 - 12x^2)x' + 4x^3 = 0$ **25.** $x'' + (x^2 - 1)x' + 2x + \sin(x) = 0$

26. $x'' + (5x^4 + 9x^2 - 4)x' + \sinh(x) = 0$ **27.** $x'' + x^3 = 0$

28. $x'' + 4x = 0$ **29.** $x'' + \dfrac{x}{1 + x^2} = 0$

30. $x'' + 2(x^2 - 1)x' + 5x^3 = 0$

31. (a) Use Bendixson's theorem to prove that the van der Pol equation does not have a closed trajectory completely contained in the infinite strip $-1 \leq x \leq 1$.

   (b) Use Bendixson's theorem to prove that the van der Pol equation does not have a closed trajectory completely contained in either the half-plane $x \geq 1$ or the half-plane $x \leq -1$.

   (c) Prove that the unique closed trajectory of van der Pol's equation, whose existence is guaranteed by Lienard's theorem, meets at least one of the lines $x = 1$ or $x = -1$ in two distinct points.

CHAPTER 9

Calculus of Variations

9.0 Introduction

The calculus of variations is an area of analysis which deals with certain kinds of *extremal problems*, in which we attempt to maximize or minimize an expression, often written as an integral. Sometimes the problem is also subject to constraints, which may also be expressed in terms of one or more integrals. The motivation for a study of problems of this form is that many processes in nature, as well as in mathematics, tend to act in such a way as to maximize or minimize some quantity which can be written as an integral.

Here are two illustrations from antiquity. Although each predates calculus by more than two thousand years, they are examples of extremal problems in interesting settings.

According to legend, an African chief offered to the Phoenician princess Dido as much land as she could enclose in an ox hide. In order to make the most of her opportunity, she cut the ox hide into thin strips and enclosed a circular region in which she eventually founded the city-state Carthage, which at one time rivaled Rome for control of the Mediterranean. In modern terminology, the problem Dido solved was one of determining a closed curve in the plane enclosing a maximum area, subject to the constraint that the length was fixed. This problem would be posed today in terms of an integral (for the area), with the constraint (fixed length of the boundary curve) as another integral.

Heron of Alexandria, an ancient Greek philosopher, was aware that the angle of incidence equals the angle of reflection for a beam of light reflecting from a polished surface. He made a more subtle observation, however, and conjectured that this law of nature actually guarantees that the beam will take the shortest path from source to reflection point by way of the mirror. The idea behind this insight is similar in spirit to

Fermat's optical principle, which states that, whenever light travels between two points in different media, the path taken will minimize the time of passage.

In the seventeenth and eighteenth centuries, the brachistochrone problem posed by James and John Bernoulli gave major impetus to the development of a calculus to deal with certain kinds of optimization problems. What path should we draw between the points P_0 and P_1 in a vertical plane so that a bead sliding along a wire from P_0 to P_1 will complete its journey in minimum time? We might be tempted to guess a straight line path (minimal length), but Galileo had already observed that some other paths yield shorter times. Using an argument from optics, the Bernoullis obtained the solution, which is a curve called a *brachistochrone* (*brachos* means "shortest" in Greek, and *chronos* means "time"). We will apply the calculus of variations to solve the brachistochrone problem in Section 9.1.

Today, it is recognized that problems involving optimization of one or more integrals abound in mathematics, physics, and engineering. The third law of thermodynamics states that the universe tends toward a state of maximum entropy, which is usually written as an integral. Hamilton's principle in mechanics states that a conservative mechanical system will move in such a way as to minimize an integral involving the potential and kinetic energies. Problems in the elastic behavior of beams, plates, and shells can be approached by observing that an integral involving the energy must be a minimum. And a soap solution stretched on a wire frame will form a surface of minimum area. With these considerations as motivation, we will now explore some introductory aspects of the calculus of variations.

9.1 *The First Problem in the Calculus of Variations*

Suppose we are given a function f of three variables and we want to choose a function $y(x)$, whose graph extends from (a, y_0) to (b, y_1), such that the integral

$$\int_a^b f(x, y(x), y'(x))\, dx \tag{9.1}$$

is a maximum or minimum. This is a bit different from the standard max/min problems seen in calculus. There, we often seek a value of some parameter, such as the diagonal of a square, which maximizes or minimizes some quantity such as area or volume, perhaps subject to some other conditions. Most max/min problems in calculus have a *number* as the solution. Here, the object sought is not a number but a function. We want a function $y(x)$ whose graph is a smooth curve extending from (a, y_0) to (b, y_1), which makes the integral (9.1) as large or as small as possible, given f.

A function y which maximizes or minimizes (9.1) is called a *stationary function*, and its graph is a *stationary curve* of this integral. In this section, we will derive a condition which enables us to reduce the problem of finding a stationary function to one of solving a related differential equation.

Suppose that $y = u(x)$ is a stationary function for (9.1). That is, $u(x)$ maximizes or minimizes $\int_a^b f(x, y(x), y'(x))\, dx$. We want some condition which will help us find u. In order to derive such a condition, we will begin by considering a special class of functions whose graphs pass through the required points (a, y_0) and (b, y_1) but are variations from u formed by adding a term of a certain form. Specifically, we will consider functions

$$y(x) = u(x) + \epsilon\eta(x), \tag{9.2}$$

in which ϵ is constant and $\eta(x)$ is any function having a continuous derivative on $[a, b]$ and satisfying the conditions

$$\eta(a) = \eta(b) = 0. \tag{9.3}$$

Such a function η is called an *admissible function*. The graph of $y(x) = u(x) + \epsilon\eta(x)$ passes through the required points (a, y_0) and (b, y_1) because of the conditions (9.3). Such curves have the appearance shown in Figure 9.1.

Substitute $y = u + \epsilon\eta$ into the integral of (9.1) to get the integral

$$\int_a^b f(x, u(x) + \epsilon\eta(x), u'(x) + \epsilon\eta'(x)) \, dx.$$

For any admissible function η, we may think of this integral as a function of ϵ, since u is also fixed (it is a solution of the problem). Thus, we will write

$$I(\epsilon) = \int_a^b f(x, u(x) + \epsilon\eta(x), u'(x) + \epsilon\eta'(x)) \, dx.$$

Since $u(x)$ is a solution of the problem, this integral has its maximum or minimum value when $\epsilon = 0$. Since $I(\epsilon)$ is a function of the single variable ϵ, we expect that $I'(\epsilon) = 0$ when $\epsilon = 0$:

$$\left.\frac{dI}{d\epsilon}\right|_{\epsilon=0} = 0.$$

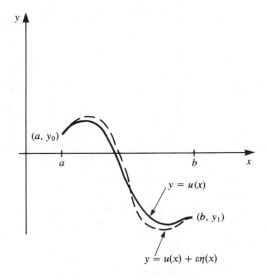

Figure 9.1

Compute $dI/d\epsilon$ and then let $\epsilon = 0$. Differentiating by chain rule under the integral sign, we have

$$\frac{dI}{d\epsilon} = \int_a^b \left[\frac{\partial f}{\partial x} \frac{\partial x}{\partial \epsilon} + \frac{\partial f}{\partial y} \frac{\partial y}{\partial \epsilon} + \frac{\partial f}{\partial y'} \frac{\partial y'}{\partial \epsilon} \right] dx. \tag{9.4}$$

Now,

$$\frac{\partial x}{\partial \epsilon} = 0, \qquad \frac{\partial y}{\partial \epsilon} = \frac{\partial}{\partial \epsilon} [u(x) + \epsilon \eta(x)] = \eta(x),$$

and

$$\frac{\partial y'}{\partial \epsilon} = \frac{\partial}{\partial \epsilon} [u'(x) + \epsilon \eta'(x)] = \eta'(x).$$

Thus, equation (9.4) becomes

$$\frac{dI}{d\epsilon} = \int_a^b \left[\frac{\partial f}{\partial y} \eta(x) + \frac{\partial f}{\partial y'} \eta'(x) \right] dx.$$

Upon setting $\epsilon = 0$, we get

$$\left. \frac{dI}{d\epsilon} \right|_{\epsilon = 0} = \int_a^b \left[\frac{\partial f}{\partial y} \eta(x) + \frac{\partial f}{\partial y'} \eta'(x) \right] dx. \tag{9.5}$$

Integrate the second term on the right of (9.5) by parts. Write

$$\int_a^b \underbrace{\frac{\partial f}{\partial y'}}_{U} \underbrace{\eta'(x)\, dx}_{dV} = \underbrace{\frac{\partial f}{\partial y'} \eta(x) \Big]_a^b}_{UV} - \int_a^b \underbrace{\frac{d}{dx}\left(\frac{\partial f}{\partial y'} \right)}_{dU} \underbrace{\eta(x)}_{V} \, dx.$$

Because $\eta(a) = \eta(b) = 0$, we have

$$\frac{\partial f}{\partial y'} \eta(x) \Big]_a^b = 0;$$

hence,

$$\int_a^b \frac{\partial f}{\partial y'} \eta'(x)\, dx = -\int_a^b \frac{d}{dx}\left(\frac{\partial f}{\partial y'} \right) \eta(x)\, dx.$$

Substitute this result into equation (9.5) to get

$$\left. \frac{dI}{d\epsilon} \right|_{\epsilon = 0} = \int_a^b \left[\frac{\partial f}{\partial y} \eta(x) - \frac{d}{dx}\left(\frac{\partial f}{\partial y'} \right) \eta(x) \right] dx = 0.$$

We can write this integral as

$$\int_a^b \left[\frac{\partial f}{\partial y} - \frac{d}{dx}\left(\frac{\partial f}{\partial y'} \right) \right] \eta(x)\, dx = 0. \tag{9.6}$$

Now, η is *any* function having a continuous derivative on $[a, b]$ and vanishing at a and at b. We claim that the only way equation (9.6) can hold for every such η is that the quantity in brackets must be zero for $a \le x \le b$. This will follow from the following lemma.

LEMMA

Suppose that g is continuous on $[a, b]$ and that $\int_a^b g(x)\eta(x)\,dx = 0$ for every function η which is continuous on $[a, b]$ and satisfies $\eta(a) = \eta(b) = 0$. Then $g(x) = 0$ for $a \le x \le b$.

Proof   Suppose that $g(x) > 0$ for some x_0 in $[a, b]$. Because g is continuous at x_0, there is some subinterval $[u, v]$ of $[a, b]$ in which $g(x) > 0$. Define

$$\eta(x) = \begin{cases} 0 & \text{if} \quad a \le x < u \\ (x - u)^2(x - v)^2 & \text{if} \quad u \le x \le v \\ 0 & \text{if} \quad v < x \le b. \end{cases}$$

Figure 9.2 shows a graph of η. It is routine to check that η is continuous on $[a, b]$. Further, $g(x)\eta(x)$ is zero on $[a, u]$ and on $[v, b]$ but positive on (u, v). Therefore,

$$\int_a^b g(x)\eta(x)\,dx = \int_u^v g(x)\eta(x)\,dx > 0,$$

a contradiction. Therefore, $g(x)$ must be zero for $a \le x \le b$. A similar argument applies if $g(x)$ is negative at any point of $[a, b]$. ∎

By applying this lemma to equation (9.6), we conclude that

$$\frac{\partial f}{\partial y} - \frac{d}{dx}\left(\frac{\partial f}{\partial y'}\right) = 0. \tag{9.7}$$

This is *Euler's equation*, and it provides the condition we want. A stationary function for $\int_a^b f(x, y, y')\,dx$ must satisfy Euler's equation (9.7).

If we use chain rule differentiation to expand the term $(d/dx)(\partial f/\partial y')$ in equation (9.7), Euler's equation can be written

$$\frac{\partial f}{\partial y} - \left[\frac{\partial}{\partial x}\left(\frac{\partial f}{\partial y'}\right) + \frac{\partial}{\partial y}\left(\frac{\partial f}{\partial y'}\right)\frac{dy}{dx} + \frac{\partial}{\partial y'}\left(\frac{\partial f}{\partial y'}\right)\frac{dy'}{dx}\right] = 0,$$

or

$$\frac{\partial f}{\partial y} - \frac{\partial^2 f}{\partial x\,\partial y'} - \frac{\partial^2 f}{\partial y\,\partial y'}\frac{dy}{dx} - \frac{\partial^2 f}{\partial y'^2}\frac{d^2 y}{dx^2} = 0. \tag{9.8}$$

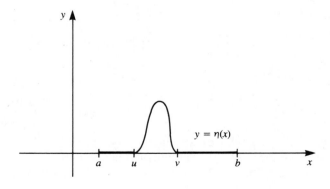

Figure 9.2

Equation (9.8) is simply a more explicit form of Euler's equation (9.7). From this formulation, it is apparent that Euler's equation is a second order ordinary differential equation for y. [Although partial derivatives appear in (9.8), these are partial derivatives of a known function and for a given f will be functions of x, y, and y'.] In Problem 13 at the end of this section, the student is asked to show that equation (9.8) may also be written

$$\frac{1}{y'}\left[\frac{d}{dx}\left(f - \frac{\partial f}{\partial y'}\frac{dy}{dx}\right) - \frac{\partial f}{\partial x}\right] = 0, \tag{9.9}$$

which is still another form of Euler's equation.

There are two special cases in which Euler's equation may be easier to solve than in others.

Case 1. If f does not explicitly depend on y, then $\partial f/\partial y = 0$, and Euler's equation (9.7) becomes simply

$$\frac{d}{dx}\left(\frac{\partial f}{\partial y'}\right) = 0.$$

We can integrate this equation once to get

$$\frac{\partial f}{\partial y'} = C, \tag{9.10}$$

a first order differential equation for a stationary function.

Case 2. If f does not explicitly depend on x, then $\partial f/\partial x = 0$, and equation (9.9) implies that

$$\frac{d}{dx}\left(f - \frac{\partial f}{\partial y'}\frac{dy}{dx}\right) = 0.$$

We can immediately carry out one integration to get

$$f - \frac{\partial f}{\partial y'}\frac{dy}{dx} = C, \tag{9.11}$$

again obtaining a first order differential equation for a stationary function.

EXAMPLE 9.1

Find the surface of revolution of least surface area passing through the points (a, y_0) and (b, y_1).

We want the equation of a curve $y = y(x)$ passing through (a, y_0) and (b, y_1) such that the surface of revolution formed by rotating the graph about the x-axis, as in Figure 9.3, has the least possible surface area. We will restrict our attention to smooth curves (so that y' must be continuous on $[a, b]$). In this case, the area of the surface of revolution is

$$\int_a^b 2\pi y \sqrt{1 + (y')^2}\, dx.$$

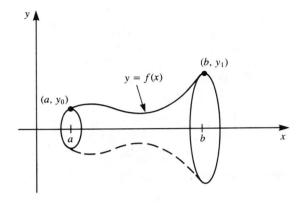

Figure 9.3

This is the integral we want to minimize. Here, $f(x, y, y') = 2\pi y \sqrt{1 + (y')^2}$, a function which does not explicitly contain x. We may therefore use Euler's equation in the form (9.11). Compute

$$\frac{\partial f}{\partial y'} = \frac{2\pi y y'}{\sqrt{1 + (y')^2}}.$$

Write y' for dy/dx. Euler's equation (9.11) for this problem is

$$2\pi y \sqrt{1 + (y')^2} - 2\pi y (y')^2 \frac{1}{\sqrt{1 + (y')^2}} = C.$$

In this equation, let $C/2\pi = K$ and multiply both sides by $\sqrt{1 + (y')^2}$ to get

$$y[1 + (y')^2] - y(y')^2 = K\sqrt{1 + (y')^2}.$$

Then

$$y = K\sqrt{1 + (y')^2}.$$

Square both sides of this equation and solve for y' to get

$$y' = \sqrt{\frac{y^2}{K^2} - 1}.$$

As we expected, Euler's equation has yielded a first order differential equation for y. In this problem, Euler's equation is separable and can be written

$$\frac{1}{\sqrt{\dfrac{y^2}{K^2} - 1}} dy = dx.$$

Integrate to obtain

$$K \cosh^{-1}\left(\frac{y}{K}\right) = x + c.$$

Solve this equation for y to get

$$y = K \cosh\left(\frac{x}{K} + \frac{c}{K}\right),$$

in which c and K are as yet unspecified constants. We must choose c and K so that this curve passes through the required points (a, y_0) and (b, y_1). This requires that

$$y_0 = K \cosh\left(\frac{a}{K} + \frac{c}{K}\right) \quad \text{and} \quad y_1 = K \cosh\left(\frac{b}{K} + \frac{c}{K}\right),$$

in which y_0, y_1, a, and b are given. The values of c and K determined by these equations give us the stationary function. The stationary graph is part of a hyperbolic cosine curve. ∎

EXAMPLE 9.2

Find the smooth curve of shortest length between points (a, y_0) and (b, y_1) in the plane.
    The length of a smooth curve $y(x)$ between these two points is

$$\int_a^b \sqrt{1 + (y')^2}\, dx.$$

We will apply Euler's equation with $f(x, y, y') = \sqrt{1 + (y')^2}$. Since f does not explicitly contain y, we may use the form (9.10) of Euler's equation to get

$$\frac{\partial}{\partial y'} \sqrt{1 + (y')^2} = C.$$

Then

$$y' \frac{1}{\sqrt{1 + (y')^2}} = C,$$

or

$$y' = C\sqrt{1 + (y')^2}.$$

Square both sides of this equation and write $K = C^2$ to get

$$(y')^2 = K[1 + (y')^2].$$

Solve this equation for y' to get

$$(y')^2 = \frac{K}{1 - K},$$

and conclude that

$$y' = \text{constant.}$$

Therefore, y must be of the form

$$y = cx + d,$$

a straight line. We must choose c and d so that this straight line passes through (a, y_0) and (b, y_1). We conclude that the smooth path of shortest length between two given points in the plane is a straight line. ∎

In concluding this section, it must be noted that Euler's equation is a necessary condition for y to be a stationary function for $\int_a^b f(x, y, y')\, dx$. That is, a stationary function must satisfy Euler's equation. This condition is not sufficient, however. Sufficient conditions for a function to optimize $\int_a^b f(x, y(x), y'(x))\, dx$ are more subtle than sufficient conditions for max/min problems of the kind we see in elementary calculus; these can be pursued in the references provided at the end of this book.

PROBLEMS FOR SECTION 9.1

In each of Problems 1 through 10, find a stationary function for the integral satisfying the given conditions at the ends of the interval.

1. $\displaystyle\int_1^2 (xy')^2\, dx, \quad y(1) = 0, \quad y(2) = 1$

2. $\displaystyle\int_1^3 [x(y')^2 - y]\, dx, \quad y(1) = 3, \quad y(3) = 4$

3. $\displaystyle\int_1^2 \frac{1}{x^3}(y')^2\, dx, \quad y(1) = 0, \quad y(2) = -3$

4. $\displaystyle\int_0^{\ln(3)} [y^2 + (y')^2 + 2ye^x]\, dx, \quad y(0) = 0, \quad y[\ln(3)] = 4$

5. $\displaystyle\int_0^4 [y^2 + (y')^2]\, dx, \quad y(0) = 0, \quad y(4) = 1$

6. $\displaystyle\int_1^3 x(y')^2\, dx, \quad y(1) = 1, \quad y(3) = 8$

7. $\displaystyle\int_1^{e^2} [x(y')^2 - yy' + y]\, dx, \quad y(1) = 1, \quad y(e^2) = 3$

8. $\displaystyle\int_1^3 [x^2(y')^2 - y^2]\, dx, \quad y(1) = 0, \quad y(3) = 1$

9. $\displaystyle\int_0^1 [\tfrac{1}{2}(y')^2 + yy' + y' + y]\, dx, \quad y(0) = 0, \quad y(1) = 4$

10. $\displaystyle\int_{-2}^3 [(y')^2 - x^2 y]\, dx, \quad y(0) = -2, \quad y(1) = 3$

11. Suppose that $f(x, y, y') = \alpha(x)(y')^2 + 2\beta(x)yy' + \gamma(x)y^2$. Show that Euler's equation is a second order linear differential equation.

12. Find a stationary function for $\int_a^b (1/x)\sqrt{1 + (y')^2}\, dx$ which passes through the points (a, y_0) and (b, y_1).

13. Derive equation (9.9) from equation (9.8).

14. The *brachistochrone problem* was formulated by John Bernoulli in 1696. Suppose a point $P_0 : (x_0, y_0)$ is joined in a vertical plane to $P_1 : (x_1, y_1)$ by a smooth curve $y = y(x)$, as in Figure (9.4). Imagine that a wire is bent into the shape of this curve and a bead slides without friction from P_0 to P_1 along the wire. How should the curve be chosen so that the time it takes the bead to slide from P_0 to P_1 is a minimum?

   As a simplification in deriving an equation for the curve, suppose that the bead starts from rest (zero initial velocity), and place P_0 at the origin so that $x_0 = y_0 = 0$. Let the mass of the bead be one unit. Let the downward direction be positive.

   (a) By considering the kinetic and potential energy of the bead at a point $P : (x, y)$ between P_0 and P_1 on the wire, show that

   $$\tfrac{1}{2}v^2 = gy$$

   and hence show that $v^2 = 2gy$, with v the velocity of the bead.

   (b) If $s(t)$ is distance traveled along the path at time t, then $v = ds/dt$. Use this fact and the result of (a) to conclude that $ds/dt = \sqrt{2gy}$.

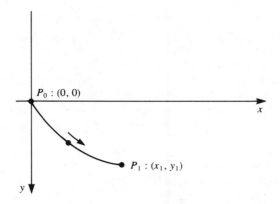

Figure 9.4

(c) Write the time it takes the bead to travel from P_0 to P_1 as

$$\text{time} = \int_0^{x_1} dt = \int_0^{x_1} \frac{1}{v}\, ds.$$

(d) If the curve is the graph of a function $y = y(x)$, then $ds = \sqrt{1 + (y')^2}\, dx$. Use this in the result of (c), together with the result of (a), to show that

$$\text{time} = \int_0^{x_1} \frac{\sqrt{1 + (y')^2}}{\sqrt{2gy}}\, dx.$$

(e) The path we want is a stationary curve for this integral subject to the conditions $y(0) = 0$ and $y(x_1) = y_1$. We have

$$f(x, y, y') = \frac{\sqrt{1 + (y')^2}}{\sqrt{2gy}}$$

for the brachistochrone problem. Use Euler's equation (9.11), together with some algebraic manipulation, to show that y is a solution of the differential equation

$$y[1 + (y')^2] = K$$

for some constant K.

(f) Use the result of (e) to show that

$$\sqrt{\frac{y}{K - y}}\, dy = dx.$$

(g) To solve for y, let $\sqrt{y/(K - y)} = \tan(\varphi)$. Show that $dy = 2K\sin(\varphi)\cos(\varphi)\, d\varphi$ and $dx = \tan(\varphi)\, dy = 2K\sin^2(\varphi)\, d\varphi = K[1 - \cos(2\varphi)]\, d\varphi$.

(h) By integrating an equation in (g), show that

$$x = \tfrac{1}{2}K[2\varphi - \sin(2\varphi)] + c_1.$$

Use the fact that the curve passes through the origin to show that $c_1 = 0$.

(i) Let $\theta = 2\varphi$ to show that

$$x = \alpha[\theta - \sin(\theta)], \qquad y = \alpha[1 - \cos(\theta)],$$

in which α must be chosen so that the curve passes through P_1.

These are parametric equations of a curve called the brachistochrone. It is the shape of the path we should form to have the bead slide from P_0 to P_1 in the least time. This curve has other remarkable properties as well. If a point is marked on the rim of a wheel and the wheel is rolled along a straight line, the point will trace out a locus called a *cycloid*. A brachistochrone is half an arc of a cycloid.

9.2 An Euler Equation for $\int_a^b f(x, y, y', y'') \, dx$

In this section, we will continue the theme of the preceding section, but now we will allow f to depend on a second derivative y'', as well as on x, y, and y'. The problem is to determine a continuous function y which minimizes or maximizes an integral $\int_a^b f(x, y, y', y'') \, dx$, in which f is a given function and is subject to the condition that the graph of y must pass through given points (a, y_0) and (b, y_1).

We will argue much as we did in Section 9.1. Suppose that $u(x)$ is a stationary function for this integral. Consider variations on u of the form

$$y(x) = u(x) + \epsilon\eta(x),$$

in which η is any function having a continuous derivative on $[a, b]$ and such that *both* $\eta(x)$ *and* $\eta'(x)$ vanish at a and b. For any such function η, think of the integral as a function of ϵ:

$$I(\epsilon) = \int_a^b f(x, u + \epsilon\eta, u' + \epsilon\eta', u'' + \epsilon\eta'') \, dx.$$

This function of ϵ has an extreme value at $\epsilon = 0$ because u is assumed to be a stationary function. Thus, we expect to have

$$I'(\epsilon)|_{\epsilon=0} = 0.$$

Using the chain rule, compute

$$I'(\epsilon) = \int_a^b \left[\frac{\partial f}{\partial x} \frac{\partial x}{\partial \epsilon} + \frac{\partial f}{\partial y} \frac{\partial y}{\partial \epsilon} + \frac{\partial f}{\partial y'} \frac{\partial y'}{\partial \epsilon} + \frac{\partial f}{\partial y''} \frac{\partial y''}{\partial \epsilon} \right] dx.$$

Now,

$$\frac{\partial x}{\partial \epsilon} = 0, \qquad \frac{\partial y}{\partial \epsilon} = \eta(x), \qquad \frac{\partial y'}{\partial \epsilon} = \eta'(x), \quad \text{and} \quad \frac{\partial y''}{\partial \epsilon} = \eta''(x).$$

Therefore,

$$I'(\epsilon)\bigg|_{\epsilon=0} = \int_a^b \left[\frac{\partial f}{\partial y} \eta(x) + \frac{\partial f}{\partial y'} \eta'(x) + \frac{\partial f}{\partial y''} \eta''(x) \right] dx = 0. \tag{9.12}$$

Integrate the second and third terms by parts. First,

$$\int_a^b \frac{\partial f}{\partial y'}\, \eta'(x)\, dx = \frac{\partial f}{\partial y'}\, \eta(x) \bigg]_a^b - \int_a^b \eta(x) \frac{d}{dx}\left(\frac{\partial f}{\partial y'}\right) dx$$

$$= -\int_a^b \eta(x) \frac{d}{dx}\left(\frac{\partial f}{\partial y'}\right) dx \tag{9.13}$$

because $\eta(a) = \eta(b) = 0$. Next,

$$\int_a^b \frac{\partial f}{\partial y''}\, \eta''(x)\, dx = \frac{\partial f}{\partial y''}\, \eta'(x) \bigg]_a^b - \int_a^b \eta'(x) \frac{d}{dx}\left(\frac{\partial f}{\partial y''}\right) dx$$

$$= -\int_a^b \eta'(x) \frac{d}{dx}\left(\frac{\partial f}{\partial y''}\right) dx \tag{9.14}$$

because $\eta'(a) = \eta'(b) = 0$. Integrate this last integral in equation (9.14) again by parts to get

$$-\int_a^b \eta'(x) \frac{d}{dx}\left(\frac{\partial f}{\partial y''}\right) dx = -\frac{d}{dx}\left(\frac{\partial f}{\partial y''}\right)\eta(x) \bigg]_a^b + \int_a^b \eta(x) \frac{d^2}{dx^2}\left(\frac{\partial f}{\partial y''}\right) dx$$

$$= \int_a^b \eta(x) \frac{d^2}{dx^2}\left(\frac{\partial f}{\partial y''}\right) dx \tag{9.15}$$

because $\eta(a) = \eta(b) = 0$. Substitute the results of these integrations into equation (9.12) to get

$$\int_a^b \left[\frac{\partial f}{\partial x} - \frac{d}{dx}\left(\frac{\partial f}{\partial y'}\right) + \frac{d^2}{dx^2}\left(\frac{\partial f}{\partial y''}\right) \right] \eta(x)\, dx = 0.$$

By a modification of the lemma of the preceding section, this equation implies that

$$\frac{\partial f}{\partial x} - \frac{d}{dx}\left(\frac{\partial f}{\partial y'}\right) + \frac{d^2}{dx^2}\left(\frac{\partial f}{\partial y''}\right) = 0. \tag{9.16}$$

This is *Euler's equation* for $\int_a^b f(x, y, y', y'')\, dx$; it provides a necessary (but not sufficient) condition for a function y to be a stationary function of this integral. Generally, Euler's equation (9.16) for the problem now being considered is a fourth order ordinary differential equation.

In Section 9.1, we used an Euler equation to solve specific problems involving curves in the plane. Often, Euler's equation is used not to obtain a solution of a specific problem but to derive a differential equation and boundary conditions modeling a physical phenomenon. In such a case, we must often be more general than in the above derivation. Recall equation (9.12), in which we had

$$\int_a^b \left[\frac{\partial f}{\partial y}\, \eta(x) + \frac{\partial f}{\partial y'}\, \eta'(x) + \frac{\partial f}{\partial y''}\, \eta''(x) \right] dx = 0.$$

We integrated the second and third terms by parts, using the facts that $\eta(a) = \eta(b) = \eta'(a) = \eta'(b) = 0$. Now suppose we omit these conditions; that is, we consider variations $y = u + \epsilon\eta$ of the form shown in Figure 9.5. The varied curve $y(x) = u(x) + \epsilon\eta(x)$ need not pass through (a, y_0) and (b, y_1), nor does it have to have the

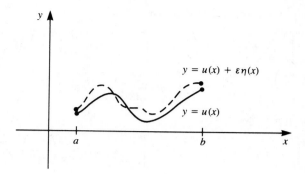

Figure 9.5

same slope as u at those points. In this event, certain terms which were zero in the integrations by parts (9.13), (9.14), and (9.15) no longer need be zero. In particular, equation (9.13) is replaced by

$$\int_a^b \frac{\partial f}{\partial y'}\, \eta'(x)\, dx = \frac{\partial f}{\partial y'}\, \eta(x)\, \bigg]_a^b - \int_a^b \eta(x)\frac{d}{dx}\left(\frac{\partial f}{\partial y'}\right) dx,$$

in which the first term on the right must now be retained. Further, the integrations by parts (9.14) and (9.15), in which certain terms were zero, are now replaced by

$$\int_a^b \frac{\partial f}{\partial y''}\, \eta''(x)\, dx = \frac{\partial f}{\partial y''}\, \eta'(x)\, \bigg]_a^b - \int_a^b \eta'(x)\frac{d}{dx}\left(\frac{\partial f}{\partial y''}\right) dx$$

$$= \frac{\partial f}{\partial y''}\, \eta'(x)\, \bigg]_a^b - \frac{d}{dx}\left(\frac{\partial f}{\partial y''}\right)\eta(x)\, \bigg]_a^b + \int_a^b \frac{d^2}{dx^2}\left(\frac{\partial f}{\partial y''}\right)\eta(x)\, dx.$$

Thus, equation (9.12) must now be written

$$\int_a^b \left[\frac{\partial f}{\partial y} - \frac{d}{dx}\left(\frac{\partial f}{\partial y'}\right) + \frac{d^2}{dx^2}\left(\frac{\partial f}{\partial y''}\right)\right]\eta(x)\, dx$$

$$+ \frac{\partial f}{\partial y'}\, \eta(x)\, \bigg]_a^b + \frac{\partial f}{\partial y''}\, \eta'(x)\, \bigg]_a^b - \frac{d}{dx}\left(\frac{\partial f}{\partial y''}\right)\eta(x)\, \bigg]_a^b = 0.$$

Write this equation as

$$\int_a^b \left[\frac{\partial f}{\partial y} - \frac{d}{dx}\left(\frac{\partial f}{\partial y'}\right) + \frac{d^2}{dx^2}\left(\frac{\partial f}{\partial y''}\right)\right]\eta(x)\, dx$$

$$+ \left[\left[\frac{\partial f}{\partial y'} - \frac{d}{dx}\left(\frac{\partial f}{\partial y''}\right)\right]\eta(x)\right]_a^b + \left[\frac{\partial f}{\partial y''}\, \eta'(x)\right]_a^b = 0. \tag{9.17}$$

We can satisfy (9.17) by having

$$\frac{\partial f}{\partial y} - \frac{d}{dx}\left(\frac{\partial f}{\partial y'}\right) + \frac{d^2}{dx^2}\left(\frac{\partial f}{\partial y''}\right) = 0. \tag{9.18}$$

and

$$\left[\frac{\partial f}{\partial y'} - \frac{d}{dx}\left(\frac{\partial f}{\partial y''}\right)\right]_a^b = 0 \quad \text{and} \quad \left[\frac{\partial f}{\partial y''}\right]_a^b = 0. \tag{9.19}$$

We attempt to solve the differential equation (9.18), which is also called Euler's equation, subject to the conditions (9.19). The two conditions (9.19) are called *natural boundary conditions* for the problem.

EXAMPLE 9.3

Suppose we have a uniform elastic cantilever beam supported as shown in Figure 9.6. Assume that the beam is subjected to a uniform external load of magnitude K. Let EI be the beam's bending rigidity and L be its length. We want a differential equation to describe the vertical deflection $y(x)$ of the beam, together with any relevant boundary conditions.

It is shown in the theory of elasticity that the strain energy caused by the bending of the beam is

$$\tfrac{1}{2}EI \int_0^L (y'')^2 \, dx.$$

Further, the gain in potential energy of the external load is

$$-\int_0^L Ky \, dx.$$

The potential energy of the beam is

$$\int_0^L [\tfrac{1}{2}EI(y'')^2 - Ky] \, dx.$$

We will assume, by the principle of elasticity, that the beam acts in such a way as to minimize this integral. Thus, we want a stationary function of this integral. Apply Euler's equation (9.18), with $f(x, y, y', y'') = \tfrac{1}{2}EI(y'')^2 - Ky$. Calculate

$$\frac{\partial f}{\partial y} = -K, \qquad \frac{\partial f}{\partial y'} = 0, \quad \text{and} \quad \frac{\partial f}{\partial y''} = EIy''.$$

The Euler equation (9.18) for this problem is

$$EIy^{(4)} - K = 0, \tag{9.20}$$

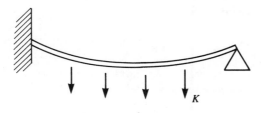

Figure 9.6

a fourth order ordinary differential equation for the displacement function y. Since the left end of the beam is clamped, we have two conditions there:

$$y(0) = y'(0) = 0. \tag{9.21}$$

Since we have a fourth order differential equation, we need two additional conditions to specify a solution uniquely. These come from the natural boundary conditions (9.19), which for the present case yield the equations

$$EIy^{(3)}(L) = 0 \quad \text{and} \quad EIy''(L) = 0. \tag{9.22}$$

Equations (9.22) state, respectively, that the shear is zero at the unclamped end and that the bending moment is zero there. The differential equation (9.20) is easily solved subject to the boundary conditions (9.21) and (9.22), yielding the deflection function

$$y(x) = \frac{K}{2EI}\left[\frac{1}{12}x^4 - \frac{L}{3}x^3 + \frac{1}{2}L^2 x^2\right]. \quad \blacksquare$$

PROBLEMS FOR SECTION 9.2

For each of Problems 1 through 5, find the general form of a stationary function for the integral.

1. $\displaystyle\int_a^b xy'y'' \, dx$

2. $\displaystyle\int_a^b x(y'')^2 \, dx$

3. $\displaystyle\int_a^b x^2 y'y'' \, dx$

4. $\displaystyle\int_a^b y^2 y'' \, dx$

5. $\displaystyle\int_a^b (xy + xy' + y'y'') \, dx$

6. Find a stationary curve for the integral $\int_0^1 [16y^2 - (y'')^2 + x^2] \, dx$ satisfying the conditions $y(0) = 0$, $y(1) = 4$, $y'(0) = -1$, $y'(1) = 2$.

7. Let $I = \int_a^b f(x, y, y', y'', y^{(3)}) \, dx$. Use the reasoning of this and the preceding section to show that a stationary function for this integral must satisfy an Euler equation of the form

$$\frac{\partial f}{\partial y} - \frac{d}{dx}\frac{\partial f}{\partial y'} + \frac{d^2}{dx^2}\frac{\partial f}{\partial y''} - \frac{d^3}{dx^3}\frac{\partial f}{\partial y^{(3)}} = 0.$$

8. For the integral $\int_a^b f(x, y, y', y'', \ldots, y^{(n)}) \, dx$, derive the Euler equation

$$\frac{\partial f}{\partial y} - \frac{d}{dx}\frac{\partial f}{\partial y'} + \frac{d^2}{dx^2}\frac{\partial f}{\partial y''} - \cdots + (-1)^n \frac{d^n}{dx^n}\frac{\partial f}{\partial y^{(n)}} = 0.$$

9. Suppose we have a uniform elastic circular plate of radius R with a hole of radius a at the center. The plate is loaded axisymmetrically, and $f(r)$ is the normal force per unit surface area at the top of the plate. Use Euler's equation and the natural boundary conditions to derive a mathematical model for the deflection function $w(r)$, which measures normal deflection of the plate at a distance r from the center.

It can be shown that the potential energy of the system is

$$\pi D \int_a^R \left[rw'' + \frac{1}{r}(w')^2 + 2vw'w''\right] dr - 2\pi \int_a^R f(r)rw \, dr + 2\pi \left[rMw' - rF(r)w\right]_{r=R},$$

where D is the flexural rigidity of the plate, M is the radial bending moment per unit of circumferential length, $F(r)$ is the radial shear per unit of circumferential length, and v is Poisson's ratio. Use Euler's equation to derive the differential equation

$$rw^{(4)} + 2w^{(3)} - \frac{1}{r}w'' + \frac{1}{r^2}w' - \frac{r}{D}f(r) = 0.$$

Also derive the natural boundary conditions.

$$\left[\frac{d}{dr}\left(\frac{1}{r}(rw')'\right)\right]_{r=a} = 0,$$

$$\left[w'' + \frac{v}{r} w'\right]_{r=a} = 0,$$

$$\left[vw'' + \frac{1}{r} w' - (rw'' + vw')'\right]_{r=R} = \frac{1}{D} F(R),$$

and

$$\left[\frac{r}{D} M + (rw'' + vw')\right]_{r=R} = 0.$$

9.3 *An Euler Equation for* $\iint_D f(x, y, w, w_x, w_y)\, dA$ _____

In this section, we will derive an Euler equation which constitutes a necessary condition for a function $w(x, y)$ of two variables to maximize or minimize a double integral $\iint_D f(x, y, w, w_x, w_y)\, dA$, in which f is a given function of five variables and D is a region of the plane bounded by a curve C. Letter subscripts denote partial derivatives: $w_x = \partial w/\partial x$ and $w_y = \partial w/\partial y$. Usually, we make the assumption that C is piecewise smooth, which means geometrically that C has a continuous tangent, except perhaps at finitely many points. Thus, for example, the curve bounding a rectangle is piecewise smooth, having a continuous tangent except at the corners.

We will proceed much as we did for the integrals involving one independent variable. Suppose that $u(x, y)$ is a stationary function for the integral so that u is a function which maximizes or minimizes $\iint_D f(x, y, w, w_x, w_y)\, dA$. Consider variations from u by letting

$$w(x, y) = u(x, y) + \epsilon\eta(x, y),$$

where $\eta(x, y)$ is continuous with continuous first and second partial derivatives at each point of D. We also assume that η vanishes on the boundary C of D. [This is analogous to having $\eta(a) = \eta(b) = 0$ in the case of one independent variable.] For a given function η having these properties, we have a function of ϵ:

$$I(\epsilon) = \iint_D f(x, y, u + \epsilon\eta, u_x + \epsilon\eta_x, u_y + \epsilon\eta_y)\, dA.$$

Since u is a stationary function, we expect $I(\epsilon)$ to have its maximum or minimum at $\epsilon = 0$; hence, we look at the equation

$$I'(\epsilon)\bigg|_{\epsilon=0} = 0.$$

This yields the equation

$$\iint_D \left[\frac{\partial f}{\partial x}\frac{\partial x}{\partial \epsilon} + \frac{\partial f}{\partial y}\frac{\partial y}{\partial \epsilon} + \frac{\partial f}{\partial w}\frac{\partial w}{\partial \epsilon} + \frac{\partial f}{\partial w_x}\frac{\partial w_x}{\partial \epsilon} + \frac{\partial f}{\partial w_y}\frac{\partial w_y}{\partial \epsilon}\right] dA = 0.$$

Now,

$$\frac{\partial x}{\partial \epsilon} = \frac{\partial y}{\partial \epsilon} = 0, \qquad \frac{\partial w}{\partial \epsilon} = \eta, \qquad \frac{\partial w_x}{\partial \epsilon} = \eta_x, \quad \text{and} \quad \frac{\partial w_y}{\partial \epsilon} = \eta_y.$$

Therefore,

$$\iint_D \left[\frac{\partial f}{\partial w} \eta + \frac{\partial f}{\partial w_x} \eta_x + \frac{\partial f}{\partial w_y} \eta_y \right] dA = 0. \tag{9.23}$$

We would like to factor η out of certain terms of this integral. In the case of one independent variable, we were able to do this after integration by parts. In the case of two independent variables, we use Green's theorem. Write

$$\oint_C \left(-\eta \frac{\partial f}{\partial w_y} \right) dx + \left(\eta \frac{\partial f}{\partial w_x} \right) dy = \iint_D \left[\frac{\partial}{\partial x} \left(\eta \frac{\partial f}{\partial w_x} \right) + \frac{\partial}{\partial y} \left(\eta \frac{\partial f}{\partial w_y} \right) \right] dA. \tag{9.24}$$

Now recall that we chose η so that $\eta(x, y) = 0$ for (x, y) on C, the boundary of D. Thus, the line integral on the left side of equation (9.24) is zero, and we have, upon carrying out the differentiations in the integral on the right side,

$$\iint_D \left[\eta \frac{\partial}{\partial x} \frac{\partial f}{\partial w_x} + \eta \frac{\partial}{\partial y} \frac{\partial f}{\partial w_y} + \frac{\partial \eta}{\partial x} \frac{\partial f}{\partial w_x} + \frac{\partial \eta}{\partial y} \frac{\partial f}{\partial w_y} \right] dA = 0.$$

Therefore,

$$\iint_D \left[\frac{\partial \eta}{\partial x} \frac{\partial f}{\partial w_x} + \frac{\partial \eta}{\partial y} \frac{\partial f}{\partial w_y} \right] dA = -\iint_D \left[\eta \frac{\partial}{\partial x} \frac{\partial f}{\partial w_x} + \eta \frac{\partial}{\partial y} \frac{\partial f}{\partial w_y} \right] dA. \tag{9.25}$$

Substitute this result into equation (9.23) to write that integral as

$$\iint_D \left[\frac{\partial f}{\partial w} - \frac{\partial}{\partial x} \frac{\partial f}{\partial w_x} - \frac{\partial}{\partial y} \frac{\partial f}{\partial w_y} \right] \eta \, dA = 0.$$

This integral can be zero for all allowable choices of η only if the quantity in square brackets is zero. Therefore,

$$\frac{\partial f}{\partial w} - \frac{\partial}{\partial x} \frac{\partial f}{\partial w_x} - \frac{\partial}{\partial y} \frac{\partial f}{\partial w_y} = 0. \tag{9.26}$$

This is the Euler equation which constitutes a necessary condition for a function w to be a stationary function for $\iint_D f(x, y, w, w_x, w_y) \, dA$.

EXAMPLE 9.4 The One-Dimensional Wave Equation

As an application of equation (9.26), we will derive the equation of motion of a vibrating string, assuming a principle of physics called Hamilton's principle. Suppose we have an elastic string with its ends fastened at the same horizontal level. At time zero, we displace the string and release it. We want an equation to describe its shape at any later time.

Imagine the string pegged along the x-axis, with one end at zero and the other at $x = L$. Let $y(x, t)$ be the vertical displacement of the particle of string, which, at time zero, was at x. If $t > 0$ is fixed, the graph of $y = y(x, t)$ is the shape of the string at time t. We will derive an equation for this displacement function y.

Let the tension in the string be T, a constant, and let the mass per unit length be $\rho(x)$. The kinetic energy of the string is

$$\tfrac{1}{2} \int_0^L \rho(x) y_t^2 \, dx.$$

We now want an expression for the potential energy. This is the work done in displacing the string from its initial length L to its length at time t. Since $y = y(x, t)$ is the function describing the shape of the string at time t, for fixed t, the length at time t is $\int_0^L \sqrt{1 + y_x^2} \, dx$. Thus, the potential energy is

$$T\left(\int_0^L \sqrt{1 + y_x^2} \, dx - L \right).$$

Use a binomial expansion to write

$$\sqrt{1 + \alpha} = 1 + \tfrac{1}{2}\alpha - \tfrac{1}{4}\alpha^2 + \cdots.$$

With $\alpha = y_x^2$, we have

$$\sqrt{1 + y_x^2} = 1 + \tfrac{1}{2}y_x^2 - \tfrac{1}{4}y_x^4 + \cdots. \tag{9.27}$$

Assume that the oscillations of the string are small in magnitude, by which we mean that $|y_x|$ is much less than 1. In this event, all terms y_x^4, and higher powers, can be neglected in equation (9.27). That is, we assume that these powers are approximately zero. This enables us to write

$$\sqrt{1 + y_x^2} = 1 + \tfrac{1}{2}y_x^2.$$

The potential energy of the string is therefore

$$T\left[\int_0^L (1 + \tfrac{1}{2}y_x^2) \, dx - L \right] = \tfrac{1}{2}T \int_0^L y_x^2 \, dx.$$

The *Lagrangian* $\mathcal{L}$ of the string is defined to be the kinetic energy minus the potential energy:

$$\mathcal{L} = \tfrac{1}{2} \int_0^L [\rho(x) y_t^2 - T y_x^2] \, dx.$$

Hamilton's principle states that the motion of the system must minimize the value of this integral over any interval of time $[t_1, t_2]$. Thus, we must choose y to minimize

$$\int_{t_1}^{t_2} \int_0^L \tfrac{1}{2}[\rho(x) y_t^2 - T y_x^2] \, dx \, dt.$$

We can think of this iterated integral as the double integral of

$$f(x, t, y, y_x, y_t) = \tfrac{1}{2}[\rho(x) y_t^2 - T y_x^2]$$

over the rectangle $0 \le x \le L$, $t_1 \le t \le t_2$, in the (x, t)-plane. The minimizing function must satisfy Euler's equation (9.26). For this problem, Euler's equation is

$$\frac{\partial f}{\partial y} - \frac{\partial}{\partial x}\left(\frac{\partial f}{\partial y_x} \right) - \frac{\partial}{\partial t}\left(\frac{\partial f}{\partial y_t} \right) = 0.$$

For the function f which we have for this problem, this equation yields

$$-\frac{\partial}{\partial x}\left(-T\frac{\partial y}{\partial x}\right) - \frac{\partial}{\partial t}\left(\rho\frac{\partial y}{\partial t}\right) = 0.$$

Assume that T is constant and recall that ρ depends only on x. We obtain

$$T\frac{\partial^2 y}{\partial x^2} = \rho(x)\frac{\partial^2 y}{\partial t^2}.$$

This is the one-dimensional wave equation. ∎

From the form of equation (9.26), it is not difficult to imagine an Euler equation for a triple integral $\iiint_M f(x, y, z, w, w_x, w_y, w_z)\, dV$ over a region M in three-space. By a derivation similar to that just done, but using Gauss's divergence theorem instead of Green's theorem, we find that the Euler equation for this triple integral is

$$\frac{\partial f}{\partial w} - \frac{\partial}{\partial x}\left(\frac{\partial f}{\partial w_x}\right) - \frac{\partial}{\partial y}\left(\frac{\partial f}{\partial w_y}\right) - \frac{\partial}{\partial z}\left(\frac{\partial f}{\partial w_z}\right) = 0. \tag{9.28}$$

As an application of equation (9.28), we will derive the two-dimensional wave equation.

EXAMPLE 9.5   The Two-Dimensional Wave Equation

Suppose we have an elastic membrane stretched over a framework that forms a piecewise smooth, simple closed curve in the xy-plane. (Think, for example, of the surface of a drum.) The membrane is displaced at time $t = 0$ and is allowed to vibrate. At time t, let $z = w(x, y, t)$ denote the z-coordinate of the particle, which, at time zero, was at the point (x, y) in the membrane. We want a differential equation for this displacement function w.

Although w is a function of three variables, we can think of t as fixed for the moment and apply Euler's equation (9.28). The argument mirrors that in Example 9.4. Let $\rho(x, y)$ be the mass per unit area of the membrane. The kinetic energy at any time t is

$$\iint_D \rho z_t^2\, dA.$$

The potential energy at any time t is the work done in stretching the membrane from its initial surface area K to its surface area at time t, and this is

$$T\left(\iint_D \sqrt{1 + z_x^2 + z_y^2}\, dA - K\right).$$

(This double integral is the surface area of the membrane at time t.) Using a binomial expansion as we did in Example 9.4, write

$$\sqrt{1 + z_x^2 + z_y^2} = 1 + \tfrac{1}{2}[z_x^2 + z_y^2] - \tfrac{1}{4}[z_x^2 + z_y^2]^2 + \cdots.$$

Assume that the magnitude of the oscillations is small, so that $|z_x|$ and $|z_y|$ are both much smaller than 1. Then the potential energy is approximately

$$T\left\{\iint_D (1 + \tfrac{1}{2}[z_x^2 + z_y^2])\, dA - K\right\},$$

or

$$T \iint_D \tfrac{1}{2}[z_x^2 + z_y^2] \, dA,$$

since $\iint_D 1 \, dA = \text{area of } D = K$.

Again using Hamilton's principle, we assert that the displacement function $z(x, t)$ must be a stationary function of the integral of the Lagrangian over any time interval $[t_1, t_2]$. That is, z must minimize

$$\int_{t_1}^{t_2} \left\{ \iint_D [\tfrac{1}{2}T(z_x^2 + z_y^2) - \tfrac{1}{2}\rho z_t^2] \, dA \right\} dt.$$

Upon applying the Euler equation (9.28) to this integral, we obtain

$$\frac{\partial^2 z}{\partial x^2} + \frac{\partial^2 z}{\partial y^2} = \frac{\rho}{T} \frac{\partial^2 z}{\partial t^2},$$

and this is the two-dimensional wave equation. ∎

PROBLEMS FOR SECTION 9.3

1. The second order partial differential equation

$$\frac{\partial^2 u}{\partial x^2} + \frac{\partial^2 u}{\partial y^2} = f(x, y),$$

in which f is a given function, is called *Poisson's equation*. Show that Poisson's equation is Euler's equation for the integral $\iint_D [u_x^2 + u_y^2 + 2f(x, y)] \, dA = 0$.

2. Derive an Euler equation for a stationary function for the triple integral

$$\iiint_M f(x, y, z, w, w_x, w_y, w_z) \, dV$$

over a region M in three-space.

3. For a function w of three variables to be a stationary function for the triple integral

$$\iiint_M f(x, y, z, w, w_x, w_y, w_z, w_{xx}, w_{yy}, w_{zz}, w_{xy}, w_{yz}, w_{zx}) \, dV,$$

derive the Euler equation

$$\frac{\partial f}{\partial x} - \frac{\partial}{\partial x}\frac{\partial f}{\partial w_x} - \frac{\partial}{\partial y}\frac{\partial f}{\partial w_y} - \frac{\partial}{\partial z}\frac{\partial f}{\partial w_z} + \frac{\partial^2}{\partial x^2}\frac{\partial^2 f}{\partial w_{xx}} + \frac{\partial^2}{\partial y^2}\frac{\partial^2 f}{\partial w_{yy}} + \frac{\partial^2}{\partial z^2}\frac{\partial^2 f}{\partial w_{zz}}$$

$$+ \frac{\partial^2}{\partial x \, \partial y}\frac{\partial^2 f}{\partial w_x \, \partial w_y} + \frac{\partial^2}{\partial y \, \partial z}\frac{\partial^2 f}{\partial w_y \, \partial w_z} + \frac{\partial^2}{\partial z \, \partial x}\frac{\partial^2 f}{\partial w_z \, \partial w_x} = 0.$$

9.4 Isoperimetric Problems _____

The term *isoperimetric problem* refers to a problem of maximizing or minimizing an integral subject to the condition that another integral is to have a specified value.

Typically, we want to find a function y whose graph passes through the points (a, y_0) and (b, y_1), which extremizes

$$\int_a^b f(x, y, y') \, dx$$

and also satisfies a condition

$$\int_a^b g(x, y, y') \, dx = K,$$

with K a given constant. In these integrals, f and g are given functions of three variables.

We will begin as before, with a function u which is assumed to be a solution of the problem. We would like to consider variations $y(x) = u(x) + \epsilon \eta(x)$, but the constraint that $\int_a^b g(x, y, y') \, dx = K$ presents a difficulty. The difficulty is that, for a given η, a change in ϵ will probably cause a change in $\int_a^b g(x, y, y') \, dx$ as well, and this integral is constrained to remain equal to the given constant K.

One way to circumvent this difficulty is to vary u in a more complicated way. Let

$$y(x) = u(x) + \epsilon_1 \eta_1(x) + \epsilon_2 \eta_2(x),$$

where η_1 and η_2 have continuous derivatives on $[a, b]$ and $\eta_1(a) = \eta_1(b) = \eta_2(a) = \eta_2(b) = 0$. In this way, we can vary not just one quantity, ϵ, but two quantities, ϵ_1 and ϵ_2, in such a way that $\int_a^b g(x, y, y') \, dx$ remains equal to K. For such functions η_1 and η_2, we have a max/min problem in two variables. We want to extremize the function

$$I(\epsilon_1, \epsilon_2) = \int_a^b f(x, u(x) + \epsilon_1 \eta_1(x) + \epsilon_2 \eta_2(x), u'(x) + \epsilon_1 \eta_1'(x) + \epsilon_2 \eta_2'(x)) \, dx$$

subject to the constraint

$$J(\epsilon_1, \epsilon_2) = \int_a^b g(x, u(x) + \epsilon_1 \eta_1(x) + \epsilon_2 \eta_2(x), u'(x) + \epsilon_1 \eta_1'(x) + \epsilon_2 \eta_2'(x)) \, dx = K.$$

We can treat this problem as a constrained max/min problem in two variables. To do this, introduce the Lagrange multiplier λ and consider the function

$$H(\epsilon_1, \epsilon_2) = I(\epsilon_1, \epsilon_2) + \lambda J(\epsilon_1, \epsilon_2).$$

Since we know that the extremum occurs when $\epsilon_1 = \epsilon_2 = 0$, the equations for the critical points of this problem are

$$\left. \frac{\partial H}{\partial \epsilon_1} \right|_{\substack{\epsilon_1 = 0 \\ \epsilon_2 = 0}} = \left. \frac{\partial H}{\partial \epsilon_2} \right|_{\substack{\epsilon_1 = 0 \\ \epsilon_2 = 0}} = 0, \tag{9.29}$$

together with the constraint condition

$$\left. \frac{\partial H}{\partial \lambda} \right|_{\substack{\epsilon_1 = 0 \\ \epsilon_2 = 0}} = \int_a^b g(x, u, u') \, dx = K.$$

Perform the chain rule differentiations under the integral sign to calculate equations

(9.29). We get

$$\int_a^b \left[\frac{\partial(f + \lambda g)}{\partial y} \eta_1 + \frac{\partial(f + \lambda g)}{\partial y'} \eta_1' \right] dx = 0$$

and

$$\int_a^b \left[\frac{\partial(f + \lambda g)}{\partial y} \eta_2 + \frac{\partial(f + \lambda g)}{\partial y'} \eta_2' \right] dx = 0.$$

Integrate the second term in each of these integrals by parts, and use the fact that η_1 and η_2 both vanish at a and at b. Since both η_1 and η_2 are arbitrary differentiable functions (vanishing at a and b), we obtain, in the usual way,

$$\frac{\partial(f + \lambda g)}{\partial y} - \frac{d}{dx} \left(\frac{\partial(f + \lambda g)}{\partial y'} \right) = 0.$$

This is the Euler equation for the isoperimetric problem. This equation must be solved subject to the constraint condition $\int_a^b g(x, y, y')\, dx = K$.

EXAMPLE 9.6

Find the curve C having given length L and passing through the points (a, y_0) and (b, y_1) such that the area enclosed by C, the x-axis, and the lines $x = a$ and $x = b$ is a maximum. Here y_0 and y_1 are given nonnegative numbers.

The area is the integral

$$I = \int_a^b y\, dx = \text{area},$$

and the constraint condition is

$$J = \int_a^b \sqrt{1 + (y')^2}\, dx = \text{length} = L.$$

Consider the function

$$H(x, y, y', \lambda) = I + \lambda J,$$

where λ is a Lagrange multiplier. Thus,

$$H = \int_a^b \left[y + \lambda \sqrt{1 + (y')^2} \right] dx.$$

Note that H does not explicitly contain x. After some routine computation, which we omit, we obtain the Euler equation for this problem. It is

$$y + \lambda \sqrt{1 + (y')^2} - \frac{\lambda(y')^2}{\sqrt{1 + (y')^2}} = C.$$

Some algebraic manipulation yields the equation

$$y - C = -\frac{\lambda}{\sqrt{1 + (y')^2}}.$$

Instead of dealing with this differential equation directly, we will use a change of variables to derive parametric equations for the curve which is the graph of the optimizing function. Let $y' = \tan(\theta)$ and use the fact that $1 + \tan^2(\theta) = \sec^2(\theta)$ to obtain

$$y - C = \frac{-\lambda}{\sqrt{1 + \tan^2(\theta)}} = -\lambda \cos(\theta). \tag{9.30}$$

Since $y' = dy/dx = \tan(\theta)$,

$$dx = \frac{1}{\tan(\theta)} \, dy. \tag{9.31}$$

But, from equation (9.30), we have $dy = \lambda \sin(\theta) \, d\theta$, so (9.31) becomes

$$dx = \frac{1}{\tan(\theta)} \lambda \sin(\theta) \, d\theta = \lambda \cos(\theta) \, d\theta.$$

This equation can be integrated to yield

$$x = \lambda \sin(\theta) + k. \tag{9.32}$$

Equations (9.30) and (9.32) represent parametric equations of the curve we want. We can write these equations in the form

$$x - k = \lambda \sin(\theta), \qquad y - C = -\lambda \cos(\theta).$$

It is now easy to eliminate θ by squaring both equations and adding to get

$$(x - k)^2 + (y - C)^2 = \lambda^2.$$

The solution is an arc of a circle of radius λ and center (k, C). The constants λ, k, and C are determined from the condition that the length of the curve must be L and the curve must pass through the two given points (a, y_0) and (b, y_1). ∎

PROBLEMS FOR SECTION 9.4

1. Show that stationary functions of $\int_a^b [p(x)(y')^2 - q(x)y^2] \, dx$, subject to the constraint $\int_a^b r(x)y^2 \, dx = 1$, are solutions of the Sturm-Liouville differential equation $(py')' + (q + \lambda p)y = 0$ for some constants λ.
2. Determine a stationary function for $\int_0^\pi (y')^2 \, dx$ if $\int_0^\pi y^2 \, dx = 1$. Assume that $y(0) = y_0$ and $y(\pi) = y_1$.
3. Determine a stationary function for $\int_0^1 [(y')^2 + x^2] \, dx$ if $\int_0^1 y^2 \, dx = 2$. Assume that $y(0) = y_0$ and $y(1) = y_1$.
4. Use the calculus of variations to determine the shape of a perfectly elastic rope of uniform density that hangs at rest. Assume that the endpoints are fixed at given locations and that the length is a given constant L. *Hint:* Use the principle of physics that states that a system in a state of stable equilibrium (as the rope is) must have minimum potential energy. If the rope is in the xy-plane, the potential energy is $\rho g \int_0^L y \, ds$, where ρ is the constant mass per unit length, L is its length, and s measures the arc length along the curve. If the curve is the graph of a function $y = y(x)$, for $a \le x \le b$, then $ds = \sqrt{1 + (y')^2} \, dx$, and the potential energy is

$$\rho g \int_a^b y\sqrt{1 + (y')^2} \, dx.$$

Find a stationary function for this integral subject to the condition that

$$\int_a^b \sqrt{1 + (y')^2} \, dx = L.$$

Let

$$H = \int_a^b \left[y\sqrt{1 + (y')^2} + \lambda\sqrt{1 + (y')^2} \right] dx.$$

Apply Euler's equation to obtain

$$y + \lambda = C\sqrt{1 + (y')^2}.$$

Change variables by putting $y' = \sinh(\theta)$. Show that $y + \lambda = C\cosh(\theta)$ and that $dx = C\,dt$. Solve for $y + \lambda$ as a function of x to get

$$y + \lambda = C\cosh\left(\frac{x - K}{C}\right),$$

where the constants C, K, and λ must be determined from the information that the curve has given length L and passes through (a, y_0) and (b, y_1).

5. Suppose we want a stationary function for $\int_a^b f(x, y, y')\,dx$ subject to m constraints

$$\int_a^b g_i(x, y, y')\,dx = K_i$$

for $i = 1, 2, \ldots, m$. Here, $K_1, \ldots, K_m$ are given constants and the functions $f, g_1, \ldots, g_m$ are given. Derive the Euler equation

$$\frac{\partial}{\partial y}\left(f + \sum_{i=1}^m \lambda_i g_i \right) - \frac{d}{dx}\left(f + \sum_{i=1}^m \lambda_i g_i \right) = 0,$$

in which $\lambda_1, \ldots, \lambda_m$ are Lagrange multipliers. *Hint:* If $u(x)$ is a stationary function for this problem, let $y(x) = u(x) + \sum_{i=1}^{m+1} \epsilon_i \eta_i(x)$, where each $\eta_i(a) = \eta_i(b) = 0$.

6. Suppose we want to find a stationary function for

$$\int_a^b f(x, y_1, y_2, \ldots, y_m, y'_1, \ldots, y'_m)\,dx,$$

in which each y_i is a function x having a continuous derivative subject to N constraints of the form

$$\int_a^b g_i(x, y_1, \ldots, y_m, y'_1, \ldots, y'_m)\,dx = K_i.$$

Show that a necessary condition for a function y to be a stationary function for this problem is that y satisfy the system of m ordinary differential equations

$$\frac{\partial}{\partial y_i}\left(f + \sum_{i=1}^N \lambda_i g_i \right) - \frac{d}{dx}\left[\frac{\partial}{\partial y'_i}\left(f + \sum_{i=1}^N \lambda_i g_i \right) \right] = 0$$

for $i = 1, 2, \ldots, N$. Here, $\lambda_1, \ldots, \lambda_N$ are Lagrange multipliers.

7. Restate the conclusion of Problem 6, with t as the independent variable and $x(t)$ and $y(t)$ as the dependent variables, instead of $y_1, \ldots, y_m$.

8. Apply the conclusion of Problem 7 to redo the isoperimetric problem of Example 9.6. *Hint:* Suppose that the curve we want has parametric equations $x = x(t)$, $y = y(t)$ for $a \le t \le b$. Use Green's theorem to write the area as $\frac{1}{2}\int_a^b (xy' - yx')\,dt$. We want to choose continuously differentiable functions $x(t)$ and $y(t)$ to minimize this integral, subject to the constraint that the length $\int_a^b [(x')^2 + (y')^2]\,dt = L$, a given number. Use the Euler equations to obtain

$$\frac{1}{2} y' - \frac{d}{dt}\left[-\frac{1}{2} y + \frac{\lambda x'}{\sqrt{(x')^2 + (y')^2}} \right] = 0$$

and

$$\frac{1}{2} x' + \frac{d}{dt}\left[\frac{1}{2} x + \frac{\lambda y'}{\sqrt{(x')^2 + (y')^2}} \right] = 0.$$

To solve these differential equations, choose $t = s$, arc length along the curve. With this choice, we have $(x')^2 + (y')^2 = 1$, and the differential equations become

$$y' - \lambda x'' = 0$$
$$x' + \lambda y'' = 0.$$

Integrate these equations and substitute the solution for y into $y' - \lambda x'' = 0$ to obtain a differential equation in y alone. Find the general solution of this differential equation, then obtain the solution for x. Show that the resulting curve is an arc of a circle.

9. Suppose we want to find continuously differentiable functions $x(t)$, $y(t)$, and $z(t)$ which extremize $\int_a^b f(t, x, y, z, x', y', z')\, dt$ subject to a constraint $g(x, y, z) = 0$. Write $x(t) = u(t) + \epsilon_1\eta_1(t)$, $y(t) = v(t) + \epsilon_2\eta_2(t)$, and $z(t) = x(t) = w(t) + \epsilon_3\eta_3(t)$, where $x = u(t)$, $y = v(t)$, and $z = w(t)$ form a solution of the problem. Derive the Euler equations

$$\frac{\partial f}{\partial x} - \frac{d}{dt}\frac{\partial f}{\partial x'} + \lambda \frac{\partial g}{\partial x} = 0, \qquad \frac{\partial f}{\partial y} - \frac{d}{dt}\frac{\partial f}{\partial y'} + \lambda \frac{\partial g}{\partial y} = 0, \qquad \frac{\partial f}{\partial z} - \frac{d}{dt}\frac{\partial f}{\partial z'} + \lambda \frac{\partial g}{\partial z} = 0,$$

in which λ is a Lagrange multiplier. These equations must be solved subject to the constraint that $g(x, y, z) = 0$.

10. A *geodesic* on a surface is a curve of shortest length on the surface between two given points. In the plane, geodesics are straight lines, but on nonplanar surfaces, they will generally be curves in three-space. A geodesic must minimize the integral for arc length

$$\int_a^b \sqrt{(x')^2 + (y')^2 + (z')^2}\, dt$$

subject to the constraint that $g(x(t), y(t), z(t)) = 0$, where $g(x, y, z) = 0$ is the equation of the surface. Use the result of Problem 9 to derive the following system of ordinary differential equations for geodesics on the surface $g(x, y, z) = 0$:

$$\frac{\frac{d}{dt}\left(\frac{1}{f} x'\right)}{\frac{\partial g}{\partial x}} = \frac{\frac{d}{dt}\left(\frac{1}{f} y'\right)}{\frac{\partial g}{\partial y}} = \frac{\frac{d}{dt}\left(\frac{1}{f} z'\right)}{\frac{\partial g}{\partial z}},$$

where $f(x, y, z, x', y', z') = \sqrt{(x')^2 + (y')^2 + (z')^2}$.

11. Use the result of Problem 10 to show that geodesics on a sphere are great circles on the sphere (that is, intersections of planes with the sphere). This is a well-known fact in navigation; airplanes flying over the pole take a great circle route. *Hint:* Write the equation of the sphere as $g(x, y, z) = x^2 + y^2 + z^2 - R^2 = 0$. (We lose no generality by placing the origin at the center of the sphere.) Show that the Euler equations in Problem 10 take the form

$$\frac{x''f - x'f'}{2xf^2} = \frac{y''f - y'f'}{2yf^2} = \frac{z''f - z'f'}{2zf^2},$$

where $f' = df/dt$, a chain rule differentiation of $\sqrt{(x')^2 + (y')^2 + (z')^2}$. Show that this equation can be written in the form

$$\frac{yx'' - xy''}{yx' - xy'} = \frac{zy'' - yz''}{zy' - yz'} = \frac{1}{f} f'.$$

Now show that

$$\frac{(yx' - xy')'}{yx' - xy'} = \frac{(zy' - yz')'}{zy' - yz'}.$$

Integrate this equation and rearrange terms to show that

$$\frac{x' + Cz'}{x + Cz} = \frac{1}{y} y',$$

in which C is the constant in the above integration. Now integrate again to show that

$$x + Cz = Ky,$$

in which K is another constant of integration. This is the equation of a plane through the origin and shows that, to lie on a geodesic, (x, y, z) must lie on the intersection of the sphere with a plane through the origin.

CHAPTER 10

Difference Equations

10.0 Introduction

We use differential equations to model processes in which we know relationships between rates of change. Sometimes, however, we know a relationship between changes, or differences, rather than rates of change. For example, suppose we are told that rabbits reproduce according to the rule that one pair of rabbits is born each month from each pair of adults of the opposite sex which are not less than 2 months old. Assume that initially one male-female pair is present. In successive months, the total number of pairs of rabbits forms the sequence

$$1, 1, 2, 3, 5, 8, 13, \ldots.$$

Each term is the sum of the preceding two terms. If r_k is the number of rabbits in month k, then

$$r_k = r_{k-1} + r_{k-2}$$

for $k = 2, 3, \ldots$, with $r_1 = r_2 = 1$. The equation $r_k = r_{k-1} + r_{k-2}$ is an example of a *difference equation*, and the conditions $r_1 = r_2 = 1$ are analogous to initial conditions for a differential equation. We will see later how to solve this difference equation initial value problem to obtain the formula

$$r_k = \frac{1}{\sqrt{5}}\left[\left(\frac{1 + \sqrt{5}}{2}\right)^k - \left(\frac{1 - \sqrt{5}}{2}\right)^k\right].$$

Difference equations arise in a variety of contexts. A mathematical model of a physical phenomenon may lead to a difference equation. An example involving a string loaded with a finite number of beads at equally spaced points is discussed in Section 10.5. Each recurrence relation obtained when we solve a differential equation by the power series method or the method of Frobenius is also a difference equation. Further, since a derivative can be approximated as a difference quotient, difference equations are sometimes used to approximate differential equations. This strategy is used in formulating numerical techniques for approximating solutions of differential equations.

Finally, we sometimes evaluate a quantity of interest by first showing that it satisfies a difference equation, then solving the difference equation. As an illustration of this, suppose we want to evaluate the integral

$$\int_0^\pi \frac{\cos(k\theta) - \cos(k\varphi)}{\cos(\theta) - \cos(\varphi)} \, d\theta,$$

in which φ is a given constant and k is a nonnegative integer. Denote the integral as I_k. The strategy is to show (see Problem 22 of Section 10.2) that I_k satisfies the difference equation

$$I_{k+2} - 2 \cos(\varphi)I_{k+1} + I_k = 0.$$

Further, by direct integration, we have the initial conditions

$$I_0 = 0 \quad \text{and} \quad I_1 = \pi.$$

Using methods we will develop in Section 10.2, this problem is not difficult to solve, and we obtain $I_k = \pi \left[\sin(k\varphi)/\sin(\varphi) \right]$, provided that $\sin(\varphi) \neq 0$.

We will now develop some introductory aspects of the theory of difference equations. Notice throughout the strong parallel which exists between both the theory and methods of differential equations and those of difference equations.

10.1 Notation, Terminology, and Linear First Order Difference Equations

In this section, we will establish some notation and terminology and discuss ways of drawing helpful analogies between difference and differential equations. We will also see how to solve a class of first order difference equations.

A *difference equation* is an equation relating various terms of a sequence $y_0, y_1, y_2, \ldots$. Examples are

$$y_{k+3} - 4y_{k+2} + y_{k+1} - 6y_k = k^2 \tag{10.1}$$

and

$$y_{k+2} + ky_{k+1} - \cos(k)y_k = 0. \tag{10.2}$$

Given a difference equation, our objective is to obtain an expression for y_k as a function of k. Sometimes we can obtain y_k as an explicit combination of elementary functions of k. For example, $y_k = 2^k$ is a solution of $y_{k+2} - 4y_k = 0$, as is $(-1)^k 2^k$. These assertions can be verified by substitution into the difference equation. Similarly, $(2\sqrt{2})^k [c_1 \cos(k\pi/4) + c_2 \sin(k\pi/4)]$ is a solution of $y_{k+2} - 4y_{k+1} + 8y_k = 0$ for $k = 0$, $1, 2, 3, \ldots$ and any numbers c_1 and c_2.

The *order* of a difference equation is the difference between the largest and smallest indices appearing in the equation. For example, equation (10.1) is of order 3 because the largest index is $k + 3$ and the smallest is k. Equation (10.2) is of order 2. The difference equation $y_{k+4} - 5y_{k+2} = k$ is of order 2 (not 4) because $(k + 4) - (k + 2) = 2$.

We can draw a parallel between difference equations and differential equations if we make the appropriate associations between continuous quantities (x varying over an interval) and discrete quantities (k taking on nonnegative integer values). With the associations we will discuss, many theorems and methods involving difference equations have recognizable analogues for differential equations.

To begin the analogy, we obviously think of $y(x)$ in a differential equation as replaced by y_k in a difference equation. Next, the *discrete derivative* Dy_k is defined by

$$Dy_k = y_{k+1} - y_k.$$

This plays the role for difference equations that $y'(x)$ does for differential equations. The second discrete derivative is

$$D^2 y_k = D[Dy_k].$$

Observe that

$$D^2 y_k = D[Dy_k] = D[y_{k+1} - y_k]$$
$$= [y_{k+2} - y_{k+1}] - [y_{k+1} - y_k] = y_{k+2} - 2y_{k+1} + y_k.$$

It is not immediately obvious how to form the discrete analogue of the power function $y(x) = x^n$, with n any positive integer. If we attempt the "natural" choice $y_k = k^n$, the discrete derivative is

$$Dy_k = y_{k+1} - y_k = (k + 1)^n - k^n,$$

a formula which bears no resemblance to the continuous derivative $y'(x) = nx^{n-1}$.

We get a better analogy if we define the discrete analogue of x^n to be

$$y_k = k(k + 1) \cdots (k + n - 1),$$

a product of n factors. Thus, the discrete analogue of x^2 is $k(k + 1)$, that of x^3 is $k(k + 1)(k + 2)$, and so on. With this definition, the discrete derivative is

$$Dy_k = y_{k+1} - y_k = (k + 1)(k + 2) \cdots (k + n - 1)(k + n) - k(k + 1) \cdots (k + n - 1)$$
$$= (k + 1) \cdots (k + n - 1)[(k + n) - k]$$
$$= n(k + 1) \cdots (k + n - 1)$$
$$= \frac{n}{k} k(k + 1) \cdots (k + n - 1),$$

which is the correct analogue of nx^{n-1} if we write $nx^{n-1} = (n/x) x^n$.

If r is not a negative integer, we can use the gamma function to write

$$\frac{\Gamma(k + r)}{\Gamma(k)} = k(k + 1) \cdots (k + r - 1).$$

Using this as a rationale, we choose $[\Gamma(k + r)]/\Gamma(k)$ to be the analogue of x^r for any number r which is not a negative integer. This choice is seen to be the appropriate one if we compute the discrete derivative. With $y_k = [\Gamma(k + r)]/\Gamma(k)$ (analogous to $y = x^r$), we get

$$Dy_k = y_{k+1} - y_k = \frac{\Gamma(k + r + 1)}{\Gamma(k + 1)} - \frac{\Gamma(k + r)}{\Gamma(k)}$$

$$= (k + 1)(k + 2) \cdots (k + r - 1)(k + r) - k(k + 1)(k + 2) \cdots (k + r - 1)$$

$$= (k + 1) \cdots (k + r - 1)[(k + r) - k]$$

$$= r(k + 1) \cdots (k + r - 1) = \frac{r}{k} k(k + 1) \cdots (k + r - 1)$$

$$= \frac{r}{k} \frac{\Gamma(k + r)}{\Gamma(k)}.$$

This is the discrete analogue of the familiar formula $(d/dx) [x^r] = (r/x) x^r$. A similar calculation yields

$$D^2 y_k = \frac{r(r - 1)}{k(k + 1)} \frac{\Gamma(k + r)}{\Gamma(k)}$$

if $y_k = [\Gamma(k + r)]/\Gamma(k)$. This is the discrete analogue of

$$\frac{d^2}{dx^2} [x^r] = \frac{r(r - 1)}{x^2} x^r.$$

We will now develop methods for solving certain difference equations. We will deal for the most part with the linear case, beginning with difference equations of order 1. A first order difference equation is linear if it has the form

$$y_{k+1} = a_k y_k + b_k, \tag{10.3}$$

in which a_k and b_k are given for $k = 0, 1, 2, 3, \ldots$. We will see how to obtain an explicit formula for the solution of this equation, assuming that each $a_k \neq 0$.

Recall that the first order linear differential equation can be solved by multiplying the equation by an integrating factor. A similar strategy can be devised for equation (10.3). Form the product

$$\prod_{j=0}^{k} a_j = a_0 a_1 \cdots a_k. \tag{10.4}$$

(The symbol $\prod$ plays the role for products that $\sum$ does for sums.) Assume that each $a_k \neq 0$ and multiply equation (10.3) by the reciprocal of the product (10.4) to get

$$\frac{1}{\prod\limits_{j=0}^{k} a_j} y_{k+1} = \frac{1}{\prod\limits_{j=0}^{k} a_j} a_k y_k + \frac{1}{\prod\limits_{j=0}^{k} a_j} b_k.$$

In the first term on the right, a_k cancels a factor of the product $a_0 a_1 \cdots a_k$ in the denominator, and we obtain

$$\frac{1}{\prod\limits_{j=0}^{k} a_j} y_{k+1} = \frac{1}{\prod\limits_{j=0}^{k-1} a_j} y_k + \frac{1}{\prod\limits_{j=0}^{k} a_j} b_k.$$

Write this equation as

$$\frac{1}{\prod\limits_{j=0}^{k} a_j} y_{k+1} - \frac{1}{\prod\limits_{j=0}^{k-1} a_j} y_k = \frac{1}{\prod\limits_{j=0}^{k} a_j} b_k.$$

Let

$$A_k = \frac{1}{\prod\limits_{j=0}^{k-1} a_j} y_k$$

to write this equation as

$$A_{k+1} - A_k = \frac{1}{\prod\limits_{j=0}^{k} a_j} b_k.$$

We can solve this equation by summing both sides and recognizing that the sum of terms on the left telescopes. We obtain

$$\sum_{j=0}^{k-1} [A_{j+1} - A_j] = [A_1 - A_0] + [A_2 - A_1] + \cdots + [A_k - A_{k-1}]$$

$$= A_k - A_0 = \sum_{j=0}^{k-1} \frac{1}{\prod\limits_{s=0}^{k} a_s} b_j.$$

Therefore,

$$y_k = \left[\prod_{j=0}^{k-1} a_j \right] A_k = \prod_{j=0}^{k-1} a_j \left[A_0 + \sum_{j=0}^{k-1} \frac{1}{\prod\limits_{s=0}^{k} a_s} b_j \right] \tag{10.5}$$

for $k = 1, 2, 3, \ldots$ and any number A_0. Formula (10.5) is an explicit formula for the solution of equation (10.3). If each $b_k = 0$, we have the homogeneous linear difference equation $y_{k+1} - a_k y_k = 0$, with solution

$$y_k = A_0 \prod_{j=0}^{k-1} a_j \tag{10.6}$$

for $k = 1, 2, 3, \ldots$ and any number A_0.

In practice, rather than memorize equations (10.5) and (10.6), we usually multiply the linear first order difference equation by $[a_0 a_1 \cdots a_k]^{-1}$ and sum the resulting equation to obtain the solution. The factor $[a_0 a_1 \cdots a_k]^{-1}$ is called a *summing factor* for the difference equation (10.3). It plays the role of an integrating factor for a differential equation, with summation considered as the discrete analogue of integration.

EXAMPLE 10.1

Solve

$$y_{k+1} = \frac{k}{k+1} y_k + 4.$$

In order for $a_k = k/(k + 1)$ to be nonzero, we will consider this equation only for $k = 1, 2, 3, \ldots$. Form the product

$$a_1 a_2 \cdots a_k = \frac{1}{2} \frac{2}{3} \frac{3}{4} \cdots \frac{k-1}{k} \frac{k}{k+1} = \frac{1}{k+1}.$$

The summing factor is the reciprocal of this product, or $k + 1$. Multiply the difference equation by $k + 1$ to get

$$(k + 1)y_{k+1} = ky_k + 4(k + 1),$$

or

$$(k + 1)y_{k+1} - ky_k = 4(k + 1).$$

In the general discussion above, we put $A_k = ky_k$. In this notation, we have

$$A_{k+1} - A_k = 4(k + 1).$$

Sum this equation from $j = 1$ to $k - 1$ to get

$$\sum_{j=1}^{k-1} [A_{j+1} - A_j] = A_k - A_1 = \sum_{j=1}^{k-1} 4(j + 1)$$

$$= 4 \sum_{j=1}^{k-1} j + 4 \sum_{j=1}^{k-1} 1.$$

Now use the standard formulas

$$\sum_{j=1}^{n} j = \tfrac{1}{2}n(n + 1) \quad \text{and} \quad \sum_{j=1}^{n} 1 = n$$

to get

$$A_k - A_1 = 4\tfrac{1}{2}(k - 1)k + 4(k - 1) = 2(k - 1)(k + 2).$$

Then

$$A_k = A_1 + 2(k - 1)(k + 2).$$

Recalling that $A_k = ky_k$, we have the general solution

$$y_k = \frac{1}{k} A_1 + \frac{2}{k}(k - 1)(k + 2),$$

in which A_1 is arbitrary and $k = 1, 2, 3, \ldots$. The arbitrary constant is A_1, not A_0, because for this problem k varies from 1, not from zero.   ■

PROBLEMS FOR SECTION 10.1

Solve each of the difference equations in Problems 1 through 10.

1. $y_{k+1} = y_k + 3^k$ 　　**2.** $y_{k+1} = y_k + 1$ 　　**3.** $y_{k+1} = ky_k$ 　　**4.** $y_{k+1} = ky_k - k$

5. $y_{k+1} = ky_k + k^2$ 　　**6.** $y_{k+1} = \dfrac{k}{k+1} y_k - 4k$ 　　**7.** $y_{k+1} = \dfrac{k}{k+1} y_k + k^2$ 　　**8.** $y_{k+1} = 3^k y_k + k$

9. $y_{k+1} = y_k + k^2 - k$ 　　**10.** $y_{k+1} + 2^k y_k = 1$

11. Determine formulas for the discrete derivatives $D^3 y_k$ and $D^4 y_k$.

12. Based on results for Dy_k through $D^4 y_k$, write a general expression for $D^n y_k$, with n any positive integer.

10.2 *Linear Homogeneous Difference Equations of Order Two*_____

A second order difference equation is called *linear* if it has the form

$$y_{k+2} + a_k y_{k+1} + b_k y_k = f_k, \tag{10.7}$$

in which a_k, b_k, and f_k are given numbers for each nonnegative integer k. If $f_k = 0$ for $k = 0, 1, 2, \ldots$, equation (10.7) is said to be *homogeneous*; if some $f_k \neq 0$, the equation is *nonhomogeneous*. For example, $y_{k+2} - k^2 y_k = 0$ is homogeneous, while $y_{k+2} - k^2 y_k = 1 + 4k$ is nonhomogeneous.

The theory and methods of solving equation (10.7) closely parallel the theory and methods for solving the differential equation $y'' + p(x)y' + q(x)y = f(x)$. As we did for this differential equation, we will begin with the homogeneous case. For the remainder of this section, consider the homogeneous second order linear difference equation

$$y_{k+2} + a_k y_{k+1} + b_k y_k = 0. \tag{10.8}$$

THEOREM 10.1

If $y_k^{(1)}$ and $y_k^{(2)}$ are solutions of equation (10.8), $c_1 y_k^{(1)} + c_2 y_k^{(2)}$ is a solution for any constants c_1 and c_2. ∎

This theorem can be proved by substituting $c_1 y_k^{(1)} + c_2 y_k^{(2)}$ into equation (10.8).

We call $c_1 y_k^{(1)} + c_2 y_k^{(2)}$ a *linear combination* of $y_k^{(1)}$ and $y_k^{(2)}$. For example, it is easy to verify that 2^k and $(-1)^k 2^k$ are solutions of $y_{k+2} - 4y_k = 0$; hence, $c_1 2^k + c_2 (-1)^k 2^k$ is also a solution for any constants c_1 and c_2.

We call two functions $y_k^{(1)}$ and $y_k^{(2)}$ *linearly dependent* if, for some constant c, $y_k^{(1)} = c y_k^{(2)}$ for $k = 0, 1, 2, \ldots$, or $y_k^{(2)} = c y_k^{(1)}$ for $k = 0, 1, 2, \ldots$. If $y_k^{(1)}$ and $y_k^{(2)}$ are not linearly dependent, they are *linearly independent*. In general, $y_k^{(1)}$ and $y_k^{(2)}$ are linearly independent if and only if the linear combination $c_1 y_k^{(1)} + c_2 y_k^{(2)}$ can be zero for $k = 0, 1, 2, 3, \ldots$, only if $c_1 = c_2 = 0$.

The *Wronskian* of two solutions $y_k^{(1)}$ and $y_k^{(2)}$ of equation (10.8) is defined to be

$$W_k = \begin{vmatrix} y_k^{(1)} & y_k^{(2)} \\ y_{k+1}^{(1)} & y_{k+1}^{(2)} \end{vmatrix}.$$

THEOREM 10.2   Wronskian Test for Linear Independence

Let $y_k^{(1)}$ and $y_k^{(2)}$ be solutions of $y_{k+2} + a_k y_{k+1} + b_k y_k = 0$. Then $y_k^{(1)}$ and $y_k^{(2)}$ are linearly dependent if and only if $W_k = 0$ for $k = 0, 1, 2, \ldots$. ∎

For a proof of Theorem 10.2, see Problems 23 and 24 at the end of this section.

THEOREM 10.3

Suppose that $y_k^{(1)}$ and $y_k^{(2)}$ are linearly independent solutions of equation (10.8). Then every solution of equation (10.8) is a linear combination of $y_k^{(1)}$ and $y_k^{(2)}$. ∎

When $y_k^{(1)}$ and $y_k^{(2)}$ are linearly independent solutions of equation (10.8), we say that they form a *fundamental set of solutions*. In this event, the linear combination $c_1 y_k^{(1)} + c_2 y_k^{(2)}$ is called the *general solution* of the difference equation.

An initial value problem for the linear homogeneous second order difference equation has the form

$$y_{k+2} + a_k y_{k+1} + b_k y_k = 0; \qquad y_0 = \alpha, \qquad y_1 = \beta,$$

with α and β given numbers. We can solve an initial value problem by finding the general solution $c_1 y_k^{(1)} + c_2 y_k^{(2)}$ of the difference equation and solving for the two constants to satisfy the initial conditions.

EXAMPLE 10.2

Solve the initial value problem

$$y_{k+2} - 4y_k = 0; \qquad y_0 = 0, \qquad y_1 = 1.$$

From previous remarks, the general solution of this difference equation is $c_1 2^k + c_2(-1)^k 2^k$. We need $y_0 = c_1 + c_2 = 0$ and $y_1 = 2c_1 - 2c_2 = 1$. Then $c_1 = \frac{1}{4}$ and $c_2 = -\frac{1}{4}$, so the solution of the initial value problem is

$$y_k = \tfrac{1}{4} 2^k [1 - (-1)^k]. \quad \blacksquare$$

For the remainder of this section, we will consider the constant coefficient linear second order difference equation

$$y_{k+2} + a y_{k+1} + b y_k = 0, \tag{10.9}$$

in which a and b are constants and $b \neq 0$ (for the equation to be of second order). The differential equation analogue $y'' + ay' + by = 0$ always has exponential solutions e^{rx}. For the difference equation (10.9), we attempt solutions of the form

$$y_k = r^k,$$

in which r is to be determined. Substitute this proposed solution into the difference equation to get

$$r^{k+2} + a r^{k+1} + b r^k = 0.$$

If $r \neq 0$, we can divide this equation by r^k to get the *characteristic equation*

$$r^2 + ar + b = 0. \tag{10.10}$$

Roots of this equation yield values of r for which $y_k = r^k$ is a solution of (10.9). There are three cases.

Case 1   If the characteristic equation has two distinct real roots β_1 and β_2, then $y_k^{(1)} = \beta_1^k$ and $y_k^{(2)} = \beta_2^k$ are solutions. Further, their Wronskian is found to be $W_k = \beta_1^k \beta_2^k (\beta_2 - \beta_1)$, which is nonzero if $\beta_1 \neq \beta_2$. Thus, β_1^k and β_2^k form a fundamental set of solutions, and the general solution of (10.9) is

$$y_k = c_1 \beta_1^k + c_2 \beta_2^k.$$

EXAMPLE 10.3

Solve $y_{k+2} + 2y_{k+1} - 3y_k = 0$.

The characteristic equation is $r^2 + 2r - 3 = 0$, with roots 1 and -3. The general solution is $y_k = c_1(1)^k + c_2(-3)^k = c_1 + c_2(-1)^k 3^k$. ∎

Case 2 Suppose the characteristic equation has one root (repeated). If β is the root, $y_k^{(1)} = \beta^k$ is one solution. We claim that $y_k^{(2)} = k\beta^k$ is also a solution. This is analogous to having e^{rx} and xe^{rx} as solutions of $y'' + ay' + by = 0$ when the characteristic equation has a repeated root r.

To prove that $k\beta^k$ is a solution, substitute this into the difference equation to get

$$y_{k+2}^{(2)} + ay_{k+1}^{(2)} + by_k^{(2)} = (k+2)\beta^{k+2} + a(k+1)\beta^{k+1} + bk\beta^k$$
$$= k[\beta^{k+2} + a\beta^{k+1} + b\beta^k] + 2\beta^{k+2} + a\beta^{k+1}.$$

Now, $\beta^{k+2} + a\beta^{k+1} + b\beta^k = 0$, because β^k is a solution of the difference equation. We will show that $2\beta^{k+2} + a\beta^{k+1} = 0$ also.

The roots of $r^2 + ar + b = 0$ are $(-a \pm \sqrt{a^2 - 4b})/2$. Assuming that there is only one root, $\beta = -a/2$ and $a^2 = 4b$. Therefore,

$$2\beta^{k+2} + a\beta^{k+1} = 2\left(-\frac{a}{2}\right)^{k+2} + a\left(-\frac{a}{2}\right)^{k+1}$$
$$= \left(-\frac{a}{2}\right)^{k+1}\left[2\left(-\frac{a}{2}\right) + a\right] = 0.$$

Thus, $y_k^{(2)} = k\beta^k$ is a second solution. Since β^k and $k\beta^k$ are linearly independent, the general solution of (10.9) in this case is

$$y = (c_1 + c_2 k)\beta^k.$$

EXAMPLE 10.4

Solve $y_{k+2} - 8y_{k+1} + 16y_k = 0$.

The characteristic equation is $r^2 - 8r + 16 = 0$, with repeated root $r = 4$. The general solution of the difference equation is $y_k = (c_1 + c_2 k)4^k$. ∎

Case 3 Suppose the characteristic equation of (10.9) has complex roots. Since a and b are real, the roots are complex conjugates $\alpha + i\beta$ and $\alpha - i\beta$. Now the general solution of equation (10.9) is

$$y_k = c_1(\alpha + i\beta)^k + c_2(\alpha - i\beta)^k.$$

We can write a real-valued solution as follows. Let (ρ, θ) be polar coordinates of the rectangular point (α, β). Then $\alpha = \rho\cos(\theta)$ and $\beta = \rho\sin\theta$. By Euler's formula,

$$\alpha + i\beta = \rho[\cos(\theta) + i\sin(\theta)] = \rho e^{i\theta};$$

hence,

$$(\alpha + i\beta)^k = \rho^k e^{ik\theta} = \rho^k[\cos(k\theta) + i\sin(k\theta)].$$

Similarly,

$$\alpha - i\beta = \rho[\cos(\theta) - i\sin(\theta)] = \rho e^{-i\theta}; \tag{10.11}$$

hence,

$$(\alpha - i\beta)^k = \rho^k e^{-ik\theta} = \rho^k[\cos(k\theta) - i\sin(k\theta)]. \qquad (10.12)$$

If we add the two solutions (10.11) and (10.12) and multiply the result by $\frac{1}{2}$, we obtain another solution,

$$y_k^{(1)} = \rho^k\cos(k\theta).$$

If we subtract (10.12) from (10.11) and multiply the result by $1/(2i)$, we obtain the solution

$$y_k^{(2)} = \rho^k\sin(k\theta).$$

We may therefore write the general solution of (10.9) in this case as

$$y_k = \rho^k[c_1\cos(k\theta) + c_2\sin(k\theta)],$$

in which ρ and θ are polar coordinates of (α, β) and $\alpha + i\beta$ is a root of the characteristic equation.

EXAMPLE 10.5

Solve $y_{k+2} + 4y_k = 0$.

The characteristic equation is $r^2 + 4 = 0$, with roots $\pm 2i$. Here, $\alpha = 0$ and $\beta = 2$. Polar coordinates of $(0, 2)$ are $\rho = 2$ and $\theta = \pi/2$. The general solution of the difference equation is

$$y = 2^k\left[c_1\cos\left(\frac{k\pi}{2}\right) + c_2\sin\left(\frac{k\pi}{2}\right)\right]. \quad \blacksquare$$

In the next section, we will treat the nonhomogeneous linear difference equation $y_{k+2} + a_k y_{k+1} + b_k y_k = f_k$.

PROBLEMS FOR SECTION 10.2

In each of Problems 1 through 10, write the general solution of the difference equation.

1. $y_{k+2} + 4y_{k+1} - 12y_k = 0$ **2.** $y_{k+2} - 6y_{k+1} + 9y_k = 0$

3. $y_{k+2} + 4y_{k+1} + 3y_k = 0$ **4.** $y_{k+2} - 2y_{k+1} + 2y_k = 0$

5. $y_{k+2} + 8y_{k+1} + 16y_k = 0$ **6.** $y_{k+2} - 36y_k = 0$

7. $y_{k+2} + 2y_{k+1} + 2y_k = 0$ **8.** $y_{k+2} - 13y_{k+1} + 42y_k = 0$

9. $y_{k+2} - 2\sqrt{3}y_{k+1} + 4y_k = 0$ **10.** $y_{k+2} + 2y_{k+1} + y_k = 0$

In each of Problems 11 through 20, solve the initial value problem.

11. $y_{k+2} - 6y_{k+1} + 8y_k = 0; \quad y_0 = 1, \quad y_1 = 0$ **12.** $y_{k+2} - 9y_{k+1} + 20y_k = 0; \quad y_0 = 0, \quad y_1 = 4$

13. $y_{k+2} + 10y_{k+1} + 25y_k = 0; \quad y_0 = y_1 = 2$ **14.** $y_{k+2} - 4y_{k+1} + 4y_k = 0; \quad y_0 = 1, \quad y_1 = \sqrt{8}$

15. $y_{k+2} + 2y_{k+1} - 8y_k = 0; \quad y_0 = 2, \quad y_1 = 1$ **16.** $y_{k+2} + 2y_{k+1} + 2y_k = 0; \quad y_0 = 6, \quad y_1 = 0$

17. $y_{k+2} + 2y_{k+1} + 2y_k = 0; \quad y_0 = y_1 = 2$ **18.** $y_{k+2} - 16y_{k+1} + 64y_k = 0; \quad y_0 = 0, \quad y_1 = 8$

19. $y_{k+2} - 9y_k = 0; \quad y_0 = 3, \quad y_1 = -3$ **20.** $y_{k+2} + 9y_k = 0; \quad y_0 = 4, \quad y_1 = 15$

21. *Fibonacci numbers* are defined to be solutions of the difference equation

$$y_{k+2} = y_{k+1} + y_k.$$

Assuming that $y_0 = y_1 = 1$, find a formula for y_k in terms of k.

22. Let

$$I_k = \int_0^\pi \frac{\cos(k\theta) - \cos(k\varphi)}{\cos(\theta) - \cos(\varphi)} \, d\theta$$

for $k = 0, 1, 2, \ldots$. Assume that φ is any number such that $\sin(\varphi) \neq 0$.
(a) Show that $I_{k+2} - 2\cos(\varphi)I_{k+1} + I_k = 0$.
(b) Show that $I_0 = 0$ and $I_1 = \pi$.
(c) Solve the initial value problem determined by results from (a) and (b) to conclude that

$$I_k = \pi \frac{\sin(k\varphi)}{\sin(\varphi)}.$$

23. Suppose that $y_k^{(1)}$ and $y_k^{(2)}$ are solutions of $y_{k+2} + a_k y_{k+1} + b_k y_k = 0$. Show that $W_k = A \prod_{j=0}^{k-1} b_j$ for some constant A. *Hint:* We know that $y_{k+2}^{(1)} + a_k y_{k+1}^{(1)} + b_k y_k^{(1)} = 0$ and $y_{k+2}^{(2)} + a_k y_{k+1}^{(2)} + b_k y_k^{(2)} = 0$. Multiply the first equation by $y_{k+1}^{(2)}$ and the second by $y_{k+1}^{(1)}$ and subtract to show that W_k satisfies a homogeneous first order difference equation. Use results from Section 10.1 to determine W_k. This formula for W_k is the difference equation analogue of Abel's formula (given in Problem 20 of Section 2.2).
24. Prove Theorem 10.2.

10.3 The Nonhomogeneous Linear Second Order Difference Equation

We will now consider methods for solving the nonhomogeneous equation

$$y_{k+2} + a_k y_{k+1} + b_k y_k = f_k. \tag{10.13}$$

First, we will state the difference equation analogue of Theorem 2.5 in Section 2.9.

THEOREM 10.4

Let $y_k^{(1)}$ and $y_k^{(2)}$ form a fundamental set of solutions of $y_{k+2} + a_k y_{k+1} + b_k y_k = 0$. Suppose that $y_k^{(p)}$ is any solution of equation (10.13). Then, given any solution y_k of equation (10.13), there are constants c_1 and c_2 such that $y_k = c_1 y_k^{(1)} + c_2 y_k^{(2)} + y_k^{(p)}$. ∎

That is, the expression $c_1 y_k^{(1)} + c_2 y_k^{(2)} + y_k^{(p)}$ contains every solution of equation (10.13), and for this reason it is called the *general solution* of equation (10.13).
This theorem reduces the problem of finding all solutions of equation (10.13) to two steps: first, find the general solution of the *associated homogeneous equation* $y_{k+2} + a_k y_{k+1} + b_k y_k = 0$, then find any particular of equation (10.13). We will now develop two methods which sometimes enable us to find $y_k^{(p)}$.

THE METHOD OF UNDETERMINED COEFFICIENTS

We want to find a particular solution of the *constant coefficient* equation

$$y_{k+2} + a y_{k+1} + b y_k = f_k, \tag{10.14}$$

assuming that a and b are constants and $b \neq 0$.

If f_k is of a certain form, we can guess a general form of a solution $y_k^{(p)}$ and determine the coefficients by substituting $y_k^{(p)}$ into equation (10.14), just as we did for certain second order differential equations. The method closely parallels undetermined coefficients for differential equations, with the same potential difficulties (see Example 10.6 below).

EXAMPLE 10.6

Solve $y_{k+2} - 7y_{k+1} + 12y_k = -16 + 12k$.

Since $f_k = -16 + 12k$ is a first degree polynomial in k, we will attempt a first degree polynomial in k for a particular solution. Thus, try $y_k^{(p)} = A + Bk$. Substitute this function into the difference equation to get

$$A + B(k + 2) - 7[A + B(k + 1)] + 12[A + Bk] = -16 + 12k.$$

Collect terms to obtain

$$6A - 5B + 6Bk = -16 + 12k.$$

Then $6A - 5B = -16$ and $6B = 12$, so $B = 2$ and $A = -1$. A particular solution is $y_k^{(p)} = -1 + 2k$.

The associated homogeneous equation $y_{k+2} - 7y_{k+1} + 12y_k = 0$ has characteristic equation $r^2 - 7r + 12 = 0$, with roots 3 and 4. Therefore, 3^k and 4^k form a fundamental set of solutions of $y_{k+2} - 7y_{k+1} + 12y_k = 0$. The general solution of $y_{k+2} - 7y_{k+1} + 12y_k = -16 + 12k$ is

$$y_k = c_1 3^k + c_2 4^k - 1 + 2k. \quad \blacksquare$$

The following example demonstrates the need for caution when the "natural" form for $y_k^{(p)}$ is a solution of the associated homogeneous equation (just as with differential equations).

EXAMPLE 10.7

Solve $y_{k+2} + 3y_{k+1} - 4y_k = 20k$.

Since $f_k = 20k$, we are tempted to try a solution $y_k^{(p)} = A + Bk$. However, if we substitute this function into the difference equation and arrange terms, we get

$$5B = 20k.$$

This forces $B = 4k$, an impossibility. The problem is that $y_k = 1$ is a solution of $y_{k+2} + 3y_{k+1} - 4y_k = 0$; hence, any constant multiple of 1—in particular, A—is a solution of the associated homogeneous equation, and A is one term of the proposed particular solution of the nonhomogeneous equation.

Analogous to the way around this difficulty for differential equations, multiply the "natural" choice of $y_k^{(p)}$ by k and instead use $y_k^{(p)} = Ak + Bk^2$. Substitute this choice into the original difference equation to get

$$A(k + 2) + B(k + 2)^2 + 3[A(k + 1) + B(k + 1)^2] - 4[Ak + Bk^2] = 20k.$$

Then

$$Ak + 2A + Bk^2 + 4Bk + 4B + 3Ak + 3A$$
$$+ 3Bk^2 + 6Bk + 3B - 4Ak - 4Bk^2 = 20k.$$

The terms involving k^2 cancel, and we are left with

$$5A + 7B + 10Bk = 20k.$$

Then $10B = 20$ (so $B = 2$) and $5A + 7B = 0$ (so $A = -\frac{14}{5}$). A particular solution is $y_k^{(p)} = -\frac{14}{5}k + 2k^2$.

We find that 1 and $(-4)^k$ form a fundamental set of solutions of $y_{k+2} + 3y_{k+1} - 4y_k = 0$. The general solution of $y_{k+2} + 3y_{k+1} - 4y_k = 20k$ is

$$y_k = c_1 + c_2(-1)^k 4^k - \frac{14}{5}k + 2k^2. \quad \blacksquare$$

EXAMPLE 10.8

Solve $y_{k+2} - 4y_k = \sin(k)$.

We find that the general solution of $y_{k+2} - 4y_k = 0$ is $y_k^{(h)} = c_1 2^k + c_2(-1)^k 2^k$. Attempt a particular solution of $y_{k+2} - 4y_k = \sin(k)$ of the form

$$y_k^{(p)} = A\sin(k) + B\cos(k).$$

Substitute this function into the difference equation to get

$$A\sin(k+2) + B\cos(k+2) - 4A\sin(k) - 4B\cos(k) = \sin(k).$$

Then

$$A[\sin(k)\cos(2) + \cos(k)\sin(2)] + B[\cos(k)\cos(2) - \sin(k)\sin(2)]$$
$$- 4A\sin(k) - 4B\cos(k) = \sin(k).$$

Group terms to get

$$[A\cos(2) - B\sin(2) - 4A]\sin(k) + [A\sin(2) + B\cos(2) - 4B]\cos(k) = \sin(k).$$

Since $\sin(k)$ and $\cos(k)$ are linearly independent,

$$A\cos(2) - B\sin(2) - 4A = 1$$

and

$$A\sin(2) + B\cos(2) - 4B = 0.$$

Solve these equations to get

$$A = \frac{\cos(2) - 4}{17 - 8\cos(2)} \quad \text{and} \quad B = \frac{-\sin(2)}{17 - 8\cos(2)}.$$

A particular solution of $y_{k+2} - 4y_k = \sin(k)$ is

$$y_k^{(p)} = \frac{\cos(2) - 4}{17 - 8\cos(2)}\sin(k) + \frac{-\sin(2)}{17 - 8\cos(2)}\cos(k).$$

The general solution is

$$y_k = 2^k[c_1 + c_2(-1)^k] + \frac{\cos(2) - 4}{17 - 8\cos(2)}\sin(k) + \frac{-\sin(2)}{17 - 8\cos(2)}\cos(k). \quad \blacksquare$$

VARIATION OF PARAMETERS

Consider the nonhomogeneous difference equation

$$y_{k+2} + a_k y_{k+1} + b_k y_k = f_k, \tag{10.15}$$

in which the coefficients need not be constant. Suppose we have a fundamental set of solutions $y_k^{(1)}$ and $y_k^{(2)}$ of the associated homogeneous equation $y_{k+2} + a_k y_{k+1} + b_k y_k = 0$. We will attempt to produce A_k and B_k such that

$$y_k^{(p)} = A_k y_k^{(1)} + B_k y_k^{(2)} \tag{10.16}$$

is a particular solution of equation (10.15). This is analogous to the method of variation of parameters for differential equations. We will produce equations for DA_k and DB_k, from which we will recover A_k and B_k by summation, the discrete analogue of integration. To do this, we must put $y_{k+1}^{(p)}$ and $y_{k+2}^{(p)}$ into a suitable form to substitute into the difference equation (10.15).

First, write

$$\begin{aligned}
y_{k+1}^{(p)} &= A_{k+1} y_{k+1}^{(1)} + B_{k+1} y_{k+1}^{(2)} \\
&= A_k y_{k+1}^{(1)} + B_k y_{k+1}^{(2)} + [y_{k+1}^{(1)}(A_{k+1} - A_k) + y_{k+1}^{(2)}(B_{k+1} - B_k)] \\
&= A_k y_{k+1}^{(1)} + B_k y_{k+1}^{(2)} + y_{k+1}^{(1)} DA_k + y_{k+1}^{(2)} DB_k.
\end{aligned}$$

To simplify this expression, impose the condition that

$$y_{k+1}^{(1)} DA_k + y_{k+1}^{(2)} DB_k = 0. \tag{10.17}$$

Now we have

$$y_{k+1}^{(p)} = A_k y_{k+1}^{(1)} + B_k y_{k+1}^{(2)}. \tag{10.18}$$

From this, compute

$$\begin{aligned}
y_{k+2}^{(p)} &= A_{k+1} y_{k+2}^{(1)} + B_{k+1} y_{k+2}^{(1)} \\
&= y_{k+2}^{(1)}[A_{k+1} - A_k] + y_{k+2}^{(2)}[B_{k+1} - B_k] + A_k y_{k+2}^{(1)} + B_k y_{k+2}^{(2)} \\
&= y_{k+2}^{(1)} DA_k + y_{k+2}^{(2)} DB_k + A_k y_{k+2}^{(1)} + B_k y_{k+2}^{(2)}.
\end{aligned}$$

We now have

$$\begin{aligned}
y_{k+2}^{(p)} + a_k y_{k+1}^{(p)} + b_k y_k^{(p)} = {} & y_{k+2}^{(1)} DA_k + y_{k+2}^{(2)} DB_k + A_k y_{k+2}^{(1)} + B_k y_{k+2}^{(2)} \\
& + a_k A_k y_{k+1}^{(1)} + a_k B_k y_{k+1}^{(2)} + b_k A_k y_k^{(1)} + b_k B_k y_k^{(2)} = f_k.
\end{aligned} \tag{10.19}$$

But

$$A_k y_{k+2}^{(1)} + B_k y_{k+2}^{(2)} + a_k A_k y_{k+1}^{(1)} + a_k B_k y_{k+1}^{(2)} + b_k A_k y_k^{(1)} + b_k B_k y_k^{(2)} = 0$$

because $y_k^{(1)}$ and $y_k^{(2)}$ satisfy $y_{k+2} + a_k y_{k+1} + b_k y_k = 0$. Therefore, equation (10.19) becomes

$$y_{k+2}^{(1)} DA_k + y_{k+2}^{(2)} DB_k = f_k. \tag{10.20}$$

Solve equations (10.17) and (10.20) for DA_k and DB_k to get

$$DA_k = A_{k+1} - A_k = -\frac{f_k y_{k+1}^{(2)}}{W_{k+1}} \quad \text{and} \quad DB_k = B_{k+1} - B_k = \frac{f_k y_{k+1}^{(1)}}{W_{k+1}},$$

where $W_{k+1} = y_{k+2}^{(2)} y_{k+1}^{(1)} - y_{k+2}^{(1)} y_{k+1}^{(2)}$ is the Wronskian. We can solve for A_k and B_k by summation:

$$\sum_{j=0}^{k-1} [A_{j+1} - A_j] = A_k - A_0 = \sum_{j=0}^{k-1} \frac{-f_j y_{j+1}^{(2)}}{W_{j+1}},$$

hence

$$A_k = A_0 + \sum_{j=0}^{k-1} \frac{-f_j y_{j+1}^{(2)}}{W_{j+1}},$$

with A_0 any constant. Similarly,

$$B_k = B_0 + \sum_{j=0}^{k-1} \frac{f_j y_{j+1}^{(1)}}{W_{j+1}},$$

with B_0 any constant. We may choose $A_0 = B_0 = 0$ and obtain the general solution of (10.15) as

$$y_k = c_1 y_k^{(1)} + c_2 y_k^{(2)} + \sum_{j=0}^{k-1} \frac{-f_j y_{j+1}^{(2)}}{W_{j+1}} y_k^{(1)} + \sum_{j=0}^{k-1} \frac{f_j y_{j+1}^{(1)}}{W_{j+1}} y_k^{(2)}.$$

EXAMPLE 10.9

Find the general solution of $y_{k+2} - 4y_k = \sin^2(k)$.

We find that $y_{k+2} - 4y_k = 0$ has fundamental set of solutions $y_k^{(1)} = 2^k$ and $y_k^{(2)} = (-2)^k$. Compute

$$W_{k+1} = \begin{vmatrix} y_{k+1}^{(1)} & y_{k+1}^{(2)} \\ y_{k+2}^{(1)} & y_{k+2}^{(2)} \end{vmatrix} = (-1)^k 4^{k+2}.$$

With $f_k = \sin^2(k)$, the general solution of the difference equation is

$$y_k = c_1 2^k + c_2(-1)^k 2^k + \sum_{j=0}^{k-1} \frac{-\sin^2(j)(-1)^{j+1} 2^{j+1}}{(-1)^j 4^{j+2}} 2^k$$

$$+ \sum_{j=0}^{k-1} \frac{\sin^2(j) 2^{j+1}}{(-1)^j 4^{j+2}} (-2)^k$$

$$= c_2 2^k + c_2(-1)^k 2^k + \sum_{j=0}^{k-1} \frac{\sin^2(j)}{2^{j+3}} 2^k + \sum_{j=0}^{k-1} \frac{(-1)^j \sin^2(j)}{2^{j+3}} (-2)^k$$

for $k = 1, 2, 3, \ldots$. ∎

PROBLEMS FOR SECTION 10.3

In each of Problems 1 through 10, find the general solution of the difference equation using the method of undetermined coefficients.

1. $y_{k+2} - 5y_{k+1} + 6y_k = 1 + k$
3. $y_{k+2} + 5y_{k+1} - 6y_k = 4$
5. $y_{k+2} + 7y_{k+1} + 12y_k = 1$
7. $y_{k+2} + 8y_{k+1} + 12y_k = e^k$
9. $y_{k+2} - 5y_{k+1} + 6y_k = 2 - k^2$

2. $y_{k+2} - 8y_{k+1} + 16y_k = k^2$
4. $y_{k+2} - 4y_k = 3\sin(k)$
6. $y_{k+2} + 4y_{k+1} - 5y_k = k$
8. $y_{k+2} - 2y_{k+1} + y_k = 1 - 4k$
10. $y_{k+2} + 4y_k = \sin(2k)$

In each of Problems 11 through 15, use variation of parameters to find the general solution of the difference equation.

11. $y_{k+2} - 9y_k = \dfrac{1}{k}$

12. $y_{k+2} + 4y_{k+1} + 4y_k = e^k$

13. $y_{k+2} + y_{k+1} - 12y_k = \sin^2(k)$

14. $y_{k+2} + 5y_{k+1} + 4y_k = \dfrac{k+1}{k}$

15. $y_{k+2} + 8y_{k+1} + 7y_k = ke^k$

16. Find the general solution of $y_{k+2} - 4y_{k+1} + 3y_k = 6 + \alpha^k$, in which α is any positive number.

10.4 *The Cauchy-Euler Difference Equation* _____

Recall the Cauchy-Euler differential equation

$$x^2 y'' + axy' + by = 0. \tag{10.21}$$

The discrete analogue of x is k, and that of x^2 is $k(k+1)$. The discrete analogue of dy/dx is $Dy_k = y_{k+1} - y_k$. We therefore define the difference equation analogue of the Cauchy-Euler differential equation to be

$$k(k+1)\, D^2 y_k + ak\, Dy_k + by = 0. \tag{10.22}$$

This is called the *Cauchy-Euler difference equation.*

Recall that solutions of (10.21) have the form $y = x^r$. We also saw in Section 10.1 that the discrete analogue of x^r is $[\Gamma(k+r)]/\Gamma(k)$, which is the product

$$k(k+1)\cdots(k+r-1).$$

This suggests that we attempt solutions of equation (10.22) of the form

$$y_k = \frac{\Gamma(k+r)}{\Gamma(k)}. \tag{10.23}$$

The idea is to substitute this proposed solution into equation (10.22) and attempt to choose r to obtain a solution. From Section 10.1,

$$Dy_k = \frac{r}{k}\frac{\Gamma(k+r)}{\Gamma(k)} \quad \text{and} \quad D^2 y_k = \frac{r(r-1)}{k(k+1)}\frac{\Gamma(k+r)}{\Gamma(k)}.$$

Substitute the proposed solution (10.23) into equation (10.22) to get

$$k(k+1)\frac{r(r-1)}{k(k+1)}\frac{\Gamma(k+r)}{\Gamma(k)} + ak\frac{r}{k}\frac{\Gamma(k+r)}{\Gamma(k)} + b\frac{\Gamma(k+r)}{\Gamma(k)} = 0.$$

After cancellations, this equation can be written

$$r(r-1) + ar + b = 0,$$

or

$$r^2 + (a-1)r + b = 0. \tag{10.24}$$

This is the *characteristic equation* for the Cauchy-Euler difference equation. It is the same as the characteristic equation of $x^2 y'' + axy' + by = 0$. The roots of equation (10.24) yield values of r such that $y_k = [\Gamma(k+r)]/\Gamma(k)$ is a solution of equation (10.21).

EXAMPLE 10.10

Solve $k(k+1)\, D^2 y_k - 5k\, Dy_k + 8y_k = 0$.

With $a = -5$ and $b = 8$, the characteristic equation is

$$r^2 - 6r + 8 = 0,$$

with roots 2 and 4. Two linearly independent solutions are $[\Gamma(k + 2)]/\Gamma(k)$ and $[\Gamma(k + 4)]/\Gamma(k)$, and the general solution is

$$y_k = c_1 \frac{\Gamma(k + 2)}{\Gamma(k)} + c_2 \frac{\Gamma(k + 4)}{\Gamma(k)}.$$

Since $[\Gamma(k + 2)]/\Gamma(k) = k(k + 1)$ and $[\Gamma(k + 4)]/\Gamma(k) = k(k + 1)(k + 2)(k + 3)$, we may write the general solution as

$$y_k = k(k + 1)[c_1 + c_2(k + 2)(k + 3)]. \quad \blacksquare$$

If the characteristic equation has repeated roots, it is possible to derive a second solution in the form of the first solution multiplied by a discrete analogue of the logarithm, similar to what we saw with the Cauchy-Euler differential equation. It is also possible to write solutions when the characteristic equation has complex roots. We will not pursue these cases.

PROBLEMS FOR SECTION 10.4

In each of Problems 1 through 10, write the general solution of the difference equation.

1. $k(k + 1) D^2 y_k - 6k \, Dy_k + 12y_k = 0$

2. $k(k + 1) D^2 y_k + 2k \, Dy_k - 2y_k = 0$

3. $k(k + 1) D^2 y_k - 3k \, Dy_k + 3y_k = 0$

4. $k(k + 1) D^2 y_k + \frac{5}{2}k \, Dy_k - y_k = 0$

5. $k(k + 1) D^2 y_k - \frac{7}{4}k \, Dy_k - \frac{3}{4}y_k = 0$

6. $k(k + 1) D^2 y_k + \frac{11}{6}k \, Dy_k + \frac{1}{6}y_k = 0$

7. $k(k + 1) D^2 y_k + \frac{4}{3}k \, Dy_k - \frac{4}{3}y_k = 0$

8. $k(k + 1) D^2 y_k + 8k \, Dy_k + 12y_k = 0$

9. $k(k + 1) D^2 y_k + 3k \, Dy_k + \frac{3}{4}y_k = 0$

10. $k(k + 1) D^2 y_k + \frac{43}{8}k \, Dy_k + \frac{3}{2}y_k = 0$

11. Derive a reduction of order method for second order linear difference equations as follows. Consider the difference equation (not necessarily Cauchy-Euler) $y_{k+2} + a_k y_{k+1} + b_k y_k = 0$. Assume that one solution $y_k^{(1)}$ is known, and look for a second solution $y_k^{(2)} = A_k y_k^{(1)}$. Derive a difference equation for A_k as follows.

Substitute $y_k^{(2)}$ into the difference equation, then substitute $a_k y_k^{(1)} = -y_{k+2}^{(1)} - b_k y_k^{(1)}$ to obtain $y_{k+2}^{(1)}(A_{k+2} - A_{k+1}) = b_k y_k^{(1)}(A_{k+1} - A_k)$. This is a first order homogeneous difference equation for A_k. Use ideas from Section 10.1 to solve this equation for A_k.

12. One solution of $k^2 y_{k+2} - k(k + 1)y_{k+1} + y_k = 0$ is $y_k^{(1)} = k$. Use reduction of order to find a second, linearly independent solution.

13. Use reduction of order to find two linearly independent solutions of the Cauchy-Euler equation $k(k + 1) D^2 y_k + \frac{1}{4}y_k = 0$.

10.5 *Difference Equations and the Loaded String*

In this section, we will apply difference equations to an analysis of a loaded string. Consider a tightly stretched elastic string with loads of magnitude f_k applied at equally spaced intervals, as in Figure 10.1. We want to describe the deflections of the string.

We will first construct a mathematical model of the system. Assume that the string has a uniform tension T which is large in magnitude when compared with each f_k. This effect is seen in the deflection of the string at each point where a load is applied. We will assume that each deflection is small. We will also neglect the weight of the string.

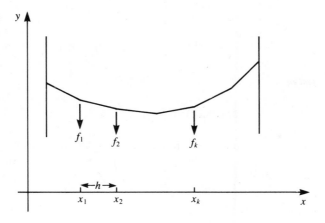

Figure 10.1

From Figure 10.2, we see that the vertical deflections y_k at x_k and y_{k+1} at x_{k+1} satisfy the difference equation

$$y_{k+1} - y_k = h \tan(\varphi_k), \tag{10.25}$$

where h is the distance between x_k and x_{k+1}. Assuming that the string is in equilibrium, the vertical components of the force at each point must balance; hence,

$$T \sin(\varphi_k) - T \sin(\varphi_{k-1}) = -f_k.$$

Assuming that each φ_k is small, we can approximate $\sin(\varphi_k) \approx \tan(\varphi_k)$ to obtain

$$T \tan(\varphi_k) - T \tan(\varphi_{k-1}) = -f_k. \tag{10.26}$$

But, from equation (10.25), $\tan(\varphi_k = (1/h)[y_{k+1} - y_k]$. Substitute this into equation

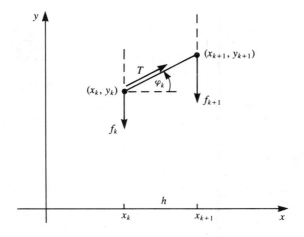

Figure 10.2

(10.26) to get

$$\frac{T}{h}[y_{k+1} - y_k] - \frac{T}{h}[y_k - y_{k-1}] = -f_k.$$

Then

$$y_{k+1} - 2y_k + y_{k-1} = -\frac{h}{T}f_k. \tag{10.27}$$

If there are N points $x_1, \ldots, x_N$ at which the loads are placed, we can consider the left end as x_0 and the right end as x_{N+1}. If the ends of the string are fixed on the x-axis, we have the boundary conditions

$$y_0 = y_{N+1} = 0. \tag{10.28}$$

That is, the deflections at the ends are zero.

We want to solve equation (10.27) subject to conditions (10.28). Equation (10.27) is not written in the form we are accustomed to dealing with. Rewrite equation (10.27) as

$$y_{k+2} - 2y_{k+1} + y_k = -\frac{h}{T}f_{k+1} \tag{10.29}$$

for $k = 0, 1, 2, \ldots, N - 1$. We will solve this equation.

The associated homogeneous equation is $y_{k+2} - 2y_{k+1} + y_k = 0$, with characteristic equation $r^2 - 2r + 1 = 0$ having repeated root 1. Thus, one homogeneous solution is $y_k^{(1)} = 1^k = 1$, and a second is $y_k^{(2)} = k$. The general homogeneous solution is $y_k^{(h)} = c_1 + c_2 k$.

For a particular solution of equation (10.29), we must have more information about f_k. This function can have different forms, depending on conditions imposed on the string. Suppose, to be specific, that the string hangs under the influence of gravity. Then $f_k = -mg$, and equation (10.29) is

$$y_{k+2} - 2y_{k+1} + y_k = \frac{mgh}{T}. \tag{10.30}$$

To use the method of undetermined coefficients, we would be tempted to try $y_k^{(p)} = $ constant. But 1 is a solution of the homogeneous equation, as is k. Thus, we will try $y_k^{(p)} = Ak^2$. Substitute this into (10.30) to get

$$A(k + 2)^2 - 2A(k + 1)^2 + Ak^2 = \frac{mgh}{T}.$$

After some cancellations, we get

$$2A = \frac{mgh}{T}$$

and hence $A = mgh/2T$. A particular solution is $y_k^{(p)} = (mgh/2T)k^2$. The general solution of (10.30) is

$$y_k = c_1 + c_2 k + \frac{mgh}{2T}k^2.$$

Now consider the boundary conditions. We need

$$y_0 = c_1 = 0$$

and

$$y_{N+1} = c_2(N+1) + \frac{mgh}{2T}(N+1)^2 = 0.$$

Solve this equation to get $c_2 = -(mgh/2T)(N+1)$. Therefore,

$$y_k = -\frac{mgh}{2T}(N+1)k + \frac{mgh}{2T}k^2 = \frac{mgh}{2T}k[k - (N+1)].$$

If the length of the unstretched string is L and we place the left end at the origin, $x_0 = 0$ and $x_k = kh$. Then $k = (1/h)x_k$ and, in terms of x_k's,

$$y(x_k) = \frac{mgh}{2T}\frac{1}{h}x_k\left(\frac{1}{h}x_k - (N+1)\right).$$

Since $(N+1)h = L$,

$$y(x_k) = \frac{mg}{2Th}x_k(x_k - L) \tag{10.31}$$

for $k = 0, 1, 2, \ldots, N+1$. We recognize from equation (10.31) that the sections of string between the beads are parabolic arcs.

It is interesting to take a limit as h approaches zero (and hence as N approaches infinity) in this solution. The points x_k move closer together, and in the limit the string loaded at discrete points becomes a uniformly loaded string. In the limit, replace mg/h by the linear intensity p of the uniformly distributed horizontal load. Equation (10.31) gives us in the limit that

$$y(x) = \frac{p}{2T}x(x - L) = -\frac{p}{2T}x(L - x),$$

the familiar result that a uniformly horizontally loaded cable hangs in the shape of a parabola. We can see this effect in cables of suspension bridges. We will pursue other aspects of the loaded string model in the problems that follow.

PROBLEMS FOR SECTION 10.5

1. Solve the difference equation for the string if $f_k = m\omega^2 y_k$. This forcing function models the string when beads are attached at the load points and the string is rotated about the x-axis with constant angular velocity of magnitude ω.

2. A uniform unloaded beam rests on N equally spaced supports. Show that the bending moment M_k at the kth support satisfies the difference equation

$$M_{k+2} + 4M_{k+1} + M_k = 0$$

for $k = 1, 2, \ldots, N - 2$.

3. Suppose that N identical objects of mass m are attached by springs, all having spring constant k, as in the accompanying diagram.

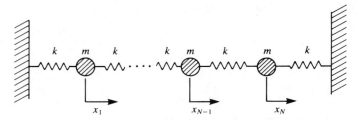

(a) If x_k is the displacement of the kth mass, show that, for small displacements,

$$m\frac{d^2x_{k+1}}{dt^2} = k[x_{k+2} - 2x_{k+1} + x_k]$$

for $k = 0, 1, 2, \ldots, N-1$. An equation of this type, containing differences and derivatives, is called a *difference-differential equation*.

(b) Assume that $x_k = A_k\cos(\omega t + \beta)$, where A_k, ω, and β are constants. Substitute x_k into the difference-differential equation of (a) to show that

$$A_{k+2} + \left(\frac{1}{k}m\omega^2 - 2\right)A_{k+1} + A_k = 0$$

for $k = 0, 1, 2, \ldots, N-1$, with $A_0 = A_{N+1} = 0$.

4. Let A, B, and C be constants, with $A \neq 0$, and let $F(t) = \sum_{k=0}^{\infty} F_k t^k$ be given. We want to determine f_k so that

$$\frac{F(t)}{A - 2Bt + Ct^2} = \sum_{k=0}^{\infty} f_k t^k. \tag{10.32}$$

Multiply both sides of (10.32) by $A - 2Bt + Ct^2$ and equate coefficients of like powers of k on both sides of the resulting equation to get

$$Af_{k+2} - 2Bf_{k+1} + Cf_k = F_{k+2}$$

for $k = 2, 3, 4, \ldots,$ and

$$f_0 = \frac{1}{A}F_0 \quad \text{and} \quad f_1 = \frac{1}{A}\left(F_1 + \frac{2}{A}Bf_0\right).$$

5. Use the result of Problem 4 to show that

$$\frac{1}{1 - 2\cos(\theta)t + t^2} = \sum_{k=0}^{\infty} \frac{\sin[(k+1)\theta]}{\sin(\theta)}t^k$$

if $|t| < 1$ and $\sin(\theta) \neq 0$.

6. Let $t = 1/r$, with $r > 1$. Use the result of Problem 5 to show that

$$\sum_{k=1}^{\infty} \frac{1}{r^k}\sin(k\theta) = \frac{r\sin(\theta)}{r^2 - 2r\cos(\theta) + 1}$$

if θ is not an integer multiple of π.

7. Let $t = -1/r$, with $r > 1$, in the result of Problem 5 to show that

$$\sum_{k=1}^{\infty} (-1)^k \frac{1}{r^k}\sin(k\theta) = \frac{-r\sin(\theta)}{r^2 + 2r\cos(\theta) + 1}$$

if θ is not an integer multiple of π.

Historical Notes on the Development of Ordinary Differential Equations

Sir Isaac Newton (1642–1727) and Gottfried Wilhelm Leibniz (1646–1716) are credited with independently discovering the fundamental ideas of the calculus (the processes of differentiation and integration and the fundamental theorem which relates them). Of course, many others made discoveries or posed questions which prepared the way, and certain ideas of the calculus may even be inherent in ancient Greek and Indian manuscripts. Particularly in Newton's case, a primary motivation for the development of the calculus was to provide a tool for solving problems involving motion and other physical phenomena. Dominating these considerations in turn were astronomy and mechanics.

In the first uses of calculus to solve problems in geometry or the physical sciences, the term *differential equation* had not yet been formalized. Rather, investigators used derivatives and differentials without distinction and derived particular equations for particular problems, which they would then attempt to solve by any method that occurred to them. The method might be analytical or geometrical or might employ reasoning combining components of analysis, geometry, algebra, and physical insight.

About the beginning of the eighteenth century, several categories of problems dominated scientific investigations and led not only to the consideration of differential equations but to methods for their solution. One class of problems concerned elasticity, which deals with properties of bodies which deform under stress of loading but return to their original configurations when the stress is removed. Analysis of the bending of beams subjected to stress was important in the practical considerations of constructing buildings and bridges and was one of the sciences discussed by Galileo in his *Dialogues Concerning Two New Sciences*.

Another problem of great interest was the shape a string or rope would take under various conditions. What shape will a flexible rope suspended from two fixed points assume? What is the shape assumed by a vertically suspended cable which is caused to vibrate? What is the shape of a rod fixed at its ends and subjected to a load?

Pendulums also attracted much attention. Galileo had conducted many experiments with pendulums and had stated a formula for the period of a simple pendulum. It soon became clear, however, that the period is not independent of the mass at the end of the pendulum, and a more careful series of investigations followed.

Another source of problems leading to differential equations was astronomy. Newton's *Principia* had answered certain questions and provided the world with the inverse square law of gravitational attraction, but it had also raised a new set of problems to be solved. Differential equations resulted from attempts to study planetary motions analytically. In particular, perturbations had been observed in planetary orbits. That is, it was observed that orbits were not quite the ellipses predicted by theory. Attempts to explain this phenomenon led to difficult problems requiring new analytical methods (a line of research which continues today).

In the late seventeenth and early eighteenth centuries, some results were published in journals (such as those sponsored by royal societies), but many ideas were communicated by private correspondence or even by word of mouth. In addition, some discoveries led to disputes over priority, so it is not always clear how to credit new methods, theory, or terminology.

James Bernoulli (1654–1705) was a member of a prominent Swiss family which produced several generations of outstanding mathematicians and scientists (as well as merchants, lawyers, and physicians). He was among the first to use calculus explicitly to solve a differential equation. In 1690, he solved the problem of finding the curve along which a pendulum takes the same time to make one complete oscillation regardless of the length of arc through which it swings. The curve is a *cycloid*, and in deriving it Bernoulli made the first use of the word "integral" in connection with the solution of a differential equation.

In the same investigation, James Bernoulli posed the problem of finding the shape taken by a flexible cord suspended freely between two points. Galileo had considered this problem and thought that the answer was an arc of a parabola. The solution (known today as a *catenary*, or part of a hyperbolic cosine curve) was derived in 1691 by Leibniz, the Dutch scientist Christian Huygens (1629–1695), and John Bernoulli (1667–1748). Huygens, who also did pioneering work in optics, based his solution on geometric reasoning, while the solutions of Bernoulli and Leibniz were more analytical.

The method of separation of variables is credited to Leibniz, who described it in a letter to Huygens in 1691. Leibniz also knew how to solve the homogeneous differential equation $y' = f(y/x)$ by reducing it to a separable differential equation using the change of variables $u = y/x$.

In 1694, Leibniz showed how to obtain the general integral solution of the linear differential equation $y' + p(x)y = q(x)$.

In 1695, James Bernoulli considered the differential equation $y' = P(x)y + Q(x)y^n$, which today bears his name. Leibniz showed in 1696 that this equation could be reduced to a linear equation by the change of variables $z = y^{1-n}$.

In 1694, James Bernoulli posed the question of determining curves intersecting other curves in given angles. The problem of determining the orthogonal trajectories of a family of curves was considered by Leibniz and John Bernoulli, and in 1715 Newton revealed his general method for finding orthogonal trajectories.

A singular solution of a first order differential equation is one which cannot be obtained from the general solution by any choice of the arbitrary constant. The existence of singular solutions had been observed by the British mathematician Brook Taylor (1685–1731). A systematic study of singular solutions was undertaken by Joseph-Louis Lagrange (1736–1813), who showed how to obtain a singular solution in certain cases by eliminating the constant from the general solution of the differential equation. Lagrange also showed that a singular solution could be interpreted as the envelope of the family of integral curves.

In 1734, the French mathematician Alexis-Claude Clairaut (1713–1765) studied the differential equation $y' = xy + f(y')$, known today as Clairaut's equation. This was a break with tradition because up to this time much of the work done had concentrated on specific differential equations arising in specific contexts. Clairaut's work was a study of a general differential equation from a general point of view.

In the early eighteenth century, second order differential equations began to attract attention in connection with a variety of problems. In 1733, Daniel Bernoulli (1700–1782), son of John, considered the problem of describing the motion of a heavy, oscillating hanging chain. The displacement $y(x)$ at a distance x from one end was found to satisfy an equation of the form

$$\alpha(xy')' + y = 0,$$

in which α is constant. Bernoulli derived a series solution which today would be written $y = kJ_0(2\sqrt{x/\alpha})$, in which J_0 is a zero order Bessel function of the first kind. If the thickness of the chain is not uniform, y satisfies

$$\alpha(g(x)y')' + yg'(x) = 0,$$

in which $g(x)$ describes the distribution of weight along the chain. Bernoulli also obtained one solution of this equation in a form we would recognize today as

$$y = k\,\frac{1}{\sqrt{x}}\,J_1\!\left(2\,\sqrt{\frac{2x}{\alpha}}\right),$$

with J_1 a Bessel function of the first kind of order 1.

John Bernoulli, in a letter to Daniel, considered a weightless elastic string loaded with equally spaced objects of equal mass. By passing to the limit as the number of masses increased, he considered a continuous string and calculated the fundamental frequencies when the string is allowed to vibrate.

About 1728, the Swiss mathematician Leonhard Euler began to study second order differential equations. At the time, Euler was a resident mathematician at the Academy of Sciences in St. Petersburg, Russia. In 1734, Daniel Bernoulli wrote to him concerning a fourth order differential equation describing the bending of a loaded beam. In response to this inquiry and on the basis of some work of his own, Euler initiated a systematic study of second and higher order differential equations, particularly linear constant coefficient equations. In the course of this study, he introduced the exponential function. Euler also observed the phenomenon of resonance while studying the second order equation $y'' + ky = \alpha\sin(\omega t)$.

Lagrange attempted to extend Euler's results to the case of linear equations with nonconstant coefficients. Closely related to Lagrange's work was a study of Riccati's equation, $y' = p(x)y + q(x)y' + r(x)y^3$. This had been introduced by Count Jacopo Francesco Riccati of Venice (1676–1754) about 1724.

Lagrange, who was of French and Italian origin, is sometimes credited with developing the method of variation of parameters. He also formalized much of Newton's work in his masterpiece *Mécanique analytique*, published in 1788.

Reduction of order is sometimes credited to Jean le Rond d'Alembert (1717–1783), who also contributed to the development of partial derivatives and partial differential equations.

Series solutions were used from Newton's time, often without regard to such details as convergence. It was Euler who systematized their use and even had in his possession an ad hoc form of the method of Frobenius. Georg Frobenius (1849–1917) formalized the method much later. Euler arrived in this way at a solution of what we now know as Bessel's equation, which is the differential equation

$$y'' + \frac{1}{x} y' + \left(1 - \frac{n^2}{x^2}\right)y = 0.$$

In addition, Euler considered the hypergeometric equation

$$x(1 - x)y'' + [c - (a + b + 1)x]y' - aby = 0,$$

although the name was bestowed by his friend and colleague Johann Pfaff (1765–1825). Solutions of this equation were studied in the early nineteenth century by the great German mathematician Carl Friedrich Gauss (1777–1855).

Systems of ordinary differential equations arise naturally by writing Newton's laws of motion (which deal with vectors) in component form. Much of the early work on systems appears in Lagrange's *Mécanique analytique*. In this work, Lagrange considered the classical three-body problem. This is the problem of describing the motion of three planets, each of which influences the other through Newton's law of gravitational attraction. This problem is still an important subject of research.

Lagrange also applied the method of variation of parameters to differential equations of order n, though the method is suggested in Newton's *Principia*. Further work on the n-body problem was done by Pierre Simon Laplace (1749–1827) in his *Mécanique celeste*.

The nineteenth century saw an expanded use of series and a recognition of special functions. The first systematic study of Bessel's equation was undertaken by Friedrich Wilhelm Bessel (1784–1846), who was director of the observatory at Königsberg, in connection with a problem in astronomy. He derived various expressions for the Bessel functions $J_n(x)$ as well as recurrence formulas and results on zeros of Bessel functions.

Legendre polynomials, which arise in connection with potential theory in spherical coordinates, appeared in the writings of Laplace and Adrien-Marie Legendre (1752–1833). Legendre was a mathematician at the École Militaire. The first systematic treatise on Legendre polynomials was written by Robert Murphy of Cambridge in 1833.

Hermite polynomials are named for Charles Hermite (1822–1901), a professor at the École Polytechnique and the Sorbonne in Paris. Laguerre polynomials are named for Edmond Laguerre (1834–1866) of the Collège de France.

Sturm-Liouville theory originated in the work of Charles Sturm (1803–1855), a professor of mechanics at the Sorbonne, and Joseph Liouville (1809–1882), professor of mathematics at the Collège de France. Problems associated with eigenvalues, and the idea of eigenfunction expansions, were first considered about 1750 and arose in connection with special functions and studies of heat conduction. Sturm and Liouville were the first to construct a general theory for a class of differential equations.

The Laplace transform can be found in writings of Laplace in the 1780s and Poisson in the 1820s and in Fourier's famous 1811 paper on heat conduction, in which he laid the foundations for Fourier series and integrals. The popularity of the Laplace transform as a computational method in elementary differential equations and electrical engineering is credited to the work of Oliver Heaviside (1850–1925), a telephone and telegraph engineer who spent the later years of his career developing vector and transform methods for use in solving engineering problems.

Stability and the qualitative theory of nonlinear systems is a relatively new area of mathematical activity. Seminal work was done by Alexander Mikhailovich Liapunov (1857–1918), a Russian mathematician and mechanical engineer. His classical work appeared in 1892 and gave birth to an approach to stability through what are known today as Liapunov functions.

Alfred Lienard (1869–1958) was director of the School of Mines in Paris and worked on spiral approaches of trajectories in the phase plane.

The Poincaré-Bendixson theorem deals with the existence of closed trajectories in the phase plane. The theorem is named for Ivar Bendixson (1861–1935), who was a professor at the University of Stockholm in Sweden, and Jules Henri Poincaré (1854–1912), who was perhaps the greatest mathematician of his time. Poincaré did important work in differential equations, dynamical systems, mathematical physics, stability, and an emerging field known today as topology.

Finally, van der Pol's equation is due to Balthasar van der Pol (1889–1959), a Dutch radio engineer who derived his equation in the late 1920s.

Biographical Sketches

SIR ISAAC NEWTON (1642–1727)

The year of Galileo's death saw the birth of the man who would establish the "system of the world." Isaac Newton was born on Christmas day in 1642 in Woolsthorpe, England, a small farming town. The English Civil War had just begun, and the Stuart king Charles I was on the throne. John Milton was 34 years of age, and in the Netherlands, Rembrandt was producing his masterpiece *The Night Watch*. René Descartes was living in the Netherlands and working on his *Principles of Philosophy*, which later influenced Newton's own physical theories.

Newton's father died before he was born, but his mother married again while he was very young. He had a lonely childhood and spent much of his time reading from a small collection of books belonging to a relative. In June 1661 he entered Trinity College, Cambridge, as a subsizar (a student working to earn tuition and lodgings). His mathematics instructor was Isaac Barrow, who held the professorship in geometry. He was to all appearances an undistinguished student.

From 1664 to 1666, Cambridge was closed because of the plague in nearby London, and Newton returned to Woolsthorpe, where he laid the foundations for his theory of light and colors, his laws of motion, and the ideas of differential and integral calculus. He returned to Cambridge in 1666 and was elected a fellow of Trinity in 1667 (in a close vote, because he had not at this time published his ideas or otherwise distinguished himself to his new colleagues). The following year, Barrow resigned his chair in favor of Newton, who thereby became Lucasian professor of mathematics.

Newton was a poor lecturer and was unsuccessful as a teacher, but he pursued research in many areas, including biblical history and alchemy. More than once, the small furnaces he used in his alchemy experiments started fires in his quarters, and there is some evidence today that during one of these he lost some papers he regarded as highly important.

Newton's unusual personality combined a desire for recognition with extreme reluctance to publish or expose himself to possible criticism. With the help of Edmund Halley, he published his great work *Mathematical Principles of Natural Philosophy*, whose original Latin title is usually abbreviated to *Principia*. In this work, Newton developed the basic notions of calculus and the physics of motion, the inverse square law of gravitational attraction, and the theory of planetary perturbations, the flattening of the earth at its poles, and the influence of the moon on the tides, as well as many other subjects. The book was widely praised for the variety and originality of the ideas it contained, and it firmly established Newton as a leading scientific figure of the period.

Newton's influence grew, and he eventually became president of the Royal Society and engaged in a wide correspondence on scientific matters. Not all of this was of a friendly nature, as Newton was a suspicious and private person. He had a particularly strong dislike for Robert Hooke, and the two engaged in many intellectual, and sometimes personal, quarrels.

In his later life, Newton moved more into the area of public service. He served for a time in Parliament, representing Cambridge, and later moved to London and became warden, then master, of the mint. He showed a talent for administration and actually brought the mint's affairs into good order.

Although his scientific research ceased during this period, Newton's mind remained sharp and he retained his mathematical power. In 1696, the Swiss mathematician John Bernoulli conspired with Leibniz to challenge the mathematical world with two problems, one of which was the brachistochrone problem. In what shape should one bend a wire so that a bead sliding down it will reach the bottom in the least time? After some of the great minds of Europe had foundered on the problem, Newton heard of it and solved it after dinner and a long day at the mint. He sent his solution to the Royal Society anonymously. According to one story, Bernoulli, upon seeing the solution, immediately recognized it as the work of Newton—"The lion is known by his claw."

Perhaps the most unfortunate aspect of Newton's last years was his acrimonious dispute with Leibniz over primacy in "discovering" the calculus. In this dispute, both men became embroiled in a quarrel which became a matter of national pride, transcending them both.

Newton died in his sleep on March 20, 1727, at the age of eighty-five, and was buried in Westminster Abbey. Another mathematician once remarked how lucky Newton was because "one could only discover the laws of the universe once." Earlier, Newton had paid tribute to his predecessors, saying, "If I have seen farther than others, it is because I have stood upon the shoulders of giants."

GOTTFRIED WILHELM LEIBNIZ (1646–1716)

Leibniz was born in Leipzig, East Germany, on July 1, 1646. His father was a professor of moral philosophy, and at an early age Leibniz distinguished himself in the classics. At fifteen he entered the University of Leipzig, and at age twenty he should have received his doctorate but was denied by a faculty already jealous of his intellectual achievements. This occurred in 1666, while Newton was in Woolsthorpe formulating his theories.

Leibniz proceeded to the University of Altdorf, where he received his degree and an offer of a professorship. The latter he declined; instead, he set out on a career as a lawyer and diplomat, pursuing philosophy, logic, and mathematics as his intellectual interests. He traveled widely, meeting many of the leading philosophers and scientists of his day (including Newton himself, during a trip to London to visit the Royal Society).

At one point, Leibniz was employed by the prominent Brunswick family as a lawyer and diplomat. One of his tasks was to produce records (by any means necessary) validating the family's claims to certain positions and fortunes. Not one to engage matters in a small way, Leibniz actually attempted to reunite the Protestant and Catholic churches, a feat which would have provided him and his clients with considerable personal gain.

By about 1676, Leibniz was certainly working on original ideas in what would become the calculus. He published some ideas on the fundamental theorem about 11 years after Newton was in possession of them but before Newton had actually published anything about them himself. Amidst all this activity, he founded the Berlin Academy of Sciences around the turn of the century.

Leibniz died in 1716 while still engaged in rewriting certain aspects of the Brunswick family history. In all his dealings, he was a curious mixture of accuracy and industrious intelligence, with a tendency to bend the rules of ethical behavior when this would serve his cause or that of his employer. He seemed to find everyone interesting, and (toward the end, with the exception of Newton) seemed to have a kind word for everyone.

THE BERNOULLI FAMILY

Switzerland in the seventeenth and eighteenth centuries saw the rise to prominence of the remarkable Bernoulli family, which produced some of the outstanding mathematicians, lawyers, merchants, and diplomats of the times. Here is part of the family tree relevant to mathematics.

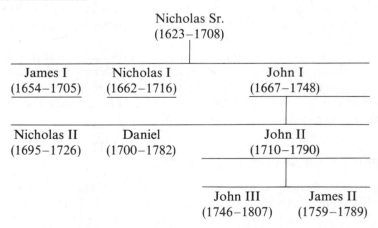

The Roman numeral designations (John I, John II, and so on) are simply a convenient notation used here to distinguish one person from another and were not parts of the family's given names.

The Bernoullis had moved from Antwerp, Belgium, in 1583 to avoid religious persecution, settling in Basel, Switzerland. Nicholas Sr. and his father and grandfather had amassed a fortune as merchants.

Nicholas Sr. had three sons of mathematical prominence. James taught himself calculus (from Leibniz's writings) and made important contributions in probability and in the calculus of variations, which deals with constrained max/min problems. James's brother John began as a medical doctor and learned mathematics from James. Together they discovered (about 1697) that an arc of the cycloid is the solution of the brachistochrone problem. John became professor of mathematics at Gröningen University in the Netherlands in 1695 and succeeded his brother at the University of Basel upon James's death.

James and John were an interesting pair, carrying on a long and at times bitter feud over mathematical discoveries. Each often accused the other of plagiarizing solutions or ideas, and these charges were sometimes accurate. The other brother, Nicholas, took a doctorate in philosophy at age sixteen and, at age twenty, added a degree in law.

John had three sons: Nicholas, Daniel, and John. Daniel became a medical doctor, then a mathematician, working in mathematical models of hydrodynamics. He was a collaborator with Euler in St. Petersburg, Russia. Today, Daniel is known as the founder of mathematical physics, and he also worked in the theory of vibrating strings and the kinetic theory of gases.

John's son John (designated John II in the table) began in law, then took up mathematics and its applications to physics. His son John (designated John III in the table) took a doctorate in philosophy at age thirteen, then studied law and finally became astronomer royal in Berlin. His brother James began with the law, then turned to experimental physics and mathematics, which he, like Daniel, pursued at the Academy of Sciences in St. Petersburg, Russia. He died at the age of thirty by drowning.

LEONHARD EULER (1707–1783)

Euler was born in Basel, Switzerland, on April 15. His father was a Calvinist minister and intended him for the ministry. He began the study of theology at Basel, but studied mathematics with John Bernoulli, the father of Daniel and Nicholas.

From about age nineteen on, Euler pursued a mathematical career. Daniel and Nicholas Bernoulli were at the Academy of Sciences in St. Petersburg, and Euler joined them in 1727. When Daniel returned to Switzerland 6 years later, Euler officially assumed the position Daniel had vacated. While in Russia, Euler married and eventually had thirteen children, only five of whom survived childhood.

Euler's powers of memory and concentration were prodigious. At one point, he solved in 3 days a problem in astronomy projected by experts to be a several-month effort. The strain apparently made him ill, and he subsequently lost the sight in his right eye.

While in Russia, Euler extended methods of the calculus and applied them to a wide variety of problems. He also assumed tasks assigned to him by the Russian government, including writing elementary mathematics textbooks and redesigning the system of weights and measures.

In 1740, Euler and his family moved to the Berlin Academy of Sciences at the invitation of Frederick the Great. Eventually he fell out of favor there, partly because of his habit of debating philosophical issues, about which he knew very little. At the invitation of Catherine the Great, he returned to Russia in 1766. Although he soon lost the sight of his other eye, he was extended every consideration by Catherine and was able to continue his mathematical work, with his sons copying formulas and text onto a slate as he dictated to them. He would often spend his evenings surrounded by his children and, later, his grandchildren, carrying out his work mentally while inventing games for them to play.

In 1776, Euler's wife died, and he soon remarried. Shortly thereafter, an operation appeared to restore his sight, but an infection set in and he again became totally blind. He died on September 7, 1783, of a stroke suffered while playing with a grandson.

CARL FRIEDRICH GAUSS (1777–1855)

Gauss was born to a poor family in Brunswick, Germany, on April 30, 1777. His brilliance was apparent at an early age, and in grammar school he astounded his teacher with his abilities at calculating. Encouraged by his teacher, the young Gauss began to study mathematics.

The wealthy Duke of Brunswick recognized the young man's talent and became his patron, sponsoring his education at the Collegium Carolinum in Brunswick. By this time, Gauss had mastered the binomial theorem and understood the idea of convergence of an infinite series.

At school, Gauss learned the classics and languages and read and understood Newton's *Principia*. His first discoveries in mathematics were in number theory and geometry, and at age twenty he decided on mathematics as a career. At this point, he began to keep a diary in which he recorded brief, often cryptic remarks indicating his thoughts on various topics in mathematics. The diary became known only after Gauss's death, when it was loaned to the Royal Society of Göttingen by his grandson. It contains 149 brief inscriptions in Latin and mathematical notation; in many instances, it provides clear proof that Gauss had in his possession many of the great mathematical discoveries which would be made by others over the next century.

Gauss earned his doctorate from Göttingen in 1799, writing on the fundamental theorem of algebra. He became interested in astronomy and calculated the orbit of Ceres. In 1809, he published a book on the motion of the planets and in the meantime married and had three children, Joseph, Minna, and Louis. Gauss's wife died after Louis's birth, and he soon remarried. Two of his sons eventually emigrated to the United States, where one became a prosperous merchant in the St. Louis area.

After the death of the Duke of Brunswick (who was mortally wounded in the war with Napoleon), Gauss was offered a position at the Royal Academy in St. Petersburg, but instead he became director of the observatory at the University of Göttingen, where he spent the rest of his life. He was a mediocre teacher but gave careful attention to students of genuine talent, and he collaborated with the physicist Wilhelm Weber in studies of electricity and magnetism. He also did important work in geometry, number theory, and complex analysis as well as on more advanced topics such as elliptic functions and differential equations. He kept some of this work to himself, even when the results were rediscovered and announced by others years later. It was his habit to publish only what he regarded as "finished products," and he often left it for others to retrace his steps and even to claim the credit for results he had kept to himself.

In 1855, Gauss began to suffer from shortness of breath, probably caused by an enlarged heart. He died on February 23, 1855, and is regarded today as the greatest mathematician of his century.

JOSEPH-LOUIS LAGRANGE (1736–1813)

Lagrange was born in Turin, Italy, and was of French and Italian extraction. While still a schoolboy, he read Edmund Halley's treatise on the merits of Newton's calculus and became interested in mathematics. By the age of nineteen, he had mastered much of what was known at that time and became a professor at the royal artillery school at Turin.

Lagrange's chief interest lay in applying Newton's law of gravitational attraction to an analysis of planetary motion. His crowning achievement in this connection was his *Mécanique analytique*, an attempt at a rigorous treatment of the mathematics of planetary motion. The book first appeared in 1788, with the second edition following in 1811.

In 1766, when Euler left Berlin to return to the St. Petersburg Academy, he recommended to Frederick the Great that Lagrange be offered his position. This was done, and Lagrange lived in Berlin for 20 years. After Frederick's death, Lagrange was invited by Louis XVI to return to Paris, where he was given apartments in the Louvre.

In 1772, Lagrange received a prize for his paper *Essai sur le problème des trois corps*, which dealt with the three-body problem. This is the problem of determining the motion of three spherical bodies, each acting on the other through gravitational attraction. Although a general solution defies a complete analysis, special cases can be successfully treated. In one special case, Lagrange analyzed the motion if the bodies began at vertices of an equilateral triangle. In 1906, it was observed that this situation actually exists in our solar system, with the three bodies being the sun, the planet Jupiter, and an asteroid called Achilles.

Lagrange also studied planetary perturbations. If two spherical bodies act on each other through gravitational attraction, they will describe orbits that are conic sections. (For example, in theory, a planet moves on an elliptical orbit with the sun at one focus.) Any deviation from this conic section motion is called a *perturbation*. In fact, planets do not describe perfectly elliptical orbits because they are influenced by the gravitational attraction of many other bodies. Lagrange applied himself to the analysis of such perturbed motion.

Lagrange's work was characterized not only by its importance in describing a physical system but also by its elegance and clarity. His masterpiece contained much that is still of value today, including Lagrange's equations of motion, which physics students see in classical mechanics.

PIERRE SIMON LAPLACE (1749–1827)

Laplace was born to moderately wealthy parents in Beaumont-en-Auge, France. At sixteen he entered the University of Caen with the intention of studying for the priesthood, but he soon became interested in mathematics. From Caen, he proceeded to Paris with a letter of introduction to Jean le Rond d'Alembert, a leading French mathematician of the day. Being a very busy person, d'Alembert paid little attention to the young, unknown Laplace. Shortly thereafter, Laplace wrote d'Alembert a letter outlining some of his work on mechanics. After receiving this, d'Alembert sent for Laplace and subsequently recommended him for a professorship at the École Militaire in Paris. It is interesting to note that, in 1783, Laplace became an examiner at the École, and posed questions to a young student named Napoleon. (Napoleon's path crossed that of more than one great mathematician of the time. Joseph Fourier, for whom Fourier analysis is named, accompanied Napoleon on his Egyptian campaign.)

During the French Revolution, Laplace was made a member of the Commission on Weights and Measures but was later removed for failing to embrace certain Republican ideals. He moved to Melun, near Paris, to work on his *Exposition du système du monde*, which appeared in 1796.

After the revolution, Laplace became a professor at the École Normale (along with Lagrange). Like Newton, he later assumed administrative positions, becoming minister of the interior and then a member and, later, chancellor, of the senate. Napoleon made him a count, but Laplace was a royalist who supported Louis XVII, and when Louis returned to power, Laplace was made a marquis and peer of France.

Over the period 1799–1825, Laplace published his five-volume work *Mécanique céleste*, in which he attempted to obtain complete analytic solutions of the equations of motion of various bodies in the solar system. He also worked on hydrodynamics, the wave propagation of sound, and the movement of the tides. His primary interest was in using mathematics to describe nature, and in this he disregarded the attempts at elegance which had distinguished Lagrange's work.

There are, of course, many others whose names figure prominently in the development of differential equations. In some instances, these were great figures of mathematics whose contributions reached into areas beyond those we have studied. Others made early contributions and have their names attached to methods or equations even though they themselves were not of the stature of a Newton or an Euler.

Jules Henri Poincaré (1854–1912) was a professor of mathematics at the University of Paris. Although much of his work was beyond the introductory level, he deserves mention as perhaps the greatest mathematician of the latter part of the nineteeth century and early part of the twentieth century. He is considered the last "universal mathematician." That is, his grasp of mathematics encompassed all of the areas known at that time, and he contributed to a wide variety of fields. In addition to many research papers of remarkable depth and originality, he wrote the three-volume work *Les Méthodes nouvelles de la mécanique céleste* (in the period 1892–1899).

Poincaré did important work on eigenvalue problems in partial differential equations and was one of the founders of a new branch of mathematics, analysis situs, which has since become known as topology. He was led to this field in his study of combinatorial structures arising in the qualitative behavior of differential equations, another area in which he helped lay foundations. After his death at the relatively early age of fifty-eight, his work in the qualitative theory of differential equations was continued by *Ivar Bendixson* (1861–1935), among others.

Georg Frobenius (1849–1917) was a German mathematician who worked in many fields. He has been mentioned in this book for his series method of solving certain types of differential equations, but he also studied linear associative algebras and group theory.

Friedrich Wilhelm Bessel (1784–1846) was director of the observatory in Königsberg. He began work around 1816 on the equation which now bears his name, and he derived many properties of its solutions, which are now called Bessel functions. His results included recurrence relations for Bessel functions, integral representations, and approximations of their zeros.

Oliver Heaviside (1850–1925) was a telegraph and telephone engineer who retired to private life in 1874, spending his time writing about electricity and magnetism. In addition to helping popularize the use of the Laplace transform in dealing with problems in electrical engineering, Heaviside had another important influence in mathematics. During his period of activity, there was division in mathematics and the sciences over the use of quaternions or vectors as a notational device for the mathematical treatment of problems in the physical sciences and engineering. Today, many students in engineering have not heard of quaternions; one reason is that Heaviside was a strong supporter of vectors and devoted a chapter of his three-volume work *Electromagnetic Theory* (published 1893–1912) to vectors and their applications. (As a proponent of vectors, Heaviside was not alone; another, even more influential figure in their eventual victory over quaternions was the Yale mathematician Josiah Willard Gibbs, for whom the Gibbs phenomenon of Fourier series is named.)

Adrien-Marie Legendre (1752–1833) was a professor of mathematics at the École Militaire and, later, was an examiner at the École Polytechnique. In addition to performing government service, he contributed to a variety of fields of mathematics, including treatment of problems involving gravitational attraction, which led him to the polynomials bearing his name. About 1790, he introduced an expression for the Legendre polynomials $P_n(x)$ as we know them today and showed that they satisfy Legendre's differential equation.

Alexis-Claude Clairaut (1713–1765) began his mathematical research when he was twelve, and at the age of sixteen he published a book on curves. At seventeen he was elected to the Paris Academy of Sciences and worked on the three-body problem, the motion of the moon, and the shape assumed by a rotating body acting under the influence of gravity. Although he is mentioned in this book for the differential equation which bears his name, and for the fact that he was among the first to study a class of differential equations in general, most of his work dealt with curves and surfaces and might today be placed in an area of mathematics called differential geometry.

CHAPTER 11

Vectors and Vector Spaces

11.0 Introduction

Many quantities require two pieces of information for their complete description. For example, force, velocity, and acceleration each have both a magnitude and a direction. Such quantities are called *vectors*. Vectors must be distinguished from *scalars*, which can be completely described by a single number. Mass, temperature, and length are examples of scalars.

In this chapter, we will develop some aspects of the algebra and geometry of vectors. In the final section of the chapter, we will place some of these ideas in the general setting of a vector space. The language of vector spaces is useful in unifying many concepts and for understanding the underlying mechanism of such processes as eigenfunction expansions, which we discussed in Chapter 7.

11.1 The Algebra and Geometry of Vectors

In the context of vectors, we will refer to real numbers as *scalars*. Thus, $\pi, 2$, and $\sqrt{41}$ are scalars. Quantities such as mass, temperature, and density, which can be measured or described by a single real number, are called *scalar quantities*.

A vector, by contrast, must be specified by giving both a direction and a magnitude. If we exert a force on an object, the effect on the object will be determined by both the magnitude of the force and its direction. Thus, force is a vector, not a scalar.

One way of efficiently conveying the magnitude and direction of a vector is to define a *vector* as an ordered triple $\langle a, b, c \rangle$ of real numbers. Given a rectangular coordinate system, as in Figure 11.1, we think of the triple $\langle a, b, c \rangle$ in two interchangeable ways: (1) as coordinates of a point (a, b, c) and (2) as an arrow from the origin to the point (a, b, c). In this way, we have a geometric way of representing a vector. The direction of the arrow gives the direction of the vector, and its length represents the magnitude. A longer arrow represents a vector of greater magnitude.

We call the number a the *first component* of the vector $\langle a, b, c \rangle$; b is the *second component*, and c is the *third component*. Two vectors are equal if and only if their respective components are equal:

$$\langle a, b, c \rangle = \langle x, y, z \rangle \quad \text{if and only if} \quad a = x, \qquad b = y, \quad \text{and} \quad c = z.$$

We will write scalars as Greek or Roman letters ($\alpha, \beta, a, b, A, B, \ldots$) and vectors as Greek or Roman letters printed in boldface type ($\boldsymbol{\alpha}, \boldsymbol{\beta}, \mathbf{a}, \mathbf{b}, \mathbf{A}, \mathbf{B}, \ldots$). The *magnitude*, or *norm*, of a vector $\mathbf{F}$ is denoted $\|\mathbf{F}\|$. If $\mathbf{F} = \langle a, b, c \rangle$, then $\|\mathbf{F}\|$ is the length of the arrow from the origin to the point (a, b, c) and hence is the distance from the origin to this point:

$$\|\mathbf{F}\| = \sqrt{a^2 + b^2 + c^2}.$$

The magnitude of a vector is a nonnegative scalar.

Figure 11.2 shows the vectors $\langle 1, -1, 2 \rangle$ and $\langle -3, 2, 4 \rangle$ as arrows from the origin to the points $(1, -1, 2)$ and $(-3, 2, 4)$, respectively. Actually, *any* arrow having a given length and direction specifies the same vector. Thus, in Figure 11.3, all of the arrows have the same length and direction and so represent the same vector. As a convenience in discussions, we often identify a vector with an arrow representing the vector and speak of the arrow and the vector interchangeably.

There are two fundamental algebraic operations involving vectors. Let

$$\mathbf{F} = \langle a_1, b_1, c_1 \rangle \quad \text{and} \quad \mathbf{G} = \langle a_2, b_2, c_2 \rangle.$$

The *vector sum*, or just sum, of $\mathbf{F}$ and $\mathbf{G}$ is

$$\mathbf{F} + \mathbf{G} = \langle a_1 + a_2, b_1 + b_2, c_1 + c_2 \rangle.$$

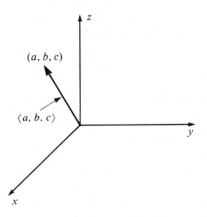

Figure 11.1. $\langle a, b, c \rangle$—represented by the arrow from $(0, 0, 0)$ to (a, b, c).

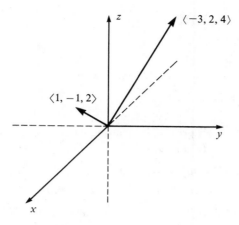

Figure 11.2

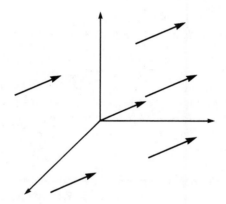

Figure 11.3

Thus, we add vectors by adding the respective components. The sum of two vectors is again a vector. For example,

$$\langle -1, 3, \pi \rangle + \langle \sqrt{2}, 1, 6 \rangle = \langle -1 + \sqrt{2}, 4, \pi + 6 \rangle.$$

The *scalar product* of a vector $\langle a, b, c \rangle$ with a scalar α is obtained by multiplying each component of the vector by the scalar:

$$\alpha \langle a, b, c \rangle = \langle \alpha a, \alpha b, \alpha c \rangle.$$

Again, this operation results in a vector.

The following theorem summarizes some properties of these two operations.

THEOREM 11.1

Let **F**, **G**, and **H** be vectors, and let α and β be scalars. Then

1. F + G = G + F.

2. $(\mathbf{F} + \mathbf{G}) + \mathbf{H} = \mathbf{F} + (\mathbf{G} + \mathbf{H})$.

3. $\mathbf{F} + \langle 0, 0, 0 \rangle = \mathbf{F}$.

4. $\alpha(\mathbf{F} + \mathbf{G}) = \alpha\mathbf{F} + \alpha\mathbf{G}$.

5. $(\alpha\beta)\mathbf{F} = \alpha(\beta\mathbf{F})$.

6. $(\alpha + \beta)\mathbf{F} = \alpha\mathbf{F} + \beta\mathbf{F}$. ∎

These properties are all proved by routine calculations. For example, to prove conclusion (1), let $\mathbf{F} = \langle a_1, b_1, c_1 \rangle$ and $\mathbf{G} = \langle a_2, b_2, c_2 \rangle$. Then

$$\mathbf{F} + \mathbf{G} = \langle a_1 + a_2, b_1 + b_2, c_1 + c_2 \rangle$$
$$= \langle a_2 + a_1, b_2 + b_1, c_2 + c_1 \rangle = \mathbf{G} + \mathbf{F},$$

by the fact that $x + y = y + x$ for any real numbers x and y. To prove (6), write

$$(\alpha + \beta)\mathbf{F} = (\alpha + \beta)\langle a, b, c \rangle = \langle (\alpha + \beta)a, (\alpha + \beta)b, (\alpha + \beta)c \rangle$$
$$= \langle \alpha a + \beta a, \alpha b + \beta b, \alpha c + \beta c \rangle$$
$$= \langle \alpha a, \alpha b, \alpha c \rangle + \langle \beta a, \beta b, \beta c \rangle$$
$$= \alpha\langle a, b, c \rangle + \beta\langle a, b, c \rangle = \alpha\mathbf{F} + \beta\mathbf{F}.$$

The other properties are proved by similar reasoning.

Conclusion (1) of Theorem 11.1 is the *commutative law* for vector addition; conclusion (2) is the *associative law*.

In view of conclusion (3) of the theorem, the vector $\langle 0, 0, 0 \rangle$ is called the *zero vector* and is denoted $\mathbf{O}$. Given $\mathbf{F} = \langle a, b, c \rangle$, there is exactly one vector $\mathbf{G}$ having the property that $\mathbf{F} + \mathbf{G} = \mathbf{O}$. In fact, $\mathbf{G} = \langle -a, -b, -c \rangle$, because

$$\langle a, b, c \rangle + \langle -a, -b, -c \rangle = \langle 0, 0, 0 \rangle = \mathbf{O}.$$

We call $\langle -a, -b, -c \rangle$ the negative of $\langle a, b, c \rangle$ and denote the negative of a vector $\mathbf{F}$ as $-\mathbf{F}$. To simplify notation, if we add a vector $\mathbf{F}$ to the negative of a vector $\mathbf{G}$, we usually write $\mathbf{F} + (-\mathbf{G})$ as just $\mathbf{F} - \mathbf{G}$. If $\mathbf{F} = \langle a_1, b_1, c_1 \rangle$, and $\mathbf{G} = \langle a_2, b_2, c_2 \rangle$, then

$$\mathbf{F} - \mathbf{G} = \langle a_1 - a_2, b_1 - b_2, c_1 - c_2 \rangle.$$

We also have the following properties of the norm of a vector.

THEOREM 11.2

Let α be a scalar and let $\mathbf{F}$ be a vector. Then

1. $\|\alpha\mathbf{F}\| = |\alpha|\,\|\mathbf{F}\|$.

2. $\|\mathbf{F}\| = 0$ if and only if $\mathbf{F} = \mathbf{O}$. ∎

Conclusion (1) states that the norm of $\alpha\mathbf{F}$ is the absolute value of α times the norm of $\mathbf{F}$. Conclusion (2) states that the only vector with zero norm is the zero vector. This is consistent with the interpretation of the norm of $\mathbf{F}$ as the length of an arrow representing $\mathbf{F}$, since any nonzero vector is represented by an arrow with positive length and hence has positive norm.

Proof of (1)   Let $\mathbf{F} = \langle a, b, c \rangle$. Then $\alpha \mathbf{F} = \langle \alpha a, \alpha b, \alpha c \rangle$, so

$$\|\alpha \mathbf{F}\| = \sqrt{(\alpha a)^2 + (\alpha b)^2 + (\alpha c)^2}$$
$$= \sqrt{\alpha^2 a^2 + \alpha^2 b^2 + \alpha^2 c^2}$$
$$= |\alpha| \sqrt{a^2 + b^2 + c^2} = |\alpha| \|\mathbf{F}\|.$$

Here we have used the fact that the square root of α^2 is $|\alpha|$, not just α. For example, the square root of $(-4)^2$, or 16, is 4, and this is exactly $|-4|$. ∎

Proof of (2)   If $\mathbf{F} = \mathbf{O}$, then $\mathbf{F} = \langle 0, 0, 0 \rangle$, so $\|\mathbf{F}\| = \sqrt{0^2 + 0^2 + 0^2} = 0$. Conversely, suppose that $\|\mathbf{F}\| = 0$. Write $\mathbf{F} = \langle a, b, c \rangle$. Then

$$\sqrt{a^2 + b^2 + c^2} = 0.$$

Upon squaring both sides of this equation, we have

$$a^2 + b^2 + c^2 = 0.$$

If any of a, b, or c is nonzero, $a^2 + b^2 + c^2$ is positive. Hence, we conclude that $a = b = c = 0$, so $\mathbf{F} = \mathbf{O}$. ∎

We have seen that a vector $\langle a, b, c \rangle$ is represented as an arrow from the origin to the point (a, b, c). In view of this, it is natural to seek some geometric interpretation of vector addition and multiplication of a vector by a scalar. For vector addition, the geometric interpretation is provided by the parallelogram law.

Parallelogram law for vector addition: The vector sum $\mathbf{F} + \mathbf{G}$ is represented by the diagonal of the parallelogram having $\mathbf{F}$ and $\mathbf{G}$ as sides, as shown in Figure 11.4.

Note in Figure 11.4 that $\mathbf{F}$ and $\mathbf{G}$ can be drawn as arrows from a common point, or either vector may be represented as an arrow drawn from the end of the other. In reading vector diagrams, the latter viewpoint is often useful; it is based on the fact that two arrows of the same length and direction represent the same vector.

Although the parallelogram law holds in three-space, it is perhaps most easily visualized when $\mathbf{F}$ and $\mathbf{G}$ have zero third component and can be represented as arrows

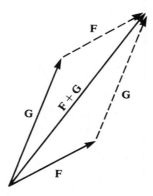

Figure 11.4. Parallelogram law for vector addition.

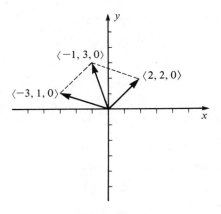

Figure 11.5. $\langle 2, 2, 0 \rangle + \langle -3, 1, 0 \rangle = \langle -1, 3, 0 \rangle$ by the parallelogram law.

in the xy-plane. Figure 11.5 shows the parallelogram law for the addition of $\langle -3, 1, 0 \rangle$ and $\langle 2, 2, 0 \rangle$.

The parallelogram law immediately suggests the following important inequality.

THEOREM 11.3

For any vectors **F** and **G**, $\|\mathbf{F} + \mathbf{G}\| \leq \|\mathbf{F}\| + \|\mathbf{G}\|$.

Proof   Represent **F** as an arrow in three-space and **G** as an arrow from the tip of **F**. The parallelogram sum **F** + **G** is shown in Figure 11.6. We may think of **F**, **G**, and **F** + **G** as forming three sides of a triangle; the lengths of these sides are $\|\mathbf{F}\|$, $\|\mathbf{G}\|$, and $\|\mathbf{F} + \mathbf{G}\|$. But the length of one side of a triangle cannot exceed the sum of the lengths of the other two sides; hence,

$$\|\mathbf{F} + \mathbf{G}\| \leq \|\mathbf{F}\| + \|\mathbf{G}\|. \quad \blacksquare$$

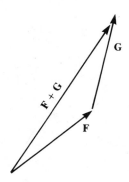

Figure 11.6

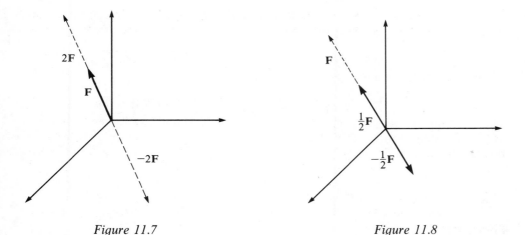

Figure 11.7   *Figure 11.8*

In view of this argument, this inequality is called the *triangle inequality*. It is also possible to prove this inequality without recourse to geometry, as we will do in *n*-dimensional space in Section 11.5.

The scalar product $\alpha\mathbf{F}$ also has a geometric interpretation, which is suggested by Figures 11.7 and 11.8. The relationship $\|\alpha\mathbf{F}\| = |\alpha|\,\|\mathbf{F}\|$ states that $\alpha\mathbf{F}$ has length equal to $|\alpha|$ times the length of $\mathbf{F}$. Thus, α acts as a scaling factor. Further, if α is positive, $\alpha\mathbf{F}$ is represented by an arrow in the same direction as $\mathbf{F}$. If α is negative, the arrow representing $\alpha\mathbf{F}$ is in the direction opposite that of $\mathbf{F}$. Thus, $\alpha\mathbf{F}$ is in the same direction as $\mathbf{F}$ if $\alpha > 0$ and is longer than $\mathbf{F}$ if $\alpha > 1$, shorter if $\alpha < 1$; $\alpha\mathbf{F}$ is in the opposite direction from $\mathbf{F}$ if $\alpha < 0$ and is longer if $|\alpha| > 1$, shorter if $|\alpha| < 1$.

Two vectors $\mathbf{F}$ and $\mathbf{G}$ have the same direction if and only if one is a positive scalar multiple of the other:

$$\mathbf{F} = \alpha\mathbf{G} \quad \text{for some} \quad \alpha > 0.$$

Of course, if this holds, we also have $\mathbf{G} = \beta\mathbf{F}$ with $\beta = 1/\alpha > 0$. In this event, $\mathbf{F}$ and $\mathbf{G}$ differ only in magnitude or length, as shown in Figure 11.9. Similarly, $\mathbf{F}$ and $\mathbf{G}$ are parallel, but in opposite directions, if and only if one is a negative scalar multiple of the other. This is indicated in Figure 11.10. Combining these two statements, we have that $\mathbf{F}$ and $\mathbf{G}$ are parallel (in the same or opposite directions) if and only if each is a nonzero scalar multiple of the other.

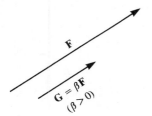

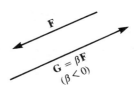

Figure 11.9   *Figure 11.10*

The operations of vector addition and scalar multiplication enable us to decompose any vector into a sum of scalar multiples of "standardized" vectors. Write

$$\mathbf{F} = \langle a, b, c \rangle = a\langle 1, 0, 0 \rangle + b\langle 0, 1, 0 \rangle + c\langle 0, 0, 1 \rangle. \tag{11.1}$$

It is customary to denote

$$\mathbf{i} = \langle 1, 0, 0 \rangle, \qquad \mathbf{j} = \langle 0, 1, 0 \rangle, \quad \text{and} \quad \mathbf{k} = \langle 0, 0, 1 \rangle.$$

These are unit vectors (having magnitude, or length, 1) and can be represented as arrows along the positive x-, y-, and z-axes, respectively, as shown in Figure 11.11. In this notation, equation (11.1) can be written

$$\mathbf{F} = a\mathbf{i} + b\mathbf{j} + c\mathbf{k}.$$

This is called the *standard form* of $\mathbf{F}$.

When a component of a vector is zero, we usually omit it in the standard form. For example,

$$\langle 0, 2, -5 \rangle = 2\mathbf{j} - 5\mathbf{k}$$

and

$$\langle 9, 0, 0 \rangle = 9\mathbf{i}.$$

Figure 11.12 shows arbitrary points (a_1, b_1, c_1) and (a_2, b_2, c_2) in three-space. We ask: What is the vector represented by the arrow extending from (a_1, b_1, c_1) to (a_2, b_2, c_2)? To determine the components of this vector, let $\mathbf{F} = a_2\mathbf{i} + b_2\mathbf{j} + c_2\mathbf{k}$ and $\mathbf{G} = a_1\mathbf{i} + b_1\mathbf{j} + c_1\mathbf{k}$. Call the unknown vector $\mathbf{H}$, representing it as the arrow from (a_1, b_1, c_1) to (a_2, b_2, c_2). From the parallelogram law in Figure 11.13, we read that

$$\mathbf{G} + \mathbf{H} = \mathbf{F}.$$

Therefore,

$$\mathbf{H} = \mathbf{F} - \mathbf{G} = (a_2 - a_1)\mathbf{i} + (b_2 - b_1)\mathbf{j} + (c_2 - c_1)\mathbf{k}.$$

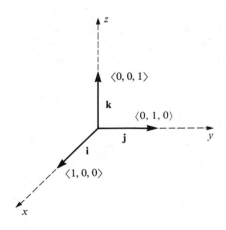

Figure 11.11. Unit vectors along the axes.

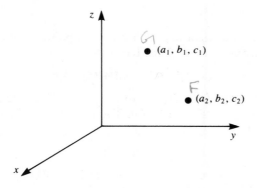

Figure 11.12

This fact is so useful that we will repeat it:
   The arrow from (a_1, b_1, c_1) to (a_2, b_2, c_2) represents the vector

$$(a_2 - a_1)\mathbf{i} + (b_2 - b_1)\mathbf{j} + (c_2 - c_1)\mathbf{k}.$$

For example, the arrow from $(0, -3, 2)$ to $(7, 5, -21)$ represents the vector

$$(7 - 0)\mathbf{i} + (5 - (-3))\mathbf{j} + (-21 - 2)\mathbf{k},$$

or $7\mathbf{i} + 8\mathbf{j} - 23\mathbf{k}$.

We will conclude this section with some examples involving vectors.

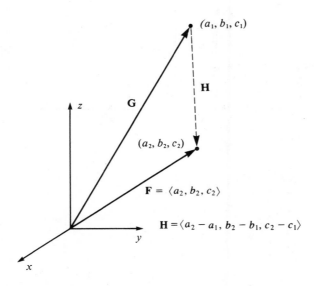

Figure 11.13. The arrow from (a_1, b_1, c_1) to (a_2, b_2, c_2) is
$(a_2 - a_1)\mathbf{i} + (b_2 - b_1)\mathbf{j} + (c_2 - c_1)\mathbf{k}$.

EXAMPLE 11.1

Find the equation of the straight line L through the points $(1, -2, 4)$ and $(6, 2, -3)$.
Let (x, y, z) be any point on L. Then the vector

$$(x - 1)\mathbf{i} + (y + 2)\mathbf{j} + (z - 4)\mathbf{k} \tag{11.2}$$

is represented by the arrow from $(1, -2, 4)$ to (x, y, z). This vector is parallel to L because (x, y, z) is on L (see Figure 11.14). The arrow from $(1, -2, 4)$ to $(6, 2, -3)$ represents the vector

$$5\mathbf{i} + 4\mathbf{j} - 7\mathbf{k} \tag{11.3}$$

and is also parallel to L. Therefore, the vectors (11.2) and (11.3) are parallel, and one is a scalar multiple of the other. For some scalar t,

$$(x - 1)\mathbf{i} + (y + 2)\mathbf{j} + (z - 4)\mathbf{k} = (5\mathbf{i} + 4\mathbf{j} - 7\mathbf{k})t.$$

Then

$$(x - 1)\mathbf{i} + (y + 2)\mathbf{j} + (z - 4)\mathbf{k} = 5t\mathbf{i} + 4t\mathbf{j} - 7t\mathbf{k}.$$

Respective components of these vectors must be equal:

$$x - 1 = 5t$$
$$y + 2 = 4t$$
$$z - 4 = -7t.$$

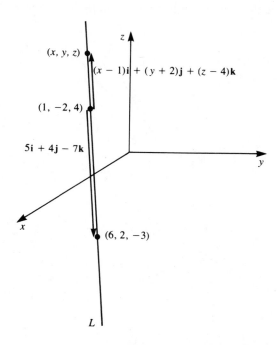

Figure 11.14

These are parametric equations of L and may be written

$$x = 1 + 5t$$
$$y = -2 + 4t$$
$$z = 4 - 7t.$$

As t takes on all real values, the point $(1 + 5t, -2 + 4t, 4 - 7t)$ varies over L. We get $(1, -2, 4)$ when $t = 0$ and $(6, 2, -3)$ when $t = 1$.

The equation of the line can be written in *normal form* by eliminating t to write

$$\frac{x - 1}{5} = \frac{y + 2}{4} = \frac{z - 4}{-7}.$$

We may also envision the line as swept out by an arrow pivoted at the origin and extending to the point $(1 + 5t, -2 + 4t, 4 - 7t)$ as t varies from $-\infty$ to ∞. ∎

EXAMPLE 11.2

Find a vector $\mathbf{F}$ of length 17 in the xy-plane making an angle of 42 degrees with the positive x-axis.

By "find a vector" we mean determine its components. Figure 11.15 shows $\mathbf{F}$ and the numbers a and b such that $\mathbf{F} = a\mathbf{i} + b\mathbf{j}$. The numbers a and b are the components of $\mathbf{F}$. From the right triangle in the diagram,

$$\cos(42°) = \frac{a}{17} \quad \text{and} \quad \sin(42°) = \frac{b}{17}.$$

Then

$$a = 17\cos(42°) \quad \text{and} \quad b = 17\sin(42°).$$

The vector is

$$\mathbf{F} = 17\cos(42°)\mathbf{i} + 17\cos(42°)\mathbf{j}.$$

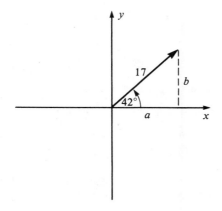

Figure 11.15   ∎

EXAMPLE 11.3

As an illustration of the efficiency of vectors in storing information and carrying out an argument, we will prove the following fact: the line segments formed by connecting successive midpoints of the sides of a quadrilateral form a parallelogram.

　　This statement is illustrated in Figure 11.16. In that diagram, we must show that the dashed lines form a parallelogram. Redraw the quadrilateral as in Figure 11.17, with arrows (vectors) as sides. The vectors **x**, **y**, **u**, and **v**, drawn with dashed lines for emphasis, connect the midpoints. We want to show that **x** and **u** are parallel and of the same length and that **y** and **v** are parallel and of the same length.

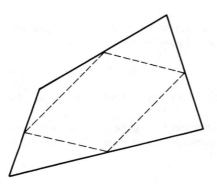

Figure 11.16

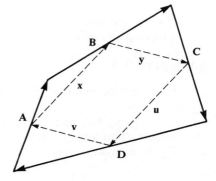

Figure 11.17

　　From the parallelogram law and the definitions of **x** and **u**, observe first that

$$\mathbf{x} = \tfrac{1}{2}\mathbf{A} + \tfrac{1}{2}\mathbf{B} = \tfrac{1}{2}(\mathbf{A} + \mathbf{B})$$

and

$$\mathbf{u} = \tfrac{1}{2}\mathbf{C} + \tfrac{1}{2}\mathbf{D} = \tfrac{1}{2}(\mathbf{C} + \mathbf{D}).$$

But

$$\mathbf{A} + \mathbf{B} + \mathbf{C} + \mathbf{D} = \mathbf{O}$$

(see Problem 36 at the end of this section). Hence,

$$\mathbf{A} + \mathbf{B} = -(\mathbf{C} + \mathbf{D}).$$

But then $\mathbf{x} = -\mathbf{u}$, so **x** and **u** are parallel. Further,

$$\|\mathbf{x}\| = \|-\mathbf{u}\| = |-1|\|\mathbf{u}\| = \|\mathbf{u}\|,$$

so **x** and **u** are also of the same length. A similar argument shows that **y** and **v** are parallel and of the same length. ∎

　　In the next two sections, we will introduce further vector operations which are used in applying vectors to a variety of problems.

PROBLEMS FOR SECTION 11.1

In Problems 1 through 5, compute $\mathbf{F} + \mathbf{G}$, $\mathbf{F} - \mathbf{G}$, $\|\mathbf{F}\|$, $\|\mathbf{G}\|$, $2\mathbf{F}$, and $3\mathbf{G}$.

1. $\mathbf{F} = 2\mathbf{i} - 3\mathbf{j} + 5\mathbf{k}$, $\mathbf{G} = \sqrt{2}\mathbf{i} + 6\mathbf{j} - 5\mathbf{k}$ **2.** $\mathbf{F} = \mathbf{i} - 3\mathbf{k}$, $\mathbf{G} = 4\mathbf{j}$

3. $\mathbf{F} = 2\mathbf{i} - 5\mathbf{j}$, $\mathbf{G} = \mathbf{i} + 5\mathbf{j} - \mathbf{k}$ **4.** $\mathbf{F} = \sqrt{2}\mathbf{i} + \mathbf{j} - 6\mathbf{k}$, $\mathbf{G} = 8\mathbf{i} + 2\mathbf{k}$

5. $\mathbf{F} = \mathbf{i} + \mathbf{j} + \mathbf{k}$, $\mathbf{G} = 2\mathbf{i} - 2\mathbf{j} + 2\mathbf{k}$

In each of Problems 6 through 10, calculate $\mathbf{F} + \mathbf{G}$ and $\mathbf{F} - \mathbf{G}$ by making a diagram and applying the parallelogram law. Each of these vectors has a zero third component, so it can be drawn as an arrow in the plane.

6. $\mathbf{F} = \mathbf{i}$, $\mathbf{G} = 6\mathbf{j}$ **7.** $\mathbf{F} = 2\mathbf{i} - \mathbf{j}$, $\mathbf{G} = \mathbf{i} - \mathbf{j}$

8. $\mathbf{F} = -3\mathbf{i} + \mathbf{j}$, $\mathbf{G} = 4\mathbf{j}$ **9.** $\mathbf{F} = \mathbf{i} - 2\mathbf{j}$, $\mathbf{G} = \mathbf{i} - 3\mathbf{j}$

10. $\mathbf{F} = -\mathbf{i} + 4\mathbf{j}$, $\mathbf{G} = -2\mathbf{i} - 3\mathbf{j}$

In each of Problems 11 through 15, calculate $\alpha\mathbf{F}$. Draw vector diagrams of $\mathbf{F}$, $\alpha\mathbf{F}$, and $-\alpha\mathbf{F}$. Again, these vectors have zero third components to make the diagrams easier to draw.

11. $\mathbf{F} = \mathbf{i} + \mathbf{j}$, $\alpha = -\frac{1}{2}$ **12.** $\mathbf{F} = 6\mathbf{i} - 2\mathbf{j}$, $\alpha = 2$

13. $\mathbf{F} = -3\mathbf{j}$, $\alpha = -4$ **14.** $\mathbf{F} = 6\mathbf{i} + 6\mathbf{j}$, $\alpha = \frac{1}{2}$

15. $\mathbf{F} = -3\mathbf{i} + 2\mathbf{j}$, $\alpha = 3$

In each of Problems 16 through 25, use vectors to find the parametric equations of the straight line containing the two given points.

16. $(1, 0, 4)$, $(2, 1, 1)$ **17.** $(3, 0, 0)$, $(-3, 1, 0)$

18. $(2, 1, 1)$, $(2, 1, -2)$ **19.** $(0, 1, 3)$, $(0, 0, 1)$

20. $(1, 0, -4)$, $(-2, -2, 5)$ **21.** $(2, -3, 6)$, $(-1, 6, 4)$

22. $(-4, -2, 5)$, $(1, 1, -5)$ **23.** $(3, 3, -5)$, $(2, -6, 1)$

24. $(0, -3, 0)$, $(1, -1, 5)$ **25.** $(4, -8, 1)$, $(-1, 0, 0)$

In each of Problems 26 through 35, find a vector $\mathbf{F}$ in the xy-plane (that is, zero third component) having the given length and making the given angle with the positive x-axis. Draw a vector diagram showing the vector.

26. $\sqrt{5}$, 45 degrees   **27.** 6, 60 degrees

28. 12, 135 degrees   **29.** 1, 315 degrees

30. 14, 90 degrees   **31.** $\sqrt{2}$, 30 degrees

32. 5, 140 degrees   **33.** 15, 175 degrees

34. 12, 225 degrees   **35.** 25, 270 degrees

36. Let P, Q, R, and S be any four points in three-space. Let $\mathbf{A}$ be the vector from P to Q, $\mathbf{B}$ be the vector from Q to R, $\mathbf{C}$ be the vector from R to S, and $\mathbf{D}$ be the vector from S to P. Prove that

$$\mathbf{A} + \mathbf{B} + \mathbf{C} + \mathbf{D} = \mathbf{O}.$$

Hint: Apply the parallelogram law to determine the vector sum $\mathbf{A} + \mathbf{B}$, and compare this with the vector sum $\mathbf{C} + \mathbf{D}$.

37. Let $\mathbf{A}$ be any vector except $\mathbf{O}$. Determine a scalar t such that $\|t\mathbf{A}\| = 1$.

38. Use vectors to prove that the altitudes of any triangle intersect in a single point.

39. Let P, Q, and R be any three points not on a straight line. Let $\mathbf{A}$ be the vector from P to Q and let $\mathbf{B}$ be the vector from P to R. Determine the vector from P to the midpoint of the segment $\overline{RQ}$. The answer should be some vector combination of scalar multiples of $\mathbf{A}$ and $\mathbf{B}$.

40. Use vectors to prove that lines drawn from a vertex of a parallelogram to the midpoints of the opposite sides trisect a diagonal of the parallelogram.

41. Suppose that $\alpha\mathbf{F} + \beta\mathbf{G} = \mathbf{O}$ and that $\mathbf{F}$ and $\mathbf{G}$ are nonzero vectors which are not parallel. Prove that α and β must both be zero.

11.2 The Dot Product of Vectors

Let

$$\mathbf{F} = a_1\mathbf{i} + b_1\mathbf{j} + c_1\mathbf{k} \quad \text{and} \quad \mathbf{G} = a_2\mathbf{i} + b_2\mathbf{j} + c_2\mathbf{k}.$$

The *dot product* of $\mathbf{F}$ with $\mathbf{G}$ is the scalar $\mathbf{F} \cdot \mathbf{G}$ defined by

$$\mathbf{F} \cdot \mathbf{G} = a_1a_2 + b_1b_2 + c_1c_2.$$

We obtain the dot product by multiplying respective components and adding the resulting numbers. Therefore, the dot product of two vectors is a scalar, not a vector. For example,

$$(\sqrt{3}\mathbf{i} + 4\mathbf{j} - \pi\mathbf{k}) \cdot (-2\mathbf{i} + 6\mathbf{j} + 3\mathbf{k}) = -2\sqrt{3} + 24 - 3\pi.$$

Before examining the geometry underlying the dot product, here are some properties of this operation.

THEOREM 11.4

Let $\mathbf{F}$, $\mathbf{G}$, and $\mathbf{H}$ be vectors and let α be a scalar. Then

1. $\mathbf{F} \cdot \mathbf{G} = \mathbf{G} \cdot \mathbf{F}$.
2. $(\mathbf{F} + \mathbf{G}) \cdot \mathbf{H} = \mathbf{F} \cdot \mathbf{H} + \mathbf{G} \cdot \mathbf{H}$.
3. $\alpha(\mathbf{F} \cdot \mathbf{G}) = (\alpha\mathbf{F}) \cdot \mathbf{G} = \mathbf{F} \cdot (\alpha\mathbf{G})$.
4. $\mathbf{F} \cdot \mathbf{F} = \|\mathbf{F}\|^2$.
5. $\mathbf{F} \cdot \mathbf{F} = 0$ if and only if $\mathbf{F} = \mathbf{O}$. ∎

The dot product is commutative [conclusion (1)] and also obeys a distributive law [conclusion (2)]. Conclusion (3) states that scalars can be factored through the dot product.

Theorem 11.4 can be proved by straightforward arguments. We will see this by proving conclusions (1) and (4).

Proof of (1)   Let $\mathbf{F} = a_1\mathbf{i} + b_1\mathbf{j} + c_1\mathbf{k}$ and $\mathbf{G} = a_2\mathbf{i} + b_2\mathbf{j} + c_2\mathbf{k}$. Then

$$\mathbf{F} \cdot \mathbf{G} = a_1a_2 + b_1b_2 + c_1c_2 = a_2a_1 + b_2b_1 + c_2c_1 = \mathbf{G} \cdot \mathbf{F}.$$

Proof of (4)   If $\mathbf{F} = a\mathbf{i} + b\mathbf{j} + c\mathbf{k}$, then

$$\mathbf{F} \cdot \mathbf{F} = (a\mathbf{i} + b\mathbf{j} + c\mathbf{k}) \cdot (a\mathbf{i} + b\mathbf{j} + c\mathbf{k}) = a^2 + b^2 + c^2$$
$$= (\sqrt{a^2 + b^2 + c^2})^2 = \|\mathbf{F}\|^2. \quad \blacksquare$$

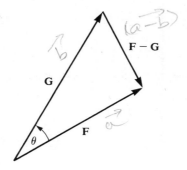

Figure 11.18

The dot product has an important geometric interpretation. Represent **F** and **G** as arrows from a common point, as in Figure 11.18, and let θ be the angle between **F** and **G**. The arrow from the tip of **G** to the tip of **F** represents **F** − **G**, by the parallelogram law. The three vectors **F**, **G**, and **F** − **G** form sides of a triangle in three-space.

For the triangle in Figure 11.19, the law of cosines states that

$$a^2 + b^2 - 2ab\cos(\theta) = c^2.$$

Apply this fact to the triangle of Figure 11.18 to conclude that

$$\|\mathbf{F}\|^2 + \|\mathbf{G}\|^2 - 2\|\mathbf{F}\|\|\mathbf{G}\|\cos(\theta) = \|\mathbf{F} - \mathbf{G}\|^2. \tag{11.4}$$

Now write

$$\mathbf{F} = a_2\mathbf{i} + b_2\mathbf{j} + c_2\mathbf{k} \quad \text{and} \quad \mathbf{G} = a_1\mathbf{i} + b_1\mathbf{j} + c_1\mathbf{k}.$$

Then

$$\mathbf{F} - \mathbf{G} = (a_2 - a_1)\mathbf{i} + (b_2 - b_1)\mathbf{j} + (c_2 - c_1)\mathbf{k}.$$

In terms of components, equation (11.4) can be written

$$a_2^2 + b_2^2 + c_2^2 + a_1^2 + b_1^2 + c_1^2 - 2\|\mathbf{F}\|\|\mathbf{G}\|\cos(\theta) = (a_2 - a_1)^2 + (b_2 - b_1)^2 + (c_2 - c_1)^2.$$

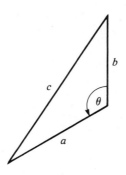

Figure 11.19. Law of cosines: $a^2 + b^2 - 2ab\cos(\theta) = c^2$.

Upon squaring the quantities on the right side of this equation and simplifying the resulting equation, we get

$$\|\mathbf{F}\|\|\mathbf{G}\|\cos(\theta) = a_1a_2 + b_1b_2 + c_1c_2 = \mathbf{F}\cdot\mathbf{G}.$$

Assuming that $\mathbf{F} \neq \mathbf{O}$ and $\mathbf{G} \neq \mathbf{O}$, this equation yields

$$\cos(\theta) = \frac{\mathbf{F}\cdot\mathbf{G}}{\|\mathbf{F}\|\|\mathbf{G}\|}, \tag{11.5}$$

with θ in radians. The dot product therefore enables us to calculate the cosine of the angle between two vectors.

EXAMPLE 11.4

Let $\mathbf{F} = -\mathbf{i} + 3\mathbf{j} + \mathbf{k}$ and $\mathbf{G} = 2\mathbf{j} - 4\mathbf{k}$. The angle θ between $\mathbf{F}$ and $\mathbf{G}$ satisfies

$$\cos(\theta) = \frac{\mathbf{F}\cdot\mathbf{G}}{\|\mathbf{F}\|\|\mathbf{G}\|} = \frac{6-4}{\sqrt{11}\,\sqrt{20}} = \frac{2}{\sqrt{220}}.$$

θ is that unique number in $[0, \pi]$ whose cosine is $2/\sqrt{220}$. Using a calculator, we find that $\theta \approx 1.436$ radians (about 82.25 degrees). ∎

EXAMPLE 11.5

Find the angle between the intersecting lines L_1 and L_2 given parametrically by

$$L_1: \quad \begin{aligned} x &= 1 + 6t \\ y &= 2 - 4t \\ z &= -1 + 3t, \qquad -\infty < t < \infty \end{aligned}$$

and

$$L_2: \quad \begin{aligned} x &= 4 - 3p \\ y &= 2p \\ z &= -5 + 4p, \qquad -\infty < p < \infty. \end{aligned}$$

In fact, if two lines intersect, there are two angles, θ and φ, "between" them, as indicated in Figure 11.20. However, $\theta + \varphi = \pi$, so either of these angles determines the other.

The strategy for solving the problem is as follows. Form a vector $\mathbf{F}$ along L_1 and a vector $\mathbf{G}$ along L_2, then use equation (11.5) to calculate the cosine of the angle between

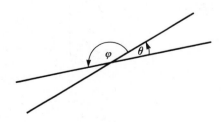

Figure 11.20

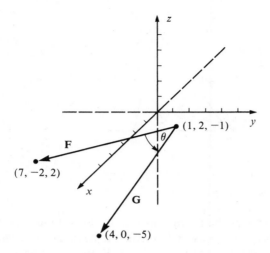

Figure 11.21

these vectors. This will give us the cosine of an angle between the lines. For **F**, take any two points on L_1, say, $(1, 2, -1)$ when $t = 0$ and $(7, -2, 2)$ when $t = 1$. Form **F** as the arrow from the first point to the second:

$$\mathbf{F} = (7\mathbf{i} - 2\mathbf{j} + 2\mathbf{k}) - (\mathbf{i} + 2\mathbf{j} - \mathbf{k}) = 6\mathbf{i} - 4\mathbf{j} + 3\mathbf{k}.$$

Now take two points on L_2, say, $(4, 0, -5)$ when $p = 0$ and $(1, 2, -1)$ when $p = 1$. Form

$$\mathbf{G} = (\mathbf{i} + 2\mathbf{j} - \mathbf{k}) - (4\mathbf{i} - 5\mathbf{k}) = -3\mathbf{i} + 2\mathbf{j} + 4\mathbf{k}.$$

F and **G** are shown in Figure 11.21. Now compute

$$\cos(\theta) = \frac{\mathbf{F} \cdot \mathbf{G}}{\|\mathbf{F}\|\|\mathbf{G}\|} = \frac{6(-3) + (-4)(2) + 3(4)}{\sqrt{6^2 + 4^2 + 3^2}\,\sqrt{3^2 + 2^2 + 4^2}} = \frac{-14}{\sqrt{1769}}.$$

Then $\theta = \cos^{-1}(-14/\sqrt{1769}) \approx 1.91$ radians, or about 109.44 degrees.

If we use $-\mathbf{G}$ in place of **G** in this calculation, we obtain $\cos(\theta) = +14/\sqrt{1769}$. Now we get $\theta = \cos^{-1}(14/\sqrt{1769}) \approx \cos^{-1}(0.3329) \approx 1.23$ radians, or about 70.56 degrees. By using $-\mathbf{G}$ in place of **G** in the calculation, we have found the complement of the angle found previously, since 109.44 degrees + 70.56 degrees = 180 degrees. ∎

In this example, we happened to find the point of intersection of L_1 and L_2. It is $(1, 2, -1)$, on L_1 when $t = 0$ and on L_2 when $p = 1$. We did not, however, need this point to solve the problem. **F** could have been any vector along L_1, and **G** could have been any vector along L_2.

EXAMPLE 11.6

The points $A : (1, -2, 1)$, $B : (0, 1, 6)$, and $C : (-3, 4, -2)$ form a triangle. Find the angle between the line $\overline{AB}$ and the line from A to the midpoint of $\overline{BC}$. Such a line is called a *median* and is shown in Figure 11.22.

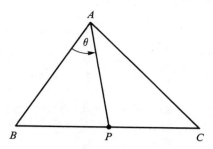

Figure 11.22

   Visualize the sides of the triangle as vectors, as in Figure 11.23. If P is the mid-point of $\overline{BC}$, then $\mathbf{H}_1 = \mathbf{H}_2$ because both vectors have the same length and direction. Further, from the coordinates of the points A, B, and C, we have

$$\mathbf{F} = -\mathbf{i} + 3\mathbf{j} + 5\mathbf{k} \quad \text{and} \quad \mathbf{G} = -4\mathbf{i} + 6\mathbf{j} - 3\mathbf{k}.$$

We want the angle between $\mathbf{F}$ and $\mathbf{K}$. We know $\mathbf{F}$. To find $\mathbf{K}$, we note by the parallelogram law that

$$\mathbf{F} + \mathbf{H}_1 = \mathbf{K} \quad \text{and} \quad \mathbf{K} + \mathbf{H}_2 = \mathbf{G}.$$

Since $\mathbf{H}_1 = \mathbf{H}_2$,

$$\mathbf{H}_1 = \mathbf{K} - \mathbf{F} = \mathbf{G} - \mathbf{K}.$$

Therefore,

$$\mathbf{K} = \tfrac{1}{2}(\mathbf{F} + \mathbf{G}) = \tfrac{1}{2}(-5\mathbf{i} + 9\mathbf{j} + 2\mathbf{k}).$$

The cosine of the angle θ between $\mathbf{F}$ and $\mathbf{K}$ is

$$\cos(\theta) = \frac{\mathbf{F} \cdot \mathbf{K}}{\|\mathbf{F}\|\|\mathbf{K}\|} = \frac{42}{\sqrt{35}\sqrt{110}} = \frac{42}{\sqrt{3850}}.$$

Then $\theta = \cos^{-1}(42/\sqrt{3850}) \approx 0.83$ radian, or about 47.4 degrees.   ∎

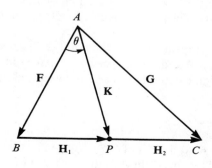

Figure 11.23

The angle between two vectors is $\pi/2$, or 90 degrees, exactly when the cosine of the angle between the vectors is zero. By equation (11.5), this occurs when $\mathbf{F} \cdot \mathbf{G} = 0$. Thus, the vanishing of the dot product is a test for two nonzero vectors to be perpendicular (that is, for arrows representing them to be perpendicular). This motivates the definition that two vectors are *orthogonal* (perpendicular) if and only if their dot product is zero. Since the dot product of any vector with $\mathbf{O}$ is zero, it is customary to say that the zero vector is orthogonal to every other vector.

EXAMPLE 11.7

Let $\mathbf{F} = -4\mathbf{i} + \mathbf{j} + 2\mathbf{k}$ and $\mathbf{G} = 2\mathbf{i} + 4\mathbf{k}$. Then $\mathbf{F} \cdot \mathbf{G} = -8 + 8 = 0$, so $\mathbf{F}$ and $\mathbf{G}$ are orthogonal.

However, vectors $\mathbf{H} = 6\mathbf{i} - \mathbf{j} - 2\mathbf{k}$ and $\mathbf{L} = 3\mathbf{i} + \mathbf{j} + 4\mathbf{k}$ are not orthogonal, because $\mathbf{H} \cdot \mathbf{L} = 9$. If θ is the angle between these vectors, $\cos(\theta) = 9/\sqrt{1066}$, so θ is approximately 1.29 radians, or approximately 74 degrees. ∎

EXAMPLE 11.8

Determine whether the lines L_1 and L_2 are perpendicular, where

$$L_1: \quad \begin{array}{l} x = 2 - 4t \\ y = 6 + t \\ z = 3t, \quad -\infty < t < \infty \end{array} \qquad \text{and} \qquad L_2: \quad \begin{array}{l} x = -2 + p \\ y = 7 + 2p \\ z = 3 - 4p, \quad -\infty < p < \infty. \end{array}$$

These lines intersect at $(-2, 7, 3)$, which is on L_1 when $t = 1$ and on L_2 when $p = 0$. Choose any other point on L_1, say, $(2, 6, 0)$ when $t = 0$. The vector $\mathbf{F} = [2 - (-2)]\mathbf{i} + (6 - 7)\mathbf{j} + (0 - 3)\mathbf{k} = 4\mathbf{i} - \mathbf{j} - 3\mathbf{k}$ is parallel to L_1.

Choose any other point on L_2, say, $(-1, 9, -1)$ when $p = 1$. Then $\mathbf{G} = [-1 - (-2)]\mathbf{i} + (9 - 7)\mathbf{j} + (-1 - 3)\mathbf{k} = \mathbf{i} + 2\mathbf{j} - 4\mathbf{k}$ is parallel to L_2.

The angle between L_1 and L_2 is the same as the angle between $\mathbf{F}$ and $\mathbf{G}$. Calculate $\mathbf{F} \cdot \mathbf{G} = 4 - 2 + 12 = 14 \neq 0$. Thus, $\mathbf{F}$ and $\mathbf{G}$ are not orthogonal, and the lines are not perpendicular. ∎

EXAMPLE 11.9

Find the equation of a plane Π passing through $(-6, 1, 1)$ and perpendicular to $-2\mathbf{i} + 4\mathbf{j} + \mathbf{k}$.

Let (x, y, z) be any point on Π (Figure 11.24). Any vector drawn as an arrow between two points in Π is also in Π. In particular, the vector

$$(x + 6)\mathbf{i} + (y - 1)\mathbf{j} + (z - 1)\mathbf{k}$$

is in the plane and hence must be perpendicular to $-2\mathbf{i} + 4\mathbf{j} + \mathbf{k}$. Then

$$[(x + 6)\mathbf{i} + (y - 1)\mathbf{j} + (z - 1)\mathbf{k}] \cdot [-2\mathbf{i} + 4\mathbf{j} + \mathbf{k}] = 0,$$

so

$$-2(x + 6) + 4(y - 1) + (z - 1) = 0,$$

or

$$-2x + 4y + z = 17. \tag{11.6}$$

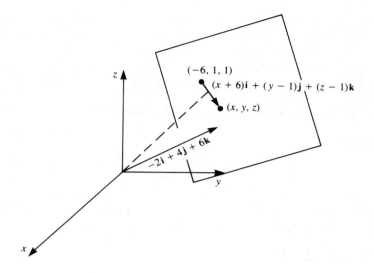

Figure 11.24

Equation (11.6) is the equation of the plane in the sense that (x, y, z) is on the plane if and only if x, y, and z satisfy equation (11.6). ∎

We will conclude this section with two additional results about the dot product. The first is called the Cauchy-Schwarz inequality.

THEOREM 11.5   Cauchy-Schwarz Inequality

For any vectors **F** and **G**,

$$|\mathbf{F} \cdot \mathbf{G}| \leq \|\mathbf{F}\|\|\mathbf{G}\|.$$

Proof   If **F** or **G** is the zero vector, $\mathbf{F} \cdot \mathbf{G} = \|\mathbf{F}\|\|\mathbf{G}\| = 0$, and the inequality becomes $0 \leq 0$, which is true. Thus, assume that $\mathbf{F} \neq \mathbf{O}$ and $\mathbf{G} \neq \mathbf{O}$. By equation (11.5), if θ is the angle between **F** and **G**, then

$$\cos(\theta) = \frac{\mathbf{F} \cdot \mathbf{G}}{\|\mathbf{F}\|\|\mathbf{G}\|}.$$

Since $-1 \leq \cos(\theta) \leq 1$,

$$-1 \leq \frac{\mathbf{F} \cdot \mathbf{G}}{\|\mathbf{F}\|\|\mathbf{G}\|} \leq 1.$$

Therefore,

$$-\|\mathbf{F}\|\|\mathbf{G}\| \leq \mathbf{F} \cdot \mathbf{G} \leq \|\mathbf{F}\|\|\mathbf{G}\|,$$

which is equivalent to the assertion that $|\mathbf{F} \cdot \mathbf{G}| \leq \|\mathbf{F}\|\|\mathbf{G}\|$. ∎

The following result is often used in performing computations with vectors.

THEOREM 11.6

Let α and β be scalars, and let $\mathbf{F}$ and $\mathbf{G}$ be vectors. Then

$$\|\alpha\mathbf{F} + \beta\mathbf{G}\|^2 = \alpha^2\|\mathbf{F}\|^2 + 2\alpha\beta\mathbf{F} \cdot \mathbf{G} + \beta^2\|\mathbf{G}\|^2.$$

Proof Using the conclusions of Theorem 11.4, we have

$$
\begin{aligned}
\|\alpha\mathbf{F} + \beta\mathbf{G}\|^2 &= (\alpha\mathbf{F} + \beta\mathbf{G}) \cdot (\alpha\mathbf{F} + \beta\mathbf{G}) \\
&= (\alpha\mathbf{F}) \cdot (\alpha\mathbf{F} + \beta\mathbf{G}) + (\beta\mathbf{G}) \cdot (\alpha\mathbf{F} + \beta\mathbf{G}) \\
&= (\alpha\mathbf{F}) \cdot (\alpha\mathbf{F}) + (\alpha\mathbf{F}) \cdot (\beta\mathbf{G}) + (\beta\mathbf{G}) \cdot (\alpha\mathbf{F}) + (\beta\mathbf{G}) \cdot (\beta\mathbf{G}) \\
&= \alpha^2\mathbf{F} \cdot \mathbf{F} + \alpha\beta\mathbf{F} \cdot \mathbf{G} + \beta\alpha\mathbf{G} \cdot \mathbf{F} + \beta^2\mathbf{G} \cdot \mathbf{G} \\
&= \alpha^2\|\mathbf{F}\|^2 + 2\alpha\beta\mathbf{F} \cdot \mathbf{G} + \beta^2\|\mathbf{G}\|^2. \quad \blacksquare
\end{aligned}
$$

The key to the proof is to compute $\|\alpha\mathbf{F} + \beta\mathbf{G}\|^2$ as $(\alpha\mathbf{F} + \beta\mathbf{G}) \cdot (\alpha\mathbf{F} + \beta\mathbf{G})$. It is often handy to remember that the dot product of a vector with itself is the square of the norm of the vector.

The Cauchy-Schwarz inequality and Theorem 11.6 yield a short proof of the triangle inequality of the preceding section. Let $\mathbf{F}$ and $\mathbf{G}$ be vectors. We want to show that $\|\mathbf{F} + \mathbf{G}\| \leq \|\mathbf{F}\| + \|\mathbf{G}\|$. With $\alpha = \beta = 1$ in Theorem 11.6, we have

$$
\begin{aligned}
\|\mathbf{F} + \mathbf{G}\|^2 &= \|\mathbf{F}\|^2 + 2\mathbf{F} \cdot \mathbf{G} + \|\mathbf{G}\|^2 \\
&\leq \|\mathbf{F}\|^2 + 2|\mathbf{F} \cdot \mathbf{G}| + \|\mathbf{G}\|^2 \\
&\leq \|\mathbf{F}\|^2 + 2\|\mathbf{F}\|\|\mathbf{G}\| + \|\mathbf{G}\|^2 \qquad \text{(by Cauchy-Schwarz)} \\
&= (\|\mathbf{F}\| + \|\mathbf{G}\|)^2;
\end{aligned}
$$

hence, $\|\mathbf{F} + \mathbf{G}\| \leq \|\mathbf{F}\| + \|\mathbf{G}\|$.

PROBLEMS FOR SECTION 11.2

In each of Problems 1 through 10, compute the dot product of the two vectors and the cosine of the angle between them, determine if they are orthogonal, and verify the Cauchy-Schwarz inequality for the two given vectors.

1. $\mathbf{i}, \quad 2\mathbf{i} - 3\mathbf{j} + \mathbf{k}$

2. $2\mathbf{i} - 6\mathbf{j} + \mathbf{k}, \quad \mathbf{i} - \mathbf{j}$

3. $-4\mathbf{i} - 2\mathbf{j} + 3\mathbf{k}, \quad 6\mathbf{i} - 2\mathbf{j} - \mathbf{k}$

4. $\mathbf{i}, \quad \mathbf{j}$

5. $-3\mathbf{i} + 2\mathbf{k}, \quad 6\mathbf{j}$

6. $8\mathbf{i} - 3\mathbf{j} + 2\mathbf{k}, \quad -8\mathbf{i} - 3\mathbf{j} + \mathbf{k}$

7. $\mathbf{i} - 2\mathbf{j}, \quad \mathbf{i} - 2\mathbf{j}$

8. $-5\mathbf{i} + 6\mathbf{j} - 3\mathbf{k}, \quad 5\mathbf{i} - 3\mathbf{j} - 3\mathbf{k}$

9. $\mathbf{i} - 3\mathbf{k}, \quad 2\mathbf{j} + 6\mathbf{k}$

10. $\mathbf{i} + \mathbf{j} + 2\mathbf{k}, \quad \mathbf{i} - \mathbf{j} + 2\mathbf{k}$

In each of Problems 11 through 20, find the equation of the plane passing through the given point and perpendicular to the given vector.

11. $(-1, 1, 2), \quad 3\mathbf{i} - \mathbf{j} + 4\mathbf{k}$

12. $(-1, 0, 0), \quad \mathbf{i} - 2\mathbf{j}$

13. $(2, -3, 4), \quad 8\mathbf{i} - 6\mathbf{j} + 4\mathbf{k}$

14. $(-1, -1, -5), \quad -3\mathbf{i} + 2\mathbf{j}$

15. $(0, -1, 4), \quad 8\mathbf{i} - 3\mathbf{j} + 4\mathbf{k}$

16. $(-2, 3, 2), \quad 4\mathbf{i} - 8\mathbf{j} + 6\mathbf{k}$

17. $(0, 2, 0), \quad -2\mathbf{i} + \mathbf{k}$

18. $(0, 0, 0), \quad 3\mathbf{i} - \mathbf{j} - 5\mathbf{k}$

19. $(1, -1, 5), \quad 6\mathbf{i} - 14\mathbf{j} - 2\mathbf{k}$

20. $(-2, 1, -1), \quad 4\mathbf{i} + 3\mathbf{j} + \mathbf{k}$

In each of Problems 21 through 25, find the cosine of the angle between the line $\overline{AB}$ and the line from A to the midpoint of the line $\overline{BC}$.

	A	B	C
21.	$(1, -2, 6)$	$(3, 0, 1)$	$(4, 2, -7)$

22. $(3, -2, -3)$ $(-2, 0, 1)$ $(1, 1, 7)$

23. $(1, -2, 6)$ $(0, 4, -3)$ $(-3, -2, 7)$

24. $(0, 5, -1)$ $(1, -2, 5)$ $(7, 0, -1)$

25. $(0, 0, -2)$ $(1, -3, 4)$ $(-2, 6, 1)$

26. Suppose that $\mathbf{F} \cdot \mathbf{X} = 0$ for *every* vector $\mathbf{X}$. What can be concluded about $\mathbf{F}$? = zero vector ✓

27. Suppose that $\mathbf{F} \cdot \mathbf{i} = \mathbf{F} \cdot \mathbf{j} = \mathbf{F} \cdot \mathbf{k} = 0$. What conclusion can be drawn about $\mathbf{F}$? *Hint:* Let $\mathbf{X} = a\mathbf{i} + b\mathbf{j} + c\mathbf{k}$ and use the result of Problem 26.

28. Let $\mathbf{F} \neq \mathbf{O}$. Prove that the unit vector $\mathbf{u}$ for which $|\mathbf{F} \cdot \mathbf{u}|$ is a maximum must be parallel to $\mathbf{F}$.

29. Prove that, for any vector $\mathbf{F}$,

$$\mathbf{F} = (\mathbf{F} \cdot \mathbf{i})\mathbf{i} + (\mathbf{F} \cdot \mathbf{j})\mathbf{j} + (\mathbf{F} \cdot \mathbf{k})\mathbf{k}.$$

30. Let $\mathbf{A}$ and $\mathbf{B}$ be vectors, with $\mathbf{B} \neq \mathbf{O}$. Prove that $\mathbf{A}$ can be written as $\mathbf{F} + \mathbf{G}$, where $\mathbf{F}$ is parallel to $\mathbf{B}$ and $\mathbf{G}$ is orthogonal to $\mathbf{B}$.

31. Let $\mathbf{F}$, $\mathbf{G}$, and $\mathbf{H}$ be nonzero vectors, each orthogonal to the other two. Show that, for any vector $\mathbf{A}$, there are unique scalars α, β, and γ such that

$$\mathbf{A} = \alpha\mathbf{F} + \beta\mathbf{G} + \gamma\mathbf{H}.$$

32. Suppose, instead of the definition we used, we defined the dot product by

$$(a_1\mathbf{i} + b_1\mathbf{j} + c_1\mathbf{k}) \cdot (a_2\mathbf{i} + b_2\mathbf{j} + c_2\mathbf{k}) = 2a_1 a_2 + b_1 b_2 + c_1 c_2.$$

Which properties of the dot product would still hold, and which would now fail?

33. Suppose we defined the dot product by putting

$$(a_1\mathbf{i} + b_1\mathbf{j} + c_1\mathbf{k}) \cdot (a_2\mathbf{i} + b_2\mathbf{j} + c_2\mathbf{k}) = a_1 a_2 - b_1 b_2 + c_1 c_2.$$

Which properties of the dot product would still hold, and which would fail?

34. Prove conclusion (2) of Theorem 11.4.

35. Prove conclusion (3) of Theorem 11.4.

36. Prove conclusion (5) of Theorem 11.4.

37. Prove that the diagonals of a rhombus must be perpendicular to each other. (Recall that a rhombus is a parallelogram whose sides have equal length.)

38. Let $\mathbf{F}$ and $\mathbf{G}$ be nonzero vectors. Prove that the vector

$$\frac{1}{\|\mathbf{F}\|\|\mathbf{G}\|}(\|\mathbf{G}\|\mathbf{F} + \|\mathbf{F}\|\mathbf{G})$$

bisects the angle between $\mathbf{F}$ and $\mathbf{G}$. *Hint:* Show that this vector makes the same angle with $\mathbf{F}$ that it does with $\mathbf{G}$.

11.3 The Cross Product of Vectors

The dot product produces a scalar from two vectors. The cross product produces a vector from two vectors. Let

$$\mathbf{F} = a_1\mathbf{i} + b_1\mathbf{j} + c_1\mathbf{k} \quad \text{and} \quad \mathbf{G} = a_2\mathbf{i} + b_2\mathbf{j} + c_2\mathbf{k}.$$

The cross product of $\mathbf{F}$ with $\mathbf{G}$ is the vector $\mathbf{F} \times \mathbf{G}$ defined by

$$\mathbf{F} \times \mathbf{G} = (b_1 c_2 - b_2 c_1)\mathbf{i} + (a_2 c_1 - a_1 c_2)\mathbf{j} + (a_1 b_2 - a_2 b_1)\mathbf{k}.$$

This vector is read "**F** cross **G**." Using this definition, we obtain, for example,

$$(\mathbf{i} + 2\mathbf{j} - 3\mathbf{k}) \times (-2\mathbf{i} + \mathbf{j} + 4\mathbf{k}) = (8 + 3)\mathbf{i} + (6 - 4)\mathbf{j} + (1 + 4)\mathbf{k} = 11\mathbf{i} + 2\mathbf{j} + 5\mathbf{k}.$$

There is a notational device which is helpful in remembering the components of $\mathbf{F} \times \mathbf{G}$. If we expand the 3×3 "determinant"

$$\begin{vmatrix} \mathbf{i} & \mathbf{j} & \mathbf{k} \\ a_1 & b_1 & c_1 \\ a_2 & b_2 & c_2 \end{vmatrix}$$

by the first row, we get exactly $\mathbf{F} \times \mathbf{G}$. For example,

$$(\mathbf{i} + 2\mathbf{j} - 3\mathbf{k}) \times (-2\mathbf{i} + \mathbf{j} + 4\mathbf{k}) = \begin{vmatrix} \mathbf{i} & \mathbf{j} & \mathbf{k} \\ 1 & 2 & -3 \\ -2 & 1 & 4 \end{vmatrix}$$

$$= (8 + 3)\mathbf{i} + (6 - 4)\mathbf{j} + (1 + 4)\mathbf{k} = 11\mathbf{i} + 2\mathbf{j} + 5\mathbf{k}.$$

The following theorem lists some properties of the cross product.

THEOREM 11.7

Let $\mathbf{F}$, $\mathbf{G}$, and $\mathbf{H}$ be vectors and let α be a scalar. Then

1. $\mathbf{F} \times \mathbf{G}$ is orthogonal to both $\mathbf{F}$ and $\mathbf{G}$.
2. $\|\mathbf{F} \times \mathbf{G}\| = \|\mathbf{F}\|\|\mathbf{G}\|\sin(\theta)$, where θ is the angle between $\mathbf{F}$ and $\mathbf{G}$.
3. If $\mathbf{F} \neq \mathbf{O}$ and $\mathbf{G} \neq \mathbf{O}$, then $\mathbf{F} \times \mathbf{G} = \mathbf{O}$ if and only if $\mathbf{F}$ and $\mathbf{G}$ are parallel.
4. $\mathbf{F} \times (\mathbf{G} + \mathbf{H}) = \mathbf{F} \times \mathbf{G} + \mathbf{F} \times \mathbf{H}$.
5. $(\alpha\mathbf{F}) \times \mathbf{G} = \alpha(\mathbf{F} \times \mathbf{G}) = \mathbf{F} \times (\alpha\mathbf{G})$.
6. $\mathbf{F} \times \mathbf{G} = -\mathbf{G} \times \mathbf{F}$. ∎

Proofs of these conclusions are, for the most part, routine calculations. We will prove (1), (2), and (3) as illustrations.

Proof of (1) Let $\mathbf{F} = a_1\mathbf{i} + b_1\mathbf{j} + c_1\mathbf{k}$ and $\mathbf{G} = a_2\mathbf{i} + b_2\mathbf{j} + c_2\mathbf{k}$. Then

$$\mathbf{F} \cdot (\mathbf{F} \times \mathbf{G}) = \mathbf{F} \cdot [(b_1c_2 - b_2c_1)\mathbf{i} + (a_2c_1 - a_1c_2)\mathbf{j} + (a_1b_2 - a_2b_1)\mathbf{k}]$$

$$= a_1(b_1c_2 - b_2c_1) + b_1(a_2c_1 - a_1c_2) + c_1(a_1b_2 - a_2b_1) = 0;$$

hence, $\mathbf{F} \times \mathbf{G}$ is orthogonal to $\mathbf{F}$. By a similar argument, $\mathbf{F} \times \mathbf{G}$ is orthogonal to $\mathbf{G}$. ∎

Proof of (2) Compute

$$\|\mathbf{F} \times \mathbf{G}\|^2 = (b_1c_2 - b_2c_1)^2 + (a_2c_1 - a_1c_2)^2 + (a_1b_2 - a_2b_1)^2$$

$$= (a_1^2 + b_1^2 + c_1^2)(a_2^2 + b_2^2 + c_2^2) - (a_1a_2 + b_1b_2 + c_1c_2)^2$$

$$= \|\mathbf{F}\|^2\|\mathbf{G}\|^2 - (\mathbf{F} \cdot \mathbf{G})^2$$

$$= \|\mathbf{F}\|^2\|\mathbf{G}\|^2 - \|\mathbf{F}\|^2\|\mathbf{G}\|^2\cos^2(\theta)$$

$$= \|\mathbf{F}\|^2\|\mathbf{G}\|^2[1 - \cos^2(\theta)]$$

$$= \|\mathbf{F}\|^2\|\mathbf{G}\|^2\sin^2(\theta).$$

Therefore, $\|\mathbf{F} \times \mathbf{G}\| = \|\mathbf{F}\|\|\mathbf{G}\|\sin(\theta)$. ∎

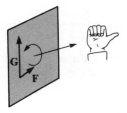

Figure 11.25. Plane determined
by **F** and **G**.

Figure 11.26. Right-hand rule for
the direction of **F** × **G**.

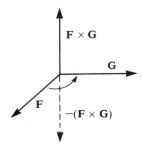

Figure 11.27. **F** × **G** = −(**G** × **F**).

Proof of (3)   Suppose that **F** × **G** = **O**. By (2), $\|\mathbf{F} \times \mathbf{G}\| = \|\mathbf{F}\|\|\mathbf{G}\|\sin(\theta) = 0$. But $\|\mathbf{F}\| \neq 0$ and $\|\mathbf{G}\| \neq 0$, so $\sin(\theta) = 0$; hence, θ must be zero or π. If $\theta = 0$, **F** and **G** are parallel and in the same direction; if $\theta = \pi$, **F** and **G** are parallel and in opposite directions.

Conversely, suppose that **F** and **G** are parallel. Then the angle θ between **F** and **G** is zero or π; hence, $\sin(\theta) = 0$. By (2), $\|\mathbf{F} \times \mathbf{G}\| = 0$; hence, **F** × **G** = **O**.   ∎

If **F** and **G** are nonzero vectors which are not parallel, **F** and **G** determine a plane, as shown in Figure 11.25. By (1), **F** × **G** must be perpendicular to this plane. In general, the direction of **F** × **G** is determined by the *right-hand rule*: if the right hand is held so that the fingers curl in the direction of orientation from **F** to **G**, the thumb points along **F** × **G** (see Figure 11.26).

The right-hand rule emphasizes the property stated as conclusion (6) of the theorem: the cross product is *anticommutative*; reversing the order of the factors changes the sign of the cross product. If **F** × **G** and **G** × **F** are computed directly from the definition, it is seen that each is the negative of the other (see Figure 11.27).

We will use the cross product later in applying vectors to problems in physics and engineering. Here is an elementary application to geometry.

EXAMPLE 11.10

Find the equation of the plane Π containing the points $(1, 2, 1)$, $(-1, 1, 3)$, and $(-2, -2, -2)$.

We can find the equation of Π if we can find a vector normal (perpendicular) to Π. To do this, first use the three given points to find two vectors lying in Π. Let $\mathbf{F}$ be the vector from $(1, 2, 1)$ to $(-1, 1, 3)$:

$$\mathbf{F} = -2\mathbf{i} - \mathbf{j} + 2\mathbf{k}.$$

Let $\mathbf{G}$ be the vector from $(1, 2, 1)$ to $(-2, -2, -2)$:

$$\mathbf{G} = -3\mathbf{i} - 4\mathbf{j} - 3\mathbf{k}.$$

Since both $\mathbf{F}$ and $\mathbf{G}$ lie in Π, their cross product $\mathbf{F} \times \mathbf{G}$ is perpendicular to Π. Compute

$$\mathbf{F} \times \mathbf{G} = \begin{vmatrix} \mathbf{i} & \mathbf{j} & \mathbf{k} \\ -2 & -1 & 2 \\ -3 & -4 & -3 \end{vmatrix} = 11\mathbf{i} - 12\mathbf{j} + 5\mathbf{k}.$$

If (x, y, z) is on Π, $(x - 1)\mathbf{i} + (y - 2)\mathbf{j} + (z - 1)\mathbf{k}$ is orthogonal to $\mathbf{F} \times \mathbf{G}$; hence,

$$[(x - 1)\mathbf{i} + (y - 2)\mathbf{j} + (z - 1)\mathbf{k}] \cdot [11\mathbf{i} - 12\mathbf{j} + 5\mathbf{k}] = 0.$$

Then

$$11(x - 1) - 12(y - 2) + 5(z - 1) = 0,$$

or

$$11x - 12y + 5z = -8.$$

This is the equation of the plane Π. ∎

Had the three given points been *collinear* in this example (lying along a straight line), they would not have determined a plane. In this event, the vectors $\mathbf{F}$ and $\mathbf{G}$ would have been parallel, and we would have gotten $\mathbf{F} \times \mathbf{G} = \mathbf{O}$. When we found that $\mathbf{F} \times \mathbf{G} \neq \mathbf{O}$, we knew that the given points were not collinear.

The cross product can also be used conveniently to calculate certain areas and volumes.

THEOREM 11.8

Let $\mathbf{F}$ and $\mathbf{G}$ be vectors represented by arrows along two incident sides of a parallelogram. Then the area of the parallelogram is $\|\mathbf{F} \times \mathbf{G}\|$. ∎

A typical parallelogram, with incident sides $\mathbf{F}$ and $\mathbf{G}$, is shown in Figure 11.28. Since $\|\mathbf{F}\|$ and $\|\mathbf{G}\|$ are the lengths of two sides of the parallelogram and θ is the

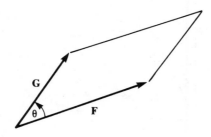

Figure 11.28. Area $= \|\mathbf{F} \times \mathbf{G}\|$.

angle between them, the area of the parallelogram is $\|\mathbf{F}\|\,\|\mathbf{G}\|\sin(\theta)$, which is exactly $\|\mathbf{F} \times \mathbf{G}\|$.

EXAMPLE 11.11

Find the area of the parallelogram having two of its incident sides extending from $(0, 1, -2)$ to $(1, 2, 2)$ and from $(0, 1, -2)$ to $(1, 4, 1)$.

Form vectors along these sides:

$$\mathbf{F} = (1 - 0)\mathbf{i} + (2 - 1)\mathbf{j} + [2 - (-2)]\mathbf{k} = \mathbf{i} + \mathbf{j} + 4\mathbf{k}$$

and

$$\mathbf{G} = (1 - 0)\mathbf{i} + (4 - 1)\mathbf{j} + [1 - (-2)]\mathbf{k} = \mathbf{i} + 3\mathbf{j} + 3\mathbf{k}.$$

Calculate

$$\mathbf{F} \times \mathbf{G} = \begin{vmatrix} \mathbf{i} & \mathbf{j} & \mathbf{k} \\ 1 & 1 & 4 \\ 1 & 3 & 3 \end{vmatrix} = -9\mathbf{i} + \mathbf{j} + 2\mathbf{k}.$$

The area of the parallelogram is $\|-9\mathbf{i} + \mathbf{j} + 2\mathbf{k}\|$, or $\sqrt{86}$. ∎

The cross product can also be used to find the volume of a rectangular parallelepiped (skewed rectangular box in three-space).

THEOREM 11.9

Let $\mathbf{F}$, $\mathbf{G}$, and $\mathbf{H}$ be vectors along incident sides of a rectangular parallelepiped. Then the volume of the parallelepiped is $|\mathbf{H} \cdot (\mathbf{F} \times \mathbf{G})|$.

Proof   Figure 11.29 shows the parallelepiped, with θ the angle between $\mathbf{H}$ and $\mathbf{F} \times \mathbf{G}$. (If $\mathbf{F}$ and $\mathbf{G}$ are in the xy-plane, θ is the angle between $\mathbf{H}$ and the z-axis.) Note that

$$|\mathbf{H} \cdot (\mathbf{F} \times \mathbf{G})| = \|\mathbf{H}\|\,\|\mathbf{F} \times \mathbf{G}\|\,|\cos(\theta)|. \tag{11.7}$$

Now, $\|\mathbf{F} \times \mathbf{G}\|$ is the area of the parallelogram forming the base of the parallelepiped, by Theorem 11.8. Further, $\|\mathbf{H}\|\,|\cos(\theta)|$ is the altitude of the parallelepiped. Since the

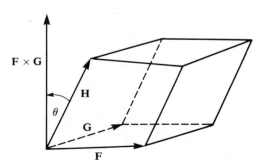

Figure 11.29. Volume $= |\mathbf{H} \cdot (\mathbf{F} \times \mathbf{G})|$.

volume is the area of the base times the altitude, equation (11.7) establishes the theorem. ∎

EXAMPLE 11.12

Suppose that one corner of a rectangular parallelepiped is at $(-1, 2, 2)$ and that three incident sides extend from this point to $(0, 1, 1), (-4, 6, 8)$, and $(-3, -2, 4)$. To find the volume, first form vectors along these sides:

$$\mathbf{F} = [0 - (-1)]\mathbf{i} + (1 - 2)\mathbf{j} + (1 - 2)\mathbf{k} = \mathbf{i} - \mathbf{j} - \mathbf{k},$$
$$\mathbf{G} = [-4 - (-1)]\mathbf{i} + (6 - 2)\mathbf{j} + (8 - 2)\mathbf{k} = -3\mathbf{i} + 4\mathbf{j} + 6\mathbf{k},$$

and

$$\mathbf{H} = [-3 - (-1)]\mathbf{i} + (-2 - 2)\mathbf{j} + (4 - 2)\mathbf{k} = -2\mathbf{i} - 4\mathbf{j} + 2\mathbf{k}.$$

Calculate $\mathbf{F} \times \mathbf{G} = -2\mathbf{i} - 3\mathbf{j} + \mathbf{k}$, and obtain

$$\text{volume} = |\mathbf{H} \cdot (\mathbf{F} \times \mathbf{G})| = |4 + 12 + 2| = 18. \quad ∎$$

The quantity $\mathbf{H} \cdot (\mathbf{F} \times \mathbf{G})$ is called a *scalar triple product*. In the next section, we will examine the scalar triple product and describe certain vector identities.

PROBLEMS FOR SECTION 11.3

In each of Problems 1 through 15, compute $\mathbf{F} \times \mathbf{G}$ and $\mathbf{G} \times \mathbf{F}$. Use the dot product to compute $\cos(\theta)$, with θ the angle between $\mathbf{F}$ and $\mathbf{G}$, and then determine $\sin(\theta)$ as $\sqrt{1 - \cos^2(\theta)}$. Using this result, calculate $\|\mathbf{F}\|\|\mathbf{G}\|\sin(\theta)$ and verify that this equals $\|\mathbf{F} \times \mathbf{G}\|$.

1. $\mathbf{F} = -3\mathbf{i} + 6\mathbf{j} + \mathbf{k}, \quad \mathbf{G} = -\mathbf{i} - 2\mathbf{j} + \mathbf{k}$
2. $\mathbf{F} = 6\mathbf{i} - \mathbf{k}, \quad \mathbf{G} = \mathbf{j} + 2\mathbf{k}$
3. $\mathbf{F} = 2\mathbf{i} - 3\mathbf{j} + 4\mathbf{k}, \quad \mathbf{G} = -3\mathbf{i} + 2\mathbf{j}$
4. $\mathbf{F} = 8\mathbf{i} + 6\mathbf{j}, \quad \mathbf{G} = 14\mathbf{j}$
5. $\mathbf{F} = 5\mathbf{i} + 3\mathbf{j} + 4\mathbf{k}, \quad \mathbf{G} = 20\mathbf{i} + 6\mathbf{k}$
6. $\mathbf{F} = 2\mathbf{k}, \quad \mathbf{G} = 8\mathbf{i} - \mathbf{j}$
7. $\mathbf{F} = 18\mathbf{i} - 3\mathbf{j} + 4\mathbf{k}, \quad \mathbf{G} = 22\mathbf{j} - \mathbf{k}$
8. $\mathbf{F} = \mathbf{i} - 3\mathbf{j} - \mathbf{k}, \quad \mathbf{G} = 18\mathbf{i} - 21\mathbf{j}$
9. $\mathbf{F} = -4\mathbf{i} + 6\mathbf{k}, \quad \mathbf{G} = \mathbf{i} - 2\mathbf{j} + 7\mathbf{k}$
10. $\mathbf{F} = -3\mathbf{i} + 2\mathbf{j} + \mathbf{k}, \quad \mathbf{G} = 8\mathbf{i} - 6\mathbf{j} + 2\mathbf{k}$
11. $\mathbf{F} = \mathbf{i} + 3\mathbf{j} - \mathbf{k}, \quad \mathbf{G} = 2\mathbf{i} + 6\mathbf{j} - \mathbf{k}$
12. $\mathbf{F} = 2\mathbf{j} - \mathbf{k}, \quad \mathbf{G} = \mathbf{i} + 6\mathbf{j} + 3\mathbf{k}$
13. $\mathbf{F} = -\mathbf{i} + 8\mathbf{j} + 3\mathbf{k}, \quad \mathbf{G} = 7\mathbf{i} + 2\mathbf{j}$
14. $\mathbf{F} = 2\mathbf{i} + 3\mathbf{j} - 5\mathbf{k}, \quad \mathbf{G} = -2\mathbf{i} + 3\mathbf{j} - 4\mathbf{k}$
15. $\mathbf{F} = 5\mathbf{i} + 8\mathbf{j} - 3\mathbf{k}, \quad \mathbf{G} = \mathbf{i} - \mathbf{j} - 3\mathbf{k}$

In each of Problems 16 through 20, (a) find a vector normal to the plane of the three points, or show that the points are collinear, and (b) in the former case, find the equation of the plane determined by the three points.

16. $(-1, 1, 6), \quad (2, 0, 1), \quad (3, 0, 0)$
17. $(4, 1, 1), \quad (-2, -2, 3), \quad (6, 0, 1)$
18. $(1, 0, -2), \quad (0, 0, 0), \quad (5, 1, 1)$
19. $(0, 0, 2), \quad (-4, 1, 0), \quad (2, -1, 1)$
20. $(-4, 2, -6), \quad (1, 1, 3), \quad (-2, 4, 5)$

In each of Problems 21 through 25, find the area of the parallelogram with incident sides $\overline{PQ}$ and $\overline{PR}$.

	P	*Q*	*R*
21.	$(1, -3, 7)$	$(2, 1, 1)$	$(6, -1, 2)$
22.	$(6, 1, 1)$	$(7, -2, 4)$	$(8, -4, 3)$
23.	$(-2, 1, 6)$	$(2, 1, -7)$	$(4, 1, 1)$
24.	$(4, 2, -3)$	$(6, 2, -1)$	$(2, -6, 4)$
25.	$(1, 1, -6)$	$(5, -3, 0)$	$(-2, 4, 1)$

In each of Problems 26 through 30, find the volume of the parallelepiped with incident sides $\overline{PQ}$, $\overline{PR}$, and $\overline{PS}$.

	P	Q	R	S
26.	$(1, 1, 1)$	$(-2, 1, 6)$	$(3, 5, 7)$	$(0, 1, 6)$
27.	$(0, 1, -6)$	$(-3, 1, 4)$	$(1, 7, 2)$	$(-3, 0, 4)$
28.	$(1, 6, 1)$	$(-2, 1, 4)$	$(3, 0, 0)$	$(2, 2, -4)$
29.	$(0, 1, 7)$	$(9, 1, 3)$	$(-2, 4, 1)$	$(3, 0, -2)$
30.	$(1, 1, 1)$	$(2, 2, 2)$	$(6, 1, 3)$	$(-2, 4, 1)$

In each of Problems 31 through 35, find a vector normal to the given plane. (There are infinitely many such vectors.)

31. $8x - y + z = 12$ **32.** $x - y + 2z = 0$ **33.** $x - 3y + 2z = 9$

34. $7x + y - 7z = 7$ **35.** $4x + 6y + 4z = -5$

36. Prove that $\mathbf{F} \times (\mathbf{G} + \mathbf{H}) = \mathbf{F} \times \mathbf{G} + \mathbf{F} \times \mathbf{H}$.

37. Prove that $(\alpha\mathbf{F}) \times \mathbf{G} = \alpha(\mathbf{F} \times \mathbf{G}) = \mathbf{F} \times (\alpha\mathbf{G})$.

38. Prove that $\mathbf{F} \times (\mathbf{F} \times \mathbf{G}) = (\mathbf{F} \cdot \mathbf{G})\mathbf{F} - (\mathbf{F} \cdot \mathbf{F})\mathbf{G}$.

39. Prove that $\mathbf{F} \cdot (\mathbf{G} \times \mathbf{H}) = \mathbf{G} \cdot (\mathbf{H} \times \mathbf{F}) = \mathbf{H} \cdot (\mathbf{F} \times \mathbf{G})$.

40. Give a geometric reason why $\mathbf{F} \times (\mathbf{G} \times \mathbf{H})$ should be in the plane determined by $\mathbf{G}$ and $\mathbf{H}$, assuming that $\mathbf{G}$ and $\mathbf{H}$ are not parallel. Also prove that $\mathbf{F} \times (\mathbf{G} \times \mathbf{H}) = (\mathbf{F} \cdot \mathbf{H})\mathbf{G} - (\mathbf{F} \cdot \mathbf{G})\mathbf{H}$.

41. Give a geometric reason why $(\mathbf{F} \times \mathbf{G}) \times (\mathbf{H} \times \mathbf{K})$ should be in the plane determined by $\mathbf{F}$ and $\mathbf{G}$, assuming that $\mathbf{F}$ and $\mathbf{G}$ are not parallel.

42. Suppose that $\mathbf{F}, \mathbf{G}, \mathbf{H}$, and $\mathbf{K}$ all lie in the same plane. What can be said about $(\mathbf{F} \times \mathbf{G}) \times (\mathbf{H} \times \mathbf{K})$?

43. Prove that $(\mathbf{F} \times \mathbf{G}) \times (\mathbf{H} \times \mathbf{K}) = [\mathbf{H} \cdot (\mathbf{K} \times \mathbf{F})]\mathbf{G} - [\mathbf{H} \cdot (\mathbf{K} \times \mathbf{G})]\mathbf{F}$.

44. Prove *Lagrange's identity*:

$$(\mathbf{F} \times \mathbf{G}) \cdot (\mathbf{H} \times \mathbf{K}) = (\mathbf{F} \cdot \mathbf{H})(\mathbf{G} \cdot \mathbf{K}) - (\mathbf{F} \cdot \mathbf{K})(\mathbf{G} \cdot \mathbf{H}).$$

45. Use vector operations to find a formula for the area of a triangle having vertices (a_1, b_1, c_1), (a_2, b_2, c_2), and (a_3, b_3, c_3). What conditions must be placed on the coordinates of these points in order to ensure that the points do not all lie on a straight line?

11.4 Scalar Triple Product and Related Vector Identities

The *scalar triple product* of vectors $\mathbf{F}, \mathbf{G}$, and $\mathbf{H}$ is a scalar denoted $[\mathbf{F}, \mathbf{G}, \mathbf{H}]$ and defined by

$$[\mathbf{F}, \mathbf{G}, \mathbf{H}] = \mathbf{F} \cdot (\mathbf{G} \times \mathbf{H}).$$

Scalar triple products arise in a variety of ways. From the preceding section, the absolute value of $[\mathbf{F}, \mathbf{G}, \mathbf{H}]$ can be interpreted as the volume of a rectangular parallelepiped having arrows representing $\mathbf{F}, \mathbf{G}$, and $\mathbf{H}$ drawn as incident sides. In the theory of electricity and magnetism, the electromotive force $d\mathbf{E}$ induced in an element $d\mathbf{L}$ of a conducting wire moving with velocity $\mathbf{v}$ through a magnetic field having flux density $\mathbf{B}$ is equal to $d\mathbf{E} = [\mathbf{v}, \mathbf{B}, d\mathbf{L}]$.

Here are some properties of the scalar triple product.

THEOREM 11.10

1. $[\mathbf{F}, \mathbf{G}, \mathbf{H}] = [\mathbf{G}, \mathbf{H}, \mathbf{F}] = [\mathbf{H}, \mathbf{F}, \mathbf{G}]$.
2. $[\mathbf{F}, \mathbf{G}, \mathbf{H}] = -[\mathbf{F}, \mathbf{H}, \mathbf{G}]$.
3. If $\mathbf{F} = a_1\mathbf{i} + b_1\mathbf{j} + c_1\mathbf{k}$, $\mathbf{G} = a_2\mathbf{i} + b_2\mathbf{j} + c_2\mathbf{k}$, and $\mathbf{H} = a_3\mathbf{i} + b_3\mathbf{j} + c_3\mathbf{k}$, then

$$[\mathbf{F}, \mathbf{G}, \mathbf{H}] = \begin{vmatrix} a_1 & b_1 & c_1 \\ a_2 & b_2 & c_2 \\ a_3 & b_3 & c_3 \end{vmatrix}.$$

If expanded, this determinant yields

$$[\mathbf{F}, \mathbf{G}, \mathbf{H}] = a_1 b_2 c_3 - a_1 c_2 b_3 + b_1 c_2 a_3 - b_1 a_2 c_3 + c_1 a_2 b_3 - c_1 b_2 a_3.$$

4. For any scalars α and β, we have

$$[\alpha\mathbf{F} + \beta\mathbf{K}, \mathbf{G}, \mathbf{H}] = \alpha[\mathbf{F}, \mathbf{G}, \mathbf{H}] + \beta[\mathbf{K}, \mathbf{G}, \mathbf{H}],$$
$$[\mathbf{F}, \alpha\mathbf{G} + \beta\mathbf{K}, \mathbf{H}] = \alpha[\mathbf{F}, \mathbf{G}, \mathbf{H}] + \beta[\mathbf{F}, \mathbf{K}, \mathbf{H}],$$

and

$$[\mathbf{F}, \mathbf{G}, \alpha\mathbf{H} + \beta\mathbf{K}] = \alpha[\mathbf{F}, \mathbf{G}, \mathbf{H}] + \beta[\mathbf{F}, \mathbf{G}, \mathbf{K}].$$

5. If any one of $\mathbf{F}$, $\mathbf{G}$, or $\mathbf{H}$ is a sum of scalar multiples of the other two, then

$$[\mathbf{F}, \mathbf{G}, \mathbf{H}] = 0. \quad \blacksquare$$

Proofs of these properties are straightforward but in some cases quite tedious, and we omit them.

Conclusion (1) is easily remembered in terms of the diagram

As long as the arrows are followed in cyclic fashion, we can write the factors starting anywhere in the diagram without changing the value of the scalar triple product. For example, starting with $\mathbf{G}$ yields $[\mathbf{G}, \mathbf{H}, \mathbf{F}]$, and starting with $\mathbf{H}$ yields $[\mathbf{H}, \mathbf{F}, \mathbf{G}]$, and these are equal. If we interchange the order of two factors, however, we change the sign, as stated in (2) of the theorem.

Combining conclusions (1) and (2) of the theorem, the absolute value of $[\mathbf{F}, \mathbf{G}, \mathbf{H}]$ is the same regardless of the order of the vectors:

$$|[\mathbf{F}, \mathbf{G}, \mathbf{H}]| = |[\mathbf{G}, \mathbf{H}, \mathbf{F}]| = |[\mathbf{H}, \mathbf{F}, \mathbf{G}]| = \cdots.$$

Conclusion (5) can be interpreted in several ways. Suppose, to be specific, that $\mathbf{H} = \alpha\mathbf{F} + \beta\mathbf{G}$. Then, row 3 in the determinant expansion of $[\mathbf{F}, \mathbf{G}, \mathbf{H}]$ is a sum of α times the first row and β times the second row, implying that this determinant is zero.

From a geometric point of view, if $\mathbf{H} = \alpha\mathbf{F} + \beta\mathbf{G}$, then $\mathbf{H}$ lies in the plane or line determined by $\mathbf{F} + \mathbf{G}$. In this case, the rectangular parallelepiped determined by $\mathbf{F}$, $\mathbf{G}$, and **H** is degenerate, lying in a plane or on a line, and has zero volume. Again, this is consistent with the conclusion that $[\mathbf{F}, \mathbf{G}, \mathbf{H}] = 0$ in such a circumstance.

Conclusion (4) of the theorem can be used to reduce calculations with a scalar triple product to calculations involving just scalar triple products involving **i**, **j**, and **k**. To illustrate, suppose we want to compute $[2\mathbf{i} - 4\mathbf{j}, 3\mathbf{i} + \mathbf{j}, \mathbf{k}]$. Of course, we can do this as a determinant:

$$[2\mathbf{i} - 4\mathbf{j}, 3\mathbf{i} + \mathbf{j}, \mathbf{k}] = \begin{vmatrix} 2 & -4 & 0 \\ 3 & 1 & 0 \\ 0 & 0 & 1 \end{vmatrix} = 14.$$

But we can also use conclusions (4) and (5). Note first that

$$[\mathbf{i}, \mathbf{i}, \mathbf{k}] = [\mathbf{j}, \mathbf{j}, \mathbf{k}] = 0$$

by (5) of the theorem. Further, by direct computation,

$$[\mathbf{i}, \mathbf{j}, \mathbf{k}] = -[\mathbf{j}, \mathbf{i}, \mathbf{k}] = 1.$$

Then

$$\begin{aligned} [2\mathbf{i} - 4\mathbf{j}, 3\mathbf{i} + \mathbf{j}, \mathbf{k}] &= 2[\mathbf{i}, 3\mathbf{i} + \mathbf{j}, \mathbf{k}] - 4[\mathbf{j}, 3\mathbf{i} + \mathbf{j}, \mathbf{k}] \\ &= 2(3)[\mathbf{i}, \mathbf{i}, \mathbf{k}] + 2[\mathbf{i}, \mathbf{j}, \mathbf{k}] - 4(3)[\mathbf{j}, \mathbf{i}, \mathbf{k}] - 4[\mathbf{j}, \mathbf{j}, \mathbf{k}] \\ &= 2(3)(0) + 2(1) - 4(3)(-1) - 4(0) = 14. \end{aligned}$$

This method of computing a scalar triple product is sometimes useful in carrying out or programming long calculations or for theoretical considerations.

PROBLEMS FOR SECTION 11.4

In each of Problems 1 through 10, compute $[\mathbf{F}, \mathbf{G}, \mathbf{H}]$, with **F** the first vector listed, **G** the second, and **H** the third.

1. $\mathbf{i} - 2\mathbf{j}$, $\mathbf{i} + 2\mathbf{j} - \mathbf{k}$, $\mathbf{i} + 3\mathbf{j} - \mathbf{k}$

2. $2\mathbf{i} + 6\mathbf{j}$, $\mathbf{i} + 2\mathbf{j} - \mathbf{k}$, $-3\mathbf{i} + 6\mathbf{k}$

3. $4\mathbf{i} + 2\mathbf{j}$, $\mathbf{i} - 2\mathbf{k}$, $2\mathbf{i} + 6\mathbf{j} + \mathbf{k}$

4. $-3\mathbf{i} + 3\mathbf{j} + 6\mathbf{k}$, $3\mathbf{i} + 4\mathbf{j}$, $8\mathbf{i} + 7\mathbf{j} - \mathbf{k}$

5. $2\mathbf{k}$, $8\mathbf{i} + 6\mathbf{j}$, $\mathbf{j} - 2\mathbf{k}$

6. $5\mathbf{i} + 3\mathbf{j} + \mathbf{k}$, $-2\mathbf{i} + 7\mathbf{k}$, $\mathbf{i} - 2\mathbf{j} + \mathbf{k}$

7. $-7\mathbf{i} + 4\mathbf{j} - \mathbf{k}$, $8\mathbf{i} + 6\mathbf{j} - \mathbf{k}$, $4\mathbf{i} + 6\mathbf{j} + 3\mathbf{k}$

8. $2\mathbf{i} - 4\mathbf{j} + 5\mathbf{k}$, $6\mathbf{i} - 3\mathbf{k}$, $\mathbf{i} + \mathbf{j} - \mathbf{k}$

9. $3\mathbf{i} + 3\mathbf{j} - 4\mathbf{k}$, $\mathbf{i} - 6\mathbf{j} + 3\mathbf{k}$, $3\mathbf{i} + 4\mathbf{k}$

10. $-10\mathbf{i} + \mathbf{j} - 2\mathbf{k}$, $8\mathbf{i} + 6\mathbf{j} - \mathbf{k}$, $8\mathbf{i} - 11\mathbf{j} + 3\mathbf{k}$

11. Suppose we want the equation of a plane containing a given point (x_0, y_0, z_0) and parallel to given vectors **F** and **G**, where $\mathbf{F} \times \mathbf{G} \neq \mathbf{O}$. Let $\mathbf{X} = x\mathbf{i} + y\mathbf{j} + z\mathbf{k}$ and $\mathbf{X}_0 = x_0\mathbf{i} + y_0\mathbf{j} + z_0\mathbf{k}$. Show that the equation of the plane can be written

$$[\mathbf{X} - \mathbf{X}_0, \mathbf{F}, \mathbf{G}] = 0.$$

12. Give a geometric interpretation of the number $[\mathbf{F}, \mathbf{G}, \mathbf{H}]/\|\mathbf{F} \times \mathbf{G}\|$.

13. Prove conclusion (1) of Theorem 11.10.

14. Prove conclusion (2) of Theorem 11.10.

15. Prove conclusion (3) of Theorem 11.10.

16. Prove conclusion (4) of Theorem 11.10.

17. Prove conclusion (5) of Theorem 11.10. *Hint:* Use conclusions (3) and (4).

11.5 *The Vector Space* R^n

Much (but not all) of what we have seen about vectors in three-space can be generalized to vectors with n components. There is a reason for doing this. Although a quantity such as a force in the real world can have only three components, many problems involve more than three variables and are treated as functions of vectors having the variables as components.

Taking a cue from vectors in three-space, we define an *n-vector* to be an *n-tuple* $\langle x_1, x_2, \ldots, x_n \rangle$ in which each x_j is a real number. The number x_j is called the *j*th *component* of $\langle x_1, x_2, \ldots, x_n \rangle$. We let R^n denote the totality of all *n*-vectors. If n is understood, we often refer to an *n*-vector as just a vector.

We use the subscript notation x_j for the *j*th component of $\langle x_1, x_2, \ldots, x_n \rangle$ because it is awkward to attempt to assign a different letter to each component if n is large. If n is small, we may drop the subscripts and use different letters. For example, a 4-vector could be conveniently denoted $\langle x, y, z, w \rangle$.

The *length*, or *norm*, of a vector $\langle x_1, x_2, \ldots, x_n \rangle$ is defined to be

$$\|\langle x_1, x_2, \ldots, x_n \rangle\| = \left[\sum_{j=1}^{n} x_j^2 \right]^{1/2} = \sqrt{x_1^2 + x_2^2 + \cdots + x_n^2}.$$

We add two vectors in R^n by adding their respective components:

$$\langle x_1, x_2, \ldots, x_n \rangle + \langle y_1, y_2, \ldots, y_n \rangle = \langle x_1 + y_1, x_2 + y_2, \ldots, x_n + y_n \rangle.$$

We multiply an *n*-vector by a scalar by multiplying each component by the scalar:

$$\alpha \langle x_1, x_2, \ldots, x_n \rangle = \langle \alpha x_1, \ldots, \alpha x_n \rangle.$$

The following properties of these operations are similar to those in R^3.

THEOREM 11.11

Let **F**, **G**, and **H** be vectors in R^n, and let α be any scalar. Then

1. $\mathbf{F} + \mathbf{G} = \mathbf{G} + \mathbf{F}$.
2. $(\mathbf{F} + \mathbf{G}) + \mathbf{H} = \mathbf{F} + (\mathbf{G} + \mathbf{H})$.
3. $\mathbf{F} + \langle 0, 0, \ldots, 0 \rangle = \mathbf{F}$.
4. $(\alpha + \beta)\mathbf{F} = \alpha\mathbf{F} + \beta\mathbf{F}$.
5. $(\alpha\beta)\mathbf{F} = \alpha(\beta\mathbf{F})$.
6. $\alpha(\mathbf{F} + \mathbf{G}) = \alpha\mathbf{F} + \alpha\mathbf{G}$.
7. $\alpha\langle 0, 0, \ldots, 0 \rangle = \langle 0, 0, \ldots, 0 \rangle$. ∎

These conclusions are proved by straightforward arguments. Because of these properties of addition of *n*-vectors and multiplication of an *n*-vector by a scalar, R^n is called a *vector space*. In Section 11.7 we will use some of the properties of vectors in R^n to give a general definition of vector space, of which R^n will be a special case.

The vector $\langle 0, 0, \ldots, 0 \rangle$ having all components equal to zero is called the *zero vector* in R^n and is denoted **O**. From conclusions (1) and (3), $\mathbf{F} + \mathbf{O} = \mathbf{O} + \mathbf{F} = \mathbf{F}$ for any *n*-vector **F**.

If $\mathbf{F} = \langle x_1, x_2, \ldots, x_n \rangle$, the vector $\langle -x_1, -x_2, \ldots, -x_n \rangle$ is called the *negative* of $\mathbf{F}$ and is denoted $-\mathbf{F}$. Then $\mathbf{F} + (-\mathbf{F}) = \mathbf{O}$. If we add $\mathbf{F}$ and $-\mathbf{G}$, we usually denote $\mathbf{F} + (-\mathbf{G})$ as $\mathbf{F} - \mathbf{G}$. If $\mathbf{G} = \langle y_1, y_2, \ldots, y_n \rangle$, then

$$\mathbf{F} - \mathbf{G} = \langle x_1, x_2, \ldots, x_n \rangle - \langle y_1, y_2, \ldots, y_n \rangle$$
$$= \langle x_1 - y_1, x_2 - y_2, \ldots, x_n - y_n \rangle.$$

The *dot product* of two vectors in R^n is defined to be the sum of the products of respective components:

$$\langle x_1, x_2, \ldots, x_n \rangle \cdot \langle y_1, y_2, \ldots, y_n \rangle = x_1 y_1 + x_2 y_2 + \cdots + x_n y_n = \sum_{j=1}^{n} x_j y_j.$$

In R^n, as in three-space, the dot product of two n-vectors is a scalar. The properties of the dot product in R^n are similar to those in R^3.

THEOREM 11.12

Let $\mathbf{F}$, $\mathbf{G}$, and $\mathbf{H}$ be vectors in R^n, and let α be any scalar. Then

1. $\mathbf{F} \cdot \mathbf{G} = \mathbf{G} \cdot \mathbf{F}$.
2. $(\mathbf{F} + \mathbf{G}) \cdot \mathbf{H} = \mathbf{F} \cdot \mathbf{H} + \mathbf{G} \cdot \mathbf{H}$.
3. $\mathbf{F} \cdot (\mathbf{G} + \mathbf{H}) = \mathbf{F} \cdot \mathbf{G} + \mathbf{F} \cdot \mathbf{H}$.
4. $\alpha(\mathbf{F} \cdot \mathbf{G}) = (\alpha\mathbf{F}) \cdot \mathbf{G} = \mathbf{F} \cdot (\alpha\mathbf{G})$.
5. $\mathbf{F} \cdot \mathbf{F} = \|\mathbf{F}\|^2$.
6. $\mathbf{F} \cdot \mathbf{F} = 0$ if and only if $\mathbf{F} = \mathbf{O}$. ∎

The proofs of these conclusions are routine and are left to the student.

THEOREM 11.13

Let $\mathbf{F}$ and $\mathbf{G}$ be in R^n, and let α and β be any scalars. Then

$$\|\alpha\mathbf{F} + \beta\mathbf{G}\|^2 = \alpha^2\|\mathbf{F}\|^2 + 2\alpha\beta\mathbf{F} \cdot \mathbf{G} + \beta^2\|\mathbf{G}\|^2. \quad ∎$$

Theorem 11.13 is simply the n-vector version of Theorem 11.6 in Section 11.2. The proof given for that theorem adapts verbatim to the present case.

Some statements look the same in R^n as they do in three-space, as we saw with the last theorem. However, when we go from R^3 to R^n, for $n > 3$, we lose some geometric intuition because we cannot draw n mutually perpendicular axes and represent an n-vector as an arrow. For example, we lose the parallelogram law for addition. Similarly, we lose the use of the law of cosines to argue that the cosine of the angle between two vectors equals $(\mathbf{F} \cdot \mathbf{G})/\|\mathbf{F}\|\|\mathbf{G}\|$, a fact we used to prove the Cauchy-Schwarz inequality (Theorem 11.5 in Section 11.2). Nevertheless, the Cauchy-Schwarz inequality does hold for n-vectors. We will now examine a proof which makes use of Theorem 11.13 without depending on geometric properties of vectors in R^3.

THEOREM 11.14   Cauchy-Schwarz Inequality

Let $\mathbf{F}$ and $\mathbf{G}$ be vectors in R^n. Then

$$|\mathbf{F} \cdot \mathbf{G}| \leq \|\mathbf{F}\|\|\mathbf{G}\|.$$

Proof   If either $\mathbf{F}$ or $\mathbf{G}$ is the zero vector, the proposed inequality becomes $0 \leq 0$, which is true. Thus, we may assume that $\mathbf{F} \neq \mathbf{O}$ and $\mathbf{G} \neq \mathbf{O}$.

For any scalars α and β, we have from Theorem 11.13 that

$$0 \leq \|\alpha\mathbf{F} + \beta\mathbf{G}\|^2 = \alpha^2\|\mathbf{F}\|^2 + 2\alpha\beta\mathbf{F} \cdot \mathbf{G} + \beta^2\|\mathbf{G}\|^2. \tag{11.8}$$

Choose $\alpha = \|\mathbf{G}\|$ and $\beta = -\|\mathbf{F}\|$. Then

$$0 \leq \|\mathbf{G}\|^2\|\mathbf{F}\|^2 - 2\|\mathbf{G}\|\|\mathbf{F}\|\mathbf{F} \cdot \mathbf{G} + \|\mathbf{F}\|^2\|\mathbf{G}\|^2.$$

Then

$$2\|\mathbf{F}\|\|\mathbf{G}\|\mathbf{F} \cdot \mathbf{G} \leq 2\|\mathbf{F}\|^2\|\mathbf{G}\|^2.$$

Upon dividing this inequality by the positive number $\|\mathbf{F}\|\|\mathbf{G}\|$, we get

$$\mathbf{F} \cdot \mathbf{G} \leq \|\mathbf{F}\|\|\mathbf{G}\|. \tag{11.9}$$

Now begin again with inequality (11.8) and let $\alpha = \|\mathbf{G}\|$ and $\beta = \|\mathbf{F}\|$. We get

$$0 \leq \|\mathbf{G}\|^2\|\mathbf{F}\|^2 + 2\|\mathbf{G}\|\|\mathbf{F}\|\mathbf{F} \cdot \mathbf{G} + \|\mathbf{G}\|^2\|\mathbf{F}\|^2$$

and hence

$$0 \leq 2\|\mathbf{G}\|^2\|\mathbf{F}\|^2 + 2\|\mathbf{G}\|\|\mathbf{F}\|\mathbf{F} \cdot \mathbf{G}.$$

Then

$$-\|\mathbf{G}\|^2\|\mathbf{F}\|^2 \leq \|\mathbf{G}\|\|\mathbf{F}\|\mathbf{F} \cdot \mathbf{G}.$$

Divide this inequality by the positive number $\|\mathbf{F}\|\|\mathbf{G}\|$ to get

$$-\|\mathbf{F}\|\|\mathbf{G}\| \leq \mathbf{F} \cdot \mathbf{G}. \tag{11.10}$$

By inequalities (11.9) and (11.10),

$$-\|\mathbf{F}\|\|\mathbf{G}\| \leq \mathbf{F} \cdot \mathbf{G} \leq \|\mathbf{F}\|\|\mathbf{G}\|$$

and hence

$$|\mathbf{F} \cdot \mathbf{G}| \leq \|\mathbf{F}\|\|\mathbf{G}\|. \quad\blacksquare$$

Of course, this reasoning holds for $n = 3$ as a special case, providing a proof of the Cauchy-Schwarz inequality in three-space, which is independent of the law of cosines.

We may now assert that $-1 \leq (\mathbf{F} \cdot \mathbf{G})/\|\mathbf{F}\|\|\mathbf{G}\| \leq 1$ for any nonzero vectors $\mathbf{F}$ and $\mathbf{G}$ in R^n. This enables us to *define* the angle between $\mathbf{F}$ and $\mathbf{G}$ to be that unique number in $[0, \pi]$ whose cosine is $(\mathbf{F} \cdot \mathbf{G})/\|\mathbf{F}\|\|\mathbf{G}\|$:

$$\cos(\theta) = \frac{\mathbf{F} \cdot \mathbf{G}}{\|\mathbf{F}\|\|\mathbf{G}\|}.$$

This suggests that we define two vectors in R^n to be *orthogonal*, or *perpendicular*, exactly when $\mathbf{F} \cdot \mathbf{G} = 0$. Since $\mathbf{F} \cdot \mathbf{O} = 0$ for every $\mathbf{F}$ in R^n, we consider the zero vector to be orthogonal to every vector in R^n.

EXAMPLE 11.13

$\langle -1, 3, 4, 6, 2 \rangle$ and $\langle 4, 2, 1, 6, -21 \rangle$ are orthogonal vectors in R^5, because

$$\langle -1, 3, 4, 6, 2 \rangle \cdot \langle 4, 2, 1, 6, -21 \rangle = -4 + 6 + 4 + 36 - 42 = 0. \quad\blacksquare$$

In R^3, it was convenient to express vectors in terms of the mutually orthogonal unit vectors $\mathbf{i}, \mathbf{j}$, and $\mathbf{k}$. We can develop a similar standard representation in R^n. Define

$$\mathbf{e}_1 = \langle 1, 0, 0, \ldots, 0 \rangle,$$
$$\mathbf{e}_2 = \langle 0, 1, 0, \ldots, 0 \rangle,$$
$$\vdots$$
$$\mathbf{e}_n = \langle 0, 0, \ldots, 0, 1 \rangle.$$

Each $\|\mathbf{e}_j\| = 1$, so $\mathbf{e}_j$ is a unit vector, and any two of these vectors are orthogonal because $\mathbf{e}_i \cdot \mathbf{e}_j = 0$ if $\mathbf{i} \neq \mathbf{j}$. Further, we can write

$$\langle x_1, x_2, \ldots, x_n \rangle = x_1 \mathbf{e}_1 + x_2 \mathbf{e}_2 + \cdots + x_n \mathbf{e}_n,$$

the *standard representation* of $\langle x_1, x_2, \ldots, x_n \rangle$. For example,

$$\langle 3, -2, \pi, 1, -7 \rangle = 3\mathbf{e}_1 - 2\mathbf{e}_2 + \pi \mathbf{e}_3 + \mathbf{e}_4 - 7\mathbf{e}_5,$$

where $\mathbf{e}_1 = \langle 1, 0, 0, 0, 0 \rangle$, $\mathbf{e}_2 = \langle 0, 1, 0, 0, 0 \rangle$, $\mathbf{e}_3 = \langle 0, 0, 1, 0, 0 \rangle$, $\mathbf{e}_4 = \langle 0, 0, 0, 1, 0 \rangle$, and $\mathbf{e}_5 = \langle 0, 0, 0, 0, 1 \rangle$.

It is obvious that a sum of vectors in R^n is in R^n and a scalar multiple of a vector in R^n is in R^n. Sometimes, however, we must deal with a set or collection S of vectors in R^n, and these two properties may not hold for S. For example, let S consist of all vectors in R^5 having first component 1. That is, S consists of all vectors $\langle 1, x, y, z, w \rangle$. A sum of two vectors in S is not in S:

$$\langle 1, x_1, y_1, z_1, w_1 \rangle + \langle 1, x_2, y_2, z_2, w_2 \rangle = \langle 2, x_1 + x_2, y_1 + y_2, z_1 + z_2, w_1 + w_2 \rangle,$$

which is not in S because the first component of the sum is 2, not 1. Further, scalar multiples of vectors in S need not be in S. For example,

$$4 \langle 1, x, y, z, w \rangle = \langle 4, 4x, 4y, 4z, 4w \rangle,$$

and this vector is not in S because the first component is 4.

If a set of vectors in R^n does have the property that the sum of any vectors in S is in S, and any scalar multiple of any vector in S is in S, then we call S a *subspace* of R^n. For example, let S consist of all vectors in R^6 having zero as first, third, and fifth components. Then a sum of any two vectors in S is in S because

$$\langle 0, x_1, 0, y_1, 0, z_1 \rangle + \langle 0, x_2, 0, y_2, 0, z_2 \rangle = \langle 0, x_1 + x_2, 0, y_1 + y_2, 0, z_1 + z_2 \rangle$$

has zero first, third, and fifth components. Further, any scalar multiple

$$\alpha \langle 0, x, 0, y, 0, z \rangle = \langle 0, \alpha x, 0, \alpha y, 0, \alpha z \rangle$$

has zero first, third, and fifth components and so is in S. Therefore, S is a subspace of R^6.

Another example of a subspace of R^n is R^n itself. Certainly, a sum of vectors in R^n is in R^n and a scalar times a vector in R^n is in R^n. Thus, R^n is a subspace of itself.

At the other extreme, let T consist of just the zero vector $\langle 0, 0, \ldots, 0 \rangle$. Then a sum of vectors in T is in T (because we are adding the zero vector to itself), and a scalar times a vector in T is in T, because $\alpha \langle 0, 0, \ldots, 0 \rangle = \langle 0, 0, \ldots, 0 \rangle$. We call T the *trivial subspace* of R^n.

If n equals 2 or 3, it is possible to characterize the subspaces of R^n geometrically.

THEOREM 11.15

The only subspaces of R^2 are (1) R^2 itself; (2) the trivial subspace consisting of just the zero vector $\langle 0, 0 \rangle$; and (3) any set consisting of vectors $\langle x, y \rangle$ represented by arrows lying along a given straight line through the origin.

That is, given any line through the origin, all vectors in R^2 which can be represented as vectors along this line constitute a subspace of R^2. Further, any subspace other than R^2 and the trivial subspace must consist of vectors along a line through the origin.

Proof   We will sketch a geometric proof of this theorem. First, suppose we have a straight line L through the origin. If $\mathbf{F}$ and $\mathbf{G}$ are vectors parallel to L, $\mathbf{F} + \mathbf{G}$ is parallel to L by the parallelogram law. Further, $\alpha\mathbf{F}$ is parallel to L for any scalar α. Thus, the set of vectors parallel to L forms a subspace of R^2.

We must now show that the theorem lists *all* subspaces of R^2. Suppose that S is a subspace of R^2. Suppose that S is not all of R^2 but that S contains vectors other than the zero vector. We will produce a line L through the origin such that S consists exactly of those vectors in R^2 parallel to L.

Let $\mathbf{F}$ be any nonzero vector in S. Represent $\mathbf{F}$ by an arrow from the origin, as in Figure 11.30. Let L be the straight line through the origin and along $\mathbf{F}$. We claim that S must consist of only those vectors represented by arrows parallel to L.

For example, suppose that S contains a nonzero vector $\mathbf{G}$ not parallel to L. Represent $\mathbf{G}$ by an arrow from the origin, as in Figure 11.31. Then $\mathbf{G}$ determines a second line L' through the origin. We will show that S must now contain every vector in the plane.

To prove this, let $\mathbf{V}$ be any vector in R^2. If $\mathbf{V}$ is along L, $\mathbf{V}$ is in S. Similarly, if $\mathbf{V}$ is along L', $\mathbf{V}$ is a scalar multiple of $\mathbf{G}$ and so is in S because $\mathbf{G}$ is in S.

If $\mathbf{V}$ is not along L or L', as in Figure 11.32, draw $\mathbf{V}$ as an arrow from the origin. Now draw parallels to L and L' from the tip of $\mathbf{V}$. This construction determines vectors $\mathbf{A}$ along L and $\mathbf{B}$ along L' such that $\mathbf{V} = \mathbf{A} + \mathbf{B}$ by the parallelogram law. But $\mathbf{A}$ is in S (being along L), and $\mathbf{B}$ is in S (being a scalar multiple of $\mathbf{G}$). Thus, V is in S because S is

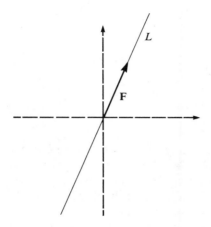

Figure 11.30. Line determined by a vector from the origin in R^2.

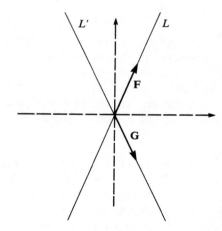

Figure 11.31

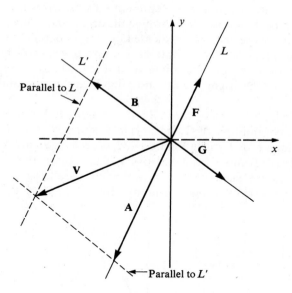

Figure 11.32

a subspace. This shows that any vector in the plane is in S, contradicting the assumption that S is not all of R^2. Therefore, there is no vector $\mathbf{G}$ in S which is not along L, and we have shown that the subspace S consists only of vectors along a line through the origin. This completes the proof of the theorem. ∎

By a similar argument, but involving more cases, we can prove the following characterization of subspaces of R^3.

THEOREM 11.16

The only subspaces of R^3 are as follows.

1. R^3.

2. The trivial subspace consisting of just the zero vector $\langle 0, 0, 0 \rangle$.

3. All vectors parallel to a given straight line through the origin.

4. All vectors lying on a given plane through the origin. ■

   In the next section, we will develop the concepts of linear independence and dimension in R^n. These ideas are important in solving systems of linear equations using matrices.

PROBLEMS FOR SECTION 11.5

In each of Problems 1 through 10, determine the sum of the given vectors and express this sum in standard form. Also determine the dot product of the vectors, and find the cosine of the angle between them.

1. $\langle -1, 6, 2, 4, 0 \rangle$, $\langle 6, -1, 4, 1, 1 \rangle$

2. $\langle 0, 1, 4, -3 \rangle$, $\langle 2, 8, 6, -4 \rangle$

3. $\langle 1, -4, 3, 2 \rangle$, $\langle 16, 0, 0, 4 \rangle$

4. $\langle 8, -3, 2 \rangle$, $\langle 9, 1, 7 \rangle$

5. $\langle -14, 6, 1, -2, 3, 2, 8 \rangle$, $\langle 9, -3, 4, 4, 1, -2, 7 \rangle$

6. $\langle 3, -5, 2, 8 \rangle$, $\langle 6, 1, -7, -2 \rangle$

7. $\langle 6, 1, 1, -1, 2 \rangle$, $\langle 4, 3, 5, 1, -2 \rangle$

8. $\langle 5, -3, 2, 1, 6 \rangle$, $\langle 1, -4, 5, 14, 12 \rangle$

9. $\langle -3, 2, 1, -4, -5 \rangle$, $\langle 18, 1, -5, 6, 7 \rangle$

10. $\langle 0, 1, 6, -4, 1, 2 \rangle$, $\langle 14, 1, -7, 6, 1, 5 \rangle$

In each of Problems 11 through 20, determine whether the set S of vectors in R^n forms a subspace of R^n.

11. S consists of all vectors in R^5 of the form $\langle x, y, z, x, x \rangle$.

12. S consists of all vectors in R^4 of the form $\langle x, 2x, 3x, y \rangle$.

13. S consists of all vectors in R^6 of the form $\langle x, 0, 0, 1, 0, y \rangle$.

14. S consists of all vectors in R^3 of the form $\langle 0, x, y \rangle$.

15. S consists of all vectors in R^4 of the form $\langle x, y, x + y, x - y \rangle$.

16. S consists of all vectors in R^7 with zero third and fifth components.

17. S consists of all vectors in R^4 with first and second components equal.

18. S consists of all vectors in R^4 with third component 2.

19. S consists of all vectors in R^7 with seventh component the sum of the first six components.

20. S consists of all vectors in R^8 with zero first, second, and fourth components and the third component equal to the sixth.

21. Prove the following theorem: If $\mathbf{F}$ and $\mathbf{G}$ are in R^n, then

$$\|\mathbf{F} + \mathbf{G}\|^2 + \|\mathbf{F} - \mathbf{G}\|^2 = 2(\|\mathbf{F}\|^2 + \|\mathbf{G}\|^2).$$

22. Prove *Pythagoras's theorem*: If $\mathbf{F}$ and $\mathbf{G}$ are orthogonal vectors in R^n, then

$$\|\mathbf{F} + \mathbf{G}\|^2 = \|\mathbf{F}\|^2 + \|\mathbf{G}\|^2.$$

23. Suppose that $\mathbf{F}$ and $\mathbf{G}$ are nonzero vectors in R^n and that

$$\|\mathbf{F} + \mathbf{G}\|^2 = \|\mathbf{F}\|^2 + \|\mathbf{G}\|^2.$$

Is it true that $\mathbf{F}$ and $\mathbf{G}$ are orthogonal?

24. Prove the *triangle inequality* in R^n. Suppose that $\mathbf{F}$ and $\mathbf{G}$ are in R^n. Then

$$\|\mathbf{F} + \mathbf{G}\| \le \|\mathbf{F}\| + \|\mathbf{G}\|.$$

25. Use the Cauchy-Schwarz inequality to prove the following fact about numbers. If $x_1, x_2, \ldots, x_n, y_1, y_2, \ldots, y_n$ are real numbers, then

$$\left| \sum_{j=1}^{n} x_j y_j \right|^2 \le \left(\sum_{j=1}^{n} x_j^2 \right)\left(\sum_{j=1}^{n} y_j^2 \right).$$

11.6 Linear Independence and Dimension

Suppose we are given m vectors $\mathbf{F}_1, \ldots, \mathbf{F}_m$ in R^n. A *linear combination* of $\mathbf{F}_1, \ldots, \mathbf{F}_m$ is a sum of scalar multiples of these vectors:

$$\alpha_1 \mathbf{F}_1 + \alpha_2 \mathbf{F}_2 + \cdots + \alpha_m \mathbf{F}_m,$$

in which each α_j is a scalar.

We call a vector $\mathbf{G}$ a *linear combination of* $\mathbf{F}_1, \ldots, \mathbf{F}_m$ if $\mathbf{G}$ can be written as a sum of scalar multiples of $\mathbf{F}_1, \ldots, \mathbf{F}_m$. That is, there must be scalars $\alpha_1, \ldots, \alpha_m$ such that

$$\mathbf{G} = \alpha_1 \mathbf{F}_1 + \alpha_2 \mathbf{F}_2 + \cdots + \alpha_m \mathbf{F}_m.$$

For example, $\langle 18, 0, -7, 18 \rangle$ is a linear combination of $\langle 4, 1, -2, 6 \rangle$, $\langle 1, 1, 0, 3 \rangle$, and $\langle 8, -1, -1, 6 \rangle$ because

$$\langle 18, 0, -7, 18 \rangle = 3\langle 4, 1, -2, 6 \rangle - 2\langle 1, 1, 0, 3 \rangle + \langle 8, -1, -1, 6 \rangle. \quad (11.11)$$

A collection of vectors $\mathbf{F}_1, \ldots, \mathbf{F}_m$ in R^n is called *linearly dependent* if one of the vectors can be written as a linear combination of the others. For example, equation (11.11) implies that the vectors $\langle 4, 1, -2, 6 \rangle$, $\langle 1, 1, 0, 3 \rangle$, $\langle 18, 0, -7, 18 \rangle$, and $\langle 8, -1, -1, 6 \rangle$ are linearly dependent. If vectors $\mathbf{F}_1, \ldots, \mathbf{F}_m$ are not linearly dependent, they are said to be *linearly independent*. For example, $\mathbf{e}_1, \mathbf{e}_2, \ldots, \mathbf{e}_n$ are linearly independent in R^n.

The following theorem provides an often-used characterization of linear dependence and linear independence.

THEOREM 11.17

Let $\mathbf{F}_1, \ldots, \mathbf{F}_m$ be vectors in R^m. Then

1. $\mathbf{F}_1, \ldots, \mathbf{F}_m$ are linearly dependent if and only if there are scalars $\alpha_1, \ldots, \alpha_m$ not all zero such that $\alpha_1 \mathbf{F}_1 + \cdots + \alpha_m \mathbf{F}_m = \mathbf{O}$.

2. $\mathbf{F}_1, \ldots, \mathbf{F}_m$ are linearly independent if and only if the equation

$$\alpha_1 \mathbf{F}_1 + \cdots + \alpha_m \mathbf{F}_m = \mathbf{O}$$

   is true only when $\alpha_1 = \alpha_2 = \cdots = \alpha_m = 0$.

Proof of (1) Suppose first that $\mathbf{F}_1, \ldots, \mathbf{F}_m$ are linearly dependent. Then one of these vectors is a linear combination of the others. Suppose, to simplify the notation, that the vectors have been labeled so that $\mathbf{F}_1$ is a linear combination of $\mathbf{F}_2, \ldots, \mathbf{F}_m$. Then

$$\mathbf{F}_1 = \alpha_2 \mathbf{F}_2 + \cdots + \alpha_m \mathbf{F}_m$$

for some scalars $\alpha_2, \ldots, \alpha_m$. But then

$$\mathbf{F}_1 - \alpha_2 \mathbf{F}_2 - \cdots - \alpha_m \mathbf{F}_m = \mathbf{O},$$

a linear combination of $\mathbf{F}_1, \ldots, \mathbf{F}_m$ which equals $\mathbf{O}$ but has at least one nonzero coefficient (the coefficient of $\mathbf{F}_1$ is 1).

Conversely, suppose that there are scalars $\alpha_1, \ldots, \alpha_m$, not all zero, such that

$$\alpha_1 \mathbf{F}_1 + \cdots + \alpha_m \mathbf{F}_m = \mathbf{O}.$$

Again, to simplify the notation, suppose that α_1 is nonzero. Then we can write

$$\mathbf{F}_1 = -\frac{1}{\alpha_1} \alpha_2 \mathbf{F}_2 - \cdots - \frac{1}{\alpha_1} \alpha_m \mathbf{F}_m,$$

and $\mathbf{F}_1$ is a linear combination of $\mathbf{F}_2, \ldots, \mathbf{F}_m$. Thus, $\mathbf{F}_1, \ldots, \mathbf{F}_m$ are linearly dependent.

A proof of (2) can be based on the conclusion of (1). ■

EXAMPLE 11.14

In the plane, two vectors are linearly dependent if they are parallel. In R^3, three vectors are linearly dependent if they all lie in the same plane or if all three are parallel. ■

EXAMPLE 11.15

Any collection of vectors $\mathbf{F}_1, \ldots, \mathbf{F}_m$ in R^n is linearly dependent if one of the vectors is the zero vector. For example, suppose that $\mathbf{F}_1 = \mathbf{O}$. Then we can write $\mathbf{F}_1$ as a linear combination of the other $\mathbf{F}_j$'s:

$$\mathbf{F}_1 = \mathbf{O} = 0\mathbf{F}_2 + 0\mathbf{F}_3 + \cdots + 0\mathbf{F}_m. ■$$

Notice that the characterization of linear dependence and independence given in Theorem 11.17 is similar to the concept of linear dependence and independence for functions given in Chapter 2. The notion of an abstract vector space, which we will discuss in the next section, provides a background against which these concepts can be unified and treated as special cases of a general definition.

There is an important special case in which it is easy to tell that a collection of m vectors $\mathbf{F}_1, \ldots, \mathbf{F}_m$ in R^n is linearly independent. This case will be stated as a lemma, anticipating its use in connection with matrices in Section 12.6.

LEMMA 11.1

Suppose that $\mathbf{F}_1, \ldots, \mathbf{F}_m$ are vectors in R^n satisfying the following condition: each vector has its first nonzero component equal to 1, and, if the first nonzero component of $\mathbf{F}_j$ is the kth component, the other vectors have k component zero. Then $\mathbf{F}_1, \ldots, \mathbf{F}_m$ are linearly independent. ■

This result is more complicated to state carefully than it is to apply, and in practice we can use it to tell at a glance that certain vectors are linearly independent. To illustrate, consider the following vectors in R^6:

$$\langle 0, 1, 0, -2, 0, 0\rangle, \quad \langle 0, 0, 1, 8, 0, 0\rangle, \quad \langle 0, 0, 0, 0, 1, 0\rangle, \quad \langle 0, 0, 0, 0, 0, 1\rangle.$$

The first nonzero component of each of these vectors is 1, and wherever any of the vectors has a first nonzero component, the others have component zero. For example, $\langle 0, 1, 0, -2, 0, 0\rangle$ has its second component as its first nonzero component, and all the other vectors in this set have second component zero.

The lemma tells us immediately that these vectors are linearly independent. We will show why this is so for these vectors. Suppose that

$$\alpha <0, 1, 0, -2, 0, 0> + \beta\langle 0, 0, 1, 8, 0, 0\rangle + \gamma\langle 0, 0, 0, 0, 1, 0\rangle \qquad (11.12)$$
$$+ \delta\langle 0, 0, 0, 0, 0, 1\rangle = \langle 0, 0, 0, 0, 0, 0\rangle.$$

Then

$$\langle 0, \alpha, \beta, -2\alpha + 8\beta, \gamma, \delta\rangle = \langle 0, 0, 0, 0, 0, 0\rangle.$$

Equating components, we have $\alpha = 0$, $\beta = 0$, $-2\alpha + 8\beta = 0$, $\gamma = 0$, and $\delta = 0$, implying that all the coefficients in (11.12) are zero. By conclusion (2) of Theorem 11.17, these vectors are linearly independent.

The fact that each vector has first nonzero component 1 and all other vectors in the list have that component zero means that each α_j will appear as some component of the linear combination $\alpha_1 F_1 + \cdots + \alpha_m F_m$. Thus, if this linear combination equals $\langle 0, 0, \ldots, 0\rangle$, each α_j must be zero, proving by conclusion (2) of Theorem 11.17 that $F_1, \ldots, F_m$ are linearly independent.

Using the concept of linear independence, we can introduce the idea of a basis. Suppose we have a subspace S of R^n. Vectors $F_1, \ldots, F_m$ in S form a *basis* for S if the following two conditions hold:

1. $F_1, \ldots, F_m$ are linearly independent.
2. Every vector in S is a linear combination of $F_1, \ldots, F_m$.

That is, every vector in S can be written in the form $\alpha_1 F_1 + \cdots + \alpha_m F_m$ for some scalars $\alpha_1, \ldots, \alpha_m$. Notice that, if we know just the basis vectors $F_1, \ldots, F_m$, we know all the vectors in S because every vector in S is a linear combination of these m vectors.

EXAMPLE 11.16

$e_1, \ldots, e_n$ form a basis for R^n. These vectors are linearly independent (by Lemma 11.1), and every n-vector is a linear combination of these vectors (in fact, $\alpha_1 e_1 + \cdots + \alpha_n e_n$ is the standard form of a vector in R^n). ∎

EXAMPLE 11.17

In R^2, let S be the subspace consisting of all vectors parallel to the line $x - y = 0$. Then a vector $\langle x, y\rangle$ is in S if and only if $x = y$. Thus, S consists of all vectors in R^2 of the form $\langle x, x\rangle$ having first and second components equal.

The single vector $\langle 1, 1\rangle$ forms a basis for S. This vector is in S and is nonzero, so it is linearly independent (see Example 11.14). Further, $\langle x, x\rangle = x\langle 1, 1\rangle$, so every vector in S is a linear combination (in this case, just a multiple) of $\langle 1, 1\rangle$. ∎

EXAMPLE 11.18

In R^3, let S be the subspace consisting of all vectors on the plane $x + y + z = 0$. A vector $\langle x, y, z \rangle$ is in S if and only if $x + y + z = 0$ or $z = -x - y$. Thus, S consists of all 3-vectors of the form $\langle x, y, -x - y \rangle$.

The vectors $\langle 1, 0, -1 \rangle$ and $\langle 0, 1, -1 \rangle$ form a basis for S. These two vectors are in S and are linearly independent because neither is a scalar multiple of the other (or apply the lemma). Further, given any vector $\langle x, y, -x - y \rangle$ in S, we can write

$$\langle x, y, -x - y \rangle = x\langle 1, 0, -1 \rangle + y\langle 0, 1, -1 \rangle,$$

a linear combination of $\langle 1, 0, -1 \rangle$ and $\langle 0, 1, -1 \rangle$. ∎

EXAMPLE 11.19

In R^5, let S consist of all vectors $\langle 0, x, y, 0, y \rangle$, with zero first and fourth components and with the third and fifth components equal. It is routine to check that S is a subspace of R^5. The vectors $\langle 0, 1, 0, 0, 0 \rangle$ and $\langle 0, 0, 1, 0, 1 \rangle$ form a basis for S. These two vectors are in S and are linearly independent (by Lemma 11.1), and

$$\langle 0, x, y, 0, y \rangle = x\langle 0, 1, 0, 0, 0 \rangle + y\langle 0, 0, 1, 0, 1 \rangle,$$

a linear combination of $\langle 0, 1, 0, 0, 0 \rangle$ and $\langle 0, 0, 1, 0, 1 \rangle$. ∎

Given a subspace of R^3, it is sometimes important to be able to find a basis. Usually, the description of the vectors in S provides a clue for doing this, as we have just seen. Here is an additional example.

EXAMPLE 11.20

Let S consist of all 6-vectors of the form $\langle x, y, 0, x - y, x + y, z \rangle$. Then S forms a subspace of R^6. Find a basis for S.

Proceed by examining a typical vector in S and attempting to write it as a linear combination of specific, linearly independent vectors in S. Write

$$\langle x, y, 0, x - y, x + y, z \rangle = x\langle 1, 0, 0, 1, 1, 0 \rangle$$
$$+ y\langle 0, 1, 0, -1, 1, 0 \rangle + z\langle 0, 0, 0, 0, 0, 1 \rangle.$$

By Lemma 11.1, $\langle 1, 0, 0, 1, 1, 0 \rangle$, $\langle 0, 1, 0, -1, 1, 0 \rangle$, and $\langle 0, 0, 0, 0, 0, 1 \rangle$ are linearly independent. Since every vector in S is a linear combination of these three vectors, they form a basis for S. ∎

In general, a subspace of R^n may have many different bases. For example, in the basis for S in Example 11.20, we could replace $\langle 0, 0, 0, 0, 0, 1 \rangle$ with $\langle 0, 0, 0, 0, 0, 2 \rangle$ and obtain another basis. It is a fact from linear algebra (which deals with vector spaces) that every basis for a given subspace of R^n has the same *number* of vectors in it. This number is called the *dimension* of the subspace. For example, the subspace S in Example 11.20 has dimension 3 because we found a basis with three vectors in it. The subspaces in Examples 11.18 and 11.19 have dimension 2. R^n has dimension n, a basis being $e_1, \ldots, e_n$.

We can think of a basis of S as a minimal set of information needed to specify S. The basis determines the subspace because every vector in the subspace is a linear combination of the vectors in the basis. But no fewer vectors can specify the subspace.

If we leave out one of the basis vectors, that vector is lost from the subspace because no vector in the basis is a linear combination of the others (remember that basis vectors are linearly independent). We saw a similar idea when we formed a fundamental set of solutions of a linear homogeneous differential equation. The fundamental set formed a "smallest" set of functions needed to specify the general solution. We will expand on this idea when we treat abstract vector spaces in the next section.

PROBLEMS FOR SECTION 11.6

In each of Problems 1 through 10, determine whether the vectors are linearly dependent or independent in the appropriate R^n.

1. $3\mathbf{i} + 2\mathbf{j}$, $\mathbf{i} - \mathbf{j}$ in R^3

2. $2\mathbf{i}$, $3\mathbf{j}$, $5\mathbf{i} - 12\mathbf{j}$ in R^3

3. $8\mathbf{e}_1 + 2\mathbf{e}_3$, $\mathbf{e}_5 - \mathbf{e}_6$ in R^7

4. $\mathbf{e}_1$, $\mathbf{e}_2 + \mathbf{e}_3$, $-4\mathbf{e}_1 + 6\mathbf{e}_2 + 6\mathbf{e}_3$ in R^4

5. $\langle 1, 2, -3, 1 \rangle$, $\langle 4, 0, 0, 2 \rangle$ in R^4

6. $\mathbf{e}_2 - \mathbf{e}_3$, $\mathbf{e}_3 - \mathbf{e}_4$, $\mathbf{e}_4 - \mathbf{e}_1$ in R^6

7. $8\mathbf{e}_1 - 6\mathbf{e}_2$, $-4\mathbf{e}_1 + 3\mathbf{e}_2$, $\mathbf{e}_1 + \mathbf{e}_4$ in R^5

8. $\mathbf{e}_2 - \mathbf{e}_1$, $\mathbf{e}_3 - \mathbf{e}_2$, $\mathbf{e}_3 - \mathbf{e}_1$ in R^8

9. $\langle -2, 0, 0, 1, 1 \rangle$, $\langle 1, 0, 0, 0, 0 \rangle$, $\langle 0, 0, 0, 0, 2 \rangle$ in R^5

10. $\langle 3, 0, 0, 4 \rangle$, $\langle 2, 0, 0, 8 \rangle$ in R^4

11. Let $\mathbf{F}$, $\mathbf{G}$ and $\mathbf{H}$ be vectors in R^3. Prove that $\mathbf{F}$, $\mathbf{G}$, and $\mathbf{H}$ are linearly dependent if and only if $[\mathbf{F}, \mathbf{G}, \mathbf{H}] = 0$.

In each of Problems 12 through 16, use the result of Problem 11 to test the three given vectors for linear independence in R^3.

12. $3\mathbf{i} + 6\mathbf{j} - \mathbf{k}$, $8\mathbf{i} + 2\mathbf{j} - 4\mathbf{k}$, $\mathbf{i} - \mathbf{j} + \mathbf{k}$

13. $\mathbf{i} + 6\mathbf{j} - 2\mathbf{k}$, $-\mathbf{i} + 4\mathbf{j} - 3\mathbf{k}$, $\mathbf{i} + 16\mathbf{j} - 7\mathbf{k}$

14. $4\mathbf{i} - 3\mathbf{j} + \mathbf{k}$, $10\mathbf{i} - 3\mathbf{j}$, $2\mathbf{i} - 6\mathbf{j} + 3\mathbf{k}$

15. $8\mathbf{i} + 6\mathbf{j}$, $2\mathbf{i} - 4\mathbf{j}$, $\mathbf{i} + \mathbf{k}$

16. $12\mathbf{i} - 3\mathbf{k}$, $\mathbf{i} + 2\mathbf{j} - \mathbf{k}$, $-3\mathbf{i} + 4\mathbf{j}$

In each of Problems 17 through 24, determine a basis for the subspace S of R^n and determine the dimension of the subspace.

17. S consists of all vectors $\langle x, y, -y, -x \rangle$ in R^4.

18. S consists of all vectors $\langle x, y, 2x, 3y \rangle$ in R^4.

19. S consists of all vectors in the plane $2x - y + z = 0$ in 3-space.

20. S consists of all vectors $\langle x, y, -y, x - y, z \rangle$ in R^5.

21. S consists of all vectors in R^4 with second component zero.

22. S consists of all vectors $\langle -x, x, y, 2y \rangle$ in R^4.

23. S consists of all vectors parallel to the line $y = 4x$ in R^2.

24. S consists of all vectors in the plane $4x + 2y - z = 0$ in R^3.

25. In R^2, all the vectors parallel to a given line L through the origin form a subspace S. Determine the dimension of S for any such line L.

26. In R^3, all the vectors in a plane Π through the origin form a subspace S of R^3. Determine the dimension of S for any such plane Π.

27. Prove that any three vectors in R^2 must be linearly dependent.

28. Prove that any four vectors in R^3 must be linearly dependent.

29. Suppose that $\mathbf{F}_1, \ldots, \mathbf{F}_m$ are m linearly independent vectors in R^n. Prove that any k of these vectors are linearly independent where $1 \le k \le m$.

30. Suppose that $\mathbf{F}_1, \ldots, \mathbf{F}_m$ are linearly dependent in R^n. Prove that any finite collection of vectors in R^n containing these m vectors must be linearly dependent.

31. Let $\mathbf{F}$, $\mathbf{G}$, and $\mathbf{H}$ be linearly independent vectors in R^3, and let $\mathbf{V}$ be any vector in R^3. Prove that

$$\mathbf{V} = \frac{[\mathbf{V}, \mathbf{G}, \mathbf{H}]}{[\mathbf{F}, \mathbf{G}, \mathbf{H}]} \mathbf{F} + \frac{[\mathbf{V}, \mathbf{H}, \mathbf{F}]}{[\mathbf{F}, \mathbf{G}, \mathbf{H}]} \mathbf{G} + \frac{[\mathbf{V}, \mathbf{F}, \mathbf{G}]}{[\mathbf{F}, \mathbf{G}, \mathbf{H}]} \mathbf{H}.$$

32. Use the discussion following Lemma 11.1 to construct a proof of the lemma.

11.7 *Abstract Vector Spaces* _____

Suppose V is a collection of objects, which may be vectors in R^n, functions, or other objects. We will assume that there are two algebraic operations called addition and scalar multiplication and denoted as follows:

>   the sum of objects **a** and **b** in V is denoted **a** + **b**;
>   the product of a scalar α with an object **a** in V is denoted α**a**.

Finally, we will assume that the objects in V, together with these two operations, satisfy the following conditions. If **a**, **b**, and **c** are in V, and α and β are scalars, then

1. **a** + **b** is in V.
2. **a** + **b** = **b** + **a**.
3. (**a** + **b**) + **c** = **a** + (**b** + **c**).
4. There is some object θ in V such that **a** + θ = **a** for every **a** in V.
5. There is some **d** in V such that **a** + **d** = θ.
6. α**a** is in V.
7. $(\alpha + \beta)$**a** = α**a** + β**a**.
8. $(\alpha\beta)$**a** = $\alpha(\beta$**a**$)$.
9. α(**a** + **b**) = α**a** + α**b**.
10. 1**a** = **a**.

When these conditions are satisfied, we call V a *vector space* relative to the addition and scalar multiplication operations. If we change V or the addition or scalar multiplication operations satisfying (1) through (10), we obtain a different vector space.

We have already seen examples of vector spaces. R^n is a vector space, with the addition and scalar multiplication we defined, and is the prototype after which conditions (1) through (10) are modeled. Any subspace of R^n is also, in its own right, a vector space, with the same addition and scalar multiplication operations.

Sometimes V is called an *abstract vector space*. The adjective "abstract" actually. contributes nothing to the discussion or theory but is sometimes used to distinguish the "concrete" vector space R^n, which has n-tuples of real numbers as vectors, from the general concept in which V might contain different kinds of objects.

Objects in a vector space are called vectors even if they are different from the vectors in R^n. When we are dealing with a vector space of familiar objects such as functions, we will often discontinue the boldface notation and adopt familiar notation customarily used with those objects.

EXAMPLE 11.21

Let V consist of all real-valued functions which are defined and continuous on $[0, 1]$. We add two functions in the usual way. If f and g are in V, $f + g$ is the function defined by $(f + g)(x) = f(x) + g(x)$. Similarly, for any scalar α, αf is the function defined by $(\alpha f)(x) = \alpha f(x)$. The vector θ in (4) of the definition is the zero function, defined by $\theta(x) = 0$ for $0 \le x \le 1$.

Our usual experience with functions enables us to verify routinely that all of the conditions (1) through (10) are satisfied. ∎

EXAMPLE 11.22

Let V consist of all real-valued functions which are differentiable for all x and which satisfy $f'(0) = 0$. With the usual addition of functions and multiplication by scalars, V is a vector space. ∎

EXAMPLE 11.23

Let V consist of all solutions of the differential equation $y'' + 2xy' + x^2 y = 0$. With the usual addition of functions and multiplication by scalars, V is a vector space. For example, sums of solutions are solutions [condition (1)], and a scalar times a solution is again a solution [condition (6)]. The zero function is a solution of this differential equation and yields the vector θ required in condition (4). ∎

The vector θ whose existence is required by condition (4) is called the *zero vector* of V. Given $\mathbf{a}$ in V, there is some $\mathbf{b}$ in V such that $\mathbf{a} + \mathbf{b} = \theta$. We denote $\mathbf{b}$ as $-\mathbf{a}$ and call it the negative of $\mathbf{a}$. It can be shown (see Problem 21 at the end of this section) that a vector space can have only one zero vector and that a vector in V can have only one negative vector (Problem 22).

Because the definition of an abstract vector space mirrors the properties we have seen for R^n, we might expect many of the concepts we have discussed in R^n to carry over naturally into the vector space setting. In fact, they do.

Suppose that $\mathbf{v}_1, \ldots, \mathbf{v}_n$ are vectors in V. A *linear combination* of $\mathbf{v}_1, \ldots, \mathbf{v}_n$ is any sum of scalars times these vectors. Such a linear combination has the form

$$\alpha_1 \mathbf{v}_1 + \cdots + \alpha_n \mathbf{v}_n.$$

If one of $\mathbf{v}_1, \ldots, \mathbf{v}_n$ is a linear combination of the others, we say that $\mathbf{v}_1, \ldots, \mathbf{v}_n$ are *linearly dependent*. Otherwise, we say that these vectors are *linearly independent*. For example, $\sin^2(x)$, $\cos^2(x)$, and $\cos(2x)$ are linearly dependent in the vector space of continuous, real-valued functions defined for all x, because

$$\cos(2x) = \cos^2(x) - \sin^2(x).$$

In the vector space of all polynomials with real coefficients, the polynomials 1, $x, x^2, \ldots, x^n$ are linearly independent for any n. The reason is that we cannot write one of these functions as a linear combination of the others valid for all x. Alternatively, we cannot find constants $\alpha_0, \alpha_1, \ldots, \alpha_n$, not all zero, such that

$$\alpha_0 + \alpha_1 x + \cdots + \alpha_n x^n = 0$$

for all x, because an nth degree polynomial with a nonzero coefficient can have at most n roots. Vectors $\mathbf{v}_1, \ldots, \mathbf{v}_n$ form a *basis* for a vector space V if they are linearly independent and every vector in V is a linear combination of these vectors.

EXAMPLE 11.24

Let V consist of all solutions of the differential equation

$$y'' + 4y = 0,$$

with the usual addition of functions and multiplication of a function by a scalar. Then V is a vector space, because a sum of solutions is a solution and a scalar times a solution is also a solution.

We know that $\cos(2x)$ and $\sin(2x)$ form a fundamental set of solutions for this equation. In current terminology, $\cos(2x)$ and $\sin(2x)$ form a basis for the vector space of solutions of this differential equation. That is, these functions are linearly independent, and every solution is of the form $c_1\cos(2x) + c_2\sin(2x)$, a linear combination of these solutions.

This vector space is also referred to as the *solution space* of the differential equation. This differential equation has order 2, and the solution space has dimension 2. This fact is not a coincidence. ∎

It is not always possible to find a basis for a vector space consisting of a finite number of functions. For example, consider the vector space of real-valued, continuous functions on $[0, 1]$. It can be shown that this vector space has a basis. It does not, however, have a finite basis. It is not possible to list a finite number of continuous functions of which every continuous function is a linear combination. For example, $1, x, x^2, \ldots, x^n$ are continuous for n as large as we like, but if we end the list at some particular n, we omit x^{n+1}, which is not a linear combination of $1, x, \ldots, x^n$. A vector space without a finite basis is said to be *infinite dimensional*.

It is sometimes possible to imitate the notions of dot product and length in an abstract vector space. As an example, consider the vector space V of all real-valued functions defined and continuous on $[0, 1]$. We can define the dot product of two functions f and g in V by setting

$$f \cdot g = \int_0^1 f(x)g(x)\, dx.$$

It is appropriate to call this integral the dot product of f and g because this operation mirrors the properties we saw for the dot product of vectors in R^3. For example,

$$
\begin{aligned}
(f + g) \cdot h &= \int_0^1 [f(x) + g(x)]h(x)\, dx \\
&= \int_0^1 f(x)h(x)\, dx + \int_0^1 g(x)h(x)\, dx = f \cdot h + g \cdot h.
\end{aligned}
$$

In terms of dot product, the length, or norm, of a vector f in V can be defined by

$$\|f\| = (f \cdot f)^{1/2} = \left[\int_0^1 [f(x)]^2\, dx \right]^{1/2}.$$

This definition yields a concept of length for a vector space of functions and has the properties we usually associate with a length. For example, $\|f\|$ is always nonnegative, and the only function in V having zero length is the zero function, $f(x) = 0$ for $0 \le x \le 1$.

One rationale for extending the concepts of vector space, dot product, and length to this abstract setting is that facts established for vector spaces in general can be applied to a wide variety of objects. For example, consider the triangle inequality

$$\|\mathbf{F} + \mathbf{G}\| \le \|\mathbf{F}\| + \|\mathbf{G}\|,$$

which we proved for vectors in R^3 and which the student was asked to prove for vectors in R^n (Problem 24 of Section 11.5). A review of the proof of this inequality for R^3, given at the end of Section 11.2, shows that it does not depend on any particular properties of R^3 other than those reflected in the general definition of vector space. In fact, the triangle inequality is true for any vectors in an abstract vector space having a norm.

Now consider what this means for the vector space of real-valued continuous functions on $[0, 1]$. With the norm defined by $\|f\| = \{\int_0^1 [f(x)]^2 \, dx\}^{1/2}$, the triangle inequality becomes the integral inequality

$$\left\{ \int_0^1 [f(x) + g(x)]^2 \, dx \right\}^{1/2} \leq \left\{ \int_0^1 [f(x)]^2 \, dx \right\}^{1/2} + \left\{ \int_0^1 [g(x)]^2 \, dx \right\}^{1/2}.$$

This inequality is not entirely obvious, and it might not be clear how to establish it. The point is that we do not have to prove it—it has already been proved as the triangle inequality for vector spaces. This integral inequality is simply a particularization of the triangle inequality to a special case.

As another example, consider the Cauchy-Schwarz inequality, $|\mathbf{F} \cdot \mathbf{G}| \leq \|\mathbf{F}\| \|\mathbf{G}\|$. If we phrase this in terms of the vector space of real-valued continuous functions on $[0, 1]$, we have another integral inequality:

$$\left| \int_0^1 f(x) g(x) \, dx \right| \leq \left\{ \int_0^1 [f(x)]^2 \, dx \right\}^{1/2} \left\{ \int_0^1 [g(x)]^2 \, dx \right\}^{1/2}.$$

Again, we do not have to discover or prove this inequality as a new result. It is simply the Cauchy-Schwarz inequality in the particular setting of continuous functions on $[0, 1]$.

These examples are only a brief introduction to the concept of a vector space. The power of the concept is that we are able to consider many apparently unrelated ideas and results as special cases of very general ones. Further, the more general result is sometimes actually easier to establish or discover than the particular one. For example, the Cauchy-Schwarz inequality was easily proved for vectors in R^n. A direct approach to the last integral inequality, using facts about integrals but none of the power of the vector space concept, can be quite demanding.

We will make use of the vector space R^n again in the next chapter when we solve systems of linear equations, and we will use the notion of an abstract vector space when we deal with matrices in that chapter as well as when we solve systems of linear differential equations using matrix methods.

PROBLEMS FOR SECTION 11.7

1. Let P_m be the set of all polynomials with real coefficients and degree $\leq m$. Show that P_m is a vector space, with the usual addition of polynomials and multiplication of a polynomial by a scalar. Determine a basis for this vector space.
2. Let V consist of all real-valued functions defined and continuous on $[0, 1]$ such that $f(1) = 0$, with the usual addition of functions and multiplication of a function by a scalar. Show that V is a vector space. Does V have finite dimension?
3. Let V consist of all real-valued functions defined and continuous on $[0, 1]$ such that $f(1) = 1$, with the usual addition of functions and multiplication of a function by a scalar. Show that V is not a vector space.
4. Let P consist of all polynomials with real coefficients. Show that P is a vector space, with the usual addition of polynomials and multiplication of a polynomial by a scalar. Show that P does not have finite dimension.

5. Let n be a positive integer. Let D_n consist of all polynomials with real coefficients and of degree n, with the usual addition of polynomials and multiplication of a polynomial by a scalar. Show that D_n is not a vector space. Which conditions in the definition fail to hold?

6. Let S consist of all solutions of $y'' - 8xy = 0$, with the usual addition of functions and multiplication of a function by a scalar. Show that S is a vector space. *Hint:* You need not (in fact, *should not*) solve the differential equation to do this problem.

7. Let T consist of all solutions of $y'' - 8xy = x$, with the usual addition of functions and multiplication of a function by a scalar. Show that T is not a vector space.

8. Let K consist of all $\langle x, y \rangle$, with x and y real numbers. Define addition in K by

$$\langle x_1, y_1 \rangle + \langle x_2, y_2 \rangle = \langle 2x_1 + 2x_2, y_1 + y_2 \rangle,$$

and define multiplication by a scalar by

$$\alpha \langle x, y \rangle = \left\langle \frac{\alpha x}{2}, \frac{\alpha y}{2} \right\rangle.$$

Is K a vector space with these operations?

9. Let W consist of all $\langle x, y, z \rangle$, with x, y, and z real numbers. Define

$$\langle x_1, y_1, z_1 \rangle + \langle x_2, y_2, z_2 \rangle = \langle 3x_1 + 3y_1, 2y_1 + 2y_2, z_1 + z_2 \rangle,$$

and define the usual multiplication by scalars

$$\alpha \langle x, y, z \rangle = \langle \alpha x, \alpha y, \alpha z \rangle.$$

Is W a vector space with these operations?

10. Let V consist of all real-valued functions continuous on $[0, 1]$ such that $\int_0^1 f(x)\, dx = 0$. With the usual addition of functions and multiplication by scalars, is V a vector space?

11. Let M consist of all polynomials with real coefficients having degree ≤ 2 and having 1 as a root. With the usual addition of polynomials and multiplication by scalars, does V form a vector space?

12. Let D consist of all real-valued functions continuous on $[0, 1]$ such that $f(0) = f(\frac{1}{2}) = f(1) = 0$. With the usual addition of functions and multiplication of a function by a scalar, is D a vector space?

In each of Problems 13 through 20, list all the conditions that are violated in the definition of vector space.

13. Q is the set of rational numbers, with the usual addition of real numbers and multiplication of a real number by another real number.

14. S is the set of all solutions of $y'' - 4y = 8$, with the usual addition of functions and multiplication of a function by a scalar.

15. P is the set of polynomials having degree either 4 or 8, with the usual addition of polynomials and multiplication of a polynomial by a scalar.

16. V consists of all $\langle x, y \rangle$, with x and y real numbers and addition and multiplication by scalars defined by

$$\langle x_1, y_1 \rangle + \langle x_2, y_2 \rangle = \langle x_1 + x_2, y_1 - y_2 \rangle$$

and

$$\alpha \langle x, y \rangle = \langle \alpha x, y \rangle.$$

17. P consists of all polynomials with real coefficients. Addition is the usual addition of polynomials, but α times a polynomial is defined to be 1.

18. V consists of all real-valued functions continuous on $[0, 1]$, with $f(\frac{1}{3}) = 1$ and the usual addition of functions and multiplication of a function by a scalar.

19. V consists of all vectors in R^2 parallel to the line $x + y = 1$, with the usual addition of 2-vectors and the usual multiplication by scalars.

20. V consists of all vectors in R^3 parallel to the plane $x + 2y - z = 4$, with the usual addition of 3-vectors and the usual multiplication by scalars.

21. Let V be a vector space. Prove that V can have only one zero element. *Hint:* The zero element of V is a vector $\boldsymbol{\theta}$ having the property that $\mathbf{x} + \boldsymbol{\theta} = \mathbf{x}$ for all $\mathbf{x}$ in V. Suppose that two vectors $\boldsymbol{\theta}$ and $\boldsymbol{\theta}'$ in V satisfy this condition. Consider $\boldsymbol{\theta} + \boldsymbol{\theta}'$.

22. Let V be a vector space, and let $\mathbf{a}$ be in V. Prove that there is only one $\mathbf{b}$ in V such that $\mathbf{a} + \mathbf{b} = \boldsymbol{\theta}$. *Hint:* Suppose that $\mathbf{a} + \mathbf{b} = \boldsymbol{\theta}$ and also $\mathbf{a} + \mathbf{b}' = \boldsymbol{\theta}$. Show that $\mathbf{b} = \mathbf{b}'$ by adding $\mathbf{b}'$ to both sides of the equation $\mathbf{a} + \mathbf{b} = \boldsymbol{\theta}$ and applying condition (3) on page 615.

ADDITIONAL PROBLEMS

In each of Problems 1 through 15, compute $\mathbf{F} \cdot \mathbf{G}$, $\mathbf{F} \times \mathbf{G}$, and the cosine of the angle θ between $\mathbf{F}$ and $\mathbf{G}$. $\mathbf{F}$ is the first vector listed; $\mathbf{G}$ is the second. If you have a calculator, determine θ in radians and degrees to two decimal places.

1. $2\mathbf{i} - 3\mathbf{j} + \mathbf{k}$, $\mathbf{i} + 6\mathbf{j} - 4\mathbf{k}$ **2.** $3\mathbf{i} + \mathbf{j}$, $\mathbf{i} - 2\mathbf{j} - \mathbf{k}$ **3.** $-2\mathbf{i} + \mathbf{j} - 6\mathbf{k}$, $4\mathbf{i} + 2\mathbf{j} + 3\mathbf{k}$

4. $7\mathbf{i} - 3\mathbf{k}$, $-\mathbf{i} + 4\mathbf{j}$ **5.** $8\mathbf{i} + 2\mathbf{j} + \mathbf{k}$, $\mathbf{i} - 2\mathbf{j} + 5\mathbf{k}$ **6.** $7\mathbf{i} + \mathbf{j} - 6\mathbf{k}$, $2\mathbf{i} - 3\mathbf{j} - 4\mathbf{k}$

7. $-8\mathbf{i} - 3\mathbf{j} + \mathbf{k}$, $\mathbf{j} - 4\mathbf{k}$ **8.** $2\mathbf{j} - 3\mathbf{k}$, $\mathbf{j} - 4\mathbf{k}$ **9.** $8\mathbf{i} - 4\mathbf{j} + 3\mathbf{k}$, $-2\mathbf{i} + 6\mathbf{j} - 2\mathbf{k}$

10. $-4\mathbf{i} + 5\mathbf{j} - \mathbf{k}$, $2\mathbf{i} + 4\mathbf{j} + 2\mathbf{k}$ **11.** $3\mathbf{i} - 4\mathbf{j}$, $\mathbf{i} - \mathbf{j} - 5\mathbf{k}$ **12.** $-3\mathbf{i} - 5\mathbf{j} + \mathbf{k}$, $8\mathbf{i} - 3\mathbf{j} + 2\mathbf{k}$

13. $5\mathbf{i} + 6\mathbf{j} - 4\mathbf{k}$, $2\mathbf{i} + 6\mathbf{j} + 3\mathbf{k}$ **14.** $\mathbf{j} - 5\mathbf{k}$, $2\mathbf{i} + 2\mathbf{j} - 3\mathbf{k}$ **15.** $14\mathbf{i} + 8\mathbf{j} + 7\mathbf{k}$, $-3\mathbf{i} + 7\mathbf{k}$

In each of Problems 16 through 21, use vectors to find the parametric equations of the straight line containing the two given points.

16. $(1, -2, 4)$, $(6, 1, 1)$ **17.** $(0, 2, 3)$, $(-2, 4, 1)$ **18.** $(6\ 5, 2)$, $(3, -5, 1)$

19. $(-2, 1, -5)$, $(6, 7, 2)$ **20.** $(0, 1, -4)$, $(8, -3, -5)$ **21.** $(2, 14, 1)$, $(7, 0, 0)$

In each of Problems 22 through 27, use vectors to find the equation of a plane containing the given point and orthogonal (normal) to the given vector.

22. $(2, 1, -4)$, $3\mathbf{i} - 2\mathbf{j} + \mathbf{k}$ **23.** $(0, 0, 0)$, $4\mathbf{i} - 5\mathbf{j} - \mathbf{k}$ **24.** $(1, 1, -3)$, $-6\mathbf{i} + \mathbf{j} - 2\mathbf{k}$

25. $(4, 4, 7)$, $-4\mathbf{i} + 2\mathbf{j} + 3\mathbf{k}$ **26.** $(1, 5, -2)$, $5\mathbf{i} + 3\mathbf{j} - \mathbf{k}$ **27.** $(-3, -7, 0)$, $\mathbf{i} + 2\mathbf{k}$

In each of Problems 28 through 33, use vectors to find the equation of the plane containing the three given points, or else show that the points are collinear.

28. $(2, 2, -6)$, $(0, 0, 0)$, $(1, 5, 3)$ **29.** $(-3, 2, 6)$, $(1, 1, -5)$, $(-4, 5, 0)$

30. $(-3, 8, 1)$, $(1, 1, -7)$, $(0, 2, 0)$ **31.** $(-5, 3, 8)$, $(1, -4, -4)$, $(6, 6, -2)$

32. $(3, 1, -7)$, $(2, 2, 4)$, $(0, 0, -2)$ **33.** $(6, 1, -3)$, $(2, 2, -6)$, $(4, 8, 2)$

In each of Problems 34 through 39, calculate $[\mathbf{F}, \mathbf{G}, \mathbf{H}]$, with $\mathbf{F}$, $\mathbf{G}$, and $\mathbf{H}$, respectively, the vectors listed.

34. $3\mathbf{i} - \mathbf{j} + 4\mathbf{k}$, $\mathbf{i} - 2\mathbf{k}$, $\mathbf{i} + \mathbf{j}$ **35.** $8\mathbf{i} + 3\mathbf{j} - \mathbf{k}$, $\mathbf{i} - 2\mathbf{j} - 5\mathbf{k}$, $3\mathbf{i} + 4\mathbf{j} - \mathbf{k}$

36. $3\mathbf{i} + \mathbf{j} - \mathbf{k}$, $2\mathbf{i} + 4\mathbf{j} - 9\mathbf{k}$, $\mathbf{j} - 4\mathbf{k}$ **37.** $\mathbf{i} + 4\mathbf{j} + 3\mathbf{k}$, $-3\mathbf{j} + 6\mathbf{k}$, $\mathbf{i} - 2\mathbf{j} + 8\mathbf{k}$

38. $\mathbf{i} - 3\mathbf{j}$, $3\mathbf{i} + \mathbf{j} + 2\mathbf{k}$, $\mathbf{i} - 2\mathbf{j} + 5\mathbf{k}$ **39.** $3\mathbf{i} - \mathbf{j} + 3\mathbf{k}$, $2\mathbf{i} + \mathbf{j} + \mathbf{k}$, $\mathbf{i} - 2\mathbf{j} + 5\mathbf{k}$

In each of Problems 40 through 44, find the volume of the rectangular parallelepiped having incident sides from the first point to each of the other three.

40. $(-3, 2, 3)$, $(1, 1, 0)$, $(0, -1, 0)$, $(4, 3, -7)$ **41.** $(-2, 4, 4)$, $(7, 2, -3)$, $(5, 5, 8)$, $(-2, 4, 1)$

42. $(6, -1, 4)$, $(0, -3, 0)$, $(-5, 7, 2)$, $(1, 1, -7)$ **43.** $(4, 4, -2)$, $(0, 0, 0)$, $(4, -2, 8)$, $(5, 7, 1)$

44. $(-2, 2, 2)$, $(0, 2, -1)$, $(5, 5, 7)$, $(-3, 5, -3)$

45. Show by an example that the cancellation law fails for cross products. That is, give an example in which $\mathbf{F} \times \mathbf{G} = \mathbf{F} \times \mathbf{H}$ but $\mathbf{G} \neq \mathbf{H}$. Do not use $\mathbf{O}$ in the example.

46. Suppose $\mathbf{F} \times \mathbf{G} = \mathbf{F} \times \mathbf{H}$ for *every* vector $\mathbf{F}$. Show that $\mathbf{G} = \mathbf{H}$.

47. Let P be a parallelogram with adjacent sides $\mathbf{F}$ and $\mathbf{G}$. Prove that the area of P is the square root of

$$\begin{vmatrix} \|\mathbf{F}\|^2 & \mathbf{F} \cdot \mathbf{G} \\ \mathbf{F} \cdot \mathbf{G} & \|\mathbf{G}\|^2 \end{vmatrix}.$$

48. Let $P_1, \ldots, P_n$ be points in three-space. Form n vectors by drawing an arrow from P_1 to P_2 to represent $\mathbf{F}_1$, an arrow from P_2 to P_3 to represent $\mathbf{F}_2, \ldots$, an arrow from P_{n-1} to P_n to represent $\mathbf{F}_{n-1}$, and an arrow from P_n to P_1 to represent $\mathbf{F}_n$ (see Figure 11.33). Show that

$$\mathbf{F}_1 + \mathbf{F}_2 + \cdots + \mathbf{F}_n = \mathbf{O}.$$

49. Show that

$$(\mathbf{F} \times \mathbf{G}) \cdot (\mathbf{H} \times \mathbf{K}) = \begin{vmatrix} \mathbf{F} \cdot \mathbf{H} & \mathbf{F} \cdot \mathbf{K} \\ \mathbf{G} \cdot \mathbf{H} & \mathbf{G} \cdot \mathbf{K} \end{vmatrix}.$$

50. Let the sides of a triangle be represented by vectors $\mathbf{F}$, $\mathbf{G}$, and $\mathbf{H}$, as in Figure 11.34. Show that

$$\mathbf{F} \times (\mathbf{G} \times \mathbf{H}) = \mathbf{G} \times (\mathbf{H} \times \mathbf{F}) = \mathbf{H} \times (\mathbf{F} \times \mathbf{G}) = \mathbf{O}.$$

Hint: Use the result of Problem 48.

51. Use vectors to derive a formula for the angle between the diagonal of a cube and an edge of the cube.

52. Use vectors to show that the distance in the xy-plane between a point (x_0, y_0) and a line $ax + by + c = 0$ (a and b not both zero) is $|ax_0 + by_0 + c|/\sqrt{a^2 + b^2}$.

53. Use vectors to show that the distance in R^3 between a point (x_0, y_0, z_0) and a plane $ax + by + cz + d = 0$ is $|ax_0 + by_0 + cz_0 + d|/\sqrt{a^2 + b^2 + c^2}$, assuming that a, b, and c are not all zero.

54. Consider an n-sided polygon drawn in the plane as shown in Figure 11.35. The polygon is assumed to be convex, which means that a line drawn between any two points inside the polygon lies entirely within the polygon. By contrast, Figure 11.36 shows a polygon which is not convex. Show that the area of the polygon is

$$\frac{1}{2} \left\{ \begin{vmatrix} x_1 & y_1 \\ x_2 & y_2 \end{vmatrix} + \begin{vmatrix} x_2 & y_2 \\ x_3 & y_3 \end{vmatrix} + \cdots + \begin{vmatrix} x_{n-1} & y_{n-1} \\ x_n & y_n \end{vmatrix} + \begin{vmatrix} x_n & y_n \\ x_1 & y_1 \end{vmatrix} \right\}.$$

What role did convexity play in the argument?

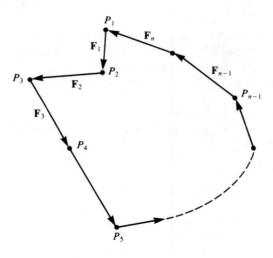

Figure 11.33

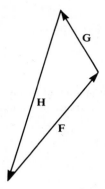

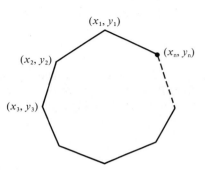

Figure 11.34

Figure 11.35. Convex *n*-sided polygon in the plane.

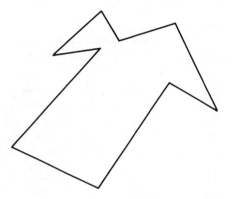

Figure 11.36. A nonconvex plane polygon.

55. Use vectors to prove that the sum of the squares of the lengths of the diagonals of any quadrilateral in R^2 is twice the sum of the squares of the lengths of the lines joining the midpoints of opposite sides.

56. Prove that $\mathbf{i} - \mathbf{j} + \mathbf{k}$, $2\mathbf{i} + 3\mathbf{j}$, and $4\mathbf{i} - 6\mathbf{k}$ form a basis for R^3.

57. Prove that e^{2x} and e^{-2x} form a basis for the solution space of the differential equation $y'' - 4y = 0$. What is the dimension of this solution space?

58. Show that the set of all vectors $\mathbf{F}$ in R^3 satisfying $\|\mathbf{F}\| = 1$ does not form a subspace of R^3.

59. Give an example of a set S of vectors in R^3 such that (a) $\mathbf{F} + \mathbf{G}$ is in S whenever $\mathbf{F}$ and $\mathbf{G}$ are in S, but (b) for some $\mathbf{F}$ in S and some scalar α, $\alpha\mathbf{F}$ is not in S.

60. Give an example of a set S of vectors in R^3 such that (a) if $\mathbf{F}$ is in S and α is any scalar, $\alpha\mathbf{F}$ is in S, but (b) for some $\mathbf{F}$ and $\mathbf{G}$ in S, $\mathbf{F} + \mathbf{G}$ is not in S.

61. Let S be a subspace of R^3, and suppose that S has dimension 3. Show that S must be all of R^3. *Hint:* Show that any vector in R^3 must be in S.

62. Let S be the subspace of R^3 consisting of all vectors parallel to a given plane through the origin. Show that S has dimension 2.

63. Show that e^x, e^{2x}, $\sin(x)$, and $\cos(x)$ are linearly independent in the vector space of all real-valued functions defined and continuous for all x.

Matrices, Determinants, and Systems of Linear Equations

12.0 Introduction _____

We will now consider the concept of a matrix. Matrices fit into the context of vector spaces, as we will see later. However, to begin, we will look at matrices and algebraic manipulations involving them without reference to the structure of a vector space. Matrices will provide a theoretically and practically useful way of approaching many types of problems, including the solution of systems of linear algebraic equations and systems of linear differential equations. Consider, for example, a system of algebraic equations

$$x + 2y - z + 4w = 0$$
$$3x - 4y + 2z - 6w = 0$$
$$x - 3y - 2z + w = 0.$$

We can solve this system by adding and subtracting appropriate multiples of selected equations to eliminate variables. Observe, however, that such manipulations really involve the coefficients; the names of the variables, or unknowns, are unimportant. If we called the unknowns by different names but kept the coefficients the same, we could carry out the same manipulations to solve the system and arrive at the same solutions.

In fact, we can carry out the manipulations without calling the unknowns anything. Consider the rectangular array

$$\begin{bmatrix} 1 & 2 & -1 & 4 \\ 3 & -4 & 2 & -6 \\ 1 & -3 & -2 & 1 \end{bmatrix}.$$

This array stores the coefficients as they appear in the system. Each row lists the coefficients of one equation, and each column lists the coefficients of a particular unknown. Thus, -4 (row 2, column 2) is the coefficient of the second unknown (y) in

623

the second equation; -6 (row 2, column 4) is the coefficient of the fourth unknown (w) in the second equation, and so on. By working with just this array of coefficients, we can carry out a method for solving the system (to be described later).

Such a rectangular array is called a *matrix*. Matrices may have numbers in them (as in this example), or functions, or other objects. Their important feature is that they provide a grid in which we can store objects in named locations (such as row 1, column 3) and from which these objects can be recalled. We will also develop an algebra enabling us to manipulate matrices in ways helpful in solving a variety of problems.

In this and the next chapter, we will concentrate on the algebraic aspects of matrices which render them useful and interesting. In Chapter 14, we will apply matrices to the solution of systems of ordinary differential equations.

12.1 *Matrices*

In this section, we will begin to develop the notation and algebra of matrices. A *matrix* is a rectangular array of objects written in rows and columns. Generally, these objects will be real or complex numbers or functions. For example,

$$\begin{bmatrix} 2 & 1 & 3 \\ 1 & 4 & 2 \end{bmatrix}, \quad \begin{bmatrix} e^{2x} & e^{3x} \\ 1 & x^2 \end{bmatrix}, \quad \text{and} \quad \begin{bmatrix} 1+i & 2 \\ 0 & -i \\ 0 & -2 \end{bmatrix}$$

are matrices.

If a matrix A has n rows and m columns, we say that A is "n by m," written $n \times m$. A is *square* if $n = m$. The matrix on the left above is 2×3 (two rows, three columns); the matrix in the middle is 2×2 and is square; and the matrix on the right is 3×2.

The entry in row i and column j of A is called the i, j *entry*, or i, j *element*, of A. It is common to see a capital letter used for the matrix and a lowercase letter for the matrix elements. Thus, the i, j element of A might be denoted a_{ij}. We also use the notation $A = [a_{ij}]$ to indicate that A is a matrix whose i, j element is a_{ij}. For example, if $H = [h_{ij}]$ is the matrix

$$\begin{bmatrix} 1+i & 2 \\ 0 & -i \\ 0 & -2 \end{bmatrix},$$

H is 3×2, and we have $h_{11} = 1 + i$, $h_{12} = 2$, $h_{21} = h_{31} = 0$, $h_{22} = -i$, and $h_{32} = -2$.

Matrices $A = [a_{ij}]$ and $B = [b_{ij}]$ are *equal* if they have the same number of rows and the same number of columns and if, for each i and j, $a_{ij} = b_{ij}$. That is, corresponding elements of both matrices are equal. If two elements in the same location are different, or if A and B have different numbers of rows or different numbers of columns, A and B are not equal.

Given two $n \times m$ matrices $A = [a_{ij}]$ and $B = [b_{ij}]$, the *sum* $A + B$ is the $n \times m$ matrix defined by

$$A + B = [a_{ij} + b_{ij}].$$

Thus, we add matrices of the same dimensions by adding corresponding elements. For example,

$$\begin{bmatrix} 1 & 2 & -3 \\ 4 & 0 & 2 \end{bmatrix} + \begin{bmatrix} -1 & 6 & 3 \\ 8 & 12 & 14 \end{bmatrix} = \begin{bmatrix} 0 & 8 & 0 \\ 12 & 12 & 16 \end{bmatrix}.$$

If A and B are $n \times m$, $A + B$ is also $n \times m$. If two matrices differ in number of rows or number of columns, they cannot be added.

We multiply a matrix by a scalar by multiplying each element of the matrix by the scalar. If $A = [a_{ij}]$ and α is any number,

$$\alpha A = [\alpha a_{ij}].$$

This is again an $n \times m$ matrix. For example,

$$3 \begin{bmatrix} 2 & 0 \\ 0 & 0 \\ 1 & 4 \\ 2 & 6 \end{bmatrix} = \begin{bmatrix} 6 & 0 \\ 0 & 0 \\ 3 & 12 \\ 6 & 18 \end{bmatrix}.$$

Some, but not all, pairs of matrices can be multiplied. If $A = [a_{ij}]$ is an $n \times r$ matrix and $B = [b_{ij}]$ is $r \times m$, the *matrix product AB* is the $n \times m$ matrix whose i, j element is

$$\begin{bmatrix} a_{i1}b_{1j} + a_{i2}b_{2j} + \cdots + a_{ir}b_{rj} \end{bmatrix}$$

for $i = 1, \ldots, n$ and $j = 1, \ldots, m$.

If we think of row i of A as the r-vector $\langle a_{i1}, a_{i2}, \ldots, a_{ir} \rangle$ and column j of B as the r-vector $\langle b_{1j}, b_{2j}, \ldots, b_{rj} \rangle$, we can recognize the i, j element of AB as the dot product of row i of A with column j of B:

$$i, j \text{ element of } AB = \langle \text{row } i \text{ of } A \rangle \cdot \langle \text{column } j \text{ of } B \rangle.$$

This product is defined only if the number of columns of A equals the number of rows of B so that row i of A and column j of B are vectors with the same number of components. When AB is defined, AB has the same number of rows as A and the same number of columns as B. At the end of this section and in the next section, we will examine how this definition of matrix product is naturally suited to certain types of calculations.

When we multiply matrices, we usually write the matrices side by side (in the correct order) and systematically take dot products of row and column vectors.

EXAMPLE 12.1

Let $A = \begin{bmatrix} 1 & 3 \\ 2 & 5 \end{bmatrix}$ and $B = \begin{bmatrix} 1 & 1 & 3 \\ 2 & 1 & 4 \end{bmatrix}$. Since A is 2×2 and B is 2×3, the matrix product AB is defined and is a 2×3 matrix. Perform the multiplication:

$$AB = \begin{bmatrix} 1 & 3 \\ 2 & 5 \end{bmatrix} \begin{bmatrix} 1 & 1 & 3 \\ 2 & 1 & 4 \end{bmatrix}$$

$$= \begin{bmatrix} \langle 1, 3 \rangle \cdot \langle 1, 2 \rangle & \langle 1, 3 \rangle \cdot \langle 1, 1 \rangle & \langle 1, 3 \rangle \cdot \langle 3, 4 \rangle \\ \langle 2, 5 \rangle \cdot \langle 1, 2 \rangle & \langle 2, 5 \rangle \cdot \langle 1, 1 \rangle & \langle 2, 5 \rangle \cdot \langle 3, 4 \rangle \end{bmatrix}$$

$$= \begin{bmatrix} 7 & 4 & 15 \\ 12 & 7 & 26 \end{bmatrix}.$$

In this example, BA is not defined. B is 2×3 and A is 2×2, and $3 \neq 2$. That is, the number of columns of B does not equal the number of rows of A. ∎

EXAMPLE 12.2

$$
\begin{bmatrix} 1 & 1 & 2 & 1 \\ 4 & 1 & 6 & 2 \end{bmatrix}
\begin{bmatrix} -1 & 8 \\ 2 & 1 \\ 1 & 1 \\ 12 & 6 \end{bmatrix}
=
\begin{bmatrix}
\langle 1, 1, 2, 1 \rangle \cdot \langle -1, 2, 1, 12 \rangle & \langle 1, 1, 2, 1 \rangle \cdot \langle 8, 1, 1, 6 \rangle \\
\langle 4, 1, 6, 2 \rangle \cdot \langle -1, 2, 1, 12 \rangle & \langle 4, 1, 6, 2 \rangle \cdot \langle 8, 1, 1, 6 \rangle
\end{bmatrix}
$$

$$
=
\begin{bmatrix} 15 & 17 \\ 28 & 51 \end{bmatrix}.
$$

Since A is 2×4 and B is 4×2, AB is defined and is 2×2. In this example, BA is also defined and is a 4×4 matrix:

$$
BA = \begin{bmatrix} -1 & 8 \\ 2 & 1 \\ 1 & 1 \\ 12 & 6 \end{bmatrix}
\begin{bmatrix} 1 & 1 & 2 & 1 \\ 4 & 1 & 6 & 2 \end{bmatrix}
$$

$$
=
\begin{bmatrix}
\langle -1, 8 \rangle \cdot \langle 1, 4 \rangle & \langle -1, 8 \rangle \cdot \langle 1, 1 \rangle & \langle -1, 8 \rangle \cdot \langle 2, 6 \rangle & \langle -1, 8 \rangle \cdot \langle 1, 2 \rangle \\
\langle 2, 1 \rangle \cdot \langle 1, 4 \rangle & \langle 2, 1 \rangle \cdot \langle 1, 1 \rangle & \langle 2, 1 \rangle \cdot \langle 2, 6 \rangle & \langle 2, 1 \rangle \cdot \langle 1, 2 \rangle \\
\langle 1, 1 \rangle \cdot \langle 1, 4 \rangle & \langle 1, 1 \rangle \cdot \langle 1, 1 \rangle & \langle 1, 1 \rangle \cdot \langle 2, 6 \rangle & \langle 1, 1 \rangle \cdot \langle 1, 2 \rangle \\
\langle 12, 6 \rangle \cdot \langle 1, 4 \rangle & \langle 12, 6 \rangle \cdot \langle 1, 1 \rangle & \langle 12, 6 \rangle \cdot \langle 2, 6 \rangle & \langle 12, 6 \rangle \cdot \langle 1, 2 \rangle
\end{bmatrix}
$$

$$
=
\begin{bmatrix}
31 & 7 & 46 & 15 \\
6 & 3 & 10 & 4 \\
5 & 2 & 8 & 3 \\
36 & 18 & 60 & 24
\end{bmatrix}. \quad ∎
$$

EXAMPLE 12.3

Let

$$
A = \begin{bmatrix} 1 & -1 & 0 & 1 & -2 & -4 \end{bmatrix} \quad \text{and} \quad B = \begin{bmatrix} 0 \\ 1 \\ 1 \\ 4 \\ 2 \\ 6 \end{bmatrix}.
$$

Then A is 1×6 and B is 6×1, so AB is defined, but BA is not. AB is a 1×1 matrix obtained by taking the dot product of the single row vector of A with the column vector of B:

$$
AB = [\langle 1, -1, 0, 1, -2, -4 \rangle \cdot \langle 0, 1, 1, 4, 2, 6 \rangle] = [-25]. \quad ∎
$$

EXAMPLE 12.4

Matrix multiplication can be used to simplify otherwise cumbersome expressions. For example, consider the system of n algebraic equations in m unknowns

$$
\begin{aligned}
a_{11}x_1 + a_{12}x_2 + \cdots + a_{1m}x_m &= b_1 \\
a_{21}x_1 + a_{22}x_2 + \cdots + a_{2m}x_m &= b_2 \\
\vdots \qquad \vdots \qquad\qquad \vdots \qquad \vdots \\
a_{n1}x_1 + a_{n2}x_2 + \cdots + a_{nm}x_m &= b_m.
\end{aligned}
\tag{12.1}
$$

The numbers $a_{11}, \ldots, a_{nm}$ are called *coefficients* of the system. We can write this system in compact matrix notation as follows. Let A be the $n \times m$ *matrix of coefficients* of the system

$$
A = \begin{bmatrix}
a_{11} & a_{12} & \cdots & a_{1m} \\
a_{21} & a_{22} & \cdots & a_{2m} \\
\vdots & \vdots & & \vdots \\
a_{n1} & a_{n2} & \cdots & a_{nm}
\end{bmatrix}.
$$

Let X be the $m \times 1$ matrix of the unknowns:

$$
X = \begin{bmatrix}
x_1 \\
x_2 \\
\vdots \\
x_m
\end{bmatrix}.
$$

Finally, let B be the $n \times 1$ matrix of constants appearing on the right side of the system (12.1):

$$
B = \begin{bmatrix}
b_1 \\
b_2 \\
\vdots \\
b_n
\end{bmatrix}.
$$

The matrix product AX is defined because A is $n \times m$ and X is $m \times 1$. Further, if we carry out the matrix product, we get

$$
\begin{aligned}
AX &= \begin{bmatrix}
a_{11} & a_{12} & \cdots & a_{1m} \\
a_{21} & a_{22} & \cdots & a_{2m} \\
\vdots & \vdots & & \vdots \\
a_{n1} & a_{n2} & \cdots & a_{nm}
\end{bmatrix}
\begin{bmatrix}
x_1 \\
x_2 \\
\vdots \\
x_m
\end{bmatrix} \\
&= \begin{bmatrix}
a_{11}x_1 + a_{12}x_2 + \cdots + a_{1m}x_m \\
a_{21}x_1 + a_{22}x_2 + \cdots + a_{2m}x_m \\
\vdots \qquad \vdots \qquad\qquad \vdots \\
a_{n1}x_1 + a_{n2}x_2 + \cdots + a_{nm}x_m
\end{bmatrix}
= \begin{bmatrix}
b_1 \\
b_2 \\
\vdots \\
b_n
\end{bmatrix}
\end{aligned}
$$

because the first row of AX equals b_1, the second row of AX equals b_2, and so on. Thus, the system of equations (12.1) becomes the matrix equation

$$
AX = B.
$$

This way of writing the system is not simply a labor-saving device. It suggests analogies with the equation $ax = b$ when a and b are numbers. More important, it enables us to bring the theory of matrices to bear on solving the system. One of the objectives of this chapter is to develop matrix methods for solving systems $AX = B$. ∎

EXAMPLE 12.5

In a manner entirely like that of Example 12.4, we can use matrices to write systems of linear differential equations in compact form. The system

$$
\begin{aligned}
a_{11}x_1' + a_{12}x_2' + \cdots + a_{1m}x_m' &= f_1(t) \\
a_{21}x_1' + a_{22}x_2' + \cdots + a_{2m}x_m' &= f_2(t) \\
&\ \ \vdots \\
a_{n1}x_1' + a_{n2}x_2' + \cdots + a_{nm}x_m' &= f_n(t)
\end{aligned}
$$

in which each a_{ij} is a function of t and $x_1, \ldots, x_n$ are functions of t, can be written as

$$AX' = F,$$

where

$$
A =
\begin{bmatrix}
a_{11}(t) & a_{12}(t) & \cdots & a_{1m}(t) \\
a_{21}(t) & a_{22}(t) & \cdots & a_{2m}(t) \\
\vdots & \vdots & & \vdots \\
a_{n1}(t) & a_{n2}(t) & \cdots & a_{nm}(t)
\end{bmatrix},
$$

$$
X =
\begin{bmatrix}
x_1 \\
x_2 \\
\vdots \\
x_m
\end{bmatrix},
\qquad
F =
\begin{bmatrix}
f_1(t) \\
f_2(t) \\
\vdots \\
f_n(t)
\end{bmatrix},
$$

and X' is obtained by differentiating each matrix element of X.

Again, the formulation $AX' = F$ is not simply easier to write. This equation suggests analogies with the first order equation $ax' = g$ and also enables us to develop very powerful matrix methods for solving certain systems of differential equations. We will pursue this theme in Chapter 14. ∎

If A is a square matrix, say, $n \times n$, then AA is defined and is also $n \times n$. We denote AA as A^2. Then $(AA)A$ is also $n \times n$ and is denoted A^3, and so on. In general, A^k denotes the product $AA \cdots A$, with k factors of A.

Some of the rules for manipulating matrices by addition and multiplication are similar to those for real numbers, except that not just any two matrices can be added or multiplied and, even when AB and BA are both defined, they may not be equal (remember Example 12.2).

THEOREM 12.1

1. $A + B = B + A$ if A and B are $n \times m$ matrices.
2. $A(B + C) = AB + AC$ if A is $n \times k$ and B and C are $k \times m$.
3. $(A + B)C = AC + BC$ if A and B are $n \times k$ and C is $k \times m$.
4. $A(BC) = (AB)C$ if A is $n \times k$, B is $k \times r$, and C is $r \times m$.

These conclusions look like the corresponding rules for real numbers, but, in each case, care must be taken that the indicated operations are defined. For example, in asserting that $A(BC) = (AB)C$, we must first check that BC and AB are defined and that we can multiply BC on the left by A and AB on the right by C. We will prove conclusions (1) and (2) to illustrate the type of reasoning involved.

Proof of (1)   Let $A = [a_{ij}]$ and $B = [b_{ij}]$. For $i = 1, \ldots, n$ and $j = 1, \ldots, m$, we have $a_{ij} + b_{ij} = b_{ij} + a_{ij}$ because we are assuming that matrix elements are objects which commute under addition (such as real numbers or real-valued functions). Thus,

$$A + B = [a_{ij} + b_{ij}] = [b_{ij} + a_{ij}] = B + A.$$

Proof of (2)   Let $A = [a_{ij}]$, $B = [b_{ij}]$, and $C = [c_{ij}]$. Since B and C are $k \times m$, $B + C$ is defined and is $k \times m$. Since A is $n \times k$, $A(B + C)$ is defined. Similarly, AB is $n \times m$ and AC is $n \times m$, so $AB + AC$ is defined.

The i, j element of $A(B + C)$ is the dot product of row i of A with column j of $B + C$:

i, j element of $A(B + C) = \langle a_{i1}, a_{i2}, \ldots, a_{ik} \rangle \cdot \langle b_{1j} + c_{1j}, b_{2j} + c_{2j}, \ldots, b_{kj} + c_{kj} \rangle.$

Using properties of the dot product, we can write the last line as

$\langle a_{i1}, a_{i2}, \ldots, a_{ik} \rangle \cdot \langle b_{1j}, b_{2j}, \ldots, b_{kj} \rangle + \langle a_{i1}, a_{i2}, \ldots, a_{ik} \rangle \cdot \langle c_{1j}, c_{2j}, \ldots, c_{kj} \rangle,$

which we recognize as

$(i, j \text{ element of } AB) + (i, j \text{ element of } AC).$

Therefore, $A(B + C) = AB + AC$. ∎

Conclusions (3) and (4) are proved by similar arguments.

With real numbers, we have $x + 0 = x$ for every x. With matrices, we have a zero matrix of every order. That is, if n and m are positive integers, the $n \times m$ zero matrix is the $n \times m$ matrix O_{nm} having each element equal to zero. For example,

$$O_{13} = \begin{bmatrix} 0 \\ 0 \\ 0 \end{bmatrix} \quad \text{and} \quad O_{22} = \begin{bmatrix} 0 & 0 \\ 0 & 0 \end{bmatrix}.$$

THEOREM 12.2

If A is an $n \times m$ matrix, $A + O_{nm} = O_{nm} + A = A$. ∎

A proof of this result is left to the student.

If A is an $n \times m$ matrix, $-A$ denotes the matrix obtained by replacing each element of A by its negative. For example,

$$-\begin{bmatrix} 2 & 1 & -6 \\ -1 & 4 & 2 \end{bmatrix} = \begin{bmatrix} -2 & -1 & 6 \\ 1 & -4 & -2 \end{bmatrix}.$$

It is obvious that $A + (-A) = O_{nm}$ for any $n \times m$ matrix A. As we did with vectors, we will usually write a sum $A + (-B)$ as just $A - B$.

Although we have seen some similarities between matrix operations and operations with real numbers, there are some significant differences as well. One obvious difference is that we can add and multiply any two numbers, but we cannot add or multiply any two matrices. There are more striking differences, of which we will discuss three.

Difference 1 Even if AB and BA are both defined, in general $AB \neq BA$. That is, matrix multiplication is not commutative.

EXAMPLE 12.6

Let

$$A = \begin{bmatrix} 1 & 0 \\ -2 & 4 \end{bmatrix} \quad \text{and} \quad B = \begin{bmatrix} -2 & 6 \\ 1 & 3 \end{bmatrix}.$$

Then

$$AB = \begin{bmatrix} -2 & 6 \\ 8 & 0 \end{bmatrix} \quad \text{and} \quad BA = \begin{bmatrix} -14 & 24 \\ -5 & 12 \end{bmatrix}. \quad \blacksquare$$

In fact, if we choose two $n \times n$ matrices A and B at random, it is likely that AB will not equal BA. Commutativity of a matrix product is the exception rather than the rule.

Difference 2 We may have $AB = AC$, with $B \neq C$ and A not a zero matrix. That is, there is in general no "cancellation" of A in an equation $AB = AC$.

EXAMPLE 12.7

$$\begin{bmatrix} 1 & 1 \\ 3 & 3 \end{bmatrix}\begin{bmatrix} 4 & 2 \\ 3 & 16 \end{bmatrix} = \begin{bmatrix} 1 & 1 \\ 3 & 3 \end{bmatrix}\begin{bmatrix} 2 & 7 \\ 5 & 11 \end{bmatrix} = \begin{bmatrix} 7 & 18 \\ 21 & 54 \end{bmatrix},$$

even though

$$\begin{bmatrix} 4 & 2 \\ 3 & 16 \end{bmatrix} \neq \begin{bmatrix} 2 & 7 \\ 5 & 11 \end{bmatrix}.$$

As might be expected, it is also possible to have $BA = CA$, with $B \neq C$ and A not a zero matrix. $\blacksquare$

Difference 3 The product AB may be a zero matrix with neither A nor B a zero matrix.

EXAMPLE 12.8

$$\begin{bmatrix} 1 & 2 \\ 0 & 0 \end{bmatrix}\begin{bmatrix} 6 & 4 \\ -3 & -2 \end{bmatrix} = \begin{bmatrix} 0 & 0 \\ 0 & 0 \end{bmatrix}. \quad \blacksquare$$

Differences 1, 2, and 3 have no analogues for ordinary real number multiplication and hence must be carefully considered in the course of doing calculations with matrices.

   We will conclude this section by studying an example that demonstrates why the definition of matrix product was chosen to have the form that it does. Suppose we want to change variables in a system of equations. As a simplification, we will consider a system of just two equations in two unknowns:

$$a_{11}x + a_{12}y = c_1$$
$$a_{21}x + a_{22}y = c_2. \tag{12.2}$$

Change variables by putting

$$x = h_{11}u + h_{12}v$$
$$y = h_{21}u + h_{22}v. \tag{12.3}$$

Substitute the equations (12.3) into the system (12.2) to get

$$a_{11}(h_{11}u + h_{12}v) + a_{12}(h_{21}u + h_{22}v) = c_1$$
$$a_{21}(h_{11}u + h_{12}v) + a_{22}(h_{21}u + h_{22}v) = c_2.$$

Rearrange terms in this system to write it as

$$(a_{11}h_{11} + a_{12}h_{21})u + (a_{11}h_{12} + a_{12}h_{22})v = c_1$$
$$(a_{21}h_{11} + a_{22}h_{21})u + (a_{21}h_{12} + a_{22}h_{22})v = c_2. \tag{12.4}$$

Now carry out this calculation in matrix notation and observe what happens. First, write the system (12.2) as

$$AX = C, \tag{12.5}$$

where

$$A = \begin{bmatrix} a_{11} & a_{12} \\ a_{21} & a_{22} \end{bmatrix}, \quad C = \begin{bmatrix} c_1 \\ c_2 \end{bmatrix}, \quad \text{and} \quad X = \begin{bmatrix} x \\ y \end{bmatrix}.$$

We can write the system (12.3) as

$$X = HU, \tag{12.6}$$

where

$$H = \begin{bmatrix} h_{11} & h_{12} \\ h_{21} & h_{22} \end{bmatrix} \quad \text{and} \quad U = \begin{bmatrix} u \\ v \end{bmatrix}.$$

Substitute HU for X in equation (12.5) and use conclusion (4) of Theorem 12.1 to write

$$A(HU) = (AH)U = C.$$

Notice that

$$AH = \begin{bmatrix} a_{11}h_{11} + a_{12}h_{21} & a_{11}h_{12} + a_{12}h_{22} \\ a_{21}h_{11} + a_{22}h_{21} & a_{21}h_{12} + a_{22}h_{22} \end{bmatrix},$$

so $(AH)U = C$ is the transformed system (12.4) after the change of variables (12.3). The definition of matrix product has given us exactly what we obtain by direct substitution of the new variables into the original system. This is no coincidence. Matrix products were defined with this result in mind because this is one type of calculation for which matrices are often used.

In the next section, we will discuss an application of matrix multiplication to a physical problem.

PROBLEMS FOR SECTION 12.1

In each of Problems 1 through 10, calculate the indicated matrix combination, with A the first matrix listed and B the second.

1. $\begin{bmatrix} 1 & -1 & 3 \\ 2 & -4 & 6 \\ -1 & 1 & 2 \end{bmatrix}$ and $\begin{bmatrix} -4 & 0 & 0 \\ -2 & -1 & 6 \\ 8 & 15 & 4 \end{bmatrix}$; $2A - 3B$

2. $\begin{bmatrix} -2 & 2 \\ 0 & 1 \\ 14 & 2 \\ 6 & 8 \end{bmatrix}$ and $\begin{bmatrix} 3 & 4 \\ 2 & 1 \\ 14 & 16 \\ 1 & 25 \end{bmatrix}$; $A + 5B$

3. $\begin{bmatrix} -22 & 1 & 6 & 4 & 5 \\ -3 & -2 & 14 & 2 & 25 \\ 18 & 1 & 16 & -4 & -6 \end{bmatrix}$ and $\begin{bmatrix} 0 & 1 & 3 & 1 & 14 \\ -8 & 6 & -10 & 4 & 10 \\ 21 & 6 & 17 & 3 & 2 \end{bmatrix}$; $-2A + 6B$

4. $[14]$ and $[-22]$; $-A - B$

5. $[-4 \quad 8 \quad 6]$ and $[22 \quad 7 \quad -3]$; $3A + 2B$

6. $\begin{bmatrix} -4 & 1 & 1 \\ 2 & 0 & 6 \end{bmatrix}$ and $\begin{bmatrix} -3 & -2 & 5 \\ 3 & 1 & 7 \end{bmatrix}$; $-A + 4B$

7. $\begin{bmatrix} 1 \\ 0 \\ -4 \end{bmatrix}$ and $\begin{bmatrix} -3 \\ 1 \\ 5 \end{bmatrix}$; $12A + 3B$

8. $\begin{bmatrix} 2 & 1 & 1 & 7 & 3 \\ 8 & 0 & 0 & 2 & 4 \end{bmatrix}$ and $\begin{bmatrix} -2 & 3 & 0 & 4 & 6 \\ -2 & 2 & 1 & 3 & 7 \end{bmatrix}$; $-4A - 6B$

9. $\begin{bmatrix} 2 & 1 \\ 0 & 0 \\ 2 & 5 \end{bmatrix}$ and $\begin{bmatrix} -2 & 4 \\ -5 & 1 \\ -6 & 3 \end{bmatrix}$; $A - B$

10. $\begin{bmatrix} 4 & -3 & 2 & 1 \\ 3 & 4 & 7 & 5 \\ 0 & 0 & 9 & 2 \end{bmatrix}$ and $\begin{bmatrix} 10 & 0 & 0 & 9 \\ -3 & 0 & 1 & 4 \\ 2 & 2 & -1 & -7 \end{bmatrix}$; $4A + B$

In each of Problems 11 through 30, compute the products AB and BA where possible. Specify any products that are not defined.

11. $A = \begin{bmatrix} -4 & 6 & 2 \\ -2 & -2 & 3 \\ 1 & 1 & 8 \end{bmatrix}$, $B = \begin{bmatrix} -2 & 4 & 6 & 12 & 5 \\ -3 & -3 & 1 & 1 & 4 \\ 0 & 0 & 1 & 6 & -9 \end{bmatrix}$

12. $A = \begin{bmatrix} -2 & -4 \\ 3 & -1 \end{bmatrix}$, $B = \begin{bmatrix} 6 & 8 \\ 1 & -4 \end{bmatrix}$

13. $A = [-1 \quad 6 \quad 2 \quad 14 \quad -22]$, $B = \begin{bmatrix} -3 \\ 2 \\ 6 \\ 0 \\ -4 \end{bmatrix}$

14. $A = \begin{bmatrix} -3 & 1 \\ 6 & 2 \\ 18 & -22 \\ 1 & 6 \end{bmatrix}$, $B = \begin{bmatrix} -16 & 0 & 0 & 28 \\ 0 & 1 & 1 & 26 \end{bmatrix}$

15. $A = \begin{bmatrix} -26 & 1 & 13 \end{bmatrix}$, $B = \begin{bmatrix} 22 \\ 1 \\ 6 \\ 14 \\ 2 \end{bmatrix}$

16. $A = \begin{bmatrix} -21 & 4 & 8 & -3 \\ 12 & 1 & 0 & 14 \\ 1 & 16 & 0 & -8 \\ 13 & 4 & 8 & 0 \end{bmatrix}$, $B = \begin{bmatrix} -9 & 16 & 3 & 2 \\ 5 & 9 & 14 & 0 \end{bmatrix}$

17. $A = \begin{bmatrix} -21 & 16 \end{bmatrix}$, $B = \begin{bmatrix} 32 & 4 & 16 \\ -8 & 7 & 0 \end{bmatrix}$ **18.** $A = \begin{bmatrix} 4 & 6 \\ -3 & 1 \end{bmatrix}$, $B = \begin{bmatrix} 1 & 16 & -7 & 9 \end{bmatrix}$

19. $A = \begin{bmatrix} 4 & 6 \\ -3 & 1 \end{bmatrix}$, $B = \begin{bmatrix} 1 & 16 & -7 & 9 \end{bmatrix}$ **20.** $A = \begin{bmatrix} 3 \\ -4 \\ 7 \\ 3 \\ 9 \end{bmatrix}$, $B = \begin{bmatrix} 2 & 7 & -5 & 6 & 10 \\ 4 & 6 & 0 & 0 & -5 \end{bmatrix}$

21. $A = \begin{bmatrix} -2 & 4 \\ 3 & 9 \end{bmatrix}$, $B = \begin{bmatrix} 1 & -3 & 7 & 2 \\ -5 & 6 & 1 & 0 \end{bmatrix}$ **22.** $A = \begin{bmatrix} -4 & -2 & 0 \\ 0 & 5 & 3 \\ -3 & 1 & 1 \end{bmatrix}$, $B = \begin{bmatrix} 1 & -3 & 4 \end{bmatrix}$

23. $A = \begin{bmatrix} 3 \\ 0 \\ -1 \\ 4 \end{bmatrix}$, $B = \begin{bmatrix} 3 & -2 & 7 \end{bmatrix}$ **24.** $A = \begin{bmatrix} 3 & -8 \\ 1 & 6 \end{bmatrix}$, $B = \begin{bmatrix} 1 & -4 & 3 \\ -4 & 7 & 0 \end{bmatrix}$

25. $A = \begin{bmatrix} -3 & 2 \\ 0 & -2 \\ 1 & 8 \\ 3 & -3 \end{bmatrix}$ and $B = \begin{bmatrix} -3 & 5 & 7 & 2 \end{bmatrix}$

26. $A = \begin{bmatrix} -4 & 2 & 8 \\ 1 & 6 & -4 \\ 2 & 2 & 0 \end{bmatrix}$, $B = \begin{bmatrix} 2 & -4 & 6 \\ 1 & 9 & -5 \\ 0 & -5 & 1 \end{bmatrix}$ **27.** $A = \begin{bmatrix} 1 \\ 0 \\ -3 \end{bmatrix}$, $B = \begin{bmatrix} 2 & -6 \end{bmatrix}$

28. $A = \begin{bmatrix} 2 & -3 & 5 \\ 0 & 0 & -4 \end{bmatrix}$, $B = \begin{bmatrix} 0 \\ 3 \\ -4 \end{bmatrix}$

29. $A = \begin{bmatrix} 1 & -2 \\ 2 & 4 \end{bmatrix}$, $B = \begin{bmatrix} -1 & 3 & 2 & 9 & -4 & 6 \\ 0 & -1 & 6 & 0 & 9 & -4 \end{bmatrix}$

30. $A = \begin{bmatrix} -2 & 1 \\ 2 & 0 \\ 0 & 9 \\ 6 & -5 \end{bmatrix}$, $B = \begin{bmatrix} 1 & 1 & -5 \\ 0 & 4 & 2 \end{bmatrix}$

In each of Problems 31 through 36, determine whether AB and BA are defined and determine how many rows and columns each product has if it is defined.

31. A is 14×21, B is 21×14
32. A is 18×4, B is 18×4
33. A is 6×2, B is 4×6
34. A is 1×3, B is 1×3
35. A is 7×6, B is 6×7
36. A is 8×6, B is 8×8
37. Find nonzero matrices A, B, and C such that $BA = CA$ but $B \neq C$.

12.2 *Matrix Multiplication and Random Walks in Crystals*

In this section, we will describe a problem of importance in physics and chemistry whose solution is obtained through matrix multiplication.

A *crystal* can be modeled as a lattice of sites occupied by atoms. In a defective crystal, one or more sites may be vacant at any time. An atom may jump from a given site to a neighboring, unoccupied one; after a time, another atom may in turn jump to the site just vacated, or the same atom may make a second jump to another empty site, and so on. The ensuing motion is called a *random walk* of the atoms, in the sense that an atom describes a path from one vacant site to another through the lattice but does so randomly. In each time interval, each atom may have various possible directions in which to move, and it may move in one of them or remain where it is.

A mathematical model for the crystal can be formed by drawing a lattice diagram in which points designate positions or sites and lines connect neighboring sites. An atom can move from one site to another only if the two are joined by a line. Such a diagram is called a *graph*, and the sites are often called *points*, or *vertices*, of the graph. Figure 12.1 shows a typical graph having six points labeled $v_1, \ldots, v_6$. In this graph, v_3 has neighbors v_2, v_1, and v_4, so an atom at v_3 can jump to any of these sites (if it is unoccupied at that time). On the other hand, v_2 and v_5 are not joined by a line, so an atom can never jump directly between v_2 and v_5 but would have to move from v_2 to v_5 through intermediate sites.

A *walk of length n* in a graph is a sequence $t_1, \ldots, t_{n+1}$ of sites (not necessarily all different) with t_{i-1} and t_i neighbors (joined by a line) for $i = 2, \ldots, n$. In Figure 12.1,

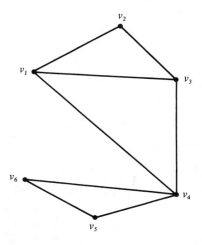

Figure 12.1. A typical graph.

$v_1, v_2, v_3, v_1, v_4, v_6$ is a walk of length 5 from v_1 to v_6 (denoted a v_1-v_6 walk). Similarly, $v_1, v_2, v_1, v_2, v_1, v_3, v_4, v_6, v_5, v_4, v_1, v_3, v_1$ is a v_1-v_1 walk of length 12. Walks represent possible paths of an atom through the crystal. We allow the possibility that $t_{n+1} = t_1$ in a walk because an atom may return to its original site.

Sometimes a physicist or chemist wants to know how many different walks of a given length there are from one site to another in a crystal. Two walks $w_1, w_2, \ldots, w_m$ and $t_1, t_2, \ldots, t_n$ are *different* if $n \neq m$ or if $m = n$ and some $t_i \neq w_i$. For example, v_1, v_3, v_2, v_1 and v_1, v_3, v_4, v_1 are different v_1-v_1 walks of length 3 in Figure 12.1.

We now look at the following problem. Given a crystal with sites labeled $v_1, \ldots, v_n$, how many different walks of length k are there between any two sites? Matrices enter the picture through the concept of an *adjacency matrix*. The adjacency matrix of a graph with n points is the $n \times n$ matrix with i, j entry 1 if v_i and v_j are neighbors and zero otherwise. For example, the graph of Figure 12.1 has adjacency matrix

$$\begin{bmatrix} 0 & 1 & 1 & 1 & 0 & 0 \\ 1 & 0 & 1 & 0 & 0 & 0 \\ 1 & 1 & 0 & 1 & 0 & 0 \\ 1 & 0 & 1 & 0 & 1 & 1 \\ 0 & 0 & 0 & 1 & 0 & 1 \\ 0 & 0 & 0 & 1 & 1 & 0 \end{bmatrix}.$$

The diagonal elements a_{ii} are all zero because we do not think of a site as adjacent to itself. The following remarkable theorem solves the problem posed above.

THEOREM 12.3

Let $A = [a_{ij}]$ be the adjacency matrix of a graph G having n points $v_1, \ldots, v_n$. Then the number of distinct v_i-v_j walks in G of length k is the i, j entry of A^k for every positive integer k.

Proof   We will prove the theorem by mathematical induction on k. Let $A = [a_{ij}]$ be the adjacency matrix of the graph. First, suppose that $k = 1$. Given any point v_i in the graph, there is no v_i-v_i walk of length 1 in the graph. Since $a_{ii} = 0$, the number of v_i-v_i walks of length 1 does equal the i, i element of A^1 (which, of course, is A). If $i \neq j$, there is a v_i-v_j walk of length 1 if v_i is adjacent to v_j (in which case $a_{ij} = 1$), and there is no v_i-v_j walk of length 1 if v_i is not adjacent to v_j (and then $a_{ij} = 0$). Therefore, a_{ij} is the number of v_i-v_j walks of length 1 in the graph, and the conclusion of the theorem holds for $k = 1$.

Now assume that the conclusion of the theorem holds for some positive integer k. We will prove that it holds for $k + 1$. Thus, we are assuming that the i, j element of A^k is the number of distinct v_i-v_j walks of length k in G. We want to show that the i, j element of A^{k+1} is the number of distinct v_i-v_j walks of length $k + 1$ in G.

Think about how we form a v_i-v_j walk of length $k + 1$ in G. We must first make a v_i-v_r walk of length 1 for some neighbor v_r of v_i and then a v_r-v_j walk of length k. This fact is illustrated in Figure 12.2. Thus,

the number of distinct v_i-v_j walks of length $k + 1$ in G	$=$	the sum of the number of distinct walks of length k from v_r to v_j for each v_r neighboring on v_i.

Figure 12.2

Now, $a_{ir} = 1$ if v_r is a neighbor of v_i, and $a_{ir} = 0$ otherwise. Further, by the inductive hypothesis, the number of distinct walks of length k from v_r to v_j is the r, j element of A^k. Denote $A^k = B$ so that the i, j element of A^k is b_{ij}. Then, for $r = 1, \ldots, n$,

$$a_{ir}b_{rj} = \begin{cases} 0 \text{ if } v_r \text{ is not a neighbor of } v_i; \\ \text{the number of distinct walks of length } k + 1 \text{ from} \\ v_i \text{ to } v_j \text{ through } v_r, \text{ if } v_r \text{ is a neighbor of } v_i. \end{cases}$$

Therefore,

$$a_{i1}b_{1j} + a_{i2}b_{2j} + \cdots + a_{in}b_{nj}$$

is the number of v_i–v_j walks of length $k + 1$ because it counts the number of walks of length k from v_r to v_j over each neighbor v_r of v_i. But this sum is exactly the i, j element of AB, which is AA^k, or A^{k+1}, and the proof is complete. ∎

Here is an illustration of the theorem.

EXAMPLE 12.9

Let G be the graph in Figure 12.3. The adjacency matrix is

$$A = \begin{bmatrix} 0 & 1 & 0 & 0 & 0 & 1 & 0 & 0 \\ 1 & 0 & 1 & 0 & 0 & 0 & 1 & 1 \\ 0 & 1 & 0 & 1 & 0 & 0 & 0 & 0 \\ 0 & 0 & 1 & 0 & 1 & 1 & 1 & 1 \\ 0 & 0 & 0 & 1 & 0 & 1 & 1 & 0 \\ 1 & 0 & 0 & 1 & 1 & 0 & 0 & 0 \\ 0 & 1 & 0 & 1 & 1 & 0 & 0 & 1 \\ 0 & 1 & 0 & 1 & 0 & 0 & 1 & 0 \end{bmatrix}$$

A straightforward calculation gives us

$$A^2 = \begin{bmatrix} 2 & 0 & 1 & 1 & 1 & 0 & 1 & 1 \\ 0 & 4 & 0 & 3 & 1 & 1 & 1 & 1 \\ 1 & 0 & 2 & 0 & 1 & 1 & 2 & 2 \\ 1 & 3 & 0 & 5 & 2 & 1 & 2 & 1 \\ 1 & 1 & 1 & 2 & 3 & 1 & 1 & 2 \\ 0 & 1 & 1 & 1 & 1 & 3 & 2 & 1 \\ 1 & 1 & 2 & 2 & 1 & 2 & 4 & 2 \\ 1 & 1 & 2 & 1 & 2 & 1 & 2 & 3 \end{bmatrix}$$

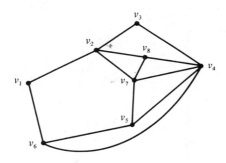

Figure 12.3

and

$$A^3 = \begin{bmatrix} 0 & 5 & 1 & 4 & 2 & 4 & 3 & 2 \\ 5 & 2 & 7 & 4 & 5 & 4 & 9 & 8 \\ 1 & 7 & 0 & 8 & 3 & 2 & 3 & 2 \\ 4 & 4 & 8 & 6 & 8 & 8 & 11 & 10 \\ 2 & 5 & 3 & 8 & 4 & 6 & 8 & 4 \\ 4 & 4 & 2 & 8 & 6 & 2 & 4 & 4 \\ 3 & 9 & 3 & 11 & 8 & 4 & 6 & 7 \\ 2 & 8 & 2 & 10 & 4 & 4 & 7 & 4 \end{bmatrix}.$$

From A^2, we can read the following information:

there are two $v_5 - v_8$ walks of length 2 (they are v_5, v_7, v_8 and v_5, v_4, v_8);

there are three $v_2 - v_4$ walks of length 2 ($v_2, v_3, v_4; v_2, v_8, v_4; v_2, v_7, v_4$);

and so on.

From A^3, we can read information about walks of length 3. For example, there are

eleven walks of length 3 from v_4 to v_7 ($v_4, v_7, v_4, v_7; v_4, v_3, v_4, v_7;$
$v_4, v_8, v_4, v_7; v_4, v_5, v_4, v_7; v_4, v_6, v_4, v_7; v_4, v_7, v_8, v_7; v_4, v_7, v_5, v_7;$
$v_4, v_7, v_2, v_7; v_4, v_3, v_2, v_7; v_4, v_8, v_2, v_7;$ and v_4, v_6, v_5, v_7); and
seven walks of length 3 from v_7 to v_8 ($v_7, v_8, v_7, v_8; v_7, v_8, v_4, v_8;$
$v_7, v_8, v_2, v_8; v_7, v_2, v_7, v_8; v_7, v_4, v_7, v_8; v_7, v_5, v_4, v_8;$ and v_7, v_5, v_7, v_8). ∎

Even for walks of length 2 or 3, it is easy to miss some if we attempt to count them by listing them all. Thus, even though computing higher powers of A is tedious, it is the most effective way of counting walks of a given length. Computer calculation of powers of the incidence matrix makes the method quite tractable in practical cases.

PROBLEMS FOR SECTION 12.2

1. Let G be the graph shown in the diagram. Determine the incidence matrix A, and compute A^3 and A^4. Determine the number of $v_1 - v_4$ walks of length 3, the number of $v_2 - v_3$ walks of length 3, the number of $v_1 - v_4$ walks of length 4, and the number of $v_2 - v_4$ walks of length 4.

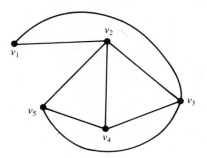

2. Let G be the graph shown. How many distinct walks of length 4 are there from v_1 to v_4 in G? How many distinct walks of length 2 are there from v_2 to v_3?

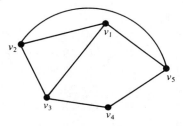

3. Let G be the graph shown. How many distinct walks of length 3 are there from v_2 to v_3; how many of length 2 are there from v_4 to v_5? How many distinct walks in G are there of length 4 from v_1 to v_2; from v_4 to v_5?

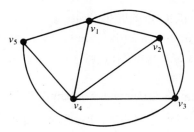

4. Let A be the adjacency matrix of a graph G. Prove that the i, i element of A^2 is the number of vertices neighboring on v_i in G.

5. Let A be the adjacency matrix of a graph G. Prove that the i, i element of A^3 is twice the number of triangles in G containing v_i as a vertex. (A triangle in G consists of three vertices, each neighboring on the other.)

12.3 *Some Special Matrices*

In this section, we will define certain matrices which occur often enough to warrant special names. We have already seen one example in the $n \times m$ zero matrix 0_{nm}. The *main diagonal* of a square matrix $A = [a_{ij}]$ consists of the elements $a_{11}, a_{22}, \ldots, a_{nn}$, which we see if we read downward from upper left to lower right in the matrix. For example, the main diagonal of

$$\begin{bmatrix} 1 & -4 & 2 \\ 0 & 1 & 6 \\ 0 & -2 & 4 \end{bmatrix}$$

has the numbers 1, 1, and 4 on it. Usually, main diagonal elements are listed in this way, from upper left to lower right. A matrix element not on the main diagonal is called an *off-diagonal element* (-4, 2, 6, -2, and 0 in the above matrix).

A square matrix $D = [d_{ij}]$ is called a *diagonal matrix* if every off-diagonal element is zero. That is, $d_{ij} = 0$ if $i \neq j$. The main diagonal may or may not have nonzero elements. For example,

$$\begin{bmatrix} 1 & 0 & 0 \\ 0 & 19 & 0 \\ 0 & 0 & 4 \end{bmatrix}$$

is a 3×3 diagonal matrix.

THEOREM 12.4

Let A and B be $n \times n$ diagonal matrices. Then

1. $A + B$ is a diagonal matrix.
2. AB is a diagonal matrix.

Proof   Conclusion (1) is obvious. To prove (2), we must show that every off-diagonal element of AB is zero. If $A = [a_{ij}]$ and $B = [b_{ij}]$, the i, j element of AB is

$$\sum_{k=1}^{n} a_{ik}b_{kj}.$$

Each a_{ik} is zero except when $k = i$, so this sum reduces to just $a_{ii}b_{ij}$. If $i \neq j$, then b_{ij} is zero, *so* $a_{ii}b_{ij} = 0$ and this sum is zero, completing the proof.   ∎

A diagonal matrix with each diagonal element equal to 1 is called an *identity matrix*. The $n \times n$ identity matrix is denoted I_n:

$$I_n = \begin{bmatrix} 1 & 0 & 0 & \cdots & 0 & 0 \\ 0 & 1 & 0 & \cdots & 0 & 0 \\ 0 & 0 & 1 & \cdots & 0 & 0 \\ \vdots & \vdots & \vdots & \cdots & \vdots & \vdots \\ 0 & 0 & 0 & \cdots & 0 & 1 \end{bmatrix}.$$

I_n derives its name, identity matrix, from the following theorem.

THEOREM 12.5

If A is $n \times m$, then $AI_m = A$ and $I_n A = A$.   ∎

Note that we must multiply A on the left by I_n (n rows and n columns) but on the right by I_m (m rows and columns) to have the products defined. Proof of Theorem 12.5 is left to the student.

EXAMPLE 12.10

Let A be the 3×2 matrix

$$A = \begin{bmatrix} 1 & 0 \\ 2 & 1 \\ -1 & 8 \end{bmatrix}.$$

Then

$$I_3 A = \begin{bmatrix} 1 & 0 & 0 \\ 0 & 1 & 0 \\ 0 & 0 & 1 \end{bmatrix} \begin{bmatrix} 1 & 0 \\ 2 & 1 \\ -1 & 8 \end{bmatrix} = \begin{bmatrix} 1 & 0 \\ 2 & 1 \\ -1 & 8 \end{bmatrix}$$

and

$$A I_2 = \begin{bmatrix} 1 & 0 \\ 2 & 1 \\ -1 & 8 \end{bmatrix} \begin{bmatrix} 1 & 0 \\ 0 & 1 \end{bmatrix} = \begin{bmatrix} 1 & 0 \\ 2 & 1 \\ -1 & 8 \end{bmatrix}.$$

In this example, $I_2 A$ and $A I_3$ are not defined. ∎

The *transpose* of an $n \times m$ matrix $A = [a_{ij}]$ is the $m \times n$ matrix A^t whose i, j element is a_{ji}. That is, the rows of A^t are the columns of A, and the columns of A^t are the rows of A. For example, if

$$A = \begin{bmatrix} 1 & -1 & 3 & 2 \\ 0 & 0 & 1 & 6 \\ 1 & -2 & 4 & 0 \end{bmatrix},$$

a 3×4 matrix, then

$$A^t = \begin{bmatrix} 1 & 0 & 1 \\ -1 & 0 & -2 \\ 3 & 1 & 4 \\ 2 & 6 & 0 \end{bmatrix},$$

a 4×3 matrix.

THEOREM 12.6

1. $I_n^t = I_n$ for any positive integer n.
2. $(A^t)^t = A$ for any matrix A.
3. Whenever AB is defined, $(AB)^t = B^t A^t$. ∎

Conclusion (1) is obvious, since row k and column k of I_n are the same; hence, interchanging them has no effect.

For (2), observe that we get A^t by interchanging the rows and columns of A and that forming $(A^t)^t$ interchanges them back again.

Conclusion (3) may be surprising the first time it is seen. The transpose of a product is the product of the transposes, *in reverse order*. Here is a proof of this fact.

Proof of (3)   Let $A = [a_{ij}]$ be $n \times k$, and let $B = [b_{ij}]$ be $k \times m$. Then AB is $n \times m$ and $(AB)^t$ is $m \times n$. B^t is $m \times k$ and A^t is $k \times n$, so $B^t A^t$ is also $m \times n$. To show that $(AB)^t = B^t A^t$, we have now to show that the i, j element of $(AB)^t$ equals the i, j element of $B^t A^t$.

Denote $A^t = [\alpha_{ij}]$ and $B^t = [\beta_{ij}]$. Note that $\alpha_{ij} = a_{ji}$ and $\beta_{ij} = b_{ji}$. The i, j element of $B^t A^t$ is

$$\sum_{s=1}^{k} \beta_{is}\alpha_{sj} = \sum_{s=1}^{k} b_{si}a_{js} = \sum_{s=1}^{k} a_{js}b_{si},$$

and the last sum is exactly the j, i element of AB. But this is the i, j element of $(AB)^t$. Thus, the i, j elements of $B^t A^t$ and $(AB)^t$ are the same; hence, $B^t A^t = (AB)^t$.   ∎

A matrix A is called *symmetric* if $A = A^t$. If A is $n \times m$, A^t is $m \times n$, and the equality $A = A^t$ forces m to equal n. Thus, a symmetric matrix must be square.

EXAMPLE 12.11

$$\begin{bmatrix} 1 & -1 & 3 \\ -1 & 0 & 2 \\ 3 & 2 & 4 \end{bmatrix} \quad \text{and} \quad \begin{bmatrix} -3 & 0 & 0 & 2 \\ 0 & 1 & -5 & 6 \\ 0 & -5 & 0 & -2 \\ 2 & 6 & -2 & 1 \end{bmatrix}$$

are symmetric matrices.   ∎

To specify a symmetric matrix A, we need only give the main diagonal elements and the elements above (or below) the main diagonal. These can be reflected across the main diagonal to fill in the rest of the matrix. This fact can be important for saving space when storing a matrix in computer memory.

PROBLEMS FOR SECTION 12.3

1. Let D be an $n \times n$ diagonal matrix with no zero elements on the main diagonal. Find a diagonal matrix E such that $DE = ED = I_n$.
2. Suppose that A and B are $n \times n$ symmetric matrices. Prove that $A + B$ is symmetric.
3. Show that $A + A^t$ is symmetric for any square matrix A.
4. Let A and B be symmetric $n \times n$ matrices.
   (a) Give an example to show that AB need not be symmetric.
   (b) Prove that AB is symmetric if and only if $AB = BA$.
5. Let A be an $n \times n$ matrix. Suppose that $AB = B$ for every $n \times n$ matrix B. Prove that $A = I_n$.
6. An $n \times n$ matrix A is said to be *upper triangular* if all elements below the main diagonal are zero (that is, $a_{ij} = 0$ if $i > j$).
   (a) Show that a sum of $n \times n$ upper triangular matrices is upper triangular.
   (b) Show that a product of $n \times n$ upper triangular matrices is upper triangular.

7. Let

$$A = \begin{bmatrix} a & b \\ c & d \end{bmatrix}$$

be any 2×2 matrix with $ad - bc \neq 0$. Find a matrix B such that $AB = BA = I_2$.

8. Let A and B be $n \times m$ matrices. Prove that $(A + B)^t = A^t + B^t$.

9. Prove Theorem 12.5.

12.4 *Elementary Row Operations and Elementary Matrices*

In solving a system of linear equations by eliminating unknowns, we generally perform three kinds of operations: we interchange equations, multiply an equation by a nonzero scalar, and add a scalar multiple of one equation to another equation.

If we write a system of equations in matrix form as $AX = B$, the coefficients and constant terms in the equation are contained in the coefficient matrix A and in B. Further, each row of A and B corresponds to an equation of the system. This suggests that we consider similar operations on rows of matrices. We will now discuss such operations, which have a variety of uses.

Given a matrix A, we define three types of row operations.

Type I: Interchange of two rows.

Type II: Multiplication of a row by a nonzero scalar.

Type III: Addition of a scalar multiple of one row to another row.

In a type II operation, we multiply a row by a scalar by multiplying each element of the row by the scalar. If we think of row i as a vector $\langle a_{i1}, a_{i1}, \ldots, a_{im} \rangle$, this corresponds to multiplying an m-vector by a scalar.

In a type III operation, we add a scalar multiple of one row to another row. If we add α times row i to row k, we replace each element of row k with that element plus α times the corresponding element of row i. Again, thinking in terms of vectors, row i is $\langle a_{i1}, a_{i2}, \ldots, a_{im} \rangle$, and row k is $\langle a_{k1}, a_{k2}, \ldots, a_{km} \rangle$. When we add α times row i to row k, we replace row k with the vector

$$\langle \alpha a_{i1} + a_{k1}, \alpha a_{i2} + a_{k2}, \ldots, \alpha a_{im} + a_{km} \rangle.$$

EXAMPLE 12.12

Let

$$A = \begin{bmatrix} -2 & 1 & 6 \\ 1 & 1 & 2 \\ 0 & 1 & 3 \\ 2 & -3 & 4 \end{bmatrix}.$$

For an example of a type I operation, interchange rows 1 and 3 to obtain the new matrix

$$\begin{bmatrix} 0 & 1 & 3 \\ 1 & 1 & 2 \\ -2 & 1 & 6 \\ 2 & -3 & 4 \end{bmatrix}.$$

For an example of a type II operation, multiply row 4 of A by 6 to get

$$\begin{bmatrix} -2 & 1 & 6 \\ 1 & 1 & 2 \\ 0 & 1 & 3 \\ 12 & -18 & 24 \end{bmatrix}.$$

For an example of a type III operation, add -5 times row 2 of A to row 3:

$$\begin{bmatrix} -2 & 1 & 6 \\ 1 & 1 & 2 \\ -5 & -4 & -7 \\ 2 & -3 & 4 \end{bmatrix}. \quad \blacksquare$$

We will now show that each row operation can be performed on A by multiplying A on the *left* by a matrix obtained by performing the operation on I_n. To this end, define an *elementary matrix* as any matrix obtained from I_n by an elementary row operation. For example,

$$\begin{bmatrix} 0 & 1 \\ 1 & 0 \end{bmatrix}$$

is obtained from I_2 by interchanging rows 1 and 2 and hence is an elementary matrix. If we multiply row 3 of I_3 by -5, we obtain the elementary matrix

$$\begin{bmatrix} 1 & 0 & 0 \\ 0 & 1 & 0 \\ 0 & 0 & -5 \end{bmatrix}.$$

And if we add -3 times row 2 to row 3 of I_3, we get the elementary matrix

$$\begin{bmatrix} 1 & 0 & 0 \\ 0 & 1 & 0 \\ 0 & -3 & 1 \end{bmatrix}.$$

We will make frequent use of the following fact.

THEOREM 12.7

Let A be any $n \times m$ matrix. Suppose that B is formed from A by an elementary row operation. Let E be the elementary matrix formed by performing the same elementary row operation on I_n. Then

$$B = EA. \quad \blacksquare$$

That is, the elementary matrix formed from I_n by an elementary row operation performs the elementary operation on A upon multiplying A on the *left* by E. We will defer a proof to the exercises and illustrate the theorem with some examples.

EXAMPLE 12.13

Let

$$A = \begin{bmatrix} -2 & 1 & 6 \\ 1 & 1 & 2 \\ 0 & 1 & 3 \\ 2 & -3 & 4 \end{bmatrix}.$$

Form B from A by the type I operation of interchanging rows 3 and 4:

$$B = \begin{bmatrix} -2 & 1 & 6 \\ 1 & 1 & 2 \\ 2 & -3 & 4 \\ 0 & 1 & 3 \end{bmatrix}.$$

Perform the same row operation on I_4 to get

$$E = \begin{bmatrix} 1 & 0 & 0 & 0 \\ 0 & 1 & 0 & 0 \\ 0 & 0 & 0 & 1 \\ 0 & 0 & 1 & 0 \end{bmatrix}.$$

Now verify that E achieves the row operation on A upon multiplying A on the left by E:

$$EA = \begin{bmatrix} 1 & 0 & 0 & 0 \\ 0 & 1 & 0 & 0 \\ 0 & 0 & 0 & 1 \\ 0 & 0 & 1 & 0 \end{bmatrix} \begin{bmatrix} -2 & 1 & 6 \\ 1 & 1 & 2 \\ 0 & 1 & 3 \\ 2 & -3 & 4 \end{bmatrix}$$

$$= \begin{bmatrix} -2 & 1 & 6 \\ 1 & 1 & 2 \\ 2 & -3 & 4 \\ 0 & 1 & 3 \end{bmatrix} = B. \quad \blacksquare$$

EXAMPLE 12.14

Again, let

$$A = \begin{bmatrix} -2 & 1 & 6 \\ 1 & 1 & 2 \\ 0 & 1 & 3 \\ 2 & -3 & 4 \end{bmatrix}.$$

Multiply row 1 of A by -4 to get C:

$$C = \begin{bmatrix} 8 & -4 & -24 \\ 1 & 1 & 2 \\ 0 & 1 & 3 \\ 2 & -3 & 4 \end{bmatrix}.$$

Perform the same operation on I_4 to get

$$E = \begin{bmatrix} -4 & 0 & 0 & 0 \\ 0 & 1 & 0 & 0 \\ 0 & 0 & 1 & 0 \\ 0 & 0 & 0 & 1 \end{bmatrix}.$$

Then

$$EA = \begin{bmatrix} -4 & 0 & 0 & 0 \\ 0 & 1 & 0 & 0 \\ 0 & 0 & 1 & 0 \\ 0 & 0 & 0 & 1 \end{bmatrix} \begin{bmatrix} -2 & 1 & 6 \\ 1 & 1 & 2 \\ 0 & 1 & 3 \\ 2 & -3 & 4 \end{bmatrix}$$

$$= \begin{bmatrix} 8 & -4 & -24 \\ 1 & 1 & 2 \\ 0 & 1 & 3 \\ 2 & -3 & 4 \end{bmatrix} = C. \ \blacksquare$$

EXAMPLE 12.15

Form D from A by adding 6 times row 1 to row 4, with A as in the last two examples. Then

$$D = \begin{bmatrix} -2 & 1 & 6 \\ 1 & 1 & 2 \\ 0 & 1 & 3 \\ -10 & 3 & 40 \end{bmatrix}.$$

Form the elementary matrix E from I_4 by adding 6 times row 1 to row 4:

$$E = \begin{bmatrix} 1 & 0 & 0 & 0 \\ 0 & 1 & 0 & 0 \\ 0 & 0 & 1 & 0 \\ 6 & 0 & 0 & 1 \end{bmatrix}.$$

Then

$$EA = \begin{bmatrix} 1 & 0 & 0 & 0 \\ 0 & 1 & 0 & 0 \\ 0 & 0 & 1 & 0 \\ 6 & 0 & 0 & 1 \end{bmatrix} \begin{bmatrix} -2 & 1 & 6 \\ 1 & 1 & 2 \\ 0 & 1 & 3 \\ 2 & -3 & 4 \end{bmatrix}$$

$$= \begin{bmatrix} -2 & 1 & 6 \\ 1 & 1 & 2 \\ 0 & 1 & 3 \\ -10 & 3 & 40 \end{bmatrix} = D. \ \blacksquare$$

To reiterate, any elementary row operation on an $n \times m$ matrix A can be achieved by multiplying A on the left by the elementary matrix formed by performing the same row operation on I_n.

Sometimes we will need to perform a sequence of row operations on a matrix A. This can be done by multiplying A on the left by a product of elementary matrices. Suppose we want to perform operation $\mathcal{O}_1$ on A to form A_1, then $\mathcal{O}_2$ on A_1 to form A_2, and so on, until finally we perform $\mathcal{O}_n$ on A_{n-1} to form A_n. Let E_j be the elementary matrix which performs $\mathcal{O}_j$ by multiplication on the left. Now,

$$A_1 = E_1 A.$$

Next,

$$A_2 = E_2 A_1 = E_2(E_1 A) = (E_2 E_1)A.$$

Next,

$$A_3 = E_3 A_2 = (E_3 E_2 E_1)A,$$

and so on, until finally

$$A_n = (E_n E_{n-1} \cdots E_1)A.$$

The sequence of elementary row operations $\mathcal{O}_1, \mathcal{O}_2, \ldots, \mathcal{O}_n$, in this order, is therefore achieved by multiplying A on the left by $E_n E_{n-1} \cdots E_2 E_1$, in this order. The order of these factors is important because we want to act first on A by E_1, then on the result by E_2, and so on.

EXAMPLE 12.16

Perform the following sequence of elementary row operations, beginning with A:

$$A = \begin{bmatrix} 2 & 1 & 0 \\ 0 & 1 & 2 \\ -1 & 3 & 2 \end{bmatrix} \xrightarrow[\text{rows 1 and 2}]{\mathcal{O}_1:\ \text{interchange}} \begin{bmatrix} 0 & 1 & 2 \\ 2 & 1 & 0 \\ -1 & 3 & 2 \end{bmatrix}$$

$$\xrightarrow[\text{3 by 2}]{\mathcal{O}_2:\ \text{multiply row}} \begin{bmatrix} 0 & 1 & 2 \\ 2 & 1 & 0 \\ -2 & 6 & 4 \end{bmatrix}$$

$$\xrightarrow[\text{to row 3}]{\mathcal{O}_3:\ \text{add 2(row 1)}} \begin{bmatrix} 0 & 1 & 2 \\ 2 & 1 & 0 \\ -2 & 8 & 8 \end{bmatrix} = B.$$

Each operation can be performed by an elementary matrix:

$$\mathcal{O}_1 \text{ by } E_1 = \begin{bmatrix} 0 & 1 & 0 \\ 1 & 0 & 0 \\ 0 & 0 & 1 \end{bmatrix} \qquad (\text{perform } \mathcal{O}_1 \text{ on } I_3);$$

$$\mathcal{O}_2 \text{ by } E_2 = \begin{bmatrix} 1 & 0 & 0 \\ 0 & 1 & 0 \\ 0 & 0 & 2 \end{bmatrix} \qquad (\text{perform } \mathcal{O}_2 \text{ on } I_3);$$

$$\mathcal{O}_3 \text{ by } E_3 = \begin{bmatrix} 1 & 0 & 0 \\ 0 & 1 & 0 \\ 2 & 0 & 1 \end{bmatrix} \qquad (\text{perform } \mathcal{O}_3 \text{ on } I_3).$$

Form

$$E_3 E_2 E_1 = \begin{bmatrix} 1 & 0 & 0 \\ 0 & 1 & 0 \\ 2 & 0 & 1 \end{bmatrix} \begin{bmatrix} 1 & 0 & 0 \\ 0 & 1 & 0 \\ 0 & 0 & 2 \end{bmatrix} \begin{bmatrix} 0 & 1 & 0 \\ 1 & 0 & 0 \\ 0 & 0 & 1 \end{bmatrix}$$

$$= \begin{bmatrix} 0 & 1 & 0 \\ 1 & 0 & 0 \\ 0 & 2 & 2 \end{bmatrix}.$$

If A is multiplied on the left by this product of elementary matrices, we obtain the final result of the above three row operations:

$$(E_3 E_2 E_1)A = \begin{bmatrix} 0 & 1 & 0 \\ 1 & 0 & 0 \\ 0 & 2 & 2 \end{bmatrix} \begin{bmatrix} 2 & 1 & 0 \\ 0 & 1 & 2 \\ -1 & 3 & 2 \end{bmatrix}$$

$$= \begin{bmatrix} 0 & 1 & 2 \\ 2 & 1 & 0 \\ -2 & 8 & 8 \end{bmatrix}. \quad \blacksquare$$

We say that A is *row equivalent* to B if we obtain B from A by a sequence of elementary row operations. This means that B must be of the form $E_n E_{n-1} \cdots E_1 A$ for some elementary matrices $E_1, E_2, \ldots, E_n$. Row equivalent matrices have the following properties.

THEOREM 12.8

1. Every matrix is row equivalent to itself.   *reflexive*

2. If A is row equivalent to B, B is row equivalent to A.   *symmetric*

3. If A is row equivalent to B, and B is row equivalent to C, A is row equivalent to C.   *Transitive*

Proof   For (1), note that A can always be obtained from A by an elementary row operation (for example, multiply row 1 by 1).

For (3), if B is obtained from A by a product of elementary matrices, and C is obtained from B by a product of elementary matrices, we have obtained C from A by a product of elementary matrices, so A is row equivalent to C.

Conclusion (2) requires an observation: we can reverse the effect of any elementary row operation by performing another elementary row operation. Consider the row operations by type.

If we obtain B from A by interchanging rows i and j of A, we obtain A from B by interchanging rows i and j of B.

If we obtain B from A by multiplying row j of A by a nonzero scalar α, we obtain A from B by multiplying row j of B by $1/\alpha$.

If we obtain B from A by adding α times row i to row j, we obtain A from B by adding $-\alpha$ times row i to row j of B.

In summary, if we obtain B from A by an elementary row operation, we can recover A from B by an elementary row operation of the same type.

$$\begin{bmatrix} B = n \leftarrow 1 A \\ A = \overset{\ast}{i} \leftarrow n \ast B \end{bmatrix}$$

Now suppose that A is row equivalent to B. Then B is obtained from A by a sequence of elementary row operations. Suppose the operations are, in turn, $\mathcal{O}_1$, then $\mathcal{O}_2, \ldots, \mathcal{O}_n$. Let $\mathcal{O}_j^*$ be the elementary row operation that reverses the effects of $\mathcal{O}_j$. Starting with B, perform $\mathcal{O}_n^*$, reversing the effect of $\mathcal{O}_n$; on the resulting matrix, perform $\mathcal{O}_{n-1}^*$, reversing the effect of $\mathcal{O}_{n-1}$, and so on, until at the last step we perform $\mathcal{O}_1^*$, reversing the effect of $\mathcal{O}_1$ and returning us to A. Thus, A is obtained from B by a sequence of elementary row operations; hence B is row equivalent to A. ∎

In view of this theorem, we will say that A and B are *row equivalent* if either is obtained from the other by a sequence of elementary row operations. In the next section, we will see that every matrix is row equivalent to a matrix having a form which is particularly simple to work with in solving equations.

PROBLEMS FOR SECTION 12.4

In each of Problems 1 through 10, perform the indicated elementary row operation on A to produce B. Then find an elementary matrix E such that $EA = B$.

1. $A = \begin{bmatrix} -2 & 1 & 4 & 2 \\ 0 & 1 & 16 & 3 \\ 1 & -2 & 4 & 8 \end{bmatrix}$; multiply row 2 by $\sqrt{3}$.

2. $A = \begin{bmatrix} 2 & 16 & -4 \\ 8 & 1 & 7 \\ 2 & 2 & 3 \end{bmatrix}$; add $\sqrt{3}$(row 3) to row 1.

3. $A = \begin{bmatrix} -2 & 1 & 1 & 7 & 13 \\ 3 & 0 & -1 & 14 & 2 \end{bmatrix}$; add 5(row 2) to row 1.

4. $A = \begin{bmatrix} 18 & 2 \\ 4 & 6 \\ -3 & 10 \end{bmatrix}$; multiply row 3 by $\sqrt{7}$.

5. $A = \begin{bmatrix} 2 & 2 & -6 \\ 1 & 0 & 3 \\ -2 & 4 & 3 \end{bmatrix}$; interchange rows 2 and 3.

6. $A = \begin{bmatrix} -3 & 5 & -7 & -8 & 5 \\ 14 & 5 & 1 & 3 & -6 \\ 2 & 0 & 4 & 3 & 9 \\ -5 & 4 & 8 & -2 & 4 \end{bmatrix}$; add 13(row 2) to row 3.

7. $A = \begin{bmatrix} -9 & 0 & 4 \\ 3 & 5 & 2 \\ 14 & 4 & 4 \end{bmatrix}$; multiply row 1 by 5.

8. $A = \begin{bmatrix} 14 & 6 & 3 & -7 \\ 3 & 2 & 7 & 0 \\ 2 & 2 & -4 & 5 \end{bmatrix}$; add 3(row 1) to row 2.

9. $A = \begin{bmatrix} 22 & 13 & -9 \\ 3 & 5 & 12 \\ -6 & 8 & 0 \end{bmatrix}$; interchange rows 1 and 3.

10. $A = \begin{bmatrix} 14 & 2 & -3 & 7 \\ 8 & 5 & 12 & -3 \end{bmatrix}$;   add 3(row 2) to row 1.

In each of Problems 11 through 20, produce a matrix B from the given matrix A by performing the sequence of operations $\mathcal{O}_1, \ldots, \mathcal{O}_k$ (k may vary from problem to problem). Then produce a matrix C such that $CA = B$.

11. $A = \begin{bmatrix} -2 & 14 & 6 \\ 8 & 1 & -3 \\ 2 & 9 & 5 \end{bmatrix}$

$\mathcal{O}_1$: add $\sqrt{13}$(row 3) to row 1; $\mathcal{O}_2$: interchange rows 2 and 3; $\mathcal{O}_3$: multiply row 1 by 5.

12. $A = \begin{bmatrix} -4 & 6 & -3 \\ 12 & 4 & -4 \\ 1 & 3 & 0 \end{bmatrix}$

$\mathcal{O}_1$: interchange rows 2 and 3; $\mathcal{O}_2$: add -1(row 1) to row 2.

13. $A = \begin{bmatrix} -3 & 15 \\ 2 & 8 \end{bmatrix}$

$\mathcal{O}_1$: add $\sqrt{3}$(row 2) to row 1; $\mathcal{O}_2$: multiply row 2 by 15; $\mathcal{O}_3$: interchange rows 1 and 2.

14. $A = \begin{bmatrix} 3 & -4 & 5 & 9 \\ 2 & 1 & 3 & -6 \\ 1 & 13 & 2 & 6 \end{bmatrix}$

$\mathcal{O}_1$: add row 1 to row 3; $\mathcal{O}_2$: add $\sqrt{3}$(row 1) to row 2; $\mathcal{O}_3$: multiply row 3 by 4; $\mathcal{O}_4$: add row 2 to row 3.

15. $A = \begin{bmatrix} -1 & 0 & 3 & 0 \\ 1 & 3 & 2 & 9 \\ -9 & 7 & -5 & 7 \end{bmatrix}$

$\mathcal{O}_1$: multiply row 3 by 4; $\mathcal{O}_2$: add 14(row 1) to row 2; $\mathcal{O}_3$: interchange rows 3 and 2.

16. $A = \begin{bmatrix} 0 & -9 & 14 \\ 1 & 5 & 2 \\ 9 & 15 & 0 \end{bmatrix}$

$\mathcal{O}_1$: interchange rows 2 and 3; $\mathcal{O}_2$: add 3(row 2) to row 3; $\mathcal{O}_3$: interchange rows 1 and 3; $\mathcal{O}_4$: multiply row 3 by 5.

17. $A = \begin{bmatrix} -3 & 7 & 1 & 1 \\ 0 & 3 & 3 & -5 \\ 2 & 1 & -5 & 3 \end{bmatrix}$

$\mathcal{O}_1$: add 2(row 1) to row 3; $\mathcal{O}_2$: multiply row 3 by -5; $\mathcal{O}_3$: interchange rows 2 and 3.

18. $A = \begin{bmatrix} 2 & -6 & 5 & 8 \\ 0 & 1 & -3 & 5 \\ 0 & -4 & 2 & -6 \\ 1 & 7 & 3 & -3 \end{bmatrix}$

$\mathcal{O}_1$: multiply row 4 by -5; $\mathcal{O}_2$: add $\sqrt{3}$(row 4) to row 1; $\mathcal{O}_3$: interchange rows 1 and 3; $\mathcal{O}_4$: multiply row 3 by -1.

19. $A = \begin{bmatrix} 2 & -3 & 1 \\ 0 & 0 & 0 \\ 1 & -5 & 0 \end{bmatrix}$

$\mathcal{O}_1$: interchange rows 1 and 2; $\mathcal{O}_2$: multiply row 2 by 5; $\mathcal{O}_3$: add -3(row 3) to row 1.

20. $A = \begin{bmatrix} -5 & 1 & -4 \\ 0 & 3 & -2 \\ 1 & 2 & 2 \end{bmatrix}$

$\mathcal{O}_1$: add 4(row 2) to row 3; $\mathcal{O}_2$: interchange rows 1 and 3; $\mathcal{O}_3$: multiply row 1 by -2.

21. Suppose that B is formed from A by a type I elementary row operation. Form E from I_n by performing the same elementary operation on I_n, where n is the number of rows of A. Prove that $EA = B$.

22. Suppose that B is formed from A by a type II elementary row operation. Form E from I_n by performing the same elementary operation on I_n, where n is the number of rows of A. Prove that $EA = B$.

23. Suppose that B is formed from A by a type III elementary row operation. Form E from I_n by performing the same elementary operation I_n, where n is the number of rows of A. Prove that $EA = B$.

24. Define the elementary column operations of types I, II, and III by substituting the word "column" for "row" in the corresponding definitions of the row operations. Prove that each elementary column operation on A can be achieved as a matrix product AE, where E is obtained from I_m (m is the number of columns of A) by performing the column operation on I_m. Notice that, to perform a column operation by matrix multiplications, we multiply on the *right* by the appropriate matrix.

12.5 *Reduced Form of a Matrix*

Many problems which can be formulated in matrix terms are easier to solve if the matrices involved have a special form. In this section, we will define one such form. Let A be an $n \times m$ matrix. If a row has a nonzero element, the *leading entry* of that row is its first nonzero entry, reading from left to right. For example, consider the matrix

$$\begin{bmatrix} 1 & 3 & 1 \\ 0 & 0 & 6 \\ 0 & -4 & 1 \\ 0 & 0 & 0 \end{bmatrix}.$$

The leading entry of row 1 is 1; the leading entry of row 2 is 6; and the leading entry of row 3 is -4. A row having all zeros is said to have no leading entry. Row 4 of the above matrix has no leading entry. We say that A is a *reduced matrix* (or is in *reduced form*) if A has the following properties.

1. The leading entry of any nonzero row is 1.
2. If a row has its leading entry in column c, all other elements of column c are zero.
3. Each row having all zero elements (if there is such a row) lies below any row having a nonzero element.
4. If the leading entry in row r_1 lies in column c_1 and the leading entry of row r_2 is in column c_2, and $r_1 < r_2$, then $c_1 < c_2$.

By requirement (2), if a column contains a leading entry of any row, all other entries of that column are zero. Put another way, all elements directly above and below any leading entry are zero. Condition (4) says that the leading entries move downward to the right as you look at the matrix.

EXAMPLE 12.17

The following four matrices are all in reduced form:

$$\begin{bmatrix} 1 & -4 & 1 & 0 \\ 0 & 0 & 0 & 1 \end{bmatrix}, \qquad \begin{bmatrix} 0 & 1 & 3 & 2 & 0 \\ 0 & 0 & 0 & 0 & 1 \\ 0 & 0 & 0 & 0 & 0 \end{bmatrix},$$

$$\begin{bmatrix} 0 & 1 & 2 & 0 & 0 \\ 0 & 0 & 0 & 1 & 0 \\ 0 & 0 & 0 & 0 & 0 \\ 0 & 0 & 0 & 0 & 0 \end{bmatrix}, \quad \text{and} \quad \begin{bmatrix} 1 & 0 & 0 & 3 & 1 \\ 0 & 1 & 0 & -2 & 4 \\ 0 & 0 & 1 & 0 & 1 \\ 0 & 0 & 0 & 0 & 0 \end{bmatrix}.$$

However, the matrix

$$\begin{bmatrix} 0 & 1 & 2 & 0 & 0 \\ 0 & 0 & 1 & 0 & 0 \\ 0 & 0 & 0 & 1 & 0 \\ 0 & 0 & 0 & 0 & 1 \end{bmatrix}$$

is not in reduced form because the leading entry of row 2 is in column 3, and this column has another nonzero element. This matrix is, however, row equivalent to a reduced matrix (add -2 times row 2 to row 1).

The matrix

$$\begin{bmatrix} 0 & 1 & 1 & 0 & 0 \\ 1 & 0 & 0 & 0 & 0 \\ 0 & 0 & 0 & 0 & 0 \end{bmatrix}$$

is not reduced because condition (4) of the definition is violated. We can obtain a reduced matrix by interchanging rows 1 and 2.

The matrix

$$\begin{bmatrix} 2 & 0 & 0 \\ 0 & 1 & 0 \\ 1 & 0 & 1 \end{bmatrix}$$

is not in reduced form. We can obtain a reduced matrix as follows. First, multiply row 1 by $\frac{1}{2}$ to get

$$\begin{bmatrix} 1 & 0 & 0 \\ 0 & 1 & 0 \\ 1 & 0 & 1 \end{bmatrix}.$$

Now add -1 times row 1 to row 3 to get

$$\begin{bmatrix} 1 & 0 & 0 \\ 0 & 1 & 0 \\ 0 & 0 & 1 \end{bmatrix},$$

which is in reduced form. ■

In each of the examples showing a matrix not in reduced form, we were able to obtain a reduced matrix by a sequence of elementary row operations. This can always be done.

THEOREM 12.9

Let A be an $n \times m$ matrix. Then A is row equivalent to a matrix in reduced form.

Proof   The proof is a constructive one in which we outline a procedure for finding a reduced matrix row equivalent to A. If A is in reduced form, we are done. If not, reading from left to right, suppose column c_1 is the first column having a nonzero entry. Let α be the top nonzero element of this column, occurring in, say, row r_1. Multiply row r_1 by $1/\alpha$ to obtain a matrix B. By choice of r_1, column c_1 of B has only zeros above the 1 in row r_1. If any row below r_1 has a nonzero element β in column c_1, add $-\beta$ times row r_1 to this row, yielding a new matrix with a zero where β was located in B. Repetition of this process results in a matrix C having zeros above and below row r_1 in column c_1. Now interchange rows 1 and r_1 of C to produce a matrix D having leading entry 1 in row 1 and column c_1 and all other elements of this column zero. Further, by choice of c_1, any column of D to the left of column c_1 has only zero elements. Finally, D is row equivalent to A, since we have arrived at D by a sequence of elementary row operations.

If D is reduced, we are done. If not, repeat this procedure, but now look for the first column, say, column c_2, to the right of column c_1 and having a nonzero element below row 1. Let r_2 be the first row below row 1 having a nonzero entry, say, γ. Divide row r_2 by γ to obtain a matrix E with 1 as its r_2, c_2 element. If column c_2 of E has a nonzero entry δ in a row above or below row r_2, add $-\delta$ times row r_2 to this row. Repetition of this process yields a matrix F with zeros in column c_2 above and below the element 1 in row r_2. Finally, interchange rows 2 and r_2 of F to form G.

If G is in reduced form, we are done. If not, locate the first column to the right of column c_2 and having a nonzero element below row r_2, and repeat the procedure we have been using. Since A has only finitely many columns, eventually we arrive at a reduced matrix, and this reduced matrix is row equivalent to A.   ∎

The process of obtaining a reduced matrix row equivalent to a given matrix A is called *reducing* A. We will discuss some examples, but first we will observe that it is usually possible to carry out several different sequences of elementary row operations on A to produce a reduced matrix. The following theorem says that, for a given A, the end result is always the same no matter what elementary row operations we use in arriving at it.

THEOREM 12.10

Let A be an $n \times m$ matrix. Suppose a sequence $\mathcal{T}_1$ of elementary row operations is applied, beginning with A and resulting in a reduced matrix R_1. Suppose that a sequence $\mathcal{T}_2$ of elementary row operations, beginning with A, also results in a reduced matrix R_2. Then $R_1 = R_2$.   ∎

Because of this theorem, we can speak of *the* reduced form of A instead of *a* reduced form of A. We denote the reduced form of A as A_R.

EXAMPLE 12.18

Reduce the matrix

$$A = \begin{bmatrix} -2 & 1 & 3 \\ 0 & 1 & 1 \\ 2 & 0 & 1 \end{bmatrix}.$$

Begin with column 1, which has a nonzero entry in row 1. Proceed as follows:

$$A \xrightarrow[\text{1 by } -\frac{1}{2}]{\text{multiply row}} \begin{bmatrix} 1 & -\frac{1}{2} & -\frac{3}{2} \\ 0 & 1 & 1 \\ 2 & 0 & 1 \end{bmatrix} \xrightarrow[\text{to row 3}]{\text{add } -2(\text{row 1})} \begin{bmatrix} 1 & -\frac{1}{2} & -\frac{3}{2} \\ 0 & 1 & 1 \\ 0 & 1 & 4 \end{bmatrix}.$$

Column 2 of the last matrix has a nonzero entry below row 1, the highest being the element 1 in the 2, 2 location. Since we want a 1 here, we need not multiply this row by any constant. However, we want zeros elsewhere in column 2, so proceed from the last matrix and add $\frac{1}{2}$ times row 2 to row 1 and -1 times row 2 to row 3 to get

$$\begin{bmatrix} 1 & 0 & -1 \\ 0 & 1 & 1 \\ 0 & 0 & 3 \end{bmatrix}.$$

Multiply row 3 by $\frac{1}{3}$ to get

$$\begin{bmatrix} 1 & 0 & -1 \\ 0 & 1 & 1 \\ 0 & 0 & 1 \end{bmatrix}.$$

Finally, add row 3 to row 1 and -1 times row 3 to row 2 to get zeros above row 3 in column 3

$$\begin{bmatrix} 1 & 0 & 0 \\ 0 & 1 & 0 \\ 0 & 0 & 1 \end{bmatrix}.$$

This is A_R, the reduced form of A. It happens that $A_R = I_3$ in this example.

To illustrate Theorem 12.10, we will use a different sequence of elementary row operations to reduce A and observe that we obtain the same reduced form A_R:

$$A = \begin{bmatrix} -2 & 1 & 3 \\ 0 & 1 & 1 \\ 2 & 0 & 1 \end{bmatrix} \xrightarrow[\text{to row 1}]{\text{add row 3}} \begin{bmatrix} 0 & 1 & 4 \\ 0 & 1 & 1 \\ 2 & 0 & 1 \end{bmatrix}$$

$$\xrightarrow[\text{to row 1}]{\text{add } -1(\text{row 2})} \begin{bmatrix} 0 & 0 & 3 \\ 0 & 1 & 1 \\ 2 & 0 & 1 \end{bmatrix}$$

$$\xrightarrow[\text{by } \frac{1}{3}]{\text{multiply row 1}} \begin{bmatrix} 0 & 0 & 1 \\ 0 & 1 & 1 \\ 2 & 0 & 1 \end{bmatrix}$$

$$\xrightarrow[\text{to rows 2 and 3}]{\text{add } -1(\text{row 1})} \begin{bmatrix} 0 & 0 & 1 \\ 0 & 1 & 0 \\ 2 & 0 & 0 \end{bmatrix}$$

$$\xrightarrow[\text{1 and 3}]{\text{interchange rows}} \begin{bmatrix} 2 & 0 & 0 \\ 0 & 1 & 0 \\ 0 & 0 & 1 \end{bmatrix}$$

$$\xrightarrow[\text{by } \frac{1}{2}]{\text{multiply row 1}} \begin{bmatrix} 1 & 0 & 0 \\ 0 & 1 & 0 \\ 0 & 0 & 1 \end{bmatrix} = A_R. \quad \blacksquare$$

EXAMPLE 12.19

Let

$$B = \begin{bmatrix} 0 & 0 & 0 & 0 & 0 \\ 0 & 0 & 2 & 0 & 0 \\ 0 & 1 & 0 & 1 & 1 \\ 0 & 4 & 3 & 4 & 0 \end{bmatrix}.$$

Reduce B as follows. The first column having a nonzero element is the second, and its first nonzero element is 1 in the 3, 2 place. Add -4 times row 3 to row 4 to get

$$\begin{bmatrix} 0 & 0 & 0 & 0 & 0 \\ 0 & 0 & 2 & 0 & 0 \\ 0 & 1 & 0 & 1 & 1 \\ 0 & 0 & 3 & 0 & -4 \end{bmatrix}.$$

Now interchange rows 3 and 1 to get

$$\begin{bmatrix} 0 & 1 & 0 & 1 & 1 \\ 0 & 0 & 2 & 0 & 0 \\ 0 & 0 & 0 & 0 & 0 \\ 0 & 0 & 3 & 0 & -4 \end{bmatrix}.$$

Multiply row 2 by $\frac{1}{2}$ to get

$$\begin{bmatrix} 0 & 1 & 0 & 1 & 1 \\ 0 & 0 & 1 & 0 & 0 \\ 0 & 0 & 0 & 0 & 0 \\ 0 & 0 & 3 & 0 & -4 \end{bmatrix}.$$

Add -3 times row 2 to row 4 to get

$$\begin{bmatrix} 0 & 1 & 0 & 1 & 1 \\ 0 & 0 & 1 & 0 & 0 \\ 0 & 0 & 0 & 0 & 0 \\ 0 & 0 & 0 & 0 & -4 \end{bmatrix}.$$

Since column 4 of this matrix has only zeros below row 2, concentrate next on column 5. Multiply row 4 by $-\frac{1}{4}$ and then add -1 times row 4 to row 1 and interchange rows 3 and 4 to get

$$\begin{bmatrix} 0 & 1 & 0 & 1 & 0 \\ 0 & 0 & 1 & 0 & 0 \\ 0 & 0 & 0 & 0 & 1 \\ 0 & 0 & 0 & 0 & 0 \end{bmatrix}.$$

This is B_R, the reduced form of B.   ∎

EXAMPLE 12.20

Let

$$D = \begin{bmatrix} 0 & 1 & 1 & 0 \\ 1 & 0 & 2 & 0 \\ 2 & 0 & 1 & 0 \end{bmatrix}.$$

Reduce D as follows:

$$D \xrightarrow[\text{to row 3}]{\text{add } -2(\text{row 2})} \begin{bmatrix} 0 & 1 & 1 & 0 \\ 1 & 0 & 2 & 0 \\ 0 & 0 & -3 & 0 \end{bmatrix}$$

$$\xrightarrow[\text{rows 1 and 2}]{\text{interchange}} \begin{bmatrix} 1 & 0 & 2 & 0 \\ 0 & 1 & 1 & 0 \\ 0 & 0 & -3 & 0 \end{bmatrix}$$

$$\xrightarrow[\text{by } -\frac{1}{3}]{\text{multiply row 3}} \begin{bmatrix} 1 & 0 & 2 & 0 \\ 0 & 1 & 1 & 0 \\ 0 & 0 & 1 & 0 \end{bmatrix}$$

$$\xrightarrow[\text{add } -1(\text{row 3}) \text{ to row 2}]{\text{add } -2(\text{row 3}) \text{ to row 1}} \begin{bmatrix} 1 & 0 & 0 & 0 \\ 0 & 1 & 0 & 0 \\ 0 & 0 & 1 & 0 \end{bmatrix} = D_R. \quad ∎$$

We obtain A_R from A by a sequence of elementary row operations. Further, we have seen that we can perform any sequence of elementary row operations on A by multiplying A on the left by a product of elementary matrices. This yields the following result, which is stated for later reference.

THEOREM 12.11

Let A be an $n \times m$ matrix. Then there exist elementary matrices $E_1, E_2, \ldots, E_n$ such that $A_R = E_n E_{n-1} \cdots E_1 A$.   ∎

In the next section, we will introduce the rank of a matrix and relate this concept to the reduced form of the matrix.

PROBLEMS FOR SECTION 12.5

In each of Problems 1 through 25, determine whether the matrix is in reduced form. If it is not, list all the conditions of the definition which are violated, and use elementary row operations to reduce the matrix.

1. $\begin{bmatrix} 1 & -1 & 3 \\ 0 & 1 & 2 \\ 0 & 0 & 0 \end{bmatrix}$

2. $\begin{bmatrix} 3 & 1 & 1 & 4 \\ 0 & 1 & 0 & 0 \end{bmatrix}$

3. $\begin{bmatrix} -1 & 4 & 1 & 1 \\ 0 & 0 & 0 & 0 \\ 0 & 0 & 0 & 0 \\ 0 & 0 & 0 & 1 \end{bmatrix}$

4. $\begin{bmatrix} 1 & 0 & 1 & 1 & -1 \\ 0 & 1 & 0 & 0 & 2 \end{bmatrix}$

5. $\begin{bmatrix} 6 & 1 \\ 0 & 0 \\ 1 & 3 \\ 0 & 1 \end{bmatrix}$

6. $\begin{bmatrix} 2 & 2 \\ 1 & 1 \end{bmatrix}$

7. $\begin{bmatrix} -1 & 4 & 6 \\ 2 & 3 & -5 \\ 7 & 1 & 1 \end{bmatrix}$

8. $\begin{bmatrix} -3 & 4 & 4 \\ 0 & 0 & 0 \end{bmatrix}$

9. $\begin{bmatrix} -1 & 2 & 3 & 1 \\ 1 & 0 & 0 & 0 \end{bmatrix}$

10. $\begin{bmatrix} 8 & 2 & 1 & 0 \\ 0 & 1 & 1 & 3 \\ 4 & 0 & 0 & -3 \end{bmatrix}$

11. $\begin{bmatrix} 4 & 1 & -7 \\ 2 & 2 & 0 \\ 0 & 1 & 0 \end{bmatrix}$

12. $\begin{bmatrix} 0 & 0 & 1 \\ 1 & 0 & 0 \\ 0 & 0 & 1 \end{bmatrix}$

13. $\begin{bmatrix} -5 & 1 & 0 & 0 \\ 2 & 0 & 0 & 0 \\ 0 & 1 & 1 & -1 \end{bmatrix}$

14. $\begin{bmatrix} 6 \\ -3 \\ 1 \\ 1 \end{bmatrix}$

15. $\begin{bmatrix} 0 & 0 & 2 & 1 & -1 \\ 1 & -1 & 3 & 0 & 0 \end{bmatrix}$

16. $\begin{bmatrix} 0 & 1 & 2 & 0 & 0 \\ 0 & 0 & 0 & 0 & 1 \end{bmatrix}$

17. $\begin{bmatrix} 1 & 0 & -4 & 0 & 6 \\ 5 & 1 & -3 & -3 & 9 \\ 6 & 3 & 7 & -3 & 1 \end{bmatrix}$

18. $\begin{bmatrix} 0 & -1 & 5 & 2 \\ 1 & 1 & -5 & 2 \\ -5 & 3 & 7 & 3 \\ 0 & 2 & 7 & 0 \end{bmatrix}$

19. $\begin{bmatrix} 1 & -5 & 3 & 0 & 8 \\ -10 & 3 & 7 & 3 & 6 \end{bmatrix}$

20. $\begin{bmatrix} -4 & 0 & 1 & 1 & -5 \end{bmatrix}$

21. $\begin{bmatrix} -12 & 9 & 1 & -2 & 4 \\ 5 & 9 & 1 & 7 & 6 \end{bmatrix}$

22. $\begin{bmatrix} 5 & -2 & 3 \\ 0 & 1 & -6 \\ -3 & 5 & 11 \end{bmatrix}$

23. $\begin{bmatrix} -3 & 6 & 1 \\ 0 & -6 & 4 \\ 1 & -1 & 7 \\ 9 & -6 & 4 \end{bmatrix}$

24. $\begin{bmatrix} -2 & 3 & 8 & 5 \\ 1 & -5 & 3 & 3 \end{bmatrix}$

25. $\begin{bmatrix} 5 & -2 & 1 & 5 \\ 0 & 3 & 3 & -7 \\ 7 & -4 & 1 & 5 \\ 9 & 5 & 3 & -8 \end{bmatrix}$

26. Prove that, for any matrix A, there is a matrix B such that $BA = A_R$.

12.6 *The Rank and Row Space of a Matrix* _____

The number of nonzero rows of the reduced form of a matrix A has a special significance, for example, in the solution of systems of linear algebraic equations and linear differential equations. This number is called the *rank* of A and is denoted rank(A):

$$\text{rank}(A) = \text{number of nonzero rows of } A_R.$$

If A is $n \times m$, obviously rank$(A) \leq n$. Further, since A_R is a reduced matrix, its rank is its number of nonzero rows; hence,

$$\text{rank}(A) = \text{rank}(A_R).$$

EXAMPLE 12.21

Let

$$A = \begin{bmatrix} 1 & -1 & 4 & 2 \\ 0 & 1 & 3 & 2 \\ 3 & -2 & 15 & 8 \end{bmatrix}.$$

We find that

$$A_R = \begin{bmatrix} 1 & 0 & 7 & 4 \\ 0 & 1 & 3 & 2 \\ 0 & 0 & 0 & 0 \end{bmatrix}.$$

Then rank$(A) = 2$, the number of nonzero rows of A_R. Of course, rank$(A_R) = 2$ also. ∎

We will find the following technical lemma useful later.

LEMMA 12.1

Let A be $n \times m$. Then rank$(A) = n$ if and only if $A_R = I_n$. That is, A has rank equal to its number of rows if and only if the reduced form of A is the identity matrix.

Proof   If $A_R = I_n$, A_R has n nonzero rows; hence, rank$(A) = n$. Conversely, suppose that rank$(A) = n$. Then A_R has exactly n nonzero rows and hence no zero rows. By condition (1) of the definition of reduced form, every row of A_R has leading entry 1. By (4) of this definition, A_R has each main diagonal element equal to 1. By (2) of the definition, all entries of column j above and below the main diagonal element are zero. Thus, $A_R = I_n$. ∎

The rank of a matrix can also be related to a vector space formed from the matrix in the following natural way. We have already observed that the rows of an $n \times m$ matrix can be thought of as vectors in R^m. For example, the rows of A in Example 12.21 form three vectors in R^4,

$$\langle 1, -1, 4, 2 \rangle, \qquad \langle 0, 1, 3, 2 \rangle, \qquad \langle 3, -2, 15, 8 \rangle.$$

Vectors formed from the rows of an $n \times m$ matrix are called the *row vectors* of the matrix.

For an arbitrary $n \times m$ matrix A, let S be the set of all vectors in R^m which are linear combinations of the row vectors of A. Then S is a subspace of R^m because (1) a sum of linear combinations of row vectors is again a linear combination of row vectors and (2) a scalar times a linear combination of row vectors is a linear combination of row vectors.

The subspace of R^m consisting of all the linear combinations of the row vectors of A is called the *row space* of A.

EXAMPLE 12.22

For the matrix

$$\begin{bmatrix} 1 & -1 & 4 & 2 \\ 0 & 1 & 3 & 2 \\ 3 & -2 & 15 & 8 \end{bmatrix}$$

of Example 12.21, the row space consists of all vectors in R^4 of the form

$$\alpha\langle 1, -1, 4, 2 \rangle + \beta\langle 0, 1, 3, 2 \rangle + \gamma\langle 3, -2, 15, 8 \rangle. \tag{12.7}$$

In this example, the row vectors are linearly dependent because the third row vector is a linear combination of the first two:

$$\langle 3, -2, 15, 8 \rangle = 3\langle 1, -1, 4, 2 \rangle + \langle 0, 1, 3, 2 \rangle.$$

Thus, any linear combination of the form (12.7) can actually be written as a linear combination of just $\langle 1, -1, 4, 2 \rangle$ and $\langle 0, 1, 3, 2 \rangle$, and the row space of A consists of all vectors in R^4 of the form

$$a\langle 1, -1, 4, 2 \rangle + b\langle 0, 1, 3, 2 \rangle.$$

Since $\langle 1, -1, 4, 2 \rangle$ and $\langle 0, 1, 3, 2 \rangle$ are linearly independent, they form a basis for the row space of A, and this row space has dimension 2. ∎

As we saw in Example 12.21, $\text{rank}(A) = 2$. At least for this matrix, then,

$$\text{rank}(A) = \text{dimension of the row space of } A.$$

It is claimed that the rank of any matrix equals the dimension of its row space. To see why, first observe that this is true for a reduced matrix.

LEMMA 12.2

If A_R is a reduced matrix, the rank of A_R equals the dimension of the row space of A_R.

Proof A_R may or may not have zero rows. In any event, let $\mathbf{F}_1, \ldots, \mathbf{F}_r$ be the nonzero row vectors of A_R. Then the row space consists of all vectors of the form

$$\alpha_1 \mathbf{F}_1 + \cdots + \alpha_r \mathbf{F}_r,$$

since zero row vectors will not contribute anything to such a linear combination.

By definition of a reduced matrix, each $\mathbf{F}_i$ has first nonzero component 1. Further, if this 1 appears in column k of A_R, all the other elements of column k are zero, so each of the other $\mathbf{F}_j$'s has k component zero. By Lemma 11.1 in Section 11.6, $\mathbf{F}_1, \ldots, \mathbf{F}_r$ are linearly independent and hence form a basis for the row space of A_R. Thus, the row space of A_R has dimension r, the number of nonzero rows of A_R, and this is the rank of A_R. ∎

We can now prove that the conclusion of the lemma holds for any matrix.

THEOREM 12.12

For any matrix A, the rank of A equals the dimension of the row space of A.

Proof   We want to show that

$$\text{rank}(A) = \text{dimension of the row space of } A.$$

By Lemma 12.2, we know that

$$\text{rank}(A) = \text{rank}(A_R) = \text{dimension of the row space of } A_R.$$

Thus, it is enough to show that the dimension of the row space of A equals the dimension of the row space of A_R. In fact, we will show that A and A_R have exactly the same row space and hence conclude that these row spaces have the same dimension.

A_R is obtained from A by a sequence of elementary row operations. We will show that each elementary row operation performed on a matrix leaves the row space unchanged. This will enable us to conclude that A and A_R have the same row space.

Suppose that A is $n \times m$, and let the row vectors of A be $\mathbf{F}_1, \ldots, \mathbf{F}_n$. The row space of A consists of all vectors in R^m of the form

$$\alpha_1 \mathbf{F}_1 + \alpha_2 \mathbf{F}_2 + \cdots + \alpha_n \mathbf{F}_n, \tag{12.8}$$

with $\alpha_1, \ldots, \alpha_n$ arbitrary constants. Observe the effects of elementary row operations on such a linear combination.

Type I operation: If we interchange two rows of A to form a matrix B, the row vectors of B are the same as those of A. Thus, A and B have the same row space.

Type II operation: If we multiply row j of A by a nonzero scalar α to form a matrix B, the row vectors of B are $\mathbf{F}_1, \ldots, \alpha \mathbf{F}_j, \ldots, \mathbf{F}_m$. Now the linear combination (12.8) is

$$\alpha_1 \mathbf{F}_1 + \cdots + \alpha \alpha_j \mathbf{F}_j + \cdots + \alpha_n \mathbf{F}_n,$$

and any vector of the form (12.8) can be written in this way (perhaps with different choices of $\alpha_1, \ldots, \alpha_n$). Thus, the row space of B is the same as that of A.

Type III operation: If we add α times row i of A to row j to form a matrix B, the row vectors of B are $\mathbf{F}_1, \mathbf{F}_2, \ldots, \mathbf{F}_{j-1}, \alpha \mathbf{F}_i + \mathbf{F}_j, \mathbf{F}_{j+1}, \ldots, \mathbf{F}_n$. A linear combination of the row vectors of B has the form

$$\alpha_1 \mathbf{F}_1 + \cdots + \alpha_i \mathbf{F}_i + \cdots + \alpha_j(\alpha \mathbf{F}_i + \mathbf{F}_j) + \cdots + \alpha_n \mathbf{F}_n,$$

which can be written

$$\alpha_1 \mathbf{F}_1 + \cdots + (\alpha_i + \alpha \alpha_j)\mathbf{F}_i + \cdots + \alpha_j \mathbf{F}_j + \cdots + \alpha_n \mathbf{F}_n, \tag{12.9}$$

which is still a linear combination of $\mathbf{F}_1, \ldots, \mathbf{F}_n$. Thus, again the row space of B is the same as the row space of A. Since elementary row operations leave the row space unchanged, they leave the dimension of the row space unchanged, and the theorem is proved.   ∎

EXAMPLE 12.23

Here is an additional illustration of these concepts. Let

$$A = \begin{bmatrix} -1 & 4 & 0 & 1 & 6 \\ -2 & 8 & 0 & 2 & 12 \end{bmatrix}.$$

The row vectors of A are

$$\mathbf{F}_1 = \langle -1, 4, 0, 1, 6 \rangle \quad \text{and} \quad \mathbf{F}_2 = \langle -2, 8, 0, 2, 12 \rangle.$$

The row space of A is the subspace of R^5 consisting of all linear combinations

$$\alpha \mathbf{F}_1 + \beta \mathbf{F}_2.$$

But $\mathbf{F}_2 = 2\mathbf{F}_1$. So $\alpha \mathbf{F}_1 + \beta \mathbf{F}_2 = \alpha \mathbf{F}_1 + 2\beta \mathbf{F}_1 = (\alpha + 2\beta)\mathbf{F}_1$. Thus, any linear combination of $\mathbf{F}_1$ and $\mathbf{F}_2$ is really just a scalar multiple of $\mathbf{F}_1$, and the row space of A consists of all scalar multiples of just $\mathbf{F}_1$. This row space has dimension 1, and this is the rank of A. We find that

$$A_R = \begin{bmatrix} 1 & -4 & 0 & -1 & -6 \\ 0 & 0 & 0 & 0 & 0 \end{bmatrix}.$$

The row space of A_R consists of all multiples of $\langle 1, -4, 0, -1, -6 \rangle$ and so has dimension 1 as well. ∎

If A is an $n \times m$ matrix, we may think of the columns as vectors in R^n. The set of all linear combinations of these column vectors forms a subspace of R^n called the *column space* of A. For example, the matrix of Example 12.23 is

$$A = \begin{bmatrix} -1 & 4 & 0 & 1 & 6 \\ -2 & 8 & 0 & 2 & 12 \end{bmatrix}.$$

The column vectors of A are vectors in R^2:

$$\langle -1, -2 \rangle, \quad \langle 4, 8 \rangle, \quad \langle 0, 0 \rangle, \quad \langle 1, 2 \rangle, \quad \text{and} \quad \langle 6, 12 \rangle.$$

The column space consists of all vectors in R^2 of the form

$$\alpha \langle -1, -2 \rangle + \beta \langle 4, 8 \rangle + \gamma \langle 0, 0 \rangle + \delta \langle 1, 2 \rangle + \epsilon \langle 6, 12 \rangle. \tag{12.10}$$

Since $\langle 0, 0 \rangle$ contributes nothing to this linear combination, we can omit this term. Further, $\langle 4, 8 \rangle$, $\langle 1, 2 \rangle$, and $\langle 6, 12 \rangle$ are all scalar multiples of $\langle -1, -2 \rangle$, so any vector written as the linear combination (12.10) is just a scalar multiple of $\langle -1, -2 \rangle$. The column space of A therefore consists of just scalar multiples of $\langle -1, -2 \rangle$. This column space has dimension 1.

Now recall that the row space of A also has dimension 1. This is no coincidence. For any matrix, the row space and column space have the same dimension. This is quite remarkable because the row space consists of vectors in R^m and the column space consists of of vectors in R^n, and these may seem unrelated.

A proof of this equality of row and column space dimensions is outlined in Problem 21. The result is stated here as a theorem for later reference.

THEOREM 12.13

For any matrix A, the row space and the column space have the same dimension. ∎

PROBLEMS FOR SECTION 12.6

In each of Problems 1 through 20, determine rank(A) first by finding A_R and counting the number of nonzero rows and second by listing the row vectors of A and determining a basis for the row space (this will consist of all the linearly independent row vectors).

1. $\begin{bmatrix} -4 & 1 & 3 \\ 2 & 2 & 0 \end{bmatrix}$

2. $\begin{bmatrix} 1 & -1 & 4 \\ 0 & 1 & 3 \\ 2 & -1 & 11 \end{bmatrix}$

3. $\begin{bmatrix} -3 & 1 \\ 2 & 2 \\ 4 & -3 \end{bmatrix}$

4. $\begin{bmatrix} 6 & 0 & 0 & 1 & 1 \\ 12 & 0 & 0 & 2 & 2 \\ 1 & -1 & 0 & 0 & 0 \end{bmatrix}$

5. $\begin{bmatrix} 8 & -4 & 3 & 2 \\ 1 & -1 & 1 & 0 \end{bmatrix}$

6. $\begin{bmatrix} 1 & 3 & 0 \\ 0 & 0 & 1 \end{bmatrix}$

7. $\begin{bmatrix} 2 & 2 & 1 \\ 1 & -1 & 3 \\ 0 & 0 & 1 \\ 4 & 0 & 7 \end{bmatrix}$

8. $\begin{bmatrix} 0 & -1 & 0 \\ 0 & 0 & -1 \\ 0 & 0 & 2 \end{bmatrix}$

9. $\begin{bmatrix} 0 & 4 & 3 \\ 6 & 1 & 0 \\ 2 & 2 & 2 \end{bmatrix}$

10. $\begin{bmatrix} 1 & 0 & 0 \\ 2 & 0 & 0 \\ 1 & 0 & -1 \\ 3 & 0 & 0 \end{bmatrix}$

11. $\begin{bmatrix} -3 & 2 & 2 \\ 1 & 0 & 5 \\ 0 & 0 & 2 \end{bmatrix}$

12. $\begin{bmatrix} -4 & -2 & 1 & 6 \\ 0 & 4 & -4 & 2 \\ 1 & 0 & 0 & 0 \end{bmatrix}$

13. $\begin{bmatrix} -2 & 5 & 7 \\ 0 & 1 & -3 \\ -4 & 11 & 11 \end{bmatrix}$

14. $\begin{bmatrix} -3 & 2 & 1 & 1 \\ 6 & -4 & -2 & -2 \end{bmatrix}$

15. $\begin{bmatrix} 7 & -2 & 1 & -2 \\ 0 & 2 & 6 & 3 \\ 7 & 2 & 13 & 4 \\ 7 & 0 & 7 & 1 \end{bmatrix}$

16. $\begin{bmatrix} -4 & 2 & 5 \\ 0 & 0 & 0 \\ 0 & 0 & 0 \end{bmatrix}$

17. $\begin{bmatrix} 4 & 1 & -3 & 5 \\ 2 & 0 & 0 & -2 \\ 13 & 2 & 0 & -1 \end{bmatrix}$

18. $\begin{bmatrix} -4 & 2 & 6 & 1 \\ 0 & 0 & 4 & 1 \\ 4 & -2 & -2 & 0 \end{bmatrix}$

19. $\begin{bmatrix} 5 & -2 & 5 & 6 & 1 \\ -2 & 0 & 1 & -1 & 3 \\ -1 & -2 & 8 & 3 & 10 \end{bmatrix}$

20. $\begin{bmatrix} 3 & -3 & 5 & 1 \\ 0 & 2 & 1 & -5 \\ 0 & 0 & 0 & 1 \end{bmatrix}$

21. Let $A = [a_{ij}]$ be an $n \times m$ matrix. Prove that the row space of A has the same dimension as the column space of A. *Hint:* Let the row vectors of A be $\mathbf{R}_1, \ldots, \mathbf{R}_n$. If the dimension of the row space is r, exactly r of these vectors are linearly independent. Suppose we have labeled these vectors so that $\mathbf{R}_1, \ldots, \mathbf{R}_r$ are linearly independent, while $\mathbf{R}_{r+1}, \ldots, \mathbf{R}_n$ are linear combinations of $\mathbf{R}_1, \ldots, \mathbf{R}_r$. Write

$$\mathbf{R}_{r+1} = \alpha_{r+1,1}\mathbf{R}_1 + \cdots + \alpha_{r+1,r}\mathbf{R}_r$$
$$\mathbf{R}_{r+2} = \alpha_{r+2,1}\mathbf{R}_1 + \cdots + \alpha_{r+2,r}\mathbf{R}_r$$
$$\vdots \qquad \vdots \qquad \qquad \vdots$$
$$\mathbf{R}_n = \alpha_{n,1}\mathbf{R}_1 \quad + \cdots + \alpha_{n,r}\mathbf{R}_r.$$

Show that column j of A can be written

$$\begin{bmatrix} a_{1j} \\ a_{2j} \\ \vdots \\ a_{nj} \end{bmatrix} = a_{1j}\begin{bmatrix} 1 \\ 0 \\ \vdots \\ \alpha_{r+1,1} \\ \vdots \\ \alpha_{n1} \end{bmatrix} + a_{2j}\begin{bmatrix} 0 \\ 1 \\ \vdots \\ \alpha_{r+1,2} \\ \vdots \\ \alpha_{n2} \end{bmatrix} + \cdots + a_{rj}\begin{bmatrix} 0 \\ 0 \\ \vdots \\ 1 \\ \alpha_{r+1,r} \\ \vdots \\ \alpha_{nr} \end{bmatrix}.$$

Conclude that every column of A is a linear combination of r vectors in R^n and hence that the dimension of the column space does not exceed r. By reversing the roles of rows and columns in this argument, show that the dimension of the column space is at least r.

In each of Problems 22 through 25, find a basis for the row space of A and also a basis for the column space of A, and show that these two spaces have the same dimension.

22. $\begin{bmatrix} -1 & 4 & 2 \\ 0 & 1 & 6 \\ 2 & 2 & 0 \\ 0 & 0 & 1 \end{bmatrix}$
23. $\begin{bmatrix} 1 & 1 & -4 & 2 \\ 0 & 1 & 1 & 3 \end{bmatrix}$
24. $\begin{bmatrix} 1 & -1 & 3 \\ 0 & -1 & 2 \\ 1 & -1 & 3 \end{bmatrix}$
25. $\begin{bmatrix} 8 & 4 \\ 2 & 1 \\ 0 & 3 \end{bmatrix}$

26. Use Theorem 12.13 to show that $\text{rank}(A) = \text{rank}(A^t)$.

27. Let V be the set of all $n \times m$ matrices having real numbers as elements. Show that V is a vector space, using the usual addition of $n \times m$ matrices and multiplication of a matrix by a scalar. Find a basis for this vector space and show that its dimension is nm. *Hint:* Look first at simple cases, such as $1 \times 2, 2 \times 2$, or 2×3 matrices, to get some idea about how to form a basis.

28. Let V be the set of all $n \times m$ matrices having complex numbers as elements. Show that V is a vector space, with the usual addition of matrices and multiplication of a matrix by a number. Find a basis for this space and determine its dimension.

12.7 *Solution of Homogeneous Systems of Linear Equations*

We have now assembled enough notation and machinery to use matrices to solve systems of linear algebraic equations. Consider a typical such system:

$$
\begin{aligned}
a_{11}x_1 + a_{12}x_2 + \cdots + a_{1m}x_m &= b_1 \\
a_{21}x_1 + a_{22}x_2 + \cdots + a_{2m}x_m &= b_2 \\
\vdots \qquad \vdots \qquad\qquad \vdots \qquad \vdots \\
a_{n1}x_1 + a_{n2}x_2 + \cdots + a_{nm}x_m &= b_n.
\end{aligned}
\tag{12.11}
$$

These equations are called linear because each unknown x_j appears to the first power only and there are no cross product terms $x_i x_j$.

The system (12.11) has n equations in m unknowns. The object is to find all values of $x_1, \ldots, x_m$ satisfying all equations of the system simultaneously, given the coefficients $a_{11}, a_{12}, \ldots, a_{nm}$ and the numbers $b_1, \ldots, b_n$.

As noted in Section 12.1, we can write the system (12.11) in matrix form as

$$AX = B,$$

in which

$$
A = \begin{bmatrix} a_{11} & a_{12} & \cdots & a_{1m} \\ a_{21} & a_{22} & \cdots & a_{2m} \\ \vdots & \vdots & & \vdots \\ a_{n1} & a_{n2} & \cdots & a_{nm} \end{bmatrix}, \quad X = \begin{bmatrix} x_1 \\ x_2 \\ \vdots \\ x_m \end{bmatrix}, \quad \text{and} \quad B = \begin{bmatrix} b_1 \\ b_2 \\ \vdots \\ b_n \end{bmatrix}.
$$

A is called the *coefficient matrix* of the system. The number of rows of A is the number of equations, while the number of columns is the number of unknowns. Row k of A and B contains the coefficients of equation k,

$$a_{k1}x_1 + \cdots + a_{km}x_m = b_k,$$

while column j of A lists the coefficients of x_j in the equations of the system.

In order to use matrix notation, we will write a solution $x_1 = \alpha_1$, $x_2 = \alpha_2, \ldots,$ $x_m = \alpha_m$ as an $m \times 1$ column matrix

$$X = \begin{bmatrix} \alpha_1 \\ \alpha_2 \\ \vdots \\ \alpha_m \end{bmatrix}.$$

This enables us to verify a solution by substituting it into the equation $AX = B$ and carrying out the matrix multiplication (A is $n \times m$, and the solution is written as an $m \times 1$ matrix, so AX is $n \times 1$, the same as B).

EXAMPLE 12.24

The system

$$x_1 - 2x_2 = 3$$
$$4x_1 + 6x_2 = -5$$

can be written

$$\begin{bmatrix} 1 & -2 \\ 4 & 6 \end{bmatrix} \begin{bmatrix} x_1 \\ x_2 \end{bmatrix} = \begin{bmatrix} 3 \\ -5 \end{bmatrix}.$$

A solution of this system is $x_1 = \frac{8}{14}$, $x_2 = -\frac{17}{14}$. We will write this solution as the column matrix

$$\begin{bmatrix} \frac{8}{14} \\ -\frac{17}{14} \end{bmatrix}.$$

In this way, we can substitute back into the matrix formulation $AX = B$ to check the solution.

In this example, the solution is easily obtained by elimination (multiply the first equation by -4, add the result to the second equation, and solve for x_2; then solve for x_1). ∎

EXAMPLE 12.25

The system

$$x_1 + 2x_2 = 1$$

has one equation in two unknowns and can be written

$$\begin{bmatrix} 1 & 2 \end{bmatrix} \begin{bmatrix} x_1 \\ x_2 \end{bmatrix} = \begin{bmatrix} 1 \end{bmatrix}.$$

We can solve this system by writing $x_1 = 1 - 2x_2$. For any choice of x_2, say, $x_2 = \alpha$, we have the solution $x_1 = 1 - 2\alpha$, $x_2 = \alpha$. In matrix notation, the solution is

$$\begin{bmatrix} 1 - 2\alpha \\ \alpha \end{bmatrix},$$

with α arbitrary.   ■

Both of these systems were simple, and matrix notation did not contribute anything to their solution. We will now consider general methods for solving linear systems with arbitrarily many equations and unknowns. To do this, we consider first the special case in which the system is homogeneous.

The system $AX = B$ is *homogeneous* if each element of B is zero. If some $b_j \neq 0$, the system is *nonhomogeneous*. For example,

$$\begin{bmatrix} 1 & -1 & 2 \\ 0 & 1 & 6 \end{bmatrix} \begin{bmatrix} x_1 \\ x_2 \\ x_3 \end{bmatrix} = \begin{bmatrix} 0 \\ 0 \end{bmatrix}$$

is a homogeneous system, while

$$\begin{bmatrix} 1 & -1 & 2 \\ 0 & 1 & 6 \end{bmatrix} \begin{bmatrix} x_1 \\ x_2 \\ x_3 \end{bmatrix} = \begin{bmatrix} -4 \\ 0 \end{bmatrix}$$

is nonhomogeneous.

For the remainder of this section, we will concentrate on homogeneous systems. In the next section, we will deal with nonhomogeneous systems. Thus, consider the system

$$AX = O, \tag{12.12}$$

where A is $n \times m$, X is $m \times 1$, and O is the $n \times 1$ zero matrix. For simplicity, we will write O for this matrix instead of O_{n1}. The key to solving the system (12.12) is the following observation. Consider the *reduced system*

$$A_R X = O, \tag{12.13}$$

in which we replace A with its reduced form. Because of the special structure of A_R, the system (12.13) is usually easier to solve than the system (12.12). We claim that

the systems (12.12) and (12.13)

have exactly the same solutions.

The reason for this can be seen in the way we obtain A_R from A. We do this by performing elementary row operations. Since rows of A correspond to equations of the system, the row operations performed on A correspond to operations performed on equations of the system. Now consider these operations in turn.

Type I: If we interchange two rows of A, we interchange two equations of the system. This has no effect on the solutions.

Type II: If we multiply a row of A by a nonzero scalar α, we multiply the corresponding equation by α. Again, this does not alter the solutions.

Type III: If we add $\alpha \cdot$ row i of A to row j, we are adding α times equation i to equation j. The new system will have the same solutions as the original system.

In fact, the row operations we have been using are exactly the operations used to solve the system by eliminating unknowns.

We can now outline a procedure called the *Gauss-Jordan reduction method* for solving the system $AX = O$.

GAUSS-JORDAN REDUCTION METHOD FOR SOLVING $AX = O$

Step 1   Reduce A to A_R. Since the reduced system has the same solutions as the original system, we will work with the reduced system $A_R X = O$.

Step 2   In the system $A_R X = O$, label each unknown as dependent or independent according to the following test. If column j contains the leading entry of any row of A, called x_j *dependent*; otherwise, x_j is *independent*.

Step 3   Express each dependent unknown in terms of the independent ones, using the rows of A_R. If, say, x_j is dependent because column j contains the leading entry of row i, we can solve for x_j in terms of independent unknowns using equation i.

Step 4   For the solution, the independent unknowns can be assigned any values; the dependent unknowns are expressed in terms of the independent ones from step 3.

This method is sometimes also known as *complete pivoting*. Although we can carry out the method by working directly with the equations, it is efficient to work with the matrix of coefficients.

EXAMPLE 12.26

Solve the system

$$x_1 - 3x_2 + 2x_3 = 0$$
$$-2x_1 + x_2 - 3x_3 = 0.$$

This system is easily solved without matrices, but we want to illustrate the matrix method. The system is $AX = O$, where

$$A = \begin{bmatrix} 1 & -3 & 2 \\ -2 & 1 & -3 \end{bmatrix}, \quad X = \begin{bmatrix} x_1 \\ x_2 \\ x_3 \end{bmatrix}, \quad \text{and} \quad O = \begin{bmatrix} 0 \\ 0 \end{bmatrix}.$$

Step 1   Reduce A. Proceed as follows:

$$A \xrightarrow[\text{to row 2}]{\text{add 2(row 1)}} \begin{bmatrix} 1 & -3 & 2 \\ 0 & -5 & 1 \end{bmatrix}$$

$$\xrightarrow[\text{by } -\frac{1}{5}]{\text{multiply row 2}} \begin{bmatrix} 1 & -3 & 2 \\ 0 & 1 & -\frac{1}{5} \end{bmatrix}$$

$$\xrightarrow[\text{to row 1}]{\text{add 3(row 2)}} \begin{bmatrix} 1 & 0 & \frac{7}{5} \\ 0 & 1 & -\frac{1}{5} \end{bmatrix} = A_R.$$

Step 2   Identify dependent and independent unknowns. The leading entry of row 1 is in column 1, so x_1 is dependent; similarly, x_2 is dependent. Finally, x_3 is independent.

Step 3   Write the dependent unknowns in terms of the independent ones. From row 1 of A_R, $x_1 + \frac{7}{5}x_3 = 0$, so $x_1 = -\frac{7}{5}x_3$. From row 2, $x_2 - \frac{1}{5}x_3 = 0$, so $x_2 = \frac{1}{5}x_3$. In these equations, x_3 is arbitrary.

Step 4   For convenience, let $x_3 = \alpha$, which can be any number. The solution of the system is $x_1 = -\frac{7}{5}\alpha$, $x_2 = \frac{1}{5}\alpha$, $x_3 = \alpha$. In matrix form, the solution is

$$\begin{bmatrix} -\dfrac{7\alpha}{5} \\ \dfrac{\alpha}{5} \\ \alpha \end{bmatrix},$$

or

$$\alpha\begin{bmatrix} -\frac{7}{5} \\ \frac{1}{5} \\ 1 \end{bmatrix}.$$

This expression is called the *general solution* of the system because we get all solutions by choosing α to have different values. ∎

The general solution of a system $AX = O$ may have more than one arbitrary constant, as the next example shows.

EXAMPLE 12.27

Solve the system

$$\begin{aligned} x_1 - 3x_2 + x_3 - 7x_4 + 4x_5 &= 0 \\ x_1 + 2x_2 - 3x_3 \qquad\qquad &= 0 \\ x_2 - 4x_3 \qquad + x_5 &= 0. \end{aligned}$$

This system can be written $AX = O$, with

$$A = \begin{bmatrix} 1 & -3 & 1 & -7 & 4 \\ 1 & 2 & -3 & 0 & 0 \\ 0 & 1 & -4 & 0 & 1 \end{bmatrix}.$$

To solve this system, first reduce A. We find (details are omitted) that

$$A_R = \begin{bmatrix} 1 & 0 & 0 & -\frac{35}{16} & \frac{13}{16} \\ 0 & 1 & 0 & \frac{28}{16} & -\frac{20}{16} \\ 0 & 0 & 1 & \frac{7}{16} & -\frac{9}{16} \end{bmatrix}.$$

From A_R, read that x_1, x_2, and x_3 are dependent (because columns 1, 2, and 3 contain leading entries of rows) and x_4 and x_5 are independent. From row 1, we have

$$x_1 - \tfrac{35}{16}x_4 + \tfrac{13}{16}x_5 = 0$$

and hence

$$x_1 = \tfrac{35}{16}x_4 - \tfrac{13}{16}x_5,$$

and we have written x_1 in terms of the independent unknowns. Similarly, from rows 2 and 3 of A_R, we read that

$$x_2 = -\tfrac{28}{16}x_4 + \tfrac{20}{16}x_5$$

and

$$x_3 = -\tfrac{7}{16}x_4 + \tfrac{9}{16}x_5.$$

With $x_4 = \alpha$ and $x_5 = \beta$, and both α and β arbitrary, the general solution is

$$\begin{bmatrix} \tfrac{35}{16}\alpha - \tfrac{13}{16}\beta \\ -\tfrac{28}{16}\alpha + \tfrac{20}{16}\beta \\ -\tfrac{7}{16}\alpha + \tfrac{9}{16}\beta \\ \alpha \\ \beta \end{bmatrix}.$$

This general solution can also be written

$$\alpha \begin{bmatrix} \tfrac{35}{16} \\ -\tfrac{28}{16} \\ -\tfrac{7}{16} \\ 1 \\ 0 \end{bmatrix} + \beta \begin{bmatrix} -\tfrac{13}{16} \\ \tfrac{20}{16} \\ \tfrac{9}{16} \\ 0 \\ 1 \end{bmatrix}.$$

Since α and β are arbitrary, we can replace them with the equally arbitrary $a = \alpha/16$ and $b = \beta/16$ to write the general solution more neatly as

$$a \begin{bmatrix} 35 \\ -28 \\ -7 \\ 16 \\ 0 \end{bmatrix} + b \begin{bmatrix} -13 \\ 20 \\ 9 \\ 0 \\ 16 \end{bmatrix}. \quad \blacksquare$$

At this point, we will make an important observation. Notice that the number of arbitrary scalars in the general solution has, in both examples, equaled the number of independent unknowns. If A is $n \times m$, the number of independent unknowns is the total number of unknowns m minus the number of dependent unknowns. But the number of dependent unknowns is the number of rows of A_R having leading entries and hence equals the number of nonzero rows of A_R, or the rank of A. This argument proves the following theorem.

THEOREM 12.14

If A is $n \times m$, the number of arbitrary scalars in the general solution of $AX = O$ is $m - \text{rank}(A)$. $\blacksquare$

As a check, in Example 12.26 we had $m = 3$ and rank$(A) = 2$, and the general solution had one arbitrary constant. In Example 12.27, we had $m = 5$ and rank$(A) = 3$, and there were two arbitrary constants in the general solution.

Theorem 12.14 enables us to tell from the rank of the coefficient matrix and the number of unknowns how many arbitrary constants to expect in the general solution.

EXAMPLE 12.28

Solve the system

$$
\begin{aligned}
-x_2 + 2x_3 + 4x_4 &= 0 \\
- x_3 + 3x_4 &= 0 \\
2x_1 + x_2 + 3x_3 + 7x_4 &= 0 \\
6x_1 + 2x_2 + 10x_3 + 28x_4 &= 0.
\end{aligned}
$$

The matrix of coefficients is

$$
A = \begin{bmatrix} 0 & -1 & 2 & 4 \\ 0 & 0 & -1 & 3 \\ 2 & 1 & 3 & 7 \\ 6 & 2 & 10 & 28 \end{bmatrix}.
$$

We find that

$$
A_R = \begin{bmatrix} 1 & 0 & 0 & 13 \\ 0 & 1 & 0 & -10 \\ 0 & 0 & 1 & -3 \\ 0 & 0 & 0 & 0 \end{bmatrix}.
$$

A is 4×4, and rank$(A) = 3$, so the general solution will have $4 - 3$, or one, arbitrary constant.

We have x_1, x_2, and x_3 dependent and x_4 independent. From rows 1, 2, and 3 of A_R, we can write

$$
\begin{aligned}
x_1 &= -13x_4 \\
x_2 &= 10x_4 \\
x_3 &= 3x_4.
\end{aligned}
$$

The general solution can be written

$$
\alpha \begin{bmatrix} -13 \\ 10 \\ 3 \\ 1 \end{bmatrix},
$$

with α arbitrary.   ■

All of these considerations can be related to vector spaces as follows. Suppose that A is $n \times m$. If X_1 and X_2 are two solutions written as $m \times 1$ matrices,

$$
A(X_1 + X_2) = AX_1 + AX_2 = O,
$$

so $X_1 + X_2$ is a solution. Further, for any constant α,

$$A(\alpha X_1) = \alpha A X_1 = \alpha O = O,$$

so αX_1 is a solution. Thus, the set of all solutions of $AX = O$ forms a vector space.

We can think of any solution as an m-vector $\langle x_1, x_2, \ldots, x_m \rangle$. Then the set of all solutions forms a subspace of R^m. This subspace is called the *solution space* of the system $AX = O$. Above, when we found the general solution of a system, we wrote it as a linear combination of vectors, each containing one arbitrary constant. These vectors are linearly independent and constitute a basis for the solution space.

To illustrate, in Example 12.26, $\langle -\frac{7}{5}, \frac{1}{5}, 1 \rangle$ forms a basis for the solution space, which has dimension 1. In Example 12.27, $\langle 35, -28, -7, 16, 0 \rangle$ and $\langle -13, 20, 9, 0, 16 \rangle$ form a basis for the solution space, which has dimension 2. And in Example 12.28, $\langle -13, 10, 3, 1 \rangle$ forms a basis for the one-dimensional solution space.

The number of arbitrary constants in the general solution is the dimension of the solution space. We have seen that this number is $m - \text{rank}(A)$ if A is $n \times m$. Thus, the dimension of the solution space is the number of columns of A minus the rank of A.

EXAMPLE 12.29

Solve the system

$$
\begin{aligned}
-x_1 \quad\quad + \quad x_3 + x_4 + 2x_5 &= 0 \\
x_2 + 3x_3 \quad\quad + 4x_5 &= 0 \\
x_1 + 2x_2 + \quad x_3 + x_4 + \quad x_5 &= 0 \\
-3x_1 + \quad x_2 \quad\quad\quad + 4x_5 &= 0.
\end{aligned}
$$

The coefficient matrix is

$$
A = \begin{bmatrix}
-1 & 0 & 1 & 1 & 2 \\
0 & 1 & 3 & 0 & 4 \\
1 & 2 & 1 & 1 & 1 \\
-3 & 1 & 0 & 0 & 4
\end{bmatrix}.
$$

We find that

$$
A_R = \begin{bmatrix}
1 & 0 & 0 & 0 & -\frac{9}{8} \\
0 & 1 & 0 & 0 & \frac{5}{8} \\
0 & 0 & 1 & 0 & \frac{9}{8} \\
0 & 0 & 0 & 1 & -\frac{2}{8}
\end{bmatrix}.
$$

Since $m = 5$ and the rank of A is 4, there is one independent unknown, x_5, and the dimension of the solution space is 1. This means that the solution space will consist of scalar multiples of a single vector in R^5.

From the rows of A_R, we read the following dependencies:

$$
\begin{aligned}
x_1 &= \tfrac{9}{8} x_5 \\
x_2 &= -\tfrac{5}{8} x_5 \\
x_3 &= -\tfrac{9}{8} x_5 \\
x_4 &= \tfrac{2}{8} x_5.
\end{aligned}
$$

With $x_5 = \alpha$, any number, the general solution is

$$\alpha \begin{bmatrix} \frac{9}{8} \\ -\frac{5}{8} \\ -\frac{9}{8} \\ \frac{2}{8} \\ 1 \end{bmatrix}.$$

The 5-vector $\langle \frac{9}{8}, -\frac{5}{8}, -\frac{9}{8}, \frac{2}{8}, 1 \rangle$ forms a basis for the solution space of this system.   ∎

A homogeneous system $AX = O$ always has at least one solution, namely,

$$x_1 = x_2 = \cdots = x_m = 0.$$

This is called the *trivial solution*. In all of the above examples, there were nontrivial solutions as well. It is possible, however, for a homogeneous system to have only the trivial solution. The next theorem tells us when this will occur for the case in which the number of unknowns equals the number of equations.

THEOREM 12.15

If A is $n \times n$, then $AX = O$ has only the trivial solution if and only if $\text{rank}(A) = n$.

Proof   Suppose that $\text{rank}(A) = n$. By Lemma 12.1 in the preceding section, $A_R = I_n$. Now, solutions of $AX = O$ are the same as solutions of $A_R X = O$, which is the system $I_n X = O$, or $X = O$. Thus, the system $AX = O$ has only the trivial solution when $\text{rank}(A) = n$.

Conversely, suppose that the system $AX = O$ has only the trivial solution. Then there can be no arbitrary constants in the general solution, since we could assign such constants nonzero values. By Theorem 12.14, $m - \text{rank}(A) = 0$. But A is square, so $m = n$ and $n - \text{rank}(A) = 0$; hence, $\text{rank}(A) = n$.   ∎

Here is a useful alternative way of stating the conclusion of this theorem.

COROLLARY TO THEOREM 12.15

Let A be $n \times n$. Then the system $AX = O$ has a nontrivial solution if and only if $\text{rank}(A) < n$.   ∎

EXAMPLE 12.30

Consider the system

$$3x_1 - 11x_2 + 5x_3 = 0$$
$$4x_1 + x_2 - 10x_3 = 0$$
$$4x_1 + 9x_2 - 6x_3 = 0.$$

The coefficient matrix is

$$A = \begin{bmatrix} 3 & -11 & 5 \\ 4 & 1 & -10 \\ 4 & 9 & -6 \end{bmatrix}.$$

We find that

$$A_R = \begin{bmatrix} 1 & 0 & 1 \\ 0 & 1 & 0 \\ 0 & 0 & 1 \end{bmatrix} = I_3,$$

and $\text{rank}(A) = 3$. The system $AX = O$ has the same solutions as the system $A_R X = I_3 X = X = O$, from which we read the trivial solution $x_1 = x_2 = x_3 = 0$ as the only solution. ∎

Here is an important circumstance in which a homogeneous system always has a nontrivial solution.

THEOREM 12.16

A homogeneous linear system with more unknowns than equations always has a nontrivial solution.

Proof   Write the system as $AX = O$, with A $n \times m$. This system has n equations and m unknowns. If there are more unknowns than equations, $m > n$.

Now, $\text{rank}(A)$ is the number of nonzero rows of A_R and cannot exceed n. Since $\text{rank}(A) \le n$, $m - \text{rank}(A) \ge m - n > 0$. Thus, there is at least one independent unknown which can be assigned any value in the general solution and hence can be given nonzero values, yielding a nontrivial solution. ∎

Alternatively, we could argue that $m - \text{rank}(A) > 0$ and hence the dimension of the solution space is positive, so the solution space has nonzero vectors in it. These nonzero vectors correspond to nontrivial solutions of the system.

We will return to homogeneous linear systems later. In the next section, we will discuss nonhomogeneous linear systems.

PROBLEMS FOR SECTION 12.7

In each of Problems 1 through 20, find the general solution of the homogeneous system by using Gauss-Jordan reduction. In each case, also determine the dimension of the solution space.

1. $x_1 + 2x_2 - x_3 + x_4 = 0$
$\quad\quad x_2 - x_3 + x_4 = 0$

2. $-3x_1 + x_2 - x_3 + x_4 + x_5 = 0$
$\quad\quad x_2 + x_3 \quad\quad + 4x_5 = 0$
$\quad\quad -3x_3 + 2x_4 + x_5 = 0$

3. $-2x_1 + x_2 + 2x_3 = 0$
$\quad\quad x_1 - x_2 \quad\quad = 0$
$\quad\quad x_1 + x_2 \quad\quad = 0$

4. $4x_1 + x_2 - 3x_3 + x_4 = 0$
$\quad\quad 2x_1 \quad\quad - x_3 \quad\quad = 0$

5. $x_1 - x_2 + 3x_3 - x_4 + 4x_5 = 0$
$\quad 2x_1 - 2x_2 + x_3 + x_4 \quad\quad = 0$
$\quad x_1 \quad\quad - 2x_3 \quad\quad + x_5 = 0$
$\quad\quad x_3 + x_4 - x_5 = 0$

6. $6x_1 - x_2 + x_3 \quad\quad\quad = 0$
$\quad x_1 \quad\quad - x_4 + 2x_5 = 0$
$\quad x_1 \quad\quad - 2x_5 = 0$

7. $-10x_1 - x_2 + 4x_3 - x_4 + x_5 - x_6 = 0$
$\quad\quad x_2 - x_3 + 3x_4 \quad\quad = 0$
$\quad 2x_1 - x_2 \quad\quad + x_5 \quad\quad = 0$
$\quad\quad x_2 \quad\quad - x_4 \quad\quad + x_6 = 0$

8. $8x_1 \quad\quad - 2x_3 \quad\quad + x_6 = 0$
$\quad 2x_1 - x_2 \quad\quad + 3x_4 \quad\quad - x_6 = 0$
$\quad\quad x_2 + x_3 \quad\quad - 2x_5 - x_6 = 0$
$\quad\quad x_4 - 3x_5 + 2x_6 = 0$

9.
$$x_2 - 3x_4 + x_5 = 0$$
$$2x_1 - x_2 + x_4 = 0$$
$$2x_1 - 3x_2 + 4x_5 = 0$$

10.
$$4x_1 - 3x_2 + x_4 + x_5 - 3x_6 = 0$$
$$2x_2 + 4x_4 - x_5 - 6x_6 = 0$$
$$3x_1 - 2x_2 + 4x_5 - x_6 = 0$$
$$2x_1 + x_2 - 3x_3 + 4x_4 = 0$$

11.
$$x_1 - 2x_2 + x_5 - x_6 + x_7 = 0$$
$$x_3 - x_4 + x_5 - 2x_6 + 3x_7 = 0$$
$$x_1 - x_5 + 2x_6 = 0$$
$$2x_1 - 3x_4 + x_5 = 0$$

12.
$$2x_1 - 4x_5 + x_7 + x_8 = 0$$
$$2x_2 - x_6 + x_7 - x_8 = 0$$
$$x_3 - 4x_4 + x_8 = 0$$
$$x_2 - x_3 + x_4 = 0$$
$$x_2 - x_5 + x_6 - x_7 = 0$$

For ease of recognition, problems up to this point have been written with the unknowns spaced so that coefficients appear almost as they do in the coefficient matrix. In Problems 13 through 20, the equations are written as they would probably appear in a normal encounter. Solve each system of equations, exercising caution in writing the coefficient matrix.

13.
$$x_1 - 4x_3 + x_5 = 0$$
$$2x_3 - 4x_4 = 0$$
$$x_2 - 5x_4 + 6x_5 = 0$$

14.
$$12x_1 + 4x_2 - x_3 + x_4 = 0$$
$$-x_1 + 2x_2 + 5x_3 - 5x_4 = 0$$

15.
$$-5x_1 + x_2 - 3x_3 + 4x_5 = 0$$
$$x_2 - 5x_3 + 7x_5 - x_4 = 0$$

16.
$$-3x_1 - 4x_2 + x_3 + x_4 = 0$$
$$x_2 - 4x_3 + 2x_4 = 0$$
$$-2x_1 + 4x_2 - 5x_3 + 2x_4 = 0$$

17.
$$9x_2 + x_3 - 5x_4 = 0$$
$$x_1 + x_2 - 4x_4 = 0$$
$$x_3 + 8x_4 = 0$$

18.
$$-3x_1 + 3x_2 - 8x_3 + x_5 = 0$$
$$x_3 + 6x_4 - 2x_5 = 0$$
$$x_2 + x_4 + 5x_5 = 0$$
$$x_1 + x_2 + x_4 + 7x_5 = 0$$

19.
$$5x_1 + x_2 - 4x_3 - x_6 = 0$$
$$x_2 + x_4 + 6x_5 - x_6 = 0$$
$$2x_1 + x_3 - 5x_4 + 11x_6 = 0$$

20.
$$x_3 + 3x_4 - x_5 = 0$$
$$2x_1 + 5x_2 + 3x_3 - x_5 = 0$$

21. Suppose we have a homogeneous system of n equations in m unknowns and $n \geq m$. Can there be nontrivial solutions?

12.8 *Solution of Nonhomogeneous Systems of Linear Equations*

We will now consider the nonhomogeneous linear system

$$AX = B, \tag{12.14}$$

in which A is $n \times m$ and B is $n \times 1$ and has at least one nonzero element. Recall that, with the homogeneous system $AX = O$, row operations on the coefficient matrix corresponded to similar operations on the equations. For example, multiplying row j of A by a nonzero scalar α corresponds to multiplying equation j by α.

Now we must be more careful. If, say, we multiply row j by nonzero α, and $b_j \neq 0$, we must multiply b_j by α also. This means that the terms $b_1, \ldots, b_n$, not all of which are zero, must be included in our calculations and row operations.

We can do this efficiently by attaching B as an extra column to the right of A, forming a new $n \times (m + 1)$ matrix, denoted $[A \mid B]$ and called the *augmented matrix* of the system (12.14). $[A \mid B]$ has the appearance

$$[A \mid B] = \begin{bmatrix} a_{11} & a_{12} & \cdots & a_{1n} & \vdots & b_1 \\ a_{21} & a_{22} & \cdots & a_{2n} & \vdots & b_2 \\ \vdots & \vdots & & \vdots & \vdots & \vdots \\ a_{n1} & a_{n2} & \cdots & a_{nm} & \vdots & b_n \end{bmatrix}.$$

For example, consider the system

$$2x_1 - x_2 + 3x_3 = 4$$
$$x_1 + 3x_2 - x_3 = -2.$$

This can be written $AX = B$, with

$$A = \begin{bmatrix} 2 & -1 & 3 \\ 1 & 3 & -1 \end{bmatrix} \quad \text{and} \quad B = \begin{bmatrix} 4 \\ -2 \end{bmatrix}.$$

The augmented matrix is

$$[A \mid B] = \begin{bmatrix} 2 & -1 & 3 & \vdots & 4 \\ 1 & 3 & -1 & \vdots & -2 \end{bmatrix}.$$

The dashed line inserted in the matrix helps us remember that the last column is really the attached matrix B of a nonhomogeneous system $AX = B$.

We will soon have occasion to find the reduced forms of both A and $[A \mid B]$. The following lemma simplifies this task.

LEMMA 12.3

Let A be $n \times m$. Then, for some $n \times 1$ column matrix C,

$$[A \mid B]_R = [A_R \mid C]. \quad \blacksquare$$

That is, the reduced form of the augmented matrix $[A \mid B]$ will have as its first m columns exactly the reduced form of A. The reason for this can be seen by reviewing the procedure for finding the reduced form of a matrix. We perform elementary row operations, beginning with the left-most column containing a leading entry, and work from the left to the right through the columns of the matrix. Thus, in finding the reduced form of $[A \mid B]$, we deal first with the first m columns, which constitute A. The elementary row operations used in reducing A will, of course, operate on elements of the last column, B, as well; hence, the reduced form of the augmented matrix will be of the general form $[A_R \mid C]$. Here is an example to clarify these ideas.

EXAMPLE 12.31

Consider the system

$$2x_1 + 4x_2 - x_3 = 8$$
$$x_1 - 2x_2 + x_3 = 1.$$

The augmented matrix is

$$[A \mid B] = \begin{bmatrix} 2 & 4 & -1 & \vdots & 8 \\ 1 & -2 & 1 & \vdots & 1 \end{bmatrix}.$$

Reduce this augmented matrix:

$$[A \mid B] \xrightarrow[\text{to row 1}]{\text{add } -2(\text{row 2})} \begin{bmatrix} 0 & 8 & -3 & \vdots & 6 \\ 1 & -2 & 1 & \vdots & 1 \end{bmatrix}$$

$$\xrightarrow[\text{1 and 2}]{\text{interchange rows}} \begin{bmatrix} 1 & -2 & 1 & \vdots & 1 \\ 0 & 8 & -3 & \vdots & 6 \end{bmatrix}$$

$$\xrightarrow{\frac{1}{8}(\text{row 2})} \begin{bmatrix} 1 & -2 & 1 & \vdots & 1 \\ 0 & 1 & -\frac{3}{8} & \vdots & \frac{6}{8} \end{bmatrix}$$

$$\xrightarrow[\text{to row 1}]{\text{add } 2(\text{row 2})} \begin{bmatrix} 1 & 0 & \frac{2}{8} & \vdots & \frac{20}{8} \\ 0 & 1 & -\frac{3}{8} & \vdots & \frac{6}{8} \end{bmatrix}.$$

The first three columns of this augmented matrix form the matrix

$$\begin{bmatrix} 1 & 0 & \frac{2}{8} \\ 0 & 1 & -\frac{3}{8} \end{bmatrix},$$

and this is A_R. Thus, the reduced form of $[A \mid B]$ has the form of the reduced form of A, augmented by an additional column formed from B during the reduction process. ∎

One dramatic difference between homogeneous and nonhomogeneous systems is that a nonhomogeneous system need not have a solution. We will state necessary and sufficient conditions for a nonhomogeneous system to have a solution. First, consider an example.

EXAMPLE 12.32

Consider the nonhomogeneous system

$$\begin{bmatrix} 2 & -3 \\ 4 & -6 \end{bmatrix} \begin{bmatrix} x_1 \\ x_2 \end{bmatrix} = \begin{bmatrix} 6 \\ 18 \end{bmatrix}.$$

Clearly, this system can have no solution. If $x_1 = \alpha$ and $x_2 = \beta$, then from the first equation, $2\alpha - 3\beta = 6$. Multiply this equation by 2 to get

$$4\alpha - 6\beta = 12.$$

But the second equation of the system requires that $4\alpha - 6\beta = 18$, a contradiction.

Now look at the coefficient matrix and the augmented matrix of the system. We have

$$A = \begin{bmatrix} 2 & -3 \\ 4 & -6 \end{bmatrix} \quad \text{and} \quad [A \mid B] = \begin{bmatrix} 2 & -3 & \vdots & 6 \\ 4 & -6 & \vdots & 18 \end{bmatrix}.$$

We find that the reduced form of the augmented matrix is

$$[A \mid B]_R = \begin{bmatrix} 1 & -\frac{3}{2} & \vdots & 0 \\ 0 & 0 & \vdots & 1 \end{bmatrix};$$

thus, $\text{rank}([A \mid B]) = 2$. However, from the first two columns of $[A \mid B]_R$, we read that

$$A_R = \begin{bmatrix} 1 & -\frac{3}{2} \\ 0 & 0 \end{bmatrix};$$

hence, $\text{rank}(A) = 1 \neq \text{rank}([A \mid B])$.

As we will now see, this system has no solution because the augmented matrix has rank greater than A. The effect is seen in the last row of $[A \mid B]_R$. It corresponds to an equation $0x_1 + 0x_2 = 1$, which can have no solution. ∎

In general, whenever the augmented matrix has rank greater than A, there is at least one equation in the reduced system of the form $0x_1 + 0x_2 + \cdots + 0x_m = \alpha \neq 0$, and the system can have no solution.

THEOREM 12.17

The nonhomogeneous system $AX = B$ has a solution if and only if A and $[A \mid B]$ have the same rank.

Proof Let A be $n \times m$ and suppose first that $\text{rank}(A) = \text{rank}([A \mid B]) = r$. By Theorems (12.12) and (12.13) in Section 12.6, the column space of $[A \mid B]$ has dimension r, and certainly $r \leq m$. Thus, column $m + 1$ of $[A \mid B]$ is a linear combination of the first m columns, and, for some constants $\alpha_1, \ldots, \alpha_m$,

$$B = \alpha_1 \begin{bmatrix} a_{11} \\ a_{21} \\ \vdots \\ a_{n1} \end{bmatrix} + \alpha_2 \begin{bmatrix} a_{12} \\ a_{22} \\ \vdots \\ a_{n2} \end{bmatrix} + \cdots + \alpha_m \begin{bmatrix} a_{1m} \\ a_{2m} \\ \vdots \\ a_{nm} \end{bmatrix}.$$

This is the same as writing

$$A \begin{bmatrix} \alpha_1 \\ \alpha_2 \\ \vdots \\ \alpha_m \end{bmatrix} = B;$$

hence,

$$\begin{bmatrix} \alpha_1 \\ \alpha_2 \\ \vdots \\ \alpha_m \end{bmatrix}$$

is a solution of $AX = B$.

Conversely, suppose that $AX = B$ has a solution

$$\begin{bmatrix} \alpha_1 \\ \alpha_2 \\ \vdots \\ \alpha_m \end{bmatrix}.$$

Then

$$B = A \begin{bmatrix} \alpha_1 \\ \alpha_2 \\ \vdots \\ \alpha_m \end{bmatrix} = \alpha_1 \begin{bmatrix} a_{11} \\ a_{21} \\ \vdots \\ a_{n1} \end{bmatrix} + \alpha_2 \begin{bmatrix} a_{12} \\ a_{22} \\ \vdots \\ a_{n2} \end{bmatrix} + \cdots + \alpha_m \begin{bmatrix} a_{1m} \\ a_{2m} \\ \vdots \\ a_{nm} \end{bmatrix};$$

hence, B is a linear combination of the columns of A. Thus, the subspace of R^n consisting of all linear combinations of column vectors of $[A \mid B]$ is the same as the subspace of R^n consisting of all linear combinations of column vectors of A. Therefore, the column spaces of A and $[A \mid B]$ have the same dimension. These dimensions are rank(A) and rank($[A \mid B]$), respectively, and we have shown that A and $[A \mid B]$ have the same rank. ∎

Now that we have an existence criterion, we will determine what we should look for in attempting to find all solutions of a system $AX = B$. The result is similar to Theorem 2.5, which dealt with the general solution of a nonhomogeneous second order linear differential equation.

THEOREM 12.18

Let U be any solution of $AX = B$. Then every solution of $AX = B$ is of the form $U + H$, where H is a solution of $AX = O$.

Note: In writing $U + H$, we are writing solutions as $m \times 1$ column matrices and performing the addition as a matrix sum.

Proof   Let W be any solution of $AX = B$. Then $W - U$ is a solution of $AX = O$ because

$$A(W - U) = AW - AU = B - B = O.$$

Let $H = W - U$. Then H is a solution of $AX = O$, and certainly $W = U + H$. ∎

Theorem 12.18 tells us that, if U is any solution of $AX = B$, the expression $U + H$ contains every solution of $AX = B$ as H varies over all solutions of $AX = O$. Thus, if H is the general solution of $AX = O$, $U + H$ contains every solution of $AX = B$, with U any particular solution of the nonhomogeneous system. For this reason, an expression $U + H$, with U any solution of $AX = B$ and H the general solution of $AX = O$, is called the *general solution* of the nonhomogeneous system.

We can now outline a procedure for finding the general solution of a nonhomogeneous linear system.

PROCEDURE FOR SOLVING $AX = B$

Step 1   Reduce $[A \mid B]$ to obtain a reduced matrix of the form $[A_R \mid C]$. The solutions of $AX = B$ are the same as solutions of $A_R X = C$, so work with this reduced system.

Step 2   If rank($[A \mid B]$) $\neq$ rank(A), the system has no solution and we are done. If these two ranks are the same, continue.

Step 3 Identify the dependent variables. If column j contains the leading entry of row i, use equation i to write x_j in terms of the independent variables *and* c_i (in the homogeneous case, each $c_i = 0$).

Step 4 Write a column matrix

$$\begin{bmatrix} x_1 \\ x_2 \\ \vdots \\ x_m \end{bmatrix}$$

with each dependent x_j written in terms of independent unknowns and c_i. The independent unknowns are arbitrary and can be assigned any values.

Step 5 To clarify the structure of the solution, write it as a sum of column matrices multiplied by the independent unknowns (arbitrary scalars), as in the homogeneous case, *plus* a column matrix containing the c_i's appearing in the expressions for the dependent unknowns. This constant column matrix is a particular solution of $A_R X = C$. We now have the general solution of $AX = B$ written as the general solution of $A_R X = O$ plus a particular solution of $A_R X = C$, yielding the general solution of the original nonhomogeneous system.

EXAMPLE 12.33

Find the general solution of

$$-x_1 + x_2 + 3x_3 = -2$$
$$x_2 + 2x_3 = 4.$$

Step 1 We have

$$[A \mid B] = \begin{bmatrix} -1 & 0 & 3 & \vdots & -2 \\ 0 & 1 & 2 & \vdots & 4 \end{bmatrix}.$$

We find that

$$[A \mid B]_R = \begin{bmatrix} 1 & 0 & -1 & \vdots & 6 \\ 0 & 1 & 2 & \vdots & 4 \end{bmatrix}.$$

Step 2 The first three columns make up A_R, and we can read that rank$(A) = 2$. Since rank$([A \mid B]_R) = 2$ also, the system has a solution.

Step 3 Read from A_R that x_1 and x_2 are dependent and x_3 is independent. From $[A \mid B]_R$, we read that

$$x_1 - x_3 = 6$$
$$x_2 + 2x_3 = 4.$$

Thus,

$$x_1 = x_3 + 6 \quad \text{and} \quad x_2 = -2x_2 + 4,$$

with x_3 arbitrary.

Step 4   Write the column matrix

$$\begin{bmatrix} x_1 \\ x_2 \\ x_3 \end{bmatrix} = \begin{bmatrix} 6 + x_3 \\ 4 - 2x_3 \\ x_3 \end{bmatrix}.$$

Step 5   Let $x_3 = \alpha$, an arbitrary constant, and write the solution of step 4 as

$$\begin{bmatrix} 6 \\ 4 \\ 0 \end{bmatrix} + \alpha \begin{bmatrix} 1 \\ -2 \\ 1 \end{bmatrix}.$$

This is the general solution of the system $AX = B$. Note that it is a sum of

$$\begin{bmatrix} 6 \\ 4 \\ 0 \end{bmatrix}, \quad \text{a particular solution of } AX = B$$

and

$$\alpha \begin{bmatrix} 1 \\ -2 \\ 1 \end{bmatrix}, \quad \text{the general solution of } AX = O.$$

You should verify by substitution that $x_1 = 6 + \alpha$, $x_2 = 4 - 2\alpha$, and $x_3 = \alpha$ satisfy the system for any α.   ∎

EXAMPLE 12.34

Solve the system

$$\begin{aligned} x_1 - \quad x_2 + \quad 2x_3 &= -1 \\ x_3 &= \quad 0 \\ 3x_1 - \quad 3x_2 + \quad 7x_3 &= \quad 1 \\ 10x_1 - 10x_2 + 24x_3 &= -2. \end{aligned}$$

Write

$$[A \mid B] = \begin{bmatrix} 1 & -1 & 2 & \vdots & -1 \\ 0 & 0 & 1 & \vdots & 0 \\ 3 & -3 & 7 & \vdots & 1 \\ 10 & -10 & 24 & \vdots & -2 \end{bmatrix}.$$

We find that

$$[A \mid B]_R = \begin{bmatrix} 1 & -1 & 0 & \vdots & 0 \\ 0 & 0 & 1 & \vdots & 0 \\ 0 & 0 & 0 & \vdots & 1 \\ 0 & 0 & 0 & \vdots & 0 \end{bmatrix}.$$

Thus, rank($[A \mid B]$) = 3 because $[A \mid B]_R$ has three nonzero rows. But we can read

A_R from the first three columns of $[A \mid B]$. Since A_R has only two nonzero rows, $\text{rank}(A) = 2$. By Theorem 12.17, this system has no solution.

Note that, from $[A \mid B]_R$, the third equation of the reduced system is $0x_1 + 0x_2 + 0x_3 = 1$, which can have no solution. ■

EXAMPLE 12.35

Solve

$$
\begin{aligned}
x_1 \quad\quad - x_3 + 2x_4 + x_5 + 6x_6 &= -3 \\
x_2 + x_3 + 3x_4 + 2x_5 + 4x_6 &= 1 \\
x_1 - 4x_2 + 3x_3 + x_4 \quad\quad + 2x_6 &= 0.
\end{aligned}
$$

Here

$$
[A \mid B] = \begin{bmatrix} 1 & 0 & -1 & 2 & 1 & 6 & \vdots & -3 \\ 0 & 1 & 1 & 3 & 2 & 4 & \vdots & 1 \\ 1 & -4 & 3 & 1 & 0 & 2 & \vdots & 0 \end{bmatrix}.
$$

We find that

$$
[A \mid B]_R = \begin{bmatrix} 1 & 0 & 0 & \frac{27}{8} & \frac{15}{8} & \frac{60}{8} & \vdots & -\frac{17}{8} \\ 0 & 1 & 0 & \frac{13}{8} & \frac{9}{8} & \frac{20}{8} & \vdots & \frac{1}{8} \\ 0 & 0 & 1 & \frac{11}{8} & \frac{7}{8} & \frac{12}{8} & \vdots & \frac{7}{8} \end{bmatrix}.
$$

Since $\text{rank}(A) = 3 = \text{rank}([A \mid B])$, a solution exists. From $[A \mid B]_R$, we identify x_1, x_2, and x_3 as dependent and x_4, x_5, and x_6 as independent. Further, from the first row of $[A \mid B]_R$, we read

$$
x_1 + \tfrac{27}{8}x_4 + \tfrac{15}{8}x_5 + \tfrac{60}{8}x_6 = -\tfrac{17}{8};
$$

hence,

$$
x_1 = -\tfrac{17}{8} - \tfrac{27}{8}x_4 - \tfrac{15}{8}x_5 - \tfrac{60}{8}x_6.
$$

Similarly, from rows 2 and 3, we read

$$
x_2 = \tfrac{1}{8} - \tfrac{13}{8}x_4 - \tfrac{9}{8}x_5 - \tfrac{20}{8}x_6
$$

and

$$
x_3 = \tfrac{7}{8} - \tfrac{11}{8}x_4 - \tfrac{7}{8}x_5 - \tfrac{12}{8}x_6.
$$

Write $x_4 = \alpha$, $x_5 = \beta$, and $x_6 = \gamma$, arbitrary constants, to write the general solution

$$
\begin{bmatrix} x_1 \\ x_2 \\ x_3 \\ x_4 \\ x_5 \\ x_6 \end{bmatrix} = \begin{bmatrix} -\frac{17}{8} \\ \frac{1}{8} \\ \frac{7}{8} \\ 0 \\ 0 \\ 0 \end{bmatrix} + \alpha \begin{bmatrix} -\frac{27}{8} \\ -\frac{13}{8} \\ -\frac{11}{8} \\ 1 \\ 0 \\ 0 \end{bmatrix} + \beta \begin{bmatrix} -\frac{15}{8} \\ -\frac{9}{8} \\ -\frac{7}{8} \\ 0 \\ 1 \\ 0 \end{bmatrix} + \gamma \begin{bmatrix} -\frac{60}{8} \\ -\frac{20}{8} \\ -\frac{12}{8} \\ 0 \\ 0 \\ 1 \end{bmatrix}.
$$

You can check by substitution that this is a solution for any choices of α, β, and γ. ■

We will conclude this section with a sufficient condition for a nonhomogeneous system to have a unique solution when the number of equations equals the number of unknowns.

THEOREM 12.19

Let A be $n \times n$. Then the nonhomogeneous system $AX = B$ has a unique solution if and only if rank$(A) = n$. Alternatively, the system has a unique solution if and only if $A_R = I_n$.

Proof   Suppose first that rank$(A) = n$. Then $A_R = I_n$ by Lemma 12.1 in Section 12.6. Then $[A \mid B]_R$ is of the form $[I_n \mid C]$ for some $n \times 1$ matrix C. The system $I_n X = C$ has exactly one solution, namely, $x_1 = c_1, x_2 = c_2, \ldots, x_n = c_n$, and this is the only solution of the original system.

Conversely, suppose that $AX = B$ has exactly one solution U. If $AX = O$ has a solution H, then by Theorem 12.18 $U + H$ is a solution of $AX = B$. But then $U = U + H$, by the assumption that $AX = B$ has only one solution U. We conclude that $H = O$ and hence that $AX = O$ has only the trivial solution. By Theorem 12.15 in Section 12.7, rank$(A) = n$. ∎

EXAMPLE 12.36

Consider the system

$$\begin{bmatrix} 2 & -1 \\ 0 & 3 \end{bmatrix} \begin{bmatrix} x_1 \\ x_2 \end{bmatrix} = \begin{bmatrix} -1 \\ 4 \end{bmatrix}.$$

We find that $A_R = I_2$; hence, rank$(A) = 2$. This system therefore has a unique solution. We find that the solution is $\begin{bmatrix} \frac{1}{6} \\ \frac{4}{3} \end{bmatrix}$. ∎

In the next section, we will discuss matrix inverses and then return to the nonhomogeneous system $AX = B$ in the case in which A is square.

PROBLEMS FOR SECTION 12.8

In each of Problems 1 through 20, find the general solution of the system or show that there is no solution.

1.
$$\begin{aligned} 3x_1 - 2x_2 + x_3 &= 6 \\ x_1 + 10x_2 - x_3 &= 2 \\ -3x_1 - 2x_2 + x_3 &= 0 \end{aligned}$$

2.
$$\begin{aligned} 4x_1 - 2x_2 + 3x_3 + 10x_4 &= 1 \\ x_1 \qquad\qquad - 3x_4 &= 8 \\ 2x_1 - 3x_2 \qquad + x_4 &= 16 \end{aligned}$$

3.
$$\begin{aligned} 2x_1 - 3x_2 \quad + x_4 \quad - x_6 &= 0 \\ 3x_1 \qquad - 2x_3 \quad + x_5 \quad &= 1 \\ x_2 \qquad - x_4 \quad + 6x_6 &= -3 \end{aligned}$$

4.
$$\begin{aligned} 2x_1 - 3x_2 &= 1 \\ -x_1 + 3x_2 &= 0 \\ x_1 - 4x_2 &= 3 \end{aligned}$$

5.
$$\begin{aligned} 3x_2 \qquad - 4x_4 \qquad\qquad &= 10 \\ x_1 - 3x_2 \qquad\qquad + 4x_5 - x_6 &= 8 \\ x_2 + x_3 - 6x_4 \qquad + x_6 &= -9 \\ x_1 - x_2 \qquad\qquad\qquad + x_6 &= 0 \end{aligned}$$

6.
$$\begin{aligned} 2x_1 - 3x_2 \qquad + x_4 &= 1 \\ 3x_2 + x_3 - x_4 &= 0 \\ 2x_1 - 3x_2 + 10x_3 \qquad &= 0 \end{aligned}$$

7. $8x_2 - 4x_3 \qquad\qquad + 10x_6 = 1$
$\qquad\qquad x_3 \quad + x_5 - \quad x_6 = 2$
$\qquad\qquad x_4 - 3x_5 + \; 2x_6 = 0$

8. $2x_1 \qquad\quad - 3x_3 = 1$
$\quad x_1 - \; x_2 + \; x_3 = 1$
$\; 2x_1 - 4x_2 + \; x_3 = 2$

9. $\qquad\qquad 14x_3 \qquad - 3x_5 \qquad + x_7 = \quad 2$
$x_1 + x_2 + \quad x_3 - x_4 \qquad\quad + x_6 \qquad = -4$

10. $3x_1 - 2x_2 = -1$
$\;\, 4x_1 + 3x_2 = \quad 4$

11. $7x_1 - 3x_2 + 4x_3 \qquad\quad = -7$
$2x_1 + \; x_2 - \; x_3 + 4x_4 = \quad 6$
$\qquad\quad x_2 \qquad - 3x_4 = -5$

12. $-4x_1 + \; 5x_2 - \; 6x_3 = \quad 2$
$\quad 2x_1 - \; 6x_2 + \quad x_3 = -5$
$-6x_1 + 16x_2 - 11x_3 = \quad 1$

13. $\quad 4x_1 - \; x_2 + 4x_3 = 1$
$\quad x_1 + x_2 - 5x_3 = 0$
$-2x_1 + x_2 + 7x_3 = 4$

14. $-6x_1 + 2x_2 - \; x_3 + \; x_4 = \quad 0$
$\quad x_1 + 4x_2 \qquad\quad - \; x_4 = -5$
$\quad x_1 + \; x_2 + x_3 - 7x_4 = \quad 0$

15. $\quad 4x_1 - 3x_2 + \quad x_3 = -1$
$-3x_1 + \; x_2 - \; 5x_3 = \quad 0$
$-5x_1 \qquad\quad - 14x_3 = -10$

16. $9x_1 + x_2 \qquad\qquad\quad - 4x_5 = -1$
$\qquad\quad x_2 + 4x_3 \qquad - 4x_5 = \quad 2$
$\qquad\quad x_2 \qquad\quad + x_4 - \; x_5 = \quad 0$

17. $-5x_1 + 3x_2 - \; x_3 \qquad\qquad = \quad 0$
$\qquad\qquad\qquad\qquad + \; x_4 = -4$
$\quad x_1 - \; x_2 + 3x_3 - 6x_4 = -11$

18. $-6x_1 + \; x_2 - 4x_3 = \quad 1$
$\quad 2x_1 - \; x_2 - \; x_3 = \quad 8$
$\quad x_1 + 6x_2 - \; x_3 = -3$

19. $-5x_1 + 3x_2 + \; x_3 - x_4 = -8$
$\quad 4x_1 + 3x_2 \qquad\quad - x_4 = \quad 9$
$\quad 2x_1 + 3x_2 - 3x_3 + x_4 = -7$

20. $\qquad\quad 3x_2 \qquad\quad + \quad x_4 - \; x_5 = 15$
$\quad x_1 + 3x_2 \qquad\quad + \quad x_4 - 7x_5 = 10$
$-5x_1 + \; x_2 - 4x_3 + \quad x_4 + 6x_5 = \quad 1$
$\;\; 2x_1 + 4x_2 - \; x_3 + 10x_4 + 8x_5 = \quad 7$

21. Let the $n \times m$ reduced matrix A have rank k. Prove that the nonhomogeneous system of equations $A_R X = B$ has a solution if and only if $b_{k+1} = \cdots = b_n = 0$.

In each of Problems 23 through 27, show that the system of equations has a unique solution, and find the solution.

22. $-2x_1 + 3x_2 \qquad\quad = -1$
$\qquad\quad - 4x_2 + x_3 = \quad 0$
$\quad x_1 + 3x_2 \qquad\quad = \quad 0$

23. $\qquad\quad 2x_2 \qquad\qquad\qquad = \quad 1$
$\qquad\quad x_2 - 2x_3 \qquad\quad = \quad 2$
$\; x_1 \qquad\qquad - 2x_4 = -10$
$\; x_1 \qquad\qquad\quad + x_4 = \quad 5$

24. $x_1 \qquad + 2x_3 \qquad\quad = -12$
$x_1 \qquad - 3x_3 + x_4 = \quad 0$
$\qquad x_2 \qquad\quad - x_4 = \quad 1$
$\qquad x_2 - \; x_3 \qquad\quad = \quad 8$

25. $4x_1 - x_2 \qquad\qquad = \quad 1$
$\qquad\quad x_2 + 3x_3 = \quad 0$
$\; x_1 \qquad\quad - 4x_3 = -2$

26. $\qquad 2x_2 \qquad\quad - x_4 = \quad 1$
$\; x_1 \qquad\quad - 3x_3 \qquad\quad = -2$
$\; x_1 - 4x_2 \qquad\qquad = \quad 0$
$\qquad 3x_2 \qquad\qquad\quad = \quad 1$

27. $\qquad\quad 2x_2 - 3x_3 = 0$
$\; 2x_1 \qquad\quad - 3x_3 = 0$
$\; x_1 - \; x_2 + \; x_3 = 1$

12.9 *Matrix Inverses* _____

If A and B are $n \times n$ matrices, we call each an *inverse* of the other if

$$AB = BA = I_n.$$

We will now examine questions about inverses. How do we know when a matrix has an inverse? How many inverses can a matrix have? How do we find an inverse? The following example shows that a matrix need not have an inverse.

EXAMPLE 12.37

We can verify easily that the 2×2 matrix $\begin{bmatrix} 1 & 0 \\ 0 & 0 \end{bmatrix}$ has no inverse. Suppose that $\begin{bmatrix} a & b \\ c & d \end{bmatrix}$ is an inverse. Then we must have

$$\begin{bmatrix} 1 & 0 \\ 0 & 0 \end{bmatrix} \begin{bmatrix} a & b \\ c & d \end{bmatrix} = \begin{bmatrix} 1 & 0 \\ 0 & 1 \end{bmatrix}.$$

Carry out the matrix product on the left to get

$$\begin{bmatrix} a & b \\ 0 & 0 \end{bmatrix} = \begin{bmatrix} 1 & 0 \\ 0 & 1 \end{bmatrix}.$$

From the 2, 2 element, we conclude that $0 = 1$, a contradiction. Therefore, this matrix has no inverse. ∎

Some matrices do have an inverse. For example,

$$\begin{bmatrix} 2 & 1 \\ 1 & 4 \end{bmatrix} \begin{bmatrix} \frac{4}{7} & -\frac{1}{7} \\ -\frac{1}{7} & \frac{2}{7} \end{bmatrix} = \begin{bmatrix} \frac{4}{7} & -\frac{1}{7} \\ -\frac{1}{7} & \frac{2}{7} \end{bmatrix} \begin{bmatrix} 2 & 1 \\ 1 & 4 \end{bmatrix} = I_2,$$

so each of the matrices in these products is an inverse of the other.

A square matrix is called *nonsingular* when it has an inverse and *singular* when it does not. Thus, $\begin{bmatrix} 2 & 1 \\ 1 & 4 \end{bmatrix}$ and $\begin{bmatrix} \frac{4}{7} & -\frac{1}{7} \\ -\frac{1}{7} & \frac{2}{7} \end{bmatrix}$ are nonsingular, while $\begin{bmatrix} 1 & 0 \\ 0 & 0 \end{bmatrix}$ is singular. We will now see that a matrix can have only one inverse.

THEOREM 12.20

A nonsingular matrix has exactly one inverse.

Proof   Suppose that B and C are inverses of the square matrix A. Then

$$AB = BA = I_n \quad \text{and} \quad AC = CA = I_n.$$

Then

$$B = BI_n = B(AC) = (BA)C = I_nC = C. \quad \blacksquare$$

In view of this result, we designate the unique inverse of A as A^{-1}. The next theorem lists additional properties of matrix inverses.

THEOREM 12.21

1. I_n is nonsingular, and $(I_n)^{-1} = I_n$.
2. If A and B are nonsingular $n \times n$ matrices, AB is also nonsingular, and

$$(AB)^{-1} = B^{-1}A^{-1}.$$

3. If A is nonsingular, so is A^{-1}, and $(A^{-1})^{-1} = A$.

4. If A is nonsingular, so is A^t, and $(A^t)^{-1} = (A^{-1})^t$.

5. If A and B are $n \times n$ matrices, and either A or B is singular, AB and BA are also singular.

Proof of (1)   We have $I_n I_n = I_n$; hence, $I_n = (I_n)^{-1}$.

Proof of (2)   Because A and B are nonsingular, A^{-1} and B^{-1} exist. Now compute

$$(AB)(B^{-1}A^{-1}) = A(BB^{-1})A^{-1} = A(I_n)A^{-1} = AA^{-1} = I_n$$

and

$$(B^{-1}A^{-1})(AB) = B^{-1}(A^{-1}A)B = B^{-1}(I_n)B = B^{-1}B = I_n.$$

Therefore, $B^{-1}A^{-1}$ is the inverse of AB.

Proofs of (3) and (4) are left to the student. Conclusion (5) will be easily proved when we have developed the concept of an $n \times n$ determinant.   ∎

The following theorem gives a necessary and sufficient condition for a matrix to have an inverse.

THEOREM 12.22

Let A be $n \times n$. Then A is nonsingular if and only if rank$(A) = n$.

Proof   Consider the equation $AB = I_n$, with B an $n \times n$ matrix of unknowns for which we want to solve. If $AB = I_n$, column j of AB equals column j of I_n. But then

$$\text{column } j \text{ of } AB = A(\text{column } j \text{ of } B) = \begin{bmatrix} 0 \\ 0 \\ \vdots \\ 1 \\ \vdots \\ 0 \end{bmatrix},$$

in which the last column matrix has 1 in row j and zeros elsewhere. Thus, column j of B is found from the system of equations

$$A \begin{bmatrix} b_{1j} \\ b_{2j} \\ \vdots \\ b_{nj} \end{bmatrix} = \begin{bmatrix} 0 \\ 0 \\ \vdots \\ 1 \\ \vdots \\ 0 \end{bmatrix} \leftarrow \text{row } j. \tag{12.15}$$

Now suppose that rank$(A) = n$. Then the system (12.15) has a unique solution by Theorem 12.19 in Section 12.8. Thus, we can find a unique matrix B such that $AB = I_n$. It is possible to show that $BA = I_n$ also; hence, B is the inverse of A.

Conversely, if A is nonsingular, the system (12.15) has a unique solution for $j = 1, 2, \ldots, n$ because these solutions form the columns of A^{-1}. By Theorem 12.19 in Section 12.8, $\text{rank}(A) = n$. ∎

With this much theory behind us, we will now turn to a method for producing A^{-1} when A is nonsingular. Let A be any $n \times n$ matrix.

A METHOD FOR FINDING A^{-1}

Step 1   Write an $n \times 2n$ matrix consisting of I_n placed to the left of A, as in Figure 12.4. This matrix is denoted $[I_n \mid A]$. The first n columns are called the *left side*, and the second n columns are called the *right side*, of this matrix.

$$
\begin{bmatrix}
1 & 0 & 0 & \cdots & 0 & A_{11} & A_{12} & A_{13} & \cdots & A_{1n} \\
0 & 1 & 0 & \cdots & 0 & A_{21} & A_{22} & A_{23} & \cdots & A_{2n} \\
0 & 0 & 1 & \cdots & 0 & A_{31} & A_{32} & A_{33} & \cdots & A_{3n} \\
\vdots & \vdots & \vdots & & \vdots & \vdots & \vdots & \vdots & & \vdots \\
0 & 0 & 0 & \cdots & 1 & A_{n1} & A_{n2} & A_{n3} & \cdots & A_{nn}
\end{bmatrix}
$$

Figure 12.4. $n \times 2n$ matrix $[I_n \mid A]$.

Step 2   Carry out elementary row operations to reduce A to A_R. If these operations are carried out on the entire matrix $[I_n \mid A]$, there results an $n \times 2n$ matrix of the form $[C \mid A_R]$, in which C is the $n \times n$ matrix obtained from I_n by carrying out the row operations used to reduce A.

Step 3   If $A_R \neq I_n$, A is singular and has no inverse. If $A_R = I_n$, $C = A^{-1}$.

Thus, in practice, we write the matrix $[I_n \mid A]$, perform elementary row operations until the right side is reduced, and read A^{-1} from whatever results on the left side, or else conclude that A is singular.

Why does this method produce A^{-1}? When we perform the elementary row operations to reduce A on the right side of $[I_n \mid A]$, we also perform them on the left side. Thus, we perform, on I_n, exactly the elementary row operations used to reduce A. This produces a matrix C which is a product of elementary matrices, each performing one of the operations used in reducing A. Thus, whether or not $A_R = I_n$, we have $CA = A_R$. When $A_R = I_n$, C must be A^{-1}.

EXAMPLE 12.38

Let

$$
A = \begin{bmatrix}
2 & -1 & 3 \\
1 & 0 & -2 \\
4 & 0 & 2
\end{bmatrix}.
$$

We want to find A^{-1} or else show that A is singular. Begin with the 3×6 matrix

$[I_3 \mid A]$, and reduce A, performing the same operations on all the rows of this matrix:

$$\begin{bmatrix} 1 & 0 & 0 & \vdots & 2 & -1 & 3 \\ 0 & 1 & 0 & \vdots & 1 & 0 & -2 \\ 0 & 0 & 1 & \vdots & 4 & 0 & 2 \end{bmatrix}$$

$\xrightarrow{\frac{1}{2}(\text{row 1})}$
$$\begin{bmatrix} \frac{1}{2} & 0 & 0 & \vdots & 1 & -\frac{1}{2} & \frac{3}{2} \\ 0 & 1 & 0 & \vdots & 1 & 0 & -2 \\ 0 & 0 & 1 & \vdots & 4 & 0 & 2 \end{bmatrix}$$

$\xrightarrow[\text{add } -4(\text{row 1}) \text{ to row 3}]{\text{add } -1(\text{row 1}) \text{ to row 2}}$
$$\begin{bmatrix} \frac{1}{2} & 0 & 0 & \vdots & 1 & -\frac{1}{2} & \frac{3}{2} \\ -\frac{1}{2} & 1 & 0 & \vdots & 0 & \frac{1}{2} & -\frac{7}{2} \\ -2 & 0 & 1 & \vdots & 0 & 2 & -4 \end{bmatrix}$$

$\xrightarrow{2(\text{row 2})}$
$$\begin{bmatrix} \frac{1}{2} & 0 & 0 & \vdots & 1 & -\frac{1}{2} & \frac{3}{2} \\ -1 & 2 & 0 & \vdots & 0 & 1 & -7 \\ -2 & 0 & 1 & \vdots & 0 & 2 & -4 \end{bmatrix}$$

$\xrightarrow[\text{and } -2(\text{row 2}) \text{ to row 3}]{\text{add } \frac{1}{2}(\text{row 2}) \text{ to row 1}}$
$$\begin{bmatrix} 0 & 1 & 0 & \vdots & 1 & 0 & -2 \\ -1 & 2 & 0 & \vdots & 0 & 1 & -7 \\ 0 & -4 & 1 & \vdots & 0 & 0 & 10 \end{bmatrix}$$

$\xrightarrow{\frac{1}{10}(\text{row 3})}$
$$\begin{bmatrix} 0 & 1 & 0 & \vdots & 1 & 0 & -2 \\ -1 & 2 & 0 & \vdots & 0 & 1 & -7 \\ 0 & -\frac{4}{10} & \frac{1}{10} & \vdots & 0 & 0 & 1 \end{bmatrix}$$

$\xrightarrow[\text{add } 7(\text{row 3}) \text{ to row 2}]{\text{add } 2(\text{row 3}) \text{ to row 1}}$
$$\begin{bmatrix} 0 & \frac{2}{10} & \frac{2}{10} & \vdots & 1 & 0 & 0 \\ -1 & -\frac{8}{10} & \frac{7}{10} & \vdots & 0 & 1 & 0 \\ 0 & -\frac{4}{10} & \frac{1}{10} & \vdots & 0 & 0 & 1 \end{bmatrix}.$$

Since I_3 has appeared on the right side of the last matrix, A is nonsingular, and the left side is A^{-1}:

$$A^{-1} = \begin{bmatrix} 0 & \frac{2}{10} & \frac{2}{10} \\ -1 & -\frac{8}{10} & \frac{7}{10} \\ 0 & -\frac{4}{10} & \frac{1}{10} \end{bmatrix}. \quad \blacksquare$$

EXAMPLE 12.39

Let

$$A = \begin{bmatrix} -3 & 1 & -1 \\ 1 & 0 & 1 \\ -2 & 2 & 2 \end{bmatrix}.$$

Find A^{-1} or show that A is singular.
    Proceed as follows:

$$\begin{bmatrix} 1 & 0 & 0 & \vdots & -3 & 1 & -1 \\ 0 & 1 & 0 & \vdots & 1 & 0 & 1 \\ 0 & 0 & 1 & \vdots & -2 & 2 & 2 \end{bmatrix}$$

$$\xrightarrow{-\frac{1}{3}(\text{row 1})}
\begin{bmatrix}
-\frac{1}{3} & 0 & 0 & \vdots & 1 & -\frac{1}{3} & \frac{1}{3} \\
0 & 1 & 0 & \vdots & 1 & 0 & 1 \\
0 & 0 & 1 & \vdots & -2 & 2 & 2
\end{bmatrix}$$

$$\xrightarrow[\text{add } 2(\text{row 1}) \text{ to row 3}]{\text{add } -1(\text{row 1}) \text{ to row 2}}
\begin{bmatrix}
-\frac{1}{3} & 0 & 0 & \vdots & 1 & -\frac{1}{3} & \frac{1}{3} \\
\frac{1}{3} & 1 & 0 & \vdots & 0 & \frac{1}{3} & \frac{2}{3} \\
-\frac{2}{3} & 0 & 1 & \vdots & 0 & \frac{4}{3} & \frac{8}{3}
\end{bmatrix}$$

$$\xrightarrow{3(\text{row 2})}
\begin{bmatrix}
-\frac{1}{3} & 0 & 0 & \vdots & 1 & -\frac{1}{3} & \frac{1}{3} \\
1 & 3 & 0 & \vdots & 0 & 1 & 2 \\
-\frac{2}{3} & 0 & 1 & \vdots & 0 & \frac{4}{3} & \frac{8}{3}
\end{bmatrix}$$

$$\xrightarrow[\text{add } -\frac{4}{3}(\text{row 2}) \text{ to row 3}]{\text{add } \frac{1}{3}(\text{row 2}) \text{ to row 1}}
\begin{bmatrix}
0 & 1 & 0 & \vdots & 1 & 0 & 1 \\
1 & 3 & 0 & \vdots & 0 & 1 & 2 \\
-\frac{6}{3} & -4 & 1 & \vdots & 0 & 0 & 0
\end{bmatrix}.$$

The right side forms the matrix A_R. Since $A_R \neq I_3$, A is singular and has no inverse. ∎

Inverses relate to the problem of solving a system in the following way.

THEOREM 12.23

Let A be an $n \times n$ matrix. Then

1. The nonhomogeneous system $AX = B$ has a unique solution if and only if A is nonsingular. In this event, the unique solution is $X = A^{-1}B$.
2. The homogeneous system $AX = O$ has a nontrivial solution if and only if A is singular.

Proof of (1)   By Theorem 12.19 in Section 12.8, the system $AX = B$ has a unique solution if and only if $\text{rank}(A) = n$, and this occurs if and only if A is nonsingular. Therefore, $AX = B$ has a unique solution if and only if A is nonsingular.

When A^{-1} exists, multiply both sides of $AX = B$ *on the left* by A^{-1} to get $X = A^{-1}B$; this is the unique solution.

Proof of (2)   By the corollary to Theorem 12.15, the system $AX = O$ has a nontrivial solution if and only if $\text{rank}(A) < n$, and this occurs if and only if A is singular. ∎

EXAMPLE 12.40

Solve the system

$$\begin{aligned}
2x_1 - x_2 + 3x_3 &= 4 \\
x_1 + 9x_2 - 2x_3 &= -8 \\
4x_1 - 8x_2 + 11x_3 &= 15
\end{aligned}$$

or show that no solution exists.

The matrix of coefficients is

$$A = \begin{bmatrix}
2 & -1 & 3 \\
1 & 9 & -2 \\
4 & -8 & 11
\end{bmatrix}.$$

We find that

$$A^{-1} = \tfrac{1}{53} \begin{bmatrix} 83 & -13 & -25 \\ -19 & 10 & 7 \\ -44 & 12 & 19 \end{bmatrix}.$$

The system has a unique solution, and this solution is

$$X = A^{-1}B = \tfrac{1}{53} \begin{bmatrix} 83 & -13 & -25 \\ -19 & 10 & 7 \\ -44 & 12 & 19 \end{bmatrix} \begin{bmatrix} 4 \\ -8 \\ 15 \end{bmatrix}$$

$$= \begin{bmatrix} \frac{61}{53} \\ -\frac{51}{53} \\ \frac{13}{53} \end{bmatrix}. \quad \blacksquare$$

PROBLEMS FOR SECTION 12.9

In each of Problems 1 through 25, find the inverse of the matrix or show that the matrix is singular.

1. $\begin{bmatrix} -1 & 2 \\ 2 & 1 \end{bmatrix}$

2. $\begin{bmatrix} 1 & 1 & -3 \\ 2 & 16 & 1 \\ 0 & 0 & 4 \end{bmatrix}$

3. $\begin{bmatrix} -3 & 4 & 1 \\ 1 & 2 & 0 \\ 1 & 1 & 3 \end{bmatrix}$

4. $\begin{bmatrix} -2 & 1 & -5 \\ 1 & 1 & 4 \\ 0 & 3 & 3 \end{bmatrix}$

5. $\begin{bmatrix} -2 & 1 & 1 \\ 0 & 1 & 1 \\ -3 & 0 & 6 \end{bmatrix}$

6. $\begin{bmatrix} -1 & 1 & 1 & 0 \\ 1 & 0 & 2 & 0 \\ 1 & 1 & 1 & 1 \\ 3 & 0 & 0 & 1 \end{bmatrix}$

7. $\begin{bmatrix} 12 & 1 & 14 \\ -3 & 2 & 0 \\ 0 & 9 & 14 \end{bmatrix}$

8. $\begin{bmatrix} 0 & 0 & -1 \\ 1 & 12 & 0 \\ 1 & -2 & 4 \end{bmatrix}$

9. $\begin{bmatrix} -1 & 1 & 16 & 2 \\ 0 & 0 & 1 & 4 \\ 0 & 0 & 1 & 6 \\ 0 & 1 & 1 & -3 \end{bmatrix}$

10. $\begin{bmatrix} -2 & 1 & 0 \\ 0 & 2 & 4 \\ -3 & -3 & -3 \end{bmatrix}$

11. $\begin{bmatrix} 1 & 14 \\ 0 & 1 \end{bmatrix}$

12. $\begin{bmatrix} 3 & 1 & 0 & 5 \\ 1 & 1 & 0 & 3 \\ 1 & 6 & 1 & 16 \\ -2 & 4 & 3 & 15 \end{bmatrix}$

13. $\begin{bmatrix} -2 & 4 & 7 \\ -6 & -3 & 4 \\ -16 & 2 & 22 \end{bmatrix}$

14. $\begin{bmatrix} 4 & -3 & 2 & 1 \\ 0 & -2 & 4 & 3 \\ 8 & 1 & -3 & 1 \\ 2 & 3 & 3 & -5 \end{bmatrix}$

15. $\begin{bmatrix} -1 & 4 & 3 \\ 2 & 0 & 0 \\ 1 & -3 & 5 \end{bmatrix}$

16. $\begin{bmatrix} -3 & -6 & 4 \\ 2 & 1 & -4 \\ 3 & -3 & -8 \end{bmatrix}$

17. $\begin{bmatrix} 1 & -1 & 1 \\ 2 & -3 & 2 \\ -4 & 6 & 1 \end{bmatrix}$

18. $\begin{bmatrix} -5 & 3 & -7 \\ 3 & 1 & 7 \\ -2 & 4 & 0 \end{bmatrix}$

19. $\begin{bmatrix} -5 & 6 & 6 \\ 2 & -1 & 3 \\ 0 & 2 & 1 \end{bmatrix}$

20. $\begin{bmatrix} -2 & 6 & 0 & 0 \\ 1 & 4 & 4 & 11 \\ 4 & -4 & -5 & 3 \\ -3 & 1 & 2 & -6 \end{bmatrix}$

21. $\begin{bmatrix} -5 & 2 & -3 \\ 1 & 0 & 0 \\ -3 & 0 & 0 \end{bmatrix}$

22. $\begin{bmatrix} 3 & -6 & 9 \\ 1 & -2 & 3 \\ 3 & 8 & -4 \end{bmatrix}$
    23. $\begin{bmatrix} 0 & 1 & 0 \\ 2 & -1 & 3 \\ 0 & 2 & 1 \end{bmatrix}$
    24. $\begin{bmatrix} -5 & 3 & 3 \\ 2 & -1 & 0 \\ 0 & -3 & 0 \end{bmatrix}$

25. $\begin{bmatrix} -4 & 3 & 3 \\ 2 & -4 & 4 \\ 9 & 0 & 1 \end{bmatrix}$

In each of Problems 26 through 31, find the unique solution $X = A^{-1}B$ of the system.

26. $3x_1 - 4x_2 - 6x_3 = 0$
  $x_1 + x_2 - 3x_3 = 4$
  $2x_1 - x_2 + 6x_3 = -1$

27. $x_1 - x_2 + 3x_3 - x_4 = 1$
  $x_2 - 3x_3 + 5x_4 = 2$
  $x_1 - x_3 + x_4 = 0$
  $x_1 + 2x_2 - x_4 = -5$

28. $8x_1 - x_2 - x_3 = 4$
  $x_1 + 2x_2 - 3x_3 = 0$
  $2x_1 - x_2 + 4x_3 = 5$

29. $2x_1 - 6x_2 + 3x_3 = -4$
  $-x_1 + x_2 + x_3 = 5$
  $2x_1 + 6x_2 - 5x_3 = 8$

30. $12x_1 + x_2 - 3x_3 = 4$
  $x_1 - x_2 + 3x_3 = -5$
  $-2x_1 + x_2 + x_3 = 0$

31. $4x_1 + 6x_2 - 3x_3 = 0$
  $2x_1 + 3x_2 - 4x_3 = 0$
  $x_1 - x_2 + 3x_3 = -7$

32. Let A be an $n \times n$ nonsingular matrix. Prove that every vector in R^n can be written as a linear combination of the column vectors of A. *Hint:* Use the fact that rank$(A) = n$.

33. Let A be an $n \times n$ matrix. Prove that A is nonsingular if and only if the row vectors of A form a basis for R^n.

34. Let A be a nonsingular matrix. Prove that A^t is nonsingular and that $(A^t)^{-1} = (A^{-1})^t$.

35. Let A be a nonsingular matrix. Prove that, for any positive integer k, A^k is nonsingular, and prove that $(A^k)^{-1} = (A^{-1})^k$.

36. Let A, B, and C be $n \times n$ matrices. Suppose that $BA = AC = I_n$. Prove that $B = C$.

12.10 Determinants

The *determinant* of a square matrix is a number produced from the matrix in a way we will now define, beginning with "small" matrices. The determinant of A will be denoted $|A|$, or det(A). In this notation, $|A|$ is not the absolute value of A (which makes no sense for a matrix) but is a number determined from A; this number might be positive, negative, or zero.

First, if A is 1×1, say, $A = [a]$, then the determinant of A is defined by

$$|A| = a.$$

Thus, for example, if $A = [-32]$, then $|A| = -32$.
  If $A = [a_{ij}]$ is 2×2, then

$$A = \begin{bmatrix} a_{11} & a_{12} \\ a_{21} & a_{22} \end{bmatrix},$$

and we define

$$|A| = a_{11}a_{22} - a_{12}a_{21}.$$

For example, if

$$A = \begin{bmatrix} 6 & -3 \\ 2 & 1 \end{bmatrix},$$

then $|A| = 6(1) - (-3)(2) = 12$.

For a 3×3 matrix

$$A = \begin{bmatrix} a_{11} & a_{12} & a_{13} \\ a_{21} & a_{22} & a_{23} \\ a_{31} & a_{32} & a_{33} \end{bmatrix},$$

we define

$$|A| = a_{11}a_{22}a_{33} - a_{11}a_{23}a_{32} - a_{12}a_{21}a_{33} + a_{12}a_{23}a_{31} + a_{13}a_{21}a_{32} - a_{13}a_{22}a_{31}.$$

For example, if

$$A = \begin{bmatrix} 8 & -2 & 4 \\ 1 & 0 & -3 \\ -2 & 8 & 3 \end{bmatrix},$$

we find that $|A| = 218$.

Before proceeding to determinants of larger matrices, we will introduce the practice of denoting the determinant of a matrix by replacing the square brackets used for matrices with vertical lines for determinants. Thus, in the last example, we would write

$$\begin{vmatrix} 8 & -2 & 4 \\ 1 & 0 & -3 \\ -2 & 8 & 3 \end{vmatrix}$$

for the determinant of

$$\begin{bmatrix} 8 & -2 & 4 \\ 1 & 0 & -3 \\ -2 & 8 & 3 \end{bmatrix}.$$

In general, the determinant of an nth order matrix is called an nth *order determinant*, or a *determinant of order n*. We will define $|A|$, if $n > 3$, as a sum of multiples of $(n-1) \times (n-1)$ determinants formed from the elements of A. In this way, we can evaluate each of the $(n-1) \times (n-1)$ determinants as a sum of multiples of $(n-2) \times (n-2)$ determinants, and so on, until we have a sum of constants times 3×3, or 2×2, determinants, which are more easily evaluated.

The following notation will help describe this process. If $A = [a_{ij}]$ is $n \times n$, and $n \geq 2$, M_{ij} denotes the determinant of the $(n-1) \times (n-1)$ matrix formed by deleting row i and column j of A. M_{ij} is called the *minor* of a_{ij} in A. The *cofactor* of a_{ij} is defined to be $(-1)^{i+j} M_{ij}$. For example, let

$$A = \begin{bmatrix} -1 & 6 & 2 \\ 0 & 1 & 4 \\ 8 & -3 & 7 \end{bmatrix}.$$

Then

$$M_{11} = \text{minor of } a_{11} \text{ in } A = \begin{vmatrix} 1 & 4 \\ -3 & 7 \end{vmatrix} = 19$$

(cover up row 1, column 1 of A);

$$M_{23} = \text{minor of } a_{23} \text{ in } A = \begin{vmatrix} -1 & 6 \\ 8 & -3 \end{vmatrix} = -45$$

(cover up row 2, column 3 of A); and

$$M_{12} = \text{minor of } a_{12} \text{ in } A = \begin{vmatrix} 0 & 4 \\ 8 & 7 \end{vmatrix} = -32$$

(cover up row 1, column 2 of A). The cofactors of -1, 4, and 6 in A are, respectively,

$$(-1)^{1+1}(19), \quad \text{or} \quad 19,$$
$$(-1)^{2+3}(-45), \quad \text{or} \quad 45,$$

and

$$(-1)^{1+2}(-32), \quad \text{or} \quad 32.$$

Now suppose that A is $n \times n$ and $n \geq 2$. The *cofactor* (or *Laplace*) *expansion of* $|A|$ *by row* k is defined to be the sum of the elements of row k, each multiplied by its cofactor:

$$\sum_{j=1}^{n} (-1)^{k+j} a_{kj} M_{kj},$$

for any $k = 1, 2, 3, \ldots, n$.

Similarly, the *cofactor* (or *Laplace*) *expansion of* $|A|$ *by column* k is defined to be the sum of the elements of column k, each multiplied by its cofactor:

$$\sum_{i=1}^{n} (-1)^{k+i} a_{ik} M_{ik},$$

for any $k = 1, 2, 3, \ldots, n$.

It is a remarkable fact that the n row cofactor expansions of $|A|$ and the n column cofactor expansions all yield exactly the same number. This is stated below as Theorem 12.24 and allows us to define $|A|$ as the cofactor expansion by any row or column.

THEOREM 12.24

The cofactor expansion of $|A|$ by any row or column yields the same number for any given $n \times n$ matrix A. ∎

We will sketch a proof of this theorem at the end of the section. To get some experience with cofactor expansions, look first at the case $n = 2$:

$$A = \begin{bmatrix} a_{11} & a_{12} \\ a_{21} & a_{22} \end{bmatrix}.$$

The four possible row and column cofactor expansions are as follows. By row 1:

$$\sum_{j=1}^{2} (-1)^{1+j} a_{1j} M_{1j} = (-1)^{1+1} a_{11} M_{11} + (-1)^{1+2} a_{12} M_{12}$$

$$= a_{11} a_{22} - a_{12} a_{21};$$

by row 2:

$$\sum_{j=1}^{2} (-1)^{2+j} a_{2j} M_{2j} = (-1)^{2+1} a_{21} M_{21} + (-1)^{2+2} a_{22} M_{22}$$

$$= -a_{12} a_{21} + a_{22} a_{11};$$

by column 1:

$$\sum_{i=1}^{2} (-1)^{i+1} a_{i1} M_{i1} = (-1)^{1+1} a_{11} M_{11} + (-1)^{2+1} a_{21} M_{21}$$

$$= a_{11} a_{22} - a_{21} a_{22};$$

and by column 2:

$$\sum_{i=1}^{2} (-1)^{i+2} a_{i2} M_{i2} = -a_{12} M_{12} + a_{22} M_{22}$$

$$= -a_{12} a_{21} + a_{22} a_{11}.$$

Notice that we obtain the same result, $a_{11} a_{22} - a_{12} a_{21}$, from all four expansions.

When $n = 3$, there are nine row and column expansions possible. To illustrate, here is the expansion of a 3×3 determinant by column 2:

$$\sum_{i=1}^{3} (-1)^{2+i} a_{i2} M_{i2} = -a_{12} M_{12} + a_{22} M_{22} - a_{32} M_{32}$$

$$= -a_{12} \begin{vmatrix} a_{21} & a_{23} \\ a_{31} & a_{33} \end{vmatrix} + a_{22} \begin{vmatrix} a_{11} & a_{13} \\ a_{31} & a_{33} \end{vmatrix} - a_{32} \begin{vmatrix} a_{11} & a_{13} \\ a_{21} & a_{23} \end{vmatrix}$$

$$= -a_{12}(a_{21} a_{33} - a_{23} a_{31}) + a_{22}(a_{11} a_{33} - a_{13} a_{31})$$

$$- a_{32}(a_{11} a_{23} - a_{13} a_{21}),$$

and this coincides with the definition given for $|A|$ when A is 3×3.

In theory, we can evaluate any determinant by a row or column expansion. In practice, this is not an efficient method if n is large (say, if $n \geq 3$). We will now develop a sequence of theorems which will help us evaluate determinants more easily. The key is to examine the effects of elementary row operations on the value of a determinant.

In fact, we will use not only row operations but the corresponding column operations as well. That is, we will interchange columns, multiply a column by a nonzero scalar, and add scalar multiples of one column to another. Although such column operations are not relevant when we are solving systems of equations (because a row corresponds to an equation, but a column does not), we will find row and column operations equally useful in evaluating determinants.

We will therefore address the following question: if B is obtained from A by an elementary row or column operation, how is $|B|$ related to $|A|$? As we proceed, we will also discuss other properties of $|A|$ that are important to know.

THEOREM 12.25

If B is formed from A by multiplying any row or column of A by a scalar α, $|B| = \alpha |A|$.

Proof   Suppose we multiply row k of A by α to form B. Expand $|B|$ by row k to get

$$|B| = \sum_{j=1}^{n} (-1)^{k+j} b_{kj} M_{kj}.$$

But $b_{kj} = \alpha a_{kj}$. Further, the minor M_{kj} is the same for B as for A, since we obtain M_{kj} by deleting row k and column j, and A and B are the same except possibly in row k. Thus,

$$|B| = \sum_{j=1}^{n} (-1)^{k+j} b_{kj} M_{kj} = \alpha \sum_{j=1}^{n} (-1)^{k+j} a_{kj} M_{kj} = \alpha |A|.$$

A similar argument holds if we use a column expansion.   ∎

THEOREM 12.26

If A has a zero row or column, $|A| = 0$.

Proof   Form B from A by multiplying the zero row or column by 2. By Theorem 12.25, $|B| = 2|A|$. But $B = A$, so $|B| = |A|$. Therefore, $|A| = 2|A|$; hence, $|A| = 0$.   ∎

THEOREM 12.27

If B is obtained from A by interchanging two rows or columns, $|B| = -|A|$.

Proof   We proceed by mathematical induction on the order of A. If A is 2×2,

$$A = \begin{bmatrix} a_{11} & a_{12} \\ a_{21} & a_{22} \end{bmatrix}.$$

If we interchange the rows of A to form B,

$$B = \begin{bmatrix} a_{21} & a_{22} \\ a_{11} & a_{12} \end{bmatrix},$$

and we have by direct computation that $|B| = a_{21}a_{12} - a_{22}a_{11} = -|A|$. If we interchange the columns of A to form B,

$$B = \begin{bmatrix} a_{12} & a_{11} \\ a_{22} & a_{21} \end{bmatrix},$$

and again we have $|B| = a_{12}a_{21} - a_{11}a_{22}$.

Now assume that the conclusion of this theorem holds for $(n-1) \times (n-1)$ matrices. Let A be $n \times n$. Suppose, to be specific, that we form B from A by interchanging rows i and j. Expand $|B|$ and $|A|$ by a different row k to get

$$|B| = \sum_{s=1}^{n} (-1)^{k+s} b_{ks} M_{ks} \quad \text{and} \quad |A| = \sum_{s=1}^{n} (-1)^{k+s} a_{ks} N_{ks},$$

where N_{ks} is the minor of A formed by deleting row k, column s of A.

For $s = 1, 2, \ldots, n$, we get N_{ks} from M_{ks} by interchanging rows i and j. By the induction hypothesis, recalling that N_{ks} and M_{ks} are $(n-1) \times (n-1)$ determinants, we have

$$N_{ks} = -M_{ks} \quad \text{for} \quad s = 1, 2, \ldots, n.$$

Hence, $|B| = -|A|$, as was to be proved. A similar argument holds if two columns are interchanged. ∎

EXAMPLE 12.41

Expanding by any row or column, we get

$$\begin{vmatrix} 4 & -1 & 6 \\ 1 & 9 & 3 \\ 2 & 1 & 4 \end{vmatrix} = 28.$$

If we interchange rows 1 and 3, we get

$$\begin{vmatrix} 2 & 1 & 4 \\ 1 & 9 & 3 \\ 4 & -1 & 6 \end{vmatrix} = -28.$$

If we interchange columns 2 and 3 of the original determinant, we get

$$\begin{vmatrix} 4 & 6 & -1 \\ 1 & 3 & 9 \\ 2 & 4 & 1 \end{vmatrix} = -28. \quad ∎$$

THEOREM 12.28

If two rows or columns of A are identical, $|A| = 0$.

Proof Form B by interchanging the identical rows or columns. Then $B = A$, so $|B| = |A|$. But also $|B| = -|A|$, so $|A| = -|A|$, and we conclude that $|A| = 0$. ∎

EXAMPLE 12.42

$$\begin{vmatrix} -1 & 3 & 4 & 2 \\ 0 & 1 & 1 & -5 \\ 2 & 6 & 17 & 3 \\ 0 & 1 & 1 & -5 \end{vmatrix} = 0$$

without calculation, because rows 2 and 4 are identical. ∎

THEOREM 12.29

If one row (or column) is a constant multiple of another, $|A| = 0$.

Proof We will give an argument for columns. A similar argument holds for rows. Suppose that column i of A is a constant multiple α of column j. If $\alpha = 0$, $|A| = 0$ because A has a zero column. If $\alpha \neq 0$, we can obtain from A a matrix B with two identical columns by multiplying column i of A by $1/\alpha$. Then $|B| = 0$. But $|B| = (1/\alpha)|A|$, so $|A| = 0$ also. ∎

EXAMPLE 12.43

$$\begin{vmatrix} -1 & 4 & 2 & 1 \\ 8 & \sqrt{2} & 4 & 0 \\ -2 & 8 & 4 & 2 \\ 1 & 19 & 0 & -4 \end{vmatrix} = 0,$$

because row 3 = 2(row 1). ∎

THEOREM 12.30

Suppose we obtain B from A by adding a constant multiple of one row (or column) to another row (or column). Then $|B| = |A|$.

Proof   Suppose, to be specific, that we add α times row i to row j of A to form B. Then

$$B = \begin{bmatrix} a_{11} & a_{12} & \cdots & a_{1n} \\ \vdots & \vdots & & \vdots \\ \alpha a_{i1} + a_{j1} & \alpha a_{i2} + a_{j2} & \cdots & \alpha a_{in} + a_{jn} \\ \vdots & \vdots & & \vdots \\ a_{n1} & a_{n2} & \cdots & a_{nn} \end{bmatrix} \longleftarrow \text{row } j \text{ of } B.$$

Expand $|B|$ by row j to get

$$|B| = \sum_{k=1}^{n} (-1)^{j+k}(\alpha a_{ik} + a_{jk})M_{jk}$$

$$= \sum_{k=1}^{n} (-1)^{j+k}\alpha a_{ik}M_{jk} + \sum_{k=1}^{n} (-1)^{j+k}a_{jk}M_{jk}.$$

Now, $\sum_{k=1}^{n} (-1)^{j+k}a_{jk}M_{jk} = |A|$ because this is a cofactor expansion of $|A|$ by row j. Here, we used the fact that we form M_{jk} by deleting row j, column k of B, and B is the same as A except for row j.

Next, $\sum_{k=1}^{n} (-1)^{j+k}\alpha a_{ik}M_{jk} = 0$ because this is the expansion by row j of the determinant formed from A by replacing row j by α times row i. But this determinant is zero by Theorem 12.29. Therefore, $|B| = |A|$. ∎

EXAMPLE 12.44

It is easy to check that

$$\begin{vmatrix} 4 & -2 & 1 \\ 3 & 0 & -5 \\ 1 & -3 & -4 \end{vmatrix} = -83.$$

If we add -3(row 2) to row 3, we get

$$\begin{vmatrix} 4 & -2 & 1 \\ 3 & 0 & -5 \\ -8 & -3 & 11 \end{vmatrix},$$

and by expanding by any row or column, we find that this determinant is also -83. ∎

THEOREM 12.31

For any square matrix A, $|A| = |A^t|$.

Proof   We form A^t from A by rewriting the rows of A as columns of A^t. Thus, the cofactor expansion of $|A^t|$ by any row is identical with the cofactor expansion of $|A|$ by the corresponding column.   ■

Incidentally, Theorem 12.31 is the reason why column and row operations are equally applicable to the evaluation of determinants. Since the rows of A are the columns of A^t, a row operation on A is the same as a column operation on A^t. Theorem 12.31 tells us that A and A^t have the same determinant.

THEOREM 12.32

If A and B are $n \times n$ matrices,

$$|AB| = |A||B|.$$

That is, the determinant of a product is the product of the determinants.

Proof   The general argument is notationally complicated, so we will develop the details for the 2×2 case first to explain the ideas involved. Suppose, then, that A and B are 2×2. First, manufacture the 4×4 matrix

$$P = \begin{bmatrix} a_{11} & a_{12} & 0 & 0 \\ a_{21} & a_{22} & 0 & 0 \\ -1 & 0 & b_{11} & b_{12} \\ 0 & -1 & b_{21} & b_{22} \end{bmatrix}.$$

We will show that $|P| = |A||B|$ and also that $|P| = |AB|$ and hence conclude that $|A||B| = |AB|$. Begin by expanding $|P|$ by row 1 to get

$$|P| = a_{11} \begin{bmatrix} a_{22} & 0 & 0 \\ 0 & b_{11} & b_{12} \\ -1 & b_{21} & b_{22} \end{bmatrix} - a_{12} \begin{bmatrix} a_{21} & 0 & 0 \\ -1 & b_{11} & b_{12} \\ 0 & b_{21} & b_{22} \end{bmatrix}.$$

Expand each of the 3×3 determinants on the right by row 1 to get

$$|P| = a_{11}a_{22} \begin{vmatrix} b_{11} & b_{12} \\ b_{21} & b_{22} \end{vmatrix} - a_{12}a_{21} \begin{vmatrix} b_{11} & b_{12} \\ b_{21} & b_{22} \end{vmatrix}$$

$$= (a_{11}a_{22} - a_{12}a_{21}) \begin{vmatrix} b_{11} & b_{12} \\ b_{21} & b_{22} \end{vmatrix} = |A||B|.$$

To show that $|P| = |AB|$, first form a new matrix P' from P by adding a_{11} times row 3 to row 1, a_{12} times row 4 to row 1, a_{21} times row 3 to row 2, and a_{22} times row 4 to row 2. We get

$$P' = \begin{bmatrix} 0 & 0 & a_{11}b_{11} + a_{12}b_{21} & a_{11}b_{12} + a_{12}b_{22} \\ 0 & 0 & a_{21}b_{11} + a_{22}b_{21} & a_{21}b_{12} + a_{22}b_{22} \\ -1 & 0 & b_{11} & b_{12} \\ 0 & -1 & b_{21} & b_{22} \end{bmatrix}.$$

Recognize the 2×2 block in the upper right corner of P' as the elements of the product AB. Letting $C = AB = [c_{ij}]$, then,

$$P' = \begin{bmatrix} 0 & 0 & c_{11} & c_{12} \\ 0 & 0 & c_{21} & c_{22} \\ -1 & 0 & b_{11} & b_{12} \\ 0 & -1 & b_{21} & b_{22} \end{bmatrix}.$$

By Theorem 12.30, $|P| = |P'|$. Expand $|P'|$ by column 1 to get

$$|P| = |P'| = (-1) \begin{vmatrix} 0 & c_{11} & c_{12} \\ 0 & c_{21} & c_{22} \\ -1 & b_{21} & b_{22} \end{vmatrix}.$$

Finally, expand this 3×3 determinant by column 1 to get

$$|P| = |P'| = \begin{vmatrix} c_{11} & c_{12} \\ c_{21} & c_{22} \end{vmatrix} = |C| = |AB|.$$

For $n \geq 3$, we proceed in much the same way. First, define the $2n \times 2n$ matrix

$$P = \left[\begin{array}{cccc|cccc} & & & & 0 & 0 & \cdots & 0 \\ & A & & & 0 & 0 & \cdots & 0 \\ & & & & \vdots & \vdots & & \vdots \\ & & & & 0 & 0 & \cdots & 0 \\ \hline -1 & 0 & \cdots & 0 & & & & \\ 0 & -1 & \cdots & 0 & & B & & \\ \vdots & \vdots & & \vdots & & & & \\ 0 & 0 & \cdots & -1 & & & & \end{array} \right],$$

with A in the upper left, B in the lower right, zeros above B, and $-I_n$ below A. The strategy is to show that $|P| = |A||B|$ and also that $|P| = |AB|$.

To show that $|P| = |A||B|$, expand $|P|$ by row 1, obtaining $|P|$ as a sum of n determinants of order $2n - 1$. Expand each of these by row 1, and so on. After n such expansions, we obtain a sum of terms which equals $|A|$, each term having a factor of $|B|$. Thus, $|P| = |A||B|$.

To show that $|P| = |AB|$, form a new matrix P' from P as follows. First, add

$$a_{11} \text{ times row } n + 1 \text{ to row 1;}$$
$$a_{12} \text{ times row } n + 2 \text{ to row 1;}$$
$$\vdots$$
$$a_{1n} \text{ times row } 2n \text{ to row 1.}$$

The first row of the matrix formed thus far is

$$0 \quad 0 \quad \cdots \quad 0 \quad a_{11}b_{11} + a_{12}b_{21} + \cdots + a_{1n}b_{n1} \cdots a_{11}b_{1n} + a_{12}b_{2n} + \cdots + a_{1n}b_{nn},$$

with zeros in the first n columns. Now add

$$a_{21} \text{ times row } n + 1 \text{ to row 2;}$$
$$a_{22} \text{ times row } n + 2 \text{ to row 2;}$$
$$\vdots$$
$$a_{2n} \text{ times row } 2n \text{ to row 2,}$$

and so on. Eventually, we obtain

$$
P' = \left[
\begin{array}{cccc:c}
0 & 0 & \cdots & 0 & \\
0 & 0 & \cdots & 0 & \\
\vdots & \vdots & & \vdots & C \\
0 & 0 & \cdots & 0 & \\
\hdashline
-1 & 0 & \cdots & 0 & \\
0 & -1 & \cdots & 0 & \\
\vdots & \vdots & & \vdots & B \\
0 & 0 & \cdots & -1 &
\end{array}
\right],
$$

in which $C = AB$. By Theorem 12.30, $|P'| = |P|$. Expand $|P'|$ by column 1; then expand the resulting $(2n - 1) \times (2n - 1)$ determinant by its first column; and so on. After n such expansions, we obtain

$$|P| = |P'| = |C| = |AB|,$$

completing the proof. ■

EXAMPLE 12.45

Let

$$
A = \begin{bmatrix} -4 & 6 & 3 \\ 8 & 1 & 1 \\ -2 & 0 & 7 \end{bmatrix} \quad \text{and} \quad B = \begin{bmatrix} 14 & 2 & -3 \\ -6 & 1 & -1 \\ 4 & 1 & 4 \end{bmatrix}.
$$

Then

$$
AB = \begin{bmatrix} -80 & 1 & 18 \\ 110 & 18 & -21 \\ 0 & 3 & 34 \end{bmatrix}.
$$

We find that $|A| = -370$, $|B| = 140$, and $|AB| = -51,800 = (-370)(140)$. ■

We will now sketch a proof of Theorem 12.24. The proof is by mathematical induction on n.

Proof of Theorem 12.24 We have already shown explicitly that when $n = 2$, both row and column cofactor expansions of $|A|$ yield the same result, namely, $a_{11}a_{22} - a_{12}a_{21}$. To complete the proof by induction, we will assume that the conclusion of the theorem holds for $n \times n$ determinants and prove that the conclusion holds for $(n + 1) \times (n + 1)$ determinants.

Let A be an $(n + 1) \times (n + 1)$ matrix. First, look at the cofactor expansion of $|A|$ by rows i and j, with $i < j$.

$$\text{By row } i: \quad \sum_{t=1}^{n+1} (-1)^{i+t} a_{it} M_{it}.$$

$$\text{By row } j: \quad \sum_{s=1}^{n+1} (-1)^{j+s} a_{js} M_{js}.$$

Now, each M_{it} in the expansion by row i is an $n \times n$ determinant and may be expanded by any row by the induction hypothesis. Expand M_{it} by its row $j - 1$, which contains the elements of row j of A (with column t deleted). The terms in the cofactor expansion of M_{it} by its row $j - 1$ are of the form

$$(-1)^{j-1+s} a_{js} M_{itjs} \quad \text{if} \quad s < t$$

and

$$(-1)^{j+1+s-1} a_{js} M_{itjs} \quad \text{if} \quad s > t,$$

where M_{itjs} is the j, s minor of M_{it}. The reason for two cases, $s < t$ or $s > t$, is that column s of M_{ij} is column s of A (with a_{is} deleted) if $s < t$; but if $s > t$, deletion of column t from A to form M_{it} makes column s of M_{ij} the $(s + 1)$ column of A. Thus, in the expansion by row i, typical terms are of the form

$$(-1)^{i+t+j-1+s} a_{it} a_{js} M_{itjs} \quad \text{if} \quad s < t$$

and

$$(-1)^{i+t+j+s-2} a_{it} a_{js} M_{itjs} \quad \text{if} \quad s > t.$$

A similar argument leads us to the conclusion that typical terms in the expansion by row j are

$$(-1)^{j+s+i+t} a_{js} a_{it} M_{jsit} \quad \text{if} \quad t < s$$

and

$$(-1)^{k+s+i+t-1} a_{js} a_{it} M_{jsit} \quad \text{if} \quad t > s.$$

Now, $M_{itjs} = M_{jsit}$, since both are formed from A by deleting rows i and j and columns s and t. The terms in the expansion by row i, with $s < t$, correspond one for one with those in the expansion by row j. Similarly, the terms match one for one for $s > t$. Thus, the expansions by rows i and j result in the same number.

A similar argument, with a term by term comparison, shows that the cofactor expansion by any column agrees with that by any row. We omit the details of this argument. ∎

It is usually possible to use elementary row and/or column operations to evaluate a determinant more efficiently than by a straightforward cofactor expansion by a row or column. Look at a cofactor expansion by any row or column. For example, expanding by column k, we get

$$|A| = \sum_{k=1}^{n} (-1)^{k+j} a_{kj} M_{kj}.$$

Each M_{kj} is an $(n - 1) \times (n - 1)$ determinant, so this sum involves the evaluation of n such determinants. However, for each j such that $a_{kj} = 0$, we are saved one such determinant and the labor is reduced by one term. This observation leads to a first stratagem:

expand by a row (or column) having as many zeros as possible.

For example, let

$$A = \begin{vmatrix} -3 & 1 & 16 & -8 \\ 0 & 1 & 14 & 0 \\ 0 & 3 & 0 & 1 \\ 0 & 14 & 6 & 0 \end{vmatrix}.$$

Since column 1 has only one nonzero element, expand $|A|$ by column 1 to get

$$|A| = -3 \begin{vmatrix} 1 & 14 & 0 \\ 3 & 0 & 1 \\ 14 & 6 & 0 \end{vmatrix}.$$

Now column 3 has only one nonzero element, so expand by column 3 to get

$$|A| = -3(-1)^{2+3}(1) \begin{vmatrix} 1 & 14 \\ 14 & 6 \end{vmatrix} = 3[6 - 14^2] = -570.$$

In this example, columns conveniently appeared having, for the most part, zero elements. What if A has no such rows or columns? Then we make some, using elementary row and column operations. These operations will transform A into a matrix B, with the relationship between $|A|$ and the intermediate determinants known at each stage from the theorems we have proved. We attempt to manufacture B so that $|B|$ is easier to evaluate than $|A|$. Following are some examples.

EXAMPLE 12.46

Evaluate $|A|$, where

$$A = \begin{bmatrix} -5 & 1 & 1 & -6 & 2 \\ 0 & 1 & 3 & 8 & -4 \\ -1 & 2 & 1 & 8 & 9 \\ 0 & 0 & 1 & 14 & 2 \\ 1 & 1 & 0 & 0 & 0 \end{bmatrix}.$$

Notice that row 5 has three zero elements. Add -1(column 1) to column 2 to produce a row with all but one element zero:

$$B = \begin{bmatrix} -5 & 6 & 1 & -6 & 2 \\ 0 & 1 & 3 & 8 & -4 \\ -1 & 3 & 1 & 8 & 9 \\ 0 & 0 & 1 & 14 & 2 \\ 1 & 0 & 0 & 0 & 0 \end{bmatrix}.$$

Now expand $|B|$ by row 5:

$$|A| = |B| = (-1)^{1+5}(1) \begin{vmatrix} 6 & 1 & -6 & 2 \\ 1 & 3 & 8 & -4 \\ 3 & 1 & 8 & 9 \\ 0 & 1 & 14 & 2 \end{vmatrix}.$$

We can get zeros in the 4, 3 and 4, 4 places if we add -14(column 2) to column 3 and -2(column 2) to column 4. Then

$$|A| = \begin{vmatrix} 6 & 1 & -20 & 0 \\ 1 & 3 & -34 & -10 \\ 3 & 1 & -6 & 7 \\ 0 & 1 & 0 & 0 \end{vmatrix} = (-1)^{4+2}(1) \begin{vmatrix} 6 & -20 & 0 \\ 1 & -34 & -10 \\ 3 & -6 & 7 \end{vmatrix}.$$

Here, we expanded the 4×4 determinant by row 4. Now we have a 3×3 determinant left to evaluate. We can do this directly, or we can add -6(row 2) to row 1, and -3(row 2) to row 3, to get

$$|A| = \begin{vmatrix} 0 & 184 & 60 \\ 1 & -34 & -10 \\ 0 & 96 & 37 \end{vmatrix} = (-1)^{1+2} \begin{vmatrix} 184 & 60 \\ 96 & 37 \end{vmatrix}$$

$$= -1[184(37) - 60(96)] = -1048. \quad \blacksquare$$

We can always evaluate a determinant in this way—apply row or column operations to get a matrix with a row or column having "mostly" zero elements. We expand by this row or column, then continue the process until eventually we obtain a determinant that is easily evaluated.

EXAMPLE 12.47

Let

$$A = \begin{bmatrix} -6 & 0 & 1 & 3 & 2 \\ -1 & 5 & 0 & 1 & 7 \\ 8 & 3 & 2 & 1 & 7 \\ 0 & 1 & 5 & -3 & 2 \\ 1 & 15 & -3 & 9 & 4 \end{bmatrix}.$$

Exploit the fact that $a_{13} = 1$ to get zeros elsewhere in column 3. Add -2(row 1) to row 3, -5(row 1) to row 4, and 3(row 1) to row 5 to get

$$B = \begin{bmatrix} -6 & 0 & 1 & 3 & 2 \\ -1 & 5 & 0 & 1 & 7 \\ 20 & 3 & 0 & -5 & 3 \\ 30 & 1 & 0 & -18 & -8 \\ -17 & 15 & 0 & 18 & 10 \end{bmatrix}.$$

Expand $|B|$ by column 3:

$$|A| = |B| = (-1)^{1+3}(1) \begin{vmatrix} -1 & 5 & 1 & 7 \\ 20 & 3 & -5 & 3 \\ 30 & 1 & -18 & -8 \\ -17 & 15 & 18 & 10 \end{vmatrix}.$$

Now exploit the -1 element in the 1, 1 place to get zeros in the other row 1 locations. Add 5(column 1) to column 2, add column 1 to column 3, and add 7(column 1) to column 4. We get

$$|A| = \begin{vmatrix} -1 & 0 & 0 & 0 \\ 20 & 103 & 15 & 143 \\ 30 & 151 & 12 & 202 \\ -17 & -70 & 1 & -109 \end{vmatrix}.$$

Now expand by row 1:

$$|A| = (-1)^{1+1}(-1)\begin{vmatrix} 103 & 15 & 143 \\ 151 & 12 & 202 \\ -70 & 1 & -109 \end{vmatrix}.$$

Now exploit the element 1 in the 3, 2 place. Add -15(row 3) to row 1, and -12(row 3) to row 2, to get

$$|A| = -\begin{vmatrix} 1153 & 0 & 1778 \\ 991 & 0 & 1510 \\ -70 & 1 & -109 \end{vmatrix} = -(-1)^{3+2}(1)\begin{vmatrix} 1153 & 1778 \\ 991 & 1510 \end{vmatrix}$$

$$= (1153)(1510) - (1778)(991) = -20,968. \quad \blacksquare$$

EXAMPLE 12.48

Let

$$A = \begin{bmatrix} 14 & 0 & 3 & 5 & -7 & 1 \\ 2 & -6 & 0 & 18 & 0 & 3 \\ -6 & -5 & 7 & 20 & 18 & 0 \\ 0 & 4 & 8 & 0 & -3 & 2 \\ 9 & -5 & 4 & 3 & 3 & 0 \\ 10 & 0 & 5 & 3 & 21 & 10 \end{bmatrix}.$$

Add -3(row 1) to row 2, -2(row 1) to row 4, and -10(row 1) to row 6, forming the matrix

$$B = \begin{bmatrix} 14 & 0 & 3 & 5 & -7 & 1 \\ -40 & -6 & -9 & 3 & 21 & 0 \\ -6 & -5 & 7 & 20 & 18 & 0 \\ -28 & 4 & 2 & -10 & 11 & 0 \\ 9 & -5 & 4 & 3 & 3 & 0 \\ -130 & 0 & -25 & -47 & 91 & 0 \end{bmatrix}.$$

We will expand this determinant by column 6 and indicate in outline form how we proceed from there eventually to find $|A|$.

$$|A| = |B| = (-1)^{1+6}(1)\begin{vmatrix} -40 & -6 & -9 & 3 & 21 \\ -6 & -5 & 7 & 20 & 18 \\ -28 & 4 & 2 & -10 & 11 \\ 9 & -5 & 4 & 3 & 3 \\ -130 & 0 & -25 & -47 & 91 \end{vmatrix}$$

(multiply row
3 by $\frac{1}{2}$)

$$= -2 \begin{vmatrix} -40 & -6 & -9 & 3 & 21 \\ -6 & -5 & 7 & 20 & 18 \\ -14 & 2 & 1 & -5 & \frac{11}{2} \\ 9 & -5 & 4 & 3 & 3 \\ -130 & 0 & -25 & -47 & 91 \end{vmatrix}$$

[add 9(row 3) to row 1,
-7(row 3) to row 2,
-4(row 3) to row 4,
25(row 3) to row 5]

$$= -2 \begin{vmatrix} -166 & 12 & 0 & -42 & \frac{141}{2} \\ 92 & -19 & 0 & 55 & -\frac{41}{2} \\ -14 & 2 & 1 & -5 & \frac{11}{2} \\ 65 & -13 & 0 & 23 & -19 \\ -480 & 50 & 0 & -172 & \frac{457}{2} \end{vmatrix}$$

(expand by column 3)

$$= -2 \begin{vmatrix} -166 & 12 & -42 & \frac{141}{2} \\ 92 & -19 & 55 & -\frac{41}{2} \\ 65 & -13 & 23 & -19 \\ -480 & 50 & -172 & \frac{457}{2} \end{vmatrix}$$

(add row 1 to row 3)

$$= -2 \begin{vmatrix} -166 & 12 & -42 & \frac{141}{2} \\ 92 & -19 & 55 & -\frac{41}{2} \\ -101 & -1 & -19 & \frac{103}{2} \\ -480 & 50 & -172 & \frac{457}{2} \end{vmatrix}$$

[add 12(row 3) to row 1,
-19(row 3) to row 2,
50(row 3) to row 4]

$$= -2 \begin{vmatrix} -1378 & 0 & -270 & \frac{1377}{2} \\ 2011 & 0 & 416 & -999 \\ -101 & -1 & -19 & \frac{103}{2} \\ -5530 & 0 & -1122 & \frac{5607}{2} \end{vmatrix}$$

(expand by column two)

$$= -2(-1)^{3+2}(-1) \begin{vmatrix} -1378 & -270 & \frac{1377}{2} \\ 2011 & 416 & -999 \\ -5530 & -1122 & \frac{5607}{2} \end{vmatrix}.$$

At this point, it will take as much arithmetic to reduce this further as it will to evaluate the 3×3 determinant directly. We choose the latter and obtain

$$|A| = -2\left[-1378 \begin{vmatrix} 416 & -999 \\ -1122 & \frac{5607}{2} \end{vmatrix} + 270 \begin{vmatrix} 2011 & -999 \\ -5530 & \frac{5607}{2} \end{vmatrix} \right.$$

$$\left. + (1377/2) \begin{vmatrix} 2011 & 416 \\ -5530 & -1122 \end{vmatrix} \right]$$

$$= -2(-1,532,376) = 3,064,752. \quad \blacksquare$$

We will conclude this section with a general result that is sometimes useful. A matrix $U = [u_{ij}]$ is called *upper triangular* if every element below the main diagonal is zero. That is, $u_{ij} = 0$ if $i > j$. Such a matrix has this appearance:

$$\begin{bmatrix} u_{11} & u_{12} & \cdots & u_{1n} \\ 0 & u_{22} & \cdots & u_{2n} \\ \vdots & \vdots & & \vdots \\ 0 & 0 & \cdots & u_{nn} \end{bmatrix}.$$

Elements above or on the main diagonal may or may not be zero. The determinant of such a matrix is easily evaluated.

THEOREM 12.33

If $U = [u_{ij}]$ is upper triangular, $|U| = u_{11}u_{22} \cdots u_{nn}$, the product of the main diagonal elements.

Proof   We can proceed by induction. If we expand $|U|$ by column 1, we obtain $|U| = u_{11}|U'|$, where U' is the $(n-1) \times (n-1)$ matrix obtained by deleting row 1 and column 1 of U. Now note that U' is also upper triangular. It is left to the student to complete the details of the proof. $\blacksquare$

EXAMPLE 12.49

$$\begin{vmatrix} -1 & -3 & 4 \\ 0 & \pi & 16 \\ 0 & 0 & 14 \end{vmatrix} = (-1)(\pi)(14) = -14\pi. \quad \blacksquare$$

The next three sections are devoted to some uses of determinants.

PROBLEMS FOR SECTION 12.10

In each of Problems 1 through 20, evaluate the determinant by a cofactor expansion (by any row or column).

1. $\begin{vmatrix} -4 & 6 \\ 1 & 7 \end{vmatrix}$

2. $\begin{vmatrix} 8 & -3 \\ -1 & 4 \end{vmatrix}$

3. $\begin{vmatrix} 16 & 2 \\ -3 & -4 \end{vmatrix}$

4. $\begin{vmatrix} -8 & -4 & 2 \\ 0 & 1 & 1 \\ 0 & 1 & -3 \end{vmatrix}$

5. $\begin{vmatrix} 2 & -2 & 1 \\ 1 & 1 & 6 \\ 3 & -1 & 4 \end{vmatrix}$

6. $\begin{vmatrix} 7 & -3 & 1 \\ 1 & -2 & 4 \\ -3 & 1 & 0 \end{vmatrix}$

7. $\begin{vmatrix} -14 & -3 & 2 \\ 1 & -1 & 1 \\ 0 & 1 & -3 \end{vmatrix}$

8. $\begin{vmatrix} 5 & -1 & 7 \\ 0 & 0 & 2 \\ 1 & -4 & 3 \end{vmatrix}$

9. $\begin{vmatrix} 8 & 8 & 8 \\ 4 & 4 & 4 \\ -1 & 3 & 0 \end{vmatrix}$

10. $\begin{vmatrix} 2 & 20 & -5 \\ 7 & -9 & 3 \\ 1 & -1 & 4 \end{vmatrix}$

11. $\begin{vmatrix} -4 & 3 & 7 \\ 0 & 1 & 4 \\ -5 & 0 & 0 \end{vmatrix}$

12. $\begin{vmatrix} -15 & 12 & 0 \\ 1 & 14 & 1 \\ -2 & 0 & 3 \end{vmatrix}$

13. $\begin{vmatrix} 8 & -5 & 4 \\ 3 & -2 & 1 \\ 1 & -1 & 4 \end{vmatrix}$

14. $\begin{vmatrix} 8 & -8 & 4 \\ 2 & 3 & -7 \\ 1 & -1 & 2 \end{vmatrix}$

15. $\begin{vmatrix} -5 & 1 & 6 \\ 1 & -1 & 1 \\ 0 & -1 & 0 \end{vmatrix}$

16. $\begin{vmatrix} 5 & -4 & 3 \\ -1 & 1 & 6 \\ -2 & 2 & 4 \end{vmatrix}$

17. $\begin{vmatrix} -5 & 0 & 1 & 6 \\ 2 & -1 & 3 & 7 \\ 4 & 4 & -5 & -8 \\ 1 & -1 & 6 & 2 \end{vmatrix}$

18. $\begin{vmatrix} 4 & 3 & -5 & 6 \\ 1 & -5 & 15 & 2 \\ 0 & -5 & 1 & 7 \\ 8 & 9 & 0 & 15 \end{vmatrix}$

19. $\begin{vmatrix} 22 & -1 & 3 & 0 & 0 \\ 1 & 4 & -5 & 0 & 2 \\ -1 & 1 & 6 & 0 & -5 \\ 4 & 7 & 9 & 1 & -7 \\ 6 & 6 & -3 & 4 & 1 \end{vmatrix}$

20. $\begin{vmatrix} 5 & -3 & 0 & 1 & -4 \\ -5 & -3 & 2 & -3 & 1 \\ 6 & -3 & 1 & 4 & 0 \\ 0 & 0 & 5 & 5 & 0 \\ 0 & 0 & -3 & -2 & 0 \end{vmatrix}$

21. Let A be an $n \times n$ matrix, and let α be any scalar. Form B from A by multiplying every element of A by α. That is, $B = \alpha A$. Prove that

$$|B| = \alpha^n |A|.$$

22. Let $(x_1, y_1), (x_2, y_2), (x_3, y_3)$ be points in the xy-plane. Prove that these points lie on a straight line if and only if

$$\begin{vmatrix} 1 & x_1 & y_1 \\ 1 & x_2 & y_2 \\ 1 & x_3 & y_3 \end{vmatrix} = 0.$$

23. An $n \times n$ matrix A is called *skew-symmetric* if $A = -A^t$. Prove that $|A| = 0$ if A is skew-symmetric and n is odd.

24. We will prove in Section 12.12 that a square matrix A is nonsingular if and only if $|A| \neq 0$. Assuming this fact for now, prove that, if A is nonsingular,

$$|A^{-1}| = \frac{1}{|A|}.$$

25. A square matrix A is called *orthogonal* if $A^t = A^{-1}$. Prove that the determinant of an orthogonal matrix must be 1 or -1.

26. Let $A = [a_{ij}]$ be a square matrix, and let α be a nonzero number. Multiply a_{ij} by α^{i-j} for $i = 1, 2, \ldots, n$ and $j = 1, 2, \ldots, n$ to form a new matrix B. How are $|A|$ and $|B|$ related? *Hint:* Look at the 2×2 and 3×3 cases to guess the general case.

In each of Problems 27 through 51, use elementary row and column operations to evaluate the determinant, as in Examples 12.47 and 12.48.

27. $\begin{vmatrix} -3 & 1 & 14 \\ 0 & 1 & 6 \\ 2 & -3 & 4 \end{vmatrix}$

28. $\begin{vmatrix} 8 & 14 & 0 \\ 1 & -2 & 3 \\ 0 & 1 & 16 \end{vmatrix}$

29. $\begin{vmatrix} -5 & 2 & 4 \\ 1 & -3 & 4 \\ 0 & 1 & 3 \end{vmatrix}$

30. $\begin{vmatrix} -8 & -5 & 2 \\ 3 & -2 & -14 \\ 3 & 4 & -7 \end{vmatrix}$

31. $\begin{vmatrix} 2 & -3 & -5 \\ -7 & 4 & -4 \\ 1 & 3 & 5 \end{vmatrix}$

32. $\begin{vmatrix} 15 & 3 & -5 \\ 1 & 7 & -4 \\ -3 & 2 & -5 \end{vmatrix}$

33. $\begin{vmatrix} 8 & 0 & 0 & 4 \\ 9 & 1 & -7 & 2 \\ -8 & 1 & 14 & 2 \\ 0 & 0 & 1 & -3 \end{vmatrix}$

34. $\begin{vmatrix} -5 & 4 & 1 & 7 \\ -9 & 3 & 2 & -5 \\ -2 & 0 & -1 & 1 \\ 1 & 14 & 0 & 3 \end{vmatrix}$

35. $\begin{vmatrix} 14 & 13 & -2 & 5 \\ 7 & 1 & 1 & 7 \\ 0 & 2 & 12 & 3 \\ 1 & -6 & 5 & 2 \end{vmatrix}$

36. $\begin{vmatrix} -3 & 0 & 1 & -5 \\ 1 & 0 & 3 & 0 \\ 1 & 7 & 6 & -1 \\ 2 & 4 & 2 & -3 \end{vmatrix}$

37. $\begin{vmatrix} -8 & 5 & 1 & 7 & 2 \\ 0 & 1 & 3 & 5 & -6 \\ 2 & 2 & 1 & 5 & 3 \\ 0 & 4 & 3 & 7 & 2 \\ 1 & 1 & -7 & -6 & 5 \end{vmatrix}$

38. $\begin{vmatrix} 5 & 15 & 3 & 1 & 7 & 2 \\ 0 & 0 & 1 & 4 & -5 & 2 \\ 1 & 7 & -1 & 3 & 1 & 9 \\ 0 & 0 & 1 & -3 & -1 & 4 \\ 1 & 1 & 7 & -4 & 1 & 6 \\ 1 & 0 & 0 & 3 & -9 & -4 \end{vmatrix}$

39. $\begin{vmatrix} 1 & 3 & -9 & 5 \\ -6 & 2 & 1 & -1 \\ 0 & 3 & 2 & -6 \\ 8 & 5 & 3 & -8 \end{vmatrix}$

40. $\begin{vmatrix} 4 & 100 & -3 \\ 296 & -4 & 10 \\ 92 & 0 & -25 \end{vmatrix}$

41. $\begin{vmatrix} 203 & 13 & 693 \\ -12 & 1 & 10 \\ 0 & -5 & 64 \end{vmatrix}$

42. $\begin{vmatrix} 10 & 1023 & -3 \\ -3 & 3 & 21 \\ 0 & -5 & 14 \end{vmatrix}$

43. $\begin{vmatrix} -54 & 23 & 15 \\ 33 & 10 & -34 \\ 21 & -32 & 15 \end{vmatrix}$

44. $\begin{vmatrix} 23 & -12 & 15 & 9 \\ 3 & 9 & 11 & 14 \\ -5 & 6 & 17 & 12 \\ 17 & 13 & -7 & 8 \end{vmatrix}$

45. $\begin{vmatrix} 12 & 15 & 283 & -45 \\ 19 & -4 & 0 & 19 \\ -13 & 10 & 4 & 15 \\ -54 & 0 & 0 & 23 \end{vmatrix}$

46. $\begin{vmatrix} -2 & 4 & 15 & -12 \\ 20 & -5 & 0 & 23 \\ 4 & 6 & -2 & 14 \\ -9 & 0 & 0 & 13 \end{vmatrix}$

47. $\begin{vmatrix} 17 & 12 & -5 & 0 \\ -1 & 0 & 12 & 0 \\ 21 & -15 & -1 & 2 \\ 13 & -12 & 9 & 6 \end{vmatrix}$

48.
$$\begin{vmatrix} 46 & 2 & 1 & 0 \\ -3 & 13 & -3 & 6 \\ -5 & 12 & 17 & -7 \\ 1 & -1 & 0 & 0 \end{vmatrix}$$

49.
$$\begin{vmatrix} 15 & -3 & 2 & 12 \\ 17 & 6 & 0 & 14 \\ -3 & 5 & 5 & -5 \\ 3 & 0 & 0 & -2 \end{vmatrix}$$

50.
$$\begin{vmatrix} 2 & 4 & -6 & 3 & 5 \\ -6 & 16 & 15 & 4 & -6 \\ 0 & 14 & 12 & 9 & 5 \\ -4 & 0 & 22 & 6 & -8 \\ 4 & -4 & 15 & 8 & 10 \end{vmatrix}$$

51.
$$\begin{vmatrix} -5 & 6 & 1 & -1 & 0 & 2 \\ 1 & 3 & -1 & -1 & 3 & 1 \\ 4 & 2 & -2 & 1 & 1 & 0 \\ 0 & 0 & 3 & 1 & -2 & 4 \\ 1 & 0 & 0 & -2 & 1 & 7 \\ 0 & 0 & 1 & 1 & -1 & 7 \end{vmatrix}$$

52. Show that

$$\begin{vmatrix} 1 & \alpha & \alpha^2 \\ 1 & \beta & \beta^2 \\ 1 & \gamma & \gamma^2 \end{vmatrix} = (\alpha - \beta)(\gamma - \alpha)(\beta - \gamma).$$

This determinant is called *Vandermonde's determinant*.

53. Show that

$$\begin{vmatrix} \alpha & \beta & \gamma & \delta \\ \beta & \gamma & \delta & \alpha \\ \gamma & \delta & \alpha & \beta \\ \delta & \alpha & \beta & \gamma \end{vmatrix} = (\alpha + \beta + \gamma + \delta)(\beta - \alpha + \delta - \gamma) \begin{vmatrix} 0 & 1 & -1 & 1 \\ 1 & \gamma & \delta & \alpha \\ 1 & \delta & \alpha & \beta \\ 1 & \alpha & \beta & \gamma \end{vmatrix}.$$

54. Show that

$$\begin{vmatrix} 1 & 1 & 1 & 1 \\ \alpha & \beta & \gamma & \delta \\ \alpha^2 & \beta^2 & \gamma^2 & \delta^2 \\ \alpha^3 & \beta^3 & \gamma^3 & \delta^3 \end{vmatrix} = (\beta - \alpha)(\beta - \gamma)(\beta - \delta)(\gamma - \alpha)(\delta - \alpha)(\delta - \gamma).$$

55. An $n \times n$ matrix A is called *block diagonal* if it has the appearance

$$A = \begin{bmatrix} [D_1] & O & \cdots & O \\ O & [D_2] & \cdots & O \\ \vdots & \vdots & & \vdots \\ O & O & \cdots & [D_r] \end{bmatrix},$$

in which each $[D_j]$ denotes a $k_j \times k_j$ matrix, $k_1 + \cdots + k_r = n$, and each O denotes a zero matrix of appropriate size (that is, all elements of A not in some $[D_j]$ block are zero). Show that $|A| = |D_1||D_2| \cdots |D_r|$.

56. Use the result of Problem 55 to evaluate the following determinants.

(a)
$$\begin{vmatrix} 22 & -4 & 3 & 0 & 0 \\ 1 & 0 & 6 & 0 & 0 \\ 2 & 2 & 0 & 0 & 0 \\ 0 & 0 & 0 & 1 & -1 \\ 0 & 0 & 0 & 0 & 4 \end{vmatrix}$$

(b)
$$\begin{vmatrix} 8 & 4 & 0 & 0 & 0 & 0 \\ -6 & 2 & 0 & 0 & 0 & 0 \\ 0 & 0 & 3 & 1 & 0 & 0 \\ 0 & 0 & 1 & 2 & 0 & 0 \\ 0 & 0 & 0 & 0 & -4 & 8 \\ 0 & 0 & 0 & 0 & 12 & 14 \end{vmatrix}$$

$$
\text{(c)}\quad
\begin{vmatrix}
16 & 14 & 0 & 2 & 0 & 0 \\
1 & 1 & 3 & 2 & 0 & 0 \\
0 & 0 & 1 & 4 & 0 & 0 \\
1 & 3 & 6 & 2 & 0 & 0 \\
0 & 0 & 0 & 0 & -5 & 3 \\
0 & 0 & 0 & 0 & 4 & -9
\end{vmatrix}
\qquad
\text{(d)}\quad
\begin{vmatrix}
5 & 3 & 0 & 0 & 0 & 0 & 0 \\
2 & -1 & 0 & 0 & 0 & 0 & 0 \\
0 & 0 & 3 & 9 & 1 & 0 & 0 \\
0 & 0 & 2 & 2 & -4 & 0 & 0 \\
0 & 0 & 1 & -8 & 6 & 0 & 0 \\
0 & 0 & 0 & 0 & 0 & -2 & -3 \\
0 & 0 & 0 & 0 & 0 & 0 & 4
\end{vmatrix}
$$

12.11 *An Application of Determinants to Electrical Circuits*

In 1847, G. R. Kirchhoff published a classic paper in which he derived many of the electrical circuit laws which bear his name. One of these is the *matrix tree theorem*, which we will discuss in this section.

Consider an electrical circuit such as the one in Figure 12.5. The underlying geometry of the circuit is shown in Figure 12.6. Such a diagram, consisting of points and connecting lines, is called a *graph*. Graphs are useful for schematic diagrams of many objects, as we saw with crystals in Section 12.2. When the points are assigned labels, we have a labeled graph, as in Figure 12.7.

Many of Kirchhoff's formulas depend upon geometric properties of the circuit's underlying graph. One such property is the arrangement of the closed loops, which do not themselves depend on whether resistors or inductors are present in the loop.

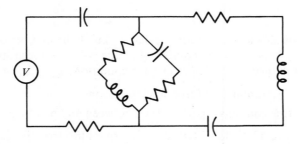

Figure 12.5

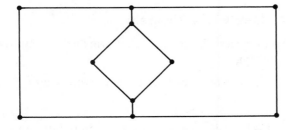

Figure 12.6

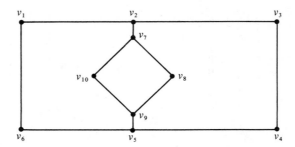

Figure 12.7

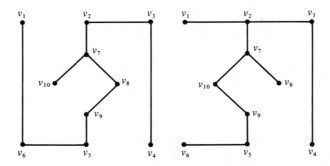

Figure 12.8

One of Kirchhoff's formulas requires that we know the number of spanning trees in a labeled graph. A *spanning tree* consists of all the points of the graph, together with some of the lines, subject to the following two conditions.

1. The lines of the spanning tree form no closed loops.

2. The spanning tree contains a path of lines between any pair of points.

For example, Figure 12.8 shows two different spanning trees in the graph of Figure 12.7.

The following result, known as the *matrix tree theorem*, tells how to compute the number of spanning trees in a labeled graph.

THEOREM 12.34   Matrix Tree Theorem

Let G be a graph with vertices labeled $1, \ldots, n$. Form an $n \times n$ matrix $T = [t_{ij}]$ (called the tree matrix) by putting

$$t_{ii} = \text{number of lines incident with point } i$$

and, for $i \neq j$,

$$t_{ij} = \begin{cases} -1 & \text{if there is a line between points } i \text{ and } j \\ 0 & \text{if there is no line between points } i \text{ and } j. \end{cases}$$

Then all cofactors of T are equal, and their common value is the number of spanning trees of G. ∎

We will not prove this result, but we will discuss an example.

EXAMPLE 12.50

Consider the labeled graph of Figure 12.9. There are seven points. The 7×7 tree matrix is

$$T = \begin{bmatrix} 3 & -1 & 0 & 0 & 0 & -1 & -1 \\ -1 & 3 & -1 & -1 & 0 & 0 & 0 \\ 0 & -1 & 3 & -1 & 0 & -1 & 0 \\ 0 & -1 & -1 & 4 & -1 & 0 & -1 \\ 0 & 0 & 0 & -1 & 3 & -1 & -1 \\ -1 & 0 & -1 & 0 & -1 & 4 & -1 \\ -1 & 0 & 0 & -1 & -1 & -1 & 4 \end{bmatrix}.$$

Evaluate any cofactor of T. For example, we find that $(-1)^{1+1}M_{11} = 386$. This is the number of spanning trees in this graph. Evaluation of any other cofactor will yield the same result. ∎

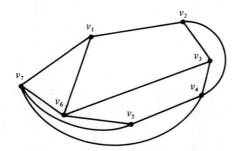

Figure 12.9

To appreciate the significance of the matrix tree theorem in calculating the number of spanning trees in a labeled graph, try determining by some other means that there are 386 spanning trees in the graph of Figure 12.9!

PROBLEMS FOR SECTION 12.11

In each of Problems 10 through 12, find the number of spanning trees in the circuit.

1.

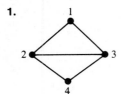

2.

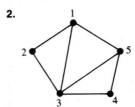

3.

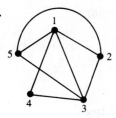

4.

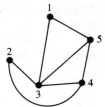

5.

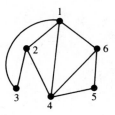

6.

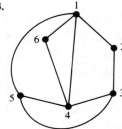

7.

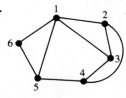

8.

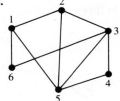

9. A complete graph on n points consists of n points and a line between each pair of distinct points. This graph is denoted K_n. With the points labeled $1, 2, \ldots, n$, use the matrix tree theorem to show that the number of spanning trees in K_n is n^{n-2}.

In each of Problems 10 through 12, find the number of spanning trees in the circuit.

10.

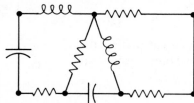

11.

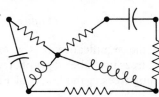

12.

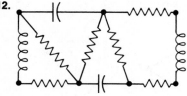

12.12 *A Determinant Formula for the Inverse of a Matrix* _____

Determinants can be used to test whether a matrix is singular or nonsingular and also to calculate the inverse of a nonsingular matrix. Recall that A has an inverse only if A is nonsingular.

THEOREM 12.35

Let A be an $n \times n$ matrix. Then A is nonsingular if and only if $|A| \neq 0$.

Proof   First, recall that elementary row operations transform A into a matrix whose determinant is a nonzero constant times $|A|$. In particular, $|A| = \alpha |A_R|$ for some $\alpha \neq 0$.

If A is nonsingular, $A_R = I_n$, by Lemma 12.1 in Section 12.6 and Theorem 12.22 in Section 12.9. Thus, for some $\alpha \neq 0$, $|A| = \alpha |A_R| = \alpha |I_n| = \alpha \neq 0$.

Conversely, if $|A| \neq 0$, $|A_R| \neq 0$. But then A_R must have no zero rows; hence, $\text{rank}(A) = n$. By Theorem 12.22 in Section 12.9, A is nonsingular.   ∎

Using this result, we can produce a determinant formula for the inverse of a nonsingular matrix.

THEOREM 12.36

Let $A = [a_{ij}]$ be $n \times n$. Define an $n \times n$ matrix $B = [b_{ij}]$ by letting

$$b_{ij} = \left(\frac{1}{|A|} \right) (\text{cofactor of } a_{ji} \text{ in } A) = \frac{1}{|A|} (-1)^{i+j} M_{ji}.$$

Then $B = A^{-1}$. (Note that the i, j element of B is defined in terms of the j, i cofactor of A, *not* the i, j cofactor.)

Proof   First, $|A| \neq 0$ by Theorem 12.35. To show that $B = A^{-1}$, we will show that $AB = I_n$. By definition of a matrix product, the i, j element of AB is

$$\sum_{k=1}^{n} a_{ik} b_{kj} = \sum_{k=1}^{n} \frac{1}{|A|} a_{ik}(-1)^{k+j} M_{jk} = \frac{1}{|A|} \sum_{k=1}^{n} (-1)^{k+j} a_{ik} M_{jk}.$$

When $i = j$, the summation on the right is the cofactor expansion of $|A|$ by row i; hence, the i, i element of AB is 1.

If $i \neq j$, $\sum_{k=1}^{n} (-1)^{k+j} a_{ik} M_{jk}$ is the cofactor expansion by row j of the determinant of the matrix formed from A by replacing row j by row i. But this matrix has two identical rows and hence has determinant zero, and the off-diagonal elements of AB are zero. This proves that $AB = I_n$. A similar argument shows that $BA = I_n$.   ∎

EXAMPLE 12.51

Let

$$A = \begin{bmatrix} 8 & 0 & 1 \\ 3 & -2 & 1 \\ 1 & 4 & 0 \end{bmatrix}.$$

We find that $|A| = -18$. We can calculate the elements of A^{-1} using the formula of Theorem 12.35. Let $A^{-1} = [c_{ij}]$. Then

$$c_{11} = \tfrac{1}{-18}(-1)^{1+1}M_{11} = -\tfrac{1}{18}\begin{vmatrix} -2 & 1 \\ 4 & 0 \end{vmatrix} = \tfrac{4}{18},$$

$$c_{12} = \tfrac{1}{-18}(-1)^{1+2}M_{21} = \tfrac{1}{18}\begin{vmatrix} 0 & 1 \\ 4 & 0 \end{vmatrix} = \tfrac{-4}{18},$$

$$c_{13} = \tfrac{1}{-18}(-1)^{1+3}M_{31} = -\tfrac{1}{18}\begin{vmatrix} 0 & 1 \\ -2 & 1 \end{vmatrix} = \tfrac{-2}{18},$$

and so on. Continuing in this way, we find that

$$A^{-1} = \tfrac{1}{18}\begin{bmatrix} 4 & -4 & -2 \\ -1 & 1 & 5 \\ -14 & 32 & 16 \end{bmatrix}. \quad \blacksquare$$

This method is not the most efficient for calculating a matrix inverse. Usually, the method of Section 12.9 is better. However, Theorem 12.35 has the value of giving an explicit formula for A^{-1} as opposed to a method for finding it. Such a formula is sometimes useful, particularly in theoretical calculations.

In concluding this section, we will observe that Theorem 12.35 gives us an easy way of proving conclusion (5) of Theorem 12.21 in Section 12.9. Suppose that A and B are $n \times n$ matrices and either A or B is singular. By Theorem 12.35 $|A|$ or $|B|$ must be zero. Then $|AB| = |A||B| = 0$; hence, AB is singular.

PROBLEMS FOR SECTION 12.12

In each of Problems 1 through 15, use Theorem 12.36 to produce the inverse of the matrix.

1. $\begin{bmatrix} 2 & -1 \\ 1 & 6 \end{bmatrix}$

2. $\begin{bmatrix} 3 & 0 \\ 1 & 4 \end{bmatrix}$

3. $\begin{bmatrix} -1 & 1 \\ 1 & 4 \end{bmatrix}$

4. $\begin{bmatrix} 2 & 5 \\ -7 & -3 \end{bmatrix}$

5. $\begin{bmatrix} -1 & 5 \\ 1 & 6 \end{bmatrix}$

6. $\begin{bmatrix} 5 & 5 \\ -2 & 1 \end{bmatrix}$

7. $\begin{bmatrix} 6 & -1 & 3 \\ 0 & 1 & -4 \\ 2 & 2 & -3 \end{bmatrix}$

8. $\begin{bmatrix} -14 & 1 & -3 \\ 2 & -1 & 3 \\ 1 & 1 & 7 \end{bmatrix}$

9. $\begin{bmatrix} 0 & -4 & 3 \\ 2 & -1 & 6 \\ 1 & -1 & -7 \end{bmatrix}$

10. $\begin{bmatrix} 11 & 0 & -5 \\ 0 & 1 & 0 \\ 4 & -7 & 9 \end{bmatrix}$

11. $\begin{bmatrix} 1 & -3 & 4 \\ 2 & 2 & -5 \\ 0 & 1 & -7 \end{bmatrix}$

12. $\begin{bmatrix} -3 & 1 & 0 \\ 2 & 0 & 0 \\ 5 & -1 & 4 \end{bmatrix}$

13. $\begin{bmatrix} 8 & -1 & 4 \\ 1 & -3 & 6 \\ 2 & 7 & 19 \end{bmatrix}$

14. $\begin{bmatrix} 3 & 1 & -2 & 1 \\ 4 & 6 & -3 & 9 \\ -2 & 1 & 7 & 4 \\ 13 & 0 & 1 & 5 \end{bmatrix}$

15. $\begin{bmatrix} 7 & -3 & -4 & 1 \\ 8 & 2 & 0 & 0 \\ 1 & 5 & -1 & 7 \\ 3 & -2 & -5 & 9 \end{bmatrix}$

12.13 *Cramer's Rule*

If A is an $n \times n$ nonsingular matrix, the system of equations $AX = B$ has a unique solution given by $X = A^{-1}B$. We will now discuss a determinant formula for this solution known as *Cramer's rule*.

THEOREM 12.37

Let A be an $n \times n$ nonsingular matrix. Then the unique solution of the nonhomogeneous system $AX = B$ is given by

$$x_k = \frac{1}{|A|}|A(k; B)| \quad \text{for} \quad k = 1, 2, \ldots, n,$$

where $A(k; B)$ is the $n \times n$ matrix obtained by replacing column k of A with B. ∎

We will not prove Cramer's rule, but we will discuss a heuristic argument to suggest how the formula arises. Begin by multiplying column k of A by x_k. This has the effect of multiplying $|A|$ by x_k, and we have

$$x_k|A| = \begin{vmatrix} a_{11} & a_{12} & \cdots & a_{1k}x_k & \cdots & a_{1n} \\ a_{21} & a_{22} & \cdots & a_{2k}x_k & \cdots & a_{2n} \\ \vdots & \vdots & & \vdots & & \vdots \\ a_{n1} & a_{n2} & \cdots & a_{nk}x_k & \cdots & a_{nn} \end{vmatrix}.$$

Next, for each $j \neq k$, add x_k times column j to column k of the last matrix. The determinant of the matrix thus formed is the same as $x_k|A|$; hence,

$$x_k|A| = \begin{vmatrix} a_{11} & a_{12} & \cdots & a_{11}x_1 + a_{12}x_2 + \cdots + a_{1n}x_n & \cdots & a_{1n} \\ a_{21} & a_{22} & \cdots & a_{21}x_1 + a_{22}x_2 + \cdots + a_{2n}x_n & \cdots & a_{2n} \\ \vdots & \vdots & & \vdots & & \vdots \\ a_{n1} & a_{n2} & \cdots & a_{n1}x_1 + a_{n2}x_2 + \cdots + a_{nn}x_n & \cdots & a_{nn} \end{vmatrix}.$$

But

$$\begin{bmatrix} a_{11}x_1 + a_{12}x_2 + \cdots + a_{1n}x_n \\ a_{21}x_1 + a_{22}x_2 + \cdots + a_{2n}x_n \\ \vdots & \vdots & \vdots \\ a_{n1}x_1 + a_{n2}x_2 + \cdots + a_{nn}x_n \end{bmatrix} = AX = B = \begin{bmatrix} b_1 \\ b_2 \\ \vdots \\ b_n \end{bmatrix}.$$

Thus, in the above expression for $x_k|A|$, we can replace column k with B, obtaining

$$x_k|A| = \begin{vmatrix} a_{11} & a_{12} & \cdots & b_1 & \cdots & a_{1n} \\ a_{21} & a_{22} & \cdots & b_2 & \cdots & a_{2n} \\ \vdots & \vdots & & \vdots & & \vdots \\ a_{n1} & a_{n2} & \cdots & b_n & \cdots & a_{nn} \end{vmatrix} = |A(k; B)|.$$

Upon dividing by $|A|$, we have $x_k = (1/|A|)|A(k; B)|$, as we wanted to show.

EXAMPLE 12.52

Solve the system

$$2x_1 - 3x_2 = 8$$
$$x_1 + 4x_2 = -3.$$

The matrix of coefficients is

$$A = \begin{bmatrix} 2 & -3 \\ 1 & 4 \end{bmatrix}.$$

We find that $|A| = 11$. By Cramer's rule,

$$x_1 = \tfrac{1}{11} \begin{bmatrix} 8 & -3 \\ -3 & 4 \end{bmatrix} = \tfrac{23}{11}$$

and

$$x_2 = \tfrac{1}{11} \begin{bmatrix} 2 & 8 \\ 1 & -3 \end{bmatrix} = -\tfrac{14}{11}.$$

Note that x_1 is $1/|A|$ times the determinant formed by replacing column 1 of A by $\begin{bmatrix} 8 \\ -3 \end{bmatrix}$, and x_2 is $1/|A|$ times the determinant formed by replacing column 2 of A by B. ■

EXAMPLE 12.53

Solve the system

$$x_1 - 3x_2 - 4x_3 = 1$$
$$-x_1 + x_2 - 3x_3 = 14$$
$$x_2 - 3x_3 = 5.$$

The matrix of coefficients is

$$A = \begin{bmatrix} 1 & -3 & -4 \\ -1 & 1 & -3 \\ 0 & 1 & -3 \end{bmatrix}.$$

We find that $|A| = 13$. By Cramer's rule,

$$x_1 = \tfrac{1}{13} \begin{bmatrix} 1 & -3 & -4 \\ 14 & 1 & -3 \\ 5 & 1 & -3 \end{bmatrix} = -\tfrac{117}{13} = -9,$$

$$x_2 = \tfrac{1}{13} \begin{bmatrix} 1 & 1 & -4 \\ -1 & 14 & -3 \\ 0 & 5 & -3 \end{bmatrix} = -\tfrac{10}{13},$$

and

$$x_3 = \tfrac{1}{13} \begin{bmatrix} 1 & -3 & 1 \\ -1 & 1 & 14 \\ 0 & 1 & 5 \end{bmatrix} = -\tfrac{25}{13}. \quad \blacksquare$$

Cramer's rule is, in general, not as efficient as Gauss-Jordan reduction. Other advantages of the Gauss-Jordan method are that it is not restricted to systems with nonsingular matrices; it produces all solutions when there is more than one; and it is easily programmed for implementation on a computer.

One value of Cramer's rule is that it presents a formula for the solution as opposed to a method for finding it. One place where this formula is commonly encountered is in the Jacobian of multivariable calculus, used in implicit differentiation and changes of variables in multiple integrals.

PROBLEMS FOR SECTION 12.13

In each of Problems 1 through 14, solve the system by using Cramer's rule or show that Cramer's rule does not apply.

1. $8x_1 - 4x_2 + 3x_3 = 0$
$x_1 + 5x_2 - x_3 = -5$
$-2x_1 + 6x_2 + x_3 = -4$

2. $15x_1 - 4x_2 = 5$
$8x_1 + x_2 = -4$

3. $x_1 + 4x_2 = 3$
$x_1 + x_2 = 0$

4. $5x_1 - 6x_2 + x_3 = 4$
$-x_1 + 3x_2 - 4x_3 = 5$
$2x_1 + 3x_2 + x_3 = -8$

5. $x_1 + x_2 - 3x_3 = 0$
$x_2 - 4x_3 = 0$
$x_1 - x_2 - x_3 = 5$

6. $-8x_1 - 6x_2 + x_3 = 4$
$x_1 - x_2 + 3x_3 = 2$
$x_1 - x_3 = 4$

7. $-5x_1 + 6x_2 - 8x_3 + x_4 = 5$
$x_1 - 3x_3 + x_4 = 6$
$x_2 + 4x_3 - x_4 = -8$
$x_1 - 8x_3 + 5x_4 = 0$

8. $6x_1 + 4x_2 - x_3 + 3x_4 - x_5 = 7$
$x_1 - 4x_2 + x_5 = -5$
$x_1 - 3x_2 + x_3 - 4x_5 = 0$
$-2x_1 + x_3 - 2x_5 = 4$
$x_3 - x_4 - x_5 = 8$

9. $2x_1 - 4x_2 + x_3 - x_4 = 6$
$x_2 - 3x_3 = 10$
$x_1 - 4x_3 = 0$
$x_2 - x_3 + 2x_4 = 4$

10. $2x_1 - 3x_2 + x_4 = 2$
$x_2 - x_3 + x_4 = 2$
$x_3 - 2x_4 = 5$
$x_1 - 3x_2 + 4x_3 = 0$

11.
$$\begin{aligned}
14x_1 \quad - 3x_3 \quad &= 5 \\
2x_1 \quad - 4x_3 + x_4 &= 2 \\
x_1 - x_2 + x_3 - 3x_4 &= 1 \\
x_3 - 4x_4 &= -5
\end{aligned}$$

12.
$$\begin{aligned}
x_2 \quad - 4x_4 &= 18 \\
x_1 - x_2 + 3x_3 \quad &= -1 \\
x_1 + x_2 - 3x_3 + x_4 &= 5 \\
x_2 \quad + 3x_4 &= 0
\end{aligned}$$

13.
$$\begin{aligned}
2x_1 - 3x_2 \quad &= 10 \\
x_1 \quad - 4x_3 + x_4 &= 5 \\
x_1 - x_2 \quad + 3x_4 &= -2 \\
x_1 - x_2 + x_3 + x_4 &= 0
\end{aligned}$$

14.
$$\begin{aligned}
x_3 - 4x_4 + x_5 &= 1 \\
x_1 - x_2 \quad + 3x_4 \quad &= 2 \\
x_2 \quad - 3x_4 + 5x_5 &= -3 \\
x_2 \quad + 3x_5 &= 1 \\
x_1 - x_2 + 3x_3 \quad &= -4
\end{aligned}$$

15. Let A be a nonsingular matrix. What happens if we attempt to apply Cramer's rule to the system $AX = O$?

ADDITIONAL PROBLEMS

In each of Problems 1 through 10, find the reduced form of the matrix and determine its rank.

1.
$$\begin{bmatrix} -4 & 5 & 8 & 1 \\ 3 & 0 & -5 & 2 \\ 1 & -1 & 5 & 8 \end{bmatrix}$$

2.
$$\begin{bmatrix} -4 & 7 & 1 & -1 \\ 2 & -3 & 6 & 2 \end{bmatrix}$$

3.
$$\begin{bmatrix} 3 & -1 \\ 2 & 5 \\ 5 & -3 \end{bmatrix}$$

4.
$$\begin{bmatrix} -1 & 1 \\ 2 & -2 \end{bmatrix}$$

5.
$$\begin{bmatrix} 8 & -3 & 2 & 1 & -5 \\ 0 & 0 & 0 & 0 & 0 \end{bmatrix}$$

6.
$$\begin{bmatrix} 6 & -2 & 3 & 1 & 1 \\ -3 & 2 & 5 & 3 & -5 \\ 0 & 2 & 13 & 7 & -9 \end{bmatrix}$$

7.
$$\begin{bmatrix} 3 & -2 & 1 \\ 0 & 0 & 0 \\ 0 & -1 & 1 \end{bmatrix}$$

8.
$$\begin{bmatrix} -7 & 0 & 0 & 0 \\ 1 & 0 & 0 & 0 \\ 0 & 0 & -2 & 4 \end{bmatrix}$$

9.
$$\begin{bmatrix} 3 & 4 & 1 & 0 \\ -3 & 2 & 2 & 0 \\ 0 & 0 & 0 & 1 \end{bmatrix}$$

10.
$$\begin{bmatrix} -5 & 7 & 7 & 0 \\ 0 & 7 & -1 & 1 \\ 0 & 0 & 0 & 0 \\ 2 & -1 & 3 & 9 \\ 1 & 0 & 0 & 1 \end{bmatrix}$$

In each of Problems 11 through 20, find the inverse of the matrix or show that the matrix is singular.

11.
$$\begin{bmatrix} -3 & 5 \\ 2 & -2 \end{bmatrix}$$

12.
$$\begin{bmatrix} 13 & 1 \\ 2 & 0 \end{bmatrix}$$

13.
$$\begin{bmatrix} 1 & -5 \\ -3 & 15 \end{bmatrix}$$

14.
$$\begin{bmatrix} -4 & 2 \\ 0 & 4 \end{bmatrix}$$

15.
$$\begin{bmatrix} -4 & 1 & -1 \\ 0 & 1 & 0 \\ 2 & -3 & 2 \end{bmatrix}$$

16.
$$\begin{bmatrix} -3 & 1 & -2 \\ 0 & -20 & 28 \\ 1 & -7 & 10 \end{bmatrix}$$

17.
$$\begin{bmatrix} 6 & -3 & 1 \\ 2 & -2 & 4 \\ 0 & 1 & -2 \end{bmatrix}$$

18.
$$\begin{bmatrix} -5 & 1 & -1 \\ 0 & 7 & 1 \\ 2 & 2 & -2 \end{bmatrix}$$

19.
$$\begin{bmatrix} -5 & 0 & 1 & 1 \\ 0 & -2 & 1 & -1 \\ 3 & -4 & 0 & -2 \\ 2 & -5 & 0 & 0 \end{bmatrix}$$

20.
$$\begin{bmatrix} -3 & 1 & 5 & 0 \\ 2 & -1 & 5 & 3 \\ 1 & -5 & 4 & 4 \\ -3 & -4 & 19 & 7 \end{bmatrix}$$

In each of Problems 21 through 30, evaluate the determinant.

21. $\begin{vmatrix} -3 & 6 \\ 2 & -4 \end{vmatrix}$

22. $\begin{vmatrix} 5 & -3 \\ 2 & 0 \end{vmatrix}$

23. $\begin{vmatrix} 3 & -2 & 4 \\ 1 & -1 & 4 \\ 3 & -5 & 6 \end{vmatrix}$

24. $\begin{vmatrix} 1 & -4 & 7 \\ 7 & -3 & 10 \\ -2 & 1 & 6 \end{vmatrix}$

25. $\begin{vmatrix} 1 & 0 & -3 & 6 \\ 2 & -1 & 5 & 3 \\ 0 & 1 & 0 & -4 \\ 1 & -4 & 6 & 5 \end{vmatrix}$

26. $\begin{vmatrix} -4 & 1 & 0 & 0 \\ 2 & -1 & 0 & -3 \\ 0 & 1 & -2 & 0 \\ 4 & 1 & -5 & 6 \end{vmatrix}$

27. $\begin{vmatrix} 6 & -2 & 1 & 0 \\ 1 & -4 & 3 & 0 \\ 2 & 0 & -6 & 2 \\ 0 & -2 & 1 & -5 \end{vmatrix}$

28. $\begin{vmatrix} 4 & -1 & 2 & 8 \\ 2 & -1 & 5 & 2 \\ 0 & 2 & 3 & -3 \\ 0 & 3 & -5 & 1 \end{vmatrix}$

29. $\begin{vmatrix} 3 & -2 & 3 & -6 & 1 \\ 0 & -2 & 0 & 0 & -2 \\ 1 & -1 & 0 & 0 & 5 \\ -1 & 4 & 0 & 1 & -5 \\ 0 & 0 & 1 & -1 & 0 \end{vmatrix}$

30. $\begin{vmatrix} 1 & -2 & 0 & 0 & 0 & 0 & 0 \\ 0 & -3 & 0 & 0 & 0 & 0 & 0 \\ 0 & 0 & -1 & 4 & 2 & 0 & 0 \\ 0 & 0 & 1 & -3 & 1 & 0 & 0 \\ 0 & 0 & 4 & -3 & 1 & 0 & 0 \\ 0 & 0 & 0 & 0 & 0 & -6 & 3 \\ 0 & 0 & 0 & 0 & 0 & 2 & 7 \end{vmatrix}$

In each of Problems 31 through 40, find the general solution of the system or show that there is no solution.

31.
$$\begin{aligned} x_1 - 4x_2 + x_3 &= 0 \\ -2x_1 + x_2 - 4x_3 &= 0 \end{aligned}$$

32.
$$\begin{aligned} x_1 + 6x_2 + x_3 &= 1 \\ -2x_1 - x_2 + 7x_3 &= -3 \\ -3x_1 + 4x_2 + 15x_3 &= 1 \end{aligned}$$

33.
$$\begin{aligned} x_1 + x_2 - x_3 + 4x_4 &= -5 \\ 2x_1 - 4x_2 + x_3 &= -2 \\ x_2 - x_3 + 7x_4 &= -1 \end{aligned}$$

34.
$$\begin{aligned} 2x_1 - x_2 + 5x_3 &= 0 \\ -2x_1 + 3x_2 + 4x_3 &= 0 \\ x_1 + 6x_2 - x_3 &= 0 \end{aligned}$$

35.
$$\begin{aligned} 2x_1 + 3x_2 - x_4 &= 5 \\ x_1 - 4x_2 - 3x_3 - x_4 &= -8 \\ x_2 - 6x_3 + x_4 &= 0 \end{aligned}$$

36.
$$\begin{aligned} -3x_1 + x_2 - 5x_4 &= 1 \\ 2x_1 + 3x_2 - 5x_3 &= 0 \\ -5x_1 + 4x_2 - 4x_3 - x_4 &= -10 \end{aligned}$$

37.
$$\begin{aligned} -4x_1 + x_2 - x_3 &= 9 \\ 2x_1 + x_2 + 7x_3 &= 1 \\ x_1 - 4x_2 + x_3 &= 0 \end{aligned}$$

38.
$$\begin{aligned} 6x_1 + x_2 - 4x_3 - x_4 &= -10 \\ x_1 - x_3 + 4x_5 &= 9 \end{aligned}$$

39.
$$\begin{aligned} 2x_1 + 7x_2 - x_3 &= 1 \\ x_2 - x_3 + 4x_4 &= 0 \\ 5x_1 + 5x_2 - 3x_3 + x_4 &= -6 \end{aligned}$$

40.
$$\begin{aligned} -3x_1 + x_2 - x_4 &= 0 \\ x_2 - x_3 + 5x_4 &= 0 \end{aligned}$$

Let A and B be $n \times n$ matrices. We say that A is *similar to* B if there is a nonsingular matrix P such that $P^{-1}AP = B$. Problems 41 through 50 deal with properties of similar matrices.

41. Let A be similar to B. Prove that B is similar to A.

42. Prove that every square matrix is similar to itself.

43. Let A be similar to B, and let B be similar to C. Prove that A is similar to C.

44. Let A be similar to B. Prove that A and B are both singular or both nonsingular.

45. Let A be similar to B, and let A be nonsingular. Prove that A^{-1} is similar to B^{-1}.

46. Let A be similar to B. Prove that, for every positive integer n, A^n is similar to B^n.

47. Let A be similar to B, and let α be any number. Prove that αA is similar to αB.

48. Let A be similar to B. Prove that A and B have the same rank.

49. Let A be similar to B. Prove that $|A| = |B|$.

50. Let A be similar to B. Prove that

$$\sum_{j=1}^{n} a_{jj} = \sum_{j=1}^{n} b_{jj}.$$

(The sum of the main diagonal elements of a matrix is called the *trace* of the matrix.)

CHAPTER 13

Eigenvalues, Eigenvectors, and Diagonalization

13.0 Introduction

In the preceding chapter, we developed some of the vocabulary and properties of matrices. In this chapter, we will continue the study of matrices with the important concepts of eigenvalues and eigenvectors and some of their ramifications.

Eigenvalues are numbers produced from a matrix in a way we will discuss in the next section. These numbers are crucial to many considerations, including representations of matrices in ways that make them easier to work with in solving problems and the solution of systems of differential equations (the subject of Chapter 14). In mathematical models of certain phenomena, the eigenvalues of a matrix may also have physical significance. For example, they may represent normal modes of vibration of a mechanical system.

13.1 Eigenvalues and Eigenvectors

Let A be an $n \times n$ matrix. A real or complex number λ is called an *eigenvalue* of A if, for some nonzero $n \times 1$ matrix X,

$$AX = \lambda X. \tag{13.1}$$

Any nonzero $n \times 1$ matrix X satisfying this equation for some number λ is called an *eigenvector* of A associated with the eigenvalue λ.

719

EXAMPLE 13.1

Let

$$
A = \begin{bmatrix} 1 & -1 & 0 \\ 0 & 1 & 1 \\ 0 & 0 & -1 \end{bmatrix}.
$$

We claim that 1 is an eigenvalue with associated

eigenvectors $\begin{bmatrix} \alpha \\ 0 \\ 0 \end{bmatrix}$ for any $\alpha \neq 0$. To verify this, calculate

$$
A\begin{bmatrix} \alpha \\ 0 \\ 0 \end{bmatrix} = \begin{bmatrix} 1 & -1 & 0 \\ 0 & 1 & 1 \\ 0 & 0 & -1 \end{bmatrix}\begin{bmatrix} \alpha \\ 0 \\ 0 \end{bmatrix} = \begin{bmatrix} \alpha \\ 0 \\ 0 \end{bmatrix} = 1\begin{bmatrix} \alpha \\ 0 \\ 0 \end{bmatrix}.
$$

Thus, $\lambda = 1$ and $X = \begin{bmatrix} \alpha \\ 0 \\ 0 \end{bmatrix}$ satisfy equation (13.1).

Another eigenvalue of A is -1, with associated eigenvectors $\begin{bmatrix} \beta \\ 2\beta \\ -4\beta \end{bmatrix}$ for any

nonzero β. Again, this can be verified by direct computation:

$$
A\begin{bmatrix} \beta \\ 2\beta \\ -4\beta \end{bmatrix} = \begin{bmatrix} 1 & -1 & 0 \\ 0 & 1 & 1 \\ 0 & 0 & -1 \end{bmatrix}\begin{bmatrix} \beta \\ 2\beta \\ -4\beta \end{bmatrix} = \begin{bmatrix} -\beta \\ -2\beta \\ 4\beta \end{bmatrix} = -1\begin{bmatrix} \beta \\ 2\beta \\ -4\beta \end{bmatrix}. \quad \blacksquare
$$

Any nonzero constant times an eigenvector is again an eigenvector.

THEOREM 13.1

Let X be an eigenvector of A associated with eigenvalue λ. Let $\alpha \neq 0$. Then αX is also an eigenvector of A associated with eigenvalue λ.

Proof. We must show that αX satisfies equation (13.1). Since $AX = \lambda X$ by assumption,

$$
A(\alpha X) = \alpha(AX) = \alpha(\lambda X) = \lambda(\alpha X),
$$

completing the proof. $\blacksquare$

We will now examine the problem of finding all the eigenvalues of a square matrix and, for each eigenvalue, all associated eigenvectors. Suppose that λ is an eigenvalue of the $n \times n$ matrix A and that X is an associated eigenvector. Then $AX = \lambda X$. Write this equation as $\lambda X - AX = O$, with O the $n \times 1$ zero matrix. This equation can in turn be written as $\lambda I_n X - AX = O$, or

$$(\lambda I_n - A)X = O. \qquad (13.2)$$

Think of this equation as the matrix formulation of a system of n equations in n unknowns, $x_1, \ldots, x_n$, the components of X. If λ is an eigenvalue, there exists a nonzero matrix X satisfying this equation, and the system (13.2) must have a nontrivial solution. Then $\lambda I_n - A$ must be singular, by conclusion (2) of Theorem 12.23 in Section 12.9. But then $|\lambda I_n - A| = 0$, by Theorem 12.35 in Section 12.12.

Conversely, if $|\lambda I_n - A| = 0$, $\lambda I_n - A$ is singular, and the system

$$(\lambda I_n - A)X = 0$$

has a nontrivial solution, implying that X is an eigenvector associated with the eigenvector λ. We will summarize these conclusions as a theorem.

THEOREM 13.2

Let A be an $n \times n$ matrix. Then

1. λ is an eigenvalue of A if and only if $|\lambda I_n - A| = 0$.
2. If λ is an eigenvalue of A, any nontrivial solution of $(\lambda I_n - A)X = O$ is an eigenvector of A associated with λ. ■

We therefore find the eigenvalues of A by solving the equation

$$|\lambda I_n - A| = 0,$$

or, more explicitly,

$$\begin{vmatrix} \lambda - a_{11} & -a_{12} & \cdots & -a_{1n} \\ -a_{21} & \lambda - a_{22} & \cdots & -a_{2n} \\ \vdots & \vdots & & \vdots \\ -a_{n1} & -a_{n2} & \cdots & \lambda - a_{nn} \end{vmatrix} = 0.$$

Given A, we expand this determinant, obtaining a polynomial of degree n in λ, with coefficients formed from the elements of A. This polynomial is called the *characteristic polynomial* of A, and its roots are the eigenvalues of A. The equation $|\lambda I_n - A| = 0$ is the *characteristic equation* of A.

Associated with each eigenvalue λ, the nontrivial solutions of $(\lambda I_n - A)X = O$ are the eigenvectors of A.

Since the eigenvalues of A are roots of an nth degree polynomial, an $n \times n$ matrix has exactly n eigenvalues, though some may be repeated. We usually list these eigenvalues $\lambda_1, \ldots, \lambda_n$, repeating in the list any multiple eigenvalues. For example, if the characteristic polynomial of a 5×5 matrix A is $(\lambda - 1)(\lambda - 3)(\lambda + 4)^3$, we would list the eigenvalues as 1, 3, -4, -4, and -4, with -4 listed three times.

Since an nth degree polynomial with real coefficients may have complex roots, a matrix may have complex eigenvalues. The solutions of $(\lambda I_n - A)X = O$ may then have complex elements, so associated eigenvectors may have complex elements.

EXAMPLE 13.2

Consider again the matrix of Example 13.1,

$$A = \begin{bmatrix} 1 & -1 & 0 \\ 0 & 1 & 1 \\ 0 & 0 & -1 \end{bmatrix}.$$

The equation $|\lambda I_3 - A| = 0$ is

$$\begin{vmatrix} \lambda - 1 & 1 & 0 \\ 0 & \lambda - 1 & -1 \\ 0 & 0 & \lambda + 1 \end{vmatrix} = 0.$$

Expand this determinant to get the third degree polynomial equation

$$(\lambda - 1)^2(\lambda + 1) = 0.$$

This is the characteristic equation of A. The eigenvalues of A are 1, 1, and -1.

Corresponding to the eigenvalue $\lambda = +1$, solve the system $(I_3 - A)X = O$, or

$$\begin{bmatrix} 0 & 1 & 0 \\ 0 & 0 & -1 \\ 0 & 0 & 2 \end{bmatrix} \begin{bmatrix} x_1 \\ x_2 \\ x_3 \end{bmatrix} = \begin{bmatrix} 0 \\ 0 \\ 0 \end{bmatrix}.$$

We obtain the general solution $X = \begin{bmatrix} \alpha \\ 0 \\ 0 \end{bmatrix}$. Any such matrix, with $\alpha \neq 0$, is an eigenvector of A associated with the eigenvalue 1.

Corresponding to the eigenvalue $\lambda = -1$, solve $(-I_3 - A)X = O$, which is the system

$$\begin{bmatrix} -2 & 1 & 0 \\ 0 & -2 & -1 \\ 0 & 0 & 0 \end{bmatrix} \begin{bmatrix} x_1 \\ x_2 \\ x_3 \end{bmatrix} = \begin{bmatrix} 0 \\ 0 \\ 0 \end{bmatrix}.$$

The general solution of this system is $X = \begin{bmatrix} \beta \\ 2\beta \\ -4\beta \end{bmatrix}$. Any such matrix with $\beta \neq 0$ is an eigenvector associated with the eigenvalue -1. These calculations explain how we got the eigenvalues and eigenvectors in Example 13.1. ■

EXAMPLE 13.3

Let $A = \begin{bmatrix} 1 & -2 \\ 2 & 0 \end{bmatrix}$. The characteristic equation of A is

$$|\lambda I_2 - A| = \begin{vmatrix} \lambda - 1 & 2 \\ -2 & \lambda \end{vmatrix} = 0,$$

or

$$\lambda(\lambda - 1) + 4 = 0.$$

The roots are $(1 \pm \sqrt{15}i)/2$, and these are the eigenvalues of A.

Corresponding to the eigenvalue $(1 + \sqrt{15}i)/2$, solve the system

$$\left[\frac{1 + \sqrt{15}i}{2} I_2 - A \right] X = O.$$

Omitting the details, we get solutions

$$\left[\begin{matrix} \alpha \\ \left(\dfrac{1 - \sqrt{15}i}{4} \right) \alpha \end{matrix} \right].$$

For any $\alpha \neq 0$, this is an eigenvector associated with eigenvalue $(1 + \sqrt{15}i)/2$. Corresponding to the eigenvalue $(1 - \sqrt{15}i)/2$, we find eigenvectors

$$\left[\begin{matrix} \beta \\ \left(\dfrac{1 + \sqrt{15}i}{4} \right) \beta \end{matrix} \right],$$

with $\beta \neq 0$. ∎

Finding the eigenvalues of an $n \times n$ matrix is equivalent to solving for the roots of an nth degree polynomial. If $n \geq 3$, the characteristic polynomial may be difficult to solve. The problem of finding the roots of an nth degree polynomial occurs throughout mathematics (we saw it previously in solving nth order differential equations). Because of the importance of eigenvalues in many applications, efficient numerical methods for approximating the eigenvalues of a matrix have been developed, as have computer programs to carry them out. Usually, such programs do not attempt to solve the characteristic polynomial directly but instead transform the matrix into one whose eigenvalues are more easily computed, noting the effect of each transformation on the eigenvalues as it is performed.

In the next section, we will see how eigenvalues can be used to transform certain matrices into new matrices having a special form that is convenient to work with.

PROBLEMS FOR SECTION 13.1

In each of Problems 1 through 30, find the characteristic polynomial and the eigenvalues of the matrix. Corresponding to each eigenvalue, find an eigenvector. (For these problems, it is possible to find all the eigenvalues.)

1. $\begin{bmatrix} 1 & 3 \\ 2 & 1 \end{bmatrix}$ 　　　 2. $\begin{bmatrix} -2 & 0 \\ 1 & 4 \end{bmatrix}$ 　　　 3. $\begin{bmatrix} -5 & 0 \\ 1 & 2 \end{bmatrix}$

4. $\begin{bmatrix} 6 & -2 \\ -3 & 4 \end{bmatrix}$ 　　　 5. $\begin{bmatrix} 1 & -6 \\ 2 & 2 \end{bmatrix}$ 　　　 6. $\begin{bmatrix} 0 & 1 \\ 0 & 0 \end{bmatrix}$

7. $\begin{bmatrix} -5 & 2 \\ 2 & -4 \end{bmatrix}$ 　　　 8. $\begin{bmatrix} 15 & 1 \\ -2 & 6 \end{bmatrix}$ 　　　 9. $\begin{bmatrix} -10 & 4 \\ 4 & -5 \end{bmatrix}$

10. $\begin{bmatrix} 5 & -4 \\ 2 & 1 \end{bmatrix}$ 　　　 11. $\begin{bmatrix} 2 & 0 & 0 \\ 1 & 0 & 2 \\ 0 & 0 & 3 \end{bmatrix}$ 　　　 12. $\begin{bmatrix} -2 & 1 & 0 \\ 1 & 3 & 0 \\ 0 & 0 & -1 \end{bmatrix}$

13. $\begin{bmatrix} -3 & 1 & 1 \\ 0 & 0 & 0 \\ 0 & 1 & 0 \end{bmatrix}$

14. $\begin{bmatrix} 0 & 0 & -1 \\ 0 & 0 & 1 \\ 2 & 0 & 0 \end{bmatrix}$

15. $\begin{bmatrix} -14 & 1 & 0 \\ 0 & 2 & 0 \\ 1 & 0 & 2 \end{bmatrix}$

16. $\begin{bmatrix} 3 & 0 & 0 \\ 1 & -2 & -8 \\ 0 & -5 & 1 \end{bmatrix}$

17. $\begin{bmatrix} 1 & -2 & 0 \\ 0 & 0 & 0 \\ -5 & 0 & 7 \end{bmatrix}$

18. $\begin{bmatrix} 9 & -7 & 1 \\ 0 & 1 & 0 \\ -1 & 0 & 1 \end{bmatrix}$

19. $\begin{bmatrix} -6 & 0 & 0 & 1 \\ 1 & 0 & 0 & 0 \\ 0 & 0 & 0 & 0 \\ 0 & 0 & 0 & 1 \end{bmatrix}$

20. $\begin{bmatrix} -2 & 1 & 0 & 0 \\ 1 & 0 & 0 & 1 \\ 0 & 0 & 0 & 0 \\ 0 & 0 & 0 & 0 \end{bmatrix}$

21. $\begin{bmatrix} -4 & 1 & 0 & 1 \\ 0 & 1 & 0 & 0 \\ 0 & 0 & 2 & 0 \\ 1 & 0 & 0 & 3 \end{bmatrix}$

22. $\begin{bmatrix} 0 & -2 & 0 & -2 & 1 \\ -1 & 0 & 0 & 0 & 0 \\ 0 & 0 & -2 & 0 & 0 \\ 0 & 0 & 0 & 0 & 0 \\ 0 & 0 & 0 & 0 & 0 \end{bmatrix}$

23. $\begin{bmatrix} 5 & 0 & 1 & 1 & 2 \\ 0 & 1 & 0 & 0 & 0 \\ 0 & 0 & 0 & -2 & 0 \\ 0 & 0 & 0 & 4 & 0 \\ 1 & 0 & 0 & 0 & 5 \end{bmatrix}$

24. $\begin{bmatrix} 6 & 0 & 0 & 1 \\ 1 & 0 & 0 & 4 \\ 0 & 0 & 0 & 1 \\ 0 & 0 & 0 & 1 \end{bmatrix}$

25. $\begin{bmatrix} 0 & -1 & 1 & 2 \\ 0 & 0 & 1 & 1 \\ 0 & 0 & 0 & 0 \\ 0 & 0 & 0 & 1 \end{bmatrix}$

26. $\begin{bmatrix} 5 & 1 & 0 & 9 \\ 0 & 1 & 0 & 9 \\ 0 & 0 & 0 & 9 \\ 0 & 0 & 0 & 0 \end{bmatrix}$

27. $\begin{bmatrix} 0 & 0 & 0 & 0 \\ 0 & 0 & 4 & 0 \\ 0 & 1 & 1 & 0 \\ 1 & 3 & 0 & 3 \end{bmatrix}$

28. $\begin{bmatrix} 1 & 0 & 0 & 2 \\ 4 & 0 & 2 & 0 \\ 0 & 2 & 0 & 0 \\ 0 & 0 & 0 & 0 \end{bmatrix}$

29. $\begin{bmatrix} -6 & 0 & 0 & 0 \\ 4 & 1 & 0 & 2 \\ 0 & 1 & 1 & 2 \\ 1 & 0 & 0 & -3 \end{bmatrix}$

30. $\begin{bmatrix} 0 & 0 & 1 & 2 \\ 1 & 0 & 1 & 3 \\ 0 & 0 & 1 & 4 \\ 0 & 0 & 0 & 1 \end{bmatrix}$

31. Show that the eigenvalues of any matrix of the form

$$\begin{bmatrix} \alpha & \beta \\ \beta & \gamma \end{bmatrix}$$

are real if α, β, and γ are real numbers.

32. Show that the eigenvalues of any matrix of the form

$$\begin{bmatrix} \alpha & \beta & \gamma \\ \beta & \epsilon & \delta \\ \gamma & \delta & \varphi \end{bmatrix}$$

are real if α, β, γ, δ, ϵ, and φ are real numbers.

33. Suppose that λ is an eigenvalue of A, with associated eigenvector X. Show that λ^2 is an eigenvalue of A^2, with associated eigenvector X. In general, show that λ^k is an eigenvalue of A^k, with associated eigenvector X, for any positive integer k.

34. Show that the eigenvalues of an upper triangular matrix are the main diagonal elements of the matrix.

35. Let λ_1 and λ_2 be distinct eigenvalues of an $n \times n$ matrix A, with associated eigenvectors X_1 and X_2, respectively. Show that X_1 and X_2 are linearly independent (that is, neither is a scalar multiple of the other).

36. Can a matrix with at least one complex, nonreal element have only real eigenvalues? If not, give a proof; if it can, give an example.

37. Find the general form of all 2×2 matrices with real elements and eigenvalues 4 and -2.

13.2 *Diagonalization*

An $n \times n$ matrix $D = [d_{ij}]$ is said to be a *diagonal matrix* if $d_{ij} = 0$ whenever $i \neq j$. Thus, D is diagonal if all the elements off the main diagonal are zero. It is convenient to write a diagonal matrix having diagonal elements $d_1, \ldots, d_n$ as

$$\begin{bmatrix} d_1 & & & \\ & d_2 & & O \\ & & \ddots & \\ & O & & \\ & & & d_n \end{bmatrix}.$$

For example, the diagonal matrix

$$\begin{bmatrix} 2 & 0 & 0 & 0 \\ 0 & -10 & 0 & 0 \\ 0 & 0 & 8 & 0 \\ 0 & 0 & 0 & \pi \end{bmatrix}$$

can be written

$$\begin{bmatrix} 2 & & & \\ & -10 & & O \\ & & 8 & \\ & O & & \\ & & & \pi \end{bmatrix}.$$

Diagonal matrices enjoy several properties which make them very pleasant to work with. Here are some of these properties.

THEOREM 13.3

1. If D and W are $n \times n$ diagonal matrices, DW is also a diagonal matrix and $DW = WD$. In fact, if the main diagonal elements of D are $d_1, \ldots, d_n$ and those of W are $w_1, \ldots, w_n$, DW is the diagonal matrix with main diagonal elements $d_1 w_1, \ldots, d_n w_n$.

2. The determinant of a diagonal matrix is the product of the elements on the main diagonal.

3. A diagonal matrix is nonsingular if and only if all the main diagonal elements are nonzero.

4. If D is a diagonal matrix with main diagonal elements $d_1, \ldots, d_n$, all nonzero, D^{-1} is the diagonal matrix with main diagonal elements $1/d_1, \ldots, 1/d_n$.

5. The eigenvalues of a diagonal matrix are its main diagonal elements.

Proof of (1)   Let $D = [d_{ij}]$ and $W = [w_{ij}]$. The i, j element of DW is

$$\sum_{k=1}^{n} d_{ik} w_{kj}.$$

A typical term $d_{ik}w_{kj}$ in this sum is zero if $i \neq k$ (because then $d_{ik} = 0$) or if $k \neq j$ (because then $w_{kj} = 0$). If $i \neq j$, every term in the sum is therefore zero. If $i = j$, the only term which is possibly nonzero is $d_{jj}w_{jj}$, and this is the j, j element of DW.

Proof of (2)   Since a diagonal matrix is upper triangular, the determinant of a diagonal matrix is the product of its main diagonal elements, by Theorem 12.33 in Section 12.10.

Proof of (3)   Recall that a square matrix is nonsingular if and only if its determinant is nonzero. If D is diagonal, $|D| = d_1 d_2 \cdots d_n$, and this is nonzero if and only if each $d_j \neq 0$.

Proof of (4)   Conclusion (4) follows immediately from (3) and (1). If each $d_i \neq 0$,

$$\begin{bmatrix} d_1 & & & \\ & d_2 & & O \\ & & \ddots & \\ & O & & \\ & & & d_n \end{bmatrix} \begin{bmatrix} 1/d_1 & & & \\ & 1/d_2 & & O \\ & & \ddots & \\ & O & & \\ & & & 1/d_n \end{bmatrix} = I_n.$$

Proof of (5)   The characteristic equation of D is

$$|\lambda I_n - D| = \begin{vmatrix} \lambda - d_1 & & & \\ & \lambda - d_2 & & O \\ & & \ddots & \\ & O & & \\ & & & \lambda - d_n \end{vmatrix}$$

$$= (\lambda - d_1)(\lambda - d_2) \cdots (\lambda - d_n) = 0,$$

which has roots $d_1, \ldots, d_n$. These are the eigenvalues of D.   ∎

Of course, "most" matrices are not diagonal. Sometimes, however, given an $n \times n$ matrix A, we can find a matrix P such that $P^{-1}AP$ is diagonal. This will be useful in several applications of matrices, among them the solution of systems of differential equations.

We say that an $n \times n$ matrix A is *diagonalizable* if there exists an $n \times n$ matrix P such that $P^{-1}AP$ is a diagonal matrix. When such a P exists, we say that P *diagonalizes* A and that A is *similar to* a diagonal matrix.

We will see in Example 13.8 that not every matrix is diagonalizable. That is, there need not exist a matrix P such that $P^{-1}AP$ is diagonal. However, the following theorem provides a sufficient condition for a matrix to be diagonalizable.

THEOREM 13.4

Let A be an $n \times n$ matrix with eigenvalues $\lambda_1, \ldots, \lambda_n$. Suppose that there exist n linearly independent eigenvectors $V_1, \ldots, V_n$ associated with these eigenvalues, respectively.

Then there exists a matrix P such that

$$P^{-1}AP = \begin{bmatrix} \lambda_1 & & & \\ & \lambda_2 & & O \\ & & \ddots & \\ O & & & \lambda_n \end{bmatrix}. \quad \blacksquare$$

That is, if A has n linearly independent eigenvectors associated, respectively, with eigenvalues $\lambda_1, \ldots, \lambda_n$, we can find a matrix P such that $P^{-1}AP$ is the diagonal matrix having these eigenvalues as main diagonal elements. In this theorem, the eigenvalues of A *need not be distinct*—all that is required is that we be able to produce n linearly independent eigenvectors.

The proof of the theorem is constructive, telling us how to find a matrix P that diagonalizes A.

Proof of Theorem 13.4   Form an $n \times n$ matrix P having V_j as its jth column. (Thus, P may have complex elements.) Since $V_1, \ldots, V_n$ are linearly independent, the dimension of the column space of P is n; hence, the rank of P is n by Theorems 12.12 and 12.13 in Section 12.6. We conclude that P is nonsingular by Theorem 12.22 in Section 12.9.

Now compute the product $P^{-1}AP$. First, observe that

$$\text{column } j \text{ of } AP = A \cdot [\text{column } j \text{ of } P]$$
$$= AV_j.$$

But V_j is an eigenvector associated with λ_j, so $AV_j = \lambda_j V_j$. Thus,

$$\text{column } j \text{ of } AP = \lambda_j V_j.$$

The columns of AP are therefore the eigenvalues times the respective eigenvectors, and AP has the form

$$AP = \begin{bmatrix} | & | & & | \\ \lambda_1 V_1 & \lambda_2 V_2 & \cdots & \lambda_n V_n \\ | & | & & | \end{bmatrix},$$

in which the vertical lines indicate that column j is the $n \times 1$ matrix $\lambda_j V_j$. Therefore,

$$P^{-1}AP = P^{-1}\begin{bmatrix} | & | & & | \\ \lambda_1 V_1 & \lambda_2 V_2 & \cdots & \lambda_n V_n \\ | & | & & | \end{bmatrix}.$$

Now, column j of this product is P^{-1} times column j of AP. Thus,

$$\text{column } j \text{ of } P^{-1}AP = P^{-1}[\text{column } j \text{ of } AP]$$
$$= P^{-1}[\lambda_j V_j]$$
$$= \lambda_j P^{-1} V_j.$$

Since V_j is column j of P,

$$P^{-1}V_j = \text{column } j \text{ of } P^{-1}P = \begin{bmatrix} 0 \\ 0 \\ \vdots \\ 1 \\ \vdots \\ 0 \end{bmatrix} \longleftarrow j\text{th element.}$$

Therefore,

$$\text{column } j \text{ of } P^{-1}AP = \lambda_j \begin{bmatrix} 0 \\ 0 \\ \vdots \\ 1 \\ \vdots \\ 0 \end{bmatrix} = \begin{bmatrix} 0 \\ 0 \\ \vdots \\ \lambda_j \\ \vdots \\ 0 \end{bmatrix}.$$

This shows that column j of $P^{-1}AP$ has all zero elements except in row j, where the element is λ_j. Thus,

$$P^{-1}AP = \begin{bmatrix} \lambda_1 & & & \\ & \lambda_2 & & O \\ & & \ddots & \\ & O & & \\ & & & \lambda_n \end{bmatrix},$$

as we wanted to prove. ∎

We will consider some examples. Take particular note of Example 13.7, which illustrates that a matrix with multiple eigenvalues may be diagonalizable.

EXAMPLE 13.4

Let $A = \begin{bmatrix} -1 & 4 \\ 0 & 3 \end{bmatrix}$. We find that the eigenvalues of A are -1 and 3. Associated eigenvectors are, respectively,

$$\begin{bmatrix} 1 \\ 0 \end{bmatrix} \quad \text{and} \quad \begin{bmatrix} 1 \\ 1 \end{bmatrix}.$$

These eigenvectors are linearly independent (as vectors in R^2). Let

$$P = \begin{bmatrix} 1 & 1 \\ 0 & 1 \end{bmatrix}.$$

From the proof of Theorem 13.4, P diagonalizes A. To show this explicitly for this example, first compute

$$P^{-1} = \begin{bmatrix} 1 & -1 \\ 0 & 1 \end{bmatrix}.$$

Now compute

$$P^{-1}AP = \begin{bmatrix} 1 & -1 \\ 0 & 1 \end{bmatrix} \begin{bmatrix} -1 & 4 \\ 0 & 3 \end{bmatrix} \begin{bmatrix} 1 & 1 \\ 0 & 1 \end{bmatrix} = \begin{bmatrix} -1 & 0 \\ 0 & 3 \end{bmatrix}.$$

This is a diagonal matrix with the eigenvalues on the main diagonal, *in the order* in which we used eigenvectors as columns of *P*.

Watch what happens if we choose different eigenvectors corresponding to -1 and 3. For example, two other associated eigenvectors are, respectively,

$$\begin{bmatrix} 3 \\ 0 \end{bmatrix} \quad \text{and} \quad \begin{bmatrix} -2 \\ -2 \end{bmatrix}.$$

Now let

$$Q = \begin{bmatrix} 3 & -2 \\ 0 & -2 \end{bmatrix}.$$

We find that

$$Q^{-1} = \begin{bmatrix} \frac{1}{3} & -\frac{1}{3} \\ 0 & -\frac{1}{2} \end{bmatrix},$$

and it is routine to verify that

$$Q^{-1}AQ = \begin{bmatrix} -1 & 0 \\ 0 & 3 \end{bmatrix}.$$

As we should have expected, *any* matrix having linearly independent eigenvectors as columns diagonalizes *A*.

What happens if we write the eigenvectors in a different order, as columns of a new matrix *S*? Then *S* will still diagonalize *A*, but the eigenvalues appearing on the main diagonal of $S^{-1}AS$ will appear in the order corresponding to the eigenvectors in the columns of *S*. For example, let

$$S = \begin{bmatrix} 1 & 1 \\ 1 & 0 \end{bmatrix}.$$

S has the columns of *P* written in reverse order. Compute

$$S^{-1}AS = \begin{bmatrix} 0 & 1 \\ 1 & -1 \end{bmatrix} \begin{bmatrix} -1 & 4 \\ 0 & 3 \end{bmatrix} \begin{bmatrix} 1 & 1 \\ 1 & 0 \end{bmatrix} = \begin{bmatrix} 3 & 0 \\ 0 & -1 \end{bmatrix},$$

with the eigenvalues appearing on the main diagonal in the order corresponding to the eigenvector columns of *S*. ∎

EXAMPLE 13.5

Let

$$A = \begin{bmatrix} -1 & 0 & 5 \\ 0 & 1 & 0 \\ 0 & 0 & -2 \end{bmatrix}.$$

We find that the eigenvalues of A are 1, -1, and -2. Associated eigenvectors are, respectively,

$$\begin{bmatrix} 0 \\ 1 \\ 0 \end{bmatrix}, \qquad \begin{bmatrix} 1 \\ 0 \\ 0 \end{bmatrix}, \quad \text{and} \quad \begin{bmatrix} 5 \\ 0 \\ -1 \end{bmatrix}.$$

Form

$$P = \begin{bmatrix} 0 & 1 & 5 \\ 1 & 0 & 0 \\ 0 & 0 & -1 \end{bmatrix}.$$

We find that

$$P^{-1} = \begin{bmatrix} 0 & 1 & 0 \\ 1 & 0 & 5 \\ 0 & 0 & -1 \end{bmatrix}.$$

It is routine to check that

$$P^{-1}AP = \begin{bmatrix} 1 & 0 & 0 \\ 0 & -1 & 0 \\ 0 & 0 & -2 \end{bmatrix}. \quad \blacksquare$$

EXAMPLE 13.6

Sometimes we must use a matrix with complex elements to diagonalize A even though A may have only real elements. For example, let

$$A = \begin{bmatrix} -1 & -4 \\ 3 & -2 \end{bmatrix}.$$

We find that the eigenvalues of A are $(-3 \pm \sqrt{47}i)/2$, with associated eigenvectors

$$\begin{bmatrix} 1 \\ \dfrac{1 - \sqrt{47}i}{8} \end{bmatrix} \quad \text{and} \quad \begin{bmatrix} 1 \\ \dfrac{1 + \sqrt{47}i}{8} \end{bmatrix}.$$

A matrix that diagonalizes A is

$$\begin{bmatrix} 1 & 1 \\ \dfrac{1 - \sqrt{47}i}{8} & \dfrac{1 + \sqrt{47}i}{8} \end{bmatrix}. \quad \blacksquare$$

EXAMPLE 13.7

It is not necessary for A to have n distinct eigenvalues in order to be diagonalizable. For example, let

$$A = \begin{bmatrix} 5 & -4 & 4 \\ 12 & -11 & 12 \\ 4 & -4 & 5 \end{bmatrix}.$$

We find that the eigenvalues of A are 1, 1, and -3. Associated with -3, we find an eigenvector $\begin{bmatrix} 1 \\ 3 \\ 1 \end{bmatrix}$.

Now consider eigenvectors associated with 1. We must solve the system $(I_3 - X) = O$, or

$$\begin{bmatrix} -4 & 4 & -4 \\ -12 & 12 & -12 \\ -4 & 4 & -4 \end{bmatrix} \begin{bmatrix} x \\ y \\ z \end{bmatrix} = \begin{bmatrix} 0 \\ 0 \\ 0 \end{bmatrix}.$$

This system has general solution

$$\begin{bmatrix} \alpha \\ \beta \\ \beta - \alpha \end{bmatrix},$$

with α and β arbitrary. Since there are two constants we can specify independently, we can find two linearly independent eigenvectors by choosing first $\alpha = 1$ and $\beta = 0$, then $\alpha = 0$ and $\beta = 1$. This gives us two linearly independent eigenvectors

$$\begin{bmatrix} 1 \\ 0 \\ -1 \end{bmatrix} \quad \text{and} \quad \begin{bmatrix} 0 \\ 1 \\ 1 \end{bmatrix}$$

associated with eigenvalue 1. We now have three linearly independent eigenvectors of A,

$$\begin{bmatrix} 1 \\ 3 \\ 1 \end{bmatrix}, \quad \begin{bmatrix} 1 \\ 0 \\ -1 \end{bmatrix}, \quad \text{and} \quad \begin{bmatrix} 0 \\ 1 \\ 1 \end{bmatrix}.$$

Form

$$P = \begin{bmatrix} 1 & 1 & 0 \\ 3 & 0 & 1 \\ 1 & -1 & 1 \end{bmatrix}.$$

This matrix diagonalizes A. We find that

$$P^{-1}AP = \begin{bmatrix} -3 & 0 & 0 \\ 0 & 1 & 0 \\ 0 & 0 & 1 \end{bmatrix}. \quad \blacksquare$$

We will now show that the condition of Theorem 13.4 is necessary as well as sufficient. That is, if A does *not* have n linearly independent eigenvectors, A is *not* diagonalizable. But much more can be said. Above, we saw that

$$P^{-1}AP = \begin{bmatrix} \lambda_1 & & & \\ & \lambda_2 & & O \\ & & \ddots & \\ & O & & \\ & & & \lambda_n \end{bmatrix}$$

if column j of P is an eigenvector of A associated with eigenvalue λ_j. We will now show that *whenever* an equation

$$Q^{-1}AQ = \begin{bmatrix} d_1 & & & & \\ & d_2 & & O & \\ & & \ddots & & \\ & O & & & \\ & & & & d_n \end{bmatrix}$$

holds, the d_j's must be the eigenvalues of A, and the columns of Q must be associated eigenvectors, respectively, of these eigenvalues.

THEOREM 13.5

Let A be an $n \times n$ diagonalizable matrix. Then A must have n linearly independent eigenvectors. Further, if

$$Q^{-1}AQ = \begin{bmatrix} d_1 & & & & \\ & d_2 & & O & \\ & & \ddots & & \\ & O & & & \\ & & & & d_n \end{bmatrix},$$

$d_1, \ldots, d_n$ are the eigenvalues of A, and column j of Q is an eigenvector of A associated with eigenvalue d_j. (We emphasize again that this theorem does *not* imply that the eigenvalues of A are distinct; remember Example 13.7.)

Proof   Since A is diagonalizable, there is a nonsingular matrix Q that diagonalizes A:

$$Q^{-1}AQ = \begin{bmatrix} d_1 & & & & \\ & d_2 & & O & \\ & & \ddots & & \\ & O & & & \\ & & & & d_n \end{bmatrix}.$$

Since Q is nonsingular, its columns are linearly independent. Denote column j of Q by V_j. We will prove the theorem by showing that V_j is an eigenvector of A associated with eigenvalue d_j. That is, we will show that $AV_j = d_jV_j$.

Let $D = Q^{-1}AQ$, as a convenience. Then $QD = AQ$. Compute both of these products in order to compare them. We have

$$QD = \begin{bmatrix} | & | & & | \\ V_1 & V_2 & \cdots & V_n \\ | & | & & | \end{bmatrix} D = \begin{bmatrix} | & | & & | \\ d_1V_1 & d_2V_2 & \cdots & d_nV_n \\ | & | & & | \end{bmatrix}$$

and

$$AQ = A \begin{bmatrix} | & | & & | \\ V_1 & V_2 & \cdots & V_n \\ | & | & & | \end{bmatrix} = \begin{bmatrix} | & | & & | \\ AV_1 & AV_2 & \cdots & AV_n \\ | & | & & | \end{bmatrix}.$$

Since $QD = AQ$, column j of QD equals column j of AQ, and $AV_j = d_jV_j$, proving that d_j is an eigenvalue of A, with associated eigenvector V_j. ∎

Theorems 13.4 and 13.5 not only say when a matrix can be diagonalized but also describe the makeup of any diagonalizing matrix and the exact form of the resulting diagonal matrix. It has been stated that not every matrix can be diagonalized. Here is an example.

EXAMPLE 13.8

Let $B = \begin{bmatrix} 1 & -1 \\ 0 & 1 \end{bmatrix}$. The eigenvalues of B are 1 and 1. By solving $(I_2 - B)X = O$, we find that all eigenvectors must have the form $\begin{bmatrix} \alpha \\ 0 \end{bmatrix}$, with $\alpha \neq 0$. We cannot find two linearly independent eigenvectors, since all eigenvectors are constant multiples of $\begin{bmatrix} 1 \\ 0 \end{bmatrix}$.

Therefore, B is not diagonalizable.

In fact, if there were P such that

$$P^{-1}AP = \begin{bmatrix} d_1 & 0 \\ 0 & d_2 \end{bmatrix},$$

then by the last part of Theorem 13.5, P would have to have eigenvectors as columns. But then P would have to be of the form $\begin{bmatrix} \alpha & \beta \\ 0 & 0 \end{bmatrix}$, and this matrix is singular because its determinant is zero no matter how we choose α and β. ∎

By reviewing examples in which eigenvalues and eigenvectors of a matrix have been found, it can be seen that eigenvectors associated with distinct eigenvalues are linearly independent. This is true in general and is important to know.

THEOREM 13.6

Eigenvectors associated with distinct eigenvalues of a matrix are linearly independent.

Proof   Proceed by mathematical induction on the number of distinct eigenvalues being considered. If we are looking at just one eigenvalue, any associated eigenvector is by itself linearly independent, thought of as a vector $\langle x_1, \ldots, x_n \rangle$.

Now suppose that any $k - 1$ eigenvectors associated with $k - 1$ distinct eigenvalues of A are linearly independent. Suppose that we have k distinct eigenvalues $\lambda_1, \ldots, \lambda_k$, with associated eigenvectors $V_1, \ldots, V_k$, respectively. We will suppose that $V_1, \ldots, V_k$ are linearly dependent and derive a contradiction.

If $V_1, \ldots, V_k$ are linearly dependent, then by Theorem 11.7 in Section 11.6, there are constants $\alpha_1, \ldots, \alpha_n$, possibly complex but not all zero, such that

$$\alpha_1 V_1 + \alpha_2 V_2 + \cdots + \alpha_k V_k = O, \tag{13.3}$$

where O is the $n \times 1$ zero matrix. By relabeling the constants and eigenvectors, if necessary, we may suppose that $\alpha_1 \neq 0$. Compute $\lambda_1 I_n - A$ times both sides of equation (13.3) to get

$$(\lambda_1 I_n - A)(\alpha_1 V_1 + \alpha_2 V_2 + \cdots + \alpha_k V_k) = O.$$

Then

$$\alpha_1(\lambda_1 V_1 - AV_1) + \alpha_2(\lambda_1 V_2 - AV_2) + \cdots + \alpha_k(\lambda_1 V_k - AV_k)$$
$$= \alpha_1(\lambda_1 V_1 - \lambda_1 V_1) + \alpha_2(\lambda_1 V_2 - \lambda_2 V_2) + \cdots + \alpha_k(\lambda_1 V_k - \lambda_k V_k)$$
$$= \alpha_2(\lambda_1 - \lambda_2)V_2 + \cdots + \alpha_k(\lambda_1 - \lambda_k)V_k = O.$$

But $V_2, \ldots, V_k$ form $k - 1$ distinct eigenvectors of A, so by the induction hypothesis, they are linearly independent. By conclusion (2) of Theorem 11.7, all of the coefficients of $V_2, \ldots, V_k$ in the last equation must be zero:

$$\alpha_2(\lambda_1 - \lambda_2) = \cdots = \alpha_k(\lambda_1 - \lambda_k) = 0.$$

But λ_1 is assumed to be distinct from $\lambda_2, \ldots, \lambda_k$, so each of $\lambda_1 - \lambda_2, \ldots, \lambda_1 - \lambda_k$ is nonzero. Therefore,

$$\alpha_2 = \cdots = \alpha_k = 0.$$

But then, from equation (13.3), $\alpha_1 V_1 = O$. Since V_1 is an eigenvector, it is not a zero vector; hence, $\alpha_1 = 0$, contradicting our assumption. Hence $\mathbf{V}_1, \ldots, \mathbf{V}_k$ are linearly independent. ■

If we combine the results of Theorems 13.4 and 13.6, we immediately have the following.

COROLLARY TO THEOREM 13.6

Any $n \times n$ matrix with n distinct eigenvalues is diagonalizable. ■

EXAMPLE 13.9

Let

$$A = \begin{bmatrix} -4 & 0 & 1 & 0 \\ 0 & -2 & 0 & 0 \\ 0 & 0 & 1 & 2 \\ 0 & 0 & 0 & 0 \end{bmatrix}.$$

The eigenvalues of A are $0, 1, -2$, and -4. Since these are distinct, A is diagonalizable. Further, by the corollary and Theorem 13.4, there is a matrix P such that

$$P^{-1}AP = \begin{bmatrix} 0 & 0 & 0 & 0 \\ 0 & 1 & 0 & 0 \\ 0 & 0 & -2 & 0 \\ 0 & 0 & 0 & -4 \end{bmatrix}.$$

This conclusion can be drawn without finding any eigenvectors or P or P^{-1}. ■

In the next chapter, we will use diagonalization to solve certain systems of differential equations.

PROBLEMS FOR SECTION 13.2

In each of Problems 1 through 12, produce a matrix that diagonalizes A or show that A is not diagonalizable. In the former case, verify explicitly that $P^{-1}AP$ is a diagonal matrix with eigenvalues of A on the main diagonal.

1. $A = \begin{bmatrix} 0 & -1 \\ 4 & 3 \end{bmatrix}$

2. $A = \begin{bmatrix} 5 & 3 \\ 1 & 3 \end{bmatrix}$

3. $A = \begin{bmatrix} 1 & 0 \\ -4 & 1 \end{bmatrix}$

4. $A = \begin{bmatrix} -5 & 3 \\ 0 & 9 \end{bmatrix}$

5. $A = \begin{bmatrix} 5 & 0 & 0 \\ 1 & 0 & 3 \\ 0 & 0 & -2 \end{bmatrix}$

6. $A = \begin{bmatrix} 0 & 0 & 0 \\ 1 & 0 & 2 \\ 0 & 1 & 3 \end{bmatrix}$

7. $A = \begin{bmatrix} -2 & 0 & 1 \\ 1 & 1 & 0 \\ 0 & 0 & -2 \end{bmatrix}$

8. $A = \begin{bmatrix} 2 & 0 & 0 \\ 0 & 2 & 1 \\ 0 & -1 & 2 \end{bmatrix}$

9. $A = \begin{bmatrix} 1 & 0 & 0 & 0 \\ 0 & 4 & 1 & 0 \\ 0 & 0 & -3 & 1 \\ 0 & 0 & 1 & -2 \end{bmatrix}$

10. $A = \begin{bmatrix} -2 & 0 & 0 & 0 \\ -4 & -2 & 0 & 0 \\ 0 & 0 & -2 & 0 \\ 0 & 0 & 0 & -2 \end{bmatrix}$

11. $A = \begin{bmatrix} 8 & -7 & 1 & 0 \\ 0 & 1 & 0 & 0 \\ 0 & 0 & 0 & 0 \\ 1 & 0 & 0 & 0 \end{bmatrix}$

12. $A = \begin{bmatrix} -7 & 0 & 1 & 0 \\ 0 & 1 & 1 & 0 \\ -4 & 0 & 2 & 0 \\ 0 & 0 & 0 & 0 \end{bmatrix}$

13. Let A be an $n \times n$ matrix. Prove that the constant term in the characteristic polynomial of A is $(-1)^n|A|$.

14. Suppose that A^2 is diagonalizable. Prove that A is diagonalizable.

15. Suppose that A has eigenvalues $\lambda_1, \ldots, \lambda_n$ and that P diagonalizes A. Prove that, for any positive integer k,

$$A^k = P \begin{bmatrix} \lambda_1^k & & & 0 \\ & \lambda_2^k & & \\ & & \ddots & \\ 0 & & & \lambda_n^k \end{bmatrix} P^{-1}.$$

In each of Problems 16 through 19, compute the indicated power of the matrix, using the idea of Problem 15.

16. $A = \begin{bmatrix} -1 & 0 \\ 1 & -5 \end{bmatrix}$; A^{18}

17. $A = \begin{bmatrix} -3 & -3 \\ -2 & 4 \end{bmatrix}$; A^{16}

18. $A = \begin{bmatrix} 0 & -2 \\ 1 & 0 \end{bmatrix}$; A^{43}

19. $A = \begin{bmatrix} -2 & 3 \\ 3 & -4 \end{bmatrix}$; A^{31}

20. Let A be any 2×2 matrix with real elements. Prove that there is some nonsingular matrix P such that $P^{-1}AP$ has one of the following forms:

$$\begin{bmatrix} \alpha & 0 \\ 0 & \beta \end{bmatrix}, \text{ with } \alpha \neq \beta; \quad \text{or} \quad \begin{bmatrix} \alpha & 0 \\ 0 & \alpha \end{bmatrix}; \quad \text{or} \quad \begin{bmatrix} \alpha & -1 \\ 0 & \alpha \end{bmatrix}.$$

21. Suppose that $L^{-1}AL = B$, for some nonsingular matrix L. Prove that A and B must both be either diagonalizable or nondiagonalizable. If both A and B are diagonalizable, and $P^{-1}AP = D_1$, a diagonal matrix, and $Q^{-1}BQ = D_2$, a diagonal matrix, how are D_1 and D_2 related?

13.3 *Eigenvalues and Eigenvectors of Real, Symmetric Matrices*

An $n \times n$ matrix A may have complex eigenvalues even if all the elements of A are real. We will now consider an important special case in which the eigenvalues of A are necessarily real numbers. An $n \times n$ matrix $A = [a_{ij}]$ is said to be *symmetric* if $a_{ij} = a_{ji}$ (or, equivalently, if $A = A^t$). This means that interchanging the rows and columns of A results in the same matrix A. Alternatively, a_{ij} is equal to its reflection a_{ji} across the main diagonal. For example,

$$\begin{bmatrix} -3 & 4 \\ 4 & 9 \end{bmatrix}$$

is a symmetric 2×2 matrix, and

$$\begin{bmatrix} 1 & 0 & -5 \\ 0 & 8 & 2 \\ -5 & 2 & -4 \end{bmatrix}$$

is a symmetric 3×3 matrix. Here, $a_{13} = a_{31} = -5$, $a_{12} = a_{21} = 0$, and $a_{23} = a_{32} = 2$.

As a convenience in terminology, we will refer to a matrix A as a *real matrix* if every element of A is real.

THEOREM 13.7

The eigenvalues of a real, symmetric matrix are real. ∎

Before proving the theorem, recall that the complex conjugate of a number $a + ib$ is the number $a - ib$ and is denoted $\overline{a + ib}$. A number z is real if and only if $z = \bar{z}$. Further,

$$(a + ib)(a - ib) = a^2 + b^2,$$

a real number. More compactly, $z\bar{z}$ is real for any number z, real or complex.

Note that the conjugate of a conjugate is the original number:

$$\overline{(\bar{z})} = z.$$

For if $z = a + ib$, $\overline{(\bar{z})} = \overline{(a - ib)} = a + ib = z$.

We take the conjugate of a (not necessarily square) matrix by taking the conjugate of each matrix element. The conjugate of a matrix A is denoted $\bar{A}$. If A is a real matrix, each element equals its own conjugate, and $\bar{A} = A$. Conversely, if $A = \bar{A}$, each element must equal its conjugate and hence must be a real number, so A is a real matrix.

The conjugate of a transpose is the transpose of the conjugate. That is,

$$\overline{(A^t)} = (\bar{A})^t. \tag{13.4}$$

To form $\overline{(A^t)}$, first interchange the rows and columns of A, then replace i with $-i$ in each matrix element. To form $(\bar{A})^t$, first replace i with $-i$ in each matrix element, then interchange rows and columns. These two operations can be performed in either order, yielding the same result. For example, let

$$A = \begin{bmatrix} 1 - i & i \\ 2 & 3 + 4i \\ 2i & 6 + i \end{bmatrix}.$$

Then

$$A^t = \begin{bmatrix} 1 - i & 2 & 2i \\ i & 3 + 4i & 6 + i \end{bmatrix},$$

so

$$\overline{(A^t)} = \begin{bmatrix} 1 + i & 2 & -2i \\ -i & 3 - 4i & 6 - i \end{bmatrix}.$$

But we also have

$$\bar{A} = \begin{bmatrix} 1 + i & -i \\ 2 & 3 - 4i \\ -2i & 6 - i \end{bmatrix},$$

so

$$(\bar{A})^t = \begin{bmatrix} 1 + i & 2 & -2i \\ -i & 3 - 4i & 6 - i \end{bmatrix} = \overline{(A^t)}.$$

In view of equation (13.4), we will write $\bar{A}^t$ for the matrix formed by taking the conjugate and the transpose of A, in either order.

With these remarks as background, we will prove the theorem.

Proof of Theorem 13.7   Let A be an $n \times n$ real, symmetric matrix. Let λ be an eigenvalue of A. We want to show that λ is real. Let

$$X = \begin{bmatrix} x_1 \\ x_2 \\ \vdots \\ x_n \end{bmatrix}$$

be an eigenvector of A associated with λ. Then $AX = \lambda X$. Multiply this equation on the left by the $1 \times n$ matrix $\bar{X}^t = [\bar{x}_1, \bar{x}_2, \ldots, \bar{x}_n]$ to get

$$\bar{X}^t A X = \bar{X}^t \lambda X = \lambda \bar{X}^t X. \tag{13.5}$$

Now, X^t is $1 \times n$, and X is $n \times 1$, so $\bar{X}^t X$ is a 1×1 matrix:

$$\bar{X}^t X = [\bar{x}_1, \bar{x}_2, \ldots, \bar{x}_n] \begin{bmatrix} x_1 \\ x_2 \\ \vdots \\ x_n \end{bmatrix} = [\bar{x}_1 x_1 + \bar{x}_2 x_2 + \cdots + \bar{x}_n x_n].$$

Since $\bar{x}_j x_j$ is real for each j, $\bar{X}^t X$ is a 1×1 real matrix.

Now look at the 1×1 matrix $\bar{X}'AX$. Its conjugate is

$$\overline{\bar{X}'AX} = (\overline{\bar{X}'})\bar{A}\bar{X} = (\overline{\bar{X}'})\bar{A}\bar{X} = X'A\bar{X}, \tag{13.6}$$

in which we have used the fact that $A = \bar{A}$ because A is a real matrix.

Now, $X'A\bar{X}$ is a 1×1 matrix, so it equals its own transpose. By conclusions (2) and (3) of Theorem 12.6 in Section 12.3, we have

$$X'A\bar{X} = (X'A\bar{X})^t = \bar{X}'A^t(X')^t = \bar{X}'AX. \tag{13.7}$$

Here, we used the fact that $A^t = A$ because A is symmetric.

From the first part of equation (13.6) and the last part of equation (13.7), we conclude that

$$\overline{\bar{X}'AX} = \bar{X}'AX.$$

Thus, $\bar{X}'AX$ is a 1×1 real matrix.

Now look at equation (13.5) again. Since $\bar{X}'AX$ is real and $\bar{X}'X$ is real, equation (13.5) implies that λ must be real, proving the theorem. ∎

Of course, with any real eigenvalue λ, we can also associate real eigenvectors because the eigenvectors are solutions of the system $(\lambda I_n - A)X = O$ and are obtained from arithmetic operations applied to the elements of the real matrix $\lambda I_n - A$. We conclude that a real, symmetric matrix has real eigenvalues and eigenvectors.

Eigenvectors of a real, symmetric matrix enjoy another property. A real eigenvector is an $n \times 1$ matrix which we can think of as a vector in R^n. This enables us to associate with real eigenvectors geometric terms commonly used in n-space.

THEOREM 13.8

Let A be a real, symmetric matrix. Then eigenvectors associated with distinct eigenvectors of A are orthogonal.

Proof   Let λ and μ be distinct eigenvalues of A, with associated eigenvectors

$$X = \begin{bmatrix} x_1 \\ x_2 \\ \vdots \\ x_n \end{bmatrix} \quad \text{and} \quad Y = \begin{bmatrix} y_1 \\ y_2 \\ \vdots \\ y_n \end{bmatrix},$$

respectively. We will show that the dot product of X with Y (thought of as vectors in R^n) is zero. That is, we will show that

$$\langle x_1, x_2, \ldots, x_n \rangle \cdot \langle y_1, y_2, \ldots, y_n \rangle = 0.$$

We have $AX = \lambda X$ and $AY = \mu Y$. Carry out the following calculation:

$$\lambda X'Y = (\lambda X)'Y = (AX)'Y$$
$$= X'A'Y = X'(AY) = X'(\mu Y) = \mu X'Y.$$

Since $\lambda - \mu \neq 0$, we conclude that $X'Y = O$ and hence that

$$x_1 y_1 + x_2 y_2 + \cdots + x_n y_n = 0.$$

But this is the dot product of the n-vectors $\langle x_1, x_2, \ldots, x_n \rangle$ and $\langle y_1, y_2, \ldots, y_n \rangle$, and therefore these two vectors are orthogonal. ∎

We call a set of vectors in R^n *mutually orthogonal* if each vector in the set is orthogonal to every other vector in the set. By Theorem 13.8, eigenvectors associated with distinct eigenvalues of a real, symmetric matrix are mutually orthogonal.

EXAMPLE 13.10

Let

$$A = \begin{bmatrix} 3 & 0 & -2 \\ 0 & 2 & 0 \\ -2 & 0 & 0 \end{bmatrix}.$$

Then A is a real, symmetric matrix. The eigenvalues of A are 2, -1, and 4. Corresponding eigenvectors are

$$\begin{bmatrix} 0 \\ 1 \\ 0 \end{bmatrix}, \quad \begin{bmatrix} 1 \\ 0 \\ 2 \end{bmatrix}, \quad \text{and} \quad \begin{bmatrix} 2 \\ 0 \\ -1 \end{bmatrix}.$$

These are mutually orthogonal, considered as vectors in R^3. ∎

We will point out a fact to be discussed in the next section. Notice that one of the eigenvectors in Example 13.10 (the first one listed) has length 1, while the others have length $\sqrt{5}$. Multiplying an eigenvector by a nonzero scalar results in another eigenvector. Multiply the other two eigenvectors by $1/\sqrt{5}$ to get the eigenvectors

$$\begin{bmatrix} 0 \\ 1 \\ 0 \end{bmatrix}, \quad \begin{bmatrix} 1/\sqrt{5} \\ 0 \\ 2/\sqrt{5} \end{bmatrix}, \quad \text{and} \quad \begin{bmatrix} 2/\sqrt{5} \\ 0 \\ -1/\sqrt{5} \end{bmatrix}.$$

These eigenvectors form mutually orthogonal *unit* vectors in R^3. Furthermore, if we form the matrix Q having these eigenvectors as columns,

$$Q = \begin{bmatrix} 0 & 1/\sqrt{5} & 2/\sqrt{5} \\ 1 & 0 & 0 \\ 0 & 2/\sqrt{5} & -1/\sqrt{5} \end{bmatrix},$$

then Q diagonalizes A but also has the remarkable property that $Q^{-1} = Q^t$. Such a matrix is called an *orthogonal matrix*.

In the next section, we will discuss how a real, symmetric matrix can always be diagonalized by an orthogonal matrix.

PROBLEMS FOR SECTION 13.3

In each of Problems 1 through 15, find the eigenvalues of the given real, symmetric matrix. For each eigenvalue, find an associated eigenvector. By taking dot products, check that eigenvectors associated with distinct eigenvalues are mutually orthogonal.

1. $\begin{bmatrix} 4 & -2 \\ -2 & 1 \end{bmatrix}$
2. $\begin{bmatrix} -3 & 5 \\ 5 & 4 \end{bmatrix}$
3. $\begin{bmatrix} 6 & 1 \\ 1 & 4 \end{bmatrix}$
4. $\begin{bmatrix} -13 & 1 \\ 1 & 4 \end{bmatrix}$

5. $\begin{bmatrix} 5 & 0 & 2 \\ 0 & 0 & 0 \\ 2 & 0 & 0 \end{bmatrix}$
6. $\begin{bmatrix} 5 & -10 \\ -10 & 3 \end{bmatrix}$
7. $\begin{bmatrix} 0 & 1 & 0 \\ 1 & -2 & 0 \\ 0 & 0 & 3 \end{bmatrix}$
8. $\begin{bmatrix} 0 & 1 & 1 \\ 1 & 2 & 0 \\ 1 & 0 & 2 \end{bmatrix}$

9. $\begin{bmatrix} -1 & 6 \\ 6 & 4 \end{bmatrix}$
10. $\begin{bmatrix} 4 & -11 \\ -11 & 12 \end{bmatrix}$
11. $\begin{bmatrix} 5 & -3 \\ -3 & -4 \end{bmatrix}$
12. $\begin{bmatrix} 9 & -15 \\ -15 & 4 \end{bmatrix}$

13. $\begin{bmatrix} 6 & -3 \\ -3 & 6 \end{bmatrix}$
14. $\begin{bmatrix} -7 & 2 \\ 2 & 9 \end{bmatrix}$
15. $\begin{bmatrix} 2 & -4 & 0 \\ -4 & 0 & 0 \\ 0 & 0 & 0 \end{bmatrix}$

16. Let A and B be $n \times n$ real, symmetric matrices. Prove that AB is symmetric if and only if $AB = BA$.

17. Let A be any $n \times n$ matrix (not necessarily symmetric). Are the eigenvalues of A in general different from those of A^t?

18. Let A be a real, symmetric matrix, and let λ be an eigenvalue of multiplicity k. Prove that there can be at most k mutually orthogonal eigenvectors associated with λ.

13.4 *Orthogonal Matrices and Diagonalization of Real, Symmetric Matrices*

We will pursue the idea mentioned at the end of the preceding section. A real, square matrix Q is called *orthogonal* if $Q^{-1} = Q^t$ (or, equivalently, if $QQ^t = I_n = Q^tQ$). The matrix Q following Example 13.10 in the preceding section is an orthogonal matrix. We will derive some facts about orthogonal matrices and then relate them to the concept of diagonalization.

THEOREM 13.9

An $n \times n$ matrix Q is orthogonal if and only if Q^t is orthogonal. ∎

This theorem is a simple consequence of the definition, and we omit the proof.
Vectors $\mathbf{F}_1, \ldots, \mathbf{F}_r$ in R^n are said to be *orthonormal* if each $\mathbf{F}_j$ has length 1 and the vectors are mutually orthogonal.

THEOREM 13.10

An $n \times n$ matrix Q is orthogonal if and only if the rows of Q are orthonormal vectors in R^n.

Proof   Suppose first that the rows of Q are orthonormal. Then

$$i, j \text{ element of } QQ^t = (\text{row } i \text{ of } Q) \cdot (\text{column } j \text{ of } Q^t)$$

$$= (\text{row } i \text{ of } Q) \cdot (\text{row } j \text{ of } Q)$$

$$= \begin{cases} 0 & \text{if } i \neq j \\ 1 & \text{if } i = j, \end{cases}$$

because the rows of Q are mutually orthogonal unit vectors and the dot product of a vector with itself is the square of its length. Therefore, $QQ^t = I_n$.

Conversely, suppose that Q is orthogonal. Then $QQ^t = I_n$, so for $i \neq j$,

$$(\text{row } i \text{ of } Q) \cdot (\text{column } j \text{ of } Q^t) = (\text{row } i \text{ of } Q) \cdot (\text{row } j \text{ of } Q)$$
$$= (i, j \text{ element of } I_n) = 0.$$

Hence, the rows of Q are mutually orthogonal vectors in R^n. Further,

$$\|\text{row } i\|^2 = (\text{row } i) \cdot (\text{row } i)$$
$$= (\text{row } i \text{ of } Q) \cdot (\text{column } i \text{ of } Q^t) = i, i \text{ element of } QQ^t$$
$$= i, i \text{ element of } I_n = 1.$$

Therefore, the row vectors of Q are unit vectors, and the proof is complete. ∎

By combining the last two theorems, we have the following.

COROLLARY TO THEOREM 13.10

Let Q be an $n \times n$ matrix. Then Q is orthogonal if and only if the columns of Q are orthonormal vectors in R^n. ∎

THEOREM 13.11

Let Q be an orthogonal matrix. Then $|Q| = 1$ or $|Q| = -1$.

Proof   Since $QQ^t = I_n$, $|QQ^t| = |Q||Q^t| = |Q|^2 = |I_n| = 1$. Thus, $|Q|$ must be either 1 or -1. ∎

EXAMPLE 13.11

We now have a great deal of information about orthogonal matrices. To illustrate how this information can be used, we will determine all orthogonal 2×2 matrices. Suppose that

$$Q = \begin{bmatrix} a & b \\ c & d \end{bmatrix}$$

is orthogonal. What can be said about a, b, c, and d?
First,

$$QQ^t = \begin{bmatrix} a & b \\ c & d \end{bmatrix} \begin{bmatrix} a & c \\ b & d \end{bmatrix}$$
$$= \begin{bmatrix} a^2 + b^2 & ac + bd \\ ac + bd & c^2 + d^2 \end{bmatrix} = \begin{bmatrix} 1 & 0 \\ 0 & 1 \end{bmatrix}.$$

Therefore,

$$\begin{aligned} a^2 + b^2 &= 1 \\ c^2 + d^2 &= 1 \\ ac + bd &= 0. \end{aligned} \tag{13.8}$$

Further, $|Q| = 1$ or $|Q| = -1$; hence, $ad - bc = \pm 1$. This leads to two cases.

Case 1: *ad* − *bc* = 1 We now have four equations for a, b, c, and d, namely, equations (13.8), together with $ad - bc = 1$. After some algebra, which we omit, we find that these equations require that $a = d$ and $b = -c$. Thus, Q has the form

$$\begin{bmatrix} a & -c \\ c & a \end{bmatrix} \quad \text{or} \quad \begin{bmatrix} a & c \\ -c & a \end{bmatrix}.$$

Further, $|a| \leq 1$ and $c^2 = 1 - a^2$. Then, for some θ, $0 \leq \theta \leq \pi$ and $a = \cos(\theta)$ and $c = \sin(\theta)$. Thus, Q must have the form

$$\begin{bmatrix} \cos(\theta) & -\sin(\theta) \\ \sin(\theta) & \cos(\theta) \end{bmatrix} \quad \text{or} \quad \begin{bmatrix} \cos(\theta) & \sin(\theta) \\ -\sin(\theta) & \cos(\theta) \end{bmatrix}.$$

Case 2: *ad* − *bc* = −1 Now we find that Q must have the form

$$\begin{bmatrix} \cos(\theta) & \sin(\theta) \\ \sin(\theta) & -\cos(\theta) \end{bmatrix}.$$

Geometrically, these 2×2 orthogonal matrices are coefficient matrices of the equations of rotation of a coordinate system about the origin in the plane. If we write

$$\begin{bmatrix} \cos(\theta) & \sin(\theta) \\ \sin(\theta) & -\cos(\theta) \end{bmatrix} \begin{bmatrix} x \\ y \end{bmatrix} = \begin{bmatrix} X \\ Y \end{bmatrix},$$

then

$$X = x \cos(\theta) + y \sin(\theta)$$

and

$$Y = x \sin(\theta) - y \cos(\theta),$$

the familiar formulas from analytic geometry for a counterclockwise rotation through an angle θ about the origin. ∎

The relationship between 2×2 orthogonal matrices and rotations of coordinate systems suggests that orthogonal matrices might have important geometric interpretations and applications. We will look at orthogonal matrices and quadratic forms in the next section. For now, we will relate orthogonal matrices to diagonalization.

THEOREM 13.12

Any real, symmetric matrix can be diagonalized by an orthogonal matrix. ∎

This is a very strong theorem, asserting two things. First, any real, symmetric matrix can be diagonalized. The implication of this is that a real, symmetric $n \times n$ matrix always has n linearly independent eigenvectors (even if it has only one eigenvalue of multiplicity n). This is quite remarkable, since a real matrix with repeated eigenvalues may not be diagonalizable.

The second conclusion of the theorem is that a real, symmetric matrix can always be diagonalized by an orthogonal matrix. That is, there is an orthogonal matrix Q such that $Q^t A Q$ is a diagonal matrix.

We will not prove Theorem 13.12. The delicate part of the proof is in showing that a real, symmetric matrix always has n linearly independent eigenvectors, even if some eigenvalues have multiplicities greater than 1.

In practice, diagonalizing a real, symmetric matrix by an orthogonal matrix is not difficult if we can find the eigenvalues.

EXAMPLE 13.12

Let

$$A = \begin{bmatrix} 1 & 0 & \sqrt{2} \\ 0 & 2 & 0 \\ \sqrt{2} & 0 & 0 \end{bmatrix}.$$

A is a real, symmetric 3×3 matrix. The eigenvalues of A are 2, 2, and -1. Corresponding to -1, we find eigenvectors

$$\begin{bmatrix} \alpha \\ 0 \\ -\sqrt{2}\alpha \end{bmatrix}, \quad \alpha \neq 0.$$

To be specific, choose $\alpha = 1$ to get the eigenvector

$$\begin{bmatrix} 1 \\ 0 \\ -\sqrt{2} \end{bmatrix}.$$

The eigenvalue 2 has multiplicity 2. Corresponding to it, we find that eigenvectors have the form

$$\begin{bmatrix} \sqrt{2}\,\gamma \\ \beta \\ \gamma \end{bmatrix},$$

with γ and β arbitrary but not both zero. The important thing to notice is that we can find two linearly independent eigenvectors associated with 2 by making different choices of β and γ. If we choose $\beta = 1$ and $\gamma = 0$, we get the eigenvector

$$\begin{bmatrix} 0 \\ 1 \\ 0 \end{bmatrix}.$$

And by choosing $\beta = 0$ and $\gamma = 1$, we get the eigenvector

$$\begin{bmatrix} \sqrt{2} \\ 0 \\ 1 \end{bmatrix}.$$

We now have three eigenvectors of A:

$$\begin{bmatrix} 1 \\ 0 \\ -\sqrt{2} \end{bmatrix}, \quad \begin{bmatrix} 0 \\ 1 \\ 0 \end{bmatrix}, \quad \text{and} \quad \begin{bmatrix} \sqrt{2} \\ 0 \\ 1 \end{bmatrix}.$$

These are mutually orthogonal vectors in R^3. We can use these eigenvectors to form the columns of a matrix P which diagonalizes A. However, if we divide each of these eigenvectors by its length, we get three mutually orthogonal unit eigenvectors,

$$\begin{bmatrix} 1/\sqrt{3} \\ 0 \\ -\sqrt{\frac{2}{3}} \end{bmatrix}, \quad \begin{bmatrix} 0 \\ 1 \\ 0 \end{bmatrix}, \quad \text{and} \quad \begin{bmatrix} \sqrt{\frac{2}{3}} \\ 0 \\ 1/\sqrt{3} \end{bmatrix}.$$

These can be used to form the columns of an orthogonal matrix Q which diagonalizes A,

$$Q = \begin{bmatrix} 1/\sqrt{3} & 0 & \sqrt{\frac{2}{3}} \\ 0 & 1 & 0 \\ -\sqrt{\frac{2}{3}} & 0 & 1/\sqrt{3} \end{bmatrix}.$$

It is routine to check that

$$Q^t A Q = \begin{bmatrix} -1 & 0 & 0 \\ 0 & 2 & 0 \\ 0 & 0 & 2 \end{bmatrix}. \quad \blacksquare$$

PROBLEMS FOR SECTION 13.4

In each of Problems 1 through 4, verify that the matrix is orthogonal.

1. $\begin{bmatrix} \sqrt{3}/2 & -\frac{1}{2} \\ \frac{1}{2} & \sqrt{3}/2 \end{bmatrix}$

2. $\begin{bmatrix} 1/\sqrt{2} & 0 & 1/\sqrt{2} \\ 0 & 1 & 0 \\ -1/\sqrt{2} & 0 & 1/\sqrt{2} \end{bmatrix}$

3. $\begin{bmatrix} 1/\sqrt{5} & 2/\sqrt{5} & 0 \\ -2/\sqrt{5} & 1/\sqrt{5} & 0 \\ 0 & 0 & 1 \end{bmatrix}$

4. $\begin{bmatrix} 1/\sqrt{3} & -\sqrt{\frac{2}{3}} & 0 \\ 1/\sqrt{3} & 1/\sqrt{6} & -1/\sqrt{2} \\ 1/\sqrt{3} & 1/\sqrt{6} & 1/\sqrt{2} \end{bmatrix}$

In each of Problems 5 through 12, find an orthogonal matrix that diagonalizes the given matrix.

5. $\begin{bmatrix} -2 & 1 \\ 1 & 3 \end{bmatrix}$

6. $\begin{bmatrix} 4 & -2 \\ -2 & 1 \end{bmatrix}$

7. $\begin{bmatrix} -2 & 6 \\ 6 & -2 \end{bmatrix}$

8. $\begin{bmatrix} 0 & 1 & 1 \\ 1 & 2 & 0 \\ 1 & 0 & 2 \end{bmatrix}$

9. $\begin{bmatrix} 0 & 0 & 0 \\ 0 & 1 & -2 \\ 0 & -2 & 0 \end{bmatrix}$

10. $\begin{bmatrix} 1 & 3 & 0 \\ 3 & 0 & 1 \\ 0 & 1 & 1 \end{bmatrix}$

11. $\begin{bmatrix} 0 & 0 & 0 & 0 \\ 0 & 1 & -2 & 0 \\ 0 & -2 & 1 & 0 \\ 0 & 0 & 0 & 0 \end{bmatrix}$

12. $\begin{bmatrix} 5 & 0 & 0 & 0 \\ 0 & 0 & -1 & 0 \\ 0 & -1 & 0 & 0 \\ 0 & 0 & 0 & 0 \end{bmatrix}$

13. A real, symmetric matrix is said to be *positive definite* if its eigenvalues are all positive. Let A be a real, symmetric matrix. Prove that A is positive definite if and only if there is a nonsingular matrix P such that $A = P^t P$.

14. Use the result of Problem 13 to show that every real, nonsingular matrix A can be written in the form $A = QS$, where S is real, symmetric and positive definite and Q is an orthogonal matrix. *Hint:* First note that $A'A$ is positive definite and symmetric. For some orthogonal matrix W, $W'(A'A)W = D$, a diagonal matrix whose main diagonal elements are the eigenvalues of $A'A$. Let D_1 be the diagonal matrix whose main diagonal elements are the square roots of these eigenvalues. Show that we can choose $S = WD_1W'$ and $Q = AS^{-1}$.

13.5 *Orthogonal Matrices and Real, Quadratic Forms* _____

A *real quadratic form* in $x_1, \ldots, x_n$ is a polynomial

$$\sum_{j=1}^{n} \sum_{i=1}^{n} \alpha_{ij} x_i x_j$$

in which the α_{ij}'s are real numbers. The general quadratic form in x_1 and x_2 ($n = 2$) has the appearance

$$\alpha_{11} x_1^2 + \alpha_{12} x_1 x_2 + \alpha_{21} x_2 x_1 + \alpha_{22} x_2^2.$$

We would normally combine the $x_1 x_2$ and $x_2 x_1$ terms (assuming that x_1 and x_2 commute) and write this form as

$$\alpha_{11} x_1^2 + (\alpha_{12} + \alpha_{21}) x_1 x_2 + \alpha_{22} x_2^2.$$

As a specific example, we might have the form

$$8x_1^2 + 4x_1 x_2 - x_2^2.$$

Similarly, if $n = 3$, we obtain the quadratic form in three variables

$$\alpha_{11} x_1^2 + \alpha_{12} x_1 x_2 + \alpha_{13} x_1 x_3 + \alpha_{21} x_2 x_1 + \alpha_{22} x_2^2 + \alpha_{23} x_2 x_3$$
$$+ \alpha_{31} x_3 x_1 + \alpha_{32} x_3 x_2 + \alpha_{33} x_3^2,$$

which we can write as

$$\alpha_{11} x_1^2 + \alpha_{22} x_2^2 + \alpha_{33} x_2^3 + (\alpha_{12} + \alpha_{21}) x_1 x_2 + (\alpha_{13} + \alpha_{31}) x_1 x_3 + (\alpha_{23} + \alpha_{32}) x_2 x_3.$$

A specific example is

$$-14x_1^2 + 8x_1 x_2 + 3x_1 x_3 - 4x_2 x_3 + 5x_3^2.$$

Quadratic forms arise in a variety of ways. In physics, the kinetic energy of a particle is a quadratic form, and in analytic geometry, a conic such as $x^2 + y^2 = 14$, or $x^2 - 4xy + 6y^2 = 9$, may be thought of as having a quadratic form on one side of the equation.

As we can see from the quadratic forms we have written out, the quadratic form $\sum_{j=1}^{n} \sum_{i=1}^{n} \alpha_{ij} x_i x_j$ will, in general, contain *square terms*

$$\alpha_{11} x_1^2, \qquad \alpha_{22} x_2^2, \qquad \ldots, \qquad \alpha_{nn} x_n^2,$$

and *mixed products* of the form

$$(\alpha_{ij} + \alpha_{ji}) x_i x_j \qquad (i \neq j).$$

We can write quadratic forms in matrix notation as follows. Let

$$X = \begin{bmatrix} x_1 \\ x_2 \\ \vdots \\ x_n \end{bmatrix},$$

an $n \times 1$ matrix. Then

$$X^t = [x_1 x_2 \cdots x_n],$$

a $1 \times n$ matrix. Now define the $n \times n$ real, symmetric matrix $A = [a_{ij}]$ by letting

$$a_{ii} = \alpha_{ii} \quad \text{for} \quad i = 1, 2, \ldots, n$$

and

$$a_{ij} = \tfrac{1}{2}(\alpha_{ij} + \alpha_{ji}) \quad \text{for} \quad i, j = 1, 2, \ldots, n \qquad (i \neq j).$$

A routine calculation shows that

$$X^t A X = \left[\sum_{j=1}^{n} \sum_{i=1}^{n} \alpha_{ij} x_i x_j \right],$$

a 1×1 matrix whose single element is the quadratic form. We will adopt the notational convenience of writing $X^t A X = \sum_{j=1}^{n} \sum_{i=1}^{n} \alpha_{ij} x_i x_j$, identifying this 1×1 matrix with its single element. For example, $8x_1^2 + 4x_1 x_2 - x_2^2$ can be written as

$$8x_1^2 + 2x_1 x_2 + 2x_2 x_1 - x_2^2,$$

and this can be thought of as

$$[x_1 x_2] \begin{bmatrix} 8 & 2 \\ 2 & -1 \end{bmatrix} \begin{bmatrix} x_1 \\ x_2 \end{bmatrix}.$$

Similarly, $-14x_1^2 + 8x_1 x_2 + 3x_1 x_3 - 4x_2 x_3 + 5x_3^2$ can be written as

$$-14x_1^2 + 4x_1 x_2 + 4x_2 x_1 + \tfrac{3}{2} x_1 x_3 + \tfrac{3}{2} x_3 x_1 - 2x_2 x_3 - 2x_3 x_2 + 0x_2^2 + 5x_3^2,$$

and this is

$$[x_1 x_2 x_3] \begin{bmatrix} -14 & 4 & \tfrac{3}{2} \\ 4 & 0 & -2 \\ \tfrac{3}{2} & -2 & 5 \end{bmatrix} \begin{bmatrix} x_1 \\ x_2 \\ x_3 \end{bmatrix}.$$

In many problems involving quadratic forms, we want to transform from the $(x_1, \ldots, x_n)$ coordinate system to a coordinate system $(y_1, \ldots, y_n)$ in which the new quadratic form (in terms of $y_1, \ldots, y_n$) has no mixed product terms. That is, we want to write

$$\sum_{j=1}^{n} \sum_{i=1}^{n} \alpha_{ij} x_i x_j = \beta_1 y_1^2 + \beta_2 y_2^2 + \cdots + \beta_n y_n^2, \tag{13.9}$$

a new quadratic form containing only square terms. This is commonly done in analytic geometry, where a rotation of axes is used to eliminate cross product xy-terms in a conic section. For example, given the quadratic form

$$x_1^2 - 2x_1x_2 + x_2^2,$$

put

$$x_1 = \frac{1}{\sqrt{2}} y_1 + \frac{1}{\sqrt{2}} y_2$$

$$x_2 = \frac{1}{\sqrt{2}} y_1 - \frac{1}{\sqrt{2}} y_2.$$

(13.10)

We obtain

$$x_1^2 - 2x_1x_2 + x_2^2 = 2y_2^2,$$

and $2y_2^2$ is a much simpler quadratic form to deal with than the original form in x_1 and x_2. For example, a conic $x_1^2 - 2x_1x_2 + x_2^2 = K$ could be analyzed in terms of the simpler rotated conic $2y_2^2 = K$.

We will now show that there always exists a change of variables transforming $\sum_{j=1}^{n} \sum_{i=1}^{n} \alpha_{ij}x_ix_j$ into a simpler quadratic form $\beta_1 y_1^2 + \beta_2 y_2^2 + \cdots + \beta_n y_n^2$. In fact, we can determine the transformation explicitly if we can find the eigenvalues of A.

THEOREM 13.13

Let A be a real, symmetric matrix with eigenvalues $\lambda_1, \ldots, \lambda_n$. Let Q be an orthogonal matrix that diagonalizes A. Then the change of coordinates

$$X = QY$$

transforms $\sum_{j=1}^{n} \sum_{i=1}^{n} \alpha_{ij}x_ix_j$ into $\lambda_1 y_1^2 + \lambda_2 y_2^2 + \cdots + \lambda_n y_n^2$.

Note that the eigenvalues of A need not be distinct, since Theorem 13.12 in the preceding section guarantees existence of an orthogonal matrix diagonalizing A as long as A is real and symmetric.

Proof   The proof is a straightforward calculation. Write

$$X^tAX = (QY)^tA(QY) = (Y^tQ^t)A(QY) = Y^t(Q^tAQ)Y$$

$$= Y^t \begin{bmatrix} \lambda_1 & & & \\ & \lambda_2 & & O \\ & & \ddots & \\ O & & & \lambda_n \end{bmatrix} Y = \lambda_1 y_1^2 + \lambda_2 y_2^2 + \cdots + \lambda_n y_n^2. \quad \blacksquare$$

Theorem 13.13 is called the *principal axis theorem* because it defines new axes (the principal axes) with respect to which the quadratic form has a particularly simple appearance. The quadratic form $\lambda_1 y_1^2 + \lambda_2 y_2^2 + \cdots + \lambda_n y_n^2$ is called the *standard form* of $\sum_{j=1}^{n} \sum_{i=1}^{n} \alpha_{ij}x_ix_j$.

EXAMPLE 13.13

Consider again the quadratic form $x_1^2 - 2x_1x_2 + x_2^2$. Previously, we defined a change of variables (13.10) which transformed this into $2y_2^2$. We will now see how we got this change of variables. Write the quadratic form as $X'AX$, with $A = \begin{bmatrix} 1 & -1 \\ -1 & 1 \end{bmatrix}$.

A is real and symmetric and has eigenvalues zero and 2. Corresponding eigenvectors are

$$\begin{bmatrix} 1/\sqrt{2} \\ 1/\sqrt{2} \end{bmatrix} \quad \text{and} \quad \begin{bmatrix} 1/\sqrt{2} \\ -1/\sqrt{2} \end{bmatrix}.$$

These vectors have already been divided by their lengths, so we have orthonormal eigenvectors. Form the orthogonal matrix

$$Q = \begin{bmatrix} 1/\sqrt{2} & 1/\sqrt{2} \\ 1/\sqrt{2} & -1/\sqrt{2} \end{bmatrix}.$$

Then Q diagonalizes A. More important for present purposes, it defines the transformation $X = QY$, which transforms $x_1^2 - 2x_1x_2 + x_2^2$ into $\lambda_1 y_1^2 + \lambda_2 y_2^2$. Since $\lambda_1 = 0$ and $\lambda_2 = 2$, this standard form is just $2y_2^2$.

If we write the matrix product $X = QY$ as

$$\begin{bmatrix} x_1 \\ x_2 \end{bmatrix} = \begin{bmatrix} 1/\sqrt{2} & 1/\sqrt{2} \\ 1/\sqrt{2} & -1/\sqrt{2} \end{bmatrix} \begin{bmatrix} y_1 \\ y_2 \end{bmatrix} = \begin{bmatrix} \dfrac{1}{\sqrt{2}} y_1 + \dfrac{1}{\sqrt{2}} y_2 \\ \dfrac{1}{\sqrt{2}} y_1 - \dfrac{1}{\sqrt{2}} y_2 \end{bmatrix},$$

then we can read the equations of transformation (13.10) stated previously without justification.

Look at this example geometrically. For any constant $K > 0$, the points (x_1, x_2) satisfying $x_1^2 - 2x_1x_2 + x_2^2 = K$ form a conic in the plane. This conic is easier to

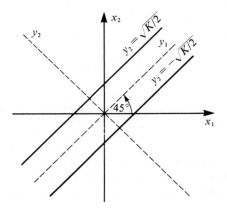

Figure 13.1

visualize when put into standard form $2y_2^2 = K$, or $y_2 = \pm(1/\sqrt{2})K$. The graph of this equation is immediately recognized as two straight lines. Figure 13.1 shows the original and rotated coordinate systems and the graph of the conic. ∎

EXAMPLE 13.14

Analyze the conic $4x_1^2 - 3x_1x_2 + 2x_2^2 = 8$.

We can write this equation as $X'AX = 8$, with

$$A = \begin{bmatrix} 4 & -\frac{3}{2} \\ -\frac{3}{2} & 2 \end{bmatrix}.$$

The eigenvalues of A are $(6 \pm \sqrt{13})/2$. Even if we do not explicitly find the eigenvectors and the diagonalizing orthogonal matrix Q, we know that Q exists so that $X = QY$ transforms the given quadratic form into standard form $\lambda_1 y_1^2 + \lambda_2 y_2^2$, or

$$\frac{6 + \sqrt{13}}{2} y_1^2 + \frac{6 - \sqrt{13}}{2} y_2^2 = 8.$$

This locus is an ellipse, whose graph is sketched in Figure 13.2.

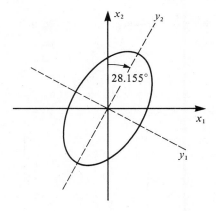

Figure 13.2 ∎

EXAMPLE 13.15

Consider the quadratic form in three variables $x_1^2 + 2x_2^2 + 2\sqrt{2}x_1x_3$. This is $X'AX$, with

$$A = \begin{bmatrix} 1 & 0 & \sqrt{2} \\ 0 & 2 & 0 \\ \sqrt{2} & 0 & 0 \end{bmatrix}.$$

Eigenvalues of A are $-1, 2,$ and 2. If Q is an orthogonal matrix that diagonalizes A, the transformation $X = QY$ puts the quadratic form into the standard form

$$-y_1^2 + 2y_2^2 + 2y_3^2.$$

In fact, we can find that

$$Q = \begin{bmatrix} 1/\sqrt{3} & 0 & \sqrt{\frac{2}{3}} \\ 0 & 1 & 0 \\ -\sqrt{\frac{2}{3}} & 0 & 1/\sqrt{3} \end{bmatrix}.$$

In terms of individual coordinates, the transformation is

$$x_1 = \frac{1}{\sqrt{3}} y_1 + \sqrt{\frac{2}{3}} y_3$$

$$x_2 = y_2$$

$$x_3 = -\sqrt{\frac{2}{3}} y_1 + \frac{1}{\sqrt{3}} y_3.$$

We can check by direct substitution that these equations transform $x_1^2 + 2x_2^2 + 2\sqrt{2}x_1x_3$ into $-y_1^2 + 2y_2^2 + 2y_3^2$. ∎

PROBLEMS FOR SECTION 13.5

In each of Problems 1 through 6, find A such that the quadratic form can be written as $X'AX$.

1. $x_1^2 + 2x_1x_2 + 6x_2^2$

2. $3x_1^2 + 3x_2^2 - 4x_1x_2 - 3x_1x_2 + 2x_1x_2 + x_3^2$

3. $x_1^2 - 4x_1x_2 + x_2^2$

4. $2x_1^2 - x_2^2 + 2x_1x_2$

5. $-x_1^2 + x_4^2 - 2x_1x_4 + 3x_2x_4 - x_1x_3 + 4x_2x_3$

6. $x_1^2 - x_2^2 - x_1x_3 + 4x_2x_3$

In each of Problems 7 through 12, write the quadratic form $X'AX$ defined by the given real, symmetric matrix.

7. $\begin{bmatrix} -2 & 1 \\ 1 & 6 \end{bmatrix}$

8. $\begin{bmatrix} 14 & -3 & 0 \\ -3 & 2 & 1 \\ 0 & 1 & 7 \end{bmatrix}$

9. $\begin{bmatrix} 6 & 1 & -7 \\ 1 & 2 & 0 \\ -7 & 0 & 1 \end{bmatrix}$

10. $\begin{bmatrix} 0 & -2 & 1 \\ -2 & 1 & 3 \\ 1 & 3 & 14 \end{bmatrix}$

11. $\begin{bmatrix} 8 & 0 & 0 \\ 0 & 2 & -4 \\ 0 & -4 & 3 \end{bmatrix}$

12. $\begin{bmatrix} 7 & 1 & -2 \\ 1 & 0 & -1 \\ -2 & -1 & 3 \end{bmatrix}$

In each of Problems 13 through 19, find the standard form of the quadratic form.

13. $-5x_1^2 + 4x_1x_2 + 3x_2^2$ **14.** $4x_1^2 - 12x_1x_2 + x_2^2$ **15.** $-3x_1^2 + 4x_1x_2 + 7x_2^2$ **16.** $4x_1^2 - 4x_1x_2 + x_2^2$

17. $-6x_1x_2 + 4x_2^2$ **18.** $5x_1^2 + 4x_1x_2 + 2x_2^2$ **19.** $-2x_1x_2 + 2x_3^2$

In each of Problems 20 through 24, obtain the standard form of the quadratic form and identify the conic as an ellipse, a parabola, a hyperbola, or a straight line or lines.

20. $x_1^2 - 2x_1x_2 + 4x_2^2 = 6$ **21.** $3x_1^2 + 5x_1x_2 - 3x_2^2 = 5$ **22.** $-2x_1^2 + 3x_2^2 + x_1x_2 = 5$

23. $4x_1^2 - 4x_2^2 + 6x_1x_2 = 8$ **24.** $6x_1^2 + 2x_1x_2 + 5x_2^2 = 14$

25. Let A be a real, symmetric matrix. Prove that there is a matrix P such that $X = PY$ transforms $X'AX$ into the form $y_1^2 + \cdots + y_p^2 - y_{p+1}^2 - \cdots - y_r^2$, where $r = \text{rank}(A)$ and p is the number of positive eigenvalues of A. This quadratic form in y is called the *canonical form* of $X'AX$. *Note:* We do not require that P be an orthogonal matrix.

26. Prove that the quadratic forms $X'AX$ and $X'BX$ have the same canonical form (see Problem 25) if and only if, for some matrix P, $B = P^{-1}AP$.

27. Let A be a real, symmetric matrix. A is *positive definite* if all the eigenvalues of A are positive (see Problem 13 of Section 13.4). Prove that A is positive definite if and only if $X^t A X > 0$ for every nonzero vector ($n \times 1$ matrix) X.

13.6 Unitary, Hermitian, and Skew-Hermitian Matrices _____

In this section, we will look at three other types of matrices which are frequently encountered. Recall that an orthogonal matrix A is a real matrix having the property that $A^{-1} = A^t$. We saw that this condition is equivalent to the assertion that the row (or column) vectors of A are orthonormal vectors in R^n. We will now consider the complex analogue of the orthogonal matrix.

Let U be a square matrix with complex elements (some or all of which may be real). As we have done before, let $\bar{U}$ denote the conjugate of U. We have noted that U is a real matrix if and only if $U = \bar{U}$; that $(\overline{\bar{U}}) = U$; and that $(\bar{U})^t = (\overline{U^t})$. We will usually write $(\bar{U})^t$ as just $\bar{U}^t$.

If U is nonsingular, it is not difficult to show that $\overline{(U)^{-1}} = (\bar{U})^{-1}$. In view of this equality, we will usually write the inverse of the conjugate as just $\bar{U}^{-1}$.

We call U a *unitary matrix* if

$$\bar{U}^{-1} = U^t.$$

That is, a complex, square matrix U is unitary if the inverse of the conjugate of U equals the transpose of U.

If U is real, $\bar{U} = U$, and a real, unitary matrix is orthogonal.

In general, the row and column vectors of an $n \times n$ unitary matrix are not in R^n because the coordinates are complex. We can, however, define a kind of dot product (often called an *inner product*) for complex vectors by putting

$$\langle x_1, \ldots, x_n \rangle \cdot \langle y_1, \ldots, y_n \rangle = \sum_{j=1}^{n} \bar{x}_j y_j.$$

This inner product reduces to the usual dot product when the vector components are real.

If we write

$$X = \begin{bmatrix} x_1 \\ x_2 \\ \vdots \\ x_n \end{bmatrix} \quad \text{and} \quad Y = \begin{bmatrix} y_1 \\ y_2 \\ \vdots \\ y_n \end{bmatrix},$$

the inner product of X with Y can be written $\bar{X}^t Y$, in which we again identify the 1×1 matrix with its single element and write it as a number.

Complex n-vectors $\mathbf{F}_1, \ldots, \mathbf{F}_n$ (that is, n-tuples $\langle x_1, \ldots, x_n \rangle$ with complex components) form a *unitary system* if

$$\mathbf{F}_i \cdot \mathbf{F}_j = \begin{cases} 0 & \text{if} \quad i \neq j \\ 1 & \text{if} \quad i = j. \end{cases}$$

Thus, a unitary system is the complex analogue of a set of n orthonormal vectors in R^n.

Here is the analogue of Theorem 13.10 and its corollary in Section 13.4.

THEOREM 13.14

Let U be an $n \times n$ complex matrix. Then U is unitary if and only if the rows (or columns) of U form a unitary system. ■

The proof is like that of Theorem 13.10 in Section 13.4, and we omit the details.

Unitary matrices generalize orthogonal matrices to the complex case. In similar fashion, hermitian matrices generalize real, symmetric matrices. Recall that a matrix A is symmetric if $A = A^t$. We define an $n \times n$ complex matrix H to be *hermitian* if

$$\bar{H} = H^t.$$

A real, hermitian matrix is symmetric because $\bar{H} = H = H^t$ if H is real.

EXAMPLE 13.16

Let

$$H = \begin{bmatrix} 15 & 8i & 6 - 2i \\ -8i & 0 & -4 + i \\ 6 + 2i & -4 - i & -3 \end{bmatrix}.$$

Then H is hermitian because

$$\bar{H} = \begin{bmatrix} 15 & -8i & 6 + 2i \\ 8i & 0 & -4 - i \\ 6 - 2i & -4 + i & -3 \end{bmatrix} = H^t. \quad ■$$

If S is an $n \times n$ complex matrix, we call S *skew-hermitian* if $\bar{S} = -S^t$. This definition generalizes the concept of skew-symmetric matrices mentioned in Problem 23 of Section 12.10.

EXAMPLE 13.17

Let

$$S = \begin{bmatrix} 0 & 8i & 2i \\ 8i & 0 & 4i \\ 2i & 4i & 0 \end{bmatrix}.$$

Then S is skew-hermitian because

$$\bar{S} = \begin{bmatrix} 0 & -8i & -2i \\ -8i & 0 & -4i \\ -2i & -4i & 0 \end{bmatrix} = -S^t. \quad ■$$

Recall from the preceding section that a real, quadratic form can be written $X^t A X$, with A a real, symmetric matrix. By the same token, we can consider the *hermitian form* $\bar{Z}^t H Z$, in which H is an hermitian matrix and

$$Z = \begin{bmatrix} z_1 \\ z_2 \\ \vdots \\ z_n \end{bmatrix}.$$

Of course, $\bar{Z}^t HZ$ is a 1×1 matrix, and we are identifying the matrix with its single element. If it is written as a summation,

$$\bar{Z}^t HZ = \sum_{j=1}^{n} \sum_{k=1}^{n} h_{jk} \bar{z}_j z_k.$$

Similarly, we can consider the *skew-hermitian form* $\bar{Z}^t SZ$, in which S is a skew-hermitian matrix.

THEOREM 13.15

Let

$$Z = \begin{bmatrix} z_1 \\ z_2 \\ \vdots \\ z_n \end{bmatrix}$$

be a complex $n \times 1$ matrix. Then

1. $\bar{Z}^t HZ$ is real if H is hermitian.
2. $\bar{Z}^t SZ$ is zero or pure imaginary if S is skew-hermitian. (By pure imaginary, we mean a complex number αi, with α real.)

Proof of (1)   Let H be hermitian. Then $\bar{H}^t = H$, so

$$(\overline{\bar{Z}^t HZ}) = \overline{\bar{Z}^t} \bar{H}\bar{Z} = Z^t \bar{H}\bar{Z}.$$

Now, $Z^t \bar{H}\bar{Z}$ is a 1×1 matrix and hence equals its own transpose. Continuing from the previous equation, we have

$$Z^t \bar{H}\bar{Z} = (Z^t \bar{H}\bar{Z})^t = \bar{Z}^t \bar{H}^t (Z^t)^t = \bar{Z}^t HZ.$$

Thus,

$$(\overline{\bar{Z}^t HZ}) = \bar{Z}^t HZ,$$

proving that $\bar{Z}^t HZ$ is real.

Proof of (2)   Let S be skew-hermitian. Then $\bar{S}^t = -S$. Imitate the argument used in proving conclusion (1) to obtain

$$(\overline{\bar{Z}^t SZ}) = -\bar{Z}^t SZ. \tag{13.11}$$

Write $\bar{Z}^t SZ = \alpha + i\beta$. Then equation (13.11) tells us that

$$\alpha - i\beta = -(\alpha + i\beta).$$

But then $\alpha = -\alpha$, so $\alpha = 0$, and therefore $\bar{Z}^t SZ$ is pure imaginary.   ∎

Using Theorem 13.15, we can draw an interesting conclusion about eigenvalues of hermitian and skew-hermitian matrices. We will also examine a result about eigenvalues of unitary matrices.

THEOREM 13.16

1. The eigenvalues of a unitary matrix have absolute value 1.

2. The eigenvalues of a hermitian matrix are real.

3. The eigenvalues of a skew-hermitian matrix must be zero or pure imaginary.

Note: In (1), there is no implication that the eigenvalues of a unitary matrix must be real—only that their absolute value must be 1. Part (2) implies, but is stronger than, the result that the eigenvalues of a real, symmetric matrix are real (since every real, symmetric matrix is hermitian).

Proof of (1)   Let U be a unitary matrix. If λ is an eigenvalue with associated eigenvector Z,

$$UZ = \lambda Z.$$

Then

$$\bar{U}\bar{Z} = \bar{\lambda}\bar{Z},$$

so

$$(\bar{U}\bar{Z})^t = \bar{\lambda}(\bar{Z}^t).$$

Then

$$\bar{Z}^t\bar{U}^t = \bar{\lambda}\bar{Z}^t.$$

But U is unitary, so $\bar{U}^t = U^{-1}$. Therefore,

$$\bar{Z}^tU^{-1} = \bar{\lambda}\bar{Z}^t.$$

Multiply both sides of this equation on the right by UZ to get

$$\bar{Z}^tU^{-1}UZ = \bar{\lambda}\bar{Z}^tUZ.$$

But $UZ = \lambda Z$, and $\bar{Z}^tUU^{-1}Z = \bar{Z}^tZ$, so we have

$$\bar{Z}^tZ = \bar{\lambda}\bar{Z}^t\lambda Z,$$

or

$$\bar{Z}^tZ = \bar{\lambda}\lambda\bar{Z}^tZ.$$

Since $\bar{Z}^tZ \neq 0$, we conclude that $\bar{\lambda}\lambda = 1$; hence, $|\lambda|^2 = 1$, so $|\lambda| = 1$.

Proof of (2) and (3)   Let A be any $n \times n$ matrix, real or complex. If λ is an eigenvalue of A, then for some nonzero $n \times 1$ matrix Z we have $AZ = \lambda Z$. Then

$$\bar{Z}^tAZ = \bar{Z}^t(\lambda Z) = \lambda\bar{Z}^tZ.$$

If

$$Z = \begin{bmatrix} z_1 \\ z_2 \\ \vdots \\ z_n \end{bmatrix},$$

then

$$\bar{Z}^tZ = \bar{z}_1z_1 + \bar{z}_2z_2 + \cdots + \bar{z}_nz_n$$
$$= |z_1|^2 + |z_2|^2 + \cdots + |z_n|^2,$$

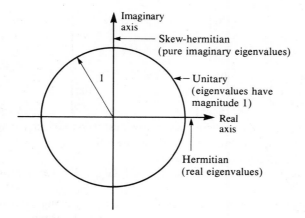

Figure 13.3. Eigenvalue locations.

which is real and nonzero. Therefore,

$$\lambda = \frac{\bar{Z}^t A Z}{Z^t Z}.$$

Now, if A is hermitian, $\bar{Z}^t A Z$ is real by Theorem 13.15; hence, λ is real, proving (2). If A is skew-hermitian, $\bar{Z}^t A Z$ is zero or pure imaginary, so λ is zero or pure imaginary, proving (3). ∎

The conclusions of Theorem 13.16 are represented graphically in Figure 13.3. Real numbers are plotted along the horizontal (real) axis, and pure imaginary numbers are plotted along the vertical (imaginary) axis. A complex number $a + ib$ is represented as a point (a, b) in the plane.

Eigenvalues of a unitary matrix may be real or complex, but they must have magnitude 1 and hence must lie on the unit circle about the origin. Eigenvalues of hermitian matrices must lie on the horizontal (real) axis, and eigenvalues of skew-hermitian matrices must lie on the vertical (imaginary) axis.

Since a real, symmetric matrix is hermitian, Theorem 13.16 again gives us the result that the eigenvalues of a real, symmetric matrix are real. Also, an orthogonal matrix is unitary, so Theorem 13.16 gives us the previously unstated result that the eigenvalues of an orthogonal matrix have magnitude 1.

PROBLEMS FOR SECTION 13.6

1. Let A be either hermitian, skew-hermitian, or unitary. Prove that $A\bar{A}^t = \bar{A}^t A$.

In each of Problems 2 through 10, determine whether the matrix is unitary, hermitian, skew-hermitian, or none of these. Find the eigenvalues of each matrix.

2. $\begin{bmatrix} 3 & -4i \\ 4i & 2 + 3i \end{bmatrix}$

3. $\begin{bmatrix} 3 & 4i \\ 4i & -5 \end{bmatrix}$

4. $\begin{bmatrix} 0 & 1 & 0 \\ -1 & 0 & 1 - i \\ 0 & -1 - i & 0 \end{bmatrix}$

5. $\begin{bmatrix} 1/\sqrt{2} & i/\sqrt{2} & 0 \\ -1/\sqrt{2} & i/\sqrt{2} & 0 \\ 0 & 0 & 1 \end{bmatrix}$ **6.** $\begin{bmatrix} 3 & 2 & 0 \\ 2 & 0 & i \\ 0 & -i & 0 \end{bmatrix}$ **7.** $\begin{bmatrix} -1 & 0 & 3-i \\ 0 & 1 & 0 \\ 3+i & 0 & 0 \end{bmatrix}$

8. $\begin{bmatrix} i & 1 & 0 \\ -1 & 0 & 2i \\ 0 & 2i & 0 \end{bmatrix}$ **9.** $\begin{bmatrix} 3i & 0 & 0 \\ -1 & 0 & i \\ 0 & -i & 0 \end{bmatrix}$ **10.** $\begin{bmatrix} 8 & -1 & i \\ -1 & 0 & 0 \\ -i & 0 & 0 \end{bmatrix}$

11. Prove that the main diagonal elements of a skew-hermitian matrix must be zero or pure imaginary.

12. Prove that the main diagonal elements of a hermitian matrix must be real.

13. Prove that the product of two $n \times n$ unitary matrices is unitary.

14. Suppose that A is both unitary and hermitian. Prove that $A = A^{-1}$. Give a 3×3 example of such a matrix (different from I_3).

15. Let A be any $n \times n$ matrix. Prove that A is the sum of a hermitian and a skew-hermitian matrix.

16. Let H be hermitian, and let α be any real number. Prove that αH is hermitian. Does this conclusion hold if α is complex but not real?

17. Let S be skew-hermitian, and let α be any real number. Prove that αS is skew-hermitian. Does this conclusion hold if α is complex but not real?

In each of Problems 18 through 22, compute $\bar{Z}^t A Z$ for the given matrices A and Z.

18. $A = \begin{bmatrix} 2 & 2-i \\ 2+i & 5 \end{bmatrix}$, $Z = \begin{bmatrix} -i \\ 2+i \end{bmatrix}$ **19.** $A = \begin{bmatrix} 5 & 3i & 4-i \\ -3i & 0 & 2-2i \\ 4+i & 2+2i & -3 \end{bmatrix}$, $Z = \begin{bmatrix} 1 \\ -5 \\ 3-i \end{bmatrix}$

20. $A = \begin{bmatrix} i & 3-i \\ -3-i & 0 \end{bmatrix}$, $Z = \begin{bmatrix} 2+5i \\ -3i \end{bmatrix}$ **21.** $A = \begin{bmatrix} 0 & -3i & 2+i \\ -3i & -4i & 0 \\ -2+i & 0 & 3i \end{bmatrix}$, $Z = \begin{bmatrix} 0 \\ -2i \\ 1-5i \end{bmatrix}$

22. $A = \begin{bmatrix} 4 & -1 & 2i \\ -1 & 0 & 6-i \\ -2i & 6+i & -5 \end{bmatrix}$, $Z = \begin{bmatrix} 3i \\ 4 \\ 0 \end{bmatrix}$

ADDITIONAL PROBLEMS

In each of Problems 1 through 10, find the eigenvalues of the matrix. Corresponding to each eigenvalue, find as many linearly independent eigenvectors as possible. (Many correct answers are possible for the eigenvectors.)

1. $\begin{bmatrix} 1 & -3 \\ -3 & 4 \end{bmatrix}$ **2.** $\begin{bmatrix} 1 & 0 & -1 \\ 0 & 0 & 0 \\ -1 & 0 & 4 \end{bmatrix}$ **3.** $\begin{bmatrix} i & 1 \\ 1 & 3i \end{bmatrix}$ **4.** $\begin{bmatrix} 2-3i & 0 \\ i & -5 \end{bmatrix}$

5. $\begin{bmatrix} 1-i & 2i \\ 0 & 4 \end{bmatrix}$ **6.** $\begin{bmatrix} i & 0 & -1 \\ -1 & 0 & 0 \\ 0 & 0 & 4i \end{bmatrix}$ **7.** $\begin{bmatrix} 3-2i & 0 \\ 1 & -4i \end{bmatrix}$ **8.** $\begin{bmatrix} -3 & 0 & 0 \\ 1 & -2 & 0 \\ 0 & 0 & -4 \end{bmatrix}$

9. $\begin{bmatrix} 1 & -5 & 3 \\ 0 & 0 & 0 \\ -2 & 0 & 0 \end{bmatrix}$ **10.** $\begin{bmatrix} -2 & 1 & -3 \\ 0 & 0 & 0 \\ 4 & 0 & 0 \end{bmatrix}$

In each of Problems 11 through 15, find an orthogonal matrix which diagonalizes the given matrix.

11. $\begin{bmatrix} 1 & -3 \\ -3 & 4 \end{bmatrix}$ **12.** $\begin{bmatrix} -2 & 0 \\ 0 & 1 \end{bmatrix}$ **13.** $\begin{bmatrix} -5 & 2 \\ 2 & 8 \end{bmatrix}$

14. $\begin{bmatrix} -1 & 2 & 0 \\ 2 & 5 & 0 \\ 0 & 0 & -4 \end{bmatrix}$
   15. $\begin{bmatrix} -4 & 0 & -3 \\ 0 & 1 & 0 \\ -3 & 0 & 0 \end{bmatrix}$

16. Let A and B be square matrices. Suppose that, for some nonsingular P, $P^{-1}AP = B$. Prove that A and B have the same eigenvalues. How are eigenvectors of A related to eigenvectors of B for a given eigenvalue?

17. Let A be a real, square matrix. Suppose that $a + ib$ is an eigenvalue of A. Prove that $a - ib$ is an eigenvalue of A also. Is this result true if A has complex elements?

18. Let A be an $n \times n$ diagonalizable matrix. Prove that there are $n \times n$ matrices $P_1, \ldots, P_n$ such that
   (a) $A = \lambda_1 P_1 + \lambda_2 P_2 + \cdots + \lambda_n P_n$, in which $\lambda_1, \ldots, \lambda_n$ are the eigenvalues of A.
   (b) $P_j^2 = P_j$ for $j = 1, 2, \ldots, n$.
   (c) $P_j P_k = O$ if $j \neq k$.
   (d) $P_1 + \cdots + P_n = I_n$.
   The sum $\lambda_1 P_1 + \lambda_2 P_2 + \cdots + \lambda_n P_n$ is called the *spectral decomposition* of A, and the matrices $P_1, \ldots, P_n$ are called *projections*. *Hint:* First write $P^{-1}AP = D$ for some nonsingular matrix P and a diagonal matrix D. Write

$$D = \lambda_1 T_{11} + \lambda_2 T_{22} + \cdots + \lambda_n T_{nn},$$

where T_{jj} is the $n \times n$ matrix with j, j element 1 and every other element zero. Then write $A = PDP^{-1}$.

19. Let A be an $n \times n$ real matrix. Prove that the coefficient of λ^n in the characteristic polynomial of A has coefficient 1.

20. Let A be an $n \times n$ diagonalizable matrix with characteristic equation

$$\lambda^n + a_{n-1}\lambda^{n-1} + \cdots + a_1\lambda + a_0 = 0.$$

Prove that

$$A^n + a_{n-1}A^{n-1} + \cdots + a_1 A + a_0 I_n = O,$$

where O is the $n \times n$ zero matrix. This result is a special case of the *Cayley-Hamilton theorem*, which is true in general for square matrices but is easier to prove in the case in which A is diagonalizable.

In each of Problems 21 through 25, find the standard form of the quadratic form.

21. $2x_1^2 - 6x_1x_2 + x_2^2$
22. $-4x_1^2 + 5x_1x_2 + x_2^2$
23. $-6x_1^2 + 4x_1x_2 + 3x_2^2$
24. $2x_1^2 - x_1x_2 + 3x_2x_3 - 4x_1x_3 + 2x_2^2 - x_3^2$
25. $-4x_1^2 + 3x_2^2 + 2x_1x_2 - x_3^2$

26. For each positive integer n, find an orthogonal matrix with determinant -1.

27. Let A and B be $n \times n$ matrices. Prove that AB and BA have the same eigenvalues. Will they have the same eigenvectors?

28. An $n \times n$ matrix A is called *nilpotent* if, for some positive integer k, $A^k = O$. Prove that a nilpotent matrix has zero as its only eigenvalue.

29. Let λ be an eigenvalue of the nonsingular matrix A.
   (a) Prove that $\lambda \neq 0$.
   (b) Prove that $1/\lambda$ is an eigenvalue of A^{-1}.

30. Suppose that zero is an eigenvalue of A. Is A necessarily singular? Can a singular matrix have all nonzero eigenvalues?

31. A square matrix A is called an *involution* if $A^2 = I_n$. What can be said about the eigenvalues of an involution?

32. Prove *Gerschgorin's theorem*: Let $A = [a_{ij}]$ be an $n \times n$ matrix, and let

$$r_k = \sum_{\substack{j=1 \\ j \neq k}}^{n} |a_{kj}|.$$

Let C_k be the circle of radius r_k centered at (α_k, β_k), where $a_{kk} = \alpha_k + i\beta_k$. Let λ be an eigenvalue of A. Then, plotted as a point in the plane, λ lies on or inside one of the circles $C_1, \ldots, C_n$. (This result can be used to obtain rough estimates on the magnitude of eigenvalues. The circles $C_1, \ldots, C_n$ are sometimes called *Gerschgorin circles* for the matrix.)

For each of the matrices of Problems 33 through 36, draw the Gerschgorin circles in the plane, and plot the eigenvalues to check that they fall within the region enclosed by the circles.

33. $\begin{bmatrix} 2 & 1 \\ 1 & 4 \end{bmatrix}$ 34. $\begin{bmatrix} i & 1 \\ 1 & 2i \end{bmatrix}$ 35. $\begin{bmatrix} 0 & 12 \\ -1 & 6 \end{bmatrix}$ 36. $\begin{bmatrix} -4 & 3 \\ 2 & 1+i \end{bmatrix}$

CHAPTER 14

Matrix Solution of Systems of Differential Equations

14.0 Introduction

Many problems of both practical and theoretical interest give rise to systems of linear differential equations. For example, mathematical models of series circuits and mechanical systems involving several springs attached in series can lead to systems of first order differential equations. It is also the case, as will be shown later, that any higher order linear differential equation can be related to a system of first order differential equations in such a way that the solution of the system enables us to find the solution of the higher order equation.

One reason for converting a higher order differential equation to a system of first order equations is that we may then bring matrix methods to bear to solve the system. The objectives of this chapter are to develop some of these methods and to show how the language and theory of matrices help us solve certain systems. It is also the case that some numerical schemes for approximating solutions of linear systems are posed in matrix terms; familiarity with this approach is a prerequisite to reading the literature and using such methods.

We will begin in the next section with the theoretical foundation, then develop methods for solving systems of first order linear differential equations.

14.1 Theory of Systems of Linear First Order Differential Equations

In this section, we will establish the theoretical basis for working with systems of first order differential equations. Systems containing higher order equations can be

759

transformed into larger systems containing just first order equations, as we will discuss in Section 14.8. Consider a system of differential equations

$$\frac{dx_1}{dt} = F_1(t, x_1, \ldots, x_n)$$

$$\frac{dx_2}{dt} = F_2(t, x_1, \ldots, x_n)$$

$$\vdots \qquad\qquad \vdots \qquad\qquad (14.1)$$

$$\frac{dx_n}{dt} = F_n(t, x_1, \ldots, x_n)$$

Here, t is the independent variable, and we want functions $x_1(t), \ldots, x_n(t)$ satisfying simultaneously the differential equations of the system. Initial conditions for this system have the form

$$x_1(t_0) = x_1^0, \qquad x_2(t_0) = x_2^0, \qquad \ldots, \qquad x_n(t_0) = x_n^0, \qquad (14.2)$$

in which $t_0, x_1^0, x_2^0, \ldots, x_n^0$ are given numbers.

The system (14.1) and the initial conditions (14.2) constitute an initial value problem. We will discuss a fundamental existence/uniqueness result for such a problem. By an open rectangular parallelepiped in the $(n + 1)$-dimensional space R^{n+1} having axes labeled $t, x_1, \ldots, x_n$, we mean the set of all points $(t, x_1, \ldots, x_n)$ whose coordinates satisfy inequalities $\alpha < t < \beta, a_1 < x_1 < b_1, \ldots, a_n < x_n < b_n$. This set generalizes the idea of an interval $a < t < b$ on the real line or an open rectangle $a < x < b, c < y < d$ in the plane. ("Open" in this context means that the endpoints or boundary points are not included.)

THEOREM 14.1 Existence/Uniqueness for the Initial Value Problem

Suppose that $F_1, \ldots, F_n$ are functions of the $n + 1$ variables $t, x_1, \ldots, x_n$. Suppose that each of $F_1, F_2, \ldots, F_n$ and their first partial derivatives are continuous in an open rectangular parallelepiped K of $(n + 1)$-dimensional space having axes $t, x_1, x_2, \ldots, x_n$. Suppose also that $(t_0, x_1^0, x_2^0, \ldots, x_n^0)$ is in K. Then there exists an interval $(t_0 - h, t_0 + h)$ in which the initial value problem consisting of the system (14.1) and the initial conditions (14.2) has a unique solution

$$x_1 = \varphi_1(t), \qquad x_2 = \varphi_2(t), \qquad \ldots, \qquad x_n = \varphi_n(t). \quad \blacksquare$$

The system (14.1) is too general to discuss in an introductory treatment, so we will immediately consider a special case which is rich in both theory and applications. Consider the *linear system*

$$x_1'(t) = a_{11}(t)x_1(t) + a_{12}(t)x_2(t) + \cdots + a_{1n}(t)x_n(t) + g_1(t)$$
$$x_2'(t) = a_{21}(t)x_1(t) + a_{22}(t)x_2(t) + \cdots + a_{2n}(t)x_n(t) + g_2(t)$$
$$\vdots \qquad\qquad\qquad \vdots$$
$$x_n'(t) = a_{n1}(t)x_1(t) + a_{n2}(t)x_2(t) + \cdots + a_{nn}(t)x_n(t) + g_n(t),$$

in which each differential equation is linear and of first order. We will write this system in matrix form. Let

$$X(t) = \begin{bmatrix} x_1(t) \\ x_2(t) \\ \vdots \\ x_n(t) \end{bmatrix}, \quad G(t) = \begin{bmatrix} g_1(t) \\ g_2(t) \\ \vdots \\ g_n(t) \end{bmatrix},$$

and

$$A(t) = \begin{bmatrix} a_{11}(t) & a_{12}(t) & \cdots & a_{1n}(t) \\ a_{21}(t) & a_{22}(t) & \cdots & a_{2n}(t) \\ \vdots & \vdots & & \vdots \\ a_{n1}(t) & a_{n2}(t) & \cdots & a_{nn}(t) \end{bmatrix}.$$

Now write the linear system as

$$X' = AX + G,$$

in which X, A, and G are matrix functions of t. For example, the system

$$x_1' = x_1 + tx_2 + \cos(t)$$
$$x_2' = t^3 x_1 - e^t x_2 + 1 - t$$

can be written as $X' = AX + G$, with

$$X = \begin{bmatrix} x_1 \\ x_2 \end{bmatrix}, \quad A = \begin{bmatrix} 1 & t \\ t^3 & -e^t \end{bmatrix}, \quad \text{and} \quad G = \begin{bmatrix} \cos(t) \\ 1 - t \end{bmatrix}.$$

The system $X' = AX + G$ is called *homogeneous* if each $g_j(t) = 0$ for all t in the relevant interval J. In this case, $G(t) = O$ for all t in J, and the system is

$$X' = AX.$$

If some $g_j(t)$ is nonzero for some t in J, the system $X' = AX + G$ is *nonhomogeneous.* Initial values can be written as an $n \times 1$ matrix:

$$X(t_0) = \begin{bmatrix} x_1^0 \\ x_2^0 \\ \vdots \\ x_n^0 \end{bmatrix}.$$

For convenience, we will usually denote the matrix on the right as X^0 and write the initial conditions as

$$X(t_0) = X^0.$$

X^0 is simply an $n \times 1$ matrix of constants giving the value of $X(t)$ at t_0.

The conclusion of Theorem 14.1 can be strengthened in the case in which the initial value problem involves a linear system.

THEOREM 14.2

Suppose that the functions a_{ij} and g are continuous on an open interval J containing t_0. Then the linear initial value problem

$$X' = AX + G; \quad X(t_0) = X^0$$

has a unique solution defined for all t in J. ∎

We will now develop a theory for the system $X' = AX + G$ which closely parallels that for nth order linear differential equations. As we did there, we will begin with homogeneous systems.

THEORY OF THE HOMOGENEOUS LINEAR SYSTEM $X' = AX$

It is straightforward to verify that a linear combination of a finite number of solutions of $X' = AX$ is also a solution.

THEOREM 14.3

Suppose that $\Phi_1, \ldots, \Phi_k$ are solutions of $X' = AX$. Then any linear combination of $\Phi_1, \ldots, \Phi_k$ is also a solution.

Proof If $\Phi_1, \ldots, \Phi_k$ are solutions of $X' = AX$, and $\alpha_1, \ldots, \alpha_k$ are any real numbers,

$$
\begin{aligned}
(\alpha_1 \Phi_1 + \alpha_2 \Phi_2 + \cdots + \alpha_k \Phi_k)' &= \alpha_1 \Phi_1' + \alpha_2 \Phi_2' + \cdots + \alpha_k \Phi_k' \\
&= \alpha_1 A\Phi_1 + \alpha_2 A\Phi_2 + \cdots + \alpha_k A\Phi_k \\
&= A(\alpha_1 \Phi_1 + \alpha_2 \Phi_2 + \cdots + \alpha_k \Phi_k),
\end{aligned}
$$

proving that $\alpha_1 \Phi_1 + \alpha_2 \Phi_2 + \cdots + \alpha_k \Phi_k$ is also a solution of $X' = AX$. ∎

From Theorem 14.3, it follows that the set of all solutions of $X' = AX$ has the structure of a vector space, with the usual addition of solutions and multiplication of a solution by a constant.

THEOREM 14.4

Let V be the set of all solutions of $X' = AX$. Then V is a vector space. ∎

A proof of Theorem 14.4 is a routine verification that V satisfies the requirements for a vector space listed in Section 11.7. The two main observations are that a sum of solutions of $X' = AX$ is a solution and that a real number times a solution is a solution.

The vector space of all solutions of $X' = AX$ is called the *solution space* of this system. If we can determine the dimension of this solution space, we will know how many linearly independent solutions to look for to form a basis. Once we have a basis, every solution is a linear combination of the solutions in the basis.

In looking for a basis for the solution space of $X' = AX$, it is important to have a test to determine when a collection of n solutions is linearly independent. The following theorem provides such a test.

THEOREM 14.5

Suppose that

$$
X^{(1)}(t) = \begin{bmatrix} x_{11}(t) \\ x_{21}(t) \\ \vdots \\ x_{n1}(t) \end{bmatrix}, \quad
X^{(2)}(t) = \begin{bmatrix} x_{12}(t) \\ x_{22}(t) \\ \vdots \\ x_{n2}(t) \end{bmatrix}, \quad \ldots, \quad
X^{(n)}(t) = \begin{bmatrix} x_{1n}(t) \\ x_{2n}(t) \\ \vdots \\ x_{nn}(t) \end{bmatrix}
$$

are solutions of $X' = AX$ on an open interval J. Then

1. $X^{(1)}, \ldots, X^{(n)}$ are linearly independent if and only if, for any t_0 in J,

$$X^{(1)}(t_0), \qquad \ldots, \qquad X^{(n)}(t_0)$$

are linearly independent when considered as vectors in R^n.

2. $X^{(1)}, \ldots, X^{(n)}$ are linearly independent if and only if, for any t_0 in J,

$$\begin{vmatrix} x_{11}(t_0) & x_{12}(t_0) & \cdots & x_{1n}(t_0) \\ x_{21}(t_0) & x_{22}(t_0) & \cdots & x_{2n}(t_0) \\ \vdots & \vdots & & \vdots \\ x_{n1}(t_0) & x_{n2}(t_0) & \cdots & x_{nn}(t_0) \end{vmatrix} \neq 0.$$

Proof of (1)   We will use Theorem 11.17 from Section 11.6. If $X^{(1)}, \ldots, X^{(n)}$ are linearly dependent, there are constants $c_1, \ldots, c_n$, not all zero, such that

$$c_1 X^{(1)}(t) + c_2 X^{(2)}(t) + \cdots + c_n X^{(n)}(t) = O$$

for all t in J. In particular,

$$c_1 X^{(1)}(t_0) + c_2 X^{(2)}(t_0) + \cdots + c_n X^{(n)}(t_0) = O,$$

and we conclude that the vectors $X^{(1)}(t_0), \ldots, X^{(n)}(t_0)$ are linearly dependent in R^n.

Conversely, suppose that, for some t_0 in J, $X^{(1)}(t_0), \ldots, X^{(n)}(t_0)$ are linearly dependent vectors in R^n. Then there are constants $c_1, \ldots, c_n$, not all zero, such that

$$c_1 X^{(1)}(t_0) + c_2 X^{(2)}(t_0) + \cdots + c_n X^{(n)}(t_0) = O$$

in R^n. Define an $n \times 1$ vector function Φ by

$$\Phi(t) = c_1 X^{(1)}(t) + c_2 X^{(2)}(t) + \cdots + c_n X^{(n)}(t)$$

for t in J. Then Φ is a solution of the initial value problem

$$X' = AX; \qquad X(t_0) = O.$$

But the zero function,

$$\psi(t) = \begin{bmatrix} 0 \\ 0 \\ \vdots \\ 0 \end{bmatrix}$$

for all t in J, is also a solution of this initial value problem. By the uniqueness conclusion of Theorem 14.1, $\Phi(t) = \psi(t)$ for all t in J. Therefore, $c_1 X^{(1)}(t) + c_2 X^{(2)}(t) + \cdots + c_n X^{(n)}(t) = O$ for all t in I. Since the coefficients $c_1, \ldots, c_n$ are not all zero, the functions $X^{(1)}, X^{(2)}, \ldots, X^{(n)}$ are linearly dependent in the solution space of $X' = AX$, completing the proof of (1).

Proof of (2)   Conclusion (2) follows from (1) and the fact that n vectors in R^n are linearly independent if and only if the determinant of the $n \times n$ matrix having these vectors as columns is nonzero.   ∎

Conclusion (2) of Theorem 14.5 provides an effective test for linear dependence or independence of n solutions of $X' = AX$ on an interval.

THEOREM 14.6

Suppose that the functions $a_{ij}(t)$ are continuous on an open interval J for $i = 1, \ldots, n$ and $j = 1, \ldots, n$, and let $A = [a_{ij}]$. Then the dimension of the solution space of $X' = AX$ is n.

Proof   We will produce a basis consisting of n solutions for this solution space. Define

$$E^{(1)} = \begin{bmatrix} 1 \\ 0 \\ 0 \\ \vdots \\ 0 \end{bmatrix}, \qquad E^{(2)} = \begin{bmatrix} 0 \\ 1 \\ 0 \\ \vdots \\ 0 \end{bmatrix}, \qquad \ldots, \qquad E^{(n)} = \begin{bmatrix} 0 \\ 0 \\ 0 \\ \vdots \\ 1 \end{bmatrix}.$$

Choose any t_0 in J. For $j = 1, 2, \ldots, n$, let $X^{(j)}$ be the unique solution of the initial value problem

$$X' = AX; \qquad X(t_0) = E^{(j)},$$

whose existence on J is guaranteed by Theorem 14.1. We claim that the solutions $X^{(1)}, X^{(2)}, \ldots, X^{(n)}$ of $X' = AX$ form a basis for the solution space of $X' = AX$ on J.

First, by conclusion (2) of Theorem 14.5, these solutions are linearly independent. Indeed, the $n \times n$ matrix having $X^{(1)}(t_0), \ldots, X^{(n)}(t_0)$ as columns is exactly the $n \times n$ identity matrix having determinant 1.

To complete the proof, we must show that any solution ξ of $X' = AX$ on J is a linear combination of $X^{(1)}, \ldots, X^{(n)}$. To do this, note that $\xi(t_0)$ is a vector in R^n, and write

$$\xi(t_0) = \begin{bmatrix} \xi_1 \\ \xi_2 \\ \vdots \\ \xi_n \end{bmatrix} = \xi_1 \begin{bmatrix} 1 \\ 0 \\ 0 \\ \vdots \\ 0 \end{bmatrix} + \xi_2 \begin{bmatrix} 0 \\ 1 \\ 0 \\ \vdots \\ 0 \end{bmatrix} + \cdots + \xi_n \begin{bmatrix} 0 \\ 0 \\ 0 \\ \vdots \\ 1 \end{bmatrix}$$

$$= \xi_1 X^{(1)}(t_0) + \xi_2 X^{(2)}(t_0) + \cdots + \xi_n X^{(n)}(t_0).$$

Thus, ξ is a solution of the initial value problem

$$X' = AX; \qquad X(t_0) = \xi_1 X^{(1)}(t_0) + \xi_2 X^{(2)}(t_0) + \cdots + \xi_n X^{(n)}(t_0).$$

But $\xi_1 X^{(1)} + \cdots + \xi_n X^{(n)}$ is also a solution of this initial value problem. By the uniqueness of solutions of this initial value problem, we conclude that

$$\xi = \xi_1 X^{(1)} + \xi_2 X^{(2)} + \cdots + \xi_n X^{(n)}.$$

Therefore, ξ is a linear combination of $X^{(1)}, X^{(2)}, \ldots, X^{(n)}$, and the proof is complete. ∎

The basis produced in proving this theorem is the simplest for the purpose of constructing a proof. However, in actually solving a given system, any set of n linearly independent solutions can be used to form a basis for the solution space.

Theorem 14.6 tells us what to look for in solving $X' = AX$. We attempt to find n linearly independent solutions. Every solution is a linear combination of these n solutions. For this reason, any set of n linearly independent solutions of $X' = AX$ is called a *fundamental set of solutions*.

If we have n linearly independent solutions $X^{(1)}, \ldots, X^{(n)}$, the general solution of $X' = AX$ has the form

$$c_1 X^{(1)} + c_2 X^{(2)} + \cdots + c_n X^{(n)},$$

in which $c_1, \ldots, c_n$ are arbitrary constants. This linear combination of $X^{(1)}$, $X^{(2)}, \ldots, X^{(n)}$ is often written in matrix form as follows. Form the $n \times 1$ matrix of arbitrary constants

$$C = \begin{bmatrix} c_1 \\ c_2 \\ \vdots \\ c_n \end{bmatrix}.$$

Now define a matrix Ω whose columns are the linearly independent solutions $X^{(1)}, X^{(2)}, \ldots, X^{(n)}$. A matrix formed in this way is called a *fundamental matrix* for the system $X' = AX$. If

$$X^{(1)}(t) = \begin{bmatrix} x_{11}(t) \\ x_{21}(t) \\ \vdots \\ x_{n1}(t) \end{bmatrix}, \qquad X^{(2)}(t) = \begin{bmatrix} x_{12}(t) \\ x_{22}(t) \\ \vdots \\ x_{n2}(t) \end{bmatrix}, \qquad \ldots, \qquad X^{(n)}(t) = \begin{bmatrix} x_{1n}(t) \\ x_{2n}(t) \\ \vdots \\ x_{nn}(t) \end{bmatrix},$$

then

$$\Omega(t) = \begin{bmatrix} x_{11}(t) & x_{12}(t) & \cdots & x_{1n}(t) \\ x_{21}(t) & x_{22}(t) & \cdots & x_{2n}(t) \\ \vdots & \vdots & & \vdots \\ x_{n1}(t) & x_{n2}(t) & \cdots & x_{nn}(t) \end{bmatrix}.$$

Now observe that

$$\Omega C = \begin{bmatrix} x_{11} & x_{12} & \cdots & x_{1n} \\ x_{21} & x_{22} & \cdots & x_{2n} \\ \vdots & \vdots & & \vdots \\ x_{n1} & x_{n2} & \cdots & x_{nn} \end{bmatrix} \begin{bmatrix} c_1 \\ c_2 \\ \vdots \\ c_n \end{bmatrix}$$

$$= \begin{bmatrix} c_1 x_{11} + c_2 x_{12} + \cdots + c_n x_{1n} \\ c_1 x_{21} + c_2 x_{22} + \cdots + c_n x_{2n} \\ \vdots \qquad \vdots \\ c_1 x_{n1} + c_2 x_{n2} + \cdots + c_n x_{nn} \end{bmatrix}$$

$$= c_1 \begin{bmatrix} x_{11} \\ x_{21} \\ \vdots \\ x_{n1} \end{bmatrix} + c_2 \begin{bmatrix} x_{12} \\ x_{22} \\ \vdots \\ x_{n2} \end{bmatrix} + \cdots + c_n \begin{bmatrix} x_{1n} \\ x_{2n} \\ \vdots \\ x_{nn} \end{bmatrix}$$

$$= c_1 X^{(1)} + c_2 X^{(2)} + \cdots + c_n X^{(n)}.$$

Thus, the general solution $c_1 X^{(1)} + c_2 X^{(2)} + \cdots + c_n X^{(n)}$ of $X' = AX$ can be written as just ΩC, where Ω is a fundamental matrix whose columns are the n linearly independent solutions $X^{(1)}, X^{(2)}, \ldots, X^{(n)}$ and C is an $n \times 1$ matrix of arbitrary constants. We will summarize this important result as a theorem.

THEOREM 14.7

Let Ω be a fundamental matrix of $X' = AX$ on J. Then the general solution of $X' = AX$ is $X = \Omega C$, where C is an $n \times 1$ matrix of arbitrary constants. ∎

We will illustrate the ideas we have just introduced.

EXAMPLE 14.1

Consider the system

$$x_1' = x_1 - 4x_2$$
$$x_2' = x_1 + 5x_2.$$

This system has general solution

$$x_1 = (-2c_1 + c_2)e^{3t} - 2c_2 t e^{3t}$$
$$x_2 = c_1 e^{3t} + c_2 t e^{3t}. \tag{14.3}$$

(One way to obtain this solution is to use the Laplace transform.) To illustrate the matrix notation we have been discussing, let

$$A = \begin{bmatrix} 1 & -4 \\ 1 & 5 \end{bmatrix}.$$

A is the matrix of coefficients of the system and in this example is a constant matrix. The system is $X' = AX$, where

$$X = \begin{bmatrix} x_1 \\ x_2 \end{bmatrix}.$$

We can write two linearly independent solutions as 2×1 matrices

$$X^{(1)} = \begin{bmatrix} -2e^{3t} \\ e^{3t} \end{bmatrix}$$

[choose $c_1 = 1$ and $c_2 = 0$ in the general solution (14.3)] and

$$X^{(2)} = \begin{bmatrix} (1 - 2t)e^{3t} \\ t e^{3t} \end{bmatrix}$$

[choose $c_1 = 0$ and $c_2 = 1$ in the general solution (14.3)].

We obtain a fundamental matrix by writing $X^{(1)}$ and $X^{(2)}$ as columns of a 2×2 matrix

$$\Omega = \begin{bmatrix} -2e^{3t} & (1 - 2t)e^{3t} \\ e^{3t} & t e^{3t} \end{bmatrix}.$$

Now the general solution (14.3) can be written in matrix form as simply ΩC, where

$$C = \begin{bmatrix} c_1 \\ c_2 \end{bmatrix}.$$

In this example, we will actually carry out this matrix product:

$$\Omega C = \begin{bmatrix} -2e^{3t} & (1 - 2t)e^{3t} \\ e^{3t} & te^{3t} \end{bmatrix} \begin{bmatrix} c_1 \\ c_2 \end{bmatrix}$$

$$= \begin{bmatrix} -2c_1 e^{3t} + (1 - 2t)c_2 e^{3t} \\ c_1 e^{3t} + c_2 te^{3t} \end{bmatrix}$$

$$= \begin{bmatrix} (-2c_1 + c_2)e^{3t} - 2c_2 te^{3t} \\ c_1 e^{3t} + c_2 te^{3t} \end{bmatrix}.$$

The general solution (14.3) is exactly reproduced in the rows of ΩC. ∎

THE NONHOMOGENEOUS LINEAR SYSTEM $X' = AX + G$

We will now discuss the nonhomogeneous linear system. Immediately we encounter a serious difference between this and the homogeneous case: the solutions of $X' = AX + G$ do not form a vector space (a sum of solutions is not a solution).

We can, however, state the following fundamental result, which is analogous to Theorem 3.5 for solutions of the nonhomogeneous nth order linear differential equation.

THEOREM 14.8

Let Ω be a fundamental matrix for the homogeneous system $X' = AX$, and let ξ be any solution of $X' = AX + G$. Then the general solution φ of $X' = AX + G$ is

$$\varphi = \Omega C + \xi,$$

in which C is an $n \times 1$ matrix of arbitrary constants. That is,

the general solution of the nonhomogeneous system

= (the general solution of the homogeneous system)

+ (any particular solution of the nonhomogeneous system).

Proof First, we will show that φ is a solution of $X' = AX + G$. Let the columns of Ω be the solutions $X^{(1)}, \ldots, X^{(n)}$ of $X' = AX$, and let

$$C = \begin{bmatrix} c_1 \\ c_2 \\ \vdots \\ c_n \end{bmatrix}.$$

Since ξ is a solution of $X' = AX + G$, $\xi' = A\xi + G$. Now calculate

$$\begin{aligned} \varphi' = (\Omega C + \xi)' &= \Omega' C + \xi' \\ &= c_1[X^{(1)}]' + c_2[X^{(2)}]' + \cdots + c_n[X^{(n)}]' + \xi' \\ &= c_1[AX^{(1)}] + c_2[AX^{(2)}] + \cdots + c_n[AX^{(n)}] + \xi' \\ &= (A\Omega)C + \xi' \\ &= (A\Omega)C + (A\xi + G) \\ &= A(\Omega C) + A\xi + G = A(\Omega C + \xi) + G = A\varphi + G. \end{aligned}$$

Therefore, φ is indeed a solution of $X' = AX + G$ for any $n \times 1$ constant matrix C. Now let ψ be any solution of $X' = AX + G$. Then

$$(\psi - \xi)' = \psi' - \xi' = (A\psi + G) - (A\xi + G) = A\psi - A\xi = A(\psi - \xi).$$

Therefore, $\psi - \xi$ is a solution of the homogeneous system $X' = AX$. Then, for some $n \times 1$ constant matrix C, $\psi - \xi = \Omega C$, and we have $\psi = \Omega C + \xi$, as was to be proved. ■

We now have the theoretical background with which to approach the system of linear differential equations $X' = AX + G$, with G zero or nonzero. In the next section, we will begin to examine methods for actually producing solutions in specific cases.

PROBLEMS FOR SECTION 14.1

For the first order initial value problem

$$y' = f(t, x); \qquad y(x_0) = y_0,$$

the existence/uniqueness theorem is proved by converting the problem to an integral equation and making use of convergence of the Picard iterates (see the remarks following Problem 7 in Section 1.10). A proof of Theorem 14.1 of this section can be developed along the same lines. While we do not intend to pursue the details of such a proof, the following problems will provide some idea of what is involved.

Consider the initial value problem

$$
\begin{aligned}
x'(t) &= f(t, x, y), \\
y'(t) &= g(t, x, y); \\
x(t_0) &= x_0, \qquad y(t_0) = y_0.
\end{aligned}
\tag{14.4}
$$

1. Prove that any solution of the problem (14.4) is also a solution of the pair of integral equations

$$
\begin{aligned}
x(t) &= x_0 + \int_{t_0}^{t} f(\xi, x(\xi), y(\xi))\, d\xi \\
y(t) &= y_0 + \int_{t_0}^{t} g(\xi, x(\xi), y(\xi))\, d\xi.
\end{aligned}
\tag{14.5}
$$

2. Prove that any solution of the system of integral equations (14.5) is also a solution of problem (14.4).

Now define Picard iterates for the system (14.5) of integral equations as follows. Let

$$
\begin{aligned}
x_0(t) &= x_0 \qquad \text{and} \qquad y_0(t) = y_0; \\
x_1(t) &= x_0 + \int_{t_0}^{t} f(\xi, x_0(\xi), y_0(\xi))\, d\xi;
\end{aligned}
$$

and

$$y_1(t) = y_0 + \int_{t_0}^{t} g(\xi, x_0(\xi), y_0(\xi))\, d\xi.$$

Next, for $n = 1, 2, 3, \ldots$, let

$$x_{n+1}(t) = x_0 + \int_{t_0}^{t} f(\xi, x_n(\xi), y_n(\xi))\, d\xi$$

and

$$y_{n+1}(t) = y_0 + \int_{t_0}^{t} g(\xi, x_n(\xi), y_n(\xi)) \, d\xi.$$

It can be shown that the sequences $\{x_n\}$ and $\{y_n\}$ converge, assuming that there is a parallelepiped centered at (t_0, x_0, y_0) throughout which $f, g, \partial f/\partial x, \partial f/\partial y, \partial g/\partial x$, and $\partial g/\partial y$ are continuous. Further, these sequences converge to the unique solution of the problem (14.4).

In each of Problems 3 through 6, find the first four Picard iterates (x_0 and y_0 through x_3 and y_3) of the system.

3. $x' = x + y$,
$\quad y' = 2x + 2$;
$\quad x(0) = 1, \quad y(0) = 0$

4. $x' = t + x$,
$\quad y' = t - 2x$;
$\quad x(0) = 2, \quad y(0) = 1$

5. $x' = 2e^t + y$,
$\quad y' = 1 + x$;
$\quad x(0) = -1, \quad y(0) = -1$

6. $x' = t^2 + 4y$,
$\quad y' = t - x$;
$\quad x(0) = 0, \qquad y(0) = 2$

7. Show that

$$x = \varphi(t) = e^{2t} + e^{-t} - 1$$
$$y = \psi(t) = e^{2t} - 2e^{-t} + 1$$

is the solution of the initial value problem given in Problem 3. Calculate the first four partial sums of the Maclaurin series of φ and ψ, and compare these with the Picard iterates found in Problem 3.

8. Construct a system of integral equations which is equivalent (in the sense of having the same solutions) to the initial value problem

$$x' = 3x - y - z,$$
$$y' = x + y - z + 2e^{2t},$$
$$z' = x - y + z;$$
$$x(0) = 4, \qquad y(0) = 4, \qquad z(0) = 3.$$

9. Define the Picard iterates of the system of Problem 8, and calculate the first four iterates $x_0, \ldots, x_3; y_0, \ldots, y_3$; and $z_0, \ldots, z_3$ of x, y, and z.

10. Show that the solution of the initial value problem of Problem 8 is

$$x = \varphi(t) = e^t + (3 - 2t)e^{2t}$$
$$y = \psi(t) = e^t + 3e^{2t}$$
$$z = \xi(t) = e^t + (2 - 2t)e^{2t}.$$

Calculate the first four partial sums of the Maclaurin series of φ, ψ, and ξ, and compare these partial sums with the Picard iterates obtained in Problem 9.

14.2 *Homogeneous Linear Systems With Constant Coefficients*

Consider the system

$$X' = AX, \tag{14.6}$$

in which A is an $n \times n$ matrix of constants.

For a single first order equation $y' = ay$, we always obtain exponential solutions $y = ce^{at}$. This suggests that we attempt solutions of the system (14.6) having the form $\xi e^{\lambda t}$, in which ξ is an $n \times 1$ constant matrix and λ is a constant to be determined.

To see how to choose ξ and λ, substitute $X = \xi e^{\lambda t}$ into the system (14.6) to get

$$(\xi e^{\lambda t})' = A\xi e^{\lambda t},$$

or

$$\lambda \xi e^{\lambda t} = A\xi e^{\lambda t}.$$

Upon dividing this equation by the scalar quantity $e^{\lambda t}$, we have

$$\lambda \xi = A\xi.$$

If we require that $\xi \neq O$, this equation is exactly the condition for λ to be an eigenvalue of A, with ξ an associated eigenvector. We will summarize this discussion as a theorem.

THEOREM 14.9

Let A be an $n \times n$ constant matrix. Then $\xi e^{\lambda t}$ is a nontrivial solution of $X' = AX$ if and only if λ is an eigenvalue of A and ξ is a corresponding eigenvector. ∎

Since A is $n \times n$, we need n linearly independent solutions of $X' = AX$ in order to write the general solution. The following theorem is useful in determining when n solutions are linearly independent.

THEOREM 14.10

Let A be an $n \times n$ matrix of constants having eigenvalues $\lambda_1, \ldots, \lambda_n$ (some of which may be repeated). Let $\xi^{(j)}$ be an eigenvector associated with λ_j.

If $\xi^{(1)}, \ldots, \xi^{(n)}$ are linearly independent as vectors in R^n, $\xi^{(1)}e^{\lambda_1 t}, \ldots, \xi^{(n)}e^{\lambda_n t}$ are linearly independent solutions of $X' = AX$.

Proof Since $\xi^{(j)}$ is an eigenvector corresponding to the eigenvalue λ_j, we know from Theorem 14.9 that $\xi^{(j)}e^{\lambda_j t}$ is a solution of $X' = AX$ for $j = 1, 2, \ldots, n$. All we must prove is that these n solutions are linearly independent.

Form the $n \times n$ matrix S whose columns are the solutions $\xi^{(1)}e^{\lambda_1 t}, \ldots, \xi^{(n)}e^{\lambda_n t}$ evaluated at $t = 0$. The columns of S are therefore just $\xi^{(1)}, \ldots, \xi^{(n)}$, which are linearly independent by assumption. Therefore, $\det(S) \neq 0$. By conclusion (2) of Theorem 14.5 in the preceding section, the solutions $\xi^{(1)}e^{\lambda_1 t}, \ldots, \xi^{(n)}e^{\lambda_n t}$ are linearly independent. ∎

It is important to understand that the theorem requires only that we be able to produce n linearly independent eigenvectors. This does not require that all of the eigenvalues of A be distinct, since repeated eigenvalues may give rise to linearly independent eigenvectors. If, however, A does have n distinct eigenvalues, we are guaranteed that corresponding eigenvectors are linearly independent, by Theorem 13.6 in Section 13.2.

EXAMPLE 14.2

Solve the system $X' = AX$, with

$$A = \begin{bmatrix} 4 & 2 \\ 3 & 3 \end{bmatrix}. \tag{14.7}$$

First, find the eigenvalues of A. The characteristic equation is

$$\det(\lambda I_2 - A) = \det\begin{bmatrix} \lambda - 4 & -2 \\ -3 & \lambda - 3 \end{bmatrix} = 0,$$

or

$$\lambda^2 - 7\lambda + 6 = 0.$$

Solve this equation to find the eigenvalues $\lambda_1 = 1$ and $\lambda_2 = 6$. Corresponding to $\lambda_1 = 1$, solve the system of algebraic equations $(I_2 - A)\xi = O$ to get

$$\xi^{(1)} = \begin{bmatrix} 2 \\ -3 \end{bmatrix}$$

(or any nonzero constant multiple of this vector). Therefore, $\xi^{(1)}e^t$ is one solution of the system (14.7). Corresponding to $\lambda_2 = 6$, solve the system $(6I_2 - A)\xi = O$ to get

$$\xi^{(2)} = \begin{bmatrix} 1 \\ 1 \end{bmatrix},$$

or any nonzero constant multiple thereof. A second solution of the system (14.7) is $\xi^{(2)}e^{6t}$. The general solution of (14.7) may be written

$$\varphi(t) = c_1\xi^{(1)}e^t + c_2\xi^{(2)}e^{6t}$$

$$= c_1\begin{bmatrix} 2 \\ -3 \end{bmatrix}e^t + c_2\begin{bmatrix} 1 \\ 1 \end{bmatrix}e^{6t}$$

$$= \begin{bmatrix} 2c_1e^t + c_2e^{6t} \\ -3c_1e^t + c_2e^{6t} \end{bmatrix}.$$

If we write the solution in terms of individual components, we have

$$x_1(t) = 2c_1e^t + c_2e^{6t}$$
$$x_2(t) = -3c_1e^t + c_2e^{6t}.$$

A fundamental matrix Ω for this system can be formed by using the linearly independent solutions $\xi^{(1)}e^t$ and $\xi^{(2)}e^{6t}$ as columns:

$$\Omega = \begin{bmatrix} 2e^t & e^{6t} \\ -3e^t & e^{6t} \end{bmatrix}.$$

In terms of Ω, the general solution we have found can be written

$$\varphi(t) = \Omega(t)C,$$

in which $C = \begin{bmatrix} c_1 \\ c_2 \end{bmatrix}$ is a matrix of arbitrary constants. ∎

EXAMPLE 14.3

Solve the system $X' = AX$, where

$$A = \begin{bmatrix} 1 & 0 & 1 \\ 0 & 1 & 1 \\ 0 & 2 & 1 \end{bmatrix}. \tag{14.8}$$

First, find the eigenvalues of A. The characteristic equation is

$$\det(\lambda I_3 - A) = \det \begin{bmatrix} \lambda - 1 & 0 & -1 \\ 0 & \lambda - 1 & -1 \\ 0 & -2 & \lambda - 1 \end{bmatrix}$$
$$= (\lambda - 1)(\lambda^2 - 2\lambda - 1) = 0.$$

The eigenvalues are $\lambda_1 = 1, \lambda_2 = 1 + \sqrt{2},$ and $\lambda_3 = 1 - \sqrt{2}$. Eigenvectors corresponding to these eigenvalues are, respectively,

$$\xi^{(1)} = \begin{bmatrix} 1 \\ 0 \\ 0 \end{bmatrix}, \quad \xi^{(2)} = \begin{bmatrix} 1 \\ 1 \\ \sqrt{2} \end{bmatrix}, \quad \text{and} \quad \xi^{(3)} = \begin{bmatrix} 1 \\ 1 \\ -\sqrt{2} \end{bmatrix}.$$

These eigenvectors are linearly independent (they correspond to distinct eigenvalues of A). Therefore, three linearly independent solutions of $X' = AX$ are

$$\xi^{(1)}e^t, \quad \xi^{(2)}e^{(1+\sqrt{2})t}, \quad \text{and} \quad \xi^{(3)}e^{(1-\sqrt{2})t}.$$

These solutions form a basis for the solution space. The general solution of the system (14.8) is

$$\varphi(t) = c_1\xi^{(1)}e^t + c_2\xi^{(2)}e^{(1+\sqrt{2})t} + c_3\xi^{(3)}e^{(1-\sqrt{2})t}.$$

As in Example 14.2, we may write this general solution in terms of a fundamental matrix

$$\varphi(t) = \Omega(t)C,$$

where

$$\Omega(t) = \begin{bmatrix} e^t & e^{(1+\sqrt{2})t} & e^{(1-\sqrt{2})t} \\ 0 & e^{(1+\sqrt{2})t} & e^{(1-\sqrt{2})t} \\ 0 & \sqrt{2}e^{(1+\sqrt{2})t} & -\sqrt{2}e^{(1-\sqrt{2})t} \end{bmatrix}.$$

In terms of individual components, the general solution of the system is

$$x_1(t) = c_1 e^t + c_2 e^{(1+\sqrt{2})t} + c_3 e^{(1-\sqrt{2})t}$$
$$x_2(t) = c_2 e^{(1+\sqrt{2})t} + c_3 e^{(1-\sqrt{2})t}$$
$$x_3(t) = \sqrt{2}c_2 e^{(1+\sqrt{2})t} - \sqrt{2}c_3 e^{(1-\sqrt{2})t}. \quad \blacksquare$$

In both of these examples, A had n distinct eigenvalues. If A does not have n distinct eigenvalues, A may or may not have n linearly independent eigenvectors. The following is an example in which A has repeated eigenvalues but still has n linearly independent eigenvectors (two corresponding to one eigenvalue).

EXAMPLE 14.4

Solve the system $X' = AX$, where

$$A = \begin{bmatrix} 5 & -4 & 4 \\ 12 & -11 & 12 \\ 4 & -4 & 5 \end{bmatrix}.$$

We find that the eigenvalues of A are -3, 1, and 1. Corresponding to the eigenvalue -3, we find an eigenvector

$$\xi^{(1)} = \begin{bmatrix} 1 \\ 3 \\ 1 \end{bmatrix}.$$

Therefore, one solution of $X' = AX$ is $\xi^{(1)}e^{-3t}$.

Now consider eigenvectors corresponding to the repeated eigenvalue 1. We must solve $(I_3 - A)\xi = O$. With

$$\xi = \begin{bmatrix} \xi_1 \\ \xi_2 \\ \xi_3 \end{bmatrix},$$

this system can be written

$$-4\xi_1 + 4\xi_2 - 4\xi_3 = 0$$
$$-12\xi_1 + 12\xi_2 - 12\xi_3 = 0$$
$$-4\xi_1 + 4\xi_2 - 4\xi_3 = 0.$$

The first and third equations are the same, while the second is the first multiplied by 3. Thus, we have, in effect, one equation with three unknowns:

$$-4\xi_1 + 4\xi_2 - 4\xi_3 = 0.$$

Divide this equation by 4 and solve for ξ_3 to get

$$\xi_3 = \xi_2 - \xi_1.$$

Thus, ξ_1 and ξ_2 are arbitrary and $\xi_3 = \xi_2 - \xi_1$. Eigenvectors corresponding to the eigenvalue 1 all have the form

$$\begin{bmatrix} \alpha \\ \beta \\ \beta - \alpha \end{bmatrix},$$

with α and β not both zero but otherwise arbitrary.

We can get two linearly independent eigenvectors corresponding to the eigenvalue 1 by choosing first $\alpha = 1$ and $\beta = 0$, then $\alpha = 0$ and $\beta = 1$. This gives us the eigenvectors

$$\xi^{(2)} = \begin{bmatrix} 1 \\ 0 \\ -1 \end{bmatrix} \quad \text{and} \quad \xi^{(3)} = \begin{bmatrix} 0 \\ 1 \\ 1 \end{bmatrix}.$$

$\xi^{(1)}$, $\xi^{(2)}$, and $\xi^{(3)}$ are linearly independent. The general solution of $X' = AX$ is therefore

$$\varphi(t) = c_1\xi^{(1)}e^{-3t} + c_2\xi^{(2)}e^t + c_3\xi^{(3)}e^t.$$

This general solution can also be written $\varphi = \Omega C$, with Ω the fundamental matrix

$$\Omega = \begin{bmatrix} e^{-3t} & e^t & 0 \\ 3e^{-3t} & 0 & e^t \\ e^{-3t} & -e^t & e^t \end{bmatrix}. \quad \blacksquare$$

In this example, A had three linearly independent eigenvectors even though it had only two distinct eigenvalues. However, an $n \times n$ matrix may not have n linearly independent eigenvectors (see Example 13.8). In this event, the method of this section will not produce the general solution. In Section 14.6, we will develop the exponential matrix and show how to proceed when this occurs.

Thus far, we have not solved a system $A' = AX$ in which A has complex eigenvalues. Although we can write a fundamental matrix with complex functions, we usually prefer to use real-valued functions. In the next section, we will see how to do this even when A has complex eigenvalues.

PROBLEMS FOR SECTION 14.2

In each of Problems 1 through 7, find a fundamental matrix for the system and write the general solution in terms of this fundamental matrix. Also write the general solution in terms of the individual functions $x_1, \ldots, x_n$.

1. $x_1' = 3x_1$
$x_2' = 5x_1 - 4x_2$

2. $x_1' = 4x_1 + 2x_2$
$x_2' = 3x_1 + 3x_2$

3. $x_1' = x_1 + x_2$
$x_2' = x_1 + x_2$

4. $x_1' = 2x_1 + x_2 - 2x_3$
$x_2' = 3x_2 - 2x_3$
$x_3' = 3x_1 + x_2 - 3x_3$

5. $x_1' = x_1 + 2x_2 + x_3$
$x_2' = 6x_1 - x_2$
$x_3' = -x_1 - 2x_2 - x_3$

6. $x_1' = 6x_1 + 2x_2$
$x_2' = 4x_1 + 4x_2$
$x_3' = 2x_3 + 2x_4$
$x_4' = x_3 + 3x_4$

7. $x_1' = x_1 - x_2 + 4x_3$
$x_2' = 3x_1 + 2x_2 - x_3$
$x_3' = 2x_1 + x_2 - x_3$

In each of Problems 8 through 13, solve the initial value problem. Write the solution first in terms of matrices, then by listing the components.

8. $x_1' = 3x_1 - 4x_2,$
$x_2' = 2x_1 - 3x_2;$
$x_1(0) = 7, \quad x_2(0) = 5$

9. $x_1' = x_1 - 2x_2,$
$x_2' = -6x_1;$
$x_1(0) = 1, \quad x_2(0) = -19$

10. $x_1' = 2x_1 - 10x_2,$
$x_2' = -x_1 - x_2;$
$x_1(0) = -3, \quad x_2(0) = 2$

11. $x_1' = 3x_1 - x_2 + x_3,$
$x_2' = x_1 + x_2 - x_3,$
$x_3' = x_1 - x_2 + x_3;$
$x_1(0) = 1, \quad x_2(0) = 5, \quad x_3(0) = 1$

12. $x_1' = 2x_1 + x_2 - 2x_3,$
$x_2' = 3x_1 - 2x_2,$
$x_3' = 3x_1 + x_2 - 3x_3;$
$x_1(0) = 1, \quad x_2(0) = 7, \quad x_3(0) = 3$

13. $x_1' = 2x_1 + 3x_2 + 3x_3,$
$x_2' = -x_2 - 3x_3,$
$x_3' = 2x_3;$
$x_1(0) = 9, \quad x_2(0) = -1, \quad x_3(0) = -3$

14. Show that the change of variables $z = \ln(t)$ transforms the system

$$tx_1' = ax_1 + bx_2$$
$$tx_2' = cx_1 + dx_2$$

into a linear system $X' = AX$, assuming that $a, b, c,$ and d are constants and $t > 0$.

15. Use the transformation given in Problem 14 to find the general solution of

$$tx_1' = 6x_1 + 2x_2$$
$$tx_2' = 4x_1 + 4x_2.$$

16. Use the transformation given in Problem 14 to find the general solution of

$$tx'_1 = -x_1 - 3x_2$$
$$tx'_2 = x_1 - 5x_2.$$

14.3 Real-Valued Solutions of $X' = AX$ When A Has Complex Eigenvalues _____

Suppose we want to solve $X' = AX$, in which A is an $n \times n$ constant matrix with real elements. If A has a complex eigenvalue $\alpha + i\beta$, then $\alpha - i\beta$ is also an eigenvalue because the characteristic equation has real coefficients (from the elements of A), and complex roots of polynomials with real coefficients occur in conjugate pairs.

Corresponding to a complex eigenvalue $\alpha + i\beta$, we expect to find a complex eigenvector $U + iV$. It is routine to show that $U - iV$ is an eigenvector corresponding to $\alpha - i\beta$. In fact, take the complex conjugate of the equation

$$A(U + iV) = (\alpha + i\beta)(U + iV)$$

to get

$$\bar{A}(U - iV) = (\alpha - i\beta)(U - iV).$$

Since A has real elements, $\bar{A} = A$, and we have

$$A(U - iV) = (\alpha - i\beta)(U - iV);$$

hence, $U - iV$ is an eigenvector of A associated with eigenvalue $\alpha - i\beta$.

Since $\alpha + i\beta$ and $\alpha - i\beta$ are distinct eigenvalues, $U + iV$ and $U - iV$ are linearly independent eigenvectors, yielding two linearly independent solutions

$$(U + iV)e^{(\alpha + i\beta)t} \quad \text{and} \quad (U - iV)e^{(\alpha - i\beta)t}.$$

These solutions, together with $n - 2$ appropriate additional solutions, form a basis for the solution space of $X' = AX$.

There is nothing wrong with these two complex-valued solutions. It is possible, however, to replace them with two linearly independent solutions involving only real-valued quantities, much as we did in solving constant coefficient linear differential equations of order 2 or higher. The following theorem tells how to do this.

THEOREM 14.11

Suppose that $\lambda = \alpha + i\beta$ is an eigenvalue of A, with a corresponding eigenvector $\xi = U + iV$. Then

$$e^{\alpha t}(U \cos(\beta t) - V \sin(\beta t))$$

and

$$e^{\alpha t}(U \sin(\beta t) + V \cos(\beta t))$$

are linearly independent solutions of $X' = AX$.   ∎

We will consider two examples before proving the theorem.

EXAMPLE 14.5

Solve the system $X' = AX$, where

$$A = \begin{bmatrix} 6 & -5 \\ 5 & -2 \end{bmatrix}.$$

The eigenvalues of A are $2 + 3i$ and $2 - 3i$. Corresponding eigenvectors are

$$\xi^{(1)} = \begin{bmatrix} 5 \\ 4 - 3i \end{bmatrix} \quad \text{and} \quad \xi^{(2)} = \begin{bmatrix} 5 \\ 4 + 3i \end{bmatrix}.$$

The general solution of $X' = AX$ is

$$\varphi(t) = c_1 \xi^{(1)} e^{(2 + 3i)t} + c_2 \xi^{(2)} e^{(2 - 3i)t}.$$

Notice that $\xi^{(2)}$ is the complex conjugate of $\xi^{(1)}$. If we let $\lambda = 2 + 3i$ and write $\xi^{(1)}$ as just ξ, this general solution can be written more neatly as

$$\varphi(t) = c_1 \xi e^{\lambda t} + c_2 \bar{\xi} e^{\bar{\lambda} t}. \tag{14.9}$$

If we wish, we can write the general solution (14.9) strictly in terms of real-valued quantities, using Theorem 14.11. Write

$$\xi = \begin{bmatrix} 5 \\ 4 - 3i \end{bmatrix} = \begin{bmatrix} 5 \\ 4 \end{bmatrix} + i \begin{bmatrix} 0 \\ -3 \end{bmatrix}.$$

To use the notation of the theorem, let $\alpha = 2$, $\beta = 3$, $U = \begin{bmatrix} 5 \\ 4 \end{bmatrix}$, and $V = \begin{bmatrix} 0 \\ -3 \end{bmatrix}$. Theorem 14.11 tells us that $X' = AX$ has linearly independent solutions

$$e^{2t} \left\{ \begin{bmatrix} 5 \\ 4 \end{bmatrix} \cos(3t) - \begin{bmatrix} 0 \\ -3 \end{bmatrix} \sin(3t) \right\}$$

and

$$e^{2t} \left\{ \begin{bmatrix} 5 \\ 4 \end{bmatrix} \sin(3t) + \begin{bmatrix} 0 \\ -3 \end{bmatrix} \cos(3t) \right\}.$$

These solutions may be written, respectively,

$$\begin{bmatrix} 5e^{2t}\cos(3t) \\ 4e^{2t}\cos(3t) + 3e^{2t}\sin(2t) \end{bmatrix}$$

and

$$\begin{bmatrix} 5e^{2t}\sin(3t) \\ 4e^{2t}\sin(3t) - 3e^{2t}\cos(3t) \end{bmatrix}.$$

The real-valued general solution of $X' = AX$ is

$$\psi(t) = c_1 \begin{bmatrix} 5e^{2t}\cos(3t) \\ 4e^{2t}\cos(3t) + 3e^{2t}\sin(3t) \end{bmatrix} + c_2 \begin{bmatrix} 5e^{2t}\sin(3t) \\ 4e^{2t}\sin(3t) - 3e^{2t}\cos(3t) \end{bmatrix}.$$

We can also form a fundamental matrix using these real-valued solutions as columns:

$$\Omega = \begin{bmatrix} 5e^{2t}\cos(3t) & 5e^{2t}\sin(3t) \\ 4e^{2t}\cos(3t) + 3e^{2t}\sin(3t) & 4e^{2t}\sin(3t) - 3e^{2t}\cos(3t) \end{bmatrix}.$$

In terms of Ω,

$$\psi = \Omega C$$

is the real-valued general solution of the system $X' = AX$. ∎

EXAMPLE 14.6

Solve the system $X' = AX$, where

$$A = \begin{bmatrix} 2 & 0 & 1 \\ 0 & -2 & -2 \\ 0 & 2 & 0 \end{bmatrix}.$$

The eigenvalues of A are 2, $-1 + \sqrt{3}i$, and $-1 - \sqrt{3}i$. Corresponding to 2, we find an eigenvector

$$\xi^{(1)} = \begin{bmatrix} 1 \\ 0 \\ 0 \end{bmatrix}.$$

This gives us one solution, $X^{(1)} = \xi^{(1)}e^{2t}$. Corresponding to $-1 + \sqrt{3}i$, we find an eigenvector

$$\begin{bmatrix} 1 \\ -2\sqrt{3}i \\ -3 + \sqrt{3}i \end{bmatrix}.$$

Write this eigenvector as

$$U + iV = \begin{bmatrix} 1 \\ 0 \\ -3 \end{bmatrix} + i \begin{bmatrix} 0 \\ -2\sqrt{3} \\ \sqrt{3} \end{bmatrix}.$$

From this and Theorem 14.11 we get two linearly independent, real-valued solutions

$$X^{(2)} = e^{-t} \left\{ \begin{bmatrix} 1 \\ 0 \\ -3 \end{bmatrix} \cos(\sqrt{3}t) - \begin{bmatrix} 0 \\ -2\sqrt{3} \\ \sqrt{3} \end{bmatrix} \sin(\sqrt{3}t) \right\}$$

and

$$X^{(3)} = e^{-t} \left\{ \begin{bmatrix} 1 \\ 0 \\ -3 \end{bmatrix} \sin(\sqrt{3}t) + \begin{bmatrix} 0 \\ -2\sqrt{3} \\ \sqrt{3} \end{bmatrix} \cos(\sqrt{3}t) \right\}.$$

The general solution of the system is

$$\varphi = c_1 X^{(1)} + c_2 X^{(2)} + c_3 X^{(3)}.$$

We can also form a fundamental matrix

$$\Omega = \begin{bmatrix} e^{2t} & e^{-t}\cos(\sqrt{3}t) & e^{-t}\sin(\sqrt{3}t) \\ 0 & 2\sqrt{3}e^{-t}\sin(\sqrt{3}t) & -2\sqrt{3}e^{-t}\cos(\sqrt{3}t) \\ 0 & e^{-t}[-3\cos(\sqrt{3}t) - \sqrt{3}\sin(\sqrt{3}t)] & e^{-t}[-3\sin(\sqrt{3}t) + \sqrt{3}\cos(\sqrt{3}t)] \end{bmatrix}.$$

In terms of Ω, the general solution is $\varphi = \Omega C$.

If we write this general solution in terms of the individual functions x_1, x_2, and x_3, we have

$$x_1(t) = c_1 e^{2t} + c_2 e^{-t}\cos(\sqrt{3}t) + c_3 e^{-t}\sin(\sqrt{3}t)$$

$$x_2(t) = 2\sqrt{3}c_2 e^{-t}\sin(\sqrt{3}t) - 2\sqrt{3}c_3 e^{-t}\cos(\sqrt{3}t)$$

$$x_3(t) = (-3c_2 + \sqrt{3}c_3)e^{-t}\cos(\sqrt{3}t) + (-\sqrt{3}c_2 - 3c_3)e^{-t}\sin(\sqrt{3}t). \quad \blacksquare$$

We will conclude this section with a proof of Theorem 14.11.

Proof of Theorem 14.11   Use Euler's formula to write

$$c_1(U + iV)e^{(\alpha + i\beta)t} + c_2(U - iV)e^{(\alpha - i\beta)t}$$

$$= c_1(U + iV)e^{\alpha t}[\cos(\beta t) + i\sin(\beta t)] + c_2(U - iV)e^{\alpha t}[\cos(\beta t) - i\sin(\beta t)]$$

$$= c_1 e^{\alpha t}\{U\cos(\beta t) - V\sin(\beta t) + i[U\sin(\beta t) + V\cos(\beta t)]\}$$

$$\quad + c_2 e^{\alpha t}\{U\cos(\beta t) - V\sin(\beta t) + i[-U\sin(\beta t) - V\cos(\beta t)]\}$$

$$= e^{\alpha t}\{c_1[U\cos(\beta t) - V\sin(\beta t)] + c_2[U\cos(\beta t) - V\sin(\beta t)]\}$$

$$\quad + ie^{\alpha t}\{c_1[U\sin(\beta t) + V\cos(\beta t)] - c_2[U\sin(\beta t) + V\cos(\beta t)]\}.$$

This is a solution for any choice of the constants c_1 and c_2. If we choose $c_1 = c_2 = \frac{1}{2}$, we get the solution

$$e^{\alpha t}[U\cos(\beta t) - V\sin(\beta t)].$$

If we choose $c_1 = -c_2 = -i/2$, we get the solution

$$e^{\alpha t}[U\sin(\beta t) + V\cos(\beta t)].$$

Finally, observe that these two solutions are linearly independent. This completes the proof of the theorem. $\blacksquare$

PROBLEMS FOR SECTION 14.3

In each of Problems 1 through 7, find a fundamental matrix in terms of real-valued functions and use this fundamental matrix to write the general solution of the system. Also write the general solution without matrix notation by listing solutions for $x_1, \ldots, x_n$.

1. $x_1' = 2x_1 - 4x_2$
$x_2' = x_1 + 2x_2$

2. $x_1' = 5x_2$
$x_2' = -x_1 - 2x_2$

3. $x_1' = 3x_1 - 5x_2$
$x_2' = x_1 - x_2$

4. $x_1' = x_1 - x_2 - x_3$
$x_2' = x_1 - x_2$
$x_3' = x_1 - x_3$

5. $x_1' = -2x_1 + x_2$
$x_2' = -5x_1$
$x_3' = 3x_2 - 2x_3$

6. $x_1' = 3x_1 + x_3$
$x_2' = 9x_1 - x_2 + 2x_3$
$x_3' = -9x_1 + 4x_2 - x_3$

7. $x_1' = 3x_1 - 2x_2$
$x_2' = 5x_1 - 3x_2$
$x_3' = 3x_3 - 2x_4$
$x_4' = 5x_3 - 3x_4$

In each of Problems 8 through 13, solve the initial value problem. The solution should be written in terms of real-valued functions.

8. $x_1' = 3x_1 + 2x_2,$
$x_2' = -5x_1 + x_2;$
$x_1(0) = 2, \quad x_2(0) = 8$

9. $x_1' = 3x_1 - 2x_2,$
$x_2' = 5x_1 - 3x_2;$
$x_1(0) = 2, \quad x_2(0) = 10$

10. $x_1' = 2x_1 - 5x_2,$
$x_2' = x_1 - 2x_2;$
$x_1(0) = 5, \quad x_2(0) = 0$

11. $x_1' = 3x_1 - 3x_2 + x_3,$
$x_2' = 2x_1 - x_2,$
$x_3' = x_1 - x_2 + x_3;$
$x_1(0) = 7, \quad x_2(0) = 4, \quad x_3(0) = 3$

12. $x_1' = 2x_1 - 5x_2,$
$x_2' = 2x_1 - 4x_2,$
$x_3' = 4x_1 - 5x_2 - 2x_3;$
$x_1(0) = 5, \quad x_2(0) = 5, \quad x_3(0) = 9$

13. $tx_1' = 5x_1 - 4x_2,$
$tx_2' = 2x_1 + x_2;$
$x_1(1) = 6, \quad x_2(1) = 5$

Hint: Note Problem 14 of Section 14.2.

14.4 *Solving X' = AX by Diagonalizing A*

A system of first order differential equations is said to be *uncoupled* if each equation of the system involves only one of the variables. For example, the system

$$x_1' = x_1$$
$$x_2' = -4x_2$$

is uncoupled. The nice feature of an uncoupled system is that it can be treated as a set of n independent differential equations, which we can attempt to solve individually.

For $X' = AX$ to be an uncoupled system, A must be a diagonal matrix. All off-diagonal elements a_{ij} with $i \neq j$ must be zero because any nonzero a_{ij} will result in x_j appearing in equation i.

EXAMPLE 14.7

Consider the system $X' = AX$, where

$$A = \begin{bmatrix} 3 & 0 & 0 & 0 \\ 0 & 4 & 0 & 0 \\ 0 & 0 & -2 & 0 \\ 0 & 0 & 0 & 6 \end{bmatrix}.$$

A is a diagonal matrix, and the system $X' = AX$ is uncoupled. It is the system

$$x_1' = 3x_1$$
$$x_2' = 4x_2$$
$$x_3' = -2x_3$$
$$x_4' = 6x_4.$$

We can solve this system by simply integrating each equation, obtaining

$$x_1(t) = c_1 e^{3t}$$
$$x_2(t) = c_2 e^{4t}$$
$$x_3(t) = c_3 e^{-2t}$$
$$x_4(t) = c_4 e^{6t}.$$

A fundamental matrix for this system is

$$\Omega = \begin{bmatrix} e^{3t} & 0 & 0 & 0 \\ 0 & e^{4t} & 0 & 0 \\ 0 & 0 & e^{-2t} & 0 \\ 0 & 0 & 0 & e^{6t} \end{bmatrix},$$

and the general solution is ΩC. ■

Observe that in this example Ω can be read directly from A as the diagonal matrix with diagonal entries $e^{a_{11}t}, e^{a_{22}t}, \ldots, e^{a_{nn}t}$. This result holds in general for systems $X' = AX$ in which A is a diagonal matrix.

THEOREM 14.12

Let $A = [a_{ij}]$ be an $n \times n$ diagonal matrix of constants. Then a fundamental matrix for the system $X' = AX$ is the diagonal matrix $\Omega = [\omega_{ij}]$, in which

$$\omega_{jj} = e^{a_{jj}t}$$

for $j = 1, 2, \ldots, n$ and $\omega_{ij} = 0$ if $i \neq j$. ■

A proof of this theorem is straightforward, and we omit the details.

Theorem 14.12 by itself is not a very powerful result, since "most" systems are not uncoupled. We will now show, however, that under certain conditions, a system $X' = AX$ can be transformed into an uncoupled system by a suitable change of variables. We attempt to solve the uncoupled system by solving each equation of the system independently of the others and then using these solutions to find a solution of the original system.

Recall from Section 13.2 that an $n \times n$ matrix A is diagonalizable if and only if A has n linearly independent eigenvectors. Further, if $\lambda_1, \ldots, \lambda_n$ are the eigenvalues of A, and $\xi^{(1)}, \ldots, \xi^{(n)}$ are associated linearly independent eigenvectors, the matrix P formed using $\xi^{(j)}$ as column j diagonalizes A. That is,

$$P^{-1}AP = \begin{bmatrix} \lambda_1 & 0 & 0 & \cdots & 0 \\ 0 & \lambda_2 & 0 & \cdots & 0 \\ \vdots & \vdots & \vdots & & \vdots \\ 0 & 0 & 0 & \cdots & \lambda_n \end{bmatrix},$$

with the eigenvalues of A down the main diagonal in order corresponding to the order in which the eigenvectors are listed.

How does this information help us solve a system $X' = AX$? If A is diagonalizable, change variables from $x_1, x_2, \ldots, x_n$ to $z_1, z_2, \ldots, z_n$ by setting

$$Z = P^{-1}X,$$

where

$$Z = \begin{bmatrix} z_1 \\ z_2 \\ \vdots \\ z_n \end{bmatrix}.$$

Since $Z = P^{-1}X$, $X = PZ$, and the system $X' = AX$ becomes

$$(PZ)' = A(PZ). \tag{14.10}$$

Since P is a matrix of constants, $(PZ)' = P'Z + PZ' = PZ'$, and equation (14.10) becomes

$$PZ' = (AP)Z. \tag{14.11}$$

Multiply equation (14.11) on the left by P^{-1} to get

$$Z' = (P^{-1}AP)Z.$$

This is an uncoupled system because we chose P in such a way that $P^{-1}AP$ is a diagonal matrix. If we can solve this uncoupled system, we can obtain X as

$$X = PZ.$$

In practice, this procedure is very straightforward. The main computation is to find n linearly independent eigenvectors of A. Once this is done, we know P. Further, we know $P^{-1}AP$ just from knowing the eigenvalues of A (simply list the eigenvalues of A down the main diagonal in the order corresponding to the eigenvectors listed as columns of P). We therefore know $P^{-1}AP$ without finding P^{-1} explicitly. If we can solve the uncoupled system $Z' = (P^{-1}AP)Z$, $X = PZ$.

Throughout this process, we never need to calculate P^{-1}. This is important, because finding the inverse of a matrix can be difficult.

EXAMPLE 14.8

Solve the system

$$x_1' = 3x_1 + 2x_2$$
$$x_2' = -3x_1 - 4x_2.$$

This system can be written $X' = AX$, with

$$A = \begin{bmatrix} 3 & 2 \\ -3 & -4 \end{bmatrix}.$$

The characteristic equation of A is

$$\lambda^2 + \lambda - 6 = 0,$$

with roots $\lambda_1 = 2$ and $\lambda_2 = -3$. Corresponding eigenvectors are

$$\xi^{(1)} = \begin{bmatrix} 2 \\ -1 \end{bmatrix} \quad \text{and} \quad \xi^{(2)} = \begin{bmatrix} 1 \\ -3 \end{bmatrix}.$$

These are linearly independent; hence, A is diagonalizable. Let

$$P = \begin{bmatrix} 2 & 1 \\ -1 & -3 \end{bmatrix}.$$

Let $X = PZ$. Then $X' = AX$ transforms into the system $Z' = (P^{-1}AP)Z$, where $P^{-1}AP$ is the diagonal matrix having the corresponding eigenvalues on the main diagonal:

$$P^{-1}AP = \begin{bmatrix} 2 & 0 \\ 0 & -3 \end{bmatrix}.$$

Note that we know $P^{-1}AP$ without calculating P^{-1} and without doing any matrix multiplications. The uncoupled system $Z' = (P^{-1}AP)Z$ is

$$z_1' = 2z_1$$
$$z_2' = -3z_2.$$

Solve these equations individually to get the general solutions

$$z_1 = c_1 e^{2t}$$
$$z_2 = c_2 e^{-3t}.$$

Then

$$Z = \begin{bmatrix} c_1 e^{2t} \\ c_2 e^{-3t} \end{bmatrix}.$$

The general solution of the original system $X' = AX$ is

$$X = PZ = \begin{bmatrix} 2 & 1 \\ -1 & -3 \end{bmatrix}\begin{bmatrix} c_1 e^{2t} \\ c_2 e^{-3t} \end{bmatrix} = \begin{bmatrix} 2c_1 e^{2t} + c_2 e^{-3t} \\ -c_1 e^{2t} - 3c_2 e^{-3t} \end{bmatrix}. \quad \blacksquare$$

In the next section, we will apply these ideas to the nonhomogeneous system $X' = AX + G$.

PROBLEMS FOR SECTION 14.4

In each of Problems 1 through 10, find the general solution of the system by diagonalizing the coefficient matrix and solving the resulting uncoupled system. If a solution contains complex-valued functions, rewrite it in terms of real-valued functions.

1. $x_1' = x_1 + x_2$
$x_2' = x_1 + x_2$

2. $x_1' = 6x_1 + 2x_2$
$x_2' = 4x_1 + 4x_2$

3. $x_1' = 2x_1 + 2x_2$
$x_2' = x_1 + 3x_2$

4. $x_1' = 2x_1 - 3x_2$
$x_2' = x_1 - 2x_2$

5. $x_1' = 5x_1 - 4x_2 + 4x_3$
$x_2' = 12x_1 - 11x_2 + 12x_3$
$x_3' = 4x_1 - 4x_2 + 5x_3$

6. $x_1' = 2x_1 + x_2 - 2x_3$
$x_2' = 3x_1 - 2x_2$
$x_3' = 3x_1 + x_2 - 3x_3$

7. $x_1' = 4x_1 + 2x_2$
$x_2' = 3x_1 + 3x_2$
$x_3' = 3x_3 - 4x_4$
$x_4' = 2x_3 - 3x_4$

8. $x_1' = 2x_1 - 5x_2$
$x_2' = x_1 - 2x_2$

9. $x_1' = 3x_1 - x_2$
$x_2' = x_1 + 3x_2$

10. $x_1' = 2x_1 - 9x_2$
$x_2' = x_1 + 2x_2$
$x_3' = 2x_1 + 6x_2 - x_3$

In each of Problems 11 through 14, use the method of this section to solve the initial value problem. If a solution contains complex terms, rewrite it in terms of real-valued quantities.

11. $x_1' = -x_1 - 3x_2,$
$x_2' = x_1 - 5x_2;$
$x_1(0) = -4, \quad x_2(0) = 0$

12. $x_1' = x_1 - 4x_2,$
$x_2' = 2x_1 - 5x_2;$
$x_1(0) = 1, \quad x_2(0) = -1$

13. $x_1' = x_1 - x_2 + 4x_3,$
$\quad x_2' = 3x_1 + 2x_2 - x_3,$
$\quad x_3' = 2x_1 + x_2 - x_3;$
$\quad x_1(0) = 7, \quad x_2(0) = -4, \quad x_3(0) = -1$

14. $x_1' = 4x_1 - 13x_2,$
$\quad x_2' = x_1 - 2x_2;$
$\quad x_1(0) = 26, \quad x_2(0) = 0$

14.5 *The Nonhomogeneous System $X' = AX + G$ When A Is Diagonalizable*

We will now consider the nonhomogeneous linear system

$$X' = AX + G, \tag{14.12}$$

restricting our attention in this section to the case in which A is an $n \times n$ diagonalizable matrix of constants. Let P be an $n \times n$ matrix whose columns are any n linearly independent eigenvectors of A (so that P diagonalizes A), and make the change of variables

$$X = PZ.$$

Upon substituting $X = PZ$ into the system (14.12), we obtain

$$(PZ)' = A(PZ) + G,$$

or, since P is a matrix of constants,

$$PZ' = (AP)Z + G.$$

Multiply this equation on the left by P^{-1}. We obtain the system

$$Z' = (P^{-1}AP)Z + P^{-1}G. \tag{14.13}$$

We know that $P^{-1}AP$ is the diagonal matrix having the eigenvalues of A along its main diagonal. If these eigenvalues are $\lambda_1, \lambda_2, \ldots, \lambda_n$, we have the system

$$\begin{bmatrix} z_1 \\ z_2 \\ \vdots \\ z_n \end{bmatrix}' = \begin{bmatrix} \lambda_1 & 0 & 0 & \cdots & 0 \\ 0 & \lambda_2 & 0 & \cdots & 0 \\ \vdots & \vdots & \vdots & & \vdots \\ 0 & 0 & 0 & \cdots & \lambda_n \end{bmatrix} \begin{bmatrix} z_1 \\ z_2 \\ \vdots \\ z_n \end{bmatrix} + \begin{bmatrix} k_1(t) \\ k_2(t) \\ \vdots \\ k_n(t) \end{bmatrix},$$

in which $K = P^{-1}G$.

This system for $z_1, \ldots, z_n$ is uncoupled and consists of n independent differential equations, each involving only one of the z_j's:

$$z_1' - \lambda_1 z_1 = k_1(t)$$
$$z_2' - \lambda_2 z_2 = k_2(t)$$
$$\vdots \qquad \vdots$$
$$z_n' - \lambda_n z_n = k_n(t).$$

We attempt to solve these n first order differential equations, obtaining Z, and then obtain $X = PZ$.

In applying this method to the nonhomogeneous system, we must actually calculate P^{-1} explicitly in order to calculate the term $K = P^{-1}G$ in the transformed system. In the homogeneous case, we never actually required P^{-1}.

EXAMPLE 14.9

Solve the system

$$x_1' = 3x_1 + 3x_2 + 8$$
$$x_2' = x_1 + 5x_2 + 4e^{3t}.$$

Write this system as $X' = AX + G$, in which

$$A = \begin{bmatrix} 3 & 3 \\ 1 & 5 \end{bmatrix} \quad \text{and} \quad G = \begin{bmatrix} 8 \\ 4e^{3t} \end{bmatrix}.$$

We find that the eigenvalues of A are $\lambda_1 = 2$ and $\lambda_2 = 6$. Corresponding eigenvectors are, respectively,

$$\xi^{(1)} = \begin{bmatrix} 3 \\ -1 \end{bmatrix} \quad \text{and} \quad \xi^{(2)} = \begin{bmatrix} 1 \\ 1 \end{bmatrix}.$$

Let

$$P = \begin{bmatrix} 3 & 1 \\ -1 & 1 \end{bmatrix}.$$

We find that

$$P^{-1} = \tfrac{1}{4}\begin{bmatrix} 1 & -1 \\ 1 & 3 \end{bmatrix}.$$

Now let $X = PZ$ and let $K = P^{-1}G$ to obtain the system

$$Z' = (P^{-1}AP)Z + K,$$

where

$$P^{-1}AP = \begin{bmatrix} 2 & 0 \\ 0 & 6 \end{bmatrix},$$

a diagonal matrix with the eigenvalues on the main diagonal. Now calculate

$$K = P^{-1}G = \tfrac{1}{4}\begin{bmatrix} 1 & -1 \\ 1 & 3 \end{bmatrix}\begin{bmatrix} 8 \\ 4e^{3t} \end{bmatrix} = \begin{bmatrix} 2 - e^{3t} \\ 2 + 3e^{2t} \end{bmatrix}.$$

The transformed system $Z' = (P^{-1}AP)Z + K$ is the uncoupled system

$$z_1' - 2z_1 = 2 - e^{3t}$$
$$z_2' - 6z_2 = 2 + 3e^{3t}.$$

Solve these differential equations individually to get

$$z_1(t) = c_1 e^{2t} - e^{3t} - 1$$
$$z_2(t) = c_2 e^{6t} - e^{3t} - \tfrac{1}{3}.$$

The general solution of the system $X' = AX + G$ is

$$X = PZ = \begin{bmatrix} 3 & 1 \\ -1 & 1 \end{bmatrix} \begin{bmatrix} c_1 e^{2t} - e^{3t} - 1 \\ c_2 e^{6t} - e^{3t} - \frac{1}{3} \end{bmatrix}$$

$$= \begin{bmatrix} 3c_1 e^{2t} + c_2 e^{6t} - 4e^{3t} - \frac{10}{3} \\ -c_1 e^{2t} + c_2 e^{6t} + \frac{2}{3} \end{bmatrix}.$$

In terms of x_1 and x_2, the general solution is

$$x_1(t) = 3c_1 e^{2t} + c_2 e^{6t} - 4e^{3t} - \frac{10}{3}$$
$$x_2(t) = -c_1 e^{2t} + c_2 e^{6t} + \frac{2}{3}. \quad \blacksquare$$

Here is an application of this method to a nontrivial initial value problem.

EXAMPLE 14.10

Solve the initial value problem

$$x_1' = 31x_1 - 21x_2 + 9x_3 - e^{-3t},$$
$$x_2' = 44x_1 - 30x_2 + 12x_3 + 2te^{-t},$$
$$x_3' = -22x_1 + 14x_2 - 8x_3 + \sin(t);$$
$$x_1(0) = -2, \qquad x_2(0) = 1, \qquad x_3(0) = 0.$$

In matrix form, we have $X' = AX + G$, with

$$A = \begin{bmatrix} 31 & -21 & 9 \\ 44 & -30 & 12 \\ -22 & 14 & -8 \end{bmatrix} \quad \text{and} \quad G = \begin{bmatrix} -e^{-3t} \\ 2te^{-t} \\ \sin(t) \end{bmatrix}.$$

The characteristic equation of A is $(\lambda + 2)^2(\lambda + 3) = 0$, and the eigenvalues are $\lambda_1 = -2$, $\lambda_2 = -2$, and $\lambda_3 = -3$. Corresponding eigenvectors are

$$\xi^{(1)} = \begin{bmatrix} 1 \\ 2 \\ 1 \end{bmatrix}, \qquad \xi^{(2)} = \begin{bmatrix} 4 \\ 5 \\ -3 \end{bmatrix}, \quad \text{and} \quad \xi^{(3)} = \begin{bmatrix} 3 \\ 4 \\ -2 \end{bmatrix}.$$

Since A has three linearly independent eigenvectors, A is diagonalizable. Let

$$P = \begin{bmatrix} 1 & 4 & 3 \\ 2 & 5 & 4 \\ 1 & -3 & -2 \end{bmatrix}.$$

We know that

$$P^{-1}AP = \begin{bmatrix} -2 & 0 & 0 \\ 0 & -2 & 0 \\ 0 & 0 & -3 \end{bmatrix}.$$

After some computation, we find that

$$P^{-1} = \begin{bmatrix} 2 & -1 & 1 \\ 8 & -5 & 2 \\ -11 & 7 & -3 \end{bmatrix}.$$

Finally, let $X = PZ$ to transform the system $X' = AX + G$ into $Z' = (P^{-1}AP)Z + K$, where

$$K = P^{-1}G = \begin{bmatrix} 2 & -1 & 1 \\ 8 & -5 & 2 \\ -11 & 7 & -3 \end{bmatrix} \begin{bmatrix} -e^{-3t} \\ 2te^{-t} \\ \sin(t) \end{bmatrix} = \begin{bmatrix} -2e^{-3t} - 2te^{-t} + \sin(t) \\ -8e^{-3t} - 10te^{-t} + 2\sin(t) \\ 11e^{-3t} + 14te^{-t} - 3\sin(t) \end{bmatrix}.$$

The uncoupled system is

$$z_1' + 2z_1 = -2e^{-3t} - 2te^{-t} + \sin(t)$$
$$z_2' + 2z_2 = -8e^{-3t} - 10te^{-t} + 2\sin(t)$$
$$z_3' + 3z_3 = 11e^{-3t} + 14te^{-t} - 3\sin(t).$$

These differential equations are not difficult to solve, although some labor is involved. Using the method of undetermined coefficients, we find that

$$z_1(t) = c_1 e^{-2t} + 2e^{-3t} - 2te^{-t} + 2e^{-t} + \tfrac{2}{5}\sin(t) - \tfrac{1}{5}\cos(t)$$
$$z_2(t) = c_2 e^{-2t} + 8e^{-3t} - 10te^{-t} + 10e^{-t} + \tfrac{4}{5}\sin(t) - \tfrac{2}{5}\cos(t)$$
$$z_3(t) = c_3 e^{-3t} + 11te^{-3t} + 7te^{-t} - \tfrac{7}{2}e^{-t} - \tfrac{9}{10}\sin(t) + \tfrac{3}{10}\cos(t).$$

Now recall that $X = PZ$ to obtain the general solution of the original system $X' = AX + G$:

$$x_1(t) = (c_1 + 4c_2)e^{-2t} + 3c_3 e^{-3t} + (34 + 33t)e^{-3t} - 21te^{-t}$$
$$+ \tfrac{63}{2}e^{-t} + \tfrac{9}{10}\sin(t) - \tfrac{9}{10}\cos(t)$$
$$x_2(t) = (2c_1 + 5c_2)e^{-2t} + 4c_3 e^{-3t} + 44(1 + t)e^{-3t} - 26te^{-t}$$
$$+ 40e^{-t} + \tfrac{6}{5}\sin(t) - \tfrac{6}{5}\cos(t)$$
$$x_3(t) = (c_1 - 3c_2)e^{-2t} - 2c_3 e^{-3t} - (22 + 22t)e^{-3t} + 14te^{-t}$$
$$- 21e^{-t} - \tfrac{1}{5}\sin(t) + \tfrac{2}{5}\cos(t).$$

We must solve for the constants to satisfy the initial conditions. Substitute the initial conditions into this general solution to obtain

$$c_1 + 4c_2 + 3c_3 = -\tfrac{333}{5}$$
$$2c_1 + 5c_2 + 4c_3 = -\tfrac{409}{5}$$
$$c_1 - 3c_2 - 2c_3 = \tfrac{213}{5}.$$

We can solve these equations directly. However, here is a useful observation. Notice that the matrix of coefficients of the c_j's is exactly the 3×3 matrix P. Therefore, this system of algebraic equations is actually the system

$$PC = \begin{bmatrix} -\tfrac{333}{5} \\ -\tfrac{409}{5} \\ \tfrac{213}{5} \end{bmatrix}.$$

We can solve for C by multiplying the last equation on the left by P^{-1}, which we have previously calculated. We get

$$C = P^{-1} \begin{bmatrix} -\tfrac{333}{5} \\ -\tfrac{409}{5} \\ \tfrac{213}{5} \end{bmatrix} = \begin{bmatrix} 2 & -1 & 1 \\ 8 & -5 & 2 \\ -11 & 7 & -3 \end{bmatrix} \begin{bmatrix} -\tfrac{333}{5} \\ -\tfrac{409}{5} \\ \tfrac{213}{5} \end{bmatrix} = \begin{bmatrix} -\tfrac{44}{5} \\ -\tfrac{193}{5} \\ \tfrac{161}{5} \end{bmatrix}.$$

The solution of the initial value problem is

$$x_1(t) = -\tfrac{816}{5}e^{-2t} + \tfrac{483}{5}e^{-3t} + (34 + 33t)e^{-3t} - 21te^{-t}$$
$$+ \tfrac{63}{2}e^{-t} + \tfrac{9}{10}\sin(t) - \tfrac{9}{10}\cos(t)$$
$$x_2(t) = -\tfrac{1053}{5}e^{-2t} + \tfrac{644}{5}e^{-3t} + 44(1 + t)e^{-3t} - 26te^{-t}$$
$$+ 40e^{-t} + \tfrac{6}{5}[\sin(t) - \cos(t)]$$
$$x_3(t) = -\tfrac{623}{5}e^{-2t} - \tfrac{322}{5}e^{-3t} - (22 + 22t)e^{-3t} + 14te^{-t} - 21e^{-t}$$
$$- \tfrac{1}{5}[\sin(t) - 2\cos(t)]. \quad \blacksquare$$

The method we have discussed in this section applies only to linear systems $X' = AX + G$ in which A is diagonalizable. In Section 14.7, we will discuss a method for solving the linear system $X' = AX + G$ in the case in which A is not diagonalizable.

PROBLEMS FOR SECTION 14.5

In each of Problems 1 through 9, find the general solution of the system. Express all solutions in terms of real-valued functions.

1. $x_1' = -2x_1 + x_2$
$x_2' = -4x_1 + 3x_2 + 10\cos(t)$

2. $x_1' = 3x_1 + 3x_2 + 8$
$x_2' = x_1 + 5x_2 + 4e^{3t}$

3. $x_1' = x_1 + x_2 + 6e^{3t}$
$x_2' = x_1 + x_2 + 4$

4. $x_1' = 6x_1 + 5x_2 - 4\cos(3t)$
$x_2' = x_1 + 2x_2 + 8$

5. $x_1' = 3x_1 - 2x_2 + 3e^{2t}$
$x_2' = 9x_1 - 3x_2 + e^{2t}$

6. $x_1' = 2x_1 + x_2 - 2x_3 - 2$
$x_2' = 3x_1 - 2x_2 + 5e^{2t}$
$x_3' = 3x_1 + x_2 - 3x_3 + 9t$

7. $x_1' = 3x_1 - x_2 + x_3 + 12e^{4t}$
$x_2' = x_1 + x_2 - x_3 + 4\cos(2t)$
$x_3' = x_1 - x_2 + x_3 + 4\cos(2t)$

8. $x_1' = x_1 - x_2 - x_3 + 4e^t$
$x_2' = x_1 - x_2 + 2e^{-3t}$
$x_3' = x_1 - x_3 - 2e^{-3t}$

9. $x_1' = x_1 + x_2 + e^{2t}$
$x_2' = x_1 + x_2 - e^{2t}$
$x_3' = 4x_3 + 2x_4 + 10e^{6t}$
$x_4' = 3x_3 + 3x_4 + 15e^{6t}$

In each of Problems 10 through 16, solve the initial value problem. Express each solution in terms of real-valued functions.

10. $x_1' = x_1 + x_2 + 6e^{2t}$,
$x_2' = x_1 + x_2 + 2e^{2t}$;
$x_1(0) = 6, \quad x_2(0) = 0$

11. $x_1' = x_1 - 2x_2 + 2t$,
$x_2' = -x_1 + 2x_2 + 5$;
$x_1(0) = 13, \quad x_2(0) = 12$

12. $x_1' = 2x_1 - 5x_2 + 5\sin(t)$,
$x_2' = x_1 - 2x_2 + 2\sin(t)$;
$x_1(0) = 10, \quad x_2(0) = 5$

13. $x_1' = -2x_2 + \tfrac{1}{2}t$,
$x_2' = x_1 + 2x_2 - \tfrac{1}{2}t$;
$x_1(0) = 0, \quad x_2(0) = 0$

14. $x_1' = 5x_1 - 4x_2 + 4x_3 - 3e^{-3t}$,
$x_2' = 12x_1 - 11x_2 + 12x_3 + t$,
$x_3' = 4x_1 - 4x_2 + 5x_3 + t$;
$x_1(0) = 1, \quad x_2(0) = -1, \quad x_3(0) = 2$

15. $x_1' = 3x_1 - x_2 - x_3$,
$x_2' = x_1 + x_2 - x_3 + t$,
$x_3' = x_1 - x_2 + x_3 + 2e^t$;
$x_1(0) = 1, \quad x_2(0) = 2, \quad x_3(0) = -2$

16. $x'_1 = 3x_1 - 4x_2 + 2,$
$x'_2 = 2x_1 - 3x_2 + 4t,$
$x'_3 = x_3 - 2x_4 + 14,$
$x'_4 = -6x_3 + 7t;$
$x_1(0) = 2, \quad x_2(0) = 0, \quad x_3(0) = 1, \quad x_4(0) = -1$

14.6 Exponential Matrix Solutions of $X' = AX$

We now have a procedure for solving $X' = AX + G$ when A is an $n \times n$ diagonalizable matrix of constants. In this and the next section, we will discuss a method to use if A is not diagonalizable (does not have n linearly independent eigenvectors). The key is to introduce a matrix called the *exponential matrix*.

As we have done before, we will look to simple cases for an idea of how to proceed. The first order equation $x' = ax$ has general solution $x(t) = ce^{at}$, with c any constant. This suggests that $X' = AX$ might have a solution of the form $e^{At}C$, in which C is an $n \times 1$ matrix of constants. We therefore seek to assign a meaning to the symbol e^{At} when A is a matrix.

In order to do this, recall the Maclaurin series

$$e^t = \sum_{j=0}^{\infty} \frac{1}{j!} t^j = 1 + t + \frac{1}{2!} t^2 + \frac{1}{3!} t^3 + \cdots$$

for all t. This suggests that we define

$$e^{At} = I + At + \frac{1}{2!} A^2 t^2 + \frac{1}{3!} A^3 t^3 + \cdots,$$

in which we have written I_n as I for convenience. It can be shown that this infinite series of matrices converges in the sense that the infinite series of constants in the i, j place converges for all t. The matrix function e^{At} is called the *matrix exponential function* and is an $n \times n$ matrix whose elements are usually infinite series in t.

Because A is a constant matrix, we can compute

$$\frac{d}{dt} e^{At} = \frac{d}{dt} \left(I + At + \frac{1}{2!} A^2 t^2 + \frac{1}{3!} A^3 t^3 + \cdots \right)$$

$$= A + A^2 t + \frac{1}{2!} A^3 t^2 + \cdots$$

$$= A \left(I + At + \frac{1}{2!} A^2 t^2 + \frac{1}{3!} A^3 t^3 + \cdots \right) = Ae^{At}.$$

The formula

$$\frac{d}{dt} e^{At} = Ae^{At}$$

is in appearance very much like the derivative formula for the real-valued exponential function $(d/dt)e^{at} = ae^{at}$. Of course, e^{At} is an $n \times n$ matrix, so Ae^{At} is a product of two $n \times n$ matrices and is itself an $n \times n$ matrix.

One must be careful in computing with exponential matrices because matrix multiplication is noncommutative. Problem 17 at the end of this section requests a proof that

$$e^{At}e^{Bt} = e^{(A+B)t} \tag{14.14}$$

if $AB = BA$. If $AB \neq BA$, equation (14.14) is not true in general.

We will now use these facts to solve $X' = AX$, considering $X' = AX + G$ in the next section. For any constant $n \times 1$ matrix C, $e^{At}C$ is a solution of $X' = AX$ (verify this by substituting $X = e^{At}C$ into the system). Further, if $C^{(1)}, C^{(2)}, \ldots, C^{(n)}$ are any linearly independent $n \times 1$ constant matrices,

$$X^{(1)} = e^{At}C^{(1)} \qquad X^{(2)} = e^{At}C^{(2)}, \qquad \ldots, \qquad X^{(n)} = e^{At}C^{(n)} \tag{14.15}$$

are n linearly independent solutions of $X' = AX$ and hence form a basis for the solution space of this system.

In theory, the solutions (14.15) form a fundamental set of solutions of the system. In practice, these solutions are difficult to work with because usually each $e^{At}C^{(j)}$ is an infinite series of matrices which we have no way of computing in closed form.

Notice, however, that $C^{(1)}, C^{(2)}, \ldots, C^{(n)}$ in (14.15) are *any* n linearly independent $n \times 1$ matrices. We will now examine a remarkable procedure for choosing n specific matrices $C^{(1)}, \ldots, C^{(n)}$ in such a way that each $e^{At}C^{(j)}$ in (14.15) is a finite sum. In this event, (14.15) will provide us with not only a fundamental set of solutions but ones which we have some realistic chance of computing. The following preliminary result will be needed in our discussion.

LEMMA 14.1

Let C be an $n \times 1$ matrix of constants. Then

1. $e^{\lambda I t}C = e^{\lambda t}C$.
2. $e^{At}C = e^{\lambda t}e^{(A-\lambda I)t}C$.

Proof of (1) I denotes the $n \times n$ identity matrix. Since $I^k = I$ for any positive integer k,

$$\begin{aligned}
e^{\lambda I t}C &= \left[I + (\lambda I t) + \frac{1}{2!}(\lambda I t)^2 + \frac{1}{3!}(\lambda I t)^3 + \cdots\right]C \\
&= \left[I + (\lambda t)I + \frac{1}{2!}(\lambda t)^2 I + \frac{1}{3!}(\lambda t)^3 I + \cdots\right]C \\
&= \left[1 + (\lambda t) + \frac{1}{2!}(\lambda t)^2 + \frac{1}{3!}(\lambda t)^3 + \cdots\right]IC \\
&= e^{\lambda t}IC = e^{\lambda t}C.
\end{aligned}$$

Proof of (2) First, observe that λI and $A - \lambda I$ commute because

$$(\lambda I)(A - \lambda I) = \lambda I(A - \lambda I) = \lambda(A - \lambda I) = (A - \lambda I)(\lambda) = (A - \lambda I)(\lambda I).$$

Using part (1) of the lemma and equation (14.14), we have

$$e^{\lambda t}e^{(A-\lambda I)t} = e^{\lambda It}e^{(A-\lambda I)t} = e^{(\lambda I + A - \lambda I)t} = e^{At}.$$

Therefore,

$$e^{\lambda t}e^{(A-\lambda I)t}C = e^{At}C,$$

proving (14.15). ■

Now return to the problem of choosing C so that $e^{At}C$ is a finite sum. The key observation is that, if $(A - \lambda I)^k C = O$ for some λ and some positive integer k, we will have $(A - \lambda I)^j C = O$ for all $j \geq k$, and the infinite series for $e^{(A-\lambda I)t}C$ will terminate after k terms. In fact, under these circumstances,

$$e^{(A-\lambda I)t}C = \left[I + (A - \lambda I)t + \frac{1}{2!}(A - \lambda I)^2 t^2 + \cdots + \frac{1}{(k-1)!}(A - \lambda I)^{k-1} t^{k-1} \right.$$
$$\left. + \frac{1}{k!}(A - \lambda I)^k t^k + \cdots \right] C$$
$$= \left[I + (A - \lambda I)t + \frac{1}{2!}(A - \lambda I)^2 t^2 + \cdots + \frac{1}{(k-1)!}(A - \lambda I)^{k-1} t^{k-1} \right] C.$$

Using the lemma, the solution $e^{At}C$ can be written as the finite sum

$$e^{At}C = e^{\lambda t}e^{(A-\lambda I)t}C$$
$$= e^{\lambda t}\left[I + (A - \lambda I)t + \frac{1}{2!}(A - \lambda I)^2 t^2 + \cdots \right.$$
$$\left. + \frac{1}{(k-1)!}(A - \lambda I)^{k-1} t^{k-1} \right] C$$
$$= e^{\lambda t}\left[C + (A - \lambda I)Ct + \frac{1}{2!}(A - \lambda I)^2 Ct^2 + \cdots \right.$$
$$\left. + \frac{1}{(k-1)!}(A - \lambda I)^{k-1} Ct^{k-1} \right].$$

Now we know what kind of C to look for: an $n \times 1$ constant matrix C such that, for some number λ and some positive integer k,

$$(A - \lambda I)^k C = O.$$

We will outline a method for producing such numbers λ and k and an $n \times 1$ matrix C satisfying these requirements. In fact, we must produce n linearly independent matrices C to use (2) to find a basis for the solution space of $X' = AX$.

Objective: Produce n linearly independent $n \times 1$ constant matrices $C^{(1)}$, $C^{(2)}, \ldots, C^{(n)}$ such that each solution $e^{At}C^{(j)}$ can be computed as a finite sum.

Procedure: Begin by finding the eigenvalues of A. If we can find n linearly independent eigenvectors, we do not need the exponential matrix, and we can write the general solution using previous methods.

Suppose, then, that A does not have n linearly independent eigenvectors. Then A cannot have n distinct eigenvalues; hence, at least one eigenvalue has multiplicity greater than 1. Suppose that the k distinct eigenvalues of A are $\lambda_1, \ldots, \lambda_k$, with multiplicities $\mu_1, \mu_2, \ldots, \mu_k$, respectively. Then

$$\mu_1 + \mu_2 + \cdots + \mu_k = n, \tag{14.16}$$

the degree of the characteristic polynomial of A. It is a fact, which we will not prove, that

>   an eigenvalue with multiplicity μ_j will provide us with
>   exactly μ_j linearly independent solutions of the system $X' = AX$.

This fact, coupled with equation (14.16), means that we will eventually obtain n linearly independent solutions from the k distinct eigenvalues $\lambda_1, \ldots, \lambda_k$. Now follow these steps.

Step 1   Produce as many linearly independent eigenvectors as possible corresponding to each eigenvalue $\lambda_1, \ldots, \lambda_k$. These eigenvectors yield at least k linearly independent solutions, each of the form $\xi e^{\lambda t}$.

Step 2   Choose an eigenvalue λ_j of multiplicity μ_j which thus far has produced fewer than μ_j linearly independent solutions. Find an $n \times 1$ matrix C which is linearly independent from the eigenvectors found in step 1 and such that

$$(A - \lambda I)^2 C = O$$

but

$$(A - \lambda I)C \neq O.$$

Then $e^{At}C$ is another solution of $X' = AX$. Further,

$$e^{At}C = e^{\lambda t}e^{(A - \lambda I)t}C = e^{\lambda t}[C + (A - \lambda I)Ct],$$

a finite sum we can compute explicitly.

If λ_j has multiplicity 2, the solution just found using λ_j and the solution just found from the eigenvector of λ_j in step 1 constitute the entire set of μ_j linearly independent solutions which can be obtained from this eigenvalue.

Step 3   If λ_j has multiplicity 3, find an $n \times 1$ matrix C, linearly independent from those already found associated with λ_j in steps 1 and 2, satisfying

$$(A - \lambda I)^3 C = O$$

but

$$(A - \lambda I)^2 C \neq O.$$

For such a matrix C, we have another solution $e^{At}C$ which is linearly independent from those already found. Further,

$$e^{At}C = e^{\lambda t}\left[C + (A - \lambda I)Ct + \frac{1}{2!}(A - \lambda I)^2 Ct^2 \right],$$

a finite sum we can compute.

If λ_j has multiplicity greater than 3, continue, this time seeking an $n \times 1$ matrix C linearly independent from those found thus far associated with λ_j and satisfying

$$(A - \lambda I)^4 C = O \quad \text{but} \quad (A - \lambda I)^3 C \neq O.$$

This yields a solution $e^{At}C$ linearly independent from those already found from λ_j. Continue this process until μ_j linearly independent solutions have been produced from λ_j.

Now repeat this process for each eigenvalue of multiplicity greater than 1. Each λ_j yields μ_j linearly independent solutions, and $\lambda_1, \ldots, \lambda_k$ in total yield $\mu_1 + \mu_2 + \cdots + \mu_k = n$ linearly independent solutions, which form a basis for the solution space of the system $X' = AX$.

EXAMPLE 14.11

Solve the system $X' = AX$, where

$$A = \begin{bmatrix} 4 & 1 & 3 \\ 0 & 4 & 1 \\ 0 & 0 & 4 \end{bmatrix}.$$

The characteristic equation of A is $(\lambda - 4)^3 = 0$, so A has eigenvalue $\lambda = 4$ with multiplicity $\mu = 3$. Corresponding to $\lambda = 4$, we find that the only eigenvectors are nonzero multiples of $\begin{bmatrix} 1 \\ 0 \\ 0 \end{bmatrix}$. Thus, A is not diagonalizable, and at this point we have only one solution, $\begin{bmatrix} 1 \\ 0 \\ 0 \end{bmatrix} e^{4t}$.

We need two additional solutions to form a basis for the solution space. Look at the equation $(A - 4I)^2 C = O$. First, we must compute

$$(A - 4I)^2 = \begin{bmatrix} 0 & 1 & 3 \\ 0 & 0 & 1 \\ 0 & 0 & 0 \end{bmatrix} \begin{bmatrix} 0 & 1 & 3 \\ 0 & 0 & 1 \\ 0 & 0 & 0 \end{bmatrix} = \begin{bmatrix} 0 & 0 & 1 \\ 0 & 0 & 0 \\ 0 & 0 & 0 \end{bmatrix}.$$

The system of equations $(A - 4I)^2 C = O$ is

$$0c_1 + 0c_2 + c_3 = 0$$
$$0c_1 + 0c_2 + 0c_3 = 0$$
$$0c_1 + 0c_2 + 0c_3 = 0.$$

Then $c_3 = 0$, but c_1 and c_2 are arbitrary, and

$$C = \begin{bmatrix} c_1 \\ c_2 \\ 0 \end{bmatrix}.$$

To be specific, choose $c_1 = 0$ and $c_2 = 1$ so that

$$C = \begin{bmatrix} 0 \\ 1 \\ 0 \end{bmatrix}.$$

We find that $(A - 4I)C \neq O$. Therefore, a second solution, linearly independent from the first, is $e^{At}C$. Further, we can compute this solution as a finite sum

$$e^{At}C = e^{4t}[C + (A - 4I)Ct]$$

$$= e^{4t}\left(\begin{bmatrix} 0 \\ 1 \\ 0 \end{bmatrix} + \begin{bmatrix} 0 & 1 & 3 \\ 0 & 0 & 1 \\ 0 & 0 & 0 \end{bmatrix}\begin{bmatrix} 0 \\ 1 \\ 0 \end{bmatrix}t\right)$$

$$= e^{4t}\left(\begin{bmatrix} 0 \\ 1 \\ 0 \end{bmatrix} + \begin{bmatrix} 1 \\ 0 \\ 0 \end{bmatrix}t\right) = e^{4t}\begin{bmatrix} t \\ 1 \\ 0 \end{bmatrix}.$$

Now seek a third solution. Look at 3×1 matrices C such that $(A - 4I)^3C = O$ but $(A - 4I)^2C \neq O$. We find that $(A - 4I)^3 = O$, so any 3×1 matrix C is a candidate. To avoid duplicating the first two solutions, choose $C = \begin{bmatrix} 0 \\ 0 \\ 1 \end{bmatrix}$. We find that C satisfies $(A - 4I)^2C \neq O$. This choice of C yields the solution $e^{At}C$, which we compute as

$$e^{At}C = e^{4t}\left[C + (A - 4I)Ct + \frac{1}{2!}(A - 4I)^2Ct^2 \right]$$

$$= e^{4t}\left(\begin{bmatrix} 0 \\ 0 \\ 1 \end{bmatrix} + \begin{bmatrix} 0 & 1 & 3 \\ 0 & 0 & 1 \\ 0 & 0 & 0 \end{bmatrix}\begin{bmatrix} 0 \\ 0 \\ 1 \end{bmatrix}t + \begin{bmatrix} 0 & 0 & 1 \\ 0 & 0 & 0 \\ 0 & 0 & 0 \end{bmatrix}\begin{bmatrix} 0 \\ 0 \\ 1 \end{bmatrix}\tfrac{1}{2}t^2\right)$$

$$= e^{4t}\left(\begin{bmatrix} 0 \\ 0 \\ 1 \end{bmatrix} + \begin{bmatrix} 3 \\ 1 \\ 0 \end{bmatrix}t + \begin{bmatrix} 1 \\ 0 \\ 0 \end{bmatrix}\tfrac{1}{2}t^2\right)$$

$$= e^{4t}\begin{bmatrix} 3t + \tfrac{1}{2}t^2 \\ t \\ 1 \end{bmatrix}.$$

This is a third solution. These three solutions are linearly independent, and the general solution is

$$\varphi(t) = c_1e^{4t}\begin{bmatrix} 1 \\ 0 \\ 0 \end{bmatrix} + c_2e^{4t}\begin{bmatrix} t \\ 1 \\ 0 \end{bmatrix} + c_3e^{4t}\begin{bmatrix} 3t + t^2/2 \\ t \\ 1 \end{bmatrix}$$

$$= \begin{bmatrix} e^{4t}[c_1 + (c_2 + 3c_3)t + c_3t^2/2] \\ e^{4t}[c_2 + c_3t] \\ c_3e^{4t} \end{bmatrix}.$$

If we wish, we can form the fundamental matrix

$$\Omega = e^{4t}\begin{bmatrix} 1 & t & (3t + t^2/2) \\ 0 & 1 & t \\ 0 & 0 & 1 \end{bmatrix}$$

and express the general solution as

$$\varphi = \Omega C. \quad \blacksquare$$

EXAMPLE 14.12

Solve the system $X' = AX$, with

$$A = \begin{bmatrix} 2 & 1 & 0 & 3 \\ 0 & 2 & 1 & 1 \\ 0 & 0 & 2 & 4 \\ 0 & 0 & 0 & 4 \end{bmatrix}.$$

The eigenvalues of A are 2, 2, 2, and 4; 2 has multiplicity 3, and 4 has multiplicity 1. Corresponding to the eigenvalue 4, we find that the eigenvectors have the form

$$\begin{bmatrix} 9\beta \\ 6\beta \\ 8\beta \\ 4\beta \end{bmatrix}, \quad \beta \neq 0.$$

If we choose $\beta = 1$, we obtain one solution of the system,

$$e^{4t} \begin{bmatrix} 9 \\ 6 \\ 8 \\ 4 \end{bmatrix}.$$

Corresponding to the eigenvalue 2, we find that the eigenvectors have the form

$$\begin{bmatrix} \alpha \\ 0 \\ 0 \\ 0 \end{bmatrix}, \quad \alpha \neq 0.$$

We may choose $\alpha = 1$ and obtain a second solution,

$$e^{2t} \begin{bmatrix} 1 \\ 0 \\ 0 \\ 0 \end{bmatrix}.$$

The two solutions found thus far are linearly independent (having been found from two distinct eigenvalues). Additional solutions will have to come from the eigenvalue 2 of multiplicity greater than 1. Compute

$$(A - 2I)^2 = \begin{bmatrix} 0 & 0 & 1 & 7 \\ 0 & 0 & 0 & 6 \\ 0 & 0 & 0 & 8 \\ 0 & 0 & 0 & 4 \end{bmatrix}.$$

Solve the system of algebraic equations $(A - 2I)^2 C = O$ to get

$$C = \begin{bmatrix} \alpha \\ \beta \\ 0 \\ 0 \end{bmatrix}.$$

We will choose $\alpha = 0$ and $\beta = 1$ to have C linearly independent from the 4×1 matrix found previously as an eigenvector associated with eigenvalue 2. Thus, choose

$$C = \begin{bmatrix} 0 \\ 1 \\ 0 \\ 0 \end{bmatrix}.$$

We find that $(A - 2I)C \neq O$. A third solution is therefore

$$e^{At}C = e^{2t}[C + (A - 2I)Ct]$$

$$= e^{2t}\left(\begin{bmatrix} 0 \\ 1 \\ 0 \\ 0 \end{bmatrix} + \begin{bmatrix} 0 & 1 & 0 & 3 \\ 0 & 0 & 1 & 1 \\ 0 & 0 & 0 & 4 \\ 0 & 0 & 0 & 2 \end{bmatrix}\begin{bmatrix} 0 \\ 1 \\ 0 \\ 0 \end{bmatrix}t\right)$$

$$= e^{2t}\left(\begin{bmatrix} 0 \\ 1 \\ 0 \\ 0 \end{bmatrix} + \begin{bmatrix} 1 \\ 0 \\ 0 \\ 0 \end{bmatrix}t\right) = e^{2t}\begin{bmatrix} t \\ 1 \\ 0 \\ 0 \end{bmatrix}.$$

The three solutions we have found up to this point are linearly independent.

We need one more solution. Since the eigenvalue 2 has multiplicity 3 and we have obtained only two linearly independent solutions from it, we will seek another solution from this eigenvalue. (Eigenvalue 4 has multiplicity 1, and we have already used it to obtain one solution.) Look at the system $(A - 2I)^3 C = O$. We find that

$$(A - 2I)^3 = \begin{bmatrix} 0 & 0 & 0 & 18 \\ 0 & 0 & 0 & 12 \\ 0 & 0 & 0 & 16 \\ 0 & 0 & 0 & 8 \end{bmatrix}$$

and that solutions of $(A - 2I)^3 C = O$ have the form

$$C = \begin{bmatrix} \alpha \\ \beta \\ \gamma \\ 0 \end{bmatrix}.$$

If we choose $\alpha = \beta = \gamma = 1$, we obtain a 4×1 matrix different from those used previously to generate solutions from the eigenvalue 2. Thus, choose

$$C = \begin{bmatrix} 1 \\ 1 \\ 1 \\ 0 \end{bmatrix}.$$

Then $(A - 2I)^2 C \neq O$, and we have a fourth solution, $e^{At}C$. Calculate this solution as

$$e^{At}C = e^{2t}\left[C + (A - 2I)Ct + (A - 2I)^2 \frac{Ct^2}{2} \right]$$

$$= e^{2t}\left(\begin{bmatrix} 1 \\ 1 \\ 1 \\ 0 \end{bmatrix} + \begin{bmatrix} 0 & 1 & 0 & 3 \\ 0 & 0 & 1 & 1 \\ 0 & 0 & 0 & 4 \\ 0 & 0 & 0 & 2 \end{bmatrix}\begin{bmatrix} 1 \\ 1 \\ 1 \\ 0 \end{bmatrix}t + \begin{bmatrix} 0 & 0 & 1 & 7 \\ 0 & 0 & 0 & 6 \\ 0 & 0 & 0 & 8 \\ 0 & 0 & 0 & 4 \end{bmatrix}\begin{bmatrix} 1 \\ 1 \\ 1 \\ 0 \end{bmatrix}\tfrac{1}{2}t^2 \right)$$

$$= e^{2t}\left(\begin{bmatrix} 1 \\ 1 \\ 1 \\ 0 \end{bmatrix} + \begin{bmatrix} 1 \\ 1 \\ 0 \\ 0 \end{bmatrix}t + \begin{bmatrix} 1 \\ 0 \\ 0 \\ 0 \end{bmatrix}\tfrac{1}{2}t^2 \right)$$

$$= e^{2t}\begin{bmatrix} 1 + t + t^2/2 \\ 1 + t \\ 1 \\ 0 \end{bmatrix}.$$

We now have four linearly independent solutions of $X' = AX$ and hence a fundamental set of solutions. The general solution is

$$\varphi(t) = c_1 \begin{bmatrix} 9 \\ 6 \\ 8 \\ 4 \end{bmatrix} e^{4t} + c_2 \begin{bmatrix} 1 \\ 0 \\ 0 \\ 0 \end{bmatrix} e^{2t} + c_3 \begin{bmatrix} t \\ 1 \\ 0 \\ 0 \end{bmatrix} e^{2t} + c_4 \begin{bmatrix} 1 + t + t^2/2 \\ 1 + t \\ 1 \\ 0 \end{bmatrix} e^{2t}$$

$$= \begin{bmatrix} 9c_1 e^{4t} + [c_2 + c_4 + (c_3 + c_4)t + \tfrac{1}{2}c_4 t^2]e^{2t} \\ 6c_1 e^{4t} + [(c_3 + c_4) + c_4 t]e^{2t} \\ 8c_1 e^{4t} + c_4 e^{2t} \\ 4c_1 e^{4t} \end{bmatrix}.$$

Again, we may also define the fundamental matrix

$$\Omega = \begin{bmatrix} 9e^{4t} & e^{2t} & te^{2t} & (1 + t + t^2/2)e^{2t} \\ 6e^{4t} & 0 & e^{2t} & (1 + t)e^{2t} \\ 8e^{4t} & 0 & 0 & e^{2t} \\ 4e^{4t} & 0 & 0 & 0 \end{bmatrix},$$

in terms of which the general solution is $\varphi = \Omega C$. ∎

As an illustration, observe what happens in the last example if we try to find more solutions from an eigenvalue than the multiplicity of that eigenvalue. We had the eigenvalue 4 of multiplicity 1 and found from it one solution. If we use $\lambda = 4$ in step 2 of the procedure to try to generate another solution, we look for a 4×1 matrix C such that $(A - 4I)^2 C = O$. After squaring $A - 4I$, we get the system

$$\begin{bmatrix} 4 & -4 & 1 & -5 \\ 0 & 4 & -4 & 2 \\ 0 & 0 & 4 & -8 \\ 0 & 0 & 0 & 0 \end{bmatrix} C = O.$$

Solve this system to conclude that C must be of the form

$$C = \begin{bmatrix} 9\alpha \\ 6\alpha \\ 8\alpha \\ 4\alpha \end{bmatrix}, \qquad \alpha \neq 0.$$

But we have already found this choice for C! In fact, it satisfies $(A - 4I)C = O$. We therefore cannot use C of this form to generate a new solution which is linearly independent from the solution already found associated with 4. In other words, the eigenvalue 4 has multiplicity 1, and one solution is all we will get from it.

EXAMPLE 14.13

Solve the system $X' = AX$, where

$$A = \begin{bmatrix} 1 & 1 & -2 & 0 & 0 \\ 0 & 1 & 0 & 0 & 0 \\ 0 & 0 & 1 & 0 & 0 \\ 0 & 0 & 0 & 2 & 1 \\ 0 & 0 & 0 & 0 & 2 \end{bmatrix}.$$

The eigenvalues of A are 1, 1, 1, 2, and 2. The eigenvalue 1 has multiplicity 3, and 2 has multiplicity 2. Corresponding to the eigenvalue 1, we find eigenvectors

$$\begin{bmatrix} \alpha \\ 2\beta \\ \beta \\ 0 \\ 0 \end{bmatrix},$$

with not both α and β zero. If we choose $\alpha = 1$ and $\beta = 0$, we obtain one solution of the system,

$$\begin{bmatrix} 1 \\ 0 \\ 0 \\ 0 \\ 0 \end{bmatrix} e^t.$$

Upon letting $\alpha = 0$ and $\beta = 1$, we obtain a second, linearly independent solution,

$$\begin{bmatrix} 0 \\ 2 \\ 1 \\ 0 \\ 0 \end{bmatrix} e^t.$$

Corresponding to the eigenvalue 2, we find eigenvectors

$$\begin{bmatrix} 0 \\ 0 \\ 0 \\ \gamma \\ 0 \end{bmatrix}, \qquad \gamma \neq 0.$$

Choosing $\gamma = 1$, we obtain a third solution, linearly independent from the preceding two:

$$\begin{bmatrix} 0 \\ 0 \\ 0 \\ 1 \\ 0 \end{bmatrix} e^{2t}.$$

We need two more linearly independent solutions. Since 1 has multiplicity 3 and thus far has produced only two linearly independent solutions, we can expect to use the eigenvalue 1 to find one more solution. Likewise, the eigenvalue 2 has multiplicity 2 and should yield one more solution, linearly independent from the others. For the eigenvalue 1, consider

$$(A - I)^2 = \begin{bmatrix} 0 & 0 & 0 & 0 & 0 \\ 0 & 0 & 0 & 0 & 0 \\ 0 & 0 & 0 & 0 & 0 \\ 0 & 0 & 0 & 1 & 2 \\ 0 & 0 & 0 & 0 & 1 \end{bmatrix}.$$

If we choose

$$C = \begin{bmatrix} 0 \\ 0 \\ 1 \\ 0 \\ 0 \end{bmatrix},$$

then

$$(A - I)C = \begin{bmatrix} -2 \\ 0 \\ 0 \\ 0 \\ 0 \end{bmatrix};$$

hence, a fourth, linearly independent solution (the third one associated with eigenvalue 1) is

$$\left(\begin{bmatrix} 0 \\ 0 \\ 1 \\ 0 \\ 0 \end{bmatrix} + \begin{bmatrix} -2 \\ 0 \\ 0 \\ 0 \\ 0 \end{bmatrix} t \right) e^t,$$

or

$$\begin{bmatrix} -2t \\ 0 \\ 1 \\ 0 \\ 0 \end{bmatrix} e^t.$$

For a second solution associated with eigenvalue 2, consider

$$(A - 2I)^2 = \begin{bmatrix} 1 & -2 & 4 & 0 & 0 \\ 0 & 1 & 0 & 0 & 0 \\ 0 & 0 & 1 & 0 & 0 \\ 0 & 0 & 0 & 0 & 0 \\ 0 & 0 & 0 & 0 & 0 \end{bmatrix}.$$

Let

$$C = \begin{bmatrix} 0 \\ 0 \\ 0 \\ 0 \\ 1 \end{bmatrix}.$$

Then

$$(A - 2I)C = \begin{bmatrix} 0 \\ 0 \\ 0 \\ 1 \\ 0 \end{bmatrix}.$$

Thus, a second, linearly independent solution corresponding to the eigenvalue 2 is

$$\left(\begin{bmatrix} 0 \\ 0 \\ 0 \\ 0 \\ 1 \end{bmatrix} + \begin{bmatrix} 0 \\ 0 \\ 0 \\ 1 \\ 0 \end{bmatrix} t \right) e^{2t},$$

or

$$\begin{bmatrix} 0 \\ 0 \\ 0 \\ t \\ 1 \end{bmatrix} e^{2t}.$$

A fundamental matrix for the system $X' = AX$ is

$$\Omega(t) = \begin{bmatrix} e^t & 0 & -2te^t & 0 & 0 \\ 0 & 2e^t & 0 & 0 & 0 \\ 0 & e^t & e^t & 0 & 0 \\ 0 & 0 & 0 & e^{2t} & te^{2t} \\ 0 & 0 & 0 & 0 & e^{2t} \end{bmatrix}.$$

The first three columns were generated from the eigenvalue 1, and the last two are from the eigenvalue 2. The general solution of the system of equations is $\Omega(t)C$, in which C is a 5×1 matrix of arbitrary constants. ∎

In the next section, we will look at the nonhomogeneous system $X' = AX + G$ for the case in which A is not diagonalizable.

PROBLEMS FOR SECTION 14.6

In each of Problems 1 through 8, use the exponential matrix to find the general solution, containing only real-valued functions, of the system.

1. $x'_1 = 3x_1 + 2x_2$
$x'_2 = 3x_2$

2. $x'_1 = 2x_1$
$x'_2 = 5x_1 + 2x_2$

3. $x'_1 = 2x_1 - 4x_2$
$x'_2 = x_1 + 6x_2$

4. $x'_1 = 5x_1 - 3x_2$
$x'_2 = 3x_1 - x_2$

5. $x'_1 = 2x_1 + 5x_2 + 6x_3$
$x'_2 = 8x_2 + 9x_3$
$x'_3 = -x_2 + 2x_3$

6. $x'_1 = x_1 + 5x_2$
$x'_2 = x_2$
$x'_3 = 4x_1 + 8x_2 + x_3$

7. $x'_1 = x_1 + 5x_2 - 2x_3 + 6x_4$
$x'_2 = 3x_2 + 4x_4$
$x'_3 = 3x_3 + 4x_4$
$x'_4 = x_4$

8. $x'_1 = x_2$
$x'_2 = x_3$
$x'_3 = x_4$
$x'_4 = -x_1 - 2x_3$

In each of Problems 9 through 14, solve the initial value problem. Each solution should be in terms of real-valued functions.

9. $x'_1 = 7x_1 - x_2,$
$x'_2 = x_1 + 5x_2;$
$x_1(0) = 5, \quad x_2(0) = 3$

10. $x'_1 = 2x_1,$
$x'_2 = 5x_1 + 2x_2;$
$x_1(0) = 4, \quad x_2(0) = 3$

11. $x'_1 = -4x_1 + x_2 + x_3,$
$x'_2 = 2x_2 - 5x_3,$
$x'_3 = -4x_3;$
$x_1(0) = 0, \quad x_2(0) = 4, \quad x_3(0) = 12$

12. $x'_1 = -5x_1 + 2x_2 + x_3,$
$x'_2 = -5x_2 + 3x_3,$
$x'_3 = -5x_3;$
$x_1(0) = 2, \quad x_2(0) = -3, \quad x_3(0) = 4$

13. $x_1' = x_1 - 2x_2,$
$x_2' = x_1 - x_2,$
$x_3' = 5x_3 - 3x_4,$
$x_4' = 3x_3 - x_4;$
$x_1(0) = 2,\quad x_2(0) = -2,\quad x_3(0) = 1,\quad x_4(0) = 4$

14. $x_1' = x_1 + 4x_2,$
$x_2' = x_2,$
$x_3' = x_3,$
$x_4' = x_1 - 3x_2 + 2x_3;$
$x_1(0) = 7,\quad x_2(0) = 1,\quad x_3(0) = -4,\quad x_4(0) = -6$

15. Suppose that A and B are $n \times n$ matrices and that $AB = BA$. Prove that

$$(A + B)^2 = A^2 + 2AB + B^2.$$

Find 2×2 matrices A and B such that $(A + B)^2 \neq A^2 + 2AB + B^2$.

16. Let A and B be $n \times n$ matrices such that $AB = BA$. Prove that

$$(A + B)^k = \sum_{j=0}^{k} \frac{k!}{j!(k-j)!} A^j B^{k-j}.$$

17. Let A and B be $n \times n$ matrices such that $AB = BA$. Prove that

$$e^{(A+B)t} = e^{At}e^{Bt}.$$

14.7 Variation of Parameters for the System $X' = AX + G$

We will now develop a method for solving a nonhomogeneous system $X' = AX + G$ that is applicable whether or not A is diagonalizable. This method is analogous to variation of parameters for nth order linear differential equations. We will also point out an interesting connection with the Laplace transform.

Recall that the general solution of $X' = AX + G$ is a sum of the general solution of the homogeneous system $X' = AX$ and any particular solution of the nonhomogeneous system. Assume that we are able to find n linearly independent solutions of $X' = AX$. These n solutions can be used to form the columns of a fundamental matrix Ω for the homogeneous system. We will look for a particular solution Ψ of the nonhomogeneous system of the form

$$\Psi(t) = \Omega(t)u(t),$$

in which $u(t)$ is an $n \times 1$ matrix of functions of t. We want to choose u so that Ψ is a solution of $X' = AX + G$. Substitute Ψ into the system to get

$$(\Omega u)' = A(\Omega u) + G.$$

The rule for differentiating a product of matrices is similar in form to the product rule for functions, and we obtain

$$\Omega'u + \Omega u' = A(\Omega u) + G,$$

or

$$\Omega'u + \Omega u' = (A\Omega)u + G. \tag{14.17}$$

Now recall that each column of Ω is a solution of $X' = AX$. The kth column of $A\Omega$ is A times the kth column of Ω and hence equals the kth column of Ω'. That is, $\Omega' = A\Omega$.

Then

$$\Omega' u = (A\Omega)u,$$

and we can write equation (14.17) as

$$(A\Omega)u + \Omega u' = (A\Omega)u + G,$$

or

$$\Omega u' = G. \tag{14.18}$$

Since the columns of Ω are linearly independent, Ω has an inverse. Multiply equation (14.18) on the left by Ω^{-1} to obtain

$$u' = \Omega^{-1} G.$$

Therefore,

$$u(t) = \int \Omega^{-1}(t)G(t) \, dt. \tag{14.19}$$

Equation (14.19) is a formula for calculating the $n \times 1$ matrix u and obtaining a solution Ωu of $X' = AX + G$. Once we find u, the general solution of $X' = AX + G$ is

$$\varphi(t) = \Omega(t)C + \Omega(t)u(t),$$

since ΩC is the general solution of $X' = AX$.

EXAMPLE 14.14

Solve the system $X' = AX + G$, where

$$A = \begin{bmatrix} 1 & -10 \\ -1 & 4 \end{bmatrix} \quad \text{and} \quad G = \begin{bmatrix} e^t \\ \sin(t) \end{bmatrix}.$$

First, we seek a fundamental matrix for $X' = AX$. The eigenvalues of A are -1 and 6. An eigenvector corresponding to -1 is $\begin{bmatrix} 5 \\ 1 \end{bmatrix}$, and an eigenvector corresponding to 6 is $\begin{bmatrix} -2 \\ 1 \end{bmatrix}$. Two linearly independent solutions of $X' = AX$ are

$$\begin{bmatrix} 5 \\ 1 \end{bmatrix} e^{-t} \quad \text{and} \quad \begin{bmatrix} -2 \\ 1 \end{bmatrix} e^{6t}.$$

Use these 2×1 matrices as columns to form a fundamental matrix

$$\Omega(t) = \begin{bmatrix} 5e^{-t} & -2e^{6t} \\ e^{-t} & e^{6t} \end{bmatrix}.$$

We find that

$$\Omega^{-1}(t) = \tfrac{1}{7}\begin{bmatrix} e^t & 2e^t \\ -e^{-6t} & 5e^{-6t} \end{bmatrix}.$$

Let

$$u'(t) = \Omega^{-1}(t)G(t)$$

$$= \tfrac{1}{7}\begin{bmatrix} e^t & 2e^t \\ -e^{-6t} & 5e^{-6t} \end{bmatrix}\begin{bmatrix} e^t \\ \sin(t) \end{bmatrix} = \tfrac{1}{7}\begin{bmatrix} e^{2t} + 2e^t\sin(t) \\ -e^{-5t} + 5e^{-6t}\sin(t) \end{bmatrix}.$$

Then

$$u(t) = \int \Omega^{-1}(t)G(t)\, dt$$

$$= \frac{1}{7}\begin{bmatrix} \int [e^{2t} + 2e^t \sin(t)]\, dt \\ \int [-e^{-5t} + 5e^{-6t}\sin(t)]\, dt \end{bmatrix}$$

$$= \begin{bmatrix} \frac{1}{14}e^{2t} + \frac{1}{7}e^t[\sin(t) - \cos(t)] \\ \frac{1}{35}e^{-5t} + \frac{5}{259}e^{-6t}[-6\sin(t) - \cos(t)] \end{bmatrix}.$$

The general solution of $X' = AX + G$ is

$$X = \Omega C + \Omega u = \begin{bmatrix} 5e^{-t} & -2e^{6t} \\ e^{-t} & e^{6t} \end{bmatrix}\begin{bmatrix} c_1 \\ c_2 \end{bmatrix}$$

$$+ \begin{bmatrix} 5e^{-t} & -2e^{6t} \\ e^{-t} & e^{6t} \end{bmatrix}\begin{bmatrix} \frac{1}{14}e^{2t} + \frac{1}{7}e^t[\sin(t) - \cos(t)] \\ \frac{1}{35}e^{-5t} + \frac{5}{259}e^{-6t}[-6\sin(t) - \cos(t)] \end{bmatrix}$$

$$= \begin{bmatrix} 5c_1e^{-t} - 2c_2e^{6t} + \frac{3}{10}e^t + \frac{35}{37}\sin(t) - \frac{25}{37}\cos(t) \\ c_1e^{-t} + c_2e^{6t} + \frac{1}{10}e^t + \frac{1}{37}\sin(t) - \frac{6}{37}\cos(t) \end{bmatrix}. \quad\blacksquare$$

In this example, A was diagonalizable, so we could also have found the general solution using the method of Section 14.5.

VARIATION OF PARAMETERS AND THE LAPLACE TRANSFORM

There is an interesting connection between the variation of parameters method for systems and the Laplace transform. We will begin with an informal argument suggesting how the connection arises. Suppose we want a particular solution Ψ of the system

$$X' = AX + G, \tag{14.20}$$

in which A is an $n \times n$ matrix of real numbers. We know from the variation of parameters method that we can choose

$$\Psi(t) = \Omega(t)u(t), \tag{14.21}$$

in which

$$u(t) = \int \Omega^{-1}(t)G(t)\, dt. \tag{14.22}$$

In equation (14.22), $\Omega(t)$ is any fundamental matrix for the associated homogeneous system $X' = AX$. We will manipulate equations (14.21) and (14.22) in such a way as to suggest the Laplace transform. First, equation (14.22) is written as an indefinite integral, in which we may choose the constant of integration in any way we like. Suppose we want a particular solution Ψ of the original system satisfying $\Psi(0) = O$. Then we can write

$$u(t) = \int_0^t \Omega^{-1}(s)G(s)\, ds, \tag{14.23}$$

in which s is the variable of integration. Now equation (14.21) becomes

$$\Psi(t) = \Omega(t) \int_0^t \Omega^{-1}(s)G(s)\,ds = \int_0^t \Omega(t)\Omega^{-1}(s)G(s)\,ds. \qquad (14.24)$$

We could bring $\Omega(t)$ inside the integral sign because the integration is with respect to s. Now recall that e^{At} is a fundamental matrix for the system $X' = AX$. Although $\Omega(t)$ can be any fundamental matrix, now make the special choice $\Omega(t) = e^{At}$. Then

$$\Omega^{-1}(s) = e^{-As},$$

and

$$\Omega(t)\Omega^{-1}(s) = e^{At}e^{-As} = e^{A(t-s)} = \Omega(t-s).$$

Now equation (14.24) for a particular solution Ψ satisfying $\Psi(0) = O$ is

$$\Psi(t) = \int_0^t \Omega(t-s)G(s)\,ds.$$

This integral has the same form as the Laplace transform convolution of two functions. We will now see that this intuition is correct if we extend the Laplace transform to matrix functions.

DERIVATION OF A LAPLACE TRANSFORM EXPRESSION FOR Ψ

If M is any $n \times n$ matrix, define the Laplace transform $\mathscr{L}[M]$ to be the matrix obtained by taking the Laplace transform of each element of M. If $M = [m_{ij}(t)]$, then $\mathscr{L}[M] = [\mathscr{L}[m_{ij}](s)]$.

We cannot use the same notation now that we did in Chapter 4 with Laplace transforms because we often use capital letters for matrices. We will denote the Laplace transform $\mathscr{L}[M]$ of a matrix $M(t)$ as $\hat{M}(s)$.

Many of the usual operational rules for the Laplace transform extend immediately to matrices because we take the Laplace transform of each matrix element. In particular, if M is a differentiable $n \times n$ matrix,

$$\mathscr{L}[M'(t)] = s\hat{M}(s) - M(0).$$

We will now look for a particular solution Ψ of the system $X' = AX + G$ satisfying $\Psi(0) = O$. Since Ψ is to be a solution of the system, we must have

$$\Psi' = A\Psi + G.$$

Take the Laplace transform of this equation to get

$$\mathscr{L}[\Psi'] = \mathscr{L}[A\Psi] + \mathscr{L}[G]. \qquad (14.25)$$

Now,

$$\mathscr{L}[\Psi'] = s\hat{\Psi}(s) - \Psi(0) = s\hat{\Psi}(s)$$

because we are seeking a solution with $\Psi(0) = O$. Further, A is an $n \times n$ matrix of constants and so can be factored through the Laplace transform. Therefore, equation (14.25) can be written

$$s\hat{\Psi}(s) = A\hat{\Psi}(s) + \hat{G}(s).$$

Write $s\hat{\Psi}(s) = sI\hat{\Psi}(s)$ so that the last equation can be written

$$sI\hat{\Psi}(s) - A\hat{\Psi}(s) = \hat{G}(s).$$

Then

$$(sI - A)\hat{\Psi}(s) = \hat{G}(s). \qquad (14.26)$$

We can solve equation (14.26) for $\hat{\Psi}(s)$ if we can find the inverse of the $n \times n$ matrix $sI - A$. To do this, remember that $\Omega(t) = e^{At}$ is a fundamental matrix for the system $X' = AX$. Hence,

$$\Omega'(t) = A\Omega(t).$$

Apply the Laplace transform to this equation to get

$$\mathscr{L}[\Omega'(t)] = s\hat{\Omega}(s) - \Omega(0) = \mathscr{L}[A\Omega(t)] = A\hat{\Omega}(s).$$

But $\Omega(t) = e^{At}$, so $\Omega(0) = I$, and the last equation yields

$$s\hat{\Omega}(s) - I = A\hat{\Omega}(s).$$

Write this equation as

$$sI\hat{\Omega}(s) - A\hat{\Omega}(s) = I,$$

or

$$(sI - A)\hat{\Omega}(s) = I.$$

This proves that $sI - A$ is nonsingular, with inverse $\hat{\Omega}(s)$:

$$(sI - A)^{-1} = \hat{\Omega}(s).$$

We can now multiply both sides of equation (14.26) by $(sI - A)^{-1}$ to obtain

$$\hat{\Psi}(s) = (sI - A)^{-1}G(s) = \hat{\Omega}(s)\hat{G}(s).$$

We now have the Laplace transform $\hat{\Psi}(s)$ of the solution we want, written as the product of two matrices we presumably know. Take the inverse Laplace transform of the last equation to obtain

$$\Psi(t) = \mathscr{L}^{-1}[\hat{\Omega}(s)\hat{G}(s)]. \qquad (14.27)$$

The convolution theorem for the Laplace transform holds for a matrix product of this form, and we obtain

$$\Psi(t) = \Omega(t) * G(t). \qquad (14.28)$$

Therefore, a particular solution Ψ of the system $X' = AX + G$ may be found as the convolution of the exponential matrix e^{At}, with the nonhomogeneous term $G(t)$ in the system. Equation (14.28) provides a formula for a particular solution of the system (14.17) whenever the elements of the matrices involved have well-defined Laplace transforms. The following is an illustration of this method.

EXAMPLE 14.15

Solve the system

$$x_1' = x_1 - 4x_2 + e^{2t}$$
$$x_2' = x_1 + 5x_2 + t.$$

Write this system as $X' = AX + G$. The coefficient matrix is

$$A = \begin{bmatrix} 1 & -4 \\ 1 & 5 \end{bmatrix}.$$

The only eigenvalue of A is 3, with eigenvector $\begin{bmatrix} -2 \\ 1 \end{bmatrix}$. We cannot generate two linearly independent eigenvectors for A; hence, A is not diagonalizable and we cannot uncouple the system and use the method of Section 14.5.

First, we need the general solution of the homogeneous system $X' = AX$. We have one solution, $\begin{bmatrix} -2 \\ 1 \end{bmatrix} e^{3t}$, of $X' = AX$. Using the exponential matrix method of the preceding section, we find that a second, linearly independent solution is

$$\left(\begin{bmatrix} 1 \\ 0 \end{bmatrix} + (A - 3I) \begin{bmatrix} 1 \\ 0 \end{bmatrix} \right) e^{3t}$$

or

$$\begin{bmatrix} 1 - 2t \\ t \end{bmatrix} e^{3t}.$$

A fundamental matrix for $X' = AX$ is

$$\begin{bmatrix} -2e^{3t} & (1 - 2t)e^{3t} \\ e^{3t} & te^{3t} \end{bmatrix}.$$

The general solution of $X' = AX$ is

$$\begin{bmatrix} -2e^{3t} & (1 - 2t)e^{3t} \\ e^{3t} & te^{3t} \end{bmatrix} \begin{bmatrix} c_1 \\ c_2 \end{bmatrix}. \tag{14.29}$$

To write the general solution of $X' = AX + G$, we need a particular solution of this system. For illustration, we will use the Laplace transform. This method requires that we use a fundamental matrix $\Omega(t)$ satisfying $\Omega(0) = I$, a property which does not hold for the fundamental matrix found above. To find such a fundamental matrix, use the general solution (14.29) of $X' = AX$ to find homogeneous solutions $X^{(1)}(t)$ and $X^{(2)}(t)$ satisfying $X^{(1)}(0) = \begin{bmatrix} 1 \\ 0 \end{bmatrix}$ and $X^{(2)}(0) = \begin{bmatrix} 0 \\ 1 \end{bmatrix}$. We can use these solutions to form columns of $\Omega(t)$.

For $X^{(1)}$, let $t = 0$ in (14.29) and solve for c_1 and c_2 so that

$$\begin{bmatrix} -2 & 1 \\ 1 & 0 \end{bmatrix} \begin{bmatrix} c_1 \\ c_2 \end{bmatrix} = \begin{bmatrix} -2c_1 + c_2 \\ c_1 \end{bmatrix} = \begin{bmatrix} 1 \\ 0 \end{bmatrix}.$$

Then $c_1 = 0$ and $c_2 = 1$, so we let

$$X^{(1)}(t) = \begin{bmatrix} (1 - 2t)e^{3t} \\ te^{3t} \end{bmatrix}.$$

For $X^{(2)}$, solve for c_1 and c_2 such that

$$\begin{bmatrix} -2 & 1 \\ 1 & 0 \end{bmatrix} \begin{bmatrix} c_1 \\ c_2 \end{bmatrix} = \begin{bmatrix} 0 \\ 1 \end{bmatrix}.$$

Then $c_1 = 1$ and $c_2 = 2$, so

$$X^{(2)}(t) = \begin{bmatrix} -4te^{3t} \\ (1 + 2t)e^{3t} \end{bmatrix}.$$

Then

$$\Omega(t) = e^{At} = \begin{bmatrix} (1 - 2t)e^{3t} & -4te^{3t} \\ te^{3t} & (1 + 2t)e^{3t} \end{bmatrix}.$$

This is a fundamental matrix for $X' = AX$ which satisfies $\Omega(0) = I$. This fundamental matrix is often called the *transition matrix* for the system $X' = AX$.

We can now find a particular solution Ψ for $X' = AX + G$. There are at least two ways to proceed. First, from equation (14.27), we have

$$\mathscr{L}[\Psi(t)] = \hat{\Psi}(s) = \hat{\Omega}(s)\hat{G}(s).$$

By calculating the Laplace transform of each matrix entry, we have

$$\hat{\Omega}(s) = \begin{bmatrix} \dfrac{1}{s - 3} - \dfrac{2}{(s - 3)^2} & \dfrac{-4}{(s - 3)^2} \\ \dfrac{1}{(s - 3)^2} & \dfrac{1}{s - 3} + \dfrac{2}{(s - 3)^2} \end{bmatrix}$$

and

$$\hat{G}(s) = \begin{bmatrix} \dfrac{1}{s - 2} \\ \dfrac{1}{s^2} \end{bmatrix}.$$

We can now form the matrix product $\hat{\Omega}(s)\hat{G}(s)$ and find the inverse Laplace transform of the resulting 2×1 matrix by calculating the inverse transform of each matrix element. This involves a good deal of labor.

Another approach is to use convolution [see equation (14.28)]. In this case, we never have to compute the Laplace transform of any function; we just make use of the observation that Ψ can be represented as a convolution:

$$\Psi(t) = \mathscr{L}^{-1}[\hat{\Omega}(s)\hat{G}(s)] = \Omega(t) * G(t) = \int_0^t \Omega(t - s)G(s) \, ds.$$

This integral is quite tedious. However (see Problem 16), it can be written as

$$\begin{aligned}
\Psi(t) &= \int_0^t \Omega(s)G(t - s) \, ds \\
&= \int_0^t \begin{bmatrix} (1 - 2s)e^{3s} & -4se^{3s} \\ se^{3s} & (1 + 2s)e^{3s} \end{bmatrix}\begin{bmatrix} e^{2(t - s)} \\ t - s \end{bmatrix} ds \\
&= \int_0^t \begin{bmatrix} e^{2t}(1 - 2s)e^s + (4s^2 - 4st)e^{3s} \\ e^{2t}se^s + (t + 2st - s - 2s^2)e^{3s} \end{bmatrix} ds \\
&= \begin{bmatrix} -\frac{8}{27} - \frac{4}{9}t - 3e^{2t} + \frac{89}{27}e^{3t} - \frac{22}{9}te^{3t} \\ \frac{1}{27} - \frac{1}{9}t + e^{2t} - \frac{28}{27}e^{3t} + \frac{11}{9}te^{3t} \end{bmatrix}.
\end{aligned}$$

The general solution of the system $X' = AX + G$ is $\Omega(t)C + \Psi(t)$, with C a 2×1 matrix of arbitrary constants. ∎

PROBLEMS FOR SECTION 14.7

In each of Problems 1 through 5, use the method of variation of parameters to find the general solution. Express all solutions in terms of real-valued functions.

1. $x_1' = 5x_1 + 2x_2 - 3e^t$
$x_2' = -2x_1 + x_2 + e^{3t}$

2. $x_1' = 2x_1 - 4x_2 + 1$
$x_2' = x_1 - 2x_2 + 3t$

3. $x_1' = 7x_1 - x_2 + 2e^{6t}$
$x_2' = x_1 + 5x_2 + 6te^{6t}$

4. $x_1' = 2x_1 + 6e^{2t}\cos(3t)$
$x_2' = 6x_2 - 4x_3 - 2$
$x_3' = 4x_2 - 2x_3 - 2$

5. $x_1' = x_1$
$x_2' = 4x_1 + 3x_2 - 2e^t$
$x_3' = 3x_3$
$x_4' = -x_1 + 2x_2 + 9x_3 + x_4 + e^t$

In each of Problems 6 through 9, solve the initial value problem. Express the solution in terms of real-valued functions.

6. $x_1' = 2x_1 + 2,$
$x_2' = 5x_1 + 2x_2 + 10t;$
$x_1(0) = 0, \quad x_2(0) = 3$

7. $x_1' = 5x_1 - 4x_2 + 2e^t,$
$x_2' = 4x_1 - 3x_2 + 2e^t;$
$x_1(0) = -1, \quad x_2(0) = 3$

8. $x_1' = 2x_1 - 3x_2 + x_3 + 10e^{2t},$
$x_2' = 2x_2 + 4x_3 + 6e^{2t},$
$x_3' = x_3 - e^{2t};$
$x_1(0) = 5, \quad x_2(0) = 11, \quad x_3(0) = -2$

9. $x_1' = x_1 - 3x_2 + 2te^{-2t},$
$x_2' = 3x_1 - 5x_2 + 2te^{-2t},$
$x_3' = 4x_1 + 7x_2 - 2x_3 + 22t^2e^{-2t};$
$x_1(0) = 6, \quad x_2(0) = 2, \quad x_3(0) = 3$

10. Suppose that A is an $m \times n$ matrix and B is an $n \times k$ matrix both of whose entries are differentiable functions. Show that $(AB)' = A'B + AB'$. *Hint:* Look at the summation formula for the i, j-entry of the matrix product AB.

11. For the transition matrix $\Omega(t)$ of Example 14.15, verify that $\Omega^{-1}(t) = \Omega(-t)$ and that $\Omega(t)\Omega(s) = \Omega(t + s)$ for real s and t.

12. Suppose that $\Omega(t)$ is any fundamental matrix of the system $X' = AX$. Prove that $\Phi(t) = \Omega(t)\Omega^{-1}(0)$ is a transition matrix of the system. That is, $\Phi(t)$ is a fundamental matrix with the property that $\Phi(0) = I$.

In each of Problems 13 through 15, verify that the given matrix is a fundamental matrix for the system, then use it to find the transition matrix $\Omega(t)$. Finally, find $\Omega^{-1}(t)$. *Hint:* Note Problem 11.

13. $x_1' = 4x_1 + 2x_2$
$x_2' = 3x_1 + 3x_2,$ $\begin{bmatrix} 2e^t & e^{6t} \\ -3e^t & e^{6t} \end{bmatrix}$

14. $x_1' = -10x_2$
$x_2' = \frac{5}{2}x_1 - 10x_2,$ $\begin{bmatrix} 2e^{-5t} & (1 + 5t)e^{-5t} \\ e^{-5t} & 5te^{-5t}/2 \end{bmatrix}$

15. $x_1' = 5x_1 - 4x_2 + 4x_3$
$x_2' = 12x_1 - 11x_2 + 12x_3$
$x_3' = 4x_1 - 4x_2 + 5x_3$ $\begin{bmatrix} e^{-3t} & e^t & 0 \\ e^{-3t} & 0 & e^t \\ e^{-3t} & -e^t & e^t \end{bmatrix}$

16. Suppose that A is a 2×2 matrix and B is a 2×1 matrix. The entries of both matrices are integrable functions. Use the fact that convolution of functions is commutative to prove that

$$\int_0^t A(t - s)B(s) \, ds = \int_0^t A(s)B(t - s) \, ds.$$

Hint: Write

$$A = \begin{bmatrix} a_{11}(t) & a_{12}(t) \\ a_{21}(t) & a_{22}(t) \end{bmatrix} \quad \text{and} \quad B = \begin{bmatrix} b_1(t) \\ b_2(t) \end{bmatrix},$$

perform the indicated multiplication in the integrands, integrate, and express the components in the result as convolutions (which we know commute).

14.8 Transforming an nth Order Differential Equation into a System _____

It is common practice to approach an nth order differential equation by converting it into a system of n first order differential equations, to which matrix methods apply.

From a computational point of view, there is a significant difference between solving an nth order differential equation and solving a system of n first order differential equations. For the nth order constant coefficient linear equation, the major problem is in finding the roots of the nth degree characteristic polynomial. If we convert the nth order equation to a system of n linear first order equations, the main problem is finding the eigenvalues of the coefficient matrix, which again involves finding the roots of an nth degree polynomial. On the face of it, then, we seem to encounter the same difficulty either way.

In fact, we do not, because numerical methods for finding eigenvalues do not depend on solving the characteristic equation of the matrix. Instead, they proceed by iterative schemes which converge to the eigenvalues through a sequence of transformations of the matrix. This method is generally faster than numerical schemes for approximating roots of polynomials.

We will now see how to convert an nth order linear differential equation to a system of n first order linear differential equations. Suppose we have an nth order linear equation

$$y^{(n)} + a_{n-1}(t)y^{(n-1)} + \cdots + a_1(t)y' + a_0(t)y = f(t). \tag{14.30}$$

Define functions $x_1, x_2, \ldots, x_n$ of t by letting

$$\begin{aligned}
x_1 &= y \\
x_2 &= x_1' = y' \\
x_3 &= x_2' = y'' \\
&\vdots \\
x_n &= x_{n-1}' = y^{(n-1)}.
\end{aligned}$$

Observe that

$$\begin{aligned}
x_n' = y^{(n)} &= -a_{n-1}y^{(n-1)} - a_{n-2}y^{(n-2)} - \cdots - a_1 y' - a_0 y + f(t) \\
&= -a_{n-1}x_n - a_{n-2}x_{n-1} - \cdots - a_1 x_2 - a_0 x_1 + f(t).
\end{aligned}$$

The nth order differential equation (14.30) therefore transforms to the system

$$\begin{aligned}
x_1' &= x_2 \\
x_2' &= x_3 \\
&\vdots \\
x_{n-1}' &= x_n \\
x_n' &= -a_{n-1}x_n - a_{n-2}x_{n-1} - \cdots - a_1 x_2 - a_0 x_1 + f(t).
\end{aligned}$$

If we can solve this system, $y = x_1$ is a solution of the nth order differential equation (14.30).

EXAMPLE 14.16

Consider the second order differential equation

$$y'' + 3y' + 2y = 6t.$$

Let $x_1 = y$ and $x_2 = x_1' = y'$ to get the system

$$x_1' = x_2$$
$$x_2' = -2x_1 - 3x_2 + 6t. \quad \blacksquare$$

This example illustrates conversion of a differential equation to a system. Of course, in this example, both the differential equation and the system are easily solved. Sometimes we encounter a system involving higher order differential equations. In such a case, we can always redefine the variables to obtain a larger system of first order equations.

EXAMPLE 14.17

Consider the system

$$x_1'' + x_2' + 3x_1 = t$$
$$x_1' - x_2' - 2x_2 = \cos(t).$$

Let

$$y_1 = x_1$$
$$y_2 = x_2$$
$$y_3 = x_1'.$$

We obtain

$$y_1' = y_3$$
$$y_2' = x_2' = x_1' - 2x_2 - \cos(t) = y_3 - 2y_2 - \cos(t)$$
$$y_3' = x_1'' = -x_2' - 3x_1 + t = -[y_3 - 2y_2 - \cos(t)] - 3y_1 + t$$
$$= -3y_1 + 2y_2 - y_3 + \cos(t) + t.$$

This is a system of three first order linear differential equations and may be written as the matrix differential equation $Y' = AY + G$, with

$$A = \begin{bmatrix} 0 & 0 & 1 \\ 0 & -2 & 1 \\ -3 & 2 & -1 \end{bmatrix} \quad \text{and} \quad G = \begin{bmatrix} 0 \\ -\cos(t) \\ t + \cos(t) \end{bmatrix}. \quad \blacksquare$$

In Example 14.17, we could have defined a fourth variable by letting

$$y_4 = x_2'.$$

This would have made y_2' easier to determine as simply y_4. However, to calculate y_4', we would have had to differentiate one of the two original equations to produce x_2''. We avoided this by not introducing y_4. In addition, introduction of y_4 would have resulted in a 4 × 4 system instead of the 3 × 3 system we actually generated. This is an example of the type of judgment we must make in deciding how many new variables to introduce when converting a differential equation to a system.

Sometimes we encounter a system of the form

$$BX' = EX + H. \tag{14.31}$$

This leads us to consider two cases. If B is nonsingular, multiply the equation (14.31) on the left by B^{-1} to get

$$X' = (B^{-1}E)X + B^{-1}H.$$

This is in the familiar form $X' = AX + G$, and we can attempt to find a solution using the techniques we have developed.

If B is singular, it has no inverse, and we cannot perform this manipulation of the system (14.31) into standard form. It can be shown, however, that when B is singular, the system (14.31) can be transformed into a new system in which at least one of the equations has no derivative terms. We use this equation to write one of the variables as a linear combination of the others and to replace this variable in the other equations of the system. This results in a system of $n - 1$ equations in $n - 1$ variables. If necessary, this process can be continued until the coefficient matrix of X' in the resulting system is nonsingular.

PROBLEMS FOR SECTION 14.8

In each of Problems 1 through 6, find a system of differential equations whose solution will yield a solution of the given *n*th order differential equation. (The actual solution is not requested.)

1. $y^{(3)} - 4y'' + 8y' - 10y = \cos(t)$

2. $y^{(5)} + 16y^{(3)} + 5y'' - 8y = t^2 - 2t$

3. $y^{(4)} - 22y'' + 8y' + 12y = 2\cos(t) - e^t$

4. $y^{(4)} + y^{(3)} + 3y'' - 8y = 1$

5. $y^{(4)} - 6y^{(3)} + 10y'' + 4y' - 9y = t^3$

6. $y^{(6)} + 5y^{(5)} + 2y^{(4)} - 17y^{(3)} + 12y'' - 8y' - y = 2t - e^t$

In each of Problems 7 through 14, convert the given system or single differential equation into a first order system of the form $X' = AX + G$. A solution is not requested.

7. $x_1' - 3x_2' - x_1 + 2x_2 = e^{5t} + 6t$
$x_2'' - 2x_2' - 3x_1 = \cos(5t)$

8. $x_1' + 3x_2' - 5x_1 + 4x_2 = 3t + \sin(2t)$
$x_2'' - 8x_1' + 3x_1 - 2x_2 = e^{6t}$

9. $x_1' - 2x_1 - x_2' + 2x_2 = 8 - 4e^{3t}$
$x_1' - 4x_1 + x_2' - 8x_2 = 8 + 4e^{3t}$

10. $5x_1' - 31x_1 + x_2' - 27x_2 = t + 2$
$x_1' - 7x_1 + x_2' - 7x_2 = t - 6$

11. $y^{(3)} - 2y'' + 3y' - 5y = 6e^{2t} - 5t$

12. $y^{(4)} + 3y^{(3)} - 4y' + 2y = 2e^{3t}$

13. $2x_1'' - 6x_1' - 4x_1 + 8x_2' - 2x_2 = 16e^{4t}$
$3x_1' + 5x_1 - x_2'' + 4x_2' + 7x_2 = 2e^{-t}$

14. $x_1' - 14x_1 + 4x_2' + 14x_2 + 3x_3' - 7x_3 = 2t$
$2x_1' - 19x_1 + 5x_2' + 17x_2 + 4x_3' - 14x_3 = 3\cos(4t) + t$
$x_1' + 7x_1 - 3x_2' - 11x_2 - 2x_3' - 4x_3 = -3t$

In each of Problems 15 through 17, determine the general solution of the system.

15. $2x_1' - 3x_1 - 7x_2' + 23x_2 = 20e^{2t}$
$x_1' - x_1 - 4x_2' + 13x_2 = 12e^{2t}$

16. $3x_1' - 8x_1 + 5x_2' - 8x_2 = 8e^{4t}$
$x_1' - 3x_1 + 2x_2' - 3x_2 = 2e^{4t}$

17. $x_1' - 5x_1 + x_2' + x_2 + x_3' + x_3 = 8e^t$
$x_1' - 4x_1 + x_2' + 2x_3 = 7e^t$
$x_1' - 4x_1 + 2x_2 + x_3' = 5e^t$

In Problems 18 and 19, solve the initial value problem.

18. $4x_1' - 9x_1 + x_2' + 18x_2 = 3t + 4,$
    $x_1'' - 7x_1 + 7x_2' + 14x_2 = 9t + 2;$
    $x_1(2) = 7, \quad x_2(2) = 7; \quad x_1'(2) = -13$

19. $x_1'' - 2x_1' + x_1 - x_2 = 5e^{-t},$
    $2x_1' - 2x_1 - x_2' = e^{-t};$
    $x_1(0) = 1, \quad x_2(0) = 1, \quad x_1'(0) = -4$

14.9 *Applications and Illustrations of Techniques*

In this section, we will discuss five applications of systems of differential equations. These examples have been chosen so that the solutions utilize various methods and techniques discussed throughout this chapter. Here is a summary of the applications and the techniques they illustrate.

Example 14.18 (mixing)—a 2×2 homogeneous system $X' = AX$, with A diagonalizable.

Example 14.19 (mass-spring system)—a 4×4 homogeneous system with complex eigenvalues and diagonalizable coefficient matrix.

Example 14.20 (electrical circuit)—a 2×2 nonhomogeneous system with complex eigenvalues and diagonalizable coefficient matrix.

Example 14.21 (electrical circuit)—a 3×3 nonhomogeneous system with real eigenvalues and a diagonalizable coefficient matrix.

Example 14.22 (electrical circuit)—a 2×2 nonhomogeneous system with real eigenvalues but coefficient matrix nondiagonalizable. This example makes use of the exponential matrix and the method of variation of parameters.

EXAMPLE 14.18   Mixing

Two tanks are connected by a series of pipes, as shown in Figure 14.1. Tank 1 initially contains 20 liters of water in which 150 grams of chlorine are dissolved. Tank 2 initially contains 50 grams of chlorine dissolved in 10 liters of water.

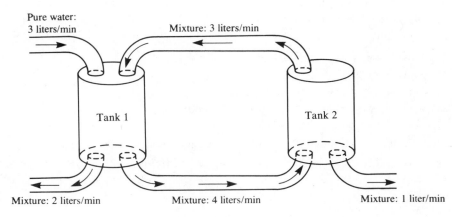

Figure 14.1

Beginning at time zero, pure water is pumped into tank 1 at a rate of 3 liters/minute, while chlorine and water solutions are interchanged between the tanks and also flow out of both tanks at the rates shown. The problem is to determine the amount of chlorine in each tank at any time $t > 0$.

At the given rates of input and discharge of solutions, the amount of solution in each tank will remain constant. Therefore, the ratio of chlorine to chlorine and water solution in each tank should, over the long run, approach that of the input, which is pure water. We will use this observation to check our final results.

Let $x_j(t)$ equal the number of grams of chlorine in tank j at time t. Reading from Figure 14.1, we have

$$x_1'(t) = \text{rate in minus rate out}$$
$$= (3 \text{ l/min}) \cdot (0 \text{ g/l}) + (3 \text{ l/min}) \cdot (x_2/10 \text{ g/l})$$
$$- (2 \text{ l/min}) \cdot (x_1/20 \text{ g/l}) - (4 \text{ l/min}) \cdot (x_1/20 \text{ g/l})$$
$$= -\tfrac{6}{20}x_1 + \tfrac{3}{10}x_2.$$

Similarly,

$$x_2'(t) = (4 \text{ l/min}) \cdot (x_1/20 \text{ g/l}) - (3 \text{ l/min}) \cdot (x_2/10 \text{ g/l})$$
$$- (1 \text{ l/min}) \cdot (x_2/10 \text{ g/l}) = \tfrac{4}{20}x_1 - \tfrac{4}{10}x_2.$$

We have the system of differential equations

$$x_1' = -\tfrac{3}{10}x_1 + \tfrac{3}{10}x_2$$
$$x_2' = \tfrac{1}{5}x_1 - \tfrac{2}{5}x_2,$$

with initial conditions

$$x_1(0) = 150, \qquad x_2(0) = 50.$$

In matrix notation, this initial value problem is

$$X' = AX; \qquad X(0) = X^0,$$

where

$$A = \begin{bmatrix} -\tfrac{3}{10} & \tfrac{3}{10} \\ \tfrac{1}{5} & -\tfrac{2}{5} \end{bmatrix} \quad \text{and} \quad X^0 = \begin{bmatrix} 150 \\ 50 \end{bmatrix}.$$

The eigenvalues of A are $\lambda_1 = -\tfrac{1}{10}$ and $\lambda_2 = -\tfrac{3}{5}$, and we find corresponding eigenvectors

$$\xi^{(1)} = \begin{bmatrix} 3 \\ 2 \end{bmatrix} \quad \text{and} \quad \xi^{(2)} = \begin{bmatrix} 1 \\ -1 \end{bmatrix}.$$

A fundamental matrix for $X' = AX$ is

$$\Omega(t) = \begin{bmatrix} 3e^{-t/10} & e^{-3t/5} \\ 2e^{-t/10} & -e^{-3t/5} \end{bmatrix}.$$

The general solution of $X' = AX$ is $\varphi = \Omega C$. To solve for C, use the initial conditions

$$\varphi(0) = X^0 = \begin{bmatrix} 150 \\ 50 \end{bmatrix} = \Omega(0)C = \begin{bmatrix} 3 & 1 \\ 2 & -1 \end{bmatrix} \begin{bmatrix} c_1 \\ c_2 \end{bmatrix}.$$

Then

$$3c_1 + c_2 = 150$$
$$2c_1 - c_2 = 50;$$

therefore, $c_1 = 40$ and $c_2 = 30$. The solution of the initial value problem is

$$\varphi(t) = \Omega(t)\begin{bmatrix} 40 \\ 30 \end{bmatrix} = \begin{bmatrix} 3e^{-t/10} & e^{-3t/5} \\ 2e^{-t/10} & -e^{-3t/5} \end{bmatrix}\begin{bmatrix} 40 \\ 30 \end{bmatrix}$$

$$= \begin{bmatrix} 120e^{-t/10} + 30e^{-3t/5} \\ 80e^{-t/10} - 30e^{-3t/5} \end{bmatrix}.$$

In terms of $x_1(t)$ and $x_2(t)$, this solution can be written

$$x_1(t) = 120e^{-t/10} + 30e^{-3t/5}$$
$$x_2(t) = 80e^{-t/10} - 30e^{-3t/5}.$$

Notice that, as $t \to \infty$, $x_1(t) \to 0$ and $x_2(t) \to 0$. We conclude that the amount of chlorine in the tanks approaches zero as time increases, as we observed at the beginning of the analysis. ∎

EXAMPLE 14.19   Mass-Spring System

Consider the mass-spring system of Figure 14.2. Assume that there is no damping and that no external force is applied to the system. Suppose that the upper weight is pulled down one unit and the lower weight is raised one unit, then both weights are released

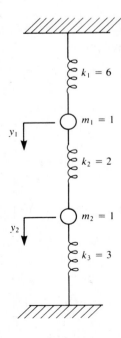

Figure 14.2

from rest simultaneously at time $t = 0$. We want to know the position of the weights relative to their equilibrium positions at any time $t > 0$.

The differential equations governing such a system are

$$m_1 y_1'' = -k_1 y_1 + k_2(y_2 - y_1)$$
$$m_2 y_2'' = -k_2(y_2 - y_1) - k_3 y_2.$$

In particular, $m_1 = m_2 = 1$, $k_1 = 6$, $k_2 = 2$, and $k_3 = 3$. Therefore, the system of differential equations to be solved is

$$y_1'' = -8y_1 + 2y_2$$
$$y_2'' = 2y_1 - 5y_2.$$

The initial conditions are

$$y_1(0) = 1, \qquad y_2(0) = -1, \qquad y_1'(0) = y_2'(0) = 0.$$

As a first step, convert the system of two second order differential equations into a system of four first order differential equations. Let

$$x_1 = y_1$$
$$x_2 = y_2$$
$$x_3 = y_1'$$
$$x_4 = y_2'.$$

The system for y_1 and y_2 transforms into the following system for x_1 through x_4:

$$x_1' = y_1' = x_3$$
$$x_2' = y_2' = x_4$$
$$x_3' = y_1'' = -8y_1 + 2y_2 = -8x_1 + 2x_2$$
$$x_4' = y_2'' = 2y_1 - 5y_2 = 2x_1 - 5x_2.$$

This system can be written $X' = AX$, with

$$A = \begin{bmatrix} 0 & 0 & 1 & 0 \\ 0 & 0 & 0 & 1 \\ -8 & 2 & 0 & 0 \\ 2 & -5 & 0 & 0 \end{bmatrix}.$$

The initial conditions are

$$X(0) = X^0 = \begin{bmatrix} 1 \\ -1 \\ 0 \\ 0 \end{bmatrix}.$$

The characteristic equation of A is $(\lambda^2 + 4)(\lambda^2 + 9) = 0$; hence, the eigenvalues are $2i$, $-2i$, $3i$, and $-3i$. Eigenvectors corresponding to $2i$ and $3i$, respectively, are

$$\xi^{(1)} = \begin{bmatrix} 1 \\ 2 \\ 2i \\ 4i \end{bmatrix} \quad \text{and} \quad \xi^{(2)} = \begin{bmatrix} 2 \\ -1 \\ 6i \\ -3i \end{bmatrix}.$$

We did not find eigenvectors for $-2i$ and $-3i$ because we will use Theorem 14.11 from Section 14.3 to write real-valued solutions. Write

$$\xi^{(1)} = \begin{bmatrix} 1 \\ 2 \\ 0 \\ 0 \end{bmatrix} + i \begin{bmatrix} 0 \\ 0 \\ 2 \\ 4 \end{bmatrix} \quad \text{and} \quad \xi^{(2)} = \begin{bmatrix} 2 \\ -1 \\ 0 \\ 0 \end{bmatrix} + i \begin{bmatrix} 0 \\ 0 \\ 6 \\ -3 \end{bmatrix}.$$

By Theorem 14.11 we obtain four linearly independent solutions, the first two from $\xi^{(1)}$ and the third and fourth from $\xi^{(2)}$. These solutions are

$$\begin{bmatrix} 1 \\ 2 \\ 0 \\ 0 \end{bmatrix} \cos(2t) - \begin{bmatrix} 0 \\ 0 \\ 2 \\ 4 \end{bmatrix} \sin(2t), \qquad \begin{bmatrix} 1 \\ 2 \\ 0 \\ 0 \end{bmatrix} \sin(2t) + \begin{bmatrix} 0 \\ 0 \\ 2 \\ 4 \end{bmatrix} \cos(2t),$$

$$\begin{bmatrix} 2 \\ -1 \\ 0 \\ 0 \end{bmatrix} \cos(3t) - \begin{bmatrix} 0 \\ 0 \\ 6 \\ -3 \end{bmatrix} \sin(3t), \quad \text{and} \quad \begin{bmatrix} 2 \\ -1 \\ 0 \\ 0 \end{bmatrix} \sin(3t) + \begin{bmatrix} 0 \\ 0 \\ 6 \\ -3 \end{bmatrix} \cos(3t).$$

These solutions may be written, respectively,

$$\begin{bmatrix} \cos(2t) \\ 2\cos(2t) \\ -2\sin(2t) \\ -4\sin(2t) \end{bmatrix}, \quad \begin{bmatrix} \sin(2t) \\ 2\sin(2t) \\ 2\cos(2t) \\ 4\cos(2t) \end{bmatrix}, \quad \begin{bmatrix} 2\cos(3t) \\ -\cos(3t) \\ -6\sin(3t) \\ 3\sin(3t) \end{bmatrix}, \quad \text{and} \quad \begin{bmatrix} 2\sin(3t) \\ -\sin(3t) \\ 6\cos(3t) \\ -3\cos(3t) \end{bmatrix}.$$

These solutions are linearly independent and can be used as columns of a fundamental matrix Ω of the system $X' = AX$:

$$\Omega = \begin{bmatrix} \cos(2t) & \sin(2t) & 2\cos(3t) & 2\sin(3t) \\ 2\cos(2t) & 2\sin(2t) & -\cos(3t) & -\sin(3t) \\ -2\sin(2t) & 2\cos(2t) & -6\sin(3t) & 6\cos(3t) \\ -4\sin(2t) & 4\cos(2t) & 3\sin(3t) & -3\cos(3t) \end{bmatrix}.$$

Notice that row 3 is the derivative of row 1 and row 4 is the derivative of row 2. This checks with the fact that $x_3 = y_1' = x_1'$ and $x_4 = y_2' = x_2'$.

The general solution for X is $\varphi = \Omega C$. To solve the initial value problem, we must choose C so that $\varphi(0) = \Omega(0)C = X^0$:

$$\begin{bmatrix} 1 & 0 & 2 & 0 \\ 2 & 0 & -1 & 0 \\ 0 & 2 & 0 & 6 \\ 0 & 4 & 0 & -3 \end{bmatrix} \begin{bmatrix} c_1 \\ c_2 \\ c_3 \\ c_4 \end{bmatrix} = \begin{bmatrix} 1 \\ -1 \\ 0 \\ 0 \end{bmatrix}.$$

Upon solving this system, we obtain $c_1 = -\frac{1}{5}$, $c_3 = \frac{3}{5}$, and $c_2 = c_4 = 0$. The unique solution of the initial value problem for X is

$$\varphi(t) = \Omega(t) \begin{bmatrix} -\frac{1}{5} \\ 0 \\ \frac{3}{5} \\ 0 \end{bmatrix}$$

$$= \begin{bmatrix} \cos(2t) & \sin(2t) & 2\cos(3t) & 2\sin(3t) \\ 2\cos(2t) & 2\sin(2t) & -\cos(3t) & -\sin(3t) \\ -2\sin(2t) & 2\cos(2t) & -6\sin(3t) & 6\cos(3t) \\ -4\sin(2t) & 4\cos(2t) & 3\sin(3t) & -3\cos(3t) \end{bmatrix} \begin{bmatrix} -\frac{1}{5} \\ 0 \\ \frac{3}{5} \\ 0 \end{bmatrix}$$

$$= \begin{bmatrix} -\frac{1}{5}\cos(2t) + \frac{6}{5}\cos(3t) \\ -\frac{2}{5}\cos(2t) - \frac{3}{5}\cos(3t) \\ \frac{2}{5}\sin(2t) - \frac{18}{5}\sin(3t) \\ \frac{4}{5}\sin(2t) + \frac{9}{5}\sin(3t) \end{bmatrix}.$$

Since $x_1 = y_1$ and $x_2 = y_2$, the solution for y_1 and y_2 is

$$y_1(t) = -\tfrac{1}{5}\cos(2t) + \tfrac{6}{5}\cos(3t)$$
$$y_2(t) = -\tfrac{2}{5}\cos(2t) - \tfrac{3}{5}\cos(3t). \quad \blacksquare$$

EXAMPLE 14.20 Electrical Circuit

Assume that all the currents and charges in the circuit of Figure 14.3 are zero until time $t = 0$, at which time the switch is closed. Determine the current in each loop for $t > 0$.

Use Kirchhoff's voltage and current laws (which are valid once the switch is closed) on the left and right loops to get

$$5i_1 + 5(i_1' - i_2') = 10 \tag{14.32}$$

$$5(i_1' - i_2') = 20i_2 + \frac{q_2}{5 \times 10^{-2}}. \tag{14.33}$$

If we use the outside loop (around the entire circuit), we get a third equation,

$$5i_1 + 20i_2 + \frac{q_2}{5 \times 10^{-2}} = 10. \tag{14.34}$$

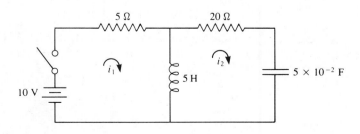

Figure 14.3

Any two of these equations contain the information needed to solve the problem; the third equation is redundant. We will use equations (14.32) and (14.34), omitting (14.33). The reason for this choice is that elimination of the q_2 term in equation (14.33) would require differentiating this equation, introducing second derivatives of i_1 and i_2. By using (14.32) and (14.34), we have a first order system rather than a second order one.

Divide equations (14.32) and (14.34) by 5 and differentiate the new equation (14.34) with respect to t to get

$$i_1' - i_2' = -i_1 + 2$$
$$i_1' + 4i_2' = -4i_2.$$

We must now determine the initial conditions. We know that

$$i_1(0-) = i_2(0-) = q_2(0-) = 0.$$

Then $(i_1 - i_2)(0-) = 0$. Since the current $i_1 - i_2$ through the inductor is continuous, we conclude that $(i_1 - i_2)(0+) = 0$ also. Therefore,

$$i_1(0+) = i_2(0+).$$

Now put this information, and the fact that the charge q_2 on the capacitor is continuous, into equation (14.34) and solve for $i_1(0+)$ to get

$$5i_1(0+) + 20i_2(0+) + 20q_2(0+) = 10,$$

or

$$25i_1(0+) + 20q_2(0+) = 25i_1(0+) = 10.$$

Then

$$i_1(0+) = i_2(0+) = \tfrac{10}{25} = \tfrac{2}{5}$$

amperes. The initial value problem for i_1 and i_2 is

$$i_1' - i_2' = -i_1 + 2$$
$$i_1' + 4i_2' = -4i_2;$$
$$i_1(0+) = \tfrac{2}{5} \quad \text{and} \quad i_2(0+) = \tfrac{2}{5}.$$

In matrix form,

$$\begin{bmatrix} 1 & -1 \\ 1 & 4 \end{bmatrix} \begin{bmatrix} i_1 \\ i_2 \end{bmatrix}' = \begin{bmatrix} -1 & 0 \\ 0 & -4 \end{bmatrix} \begin{bmatrix} i_1 \\ i_2 \end{bmatrix} + \begin{bmatrix} 2 \\ 0 \end{bmatrix}, \tag{14.35}$$

or $Bi' = Ai + K$. Because of the factor B, this is not quite in the form we have studied. However, notice that $|B| = 4 + 1 \neq 0$, so B has an inverse. We find that

$$B^{-1} = \begin{bmatrix} 1 & -1 \\ 1 & 4 \end{bmatrix}^{-1} = \tfrac{1}{5} \begin{bmatrix} 4 & 1 \\ -1 & 1 \end{bmatrix}.$$

Multiply equation (14.35) on the left by B^{-1} to obtain

$$\begin{bmatrix} i_1 \\ i_2 \end{bmatrix}' = \tfrac{1}{5} \begin{bmatrix} 4 & 1 \\ -1 & 1 \end{bmatrix} \begin{bmatrix} -1 & 0 \\ 0 & -4 \end{bmatrix} \begin{bmatrix} i_1 \\ i_2 \end{bmatrix} + \tfrac{1}{5} \begin{bmatrix} 4 & 1 \\ -1 & 1 \end{bmatrix} \begin{bmatrix} 2 \\ 0 \end{bmatrix}.$$

Upon carrying out the matrix multiplications, we obtain

$$\begin{bmatrix} i_1 \\ i_2 \end{bmatrix}' = \begin{bmatrix} -\frac{4}{5} & -\frac{4}{5} \\ \frac{1}{5} & -\frac{4}{5} \end{bmatrix} \begin{bmatrix} i_1 \\ i_2 \end{bmatrix} + \begin{bmatrix} \frac{8}{5} \\ -\frac{2}{5} \end{bmatrix}.$$

This system is in the standard form $i' = Ai + G$. The eigenvalues of A are $-\frac{4}{5} + \frac{2}{5}i$ and $-\frac{4}{5} - \frac{2}{5}i$, and corresponding eigenvectors are

$$\xi^{(1)} = \begin{bmatrix} 2 \\ -i \end{bmatrix} \quad \text{and} \quad \xi^{(2)} = \begin{bmatrix} 2 \\ i \end{bmatrix} = \text{conjugate of } \xi^{(1)}.$$

Since A has two linearly independent eigenvectors, A is diagonalizable. We will transform this system into an uncoupled one. Form the matrix P whose columns are the eigenvectors of A:

$$P = \begin{bmatrix} 2 & 2 \\ -i & i \end{bmatrix}.$$

Now change variables by letting $i = PZ$. The system $i' = Ai + G$ becomes

$$PZ' = (AP)Z + G.$$

Multiply this equation on the left by P^{-1} to get

$$Z' = (P^{-1}AP)Z + P^{-1}G.$$

We know that $P^{-1}AP$ is the 2×2 diagonal matrix with the eigenvalues of A on the main diagonal

$$P^{-1}AP = \tfrac{1}{5}\begin{bmatrix} -4 + 2i & 0 \\ 0 & -4 - 2i \end{bmatrix}.$$

However, because this system is nonhomogeneous, we must actually compute P^{-1} to determine $P^{-1}G$. We find that

$$P^{-1} = \tfrac{1}{4}\begin{bmatrix} 1 & 2i \\ 1 & -2i \end{bmatrix}.$$

Now we can calculate

$$P^{-1}G = \tfrac{1}{4}\begin{bmatrix} 1 & 2i \\ 1 & -2i \end{bmatrix}\begin{bmatrix} \frac{8}{5} \\ -\frac{2}{5} \end{bmatrix} = \tfrac{1}{5}\begin{bmatrix} 2 - i \\ 2 + i \end{bmatrix}.$$

The transformed system is

$$\begin{bmatrix} z_1 \\ z_2 \end{bmatrix}' = \tfrac{1}{5}\begin{bmatrix} -4 + 2i & 0 \\ 0 & -4 - 2i \end{bmatrix}\begin{bmatrix} z_1 \\ z_2 \end{bmatrix} + \tfrac{1}{5}\begin{bmatrix} 2 - i \\ 2 + i \end{bmatrix}.$$

This is the uncoupled system

$$z_1' + \tfrac{2}{5}(2 - i)z_1 = \tfrac{1}{5}(2 - i)$$
$$z_2' + \tfrac{2}{5}(2 + i)z_2 = \tfrac{1}{5}(2 + i).$$

Solve these constant coefficient linear first order differential equations to get

$$z_1(t) = \tfrac{1}{2} + c_1 e^{-(4 - 2i)t/5}$$
$$z_2(t) = \tfrac{1}{2} + c_2 e^{-(4 + 2i)t/5}.$$

Then

$$\begin{bmatrix} i_1 \\ i_2 \end{bmatrix} = \begin{bmatrix} 2 & 2 \\ -i & i \end{bmatrix} \begin{bmatrix} z_1 \\ z_2 \end{bmatrix} = \begin{bmatrix} 2z_1 + 2z_2 \\ -iz_1 + iz_2 \end{bmatrix}$$

$$= \begin{bmatrix} 2 + 2c_1 e^{-(4-2i)t/5} + 2c_2 e^{-(4+2i)t/5} \\ -ic_1 e^{-(4-2i)t/5} + ic_2 e^{-(4+2i)t/5} \end{bmatrix}.$$

Since $i_1(0+) = i_2(0+) = \frac{2}{5}$, we have the equations

$$2 + 2c_1 + 2c_2 = \frac{2}{5}$$
$$-ic_1 + ic_2 = \frac{2}{5}.$$

Solve these equations to get

$$c_1 = \tfrac{1}{5}(-2 + i) \quad \text{and} \quad c_2 = \tfrac{1}{5}(-2 - i).$$

The unique solution of the initial value problem is

$$\begin{bmatrix} i_1 \\ i_2 \end{bmatrix} = \frac{1}{5} \begin{bmatrix} 10 + 2(-2 + i)e^{-(4-2i)t/5} + 2(-2 - i)e^{-(4+2i)t/5} \\ (1 + 2i)e^{-(4-2i)t/5} + (1 - 2i)e^{-(4+2i)t/5} \end{bmatrix}.$$

If we wish, we can use Euler's formula to express this solution in terms of real-valued functions. Write

$$e^{-(4-2i)t/5} = e^{-4t/5}\left[\cos\left(\frac{2t}{5}\right) + i\sin\left(\frac{2t}{5}\right)\right]$$

and

$$e^{-(4+2i)t/5} = e^{-4t/5}\left[\cos\left(\frac{2t}{5}\right) - i\sin\left(\frac{2t}{5}\right)\right].$$

Then

$$i_1(t) = 2 + \frac{2}{5}(-2 + i)e^{-(4-2i)t/5} + \frac{2}{5}(-2 - i)e^{-(4+2i)t/5}$$

$$= 2 + \frac{2}{5}e^{-4t/5}\left\{(-2 + i)\left[\cos\left(\frac{2t}{5}\right) + i\sin\left(\frac{2t}{5}\right)\right]\right.$$

$$\left. + (-2 - i)\left[\cos\left(\frac{2t}{5}\right) - i\sin\left(\frac{2t}{5}\right)\right]\right\}$$

$$= 2 + \frac{2}{5}e^{-4t/5}\left[-2\cos\left(\frac{2t}{5}\right) - \sin\left(\frac{2t}{5}\right) - 2\cos\left(\frac{2t}{5}\right) - \sin\left(\frac{2t}{5}\right)\right],$$

with all the terms involving i canceling. Similarly,

$$i_2(t) = \frac{1}{5}[(1 + 2i)e^{-(4-2i)t/5} + (1 - 2i)e^{-(4+2i)t/5}]$$

$$= \frac{1}{5}e^{-4t/5}\left\{(1 + 2i)\left[\cos\left(\frac{2t}{5}\right) + i\sin\left(\frac{2t}{5}\right)\right] + (1 - 2i)\left[\cos\left(\frac{2t}{5}\right) - i\sin\left(\frac{2t}{5}\right)\right]\right\}$$

$$= \frac{1}{5}e^{-4t/5}\left[\cos\left(\frac{2t}{5}\right) - 2\sin\left(\frac{2t}{5}\right) + \cos\left(\frac{2t}{5}\right) - 2\sin\left(\frac{2t}{5}\right)\right],$$

again with all terms involving i canceling. Finally, we can write the solution as

$$i_1(t) = 2 - \frac{4}{5} e^{-4t/5} \left[2 \cos\left(\frac{2t}{5}\right) + \sin\left(\frac{2t}{5}\right) \right]$$

$$i_2(t) = \frac{2}{5} e^{-4t/5} \left[\cos\left(\frac{2t}{5}\right) - 2 \sin\left(\frac{2t}{5}\right) \right].$$

Units for both $i_1(t)$ and $i_2(t)$ are amperes. ■

EXAMPLE 14.21   Electrical Circuit

The circuit shown in Figure 14.4 has three connected loops. Assume that all three loop currents are zero prior to the time the switch is closed at $t = 0$. Also assume that the capacitor is in a discharged state. We want to determine the current in each loop for $t > 0$.

Apply Kirchhoff's current and voltage laws to get

$$4i_1 + 2i_1' - 2i_2' = 36 \tag{14.36}$$
$$2i_1' - 2i_2' = 5i_2 + 10q_2 - 10q_3 \tag{14.37}$$
$$10q_2 - 10q_3 = 5i_3 \tag{14.38}$$
$$4i_1 + 5i_2 + 10q_2 - 10q_3 = 36 \tag{14.39}$$
$$4i_1 + 5i_2 + 5i_3 = 36 \tag{14.40}$$
$$2i_1' - 2i_2' = 5i_2 + 5i_3. \tag{14.41}$$

We may use any three of these equations to determine i_1, i_2, and i_3. The other three equations are redundant. Our first task is to choose the three to use. Since equation (14.37) involves both charge and current terms, we will not use it. If we use equation (14.40) to eliminate one variable, reducing the problem to 2×2 size, we have to do a lot of algebra. Hence, we will retain the 3×3 size. Equation (14.38) is a likely one to retain.

If we use equations (14.36) and (14.41) together, we obtain a system of the form $Bi' = Di + F$ but with B singular. We would then be unable to multiply by B^{-1} and put the system into the standard form $i' = Ai + G$. We will therefore use equations (14.36), (14.38), and (14.39). These equations can be written

$$i_1' - i_2' = -2i_1 + 18$$
$$4i_1' + 5i_2' = -10i_2 + 10i_3 \tag{14.42}$$
$$i_3' = 2i_2 - 2i_3.$$

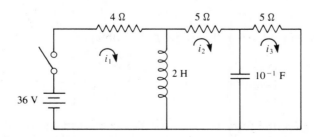

Figure 14.4

Now determine the initial conditions. The conductor current $i_1 - i_2$ is continuous, and $i_1(0-) = i_2(0-) = 0$. Therefore,

$$(i_1 - i_2)(0+) = (i_1 - i_2)(0-) = 0.$$

Then

$$i_1(0+) = i_2(0+).$$

Further, the capacitor charge, $q_2 - q_3$, is continuous. Since $(q_2 - q_3)(0-) = 0$, $q_2(0+) = q_3(0+)$. This condition and equation (14.38) require that

$$i_3(0+) = 0.$$

Now use equation (14.40) to get

$$4i_1(0+) + 5i_2(0+) + 5i_3(0+) = 36.$$

Since $i_3(0+) = 0$ and $i_1(0+) = i_2(0+)$,

$$9i_1(0+) = 36.$$

Then

$$i_1(0+) = i_2(0+) = 4.$$

Now summarize the information we have accumulated. The system (14.42) can be written in matrix form $Bi' = Di + F$:

$$\begin{bmatrix} 1 & -1 & 0 \\ 4 & 5 & 0 \\ 0 & 0 & 1 \end{bmatrix} \begin{bmatrix} i_1 \\ i_2 \\ i_3 \end{bmatrix}' = \begin{bmatrix} -2 & 0 & 0 \\ 0 & -10 & 10 \\ 0 & 2 & -2 \end{bmatrix} \begin{bmatrix} i_1 \\ i_2 \\ i_3 \end{bmatrix} + \begin{bmatrix} 18 \\ 0 \\ 0 \end{bmatrix}. \tag{14.43}$$

The initial conditions are

$$i(0+) = \begin{bmatrix} 4 \\ 4 \\ 0 \end{bmatrix}.$$

Finally, we have an initial value problem for the currents in the circuit. Compute

$$B^{-1} = \begin{bmatrix} 1 & -1 & 0 \\ 4 & 5 & 0 \\ 0 & 0 & 1 \end{bmatrix}^{-1} = \tfrac{1}{9} \begin{bmatrix} 5 & 1 & 0 \\ -4 & 1 & 0 \\ 0 & 0 & 9 \end{bmatrix}.$$

Multiply the system (14.43) on the left by B^{-1} to obtain

$$\begin{bmatrix} i_1 \\ i_2 \\ i_3 \end{bmatrix}' = \begin{bmatrix} -\frac{10}{9} & -\frac{10}{9} & \frac{10}{9} \\ \frac{8}{9} & -\frac{10}{9} & \frac{10}{9} \\ 0 & 2 & -2 \end{bmatrix} \begin{bmatrix} i_1 \\ i_2 \\ i_3 \end{bmatrix} + \begin{bmatrix} 10 \\ -8 \\ 0 \end{bmatrix}.$$

This equation is of the form $i' = Ai + G$. The coefficient matrix A has characteristic equation $\lambda(\lambda + 2)(\lambda + \frac{20}{9}) = 0$ and hence has eigenvalues $\lambda_1 = 0$, $\lambda_2 = -2$, and

$\lambda_3 = -\frac{20}{9}$. Corresponding eigenvectors are

$$\xi^{(1)} = \begin{bmatrix} 0 \\ 1 \\ 1 \end{bmatrix}, \qquad \xi^{(2)} = \begin{bmatrix} 5 \\ 0 \\ -4 \end{bmatrix}, \quad \text{and} \quad \xi^{(3)} = \begin{bmatrix} 10 \\ 1 \\ -9 \end{bmatrix}.$$

These are linearly independent (they correspond to distinct eigenvalues), so A is diagonalizable. Let

$$P = \begin{bmatrix} 0 & 5 & 10 \\ 1 & 0 & 1 \\ 1 & -4 & -9 \end{bmatrix}.$$

Let $i = PZ$ to obtain

$$PZ' = (AP)Z + G.$$

Multiply this equation on the left by P^{-1} to get

$$Z' = (P^{-1}AP)Z + P^{-1}G.$$

We find that

$$P^{-1} = \frac{1}{10} \begin{bmatrix} 4 & 5 & 5 \\ 10 & -10 & 10 \\ -4 & 5 & -5 \end{bmatrix}.$$

Calculate

$$P^{-1}G = \frac{1}{10} \begin{bmatrix} 4 & 5 & 5 \\ 10 & -10 & 10 \\ -4 & 5 & -5 \end{bmatrix} \begin{bmatrix} 10 \\ -8 \\ 0 \end{bmatrix} = \begin{bmatrix} 0 \\ 18 \\ -8 \end{bmatrix}.$$

The uncoupled system for Z is

$$\begin{bmatrix} z_1 \\ z_2 \\ z_3 \end{bmatrix}' = \begin{bmatrix} 0 & 0 & 0 \\ 0 & -2 & 0 \\ 0 & 0 & -\frac{20}{9} \end{bmatrix} \begin{bmatrix} z_1 \\ z_2 \\ z_3 \end{bmatrix} + \begin{bmatrix} 0 \\ 18 \\ -8 \end{bmatrix},$$

or

$$z_1' = 0$$
$$z_2' + 2z_2 = 18$$
$$z_3' + \frac{20}{9}z_3 = -8.$$

Solve these differential equations individually to get

$$z_1 = c_1$$
$$z_2 = c_2 e^{-2t} + 9$$
$$z_3 = c_3 e^{-20t/9} - \frac{18}{5}.$$

Since $i = PZ$,

$$\begin{bmatrix} i_1 \\ i_2 \\ i_3 \end{bmatrix} = \begin{bmatrix} 0 & 5 & 10 \\ 1 & 0 & 1 \\ 1 & -4 & -9 \end{bmatrix} \begin{bmatrix} z_1 \\ z_2 \\ z_3 \end{bmatrix}$$

$$= \begin{bmatrix} 0 & 5 & 10 \\ 1 & 0 & 1 \\ 1 & -4 & -9 \end{bmatrix} \begin{bmatrix} c_1 \\ c_2 e^{-2t} + 9 \\ c_3 e^{-20t/9} - \frac{18}{5} \end{bmatrix}$$

$$= \begin{bmatrix} 9 + 5c_2 e^{-2t} + 10c_3 e^{-20t/9} \\ -\frac{18}{5} + c_1 + c_3 e^{-20t/9} \\ -\frac{18}{5} + c_1 - 4c_2 e^{-2t} - 9c_3 e^{-20t/9} \end{bmatrix}.$$

Now use the initial conditions to solve for c_1, c_2, and c_3. We get

$$i(0+) = \begin{bmatrix} 4 \\ 4 \\ 0 \end{bmatrix} = \begin{bmatrix} 9 + 5c_2 + 10c_3 \\ -\frac{18}{5} + c_1 + c_3 \\ -\frac{18}{5} + c_1 - 4c_2 - 9c_3 \end{bmatrix}.$$

This system of algebraic equations can be written

$$\begin{bmatrix} 0 & 5 & 10 \\ 1 & 0 & 1 \\ 1 & -4 & -9 \end{bmatrix} C = \begin{bmatrix} -5 \\ \frac{38}{5} \\ \frac{18}{5} \end{bmatrix}.$$

The matrix of coefficients on the left is exactly P. We can therefore use P^{-1} (already calculated) to solve the system. We get

$$C = P^{-1} \begin{bmatrix} -5 \\ \frac{38}{5} \\ \frac{18}{5} \end{bmatrix} = \frac{1}{10} \begin{bmatrix} 4 & 5 & 5 \\ 10 & -10 & 10 \\ -4 & 5 & -5 \end{bmatrix} \begin{bmatrix} -5 \\ \frac{38}{5} \\ \frac{18}{5} \end{bmatrix} = \begin{bmatrix} \frac{18}{5} \\ -9 \\ 4 \end{bmatrix}.$$

Thus, the loop currents are

$$i_1(t) = 9 - 45e^{-2t} + 40e^{-20t/9}$$

$$i_2(t) = 4e^{-20t/9}$$

$$i_3(t) = 36e^{-2t} - 36e^{-20t/9}.$$

All of these currents are in amperes.   ∎

EXAMPLE 14.22   Electrical Circuit

Consider the circuit shown in Figure 14.5. Assume that all currents and charges are zero until time $t = 0$, when the switch is closed. We want to know the loop currents for any time $t > 0$.

Kirchhoff's laws give us

$$10i_1 + 4(i_1' - i_2') = 4 \tag{14.44}$$

$$4(i_1' - i_2') = 100q_2 \tag{14.45}$$

$$10i_1 + 100q_2 = 4. \tag{14.46}$$

We may use any two of these equations. We will use equations (14.44) and (14.46) to avoid a mixture of terms involving i and q. Thus, consider the system

$$i'_1 = -10i_2$$
$$2i'_1 - 2i'_2 = -5i_1 + 2.$$

The current $i_1 - i_2$ through the inductor is zero prior to switch closure and hence is zero "immediately" afterward as well. Therefore, $i_1(0+) = i_2(0+)$. Since the capacitor charge is continuous and zero prior to $t = 0$, we also have from equation (14.46) that

$$10i_1(0+) + 100q_2(0+) = 10i_1(0+) + 100q_2(0-)$$
$$= 10i_1(0+) + 0 = 10i_1(0+) = 4.$$

We conclude that

$$i_1(0+) = \tfrac{2}{5} \quad \text{and} \quad i_2(0+) = \tfrac{2}{5}.$$

The system to be solved is

$$\begin{bmatrix} 1 & 0 \\ 2 & -2 \end{bmatrix} \begin{bmatrix} i_1 \\ i_2 \end{bmatrix}' = \begin{bmatrix} 0 & -10 \\ -5 & 0 \end{bmatrix} \begin{bmatrix} i_1 \\ i_2 \end{bmatrix} + \begin{bmatrix} 0 \\ 2 \end{bmatrix}.$$

Calculate

$$\begin{bmatrix} 1 & 0 \\ 2 & -2 \end{bmatrix}^{-1} = \begin{bmatrix} 1 & 0 \\ 1 & -\tfrac{1}{2} \end{bmatrix}.$$

Multiply the system on the left by this matrix to obtain

$$\begin{bmatrix} i_1 \\ i_2 \end{bmatrix}' = \begin{bmatrix} 1 & 0 \\ 1 & -\tfrac{1}{2} \end{bmatrix} \begin{bmatrix} 0 & -10 \\ -5 & 0 \end{bmatrix} \begin{bmatrix} i_1 \\ i_2 \end{bmatrix} + \begin{bmatrix} 1 & 0 \\ 1 & -\tfrac{1}{2} \end{bmatrix} \begin{bmatrix} 0 \\ 2 \end{bmatrix},$$

or

$$\begin{bmatrix} i_1 \\ i_2 \end{bmatrix}' = \begin{bmatrix} 0 & -10 \\ \tfrac{5}{2} & -10 \end{bmatrix} \begin{bmatrix} i_1 \\ i_2 \end{bmatrix} + \begin{bmatrix} 0 \\ -1 \end{bmatrix}.$$

This system is in the standard form $i' = Ai + G$. The initial condition is

$$i(0) = \begin{bmatrix} \tfrac{2}{5} \\ \tfrac{2}{5} \end{bmatrix}.$$

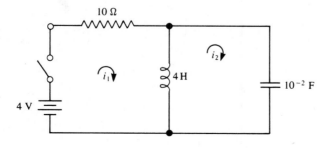

Figure 14.5

The characteristic equation of A is $(\lambda + 5)^2 = 0$, with repeated root $\lambda_1 = \lambda_2 = -5$. All eigenvectors are nonzero multiples of $\xi = \begin{bmatrix} 2 \\ 1 \end{bmatrix}$, and A does not have two linearly independent eigenvectors. Thus, A is not diagonalizable. We will use the exponential matrix together with the method of variation of parameters.

First, we must find the general solution of $i' = Ai$. Immediately, one solution is ξe^{-5t}. To find a second, linearly independent solution, we will look for a 2×1 constant matrix C such that

$$(A + 5I)^2 C = O \quad \text{but} \quad (A + 5I)C \neq O.$$

Calculate

$$(A + 5I)^2 = \begin{bmatrix} 5 & -10 \\ \frac{5}{2} & -5 \end{bmatrix}^2 = \begin{bmatrix} 0 & 0 \\ 0 & 0 \end{bmatrix}.$$

Any nonzero 2×1 matrix, other than a multiple of ξ, will do. For simplicity, choose $C = \begin{bmatrix} 1 \\ 0 \end{bmatrix}$. A second solution of $i' = Ai$ is

$$\begin{aligned}
e^{At}C &= e^{-5t}e^{(A + 5I)t}C \\
&= e^{-5t}[C + (A + 5I)Ct] \\
&= e^{-5t}\left(\begin{bmatrix} 1 \\ 0 \end{bmatrix} + \begin{bmatrix} 5 & 0 \\ \frac{5}{2} & -5 \end{bmatrix} \begin{bmatrix} 1 \\ 0 \end{bmatrix} t \right) \\
&= e^{-5t} \begin{bmatrix} 1 + 5t \\ 5t/2 \end{bmatrix}.
\end{aligned}$$

This solution and ξe^{-5t} form a basis for the solution space of $i' = Ai$. A fundamental matrix is

$$\Omega(t) = e^{-5t} \begin{bmatrix} 2 & (1 + 5t) \\ 1 & 5t/2 \end{bmatrix}.$$

Now we need a particular solution of $i' = Ai + G$. We know that such a solution can be written Ωu, where $u = \int \Omega^{-1}(t)G(t)\,dt$. Calculate

$$\Omega^{-1}(t) = e^{5t} \begin{bmatrix} -5t/2 & 1 + 5t \\ 1 & -2 \end{bmatrix}.$$

Then

$$\begin{aligned}
\Omega^{-1}(t)G(t) &= e^{5t} \begin{bmatrix} -5t/2 & 1 + 5t \\ 1 & -2 \end{bmatrix} \begin{bmatrix} 0 \\ -1 \end{bmatrix} \\
&= e^{5t} \begin{bmatrix} -1 - 5t \\ 2 \end{bmatrix}.
\end{aligned}$$

Then

$$u(t) = \begin{bmatrix} \int (-1 - 5t)e^{5t}\,dt \\ \int 2e^{5t}\,dt \end{bmatrix} = \begin{bmatrix} -te^{5t} \\ \frac{2}{5}e^{5t} \end{bmatrix}.$$

A particular solution of $i' = Ai + G$ is

$$\psi(t) = \Omega(t)u(t) = e^{-5t}\begin{bmatrix} 2 & (1 + 5t) \\ 1 & 5t/2 \end{bmatrix}\begin{bmatrix} -te^{5t} \\ \frac{2}{5}e^{5t} \end{bmatrix} = \begin{bmatrix} \frac{2}{5} \\ 0 \end{bmatrix}.$$

(This problem happens to have a constant solution.) The general solution of $i' = Ai + G$ is

$$i(t) = \Omega C + \psi = e^{-5t}\begin{bmatrix} 2 & (1 + 5t) \\ 1 & 5t/2 \end{bmatrix}\begin{bmatrix} c_1 \\ c_2 \end{bmatrix} + \begin{bmatrix} \frac{2}{5} \\ 0 \end{bmatrix}.$$

Use the initial conditions to solve for C:

$$i(0) = \begin{bmatrix} \frac{2}{5} \\ \frac{2}{5} \end{bmatrix} = \begin{bmatrix} 2 & 1 \\ 1 & 0 \end{bmatrix}C + \begin{bmatrix} \frac{2}{5} \\ 0 \end{bmatrix}.$$

This system of equations can be written

$$\frac{2}{5} = 2c_1 + c_2 + \frac{2}{5}$$
$$c_1 = \frac{2}{5};$$

hence, $c_1 = \frac{2}{5}$ and $c_2 = -\frac{4}{5}$. The solution for the current is

$$i(t) = e^{-5t}\begin{bmatrix} 2 & (1 + 5t) \\ 1 & 5t/2 \end{bmatrix}\begin{bmatrix} \frac{2}{5} \\ -\frac{4}{5} \end{bmatrix} + \begin{bmatrix} \frac{2}{5} \\ 0 \end{bmatrix}.$$

We can also write the solution as

$$i_1(t) = -4te^{-5t} + \frac{2}{5}$$
$$i_2(t) = (\frac{2}{5} - 2t)e^{-5t}.$$

As usual, these currents are in amperes. ∎

PROBLEMS FOR SECTION 14.9

1. Determine the amount of chlorine in each of the tanks of Figure 14.1 if the input is at the same rate of 3 liters/minute but consists of a mixture of 4 grams of chlorine per liter rather than pure water, as in that example.
2. Referring to the circuit depicted in Figure 14.4, determine how much time elapses between the time the switch is closed and the time the charge on the capacitor is a maximum. What is the maximum voltage on the capacitor?

Each of Problems 3 through 5 deals with the system of tanks and interconnecting pipes shown in the diagram.

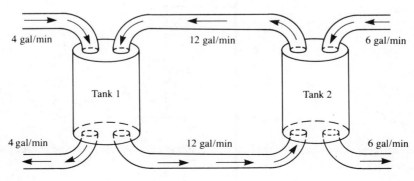

Tank-1 initially contains 200 gallons of saltwater mixture (brine), while tank 2 initially contains 300 gallons of brine. Beginning at time zero, a brine solution is pumped into tank 1 at the rate of 4 gallons/minute, pure water is pumped into tank 2 at the rate of 6 gallons/minute, and the brine solutions are interchanged between the tanks and also flow out of both tanks at the rates shown. Determine the amount of salt in each tank for $t \geq 0$ in each of the following cases.

3. Assume that the input to tank 1 is pure water, that tank 1 initially contains 80 pounds of salt, and that tank 2 initially contains 10 pounds of salt.

4. Assume that the input solution to tank 1 contains $\frac{1}{4}$ pound of salt per gallon, that tank 1 has 200 pounds of salt at time zero, and that tank 2 initially contains 150 pounds of salt.

5. Assume that tank 1 contains pure water at time zero, that tank 2 initially contains 100 pounds of salt, and that the input to tank 1 contains $\frac{1}{6}$ pound of salt per gallon.

6. Two tanks are connected by a series of pipes as shown. Tank 1 initially contains 100 gallons of water in which 40 pounds of salt are dissolved. Tank 2 initially contains 150 gallons of pure water. Beginning at time zero, a brine solution containing $\frac{1}{5}$ pound of salt per gallon is pumped into tank 1 at the rate of 5 gallons/minute. At the same time, a solution which also contains $\frac{1}{5}$ pound of salt per gallon is pumped into tank 2 at the rate of 10 gallons/minute. The brine solutions are interchanged between the tanks and also flow out of both tanks at the rates shown in the diagram.

   Determine the amount of salt in each tank for $t \geq 0$. Also calculate the time at which the brine solution in tank 1 reaches its minimum salinity (concentration of salt), and determine how much salt is in tank 1 at that time.

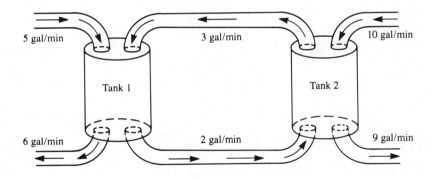

7. Suppose that all of the currents in the circuit shown in the diagram are zero prior to the time the switch is closed at time zero. Determine the currents i_1 and i_2 for $t > 0$.

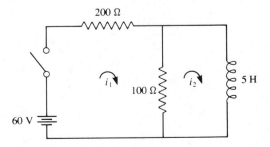

8. Find the loop currents in the circuit shown for $t > 0$, assuming that the currents and charge are all zero prior to the time the switch is closed at $t = 0$.

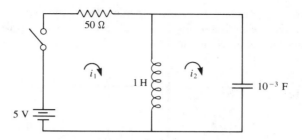

9. Determine the steady-state currents in the circuit shown in Figure 14.4 (Example 14.21) if the battery is replaced with a sine wave generator which has an output of $3\cos(3t)$ volts.

Each of Problems 10 through 12 deals with the mass-spring system shown. In each problem, derive and solve the system of differential equations for the motion of the weights under the assumption that there is no damping.

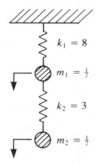

10. Each weight is pulled downward one unit and is then released from rest, with no external driving forces.
11. The upper weight is pulled downward one unit and the lower one is raised one unit, then both are released from rest. Assume that no external driving forces are present.
12. The weights have zero initial displacement and velocity. The lower weight is subjected to an external driving force $F(t) = 2\sin(3t)$, and the upper weight has no external driving force applied to it.

Each of Problems 13 through 15 deals with the mass-spring system shown. In each problem, derive and solve the system of differential equations for the displacement functions of the weights under the assumption that there is no damping. Denote left to right as the positive direction.

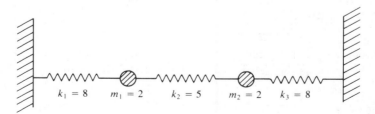

13. Each weight is pulled one unit to the left and then released, with no external driving forces. In addition to solving for the displacement functions of the weights, calculate the distance between the two weights and determine how rapidly this distance is changing for $t \geq 0$.
14. The left weight is pushed to the right one unit, and the right weight is pushed to the left one unit. Both are released from rest. Assume that there are no external driving forces.

15. The weights have zero initial displacements and velocities. The left weight is subjected to an external force $F(t) = 4\sin(t)$, and the right weight is not subjected to any external driving force.

16. Find the currents in each loop of the circuit shown. Assume that the currents and charge are all zero prior to the time the switch is closed ($t = 0$).

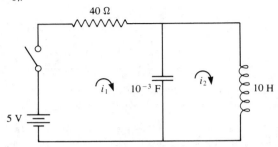

17. Find the loop currents in the circuit shown below for $t > 0$ under the assumption that the currents and charges are all zero prior to the time the switch is closed (at $t = 0$).

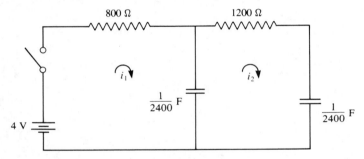

18. In the circuit shown, assume that the currents and charge are all zero prior to the time the switch is closed ($t = 0$). Find the loop currents for $t > 0$.

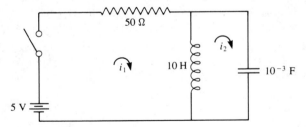

19. Find the output voltage E_{out} in the circuit shown for $t > 0$, assuming that the currents and charge are all zero prior to the time the switch is closed ($t = 0$) and that $V(t) = 20[1 - H(t - 3)]$.

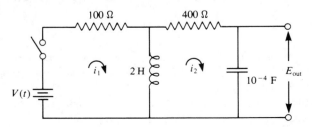

20. Find the loop currents in the circuit shown below for $t > 0$, assuming that the currents and charge are all zero prior to the time the switch is closed ($t = 0$). Also determine how much time elapses after the switch is closed until E_{out} assumes its maximum value. What is this maximum value?

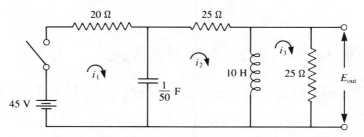

21. Derive and solve the system of differential equations for the displacement functions for the weights in the mass-spring system shown. Assume that the top weight is lowered one unit and the bottom weight is raised one unit, then both are released from rest. Assume that there are no external driving forces acting on the weights. The value of the spring constant in the middle spring is $\alpha = 10\sqrt{26}$. *Hint:* One root of the characteristic polynomial is $-1 + 2i$.

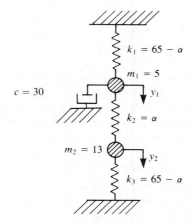

22. Consider the system of Problem 21, with the assumption that the weights have zero initial displacements and velocities and that the bottom weight is subjected to an external force of $F(t) = 39 \sin(t)$ while the top weight is free of (nongravitational) external forces.

Notes on the History of Matrices and Determinants

Matrices and determinants developed out of the mathematics of the eighteenth and nineteenth centuries, particularly that dealing with transformations of geometric objects and the solution of systems of linear equations.

Perhaps surprisingly, the early emphasis was on the determinant, not the matrix, although we usually consider matrices first in modern treatments of linear algebra. While many properties of determinants were discovered by talented amateurs and figures of minor importance in the overall history of mathematics, such mathematical giants as Gauss and Cauchy also contributed to their development. Cauchy stated the correct definition of the product of two matrices and showed that the determinant of a product equals the product of the determinants. Cauchy also proved Laplace's formula for the cofactor expansions of a determinant. Gauss's main interest in determinants was in the study of quadratic forms.

A leading figure in the history of determinants was James Joseph Sylvester (1814–1897), who was denied a teaching post at Cambridge because of his Jewish ancestry. His varied career included periods at the University of Virginia and Johns Hopkins University in the United States as well as positions as a lawyer and actuary in his native England. Sylvester's work touched upon determinant applications to the solution of polynomial equations as well as geometric aspects of quadratic forms. Out of this, and Lagrange's work on planetary motion, came the determinant formulation of the characteristic equation of a matrix (which was then called the "secular equation").

Carl Gustav Jacob Jacobi (1804–1851) and Eugène Catalan (1814–1894) are credited with discovering the determinant used in transforming multiple integrals. Nowadays, this determinant is called the *Jacobian*. The *Wronskian*, a determinant involving solutions of differential equations, is named for the Polish mathematician H. Wronski (1778–1853).

In connection with determinants, it is interesting to note that Lewis Carroll was the pseudonym of the Reverend C. L. Dodgson (1832–1898), the creator of *Alice's Adventures in Wonderland*. Dodgson was a fellow of Christ Church College, Oxford, an Anglican clergyman, and Tutor in Mathematics. His popular stories about Alice so delighted the queen that she commanded him to send her his next work, which happened to be a treatise on determinants. History has not recorded her reaction.

The term *matrix* was apparently coined by Sylvester about 1850. Arthur Cayley (1821–1895) pointed out the rationale for thinking of matrices as preceding determinants in the logical order of linear algebra, but in fact the historical development was the reverse, and many properties of determinants were known before matrices were formalized. Cayley collaborated with Sylvester in the development of matrices, with emphasis on the geometry of R^n, invariants of quadratic forms, and linear transformations. Cayley and Sylvester must have made an interesting pair. The latter was a lively, well-traveled man of many accomplishments, while the former was as calm and even as Sylvester was excitable. Cayley spent his entire career at Cambridge except for the year 1892, when he visited his friend at Johns Hopkins.

The fact that real, symmetric matrices have real eigenvalues was known to Cauchy and was established later by Charles Hermite, who developed hermitian matrices and is also known for Hermite polynomials. Rudolf Friedrich Alfred Clebsch (1833–1872) showed that the nonzero eigenvalues of a real, skew-hermitian matrix are pure imaginary. Hermite had the idea of orthogonal matrices, but Georg Frobenius (1849–1917) gave the definition we use today. Frobenius is also credited with a method for solving differential equations by infinite series.

Of the topics we have omitted in our brief introduction to linear algebra, two are the Cayley-Hamilton theorem, which states that a square matrix satisfies its own characteristic equation, and the Jordan canonical form. Sir William Rowan Hamilton (1805–1865) was a major figure in the development of quaternions and vectors. By age thirteen Hamilton had mastered thirteen languages, including Persian and Sanskrit. Camille Jordan (1838–1922) was a French mathematician who made fundamental contributions to the growing fields of modern algebra and topology.

Vector Calculus _____

_____

*Vector Differential Calculus*_____

_____

_____

_____

_____

15.0 *Introduction* _____

In Chapter 11, we developed some of the algebra and geometry of vectors in the plane, in three-space, and eventually in R^n. In this and the next chapter, we will combine vector algebra and geometry with the processes of calculus to develop what is known as vector calculus. This chapter deals primarily with derivatives and is called vector differential calculus; the next chapter deals with various kinds of integrals of vector functions, which come under the heading of vector integral calculus.

For the most part, we will be dealing with vectors in the plane and in three-space, where we have geometric intuition available to support the concepts we will be studying. It is assumed that the reader is familiar with the material of Sections 11.1 through 11.3, inclusive. We will not require concepts from the vector space structure of R^n in this chapter, nor will we need the idea of a scalar triple product.

15.1 *Vector Functions of One Variable* _____

It is often the case that we are dealing with a vector whose components are functions of either one or several variables. In this section, we will treat the one-variable case, in which we have a vector function of the form

$$\mathbf{F}(t) = x(t)\mathbf{i} + y(t)\mathbf{j} + z(t)\mathbf{k}.$$

Such a function is referred to as a *vector function of one real variable*, since the variable is a scalar t and the function value $\mathbf{F}(t)$ is a vector. For example, we might have

$$\mathbf{F}(t) = \cos(t)\mathbf{i} - 2t^2\mathbf{j} + e^t\mathbf{k}$$

for all t. Given any value of t, $\mathbf{F}(t)$ is a vector. For example, $\mathbf{F}(\pi) = -\mathbf{i} - 2\pi^2\mathbf{j} + e^\pi\mathbf{k}$ and $\mathbf{F}(0) = \mathbf{i} + \mathbf{k}$.

The functions $x(t)$, $y(t)$, and $z(t)$ are the *component functions*, or *components*, of $\mathbf{F}(t)$.

We call $\mathbf{F}$ *continuous* if each component function is continuous. For example, $\cos(t)\mathbf{i} - 2t^2\mathbf{j} + e^t\mathbf{k}$ is continuous for all t, while $(1/t)\mathbf{i} - \ln(t)\mathbf{k}$ is continuous for $t > 0$.

We say that $\mathbf{F}$ is *differentiable* if each component function is differentiable; in this case, we define $\mathbf{F}'(t) = x'(t)\mathbf{i} + y'(t)\mathbf{j} + z'(t)\mathbf{k}$. The derivative of $\mathbf{F}(t)$, evaluated at a specific point t_0, is denoted $\mathbf{F}'(t_0)$, or $d\mathbf{F}/dt\big|_{t_0}$. Note that the derivative of a vector function is again a vector function. For example, if $\mathbf{F}(t) = (1/t)\mathbf{i} - \ln(t)\mathbf{k}$ for $t > 0$, then

$$\mathbf{F}'(t) = -\frac{1}{t^2}\mathbf{i} - \frac{1}{t}\mathbf{k}$$

and

$$\mathbf{F}'(1) = -\mathbf{i} - \mathbf{k}.$$

As with real-valued functions of a single variable, a vector function need not be differentiable. For example, consider

$$\mathbf{K}(t) = |t|\mathbf{i} - 3t^2\mathbf{j}.$$

Then $\mathbf{K}$ is continuous for all t. However, the first component function $x(t) = |t|$ is not differentiable at $t = 0$; hence, $\mathbf{K}$ is not differentiable at zero.

Geometrically, we may envision a vector function as an adjustable arrow pivoted at the origin. $\mathbf{F}(t)$ is represented as the arrow from the origin to the point $(x(t), y(t), z(t))$; as t varies, the arrow sweeps out a curve in space. This curve is the locus of points $(x(t), y(t), z(t))$, typically shown as in Figure 15.1. The curve is also called a *trajectory*. The vector $\mathbf{F}(t) = x(t)\mathbf{i} + y(t)\mathbf{j} + z(t)\mathbf{k}$ is called the *position vector* of this curve.

The vector $\mathbf{F}'(t_0)$ can be interpreted as a tangent vector to this curve at the point $(x(t_0), y(t_0), z(t_0))$. To understand this, note that, from the parallelogram law, the vector

$$\mathbf{F}(t_0 + \Delta t) - \mathbf{F}(t_0)$$

is represented by the arrow from the point $(x(t_0), y(t_0), z(t_0))$ to the point $(x(t_0 + \Delta t),$ $y(t_0 + \Delta t), z(t_0 + \Delta t))$. This vector is shown in Figure 15.2. Since Δt is a nonzero scalar, the vector

$$\frac{1}{\Delta t}[\mathbf{F}(t_0 + \Delta t) - \mathbf{F}(t_0)]$$

is also along the line from $(x(t_0), y(t_0), z(t_0))$ to $(x(t_0 + \Delta t), y(t_0 + \Delta t), z(t_0 + \Delta t))$. In terms of components,

$$\frac{1}{\Delta t}[\mathbf{F}(t_0 + \Delta t) - \mathbf{F}(t_0)] = \frac{x(t_0 + \Delta t) - x(t_0)}{\Delta t}\mathbf{i} + \frac{y(t_0 + \Delta t) - y(t_0)}{\Delta t}\mathbf{j} \quad (15.1)$$

$$+ \frac{z(t_0 + \Delta t) - z(t_0)}{\Delta t}\mathbf{k}.$$

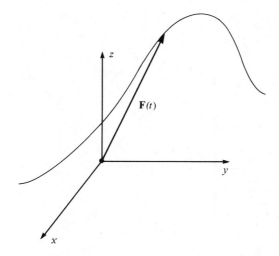

Figure 15.1

In the limit as $\Delta t \to 0$, the vector on the left side of equation (15.1) moves into a position tangent to the curve at $(x(t_0), y(t_0), z(t_0))$, while the vector on the right side approaches $x'(t_0)\mathbf{i} + y'(t_0)\mathbf{j} + z'(t_0)\mathbf{k}$. This limit process is illustrated in Figure 15.3 and provides the rationale for interpreting $\mathbf{F}'(t_0)$ as a tangent to the curve at $(x(t_0), y(t_0), z(t_0))$.

Henceforth, when we speak of the tangent to a curve $x = x(t)$, $y = y(t)$, $z = z(t)$ at a point t_0, we mean the vector $\mathbf{F}'(t_0)$, where $\mathbf{F}(t) = x(t)\mathbf{i} + y(t)\mathbf{j} + z(t)\mathbf{k}$ is the position vector of the curve.

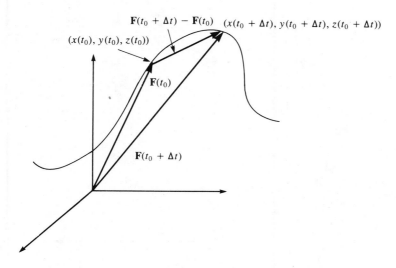

Figure 15.2

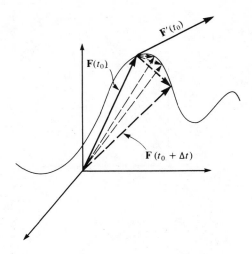

Figure 15.3

It is natural to draw a tangent vector $\mathbf{F}'(t)$ as an arrow from the point $(x(t),$ $y(t), z(t))$ on the curve rather than as an arrow from the origin. This associates the vector $\mathbf{F}'(t)$ more clearly with its role as a tangent vector to a particular curve. A typical curve and tangent vector are shown in Figure 15.4.

In calculus, it is shown that the length of a curve given parametrically by

$$x = x(t), \qquad y = y(t), \qquad z = z(t); \qquad a \leq t \leq b$$

is given by

$$\text{length} = \int_a^b \sqrt{[x'(t)]^2 + [y'(t)]^2 + [z'(t)]^2}\, dt$$

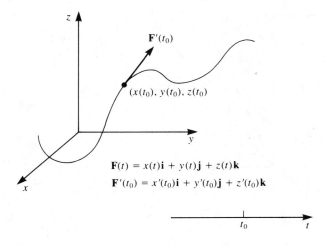

Figure 15.4

if x', y', and z' are continuous on $[a, b]$. The quantity under the radical is exactly the length of the tangent vector

$$\|\mathbf{F}'(t)\| = \sqrt{[x'(t)]^2 + [y'(t)]^2 + [z'(t)]^2},$$

where $\mathbf{F}(t) = x(t)\mathbf{i} + y(t)\mathbf{j} + z(t)\mathbf{k}$. Thus, we may write the length of the curve as

$$\text{length} = \int_a^b \|\mathbf{F}'(t)\| \, dt.$$

The length of a curve is the integral of the length of the tangent vector.

EXAMPLE 15.1

Consider the curve C in three-space given parametrically by

$$x = 2\cos(t), \qquad y = 2\sin(t), \qquad z = t; \qquad 0 \le t \le 2\pi.$$

As t varies from zero to 2π, C begins at $(x(0), y(0), z(0)) = (2, 0, 0)$ and ends at $(x(2\pi), y(2\pi), z(2\pi)) = (2, 0, 2\pi)$. The curve winds around the cylinder $x^2 + y^2 = 4$, since $[2\cos(t)]^2 + [2\sin(t)]^2 = 4$, as shown in Figure 15.5.

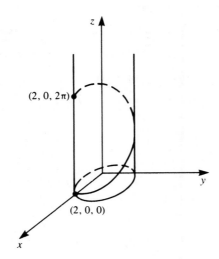

Figure 15.5

The position vector of this curve is

$$\mathbf{F}(t) = 2\cos(t)\mathbf{i} + 2\sin(t)\mathbf{j} + t\mathbf{k}.$$

The tangent vector is

$$\mathbf{F}'(t) = -2\sin(t)\mathbf{i} + 2\cos(t)\mathbf{j} + \mathbf{k},$$

and the length of the tangent vector is

$$\|\mathbf{F}'(t)\| = \sqrt{4\sin^2(t) + 4\cos^2(t) + 1} = \sqrt{5}.$$

The length of the curve is

$$\text{length} = \int_0^{2\pi} \sqrt{5}\, dt = 2\pi\sqrt{5}. \quad \blacksquare$$

It is often convenient to deal with a unit tangent vector to a curve. In general, the tangent vector $\mathbf{F}'(t)$ will not have length 1 for all t. However, if we change the parameter and use arc length s along the curve to describe the curve, we obtain a position function whose derivative has length 1, yielding a unit tangent. Here is how this is done.

Suppose we are given a position function $\mathbf{F}(t) = x(t)\mathbf{i} + y(t)\mathbf{j} + z(t)\mathbf{k}$ for a curve C given parametrically by $x = x(t)$, $y = y(t)$, $z = z(t)$ for $a \le t \le b$. Let P_0 be the point $(x(a), y(a), z(a))$ on the curve, as shown in Figure 15.6. We may think of this as the initial, or starting, point of C. For $a \le t \le b$, define a function $s(t)$ by

$$s(t) = \int_a^t \|\mathbf{F}'(\xi)\|\, d\xi. \tag{15.2}$$

Thus, $s(t)$ is the length of the part of C from P_0 to $P:(x(t), y(t), z(t))$. Obviously, $s(t)$ is an increasing function of t. As t increases, the distance along C from P_0 to P increases. At least in theory, the equation $s = s(t)$ can be solved for t as a function of s, $t = t(s)$. Substitute $t = t(s)$ into the parametric equations of C to get

$$x = x(t(s)), \qquad y = y(t(s)), \qquad z = z(t(s)); \qquad 0 \le s \le L,$$

in which L is the length of the curve. In terms of s, the position vector is

$$\mathbf{F}(t(s)) = x(t(s))\mathbf{i} + y(t(s))\mathbf{j} + z(t(s))\mathbf{k}.$$

By the chain rule, the tangent vector, in terms of s, is

$$\begin{aligned}
\mathbf{F}'(t(s)) &= \frac{d}{ds}[x(t(s))]\mathbf{i} + \frac{d}{ds}[y(t(s))]\mathbf{j} + \frac{d}{ds}[z(t(s))]\mathbf{k} \\
&= x'(t)t'(s)\mathbf{i} + y'(t)t'(s)\mathbf{j} + z'(t)t'(s)\mathbf{k} \\
&= t'(s)[x'(t)\mathbf{i} + y'(t)\mathbf{j} + z'(t)\mathbf{k}] = t'(s)\mathbf{F}'(t).
\end{aligned}$$

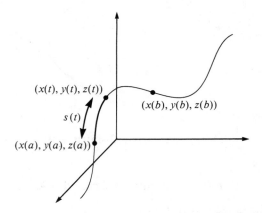

Figure 15.6. $s(t) = $ length of C from $(x(a), y(a), z(a))$ to $(x(t), y(t), z(t))$.

But, from equation (15.2) and the rule for differentiating an inverse function,

$$t'(s) = \frac{1}{s'(t)} = \frac{1}{\|\mathbf{F}'(t)\|},$$

so

$$\|\mathbf{F}'(t(s))\| = \|t'(s)\mathbf{F}'(t)\| = |t'(s)|\,\|\mathbf{F}'(t)\|$$

$$= \frac{1}{\|\mathbf{F}'(t)\|}\|\mathbf{F}'(t)\| = 1.$$

This shows that, when arc length s is used as a parameter to describe the curve, the tangent vector obtained by taking the derivative of the position vector has length 1. Thus, when we wish to have a unit tangent vector to a curve, we choose arc length as parameter, or variable, in the coordinate functions of the curve.

EXAMPLE 15.2

Consider again the curve of Example 15.1. We had

$$\mathbf{F}(t) = 2\cos(t)\mathbf{i} + 2\sin(t)\mathbf{j} + t\mathbf{k}; \qquad 0 \le t \le 2\pi.$$

Here P_0 is the point $(x(0), y(0), z(0)) = (2, 0, 0)$. The length along the curve from P_0 to an arbitrary point $(x(t), y(t), z(t))$ is

$$s(t) = \int_0^t \|\mathbf{F}'(\xi)\|\, d\xi = \int_0^t \sqrt{5}\, d\xi = \sqrt{5}t.$$

In this simple example, we can explicitly solve for t in terms of s:

$$t = \frac{1}{\sqrt{5}}s.$$

In terms of s, the position vector of the curve is

$$\mathbf{F}(s) = 2\cos\left(\frac{1}{\sqrt{5}}s\right)\mathbf{i} + 2\sin\left(\frac{1}{\sqrt{5}}s\right)\mathbf{j} + \frac{1}{\sqrt{5}}s\mathbf{k}.$$

It is routine to calculate $\mathbf{F}'(s)$ and check that $\|\mathbf{F}'(s)\| = 1$. ∎

Henceforth, we will always use s to denote arc length when discussing curves and vector functions of a single variable.

We will conclude this section with some rules for differentiating various combinations of vector functions. If $\mathbf{F}$ and $\mathbf{G}$ are differentiable vector functions of a single variable, then

$$(\mathbf{F} + \mathbf{G})' = \mathbf{F}' + \mathbf{G}',$$

exactly as with real-valued differentiable functions of one variable. This is obvious because we add vectors by adding respective components, and we differentiate by differentiating each component.

Further, if f is a differentiable real-valued function, then

$$[f(t)\mathbf{F}(t)]' = f'(t)\mathbf{F}(t) + f(t)\mathbf{F}'(t).$$

To derive this rule, write

$$f(t)\mathbf{F}(t) = f(t)x(t)\mathbf{i} + f(t)y(t)\mathbf{j} + f(t)z(t)\mathbf{k}.$$

Then

$$
\begin{aligned}
[f(t)\mathbf{F}(t)]' &= \frac{d}{dt}[f(t)x(t)]\mathbf{i} + \frac{d}{dt}[f(t)y(t)]\mathbf{j} + \frac{d}{dt}[f(t)z(t)]\mathbf{k} \\
&= [f'(t)x(t) + f(t)x'(t)]\mathbf{i} + [f'(t)y(t) + f(t)y'(t)]\mathbf{j} \\
&\quad + [f'(t)z(t) + f(t)z'(t)]\mathbf{k} \\
&= f'(t)[x(t)\mathbf{i} + y(t)\mathbf{j} + z(t)\mathbf{k}] + f(t)[x'(t)\mathbf{i} + y'(t)\mathbf{j} + z'(t)\mathbf{k}] \\
&= f'(t)\mathbf{F}(t) + f(t)\mathbf{F}'(t).
\end{aligned}
$$

This rule is similar in form to the rule for differentiating a product $f(t)g(t)$, but we must remember that one factor here is a vector function. For each t, $f'(t)\mathbf{F}(t)$ and $f(t)\mathbf{F}'(t)$ are products of a scalar with a vector.

The rules for differentiating dot and cross products of vectors are also very similar in form to the usual product rule. The rules are

$$(\mathbf{F} \cdot \mathbf{G})' = \mathbf{F}' \cdot \mathbf{G} + \mathbf{F} \cdot \mathbf{G}' \tag{15.3}$$

and

$$(\mathbf{F} \times \mathbf{G})' = \mathbf{F}' \times \mathbf{G} + \mathbf{F} \times \mathbf{G}'. \tag{15.4}$$

These can be proved by writing $\mathbf{F} \cdot \mathbf{G}$ and $\mathbf{F} \times \mathbf{G}$ in terms of the component functions of $\mathbf{F}$ and $\mathbf{G}$ and differentiating the components.

The formulas (15.3) and (15.4) are important in general calculations with vectors. In specific instances, it is usually just as easy to carry out the dot or cross product first and differentiate the resulting scalar or vector function.

Finally, the chain rule applies to vector functions in the following form:

$$\frac{d}{dt}\mathbf{F}[f(t)] = f'(t)\mathbf{F}'(f(t)),$$

assuming that f and $\mathbf{F}$ are differentiable. This fact can be established by writing

$$\mathbf{F}[f(t)] = x(f(t))\mathbf{i} + y(f(t))\mathbf{j} + z(f(t))\mathbf{k}$$

and applying the chain rule to each component.

In the next section, we will develop the concepts of velocity and acceleration and use them to discuss some aspects of the geometry of curves in three-space.

PROBLEMS FOR SECTION 15.1

In each of Problems 1 through 5, (a) calculate the vector $f(t)\mathbf{F}(t)$ and differentiate this vector by differentiating each component and (b) compute $[f(t)\mathbf{F}(t)]'$ using the formula for the derivative of this type of product. These should yield the same result.

1. $f(t) = 4\cos(3t)$, $\mathbf{F}(t) = \mathbf{i} + 3t^2\mathbf{j} + 2t\mathbf{k}$

2. $f(t) = 1 - 2t^3$, $\mathbf{F}(t) = t\mathbf{i} - \cosh(t)\mathbf{j} + e^t\mathbf{k}$

3. $f(t) = \sin^2(t)$, $\mathbf{F}(t) = 4\mathbf{i} + t^5\mathbf{k}$

4. $f(t) = t^2 - 2t + 3$, $\mathbf{F}(t) = \ln(t)\mathbf{i} + e^t\mathbf{j} - t^2\mathbf{k}$

5. $f(t) = 2t + 3$, $\mathbf{F}(t) = (1 - 3t)\mathbf{i} + t^4\mathbf{j} - t\mathbf{k}$

In each of Problems 6 through 10, (a) compute $\mathbf{F} \cdot \mathbf{G}$ and differentiate the resulting real-valued function and (b) compute $(\mathbf{F} \cdot \mathbf{G})'$ using equation (15.3). The answers in (a) and (b) should agree.

6. $\mathbf{F}(t) = t\mathbf{i} - 3t^2\mathbf{k}, \quad \mathbf{G}(t) = \mathbf{i} + \cos(t)\mathbf{k}$ **7.** $\mathbf{F}(t) = e^t\mathbf{j} + 2\mathbf{k}, \quad \mathbf{G}(t) = \cos(t)\mathbf{i} + 2\mathbf{j} + t^2\mathbf{k}$

8. $\mathbf{F}(t) = t\mathbf{i} + t\mathbf{j} - t^2\mathbf{k}, \quad \mathbf{G}(t) = \sinh(t)\mathbf{i} - 4\mathbf{j} - t^3\mathbf{k}$ **9.** $\mathbf{F}(t) = -4\cos(t)\mathbf{k}, \quad \mathbf{G}(t) = -t^2\mathbf{i} + 4\sin(t)\mathbf{k}$

10. $\mathbf{F}(t) = \sqrt{t}\,\mathbf{i} + 2t\mathbf{j} - \ln(t)\mathbf{k}, \quad \mathbf{G}(t) = t^2\mathbf{i} - 4t\mathbf{j}$

In each of Problems 11 through 15, (a) compute $\mathbf{F} \times \mathbf{G}$ and differentiate the resulting vector function by differentiating each component and (b) use equation (15.4) to compute $(\mathbf{F} \times \mathbf{G})'$. The answers to (a) and (b) should agree.

11. $\mathbf{F}(t) = t\mathbf{i} + \mathbf{j} + 4\mathbf{k}, \quad \mathbf{G}(t) = \mathbf{i} - \cos(t)\mathbf{j} + t\mathbf{k}$ **12.** $\mathbf{F}(t) = -2t\mathbf{i} + t^2\mathbf{j}, \quad \mathbf{G}(t) = t\mathbf{j} - 4t^3\mathbf{k}$

13. $\mathbf{F}(t) = -9\mathbf{i} + t^2\mathbf{j} + t^2\mathbf{k}, \quad \mathbf{G}(t) = e^t\mathbf{i}$ **14.** $\mathbf{F}(t) = \sinh(t)\mathbf{j} - t\mathbf{k}, \quad \mathbf{G}(t) = t\mathbf{i} + t^2\mathbf{j} - t^2\mathbf{k}$

15. $\mathbf{F}(t) = -4t\mathbf{i} + \cos(t)\mathbf{k}, \quad \mathbf{G}(t) = t^2\mathbf{i} + t\mathbf{j}$

In each of Problems 16 through 25, parametric equations of a curve in three-space are given. Write the position vector for this curve and find the tangent vector to the curve.

16. $x = 1, \quad y = 2t, \quad z = 1; \quad 0 \le t \le 4$

17. $x = t, \quad y = \sin(2\pi t), \quad z = \cos(2\pi t); \quad 0 \le t \le 1$

18. $x = \cosh(t), \quad y = \sinh(t), \quad z = 4t; \quad 0 \le t \le 2$

19. $x = e^t\cos(t), \quad y = e^t\sin(t), \quad z = e^t; \quad 0 \le t \le \pi$

20. $x = y = t^2, \quad z = 3t^2; \quad 0 \le t \le 1$

21. $x = y = z = t^3; \quad 1 \le t \le 2$

22. $x = 2\cos(2t), \quad y = 2\sin(2t), \quad z = 1 - 3t; \quad 0 \le t \le 2\pi$

23. $x = 4\ln(2t + 1), \quad y = 4\sinh(3t), \quad z = 2; \quad 3 \le t \le 9$

24. $x = 2 - \sinh(t), \quad y = \cosh(t), \quad z = \ln(t); \quad 1 \le t \le 2$

25. $x = 4t^2 - 5, \quad y = 3t, \quad z = t; \quad -1 \le t \le 5$

In each of Problems 26 through 30, the position vector $\mathbf{F}(t)$ of a curve in three-space is given. (a) Compute $\mathbf{F}'(t)$. (b) Write an integral for the length function $s(t)$ along the curve (do not attempt to evaluate this integral explicitly). (c) Compute $s'(t)$. (d) Use the chain rule to compute $d/ds\,[\mathbf{F}(t(s))]$, and show that the resulting tangent vector has length 1.

26. $\mathbf{F} = 3t^2\mathbf{i} - 2t\mathbf{j} + t\mathbf{k}$ **27.** $\mathbf{F} = 4\cos(t)\mathbf{i} + 2\sin(3t)\mathbf{j} - \mathbf{k}$ **28.** $\mathbf{F} = (3 - t)\mathbf{i} + 6t^3\mathbf{j} + t\mathbf{k}$

29. $\mathbf{F} = e^t\sin(t)\mathbf{i} + t\mathbf{j} + e^t\cos(t)\mathbf{k}$ **30.** $\mathbf{F} = (3t^2 - 1)\mathbf{i} + 2t\mathbf{j} - t^2\mathbf{k}$

In each of Problems 31 through 34, explicitly determine arc length s along the curve as a function of t, and solve for t as a function of s. Write the position vector $\mathbf{F}$ in terms of s, and show that $\|\mathbf{F}'(s)\| = 1$.

31. $x = t, \quad y = \cosh(t), \quad z = 1; \quad 0 \le t \le \pi$ **32.** $x = \sin(t), \quad y = \cos(t), \quad z = 45; \quad 0 \le t \le \pi$

33. $x = y = z = t^3; \quad -1 \le t \le 1$ **34.** $x = 2t^2, \quad y = 3t^2, \quad z = 4t^2; \quad 1 \le t \le 3$

35. Let $\mathbf{F}(t) = x(t)\mathbf{i} + y(t)\mathbf{j} + z(t)\mathbf{k}$, with $x, y,$ and z twice-differentiable functions. Prove that $(\mathbf{F} \times \mathbf{F}')' = \mathbf{F} \times \mathbf{F}''$.

36. Let $\mathbf{F}(t) = x(t)\mathbf{i} + y(t)\mathbf{j} + z(t)\mathbf{k}$, with $x, y,$ and z differentiable. Consider $\mathbf{F}(t)$ as the position vector of a particle at time t moving along a curve C. Suppose that $\mathbf{F} \times \mathbf{F}' = \mathbf{O}$. Prove that the particle always moves in the same direction.

37. Let $\mathbf{F}(t) = x(t)\mathbf{i} + y(t)\mathbf{j} + z(t)\mathbf{k}$, with $x, y,$ and z twice-differentiable. Think of $\mathbf{F}(t)$ as the position of a particle moving along a curve in three-space.

   (a) Prove that $\|\mathbf{F}(t) \times [\mathbf{F}(t + \Delta t) - \mathbf{F}(t)]\|$ is twice the area of the sector bounded by $C, \mathbf{F}(t),$ and $\mathbf{F}(t + \Delta t)$, with these vectors represented as arrows from the origin to points of C.

   (b) Prove that, if $\mathbf{F} \times \mathbf{F}'' = \mathbf{O}$, the particle moves so that equal areas are swept out by the position vector in equal times. (Compare this result with one of Kepler's laws of planetary motion.)

38. Assuming that $\mathbf{F}, \mathbf{G},$ and $\mathbf{H}$ are differentiable, derive a formula for $d/dt\,[\mathbf{F}, \mathbf{G}, \mathbf{H}]$, the derivative of the scalar triple product of $\mathbf{F}, \mathbf{G},$ and $\mathbf{H}$.

15.2 *Velocity, Acceleration, Curvature, and Torsion*

We will now use vector functions to analyze the motion of a particle moving along a trajectory in three-space. In doing this, we will encounter certain concepts from the differential geometry of curves.

Imagine a particle moving in three-space (or, as a special case, in a plane). At time t, the particle is at a point $(x(t), y(t), z(t))$. The position vector

$$\mathbf{F}(t) = x(t)\mathbf{i} + y(t)\mathbf{j} + z(t)\mathbf{k}$$

can be thought of as an arrow from the origin to the particle at time t. We can also think of the coordinate functions $x = x(t)$, $y = y(t)$, $z = z(t)$ as describing the trajectory of the particle as t (usually thought of as time) increases.

For the calculations we want to carry out in this section, we will assume that x, y, and z are twice-differentiable functions of t. In any time interval $[t_1, t_2]$, the particle moves a distance along its trajectory given by

$$\int_{t_1}^{t_2} \|\mathbf{F}'(t)\| \, dt.$$

The *velocity* of the particle at time t is defined to be

$$\mathbf{v}(t) = \mathbf{F}'(t) = x'(t)\mathbf{i} + y'(t)\mathbf{j} + z'(t)\mathbf{k}.$$

The vector $\mathbf{v}(t)$ is in the direction of the tangent to the trajectory at any time t. The magnitude of the velocity is the *speed* and is a scalar quantity denoted v:

$$v(t) = \|\mathbf{v}(t)\| = \|\mathbf{F}'(t)\|.$$

Consistent with intuition, the speed of the particle is the rate of change with respect to time of distance traveled along the curve, since

$$\frac{d}{dt} \int_{t_0}^{t} \|\mathbf{F}'(\xi)\| \, d\xi = \|\mathbf{F}'(t)\| = \|\mathbf{v}(t)\| = v(t).$$

The acceleration of the particle is the rate of change of the velocity with respect to time,

$$\mathbf{a}(t) = \mathbf{v}'(t) = x''(t)\mathbf{i} + y''(t)\mathbf{j} + z''(t)\mathbf{k}.$$

EXAMPLE 15.3

Suppose that

$$\mathbf{F}(t) = \sin(t)\mathbf{i} + 2e^{-t}\mathbf{j} + t^2\mathbf{k}.$$

The particle is moving along the trajectory given by

$$x = \sin(t), \qquad y = 2e^{-t}, \qquad z = t^2.$$

The velocity is

$$\mathbf{v}(t) = \cos(t)\mathbf{i} - 2e^{-t}\mathbf{j} + 2t\mathbf{k},$$

and the acceleration is

$$\mathbf{a}(t) = -\sin(t)\mathbf{i} + 2e^{-t}\mathbf{j} + 2\mathbf{k}.$$

The speed is

$$v(t) = \|\mathbf{v}(t)\| = \sqrt{\cos^2(t) + 4e^{-2t} + 4t^2}. \quad \blacksquare$$

We have seen that $\mathbf{F}'(t)$ is in the direction of the tangent to the trajectory at $(x(t), y(t), z(t))$. If we parametrize the curve in terms of length s along the curve from some initial point, then $\mathbf{F}'(s)$ has length 1 and is a unit tangent. If it is inconvenient to introduce s, we can obtain a unit tangent vector by dividing $\mathbf{F}'(t)$ by its length. This leads us to define the *unit tangent* vector $\mathbf{T}(t)$ as

$$\mathbf{T}(t) = \frac{1}{\|\mathbf{F}'(t)\|} \mathbf{F}'(t),$$

provided that $\mathbf{F}'(t) \neq \mathbf{O}$. If $\mathbf{F}'(t) = \mathbf{O}$, we do not define a tangent vector to the curve at $(x(t), y(t), z(t))$. If $t = s$, $\|\mathbf{F}'(s)\| = 1$, and this equation just gives us $\mathbf{T}(s) = \mathbf{F}'(s)$.

The *curvature κ* of a curve in three-space is defined to be the magnitude of the rate of change of the tangent vector with respect to s,

$$\kappa = \left\| \frac{d\mathbf{T}}{ds} \right\|.$$

Intuitively, the greater the magnitude $\mathbf{T}'(s)$, the sharper the curve bends at that point. We use this magnitude to quantify the "amount of bending," or curvature. A straight line has constant tangent vector and hence zero curvature, as we would expect.

It is shown in calculus that , in terms of the position vector $\mathbf{F}(t)$,

$$\kappa = \frac{\|\mathbf{F}' \times \mathbf{F}''\|}{\|\mathbf{F}'\|^3},$$

assuming that $\mathbf{F}'(t) \neq \mathbf{O}$.

The quantity $\rho = 1/\kappa$, the reciprocal of the curvature, is called the *radius of curvature* of a curve, provided that $\kappa \neq 0$. The circle which best approximates C near P on its concave side can be shown to have radius ρ. Such a circle is called the *osculating circle* to the curve at P. A typical osculating circle is shown in Figure 15.7.

In terms of the scalar quantity ρ and the unit tangent vector, it is possible to define a *unit normal* vector to the curve at a point P. This is a unit vector orthogonal (perpendicular) to P defined by

$$\mathbf{N}(s) = \rho \frac{d\mathbf{T}}{ds}.$$

To verify that this is indeed a unit vector, compute

$$\|\mathbf{N}\| = \rho \left\| \frac{d\mathbf{T}}{ds} \right\| = \frac{1}{\kappa} \left\| \frac{d\mathbf{T}}{ds} \right\| = 1,$$

in view of the definition of κ. Further, $\mathbf{N}$ is orthogonal to $\mathbf{T}$. To prove this, begin by recalling that $\|\mathbf{T}\| = 1$. Thus, $\|\mathbf{T}\|^2 = \mathbf{T} \cdot \mathbf{T} = 1$. Differentiate this equation with respect to s:

$$\frac{d}{ds}(\mathbf{T} \cdot \mathbf{T}) = \mathbf{T}'(s) \cdot \mathbf{T}(s) + \mathbf{T}(s) \cdot \mathbf{T}'(s) = \frac{d}{ds}[1] = 0.$$

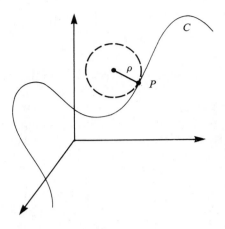

Figure 15.7

Therefore,

$$2\mathbf{T}'(s) \cdot \mathbf{T}(s) = 0;$$

hence, $\mathbf{T}'(s) \cdot \mathbf{T}(s) = 0$, and $\mathbf{T}'(s)$ is orthogonal to $\mathbf{T}(s)$. But $\mathbf{N}$ is a positive scalar multiple of $\mathbf{T}'(s)$ and hence is in the same direction as $\mathbf{T}'(s)$. Therefore, $\mathbf{N}$ is also orthogonal to $\mathbf{T}$.

The unit tangent and normal vectors $\mathbf{T}$ and $\mathbf{N}$ can be used to decompose the acceleration into a tangential component and a centripetal component as follows.

THEOREM 15.1

$$\mathbf{a} = \frac{dv}{dt}\,\mathbf{T} + \frac{1}{\rho}\,v^2\mathbf{N}.$$

Proof   Notice first that, in terms of an arbitrary parameter t,

$$\mathbf{T}(t) = \frac{1}{\|\mathbf{F}'(t)\|}\,\mathbf{F}'(t) = \frac{1}{v}\,\mathbf{v},$$

since $v = \|\mathbf{v}\|$ and $\mathbf{v} = \mathbf{F}'(t)$. Then

$$\mathbf{v} = v\mathbf{T}.$$

Then

$$\begin{aligned}
\mathbf{a} &= \frac{d\mathbf{v}}{dt} = \frac{d}{dt}[v\mathbf{T}] = \frac{dv}{dt}\,\mathbf{T} + v\,\frac{d\mathbf{T}}{dt} \\
&= \frac{dv}{dt}\,\mathbf{T} + v\,\frac{ds}{dt}\,\frac{d\mathbf{T}}{ds} \\
&= \frac{dv}{dt}\,\mathbf{T} + v^2\,\frac{d\mathbf{T}}{ds},
\end{aligned}$$

since $v = ds/dt$. But we defined the unit normal vector by $\mathbf{N} = \rho(d\mathbf{T}/ds)$; hence, $d\mathbf{T}/ds = (1/\rho)\mathbf{N}$, and we have

$$\mathbf{a} = \frac{dv}{dt}\,\mathbf{T} + \frac{1}{\rho}\,v^2\mathbf{N},$$

as we wanted to show.   ■

The scalar dv/dt is called the *tangential component* of the acceleration and is denoted a_T. The scalar $(1/\rho)v^2$ is called the *centripetal*, or *normal*, *component* of the acceleration and is denoted a_N. We have therefore decomposed the acceleration vector into the sum

$$\mathbf{a} = a_T\mathbf{T} + a_N\mathbf{N}.$$

EXAMPLE 15.4

Suppose the position vector of a particle at time $t > 0$ is

$$\mathbf{F}(t) = [\cos(t) + t\sin(t)]\mathbf{i} + [\sin(t) - t\cos(t)]\mathbf{j} + t^2\mathbf{k}.$$

Then

$$\mathbf{v}(t) = \mathbf{F}'(t) = t\cos(t)\mathbf{i} + t\sin(t)\mathbf{j} + 2t\mathbf{k},$$
$$v = \|\mathbf{v}\| = \sqrt{t^2\cos^2(t) + t^2\sin^2(t) + 4t^2} = \sqrt{5t^2} = \sqrt{5}t \qquad \text{(because } t > 0\text{)},$$

and

$$\mathbf{a} = [\cos(t) - t\sin(t)]\mathbf{i} + [\sin(t) + t\cos(t)]\mathbf{j} + 2\mathbf{k}.$$

We also have

$$\|\mathbf{a}\| = \sqrt{[\cos(t) - t\sin(t)]^2 + [\sin(t) + t\cos(t)]^2 + 4},$$

which simplifies to

$$\|\mathbf{a}\| = \sqrt{5 + t^2}.$$

The tangential component of the acceleration is

$$a_T = \frac{dv}{dt} = \sqrt{5}.$$

To compute the normal component with a minimum of effort, apply the Pythagorean theorem to the parallelogram sum $\mathbf{a} = a_T\mathbf{T} + a_N\mathbf{N}$ shown in Figure 15.8. Since $\mathbf{T}$ and $\mathbf{N}$ are unit vectors, two sides of the triangle have lengths a_T and a_N, and the third side has length $\|\mathbf{a}\|$. Thus,

$$a_N^2 = \|\mathbf{a}\|^2 - a_T^2 = 5 + t^2 - 5 = t^2.$$

Then

$$a_N = |t| = t,$$

since $t > 0$. The acceleration vector may therefore be written

$$\mathbf{a} = \sqrt{5}\mathbf{T} + t\mathbf{N}.$$

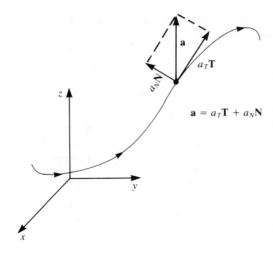

$$\mathbf{a} = a_T\mathbf{T} + a_N\mathbf{N}$$

Figure 15.8

From a_N, we easily obtain the curvature of the trajectory. We have

$$a_N = t = \frac{1}{\rho}v^2 = \kappa v^2 = \kappa(5t^2),$$

so

$$\kappa = \frac{1}{5t}.$$

Notice that $\kappa \to 0$ as $t \to \infty$; hence, the trajectory tends to straighten out as time increases. As $t \to 0+$, $\kappa \to \infty$, so the curve begins at $(1,0,0)$ with very large curvature.

We have not computed $\mathbf{T}$ or $\mathbf{N}$ in this example, but we can easily do so. We have

$$\mathbf{T} = \frac{d\mathbf{R}}{ds} = \frac{dt}{ds}\frac{d\mathbf{R}}{dt} = \frac{1}{v}\mathbf{v}$$

$$= \frac{1}{\sqrt{5}}[\cos(t)\mathbf{i} + \sin(t)\mathbf{j} + 2\mathbf{k}]$$

and

$$\mathbf{N} = \rho\frac{d\mathbf{T}}{ds} = \rho\frac{dt}{ds}\frac{d\mathbf{T}}{dt} = \frac{1}{v}\rho\frac{d\mathbf{T}}{dt}$$

$$= \frac{5t}{\sqrt{5t}}\frac{1}{\sqrt{5}}[-\sin(t)\mathbf{i} + \cos(t)\mathbf{j}] = -\sin(t)\mathbf{i} + \cos(t)\mathbf{j}.$$

Notice that $\mathbf{T}$ and $\mathbf{N}$ are orthogonal unit vectors. ∎

The vector $\mathbf{B} = \mathbf{T} \times \mathbf{N}$ is called the *binormal* of the curve and is a unit vector orthogonal to the plane of $\mathbf{T}$ and $\mathbf{N}$. The vectors $\mathbf{T}$, $\mathbf{N}$, and $\mathbf{B}$ form a right-handed

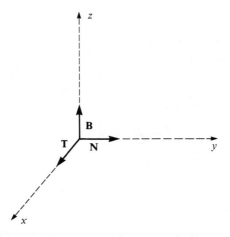

Figure 15.9. Right-handed coordinate system formed by tangent,
normal, and binormal vectors.

coordinate system at each point on the curve. This means that, if we take the positive
x-axis along $\mathbf{T}$, the positive y-axis along $\mathbf{N}$, and the positive z-axis along $\mathbf{B}$, we obtain
a coordinate system as shown in Figure 15.9. By contrast, a left-handed coordinate
system is shown in Figure 15.10.

Since $\mathbf{N}$ is defined to be $\rho(d\mathbf{T}/ds)$, and $\rho = 1/\kappa$, $\mathbf{N} = (1/\kappa)d\mathbf{T}/ds$; hence,

$$\frac{d\mathbf{T}}{ds} = \kappa\mathbf{N}. \tag{15.5}$$

It can be shown that, for some scalar function $\tau(t)$,

$$\frac{d\mathbf{N}}{ds} = -\kappa\mathbf{T} + \tau\mathbf{B} \tag{15.6}$$

and

$$\frac{d\mathbf{B}}{ds} = -\tau\mathbf{N}. \tag{15.7}$$

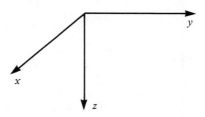

Figure 15.10. Left-handed coordinate system.

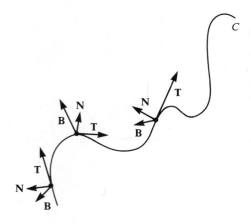

Figure 15.11. Torsion as a measure of twisting along a curve.

The scalar $\tau(s)$ is called the *torsion* of the curve at the point $(x(s), y(s), z(s))$. The three formulas (15.5), (15.6), and (15.7) are called the *Frenet formulas*. Intuitively, torsion measures the amount a curve twists in the sense that the coordinate system formed at each point by $\mathbf{T}$, $\mathbf{N}$, and $\mathbf{B}$ seems to twist about C as one moves along the curve. This concept is illustrated in Figure 15.11.

Up to this point, we have concentrated on the idea of a vector-valued function of a single real variable, the trajectory defined in three-space by its coordinate functions, and some geometric concepts which describe the behavior of the trajectory. In the next section, we will turn our attention to vector functions of two and three variables and the three fundamental operations of vector differential calculus: gradient, divergence, and curl.

PROBLEMS FOR SECTION 15.2

In each of Problems 1 through 15, determine $\mathbf{v}$, v, $\mathbf{a}$, a_T, a_N, κ, $\mathbf{T}$, $\mathbf{N}$, and $\mathbf{B}$, given the position vector $\mathbf{F}$.

1. $\mathbf{F} = 3t\mathbf{i} - 2\mathbf{j} + t^2\mathbf{k}$

2. $\mathbf{F} = t \sin(t)\mathbf{i} + t \cos(t)\mathbf{j} + \mathbf{k}$

3. $\mathbf{F} = 2t\mathbf{i} - 2t\mathbf{j} + t\mathbf{k}$

4. $\mathbf{F} = e^t\sin(t)\mathbf{i} - \mathbf{j} + e^t\cos(t)\mathbf{k}$

5. $\mathbf{F} = 2 \sin(t)\mathbf{i} + t\mathbf{j} + 2 \cos(t)\mathbf{k}$

6. $\mathbf{F} = 2t^2\mathbf{i} + t\mathbf{j} + t\mathbf{k}$

7. $\mathbf{F} = e^{-t}(\mathbf{i} + \mathbf{j} - 2\mathbf{k})$

8. $\mathbf{F} = \alpha \cos(t)\mathbf{i} + \alpha \sin(t)\mathbf{j} + \beta t\mathbf{k}$

9. $\mathbf{F} = 2 \sinh(t)\mathbf{j} - 2 \cosh(t)\mathbf{k}$

10. $\mathbf{F} = \alpha \cos(t)\mathbf{i} + \beta \sin(t)\mathbf{k}$

11. $\mathbf{F} = 2t\mathbf{i} - \cos(t)\mathbf{j} - \sin(t)\mathbf{k}$

12. $\mathbf{F} = e^{-t}(\mathbf{i} - \mathbf{j} + t\mathbf{k})$

13. $\mathbf{F} = t^2\mathbf{i} + t^2\mathbf{j} - 2t\mathbf{k}$

14. $\mathbf{F} = 3t \cos(t)\mathbf{j} - 3t \sin(t)\mathbf{k}$

15. $\mathbf{F} = \alpha t^2\mathbf{i} + \beta t^2\mathbf{j} + \gamma t^2\mathbf{k}$

16. Let $\mathbf{F}$ be the position vector of a curve C. Suppose that the unit tangent $\mathbf{T}$ is a constant vector. Prove that C is a straight line.

17. Prove that $\tau(s) = -\mathbf{N}(s) \cdot \mathbf{B}'(s)$.

18. Show that $\tau(s) = [\mathbf{T}(s), \mathbf{N}(s), \mathbf{N}'(s)]$, a scalar triple product.

19. Show that $\tau(s) = (1/\kappa^2)[\mathbf{F}'(s), \mathbf{F}''(s), \mathbf{F}^{(3)}(s)]$.

20. Let $\mathbf{F}(t) = \alpha[\cos(\omega t)\mathbf{i} + \sin(\omega t)\mathbf{j}]$ be the position vector of a particle moving in the xy-plane.

(a) Show that the angular speed (speed divided by distance α from the center of the circular path) is ω.

(b) Show that $\mathbf{a}$ is directed toward the origin, with constant magnitude $\alpha\omega^2$. (This is the *centripetal acceleration*. The *centripetal force* is $m\mathbf{a}$, and the *centrifugal force* is $-m\mathbf{a}$, where m is the mass of the particle.)

15.3 *Vector Fields and Lines of Force*

Until now, we have dealt with vector-valued functions of a single real variable. We also frequently encounter vector functions whose components are functions of two or three real variables. A typical vector function of two variables has the appearance

$$\mathbf{G}(x, y) = g_1(x, y)\mathbf{i} + g_2(x, y)\mathbf{j},$$

while a typical vector function of three variables has the form

$$\mathbf{F}(x, y, z) = f_1(x, y, z)\mathbf{i} + f_2(x, y, z)\mathbf{j} + f_3(x, y, z)\mathbf{k}.$$

Such functions are called *vector fields*. $\mathbf{G}$ is a vector field in the plane, and $\mathbf{F}$ is a vector field in three-space. At each point (x, y) at which a vector field $\mathbf{G}(x, y)$ in the plane is defined, we can draw an arrow representing the vector $\mathbf{G}(x, y)$. Similarly, at each point (x, y, z) in three-space, we can draw an arrow representing the vector $\mathbf{F}(x, y, z)$. As a specific example, let

$$\mathbf{G}(x, y) = xy\mathbf{i} + (x - y)\mathbf{j}.$$

Then

$$\mathbf{G}(1, 1) = \mathbf{i}, \qquad \mathbf{G}(1, 4) = 4\mathbf{i} - 3\mathbf{j}, \qquad \mathbf{G}(-1, 0) = -\mathbf{j},$$

and so on. These vectors are drawn as arrows from the points $(1, 1)$, $(1, 4)$, and $(-1, 0)$, respectively, in Figure 15.12. By drawing enough vectors, we get a pictorial representation of the vector field. Such diagrams are effective in representing such quantities as magnetic or electrical fields. A vector field is *continuous* if each of the component functions is continuous.

We define the partial derivative of a vector field to be the vector field obtained by taking the partial derivative of each of the components. With $\mathbf{G}(x, y)$ and $\mathbf{F}(x, y, z)$

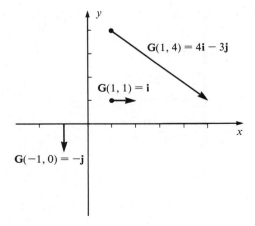

Figure 15.12

as defined above,

$$\frac{\partial \mathbf{G}}{\partial x} = \frac{\partial g_1}{\partial x}\mathbf{i} + \frac{\partial g_2}{\partial x}\mathbf{j} \quad \text{and} \quad \frac{\partial \mathbf{G}}{\partial y} = \frac{\partial g_1}{\partial y}\mathbf{i} + \frac{\partial g_2}{\partial y}\mathbf{j};$$

and

$$\frac{\partial \mathbf{F}}{\partial x} = \frac{\partial f_1}{\partial x}\mathbf{i} + \frac{\partial f_2}{\partial x}\mathbf{j} + \frac{\partial f_3}{\partial x}\mathbf{k}, \qquad \frac{\partial \mathbf{F}}{\partial y} = \frac{\partial f_1}{\partial y}\mathbf{i} + \frac{\partial f_2}{\partial y}\mathbf{j} + \frac{\partial f_3}{\partial y}\mathbf{k};$$

and

$$\frac{\partial \mathbf{F}}{\partial z} = \frac{\partial f_1}{\partial z}\mathbf{i} + \frac{\partial f_2}{\partial z}\mathbf{j} + \frac{\partial f_3}{\partial z}\mathbf{k}.$$

For example, let $\mathbf{F}(x, y, z) = 3xz\mathbf{i} + 2e^z\cos(x)\mathbf{j} - y^2\mathbf{k}$. Then

$$\frac{\partial \mathbf{F}}{\partial x} = 3z\mathbf{i} - 2e^z\sin(x)\mathbf{j}, \qquad \frac{\partial \mathbf{F}}{\partial y} = -2y\mathbf{k}, \quad \text{and} \quad \frac{\partial \mathbf{F}}{\partial z} = 3x\mathbf{i} + 2e^z\cos(x)\mathbf{j}.$$

A partial derivative of a vector field is again a vector field.

As we have noted, we can represent a vector field $\mathbf{F}(x, y, z)$ by drawing from each point (x, y, z) an arrow representing the vector $\mathbf{F}(x, y, z)$. In some cases, we can envision these arrows as tangents to curves in space. Such curves are called *streamlines, flow lines*, or *lines of force*, of the vector field. Terminology varies according to context. For example, if $\mathbf{F}$ is the velocity of a fluid, the curves are often called streamlines or flow lines; if $\mathbf{F}$ is a magnetic field, we usually speak of lines of force. If iron filings are placed on a piece of cardboard and a magnet is held underneath, the filings will arrange themselves along the lines of force of the magnetic field, with a greater concentration of filings where the field is stronger.

Given a vector field $\mathbf{F}$ in three-space, how do we construct these lines of force? This is a problem of constructing a curve given its tangent at each point. Suppose that C is such a curve, with parametric equations

$$x = x(s), \qquad y = y(s), \qquad z = z(s),$$

parametrized in terms of arc length along the curve. Let the position vector for this trajectory be

$$\mathbf{R}(s) = x(s)\mathbf{i} + y(s)\mathbf{j} + z(s)\mathbf{k}.$$

Now,

$$\mathbf{R}'(s) = x'(s)\mathbf{i} + y'(s)\mathbf{j} + z'(s)\mathbf{k}$$

is a unit tangent vector to the curve. But $\mathbf{F}(x(s), y(s), z(s))$ is tangent to the curve at $(x(s), y(s), z(s))$ and hence is parallel to $\mathbf{R}'(s)$. Thus, for any given s, there is a scalar t such that

$$\mathbf{R}'(s) = t\mathbf{F}(x(s), y(s), z(s)).$$

Then

$$\frac{dx}{ds}\mathbf{i} + \frac{dy}{ds}\mathbf{j} + \frac{dz}{ds}\mathbf{k} = t\mathbf{F}(x(s), y(s), z(s)).$$

Let $\mathbf{F} = f_1\mathbf{i} + f_2\mathbf{j} + f_3\mathbf{k}$. Then

$$\frac{dx}{ds} = tf_1, \qquad \frac{dy}{ds} = tf_2, \quad \text{and} \quad \frac{dz}{ds} = tf_3. \tag{15.8}$$

Since f_1, f_2, and f_3 are known functions, this is a system of differential equations whose solution gives the coordinate functions of the lines of force. If f_1, f_2, and f_3 are nonzero, we can eliminate t and write the system in differential form as

$$\frac{dx}{f_1} = \frac{dy}{f_2} = \frac{dz}{f_3}.$$

EXAMPLE 15.5

Let $\mathbf{F}(x, y, z) = x^2\mathbf{i} + 2y\mathbf{j} - \mathbf{k}$. We will find the lines of force of this vector field.

The lines of force satisfy the system (15.8) of differential equations, which for this vector field are

$$\frac{dx}{ds} = tx^2, \qquad \frac{dy}{ds} = 2ty, \qquad \frac{dz}{ds} = -t.$$

Alternatively, if $x \neq 0$ and $y \neq 0$, we can write the system in the form

$$\frac{dx}{x^2} = \frac{dy}{2y} = \frac{dz}{-1}.$$

We can solve these equations in pairs. Begin, for example, with $dx/x^2 = dz/-1$, which we can write as $dz = -(1/x^2)\,dx$. Integrate to obtain

$$z = \frac{1}{x} + c_1,$$

in which c_1 is a constant of integration.

Next, consider $dy/2y = dz/-1$, which we write as $dz = -(1/2y)\,dy$. Integrate to get

$$z = -\tfrac{1}{2}\ln(y) + c_2.$$

In this example, it seems easiest to express x and y in terms of z. We get

$$x = \frac{1}{z - c_1}, \qquad y = c_3 e^{-2z}, \qquad z = z,$$

in which $c_3 = e^{2c_2}$. These are parametric equations of the lines of force. These curves will pass through different points depending on the choices of the arbitrary constants c_1 and c_2. For example, suppose we want the line of force passing through $(-1, 6, 2)$. Let $x = -1$, $y = 6$, and $z = 2$ in the parametric equations to get

$$-1 = \frac{1}{2 - c_1} \quad \text{and} \quad 6 = c_3 e^{-4}.$$

Then $c_1 = 3$ and $c_3 = 6e^4$. The line of force passing through $(-1, 6, 2)$ has parametric equations

$$x = \frac{1}{z-3}, \qquad y = 6e^{4-2z}, \qquad z = z. \quad \blacksquare$$

EXAMPLE 15.6

Find the lines of force of $\mathbf{F}(x, y, z) = -y\mathbf{j} + z\mathbf{k}$.

Here, $f_1 = 0$, and the lines of force satisfy the equations

$$\frac{dx}{ds} = 0, \qquad \frac{dy}{ds} = -ty, \qquad \frac{dz}{ds} = tz.$$

The first equation implies that $\dot{x} = \text{constant} = c_1$. Thus, each line of force is in a plane parallel to the yz-plane. The other two equations can be combined to write

$$\frac{dy}{-y} = \frac{dz}{z}.$$

Integrate this equation to obtain

$$-\ln(y) = \ln(z) + c_2,$$

or

$$\frac{1}{y} = c_3 z,$$

where $c_3 = e^{c_2}$. The lines of force are given parametrically by the equations

$$x = c_1, \qquad \frac{1}{y} = c_3 z.$$

Suppose, to be specific, that we want the line of force passing through the point $(-4, 1, 7)$. Then we must choose $c_1 = -4$. Further, we need

$$1 = 7c_3,$$

so $c_3 = \frac{1}{7}$. The line of force passing through $(-4, 1, 7)$ is the curve given by

$$x = -4, \qquad \frac{1}{y} = \frac{1}{7} z.$$

This is the hyperbola $z = 7/y$ in the plane $x = -4$. $\blacksquare$

In the next section, we will define the gradient operator, which produces a vector field from a real-valued function of two or three variables.

PROBLEMS FOR SECTION 15.3

In each of Problems 1 through 10, (a) find $\partial \mathbf{G}/\partial x$ and $\partial \mathbf{G}/\partial y$ and (b) make a diagram in which each indicated vector $\mathbf{G}(x, y)$ is drawn as an arrow in the xy-plane from the point (x, y).

1. $\mathbf{G}(x, y) = 3x\mathbf{i} - 4xy\mathbf{j}$; $\mathbf{G}(0, 1)$, $\mathbf{G}(1, 3)$, $\mathbf{G}(1, 4)$, $\mathbf{G}(-1, 2)$
2. $\mathbf{G}(x, y) = e^x\mathbf{i} - 2x^2 y\mathbf{j}$; $\mathbf{G}(0, 0)$, $\mathbf{G}(1, 0)$, $\mathbf{G}(0, 1)$, $\mathbf{G}(3, -2)$

3. $\mathbf{G}(x, y) = 4xy\mathbf{i} - y\mathbf{j}$; $\mathbf{G}(0, 0)$, $\mathbf{G}(-1, 0)$, $\mathbf{G}(0, 1)$, $\mathbf{G}(1, 3)$

4. $\mathbf{G}(x, y) = 3\mathbf{i} - 4x\mathbf{j}$; $\mathbf{G}(1, 0)$, $\mathbf{G}(2, 0)$, $\mathbf{G}(0, 1)$, $\mathbf{G}(3, -1)$

5. $\mathbf{G}(x, y) = 2xy\mathbf{i} + \cos(x)\mathbf{j}$; $\mathbf{G}\left(\dfrac{\pi}{2}, 0\right)$, $\mathbf{G}(0, 0)$, $\mathbf{G}(\pi, 1)$, $\mathbf{G}(3\pi, 0)$

6. $\mathbf{G}(x, y) = (-5x + 1)\mathbf{i} + 2x^2\mathbf{j}$; $\mathbf{G}(0, 2)$, $\mathbf{G}(-1, 1)$, $\mathbf{G}(3, 1)$, $\mathbf{G}(1, -2)$

7. $\mathbf{G}(x, y) = e^{-x}y\mathbf{i} + 8xy\mathbf{j}$; $\mathbf{G}(1, -3)$, $\mathbf{G}(0, 0)$, $\mathbf{G}(-1, -2)$, $\mathbf{G}(0, 2)$

8. $\mathbf{G}(x, y) = \sin(2x)y\mathbf{i} + x^2\mathbf{j}$; $\mathbf{G}(0, 1)$, $\mathbf{G}(-3, 1)$, $\mathbf{G}(0, 0)$, $\mathbf{G}(1, -4)$

9. $\mathbf{G}(x, y) = 2\ln(x + 1)y^2\mathbf{i} + 8xy^3\mathbf{j}$; $\mathbf{G}(0, 1)$, $\mathbf{G}(0, -2)$, $\mathbf{G}(1, 1)$, $\mathbf{G}(1, -1)$

10. $\mathbf{G}(x, y) = 3x^2\mathbf{i} + 11xy\mathbf{j}$; $\mathbf{G}(1, -2)$, $\mathbf{G}(0, -2)$, $\mathbf{G}(3, 0)$, $\mathbf{G}(-1, 2)$

In each of Problems 11 through 15, calculate $\partial\mathbf{F}/\partial x$, $\partial\mathbf{F}/\partial y$, and $\partial\mathbf{F}/\partial z$.

11. $\mathbf{F}(x, y, z) = e^{xy}\mathbf{i} - 2x^2y\mathbf{j} + \cosh(z + y)\mathbf{k}$

12. $\mathbf{F}(x, y, z) = 4z^2\cos(x)\mathbf{i} - x^3yz\mathbf{j} + x^3y\mathbf{k}$

13. $\mathbf{F}(x, y, z) = 3xy^3\mathbf{i} + \ln(x + y + z)\mathbf{j} + \cosh(xyz)\mathbf{k}$

14. $\mathbf{F}(x, y, z) = -\sin(xy)z^2\mathbf{i} + 3xy^4z\mathbf{j} + \cosh(z - x)\mathbf{k}$

15. $\mathbf{F}(x, y, z) = (14x - 2y)\mathbf{i} + (x^2 - y^2 - z^2)\mathbf{j} + 5xy\mathbf{k}$

In each of Problems 16 through 25, find the lines of force for the given vector field, then find the particular line of force passing through the given point.

16. $\mathbf{F} = \mathbf{i} - y^2\mathbf{j} + z\mathbf{k}$; $(2, 1, 1)$

17. $\mathbf{F} = \mathbf{i} - 2\mathbf{j} + \mathbf{k}$; $(0, 1, 1)$

18. $\mathbf{F} = \dfrac{1}{x}\mathbf{i} + e^y\mathbf{j} - \mathbf{k}$; $(2, 0, 4)$

19. $\mathbf{F} = \cos(y)\mathbf{i} + \sin(x)\mathbf{j}$; $\left(\dfrac{\pi}{2}, 0, -4\right)$

20. $\mathbf{F} = 2e^z\mathbf{j} - \cos(y)\mathbf{k}$; $\left(3, \dfrac{\pi}{4}, 0\right)$

21. $\mathbf{F} = 3x^2\mathbf{i} - y\mathbf{j} + z^3\mathbf{k}$; $(2, 1, 6)$

22. $\mathbf{F} = e^z\mathbf{i} - x^2\mathbf{k}$; $(4, 2, 0)$

23. $\mathbf{F} = x^2\mathbf{i} + y^2\mathbf{j} - z^2\mathbf{k}$; $(1, 1, 1)$

24. $\mathbf{F} = \sec(x)\mathbf{i} - \cot(y)\mathbf{j} + \mathbf{k}$; $\left(\dfrac{\pi}{4}, 0, 1\right)$

25. $\mathbf{F} = -3\mathbf{i} + 2e^z\mathbf{j} - \cos(y)\mathbf{k}$; $\left(2, \dfrac{\pi}{4}, 0\right)$

26. Construct a vector field in three-space whose lines of force are straight lines.

27. Construct a vector field in the plane whose lines of force are circles about the origin.

28. Is it possible for a vector field in three-space to have lines of force lying only in the xy-plane?

15.4 *The Gradient Vector Field* _____

Let $\varphi(x, y, z)$ be a real-valued function of three variables. In the context of vectors, such a function is called a *scalar field*. The gradient of φ, denoted grad φ, or $\nabla\varphi$, is a vector field manufactured from φ according to the rule

$$\nabla\varphi = \frac{\partial\varphi}{\partial x}\mathbf{i} + \frac{\partial\varphi}{\partial y}\mathbf{j} + \frac{\partial\varphi}{\partial z}\mathbf{k}$$

wherever these partial derivatives are defined. The symbol $\nabla\varphi$ is read "del phi," and ∇ is called the *del operator*. It operates on a scalar field to produce a vector field.

If φ is a function of just two variables x and y, the gradient of φ is a vector field in the plane:

$$\nabla\varphi = \frac{\partial\varphi}{\partial x}\mathbf{i} + \frac{\partial\varphi}{\partial y}\mathbf{j}.$$

$\nabla\varphi(P_0)$ denotes the gradient of φ evaluated at P_0.

EXAMPLE 15.7

Let $\varphi = 2xy + e^z - x^2 z$. Then

$$\nabla \varphi = (2y - 2xz)\mathbf{i} + 2x\mathbf{j} + (e^z - x^2)\mathbf{k},$$

and

$$\nabla \varphi(1, 2, -1) = 6\mathbf{i} + 2\mathbf{j} + (e^{-1} - 1)\mathbf{k}. \quad \blacksquare$$

The gradient is related to the directional derivative of φ. Suppose we are given a point P_0 and a unit vector $\mathbf{u}$ specifying a direction from P_0. The directional derivative of φ at P_0 in the direction of $\mathbf{u}$ is the rate of change of φ with respect to (x, y, z) as (x, y, z) varies in the direction of $\mathbf{u}$ from P_0. This directional derivative is denoted $D_{\mathbf{u}}\varphi(P_0)$. Although this concept is usually taught in calculus, we will review it here.

Let P_0 have coordinates (x_0, y_0, z_0). In Figure 15.13, (x, y, z) is any point on the line from P_0 in the direction of $\mathbf{u}$. The vector $(x - x_0)\mathbf{i} + (y - y_0)\mathbf{j} + (z - z_0)\mathbf{k}$ is parallel to $\mathbf{u}$ and in the same direction. Hence, for some $t > 0$,

$$(x - x_0)\mathbf{i} + (y - y_0)\mathbf{j} + (z - z_0)\mathbf{k} = t\mathbf{u}.$$

Let $\mathbf{u} = u_1\mathbf{i} + u_2\mathbf{j} + u_3\mathbf{k}$. Then

$$x - x_0 = tu_1, \qquad y - y_0 = tu_2, \qquad z - z_0 = tu_3,$$

so

$$x = x_0 + tu_1, \qquad y = y_0 + tu_2, \qquad z = z_0 + tu_3.$$

Along this direction from P_0,

$$\varphi(x, y, z) = \varphi(x_0 + tu_1, y_0 + tu_2, z_0 + tu_3),$$

a function of the nonnegative parameter t. The rate of change of φ at P_0 along this direction is the derivative of $\varphi(x_0 + tu_1, y_0 + tu_2, z_0 + tu_3)$ with respect to t, evaluated at $t = 0$. Thus,

$$D_{\mathbf{u}}\varphi(P_0) = \frac{d}{dt}\left[\varphi(x_0 + tu_1, y_0 + tu_2, z_0 + tu_3)\right]\Big|_{t=0}.$$

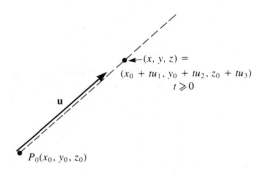

Figure 15.13

By the chain rule,

$$\frac{d}{dt}[\varphi(x_0 + tu_1, y_0 + tu_2, z_0 + tu_3)] = \frac{\partial \varphi}{\partial x}\frac{\partial x}{\partial t} + \frac{\partial \varphi}{\partial y}\frac{\partial y}{\partial t} + \frac{\partial \varphi}{\partial z}\frac{\partial z}{\partial t}$$

$$= \frac{\partial \varphi}{\partial x}\frac{\partial}{\partial t}[x_0 + tu_1] + \frac{\partial \varphi}{\partial y}\frac{\partial}{\partial t}[y_0 + tu_2]$$

$$+ \frac{\partial \varphi}{\partial z}\frac{\partial}{\partial t}[z_0 + tu_3]$$

$$= \frac{\partial \varphi}{\partial x}u_1 + \frac{\partial \varphi}{\partial y}u_2 + \frac{\partial \varphi}{\partial z}u_3.$$

We get the directional derivative by evaluating these partial derivatives at $t = 0$ [that is, at $(x, y, z) = (x_0, y_0, z_0)$]. Then

$$D_{\mathbf{u}}\varphi(P_0) = \frac{\partial \varphi}{\partial x}(P_0)u_1 + \frac{\partial \varphi}{\partial y}(P_0)u_2 + \frac{\partial \varphi}{\partial z}(P_0)u_3.$$

Now recognize that this is the dot product of the gradient of φ at P_0 with $\mathbf{u}$:

$$D_{\mathbf{u}}\varphi(P_0) = \nabla\varphi(P_0) \cdot \mathbf{u}.$$

In applying this formula to calculate a directional derivative, $\mathbf{u}$ must be a unit vector. If the direction is specified by a vector $\mathbf{v}$ of length different from 1, let $\mathbf{u} = \mathbf{v}/\|\mathbf{v}\|$ in this formula.

EXAMPLE 15.8

Let $\varphi(x, y, z) = x^2 y - xe^z$. Then

$$\nabla\varphi = (2xy - e^z)\mathbf{i} + x^2\mathbf{j} - xe^z\mathbf{k}.$$

Let P_0 be $(2, -1, 0)$. Then

$$\nabla\varphi(P_0) = -5\mathbf{i} + 4\mathbf{j} - 2\mathbf{k}.$$

Let

$$\mathbf{u} = \frac{1}{\sqrt{21}}(2\mathbf{i} - 4\mathbf{j} + \mathbf{k}).$$

Then $\mathbf{u}$ is a unit vector, and the directional derivative of φ at P_0 in the direction of $\mathbf{u}$ is

$$D_{\mathbf{u}}\varphi(2, -1, 0) = \nabla\varphi(2, -1, 0) \cdot \mathbf{u}$$

$$= [-5\mathbf{i} + 4\mathbf{j} - 2\mathbf{k}] \cdot \left[\frac{1}{\sqrt{21}}(2\mathbf{i} - 4\mathbf{j} + \mathbf{k})\right] = \frac{-28}{\sqrt{21}}. \quad \blacksquare$$

Suppose $\varphi(x, y, z)$ is defined for all (x, y, z) in some sphere about P_0. If we look in various directions from P_0, we may see φ increasing in some directions, decreasing in others, and perhaps remaining constant in some directions. We can ask: in what direction from P_0 does $\varphi(x, y, z)$ increase (or decrease) at the greatest rate? The answer is provided by the following theorem.

THEOREM 15.2

Let $\varphi(x, y, z)$ and its first partial derivatives be continuous in some sphere about P_0, and assume that $\nabla\varphi(P_0) \neq \mathbf{O}$.

1. The direction from P_0 in which φ has its maximum rate of change is the direction of the gradient vector at P_0. Further, this maximum rate of change is $\|\nabla\varphi(P_0)\|$.
2. The direction from P_0 in which φ has its minimum rate of change is the direction of $-\nabla\varphi(P_0)$, and this minimum rate of change is $-\|\nabla\varphi(P_0)\|$.

Proof   Let $\mathbf{u}$ be any unit vector drawn as an arrow from P_0. Then

$$D_{\mathbf{u}}\varphi(P_0) = \nabla\varphi(P_0) \cdot \mathbf{u}.$$

We want to choose $\mathbf{u}$ so that this directional derivative is as large as possible. If θ is the angle between $\mathbf{u}$ and $\nabla\varphi(P_0)$,

$$\nabla\varphi(P_0) \cdot \mathbf{u} = \|\nabla\varphi(P_0)\|\,\|\mathbf{u}\|\cos(\theta) = \|\nabla\varphi(P_0)\|\cos(\theta),$$

since $\|\mathbf{u}\| = 1$. This directional derivative has its maximum value when $\cos(\theta) = 1$, or $\theta = 0$, in which case $\mathbf{u}$ is in the same direction as $\nabla\varphi(P_0)$. Further, in this case, the directional derivative equals $\|\nabla\varphi(P_0)\|$.

The directional derivative $D_{\mathbf{u}}\varphi(P_0)$ has its minimum value when $\cos(\theta) = -1$. In this event, $\theta = \pi$, and $\mathbf{u}$ is in the direction of $-\mathbf{V}(P_0)$. Further, $D_{\mathbf{u}}\varphi(P_0) = \|\nabla\varphi(P_0)\|\cos(\pi) = -\|\nabla\varphi(P_0)\|$ in this event.   ∎

Depending on φ and c, the locus of points satisfying an equation $\varphi(x, y, z) = c$, for c constant, may form a surface in three-space. For example, if $\varphi(x, y, z) = x^2 + y^2 + z^2$, the equation $\varphi(x, y, z) = 16$ has as its graph a sphere of radius 4 about the origin. By contrast, the locus $\varphi(x, y, z) = 0$ consists of a single point, the origin; and the locus $\varphi(x, y, z) = -4$ is empty, containing no points.

A surface $\varphi(x, y, z) = c$ is called a *level surface* of the function φ. We claim that the gradient is perpendicular to this surface at any point where the gradient is not zero, in the sense that $\nabla\varphi$ is perpendicular to the tangent plane to this surface there.

THEOREM 15.3

Let φ and its first partial derivatives be continuous for (x, y, z) on the level surface $\varphi(x, y, z) = c$. Then the gradient vector is normal (perpendicular) to this surface at any point where the gradient vector is not the zero vector.   ∎

This conclusion is illustrated in Figure 15.14. We will sketch the idea behind the theorem. Suppose we have a curve through P_0 and lying on the surface $\varphi(x, y, z) = c$, as in Figure 15.15. Let the curve have parametric equations

$$x = x(t), \qquad y = y(t), \qquad z = z(t)$$

for t in some interval I, and let $P_0 = (x(t_0), y(t_0), z(t_0))$. Since the curve is on the surface,

$$\varphi(x(t), y(t), z(t)) = c$$

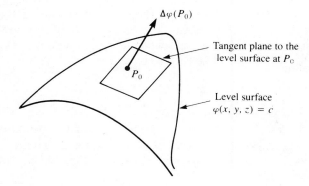

Figure 15.14. $\mathbf{V}\varphi(P_0)$ is normal to the level surface at P_0.

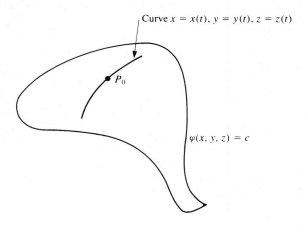

Figure 15.15

for t in I. Therefore, for t in I,

$$\frac{d}{dt}\left[\varphi(x(t), y(t), z(t))\right] = 0.$$

By the chain rule,

$$\frac{\partial \varphi}{\partial x}\frac{dx}{dt} + \frac{\partial \varphi}{\partial y}\frac{dy}{dt} + \frac{\partial \varphi}{\partial z}\frac{dz}{dt} = 0.$$

This equation can be written

$$\mathbf{V}\varphi \cdot \left[x'(t)\mathbf{i} + y'(t)\mathbf{j} + z'(t)\mathbf{k}\right] = 0.$$

In particular,

$$\mathbf{V}\varphi(P_0) \cdot \left[x'(t_0)\mathbf{i} + y'(t_0)\mathbf{j} + z'(t_0)\mathbf{k}\right] = 0.$$

But $x'(t_0)\mathbf{i} + y'(t_0)\mathbf{j} + z'(t_0)\mathbf{k}$ is tangent to the curve at P_0, so $\nabla\varphi(P_0)$ must be perpendicular to the (tangent to the) curve there. Since this holds for any smooth curve on the surface through P_0, $\nabla\varphi(P_0)$ is perpendicular to the surface at P_0.

We will illustrate Theorem 15.3 with some examples.

EXAMPLE 15.9

Let $\varphi(x, y, z) = x^2 + y^2 + z^2$. Then $\nabla\varphi = 2x\mathbf{i} + 2y\mathbf{j} + 2z\mathbf{k}$.

If $c > 0$, the level surface $\varphi(x, y, z) = c$ is a sphere of radius $\sqrt{c}$ about the origin. At any point $P_0 : (x_0, y_0, z_0)$ on this sphere, $\nabla\varphi(P_0) = 2x_0\mathbf{i} + 2y_0\mathbf{j} + 2z_0\mathbf{k}$ can be represented as an arrow from the origin through P_0, as in Figure 15.16. This vector is normal to the sphere at P_0.

For example, if P_0 is the "north pole" $(0, 0, \sqrt{c})$, then $\nabla\varphi(P_0) = 2\sqrt{c}\,\mathbf{k}$, normal to the surface and pointing along the z-axis.

Usually we draw a normal vector as an arrow from the point on the surface to clarify its relationship to the surface. Figure 15.17 shows a normal vector as an arrow from a point on the surface. ∎

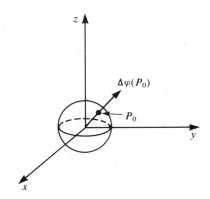

Figure 15.16

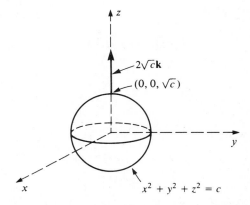

Figure 15.17. Normal to a sphere at $(0, 0, r)$.

EXAMPLE 15.10

Let $\varphi(x, y, z) = z - \sqrt{x^2 + y^2}$. Then the surface $\varphi(x, y, z) = 0$ is the cone

$$z = \sqrt{x^2 + y^2}$$

shown in Figure 15.18. Calculate

$$\nabla\varphi = \frac{-x}{\sqrt{x^2 + y^2}}\,\mathbf{i} + \frac{-y}{\sqrt{x^2 + y^2}}\,\mathbf{j} + \mathbf{k} = -\frac{x}{z}\,\mathbf{i} - \frac{y}{z}\,\mathbf{j} + \mathbf{k},$$

provided that $x^2 + y^2 \neq 0$. At any point on the cone (other than the origin), the gradient $\nabla\varphi$ can be drawn as an arrow pointing into the cone and perpendicular to the side of the cone at that point. Put another way, $\nabla\varphi(P)$ is perpendicular to the position vector $\mathbf{R}(x, y, z) = x\mathbf{i} + y\mathbf{j} + z\mathbf{k}$ at (x, y, z). $\mathbf{R}$ and $\nabla\varphi(P)$ are shown in Figure 15.19. To

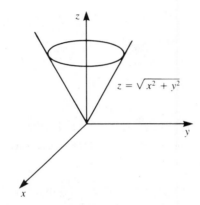

Figure 15.18

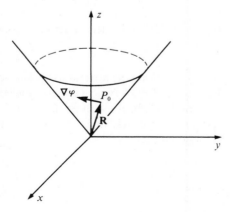

Figure 15.19. Cone $z = \sqrt{x^2 + y^2}$ and normal at P_0.

show that these two vectors are indeed orthogonal, calculate their dot product

$$\mathbf{R} \cdot \nabla \varphi = [x\mathbf{i} + y\mathbf{j} + z\mathbf{k}] \cdot \left[-\frac{x}{z}\mathbf{i} - \frac{y}{z}\mathbf{j} + \mathbf{k} \right]$$

$$= -\frac{1}{z}x^2 - \frac{1}{z}y^2 + z = \frac{-x^2 - y^2 + z^2}{z} = 0$$

because, on the surface, $z^2 = x^2 + y^2$. ∎

In view of Theorem 15.3, we can use the gradient vector to find the tangent plane to a surface $\varphi(x, y, z) = c$ at any point P_0 where the gradient is defined and is not the zero vector. If (x, y, z) is on the tangent plane, $(x - x_0)\mathbf{i} + (y - y_0)\mathbf{j} + (z - z_0)\mathbf{k}$ is a vector parallel to the tangent plane and hence is orthogonal to $\nabla \varphi(P_0)$. Therefore,

$$\nabla \varphi(P_0) \cdot [(x - x_0)\mathbf{i} + (y - y_0)\mathbf{j} + (z - z_0)\mathbf{k}] = 0.$$

This yields the equation

$$\frac{\partial \varphi}{\partial x}(P_0)(x - x_0) + \frac{\partial \varphi}{\partial y}(y - y_0) + \frac{\partial \varphi}{\partial z}(P_0)(z - z_0) = 0$$

for the tangent plane at P_0.

EXAMPLE 15.11

Consider again the cone $z = \sqrt{x^2 + y^2}$. This is the locus of $\varphi(x, y, z) = 0$ if $\varphi(x, y, z) = z - \sqrt{x^2 + y^2}$. The point $(1, 1, \sqrt{2})$ is on this surface. From Example 15.10, the gradient at this point is

$$\nabla \varphi(P_0) = \frac{-1}{\sqrt{2}}\mathbf{i} + \frac{-1}{\sqrt{2}}\mathbf{j} + \mathbf{k}.$$

The tangent plane at P_0 has equation

$$\frac{-1}{\sqrt{2}}(x - 1) - \frac{1}{\sqrt{2}}(y - 1) + (z - \sqrt{2}) = 0.$$

Notice that the gradient of φ is not defined at the origin. The cone has no tangent plane at the origin. ∎

A straight line through P_0 and along the normal to the surface $\varphi(x, y, z) = c$ at P_0 is called the *normal line* to the surface at P_0. A typical normal line is shown in Figure 15.20. To find the equation of a normal line, begin with the fact that $\nabla \varphi(P_0)$ is normal to the surface at P_0. If (x, y, z) is on the normal line, $(x - x_0)\mathbf{i} + (y - y_0)\mathbf{j} + (z - z_0)\mathbf{k}$ is parallel to $\nabla \varphi(P_0)$; hence, for some scalar t,

$$(x - x_0)\mathbf{i} + (y - y_0)\mathbf{j} + (z - z_0)\mathbf{k} = t\nabla \varphi(P_0).$$

Equate respective components to get

$$x - x_0 = t\frac{\partial \varphi}{\partial x}(P_0),$$

$$y - y_0 = t \, \frac{\partial \varphi}{\partial y}(P_0),$$

$$z - z_0 = t \, \frac{\partial \varphi}{\partial z}(P_0).$$

These are the parametric equations of the normal line. They may also be written

$$x = x_0 + t \, \frac{\partial \varphi}{\partial x}(P_0), \qquad y = y_0 + t \, \frac{\partial \varphi}{\partial x}(P_0), \qquad z = z_0 + t \, \frac{\partial \varphi}{\partial x}(P_0).$$

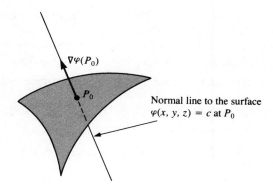

Normal line to the surface
$\varphi(x, y, z) = c$ at P_0

Figure 15.20

EXAMPLE 15.12

For the cone of Examples 15.10 and 15.11, the normal line at $(1, 1, \sqrt{2})$ has parametric equations

$$x = 1 - \frac{1}{\sqrt{2}} \, t, \qquad y = 1 - \frac{1}{\sqrt{2}} \, t, \qquad z = \sqrt{2} + t,$$

$-\infty < t < \infty$. The point $(1, 1, \sqrt{2})$ is on this line when $t = 0$. ∎

We will conclude this section with two additional examples.

EXAMPLE 15.13

Let $\varphi(x, y, z) = 2xz + e^y z^2$.

1. Find the rate of change of φ in the direction of $2\mathbf{i} + 3\mathbf{j} - \mathbf{k}$ at $(2, 1, 1)$.

The rate of change we want is $D_{\mathbf{u}}\varphi(2, 1, 1)$, the directional derivative of φ in the direction given. Since the vector specifying the direction is not a unit vector, divide it by its length to let

$$\mathbf{u} = \frac{1}{\sqrt{14}}(2\mathbf{i} + 3\mathbf{j} - \mathbf{k}).$$

Now calculate

$$\nabla\varphi = 2z\mathbf{i} + e^y z^2 \mathbf{j} + (2x + 2ze^y)\mathbf{k}.$$

Then

$$\nabla\varphi(2, 1, 1) = 2\mathbf{i} + e\mathbf{j} + (4 + 2e)\mathbf{k},$$

so

$$D_\mathbf{u}\varphi(2, 1, 1) = \nabla\varphi(2, 1, 1) \cdot \mathbf{u} = \frac{1}{\sqrt{14}}(4 + 3e - 4 - 2e) = \frac{e}{\sqrt{14}}.$$

2. Find the direction from $(2, 1, 1)$ in which φ has its greatest rate of change, and calculate this maximum rate of change.

   The direction is that of the gradient at $(2, 1, 1)$, and this is $2\mathbf{i} + e\mathbf{j} + (4 + 2e)\mathbf{k}$. The maximum rate of change is $\|2\mathbf{i} + e\mathbf{j} + (4 + 2e)\mathbf{k}\|$, or $\sqrt{20 + 16e + 5e^2}$.

3. Find the tangent plane and normal vector to the surface $\varphi(x, y, z) = 4 + e$ at $(2, 1, 1)$.

   The normal vector to the surface at $(2, 1, 1)$ is $\nabla\varphi(2, 1, 1) = 2\mathbf{i} + e\mathbf{j} + (4 + 2e)\mathbf{k}$. The tangent plane has equation

$$2(x - 2) + e(y - 1) + (4 + 2e)(z - 1) = 0,$$

or

$$2x + ey + (4 + 2e)z = 8 + 3e.$$

The normal line has parametric equations

$$x = 2 + 2t, \qquad y = 1 + et, \qquad z = 1 + (4 + 2e)t.$$

By eliminating t, the equation of this line can also be written

$$\frac{x - 2}{2} = \frac{y - 1}{e} = \frac{z - 1}{4 + 2e}. \quad \blacksquare$$

EXAMPLE 15.14

Find the tangent plane and normal line to the surface $z = x^2 + y^2$ at $(2, -2, 8)$.

The point is on the surface because $2^2 + (-2)^2 = 8$. Let $\varphi(x, y, z) = z - x^2 - y^2$. The surface is the graph of $\varphi(x, y, z) = 0$. A normal vector at $(2, -2, 8)$ is $\nabla\varphi(2, -2, 8)$. Since $\nabla\varphi = -2x\mathbf{i} - 2y\mathbf{j} + \mathbf{k}$, $\nabla\varphi(2, -2, 8) = -4\mathbf{i} + 4\mathbf{j} + \mathbf{k}$. The tangent plane has equation

$$-4(x - 2) + 4(y + 2) + (z - 8) = 0,$$

or

$$-4x + 4y + z = -8.$$

The normal line has parametric equations

$$x = 2 - 4t, \qquad y = -2 + 4t, \qquad z = 8 + t,$$

or

$$\frac{x-2}{-4} = \frac{y+2}{4} = \frac{z-8}{1}. \quad \blacksquare$$

PROBLEMS FOR SECTION 15.4

In each of Problems 1 through 15, compute $\nabla\varphi(x, y, z)$, then determine $\nabla\varphi(P_0)$ for the given point.

1. $\varphi = xyz$; $(1, 1, 1)$

2. $\varphi = x^2y - \sin(zx)$; $\left(1, -1, \dfrac{\pi}{4}\right)$

3. $\varphi = 2xy + xe^z$; $(-2, 1, 6)$

4. $\varphi = \cos(xyz)$; $\left(-1, 1, \dfrac{\pi}{2}\right)$

5. $\varphi = \cosh(2xy) - \sinh(z)$; $(0, 1, 1)$

6. $\varphi = \sqrt{x^2 + y^2 + z^2}$; $(2, 2, 2)$

7. $\varphi = \ln(x + y + z)$; $(3, 1, -2)$

8. $\varphi = \dfrac{-1}{\|\mathbf{R}\|}$,   where $\mathbf{R} = x\mathbf{i} + y\mathbf{j} + z\mathbf{k}$; $(-2, 1, 1)$

9. $\varphi = e^x\cos(y)\cos(z)$; $\left(0, \dfrac{\pi}{4}, \dfrac{\pi}{4}\right)$

10. $\varphi = 2x^3y + ze^y$; $(1, 1, 2)$

11. $\varphi = x^2y \cosh(xz)$; $(0, 0, 1)$

12. $\varphi = e^{xy} + xz^2$; $(0, 0, 4)$

13. $\varphi = x - 2 \cosh(y + z)$; $(1, -1, 0)$

14. $\varphi = x^2 + 2y + z \ln(x)$; $(1, 3, 1)$

15. $\varphi = \cosh(x - y + 2z)$; $(2, 0, -1)$

In each of Problems 16 through 25, find the tangent plane and normal line to the surface at the point.

16. $x^2 + y^2 + z^2 = 4$; $(1, 1, \sqrt{2})$

17. $z = x^2 + y$; $(-1, 1, 2)$

18. $z^2 = x^2 - y^2$; $(1, 1, 0)$

19. $\sinh(x + y + z) = 0$; $(0, 0, 0)$

20. $2x - 4y^2 + z^3 = 0$; $(-4, 0, 2)$

21. $x^2 - y^2 + z^2 = 0$; $(1, 1, 0)$

22. $2x - \cos(xyz) = 3$; $(1, \pi, 1)$

23. $3x^4 + 3y^4 + 6z^4 = 12$; $(1, 1, 1)$

24. $x^2 - 2y^2 + z^4 = 0$; $(1, 1, 1)$

25. $\cos(x) - \sin(y) + z = 1$; $(0, \pi, 0)$

In each of Problems 26 through 29, find the angle between the two surfaces at the given point of intersection. (This angle is the smaller of the two angles between normal vectors to the surfaces at the given points.)

26. $z = 3x^2 + 2y^2$, $-2x + 7y^2 - z = 0$; $(1, 1, 5)$

27. $x^2 + y^2 + z^2 = 4$, $z^2 + x^2 = 2$; $(1, \sqrt{2}, 1)$

28. $z = \sqrt{x^2 + y^2}$, $x^2 + y^2 = 4$; $(2, 2, \sqrt{8})$

29. $\frac{1}{2}x^2 + \frac{1}{2}y^2 + z^2 = 5$, $x + y + z = 5$; $(2, 2, 1)$

In each of Problems 30 through 40, find a unit vector in the direction from P_0 in which the function has its maximum rate of change, and find this maximum rate of change.

30. $\varphi = x^2 - 3xy + 2y^2$; $(0, 0, 1)$

31. $\varphi = e^x\cos(yz)$; $(1, 1, \pi)$

32. $\varphi = 2xy - 3xz^2$; $(1, 2, 1)$

33. $\varphi = \dfrac{1}{x} - 3yz$; $(2, 1, 1)$

34. $\varphi = 3x^2y - 2 \sin(z)$; $(1, 3, \pi)$

35. $\varphi = 14x - 3y^2 + 2xye^z$; $(0, 1, 0)$

36. $\varphi = \sin(xyz)$; $(1, \pi, 1)$

37. $\varphi = -3z^3 + e^x\sin(y)$; $(0, 1, -2)$

38. $\varphi = x \ln(y + z)$; $(1, 3, -2)$

39. $\varphi = \tan^{-1}\left(\dfrac{y}{x}\right)$; $(1, -1, 3)$

40. $\varphi = 4x^3y^2z^2$; $(1, -1, 1)$

41. Determine a function φ, not identically zero, such that $\nabla\varphi = \mathbf{O}$. Can the surface $\varphi = $ constant have a normal vector?

42. Suppose that $\nabla\varphi = \mathbf{i} + \mathbf{k}$. What can be said about the surfaces $\varphi = $ constant? Prove that the lines of force of $\nabla\varphi$ are orthogonal to the surfaces $\varphi = $ constant.

15.5 *Divergence and Curl*

The gradient operation produces a vector field from a scalar field. We will now develop two additional vector operations. The divergence produces a scalar field from a vector field, and the curl produces a vector field from a vector field. Suppose we are given

$$\mathbf{F}(x, y, z) = f_1(x, y, z)\mathbf{i} + f_2(x, y, z)\mathbf{j} + f_3(x, y, z)\mathbf{k}.$$

The divergence of this vector field is the scalar field

$$\text{div } \mathbf{F} = \frac{\partial f_1}{\partial x} + \frac{\partial f_2}{\partial y} + \frac{\partial f_3}{\partial z}.$$

Read div $\mathbf{F}$ as "divergence of $\mathbf{F}$." The curl of this vector field is the vector field

$$\text{curl } \mathbf{F} = \left(\frac{\partial f_3}{\partial y} - \frac{\partial f_2}{\partial z}\right)\mathbf{i} + \left(\frac{\partial f_1}{\partial z} - \frac{\partial f_3}{\partial x}\right)\mathbf{j} + \left(\frac{\partial f_2}{\partial x} - \frac{\partial f_1}{\partial y}\right)\mathbf{k}.$$

This is read "curl of $\mathbf{F}$," or "curl $\mathbf{F}$." For example, if $\mathbf{F}(x, y, z) = y\mathbf{i} + 2xz\mathbf{j} + ze^x\mathbf{k}$,

$$\text{div } \mathbf{F} = 0 + 0 + e^x = e^x,$$

and

$$\text{curl } \mathbf{F} = -2x\mathbf{i} - ze^x\mathbf{j} + (2z - 1)\mathbf{k}.$$

Both divergence and curl have physical interpretations which we will discuss at the end of this section and again when we have the vector integral theorems of Gauss and Stokes in the next chapter.

Divergence and curl can be written in terms of the vector operations of dot and cross product and the gradient. First, define the *del operator*

$$\nabla = \frac{\partial}{\partial x}\mathbf{i} + \frac{\partial}{\partial y}\mathbf{j} + \frac{\partial}{\partial z}\mathbf{k}.$$

The symbol ∇ is read "del." We will treat the del operator as a vector in carrying out calculations. The "product" of the scalar φ with the "vector" ∇ is the gradient of φ:

$$\nabla\varphi = \frac{\partial\varphi}{\partial x}\mathbf{i} + \frac{\partial\varphi}{\partial y}\mathbf{j} + \frac{\partial\varphi}{\partial z}\mathbf{k}.$$

Here, we interpret φ times $(\partial/\partial x)\mathbf{i}$ as $(\partial\varphi/\partial x)\mathbf{i}$, and similarly for the $\mathbf{j}$ and $\mathbf{k}$ terms.

We can interpret the divergence as the dot product of ∇ with a vector field $\mathbf{F}$:

$$\nabla \cdot \mathbf{F} = \left[\frac{\partial}{\partial x}\mathbf{i} + \frac{\partial}{\partial y}\mathbf{j} + \frac{\partial}{\partial z}\mathbf{k}\right] \cdot [f_1(x, y, z)\mathbf{i} + f_2(x, y, z)\mathbf{j} + f_3(x, y, z)\mathbf{k}]$$

$$= \frac{\partial f_1}{\partial x} + \frac{\partial f_2}{\partial y} + \frac{\partial f_3}{\partial z}.$$

Finally, we can interpret the curl of **F** as the cross product of **∇** with **F**:

$$\text{curl } \mathbf{F} = \left[\frac{\partial}{\partial x}\mathbf{i} + \frac{\partial}{\partial y}\mathbf{j} + \frac{\partial}{\partial z}\mathbf{k} \right] \times [f_1(x, y, z)\mathbf{i} + f_2(x, y, z)\mathbf{j} + f_3(x, y, z)\mathbf{k}]$$

$$= \begin{vmatrix} \mathbf{i} & \mathbf{j} & \mathbf{k} \\ \dfrac{\partial}{\partial x} & \dfrac{\partial}{\partial y} & \dfrac{\partial}{\partial z} \\ f_1 & f_2 & f_3 \end{vmatrix}$$

$$= \left(\frac{\partial f_3}{\partial y} - \frac{\partial f_2}{\partial z} \right)\mathbf{i} + \left(\frac{\partial f_1}{\partial z} - \frac{\partial f_3}{\partial x} \right)\mathbf{j} + \left(\frac{\partial f_2}{\partial x} - \frac{\partial f_1}{\partial y} \right)\mathbf{k}.$$

Thus

$$\mathbf{\nabla} \cdot \mathbf{F} = \text{div } \mathbf{F}$$

(divergence = del dot), and

$$\mathbf{\nabla} \times \mathbf{F} = \text{curl } \mathbf{F}$$

(curl = del cross).

The del operator provides a compact and efficient way of writing and computing with divergence, curl, and the identities involving them. It is also used to define the *Laplacian* of a scalar field, which is defined to be the scalar field

$$\mathbf{\nabla} \cdot (\mathbf{\nabla}\varphi) = \frac{\partial^2 \varphi}{\partial x^2} + \frac{\partial^2 \varphi}{\partial y^2} + \frac{\partial^2 \varphi}{\partial z^2}.$$

The quantity $\mathbf{\nabla} \cdot (\mathbf{\nabla}\varphi)$ is often denoted ∇^2 (read "del squared"), and the partial differential equation $\nabla^2\varphi = 0$ is called *Laplace's equation*. We will encounter this equation in many applications of mathematics to physics and engineering.

We will now look at two relationships among gradient, divergence, and curl.

THEOREM 15.4

If φ is continuous with continuous first and second partial derivatives, then

$$\mathbf{\nabla} \times (\mathbf{\nabla}\varphi) = \mathbf{O}.$$

(That is, the curl of a gradient vector is the zero vector.)

Proof   First, $\mathbf{\nabla}\varphi = (\partial\varphi/\partial x)\mathbf{i} + (\partial\varphi/\partial y)\mathbf{j} + (\partial\varphi/\partial z)\mathbf{k}$, so

$$\mathbf{\nabla} \times (\mathbf{\nabla}\varphi) = \begin{vmatrix} \mathbf{i} & \mathbf{j} & \mathbf{k} \\ \dfrac{\partial}{\partial x} & \dfrac{\partial}{\partial y} & \dfrac{\partial}{\partial z} \\ \dfrac{\partial\varphi}{\partial x} & \dfrac{\partial\varphi}{\partial y} & \dfrac{\partial\varphi}{\partial z} \end{vmatrix}$$

$$= \left(\frac{\partial^2 \varphi}{\partial y\,\partial z} - \frac{\partial^2 \varphi}{\partial z\,\partial y} \right)\mathbf{i} + \left(\frac{\partial^2 \varphi}{\partial z\,\partial x} - \frac{\partial^2 \varphi}{\partial x\,\partial z} \right)\mathbf{j} + \left(\frac{\partial^2 \varphi}{\partial y\,\partial x} - \frac{\partial^2 \varphi}{\partial x\,\partial y} \right)\mathbf{k}.$$

Each component of this vector is zero because the mixed partial derivatives within each pair of brackets are equal. ∎

THEOREM 15.5

Let **F** be a continuous vector field with continuous first and second partial derivatives. Then

$$\mathbf{V} \cdot (\mathbf{V} \times \mathbf{F}) = 0.$$

That is, the divergence of a curl is zero.

Proof   Let $\mathbf{F} = f_1\mathbf{i} + f_2\mathbf{j} + f_3\mathbf{k}$. Then

$$\mathbf{V} \cdot (\mathbf{V} \times \mathbf{F}) = \frac{\partial}{\partial x}\left(\frac{\partial f_3}{\partial y} - \frac{\partial f_2}{\partial z}\right) + \frac{\partial}{\partial y}\left(\frac{\partial f_1}{\partial z} - \frac{\partial f_3}{\partial x}\right) + \frac{\partial}{\partial z}\left(\frac{\partial f_2}{\partial x} - \frac{\partial f_1}{\partial y}\right)$$

$$= \frac{\partial^2 f_3}{\partial x\,\partial y} - \frac{\partial^2 f_2}{\partial x\,\partial z} + \frac{\partial^2 f_1}{\partial y\,\partial z} - \frac{\partial^2 f_3}{\partial y\,\partial x} + \frac{\partial^2 f_2}{\partial z\,\partial x} - \frac{\partial^2 f_1}{\partial z\,\partial y} = 0$$

because the mixed partials cancel each other in pairs. ∎

To illustrate how the del operator suggests relationships, observe that Theorem 15.5 can be "proved" by noting that $\mathbf{V} \cdot (\mathbf{V} \times \mathbf{F})$ is the scalar triple product $[\mathbf{V}, \mathbf{V}, \mathbf{F}]$, and this is zero because two of its factors are equal.

We will now discuss a physical model for the divergence of a vector field. Imagine that $\mathbf{F}(x, y, z, t)$ is the velocity of a fluid at a point (x, y, z) and time t. We have included a time dependence to be realistic, but it plays no role in calculating the divergence of **F**.

Now imagine a small rectangular box in the fluid, as in Figure 15.21. We want some measure of the rate per unit volume of fluid flow out of the box across its faces at time t. To do this, look at pairs of opposite faces of the box.

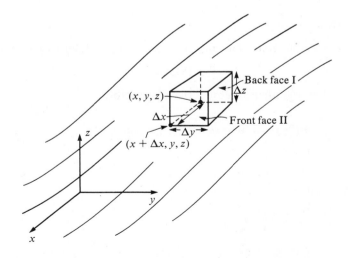

Figure 15.21

First, look at the front and back faces, labeled II and I, parallel to the yz-plane in Figure 15.21. If Δx is small, then $\mathbf{F}$ on face II is approximately $\mathbf{F}(x + \Delta x, y, z, t)$. The outer normal vector to this face is $\mathbf{i}$, and the area of the face is $\Delta x \, \Delta y$. Therefore,

$\mathbf{F}(x + \Delta x, y, z, t) \cdot \mathbf{i} \, \Delta y \, \Delta z$

$= f_1(x + \Delta x, y, z, t) \, \Delta y \, \Delta z$

$=$ (normal component of fluid velocity outward across face II) $\cdot$ (area of face II)

$=$ flux of flow out of the box across face II.

On face I, $\mathbf{F}$ is approximately $\mathbf{F}(x, y, z, t)$, and the outer normal is $-\mathbf{i}$. Hence, the flux out of the box across face I is approximately $\mathbf{F}(x, y, z, t) \cdot (-\mathbf{i}) \Delta y \, \Delta z$, or

$$-f_1(x, y, z, t) \, \Delta y \, \Delta z.$$

A similar calculation can be done for faces III and IV and faces V and VI. Summing over these pairs of faces, the total flux out of the box is

$$[f_1(x + \Delta x, y, z, t) - f_1(x, y, z, t)] \, \Delta y \, \Delta z$$
$$+ [f_2(x, y + \Delta y, z, t) - f_2(x, y, z, t)] \, \Delta x \, \Delta z$$
$$+ [f_3(x, y, z + \Delta z, t) - f_3(x, y, z, t)] \, \Delta x \, \Delta y.$$

The outward flux per unit volume is the outward flux divided by the volume of the box, $\Delta x \, \Delta y \, \Delta z$. This outward flux per unit volume is

$$\frac{f_1(x + \Delta x, y, z, t) - f_1(x, y, z, t)}{\Delta x} + \frac{f_2(x, y + \Delta y, z, t) - f_2(x, y, z, t)}{\Delta y}$$
$$+ \frac{f_3(x, y, z + \Delta z, t) - f_3(x, y, z, t)}{\Delta z}.$$

Take the limit as Δx, Δy, and Δz go to zero. The quantity we have just calculated has limit

$$\frac{\partial f_1}{\partial x} + \frac{\partial f_2}{\partial y} + \frac{\partial f_3}{\partial z},$$

which is div $\mathbf{F}$. The divergence may be interpreted as the outward flux per unit volume of the flow at (x, y, z) and time t. This measures the expansion, or "divergence," of the fluid away from this point.

In this context of fluid flow, the curl of the velocity field measures the degree to which the fluid swirls, or rotates, about a given direction. We will develop this interpretation when we have Stokes's theorem at our disposal.

For now, here is an interpretation of curl in terms of a rotating object. Suppose a body rotates with uniform angular speed ω about a line L, as in Figure 15.22. The angular velocity vector Ω has magnitude ω and direction along L in the direction a right-handed screw would progress if given the same sense of rotation as the object.

Suppose that L goes through the origin of the coordinate system, and let $\mathbf{R} = x\mathbf{i} + y\mathbf{j} + z\mathbf{k}$ for any point (x, y, z) on the rotating object. Let $\mathbf{T}$ be the tangential (linear) velocity at any point. Then

$$\|\mathbf{T}\| = \omega\|\mathbf{R}\|\,|\sin(\theta)| = \|\Omega \times \mathbf{R}\|.$$

Since $\mathbf{T}$ is also in the direction of $\Omega \times \mathbf{R}$, we conclude that $\mathbf{T} = \Omega \times \mathbf{R}$.

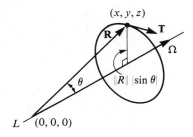

Figure 15.22. Angular velocity as the curl of the linear velocity.

Now write $\mathbf{\Omega} = \Omega_1\mathbf{i} + \Omega_2\mathbf{j} + \Omega_3\mathbf{k}$. Then

$$\mathbf{\Omega} \times \mathbf{R} = (\Omega_2 z - \Omega_3 y)\mathbf{i} + (\Omega_3 x - \Omega_1 z)\mathbf{j} + (\Omega_1 y - \Omega_2 x)\mathbf{k}.$$

But we can also compute

$$\mathbf{V} \times \mathbf{T} = \begin{vmatrix} \mathbf{i} & \mathbf{j} & \mathbf{k} \\ \dfrac{\partial}{\partial x} & \dfrac{\partial}{\partial y} & \dfrac{\partial}{\partial z} \\ \Omega_2 z - \Omega_3 y & \Omega_3 x - \Omega_1 z & \Omega_1 y - \Omega_2 x \end{vmatrix}$$

$$= 2\Omega_1\mathbf{i} + 2\Omega_2\mathbf{j} + 2\Omega_3\mathbf{k} = 2\mathbf{\Omega}$$

because Ω_1, Ω_2, and Ω_3 are constants. We conclude that

$$\mathbf{\Omega} = \tfrac{1}{2}\mathbf{V} \times \mathbf{T}.$$

This equation states that the angular velocity of the uniformly rotating body is a scalar multiple of the curl of the linear velocity. This fact is one motivation for the term *curl*. In fact, curl was once commonly known as rotation, and curl **F** was written rot **F**, especially in British publications. It is also motivation for the term *irrotational* to describe a vector field whose curl is the zero vector.

PROBLEMS FOR SECTION 15.5

In each of Problems 1 through 10, compute $\mathbf{V} \cdot \mathbf{F}$ and $\mathbf{V} \times \mathbf{F}$ and verify explicitly that $\mathbf{V} \cdot (\mathbf{V} \times \mathbf{F}) = 0$.

1. $\mathbf{F} = x\mathbf{i} + y\mathbf{j} + 2z\mathbf{k}$

2. $\mathbf{F} = \sinh(xyz)\mathbf{j}$

3. $\mathbf{F} = 2xy\mathbf{i} + e^y\mathbf{j} + 2z\mathbf{k}$

4. $\mathbf{F} = zx^2\mathbf{i} - yj + z^3\mathbf{k}$

5. $\mathbf{F} = -2e^z\mathbf{i} - zy^2\mathbf{j} + 2\mathbf{k}$

6. $\mathbf{F} = -yz\mathbf{j} - 6x^3\mathbf{k}$

7. $\mathbf{F} = 2x\mathbf{i} - 3y\mathbf{j} + \mathbf{k}$

8. $\mathbf{F} = \sinh(x)\mathbf{i} + \cosh(y)\mathbf{j} - xyz\mathbf{k}$

9. $\mathbf{F} = x^2\mathbf{i} + y^2\mathbf{j} + z^2\mathbf{k}$

10. $\mathbf{F} = \sinh(x - z)\mathbf{i} + 2y\mathbf{j} + z^2\mathbf{k}$

In each of Problems 11 through 20, compute $\mathbf{V}\varphi$ and verify explicitly that $\mathbf{V} \times (\mathbf{V}\varphi) = \mathbf{O}$.

11. $\varphi = x - y + 2z^2$

12. $\varphi = 18xyz + e^x$

13. $\varphi = -2x^3yz^2$

14. $\varphi = \sin(xz)$

15. $\varphi = x^3y^2e^z$

16. $\varphi = 2xy - 3z^2$

17. $\varphi = 2e^x\ln(yz)$

18. $\varphi = -4xy^3 + xz^2$

19. $\varphi = \cos(x + y + z)$

20. $\varphi = e^{x+y+z}$

21. Find a vector field **F** such that $\mathbf{V} \times \mathbf{F} = 3\mathbf{k}$.

22. Find a vector field **F** such that $\mathbf{V} \cdot \mathbf{F} = 14xyz$.

23. Let φ be a scalar field, and let $\mathbf{F}$ be a vector field. Derive expressions for $\mathbf{V} \cdot (\varphi\mathbf{F})$ and $\mathbf{V} \times (\varphi\mathbf{F})$ in terms of vector operations applied to φ and $\mathbf{F}$.

24. Let $\mathbf{F}$ and $\mathbf{G}$ be vector fields. Prove that

$$\mathbf{V}(\mathbf{F} \cdot \mathbf{G}) = (\mathbf{F} \cdot \mathbf{V})\mathbf{G} + (\mathbf{G} \cdot \mathbf{V})\mathbf{F} + \mathbf{F} \times (\mathbf{V} \times \mathbf{G}) + \mathbf{G} \times (\mathbf{V} \times \mathbf{F}),$$

in which

$$\mathbf{F} \cdot \mathbf{V} = \mathbf{F} \cdot \left[\frac{\partial}{\partial x}\mathbf{i} + \frac{\partial}{\partial y}\mathbf{j} + \frac{\partial}{\partial z}\mathbf{k}\right] = f_1\frac{\partial}{\partial x} + f_2\frac{\partial}{\partial y} + f_3\frac{\partial}{\partial z}.$$

25. Let $\mathbf{F}$ and $\mathbf{G}$ be vector fields. Prove that $\mathbf{V} \cdot (\mathbf{F} \times \mathbf{G}) = \mathbf{G} \cdot (\mathbf{V} \times \mathbf{F}) - \mathbf{F} \cdot (\mathbf{V} \times \mathbf{G})$.

26. Let φ and ψ be scalar fields. Prove that $\mathbf{V} \cdot (\mathbf{V}\varphi \times \mathbf{V}\psi) = 0$.

27. Let $\mathbf{R} = x\mathbf{i} + y\mathbf{j} + z\mathbf{k}$, and let $\|\mathbf{R}\| = r$.
    (a) Prove that $\mathbf{V}r^n = nr^{n-2}\mathbf{R}$ for $n = 1, 2, \ldots$.
    (b) Let φ be a real-valued function of one variable. Prove that $\mathbf{V} \times (\varphi(r)\mathbf{R}) = \mathbf{O}$.

28. Let $\mathbf{F}$ be a vector field. Prove that $\mathbf{V} \times (\mathbf{V} \times \mathbf{F}) = \mathbf{V}(\mathbf{V} \cdot \mathbf{F}) - \mathbf{V}^2\mathbf{F}$, where

$$\mathbf{V}^2\mathbf{F} = \frac{\partial^2\mathbf{F}}{\partial x^2} + \frac{\partial^2\mathbf{F}}{\partial y^2} + \frac{\partial^2\mathbf{F}}{\partial z^2}.$$

29. Let $\mathbf{A}$ be a constant vector, and let $\mathbf{R} = x\mathbf{i} + y\mathbf{j} + z\mathbf{k}$.
    (a) Prove that $\mathbf{V}(\mathbf{R} \cdot \mathbf{A}) = \mathbf{A}$.
    (b) Prove that $\mathbf{V} \cdot (\mathbf{R} - \mathbf{A}) = 3$.
    (c) Prove that $\mathbf{V} \times (\mathbf{R} - \mathbf{A}) = \mathbf{O}$.

30. Let $\mathbf{F}$ and $\mathbf{G}$ be vector fields. Prove that

$$\mathbf{V} \times (\mathbf{F} \times \mathbf{G}) = (\mathbf{G} \cdot \mathbf{V})\mathbf{F} - (\mathbf{F} \cdot \mathbf{V})\mathbf{G} + (\mathbf{V} \cdot \mathbf{G})\mathbf{F} - (\mathbf{V} \cdot \mathbf{F})\mathbf{G}.$$

ADDITIONAL PROBLEMS

In each of Problems 1 through 10, compute $(\mathbf{F} \cdot \mathbf{G})'$ and $(\mathbf{F} \times \mathbf{G})'$.

1. $\mathbf{F} = 4t\mathbf{i} + t^2\mathbf{j} - \mathbf{k}, \quad \mathbf{G} = \cos(2t)\mathbf{i} - 4t\mathbf{k}$
2. $\mathbf{F} = -\sin(t)\mathbf{j} + 3\mathbf{k}, \quad \mathbf{G} = 2\mathbf{i} + t^2\mathbf{j} - \ln(t)\mathbf{k}$
3. $\mathbf{F} = 2t^{1/2}\mathbf{i} + \sinh(t)\mathbf{j} + t\mathbf{k}, \quad \mathbf{G} = e^{-t}\mathbf{i} + \mathbf{j} - 2t\mathbf{k}$
4. $\mathbf{F} = e^t\mathbf{i} - \cos(t)\mathbf{j} + t\mathbf{k}, \quad \mathbf{G} = 2\mathbf{i} - t^3\mathbf{j} + \mathbf{k}$
5. $\mathbf{F} = \sinh(3t)\mathbf{j}, \quad \mathbf{G} = t^2\mathbf{i} - t\mathbf{j} + \mathbf{k}$
6. $\mathbf{F} = (1 - 2t)\mathbf{i} + t^3\mathbf{j} - 3t^2\mathbf{k}, \quad \mathbf{G} = 4t\mathbf{i} + \frac{1}{t}\mathbf{k}$
7. $\mathbf{F} = t^2\mathbf{i} - \mathbf{j} - 4\mathbf{k}, \quad \mathbf{G} = \mathbf{i} + \cosh(t)\mathbf{j} + t\mathbf{k}$
8. $\mathbf{F} = e^t\mathbf{i} + t\mathbf{j} - 4t\mathbf{k}, \quad \mathbf{G} = \cosh(t)\mathbf{j} - t^3\mathbf{k}$
9. $\mathbf{F} = \ln(t)\mathbf{i} - t\mathbf{j} + \mathbf{k}, \quad \mathbf{G} = \mathbf{i} + \mathbf{j} + t^3\mathbf{k}$
10. $\mathbf{F} = 3t\mathbf{j} + \sin(t)\mathbf{k}, \quad \mathbf{G} = -t^2\mathbf{i} + t\mathbf{k}$

In each of Problems 11 through 15, compute the velocity, speed, acceleration, and tangential and centripetal components of the vector $\mathbf{R} = x\mathbf{i} + y\mathbf{j} + z\mathbf{k}$. Also compute the curvature, radius of curvature, and torsion of the trajectory.

11. $x = y = 2t^2, \quad z = \cos(t); \quad 0 \le t \le \pi$
12. $x = 2 - \sin(t), \quad y = 2 - \cos(t), \quad z = 1; \quad 0 \le t \le \pi$
13. $x = t^2, \quad y = 2 + t, \quad z = e^t; \quad 2 \le t \le 5$
14. $x = 4t, \quad y = 2 + t^2, \quad z = \sin(2t); \quad -1 \le t \le 1$
15. $x = 3t^3, \quad y = 2t = z; \quad 0 \le t \le 4$

In each of Problems 16 through 20, find the lines of force of the vector field.

16. $\mathbf{F} = \mathbf{i} - zy\mathbf{j} + z\mathbf{k}$
17. $\mathbf{F} = 4x\mathbf{i} - x^2\mathbf{j} + z\mathbf{k}$
18. $\mathbf{F} = y\mathbf{i} - x\mathbf{j} + yx\mathbf{k}$
19. $\mathbf{F} = z\mathbf{i} - 3y\mathbf{j} + z^3\mathbf{k}$
20. $\mathbf{F} = \mathbf{i} - y\mathbf{j} + z^2\mathbf{k}$

In each of Problems 21 through 25, find the direction in which φ has its maximum rate of change at the given point, and find the value of this maximum rate of change.

21. $\varphi = x^2y - \cos(zx); \quad (1, 1, \pi)$
22. $\varphi = 2yz^3 + xy\cos(z); \quad (2, -1, 0)$

23. $\varphi = -\ln(x + y + z)$; $(-2, 2, e)$ **24.** $\varphi = e^{x-z}\sin(y - z)$; $(2, 2, 2)$

25. $\varphi = \cos(zx)$; $\left(1, 0, \dfrac{\pi}{4}\right)$

In each of Problems 26 through 30, find the tangent plane and normal line to the surface at the given point.

26. $z = x^2 + y^2 - 2$; $(2, 2, 6)$ **27.** $z^2 = x^2 + y^2$; $(1, 1, \sqrt{2})$

28. $x^2 + (y - 2)^2 + z^2 = 11$; $(1, 1, 3)$ **29.** $2x^2 - y^2 - z^2 = 0$; $(1, 1, 1)$

30. $4xyz + x^2 - y^2 = 0$; $(2, 4, \frac{3}{8})$

In each of Problems 31 through 35, calculate $\mathbf{V} \cdot \mathbf{F}$ and $\mathbf{V} \times \mathbf{F}$.

31. $\mathbf{F} = xy\mathbf{i} - z^3\mathbf{j} + x\mathbf{k}$ **32.** $\mathbf{F} = (1 - z)\mathbf{i} + x^3\mathbf{j} - \cos(yz)\mathbf{k}$ **33.** $\mathbf{F} = \mathbf{i} - z^3\mathbf{j} + e^{xy}\mathbf{k}$

34. $\mathbf{F} = 2xy\mathbf{i} - z^2\mathbf{j} + x\mathbf{k}$ **35.** $\mathbf{F} = \sin(y - z)\mathbf{i} + e^z\mathbf{j} - x\mathbf{k}$

CHAPTER 16

Vector Integral Calculus

16.1 Line Integrals

Thus far, we have dealt with derivative operations on vector fields. We will now develop the concept of an integral of a vector field over a curve in three-space. To begin, we need some terminology about curves. A curve in three-space is usually given parametrically by specifying the *coordinate functions*

$$x = x(t), \qquad y = y(t), \qquad z = z(t)$$

for t in some interval I (which may be the entire real line). The *graph* of a curve consists of the locus of points $(x(t), y(t), z(t))$ as t varies over I. The coordinate functions of the curve may also be specified by giving the *position vector*

$$\mathbf{R}(t) = x(t)\mathbf{i} + y(t)\mathbf{j} + z(t)\mathbf{k} \quad \text{for} \quad t \text{ in } I.$$

Thus, when we speak of the curve given by $\mathbf{R}(t)$, we mean the curve whose coordinate functions are the components of $\mathbf{R}(t)$. We may think of the position vector $\mathbf{R}(t)$ as an arrow from the origin to the point $(x(t), y(t), z(t))$, tracing out the graph as t varies over I. This interpretation is illustrated in Figure 16.1.

Suppose the parameter t is defined over a closed interval $[a, b]$. We call the point $(x(a), y(a), z(a))$ the *initial point* of the curve and $(x(b), y(b), z(b))$ the *terminal point*. We obtain a sense of direction or orientation along the curve by thinking of the point $(x(t), y(t), z(t))$ as moving from the initial to the terminal point as t varies from a to b. This orientation is often indicated by putting an arrow on the graph, as in Figure 16.2. If we let t vary from b to a, we reverse orientation on the curve, moving from the new initial point $(x(b), y(b), z(b))$ to what is now the terminal point $(x(a), y(a), z(a))$, as indicated in Figure 16.3. Unless otherwise specified, we understand the orientation of a curve to be that obtained by letting t increase over its interval of definition.

875

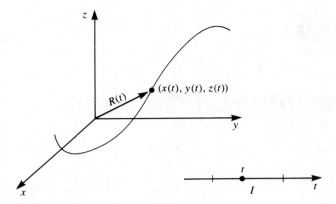

Figure 16.1

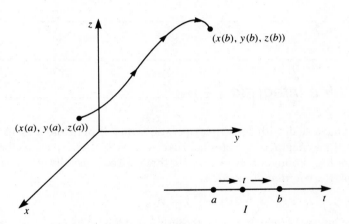

Figure 16.2

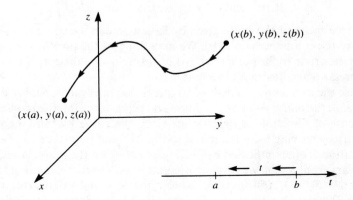

EXAMPLE 16.1

Let C be given by

$$x = \cos(t), \qquad y = \sin(t), \qquad z = 1; \qquad 0 \le t \le 2\pi.$$

Since $x^2 + y^2 = 1$, the graph of C is a circle of radius 1 about the origin in the plane $z = 1$. The initial point is $(1, 0, 1)$. As t increases from zero to 2π, the point $(x(t), y(t), z(t))$ moves around the circle, through $(0, 1, 1)$ when $t = \pi/2$, through $(-1, 0, 1)$ when $t = \pi$, through $(0, -1, 1)$ when $t = 3\pi/2$, and returning to $(1, 0, 1)$ when $t = 2\pi$. In this case, the initial and terminal points are the same. The graph, with the orientation from initial to terminal point, is shown in Figure 16.4. ∎

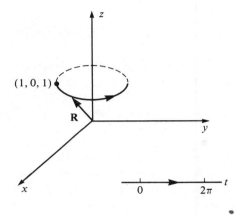

Figure 16.4

We call C a *closed curve* if the initial and terminal points are the same. The curve in Example 16.1 is a closed curve. We cannot always tell from the graph whether or not a curve is closed. Look at the following variation on Example 16.1.

EXAMPLE 16.2

Let K be the curve given by

$$x = \cos(t), \qquad y = \sin(t), \qquad z = 1; \qquad 0 \le t \le 3\pi.$$

The graph of K looks just like the graph of C, but there is a difference. As t increases from zero to 3π, the point $(\cos(t), \sin(t), 1)$ moves around the circle of Figure 16.4 from $(1, 0, 1)$ at $t = 0$ to $(1, 0, 1)$ at $t = 2\pi$, then starts around the circle again, ending at $(-1, 0, 1)$ when $t = 3\pi$. Thus, K has initial point $(1, 0, 1)$ and terminal point $(-1, 0, 1)$ and is not a closed curve, even though its graph appears to be a circle. ∎

We can dramatize the difference between Examples 16.1 and 16.2 by imagining the graph as a running track. If we run the entire course once in Example 16.1, we go around the circle once; if we run the entire course in Example 16.2, we complete one circuit, then go around half of the circle again. This makes a difference in energy expended and work done.

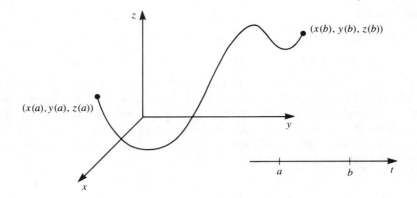

Figure 16.5. Smooth curve (continuous tangent vector).

Examples 16.1 and 16.2 emphasize that a curve is not the same as its graph. The curve consists of the three coordinate functions and the interval over which they are defined, together with a sense of direction on the curve. The graph is a locus of points in three-space (or the plane). As Examples 16.1 and 16.2 show, two different curves may have the same graph.

If the coordinate functions are continuous on $[a, b]$, we call the curve *continuous*; if they are differentiable, we call the curve *differentiable*.

If $x'(t)$, $y'(t)$, and $z'(t)$ are continuous on $[a, b]$ and are not all zero for any value of t [so that $\mathbf{R}'(t) \neq \mathbf{O}$], $\mathbf{R}'(t)$ is the tangent vector to the curve and is continuous as well. A curve with a continuous tangent vector at each point is called *smooth* and typically has the appearance of Figure 16.5, with no sharp points or spikes. For example, roller coaster paths are smooth, and the curves of Examples 16.1 and 16.2 are smooth.

We are now ready to define the line integral of a vector field over a smooth curve. Suppose we are given a smooth curve C and a vector field

$$\mathbf{F}(x, y, z) = f(x, y, z)\mathbf{i} + g(x, y, z)\mathbf{j} + h(x, y, z)\mathbf{k},$$

with component functions which are continuous over the graph of C. The *line integral of $\mathbf{F}$ over C*, denoted $\int_C \mathbf{F} \cdot d\mathbf{R}$, is defined to be

$$\int_C \mathbf{F} \cdot d\mathbf{R} = \int_a^b \mathbf{F}(x(t), y(t), z(t)) \cdot \mathbf{R}'(t) \, dt.$$

To evaluate $\int_C \mathbf{F} \cdot d\mathbf{R}$,

1. Form the dot product $\mathbf{F} \cdot \mathbf{R}'$.

2. Replace x, y, and z in $\mathbf{F} \cdot \mathbf{R}'$ with the coordinate functions $x(t)$, $y(t)$, and $z(t)$ of C.

3. Integrate the resulting function of t from a to b.

EXAMPLE 16.3

Evaluate $\int_C \mathbf{F} \cdot d\mathbf{R}$ if $\mathbf{F}(x, y, z) = x\mathbf{i} - yz\mathbf{j} + e^z\mathbf{k}$ and C is specified by $\mathbf{R}(t) = t^3\mathbf{i} - t\mathbf{j} + t\mathbf{k}$ for $0 \leq t \leq 1$.

First, $\mathbf{R}'(t) = 3t^2\mathbf{i} - \mathbf{j} + \mathbf{k}$, a continuous tangent vector, so C is smooth. Compute

$$\mathbf{F} \cdot \mathbf{R}' = x(3t^2) - (yz)(-1) + e^z(1) = 3xt^2 + yz + e^z.$$

Next, on C, $x(t) = t^3$, $y(t) = -t$, and $z(t) = t$. Substitute these functions into $\mathbf{F} \cdot \mathbf{R}'$ to get

$$\mathbf{F} \cdot \mathbf{R}' = 3t^3t^2 + (-t)(t) + e^t = 3t^5 - t^2 + e^t.$$

The line integral of $\mathbf{F}$ over C is the integral of this function of t from zero to 1:

$$\int_C \mathbf{F} \cdot d\mathbf{R} = \int_0^1 (3t^5 - t^2 + e^t)\, dt$$

$$= \tfrac{1}{2}t^6 - \tfrac{1}{3}t^3 + e^t \bigg]_0^1 = \tfrac{1}{2} - \tfrac{1}{3} + e^1 - e^0 = -\tfrac{5}{6} + e. \quad\blacksquare$$

Often, C is described in words, and we must construct the parametric equations and position vector of C.

EXAMPLE 16.4

Evaluate $\int_C \mathbf{F} \cdot d\mathbf{R}$ if $\mathbf{F} = x\mathbf{i} - y\mathbf{j} + z\mathbf{k}$ and C is the straight line segment from $(1, 1, 1)$ to $(-2, 1, 3)$.

First, find the parametric equations of the line through $(1, 1, 1)$ and $(-2, 1, 3)$. They are

$$x = 1 - 3t, \qquad y = 1, \qquad z = 1 + 2t; \qquad 0 \le t \le 1.$$

Since $(1, 1, 1)$ is on this line for $t = 0$ and $(-2, 1, 3)$ is on the line for $t = 1$, we must let t vary from zero to 1. Thus, let

$$\mathbf{R}(t) = (1 - 3t)\mathbf{i} + \mathbf{j} + (1 + 2t)\mathbf{k}; \qquad 0 \le t \le 1.$$

$\mathbf{R}'(t) = -3\mathbf{i} + 2\mathbf{k}$ is a continuous tangent vector, and C is a smooth curve. Next,

$$\mathbf{F} \cdot \mathbf{R}' = x(-3) - y(0) + z(2) = -3x + 2z.$$

Substitute the parametric equations of C into $\mathbf{F} \cdot \mathbf{R}'$ to get

$$\mathbf{F} \cdot \mathbf{R}' = (-3)(1 - 3t) + (2)(1 + 2t) = -1 + 13t.$$

Then

$$\int_C \mathbf{F} \cdot d\mathbf{R} = \int_0^1 (-1 + 3t)\, dt = -t + \tfrac{3}{2}t^2 \bigg]_0^1 = \tfrac{1}{2}. \quad\blacksquare$$

A curve C having continuous position vector $\mathbf{R}(t)$ is *piecewise smooth* if $\mathbf{R}'(t)$ is continuous and different from $\mathbf{O}$ at all but possibly a finite number of values of t. A typical piecewise-smooth curve is shown in Figure 16.6; it consists of smooth curves connected at points where the curve has no tangent vector.

If C is piecewise smooth, consisting of smooth curves $C_1, C_2, \ldots, C_n$ as in Figure 16.6, the line integral of $\mathbf{F}$ over C is defined to be the sum of the line integrals of $\mathbf{F}$ over each of the smooth curves making up C:

$$\int_C \mathbf{F} \cdot d\mathbf{R} = \int_{C_1} \mathbf{F} \cdot d\mathbf{R} + \int_{C_2} \mathbf{F} \cdot d\mathbf{R} + \cdots + \int_{C_n} \mathbf{F} \cdot d\mathbf{R}.$$

In this sum, the orientation along C must be maintained over the curves $C_1, \ldots, C_n$. That is, the initial point of C_j is the terminal point of C_{j-1}. This requirement is indicated by the arrows in Figure 16.6.

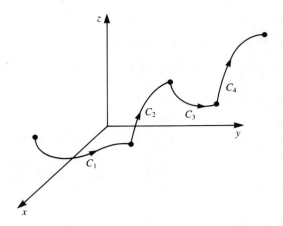

Figure 16.6

EXAMPLE 16.5

Let C be the curve traversing the quarter-circle $x^2 + y^2 = 1$ from $(1, 0)$ to $(0, 1)$ in the plane, then moving along the horizontal line from $(0, 1)$ to $(2, 1)$. Let $\mathbf{F}(x, y, z) = 4x\mathbf{i}$. Compute $\int_C \mathbf{F} \cdot d\mathbf{R}$.

Here we have a line integral in the plane because $\mathbf{F}$ is independent of z and has zero $\mathbf{k}$-component, and C is in the plane. The graph of C is shown in Figure 16.7; it consists of smooth curves C_1 and C_2. We can parametrize these individually as

$$C_1: \quad x = \cos(t), \qquad y = \sin(t); \qquad 0 \le t \le \frac{\pi}{2}$$

$$C_2: \quad x = t, \qquad y = 1; \qquad 0 \le t \le 2$$

We will evaluate $\int_{C_1} \mathbf{F} \cdot d\mathbf{R}$ and $\int_{C_2} \mathbf{F} \cdot d\mathbf{R}$ independently. For C_1, $\mathbf{R} = \cos(t)\mathbf{i} + \sin(t)\mathbf{j}$,

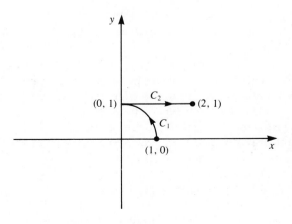

Figure 16.7

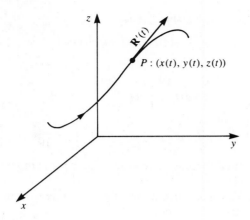

Figure 16.8

and $\mathbf{R}' = -\sin(t)\mathbf{i} + \cos(t)\mathbf{j}$. Then

$$\mathbf{F} \cdot \mathbf{R}' = 4x[-\sin(t)] = -4\cos(t)\sin(t),$$

and

$$\int_{C_1} \mathbf{F} \cdot d\mathbf{R} = \int_0^{\pi/2} -4\sin(t)\cos(t)\, dt = -2\sin^2(t)\Big]_0^{\pi/2} = -2.$$

On C_2, $\mathbf{R} = t\mathbf{i} + \mathbf{j}$ and $\mathbf{R}' = \mathbf{i}$. Further, $\mathbf{F} \cdot \mathbf{R}' = 4x(1) = 4x = 4t$. Then

$$\int_{C_2} \mathbf{F} \cdot d\mathbf{R} = \int_0^2 4t\, dt = 2t^2\Big]_0^2 = 8.$$

Then $\int_C \mathbf{F} \cdot d\mathbf{R} = -2 + 8 = 6.$ ∎

Here is one physical interpretation of the line integral $\int_C \mathbf{F} \cdot d\mathbf{R}$. Think of $\mathbf{F}$ as a force moving an object along C from its initial point $\mathbf{R}(a)$ to its terminal point $\mathbf{R}(b)$. We will argue that $\int_C \mathbf{F} \cdot d\mathbf{R}$ is the work done by this force.

Recall that a constant force $\mathbf{F}$ acting along a constant vector $\mathbf{D}$ does $\mathbf{F} \cdot \mathbf{D}$ units of work. In the case of a variable force acting along a curved path C, we can derive an expression for the work as follows. Choose any point $P:(x(t), y(t), z(t))$ on C, as in Figure 16.8. At P, the object being moved may be thought of as having direction $\mathbf{R}'(t)$, which is tangent to C at P. We can "sum" $\mathbf{F} \cdot \mathbf{R}'$ over the entire curve by integrating, leading us to interpret $\int_a^b \mathbf{F}(x(t), y(t), z(t)) \cdot \mathbf{R}'(t)\, dt$ as the total work done. This is exactly $\int_C \mathbf{F} \cdot d\mathbf{R}$.

EXAMPLE 16.6

Calculate the work done by $\mathbf{F} = \mathbf{i} - y\mathbf{j} + xyz\mathbf{k}$ as it moves an object along the curve $\mathbf{R}(t) = t\mathbf{i} - t^2\mathbf{j} + t\mathbf{k}$ from $(0, 0, 0)$ to $(1, -1, 1)$.

The work done is $\int_C \mathbf{F} \cdot d\mathbf{R} = \int_0^1 \mathbf{F}(x(t), y(t), z(t)) \cdot \mathbf{R}'(t)\, dt$. Now, $\mathbf{R}' = \mathbf{i} - 2t\mathbf{j} + \mathbf{k}$, so

$$\mathbf{F} \cdot \mathbf{R}' = 1(1) - y(-2t) + xyz(1) = 1 + 2yt + xyz.$$

On C, $x(t) = t$, $y(t) = -t^2$, and $z(t) = t$. Then

$$\mathbf{F} \cdot \mathbf{R}' = 1 + 2(-t^2)t + (t)(-t^2)(t) = 1 - 2t^3 - t^4,$$

so

$$\int_C \mathbf{F} \cdot d\mathbf{R} = \int_0^1 (1 - 2t^3 - t^4)\, dt$$

$$= t - \tfrac{1}{2}t^4 - \tfrac{1}{5}t^5 \bigg]_0^1 = \frac{3}{10}.$$

If distance is in meters and force is in newtons, the work is in newton-meters. ∎

Line integrals have the properties we usually associate with integrals.

THEOREM 16.1

Let $\mathbf{R}(t)$ be the position vector of a piecewise-smooth curve C, and let $\mathbf{F}$ and $\mathbf{G}$ be vector fields with continuous component functions over C. Then

1. $\int_C (\mathbf{F} + \mathbf{G}) \cdot d\mathbf{R} = \int_C \mathbf{F} \cdot d\mathbf{R} + \int_C \mathbf{G} \cdot d\mathbf{R}$.
2. If β is any number, $\int_C (\beta \mathbf{F}) \cdot d\mathbf{R} = \beta \int_C \mathbf{F} \cdot d\mathbf{R}$. ∎

In addition, the sign of a line integral is determined by the orientation, or direction traveled along the curve. To illustrate, suppose in Example 16.4 we had wanted to go from $(-2, 1, 3)$ to $(1, 1, 1)$, the direction opposite that used in the example. Then we would have integrated from $t = 1$ to $t = 0$, giving us the negative of the line integral from $(1, 1, 1)$ to $(-2, 1, 3)$.

Given any curve C, we let $-C$ denote the curve with direction reversed. Changing direction on C changes the sign of the line integral.

THEOREM 16.2

$$\int_C \mathbf{F} \cdot d\mathbf{R} = -\int_{-C} \mathbf{F} \cdot d\mathbf{R}. \quad ∎$$

There is a differential notation for line integrals which is commonly seen. If we write $\mathbf{R} = x\mathbf{i} + y\mathbf{j} + z\mathbf{k}$,

$$d\mathbf{R} = dx\mathbf{i} + dy\mathbf{j} + dz\mathbf{k}.$$

If $\mathbf{F}(x, y, z) = f(x, y, z)\mathbf{i} + g(x, y, z)\mathbf{j} + h(x, y, z)\mathbf{k}$,

$$\mathbf{F} \cdot d\mathbf{R} = (f\mathbf{i} + g\mathbf{j} + h\mathbf{k}) \cdot (dx\mathbf{i} + dy\mathbf{j} + dz\mathbf{k})$$

$$= f(x, y, z)\, dx + g(x, y, z)\, dy + h(x, y, z)\, dz.$$

For this reason, we often see $\int_C \mathbf{F} \cdot d\mathbf{R}$ written as

$$\int_C f(x, y, z)\, dx + g(x, y, z)\, dy + h(x, y, z)\, dz. \tag{16.1}$$

We evaluate this expression by substituting the parametric functions $x(t)$, $y(t)$, and $z(t)$

from the curve into equation (16.1) and obtaining a definite integral in terms of t (or whatever parameter is used to describe C).

EXAMPLE 16.7

Evaluate $\int_C x^2 \, dx - y \, dz$, with C the curve whose parametric functions are

$$x(t) = 2t, \qquad y(t) = t^2, \qquad z(t) = -t; \qquad 1 \leq t \leq 2.$$

Using the parametric functions for C, we have

$$dx = 2 \, dt, \qquad dy = 2t \, dt, \qquad dz = -dt.$$

By direct substitution, we have

$$x^2 \, dx - y \, dz = (2t)^2 (2) \, dt - t^2 (-1) \, dt = 9t^2 \, dt.$$

Then

$$\int_C x^2 \, dx - y \, dz = \int_1^2 9t^2 \, dt = 21.$$

It is routine to check that this is the same as $\int_C \mathbf{F} \cdot d\mathbf{R}$ if we choose $\mathbf{F}(x, y, z) = x^2 \mathbf{i} - y \mathbf{k}$ and $\mathbf{R} = 2t\mathbf{i} + t^2 \mathbf{j} - t\mathbf{k}$. ∎

EXAMPLE 16.8

Evaluate $\int_C xz \, dy$ on the line segment from $(0, 0, 0)$ to $(1, 3, 2)$.

The parametric equations of the line through these points are

$$x = t, \qquad y = 3t, \qquad z = 2t; \qquad 0 \leq t \leq 1.$$

The origin is on C when $t = 0$, and $(1, 3, 2)$ is on C when $t = 1$. Thus, t will vary from zero to 1 in the integral. On C, $xz = t(2t) = 2t^2$, and $dy = 3 \, dt$, so

$$xz \, dy = t(2t)(3) \, dt = 6t^2 \, dt.$$

Then

$$\int_C xy \, dy = \int_0^1 6t^2 \, dt = 2t^3 \Big]_0^1 = 2. \quad ∎$$

EXAMPLE 16.9

Evaluate $\int_C xyz^2 \, dz$ over the curve given by $x = t^2, y = t^3, z = 2t^2$ for t varying from zero to 1.

We have $xyz^2 = (t^2)(t^3)(2t^2)^2 = 4t^9$, and $dz = 4t \, dt$, so

$$xyz^2 \, dz = 4t^9(4t) \, dt = 16t^{10} \, dt.$$

Then

$$\int_C xyz^2 \, dz = 16 \int_0^1 t^{10} \, dt = \tfrac{16}{11}. \quad ∎$$

Thus far, we have considered line integrals of the form $\int_C \mathbf{F} \cdot d\mathbf{R}$ or, equivalently, $\int_C f(x, y, z) \, dx + g(x, y, z) \, dy + h(x, y, z) \, dz$. There is another type of line integral which is often used. The line integral of $f(x, y, z)$ over a smooth curve C with respect to arc length is denoted $\int_C f(x, y, z)\, ds$ and is defined by

$$\int_C f(x, y, z)\, ds = \int_a^b f(x(t), y(t), z(t))\sqrt{[x'(t)]^2 + [y'(t)]^2 + [z'(t)]^2}\, dt.$$

This definition is motivated by the fact that

$$ds = \sqrt{[x'(t)]^2 + [y'(t)]^2 + [z'(t)]^2}\, dt.$$

We assume here that C is smooth. If C is piecewise smooth, sum the line integrals over the individual smooth curves comprising C.

EXAMPLE 16.10

Evaluate $\int_C xy\, ds$, with C given by

$$x = 4\cos(t), \qquad y = 4\sin(t), \qquad z = -3; \qquad 0 \le t \le \frac{\pi}{2}.$$

The curve is shown in Figure 16.9; it is an arc of the circle $x^2 + y^2 = 16$, $z = -3$, beginning at $(4, 0, -3)$ and ending at $(0, 4, -3)$. Compute

$$ds = \sqrt{[x'(t)]^2 + [y'(t)]^2 + [z'(t)]^2}\, dt$$
$$= \sqrt{[-4\sin(t)]^2 + [4\cos(t)]^2}\, dt = 4\, dt.$$

Then

$$xy\, ds = [4\cos(t)][4\sin(t)](4)\, dt = 64\sin(t)\cos(t)\, dt,$$

and

$$\int_C xy\, ds = \int_0^{\pi/2} 64\sin(t)\cos(t)\, dt$$
$$= 32\sin^2(t)\Big]_0^{\pi/2} = 32. \quad\blacksquare$$

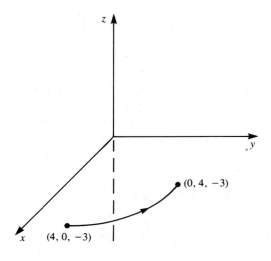

Figure 16.9

To illustrate how $\int_C f(x, y, z)\, ds$ can arise in a practical problem, suppose we want the mass and center of mass of a thin wire occupying the graph of a curve C in three-space. Let the position vector of the wire be $\mathbf{R}(t)$ for $a \leq t \leq b$, and let the density of the wire at (x, y, z) be $\delta(x, y, z)$.

To compute the mass, begin by partitioning $[a, b]$, and look at the piece of wire between $P_{j-1}:(x(t_{j-1}), y(t_{j-1}), z(t_{j-1}))$ and $P_j:(x(t_j), y(t_j), z(t_j))$ shown in Figure 16.10. Choose t_j^* between t_{j-1} and t_j. If $\Delta t_j = t_j - t_{j-1}$ is small, δ is approximately constant on the segment between P_{j-1} and P_j, and we can consider

$$\delta(x, y, z) \approx \delta(x(t_j^*), y(t_j^*), z(t_j^*))$$

on this segment. The length of this segment is ds, where

$$ds \approx \sqrt{[x'(t_j)]^2 + [y'(t_j)]^2 + [z'(t_j)]^2}\, \Delta t_j.$$

The mass of this segment of wire is therefore approximated by

$$\delta(x(t_j^*), y(t_j^*), z(t_j^*))\sqrt{[x'(t_j)]^2 + [y'(t_j)]^2 + [z'(t_j)]^2}\, \Delta t_j.$$

The total mass m of the wire is approximated by the sum of the masses of the segments,

$$m \approx \sum_{j=1}^n \delta(x(t_j^*), y(t_j^*), z(t_j^*))\sqrt{[x'(t_j)]^2 + [y'(t_j)]^2 + [z'(t_j)]^2}\, \Delta t_j.$$

As $n \to \infty$ and each $\Delta t_j \to 0$, we get the exact value for the mass:

$$m = \int_a^b \delta(x(t), y(t), z(t))\sqrt{[x'(t)]^2 + [y'(t)]^2 + [z'(t)]^2}\, dt$$

$$= \int_C \delta(x, y, z)\, ds.$$

The mass of the wire is the line integral with respect to arc length of the density function over the wire. A similar type of reasoning yields the result that the center of mass of the

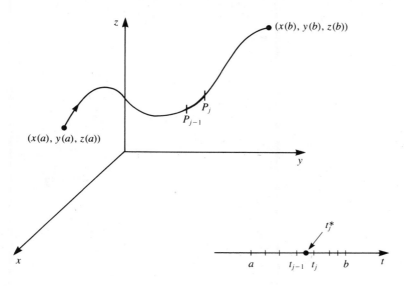

Figure 16.10

wire has coordinates

$$\bar{x} = \frac{1}{m} \int_C x \, \delta(x, y, z) \, ds \qquad \bar{y} = \frac{1}{m} \int_C y \, \delta(x, y, z) \, ds, \qquad \bar{z} = \frac{1}{m} \int_C z \, \delta(x, y, z) \, ds.$$

EXAMPLE 16.11

A wire is bent into the shape of a quarter-circle whose parametric equations are $x = 2 \cos(t)$, $y = 2 \sin(t)$, $z = 3$ for $0 \le t \le \pi/2$. The wire is shown in Figure 16.11. The density function is $\delta(x, y, z) = xy$ grams/centimeter. Find the mass and center of mass.

For this curve, we have

$$\begin{aligned}
ds &= \sqrt{[x'(t)]^2 + [y'(t)]^2 + [z'(t)]^2} \, dt \\
&= \sqrt{[-2 \sin(t)]^2 + [2 \cos(t)]^2 + 0^2} \, dt = 2 \, dt.
\end{aligned}$$

The mass is

$$\begin{aligned}
m &= \int_C \delta(x, y, z) \, ds \\
&= \int_0^{\pi/2} [2 \cos(t)][2 \sin(t)] 2 \, dt \\
&= \int_0^{\pi/2} 8 \sin(t)\cos(t) \, dt = 4 \sin^2(t) \Big]_0^{\pi/2} = 4 \text{ grams.}
\end{aligned}$$

The x-coordinate of the center of mass is

$$\begin{aligned}
\bar{x} &= \tfrac{1}{4} \int_C x \, \delta(x, y, z) \, ds = \tfrac{1}{4} \int_0^{\pi/2} [2 \cos(t)][4 \sin(t)\cos(t)] 2 \, dt \\
&= 4 \int_0^{\pi/2} \cos^2(t)\sin(t) \, dt = -\tfrac{4}{3} \cos^3(t) \Big]_0^{\pi/2} = \tfrac{4}{3}.
\end{aligned}$$

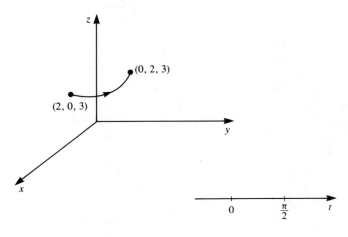

Figure 16.11

By symmetry of the density function and the curve, $\bar{y} = \bar{x}$. Finally,

$$\bar{z} = \tfrac{1}{4} \int_C z\, \delta(x, y, z)\, ds = \tfrac{1}{4} \int_0^{\pi/2} 3[4\cos(t)\sin(t)]2\, dt$$

$$= 6 \int_0^{\pi/2} \cos^2(t)\sin(t)\, dt = 3\sin^2(t)\Big]_0^{\pi/2} = 3,$$

hardly a surprising result. The center of mass is $(\tfrac{4}{3}, \tfrac{4}{3}, 3)$. ∎

In the next section, we will develop a relationship between double integrals and line integrals over closed curves in the plane.

PROBLEMS FOR SECTION 16.1

In each of Problems 1 through 25, compute the line integral.

1. $\int_C \mathbf{F} \cdot d\mathbf{R}$, $\mathbf{F} = x\mathbf{i} - \mathbf{j} + z\mathbf{k}$, $\mathbf{R} = t\mathbf{i} + t\mathbf{j} + t^3\mathbf{k}$ for $1 \le t \le 2$

2. $\int_C \mathbf{F} \cdot d\mathbf{R}$, $\mathbf{F} = -4x\mathbf{i} + y^2\mathbf{j} - yz\mathbf{k}$, $\mathbf{R} = t^2\mathbf{i} - 3t\mathbf{k}$ for $0 \le t \le 1$

3. $\int_C \mathbf{F} \cdot d\mathbf{R}$, $\mathbf{F} = \cos(x)\mathbf{i} - y\mathbf{j} + xz\mathbf{k}$, $\mathbf{R} = t\mathbf{i} - t^2\mathbf{j} + \mathbf{k}$ for $0 \le t \le 1$

4. $\int_C x^2 z\, dz$, C the line segment from $(0, 1, 1)$ to $(1, 2, -1)$

5. $\int_C x\, dy - yz\, dz$, C the parabola $y = z^2$, $x = 2$ from $(2, 1, 1)$ to $(2, 9, 3)$

6. $\int_C x\, ds$, C the line segment from $(0, 0, 0)$ to $(2, 1, 1)$

7. $\int_C \mathbf{F} \cdot d\mathbf{R}$, $\mathbf{F} = x\mathbf{i} + y\mathbf{j} - z\mathbf{k}$, C the circle $x^2 + y^2 = 4$, $z = 0$, going around once counterclockwise

8. $\int_C yz\, ds$, C the parabola $z = y^2$, $x = 1$ for $0 \le y \le 2$

9. $\int_C -xyz\, dz$, C the curve $y = \sqrt{z}$, $x = 1$ for $4 \le z \le 9$

10. $\int_C xz\, dy$, C the curve $x = y = t$, $z = -4t^2$, $1 \le t \le 3$

11. $\int_C 8z^2\, ds$, C the curve $x = y = 2t^2$, $z = 1$ for $1 \le t \le 2$

12. $\int_C \mathbf{F} \cdot d\mathbf{R}$, $\mathbf{F} = \mathbf{i} - x\mathbf{j} + \mathbf{k}$, $\mathbf{R} = \cos(t)\mathbf{i} - \sin(t)\mathbf{j} + t\mathbf{k}$, $0 \le t \le \pi$

13. $\int_C \mathbf{F} \cdot d\mathbf{R}$, $\mathbf{F} = 8x^2\mathbf{j}$, $\mathbf{R} = e^t\mathbf{i} - t^2\mathbf{j} + t\mathbf{k}$ for $1 \le t \le 2$

14. $\int_C x\, dy - y\, dz$ for C the curve $x = y = 2t$, $z = e^{-t}$, $0 \le t \le 3$

15. $\int_C \mathbf{F} \cdot d\mathbf{R}$, $\mathbf{F} = x\mathbf{i} + y\mathbf{j} - xyz\mathbf{k}$, C the curve $x = y = t$, $z = -3t^2$, $-1 \le t \le 3$

16. $\int_C \mathbf{F} \cdot d\mathbf{R}$, $\mathbf{F} = 3x\mathbf{i} - y^2\mathbf{j} + \mathbf{k}$, C the curve $x = 2t$, $y = 1 - t$, $z = t^2 + 2$, $1 \le t \le 3$

17. $\int_C \mathbf{F} \cdot d\mathbf{R}$, $\mathbf{F} = \cos(xy)\mathbf{j}$, C the curve $x = 1$, $y = 2t - 1$, $z = t$, $0 \le t \le \pi$

18. $\int_C (x + y + z^2)\, dx$, C given by $x = 2y = z$, $4 \le x \le 8$

19. $\int_C (x^2 - yz)\, dy$, C given by $x = t$, $y = z = \sqrt{t}$, $1 \le t \le 4$

20. $\int_C \mathbf{F} \cdot d\mathbf{R}$, $\mathbf{F} = -y\mathbf{i} + xy\mathbf{j} + x^2\mathbf{k}$, C given by $x = \sqrt{t}$, $y = 2t$, $z = t$, $1 \le t \le 4$

21. $\int_C \mathbf{F} \cdot d\mathbf{R}$, $\mathbf{F} = \mathbf{j} - 3x\mathbf{k}$, C given by $x = 1 + t^2$, $y = -t$, $z = 1 + t$, $2 \le t \le 5$

22. $\int_C \sin(z)\, dy$, C given by $x = 1 - t$, $y = 1 + t$, $z = 2t$, $0 \le t \le 1$

23. $\int_C \mathbf{F} \cdot d\mathbf{R}$, $\mathbf{F} = \sin(x)\mathbf{i} + 2z\mathbf{j} - \mathbf{k}$, C given by $x = 1$, $y = 3t^2$, $z = 4t$, $0 \le t \le 5$

24. $\int_C xyz\, dz$, C given by $x = t$, $y = t^2$, $z = t^3$, $0 \le t \le 2$

25. $\int_C \mathbf{F} \cdot d\mathbf{R}$, $\mathbf{F} = -3xy\mathbf{i} + 2y\mathbf{k}$, C the semicircle $x^2 + z^2 = 4$, $y = 1$, $z \ge 0$, oriented from $(2, 1, 0)$ to $(-2, 1, 0)$

In each of Problems 26 through 30, evaluate $\int_C f(x, y, z)\, ds$.

26. $f(x, y, z) = 4xy$, C given by $x = y = t$, $z = 2t$, $1 \le t \le 2$

27. $f(x, y, z) = x + y$, C given by $x = y = t$, $z = t^2$, $0 \le t \le 2$

28. $f(x, y, z) = \sin(x)$, C given by $x = t$, $y = 2t$, $z = 3t$, $1 \le t \le 3$

29. $f(x, y, z) = 3y^3$, C given by $x = z = t^2$, $y = 1$, $0 \le t \le 3$

30. $f(x, y, z) = x - y + 3z$, C given by $x = 3\cos(t)$, $y = 2$, $z = 3\sin(t)$, $0 \le t \le \pi$

31. Find the mass and center of mass of a thin wire stretched from $(0, 0, 0)$ to $(3, 3, 3)$ if $\delta(x, y, z) = x + y + z$ grams.

32. Find the mass of a thin wire in the shape of the circle $x^2 + z^2 = 4$, $y = 2$ if $\delta(x, y, z) = yz$.

33. Find the work done if an object is moved along the straight line from $(1, 1, 1)$ to $(4, 4, 4)$ by a force $\mathbf{F} = x^2\mathbf{i} - 2yz\mathbf{j} + z\mathbf{k}$.

34. Find the mass and center of mass of a wire triangle with vertices at the origin, $(0, 1, 0)$, and $(1, 1, 1)$ if $\delta(x, y, z) = 4$ on the side from $(0, 0, 0)$ to $(0, 1, 0)$ and $\delta(x, y, z) = 2$ on the other two sides.

35. Find the mass and center of mass of a wire bent into a rectangle having vertices $(1, 1, 3)$, $(1, 4, 3)$, $(6, 1, 3)$, and $(6, 4, 3)$ if $\delta(x, y, z) = 3$ on the sides from $(1, 1, 3)$ to $(1, 4, 3)$ and from $(1, 4, 3)$ to $(6, 1, 3)$ and $\delta = 5$ on the other two sides.

36. Suppose that $\mathbf{F}(x, y, z) = \nabla\varphi(x, y, z)$.

(a) Let C be a piecewise-smooth curve from P_0 to P_1, and suppose that the first partial derivatives of φ are continuous on C. Show that

$$\int_C \mathbf{F} \cdot d\mathbf{R} = \varphi(P_1) - \varphi(P_0).$$

(b) Show that $\int_C \mathbf{F} \cdot d\mathbf{R} = 0$ if C is a closed curve (that is, if $P_0 = P_1$).

37. Show that any ordinary Riemann integral $\int_a^b f(x)\,dx$ is a line integral $\int_C \mathbf{F} \cdot d\mathbf{R}$ for appropriate choices of $\mathbf{F}$ and C.

16.2 Green's Theorem

In this section, we will consider vector functions of just x and y and derive a relationship between a line integral around a closed curve and a double integral over the part of the plane enclosed by the curve.

Suppose we have a curve C in the plane given by the position vector $\mathbf{R}(t) = x(t)\mathbf{i} + y(t)\mathbf{j}, a \leq t \leq b$. Recall that C is closed if its initial point $(x(a), y(a))$ and terminal point $(x(b), y(b))$ are the same.

We call a closed curve C *positively oriented* if $(x(t), y(t))$ moves around C counterclockwise as t increases from a to b. For example, $\mathbf{R}(t) = \cos(t)\mathbf{i} + \sin(t)\mathbf{j}, 0 \leq t \leq 2\pi$ is positively oriented (Figure 16.12). In this case, the set D enclosed by C is over a person's left shoulder as he walks around C in the positive sense.

We call a nonclosed curve C *simple* if $\mathbf{R}(t_1) \neq \mathbf{R}(t_2)$ if $t_1 \neq t_2$. Thus, a curve is simple if it does not cross itself. Figure 16.13 shows the graph of a nonsimple curve. If t is time and the graph of C is a railroad track, a curve is simple if the train does not return to the same point at two different times.

If C is closed, $\mathbf{R}(a) = \mathbf{R}(b)$, so C cannot be simple because $a \neq b$. Nevertheless, we will still refer to a closed curve as simple if the initial and terminal points are the only

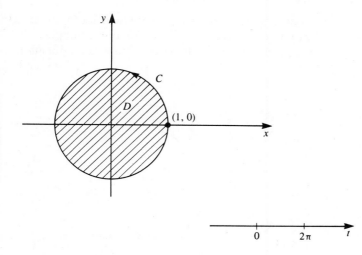

Figure 16.12

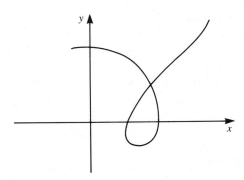

Figure 16.13. A nonsimple curve.

points which coincide for different values of the parameter. Figure 16.14 shows the graph of a nonsimple closed curve.

It is common to denote $\int_C \mathbf{F} \cdot d\mathbf{R}$ as $\oint_C \mathbf{F} \cdot d\mathbf{R}$ when C is a closed curve. The oval on the integral sign is simply a reminder that the curve is closed; it does not play any role in the way we define or evaluate the integral. Sometimes an arrow is inserted on the oval to remind us which way we are integrating around the curve (usually counterclockwise).

A piecewise-smooth simple closed curve C in the plane partitions the plane into two sets, as shown in Figure 16.15, with C their common boundary. One set contains points arbitrarily far from the origin and is called the *exterior* of C. The other set is called the *interior* of C. One way to think of this is as follows. If we took scissors and cut along C, two pieces of the plane would fall out. One is the interior and has finite area, while the other is the exterior. C itself does not belong to either of these sets but forms the common boundary between them.

THEOREM 16.3　Green's Theorem in the Plane

Let C be a simple closed positively oriented piecewise-smooth curve in the plane. Let D consist of all points on C and in the interior of C. Suppose that $\mathbf{F}(x, y) = f(x, y)\mathbf{i} + g(x, y)\mathbf{j}$ is a continuous vector function whose components have contin-

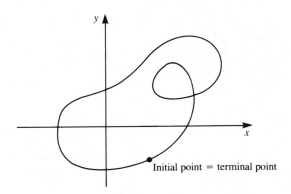

Figure 16.14. A nonsimple closed curve.

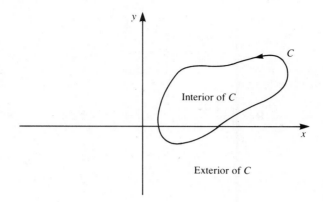

Figure 16.15

uous first partial derivatives throughout D. Then

$$\oint_C \mathbf{F} \cdot d\mathbf{R} = \iint_D \left[\frac{\partial g}{\partial x} - \frac{\partial f}{\partial y} \right] dA.$$

This conclusion can also be written

$$\oint_C f(x, y)\, dx + g(x, y)\, dy = \iint_D \left[\frac{\partial g}{\partial x} - \frac{\partial f}{\partial y} \right] dA. \quad \blacksquare$$

We will prove the theorem under special circumstances at the end of this section. Here are three examples.

EXAMPLE 16.12

Evaluate $\oint_C \mathbf{F} \cdot d\mathbf{R}$ if $\mathbf{F}(x, y) = [y - x^2 e^x]\mathbf{i} + [\cos(2y^2) - x]\mathbf{j}$ and C is the rectangle with vertices $(1, 1)$, $(0, 1)$, $(1, 3)$, and $(0, 3)$ oriented counterclockwise.

The curve is shown in Figure 16.16. It is routine but tedious to evaluate $\oint \mathbf{F} \cdot d\mathbf{R}$ directly. To use Green's theorem, let $f(x, y) = [y - x^2 e^x]$ and $g(x, y) = \cos(2y^2) - x$.

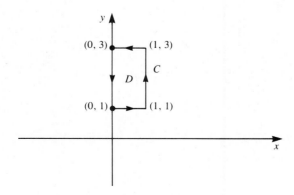

Figure 16.16

Then

$$\frac{\partial g}{\partial x} - \frac{\partial f}{\partial y} = -1 - (1) = -2,$$

and we have immediately that

$$\oint_C \mathbf{F} \cdot d\mathbf{R} = \iint_D \left[\frac{\partial g}{\partial x} - \frac{\partial f}{\partial y} \right] dA = \iint_D -2 \, dA$$

$$= -2(\text{area of } D) = -2(1)(2) = -4. \quad \blacksquare$$

EXAMPLE 16.13

Evaluate $\oint_C 2x \cos(2y) \, dx - 2x^2 \sin(2y) \, dy$ for *every* positively oriented piecewise-smooth simple closed curve in the plane.

If C is any such curve, then

$$\oint_C 2x \cos(2y) \, dx - 2x^2 \sin(2y) \, dy = \iint_D \left[\frac{\partial g}{\partial x} - \frac{\partial f}{\partial y} \right] dA$$

$$= \iint_D \left[-4x \sin(2y) - [-4x \sin(2y)] \right] dA$$

$$= \iint_D 0 \, dA = 0. \quad \blacksquare$$

EXAMPLE 16.14

A particle moves counterclockwise around the rectangle having vertices $(0, 0)$, $(6, 0)$, $(0, 4)$, $(6, 4)$ under the influence of the force $\mathbf{F} = x^2\mathbf{i} + 2xy\mathbf{j}$. Calculate the work done by $\mathbf{F}$ after one complete circuit.

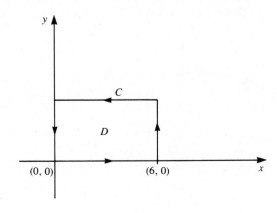

Figure 16.17

The curve is shown in Figure 16.17. Compute

$$\text{work done} = \oint_C \mathbf{F} \cdot d\mathbf{R} = \iint_D \left[\frac{\partial g}{\partial x} - \frac{\partial f}{\partial y} \right] dA$$

$$= \iint_D \left[\frac{\partial}{\partial x}(2xy) - \frac{\partial}{\partial y}(x^2) \right] dA = \iint_D 2y \, dA$$

$$= \int_0^6 dx \int_0^4 2y \, dy = 6 \cdot y^2 \Big]_0^4 = 6(16) = 96. \quad \blacksquare$$

We will conclude this section with a proof of a special case of Green's theorem. Assume that D can be described in two ways. First, suppose that the boundary has an upper portion which is the graph of $y = k(x)$ and a lower portion which is the graph of $y = h(x)$, as in Figure 16.18. Then D consists of all points (x, y) with

$$h(x) \le y \le k(x), \qquad a \le x \le b.$$

Second, suppose that the boundary of D also has a description in terms of a left portion [graph of $x = F(y)$] and a right portion [graph of $x = G(y)$], as in Figure 16.19. Then D consists of all (x, y) such that

$$F(y) \le x \le G(y), \qquad c \le y \le d.$$

We will use these descriptions to show explicitly that

$$\oint_C f(x, y) \, dx = - \iint_D \frac{\partial f}{\partial y} \, dA$$

and

$$\oint_C g(x, y) \, dy = \iint_D \frac{\partial g}{\partial x} \, dA.$$

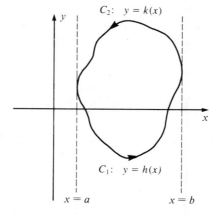

Figure 16.18

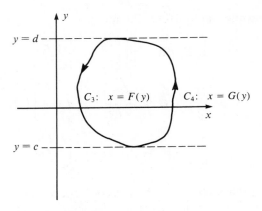

Figure 16.19

We get the conclusion of Green's theorem upon adding these equations. Begin with $\oint_C f(x, y)\, dx$. From the first description of D (Figure 16.18), we can think of C as consisting of C_1: $y = h(x)$, with x varying from a to b, and C_2: $y = k(x)$, with x varying from b to a (to maintain counterclockwise orientation). Using x as the parameter, we have

$$\oint_C f(x, y)\, dx = \int_{C_1} f(x, y)\, dx + \int_{C_2} f(x, y)\, dx$$

$$= \int_a^b f(x, h(x))\, dx + \int_b^a f(x, k(x))\, dx$$

$$= \int_a^b f(x, h(x))\, dx - \int_a^b f(x, k(x))\, dx$$

$$= \int_a^b - [f(x, k(x)) - f(x, h(x))]\, dx. \tag{16.2}$$

Next,

$$\iint_D \frac{\partial f}{\partial y}\, dA = \int_a^b \int_{h(x)}^{k(x)} \frac{\partial f}{\partial y}\, dy\, dx$$

$$= \int_a^b \left[f(x, y) \right]_{h(x)}^{k(x)} dx$$

$$= \int_a^b [f(x, k(x)) - f(x, h(x))]\, dx. \tag{16.3}$$

Comparing equations (16.2) and (16.3), we conclude that

$$\oint_C f(x, y)\, dx = - \iint_D \frac{\partial f}{\partial y}\, dA.$$

By using the other description of D (Figure 16.19), we can show by a similar argument that

$$\oint_C g(x, y) \, dy = \iint_D \frac{\partial g}{\partial x} \, dA.$$

Adding the last two equations gives us the conclusion of Green's theorem.

In the next section, we will examine circumstances in which the value of $\int_C \mathbf{F} \cdot d\mathbf{R}$ is independent of the curve C between two given points.

PROBLEMS FOR SECTION 16.2

In Each of Problems 1 through 15, use Green's theorem to evaluate $\oint_C \mathbf{F} \cdot d\mathbf{R}$. All curves are oriented counterclockwise.

1. $\mathbf{F} = 2y\mathbf{i} - x\mathbf{j}$, C the circle of radius 4 about $(1, 3)$
2. $\mathbf{F} = x^2\mathbf{i} - 2xy\mathbf{j}$, C the triangle with vertices $(1, 1), (4, 1), (2, 6)$
3. $\mathbf{F} = (x + y)\mathbf{i} + (x - y)\mathbf{j}$, C the ellipse $x^2 + 4y^2 = 1$
4. $\mathbf{F} = 8xy^2\mathbf{j}$, C the circle of radius 4 about the origin
5. $\mathbf{F} = (x^2 - y)\mathbf{i} + [\cos(2y) - e^{3y} + 4x]\mathbf{j}$, C any square with sides of length 5
6. $\mathbf{F} = (x^3 + y^2)\mathbf{i} + 4xy\mathbf{j}$, C the circle of radius 3 about $(4, 0)$. *Hint:* The resulting double integral can be evaluated by inspection if it is recognized as the y-coordinate of the centroid of D, which can be determined by symmetry.
7. $\mathbf{F} = e^x\cos(y)\mathbf{i} - e^x\sin(y)\mathbf{j}$, C any simple closed curve in the plane
8. $\mathbf{F} = (e^{x^3} - 4y)\mathbf{i} - [\cos(y^2) + 6x]\mathbf{j}$, C the square with vertices $(1, 0), (3, 0), (1, 3), (3, 3)$
9. $\oint_C x^2y \, dx - xy^2 \, dy$, C the boundary of the region $x^2 + y^2 \le 4, x \ge 0, y \ge 0$
10. $\mathbf{F} = xy\mathbf{i} + xy^2\mathbf{j}$, C the triangle with vertices $(0, 0), (3, 0), (0, 5)$
11. $\oint_C x \, dx - 4xy \, dy$, C the boundary of the region bounded by $y = x^2$ and $y = x$
12. $\oint_C \cos(x)y \, dx - y^3 \, dy$, C the square with vertices $(-1, 0), (0, 0), (0, 1), (-1, 1)$
13. $\mathbf{F} = xe^y\mathbf{i} - \sin(2y)\mathbf{j}$, C the triangle with vertices $(1, 1), (1, 3), (4, 1)$
14. $\mathbf{F} = (x^2 + y^2)\mathbf{i} + (x^2 - y^2)\mathbf{j}$, C the ellipse $4x^2 + y^2 = 16$
15. $\mathbf{F} = [e^{\sin(x)} - y]\mathbf{i} + [\sinh(y^3) - 4x]\mathbf{j}$, C the circle of radius 2 about $(-8, 0)$
16. Let C be a piecewise-smooth simple closed curve with interior D. Show that

$$\text{area of } D = \oint_C - y \, dx.$$

17. Under the conditions of Problem 16, show that

$$\text{area of } D = \oint_C x \, dy.$$

18. Under the conditions of Problem 16, show that

$$\text{area of } D = \frac{1}{2}\oint_C - y \, dx + x \, dy.$$

19. Under the conditions of Problem 16, show that the centroid of D has coordinates

$$\bar{x} = \frac{1}{2A(D)} \oint_C x^2 \, dy, \qquad \bar{y} = \frac{-1}{2A(D)} \oint_C y^2 \, dx,$$

in which $A(D) =$ area of D.

20. Suppose that $u(x, y)$ is continuous with continuous first and second partial derivatives on a piecewise-smooth simple closed curve C and throughout the interior of C. Show that

$$\iint_D \left[\frac{\partial^2 u}{\partial x^2} + \frac{\partial^2 u}{\partial y^2} \right] dA = \oint_C -\frac{\partial u}{\partial y} \, dx + \frac{\partial u}{\partial x} \, dy.$$

21. Let C be the piecewise-smooth simple closed curve in the plane made up of the graph of $r = f(\theta)$ for $\alpha \le \theta \le \beta$ (polar coordinates) and parts of the lines $\theta = \alpha$ and $\theta = \beta$. Assume that $f'(\theta)$ is continuous. Let D be the interior of C. Use the result of Problem 18 to show that

$$A(D) = \tfrac{1}{2} \int_\alpha^\beta f(\theta)^2 \, d\theta.$$

Hint: Calculate the line integral in Problem 18, using r as parameter on the straight line segments and θ as parameter on the portion of C given by the graph of $r = f(\theta)$

22. A particle moves once counterclockwise about the triangle with vertices $(0, 0), (4, 0), (1, 6)$. Find the work done if the force acting on the particle is $\mathbf{F} = xy\mathbf{i} + x\mathbf{j}$.

23. A particle moves once counterclockwise about the circle of radius 6 about the origin. Find the work done if the force acting on the particle is

$$\mathbf{F} = [e^x - y + x \cosh(x)]\mathbf{i} + (y^{3/2} + x)\mathbf{j}.$$

24. A particle moves once counterclockwise around the rectangle with vertices $(1, 1), (1, 7), (3, 1), (3, 7)$ under the influence of the force

$$\mathbf{F} = [-\cosh(4x^4) + xy]\mathbf{i} + (e^{-y} + x)\mathbf{j}.$$

Find the work done.

16.3 *Independence of Path and Potential Theory in the Plane* _____

Suppose we have a vector function $\mathbf{F}(x, y)$ defined for all (x, y) in some set D of points in the plane. We call $\int_C \mathbf{F} \cdot d\mathbf{R}$ *independent of path* in D if, for any two points P_0 and P_1 in D, $\int_C \mathbf{F} \cdot d\mathbf{R}$ has the same value for any piecewise-smooth curve C in D from P_0 to P_1. In this event, $\int_C \mathbf{F} \cdot d\mathbf{R}$ depends on only the endpoints of C, not C itself. (Of course, $\int_{C_1} \mathbf{F} \cdot d\mathbf{R}$ may still differ from $\int_{C_2} \mathbf{F} \cdot d\mathbf{R}$ if C_1 and C_2 have different endpoints or orientations.) In this context, a piecewise-smooth curve from P_0 to P_1 is often called a *path* from P_0 to P_1.

Independence of path has been observed in physics. It is customary to call φ a *potential function* for $\mathbf{F}$ if $\mathbf{F} = \nabla\varphi$ (some authors write $\mathbf{F} = -\nabla\varphi$). If a particle moves under the influence of $\mathbf{F}$ along a path C from P_0 to P_1, the work done is equal to the difference in the potential energy at the endpoints of C. If $\varphi(P)$ is the potential energy at P, this statement can be phrased

$$\text{work done} = \varphi(P_1) - \varphi(P_0).$$

Since the work done is also the line integral of $\mathbf{F}$ along C, this result suggests that

$$\int_C \mathbf{F} \cdot d\mathbf{R} = \varphi(P_1) - \varphi(P_0)$$

when $\mathbf{F} = \nabla\varphi$. This result depends on only the endpoints of C, not C itself.

We will discuss this result in a general context, then explore some of its ramifications.

THEOREM 16.4

Suppose that φ and its first partial derivatives are continuous for all (x, y) in a set D of the plane and that $\mathbf{F} = \nabla\varphi$ in D. Then $\int_C \mathbf{F} \cdot d\mathbf{R}$ is independent of path in D.

Proof   We are assuming that

$$\mathbf{F} = \nabla\varphi = \frac{\partial\varphi}{\partial x}\,\mathbf{i} + \frac{\partial\varphi}{\partial y}\,\mathbf{j}.$$

Suppose first that C is a smooth curve parametrized by $x = x(t)$, $y = y(t)$ for $a \le t \le b$. Then $x'(t)$ and $y'(t)$ are continuous on $[a, b]$. A position vector for C is $\mathbf{R}(t) = x(t)\mathbf{i} + y(t)\mathbf{j}$, and we can write

$$\mathbf{F} \cdot \mathbf{R}' = \left[\frac{\partial\varphi}{\partial x}\,\mathbf{i} + \frac{\partial\varphi}{\partial y}\,\mathbf{j}\right]\left[\frac{dx}{dt}\,\mathbf{i} + \frac{dy}{dt}\,\mathbf{j}\right]$$

$$= \frac{\partial\varphi}{\partial x}\frac{dx}{dt} + \frac{\partial\varphi}{\partial y}\frac{dy}{dt}.$$

Now recognize the quantity in the last line as a chain rule calculation of $(d/dt)\varphi(x(t), y(t))$. Therefore,

$$\int_C \mathbf{F} \cdot d\mathbf{R} = \int_a^b \frac{d}{dt}[\varphi(x(t), y(t))]\,dt$$

$$= \varphi(x(t), y(t))\Big]_a^b = \varphi(x(b), y(b)) - \varphi(x(a), y(a))$$

$$= \varphi(P_1) - \varphi(P_0).$$

If C is piecewise smooth, write $\int_C \mathbf{F} \cdot d\mathbf{R}$ as a sum of line integrals over the smooth curves comprising C and apply the preceding argument to each curve to complete the proof. ∎

Thus, when $\mathbf{F} = \nabla\varphi$, the line integral $\int_C \mathbf{F} \cdot d\mathbf{R}$ is obtained by subtracting the value of φ at the terminal point of C from its value at the initial point.

EXAMPLE 16.15

Consider $\int_C \mathbf{F} \cdot d\mathbf{R}$, with $\mathbf{F} = 2x\cos(2y)\mathbf{i} - 2x^2\sin(2y)\mathbf{j}$. It is easy to check that $\mathbf{F} = \mathbf{V}[x^2\cos(2y)]$. Further, $\varphi(x, y) = x^2\cos(2y)$ and its partial derivatives are continuous in the entire plane. Therefore, $\int_C \mathbf{F} \cdot d\mathbf{R}$ is independent of path in the entire plane.

For example, if C is any curve from $(0, 0)$ to $(-3, \pi/8)$, as in Figure 16.20,

$$\int_C \mathbf{F} \cdot d\mathbf{R} = \varphi\left(-3, \frac{\pi}{8}\right) - \varphi(0, 0) = (-3)^2\cos\left(\frac{2\pi}{8}\right) = \frac{9\sqrt{2}}{2}.$$

Similarly, if K is any curve from $(1, 1)$ to $(2, \pi)$, also shown in Figure 16.20,

$$\int_K \mathbf{F} \cdot d\mathbf{R} = \varphi(2, \pi) - \varphi(1, 1) = 4\cos(2\pi) - \cos(2) = 4 - \cos(2).$$

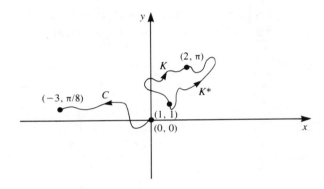

Figure 16.20

If we choose another path K^* from $(1, 1)$ to $(2, \pi)$, we get the same value for the line integral: $\int_{K^*} \mathbf{F} \cdot d\mathbf{R} = 4 - \cos(2)$. ■

A vector field $\mathbf{F}(x, y)$ is called *conservative* if $\mathbf{F} = \mathbf{V}\varphi$ for some function φ. As already noted, we call φ a *potential function*, or *potential*, for $\mathbf{F}$. Theorem 16.4 may be restated as follows: If $\mathbf{F}$ is conservative over D, the line integral of $\mathbf{F}$ is independent of path in D.

One ramification of Theorem 16.4 is that the line integral of any conservative force around a closed path is zero.

COROLLARY TO THEOREM 16.4

Suppose that $\mathbf{F} = \mathbf{V}\varphi$ for (x, y) in a set D of the plane and that φ and its first partial derivatives are continuous in D. Then

$$\oint_C \mathbf{F} \cdot d\mathbf{R} = 0$$

for any piecewise-smooth simple closed curve C in D.

Proof   Since C is a closed curve, the initial point P_0 and the terminal point P_1 are the same. By Theorem 16.4,

$$\oint_C \mathbf{F} \cdot d\mathbf{R} = \varphi(P_1) - \varphi(P_0) = 0. \quad \blacksquare$$

The converse of Theorem 16.4 is, in general, false unless we place additional restrictions on D. A set Ω of points in the plane is called a *domain* if it has the following two properties.

1. Given any point P_0 in Ω, there is a circle centered at P_0 such that every point inside the circle is also in Ω.

2. Given any points P_0 and P_1 in Ω, there is a piecewise-smooth curve C from P_0 to P_1 whose graph lies entirely in Ω.

For example, let Ω consist of all points (x, y) with $x > 0$ and $y > 0$. Then Ω is a domain. Check this as follows.

1. If P_0 is any point of Ω, we can draw a small enough circle about P_0 that every point inside the circle has positive coordinates and hence is in Ω (see Figure 16.21).

2. Certainly any two points in Ω are connected by a piecewise-smooth curve in Ω (in fact, by the straight line between the points). This is indicated in Figure 16.22.

If we let Ω consist of all (x, y) with $x \ge 0$ and $y \ge 0$, Ω is not a domain because condition (1) fails to hold. In particular, we cannot draw a circle about $(0, 0)$ (or any point on the positive x- or y-axis) containing only points with nonnegative coordinates (see Figure 16.23).

The set shown in Figure 16.24 fails to be a domain because condition (2) fails to hold; we cannot connect P_0 and P_1 with a curve lying entirely in Ω.

We can now state an improvement of Theorem 16.4.

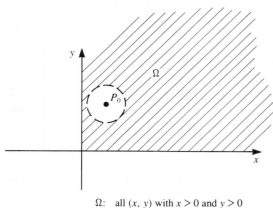

Ω: all (x, y) with $x > 0$ and $y > 0$

Figure 16.21

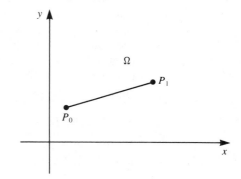

Figure 16.22

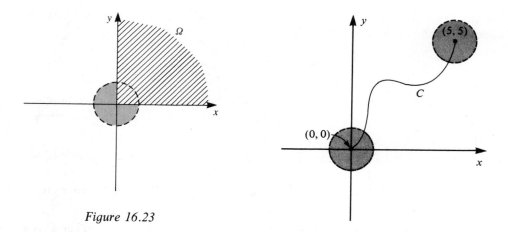

Figure 16.23

Figure 16.24. Ω (shaded) is not a domain.

THEOREM 16.5

Let $\mathbf{F}(x, y)$ be a vector function which is continuous in a domain Ω of the plane. Then $\int_C \mathbf{F} \cdot d\mathbf{R}$ is independent of path in Ω if and only if $\mathbf{F}$ is conservative.

Proof   If $\mathbf{F}$ is conservative, $\mathbf{F} = \nabla\varphi$, and we have already seen by Theorem 16.4 that $\int_C \mathbf{F} \cdot d\mathbf{R}$ is independent of path in Ω. We do not need Ω to be a domain for this to hold.

The proof of the converse is more difficult, exploiting properties (1) and (2) of the definition of a domain. We will examine this part of the proof at the end of this section. ∎

By Theorem 16.5, if we restrict ourselves to domains, the line integral is independent of path exactly when the vector field is conservative. We would now like a convenient test to determine when a vector field $\mathbf{F}$ is conservative. The next theorem provides such a test but requires still another condition on the set of points in which the vector field is defined.

We call a domain Ω *simply connected* if every simple closed curve in Ω encloses only points of Ω. Otherwise, Ω is said to be *not simply connected*. Figure 16.25 shows some domains, indicating which are simply connected and which are not. For example, the domain Ω consisting of all (x, y) with $4 < x^2 + y^2 < 6$ is not simply connected (Figure 16.26) because the simple closed curve C shown in Ω encloses points not in Ω (namely, the points on and inside the circle $x^2 + y^2 = 4$). Intuitively, Ω is simply connected if it has no "holes" in it, a test failed by this domain.

THEOREM 16.6

Let $\mathbf{F}(x, y) = f(x, y)\mathbf{i} + g(x, y)\mathbf{j}$ have continuous components, with continuous first partial derivatives, in a simply connected domain Ω. Then

$$\mathbf{F} \text{ is conservative if and only if } \frac{\partial f}{\partial y} = \frac{\partial g}{\partial x} \text{ at each point of } \Omega.$$

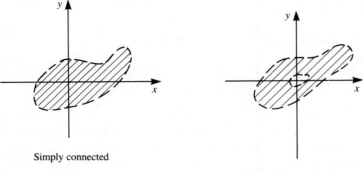

Simply connected

Not simply connected

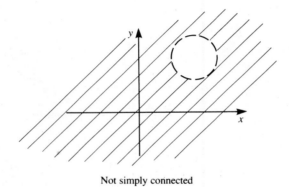

Not simply connected

Figure 16.25

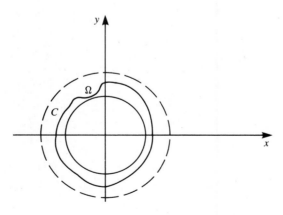

Figure 16.26

Proof   Suppose first that **F** is conservative. For some φ, $\mathbf{F} = \mathbf{V}\varphi$. Then

$$f(x, y) = \frac{\partial \varphi}{\partial x} \quad \text{and} \quad g(x, y) = \frac{\partial \varphi}{\partial y}.$$

Since the mixed partial derivatives of φ are equal, we have

$$\frac{\partial f}{\partial y} = \frac{\partial^2 \varphi}{\partial y \, \partial x} = \frac{\partial^2 \varphi}{\partial x \, \partial y} = \frac{\partial}{\partial x} \frac{\partial \varphi}{\partial y} = \frac{\partial g}{\partial x}.$$

Conversely, suppose that $\partial f / \partial y = \partial g / \partial x$ throughout Ω. We will show that $\int_C \mathbf{F} \cdot d\mathbf{R}$ is independent of path in Ω. Let P_0 and P_1 be points in Ω. Let C_1 and C_2 be paths from P_0 to P_1 in Ω, as shown in Figure 16.27. Form a closed path C using C_1 from P_0 to P_1 and

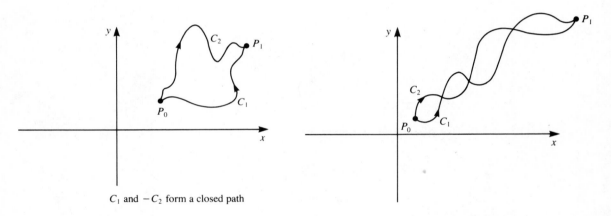

C_1 and $-C_2$ form a closed path

Figure 16.27

Figure 16.28

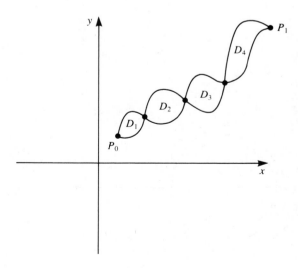

Figure 16.29

$-C_2$ from P_1 to P_0. Since Ω is simply connected, C encloses only points of Ω. By Green's theorem, $\oint_C \mathbf{F} \cdot d\mathbf{R} = 0$. But then $\int_{C_1} \mathbf{F} \cdot d\mathbf{R} + \int_{-C_2} \mathbf{F} \cdot d\mathbf{R} = 0$, implying that $\int_{C_1} \mathbf{F} \cdot d\mathbf{R} = \int_{C_2} \mathbf{F} \cdot d\mathbf{R}$; hence, $\int_C \mathbf{F} \cdot d\mathbf{R}$ is independent of path in Ω.

There is a technical difficulty if C_1 and C_2 cross each other, as in Figure 16.28. We can still prove the theorem in such a case by applying Green's theorem as indicated in Figure 16.29, but we will not pursue all of these details. ■

Problem 21 at the end of this section gives an illustration of how the conclusion of Theorem 16.6 can fail to hold if Ω is not simply connected.

Now that we can determine quite easily, under some conditions, whether $\mathbf{F}$ is conservative, how do we find a potential function φ for $\mathbf{F}$? The problem of producing a potential function for $\mathbf{F}(x, y) = f(x, y)\mathbf{i} + g(x, y)\mathbf{j}$ is equivalent to finding a potential function for the exact differential equation $dy/dx = -[f(x, y)]/[g(x, y)]$, a topic discussed in Section 1.4. Here is an example.

EXAMPLE 16.16

Evaluate $\int_C (6xy - 4e^x)\,dx + 3x^2\,dy$, with C any piecewise-smooth curve from $(0, 0)$ to $(-2, 1)$.

The component functions of $\mathbf{F}$ are continuous for all (x, y), and we can choose Ω as the entire xy-plane, which is simply connected. Compute

$$\frac{\partial}{\partial y}[6xy - 4e^x] = 6x = \frac{\partial}{\partial x}[3x^2],$$

so $\mathbf{F}$ is conservative. A potential function φ must satisfy

$$\frac{\partial \varphi}{\partial x} = 6xy - 4e^x \quad \text{and} \quad \frac{\partial \varphi}{\partial y} = 3x^2.$$

Choose one of these, say, $\partial\varphi/\partial y = 3x^2$. Integrate with respect to y, thinking of x as fixed, to get

$$\varphi(x, y) = 3x^2 y + K(x).$$

The constant of integration may involve x because this will be zero if we differentiate φ partially with respect to y. Then

$$\frac{\partial \varphi}{\partial x} = 6xy - 4e^x = \frac{\partial}{\partial x}[3x^2 y + K(x)] = 6xy + K'(x).$$

Then $K'(x) = -4e^x$, so $K(x) = -4e^x$ (letting the constant of integration be zero). Therefore,

$$\varphi(x, y) = 3x^2 y - 4e^x$$

is a potential function for $\mathbf{F}$, and

$$\int_C \mathbf{F} \cdot d\mathbf{R} = \varphi(-2, 1) - \varphi(0, 0) = (12 - 4e^{-2}) - (-4) = 16 - \frac{4}{e^2}. \quad ■$$

Many of the ideas we have discussed extend to vector functions and line integrals in three-space. An existence theorem for a potential function in three variables is more complicated than Theorem 16.6, but the mechanics of finding φ when it exists are virtually the same. Here is an example.

EXAMPLE 16.17

Find a potential function for

$$\mathbf{F}(x, y, z) = (yz^2 - 1)\mathbf{i} + (xz^2 + e^y)\mathbf{j} + (2xyz + 1)\mathbf{k}.$$

If $\mathbf{F} = \nabla\varphi = (\partial\varphi/\partial x)\mathbf{i} + (\partial\varphi/\partial y)\mathbf{j} + (\partial\varphi/\partial z)\mathbf{k}$, we must have

$$\frac{\partial\varphi}{\partial x} = yz^2 - 1, \qquad \frac{\partial\varphi}{\partial y} = xz^2 + e^y \quad \text{and} \quad \frac{\partial\varphi}{\partial z} = 2xyz + 1.$$

Pick one of these equations and integrate it. Starting with $\partial\varphi/\partial x = yz^2 - 1$, integrate with respect to x, holding y and z fixed. Then

$$\varphi(x, y, z) = xyz^2 - x + K(y, z),$$

with the "constant" of integration possibly containing y and z because the partial derivative of such a function with respect to x is zero. Then

$$\frac{\partial\varphi}{\partial y} = xz^2 + e^y = \frac{\partial}{\partial y}[xyz^2 - x + K(y, z)] = xz^2 + \frac{\partial K}{\partial y}.$$

Then

$$\frac{\partial K}{\partial y} = e^y.$$

Integrate this equation with respect to y, holding z fixed (K cannot involve x):

$$K(y, z) = e^y + A(z).$$

The constant of integration $A(z)$ may involve z because we will still have $\partial K/\partial y = e^y$. Thus far,

$$\varphi(x, y) = xyz^2 - x + e^y + A(z).$$

Finally,

$$\frac{\partial\varphi}{\partial z} = 2xyz + 1 = \frac{\partial}{\partial z}[xyz^2 - x + e^y + A(z)] = 2xyz + A'(z).$$

Then $A'(z) = 1$, so $A(z) = z$ (let the constant of integration be zero). Then $\varphi(x, y, z) = xyz^2 - x + e^y + z$ is a potential function for $\mathbf{F}$, as can be easily checked.

As a specific example, if we want $\int_C \mathbf{F} \cdot d\mathbf{R}$, with C a piecewise-smooth curve from $(1, 1, 1)$ to $(-2, 1, 3)$, then

$$\int_C \mathbf{F} \cdot d\mathbf{R} = \varphi(-2, 1, 3) - \varphi(1, 1, 1)$$

$$= -18 - (-2) + e + 3 - [1 - 1 + e + 1] = -14. \quad \blacksquare$$

We conclude this section with the rest of the proof of Theorem 16.5.

Proof of the Converse of Theorem 16.5   Suppose that $\int_C \mathbf{F} \cdot d\mathbf{R}$ is independent of path in Ω. Choose a point (x_0, y_0) in Ω. If (x, y) is any point of Ω, select any piecewise-smooth curve C from (x_0, y_0) to (x, y) and define

$$\varphi(x, y) = \int_C \mathbf{F} \cdot d\mathbf{R}.$$

This is a function of x and y because this line integral depends only on the endpoints (x, y) and (x_0, y_0) of C, and (x_0, y_0) is held constant.

We will show that $\mathbf{F} = \nabla\varphi$. To do this, let $\mathbf{F}(x, y) = f(x, y)\mathbf{i} + g(x, y)\mathbf{j}$. It will make the notation easier to follow if we choose any point (a, b) in Ω and show that

$$\frac{\partial\varphi}{\partial x}(a, b) = f(a, b) \quad \text{and} \quad \frac{\partial\varphi}{\partial y}(a, b) = g(a, b).$$

This will show that $\mathbf{F}(a, b) = \nabla\varphi(a, b)$ and complete the proof. First, we will show that $(\partial\varphi/\partial x)(a, b) = f(a, b)$. We have

$$\frac{\partial\varphi}{\partial x}(a, b) = \lim_{\Delta x \to 0} \frac{\varphi(a + \Delta x, b) - \varphi(a, b)}{\Delta x}.$$

Since Ω is a domain, there is some positive number r such that the circle of radius r about (a, b) contains only points of Ω. Choose Δx so that $0 < \Delta x < r$; let C_1 be any piecewise-smooth curve in Ω from (x_0, y_0) to (a, b), and let C_2 be the horizontal line segment from (a, b) to $(a + \Delta x, b)$, as shown in Figure 16.30. Then C, consisting of C_1 and C_2, is a piecewise-smooth curve from (x_0, y_0) to $(a + \Delta x, b)$. Further,

$$\varphi(a + \Delta x, b) - \varphi(a, b) = \int_{C_1 + C_2} \mathbf{F} \cdot d\mathbf{R} - \int_{C_1} \mathbf{F} \cdot d\mathbf{R} = \int_{C_2} \mathbf{F} \cdot d\mathbf{R}$$

$$= \int_{C_2} f(x, y)\, dx + g(x, y)\, dy.$$

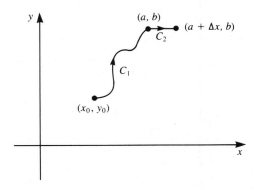

Figure 16.30

Parametrize C_2 by

$$x = a + t\,\Delta x, \qquad y = b \quad \text{for} \quad 0 \le t \le 1.$$

Then $dx = (\Delta x)\,dt$ and $dy = 0$, so

$$\varphi(a + \Delta x, b) - \varphi(a, b) = \int_0^1 f(a + t\,\Delta x, b)(\Delta x)\,dt$$

$$= \Delta x \int_0^1 f(a + t\,\Delta x, b)\,dt.$$

In this integral, we consider $f(a + t\,\Delta x, b)$ as a function of t. By the mean value theorem for integrals, there exists some number c, with $0 < c < 1$, such that

$$\int_0^1 f(a + t\,\Delta x, b)\,dt = f(a + c\,\Delta x, b)(1 - 0) = f(a + c\,\Delta x, b).$$

Therefore,

$$\varphi(a + \Delta x, b) - \varphi(a, b) = \Delta x f(a + c\,\Delta x, b).$$

Then

$$\frac{\varphi(a + \Delta x, b) - \varphi(a, b)}{\Delta x} = f(a + c\,\Delta x, b).$$

In the limit as $\Delta x \to 0$, we have $a + c\,\Delta x \to a$, and by continuity of f, we have $f(a + c\,\Delta x, b) \to f(a, b)$. Since we have $\Delta x > 0$ here, we have shown that

$$\lim_{\Delta x \to 0+} \frac{\varphi(a + \Delta x, b) - \varphi(a, b)}{\Delta x} = f(a, b).$$

By a similar argument, using a curve such as that shown in Figure 16.31, we can show that $\lim_{\Delta x \to 0-} [\varphi(a + \Delta x, b) - \varphi(a, b)]/\Delta x = f(a, b)$. We conclude that $(\partial \varphi/\partial x)(a, b) = f(a, b)$. Similarly, by using curves as shown in Figures 16.32 and 16.33, we can show that $(\partial \varphi/\partial y)(a, b) = g(a, b)$, completing the proof.

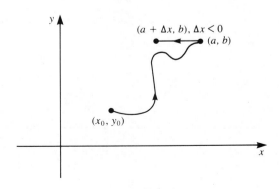

Figure 16.31

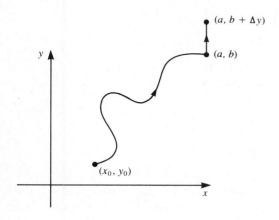

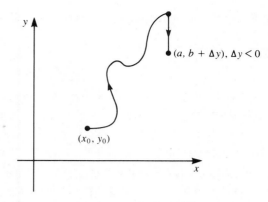

Figure 16.32

Figure 16.33

PROBLEMS FOR SECTION 16.3

In each of Problems 1 through 10, use Theorem 16.6 to determine whether **F** is conservative in the given domain Ω. A potential function need not be found.

1. $\mathbf{F} = y^3\mathbf{i} + (3xy^2 - 4)\mathbf{j}$, Ω the entire plane

2. $\mathbf{F} = (6y + ye^{xy})\mathbf{i} + (6x + xe^{xy})\mathbf{j}$, Ω the entire plane

3. $\mathbf{F} = 16x\mathbf{i} + (2 - y^2)\mathbf{j}$, Ω the entire plane

4. $\mathbf{F} = 2xy\cos(x^2)\mathbf{i} + \sin(x^2)\mathbf{j}$, Ω the domain $x^2 + y^2 < 1$

5. $\mathbf{F} = \dfrac{2x}{x^2 + y^2}\mathbf{i} + \dfrac{2y}{x^2 + y^2}\mathbf{j}$, Ω the domain enclosed by the circle of radius 2 about $(9, 0)$

6. $\mathbf{F} = \sinh(x + y)(\mathbf{i} + \mathbf{j})$, Ω the entire plane

7. $\mathbf{F} = 2\cos(2x)e^y\mathbf{i} + [e^y\sin(2x) - y]\mathbf{j}$, Ω the entire plane

8. $\mathbf{F} = [3x^2y - \sin(x) + 1]\mathbf{i} + (x^3 - e^y)\mathbf{j}$, Ω the entire plane

9. $\mathbf{F} = (y^2 + 3)\mathbf{i} + (2xy + 3x)\mathbf{j}$, Ω the entire plane

10. $\mathbf{F} = (3x^2 - 2y)\mathbf{i} + (12y - 2x)\mathbf{j}$, Ω the entire plane

In each of Problems 11 through 20, evaluate $\int_C \mathbf{F} \cdot d\mathbf{R}$ for C any piecewise-smooth curve from the first point to the second.

11. $\mathbf{F} = 3x^2(y^2 - 4y)\mathbf{i} + (2x^3y - 4x^3)\mathbf{j}$; $(1, 1)$ to $(2, 3)$

12. $\mathbf{F} = e^x\cos(y)\mathbf{i} - e^x\sin(y)\mathbf{j}$; $(0, 0)$ to $\left(2, \dfrac{\pi}{4}\right)$

13. $\mathbf{F} = 2xy\mathbf{i} + \left(x^2 - \dfrac{1}{y}\right)\mathbf{j}$; $(1, 3)$ to $(2, 2)$

14. $\mathbf{F} = \mathbf{i} + [6y + \sin(y)]\mathbf{j}$; $(0, 0)$ to $(-1, 3)$

15. $\mathbf{F} = (3x^2y^2 - 6y^3)\mathbf{i} + (2x^3y - 18xy^2)\mathbf{j}$; $(0, 0)$ to $(1, 1)$

16. $\mathbf{F} = \dfrac{y}{x}\mathbf{i} + \ln(x)\mathbf{j}$; $(1, 1)$ to $(2, 2)$ in the half-plane $x > 0$

17. $\mathbf{F} = (-8e^y + e^x)\mathbf{i} - 8xe^y\mathbf{j}$; $(-1, -1)$ to $(3, 1)$

18. $\mathbf{F} = \left(4xy + \dfrac{3}{x^2}\right)\mathbf{i} + 2x^2\mathbf{j}$; $(1, 2)$ to $(3, 3)$ in the half-plane $x > 0$

19. $\mathbf{F} = [-4\cosh(xy) - 4xy\sinh(xy)]\mathbf{i} - 4x^2\sinh(xy)\mathbf{j}$; $(1, 0)$ to $(2, 1)$

20. $\mathbf{F} = [3y^2 + 3\sin(y)]\mathbf{i} + [6xy + 3x\cos(y)]\mathbf{j}$; (0, 0) to $(-3, \pi)$

21. (a) Evaluate

$$\oint_C \left[\frac{-y}{x^2 + y^2} \mathbf{i} + \frac{x}{x^2 + y^2} \mathbf{j} \right] \cdot d\mathbf{R},$$

with C the circle of radius 4 about the origin, oriented positively.

(b) Show that

$$\frac{\partial}{\partial y} \frac{-y}{x^2 + y^2} = \frac{\partial}{\partial x} \frac{x}{x^2 + y^2}.$$

(c) Explain why the results of (a) and (b) do not contradict Theorem 16.6.

22. Prove the law of conservation of energy of physics. This states that the sum of the kinetic and potential energies of an object acted on by a conservative force $\mathbf{F}$ is a constant. *Hint:* The kinetic energy is $(m/2)\|\mathbf{R}'(t)\|^2$, with m the mass and $\mathbf{R}(t)$ the position vector of the particle. The potential energy is $-\varphi$, with φ a potential function for $\mathbf{F}$. To derive the theorem, show that the derivative of the sum of the kinetic and potential energies is zero along the trajectory.

23. Find a potential function for $\mathbf{F} = yz^3\mathbf{i} + xz^3\mathbf{j} + 3xyz^2\mathbf{k}$. Use this result to evaluate $\int_C \mathbf{F} \cdot d\mathbf{R}$ for any piecewise-smooth curve from (0, 1, 1) to (3, −1, 2).

24. Find a potential function for $\mathbf{F} = 2xy\cos(z)\mathbf{i} + x^2\cos(z)\mathbf{j} - x^2y\sin(z)\mathbf{k}$. Use this to evaluate $\int_C \mathbf{F} \cdot d\mathbf{R}$ for any piecewise-smooth curve from (1, 1, −1) to (2, 0, 5).

25. Find a potential function for $\mathbf{F} = e^{xyz}[yz\mathbf{i} + xz\mathbf{j} + xy\mathbf{k}]$. Use this to evaluate $\int_C \mathbf{F} \cdot d\mathbf{R}$, with C any piecewise-smooth curve from (1, 1, −2) to (3, 1, 0).

26. Find a potential function for $\mathbf{F} = (3x^2yz^2 + e^x)\mathbf{i} + x^3z^2\mathbf{j} + 2x^3yz\mathbf{k}$. Use this to evaluate $\int_C \mathbf{F} \cdot d\mathbf{R}$ along any piecewise-smooth curve from the origin to (−1, 3, 1).

27. Show that there is no potential function for $\mathbf{F} = x^2\mathbf{i} - y\mathbf{j} + z\cos(x)\mathbf{k}$. *Hint:* Try to find one and reach a contradiction.

28. Show that there is no potential function for $\mathbf{F} = e^{xyz}\mathbf{i} - \mathbf{j} + z\mathbf{k}$.

16.4 Surfaces and Surface Integrals

Intuitively, we think of a thin wire or curve as one-dimensional, a flat plate or sheet of metal as two-dimensional, and a solid object occupying a volume in three-space as three-dimensional. We have integrals tailored to solving certain kinds of problems involving such objects. For the mass of a wire, we use a line integral; for a thin, flat plate, a double integral; and for a solid object in three-space, a triple integral.

But there are other kinds of objects we are as yet unequipped to deal with. For example, a thin shell in three-space is two-dimensional (think, for example, of an eggshell or a geodesic dome). Such an object exists in three-space and may even enclose a volume, even though it is only two-dimensional. To find the mass of such a shell, given the density function, we must use a new kind of integral called a *surface integral*, which we will now define. We will begin with some preliminary remarks about surfaces.

SURFACES IN THREE-SPACE

Often a surface is thought of as a locus of points (x, y, z) with x, y, and z specified by parametric functions of two variables:

$$x = x(u, v), \qquad y = y(u, v), \qquad z = z(u, v) \tag{16.4}$$

for (u, v) in some set D in the uv-plane. For example, using spherical coordinates θ and φ (see Appendix E at the end of this part), a sphere of radius 1 about the origin is given by

$$x = \cos(\theta)\sin(\varphi), \qquad y = \sin(\theta)\sin(\varphi), \qquad z = \cos(\varphi); \qquad 0 \le \theta \le 2\pi, \qquad 0 \le \varphi \le \pi.$$

Figure 16.34 shows a graph of this surface together with the parameter domain (the rectangle $0 \le \theta \le 2\pi, 0 \le \varphi \le \pi$ in the $\theta\varphi$-plane).

A surface might also be specified by an equation $z = S(x, y)$. Here, we can think of x and y as parameters so that the parametric equations are simply

$$x = x, \qquad y = y, \qquad z = S(x, y)$$

for (x, y) in some given set D. For example, the equation $z = \sqrt{x^2 + y^2}$ for all (x, y) has as its graph the cone of Figure 16.35.

Another common way of specifying a surface is as a locus of points satisfying an equation $f(x, y, z) = 0$. In this formulation, we can think of any two of the variables x, y, and z as parameters, with the third defined implicitly by the equation. For example, the sphere specified above in spherical coordinates is also given by the equation $f(x, y, z) = 0$, with $f(x, y, z) = x^2 + y^2 + z^2 - 1$.

Associated with the parametric equations (16.4), we can define a position vector $\mathbf{R}(u, v) = x(u, v)\mathbf{i} + y(u, v)\mathbf{j} + z(u, v)\mathbf{k}$. This is like the position vector for a curve except that now $\mathbf{R}(u, v)$ depends on two variables, not just one. For each (u, v) in the parameter domain, $\mathbf{R}(u, v)$ can be represented as an arrow from the origin to a point on the surface, as in Figure 16.36. As (u, v) moves over the parameter domain, the position vector $\mathbf{R}(u, v)$ sweeps out the surface in three-space.

We call a surface *simple* if $\mathbf{R}(u, v)$ does not return to the same point for different values of (u, v). For example, the surface of Figure 16.37 folds across itself and is not simple. This concept is analogous to a simple curve, which does not cross itself.

On a surface, there is no unambiguous way to specify a tangent vector at a point. A surface may, however, have a tangent plane at a point P_0 in which all vectors are tangent

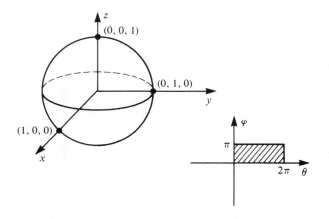

Figure 16.34

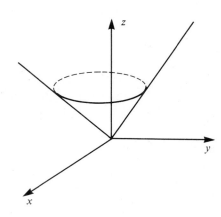

Figure 16.35. Cone $z = \sqrt{x^2 + y^2}$.

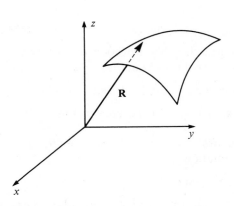

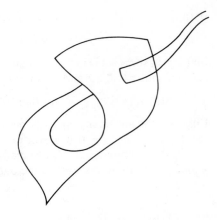

Figure 16.36. Position vector $\mathbf{R}(u, v)$ sweeping out a surface.

Figure 16.37. A nonsimple surface.

to the surface at P_0. In this event, we can define a normal vector at P_0 as a vector normal (orthogonal) to the tangent plane there, as shown in Figure 16.38.

To determine the normal vector to a surface given by (16.4), consider a point P_0 on the surface, as in Figure 16.39. P_0 has coordinates $(x(u_0, v_0), y(u_0, v_0), z(u_0, v_0))$. If we fix $v = v_0$ and vary u in equations (16.4), we obtain a curve

$$C_u: \quad x = x(u, v_0), \qquad y = y(u, v_0), \qquad z = z(u, v_0)$$

on the surface through P_0. Similarly, if we fix $u = u_0$, we obtain another curve

$$C_v: \quad x = x(u_0, v), \qquad y = y(u_0, v), \qquad z = z(u_0, v)$$

on the surface through P_0. These curves are shown in Figure 16.39. The curves have position vectors

$$\mathbf{R}_u(u, v_0) = x(u, v_0)\mathbf{i} + y(u, v_0)\mathbf{j} + z(u, v_0)\mathbf{k}$$

and

$$\mathbf{R}_v(u_0, v) = x(u_0, v)\mathbf{i} + y(u_0, v)\mathbf{j} + z(u_0, v)\mathbf{k},$$

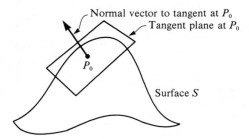

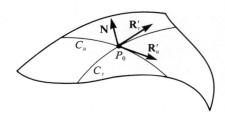

Figure 16.38

Figure 16.39. Normal vector $\mathbf{N} = \mathbf{R}'_u \times \mathbf{R}'_v$.

respectively. The tangents to these curves are

$$\mathbf{R}'_u = \frac{\partial x}{\partial u}\mathbf{i} + \frac{\partial y}{\partial u}\mathbf{j} + \frac{\partial z}{\partial u}\mathbf{k}$$

and

$$\mathbf{R}'_v = \frac{\partial x}{\partial v}\mathbf{i} + \frac{\partial y}{\partial v}\mathbf{j} + \frac{\partial z}{\partial v}\mathbf{k},$$

with the partial derivatives evaluated at (u_0, v_0).

Both of these vectors are in the tangent plane to the surface at P_0 because the curves C_u and C_v are on the surface through P_0. Therefore, their cross product is orthogonal to the tangent plane at P_0. We define the vector

$$\mathbf{N} = \mathbf{R}'_u \times \mathbf{R}'_v$$

to be the *normal to the surface at* P_0, provided that it is not the zero vector; if $\mathbf{N} = \mathbf{O}$, we say that the surface has no normal vector at the point.

Calculate

$$\mathbf{N} = \begin{vmatrix} \mathbf{i} & \mathbf{j} & \mathbf{k} \\ \dfrac{\partial x}{\partial u} & \dfrac{\partial y}{\partial u} & \dfrac{\partial z}{\partial u} \\ \dfrac{\partial x}{\partial v} & \dfrac{\partial y}{\partial v} & \dfrac{\partial z}{\partial v} \end{vmatrix}$$

$$= \begin{vmatrix} \dfrac{\partial y}{\partial u} & \dfrac{\partial y}{\partial v} \\ \dfrac{\partial z}{\partial u} & \dfrac{\partial z}{\partial v} \end{vmatrix}\mathbf{i} + \begin{vmatrix} \dfrac{\partial z}{\partial u} & \dfrac{\partial z}{\partial v} \\ \dfrac{\partial x}{\partial u} & \dfrac{\partial x}{\partial v} \end{vmatrix}\mathbf{j} + \begin{vmatrix} \dfrac{\partial x}{\partial u} & \dfrac{\partial x}{\partial v} \\ \dfrac{\partial y}{\partial u} & \dfrac{\partial y}{\partial v} \end{vmatrix}\mathbf{k}.$$

For those familiar with Jacobian notation from multivariable calculus, this vector can be written

$$\mathbf{N} = \frac{\partial(y, z)}{\partial(u, v)}\mathbf{i} + \frac{\partial(z, x)}{\partial(u, v)}\mathbf{j} + \frac{\partial(x, y)}{\partial(u, v)}\mathbf{k}.$$

In the special case in which the surface is the locus of points satisfying $z = S(x, y)$, we use x and y as parameters, and the coordinate functions are

$$x = u, \qquad y = v, \qquad z = S(x, y).$$

In this event, the normal vector is

$$\mathbf{N} = \begin{vmatrix} \dfrac{\partial y}{\partial x} & \dfrac{\partial y}{\partial y} \\ \dfrac{\partial S}{\partial x} & \dfrac{\partial S}{\partial y} \end{vmatrix}\mathbf{i} + \begin{vmatrix} \dfrac{\partial S}{\partial x} & \dfrac{\partial S}{\partial y} \\ \dfrac{\partial x}{\partial x} & \dfrac{\partial x}{\partial y} \end{vmatrix}\mathbf{j} + \begin{vmatrix} \dfrac{\partial x}{\partial x} & \dfrac{\partial x}{\partial y} \\ \dfrac{\partial y}{\partial x} & \dfrac{\partial y}{\partial y} \end{vmatrix}\mathbf{k}$$

$$= -\frac{\partial S}{\partial x}\mathbf{i} - \frac{\partial S}{\partial y}\mathbf{j} + \mathbf{k}$$

because $\partial x/\partial x = \partial y/\partial y = 1$ and $\partial x/\partial y = \partial y/\partial x = 0$.

Often, we write

$$\mathbf{N} = -\frac{\partial z}{\partial x}\,\mathbf{i} - \frac{\partial z}{\partial y}\,\mathbf{j} + \mathbf{k}.$$

We call a surface Σ *smooth* if it is simple and $\mathbf{N}$ is continuous and nonzero over the entire surface. If Σ consists of finitely many smooth pieces, we say that Σ is *piecewise smooth*. For example, a sphere is smooth, while the surface of a cube is piecewise smooth, consisting of six smooth faces.

EXAMPLE 16.18

Let $z = S(x, y) = \sqrt{x^2 + y^2}$ for $x^2 + y^2 \le 1$. The graph of this surface is the cone of Figure 16.40. The normal vector is

$$\mathbf{N} = -\frac{\partial z}{\partial x}\,\mathbf{i} - \frac{\partial z}{\partial y}\,\mathbf{j} + \mathbf{k}$$

$$= \frac{-x}{\sqrt{x^2 + y^2}}\,\mathbf{i} + \frac{-y}{\sqrt{x^2 + y^2}}\,\mathbf{j} + \mathbf{k}$$

$$= -\frac{x}{z}\,\mathbf{i} - \frac{y}{z}\,\mathbf{j} + \mathbf{k}$$

if $z \ne 0$. Technically, this surface is not smooth because it has no normal vector at the origin, where there is no tangent plane. For some types of calculations, this single "exceptional point" plays no significant role. ■

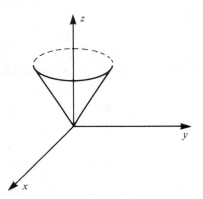

Figure 16.40. Cone $z = \sqrt{x^2 + y^2}, 0 \le x^2 + y^2 \le 1$.

EXAMPLE 16.19

The function $z = x^2 + y^2$, for $x^2 + y^2 \le 4$, has the paraboloid of Figure 16.41 as its graph. The normal vector is

$$\mathbf{N} = -2x\mathbf{i} - 2y\mathbf{j} + \mathbf{k}.$$

This surface is smooth, having a continuous nonzero normal at each point. ■

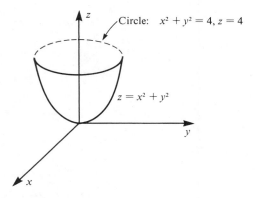

Figure 16.41. Paraboloid $z = x^2 + y^2$.

Recall the formula for surface area from calculus. If the surface Σ is given by $z = S(x, y)$, for (x, y) in a set D of the plane, the area of Σ is

$$A(\Sigma) = \iint_D \sqrt{1 + \left(\frac{\partial z}{\partial x}\right)^2 + \left(\frac{\partial z}{\partial y}\right)^2} \, dA,$$

assuming that these partial derivatives are continuous for (x, y) in D. This is the double integral of the length of the normal vector

$$A(\Sigma) = \iint_D \|\mathbf{N}(x, y)\| \, dA.$$

This formula is the analogue, for surfaces, of the fact that the length of a curve is the integral of the length of the tangent vector.

SURFACE INTEGRALS

We are now prepared to define the concept of a surface integral. Suppose that $f(x, y, z)$ is defined at least for all points on a surface Σ, which we will think of as the graph of $z = S(x, y)$ for (x, y) in some set D of points in the plane. We will suppose that Σ is smooth and that the parameter domain D is bounded. This means that it is possible to draw a rectangle containing the set D in the xy-plane. We now proceed to define an integral of $f(x, y, z)$ over Σ.

Step 1   Form a rectangular grid over D by lines $x = x_j$, $y = y_k$, retaining only those rectangles $R_1, \ldots, R_N$ containing points of D. The straight lines $x = x_j$, $y = y_k$ project onto curves $z = S(x_j, y)$ and $z = S(x, y_k)$ on Σ, forming a grid of "curved parallelograms" $\Sigma_1, \Sigma_2, \ldots, \Sigma_N$ over the surface, as shown in Figure 16.42.

Step 2   Choose a point (x_j^*, y_j^*) in each R_j, and let $z_j^* = S(x_j^*, y_j^*)$. The point $P_j^*:(x_j^*, y_j^*, z_j^*)$ is on Σ_j, as in Figure 16.43.

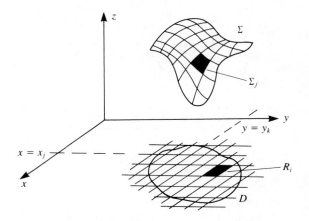

Figure 16.42

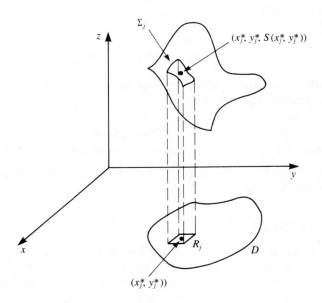

Figure 16.43

Step 3   Let $A(\Sigma_j)$ be the area of Σ_j. Form the sum

$$\sum_{j=1}^{N} f(x_j^*, y_j^*, z_j^*) A(\Sigma_j). \qquad (16.5)$$

If these sums approach a number L as $N \to \infty$ and as the dimensions of R_j tend to zero, we call L the *surface integral* of $f(x, y, z)$ over Σ and denote this integral $\iint_{\Sigma} f(x, y, z)\, d\sigma$. Surface integrals have some of the familiar properties of integrals.

THEOREM 16.7

If Σ is a smooth surface and $f(x, y, z)$ is a bounded function which is continuous except possibly at finitely many points of Σ, then $\iint_\Sigma f(x, y, z)\, d\sigma$ exists. ∎

THEOREM 16.8

1. If $\iint_\Sigma f(x, y, z)\, d\sigma$ and $\iint_\Sigma g(x, y, z)\, d\sigma$ exist, then

$$\iint_\Sigma [f(x, y, z) + g(x, y, z)]\, d\sigma = \iint_\Sigma f(x, y, z)\, d\sigma + \iint_\Sigma g(x, y, z)\, d\sigma.$$

2. If $\iint_\Sigma f(x, y, z)\, d\sigma$ exists, then for any constant β,

$$\iint_\Sigma \beta f(x, y, z)\, d\sigma = \beta \iint_\Sigma f(x, y, z)\, d\sigma.$$

3. If Σ_1 and Σ_2 are smooth surfaces having at most a curve in common (as in Figure 16.44), and Σ is a surface formed from Σ_1 and Σ_2, then

$$\iint_\Sigma f(x, y, z)\, d\sigma = \iint_{\Sigma_1} f(x, y, z)\, d\sigma + \iint_{\Sigma_2} f(x, y, z)\, d\sigma. \quad ∎$$

We want to be able to evaluate $\iint_\Sigma f(x, y, z)\, d\sigma$ explicitly for specific surfaces Σ and functions f. We will now see how to do this by converting $\iint_\Sigma f(x, y, z)\, d\sigma$ to a double integral over a set in the plane.

The key lies in approximating the factor $A(\Sigma_j)$ in the summation (2) in terms of known quantities. $A(\Sigma_j)$ is the area of the projection of the rectangle R_j onto Σ. Draw the tangent plane to the surface at P_j^*, which is a point of Σ_j. The rectangle R_j projects onto a rectangle R_j^* on the tangent plane to Σ at P_j^*, as shown in Figure 16.45. If R_j is

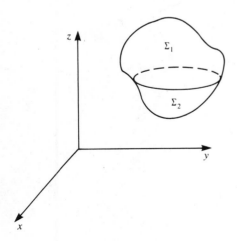

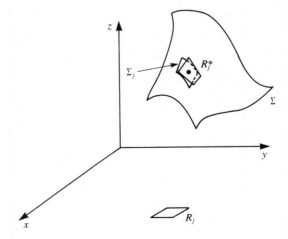

Figure 16.44   Figure 16.45

small, the area $A(R_j^*)$ of this rectangle approximates $A(\Sigma_j)$. We therefore want to relate the area of a rectangle in the xy-plane to the area of its projection onto a plane in three-space. The following lemma does this.

LEMMA 16.1

Let R be a rectangle in the xy-plane having sides of length L_1 and L_2. Let T be a projection of R onto the plane $z = ax + by + c$, as in Figure 16.46. Then the areas $A(T)$ of T and $A(R)$ of R are related by

$$A(T) = \sqrt{a^2 + b^2 + 1}\,A(R).$$

Proof   Recall that the area of a parallelogram having vectors $\mathbf{u}$ and $\mathbf{v}$ along two incident sides is $\|\mathbf{u} \times \mathbf{v}\|$. All we have to do is find $\mathbf{u}$ and $\mathbf{v}$ for the parallelogram T. To do this, consider Figure 16.47. With the vertices of R as labeled, the vertices of T are

$$(x_0, y_0, ax_0 + by_0 + c), \qquad (x_0 + L_1, y_0, a(x_0 + L_1) + by_0 + c),$$
$$(x_0, y_0 + L_2, ax_0 + b(y_0 + L_2) + c),$$

and

$$(x_0 + L_1, y_0 + L_2, a(x_0 + L_1) + b(y_0 + L_2) + c).$$

With $\mathbf{u}$ and $\mathbf{v}$ as shown in Figure 16.47, we have

$$\mathbf{u} = L_1\mathbf{i} + 0\mathbf{j} + aL_1\mathbf{k} = L_1(\mathbf{i} + a\mathbf{k})$$

and

$$\mathbf{v} = 0\mathbf{i} + L_2\mathbf{j} + bL_2\mathbf{k} = L_2(\mathbf{j} + b\mathbf{k}).$$

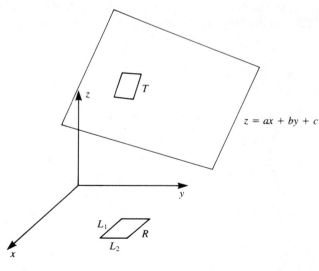

$$A(T) = \sqrt{a^2 + b^2 + 1}\,A(R)$$

Figure 16.46

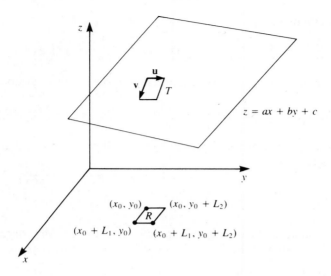

Figure 16.47

Compute

$$\mathbf{u} \times \mathbf{v} = \begin{vmatrix} \mathbf{i} & \mathbf{j} & \mathbf{k} \\ L_1 & 0 & aL_1 \\ 0 & L_2 & bL_2 \end{vmatrix}$$

$$= L_1 L_2(-a\mathbf{j} - b\mathbf{j} + \mathbf{k}) = (-a\mathbf{i} - b\mathbf{j} + \mathbf{k})A(R).$$

Then

$$A(T) = \|\mathbf{u} \times \mathbf{v}\| = \|-a\mathbf{i} - b\mathbf{j} + \mathbf{k}\|A(R)$$
$$= \sqrt{a^2 + b^2 + 1}\,A(R),$$

proving the lemma. ∎

Now relate this lemma to Figure 16.45. We have R_j^*, a rectangle on the tangent plane at P_j^*. This tangent plane has equation

$$-\frac{\partial S}{\partial x}(P_j^*)(x - x_j^*) - \frac{\partial S}{\partial y}(P_j^*)(y - y_j^*) + (z - z_j^*) = 0.$$

Thus, apply the lemma with

$$a = -\frac{\partial S}{\partial x}(P_j^*) \quad \text{and} \quad b = -\frac{\partial S}{\partial y}(P_j^*).$$

We get

$$A(R_j^*) = \sqrt{1 + \frac{\partial S}{\partial x}(x_j^*, y_j^*)^2 + \frac{\partial S}{\partial y}(x_j^*, y_j^*)^2}\,A(R_j).$$

If R_j is small, $A(\Sigma_j)$ is approximately $A(R_j^*)$, and summation (2) of Theorem 16.8

becomes approximately

$$\sum_{j=1}^{N} f(x_j^*, y_j^*, z_j^*)\sqrt{1 + \frac{\partial S}{\partial x}(x_j^*, y_j^*)^2 + \frac{\partial S}{\partial y}(x_j^*, y_j^*)^2}A(R_j).$$

Since $z = S(x, y)$, this sum can be written in terms of the x and y variables as

$$\sum_{j=1}^{N} f(x_j^*, y_j^*, S(x_j^*, y_j^*))\sqrt{1 + \frac{\partial S}{\partial x}(x_j^*, y_j^*)^2 + \frac{\partial S}{\partial y}(x_j^*, y_j^*)^2}A(R_j).$$

Recognize this as a Riemann sum for a double integral; we conclude that, in the limit as $N \to \infty$, we get

$$\iint_{\Sigma} f(x, y, z)\, d\sigma = \iint_{D} f(x, y, S(x, y))\sqrt{1 + \left(\frac{\partial S}{\partial x}\right)^2 + \left(\frac{\partial S}{\partial y}\right)^2}\, dA.$$

This is how we evaluate surface integrals in practice: we convert the surface integral to a double integral over a set in the plane. This result is often written as

$$\iint_{\Sigma} f(x, y, z)\, d\sigma = \iint_{D} f(x, y, z(x, y))\sqrt{1 + \left(\frac{\partial z}{\partial x}\right)^2 + \left(\frac{\partial z}{\partial y}\right)^2}\, dA,$$

in which the surface is given by $z = z(x, y)$.

EXAMPLE 16.20

Evaluate $\iint_{\Sigma} z\, d\sigma$, with Σ that part of the plane $x + y + z = 4$ lying above the rectangle $0 \le x \le 2, 0 \le y \le 1$.

The surface is shown in Figure 16.48. We have $f(x, y, z) = z$ and $S(x, y) = z = 4 - x - y$. On Σ, $\partial z/\partial x = \partial z/\partial y = -1$, and

$$\iint_{\Sigma} z\, d\sigma = \iint_{D} (4 - x - y)\sqrt{1 + (-1)^2 + (-1)^2}\, dA$$

$$= \sqrt{3} \iint_{D} (4 - x - y)\, dA$$

$$= \sqrt{3} \int_{0}^{2} \int_{0}^{1} (4 - x - y)\, dy\, dx.$$

The y-integration is

$$\int_{0}^{1} (4 - x - y)\, dy = 4y - xy - \tfrac{1}{2}y^2 \Big]_{y=0}^{1} = 4 - x - \tfrac{1}{2} = \tfrac{7}{2} - x.$$

Then

$$\iint_{\Sigma} z\, d\sigma = \sqrt{3} \int_{0}^{2} \left(\frac{7}{2} - x\right) dx$$

$$= \sqrt{3} \left[\frac{7}{2}x - \frac{1}{2}x^2\right]_{0}^{2} = \frac{\sqrt{3}}{2}(14 - 4) = 5\sqrt{3}. \quad \blacksquare$$

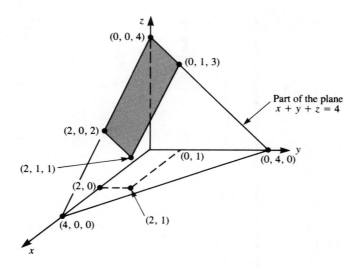

Figure 16.48

EXAMPLE 16.21

Compute $\iint_{\Sigma} (xy/z)\, d\sigma$ over that part of the paraboloid $z = x^2 + y^2$ lying in the first octant for $4 \le x^2 + y^2 \le 9$.

The surface is shown in Figure 16.49. Here, $f(x, y, z) = (1/z)xy$, and the surface is given as part of $z = S(x, y) = x^2 + y^2$. Now, $\partial z/\partial x = 2x$ and $\partial z/\partial y = 2y$, so

$$\iint_{\Sigma} \frac{xy}{z}\, d\sigma = \iint_{D} \frac{xy}{x^2 + y^2} \sqrt{1 + 4x^2 + 4y^2}\, dA.$$

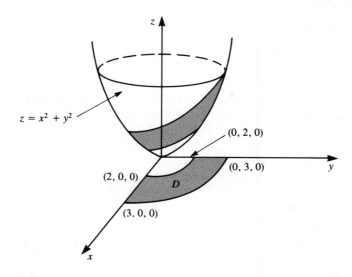

Figure 16.49

Since D is a part of the plane between two circles, it is convenient to transform this double integral to polar coordinates (see Appendix E). In terms of r and θ, D consists of all points (r, θ) with $2 \le r \le 3, 0 \le \theta \le \pi/2$. Therefore,

$$\iint_\Sigma \frac{xy}{z}\, d\sigma = \int_0^{\pi/2} \int_2^3 \frac{r^2\cos(\theta)\sin(\theta)}{r^2} \sqrt{1 + 4r^2}\, r\, dr\, d\theta$$

$$= \int_0^{\pi/2} \cos(\theta)\sin(\theta)\, d\theta \int_2^3 r\sqrt{1 + 4r^2}\, dr.$$

The θ-integration is

$$\int_0^{\pi/2} \cos(\theta)\sin(\theta)\, d\theta = \left[\tfrac{1}{2}\sin^2(\theta) \right]_0^{\pi/2} = \tfrac{1}{2}.$$

For the r-integration, let $u = 1 + 4r^2$. Then $du = 8r\, dr$, so $r\, dr = \tfrac{1}{8}\, du$. Further, when $u = 2, r = 17$; when $u = 3, r = 37$. Then

$$\int_2^3 r\sqrt{1 + 4r^2}\, dr = \int_{17}^{37} u^{1/2} \tfrac{1}{8}\, du$$

$$= \tfrac{1}{12} u^{3/2} \bigg]_{17}^{37} = \tfrac{1}{12}[37^{3/2} - 17^{3/2}].$$

Finally, we have

$$\iint_\Sigma \frac{xy}{z}\, d\sigma = \frac{1}{24}[37^{3/2} - 17^{3/2}]. \quad\blacksquare$$

Here are some applications of surface integrals.

SURFACE AREA

With $f(x, y, z) = 1$, we get

$$\iint_\Sigma 1\, d\sigma = \iint_D \sqrt{1 + \left(\frac{\partial z}{\partial x}\right)^2 + \left(\frac{\partial z}{\partial y}\right)^2}\, dA = \text{area of } \Sigma.$$

This result is analogous to $\iint_D 1\, dA$ being the area of a plane region A and $\iiint_M 1\, dV$ being the volume of a solid region M in three-space. We usually write $\iint_\Sigma 1\, d\sigma$ as just $\iint_\Sigma d\sigma$.

MASS OF A THIN SHELL

Imagine a thin shell of negligible thickness in three-space, taking the shape of a surface Σ. Let $\delta(x, y, z)$ be the density at (x, y, z). We want the mass of the shell.

Imagine that Σ is given by $z = S(x, y)$ for (x, y) in D. Form a grid over D, and look at a typical rectangle R_j, which projects onto a piece of surface Σ_j. Choose (x_j^*, y_j^*) in R_j and approximate the density on Σ_j by the constant $\delta(x_j^*, y_j^*, z_j^*)$, where $z_j^* = S(x_j^*, y_j^*)$. The mass of Σ_j is approximately $\delta(x_j^*, y_j^*, z_j^*)A(\Sigma_j)$, and the mass of the entire shell is approximated by

$$m \approx \sum_{j=1}^{N} \delta(x_j^*, y_j^*, z_j^*) A(\Sigma_j).$$

Taking the limit as $N \to \infty$ and the dimensions of each R_j approach zero, we get

$$m = \iint_{\Sigma} \delta(x, y, z) \, d\sigma.$$

The mass of the shell is the surface integral of the density function over the surface formed by the shell. This should not be surprising in view of the facts that the double integral of a density function $\delta(x, y)$ over a set D in the plane yields the mass of a flat plate occupying D and the triple integral of a density function $\delta(x, y, z)$ over a set M in three-space gives the mass of a solid object occupying M.

CENTER OF MASS OF A SHELL

Following the line of reasoning just employed for mass, it is not difficult to show that the center of mass of a shell Σ has coordinates

$$\bar{x} = \frac{1}{m} \iint_{\Sigma} x \, \delta(x, y, z) \, d\sigma, \qquad \bar{y} = \frac{1}{m} \iint_{\Sigma} y \, \delta(x, y, z) \, d\sigma, \qquad \bar{z} = \frac{1}{m} \iint_{\Sigma} z \, \delta(x, y, z) \, d\sigma.$$

EXAMPLE 16.22

Find the mass and center of mass of the cone $z = \sqrt{x^2 + y^2}$, $0 \le x^2 + y^2 \le 4$ if $\delta(x, y, z) = x^2 + y^2$.

The cone is shown in Figure 16.50. Here, $z = S(x, y) = \sqrt{x^2 + y^2}$. To compute $\partial z/\partial x$ and $\partial z/\partial y$ efficiently, write $z^2 = x^2 + y^2$ and differentiate both sides with respect

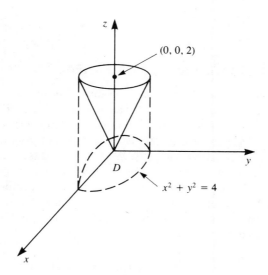

Figure 16.50

to x, recalling that $\partial y/\partial x = 0$ because x and y are independent. We get

$$2z \frac{\partial z}{\partial x} = 2x.$$

Then $\partial z/\partial x = x/z$. Similarly, $\partial z/\partial y = y/z$. Then

$$\delta(x, y, z(x, y)) \sqrt{1 + \left(\frac{\partial z}{\partial x}\right)^2 + \left(\frac{\partial z}{\partial y}\right)^2} = z^2 \sqrt{1 + \frac{x^2 + y^2}{z^2}} = z\sqrt{z^2 + x^2 + y^2}$$

$$= \sqrt{x^2 + y^2} \sqrt{2(x^2 + y^2)} = \sqrt{2}(x^2 + y^2).$$

Then

$$m = \iint_{\Sigma} \delta(x, y, z) \, d\sigma = \iint_{D} \sqrt{2}(x^2 + y^2) \, dA$$

$$= \sqrt{2} \int_0^{2\pi} \int_0^2 r^2 r \, dr \, d\theta = \sqrt{2} \int_0^{2\pi} d\theta \int_0^2 r^3 \, dr = \sqrt{2}(2\pi)\tfrac{1}{4}r^4 \Big]_0^2 = 8\pi\sqrt{2}.$$

By symmetry of the shell and of the density function, we expect the center of mass to lie on the z-axis; hence, $\bar{x} = \bar{y} = 0$. Calculate

$$\bar{z} = \frac{1}{m} \iint_{\Sigma} z \, \delta(x, y, z) \, d\sigma = \frac{1}{m} \iint_{D} z(x^2 + y^2) \sqrt{\frac{z^2 + x^2 + y^2}{z^2}} \, dA$$

$$= \frac{1}{m} \iint_{D} \sqrt{2}(x^2 + y^2)^{3/2} \, dA = \frac{\sqrt{2}}{8\pi\sqrt{2}} \int_0^{2\pi} \int_0^2 r^3 r \, dr \, d\theta$$

$$= \frac{1}{8\pi} (2\pi) \frac{1}{5} r^5 \Big]_0^2 = \frac{1}{20} 2^5 = \frac{8}{5}.$$

The center of mass is $(0, 0, \tfrac{8}{5})$. ∎

FLUX OF A VECTOR FIELD ACROSS A SURFACE

A surface integral can be used to measure the flux of a vector across a surface. To illustrate, suppose a fluid moves in three-space with velocity $\mathbf{F}(x, y, z)$ at (x, y, z). Imagine a surface Σ in the fluid with continuous unit normal $\mathbf{N}(x, y, z)$, as in Figure 16.51. The flux of $\mathbf{F}$ across Σ is defined to be the net volume per unit time of fluid flowing across Σ in the direction of $\mathbf{N}$. We want to measure this quantity.

In a small time interval Δt, the volume of fluid flowing across a small piece of surface Σ_j equals the volume of the cylinder with base Σ_j and altitude $F_{\mathbf{N}} \Delta t$, in which $F_{\mathbf{N}}$ is the component of $\mathbf{F}$ in the direction of $\mathbf{N}$, evaluated at some point of Σ_j. This volume is $F_{\mathbf{N}} \Delta t \cdot A(\Sigma_j)$, shown in Figure 16.52. Because $\|\mathbf{N}\| = 1$, $F_{\mathbf{N}} = \mathbf{F} \cdot \mathbf{N}$. The volume of fluid flowing across Σ_j per unit time is

$$\frac{F_{\mathbf{N}} \Delta t \cdot A(\Sigma_j)}{\Delta t} = F_{\mathbf{N}} A(\Sigma_j) = \mathbf{F} \cdot \mathbf{N} A(\Sigma_j).$$

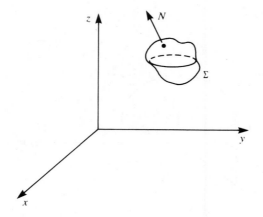

Figure 16.51

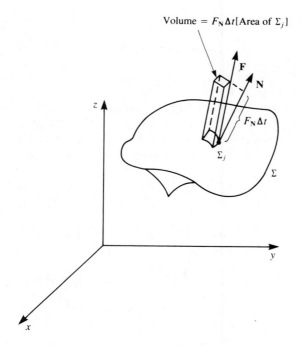

Figure 16.52

Sum these quantities over all the pieces of surface and take a limit as these pieces are chosen smaller. We get

$$\text{flux of } \mathbf{F} \text{ across } \Sigma \text{ in the direction of } \mathbf{N} = \iint_{\Sigma} \mathbf{F} \cdot \mathbf{N} \, d\sigma.$$

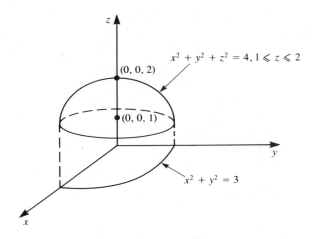

$x^2 + y^2 + z^2 = 4, 1 \leqslant z \leqslant 2$

$(0, 0, 2)$

$(0, 0, 1)$

$x^2 + y^2 = 3$

Figure 16.53

EXAMPLE 16.23

Find the flux of $\mathbf{F} = x\mathbf{i} + y\mathbf{j} + z\mathbf{k}$ across the part of the sphere $x^2 + y^2 + z^2 = 4$ lying between the planes $z = 1$ and $z = 2$.

The surface Σ is shown in Figure 16.53. Since the flux is $\iint_\Sigma \mathbf{F} \cdot \mathbf{N} \, d\sigma$, we must first find $\mathbf{N}$ in order to compute $\mathbf{F} \cdot \mathbf{N}$.

A normal vector to Σ is $-(\partial S/\partial x)\mathbf{i} - (\partial S/\partial y)\mathbf{j} + \mathbf{k}$. For this surface, this vector is $(x/z)\mathbf{i} + (y/z)\mathbf{j} + \mathbf{k}$. This vector does point out of the sphere, a necessity for this application because we are considering flow outward across the surface from the interior to the exterior of Σ. However, this is not a unit vector, so divide the vector by its length

$$\left\| \frac{x}{z}\mathbf{i} + \frac{y}{z}\mathbf{j} + \mathbf{k} \right\| = \sqrt{\frac{x^2}{z^2} + \frac{y^2}{z^2} + 1} = \sqrt{\frac{x^2 + y^2 + z^2}{z^2}} = \frac{2}{z}$$

(recall that $x^2 + y^2 + z^2 = 4$ on S). The unit normal vector out of the surface, pointing away from the enclosed volume, is

$$\mathbf{N} = \frac{z}{2}\left[\frac{x}{z}\mathbf{i} + \frac{y}{z}\mathbf{j} + \mathbf{k} \right] = \tfrac{1}{2}(x\mathbf{i} + y\mathbf{j} + z\mathbf{k}).$$

Now compute

$$\mathbf{F} \cdot \mathbf{N} = \tfrac{1}{2}(x^2 + y^2 + z^2).$$

The flux of $\mathbf{F}$ across Σ outward from the origin is

$$\iint_\Sigma \mathbf{F} \cdot \mathbf{N} \, d\sigma = \iint_D \tfrac{1}{2}(x^2 + y^2 + z^2)\sqrt{1 + \frac{x^2}{z^2} + \frac{y^2}{z^2}} \, dA$$

$$= \iint_D \tfrac{1}{2}[4]\sqrt{\frac{4}{4 - x^2 - y^2}} \, dA = 4\iint \frac{1}{\sqrt{4 - x^2 - y^2}} \, dA.$$

Now, Σ is the portion of the sphere of radius 2 about the origin lying between the planes $z = 1$ and $z = 2$. When $z = 1$, $x^2 + y^2 + 1^2 = 4$; hence, $x^2 + y^2 = 3$. The plane

$z = 2$ intersects the sphere in the circle $x^2 + y^2 = 3, z = 1$. When $z = 2, x^2 + y^2 + 2^2 = 4$, so $x = y = 0$, and the plane $z = 2$ intersects the sphere at the single point $(0, 0, 2)$, the "north pole" of the sphere. Thus, the portion of the sphere that constitutes the surface projects onto the xy-plane to the set D given by $0 \leq x^2 + y^2 \leq 3$, the interior of the circle of radius $\sqrt{3}$ about the origin. This set is also shown in Figure 16.53.

Convert the double integral over D to polar coordinates to get

$$\iint_{\Sigma} \mathbf{F} \cdot \mathbf{N} \, d\sigma = 4 \int_0^{2\pi} \int_0^{\sqrt{3}} \frac{r}{\sqrt{4 - r^2}} \, dr \, d\theta$$

$$= 8\pi \int_0^{\sqrt{3}} \frac{r}{\sqrt{4 - r^2}} \, dr = 8\pi \left[-(4 - r^2)^{1/2} \right]_0^{\sqrt{3}}$$

$$= 8\pi[-1 + 2] = 8\pi. \quad \blacksquare$$

This flux interpretation of $\iint_{\Sigma} \mathbf{F} \cdot \mathbf{N} \, d\sigma$ will be useful later in understanding Gauss's divergence theorem, which is one generalization of Green's theorem to three-space.

PROBLEMS FOR SECTION 16.4

In each of Problems 1 through 10, evaluate the surface integral.

1. $\iint_{\Sigma} x \, d\sigma$, with Σ the part of the plane $x + 4y + z = 10$ in the first octant

2. $\iint_{\Sigma} y^2 \, d\sigma$, with Σ the part of the plane $z = x$ for $0 \leq x \leq 2, 0 \leq y \leq 4$

3. $\iint_{\Sigma} d\sigma$, with Σ the part of the paraboloid $z = x^2 + y^2$ lying between the planes $z = 2$ and $z = 7$

4. $\iint_{\Sigma} (x + y) \, d\sigma$, with Σ the part of the plane $4x + 8y + 10z = 25$ lying above the triangle in the xy-plane having vertices $(0, 0), (1, 1), (1, 0)$

5. $\iint_{\Sigma} z \, d\sigma$, with Σ the part of the cone $z = \sqrt{x^2 + y^2}$ lying in the first octant and between the planes $z = 2$ and $z = 4$

6. $\iint_{\Sigma} xyz \, d\sigma$, with Σ the part of the plane $z = x + y$ for (x, y) in the square with vertices $(0, 0), (1, 0), (0, 1), (1, 1)$

7. $\iint_{\Sigma} y \, d\sigma$, with Σ the part of the cylinder $z = x^2$ for $0 \leq x \leq 2$ and $0 \leq y \leq 3$

8. $\iint_{\Sigma} x^2 \, d\sigma$, with Σ the part of the paraboloid $z = 4 - x^2 - y^2$ lying above the xy-plane

9. $\iint_{\Sigma} z \, d\sigma$, with Σ the part of the plane $z = x - y$ for $0 \leq x \leq 1, 0 \leq y \leq 5$

10. $\iint_{\Sigma} xyz \, d\sigma$, with Σ the part of the cylinder $z = 1 + y^2$ for $0 \leq y \leq 1$ and $0 \leq x \leq 1$

11. Find the mass of the triangular shell having vertices $(1, 0, 0), (0, 3, 0), (0, 0, 2)$ if $\delta(x, y, z) = xz + 1$.

12. Find the center of mass of the portion of the homogeneous sphere $x^2 + y^2 + z^2 = 9$ lying above the plane $z = 1$.

13. Find the center of mass of the homogeneous cone $z = \sqrt{x^2 + y^2}$ for $0 \leq x^2 + y^2 \leq 9$.

14. Find the mass and center of mass of the part of the paraboloid $z = 16 - x^2 - y^2$ lying in the first octant and between the cylinders $x^2 + y^2 = 1$ and $x^2 + y^2 = 9$ if $\delta(x, y, z) = xy/\sqrt{1 + 4x^2 + 4y^2}$.

15. Find the mass and center of mass of the paraboloid $z = 6 - x^2 - y^2$ lying above the xy-plane if $\delta(x, y, z) = \sqrt{1 + 4x^2 + 4y^2}$.

16. Find the center of mass of the part of the homogeneous sphere $x^2 + y^2 + z^2 = 1$ lying in the first octant.

17. Find the flux of $\mathbf{F} = x\mathbf{i} + y\mathbf{j} - z\mathbf{k}$ across the part of the plane $x + 2y + z = 8$ lying in the first octant.

18. Find the flux of $\mathbf{F} = xz\mathbf{i} - y\mathbf{k}$ across the part of the sphere $x^2 + y^2 + z^2 = 4$ lying above the plane $z = 1$.

If a surface is given as $x = S(y, z)$ for (y, z) in a region D of the yz-plane, then

$$\iint\limits_{\Sigma} f(x, y, z) \, d\sigma = \iint\limits_{D} f(S(y, z), y, z) \sqrt{1 + \left(\frac{\partial x}{\partial y}\right)^2 + \left(\frac{\partial x}{\partial z}\right)^2} \, dA.$$

If the surface is given as $y = S(x, z)$ for (x, z) in a region D of the xz-plane, then

$$\iint\limits_{\Sigma} f(x, y, z) \, d\sigma = \iint\limits_{D} f(x, S(x, z), z) \sqrt{1 + \left(\frac{\partial y}{\partial x}\right)^2 + \left(\frac{\partial y}{\partial z}\right)^2} \, dA.$$

Use these formulations to evaluate the following surface integrals.

19. $\iint_{\Sigma} z \, d\sigma$, Σ the cone $x = \sqrt{y^2 + z^2}$ for $0 \le y^2 + z^2 \le 9$

20. $\iint_{\Sigma} x \, d\sigma$,   with Σ the parabolic bowl $y = x^2 + z^2$ for $0 \le x^2 + z^2 \le 4$

21. $\iint_{\Sigma} y \, d\sigma$,   with Σ the square $0 \le x \le 3, 0 \le z \le 4$ in the plane $y = 5$

22. $\iint_{\Sigma} x \, d\sigma$, Σ bounded by $x = y^3$ for $0 \le y \le 2$ and $0 \le z \le 4$

23. $\iint_{\Sigma} z^2 \, d\sigma$,   with Σ the surface $x = y + z, 0 \le y \le 1, 0 \le z \le 4$

24. $\iint_{\Sigma} xyz \, d\sigma$,   with Σ the part of the plane $x = 4z + y$ determined by $0 \le y \le 2, 0 \le z \le 6$

25. $\iint_{\Sigma} xz \, d\sigma$,   with Σ the part of the surface $y = x^2 + z^2$ in the first octant, with $1 \le x^2 + z^2 \le 4$

16.5 Preparation for the Theorems of Gauss and Stokes

In this section, we will prepare for the two main theorems of vector integral calculus, the divergence theorem of Gauss and Stokes's theorem. This preparation will consist of rewriting the conclusion of Green's theorem in ways which suggest generalizations to three-space.

The conclusion of Green's theorem can be written

$$\oint_{C} f(x, y) \, dx + g(x, y) \, dy = \iint\limits_{D} \left[\frac{\partial g}{\partial x} - \frac{\partial f}{\partial y}\right] dA, \qquad (16.6)$$

with appropriate conditions on C, f, and g. We can write this conclusion in vector form as follows. Let

$$\mathbf{F}(x, y) = g(x, y)\mathbf{i} - f(x, y)\mathbf{j}$$

and observe that

$$\nabla \cdot \mathbf{F} = \frac{\partial g}{\partial x} - \frac{\partial f}{\partial y},$$

which we recognize as the quantity in the double integral in Green's theorem.

Now consider the line integral. Parametrize C by arc length $x = x(s)$ and $y = y(s)$. The unit tangent vector to C is $\mathbf{T}(s) = x'(s)\mathbf{i} + y'(s)\mathbf{j}$, and the unit normal is $\mathbf{N}(s) = y'(s)\mathbf{i} - x'(s)\mathbf{j}$, pointing away from D when drawn as an arrow from (x, y) on C, as in

Figure 16.54. Now,

$$\mathbf{F} \cdot \mathbf{N} = g(x, y)\frac{dy}{ds} + f(x, y)\frac{dx}{ds}.$$

Then

$$\oint_C f(x, y)\, dx + g(x, y)\, dy = \oint_C \left[f(x, y)\frac{dx}{ds} + g(x, y)\frac{dy}{ds} \right] ds = \oint_C \mathbf{F} \cdot \mathbf{N}\, ds.$$

Thus, the conclusion of Green's theorem may be written in vector form as

$$\oint_C \mathbf{F} \cdot \mathbf{N}\, ds = \iint_D \mathbf{\nabla} \cdot \mathbf{F}\, dA.$$

This suggests a generalization to three-space. Replace C with a surface Σ, enclosing a set M in three-space, and replace the line integral with a surface integral and the double integral with a triple integral. If $\mathbf{F}$ is a vector field in three-space, we conjecture that

$$\iint_\Sigma \mathbf{F} \cdot \mathbf{N}\, d\sigma = \iiint_M \mathbf{\nabla} \cdot \mathbf{F}\, dV,$$

with $\mathbf{N}$ the unit outer normal to Σ, as in Figure 16.55. Under appropriate conditions on $\mathbf{F}$ and Σ, this is the conclusion of Gauss's divergence theorem, which will be stated in the next section.

We can arrive at another vector integral theorem by writing equation (16.6) in a different way. Let $\mathbf{F}(x, y, z) = f(x, y)\mathbf{i} + g(x, y)\mathbf{j} + 0\mathbf{k}$. We have added the third component so we can think of $\mathbf{F}$ as a vector field in three-space and compute the curl of $\mathbf{F}$:

$$\mathbf{\nabla} \times \mathbf{F} = \begin{vmatrix} \mathbf{i} & \mathbf{j} & \mathbf{k} \\ \dfrac{\partial}{\partial x} & \dfrac{\partial}{\partial y} & \dfrac{\partial}{\partial z} \\ f & g & 0 \end{vmatrix} = \left[\frac{\partial g}{\partial x} - \frac{\partial f}{\partial y} \right]\mathbf{k}.$$

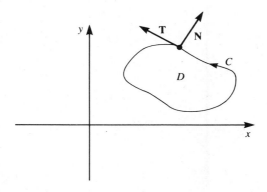

Figure 16.54

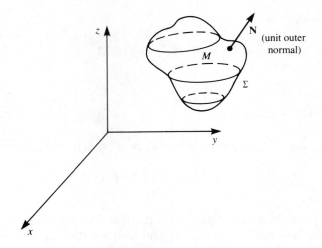

Figure 16.55

Then

$$(\mathbf{\nabla} \times \mathbf{F}) \cdot \mathbf{k} = \frac{\partial g}{\partial x} - \frac{\partial f}{\partial y}.$$

Further, with unit tangent $\mathbf{T} = x'(s)\mathbf{i} + y'(s)\mathbf{j}$ to C, we have

$$\mathbf{F} \cdot \mathbf{T} \, ds = [f\mathbf{i} + g\mathbf{j}] \cdot \left[\frac{dx}{ds}\mathbf{i} + \frac{dy}{ds}\mathbf{j}\right] ds = f(x, y) \, dx + g(x, y) \, dy.$$

Now equation (16.6) can be written

$$\oint_C \mathbf{F} \cdot \mathbf{T} \, ds = \iint_D (\mathbf{\nabla} \times \mathbf{F}) \cdot \mathbf{k} \, dA. \tag{16.7}$$

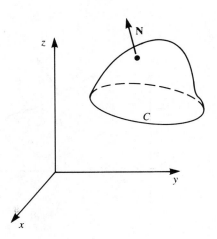

Figure 16.56

Now, we can think of D as a surface (in the xy-plane) with unit normal vector $\mathbf{k}$ and bounded by a closed curve C. This suggests a second generalization of equation (16.6) to three-space. Think of C as a closed curve in three-space, bounding a surface Σ having unit normal $\mathbf{N}$, as in Figure 16.56. Equation (16.7), which holds for a vector field and closed curve in the plane, suggests that in three-space we might have

$$\oint_C \mathbf{F} \cdot \mathbf{T} \, ds = \iint_\Sigma (\nabla \times \mathbf{F}) \cdot \mathbf{N} \, d\sigma.$$

We will see in Section 16.8 that this is the conclusion of Stokes's theorem under conditions we will state on $\mathbf{F}$, C and Σ.

PROBLEMS FOR SECTION 16.5

1. Let $\varphi(x, y)$ and $\psi(x, y)$ be scalar fields in the plane. Let C be a simple closed piecewise-smooth curve in the plane enclosing a region D over which φ, ψ, and their first and second partial derivatives are continuous. Show that

$$\iint_D \varphi \nabla^2 \psi \, dA = \oint_C - \varphi \frac{\partial \psi}{\partial y} \, dx + \varphi \frac{\partial \psi}{\partial x} \, dy - \iint_D \nabla \varphi \cdot \nabla \psi \, dA.$$

Hint: Apply Green's theorem to the line integral on the right.

2. Use the result of Problem 1 to show that, under the conditions stated there,

$$\iint_D (\varphi \nabla^2 \psi - \psi \nabla^2 \varphi) \, dA = \oint_C \left[\psi \frac{\partial \varphi}{\partial y} - \varphi \frac{\partial \psi}{\partial y} \right] dx + \left[\varphi \frac{\partial \psi}{\partial x} - \psi \frac{\partial \varphi}{\partial x} \right] dy.$$

Hint: Interchange φ and ψ in the result of Problem 1, then subtract.

3. Let $\varphi(x, y)$ be a continuous scalar field in the plane, with continuous first and second partial derivatives on a simple closed piecewise-smooth curve C and in the set D enclosed by C. Let $\mathbf{N}(x, y)$ be the unit normal to C. Show that

$$\oint_C D_\mathbf{N} \varphi(x, y) \, ds = \iint_D \nabla^2 \varphi \, dA.$$

$[D_\mathbf{N} \varphi(x, y)$ is the directional derivative of φ in the direction of $\mathbf{N}$.]

16.6 *The Divergence Theorem of Gauss*

We will now develop the first of two vector integral theorems which are generalizations of Green's theorem to three-space. Recall from Section 16.5 that one vector form of Green's theorem states that

$$\oint_C \mathbf{F} \cdot \mathbf{N} \, ds = \iint_D \nabla \cdot \mathbf{F} \, dA$$

under appropriate conditions on the two-dimensional vector field $\mathbf{F}$ and the closed curve C in the plane. We will generalize this formula as follows. List the important components of Green's theorem and look at their three-dimensional analogues.

In the Plane	*In Three-Space*
A set D	A three-dimensional solid M
A curve C bounding D	A surface Σ enclosing M
A unit outer normal $\mathbf{N}$ to C	A unit outer normal $\mathbf{N}$ to Σ
A vector field $\mathbf{F}$ in the plane	A vector field $\mathbf{F}$ in three-space
A line integral $\oint_C \mathbf{F} \cdot \mathbf{N} \, ds$	A surface integral $\iint_\Sigma \mathbf{F} \cdot \mathbf{N} \, d\sigma$
A double integral $\iint_D \nabla \cdot \mathbf{F} \, dA$	A triple integral $\iiint_M \nabla \cdot \mathbf{F} \, dV.$

With some preparation in terminology, these natural matchings suggest Gauss's theorem, named after the great nineteenth-century German mathematician and scientist Carl Friedrich Gauss.

We call a surface Σ a *closed surface* if it encloses a three-dimensional volume. For example, a sphere is closed, as is a cube; a hemisphere is not closed. A hemispherical bowl will spill water if it is tipped.

A normal vector $\mathbf{N}$ is called an *outer normal* to Σ if, when drawn as an arrow from a point on Σ, it points outward from the region enclosed by the surface (see Figure 16.57). If $\mathbf{N}$ is also a unit vector, $\mathbf{N}$ is a *unit outer normal*.

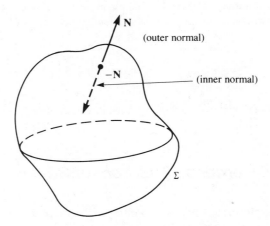

Figure 16.57

THEOREM 16.9   Gauss's Divergence Theorem

Let Σ be a piecewise-smooth closed surface enclosing a set M in three-space. Let Σ have unit outer normal $\mathbf{N}$. Let $\mathbf{F}(x, y, z)$ be a vector field whose component functions are continuous with continuous first and second partial derivatives on Σ and throughout M. Then

$$\iint_{\Sigma} \mathbf{F} \cdot \mathbf{N} \, d\sigma = \iiint_{M} \nabla \cdot \mathbf{F} \, dV. \quad \blacksquare$$

This conclusion may also be written

$$\iint_{\Sigma} \mathbf{F} \cdot \mathbf{N} \, d\sigma = \iiint_{M} \operatorname{div} \mathbf{F} \, dV.$$

We will sketch a proof of a special case in Problem 17 at the end of this section.

EXAMPLE 16.24

Let Σ be the closed surface consisting of the surface Σ_1 of the cone $z^2 = x^2 + y^2$ for $0 \leq x^2 + y^2 \leq 1$ and the flat cap Σ_2 consisting of the disk $x^2 + y^2 \leq 1, z = 1$ as shown in Figure 16.58. Let $\mathbf{F}(x, y, z) = x\mathbf{i} + y\mathbf{j} + z\mathbf{k}$. We will illustrate Gauss's theorem by computing both sides of the equation separately.

First, compute $\iint_{\Sigma} \mathbf{F} \cdot \mathbf{N} \, d\sigma$ as $\iint_{\Sigma_1} \mathbf{F} \cdot \mathbf{N} \, d\sigma + \iint_{\Sigma_2} \mathbf{F} \cdot \mathbf{N} \, d\sigma$. For $\iint_{\Sigma_1} \mathbf{F} \cdot \mathbf{N} \, d\sigma$, we first must find the unit outer normal vector $\mathbf{N}$. Let $z = \sqrt{x^2 + y^2}$. We found in Example 16.18 that the normal vector $-(\partial z/\partial x)\mathbf{i} - (\partial z/\partial y)\mathbf{j} + \mathbf{k}$ is $-(x/z)\mathbf{i} - (y/z)\mathbf{j} + \mathbf{k}$. However, this normal points into the cone (Figure 16.59) because of the positive $\mathbf{k}$-component. Thus, take the negative of this, or $(x/z)\mathbf{i} + (y/z)\mathbf{j} - \mathbf{k}$, which we write as $(1/z)(x\mathbf{i} + y\mathbf{j} - z\mathbf{k})$. This is an outer normal but does not in general have length 1, so divide by its length to obtain the unit outer normal

$$\mathbf{N} = \frac{1}{\sqrt{z^2 + x^2 + y^2}} (x\mathbf{i} + y\mathbf{j} - z\mathbf{k}).$$

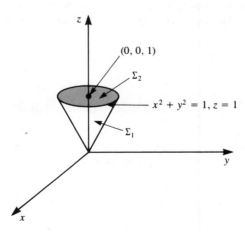

Figure 16.58

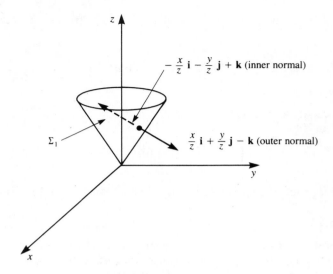

Figure 16.59

On Σ_1,

$$\mathbf{F} \cdot \mathbf{N} = \frac{1}{\sqrt{z^2 + x^2 + y^2}}(x^2 + y^2 - z^2) = 0$$

because on Σ_1, $z^2 = x^2 + y^2$. Therefore,

$$\iint\limits_{\Sigma_1} \mathbf{F} \cdot \mathbf{N}\, d\sigma = 0.$$

Next, Σ_2 is the disk forming the cap of the cone. We can write Σ_2 as $z = 1$ for $0 \le x^2 + y^2 \le 1$. Here, the unit outer normal is obviously $\mathbf{N} = \mathbf{k}$, and $\mathbf{F} \cdot \mathbf{N} = z$. Since $z = 1$ on Σ_2,

$$\iint\limits_{\Sigma_2} \mathbf{F} \cdot \mathbf{N}\, d\sigma = \iint\limits_{\Sigma_2} 1\, d\sigma = \text{area of } \Sigma_2$$

$$= \text{area of the circle of radius 1 about the origin} = \pi.$$

Therefore,

$$\iint\limits_{\Sigma} \mathbf{F} \cdot \mathbf{N}\, d\sigma = \pi.$$

Now compute $\iiint_M \nabla \cdot \mathbf{F}\, dV$. First,

$$\nabla \cdot \mathbf{F} = \frac{\partial}{\partial x}[x] + \frac{\partial}{\partial y}[y] + \frac{\partial}{\partial z}[z] = 3.$$

Then

$$\iiint_M \nabla \cdot \mathbf{F} \, dV = \iiint_M 3 \, dV$$

$$= 3[\text{volume of the cone of height 1 and radius 1}] = 3\tfrac{1}{3}\pi(1)^2(1) = \pi.$$

In this example, $\iint_\Sigma \mathbf{F} \cdot \mathbf{N} \, d\sigma = \iiint_M \nabla \cdot \mathbf{F} \, dV = \pi.$ ∎

EXAMPLE 16.25

Let Σ be the surface of the cube with vertices $(0, 0, 0), (1, 0, 0), (0, 1, 0), (0, 0, 1), (1, 1, 0),$ $(1, 0, 1), (0, 1, 1), (1, 1, 1).$ Let $\mathbf{F}(x, y, z) = x^2\mathbf{i} + y^2\mathbf{j} + z^2\mathbf{k}.$ We will illustrate Gauss's theorem by computing the sum of the surface integrals of $\mathbf{F} \cdot \mathbf{N}$ over each of the six faces of the cube (using the appropriate unit outer normal for each face), then comparing this result with the integral of $\nabla \cdot \mathbf{F}$ over the region enclosed by the cube.

Figure 16.60 shows the surface with the faces labeled. On the top face Σ_1, we have $0 \le x \le 1, 0 \le y \le 1,$ and $z = 1,$ and $\mathbf{N} = \mathbf{k}.$ Here, $\mathbf{F} \cdot \mathbf{N} = z^2 = 1^2 = 1.$ Further, $\partial z/\partial x = \partial z/\partial y = 0$ on Σ_1, so

$$\iint_{\Sigma_1} \mathbf{F} \cdot \mathbf{N} \, d\sigma = \iint_{D_1} dA = \text{area of } D_1 = 1,$$

since D_1 is the square $0 \le x \le 1, 0 \le y \le 1.$

On the front face Σ_2, we have $x = 1$ and $0 \le y \le 1, 0 \le z \le 1.$ Here, $\mathbf{N} = \mathbf{i}$, so $\mathbf{F} \cdot \mathbf{N} = x^2 = 1.$ Further, on Σ_2, $\partial x/\partial y = \partial x/\partial z = 0$, so

$$\iint_{\Sigma_2} \mathbf{F} \cdot \mathbf{N} \, d\sigma = \iint_{D_2} dA = \text{area of } D_2 = 1.$$

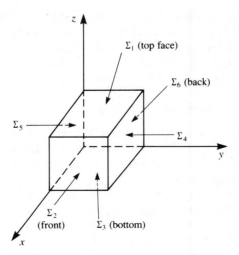

Figure 16.60

In similar fashion (details omitted), we obtain

$$\iint\limits_{\Sigma_3} \mathbf{F} \cdot \mathbf{N} \, d\sigma = \iint\limits_{\Sigma_5} \mathbf{F} \cdot \mathbf{N} \, d\sigma = \iint\limits_{\Sigma_6} \mathbf{F} \cdot \mathbf{N} \, d\sigma = 0 \quad \text{and} \quad \iint\limits_{\Sigma_4} \mathbf{F} \cdot \mathbf{N} \, d\sigma = 1.$$

Therefore,

$$\iint\limits_{\Sigma} \mathbf{F} \cdot \mathbf{N} \, d\sigma = 3.$$

Now compute $\mathbf{V} \cdot \mathbf{F} = 2x + 2y + 2z$. Since M is the region $0 \le x \le 1, 0 \le y \le 1,$ $0 \le z \le 1$,

$$\iiint\limits_{M} \mathbf{V} \cdot \mathbf{F} \, dV = \int_0^1 \int_0^1 \int_0^1 (2x + 2y + 2z) \, dz \, dy \, dx.$$

The z-integration is

$$\int_0^1 (2x + 2y + 2z) \, dz = (2x + 2y)z + z^2 \Big]_{z=0}^1 = 2x + 2y + 1.$$

The y-integration is

$$\int_0^1 (2x + 2y + 1) \, dy = (2x + 1)y + y^2 \Big]_{y=0}^1 = 2x + 2.$$

Finally,

$$\iiint\limits_{M} \mathbf{V} \cdot \mathbf{F} \, dV = \int_0^1 (2x + 2) \, dx = x^2 + 2x \Big]_0^1 = 3,$$

in agreement with $\iint_\Sigma \mathbf{F} \cdot \mathbf{N} \, d\sigma$. ■

These examples are given simply to illustrate the equation in Gauss's theorem. Usually, we would not compute both sides of this equation but would use one side to evaluate the other. In Example 16.25, the triple integral is easier to evaluate than the surface integral. The theorem also provides a tool for approaching significant real-world problems. Following are some examples of this.

EXAMPLE 16.26 Archimedes' Principle

Archimedes' principle states that the buoyant force on a solid immersed in a fluid of constant density equals the weight of the fluid displaced. For example, a metal ship will float if it displaces a volume of water weighing more than the ship. We will derive this principle using Gauss's theorem.

Consider a piecewise-smooth surface Σ in three-space with interior M. Let ρ be the constant density of the fluid. Draw a coordinate system as in Figure 16.61, with M lying below the surface of the fluid. We will assume that the pressure p on Σ equals the density of the fluid times the depth of Σ. The pressure at (x, y, z) on Σ is then $p = -\rho z$; the minus is introduced because z is negative in the downward direction and we want the pressure itself to be positive.

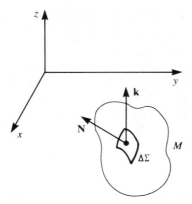

Pressure force on $\Delta \approx \varrho z \mathbf{N} \cdot A(\Delta \Sigma)$
Vertical component $= \varrho z \mathbf{N} \cdot \mathbf{k} A(\Delta \Sigma)$

Figure 16.61

Consider a small piece $\Delta \Sigma$ of surface, also shown in Figure 16.61. The force caused by the pressure on $\Delta \Sigma$ has magnitude approximately $-\rho z$ times the area $A(\Delta \Sigma)$ of $\Delta \Sigma$. If $\mathbf{N}$ is the unit outer normal to Σ, the force caused by the pressure on $\Delta \Sigma$ is approximately $\rho z \mathbf{N}[A(\Delta \Sigma)]$. The vertical component of this force is the magnitude of the buoyant force acting upward on $\Delta \Sigma$. This vertical component is $\rho z \mathbf{N} \cdot \mathbf{k} A(\Delta \Sigma)$. Sum these vertical components over the entire surface to obtain (approximately) the net buoyant force on the object, then take the limit as the surface elements $\Delta \Sigma$ are chosen smaller. We obtain in the limit that

$$\text{the net buoyant force on } \Sigma = \iint_{\Sigma} \rho z \mathbf{N} \cdot \mathbf{k} \, d\sigma = \iint_{\Sigma} \rho z \mathbf{k} \cdot \mathbf{N} \, d\sigma.$$

Now apply Gauss's theorem. With $\mathbf{F} = \rho z \mathbf{k}$, we obtain

$$\text{the net buoyant force} = \iint_{\Sigma} \rho z \mathbf{k} \cdot \mathbf{N} \, d\sigma$$

$$= \iiint_{M} \nabla \cdot (\rho z \mathbf{k}) \, dV = \iiint_{M} \rho \nabla \cdot (z \mathbf{k}) \, dV$$

$$= \iiint_{M} \rho \, dV = \rho (\text{volume of } M).$$

Since ρ is constant and equals the weight per unit volume, this equation states that the net buoyant force equals the weight of fluid displaced, which is what we wanted to show. ∎

Here is an application of Gauss's theorem to a physical interpretation of the divergence of a vector.

EXAMPLE 16.27   An Interpretation of Divergence

Suppose that $\mathbf{F}(x, y, z)$ is the velocity of a fluid moving in three-space. Let Σ_r be an imaginary sphere in the fluid bounding M_r, as in Figure 16.62. Σ_r is a purely mathematical construction whose purpose is to help study the fluid's motion. We imagine that the fluid can flow through Σ_r into and out of M_r.

We have already seen that $\iint_{\Sigma_r} \mathbf{F} \cdot \mathbf{N} \, d\sigma$ is the flux of $\mathbf{F}$ out of M_r across Σ_r. Let $P_0 : (x_0, y_0, z_0)$ be the center of Σ_r. If r is small, $\boldsymbol{\nabla} \cdot \mathbf{F}(x, y, z) \approx \boldsymbol{\nabla} \cdot \mathbf{F}(P_0)$ for all (x, y, z) in M_r. Then

$$\iiint_{M_r} \boldsymbol{\nabla} \cdot \mathbf{F} \, dV \approx \iiint_{M_r} \boldsymbol{\nabla} \cdot \mathbf{F}(P_0) \, dV$$

$$= [\boldsymbol{\nabla} \cdot \mathbf{F}(P_0)][\text{volume of } M_r] = \tfrac{4}{3}\pi r^3 \boldsymbol{\nabla} \cdot \mathbf{F}(P_0).$$

Then

$$\boldsymbol{\nabla} \cdot \mathbf{F}(P_0) \approx \frac{3}{4\pi r^3} \iiint_{M_r} \boldsymbol{\nabla} \cdot \mathbf{F} \, dV = \frac{3}{4\pi r^3} \iint_{\Sigma_r} \mathbf{F} \cdot \mathbf{N} \, d\sigma.$$

Let $r \to 0$. Then Σ_r collapses to its center P_0, and this approximation becomes more accurate, approaching in the limit an equality:

$$\boldsymbol{\nabla} \cdot \mathbf{F}(P_0) = \lim_{r \to 0} \frac{3}{4\pi r^3} \iint_{\Sigma} \mathbf{F} \cdot \mathbf{N} \, d\sigma.$$

On the right is the limit, as $r \to 0$, of the flux divided by the volume of the sphere. This is the limit, as $r \to 0$, of the flux per unit volume of fluid flowing out of M_r across its boundary Σ_r. Since the sphere contracts to P_0 as $r \to 0$, we interpret the right side, hence also $\boldsymbol{\nabla} \cdot \mathbf{F}(P_0)$, as a measure of fluid divergence from, or flow away from, P_0. This provides a physical interpretation of the divergence of a vector.

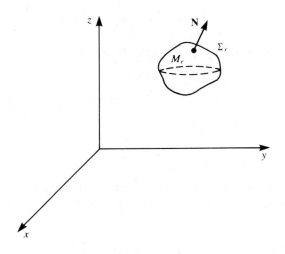

Figure 16.62

Now consider again the equation

$$\iint_{\Sigma} \mathbf{F} \cdot \mathbf{N} \, d\sigma = \iiint_{M} \mathbf{V} \cdot \mathbf{F} \, dV.$$

This equation states that the flux of $\mathbf{F}$ out of M across its bounding surface Σ exactly balances the divergence of fluid away from points of M. This is a conservation of mass statement, in the absence of fluid produced or destroyed within M, and gives us a physical model for the divergence theorem. ■

Gauss's theorem has wide application in mathematics, physics, and engineering and is used to derive equations governing the flow of fluids, heat conduction, and wave propagation. We will discuss some of these applications in the next section.

PROBLEMS FOR SECTION 16.6

In each of Problems 1 through 10, use Gauss's theorem to evaluate $\iint_{\Sigma} \mathbf{F} \cdot \mathbf{N} \, d\sigma$.

1. $\mathbf{F} = x\mathbf{i} + y\mathbf{j} - z\mathbf{k}$, Σ the sphere of radius 4 about $(1, 1, 1)$
2. $\mathbf{F} = 4x\mathbf{i} - 6y\mathbf{j} + \mathbf{k}$, Σ the surface of the solid cylinder $x^2 + y^2 \leq 4$, $0 \leq z \leq 2$
3. $\mathbf{F} = 2yz\mathbf{i} - 4xz\mathbf{j} + xy\mathbf{k}$, Σ the sphere of radius 4 about $(-1, 3, 1)$
4. $\mathbf{F} = x^3\mathbf{i} + y^3\mathbf{j} + z^3\mathbf{k}$, Σ the sphere of radius 1 about the origin
5. $\mathbf{F} = 4x\mathbf{i} - z\mathbf{j} + x\mathbf{k}$, Σ the surface bounding the hemisphere $x^2 + y^2 + z^2 \leq 1, z \geq 0$; together with the bottom cap $x^2 + y^2 \leq 1$ in the xy-plane
6. $\mathbf{F} = (x - y)\mathbf{i} + (y - 4xz)\mathbf{j} + xz\mathbf{k}$, Σ the rectangular box bounded by the coordinate planes $x = 0, y = 0, z = 0$ and the planes $x = 4, y = 2, z = 3$
7. $\mathbf{F} = x^2\mathbf{i} + y^2\mathbf{j} + z^2\mathbf{k}$, Σ the surface bounding the cone $z = \sqrt{x^2 + y^2}$ for $0 \leq z \leq 2$
8. $\mathbf{F} = x^2\mathbf{i} - e^z\mathbf{j} + z\mathbf{k}$, Σ the surface bounding the circular cylinder $x^2 + y^2 \leq 4, 0 \leq z \leq 2$, including top and bottom caps
9. $\mathbf{F} = 3xy\mathbf{i} + z^2\mathbf{k}$, Σ the sphere of radius 1 about the origin
10. $\mathbf{F} = x^2\mathbf{i} + y^2\mathbf{j} + z^2\mathbf{k}$, Σ the rectangular box bounded by the coordinate planes and the planes $x = 6, y = 2,$ $z = 7$
11. Suppose that Σ is a smooth closed surface and that $\mathbf{F}$ is a vector field with continuous components having continuous first and second partial derivatives on Σ and throughout the interior M of Σ. Evaluate $\iint_{\Sigma} (\mathbf{V} \times \mathbf{F}) \cdot \mathbf{N} \, d\sigma$.
12. Suppose that $\varphi(x, y, z)$ and $\psi(x, y, z)$ are continuous, with continuous first and second partial derivatives in the interior K of a smooth closed surface Σ. Show that

$$\iint_{\Sigma} (\varphi \mathbf{V} \psi) \cdot \mathbf{N} \, d\sigma = \iiint_{M} (\varphi \nabla^2 \psi + \mathbf{V}\varphi \cdot \mathbf{V}\psi) \, dV.$$

  Hint: Let $\mathbf{F} = \varphi \mathbf{V}\psi$ in Gauss's theorem.
13. Under the conditions of Problem 12, show that

$$\iint_{\Sigma} (\varphi \mathbf{V}\psi - \psi \mathbf{V}\varphi) \cdot \mathbf{N} \, d\sigma = \iiint_{M} (\varphi \nabla^2 \psi - \psi \nabla^2 \varphi) \, dV.$$

14. Suppose that Σ is a smooth closed surface with interior M in three-space. Show that

$$\text{volume of } M = \frac{1}{3} \iint_{\Sigma} \mathbf{R} \cdot \mathbf{N} \, d\sigma,$$

where $\mathbf{R}$ is the position vector of the surface.

15. Show that, for any constant vector $\mathbf{K}$ and any smooth closed surface Σ,

$$\iint_{\Sigma} \mathbf{K} \cdot \mathbf{N} \, d\sigma = 0.$$

16. Find the flux of the vector field $\mathbf{F} = xy^2\mathbf{i} + yz^2\mathbf{j} + zx^2\mathbf{k}$ across the surface Σ bounding the cylinder $2 \leq x^2 + y^2 \leq 4, 0 \leq z \leq 7$.

17. Fill in the details of this proof of a special case of Gauss's theorem. Assume that Σ is made up of an upper surface Σ_2, a lower surface Σ_1, and possibly a lateral surface Σ_3, as shown in Figure (a). If Σ_1 and Σ_2 meet as in Figure (b), omit Σ_3. Suppose that Σ_2 is given by $z = K(x, y)$ and Σ_1 is given by $z = H(x, y)$ for (x, y) in a region D of the plane. Let $\mathbf{F}(x, y, z) = f(x, y, z)\mathbf{i} + g(x, y, z)\mathbf{j} + h(x, y, z)\mathbf{k}$. Show, by considering the surface integrals on each of Σ_1, Σ_2, and Σ_3, that

$$\iint_{\Sigma} f(x, y, z)\mathbf{i} \cdot \mathbf{N} \, d\sigma = \iiint_{M} \frac{\partial f}{\partial x} \, dV,$$

$$\iint_{\Sigma} g(x, y, z)\mathbf{j} \cdot \mathbf{N} \, d\sigma = \iiint_{M} \frac{\partial g}{\partial y} \, dV,$$

and

$$\iint_{\Sigma} h(x, y, z)\mathbf{k} \cdot \mathbf{N} \, d\sigma = \iiint_{M} \frac{\partial h}{\partial z} \, dV.$$

Add these equations to establish the theorem for this case.

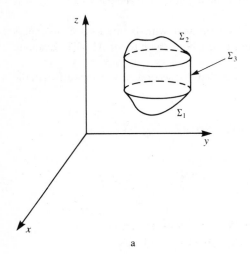

a

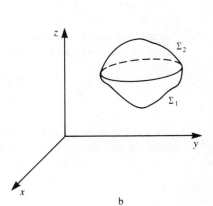

b

16.7 *Some Applications of the Divergence Theorem*

In this section, we will show how the divergence theorem can be used to develop mathematical models of various physical phenomena, including heat conduction, flow of fluids, and electrical fields.

EQUATION OF HEAT CONDUCTION

We will derive a partial differential equation governing heat conduction. Suppose that a medium has density $\rho(x, y, z)$, specific heat $\sigma(x, y, z)$, and coefficient of thermal conductivity $K(x, y, z)$. Let $u(x, y, z, t)$ be the temperature at time t and point (x, y, z) in the medium. (The medium might be the air in a room, or we might be considering heat conduction through a metal bar, in which case the metal is the medium.) We want to derive an equation for u.

Imagine a smooth closed surface Σ within the medium enclosing a set of points M. We will assume Fourier's law, which states that the amount of heat energy leaving M across Σ in a time interval Δt equals $\iint_\Sigma (K \nabla u) \cdot \mathbf{N} \, d\sigma \, \Delta t$.

The change in temperature at (x, y, z) in M in the interval Δt is approximately $(\partial u / \partial t) \, \Delta t$, and the resulting heat loss in M is

$$\left(\iiint_M \sigma \rho \frac{\partial u}{\partial t} \, dV \right) \Delta t.$$

In the absence of sources of heat energy within M, the change in heat energy in M must equal the amount of heat energy lost across Σ; hence,

$$\iint_\Sigma (K \nabla u) \cdot \mathbf{N} \, d\sigma \, \Delta t = \left(\iiint_M \sigma \rho \frac{\partial u}{\partial t} \, dV \right) \Delta t.$$

Upon dividing this equation by Δt, we get

$$\iint_\Sigma (K \nabla u) \cdot \mathbf{N} \, d\sigma = \iiint_M \sigma \rho \frac{\partial u}{\partial t} \, dV. \tag{16.8}$$

Apply Gauss's theorem to the surface integral in equation (16.8) to get

$$\iiint_M \nabla \cdot (K \nabla u) \, dV = \iiint_M \sigma \rho \frac{\partial u}{\partial t} \, dV,$$

or

$$\iiint_M \left(\nabla \cdot (K \nabla u) - \sigma \rho \frac{\partial u}{\partial t} \right) dV = 0. \tag{16.9}$$

But Σ is *any* smooth closed surface in the medium. If the integrand in equation (16.9) were nonzero at some point P_0, we could by continuity choose a surface about P_0 so that the integrand has the same sign throughout the interior of Σ and obtain a nonzero value for the integral, a contradiction. Therefore, the integrand must be identically zero throughout the medium, and we have

$$\mathbf{V} \cdot (K\mathbf{V}u) - \sigma\rho \frac{\partial u}{\partial t} = 0.$$

This partial differential equation may be written

$$\sigma\rho \frac{\partial u}{\partial t} = \mathbf{V} \cdot (K\mathbf{V}u).$$

This equation is a very general form of the three-dimensional heat equation. To write this equation more explicitly, compute

$$\mathbf{V} \cdot (K\mathbf{V}u) = \mathbf{V} \cdot \left(K \frac{\partial u}{\partial x} \mathbf{i} + K \frac{\partial u}{\partial y} \mathbf{j} + K \frac{\partial u}{\partial z} \mathbf{k} \right)$$

$$= \frac{\partial}{\partial x} \left(K \frac{\partial u}{\partial x} \right) + \frac{\partial}{\partial y} \left(K \frac{\partial u}{\partial y} \right) + \frac{\partial}{\partial z} \left(K \frac{\partial u}{\partial z} \right)$$

$$= K \left(\frac{\partial^2 u}{\partial x^2} + \frac{\partial^2 u}{\partial y^2} + \frac{\partial^2 u}{\partial z^2} \right) + \mathbf{V}K \cdot \mathbf{V}u.$$

The sum $\partial^2 u/\partial x^2 + \partial^2 u/\partial y^2 + \partial^2 u/\partial z^2$ is called the *Laplacian* of u and is denoted $\nabla^2 u$ (read "del squared u"). The heat equation may now be written

$$\sigma\rho \frac{\partial u}{\partial t} = K\nabla^2 u + \mathbf{V}K \cdot \mathbf{V}u.$$

If K is constant, then $\mathbf{V}K = \mathbf{O}$, and this equation is

$$\frac{\partial u}{\partial t} = \frac{K}{\sigma\rho} \nabla^2 u.$$

The steady-state case occurs when u does not change with time. Then $\partial u/\partial t = 0$, and the heat equation becomes

$$\nabla^2 u = 0,$$

a partial differential equation known as *Laplace's equation*. Laplace's equation arises in many contexts, including fluid flow, electricity and magnetism, and astronomy.

We will solve the heat equation under various conditions on the medium in Chapter 18. There we will also solve Laplace's equation under a variety of conditions which apply to many models of physical phenomena.

EQUATIONS OF HYDRODYNAMICS

Consider a fluid (for example, water, air, or oil) flowing in some region of three-space. The standard device we have been using to derive equations for a mathematical model has been to insert an imaginary surface in the medium, in this case, the fluid. We will use this strategy again. Imagine a smooth surface Σ in the fluid and observe the flow of fluid through the (presumably perfectly porous) surface. Let $\rho(x, y, z, t)$ be the density of the fluid at time t and point (x, y, z) and let $\mathbf{F}(x, y, z, t)$ be the velocity field. We want to derive equations which give us information about the movement of the fluid.

Suppose Σ bounds a region M in three-space. The amount of fluid flowing out of M across Σ in time Δt is $\left(\iint_{\Sigma} \rho \mathbf{F} \cdot \mathbf{N} \, d\sigma\right) \Delta t$. This requires that the quantity of fluid within Σ be changed by the amount

$$-\left(\iiint_{M} \frac{\partial \rho}{\partial t} \, dV\right) \Delta t,$$

with the minus sign indicating a decrease. This statement assumes that fluid is neither created nor destroyed within Σ. If we assume that fluid is created or destroyed at a rate proportional to ρ, with constant of proportionality K, we must add to the last triple integral a term

$$\left(\iiint_{M} K\rho \, dV\right) \Delta t,$$

with K positive if fluid is created and negative if it is destroyed.

With this term added, the change in fluid within Σ in time Δt is

$$-\left(\iiint_{M} \frac{\partial \rho}{\partial t} \, dV\right) \Delta t + \left(\iiint_{M} K\rho \, dV\right) \Delta t.$$

This sum of terms must equal the surface integral for the amount of fluid flowing out of M across Σ in time Δt. Therefore,

$$\left(\iint_{\Sigma} \rho \mathbf{F} \cdot \mathbf{N} \, d\sigma\right) \Delta t = -\left(\iiint_{M} \frac{\partial \rho}{\partial t} \, dV\right) \Delta t + \left(\iiint_{M} K\rho \, dV\right) \Delta t.$$

Divide this equation by Δt to get

$$\iint_{\Sigma} \rho \mathbf{F} \cdot \mathbf{N} \, d\sigma = -\iiint_{M} \frac{\partial \rho}{\partial t} \, dV + \iiint_{M} K\rho \, dV.$$

Now apply Gauss's theorem to the surface integral on the left to obtain

$$\iiint_{M} (\mathbf{\nabla} \cdot \rho \mathbf{F}) \, dV = \iiint_{M} \left(K\rho - \frac{\partial \rho}{\partial t}\right) dV,$$

or

$$\iiint_{M} \left(\mathbf{\nabla} \cdot (\rho F) + \frac{\partial \rho}{\partial t} - K\rho\right) dV = 0.$$

Again, the only way this triple integral can be zero for any smooth surface Σ within the medium is for the integrand to be identically zero. Therefore,

$$\mathbf{\nabla} \cdot (\rho \mathbf{F}) + \frac{\partial \rho}{\partial t} - K\rho = 0. \qquad (16.10)$$

If $K = 0$, there are no sources or sinks within the medium, and we obtain

$$\mathbf{V} \cdot (\rho \mathbf{F}) + \frac{\partial \rho}{\partial t} = 0. \tag{16.11}$$

This is the *continuity equation* of fluid dynamics. It is a law of conservation of matter.

In equation (16.10), $\partial \rho / \partial t$ is the rate of change of ρ with respect to time at a given point in the medium. If we imagine a particle of fluid moving along a path in the medium parametrized by

$$x = x(t), \qquad y = y(t), \qquad z = z(t),$$

then $\rho = \rho(x(t), y(t), z(t), t)$, and by the chain rule we have

$$\frac{d\rho}{dt} = \frac{\partial \rho}{\partial x}\frac{dx}{dt} + \frac{\partial \rho}{\partial y}\frac{dy}{dt} + \frac{\partial \rho}{\partial z}\frac{dz}{dt} + \frac{\partial \rho}{\partial t}.$$

This is the rate of change of ρ with respect to time as the particle of fluid moves along the given path in the fluid. Now,

$$\mathbf{F} = \frac{dx}{dt}\mathbf{i} + \frac{dy}{dt}\mathbf{j} + \frac{dz}{dt}\mathbf{k}$$

is the velocity of the fluid along the path, and

$$\mathbf{V}\rho = \frac{\partial \rho}{\partial x}\mathbf{i} + \frac{\partial \rho}{\partial y}\mathbf{j} + \frac{\partial \rho}{\partial z}\mathbf{k}.$$

Therefore,

$$\frac{d\rho}{dt} = \mathbf{F} \cdot \mathbf{V}\rho + \frac{\partial \rho}{\partial t}.$$

Solve this equation for $\partial \rho / \partial t$ and substitute the result into the continuity equation (16.11) to get

$$\mathbf{V} \cdot (\rho \mathbf{F}) + \frac{d\rho}{dt} - \mathbf{F} \cdot \mathbf{V}\rho = 0,$$

or

$$\frac{d\rho}{dt} = \mathbf{F} \cdot \mathbf{V}\rho - \mathbf{V} \cdot (\rho F). \tag{16.12}$$

Now,

$$\mathbf{V} \cdot (\rho \mathbf{F}) = \mathbf{F} \cdot \mathbf{V}\rho + \rho \mathbf{V} \cdot \mathbf{F}, \tag{16.13}$$

an identity we derived in detail when treating the heat equation previously. Upon substituting (16.13) into (16.12), we get

$$\frac{d\rho}{dt} = \mathbf{F} \cdot \mathbf{V}\rho - \mathbf{F} \cdot \mathbf{V}\rho - \rho \mathbf{V} \cdot \mathbf{F} = -\rho \mathbf{V} \cdot \mathbf{F},$$

or

$$-\frac{1}{\rho}\frac{d\rho}{dt} = \mathbf{V} \cdot \mathbf{F}.$$

The fluid is called *incompressible* if $d\rho/dt = 0$ (for example, water is incompressible under normal conditions). In this case, the fluid flow is governed by the simple equation

$$\mathbf{V} \cdot \mathbf{F} = 0.$$

If the velocity field is conservative, $\mathbf{F}$ has a potential function φ, and we can write $\mathbf{F} = \mathbf{V}\varphi$. In this case, the equation $\mathbf{V} \cdot \mathbf{F} = 0$ becomes $\mathbf{V} \cdot (\mathbf{V}\varphi) = 0$, or

$$\frac{\partial^2 \varphi}{\partial x^2} + \frac{\partial^2 \varphi}{\partial y^2} + \frac{\partial^2 \varphi}{\partial z^2} = 0,$$

which is Laplace's equation again.

GAUSS'S LAW

An electrical charge q at the origin produces an electric field

$$\mathbf{E} = \frac{q}{4\pi\epsilon \|\mathbf{R}\|^3} \, \mathbf{R},$$

where $\mathbf{R} = x\mathbf{i} + y\mathbf{j} + z\mathbf{k}$ and ϵ is the electric permittivity of the medium.

Gauss's law states that, if Σ is a smooth closed surface in the medium, then

$$\iint\limits_{\Sigma} \mathbf{E} \cdot \mathbf{N} \, d\sigma = \begin{cases} \dfrac{q}{\epsilon} & \text{if} \quad \Sigma \text{ encloses the origin} \\[2mm] 0 & \text{if} \quad \Sigma \text{ does not enclose the origin.} \end{cases}$$

We will use Gauss's theorem to establish this result. Consider two cases.

Case 1 Σ does not enclose the origin.
Now we can apply Gauss's theorem to obtain

$$\iint\limits_{\Sigma} \mathbf{E} \cdot \mathbf{N} \, d\sigma = \iiint\limits_{M} \mathbf{V} \cdot \mathbf{E} \, dV.$$

A routine calculation shows that $\mathbf{V} \cdot \mathbf{E} = 0$; hence, $\iint_{\Sigma} \mathbf{E} \cdot \mathbf{N} \, d\sigma = 0$.

Case 2 Σ encloses the origin.
Now Gauss's theorem cannot be used because $\mathbf{E}$ is undefined at the origin, interior to Σ. Enclose the origin in a sphere Σ_r of radius r, with r sufficiently small that Σ_r lies entirely within Σ, as in Figure 16.63. Cut small disks out of Σ and Σ_r and connect Σ to Σ_r with a tube T, as shown in Figure 16.64. Consider the new surface Σ^* consisting of Σ' (which is Σ minus the small excised disk), Σ_r' (which is Σ_r minus its excised disk), and the connecting tube T. The interesting thing about Σ^* is that it does not enclose the origin! By Gauss's theorem,

$$\iint\limits_{\Sigma^*} \mathbf{E} \cdot \mathbf{N} \, d\sigma = \iiint\limits_{M^*} \mathbf{V} \cdot \mathbf{E} \, dV = 0$$

because $\mathbf{V} \cdot \mathbf{E} = 0$. Now,

$$\iint\limits_{\Sigma^*} \mathbf{E} \cdot \mathbf{N} \, d\sigma = \iint\limits_{\Sigma'} \mathbf{E} \cdot \mathbf{N} \, d\sigma + \iint\limits_{\Sigma_r'} \mathbf{E} \cdot \mathbf{N} \, d\sigma + \iint\limits_{T} \mathbf{E} \cdot \mathbf{N} \, d\sigma.$$

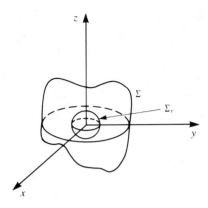

Figure 16.63

Therefore,

$$\iint\limits_{\Sigma'} \mathbf{E} \cdot \mathbf{N} \, d\sigma + \iint\limits_{\Sigma_r'} \mathbf{E} \cdot \mathbf{N} \, d\sigma + \iint\limits_{T} \mathbf{E} \cdot \mathbf{N} \, d\sigma = 0. \tag{16.14}$$

In this equation, take a limit as the connecting tube T is chosen thinner, approaching a curve (with no thickness). In this limit,

$$\iint\limits_{T} \mathbf{E} \cdot \mathbf{N} \, d\sigma \to 0$$

because the surface integral over a curve is zero. Further, in this limit, the disks excised from Σ and Σ_r to form the connecting ends of T go to zero radius, and $\Sigma' \to \Sigma$ and $\Sigma_r' \to \Sigma_r$. Equation (16.14) gives us in the limit that

$$\iint\limits_{\Sigma} \mathbf{E} \cdot \mathbf{N} \, d\sigma + \iint\limits_{\Sigma_r} \mathbf{E} \cdot \mathbf{N} \, d\sigma = 0,$$

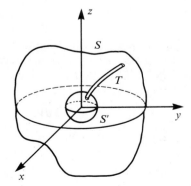

Figure 16.64

or

$$\iint\limits_{\Sigma} \mathbf{E} \cdot \mathbf{N} \, d\sigma = -\iint\limits_{\Sigma_r} \mathbf{E} \cdot \mathbf{N} \, d\sigma.$$

But we can evaluate $\iint_{\Sigma_r} \mathbf{E} \cdot \mathbf{N} \, d\sigma$ explicitly because Σ_r is a sphere of radius r about the origin and $\mathbf{E}$ is a given vector field. Parametrize Σ_r by spherical coordinates:

$$x = r \cos(\theta)\sin(\varphi), \qquad y = r \sin(\theta)\sin(\varphi), \qquad z = r \cos(\varphi),$$

$0 \le \theta \le 2\pi$, $0 \le \varphi \le \pi$. Using the determinant expression for the normal vector to a surface given in Section 16.4, we find that

$$\mathbf{N} = -\sin(\varphi)[r^2\cos(\theta)\sin(\varphi)\mathbf{i} + r^2\sin(\theta)\sin(\varphi)\mathbf{j} + r^2\cos(\varphi)\mathbf{k}].$$

(This is an *inner* normal on Σ_r considered by itself but is an *outer* normal on Σ_r considered as a part of Σ^* because $\mathbf{N}$ points *away from* the region bounded by Σ^*.)

After a routine calculation, we find that

$$\mathbf{E} \cdot \mathbf{N} = -\frac{q}{4\pi\epsilon}\sin(\varphi).$$

We now have

$$\iint\limits_{\Sigma_r} \mathbf{E} \cdot \mathbf{N} \, d\sigma = \int_0^\pi \int_0^{2\pi} \mathbf{E} \cdot \mathbf{N} \, d\theta \, d\varphi$$

$$= \int_0^\pi \int_0^{2\pi} -\frac{q}{4\pi\epsilon}\sin(\varphi) \, d\theta \, d\varphi = -\frac{q}{\epsilon}.$$

We conclude that $\iint_{\Sigma} \mathbf{E} \cdot \mathbf{N} \, d\sigma = -(-q/\epsilon) = q/\epsilon$ when Σ encloses the origin.

This completes the derivation of Gauss's law. The reasonsing was intuitive, since the construction of Σ^* from Σ and the ensuing limiting process require delicate arguments from topology if we wish to justify every detail.

We will see this kind of reasoning again when we treat singularities of complex functions in Part VI of this book, as well as when we derive an extension of the divergence theorem in Appendix B to this part.

POISSON'S EQUATION

We can use Gauss's law to draw further conclusions about electrical fields. Let Q be the charge density of the field. If Σ is a smooth closed surface bounding a set M in three-space, the distributed charge over Σ is

$$q = \iiint\limits_{M} Q(x, y, z) \, dV.$$

If Σ encloses the origin, then by Gauss's law,

$$\iint\limits_{\Sigma} \mathbf{E} \cdot \mathbf{N} \, d\sigma = \frac{q}{\epsilon} = \frac{1}{\epsilon}\iiint\limits_{M} Q \, dV.$$

Apply Gauss's theorem to the surface integral on the left to get

$$
\iiint_M \mathbf{V} \cdot \mathbf{E} \, dV = \frac{1}{\epsilon} \iiint_M Q \, dV,
$$

or

$$
\iiint_M \left(\mathbf{V} \cdot \mathbf{E} - \frac{1}{\epsilon} Q \right) dV = 0.
$$

As usual, since Σ is any smooth closed surface enclosing the origin, we conclude that

$$
\mathbf{V} \cdot \mathbf{E} - \frac{1}{\epsilon} Q = 0.
$$

This is one of Maxwell's equations for an electrical field.

If $\mathbf{E}$ is conservative, $\mathbf{E} = \mathbf{V}\varphi$ for some potential function φ, and the last equation becomes

$$
\frac{\partial^2 \varphi}{\partial x^2} + \frac{\partial^2 \varphi}{\partial y^2} + \frac{\partial^2 \varphi}{\partial z^2} = \frac{1}{\epsilon} Q.
$$

This partial differential equation is called *Poisson's equation*. Notice that, if $Q = 0$, Poisson's equation becomes Laplace's equation.

GREEN'S IDENTITIES

There are several integral identities which are used in treating partial differential equations. Green's first identity is

$$
\iint_\Sigma f \mathbf{V}g \cdot \mathbf{N} \, d\sigma = \iiint_M (f \nabla^2 g + \mathbf{V}f \cdot \mathbf{V}g) \, dV, \tag{16.15}
$$

in which Σ is a piecewise-smooth closed surface bounding M, and f and g are continuous with continuous first and second partial derivatives on Σ and throughout M. To derive equation (16.15), use Gauss's theorem to write

$$
\iint_\Sigma f \mathbf{V}g \cdot \mathbf{N} \, d\sigma = \iiint_M \mathbf{V} \cdot (f \mathbf{V}g) \, dV.
$$

But

$$
\mathbf{V} \cdot (f \mathbf{V}g) = f \nabla^2 g + \mathbf{V}f \cdot \mathbf{V}g,
$$

yielding equation (16.15).

Green's second identity is

$$
\iint_\Sigma (f \mathbf{V}g - g \mathbf{V}f) \cdot \mathbf{N} \, d\sigma = \iiint_M (f \nabla^2 g - g \nabla^2 f) \, dV.
$$

To prove this identity, interchange f with g in Green's first identity, equation (16.15), to write

$$\iint_{\Sigma} g\nabla f \cdot \mathbf{N} \, d\sigma = \iiint_{M} (g\nabla^2 f + \nabla g \cdot \nabla f) \, dV. \qquad (16.16)$$

Now subtract equation (16.16) from equation (16.15) to get

$$\iint_{\Sigma} (f\nabla g - g\nabla f) \cdot \mathbf{N} \, d\sigma = \iiint_{M} [f\nabla^2 g + \nabla f \cdot \nabla g - (g\nabla^2 f + \nabla g \cdot \nabla f)] \, dV.$$

but $\nabla f \cdot \nabla g - \nabla g \cdot \nabla f = 0$, so the last equation yields Green's second identity. If we put $f = 1$ into Green's second identity, we obtain

$$\iint_{\Sigma} \nabla g \cdot \mathbf{N} \, d\sigma = \iiint_{M} \nabla^2 g \, dV.$$

If g satisfies Laplace's equation, $\nabla^2 g = 0$, and we conclude that

$$\iint_{\Sigma} \nabla g \cdot \mathbf{N} \, d\sigma = 0.$$

Therefore, if g is a solution of Laplace's equation in a region bounded by a surface Σ, the surface integral over Σ of the normal component of the gradient of g must be zero.

PROBLEMS FOR SECTION 16.7

1. Show that

$$\nabla^2 f(P_0) = \lim_{\mathscr{V} \to 0} \frac{1}{\mathscr{V}} \iint_{\Sigma} \nabla f \cdot \mathbf{N} \, d\sigma,$$

in which Σ is a piecewise-smooth closed surface containing P_0 and bounding a region M having volume $\mathscr{V}$, and $\lim_{\mathscr{V} \to 0}$ means that Σ contracts to P_0. Use this formula to give a physical interpretation of $\nabla^2 f$.

2. Let Σ be a piecewise-smooth closed surface, and let $f(x, y, z)$ be a given continuous function for (x, y, z) on Σ. A *Dirichlet problem* consists of finding a function satisfying Laplace's equation $\nabla^2 u = 0$ for (x, y, z) in the region M bounded by Σ and satisfying the boundary condition $u(x, y, z) = f(x, y, z)$ for (x, y, z) on Σ. Prove that a Dirichlet problem can have only one continuous solution having continuous first and second partial derivatives in M. *Hint:* Suppose that F and G are solutions. Let $w(x, y, z) = F(x, y, z) - G(x, y, z)$. Show that $\nabla^2 w = 0$ for all (x, y, z) in M and that $w(x, y, z) = 0$ for all (x, y, z) on Σ. Use Gauss's theorem to show that $w(x, y, z) = 0$ for all (x, y, z) in M.

3. The partial differential equation for the temperature function $u(x, y, z, t)$ in a solid M bounded by a piecewise-smooth surface Σ can be written

$$\frac{\partial u}{\partial t} = k\nabla^2 u + \varphi(x, y, z, t).$$

Suppose that the temperature on the boundary Σ of M is given by

$$u(x, y, z, t) = f(x, y, z) \quad \text{for all } (x, y, z) \text{ on } \Sigma \quad \text{and} \quad t > 0.$$

The initial temperature (at time $t = 0$) is given throughout M by

$$u(x, y, z, 0) = g(x, y, z) \quad \text{for } (x, y, z) \text{ in } M.$$

Show that there can be at most one temperature function $u(x, y, z, t)$ satisfying these equations, continuous throughout M, with continuous first and second partial derivatives in M. *Hint:* Suppose that F and G are solutions. Let $w(x, y, z, t) = F(x, y, z, t) - G(x, y, z, t)$. Show that w satisfies the partial differential equation $\partial w/\partial t = k\nabla^2 w$, that $w(x, y, z, t) = 0$ for (x, y, z) on Σ and $t > 0$, and that $w(x, y, z, 0) = 0$ for (x, y, z) in M. Now let

$$I(t) = \tfrac{1}{2} \iiint\limits_M w^2(x, y, z, t)\, dV.$$

Show that

$$I'(t) = -k \iiint\limits_M \left[\left(\frac{\partial w}{\partial x}\right)^2 + \left(\frac{\partial w}{\partial y}\right)^2 + \left(\frac{\partial w}{\partial z}\right)^2\right] dV.$$

Conclude that $I'(t) \leq 0$ for $t > 0$. Next, apply the mean value theorem to $I(t)$ on the interval $[0, t]$ to show that $I(t) \leq 0$ for $t > 0$. Thus, show that $I(t) = 0$ for $t > 0$. Finally, show that $w(x, y, z, t) = 0$ for $t > 0$ and (x, y, z) in M.

4. Prove that the following problem can have only one solution which is continuous and has continuous first and second partial derivatives:

$$\frac{\partial u}{\partial t} = k\nabla^2 u + \varphi(x, y, z, t) \quad \text{for} \quad (x, y, z) \text{ in } M, \qquad t > 0,$$

$$\frac{\partial u}{\partial \eta} + hu = f(x, y, z, t) \quad \text{for} \quad (x, y, z) \text{ in } \Sigma \quad \text{and} \quad t > 0,$$

$$u(x, y, z, 0) = g(x, y, z) \quad \text{for} \quad (x, y, z) \text{ in } M.$$

Here, f and g are given, h is a positive constant, and $\partial u/\partial \eta = \nabla u \cdot \mathbf{N}$ is the normal derivative of u, with $\mathbf{N}$ the normal vector to Σ.

5. Let f and g satisfy Laplace's equation in a region M of three-space bounded by a piecewise-smooth surface Σ. Suppose that $\partial f/\partial \eta = \partial g/\partial \eta$ at all points of Σ. Prove that, for some constant k, $f(x, y, z) = g(x, y, z) + k$ for all (x, y, z) in M.

6. Express the moment of inertia I of a uniform solid object about the z-axis as the flux of a vector field across the surface of the object. [Recall that

$$I = \iiint\limits_M \rho(x^2 + y^2)\, dV,$$

where ρ is the constant density of the object.]

16.8 Stokes's Theorem

From Section 16.5, the conclusion of Green's theorem can be written

$$\oint_C \mathbf{F} \cdot \mathbf{T}\, ds = \iint\limits_D (\nabla \times \mathbf{F}) \cdot \mathbf{k}\, dA.$$

Here, $\mathbf{T}$ is the unit tangent to C, and C is a simple closed piecewise-smooth counterclockwise-oriented curve with interior D.

We may think of D as a flat surface with unit normal vector $\mathbf{k}$ and bounded by C. To generalize Green's theorem to three-space, allow C to be a closed curve in three-

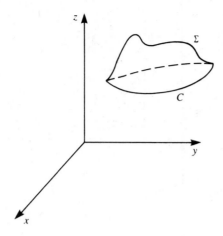

Figure 16.65

space, bounding a smooth surface Σ as in Figure 16.65. Finally, let **N** be a unit normal vector to Σ.

This raises a difficulty. At any point of Σ, there are exactly two normal vectors, shown in Figure 16.66. Which should we choose? In addition, we must choose a direction on C. In the plane, this is done by parametrizing C so that if we walk around C counterclockwise, D is over our left shoulder. In three-space, however, "counterclockwise" has no unambiguous meaning. We must therefore develop a concept of positive orientation for a surface in three-space.

We will begin by choosing a unit vector **N** (not necessarily outer) and use this in turn to select a direction along C. Remember that, for a surface $z = S(x, y)$, a normal vector is

$$-\frac{\partial S}{\partial x}\mathbf{i} - \frac{\partial S}{\partial y}\mathbf{j} + \mathbf{k}.$$

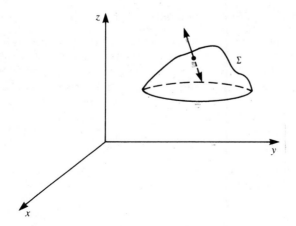

Figure 16.66

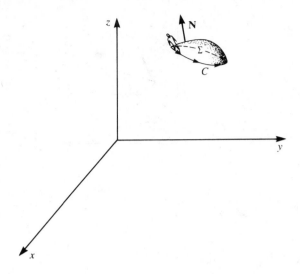

Figure 16.67

We can choose this vector or its negative, but, whichever we choose, we retain it for all points of Σ. Divide the vector by its length to produce a unit normal **N**.

Having chosen **N**, we want to determine a direction on C to call the positive direction. We will use the intuitive rule illustrated in Figure 16.67. If a person stands on C so that **N** points from his feet toward his head when drawn as an arrow from C, then the positive direction on C is the one he must walk to have Σ over his left shoulder. When the positive direction on C is chosen in this way, we say that C has been *oriented coherently* with Σ.

The choice of **N** therefore determines a direction on C according to the above rule; then the surface and curve are oriented coherently. We can now state Stokes's theorem.

THEOREM 16.10 Stokes's Theorem

Let Σ be a piecewise-smooth surface bounded by a piecewise-smooth closed curve C. Suppose that a unit normal **N** has been chosen on Σ and that C has been oriented coherently with Σ. Let $\mathbf{F}(x, y, z)$ be a vector field whose component functions are continuous with continuous first and second partial derivatives on Σ. Then

$$\oint_C \mathbf{F} \cdot d\mathbf{R} = \iint_\Sigma (\nabla \times \mathbf{F}) \cdot \mathbf{N} \, d\sigma. \quad \blacksquare$$

This equation is often written as

$$\oint_C \mathbf{F} \cdot d\mathbf{R} = \iint_\Sigma (\text{curl } \mathbf{F}) \cdot \mathbf{N} \, d\sigma.$$

Here is a computational example to illustrate Stokes's theorem.

EXAMPLE 16.28

Let $\mathbf{F} = -y\mathbf{i} + x\mathbf{j} - xyz\mathbf{k}$, and consider the surface Σ consisting of the cone $z = \sqrt{x^2 + y^2}$ for $x^2 + y^2 \le 9$. We will compute both sides of the equation in Stokes's theorem as an illustration.

Σ and its boundary curve C are shown in Figure 16.68. C is the circle $x^2 + y^2 = 9$ in the plane $z = 3$. (Σ is "almost" smooth, having a continuous normal at each point except the origin. The student is asked to believe that the conclusion of Stokes's theorem holds in this example despite this exceptional point.)

First, we need the normal vector. We find that

$$-\frac{\partial z}{\partial x}\mathbf{i} - \frac{\partial z}{\partial y}\mathbf{j} + \mathbf{k} = -\frac{x}{z}\mathbf{i} - \frac{y}{z}\mathbf{j} + \mathbf{k}$$

at each point except the origin. This vector has magnitude

$$\sqrt{\frac{x^2}{z^2} + \frac{y^2}{z^2} + 1} = \sqrt{\frac{x^2 + y^2 + z^2}{z^2}} = \sqrt{2}$$

because $z^2 = x^2 + y^2$. Thus, a unit normal to Σ is

$$\mathbf{N} = \frac{1}{\sqrt{2}}\left[-\frac{x}{z}\mathbf{i} - \frac{y}{z}\mathbf{j} + \mathbf{k}\right] = \frac{1}{\sqrt{2}z}(-x\mathbf{i} - y\mathbf{j} + z\mathbf{k})$$

if $z \neq 0$. For example, at $(3, 0, 3)$,

$$\mathbf{N}(3, 0, 3) = \frac{1}{\sqrt{2}}(-\mathbf{i} + \mathbf{k}),$$

shown in Figure 16.69. This normal vector points "into" the cone.

If we are on C and walking in the direction of the arrow in Figure 16.69 with our head in the direction of $\mathbf{N}$, the surface is over our left shoulder. Thus, the coherent orientation of C for this choice of $\mathbf{N}$ is the direction of the arrow in Figure 16.69.

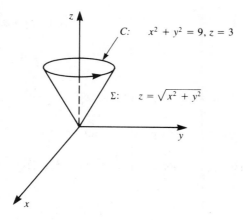

Figure 16.68

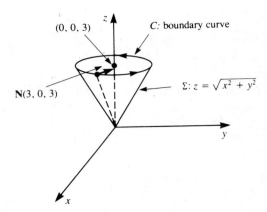

Figure 16.69

Parametrize C by

$$x = 3\cos(t), \qquad y = 3\sin(t), \qquad z = 3; \qquad 0 \le t \le 2\pi.$$

The point (x, y, z) traverses C once in the positive direction as t increases from 0 to 2π.
Now compute

$$\oint_C \mathbf{F} \cdot d\mathbf{R} = \oint_C (-y\, dx + x\, dy - xyz\, dz)$$

$$= \int_0^{2\pi} \{-3\sin(t)[-3\sin(t)] + 3\cos(t)[3\cos(t)] - 27\sin(t)\cos(t)[0]\}\, dt$$

$$= \int_0^{2\pi} 9\, dt = 18\pi.$$

For $\iint_\Sigma (\nabla \times \mathbf{F}) \cdot \mathbf{N}\, d\sigma$, first calculate

$$\nabla \times \mathbf{F} = \begin{vmatrix} \mathbf{i} & \mathbf{j} & \mathbf{k} \\ \dfrac{\partial}{\partial x} & \dfrac{\partial}{\partial y} & \dfrac{\partial}{\partial z} \\ -y & x & -xyz \end{vmatrix} = -xz\mathbf{i} + yz\mathbf{j} + 2\mathbf{k}.$$

Then

$$(\nabla \times \mathbf{F}) \cdot \mathbf{N} = \frac{1}{\sqrt{2}z}[x^2 z - y^2 z + 2z] = \frac{1}{\sqrt{2}}(x^2 - y^2 + 2).$$

Remember that $1 + (\partial z/\partial x)^2 + (\partial z/\partial y)^2 = 2$ (from computing $\|\mathbf{N}\|$). Then

$$\iint_\Sigma (\nabla \times \mathbf{F}) \cdot \mathbf{N}\, d\sigma = \iint_D \frac{1}{\sqrt{2}}(x^2 - y^2 + 2)\sqrt{2}\, dA$$

$$= \iint_D (x^2 - y^2 + 2)\, dA.$$

D is the disk $x^2 + y^2 \leq 9$ in the xy-plane. Transform this integral to polar coordinates:

$$\iint\limits_{D} (x^2 - y^2 + 2)\, dA = \int_0^{2\pi} \int_0^3 [r^2\cos^2(\theta) - r^2\sin^2(\theta) + 2]r\, dr\, d\theta.$$

Since $\cos^2(\theta) - \sin^2(\theta) = \cos(2\theta)$, this is

$$\int_0^{2\pi} \int_0^3 [r^3\cos(2\theta) + 2r]\, dr\, d\theta = \int_0^{2\pi} \cos(2\theta)\, d\theta \int_0^3 r^3\, dr + \int_0^{2\pi} d\theta \int_0^3 2r\, dr$$

$$= 0 + (2\pi)\left[r^2\right]_0^3 = 18\pi,$$

in agreement with the previous calculation of $\oint_C \mathbf{F} \cdot d\mathbf{R}$. ∎

We will use Stokes's theorem to yield a physical interpretation of the curl of a vector.

EXAMPLE 16.29 Interpretation of Curl

Think of $\mathbf{F}$ as the velocity of a fluid, and let $P_0: (x_0, y_0, z_0)$ be any point. Consider a disk Σ_r of radius r about P_0, with boundary C_r coherently oriented with Σ_r, as in Figure 16.70. Here, $\mathbf{N}$ is a constant vector. Then

$$\oint_{C_r} \mathbf{F} \cdot d\mathbf{R} = \iint\limits_{\Sigma_r} (\nabla \times \mathbf{F}) \cdot \mathbf{N}\, d\sigma.$$

Since $\mathbf{F} \cdot \mathbf{R}'$ is the tangential component of $\mathbf{F}$ about C_r, $\oint_C \mathbf{F} \cdot d\mathbf{R}$ measures the circulation of the fluid about C_r. If r is small, $\nabla \times \mathbf{F}(x, y, z) \approx \nabla \times \mathbf{F}(P_0)$ on Σ_r. Since $\mathbf{N}$

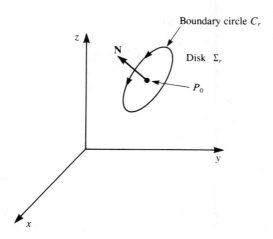

Figure 16.70

is constant,

$$\text{circulation of } \mathbf{F} \text{ about } C_r \approx \iint\limits_{\Sigma_r} (\mathbf{V} \times \mathbf{F})(P_0) \cdot \mathbf{N} \, d\sigma$$

$$= (\mathbf{V} \times \mathbf{F})(P_0) \cdot \mathbf{N}(\text{area of } \Sigma_r)$$

$$= \pi r^2 \mathbf{V} \times \mathbf{F}(P_0) \cdot \mathbf{N}.$$

Then

$$\mathbf{V} \times \mathbf{F}(P_0) \cdot \mathbf{N} \approx \frac{1}{\pi r^2}(\text{circulation of } \mathbf{F} \text{ about } C_r).$$

As $r \to 0$, Σ_r contracts to P_0, and we get

$$\mathbf{V} \times \mathbf{F}(P_0) \cdot \mathbf{N} = \lim_{r \to 0} \frac{1}{\pi r^2}(\text{circulation of } \mathbf{F} \text{ about } P_0).$$

Since $\mathbf{N}$ is normal to the plane of C_r, this equation can be read

$$\mathbf{V} \times \mathbf{F}(P_0) \cdot \mathbf{N} = \text{circulation of } \mathbf{F} \text{ per unit area in the plane normal to } \mathbf{N}.$$

Thus, the curl of $\mathbf{F}$ is a measure of rotation of the fluid at a point. In particular, $\mathbf{F}$ is called *irrotational* if its curl is the zero vector. For example, any conservative field is irrotational, since if $\mathbf{F} = \mathbf{V}\varphi$, then $\mathbf{V} \times \mathbf{F} = \mathbf{V} \times (\mathbf{V}\varphi) = \mathbf{O}$. ∎

Stokes's theorem is also important in the analysis of Maxwell's equations for electrical and magnetic fields. We will discuss Maxwell's equations and potential theory in three-space in the next section.

PROBLEMS FOR SECTION 16.8

In each of Problems 1 through 5, evaluate both $\oint_C \mathbf{F} \cdot d\mathbf{R}$ and $\iint_\Sigma (\mathbf{V} \times \mathbf{F}) \cdot \mathbf{N} \, d\sigma$.

1. $\mathbf{F} = yx^2\mathbf{i} - xy^2\mathbf{j} + z^2\mathbf{k}$; Σ is the hemisphere $x^2 + y^2 + z^2 = 4$, $z \geq 0$.
2. $\mathbf{F} = xy\mathbf{i} + yz\mathbf{j} + xz\mathbf{k}$; Σ is the paraboloid $z = x^2 + y^2$ for $x^2 + y^2 \leq 9$.
3. $\mathbf{F} = z\mathbf{i} + x\mathbf{j} + y\mathbf{k}$; Σ is the cone $z = \sqrt{x^2 + y^2}$ for $0 \leq z \leq 4$.
4. $\mathbf{F} = z^2\mathbf{i} + x^2\mathbf{j} + y^2\mathbf{k}$; Σ is the part of the parabolid $z = 6 - x^2 - y^2$ lying above the xy-plane.
5. $\mathbf{F} = xy\mathbf{i} + yz\mathbf{j} + xy\mathbf{k}$; Σ is the part of the plane $2x + 4y + z = 8$ lying in the first octant.
6. Calculate the circulation of $\mathbf{F} = (x - y)\mathbf{i} + x^2 y\mathbf{j} + xza\mathbf{k}$ counterclockwise about the circle $x^2 + y^2 = 1$. *Hint:* Use Stokes's theorem, letting Σ be any smooth surface having C as boundary.
7. Use Stokes's theorem to evaluate $\oint_C \mathbf{F} \cdot \mathbf{T} \, ds$, where C is the boundary of the part of the plane $x + 4y + z = 12$ lying in the first octant and $\mathbf{F} = (x - z)\mathbf{i} + (y - x)\mathbf{j} + (z - y)\mathbf{k}$.
8. Let $\mathbf{E}(x, y, z, t)$ be the electrical field at (x, y, z) and time t, and let $\mathbf{B}(x, y, z, t)$ be the magnetic field induced by $\mathbf{E}$. According to Maxwell's equations, $\mathbf{V} \times \mathbf{E} = -\partial \mathbf{B}/\partial t$, where the partial derivative is calculated by taking the partial derivative of each component. Now suppose that C is a smooth simple closed curve in three-space. The voltage drop around C is defined to be $\oint_C \mathbf{E} \cdot d\mathbf{R}$.

    Use Stokes's theorem to prove Faraday's law, which states that, if Σ is a smooth surface bounded by C, the voltage drop around C equals the rate of decrease, with respect to time, of the magnetic flux through Σ. *Hint:* The magnetic flux is a surface integral of the magnetic field over Σ. Compute the partial derivative of this surface integral with respect to t by interchanging $\iint_\Sigma$ and $\partial/\partial t$, then apply Stokes's theorem.

9. Fill in the details of this proof of a special case of Stokes's theorem. Assume that Σ is the graph of $z = S(x, y)$. Let $\mathbf{F}(x, y, z) = f(x, y, z)\mathbf{i} + g(x, y, z)\mathbf{j} + h(x, y, z)\mathbf{k}$. First, show that

$$\iint_{\Sigma} (\nabla \times \mathbf{F}) \cdot \mathbf{N} \, d\sigma = \iint_{D} \left(\left[\frac{\partial h}{\partial y} - \frac{\partial g}{\partial z} \right] \left[-\frac{\partial z}{\partial x} \right] + \left[\frac{\partial f}{\partial z} - \frac{\partial h}{\partial x} \right] \left[-\frac{\partial z}{\partial y} \right] + \left[\frac{\partial g}{\partial x} - \frac{\partial f}{\partial y} \right] \right) dA.$$

Next, parametrize the boundary C^* of D in the xy-plane by $x = x(t)$, $y = y(t)$ for $a \le t \le b$, oriented counterclockwise. Show that the boundary C of Σ can be parametrized as $x = x(t)$, $y = y(t)$, $z = S(x(t), y(t))$ for $a \le t \le b$. Next, show that

$$\oint_C \mathbf{F} \cdot d\mathbf{R} = \oint_{C^*} \left[f + h \frac{\partial z}{\partial x} \right] dx + \left[g + h \frac{\partial z}{\partial y} \right] dy.$$

Hint:

$$\frac{\partial z}{\partial t} = \frac{\partial z}{\partial x} \frac{dx}{dt} + \frac{\partial z}{\partial y} \frac{dy}{dt}$$

in evaluating $\oint_C \mathbf{F} \cdot \mathbf{T} \, ds$.

Finally, apply Green's theorem to the line integral over C^* and use the first step of the proof to conclude the argument.

16.9 Maxwell's Equations and Potential Theory in Three-Space_____

In this section, we will discuss two applications of Stokes's theorem.

MAXWELL'S EQUATIONS

We will apply the theorems of both Gauss and Stokes, along with some vector analysis, to a mathematical treatment of electrical and magnetic fields, deriving the classical equations of Maxwell. We will use the following standard symbols:

$\mathbf{E}$ is the electrical intensity of the field.

$\mathbf{H}$ is the magnetic intensity of the field.

$\mathbf{J}$ is the current density.

ϵ is the permittivity of the medium.

μ is the permeability of the medium.

σ is the conductivity of the medium.

Q is the charge density.

$\mathbf{D} = \epsilon \mathbf{E}$ is the electrical flux density.

$\mathbf{B} = \mu \mathbf{H}$ is the magnetic flux density.

$q = \iiint_M Q \, dV$ is the total charge in a region M of three-space.

$\varphi = \iint_{\Sigma} \mathbf{B} \cdot \mathbf{N} \, d\sigma$ is the total outward magnetic flux through a closed surface Σ.

$i = \iint_{\Sigma} \mathbf{J} \cdot \mathbf{N} \, d\sigma$ is the total current flowing outward from M across Σ.

As with any real-world phenomenon, we must begin with experimental observations and empirically derived relationships. We will use the following laws from physics.

Faraday's law states that

$$\int_C \mathbf{E} \cdot d\mathbf{R} = -\frac{\partial}{\partial t} \iint_\Sigma \mathbf{B} \cdot \mathbf{N} \, d\sigma \qquad \left(= -\frac{\partial \varphi}{\partial t} \right)$$

around any closed curve C bounding a closed surface Σ. That is, the tangential component of $\mathbf{E}$ measured around C is the negative of the rate of change with respect to time of the magnetic flux through any surface bounded by C.

Ampere's law states that, for a closed curve C,

$$\int_C \mathbf{H} \cdot d\mathbf{R} = \iint_\Sigma \mathbf{J} \cdot \mathbf{N} \, d\sigma \qquad (=i).$$

That is, the tangential component of magnetic intensity over any closed curve C equals the current flowing through any surface bounded by C.

Finally, *Gauss's laws* state that

$$\iint_\Sigma \mathbf{D} \cdot \mathbf{N} \, d\sigma = q \quad \text{and} \quad \iint_\Sigma \mathbf{B} \cdot \mathbf{N} \, d\sigma = 0.$$

We can now begin our analysis. Apply Stokes's theorem to Faraday's law to get

$$\int_C \mathbf{E} \cdot d\mathbf{R} = \iint_\Sigma (\mathbf{\nabla} \times \mathbf{E}) \cdot \mathbf{N} \, d\sigma = -\frac{\partial \varphi}{\partial t} = -\frac{\partial}{\partial t} \iint_\Sigma \mathbf{B} \cdot \mathbf{N} \, d\sigma = \iint_\Sigma -\frac{\partial \mathbf{B}}{\partial t} \cdot \mathbf{N} \, d\sigma.$$

Then

$$\iint_\Sigma \left(\mathbf{\nabla} \times \mathbf{E} + \frac{\partial \mathbf{B}}{\partial t} \right) \cdot \mathbf{N} \, d\sigma = 0.$$

Since Σ is any smooth closed surface in the medium having C as boundary,

$$\mathbf{\nabla} \times \mathbf{E} + \frac{\partial \mathbf{B}}{\partial t} = \mathbf{O}.$$

This equation is often written

$$\mathbf{\nabla} \times \mathbf{E} = -\frac{\partial \mathbf{B}}{\partial t}. \tag{16.17}$$

A similar argument, beginning with Ampere's law, leads to the equation

$$\mathbf{\nabla} \times \mathbf{H} = \mathbf{J}.$$

Maxwell observed that

$$\mathbf{J} = \sigma \mathbf{E} + \epsilon \frac{\partial \mathbf{E}}{\partial t}.$$

Combining the last two equations, we have

$$\mathbf{V} \times \mathbf{H} = \sigma \mathbf{E} + \epsilon \frac{\partial \mathbf{E}}{\partial t}. \tag{16.18}$$

Now start again by applying Gauss's theorem to Gauss's law to get

$$\iint_{\Sigma} \mathbf{D} \cdot \mathbf{N} \, d\sigma = \iiint_{M} (\mathbf{V} \cdot \mathbf{D}) \, dV = q = \iiint_{M} Q \, dV.$$

Then

$$\iiint_{M} (\mathbf{V} \cdot \mathbf{D} - Q) \, dV = 0.$$

Again, by the arbitrary nature of Σ, we conclude that

$$\mathbf{V} \cdot \mathbf{D} = Q.$$

Now take the curl of both sides of equation (16.17) to get

$$\mathbf{V} \times (\mathbf{V} \times \mathbf{E}) = \mathbf{V} \times \left(-\frac{\partial \mathbf{B}}{\partial t} \right) = -\frac{\partial}{\partial t} (\mathbf{V} \times \mathbf{B}).$$

We can interchange $\mathbf{V} \times$ and $\partial/\partial t$ because the curl operation involves only the space variables and the partial derivative involves only the time variable. Since $\mathbf{B} = \mu \mathbf{H}$,

$$\mathbf{V} \times (\mathbf{V} \times \mathbf{E}) = -\frac{\partial}{\partial t} (\mathbf{V} \times (\mu \mathbf{H})) = -\mu \frac{\partial}{\partial t} (\mathbf{V} \times \mathbf{H}).$$

It is a routine calculation to verify that $\mathbf{V} \times (\mathbf{V} \times \mathbf{E}) = \mathbf{V}(\mathbf{V} \cdot \mathbf{E}) - (\mathbf{V} \cdot \mathbf{V})\mathbf{E}$; hence,

$$\mathbf{V}(\mathbf{V} \cdot \mathbf{E}) - (\mathbf{V} \cdot \mathbf{V})\mathbf{E} = -\mu \frac{\partial}{\partial t} (\mathbf{V} \times \mathbf{H}). \tag{16.19}$$

In this equation,

$$\mathbf{V} \cdot \mathbf{V} = \frac{\partial^2}{\partial x^2} + \frac{\partial^2}{\partial y^2} + \frac{\partial^2}{\partial z^2}.$$

From equation (16.18), $\mathbf{V} \times \mathbf{H} = \sigma \mathbf{E} + \epsilon(\partial \mathbf{E}/\partial t)$. Substitute this result into the right side of equation (16.19) to get

$$\mathbf{V}(\mathbf{V} \cdot \mathbf{E}) - (\mathbf{V} \cdot \mathbf{V})\mathbf{E} = -\mu \frac{\partial}{\partial t} \left(\sigma \mathbf{E} + \epsilon \frac{\partial \mathbf{E}}{\partial t} \right). \tag{16.20}$$

In practice, we often encounter the case in which $Q = 0$. Then

$$Q = \mathbf{V} \cdot \mathbf{D} = \mathbf{V} \cdot (\epsilon \mathbf{E}) = \epsilon \mathbf{V} \cdot \mathbf{E} = 0;$$

hence, in this case,

$$\mathbf{V} \cdot \mathbf{E} = 0.$$

In the case in which $Q = 0$, equation (16.20) becomes

$$(\mathbf{V} \cdot \mathbf{V})\mathbf{E} = \mu\sigma \frac{\partial \mathbf{E}}{\partial t} + \mu\epsilon \frac{\partial^2 \mathbf{E}}{\partial t^2}. \tag{16.21}$$

This is Maxwell's equation for $\mathbf{E}$. By a similar analysis, we can obtain Maxwell's equation for $\mathbf{H}$:

$$(\mathbf{V} \cdot \mathbf{V})\mathbf{H} = \mu\sigma \frac{\partial \mathbf{H}}{\partial t} + \mu\epsilon \frac{\partial^2 \mathbf{H}}{\partial t^2}. \tag{16.22}$$

If $\sigma = 0$, the medium is a perfect dielectric, and Maxwell's equations (16.21) and (16.22) become

$$(\mathbf{V} \cdot \mathbf{V})\mathbf{E} = \mu\epsilon \frac{\partial^2 \mathbf{E}}{\partial t^2} \quad \text{and} \quad (\mathbf{V} \cdot \mathbf{V})\mathbf{H} = \mu\epsilon \frac{\partial^2 \mathbf{H}}{\partial t^2}.$$

Each of these equations is a vector form of the three-dimensional wave equation.

If $\sigma \neq 0$ but $\epsilon = 0$, Maxwell's equations (16.21) and (16.22) become

$$(\mathbf{V} \cdot \mathbf{V})\mathbf{E} = \mu\sigma \frac{\partial \mathbf{H}}{\partial t} \quad \text{and} \quad (\mathbf{V} \cdot \mathbf{V})\mathbf{H} = \mu\epsilon \frac{\partial \mathbf{H}}{\partial t}.$$

These are vector forms of the three-dimensional heat equation.

POTENTIAL THEORY IN THREE-SPACE

In Section 16.3, we discussed potential theory in the plane, making use of Green's theorem. We can develop similar ideas in three-space, where Stokes's theorem plays the role of Green's theorem.

Suppose that $\mathbf{F}$ is a vector field defined and continuous over a set Ω in three-space. Following the lead from the two-dimensional case, we say that

1. $\mathbf{F}$ is *conservative* if there exists a potential function φ for $\mathbf{F}$ (that is, a scalar field φ such that $\mathbf{F} = \mathbf{V}\varphi$).
2. The line integral $\int_C \mathbf{F} \cdot d\mathbf{R}$ is *independent of path* in Ω if $\int_C \mathbf{F} \cdot d\mathbf{R} = \int_K \mathbf{F} \cdot d\mathbf{R}$ for any two piecewise-smooth curves C and K in Ω having the same initial points and the same terminal points.

Notice that, if $\int_C \mathbf{F} \cdot d\mathbf{R}$ is independent of path in Ω, $\oint_C \mathbf{F} \cdot d\mathbf{R} = 0$ for every piecewise-smooth closed curve in Ω. Conversely, if the line integral about every closed curve in Ω is zero, the line integral is independent of path in Ω.

If $\mathbf{F} = \mathbf{V}\varphi$, then, for any piecewise-smooth curve C in Ω from P_0 to P_1,

$$\int_C \mathbf{F} \cdot d\mathbf{R} = \varphi(P_1) - \varphi(P_0).$$

The argument is similar to that for functions of two variables.

We will now provide a simple test to determine whether a vector field is conservative. A hint of a test is provided by the vector identity

$$\mathbf{V} \times (\mathbf{V}\varphi) = \mathbf{O}.$$

If $\mathbf{F} = \mathbf{V}\varphi$, then certanly $\mathbf{V} \times \mathbf{F} = \mathbf{O}$. Is the converse true? That is, if $\mathbf{V} \times \mathbf{F} = \mathbf{O}$, does

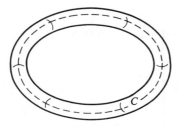

Figure 16.71. A nonsimply connected region in R^3.

$\mathbf{F} = \nabla\varphi$ for some scalar field φ? The answer is no in general but yes with two additional assumptions on the set Ω of points in three-space.

We call Ω a *domain* if (1) around any point in Ω, there is a sphere containing only points of Ω and (2) if P_0 and P_1 are in Ω, there is a piecewise-smooth curve in Ω having P_0 as initial point and P_1 as terminal point. This is a straightforward generalization of the idea of a domain in the plane (Section 16.3).

For example, if Ω consists of all points (x, y, z) with positive coordinates, Ω is a domain. But if Ω consists of all points (x, y, z) with nonnegative coordinates, Ω is not a domain. For example, the origin has no sphere about it containing only points with positive coordinates.

We call Ω *simply connected* if every simple piecewise-smooth closed curve in Ω is the boundary of a piecewise-smooth surface in Ω. To see how this condition might fail to be satisfied, let Ω be the set of points inside the doughnut-shaped surface of Figure 16.71. By choosing C as shown, enclosing the "hole" in the doughnut, we produce a simple closed curve in Ω which is not the boundary of a surface completely contained in Ω.

By contrast, the first octant, consisting of all (x, y, z) with positive coordinates, is simply connected. Figure 16.72 shows a simple piecewise-smooth closed curve C and a surface Σ constructed within Ω so that C is the boundary of Σ.

We can now state a necessary and sufficient condition for a vector field to be conservative over a set of points in three-space.

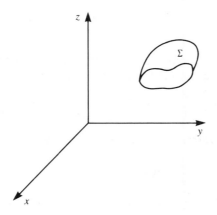

Figure 16.72

THEOREM 16.11

Let Ω be a simply connected domain in three-space. If $\mathbf{F}$ and $\nabla \times \mathbf{F}$ are continuous on Ω, $\mathbf{F}$ is conservative if and only if $\nabla \times \mathbf{F} = \mathbf{O}$ in Ω.

Proof   If $\mathbf{F}$ is conservative, $\mathbf{F} = \nabla\varphi$ for some potential function φ. Then

$$\nabla \times \mathbf{F} = \nabla \times (\nabla\varphi) = \mathbf{O}.$$

Conversely, suppose that $\nabla \times \mathbf{F} = \mathbf{O}$ in Ω. Let C be any piecewise-smooth simple closed curve in Ω. Since Ω is simply connected, C is the boundary of a piecewise-smooth surface Σ lying in Ω. Then, by Stokes's theorem,

$$\int_C \mathbf{F} \cdot d\mathbf{R} = \iint_\Sigma (\nabla \times \mathbf{F}) \cdot \mathbf{N} \, d\sigma = \iint_\Sigma \mathbf{O} \cdot \mathbf{N} \, d\sigma = 0.$$

Thus, the line integral of $\mathbf{F}$ about any simple piecewise-smooth closed curve in Ω is zero; hence, $\int_C \mathbf{F} \cdot d\mathbf{R}$ is independent of path in Ω. ∎

In practice, we find potential functions for vector functions of three variables just as we did for functions of two variables.

EXAMPLE 16.30

Let $\mathbf{F} = 2xy\mathbf{i} + z^2\mathbf{j} + (x - y + z)\mathbf{k}$ for all x, y, and z.

Here, Ω consists of all points (x, y, z). Certainly, Ω is a simply connected domain. But we find that $\nabla \times \mathbf{F} = (-2z - 1)\mathbf{i} - \mathbf{j} - 2x\mathbf{k} \neq \mathbf{O}$; hence, $\mathbf{F}$ is not conservative and has no potential function. ∎

EXAMPLE 16.31

Let $\mathbf{F} = (yze^{xyz} - 4x)\mathbf{i} + (xze^{xyz} + z)\mathbf{j} + (xye^{xyz} + y)\mathbf{k}$ for all x, y and z.

As in Example 16.30, we can choose Ω to be all of three-space. We find that $\nabla \times \mathbf{F} = \mathbf{O}$; hence, $\mathbf{F}$ is conservative. We will find a potential function for $\mathbf{F}$.

To have $\mathbf{F} = \nabla\varphi$, we require that

$$\frac{\partial\varphi}{\partial x} = yze^{xyz} - 4x, \quad \frac{\partial\varphi}{\partial y} = xze^{xyz} + z, \quad \text{and} \quad \frac{\partial\varphi}{\partial z} = xye^{xyz} + y.$$

Choose one of these equations, say, the first one, and integrate with respect to x, treating y and z as constants in the integration. We obtain

$$\varphi(x, y, z) = \int (yze^{xyz} - 4x) \, dx = e^{xyz} - 2x^2 + k(y, z).$$

The "constant" of integration may involve y and z. We now require that

$$\frac{\partial\varphi}{\partial y} = xze^{xyz} + \frac{\partial k}{\partial y} = xze^{xyz} + z.$$

We must choose k so that $\partial k/\partial y = z$. Integrate this equation:

$$\int \frac{\partial k}{\partial y} \, dy = \int z \, dy = zy + c(z),$$

treating z as constant in this integration. Thus far, we have

$$\varphi(x, y, z) = e^{xyz} - 2x^2 + zy + c(z).$$

Finally, we must have

$$\frac{\partial \varphi}{\partial z} = xye^{xyz} + y + c'(z) = xye^{xyz} + y.$$

Thus, we must have $c'(z) = 0$, and we may choose $c(z) = 0$. A potential function for **F** is

$$\varphi(x, y, z) = e^{xyz} - 2x^2 + zy. \quad \blacksquare$$

PROBLEMS FOR SECTION 16.9

In each of Problems 1 through 10, take Ω as all of three-space. Use Theorem 16.11 to determine whether **F** is conservative. If it is, find a potential function.

1. $\mathbf{F} = \cosh(x + y)(\mathbf{i} + \mathbf{j})$
2. $\mathbf{F} = 2x\mathbf{i} - 2y\mathbf{j} + 2z\mathbf{k}$
3. $\mathbf{F} = \mathbf{i} - 2\mathbf{j} + 2z\mathbf{k}$
4. $\mathbf{F} = z\mathbf{i} + \mathbf{j} + x\mathbf{k}$
5. $\mathbf{F} = 2x\mathbf{i} - 2\mathbf{j} + \mathbf{k}$
6. $\mathbf{F} = yz\cos(x)\mathbf{i} + [z\sin(x) + 1]\mathbf{j} + y\sin(x)\mathbf{k}$
7. $\mathbf{F} = (x^2 - 2)\mathbf{i} + xyz\mathbf{j} - yz^2\mathbf{k}$
8. $\mathbf{F} = e^{xyz}[(1 + xyz)\mathbf{i} + x^2z\mathbf{j} + x^2y\mathbf{k}]$
9. $\mathbf{F} = [\cos(x) + y\sin(xy)]\mathbf{i} + x\sin(xy)\mathbf{j} + \mathbf{k}$
10. $\mathbf{F} = (2x^2 + 3y^2z)\mathbf{i} + 6xyz\mathbf{j} + 3xy^2\mathbf{k}$

In each of Problems 11 through 20, find a potential function for **F** and use it to evaluate $\int_C \mathbf{F} \cdot d\mathbf{R}$ for C any piecewise-smooth curve from the first point to the second.

11. $\mathbf{F} = \mathbf{i} - 9y^2z\mathbf{j} - 3y^3\mathbf{k}$; $(1, 1, 1)$ to $(0, 3, 5)$
12. $\mathbf{F} = [y\cos(xz) - xyz\sin(xz)]\mathbf{i} + x\cos(xz)\mathbf{j} - x^2y\sin(xz)\mathbf{k}$; $(1, 0, \pi)$ to $(1, 1, 7)$
13. $\mathbf{F} = 6x^2e^{yz}\mathbf{i} + 2x^3ze^{yz}\mathbf{j} + 2x^3ye^{yz}\mathbf{k}$; $(0, 0, 0)$ to $(1, 2, -1)$
14. $\mathbf{F} = -8y^2\mathbf{i} - (16xy + 4z)\mathbf{j} - 4y\mathbf{k}$; $(-2, 1, 1)$ to $(1, 3, 2)$
15. $\mathbf{F} = -\mathbf{i} + 2z^2\mathbf{j} + 4yz\mathbf{k}$; $(0, 0, -4)$ to $(1, 1, 6)$
16. $\mathbf{F} = (y - 4xz)\mathbf{i} + x\mathbf{j} + (3z^2 - 2x^2)\mathbf{k}$; $(1, 1, 1)$ to $(3, 1, 4)$
17. $\mathbf{F} = (4y^3 - 8x)\mathbf{i} + 12xy^2\mathbf{j} - 8z\mathbf{k}$; $(-1, 2, 2)$ to $(0, 1, 6)$
18. $\mathbf{F} = z\sin(yz)\mathbf{i} + xz^2\cos(yz)\mathbf{j} + [xyz\cos(yz) + x\sin(yz)]\mathbf{k}$; $(0, 1, 1)$ to $(1, 7, -2)$
19. $\mathbf{F} = yz\cosh(xy)\mathbf{i} + xz\cosh(xy)\mathbf{j} + \sinh(xy)\mathbf{k}$; $(1, 1, 3)$ to $(1, -1, 4)$
20. $\mathbf{F} = (yz^2 - 2xz)\mathbf{i} + xz^2\mathbf{j} + (2xyz - x^2)\mathbf{k}$; $(-1, 4, 4)$ to $(0, 0, 0)$
21. Prove that

$$\iint_\Sigma [(\nabla f) \times (\nabla g)] \cdot \mathbf{N} \, d\sigma = \oint_C (f\nabla g) \cdot d\mathbf{R}$$

if Σ is a piecewise-smooth surface bounded by the piecewise-smooth closed curve C. Here, $\mathbf{R} = x\mathbf{i} + y\mathbf{j} + z\mathbf{k}$.

22. Use Maxwell's equation $\nabla \times \mathbf{H} = \mathbf{O}$ to show that the net electrical current enclosed by a piecewise-smooth simple closed curve C is

$$\oint_C \mathbf{H} \cdot d\mathbf{R},$$

under the assumption that $\mathbf{E} = \mathbf{O}$.

23. Assume that $\nabla \times \mathbf{B} = \mu_0 \mathbf{J}$. Show that

$$\oint_C \mathbf{B} \cdot d\mathbf{R} = \mu_0 i$$

for any piecewise-smooth simple closed curve C.

ADDITIONAL PROBLEMS

In each of Problems 1 through 10, compute the line integral.

1. $\int_C x^2 y \, dx - y \, dy$, $\quad C$ the straight line from $(-2, 3)$ to $(1, 1)$

2. $\int_C (xz\mathbf{i} - y\mathbf{j} + xz\mathbf{k}) \cdot d\mathbf{R}$, $\quad C$ the straight line from $(1, 1, 0)$ to $(-2, 3, -1)$

3. $\int_C xy \, dz$, $\quad C$ the straight line from the origin to $(1, 1, -2)$

4. $\int_C -4xy \, dx + zx \, dy - x \, dz$, $\quad C$ given by $x = 2t$, $\quad y = 3t^2$, $\quad z = t$; $\quad 1 \le t \le 2$

5. $\int_C y \cos(z) \, dx$, $\quad C$ the straight line from $(1, 1, 1)$ to $(3, 1, 1)$

6. $\int_C (x\mathbf{i} - y\mathbf{j} + xyz\mathbf{k}) \cdot d\mathbf{R}$, $\quad C$ the parabola $y = x^2$, $\quad z = 4$, with $\quad 1 \le x \le 3$

7. $\int_C 4xy \, dx - xz \, dy + e^{-z} \, dz$, $\quad C$ given by $x = y = t$, $\quad z = -t$, with $-1 \le t \le 1$

8. $\int_C -5yz \, dy - z^3 \, dz$, $\quad C$ the straight line from $(1, 1, 1)$ to $(-4, 2, 5)$

9. $\int_C xz \, ds$, $\quad C$ the circle $x = 2\cos(t)$, $\quad y = 2$, $\quad z = 2\sin(t)$; $\quad 0 \le t \le 2\pi$

10. $\int_C x^3 \, dx - 4yz \, dz$, $\quad C$ given by $x = 1$, $\quad y = 2\cos(t)$, $z = 3\sin(t)$; $\quad 0 \le t \le \pi$

In each of Problems 11 through 15, verify Green's theorem for the given vector field and curve. All curves are oriented counterclockwise.

11. $\mathbf{F} = xy\mathbf{i} - y\mathbf{j}$, $\quad C$ the circle $x^2 + y^2 = 9$
12. $\mathbf{F} = -x\mathbf{j}$, $\quad C$ the triangle with vertices $(1, 0)$, $(2, 0)$, $(2, 4)$
13. $\mathbf{F} = xy^2\mathbf{i}$, $\quad C$ the square with vertices $(0, 0)$, $(1, 0)$, $(0, 1)$, $(1, 1)$
14. $\mathbf{F} = -2y\mathbf{i} + x\mathbf{j}$, $\quad C$ the semicircle $x^2 + y^2 = 4$, $y \ge 0$ and the segment of the x-axis from -2 to 2
15. $\mathbf{F} = \mathbf{i} - y\mathbf{j}$, $\quad C$ the triangle with vertices $(1, 2)$, $(1, 4)$, $(2, 2)$

In each of Problems 16 through 20, verify Gauss's theorem for the given vector field and surface.

16. $\mathbf{F} = -5x\mathbf{i} + y\mathbf{j} - z\mathbf{k}$, $\quad \Sigma$ the surface of the cube with four of its vertices at $(0, 0, 0), (1, 0, 0), (0, 1, 0)$ and $(0, 0, 1)$
17. $\mathbf{F} = x^2\mathbf{i}$, $\quad \Sigma$ the surface consisting of the hemisphere $x^2 + y^2 + z^2 = 1$, $z \ge 0$ and the disk $x^2 + y^2 \le 1$, $z = 0$ in the xy-plane
18. $\mathbf{F} = x\mathbf{i} + y\mathbf{j} + z\mathbf{k}$, $\quad \Sigma$ the surface consisting of the cylinder $x^2 + y^2 = 1$, $0 \le z \le 4$ together with the top and bottom disks, $x^2 + y^2 \le 1$, $z = 0$; and $x^2 + y^2 \le 1$, $z = 4$
19. $\mathbf{F} = z\mathbf{k}$, $\quad \Sigma$ the surface consisting of the cone $z = \sqrt{x^2 + y^2}$ for $0 \le x^2 + y^2 \le 4$ together with the disk $x^2 + y^2 \le 4$, $z = 2$
20. $\mathbf{F} = yz\mathbf{j}$, $\quad \Sigma$ the surface consisting of the paraboloid $z = x^2 + y^2$ for $0 \le z \le 9$ and the disk $x^2 + y^2 \le 9$, $z = 9$

In each of Problems 21 through 25, verify Stokes's theorem for the given vector field and surface.

21. $\mathbf{F} = y\mathbf{i}$, $\quad \Sigma$ the hemisphere $x^2 + y^2 + z^2 = 4$, $z \ge 0$
22. $\mathbf{F} = -x\mathbf{j} + y\mathbf{k}$, $\quad \Sigma$ the cone $z = \sqrt{x^2 + y^2}$ for $0 \le z \le 4$
23. $\mathbf{F} = xz\mathbf{i}$, $\quad \Sigma$ the frustrum of the cone $z = \sqrt{x^2 + y^2}$ given by $1 \le z \le 9$
24. $\mathbf{F} = yz\mathbf{i} + x\mathbf{j}$, $\quad \Sigma$ the parabolic bowl $z = x^2 + y^2$, $0 \le z \le 9$
25. $\mathbf{F} = x\mathbf{i} - y\mathbf{j} + z\mathbf{k}$, $\quad \Sigma$ the disk $x^2 + y^2 \le 5$ in the xy-plane

In each of Problems 26 through 30, find a potential function for $\mathbf{F}$ and use this to evaluate the line integral of $\mathbf{F}$ over any piecewise-smooth curve from the first given point to the second.

26. $\mathbf{F} = (z^2 - 2)\mathbf{i} - 2x\mathbf{j} + 2xz\mathbf{k}$; from $(0, 6, -2)$ to $(1, 1, -4)$
27. $\mathbf{F} = 2x\mathbf{i} - 3z\mathbf{j} + [-3y - \sin(z)]\mathbf{k}$; from $(0, 0, 0)$ to $(1, 1, 4)$
28. $\mathbf{F} = -z \sin(xz)\mathbf{i} - ze^{yz}\mathbf{j} + [-x\sin(xz) - ye^{yz}]\mathbf{k}$; from $(1, 0, 0)$ to $(-1, 0, -2)$
29. $\mathbf{F} = \mathbf{i} - 2y\mathbf{j} + \mathbf{k}$; from $(-2, 2, 5)$ to $(0, 0, 0)$
30. $\mathbf{F} = 3x^2\mathbf{i} - z^2\mathbf{j} - 2yz\mathbf{k}$; from $(-5, 2, 2)$ to $(1, -1, 3)$
31. Let C be a smooth simple closed curve in the xy-plane, and let f and g be continuous functions on C and its

interior D. Suppose that the partial derivatives of f and g are also continuous on C and throughout D. Let $\mathbf{T}$ be the unit tangent to C, and let $\mathbf{N}$ be the unit outer normal.

(a) Prove that

$$\int_C \left(f \frac{\partial g}{\partial \eta} - g \frac{\partial f}{\partial \eta} \right) ds = \iint_D \left[f(\mathbf{V} \cdot \mathbf{V}g) - g(\mathbf{V} \cdot \mathbf{V}f) \right] dA.$$

Here, $\mathbf{V} = (\partial/\partial x)\mathbf{i} + (\partial/\partial y)\mathbf{j}$, and $\partial/\partial \eta$ is the directional derivative in the direction of the unit outer normal to C.

(b) Show that, if $\mathbf{V} \cdot \mathbf{V}g = 0$ in D, $\int_C (\partial g/\partial \eta)\, ds = 0$.

(c) Show that, if $\mathbf{V} \cdot \mathbf{V}f = \mathbf{V} \cdot \mathbf{V}g = 0$ for (x, y) in D,

$$\int_C f \frac{\partial g}{\partial \eta}\, ds = \int_C g \frac{\partial f}{\partial \eta}\, ds.$$

APPENDIX A: *Orthogonal Curvilinear Coordinates*

Often, we use different coordinate systems depending upon the setting in which we are doing our computations. Familiar systems include rectangular, cylindrical, and spherical coordinates. We will now see that each of these coordinate systems shares an important property which forms the basis for a very general theory of coordinate systems. This theory is used to construct specialized systems for use in specialized settings.

Two surfaces are said to be *orthogonal* if their normal vectors are orthogonal at any points where the surfaces intersect. We will begin by relating this concept to familiar coordinate systems.

RECTANGULAR COORDINATES

The rectangular coordinates of a point in three-space are usually given as an ordered triple (x, y, z) with x, y, and z measured against three mutually perpendicular axes, as shown in Figure A.1. This coordinate system is very familiar. However, we will look at a feature which sometimes goes unnoticed.

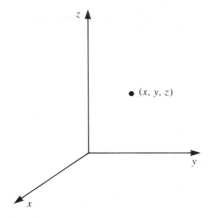

Figure A.1

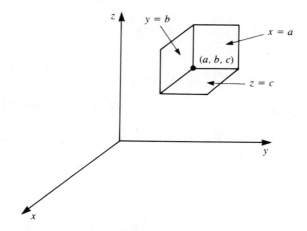

Figure A.2

We obtain a *coordinate surface* by letting one coordinate equal a constant. The three coordinate surfaces $x = a$, $y = b$, and $z = c$ are mutually orthogonal surfaces in the sense that any two of them intersect at right angles. Further, each point (a, b, c) in three-space is the intersection of three coordinate surfaces, namely, the planes $x = a$, $y = b$, and $z = c$. This fact is indicated in Figure A.2.

These coordinate surfaces (planes, in this case) are mutually orthogonal. Thus, at least for rectangular coordinates, each point in three-space is the point of intersection of three mutually orthogonal coordinate surfaces. We will now observe that this is true for cylindrical and spherical coordinates as well.

CYLINDRICAL COORDINATES

Any point (x, y, z) in three-space has cylindrical coordinates (r, θ, z), where (r, θ) are polar coordinates of the point (x, y) in the xy-plane. These coordinates are shown in Figure A.3.

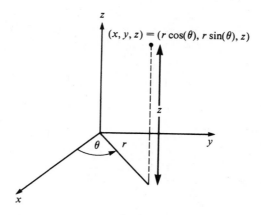

Figure A.3. Cylindrical coordinates.

   Now look at the coordinate surfaces obtained by letting each coordinate equal a constant. The surface $r = a$ in three-space is a cylinder of radius a about the origin (Figure A.4). The surface $\theta = b$ is a plane through the z-axis making an angle θ with the positive x-axis, as shown in Figure A.5. And the surface $z = c$ is a plane parallel to the xy-plane, as in Figure A.6. Any point in three-space not on the z-axis with cylindrical coordinates (a, b, c) is the intersection of the cylinder $r = a$ with the planes $\theta = b$ and $z = c$. This is shown in Figure A.7.

   Further, the cylinder $z = a$ and the planes $\theta = b$ and $z = c$ are mutually orthogonal surfaces; any two of them intersect at right angles. Thus, each point in three-space is the intersection of three mutually orthogonal (cylindrical coordinate) surfaces.

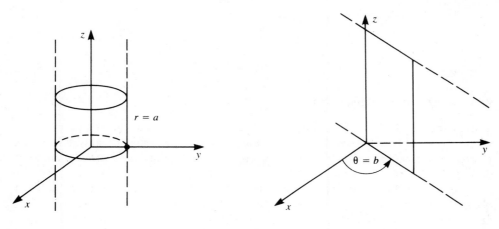

Figure A.4   Figure A.5

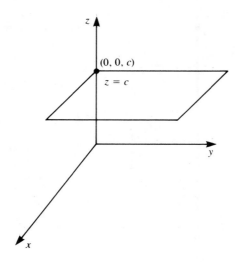

Figure A.6

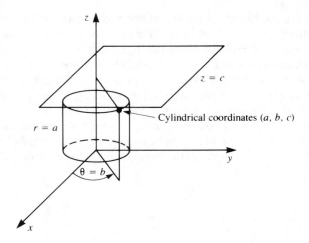

Figure A.7

SPHERICAL COORDINATES

The spherical coordinates of a point in three-space are (ρ, θ, φ), as shown in Figure A.8, where ρ is the distance from the origin to the point, θ is the angle of rotation from the positive x-axis to the projection of the point into the xy-plane, and φ is the angle of declination from the positive z-axis.

Figure A.9 shows the coordinate surfaces. A surface $\rho = a$ is a sphere of radius a about the origin; a surface $\theta = b$ is a plane making an angle b with the positive x-axis; and a surface $\varphi = c$ is a cone whose slanted side makes an angle c with the positive z-axis. Any two of these coordinate surfaces intersect at right angles; hence, these coordinate surfaces are mutually orthogonal. Further, any point in three-space is the intersection of three coordinate surfaces.

There is nothing unique about planes, spheres, and cones in their forming mutually orthogonal surfaces whose intersections can be used to specify points in three-

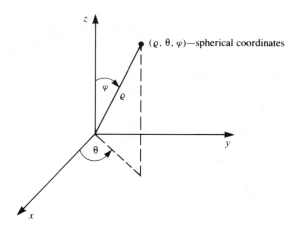

Figure A.8

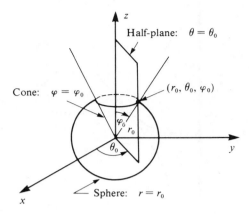

Figure A.9. Spherical coordinates of a point at the intersection of mutually orthogonal coordinate surfaces.

space. In fact, we can construct a coordinate system by using any three types of mutually orthogonal surfaces, with the property that each point in three-space can be specified as the point of intersection of three of these surfaces.

To describe this idea mathematically, suppose that x, y, and z are functions of variables q_1, q_2, and q_3:

$$x = x(q_1, q_2, q_3), \qquad y = y(q_1, q_2, q_3), \qquad z = z(q_1, q_2, q_3).$$

We will assume that

1. The equations can be solved for q_1, q_2, and q_3 in terms of x, y, and z.

2. Each point in space can be described by exactly one triple of numbers (q_1, q_2, q_3).

In this event, we say that q_1, q_2, and q_3 form a system of *curvilinear coordinates*. If, in addition, the coordinate surfaces $q_1 = k_1$, $q_2 = k_2$, and $q_3 = k_3$ are mutually orthogonal for every triple (k_1, k_2, k_3) of real numbers, we say that q_1, q_2, and q_3 constitute a system of *orthogonal curvilinear coordinates*.

At any point P_0 where two of these coordinate surfaces intersect, their normal vectors are orthogonal. Since the gradient vector is normal to a surface, this means that $\nabla q_1(P_0)$, $\nabla q_2(P_0)$, and $\nabla q_3(P_0)$ are mutually orthogonal if the three coordinate surfaces $q_1 = k_1$, $q_2 = k_2$, and $q_3 = k_3$ intersect at P_0. This is illustrated in Figure A.10.

We will now consider the problem of computing various quantities such as arc length, gradient, divergence, and curl in terms of general orthogonal curvilinear coordinates. In rectangular coordinates, the differential element of arc length along a curve is

$$(ds)^2 = (dx)^2 + (dy)^2 + (dz)^2. \tag{A.1}$$

We will assume that this differential, in terms of x, y, and z, is related to differentials in terms of q_1, q_2, and q_3 by a quadratic form

$$(ds)^2 = \sum_{i=1}^{3} \sum_{j=1}^{3} h_{ij}^2 \, dq_i \, dq_j,$$

in which $h_{ij} = h_{ji}$. The numbers h_{ij} are called *scale factors*. Given an orthogonal

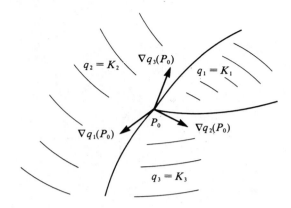

Figure A.10. Orthogonal curvilinear coordinates.

curvilinear coordinate system, we attempt to determine the scale factors in order to calculate such quantities as area, volume, and arc length in the new coordinate system.

To see how to do this, begin with the equations

$$x = x(q_1, q_2, q_3), \qquad y = y(q_1, q_2, q_3), \qquad z = z(q_1, q_2, q_3).$$

Compute the total differentials

$$dx = \frac{\partial x}{\partial q_1} \, dq_1 + \frac{\partial x}{\partial q_2} \, dq_2 + \frac{\partial x}{\partial q_3} \, dq_3,$$

$$dy = \frac{\partial y}{\partial q_1} \, dq_1 + \frac{\partial y}{\partial q_2} \, dq_2 + \frac{\partial y}{\partial q_3} \, dq_3,$$

and

$$dz = \frac{\partial z}{\partial q_1} \, dq_1 + \frac{\partial z}{\partial q_2} \, dq_2 + \frac{\partial z}{\partial q_3} \, dq_3.$$

Substitute these expressions into equation (A.1) to obtain

$$
\begin{aligned}
(ds)^2 &= \left(\frac{\partial x}{\partial q_1}\right)^2 (dq_1)^2 + \left(\frac{\partial x}{\partial q_2}\right)^2 (dq_2)^2 + \left(\frac{\partial x}{\partial q_3}\right)^2 (dq_3)^2 \\
&\quad + 2\frac{\partial x}{\partial q_1}\frac{\partial x}{\partial q_2} \, dq_1 \, dq_2 + 2\frac{\partial x}{\partial q_1}\frac{\partial x}{\partial q_3} \, dq_1 \, dq_3 + 2\frac{\partial x}{\partial q_2}\frac{\partial x}{\partial q_3} \, dq_2 \, dq_3 \\
&\quad + \left(\frac{\partial y}{\partial q_1}\right)^2 (dq_1)^2 + \left(\frac{\partial y}{\partial q_2}\right)^2 (dq_2)^2 + \left(\frac{\partial y}{\partial q_3}\right)^2 (dq_3)^2 \\
&\quad + 2\frac{\partial y}{\partial q_1}\frac{\partial y}{\partial q_2} \, dq_1 \, dq_2 + 2\frac{\partial y}{\partial q_1}\frac{\partial y}{\partial q_3} \, dq_1 \, dq_3 + 2\frac{\partial y}{\partial q_2}\frac{\partial y}{\partial q_3} \, dq_2 \, dq_3 \\
&\quad + \left(\frac{\partial z}{\partial q_1}\right)^2 (dq_1)^2 + \left(\frac{\partial z}{\partial q_2}\right)^2 (dq_2)^2 + \left(\frac{\partial z}{\partial q_3}\right)^2 (dq_3)^2 \\
&\quad + 2\frac{\partial z}{\partial q_1}\frac{\partial z}{\partial q_2} \, dq_1 \, dq_2 + 2\frac{\partial z}{\partial q_1}\frac{\partial z}{\partial q_3} \, dq_1 \, dq_3 + 2\frac{\partial z}{\partial q_2}\frac{\partial z}{\partial q_3} \, dq_2 \, dq_3 \\
&= \sum_{i=1}^{3} \sum_{j=1}^{3} h_{ij}^2 \, dq_i \, dq_j.
\end{aligned}
$$

By equating like coefficients of $dq_i\, dq_j$ on both sides of this equation, we obtain

$$h_{11}^2 = \left(\frac{\partial x}{\partial q_1}\right)^2 + \left(\frac{\partial y}{\partial q_1}\right)^2 + \left(\frac{\partial z}{\partial q_1}\right)^2,$$

$$h_{22}^2 = \left(\frac{\partial x}{\partial q_2}\right)^2 + \left(\frac{\partial y}{\partial q_2}\right)^2 + \left(\frac{\partial z}{\partial q_2}\right)^2,$$

and

$$h_{33}^2 = \left(\frac{\partial x}{\partial q_3}\right)^2 + \left(\frac{\partial y}{\partial q_3}\right)^2 + \left(\frac{\partial z}{\partial q_3}\right)^2.$$

For $i \neq j$, we get $h_{ij} = 0$, assuming that we have an orthogonal curvilinear coordinate system. For example,

$$h_{12}^2 = \frac{\partial x}{\partial q_1}\frac{\partial x}{\partial q_2} + \frac{\partial y}{\partial q_1}\frac{\partial y}{\partial q_2} + \frac{\partial z}{\partial q_1}\frac{\partial z}{\partial q_2}$$

$$= \left(\frac{\partial x}{\partial q_1}\mathbf{i} + \frac{\partial y}{\partial q_1}\mathbf{j} + \frac{\partial z}{\partial q_1}\mathbf{k}\right) \cdot \left(\frac{\partial x}{\partial q_2}\mathbf{i} + \frac{\partial y}{\partial q_2}\mathbf{j} + \frac{\partial z}{\partial q_2}\mathbf{k}\right) = 0.$$

under the assumption that these two vectors are orthogonal.

As a convenience for the rest of this section, we will denote h_{jj} as just h_j.

EXAMPLE A.1

Consider spherical coordinates ρ, θ, and φ. In the above notation, we have $q_1 = \rho$, $q_2 = \theta$, and $q_3 = \varphi$. Spherical and rectangular coordinates are related by the equations

$$x = \rho\, \cos(\theta)\sin(\varphi), \qquad y = \rho\, \sin(\theta)\sin(\varphi), \qquad z = \rho\, \cos(\varphi),$$

$0 \leq \theta \leq 2\pi, 0 \leq \varphi \leq \pi$.

Compute

$$dx = \frac{\partial x}{\partial \rho}\, d\rho + \frac{\partial x}{\partial \theta}\, d\theta + \frac{\partial x}{\partial \varphi}\, d\varphi$$

$$= \cos(\theta)\sin(\varphi)\, d\rho - \rho\, \sin(\theta)\sin(\varphi)\, d\theta + \rho\, \cos(\theta)\cos(\varphi)\, d\varphi.$$

Similarly,

$$dy = \sin(\theta)\sin(\varphi)\, d\rho + \rho\, \cos(\theta)\sin(\varphi)\, d\theta + \rho\, \sin(\theta)\cos(\varphi)\, d\varphi$$

and

$$dz = \cos(\varphi)\, d\rho - \rho\, \sin(\varphi)\, d\varphi.$$

Compute

$$(ds)^2 = (dx)^2 + (dy)^2 + (dz)^2$$
$$= [\cos^2(\theta)\sin^2(\varphi) + \sin^2(\theta)\sin^2(\varphi) + \cos^2(\varphi)](d\rho)^2$$
$$+ [\rho^2\sin^2(\theta)\sin^2(\varphi) + \rho^2\cos^2(\theta)\sin^2(\varphi)](d\theta)^2$$
$$+ [\rho^2\cos^2(\theta)\cos^2(\varphi) + \rho^2\sin^2(\theta)\cos^2(\varphi) + \rho^2\sin^2(\varphi)](d\varphi)^2$$
$$+ [-2\rho\, \sin(\theta)\cos(\theta)\sin^2(\varphi) + 2\rho\, \sin(\theta)\cos(\theta)\sin^2(\varphi)]\, d\rho\, d\theta$$
$$+ [-2\rho^2\cos(\theta)\sin(\theta)\cos(\varphi)\sin(\varphi) + 2\rho^2\cos(\theta)\sin(\theta)\cos(\varphi)\sin(\varphi)]\, d\theta\, d\varphi$$
$$+ [2\rho\, \sin^2(\theta)\sin(\varphi)\cos(\varphi) - 2\rho\, \cos(\varphi)\sin(\varphi) + 2\rho\, \cos^2(\theta)\sin(\varphi)\cos(\varphi)]\, d\rho\, d\varphi$$
$$= (d\rho)^2 + \rho^2\sin^2(\varphi)(d\theta)^2 + \rho^2(d\varphi)^2.$$

The scale factors relating spherical and rectangular coordinates are

$$h_\rho^2 = \text{coefficient of } (d\rho)^2 = 1,$$
$$h_\theta^2 = \text{coefficient of } (d\theta)^2 = \rho^2 \sin^2(\varphi),$$

and

$$h_\varphi^2 = \text{coefficient of } (d\varphi)^2 = \rho^2.$$

Here we have used the names of the coordinates in the subscripts, rather than subscripts 1, 2, and 3, for ease in relating the scale factors to the coordinates. We therefore have, in spherical coordinates, the differential element of arc length

$$ds = \sqrt{(d\rho)^2 + \rho^2 \sin^2(\varphi)(d\theta)^2 + \rho^2(d\varphi)^2}.$$

For example, if a curve is given in spherical coordinates by parametric equations

$$\rho = \rho(t), \qquad \theta = \theta(t), \qquad \varphi = \varphi(t)$$

for $a \leq t \leq b$, then

$$ds = \sqrt{[\rho'(t)]^2 + \rho^2 \sin^2(\varphi)[\theta'(t)]^2 + \rho^2[\varphi'(t)]^2}\; dt,$$

and the length of the curve is

$$\int_a^b ds = \int_a^b \sqrt{[\rho'(t)]^2 + \rho^2 \sin^2(\varphi)[\theta'(t)]^2 + \rho^2[\varphi'(t)]^2}\; dt.$$

If we write $ds_\rho = h_\rho\, d\rho$, $ds_\theta = h_\theta\, d\theta$, and $ds_\varphi = h_\varphi\, d\varphi$, we have the differential elements of arc length in the ρ, θ, and φ directions, respectively, in the spherical coordinate system. The differential elements of area are

$$ds_\rho\, ds_\theta = h_\rho h_\theta\, d\rho\, d\theta = \rho\, \sin(\varphi)\, d\rho\, d\theta,$$
$$ds_\rho\, ds_\varphi = h_\rho h_\varphi\, d\rho\, d\varphi = \rho\, d\rho\, d\varphi,$$

and

$$ds_\theta\, ds_\varphi = \rho^2 \sin(\varphi)\, d\theta\, d\varphi.$$

Thus, $ds_\rho\, ds_\theta$ is the area of a "rectangle" having sides of the form $\rho = $ constant, $\theta = $ constant, with similar interpretations for the other two area elements. These area elements correspond to $dx\, dy$, $dx\, dz$, and $dy\, dz$ in rectangular coordinates.

The differential element of volume in spherical coordinates is

$$ds_\rho\, ds_\theta\, ds_\varphi = \rho^2 \sin(\varphi)\, d\rho\, d\theta\, d\varphi.$$

This corresponds to the differential element of volume $dV = dx\, dy\, dz$ in rectangular coordinates. We can recognize $\rho^2 \sin(\varphi)$ as the factor which appears in a triple integral upon changing variables from rectangular to spherical coordinates. ∎

As noted at the end of this example, expressions for the differential elements of area and volume in curvilinear coordinates are involved in formulas for change of variables in double and triple integrals from one coordinate system to another.

Now let $\mathbf{u}_i$ be a unit vector in the direction of increasing q_i at any point $(x(q_1, q_2, q_3), y(q_1, q_2, q_3), z(q_1, q_2, q_3))$. For example, in spherical coordinates, these vectors can be written in terms of the rectangular $\mathbf{i}$, $\mathbf{j}$, and $\mathbf{k}$ by

$$\mathbf{u}_\rho = \cos(\theta)\sin(\varphi)\mathbf{i} + \sin(\theta)\sin(\varphi)\mathbf{j} + \cos(\varphi)\mathbf{k},$$
$$\mathbf{u}_\theta = -\sin(\theta)\mathbf{i} + \cos(\theta)\mathbf{j},$$
$$\mathbf{u}_\varphi = \cos(\theta)\cos(\varphi)\mathbf{i} + \sin(\theta)\cos(\varphi)\mathbf{j} - \sin(\varphi)\mathbf{k}.$$

Unlike rectangular coordinates, where $\mathbf{i}$, $\mathbf{j}$, and $\mathbf{k}$ are fixed, in general $\mathbf{u}_1$, $\mathbf{u}_2$, and $\mathbf{u}_3$ will vary from point to point.

A vector field in curvilinear coordinates will have the form

$$\mathbf{F}(q_1, q_2, q_3) = F_1(q_1, q_2, q_3)\mathbf{u}_1 + F_2(q_1, q_2, q_3)\mathbf{u}_2 + F_3(q_1, q_2, q_3)\mathbf{u}_3.$$

A scalar field has the form $\psi(q_1, q_2, q_3)$, a real-valued function of the three coordinates.

We want to write expressions for the gradient of a scalar field and the divergence and curl of a vector field in curvilinear coordinates.

For the gradient, argue that at any point $P:(q_1, q_2, q_3)$, $\nabla\psi$ should be a vector normal to the surface $\psi = $ constant passing through P and that $\nabla\psi$ should have magnitude equal to the maximum rate of change of ψ from that point. Thus, the component of $\nabla\psi$ normal to the coordinate surface $q_1 = $ constant is $\partial\psi/\partial s_1$, or $(1/h_1)(\partial\psi/\partial q_1)$. Arguing in like fashion for the other components, we have

$$\nabla\psi(q_1, q_2, q_3) = \frac{1}{h_1}\frac{\partial\psi}{\partial q_1}\mathbf{u}_1 + \frac{1}{h_2}\frac{\partial\psi}{\partial q_2}\mathbf{u}_2 + \frac{1}{h_3}\frac{\partial\psi}{\partial q_3}\mathbf{u}_3.$$

For the divergence of a vector field $\mathbf{F}$ in curvilinear coordinates, we will use the flux interpretation of divergence. Write

$$\mathbf{F} = F_1\mathbf{u}_1 + F_2\mathbf{u}_2 + F_3\mathbf{u}_3.$$

Refer to Figure A.11. The flux across face $abcd$ is

$$\mathbf{F}(q_1 + dq_1, q_2, q_3) \cdot \mathbf{u}_1 h_2(q_1 + dq_1, q_2, q_3)h_3(q_1 + dq_1, q_2, q_3)\, dq_2\, dq_3.$$

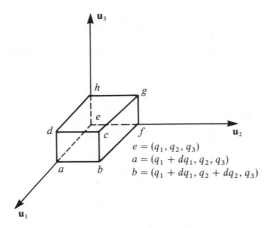

$$e = (q_1, q_2, q_3)$$
$$a = (q_1 + dq_1, q_2, q_3)$$
$$b = (q_1 + dq_1, q_2 + dq_2, q_3)$$

Figure A.11

Across face *efgh*, the flux is

$$-\mathbf{F}(q_1, q_2, q_3) \cdot \mathbf{u}_1 h_2(q_1, q_2, q_3) h_3(q_1, q_2, q_3) \, dq_2 \, dq_3.$$

Across both faces, the flux is approximately

$$\frac{\partial}{\partial q_1} [F_1 h_2 h_3] \, dq_1 \, dq_2 \, dq_3.$$

Similarly, the fluxes across the other pairs of opposite faces are

$$\frac{\partial}{\partial q_2} [F_2 h_1 h_3] \, dq_1 \, dq_2 \, dq_3 \quad \text{and} \quad \frac{\partial}{\partial q_3} [F_3 h_1 h_2] \, dq_1 \, dq_2 \, dq_3.$$

Upon adding these results, we conclude that the flux per unit volume at P, or the divergence of $\mathbf{F}$ at P, is

$$\mathbf{V} \cdot \mathbf{F}(P) = \frac{1}{h_1 h_2 h_3} \left(\frac{\partial}{\partial q_1} [F_1 h_2 h_3] + \frac{\partial}{\partial q_2} [F_2 h_1 h_3] + \frac{\partial}{\partial q_3} [F_3 h_1 h_2] \right).$$

From this, we obtain an expression for the Laplacian in curvilinear coordinates with very little additional effort. In rectangular coordinates,

$$\nabla^2 f(x, y, z) = \mathbf{V} \cdot \mathbf{V} f = \frac{\partial^2 f}{\partial x^2} + \frac{\partial^2 f}{\partial y^2} + \frac{\partial^2 f}{\partial z^2}.$$

In orthogonal curvilinear coordinates,

$$\nabla^2 \psi(q_1, q_2, q_3) = \mathbf{V} \cdot \mathbf{V} \psi(q_1, q_2, q_3)$$
$$= \frac{1}{h_1 h_2 h_3} \left[\frac{\partial}{\partial q_1} \left(\frac{h_2 h_3}{h_1} \frac{\partial \psi}{\partial q_1} \right) + \frac{\partial}{\partial q_2} \left(\frac{h_1 h_3}{h_2} \frac{\partial \psi}{\partial q_2} \right) + \frac{\partial}{\partial q_3} \left(\frac{h_1 h_2}{h_3} \frac{\partial \psi}{\partial q_3} \right) \right].$$

Finally, we want to express the curl of a vector in curvilinear coordinates. Here, we will use the interpretation of the normal component of the curl, $(\mathbf{V} \times \mathbf{F}) \cdot \mathbf{N}$, as the swirl of a fluid with velocity $\mathbf{F}$ about a point in a plane normal to $\mathbf{N}$. At point P, the component of $\mathbf{V} \times \mathbf{F}$ in the direction $\mathbf{u}_1$ is

$$\lim_{A \to 0} \frac{1}{A} \oint_C \mathbf{F} \cdot \mathbf{N} \, ds,$$

where C may be taken as a curvilinear rectangle about P in the $\mathbf{u}_2 \mathbf{u}_3$-plane, as in Figure A.12, and A is the area of this rectangle. Now,

$$A = ds_2 \, ds_3 = h_2 h_3 \, dq_2 \, dq_3.$$

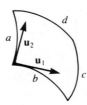

Figure A.12

To compute the line integral, look at the line integral over each side a, b, c, and d of the rectangle. On the side labeled a, $\int_c \mathbf{F} \cdot \mathbf{N}\, ds$ is approximately equal to $F_2(q_1, q_2, q_3)h_2(q_1, q_2, q_3)\, dq_2$ because $\mathbf{u}_2$ is tangent to this side. On side c, $\int_c \mathbf{F} \cdot \mathbf{N}\, ds$ is approximately $-F_2(q_1, q_2, q_3 + dq_3)h_2(q_1, q_2, q_3 + dq_3)\, dq_2$. The net contribution from sides a and c is approximately $-(\partial/\partial q_3)[F_2 h_2]\, dq_2\, dq_3$. Similarly, the contribution to the line integral from sides b and d is approximately $(\partial/\partial q_2)[F_3 h_3]\, dq_2\, dq_3$. Thus,

$$(\nabla \times \mathbf{F}) \cdot \mathbf{u}_1 = \frac{1}{h_2 h_3\, dq_2\, dq_3}\left(\frac{\partial}{\partial q_2}[F_3 h_3] - \frac{\partial}{\partial q_3}[F_2 h_2]\right) dq_2\, dq_3.$$

We obtain the other components $(\nabla \times \mathbf{F}) \cdot \mathbf{u}_2$ and $(\nabla \times \mathbf{F}) \cdot \mathbf{u}_3$ of the curl in similar fashion. We obtain

$$\begin{aligned}
\nabla \times \mathbf{F}(q_1, q_2, q_3) = {} & \frac{1}{h_2 h_3}\left(\frac{\partial}{\partial q_2}[F_3 h_3] - \frac{\partial}{\partial q_3}[F_2 h_2]\right)\mathbf{u}_1 \\
& + \frac{1}{h_1 h_3}\left(\frac{\partial}{\partial q_3}[F_1 h_1] - \frac{\partial}{\partial q_1}[F_3 h_3]\right)\mathbf{u}_2 \\
& + \frac{1}{h_1 h_2}\left(\frac{\partial}{\partial q_1}[F_2 h_2] - \frac{\partial}{\partial q_2}[F_1 h_1]\right)\mathbf{u}_3.
\end{aligned}$$

This expression can be written in determinant form as

$$\nabla \times \mathbf{F} = \frac{1}{h_1 h_2 h_3}\begin{vmatrix} h_1 \mathbf{u}_1 & h_2 \mathbf{u}_2 & h_3 \mathbf{u}_3 \\ \dfrac{\partial}{\partial q_1} & \dfrac{\partial}{\partial q_2} & \dfrac{\partial}{\partial q_3} \\ F_1 h_1 & F_2 h_2 & F_3 h_3 \end{vmatrix}.$$

In the problems that follow, several curvilinear coordinate systems are defined; the student is asked to compute the scale factors, gradient, divergence, and curl for these systems.

PROBLEMS FOR APPENDIX A

1. Compute the scale factors for cylindrical coordinates. Using these, write expressions for $\nabla \cdot \mathbf{F}$ and $\nabla \times \mathbf{F}$ if $\mathbf{F}(r, \theta, z)$ is a vector field in cylindrical coordinates. Write an expression for $\nabla \psi$ if $\psi(r, \theta, z)$ is a scalar field in cylindrical coordinates. Also write the Laplacian of a scalar field in cylindrical coordinates.

2. The scale factors for spherical coordinates were computed in this section. Using these, show that the divergence of a vector field $\mathbf{F}(\rho, \theta, \varphi)$ in spherical coordinates is

$$\nabla \cdot \mathbf{F} = \frac{1}{\rho^2}\frac{\partial}{\partial \rho}(\rho^2 F_\rho) + \frac{1}{\rho \sin(\varphi)}\frac{\partial}{\partial \theta}F_\theta + \frac{1}{\rho \sin(\varphi)}\frac{\partial}{\partial \varphi}[F_\varphi \sin(\varphi)].$$

Also compute the curl of $\mathbf{F}$ and the gradient and Laplacian of a scalar field $\psi(\rho, \theta, \varphi)$.

3. Elliptic cylindrical coordinates are defined by

$$x = \alpha \cosh(u)\cos(v), \qquad y = \alpha \sinh(u)\sin(v), \qquad z = z,$$

with $0 \le u < \infty$, $0 \le v < 2\pi$, $-\infty < z < \infty$, and α a positive constant.

(a) Sketch some of the coordinate surfaces.

(b) Determine the scale factors h_u, h_v, h_z.

(c) Determine $\nabla f(u, v, z)$.

(d) Determine $\nabla \cdot \mathbf{F}(u, v, z)$ and $\nabla \times \mathbf{F}(u, v, z)$.

(e) Determine $\nabla^2 f(u, v, z)$.

4. Bipolar coordinates are defined by

$$x = \frac{\alpha \sinh(v)}{\cosh(v) - \cos(u)}, \qquad y = \frac{\alpha \sin(u)}{\cosh(v) - \cos(u)}, \qquad z = z.$$

(a) Sketch some of the coordinate surfaces.

(b) Determine the scale factors h_u, h_v, h_z.

(c) Determine $\nabla f(u, v, z)$.

(d) Determine $\nabla \cdot \mathbf{F}(u, v, z)$ and $\nabla \times \mathbf{F}(u, v, z)$.

(e) Determine $\nabla^2 f(u, v, z)$.

5. Parabolic cylindrical coordinates are defined by

$$x = uv, \qquad y = \tfrac{1}{2}(u^2 - v^2), \qquad z = z,$$

where $-\infty < v < \infty, 0 \leq u < \infty$, and $-\infty < z < \infty$.

(a) Sketch some of the coordinate surfaces.

(b) Determine the scale factors h_u, h_v, h_z.

(c) Determine $\nabla f(u, v, z)$.

(d) Determine $\nabla \cdot \mathbf{F}(u, v, z)$ and $\nabla \times \mathbf{F}(u, v, z)$.

(e) Determine $\nabla^2 f(u, v, z)$.

APPENDIX B: An Extension of Green's Theorem

Let $\mathbf{F}(x, y) = f(x, y)\mathbf{i} + g(x, y)\mathbf{j}$. We have seen in Green's theorem that, under certain conditions,

$$\oint_C \mathbf{F} \cdot d\mathbf{R} = \iint_D \left(\frac{\partial g}{\partial x} - \frac{\partial f}{\partial y} \right) dA.$$

In writing this equation, we assume that f, g, $\partial f/\partial y$, and $\partial g/\partial x$ are continuous over D and on C. We will now examine to what extent these conditions can be relaxed.

Suppose that f, g, $\partial f/\partial y$, and $\partial g/\partial x$ are continuous on C and at all points of D except $P_1, \ldots, P_n$, where one or more of these functions may not be defined at all. Green's theorem no longer applies, but we can still derive an interesting relationship which, in a sense, extends Green's theorem.

Enclose each P_j in a circle K_j of radius r_j. Choose these radii sufficiently small that no two of these circles intersect and no one of these circles intersects C. This is illustrated in Figure B.1. Now cut a "channel" from C to K_1, from K_1 to K_2, from K_2 to $K_3, \ldots$, and finally from K_{n-1} to K_n, as in Figure B.2. Assuming that C is piecewise smooth, this enables us to form the piecewise-smooth curve C^* shown in Figure B.3, consisting of "most" of C, $K_1, \ldots, K_n$ and the segments forming the channels we inserted in Figure B.2.

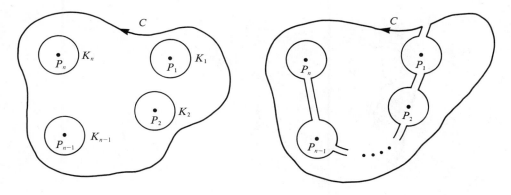

Figure B.1

Figure B.2

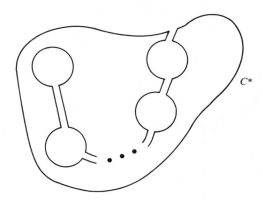

Figure B.3

Observe now that $P_1, \ldots, P_n$ are *exterior* to the region enclosed by C^* and that f, g, $\partial g/\partial x$, and $\partial f/\partial y$ are continuous in the region enclosed by C^*. By Green's theorem,

$$\oint_{C^*} \mathbf{F} \cdot d\mathbf{R} = \iint_{D^*} \left(\frac{\partial g}{\partial x} - \frac{\partial f}{\partial y} \right) dA,$$

where D^* is the region enclosed by C^* and is shaded in Figure B.4.

Now take a limit as the channels become narrower, merging to form curves between the circles $K_1, \ldots, K_n$. The curve C^* approaches the curve shown in Figure B.5. The line integrals over the segments between the K_j's are zero because we integrate in both directions over these segments, and the integrals cancel. Further, as we shrink the channels, we restore the small pieces of C and the K_j's which were previously excised. Finally, D^* approaches D, which is all of D with the interiors of

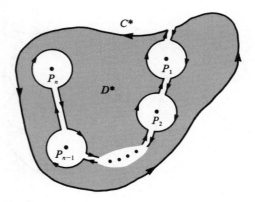

Figure B.4

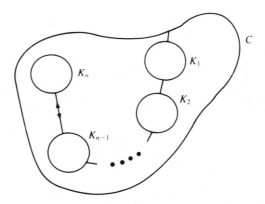

Figure B.5

the K_j's excised. We obtain in this limit

$$\oint_{C^*} \mathbf{F} \cdot d\mathbf{R} = \oint_C \mathbf{F} \cdot dR + \sum_{j=1}^{n} \oint_{K_j} \mathbf{F} \cdot d\mathbf{R} = \iint_{D^*} \left(\frac{\partial g}{\partial x} - \frac{\partial f}{\partial y} \right) dA.$$

The line integrals about the circles K_j are taken in a clockwise sense. Replace these line integrals with integrals going around the circles in the usual counterclockwise sense, introducing a minus sign, to get

$$\oint_C \mathbf{F} \cdot d\mathbf{R} - \sum_{j=1}^{n} \oint_{K_j} \mathbf{F} \cdot d\mathbf{R} = \iint_{D^*} \left(\frac{\partial g}{\partial x} - \frac{\partial f}{\partial y} \right) dA.$$

This equation may be written

$$\oint_C \mathbf{F} \cdot d\mathbf{R} = \sum_{j=1}^{n} \oint_{K_j} \mathbf{F} \cdot d\mathbf{R} + \iint_{D^*} \left(\frac{\partial g}{\partial x} - \frac{\partial f}{\partial y} \right) dA.$$

This is an extended form of Green's theorem. The line integrals about the circles on the right account for the points within D at which the components of $\mathbf{F}$ do not satisfy the hypotheses of Green's theorem.

EXAMPLE B.1

Evaluate

$$\oint_C \frac{-y}{x^2 + y^2} \, dx + \frac{x}{x^2 + y^2} \, dy,$$

with C any piecewise-smooth simple closed curve in the plane. First, notice that

$$\frac{\partial}{\partial x}\left(\frac{x}{x^2 + y^2}\right) = \frac{\partial}{\partial y}\left(\frac{-y}{x^2 + y^2}\right).$$

Now consider two cases.

Case 1 C does not enclose the origin.

Then the components of

$$\frac{-y}{x^2 + y^2}\,\mathbf{i} + \frac{x}{x^2 + y^2}\,\mathbf{j}$$

satisfy the hypotheses of Green's theorem, and

$$\oint_C \mathbf{F} \cdot d\mathbf{R} = \iint_D \left[\frac{\partial}{\partial x}\left(\frac{x}{x^2 + y^2}\right) - \frac{\partial}{\partial y}\left(\frac{-y}{x^2 + y^2}\right)\right] dA = \iint_D 0 \, dA = 0.$$

Case 2 C does enclose the origin.

Now enclose the origin with a circle K of sufficiently small radius r that K does not intersect C (Figure B.6). Parametrize K: $x = r\cos(\theta)$, $y = r\sin(\theta)$ for $0 \le \theta \le 2\pi$. By

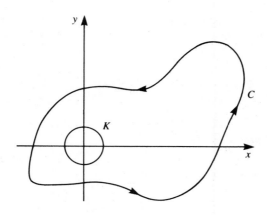

Green's extended theorem,

$$\oint_C \mathbf{F} \cdot d\mathbf{R} = \oint_K \mathbf{F} \cdot d\mathbf{R} + \iint_{D^*} \left[\frac{\partial}{\partial x} \left(\frac{x}{x^2 + y^2} \right) - \frac{\partial}{\partial y} \left(\frac{-y}{x^2 + y^2} \right) \right] dA$$

$$= \oint_K \mathbf{F} \cdot d\mathbf{R} = \oint_K \frac{-y}{x^2 + y^2} \, dx + \frac{x}{x^2 + y^2} \, dy$$

$$= \int_0^{2\pi} \left\{ \frac{-r \sin(\theta)}{r^2} [-r \sin(\theta)] + \frac{r \cos(\theta)}{r^2} [r \cos(\theta)] \right\} d\theta$$

$$= \int_0^{2\pi} d\theta = 2\pi.$$

We have therefore evaluated $\oint_C \mathbf{F} \cdot d\mathbf{R}$ explicitly for any piecewise-smooth simple closed curve in the plane. Its value is either zero (if C does not enclose the origin) or 2π (if C does enclose the origin). ∎

A similar extension can be derived for Gauss's theorem (note the argument used to derive Gauss's law in Section 16.7). We will leave this for Problem 6.

PROBLEMS FOR APPENDIX B

In each of Problems 1 through 5, evaluate $\oint_C \mathbf{F} \cdot d\mathbf{R}$ for any piecewise-smooth simple closed curve in the plane not passing through the origin.

1. $\mathbf{F} = \dfrac{x}{x^2 + y^2} \mathbf{i} + \dfrac{y}{x^2 + y^2} \mathbf{j}$

2. $\mathbf{F} = \left(\dfrac{1}{x^2 + y^2} \right)^{3/2} (x\mathbf{i} + y\mathbf{j})$

3. $\mathbf{F} = \left(\dfrac{-y}{x^2 + y^2} + x^2 \right) \mathbf{i} + \left(\dfrac{x}{x^2 + y^2} - 2y \right) \mathbf{j}$

4. $\mathbf{F} = \left(\dfrac{-y}{x^2 + y^2} + 2x^2 \right) \mathbf{i} + \left(\dfrac{x}{x^2 + y^2} - y \right) \mathbf{j}$

5. $\mathbf{F} = \left(\dfrac{x}{\sqrt{x^2 + y^2}} + 2x \right) \mathbf{i} + \left(\dfrac{y}{\sqrt{x^2 + y^2}} - 3y^2 \right) \mathbf{j}$

6. Suppose that $\mathbf{F}$ is a continuous vector field whose components are continuous with continuous first partial derivatives on and inside the piecewise-smooth closed surface Σ, except at points $P_1, \ldots, P_n$ enclosed by Σ. Enclose P_j with a sphere S_j of radius r_j, with $r_1, \ldots, r_n$ chosen small enough that no two spheres intersect each other and no sphere intersects Σ. Assuming that $\nabla \cdot \mathbf{F} = 0$ on the region inside Σ and outside the spheres, show that

$$\iint_\Sigma \mathbf{F} \cdot \mathbf{N} \, d\sigma = \sum_{j=1}^n \iint_{S_j} \mathbf{F} \cdot \mathbf{N} \, d\sigma.$$

Hint: Use an argument like that just used for the extended Green's theorem, but use tubes between successive spheres, as we did in deriving Gauss's law in Section 16.7.

In each of Problems 7 through 10, evaluate $\iint_\Sigma \mathbf{F} \cdot \mathbf{N} \, d\sigma$, in which Σ may be any piecewise-smooth closed surface in three-space not passing through the origin.

7. $\mathbf{F} = \dfrac{1}{x^2 + y^2 + z^2} [(y - z)\mathbf{i} + (z - x)\mathbf{j} + (x - y)\mathbf{k}]$

8. $\mathbf{F} = \dfrac{1}{y^2 + z^2} \mathbf{i} + \dfrac{1}{x^2 + z^2} \mathbf{j} + \dfrac{1}{y^2 + x^2} \mathbf{k}$

9. $\mathbf{F} = y\mathbf{i} + z\mathbf{j} + x\mathbf{k}$

10. $\mathbf{F} = \dfrac{1}{\sqrt{y^2 + z^2}} \mathbf{i} + \dfrac{1}{\sqrt{x^2 + z^2}} \mathbf{j} + \dfrac{1}{\sqrt{x^2 + y^2}} \mathbf{k}$

APPENDIX C: *Review of Double Integrals*

It is assumed that the student has seen double integrals before. This section is a review of the definition of the double integral and evaluation by iterated integrals.

DEFINITION OF THE DOUBLE INTEGRAL

Suppose we have a set D of points in the xy-plane. Assume that for any (x, y) in D, we have $a \leq x \leq b$ and $c \leq y \leq d$, as in Figure C.1. Let $f(x, y)$ be a function of two variables defined for all (x, y) in D.

Partition the intervals $[a, b]$ and $[c, d]$ on the x- and y-axes, respectively, by inserting points $a = x_0 < x_1 < \cdots < x_{n-1} < x_n = b$ and $c = y_0 < y_1 < \cdots < y_m = d$. The vertical lines $x = x_j$ and horizontal lines $y = y_j$ form a grid over D (Figure C.2).

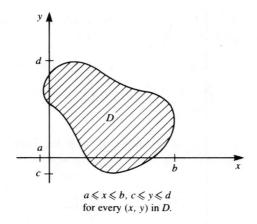

$a \leqslant x \leqslant b, \ c \leqslant y \leqslant d$
for every (x, y) in D.

Figure C.1

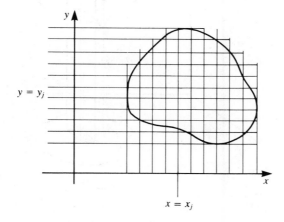

Figure C.2

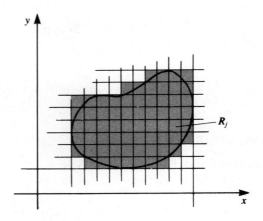

Figure C.3

Retain the rectangles of this grid which have points in common with D or its boundary. Label these rectangles $R_1, \ldots, R_N$. These are the shaded rectangles of Figure C.3.

In each rectangle R_j, choose a point (x_j^*, y_j^*). Form the sum

$$\sum_{j=1}^{N} f(x_j^*, y_j^*) A(R_j), \tag{C.1}$$

where $A(R_j)$ is the area of R_j.

If all the sums (C.1) can be made arbitrarily close to a number L by choosing n and m large enough and each $x_j - x_{j-1}$ and $y_i - y_{i-1}$ sufficiently small, we say that f

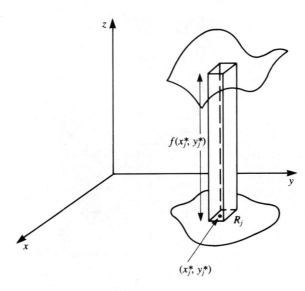

Figure C.4

is *integrable over D* and call the number L the *double integral of f over D*. We denote this double integral $\iint_D f(x, y) \, dA$.

If $f(x, y) \geq 0$ for all (x, y) in D, we can envision the graph of $z = f(x, y)$ as a surface above the xy-plane, as in Figure C.4. For each rectangle R_j, the number $f(x_j^*, y_j^*)A(R_j)$ is the area of a rectangular parallelepiped, with $A(R_j)$ the area of the base and $f(x_j^*, y_j^*)$ its height. The sum (C.1) of the volumes of these parallelepipeds approximates the volume of the solid with D as base and capped by the surface $z = f(x, y)$. In the limit, these parallelepipeds fit closer to this solid, and the double integral is the volume of this solid.

EVALUATION OF DOUBLE INTEGRALS BY ITERATED INTEGRALS

Usually, we evaluate $\iint_D f(x, y) \, dA$ by evaluating two single, or iterated, integrals.

THEOREM C.1   Evaluation of $\iint_D f(x, y) \, dA$ as an Iterated Integral

Suppose that $h(x)$ and $k(x)$ are continuous on $[a, b]$ and that $f(x, y)$ is continuous on the region D of the plane, consisting of all (x, y) such that

$$h(x) \leq y \leq k(x), \qquad a \leq x \leq b.$$

Then

$$\iint\limits_D f(x, y) \, dA = \int_a^b \int_{h(x)}^{k(x)} f(x, y) \, dy \, dx. \quad \blacksquare$$

EXAMPLE C.1

Evaluate $\iint_D xy \, dA$ if D is the triangular region described by $0 \leq y \leq x$, $0 \leq x \leq 2$ shown in Figure C.5.

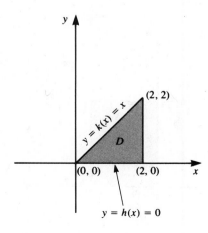

Figure C.5

The lower boundary of D is described by $h(x) = 0$ (part of the x-axis), and the upper boundary is described by $k(x) = x$ for $0 \le x \le 2$. Then

$$\iint_D xy \, dA = \int_0^2 \int_0^x xy \, dy \, dx.$$

The inner integration is

$$\int_0^x xy \, dy = \frac{x}{2} y^2 \bigg]_{y=0}^x = \frac{x}{2} [x^2 - 0^2] = \frac{1}{2} x^3.$$

Then

$$\iint_D xy \, dA = \int_0^2 \tfrac{1}{2} x^3 \, dx = \tfrac{1}{2} \tfrac{1}{4} x^4 \bigg]_0^2 = \tfrac{1}{8}[2^4 - 0^4] = \tfrac{1}{8}(16) = 2. \quad \blacksquare$$

EXAMPLE C.2

Evaluate $\iint_D x \cos(y) \, dA$, with D the rectangle $0 \le x \le 1, 0 \le y \le \pi/2$.

    D is shown in Figure C.6, with upper boundary curve $y = \pi/2$ and lower boundary curve $y = 0$ for $0 \le x \le 1$. Then

$$\iint_D x \cos(y) \, dA = \int_0^1 \int_0^{\pi/2} x \cos(y) \, dy \, dx.$$

The inner integration is

$$\int_0^{\pi/2} x \cos(y) \, dy = x \sin(y) \bigg]_{y=0}^{\pi/2} = x[1 - 0] = x.$$

Then

$$\iint_D x \cos(y) \, dA = \int_0^1 x \, dx = \tfrac{1}{2} x^2 \bigg]_0^1 = \tfrac{1}{2}. \quad \blacksquare$$

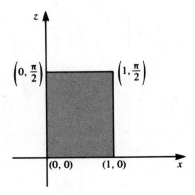

Figure C.6

EXAMPLE C.3

Evaluate $\iint_D (x + y)\, dA$, with D bounded below by $y = x^2$ and above by $y = x$ for $0 \le x \le 1$.

    D is shown in Figure C.7. We have

$$\iint_D (x + y)\, dA = \int_0^1 \int_{x^2}^x (x + y)\, dy\, dx.$$

The inner integration is

$$\int_{x^2}^x (x + y)\, dy = xy + \tfrac{1}{2}y^2 \Big]_{y=x^2}^x$$
$$= x(x) + \tfrac{1}{2}x^2 - x(x^2) - \tfrac{1}{2}(x^2)^2$$
$$= \tfrac{3}{2}x^2 - x^3 - \tfrac{1}{2}x^4.$$

Then

$$\iint_D (x + y)\, dA = \int_0^1 (\tfrac{3}{2}x^2 - x^3 - \tfrac{1}{2}x^4)\, dx$$

$$= \tfrac{1}{2}x^3 - \tfrac{1}{4}x^4 - \tfrac{1}{10}x^5 \Big]_0^1 = \tfrac{1}{2} - \tfrac{1}{4} - \tfrac{1}{10} = \tfrac{3}{20}. \quad \blacksquare$$

    We can also evaluate $\iint_D f(x, y)\, dA$ using the reverse order of integration (with respect to x first, then y), provided that the region D can be described as having a left and right boundary written as functions of y.

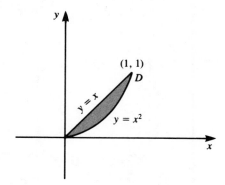

Figure C.7

THEOREM C.2

Suppose that $p(y)$ and $q(y)$ are continuous on $[c, d]$ and that $f(x, y)$ is continuous on the region D (Figure C.8) consisting of all (x, y) such that

$$p(y) \le x \le q(y), \qquad c \le y \le d.$$

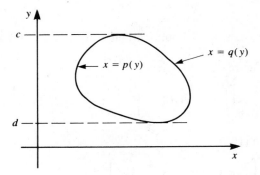

Figure C.8

Then

$$\iint\limits_{D} f(x,\ y)\ dA = \int_{c}^{d} \int_{p(y)}^{q(y)} f(x,\ y)\ dx\ dy. \quad \blacksquare$$

EXAMPLE C.4

Evaluate $\iint_{D} x^2 y\ dA$, with D bounded by the y-axis and the graph of $x = y^2$, for $1 \leq y \leq 3$ (Figure C.9).

The left boundary is $x = p(y) = 0$ (part of the y-axis), and the right boundary is $x = q(y) = y^2$ for $1 \leq y \leq 3$. Then

$$\iint\limits_{D} x^2 y\ dA = \int_{1}^{3} \int_{0}^{y^2} x^2 y\ dx\ dy.$$

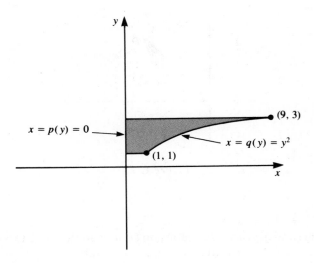

Figure C.9

The inner integration is

$$\int_0^{y^2} x^2 y \, dx = \tfrac{1}{3} x^3 y \Big]_{x=0}^{y^2} = \tfrac{1}{3}(y^2)^3 y = \tfrac{1}{3} y^7.$$

In this integration, y was thought of as fixed. Next,

$$\iint_D x^2 y \, dA = \int_1^3 \tfrac{1}{3} y^7 \, dy = \tfrac{1}{24} y^8 \Big]_1^3 = \tfrac{1}{24}[3^8 - 1] = \tfrac{6560}{24} = \tfrac{820}{3}. \quad \blacksquare$$

EXAMPLE C.5

Evaluate $\iint_D xy \, dA$ over the triangle $0 \le y \le 2x$, $0 \le x \le 2$ shown in Figure C.10. As an illustration, we will evaluate $\iint_D xy \, dA$ in two ways.

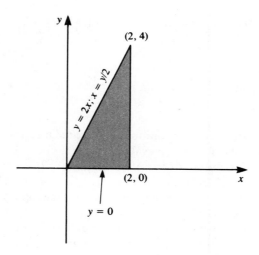

Figure C.10

Method 1 Integrate first with respect to x. Consider D as having left boundary $x = y/2$ and right boundary $x = 2$ for $0 \le y \le 4$. Then

$$\iint_D xy \, dA = \int_0^4 \int_{y/2}^2 xy \, dx \, dy.$$

Now,

$$\int_{y/2}^2 xy \, dx = \frac{1}{2} yx^2 \Big]_{x=y/2}^2$$

$$= \frac{1}{2} y \left[2^2 - \left(\frac{y}{2}\right)^2 \right] = \frac{1}{2} y \left(\frac{16 - y^2}{4}\right) = 2y - \frac{1}{8} y^3.$$

Then

$$
\iint\limits_{D} xy\, dA = \int_{0}^{4} \left(2y - \frac{1}{8} y^3 \right) dy = y^2 - \frac{1}{32} y^4 \bigg]_{0}^{4}
$$

$$
= 4^2 - \frac{4^4}{32} = 16 - 8 = 8.
$$

Method 2 Consider D as having lower boundary $y = 0$ and upper boundary $y = 2x$ for $0 \le x \le 2$. Then

$$
\iint\limits_{D} xy\, dA = \int_{0}^{2} \int_{0}^{2x} xy\, dy\, dx.
$$

The inner integration is

$$
\int_{0}^{2x} xy\, dy = \tfrac{1}{2}xy^2 \bigg]_{y=0}^{2x} = \tfrac{1}{2}x(2x)^2 = 2x^3.
$$

Then

$$
\iint\limits_{D} xy\, dA = \int_{0}^{2} 2x^3\, dx = \tfrac{1}{2}x^4 \bigg]_{0}^{2} = \tfrac{1}{2}16 = 8. \quad \blacksquare
$$

In determining the limits of integration in an iterated integral, it is often helpful to adopt the following points of view. To evaluate $\iint_{D} f(x, y)\, dA$ as $\int_{a}^{b} \int_{h(x)}^{k(x)} f(x, y)\, dy\, dx$, think of a vertical line extending from $h(x)$ to $k(x)$, moving left to right over D as x moves from a to b (Figure C.11). To evaluate $\iint_{D} f(x, y)\, dA$ as $\int_{c}^{d} \int_{p(y)}^{q(y)} f(x, y)\, dx\, dy$, think of a horizontal line extending from $p(y)$ to $q(y)$, sweeping out D as y moves upward from c to d (Figure C.12).

Sometimes the order in which the iterated integrals are written makes a great deal of difference in the amount of work involved in evaluating $\iint_{D} f(x, y)\, dA$.

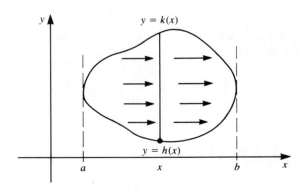

Figure C.11

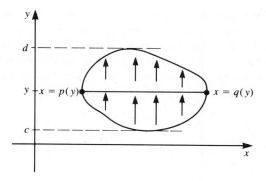

Figure C.12

EXAMPLE C.6

Evaluate $\iint_D 2x \, dA$, with D given by $y \le x \le 4y$, $0 \le y \le 2$.

   D is drawn in Figure C.13. Think of D as swept out by a horizontal line, which extends from y to $4y$, and moving upward from $y = 0$ to $y = 2$. Then

$$\iint\limits_D 2x \, dA = \int_0^2 \int_y^{4y} 2x \, dx \, dy.$$

The inner integral is

$$\int_y^{4y} 2x \, dx = x^2 \Big]_{x=y}^{4y} = [16y^2 - y^2] = 15y^2.$$

Then

$$\iint\limits_D 2x \, dA = \int_0^2 15y^2 \, dy = 5y^3 \Big]_0^2 = 5(2^3) = 40.$$

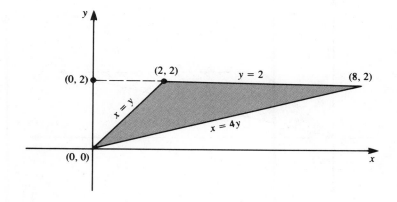

Figure C.13

Now use the reverse order of integration. Think of D as being swept out by a vertical line moving right, extending from a lower boundary of D to an upper boundary. Figure C.14 shows the difficulty with this: as x moves from zero to 2, y must extend from $y = x/4$ as lower boundary to $y = x$ as upper boundary; but as x moves from 2 to 8, y extends from $y = x/4$ on the bottom to $y = 2$ on top. This requires that we think of D as being composed of D_1 and D_2, as shown, with

$$D_1: \begin{array}{c} \dfrac{x}{4} \le y \le x \\[4pt] 0 \le x \le 2 \end{array} \quad \text{and} \quad D_2: \begin{array}{c} \dfrac{x}{4} \le y \le 2 \\[4pt] 2 \le x \le 8. \end{array}$$

Then

$$\iint\limits_{D} 2x \, dA = \iint\limits_{D_1} 2x \, dA + \iint\limits_{D_2} 2x \, dA$$

$$= \int_0^2 \int_{x/4}^x 2x \, dy \, dx + \int_2^8 \int_{x/4}^2 2x \, dy \, dx.$$

We will do the iterated integrals individually. For the left-most one, first compute the inner integral:

$$\int_{x/4}^x 2x \, dy = 2xy \Big]_{y=x/4}^x = 2x\left[x - \frac{x}{4}\right] = \frac{3}{2} x^2.$$

Then

$$\int_0^2 \int_{x/4}^x 2x \, dy \, dx = \int_0^2 \frac{3}{2} x^2 \, dx = \frac{1}{2} x^3 \Big]_0^2 = 4.$$

For the right-most iterated integral, first evaluate

$$\int_{x/4}^2 2x \, dy = 2xy \Big]_{y=x/4}^2 = 2x\left[2 - \frac{x}{4}\right] = 4x - \frac{1}{2} x^2.$$

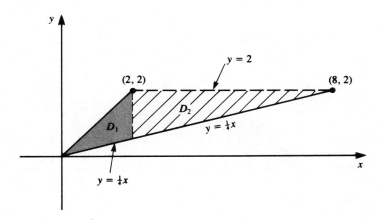

Figure C.14

Then

$$\int_2^8 \int_{x/4}^2 2x \, dy \, dx = \int_2^8 (4x - \tfrac{1}{2}x^2) \, dx$$

$$= 2x^2 - \tfrac{1}{6}x^3 \Big]_2^8 = 128 - \tfrac{512}{6} - 8 + \tfrac{8}{6} = 36.$$

Adding these gives us $\iint_D 2x \, dA = 4 + 36 = 40$, as we got before. ∎

Sometimes there is a great practical advantage to choosing one order of integration over the other, as the following example suggests.

EXAMPLE C.7

Evaluate $\iint_D e^{x^2} \, dA$, with D the triangle with vertices at $(0, 0)$, $(1, 0)$, and $(1, 1)$ shown in Figure C.15.

If we want to do the x-integration first, we write D as $y \le x \le 1, 0 \le y \le 1$ to get

$$\iint_D e^{x^2} \, dA = \int_0^1 \int_y^1 e^{x^2} \, dx \, dy.$$

The integral $\int_y^1 e^{x^2} \, dx$ can be evaluated as an infinite series but not as a finite combination of elementary functions of y. The evaluation of $\iint_D e^{x^2} \, dA$ using iterated integrals in this order will lead to an infinite series whose sum is not obvious.

Try the other order of integration. Write D as $0 \le y \le x, 0 \le x \le 1$. Then

$$\iint_D e^{x^2} \, dA = \int_0^1 \int_0^x e^{x^2} \, dy \, dx.$$

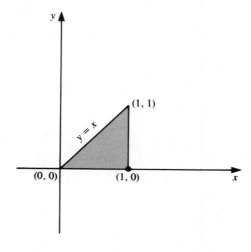

Figure C.15

First,

$$\int_0^x e^{x^2}\, dy = e^{x^2}y \Big]_{y=0}^x = xe^{x^2}.$$

Then

$$\iint_D e^{x^2}\, dA = \int_0^1 xe^{x^2}\, dx = \tfrac{1}{2}e^{x^2}\Big]_0^1 = \tfrac{1}{2}[e - 1].$$

The integration in this order yielded a simple solution in terms of familiar numbers.   ∎

We will conclude this section with a result that is often convenient in evaluating a double integral when D is a rectangle and $f(x, y)$ is a product of a function of x with a function of y.

THEOREM C.3

Let $F(x)$ be continuous on $[a, b]$, and let $G(y)$ be continuous on $[c, d]$. Let D be the rectangle $a \le x \le b$ and $c \le y \le d$, and let $f(x, y) = F(x)G(y)$. Then

$$\iint_D f(x, y)\, dA = \int_a^b F(x)\, dx \int_c^d G(y)\, dy.$$

That is, when $f(x, y)$ is a product of a function of x and a function of y, and the set to be integrated over is a rectangle, $\iint_D f(x, y)\, dA$ is a *product of two single integrals*. We evaluate $\int_a^b F(x)\, dx$ and $\int_c^d G(y)\, dy$ independently and multiply the results to obtain $\iint_D f(x, y)\, dA$.

Proof   Calculate

$$\iint_D f(x, y)\, dA = \int_a^b \int_c^d F(x)G(y)\, dy\, dx = \int_a^b F(x)\left[\int_c^d G(y)\, dy\right] dx$$

because $F(x)$ is constant in the inner integration. But $\int_c^d G(y)\, dy$ is constant, so it can be factored out of the x-integration, giving us

$$\iint_D f(x, y)\, dA = \int_a^b F(x)\, dx \int_c^d G(y)\, dy. \quad ∎$$

EXAMPLE C.8

Evaluate $\iint_D x \cos(y)\, dA$ over D: $1 \le x \le 3, 0 \le y \le \pi/4$.
    Write

$$\iint_D x \cos(y)\, dA = \int_1^3 x\, dx \int_0^{\pi/4} \cos(y)\, dy$$

$$= \left[\frac{1}{2}x^2\right]_1^3 \cdot \left[\sin(y)\right]_0^{\pi/4} = \frac{1}{2}(9 - 1)\sin\left(\frac{\pi}{4}\right) = 2\sqrt{2}. \quad ∎$$

PROBLEMS FOR APPENDIX C

In each of Problems 1 through 15, evaluate $\iint_D f(x, y)\, dA$ by means of Theorem C.3, if it applies; if it does not, use both orders of integration to evaluate $\iint_D f(x, y)$ by iterated integrals.

1. $\iint_D xy\, dA$, D the rectangle $0 \le x \le 2, 0 \le y \le 5$

2. $\iint_D (x + y)\, dA$, D the rectangle $-1 \le x \le 4, 1 \le y \le 3$

3. $\iint_D 2y\, dA$, D the triangle with vertices $(0, 0), (1, 0), (1, 1)$

4. $\iint_D (x - y)\, dA$, D the triangle with vertices $(0, 0), (2, 0), (0, 2)$

5. $\iint_D x^2 y\, dA$, D the region $1 \le x \le 3, 2 \le y \le 3$

6. $\iint_D x \sin(y)\, dA$, D the region $-2 \le x \le 4, 0 \le y \le \pi/2$

7. $\iint_D (x^2 + y)\, dA$, D the region bounded by the graphs of $y = x$ and $y = x^2$

8. $\iint_D x\, dA$, D bounded by the graphs of $y = 4 - x^2$, $y = 3x$ and the y-axis

9. $\iint_D xe^y\, dA$, D bounded by the graphs of $y = x$ and $y = 2x$ for $0 \le x \le 1$

10. $\iint_D \cos(x)\cos(y)\, dA$, D the rectangle $0 \le x \le \pi/4, 0 \le y \le \pi/3$

11. $\iint_D (x - 4y^2)\, dA$, D bounded by the graphs of $y = x^2$ and $y = 2$

12. $\iint_D \sin(x)e^{-y}\, dA$, D bounded by the graphs of $y = x$, the y-axis, and $y = 4$

13. $\iint_D (x + 4)\, dA$, D bounded by the graphs of $y = x$, $y = 2x$, and $y = 5$

14. $\iint_D -x^2 y^2\, dA$, D bounded by the graphs of $y = 2x$, the x-axis, and $x = 3$

15. $\iint_D 2xy\, dA$, D bounded by the graphs of $y = x^2$ and $y = x^3$

In each of Problems 16 through 20, evaluate $\iint_D f(x, y)\, dA$ as a product of single integrals by means of Theorem C.3.

16. $\iint_D x^2 y\, dA$, $D: -1 \le x \le 5, 3 \le y \le 4$ **17.** $\iint_D xe^{-y}\, dA$, $D: 0 \le x \le 3, 1 \le y \le 2$

18. $\iint_D \cos(x)\sin(y)\, dA$, $D: 0 \le x \le \pi/4, 0 \le y \le \pi/3$ **19.** $\iint_D x^2\sin(y)\, dA$, $D: 0 \le x \le 4, \pi/6 \le y \le \pi/3$

20. $\iint_D 4xy^2\, dA$, $D: -2 \le x \le 3, 1 \le y \le 5$

APPENDIX D: *Review of Triple Integrals*

Suppose that M is a bounded set in three-space. That is, we can define a sphere which encloses M. Suppose that $f(x, y, z)$ is defined for (x, y, z) in M. To define the triple integral of f over M, proceed as follows.

1. Use planes parallel to the xy-, xz-, and yz-planes to subdivide M into small rectangular boxes (Figure D.1).

2. Retain only those boxes $B_1, \ldots, B_N$ containing points of M.

3. Choose a point (x_j^*, y_j^*, z_j^*) in each B_j.

4. Form the sum $\sum_{j=1}^{N} f(x_j^*, y_j^*, z_j^*)V(B_j)$, in which $V(B_j)$ is the volume of B_j.

5. Let $N \to \infty$, and suppose that the dimensions of all the boxes approach zero. If the sums approach a number L, we say that $f(x, y, z)$ is *integrable* over M and call L the *triple integral* of $f(x, y, z)$ over M. In this event, we denote L as $\iiint_M f(x, y, z)\, dV$.

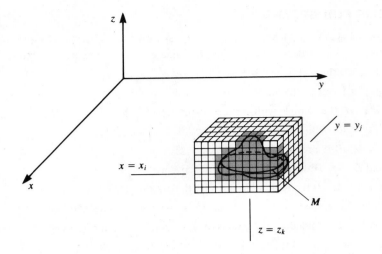

Figure D.1

The process of evaluating a triple integral is like that for double integrals, except that the iterated integral now involves three single integrations.

THEOREM D.1   Evaluation of $\iiint_M f(x, y, z)\, dV$ as an Iterated Integral

Suppose that f is continuous on the bounded set M, which consists of all (x, y, z) satisfying

$$g_1(x, y) \leq z \leq g_2(x, y) \quad \text{for} \quad h(x) \leq y \leq k(x) \quad \text{for} \quad a \leq x \leq b.$$

Suppose that $g_1(x, y)$, $g_2(x, y)$, $h(x)$, and $k(x)$ are continuous. Then

$$\iiint_M f(x, y, z)\, dV = \int_a^b \int_{h(x)}^{k(x)} \int_{g_1(x, y)}^{g_2(x, y)} f(x, y, z)\, dz\, dy\, dx. \quad \blacksquare$$

EXAMPLE D.1

Evaluate $\iiint_M 8xz\, dV$ if M consists of (x, y, z) such that

$$0 \leq z \leq \sqrt{x + y}, \qquad x^2 \leq y \leq x, \qquad 0 \leq x \leq 1.$$

Figure D.2 shows M. Write

$$\iiint_M 8xz\, dV = \int_0^1 \int_{x^2}^x \int_0^{\sqrt{x+y}} 8xz\, dz\, dy\, dx.$$

The inner integral is

$$\int_0^{\sqrt{x+y}} 8xz\, dz = 4xz^2 \bigg]_{z=0}^{\sqrt{x+y}} = 4x(x + y) = 4x^2 + 4xy.$$

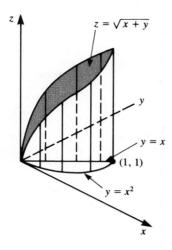

Figure D.2

The y-integration is

$$\int_{x^2}^{x} (4x^2 + 4xy) \, dy = 4x^2y + 2xy^2 \Big]_{y=x^2}^{x}$$

$$= 4x^2(x) + 2x(x)^2 - 4x^2(x^2) - 2x(x^2)^2$$

$$= 6x^3 - 4x^4 - 2x^5.$$

Finally,

$$\iiint_{M} 8xz \, dV = \int_{0}^{1} (6x^3 - 4x^4 - 2x^5) \, dx$$

$$\tfrac{3}{2}x^4 - \tfrac{4}{5}x^5 - \tfrac{1}{3}x^6 \Big]_{0}^{1} = \tfrac{3}{2} - \tfrac{4}{5} - \tfrac{1}{3} = \tfrac{11}{30}. \quad \blacksquare$$

EXAMPLE D.2

Evaluate $\iiint_{M} 2xe^y \, dV$ if M consists of (x, y, z) such that

$$0 \le z \le x, \qquad 0 \le y \le 1, \qquad 0 \le x \le 2.$$

Write

$$\iiint_{M} 2xe^y \, dV = \int_{0}^{2} \int_{0}^{1} \int_{0}^{x} 2xe^y \, dz \, dy \, dx.$$

The z-integration is

$$\int_{0}^{x} 2xe^y \, dz = 2xe^y z \Big]_{z=0}^{x} = 2xe^y[x - 0] = 2x^2e^y.$$

The y-integration is

$$\int_0^1 2x^2 e^y \, dy = 2x^2 e^y \Big]_{y=0}^{1} = 2x^2(e^1 - e^0) = 2(e-1)x^2.$$

Finally,

$$\iiint_M 2x^2 e^y \, dV = \int_0^2 2(e-1)x^2 \, dx$$

$$= \frac{2(e-1)}{3} x^3 \Big]_0^2 = \frac{16(e-1)}{3}. \quad \blacksquare$$

If $f(x, y, z) = F(x)G(y)K(z)$, and M is a rectangular box $a \le x \le b$, $c \le y \le d$, $u \le z \le v$, the triple integral of $f(x, y, z)$ over M is a product of three single integrals

$$\iiint_M f(x, y, z) \, dV = \int_a^b F(x) \, dx \int_c^d G(y) \, dy \int_u^v K(z) \, dz.$$

This is the analogue of Theorem C.3 in Appendix C.

EXAMPLE D.3

Evaluate $\iiint_M xe^y \sin(z) \, dV$ over the region $0 \le x \le 3$, $1 \le y \le 2$, $0 \le z \le \pi$.

Here, M is a rectangular box and $xe^y \sin(z)$ is a product of individual functions of x, y, and z. We can therefore write

$$\iiint_M xe^y \sin(z) \, dV = \int_0^3 x \, dx \int_1^2 e^y \, dy \int_0^\pi \sin(z) \, dz$$

$$= \tfrac{9}{2}(e^2 - e)(2) = 9(e^2 - e). \quad \blacksquare$$

We have evaluated $\iiint_M f(x, y, z) \, dV$ using a threefold iterated integral with the order of integration z, then y, then x. There are six different orders of integration we can use: z, then y, then x; z, then x, then y; x, then y, then z; x, then z, then y; y, then z, then x; y, then x, then z. We can choose any of these orders, provided that M is described appropriately to obtain the correct limits of integration.

EXAMPLE D.4

Change the order of integration in

$$\int_0^3 \int_{x^2}^9 \int_0^{9-y} f(x, y, z) \, dz \, dy \, dx.$$

From the limits and order of integration, we infer that M is given by

$$0 \le z \le 9 - y, \qquad x^2 \le y \le 9, \qquad 0 \le x \le 3.$$

First, z must extend from zero to the plane $z = 9 - y$. Then y must vary from x^2 to 9 in the xy-plane. Finally, x must vary from zero to 3. M is the solid region shown in Figure D.3; it is bounded below by the xy-plane, above by the plane $z = 9 - y$, on one side by the yz-plane, and on the other side by the cylinder $y = x^2$.

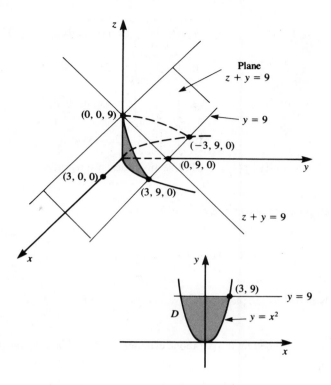

Figure D.3

Suppose we want to integrate x first, then y, then z. Now describe M as

$$0 \le x \le \sqrt{y}, \qquad 0 \le y \le 9 - z, \qquad 0 \le z \le 9.$$

Then

$$\int_0^3 \int_{x^2}^9 \int_0^{9-y} f(x, y, z)\, dz\, dy\, dx = \int_0^9 \int_0^{9-z} \int_0^{\sqrt{y}} f(x, y, z)\, dx\, dy\, dz.$$

If we want to integrate in the order x, then z, then y, we write

$$M: \quad 0 \le x \le \sqrt{y}, \qquad 0 \le z \le 9 - y, \qquad 0 \le y \le 9$$

to get

$$\int_0^3 \int_{x^2}^9 \int_0^{9-y} f(x, y, z)\, dz\, dy\, dx \int_0^9 \int_0^{9-y} \int_0^{\sqrt{y}} f(x, y, z)\, dx\, dz\, dy.$$

Similarly, we could write iterated integrals in three other orders of integration. ∎

PROBLEMS FOR APPENDIX D

In each of Problems 1 through 10, evaluate the iterated integral.

1. $\displaystyle\int_{-1}^1 \int_0^2 \int_1^4 xz\, dz\, dy\, dx$

2. $\displaystyle\int_1^3 \int_1^2 \int_0^1 (y - z)\, dz\, dy\, dx$

3. $\int_{-2}^{1} \int_{0}^{\pi} \int_{1}^{3} \cos\left(\frac{y}{3}\right) dz\, dy\, dx$

4. $\int_{-1}^{2} \int_{1}^{2} \int_{2}^{4} (x + y)\, dz\, dx\, dy$

5. $\int_{0}^{2} \int_{0}^{x} \int_{0}^{x+y} 4y\, dz\, dy\, dx$

6. $\int_{-2}^{2} \int_{x}^{x^2} \int_{0}^{x+y} 2z\, dz\, dy\, dx$

7. $\int_{1}^{3} \int_{-x}^{2x} \int_{0}^{x+y^2} x\, dz\, dy\, dx$

8. $\int_{-1}^{0} \int_{y}^{3y} \int_{1}^{x-y} 4xz\, dz\, dx\, dy$

9. $\int_{0}^{2} \int_{2z}^{3z} \int_{1}^{y+z} (y + 2z)\, dx\, dy\, dz$

10. $\int_{0}^{1} \int_{-1}^{x^2} \int_{0}^{xy} (2z - x)\, dz\, dy\, dx$

In each of Problems 11 through 20, evaluate $\iiint_{M} f(x, y, z)\, dV$. Use any order of integration in writing the iterated integrals.

11. $\iiint_{M} x\, dV$, M in the first octant bounded above by $3x + y + 2z = 6$

12. $\iiint_{M} xz\, dV$, M bounded by $x^2 + y^2 = 1$ and the planes $z = 0$ and $z = 1$

13. $\iiint_{M} (y - 2z)\, dV$, M the box bounded by the planes $x = 0$, $x = 4$, $y = 1$, $y = 6$, $z = 0$, and $z = 2$

14. $\iiint_{M} xyz\, dV$, M in the first octant, bounded by the parabolic cylinder $y = 1 - x^2$ and by the plane $z = 2$

15. $\iiint_{M} xy \cos(z)\, dV$, M bounded by the planes $z = 0$ and $z = \pi$ and the cylinder $y = x^2$ and the plane $y = 2$

16. $\iiint_{M} (x + y + z)\, dV$, M in the first octant bounded by the xy-plane and the cone $z = \sqrt{x^2 + y^2}$ for $x^2 + y^2 \le 1$

17. $\iiint_{M} (x - z)\, dV$, M in the first octant above the xy-plane, inside the cylinder $x^2 + y^2 = 4$ and below the plane $z + 2x + 6y = 12$

18. $\iiint_{M} 6z\, dV$, M bounded by the xy-plane, the xz-plane, the planes $z = 0$ and $z = 2$, and the plane $3x + 4y + z = 8$

In each of Problems 19 through 23, evaluate $\iiint_{M} f(x, y, z)\, dV$ as a product of three single integrals.

19. $\iiint_{M} xy^2\, dV$, $M: 0 \le x \le 1, 0 \le y \le 1, 2 \le z \le 5$

20. $\iiint_{M} xyz\, dV$, $M: -1 \le x \le 2, 0 \le y \le 2, 1 \le z \le 4$

21. $\iiint_{M} e^x yz^2\, dV$, $M: 0 \le x \le 3, -1 \le y \le 0, 0 \le z \le 2$

22. $\iiint\limits_M \sin(x)\cos(y)z \, dV, \quad M: 0 \le x \le \dfrac{\pi}{4}, 0 \le y \le \dfrac{\pi}{3}, 0 \le z \le 2$

23. $\iiint\limits_M xy \ln(z) \, dV, \quad M: 1 \le x \le 2, 1 \le y \le 2, 1 \le z \le 2$

APPENDIX E: Changes of Variables in Multiple Integrals _____

In this section, we will review the evaluation of double integrals using polar coordinates and that of triple integrals using cylindrical and spherical coordinates.

POLAR COORDINATES IN DOUBLE INTEGRALS

THEOREM E.1 $\iint_D f(x, y) \, dA$ in Polar Coordinates

Suppose that $f(x, y)$ is continuous on a set D which, in polar coordinates, consists of all points (r, θ) such that

$$g_1(\theta) \le r \le g_2(\theta) \quad \text{for} \quad \alpha \le \theta \le \beta,$$

as in Figure E.1. Then

$$\iint\limits_D f(x, y) \, dA = \int_\alpha^\beta \int_{g_1(\theta)}^{g_2(\theta)} f(r \cos(\theta), r \sin(\theta)) r \, dr \, d\theta. \quad \blacksquare$$

That is, to evaluate $\iint_D f(x, y) \, dA$ in polar coordinates,

1. Replace x by $r \cos(\theta)$ and y by $r \sin(\theta)$ in $f(x, y)$.
2. Multiply $f(r \cos(\theta), r \sin(\theta))$ by r.
3. Integrate $r \cdot f(r \cos(\theta), r \sin(\theta))$ with respect to r (thinking of θ as fixed) between $g_1(\theta)$ and $g_2(\theta)$.
4. Integrate the resulting function of θ between α and β.

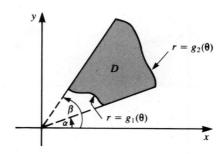

Figure E.1

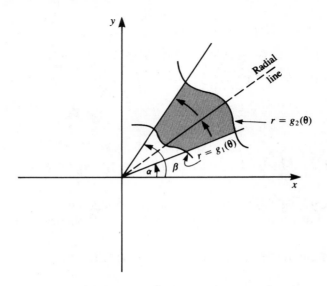

Figure E.2

In determining the limits of integration in terms of r and θ, it is useful to think of D as swept out by a radial line extending from the origin outward, as in Figure E.2. Then r extends from an inner boundary $g_1(\theta)$ to an outer one $g_2(\theta)$ as it rotates counterclockwise from $\theta = \alpha$ to $\theta = \beta$.

EXAMPLE E.1

Evaluate $\iint_D e^{-x^2 - y^2}\, dA$, with D the quarter-circle of Figure E.3.

Although this integral is difficult in rectangular coordinates, it is easily done in polar coordinates. Write D as the set of points (r, θ) with

$$0 \le r \le R, \qquad 0 \le \theta \le \frac{\pi}{2}.$$

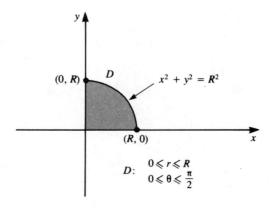

Figure E.3

Further, $-x^2 - y^2 = -r^2\cos^2(\theta) - r^2\sin^2(\theta) = -r^2$, so

$$rf(r\cos(\theta), r\sin(\theta)) = re^{-r^2}$$

Then

$$\iint_D e^{-x^2-y^2}\, dA = \int_0^{\pi/2} \int_0^R re^{-r^2}\, dr\, d\theta$$

$$= \int_0^{\pi/2} d\theta \int_0^R re^{-r^2}\, dr = \frac{\pi}{2}\left[-\frac{1}{2}e^{-r^2}\right]_0^R = \frac{\pi}{4}[1 - e^{-R^2}]. \quad \blacksquare$$

EXAMPLE E.2

Find the volume bounded above by the surface $z = 1/\sqrt{x^2 + y^2}$ and below by the set in the first quadrant inside the cardioid $r = 2[1 + \cos(\theta)]$ and outside the circle $r = 2$.

The set D is shown in Figure E.4, and the surface is shown in Figure E.5. D can be described as consisting of all (r, θ) such that

$$2 \le r \le 2[1 + \cos(\theta)], \qquad 0 \le \theta \le \frac{\pi}{2}.$$

Then

$$V = \iint_D \frac{1}{\sqrt{x^2 + y^2}}\, dA = \int_0^{\pi/2} \int_2^{2[1+\cos(\theta)]} r\frac{1}{r}\, dr\, d\theta$$

$$= \int_0^{\pi/2} \int_2^{2[1+\cos(\theta)]} dr\, d\theta.$$

The r-integration is

$$\int_2^{2[1+\cos(\theta)]} dr = 2[1 + \cos(\theta)] - 2 = 2\cos(\theta).$$

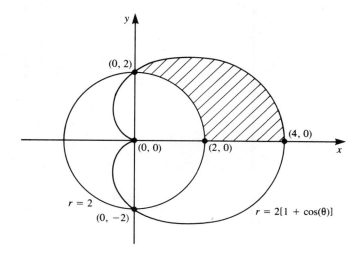

Figure E.4

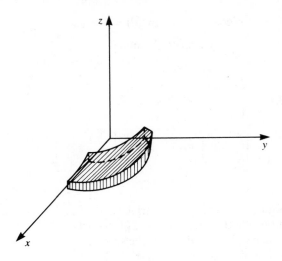

Figure E.5

Then

$$V = \int_2^{\pi/2} 2\cos(\theta)\, d\theta = 2\sin(\theta) \Big]_0^{\pi/2} = 2. \quad \blacksquare$$

Thus far, we have shown how to convert a double integral in x and y into an iterated integral in r and θ. Sometimes we also use this method to convert an iterated integral in terms of x and y into one in terms of r and θ.

EXAMPLE E.3

Evaluate $\int_0^2 \int_0^{\sqrt{4-x^2}} (x^2 + y^2)^{4/3}\, dy\, dx$.

    If done directly, this iterated integral involves a good deal of work. However, the limits of integration suggest the region D: $0 \le y \le \sqrt{4 - x^2}$, $0 \le x \le 2$ shown

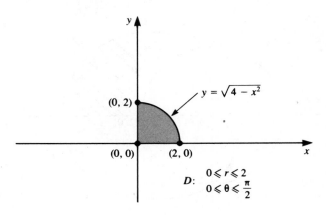

Figure E.6

in Figure E.6, and we recognize this iterated integral as exactly what we would get in evaluating the double integral $\iint_D (x^2 + y^2)^{4/3} \, dA$. Transform this double integral to polar coordinates, writing D as $0 \le r \le 2, 0 \le \theta \le \pi/2$ to get

$$\int_0^2 \int_0^{\sqrt{4-x^2}} (x^2 + y^2)^{4/3} \, dy \, dx = \iint_D (x^2 + y^2)^{4/3} \, dA$$

$$\int_0^{\pi/2} \int_0^2 r(r^2)^{4/3} \, dr \, d\theta = \int_0^{\pi/2} d\theta \int_0^2 r^{11/3} \, dr$$

$$\frac{\pi}{2} \left. \frac{3}{14} r^{14/3} \right]_0^2 = \frac{3\pi}{28} 2^{14/3}$$

$$= \frac{3\pi}{7} 2^{8/3},$$

which is approximately 8.55.   ∎

TRIPLE INTEGRALS IN CYLINDRICAL COORDINATES

Cylindrical and rectangular coordinates are related by

$$x = r \cos(\theta), \qquad y = r \sin(\theta), \qquad z = z.$$

We transform a triple integral $\iiint_M f(x, y, z) \, dV$ from rectangular to cylindrical coordinates as follows.

THEOREM E.2 $\iiint_M f(x, y, z) \, dV$ in Cylindrical Coordinates

Suppose that $f(x, y, z)$ is continuous over a set M described in cylindrical coordinates by

$$g_1(r, \theta) \le z \le g_2(r, \theta), \qquad r_1(\theta) \le r \le r_2(\theta), \qquad \alpha \le \theta \le \beta.$$

Then

$$\iiint_M f(x, y, z) \, dV = \int_\alpha^\beta \int_{r_1(\theta)}^{r_2(\theta)} \int_{g_1(r, \theta)}^{g_2(r, \theta)} f(r \cos(\theta), r \sin(\theta), z) r \, dz \, dr \, d\theta.$$ ∎

EXAMPLE E.4

Evaluate $\int_0^2 \int_0^{\sqrt{4-x^2}} \int_0^8 2yz \, dz \, dy \, dx$.

We could do this in rectangular coordinates, but the integration is easier in cylindrical coordinates. Recognize the iterated integral as an evaluation of $\iiint_M 2yz \, dV$, with M the set

$$0 \le z \le 8, \qquad 0 \le y \le \sqrt{4 - x^2}, \qquad 0 \le x \le 2$$

shown in Figure E.7. Thus, M is part of a circular cylinder. Changing to cylindrical coordinates, M has the simple description

$$0 \le z \le 8, \qquad 0 \le r \le 2, \qquad 0 \le \theta \le \frac{\pi}{2}.$$

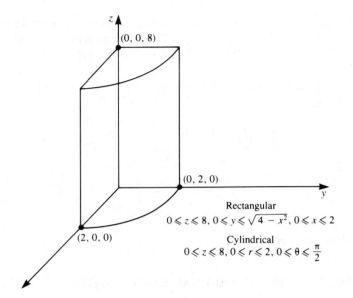

Rectangular
$$0 \leqslant z \leqslant 8,\, 0 \leqslant y \leqslant \sqrt{4 - x^2},\, 0 \leqslant x \leqslant 2$$

Cylindrical
$$0 \leqslant z \leqslant 8,\, 0 \leqslant r \leqslant 2,\, 0 \leqslant \theta \leqslant \frac{\pi}{2}$$

Figure E.7

Then

$$\int_0^2 \int_0^{\sqrt{4 - x^2}} \int_0^8 2yz\, dz\, dy\, dx = \iiint_M 2yz\, dV$$

$$= \int_0^{\pi/2} \int_0^2 \int_0^8 2[r\sin(\theta)]zr\, dz\, dr\, d\theta$$

$$= \int_0^{\pi/2} 2\sin(\theta)\, d\theta \int_0^2 r^2\, dr \int_0^8 z\, dz$$

$$= 2\left[-\cos\left(\frac{\pi}{2}\right) + \cos(0)\right]\frac{1}{3}2^3\frac{1}{2}8^2 = \frac{512}{3}.$$

We evaluated this iterated integral in r, θ, and z as a product of three integrals. ∎

TRIPLE INTEGRALS IN SPHERICAL COORDINATES

Spherical coordinates (ρ, θ, φ) and rectangular coordinates (x, y, z) of a point P are related by the equations

$$x = \rho\sin(\varphi)\cos(\theta), \qquad y = \rho\sin(\varphi)\sin(\theta), \qquad z = \rho\cos(\varphi).$$

In particular, the relationship

$$x^2 + y^2 + z^2 = \rho^2$$

arises often. Here is the formula to change variables to spherical coordinates in a triple integral $\iiint_M f(x, y, z)\, dV$.

THEOREM E.3 $\iiint_M f(x, y, z)\, dV$ in Spherical Coordinates

Suppose that $f(x, y, z)$ is continuous on M, which is described by continuous functions of ρ, θ, and φ. Then

$$\iiint_M f(x, y, z)\, dV$$

$$= \iiint_{M_{\rho\varphi\theta}} f(\rho\,\sin(\varphi)\cos(\theta),\ \rho\,\sin(\varphi)\sin(\theta),\ \rho\,\cos(\varphi))\rho^2\sin(\varphi)\, d\rho\, d\varphi\, d\theta,$$

in which $\iiint_{M_{\rho\varphi\theta}}$ denotes an iterated integral over M, with the limits of integration describing M given in terms of ρ, φ, θ. ∎

Notice the factor $\rho^2\sin(\varphi)$ in the integral after transformation to spherical coordinates. It is analogous to the factor r which appeared in the transformation to cylindrical (and polar) coordinates. See Appendix A for a discussion of this factor in the context of orthogonal curvilinear coordinates.

EXAMPLE E.5

Find the volume of the region between the cones $z = \sqrt{x^2 + y^2}$ and $z = \sqrt{3(x^2 + y^2)}$ and bounded above by the sphere $x^2 + y^2 + z^2 = 16$.

M is sketched in Figure E.8. The cone $z = \sqrt{x^2 + y^2}$ makes an angle $\pi/4$ with the positive z-axis and hence has equation $\varphi = \pi/4$ in spherical coordinates. Similarly, $z = \sqrt{3(x^2 + y^2)}$ has equation $\varphi = \pi/6$. The sphere is $\rho = 4$. Then M has the following description in spherical coordinates. It consists of points (ρ, θ, φ) such that

$$0 \le \rho \le 4, \qquad \frac{\pi}{6} \le \varphi \le \frac{\pi}{4}, \qquad 0 \le \theta \le 2\pi.$$

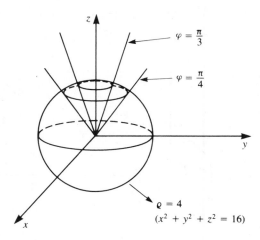

Figure E.8

The volume is

$$\iiint_M dV = \int_0^{2\pi} \int_{\pi/6}^{\pi/4} \int_0^4 \rho^2 \sin(\varphi)\, d\rho\, d\varphi\, d\theta = \int_0^{2\pi} d\theta \int_{\pi/6}^{\pi/4} \sin(\varphi)\, d\varphi \int_0^4 \rho^2\, d\rho$$

$$= 2\pi\left[-\cos\left(\frac{\pi}{4}\right) + \cos\left(\frac{\pi}{6}\right)\right] \frac{1}{3}(4^3) = \frac{64\pi}{3}[\sqrt{3} - \sqrt{2}],$$

or about 27.76. ∎

The iterated integral in this example could be evaluated as a product of three single integrals because the limits of integration were all constant, and the function of (ρ, θ, φ) was a product of functions of ρ, θ, and φ individually.

EXAMPLE E.6

Evaluate

$$\iiint_M \frac{1}{x^2 + y^2 + z^2}\, dV,$$

with M the region between the spheres $x^2 + y^2 + z^2 = 9$ and $x^2 + y^2 + z^2 = 36$.
M is shown in Figure E.9. In spherical coordinates, M is described by

$$3 \le \rho \le 6, \qquad 0 \le \varphi \le \pi, \qquad 0 \le \theta \le 2\pi.$$

Then

$$\iiint_M \frac{1}{x^2 + y^2 + z^2}\, dV = \int_0^{2\pi} \int_0^\pi \int_3^6 \frac{1}{\rho^2} \rho^2 \sin(\varphi)\, d\rho\, d\varphi\, d\theta$$

$$= \int_0^{2\pi} d\theta \int_0^\pi \sin(\varphi)\, d\varphi \int_3^6 d\rho$$

$$= (2\pi)[-\cos(\pi) + \cos(0)](6 - 3) = 12\pi. \quad \blacksquare$$

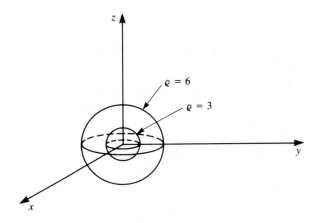

$\varrho = 6$

$\varrho = 3$

Figure E.9

PROBLEMS FOR APPENDIX E

In each of Problems 1 through 5, use polar coordinates to evaluate the iterated integral.

1. $\displaystyle\int_{-2}^{2}\int_{-\sqrt{4-x^2}}^{\sqrt{4-x^2}} e^{x^2+y^2}\, dy\, dx$

2. $\displaystyle\int_{0}^{1}\int_{0}^{\sqrt{1-x^2}}\cos(x^2+y^2)\, dy\, dx$

3. $\displaystyle\int_{-1}^{1}\int_{-\sqrt{1-y^2}}^{\sqrt{1-y^2}}(x^2+y^2)^{3/2}\, dx\, dy$

4. $\displaystyle\int_{1}^{4}\int_{0}^{x}\frac{1}{\sqrt{x^2+y^2}}\, dy\, dx$

5. $\displaystyle\int_{0}^{4}\int_{0}^{\sqrt{16-y^2}} x\cos(x^2+y^2)\, dx\, dy$. *Hint:* Use integration by parts for the *r*-integration.

In each of Problems 6 through 10, evaluate the triple integral using cylindrical coordinates.

6. $\displaystyle\iiint_M z\, dV,$ *M* bounded by $z = 12 - x^2 - y^2$ and the *xy*-plane

7. $\displaystyle\iiint_M x\, dV,$ *M* lying in the first octant and bounded by the cylinder $x^2 + y^2 = 8$ and the planes $z = 0$ and $z = 6$

8. $\displaystyle\iiint_M xy\, dV,$ *M* in the first octant bounded by $z = 8 - x^2 - y^2$

9. $\displaystyle\iiint_M \sqrt{x^2+y^2}\, dV,$ *M* bounded above by the cone $z = \sqrt{x^2+y^2}$ for $x^2 + y \le 4$ and below by the *xy*-plane

10. $\displaystyle\iiint_M z(x^2+y^2)\, dV,$ *M* in the first octant lying between the cone $z = \sqrt{x^2+y^2}$ and the sphere $x^2 + y^2 + z^2 = 8$

In each of Problems 11 through 14, evaluate the iterated integral by changing to cylindrical coordinates.

11. $\displaystyle\int_{0}^{2}\int_{0}^{\sqrt{4-x^2}}\int_{0}^{\sqrt{x^2+y^2}} 2z\, dz\, dy\, dx$

12. $\displaystyle\int_{0}^{3}\int_{-\sqrt{9-x^2}}^{\sqrt{9-x^2}}\int_{0}^{y^2+x^2}\frac{z}{\sqrt{x^2+y^2}}\, dz\, dy\, dx$

13. $\displaystyle\int_{0}^{\sqrt{2}}\int_{0}^{\sqrt{2-x^2}}\int_{0}^{6} e^{\sqrt{x^2+y^2}}\, dz\, dy\, dx$

14. $\displaystyle\int_{0}^{2}\int_{0}^{\sqrt{4-x^2}}\int_{0}^{\sqrt{4-x^2-y^2}}\frac{x}{\sqrt{x^2+y^2}}\, dz\, dy\, dx$

In each of Problems 15 through 20, evaluate the integral using spherical coordinates.

15. $\displaystyle\iiint_M (x^2+y^2+z^2)^{1/2}\, dV,$ *M* bounded by the *xy*-plane and the hemisphere $x^2 + y^2 + z^2 = 9,\ z \ge 0$

16. $\displaystyle\iiint_M x^2\, dV,$ *M* bounded by the cone $z = \sqrt{x^2+y^2}$ and the sphere $x^2 + y^2 + z^2 = 1$

17. $\displaystyle\iiint_M y\, dV,$ *M* in the first octant, bounded by $x^2 + y^2 + z^2 = 16$ and the planes $z = 0$ and $z = 2$. *Hint:* Be careful in writing the plane $z = 2$ in terms of spherical coordinates.

18. $\displaystyle\iiint_M \frac{1}{\sqrt{x^2+y^2+z^2}}\, dV,$ *M* between the spheres of radii 2 and 7 about the origin

19. $\iiint\limits_{M} z^2 \, dV$, M between the cones $\varphi = \pi/4$ and $\varphi = \pi/6$ and inside the sphere $x^2 + y^2 + z^2 = 8$

20. $\iiint\limits_{M} 2z \, dV$, M in the first octant, inside the sphere $\rho = 8$ and outside the cone $\varphi = \pi/6$

In each of Problems 21 through 23, use spherical coordinates to evaluate the iterated integral.

21. $\displaystyle\int_{0}^{4} \int_{0}^{\sqrt{16-x^2}} \int_{-\sqrt{16-x^2-y^2}}^{\sqrt{16-x^2-y^2}} z^2 y \, dz \, dy \, dx$

22. $\displaystyle\int_{0}^{\sqrt{2}} \int_{0}^{\sqrt{12-x^2}} \int_{0}^{\sqrt{12-x^2-y^2}} \frac{z}{\sqrt{x^2 + y^2 + z^2}} \, dz \, dy \, dx$

23. $\displaystyle\int_{0}^{2} \int_{0}^{\sqrt{4-x^2}} \int_{0}^{\sqrt{4-x^2-y^2}} \frac{1}{1 + x^2 + y^2 + z^2} \, dz \, dy \, dx$

APPENDIX F: Notes on the History of Vector Calculus_____

The origins of vectors can be traced to considerations of problems connected with complex numbers. In the development of mathematics, complex numbers played a significant role, presenting philosophical as well as mathematical difficulties. Their necessity arose in solving polynomial equations such as $x^2 + 1 = 0$. However, the idea of a "number" whose square is negative was very distasteful to early mathematicians.

Complex numbers became more acceptable when they were interpreted geometrically. Caspar Wessel (1745–1818), a Norwegian surveyor, Jean-Robert Argand (1768–1822), a Swiss bookkeeper, and Carl Friedrich Gauss (1777–1855), one of the great mathematicians, endowed complex numbers with a geometric interpretation. A complex number $a + ib$ can be thought of as a point (a, b) in the plane (sometimes called the Argand plane). Multiplication by i corresponds to a 90 degree rotation counterclockwise, since $(a + ib)i = ai + i^2 b = -b + ai$, and this complex number is represented by the point $(-b, a)$, as shown in Figure F.1. Further, the arrow from the origin

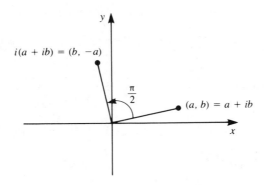

Figure F.1

to the point (a, b) represents a vector; in the early nineteenth century, the vector and the complex number $a + ib$ were sometimes used interchangeably. Addition of complex numbers

$$(a + ib) + (c + id) = (a + c) + i(b + d)$$

is consistent with the parallelogram law for vector addition, as shown in Figure F.2.

   The practice of thinking of vectors geometrically as complex numbers has a difficulty, however. Vectors also represent forces, and forces are not necessarily confined to a plane. Therefore, the problem arose of representing vectors and forces in three-space. Unfortunately, because of the established practice of representing two-dimensional vectors as complex numbers, the problem became one of generalizing complex numbers to three dimensions. This problem became the obsession of William Rowan Hamilton (1805–1865), who proposed a solution in what are known as quaternions.

   A *quaternion* is a quantity of the form

$$a + bi + cj + dk$$

(using modern notation), in which $a, b, c,$ and d are real numbers. Thus, a typical quaternion has a numerical part, a, and a "vector" part. Quaternions are added in the obvious way. For example,

$$(2 + 3i + 4j - 6k) + (-1 + 2i - 8j + 16k) = 1 + 5i - 4j + 10k.$$

However, in "multiplying" quaternions, Hamilton defined the rules

$$jk = i, \quad ki = j, \quad ij = k, \quad kj = -i, \quad ik = -j, \quad ji = -k, \quad \text{(F.1)}$$

and

$$i^2 = j^2 = k^2 = -1. \tag{F.2}$$

Today, we recognize the products in line (F.1) as the cross products of the unit vectors. Thus $i, j,$ and k behaved as do our unit vectors **i**, **j**, and **k**. However, in line (F.2),

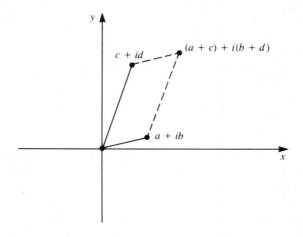

Figure F.2

Hamilton treated i, j, and k as the square root of -1 (which we denote i). This was in keeping with his intention to extend complex numbers to three dimensions. The rules (F.2) have no analogues in modern vector "multiplication."

Hamilton believed that eventually quaternions would provide the natural vehicle for the mathematical expression of physical laws. In this he was wrong; today, quaternions are used primarily in abstract algebra.

Soon after Hamilton began to publicize his views on quaternions, a debate developed among scientists over Hamilton's claims for them. James Clerk Maxwell (1831–1879), a professor of physics at Cambridge who derived the fundamental field equations of electricity and magnetism, separated in his thinking the scalar and vector parts of quaternions. Maxwell was an important and influential figure, and he helped forcus attention on expressions $ai + bj + ck$, or *vectors*, as we know them today (using **i**, **j**, and **k** in place of i, j, and k).

Maxwell was the first to use the term *rotation* (curl) of a vector in connection with swirl in fluid motion. He also introduced what he called *Laplace's operator*,

$$\nabla^2 = \frac{\partial^2}{\partial x^2} + \frac{\partial^2}{\partial y^2} + \frac{\partial^2}{\partial z^2}.$$

Maxwell further noted the identities div(curl) = 0 and curl(grad) = **O**.

William Kingdon Clifford (1845–1879), of Queen's College, London, apparently originated the term *divergence of a vector*.

The outcome of the debate on quaternions versus vectors was in doubt until the work of Yale's Josiah Willard Gibbs (1839–1903). The book *Vector Analysis*, written in 1901 by E. B. Wilson but based on Gibbs's lectures, helped to popularize vector notation and algebra. Also influential in establishing the use of vectors was Oliver Heaviside's *Electromagnetic Theory*, the first volume of which contained vector algebra. Heaviside (1850–1925) was an electrical engineer who helped develop the Laplace transform and introduce its use, as well as vector algebra, into the analysis of engineering problems.

Today, we recognize the main results of vector calculus in the integral theorems named for Gauss, Green, and Stokes. The divergence theorem, popularly named after Gauss, was discovered independently by the Ukrainian mathematician Michel Ostrogradsky (1801–1861), who published his version in 1831. In Russia, the theorem is known today as Ostrogradsky's theorem.

Stokes's theorem can be traced to a letter dated July 2, 1850, sent to Stokes by Lord Kelvin (Sir William Thomson, 1824–1907). Sir George Gabriel Stokes (1819–1903) was a professor of mathematics at Cambridge, and he presented the formula which bears his name as question 8 on the 1854 examination for the Smith Prize, a competition held annually for the leading mathematics students at Cambridge. The result soon became widely known as Stokes's theorem.

George Green (1793–1841) was a self-taught mathematician primarily interested in electricity and magnetism. In the course of applying potential theory to this subject, he developed what are today known as Green's formulas, or Green's identities (the three-dimensional versions of these identities are given in Section 16.7). These identities were also presented to the St. Petersburg Academy of Sciences in 1828 by Ostrogradsky. The theorem which today bears the name Green's theorem was originally referred to as Green's lemma.

PART **V**

Fourier Analysis and Boundary Value Problems

Fourier Analysis _____

17.0 Introduction _____

In 1807, the French mathematician Joseph Fourier (1768–1830) submitted a paper to the Academy of Sciences of Paris. In it, he presented a mathematical description of problems involving heat conduction. Although the paper was at first rejected for lack of mathematical rigor, it contained ideas which would develop into an important area of mathematics named in his honor, Fourier analysis.

One surprising ramification of Fourier's work was that many familiar functions can be expanded in infinite series and integrals involving trigonometric functions. This idea is important in modeling many phenomena in physics and engineering, as well as in areas undreamed of in Fourier's day, such as computing and the CAT scan (computer-assisted tomography) diagnostic technique of modern medicine.

In this chapter, we will develop some of the ideas of Fourier analysis and its applications, particularly in electrical engineering (for example, analyzing the frequency response of a filter). The next chapter will be devoted to the use of Fourier analysis in solving partial differential equations.

17.1 Fourier Series _____

Suppose f is integrable on $[-\pi, \pi]$. We want to entertain the possibility of choosing constants $a_0, a_1, \ldots, b_1, b_2, \ldots$ so that

$$f(x) = a_0 + \sum_{n=1}^{\infty} [a_n\cos(nx) + b_n\sin(nx)].$$

In this and the next section, we will see that, under fairly general conditions on f, we can obtain such an equality, except possibly at finitely many points of $[-\pi, \pi]$.

We will begin with an informal argument to suggest how to choose the a_n's and b_n's. The key lies in the following facts.

LEMMA 17.1

1. If n and m are distinct nonnegative integers,

$$\int_{-\pi}^{\pi} \cos(mx)\cos(nx)\,dx = \int_{-\pi}^{\pi} \sin(mx)\sin(nx)\,dx = 0.$$

2. For any positive integers m and n,

$$\int_{-\pi}^{\pi} \cos(mx)\sin(nx)\,dx = 0.$$

Proof of (1)   Both parts of the lemma are proved by integration. If $n \neq m$, both nonzero,

$$\int_{-\pi}^{\pi} \cos(nx)\cos(mx)\,dx = \left[\frac{\sin[(n-m)x]}{2(n-m)} + \frac{\sin[(n+m)x]}{2(n+m)}\right]_{-\pi}^{\pi} = 0$$

and

$$\int_{-\pi}^{\pi} \sin(nx)\sin(mx)\,dx = \left[\frac{\sin[(n-m)x]}{2(n-m)} - \frac{\sin[(n+m)x]}{2(n+m)}\right]_{-\pi}^{\pi} = 0.$$

If $m = 0$ and $n \neq 0$, $\int_{-\pi}^{\pi} \cos(nx)\cos(mx)\,dx = \int_{-\pi}^{\pi} \cos(nx)\,dx = \dfrac{1}{n}\sin(nx)\bigg]_{-\pi}^{\pi} = 0.$

Proof of (2)   If n and m are positive integers and $n \neq m$,

$$\int_{-\pi}^{\pi} \cos(mx)\sin(nx)\,dx = \left[-\frac{\cos[(n-m)x]}{2(n-m)} - \frac{\cos[(n+m)x]}{2(n+m)}\right]_{-\pi}^{\pi} = 0$$

because $\cos(A) = \cos(-A)$ for any A. If n and m are positive integers and $n = m$,

$$\int_{-\pi}^{\pi} \cos(nx)\sin(nx)\,dx = \frac{1}{2n}\sin^2(nx)\bigg]_{-\pi}^{\pi} = 0.$$

Finally, $\int_{-\pi}^{\pi} \cos(mx)\,dx = \int_{-\pi}^{\pi} \sin(mx)\,dx = 0$ for $m = 0, 1, 2, \ldots.$ ∎

The integral formulas in Lemma 17.1 are called *orthogonality relationships*, and the functions $\cos(nx)$ for $n = 0, 1, 2, \ldots,$ and $\sin(nx)$ for $n = 1, 2, \ldots$ are said to be *orthogonal* on $[-\pi, \pi]$. Lemma 17.1 states that the integral from $-\pi$ to π of a product of any two of these functions is zero.

LEMMA 17.2

For any positive integer n,

$$\int_{-\pi}^{\pi} \cos^2(nx)\,dx = \int_{-\pi}^{\pi} \sin^2(nx)\,dx = \pi.$$

Proof As with Lemma 17.1, the proof is by routine integration:

$$\int_{-\pi}^{\pi} \cos^2(nx)\, dx = \int_{-\pi}^{\pi} \frac{1}{2}[1 + \cos(2nx)]\, dx = \frac{1}{2}x + \frac{1}{2n}\sin(2nx)\Big]_{-\pi}^{\pi} = \pi.$$

Similarly,

$$\int_{-\pi}^{\pi} \sin^2(nx)\, dx = \int_{-\pi}^{\pi} \tfrac{1}{2}[1 - \cos(2nx)]\, dx = \pi. \quad\blacksquare$$

Now return to the question of how f can be written as a series of sines and cosines together with a constant term

$$f(x) = a_0 + \sum_{n=1}^{\infty} [a_n\cos(nx) + b_n\sin(nx)] \tag{17.1}$$

for $-\pi \le x \le \pi$. We will explore an informal argument suggesting how we should choose the a_n's and b_n's. Integrate both sides of equation (17.1) from $-\pi$ to π, and assume for the moment that we can interchange the summation and the integral. We get

$$\int_{-\pi}^{\pi} f(x)\, dx = a_0 \int_{-\pi}^{\pi} dx + \sum_{n=1}^{\infty} \left[a_n \int_{-\pi}^{\pi} \cos(nx)\, dx + b_n \int_{-\pi}^{\pi} \sin(nx)\, dx \right] = 2\pi a_0$$

because all of the integrals in the summation are zero. Solve this equation for a_0 to get

$$a_0 = \frac{1}{2\pi} \int_{-\pi}^{\pi} f(x)\, dx.$$

Now let k be any positive integer. We will "determine" a_k. Multiply equation (17.1) by $\cos(kx)$ to get

$$f(x)\cos(kx) = a_0\cos(kx) + \sum_{n=1}^{\infty} [a_n\cos(nx)\cos(kx) + b_n\sin(nx)\cos(kx)].$$

Integrate both sides of this equation from $-\pi$ to π, and again interchange the integral and the summation. We get

$$\int_{-\pi}^{\pi} f(x)\cos(kx)\, dx = a_0 \int_{-\pi}^{\pi} \cos(kx)\, dx$$

$$+ \sum_{n=1}^{\infty} \left[a_n \int_{-\pi}^{\pi} \cos(nx)\cos(kx)\, dx + b_n \int_{-\pi}^{\pi} \sin(nx)\cos(kx)\, dx \right].$$

By Lemma 17.1, all of the integrals on the right are zero except the one involving $\cos(nx)\cos(kx)$ when $n = k$. The last equation therefore collapses to just

$$\int_{-\pi}^{\pi} f(x)\cos(kx)\, dx = a_k \int_{-\pi}^{\pi} \cos(kx)\cos(kx)\, dx = a_k\pi,$$

by Lemma 17.2. Solve this equation for a_k to get

$$a_k = \frac{1}{\pi} \int_{-\pi}^{\pi} f(x)\cos(kx)\, dx$$

for $k = 1, 2, 3, \ldots$.

We can use the same idea to solve for b_k. Now multiply equation (17.1) by $\sin(kx)$:

$$f(x)\sin(kx) = a_0\sin(kx) + \sum_{n=1}^{\infty} [a_n\cos(nx)\sin(kx) + b_n\sin(nx)\sin(kx)].$$

Integrate this equation from $-\pi$ to π and interchange the integral and the summation:

$$\int_{-\pi}^{\pi} f(x)\sin(kx)\,dx = \int_{-\pi}^{\pi} a_0\sin(kx)\,dx$$

$$+ \sum_{n=1}^{\infty} \left[a_n \int_{-\pi}^{\pi} \cos(nx)\sin(kx)\,dx + b_n \int_{-\pi}^{\pi} \sin(nx)\sin(kx)\,dx \right].$$

Again using the orthogonality relationships, all the integrals on the right side are zero, except for $\int_{-\pi}^{\pi} \sin(nx)\sin(kx)\,dx$ when $n = k$ in the summation. The last equation reduces to

$$\int_{-\pi}^{\pi} f(x)\sin(kx)\,dx = b_k \int_{-\pi}^{\pi} \sin^2(kx)\,dx = b_k\pi,$$

from which we conclude that

$$b_k = \frac{1}{\pi} \int_{-\pi}^{\pi} f(x)\sin(kx)\,dx$$

for $k = 1, 2, 3, \ldots$.

This "derivation" of the constants in equation (17.1) is flawed by the interchange of the summation $\sum_{k=1}^{\infty}$ and the integral $\int_{-\pi}^{\pi} \cdots dx$. This step is not justified without certain assumptions about f. Nevertheless, the argument does serve as a motivation for the definition of the Fourier coefficients and Fourier series which we will now make.

DEFINITION   Fourier Series and Coefficients on $[-\pi, \pi]$

Let f be integrable on $[-\pi, \pi]$.

1. The *Fourier coefficients of f on* $[-\pi, \pi]$ are the numbers

$$a_0 = \frac{1}{2\pi} \int_{-\pi}^{\pi} f(x)\,dx,$$

$$a_n = \frac{1}{\pi} \int_{-\pi}^{\pi} f(x)\cos(nx)\,dx \quad \text{for} \quad n = 1, 2, 3, \ldots,$$

and

$$b_n = \frac{1}{\pi} \int_{-\pi}^{\pi} f(x)\sin(nx)\,dx \quad \text{for} \quad n = 1, 2, 3, \ldots.$$

2. The *Fourier series of f on* $[-\pi, \pi]$ is the series

$$a_0 + \sum_{n=1}^{\infty} [a_n\cos(nx) + b_n\sin(nx)],$$

in which the coefficients are the Fourier coefficients of f on $[-\pi, \pi]$. ∎

EXAMPLE 17.1

Write the Fourier series of $f(x) = x$ on $[-\pi, \pi]$.

Compute

$$a_0 = \frac{1}{2\pi} \int_{-\pi}^{\pi} x \, dx = \frac{1}{4\pi} x^2 \Big]_{-\pi}^{\pi} = 0,$$

$$a_n = \frac{1}{\pi} \int_{-\pi}^{\pi} x \cos(nx) \, dx = \frac{1}{n^2\pi} \cos(nx) + \frac{x}{n\pi} \sin(nx) \Big]_{-\pi}^{\pi} = 0,$$

and

$$b_n = \frac{1}{\pi} \int_{-\pi}^{\pi} x \sin(nx) \, dx = \frac{1}{n^2\pi} \sin(nx) - \frac{x}{n\pi} \cos(nx) \Big]_{-\pi}^{\pi}$$

$$= -\frac{1}{n} \cos(n\pi) - \frac{1}{n} \cos(-n\pi) = -\frac{2}{n} \cos(n\pi) = \frac{2}{n}(-1)^{n+1}$$

because $\cos(n\pi) = (-1)^n$. The Fourier series of x on $[-\pi, \pi]$ is

$$\sum_{n=1}^{\infty} \frac{2}{n}(-1)^{n+1} \sin(nx).$$

In this example, the constant term a_0 and all the a_n's are zero, and only sine terms remain in the Fourier series. ∎

We can extend this discussion to functions defined on an interval $[-L, L]$ simply by changing scale. The change of variables $t = \pi x/L$ converts the interval $[-L, L]$ to $[-\pi, \pi]$. If we apply this scaling to the definition of Fourier series and Fourier coefficients, we obtain the following, which we also state as a definition.

DEFINITION   Fourier Series and Coefficients on $[-L, L]$

Let f be integrable on $[-L, L]$.

1. The *Fourier coefficients of f on* $[-L, L]$ are

$$a_0 = \frac{1}{2L} \int_{-L}^{L} f(x) \, dx,$$

$$a_n = \frac{1}{L} \int_{-L}^{L} f(x)\cos\left(\frac{n\pi x}{L}\right) dx, \quad \text{and} \quad b_n = \frac{1}{L} \int_{-L}^{L} f(x)\sin\left(\frac{n\pi x}{L}\right) dx$$

for $n = 1, 2, 3, \ldots$.

2. The *Fourier series of f on* $[-L, L]$ is

$$a_0 + \sum_{n=1}^{\infty} \left[a_n\cos\left(\frac{n\pi x}{L}\right) + b_n\sin\left(\frac{n\pi x}{L}\right) \right],$$

in which the numbers $a_0, a_1, \ldots, b_1, \ldots$ are the Fourier coefficients of f on $[-L, L]$. ∎

EXAMPLE 17.2

Let

$$f(x) = \begin{cases} 0 & \text{for} \quad -3 \le x \le 0 \\ x & \text{for} \quad 0 \le x \le 3. \end{cases}$$

Find the Fourier series of f on $[-3, 3]$.

With $L = 3$, compute

$$a_0 = \frac{1}{6} \int_{-3}^{3} f(x)\, dx = \frac{1}{6} \int_{0}^{3} x\, dx = \frac{3}{4},$$

$$a_n = \frac{1}{3} \int_{-3}^{3} f(x)\cos\left(\frac{n\pi x}{3}\right) dx$$

$$= \frac{1}{3} \int_{0}^{3} x \cos\left(\frac{n\pi x}{3}\right) dx = \frac{3}{n^2\pi^2}[\cos(n\pi) - 1],$$

and

$$b_n = \frac{1}{3} \int_{-3}^{3} f(x)\sin\left(\frac{n\pi x}{3}\right) dx = \frac{1}{3} \int_{0}^{3} x \sin\left(\frac{n\pi x}{3}\right) dx = -\frac{3}{n\pi}\cos(n\pi).$$

Since $\cos(n\pi) = (-1)^n$, the Fourier series of f on $[-3, 3]$ is

$$\frac{3}{4} + \sum_{n=1}^{\infty}\left[\frac{3}{n^2\pi^2}[(-1)^n - 1]\cos\left(\frac{n\pi x}{3}\right) - \frac{3}{n\pi}(-1)^n\sin\left(\frac{n\pi x}{3}\right)\right]. \quad \blacksquare$$

In these two examples, we have been careful not to claim that the Fourier series of the function actually equals the function on the interval. The reason is that a function and its Fourier series need not agree. In Example 17.1, we found that the Fourier series of x on $[-\pi, \pi]$ is $\sum_{n=1}^{\infty} (2/n)(-1)^{n+1}\sin(nx)$. This series is zero at π and at $-\pi$, while $f(x) = x$ does not equal zero at either endpoint. At other points, the relationship between the Fourier series and the function is not obvious. For example, $f(\pi/2) = \pi/2$, while the Fourier series at $\pi/2$ is $\sum_{n=1}^{\infty} (2/n)(-1)^{n+1}\sin(n\pi/2)$, and the sum of this series is not apparent. We will examine convergence of Fourier series in the next section.

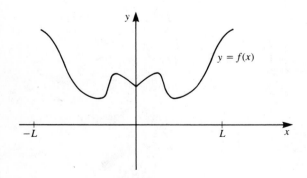

Figure 17.1. A typical even function on $[-L, L]$.

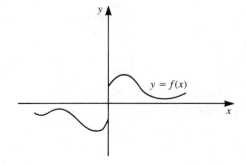

Figure 17.2. Graph of a typical odd function on $[-L, L]$.

In some cases, we can save some work in computing Fourier coefficients by observing special features of f. We say that f is *even* on $[-L, L]$ if $f(-x) = f(x)$ for $0 \leq x \leq L$. This means that the graph of f from $-L$ to zero is the reflection across the y-axis of the graph from zero to L, as in Figure 17.1. For example, $\cos(n\pi x/L)$, x^2, x^4, $x^6 - 2x^2 + \cos(x)$, and $|x|$ are even functions. Replacing x by $-x$ has no effect on $f(x)$.

A function f is *odd* on $[-L, L]$ if $f(-x) = -f(x)$ for $0 \leq x \leq L$. In this case, the graph from $-L$ to zero is the reflection across the y-axis, and then across the x-axis, of the graph from zero to L. A typical odd function is shown in Figure 17.2. Examples of odd functions are $\sin(n\pi x/L)$, x, x^3, and $x^5 - \tan(x)$. For these functions, replacing x by $-x$ results in obtaining the negative of the value obtained for x.

"Most" functions are neither even nor odd. For example, e^x and $x^2 - x$ are neither even nor odd. However, a product of two even or two odd functions is even, and a product of an even function with an odd function is odd:

$$\text{odd} \cdot \text{odd} = \text{even};$$
$$\text{even} \cdot \text{even} = \text{even};$$
$$\text{odd} \cdot \text{even} = \text{odd}.$$

For example, x^2 is even on $[-L, L]$, while $\sin(x)$ is odd. The product $x^2\sin(x)$ is odd because $x^2\sin(x)$ changes sign if x is replaced by $-x$.

We will exploit the following facts from calculus.

LEMMA 17.3

Let f be integrable on $[-L, L]$.

1. If f is even on $[-L, L]$, then $\int_{-L}^{L} f(x)\, dx = 2 \int_{0}^{L} f(x)\, dx$.
2. If f is odd on $[-L, L]$, then $\int_{-L}^{L} f(x)\, dx = 0$. ∎

A proof is outlined in Problem 17 at the end of this section. In terms of areas, these results are suggested by Figures 17.3 and 17.4.

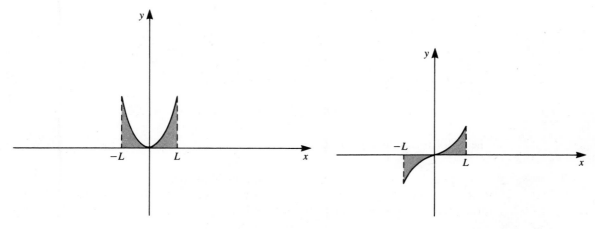

Figure 17.3. $\int_{-L}^{L} f(x)\, dx = 2 \int_{0}^{L} f(x)\, dx.$ *Figure 17.4.* $\int_{-L}^{L} f(x)\, dx = 0.$

Lemma 17.3 makes computation of the Fourier coefficients of f easier when f is even or odd. If f is even on $[-L, L]$, $f(x)\sin(n\pi x/L)$ is odd for any positive integer n; hence, all the b_n's are zero by conclusion (2) of Lemma 17.3, and we can write a somewhat simpler formula for the a_n's by using conclusion (1). If f is odd, $f(x)\cos(n\pi x/L)$ is odd and each a_n is zero. The following theorem gives the specific details.

THEOREM 17.1

Let f be integrable on $[-L, L]$.

1. If f is even, the Fourier series of f on $[-L, L]$ is

$$a_0 + \sum_{n=1}^{\infty} a_n \cos\left(\frac{n\pi x}{L}\right),$$

in which

$$a_0 = \frac{1}{L} \int_0^L f(x)\, dx \quad \text{and} \quad a_n = \frac{2}{L} \int_0^L f(x)\cos\left(\frac{n\pi x}{L}\right) dx.$$

2. If f is odd, its Fourier series on $[-L, L]$ is

$$\sum_{n=1}^{\infty} b_n \sin\left(\frac{n\pi x}{L}\right),$$

in which

$$b_n = \frac{2}{L} \int_0^L f(x)\sin\left(\frac{n\pi x}{L}\right) dx.$$

Proof   If f is even,

$$a_0 = \frac{1}{2L} \int_{-L}^L f(x)\, dx = \frac{1}{L} \int_0^L f(x)\, dx$$

and

$$a_n = \frac{1}{L} \int_{-L}^L f(x)\cos\left(\frac{n\pi x}{L}\right) dx = \frac{2}{L} \int_0^L f(x)\cos\left(\frac{n\pi x}{L}\right) dx,$$

by conclusion (1) of Lemma 17.3. By conclusion (2),

$$b_n = \frac{1}{L} \int_{-\pi}^{\pi} f(x)\sin\left(\frac{n\pi x}{L}\right) dx = 0 \quad \text{for} \quad n = 1, 2, 3, \ldots.$$

The proof is similar if f is odd.   ■

Example 17.1 illustrates conclusion (2) of the theorem. There, we had $f(x) = x$, which is odd on $[-\pi, \pi]$, and the Fourier series contained only sine terms. All of the integrals for the cosine coefficients were zero.

EXAMPLE 17.3

Find the Fourier series of $f(x) = x^4$ on $[-\pi, \pi]$.

Since x^4 is even on $[-\pi, \pi]$, each $b_n = 0$. Further,

$$a_0 = \frac{1}{\pi} \int_0^\pi x^4 \, dx = \frac{1}{\pi} \frac{1}{5} x^5 \Big]_0^\pi = \frac{1}{5} \pi^4$$

and, for $n = 1, 2, 3, \ldots$,

$$a_n = \frac{2}{\pi} \int_0^\pi x^4 \cos(nx) \, dx$$

$$= \frac{2}{n\pi} \left[x^4 \sin(nx) \right]_0^\pi - \frac{8}{n\pi} \left[\frac{3n^2 x^2 - 6}{n^4} \sin(nx) - \frac{n^2 x^3 - 6x}{n^3} \cos(nx) \right]_0^\pi$$

$$= \frac{8}{\pi n^4} (n^2 \pi^3 - 6\pi) \cos(n\pi)$$

$$= \frac{8}{n^4} (n^2 \pi^2 - 6)(-1)^n,$$

since $\cos(n\pi) = (-1)^n$. The Fourier series of x^4 on $[-\pi, \pi]$ is

$$\frac{1}{5} \pi^4 + \sum_{n=1}^\infty \frac{8}{n^4} (n^2 \pi^2 - 6)(-1)^n \cos(nx). \quad \blacksquare$$

In sum, then, if f is even on $[-L, L]$, its Fourier series contains only cosine terms (which are even) and perhaps a constant; if f is odd, its Fourier series contains only sine terms (which are odd).

PROBLEMS FOR SECTION 17.1

In each of Problem 1 through 15, find the Fourier series of f on the interval. If f is even or odd, exploit Theorem 17.1 to simplify the calculations. Use a table of integrals where convenient.

1. $f(x) = 4, \quad -3 \le x \le 3$ **2.** $f(x) = -x, \quad -1 \le x \le 1$ **3.** $f(x) = \cos(\pi x), \quad -1 \le x \le 1$

4. $f(x) = \cos(3x), \quad -2 \le x \le 2$ **5.** $f(x) = \frac{1}{2} e^{2x}, \quad -1 \le x \le 1$ **6.** $f(x) = \sinh(x), \quad -4 \le x \le 4$

7. $f(x) = x^2, \quad -3 \le x \le 3$ **8.** $f(x) = \begin{cases} 0, & -4 \le x \le 0 \\ 1, & 0 < x \le 4 \end{cases}$ **9.** $f(x) = 1 - |x|, \quad -2 \le x \le 2$

10. $f(x) = \begin{cases} 1, & -\pi \le x < 0 \\ 2, & 0 \le x \le \pi \end{cases}$ **11.** $f(x) = \begin{cases} 1 - x, & -1 \le x < 0 \\ 0, & 0 \le x \le 1 \end{cases}$ **12.** $f(x) = \begin{cases} -4, & -\pi \le x \le 0 \\ 4, & 0 < x < \pi \end{cases}$

13. $f(x) = \sin(2x), \quad -\pi \le x \le \pi$ **14.** $f(x) = \begin{cases} -1, & -3 \le x \le -1 \\ 0, & -1 < x \le 3 \end{cases}$

15. $f(x) = \cos\left(\frac{x}{2}\right) - \sin(x), \quad -\pi \le x \le \pi$

16. Suppose that f and g are integrable on $[-L, L]$ and that x_0 is in $(-L, L)$. Suppose that $f(x) = g(x)$ if $x \ne x_0$ but that $f(x_0) \ne g(x_0)$. How do the Fourier series of f and g on $[-L, L]$ differ? What does this suggest about the relationship between a function and its Fourier series on an interval?

17. Prove Lemma 17.3. *Hint:* Write $\int_{-L}^L f(x) \, dx = \int_{-L}^0 f(x) \, dx + \int_0^L f(x) \, dx$. Let $t = -x$ to write $\int_{-L}^0 f(x) \, dx = \int_L^0 f(-t)(-1) \, dt = \int_0^L f(-x) \, dx$ and hence $\int_{-L}^L f(x) \, dx = \int_0^L f(x) \, dx + \int_0^L f(-x) \, dx$. Now consider separately the cases in which f is even or f is odd.

POSTSCRIPT

We will place the discussion of Fourier series in the context of eigenfunction expansion and Sturm-Liouville theory from Sections 7.6 and 7.7. This discussion may be omitted without compromising the rest of this chapter.

Consider the periodic Sturm-Liouville problem

$$y'' + \lambda y = 0; \qquad y(-\pi) = y(\pi), \qquad y'(-\pi) = y'(\pi).$$

In Example 7.7, we showed that the eigenvalues of this problem are $\lambda = n^2$, with $n = 0, 1, 2, 3, \ldots$. Further, the corresponding eigenfunctions are 1 for $\lambda = 0$ and $\cos(nx)$ and $\sin(nx)$ for $\lambda = n^2$, with n a positive integer. The Sturm-Liouville theory tells us that these functions are orthogonal on $[-\pi, \pi]$ with weight function 1. Since the inner product defining orthogonality is the integral from $-\pi$ to π, this simply says that

$$\int_{-\pi}^{\pi} (\text{product of two distinct eigenfunctions}) \, dx = 0,$$

a conclusion we saw in Lemma 17.1.

In Sections 7.6 and 7.8, we also discussed the possibility of expanding a function g on an interval $[a, b]$ in a series of the form

$$g(x) = \sum_{n=0}^{\infty} c_n f_n(x).$$

If $f_0, f_1, \ldots$ are orthogonal on $[a, b]$, we argued that the coefficients are given by

$$c_n = \frac{\int_a^b g(x) f_n(x) \, dx}{\int_a^b f_n^2(x) \, dx}.$$

For the Fourier series of g on $[-\pi, \pi]$, the f_n's are the functions $1, \cos(x), \cos(2x), \ldots$, $\sin(x), \sin(2x), \ldots$, which are orthogonal on $[-\pi, \pi]$. Further,

$$a_0 = \frac{\int_{-\pi}^{\pi} g(x) \, dx}{\int_{-\pi}^{\pi} 1^2 \, dx} = \frac{1}{2\pi} \int_{-\pi}^{\pi} g(x) \, dx,$$

$$a_n = \frac{\int_{-\pi}^{\pi} g(x)\cos(nx) \, dx}{\int_{-\pi}^{\pi} \cos^2(nx) \, dx} = \frac{1}{\pi} \int_{-\pi}^{\pi} g(x)\cos(nx) \, dx,$$

and

$$b_n = \frac{\int_{-\pi}^{\pi} g(x)\sin(nx) \, dx}{\int_{-\pi}^{\pi} \sin^2(nx) \, dx} = \frac{1}{\pi} \int_{-\pi}^{\pi} g(x)\sin(nx) \, dx.$$

Thus, the Fourier series of g on $[-\pi, \pi]$ is exactly the expansion of g in a series of eigenfunctions of a periodic Sturm-Liouville problem. The formula for the coefficients in the eigenfunction expansion gives us the Fourier coefficients.

We can study and use Fourier series without having seen the broader context of Sturm-Liouville theory, and this is the approach we will take in the following sections. However, the Sturm-Liouville background makes it easier to understand other types of series expansions we will encounter later and to see how they are all part of a general theory of eigenfunction expansions.

17.2 *Convergence, Differentiation, and Integration of Fourier Series*

We observed in Section 17.1 that the Fourier series of $f(x) = x$ on $[-\pi, \pi]$ does not converge to x, at least for some points in $[-\pi, \pi]$. We also noted that the motivational argument for the definition of the Fourier coefficients depended on interchanging a series and an integral, a step which is not valid in general. In this section, we will state conditions on f which enable us to determine the sum of the Fourier series of f on an interval.

In treating the Laplace transform, we encountered the concept of a piecewise-continuous function. Recall that f is *piecewise continuous* on $[a, b]$ if

1. $\lim_{x \to a+} f(x)$ and $\lim_{x \to b-} f(x)$ are finite.
2. f is continuous at all but possibly finitely many points of $[a, b]$.
3. At any point in (a, b) where f is discontinuous, the function has finite left and right limits.

Figure 17.5 shows the graph of a piecewise-continuous function. At points in (a, b) where the function is discontinuous, the graph has jump discontinuities. The magnitude of the jump at x_0 is the difference between the left and right limits there:

$$\left| \lim_{h \to 0+} f(x_0 + h) - \lim_{h \to 0-} f(x_0 + h) \right|.$$

This is the distance between left and right "ends" of the graph at x_0.

We will denote left and right limits as follows:

$$f(x_0 +) = \lim_{h \to 0+} f(x_0 + h) \quad \text{and} \quad f(x_0 -) = \lim_{h \to 0+} f(x_0 - h).$$

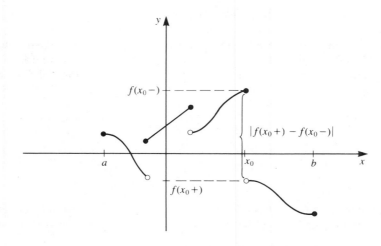

Figure 17.5

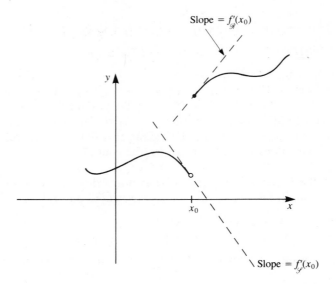

Figure 17.6

If f has a right limit at x_0, the *right derivative* of f at x_0 is defined to be

$$f'_{\mathcal{R}}(x_0) = \lim_{h \to 0+} \frac{f(x_0 + h) - f(x_0+)}{h}$$

when this limit exists and is finite. We interpret $f'_{\mathcal{R}}(x_0)$ as the slope of the tangent to the graph at x_0 if we look at only the portion of the graph to the right of x_0. The *left derivative* of f at x_0 is

$$f'_{\mathcal{L}}(x_0) = \lim_{h \to 0-} \frac{f(x_0 + h) - f(x_0-)}{h}$$

when this limit exists and is finite. This number is the slope of the graph of f at x_0, looking at only that portion to the left of x_0.

Figure 17.6 illustrates the left and right derivatives of a function. At a jump discontinuity, the left and right derivatives may exist, while the derivative does not exist. That is, a graph may have a tangent "from the left" and "from the right" at x_0 while having no tangent line at x_0.

EXAMPLE 17.4

Let

$$f(x) = \begin{cases} 1 + x & \text{for} \quad -\pi \le x < 0 \\ x^2 & \text{for} \quad 0 \le x \le \pi. \end{cases}$$

f is piecewise continuous on $[-\pi, \pi]$, with a jump discontinuity at zero. The graph is shown in Figure 17.7. The right and left limits of f at zero are, respectively,

$$f(0+) = \lim_{h \to 0+} f(0 + h) = \lim_{h \to 0+} h^2 = 0$$

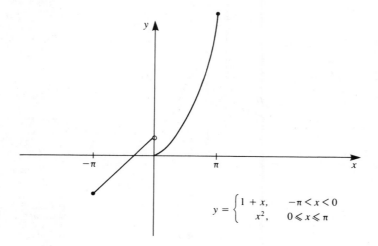

$$y = \begin{cases} 1 + x, & -\pi < x < 0 \\ x^2, & 0 \leqslant x \leqslant \pi \end{cases}$$

Figure 17.7

and

$$f(0-) = \lim_{h \to 0+} f(0 - h) = \lim_{h \to 0+} (1 - h) = 1.$$

On $(-\pi, \pi)$, f is differentiable except at zero. The right derivative at zero is

$$\mathrm{f}'_{\mathscr{R}}(0) = \lim_{h \to 0+} \frac{f(0 + h) - f(0+)}{h} = \lim_{h \to 0+} \frac{h^2 - 0}{h} = 0.$$

This is consistent with Figure 17.7. To the right of the origin, $f(x) = x^2$, with slope zero at the origin from the right. The left derivative at zero is

$$f'_{\mathscr{L}}(0) = \lim_{h \to 0-} \frac{f(0 + h) - f(0-)}{h} = \lim_{h \to 0-} \frac{(1 + h) - 1}{h} = 1.$$

This result is consistent with the fact that, to the left of the origin, $f(x) = 1 + x$, with slope 1. ∎

We can now state a result on the convergence of Fourier series.

THEOREM 17.2 Convergence of Fourier Series

Let f be piecewise continuous on $[-L, L]$.

1. If $-L < x_0 < L$ and both left and right derivatives of f exist at x_0, the Fourier series of f on $[-L, L]$ converges at x_0 to

$$\tfrac{1}{2}[f(x_0+) + f(x_0-)].$$

2. If both $f'_{\mathscr{R}}(-L)$ and $f'_{\mathscr{L}}(L)$ exist, the Fourier series converges at both L and $-L$ to

$$\tfrac{1}{2}[f(-L+) + f(L-)]. \quad \blacksquare$$

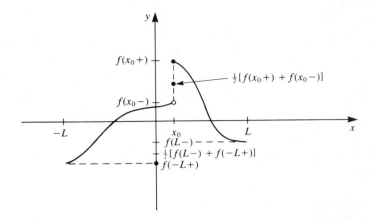

Figure 17.8

Therefore, at each point between $-L$ and L where the right and left derivatives exist, the Fourier series converges to the average of the left and right limits of f. This fact is illustrated in Figure 17.8. In particular, *if f is continuous at x_0*, the left and right limits both equal $f(x_0)$ and the series converges at x_0 to $\frac{1}{2}[f(x_0) + f(x_0)]$, which equals $f(x_0)$.

At *both* $-L$ and L, the Fourier series converges to the average of the right limit at $-L$ and the left limit at L, assuming that the right derivative exists at $-L$ and the left derivative exists at L. It should not be surprising that the Fourier series converges to the same value at L and at $-L$. Upon letting $x = L$ in the Fourier series of f, we get

$$a_0 + \sum_{n=1}^{\infty} a_n \cos(n\pi)$$

because $\sin(n\pi) = 0$. If we let $x = -L$ in the Fourier series, we get exactly the same series because $\sin(-n\pi) = 0$ and $\cos(-n\pi) = \cos(n\pi)$.

In terms of the graph, the Fourier series converges at $-L$ and at L to the point midway between the ends of the graph at $-L$ and L. This result is also indicated in Figure 17.8.

In view of Theorem 17.2, we can often tell what the Fourier series of f converges to just by looking at the graph of f.

EXAMPLE 17.5

Let $f(x) = 2x$ for $-\pi \le x \le \pi$.

Observe that f is continuous on $(-\pi, \pi)$ and has left and right derivatives at every point (in fact, f is differentiable). Thus, the Fourier series of f on $[-\pi, \pi]$ converges to $2x$ for $-\pi < x < \pi$.

At both π and $-\pi$, the Fourier series converges to the average of the left limit of f at π and the right limit at $-\pi$. Compute

$$f(\pi-) = \lim_{h \to 0+} f(\pi - h) = \lim_{h \to 0+} 2(\pi - h) = 2\pi$$

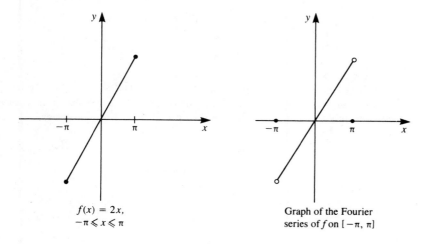

$f(x) = 2x,$
$-\pi \leqslant x \leqslant \pi$

Graph of the Fourier
series of f on $[-\pi, \pi]$

Figure 17.9

and

$$f(-\pi+) = \lim_{h \to 0+} f(-\pi + h) = \lim_{h \to 0+} 2(-\pi + h) = -2\pi.$$

The average of 2π and -2π is zero, so the Fourier series converges to zero at both π and $-\pi$. From Example 17.1, the Fourier series of f on $[-\pi, \pi]$ is

$$\sum_{n=1}^{\infty} \frac{4}{n}(-1)^{n+1}\sin(nx),$$

from which we can verify directly that the series converges to zero at π and at $-\pi$.

The graph of the Fourier series compared with that of f is shown in Figure 17.9. In this example, the Fourier series converges to the function on $(-\pi, \pi)$ but differs from function values at the endpoints. We can write

$$2x = \sum_{n=1}^{\infty} \frac{4}{n}(-1)^{n+1}\sin(nx) \quad \text{for} \quad -\pi < x < \pi. \quad \blacksquare$$

EXAMPLE 17.6

Let

$$f(x) = \begin{cases} 0 & \text{for} & -\pi \leq x < 1 \\ 1 & \text{for} & 1 \leq x < 2 \\ 3 & \text{for} & 2 \leq x \leq \pi. \end{cases}$$

The graph of f is shown in Figure 17.10. From it, we read that the Fourier series of f on $[-\pi, \pi]$ converges to

$$
\begin{array}{lll}
0 & \text{for} & -\pi < x < 1 \\
1 & \text{for} & 1 < x < 2 \\
3 & \text{for} & 2 < x < \pi
\end{array}
$$

$$\tfrac{1}{2}(0 + 3), \quad \text{or} \quad \tfrac{3}{2}, \quad \text{at} \quad -\pi \quad \text{and at} \quad \pi$$

$$\tfrac{1}{2}(0 + 1), \quad \text{or} \quad \tfrac{1}{2}, \quad \text{at} \quad 1$$

$$\tfrac{1}{2}(1 + 3), \quad \text{or} \quad 2, \quad \text{at} \quad 2. \quad \blacksquare$$

It is interesting to examine the convergence of a Fourier series by comparing the graph of the function with graphs of partial sums of the Fourier series. This is done in Figures 17.11 through 17.22 for various functions and their Fourier series.

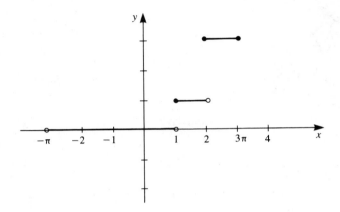

$$\text{Figure 17.10.} \quad f(x) = \begin{cases} 0, & -\pi \leq x < 1 \\ 1, & 1 \leq x < 2 \\ 2, & 2 \leq x \leq \pi. \end{cases}$$

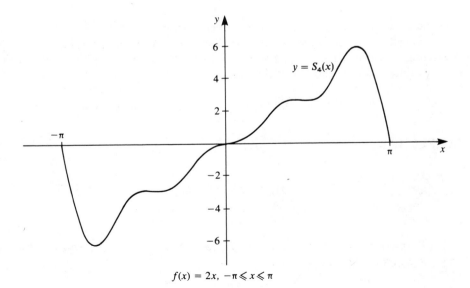

$$f(x) = 2x, \quad -\pi \leqslant x \leqslant \pi$$

$$\text{Figure 17.11.} \quad S_4(x) = \sum_{n=1}^{4} \frac{4}{n}(-1)^{n+1}\sin(nx).$$

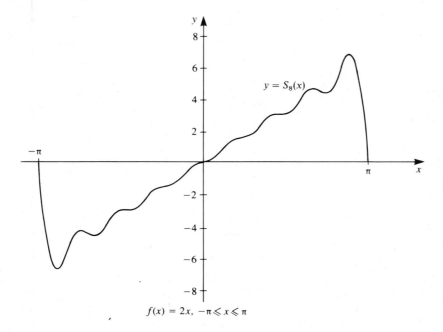

$$f(x) = 2x, \quad -\pi \leqslant x \leqslant \pi$$

Figure 17.12. $S_8(x) = \displaystyle\sum_{n=1}^{8} \frac{4}{n}(-1)^{n+1}\sin(nx).$

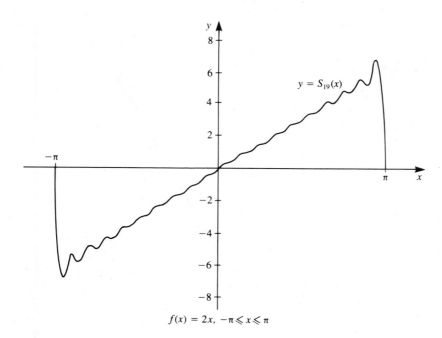

$$f(x) = 2x, \quad -\pi \leqslant x \leqslant \pi$$

Figure 17.13. $S_{19}(x) = \displaystyle\sum_{n=1}^{19} \frac{4}{n}(-1)^{n+1}\sin(nx).$

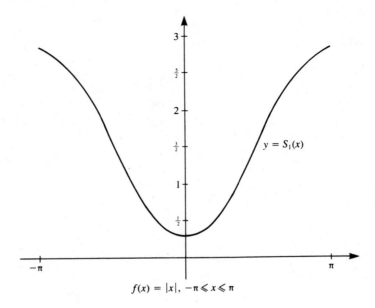

$$f(x) = |x|, \quad -\pi \leqslant x \leqslant \pi$$

Figure 17.14. $S_1(x) = \dfrac{\pi}{2} - \dfrac{4}{\pi}\cos(x).$

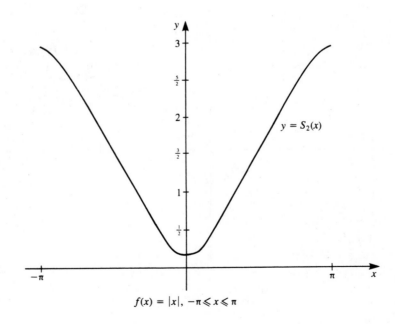

$$f(x) = |x|, \quad -\pi \leqslant x \leqslant \pi$$

Figure 17.15. $S_2(x) = \dfrac{\pi}{2} - \dfrac{4}{\pi}\cos(x) - \dfrac{4}{9\pi}\cos(3x).$

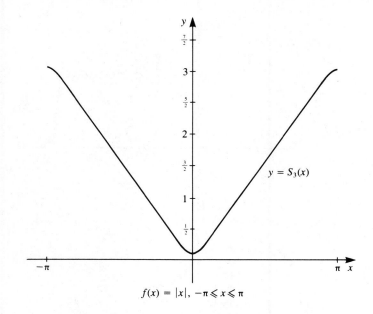

$$f(x) = |x|, \quad -\pi \leqslant x \leqslant \pi$$

Figure 17.16. $S_3(x) = \dfrac{\pi}{2} - \dfrac{4}{\pi}\cos(x) - \dfrac{4}{9\pi}\cos(3x) - \dfrac{4}{25\pi}\cos(5x).$

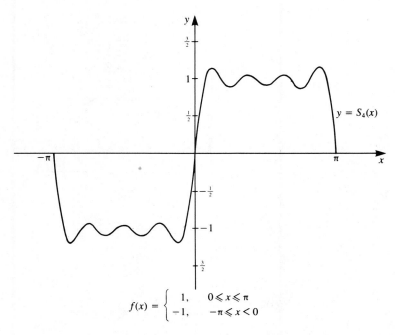

$$f(x) = \begin{cases} 1, & 0 \leqslant x \leqslant \pi \\ -1, & -\pi \leqslant x < 0 \end{cases}$$

Figure 17.17. $S_4(x) = \dfrac{4}{\pi}\sin(x) + \dfrac{4}{3\pi}\sin(3x) + \dfrac{4}{5\pi}\sin(5x) + \dfrac{4}{7\pi}\sin(7x).$

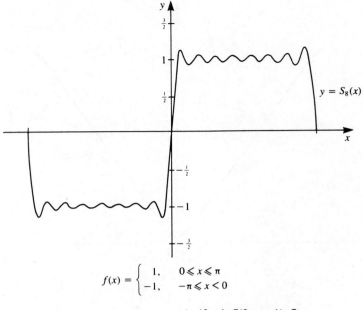

$$f(x) = \begin{cases} 1, & 0 \leqslant x \leqslant \pi \\ -1, & -\pi \leqslant x < 0 \end{cases}$$

Figure 17.18. $S_8(x) = \dfrac{4}{\pi} \displaystyle\sum_{n=1}^{15} \dfrac{\sin[(2n-1)x]}{2n-1}.$

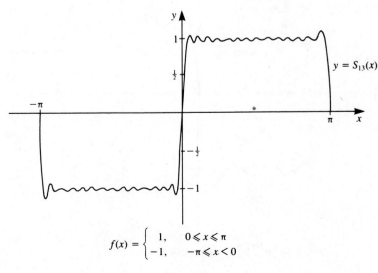

$$f(x) = \begin{cases} 1, & 0 \leqslant x \leqslant \pi \\ -1, & -\pi \leqslant x < 0 \end{cases}$$

Figure 17.19. $S_{13}(x) = \dfrac{4}{\pi} \displaystyle\sum_{n=1}^{25} \dfrac{1}{2n-1} \sin[(2n-1)x].$

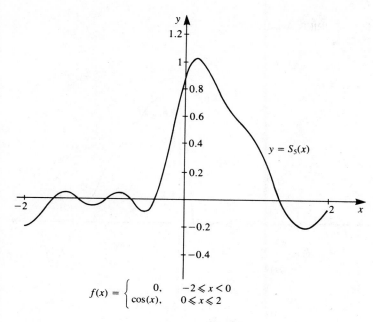

$$f(x) = \begin{cases} 0, & -2 \leqslant x < 0 \\ \cos(x), & 0 \leqslant x \leqslant 2 \end{cases}$$

Figure 17.20. $S_5(x) = \dfrac{1}{4}\sin(2) + \displaystyle\sum_{n=1}^{5} \dfrac{1}{n^2\pi^2 - 4}$

$$\left\{ (-1)^{n+1} 2\sin(2)\cos\left(\dfrac{n\pi x}{2}\right) + n\pi[1 - (-1)^n\cos(2)]\sin\left(\dfrac{n\pi x}{2}\right) \right\}.$$

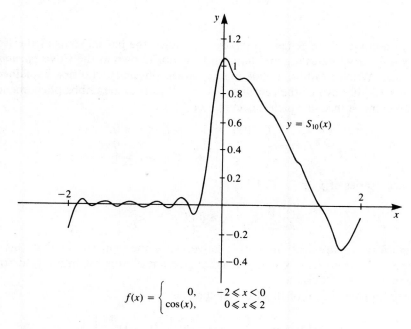

$$f(x) = \begin{cases} 0, & -2 \leqslant x < 0 \\ \cos(x), & 0 \leqslant x \leqslant 2 \end{cases}$$

Figure 17.21. $S_{10}(x) = \dfrac{1}{4}\sin(2) + \displaystyle\sum_{n=1}^{10} \dfrac{1}{n^2\pi^2 - 4}$

$$\left\{ (-1)^{n+1} 2\sin(2)\cos\left(\dfrac{n\pi x}{2}\right) + n\pi[1 - (-1)^n\cos(2)]\sin\left(\dfrac{n\pi x}{2}\right) \right\}.$$

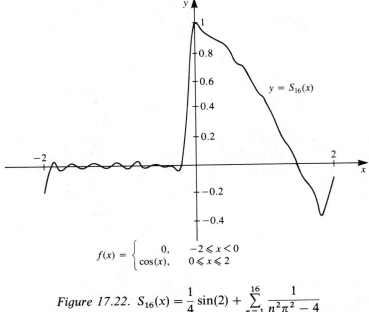

$$f(x) = \begin{cases} 0, & -2 \leqslant x < 0 \\ \cos(x), & 0 \leqslant x \leqslant 2 \end{cases}$$

Figure 17.22. $S_{16}(x) = \dfrac{1}{4}\sin(2) + \displaystyle\sum_{n=1}^{16} \dfrac{1}{n^2\pi^2 - 4}$

$$\left\{ (-1)^{n+1} 2\sin(2)\cos\left(\frac{n\pi x}{2}\right) + n\pi[1 - (-1)^n \cos(2)]\sin\left(\frac{n\pi x}{2}\right) \right\}.$$

Near a point where f has a jump discontinuity, the partial sums of the Fourier series exhibit very interesting and unusual behavior known as the *Gibbs phenomenon* (after Josiah Willard Gibbs, a Yale mathematical physicist who first explained this behavior near the turn of the century). We will illustrate the Gibbs phenomenon by examining an instance in which it occurs. Let

$$f(x) = \begin{cases} -100 & \text{if} & -L < x < 0 \\ 0 & \text{if} & x = 0 \\ 100 & \text{if} & 0 < x \leq L. \end{cases}$$

The Fourier series of f on $[-L, L]$ is

$$\frac{400}{\pi} \sum_{n=1}^{\infty} \frac{1}{2n-1} \sin\left[(2n-1)\frac{\pi x}{L}\right].$$

A straightforward application of the convergence theorem shows that this series converges to $f(x)$ for $-L < x < L$. Even though f has a jump discontinuity at zero, the series converges to $f(0)$ there.

The Nth partial sum of this Fourier series is

$$S_N(x) = \frac{400}{\pi} \sum_{n=1}^{N} \frac{1}{2n-1} \sin\left[(2n-1)\frac{\pi x}{L}\right].$$

Figure 17.23 shows graphs of S_N for various values of N and just $0 \leq x \leq L$. The graphs of $S_4(x)$ and $S_8(x)$ exhibit relatively high peaks near zero. Intuition might suggest that these peaks will decrease as N is chosen larger. In fact, they do not. The

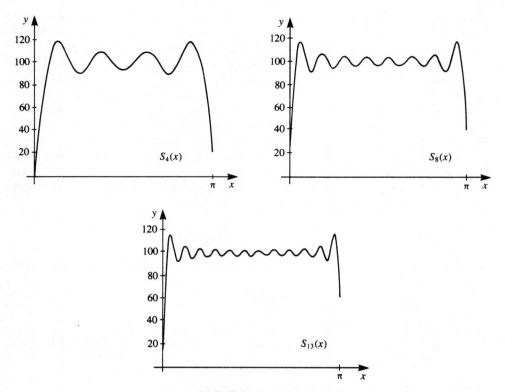

Figure 17.23. Gibbs phenomenon.

graph of S_{13} exhibits a peak of about the same size but occurring closer to zero. The partial sums will continue to show the peaks seen in Figure 17.23, but as N increases, these peaks occur nearer the jump discontinuity. This behavior is the *Gibbs phenomenon*, and it occurs in the partial sum of the Fourier series of every function having a jump discontinuity.

For this example, the maximum values of $S_N(x)$ on $[0, L]$ occur at $L/2N$ and at $[(2N - 1)/2N]L$. Some approximate values for this maximum for the first few values of N are as follows.

N	Maximum Value of $S_N(x)$	Maximum Occurs at
1	127.3240	$L/2$
2	120.0421	$L/4$ and $3L/4$
3	118.8357	$L/6$ and $5L/6$
4	118.4225	$L/8$ and $7L/8$
5	118.2328	$L/10$ and $9L/10$
6	118.1302	$L/12$ and $11L/12$

The amplitudes of these peaks approach a limit which is approximately 9% more than the magnitude of the jump at the discontinuity. Some of the mathematical details explaining this behavior are outlined in Problem 29 at the end of this section.

The Gibbs phenomenon is also a commonplace occurrence in the real world. Very high frequencies (harmonics) may leak from a circuit or wire due to stray capacitance. Further, every (well-behaved) function is really a sum of sine waves. If the high-frequency components of the function are attenuated, there remains a finite sum approximation of the function. If the function also has a jump discontinuity (as with a switched voltage), behavior very like the Gibbs phenomenon will occur.

We will conclude this section with a brief discussion of term by term differentiation and integration of Fourier series. In general, differentiation of a Fourier series term by term will result in a series which is unrelated to the derivative of the function. For example, the Fourier series of $f(x) = x$ on $[-\pi, \pi]$ is

$$\sum_{n=1}^{\infty} \frac{2}{n}(-1)^{n+1}\sin(nx),$$

and this converges to x for $-\pi < x < \pi$. Differentiate this series term by term to get

$$\sum_{n=1}^{\infty} 2(-1)^{n+1}\cos(nx),$$

which does not even converge on $(-\pi, \pi)$, much less to $f'(x) = 1$.

Quite restrictive hypotheses are needed to allow term by term differentiation of a Fourier series. Here is one such result.

THEOREM 17.3   Differentiation of a Fourier Series

Suppose that f is continuous on $[-L, L]$ and that $f(L) = f(-L)$. Suppose also that f' is piecewise continuous on $[-L, L]$. Then, for any x in $(-L, L)$ at which f' is continuous,

$$f'(x) = \frac{\pi}{L} \sum_{n=1}^{\infty}\left[-na_n\sin\left(\frac{n\pi x}{L}\right) + nb_n\cos\left(\frac{n\pi x}{L}\right)\right]. \quad \blacksquare$$

This is exactly the series obtained by differentiating the Fourier series of f term by term. A proof of Theorem 17.3 is outlined in Problem 20.

Integration of a Fourier series is an entirely different story.

THEOREM 17.4   Integration of a Fourier Series

Let f be piecewise continuous on $[-L, L]$, with Fourier series

$$a_0 + \sum_{n=1}^{\infty}\left[a_n\cos\left(\frac{n\pi x}{L}\right) + b_n\sin\left(\frac{n\pi x}{L}\right)\right].$$

Then, for any x in $[-L, L]$,

$$\int_{-L}^{x} f(t)\, dt = a_0(x + L)$$

$$+ \frac{L}{\pi} \sum_{n=1}^{\infty} \frac{1}{n}\left\{ a_n\sin\left(\frac{n\pi x}{L}\right) - b_n\left[\cos\left(\frac{n\pi x}{L}\right) - \cos(n\pi)\right]\right\}. \quad \blacksquare$$

This is exactly the result we get by integrating the Fourier series of f term by term from $-\pi$ to x. A proof of Theorem 17.4 is outlined in Problem 22.

EXAMPLE 17.7

From Example 17.5, the Fourier series of $f(x) = 2x$ on $[-\pi, \pi]$ is

$$\sum_{n=1}^{\infty} \frac{4}{n}(-1)^{n+1}\sin(nx).$$

Since f is continuous on $[-\pi, \pi]$, we can apply Theorem 17.4 to write

$$\int_{-\pi}^{x} 2t\, dt = x^2 - \pi^2$$

$$= \sum_{n=1}^{\infty} \frac{4}{n}(-1)^{n+1} \int_{-\pi}^{x} \sin(nt)\, dt$$

$$= \sum_{n=1}^{\infty} \frac{4}{n}(-1)^{n+1} \left\{ -\frac{1}{n}[\cos(nx) - \cos(n\pi)] \right\}$$

$$= \sum_{n=1}^{\infty} \frac{4(-1)^n}{n^2}[\cos(nx) + (-1)^n]. \quad \blacksquare$$

In the next section, we will discuss how to expand a function on an interval $[0, L]$ in a series containing just sines or just cosines, whichever we choose.

PROBLEMS FOR SECTION 17.2

In each of Problems 1 through 15, calculate the left and right limits and the left and right derivatives at each jump discontinuity of the function (if it has any). Also compute the left limit and left derivative at L and the right limit and right derivative at $-L$. Finally, determine what the Fourier series of f on $[-L, L]$ converges to. Sketch a graph of the function and of its Fourier series. It is not necessary to compute the Fourier series itself.

1. $f(x) = \begin{cases} x^2, & 1 \le x \le 3 \\ 0, & -2 \le x < 1 \\ 2x, & -3 \le x < -2 \end{cases}$

2. $f(x) = x^4, \quad -2 \le x \le 2$

3. $f(x) = x^2 e^{-x}, \quad -3 \le x \le 3$

4. $f(x) = \begin{cases} 2x - 2, & -\pi < x \le 1 \\ 3, & 1 < x \le \pi \end{cases}$

5. $f(x) = \begin{cases} x^2, & -\pi \le x < 0 \\ 2, & 0 \le x \le \pi \end{cases}$

6. $f(x) = \begin{cases} 1/x, & -3 \le x \le -1 \\ 2, & -1 < x \le 2 \\ 3, & 2 < x \le 3 \end{cases}$

7. $f(x) = \begin{cases} \cos(x), & -\pi \le x < 0 \\ \sin(x), & 0 \le x \le \pi \end{cases}$

8. $f(x) = \begin{cases} 0, & -1 \le x < \frac{1}{2} \\ 1, & \frac{1}{2} \le x \le \frac{3}{4} \\ 2, & \frac{3}{4} < x \le 1 \end{cases}$

9. $f(x) = e^{-|x|}, \quad -2 \le x \le 2$

10. $f(x) = \begin{cases} -2, & -4 \le x \le -2 \\ 1 + x^2, & -2 < x \le 2 \\ x^3, & 2 < x \le 4 \end{cases}$

11. $f(x) = \begin{cases} x, & -2 \leq x < 1 \\ e^x, & 1 \leq x \leq 2 \end{cases}$

12. $f(x) = \begin{cases} -1, & -1 \leq x < 0 \\ 4, & 0 \leq x \leq 1 \end{cases}$

13. $f(x) = \begin{cases} 1, & -5 \leq x \leq -1 \\ 0, & -1 < x < 3 \\ x^2, & 3 \leq x \leq 5 \end{cases}$

14. $f(x) = \begin{cases} \cos(\pi x), & -1 \leq x \leq 0 \\ 3, & 0 < x \leq 1 \end{cases}$

15. $f(x) = \begin{cases} |x|, & -1 \leq x \leq \frac{1}{2} \\ 2x, & \frac{1}{2} < x \leq \frac{2}{3} \\ 4x, & \frac{2}{3} < x \leq 1 \end{cases}$

16. Determine the Fourier series of $\frac{1}{4}x^2$ on $[-\pi, \pi]$. Show that this series converges to $\frac{1}{4}x^2$ for $-\pi < x < \pi$. By choosing x appropriately, sum the series $\sum_{n=1}^{\infty} 1/n^2$.

17. Use the result of Problem 16 to sum the series $\sum_{n=1}^{\infty} (-1)^n 1/n^2$.

18. What is wrong with the following argument? Theorem 17.2 cannot be true because the sine and cosine terms in the Fourier series are periodic, while f need not be periodic. Hence, whenever f is not periodic, its Fourier series cannot converge to f.

19. Let $f(x) = x$ for $-\pi \leq x \leq \pi$.
(a) Write the Fourier series of f on $[-\pi, \pi]$.
(b) Use Theorem 17.2 to show that the Fourier series in (a) converges to f on $(-\pi, \pi)$.
(c) Note that $f'(x) = 1$ for $-\pi < x < \pi$. Differentiate the Fourier series of f term by term and show that the resulting series diverges for x in $(-\pi, \pi)$. Thus, conclude that, in general, the Fourier series of f cannot be differentiated term by term to yield the Fourier series of f'.

20. Prove Theorem 17.3. *Hint:* Begin by writing the Fourier series of f' on $[-L, L]$. Suppose that this series is

$$A_0 + \sum_{n=1}^{\infty} \left[A_n \cos\left(\frac{n\pi x}{L}\right) + B_n \sin\left(\frac{n\pi x}{L}\right) \right].$$

Write the integral formulas for the Fourier coefficients $A_0, A_1, \ldots, B_1, \ldots$. Integrate by parts to relate these coefficients to the Fourier coefficients of f. Also show that $A_0 = 0$.

21. Let $f(x) = |x|$ for $-1 \leq x \leq 1$.
(a) Show that f satisfies the hypotheses of Theorem 17.3.
(b) Write the Fourier series of f on $[-1, 1]$.
(c) Differentiate the Fourier series found in (b) term by term to obtain a Fourier series of f', where

$$f'(x) = \begin{cases} -1 & \text{if} \quad -1 < x < 0 \\ 1 & \text{if} \quad 0 < x < 1. \end{cases}$$

22. Prove Theorem 17.4. *Hint:* Let $F(x) = \int_{-L}^{x} f(t)\, dt - a_0 x$. Show that the Fourier series of f on $[-L, L]$ converges to F on the entire interval. Let the Fourier series of F on $[-L, L]$ be

$$A_0 + \sum_{n=1}^{\infty} \left[A_n \cos\left(\frac{n\pi x}{L}\right) + B_n \sin\left(\frac{n\pi x}{L}\right) \right].$$

Integrate the formulas for the Fourier coefficients by parts to show that $A_n = -Lb_n/n\pi$ and $B_n = La_n/n\pi$.

23. Let

$$f(x) = \begin{cases} 0, & -\pi \leq x \leq 0 \\ x, & 0 < x \leq \pi. \end{cases}$$

(a) Write the Fourier series of f on $[-\pi, \pi]$.
(b) Show that the Fourier series in (1) converges to f on $(-\pi, \pi)$.
(c) Use Theorem 17.4 and the results of (a) and (b) to obtain a trigonometric series for $\int_{-\pi}^{x} f(t)\, dt$ for any x in $[-\pi, \pi]$.

24. Suppose that f satisfies the hypotheses of Theorem 17.2 on $[-L, L]$. Let

$$S_N(x) = \alpha_0 + \sum_{n=1}^{N} \left[\alpha_n \cos\left(\frac{n\pi x}{L}\right) + \beta_n \sin\left(\frac{n\pi x}{L}\right) \right].$$

How should we choose the α_n's and β_n's to minimize the integral

$$I_N = \int_{-L}^{L} [f(x) - S_N(x)]^2 \, dx?$$

(With this choice of the coefficients, $S_N(x)$ is called a best *mean square approximation* of $f(x)$ on $[-L, L]$.) *Hint:* First, perform the indicated multiplications to show that

$$[f(x) - S_N(x)]^2 = [f(x)]^2 + \left\{ \alpha_0 + \sum_{n=1}^{N} \left[\alpha_n \cos\left(\frac{n\pi x}{L}\right) + \beta_n \sin\left(\frac{n\pi x}{L}\right) \right] \right\}^2$$
$$- 2 \left\{ \alpha_0 f(x) + \sum_{n=1}^{N} \left[\alpha_n f(x) \cos\left(\frac{n\pi x}{L}\right) + \beta_n f(x) \sin\left(\frac{n\pi x}{L}\right) \right] \right\}.$$

Show from this that

$$I_N(x) = \int_{-L}^{L} [f(x)]^2 \, dx + L \left[2\alpha_0^2 + \sum_{n=1}^{N} (\alpha_n^2 + \beta_n^2) \right] - 2L \left[2\alpha_0 a_0 + \sum_{n=1}^{N} (\alpha_n a_n + \beta_n b_n) \right],$$

in which the a_n's and b_n's are the Fourier coefficients of f on $[-L, L]$. Think of the stated objective as one of choosing the α_n's and β_n's to minimize the right side. Let $\partial I_N / \partial \alpha_i = 0$ for $i = 0, 1, \ldots, N$, and let $\partial I_N / \partial \beta_i = 0$ for $i = 1, 2, \ldots, N$. Solve the resulting equations.

25. Prove *Bessel's inequality* for the Fourier coefficients of f on $[-L, L]$

$$a_0^2 + \sum_{n=1}^{\infty} (a_n^2 + b_n^2) \le \frac{1}{L} \int_{-L}^{L} [f(x)]^2 \, dx$$

if f is piecewise continuous on $[-L, L]$. *Hint:* Let $\alpha_i = a_i$ and $\beta_i = b_i$ in the last line of the hint for Problem **24**, then let $N \to \infty$.

26. Suppose that f satisfies the hypotheses of Theorem 17.2 on $[-L, L]$. Show that

$$\sum_{n=1}^{\infty} a_n^2 \quad \text{and} \quad \sum_{n=1}^{\infty} b_n^2$$

both converge if the a_n's and b_n's are the Fourier coefficients of f on $[-L, L]$.

27. Use the result of Problem 26 to show that

$$\lim_{n \to \infty} \int_{-L}^{L} f(x) \cos\left(\frac{n\pi x}{L}\right) dx = \lim_{n \to \infty} \int_{-L}^{L} f(x) \sin\left(\frac{n\pi x}{L}\right) dx = 0,$$

assuming that f satisfies the hypotheses of Theorem 17.2 on $[-L, L]$. This result is known as *Riemann's lemma.*

28. Suppose that f satisfies the hypotheses of Theorem 17.3 on $[-L, L]$.
  (a) Prove that

$$\sum_{n=1}^{\infty} \sqrt{a_n^2 + b_n^2}$$

converges. *Hint:* Integrate by parts to relate the Fourier coefficients of f' to those of f on $[-L, L]$. Use may also be made of Cauchy's inequality, which states that

$$\left(\sum_{n=1}^{N} A_n B_n \right)^2 \le \left(\sum_{n=1}^{N} A_n^2 \right) \left(\sum_{n=1}^{N} B_n^2 \right)$$

for any real numbers $A_1, \ldots, A_N, B_1, \ldots, B_N$.

(b) Use the result of (a) to show that, under the given conditions, the Fourier series of f on $[-L, L]$ converges uniformly on $[-L, L]$.

(c) Prove that $\lim_{n \to \infty} n a_n = \lim_{n \to \infty} n b_n = 0$.

(d) Give an example to show that the conclusion of (c) need not hold for the Fourier coefficients of a function which does not satisfy the hypotheses of this problem.

(e) Show that, under the conditions of this problem,

$$\frac{1}{L} \int_{-L}^{L} [f(x)]^2 \, dx = \frac{1}{2} a_0^2 + \sum_{n=1}^{\infty} (a_n^2 + b_n^2).$$

This formula is known as *Parseval's identity*.

29. *Gibbs phenomenon*   This problem deals with the Gibbs phenomenon, which states that, for large N, the graph of the Nth partial sum of the Fourier series of f on $[-L, L]$ overshoots the graph of the function at a jump discontinuity by approximately 9% of the magnitude of the jump.

(a) Show that the Nth partial sum of the Fourier series of f on $[-L, L]$ can be written

$$S_N(x) = \frac{1}{L} \int_{-L}^{L} f(t) \left\{ \frac{1}{2} + \sum_{n=1}^{N} \cos\left[\frac{n\pi(t - x)}{L} \right] \right\} dt.$$

Hint: Begin with

$$S_N(x) = a_0 + \sum_{n=1}^{N} \left[a_n \cos\left(\frac{n\pi x}{L} \right) + b_n \sin\left(\frac{n\pi x}{L} \right) \right].$$

Substitute into the right side the integral formulas for the Fourier coefficients, and use a trigonometric identity to combine terms.

(b) Derive the trigonometric identity

$$\frac{1}{2} + \sum_{n=1}^{N} \cos(n\xi) = \frac{\sin[(N + \frac{1}{2})\xi]}{2 \sin\left(\frac{\xi}{2} \right)},$$

provided that $\xi/2$ is not an integer multiple of π. *Hint:* Use the fact (from Euler's equation) that $\cos(A) = \frac{1}{2}(e^{iA} + e^{-iA})$ and $\sin(A) = (1/2i)(e^{iA} - e^{-iA})$. Replace each trigonometric function in the proposed identity with its complex exponential form, and use the geometric series relation

$$\sum_{n=1}^{N} r^n = \frac{r}{1 - r}(1 - r^N).$$

(c) Use the fact that, for small ξ, $\sin(\xi) \approx \xi$; together with the results of (a) and (b), to conclude that, if t is approximately equal to x,

$$S_N(x) \approx \frac{1}{2L} \int_{-L}^{L} f(t) \frac{\sin\left[\frac{\pi}{L}(N + \frac{1}{2})(t - x) \right]}{\frac{\pi}{2L}(t - x)} \, dt.$$

(d) Consider the function f defined by

$$f(x) = \begin{cases} 0 & \text{if} \quad -L < x \le x_0 \\ 1 & \text{if} \quad x_0 < x < L. \end{cases}$$

Note that f has a jump discontinuity at x_0. Put this function into the conclusion of (c) and make the change of variables $s = (\pi/L)(N + \frac{1}{2})(t - x)$ to show that an approximate value for the Nth partial sum of the Fourier series of f on $[-L, L]$ for x near x_0 is

$$S_N(x) \approx \frac{1}{\pi} \int_{\pi(N + 1/2)(x_0 - x)/L}^{\pi(N + 1/2)(L - x)/L} \frac{\sin(s)}{s} \, ds.$$

(e) Notice that

$$\lim_{N \to \infty} \frac{\pi}{L}(N + \tfrac{1}{2})(L - x) = \infty$$

and

$$\lim_{N \to \infty} \frac{\pi}{L}(N + \tfrac{1}{2})(x_0 - x) = \begin{cases} \infty & \text{if} \quad x < x_0 \\ 0 & \text{if} \quad x = x_0 \\ -\infty & \text{if} \quad x > x_0. \end{cases}$$

Use these observations, together with the conclusion of part (d) and Theorem 17.2, to show that

$$\lim_{N \to \infty} S_N(x) = \begin{cases} \dfrac{1}{\pi} \displaystyle\int_{\infty}^{\infty} \dfrac{\sin(s)}{s} \, ds = 0 & \text{if} \quad x < x_0 \\[2ex] \dfrac{1}{\pi} \displaystyle\int_{0}^{\infty} \dfrac{\sin(s)}{s} \, ds = \dfrac{1}{2} & \text{if} \quad x = x_0 \\[2ex] \dfrac{1}{\pi} \displaystyle\int_{-\infty}^{\infty} \dfrac{\sin(s)}{s} \, ds = 1 & \text{if} \quad x > x_0. \end{cases}$$

(f) If N is very large and x is near x_0, show that

$$S_N(x) \approx \frac{1}{\pi} \int_{\pi(N + 1/2)(x_0 - x)/L}^{\infty} \frac{\sin(s)}{s} \, ds$$

$$= \frac{1}{\pi} \int_{0}^{\infty} \frac{\sin(s)}{s} \, ds - \frac{1}{\pi} \int_{0}^{\pi(N + 1/2)(x_0 - x)/L} \frac{\sin(s)}{s} \, ds$$

$$= \frac{1}{2} - \frac{1}{\pi} \int_{0}^{\pi(N + 1/2)(x_0 - x)/L} \frac{\sin(s)}{s} \, ds.$$

Now consider the integral

$$\mathrm{Si}(x) = \int_{0}^{x} \frac{\sin(s)}{s} \, ds.$$

This function is called a *sine integral*, and its values have been tabulated for various values of x. The graphs of $[\sin(x)]/x$ and $\mathrm{Si}(x)$ are shown on page 1040. Use these to determine the value of x which maximizes $S_N(x)$. Find a table of values for $\mathrm{Si}(x)$ (for example, from a *CRC Handbook*) and use it to show that, for large N, the maximum value of $S_N(x)$ is very near 1.089 times the magnitude of the jump, or approximately 9% more than the jump magnitude.

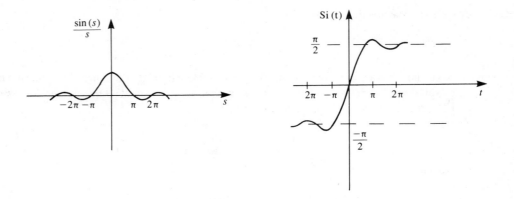

17.3 *Fourier Sine and Cosine Series*

We will now see how to write a Fourier series of f on $[0, L]$ containing either just sine terms or just cosine terms, whichever we choose. Such series are called *half-range expansions*. The key lies in Lemma 17.3.

Suppose that f is integrable on $[0, L]$ and we want to expand f in a Fourier series on $[0, L]$ containing just cosine terms. The idea is to extend f to a new function g defined on $[-L, L]$ in such a way that g is an even function. The Fourier series of g on $[-L, L]$ contains only cosine terms. Since g and f agree on $[0, L]$, this gives a Fourier cosine series of f on $[0, L]$. To do this, define

$$g(x) = \begin{cases} f(x) & \text{for} \quad 0 \le x \le L \\ f(-x) & \text{for} \quad -L \le x \le 0. \end{cases}$$

Figure 17.24 shows a typical graph of g obtained by folding the graph of f over the y-axis. Since g is an even function, its Fourier series on $[-L, L]$ is

$$a_0 + \sum_{n=1}^{\infty} a_n \cos\left(\frac{n\pi x}{L}\right), \tag{17.2}$$

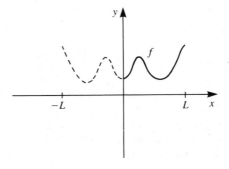

Figure 17.24. Even extension of f to $[-L, L]$.

in which

$$a_0 = \frac{1}{L} \int_0^L g(x)\, dx \quad \text{and} \quad a_n = \frac{2}{L} \int_0^L g(x)\cos\left(\frac{n\pi x}{L}\right) dx. \tag{17.3}$$

Since $g(x) = f(x)$ for $0 \le x \le L$, we may think of the series (17.2) as a Fourier cosine series of $f(x)$ on $[0, L]$. Further, the coefficients (17.3) can be written in terms of f. Because $g(x) = f(x)$ for $0 \le x \le L$, these coefficients are

$$a_0 = \frac{1}{L} \int_0^L f(x)\, dx \quad \text{and} \quad a_n = \frac{2}{L} \int_0^L f(x)\cos\left(\frac{n\pi x}{L}\right) dx.$$

This informal reasoning suggests the following definition.

DEFINITION   Fourier Cosine Series on [0, L]

If f is integrable on $[0, L]$, the *Fourier cosine series* of f on $[0, L]$ is

$$a_0 + \sum_{n=1}^{\infty} a_n \cos\left(\frac{n\pi x}{L}\right),$$

in which

$$a_0 = \frac{1}{L} \int_0^L f(x)\, dx \quad \text{and} \quad a_n = \frac{2}{L} \int_0^L f(x)\cos\left(\frac{n\pi x}{L}\right) dx. \quad \blacksquare$$

The function g we introduced had the sole purpose of suggesting how to define a pure cosine series on $[0, L]$. In computing a Fourier cosine series, we never mention g but simply calculate the coefficients directly from the definition.

EXAMPLE 17.8

Let $f(x) = e^{2x}$. Write the Fourier cosine series of f on $[0, 1]$.
   The coefficients are

$$a_0 = \frac{1}{1} \int_0^1 e^{2x}\, dx = \frac{1}{2}(e^2 - 1)$$

and

$$a_n = \frac{2}{1} \int_0^1 e^{2x}\cos(n\pi x)\, dx = \frac{4}{4 + n^2\pi^2} \left[e^2\cos(n\pi) - 1 \right].$$

The Fourier cosine series of e^{2x} on $[0, 1]$ is

$$\frac{1}{2}(e^2 - 1) + \sum_{n=1}^{\infty} \frac{4}{4 + n^2\pi^2} \left[e^2\cos(n\pi) - 1 \right]\cos(n\pi x),$$

or, because $\cos(n\pi) = (-1)^n$,

$$\frac{1}{2}(e^2 - 1) + \sum_{n=1}^{\infty} \frac{4}{4 + n^2\pi^2} \left[e^2(-1)^n - 1 \right]\cos(n\pi x). \quad \blacksquare$$

Convergence of a Fourier cosine series on $[0, L]$ can be determined from the convergence theorem of Section 17.2 if we remember that we obtained the cosine series by extending f to an even function g defined on $[-L, L]$. The convergence theorem applies to g, and, of course, $f(x) = g(x)$ for $0 \leq x \leq L$. This gives us the following theorem.

THEOREM 17.5   Convergence of a Fourier Cosine Series

Let f be piecewise continuous on $[0, L]$.

1. If $0 < x < L$, and f has both right and left derivatives at x, then at x the Fourier cosine series of f on $[0, L]$ converges to

$$\tfrac{1}{2}[f(x+) + f(x-)],$$

the average of the left and right limits of f at x. In particular, if f is also continuous at x, the series converges to $f(x)$.

2. If $f'_{\mathscr{R}}(0)$ exists, the cosine series converges at zero to $f(0+)$.

3. If $f'_{\mathscr{L}}(L)$ exists, the cosine series converges at L to $f(L-)$.   ∎

Conclusion (1) should not be surprising. To understand conclusion (2), consider convergence at zero of the Fourier series of g on $[-L, L]$. At zero, this series converges to $\tfrac{1}{2}[g(0+) + g(0-)]$. But

$$g(0+) = \lim_{h \to 0+} g(0 + h) = \lim_{h \to 0+} g(h) = \lim_{h \to 0+} f(h) = f(0+)$$

and

$$g(0-) = \lim_{h \to 0+} g(0 - h) = \lim_{h \to 0+} g(-h) = \lim_{h \to 0+} f(h) = f(0+).$$

Therefore,

$$\tfrac{1}{2}[g(0+) + g(0-)] = \tfrac{1}{2}[f(0+) + f(0+)] = f(0+),$$

and the cosine series of f converges to $f(0+)$ at zero. A similar argument establishes convergence of the series to $f(L-)$ at L.

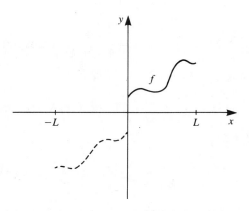

Figure 17.25. Odd extension of f to $[-L, L]$.

We will now consider the problem of writing a Fourier sine series of f on $[0, L]$. The strategy is similar to that just used. This time, however, extend f to an odd function w defined on $[-L, L]$ by letting

$$w(x) = \begin{cases} f(x) & \text{if} & 0 \le x \le L \\ -f(-x) & \text{if} & -L \le x < 0. \end{cases}$$

Figure 17.25 shows a typical graph of w. Since w is an odd function, its Fourier series on $[-L, L]$ contains only sine terms and is

$$\sum_{n=1}^{\infty} b_n \sin\left(\frac{n\pi x}{L}\right),$$

where

$$b_n = \frac{2}{L} \int_0^L w(x)\sin\left(\frac{n\pi x}{L}\right) dx.$$

Since $w(x) = f(x)$ if $0 \le x \le L$,

$$b_n = \frac{2}{L} \int_0^L f(x)\sin\left(\frac{n\pi x}{L}\right) dx.$$

Further, on the interval $[0, L]$, we may take the Fourier series of w to be a Fourier sine series of f. This suggests the following definition.

DEFINITION   Fourier Sine Series on [0, L]

If f is integrable on $[0, L]$, the *Fourier sine series* of f on $[0, L]$ is

$$\sum_{n=1}^{\infty} b_n \sin\left(\frac{n\pi x}{L}\right),$$

in which

$$b_n = \frac{2}{L} \int_0^L f(x)\sin\left(\frac{n\pi x}{L}\right) dx. \quad \blacksquare$$

EXAMPLE 17.9

Expand e^{2x} in a Fourier sine series on $[0, 1]$.
     The coefficients are

$$b_n = \frac{2}{1} \int_0^1 e^{2x}\sin(n\pi x)\, dx = \frac{2n\pi[1 - e^2\cos(n\pi)]}{4 + n^2\pi^2} = \frac{2n\pi[1 - e^2(-1)^n]}{4 + n^2\pi^2}.$$

The sine series of e^{2x} on $[0, 1]$ is

$$\sum_{n=1}^{\infty} \frac{2n\pi[1 - e^2(-1)^n]}{4 + n^2\pi^2} \sin(n\pi x). \quad \blacksquare$$

As with cosine series, we obtain a convergence theorem by recalling that the sine series of f on $[0, L]$ is derived from the Fourier series of w on $[-L, L]$ and then applying the Fourier convergence theorem to w.

THEOREM 17.6 Convergence of Fourier Sine Series

Let f be piecewise continuous on $[0, L]$.

1. At any point x in $(0, L)$ at which f has left and right derivatives, the Fourier sine series on $[0, L]$ converges to

$$\tfrac{1}{2}[f(x+) + f(x-)],$$

the average of the left and right limits at x. In particular, if f is also continuous at x, the sine series converges to $f(x)$.

2. At both zero and L, the sine series of f converges to zero. ∎

Conclusion (2) is immediate upon letting $x = 0$ or $x = L$ in the sine series; all of the terms vanish because $\sin(0) = 0$ and $\sin(n\pi) = 0$ for any integer n.

EXAMPLE 17.10

Let

$$f(x) = \begin{cases} 1 & \text{if} \quad 0 \le x \le \pi/2 \\ 2 & \text{if} \quad \pi/2 < x \le \pi. \end{cases}$$

We will write both the Fourier cosine and Fourier sine series of f on $[0, \pi]$. For the cosine series, compute

$$a_0 = \frac{1}{\pi} \int_0^\pi f(x)\, dx = \frac{1}{\pi} \int_0^{\pi/2} dx + \frac{1}{\pi} \int_{\pi/2}^\pi 2\, dx = \frac{3}{2}$$

and, for $n = 1, 2, 3, \ldots,$

$$a_n = \frac{2}{\pi} \int_0^\pi f(x)\cos(nx)\, dx$$

$$= \frac{2}{\pi} \int_0^{\pi/2} \cos(nx)\, dx + \frac{2}{\pi} \int_{\pi/2}^\pi 2\cos(nx)\, dx = \frac{-2}{n\pi} \sin\left(\frac{n\pi}{2}\right).$$

The cosine series of f on $[0, \pi]$ is

$$\frac{3}{2} + \sum_{n=1}^\infty \frac{-2}{n\pi} \sin\left(\frac{n\pi}{2}\right)\cos(nx).$$

Since $\sin(n\pi/2) = 0$ if n is an even positive integer, we need retain only the terms involving odd n. This series may therefore be written

$$\frac{3}{2} - \sum_{n=1}^\infty \frac{2}{(2n-1)\pi} \sin\left[(2n-1)\frac{\pi}{2}\right]\cos[(2n-1)x].$$

Further, $\sin[(2n-1)\pi/2] = (-1)^{n+1}$, and $-(-1)^{n+1} = (-1)^n$, so we can further simplify this cosine series and write it as

$$\frac{3}{2} + \frac{2}{\pi} \sum_{n=1}^\infty \frac{(-1)^n}{2n-1} \cos[(2n-1)x].$$

By Theorem 17.5, this series converges to

$$1 \quad \text{if} \quad 0 < x < \frac{\pi}{2};$$

$$2 \quad \text{if} \quad \frac{\pi}{2} < x < \pi;$$

$$\frac{3}{2} \quad \text{at} \quad \frac{\pi}{2} \quad \text{(the jump discontinuity)};$$

$$1 \quad \text{at} \quad 0;$$

$$2 \quad \text{at} \quad \pi.$$

For the sine series on $[0, \pi]$, compute

$$b_n = \frac{2}{\pi} \int_0^\pi f(x)\sin(nx)\, dx = \frac{2}{\pi} \int_0^{\pi/2} \sin(nx)\, dx + \frac{2}{\pi} \int_{\pi/2}^\pi 2 \sin(nx)\, dx$$

$$= \frac{2}{\pi n}\left[\cos\left(\frac{n\pi}{2}\right) + 1\right] - \frac{4}{\pi n}\cos(n\pi).$$

Since $\cos(n\pi) = (-1)^n$, the sine series of f on $[0, \pi]$ is

$$\frac{2}{\pi} \sum_{n=1}^\infty \frac{1}{n}\left[\cos\left(\frac{n\pi}{2}\right) + 1 - 2(-1)^n\right] \sin(nx).$$

By Theorem 17.6, this series converges to

$$1 \quad \text{if} \quad 0 < x < \frac{\pi}{2}$$

$$2 \quad \text{if} \quad \frac{\pi}{2} < x < \pi$$

$$\frac{3}{2} \quad \text{at} \quad \frac{\pi}{2}$$

$$0 \quad \text{at} \quad 0 \quad \text{and at} \quad \pi. \quad \blacksquare$$

In the next section, we will consider multiple Fourier series.

PROBLEMS FOR SECTION 17.3

In each of Problems 1 through 15, write the Fourier sine and cosine series of f on $[0, L]$ and determine what each series converges to.

1. $f(x) = 4, \quad 0 \le x \le 3$

2. $f(x) = \begin{cases} 1, & 0 \le x \le 1 \\ -1, & 1 < x \le 2 \end{cases}$

3. $f(x) = \begin{cases} 0, & 0 \le x < \pi \\ \cos(x), & \pi \le x \le 2\pi \end{cases}$

4. $f(x) = 2x, \quad 0 \le x \le 1$

5. $f(x) = 1 - x, \quad 0 \le x \le 5$

6. $f(x) = \begin{cases} 0, & 0 \le x < 1 \\ x, & 1 \le x \le 4 \end{cases}$

7. $f(x) = x^2, \quad 0 \le x \le 2$

8. $f(x) = e^{-x}, \quad 0 \le x \le 1$

9. $f(x) = \begin{cases} x, & 0 \le x \le 2 \\ 2 - x, & 2 \le x \le 3 \end{cases}$

10. $f(x) = \begin{cases} 1, & 0 \le x < 1 \\ 0, & 1 \le x \le 3 \\ -1, & 3 < x \le 5 \end{cases}$

11. $f(x) = 1 - \sin(\pi x), \qquad 0 \le x \le 2$

12. $f(x) = 1 - x^3, \quad 0 \le x \le 2$

13. $f(x) = \begin{cases} x^2, & 0 \le x < 1 \\ 1, & 1 \le x \le 4 \end{cases}$

14. $f(x) = \begin{cases} 4x, & 0 \le x \le 2 \\ -3, & 2 < x < 4 \\ 1, & 4 \le x \le 7 \end{cases}$

15. $f(x) = \sin(3x), \quad 0 \le x \le \pi$

16. Let f be defined on $[-L, L]$. Prove that f can be written as a sum of an even and an odd function on $[-L, L]$.

17. Find all functions defined on $[-L, L]$ which are both even and odd.

18. Find the sum of the series

$$\sum_{n=1}^{\infty} \frac{(-1)^n}{4n^2 - 1}.$$

Hint: Expand $\sin(x)$ in a Fourier cosine series on $[0, \pi]$ and let $x = \pi/2$.

17.4 Multiple Fourier Series

Problems involving several independent variables sometimes require that we expand a function of two or more variables in a series of sines and/or cosines. In this section, we will briefly discuss the idea behind such an expansion.

Suppose that f is defined on a rectangle $0 \le x \le L, 0 \le y \le K$. We will describe how we might expand f in a series of sines, analogous to a half-range sine series of a function defined on $[0, L]$.

First, think of y as fixed, and let $g(x) = f(x, y)$ for $0 \le x \le L$. Expand g in a Fourier sine series on $[0, L]$

$$g(x) = \sum_{n=1}^{\infty} b_n(y)\sin\left(\frac{n\pi x}{L}\right), \qquad (17.4)$$

where

$$b_n(y) = \frac{2}{L} \int_0^L g(x)\sin\left(\frac{n\pi x}{L}\right) dx = \frac{2}{L} \int_0^L f(x, y)\sin\left(\frac{n\pi x}{L}\right) dx. \qquad (17.5)$$

We have written $g(x)$ equal to this Fourier series as a notational convenience in carrying out the discussion. In general, the series on the right need not converge to $g(x)$ for all x in $[0, L]$. The Fourier coefficient is written $b_n(y)$ because y is fixed inside $f(x, y)$ and we may obtain a different Fourier coefficient if y is changed.

Now assume that we can expand each $b_n(y)$ in a Fourier sine series on $[0, K]$

$$b_n(y) = \sum_{m=1}^{\infty} b_{nm}\sin\left(\frac{m\pi y}{K}\right), \qquad (17.6)$$

in which

$$b_{nm} = \frac{2}{K} \int_0^K b_n(y)\sin\left(\frac{m\pi y}{K}\right) dy. \qquad (17.7)$$

In (17.7), b_{nm} is the coefficient of $\sin(m\pi y/K)$ in the Fourier expansion of $b_n(y)$ on $[0, K]$.

We obtain the double Fourier sine series of f on $0 \le x \le L$, $0 \le y \le K$ by substituting the series (17.7) into the Fourier sine series (17.6) to get

$$g(x) = f(x, y) = \sum_{n=1}^{\infty} \left[\sum_{m=1}^{\infty} b_{nm} \sin\left(\frac{m\pi y}{K}\right) \right] \sin\left(\frac{n\pi x}{L}\right)$$

$$= \sum_{n=1}^{\infty} \sum_{m=1}^{\infty} b_{nm} \sin\left(\frac{n\pi x}{L}\right) \sin\left(\frac{m\pi y}{K}\right).$$

We get a formula for the coefficients by substituting the integral (17.5) for $b_n(y)$ into the expression (17.7) for b_{nm}:

$$b_{nm} = \frac{2}{K} \int_0^K \left[\frac{2}{L} \int_0^L f(x, y) \sin\left(\frac{n\pi x}{L}\right) dx \right] \sin\left(\frac{m\pi y}{K}\right) dy$$

$$= \frac{4}{LK} \int_0^K \int_0^L f(x, y) \sin\left(\frac{n\pi x}{L}\right) \sin\left(\frac{m\pi y}{K}\right) dx \, dy.$$

This discussion suggests the following definition.

DEFINITION   Double Fourier Sine Series

Let f be integrable on a rectangle R defined by $0 \le x \le L$, $0 \le y \le K$. The Fourier sine series of f on R is

$$\sum_{n=1}^{\infty} \sum_{m=1}^{\infty} b_{nm} \sin\left(\frac{n\pi x}{L}\right) \sin\left(\frac{m\pi y}{K}\right),$$

in which

$$b_{nm} = \frac{4}{LK} \int_0^K \int_0^L f(x, y) \sin\left(\frac{n\pi x}{L}\right) \sin\left(\frac{m\pi y}{K}\right) dx \, dy$$

for $n = 1, 2, 3, \ldots$, $m = 1, 2, 3, \ldots$. ∎

In the discussion, we took some liberties and wrote $f(x, y)$ equal to this Fourier series. When dealing with a specific function, however, we must use convergence theorems to determine where the series and the function agree.

EXAMPLE 17.11

Let $f(x, y) = xy$ for $0 \le x \le 1$ and $0 \le y \le 2$. Compute the double Fourier sine series of f.

The coefficients are

$$b_{nm} = \frac{4}{2} \int_0^2 \int_0^1 xy \sin(n\pi x) \sin\left(\frac{m\pi y}{2}\right) dx \, dy$$

$$= 2 \int_0^2 y \sin\left(\frac{m\pi y}{2}\right) dy \int_0^1 x \sin(n\pi x) \, dx$$

$$= 2 \left[\frac{-4 \cos(m\pi)}{m\pi} \right] \left[\frac{-\cos(n\pi)}{n\pi} \right] = \frac{8}{nm\pi^2} (-1)^{n+m}.$$

The sine expansion of xy on $0 \leq x \leq 1, 0 \leq y \leq 2$ is

$$\frac{8}{\pi^2} \sum_{n=1}^{\infty} \sum_{m=1}^{\infty} \frac{(-1)^{n+m}}{nm} \sin(n\pi x)\sin\left(\frac{m\pi y}{2}\right). \quad \blacksquare$$

It is also possible to develop double Fourier cosine series on a rectangle $0 \leq x \leq L, 0 \leq y \leq K$ as well as a "full" double Fourier series on $-L \leq x \leq L$, $-K \leq y \leq K$. We will outline the cosine expansion in Problem 11.

PROBLEMS FOR SECTION 17.4

In each of Problems 1 through 10, write the double Fourier sine series of f on the given rectangle.

1. $xy; \quad 0 \leq x \leq \pi, \quad 0 \leq y \leq \pi$

2. $x - y; \quad 0 \leq x \leq 2, \quad 0 \leq y \leq 2$

3. $3x + y^2; \quad 0 \leq x \leq 1, \quad 0 \leq y \leq 2$

4. $\sin(2x - y); \quad 0 \leq x \leq 2, \quad 0 \leq y \leq 2$

5. $e^{x+y}; \quad 0 \leq x \leq 4, \quad 0 \leq y \leq 2$

6. $xy^2; \quad 0 \leq x \leq 2, \quad 0 \leq y \leq 2$

7. $x \sinh(y); \quad 0 \leq x \leq 4, \quad 0 \leq y \leq 2$

8. $x^2 + y^2; \quad 0 \leq x \leq 2, \quad 0 \leq y \leq 1$

9. $4y \sin(2x); \quad 0 \leq x \leq \pi, \quad 0 \leq y \leq 4$

10. $x - y^2; \quad 0 \leq x \leq 1, \quad 0 \leq y \leq 2$

11. The Fourier cosine expansion of f on $0 \leq x \leq L, 0 \leq y \leq K$ is

$$\frac{1}{4} a_{00} + \sum_{m=1}^{\infty} \frac{1}{2} a_{0m}\cos\left(\frac{m\pi y}{K}\right) + \sum_{n=1}^{\infty} \frac{1}{2} a_{n0}\cos\left(\frac{n\pi x}{L}\right) + \sum_{n=1}^{\infty} \sum_{m=1}^{\infty} a_{nm}\cos\left(\frac{n\pi x}{L}\right)\cos\left(\frac{m\pi y}{K}\right).$$

Give an informal argument to support the choice of the coefficients as

$$a_{nm} = \frac{4}{LK} \int_0^K \int_0^L f(x, y)\cos\left(\frac{n\pi x}{L}\right)\cos\left(\frac{m\pi y}{K}\right) dx\, dy.$$

In each of Problems 12 through 15, use the result of Problem 11 to expand the function in a Fourier cosine series on the rectangle.

12. $xy; \quad 0 \leq x \leq 1, \quad 0 \leq y \leq 3$

13. $x^2 y; \quad 0 \leq x \leq 1, \quad 0 \leq y \leq 1$

14. $xe^y; \quad 0 \leq x \leq 2, \quad 0 \leq y \leq 2$

15. $x - y^2; \quad 0 \leq x \leq 3, \quad 0 \leq y \leq 1$

17.5 *The Finite Fourier Sine and Cosine Transforms*

We have been exposed to the idea of a transform with the Laplace transform. Usually, a transform is developed in response to a class of problems, and there are many different transforms which are defined in terms of integrals or series. Throughout the remainder of this chapter, we will develop several transforms which are related to Fourier series and to the Fourier integral we have yet to discuss. In this section, our objective is a type of transform whose definition is based on Fourier sine and cosine series.

THE FINITE FOURIER SINE TRANSFORM

Suppose that f is piecewise continuous on $[0, \pi]$. The *finite Fourier sine transform* F_S of f is defined by

$$F_S(n) = \int_0^\pi f(x)\sin(nx)\, dx$$

for $n = 1, 2, 3, \ldots$. F_S is defined over the positive integers rather than over an interval, as are most functions we are accustomed to dealing with. Thus, $F_S(1)$, $F_S(2)$, $\ldots$ are defined, but $F_S(\frac{1}{2})$ and $F_S(\pi)$ are not.

The motivation for the definition comes from the Fourier sine series of f on $[0, \pi]$. The sine series of f on $[0, \pi]$ is

$$\frac{2}{\pi} \sum_{n=1}^{\infty} \left[\int_0^{\pi} f(\xi)\sin(n\xi)\, d\xi \right] \sin(nx).$$

In terms of the finite Fourier sine transform, this series is

$$\frac{2}{\pi} \sum_{n=1}^{\infty} F_S(n)\sin(nx).$$

We may think of this expression as an inversion formula for the finite Fourier sine transform. If we know F_S, we can recover a function f whose finite Fourier transform is F_S by letting

$$f(x) = \frac{2}{\pi} \sum_{n=1}^{\infty} F_S(n)\sin(nx)$$

for $0 \le x \le \pi$. This formula will not generally yield f in closed form, but it gives us the Fourier sine expansion of f on $[0, \pi]$.

This inversion formula will be important later when we solve problems by applying the finite Fourier sine transform, yielding the transform of the solution. We must then invert this transform to produce the solution itself, just as we did in solving initial value problems using the Laplace transform.

It is sometimes convenient to denote $F_S(n)$ by the alternate notation $S_n\{f(x)\}$. This enables us to reference the function in the symbol.

EXAMPLE 17.12

Let $f(x) = x^2$ for $0 \le x \le \pi$. Then

$$F_S(n) = \int_0^{\pi} x^2 \sin(nx)\, dx$$

$$= \frac{1}{n^3}[(n^2\pi^2 - 2)(-1)^{n+1} - 2]$$

for $n = 1, 2, 3, \ldots$. We could also denote this expression as $S_n\{x^2\}$. ∎

It is obvious from the definition that the finite Fourier sine transform is a linear operator. This means that

$$S_n\{\alpha f(x) + \beta g(x)\} = \alpha S_n\{f(x)\} + \beta S_n\{g(x)\}$$

if f and g are piecewise continuous on $[0, \pi]$.

In applying the finite Fourier transform to boundary value problems, the following operational formula is important. It is analogous to the formula expressing the Laplace transform of a derivative in terms of the Laplace transform of the function and its initial values (Theorems 4.6 and 4.7).

THEOREM 17.7

Let f and f' be continuous on $[0, \pi]$, and suppose that f'' is piecewise continuous on $[0, \pi]$. Then

$$S_n\{f''(x)\} = -n^2 F_S(n) + nf(0) - n(-1)^n f(\pi)$$

for $n = 1, 2, 3, \ldots$.

Proof   Integrate by parts twice as follows:

$$S_n\{f''(x)\} = \int_0^\pi f''(x)\sin(nx)\,dx \qquad [u = \sin(nx), \, dv = f''(x)\,dx]$$

$$= \left[\sin(nx)f'(x)\right]_0^\pi - \int_0^\pi f'(x)n\cos(nx)\,dx$$

$$= -n \int_0^\pi f'(x)\cos(nx)\,dx \qquad [u = \cos(nx), \, dv = f'(x)\,dx]$$

$$= -n\left\{\left[\cos(nx)f(x)\right]_0^\pi - \int_0^\pi f(x)[-n\sin(nx)]\,dx\right\}$$

$$= -n[\cos(n\pi)f(\pi) - f(0) + nF_S(n)]$$

$$= -n^2 F_S(n) + nf(0) - n(-1)^n f(\pi).$$

In the last line, we used the fact that $\cos(n\pi) = (-1)^n$. ∎

With corresponding assumptions about $f^{(3)}$ and $f^{(4)}$ on $[0, \pi]$, we can apply this result to the fourth derivative $f^{(4)}$ to get

$$S_n\{f^{(4)}(x)\} = -n^2 S_n\{f''(x)\} + nf''(0) - n(-1)^n f''(\pi)$$

$$= -n^2[-n^2 F_S(n) + nf(0) - n(-1)^n f(\pi)] + nf''(0) - n(-1)^n f''(\pi)$$

$$= n^4 F_S(n) - n^3[f(0) - (-1)^n f(\pi)] + n[f''(0) - (-1)^n f''(\pi)].$$

THE FINITE FOURIER COSINE TRANSFORM

The definition of the finite Fourier cosine transform is analogous to that of the finite Fourier sine transform. If f is piecewise continuous on $[0, \pi]$, the *finite Fourier cosine transform* F_C of f is defined by

$$F_C(n) = \int_0^\pi f(x)\cos(nx)\,dx$$

for $n = 0, 1, 2, \ldots$. This transform is often denoted $C_n\{f(x)\}$.

The motivation for this transform comes from the Fourier cosine expansion of f on $[0, \pi]$. With the integral formulas for the coefficients inserted, this series is

$$\frac{1}{\pi}\int_0^\pi f(\xi)\,d\xi + \sum_{n=1}^\infty \left[\frac{2}{\pi}\int_0^\pi f(\xi)\cos(n\xi)\,d\xi\right]\cos(nx).$$

In terms of the finite Fourier cosine transform, this series is

$$\frac{2}{\pi}\left[\frac{1}{2}F_C(0) + \sum_{n=1}^\infty F_C(n)\cos(nx)\right].$$

This series is an inversion formula for the finite Fourier cosine transform. Given the finite Fourier cosine transform F_C, the inversion formula yields a Fourier cosine expansion on $[0, \pi]$ of a function f whose finite Fourier cosine transform is F_C.

The following theorem provides the fundamental operational formula for the finite Fourier cosine transform.

THEOREM 17.8

Let f and f' be continuous on $[0, \pi]$, and suppose that f'' is piecewise continuous on $[0, \pi]$. Then

$$C_n\{f''(x)\} = -n^2 F_C(n) - f'(0) + (-1)^n f'(\pi)$$

for $n = 1, 2, \ldots$. ∎

The proof is by integration by parts and is similar to that of Theorem 17.7.

A table at the end of Chapter 18 gives the finite Fourier cosine transform of some familiar functions. In Chapter 18, we will use the finite Fourier cosine transform to solve boundary value problems involving wave transmission and heat conduction.

PROBLEMS FOR SECTION 17.5

In each of Problems 1 through 7, find the finite Fourier sine transform.

1. K (any constant)   **2.** x **3.** x^3 **4.** x^5

5. $\sin(ax)$ **6.** $\cos(ax)$ **7.** e^{-x}

In each of Problems 8 through 14, find the finite Fourier cosine transform.

8. $f(x) = \begin{cases} 1, & 0 \le x \le \frac{1}{2} \\ -1, & \frac{1}{2} < x \le \pi \end{cases}$ **9.** x **10.** x^2

11. e^x **12.** $\sin(ax)$ **13.** x^3 **14.** $\cosh(ax)$

15. Suppose that f is continuous and f' is piecewise continuous on $[0, \pi]$. Prove that

$$S_n\{f'(x)\} = -nC_n\{f(x)\}$$

for $n = 1, 2, 3, \ldots$.

16. Suppose that f is continuous and f' is piecewise continuous on $[0, \pi]$. Prove that

$$C_n\{f'(x)\} = nS_n\{f(x)\} - f(0) + (-1)^n f(\pi)$$

for $n = 0, 1, 2, \ldots$.

In Problems 17 through 20, n and m are positive integers and f is piecewise continuous on $[0, \pi]$.

17. Prove that $F_S(n + m) = S_n\{f(x)\cos(mx)\} + C_n\{f(x)\sin(mx)\}$.

18. Prove that $F_C(n + m) = C_n\{f(x)\cos(mx)\} - S_n\{f(x)\sin(mx)\}$.

19. Suppose that $m < n$. Prove that

$$C_n\{f(x)\cos(mx)\} = \tfrac{1}{2}[F_C(n - m) + F_C(n + m)].$$

20. Suppose that $m < n$. Prove that

$$C_n\{f(x)\sin(mx)\} = \tfrac{1}{2}[F_S(n + m) - F_S(n - m)].$$

17.6 *Periodic Functions and the Amplitude Spectrum*

A function f is said to be *periodic* if it is defined for all real t and for some positive number T, $f(t + T) = f(t)$ for all t. The number T is called a *period* of f. (We are using t as the variable here because many applications of periodic functions have t for time.)

The smallest period of f is called the *principal period*, or *fundamental period*, of f. For example, $\sin(t + 6\pi) = \sin(t)$ for all t. Thus, 6π is a period of the sine function. However, the principal period of the sine function is 2π because $\sin(t + 2\pi) = \sin(t)$ for all t and 2π is the smallest number with this property.

If f and g are periodic with period T, $\alpha f + \beta g$ is also periodic with period T for any numbers α and β.

We have seen that, under certain conditions, a function f defined on $[-L, L]$ can be expanded in a Fourier series:

$$f(t) = a_0 + \sum_{n=1}^{\infty} \left[a_n \cos\left(\frac{n\pi t}{L}\right) + b_n \sin\left(\frac{n\pi t}{L}\right) \right].$$

It is interesting to observe that, even though f may be defined on only $[-L, L]$, the Fourier series is defined for every real number t. Further, each function in the Fourier series has period $2L$; therefore, the Fourier series has period $2L$. This means that we can use the series on the right to define $f(t)$ for all real t. In effect, we take the graph of f over $[-L, L]$ and duplicate it over each interval $[L, 3L], [3L, 5L], \ldots, [-3L, -L]$, $[-5L, -3L], \ldots$. This process is illustrated in Figure 17.26. The resulting function is defined for all t, has period $2L$, and agrees with f on $[-L, L]$. This extended function is called a *periodic extension of f* (with period $2L$) to the entire real line.

If we begin with a function g which is periodic with period T, we can construct the entire graph of g by taking the graph over any interval of length T and duplicating it over subsequent intervals of length T. This is illustrated in Figure 17.27.

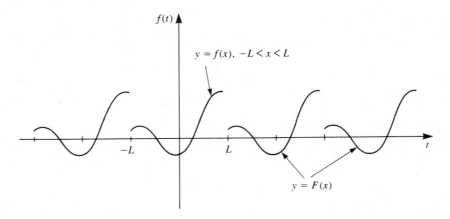

Figure 17.26. F = periodic extension of f.

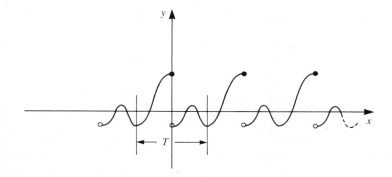

Figure 17.27

Now suppose that g is periodic of period T and we expand g in a Fourier series on $[-T/2, T/2]$. Assuming for convenience that g equals its Fourier series, we have

$$g(t) = a_0 + \sum_{n=1}^{\infty} \left[a_n \cos\left(\frac{2n\pi t}{T}\right) + b_n \sin\left(\frac{2n\pi t}{T}\right) \right],$$

obtained by putting $L = T/2$ in the usual Fourier series. The coefficients are

$$a_0 = \frac{1}{T} \int_{-T/2}^{T/2} g(t)\, dt, \qquad a_n = \frac{2}{T} \int_{-T/2}^{T/2} g(t) \cos\left(\frac{2n\pi t}{T}\right) dt$$

and

$$b_n = \frac{2}{T} \int_{-T/2}^{T/2} g(t) \sin\left(\frac{2n\pi t}{T}\right) dt.$$

Now let $\omega_0 = 2\pi/T$. The Fourier series of g becomes

$$g(t) = a_0 + \sum_{n=1}^{\infty} [a_n \cos(n\omega_0 t) + b_n \sin(n\omega_0 t)]. \qquad (17.8)$$

The coefficients can be written

$$a_0 = \frac{1}{T} \int_{-T/2}^{T/2} g(t)\, dt, \qquad a_n = \frac{2}{T} \int_{-T/2}^{T/2} g(t) \cos(n\omega_0 t)\, dt,$$

and

$$b_n = \frac{2}{T} \int_{-T/2}^{T/2} g(t) \sin(n\omega_0 t)\, dt.$$

In fact, because g is periodic of period T, these integrals can be taken over any interval of length T. We may therefore choose any real number a and write the coefficients

$$a_0 = \frac{1}{T} \int_{a}^{a+T} g(t)\, dt, \qquad a_n = \frac{2}{T} \int_{a}^{a+T} g(t) \cos(n\omega_0 t)\, dt,$$

and

$$b_n = \frac{2}{T} \int_{a}^{a+T} g(t) \sin(n\omega_0 t)\, dt.$$

The Fourier expansion (17.8) is often written in *phase angle*, or *harmonic, form*, as

$$g(t) = a_0 + \sum_{n=1}^{\infty} c_n \cos(n\omega_0 t + \delta_n),$$

in which

$$c_n = \sqrt{a_n^2 + b_n^2} \quad \text{and} \quad \delta_n = \tan^{-1}\left(-\frac{b_n}{a_n}\right).$$

This series for g emphasizes that any periodic function (with certain continuity properties) can be written as a sum of cosine waves (or, equivalently, sine waves). The sinusoidal component of frequency $n\omega_0/2\pi$ is called the *nth harmonic* of g, and the coefficients c_n and δ_n are called the *harmonic amplitudes* and *phase angles*, respectively.

The *amplitude spectrum* of g is the plot of c_0 at 0 (with $c_0 = |a_0|$) and, for $n \geq 1$, $c_n/2$ versus the angular frequency $n\omega_0$. (The factor $\frac{1}{2}$ with c_n is simply a convention; its purpose will be seen in the next section.) Since the summation index n assumes only positive integer values, the amplitude spectrum is not a continuous curve but appears as points at the discrete frequencies $n\omega_0$, prompting the name *discrete frequency*, or *line spectrum*. Here is an illustration of these terms.

EXAMPLE 17.13

Let g be the periodic function defined by

$$g(t) = t^2 \quad \text{for} \quad 0 < t < 3 \quad \text{and} \quad g(t + 3) = g(t) \quad \text{for all } t.$$

The graph of g is shown in Figure 17.28. It consists of the graph of $y = t^2$ for $0 < t < 3$, with this graph repeated over the intervals $(3, 6)$, $(6, 9)$, ..., $(-3, 0)$, $(-6, -3)$, ... Observe that g is periodic of period 3.

First, compute the Fourier series of g. In the integral formulas for the coefficients, we may integrate over any interval of length 3. For convenience, integrate from zero to 3:

$$a_0 = \frac{1}{3}\int_0^3 t^2\,dt = 3, \qquad a_n = \frac{2}{3}\int_0^3 t^2\cos\left(\frac{2n\pi t}{3}\right)dt = \frac{9}{n^2\pi^2},$$

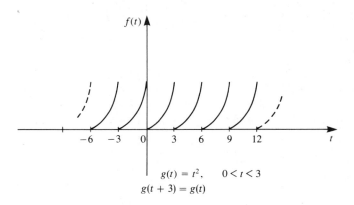

$$g(t) = t^2, \quad 0 < t < 3$$
$$g(t + 3) = g(t)$$

Figure 17.28

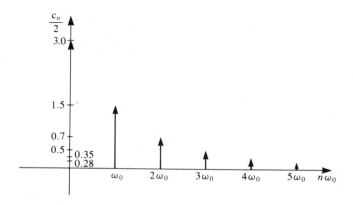

Figure 17.29. Amplitude spectrum of g of Figure 17.26.

and

$$b_n = \frac{2}{3} \int_0^3 t^2 \sin\left(\frac{2n\pi t}{3}\right) dt = -\frac{9}{n\pi}.$$

The Fourier series representation of g is

$$g(t) = 3 + \sum_{n=1}^{\infty} \left[\frac{9}{n^2\pi^2} \cos\left(\frac{2n\pi t}{3}\right) - \frac{9}{n\pi} \sin\left(\frac{2n\pi t}{3}\right) \right].$$

We have

$$\omega_0 = \frac{2\pi}{3},$$

$$c_n = \sqrt{a_n^2 + b_n^2} = \frac{9}{n^2\pi^2} \sqrt{n^2\pi^2 + 1},$$

and

$$\delta_n = \tan^{-1}\left(-\frac{b_n}{a_n}\right) = \tan^{-1}(n\pi).$$

The harmonic form of the Fourier representation of g is therefore

$$g(t) = 3 + \sum_{n=1}^{\infty} \frac{9}{n^2\pi^2} \sqrt{n^2\pi^2 + 1} \, \cos\left[\frac{2n\pi t}{3} + \tan^{-1}(n\pi)\right].$$

The amplitude spectrum for this function is the plot of the numbers $c_n/2$ against the numbers $n\omega_0$, or $2\pi n/3$, and is shown in Figure 17.29. This plot enables us to visualize the magnitude of each of the harmonics of which the periodic function is composed. ∎

The idea that a periodic function is a sum of sine waves has important applications in settings where the superposition of solutions applies. For example, the response of a circuit or mechanical system to a sum of stimuli is the sum of the responses to the individual stimuli. Here is an example involving a circuit in which the forcing function is a sum of sine waves. We will apply the method of undetermined coefficients to one harmonic at a time, determining the particular solution resulting from forcing the circuit with that particular harmonic. The sum of these functions is the Fourier series of the solution.

EXAMPLE 17.14

In some telemetry systems, it is necessary to produce a sine wave with a frequency that is related to a direct-current voltage in a linear manner. For example, we may need the oscillator output to be 370 Hz (hertz, or cycles per second) when 0 volts DC is applied and then have it vary linearly up to 430 Hz as the input is raised to 5 VDC (see Figure 17.30).

The problem is that the frequency of a sine wave oscillator cannot be changed much with just an electrical signal. Usually, these oscillators operate by picking up a natural sine wave oscillation from an inductor-capacitor circuit or from a crystal. The frequency of oscillation can then be changed over a large spectrum by physically changing the value of one of the two inductor-capacitor components. (For example, rotating the tuning knob of a radio actually changes the value of a capacitor in an inductor-capacitor oscillator circuit.) What is needed in this case is the capability of changing the output frequency by as much as 7.5% to either side of the fundamental frequency, using just an electrical signal.

We will now study how this is done. Begin with a pulse generator capable of producing pulses periodically (equal time between pulses) and with the additional capability of having this period changed significantly with a change in input voltage. This is easy to do if we do not care about the shape of the pulses. Now take the pulse generator output and put it into a flip-flop with an output that is always one of two voltages, say, k and $-k$.

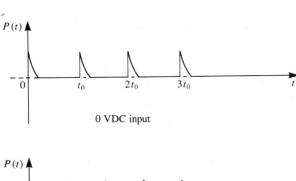

0 VDC input

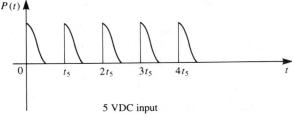

5 VDC input

Figure 17.30. Pulse generator output.

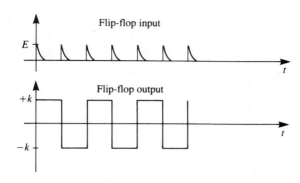

Figure 17.31

If the output is at k volts, it will stay there until a pulse is received at its input. At that time, it changes state and its output drops to $-k$ volts until the next pulse arrives. Since the time between pulses is constant, the flip-flop output will be a square wave of amplitude k and frequency one-half that of the pulse generator (see Figure 17.31). An increase of voltage into the pulse generator increases its frequency, thus increasing the square wave frequency. Therefore, it is possible to produce a square wave with a frequency that can be changed with a change in input voltage.

What is really needed is a sine wave with a frequency that can be changed in this way. Since the square wave is periodic and satisfies the conditions of Theorem 17.2, we know that this square wave is a sum of sine waves, one of which is the one we want.

Suppose we want a sine wave of frequency $\omega_0/2\pi$. Construct a square wave with frequency $\omega_0/2\pi$. The output square wave of the flip-flop is

$$f(t) = \begin{cases} k & \text{if} \quad 0 < t < \pi/\omega_0 \\ -k & \text{if} \quad \pi/\omega_0 < t < 2\pi/\omega_0 \end{cases}$$

and

$$f\left(t + \frac{2\pi}{\omega_0}\right) = f(t).$$

The Fourier series representation of f is

$$f(t) = \frac{4k}{\pi} \sum_{n=1}^{\infty} \frac{\sin[(2n-1)\omega_0 t]}{2n-1}$$

$$= \frac{4k}{\pi} \sin(\omega_0 t) + \frac{4k}{3\pi} \sin(3\omega_0 t) + \frac{4k}{5\pi} \sin(5\omega_0 t) + \cdots.$$

For present purposes, we want only the first sine term, so we remove the higher harmonics by putting the square wave signal through a low-pass filter. This is just a circuit which allows low frequencies to pass through practically unscathed while attenuating higher frequencies. To see how this is done, consider a specific example. Suppose we want to produce a 400-hertz sine wave and to be able to change this frequency electrically from 370 to 430 hertz without distorting the signal. We will assume that the value of the output amplitude is not critical (within reasonable limits); this is true in practice. What is crucial here is the output frequency.

Begin with a pulse generator with a frequency that can be varied from 740 to 860 pulses per second with a change in input from 0 VDC to 5 VDC. Make a square wave by putting these pulses into a flip-flop. The square wave frequency will be half that of the pulse generator and hence will vary from 370 to 430 hertz, as we want. We now have the above square wave with $\omega_0 = 800\pi$. That is,

$$f(t) = \begin{cases} k & \text{if} \quad 0 < t < \frac{1}{800} \\ -k & \text{if} \quad \frac{1}{800} < t < \frac{1}{400}, \end{cases}$$

and $f(t + \frac{1}{400}) = f(t)$. This function has Fourier series

$$f(t) = \frac{4k}{\pi} \sum_{n=1}^{\infty} \frac{\sin[(2n-1)800\pi t]}{2n-1}.$$

Take this signal as the input of the filter of Figure 17.32. What is the output signal? To answer this, we must calculate $q(t)/C$, which in turn requires that we know $q(t)$. The differential equation for the circuit is

$$Li' + Ri + \frac{q}{C} = f(t).$$

Recall that $i = q'$ and put $f(t)$ and the values of L, R, and C into this equation to get

$$q'' + 600q' + 6.25 \times 10^6 q = \frac{4k}{\pi} \sum_{n=1}^{\infty} \frac{\sin[(2n-1)800\pi t]}{2n-1}.$$

Since we want only the steady-state solution, disregard the homogeneous solution (in which the initial conditions play their role) and consider the nonhomogeneous differential equations

$$q'' + 600q' + 6.25 \times 10^6 q = \frac{4k}{\pi} \sin(800\pi t)$$

$$q'' + 600q' + 6.25 \times 10^6 q = \frac{4k}{3\pi} \sin(2400\pi t)$$

$$q'' + 600q' + 6.25 \times 10^6 q = \frac{4k}{5\pi} \sin(4000\pi t)$$

$$\vdots$$

$$q'' + 600q' + 6.25 \times 10^6 q = \frac{4}{(2n-1)\pi} \sin[(2n-1)800\pi t].$$

$$\vdots$$

The strategy is to solve each of these differential equations and sum the solutions to find the Fourier series representation of the capacitor charge q. Thus, consider

$$q_n'' + 600q_n' + 6.25 \times 10^6 q_n = \frac{4k}{n\pi} \sin(n\omega_0 t),$$

in which n is any odd positive integer and $\omega_0 = 800\pi$. The subscript n is introduced to remind us that this is the nth equation and we must sum the solutions.

To use the method of undetermined coefficients, attempt a solution

$$q_n(t) = A_n\cos(n\omega_0 t) + B_n\sin(n\omega_0 t).$$

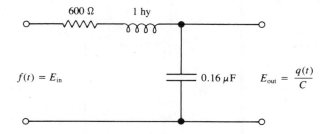

Figure 17.32. Low-pass filter (600 Ω resistance used to construct the 1 hy inductor).

We find that, if n is odd,

$$A_n = \frac{-4k(600)n\omega_0}{n\pi \, \Delta_n} = -\frac{(600)4k\omega_0}{\pi \, \Delta_n}$$

and

$$B_n = \frac{4k(6.25 \times 10^6 - n^2\omega_0^2)}{n\pi \, \Delta_n},$$

where

$$\Delta_n = (600)^2 n^2 \omega_0^2 + [6.25 \times 10^6 - n^2\omega_0^2]^2.$$

To write the solution in harmonic form, calculate

$$C_n = \sqrt{A_n^2 + B_n^2} = \frac{4k}{n\pi\sqrt{\Delta_n}}.$$

The steady-state output voltage is $q(t)/C$, or $6.25 \times 10^6 q(t)$. This has Fourier series representation (in harmonic form)

$$E_{\text{out}} = \frac{2.5 \times 10^7 k}{\pi} \sum_{n=1}^{\infty} \frac{\cos[(2n - 1)800\pi t + \delta_{2n-1}]}{(2n - 1)\sqrt{\Delta_{2n-1}}}.$$

Here are some values for these coefficients with $k = 1$.

n	A_n/C	B_n/C	C_n/C	δ_n
1	5.2669	−0.2324	5.2720	−0.0441
3	0.0015	−0.0522	0.0522	−1.4816
5	0.0001	−0.0105	0.0105	−1.5211
7	0.0000	−0.0037	0.0037	−1.5360
9	0.0000	−0.0017	0.0017	−1.5439
11	0.0000	−0.0001	0.0001	−1.5489

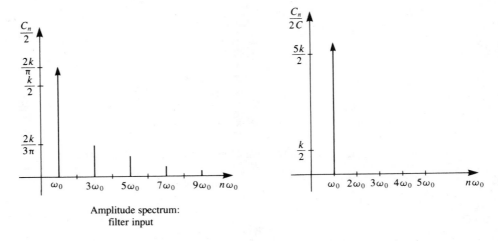

Amplitude spectrum:
filter input

Figure 17.33. Low-pass filter removes higher harmonics of input.

Notice that the output signal is dominated by a cosine term even though no such term appears in the forcing function. Further, the output (which is essentially just a sine wave) is about four times as large as the amplitude of the input.

The frequency spectra of the input and output are shown in Figure 17.33; they provide a clear picture of what has happened. The amplitude of the third harmonic is only about 1% that of the first, or fundamental, frequency. This is because it was only one-third as large to begin with and also because the filter attenuated it, since it was of higher frequency than the fundamental frequency. The subsequent harmonics are even smaller in magnitude for the same reasons.

At the end of this chapter, we will discuss filters and see how we selected the filter component values used in this example.

To complete the original project, which was to make a sine wave generator with a frequency that could be varied by a change in direct-current input voltage, simply group the pulse generator, flip-flop, and low-pass filter into one package, as shown in Figure 17.34. ∎

There is another useful function served by the Fourier series representation of a periodic function. The orthogonality of the sine and cosine functions provides a convenient means of calculating the root mean square (RMS) value of a periodic function. This discloses the power content of the signal, a difficult calculation by other means. This idea is contained in Parseval's identity (see Problem 28 of Section 17.2).

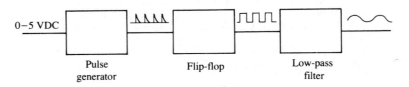

Figure 17.34. Variable-frequency oscillator.

In the next section, we will continue the study of periodic functions by introducing the concept of a complex Fourier series representation of a function.

PROBLEMS FOR SECTION 17.6

1. Suppose that f and g have period T. Show that $\alpha f + \beta g$ has period T for any numbers α and β.

2. Let f be periodic with period T, and let a and b be any nonzero constants. Show that $g(t) = f(at)$ is periodic with period T/a and that $h(t) = f(t/b)$ is periodic with period bT. Verify these results for $\cos(t)$ with $a = b = 2$. Determine the value of a so that $f(at)$ has period 2π.

3. Show that f' is periodic with period T if f is a differentiable periodic function with period T.

4. Suppose that f is periodic with period T. Show that, for any real number a,

$$\int_{a}^{a+T} f(t)\, dt = \int_{-T/2}^{T/2} f(t)\, dt.$$

Hint: Show that $\int_{\alpha}^{\beta} f(t)\, dt = \int_{\alpha+T}^{\beta+T} f(t)\, dt$ for any α and β. Use this to show that $\int_{a}^{a+T} f(t)\, dt = \int_{0}^{T} f(t)\, dt$. Finally, use this relationship with $a = -T/2$.

5. Determine the Fourier series of the function whose graph is shown below, assuming a period of 2π.

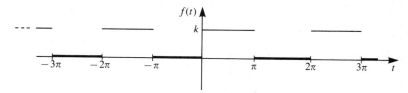

6. Determine the Fourier series of the function whose graph is shown below, assuming a period of 2π.

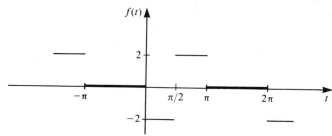

7. Suppose that f has Fourier coefficients a_n and b_n on $[-L, L]$. Show that kf has Fourier coefficients ka_n and kb_n for any constant k.

8. Suppose that f has Fourier coefficients a_n and b_n on $[-L, L]$ and g has Fourier coefficients a_n^* and b_n^* on $[-L, L]$. Show that $f + g$ has Fourier coefficients $a_n + a_n^*$ and $b_n + b_n^*$ on $[-L, L]$.

9. Use the results of Problems 5, 6, 7, and 8 to determine the Fourier series of the function whose graph is shown below. Assume that f has period 2π.

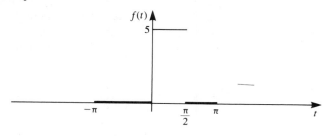

In each of Problems 10 through 21, find the phase angle form of the Fourier series of the periodic function and plot some points of the amplitude spectrum. In Problems 15 through 21, the function is specified by its graph.

10. $f(t) = t \quad 0 < t < 2; \quad f(t + 2) = f(t)$

11. $f(t) = \begin{cases} 1, & 0 < t < 1 \\ 0, & 1 \le t < 2; \end{cases}$
$f(t + 2) = f(t)$

12. $f(t) = 3t^2, \quad 0 < t < 4; \quad f(t + 4) = f(t)$

13. $f(t) = \begin{cases} 1 + t, & 0 < t < 3 \\ 2, & 3 < t < 4; \end{cases}$
$f(t + 4) = f(t)$

14. $f(t) = \cos(\pi t), \quad 0 < t < 1; \quad f(t + 1) = f(t)$

15.

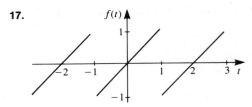

16.

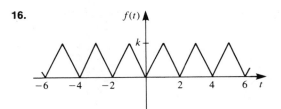

17.

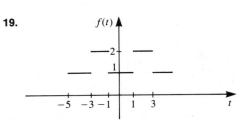

18.

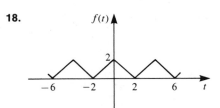

19.

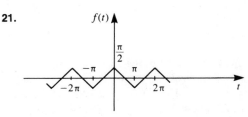

20.

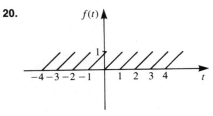

21.

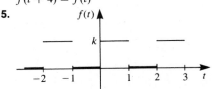

22. Find the Fourier series representation of f if $f(t) = e^t$ for $-2 < t < 2$ and $f(t + 4) = f(t)$. Be careful! Even though it is true that $a_n = \frac{1}{2} \int_0^4 f(t)\cos(n\pi t/2)\, dt$, it is *not* true that $a_n = \frac{1}{2} \int_0^4 e^t\cos(n\pi t/2)\, dt$.

23. Calculate the ratio of the amplitude of the third harmonic to the amplitude of the first (fundamental) harmonic in the output of the filter in Example 17.14.

24. The RMS (*root mean square*) value of a function f over an interval $[a, b]$ is defined to be

$$\text{RMS}(f) = \sqrt{\frac{\int_a^b [f(t)]^2\, dt}{b - a}}.$$

If f is periodic, RMS(f) is calculated by using an interval of length one period. Determine RMS$[E \sin(\omega t)]$, in which E and ω are constants.

25. Assume that f is periodic of period T and that f satisfies the hypotheses of Theorem 17.2 (Section 17.2) on $[-T/2, T/2]$. Assume also that f' is piecewise continuous and that $f(T/2) = f(-T/2)$. Prove that, for any number a,

$$\frac{1}{T} \int_{a-T/2}^{a+T/2} [f(t)]^2 \, dt = a_0^2 + \frac{1}{2} \sum_{n=1}^{\infty} (a_n^2 + b_n^2).$$

(See also Problem 28 of Section 17.2.)

26. The *power content* of a periodic signal f of period T is defined to be

$$\frac{1}{T} \int_{-T/2}^{T/2} [f(t)]^2 \, dt.$$

This definition is motivated by the fact that the power dissipation in a resistor is $i^2 R$, where i is the RMS value of the current (see Problem 24). The RMS value of a signal is that value of direct current which would dissipate the same power in a resistor. Determine the power content of the function defined in Problem 17.

27. Find the steady-state solution of $y'' + 0.04y' + 25y = f(t)$, with f as given in Problem 15.

28. Find the steady-state solution of $y'' + 0.02y' + 12y = f(t)$, with f as given in Problem 17.

29. Find the steady-state solution of $y'' + 25y = f(t)$, with f as given in Problem 19.

30. Find the steady-state solution of $y'' + 8y = f(t)$, with f as given in Problem 21.

31. Determine the Fourier series representation of the steady-state current in the circuit shown below, with $E(t) = 100t(\pi^2 - t^2)$ for $-\pi < t < \pi$, and $E(t + 2\pi) = E(t)$.

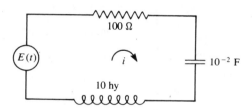

32. Determine the Fourier series representation of the steady-state current in the circuit shown below if $E(t) = |10 \sin(800\pi t)|$. *Hint:* First show that

$$E(t) = \frac{20}{\pi} \left[1 - 2 \sum_{n=1}^{\infty} \frac{\cos(1600 n \pi t)}{4n^2 - 1} \right].$$

33. Use the first two nonzero terms of the Fourier series representation of the current, together with Parseval's identity, to approximate the power dissipation in the resistor in the circuit of Problem 31 due to the steady-state current. (See Problems 24, 25, and 26.)

34. Approximate the power dissipation in the resistor in the circuit of Problem 32 due to the steady-state current. (See Problem 33.)

17.7 *Complex Fourier Series and Frequency Spectra*

We will now develop the complex form of the Fourier series. We could do without this form (except for an application to electrical circuits in Section 17.11). However, it does provide a natural setting from which to view many results of Fourier analysis, including the Fourier transform, which we will develop soon.

We will assume familiarity with the elementary arithmetic of complex numbers. The complex conjugate of $a + ib$ is denoted $\overline{a + ib}$ and equals $a - ib$. If we represent the complex number $a + ib$ as the point (a, b) in the plane, as in Figure 17.35, $\overline{a + ib}$ is represented by the point $(a, -b)$, the reflection of (a, b) across the x-axis (real axis).

If $z = a + ib$, the *modulus* of z is $|z| = \sqrt{a^2 + b^2}$. This is the distance from the origin to the point (a, b) representing z.

If we transform from rectangular to polar coordinates by writing

$$x = r\cos(\theta), \qquad y = r\sin(\theta),$$

the point $z = x + iy$ can be written

$$z = r\cos(\theta) + ir\sin(\theta) = r[\cos(\theta) + i\sin(\theta)].$$

Then $r = |z|$. The angle θ is called an *argument* of z. For example, $2 + 2i$, shown in Figure 17.36, has argument $\theta = \pi/4$, or $\pi/4 + 2k\pi$ for any integer k.

In Section 2.5, we stated Euler's formula,

$$e^{i\theta} = \cos(\theta) + i\sin(\theta).$$

In terms of this complex exponential function, we can write

$$z = x + iy = r[\cos(\theta) + i\sin(\theta)] = re^{i\theta},$$

which is called the *polar form* of z. Euler's formula enables us to write $\cos(t)$ and $\sin(t)$ in terms of complex exponentials. Replace t with $-t$ in Euler's formula

$$e^{it} = \cos(t) + i\sin(t)$$

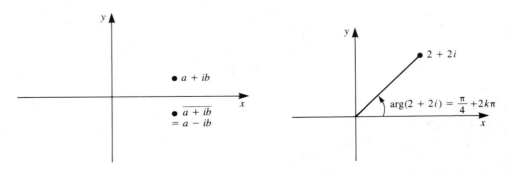

Figure 17.35   *Figure 17.36*

to get

$$e^{-it} = \cos(t) - i\sin(t).$$

Add these equations for e^{it} and e^{-it} and solve for $\cos(t)$ to get

$$\cos(t) = \tfrac{1}{2}(e^{it} + e^{-it}).$$

Similarly, subtract these equations and solve for $\sin(t)$ to get

$$\sin(t) = \frac{1}{2i}(e^{it} - e^{-it}).$$

We will use these complex expressions for $\sin(t)$ and $\cos(t)$ to derive the complex Fourier series representation of a function. Suppose that f is periodic of period T and satisfies the hypotheses of Theorem 17.2. As we did in the preceding section, we can write the Fourier series representation of f,

$$a_0 + \sum_{n=1}^{\infty} [a_n\cos(n\omega_0 t) + b_n\sin(n\omega_0 t)],$$

in which $\omega_0 = 2\pi/T$.

Substitute into this series the expressions

$$\cos(n\omega_0 t) = \frac{e^{in\omega_0 t} + e^{-in\omega_0 t}}{2} \quad \text{and} \quad \sin(n\omega_0 t) = \frac{e^{in\omega_0 t} - e^{-in\omega_0 t}}{2i}.$$

We obtain

$$a_0 + \sum_{n=1}^{\infty} \left\{ a_n\left[\frac{e^{in\omega_0 t} + e^{-in\omega_0 t}}{2}\right] + b_n\left[\frac{e^{in\omega_0 t} - e^{-in\omega_0 t}}{2i}\right]\right\},$$

or

$$a_0 + \sum_{n=1}^{\infty} [\tfrac{1}{2}(a_n - ib_n)e^{in\omega_0 t} + \tfrac{1}{2}(a_n + ib_n)e^{-in\omega_0 t}]. \qquad (17.9)$$

Now let

$$c_0 = a_0$$

and, for $n = 1, 2, 3, \ldots,$

$$c_n = \tfrac{1}{2}(a_n - ib_n).$$

Since $f(t)$ is real for each real t, a_n and b_n are themselves real, and

$$\tfrac{1}{2}(a_n + ib_n) = \bar{c}_n.$$

Therefore, the Fourier series (17.9) for $f(t)$ can be written

$$c_0 + \sum_{n=1}^{\infty} [c_n e^{in\omega_0 t} + \bar{c}_n e^{-in\omega_0 t}] = c_0 + \sum_{n=1}^{\infty} c_n e^{in\omega_0 t} + \sum_{n=1}^{\infty} \bar{c}_n e^{-in\omega_0 t}. \qquad (17.10)$$

Now use the integral formulas for the a_n's and b_n's. We have

$$c_0 = a_0 = \frac{1}{T} \int_{-T/2}^{T/2} f(t)\,dt$$

and, for $n = 1, 2, 3, \ldots$,

$$c_n = \frac{1}{2}(a_n - ib_n)$$

$$= \frac{1}{2}\frac{2}{T}\int_{-T/2}^{T/2} f(t)\cos(n\omega_0 t)\, dt - \frac{i}{2}\frac{2}{T}\int_{-T/2}^{T/2} f(t)\sin(n\omega_0 t)\, dt$$

$$= \frac{1}{T}\int_{-T/2}^{T/2} f(t)[\cos(n\omega_0 t) - i\sin(n\omega_0 t)]\, dt$$

$$= \frac{1}{T}\int_{-T/2}^{T/2} f(t)e^{-in\omega_0 t}\, dt.$$

By taking the complex conjugate of this integral, we obtain

$$\bar{c}_n = \frac{1}{T}\int_{-T/2}^{T/2} f(t)e^{in\omega_0 t}\, dt = c_{-n}.$$

Put these results into equation (17.10) to obtain

$$c_0 + \sum_{n=1}^{\infty} c_n e^{in\omega_0 t} + \sum_{n=1}^{\infty} c_{-n}e^{-in\omega_0 t} = c_0 + \sum_{\substack{n=-\infty \\ n \neq 0}}^{\infty} c_n e^{in\omega_0 t} = \sum_{n=-\infty}^{\infty} c_n e^{in\omega_0 t}.$$

This leads us to define the *complex Fourier series* of f (having period T) to be

$$\sum_{n=-\infty}^{\infty} c_n e^{in\omega_0 t},$$

in which $\omega_0 = 2\pi/T$ and

$$c_n = \frac{1}{T}\int_{-T/2}^{T/2} f(t)e^{-in\omega_0 t}\, dt$$

for $n = 0, \pm 1, \pm 2, \pm 3, \ldots$. The numbers c_n are the *complex Fourier coefficients* of f.

THEOREM 17.9   Convergence of Complex Fourier Series

Suppose that f satisfies the hypotheses of Theorem 17.2 and that f has period T. Then

$$\sum_{n=-\infty}^{\infty} c_n e^{in\omega_0 t} = \tfrac{1}{2}[f(t+) + f(t-)]$$

for every real number t at which f has a left and right derivative, provided that

$$c_n = \frac{1}{T}\int_{-T/2}^{T/2} f(t)e^{-in\omega_0 t}\, dt$$

for $n = 0, \pm 1, \pm 2, \ldots$. ∎

We can write the polar form of each c_n:

$$c_n = r_n e^{i\theta_n}.$$

The *frequency spectrum (amplitude spectrum)* of a periodic function f is a plot of the magnitudes of the c_n's versus the angular frequency. Since $|c_n| = r_n$, this is a plot of the numbers r_n versus the numbers $n\omega_0$. Notice that $|c_n|$ is one-half the value of the Fourier coefficients of the same function expressed in harmonic form.

EXAMPLE 17.15

We will determine the complex Fourier series, and plot the frequency spectrum of the sawtooth function f of period 8 defined by

$$f(t) = \tfrac{3}{4}t \quad \text{for} \quad 0 < t < 8$$

$$f(t + 8) = f(t).$$

Here, $T = 8$ and $\omega_0 = 2\pi/8 = \pi/4$. If $n \neq 0$,

$$
\begin{aligned}
c_n &= \frac{1}{8} \int_0^8 \frac{3}{4} t e^{-n\pi i t/4} \, dt \\
&= \frac{3}{32} \left[\frac{t e^{-n\pi i t/4}}{-\dfrac{n\pi i}{4}} - \frac{e^{-n\pi i t/4}}{\left(\dfrac{n\pi i}{4}\right)^2} \right]_0^8 \\
&= \frac{3}{32} \left[\frac{8 e^{-2n\pi i}}{-\dfrac{n\pi i}{4}} + \frac{e^{-2n\pi i}}{\dfrac{n^2\pi^2}{16}} - \frac{1}{\dfrac{n^2\pi^2}{16}} \right] = \frac{3i}{n\pi}.
\end{aligned}
$$

In this calculation, we used the fact that $e^{-2n\pi i} = \cos(2n\pi) - i \sin(2n\pi) = 1$. Finally,

$$c_0 = \frac{1}{8} \int_0^8 \frac{3}{4} t \, dt = \frac{3t^2}{64} \Big]_0^8 = 3.$$

The complex Fourier series of f is

$$3 + \frac{3}{\pi} i \sum_{\substack{n=-\infty \\ n \neq 0}}^{\infty} \frac{1}{n} e^{n\pi i t/4}.$$

The graph of f and its frequency spectrum are shown in Figure 17.37. Notice that the spectrum plot extends in the negative as well as the positive direction.

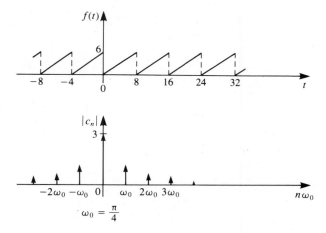

Figure 17.37. The graph of the periodic function f and its amplitude spectrum.   ∎

EXAMPLE 17.16

Determine the complex Fourier series of the full-wave rectification of $E \sin(\omega t)$.

We want to expand $|E \sin(\omega t)|$, the graph of which is shown in Figure 17.38. We will assume that E is a positive constant. Since f has period $T = \pi/\omega$, calculate

$$c_n = \frac{\omega}{\pi} \int_0^{\pi/\omega} |E \sin(\omega t)| e^{-2n\omega it} \, dt = \frac{E\omega}{\pi} \int_0^{\pi/\omega} \sin(\omega t) e^{-2n\omega it} \, dt.$$

We can simplify this integration if we write $\sin(\omega t) = (e^{i\omega t} - e^{-i\omega t})/2i$ to get

$$
\begin{aligned}
c_n &= \frac{E\omega}{\pi} \int_0^{\pi/\omega} \left[\frac{e^{i\omega t} - e^{-i\omega t}}{2i} \right] e^{-2n\omega it} \, dt \\
&= \frac{E\omega}{2\pi i} \int_0^{\pi/\omega} e^{-(2n-1)\omega it} \, dt - \frac{E\omega}{2\pi i} \int_0^{\pi/\omega} e^{-(2n+1)\omega it} \, dt \\
&= \frac{E\omega}{2\pi i} \left[-\frac{1}{(2n-1)\omega i} e^{-(2n-1)\omega it} + \frac{1}{(2n+1)\omega i} e^{-(2n+1)\omega it} \right]_0^{\pi/\omega} \\
&= \frac{E\omega}{2\pi i} \left[-\frac{1}{(2n-1)\omega i} (e^{-(2n-1)\pi i} - 1) + \frac{1}{(2n+1)\omega i} (e^{-(2n+1)\pi i} - 1) \right].
\end{aligned}
$$

Now,

$$e^{-(2n-1)\pi i} = \cos[(2n-1)\pi] - i \sin[(2n-1)\pi] = (-1)^{2n-1} = -1$$

and

$$e^{-(2n+1)\pi i} = \cos[(2n+1)\pi] + i \sin[(2n+1)\pi] = (-1)^{2n+1} = -1.$$

Therefore,

$$c_n = \frac{E\omega}{2\pi i} \left[-\frac{-2}{(2n-1)\omega i} + \frac{-2}{(2n+1)\omega i} \right] = \frac{-2E}{(4n^2 - 1)\pi}$$

for $n = 0, \pm 1, \pm 2, \ldots$. The complex Fourier series of $|E \sin(\omega t)|$ is

$$-2 \frac{E}{\pi} \sum_{n=-\infty}^{\infty} \frac{1}{4n^2 - 1} e^{2n\omega it}.$$

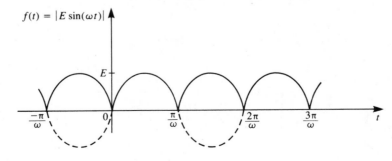

Figure 17.38. Full-wave rectification of $E \sin(\omega t)$. ∎

PROBLEMS FOR SECTION 17.7

1. Use Euler's formula to show that $\overline{e^{i\theta}} = e^{-i\theta}$ if θ is real. Does this equation still hold if θ is complex?

In each of Problems 2 through 10, find the complex Fourier series of the given periodic function and plot some points of the frequency spectrum.

2. $f(t) = 2t$ for $0 \le t < 3$; $f(t+3) = f(t)$

3. $f(t) = t^2$ for $0 \le t < 2$; $f(t+2) = f(t)$

4. $f(t) = \begin{cases} 0, & 0 \le t < 1 \\ 1, & 1 < t < 4; \end{cases}$
$f(t+4) = f(t)$

5. $f(t) = 1 - t$ for $0 \le t < 6$; $f(t+6) = f(t)$

6. $f(t) = \begin{cases} -1, & 0 < t < 2 \\ 2, & 2 < t < 4; \end{cases}$
$f(t+4) = f(t)$

7. $f(t) = e^{-t}$, $0 < t < 5$; $f(t+5) = f(t)$

8. $f(t) = \begin{cases} t, & 0 < t < 1 \\ 2 - t, & 1 < t < 2; \end{cases}$
$f(t+2) = f(t)$

9. $f(t) = \cos(t)$, $0 < t < 1$; $f(t+1) = f(t)$

10. $f(t) = \begin{cases} t, & 0 < t < 2 \\ 0, & 2 < t < 3; \end{cases}$
$f(t+3) = f(t)$

11. Determine the complex Fourier series of f and graph its frequency spectrum if f is the periodic function whose graph is shown below.

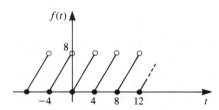

12. Graph the frequency spectrum of the function f defined by

$$f(t) = 4 + \sum_{\substack{n=-\infty \\ n \neq 0}}^{\infty} \frac{26}{n(12 - 5i)} e^{2nit}.$$

The next two problems involve the notion of a *phase spectrum*. The phase spectrum of f is a plot of the phase angle $\varphi_n = \tan^{-1}(b_n/a_n)$ of c_n versus $n\omega_0$. Thus, in Example 17.15, $\varphi_0 = 0$ and $\varphi_n = \pi/2$ for $n = 1, 2, 3, \ldots$. In Example 17.16, $\varphi_n = 0$ for $n = 0, 1, 2, \ldots$.

13. Determine φ_n for each n and plot the phase spectrum of the function of Problem 6.

14. Calculate the complex Fourier series representation of the two periodic functions f and g whose graphs are shown below. Determine a relationship between the frequency spectra of these two functions and also between their phase spectra.

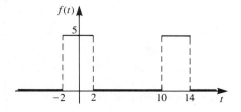

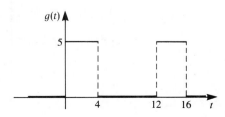

15. A set of complex functions $\{\varphi_n\}$ is said to be *orthogonal* over an interval $[a, b]$ if

$$\int_a^b \varphi_n(t)\overline{\varphi_m(t)}\, dt = 0 \quad \text{if} \quad n \neq m.$$

Show that $\{e^{in\omega_0 t}\}$ is an orthogonal set of functions on $[-T/2, T/2]$ if $\omega_0 = 2\pi/T$.

16. Use the result of Problem 15 to determine the c_n's so that a periodic function f with period T can be represented in a series

$$f(t) = \sum_{n=-\infty}^{\infty} c_n e^{in\omega_0 t}.$$

Hint: Multiply both sides of this equation by $e^{ik\omega_0 t}$, with k any integer, and integrate the resulting expression from $-T/2$ to $T/2$, taking advantage of the orthogonality.

17. Prove *Parseval's identity* in the form

$$\frac{1}{T}\int_{-T/2}^{T/2} [f(t)]^2\, dt = \sum_{n=-\infty}^{\infty} |c_n|^2,$$

in which

$$c_n = \frac{1}{T}\int_{-T/2}^{T/2} f(t)e^{-in\omega_0 t}\, dt \quad \text{and} \quad \omega_0 = \frac{2\pi}{T}.$$

[See Problem 28(e) of Section 17.2.]

17.8 *The Fourier Integral*

If f is defined on $[-L, L]$ or on $[0, L]$, it is sometimes possible to represent f in a Fourier series. If f is defined on the entire real line but is periodic, it may still be possible to represent f as a Fourier series. We will now consider the possibility of representing a function which is defined on the entire real line or on the half-line $[0, \infty)$ but is not necessarily periodic. In this circumstance, the Fourier series is replaced by a Fourier integral.

Suppose that f is defined for all t and we want to represent f in some way in terms of sines and/or cosines. One approach is to choose $T > 0$ and devise a function f_T having period T and having the property that $\lim_{T \to \infty} f_T(t) = f(t)$ for all t. On $[-T/2, T/2]$, f_T can be represented by a Fourier series. Letting $T \to \infty$, $f_T \to f$ and the Fourier series approaches a representation of f in terms of sines and cosines.

How do we devise such a periodic function f_T which approaches f as T is chosen larger? An example will clarify this process. Let, for example,

$$f(t) = \begin{cases} 1 & \text{if} \quad -1 < t < 1 \\ 0 & \text{if} \quad t \geq 1 \quad \text{or} \quad t \leq -1. \end{cases}$$

Now choose any $T > 2$ and define f_T by letting

$$f_T(t) = \begin{cases} 1 & \text{if} \quad -1 < t < 1, \\ 0 & \text{if} \quad 1 < t < T/2 \quad \text{or} \quad -T/2 < t < -1, \end{cases}$$

with $f_T(t + T) = f_T(t)$. Then f_T is periodic with period T.

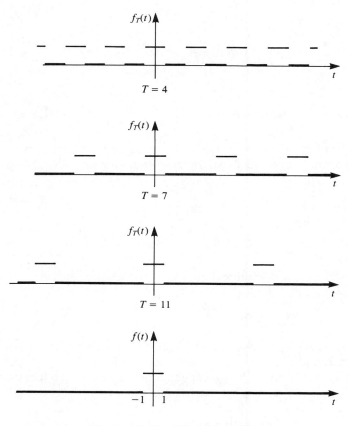

Figure 17.39. $f(t) = \lim_{T \to \infty} f_T(t)$.

The graphs of f and f_T are shown in Figure 17.39. It is apparent that, as $T \to \infty$, the graph of f_T more nearly approximates that of f, prompting us to write $f_T \to f$ as $T \to \infty$. We can attempt a Fourier representation of f by writing the Fourier series of f_T on $[-T/2, T/2]$ and letting $T \to \infty$.

We will now carry out this strategy for an arbitrary f. Suppose that f_T is periodic with period T and that $f_T \to f$ as $T \to \infty$. This means that, for any real number t, $\lim_{T \to \infty} f_T(t) = f(t)$. Assume that we can expand f_T in a Fourier series on $[-T/2, T/2]$

$$f_T(t) = a_0 + \sum_{n=1}^{\infty} [a_n \cos(n\omega_0 t) + b_n \sin(n\omega_0 t)],$$

in which $\omega_0 = 2\pi/T$. The Fourier coefficients are

$$a_0 = \frac{1}{T} \int_{-T/2}^{T/2} f_T(\xi)\, d\xi, \qquad a_n = \frac{2}{T} \int_{-T/2}^{T/2} f_T(\xi)\cos(n\omega_0 \xi)\, d\xi,$$

and

$$b_n = \frac{2}{T} \int_{-T/2}^{T/2} f_T(\xi)\sin(n\omega_0 \xi)\, d\xi.$$

Let

$$\omega_n = n\omega_0 = \frac{2n\pi}{T} \quad \text{and} \quad \Delta\omega_n = \omega_{n+1} - \omega_n = \frac{2\pi}{T}.$$

Then

$$a_n = \frac{1}{\pi}\,\omega_0 \int_{-T/2}^{T/2} f_T(\xi)\cos(\omega_n\xi)\,d\xi \quad \text{and} \quad b_n = \frac{1}{\pi}\,\omega_0 \int_{-T/2}^{T/2} f_T(\xi)\sin(\omega_n\xi)\,d\xi.$$

Replace $2/T$ with $\Delta\omega_n/\pi$ in these integral formulas for the coefficients, and substitute the resulting formulas into the Fourier series for $f_T(t)$ to get

$$f_T(t) = \frac{1}{T}\int_{-T/2}^{T/2} f_T(\xi)\,d\xi + \frac{1}{\pi}\sum_{n=1}^{\infty}\left\{\left[\int_{-T/2}^{T/2} f_T(\xi)\cos(\omega_n\xi)\,d\xi\right]\cos(\omega_n t)\right.$$
$$\left. + \left[\int_{-T/2}^{T/2} f_T(\xi)\sin(\omega_n\xi)\,d\xi\right]\sin(\omega_n t)\right\}\Delta\omega_n.$$

Now let $T \to \infty$. We are assuming that $f_T(t) \to f(t)$ for all t. Further,

$$\int_{-T/2}^{T/2} f_T(\xi)\,d\xi \to \int_{-T/2}^{T/2} f(\xi)\,d\xi.$$

Assume that $\int_{-\infty}^{\infty}|f(\xi)|\,d\xi$ converges. This implies convergence not only of $\int_{-\infty}^{\infty} f(\xi)\,d\xi$ but also of the integrals appearing in the last summation as $T \to \infty$. In fact, as $T \to \infty$, $\Delta\omega_n \to 0$ and $a_0 \to 0$. Further, the last summation is suggestive of a Riemann sum for an improper integral, and, as $\Delta\omega_n \to 0$, we obtain

$$\frac{1}{\pi}\sum_{n=1}^{\infty}\left\{\left[\int_{-T/2}^{T/2} f_T(\xi)\cos(\omega_n\xi)\,d\xi\right]\cos(\omega_n t) + \left[\int_{-T/2}^{T/2} f_T(\xi)\sin(\omega_n\xi)\,d\xi\right]\sin(\omega_n t)\right\}\Delta\omega_n$$
$$\to \frac{1}{\pi}\int_0^{\infty}\left\{\left[\int_{-\infty}^{\infty} f(\xi)\cos(\omega\xi)\,d\xi\right]\cos(\omega t) + \left[\int_{-\infty}^{\infty} f(\xi)\sin(\omega\xi)\,d\xi\right]\sin(\omega t)\right\}d\omega.$$

This suggests that the Fourier series for f_T on $[-T/2,\ T/2]$ approaches an integral representation of f on the entire real line as $T \to \infty$. The form of this integral motivates the following definition.

DEFINITION   The Fourier Integral

Let f be defined for all real t, and suppose that $\int_{-\infty}^{\infty}|f(t)|\,dt$ converges. Then the *Fourier integral*, or *Fourier integral representation*, of f is defined to be the integral

$$\frac{1}{\pi}\int_0^{\infty}\left[A(\omega)\cos(\omega t) + B(\omega)\sin(\omega t)\right]d\omega \qquad (-\infty < t < \infty),$$

in which

$$A(\omega) = \int_{-\infty}^{\infty} f(\xi)\cos(\omega\xi)\,d\xi \quad \text{and} \quad B(\omega) = \int_{-\infty}^{\infty} f(\xi)\sin(\omega\xi)\,d\xi. \quad \blacksquare$$

Fourier integrals have convergence properties similar to those of Fourier series.

THEOREM 17.10   Convergence of a Fourier Integral

Suppose that f satisfies the hypotheses of Theorem 17.2 on every finite interval $[-a, a]$. Further, suppose that $\int_{-\infty}^{\infty} |f(t)|\, dt$ converges. Let

$$A(\omega) = \int_{-\infty}^{\infty} f(\xi)\cos(\omega\xi)\, d\xi \quad \text{and} \quad B(\omega) = \int_{-\infty}^{\infty} f(\xi)\sin(\omega\xi)\, d\xi$$

for every $\omega \geq 0$. Then the Fourier integral

$$\frac{1}{\pi} \int_{0}^{\infty} [A(\omega)\cos(\omega t) + B(\omega)\sin(\omega t)]\, d\omega \qquad -\infty < t < \infty$$

converges to $\frac{1}{2}[f(t+) + f(t-)]$ at any t at which f has a jump discontinuity and at which the left and right derivatives of f exist. In particular, at any such t at which f is continuous, the Fourier integral converges to $f(t)$. ∎

As with Fourier series, we can usually tell what a Fourier integral converges to just by examining the graph of f.

EXAMPLE 17.17

Let

$$f(t) = \begin{cases} 1 & \text{if} & -1 < t < 1 \\ 0 & \text{if} & t \geq 1 \quad \text{or} \quad t \leq -1. \end{cases}$$

We will write the Fourier integral representation of f. First,

$$\int_{-\infty}^{\infty} |f(t)|\, dt = \int_{-1}^{1} dt = 2.$$

For $\omega > 0$, calculate the coefficients $A(\omega)$ and $B(\omega)$

$$A(\omega) = \int_{-\infty}^{\infty} f(\xi)\cos(\omega\xi)\, d\xi = \int_{-1}^{1} \cos(\omega\xi)\, d\xi = \frac{\sin(\omega\xi)}{\omega}\bigg]_{-1}^{1} = \frac{2\sin(\omega)}{\omega}$$

and

$$B(\omega) = \int_{-\infty}^{\infty} f(\xi)\sin(\omega\xi)\, d\xi = \int_{-1}^{1} \sin(\omega\xi)\, d\xi = -\frac{\cos(\omega\xi)}{\omega}\bigg]_{-1}^{1} = 0.$$

In fact, $A(0)$ can be assigned the value 2, since

$$\lim_{\omega \to 0} \frac{2\sin(\omega)}{\omega} = 2.$$

The Fourier integral representation of f is

$$\frac{2}{\pi} \int_{0}^{\infty} \frac{\sin(\omega)}{\omega} \cos(\omega t)\, d\omega.$$

By Theorem 17.10, this integral representation converges to

$$\begin{cases} 1 & \text{if} & -1 < t < 1 \\ 0 & \text{if} & t > 1 \quad \text{or} \quad t < -1 \\ \frac{1}{2} & \text{if} & t = 1 \quad \text{or} \quad t = -1. \end{cases}$$

Therefore, for all t except $t = \pm 1$, we have

$$f(t) = \frac{2}{\pi} \int_0^\infty \frac{\sin(\omega)}{\omega} \cos(\omega t)\, d\omega. \quad \blacksquare$$

FOURIER SINE AND COSINE INTEGRALS

Analogous to Fourier sine and cosine series, we can develop sine and cosine integrals for functions defined on the half-line $[0, \infty)$. The development mirrors that for series.

Suppose that f satisfies the hypotheses of Theorem 17.2 on every interval $[0, a]$ and that $\int_0^\infty |f(t)|\, dt$ converges. To obtain a Fourier cosine representation of f, extend f to an even function g on the entire real line by setting

$$g(t) = \begin{cases} f(t) & \text{if} \quad t \geq 0 \\ f(-t) & \text{if} \quad t < 0. \end{cases}$$

Expand g in a Fourier integral on $(-\infty, \infty)$. The coefficients are

$$A(\omega) = \int_{-\infty}^\infty g(\xi)\cos(\omega\xi)\, d\xi = 2 \int_0^\infty f(\xi)\cos(\omega\xi)\, d\xi,$$

because g is even and agrees with f on $[0, \infty)$, and

$$B(\omega) = \int_{-\infty}^\infty g(\xi)\sin(\omega\xi)\, d\xi = 0,$$

because $g(\xi)\sin(\omega\xi)$ is an odd function. This discussion suggests the following definition.

DEFINITION   Fourier Cosine Integral

Let f be integrable on $[0, \infty)$. The Fourier cosine integral of f on $[0, \infty)$ is

$$\frac{1}{\pi} \int_0^\infty A(\omega)\cos(\omega t)\, d\omega,$$

in which

$$A(\omega) = 2 \int_0^\infty f(\xi)\cos(\omega\xi)\, d\xi. \quad \blacksquare$$

Using Theorem 17.10, it can be shown that this Fourier cosine integral converges to $f(0+)$ at zero and to the average of the left and right limits of f at any point where f has left and right derivatives. By extending f to an odd function defined over $(-\infty, \infty)$, we arrive at the concept of a Fourier sine integral representation of f.

DEFINITION   Fourier Sine Integral

The Fourier sine integral representation of f on $[0, \infty)$ is

$$\frac{1}{\pi} \int_0^\infty B(\omega)\sin(\omega t)\, d\omega,$$

in which

$$B(\omega) = 2 \int_0^\infty f(\xi)\sin(\omega\xi)\, d\xi. \quad \blacksquare$$

We can check, using Theorem 17.10, that this representation converges to zero at $t = 0$ and to the average of the left and right limits of f at any point where f has a jump discontinuity and both left and right derivatives of f exist.

EXAMPLE 17.18   The Laplace Integrals

Let $f(t) = e^{-kt}$ for $k > 0$ and $t \geq 0$. Then

$$\int_0^\infty |f(\xi)|\, d\xi = \frac{1}{k};$$

hence, this integral converges. For the Fourier cosine representation of f on $[0, \infty)$, compute

$$A(\omega) = 2 \int_0^\infty e^{-k\xi}\cos(\omega\xi)\, d\xi = \frac{2k}{k^2 + \omega^2}.$$

Further, observe that e^{-kt} is continuous for all $t \geq 0$ and has a left and right derivative (in fact, is differentiable) at each $t > 0$. Then, for $t \geq 0$ and k any positive constant,

$$e^{-kt} = 2\frac{k}{\pi} \int_0^\infty \frac{\cos(\omega t)}{k^2 + \omega^2}\, d\omega.$$

For the Fourier sine representation of f, compute

$$B(\omega) = 2 \int_0^\infty e^{-k\xi}\sin(\omega\xi)\, d\xi = \frac{2\omega}{k^2 + \omega^2}.$$

The sine integral converges to zero at $t = 0$ and to e^{-kt} if $t > 0$. Thus,

$$\frac{2}{\pi} \int_0^\infty \frac{\omega \sin(\omega t)}{k^2 + \omega^2}\, d\omega = \begin{cases} e^{-kt} & \text{if} \quad t > 0. \\ 0 & \text{if} \quad t = 0. \end{cases}$$

These integral representations are called *Laplace's integrals*. The rationale for this name is that the coefficient in the cosine representation of e^{-kt} is $A(\omega) = 2\mathscr{L}[\cos(\omega t)]$, while the coefficient in the sine representation is $B(\omega) = 2\mathscr{L}[\sin(\omega t)]$. $\quad \blacksquare$

In the next section, we will begin the first of several sections devoted to the Fourier transform and some of its applications.

PROBLEMS FOR SECTION 17.8

In each of Problems 1 through 10, write the Fourier integral representation of f and determine what this representation converges to.

1. $f(t) = \begin{cases} t, & -\pi \leq t \leq \pi \\ 0, & |t| > \pi \end{cases}$

2. $f(t) = \begin{cases} C, & -10 \leq t \leq 10 \\ 0, & |t| > 10 \end{cases}$

3. $f(t) = \begin{cases} -1, & -\pi \le t \le 0 \\ 1, & 0 < t \le \pi \\ 0, & |t| > \pi \end{cases}$

4. $f(t) = \begin{cases} \sin(t), & -4 \le t \le 0 \\ \cos(t), & 0 < t \le 4 \\ 0, & |t| > 4 \end{cases}$

5. $f(t) = \begin{cases} t^2, & -100 \le t \le 100 \\ 0, & |t| > 100 \end{cases}$

6. $f(t) = \begin{cases} |t|, & -\pi \le t \le \pi \\ 0, & |t| > \pi \end{cases}$

7. $f(t) = \begin{cases} \sin(t), & -\pi \le t \le \pi \\ 0, & |t| > \pi \end{cases}$

8. $f(t) = \begin{cases} 0, & |t| > 5 \\ \frac{1}{2}, & -5 \le t < 1 \\ 1, & 1 \le t \le 5 \end{cases}$

9. $f(t) = \begin{cases} \sin(t)/t, & t \ne 0 \\ 1, & t = 0 \end{cases}$

10. $f(t) = \begin{cases} t^2 - 1, & 0 < t < 4 \\ 0, & t > 4 \quad \text{or} \quad t < 0 \end{cases}$

11. Show that the Fourier integral of f can be written

$$\frac{1}{\pi} \int_0^\infty \int_{-\infty}^\infty f(\xi) \cos[\omega(t - \xi)] \, d\xi \, d\omega.$$

12. Show that the Fourier integral of f can be written

$$\lim_{\omega \to \infty} \frac{1}{\pi} \int_{-\infty}^\infty f(\xi) \frac{\sin[\omega(\xi - t)]}{\xi - t} \, d\xi.$$

13. Use the result of Problem 11 to show that the Fourier integral of f can be written

$$\frac{1}{2\pi} \int_{-\infty}^\infty \int_{-\infty}^\infty f(\xi) \cos[\omega(t - \xi)] \, d\xi \, d\omega.$$

14. Use the result of Problem 13 to show that the Fourier integral of f can be written in complex exponential form as

$$\frac{1}{2\pi} \int_{-\infty}^\infty \int_{-\infty}^\infty e^{i\omega(t - \xi)} f(\xi) \, d\xi \, d\omega.$$

In each of Problems 15 through 25, write the Fourier cosine integral and Fourier sine integral representations of the function and determine what these integrals converge to for $t \ge 0$.

15. $f(t) = \begin{cases} t^2, & 0 \le t \le 10 \\ 0, & t > 10 \end{cases}$

16. $f(t) = \begin{cases} \sin(t), & 0 \le t \le 2\pi \\ 0, & t > 2\pi \end{cases}$

17. $f(t) = \begin{cases} 1, & 0 \le t \le 1 \\ 2, & 1 \le t \le 4 \\ 0, & t > 4 \end{cases}$

18. $f(t) = \begin{cases} \cosh(t), & 0 \le t \le 5 \\ 0, & t > 5 \end{cases}$

19. $f(t) = \begin{cases} 2t + 1, & 0 \le t \le \pi \\ 2, & \pi < t \le 3\pi \\ 1, & 3\pi < t \le 10\pi \\ 0, & t > 10\pi \end{cases}$

20. $f(t) = \begin{cases} t, & 0 \le t \le 1 \\ 1 + t, & 1 < t \le 2 \\ 0, & t > 2 \end{cases}$

21. $f(t) = e^{-t}\cos(t)$ for $t \ge 0$

22. $f(t) = te^{-t}$ for $t \ge 0$

23. $f(t) = \begin{cases} 1, & 0 \le t \le 10 \\ 0, & t > 10 \end{cases}$

24. $f(t) = \begin{cases} \sin(\pi t), & 0 \le t \le 1 \\ 0, & t > 1 \end{cases}$

25. Use the Laplace integrals from Example 17.18 to compute (a) the Fourier cosine integral of $f(t) = \dfrac{1}{1 + t^2}$, and

(b) the Fourier sine integral of $f(t) = \dfrac{t}{1 + t^2}$.

26. Suppose that $f'(0) = 0$, that $\lim_{t \to \infty} f'(t) = \lim_{t \to \infty} f(t) = 0$, and that $f''(t)$ exists for $t \geq 0$. Show that the integrand of the Fourier cosine integral representation of $f''(t)$ is $-\omega^2$ times the integrand of the Fourier cosine integral representation of $f(t)$.

In Problems 27 and 28, assume that $f(t) = (1/\pi) \int_0^\infty A(\omega)\cos(\omega t)\, d\omega$, where $A(\omega) = \int_{-\infty}^\infty f(\xi)\cos(\omega \xi)\, d\xi$.

27. Show that $t^2 f(t) = (1/\pi) \int_0^\infty A^*(\omega)\cos(\omega t)\, d\omega$, where $A^*(\omega) = -d^2 A/d\omega^2$.

28. Show that

$$tf(t) = \frac{1}{\pi} \int_0^\infty B^*(\omega)\sin(\omega t)\, d\omega,$$

where

$$B^*(\omega) = -\frac{dA}{d\omega}.$$

29. Find the Fourier cosine integral representation of f, where

$$f(t) = \begin{cases} 1 & \text{if} \quad 0 < t < 1 \\ 0 & \text{if} \qquad\quad t > 1. \end{cases}$$

Hint: Look at Example 17.17.

30. Use the results of Problems 27 and 29 to find the Fourier cosine integral representation of f, where

$$f(t) = \begin{cases} t^2 & \text{if} \quad 0 \leq t < 1 \\ 0 & \text{if} \qquad\quad t \geq 1. \end{cases}$$

17.9 *The Fourier Transform*

In this section, we will introduce the Fourier transform. As a starting point, we will derive a complex form of the Fourier integral. Suppose we have a real-valued function f represented by a Fourier integral

$$f(t) = \frac{1}{\pi} \int_0^\infty [A(\omega)\cos(\omega t) + B(\omega)\sin(\omega t)]\, d\omega.$$

Replace $\cos(\omega t)$ and $\sin(\omega t)$ by their complex exponential forms to get

$$f(t) = \frac{1}{\pi} \int_0^\infty \left\{ A(\omega)\frac{1}{2}[e^{i\omega t} + e^{-i\omega t}] + B(\omega)\frac{1}{2i}[e^{i\omega t} - e^{-i\omega t}] \right\} d\omega$$

$$= \frac{1}{\pi} \int_0^\infty \left\{ \frac{1}{2}[A(\omega) - iB(\omega)]e^{i\omega t} + \frac{1}{2}[A(\omega) + iB(\omega)]e^{-i\omega t} \right\} d\omega.$$

Now define

$$\overline{C(\omega)} = \tfrac{1}{2}[A(\omega) - iB(\omega)].$$

Since $A(\omega)$ and $B(\omega)$ are real valued for ω real, the complex conjugate of $C(\omega)$ is

$$\overline{C(\omega)} = \tfrac{1}{2}[A(\omega) + iB(\omega)].$$

Therefore,

$$f(t) = \frac{1}{\pi} \int_0^\infty [C(\omega)e^{i\omega t} + \overline{C(\omega)}e^{-i\omega t}]\, d\omega$$

$$= \frac{1}{\pi} \int_0^\infty C(\omega)e^{i\omega t}\, d\omega + \frac{1}{\pi} \int_0^\infty \overline{C(\omega)}e^{-i\omega t}\, d\omega. \qquad (17.11)$$

Now notice that

$$C(\omega) = \frac{1}{2}[A(\omega) - iB(\omega)]$$

$$= \frac{1}{2} \int_{-\infty}^\infty f(\xi)\cos(\omega\xi)\, d\xi - \frac{i}{2} \int_{-\infty}^\infty f(\xi)\sin(\omega\xi)\, d\xi$$

$$= \frac{1}{2} \int_{-\infty}^\infty f(\xi)\left(\frac{1}{2}[e^{i\xi} + e^{-i\xi}] - i\frac{1}{2i}[e^{i\omega\xi} - e^{-i\omega\xi}]\right) d\xi$$

$$= \frac{1}{2} \int_{-\infty}^\infty f(\xi)e^{-i\omega\xi}\, d\xi.$$

But

$$\overline{e^{-i\omega\xi}} = \overline{[\cos(\omega\xi) - i\sin(\omega\xi)]} = \cos(\omega\xi) + i\sin(\omega\xi) = e^{i\omega\xi}.$$

Further, $f(\xi)$ is real if ξ is real, so $\overline{f(\xi)} = f(\xi)$. Therefore,

$$\overline{C(\omega)} = \frac{1}{2} \int_{-\infty}^\infty \overline{f(\xi)}\,\overline{e^{-i\omega\xi}}\, d\xi = \frac{1}{2} \int_{-\infty}^\infty f(\xi)e^{i\omega\xi}\, d\xi = C(-\omega).$$

Upon substituting this conclusion into equation (17.11), we get

$$f(t) = \frac{1}{\pi} \int_0^\infty C(\omega)e^{i\omega t}\, d\omega + \frac{1}{\pi} \int_0^\infty C(-\omega)e^{-i\omega t}\, d\omega \qquad (\text{let } u = -\omega)$$

$$= \frac{1}{\pi} \int_0^\infty C(\omega)e^{i\omega t}\, d\omega - \frac{1}{\pi} \int_0^{-\infty} C(u)e^{iut}\, du$$

$$= \frac{1}{\pi} \int_0^\infty C(\omega)e^{i\omega t}\, d\omega + \frac{1}{\pi} \int_{-\infty}^0 C(\omega)e^{i\omega t}\, d\omega$$

$$= \frac{1}{\pi} \int_{-\infty}^\infty C(\omega)e^{i\omega t}\, d\omega.$$

The integral in the last line is the complex form of the Fourier integral of f. We will summarize this discussion in a definition.

DEFINITION   Complex Fourier Integral

The *complex Fourier integral* of f on $(-\infty, \infty)$ is

$$\frac{1}{\pi} \int_{-\infty}^\infty C(\omega)e^{i\omega t}\, d\omega,$$

in which

$$C(\omega) = \tfrac{1}{2} \int_{-\infty}^{\infty} f(\xi)e^{-i\omega\xi}\,d\xi. \quad \blacksquare$$

This complex Fourier integral representation will converge to the same values as the Fourier integral under the conditions of Theorem 17.10. This is due to the fact that we obtained the complex Fourier integral from the Fourier integral by manipulations which do not affect the convergence.

EXAMPLE 17.19

Find the complex Fourier integral of f if

$$f(t) = \begin{cases} e^t, & t < 0 \\ e^{-t}, & t \ge 0. \end{cases}$$

First, $\int_{-\infty}^{\infty} |f(t)|\,dt = 2$, and f is piecewise continuous on every finite interval. Calculate the coefficients:

$$\begin{aligned}
C(\omega) &= \frac{1}{2}\int_{-\infty}^{\infty} f(\xi)e^{-i\omega\xi}\,d\xi \\
&= \frac{1}{2}\int_{-\infty}^{0} e^{\xi}e^{-i\omega\xi}\,d\xi + \frac{1}{2}\int_{0}^{\infty} e^{-\xi}e^{-i\omega\xi}\,d\xi \\
&= \frac{1}{2}\int_{-\infty}^{0} e^{(1-i\omega)\xi}\,d\xi + \frac{1}{2}\int_{0}^{\infty} e^{-(1+i\omega)\xi}\,d\xi \\
&= \frac{\tfrac{1}{2}}{1-i\omega} e^{(1-i\omega)\xi}\Big]_{-\infty}^{0} - \frac{\tfrac{1}{2}}{1+i\omega} e^{-(1+i\omega)\xi}\Big]_{0}^{\infty}. \qquad (17.12)
\end{aligned}$$

At the infinite limits of integration, both of these quantities are zero. For example,

$$\begin{aligned}
e^{-(1+i\omega)\xi}\Big|_{\xi=\infty} &= \lim_{k\to\infty} e^{-(1+i\omega)k} \\
&= \lim_{k\to\infty} e^{-k}e^{-i\omega k} = \lim_{k\to\infty} e^{-k}[\cos(\omega k) - i\sin(\omega k)] \\
&= \lim_{k\to\infty} e^{-k}\cos(\omega k) - i\lim_{k\to\infty} e^{-k}\sin(\omega k) = 0,
\end{aligned}$$

and we can treat $e^{(1-i\omega)\xi}]_{-\infty}^{0}$ similarly. Continuing from equation (17.12), we have

$$C(\omega) = \frac{\tfrac{1}{2}}{1-i\omega} + \frac{\tfrac{1}{2}}{1+i\omega} = \frac{1}{1+\omega^2}.$$

The complex Fourier integral of f is

$$\frac{1}{\pi}\int_{-\infty}^{\infty} \frac{1}{1+\omega^2} e^{i\omega t}\,d\omega.$$

This integral converges to $f(t)$ for all t. $\blacksquare$

We are now ready to define the Fourier transform of a function f.

DEFINITION   Fourier Transform

Suppose that f satisfies the hypotheses of Theorem 17.10. Then the Fourier transform $\mathcal{F}$ of f is defined by

$$\mathcal{F}\{f(t)\} = \int_{-\infty}^{\infty} f(t)e^{-i\omega t}\, dt. \quad \blacksquare$$

Observe that $\mathcal{F}\{f(t)\}$ is a function of ω. As with the Laplace transform, we will usually denote the Fourier transform of a lowercase function by using the uppercase form of the same letter:

$$\mathcal{F}\{f(t)\} = F(\omega).$$

$F(\omega)$ may be complex valued because of the factor $e^{-i\omega t}$ in the integral defining the transform. For the Fourier transform, there is an inverse transform which is easily expressed as an integral similar to that defining $\mathcal{F}$. In fact, from the Fourier integral representation of f, we have

$$f(t) = \frac{1}{\pi}\int_{-\infty}^{\infty} C(\omega)e^{i\omega t}\, d\omega = \frac{1}{\pi}\int_{-\infty}^{\infty}\frac{1}{2}\int_{-\infty}^{\infty} f(\xi)e^{-i\omega\xi}\, d\xi\, e^{i\omega t}\, d\omega$$

$$= \frac{1}{2\pi}\int_{-\infty}^{\infty} F(\omega)e^{i\omega t}\, d\omega.$$

That is,

$$f(t) = \mathcal{F}^{-1}\{F(\omega)\} = \frac{1}{2\pi}\int_{-\infty}^{\infty} F(\omega)e^{i\omega t}\, d\omega.$$

The two transformations

$$\mathcal{F}\{f(t)\} = F(\omega) = \int_{-\infty}^{\infty} f(t)e^{-i\omega t}\, dt \qquad \text{(Fourier transform)}$$

and

$$\mathcal{F}^{-1}\{F(\omega)\} = f(t) = \frac{1}{2\pi}\int_{-\infty}^{\infty} F(\omega)e^{i\omega t}\, d\omega \qquad \text{(inverse Fourier transform)}$$

are called a *transform pair*.

Some care must be taken in using tables to evaluate these transforms. Some authors define the transform pair so that each has a coefficient of $\sqrt{\frac{1}{2\pi}}$. With this convention, and with the way we have written the definitions, the product of the coefficients of the Fourier transform and its inverse is $\frac{1}{2\pi}$.

The *amplitude spectrum* of f is the graph of the magnitude $|F(\omega)|$ of its Fourier transform versus the frequency variable ω. This amplitude spectrum is not the discrete set of lines we saw associated with the Fourier series representation of a function but is instead a continuous curve whose amplitude at any frequency ω measures the contribution of that frequency to the function. Just as the Fourier series decomposes a periodic function into a discrete set of contributions of various frequencies (all multiples of one fundamental frequency), the Fourier transform provides a continuous frequency resolution of a (possibly nonperiodic) function. We will exploit this fact in various uses of the Fourier transform in Section 17.11. Here is a computational example to illustrate these terms and concepts.

EXAMPLE 17.20

Let $a > 0$, and let

$$f(t) = \begin{cases} 0 & \text{if} \quad t < 0 \\ e^{-at} & \text{if} \quad t \geq 0. \end{cases}$$

This function is conveniently written in terms of the unit step function (Heaviside function) defined in Section 4.4

$$H(t) = \begin{cases} 0 & \text{if} \quad t < 0 \\ 1 & \text{if} \quad t \geq 0. \end{cases}$$

A graph of this function is shown in Figure 4.4. We may write

$$f(t) = H(t)e^{-at}.$$

First, we find that

$$\int_{-\infty}^{\infty} |f(t)|\, dt = \int_{0}^{\infty} e^{-at}\, dt = \frac{1}{a}.$$

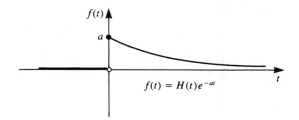

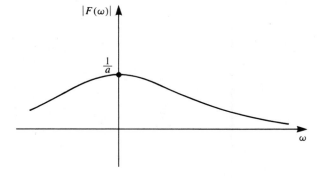

Figure 17.40. $|F(\omega)| = \dfrac{1}{\sqrt{a^2 + \omega^2}}$ [frequency spectrum of $f(t)$].

It is routine to verify that the Fourier integral of f converges to f. Therefore,

$$\mathcal{F}\{f(t)\} = \int_{-\infty}^{\infty} f(t)e^{-i\omega t}\, dt = \int_{0}^{\infty} e^{-at}e^{-i\omega t}\, dt$$

$$= \int_{0}^{\infty} e^{-(a+i\omega)t}\, dt = \frac{-1}{a+i\omega}\, e^{-(a+i\omega)t}\Big]_{0}^{\infty} = \frac{1}{a+i\omega} = F(\omega).$$

To determine the amplitude spectrum, calculate

$$|F(\omega)| = \frac{1}{|a+i\omega|} = \frac{1}{\sqrt{a^2+\omega^2}}.$$

The graph of f and its frequency spectrum are shown in Figure 17.40. ∎

EXAMPLE 17.21

Let a and k be positive numbers, and let

$$f(t) = \begin{cases} k & \text{if} \quad\quad -a < t < \quad a \\ 0 & \text{if} \quad t > a \quad \text{or} \quad t < -a. \end{cases}$$

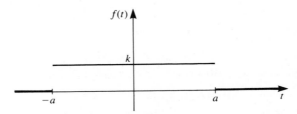

Pulse function:
$$f(t) = k[H(t+a) - H(t-a)]$$

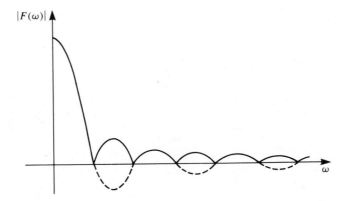

Frequency spectrum:
$$|F(\omega)| = \left|\frac{2k\sin(a\omega)}{\omega}\right|$$

Figure 17.41

In terms of the Heaviside function,

$$f(t) = kH(t + a) - kH(t - a)$$

except at $t = \pm a$. Such a function is called a *pulse function*. (Also note Example 4.19).

We find that f satisfies the requirements to have a Fourier transform. Compute

$$\mathscr{F}\{f(t)\} = \int_{-\infty}^{\infty} f(t)e^{-i\omega t}\, dt = k \int_{-a}^{a} e^{-i\omega t}\, dt = \frac{ke^{-i\omega t}}{-i\omega}\bigg]_{-a}^{a}$$

$$= \frac{k}{\omega}\frac{e^{-i\omega a} - e^{i\omega a}}{-i} = \frac{2k}{\omega}\frac{e^{i\omega a} - e^{-i\omega a}}{2i} = \frac{2k\,\sin(\omega a)}{\omega} = F(\omega).$$

Since ω is real,

$$|F(\omega)| = \left|\frac{2k\,\sin(\omega a)}{\omega}\right| = \frac{2k}{\omega}|\sin(\omega a)|.$$

This function is the amplitude spectrum of f. A graph of f and its amplitude spectrum are shown in Figure 17.41. ■

We will now develop some of the properties needed to use the Fourier transform.

THEOREM 17.11   Linearity

If α and β are real numbers,

$$\mathscr{F}\{\alpha f(t) + \beta g(t)\} = \alpha F(\omega) + \beta G(\omega),$$

assuming that the Fourier transforms of f and g exist. Similarly,

$$\mathscr{F}^{-1}\{\alpha F(\omega) + \beta G(\omega)\} = \alpha\mathscr{F}^{-1}\{F(\omega)\} + \beta\mathscr{F}^{-1}\{G(\omega)\}. \quad ■$$

THEOREM 17.12   Time Shifting

If $\mathscr{F}\{f(t)\} = F(\omega)$, then $\mathscr{F}\{f(t - t_0)\} = e^{-i\omega t_0}F(\omega)$.

That is, the Fourier transform of the translation of f by t_0 is obtained by multiplying the Fourier transform of f by $e^{-i\omega t_0}$. This result is similar to shifting in the time variable with the Laplace transform (Theorem 4.11).

Proof

$$\mathscr{F}\{f(t - t_0)\} = \int_{-\infty}^{\infty} f(t - t_0)e^{-i\omega t}\, dt$$

$$= e^{-i\omega t_0} \int_{-\infty}^{\infty} f(t - t_0)e^{-i\omega(t - t_0)}\, dt$$

$$= e^{-i\omega t_0} \int_{-\infty}^{\infty} f(x)e^{-i\omega x}\, dx \qquad (x = t - t_0)$$

$$= e^{-i\omega t_0}F(\omega). \quad ■$$

As with time-shifting with the Laplace transform, the key to using Theorem 17.12 usually lies in writing the functions involved in terms of $t - t_0$.

EXAMPLE 17.22

Determine the Fourier transform of the pulse which turns on at time 3, maintains an amplitude of 6, and shuts off at time 7.

Call this pulse g. The graph of g is shown in Figure 17.42. Notice that g is similar in form to the pulse f of Example 17.21 (shown in Figure 17.41). Letting $k = 6$ and $a = 2$ in Example 17.21, the pulse of Figure 17.41 is exactly the pulse of Figure 17.42 shifted five units to the right. That is,

$$g(t) = f(t - 5).$$

By the time-shifting theorem and the result of Example 17.21,

$$\mathscr{F}\{g(t)\} = \mathscr{F}\{f(t - 5)\}$$

$$= e^{-5i\omega}\mathscr{F}\{f(t)\} = e^{-5i\omega}\frac{12\sin(2\omega)}{\omega}.$$

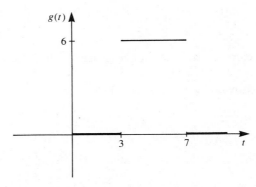

Figure 17.42   ■

The next result is reminiscent of the first shifting thoerem (Theorem 4.9) for the Laplace transform.

THEOREM 17.13   Frequency Shifting

If $\mathscr{F}\{f(t)\} = F(\omega)$, then $\mathscr{F}\{e^{i\omega_0 t}f(t)\} = F(\omega - \omega_0)$.

Proof

$$\mathscr{F}\{e^{i\omega_0 t}f(t)\} = \int_{-\infty}^{\infty} f(t)e^{i\omega_0 t}e^{-i\omega t}\,dt$$

$$= \int_{-\infty}^{\infty} f(t)e^{i(\omega - \omega_0)t}\,dt = F(\omega - \omega_0). \quad ■$$

EXAMPLE 17.23

Find

$$\mathscr{F}^{-1}\left\{\frac{4e^{(2\omega - 6)i}}{5 - (3 - \omega)i}\right\}.$$

Using both shifting theorems, we can write

$$\mathscr{F}^{-1}\left\{\frac{4e^{(2\omega-6)i}}{5-(3-\omega)i}\right\} = e^{3it}\mathscr{F}^{-1}\left\{\frac{4e^{2i\omega}}{5+i\omega}\right\}$$

$$= e^{3it}\left[\mathscr{F}^{-1}\left\{\frac{4}{5+i\omega}\right\}\right]_{t\to t+2} = e^{3it}[4H(t)e^{-5t}]_{t\to t+2}$$

$$= e^{3it}4H(t+2)e^{-5(t+2)}$$

$$= 4H(t+2)e^{-10-(5-3i)t}.$$

Without the Heaviside notation, this result is

$$\begin{cases} 4e^{-10-(5-3i)t} & \text{if } t \geq -2 \\ 0 & \text{if } t < -2. \end{cases} \blacksquare$$

THEOREM 17.14 Scaling

If $\mathscr{F}\{f(t)\} = F(\omega)$ and a is a real, nonzero constant,

$$\mathscr{F}\{f(at)\} = \frac{1}{|a|}F\left(\frac{\omega}{a}\right). \blacksquare$$

The proof is a straightforward calculation, which we omit.

The scaling theorem points out an interesting fact about signals or functions in general. Those functions which change rapidly, such as the Dirac delta function or the Heaviside step function, contain a wide spectrum of especially high frequencies, while the smooth functions contain a narrow band of frequencies. We will return to this idea in Section 17.11.

THEOREM 17.15 Time Reversal

If $\mathscr{F}\{f(t)\} = F(\omega)$, then $\mathscr{F}\{f(-t)\} = F(-\omega)$. $\blacksquare$

This conclusion follows from the scaling property upon putting $a = -1$.

THEOREM 17.16 Symmetry

If $\mathscr{F}\{f(t)\} = F(\omega)$, then $\mathscr{F}\{F(t)\} = 2\pi f(-\omega)$.

That is, if we replace ω by t in the Fourier transform F, forming $F(t)$, the Fourier transform of this function is 2π times the original function, with t replaced by $-\omega$.

Proof Working directly from the integral for the inverse Fourier transform, we have

$$f(t) = \frac{1}{2\pi}\int_{-\infty}^{\infty} F(\omega)e^{i\omega t}\,d\omega = \frac{1}{2\pi}\int_{-\infty}^{\infty} F(x)e^{ixt}\,dx.$$

Then

$$2\pi f(-\omega) = \int_{-\infty}^{\infty} F(x)e^{-ix\omega}\,dx = \int_{-\infty}^{\infty} F(t)e^{-it\omega}\,dt = \mathscr{F}\{F(t)\}. \blacksquare$$

Symmetry is an extremely useful property of the Fourier transform. Here is an illustration.

EXAMPLE 17.24

Find the Fourier transform of f, where $f(t) = 5/(4 + it)$.

This is difficult to do directly from the integral defining the Fourier transform. However, this function looks like the transformation of $5H(t)e^{-4t}$ found in Example 17.20, with ω replaced by t and $a = 4$. Thus, let $g(t) = 5H(t)e^{-4t}$. From Example 17.20, we have

$$\mathscr{F}\{g(t)\} = \frac{5}{4 + i\omega} = G(\omega).$$

From the symmetry property,

$$\mathscr{F}\{G(t)\} = \mathscr{F}\left\{\frac{5}{4 + it}\right\} = 2\pi g(-\omega) = 10\pi H(-\omega)e^{4\omega}.$$

We conclude that

$$\mathscr{F}\{f(t)\} = \begin{cases} 10\pi e^{4\omega} & \text{if} \quad \omega \leq 0 \\ 0 & \text{if} \quad \omega > 0. \end{cases} \quad \blacksquare$$

THEOREM 17.17   Modulation

If $\mathscr{F}\{f(t)\} = F(\omega)$,

$$\mathscr{F}\{f(t)\cos(\omega_0 t)\} = \tfrac{1}{2}[F(\omega - \omega_0) + F(\omega + \omega_0)]$$

and

$$\mathscr{F}\{f(t)\sin(\omega_0 t)\} = \frac{i}{2}[F(\omega + \omega_0) - F(\omega - \omega_0)].$$

Proof   Use the frequency-shifting theorem, together with the exponential form of the cosine function, to write

$$\begin{aligned}
\mathscr{F}\{f(t)\cos(\omega_0 t)\} &= \mathscr{F}\{f(t)\tfrac{1}{2}(e^{i\omega_0 t} + e^{-i\omega_0 t})\} \\
&= \tfrac{1}{2}\mathscr{F}\{e^{i\omega_0 t}f(t)\} + \tfrac{1}{2}\mathscr{F}\{e^{-i\omega_0 t}f(t)\} \\
&= \tfrac{1}{2}F(\omega - \omega_0) + \tfrac{1}{2}F(\omega + \omega_0).
\end{aligned}$$

A proof of the second equation follows in like fashion.   ■

The conclusion of the modulation theorem is similar in form to formulas for the finite Fourier sine and cosine transforms (Problems 19 and 20 of Section 17.5). Here is an example of the use of this theorem.

EXAMPLE 17.25

Consider a pulse-modulated radar in which a high-frequency signal is switched on for a short time (relative to the time between pulses). Find the amplitude spectrum of the pulse.

The pulse is modeled by the function

$$f(t) = \begin{cases} k\cos(\omega_0 t) & \text{if} \qquad -a < t < \quad a \\ 0 & \text{if} \quad t > a \quad \text{or} \quad t < -a. \end{cases}$$

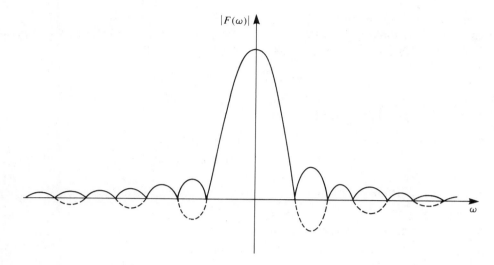

Figure 17.43. $|F(\omega)| = |k| \left| \dfrac{\sin[a(\omega - \omega_0)]}{\omega - \omega_0} + \dfrac{\sin[a(\omega + \omega_0)]}{\omega + \omega_0} \right|$.

(The actual radar output is a periodic function, but we can make the assumption that the period is much larger than a.)

First, notice that f is the product of the pulse function of Example 17.21, with $\cos(\omega_0 t)$. Hence, use the result of Example 17.21, in conjunction with the modulation theorem, to get

$$\mathscr{F}\{f(t)\} = \tfrac{1}{2}F(\omega - \omega_0) + \tfrac{1}{2}F(\omega + \omega_0).$$

Since $F(\omega) = [2k \sin(\omega a)]/\omega$,

$$\mathscr{F}\{f(t)\} = F(\omega) = \frac{k \sin[a(\omega - \omega_0)]}{\omega - \omega_0} + \frac{k \sin[a(\omega + \omega_0)]}{\omega + \omega_0}.$$

The amplitude spectrum of this modulation is shown in Figure 17.43. ■

We will conclude this section with a list of additional results which are sometimes of use. The derivations are left to the student. Assume that $a > 0$. Then

1. $\mathscr{F}\{e^{-a|t|}\} = \dfrac{2a}{a^2 + \omega^2}$.

2. $\mathscr{F}\{e^{-at^2}\} = \sqrt{\dfrac{\pi}{a}}\, e^{-\omega^2/4a}$.

3. $\mathscr{F}\left\{\dfrac{1}{a^2 + t^2}\right\} = \dfrac{\pi}{a} e^{-a|\omega|}$ [use (1) and the symmetry property].

4. $\mathscr{F}\{e^{-t^2/a}\} = \sqrt{\pi a}\, e^{-a\omega^2/4}$ [use (2) and the symmetry property].

For the remaining properties, let f be a real-valued function. Define

$$f_e(t) = \tfrac{1}{2}[f(t) + f(-t)] \quad \text{and} \quad f_o(t) = \tfrac{1}{2}[f(t) - f(-t)].$$

Then f_e is an even function and f_o is an odd function. Further, $f = f_e + f_o$. Hence, any real-valued function can be written as a sum of an even function and an odd function. In general, $F(\omega) = \mathscr{F}\{f(t)\}$ will be complex valued. Write

$$F(\omega) = R(\omega) + iX(\omega),$$

with $R(\omega)$ and $X(\omega)$ both real-valued functions of ω.

We can now continue the list of results.

5. $R(\omega) = \int_{-\infty}^{\infty} f(t)\cos(\omega t)\, dt$ $[= A(\omega)$, a Fourier integral coefficient].

6. $X(\omega) = -\int_{-\infty}^{\infty} f(t)\sin(\omega t)\, dt$ $[= -B(\omega)$, the negative of a Fourier integral coefficient].

7. $\mathscr{F}\{f_e(t)\} = R(\omega)$.

8. $\mathscr{F}\{f_o(t)\} = iX(\omega)$.

9. $R(-\omega) = R(\omega)$ (R is an even function).

10. $X(-\omega) = -X(\omega)$ (X is an odd function).

11. $F(-\omega) = \overline{F(\omega)}$.

In the next section, we will develop operational properties of the Fourier transform which we will use in solving boundary value problems.

PROBLEMS FOR SECTION 17.9

In each of Problems 1 through 16, find the Fourier transform of the function and graph the frequency spectrum.

1. $f(t) = \begin{cases} 1, & 0 \le t \le 1 \\ -1, & -1 \le t < 0 \\ 0, & |t| > 1 \end{cases}$

2. $f(t) = \begin{cases} \sin(t), & -K \le t \le K \\ 0, & |t| > K \end{cases}$

3. $f(t) = 5[H(t-3) - H(t-11)]$

4. $f(t) = 5e^{-3(t-5)^2}$

5. $f(t) = \begin{cases} \cosh(t), & -K \le t \le K \\ 0, & |t| > K \end{cases}$

6. $f(t) = \begin{cases} t^2, & -K \le t \le K \\ 0, & |t| > K \end{cases}$

7. $f(t) = \dfrac{1}{1+t^2}$

8. $f(t) = 3H(t-2)e^{-3t}$

9. $f(t) = 3e^{-4|t+2|}$

10. $f(t) = 4H(t-3)e^{-2t}$

11. $f(t) = 3e^{-4|t|}\cos(2t)$

12. $f(t) = 8e^{-2t^2}\sin(3t)$

13. $f(t) = \dfrac{\sin(3t)}{4+t^2}$

14. $f(t) = \dfrac{1}{t^2 + 6t + 13}$

15. $f(t) = k[H(t-a) - H(t-b)], \quad a < b$

16. $f(t) = 4H(t-2)e^{-3t}\cos(t-2)$

In each of Problems 17 through 22, find the inverse Fourier transform of the function.

17. $9e^{-(\omega+4)^2/32}$

18. $\dfrac{e^{(20-4\omega)i}}{3 - (5-\omega)i}$

19. $\dfrac{4e^{(2\omega-6)i}}{5 - (3-\omega)i}$

20. $\dfrac{10\sin(3\omega)}{\omega + \pi}$

21. $\dfrac{1+\omega i}{6 - \omega^2 + 5\omega i}$.   Hint: Factor the denominator and use partial fractions.

22. $\dfrac{10(4+\omega i)}{9 - \omega^2 + 8\omega i}$.   Hint: Factor the denominator.

23. Derive equation (1) at the end of this section. *Hint:* Write $e^{-a|t|} = H(t)e^{-at} + H(-t)e^{at}$. To evaluate $\mathcal{F}\{H(-t)e^{at}\}$, notice that, if $f(t) = H(t)e^{-at}$, $H(-t)e^{at} = f(-t)$, and use the time reversal property of the Fourier transform.

24. Derive equation (2). *Hint:* Let $F(\omega) = \mathcal{F}\{e^{-at^2}\}$ and compute $F'(\omega) = -i\int_{-\infty}^{\infty} te^{-at^2}e^{-i\omega t}\, dt$. Integrate by parts using $u = e^{-i\omega t}$ and $dv = te^{-at^2}\, dt$. Use the fact that a is positive to obtain the differential equation $F'(\omega) = -(\omega/2a)F(\omega)$. Solve this separable differential equation and use the fact that $\int_0^{\infty} e^{-x^2}\, dx = \frac{1}{2}\sqrt{\pi}$ to determine the constant of integration.

17.10 *Additional Properties of the Fourier Transform*

In this section, we will derive some of the operational formulas needed to use the Fourier transform to solve differential equations. As usual, $f^{(n)}$ denotes the nth derivative of f, with the convention that $f^{(0)} = f$.

THEOREM 17.18   Differentiation in the Time Variable

Let n be a positive integer. Suppose that $f^{(n)}$ is piecewise continuous on every interval $[-a, a]$ and that $\int_{-\infty}^{\infty} |f^{(n-1)}(t)|\, dt$ converges. Assume that

$$\lim_{t \to \infty} f^{(k)}(t) = \lim_{t \to -\infty} f^{(k)}(t) = 0$$

for $k = 0, 1, 2, \ldots, n - 1$. Finally, let $\mathcal{F}\{f(t)\} = F(\omega)$. Then

$$\mathcal{F}\{f^{(n)}(t)\} = (i\omega)^n F(\omega).$$

In particular,

$$\mathcal{F}\{f'(t)\} = i\omega F(\omega) \quad \text{and} \quad \mathcal{F}\{f''(t)\} = -\omega^2 F(\omega).$$

Proof   We will prove the theorem for the case $n = 1$; the general result can be proved by mathematical induction. The hypotheses of the theorem ensure existence of the Fourier transform of f'. Integrate by parts to get

$$\mathcal{F}\{f'(t)\} = \int_{-\infty}^{\infty} f'(t)e^{-i\omega t}\, dt \qquad [dv = f'(t)\, dt, u = e^{-i\omega t}]$$

$$= f(t)e^{-i\omega t}\Big]_{-\infty}^{\infty} - \int_{-\infty}^{\infty} f(t)(-i\omega)e^{-i\omega t}\, dt.$$

Since $f^{(0)} = f$, we have by assumption that

$$\lim_{t \to \infty} f(t) = \lim_{t \to -\infty} f(t) = 0.$$

Further, $|e^{-i\omega t}| = |\cos(\omega t) - i\sin(\omega t)| = 1$ for all real ω and t. Thus,

$$f(t)e^{-i\omega t}\Big]_{-\infty}^{\infty} = 0,$$

and the integration by parts yields

$$\mathscr{F}\{f'(t)\} = i\omega \int_{-\infty}^{\infty} f(t)e^{-i\omega t}\, dt = i\omega F(\omega),$$

as was to be proved.   ∎

If f has a jump discontinuity, as often occurs in applications, additional terms must be added to the formula in this theorem. The next theorem specifies what these terms are if we are dealing with just the first derivative f'.

THEOREM 17.19   Differentiation in the Time Variable for Discontinuous f

Suppose that f satisfies the hypotheses of Theorem 17.10 and that $f(t) \to 0$ as $t \to \infty$ and also as $t \to -\infty$. Suppose that f is continuous except at finitely many points t_k for $k = 1, 2, \ldots, M$ and that f has jump discontinuities at these points. Let

$$a_k = f(t_k+) - f(t_k-)$$

(this is the size of the jump at t_k). Finally, let $\mathscr{F}\{f(t)\} = F(\omega)$. Then

$$\mathscr{F}\{f'(t)\} = i\omega F(\omega) - \sum_{k=1}^{M} a_k e^{-i\omega t_k}.$$

Proof   We will suppose that f has just one jump discontinuity, at t_0, with $a = f(t_0+) - f(t_0-)$. In the case in which f has more discontinuities, the argument simply involves more of the kind of calculation we will now do. Integrate by parts to get

$$\mathscr{F}\{f'(t)\} = \int_{-\infty}^{\infty} f'(t)e^{-i\omega t}\, dt = \int_{-\infty}^{t_0} f'(t)e^{-i\omega t}\, dt + \int_{t_0}^{\infty} f'(t)e^{-i\omega t}\, dt$$

$$= f(t)e^{-i\omega t}\Big]_{-\infty}^{t_0} + i\omega \int_{-\infty}^{t_0} f(t)e^{-i\omega t}\, dt$$

$$+ f(t)e^{-i\omega t}\Big]_{t_0}^{\infty} + i\omega \int_{t_0}^{\infty} f(t)e^{-i\omega t}\, dt$$

$$= \lim_{t \to t_0-} f(t)e^{-i\omega t} + i\omega \int_{-\infty}^{t_0} f(t)e^{-i\omega t}\, dt$$

$$- \lim_{t \to t_0+} f(t)e^{-i\omega t} + i\omega \int_{t_0}^{\infty} f(t)e^{-i\omega t}\, dt$$

$$= f(t_0-)e^{-i\omega t_0} - f(t_0+)e^{-i\omega t_0} + i\omega \int_{-\infty}^{\infty} f(t)e^{-i\omega t}\, dt$$

$$= -[f(t_0+) - f(t_0-)]e^{-i\omega t_0} + i\omega F(\omega)$$

$$= i\omega F(\omega) - ae^{-i\omega t_0}.   ∎$$

EXAMPLE 17.26

Solve the first order differential equation

$$y' - 4y = H(t)e^{-4t}, \qquad -\infty < t < \infty.$$

Apply the Fourier transform to the differential equation to get

$$\mathscr{F}\{y'\} - 4\mathscr{F}\{y\} = \mathscr{F}\{H(t)e^{-4t}\}.$$

Write $\mathscr{F}\{y\} = Y(\omega)$ to obtain

$$i\omega Y(\omega) - 4Y(\omega) = \frac{1}{4 + i\omega}.$$

Solve this equation for $Y(\omega)$ to get

$$Y(\omega) = \frac{-1}{(4 + i\omega)(4 - i\omega)} = \frac{-1}{16 + \omega^2} = \mathscr{F}\left\{-\frac{1}{8}e^{-4|t|}\right\},$$

by equation (3) at the end of the preceding section. We conclude that

$$y(t) = -\tfrac{1}{8}e^{-4|t|}$$

is a solution for $-\infty < t < \infty$. We can also write this solution as

$$y(t) = \begin{cases} -\tfrac{1}{8}e^{-4t} & \text{if } t \geq 0 \\ -\tfrac{1}{8}e^{4t} & \text{if } t < 0. \end{cases} \quad \blacksquare$$

In this example, we obtained a solution with no arbitrary constant even though the problem had no initial condition. The reason is that the transform has produced the only solution which is *continuous and bounded* on the whole real line. If we solve the differential equation using an integrating factor, we obtain two arbitrary constants; one of these must be set equal to zero to get a bounded solution, and the other must be set equal to $-\tfrac{1}{8}$ to get a continuous solution. In applying the Fourier transform, we tacitly assume a bounded solution, because we assume that $\int_{-\infty}^{\infty} |y(t)|\, dt$ converges.

There are other solutions of $y' - 4y = H(t)e^{-4t}$. For example,

$$y(t) = -\tfrac{1}{8}H(t)e^{-4t}$$

is a solution, but not a continuous one.

THEOREM 17.20   Differentiation in the Frequency Variable

Let n be a positive integer. Suppose that f satisfies the hypotheses of Theorem 17.10 and that $\int_{-\infty}^{\infty} |t^n f(t)|\, dt$ converges. Let $\mathscr{F}\{f(t)\} = F(\omega)$. Then

$$\mathscr{F}\{t^n f(t)\} = i^n F^{(n)}(\omega).$$

In particular,

$$\mathscr{F}\{tf(t)\} = iF'(\omega) \quad \text{and} \quad \mathscr{F}\{t^2 f(t)\} = -F''(\omega).$$

Proof   We will prove the theorem for $n = 1$; we essentially repeat this argument for larger n. Apply Leibniz's rule to differentiate through the integral sign, obtaining

$$F'(\omega) = \frac{d}{d\omega} \int_{-\infty}^{\infty} f(t)e^{-i\omega t}\, dt = \int_{-\infty}^{\infty} \frac{\partial}{\partial\omega}[f(t)e^{-i\omega t}]\, dt$$

$$= \int_{-\infty}^{\infty} f(t)(-it)e^{-i\omega t}\, dt = -i\int_{-\infty}^{\infty} [tf(t)]e^{-i\omega t}\, dt = -i\mathscr{F}\{tf(t)\}.$$

To complete the proof for $n = 1$, multiply both sides of the equation $F'(\omega) = -i\mathscr{F}\{tf(t)\}$ by i. ∎

EXAMPLE 17.27

Let $a > 0$. Determine $\mathscr{F}\{te^{-at^2}\}$.

Theorem 17.20 applies. From equation (4) at the end of the preceding section, we know that

$$\mathscr{F}\{e^{-at^2}\} = \sqrt{\frac{\pi}{a}}\, e^{-\omega^2/4a}.$$

Therefore,

$$\mathscr{F}\{te^{-at^2}\} = i\frac{d}{d\omega}\sqrt{\frac{\pi}{a}}\, e^{-\omega^2/4a} = -\frac{i\omega}{2a}\sqrt{\frac{\pi}{a}}\, e^{-\omega^2/4a}. \quad ∎$$

THEOREM 17.21

Suppose that f satisfies the hypotheses of Theorem 17.10 and that $F(0) = 0$, where $F(\omega) = \mathscr{F}\{f(t)\}$. Then

$$\mathscr{F}\left\{\int_{-\infty}^{t} f(\tau)\, d\tau\right\} = \frac{1}{i\omega} F(\omega).$$

Proof   Let $g(t) = \int_{-\infty}^{t} f(\tau)\, d\tau$. Then $g'(t) = f(t)$ wherever f is continuous. By Theorem 17.10,

$$F(\omega) = \mathscr{F}\{f(t)\} = \mathscr{F}\{g'(t)\} = i\omega\mathscr{F}\{g(t)\} = i\omega\mathscr{F}\left\{\int_{-\infty}^{t} f(\tau)\, d\tau\right\},$$

provided that $g(t) \to 0$ as $t \to \pm\infty$. Then

$$\lim_{t\to-\infty} g(t) = \lim_{t\to-\infty}\int_{-\infty}^{t} f(\tau)\, d\tau = \int_{-\infty}^{-\infty} f(\tau)\, d\tau = 0.$$

However, to show that $g(t) \to 0$ as $t \to +\infty$, we will use the assumption that $F(0) = 0$. In fact,

$$\lim_{t\to\infty} g(t) = \lim_{t\to\infty}\int_{-\infty}^{t} f(\tau)\, d\tau = \int_{-\infty}^{\infty} f(\tau)\, d\tau = F(0) = 0.$$

This completes the proof of the theorem. ∎

The assumption that $F(0) = 0$ can be omitted in this theorem, but then we must use the Dirac delta function. We will now consider a convolution operation for the Fourier transform which is reminiscent of the convolution for the Laplace transform.

DEFINITION

If f and g both satisfy the hypotheses of Theorem 17.10, the convolution of f and g is denoted $f * g$ and is defined by

$$[f * g](t) = \int_{-\infty}^{\infty} f(\tau)g(t - \tau)\, d\tau. \quad ∎$$

The change of variables $z = t - \tau$ can be used to show that the convolution is commutative:

$$f * g = g * f.$$

Here is the fundamental theorem on convolution for the Fourier transform.

THEOREM 17.22

Suppose that f and g both satisfy the hypotheses of Theorem 17.10, and let $\mathscr{F}\{f(t)\} = F(\omega)$ and $\mathscr{F}\{g(t)\} = G(\omega)$. Then

$$\mathscr{F}\{[f * g](t)\} = F(\omega)G(\omega) \qquad \text{(time convolution)}$$

and

$$\mathscr{F}\{f(t)g(t)\} = \frac{1}{2\pi}[F * G](\omega) \qquad \text{(frequency convolution)}.$$

Proof   We will prove the time convolution formula. The frequency convolution is left to the student. Compute

$$\mathscr{F}\{[f * g](t)\} = \int_{-\infty}^{\infty} \left[\int_{-\infty}^{\infty} f(\tau)g(t - \tau) \, d\tau \right] e^{-i\omega t} \, dt$$

$$= \int_{-\infty}^{\infty} \int_{-\infty}^{\infty} f(\tau)g(t - \tau)e^{-i\omega t} \, d\tau \, dt$$

$$= \int_{-\infty}^{\infty} \int_{-\infty}^{\infty} f(\tau)g(t - \tau)e^{-i\omega t} \, dt \, d\tau$$

$$= \int_{-\infty}^{\infty} f(\tau) \left[\int_{-\infty}^{\infty} g(t - \tau)e^{-i\omega t} \, dt \right] d\tau.$$

The bracketed term in the last line is the Fourier transform of $g(t - \tau)$. Apply the time-shifting theorem to this term to get

$$\int_{-\infty}^{\infty} g(t - \tau)e^{-i\omega t} \, dt = \mathscr{F}\{g(t - \tau)\} = e^{-i\omega\tau}\mathscr{F}\{g(t)\} = e^{-i\omega\tau}G(\omega).$$

Therefore,

$$\mathscr{F}\{[f * g](t)\} = \int_{-\infty}^{\infty} f(\tau)[G(\omega)e^{-i\omega\tau}] \, d\tau$$

$$= G(\omega) \int_{-\infty}^{\infty} f(\tau)e^{-i\omega\tau} \, d\tau = F(\omega)G(\omega). \quad \blacksquare$$

EXAMPLE 17.28

Determine

$$\mathscr{F}^{-1}\left\{ \frac{5}{2 - \omega^2 + 3i\omega} \right\}.$$

Begin by factoring

$$2 - \omega^2 + 3i\omega = (2 + i\omega)(1 + i\omega).$$

Then

$$\mathcal{F}^{-1}\left\{\frac{5}{2 - \omega^2 + 3i\omega}\right\} = \mathcal{F}^{-1}\left\{\left[\frac{5}{2 + i\omega}\right]\left[\frac{1}{1 + i\omega}\right]\right\}$$

$$= \mathcal{F}^{-1}\{5\mathcal{F}\{H(t)e^{-2t}\}\mathcal{F}\{H(t)e^{-t}\}\}$$

$$= \mathcal{F}^{-1}\{\mathcal{F}\{5H(t)e^{-2t} * H(t)e^{-t}\}\}$$

$$= 5H(t)e^{-2t} * H(t)e^{-t}$$

$$= 5\int_{-\infty}^{\infty} H(\tau)e^{-2\tau}H(t - \tau)e^{-(t-\tau)}\,d\tau$$

$$= 5\int_{-\infty}^{\infty} e^{-t}e^{-\tau}H(\tau)H(t - \tau)\,d\tau$$

$$= 5e^{-t}\int_{-\infty}^{\infty} e^{-\tau}H(\tau)H(t - \tau)\,d\tau.$$

Now,

$$H(\tau)H(t - \tau) = \begin{cases} 0 & \text{if} \quad \tau < 0 \quad \text{or} \quad \tau > t \\ 1 & \text{if} \qquad\qquad 0 < \tau < t. \end{cases}$$

Continuing the above calculation, we have

$$\mathcal{F}^{-1}\left\{\frac{5}{2 - \omega^2 + 3i\omega}\right\} = \begin{cases} 5e^{-t}\displaystyle\int_0^t e^{-\tau}\,d\tau, \text{ if } t \geq 0 \\ 0, \quad t < 0 \end{cases}$$

$$= 5e^{-t}H(t)\left[-e^{-\tau}\right]_0^t = 5e^{-t}H(t)[1 - e^{-t}]$$

$$= 5[e^{-t} - e^{-2t}]H(t). \quad \blacksquare$$

The Dirac delta function plays an important role in circuit analysis, so we will discuss some facts about its Fourier transform. We will omit formal proofs. Recall from Section 4.7 that, if $a > 0$,

$$\delta(t) = \lim_{a \to 0} \frac{1}{2a}[H(t + a) - H(t - a)].$$

Assuming that we can interchange this limit and the Fourier transform, we have

$$\mathcal{F}\{\delta(t)\} = \mathcal{F}\left\{\lim_{a \to 0} \frac{1}{2a}[H(t + a) - H(t - a)]\right\}$$

$$= \lim_{a \to 0} \frac{1}{2a}\,\mathcal{F}\{H(t + a) - H(t - a)\} = \lim_{a \to 0} \frac{1}{2a}\frac{2\sin(a\omega)}{\omega}$$

$$= \lim_{a \to 0} \frac{\sin(a\omega)}{a\omega} = 1.$$

Thus, the Fourier transform of the Dirac delta function is the same as the Laplace transform of this function.

The Dirac function plays the role of a multiplicative identity with respect to convolution. That is,

$$\delta * f = f * \delta = f.$$

To prove this, write

$$\mathscr{F}\{[\delta * f](t)\} = \mathscr{F}\{\delta(t)\}\mathscr{F}\{f(t)\} = 1 \cdot \mathscr{F}\{f(t)\} = \mathscr{F}\{f(t)\}.$$

This implies that $\delta * f = f$. Since $*$ is commutative, $f * \delta = f$ also.

The Dirac function also has a filtering property. If $\mathscr{F}\{f(t)\}$ exists and f is continuous at t_0, a routine calculation gives us

$$f(t_0) = \int_{-\infty}^{\infty} \delta(\tau - t_0)f(\tau)\,d\tau.$$

If f has a jump discontinuity at t_0, this filtering formula must be modified. Assuming that f has left and right derivatives at t_0, we obtain

$$\tfrac{1}{2}[f(t_0+) - f(t_0-)] = \int_{-\infty}^{\infty} \delta(\tau - t_0)f(\tau)\,d\tau.$$

If we apply the time-shifting theorem to δ, we get

$$\mathscr{F}\{\delta(t - t_0)\} = e^{-i\omega t_0}\mathscr{F}\{\delta(t)\} = e^{-i\omega t_0}.$$

From the symmetry property of the Fourier transform and the fact that $\mathscr{F}\{\delta(t)\} = 1$, we also have

$$\mathscr{F}\{1\} = 2\pi\,\delta(-\omega) = 2\pi\,\delta(\omega). \tag{17.13}$$

Here, we have ignored the integrability hypothesis, since $\int_{-\infty}^{\infty} 1\,dt$ does not converge.

Using (17.13) and the frequency-shifting theorem, we get

$$\mathscr{F}\{e^{i\omega_0 t}\} = \mathscr{F}\{1\}_{\omega \to \omega - \omega_0} = 2\pi\,\delta(\omega)_{\omega \to \omega - \omega_0} = 2\pi\,\delta(\omega - \omega_0).$$

If we combine equation (17.13) with the modulation properties of the Fourier transform, we get

$$\mathscr{F}\{\cos(\omega_0 t)\} = \pi[\delta(\omega + \omega_0) + \delta(\omega - \omega_0)]$$

and

$$\mathscr{F}\{\sin(\omega_0 t)\} = \pi i[\delta(\omega + \omega_0) - \delta(\omega - \omega_0)].$$

In the next section, we will use some of these ideas in a variety of applications of the Fourier transform.

PROBLEMS FOR SECTION 17.10

In each of Problems 1 through 10, determine the Fourier transform of the function.

1. $\dfrac{t}{9 + t^2}$

2. $\dfrac{d}{dt}[e^{-3t^2}]$

3. $26H(t)te^{-2t}$

4. $\dfrac{t}{a^2 + t^2}$,   with $a > 0$; express the answer in terms of $\mathrm{sgn}(\omega)$, where

$$\mathrm{sgn}(\omega) = \begin{cases} 1 & \text{if} \quad \omega > 0 \\ -1 & \text{if} \quad \omega < 0. \end{cases}$$

5. $H(t - 3)(t - 3)e^{-4t}$

6. $3te^{-9t^2}$

7. $\dfrac{d}{dt}[H(t)e^{-3t}]$

8. $\delta(t - t_0)$

9. $t[H(t + 1) - H(t - 1)]$

10. $\dfrac{5e^{3it}}{t^2 - 4t + 13}$

In each of Problems 11 through 13, use the convolution theorem to find the inverse Fourier transform of the function.

11. $\dfrac{1}{(1 + i\omega)^2}$

12. $\dfrac{1}{(1 + i\omega)(2 + i\omega)}$

13. $\dfrac{\sin(3\omega)}{\omega(2 + i\omega)}$

In each of Problems 14 through 16, find the inverse Fourier transform of the function.

14. $\dfrac{6e^{4i\omega}\sin(2\omega)}{9 + \omega^2}$

15. $e^{-3|\omega + 4|}\cos(2\omega + 8)$

16. $e^{-\omega^2/9}\sin(8\omega)$

17. Suppose that $\mathscr{F}\{f(t)\} = F(\omega)$ and $\mathscr{F}\{g(t)\} = G(\omega)$. Prove that

$$\int_{-\infty}^{\infty} f(t)g(t)\,dt = \frac{1}{2\pi}\int_{-\infty}^{\infty} F(w)G(-w)\,dw.$$

Hint: Use the frequency convolution theorem to get

$$\mathscr{F}\{f(t)g(t)\} = \frac{1}{2\pi}\int_{-\infty}^{\infty} F(w)G(\omega - w)\,dw.$$

Conclude that

$$\int_{-\infty}^{\infty} f(t)g(t)e^{-i\omega t}\,dt = \frac{1}{2\pi}\int_{-\infty}^{\infty} F(w)G(\omega - w)\,dw,$$

and let $\omega = 0$.

18. Suppose that f and g are real valued. Show that

$$\int_{-\infty}^{\infty} f(t)g(t)\,dt = \frac{1}{2\pi}\int_{-\infty}^{\infty} F(\omega)\overline{G(\omega)}\,d\omega,$$

assuming that the integral on the left converges. *Hint:* Use formula (11) at the end of the preceding section.

19. Prove *Parseval's identity* in the form

$$\int_{-\infty}^{\infty} [f(t)]^2\,dt = \frac{1}{2\pi}\int_{-\infty}^{\infty} |F(\omega)|^2\,d\omega.$$

20. The *power content* of a nonperiodic signal f is defined to be $\int_{-\infty}^{\infty}|f(t)|^2\,dt$. Determine the power content of $H(t)e^{-2t}$.

21. Determine the power content of $(1/t)\sin(3t)$. *Hint:* Use Parseval's identity from Problem 19.

22. Solve the differential equation

$$y'' + 6y' + 5y = \delta(t - 3).$$

23. Suppose that f is a solution of $y'' - t^2y = y$. Show that $\mathscr{F}\{f(t)\}$ is also a solution.

24. Prove that

$$|F(\omega)| \le \int_{-\infty}^{\infty} |f(t)| \, dt,$$

in which $\mathcal{F}\{f(t)\} = F(\omega)$. *Hint:*

$$|F(\omega)| \le \frac{1}{|\omega|} \int_{-\infty}^{\infty} |f'(t)| \, dt \quad \text{and} \quad |F(\omega)| \le \frac{1}{\omega^2} \int_{-\infty}^{\infty} |f''(t)| \, dt.$$

25. Prove the frequency convolution theorem.

26. The *phase spectrum* of a function is the graph of the argument $\varphi(\omega)$ of the Fourier transformation of f. That is, $\mathcal{F}\{f(t)\} = R(\omega) + iX(\omega) = |F(\omega)|e^{i\varphi(\omega)}$, in which

$$\varphi(\omega) = \tan^{-1}\left[-\frac{X(\omega)}{R(\omega)} \right].$$

Compute the Fourier transforms of f and g, where

$$f(t) = 5[H(t + 2) - H(t - 2)] \quad \text{and} \quad g(t) = 5[H(t - 3) - H(t - 7)].$$

Compare the amplitude and phase spectra of f and g.

27. Determine the value of the amplitude spectrum at 3 of the function defined in Problem 3.

28. Determine the value of the amplitude spectrum at 6 of the function defined in Problem 6.

29. Evaluate $\int_{-\infty}^{\infty} 8\delta(t - 3)H(t - 3)e^{-5t} \, dt$.

30. *Strange bedfellows*   Define

$$f_n(t) = P_n(t)[H(t + 1) - H(t - 1)] = \begin{cases} P_n(t) & \text{if} \quad |t| < 1 \\ 0 & \text{if} \quad |t| > 1, \end{cases}$$

in which P_n is the nth Legendre polynomial. Prove that

$$\mathcal{F}\{f_n(t)\} = (-i)^n \sqrt{\frac{2\pi}{\omega}} \, J_{n+1/2}(\omega) \tag{17.14}$$

for $n = 0, 1, 2, \ldots$, with J_ν the Bessel function of the first kind of order ν. *Hint:* For $n = 0$, use the formula established in Example 17.21 and the fact that $J_{1/2}(\omega) = \sqrt{2/\pi\omega} \sin(\omega)$. For the case in which $n = 1$, note that $P_1(t) = tP_0(t)$, and use Theorem 17.20, the fact that (17.14) has been shown true for $n = 0$, and the result of Problem 13 of Section 7.1, which states that $[x^{-\nu}J_\nu(x)]' = -x^{-\nu}J_{\nu+1}(x)$.

Now proceed by mathematical induction. For $n \ge 2$, assume that (17.14) has been proved for $0, 1, \ldots, n$. Write

$$P_{n+1}(t) = \frac{2n + 1}{n + 1} tP_n(t) - \frac{n}{n + 1} P_{n-1}(t)$$

from Theorem 7.4, and apply the Fourier transform to this equation, using Theorem 17.20 to obtain

$$\frac{(-i)^{n+1}}{n + 1} \sqrt{2\pi} \left[n\omega^{-1/2}J_{n-1/2}(\omega) + \frac{2n + 1}{2} \omega^{-3/2}J_{n+1/2}(\omega) - (2n + 1)\omega^{-1/2}J'_{n+1/2}(\omega) \right].$$

Complete the proof by using the identities

$$J'_\nu = \frac{\nu}{\omega} J_\nu - J_{\nu+1} \quad \text{and} \quad \frac{2\nu}{\omega} J_\nu - J_{\nu-1} = J_{\nu+1}$$

from Problem 16 of Section 7.1. Choose $\nu = n + \frac{1}{2}$ in these identities.

17.11 *Some Applications of the Fourier Transform*

This section is devoted to three applications of the Fourier transform.

FREQUENCY RESPONSE OF A FILTER

Recall that the amplitude spectrum provided by the magnitude of the Fourier transform of a function gives us a measure of how much of a given frequency is contained in the function. We will look at some special cases of this and then see how we can use this aspect of the Fourier transform to display the frequency response of a filter. Consider the square wave f of period T defined by

$$f(t) = \begin{cases} -k & \text{if} \quad -T/2 < t < 0 \\ k & \text{if} \quad 0 < t < T/2 \end{cases}$$

and $f(t + T) = f(t)$. This function has complex Fourier representation

$$f(t) = \sum_{n=-\infty}^{\infty} \frac{-2ki}{(2n-1)\pi} e^{(2n-1)\omega_0 it},$$

in which $\omega_0 = 2\pi/T$. The amplitude spectrum of f is shown in Figure 17.44; it clearly indicates that the square wave is composed of odd multiples of the fundamental frequency $\omega_0/2\pi$.

Now consider the function having constant value 1. We would not expect this function to contain much of any one frequency. We found [Equation (17.13), at the end of the preceding section] that the Fourier transform of this function is $2\pi\,\delta(\omega)$. Figure 17.45 shows that our intuition is correct and that the spectrum of this function is nonzero only at $\omega = 0$. The function $\cos(\omega_0 t)$ has Fourier transform $\pi[\delta(\omega + \omega_0) + \delta(\omega - \omega_0)]$ and has a spectrum which rises above zero in only two places, indicating that it contains only one harmonic, which of course is exactly the case. This is shown in Figure 17.46.

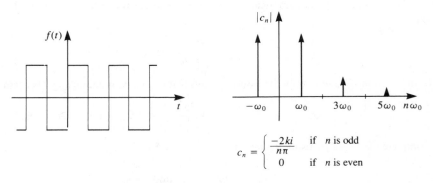

$$c_n = \begin{cases} \dfrac{-2ki}{n\pi} & \text{if} \quad n \text{ is odd} \\ 0 & \text{if} \quad n \text{ is even} \end{cases}$$

Figure 17.44. From the amplitude spectrum, the square wave is composed of only odd harmonics of the fundamental ω_0.

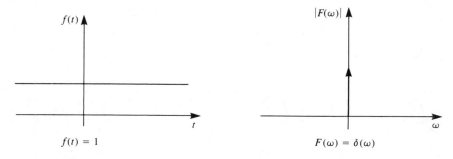

$$f(t) = 1 \qquad\qquad F(\omega) = \delta(\omega)$$

Figure 17.45. Constant functions contain no frequencies.

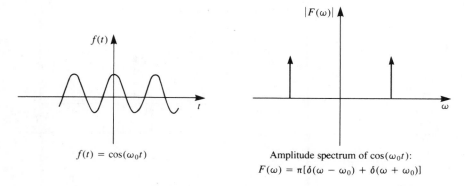

$$f(t) = \cos(\omega_0 t)$$

Amplitude spectrum of $\cos(\omega_0 t)$:
$$F(\omega) = \pi[\delta(\omega - \omega_0) + \delta(\omega + \omega_0)]$$

Figure 17.46. Cosine (sine) waves contain only one frequency.

Now look at the case of a periodic function. Suppose that f has period T and that f has complex Fourier series representation

$$f(t) = \sum_{n=-\infty}^{\infty} c_n e^{n\omega_0 it}.$$

Apply the Fourier transform to both sides of this equation, interchanging $\mathscr{F}$ and $\sum_{n=-\infty}^{\infty}$, to get

$$
\begin{aligned}
F(\omega) &= \mathscr{F}\{f(t)\} \\
&= \sum_{n=-\infty}^{\infty} c_n \mathscr{F}\{e^{n\omega_0 it}\} = \sum_{n=-\infty}^{\infty} 2\pi c_n \delta(\omega - n\omega_0).
\end{aligned}
$$

This indicates a discrete amplitude spectrum, in agreement with the fact that f is periodic. A typical example is shown in Figure 17.47.

At the other extreme from a constant function, which contains no frequencies, we have the Dirac delta function, which, with its transform having constant value 1, must in some sense contain all frequencies, each of the same amplitude (Figure 17.48). This and the fact that $\mathscr{F}\{1\} = 2\pi\,\delta(\omega)$ are the two extreme examples of the scaling property, which says, in essence, that the more passive the function, the narrower its spectrum.

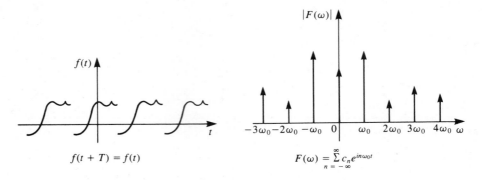

Figure 17.47. Frequency spectrum of a typical periodic function: equally spaced lines.

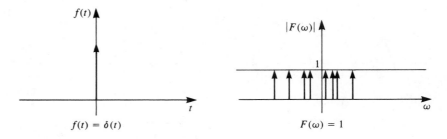

Figure 17.48. The unit impulse contains all frequencies at equal energy levels.

We can exploit the fact that $\mathscr{F}\{\delta(t)\} = 1$ to analyze the output frequency of a filter. Suppose we have a crude low-pass filter consisting of a resistor and a capacitor, as in Figure 17.49. One method of determining the output frequency response is to sweep the input with a sine wave generator, varying the frequency and maintaining a constant input amplitude. Meanwhile, measure the output amplitude at the various frequencies with a voltmeter and then graph these results.

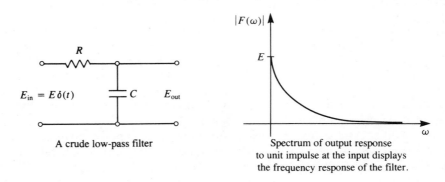

Figure 17.49

We can also determine the filter frequency response by using a differential equation. First, model the system with the equation

$$Rq' + \frac{1}{C} q = E_{in} = E \sin(\omega t).$$

Since the desired output is $q(t)/C$, we want to solve for q. Multiply the differential equation by the integrating factor $e^{t/RC}$ and integrate by parts, discarding the transient part of the solution, to obtain

$$E_{out} = \frac{1}{C} q(t) = \frac{E}{RC} \left[\frac{\frac{1}{RC} \sin(\omega t) - \omega \cos(\omega t)}{\omega^2 + \left(\frac{1}{RC}\right)^2} \right] = \frac{E \cos(\omega t + \delta_\omega)}{\sqrt{1 + R^2 C^2 \omega^2}}.$$

Using this explicit expression, we can easily graph $|E_{out}|$.

Here is an alternative method of determining the filter frequency response, using the concept of the amplitude spectrum and the idea that the unit impulse contains all frequencies, all of equal amplitude. Act on the input of the filter with the signal $E \, \delta(t)$. The resulting system is modeled by the differential equation

$$Rq' + \frac{1}{C} q = E \, \delta(t).$$

Apply the Fourier transform to the differential equation, with $\mathscr{F}\{q(t)\} = Q(\omega)$, to get

$$i\omega R Q(\omega) + \frac{1}{C} Q(\omega) = \mathscr{F}\{E \, \delta(t)\} = E,$$

since $\mathscr{F}\{\delta(t)\} = 1$. Solve this equation for $Q(\omega)$ to get

$$Q(\omega) = \frac{E}{\left(\frac{1}{C}\right) + i\omega R} = \frac{EC}{1 + i\omega RC}.$$

Since $E_{out} = [q(t)]/C$, we have the amplitude spectrum of the output signal as

$$\left| \frac{1}{C} Q(\omega) \right| = \left| \frac{E}{1 + i\omega RC} \right| = \frac{E}{\sqrt{1 + R^2 C^2 \omega^2}},$$

essentially the same result we obtained before, but with less effort.

The inclusion of an inductor in the circuit, as in Figure 17.50, results in a higher quality (sharper) low-pass filter. To see this, write the differential equation for this circuit:

$$Lq'' + Rq' + \frac{1}{C} q = E_{in}.$$

Write $E_{in} = E \, \delta(t)$ and apply the Fourier transform to this differential equation to get

$$\left[-L\omega^2 + iR\omega + \frac{1}{C} \right] Q(\omega) = E.$$

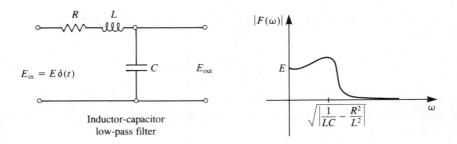

Figure 17.50. Introduction of an inductor into a low-pass filter makes it sharper at the high-pass end; the degree of sharpness depends on the inductor resistance.

Solve for $Q(\omega)/C$ to get

$$\frac{1}{C} Q(\omega) = \frac{E}{1 - \omega^2 LC + i\omega RC}.$$

Then

$$\left| \frac{1}{C} Q(\omega) \right| = \frac{E}{\sqrt{(1 - \omega^2 LC)^2 + \omega^2 R^2 C^2}}.$$

This is the amplitude spectrum of the output voltage, which in turn is the frequency response of the filter. We find that $|(1/C)Q(\omega)|$ has local extreme values equal to E when $\omega = 0$ and equal to EL/R^2C when

$$\omega = \pm\sqrt{\left| \frac{1}{LC} - \frac{R^2}{L^2} \right|}.$$

As can be seen from the amplitude spectrum, this filter has a much faster rate of fall on the pass side than does the resistor-capacitor filter. Notice from the equation of the output amplitude spectrum that the smaller the resistance in the circuit, the sharper the curve will be. The restriction is that there is always some resistance in the circuit due to the resistance of the wire in the inductor.

This inductor-capacitor filter (along with the inductor's internal resistance) is the type we used in Example 17.14. There, it took some effort to determine the amplitudes of the various harmonics at the output when we had a square wave input. If we have the amplitude spectrum of the filter output with the unit impulse input at our disposal, we can quickly determine these same values with much less effort.

To illustrate, in that example we had an input signal with component values of $4k/\pi$ volts when $\omega = 800\pi$, $4k/3\pi$ volts at $\omega = 2400\pi$, and so on. Put $L = 1$, $R = 600$, and $C = 0.16 \times 10^{-6}$ into the differential equation $Li' + Ri + q/C = f(t)$ of that example, with $\omega = 800\pi$ and $E = 4k/\pi$, then calculate the output value. Next, let $\omega = 2400\pi$ and $E = 4k/3\pi$ and recalculate. The ratio of these two values is precisely the same as the ratio of the output values of the first and third harmonics that we calculated using the method of undetermined coefficients and the Fourier series.

The circuit in Figure 17.51 is a band-pass filter, which allows frequencies only within a certain range to pass through. If the filter is to be sharp, it must be made with

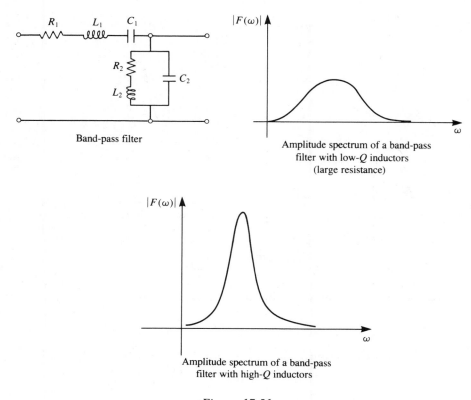

Band-pass filter

Amplitude spectrum of a band-pass
filter with low-Q inductors
(large resistance)

Amplitude spectrum of a band-pass
filter with high-Q inductors

Figure 17.51

components having a "high Q" (a high inductance-to-resistance ratio). The filter will
then necessarily have a narrow bandwidth. This can be seen by observing the change in
the spectrum of the output of such a filter as the inductor resistance is changed. In order
to construct a wide, sharp band-pass filter, several sharp narrow-band filters can be
placed in parallel to make what is called a *Tchebycheff filter*. Figure 17.52 shows a
typical Tchebycheff filter. Graphing the spectra of these filters can be beneficial in their
design.

THE SAMPLING THEOREM

We will use the Fourier transform to derive a result called the *sampling theorem*, which
states that an entire band-limited signal can be reconstructed from certain sampled
values. If we think of a signal as a function, f is said to be *band limited* if its Fourier
transform $F(\omega)$ vanishes outside of some finite interval. That is, for some number L,

$$F(\omega) = 0 \quad \text{if} \quad |\omega| > L.$$

In this event,

$$f(t) = \frac{1}{2\pi} \int_{-\infty}^{\infty} F(\omega)e^{i\omega t} \, d\omega = \frac{1}{2\pi} \int_{-L}^{L} F(\omega)e^{i\omega t} \, d\omega. \tag{17.15}$$

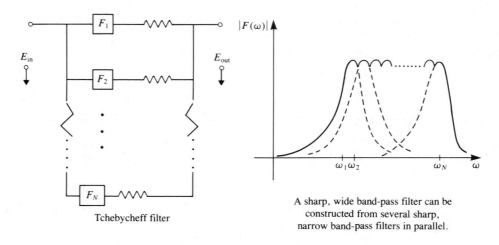

Tchebycheff filter

A sharp, wide band-pass filter can be constructed from several sharp, narrow band-pass filters in parallel.

Figure 17.52

To derive the result we want, we begin with the complex Fourier series representation

$$F(\omega) = \sum_{n=-\infty}^{\infty} c_n e^{n\pi i \omega/L} \tag{17.16}$$

for $-L \le \omega \le L$. Here,

$$c_n = \frac{1}{2L} \int_{-L}^{L} F(\omega) e^{-n\pi i \omega/L} \, d\omega.$$

Now compare this formula for the c_n's with equation (17.15) to conclude that

$$c_n = \frac{\pi}{L} F\left(\frac{-n\pi}{L}\right).$$

We can now write equation (17.16) as

$$F(\omega) = \sum_{n=-\infty}^{\infty} c_n e^{n\pi i \omega/L} = \sum_{n=-\infty}^{\infty} \frac{\pi}{L} f\left(\frac{-n\pi}{L}\right) e^{n\pi i \omega/L} = \frac{\pi}{L} \sum_{n=-\infty}^{\infty} f\left(\frac{n\pi}{L}\right) e^{-n\pi i \omega/L}.$$

We obtained the last term from the preceding one by replacing n by $-n$. This is a valid substitution because the summation is over all integers n, positive, negative, and zero.

Put the last summation for $F(\omega)$ into (17.15) to get

$$f(t) = \frac{1}{2\pi} \int_{-\infty}^{\infty} F(\omega) e^{i\omega t} \, d\omega = \frac{1}{2\pi} \int_{-\infty}^{\infty} \frac{\pi}{L} \left[\sum_{n=-\infty}^{\infty} f\left(\frac{n\pi}{L}\right) e^{-n\pi i \omega/L} \right] e^{i\omega t} \, d\omega$$

$$= \frac{1}{2L} \sum_{n=-\infty}^{\infty} f\left(\frac{n\pi}{L}\right) \int_{-L}^{L} e^{i\omega[t - (n\pi/L)]} \, d\omega$$

$$= \frac{1}{2} \sum_{n=-\infty}^{\infty} f\left(\frac{n\pi}{L}\right) \frac{e^{(Lt - n\pi)i} - e^{-(Lt - n\pi)i}}{i(Lt - n\pi)}$$

$$= \sum_{n=-\infty}^{\infty} f\left(\frac{n\pi}{L}\right) \frac{\sin(Lt - n\pi)}{Lt - n\pi}.$$

This remarkable expression implies that we know $f(t)$ for *all* t if we know just $f(n\pi/L)$ for $n = 0, \pm 1, \pm 2, \ldots$. That is, the values $f(0)$, $f(\pi/L)$, $f(-\pi/L)$, $f(2\pi/L)$, $f(-2\pi/L), \ldots$ determine $f(t)$ for all t. This conclusion is called the *sampling theorem*. It is important in electrical engineering, where a signal can be reconstructed by sampling specific data points.

There is a similar theorem, which applies to periodic functions, in which we construct a periodic function to fit data given at a finite number of points in an interval. These results usually come under the heading of *harmonic analysis*.

FOURIER TRANSFORM SOLUTION OF PARTIAL DIFFERENTIAL EQUATIONS

Fourier transforms are a powerful tool for solving certain kinds of problems involving partial differential equations. Although we will discuss this in more detail in the next chapter, we will examine the idea here.

Suppose we want to solve the partial differential equation

$$\frac{\partial^2 u}{\partial t^2} = 4 \frac{\partial^2 u}{\partial x^2} \qquad (-\infty < x < \infty, t > 0)$$

subject to the conditions

$$u(x, 0) = 0, \qquad \frac{\partial u}{\partial x}(x, 0) = 64xe^{-4x^2}.$$

The fact that x varies over the entire real line suggests the possibility of using the Fourier transform. The strategy is to apply the Fourier transform to the partial differential equation, with x, not t, as the variable (because t varies only over $[0, \infty)$). Thus, t is carried through the transform as simply a symbol, and we transform in the x-variable. For $u(x, t)$, then, the Fourier transform is

$$\mathscr{F}\{u(x, t)\} = \int_{-\infty}^{\infty} u(x, t)e^{-i\omega x}\, dx = U(\omega, t).$$

Notice that this Fourier transform is a function of ω and t, since t is unaffected by the transform with respect to x.

Now apply the Fourier transform to the partial differential equation, using subscripts to denote partial derivatives for ease in writing. We have

$$\mathscr{F}\{u_{tt}(x, t)\} = 4\mathscr{F}\{u_{xx}(x, t)\}. \tag{17.17}$$

We must compute the Fourier transforms on both sides of (17.17). For the left side,

$$\mathscr{F}\{u_{tt}(x, t)\} = \int_{-\infty}^{\infty} u_{tt}(x, t)e^{-i\omega x}\, dx = \frac{\partial^2}{\partial t^2} \int_{-\infty}^{\infty} u(x, t)e^{-i\omega x}\, dx.$$

The differentiation with respect to t can be brought outside $\int_{-\infty}^{\infty} \cdots dx$ because the integration is with respect to x, and x and t are independent. From the last equation,

$$\mathscr{F}\{u_{tt}(x, t)\} = U_{tt}(\omega, t).$$

On the right side of (17.17), apply Theorem 17.18 to get

$$\mathscr{F}\{u_{xx}(x, t)\} = (i\omega)^2 \mathscr{F}\{u(x, t)\} = -\omega^2 U(\omega, t).$$

Thus, equation (17.17) becomes

$$U_{tt}(\omega, t) = -4\omega^2 U(\omega, t),$$

or

$$U_{tt} + 4\omega^2 U = 0.$$

Think of this as an ordinary differential equation in t, with ω carried along as a constant. The general solution is

$$U(\omega, t) = A(\omega)\cos(2\omega t) + B(\omega)\sin(2\omega t).$$

The "arbitrary constants" may be functions of ω, which was held fixed to obtain this general solution for U. Now we must determine $A(\omega)$ and $B(\omega)$. We will do this using the conditions given with the partial differential equation. First, we know that $u(x, 0) = 0$. Apply the Fourier transform to this condition to get

$$\mathcal{F}\{u(x, 0)\} = U(\omega, 0) = \mathcal{F}\{0\} = 0.$$

Then

$$U(\omega, 0) = A(\omega)\cos(0) + B(\omega)\sin(0) = A(\omega) = 0.$$

Since $A(\omega) = 0$, U must be of the form

$$U(\omega, t) = B(\omega)\sin(2\omega t).$$

To solve for $B(\omega)$, recall the other condition

$$u_t(x, 0) = 64xe^{-4x^2}.$$

Since $\partial/\partial t$ and $\int_{-\infty}^{\infty} \cdots dx$ can be interchanged, we have, upon applying $\mathcal{F}$ to this condition, that

$$\mathcal{F}\{u_t(x, 0)\} = U_t(\omega, 0) = \mathcal{F}\{64xe^{-4x^2}\}.$$

By Theorem 17.20 and the table of Fourier transforms at the end of Chapter 18, we have

$$\mathcal{F}\{64xe^{-4x^2}\} = i\frac{d}{d\omega}\mathcal{F}\{64e^{-4x^2}\} = 64i\frac{d}{d\omega}\left[\frac{\sqrt{\pi}}{2}e^{-\omega^2/16}\right] = -4i\sqrt{\pi}\omega e^{-\omega^2/16}.$$

Therefore,

$$U_t(\omega, 0) = -4i\sqrt{\pi}\omega e^{-\omega^2/16}.$$

But $U(\omega, t) = B(\omega)\sin(2\omega t)$, so

$$U_t(\omega, 0) = 2\omega B(\omega)\cos(2\omega t)\Big|_{t=0} = 2\omega B(\omega) = -4i\sqrt{\pi}\omega e^{-\omega^2/16}.$$

Solve this equation for $B(\omega)$ to get

$$B(\omega) = -2i\sqrt{\pi}e^{-\omega^2/16}.$$

Finally,

$$U(\omega, t) = -2i\sqrt{\pi}e^{-\omega^2/16}\sin(2\omega t).$$

This is the Fourier transform of the solution. We must invert it to find the solution. There are several ways to do this. We could use the convolution theorem, or we could use the symmetry property in conjunction with one of the modulation theorems. A third method, which we will pursue, is to rewrite the sine term in complex exponential form and use the time-shifting theorem. We get

$$u(x, t) = \mathcal{F}^{-1}\{-2i\sqrt{\pi}e^{-\omega^2/16}\sin(2\omega t)\}$$

$$= \mathcal{F}^{-1}\left\{-2i\sqrt{\pi}e^{-\omega^2/16}\left[\frac{e^{2i\omega t} - e^{-2i\omega t}}{2i}\right]\right\}$$

$$= \mathcal{F}^{-1}\{\sqrt{\pi}e^{-\omega^2/16}e^{-i\omega(2t)} - \sqrt{\pi}e^{-\omega^2/16}e^{i\omega(2t)}\}$$

$$= \mathcal{F}^{-1}\{\sqrt{\pi}e^{-\omega^2/16}\}_{x\to x-2t} - \mathcal{F}^{-1}\{\sqrt{\pi}e^{-\omega^2/16}\}_{x\to x+2t}$$

$$= 2e^{-4(x-2t)^2} - 2e^{-4(x+2t)^2}.$$

This function satisfies the partial differential equation and the conditions imposed in the problem.

PROBLEMS FOR SECTION 17.11

1. The frequency translation (modulation) theorem (Theorem 17.17) states that the multiplication of a signal f by a sinusoidal signal of frequency ω_0 translates its spectrum by $\pm i\omega$. Suppose that m is a band-limited signal. The signal

$$f(t) = m(t)\cos(\omega_0 t)$$

is called the *modulated signal*; $\cos(\omega_0 t)$ is called the *carrier*, and m is called the *modulating* signal. Derive an expression for the spectrum of the modulated signal.

2. The process of separating a modulating signal from a modulated signal is called *demodulation*. Show that the spectrum of the modulated signal can be retranslated to its original position by multiplying the modulated signal by $\cos(\omega_0 t)$ at the receiving end. *Hint:* Write

$$f(t)\cos(\omega_0 t) = m(t)\cos^2(\omega_0 t) = m(t)\tfrac{1}{2}[1 + \cos(2\omega_0 t)] = \tfrac{1}{2}m(t) + \tfrac{1}{2}m(t)\cos(2\omega_0 t).$$

Now show that

$$\mathcal{F}\{f(t)\cos(\omega_0 t)\} = \tfrac{1}{2}M(\omega) + \tfrac{1}{4}M(\omega - 2\omega_0) + \tfrac{1}{4}M(\omega + 2\omega_0).$$

If $M(\omega) = 0$ for $|\omega| > |\omega_M|$ for some positive constant ω_M, the last two terms can be removed by putting the product $f(t)\cos(\omega_0 t)$ through a low-pass filter, which will pass through the spectrum only up to ω_M, as the following diagram illustrates.

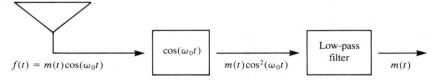

$f(t) = m(t)\cos(\omega_0 t)$ → $\cos(\omega_0 t)$ → $m(t)\cos^2(\omega_0 t)$ → Low-pass filter → $m(t)$

3. Show that demodulation can be accomplished by multiplying the modulated signal $f(t) = m(t)\cos(\omega_0 t)$ by any periodic signal of frequency $\dfrac{\omega_0}{2\pi}$. *Hint:* If p is a periodic signal of frequency $\dfrac{\omega_0}{2\pi}$ and has the form

$$p(t) = \sum_{n=-\infty}^{\infty} c_n e^{in\omega_0 t},$$

then from the Fourier transform of a complex exponential function, the Fourier transform of p can be written

$$\mathcal{F}\{p(t)\} = P(\omega) = \sum_{n=-\infty}^{\infty} c_n \delta(\omega - n\omega_0).$$

Now recall from Problem 2 that

$$\mathcal{F}\{f(t)\} = \tfrac{1}{2}M(\omega - \omega_0) + \tfrac{1}{2}M(\omega + \omega_0).$$

Therefore,

$$\mathcal{F}\{f(t)p(t)\} = F(\omega) * P(\omega)$$

$$= \tfrac{1}{2}[M(\omega - \omega_0) + M(\omega + \omega_0)] * \sum_{n=-\infty}^{\infty} c_n \delta(\omega - n\omega_0)$$

$$= \tfrac{1}{2} \sum_{n=-\infty}^{\infty} c_n\{[M(\omega - \omega_0) + M(\omega + \omega_0)] * \delta(\omega - n\omega_0)\}$$

$$= \tfrac{1}{2} \sum_{n=-\infty}^{\infty} c_n\{M[\omega - (n + 1)\omega_0] + M[(\omega - (n - 1)\omega_0]\}.$$

Finally, show that this spectrum contains a term $M(\omega)$, the spectrum of $m(t)$. This can be recovered by using a low-pass filter.

4. Suppose we have some periodic phenomenon for which data are available over one period at equally spaced points. Then we can construct a function defined for all real numbers which approximately fits the data, as follows. Assume that the function f is periodic of period 2π and that its values are known for the $N + 1$ points $-\pi, -\pi + 2\pi/N, \ldots, \pi - (2\pi/N), \pi$. Assume that f has a Fourier series representation

$$f(x) = a_0 + \sum_{n=1}^{\infty} [a_n\cos(nx) + b_n\sin(nx)].$$

We will know the value of $f(x)$ for all x if we can determine the Fourier coefficients in this expansion. The strategy is to use the trapezoidal rule to approximate these Fourier coefficients. The trapezoidal rule formula is

$$\int_a^b f(x)\, dx \approx \frac{b - a}{2N}\left[f(a) + 2f\left(a + \frac{b - a}{N}\right) + \cdots + 2f\left(b - \frac{b - a}{N}\right) + f(b)\right].$$

Use this to show that

$$a_0 \approx \frac{1}{2N}[f(-\pi) + f(\pi)] + \frac{1}{N}\sum_{k=1}^{N-1} f\left[\frac{\pi}{N}(2k - N)\right],$$

$$a_n \approx \frac{(-1)^n}{N}[f(-\pi) + f(\pi)] + \frac{2}{N}\sum_{k=1}^{N-1} f\left[\frac{\pi}{N}(2k - N)\right]\cos\left[\frac{n\pi}{n}(2k - N)\right],$$

and

$$b_n \approx \frac{2}{N}\sum_{k=1}^{N-1} f\left[\frac{\pi}{N}(2k - N)\right]\sin\left[\frac{n\pi}{N}(2k - N)\right].$$

In order to make use of this method, we can employ tables of values of $\cos[n\pi(2k - N)/N]$ and $\sin[n\pi(2k - N)/N]$ for various values of k, n, and N. As an illustration, Tables 17.1 and 17.2 give values of $\cos[n\pi(2k - N)/N]$ and $\sin[n\pi(2k - N)/N]$ for $N = 12$, $k = 1, \ldots, 11$, and $n = 1, \ldots, 6$.

5. Use the idea developed in Problem 4 to approximate the pressure in the cylinder above the piston shown in the

TABLE 17.1 $\cos[n\pi(2k - N)/N]$ for $N = 12$;
$k = 1, \ldots, 11$; $n = 1, \ldots, 6$

k	$n = 1$	2	3	4	5	6
1	−0.866	0.500	0.000	−0.500	0.866	−1.000
2	−0.400	−0.500	1.000	−0.500	−0.500	1.000
3	0.000	−1.000	0.000	1.000	0.000	−1.000
4	0.500	−0.500	−1.000	−0.500	0.500	1.000
5	0.866	0.500	0.000	−0.500	−0.866	−1.000
6	1.000	1.000	1.000	1.000	1.000	1.000
7	0.866	0.500	0.000	−0.500	−0.866	−1.000
8	0.500	−0.500	−1.000	−0.500	0.500	1.000
9	0.000	−1.000	0.000	1.000	0.000	−1.000
10	−0.500	−0.500	1.000	−0.500	−0.500	1.000
11	−0.866	0.500	0.000	−0.500	0.866	−1.000

TABLE 17.2 $\sin[n\pi(2k - N)/N]$ for $N = 12$;
$k = 1, \ldots, 11$; $n = 1, \ldots, 6$

k	$n = 1$	2	3	4	5	6
1	−0.500	0.866	−1.000	0.866	−0.500	0.000
2	−0.866	0.866	0.000	−0.866	0.866	0.000
3	−1.000	0.000	1.000	0.000	−1.000	0.000
4	−0.866	−0.866	0.000	0.866	0.866	0.000
5	−0.500	−0.866	−1.00	−0.866	−0.500	0.000
6	0.000	0.000	0.000	0.000	0.000	0.000
7	0.500	0.866	1.000	0.866	0.500	0.000
8	0.866	0.866	0.000	−0.866	−0.866	0.000
9	1.000	0.000	−1.000	0.000	1.000	0.000
10	0.866	−0.866	0.000	0.866	−0.866	0.000
11	0.500	−0.866	1.000	−0.866	0.500	0.000

**TABLE 17.3   Cylinder
Pressure $p(\theta)$ versus
Crankshaft Angle θ**

θ	$p(\theta)$
$-\pi$	22.5
$-5\pi/6$	17.3
$-2\pi/3$	15.3
$-\pi/2$	14.7
$-\pi/3$	15.3
$-\pi/6$	17.3
0	22.5
$\pi/6$	35.8
$\pi/3$	71.6
$\pi/2$	117.9
$2\pi/3$	71.6
$5\pi/6$	35.8
π	22.5

diagram for all crankshaft angles θ, using the data in Table 17.3 and the sixth partial sum of the Fourier series.

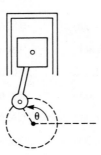

6. Suppose that g is defined on $[a, b]$. Then the change of variables

$$t = \frac{2\pi(x - a)}{b - a} - \pi$$

transforms g to a new function defined on $[-\pi, \pi]$ by

$$f(t) = g\left[a + \frac{b - a}{2\pi}(\pi + t)\right].$$

**TABLE 17.4   Average Noon
Temperature, Big Sandy,
Wyoming, Airport**

January 1	-16.7
February 1	-19.4
March 1	-8.9
April 1	12.1
May 1	37.8
June 1	61.5
July 1	76.9
August 1	79.4
September 1	68.9
October 1	47.9
November 1	22.2
December 1	-1.5

Show that if g is periodic of period $b - a$, f is periodic of period 2π, and

$$g\left[a + \frac{k(b-a)}{N}\right] = f\left(-\pi + \frac{2k\pi}{N}\right).$$

7. Use the results of Problems 4 and 6 to approximate the average noon temperature on January 21 at the Big Sandy, Wyoming, airport. Use the third partial sum of the series and assume that the data given in Table 17.4 are equally spaced (at 30-day intervals).

8. Suppose that a periodic, band-limited signal has a Fourier series representation. Prove that the series must be finite.

9. Solve

$$u_{tt}(x, t) = 16u_{xx}(x, t) \quad (-\infty < x < \infty, t > 0);$$
$$u(x, 0) = 8e^{-3x^2}, \quad u_t(x, 0) = 0.$$

10. Solve

$$u_t(x, t) = u_{xx}(x, t) \quad (-\infty < x < \infty, \quad t > 0);$$
$$u(x, 0) = 3e^{-x^2/4} \quad \text{for} \quad x > 0,$$

in which $H(x)$ is the Heaviside function.

11. Solve

$$u_{tt}(x, t) = 4u_{xx}(x, t) \quad (-\infty < x < \infty, t > 0);$$
$$u(x, 0) = 0, \quad u_t(x, 0) = \frac{12}{x^2 + 9}.$$

12. Solve

$$u_{tt}(x, t) = 9u_{xx}(x, t) \qquad (-\infty < x < \infty, t > 0);$$

$$u(x, 0) = \begin{cases} x + 4 & \text{if } -4 < x < 0 \\ 4 - x & \text{if } \quad 0 < x < 4 \\ 0 & \text{if } \quad |x| > 4, \qquad u_t(x, 0) = 0 \end{cases}$$

$$u_t(x, 0) = 0$$

13. Graph $\frac{1}{2}[f(x - 3t) + f(x + 3t)]$ for $-8 < x < 8$ and for $t = 0, \frac{1}{3}, \frac{2}{3}, 1, \frac{4}{3}, \frac{5}{3}$, and 2, where

$$f(x) = \begin{cases} x + 4 & \text{if } -4 < x < 0 \\ 4 - x & \text{if } \quad 0 < x < 4 \\ 0 & \text{if } \quad |x| > 4. \end{cases}$$

14. Solve

$$u_{tt} = c^2 u_{xx} \qquad (-\infty < x < \infty, t > 0);$$
$$u(x, 0) = f(x), \qquad u_t(x, 0) = 0.$$

Hint: Find that

$$\begin{aligned} u(x, t) &= \mathscr{F}^{-1}\{F(\omega)\cos(c\omega t)\} \\ &= f(x) * \tfrac{1}{2}[\delta(x - ct) + \delta(x + ct)] \\ &= \tfrac{1}{2}\int_{-\infty}^{\infty} f(x - \tau)\delta(\tau - ct)\, d\tau + \tfrac{1}{2}\int_{-\infty}^{\infty} f(x - \tau)\delta(\tau + ct)\, d\tau \\ &= \tfrac{1}{2}[f(x - ct) + f(x + ct)]. \end{aligned}$$

(This is part of d'Alembert's solution of the wave equation; see Problem 21 of Section 18.0.)

15. Solve

$$u_{tt} = c^2 u_{xx} \qquad (-\infty < x < \infty, t > 0);$$
$$u(x, 0) = f(x), \qquad u_t(x, 0) = g(x).$$

Hint: Find that

$$u(x, t) = \frac{1}{2}[f(x + ct) + f(x - ct)] + \frac{1}{2c}\int_{x-ct}^{x+ct} g(\xi)\, d\xi.$$

[This is d'Alembert's solution of the wave equation, containing the solution of Problem 14 as the special case in which $g(x)$ is identically zero.]

17.12 *Fourier Sine and Cosine Transforms*

In this section, we will modify the Fourier transform to obtain a transform for functions defined on only the half-line $[0, \infty)$. The idea is analogous to expanding a function defined on only $[0, L]$ in a Fourier cosine or sine series.

   Suppose first that f is defined on the entire real line. Use Euler's equation in the definition of the Fourier transform of f to write

$$\mathscr{F}\{f(t)\} = \int_{-\infty}^{\infty} f(t)e^{-i\omega t}\, dt = \int_{-\infty}^{\infty} f(t)[\cos(\omega t) - i\,\sin(\omega t)]\, dt$$

$$= \int_{-\infty}^{\infty} f(t)\cos(\omega t)\, dt - i\int_{-\infty}^{\infty} f(t)\sin(\omega t)\, dt.$$

If f is a real-valued, even function on $(-\infty, \infty)$, $\int_{-\infty}^{\infty} f(t)\sin(\omega t)\, dt = 0$, because $f(t)\sin(\omega t)$ is odd. Further, $f(t)\cos(\omega t)$ is even, so

$$\int_{-\infty}^{\infty} f(t)\cos(\omega t)\, dt = 2\int_{0}^{\infty} f(t)\cos(\omega t)\, dt.$$

When f is even, then,

$$\mathscr{F}\{f(t)\} = 2\int_{-\infty}^{\infty} f(t)\cos(\omega t)\, dt.$$

Similarly, if f is odd, $\int_{-\infty}^{\infty} f(t)\cos(\omega t)\, dt = 0$, because $f(t)\cos(\omega t)$ is odd. Further, $\int_{-\infty}^{\infty} f(t)\sin(\omega t)\, dt = 2\int_{0}^{\infty} f(t)\sin(\omega t)\, dt$. In this case,

$$\mathscr{F}\{f(t)\} = -2i\int_{0}^{\infty} f(t)\sin(\omega t)\, dt.$$

These observations lead us to define the *Fourier cosine transform* of a function f defined on $[0, \infty)$ as

$$\mathscr{F}_C\{f(t)\} = \int_{0}^{\infty} f(t)\cos(\omega t)\, dt = F_C(\omega)$$

and the *Fourier sine transform* of f as

$$\mathscr{F}_S\{f(t)\} = \int_{0}^{\infty} f(t)\sin(\omega t)\, dt = F_S(\omega).$$

Notice that the Fourier cosine transform is the coefficient $A(\omega)$ in the Fourier cosine integral representation of f. If this cosine integral equals f on $[0, \infty)$, we can write

$$f(t) = \frac{2}{\pi}\int_{0}^{\infty} F_C(\omega)\cos(\omega t)\, d\omega.$$

This is an *inversion formula for the Fourier cosine transform*. Given F_C, it enables us to recover a function f whose Fourier cosine transform is F_C.

   Similarly, the Fourier sine transform is the coefficient $B(\omega)$ in the Fourier sine integral representation of f; hence,

$$f(t) = \frac{2}{\pi}\int_{0}^{\infty} F_S(\omega)\sin(\omega t)\, d\omega.$$

This is an *inversion formula for the Fourier sine transform*. Given F_S, it enables us to recover a function f whose Fourier sine transform is F_S.

EXAMPLE 17.29

Let

$$f(t) = \begin{cases} 1, & 0 \le t \le K \\ 0, & t > K. \end{cases}$$

Then

$$\mathcal{F}_C\{f(t)\} = \int_0^\infty f(t)\cos(\omega t)\, dt = \int_0^K \cos(\omega t)\, dt = \frac{\sin(\omega K)}{\omega} = F_C(\omega)$$

and

$$\mathcal{F}_S\{f(t)\} = \int_0^\infty f(t)\sin(\omega t)\, dt = \int_0^K \sin(\omega t)\, dt = \frac{1}{\omega}[1 - \cos(\omega K)] = F_S(\omega). \quad \blacksquare$$

As a second example, a straightforward calculation shows that

$$\mathcal{F}_C\{e^{-t}\} = \frac{1}{1 + \omega^2} \quad \text{and} \quad \mathcal{F}_S\{e^{-t}\} = \frac{\omega}{1 + \omega^2}.$$

It follows immediately from the definition that the Fourier cosine and sine transforms are linear:

$$\mathcal{F}_C\{\alpha f(t) + \beta g(t)\} = \alpha\mathcal{F}_C\{f(t)\} + \beta\mathcal{F}_C\{g(t)\}$$

and

$$\mathcal{F}_S\{\alpha f(t) + \beta g(t)\} = \alpha\mathcal{F}_S\{f(t)\} + \beta\mathcal{F}_S\{g(t)\}$$

whenever the transforms on the right exist. These relationships may also be written

$$\mathcal{F}_C\{\alpha f(t) + \beta g(t)\} = \alpha F_C(\omega) + \beta G_C(\omega)$$

and

$$\mathcal{F}_S\{\alpha f(t) + \beta g(t)\} = \alpha F_S(\omega) + \beta G_S(\omega).$$

The following theorems provide the operational formulas which are used to apply the Fourier sine and cosine transforms to the solution of boundary value problems.

THEOREM 17.23

Suppose that f and f' are continuous on $[0, \infty)$ and that $f(t) \to 0$ and $f'(t) \to 0$ as $t \to \infty$. Assume that f'' is piecewise continuous on $[0, k]$ for every $k > 0$. Then

$$\mathcal{F}_S\{f''(t)\} = -\omega^2 F_S(\omega) + \omega f(0).$$

Proof Integrate by parts:

$$\mathcal{F}_S\{f''(t)\} = \int_0^\infty f''(t)\sin(\omega t)\, dt = \left[f'(t)\sin(\omega t) \right]_0^\infty - \int_0^\infty f'(t)[\omega\cos(\omega t)]\, dt$$

$$= -\omega\left\{ \left[\cos(\omega t)f(t) \right]_0^\infty - \int_0^\infty f(t)[-\omega\sin(\omega t)]\, dt \right\}$$

$$= -\omega^2 F_S(\omega) + \omega f(0). \quad \blacksquare$$

THEOREM 17.24

Assume that f and f' are continuous on $[0, \infty)$ and that $f'(t) \to 0$ as $t \to \infty$. Assume that f'' is piecewise continuous on $[0, k]$ for every $k > 0$. Then

$$\mathscr{F}_C\{f''(t)\} = -\omega^2 F_C(\omega) - f'(0). \quad \blacksquare$$

The proof is by integration by parts, as in Theorem 17.23.

Theorems 17.23 and 17.24 provide the key to deciding which transform to use to solve a boundary value problem. If the problem requires a solution defined on the entire real line, we might employ a Fourier transform. If the problem is defined on only $[0, \infty)$, a Fourier sine or cosine transform might be used. If data are provided in terms of $f(0)$, we would try the Fourier sine transform because the formula in Theorem 17.23 contains $f(0)$. If, however, data are given in terms of $f'(0)$, we would use the Fourier cosine transform because the formula in Theorem 17.24 contains $f'(0)$.

In Section 18.11, we will illustrate the use of Fourier sine and cosine transforms in solving partial differential equaions. Short tables of sine and cosine transforms are provided at the end of Chapter 18.

PROBLEMS FOR SECTION 17.12

In each of Problems 1 through 10, find the Fourier cosine transform of f.

1. $f(t) = e^{-t}$

2. $f(t) = \begin{cases} \cos(t), & 0 \le t \le K \\ 0, & t > K \end{cases}$

3. $f(t) = \begin{cases} t^2, & 0 \le t \le K \\ 0, & t > K \end{cases}$

4. $f(t) = \begin{cases} 1, & 0 \le t \le K \\ -1, & K < t \le 2K \\ 0, & t > 2K \end{cases}$

5. $f(t) = \begin{cases} 2t, & 0 \le t \le 5 \\ 0, & t > 5 \end{cases}$

6. $f(t) = \begin{cases} 0, & 0 \le t \le K \\ \sinh(t), & K < t < 2K \\ 0, & t \ge 2K \end{cases}$

7. $f(t) = te^{-at} \quad (a > 0)$

8. $f(t) = t^2 e^{-at} \quad (a > 0)$

9. $F(t) = e^{-t}\cos(t)$

10. $f(t) = \dfrac{1}{1 + t^2}$

In each of Problems 11 through 18, find the Fourier sine transform of f.

11. $f(t) = \begin{cases} 2, & 0 \le t \le K \\ -1, & K < t \le 2K \\ 0, & t > 2K \end{cases}$

12. $f(t) = e^{-4t}$

13. $f(t) = te^{-at} \quad (a > 0)$

14. $f(t) = te^{-at^2} \quad (a > 0)$

15. $f(t) = e^{-t}\sin(t)$

16. $f(t) = e^{-t}\cos(t)$

17. $f(t) = \begin{cases} \sin(Kt), & 0 \le t \le 1 \\ 0, & t > 1 \end{cases}$

18. $f(t) = \begin{cases} \cosh(Kt), & 0 \le t \le 3 \\ 0, & t > 3 \end{cases}$

19. Show that, under appropriate conditions,

$$\mathscr{F}_S\{f^{(4)}(t)\} = \omega^4 F_S(\omega) - \omega^3 f(0) + \omega f''(0).$$

State conditions on f and its derivatives which allow the use of Theorem 17.23 to derive this formula.

20. Show that, under appropriate conditions,

$$\mathscr{F}_C\{f^{(4)}(t)\} = \omega^4 F_C(\omega) + \omega^2 f'(0) - f^{(3)}(0).$$

State conditions on f and its derivatives which allow the use of Theorem 17.24 to derive this formula.

17.13 *The Discrete Fourier Transform*

The fact that we can determine the frequency content of a signal by analyzing its Fourier transform makes this transform a valuable tool in signal analysis. The major problem encountered is in evaluating the continuous spectrum defined as an integral over an infinite interval. Modern computing power can be brought to bear on this problem *if* a method can be found to approximate the integral using only a finite number of operations while still salvaging the crucial information that it contains.

The discrete Fourier transform provides just such a method. It is a scheme for numerically approximating the Fourier transform $\mathscr{F}\{f\}$ of a function f. Recall that, under appropriate conditions on f, its Fourier transform is

$$\mathscr{F}\{f\}(\omega) = F(\omega) = \int_{-\infty}^{\infty} f(t)e^{-i\omega t}\, dt.$$

We will now define the *discrete Fourier transform G* of f by setting

$$G\left(\frac{2n\pi}{NT}\right) = T\sum_{k=0}^{N-1} f(kT)e^{-2nk\pi i/N} \quad \text{for} \quad n = 0, 1, 2, \ldots, N-1,$$

in which $T > 0$ (T is the *sampling period*) and N is a positive integer. Note that G is a periodic function of period $2\pi/T$ which approximates the function F on the interval $[-\pi/T, \pi/T]$.

If F and G "nearly" agree at the points $2n\pi/NT$, this transformation yields approximate values of F at N equally spaced points by evaluating f at N equally spaced points. This type of calculation is quite suitable for implementation on a computer. The accuracy of this approximation to the continuous Fourier transform is a function of both the sampling period T and the number N of samples used.

A careful investigation of the mathematics behind the discrete Fourier transform would involve us in the theory of distributions, which is a level above our treatment in this book. Nevertheless, even without this theory, we can indicate how the discrete Fourier transform arises and how its accuracy can be improved. We will outline the idea graphically and then deal with some of the theory behind the transform.

First, we need some preliminary material. Begin by defining an *infinite train of impulses* as

$$d(t) = \sum_{n=-\infty}^{\infty} \delta(t - nT),$$

in which δ is the Dirac delta function. A graph of $d(t)$ is shown in Figure 17.53. Note that d is periodic of period T. This function and its Fourier transform will play important roles in the sampling process needed to approximate $F(\omega)$.

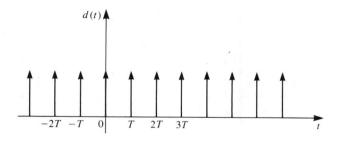

Figure 17.53

THEOREM 17.25

$$\mathscr{F}\{d(t)\} = \frac{2\pi}{T} \sum_{n=-\infty}^{\infty} \delta\left(\omega - \frac{2n\pi}{T}\right).$$

That is, the Fourier transform of an infinite train of impulses is again an infinite train of impulses.

Proof   Since d is periodic of periodic T, we can express it with a complex Fourier series

$$d(t) = \sum_{n=-\infty}^{\infty} c_n e^{2n\pi it/T},$$

where

$$c_n = \frac{1}{T} \int_0^T \delta(t) e^{-2n\pi it/T} \, dt.$$

By the filtering property of the Dirac delta function,

$$c_n = \frac{1}{T} e^0 = \frac{1}{T};$$

hence,

$$d(t) = \sum_{n=-\infty}^{\infty} \delta(t - nT) = \sum_{n=-\infty}^{\infty} \frac{1}{T} e^{2n\pi it/T}. \tag{17.18}$$

But we also know that $\mathscr{F}\{1\} = 2\pi\delta(\omega)$. Use this and the frequency-shifting property of the Fourier transform [which states that $\mathscr{F}\{e^{i\omega_0 t}f(t)\} = F(\omega - \omega_0)$] in equation (17.18) to obtain

$$\mathscr{F}\{d(t)\} = \frac{2\pi}{T} \sum_{n=-\infty}^{\infty} \delta\left(\omega - \frac{2n\pi}{T}\right),$$

as was to be proved.   ∎

As Figure 17.54 indicates, this is a good example of the scaling property of the Fourier transform. Similarly, we can show that, if f is periodic of period T, then F, the Fourier transform of f, is an infinite train of impulses. We will state this result as Theorem 17.26 and leave the proof to the problems.

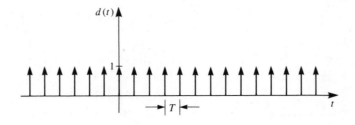

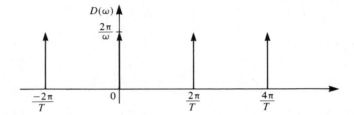

Figure 17.54

THEOREM 17.26

Suppose that f is periodic of period T. Then

$$\mathscr{F}\{f(t)\} = 2\pi \sum_{n=-\infty}^{\infty} c_n \,\delta\!\left(\omega - \frac{2n\pi}{T}\right),$$

where

$$c_n = \frac{1}{T} \int_0^T f(t) e^{-2\pi int/T}\, dt. \quad \blacksquare$$

One of the conclusions of distribution theory is the "product" formula

$$f(t)\,\delta(t - t_0) = f(t_0)\,\delta(t - t_0).$$

Using this, we may write the product of a function f and an infinite train d of impulses as

$$f_s(t) = f(t)\,d(t) = f(t) \sum_{n=-\infty}^{\infty} \delta(t - nT)$$

$$= \sum_{n=-\infty}^{\infty} f(t)\,\delta(t - nT)$$

$$= \sum_{n=-\infty}^{\infty} f(nT)\,\delta(t - nT).$$

This product function f_s is referred to as a *sampled function*, or a *sampling of f*. It is a sequence of impulses located T units apart having strength $f(t)$ at the sample point t as depicted in Figure 17.55. We will also need the following fact about the convolution of a function with a shifted Dirac delta function.

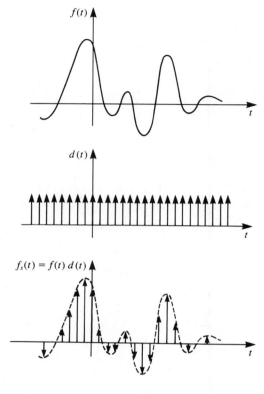

Figure 17.55

THEOREM 17.27

$$g(t) * \delta(t - a) = g(t - a).$$

Proof   Use the filtering property of the Dirac delta function to write

$$g(t) * \delta(t - a) = \int_{-\infty}^{\infty} g(\tau)\,\delta(t - \tau - a)\,d\tau = g(t - a). \quad \blacksquare$$

Theorem 17.27 can be used to find the convolution of a function g with an infinite train of impulses. For future reference, this result is listed as a theorem.

THEOREM 17.28

$$g(t) * d(t) = \sum_{n=-\infty}^{\infty} g(t - nT).$$

Proof   Write

$$g(t) * d(t) = g(t) * \sum_{n=-\infty}^{\infty} \delta(t - nT) = \sum_{n=-\infty}^{\infty} g(t) * \delta(t - nT) = \sum_{n=-\infty}^{\infty} g(t - nT). \quad \blacksquare$$

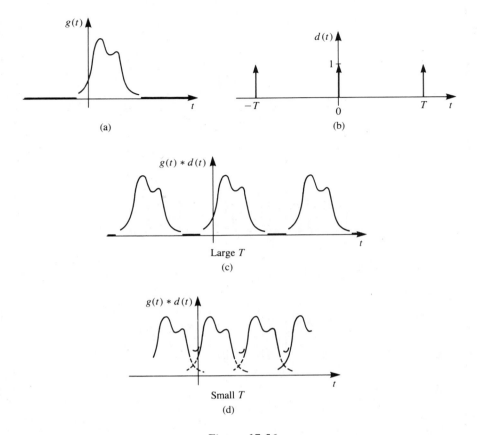

Figure 17.56

Figures 17.56(c) and (d) point out that, if g is time limited (that is, if $g(t) = 0$ for t outside some interval $[a, b]$), then $g(t) * d(t)$ produces periodic copies of g with period T, unless T is too small, in which case we have some overlapping in the copies of g. If g is not time limited, overlapping will occur regardless of the size of T.

We can now describe graphically how the discrete Fourier transform is related to the Fourier transform. Figure 17.57 gives the complete picture. The left column contains a sequence of functions, and the right column, their Fourier transforms. Begin with a function f and the infinite train of unit impulses d_1 defined by

$$d_1(t) = T \sum_{n=-\infty}^{\infty} \delta(t - nT).$$

Notice that d_1 is periodic with the sampling period T.

Now multiply f and d_1 to obtain the infinite sequence of impulses shown in Figure 17.57(c). The Fourier transform of this product is the convolution of $F(\omega)$ and $D_1(\omega)$, the Fourier transform of d_1. This creates the first deviation away from the true value of F.

The new function $F(\omega) * D_1(\omega)$ is periodic of period $2\pi/T$, and unless f is band limited [that is, for some number M, $F(\omega) = 0$ for $\omega > M$], overlapping of the copies of

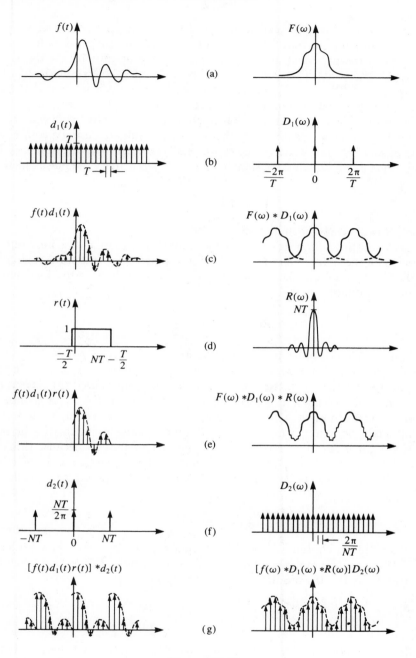

Figure 17.57

F occurs. This phenomenon is called *aliasing*. The amount of frequency aliasing can be reduced by sampling the time function at a faster rate (choosing T smaller), which in turn increases the period of D_1. Engineers use a rule of thumb that the sampling frequency should be at least as large as twice the highest frequency contained in f. This value is called the *Nyquist sampling rate*. Of course, unless f is band limited, it does not have a highest frequency component and some aliasing has to be accepted.

Next, note that, even after sampling, there is an infinite number of values of f to calculate. If the final approximating scheme is to be evaluated using a computer, the number of samples must be decreased substantially (at least to a finite number). This is accomplished by multiplying the sampled function $f(t)\,d_1(t)$ by the pulse function

$$r(t) = H\left(t + \frac{T}{2}\right) - H\left(t - NT + \frac{T}{2}\right),$$

in which H is the Heaviside function.

This truncating action reduces to N the number of values of f we must determine. The reason for the shift of half of one sampling period to the right of zero is to prevent time aliasing by adding a sampled value twice when evaluating the final sum. The Fourier transform of r is

$$\mathscr{F}\{r(t)\} = R(\omega) = \frac{2}{\omega} \sin\left(\frac{NT\omega}{2}\right) e^{-i\omega(N+1)T/2},$$

the graph of which is shown in Figure 17.57(d).

The convolution of R with $F * D_1$ introduces a ripple in the function. This ripple effect can be decreased by increasing the number of samples used (by increasing N). This sharpens R, making it look more like a Dirac delta function. This is done at the expense of more sampling, which translates into an increase in computing time.

Although the function $G = F * D_1 * R$ approximates F, at least in the interval $(-\pi/T, \pi/T)$, it is still not acceptable for computer evaluation because it is continuous and we need discrete values. The final move is to sample G. This is done by taking the convolution of $f(t)\,d_1(t)r(t)$ with another train of impulses, specifically

$$d_2(t) = \frac{NT}{2\pi} \sum_{n=-\infty}^{\infty} \delta(t - nNT).$$

Here, the period is chosen as NT so that the frequency-sampling interval length is $2\pi/NT$. This way, N samples will exactly cover one period of G. The transformation of $f(t)\,d_1(t)r(t) * d_2(t)$ is the product of $G(\omega)$ and $D_2(\omega)$, where

$$D_2(\omega) = \sum_{n=-\infty}^{\infty} \delta\left(\omega - \frac{2n\pi}{NT}\right).$$

The discrete function $G \cdot D_2 = [F(\omega) * D_1(\omega) * R(\omega)]D_2(\omega)$ has values at the points $2n\pi/NT$ for $n = 0, 1, 2, \ldots, N - 1$, which are nearly the values of $F(\omega)$ at these points.

It must be remembered that the function G is periodic, so the approximation to F is valid throughout only one period of this function on the interval $[-\pi/T, \pi/T]$. In practice, it is much easier to evaluate G on the interval $[0, 2\pi/T]$ than on $[-\pi/T, \pi/T]$. Since G has period $2\pi/T$, the values attained in $[\pi/T, 2\pi/T]$ agree with those in $[-\pi/T, 0]$, so approximations of F in $[-\pi/T, 0]$ can be obtained from the values of

G in $[\pi/T, 2\pi/T]$. To estimate the values of F outside of this range, we would have to increase the sample rate (that is, decrease the value of T).

We will now mathematically duplicate the procedure outlined by the graphs of Figure 17.57 to derive the discrete Fourier transform. That is, we will attempt to clarify the relationship between the discrete values of G at the points $2n\pi/T$ for $n = 0, 1, 2, \ldots,$ $N - 1$ and the discrete values of f at the points kT for $k = 0, 1, 2, \ldots, N - 1$.

Begin by sampling f at a sampling rate T by multiplying f and d_1, where

$$d_1(t) = T \sum_{k=-\infty}^{\infty} \delta(t - kT).$$

This sampling of f is expressible as

$$f(t)\, d_1(t) = T \sum_{k=-\infty}^{\infty} f(kT)\, \delta(t - kT).$$

Next, truncate this product by multiplying it by $r(t)$, as defined previously. This reduces the sum to a finite number of terms:

$$f(t)\, d_1(t) r(t) = \left[T \sum_{k=-\infty}^{\infty} f(kT)\, \delta(t - kT) \right] r(t) = T \sum_{k=0}^{N-1} f(kT)\, \delta(t - kT).$$

Now finish conditioning the signal by forming the convolution of this finite sum with the infinite train of impulses d_2, of period NT, where

$$d_2(t) = \frac{NT}{2\pi} \sum_{m=-\infty}^{\infty} \delta(t - mNT).$$

By Theorem 17.28, we obtain

$$[f(t)\, d_1(t) r(t)] * d_2(t) = \left[T \sum_{k=0}^{N-1} f(kT)\, \delta(t - kT) \right] * \left[\frac{NT}{2\pi} \sum_{m=-\infty}^{\infty} \delta(t - mNT) \right]$$

$$= \frac{NT^2}{2\pi} \sum_{m=-\infty}^{\infty} \sum_{k=0}^{N-1} f(kT)\, \delta(t - kT - mNT). \tag{17.19}$$

Denote this as $g(t)$. Then $G(\omega) = \mathcal{F}\{g(t)\}$ is a discrete approximation of $F(\omega)$. To see what it looks like, first notice that g is periodic of period NT; hence,

$$g(t) = \sum_{n=-\infty}^{\infty} c_n e^{2n\pi it/NT},$$

where

$$c_n = \frac{1}{NT} \int_{-T/2}^{NT - T/2} g(t) e^{-2n\pi it/NT}\, dt$$

for $n = 0, \pm 1, \pm 2, \ldots$. Hence, by Theorem 17.26, we know that the Fourier transform $G(\omega)$ of $g(t)$ will be an infinite train of impulses

$$G(\omega) = 2\pi \sum_{n=-\infty}^{\infty} c_n\, \delta\!\left(\omega - \frac{2n\pi}{NT} \right),$$

with the c_n's defined above by integrals involving g. If we can find the c_n's, we will know G.

Put the value of $g(t)$ from equation (17.19) in the integral for c_n to get

$$c_n = \frac{1}{NT} \int_{-T/2}^{NT-T/2} \left[\frac{NT^2}{2\pi} \sum_{m=-\infty}^{\infty} \sum_{k=0}^{N-1} f(kT)\, \delta(t - kT - mNT) \right] e^{-2n\pi it/NT} \, dt.$$

The only value of m for which $t - kT - mNT$ is in the interval $[-T/2, NT - T/2]$ is zero. Hence, this expression for c_n reduces to

$$c_n = \frac{T}{2\pi} \int_{-T/2}^{NT-T/2} \sum_{k=0}^{N-1} f(kT)\, \delta(t - kT) e^{-2n\pi it/NT} \, dt.$$

Now use the filtering property of the Dirac delta function in this expression to obtain

$$c_n = \frac{T}{2\pi} \sum_{k=0}^{N-1} f(kT) e^{-2n\pi ikT/NT} = \frac{T}{2\pi} \sum_{k=0}^{N-1} f(kT) e^{-2n\pi ik/N}.$$

Put this into the equation for $G(\omega)$ to get

$$G(\omega) = 2\pi \sum_{n=-\infty}^{\infty} \frac{T}{2\pi} \sum_{k=0}^{N-1} f(kT) e^{-2n\pi ik/N} \delta\left(\omega - \frac{2n\pi}{NT} \right).$$

Therefore,

$$g(\omega) = \begin{cases} 0 & \text{if } \omega \neq 2n\pi/NT \\ \sum\limits_{k=0}^{N-1} f(kT) e^{-2n\pi ik/N} & \text{if } \omega = 2n\pi/NT \end{cases}$$

for $n = 0, 1, 2, \ldots, N - 1$ and $G(\omega + 2\pi/T) = G(\omega)$. The nonzero values of G are approximations of F at these points. We then have

$$G\left(\frac{2n\pi}{NT} \right) = T \sum_{k=0}^{N-1} f(kT) e^{-2n\pi ik/N}$$

for $n = 0, 1, 2, \ldots, N - 1$. Further,

$$F\left(\frac{2n\pi}{NT} \right) \approx G\left(\frac{2n\pi}{NT} \right)$$

if $0 \leq 2n\pi/NT \leq \pi/T$, and

$$F\left(\frac{2n\pi}{NT} \right) \approx G\left(\frac{2n\pi}{NT} \right) = G\left(\frac{2n\pi}{NT} + \frac{2\pi}{T} \right)$$

if $-\pi/T \leq 2n\pi/NT \leq 0$.

The discrepancy between the values of F and G can be made small at the expense of additional computation, but it can never be reduced to zero unless the signal function f is band limited and periodic, the sampling rate exceeds twice the highest frequency contained in f, and the truncation function is chosen to match one period of f. The only functions for which all of these conditions hold are finite sums of sines and cosines, which are exactly the functions with finite Fourier series.

The *inverse discrete Fourier transform* is defined by

$$g(kT) = \frac{1}{NT} \sum_{k=0}^{N-1} F\left(\frac{2n\pi}{NT} \right) e^{2\pi ink/N}$$

for $k = 0, 1, 2, \ldots, N - 1$. Here, $g(kT)$ gives approximate values of $f(kT)$, where $F(\omega) = \mathscr{F}\{f(t)\}$.

In Problem 16 at the end of this section, the student is asked to verify that the inverse discrete Fourier transform, as we have defined it, is indeed the inverse of the discrete Fourier transform.

We will explore the use of the discrete Fourier transform to approximate the Fourier transform of a function.

EXAMPLE 17.30

Let

$$f(t) = \begin{cases} 0 & \text{if } \quad t < 0 \\ e^{-t} & \text{if } \quad t \geq 0. \end{cases}$$

First, we must choose the sampling period T and the number of samples N. We will choose $T = 0.5$ and $N = 8$. We will also redefine $f(0)$ to be the average of 1 [which is $\lim_{t \to 0+} f(t)$] and zero [which is approximately $\lim_{t \to 4-} f(t)$]. As with every other Fourier transform, series, or integral we have encountered, the transform pair will converge to the average of the left and right limits of the function at points of discontinuity. When integrating to find the transform, the value of the function at one point is insignificant. Here, however, in the finite case, every value is important. Figure 17.58 shows a graph of f and its sampled values.

The Fourier transform of f is

$$F(\omega) = \frac{1}{1 - i\omega} = \frac{1}{1 + \omega^2} + \frac{i\omega}{1 + \omega^2}.$$

Since f is real valued, the real part of $F(\omega)$ is an even function and the imaginary part is odd. Further,

$$|F(\omega)| = \frac{1}{\sqrt{1 + \omega^2}}.$$

Recall that the amplitude spectrum of f is the absolute value of its Fourier transform. This is the function which contains the information about the frequency content of f, so the absolute values of the values of G can be used to study the spectrum of f.

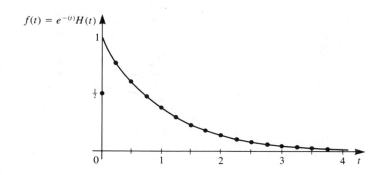

Figure 17.58

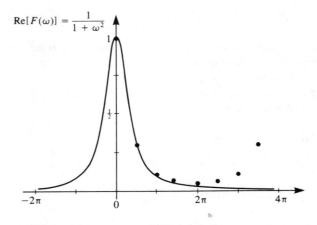

$$\mathrm{Re}[F(\omega)] = \frac{1}{1 + \omega^2}$$

Figure 17.59

Figure 17.59 shows the graph of the real part of F and values of the real part of its approximating function G at points $2\pi/NT$. Here, we can see that the periodicity of G can be used to approximate values of F at negative multiples of $2\pi/NT$. The approximations are surprisingly good, considering that we used a small number of data points.

Figure 17.60 shows a graph of the imaginary part of F in addition to the values of the imaginary values of G at the frequency domain sampling points. Here again, we can see that knowing the values of G in $[\pi/T, 2\pi/T]$ is sufficient to find the values of F in $[-\pi/T, 0]$. In this case, the approximations do not appear to be very accurate. In this example, there is not much frequency aliasing because F decreases rapidly with ω. Consequently, to improve the accuracy of the approximations, increasing the value of N would be more efficient than increasing T. Here, we have the advantage of knowing the Fourier transform of the signal we are studying, a luxury we would not normally have. In practice, we usually decrease T until further reductions result in very little

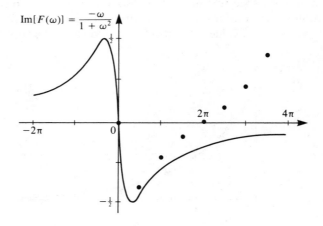

$$\mathrm{Im}[F(\omega)] = \frac{-\omega}{1 + \omega^2}$$

Figure 17.60

change in discrete transform values. Then the truncation size is increased until only small changes in output occur with changes in N. ∎

In the next section, we will discuss the fast Fourier transform, which is a numerical procedure for reducing the computation time used when evaluating the discrete Fourier transform.

PROBLEMS FOR SECTION 17.13

In each of Problems 1 through 6, use the given N discrete values of the function f to determine $G(2n\pi/NT)$ for $n = 0, 1, 2, \ldots, N - 1$. Euler's formula $e^{ix} = \cos(x) + i \sin(x)$ may prove useful.

1. $f(0) = 3$, $f(1) = 2$, $f(2) = 1$, $f(3) = 0$

2. $f(0) = 2$, $f(1) = -3$, $f(2) = -1$, $f(3) = 4$

3. $f(0) = 2$, $f(1/3) = 1$, $f(2/3) = 0$, $f(1) = 2$, $f(4/3) = 1$, $f(5/3) = 0$

4. $f(0) = 4$, $f(1/3) = 1$, $f(2/3) = -3$, $f(1) = 2$, $f(4/3) = -1$, $f(5/3) = 5$

5. $f(0) = 1/2$, $f(1/2) = 1$, $f(1) = 1$, $f(3/2) = 1$, $f(2) = 1/2$, $f(5/2) = 0$, $f(3) = 0$, $f(7/2) = 0$

6. $f(0) = 0$, $f(1/2) = 1/2$, $f(1) = 1$, $f(3/2) = 1/2$, $f(2) = 0$, $f(5/2) = 0$, $f(3) = 0$, $f(7/2) = 0$

7. Suppose that $N = 8$, $T = \frac{1}{2}$, and

$$f(t) = \begin{cases} 1 & \text{if} & 0 \le t \le 2 \\ 0 & \text{if} & t < 0 \quad \text{or} \quad t > 2. \end{cases}$$

Determine $G(2n\pi/NT)$ for $n = 0, 1, 2, \ldots, N - 1$. *Hint:* Redefine $f(0)$ and $f(2)$.

8. Suppose $N = 8$, $T = \frac{1}{2}$, and

$$f(t) = \begin{cases} t & \text{if} & 0 \le t \le 1 \\ 2 - t & \text{if} & 1 \le t \le 2 \\ 0 & \text{if} & t < 0 \quad \text{or} \quad t > 2. \end{cases}$$

Find $G(2n\pi/NT)$ for $n = 0, 1, 2, \ldots, N - 1$.

In each of Problems 9 and 10, use the given function d to graph $g(t) * d(t)$, where

$$g(t) = \begin{cases} t & \text{if} & 0 \le t \le 2 \\ 4 - t & \text{if} & 2 \le t \le 4 \\ 0 & \text{if} & t < 0 \quad \text{or} \quad t > 4. \end{cases}$$

9. $d(t) = \displaystyle\sum_{n=-\infty}^{\infty} \delta(t - 6n)$ **10.** $d(t) = \displaystyle\sum_{n=-\infty}^{\infty} \delta(t - 3n)$

11. Prove Theorem 17.26.

12. Prove that, if f is a real-valued function with Fourier transform F, $\text{Re}[F(\omega)]$, the real part of $F(\omega)$, is an even function and $\text{Im}[F(\omega)]$, the imaginary part of $F(\omega)$, is an odd function.

13. Suppose that f is a real-valued function, and let

$$G\left(\frac{2n\pi}{NT}\right) = T \sum_{k=0}^{N-1} f(kT)e^{-2n\pi ik/N} = \alpha(n) + i\beta(n).$$

Prove that $G[2(N - n)\pi/NT] = \alpha(n) - i\beta(n)$, the complex conjugate of $G(2n\pi/NT)$.

14. Derive the discrete Fourier transform as a Riemann sum as follows. Begin by approximating $F(\omega)$ with a finite integral:

$$F(\omega) = \int_{-\infty}^{\infty} f(t)e^{-i\omega t}\, dt \approx \int_{0}^{NT} f(t)e^{-i\omega t}\, dt.$$

Recall that a Riemann sum of a function h relative to a partition $a = x_0 < x_1 < \cdots < x_n = b$ of $[a, b]$ is a sum $\sum_{k=0}^{n} h(\xi_k)(x_k - x_{k-1})$, with ξ_k any point of $[x_{k-1}, x_k]$ for $k = 1, 2, \ldots, n$. This sum is an approximation of the integral $\int_a^b h(t)\, dt$.

Now choose a partition of $[0, NT]$ using N subintervals of equal length. In the kth subinterval, choose ξ_k as the left endpoint of that interval. The resulting Riemann sum, for the function $f(t)e^{-i\omega t}$, approximates $F(\omega)$. Complete the derivation by evaluating this finite sum at $\omega = 2n\pi/NT$.

15. Prove that

$$\sum_{k=0}^{N-1} e^{2\pi i k(m-n)/N} = \begin{cases} N & \text{if} \quad m = n \\ 0 & \text{if} \quad m \neq n. \end{cases}$$

Hint: Write the series as a finite geometric series with common ratio $[e^{2\pi i/N}]^{m-n}$.

16. Use the result of Problem 15 to prove that the quantity we called the inverse discrete Fourier transform is indeed the inverse of the discrete Fourier transform. That is, show that, if

$$f(kT) = \frac{1}{NT} \sum_{m=0}^{N-1} G\left(\frac{2m\pi}{NT}\right) e^{2\pi i m k/N}$$

for $k = 0, 1, 2, \ldots, N - 1$, then

$$G\left(\frac{2n\pi}{NT}\right) = T \sum_{k=0}^{N-1} f(kT) e^{-2\pi i n k/N}$$

for $n = 0, 1, 2, \ldots, N - 1$.

17.14 *The Fast Fourier Transform*

The fast Fourier transform (FFT) is not itself a transform. Rather, it is an algorithm which helps reduce the computation time required to evaluate the discrete Fourier transform. In fact, there are several such algorithms to which the name "fast Fourier transform" is routinely applied. We will explore the idea behind these algorithms by developing specific ones.

Consider the discrete Fourier transform of f, given by

$$F(n) = \sum_{k=0}^{N-1} f(k) e^{-2nk\pi i/N},$$

in which we have replaced n/NT by n and kT by k and removed the factor T to simplify the discussion. If we let $W = e^{-2\pi i/N}$, then

$$F(n) = \sum_{k=0}^{N-1} f(k) W^{nk}$$

for $n = 0, 1, 2, \ldots, N - 1$.

We want a procedure for reducing the labor needed to compute $F(n)$. The particular procedure we will describe (a fast Fourier transform) is called the *base-2 FFT*. In this method, we assume that N is a power of 2, say $N = 2^\gamma$ for some positive integer γ. To begin with a simple case, we will let $\gamma = 2$ and choose simply $N = 4$.

We therefore wish to determine the values of $F(n)$ for $n = 0, 1, 2, 3$. We have

$$F(0) = f(0)W^0 + f(1)W^0 + f(2)W^0 + f(3)W^0,$$
$$F(1) = f(0)W^0 + f(1)W^1 + f(2)W^2 + f(3)W^3,$$
$$F(2) = f(0)W^0 + f(1)W^2 + f(2)W^4 + f(3)W^6,$$

and

$$F(3) = f(0)W^0 + f(1)W^3 + f(2)W^6 + f(3)W^9.$$

Write this system of equations in matrix form as

$$\begin{bmatrix} F(0) \\ F(1) \\ F(2) \\ F(3) \end{bmatrix} = \begin{bmatrix} W^0 & W^0 & W^0 & W^0 \\ W^0 & W^1 & W^2 & W^3 \\ W^0 & W^2 & W^4 & W^6 \\ W^0 & W^3 & W^6 & W^9 \end{bmatrix} \begin{bmatrix} f(0) \\ f(1) \\ f(2) \\ f(3) \end{bmatrix}. \tag{17.20}$$

Since W and each $f(k)$ may be complex, it will take in the neighborhood of sixteen, or N^2, complex multiplications and twelve, or $N(N-1)$, complex additions to find the N values of F that we seek. Generally, when estimating computer time spent on arithmetic operations, the major consideration is the number of multiplications required to complete the job. Operations such as addition, storing, and retrieving consume much less time than multiplication.

Notice that, since $W^0 = 1$, we can calculate the product of W^0 and any number without using any multiplication at all. If we are willing to keep track of the locations of the W^0 entries of the matrix, we can reduce the number of complex multiplications to no more than $(N-1)^2$. Fast Fourier transforms are methods which exploit ideas such as this to reduce the number of multiplications required to find $F(0), F(1), \ldots, F(N-1)$. To illustrate, we will rewrite equation (17.20) in such a way that fewer computer multiplications will be required to perform the indicated matrix multiplication.

In Problem 1 at the end of this section, the student is asked to show that $W^N = W^0 = 1$ and, if k and m are integers, that $W^{mN+k} = W^k$. For example, when $N = 4$, $W^6 = W^2$ and $W^9 = W^1$. Using this fact, rewrite equation (17.20) as

$$\begin{bmatrix} F(0) \\ F(1) \\ F(2) \\ F(3) \end{bmatrix} = \begin{bmatrix} 1 & 1 & 1 & 1 \\ 1 & W & W^2 & W^3 \\ 1 & W^2 & W^0 & W^2 \\ 1 & W^3 & W^2 & W^1 \end{bmatrix} \begin{bmatrix} f(0) \\ f(1) \\ f(2) \\ f(3) \end{bmatrix}. \tag{17.21}$$

Now factor the $N \times N$ matrix into two $N \times N$ matrices and write this equation as

$$\begin{bmatrix} F(0) \\ F(2) \\ F(1) \\ F(3) \end{bmatrix} = \begin{bmatrix} 1 & W^0 & 0 & 0 \\ 1 & W^2 & 0 & 0 \\ 0 & 0 & 1 & W^1 \\ 0 & 0 & 1 & W^3 \end{bmatrix} \begin{bmatrix} 1 & 0 & W^0 & 0 \\ 0 & 1 & 0 & W^0 \\ 1 & 0 & W^2 & 0 \\ 0 & 1 & 0 & W^1 \end{bmatrix} \begin{bmatrix} f(0) \\ f(1) \\ f(2) \\ f(3) \end{bmatrix}. \tag{17.22}$$

The role of the FFT is actually to provide this factorization, which at this point must seem to have come from nowhere. Soon we will see how this is done. For now, however, it is easy to check that the product of the two 4×4 matrices in equation (17.22) equals the 4×4 matrix of equation (17.21), with the second and third rows interchanged. We have compensated for this in equation (17.22) by interchanging $F(1)$ and $F(2)$ on the left.

Now let us see what we have gained. Define

$$\begin{bmatrix} f_1(0) \\ f_1(1) \\ f_1(2) \\ f_1(3) \end{bmatrix} = \begin{bmatrix} 1 & 0 & W^0 & 0 \\ 0 & 1 & 0 & W^0 \\ 1 & 0 & W^2 & 0 \\ 0 & 1 & 0 & W^1 \end{bmatrix} \begin{bmatrix} f(0) \\ f(1) \\ f(2) \\ f(3) \end{bmatrix}.$$

To determine $f_1(0)$, we must perform one complex multiplication and one complex addition:

$$f_1(0) = f(0) + W^0 f(2).$$

(We have not reduced W^0 to 1 here because when we generalize these ideas, this term will usually have a different exponent.) It also takes one complex multiplication and one complex addition to compute $f_1(1)$:

$$f_1(1) = f(1) + W^0 f(3).$$

However, it takes only one complex addition (actually a subtraction) to determine $f_1(2)$, since $W^2 = -W^0$ and we have already performed the needed multiplication in finding $f_1(0)$. We have

$$f_1(2) = f(0) + W^2 f(2) = f(0) - W^0 f(2).$$

Finally, $f_1(3)$ can similarly be calculated with no multiplications and one addition:

$$f_1(3) = f(1) + W^2 f(3) = f(1) - W^0 f(3).$$

Thus far, in determining the intermediate matrix

$$\begin{bmatrix} f_1(0) \\ f_1(1) \\ f_1(2) \\ f_1(3) \end{bmatrix},$$

we have invested two complex multiplications and four complex additions. Pressing on, we have from equation (17.22) that

$$\begin{bmatrix} F(0) \\ F(2) \\ F(1) \\ F(3) \end{bmatrix} = \begin{bmatrix} f_2(0) \\ f_2(1) \\ f_2(2) \\ f_2(3) \end{bmatrix} = \begin{bmatrix} 1 & W^0 & 0 & 0 \\ 1 & W^2 & 0 & 0 \\ 0 & 0 & 1 & W^1 \\ 0 & 0 & 1 & W^3 \end{bmatrix} \begin{bmatrix} f_1(0) \\ f_1(1) \\ f_1(2) \\ f_1(3) \end{bmatrix}.$$

We can find $f_2(0)$ with one complex multiplication and one addition:

$$f_2(0) = f_1(0) + W^0 f_1(1).$$

We get $f_2(1)$ with one addition:

$$f_2(1) = f_1(0) - W^0 f_1(1).$$

In similar fashion, $f_2(2)$ will require one multiplication and one addition, and $f_2(3)$ one addition. The entire computation of $F(0), \ldots, F(3)$ from $f(0), \ldots, f(3)$ in this way costs a total of four complex multiplications and eight complex additions. Recall that, without the above factorization, the same result would have required nine multiplications and twelve additions.

In this discussion, we have let $\gamma = 2$. If $N = 2^\gamma$, the FFT is a method of factoring the appropriate $N \times N$ matrix into γ $N \times N$ matrices, each containing enough entries equal to zero or 1 to reduce substantially the total number of multiplications required to evaluate $F(0), F(1), \ldots, F(N-1)$.

In general, the number of multiplications required in the factored matrices is about $N\gamma/2$, with about $N\gamma$ additions needed. Without the factorization, about $(N-1)^2$ multiplications are needed, and $N(N-1)$ additions. If we assume that computing time is proportional to the number of multiplications, we conclude that FFT algorithms can produce a significant savings in computing time when N is large.

One issue we have slighted [in addition to not showing where we obtained the factorization in equation (17.22)] is the fact that two of the rows of the product of the factorization had to be interchanged to recover the original matrix. In fact, this is really a minor problem which is easily treated, even in the case in which γ is large. All that is required is to determine which rows are interchanged and then interchange those components of the F matrix.

We will now derive the matrix factorization (for $\gamma = 2$) and in the process determine which rows become interchanged. Associate with each of the integers $n = 0$, 1, 2, 3 the ordered pair (n_1, n_0), where $n = 2n_1 + n_0$ and n_1 is either zero or 1 and n_0 is zero or 1. Thus, we have the associations

$$0 \sim (0, 0), \quad 1 \sim (0, 1), \quad 2 \sim (1, 0), \quad \text{and} \quad 3 \sim (1, 1).$$

To determine which rows are interchanged in the case $N = 2^2$, use the *bit reversal* process of writing the integers in the ordered pairs in reverse order, which in the case of just two coordinates involves simply interchanging the coordinates:

$$0 \sim (0, 0) \quad \text{becomes} \quad (0, 0) \sim 0;$$
$$1 \sim (0, 1) \quad \text{becomes} \quad (1, 0) \sim 2;$$
$$2 \sim (1, 0) \quad \text{becomes} \quad (0, 1) \sim 1;$$
$$3 \sim (1, 1) \quad \text{becomes} \quad (1, 1) \sim 3.$$

Thus, row 0 is interchanged with row 0 (no change); row 1 is interchanged with row 2 and row 2 with row 1 (interchange rows 1 and 2); and row 3 is interchanged with row 3 (no change). This tells us how to decide the order in which to write the entries of the F matrix.

We will now see how the matrix factorization is determined for the base-2 FFT. Begin again with the discrete Fourier transform

$$F(n) = \sum_{k=0}^{N-1} f(k) W^{nk}$$

for $n = 0, 1, 2, \ldots, N - 1$. Here, $W = e^{-2\pi i/N}$. We are dealing with the case in which $N = 4$ and $\gamma = 2$. Define

$$\hat{F}(n_1, n_0) = F(2n_1 + n_0) = F(n)$$

and

$$\hat{f}(k_1, k_0) = f(2k_1 + k_0) = f(k).$$

In this notation, we have

$$F(n) = F(2n_1 + n_0) = \hat{F}(n_1, n_0)$$

$$= \sum_{k=0}^{3} f(k)W^{(2n_1 + n_0)k}$$

$$= \sum_{k_0=0}^{1} \sum_{k_1=0}^{1} f(2k_1 + k_0)W^{(2n_1 + n_0)(2k_1 + k_0)}$$

$$= \sum_{k_0=0}^{1} \sum_{k_1=0}^{1} \hat{f}(k_1, k_0)W^{(2n_1 + n_0)(2k_1 + k_0)}$$

for $n_0 = 0, 1; n_1 = 0, 1$. Now write

$$W^{(2n_1 + n_0)(2k_1 + k_0)} = W^{(2n_1 + n_0)2k_1}W^{(2n_1 + n_0)k_0}$$

$$= W^{4n_1k_1}W^{2n_0k_1}W^{(2n_1 + n_0)k_0}$$

$$= W^{2n_0k_1}W^{(2n_1 + n_0)k_0},$$

since $W^{4n_1k_1} = 1$. Therefore,

$$\hat{F}(n_1, n_0) = \sum_{k_0=0}^{1} \left[\sum_{k_1=0}^{1} \hat{f}(k_1, k_0)W^{2n_0k_1} \right] W^{(2n_1 + n_0)k_0} \qquad (17.23)$$

for $n_0 = 0, 1; n_1 = 0, 1$. Denote the inner sum as

$$\hat{f}_1(n_0, k_0) = \sum_{k_1=0}^{1} \hat{f}(k_1, k_0)W^{2n_0k_1}$$

for $n_0 = 0, 1; n_1 = 0, 1$. Then

$$\hat{f}_1(0, 0) = \hat{f}(0, 0)W^0 + \hat{f}(1, 0)W^0,$$
$$\hat{f}_1(0, 1) = \hat{f}(0, 1)W^0 + \hat{f}(1, 1)W^0,$$
$$\hat{f}_1(1, 0) = \hat{f}(0, 0)W^0 + \hat{f}(1, 0)W^2,$$

and

$$\hat{f}_1(0, 1) = \hat{f}(0, 1)W^0 + \hat{f}(1, 1)W^2.$$

In matrix form,

$$\begin{bmatrix} \hat{f}_1(0, 0) \\ \hat{f}_1(0, 1) \\ \hat{f}_1(1, 0) \\ \hat{f}_1(1, 1) \end{bmatrix} = \begin{bmatrix} 1 & 0 & 1 & 0 \\ 0 & 1 & 0 & 1 \\ 1 & 0 & W^2 & 0 \\ 0 & 1 & 0 & W^2 \end{bmatrix} \begin{bmatrix} f(0, 0) \\ f(0, 1) \\ f(1, 0) \\ f(1, 1) \end{bmatrix}.$$

Thus, the inner sum of equation (17.23) determines the first factor of the matrix factorization we are attempting to obtain. Next, define

$$\hat{f}_2(n_0, n_1) = \sum_{k_0=0}^{1} \hat{f}_1(n_0, k_0)W^{(2n_1 + n_0)k_0}$$

for $n_0 = 0, 1; n_1 = 0, 1$. Then

$$\begin{bmatrix} \hat{f}_2(0, 0) \\ \hat{f}_2(0, 1) \\ \hat{f}_2(1, 0) \\ \hat{f}_2(1, 1) \end{bmatrix} = \begin{bmatrix} 1 & 1 & 0 & 0 \\ 1 & W^2 & 0 & 0 \\ 0 & 0 & 1 & W^1 \\ 0 & 0 & 1 & W^3 \end{bmatrix} \begin{bmatrix} \hat{f}_1(0, 0) \\ \hat{f}_1(0, 1) \\ \hat{f}_1(1, 0) \\ \hat{f}_1(1, 1) \end{bmatrix}.$$

Therefore, the outer sum in equation (17.23) determines the second factor of the matrix factorization.

To summarize this information, we have written

$$\hat{f}_1(n_0, k_0) = \sum_{k_1=0}^{1} \hat{f}(k_1, k_0) W^{2n_0 k_1} \quad \text{for} \quad n_0 = 0, 1; k_0 = 0, 1;$$

and

$$\hat{f}_2(n_0, n_1) = \sum_{k_0=0}^{1} \hat{f}_1(n_0, k_0) W^{(2n_1 + n_0)k_0} \quad \text{for} \quad n_0 = 0, 1; k_0 = 0, 1;$$

and

$$F(n) = F(2n_1 + n_0) = \hat{F}(n_1, n_0) = \hat{f}_2(n_0, n_1),$$

where we have compensated for the interchange of rows in the matrix (seen previously) by reversing the coordinates in $\hat{F}(n_1, n_0)$ and $\hat{f}_2(n_0, n_1)$.

By writing out all of the details for the case $N = 2^2$, we can see how to extend the idea to $N = 2^3$. Now we write

$$n = 4n_2 + 2n_1 + n_0 \qquad (n_0, n_1, n_2 = 0, 1)$$

and

$$k = 4k_2 + 2k_1 + k_0 \qquad (k_0, k_1, k_2 = 0, 1).$$

Since $W^{8m} = 1$, we have

$$F(n) = \hat{F}(n_2, n_1, n_0) = \sum_{k=0}^{7} f(k) W^{nk}$$

$$= \sum_{k_0=0}^{1} \sum_{k_1=0}^{1} \sum_{k_2=0}^{1} \hat{f}(k_2, k_1, k_0) W^{(4n_2 + 2n_1 + n_0)(4k_2 + 2k_1 + k_0)}$$

$$= \sum_{k_0=0}^{1} \left[\sum_{k_1=0}^{1} \left[\sum_{k_2=0}^{1} \hat{f}(k_2, k_1, k_0) W^{4n_0 k_2} \right] W^{(2n_1 + n_0)2k_1} \right] W^{(4n_2 + 2n_1 + n_0)k_0}$$

with $n_0 = 0, 1; n_1 = 0, 1; n_2 = 0, 1$. Then

$$\hat{f}_1(n_0, k_1, k_0) = \sum_{k_2=0}^{1} \hat{f}(k_2, k_1, k_0) W^{4n_0 k_2} \qquad (n_0, k_0, k_1 = 0, 1);$$

$$\hat{f}_2(n_0, n_1, k_0) = \sum_{k_1=0}^{1} \hat{f}_1(n_0, k_1, k_0) W^{(2n_1 + n_0)2k_1} \qquad (n_0, n_1, k_0 = 0, 1);$$

$$\hat{f}_3(n_0, n_1, n_2) = \sum_{k_0=0}^{1} \hat{f}_2(n_0, n_1, k_0) W^{(4n_2 + 2n_1 + n_0)k_0} \qquad (n_0, n_1, n_2 = 0, 1);$$

$$F(n) = F(4n_2 + 2n_1 + n_0) = \hat{F}(n_2, n_1, n_0) = \hat{f}_2(n_0, n_1, n_2).$$

For completeness, we will also include the equations for the general case $N = 2^{\gamma}$.

Now denote

$$n = 2^{\gamma-1}n_{\gamma-1} + 2^{\gamma-2}n_{\gamma-2} + \cdots + 2n_1 + n_0 \qquad (n_j = 0, 1)$$

and

$$k = 2^{\gamma-1}k_{\gamma-1} + 2^{\gamma-2}k_{\gamma-2} + \cdots + 2k_1 + k_0 \qquad (k_j = 0, 1).$$

We get

$$\hat{f}_1(n_0, k_{\gamma-2}, k_{\gamma-3}, \ldots, k_0) = \sum_{k_{\gamma-1}=0}^{1} \hat{f}(k_{\gamma-1}, k_{\gamma-2}, \ldots, k_0)W^{2^{\gamma-1}n_0 k_{\gamma-1}};$$

$$\hat{f}_2(n_0, n_1, k_{\gamma-3}, \ldots, k_0) = \sum_{k_{\gamma-2}=0}^{1} \hat{f}_1(n_0, k_{\gamma-2}, \ldots, k_0)W^{(2n_1+n_0)2^{\gamma-2}k_{\gamma-2}};$$

$$\hat{f}_3(n_0, n_1, n_2, k_{\gamma-4}, \ldots, k_0)$$
$$= \sum_{k_{\gamma-3}=0}^{1} \hat{f}_2(n_0, n_1, k_{\gamma-3}, \ldots, k_0)W^{(4n_2+2n_1+n_0)2^{\gamma-3}k_{\gamma-3}};$$

$$\vdots$$

$$\hat{f}_\gamma(n_0, n_1, n_2, \ldots, n_{\gamma-1})$$
$$= \sum_{k_0=0}^{1} \hat{f}_{\gamma-1}(n_0, n_1, \ldots, n_{\gamma-2}, k_0)W^{(2^{\gamma-1}n_{\gamma-1}+\cdots+2n_1+n_0)k_0};$$

and

$$F(n) = F(2^{\gamma-1}n_{\gamma-1} + \cdots + 2n_1 + n_0) = \hat{F}(n_{\gamma-1}, \ldots, n_1, n_0)$$
$$= \hat{f}_\gamma(n_0, n_1, \ldots, n_{\gamma-1}).$$

As in the case $N = 2^2$, these functions generate a sequence of matrices which can be used to evaluate F from f. This involves γ summations, each of which consists of N equations. Each of these equations involves two multiplications, one of which always has $W^0 = 1$ as a factor times some value of an $\hat{f}_i$ and hence does not really cost another multiplication. This leaves $N\gamma$ multiplications to do. However, half of these have a factor of the form $W^{N/2} = -1$, and multiplications of this form can be observed as subtractions of values which have already been calculated. Reasoning in this way and doing some careful bookkeeping, we find that the total number of multiplications has been reduced to $N\gamma/2$. There are also $N\gamma$ additions to perform.

In the problems, a procedure will be outlined for factoring the matrix in the case in which $N = ab$, a product of two integers. This technique can be extended to integers N which are products of finitely many integers.

PROBLEMS FOR SECTION 17.14

1. Let $W = e^{2\pi i/N}$. Use Euler's equation to prove that $W^{mN+k} = W^k$ if m and n are integers. N is assumed to be a positive integer.

2. Verify that

$$\begin{bmatrix} 1 & 1 & 1 & 1 \\ 1 & W^2 & 1 & W^2 \\ 1 & W^1 & W^2 & W^3 \\ 1 & W^3 & W^2 & W^1 \end{bmatrix} = \begin{bmatrix} 1 & 1 & 0 & 0 \\ 1 & W^2 & 0 & 0 \\ 0 & 0 & 1 & W^1 \\ 0 & 0 & 1 & W^3 \end{bmatrix} \begin{bmatrix} 1 & 0 & 1 & 0 \\ 0 & 1 & 0 & 1 \\ 1 & 0 & W^2 & 0 \\ 0 & 1 & 0 & W^2 \end{bmatrix}.$$

3. Let $F(n) = \sum_{k=0}^{7} f(k)W^{nk}$ for $n = 0, 1, \ldots, 7$. Determine which rows of the product of the three matrices in the base-2 FFT factorization have to be interchanged.

4. Referring to Problem 3, which rows have to be interchanged in the case $N = 16$? Which rows must be interchanged in the case $N = 32$?

5. Write the equation $F(n) = \sum_{k=0}^{7} f(k)W^{nk}$ in matrix form.

6. Determine the base-2 FFT factorization of the matrix requested in Problem 5.

7. Let $F(n) = \sum_{k=0}^{5} f(k)W^{nk}$, $n = 0, 1, \ldots, 5$. Express this equation in matrix form.

As background for the next problem, suppose that $N = ab$ and let $F(n) = \sum_{k=0}^{N-1} f(k)W^{nk}$. We can write this equation in matrix form. We will examine a procedure for factoring the $N \times N$ coefficient matrix into a product of two $N \times N$ matrices. (If $N = abc$, we would write a product of three $N \times N$ matrices.) Write $n = n_1 a + n_0$, with $n_0 = 0, 1, 2, \ldots, a - 1$ and $n_1 = 0, 1, 2, \ldots, b - 1$. Next, write $k = k_1 b + k_0$, with $k_0 = 0, 1, 2, \ldots, b - 1$ and $k_1 = 0, 1, 2, \ldots, a - 1$. Now write

$$F(n) = \hat{F}(n_1, n_0) = \sum_{k_0=0}^{b-1} \left[\sum_{k_1=0}^{a-1} \hat{f}(k_1, k_0)W^{nk_1 b} \right] W^{nk_0}$$

Define

$$\hat{f}_1(n_0, k_0) = \sum_{k_1=0}^{a-1} \hat{f}(k_1, k_0)W^{nk_1 b}$$

and

$$\hat{f}_2(n_0, n_0) = \sum_{k_0=0}^{b-1} \hat{f}_1(n_0, k_0)W^{(n_1 a + n_0)k_0}.$$

Then

$$\hat{F}(n_1, n_0) = \hat{f}_2(n_0, n_1).$$

8. Use the factorization described above to factor the matrix obtained in Problem 7.

CHAPTER 18

Boundary Value Problems
in Partial Differential
Equations

18.0 Introduction

Mathematical models of various phenomena may lead to partial differential equations because often more than one independent variable is involved. A *partial differential equation* is an equation which contains one or more partial derivatives. For example,

$$\frac{\partial u}{\partial t} = \frac{\partial^2 u}{\partial x^2}$$

is a partial differential equation in which we seek a function u dependent upon the independent variables x and t. A *solution* of a partial differential equation is a function which satisfies the equation. For example,

$$u(x, t) = \cos(2x)e^{-4t}$$

is a solution of the above equation, as can be checked by substitution.

Occasionally, a partial differential equation can be solved by inspection, though this is rare. For example, consider

$$\frac{\partial u}{\partial x} = -4\,\frac{\partial u}{\partial y}.$$

There may be many solutions of this equation, but we can guess one by reasoning as follows. If $\partial u/\partial x$ and $\partial u/\partial y$ were both constant, we could find a solution easily. Try

$$u(x, y) = Ax + By,$$

with A and B constant. Then $\partial u/\partial x = A$ and $\partial u/\partial y = B$, and substitution of u into the differential equation shows that u is a solution if and only if

$$A = -4B.$$

Therefore, any function defined by $u(x, y) = B(-4x + y)$, with B any constant, is a solution of this partial differential equation.

A partial differential equation is *of order n* if it contains an nth partial derivative but none of higher order. For example, *Laplace's equation*

$$\frac{\partial^2 u}{\partial x^2} + \frac{\partial^2 u}{\partial y^2} + \frac{\partial^2 u}{\partial z^2} = 0,$$

which is often written as $\nabla^2 u = 0$, is of order 2. We also say that this equation is *second order*. The partial differential equation

$$\frac{\partial^2 u}{\partial x^2} = \frac{\partial^5 u}{\partial t^5} - \frac{\partial u}{\partial t}$$

is fifth order.

Just as with ordinary differential equations of order 2 or higher, we usually begin the study of partial differential equations by concentrating on the linear case. The *general linear first order partial differential equation* in three variables (with u a function of the independent variables x and y) is

$$a(x, y)\frac{\partial u}{\partial x} + b(x, y)\frac{\partial u}{\partial y} + f(x, y)u + g(x, y) = 0.$$

The *general second order linear partial differential equation* in three variables has the form

$$a(x, y)\frac{\partial^2 u}{\partial x^2} + b(x, y)\frac{\partial^2 u}{\partial x \, \partial y} + c(x, y)\frac{\partial^2 u}{\partial y^2}$$

$$+ d(x, y)\frac{\partial u}{\partial x} + e(x, y)\frac{\partial u}{\partial y} + f(x, y)u + g(x, y) = 0.$$

Many of the equations we encounter will be in one of these two forms. In both cases, the equation is *homogeneous* if $g(x, y) = 0$ for all (x, y) under consideration and *nonhomogeneous* if some $g(x, y) \neq 0$.

One reason for restricting our attention to linear first and second order equations is that they occur frequently in important contexts. Even these cases require a vast theory when treated very generally, so we will devote most of our time to equations governing vibration and heat conduction phenomena encountered in physics and engineering. The main tools we will use are Fourier series, integrals, and transforms, and we will explore the use of the Laplace transform as well. Throughout much of the chapter, we will concentrate on techniques for producing solutions, with a consideration of existence and uniqueness in Section 18.12.

As we saw with various applications of ordinary differential equations, a mathematical model of a physical phenomenon usually leads to a differential equation accompanied by other conditions, which may be initial conditions (specifying the state of the system at some initial time) and/or boundary conditions (specifying the state of the system at ends of an interval over which the solution is to be defined).

Similar remarks hold for models leading to partial differential equations. The phenomenon being modeled will lead not only to one or more partial differential equations but also to initial and boundary conditions which specify the state of the system at certain times and at certain points.

We will illustrate these remarks by deriving models for wave motion and heat conduction.

THE WAVE EQUATION

Suppose we have a flexible elastic string stretched between two pegs. We want to describe the ensuing motion if the string is lifted and then released to vibrate in a vertical plane. This models, for example, the motion of a strummed guitar string.

Place the x-axis along the length of the string at rest. At any time t and horizontal coordinate x, let $y(x, t)$ be the vertical displacement of the string. The graph of $y = y(x, t)$ at any time shows the shape of the string at that time as it vibrates in a vertical plane. A typical configuration is shown in Figure 18.1. We want to determine equations which will enable us to solve for y, thus obtaining a description of the shape of the string at any time.

We will begin by modeling a simplified case. Neglect damping forces such as air resistance and the weight of the string, and assume that the tension $T(x, t)$ in the string always acts tangentially to the string. Also assume that the mass ρ per unit length is a constant. Apply Newton's second law of motion to the segment of the string between x and $x + \Delta x$, shown in Figure 18.2, to write

$$\text{net force due to tension} \tag{18.1}$$

$$= [\text{segment mass}] \times [\text{acceleration of the center of mass of the segment}].$$

For small Δx, consideration of the vertical component of this equation gives us approximately

$$T(x + \Delta x, t)\sin(\theta + \Delta\theta) - T(x, t)\sin(\theta) = \rho \, \Delta x \, \frac{\partial^2 y}{\partial t^2}(\bar{x}, t),$$

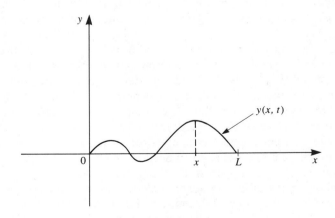

Figure 18.1. String profile at time t.

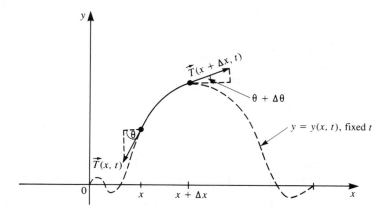

Figure 18.2

where $\bar{x}$ is the coordinate of the center of mass of the segment. Then

$$\frac{T(x + \Delta x, t)\sin(\theta + \Delta\theta) - T(x, t)\sin(\theta)}{\Delta x} = \rho \frac{\partial^2 y}{\partial t^2}(\bar{x}, t). \tag{18.2}$$

As a convenience, write

$$v(x, t) = T(x, t)\sin(\theta).$$

This is the vertical component of the tension. Write equation (18.2) as

$$\frac{v(x + \Delta x, t) - v(x, t)}{\Delta x} = \rho \frac{\partial^2 y}{\partial t^2}(\bar{x}, t).$$

Let $\Delta x \to 0$. Then $\bar{x} \to x$, and we obtain

$$\frac{\partial v}{\partial x} = \rho \frac{\partial^2 y}{\partial t^2}. \tag{18.3}$$

Write $h(x, t) = T(x, t)\cos(\theta)$, the horizontal component of the tension at (x, t). Then

$$v(x, t) = h(x, t)\tan(\theta) = h(x, t)\frac{\partial y}{\partial x}.$$

Substitute this for v in equation (18.3) to obtain

$$\frac{\partial}{\partial x}\left[h\frac{\partial y}{\partial x}\right] = \rho \frac{\partial^2 y}{\partial t^2}.$$

To compute the partial derivative on the left, recall that the horizontal component of the tension of the segment is zero; hence,

$$h(x + \Delta x, t) - h(x, t) = 0.$$

Therefore, h is independent of x; hence,

$$\frac{\partial}{\partial x}\left[h\frac{\partial y}{\partial x}\right] = h\frac{\partial^2 y}{\partial x^2}.$$

Finally, we have

$$h \frac{\partial^2 y}{\partial x^2} = \rho \frac{\partial^2 y}{\partial t^2}.$$

Let $a^2 = h/\rho$ to write this equation in standard form as

$$\frac{\partial^2 y}{\partial t^2} = a^2 \frac{\partial^2 y}{\partial x^2}.$$

This is the *one-dimensional* (one space variable) *wave equation*.

The motion of the string will be influenced by both the initial position and the initial velocity of the string. Therefore, we must specify *initial conditions*

$$y(x, 0) = f(x) \qquad \text{(initial position)}$$

and

$$\frac{\partial y}{\partial t}(x, 0) = g(x) \qquad \text{(initial velocity)},$$

with f and g given functions defined on $[0, L]$. These conditions must hold for $0 \le x \le L$. In addition, we must incorporate into the model the information that the ends of the string are fixed. Certainly the motion would be different if one end were free. Thus, we must state boundary conditions (conditions at the ends of the string)

$$y(0, t) = y(L, t) = 0 \qquad (t \ge 0).$$

The wave equation, together with the initial and boundary conditions, is an example of a *boundary value problem*. This boundary value problem is

$$\frac{\partial^2 y}{\partial t^2} = a^2 \frac{\partial^2 y}{\partial x^2} \qquad\qquad (0 < x < L, t > 0),$$

$$y(0, t) = y(L, t) = 0 \qquad\qquad (t > 0),$$

$$y(x, 0) = f(x), \quad \frac{\partial y}{\partial t}(x, 0) = g(x) \qquad\qquad (0 < x < L).$$

We expect on physical grounds that this problem will have a unique solution. We will discuss a uniqueness argument in Section 18.12.

We can also include in the model additional forces acting on the string. If an external force of magnitude F units per unit length acts parallel to the y-axis, the wave equation derived above must be adjusted by addition of a term $(1/\rho)F$. The boundary value problem for the displacement function y is

$$\frac{\partial^2 y}{\partial t^2} = a^2 \frac{\partial^2 y}{\partial x^2} + \frac{1}{\rho} F \qquad\qquad (0 < x < L, t > 0),$$

$$y(0, t) = y(L, t) = 0 \qquad\qquad (t > 0),$$

$$y(x, 0) = f(x), \quad \frac{\partial y}{\partial t}(x, 0) = g(x) \qquad\qquad (0 < x < L).$$

In particular, if F is the weight of the string, replace $(1/\rho)F$ by $-g$ in this model.

In two dimensions, we might have a membrane covering a region R in the plane and fixed on a frame forming the boundary of R. The membrane is set in motion, with vibrations occurring vertical to the plane of the membrane. If $z(x, y, t)$ is the vertical coordinate at time t of the particle at point (x, y) in the membrane, the partial differential equation for z is

$$\frac{\partial^2 z}{\partial t^2} = a^2 \left[\frac{\partial^2 z}{\partial x^2} + \frac{\partial^2 z}{\partial y^2} \right]$$

for (x, y) in R. This is the *two-dimensional wave equation*. To determine z uniquely, we must first include conditions, which specify the initial position and velocity of the membrane,

$$z(x, y, 0) = f(x, y) \quad \text{for} \quad (x, y) \text{ in } R$$

and

$$\frac{\partial z}{\partial t}(x, y, 0) = g(x, y) \quad \text{for} \quad (x, y) \text{ in } R.$$

Finally, the condition that the membrane is fixed to the frame means that points on the border of the membrane do not move, a condition we write as

$$z(x, y, t) = 0 \quad \text{for} \quad t > 0 \quad \text{and} \quad (x, y) \text{ on the boundary of } R.$$

THE HEAT EQUATION

We will now derive a boundary value problem which models the temperature distribution in a straight, thin bar under simple circumstances.

Suppose we have a straight, thin (compared with length) bar of constant density ρ and constant cross-sectional area A placed along the x axis from zero to L. Assume that the sides of the bar are insulated and do not allow heat loss and that the temperature on the cross-section of the bar perpendicular to the x-axis at x is a function $u(x, t)$ of x and t, independent of y. Let the specific heat of the bar be c, and let the thermal conductivity be k, both constant.

Now consider a typical segment of the bar between $x = \alpha$ and $x = \beta$, as in Figure 18.3. By the definition of specific heat, the rate at which heat energy accumulates in this segment of the bar is

$$\int_\alpha^\beta c\rho A \frac{\partial u}{\partial t} \, dx.$$

By Newton's law of cooling, heat energy flows within this segment from the warmer to the cooler end at a rate equal to k times the negative of the temperature gradient. Therefore, the net rate at which heat energy enters the segment of bar between α and β at time t is

$$kA \frac{\partial u}{\partial x}(\beta, t) - kA \frac{\partial u}{\partial x}(\alpha, t).$$

In the absence of heat production within the segment, the rate at which heat energy accumulates within the segment must balance the rate at which heat energy enters the

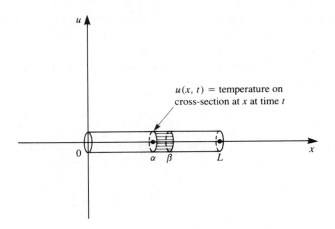

Figure 18.3

segment. Therefore,

$$\int_\alpha^\beta c\rho A \frac{\partial u}{\partial t}\, dx = kA \frac{\partial u}{\partial x}(\beta, t) - kA \frac{\partial u}{\partial x}(\alpha, t).$$

Now recognize that the right side of this equation can be written

$$kA \frac{\partial u}{\partial x}(\beta, t) - kA \frac{\partial u}{\partial x}(\alpha, t) = kA \int_\alpha^\beta \frac{\partial^2 u}{\partial x^2}\, dx.$$

Combining the last two equations, we obtain

$$\int_\alpha^\beta \left[c\rho A \frac{\partial u}{\partial t} - kA \frac{\partial^2 u}{\partial x^2} \right] dx = 0.$$

This must hold for every α and β with $0 \le \alpha < \beta \le L$. If $c\rho A(\partial u/\partial t) - kA(\partial^2 u/\partial x^2)$ were nonzero for any $t > 0$ and some x_0 in $[0, L]$, we could choose an interval $[\alpha, \beta]$ about x_0 in which $c\rho A(\partial u/\partial t) - kA(\partial^2 u/\partial x^2)$ is strictly positive or strictly negative, and we would then have $\int_\alpha^\beta [c\rho A(\partial u/\partial t) - kA(\partial^2 u/\partial x^2)]\, dx \ne 0$, a contradiction. We conclude that $c\rho A(\partial u/\partial t) - kA(\partial^2 u/\partial x^2) = 0$ for all x in $[0, L]$ and $t \ge 0$. This yields the heat equation

$$c\rho A \frac{\partial u}{\partial t} - kA \frac{\partial^2 u}{\partial x^2} = 0,$$

or, as it is more customarily written,

$$\frac{\partial u}{\partial t} = a^2 \frac{\partial^2 u}{\partial x^2},$$

in which $a^2 = k/c\rho$ is called the *thermal diffusivity* of the bar.

   To determine u uniquely, we need boundary conditions (information at the ends of the bar) and initial data (temperature throughout the bar at time zero). The differential equation, together with these pieces of information, constitutes a boundary value problem for the temperature function u. As a specific example, we might have the boundary value problem

$$\frac{\partial u}{\partial t} = a^2 \frac{\partial^2 u}{\partial x^2} \qquad (0 < x < L, t > 0),$$

$$u(0, t) = u(L, t) = T \qquad (t > 0),$$

$$u(x, 0) = f(x) \qquad (0 < x < L).$$

This boundary value problem models the temperature distribution in a bar of length L, whose ends are kept at constant temperature T, if the initial temperature in the cross-section at x is a given function $f(x)$.

As a second example, we might have a perfectly insulated bar, in which case we replace the boundary conditions $u(0, t) = u(L, t) = T$ of the above model with the conditions

$$\frac{\partial u}{\partial x}(0, t) = \frac{\partial u}{\partial x}(L, t) = 0 \qquad (t > 0).$$

These conditions specify that there is no heat flow across the ends of the bar.

As a third example, we can consider free radiation (convection), in which the bar loses heat by radiation from its ends into the surrounding medium, which is assumed to be maintained at a constant temperature T_a. In this case, the boundary conditions are (by Newton's law of cooling)

$$\frac{\partial u}{\partial x}(0, t) = A[u(0, t) - T_a] \quad \text{and} \quad \frac{\partial u}{\partial x}(L, t) = -A[u(L, t) - T_a]$$

for $t > 0$. Notice that, if the bar is hotter than the surrounding medium, the heat flow [change in temperature per unit length, or $(\partial u/\partial x)$] must be positive at the left end of the bar and negative at the right end.

We can also have a combination of these different types of conditions occurring in a boundary value problem. For example, suppose we have a bar with its left end maintained at a constant temperature T_1 and its right end radiating into a medium of temperature T_2, with an initial temperature distribution given by $f(x)$. The boundary value problem modeling the temperature distribution u is

$$\frac{\partial u}{\partial t} = a^2 \frac{\partial^2 u}{\partial x^2} \qquad (0 < x < L, t > 0),$$

$$u(0, t) = T_1, \qquad \frac{\partial u}{\partial x}(L, t) = -A[u(L, t) - T_2] \qquad (t > 0),$$

$$u(x, 0) = f(x) \qquad (0 < x < L).$$

In two dimensions, the heat equation is

$$\frac{\partial u}{\partial t} = a^2 \left[\frac{\partial^2 u}{\partial x^2} + \frac{\partial^2 u}{\partial y^2} \right],$$

and in three dimensions, it is

$$\frac{\partial u}{\partial t} = a^2 \left[\frac{\partial^2 u}{\partial x^2} + \frac{\partial^2 u}{\partial y^2} + \frac{\partial^2 u}{\partial z^2} \right].$$

Corresponding boundary and initial conditions must be specified to determine unique solutions of these partial differential equations.

LAPLACE'S EQUATION; POISSON'S EQUATION; DIRICHLET AND NEUMANN PROBLEMS

Another important partial differentiation equation is *Laplace's equation*

$$\frac{\partial^2 u}{\partial x^2} + \frac{\partial^2 u}{\partial y^2} + \frac{\partial^2 u}{\partial z^2} = 0 \quad \text{(in three dimensions)}$$

or

$$\frac{\partial^2 u}{\partial x^2} + \frac{\partial^2 u}{\partial y^2} = 0 \quad \text{(in two dimensions)}.$$

We have seen Laplace's equation in Section 16.7, in connection with heat conduction and fluid flow. In the context of heat conduction, Laplace's equation is called the *steady-state heat equation*; it is the heat equation when $\partial u/\partial t = 0$. Laplace's equation also occurs in the study of electrical field potentials. Often, Laplace's equation is written $\nabla^2 u = 0$ (in which ∇^2 is read "del squared"). A function satisfying $\nabla^2 u = 0$ is called a *harmonic function*.

There are two types of problems commonly associated with Laplace's equation. A *Dirichlet problem* consists of finding a function which is harmonic in a given set M and takes on predetermined values on the boundary of M. For example, in three-space, we could have the boundary value problem

$$\nabla^2 u = 0 \quad \text{in } M,$$
$$u(x, y, z) = f(x, y, z) \quad \text{for} \quad (x, y, z) \quad \text{on } \Sigma,$$

where Σ is a piecewise-smooth surface bounding M and f is a given function. A typical Dirichlet problem in the plane would be

$$\nabla^2 u = 0 \quad \text{in } D,$$
$$u(x, y) = g(x, y) \quad \text{for} \quad (x, y) \quad \text{on } C,$$

where C is a piecewise-smooth curve bounding the set D in the plane.

A *Neumann problem* in three-space consists of finding a function u such that $\nabla^2 u = 0$ for (x, y, z) in a region M, subject to the condition that the normal derivative of u takes on prescribed values on the boundary of the region. In three-space, this boundary value problem has the form

$$\nabla^2 u = 0 \quad \text{in } D,$$
$$\frac{\partial u}{\partial \eta} = f(x, y, z) \quad \text{for} \quad (x, y, z) \quad \text{on } \Sigma,$$

where Σ is the surface bounding M and $\partial u/\partial \eta$ denotes the directional derivative in the direction of the normal to Σ.

In Section 16.7, we encountered *Poisson's equation* in looking at potentials for electrical fields. Poisson's equation has the form

$$\nabla^2 u = f,$$

with f a given function. When f is identically zero, Poisson's equation becomes Laplace's equation.

LAPLACE'S EQUATION IN CYLINDRICAL AND SPHERICAL COORDINATES

Some problems involve the Laplace operator ∇^2 written in cylindrical or spherical coordinates. We touched on this issue briefly in Appendix A to Part IV, where we treated orthogonal curvilinear coordinates. For reference, we will derive Laplace's equation in cylindrical and spherical coordinates.

First, consider cylindrical coordinates (r, θ, z), which are related to rectangular coordinates by

$$x = r\cos(\theta), \qquad y = r\sin(\theta), \qquad z = z. \qquad (18.4)$$

We can also write

$$r = \sqrt{x^2 + y^2}, \qquad \theta = \tan^{-1}\left(\frac{y}{x}\right) \qquad (18.5)$$

if $x \neq 0$. Here, the value of θ is determined by the signs of x and y [for example, if x and y are both negative, (x, y) is in the third quadrant, so we must have $\pi < \theta < 3\pi/2$, even though the ratio y/x is positive].

Now suppose that u is a function of (x, y, z) and that u and its first and second partial derivatives are continuous throughout some set M of three-space. Since x, y, and z are functions of r, θ, and z, we may think of u as a function of r, θ, and z. By the chain rule,

$$\frac{\partial u}{\partial x} = \frac{\partial u}{\partial r}\frac{\partial r}{\partial x} + \frac{\partial u}{\partial \theta}\frac{\partial \theta}{\partial x} + \frac{\partial u}{\partial z}\frac{\partial z}{\partial x} = \frac{\partial u}{\partial r}\frac{\partial r}{\partial x} + \frac{\partial u}{\partial \theta}\frac{\partial \theta}{\partial x}$$

$$= \frac{\partial u}{\partial r}\frac{x}{\sqrt{x^2 + y^2}} - \frac{\partial u}{\partial \theta}\frac{y}{x^2 + y^2} \qquad (18.6)$$

$$= \frac{\partial u}{\partial r}\frac{x}{r} - \frac{\partial u}{\partial \theta}\frac{y}{r^2}.$$

Here, $\partial z/\partial x = 0$ because x and z are independent variables. By a similar calculation,

$$\frac{\partial u}{\partial y} = \frac{\partial u}{\partial r}\frac{y}{\sqrt{x^2 + y^2}} + \frac{\partial u}{\partial \theta}\frac{x}{x^2 + y^2} = \frac{\partial u}{\partial r}\frac{y}{r} + \frac{\partial u}{\partial \theta}\frac{x}{r^2}. \qquad (18.7)$$

Calculate $\partial^2 u/\partial x^2$ by chain rule differentiation of equation (18.6):

$$\frac{\partial^2 u}{\partial x^2} = \frac{\partial}{\partial x}\left(\frac{\partial u}{\partial r}\right)\frac{x}{r} + \frac{\partial u}{\partial r}\frac{\partial}{\partial x}\left(\frac{x}{r}\right) - \frac{\partial}{\partial x}\left(\frac{\partial u}{\partial \theta}\right)\frac{y}{r^2} - \frac{\partial u}{\partial \theta}\frac{\partial}{\partial x}\left(\frac{y}{r^2}\right)$$

$$= \frac{x}{r}\left[\frac{\partial^2 u}{\partial r^2}\frac{x}{r} - \frac{\partial^2}{\partial \theta \, \partial r}\left(\frac{y}{r^2}\right)\right] - \frac{y}{r^2}\left[\frac{\partial^2 u}{\partial r \, \partial \theta}\frac{x}{r} - \frac{\partial^2 u}{\partial \theta^2}\frac{y}{r^2}\right]$$

$$+ \frac{\partial u}{\partial r}\frac{\partial}{\partial x}\left(\frac{x}{r}\right) - \frac{\partial u}{\partial \theta}\frac{\partial}{\partial x}\left(\frac{y}{r^2}\right)$$

$$= \frac{y^2}{r^3}\frac{\partial u}{\partial r} + \frac{2xy}{r^4}\frac{\partial u}{\partial \theta} + \frac{x^2}{r^2}\frac{\partial^2 u}{\partial r^2} - \frac{2xy}{r^3}\frac{\partial^2 u}{\partial r \, \partial \theta} + \frac{y^2}{r^4}\frac{\partial^2 u}{\partial \theta^2}. \qquad (18.8)$$

Similarly, from equation (18.7), we obtain by chain rule differentiation that

$$\frac{\partial^2 u}{\partial y^2} = \frac{x^2}{r^3}\frac{\partial u}{\partial r} - \frac{2xy}{r^4}\frac{\partial u}{\partial \theta} + \frac{y^2}{r^2}\frac{\partial^2 u}{\partial r^2} + \frac{2xy}{r^3}\frac{\partial^2 u}{\partial r \partial \theta} + \frac{x^2}{r^4}\frac{\partial^2 u}{\partial \theta^2}. \tag{18.9}$$

Upon adding equations (18.8) and (18.9), we obtain

$$\frac{\partial^2 u}{\partial x^2} + \frac{\partial^2 u}{\partial y^2} = \frac{\partial^2 u}{\partial r^2} + \frac{1}{r}\frac{\partial u}{\partial r} + \frac{1}{r^2}\frac{\partial^2 u}{\partial \theta^2}. \tag{18.10}$$

Therefore, in two variables r and θ, Laplace's equation in polar coordinates is

$$\frac{\partial^2 u}{\partial r^2} + \frac{1}{r}\frac{\partial u}{\partial r} + \frac{1}{r^2}\frac{\partial^2 u}{\partial \theta^2} = 0. \tag{18.11}$$

In three variables r, θ, and z, Laplace's equation in cylindrical coordinates is

$$\frac{\partial^2 u}{\partial r^2} + \frac{1}{r}\frac{\partial u}{\partial r} + \frac{1}{r^2}\frac{\partial^2 u}{\partial \theta^2} + \frac{\partial^2 u}{\partial z^2} = 0. \tag{18.12}$$

We may also group terms and write

$$\frac{\partial^2 u}{\partial r^2} + \frac{1}{r}\frac{\partial u}{\partial r} + \frac{1}{r^2}\frac{\partial^2 u}{\partial \theta^2} = \frac{1}{r}\frac{\partial}{\partial r}\left[r\frac{\partial u}{\partial r} \right] + \frac{1}{r^2}\frac{\partial^2 u}{\partial \theta^2}$$

to write Laplace's equation as

$$\nabla^2 u(r, \theta) = \frac{1}{r}\frac{\partial}{\partial r}\left[r\frac{\partial u}{\partial r} \right] + \frac{1}{r^2}\frac{\partial^2 u}{\partial \theta^2} = 0 \qquad \text{(polar coordinates)} \tag{18.13}$$

and

$$\nabla^2 u(r, \theta, z) = \frac{1}{r}\frac{\partial}{\partial r}\left[r\frac{\partial u}{\partial r} \right] + \frac{1}{r^2}\frac{\partial^2 u}{\partial \theta^2} + \frac{\partial^2 u}{\partial z^2} = 0 \qquad \text{(cylindrical coordinates).} \tag{18.14}$$

We will now consider Laplace's equation in spherical coordinates. The spherical coordinates (ρ, θ, φ) of a point are related to the rectangular coordinates (x, y, z) by

$$x = \rho \cos(\theta)\sin(\varphi), \qquad y = \rho \sin(\theta)\sin(\varphi), \qquad z = \rho \cos(\varphi), \tag{18.15}$$

$0 \le \theta \le 2\pi$, $0 \le \varphi \le \pi$. The transformation is easy to carry out if we make the observation that

$$z = \rho \cos(\varphi), \qquad r = \rho \sin(\varphi), \qquad \theta = \theta. \tag{18.16}$$

[This observation appears in Ruel V. Churchill and James Ward Brown, *Fourier Series and Boundary Value Problems* (New York: McGraw-Hill, 1978).]

Equations (18.16) are identical in form with equations (18.4), with different names assigned to the variables. We can therefore use equation (18.10), with the appropriate changes in the names of the variables, to conclude immediately that

$$\frac{\partial^2 u}{\partial z^2} + \frac{\partial^2 u}{\partial r^2} = \frac{\partial^2 u}{\partial \rho^2} + \frac{1}{\rho}\frac{\partial u}{\partial \rho} + \frac{1}{\rho^2}\frac{\partial^2 u}{\partial \varphi^2}. \tag{18.17}$$

Similarly, using equations (18.16), equation (18.7) becomes

$$\frac{\partial u}{\partial r} = \frac{r}{\rho}\frac{\partial u}{\partial \rho} + \frac{z}{\rho^2}\frac{\partial u}{\partial \varphi}. \tag{18.18}$$

Using equation (18.18) and the first two of equations (18.16), we obtain

$$\frac{1}{r}\frac{\partial u}{\partial r} + \frac{1}{r^2}\frac{\partial^2 u}{\partial \theta^2} = \frac{1}{\rho}\frac{\partial u}{\partial \rho} + \frac{\cot(\varphi)}{\rho^2}\frac{\partial u}{\partial \varphi} + \frac{1}{\rho^2 \sin^2(\varphi)}\frac{\partial^2 u}{\partial \theta^2}.$$

From equations (18.17) and (18.18), we obtain $\nabla^2 u$ in spherical coordinates:

$$\nabla^2 u(\rho, \theta, \varphi) = \frac{\partial^2 u}{\partial \rho^2} + \frac{2}{\rho}\frac{\partial u}{\partial \rho} + \frac{1}{\rho^2 \sin^2(\varphi)}\frac{\partial^2 u}{\partial \theta^2} + \frac{1}{\rho^2}\frac{\partial^2 u}{\partial \varphi^2} + \frac{\cot(\varphi)}{\rho^2}\frac{\partial u}{\partial \varphi}. \quad (18.19)$$

We also see this expression written in the equivalent form

$$\nabla^2 u(\rho, \theta, \varphi) = \frac{1}{\rho^2}\frac{\partial^2}{\partial \rho^2}[\rho u] + \frac{1}{\rho^2 \sin^2(\varphi)}\frac{\partial^2 u}{\partial \theta^2} + \frac{1}{\rho^2 \sin(\varphi)}\frac{\partial}{\partial \varphi}\left[\sin(\varphi)\frac{\partial u}{\partial \varphi}\right]. \quad (18.20)$$

For the remainder of this chapter, we will discuss techniques and derive solutions of various boundary value problems using Fourier methods.

PROBLEMS FOR SECTION 18.0

In each of Problems 1 through 5, find a nontrivial solution of the partial differential equation by inspection and integration. Many solutions are possible.

1. $\dfrac{\partial u}{\partial x} = 0$ **2.** $\dfrac{\partial u}{\partial x} = 2y^2$ **3.** $\dfrac{\partial^2 u}{\partial x^2} = 0$

4. $\dfrac{\partial^2 u}{\partial x \partial y} = 0$ **5.** $\dfrac{\partial u}{\partial x} + \dfrac{\partial u}{\partial y} = 0$

In each of Problems 6 through 10, find a solution of the partial differential equation satisfying the given condition.

6. $\dfrac{\partial u}{\partial x} - \dfrac{\partial u}{\partial y} = 0; \quad u(0, 1) = 2$ **7.** $\dfrac{\partial^2 u}{\partial x \partial y} = 1; \quad u(0, 0) = 0$ **8.** $\dfrac{\partial^2 u}{\partial x^2} = 0; \quad u(1, 1) = 1$

9. $2\dfrac{\partial u}{\partial x} + 3\dfrac{\partial u}{\partial y} = 0; \quad u(1, 4) = 7$ **10.** $y\dfrac{\partial u}{\partial y} = 2; \quad u(1, 2) = 4$

In each of Problems 11 through 15, determine the order of the partial differential equation and determine whether it is linear; if not, explain why.

11. $\dfrac{\partial^2 u}{\partial x^2} + \dfrac{\partial^2 u}{\partial y^2} + \dfrac{\partial u}{\partial x}\dfrac{\partial u}{\partial y} = 0$ **12.** $\dfrac{\partial^2 u}{\partial x^2} + \left(\dfrac{\partial u}{\partial x}\right)^2 - 2e^{xy} = 0$

13. $\dfrac{\partial u}{\partial x} + 3\dfrac{\partial u}{\partial y} - y^2\dfrac{\partial^2 u}{\partial x^2} + x^2\dfrac{\partial^2 u}{\partial x \partial y} = 3x\cos(xy)$

14. $8x\dfrac{\partial u}{\partial y} - \dfrac{\partial u}{\partial x}\dfrac{\partial u}{\partial y} = 4y^2$ **15.** $x^2\dfrac{\partial^2 u}{\partial x^2} - 4\dfrac{\partial u}{\partial x} = 0$

In each of Problems 16 through 20, verify that the given function is a solution of the partial differential equation.

16. $\dfrac{\partial^2 u}{\partial x^2} + \dfrac{\partial^2 u}{\partial y^2} = 0; \quad u(x, y) = e^{-y}\cos(x)$

17. $\dfrac{\partial^2 y}{\partial t^2} = a^2\dfrac{\partial^2 y}{\partial x^2}; \quad y(x, t) = \sin\left(\dfrac{n\pi x}{L}\right)\cos\left(\dfrac{n\pi at}{L}\right)$ for $n = 1, 2, 3, \ldots, L$ any positive constant, n any positive integer

18. $\dfrac{\partial^2 u}{\partial x^2} + \dfrac{\partial^2 u}{\partial y^2} = 0; \quad u(x, y) = \tan^{-1}\left(\dfrac{y}{x}\right) \quad$ for $x > 0, y > 0$

19. $\dfrac{\partial^2 y}{\partial t^2} = a^2 \dfrac{\partial^2 y}{\partial x^2}; \quad y(x, t) = \dfrac{1}{2}[f(x + at) + f(x - at)] \quad$ for any twice-differentiable function f of a single variable

20. $\dfrac{\partial^2 u}{\partial t^2} = a^2 \left[\dfrac{\partial^2 u}{\partial x^2} + \dfrac{\partial^2 u}{\partial y^2}\right]; \quad u(x, y, t) = \sin(nx)\sin(my)\cos(\sqrt{n^2 + m^2}\,at) \quad$ for any positive integers n and m

21. Consider the boundary value problem for a vibrating string

$$\frac{\partial^2 y}{\partial t^2} = a^2 \frac{\partial^2 y}{\partial x^2} \qquad\qquad (0 < x < L, t > 0),$$

$$y(0, t) = y(L, t) = 0 \qquad\qquad (t > 0),$$

$$y(x, 0) = f(x), \qquad \frac{\partial y}{\partial t}(x, 0) = g(x) \qquad (0 < x < L).$$

(a) Show that the change of variables $x = \frac{1}{2}(\xi + \eta)$ and $t = (1/2a)(\xi - \eta)$ transforms the partial differential equation into one of the form

$$\frac{\partial^2 Y}{\partial \xi\, \partial \eta} = 0,$$

in which $Y(\xi, \eta) = y[(\xi + \eta)/2, (\xi - \eta)/2a]$.

(b) Show that $\partial^2 Y/\partial \xi\, \partial \eta = 0$ is satisfied by $Y(\xi, \eta) = F(\xi) + G(\eta)$ for any twice-differentiable functions F and G of one variable defined for all real ξ and η.

(c) Conclude that $y(x, t) = F(x + at) + G(x - at)$ satisfies the wave equation for any twice-differentiable functions F and G.

(d) Use the initial conditions to show that

$$F(\xi) = \frac{1}{2}f(\xi) + \frac{1}{2a}\int_0^\xi g(\alpha)\, d\alpha + k$$

and

$$G(\eta) = \frac{1}{2}f(\eta) - \frac{1}{2a}\int_0^\eta g(\alpha)\, d\alpha - k$$

for some constant k.

(e) Derive *d'Alembert's solution* of the vibrating string problem

$$y(x, t) = \frac{1}{2}[f(x + at) + f(x - at)] + \frac{1}{2a}\int_{x-at}^{x+at} g(\alpha)\, d\alpha.$$

22. Formulate a boundary value problem modeling heat conduction in a bar of length L, with the left end kept at temperature zero and the right end insulated, if the initial temperature in the cross-section of the bar at x is $f(x)$.

23. Formulate a boundary value problem for steady-state heat conduction in a thin plate placed over the rectangle $0 \le x \le a, 0 \le y \le b$. On the boundary of this rectangle, the top and bottom sides are kept at zero temperature, while the vertical sides are kept at constant temperature T. (Steady-state means that $\partial u/\partial t = 0$ in the heat equation; use the two-dimensional heat equation here.)

24. Formulate a boundary value problem for vibration of a rectangular membrane occupying a region $0 \le x \le a$, $0 \le y \le b$ if the initial position is $f(x, y)$ and the initial velocity is $g(x, y)$. The membrane is fastened to a frame along the rectangle.

25. Formulate a boundary value problem for vibration of an elastic string of length L which is fastened at both ends and released from rest at an initial position given by $y = f(x)$. The string vibrates in a vertical plane. Its motion is opposed by air resistance, with a force at each point of magnitude proportional to the square of the velocity at that point.

26. Suppose that $Y_1(x, t)$ is a solution of the boundary value problem

$$\frac{\partial^2 y}{\partial t^2} = a^2 \frac{\partial^2 y}{\partial x^2} \qquad\qquad (0 < x < L, t > 0),$$

$$y(0, t) = y(L, t) = 0 \qquad\qquad (t > 0),$$

$$y(x, 0) = f(x), \qquad \frac{\partial y}{\partial t}(x, 0) = 0 \qquad\qquad (0 < x < L).$$

Let $Y_2(x, t)$ be a solution of the boundary value problem

$$\frac{\partial^2 y}{\partial t^2} = a^2 \frac{\partial^2 y}{\partial x^2} \qquad\qquad (0 < x < L, t > 0),$$

$$y(0, t) = y(L, t) = 0 \qquad\qquad (t > 0),$$

$$y(x, 0) = 0, \qquad \frac{\partial y}{\partial t}(x, 0) = g(x) \qquad\qquad (0 < x < L).$$

Show that $Y_1(x, t) + Y_2(x, t)$ is a solution of

$$\frac{\partial^2 y}{\partial t^2} = a^2 \frac{\partial^2 y}{\partial x^2} \qquad\qquad (0 < x < L, t > 0),$$

$$y(0, t) = y(L, t) = 0 \qquad\qquad (t > 0),$$

$$y(x, 0) = f(x), \qquad \frac{\partial y}{\partial t}(x, 0) = g(x) \qquad\qquad (0 < x < L).$$

27. Show that

$$y(x, t) = \sin(x)\cos(at) + \frac{1}{a}\cos(x)\sin(at)$$

is a solution of

$$\frac{\partial^2 y}{\partial t^2} = a^2 \frac{\partial^2 y}{\partial x^2} \qquad\qquad (0 < x < \pi, t > 0),$$

$$y(x, 0) = \sin(x), \qquad \frac{\partial y}{\partial t}(x, 0) = \cos(x) \qquad\qquad (0 < x < \pi).$$

18.1 *Fourier Series Solutions of the Wave Equation*

We will now consider the solution of several boundary value problems modeling the motion of a vibrating string. We will use the method of *separation of variables,* or the *Fourier method.* It consists of attempting a solution of the form $y(x, t) = X(x)T(t)$ and then solving for X and T.

INITIALLY DISPLACED VIBRATING STRING WITH ZERO INITIAL VELOCITY

Consider the boundary value problem

$$\frac{\partial^2 y}{\partial t^2} = a^2 \frac{\partial^2 y}{\partial x^2} \qquad (0 < x < L, t > 0),$$

$$y(0, t) = y(L, t) = 0 \qquad\qquad (t > 0),$$

$$y(x, 0) = f(x) \qquad\qquad (0 < x < L),$$

$$\frac{\partial y}{\partial t}(x, 0) = 0 \qquad\qquad (0 < x < L).$$

This problem models the vibration of an elastic string of length L, fastened at the ends, picked up at time zero to assume the configuration of the graph of $y = f(x)$, and released from rest. At time t, $y(x, t)$ is the vertical displacement of the particle of string having coordinate x, as shown in Figure 18.1.

Attempt a solution

$$y(x, t) = X(x)T(t).$$

Substitute this function into the partial differential equation to get

$$XT'' = a^2 X''T,$$

or

$$\frac{X''}{X} = \frac{T''}{a^2 T}.$$

We have "separated" x and t; the left side is a function of x alone, and the right side is a function of t. Since x and t are independent, we can fix the right side by choosing $t = t_0$, and the left side must equal $T''(t_0)/a^2 T(t_0)$ for all x in $(0, L)$. Thus, X''/X must be constant. But then $T''/a^2 T$ must equal the same constant. As a notational convenience which will become apparent later, denote this constant $-\lambda$. (If we denote it λ, without the minus sign, we will arrive at the same solution.) Then

$$\frac{X''}{X} = \frac{T''}{a^2 T} = -\lambda.$$

The quantity λ is called a *separation constant*. We now have

$$X'' + \lambda X = 0 \quad \text{and} \quad T'' + \lambda a^2 T = 0.$$

These are two *ordinary* differential equations for X and T.

Next, look at the boundary conditions and relate them to X and T. From the condition that the left end of the string is fixed, we have

$$y(0, t) = X(0)T(t) = 0$$

for $t > 0$. Since $T(t)$ cannot be zero for all $t > 0$ (if the string is to move),

$$X(0) = 0.$$

Similarly,

$$y(L, t) = X(L)T(t) = 0$$

for $t > 0$; hence,

$$X(L) = 0.$$

Next, the initial condition $(\partial y/\partial t)(x, 0) = 0$ requires that

$$X(x)T'(0) = 0$$

for $0 < x < L$. Therefore,

$$T'(0) = 0.$$

At this point, we have two problems for X and T:

$$X'' + \lambda X = 0, \qquad\qquad T'' + \lambda a^2 T = 0,$$
$$X(0) = X(L) = 0; \qquad\qquad T'(0) = 0.$$

Consider first the problem for X, about which we have the most information, and consider cases on λ. From Section 7.7, this is a Sturm-Liouville problem for X, and solutions for X are eigenfunctions with corresponding eigenvalues λ. The Sturm-Liouville theorem tells us that these eigenvalues will be real and that the eigenfunctions are orthogonal on $[0, L]$ with weight function 1. We will, however, proceed from first principles and not assume this material as background.

Consider cases on λ, assuming that λ is real, as we might expect from the physics of the problem.

Case 1: $\lambda = 0$. Then $X'' = 0$, so $X(x) = cx + d$. Now $X(0) = d = 0$, and $X(L) = cL = 0$ implies that $c = 0$. Then $X(x) = 0$ for $0 \le x \le L$, and we will have $y(x, t) = 0$ as the solution. This is the case if $f(x) = 0$ for $0 \le x \le L$ because then the string was not displaced initially and simply remains stationary. If, however, $f(x) \ne 0$ for some x, $X(x)$ cannot be identically zero, and we must discard this case.

Case 2: $\lambda < 0$. Write $\lambda = -k^2$, with $k > 0$. The equation for X is

$$X'' - k^2 X = 0,$$

with general solution $X = ce^{kx} + de^{-kx}$. Since $X(0) = 0 = c + d$, $d = -c$, and

$$X(x) = c(e^{kx} - e^{-kx}) = 2c \sinh(kx).$$

Then $X(L) = 2c \sinh(kL) = 0$. Since k and L are positive, $\sinh(kL) > 0$, and we conclude that $c = 0$. But then $d = 0$, and, as in case 1, we obtain $X(x) = 0$ for $0 \le x \le L$. This case does not yield a solution (unless, again, the string is not moved).

Case 3: $\lambda > 0$. Write $\lambda = k^2$. Then $X'' + k^2 X = 0$, with general solution

$$X(x) = c \cos(kx) + d \sin(kx).$$

Since $X(0) = c = 0$, $X = d \sin(kx)$. We require also that $X(L) = d \sin(kL) = 0$. In order to be able to choose $d \ne 0$, we must have $\sin(kL) = 0$. This holds if kL is a positive integer multiple of π:

$$kL = n\pi, \qquad n = 1, 2, 3, \ldots$$

(recall that k and L are positive). Then

$$\lambda = k^2 = \frac{n^2 \pi^2}{L^2}$$

for $n = 1, 2, 3, \ldots$. Corresponding to each positive integer n, we therefore have a solution for X:

$$X_n(x) = d_n \sin\left(\frac{n\pi x}{L}\right),$$

with d_n a constant yet to be determined.

Now look at the problem for T. With $\lambda = n^2\pi^2/L^2$, we have

$$T'' + \frac{n^2 \pi^2 a^2}{L^2} T = 0; \qquad T'(0) = 0.$$

The general solution for T is

$$T = \alpha \cos\left(\frac{n\pi at}{L}\right) + \beta \sin\left(\frac{n\pi at}{L}\right).$$

Since $T'(0) = (n\pi a/L)\beta = 0$, $\beta = 0$. For each positive integer n, we have a solution for T:

$$T_n(t) = \alpha_n \cos\left(\frac{n\pi at}{L}\right).$$

We now have, for each positive integer n, a function

$$y_n(x, t) = X_n(x)T_n(t) = A_n \sin\left(\frac{n\pi x}{L}\right)\cos\left(\frac{n\pi at}{L}\right),$$

in which $A_n = d_n \alpha_n$ is a constant yet to be determined. Each of these functions satisfies the wave equation

$$\frac{\partial^2 y}{\partial t^2} = a^2 \frac{\partial^2 y}{\partial x^2} \qquad (0 < x < L, t > 0)$$

together with the boundary conditions

$$y(0, t) = y(L, t) = 0 \quad \text{for} \quad t > 0.$$

The y_n's also satisfy one initial condition:

$$\frac{\partial y}{\partial t}(x, 0) = 0 \qquad (0 < x < L).$$

We must choose n and A_n to satisfy the remaining condition, $y(x, 0) = f(x)$. Depending on $f(x)$, this may be possible. For example, if $f(x) = 8 \sin(5\pi x/L)$, we can choose $n = 5$ and $A_n = 8$. The solution of the boundary value problem for this initial displacement function f is

$$y(x, t) = 8 \sin\left(\frac{5\pi x}{L}\right)\cos\left(\frac{5\pi at}{L}\right).$$

However, if $f(x)$ is not a constant multiple of a sine function, we cannot choose any one integer n and constant A_n so that $y(x, 0) = f(x)$. In this event, we attempt an infinite superposition of the y_n's and write

$$y(x, t) = \sum_{n=1}^{\infty} y_n(x, t) = \sum_{n=1}^{\infty} A_n \sin\left(\frac{n\pi x}{L}\right) \cos\left(\frac{n\pi at}{L}\right).$$

The condition $y(x, 0) = f(x)$ requires that

$$f(x) = \sum_{n=1}^{\infty} A_n \sin\left(\frac{n\pi x}{L}\right).$$

Recognize this equation as the Fourier sine expansion of f on $[0, L]$. Thus, we should choose the A_n's as the Fourier coefficients in this expansion:

$$A_n = \frac{2}{L} \int_0^L f(x) \sin\left(\frac{n\pi x}{L}\right) dx.$$

With this choice of the constants, we have the formal solution

$$y(x, t) = \frac{2}{L} \sum_{n=1}^{\infty} \left[\int_0^L f(\xi) \sin\left(\frac{n\pi \xi}{L}\right) d\xi \right] \sin\left(\frac{n\pi x}{L}\right) \cos\left(\frac{n\pi at}{L}\right). \tag{18.21}$$

The integral in brackets is part of the Fourier coefficient, and ξ is used as the dummy variable of integration to avoid confusion with x.

As a specific illustration, suppose that initially the string is picked up $L/2$ units at its center point (Figure 18.4) and then released from rest. This initial position function is

$$f(x) = \begin{cases} x, & 0 \le x \le L/2 \\ L - x, & L/2 \le x \le L. \end{cases}$$

A routine integration gives us

$$A_n = \frac{2}{L} \int_0^{L/2} x \sin\left(\frac{n\pi x}{L}\right) dx + \frac{2}{L} \int_{L/2}^L (L - x) \sin\left(\frac{n\pi x}{L}\right) dx = \frac{4L}{n^2 \pi^2} \sin\left(\frac{n\pi}{2}\right).$$

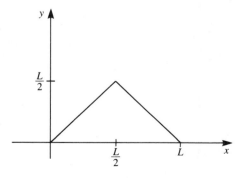

Figure 18.4. $y(x, 0) = \begin{cases} x, & 0 \le x \le \dfrac{L}{2} \\ L - x, & \dfrac{L}{2} \le x \le L. \end{cases}$

Since

$$\sin\left(\frac{n\pi}{2}\right) = \begin{cases} 0 & \text{if} \quad n \text{ is even} \\ (-1)^{k+1} & \text{if} \quad n = 2k-1, \end{cases}$$

we have

$$A_{2n} = 0 \quad \text{and} \quad A_{2n-1} = \frac{4L}{(2n-1)^2\pi^2}(-1)^{n+1}$$

for $n = 1, 2, 3, \ldots$, and the solution is

$$y(x, t) = \frac{4L}{\pi^2} \sum_{n=1}^{\infty} \frac{(-1)^{n+1}}{(2n-1)^2} \sin\left[(2n-1)\frac{\pi x}{L}\right] \cos\left[(2n-1)\frac{\pi a t}{L}\right].$$

Figure 18.5 shows the shape of the string at different times.

The numbers $n^2\pi^2/L^2$ which we found for λ are called *eigenvalues* of this problem. The functions $\sin(n\pi x/L)$, or nonzero constant multiples thereof, are corresponding *eigenfunctions*. In this physically motivated problem, the eigenvalues carry information about the frequencies of the individual sine waves which are superposed to form the final solution.

Equation (18.21) gives the solution of the wave equation with zero initial velocity and initial displacement function f. Although the reasoning leading to it was plausible, how do we know that this infinite series is indeed the solution? We will consider this question later in this section. First, we will look at another example of separation of variables applied to the wave equation.

THE WAVE EQUATION WITH ZERO INITIAL DISPLACEMENT

Now consider the case in which the string is released from its horizontal stretched position (zero initial displacement) but with a nonzero initial velocity. The boundary value problem modeling this phenomenon is

$$\frac{\partial^2 y}{\partial t^2} = a^2 \frac{\partial^2 y}{\partial x^2} \qquad (0 < x < L, t > 0),$$

$$y(0, t) = y(L, t) = 0 \qquad (t > 0),$$

$$y(x, 0) = 0 \qquad (0 < x < L),$$

$$\frac{\partial y}{\partial t}(x, 0) = g(x) \qquad (0 < x < L).$$

As before, begin by letting $y(x, t) = X(x)T(t)$. Exactly as before, we get

$$X'' + \lambda X = 0; \qquad\qquad T'' + \lambda a^2 T = 0;$$

$$X(0) = X(L) = 0.$$

The problem for X is the same as before, and we have from previous work that

$$\lambda_n = \frac{n^2\pi^2}{L^2} \quad \text{and} \quad X_n(x) = d_n \sin\left(\frac{n\pi x}{L}\right)$$

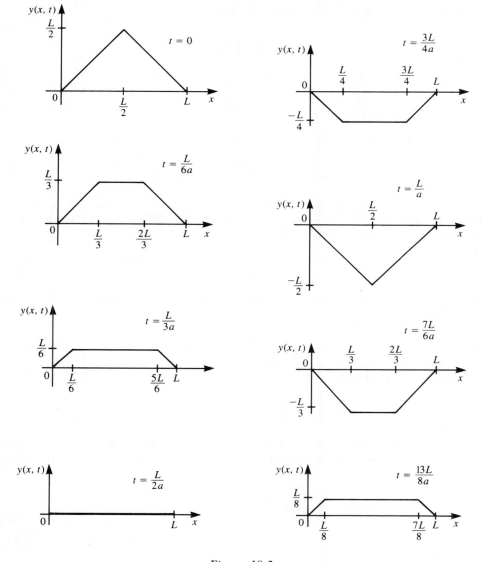

Figure 18.5

for $n = 1, 2, 3, \ldots$. The λ's are the eigenvalues of this problem, and the functions $\sin(n\pi x/L)$ are the eigenfunctions.

The problem for T, however, is not the same as before. The condition that $y(x, 0) = 0$ requires that

$$X(x)T(0) = 0$$

for $0 < x < L$ and hence that $T(0) = 0$. (In the preceding problem, with zero initial velocity, we had $T'(0) = 0$.) The general solution of $T'' + [(n^2\pi^2 a^2)/L^2]T = 0$ is

$$T(t) = \alpha_n \cos\left(\frac{n\pi a t}{L}\right) + \beta_n \sin\left(\frac{n\pi a t}{L}\right).$$

Since $T(0) = 0$, $\alpha_n = 0$, leaving us with

$$T_n(t) = \beta_n \sin\left(\frac{n\pi a t}{L}\right)$$

for $n = 1, 2, 3, \ldots$. For each positive integer n, we have a function

$$y_n(x, t) = B_n \sin\left(\frac{n\pi x}{L}\right)\sin\left(\frac{n\pi a t}{L}\right)$$

satisfying the wave equation, the boundary conditions $y(0, t) = y(L, t) = 0$, and the initial condition $y(x, 0) = 0$. We must now satisfy the initial condition $(\partial y/\partial t)(x, 0) = g(x)$. To do this, write a superposition

$$y(x, t) = \sum_{n=1}^{\infty} y_n(x, t) = \sum_{n=1}^{\infty} B_n \sin\left(\frac{n\pi x}{L}\right)\sin\left(\frac{n\pi a t}{L}\right),$$

and attempt to choose the constants so that $(\partial y/\partial t)(x, 0) = g(x)$. Differentiate the series term by term to get

$$\frac{\partial y}{\partial t}(x, t) = \sum_{n=1}^{\infty} \frac{n\pi a}{L} B_n \sin\left(\frac{n\pi x}{L}\right)\cos\left(\frac{n\pi a t}{L}\right).$$

Now we require that

$$\frac{\partial y}{\partial t}(x, 0) = \sum_{n=1}^{\infty} \frac{n\pi a}{L} B_n \sin\left(\frac{n\pi x}{L}\right) = g(x).$$

This is a Fourier sine expansion of g on $[0, L]$. The *entire coefficient* $[(n\pi a)/L]B_n$ (not just B_n) must be chosen equal to the Fourier coefficient in the sine expansion of g on $[0, L]$. Therefore,

$$\frac{n\pi a}{L} B_n = \frac{2}{L}\int_0^L g(x)\sin\left(\frac{n\pi x}{L}\right) dx \quad \text{for} \quad n = 1, 2, 3, \ldots;$$

hence,

$$B_n = \frac{2}{n\pi a}\int_0^L g(x)\sin\left(\frac{n\pi x}{L}\right) dx \quad \text{for} \quad n = 1, 2, 3, \ldots.$$

Formally, the solution of the boundary value problem is

$$y(x, t) = \sum_{n=1}^{\infty} \frac{2}{n\pi a}\left[\int_0^L g(\xi)\sin\left(\frac{n\pi \xi}{L}\right) d\xi\right]\sin\left(\frac{n\pi x}{L}\right)\sin\left(\frac{n\pi a t}{L}\right).$$

This is a "formal" solution because we have not justified all of the steps used to derive it. In particular, we did not justify differentiating the series term by term, an operation which is usually not valid without conditions on the terms of the series. Again, we will deal with this issue shortly.

As a specific example, suppose that the string is driven from its horizontal rest position with an initial velocity of one unit per second. Then $g(x) = 1$, and

$$B_n = \frac{2}{n\pi a} \int_0^L 1 \cdot \sin\left(\frac{n\pi x}{L}\right) dx = \frac{-2L}{n^2\pi^2 a}[\cos(n\pi) - 1]$$

$$= \frac{2L}{n^2\pi^2 a}[1 - (-1)^n] = \begin{cases} 0 & \text{if } n \text{ is even} \\ 4L/(n^2\pi^2 a) & \text{if } n \text{ is odd.} \end{cases}$$

In this case, the formal solution is

$$y(x, t) = \frac{4L}{\pi^2 a} \sum_{n=1}^{\infty} \frac{1}{(2n-1)^2} \sin\left[(2n-1)\frac{\pi x}{L}\right]\sin\left[(2n-1)\frac{\pi a t}{L}\right].$$

Figure 18.6 shows the position of the string at different times.

VERIFICATION OF SOLUTIONS

We have now "solved" two boundary value problems involving the wave equation. However, the proposed solutions are infinite series and were obtained without justification of critical steps. How do we know that these expressions are indeed solutions?
    Consider again the expression

$$y(x, t) = \sum_{n=1}^{\infty} \frac{2}{L}\left[\int_0^L f(\xi)\sin\left(\frac{n\pi\xi}{L}\right) d\xi\right]\sin\left(\frac{n\pi x}{L}\right)\cos\left(\frac{n\pi a t}{L}\right)$$

for the problem with initial position f and zero initial velocity. By direct substitution, we can check that $y(0, t) = y(L, t) = 0$. From the Fourier convergence theorem, we can check that $y(x, 0) = f(x)$ if f satisfies certain conditions (which are quite natural in this physical setting).
    However, this series does not converge uniformly in either x or t, so there is no evidence that we can differentiate it term by term to check that y satisfies the wave equation (although each y_n does) or that $(\partial y/\partial t)(x, 0) = 0$.
    We will now demonstrate a remarkable fact: we can sum this series solution in closed form! The key is to recognize that we can write

$$2 \sin\left(\frac{n\pi x}{L}\right)\cos\left(\frac{n\pi a t}{L}\right) = \sin\left[\frac{n\pi(x + at)}{L}\right] + \sin\left[\frac{n\pi(x - at)}{L}\right].$$

Substitute this into the series solution (with the Fourier coefficients called A_n) to get

$$\sum_{n=1}^{\infty} A_n\sin\left(\frac{n\pi x}{L}\right)\cos\left(\frac{n\pi a t}{L}\right) = \frac{1}{2}\left\{\sum_{n=1}^{\infty} A_n\sin\left[\frac{n\pi(x + at)}{L}\right]\right.$$
$$\left. + \sum_{n=1}^{\infty} A_n\sin\left[\frac{n\pi(x - at)}{L}\right]\right\}.$$

Assume that $f(x) = \sum_{n=1}^{\infty} A_n\sin(n\pi x/L)$ for $0 \leq x \leq L$. Then the series solution is exactly $\frac{1}{2}[f(x + at) + f(x - at)]$. This suggests that the solution can be written in terms of the initial position function f as

$$y(x, t) = \frac{1}{2}[f(x + at) + f(x - at)].$$

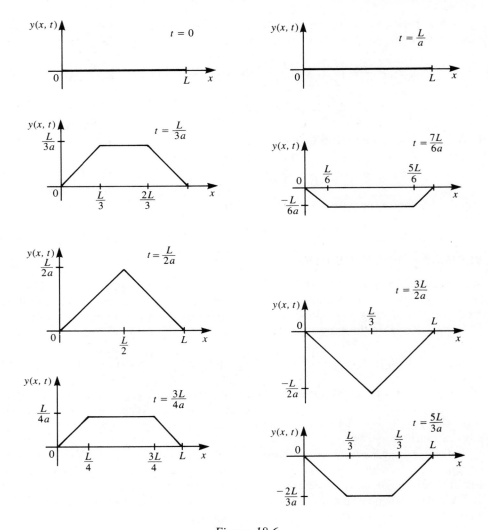

Figure 18.6

The drawback to this is that $f(x)$ is defined only for $0 \leq x \leq L$, while $x + at$ and $x - at$ take on all real values as x varies over $[0, L]$ and t varies over $[0, \infty)$. This problem can be met by extending f as follows. First, extend f to an odd function $\hat{f}$ defined on $[-L, L]$ by defining

$$\hat{f}(x) = \begin{cases} f(x) & \text{for} \quad 0 \leq x \leq L \\ -f(-x) & \text{for} \quad -L \leq x \leq 0. \end{cases}$$

We choose an odd extension because the series representing f on $[0, L]$ is a sine series. This extension is shown in Figure 18.7.

Now extend $\hat{f}$ to a periodic function F defined over the entire real line by repeating the graph of $\hat{f}$ over each interval $[L, 3L]$, $[3L, 5L], \ldots, [-3L, -L]$,

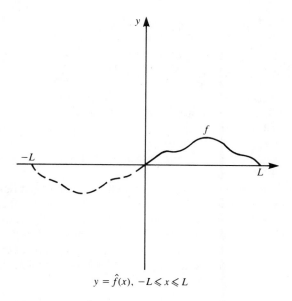

$$y = \hat{f}(x),\ -L \leqslant x \leqslant L$$

Figure 18.7. $\hat{f}$ = odd extension of f to $[-L, L]$.

$[-5L, -3L], \ldots$, as shown in Figure 18.8. We claim that the solution of the boundary value problem for y is

$$y(x, t) = \tfrac{1}{2}[F(x + at) + F(x - at)]. \tag{18.22}$$

(In this connection, note Problem 21 of Section 18.0.)

Certainly this function is continuous if F is continuous. Assuming that f is continuous on $[0, L]$, a reasonable assumption for an initial configuration of a string, the only places where F can be discontinuous are those points where we extended the original function f to $\hat{f}$ and then extended $\hat{f}$ to F. Thus, the only places where F could

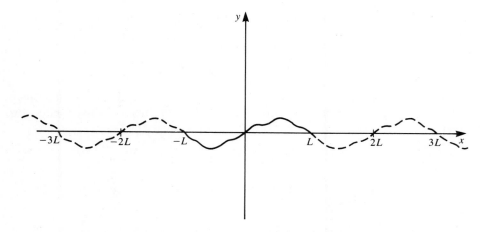

Figure 18.8. F = periodic extension of $\hat{f}$ to the entire real line.

conceivably be discontinuous are $0, \pm L, \pm 2L, \pm 3L, \ldots$. However, observe that $f(0) = f(L) = 0$ because the ends of the string are fastened on the x-axis. Then (note Figures 18.7 and 18.8) the extended graphs join with the original ones, and F is continuous even at the points $0, \pm L, \pm 2L, \ldots$. Further, we can differentiate y twice by the chain rule (something the student should do) and verify that $\frac{1}{2}[F(x + at) + F(x - at)]$ satisfies the wave equation if F is twice continuously differentiable. This will hold if F is sufficiently smooth at the places where f and $\hat{f}$ were extended in Figures 18.7 and 18.8. In particular, it is sufficient that $f''(0) = f''(L) = 0$. Thus, by assuming physically reasonable conditions on f, we can obtain a twice continuously differentiable function F such that the solution of this boundary value problem is given by equation (18.22).

This discussion applies to the case in which the string is initially displaced and released with zero initial velocity. In the case of zero initial displacement but initial velocity function g, we differentiated the proposed series solution term by term to solve for the coefficients. In fact, this device was useful in suggesting how the coefficients should be chosen, but this solution, once derived, can also be justified by an argument similar to that just carried out. It is left for the student to show that the solution of this boundary value problem is

$$y(x, t) = \frac{1}{2a} \int_{x - at}^{x + at} G(\xi) \, d\xi,$$

in which G is obtained by first extending g to an *even* function $\hat{g}$ defined on $[-L, L]$ and then extending $\hat{g}$ to a periodic function G defined on the whole real line. Again, note the formula in Problem 21(e) of Section 18.0.

THE WAVE EQUATION WITH INITIAL DISPLACEMENT AND VELOCITY

Consider the boundary value problem

$$\frac{\partial^2 y}{\partial t^2} = a^2 \frac{\partial^2 y}{\partial x^2} \qquad (0 < x < L, t > 0),$$

$$y(0, t) = y(L, t) = 0 \qquad (t > 0),$$

$$y(x, 0) = f(x) \qquad (0 < x < L),$$

$$\frac{\partial y}{\partial t}(x, 0) = g(x) \qquad (0 < x < L).$$

This models a vibrating string in which we have possibly nonzero initial displacement and velocity functions. Instead of approaching this problem directly, define two problems:

Problem 1:
$$\begin{cases} \dfrac{\partial^2 y}{\partial t^2} = a^2 \dfrac{\partial^2 y}{\partial x^2} & (0 < x < L, t > 0), \\[2mm] y(0, t) = y(L, t) = 0 & (t > 0), \\[2mm] y(x, 0) = f(x) & (0 < x < L), \\[2mm] \dfrac{\partial y}{\partial t}(x, 0) = 0 & (0 < x < L) \end{cases}$$

and

$$
\text{Problem 2:} \quad
\begin{cases}
\dfrac{\partial^2 y}{\partial t^2} = a^2 \dfrac{\partial^2 y}{\partial x^2} & (0 < x < L, t > 0), \\[2mm]
y(0, t) = y(L, t) = 0 & (t > 0), \\[2mm]
y(x, 0) = 0 & (0 < x < L), \\[2mm]
\dfrac{\partial y}{\partial t}(x, 0) = g(x) & (0 < x < L).
\end{cases}
$$

Let y_1 be the solution of problem 1, and let y_2 be the solution of problem 2. It is routine to check that $y_1 + y_2$ is the solution of the problem with initial displacement and velocity (see Problem 26 of Section 18.0).

As an example, suppose that the initial displacement is given by

$$
f(x) = \begin{cases} x, & 0 \le x \le L/2 \\ L - x, & L/2 \le x \le L \end{cases}
$$

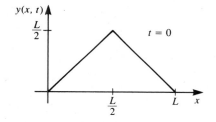

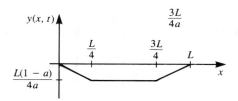

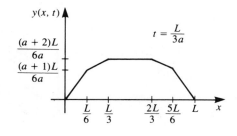

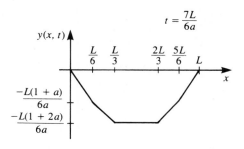

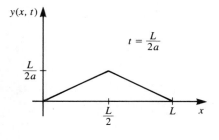

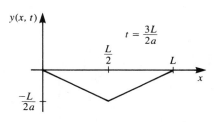

Figure 18.9

and the initial velocity by

$$g(x) = 1, \ 0 \le x \le L.$$

The displacement function for the motion of the string is

$$y(x, t) = \frac{4L}{\pi^2} \sum_{n=1}^{\infty} \frac{1}{(2n-1)^2} \sin\left[(2n-1)\frac{\pi x}{L}\right]$$

$$\times \left\{(-1)^{n+1}\cos\left[(2n-1)\frac{\pi at}{L}\right] + \frac{1}{a}\sin\left[(2n-1)\frac{\pi at}{L}\right]\right\}.$$

We may also write this solution as

$$y(x, t) = \frac{1}{2}[F(x + at) + F(x - at)] + \frac{1}{2a}\int_{x-at}^{x+at} G(\xi)\, d\xi,$$

with F obtained as an odd periodic extension of f and G as an even periodic extension of g to the whole real line.

Graphs of $y(x, t)$ for $0 \le x \le L$ and for various values of t are shown in Figure 18.9. These show the shape of the string at the indicated times.

PROBLEMS FOR SECTION 18.1

In each of Problems 1 through 10, write a series solution of the boundary value problem.

1. $\dfrac{\partial^2 y}{\partial t^2} = a^2 \dfrac{\partial^2 y}{\partial x^2}$ $(0 < x < 2, t > 0)$,

$y(0, t) = y(2, t) = 0$ $(t > 0)$,
$y(x, 0) = 0$ $(0 < x < 2)$,
$\dfrac{\partial y}{\partial t}(x, 0) = 2x$ $(0 < x < 2)$

2. $\dfrac{\partial^2 y}{\partial t^2} = 9 \dfrac{\partial^2 y}{\partial x^2}$ $(0 < x < 4, t > 0)$,

$y(0, t) = y(4, t) = 0$ $(t > 0)$,
$y(x, 0) = 2 \sin(\pi x)$ $(0 < x < 4)$,
$\dfrac{\partial y}{\partial t}(x, 0) = 0$ $(0 < x < 4)$

3. $\dfrac{\partial^2 y}{\partial t^2} = 4 \dfrac{\partial^2 y}{\partial x^2}$ $(0 < x < 3, t > 0)$,

$y(0, t) = y(3, t) = 0$ $(t > 0)$,
$y(x, 0) = 0$ $(0 < x < 3)$,
$\dfrac{\partial y}{\partial t}(x, 0) = x$ $(0 < x < 3)$

4. $\dfrac{\partial^2 y}{\partial t^2} = 9 \dfrac{\partial^2 y}{\partial x^2}$ $(0 < x < \pi, t > 0)$,

$y(0, t) = y(\pi, t) = 0$ $(t > 0)$,
$y(x, 0) = \sin(x)$ $(0 < x < \pi)$,
$\dfrac{\partial y}{\partial t}(x, 0) = 1$ $(0 < x < \pi)$

5. $\dfrac{\partial^2 y}{\partial t^2} = 8 \dfrac{\partial^2 y}{\partial x^2}$ $(0 < x < 2\pi, t > 0)$,

$y(0, t) = y(2\pi, t) = 0$ $(t > 0)$,

$y(x, 0) = \begin{cases} x, & 0 < x < \pi \\ 2\pi - x, & \pi < x < 2\pi, \end{cases}$

$\dfrac{\partial y}{\partial t}(x, 0) = 0$ $(0 < x < 2\pi)$

6. $\dfrac{\partial^2 y}{\partial t^2} = 4 \dfrac{\partial^2 y}{\partial x^2}$ $(0 < x < 5, t > 0)$,

$y(0, t) = y(5, t) = 0$ $(t > 0)$,
$y(x, 0) = 0$ $(0 < x < 5)$,

$\dfrac{\partial y}{\partial t}(x, 0) = \begin{cases} x, & 0 \le x \le \frac{5}{2} \\ 5 - x, & \frac{5}{2} \le x \le 5 \end{cases}$

7. $\dfrac{\partial^2 y}{\partial t^2} = 9 \dfrac{\partial^2 y}{\partial x^2}$ $\qquad (0 < x < 2, t > 0)$,

$y(0, t) = y(2, t) = 0$ $\qquad (t > 0)$,

$y(x, 0) = x(2 - x)$ $\qquad (0 < x < 2)$,

$\dfrac{\partial y}{\partial t}(x, 0) = 4$ $\qquad (0 < x < 2)$

8. $\dfrac{\partial^2 y}{\partial t^2} = 16 \dfrac{\partial^2 y}{\partial x^2}$ $\qquad (0 < x < 5, t > 0)$,

$y(0, t) = y(5, t) = 0$ $\qquad (t > 0)$,

$y(x, 0) = x(x - 5)$ $\qquad (0 < x < 5)$,

$\dfrac{\partial y}{\partial t}(x, 0) = 0$ $\qquad (0 < x < 5)$

9. $\dfrac{\partial^2 y}{\partial t^2} = 8 \dfrac{\partial^2 y}{\partial x^2}$ $\qquad (0 < x < 4, t > 0)$,

$y(0, t) = y(4, t) = 0$ $\qquad (t > 0)$,

$y(x, 0) = x^2(x - 4)$ $\qquad (0 < x < 4)$,

$\dfrac{\partial y}{\partial t}(x, 0) = 1$ $\qquad (0 < x < 4)$

10. $\dfrac{\partial^2 y}{\partial t^2} = 25 \dfrac{\partial^2 y}{\partial x^2}$ $\qquad (0 < x < \pi, t > 0)$,

$y(0, t) = y(\pi, t) = 0$ $\qquad (t > 0)$,

$y(x, 0) = \sin(2x)$ $\qquad (0 < x < \pi)$,

$\dfrac{\partial y}{\partial t}(x, 0) = \pi - x$ $\qquad (0 < x < \pi)$

11. Solve the boundary value problem

$$\frac{\partial^2 y}{\partial t^2} = 3 \frac{\partial^2 y}{\partial x^2} + 2x \qquad (0 < x < 2, t > 0),$$

$$y(0, t) = y(2, t) = 0 \qquad (t > 0),$$

$$y(x, 0) = 0 \qquad (0 < x < 2),$$

$$\frac{\partial y}{\partial t}(x, 0) = 0 \qquad (0 < x < 2).$$

Hint: Putting $y(x, t) = X(x)T(t)$ into the differential equation results in an equation in which X and T cannot be separated on opposite sides. Therefore, put $Y(x, t) = y(x, t) + h(x)$ and substitute $y = Y - h$ into the boundary value problem to obtain a boundary value problem for Y. Now try to choose h so that the problem for Y is like one we have already solved.

12. Solve the boundary value problem

$$\frac{\partial^2 y}{\partial t^2} = 9 \frac{\partial^2 y}{\partial x^2} + x^2 \qquad (0 < x < 4, t > 0),$$

$$y(0, t) = y(4, t) = 0 \qquad (t > 0),$$

$$y(x, 0) = 0 \qquad (0 < x < 4),$$

$$\frac{\partial y}{\partial t}(x, 0) = 0 \qquad (0 < x < 4).$$

(Use the hint with Problem 11.)

13. Solve the boundary value problem

$$\frac{\partial^2 y}{\partial t^2} = \frac{\partial^2 y}{\partial x^2} - \cos(x) \qquad (0 < x < 2\pi, t > 0),$$

$$y(0, t) = y(2\pi, t) = 0 \qquad (t > 0),$$

$$y(x, 0) = 0 \qquad (0 < x < 2\pi),$$

$$\frac{\partial y}{\partial t}(x, 0) = 0 \qquad (0 < x < 2\pi).$$

14. Solve the vibrating string problem in which there is a damping force proportional to the velocity. With zero initial velocity and with initial position given as the graph of $y = f(x)$, the boundary value problem is

$$\frac{\partial^2 y}{\partial t^2} = a^2 \frac{\partial^2 y}{\partial x^2} - \alpha \frac{\partial y}{\partial t} \qquad (0 < x < L, t > 0),$$

$$y(0, t) = y(L, t) = 0 \qquad\qquad (t > 0),$$

$$y(x, 0) = f(x) \qquad\qquad (0 < x < L),$$

$$\frac{\partial y}{\partial t}(x, 0) = 0 \qquad\qquad (0 < x < L).$$

Assume that $\alpha < 2\pi a/L$.

15. Solve the boundary value problem which models the longitudinal displacements in a cylindrical bar of length L if, at time zero, the bar is stretched by an amount AL and then released. The boundary value problem is

$$\frac{\partial^2 y}{\partial t^2} = a^2 \frac{\partial^2 y}{\partial x^2} \qquad (0 < x < L, t > 0),$$

$$\frac{\partial y}{\partial x}(0, t) = \frac{\partial y}{\partial x}(L, t) \qquad\qquad (t > 0),$$

$$y(x, 0) = (A + 1)x \qquad\qquad (0 < x < L),$$

$$\frac{\partial y}{\partial t}(x, 0) = 0 \qquad\qquad (0 < x < L).$$

Here, $a^2 = E/\delta$, where E is the modulus of elasticity of the bar and δ is the mass per unit volume (assumed constant).

16. Transverse vibrations in a homogeneous (constant-density) rod of length π are modeled by the partial differential equation

$$a^2 \frac{\partial^4 u}{\partial x^4} + \frac{\partial^2 u}{\partial t^2} = 0 \qquad (0 < x < \pi, t > 0).$$

Here, $u(x, t)$ is the displacement at time t of the cross-section through x perpendicular to the x-axis; a^2 is $EI/\rho A$, where E is Young's modulus; I is the moment of inertia of a cross-section with respect to the x-axis; ρ is the density; and A is the cross-sectional area. Assume that both ρ and A are constant.

(a) Let $u(x, t) = X(x)T(t)$ to obtain differential equations for X and T.

(b) Solve for values of the separation constant and for X and T in the case of free ends:

$$\frac{\partial^2 u}{\partial x^2}(0, t) = \frac{\partial^2 u}{\partial x^2}(\pi, t) = \frac{\partial^3 u}{\partial x^3}(0, t) = \frac{\partial^3 u}{\partial x^3}(\pi, t) = 0.$$

(c) Solve for values of the separation constant and for X and T in the case of supported ends:

$$u(0, t) = u(\pi, t) = \frac{\partial^2 u}{\partial x^2}(0, t) = \frac{\partial^2 u}{\partial x^2}(\pi, t) = 0 \qquad (t > 0).$$

17. Solve the *telegraph equation*

$$\frac{\partial^2 u}{\partial t^2} + A \frac{\partial u}{\partial t} + Bu = a^2 \frac{\partial^2 u}{\partial x^2} \qquad (0 < x < L, t > 0),$$

in which A and B are positive constants. The boundary conditions are

$$u(0, t) = u(L, t) = 0 \quad \text{for} \quad t > 0,$$

and the initial conditions are

$$u(x, 0) = f(x) \quad \text{and} \quad \frac{\partial u}{\partial t}(x, 0) = 0 \quad \text{for} \quad 0 < x < L.$$

Assume that $L^2A^2 < 4(L^2B + a^2\pi^2)$.

18. The current $i(x, t)$ and voltage $v(x, t)$ at time t and distance x from one end of a transmission line satisfy the system

$$-\frac{\partial v}{\partial x} = Ri + L\frac{\partial i}{\partial t},$$

$$-\frac{\partial i}{\partial x} = Sv + K\frac{\partial v}{\partial t},$$

in which R is the resistance, L is the inductance, S is the leakage conductance, and K is the capacitance to ground, all per unit length and assumed constant. By differentiating appropriately and eliminating i and its partial derivatives (or v and its partials), show that v (or i) satisfies the telegraph equation (see the preceding problem).

19. Consider torsional oscillations of a homogeneous cylindrical shaft. If $\theta(x, t)$ is the angular displacement at time t of the cross-section at x,

$$\frac{\partial^2\theta}{\partial t^2} = a^2\frac{\partial^2\theta}{\partial x^2}.$$

Solve this equation if $\theta(x, 0) = f(x)$ and $(\partial\theta/\partial t)(x, 0) = 0$ for $0 < x < L$ and the ends of the shaft are fixed elastically:

$$\frac{\partial\theta}{\partial x}(0, t) - \alpha\theta(0, t) = 0,$$

$$\frac{\partial\theta}{\partial x}(L, t) + \alpha\theta(L, t) = 0$$

for $t > 0$. Here, α is a positive constant.

18.2 Fourier Series Solutions of the Heat Equation

We will now apply the separation of variables method to typical boundary value problems associated with the heat equation.

ENDS OF THE BAR KEPT AT ZERO TEMPERATURE

We want to determine the temperature distribution $u(x, t)$ in a thin homogeneous bar of length L, given the initial temperature distribution throughout the bar at time zero, if the ends are maintained at zero temperature for all time. The boundary value problem is

$$\frac{\partial u}{\partial t} = a^2 \frac{\partial^2 u}{\partial x^2} \qquad (0 < x < L, t > 0),$$

$$u(0, t) = u(L, t) = 0 \qquad\qquad (t > 0),$$

$$u(x, 0) = f(x) \qquad\qquad (0 < x < L).$$

We will apply the separation of variables method and seek a solution $u(x, t) = X(x)T(t)$. Substitute this into the heat equation to get

$$XT' = a^2 X''T,$$

or

$$\frac{T'}{a^2 T} = \frac{X''}{X}.$$

Since x and t are independent, both sides of this equation must equal the same constant. For some λ,

$$\frac{T'}{a^2 T} = \frac{X''}{X} = -\lambda.$$

As with the wave equation, we call the constant $-\lambda$ simply to make the differential equation for X of Sturm-Liouville type. We obtain the same solutions if the constant is given another name. Then

$$X'' + \lambda X = 0 \quad \text{and} \quad T' + \lambda a^2 T = 0.$$

The condition that $u(0, t) = 0$ implies that $X(0)T(t) = 0$ for $t > 0$ and hence that

$$X(0) = 0,$$

assuming that $T(t)$ is not identically zero. Similarly, $u(L, t) = 0 = X(L)T(t)$ implies that

$$X(L) = 0.$$

We now have

$$T' + \lambda a^2 T = 0 \quad \text{and} \quad X'' + \lambda^2 X = 0,$$
$$X(0) = X(L) = 0.$$

Unlike that with the wave equation, this equation for T is first order, with no boundary condition. The boundary value problem for X, however, is identical with that encountered with the wave equation, so we have eigenvalues

$$\lambda_n = \frac{n^2 \pi^2}{L^2}$$

and corresponding eigenfunctions

$$X_n(x) = d_n \sin\left(\frac{n\pi x}{L}\right)$$

for $n = 1, 2, 3, \ldots$. The constants d_n will be determined later. With the values we have for λ, the differential equation for T is

$$T' + \frac{a^2 n^2 \pi^2}{L^2} T = 0,$$

with general solution

$$T_n(t) = \alpha_n e^{-n^2 \pi^2 a^2 t/L^2}$$

for $n = 1, 2, 3, \ldots$. Now let

$$u_n(x, t) = X_n(x)T_n(t) = A_n \sin\left(\frac{n\pi x}{L}\right) e^{-n^2 \pi^2 a^2 t/L^2},$$

with $A_n = d_n \alpha_n$. Each u_n satisfies the heat equation and both boundary conditions $u(0, t) = u(L, t) = 0$. There remains to satisfy the initial condition $u(x, 0) = f(x)$ for $0 < x < L$.

If we can choose some positive integer n and a constant A_n such that $u_n(x, 0) = f(x)$, we will have the solution. For example, if $f(x) = 4 \sin(3\pi x/L)$, we choose $n = 3$ and $A_3 = 4$ to obtain the solution

$$u(x, t) = u_3(x, t) = 4 \sin\left(\frac{3\pi x}{L}\right) e^{-9\pi^2 a^2 t/L^2}.$$

Usually, we cannot choose any n and A_n such that $u_n(x, 0) = f(x)$, and we attempt an infinite superposition

$$u(x, t) = \sum_{n=1}^{\infty} u_n(x, t) = \sum_{n=1}^{\infty} A_n \sin\left(\frac{n\pi x}{L}\right) e^{-n^2 \pi^2 a^2 t/L^2}.$$

The condition $u(x, 0) = f(x)$ requires that

$$u(x, 0) = f(x) = \sum_{n=1}^{\infty} A_n \sin\left(\frac{n\pi x}{L}\right).$$

This is the Fourier sine expansion of f on $[0, L]$; hence, choose the A_n's as the Fourier coefficients:

$$A_n = \frac{2}{L} \int_0^L f(x) \sin\left(\frac{n\pi x}{L}\right) dx.$$

Using ξ as the variable of integration, the proposed solution is

$$u(x, t) = \frac{2}{L} \sum_{n=1}^{\infty} \left[\int_0^L f(\xi) \sin\left(\frac{n\pi \xi}{L}\right) d\xi\right] \sin\left(\frac{n\pi x}{L}\right) e^{-n^2 \pi^2 a^2 t/L^2}.$$

As a specific example, suppose that the bar has length π and the initial temperature function is $f(x) = 2$ for $0 < x < \pi$. Compute

$$\int_0^\pi 2 \sin(n\xi) \, d\xi = -\frac{2}{n} \cos(n\xi) \Big]_0^\pi$$

$$= -\frac{2}{n}[\cos(n\pi) - 1] = \frac{2}{n}[1 - (-1)^n].$$

The proposed solution is

$$u(x, t) = \frac{2}{\pi} \sum_{n=1}^{\infty} \frac{2}{n} [1 - (-1)^n] \sin(nx) e^{-n^2 a^2 t}$$

$$= \frac{8}{\pi} \sum_{n=1}^{\infty} \frac{1}{2n - 1} \sin[(2n - 1)x] e^{-(2n-1)^2 a^2 t}.$$

We will examine this proposed solution more carefully. Each term satisfies the heat equation and the boundary conditions $u(0, t) = u(\pi, t) = 0$. None of them individually satisfies the initial condition $u(x, 0) = 2$. The entire sum satisfies the boundary conditions and the initial condition $u(x, 0) = 2$ on $0 < x < \pi$. Finally, we must verify that $u(x, t)$ satisfies the differential equation.

In order to do this, we would like to be able to differentiate the series term by term. This is certainly valid if the appropriate series converge uniformly. We will now show that they do. If $t_0 > 0$ and $0 \leq x \leq \pi$,

$$\left| \frac{1}{2n - 1} \sin[(2n - 1)x] e^{-(2n-1)^2 a^2 t} \right| \leq \frac{1}{2n - 1} e^{-(2n-1)^2 a^2 t_0}.$$

Further, the series

$$\sum_{n=1}^{\infty} \frac{1}{2n - 1} e^{-(2n-1)^2 a^2 t_0}$$

converges. By a theorem of Weierstrass (sometimes called the M-test), the series

$$\sum_{n=1}^{\infty} \frac{1}{2n - 1} \sin[(2n - 1)x] e^{-(2n-1)^2 a^2 t_0}$$

converges uniformly with respect to x and t for $0 \leq x \leq \pi$ and $t \geq t_0$.

By a similar argument, the series obtained by differentiating this series term by term with respect to t, or twice with respect to x, also converge uniformly for $0 \leq x \leq \pi$ and $t \geq t_0$. We may therefore differentiate the proposed series solution term by term once with respect to t and twice with respect to x and verify that $\partial u / \partial t = a^2 (\partial^2 u / \partial x^2)$, completing the verification of u as the solution of the boundary value problem. It can be shown that this solution is unique, as might be expected from the physical setting of the problem.

TEMPERATURE IN A BAR WITH INSULATED ENDS

Consider the problem of heat conduction in a bar with no energy loss across the ends. The boundary value problem for the temperature distribution $u(x, t)$ is

$$\frac{\partial u}{\partial t} = a^2 \frac{\partial^2 u}{\partial x^2} \qquad (0 < x < L, t > 0),$$

$$\frac{\partial u}{\partial x}(0, t) = \frac{\partial u}{\partial x}(L, t) = 0 \qquad (t > 0),$$

$$u(x, 0) = f(x) \qquad (0 < x < L).$$

As before, let $u(x, t) = X(x)T(t)$ to obtain

$$T' + \lambda a^2 T = 0, \qquad X'' + \lambda X = 0.$$

The insulation condition at the left end is

$$\frac{\partial u}{\partial x}(0, t) = X'(0)T(t) = 0,$$

implying that $X'(0) = 0$. Similarly,

$$\frac{\partial u}{\partial x}(L, t) = X'(L)T(t) = 0;$$

hence, $X'(L) = 0$. We therefore have the following problems for X and T:

$$T' + \lambda a^2 T = 0 \quad \text{and} \quad X'' + \lambda X = 0,$$
$$X'(0) = X'(L) = 0.$$

The boundary value problem for X is different from that encountered previously, so we will carry out the analysis in detail. Consider cases on λ.

Case 1: $\lambda = 0$ Now $X = cx + d$. Then $X'(0) = c = 0$, so $X(x) = d$, a constant. This does satisfy the condition $X'(L) = 0$. Thus, $\lambda = 0$ is an eigenvalue of this problem, with corresponding eigenfunctions $X(x) = $ constant.

Case 2: $\lambda < 0$ Let $\lambda = -k^2$, with $k > 0$. The general solution of $X'' - k^2 X = 0$ is

$$X = ce^{kx} + de^{-kx}.$$

Since $X'(0) = 0$,

$$kc - kd = 0.$$

But $k > 0$; hence, $c = d$. Then

$$X(x) = 2c \cosh(kx).$$

Next,

$$X'(L) = 2kc \sinh(kL) = 0.$$

Since $kL > 0$, $\sinh(kL) > 0$, and this equation forces us to choose $c = 0$. Then $c = d = 0$, and we obtain only the trivial solution for X. Therefore, this problem has no negative eigenvalue.

Case 3: $\lambda > 0$ Write $\lambda = k^2$. The general solution of $X'' + k^2 X = 0$ is

$$X = c \cos(kx) + d \sin(kx).$$

Now $X'(0) = kd = 0$ forces us to choose $d = 0$, so $X = c \cos(kx)$. Further,

$$X'(L) = -kc \sin(kL) = 0.$$

If $c = 0$, we again have only the trivial solution for X. If $c \neq 0$, $\sin(kL) = 0$, implying that kL is a positive integer multiple of π (recall that $kL > 0$). Thus, choose $kL = n\pi$, or

$k = n\pi/L$. Then

$$\lambda = \frac{n^2\pi^2}{L^2} \quad \text{for} \quad n = 1, 2, 3, \ldots$$

are eigenvalues of this problem. Corresponding to the eigenvalue $n^2\pi^2/L^2$, we have eigenfunctions

$$X_n(x) = \cos\left(\frac{n\pi x}{L}\right),$$

or constant multiples of these cosine functions.

We can combine cases 1 and 3, in which we found eigenvalues and eigenfunctions, by writing the eigenvalues as

$$\lambda_n = \frac{n^2\pi^2}{L^2} \quad \text{for} \quad n = 0, 1, 2, 3, \ldots$$

are corresponding eigenfunctions

$$X_n(x) = c_n\cos\left(\frac{n\pi x}{L}\right),$$

in which c_n is to be determined. When $n = 0$, we get the conclusion of case 1; when n is a positive integer, we obtain case 3.

The equation for T is

$$T' + \frac{a^2 n^2 \pi^2}{L^2} T = 0,$$

with solutions

$$T_0(t) = \alpha_0 = \text{constant} \quad \text{if} \quad n = 0$$

and

$$T_n(t) = \alpha_n e^{-n^2\pi^2 a^2 t/L^2} \quad \text{for} \quad n = 1, 2, 3, \ldots.$$

For $n = 0, 1, 2, \ldots$, we now have a function

$$u_n(x, t) = A_n\cos\left(\frac{n\pi x}{L}\right)e^{-n^2\pi^2 a^2 t/L^2}$$

which satisfies the heat equation and the insulation conditions at the ends of the bar. For the initial condition $u(x, 0) = f(x)$, we try a superposition

$$u(x, t) = \sum_{n=0}^{\infty} u_n(x, t) = \sum_{n=0}^{\infty} A_n\cos\left(\frac{n\pi x}{L}\right)e^{-n^2\pi^2 a^2 t/L^2}.$$

The initial condition requires that we choose the A_n's so that

$$u(x, 0) = f(x) = \sum_{n=0}^{\infty} A_n\cos\left(\frac{n\pi x}{L}\right)$$

for $0 < x < L$. This is a Fourier cosine expansion of f on $[0, L]$, so we choose

$$A_0 = \frac{1}{L}\int_0^L f(x)\,dx \quad \text{and} \quad A_n = \frac{2}{L}\int_0^L f(x)\cos\left(\frac{n\pi x}{L}\right)dx.$$

The proposed solution is

$$u(x, t) = \frac{1}{L} \int_0^L f(\xi) \, d\xi + \frac{2}{L} \sum_{n=1}^{\infty} \left[\int_0^L f(\xi) \cos\left(\frac{n\pi\xi}{L}\right) d\xi \right] \cos\left(\frac{n\pi x}{L}\right) e^{-n^2\pi^2 a^2 t/L^2}.$$

As in the case in which the ends of the bar are kept at zero temperature, this series can be differentiated term by term with respect to t, and twice with respect to x, for $0 \leq x \leq L$ and $t \geq t_0 > 0$. It therefore satisfies the heat equation and the boundary and initial conditions of the problem and is the solution of the problem.

As a specific example, suppose that $f(x) = x(x - L)$. Compute

$$\int_0^L \xi(L - \xi) \, d\xi = \tfrac{1}{6} L^3$$

and

$$\int_0^L \xi(L - \xi) \cos\left(\frac{n\pi\xi}{L}\right) d\xi = -\frac{L^3}{n^2\pi^2} [(-1)^n + 1].$$

The solution for this case is

$$u(x, t) = \frac{1}{6} L^2 - 2L^2 \sum_{n=1}^{\infty} \frac{1}{n^2\pi^2} [(-1)^n + 1] \cos\left(\frac{n\pi x}{L}\right) e^{-n^2\pi^2 a^2 t/L^2}.$$

Since

$$(-1)^n + 1 = \begin{cases} 2 & \text{if } n \text{ is even} \\ 0 & \text{if } n \text{ is odd}, \end{cases}$$

the solution may be written

$$u(x, t) = \frac{1}{6} L^2 - 4L^2 \sum_{n=1}^{\infty} \frac{1}{(2n)^2\pi^2} \cos\left(\frac{2n\pi x}{L}\right) e^{-4n^2\pi^2 a^2 t/L^2}$$

$$= \frac{1}{6} L^2 - \frac{L^2}{\pi^2} \sum_{n=1}^{\infty} \frac{1}{n^2} \cos\left(\frac{2n\pi x}{L}\right) e^{-4n^2\pi^2 a^2 t/L^2}.$$

TEMPERATURE IN A BAR WITH A RADIATING END

In this example, we will consider the heat equation for a bar of length L with its left end maintained at a constant zero temperature and its right end radiating energy into the surrounding medium. We will assume that this medium is also at zero temperature and that the initial temperature at any point x in the bar is $f(x)$. The boundary value problem is

$$\frac{\partial u}{\partial t} = a^2 \frac{\partial^2 u}{\partial x^2} \qquad (0 < x < L, t > 0),$$

$$u(0, t) = 0 \qquad (t > 0),$$

$$\frac{\partial u}{\partial x}(L, t) = -Au(L, t) \qquad (t > 0),$$

$$u(x, 0) = f(x) \qquad (0 < x < L),$$

in which A is a positive constant. Let $u(x, t) = X(x)T(t)$. Then

$$T' + \lambda a^2 T = 0 \quad \text{and} \quad X'' + \lambda X = 0.$$

Since $u(0, t) = X(0)T(t)$, we must have $X(0) = 0$. At the other end of the bar, the condition $(\partial u/\partial x)(L, t) = -Au(L, t)$ implies that

$$X'(L)T(t) + AX(L)T(t) = [X'(L) + AX(L)]T(t) = 0;$$

hence, $X'(L) + AX(L) = 0$. The equations for T and X are

$$T' + \lambda a^2 T = 0 \quad \text{and} \quad X'' + \lambda X = 0,$$
$$X(0) = 0, \qquad X'(L) + AX(L) = 0.$$

As before, consider cases on λ.

Case 1: $\lambda = 0$ Then $X = cx + d$. Since $X(0) = d = 0$, $X = cx$. Then

$$X'(L) + AX(L) = c + AcL = 0.$$

Since $1 + AL > 0$, we must have $c = 0$, leading to the trivial solution $X(x) = 0$. Therefore, zero is not an eigenvalue of this problem.

Case 2: $\lambda < 0$ Write $\lambda = -k^2$, with $k > 0$. Then $X'' - k^2 X = 0$, with general solution

$$X = ce^{kx} + de^{-kx}.$$

Since $X(0) = c + d = 0$, $d = -c$, and

$$X = 2c \sinh(kx).$$

Then

$$X'(L) + AX(L) = 2ck \cosh(kL) + 2Ac \sinh(kL) = 0.$$

Then

$$2c[k \cosh(kL) + A \sinh(kL)] = 0.$$

Now, $kL > 0$, so $k \cosh(kL) + A \sinh(kL) > 0$. Therefore, $2c = 0$, so $c = d = 0$, and we obtain only the trivial solution in this case also. This problem has no negative eigenvalue.

Case 3: $\lambda > 0$ Write $\lambda = k^2$, with $k > 0$. Then $X'' + k^2 X = 0$, with general solution

$$X = c \cos(kx) + d \sin(kx).$$

Now $X(0) = c = 0$, so $X = d \sin(kx)$. Further,

$$X'(L) + AX(L) = dk \cos(kL) + Ad \sin(kL) = 0.$$

Then

$$d[k \cos(kL) + A \sin(kL)] = 0.$$

In order to have a nontrivial solution, we must choose $d \neq 0$. This forces us to choose k so that

$$k \cos(kL) + A \sin(kL) = 0,$$

or

$$\tan(kL) = -\frac{k}{A}.$$

Let $z = kL$. We must solve for z in the equation

$$\tan(z) = -\frac{z}{AL}.$$

This problem is very similar to one we encountered in Example 7.9. Although this equation cannot be solved algebraically for z, we can show that there are infinitely many positive solutions. Consider the graphs of $y = \tan(z)$ and $y = -z/AL$ in Figure 18.10. There are infinitely many points of intersection of these graphs with $z > 0$. At each point of intersection, the horizontal coordinate yields a value of z satisfying $\tan(z) = -z/AL$, and $k = z/L$ is an eigenvalue of the problem. If we label these values of z as $z_1, z_2, \ldots$ in increasing order (as in Figure 18.10), the eigenvalues are

$$\lambda_n = k_n^2 = \frac{z_n^2}{L^2}, \qquad n = 1, 2, 3, \ldots.$$

Corresponding eigenfunctions are $X_n(x) = d_n \sin(k_n x) = d_n \sin(z_n x/L)$. (Although we have shown from graphs that this problem has infinitely many positive eigenvalues, we could also conclude this from the Sturm-Liouville theorem because the problem for X is a regular Sturm-Liouville problem.)

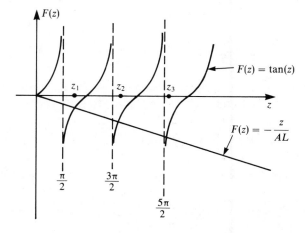

Figure 18.10

Using these values for λ, the equation for T is

$$T' + \frac{a^2 z_n^2}{L^2} T = 0,$$

with general solution

$$T_n(t) = \alpha_n e^{-a^2 z_n^2 t/L^2}.$$

For $n = 1, 2, 3, \ldots$, let

$$u_n(x, t) = A_n \sin\left(\frac{z_n x}{L}\right) e^{-a^2 z_n^2 t/L^2},$$

in which z_n is the nth positive solution of

$$\tan(z) = -\frac{z}{AL}.$$

Each of these functions satisfies the heat equation and the boundary conditions. To satisfy the initial condition, we attempt a superposition

$$u(x, t) = \sum_{n=1}^{\infty} A_n \sin\left(\frac{z_n x}{L}\right) e^{-a^2 z_n^2 t/L^2}.$$

We must choose the coefficients to satisfy

$$u(x, 0) = f(x) = \sum_{n=1}^{\infty} A_n \sin\left(\frac{z_n x}{L}\right), \qquad 0 < x < L. \tag{18.23}$$

This is like a Fourier sine expansion but is not, because $z_n \neq n\pi$. Therefore, we cannot choose the A_n's as the usual Fourier sine coefficients. However, we can solve for the A_n's in several ways.

Method 1   We can imitate the reasoning of Section 17.1 in motivating the definition of Fourier series. At the conclusion of this example, we will show that

$$\int_0^L \sin\left(\frac{z_n x}{L}\right) \sin\left(\frac{z_m x}{L}\right) dx = 0 \quad \text{if} \quad m \neq n. \tag{18.24}$$

Therefore, the functions $\sin(z_n x/L)$ are orthogonal on $[0, L]$. Now let k be a positive integer, and multiply equation (18.23) by $\sin(z_k x/L)$ to get

$$f(x)\sin\left(\frac{z_k x}{L}\right) = \sum_{n=1}^{\infty} A_n \sin\left(\frac{z_n x}{L}\right) \sin\left(\frac{z_k x}{L}\right).$$

Integrate from zero to L, interchanging the series and integral. We get

$$\int_0^L f(x)\sin\left(\frac{z_k x}{L}\right) dx = \sum_{n=1}^{\infty} A_n \int_0^L \sin\left(\frac{z_n x}{L}\right) \sin\left(\frac{z_k x}{L}\right) dx$$

$$= A_k \int_0^L \sin^2\left(\frac{z_k x}{L}\right) dx$$

because all terms of the series are zero except the term $n = k$. Then

$$A_k = \frac{\int_0^L f(x)\sin\left(\frac{z_k x}{L}\right) dx}{\int_0^L \sin^2\left(\frac{z_k x}{L}\right) dx}. \tag{18.25}$$

With this choice of the coefficients, the solution of the boundary value problem is

$$u(x, t) = \sum_{n=1}^{\infty} \left[\frac{\int_0^L f(\xi)\sin\left(\frac{z_n \xi}{L}\right) d\xi}{\int_0^L \sin^2\left(\frac{z_n \xi}{L}\right) d\xi} \right] \sin\left(\frac{z_n x}{L}\right) e^{-a^2 z_n^2 t / L^2}.$$

In actual practice, we would have to use a numerical method (such as Newton's method) to approximate the first few z_n's, then approximate the integrals to obtain the coefficients from formula (18.25). For example, if $A = L = 1$ and $f(x) = 1$, Newton's method yields

$$z_1 \approx 2.0288$$
$$z_2 \approx 4.9132$$
$$z_3 \approx 7.9787$$
$$z_4 \approx 11.0855.$$

Using these values, carry out the integrations in (18.25) to get

$$A_1 \approx 1.9207$$
$$A_2 \approx 2.6593$$
$$A_3 \approx 4.1457$$
$$A_4 \approx 5.6329.$$

In this case,

$$u(x, t) \approx 1.9207 \sin(2.0288x)e^{-4.1160a^2 t} + 2.6593 \sin(4.9132x)e^{-24.1395a^2 t}$$
$$+ 4.1457 \sin(7.9787x)e^{-63.6597a^2 t} + 5.6329 \sin(11.0855)e^{-122.8883a^2 t} + \cdots,$$

and we can include more terms by approximating $A_5, A_6, \ldots$.

Method 2   Formula (18.24) is an immediate consequence of the Sturm-Liouville theorem, and formula (18.25) is a special case of formula (7.37). Therefore, we have formulas (18.24) and (18.25) with none of the work of method 1 if these ideas have been studied previously.

If Sturm-Liouville theory has not been discussed, formula (18.24) can nevertheless be derived by integration, but we must invest some effort, which we will now outline. We want to show that

$$\int_0^L \sin\left(\frac{z_n x}{L}\right)\sin\left(\frac{z_m x}{L}\right) dx = 0$$

if $n \neq m$. We will use the fact that $\sin(z_n) = (-z_n/AL)\cos(z_n)$ (and similarly with m

replacing n). Write

$$\int_0^L \sin\left(\frac{z_n x}{L}\right)\sin\left(\frac{z_m x}{L}\right) dx = \frac{\sin\left[\dfrac{(z_m - z_n)x}{L}\right]}{\dfrac{2(z_m - z_n)}{L}} - \frac{\sin\left[\dfrac{(z_m + z_n)x}{L}\right]}{\dfrac{2(z_m + z_n)}{L}}\Bigg|_0^L$$

$$= \frac{\sin(z_m - z_n)}{\dfrac{2(z_m - z_n)}{L}} - \frac{\sin(z_m + z_n)}{\dfrac{2(z_m + z_n)}{L}}$$

$$= \frac{1}{\dfrac{2(z_m - z_n)}{L}}[\sin(z_m)\cos(z_n) - \cos(z_m)\sin(z_n)]$$

$$- \frac{1}{\dfrac{2(z_m + z_n)}{L}}[\sin(z_m)\cos(z_n) + \cos(z_m)\sin(z_n)]$$

$$= \frac{1}{\dfrac{2(z_m - z_n)}{L}}\left[-\frac{z_m}{AL}\cos(z_m)\cos(z_n) + \frac{z_n}{AL}\cos(z_m)\cos(z_n)\right]$$

$$- \frac{1}{\dfrac{2(z_m + z_n)}{L}}\left[-\frac{z_m}{AL}\cos(z_m)\cos(z_n) - \frac{z_n}{AL}\cos(z_m)\cos(z_n)\right]$$

$$= \frac{1}{\dfrac{2(z_m - z_n)}{L}}\cos(z_m)\cos(z_n)\left(\frac{z_n}{AL} - \frac{z_m}{AL}\right)$$

$$+ \frac{1}{\dfrac{2(z_m + z_n)}{L}}\cos(z_m)\cos(z_n)\left(\frac{z_m}{AL} + \frac{z_n}{AL}\right) = 0.$$

This proves the orthogonality of the functions $\sin(z_n x/L)$ on $[0, L]$.

PROBLEMS FOR SECTION 18.2

In each of Problems 1 through 5, solve the boundary value problem.

1. $\dfrac{\partial u}{\partial t} = a^2 \dfrac{\partial^2 u}{\partial x^2}$ $(0 < x < L, t > 0)$,

$u(0, t) = u(L, t) = 0$ $(t > 0)$,

$u(x, 0) = x(L - x)$ $(0 < x < L)$

2. $\dfrac{\partial u}{\partial t} = 4\dfrac{\partial^2 u}{\partial x^2}$ $(0 < x < L, t > 0)$,

$u(0, t) = u(L, t) = 0$ $(t > 0)$,

$u(x, 0) = x^2(L - x)$ $(0 < x < L)$

3. $\dfrac{\partial u}{\partial t} = 3\dfrac{\partial^2 u}{\partial x^2}$ $(0 < x < L, t > 0)$,

$u(0, t) = u(L, t) = 0$ $(t > 0)$,

$u(x, 0) = L\left[1 - \cos\left(\dfrac{2\pi x}{L}\right)\right]$ $(0 < x < L)$

4. $\dfrac{\partial u}{\partial t} = \dfrac{\partial^2 u}{\partial x^2}$ $(0 < x < \pi, t > 0)$,

$\dfrac{\partial u}{\partial x}(0, t) = \dfrac{\partial u}{\partial x}(\pi, t) = 0$ $(t > 0)$,

$u(x, 0) = \sin(x)$ $(0 < x < \pi)$

5. $\dfrac{\partial u}{\partial t} = 4\,\dfrac{\partial^2 u}{\partial x^2}$ $\qquad\qquad$ $(0 < x < 2\pi,\ t > 0)$,

$\dfrac{\partial u}{\partial x}(0, t) = \dfrac{\partial u}{\partial x}(2\pi, t) = 0$ $\qquad\qquad$ $(t > 0)$,

$u(x, 0) = x(2\pi - x)$ $\qquad\qquad$ $(0 < x < 2\pi)$

6. A thin bar of length L has insulated ends and initial temperature equal to a constant B. Find the temperature function $u(x, t)$.

7. A thin bar of length L has initial temperature equal to a constant B, and the right end (at $x = L$) is insulated, while the left end is kept at zero temperature. Find the temperature distribution $u(x, t)$.

8. A thin bar of thermal diffusivity 9, length 2, and insulated sides has its left end maintained at temperature zero. Its right end is perfectly insulated. The bar has an initial temperature distribution given by $f(x) = x^2$ for $0 < x < 2$. Determine the temperature distribution $u(x, t)$ for $t > 0$. What is $\lim_{t \to \infty} u(x, t)$?

9. A thin bar of thermal diffusivity 4 and length 6, with insulated sides, has its left end maintained at temperature zero. Its right end is radiating (with transfer coefficient $\frac{1}{2}$) into the surrounding medium, which has temperature zero. The bar has an initial temperature distribution $f(x) = x(6 - x)$ for $0 < x < 6$. Approximate the temperature distribution $u(x, t)$ for any time $t > 0$ by finding the fifth partial sum of the series representation for u. *Hint:* See the third example of this section and use the results of Example 7.9.

10. Imagine a long, thin bar of length L whose sides are poorly insulated. Heat radiates freely from the bar along its length into the surrounding medium. Assuming a positive transfer coefficient A and a medium temperature T_a, the resulting heat equation modeling this setting is

$$\frac{\partial u}{\partial t} = a^2\,\frac{\partial^2 u}{\partial x^2} - A(u - T_a).$$

Assuming an initial temperature distribution $f(x)$ and insulated ends, solve for $u(x, t)$. *Hint:* Transform the boundary value problem into one we have seen before by letting $v(x, t) = e^{At}u(x, t)$.

11. Show that the partial differential equation

$$\frac{\partial u}{\partial t} = k\left[\frac{\partial^2 u}{\partial x^2} + A\,\frac{\partial u}{\partial x} + Bu\right]$$

can be transformed into a partial differential equation of the form

$$\frac{\partial v}{\partial t} = k\,\frac{\partial^2 v}{\partial x^2}$$

by choosing α and β appropriately and using the change of variables

$$u(x, t) = e^{\alpha x + \beta t}v(x, t)$$

Hint: Substitute $u(x, t) = e^{\alpha x + \beta t}v(x, t)$ into the partial differential equation and determine how to choose α and β.

12. Use the method of Problem 11 to solve the boundary value problem

$$\frac{\partial u}{\partial t} = k\left[\frac{\partial^2 u}{\partial x^2} - \frac{a}{L}\,\frac{\partial u}{\partial x}\right] \qquad (0 < x < L,\ t > 0),$$

$$u(0, t) = u(L, t) = 0 \qquad\qquad (t > 0),$$

$$u(x, 0) = \frac{L}{ak}\left[1 - e^{-a(1 - x/L)}\right] \qquad (0 < x < L).$$

Here, $u(x, t)$ measures the concentration of positive charge carriers on the base of length L of a transistor; k and a are positive constants.

18.3 *Steady-State Temperatures* in a Flat Plate

A steady-state temperature distribution is achieved in the limit as $t \to \infty$ and is independent of time and of the initial conditions. In this section, we will consider the problem of determining the steady-state temperature distribution for a flat plate. The two-dimensional heat equation is

$$\frac{\partial u}{\partial t} = a^2 \left[\frac{\partial^2 u}{\partial x^2} + \frac{\partial^2 u}{\partial y^2} \right].$$

In the steady-state case, $\partial u/\partial t = 0$, and this becomes Laplace's equation

$$\frac{\partial^2 u}{\partial x^2} + \frac{\partial^2 u}{\partial y^2} = 0.$$

This is often written $\nabla^2 u = 0$, and $\nabla^2 u = \partial^2 u/\partial x^2 + \partial^2 u/\partial y^2$ is called the *Laplacian* of u.

A Dirichlet problem consists of Laplace's equation $\nabla^2 u = 0$ for (x, y) in a set M, together with values the solution is to have over the boundary of M. With certain conditions on M and the values of the solution on the boundary of M, a Dirichlet problem has a unique solution.

LAPLACE'S EQUATION FOR A RECTANGLE

As an example of a Dirichlet problem, consider a flat rectangular plate with u equal to zero on three sides and constant on the fourth side, as shown in Figure 18.11. The plate occupies the rectangle $0 \le x \le a$, $0 \le y \le b$. The boundary value problem for the steady-state temperature distribution over this plate is

$$\nabla^2 u = 0 \qquad\qquad (0 < x < a, 0 < y < b),$$
$$u(x, 0) = u(x, b) = 0 \qquad (0 < x < a),$$
$$u(0, y) = 0 \qquad\qquad (0 < y < b),$$
$$u(a, y) = T = \text{constant} \qquad (0 < y < b).$$

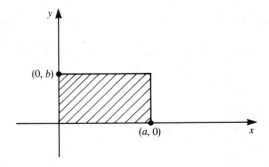

Figure 18.11

The boundary values do not agree at the corners of the rectangle. Nevertheless, we will still be able to obtain a solution which is continuous in the interior of the rectangle (that is, for $0 < x < a, 0 < y < b$) and which meets the stated conditions on the sides of the rectangle.

To separate the variables, substitute $u(x, y) = X(x)Y(y)$ into Laplace's equation. We get

$$X''Y + XY'' = 0,$$

or

$$\frac{Y''}{Y} = -\frac{X''}{X}.$$

The left side depends only on y and the right side only on x. Since x and y are independent, both sides must equal the same constant:

$$\frac{Y''}{Y} = -\frac{X''}{X} = -\lambda.$$

Then

$$Y'' + \lambda Y = 0 \quad \text{and} \quad X'' - \lambda X = 0.$$

Now use the boundary conditions:

$$u(x, 0) = X(x)Y(0) = 0 \quad \text{implies that} \quad Y(0) = 0;$$
$$u(x, b) = X(x)Y(b) = 0 \quad \text{implies that} \quad Y(b) = 0;$$
$$u(0, y) = X(0)Y(y) = 0 \quad \text{implies that} \quad X(0) = 0.$$

Therefore, X and Y must satisfy

$$Y'' + \lambda Y = 0, \qquad\qquad X'' - \lambda X = 0,$$
$$Y(0) = Y(b) = 0; \qquad\qquad X(0) = 0.$$

The problem for Y is the same as that solved for X in Section 18.1, so the eigenvalues are

$$\lambda_n = \frac{n^2 \pi^2}{b^2}$$

and corresponding eigenfunctions are

$$Y_n(y) = a_n \sin\left(\frac{n\pi y}{b}\right)$$

for $n = 1, 2, 3, \ldots.$

For any positive integer n, the problem for X is

$$X'' - \frac{n^2 \pi^2}{b^2} X = 0; \qquad X(0) = 0.$$

The general solution of this differential equation is

$$X(x) = ce^{n\pi x/b} + de^{-n\pi x/b}.$$

Since $X(0) = c + d = 0$, we must have $d = -c$, and

$$X = c(e^{n\pi x/b} - e^{-n\pi x/b}) = 2c \sinh\left(\frac{n\pi x}{b}\right).$$

Let

$$X_n(x) = 2c_n \sinh\left(\frac{n\pi x}{b}\right).$$

For each positive integer n, let

$$u_n(x, y) = X_n(x)Y_n(y) = A_n \sinh\left(\frac{n\pi x}{b}\right)\sin\left(\frac{n\pi y}{b}\right),$$

in which $A_n = 2a_n c_n$ is an arbitrary constant.

For each n, this function satisfies Laplace's equation and the boundary conditions on three sides of the rectangle. To satisfy the condition $u(a, y) = T$ on the fourth side, we must attempt a superposition

$$u(x, y) = \sum_{n=1}^{\infty} u_n(x, y) = \sum_{n=1}^{\infty} A_n \sinh\left(\frac{n\pi x}{b}\right)\sin\left(\frac{n\pi y}{b}\right).$$

We must choose the A_n's so that

$$u(a, y) = T = \sum_{n=1}^{\infty} A_n \sinh\left(\frac{n\pi a}{b}\right)\sin\left(\frac{n\pi y}{b}\right)$$

for $0 < y < b$. The series on the right is the Fourier sine expansion of T on $[0, b]$, with the *entire expression* $A_n\sinh(n\pi a/b)$ as the Fourier coefficient in this expansion. Therefore, choose

$$A_n \sinh\left(\frac{n\pi a}{b}\right) = \frac{2}{b}\int_0^b T \sin\left(\frac{n\pi y}{b}\right) dy$$

$$= \frac{2}{b}\frac{bT}{n\pi}[1 - (-1)^n] = \frac{2T}{n\pi}[1 - (-1)^n].$$

Then

$$A_n = \frac{2T}{n\pi}\frac{[1 - (-1)^n]}{\sinh\left(\dfrac{n\pi a}{b}\right)} = \begin{cases} 4T/[n\pi \sinh(n\pi a/b)] & \text{if } n \text{ is odd} \\ 0 & \text{if } n \text{ is even.} \end{cases}$$

The steady-state temperature distribution is

$$u(x, y) = \sum_{n=1}^{\infty} \frac{4T}{(2n-1)\pi \sinh\left[(2n-1)\dfrac{\pi a}{b}\right]} \sinh\left[(2n-1)\frac{\pi x}{b}\right]\sin\left[(2n-1)\frac{\pi y}{b}\right].$$

If nonzero boundary conditions are specified on each side of the rectangle, we split the problem into four simpler ones, in each of which the boundary condition is nonzero on one side and zero on the other three. The solution of the original problem is the sum of the solutions of these four problems.

LAPLACE'S EQUATION FOR A DISK

As a second example of a steady-state heat problem for a flat plate, suppose the plate is a thin disk of radius R. For a disk, it is convenient to use polar coordinates. Place the origin at the center of the disk. By equation (18.13), Laplace's equation in polar coordinates is

$$\frac{\partial^2 u}{\partial r^2} + \frac{1}{r} \frac{\partial u}{\partial r} + \frac{1}{r^2} \frac{\partial^2 u}{\partial \theta^2} = 0.$$

Assume that the temperature is known on the boundary of the disk:

$$u(R, \theta) = f(\theta) \qquad (-\pi \le \theta \le \pi).$$

We will impose two additional conditions in order to obtain a unique solution. First, we will assume that the solution is bounded, a physically realistic condition when dealing with a temperature distribution function. Second, we will assume periodicity conditions:

$$u(r, -\pi) = u(r, \pi) \quad \text{and} \quad \frac{\partial u}{\partial \theta}(r, -\pi) = \frac{\partial u}{\partial \theta}(r, \pi).$$

These conditions are motivated by the fact that $(r, -\pi)$ and (r, π) are different polar coordinates of the same point.

Attempt a solution

$$u(r, \theta) = F(r)G(\theta).$$

Substitute this expression into Laplace's equation to get

$$F''(r)G(\theta) + \frac{1}{r} F'(r)G(\theta) + \frac{1}{r^2} F(r)G''(\theta) = 0.$$

If $F(r)G(\theta) \ne 0$, we can write this equation as

$$-\frac{r^2 F''(r) + r F'(r)}{F(r)} = \frac{G''(\theta)}{G(\theta)}.$$

Since the left side depends only on r and the right side only on θ, and r and θ are independent, both sides must equal the same constant:

$$-\frac{r^2 F''(r) + r F'(r)}{F(r)} = \frac{G''(\theta)}{G(\theta)} = -\lambda.$$

Then

$$r^2 F''(r) + r F'(r) - \lambda F(r) = 0 \quad \text{and} \quad G''(\theta) + \lambda G(\theta) = 0.$$

Now look at the periodicity conditions:

$$u(r, \pi) = F(r)G(\pi) = u(r, -\pi)^{\boldsymbol{\cdot}} = F(r)G(-\pi)$$

and

$$\frac{\partial u}{\partial \theta}(r, \pi) = F(r)G'(\pi) = \frac{\partial u}{\partial \theta}(r, -\pi) = F(r)G'(-\pi).$$

Assuming that $F(r)$ is not identically zero, these equations imply that

$$G(\pi) = G(-\pi) \quad \text{and} \quad G'(\pi) = G'(-\pi).$$

The problem to solve for G is therefore

$$G'' + \lambda G = 0; \qquad G(\pi) = G(-\pi), \qquad G'(\pi) = G'(-\pi).$$

This is a periodic Sturm-Liouville problem and was solved in Example 7.7. There we found that the eigenvalues are

$$\lambda = n^2 \quad \text{for} \quad n = 0, 1, 2, 3, \ldots,$$

with associated eigenfunctions

$$G_n(\theta) = c_n \cos(n\theta) + d_n \sin(n\theta).$$

Now let $\lambda = n^2$ in the differential equation for F to get

$$r^2 F''(r) + r F'(r) - n^2 F(r) = 0.$$

This is a Cauchy-Euler equation and has general solution

$$F(r) = \begin{cases} A r^n + B r^{-n} & \text{if} \quad n = 1, 2, 3, \ldots \\ A & \text{if} \quad n = 0. \end{cases}$$

We want a bounded solution. Since r^{-n} is unbounded as $r \to 0$ (the center of the disk), we must choose $B = 0$. We can combine cases and write

$$F(r) = A r^n \quad \text{for} \quad n = 0, 1, 2, \ldots,$$

with A nonzero but otherwise arbitrary. For each $n = 0, 1, 2, 3, \ldots$, we now have functions of the form

$$u_n(x, y) = r^n [a_n \cos(n\theta) + b_n \sin(n\theta)].$$

These functions satisfy Laplace's equation and the stated conditions, except perhaps the boundary condition $u(R, \theta) = f(\theta)$. To satisfy this condition, we use a super-position

$$u(r, \theta) = \sum_{n=1}^{\infty} u_n(r, \theta) = a_0 + \sum_{n=1}^{\infty} r^n [a_n \cos(n\theta) + b_n \sin(n\theta)].$$

We must choose the coefficients to satisfy

$$u(R, \theta) = f(\theta) = a_0 + \sum_{n=1}^{\infty} [R^n a_n \cos(n\theta) + R^n b_n \sin(n\theta)].$$

This is a Fourier series expansion of f on $[-\pi, \pi]$. Therefore, choose

$$a_0 = \frac{1}{2\pi} \int_{-\pi}^{\pi} f(\theta) \, d\theta,$$

$$R^n a_n = \frac{1}{\pi} \int_{-\pi}^{\pi} f(\theta) \cos(n\theta) \, d\theta,$$

and

$$R^n b_n = \frac{1}{\pi} \int_{-\pi}^{\pi} f(\theta) \sin(n\theta) \, d\theta$$

for $n = 1, 2, 3, \ldots$. These equations can be solved for the a_n's and b_n's, yielding the coefficients to put into the series for $u(r, \theta)$ to obtain the steady-state temperature distribution.

As a specific example, suppose that the disk has radius 3 and that $f(\theta) = 10 + \pi^2\theta - \theta^3$ for $-\pi \le \theta \le \pi$. Then

$$a_0 = \frac{1}{2\pi} \int_{-\pi}^{\pi} (10 + \pi^2\theta - \theta^3) \, d\theta = 10,$$

$$a_n = \frac{1}{3^n\pi} \int_{-\pi}^{\pi} (10 + \pi^2\theta - \theta^3)\cos(n\theta) \, d\theta = 0,$$

and

$$b_n = \frac{1}{3^n\pi} \int_{-\pi}^{\pi} (10 + \pi^2\theta - \theta^3)\sin(n\theta) \, d\theta = \frac{12(-1)^{n+1}}{n^3 3^n}.$$

The steady-state temperature for the disk with the given boundary data is

$$u(r, \theta) = 10 + 12 \sum_{n=1}^{\infty} \left(\frac{r}{3}\right)^n \frac{(-1)^{n+1}}{n^3} \sin(n\theta)$$

for $0 \le r < 3, -\pi \le \theta \le \pi$.

PROBLEMS FOR SECTION 18.3

1. Solve for the steady-state temperature distribution in a flat plate placed over $0 \le x \le a, 0 \le y \le b$ if the temperature on the vertical and bottom sides is kept at zero and the temperature on the top side is kept at a constant K.

2. Solve for the steady-state temperature in a flat plate placed over $0 \le x \le a, 0 \le y \le b$ if the temperature on the left side is a constant T_1 and that on the right side is a constant T_2, while the top and bottom sides are kept at zero temperature.

3. Solve for the steady-state temperature in a semi-infinite strip $x \ge 0, 0 \le y \le 1$ if the temperature on the top and bottom sides is kept at zero and that on the left side (the portion of the y-axis between zero and 1) is kept at constant temperature T. *Hint:* There are only three boundary conditions, as the strip is open to the right with no boundary there. Impose the condition that the solution must be bounded for $x \ge 0$.

4. Find the steady-state temperature in the semi-infinite region $0 \le x \le a, y \ge 0$ if the temperature on the bottom and left sides is kept at zero and the temperature on the right side is kept at constant T. *Hint:* Look for a bounded solution.

5. Find the steady-state temperature in the semi-infinite region $0 \le x \le 4, y \ge 0$ if the temperature on the vertical sides is kept at constant T and the temperature on the bottom side is kept at zero. Look for a bounded solution.

6. Guess the steady-state temperature in a thin rod of length L if the ends are perfectly insulated and the initial temperature distribution is $f(x), 0 < x < L$. Solve for the temperature distribution $u(x, t)$, and compare the steady-state solution with $\lim_{t \to \infty} u(x, t)$.

7. Find the steady-state temperature for a thin disk of radius R if the temperature on the boundary is $f(\theta) = \cos^2(\theta)$ for $-\pi \le \theta \le \pi$.

8. Find the steady-state temperature for a thin disk of radius 1 if the temperature on the boundary is $f(\theta) = \sin^3(\theta)$ for $-\pi \le \theta \le \pi$.

9. Find the steady-state temperature for a thin disk of radius R if the temperature on the boundary is a constant T.

10. Find the steady-state temperature in a wedge-shaped flat plate occupying the region $0 \le r \le k, 0 \le \theta \le \alpha$ (polar

coordinates). The edges $\theta = 0$ and $\theta = \alpha$ are kept at zero temperature, and the arc $r = k$ $(0 \le \theta \le \alpha)$ is kept at constant temperature T.

11. Derive Poisson's solution for the steady-state temperature distribution in a flat, circular disk of radius 1 by carrying out the following steps. The boundary value problem is

$$\frac{1}{r} \frac{\partial}{\partial r}\left(r \frac{\partial u}{\partial r}\right) + \frac{1}{r^2} \frac{\partial^2 u}{\partial \theta^2} = 0 \qquad (0 \le r < 1, 0 \le \theta \le 2\pi),$$

$$u(1, \theta) = f(\theta) \qquad (0 \le \theta \le 2\pi).$$

(a) Assume a solution of the form

$$u(r, \theta) = a_0 + \sum_{n=1}^{\infty} r^n[a_n\cos(n\theta) + b_n\sin(n\theta)].$$

Substitute the integral expressions for the Fourier coefficients into this series, interchange the integral and the summation, and use a trigonometric identity to obtain

$$u(r, \theta) = \frac{1}{2\pi} \int_0^{2\pi} f(\xi)\left\{1 + 2 \sum_{n=1}^{\infty} r^n\cos[n(\theta - \xi)]\right\} d\xi.$$

(b) Use Euler's formula $e^{i\xi} = \cos(\xi) + i \sin(\xi)$, and let $z = re^{i\xi}$, to conclude that

$$z^n = r^n[\cos(n\xi) + i \sin(n\xi)].$$

Hence, conclude from (a) that

$$1 + 2 \sum_{n=1}^{\infty} r^n\cos(n\xi) = \operatorname{Re}\left(1 + 2 \sum_{n=1}^{\infty} z^n\right),$$

in which $\operatorname{Re}(A + iB) = A$.

(c) Use the geometric series $1/(1 - z) = \sum_{n=0}^{\infty} z^n$ for $|z| < 1$ to show that

$$1 + 2 \sum_{n=1}^{\infty} r^n\cos(n\xi) = \operatorname{Re}\left(\frac{1 + re^{i\xi}}{1 - re^{i\xi}}\right).$$

(d) Use the result of (c) to show that

$$1 + 2 \sum_{n=1}^{\infty} r^n\cos(n\xi) = \frac{1 - r^2}{1 + r^2 - 2r \cos(\xi)}.$$

(e) Derive the solution of the boundary value problem in the form

$$u(r, \theta) = \frac{1}{2\pi} \int_0^{2\pi} \frac{(1 - r^2)f(\xi)}{1 + r^2 - 2r \cos(\theta - \xi)} d\xi.$$

This is *Poisson's integral formula* for the solution of Laplace's equation for a disk.

12. Using the result of Problem 11, write an integral for the solution of the steady-state heat equation for a disk of radius 1 if $u(1, \theta) = \sin^2(\theta)$. Use an integral table to evaluate $u(r, \theta)$, or use a numerical integration program on a microcomputer to approximate $u(r, \theta)$ to four decimal places, for the following points (r, θ):
    (a) $(0, 0)$ (b) $(\frac{1}{2}, \frac{\pi}{2})$ (c) $(\frac{1}{4}, \pi)$ (d) $(\frac{3}{4}, 0)$

13. In this section, we derived a general solution for the steady-state temperature distribution in a thin, flat disk. Use this solution to show that the temperature at the center of the disk is the average value of the temperature on the circumference of the disk.

18.4 *Some Problems in Which Separation of Variables Fails*

The separation of variables method may fail if either the differential equation or the boundary conditions are not of the right form. In such cases, we can sometimes still apply the method after some preliminary transformations. Here are two examples.

EXAMPLE 18.1

Consider the problem of describing vibrations of a stretched string released from rest from its natural stretched position but with an external force acting on it. Suppose, to be specific, that the external force acts parallel to the y-axis in the plane of the motion and is proportional to the distance from the left end of the string. Then, for some constant A, the force per unit length acting on the particle of string at coordinate x is Ax, and we must add the forcing term Ax to the usual wave equation, obtaining

$$\frac{\partial^2 y}{\partial t^2} = \frac{\partial^2 y}{\partial x^2} + Ax \qquad (0 < x < L, t > 0).$$

Here, we have let $a = 1$ for convenience. The boundary and initial conditions are

$$y(0, t) = y(L, t) = 0 \qquad (t > 0),$$
$$y(x, 0) = 0 \qquad (0 < x < L),$$
$$\frac{\partial y}{\partial t}(x, 0) = 0 \qquad (0 < x < L).$$

The last two conditions specify zero initial displacement and velocity.

If we attempt a separation of variables with $y(x, t) = X(x)T(t)$, we obtain

$$XT'' = X''T + Ax,$$

and we cannot separate functions of x and t on opposite sides of this equation. In such a case, it is sometimes helpful to introduce new functions Y and ψ, related to y by

$$y(x, t) = Y(x, t) + \psi(x).$$

The idea is to substitute this expression into the wave equation and boundary and initial conditions and attempt to choose ψ so that the resulting boundary value problem for Y can be done by separation of variables. Substitute $y = Y + \psi$ into the partial differential equation to get

$$\frac{\partial^2 Y}{\partial t^2} = \frac{\partial^2 Y}{\partial x^2} + \psi''(x) + Ax,$$

since $\partial \psi / \partial t = 0$. Immediately, this differential equation is simplified if

$$\psi''(x) + Ax = 0.$$

This is a second order ordinary differential equation for ψ with many solutions. Before choosing one, consider the boundary conditions. We need

$$y(0, t) = Y(0, t) + \psi(0) = 0$$

and

$$y(L, t) = Y(L, t) + \psi(L) = 0.$$

Both of these conditions for Y will be simplified if we choose ψ so that

$$\psi(0) = \psi(L) = 0.$$

Therefore, choose ψ to be a solution of the boundary value problem

$$\psi'' + Ax = 0; \qquad \psi(0) = \psi(L) = 0.$$

This is a routine problem to solve. Write $\psi'' = -Ax$ and integrate twice to get

$$\psi(x) = -\tfrac{1}{6}Ax^3 + Bx + C.$$

Then

$$\psi(0) = C = 0 \quad \text{and} \quad \psi(L) = -\tfrac{1}{6}AL^3 + BL = 0.$$

From these equations, we obtain

$$B = \tfrac{1}{6}AL^2.$$

Thus, choose

$$\psi(x) = -\frac{1}{6}Ax^3 + \frac{1}{6}AL^2x = \frac{A}{6}x(L^2 - x^2).$$

With this choice of ψ, Y satisfies the equations

$$\frac{\partial^2 Y}{\partial t^2} = \frac{\partial^2 Y}{\partial x^2} \qquad\qquad (0 < x < L, t > 0),$$

$$Y(0, t) = Y(L, t) = 0 \qquad\qquad (t > 0).$$

Now recall the initial conditions. We have

$$y(x, 0) = 0 = Y(x, 0) + \psi(x).$$

Therefore,

$$Y(x, 0) = -\psi(x) = \frac{A}{6}x(x^2 - L^2).$$

Finally,

$$\frac{\partial y}{\partial t}(x, 0) = 0 = \frac{\partial Y}{\partial t}(x, 0).$$

The boundary value problem for Y is

$$\frac{\partial^2 Y}{\partial t^2} = \frac{\partial^2 Y}{\partial x^2} \qquad\qquad (0 < x < L, t > 0),$$

$$Y(0, t) = Y(L, t) = 0 \qquad\qquad (t > 0),$$

$$Y(x, 0) = \frac{A}{6}x(x^2 - L^2) \qquad\qquad (t > 0),$$

$$\frac{\partial Y}{\partial t}(x, 0) = 0 \qquad\qquad (t > 0).$$

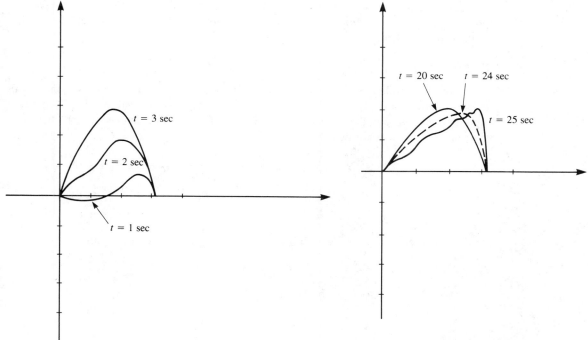

Figure 18.12. String profiles at $t = 1, 2, 3$ seconds.

Figure 18.13. String profiles at $t = 20, 24, 25$ seconds.

We have already solved this problem for Y in Section 18.1. The solution is

$$Y(x, t) = \frac{2}{L} \sum_{n=1}^{\infty} \left[\int_0^L \psi(\xi) \sin\left(\frac{n\pi\xi}{L}\right) d\xi \right] \sin\left(\frac{n\pi x}{L}\right) \cos\left(\frac{n\pi t}{L}\right).$$

Compute the integral:

$$\int_0^L \psi(\xi) \sin\left(\frac{n\pi\xi}{L}\right) d\xi = \int_0^L \frac{1}{6} A\xi(\xi^2 - L^2) \sin\left(\frac{n\pi\xi}{L}\right) d\xi$$

$$= \frac{AL^4}{n^3\pi^3} \cos(n\pi) = \frac{AL^4(-1)^n}{n^3\pi^3}.$$

Therefore,

$$Y(x, t) = \frac{2AL^3}{\pi^3} \sum_{n=1}^{\infty} \frac{(-1)^n}{n^3} \sin\left(\frac{n\pi x}{L}\right) \cos\left(\frac{n\pi t}{L}\right).$$

The original problem has solution

$$y(x, t) = \frac{2AL^3}{\pi^3} \sum_{n=1}^{\infty} \frac{(-1)^n}{n^3} \sin\left(\frac{n\pi x}{L}\right) \cos\left(\frac{n\pi t}{L}\right) + \frac{1}{6} Ax(L^2 - x^2).$$

Figures 18.12 and 18.13 show the string profile at different times. ■

In this example, separation of variables failed because of the form of the partial differential equation. Here is an example in which the method fails because of the boundary conditions.

EXAMPLE 18.2

Consider the problem of describing heat conduction in a thin bar extending from $x = 0$ to $x = L$, but suppose that the left end is maintained at constant temperature T_1 and the right end is maintained at constant temperature T_2. The boundary value problem for the temperature distribution u is

$$\frac{\partial u}{\partial t} = a^2 \frac{\partial^2 u}{\partial x^2} \qquad (0 < x < L, t > 0),$$

$$u(0, t) = T_1, \qquad u(L, t) = T_2 \qquad (t > 0),$$

$$u(x, 0) = f(x) \qquad (0 < x < L).$$

If we let $u(x, t) = X(x)T(t)$, we obtain from the differential equation that

$$X'' + \lambda X = 0 \quad \text{and} \quad T' + \lambda a^2 T = 0,$$

as before. However, look at the boundary conditions. We must have

$$u(0, t) = T_1 = X(0)T(t).$$

Whereas previously we had $T_1 = 0$ and concluded that $X(0) = 0$, here we cannot draw any meaningful conclusion. We cannot conclude that $X(0) = T_1/T(t)$ because then $X(0)$ would be a function of t; we cannot conclude that $T(t) = T_1/X(0)$ because then $T(t)$ must be constant. In the same way, the boundary condition $u(0, L) = T_2$ presents us with a condition we cannot meet.

As before, we will try to eliminate the difficulty by perturbing u. Let

$$u(x, t) = U(x, t) + \psi(x).$$

We will try to choose ψ to obtain a problem for U which we can solve. The partial differential equation transforms to

$$\frac{\partial U}{\partial t} = a^2 \frac{\partial^2 U}{\partial x^2} + a^2 \psi''(x).$$

This will simplify to the standard heat equation if we choose ψ so that $\psi''(x) = 0$. Now look at the boundary conditions. We have

$$u(0, t) = T_1 = U(0, t) + \psi(0).$$

This will simplify to $U(0, t) = 0$ if $\psi(0) = T_1$. Similarly, $u(L, t) = T_2 = U(L, t) + \psi(L)$ will simplify to $U(L, t) = 0$ if $\psi(L) = T_2$. Thus, choose ψ as a solution of the problem

$$\psi''(x) = 0; \qquad \psi(0) = T_1, \qquad \psi(L) = T_2.$$

It is routine to solve this problem by integration to obtain

$$\psi(x) = \frac{1}{L}(T_2 - T_1)x + T_1.$$

With this choice of ψ, the boundary value problem for U is

$$\frac{\partial U}{\partial t} = a^2 \frac{\partial^2 U}{\partial x^2} \qquad\qquad (0 < x < L, t > 0),$$

$$U(0, t) = U(L, t) = 0 \qquad\qquad (t > 0),$$

$$U(x, 0) = u(x, 0) - \psi(x) = f(x) - \psi(x) \qquad (0 < x < L).$$

We have again arrived at a problem we have solved previously. From Section 18.2, the solution is

$$U(x, t) = \frac{2}{L} \sum_{n=1}^{\infty} \left[\int_0^L [f(\xi) - \psi(\xi)] \sin\left(\frac{n\pi\xi}{L}\right) d\xi \right] \sin\left(\frac{n\pi x}{L}\right) e^{-n^2\pi^2 a^2 t/L^2}.$$

Once we know U, we obtain u as $u = U + \psi$. ∎

Although the technique we have just illustrated has its limitations, there are physical as well as mathematical reasons why it works in some problems. Consider again the heat conduction problem of Example 18.2. As t increases, the bar should evolve toward a steady-state temperature function $\psi(x)$ which depends only on x. This suggests that the temperature distribution $u(x, t)$ should be a sum of a transient part $U(x, t)$, which decays to zero as time increases, and a steady-state part $\psi(x)$. Thus, we should expect that $u(x, t) = U(x, t) + \psi(x)$.

This decomposition of u into a steady-state part and a transient part is analogous to the behavior of current in a circuit. As we have seen, under certain conditions, the solution for the current can be split into a steady-state current and a transient current.

PROBLEMS FOR SECTION 18.4

In each of Problems 1 through 8, solve the boundary value problem.

1. $\dfrac{\partial^2 y}{\partial t^2} = 9 \dfrac{\partial^2 y}{\partial x^2} + 4x \qquad (0 < x < 1, t > 0),$

$y(0, t) = y(1, t) = 0 \qquad\qquad (t > 0),$

$y(x, 0) = 0, \quad \dfrac{\partial y}{\partial t}(x, 0) = 1 \qquad (0 < x < 1)$

2. $\dfrac{\partial^2 y}{\partial t^2} = 16 \dfrac{\partial^2 y}{\partial x^2} \qquad\qquad (0 < x < 4, t > 0),$

$y(0, t) = 0, \quad y(4, t) = 1 \qquad\qquad (t > 0),$

$y(x, 0) = x(4 - x), \quad \dfrac{\partial y}{\partial t}(x, 0) = 0 \qquad (0 < x < 1)$

3. $\dfrac{\partial^2 y}{\partial t^2} = 4 \dfrac{\partial^2 y}{\partial x^2} \qquad\qquad (0 < x < 9, t > 0),$

$y(0, t) = 0, \quad y(9, t) = 1 \qquad\qquad (t > 0),$

$y(x, 0) = 0, \quad \dfrac{\partial y}{\partial t}(x, 0) = x \qquad (0 < x < 9)$

4. $\dfrac{\partial u}{\partial t} = a^2 \dfrac{\partial^2 u}{\partial x^2} \qquad\qquad (0 < x < L, t > 0),$

$u(0, t) = T, \quad u(L, t) = 0 \qquad\qquad (t > 0),$

$u(x, 0) = x(L - x) \qquad\qquad (0 < x < L)$

5. $\dfrac{\partial u}{\partial t} = 9 \dfrac{\partial^2 u}{\partial x^2} \qquad\qquad (0 < x < 5, t > 0),$

$u(0, t) = 0, \quad u(5, t) = 3 \qquad\qquad (t > 0),$

$u(x, 0) = 0 \qquad\qquad (0 < x < 5)$

6. $\dfrac{\partial u}{\partial t} = a^2 \dfrac{\partial^2 u}{\partial x^2} \qquad\qquad (0 < x < 9, t > 0),$

$u(0, t) = T_1, \quad u(9, t) = T_2 \qquad\qquad (t > 0),$

$u(x, 0) = x^2 \qquad\qquad (0 < x < 9)$

7. $\dfrac{\partial u}{\partial t} = 4\dfrac{\partial^2 u}{\partial x^2} - Au \qquad (0 < x < 9, t > 0),$

$u(0, t) = u(9, t) = 0 \qquad\qquad (t > 0),$
$u(x, 0) = 0 \qquad\qquad\qquad (0 < x < 9)$

In this problem, A is a positive constant. *Hint:* Let $w(x, t) = e^{\alpha t}u(x, t)$ and choose α to obtain a problem for w that can be solved by separation of variables.

8. $\dfrac{\partial u}{\partial t} = 9\dfrac{\partial^2 u}{\partial x^2} \qquad\qquad (0 < x < L, t > 0),$

$u(0, t) = T, \quad u(L, t) = 0 \qquad (t > 0),$
$u(x, 0) = 0 \qquad\qquad\qquad (0 < x < L)$

9. A thin bar of thermal diffusivity 9 and length 2 and insulated sides has its left end maintained at a temperature of 100 and its right end maintained at 200. The bar has an initial temperature distribution given by $f(x) = x$, $0 < x < 2$. Determine the temperature distribution $u(x, t)$ for $0 < x < 2$ and $t > 0$. What is the steady-state temperature distribution in the bar?

10. Determine the temperature distribution in a long, thin rod of length L which has insulated ends but radiates heat along its length into the surrounding atmosphere with transfer coefficient A. The surrounding medium has temperature T_a, and the initial temperature is given by a function f. (See Problem 10 of Section 18.2.)

18.5 *The Heat Equation in an Infinite Cylinder*

Suppose we want the temperature distribution in a solid, infinitely long, homogeneous circular cylinder of radius R. It is convenient to use cylindrical coordinates, in terms of which the three-dimensional heat equation [see equation (18.14)] is

$$\frac{\partial u}{\partial t} = a^2\left[\frac{\partial^2 u}{\partial r^2} + \frac{1}{r}\frac{\partial u}{\partial r} + \frac{1}{r^2}\frac{\partial^2 u}{\partial \theta^2} + \frac{\partial^2 u}{\partial z^2}\right].$$

To simplify the problem, we will (for now) assume that the temperature at any point in the cylinder depends on only the time t and the distance r from the axis of the cylinder. The z-axis is the axis of the cylinder. Then

$$\frac{\partial u}{\partial \theta} = \frac{\partial u}{\partial z} = 0,$$

and the heat equation is

$$\frac{\partial u}{\partial t} = a^2\left[\frac{\partial^2 u}{\partial r^2} + \frac{1}{r}\frac{\partial u}{\partial r}\right] \qquad (18.26)$$

for $0 \le r < R$ and $t > 0$. We will assume in this example that

$$u(R, t) = 0 \qquad (t > 0).$$

This means that the outer surface of the cylinder is kept at temperature zero. Further, assume that

$$u(r, 0) = f(r) \qquad (0 \le r < R).$$

This is the initial temperature distribution in the interior of the cylinder.

To separate variables in the heat equation, let

$$u(r, t) = F(r)T(t)$$

in equation (18.26). After some algebra, we obtain

$$\frac{T'}{a^2 T} = \frac{F'' + \dfrac{1}{r} F'}{F} = -\lambda$$

for some constant λ. Then

$$F'' + \frac{1}{r} F' + \lambda F = 0 \quad \text{and} \quad T' + \lambda a^2 T = 0.$$

Now consider cases on λ.

Case 1: $\lambda = 0$ The differential equation for F is $F'' + (1/r)F' = 0$, with general solution

$$F(r) = C \ln(r) + K.$$

But $\ln(r) \to -\infty$ as $r \to 0$ (at the center of the cylinder), so we must choose $C = 0$ to have a bounded solution. Then $F(r) = K$.

If $\lambda = 0$, the differential equation for T is

$$T' = 0,$$

so $T = $ constant also. Hence, when $\lambda = 0$, $u = $ constant. The function $u = $ constant will satisfy $u(R, t) = 0$ for $t > 0$ only if the constant is zero. Thus, we are led to the trivial solution $u(r, t) = 0$ in this case. This is, in fact, the solution if $f(r) = 0$ for $0 \le r < R$. If, however, $f(r) \ne 0$ for some r, $\lambda = 0$ does not yield a solution, so zero is not an eigenvalue of this problem.

Case 2: $\lambda < 0$ Write $\lambda = -k^2$, with k positive. Then $T' - a^2 k^2 T = 0$ has general solution $T = c e^{a^2 k^2 t}$, which is unbounded if $c \ne 0$. Thus, we have no (bounded) solution with $\lambda < 0$.

Case 3: $\lambda > 0$ Write $\lambda = k^2$, with k positive. Then $T' + a^2 k^2 T = 0$, so $T = c e^{-a^2 k^2 t}$. The equation for F is

$$r^2 F'' + r F' + k^2 r^2 F = 0.$$

In Problem 7 at the end of this section, the student is asked to show that this equation has general solution

$$F(r) = A J_0(kr) + B Y_0(kr),$$

in which J_0 and Y_0 are Bessel functions of order zero of the first and second kind, respectively.

As $r \to 0$, $Y_0(kr) \to -\infty$, again leading to an unbounded solution. Thus, we must choose $B = 0$, leaving us with

$$F(r) = A J_0(kr).$$

Therefore, for every $k > 0$, we have a function

$$u_k(r, t) = A_k J_0(kr)e^{-k^2 a^2 t}$$

which satisfies the heat equation in cylindrical coordinates.

Now we must satisfy the other conditions. First, for $t > 0$, we need

$$u_k(R, t) = A_k J_0(kR)e^{-k^2 a^2 t} = 0.$$

To satisfy this with $A_k \neq 0$, we need $J_0(kR) = 0$. We know (see Problem 27 of Section 7.1 and Table 7.3) that the graph of $J_0(x)$ crosses the x-axis at infinitely many positive values. If we call these values $z_1, z_2, \ldots$, kR must take on the value z_n for $n = 1, 2, 3, \ldots$. The values k can have are

$$\frac{z_1}{R}, \frac{z_2}{R}, \ldots, \frac{z_n}{R}, \ldots.$$

Corresponding to each positive integer n, we have

$$u_n(r, t) = A_n J_0\left(\frac{z_n r}{R}\right)e^{-z_n^2 a^2 t/R^2}.$$

Each u_n satisfies the heat equation and the condition that $u_n(R, t) = 0$. We must now satisfy the condition that $u(r, 0) = f(r)$. In general, we cannot choose a single A_n so that $u_n(r, t)$ satisfies this condition. As we have done before, we attempt a superposition

$$u(r, t) = \sum_{n=1}^{\infty} u_n(r, t) = \sum_{n=1}^{\infty} A_n J_0\left(\frac{z_n r}{R}\right)e^{-z_n^2 a^2 t/R^2}. \tag{18.27}$$

We must choose the A_n's to satisfy $u(x, 0) = f(r)$. We require that

$$u(r, 0) = f(r) = \sum_{n=1}^{\infty} A_n J_0\left(\frac{z_n r}{R}\right). \tag{18.28}$$

This is like a Fourier series except that we have functions defined in terms of the zero order Bessel function instead of sines and cosines. As with Fourier series of sines and cosines, the key to choosing the A_n's lies in an *orthogonality relationship* for the functions $J_0(z_n r/R)$. This relationship is

$$\int_0^R r J_0\left(\frac{z_n r}{R}\right) J_0\left(\frac{z_m r}{R}\right) dr = 0 \quad \text{if} \quad n \neq m. \tag{18.29}$$

This equation follows from Lommel's integrals [Problem 24(c) of Section 7.1] upon letting $x = r$, $\alpha = z_n/R$, and $\beta = z_m/R$ and inserting the limits zero and R on the integral. It is also a consequence of Sturm-Liouville theory (see Section 7.8).

We will exploit this relationship to find the A_n's in equation (18.28) much as we used the orthogonality of sines and cosines in Section 17.1 to find Fourier coefficients. Multiply both sides of equation (18.28) by $r J_0(z_k r/R)$, with k any positive integer. We get

$$r f(r) J_0\left(\frac{z_k r}{R}\right) = \sum_{n=1}^{\infty} A_n r J_0\left(\frac{z_n r}{R}\right) J_0\left(\frac{z_k r}{R}\right).$$

Integrate both sides from zero to R, interchanging the summation and the integral:

$$\int_0^R r f(r) J_0\left(\frac{z_k r}{R}\right) dr = \sum_{n=1}^{\infty} A_n \int_0^R r J_0\left(\frac{z_n r}{R}\right) J_0\left(\frac{z_k r}{R}\right) dr.$$

By the orthogonality relationship (18.29), all of the integrals on the right are zero except the one in which $n = k$. The last equation therefore reduces to

$$\int_0^R rf(r)J_0\left(\frac{z_k r}{R}\right) dr = A_k \int_0^R r\left[J_0\left(\frac{z_k r}{R}\right)\right]^2 dr.$$

Solve this equation for A_k to obtain

$$A_k = \frac{\displaystyle\int_0^R rf(r)J_0\left(\frac{z_k r}{R}\right) dr}{\displaystyle\int_0^R r\left[J_0\left(\frac{z_k r}{R}\right)\right]^2 dr}$$

for $k = 1, 2, 3, \ldots$. These numbers are the *Fourier-Bessel coefficients* of f. With this choice of the A_n's, the series (18.28) is called a *Fourier-Bessel expansion* of f. We obtain the solution of the heat conduction problem by using the Fourier-Bessel coefficients of f in the proposed solution (18.27):

$$u(r, t) = \sum_{n=1}^{\infty} \left[\frac{\displaystyle\int_0^R \rho f(\rho)J_0\left(\frac{z_n \rho}{R}\right) d\rho}{\displaystyle\int_0^R \rho\left[J_0\left(\frac{z_n \rho}{R}\right)\right]^2 d\rho}\right] J_0\left(\frac{z_n r}{R}\right) e^{-z_n^2 a^2 t/R^2}.$$

With a change in notation, it follows from Theorem 7.13 that

$$\int_0^R r\left[J_0\left(\frac{z_n r}{R}\right)\right]^2 dr = \frac{1}{2} R^2 [J_1(z_n)]^2.$$

Using this equation, the solution of the boundary value problem can be written more neatly as

$$u(r, t) = \frac{2}{R^2} \sum_{n=1}^{\infty} \frac{1}{[J_1(z_n)]^2} \left[\int_0^R \rho f(\rho)J_0\left(\frac{z_n \rho}{R}\right) d\rho\right] J_0\left(\frac{z_n r}{R}\right) e^{-z_n^2 a^2 t/R^2}.$$

For those familiar with the Sturm-Liouville theorem, we will fit this discussion into that context. The functions $J_0(z_n r/R)$ are eigenfunctions of the Sturm-Liouville problem for F and are orthogonal on $[0, R]$ with weight function r (this accounts for the factor r in the orthogonality relationship). The series (18.28) is an expansion of f in a series of these eigenfunctions, and the coefficients are given by equation (7.37), which yields the quotient of integrals obtained above for A_k.

In the next section, we will apply Fourier-Legendre series to the solution of a boundary value problem modeling heat conduction in a sphere.

PROBLEMS FOR SECTION 18.5

1. A homogeneous circular cylinder of radius 2 and semi-infinite length has its base, which is sitting on the polar plane $z = 0$, maintained at a constant positive temperature K. The lateral surface is kept at zero temperature. Determine the steady-state temperature of the cylinder if it has a thermal diffusivity of a^2, under the assumption that the temperature at any point depends on only the height z above the base and the distance r from the axis of the cylinder. *Hint:* Set up the boundary value problem and separate the variables; then use the zero boundary condition.

2. Redo Problem 1 with the assumption that the lateral surface is maintained at a constant positive temperature $L < K$.

3. A thin homogeneous cylindrical plate with thermal diffusivity a^2 and radius R is initially heated throughout its interior to a positive temperature K. Determine the temperature distribution for any time $t > 0$ under the assumption that the cylinder is radiating heat from its surface into the surrounding medium, which has temperature zero, with a transfer coefficient of A. Assume that the boundary of the disk is maintained at a constant temperature zero. (See Problem 10 of Section 18.2.)

4. Redo Problem 3 assuming that the surrounding medium is at a positive temperature T_a rather than zero, and the edge of the plate is insulated.

5. A solid cylinder is bounded by $z = 0$ and $z = L$, and $r = R$. The cylinder's lateral surface is insulated, while the top is kept at temperature 2 and the bottom is kept at temperature zero. Find the steady-state temperature $u(r, z)$ in the cylinder.

6. Determine the temperature distribution in a circular cylinder of radius R with insulated top and bottom under the assumption that the temperature is independent of both angle and height. Assume that heat is radiating from the lateral surface into the surrounding medium, which has temperature zero, with transfer coefficient A. The initial temperature distribution in the cylinder is $f(r)$. *Hint:* See the third example of Section 18.2. Use the fact that the boundary value problem has an infinite number of positive eigenvalues. This involves assuming that an equation of the form

$$kJ'_0(kR) + AJ_0(kR) = 0$$

has infinitely many positive solutions for k. This is true, but we will not prove it.

7. Find the general solution of the differential equation

$$r^2 F''(r) + rF(r) + \lambda^2 r^2 F(r) = 0.$$

Hint: Transform the equation by letting $x = \lambda r$. The transformed equation will be a Bessel equation of order zero, which has general solution $\varphi(x) = aJ_0(x) + bY_0(x)$. Now transform this solution back in terms of r.

18.6 *The Heat Equation in a Solid Sphere*

We will now consider a problem involving temperature distribution in a solid sphere. This will lead us to a Fourier type expansion of an initial temperature function, but this time the expansion will be in terms of functions obtained using Legendre polynomials. The idea behind choosing the coefficients will be similar to that used in Section 17.1 for Fourier coefficients and in the preceding section for Fourier-Bessel coefficients and will involve an orthogonality relationship involving Legendre polynomials.

Consider a solid sphere of radius R centered at the origin. We want to solve for the steady-state temperature distribution in the sphere, given the temperature on the surface. In three dimensions, the heat equation is

$$\frac{\partial u}{\partial t} = a^2 \left[\frac{\partial^2 u}{\partial x^2} + \frac{\partial^2 u}{\partial y^2} + \frac{\partial^2 u}{\partial z^2} \right].$$

For the steady-state case, $\partial u / \partial t = 0$, and the steady-state heat equation in three dimensions is Laplace's equation

$$\nabla^2 u = \frac{\partial^2 u}{\partial x^2} + \frac{\partial^2 u}{\partial y^2} + \frac{\partial^2 u}{\partial z^2} = 0.$$

For a sphere, we will use spherical coordinates ρ, θ, and φ. From equation (18.20), Laplace's equation in spherical coordinates is

$$\frac{1}{\rho}\frac{\partial^2}{\partial\rho^2}[\rho u] + \frac{1}{\rho^2\sin^2(\varphi)}\frac{\partial^2 u}{\partial\theta^2} + \frac{1}{\rho^2\sin(\varphi)}\frac{\partial}{\partial\varphi}\left[\frac{\partial u}{\partial\varphi}\sin(\varphi)\right] = 0.$$

We will consider the special case in which u is independent of θ. Then $\partial u/\partial\theta = 0$, and the partial differential equation for u is

$$\frac{1}{\rho}\frac{\partial^2}{\partial\rho^2}[\rho u] + \frac{1}{\rho^2\sin(\varphi)}\frac{\partial}{\partial\varphi}\left[\frac{\partial u}{\partial\varphi}\sin(\varphi)\right] = 0 \qquad (18.30)$$

for $0 \le \rho < R$ and $0 \le \varphi \le \pi$. Assume that u is given on the surface of the sphere:

$$u(R, \varphi) = f(\varphi) \qquad (0 \le \varphi \le \pi).$$

Thus, we have a Dirichlet problem in which we want to solve Laplace's equation in a set (the interior of a sphere) subject to the condition that the solution must assume given values on the boundary (the surface of the sphere). This boundary value problem also models the electrostatic potential inside the sphere, if there are no charges inside, under the assumption that $u = f(\varphi)$ on the surface.

In order to separate the variables in equation (18.30), let

$$u(\rho, \varphi) = F(\rho)\Phi(\varphi)$$

to get

$$\frac{\rho[\rho F]''}{F} + \frac{1}{\sin(\varphi)}\frac{[\Phi'(\varphi)\sin(\varphi)]'}{\Phi} = 0.$$

For some constant λ to be determined,

$$-\frac{\rho[\rho F]''}{F} = \frac{1}{\sin(\varphi)}\frac{[\Phi'(\varphi)\sin(\varphi)]'}{\Phi} = -\lambda.$$

After carrying out the indicated differentiations on $[\rho F]''$ and rearranging terms, we get

$$\rho^2 F''(\rho) + 2\rho F'(\rho) - \lambda F(\rho) = 0 \qquad (0 \le \rho < R)$$

and

$$\frac{1}{\sin(\varphi)}[\Phi'(\varphi)\sin(\varphi)]' + \lambda\Phi(\varphi) = 0 \qquad (0 < \varphi < \pi).$$

In the differential equation for Φ, change variables by letting

$$x = \cos(\varphi).$$

Then $\varphi = \cos^{-1}(x)$. Since $0 < \varphi < \pi$, $-1 < x < 1$. Let

$$G(x) = \Phi[\cos^{-1}(x)].$$

Calculate

$$\Phi'(\varphi)\sin(\varphi) = \sin(\varphi)\frac{d\Phi}{dx}\frac{dx}{d\varphi} = \frac{1 - \cos^2(\varphi)}{\sin(\varphi)}G'(x)[-\sin(\varphi)]$$

$$= -(1 - x^2)G'(x).$$

In terms of x and G, the differential equation for Φ becomes

$$\frac{d}{dx}[(1-x^2)G'(x)] + \lambda G(x) = 0 \qquad (-1 < x < 1).$$

This is Legendre's differential equation (Sections 7.4 and 7.5). This equation has bounded solutions on $(-1, 1)$ only if we choose

$$\lambda = n(n+1), \qquad n = 0, 1, 2, 3, \ldots.$$

These are the eigenvalues of the problem for G. Corresponding eigenfunctions are $G_n(x) = A_n P_n(x)$, with P_n the nth Legendre polynomial and A_n a constant to be determined. Since $x = \cos(\varphi)$, these eigenfunctions, in terms of φ, are $A_n P_n[\cos(\varphi)]$.

Now let $\lambda = n(n+1)$ in the differential equation for F to get

$$\rho^2 F''(\rho) + 2\rho F'(\rho) - n(n+1)F(\rho) = 0.$$

This is a Cauchy-Euler equation with general solution

$$F_n(\rho) = a\rho^n + b\rho^{-n-1}.$$

In order to have a solution which remains bounded throughout the sphere, we must let $b = 0$ because $\lim_{\rho \to 0+} \rho^{-n-1} = \infty$. Thus, $F_n(\rho)$ must have the form of a nonzero constant times ρ^n for any given nonnegative integer n.

We now have, in terms of ρ and φ, functions

$$u_n(\rho, \varphi) = C_n \rho^n P_n[\cos(\varphi)]$$

for $n = 0, 1, 2, 3, \ldots$. Each of these functions satisfies Laplace's equation.

In order to satisfy the boundary condition, we require that $u(R, \varphi) = f(\varphi)$ for $0 \le \varphi \le \pi$. Depending on f, this condition is usually impossible to satisfy with a particular choice of n, and we must generally use a superposition

$$u(\rho, \varphi) = \sum_{n=0}^{\infty} u_n(\rho, \varphi) = \sum_{n=0}^{\infty} C_n \rho^n P_n[\cos(\varphi)]. \qquad (18.31)$$

We must choose the C_n's to satisfy the condition

$$u(R, \varphi) = f(\varphi) = \sum_{n=0}^{\infty} C_n R^n P_n[\cos(\varphi)]. \qquad (18.32)$$

This is very much the position we reached in the preceding section, but there we had a Fourier-Bessel expansion of f, while now we have a series we will call a *Fourier-Legendre* expansion of f, in view of the appearance of P_n in the series. As we have seen, such an expansion is often possible if we have an orthogonality relationship for the functions involved. Here, the orthogonality relationship is

$$\int_{-1}^{1} P_n(x) P_m(x) \, dx = 0$$

if n and m are distinct nonnegative integers. This relationship was proved in Theorem 7.3 and was observed again in Section 7.8 as a consequence of the Sturm-Liouville theorem of Section 7.7.

In order to make use of this orthogonality relationship to choose the C_n's, write the expansion (18.32) in terms of x by recalling that $\varphi = \cos^{-1}(x)$:

$$f[\cos^{-1}(x)] = \sum_{n=0}^{\infty} C_n R^n P_n(x).$$

Choose a nonnegative integer k and multiply both sides of the last equation by $P_k(x)$:

$$f[\cos^{-1}(x)] P_k(x) = \sum_{n=0}^{\infty} C_n R^n P_n(x) P_k(x).$$

Integrate this equation from -1 to 1, and interchange the summation and the integral:

$$\int_{-1}^{1} f[\cos^{-1}(x)] P_k(x)\, dx = \sum_{n=0}^{\infty} C_n R^n \int_{-1}^{1} P_n(x) P_k(x)\, dx$$

$$= C_k R^k \int_{-1}^{1} [P_k(x)]^2\, dx,$$

with all the terms in the series vanishing if $n \neq k$ by the orthogonality relationship. Solve the last equation for C_k to get

$$C_k = \frac{1}{R^k} \frac{\int_{-1}^{1} f[\cos^{-1}(x)] P_k(x)\, dx}{\int_{-1}^{1} [P_k(x)]^2\, dx}.$$

By Theorem 7.9,

$$\int_{-1}^{1} [P_k(x)]^2\, dx = \frac{2}{2k+1} \quad \text{for} \quad k = 0, 1, 2, 3, \ldots.$$

Therefore,

$$C_k = \frac{1}{R^k} \frac{2k+1}{2} \int_{-1}^{1} f[\cos^{-1}(x)] P_k(x)\, dx,$$

and, at least formally, the solution of the boundary value problem is

$$u(\rho, \varphi) = \sum_{n=0}^{\infty} \left[\frac{2n+1}{2} \int_{-1}^{1} f[\cos^{-1}(\xi)] P_n(\xi)\, d\xi \right] \left(\frac{\rho}{R} \right)^n P_n[\cos(\varphi)].$$

Assuming certain continuity conditions on f (for example, see Theorem 7.11), this series converges to the solution of the problem.

PROBLEMS FOR SECTION 18.6

1. Consider heat conduction in a solid hemisphere (spherical coordinates $0 \leq r \leq R$, $0 \leq \varphi \leq \pi/2$, $0 \leq \theta \leq 2\pi$). Solve for the steady-state temperature function if the base is kept at zero temperature and the hemispherical portion is kept at constant temperature A. Assume that the distribution is independent of θ.

2. Redo Problem 1 with the base insulated instead of being kept at zero temperature.

3. Redo Problem 1 for the case in which the temperature on the hemispherical surface is $u(R, \varphi) = f[\cos(\varphi)]$, not necessarily the constant A.

4. Solve the boundary value problem

$$\frac{\partial u}{\partial t} = \frac{\partial}{\partial x}\left[(1 - x^2)\frac{\partial u}{\partial x} \right] \qquad (-1 < x < 1, t > 0),$$

$$u(x, 0) = f(x) \qquad\qquad (-1 < x < 1).$$

This models the temperature in a bar extending from -1 to 1 if the thermal conductivity is $1 - x^2$. Assume that the lateral surface of the bar is insulated.

5. Solve for the steady-state temperature in a hollowed-out sphere given in spherical coordinates by $R_1 \leq \rho \leq R_2$. The inner surface $\rho = R_1$ is kept at constant temperature T, and the outer surface $\rho = R_2$ is kept at temperature zero. Assume that the temperature distribution depends on only ρ and φ.

18.7 Multiple Fourier Series Solutions of Boundary Value Problems

In several instances, we have simplified a boundary value problem modeling a physical phenomenon by assuming that the solution was independent of one of the variables. We will now consider problems like those done previously but in which we retain dependence of the solution on more of the variables present in the model. Such problems may require the use of multiple Fourier series.

EXAMPLE 18.3

Consider an elastic membrane stretched across a rectangular frame. If the coordinate system is defined as in Figure 18.14, the membrane occupies a rectangle $0 \leq x \leq L$, $0 \leq y \leq K$. The membrane is given an initial displacement and velocity, and we want to determine the vertical displacement function $z(x, y, t)$ for all $t > 0$ and at all points (x, y) in the membrane.

The mathematical model consists of the two-dimensional wave equation for z, together with the following boundary conditions (the membrane is fixed at the edges on the rectangular frame) and information about the initial velocity and displacement:

$$\frac{\partial^2 z}{\partial t^2} = a^2 \left[\frac{\partial^2 z}{\partial x^2} + \frac{\partial^2 z}{\partial y^2} \right] \qquad (0 < x < L, 0 < y < K, t > 0),$$

$$z(x, 0, t) = z(x, K, t) = 0 \qquad\qquad (0 < x < L, t > 0),$$

$$z(0, y, t) = z(L, y, t) = 0 \qquad\qquad (0 < y < K, t > 0),$$

$$z(x, y, 0) = f(x, y)$$

$$\frac{\partial z}{\partial t}(x, y, 0) = g(x, y) \qquad\qquad (0 < x < L, 0 < y < K).$$

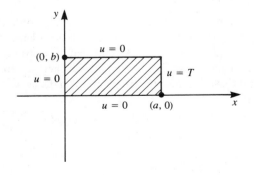

Figure 18.14

As an illustration, we will solve this problem when the membrane is stretched to an initial position and then released from rest, so that $g(x, y) = 0$.

Attempt a separation of variables by putting

$$z(x, y, t) = X(x)Y(y)T(t).$$

Substitute this expression into the wave equation to get

$$XYT'' = X''YT + XY''T,$$

which we write as

$$\frac{T''}{a^2 T} - \frac{Y''}{Y} = \frac{X''}{X}.$$

We cannot separate three variables on different sides of a single equation. However, we have managed to write the left side as a function of y and t alone and the right side as a function of just x. Since these variables are independent, we can fix y and t on the left, implying that the right side has a constant value for $0 < x < L$. Hence, the left side is also constant for $0 < y < K$ and $t > 0$. We will call this constant $-\lambda$. (We could call it λ and achieve the same results.)

We now have

$$\frac{T''}{a^2 T} - \frac{Y''}{Y} = \frac{X''}{X} = -\lambda.$$

Then $X'' + \lambda X = 0$. Further,

$$\frac{T''}{a^2 T} - \frac{Y''}{Y} = -\lambda.$$

Write this equation as

$$\frac{T''}{a^2 T} + \lambda = \frac{Y''}{Y}.$$

Since the left side depends only on t and the right side only on y, and t and y are independent, both sides must equal some constant, which we will call $-\mu$. Then

$$\frac{T''}{a^2 T} + \lambda = \frac{Y''}{Y} = -\mu;$$

hence,

$$T'' + (\lambda + \mu)a^2 T = 0 \quad \text{and} \quad Y'' + \mu Y = 0.$$

We therefore have separate differential equations for X, Y, and T involving *two* separation constants λ and μ (because there are three independent variables). These differential equations are

$$X'' + \lambda X = 0, \qquad Y'' + \mu Y = 0, \quad \text{and} \quad T'' + (\lambda + \mu)a^2 T = 0.$$

Now consider the boundary conditions. Since

$$z(0, y, t) = X(0)Y(y)T(t) = 0,$$

$X(0) = 0$. Similarly,

$$z(L, y, t) = X(L)Y(y)T(t) = 0,$$

so $X(L) = 0$.

The problem for X is

$$X'' + \lambda X = 0 \qquad (0 < x < L),$$
$$X(0) = X(L) = 0.$$

We have solved this problem before. The eigenvalues are

$$\lambda_n = \frac{n^2 \pi^2}{L^2},$$

with corresponding eigenfunctions

$$X_n(x) = A_n \sin\left(\frac{n\pi x}{L}\right)$$

for $n = 1, 2, 3, \ldots$ Similarly, $z(x, 0, t) = z(x, K, t) = 0$ implies that

$$Y(0) = Y(K) = 0.$$

The problem for Y is

$$Y'' + \mu Y = 0,$$
$$Y(0) = Y(L) = 0.$$

This problem has eigenvalues

$$\mu_m = \frac{m^2 \pi^2}{K^2}$$

and eigenfunctions

$$Y_m(y) = B_m \sin\left(\frac{m\pi y}{K}\right)$$

for $m = 1, 2, 3, \ldots$ Note that m and n *independently* take on positive integer values.
Now that we know the eigenvalues λ and μ, the equation for T is

$$T'' + \left[\frac{n^2 \pi^2}{L^2} + \frac{m^2 \pi^2}{K^2}\right] a^2 T = 0.$$

This differential equation has general solution

$$T(t) = C \cos\left(\sqrt{\frac{n^2}{L^2} + \frac{m^2}{K^2}} \, \pi a t\right) + D \sin\left(\sqrt{\frac{n^2}{L^2} + \frac{m^2}{K^2}} \, \pi a t\right).$$

Since $(\partial z/\partial t)(x, y, 0) = X(x)Y(y)T'(0) = 0$, $T'(0) = 0$, and we must choose $D = 0$. Then, for $n = 1, 2, 3, \ldots$ and $m = 1, 2, 3, \ldots$, we have

$$T_{nm}(t) = C_{nm} \cos\left(\sqrt{\frac{n^2}{L^2} + \frac{m^2}{K^2}} \, \pi a t\right).$$

For each positive integer n and m, we have

$$z_{nm}(x, y, t) = X_n(x) Y_m(y) T_{nm}(t)$$

$$= k_{nm} \sin\left(\frac{n\pi x}{L}\right) \sin\left(\frac{m\pi y}{K}\right) \cos\left(\sqrt{\frac{n^2}{L^2} + \frac{m^2}{K^2}} \, \pi a t\right),$$

in which we have aggregated all of the constants into k_{nm}. These functions all satisfy the wave equation, together with the boundary conditions and the initial condition $(\partial z/\partial t)(x, y, 0) = 0$. In order to satisfy the initial condition $z(x, y, 0) = f(x, y)$, we must generally form a superposition of these functions, which takes the form of a double summation:

$$
\begin{aligned}
z(x, y, t) &= \sum_{n=1}^{\infty} \sum_{m=1}^{\infty} z_{nm}(x, y, t) \\
&= \sum_{n=1}^{\infty} \sum_{m=1}^{\infty} k_{nm}\sin\left(\frac{n\pi x}{L}\right)\sin\left(\frac{m\pi y}{K}\right)\cos\left(\sqrt{\frac{n^2}{L^2} + \frac{m^2}{K^2}}\,\pi a t\right). \quad (18.33)
\end{aligned}
$$

The initial condition $z(x, y, 0) = f(x, y)$ requires that

$$
z(x, y, 0) = f(x, y) = \sum_{n=1}^{\infty} \sum_{m=1}^{\infty} k_{nm}\sin\left(\frac{n\pi x}{L}\right)\sin\left(\frac{m\pi y}{K}\right).
$$

This is a double Fourier sine representation of f on $0 < x < L, 0 < y < K$. Hence, choose the Fourier coefficients

$$
k_{nm} = \frac{4}{LK} \int_0^K \int_0^L f(x, y)\sin\left(\frac{n\pi x}{L}\right)\sin\left(\frac{m\pi y}{K}\right) dx\, dy.
$$

With this choice of the coefficients, equation (18.33) provides a formal solution of the problem. Assuming that f satisfies certain continuity conditions, f is represented by its double Fourier sine series on the interior of the rectangle, and (18.33) is a solution of the boundary value problem.

As a specific example, suppose that

$$
f(x, y) = x(x - L)y(y - K) \quad \text{for} \quad 0 \le x \le L, 0 \le y \le K.
$$

Then

$$
\begin{aligned}
k_{nm} &= \frac{4}{LK} \int_0^L \int_0^K x(x - L)y(y - K)\sin\left(\frac{n\pi x}{L}\right)\sin\left(\frac{m\pi y}{K}\right) dx\, dy \\
&= \frac{16L^2K^2}{(n\pi)^3(m\pi)^3}[\cos(n\pi) - 1][\cos(m\pi) - 1] \\
&= \begin{cases} \dfrac{64L^2K^2}{n^3m^3\pi^6} & \text{if} \quad n \text{ and } m \text{ are both odd} \\[2mm] 0 & \text{if} \quad n \text{ or } m \text{ is even.} \end{cases}
\end{aligned}
$$

The solution for this choice of f is

$$
\begin{aligned}
z(x, y, t) &= \frac{64L^2K^2}{\pi^6} \sum_{n=1}^{\infty} \sum_{m=1}^{\infty} \frac{1}{(2n - 1)^3(2m - 1)^3} \\
&\quad \times \sin\left[(2n - 1)\frac{\pi x}{L}\right]\sin\left[(2m - 1)\frac{\pi y}{K}\right] \\
&\quad \times \cos\left(\sqrt{\frac{(2n - 1)^2}{L^2} + \frac{(2m - 1)^2}{K^2}}\,\pi a t\right). \quad \blacksquare
\end{aligned}
$$

EXAMPLE 18.4

Consider the temperature distribution in a square plate of side length 1 if the sides are kept at temperature zero and the interior temperature at time zero is given by $f(x, y) = x(1 - x)y(1 - y)$.

The boundary value problem modeling this setting is

$$\frac{\partial u}{\partial t} = a^2 \left[\frac{\partial^2 u}{\partial x^2} + \frac{\partial^2 u}{\partial y^2} \right] \qquad (0 < x < 1, 0 < y < 1, t > 0),$$

$$u(x, 0, t) = u(x, 1, t) = 0 \qquad\qquad\qquad (0 < x < 1),$$

$$u(0, y, t) = u(1, y, t) = 0 \qquad\qquad\qquad (0 < y < 1),$$

$$u(x, y, 0) = x(1 - x)y(1 - y) \qquad\qquad (0 < x < 1, 0 < y < 1).$$

Let $u(x, y, z) = X(x)Y(y)T(t)$. After a calculation like that of Example 18.3, we obtain

$$\frac{X''}{X} = \frac{T'}{a^2 T} - \frac{Y''}{Y}.$$

Since x, y, and t are independent, for some constant λ,

$$\frac{X''}{X} = \frac{T'}{a^2 T} - \frac{Y''}{Y} = -\lambda.$$

Then $X'' + \lambda X = 0$, and, for some constant μ,

$$\frac{T'}{a^2 T} + \lambda = \frac{Y''}{Y} = -\mu$$

because t and y are independent. Then

$$T' + (\lambda + \mu)a^2 T = 0 \quad \text{and} \quad Y'' + \mu Y = 0.$$

Now consider the boundary conditions. First,

$$u(0, y, t) = X(0)Y(y)T(t) = 0 \quad \text{implies that} \quad X(0) = 0$$

and

$$u(1, y, t) = X(1)Y(y)T(t) = 0 \quad \text{implies that} \quad X(1) = 0.$$

Similarly, the other two boundary conditions imply that

$$Y(0) = Y(1) = 0.$$

The problems for X and Y are therefore

$$X'' + \lambda X = 0, \qquad\qquad Y'' + \mu Y = 0,$$
$$X(0) = X(1) = 0; \qquad\qquad Y(0) = Y(1) = 0.$$

These problems have solutions

$$\lambda_n = n^2\pi^2 \quad \text{and} \quad X_n(x) = a_n\sin(n\pi x) \quad \text{for} \quad n = 1, 2, 3, \ldots$$

and

$$\mu_m = m^2\pi^2 \quad \text{and} \quad Y_m(y) = b_m\sin(m\pi y) \quad \text{for} \quad m = 1, 2, 3, \ldots.$$

The problem for T becomes

$$T' + (\lambda + \mu)a^2 T = 0,$$

or

$$T' + (n^2 + m^2)a^2\pi^2 T = 0.$$

This differential equation has general solution

$$T_{nm}(t) = c_{nm}e^{-(n^2 + m^2)a^2\pi^2 t}.$$

For any positive integers n and m, let

$$u_{nm}(x, y, t) = d_{nm}\sin(n\pi x)\sin(m\pi y)e^{-(n^2 + m^2)a^2\pi^2 t}.$$

To satisfy the initial condition, we must use a superposition

$$u(x, y, t) = \sum_{n=1}^{\infty} \sum_{m=1}^{\infty} d_{nm}\sin(n\pi x)\sin(m\pi y)e^{-(n^2 + m^2)a^2\pi^2 t}.$$

We must choose the coefficients so that

$$u(x, y, 0) = f(x, y) = x(1 - x)y(1 - y) = \sum_{n=1}^{\infty} \sum_{m=1}^{\infty} d_{nm}\sin(n\pi x)\sin(m\pi y).$$

This is a Fourier sine expansion of f over $0 < x < 1, 0 < y < 1$, and the coefficients are

$$d_{nm} = 4 \int_0^1 \int_0^1 x(1 - x)y(1 - y)\sin(n\pi x)\sin(m\pi y)\, dx\, dy$$

$$= \begin{cases} \dfrac{64}{n^3 m^3 \pi^6} & \text{if } n \text{ and } m \text{ are both odd} \\ 0 & \text{if } n \text{ or } m \text{ is even.} \end{cases}$$

The solution is

$$u(x, y, t) = \frac{64}{\pi^6} \sum_{n=1}^{\infty} \sum_{m=1}^{\infty} \frac{1}{(2n - 1)^3(2m - 1)^3} \sin[(2n - 1)\pi x]\sin[(2m - 1)\pi y]$$
$$\times e^{-[(2n-1)^2 + (2m-1)^2]\pi^2 a^2 t}. \blacksquare$$

In the next section, we will consider a boundary value problem whose solution · requires a variety of the techniques developed up to this point.

PROBLEMS FOR SECTION 18.7

In each of Problems 1 through 8, solve the boundary value problem.

1. $\dfrac{\partial^2 u}{\partial x^2} + \dfrac{\partial^2 u}{\partial y^2} + \dfrac{\partial^2 u}{\partial z^2} = 0 \quad (0 < x < 1, 0 < y < 1, 0 < z < 1),$

   $u(0, y, z) = u(1, y, z) = u(x, 0, z) = u(x, 1, z) = u(x, y, 0) = 0,$
   $u(x, y, 1) = xy$

2. $\dfrac{\partial u}{\partial t} = \dfrac{\partial^2 u}{\partial x^2} + \dfrac{\partial^2 u}{\partial y^2} \quad (0 < x < \pi, 0 < y < \pi, t > 0),$

   $u(0, y, t) = u(\pi, y, t) = u(x, 0, t) = u(x, \pi, t) = 0,$
   $u(x, y, 0) = x(\pi - x)(\pi - y)^2$

3. $\dfrac{\partial^2 u}{\partial t^2} = \dfrac{\partial^2 u}{\partial x^2} + \dfrac{\partial^2 u}{\partial y^2}$ $(0 < x < 2\pi, 0 < y < 2\pi, t > 0)$,

$u(0, y, t) = u(2\pi, y, t) = u(x, 0, t) = u(x, 2\pi, t) = 0$,

$u(x, y, 0) = x^2 \sin(y)$,

$\dfrac{\partial u}{\partial t}(x, y, 0) = 0$

4. $\dfrac{\partial^2 u}{\partial x^2} + \dfrac{\partial^2 u}{\partial y^2} + \dfrac{\partial^2 u}{\partial z^2} = 0$ $(0 < x < 2\pi, 0 < y < 2\pi, 0 < z < 1)$,

$u(0, y, z) = u(x, 0, z) = u(x, 2\pi, z) = u(x, y, 0) = u(x, y, 1) = 0$,

$u(2\pi, y, z) = 2$

5. $\dfrac{\partial^2 u}{\partial t^2} = 4\left[\dfrac{\partial^2 u}{\partial x^2} + \dfrac{\partial^2 u}{\partial y^2}\right]$ $(0 < x < 2\pi, 0 < y < 2\pi, t > 0)$,

$u(0, y, t) = u(2\pi, y, t) = u(x, 0, t) = u(x, 2\pi, t) = 0$,

$u(x, y, 0) = 0, \quad \dfrac{\partial u}{\partial t}(x, y, 0) = 1$

6. $\dfrac{\partial^2 u}{\partial t^2} = 9\left[\dfrac{\partial^2 u}{\partial x^2} + \dfrac{\partial^2 u}{\partial y^2}\right]$ $(0 < x < \pi, 0 < y < \pi, t > 0)$,

$u(0, y, t) = u(\pi, y, t) = u(x, 0, t) = u(x, \pi, t) = 0$,

$u(x, y, 0) = \sin(x)\cos(y), \quad \dfrac{\partial u}{\partial t}(x, y, 0) = xy$

7. $\dfrac{\partial^2 u}{\partial x^2} + \dfrac{\partial^2 u}{\partial y^2} + \dfrac{\partial^2 u}{\partial z^2} = 0$ $(0 < x < 1, 0 < y < 2\pi, 0 < z < \pi)$,

$u(0, y, z) = u(1, y, z) = u(x, 0, z) = u(x, y, 0) = 0$,

$u(x, 2\pi, z) = 2, \quad u(x, y, \pi) = 1$

8. $\dfrac{\partial^2 u}{\partial x^2} + \dfrac{\partial^2 u}{\partial y^2} + \dfrac{\partial^2 u}{\partial z^2} = 0$ $(0 < x < 1, 0 < y < 1, 0 < z < 1)$,

$u(0, y, z) = u(1, y, z) = u(x, 0, z) = u(x, 1, z) = 0$,

$u(x, y, 0) = -1, \quad u(x, y, 1) = 1$

9. Find the temperature distribution for a solid cube of side length L if all sides are insulated and the initial temperature is $f(x, y, z)$ at (x, y, z) in the interior of the cube.

10. Solve the boundary value problem

$$\dfrac{\partial^2 u}{\partial t^2} = \alpha^2\left[\dfrac{\partial^2 u}{\partial x^2} + \dfrac{\partial^2 u}{\partial y^2} + \dfrac{\partial^2 u}{\partial z^2}\right]$$

$$u(0, y, z, t) = u(a, y, z, t) = u(x, 0, z, t) = u(x, b, z, t) = u(x, y, 0, t) = u(x, y, c, t) = 0$$

$$u(x, y, z, 0) = f(x, y, z)$$

$$\dfrac{\partial u}{\partial t}(x, y, z, 0) = 0,$$

in which $0 < x < a, 0 < y < b, 0 < z < c$, and $t > 0$. This boundary value problem models sound waves in a cubical room, assuming a given initial disturbance and zero initial velocity.

11. Find an expression for the motion of sound waves in a cubical room, assuming an initial disturbance $f(x, y, z)$ and an initial velocity $g(x, y, z)$. *Hint:* Half the work was done in Problem 10.

12. Solve for the displacement function for a rectangular membrane stretched onto a fixed frame if the initial position is given by a function f and the initial velocity is given by a function g.

13. Solve the boundary value problem

$$\frac{\partial^2 u}{\partial x^2} + \frac{\partial^2 u}{\partial y^2} + \frac{\partial^2 u}{\partial z^2} = 0 \qquad (0 < x < 1, 0 < y < 1, 0 < z < 2)$$

if $u = 0$ on all faces of the cube except the top face, where $u(x, y, 2) = xy^2$.

14. Solve the boundary value problem

$$\frac{\partial^2 u}{\partial x^2} + \frac{\partial^2 u}{\partial y^2} + \frac{\partial^2 u}{\partial z^2} = 0 \qquad (0 < x < 1, 0 < y < 4, 0 < z < 2)$$

if $u = 0$ on the faces $x = 0$, $x = 1$, $z = 0$, and $z = 2$ and $u(x, 0, z) = xz$ and $u(x, 4, z) = \sin(z)$.

15. Find the temperature distribution in a plate placed over the rectangle $0 \le x \le a$, $0 \le y \le b$. The plate has insulated edges and is radiating freely off its surface with transfer coefficient A into a surrounding medium, which is kept at temperature zero. The initial temperature distribution on the plate is $f(x, y) = x(x - a)$. *Hint:* See Problem 10 of Section 18.2. Transform the partial differential equation into one of the form

$$\frac{\partial v}{\partial t} = a^2 \left[\frac{\partial^2 v}{\partial x^2} + \frac{\partial^2 v}{\partial y^2} \right]$$

by letting

$$v(x, y, t) = e^{At} u(x, y, t).$$

18.8 *Vibrations of a Circular Elastic Membrane*

We will now consider the problem of describing the vibrations of an elastic membrane fixed to a circular frame. For example, we might imagine a circular drum which has been struck and attempt to describe how the drum surface moves. To formulate a boundary value problem model, consider an elastic membrane stretched over a fixed circular frame of radius R, given an initial displacement described by a function f and then released from rest. Since the frame is circular, we will use cylindrical coordinates, placing the frame in the plane $z = 0$ with the center at the origin. The vertical displacements will be given by a continuous function $z = u(r, \theta, t)$.

The two-dimensional wave equation in cylindrical coordinates is

$$\frac{\partial u}{\partial t} = a^2 \left[\frac{\partial^2 u}{\partial r^2} + \frac{1}{r} \frac{\partial u}{\partial r} + \frac{1}{r^2} \frac{\partial^2 u}{\partial \theta^2} \right].$$

We treated a simplified version of this problem, in which u was independent of θ, in Section 18.5. Now we will retain the θ-dependence and hence the term $(1/r^2)(\partial^2 u / \partial \theta^2)$. We will still, however, impose the periodicity conditions

$$u(r, -\pi, t) = u(r, \pi, t) \quad \text{and} \quad \frac{\partial u}{\partial \theta}(r, -\pi, t) = \frac{\partial u}{\partial \theta}(r, \pi, t).$$

As before, this condition is motivated by the fact that $(r, -\pi)$ and (r, π) are polar coordinate representations of the same point in the membrane. The initial displacement is

$$u(r, \theta, 0) = f(r, \theta).$$

We will assume that f is continuous and has piecewise-continuous first partial derivatives.

The boundary value problem is

$$\frac{\partial^2 u}{\partial t^2} = a^2 \left[\frac{\partial^2 u}{\partial r^2} + \frac{1}{r} \frac{\partial u}{\partial r} + \frac{1}{r^2} \frac{\partial^2 u}{\partial \theta^2} \right] \qquad (0 \le r < R, \ -\pi \le \theta \le \pi, \ t > 0),$$

$$u(R, \theta, t) = 0 \qquad (-\pi \le \theta \le \pi, \ t > 0),$$

$$u(r, -\pi, t) = u(r, \pi, t), \qquad \frac{\partial u}{\partial \theta}(r, -\pi, t) = \frac{\partial u}{\partial \theta}(r, \pi, t) \qquad (0 \le r < R, \ t > 0),$$

$$u(r, \theta, 0) = f(r, \theta) \qquad (0 \le r < R, \ -\pi \le \theta \le \pi),$$

$$\frac{\partial u}{\partial t}(r, \theta, 0) = 0 \qquad (0 \le r < R, \ -\pi \le \theta \le \pi).$$

To separate the variables, let

$$u(r, \theta, t) = F(r)G(\theta)T(t).$$

Substitute this into the wave equation to obtain

$$\frac{T''}{a^2 T} = \frac{F'' + \frac{1}{r} F'}{F} + \frac{1}{r^2} \frac{G''}{G}.$$

Since the left side depends only on t and the right side on r and θ, both sides must equal a constant:

$$\frac{T''}{a^2 T} = \frac{F'' + \frac{1}{r} F'}{F} + \frac{1}{r^2} \frac{G''}{G} = -\lambda.$$

Then

$$T'' + \lambda a^2 T = 0$$

and

$$\frac{r^2 F''(r) + r F'(r)}{F(r)} + \lambda r^2 = -\frac{G''(\theta)}{G(\theta)}.$$

Since r and θ are independent, both sides of this equation must equal some constant, which we will call μ. We obtain

$$G''(\theta) + \mu G(\theta) = 0$$

and

$$r^2 F''(r) + r F'(r) + (\lambda^2 r^2 - \mu) F(r) = 0.$$

We now have ordinary differential equations for F, G, and T together with separation constants λ and μ for which we must find values. The boundary and periodicity conditions give us

$$G(-\pi) = G(\pi), \qquad G'(-\pi) = G'(\pi), \quad \text{and} \quad F(R) = 0.$$

The fact that the initial velocity is zero requires that

$$T'(0) = 0.$$

The problem for G is a periodic Sturm-Liouville problem and was solved in Example 7.7. The eigenvalues and eigenfunctions are

$$\mu = n^2 \quad \text{and} \quad G_n(\theta) = A_n\cos(n\theta) + B_n\sin(n\theta)$$

for $n = 0, 1, 2, \ldots$.

The problem for F is a singular Sturm-Liouville problem. Let $\mu = n^2$ to get

$$r^2F''(r) + rF'(r) + (\lambda^2r^2 - n^2)F(r) = 0; \qquad F(R) = 0.$$

We can transform this differential equation into a Bessel equation by letting

$$z = \sqrt{|\lambda|}r \quad \text{and} \quad W(z) = F\left(\frac{z}{\sqrt{|\lambda|}}\right).$$

Then

$$F'(r) = \frac{dW}{dz}\frac{dz}{dr} = W'(z)\sqrt{|\lambda|}$$

and

$$F''(r) = \frac{d^2W}{dz^2}\left(\frac{dz}{dr}\right)^2 + \frac{dW}{dz}\frac{d^2z}{dr^2} = \lambda W''(z).$$

With these substitutions, the differential equation for $F(r)$ transforms into

$$z^2W''(z) + zW'(z) + (z^2 - n^2)W(z) = 0.$$

This is Bessel's equation of order n, with general solution

$$W(z) = cJ_n(z) + dY_n(z).$$

In terms of r,

$$F(r) = cJ_n(\sqrt{|\lambda|}r) + dY_n(\sqrt{|\lambda|}r).$$

The center of the disk has r coordinate zero. But $Y_n(\sqrt{|\lambda|}r)$ is unbounded as $r \to 0$. To have a bounded solution, we must choose $d = 0$. This leaves us with

$$F_n(r) = c_nJ_n(\sqrt{|\lambda|}r)$$

for $n = 0, 1, 2, \ldots$.

In order to satisfy $F_n(R) = 0$ for each n, we must choose λ so that $J_n\sqrt{|\lambda|}R) = 0$. Thus, $\sqrt{|\lambda|}R$ must be a positive zero of the nth order Bessel function of the first kind J_n. For each nonnegative integer n, J_n has infinitely many positive zeros. Denote the kth positive zero (designated in increasing order) of J_n as z_{nk}. Then we must choose λ so that

$$\sqrt{|\lambda|}R = z_{nk}.$$

Thus, choose

$$\lambda_{nk} = \frac{z_{nk}^2}{R^2}.$$

For convenience, write

$$\omega_{nk} = \frac{z_{nk}}{R}.$$

The eigenvalues of this boundary value problem for F are

$$\lambda_{nk} = \omega_{nk}^2,$$

with associated eigenfunctions $J_n(\omega_{nk}r)$ for $k = 1, 2, 3, \ldots$ and $n = 0, 1, 2, \ldots$. Put these eigenvalues into the equation for T to get

$$T''(t) + \omega_{nk}^2 a^2 T(t) = 0; \qquad T'(0) = 0.$$

The general solution of this problem has the form

$$T_{nk}(t) = c_{nk}\cos(a\omega_{nk}t).$$

Finally, we can put the solutions for F, G, and T together to get a function

$$u_{nk}(r, \theta, t) = [a_{nk}\cos(n\theta) + b_{nk}\sin(n\theta)]J_n(\omega_{nk}r)\cos(a\omega_{nk}t),$$

in which we have absorbed all of the arbitrary constants into the symbols a_{nk} and b_{nk}. Here, n can be any nonnegative integer and k any positive integer.

Each of these functions satisfies the partial differential equation and the boundary conditions, together with the condition that the initial velocity is zero. In order to satisfy the requirement that the initial position be given by f, we must, in general, use a superposition

$$u(r, \theta, t) = \sum_{n=0}^{\infty} \sum_{k=1}^{\infty} [a_{nk}\cos(n\theta) + b_{nk}\sin(n\theta)]J_n(\omega_{nk}r)\cos(\omega_{nk}at).$$

The initial displacement condition requires that

$$u(r, \theta, 0) = \sum_{n=0}^{\infty} \sum_{k=1}^{\infty} [a_{nk}\cos(n\theta) + b_{nk}\sin(n\theta)]J_n(\omega_{nk}r) = f(r, \theta).$$

This is a representation of the function f of two variables in a series of sines and cosines multiplied by functions formed from nth order Bessel functions. In order to choose the coefficients, write this series representation of f as

$$f(r, \theta) = \sum_{n=0}^{\infty} \left(\left[\sum_{k=1}^{\infty} a_{nk}J_n(\omega_{nk}r) \right] \cos(n\theta) + \left[\sum_{k=1}^{\infty} b_{nk}J_n(\omega_{nk}r) \right] \sin(n\theta) \right)$$

$$= \sum_{k=1}^{\infty} a_{0k}J_0(\omega_{0k}r) + \sum_{n=1}^{\infty} \left(\left[\sum_{k=1}^{\infty} a_{nk}J_n(\omega_{nk}r) \right] \cos(n\theta) \right.$$

$$\left. + \left[\sum_{k=1}^{\infty} b_{nk}J_n(\omega_{nk}r) \right] \sin(n\theta) \right).$$

For a given r, this is a Fourier series expansion of the function $f(r, \theta)$ of θ on $[-\pi, \pi]$. The terms in square brackets (themselves summations), as well as the "constant" term (for a given r), are the Fourier coefficients. Thus, for each r, with $0 \le r < R$, choose

$$\sum_{k=1}^{\infty} a_{0k}J_0(\omega_{0k}r) = \frac{1}{2\pi}\int_{-\pi}^{\pi} f(r, \theta)\,d\theta = \alpha_0(r),$$

$$\sum_{k=1}^{\infty} a_{nk}J_n(\omega_{nk}r) = \frac{1}{\pi}\int_{-\pi}^{\pi} f(r, \theta)\cos(n\theta)\,d\theta = \alpha_n(r),$$

and

$$\sum_{k=1}^{\infty} b_{nk} J_n(\omega_{nk} r) = \frac{1}{\pi} \int_{-\pi}^{\pi} f(r, \theta) \sin(n\theta) \, d\theta = \beta_n(r)$$

for $n = 1, 2, 3, \ldots$.

Next, for each n, we must choose a_{nk} and b_{nk} as the coefficient in the Fourier-Bessel expansion of the functions $\alpha_n(r)$ and $\beta_n(r)$ which are given by the above integrals with respect to θ. Thus, choose

$$a_{nk} = \frac{\int_0^R r \alpha_n(r) J_n(\omega_{nk} r) \, dr}{\int_0^R r [J_n(\omega_{nk} r)]^2 \, dr} \qquad (n = 0, 1, 2, \ldots, k = 1, 2, 3, \ldots)$$

and

$$b_{nk} = \frac{\int_0^R r \beta_n(r) J_n(\omega_{nk} r) \, dr}{\int_0^R r [J_n(\omega_{nk} r)]^2 \, dr} \qquad (n = 0, 1, 2, \ldots, k = 1, 2, 3, \ldots).$$

From Theorem 7.13,

$$\int_0^R r [J_n(\omega_{nk} r)]^2 \, dr = \tfrac{1}{2} R^2 [J_{n+1}(\omega_{nk} R)]^2.$$

We can therefore write the coefficients more simply as

$$a_{0k} = \frac{1}{\pi R^2 [J_1(\omega_{0k} R)]^2} \int_0^R \int_{-\pi}^{\pi} r J_0(\omega_{0k} r) f(r, \theta) \, d\theta \, dr$$

for $k = 1, 2, 3, \ldots$, and, for $n = 1, 2, 3, \ldots$ and $k = 1, 2, 3, \ldots$,

$$a_{nk} = \frac{2}{\pi R^2 [J_{n+1}(\omega_{nk} R)]^2} \int_0^R \int_{-\pi}^{\pi} r J_n(\omega_{nk} r) f(r, \theta) \cos(n\theta) \, d\theta \, dr$$

and

$$b_{nk} = \frac{2}{\pi R^2 [J_{n+1}(\omega_{nk} R)]^2} \int_0^R \int_{-\pi}^{\pi} r J_n(\omega_{nk} r) f(r, \theta) \sin(n\theta) \, d\theta \, dr.$$

With these coefficients, the vertical displacement function is

$$u(r, \theta, t) = \sum_{n=0}^{\infty} \sum_{k=1}^{\infty} [a_{nk} \cos(n\theta) + b_{nk} \sin(n\theta)] J_n(\omega_{nk} r) \cos(\omega_{nk} at).$$

PROBLEMS FOR SECTION 18.8

1. Solve the boundary value problem for the vertical displacement of the membrane for the case in which the membrane is not initially displaced but has an initial velocity given by $g(r, \theta)$.

2. Use the solution of the boundary value problem derived in this section to prove the plausible fact that the center of the membrane remains undeflected for all time if the initial displacement is an odd function of θ. That is, show that the deflection at the center is zero if $f(r, \theta) = -f(r, -\theta)$. *Hint:* The only Bessel function of integer order which is different from zero at zero in this case is J_0.

3. Approximate the vertical deflection of the center of a circular membrane of radius 2 for any time $t > 0$ by computing the first three nonzero terms of the series solution of the boundary value problem. Assume that the initial displacement is

$$f(r, \theta) = (4 - r^2) \sin^2(\theta)$$

and that the initial velocity is zero. Use $a = 2$ in the boundary value problem solved in this section.

18.9 *Solution of the Heat and Wave Equations in Unbounded Domains* _____

We will now consider problems involving the heat and wave equations when x is allowed to vary either over the entire real line or over the half-line $[0, \infty)$. Although we do not encounter elastic strings or bars of material of infinite length in reality, it has been found useful to model extremely long objects (as compared with width) in this way. For example, models of submarine activity in the ocean often assume an infinite environment, a good approximation in the sense that the environment is ocean for as far as the submarine can effectively interact with it.

When the entire line is involved, there will be no boundary conditions, and for the half-line $[0, \infty)$, there will be one boundary condition, at zero. To compensate for this absence of boundary data, we will impose the physically appealing condition that solutions must remain bounded as x and t vary over their respective intervals. This will eliminate some functions which are mathematical solutions of the boundary value problem but which cannot be solutions modeling a real-world phenomenon.

THE WAVE EQUATION FOR AN INFINITE STRING

Imagine a string stretched over the entire real line and set in motion with a given initial displacement and velocity. We want to describe the ensuing displacements of the string. The vibrating string model developed for a finite interval $[0, L]$ is easily adapted to this circumstance. The boundary value problem is

$$\frac{\partial^2 y}{\partial t^2} = a^2 \frac{\partial^2 y}{\partial x^2} \qquad (-\infty < x < \infty, t > 0),$$

$$y(x, 0) = f(x) \qquad (-\infty < x < \infty),$$

$$\frac{\partial y}{\partial t}(x, 0) = g(x) \qquad (-\infty < x < \infty).$$

To be specific, suppose we release the string with initial velocity $g(x) = e^{-|x|}$ but without initial displacement $[f(x) = 0]$. We will also impose the condition that the solution $y(x, t)$ be bounded for $-\infty < x < \infty$ and $t > 0$. This condition takes the place of the usual boundary conditions $y(0, t) = y(L, t) = 0$, which do not apply here.

Let $y(x, t) = X(x)T(t)$ to obtain, as in the finite interval case,

$$X'' + \lambda X = 0 \quad \text{and} \quad T'' + \lambda a^2 T = 0,$$

in which λ is the separation constant. Consider cases on λ and attempt to solve for X.

Case 1: $\lambda = 0$ Then $X = cx + d$. This is unbounded unless we choose $c = 0$. Then $X = \text{constant}$, a possible solution for X; hence, $\lambda = 0$ is an eigenvalue of this problem for X.

Case 2: $\lambda < 0$ Write $\lambda = -k^2$, with k positive. Then $X'' - k^2 X = 0$, with general solution

$$X(x) = ce^{kx} + de^{-kx}.$$

Notice that ce^{kx} is unbounded on $[0, \infty)$ unless we choose $c = 0$, and de^{-kx} is unbounded on $(-\infty, 0]$ unless we choose $d = 0$. We therefore have only the trivial solution in this case, and there is no negative eigenvalue of this problem.

Case 3: $\lambda > 0$ Write $\lambda = k^2$, with k positive. Now $X'' + k^2 X = 0$, with general solution

$$X(x) = c \cos(kx) + d \sin(kx).$$

This is a bounded solution for any choices of c and d. Thus, $\lambda = k^2$ is an eigenvalue, with corresponding eigenfunctions $X_k(x) = c_k \cos(kx) + d_k \sin(kx)$.

Corresponding to $\lambda = 0$, the general solution for T is

$$T(t) = \alpha t + \beta.$$

This is unbounded for $t > 0$ unless we choose $\alpha = 0$. Further, $y(x, 0) = X(x)T(0) = 0$ requires that $T(0) = \beta = 0$. Thus, $T = 0$ is the only solution for T corresponding to $\lambda = 0$, and zero is not an eigenvalue. With $\lambda = k^2$ ($k > 0$), the general solution for T is

$$T(t) = \alpha \cos(kat) + \beta \sin(kat).$$

Again, the condition that $T(0) = 0$ forces us to choose $\alpha = 0$. Therefore, T is of the form

$$T_k(t) = \beta_k \sin(kat)$$

for each $k > 0$ and β_k constant. We can combine the information up to this point and write

$$y_k(x, t) = X_k(x)T_k(t) = [A_k \cos(kx) + B_k \sin(kx)] \sin(kat)$$

for every $k > 0$. For each $k > 0$ and any choices of the constants A_k and B_k, y_k is a bounded function which satisfies the wave equation and the condition that $y(x, 0) = 0$. All that remains is to satisfy the condition that $(\partial y / \partial t)(x, 0) = e^{-|x|}$.

For the wave equation on an interval $[0, L]$, we obtained a discrete set of eigenvalues, with a λ_n associated with each positive integer n. By contrast, this problem has eigenvalues $\lambda = k^2$, with $k > 0$. Every nonnegative real number is an eigenvalue.

This difference has quite an impact when we attempt a superposition to obtain a solution satisfying $(\partial y / \partial t)(x, 0) = e^{-|x|}$. In the finite interval case, we summed the solutions y_n for $n = 1, 2, 3, \ldots$. In the present case, however, we must "sum" over k for $k > 0$. This is done by replacing $\sum_{n=1}^{\infty}$ with $\int_0^{\infty} \cdots dk$ and letting

$$y(x, t) = \int_0^{\infty} y_k(x, t) \, dk = \int_0^{\infty} [A_k \cos(kx) + B_k \sin(kx)] \sin(kat) \, dk.$$

We must choose the A_k's and B_k's to satisfy $(\partial y / \partial t)(x, 0) = e^{-|x|}$. Interchange $\int_0^{\infty} \cdots dk$ and $\partial / \partial t$ to write

$$\frac{\partial y}{\partial t} = \int_0^{\infty} [A_k \cos(kx) + B_k \sin(kx)] ka \cos(kat) \, dk.$$

The condition $(\partial y/\partial t)(x, 0) = e^{-|x|}$ is

$$e^{-|x|} = \int_0^\infty [kaA_k\cos(kx) + kaB_k\sin(kx)] \, dk.$$

This is a Fourier integral expansion of $e^{-|x|}$ on $(-\infty, \infty)$. From Section 17.8, the coefficients are

$$kaA_k = \frac{1}{\pi} \int_{-\infty}^\infty e^{-|\xi|}\cos(k\xi) \, d\xi = \frac{2}{\pi} \frac{1}{1 + k^2}$$

and

$$kaB_k = \frac{1}{\pi} \int_{-\infty}^\infty e^{-|\xi|}\sin(k\xi) \, d\xi = 0.$$

[Since $e^{-|\xi|}\sin(k\xi)$ is an odd function on $(-\infty, \infty)$, we conclude that $B_k = 0$ without having to do the integration.] Thus, let

$$A_k = \frac{2}{\pi ka} \frac{1}{1 + k^2} \quad \text{and} \quad B_k = 0 \quad \text{for} \quad k > 0.$$

The solution of the boundary value problem is

$$y(x, t) = \frac{2}{\pi a} \int_0^\infty \frac{1}{k(1 + k^2)} \cos(kx)\sin(kat) \, dk.$$

THE WAVE EQUATION FOR A SEMI-INFINITE STRING

We will now consider wave motion of a semi-infinite string. Consider a problem like the above except that now the string extends from $[0, \infty)$ and is fastened at the left end, $x = 0$. The boundary value problem for the vertical displacement function takes the general form

$$\frac{\partial^2 y}{\partial t^2} = a^2 \frac{\partial^2 y}{\partial x^2} \qquad (0 < x < \infty, t > 0),$$

$$y(0, t) = 0 \qquad\qquad\qquad (t > 0),$$

$$y(x, 0) = f(x) \qquad\qquad (0 < x < \infty),$$

$$\frac{\partial y}{\partial t}(x, 0) = g(x) \qquad\quad (0 < x < \infty).$$

We have one boundary condition, reflecting the fact that the string is fixed at one end. We will also assume that $y(x, t)$ must be bounded for $0 \leq x < \infty, t > 0$.

To be specific, suppose that the string is released from rest $[g(x) = 0]$ and that it initially has the configuration of Figure 18.15, which is the graph of

$$f(x) = \begin{cases} x, & 0 \leq x \leq 1 \\ 1, & 1 \leq x \leq 4 \\ 5 - x, & 4 \leq x \leq 5 \\ 0, & x \geq 5. \end{cases}$$

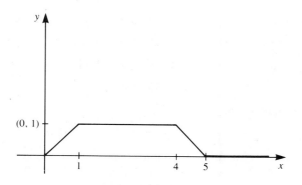

Figure 18.15. $f(x) = \begin{cases} x, & 0 \le x \le 1 \\ 1, & 1 \le x \le 4 \\ 5 - x, & 4 \le x \le 5 \\ 0, & x \ge 5. \end{cases}$

Let $y(x, t) = X(x)T(t)$ to obtain in the usual way

$$X'' + \lambda X = 0, \qquad T'' + \lambda a^2 T = 0.$$

The initial condition that $(\partial y/\partial t)(x, 0) = 0$ requires that $T'(0) = 0$. The boundary condition that $y(0, t) = 0$ requires that $X(0) = 0$. [In the case of a string on $(-\infty, \infty)$, this condition did not apply.] Now consider cases on λ.

Case 1: $\lambda = 0$ Then $X = cx + d$. For a bounded solution, we must have $c = 0$. Then $X(0) = d = 0$, and we have only the trivial solution corresponding to $\lambda = 0$. We conclude that zero is not an eigenvalue of this problem.

Case 2: $\lambda < 0$ Write $\lambda = -k^2$, with k positive. The general solution for X is

$$X(x) = ce^{kx} + de^{-kx}.$$

Since $X(0) = c + d = 0$, we have $d = -c$, and

$$X(x) = 2c \sinh(kx).$$

$X(x)$ is unbounded on $[0, \infty)$ unless we choose $c = 0$. This leads to the trivial solution; therefore, this problem has no negative eigenvalues.

Case 3: $\lambda > 0$ Write $\lambda = k^2$, with k positive. The general solution for X is

$$X(x) = c \cos(kx) + d \sin(kx).$$

Since $X(0) = c = 0$, we have $X(x) = d \sin(kx)$. This function is bounded on $[0, \infty)$ for any constant d. Therefore, every positive number $\lambda = k^2$ is an eigenvalue, with corresponding eigenfunction of the form $X_k(x) = d_k \sin(kx)$.

Now look at the equation $T'' + \lambda a^2 T = 0$. With $\lambda = k^2$, this is

$$T'' + k^2 a^2 T = 0,$$

with general solution

$$T(t) = \alpha \cos(kat) + \beta \sin(kat).$$

Since $T'(0) = ka\beta = 0$, and k and a are positive, $\beta = 0$. Thus, for $k > 0$, we must choose

$$T_k(t) = \alpha_k \cos(kat).$$

For each $k > 0$, we have a function

$$y_k(x, t) = X_k(x)T_k(t) = B_k \sin(kx)\cos(kat)$$

which satisfies the wave equation, the boundary condition $y(x, 0) = 0$, and the initial condition $(\partial y/\partial t)(x, 0) = 0$. To satisfy the initial condition $y(x, 0) = f(x)$, consider a superposition of these functions over all positive k:

$$y(x, t) = \int_0^\infty B_k \sin(kx)\cos(kat) \, dk.$$

We must choose the coefficients to satisfy

$$y(x, 0) = f(x) = \int_0^\infty B_k \sin(kx) \, dk.$$

This is a Fourier sine expansion of f on $[0, \infty)$, and we accordingly choose

$$\begin{aligned} B_k &= \frac{2}{\pi} \int_0^\infty f(\xi)\sin(k\xi) \, d\xi \\ &= \frac{2}{\pi}\left[\int_0^1 \xi \sin(k\xi) \, d\xi + \int_1^4 \sin(k\xi) \, d\xi + \int_4^5 (5 - \xi)\sin(k\xi) \, d\xi \right] \\ &= \frac{2}{\pi k^2} [\sin(k) + \sin(4k) - \sin(5k)]. \end{aligned}$$

The solution of the boundary value problem is

$$y(x, t) = \frac{2}{\pi} \int_0^\infty \frac{1}{k^2} [\sin(k) + \sin(4k) - \sin(5k)]\sin(kx)\cos(kat) \, dk.$$

HEAT CONDUCTION IN A SEMI-INFINITE BAR

Consider the problem of determining the temperature distribution in a bar extending from zero to infinity if the left end is kept at zero temperature and the initial temperature in the cross-section at x is $f(x)$. The mathematical model is

$$\frac{\partial u}{\partial t} = a^2 \frac{\partial^2 u}{\partial x^2} \qquad (x > 0, t > 0),$$

$$u(x, 0) = f(x) \qquad\qquad (x > 0),$$

$$u(0, t) = 0 \qquad\qquad (t > 0).$$

As usual, we will require that u be a bounded function of x and t. Let $u(x, t) = X(x)T(t)$ and obtain

$$X'' + \lambda X = 0 \quad \text{and} \quad T' + \lambda a^2 T = 0.$$

The information that $u(0, t) = 0$ implies that $X(0) = 0$. As we have done before, we find that

$$X_k(x) = A_k\sin(kx)$$

for every $k > 0$, with $\lambda = k^2$. The equation for T is

$$T' + k^2a^2T = 0,$$

with general solution

$$T_k(t) = \alpha_k e^{-k^2a^2t}.$$

For each $k > 0$, the function

$$u_k(x, t) = B_k\sin(kx)e^{-k^2a^2t}$$

satisfies the heat equation and the condition that $u(x, 0) = 0$. Further, this function is bounded for $x > 0$ and $t > 0$. In order to satisfy $u(x, 0) = f(x)$, we try a superposition

$$u(x, t) = \int_0^\infty B_k\sin(kx)e^{-k^2a^2t}\,dk.$$

We must choose the coefficients so that

$$u(x, 0) = f(x) = \int_0^\infty B_k\sin(kx)\,dk.$$

This is a Fourier sine integral expansion of f on $[0, \infty)$, leading us to choose

$$B_k = \frac{2}{\pi}\int_0^\infty f(\xi)\sin(k\xi)\,d\xi.$$

For example, suppose that

$$f(x) = \begin{cases} \pi - x, & 0 \le x \le \pi \\ 0, & x \ge \pi. \end{cases}$$

Then

$$B_k = \frac{2}{\pi}\int_0^\pi (\pi - \xi)\sin(k\xi)\,d\xi = \frac{2}{k}\left[1 - \frac{\sin(k\pi)}{k\pi}\right].$$

In this case, the temperature distribution, for $x > 0$ and $t > 0$, is

$$u(x, t) = \frac{2}{\pi}\int_0^\infty \frac{1}{k}\left[1 - \frac{\sin(k\pi)}{k\pi}\right]\sin(kx)e^{-k^2a^2t}\,dk.$$

STEADY-STATE TEMPERATURES IN AN INFINITE FLAT PLATE

Suppose we want to find the steady-state temperature distribution for a plate extending over the right quarter-plane $x \ge 0$, $y \ge 0$. Assume that the temperature on the vertical side $x = 0$ is kept at zero, while the bottom side $y = 0$ is kept at a constant 4 degrees

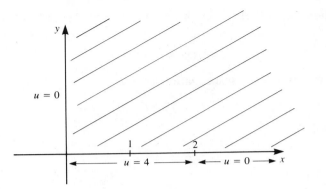

Figure 18.16

Celsius for $0 \leq x \leq 2$ and at zero for $x > 2$. (These boundary conditions are illustrated in Figure 18.16.) We want a bounded solution.

The boundary value problem is

$$\frac{\partial^2 u}{\partial x^2} + \frac{\partial^2 u}{\partial y^2} = 0 \qquad (x > 0, y > 0),$$

$$u(0, y) = 0 \qquad\qquad (y > 0),$$

$$u(x, 0) = f(x) = \begin{cases} 4, & 0 \leq x \leq 2 \\ 0, & x > 2. \end{cases}$$

Let $u(x, y) = X(x)Y(y)$ in the partial differential equation (Laplace's equation) to get

$$\frac{X''}{X} = -\frac{Y''}{Y} = -\lambda$$

for some constant λ. Then

$$X'' + \lambda X = 0 \quad \text{and} \quad Y'' - \lambda Y = 0.$$

Since $u(0, y) = X(0)Y(y) = 0$, $X(0) = 0$.

The problem for X is one we have solved before. We get eigenvalues $\lambda = k^2$, for any $k > 0$, and corresponding eigenfunctions $X_k(x) = a_k \sin(kx)$.

The problem for Y, with $\lambda = k^2$, is

$$Y'' - k^2 Y = 0,$$

with general solution

$$Y = ce^{ky} + de^{-ky}.$$

For $y > 0$, e^{ky} is unbounded; hence, we must choose $c = 0$. But e^{-ky} is bounded on $(0, \infty)$ so we have solutions

$$Y_k(y) = d_k e^{-ky}.$$

For each $k > 0$, then, we have

$$u_k(x, y) = A_k \sin(kx) e^{-ky}.$$

These functions are bounded for $x > 0$ and $y > 0$ and satisfy the steady-state heat equation. Further, they satisfy the condition $u(0, y) = 0$. To satisfy the boundary condition on the x-axis, let

$$u(x, y) = \int_0^\infty u_k(x, y)\, dk = \int_0^\infty A_k \sin(kx) e^{-ky}\, dy.$$

We now require that

$$u(x, 0) = f(x) = \int_0^\infty A_k \sin(kx)\, dk.$$

Choose A_k as the coefficient in the Fourier sine expansion of f on $[0, \infty)$:

$$A_k = \frac{2}{\pi} \int_0^\infty f(\xi) \sin(k\xi)\, d\xi = \frac{2}{\pi} \int_0^2 4 \sin(k\xi)\, d\xi$$

$$= \frac{8}{\pi}[1 - \cos(2k)].$$

The steady-state temperature distribution is

$$u(x, y) = \frac{8}{\pi} \int_0^\infty [1 - \cos(2k)] \sin(kx) e^{-ky}\, dk.$$

In the next section, we will apply the Laplace transform to the solution of boundary value problems.

PROBLEMS FOR SECTION 18.9

In each of Problems 1 through 6, solve the following boundary value problem for the given initial position and velocity functions:

$$\frac{\partial^2 y}{\partial t^2} = a^2 \frac{\partial^2 y}{\partial x^2} \qquad (0 < x < \infty, t > 0),$$

$$y(0, t) = 0 \qquad\qquad (t > 0),$$

$$y(x, 0) = f(x) \qquad\quad (0 < x < \infty),$$

$$\frac{\partial y}{\partial t}(x, 0) = g(x) \qquad (0 < x < \infty).$$

1. $f(x) = \begin{cases} x(1 - x), & 0 \le x \le 1 \\ 0, & x > 1 \end{cases}$ and $g(x) = 0$ **2.** $f(x) = e^{-|x|}$ and $g(x) = 0$

3. $f(x) = \begin{cases} \sin(x), & 0 \le x \le \pi \\ 0, & x > \pi \end{cases}$ and $g(x) = 0$ **4.** $f(x) = 0$, $g(x) = \begin{cases} 0, & 0 \le x \le 1 \\ 2, & 1 \le x \le 4 \\ 0, & x > 4 \end{cases}$

5. $f(x) = 0$ $g(x) = \begin{cases} x, & 0 \le x \le 1 \\ 0, & x > 1 \end{cases}$

6. $f(x) = \begin{cases} 3 \cos(x), & 0 \le x \le \pi/2 \\ 0, & x > \pi/2 \end{cases}$, $g(x) = \begin{cases} \sin(x), & 0 \le x \le \pi \\ 0, & x > \pi \end{cases}$

7. Determine and solve a boundary value problem modeling an infinite string stretched from $-\infty$ to ∞ and released from rest with initial displacement given by

$$f(x, 0) = \begin{cases} 1 + x, & -1 \le x \le 0 \\ 1 - x, & 0 \le x \le 1 \\ 0, & |x| > 1. \end{cases}$$

8. Set up and solve the boundary value problem describing displacements of an infinite string released from rest and given the initial displacement

$$f(x, 0) = \begin{cases} \sin(x), & -\pi \le x \le \pi \\ 0, & |x| > \pi. \end{cases}$$

9. Set up and solve the boundary value problem for an infinite string stretched along the x-axis and released from its stretched position with an initial velocity

$$g(x) = \begin{cases} 1, & -2 \le x \le 2 \\ 0, & |x| > 2. \end{cases}$$

10. Set up and solve the boundary value problem for an infinite string having initial position

$$f(x) = \begin{cases} \sin(2x), & |x| \le 2\pi \\ 0, & |x| > 2\pi \end{cases}$$

and initial velocity

$$g(x) = \begin{cases} e^{-x}, & -1 \le x \le 1 \\ 0, & |x| > 1. \end{cases}$$

11. Imagine a string stretched from zero to infinity and released from initial position $y(x, 0) = f(x)$ with initial velocity $(\partial y/\partial t)(x, 0) = g(x)$. Show how the solution for the displacement function can be obtained from the displacement function for an infinite string $(-\infty < x < \infty)$ having initial position $y(x, 0) = F(x)$ and initial velocity $(\partial y/\partial t)(x, 0) = G(x)$, where

$$F(x) + F(-x) = -\frac{1}{a} \int_{-x}^{x} G(\xi)\, d\xi.$$

In each of Problems 12 through 16, solve the boundary value problem.

12. $\dfrac{\partial u}{\partial t} = 4\dfrac{\partial^2 u}{\partial x^2}$ $(-\infty < x < \infty, t > 0)$,

$$u(x, 0) = \begin{cases} -2, & -1 \le x < 0 \\ x, & 0 \le x \le 1 \\ 0, & |x| > 1 \end{cases}$$

13. $\dfrac{\partial u}{\partial t} = 16\dfrac{\partial^2 u}{\partial x^2}$ $(0 < x < \infty, t > 0)$,

$u(0, t) = 0$ $(t > 0)$,
$u(x, 0) = e^{-x}\sin(x)$ $(x > 0)$

14. $\dfrac{\partial^2 u}{\partial x^2} + \dfrac{\partial^2 u}{\partial y^2} = 0$ $(-\infty < x < \infty, y > 0)$,

$$u(x, 0) = \begin{cases} -1, & -4 \le x < 0 \\ 0, & 0 \le x \le 4 \\ e^{-2|x|}, & |x| > 4 \end{cases}$$

15. $\dfrac{\partial^2 u}{\partial x^2} + \dfrac{\partial^2 u}{\partial y^2} = 0$ $(x > 0, y < 0)$,

$u(x, 0) = 0$ $(x > 0)$,

$$u(0, y) = \begin{cases} 0, & -5 \le y \le 0 \\ 2, & -7 \le y < -5 \\ 0, & y < -7 \end{cases}$$

16. $\dfrac{\partial^2 u}{\partial x^2} + \dfrac{\partial^2 u}{\partial y^2} = 0 \quad (x > 0,\ -\infty < y < \infty),$

$u(0, y) = \begin{cases} \cos(y), & -\pi \le y \le \pi \\ 0, & |y| > \pi \end{cases}$

17. Find the temperature distribution for a thin semi-infinite bar with temperature at $x = 0$ kept at zero and the initial temperature given by

$$f(x) = \begin{cases} x, & 0 \le x \le 4 \\ 0, & x > 4. \end{cases}$$

18. Solve for the temperature distribution in a thin bar extending from $-\infty$ to ∞ on the x-axis. The initial temperature is given by a function f.

19. Solve for the temperature distribution in a thin semi-infinite bar which is initially at zero temperature if, for $t > 0$, the left end is kept at constant temperature A.

20. Find the steady-state temperature distribution in a thin, flat plate extending over the set $x \ge 0$, $y \ge 0$ if the temperature on the side $x = 0$ is kept at e^{-y} and the temperature on the side $y = 0$ is kept at zero.

21. Solve for the steady-state temperatures in the infinite flat plate covering the half-plane $x \ge 0$ if the temperature on the boundary $x = 0$ is kept at $f(y)$, where

$$f(y) = \begin{cases} 1, & |y| \le 1 \\ 0, & |y| > 1. \end{cases}$$

22. Solve for the steady-state temperature function in the infinite flat plate covering the half-plane $y \ge 0$ if the temperature on the boundary $y = 0$ is kept at zero for $x < 4$, at constant A for $4 \le x \le 8$, and at zero for $x > 8$.

23. Find the steady-state temperature in the infinite flat plate covering the strip $0 \le y \le 1$, $x \ge 0$ if the temperature on the left boundary and bottom boundary is kept at zero and the temperature on the top side is given by $f(x)$.

24. Find the steady-state temperature in a flat plate covering the strip $-1 \le x \le 1$, $y \ge 0$ if the temperature on the bottom side is 2, that on the left side is e^{-y}, and that on the right side is 3. *Hint:* Separate this into three boundary value problems and add the solutions.

25. Find the temperature distribution for a semi-infinite bar if the left end $(x = 0)$ is insulated and the initial temperature is $f(x)$.

26. Find the steady-state temperature in a flat plate extending over the set $x \ge 0$, $0 \le y \le 2$ if the left and bottom sides are insulated and the top side has temperature $f(x)$.

27. Find the steady-state temperature distribution in a flat plate extending over the quarter-plane $x \ge 0$, $y \ge 0$ if the temperature on the vertical side is e^{-y} and the bottom side is insulated.

28. Find the steady-state temperature distribution in a flat strip extending over the set $x \ge 0$, $0 \le y \le 1$ if the temperature on the horizontal sides is kept at zero and that on the vertical side is 2.

18.10 *Laplace Transform Solution of Boundary Value Problems* _____

In this section, we will illustrate the use of the Laplace transform in the solution of boundary value problems. Consider a semi-infinite bar of metal with one end at $x = 0$ and extending to infinity along the positive x-axis. At time zero, the temperature in the bar is a constant A. From time zero to t_0, the left end is kept at constant temperature B. After time t_0, the left end is kept at temperature zero. We want the temperature distribution $u(x, t)$ for $x > 0$ and $t > 0$.

The boundary value problem for u is

$$\frac{\partial u}{\partial t} = a^2 \frac{\partial^2 u}{\partial x^2} \qquad (x > 0, t > 0),$$

$$u(x, 0) = A,$$

$$u(0, t) = \begin{cases} B, & 0 < t < t_0 \\ 0, & t > t_0. \end{cases}$$

The boundary condition at the left end of the bar involves a function of time which has a jump discontinuity at t_0. The Laplace transform is well suited to dealing with such functions. Further, t varies from zero to infinity, an appropriate range of values for the Laplace transform. This suggests that we attempt to apply the Laplace transform *with respect to* t, thinking of x as fixed in the process.

Apply the Laplace transform to the heat equation to get

$$\mathscr{L}\left\{\frac{\partial u}{\partial t}\right\} = a^2 \mathscr{L}\left\{\frac{\partial^2 u}{\partial x^2}\right\}.$$

By the operational formula for the Laplace transform of a derivative,

$$\mathscr{L}\left\{\frac{\partial u}{\partial t}\right\} = s\mathscr{L}\{u\} - u(x, 0).$$

Further,

$$\mathscr{L}\left\{\frac{\partial^2 u}{\partial x^2}\right\} = \int_0^\infty e^{-st} \frac{\partial^2 u}{\partial x^2}\, dt = \frac{\partial^2}{\partial x^2} \int_0^\infty e^{-st} u(x, t)\, dt = \frac{\partial^2}{\partial x^2}[\mathscr{L}\{u\}].$$

We interchanged $\partial^2/\partial x^2$ and $\int_0^\infty \cdots dt$ because the partial derivative is with respect to x, the integral is with respect to t, and x and t are independent. As usual, denote

$$U(x, s) = \mathscr{L}\{u(x, t)\},$$

remembering that the transform is with respect to t, leaving x fixed. The transformed heat equation therefore has the form

$$sU(x, s) - u(x, 0) = a^2 \frac{d^2 U}{dx^2}.$$

Since $u(x, 0) = A$, this equation may be written

$$\frac{d^2 U}{dx^2} - \frac{s}{a^2} U = -\frac{A}{a^2}.$$

This is a differential equation for U, with x as the variable and s thought of as a parameter. The general solution of this second order linear differential equation is

$$U(x, s) = \alpha(s)e^{\sqrt{s}x/a} + \beta(s)e^{-\sqrt{s}x/a} + \frac{A}{s}.$$

The "arbitrary constants" $\alpha(s)$ and $\beta(s)$ are functions of s, which is held fixed in solving this differential equation in x. We assume, as usual, that we want a bounded solution.

Since $e^{\sqrt{s}x/a}$ is unbounded for $x > 0$, we must choose $\alpha(s) = 0$. This leaves us with

$$U(x, s) = \beta(s)e^{-\sqrt{s}x/a} + \frac{A}{s}. \tag{18.34}$$

Now recall the condition that $u(0, t)$ be a given function with a jump discontinuity at t_0. In terms of the Heaviside unit step function,

$$u(0, t) = B[1 - H(t - t_0)],$$

where

$$H(t - t_0) = \begin{cases} 0 & \text{if} \quad t < t_0 \\ 1 & \text{if} \quad t \geq t_0. \end{cases}$$

Therefore,

$$\mathscr{L}\{u(0, t)\} = B\mathscr{L}\{1\} - B\mathscr{L}\{H(t - t_0)\}$$

$$= \frac{B}{s} - \frac{B}{s} e^{-t_0 s} = U(0, s). \tag{18.35}$$

But, from equation (18.34),

$$U(0, s) = \beta(s) + \frac{A}{s}.$$

From equations (18.34) and (18.35) for $U(0, s)$, we conclude that

$$\beta(s) + \frac{A}{s} = \frac{B}{s} - \frac{B}{s} e^{-t_0 s}.$$

Therefore,

$$\beta(s) = \frac{B - A}{s} - \frac{B}{s} e^{-t_0 s}.$$

Finally, we have

$$U(x, s) = \mathscr{L}\{u(x, t)\} = \left[\frac{B - A}{s} - \frac{B}{s} e^{-t_0 s}\right] e^{-\sqrt{s}x/a} + \frac{A}{s}.$$

This is the Laplace transform of the temperature distribution function u. The solution of the boundary value problem is

$$u(x, t) = \mathscr{L}^{-1}\{U(x, s)\}.$$

This inverse can be read from a table [for example, the *CRC Standard Mathematical Tables*, 25th ed. (Boca Raton, FL: CRC Press, 1981)]. The inverse involves two functions called the *error function*, erf(x), and the *complementary error function*, erfc(x), which arise, for example, in probability and statistics. They are defined by

$$\text{erf}(x) = \frac{2}{\sqrt{\pi}} \int_0^x e^{-\xi^2}\, d\xi$$

and

$$\text{erfc}(x) = 1 - \text{erf}(x) = \frac{2}{\sqrt{\pi}} \int_x^{\infty} e^{-\xi^2} \, d\xi.$$

In terms of these functions, we find that

$$u(x, t) = \begin{cases} A \, \text{erf}\left(\dfrac{x}{2a\sqrt{t}}\right) + B \, \text{erfc}\left(\dfrac{x}{2a\sqrt{t}}\right) & \text{for} \quad x > 0, 0 < t < t_0 \\[3mm] A \, \text{erf}\left(\dfrac{x}{2a\sqrt{t}}\right) + B \, \text{erfc}\left(\dfrac{x}{2a\sqrt{t}}\right) - B \, \text{erfc}\left(\dfrac{x}{2a\sqrt{t - t_0}}\right) & \text{for} \quad t > t_0. \end{cases}$$

Having solved the problem, it is worthwhile for us to review the process. The Laplace transform in t converted a partial differential equation into an ordinary differential equation in x containing s as a parameter. The solution of the ordinary differential equation involved "constants" which were functions of s. By using the Laplace transform of the boundary data and the requirement that the solution be bounded, we solved for these constants as functions of s. This yielded the Laplace transform of the solution of the boundary value problem. We then applied the inverse Laplace transform to find the solution.

PROBLEMS FOR SECTION 18.10

1. Use the Laplace transform to solve the boundary value problem

$$\frac{\partial^2 y}{\partial t^2} = a^2 \frac{\partial^2 y}{\partial x^2} \qquad (x > 0, t > 0),$$

$$y(0, t) = t \qquad\qquad (t > 0),$$

$$y(x, 0) = 0 \qquad\qquad (x > 0),$$

$$\frac{\partial y}{\partial t}(x, 0) = A \qquad\qquad (x > 0).$$

Assume that A is a positive constant. Also solve this problem by Fourier methods, and compare the solutions.

2. Use the Laplace transform to solve the boundary value problem

$$\frac{\partial^2 y}{\partial t^2} = a^2 \frac{\partial^2 y}{\partial x^2} \qquad (x > 0, t > 0)$$

$$y(0, t) = \begin{cases} \sin(2\pi t), & 0 \le t \le 1 \\ 0, & t \ge 1 \end{cases}$$

$$y(x, 0) = \frac{\partial y}{\partial t}(x, 0) = 0 \qquad (x > 0).$$

This problem models the displacement of a string initially at rest along the positive x-axis, with the left end moving in the prescribed fashion. Imagine, for example, a long jump rope, with a child at one end moving the rope in the given way. Look for a solution which decays to zero with increasing x for all time [that is, $y(x, t) \to 0$ as $x \to \infty$ for all t].

3. Use the Laplace transform to solve

$$\frac{\partial u}{\partial t} = a^2 \frac{\partial^2 u}{\partial x^2} \qquad (x > 0, t > 0),$$

$$u(0, t) = t^2 \qquad (t > 0),$$

$$u(x, 0) = 0 \qquad (x > 0).$$

4. Use the Laplace transform to solve

$$\frac{\partial u}{\partial t} = a^2 \frac{\partial^2 u}{\partial x^2} \qquad (x > 0, t > 0),$$

$$u(0, t) = 0 \qquad (t > 0),$$

$$u(x, 0) = e^{-x} \qquad (x > 0).$$

5. Use the Laplace transform to solve

$$a \frac{\partial u}{\partial x} + b \frac{\partial u}{\partial y} = y \qquad (x > 0, y > 0),$$

$$u(x, 0) = 0 \qquad (x > 0),$$

$$u(0, y) = 0 \qquad (y > 0).$$

6. Use the Laplace transform to solve

$$\frac{\partial u}{\partial x} + \frac{\partial u}{\partial y} = 3y^2 \qquad (x > 0, y > 0),$$

$$u(x, 0) = 0 \qquad (x > 0),$$

$$u(0, y) = -y \qquad (y > 0).$$

7. Use the Laplace transform to solve

$$\frac{\partial^2 y}{\partial t^2} = 4 \frac{\partial^2 y}{\partial x^2} \qquad (x > 0, y > 0),$$

$$y(x, 0) = 0 \qquad (x > 0),$$

$$y(0, t) = \sin(t) \qquad (t > 0),$$

$$\frac{\partial y}{\partial t}(x, 0) = e^{-x} \qquad (x > 0).$$

8. Use the Laplace transform to solve

$$\frac{\partial u}{\partial t} = 16 \frac{\partial^2 u}{\partial x^2} \qquad (0 < x < 1, t > 0),$$

$$u(0, t) = 0 \qquad (t > 0),$$

$$u(x, 0) = K \qquad (0 < x < 1),$$

$$u(1, t) = 0 \qquad (t > 0).$$

Assume that K is a positive constant.

9. Use the Laplace transform to solve the system

$$\frac{\partial u}{\partial t} + a \frac{\partial v}{\partial x} = 0$$

$$\frac{\partial v}{\partial t} + b \frac{\partial u}{\partial x} = 0 \qquad (-\infty < x < \infty, t > 0),$$

$$u(x, 0) = 1, \qquad v(x, 0) = x \qquad (-\infty < x < \infty),$$

$$v(0, t) = t \qquad\qquad\qquad (t > 0).$$

Assume that a and b are positive constants. Seek a solution having a bounded Laplace transform for $s > 0$.

10. Solve the electrical circuit problem

$$\frac{\partial i}{\partial t} + \frac{1}{L} \frac{\partial E}{\partial x} + \frac{R}{L} i = 0$$

$$\frac{\partial E}{\partial t} + \frac{1}{C} \frac{\partial i}{\partial x} + \frac{1}{RC} E = 0 \qquad (-\infty < x < \infty, t > 0),$$

$$E(x, 0) = f(x) \qquad (-\infty < x < \infty),$$

$$i(x, 0) = g(x) \qquad (-\infty < x < \infty).$$

18.11 *Fourier Transform Solution of Boundary Value Problems*

We will now explore the use of various Fourier transforms to solve boundary value problems. Here is an outline of the strategy.

1. Choose a transform appropriate for the range of variables in the problem. For example, if $-\infty < x < \infty$, a Fourier transform in x may be appropriate; if $y > 0$, a Fourier sine or cosine transform may apply. If $0 < x < \pi$, neither of these types of transforms can be applied to the x-variable, but we may be able to solve the problem using a finite Fourier sine or cosine transform in x. We may also consider a transform in one of the other variables.

   The choice of transform is also partly dictated by the form of the boundary conditions. The operational formula for the transform, used to transform derivative terms in the differential equation, will contain terms for which data must be supplied by, or obtained from, the initial and boundary conditions of the problem. We choose a transform so that the initial or boundary data of the problem can be used in the operational formula for that transform.

2. The transform used will often convert a partial differential equation into an ordinary differential equation (as we saw in the preceding section). The boundary conditions must be incorporated into this ordinary differential equation through the appropriate operational formulas.

3. This ordinary differential equation must be solved to obtain the transform of the solution of the boundary value problem.

4. The transformed solution must be inverted to solve the boundary value problem.

In problems having several independent variables, we may be able to use different transforms on different variables. For convenience in using Fourier transforms, below is a table summarizing definitions, operational formulas, and inversion formulas.

	Transform	Inversion Formula	Operational Formula
Finite Fourier cosine	$C_n\{f(x)\} = F_C(n)$ $= \int_0^\pi f(x)\cos(nx)\, dx$	$f(x) = \dfrac{1}{\pi} F_C(0)$ $+ \dfrac{2}{\pi} \sum_{n=1}^\infty F(n)\cos(nx)$	$C_n\{f''(x)\} = -n^2 F_C(n) - f'(0)$ $+ (-1)^n f'(\pi)$
Finite Fourier sine	$S_n\{f(x)\} = F_S(n)$ $= \int_0^\pi f(x)\sin(nx)\, dx$	$f(x) = \dfrac{2}{\pi} \sum_{n=1}^\infty F_S(n)\sin(nx)$	$S_n\{f''(x)\} = -n^2 F_S(n) + nf(0)$ $-n(-1)^n f(\pi)$
Fourier cosine	$\mathscr{F}_C\{f(x)\} = F_C(\omega)$ $= \int_0^\infty F(x)\cos(\omega x)\, dx$	$f(x) = \dfrac{2}{\pi} \int_0^\infty F_C(\omega)\cos(\omega x)\, d\omega$	$\mathscr{F}_C\{f''(x)\} = -\omega^2 F_C(\omega)$ $- f'(0)$
Fourier sine	$\mathscr{F}_S\{f(x)\} = F_S(\omega)$ $= \int_0^\infty f(x)\sin(\omega x)\, dx$	$f(x) = \dfrac{2}{\pi} \int_0^\infty F_S(\omega)\sin(\omega x)\, d\omega$	$\mathscr{F}_S\{f''(x)\} = -\omega^2 F_S(\omega)$ $+ \omega f(0)$
Fourier	$\mathscr{F}\{f(x)\} = F(\omega)$ $= \int_{-\infty}^\infty f(x)e^{-i\omega x}\, dx$	$f(x) = \dfrac{1}{2\pi} \int_{-\infty}^\infty F(\omega)e^{-i\omega x}\, d\omega$	$\mathscr{F}\{f''(x)\} = -\omega^2 F(\omega)$

EXAMPLE 18.5

Solve the boundary value problem

$$\frac{\partial u}{\partial t} = \frac{\partial^2 u}{\partial x^2} \qquad (-\infty < x < \infty, t > 0),$$

$$u(x, 0) = f(x) \qquad (-\infty < x < \infty).$$

This problem models heat conduction in an infinite slab with initial temperature function f. We can certainly solve this problem by separation of variables, using the Fourier integral. We will derive a solution using the Fourier transform. Since x ranges from $-\infty$ to ∞, a Fourier transform in x is a candidate for use. Denote $\mathscr{F}\{u(x, t)\} = U(\omega, t)$, taking the transform in the x-variable. This leaves t unchanged. From the operational formula for the Fourier transform, we obtain

$$\mathscr{F}\left\{\frac{\partial^2 u}{\partial x^2}\right\} = (i\omega)^2 U(\omega, t) = -\omega^2 U(\omega, t).$$

For the other side of the heat equation, compute

$$\mathscr{F}\left\{\frac{\partial u}{\partial t}\right\} = \int_{-\infty}^{\infty} \frac{\partial u}{\partial t} e^{-i\omega x} \, dx = \frac{\partial}{\partial t} \int_{-\infty}^{\infty} u(x, t)e^{-i\omega x} \, dx = \frac{d}{dt} U(\omega, t),$$

in which we treat ω as a parameter and t as the variable. We were able to interchange $\int_{-\infty}^{\infty} \cdots dx$ and $\partial/\partial t$ because x and t are independent. The Fourier transform of the heat equation therefore yields

$$\frac{d}{dt}[U(\omega, t)] + \omega^2 U(\omega, t) = 0.$$

This is a separable ordinary differential equation for U with general solution

$$U(\omega, t) = A(\omega)e^{-\omega^2 t}.$$

The "constant" of integration may be a function of ω, which was thought of as a constant in solving for $U(\omega, t)$.

To find $A(\omega)$, recall that $U(\omega, 0) = f(x)$ for $-\infty < x < \infty$. Therefore,

$$\mathscr{F}\{u(x, 0)\} = U(\omega, 0) = F\{\omega\} = A(\omega).$$

Then

$$U(\omega, t) = F(\omega)e^{-\omega^2 t}.$$

We now have the Fourier transform of the solution of the boundary value problem. To recover $u(x, t)$, we must invert the last equation. There are several ways to do this. One is to use the convolution theorem. Write

$$u(x, t) = \mathscr{F}^{-1}\{F(\omega)e^{-\omega^2 t}\} = \mathscr{F}^{-1}\{F(\omega)\} * \mathscr{F}^{-1}\{e^{-\omega^2 t}\}$$

$$= f(x) * \frac{1}{2\sqrt{\pi t}} e^{-x^2/4t}$$

$$= \frac{1}{2\sqrt{\pi t}} \int_{-\infty}^{\infty} f(x - \xi)e^{-\xi^2/4t} \, d\xi.$$

This is an integral formula for the solution of the initial value problem.

There is another way to invert $U(\omega, t)$ which employs a technique worth seeing. Use the inversion integral for the Fourier transform, substitute the definition of $F(\omega)$ into it, and obtain a complex quantity, which can be separated into its real and imaginary parts. The real part is the solution. Thus, write

$$u(x, t) = \frac{1}{2\pi} \int_{-\infty}^{\infty} U(\omega, t)e^{i\omega x} \, d\omega = \frac{1}{2\pi} \int_{-\infty}^{\infty} F(\omega)e^{-\omega^2 t}e^{i\omega x} \, d\omega$$

$$= \frac{1}{2\pi} \int_{-\infty}^{\infty} \left[\int_{-\infty}^{\infty} f(\xi)e^{-i\omega\xi} \, d\xi\right] e^{-\omega^2 t}e^{i\omega x} \, d\omega$$

$$= \frac{1}{2\pi} \int_{-\infty}^{\infty} \int_{-\infty}^{\infty} f(\xi)e^{-i\omega\xi}e^{-\omega^2 t}e^{i\omega x} \, d\xi \, d\omega$$

$$= \frac{1}{2\pi} \int_{-\infty}^{\infty} \int_{-\infty}^{\infty} f(\xi)e^{-i\omega(\xi - x)}e^{-\omega^2 t} \, d\xi \, d\omega.$$

To obtain the real part of this expression, use Euler's formula

$$e^{-i\omega(\xi - x)} = \cos[\omega(\xi - x)] - i\sin[\omega(\xi - x)].$$

Upon substituting this into the integral, we obtain

$$u(x, t) = \frac{1}{2\pi} \int_{-\infty}^{\infty} \int_{-\infty}^{\infty} f(\xi)\cos[\omega(\xi - x)]e^{-\omega^2 t}\, d\xi\, d\omega$$

$$- \frac{i}{2\pi} \int_{-\infty}^{\infty} \int_{-\infty}^{\infty} f(\xi)\sin[\omega(\xi - x)]e^{-\omega^2 t}\, d\xi\, d\omega.$$

Since $u(x, t)$ is real valued, the coefficient of i must vanish in the last equation, and we obtain

$$u(x, t) = \frac{1}{2\pi} \int_{-\infty}^{\infty} \int_{-\infty}^{\infty} f(\xi)\cos[\omega(\xi - x)]e^{-\omega^2 t}\, d\xi\, d\omega.$$

This is an integral expression for the solution (although different in form from the one derived previously using the convolution theorem). ∎

EXAMPLE 18.6

Solve the boundary value problem

$$\frac{\partial u}{\partial t} = \frac{\partial^2 u}{\partial x^2} \qquad (0 < x < \infty, t > 0),$$

$$u(0, t) = 0 \qquad\qquad (t > 0),$$

$$u(x, 0) = f(x) \qquad\quad (0 < x < \infty).$$

One difference between this problem and Example 18.5 is that here, x varies from zero to infinity instead of over the entire real line. This problem models the temperature distribution in a semi-infinite bar with left end at $x = 0$. We have the boundary condition that the temperature is zero at the left end.

Because both x and t vary from zero to infinity, a Fourier transform is inappropriate. We can try a Fourier sine or cosine transform in either x or t. Since the operational formula for either of these transforms contains the second derivative, we are led to take the transform in x, not t (the heat equation contains a second derivative $\partial^2 u/\partial x^2$ with respect to x but only a first derivative $\partial u/\partial t$ with respect to t). Further, for this problem, the boundary data are given in terms of the function value at $x = 0$, not its derivative, suggesting that we use the sine transform, not the cosine transform.

Having decided upon strategy, write

$$\mathscr{F}_S\left\{\frac{\partial u}{\partial t}\right\} = \mathscr{F}_S\left\{\frac{\partial^2 u}{\partial x^2}\right\}.$$

Let $\mathscr{F}_S\{u(x, t)\} = U_S(\omega, t)$. Since $\partial/\partial t$ and $\mathscr{F}_S$ are with respect to independent variables, these operations can be interchanged, and we have

$$\mathscr{F}_S\left\{\frac{\partial u}{\partial t}\right\} = \frac{d}{dt}[U(\omega, t)].$$

Using the operational formula for $\mathscr{F}_S$, we have

$$\mathscr{F}_S\left\{\frac{\partial^2 u}{\partial x^2}\right\} = -\omega^2 U_S(\omega, t) + \omega u(0, t) = -\omega^2 U_S(\omega, t).$$

Then

$$\frac{d}{dt}[U_S(\omega, t)] = -\omega^2 U_S(\omega, t).$$

This first order equation has general solution

$$U_S(\omega, t) = A(\omega)e^{-\omega^2 t},$$

in which the "constant" may be a function of ω. Since $u(x, 0) = f(x)$, we have $U_S(\omega, 0) = A(\omega) = F_S(\omega)$, and

$$U_S(\omega, t) = F_S(\omega)e^{-\omega^2 t}.$$

By the inversion formula for the Fourier sine transform, the solution of the boundary value problem is

$$u(x, t) = \frac{2}{\pi}\int_0^\infty F_S(\omega)e^{-\omega^2 t}\sin(\omega x)\, d\omega.$$

Upon inserting the integral for $F_S(\omega)$ into this expression, we obtain the solution

$$u(x, t) = \frac{2}{\pi}\int_0^\infty\int_0^\infty f(\xi)\sin(\omega\xi)e^{-\omega^2 t}\sin(\omega x)\, d\xi\, d\omega. \quad\blacksquare$$

EXAMPLE 18.7

Consider Poisson's equation

$$\frac{\partial^2 u}{\partial x^2} + \frac{\partial^2 u}{\partial y^2} = -h \qquad (0 < x < \pi, y > 0),$$

in which h is a constant. The strip $0 < x < \pi$, $y > 0$ is shown in Figure 18.17. We may think of $u(x, y)$ as the electrostatic potential in a space bounded by the planes $x = 0$, $x = \pi$, and $y = 0$, with a uniform distribution of charge having density $h/4\pi$.

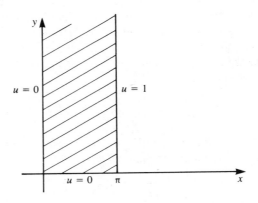

Figure 18.17

We will solve Poisson's equation subject to the conditions

$$u(0, y) = 0 \qquad (y > 0),$$
$$u(x, 0) = 0 \qquad (0 < x < \pi),$$
$$u(\pi, y) = 1 \qquad (y > 0).$$

Since $0 < x < \pi$, we can try a finite Fourier sine or cosine transform in the x-variable. Or, since $y > 0$, we can try a Fourier sine or cosine transform in t. Which should we choose? Notice that the operational formulas for the Fourier cosine and finite Fourier cosine transforms involve information about the derivative of a data function at zero. We have no such data here, so we discard these two possibilities.

If we use a Fourier sine transform, it must be in the y-variable, since $y > 0$. Compute

$$\mathscr{F}_S\left\{\frac{\partial^2 u}{\partial x^2}\right\} + \mathscr{F}_S\left\{\frac{\partial^2 u}{\partial y^2}\right\} = \frac{d^2}{dx^2}[U_S(x, \omega)] - \omega^2 U_S(x, \omega) + \omega u(x, 0) = 0.$$

Since $u(x, 0) = 0$, this gives us

$$\frac{d^2}{dx^2}[U_S(x, \omega)] - \omega^2 U_S(x, \omega) = 0,$$

with general solution

$$U_S(x, \omega) = A(\omega)e^{\omega x} + B(\omega)e^{-\omega x}.$$

Now we must use the other boundary data to solve for $A(\omega)$ and $B(\omega)$. Since $u(0, y) = 0$, $U_S(0, \omega) = A(\omega) + B(\omega) = 0$. Then $B(\omega) = -A(\omega)$, and

$$U_S(x, \omega) = 2A(\omega)\sinh(\omega x).$$

Finally, $u(\pi, y) = 1$, so $U_S(\pi, \omega) = \mathscr{F}_S\{1\} = \int_0^\infty \sin(\omega\xi)\, d\xi$, a divergent improper integral (because 1 does not have a Fourier sine transform). We have reached an impasse and cannot obtain a solution using the Fourier sine transform in y.

Try a finite Fourier sine transform in x. Write $S_n\{u(x, y)\} = U_S(n, y)$. The partial differential equation transforms to

$$-n^2 U_S(n, y) - n(-1)^n u(\pi, y) + nu(0, y) + \frac{d^2}{dy^2}[U_S(n, y)] = -hS_n\{1\}.$$

(Again, we have used the fact that the transform with respect to one variable and the derivative with respect to the other variable commute.) Since $u(\pi, y) = 1$ and $u(0, y) = 0$, the differential equation for $U_S(n, y)$ becomes

$$\frac{d^2}{dy^2}[U_S] - n^2 U_S = n(-1)^n - hS_n\{1\} = n(-1)^n - h\left[\frac{1 - (-1)^n}{n}\right].$$

This equation is easy to solve because the right side is constant as far as y is concerned. The general solution is

$$U_S(n, y) = \alpha(n)e^{ny} + \beta(n)e^{-ny} + \frac{(-1)^{n+1}}{n} + \frac{h}{n^3}[1 - (-1)^n].$$

In order for $U_S(n, y)$ to remain bounded for $y > 0$, we must choose $\alpha(n) = 0$. Thus,

$$U_S(n, y) = \beta(n)e^{-ny} + \frac{(-1)^{n+1}}{n} + \frac{h}{n^3}[1 - (-1)^n].$$

Now transform the boundary condition $u(x, 0) = 0$ to get

$$U_S(n, 0) = S_n\{0\} = \beta(n) + \frac{(-1)^{n+1}}{n} + \frac{h}{n^3}[1 - (-1)^n] = 0.$$

Therefore,

$$\beta(n) = -\left\{\frac{(-1)^{n+1}}{n} + \frac{h}{n^3}[1 - (-1)^n]\right\}.$$

The unique bounded solution of the boundary value problem has finite Fourier sine transform

$$U_S(n, y) = \left\{\frac{(-1)^{n+1}}{n} + \frac{h}{n^3}[1 - (-1)^n]\right\}(1 - e^{-ny}).$$

By the inversion formula, the solution of the boundary value problem is

$$u(x, y) = \frac{2}{\pi}\sum_{n=1}^{\infty} U_S(n, y)\sin(n\pi x)$$

$$= \sum_{n=1}^{\infty}\left\{\frac{(-1)^{n+1}}{n} + \frac{h}{n^3}[1 - (-1)^n]\right\}(1 - e^{-ny})\sin(n\pi x).$$

It is perhaps surprising that we can obtain a closed-form expression for $u(x, y)$. Using results from tables and a theorem we have not discussed, it can be shown that

$$u(x, y) = \frac{1}{2}hx(\pi - x) + \frac{1}{\pi}x - \frac{2}{\pi}\tan^{-1}\left[\frac{\sin(x)}{e^y + \cos(x)}\right]$$

$$- \frac{2}{\pi^2}hx\int_0^\pi (\pi - \xi)^2\tan^{-1}\left[\frac{\sin(\xi)}{\sinh(y)}\right]d\xi$$

$$+ \frac{2}{\pi}h\int_0^x (x - \xi)\tan^{-1}\left[\frac{\sin(\xi)}{\sinh(y)}\right]d\xi. \quad\blacksquare$$

EXAMPLE 18.8

Solve the boundary value problem

$$\frac{\partial^2 u}{\partial t^2} = 9\frac{\partial^2 u}{\partial x^2} \qquad (-\infty < x < \infty, t > 0),$$

$$u(x, 0) = 4e^{-5|x|} \qquad (-\infty < x < \infty),$$

$$\frac{\partial u}{\partial t}(x, 0) = 0 \qquad (-\infty < x < \infty).$$

Since x varies over the entire real line, we can try a Fourier transform in x. Denote $\mathscr{F}\{u(x, t)\} = U(\omega, t)$. From the partial differential equation, we obtain

$$\frac{d^2 U}{dt^2} = 9\mathscr{F}\left\{\frac{\partial^2 u}{\partial x^2}\right\} = 9(i\omega)^2 U(\omega, t) = -9U(\omega, t).$$

Then

$$\frac{d^2 U}{dt^2} + 9U = 0.$$

The general solution for U is

$$U(\omega, t) = A(\omega)\cos(3\omega t) + B(\omega)\sin(3\omega t).$$

Now use the initial conditions. Since $u(x, 0) = 4e^{-5|x|}$,

$$\mathcal{F}\{u(x, 0)\} = U(\omega, 0) = A(\omega) = \mathcal{F}\{4e^{-5|x|}\} = \frac{40}{25 + \omega^2}.$$

Next, we know that $(\partial u/\partial t)(x, 0) = 0$, so

$$\mathcal{F}\left\{\frac{\partial u}{\partial t}(x, 0)\right\} = \frac{dU}{dt}(\omega, 0) = \mathcal{F}\{0\} = 0.$$

But

$$\frac{dU}{dt}(\omega, 0) = -3\omega A(\omega)\sin(3\omega t) + 3\omega B(\omega)\cos(3\omega t)\bigg|_{t=0} = 3\omega B(\omega).$$

Therefore,

$$3\omega B(\omega) = 0,$$

from which we conclude that $B(\omega) = 0$. Then

$$U(\omega, t) = A(\omega)\cos(3\omega t) = \frac{40}{25 + \omega^2}\cos(3\omega t).$$

The solution of the boundary value problem is

$$u(x, t) = \mathcal{F}^{-1}\left\{\frac{40}{25 + \omega^2}\cos(3\omega t)\right\}.$$

There are several ways to determine this inverse Fourier transform. We will use the convolution theorem in conjunction with the symmetry property applied to the cosine term and then evaluate the convolution integral through the filtering property of the Dirac delta function. To begin, write

$$u(x, t) = \mathcal{F}^{-1}\left\{\frac{40}{25 + \omega^2}\right\} * \mathcal{F}^{-1}\{\cos(3\omega t)\}. \tag{18.36}$$

We know that the inverse Fourier transform of $40/(25 + \omega^2)$ is $4e^{-5|x|}$, from an initial condition of the problem. To determine $\mathcal{F}^{-1}\{\cos(3\omega t)\}$, first recall the formula

$$\mathcal{F}\{\cos(\omega_0 t)\} = \pi[\delta(\omega + \omega_0) + \delta(\omega - \omega_0)].$$

Apply the symmetry property of the Fourier transform to this equation. [The symmetry property states that $\mathcal{F}\{F(t)\} = 2\pi f(-\omega)$ if $\mathcal{F}\{f(t)\} = F(\omega)$.] We get

$$\mathcal{F}^{-1}\{\cos(\omega_0 t)\} = \tfrac{1}{2}[\delta(t + t_0) + \delta(t - t_0)].$$

If we replace t with x and let $t_0 = 3t$ in this equation, we have

$$\mathcal{F}^{-1}\{\cos(3\omega x)\} = \tfrac{1}{2}[\delta(x + 3t) + \delta(x - 3t)].$$

Using the definition of convolution and the information we have just found, we have from equation (18.36) that

$$u(x, t) = 2 \int_{-\infty}^{\infty} e^{-5|x - \xi|} [\delta(\xi + 3t) + \delta(\xi - 3t)] \, d\xi$$

$$= 2 \int_{-\infty}^{\infty} e^{-5|x - \xi|} \delta(\xi + 3t) \, d\xi + 2 \int_{-\infty}^{\infty} e^{-5|x - \xi|} \delta(\xi - 3t) \, d\xi.$$

Finally, use the filtering property of the delta function (Theorem 4.17) to get

$$u(x, t) = 2e^{-5|x + 3t|} + 2e^{-5|x - 3t|}.$$

It is routine to check by substitution that this is the solution of the boundary value problem. ∎

PROBLEMS FOR SECTION 18.11

In the following problems, $\nabla^2 u = 0$ denotes Laplace's equation

$$\frac{\partial^2 u}{\partial x^2} + \frac{\partial^2 u}{\partial y^2} = 0.$$

1. Solve

$$\nabla^2 u = 0 \qquad (-\infty < x < \infty, y > 0),$$
$$u(x, 0) = f(x) \qquad (-\infty < x < \infty)$$

by using the Fourier transform in x. Compare the solution with that obtained previously using separation of variables.

2. Attempt to solve the boundary value problem in Problem 1 by using the Laplace transform in y.

3. Use an appropriate Fourier transform to solve

$$\nabla^2 u = 0 \qquad (0 < x < \pi, y > 0),$$
$$u(0, y) = u(\pi, y) = 0 \qquad (y > 0),$$
$$u(x, 0) = B \sin(x) \qquad (0 < x < \pi).$$

B is a constant.

4. Use an appropriate Fourier transform to solve

$$\frac{\partial^2 y}{\partial t^2} = a^2 \frac{\partial^2 y}{\partial x^2} \qquad (0 < x < \pi, t > 0),$$

$$y(x, 0) = \frac{\partial y}{\partial t}(x, 0) = 0 \qquad (0 < x < \pi),$$

$$y(0, t) = A \sin(\omega t) \qquad (t > 0),$$
$$y(\pi, t) = 0 \qquad (t > 0).$$

Hint: There are two cases, depending on whether or not ω is an integer multiple of π.

5. Use a Fourier transform to solve

$$\nabla^2 u = 0 \qquad (0 < x < \pi, y > 0),$$
$$u(0, y) = 0 \qquad (y > 0),$$
$$u(\pi, y) = 2 \qquad (y > 0),$$
$$u(x, 0) = -4 \qquad (0 < x < \pi).$$

First, try a finite Fourier sine transform in x, then try a Fourier sine or cosine transform in y.

6. Use a Fourier transform to solve

$$\frac{\partial u}{\partial t} = \nabla^2 u \qquad (0 < x < \pi, \, y > 0, \, t > 0),$$

$$u(x, y, 0) = 0 \qquad\qquad (0 < x < \pi, \, y > 0),$$

$$\frac{\partial u}{\partial x}(0, y, t) = -5 \qquad\qquad (y > 0, \, t > 0),$$

$$\frac{\partial u}{\partial x}(\pi, y, t) = 0 \qquad\qquad (y > 0, \, t > 0),$$

$$u(x, 0, t) = 0 \qquad\qquad (0 < x < \pi, \, t > 0).$$

First, use a finite Fourier cosine transform in x. To solve the resulting ordinary differential equation, use a Laplace transform in t.

7. Use a Fourier transform to solve

$$\nabla^2 u = 0 \qquad\qquad (0 < x < \pi, \, 0 < y < 2),$$

$$u(0, y) = 0 \qquad\qquad (0 < y < 2),$$

$$u(\pi, y) = 4 \qquad\qquad (0 < y < 2),$$

$$\frac{\partial u}{\partial y}(x, 0) = u(x, 2) = 0 \qquad\qquad (0 < x < 2).$$

8. Use a Fourier transform to solve

$$\frac{\partial u}{\partial t} = 9 \frac{\partial^2 u}{\partial x^2} \qquad (x > 0, \, t > 0),$$

$$u(x, 0) = 0 \qquad\qquad (x > 0),$$

$$u(0, t) = f(t) \qquad\qquad (t > 0).$$

9. Use a Fourier transform to solve

$$\nabla^2 u = 0 \qquad\qquad (x > 0, \, 0 < y < 1),$$

$$u(0, y) = y^2(1 - y) \qquad\qquad (0 < y < 1),$$

$$u(x, 0) = u(x, 1) = 0 \qquad\qquad (x > 0).$$

10. Use a Fourier transform to solve

$$\frac{\partial u}{\partial t} = \frac{\partial^2 u}{\partial x^2} - u \qquad (x > 0, \, t > 0),$$

$$\frac{\partial u}{\partial x}(0, t) = f(t) \qquad\qquad (t > 0),$$

$$u(x, 0) = 0 \qquad\qquad (x > 0).$$

11. Use a Fourier transform to solve

$$\frac{\partial u}{\partial t} - \frac{\partial^2 u}{\partial x^2} + tu = 0 \qquad (x > 0, \, t > 0),$$

$$u(x, 0) = xe^{-x} \qquad\qquad (x > 0),$$

$$\frac{\partial u}{\partial t}(0, t) = 0 \qquad\qquad (t > 0).$$

12. Use a Fourier transform to solve

$$\nabla^2 u = 0 \qquad (-\infty < x < \infty, 0 < y < 1),$$

$$u(x, 0) = \begin{cases} 0, & x < 0 \\ e^{-ax}, & x > 0 \end{cases}$$

$$u(x, 1) = 0 \qquad (-\infty < x < \infty).$$

Here, a is a positive constant.

13. Use a Fourier transform to solve

$$\nabla^2 u = 0 \qquad (-\infty < x < \infty, 0 < y < 1),$$

$$\frac{\partial u}{\partial t}(x, 0) = 0 \qquad (-\infty < x < \infty),$$

$$u(x, 1) = e^{-x^2} \qquad (-\infty < x < \infty).$$

14. Use the method of Example 18.8 to derive d'Alembert's solution of the wave equation (see Problem 21, Section 18.0).

18.12 Existence, Uniqueness, Classification, and Well-Posed Problems

We will conclude this chapter with some remarks about theoretical aspects of partial differential equations. Using subscripts to denote partial derivatives, the second order linear partial differential equation has the form

$$au_{xx} + 2bu_{xy} + cu_{yy} + du_x + eu_y + fu + g = 0,$$

in which $a, b, \ldots, g$ are functions of x and y. The quantity

$$\Delta(x, y) = b^2(x, y) - a(x, y)c(x, y)$$

is called the *discriminant* of this partial differential equation. The equation is classified as being

hyperbolic at (x_0, y_0) if $\Delta(x_0, y_0) > 0$;
elliptic at (x_0, y_0) if $\Delta(x_0, y_0) < 0$;
parabolic at (x_0, y_0) if $\Delta(x_0, y_0) = 0$.

For example,

The wave equation $u_{xx} - u_{yy} = 0$ (with y in place of t) has positive discriminant at every point and hence is hyperbolic everywhere.

The heat equation $u_x - u_{yy} = 0$ has zero discriminant at every point and hence is parabolic everywhere.

Laplace's equation $u_{xx} + u_{yy} = 0$ has negative discriminant at every point and hence is elliptic everywhere.

The partial differential equation $yu_{xx} + u_{yy} = 0$ is called *Tricomi's equation* and has discriminant $-y$ at (x, y). This equation is therefore hyperbolic in the half-plane $y < 0$, elliptic in the half-plane $y > 0$, and parabolic on the x-axis, where $y = 0$.

Classification has proved important for an understanding of the kinds of initial and boundary data that must be furnished with the partial differential equation in order to determine a unique bounded solution. For example, the data given with the hyperbolic wave equation (boundary conditions at the ends of the interval, initial position, initial velocity) are of a different kind from those given with the parabolic heat equation (boundary conditions at the ends of the interval but just one initial condition, the initial temperature distribution). The type of equation also influences the choice of technique to use when attempting a numerical approximation of a solution. Methods suited to parabolic equations are not necessarily well suited to hyperbolic or elliptic ones.

Given a problem involving a partial differential equation and boundary and/or initial conditions, we usually ask three questions.

1. Does a solution exist?

2. Is the solution unique?

3. Does the solution depend continuously on the data?

If the answer to all three questions is affirmative, the problem is said to be *well posed*. Otherwise, the problem is *ill posed*.

While the meanings of (1) and (2) are clear, (3) requires some explanation. A solution is said to *depend continuously on the data* of the problem if small changes in the data induce correspondingly small changes in the solution. For example, consider the problem

$$u_{xx} + u_{yy} = 0 \quad \text{in a rectangle } R,$$
$$u(x, y) = f(x, y) \quad \text{on the boundary of } R.$$

For this problem, the data consist of the function f specified on the boundary of R. Suppose that this problem has solution F.

Now suppose that the problem

$$u_{xx} + u_{yy} = 0 \quad \text{in } R,$$
$$u(x, y) = g(x, y) \quad \text{on the boundary of } R$$

has solution G. We say that this problem *depends continuously on the data* if, given $\epsilon > 0$, there is some $\delta > 0$ such that

$$|F(x, y) - G(x, y)| < \epsilon \quad \text{for} \quad (x, y) \text{ in } R$$

if

$$|f(x, y) - g(x, y)| < \delta \quad \text{for} \quad (x, y) \quad \text{on the boundary of } R.$$

That is, we can make the solutions as close to one another as we like by choosing the data functions close enough. Similar definitions of continuous dependence on data can be written for other kinds of boundary value problems we have discussed.

Continuous dependence on the data is something we expect to occur, at least when the problem models a physical phenomenon. However, an example due to Jacques Hadamard shows that it need not hold even in a simple problem in which the coefficients are analytic. This example is given in Problem 3 at the end of this section.

Concerning existence, theorems have been proved for various classes of problems. As an example, we will state one of the first existence theorems, which deals with a first order problem called the *Cauchy problem*. The Cauchy problem consists of solving the partial differential equation

$$u_t(x_1, \ldots, x_n, t) = F(x_1, \ldots, x_n, t, u, u_{x_1}, \ldots, u_{x_n})$$

subject to the condition

$$u(x_1, \ldots, x_n, 0) = f(x_1, \ldots, x_n),$$

with f a given function.

THEOREM 18.1   Cauchy-Kovalevski

Suppose that F is analytic in a $(2n + 2)$-dimensional sphere about the point

$$(0, \ldots, 0, f(0, \ldots, 0), f_{x_1}(0, \ldots, 0), \ldots, f_{x_n}(0, \ldots, 0))$$

and that f is analytic in an n-dimensional sphere about $(0, \ldots, 0)$. Then the Cauchy problem has a unique solution in some $(n + 1)$-dimensional sphere about the origin. ∎

In many cases, proving uniqueness is easier than proving existence. One method is to assume that u_1 and u_2 are solutions of the problem, then show that $u_1 - u_2$ must be identically zero and hence that $u_1 = u_2$. Here is an example of this type of reasoning with a problem involving the heat equation. Consider the very general boundary value problem

$$\frac{\partial u}{\partial t} = a^2 \left(\frac{\partial^2 u}{\partial x^2} + \frac{\partial^2 u}{\partial y^2} + \frac{\partial^2 u}{\partial z^2} \right) + \varphi(x, y, z, t) \quad \text{for} \quad (x, y, z) \quad \text{in } M \text{ and } t > 0,$$

$$u(x, y, z, 0) = f(x, y, z) \quad \text{for} \quad (x, y, z) \quad \text{in } M,$$

$$u(x, y, z, t) = g(x, y, z, t) \quad \text{for} \quad t > 0 \quad \text{and} \quad (x, y, z) \quad \text{on the boundary of } M,$$

in which M is some set in three-space bounded by a piecewise-smooth surface Σ, and f, g, and φ are continuous.

We may think of $u(x, y, z, t)$ as the temperature at time t and point (x, y, z) in M, with initial temperature given by f and the temperature on the boundary given for all positive t by g. The function φ in the partial differential equation allows for the generation of heat energy within M.

Assume that this problem has a solution u which is continuous, with continuous first and second partial derivatives with respect to x, y, and z and with a continuous partial derivative with respect to t for $t > 0$ and for (x, y, z) in M and on its boundary. We will show that there is only one solution having these properties.

Suppose that u_1 and u_2 are two such solutions. Let $w = u_1 - u_2$. Then w, its first and second partial derivatives with respect to x, y, and z, and $\partial w / \partial t$ are continuous for $t > 0$ and for (x, y, z) in M and on its boundary. Further, it is routine to check by substitution that w is a solution of the problem

$$\frac{\partial w}{\partial t} = a^2 \nabla^2 w \qquad\qquad (x, y, z) \text{ in } M \text{ and } t > 0,$$

$$w(x, y, z, 0) = 0 \qquad\qquad (x, y, z) \text{ in } M,$$

$$w(x, y, z, t) = 0 \qquad (x, y, z) \text{ on the boundary surface } \Sigma \text{ of } M.$$

Define

$$I(t) = \tfrac{1}{2} \iiint\limits_M w^2 \, dx \, dy \, dz.$$

Then

$$I'(t) = \iiint_M w \frac{\partial w}{\partial t} \, dx \, dy \, dz = a^2 \iiint_M w \, \nabla^2 w \, dx \, dy \, dz.$$

Now apply Gauss's divergence theorem to the last integral to get

$$I'(t) = a^2 \iint_\Sigma w \frac{\partial w}{\partial \eta} \, d\sigma - a^2 \iiint_M (w_x^2 + w_y^2 + w_z^2) \, dx \, dy \, dz,$$

where $\partial w / \partial \eta$ is the directional derivative of w in the direction of the outer normal to Σ. This surface integral is zero because w is identically zero at points of Σ. Therefore,

$$I'(t) = -a^2 \iiint_M (w_x^2 + w_y^2 + w_z^2) \, dx \, dy \, dz$$

for $t > 0$. We conclude that

$$I'(t) \le 0 \quad \text{for} \quad t > 0.$$

Now apply the mean value theorem to I on $[0, t]$ for any positive t. There exists some ξ_t in $(0, t)$ such that

$$I'(\xi_t) = \frac{I(t) - I(0)}{t}.$$

Now,

$$I(0) = \tfrac{1}{2} \iiint_M w(x, y, z, 0)^2 \, dx \, dy \, dz = 0$$

because $w(x, y, z, 0) = 0$ for all (x, y, z) in M. Therefore,

$$I(t) = t I'(\xi_t) \le 0.$$

Since $I(t) \ge 0$ and also $I(t) \le 0$ for all $t > 0$, we conclude that $I(t) = 0$ for all $t > 0$. But then $w(x, y, z)^2 = 0$ throughout M because w is continuous and nonnegative. Then $w(x, y, z, t) = 0$ for all (x, y, z) in M and all $t > 0$, and therefore $u_1 = u_2$. We will pursue some of these ideas further in the problems.

PROBLEMS FOR SECTION 18.12

1. Devise a suitable definition of continuous dependence on data for the vibrating string problem

$$\frac{\partial^2 y}{\partial t^2} = a^2 \frac{\partial^2 y}{\partial x^2} \qquad (0 < x < L, t > 0),$$

$$y(0, t) = y(L, t) = 0 \qquad (t > 0),$$

$$y(x, 0) = f(x) \qquad (0 < x < L),$$

$$\frac{\partial y}{\partial t}(x, 0) = g(x) \qquad (0 < x < L).$$

Prove that this problem depends continuously on the data if f and g are continuous on $[0, L]$. *Hint:* Use d'Alembert's solution of the wave equation (Problem 21, Section 18.0).

2. Write a suitable definition of continuous dependence on data for the problem

$$\frac{\partial u}{\partial t} = a^2 \frac{\partial^2 u}{\partial x^2} \qquad (0 < x < L, t > 0),$$

$$u(0, t) = u(L, t) = 0 \qquad (t > 0),$$

$$u(x, 0) = f(x) \qquad (0 < x < L).$$

3. This problem deals with Hadamard's example of a problem which does not continuously depend on the data. Consider the boundary value problem

$$\frac{\partial^2 u}{\partial x^2} + \frac{\partial^2 u}{\partial y^2} = 0 \qquad (-\infty < x < \infty, -\infty < y < \infty),$$

$$u(x, 0) = 0 \qquad (-\infty < x < \infty),$$

$$\frac{\partial u}{\partial y}(x, 0) = \frac{1}{n} \sin(nx) \qquad (-\infty < x < \infty),$$

in which n is any positive integer.

(a) Show that $u(x, y) = (1/n^2)\sinh(ny)\sin(nx)$ is a solution.

(b) Let $A(x, y)$ be a solution of the problem

$$\frac{\partial^2 u}{\partial x^2} + \frac{\partial^2 u}{\partial y^2} = 0 \qquad (-\infty < x < \infty, -\infty < y < \infty),$$

$$u(x, 0) = f(x) \qquad (-\infty < x < \infty),$$

$$\frac{\partial u}{\partial y}(x, 0) = g(x) \qquad (-\infty < x < \infty),$$

and let $B(x, y)$ be a solution of the problem

$$\frac{\partial^2 u}{\partial x^2} + \frac{\partial^2 u}{\partial y^2} = 0 \qquad (-\infty < x < \infty, -\infty < y < \infty),$$

$$u(x, 0) = f(x) \qquad (-\infty < x < \infty),$$

$$\frac{\partial u}{\partial y}(x, 0) = g(x) + \frac{1}{n} \sin(nx) \qquad (-\infty < x < \infty).$$

Prove that $B(x, y) - A(x, y) = (1/n^2)\sinh(ny)\sin(nx)$.

(c) Use these results to show that the problem

$$\frac{\partial^2 u}{\partial x^2} + \frac{\partial^2 u}{\partial y^2} = 0 \qquad (-\infty < x < \infty, -\infty < y < \infty),$$

$$u(x, 0) = f(x) \qquad (-\infty < x < \infty),$$

$$\frac{\partial u}{\partial y}(x, 0) = g(x) \qquad (-\infty < x < \infty)$$

does not depend continuously on the data.

4. Consider the very general vibrating string problem

$$\frac{\partial^2 y}{\partial t^2} = a^2 \frac{\partial^2 y}{\partial x^2} + F(x) \qquad (0 < x < L, t > 0),$$

$$y(0, t) = \alpha(t), \quad y(L, t) = \beta(t) \qquad (0 < x < L),$$

$$y(x, 0) = f(x) \qquad\qquad\qquad (0 < x < L),$$

$$\frac{\partial y}{\partial t}(x, 0) = g(x) \qquad\qquad\qquad (0 < x < L).$$

Assume that α and β are continuous on $[0, \infty)$ and that f and g are continuous on $[0, L]$. Prove that there can be at most one continuous solution having continuous first and second partial derivatives. *Hint:* Suppose that y_1 and y_2 are solutions. Let $u = y_1 - y_2$. Derive a boundary value problem for which u is a solution. Then define

$$I(t) = \frac{1}{2} \int_0^L \left[\left(\frac{\partial u}{\partial x}\right)^2 + \frac{1}{a^2}\left(\frac{\partial u}{\partial t}\right)^2 \right] dx.$$

Prove that $I'(t) = 0$ for $t \geq 0$ and hence conclude that $I(t) = $ constant. By letting $t = 0$, prove that this constant is zero. From this, show that $u(x, t) = 0$ for $0 \leq x \leq L, t \geq 0$.

5. In the boundary value problem of Problem 4, replace the conditions

$$y(0, t) = \alpha(t), \qquad y(L, t) = \beta(t) \quad \text{for} \quad t > 0$$

with the conditions

$$\frac{\partial y}{\partial t}(0, t) = \alpha(t), \qquad \frac{\partial y}{\partial t}(L, t) = \beta(t) \quad \text{for} \quad t > 0.$$

Prove that there is still at most one solution which is continuous and has continuous first and second partial derivatives.

6. Consider the Dirichlet problem

$$\frac{\partial^2 u}{\partial x^2} + \frac{\partial^2 u}{\partial y^2} + \frac{\partial^2 u}{\partial z^2} = 0 \quad \text{for} \quad (x, y, z) \text{ in a region } M \text{ bounded by } \Sigma,$$

$$u(x, y, z) = f(x, y, z) \quad \text{for} \quad (x, y, z) \text{ on } \Sigma,$$

in which the surface Σ bounding M is piecewise smooth. Prove that there is at most one solution which is continuous and has continuous first partial derivatives throughout M. *Hint:* Suppose that u_1 and u_2 are solutions having these continuity properties, and let $w = u_1 - u_2$. Find a Dirichlet problem satisfied by w. Use Gauss's divergence theorem in the form

$$\iint_\Sigma w \frac{\partial w}{\partial \eta}\, d\sigma = \iiint_M [w_x^2 + w_y^2 + w_z^2 + w\, \nabla^2 w]\, dx\, dy\, dz$$

to conclude that $w(x, y, z) = 0$ for all (x, y, z) in D.

7. Generalize Problem 6 as follows. Consider the problem

$$\frac{\partial^2 u}{\partial x^2} + \frac{\partial^2 u}{\partial y^2} + \frac{\partial^2 u}{\partial z^2} = 0 \quad \text{for} \quad (x, y, z) \text{ in a region } M \text{ bounded by a surface } \Sigma,$$

$$a\frac{\partial u}{\partial \eta}(x, y, z) + bu(x, y, z) = f(x, y, z) \quad \text{for} \quad (x, y, z) \text{ on } \Sigma.$$

Show that this problem has at most one solution which is continuous and has continuous first partial derivatives in M. Here, a and b are positive constants and Σ is the piecewise-smooth surface bounding M.

TABLE 18.1   Finite Fourier Sine Transforms

$f(x)$	$S_n\{f(x)\} = F_S(n)$
$\dfrac{\pi - x}{\pi}$	$\dfrac{1}{n}$
$\dfrac{x}{\pi}$	$\dfrac{(-1)^{n+1}}{n}$
1	$\dfrac{1 - (-1)^n}{n}$
$\dfrac{x(\pi^2 - x^2)}{6}$	$\dfrac{(-1)^{n+1}}{n^3}$
$\dfrac{x(\pi - x)}{2}$	$\dfrac{1 - (-1)^n}{n^3}$
x^2	$\dfrac{\pi^2(-1)^{n+1}}{n} - \dfrac{2[1 - (-1)^n]}{n^3}$
x^3	$\pi(-1)^n\left(\dfrac{6}{n^3} - \dfrac{\pi^2}{n}\right)$
e^{ax}	$\dfrac{n}{n^2 + a^2}[1 - (-1)^n e^{an}]$
$\sin(kx), \quad k = 1, 2, 3, \ldots$	$\begin{cases} 0 & \text{if} \quad n \neq k \\ \pi/2 & \text{if} \quad n = k \end{cases}$
$\cos(ax), \quad a \neq \text{integer}$	$\dfrac{n}{n^2 - a^2}[1 - (-1)^n \cos(a\pi)]$
$\cos(kx), \quad k = 1, 2, 3, \ldots$	$\begin{cases} \dfrac{n}{n^2 - a^2}[1 - (-1)^{n+k}], & n \neq k \\ \dfrac{\pi}{2}, & n = k \end{cases}$

TABLE 18.1 Finite Fourier Sine Transforms (cont.)

$f(x)$	$S_n\{f(x)\} = F_S(n)$
$\dfrac{2}{\pi}\tan^{-1}\left(\dfrac{2a\sin(x)}{1-a^2}\right) \quad (\lvert a\rvert < 1)$	$\dfrac{1-(-1)^n}{n}a^n$
$f(\pi - x)$	$(-1)^n F_S(n)$
$\dfrac{x}{n}(\pi - x)(2\pi - x)$	$\dfrac{6}{n^3}$
$\cosh(kx)$	$\dfrac{n}{n^2 + k^2}\left[1 - (-1)^n\cosh(k\pi)\right]$
$\dfrac{\sin[k(\pi - x)]}{\sin(k\pi)}$	$\dfrac{n}{n^2 - k^2} \quad (k \neq \text{integer})$
$\dfrac{\sinh[k(\pi - x)]}{\sinh(k\pi)}$	$\dfrac{n}{n^2 + k^2} \quad (k \neq 0)$
$\dfrac{2}{\pi}\dfrac{k\sin(x)}{1 + k^2 - 2k\cos(x)}$	$k^n \quad (\lvert k\rvert < 1)$
$\dfrac{2}{\pi}\tan^{-1}\left[\dfrac{k\sin(x)}{1 - k\cos(x)}\right]$	$\dfrac{1}{n}k^n \quad (\lvert k\rvert < 1,\, n \neq 0)$
$\dfrac{1}{\pi}\dfrac{\sin(x)}{\cosh(y) - \cos(x)}$	$e^{-ny} \quad (y > 0)$
$\dfrac{1}{\pi}\tan^{-1}\left[\dfrac{\sin(x)}{e^y - \cos(x)}\right]$	$\dfrac{1}{n}e^{-ny} \quad (y > 0)$

TABLE 18.2 Finite Fourier Cosine Transforms

$f(x)$	$C_n\{f(x)\} = F_C(n)$
1	$\begin{cases} 0 & \text{if } n = 1, 2, 3, \ldots \\ \pi & \text{if } n = 0 \end{cases}$
$\begin{cases} 1, & \text{if } 0 \le x \le k \\ -1, & \text{if } k < x < \pi \end{cases}$	$\begin{cases} \dfrac{2}{n}\sin(nk) & \text{if } n = 1, 2, 3, \ldots \\ 2k - \pi & \text{if } n = 0 \end{cases}$
x	$\begin{cases} \dfrac{-[1 - (-1)^n]}{n^2} & \text{if } n = 1, 2, 3, \ldots \\ \dfrac{\pi^2}{2} & \text{if } n = 0 \end{cases}$
$\dfrac{x^2}{2\pi}$	$\begin{cases} \dfrac{(-1)^n}{n} & \text{if } n = 1, 2, 3, \ldots \\ \dfrac{\pi^2}{6} & \text{if } n = 0 \end{cases}$
x^3	$\begin{cases} 3\pi^2 \dfrac{(-1)^n}{n^2} + 6\dfrac{[1 - (-1)^n]}{n^4} & \text{if } n = 1, 2, 3, \ldots \\ \dfrac{\pi^2}{4} & \text{if } n = 0 \end{cases}$
e^{ax}	$\left(\dfrac{(-1)^n e^{a\pi} - 1}{n^2 + a^2}\right)a$
$\sin(ax), \quad a \ne \text{integer}$	$\left(\dfrac{(-1)^n \cos(a\pi) - 1}{n^2 - a^2}\right)a$
$\sin(kx), \quad k = 1, 2, \ldots$	$\begin{cases} \dfrac{(-1)^{n+k} - 1}{n^2 - k^2}k & \text{if } n \ne k \\ 0 & \text{if } n = k \end{cases}$
$\cos(kx), \quad k = 1, 2, 3, \ldots$	$\begin{cases} 0 & \text{if } n \ne k \\ \dfrac{\pi}{2} & \text{if } n = k \end{cases}$
$f(\pi - x)$	$(-1)^n F_C(n)$
$\dfrac{1}{\pi}\dfrac{1 - k^2}{1 + k^2 - 2k\cos(x)}$	$k^n \quad (\lvert k \rvert < 1)$
$-\dfrac{1}{\pi}\ln[1 + k^2 - 2k\cos(x)]$	$\dfrac{1}{n}k^n \quad (\lvert k \rvert < 1)$
$\dfrac{\cosh[k(\pi - x)]}{k\sinh(k\pi)}$	$\dfrac{1}{k^2 + n^2} \quad (k \ne 0)$
$\dfrac{1}{\pi}\dfrac{\sinh(y)}{\cosh(y) - \cos(x)}$	$e^{-ny} \quad (y > 0)$
$\dfrac{1}{\pi}[y - \ln(2\cosh(y) - 2\cos(x))]$	$\dfrac{1}{n}e^{-ny} \quad (y > 0)$

TABLE 18.3 Fourier Sine Transforms

$f(x)$	$\mathscr{F}_S\{f(x)\} = F_S(\omega)$
$\dfrac{1}{x}$	$\begin{cases} \dfrac{\pi}{2} & \text{if} \quad \omega > 0 \\[2mm] -\dfrac{\pi}{2} & \text{if} \quad \omega < 0 \end{cases}$
$x^{r-1} \qquad (0 < r < 1)$	$\Gamma(r)\sin\left(\dfrac{\pi r}{2}\right)\omega^{-r}$
$\dfrac{1}{\sqrt{x}}$	$\sqrt{\dfrac{\pi}{2}}\,\omega^{-1/2}$
$e^{-ax} \qquad (a > 0)$	$\dfrac{\omega}{a^2 + \omega^2}$
$xe^{-ax} \qquad (a > 0)$	$\dfrac{2a\omega}{(a^2 + \omega^2)^2}$
$xe^{-a^2x^2} \qquad (a > 0)$	$\dfrac{\sqrt{\pi}}{4}\,a^{-3}\omega e^{-\omega^2/4a^2}$
$x^{-1}e^{-ax} \qquad (a > 0)$	$\tan^{-1}\left(\dfrac{\omega}{a}\right)$
$\dfrac{x}{a^2 + x^2} \qquad (a > 0)$	$\dfrac{\pi}{2}\,e^{-a\omega}$
$\dfrac{x}{(a^2 + x^2)^2} \qquad (a > 0)$	$2^{-3/2}a^{-1}\omega e^{-a\omega}$
$\dfrac{1}{x(a^2 + x^2)} \qquad (a > 0)$	$\dfrac{\pi}{2}\,a^{-2}(1 - e^{-a\omega})$
$e^{-x/\sqrt{2}}\sin\left(\dfrac{x}{\sqrt{2}}\right)$	$\dfrac{\omega}{1 + \omega^4}$
$\dfrac{2}{\pi}\dfrac{x}{a^2 + x^2}$	$e^{-a\omega}$
$\dfrac{2}{\pi}\tan^{-1}\left(\dfrac{a}{x}\right)$	$\dfrac{1}{\omega}(1 - e^{-a\omega}) \qquad (a > 0)$
$\text{erfc}\left(\dfrac{x}{2\sqrt{a}}\right)$	$\dfrac{1}{\omega}(1 - e^{-a\omega^2})$
$\dfrac{4}{\pi}\dfrac{x}{4 + x^4}$	$e^{-\omega}\sin(\omega)$
$\sqrt{\dfrac{2}{\pi x}}$	$\dfrac{1}{\sqrt{\omega}}$

TABLE 18.4 Fourier Cosine Transforms

$f(x)$	$\mathscr{F}_C\{f(x)\} = F_C(\omega)$
$x^{r-1} \qquad (0 < r < 1)$	$\Gamma(r)\cos\left(\dfrac{\pi r}{2}\right)\omega^{-r}$
$e^{-ax} \qquad (a > 0)$	$\dfrac{a}{a^2 + \omega^2}$
$xe^{-ax} \qquad (a > 0)$	$\dfrac{a^2 - \omega^2}{(a^2 + \omega^2)^2}$
$e^{-a^2 x^2} \qquad (a > 0)$	$\dfrac{\sqrt{\pi}}{2}\, a^{-1}\omega e^{-\omega^2/4a^2}$
$\dfrac{1}{a^2 + x^2} \qquad (a > 0)$	$\dfrac{\pi}{2}\dfrac{1}{a}e^{-a\omega}$
$\dfrac{1}{(a^2 + x^2)^2} \qquad (a > 0)$	$\dfrac{\pi}{4}\, a^{-3}e^{-a\omega}(1 + a\omega)$
$\cos\left(\dfrac{x^2}{2}\right)$	$\dfrac{\sqrt{\pi}}{2}\left[\cos\left(\dfrac{\omega^2}{2}\right) + \sin\left(\dfrac{\omega^2}{2}\right)\right]$
$\sin\left(\dfrac{x^2}{2}\right)$	$\dfrac{\sqrt{\pi}}{2}\left[\cos\left(\dfrac{\omega^2}{2}\right) - \sin\left(\dfrac{\omega^2}{2}\right)\right]$
$\dfrac{1}{2}(1 + x)e^{-x}$	$\dfrac{1}{(1 + \omega^2)^2}$
$\sqrt{\dfrac{2}{\pi x}}$	$\dfrac{1}{\sqrt{\omega}}$
$e^{-x/\sqrt{2}}\sin\left(\dfrac{\pi}{4} + \dfrac{x}{\sqrt{2}}\right)$	$\dfrac{1}{1 + \omega^2}$
$e^{-x/\sqrt{2}}\cos\left(\dfrac{\pi}{4} + \dfrac{x}{\sqrt{2}}\right)$	$\dfrac{\omega^2}{1 + \omega^4}$
$\dfrac{2}{x}\, e^{-x}\sin(x)$	$\tan^{-1}\left(\dfrac{2}{\omega^2}\right)$
$H(x) - H(x - a)$	$\dfrac{1}{\omega}\sin(a\omega)$

TABLE 18.5   Fourier Transforms

$f(x)$	$\mathscr{F}\{f(x)\} = F(\omega)$
1	$2\pi\delta(\omega)$
$\dfrac{1}{x}$	$\begin{cases} i & \text{if} \quad \omega > 0 \\ 0 & \text{if} \quad \omega = 0 \\ -i & \text{if} \quad \omega < 0 \end{cases}$
$e^{-a\|x\|} \quad (a > 0)$	$\dfrac{2a}{a^2 + \omega^2}$
$xe^{-a\|x\|} \quad (a > 0)$	$\dfrac{4ai}{(a^2 + \omega^2)^2}$
$\|x\|e^{-a\|x\|} \quad (a > 0)$	$\dfrac{2(a^2 - \omega^2)}{(a^2 + \omega^2)^2}$
$e^{-a^2x^2} \quad (a > 0)$	$\dfrac{\sqrt{\pi}}{a} e^{-\omega^2/4a^2}$
$\dfrac{1}{a^2 + x^2} \quad (a > 0)$	$\dfrac{\pi}{a} e^{-a\|\omega\|}$
$\dfrac{x}{a^2 + x^2} \quad (a > 0)$	$-\dfrac{i}{2}\dfrac{\pi}{a}\omega e^{-a\|\omega\|}$
$H(x + a) - H(x - a)$	$\dfrac{2\sin(a\omega)}{\omega}$

Notes on the History of Fourier Analysis

Fourier analysis is named after Joseph Fourier (1768–1830), a French mathematician who lived during the Napoleonic era (and in fact accompanied Napoleon on his Egyptian campaign). Fourier is justly honored by having his name attached to this important branch of analysis. As we might expect, however, many of Fourier's contemporaries and immediate predecessors contributed to his great achievement.

Leonhard Euler (1707–1783) was a Swiss mathematician who is sometimes called the greatest analyst who ever lived. In the approximate period 1729–1753, he considered the following problem of interpolation:

Find a continuous function f defined on $[1, n]$ for some positive integer n, given the values $f(1), f(2), \ldots, f(n)$ which f assumes at the integers.

The problem arose in connection with calculations Euler was doing in studying planetary perturbations, and it led him to a trigonometric series. He found that the solution of an equation of the form

$$f(x) = f(x - 1) + F(x)$$

can be written

$$f(x) = \int_0^x F(t)\, dt + 2 \sum_{n=1}^{\infty} \int_0^x [F(t)\cos(2n\pi t)\, dt]\cos(2n\pi x)$$
$$+ 2 \sum_{n=1}^{\infty} \int_0^x [F(t)\sin(2n\pi t)\, dt]\sin(2n\pi x).$$

Soon after this observation, in 1754, Jean le Rond d'Alembert (1717–1783) obtained a trigonometric cosine expansion for the reciprocal of the distance between two planets in terms of the angle between the vectors from the origin to the planets. The integral formulas for the Fourier coefficients appear in d'Alembert's work.

Other expressions involving series and cosines soon began to appear in a variety of connections. Using geometric series, Euler found that

$$\frac{a\cos(x) - a^2}{1 - 2a\cos(x) + a^2} = \sum_{n=1}^{\infty} a^n\cos(nx)$$

and

$$\frac{a\sin(x)}{1 - 2a\cos(x) + a^2} = \sum_{n=1}^{\infty} a^n\sin(nx).$$

These expansions are valid if $|a| < 1$, a detail which was not fully appreciated at the time. Euler put $a = 1$ to conclude incorrectly that

$$\sum_{n=1}^{\infty} \cos(nx) = -\tfrac{1}{2}.$$

In fact, this series diverges. If this "equation" is integrated term by term from x to π, we obtain

$$\sum_{n=1}^{\infty} \frac{1}{n}\sin(nx) = \frac{1}{2}(\pi - x),$$

which is the correct Fourier sine expansion of $\frac{1}{2}(\pi - x)$ on $(0, \pi)$! The "result" that $\sum_{n=1}^{\infty} \cos(nx) = -\frac{1}{2}$ also appeared in a study of sound waves by Joseph-Louis Lagrange (1736–1813).

Thus, as early as the middle of the eighteenth century, trigonometric series were beginning to appear. Leading mathematicians encountered them and calculated with them. Their significance was not fully understood at that time, and some of the calculations were simply incorrect. Nevertheless, their appearance in connection with important problems raised questions and led to their further investigation. One question, for example, was how a nonperiodic function might be represented by a series of sines and cosines, which are periodic.

By the 1750s, the integral formulas for the Fourier coefficients were known but were not always trusted. Mathematicians such as Euler preferred deriving trigonometric series in other ways.

Eventually, there arose a long and heated debate involving the leading mathematicians of the day. Included were Euler, Lagrange, d'Alembert, Daniel Bernoulli (1700–1782), and, later, Pierre Simon Laplace (1749–1827). The critical issue was to specify which functions could be expanded in trigonometric series. In the course of the debate, it became clear that the term "function" itself was little understood and not generally agreed upon. The term had usually been taken for granted; upon closer investigation, it was found to have different meanings for different mathematicians. To some, it meant what we would today call a continuous function; to others, a differentiable function. To some, it carried the requirement that the function must be specified by an analytic expression, such as $x^2 - \cos(x)$. The argument continued for many years. The questions were difficult, and the tools for resolving them had yet to be developed.

Joseph Fourier was born in 1768, at the height of this debate. As a student, he showed a talent for mathematics, but he turned to it as a profession only when his common heritage (he was the son of a tailor) made it impossible for him to attain a military commission.

An outstanding problem at the turn of the nineteenth century was the mathematical description of heat conduction in various media. In 1807, Fourier submitted a paper on this subject to the prestigious Academy of Sciences of Paris, in competition for a prize which had been offered for the most successful attack on this problem. Such mathematical giants as Laplace, Lagrange, and Legendre refereed the work and rejected it for lack of rigor. They did, however, encourage Fourier to continue his research and attempt to supply some of the details he had omitted. In 1811, Fourier submitted a revised version of his paper and was awarded the prize by the academy. The academy still refused to publish the paper, however, because many details were still unclear. Finally, in 1822, Fourier published his now classic work *Théorie analytique de la chaleur*, incorporating most of his results of 1811 together with some new ones.

In his paper, Fourier considered the three-dimensional heat equation

$$\frac{\partial u}{\partial t} = k^2 \left[\frac{\partial^2 u}{\partial x^2} + \frac{\partial^2 u}{\partial y^2} + \frac{\partial^2 u}{\partial z^2} \right],$$

in which $u(x, y, z, t)$ is the temperature of an object at time t and point (x, y, z). By what is now known as separation of variables, or the Fourier method, he derived trigonometric series representations of solutions. In these derivations, Fourier made explicit use of the formulas for the coefficients which today are known as Fourier coefficients.

It is interesting to ask how Fourier's work differed from that of Euler and other predecessors who also worked with trigonometric series. First, Fourier's general methods for attacking the heat equation have proved to be of fundamental and lasting importance in solving boundary value problems of partial differential equations.

Second, while others occasionally used Fourier coefficients, none really appreciated their importance. Fourier was the first (albeit not for entirely rigorous reasons) to use them confidently and extensively and to assert convincingly the generality with which an "arbitrary" function could be represented by a Fourier series. In certain technical details, Fourier was sometimes wrong. He did not have the advantage of convergence theorems, which were developed later. Nevertheless, he took the major steps on which others built, and for this he deserves the credit inherent in the designations Fourier series, Fourier integral, Fourier coefficients, and Fourier transforms.

Finally, the importance of a piece of mathematics is often judged by its ramifications as seen later from a historical perspective. Questions and results arising from the study of Fourier series have had a profound impact on the development of mathematics. Some areas influenced by research on Fourier series include the convergence of series; solution of partial differential equations; critical examination of the concepts of function, continuity, and differentiability; properties of the real number system; set theory; measure theory; and harmonic analysis.

Sufficient conditions for the convergence of a Fourier series were first established by Peter Gustav Lejeune-Dirichlet (1805–1859) about 1829. These conditions were essentially those given in Theorem 17.2. The fact that the Fourier coefficients have limit zero as $n \to \infty$ if f is integrable on $[-L, L]$ is due to Georg Friedrich Bernhard Riemann (1826–1866), a student of Gauss who worked with Dirichlet for a time. Riemann's name is attached to many concepts and theorems in mathematics, including the Riemann integral and Riemann sums.

In 1873, Paul Du Bois-Reymond (1831–1889) gave an example of a function which is continuous on $(-\pi, \pi)$ but whose Fourier series fails to converge at any point of this interval. Surprising examples of this type led to the recognition of deep properties of the structure of the real number system and were influential in the development of measure theory and real analysis.

The Fourier integral appeared near the end of Fourier's 1811 paper, when he attempted to extend his results to functions defined over the half-line $[0, \infty)$. A separate treatment appeared in Cauchy's 1816 prize-winning paper on surface waves in a fluid. Augustin-Louis Cauchy (1789–1857) was a leading French mathematician who established the basis for much of the calculus of complex-valued functions. Still a third treatment appeared in 1816 in Poisson's paper on water waves. Siméon-Denis Poisson (1781–1840) was a leading French mathematician and mathematical physicist. The differential equation $\nabla^2 u = f$ is known as Poisson's equation; it arises, for example, in connection with electrical potential (see Section 16.7).

The Fourier transform is found in early writings of Cauchy and Laplace (from about 1782 on). Finally, a formula for the Fourier cosine transform and its inverse appears in Fourier's 1811 paper.

Notes on the History of Partial Differential Equations

Since we have restricted our attention primarily to the wave and heat equations and Laplace's equation, we will center our historical remarks on these equations. Certainly they played a central role in early interest in partial differential equations.

Perhaps surprisingly, we may think of the wave equation as belonging to the eighteenth century and the heat equation to the nineteenth century. In 1727, the Swiss mathematician John Bernoulli (1667–1748) treated the vibrating string problem by imagining the string as a flexible thread having a finite number of equally spaced weights, or beads, placed along it. The differential equation he derived was time independent, however, and so was actually an ordinary differential equation. The French mathematician Jean le Rond d'Alembert (1717–1783) introduced the time variable into the equation and let the number of weights in Bernoulli's treatment go to infinity to derive the one-dimensional wave equation as we know it today. His work, which appeared about 1746, included the solution known today as d'Alembert's solution (see Problem 21 of Section 18.0).

In the 1750s, the Swiss mathematician Leonhard Euler (1707–1783) considered what we would today call Fourier series solutions of the wave equation. In 1781, he used what was essentially a Fourier-Bessel series to solve the vibrating membrane problem for a circular drum.

Laplace's equation, also known as the potential equation, arose in the study of gravitational attraction. The main figures in its early development and solution in various special cases were Pierre Simon Laplace (1749–1827) and Adrien-Marie Legendre (1752–1833). Both were professors of mathematics at the École Militaire in France. Legendre also developed what are today known as Legendre polynomials.

In 1807, Joseph Fourier (1768–1830) submitted a paper to the Academy of Sciences of Paris. In this paper, he derived the heat equation and proposed his separation of variables method of solution. The paper, refereed by Laplace, Lagrange, and Legendre, was rejected due to lack of rigor. However, the results were promising enough for the academy to include the problem of describing heat conduction in a prize competition in 1812. Fourier's 1811 revision of his earlier paper won the prize but suffered the same criticism of lack of rigor. In 1822, Fourier finally published his classic *Théorie analytique de la chaleur*, laying the foundations not only for the separation of variables method and Fourier series but for the Fourier integral and transform as well.

Pioneering work on properties of harmonic functions (solutions of Laplace's equation) was done by George Green (1793–1841), a self-taught British mathematician. Green's main interest was in the mathematical treatment of electricity and magnetism. Much of Green's mathematics was discovered independently by Michel Ostrogradsky (1801–1861), a Ukrainian mathematician.

PART *VI*

Complex Analysis

CHAPTER 19

Complex Numbers and Complex Functions

19.1 Complex Numbers

We have already used complex numbers in, for example, Euler's formula to write solutions of certain second order linear ordinary differential equations. We have also seen complex eigenvalues and eigenvectors of matrices. In this chapter, we will concentrate on properties of complex numbers and lay the foundations for the study of functions which produce complex numbers from complex numbers. Such functions are called *complex-valued functions*, or just *complex functions*.

The *imaginary unit i* is defined by the equation

$$i^2 + 1 = 0.$$

That is, $i^2 = -1$. Clearly, i is not a real number, and we define a *complex number* to be a symbol of the form

$$a + bi \quad (\text{or } a + ib),$$

in which a and b are real numbers.

We call a the *real part* of $a + bi$ and b the *complex part*. These are denoted $\text{Re}(a + bi)$ and $\text{Im}(a + bi)$, respectively. Thus,

$$\text{Re}(a + bi) = a \quad \text{and} \quad \text{Im}(a + bi) = b.$$

$\text{Re}(z)$ and $\text{Im}(z)$ are both real numbers for any complex number z. For example,

$$\text{Re}(2 - 3i) = 2, \qquad \text{Im}(2 - 3i) = -3,$$
$$\text{Re}(\pi + \sqrt{3}i) = \pi \quad \text{and} \quad \text{Im}(\pi + \sqrt{3}i) = \sqrt{3}.$$

Two complex numbers are equal if and only if their real parts are equal and their

1255

imaginary parts are equal. That is,

$$a + bi = c + di \quad \text{if and only if} \quad a = c \quad \text{and} \quad b = d.$$

We can also write

$$z = w \quad \text{if and only if} \quad \text{Re}(z) = \text{Re}(w) \quad \text{and} \quad \text{Im}(z) = \text{Im}(w).$$

This fact is often used in solving equations involving complex numbers. For example, suppose we want to solve the equation

$$x^2 + (2x + y)i = 9 - 4i. \tag{19.1}$$

We must have

$$x^2 = 9 \quad \text{and} \quad 2x + y = -4.$$

Then $x = \pm 3$ and $y = -4 - 2x$. If $x = 3$, $y = -4 - 6 = -10$; if $x = -3$, $y = -4 + 6 = 2$. The solutions of equation (19.1) are $x = 3$, $y = -10$, and $x = -3$, $y = 2$.

The operations of addition, subtraction, and multiplication of complex numbers are defined as follows.

Addition:

$$(a + bi) + (c + di) = (a + c) + (b + d)i.$$

Subtraction:

$$(a + bi) - (c + di) = (a - c) + (b - d)i.$$

Multiplication:

$$(a + bi)(c + di) = ac - bd + (ad + bc)i. \tag{19.2}$$

We multiply two complex numbers just as we do first degree polynomials, with i in place of x. However, in multiplying complex numbers, remember that $i^2 = -1$. To see this, compare equation (19.2) with the following:

$$(a + bx)(c + dx) = ac + bdx^2 + (ad + bc)x. \tag{19.3}$$

Equations (19.2) and (19.3) are identical if $x = i$ because then $x^2 = -1$. For example,

$$(3 + 6i)(2 - i) = 3 \cdot 2 - 3 \cdot i + 2 \cdot 6i + 6i(-i)$$
$$= 6 + 6 + 9i = 12 + 9i.$$

The *complex conjugate* of $a + bi$ is defined to be $a - bi$, the complex number obtained by changing the sign of the imaginary part. We denote the operation of taking the complex conjugate by putting a bar over the number:

$$\overline{a + bi} = a - bi.$$

For example,

$$\overline{10 - 8i} = 10 + 8i.$$

Often, the letter z is used to denote a complex number, just as we often use x as a real variable in calculus. If $z = a + bi$, $\bar{z} = a - bi$. Notice that

$$\text{Re}(z) = \text{Re}(\bar{z}) \quad \text{and} \quad \text{Im}(z) = -\text{Im}(\bar{z}).$$

Further,

$$z\bar{z} = (a + bi)(a - bi) = a^2 + b^2.$$

Thus, for any complex number z, $z\bar{z}$ is a real number.

For any complex number $z = a + bi$, the *magnitude*, or *modulus*, of z, denoted $|z|$, is defined by

$$|z| = \sqrt{a^2 + b^2}.$$

For example,

$$|5 - i| = \sqrt{5^2 + 1^2} = \sqrt{26}.$$

Notice that $z = 0$ if and only if $|z| = 0$, because $z = 0$ if and only if $a = b = 0$. Magnitude is related to conjugacy by the relationship

$$z\bar{z} = |z|^2.$$

If $z = a + bi$, both sides of this equation are simply $a^2 + b^2$.

We will now define the quotient z/w of two complex numbers if $w \neq 0$. We would like to be able to compute this quotient so that it is clear what its real and imaginary parts are. To do this, write

$$\frac{z}{w} = \frac{z}{w} \cdot \frac{\bar{w}}{\bar{w}} = \frac{z\bar{w}}{w\bar{w}}.$$

This is how we actually carry out complex division. This equation gives us the quotient as a product $z\bar{w}$, which we know how to compute, divided by the *real number* $w\bar{w}$. If $z = a + bi$ and $w = c + di$, this quotient is

$$\frac{z}{w} = \frac{a + bi}{c + di} = \frac{a + bi}{c + di} \frac{c - di}{c - di} = \frac{ac + bd + (bc - ad)i}{c^2 + d^2}$$

$$= \frac{ac + bd}{c^2 + d^2} + \frac{bc - ad}{c^2 + d^2} i.$$

For example,

$$\frac{1 + 4i}{2 - 3i} = \frac{1 + 4i}{2 - 3i} \frac{2 + 3i}{2 + 3i} = \frac{2 - 12 + (3 + 8)i}{2^2 + 3^2}$$

$$= \frac{-10 + 11i}{13} = -\frac{10}{13} + \frac{11}{13} i.$$

By carrying out the division in this way, we easily read the real and imaginary parts of the quotient:

$$\mathrm{Re}\left(\frac{1 + 4i}{2 - 3i}\right) = -\frac{10}{13} \quad \text{and} \quad \mathrm{Im}\left(\frac{1 + 4i}{2 - 3i}\right) = \frac{11}{13}.$$

Complex numbers admit a useful geometric interpretation. Draw a rectangular coordinate system as in Figure 19.1, and label the horizontal axis as the real axis and the vertical axis as the imaginary axis. A complex number $a + ib$ may be identified with the point (a, b), which is plotted in the usual way. The real part of $a + ib$ is a and is measured along the real axis, and the imaginary part, b, is measured along

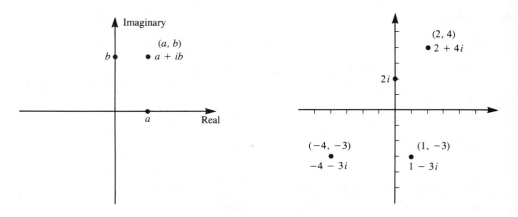

Figure 19.1   *Figure 19.2*

the imaginary axis. As specific examples, Figure 19.2 shows the points $(2, 4)$, $(1, -3)$, and $(-4, -3)$ associated, respectively, with the complex numbers $2 + 4i$, $1 - 3i$, and $-4 - 3i$. The xy-plane, with points thought of as complex numbers, is called the *complex plane*.

Every real number a may be thought of as a special case of a complex number by writing a as $a + 0i$. Thus, a real number a is associated with the point $(a, 0)$ on the real axis. A number bi, with zero real part, is called *pure imaginary* and is associated with the point $(0, b)$ on the imaginary axis.

We can also associate the complex number $a + bi$ with the vector $a\mathbf{i} + b\mathbf{j}$, represented as an arrow from the origin to the point (a, b) as in Figure 19.3. Using this interpretation, the parallelogram law holds for addition of complex numbers, as illustrated in Figure 19.4. Adding $a + bi$ to $c + di$ to obtain $a + c + (b + d)i$ exactly corresponds to the vector sum $(a\mathbf{i} + b\mathbf{j}) + (c\mathbf{i} + d\mathbf{j}) = (a + c)\mathbf{i} + (b + d)\mathbf{j}$.

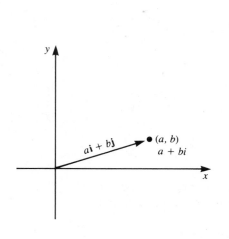

Figure 19.3

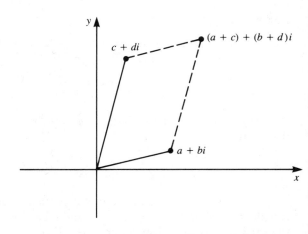

Figure 19.4. Parallelogram law for addition of complex numbers.

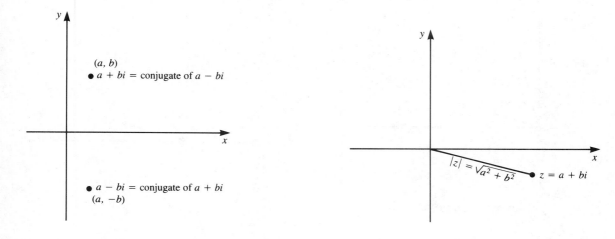

Figure 19.5   Figure 19.6

The complex conjugate $a - bi$ of $a + bi$ can be associated with the point $(a, -b)$, which is the reflection of (a, b) across the real axis (Figure 19.5).

The magnitude $|a + bi| = \sqrt{a^2 + b^2}$ is the distance from the origin to the point (a, b) representing the complex number $a + bi$. Thus, the magnitude of z can be interpreted as the distance from the origin to a point representing z in the complex plane (Figure 19.6). It is also the length of the vector (arrow) from the origin to the point representing z.

As a useful convention, we will speak of a complex number $z = a + bi$ and the point (a, b) representing it interchangeably. For example, we can refer to the point $(0, 1)$ as the number i (because $0 + 1i = i$), or we can refer to the complex number i as the point $(0, 1)$.

The geometric representation of complex numbers suggests the triangle inequality: If z and w are complex numbers,

$$|z + w| \le |z| + |w|.$$

That is, the magnitude of a sum of two complex numbers cannot exceed the sum of the magnitudes of these numbers. If $z = a + bi$ and $w = c + di$, $z + w = (a + c) + (b + d)i$, and the triangle inequality is

$$\sqrt{(a + c)^2 + (b + d)^2} \le \sqrt{a^2 + b^2} + \sqrt{c^2 + d^2}.$$

This inequality can be proved algebraically. We can also observe (see Figure 19.7) that $|z|, |w|$, and $|z + w|$ are lengths of sides of a triangle formed by vectors representing z, w, and $z + w$, respectively. With this interpretation, the triangle inequality follows from the fact that one side of a triangle cannot be longer than the sum of the lengths of the other two sides.

The number $|z - w|$ may be interpreted as the distance between z and w when z and w are represented as points in the plane (Figure 19.8). This also follows from the parallelogram law for vectors. For example, the distance between $1 - 4i$ and $-2 - 6i$

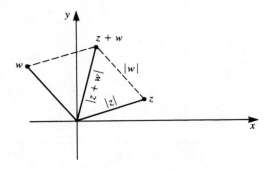

Figure 19.7. $|z + w| \leq |z| + |w|$
(triangle inequality).

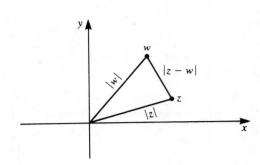

Figure 19.8

is the magnitude of their difference:

$$|(1 - (-2)) + (-4 - (-6))i| = |3 + 2i| = \sqrt{13}.$$

This is also the distance between the points $(1, -4)$ and $(-2, -6)$ in the plane. It is also the length of the vector $3\mathbf{i} + 2\mathbf{j}$ represented by the arrow from $(-2, -6)$ to $(1, -4)$.

The geometric interpretation of complex numbers as points in the plane enables us to represent equalities or inequalities involving complex numbers as loci of points in the plane. This gives us a pictorial representation of those complex numbers satisfying the equality or inequality. Here are some illustrations of this idea.

EXAMPLE 19.1

Determine all z such that $|z| = 4$.

Since $|z|$ is the distance from the origin to the complex number z in the plane, $|z| = 4$ if and only if z is at distance 4 from the origin. Thus, z must lie on the circle of radius 4 about the origin, as shown in Figure 19.9.

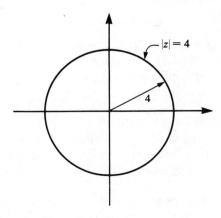

Figure 19.9. Locus of z with $|z| = 4$.

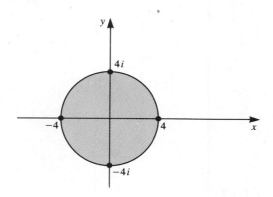

Figure 19.10. $|z| \le 4$.

We can also look at the equation $|z| = 4$ algebraically. If $z = x + iy$,

$$|z| = \sqrt{x^2 + y^2} = 4$$

if and only if

$$x^2 + y^2 = 16,$$

which means that (x, y) must lie on the circle of radius 4 about the origin.

If we want all complex numbers z with $|z| \le 4$, we require that z lie on or inside the circle of radius 4 about the origin. This set is shaded in Figure 19.10. ■

EXAMPLE 19.2

Find all complex numbers z such that $|z + 2i| = |1 + i|$.

Since $|1 + i| = \sqrt{2}$, we require that $|z + 2i| = \sqrt{2}$. Think of $|z + 2i|$ as $|z - (-2i)|$, which is the distance from z to $-2i$. The equation $|z + 2i| = \sqrt{2}$ may be read "The distance between z and $-2i$ must be $\sqrt{2}$." This puts z on the circle of radius $\sqrt{2}$ about the point $-2i$. This locus is shown in Figure 19.11.

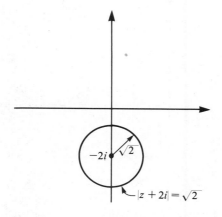

Figure 19.11. Locus of points z with $|z + 2i| = |1 + i|$. ■

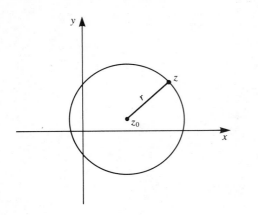

Figure 19.12. Locus of z satisfying $|z - z_0| = r$.

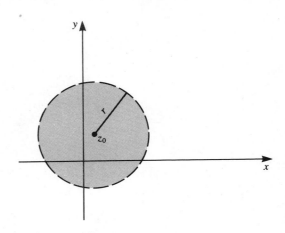

Figure 19.13. Locus of z satisfying $|z - z_0| < r$.

Given any complex number z_0 and a positive number r, the equation

$$|z - z_0| = r$$

restricts z to lie on the circle of radius r about z_0 (because the distance between z and z_0 is fixed at r). This fact is recorded in Figure 19.12. We will often write the equation of a circle in this way.

The inequality

$$|z - z_0| < r$$

is satisfied by all z within the circle of radius r about z_0 (not including points on the boundary circle). This set is shaded in Figure 19.13. We call the set of z satisfying $|z - z_0| < r$ the *open disk* of radius r about z_0. When we speak of a disk, we will always mean an open disk about a point.

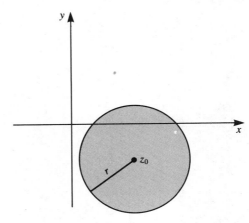

Figure 19.14. Locus of z satisfying $|z - z_0| \leq r$ (includes points on the boundary circle $|z - z_0| = r$.)

The inequality

$$|z - z_0| \le r$$

specifies all z lying on or within the circle of radius r about z_0. This set consists of the disk of radius r about z_0 together with points on the boundary circle and is called the *closed disk* of radius r about z_0. It differs from the open disk $|z - z_0| < r$ by its inclusion of the points on the boundary circle and is illustrated in Figure 19.14.

EXAMPLE 19.3

Sketch the locus of points satisfying $|z - 2 + 4i| < 3$.
    Write

$$|z - 2 + 4i| = |z - (2 - 4i)| < 3.$$

This inequality defines the open disk of radius 3 about $2 - 4i$ consisting of all points within the circle of radius 3 about $2 - 4i$. The set is shaded in Figure 19.15, where the dashed boundary circle indicates that points on the circle are not in the set.

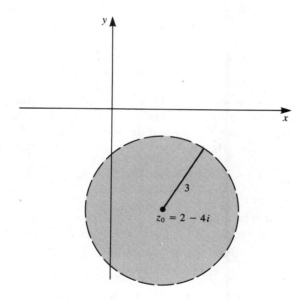

Figure 19.15. Locus of points satisfying $|z - 2 + 4i| < 3$. ∎

EXAMPLE 19.4

Determine all z such that $|z|^2 + 3 \operatorname{Re}(z^2) = 4$.
    Write $z = x + iy$. Then $z^2 = x^2 - y^2 + 2ixy$, so

$$|z|^2 + 3 \operatorname{Re}(z^2) = x^2 + y^2 + 3(x^2 - y^2) = 4$$

if and only if

$$4x^2 - 2y^2 = 4,$$

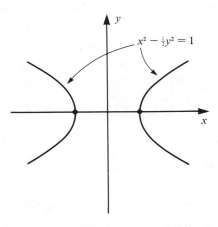

Figure 19.16. Locus of points z with $|z|^2 + 3 \operatorname{Re}(z^2) = 4$.

or

$$x^2 - \tfrac{1}{2}y^2 = 1.$$

The graph of this locus is the hyperbola shown in Figure 19.16. A complex number $x + iy$ satisfies $|z|^2 + 3 \operatorname{Re}(z^2) = 4$ exactly when (x, y) is on this hyperbola. ■

In some cases, we determine a locus from an equation or inequality by reasoning geometrically (as in Example 19.1, 19.2, and 19.3); in some instances we use algebra, writing z as $x + iy$ and attempting to find an equation in x and y. We did this in Example 19.4. Usually, we try the geometric approach first and resort to algebra if this approach fails.

EXAMPLE 19.5

Determine all z such that $|z + 2i| = |z + 1 - 3i|$.
    Write this equation as

$$|z - (-2i)| = |z - (-1 + 3i)|. \tag{19.4}$$

Now verbalize the equation, using the interpretation that $|z - w|$ is the distance between z and w. The equation reads "The distance between z and $-2i$ equals the distance between z and $-1 + 3i$." Figure 19.17 shows $-2i$ and $-1 + 3i$ as points in the complex plane. It is apparent that z can be equidistant from both points only if z lies on a straight line perpendicular to the line segment between these points and bisecting this segment.
    This geometric reasoning has given us an idea of what the locus looks like. For specific details, we must do some algebra. Write $z = x + iy$. Then equation (19.4) reads

$$|x + (y + 2)i| = |(x + 1) + (y - 3)i|.$$

Then

$$\sqrt{x^2 + (y + 2)^2} = \sqrt{(x + 1)^2 + (y - 3)^2}.$$

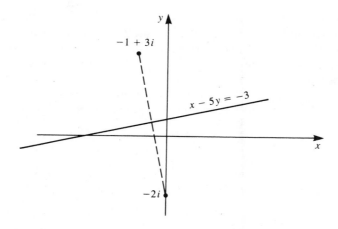

Figure 19.17. Locus of points z satisfying $|z + zi| = |z + 1 - 3i|$.

Square both sides of this equation to obtain

$$x^2 + y^2 + 4y + 4 = x^2 + 2x + 1 + y^2 - 6y + 9,$$

which can be written

$$x - 5y = -3,$$

the equation of a straight line. The complex numbers $x + iy$, with $x - 5y = -3$, are exactly those satisfying equation (19.4).

Complex numbers obey many of the rules we are accustomed to using with real numbers. In particular,

$z + w = w + z$	(commutativity of addition),
$zw = wz$	(commutativity of multiplication),
$z + (w + u) = (z + w) + u$	(associative law for addition),
$z(wu) = (zw)u$	(associative law for multiplication),
$z(u + w) = zu + zw$	(distributive law),
$z + 0 = 0 + z = z,$	

and

$$z \cdot 1 = 1 \cdot z = z.$$

An important difference between real and complex numbers is that the real numbers are ordered while the complex numbers are not. The relation $<$ ("less than") orders the real numbers. Given any two real numbers a and b, exactly one of the following must be true: $a < b$, $b < a$, or $a = b$. Geometrically, $a < b$ for a and b real means that the point representing a on the real line is to the left of the point representing b.

There is no comparable way of defining a "less than" relationship for complex numbers. We cannot order the complex numbers (or, equivalently, points in the plane) so that, given any two complex numbers z and w, we must have one of $z < w$, $w < z$, or $w = z$. For example, if we attempt to define $z < w$ to mean $\text{Re}(z) < \text{Re}(w)$ and $\text{Im}(z) < \text{Im}(w)$, the numbers i and 1 are not related, since we do not have $i = 1$, $i < 1$, or $1 < i$. It can be shown that there is no order relationship for the complex numbers, which is consistent with complex arithmetic.

In the next section, we will define the polar form of a complex number, which is obtained by writing the point representing the number in polar coordinates.

PROBLEMS FOR SECTION 19.1

In each of Problems 1 through 20, carry out the indicated numerical computation. All answers should be written in the form $a + bi$, so that the real and imaginary parts of the answer are explicitly given.

1. $(3 - 4i)(6 + 2i)$

2. $(1 - i) + (2 + 4i)$

3. $i(6 - 2i)$

4. $\dfrac{1}{i}$

5. $\dfrac{2 - i}{4 + i}$

6. $\left[2i + \dfrac{3 - i}{2i}\right](1 - i)$

7. $\dfrac{1 + i}{1 - i} + \dfrac{2}{i}$

8. $\dfrac{(2 + i) - (3 - 4i)}{(2 - i)(3 + i)}$

9. $(2 + 4i)\overline{(6 - 3i)}$

10. $\dfrac{(-4 - 5i)\overline{(8 - 4i)}}{6 + 2i}$

11. $\dfrac{8i}{6 - i}$

12. $i^3 - 4i^2 + 2$

13. $(3 + i)^2$

14. $\dfrac{17 - 3i}{2 + 4i}$

15. $\dfrac{(7 + i)(1 - 5i)}{(4 - i)(6 + i)}$

16. $\dfrac{(3 - 5i)(3 - 7i)}{(3 + i)i^3}$

17. $(1 + i)(2 + i)(1 + 5i)$

18. $(2 - i)^3$

19. $\left[\dfrac{3 + 2i}{-1 + i}\right]^2$

20. $(-3 - 8i)(2i)(3 + 2i)$

In each of Problems 21 through 25, $z = a + bi$. The answer should be in terms of a and b.

21. Find $\text{Re}(z^2)$ and $\text{Im}(z^2)$.

22. Find $|z + 2|$ and $|z - i|$.

23. Find $\text{Re}(2z - 3\bar{z} + 4)$.

24. Find $\text{Im}(z^2 + z)$.

25. Find $\text{Im}(2\bar{z}/|z|)$.

In each of Problems 26 through 32, graph the locus of points in the complex plane satisfying the equation.

26. $z^2 + \bar{z}^2 = 4$

27. $|z| = |z - i|$

28. $|z|^2 + \text{Im}(z) = 16$

29. $|z - 8 + 4i| = 9$

30. $|z| + \text{Re}(z) = 0$

31. $\text{Im}(z - i) = \text{Re}(z + 1)$

32. $|z - i| + |z| = 9$

33. Prove that z is real or pure imaginary if and only if $z^2 = (\bar{z})^2$.

34. Let z, w, and u be complex numbers. Prove that z, w, and u, when represented as points in the plane, form the vertices of an equilateral triangle if and only if

$$z^2 + w^2 + u^2 = zw + zu + wu.$$

35. Prove that $\text{Re}(iz) = -\text{Im}(z)$ and $\text{Im}(iz) = \text{Re}(z)$ for any complex number z.

36. Prove that

$$\left|\frac{z - w}{1 - \bar{z}w}\right| = 1$$

if either $|z| = 1$ or $|w| = 1$ but $\bar{z}w \neq 1$.

37. Describe geometrically the locus of points z satisfying

$$|z - i| = 2|z + 2i|.$$

38. Prove that, for any complex numbers z and w,

$$|z + w|^2 + |z - w|^2 = 2(|z|^2 + |w|^2).$$

Hint: Remember that $|z|^2 = z\bar{z}$, and similarly for $|w|^2$.

39. Prove that, for any complex numbers z and w,

$$||z| - |w|| \le |z - w|.$$

40. Prove that, for any complex numbers z and w and any positive integer n,

$$(z + w)^n = \sum_{j=0}^{n} \binom{n}{j} z^{n-j} w^j,$$

where

$$\binom{n}{j} = \frac{n!}{j!(n - j)!}.$$

For any positive integer k, $k!$ is the product of the integers from 1 to k, inclusive (for example, $4! = 1 \cdot 2 \cdot 3 \cdot 4 = 24$), and $0! = 1$ by definition. *Hint:* One way to do this is by mathematical induction.

41. Prove that, for any complex numbers a and b,

$$\left| \frac{az + b}{\bar{b}z + \bar{a}} \right| = 1$$

if z is any complex number with $|z| = 1$.

19.2 *The Polar Form of a Complex Number*

Let $z = a + bi$ be a complex number. We can associate with z the point (a, b) in the complex plane, as in Figure 19.18. Let (r, θ) be polar coordinates of (a, b). Then

$$a = r \cos(\theta) \quad \text{and} \quad b = r \sin(\theta), \tag{19.5}$$

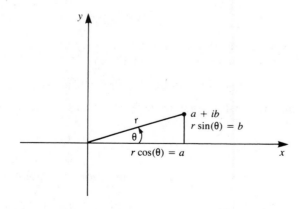

Figure 19.18

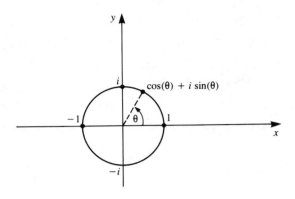

Figure 19.19

and we can write

$$z = r[\cos(\theta) + i\sin(\theta)].$$

The expression on the right is called the *polar form* of z. We have

$$r = |z|,$$

the magnitude of z.

Any number θ (in radians) satisfying equations (19.5) is called an *argument of z*. If we have one argument of z, this argument plus any integer multiple of 2π is also an argument of z. The symbol $\arg(z)$ denotes the set of all arguments of z. Thus, $\arg(z)$ is not itself a number but denotes an infinite set of numbers. Given z, any two arguments of z differ by an integer multiple of 2π. There is always exactly one argument of z lying in the interval $-\pi < \theta \leq \pi$. This number, which is uniquely determined by z, is called the *principal argument* of z and is denoted $\text{Arg}(z)$.

Geometrically, if $z \neq 0$, we get values of $\arg(z)$ by rotating the positive real axis to the line from the origin through (a, b). A counterclockwise rotation corresponds to positive values of $\arg(z)$, and a clockwise rotation corresponds to negative values.

The polar form of z,

$$z = r[\cos(\theta) + i\sin(\theta)],$$

represents z as a product of a real number, r, the magnitude of z, with a complex number $\cos(\theta) + i\sin(\theta)$ of magnitude 1, since for any θ,

$$|\cos(\theta) + i\sin(\theta)| = \sqrt{\cos^2(\theta) + \sin^2(\theta)} = 1.$$

Any number of the form $\cos(\theta) + i\sin(\theta)$ is represented by a point on the unit circle about the origin (Figure 19.19).

EXAMPLE 19.6

Let $z = i$. Then $a = 0$ and $b = 1$ in the above notation. The point $(0, 1)$ is shown in Figure 19.20.

The point $(0, 1)$ has polar coordinates $(1, \pi/2), (1, 5\pi/2), (1, 9\pi/2), \dots, (1, -3\pi/2),$ $(1, -7\pi/2), \dots$. Values of $\arg(i)$ are $\pi/2, 5\pi/2, 9\pi/2, \dots, -3\pi/2, -7\pi/2, \dots$. In this

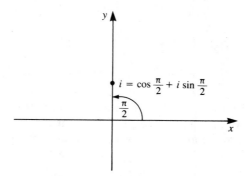

Figure 19.21

example, arg(i) consists of all numbers

$$\frac{\pi}{2} + 2n\pi \quad \text{and} \quad \frac{-3\pi}{2} + 2n\pi,$$

in which n is any integer. Arg(i) $= \pi/2$, the one argument of i lying in $(-\pi, \pi]$.
The polar form of i is

$$i = 1\left[\cos\left(\frac{\pi}{2}\right) + i \sin\left(\frac{\pi}{2}\right) \right]. \quad \blacksquare$$

EXAMPLE 19.7

Let $z = 1 + i$. Then $a = b = 1$ in the preceding discussion. The point $(1, 1)$ is shown in Figure 19.21. The principal argument of $1 + i$ is Arg($1 + i$) $= \pi/4$. Values of arg($1 + i$) are $\pi/4 + 2n\pi$ and $-7\pi/4 + 2n\pi$, as n varies over all the integers. Since $|1 + i| = \sqrt{2}$,

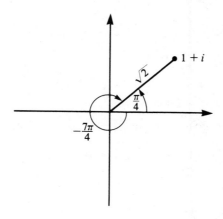

Figure 19.20

the polar form of $1 + i$ is

$$1 + i = \sqrt{2}\left[\cos\left(\frac{\pi}{4}\right) + i\sin\left(\frac{\pi}{4}\right)\right]. \quad \blacksquare$$

EXAMPLE 19.8

Let $z = 17 - 3i$. The point $(17, -3)$ is shown in Figure 19.22. With α as indicated, $\tan(\alpha) = \frac{3}{17}$, so $\alpha = \tan^{-1}(\frac{3}{17})$. Therefore, one argument of $17 - 3i$ is $-\tan^{-1}(\frac{3}{17})$, with the minus sign included because the positive real axis must be rotated clockwise through this angle to reach $(17, -3)$. Other arguments of $17 - 3i$ are

$$-\tan^{-1}(\tfrac{3}{17}) + 2n\pi, \qquad n \text{ any integer.}$$

We can also use the angle β of Figure 19.23. Since $\tan(\beta) = \frac{17}{3}$, $\beta = \tan^{-1}(\frac{17}{3})$, and another argument of $17 - 3i$ is $3\pi/2 + \tan^{-1}(\frac{17}{3})$. We reach $(17, -3)$ by rotating the positive real axis counterclockwise through this angle. Therefore, additional arguments of $17 - 3i$ are

$$\tan^{-1}\left(\frac{17}{3}\right) + \frac{3\pi}{2} + 2n\pi, \qquad n \text{ any integer.}$$

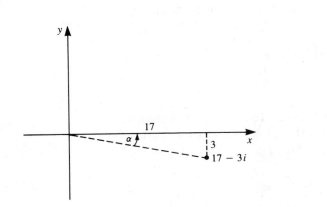

Figure 19.22. $\alpha = \tan^{-1}(\frac{3}{17})$.

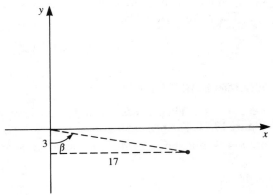

Figure 19.23. $\beta = \tan^{-1}(\frac{17}{3})$.

The principal argument of $17 - 3i$ is $\text{Arg}(17 - 3i) = -\tan^{-1}(\frac{3}{17})$. Since $|17 - 3i| = \sqrt{318}$, the polar form of $17 - 3i$ is

$$\sqrt{318}\{\cos[-\tan^{-1}(\tfrac{3}{17})] + i\sin[-\tan^{-1}(\tfrac{3}{17})]\}. \quad \blacksquare$$

Here are some properties of modulus and argument.

THEOREM 19.1

Let z and w be complex numbers. Then

1. $|zw| = |z||w|$.

2. If $w \neq 0$,

$$\left| \frac{z}{w} \right| = \frac{|z|}{|w|}.$$

3. $\arg(zw) = \arg(z) + \arg(w)$.

4. $\arg(z/w) = \arg(z) - \arg(w)$.

5. If r is a positive number, z and rz have the same arguments.   ∎

Conclusion (1) states that the magnitude of a product is the product of the magnitudes; (2) states that the magnitude of a quotient is the quotient of the magnitudes.

Conclusions (3) and (4) must be interpreted carefully, since the symbol $\arg(z)$ stands for an infinite set of numbers. Read

$$\arg(zw) = \arg(z) + \arg(w)$$

to mean that any argument of $\arg(zw)$ can be written as a sum of arguments of z and of w; in the same way, the statement

$$\arg\left(\frac{z}{w}\right) = \arg(z) - \arg(w)$$

means that any argument of z/w can be written as an argument of z minus an argument of w. These rules are reminiscent of familiar properties of logarithms.

Proof of Theorem 19.1   We will prove the theorem using trigonometric identities. Write the polar forms of z and w:

$$z = r_1[\cos(\theta_1) + i \sin(\theta_1)]$$

and

$$w = r_2[\cos(\theta_2) + i \sin(\theta_2)].$$

Then

$$zw = r_1[\cos(\theta_1) + i \sin(\theta_1)]r_2[\cos(\theta_2) + i \sin(\theta_2)]$$
$$= r_1r_2\{[\cos(\theta_1)\cos(\theta_2) - \sin(\theta_1)\sin(\theta_2)] + i[\sin(\theta_1)\cos(\theta_2) + \cos(\theta_1)\sin(\theta_2)]\}$$
$$= r_1r_2[\cos(\theta_1 + \theta_2) + i \sin(\theta_1 + \theta_2)].$$

This is the polar form of zw. From it, we conclude immediately that

$$|zw| = r_1r_2 = |z||w|.$$

Further, $\theta_1 + \theta_2$, the sum of an argument of z and an argument of w is an argument of zw.

Conclusions (2) and (4) can be proved by a similar argument (divide the polar form of z by the polar form of w); details are left to the student.

Conclusion (5) is intuitively obvious. If we represent $z = x + iy$ as a vector from the origin to the point (x, y), as in Figure 19.24, multiplying this vector by a positive number does not change the angle θ.   ∎

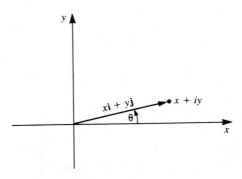

Figure 19.24

Conclusions (3) and (4) of Theorem 19.1 are not true if arg is replaced by Arg. For example, $\text{Arg}(-1) = \pi$ and $\text{Arg}(i) = \pi/2$, so

$$\text{Arg}(-1) + \text{Arg}(i) = \frac{3\pi}{2}.$$

But $\text{Arg}(-1 \cdot i) = \text{Arg}(-i) = -\pi/2$.

Here is an illustration of Theorem 19.1 for two specific complex numbers.

EXAMPLE 19.9

Let $z = 2i$ and $w = 3 + 3i$. Then $|z| = 2$, $|w| = \sqrt{18}$, and

$$\arg(z) = \frac{\pi}{2} + 2k\pi \quad \text{and} \quad \arg(w) = \frac{\pi}{4} + 2j\pi,$$

with k and j any integers. Now compute

$$zw = 2i(3 + 3i) = 6i - 6 = -6 + 6i.$$

Then

$$|zw| = |-6 + 6i| = \sqrt{36 + 36} = \sqrt{72} = 2\sqrt{18} = |z||w|.$$

Further (note Figure 19.25),

$$\arg(zw) = \arg(-6 + 6i) = \tfrac{3}{4}\pi + 2m\pi$$

and

$$\arg(z) + \arg(w) = \frac{\pi}{2} + 2k\pi + \frac{\pi}{4} + 2j\pi = \frac{3}{4}\pi + 2(k + j)\pi.$$

Comparing these two lines, we see that any value of $\arg(zw)$ does indeed equal a value of $\arg(z)$ plus a value of $\arg(w)$. Now look at quotients:

$$\frac{z}{w} = \frac{2i}{3 + 3i} = \frac{2}{3}\frac{i}{1 + i}\frac{1 - i}{1 - i} = \frac{1}{3}(1 + i).$$

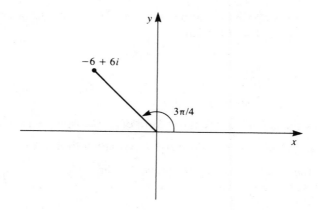

Figure 19.25

Then

$$\left|\frac{z}{w}\right| = \sqrt{\frac{1}{9} + \frac{1}{9}} = \frac{\sqrt{2}}{3} = \frac{|z|}{|w|}.$$

Now

$$\arg\left(\frac{z}{w}\right) = \frac{\pi}{4} + 2n\pi$$

with n any integer. Compare this result with

$$\arg(z) - \arg(w) = \frac{\pi}{2} + 2k\pi - \left(\frac{\pi}{4} + 2j\pi\right)$$

$$= \frac{\pi}{4} + 2(k - j)\pi.$$

Any value of $\arg(z/w)$ is a value of $\arg(z) - \arg(w)$. ∎

By a simple induction from Theorem 19.1, we have, for any positive integer n,

$$|z^n| = |z|^n$$

and

$$\arg(z^n) = \arg(z \cdot z \cdots z) = \arg(z) + \arg(z) + \cdots + \arg(z)$$
$$= n \arg(z).$$

Using these results, we can prove de Moivre's theorem.

THEOREM 19.2   de Moivre

For any integer n and any real number θ,

$$[\cos(\theta) + i \sin(\theta)]^n = \cos(n\theta) + i \sin(n\theta).$$

Proof Clearly, the proposed equation holds if $n = 0$. Now let n be any positive integer. Let $z = \cos(\theta) + i \sin(\theta)$. Then

$$|z^n| = |z|^n = 1^n = 1$$

and

$$\arg(z^n) = n \arg(z) = n\theta + 2k\pi,$$

in which k is any integer. Thus, the polar form of z^n is $\cos(n\theta) + i \sin(n\theta)$. Therefore,

$$z^n = [\cos(\theta) + i \sin(\theta)]^n = \cos(n\theta) + i \sin(n\theta).$$

This completes the proof if n is a nonnegative integer.

Now suppose that n is a negative integer. Then $m = -n$ is a positive integer, and we have

$$
\begin{aligned}
[\cos(\theta) + i \sin(\theta)]^n &= \frac{1}{[\cos(\theta) + i \sin(\theta)]^m} \\
&= \frac{1}{\cos(m\theta) + i \sin(m\theta)} \\
&= \frac{1}{\cos(m\theta) + i \sin(m\theta)} \frac{\cos(m\theta) - i \sin(m\theta)}{\cos(m\theta) - i \sin(m\theta)} \\
&= \cos(m\theta) - i \sin(m\theta) = \cos(-n\theta) - i \sin(-n\theta) \\
&= \cos(n\theta) + i \sin(n\theta)
\end{aligned}
$$

because $\cos(-n\theta) = \cos(n\theta)$ and $\sin(-n\theta) = -\sin(n\theta)$. This completes the proof of the theorem. ∎

The following calculation illustrates how a fact about complex numbers can be used to prove a fact about real-valued functions.

EXAMPLE 19.10

Express $\cos(3\theta)$ and $\sin(3\theta)$ in terms of $\sin(\theta)$ and $\cos(\theta)$.

By de Moivre's theorem,

$$[\cos(\theta) + i \sin(\theta)]^3 = \cos(3\theta) + i \sin(3\theta).$$

But

$$[\cos(\theta) + i \sin(\theta)]^3 = \cos^3(\theta) - 3 \cos(\theta)\sin^2(\theta) + i[3 \sin(\theta)\cos^2(\theta) - \sin^3(\theta)].$$

Compare the real and imaginary parts of the last two equations to get

$$\cos(3\theta) = \cos^3(\theta) - 3 \cos(\theta)\sin^2(\theta)$$

and

$$\sin(3\theta) = 3 \sin(\theta)\cos^2(\theta) - \sin^3(\theta). \quad ∎$$

The theme of this example (deriving a result about real functions by separating real and imaginary parts of a complex quantity) is a useful idea which we will exploit many times.

PROBLEMS FOR SECTION 19.2

In each of Problems 1 through 10, determine $|z|$, $\arg(z)$, and $\text{Arg}(z)$ and write the polar form of the complex number. [In some problems, the solution for $\arg(z)$ may involve $\tan^{-1}(\alpha)$ for an appropriate α.]

1. $1 + 4i$ **2.** $2 - 6i$ **3.** $8 - 2i$ **4.** $-3 - 6i$ **5.** $-14i$

6. $-2 - 12i$ **7.** $3 + 9i$ **8.** $-4 - i$ **9.** $-8 - 3i$ **10.** $5 + i$

In each of Problems 11 through 20, a complex number is given in polar form. Write the number in the form $a + bi$.

11. $3\left[\cos\left(\dfrac{\pi}{4}\right) + i\sin\left(\dfrac{\pi}{4}\right)\right]$
    12. $9\left[\cos\left(\dfrac{7\pi}{4}\right) + i\sin\left(\dfrac{7\pi}{4}\right)\right]$
    13. $8\left[\cos\left(\dfrac{2\pi}{3}\right) + i\sin\left(\dfrac{2\pi}{3}\right)\right]$

14. $14\left[\cos\left(\dfrac{7\pi}{6}\right) + i\sin\left(\dfrac{7\pi}{6}\right)\right]$
    15. $4\left[\cos\left(\dfrac{11\pi}{4}\right) + i\sin\left(\dfrac{11\pi}{4}\right)\right]$
    16. $15\left[\cos\left(\dfrac{5\pi}{6}\right) + i\sin\left(\dfrac{5\pi}{6}\right)\right]$

17. $5\left[\cos\left(\dfrac{5\pi}{4}\right) + i\sin\left(\dfrac{5\pi}{4}\right)\right]$
    18. $14\left[\cos\left(\dfrac{\pi}{6}\right) + i\sin\left(\dfrac{\pi}{6}\right)\right]$
    19. $7\left[\cos\left(\dfrac{8\pi}{3}\right) + i\sin\left(\dfrac{8\pi}{3}\right)\right]$

20. $15\left[\cos\left(\dfrac{15\pi}{4}\right) + i\sin\left(\dfrac{15\pi}{4}\right)\right]$

21. Prove *Lagrange's trigonometric identity*

$$\sum_{j=0}^{n} \cos(j\theta) = \frac{1}{2} + \frac{\sin\left[(n + \frac{1}{2})\theta\right]}{2\sin\left(\dfrac{\theta}{2}\right)}$$

if $\sin(\theta/2) \neq 0$. *Hint:* Begin with the algebraic identity

$$\sum_{j=0}^{n} z^j = \frac{1 - z^{n+1}}{1 - z} \quad \text{if} \quad z \neq 1,$$

which may be assumed. Let $z = \cos(\theta) + i\sin(\theta)$, use de Moivre's theorem, and separate real and imaginary parts in this algebraic identity.

22. Write $\cos(4\theta)$ and $\sin(4\theta)$ as a sum of products of powers of $\sin(\theta)$ and $\cos(\theta)$. *Hint:* Use the binomial theorem (Problem 40 of Section 19.1) to expand

$$[\cos(\theta) + i\sin(\theta)]^4,$$

then use de Moivre's theorem.

19.3 *Limits and Derivatives of Complex-Valued Functions* _____

Thus far, we have dealt with elementary properties of complex numbers. We will now introduce complex functions and develop first concepts of the calculus of such functions.

In a first calculus course, the emphasis is on real functions, by which we mean any function which operates on real numbers and produces real numbers. For example, the function defined by $f(x) = x^2$, for all real x, takes any real number x and produces the nonnegative real number x^2. The function defined by $g(x) = \ln(x)$ takes any positive number x and produces its natural logarithm, which is a real number.

The idea of a complex function is similar except that now we allow the function to operate on complex numbers and produce complex numbers (with the understanding that a real number x is a complex number of the form $x + 0i$). For example, we might have

$$f(z) = z^2 + 1,$$

which takes the complex number z and produces the complex number $z^2 + 1$. For example, $f(1 + i) = (1 + i)^2 + 1 = 1 + 2i$, and $f(i) = i^2 + 1 = 0$.

We say that $f(z)$ has *limit L* as z approaches z_0, and write

$$\lim_{z \to z_0} f(z) = L,$$

if the following two conditions hold:

1. $f(z)$ is defined for all z in some disk $|z - z_0| < r$ about z_0, except possibly at z_0 itself.
2. Given any $\epsilon > 0$, there exists some $\delta > 0$ such that

$$|f(z) - L| < \epsilon \quad \text{if} \quad 0 < |z - z_0| < \delta.$$

In this definition, L is complex, although in specific cases a limit might be real. For example, if $f(z) = z^2$, it is intuitively apparent that

$$\lim_{z \to 1+i} f(z) = (1 + i)^2 = 2i \qquad \text{(complex)},$$

while

$$\lim_{z \to i} f(z) = i^2 = -1 \qquad \text{(real)}.$$

In terms of disks, we can state the definition of limit as follows: $\lim_{z \to z_0} f(z) = L$ if, given any disk D_ϵ of radius ϵ about L, there exists a disk D_δ of radius δ about z_0 such that, for every z in D_δ, except possibly its center z_0, $f(z)$ lies in D_ϵ. Figure 19.26 illustrates this concept.

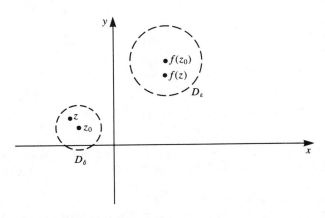

Figure 19.26. $|f(z) - f(z_0)| < \varepsilon$ if $0 < |z - z_0| < \delta$.

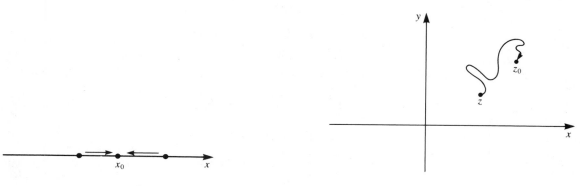

Figure 19.27   Figure 19.28

The definition of limit for complex functions is similar in form to the definition of limit for real-valued functions of a real variable. The main difference is that now we speak of a number lying in a disk about a point in the plane, while in the real function case we speak of a number lying in an interval on the real line.

On the real line, there are only two ways x can approach x_0: from the left or from the right (Figure 19.27). In the plane, however, z may approach z_0 along infinitely many different paths (Figure 19.28). In order for $f(z)$ to have the limit L as z approaches z_0, it is necessary (and sufficient) that $f(z)$ approach L along *all* of these paths.

In some instances, limits are quite obvious. For example,

$$\lim_{z \to 2i} (z^2 + z) = (2i)^2 + 2i = -4 + 2i$$

and

$$\lim_{z \to 1+i} \frac{1}{z} = \frac{1}{1+i} = \frac{1}{1+i} \frac{1-i}{1-i} = \frac{1}{2} - \frac{1}{2} i.$$

For some time, we will not see limits much more complicated than this because we have not as yet defined complex exponential, logarithm, and power functions.

Some rules for computing limits of complex functions are similar in form to rules for limits of real functions. Assume that $\lim_{z \to z_0} f(z)$ and $\lim_{z \to z_0} g(z)$ exist. Then

1. $\lim_{z \to z_0} [f(z) + g(z)] = \lim_{z \to z_0} f(z) + \lim_{z \to z_0} g(z).$

2. $\lim_{z \to z_0} \alpha f(z) = \alpha \lim_{z \to z_0} f(z)$ for any number α.

3. $\lim_{z \to z_0} f(z)g(z) = \left[\lim_{z \to z_0} f(z) \right]\left[\lim_{z \to z_0} f(z) \right].$

4. $\lim_{z \to z_0} \dfrac{f(z)}{g(z)} = \dfrac{\lim_{z \to z_0} f(z)}{\lim_{z \to z_0} g(z)},$

provided the denominator is not zero.

We say that f is *continuous* at z_0 if $\lim_{z \to z_0} f(z) = f(z_0)$. This requires that $f(z)$ be defined for z in some disk about z_0, including at z_0 itself. As in the real case, complex polynomials

$$\alpha_0 + \alpha_1 z + \cdots + \alpha_n z^n$$

are continuous for all z. Here, $\alpha_0, \ldots, \alpha_n$ are complex numbers. A rational function (quotient of polynomials) has the form

$$\frac{\alpha_0 + \alpha_1 z + \cdots + \alpha_n z^n}{\beta_0 + \beta_1 z + \cdots + \beta_m z^m}$$

and is continuous wherever the denominator does not vanish.

We say that f is *differentiable* at z_0 if

$$\lim_{z \to z_0} \frac{f(z) - f(z_0)}{z - z_0}$$

exists and is finite. An equivalent statement is that f is differentiable at z_0 if

$$\lim_{h \to 0} \frac{f(z_0 + h) - f(z_0)}{h}$$

exists and is finite. In this limit, h approaches the complex number zero through complex values. That is, think of a disk about the origin and imagine h as a complex number approaching the origin along an arbitrary path. The difference quotient

$$\frac{f(z_0 + h) - f(z_0)}{h}$$

must approach the same value as h approaches the origin along any such path. When the limit of this difference quotient exists as $h \to 0$, we call this limit the *derivative of f at z_0* and write it as $f'(z_0)$, or $df/dz|_{z = z_0}$. Sometimes we use the delta notation, with Δz in place of h:

$$f'(z_0) = \lim_{\Delta z \to 0} \frac{f(z_0 + \Delta z) - f(z_0)}{\Delta z}.$$

Rules for differentiating complex functions are similar in form to rules for real functions. If f and g are differentiable at z_0,

1. $(f + g)'(z_0) = f'(z_0) + g'(z_0)$.
2. $(\alpha f)'(z_0) = \alpha f'(z_0)$ for any complex number α.
3. $(fg)'(z_0) = f'(z_0)g(z_0) + f(z_0)g'(z_0)$.
4. $\left(\dfrac{f}{g}\right)'(z_0) = \dfrac{g(z_0)f'(z_0) - f(z_0)g'(z_0)}{[g(z_0)]^2}$ if $g(z_0) \neq 0$.

In addition, there is a complex version of the chain rule. If $w = g(z)$ and $f'(w)$ and $g'(z)$ exist, then

$$\frac{d}{dz}[f(g(z)] = f'(w)g'(z) = f'(g(z))g'(z).$$

In Leibniz notation,

$$\frac{d}{dw}[f(g(w))] = \frac{df}{dz}\frac{dg}{dw} = \frac{df}{dz}\frac{dz}{dw}.$$

In the calculus of complex functions, as in the calculus of real functions, we rarely compute a derivative using the limit definition. Instead, we use differentiation rules together with rules for differentiating specific functions, which we will develop throughout this chapter. However, the limit definition is needed in instances in which we do not know that the function is differentiable or the usual derivative rules do not apply. Here are two examples.

EXAMPLE 19.11

Let $f(z) = \bar{z}$. Show that $f'(i)$ does not exist.
   Consider

$$\lim_{z \to i} \frac{f(z) - f(i)}{z - i} = \lim_{z \to i} \frac{\bar{z} - (-i)}{z - i} = \lim_{z \to i} \frac{\bar{z} + i}{z - i}.$$

Consider different paths of approach of z toward i. In Figure 19.29, z approaches i along the imaginary axis. On this path, $z = \alpha i$, with α real, and to have $z \to i$, we must have $\alpha \to 1$. The difference quotient along this path is

$$\frac{\bar{z} + i}{z - i} = \frac{-\alpha i + i}{\alpha i - i} = \frac{1 - \alpha}{\alpha - 1} = -1,$$

a constant which remains at -1 as $\alpha \to 1$.

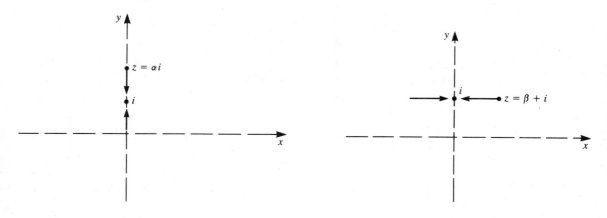

Figure 19.29   Figure 19.30

Now examine the case in which $z \to i$ along a horizontal path parallel to the real axis, as in Figure 19.30. Along this path, we have $z = \beta + i$, with β real. To have $z \to i$, we must have $\beta \to 0$. Along this path, the difference quotient is

$$\frac{\bar{z} + i}{z - i} = \frac{(\overline{\beta + i}) + i}{\beta + i - i} = \frac{\beta - i + i}{\beta} = 1.$$

Along this path, the difference quotient is identically 1 and approaches 1 as $\beta \to 0$.
   Because the difference quotient approaches two different values along two different paths of approach of z to i, we conclude that it has no limit as $z \to i$ and that $f'(i)$ does not exist. ■

EXAMPLE 19.12

Let $f(z) = |z|^2$. We claim that f is differentiable at zero and $f'(0) = 0$, but f is not differentiable at any point $z \neq 0$.

To show that $f'(0) = 0$, compute

$$\lim_{z \to 0} \frac{f(z) - f(0)}{z - 0} = \lim_{z \to 0} \frac{|z|^2}{z}.$$

Since $|z|^2 = z\bar{z}$, $|z|^2/z = \bar{z}$ if $z \neq 0$. If $z \to 0$, $\bar{z} \to 0$. Therefore,

$$f'(0) = \lim_{z \to 0} \bar{z} = 0.$$

Now we will show that $f'(z_0)$ does not exist if $z_0 \neq 0$. Write $z_0 = a + ib$. We will consider the limit of the difference quotient along two different paths of approach of z to z_0. First, consider (Figure 19.31) a vertical path of approach. Along this path, $z = a + it$, and we want to have t approach b. On this path, the difference quotient is

$$\frac{|a + it|^2 - |a + ib|^2}{(a + it) - (a + ib)} = \frac{t^2 - b^2}{i(t - b)} = \frac{t + b}{i} \to \frac{2b}{i} = -2bi$$

as $t \to b$.

Next, consider a horizontal path of approach (Figure 19.32). Along this path, $z = t + ib$, and we want $t \to a$. We have

$$\frac{|t + ib|^2 - |a + ib|^2}{(t + ib) - (a + ib)} = \frac{t^2 - a^2}{t - a} = t + a \to 2a \quad \text{as} \quad t \to a.$$

The limits along these paths are the same if and only if $2a = -2bi$, which requires that $a = b = 0$. If $z_0 \neq 0$, $a \neq 0$ or $b \neq 0$, and $f'(z_0)$ does not exist.

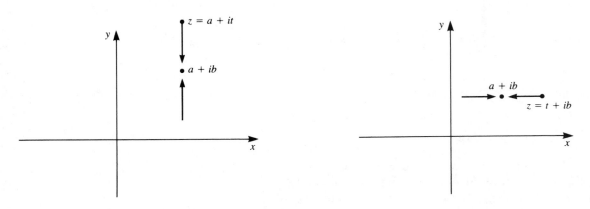

Figure 19.31   Figure 19.32   ∎

If f is differentiable at z_0, f must be continuous at z_0, just as with real-valued functions. To see this, observe that f is continuous at z_0 if and only if

$$\lim_{h \to 0} f(z_0 + h) = f(z_0),$$

or, equivalently,

$$\lim_{h \to 0} [f(z_0 + h) - f(z_0)] = 0.$$

Now write

$$\lim_{h \to 0} [f(z_0 + h) - f(z_0)] = \lim_{h \to 0} \frac{f(z_0 + h) - f(z_0)}{h} h$$

$$= \left[\lim_{h \to 0} \frac{f(z_0 + h) - f(z_0)}{h} \right] \left[\lim_{h \to 0} h \right] = f'(z_0) \cdot h = 0.$$

A complex function f is said to be *analytic* at z_0 if f is differentiable for all z in some disk about z_0. That is, there must exist a number r such that $f'(z)$ exists for all z in $|z - z_0| < r$. For example, $f(z) = z^2$ is differentiable for all z and so is analytic at each complex number. By contrast, $f(z) = |z|^2$ is differentiable at the origin and nowhere else. Thus, f is not analytic at *any* point. In particular, f is not analytic at the origin, because we cannot find a disk about the origin throughout which f is differentiable.

In the next section, we will derive an important necessary and sufficient condition for a complex function to be analytic over a given set of complex numbers.

PROBLEMS FOR SECTION 19.3

In each of Problems 1 through 10, use the limit definition of derivative to evaluate $f'(z_0)$ or to show that $f'(z_0)$ does not exist.

1. $f(z) = z^2$; $z_0 = 1 + i$

2. $f(z) = z + 2\bar{z}$; $z_0 = 3i$

3. $f(z) = \dfrac{z}{1 + z}$; $z_0 = 2$

4. $f(z) = \text{Im}(z)$; z_0 any complex number

5. $f(z) = |z|$; $z_0 = i$

6. $f(z) = \text{Re}(z)$; z_0 any complex number

7. $f(z) = \text{Re}(z) + i \, \text{Im}(z)$; z_0 any complex number

8. $f(z) = (\bar{z})^2$; $z_0 = 2 - i$

9. $f(z) = i + \text{Re}(z)$; $z_0 = 4 + 7i$

10. $f(z) = \dfrac{2}{1 + z}$; $z_0 = -1 + 4i$

11. Give an example of a complex-valued function which is continuous for all z but differentiable for no z.

12. Suppose that $\lim_{z \to z_0} f(z) = L$. Prove that $\lim_{z \to z_0} \text{Re}[f(z)] = \text{Re}(L)$ and $\lim_{z \to z_0} \text{Im}[f(z)] = \text{Im}(L)$.

19.4 *The Cauchy-Riemann Equations* _____

There is an important test for analyticity which can be stated in terms of partial derivatives of the real and imaginary parts of a complex function. Write $z = x + iy$ and $f(z) = u(x, y) + iv(x, y)$. Then

$$\text{Re}(z) = x, \qquad \text{Im}(z) = y,$$
$$\text{Re}(f(z)) = u(x, y), \quad \text{and} \quad \text{Im}(f(z)) = v(x, y).$$

For example, if $f(z) = z^2$, $f(z) = (x + iy)^2 = x^2 - y^2 + 2ixy$, and we have $u(x, y) = x^2 - y^2$ and $v(x, y) = 2xy$ *(not 2ixy)*.

THEOREM 19.3   Cauchy-Riemann Equations

Let f be continuous in a disk $|z - z_0| < r$ about $z_0 = x_0 + iy_0$, and assume that f is differentiable at z_0. Then

$$\frac{\partial u}{\partial x}(x_0, y_0) = \frac{\partial v}{\partial y}(x_0, y_0) \quad \text{and} \quad \frac{\partial u}{\partial y}(x_0, y_0) = -\frac{\partial v}{\partial x}(x_0, y_0).$$

With the understanding that these partial derivatives are evaluated at (x_0, y_0), these equations may be written

$$\frac{\partial u}{\partial x} = \frac{\partial v}{\partial y} \quad \text{and} \quad \frac{\partial u}{\partial y} = -\frac{\partial v}{\partial x}.$$

These equations are called the *Cauchy-Riemann equations* for u and v.

Proof   Since $f'(z_0)$ exists by assumption,

$$f'(z_0) = \lim_{\Delta z \to 0} \frac{f(z_0 + \Delta z) - f(z_0)}{\Delta z}.$$

Write $\Delta z = \Delta x + i \, \Delta y$, and manipulate the difference quotient as follows:

$$\frac{f(z_0 + \Delta z) - f(z_0)}{\Delta z}$$

$$= \frac{[u(x_0 + \Delta x, y_0 + \Delta y) + iv(x_0 + \Delta x, y_0 + \Delta y)] - [u(x_0, y_0) + iv(x_0, y_0)]}{\Delta x + i \, \Delta y}$$

$$= \frac{u(x_0 + \Delta x, y_0 + \Delta y) - u(x_0, y_0)}{\Delta x + i \, \Delta y} + i \, \frac{v(x_0 + \Delta x, y_0 + \Delta y) - v(x_0, y_0)}{\Delta x + i \, \Delta y}.$$

Since $f'(z_0)$ exists, we must obtain the same limit from the difference quotient along each path of approach of Δz to the origin. Choose two specific paths as follows.

Path I: Let $\Delta z \to 0$ along the real axis. Then $\Delta z = \Delta x \to 0$ and $\Delta y = 0$. We get

$$f'(z_0) = \lim_{\Delta x \to 0} \left[\frac{u(x_0 + \Delta x, y_0) - u(x_0, y_0)}{\Delta x} + i \, \frac{v(x_0 + \Delta x, y_0) - v(x_0, y_0)}{\Delta x} \right]$$

$$= \frac{\partial u}{\partial x}(x_0, y_0) + i \, \frac{\partial v}{\partial x}(x_0, y_0).$$

Path II: Let $\Delta z \to 0$ along the imaginary axis. Then $\Delta x = 0$ and $\Delta z = i \, \Delta y \to 0$. We have

$$f'(z_0) = \lim_{\Delta y \to 0} \left[\frac{u(x_0, y_0 + \Delta y) - u(x_0, y_0)}{i \, \Delta y} + i \, \frac{v(x_0, y_0 + \Delta y) - v(x_0, y_0)}{i \, \Delta y} \right]$$

$$= \frac{1}{i} \frac{\partial u}{\partial y}(x_0, y_0) + \frac{\partial v}{\partial y}(x_0, y_0)$$

$$= \frac{\partial v}{\partial y}(x_0, y_0) - i \, \frac{\partial u}{\partial y}(x_0, y_0),$$

since $1/i = -i$.

We have shown that

$$f'(z_0) = \frac{\partial u}{\partial x}(x_0, y_0) + i\frac{\partial v}{\partial x}(x_0, y_0) = \frac{\partial v}{\partial y}(x_0, y_0) - i\frac{\partial u}{\partial y}(x_0, y_0).$$

Therefore,

$$\frac{\partial u}{\partial x}(x_0, y_0) = \frac{\partial v}{\partial y}(x_0, y_0) \quad \text{and} \quad \frac{\partial v}{\partial x}(x_0, y_0) = -\frac{\partial u}{\partial y}(x_0, y_0),$$

as we wanted to show. ∎

We will now discuss some results which depend not only on properties of the function under consideration but also on properties of the set of points in the complex plane where the function is defined. We have seen this type of dependence before, in Section 16.3, when we discussed potential theory in the plane. Those ideas are intimately related to the complex function concepts we will now introduce.

Suppose we have a set D of points in the complex plane. We call D a *domain* if D satisfies the following two conditions:

1. About every point in D, we can place an open disk containing only points of D.
2. Between any two points of D, there is a piecewise-smooth curve whose graph lies entirely in D.

As an example, an open disk is a domain, as is the set of all complex numbers with positive real and imaginary parts. Similarly, the left-plane consisting of all z with $\text{Im}(z) < 0$ is a domain. A closed disk $|z - z_0| \le r$ is not a domain. Any open disk about a point on the boundary circle will extend outside the closed disk, violating part (1) of the definition of domain. We say that a complex function f is *analytic in a domain* D if f is analytic at each $z = x + iy$, where (x, y) is a point in D.

We can now state a corollary to Theorem 19.3.

Corollary to Theorem 19.3

Let $f(z) = u(x, y) + iv(x, y)$ be analytic in a domain D. Then u and v satisfy the Cauchy-Riemann equations for all (x, y) in D.

Proof   Let $z = x + iy$ be in D. Then $f'(z)$ exists at each z in D; hence, f is also continuous at each point of D. Now let $z = x + iy$ be in D. Since D is a domain, there is an open disk about z containing only points of D. Therefore, f is continuous in this disk. Since f is differentiable at z, by Theorem 19.3, u and v satisfy the Cauchy-Riemann equations at (x, y). ∎

EXAMPLE 19.13

Let $f(z) = z^3$ for all x. By cubing $x + iy$, we find that

$$f(z) = (x^3 - 3xy^2) + i(3x^2y - y^3).$$

Thus, $f = u + iv$, with

$$u(x, y) = x^3 - 3xy^2 \quad \text{and} \quad v(x, y) = 3x^2y - y^3.$$

We can let D be the entire complex plane, which is certainly a domain. By the corollary, u and v satisfy the Cauchy-Riemann equations at each point of the plane. In this example,

$$\frac{\partial u}{\partial x} = 3x^2 - 3y^2 = \frac{\partial v}{\partial y}$$

and

$$\frac{\partial v}{\partial x} = 6xy = -\frac{\partial u}{\partial y}. \quad \blacksquare$$

The converse of the corollary to Theorem 19.3 is not true. The example given in Problem 13 at the end of this section shows that it is possible for u and v to satisfy the Cauchy-Riemann equations throughout a domain D even though $f = u + iv$ is not analytic throughout D. With an additional hypothesis, however, the Cauchy-Riemann equations do imply analyticity in a domain.

To derive this result, we will need the following fact from the calculus of real-valued functions of two real variables.

LEMMA 19.1

If $F(x, y)$ is continuous with continuous first partial derivatives in some disk about (x_0, y_0), there exists a function φ of two variables such that

$$\lim_{(\Delta x, \Delta y) \to (0,0)} \varphi(\Delta x, \Delta y) = 0$$

and

$$F(x_0 + \Delta x, y_0 + \Delta y) - F(x_0, y_0) = \frac{\partial F}{\partial x}(x_0, y_0) \Delta x + \frac{\partial F}{\partial y}(x_0, y_0) \Delta y$$
$$+ \varphi(\Delta x, \Delta y)\sqrt{(\Delta x)^2 + (\Delta y)^2}. \quad \blacksquare$$

We can now state the result we want.

THEOREM 19.4

Let u and v be continuous, with continuous first partial derivatives satisfying the Cauchy-Riemann equations, in a domain D. Define a complex function f by putting $f(z) = u(x, y) + iv(x, y)$ for $z = x + iy$ in D. Then f is analytic in D.

Proof   Let z_0 be in D. Apply Lemma 19.1 to each of u and v to write

$$f(z_0 + \Delta z) - f(z_0) = u(x_0 + \Delta x, y_0 + \Delta y) - u(x_0, y_0)$$
$$+ i[v(x_0 + \Delta x, y_0 + \Delta y) - v(x_0, y_0)]$$
$$= \frac{\partial u}{\partial x}(x_0, y_0) \Delta x + \frac{\partial u}{\partial y}(x_0, y_0) \Delta y$$
$$+ \varphi_1(\Delta x, \Delta y)\sqrt{(\Delta x)^2 + (\Delta y)^2}$$
$$+ i\left[\frac{\partial v}{\partial x}(x_0, y_0) \Delta x + \frac{\partial v}{\partial y}(x_0, y_0) \Delta y\right.$$
$$\left.+ \varphi_2(\Delta x, \Delta y)\sqrt{(\Delta x)^2 + (\Delta y)^2}\right]$$

for some functions φ_1 and φ_2 having limit zero as $(\Delta x, \Delta y) \to (0, 0)$.

Use the Cauchy-Riemann equations and some algebraic manipulation in the last equation to write

$$f(z_0 + \Delta z) - f(z_0) = \left[\frac{\partial u}{\partial x}(x_0, y_0) + i \frac{\partial v}{\partial x}(x_0, y_0) \right](\Delta x + i \, \Delta y)$$
$$+ \left[\varphi_1(\Delta x, \Delta y) + i\varphi_2(\Delta x, \Delta y) \right] \sqrt{(\Delta x)^2 + (\Delta y)^2}.$$

Then

$$\frac{f(z_0 + \Delta z) - f(z_0)}{\Delta z} = \frac{\partial u}{\partial x}(x_0, y_0) + i \frac{\partial v}{\partial x}(x_0, y_0)$$
$$+ \left[\varphi_1(\Delta x, \Delta y) + i\varphi_2(\Delta x, \Delta y) \right] \frac{\sqrt{(\Delta x)^2 + (\Delta y)^2}}{\Delta x + i \, \Delta y}.$$

Now observe that

$$\left| \frac{\sqrt{(\Delta x)^2 + (\Delta y)^2}}{\Delta x + i \, \Delta y} \right| = \frac{\sqrt{(\Delta x)^2 + (\Delta y)^2}}{\sqrt{(\Delta x)^2 + (\Delta y)^2}} = 1.$$

As $\Delta z \to 0$, we have $(\Delta x, \Delta y) \to (0, 0)$; then

$$\left[\varphi_1(\Delta x, \Delta y) + i\varphi_2(\Delta x, \Delta y) \right] \frac{\sqrt{(\Delta x)^2 + (\Delta y)^2}}{\Delta x + i \, \Delta y} \to 0$$

and

$$\frac{f(z_0 + \Delta z) - f(z_0)}{\Delta z} \to \frac{\partial u}{\partial x}(x_0, y_0) + i \frac{\partial v}{\partial x}(x_0, y_0).$$

This shows that $f'(z_0)$ exists and equals $(\partial u/\partial x)(x_0, y_0) + i(\partial v/\partial x)(x_0, y_0)$.

Thus far, we have proved that f is differentiable at each point of D. To prove that f is analytic in D, let z_0 be any point of D. Since D is a domain, there is a disk about z_0 containing only points of D. Since f is differentiable at each point in this disk, f is analytic at z_0. ∎

By means of the Cauchy-Riemann equations, we can establish a relationship between analytic functions and Laplace's equation. Recall that Laplace's equation for a function φ of two variables is

$$\nabla^2 \varphi = \frac{\partial^2 \varphi}{\partial x^2} + \frac{\partial^2 \varphi}{\partial y^2} = 0.$$

This equation arises in potential theory, fluid flow, study of electrical and magnetic fields, heat conduction, and wave propagation. A function satisfying Laplace's equation in a domain is said to be *harmonic* in that domain.

THEOREM 19.5

Let $f(z) = u(x, y) + iv(x, y)$ be analytic in a domain D. Then u and v are harmonic in D.

Proof We will see later that analyticity of f in D implies that u and v have continuous second partial derivatives in D. Assuming this for now, we can conclude that the mixed partial derivatives of u are equal, and similarly for v:

$$\frac{\partial^2 u}{\partial x \, \partial y} = \frac{\partial^2 u}{\partial y \, \partial x} \quad \text{and} \quad \frac{\partial^2 v}{\partial x \, \partial y} = \frac{\partial^2 v}{\partial y \, \partial x}.$$

We can now prove the theorem. Since $\partial u / \partial x = \partial v / \partial y$,

$$\frac{\partial^2 u}{\partial x^2} = \frac{\partial}{\partial x}\left(\frac{\partial u}{\partial x}\right) = \frac{\partial}{\partial x}\left(\frac{\partial v}{\partial y}\right) = \frac{\partial}{\partial y}\left(\frac{\partial v}{\partial x}\right) = \frac{\partial}{\partial y}\left(-\frac{\partial u}{\partial y}\right) = -\frac{\partial^2 u}{\partial y^2}.$$

Therefore,

$$\frac{\partial^2 u}{\partial x} + \frac{\partial^2 u}{\partial y^2} = 0.$$

A similar argument shows that $\nabla^2 v = 0$. ∎

EXAMPLE 19.14

Let $f(x) = z^2 + 2iz$. Then $f(z) = x^2 - y^2 - 2y + i(2xy + 2x)$. Here,

$$u(x, y) = x^2 - y^2 - 2y \quad \text{and} \quad v(x, y) = 2xy + 2x.$$

It is routine to check that $\nabla^2 u = 0$ and $\nabla^2 v = 0$. ∎

In concluding this section, we will note one extra dividend of the proof of Theorem 19.4. There, we showed that

$$f'(z_0) = \frac{\partial u}{\partial x}(x_0, y_0) + i \frac{\partial v}{\partial x}(x_0, y_0)$$

if f is analytic at z_0. By using the Cauchy-Riemann equations, we obtain the following equivalent expressions for this derivative:

$$f'(z_0) = \frac{\partial u}{\partial x}(x_0, y_0) - i \frac{\partial u}{\partial y}(x_0, y_0)$$

$$= \frac{\partial v}{\partial y}(x_0, y_0) + i \frac{\partial v}{\partial x}(x_0, y_0)$$

$$= \frac{\partial v}{\partial y}(x_0, y_0) - i \frac{\partial u}{\partial y}(x_0, y_0).$$

In some instances, one of these expressions may be more convenient to use than the others in computing a derivative.

For the remainder of this chapter, we will develop complex analogues of familiar functions such as the natural logarithm, exponential, and trigonometric functions.

PROBLEMS FOR SECTION 19.4

In each of Problems 1 through 12, find u and v so that $f(z) = u(x, y) + iv(x, y)$. Determine whether the Cauchy-Riemann equations hold in the given domain, and use Theorems 19.3 and 19.4 to determine whether f is analytic in the domain. If f is analytic, verify that u and v are harmonic in D.

1. $f(z) = z - i$; D is the entire complex plane

2. $f(z) = z^2 - iz$; D is the complex plane

3. $f(z) = \dfrac{2z + 1}{z}$; D consists of all $z \neq 0$

4. $f(z) = i|z|^2$; D is the complex plane

5. $f(z) = \dfrac{z}{\text{Re}(z)}$; D consists of all z with $\text{Re}(z) > 0$

6. $f(z) = \dfrac{2 + \text{Im}(z)}{|z|^2}$; D consists of all $z \neq 0$

7. $f(z) = (z + 2)^2$; D is the complex plane

8. $f(z) = \bar{z} + \text{Im}(3z)$; D is the complex plane

9. $f(z) = iz + |z|$; D is the complex plane

10. $f(z) = z^3 + 2z + 1$; D is the complex plane

11. $f(z) = \dfrac{z - 2i}{z}$; D consists of all $z \neq 0$

12. $f(z) = |z| + z$; D is the complex plane

13. Let

$$f(z) = \begin{cases} z^5/|z|^4 & \text{if} \quad z \neq 0 \\ 0 & \text{if} \quad z = 0. \end{cases}$$

Compute $u(x, y) = \text{Re}(f(z))$ and $v(x, y) = \text{Im}(f(z))$, and show that these functions satisfy the Cauchy-Riemann equations at the origin. Show also that $f'(0)$ does not exist. Why does this not contradict Theorem 19.4?

14. Let $f(z) = |z|^2$.

(a) Show that the real and imaginary parts of f satisfy the Cauchy-Riemann equations at $z = 0$.

(b) Explain why f is not analytic at zero even though f is differentiable at zero.

15. Suppose that f is analytic in a domain D and that $f'(z) = 0$ for all z in D. Prove that, for some constant k, $f(z) = k$ for all z in D.

16. Let M be the set of all complex numbers with nonzero real parts. Give an example of a function which is analytic on M and such that $f'(z) = 0$ for all z in M but $f(z)$ is not constant on M.

19.5 *Rational Powers and Roots*

We would like to define what z^w means when z and w are complex numbers and $z \neq 0$. As a starting point, we will consider the case in which z is complex and w is a rational number (quotient of integers). Thus, consider z^r, with z complex and nonzero and r rational.

If $r = 0$, we define $z^0 = 1$.

If r is a positive integer, say, $r = n$, we define z^n as the product of z with itself n times. That is, $z^2 = z \cdot z$ and, proceeding by induction, $z^n = z^{n-1} z$ if $n \geq 2$.

If n is a positive integer and $z \neq 0$, we define

$$z^{-n} = \frac{1}{z^n}.$$

These natural definitions should hold no surprises. As examples,

$$(2 + i)^3 = (2 + i)^2 (2 + i) = (3 + 4i)(2 + i) = 2 + 11i$$

and

$$(2 + i)^{-3} = \frac{1}{(2 + i)^3} = \frac{1}{2 + 11i}$$

$$= \frac{1}{2 + 11i} \frac{2 - 11i}{2 - 11i} = \frac{2}{125} - \frac{11}{125} i.$$

We carried out this calculation until we had the power in the form $a + bi$ in order to be able to recognize the real and imaginary parts.

We will next look at $z^{1/n}$ when n is a positive integer; $z^{1/n}$ is called an nth *root* of z. Let

$$w = z^{1/n}.$$

Then

$$z = w^n.$$

Write z and w in polar form:

$$z = r[\cos(\theta) + i \sin(\theta)]$$

and

$$w = R[\cos(\varphi) + i \sin(\varphi)].$$

By de Moivre's theorem,

$$z = r[\cos(\theta) + i \sin(\theta)] = w^n = R^n[\cos(n\varphi) + i \sin(n\varphi)].$$

Therefore,

$$r = R^n$$

and

$$n\varphi = \theta + 2k\pi,$$

with k any integer. Then

$$|w| = R = r^{1/n},$$

the nonnegative nth root of the nonnegative real number r; and

$$\arg(w) = \varphi = \frac{\theta + 2k\pi}{n}$$

for $k = 0, \pm 1, \pm 2, \ldots$. The nth roots of z are therefore the complex numbers

$$w = z^{1/n} = r^{1/n}\left[\cos\left(\frac{\theta + 2k\pi}{n}\right) + i \sin\left(\frac{\theta + 2k\pi}{n}\right)\right] \tag{19.6}$$

for $k = 0, \pm 1, \pm 2, \ldots$.

In fact, equation (19.6) yields exactly n distinct values of $z^{1/n}$, and we get them by letting $k = 0, 1, \ldots, n - 1$. Any other integer value of k produces one of these n values again. Note also that in equation (19.6), θ may be any value of $\arg(z)$.

EXAMPLE 19.15

Find the cube roots of 8.

Let $z = 8$. Then $r = |z| = 8$, and one value of $\arg(z)$ is zero. Since $r^{1/3} = 2$, equation (19.6) yields the following cube roots of 8:

$$2\left[\cos\left(\frac{2k\pi}{3}\right) + i \sin\left(\frac{2k\pi}{3}\right)\right], \qquad k = 0, 1, 2.$$

These numbers are

$$2 \qquad\qquad (k = 0),$$

$$2\left[-\frac{1}{2} + \frac{\sqrt{3}}{2}i\right], \quad \text{or} \quad -1 + \sqrt{3}i \qquad (k = 1),$$

and

$$2\left[-\frac{1}{2} + \frac{\sqrt{3}}{2}i\right], \quad \text{or} \quad -1 - \sqrt{3}i \qquad (k = 2).$$

If we use other values of k, we simply repeat the roots we have already found. For example, with $k = 3$, we get the root

$$2[\cos(2\pi) + i\sin(2\pi)],$$

or 2, which we have already found. If $k = 4$, we get

$$2\left[\cos\left(\frac{8\pi}{3}\right) + i\sin\left(\frac{8\pi}{3}\right)\right].$$

But $\cos(8\pi/3) = \cos(2\pi/3)$ and $\sin(8\pi/3) = \sin(2\pi/3)$, so we get the same root obtained when $k = 1$, and so on. ∎

EXAMPLE 19.16

Find the fourth roots of $1 - i$.
   We have $r = |1 - i| = \sqrt{2}$, and one value of $\arg(1 - i)$ is $-\pi/4$. The four fourth roots of $1 - i$ are contained in the expression

$$(\sqrt{2})^{1/4}\left[\cos\left(\frac{\frac{-\pi}{4} + 2k\pi}{4}\right) + i\sin\left(\frac{\frac{-\pi}{4} + 2k\pi}{4}\right)\right], \qquad k = 0, 1, 2, 3.$$

These roots are

$$2^{1/8}\left[\cos\left(\frac{-\pi}{16}\right) + i\sin\left(\frac{-\pi}{16}\right)\right],$$

$$2^{1/8}\left[\cos\left(\frac{7\pi}{16}\right) + i\sin\left(\frac{7\pi}{16}\right)\right],$$

$$2^{1/8}\left[\cos\left(\frac{15\pi}{16}\right) + i\sin\left(\frac{15\pi}{16}\right)\right],$$

and

$$2^{1/8}\left[\cos\left(\frac{23\pi}{16}\right) + i\sin\left(\frac{23\pi}{16}\right)\right].$$

These roots are approximately

$$1.07 - 0.213i, \qquad 0.213 + 1.07i, \qquad -1.07 + 0.213i, \quad \text{and} \quad -0.213 - 1.07i.$$

A sketch of these roots as points in the plane is shown in Figure 19.33. ∎

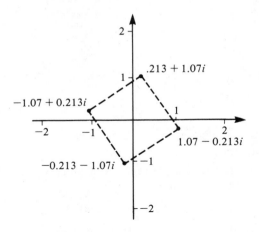

Figure 19.33. Fourth roots of $1 - i$ (forming vertices of a square).

EXAMPLE 19.17

The nth roots of 1 are called the nth *roots of unity.* They are

$$\cos\left(\frac{2k\pi}{n}\right) + i \sin\left(\frac{2k\pi}{n}\right)$$

for $k = 0, 1, 2, \ldots, n - 1$.

These roots form the vertices of a regular polygon of n sides, inscribed in the unit circle about the origin and positioned so that one vertex is at 1. Figure 19.34 shows the fifth roots of unity ($n = 5$), which are

$$\cos\left(\frac{2k\pi}{5}\right) + i \sin\left(\frac{2k\pi}{5}\right)$$

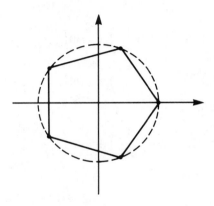

Figure 19.34. Fifth roots of unity.

for $k = 0, 1, 2, 3, 4$. The numbers are

$$1, \quad \cos\left(\frac{2\pi}{5}\right) + i \sin\left(\frac{2\pi}{5}\right), \quad \cos\left(\frac{4\pi}{5}\right) + i \sin\left(\frac{4\pi}{5}\right),$$

$$\cos\left(\frac{6\pi}{5}\right) + i \sin\left(\frac{6\pi}{5}\right), \quad \text{and} \quad \cos\left(\frac{8\pi}{5}\right) + i \sin\left(\frac{8\pi}{5}\right).$$

These roots are approximately $1, 0.309 + 0.951i, -0.809 + 0.588i, -0.809 - 0.588i,$ and $0.309 - 0.951i.$ ∎

Once we know how to compute the nth roots of z, we can compute any rational power $z^{m/n}$. We can first raise z to the power m, then take the nth roots of this number. That is,

$$z^{m/n} = (z^m)^{1/n}.$$

Equivalently, we could first compute the nth roots of z, then raise each to the power m:

$$z^{m/n} = (z^{1/n})^m.$$

EXAMPLE 19.18

Find all values of $(2 - 2i)^{3/5}$.

First, calculate $(2 - 2i)^3 = -16 - 16i$. Now take the fifth roots of this number. Calculate $|-16 - 16i| = \sqrt{512}$, and an argument of $-16 - 16i$ is $5\pi/4$. Thus, the fifth roots of $(2 - 2i)^3$ are

$$(\sqrt{512})^{1/5}\left[\cos\left(\frac{\frac{5\pi}{4} + 2k\pi}{5}\right) + i \sin\left(\frac{\frac{5\pi}{4} + 2k\pi}{5}\right)\right],$$

or

$$(512)^{1/10}\left[\cos\left(\frac{5\pi + 8k\pi}{20}\right) + \sin\left(\frac{5\pi + 8k\pi}{20}\right)\right]$$

for $k = 0, 1, 2, 3, 4$.

Specifically, the $\frac{3}{5}$ powers of $2 - 2i$ are

$$(512)^{1/10}\left[\cos\left(\frac{\pi}{4}\right) + i \sin\left(\frac{\pi}{4}\right)\right],$$

$$(512)^{1/10}\left[\cos\left(\frac{13\pi}{20}\right) + i \sin\left(\frac{13\pi}{20}\right)\right],$$

$$(512)^{1/10}\left[\cos\left(\frac{21\pi}{20}\right) + i \sin\left(\frac{21\pi}{20}\right)\right],$$

$$(512)^{1/10}\left[\cos\left(\frac{29\pi}{20}\right) + i \sin\left(\frac{29\pi}{20}\right)\right],$$

and

$$(512)^{1/10}\left[\cos\left(\frac{37\pi}{20}\right) + i \sin\left(\frac{37\pi}{20}\right)\right].$$

To four decimal places, the $\frac{3}{5}$ powers of $2 - 2i$ are approximately

$$1.3195 + 1.3195i, \qquad -0.8472 + 1.6627i, \qquad -1.8431 - 0.2919i,$$
$$-0.2919 - 1.8431i, \quad \text{and} \quad 1.6627 - 0.8472i. \quad \blacksquare$$

EXAMPLE 19.19

Calculate all values of $(2 - i)^{-2/3}$.
    First, compute

$$\frac{1}{2 - i} = \frac{1}{2 - i} \frac{2 + i}{2 + i} = \frac{2}{5} + \frac{1}{5} i.$$

Now write

$$(2 - i)^{-2/3} = \left(\frac{1}{2 - i}\right)^{2/3} = \left(\frac{2}{5} + \frac{1}{5} i\right)^{2/3}.$$

One way to proceed now is first to square $\frac{2}{5} + \frac{1}{5}i$, then to take the cube roots of the resulting number. If we do this, we first compute

$$(\tfrac{2}{5} + \tfrac{1}{5}i)^2 = \tfrac{3}{25} + \tfrac{4}{25}i.$$

Then

$$(2 - i)^{-2/3} = (\tfrac{3}{25} + \tfrac{4}{25}i)^{1/3}.$$

Now,

$$|\tfrac{3}{25} + \tfrac{4}{25}i| = \tfrac{1}{25}\sqrt{3^2 + 4^2} = \tfrac{1}{5},$$

and an argument of $\frac{3}{25} + \frac{4}{25}i$ is $\tan^{-1}(\frac{4}{3})$, as indicated in Figure 19.35. The $-\frac{2}{3}$ powers

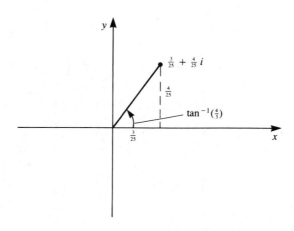

Figure 19.35

of $2 - i$ are

$$\left(\frac{1}{5}\right)^{1/3}\left[\cos\left(\frac{\tan^{-1}(\frac{4}{3}) + 2k\pi}{3}\right) + i\sin\left(\frac{\tan^{-1}(\frac{4}{3}) + 2k\pi}{3}\right)\right]$$

for $k = 0, 1, 2$. To four decimal places, these numbers are

$$0.5571 + 0.1179i, \qquad -0.4326 + 0.3935i, \quad \text{and} \quad -0.1245 - 0.5714i. \quad \blacksquare$$

In order to define z^w when w is irrational or complex, we must have the complex exponential and logarithm functions. We will define these in the next two sections.

PROBLEMS FOR SECTION 19.5

In each of Problems 1 through 15, find an expression containing all values of the given number. In some cases, the argument must be left in terms of an inverse tangent, as in Example 19.19. After obtaining an exact expression, calculate each power to four decimal places.

1. $(1 - i)^{1/2}$ **2.** $i^{1/4}$ **3.** $16^{1/4}$

4. $(1 + i)^{3/2}$ **5.** $(-16)^{1/4}$ **6.** $\left(\dfrac{1 + i}{1 - i}\right)^{1/3}$

7. $1^{1/6}$ **8.** $(-1)^{1/5}$ **9.** $(-2 - 2i)^{1/4}$

10. $(-i)^{1/3}$ **11.** $i^{3/5}$ **12.** $(3 + 3i)^{4/5}$

13. $(4i)^{-1/2}$ **14.** $(1 - i)^{-3/7}$ **15.** $(1 + 4i)^{1/3}$

16. Let $w_0, w_1, \ldots, w_{n-1}$ be the nth roots of unity. Prove that

$$w_0 + w_1 + \cdots + w_{n-1} = 0.$$

17. Show that the numbers

$$\frac{-b \pm \sqrt{b^2 - 4ac}}{2a}$$

satisfy the equation $az^2 + bz + c = 0$ for complex a, b, and c $(a \neq 0)$.

18. Solve the equation

$$z^2 + iz - 2 = 0.$$

19. Solve the equation

$$z^2 + (1 - i)z + i = 0.$$

20. Show that the roots of an equation $az^2 + bz + c = 0$ are complex conjugates if a, b, and c are real numbers and the roots are not equal.

21. The roots in Problems 18 and 19 are not complex conjugates. Why does this not contradict the result stated in Problem 20?

22. Solve the equation

$$z^4 - 2z^2 = -2.$$

Hint: Let $t = z^2$, solve for t, then solve for z.

23. Show that $(z^m)^{1/n} = (z^{1/n})^m$ for any positive integers n and m and complex z. (The equality here means that we obtain the same collection of numbers when the indicated operations are carried out on both sides of the equation.)

19.6 *The Complex Exponential Function*

We would like to define e^z for complex z in such a way that we have the usual real exponential function when z is real. If a and b are real numbers, we know that

$$e^{a+b} = e^a e^b.$$

Proceed formally to write

$$e^{x+iy} = e^x e^{iy}. \tag{19.7}$$

We know what e^x means, but we have to assign a meaning to e^{iy}. We did this in Section 2.4 when we discussed Euler's formula, but we will review the idea here.

Recall that

$$e^a = \sum_{n=0}^{\infty} \frac{1}{n!} a^n = 1 + a + \frac{1}{2!} a^2 + \frac{1}{3!} a^3 + \cdots.$$

Put $a = iy$ to obtain

$$
\begin{aligned}
e^{iy} &= \sum_{n=0}^{\infty} \frac{1}{n!} (iy)^n = 1 + iy + \frac{1}{2!} (iy)^2 + \frac{1}{3!} (iy)^3 + \frac{1}{4!} (iy)^4 + \cdots \\
&= 1 + \frac{1}{2!} (iy)^2 + \frac{1}{4!} (iy)^4 + \frac{1}{6!} (iy)^6 + \cdots + \left[iy + \frac{1}{3!} (iy)^3 + \frac{1}{5!} (iy)^5 + \cdots \right] \\
&= 1 - \frac{1}{2!} y^2 + \frac{1}{4!} y^4 - \frac{1}{6!} y^6 + \cdots + i\left[y - \frac{1}{3!} y^3 + \frac{1}{5!} y^5 - \cdots \right] \\
&= \cos(y) + i \sin(y).
\end{aligned}
$$

This is *Euler's formula*. Use this in equation (19.7) to write

$$e^{x+iy} = e^x [\cos(y) + i \sin(y)]. \tag{19.8}$$

This is the way we will define the *complex exponential function* e^{x+iy}. In fact, equation (19.8) defines e^z by specifying its polar form: the magnitude of e^z is e^x, and an argument of e^z is y.

EXAMPLE 19.20

As numerical examples, we have

$$e^{\pi i} = e^0 [\cos(\pi) + i \sin(\pi)] = -1,$$
$$e^{2-3i} = e^2 [\cos(-3) + i \sin(-3)] = e^2 [\cos(3) - i \sin(3)],$$

and

$$e^{4+i\pi/2} = e^4 \left[\cos\left(\frac{\pi}{2}\right) + i \sin\left(\frac{\pi}{2}\right) \right] = ie^4. \quad \blacksquare$$

The following theorem gives some properties of the complex exponential function.

THEOREM 19.6

1. e^z is analytic in the entire complex plane, and $(d/dz)e^z = e^z$.
2. $e^z e^w = e^{z+w}$.
3. $e^z \neq 0$ for all z.
4. $e^{-z} = 1/e^z$ for all z.
5. $e^z/e^w = e^{z-w}$.
6. If x is real, $|e^{ix}| = 1$.
7. $e^z = 1$ if and only if $z = 2n\pi i$ for some integer n.
8. $e^z = e^w$ if and only if $z = w + 2n\pi i$ for some integer n.

Proof of (1)   Write $e^z = e^x\cos(y) + ie^x\sin(y) = u(x, y) + iv(x, y)$, with

$$u(x, y) = e^x\cos(y) \quad \text{and} \quad v(x, y) = e^x\sin(y).$$

Then u and v are continuous, with continuous first partial derivatives in the entire complex plane, which is a domain. Further,

$$\frac{\partial u}{\partial x} = e^x\cos(y) = \frac{\partial v}{\partial y} \quad \text{and} \quad \frac{\partial v}{\partial x} = e^x\sin(y) = -\frac{\partial u}{\partial y}.$$

By Theorem 19.4, e^z is analytic for all z.

Proof of (2)   Write $z = a + bi$ and $w = c + di$. Then

$$
\begin{aligned}
e^z e^w &= e^a[\cos(b) + i\sin(b)]e^c[\cos(d) + i\sin(d)] \\
&= e^a e^c[\cos(b)\cos(d) - \sin(b)\sin(d) + i\{\sin(b)\cos(d) + \cos(b)\sin(d)\}] \\
&= e^{a+c}[\cos(b+d) + i\sin(b+d)] \\
&= e^{z+w}.
\end{aligned}
$$

Proof of (3)   Suppose that $e^z = 0$. Let $z = a + bi$. Then

$$e^a\cos(b) + ie^a\sin(b) = 0.$$

Then

$$e^a\cos(b) = 0 \quad \text{and} \quad e^a\sin(b) = 0.$$

Since $e^a \neq 0$, $\cos(b) = \sin(b) = 0$, which is impossible if a and b are real numbers. Thus, $e^z \neq 0$ for all z.

Proofs of (4) and (5) are left to the student.

Proof of (6)   If x is real,

$$|e^{ix}| = |\cos(x) + i\sin(x)| = \sqrt{\cos^2(x) + \sin^2(x)} = 1.$$

Proof of (7)   First, if n is any integer,

$$e^{2n\pi i} = \cos(2n\pi) + i\sin(2n\pi) = 1$$

because $\cos(2n\pi) = (-1)^{2n} = 1$ and $\sin(2n\pi) = 0$ for any integer n.

Conversely, suppose that $e^z = 1$. Write $z = a + bi$. Then

$$e^a \cos(b) + ie^a \sin(b) = 1.$$

Then

$$e^a \cos(b) = 1 \quad \text{and} \quad e^a \sin(b) = 0.$$

Since $e^a \neq 0$, $\sin(b) = 0$, so $b = k\pi$ for some integer k. Then

$$e^a \cos(k\pi) = 1.$$

But $\cos(k\pi) = (-1)^k$, so

$$e^a (-1)^k = 1.$$

Since $e^a > 0$ for any real number a, k must be even, say, $k = 2n$, for some integer n. Then

$$e^a (-1)^{2n} = e^a = 1.$$

But then $a = 0$, so $z = 2n\pi i$, as was to be proved.

Proof of (8)   If $e^z = e^w$, $e^z/e^w = e^{z-w} = 1$, so $z - w = 2n\pi i$ for some integer n. Conversely, if $z - w = 2n\pi i$ for some integer n,

$$e^{z-w} = \frac{e^z}{e^w} = e^{2n\pi i} = 1,$$

so $e^z = e^w$. ∎

We will conclude this section with the observation that the polar form of z can now be written

$$z = re^{i\theta},$$

where $r = |z|$ and θ is any argument of z. In this notation, de Moivre's theorem is particularly easy to prove. If n is an integer,

$$[\cos(\theta) + i\sin(\theta)]^n = (e^{i\theta})^n = e^{in\theta} = \cos(n\theta) + i\sin(n\theta).$$

Use of the complex exponential function in writing the polar form of a complex number is very convenient in parametrizing circles in the complex plane. Recall that a circle of radius r about z_0 is the locus of points z satisfying

$$|z - z_0| = r.$$

For z to satisfy this equation, $z - z_0$ must have magnitude r and so must be of the form $re^{i\theta}$. Thus, any point on this circle can also be written

$$z = z_0 + re^{i\theta}.$$

As θ varies from zero to 2π, $z_0 + re^{i\theta}$ varies once around this circle in a counterclockwise direction. Writing $z = z_0 + re^{i\theta}$ is equivalent to translating the origin to z_0 and writing the polar coordinates of the point. In later work on integrals of complex functions, we will often parametrize points on a circle in this way.

PROBLEMS FOR SECTION 19.6

In each of Problems 1 through 10, evaluate the exponential as a complex number in the form $a + bi$.

1. e^i

2. e^{1-i}

3. $e^{\pi - i}$

4. $e^{3\pi + 4i}$

5. e^{2-2i}

6. $e^{2 - \pi i/2}$

7. $e^{\pi(1+i)/4}$

8. $e^{2 - \pi i/6}$

9. $e^{-5 + 7i}$

10. $e^{9\pi i}$

In each of Problems 11 through 16, write the complex number in the form $re^{i\theta}$.

11. $3i$

12. $2 + i$

13. $1 - i$

14. $-1 - 2i$

15. $3 + i$

16. $-3 - 9i$

17. Prove that $e^{-z} = 1/e^z$ for any complex number z.

18. Prove that $e^{z-w} = e^z/e^w$ for any complex numbers z and w.

19. Show that, for any positive integer n, the nth roots of unity are $e^{2k\pi i/n}$, with $k = 0, 1, 2, \ldots, n - 1$.

20. Write e^{z^2} in the form $u(x, y) + iv(x, y)$, and show that u and v satisfy the Cauchy-Riemann equations in the entire complex plane.

21. Write $e^{1/z}$ in the form $u(x, y) + iv(x, y)$, and show that u and v satisfy the Cauchy-Riemann equations for $z \neq 0$.

19.7 *The Complex Logarithm Function* _____

There are several ways to approach the real-valued natural logarithm function. Whichever way we choose, one of the main properties is that, for any y, and for $x > 0$, $y = \ln(x)$ if and only if $x = e^y$. We will use this relationship to define the complex natural logarithm function, which we will denote as log to distinguish it from the real natural logarithm function ln. Define, for $z \neq 0$,

$$w = \log(z) \quad \text{if and only if} \quad z = e^w.$$

We would now like some way of explicitly calculating $\log(z)$, given $z \neq 0$.

Write z in polar form, $z = re^{i\theta}$, and let $w = u + iv$. Consider the equation $z = e^w$, which we want to solve for w. We have

$$z = re^{i\theta} = e^w = e^u e^{iv}. \tag{19.9}$$

By taking the magnitude of both sides of equation (19.9), we get

$$|r||e^{i\theta}| = |e^u||e^{iv}|.$$

But $|e^{i\theta}| = |e^{iv}| = 1$ because θ and v are real. Therefore,

$$r = e^u.$$

Since both r and e^u are real numbers, we conclude that

$$u = \ln(r) = \ln(|z|),$$

in which $\ln(r)$ is the real natural logarithm function evaluated at r.

Now we must solve for v to obtain w. Since we know that $r = e^u$, equation (19.9) yields

$$e^{i\theta} = e^{iv},$$

or

$$e^{i(v-\theta)} = 1.$$

By conclusion (7) of Theorem 19.6, $v - \theta$ must be an integer multiple of 2π. Therefore, for some integer n,

$$v = \theta + 2n\pi.$$

We now have, for $z \neq 0$,

$$w = \log(z) = \log(|z|) + i(\theta + 2n\pi).$$

Since θ is any argument of z, the symbol arg(z) contains all numbers of the form $\theta + 2n\pi$, and we can write

$$\log(z) = \ln|z| + i \arg(z).$$

Notice that there are infinitely many different natural logarithms of any nonzero complex number z. Given z, these logarithms differ by integer multiples of $2\pi i$.

EXAMPLE 19.21

$$\log(i) = \ln|i| + i \arg(i) = i\left(\frac{\pi}{2} + 2n\pi\right);$$

$$\log(1 - i) = \ln|1 - i| + i \arg(1 - i) = \ln(\sqrt{2}) + i\left(-\frac{\pi}{4} + 2n\pi\right);$$

and

$$\log(-4) = \ln|-4| + i \arg(-4) = \ln(4) + i(\pi + 2n\pi) = \ln(4) + (2n + 1)\pi i.$$

In these calculations, n can be any integer. ∎

Note that we can now take the logarithm of a negative number, something we cannot do with the real-valued natural logarithm.

Since the symbol log(z) denotes an infinite set of different complex numbers, log(z) is not a function as we usually define the term. For example, log(-4) actually stands for the set of complex numbers $\ln(4), \ln(4) + 3\pi i, \ln(4) + 5\pi i, \ldots, \ln(4) - 3\pi i, \ln(4) - 5\pi i, \ldots$. In order to have a complex logarithm *function*, we define the *principal logarithm* of $z \neq 0$ by

$$\text{Log}(z) = \ln(|z|) + i \text{Arg}(z),$$

using the principal value of the argument. For example,

$$\text{Log}(i) = \ln(|i|) + i \text{Arg}(i) = \frac{i\pi}{2},$$

$$\text{Log}(1 - i) = \ln(|1 - i|) + i \text{Arg}(1 - i) = \ln(\sqrt{2}) - \frac{i\pi}{4},$$

and

$$\text{Log}(-4) = \ln(4) + i \text{Arg}(-4) = \ln(4) + \pi i.$$

Here are some properties of the complex logarithm. Remember in this statement that Log(z) has a uniquely determined value for any nonzero complex number, while log(z) denotes a set of complex numbers, differing, for a given z, by integer multiples of $2n\pi i$.

THEOREM 19.7

Let z and w be nonzero complex numbers. Then

1. $e^{\log(z)} = z$, and $\log(e^z) = z + 2n\pi i$, with n any integer.
2. $\log(zw) = \log(z) + \log(w)$.
3. $\log(z/w) = \log(z) - \log(w)$.
4. For any rational number r, $\log(z^r) = r \log(z)$.
5. Let D consist of all complex numbers $x + iy$ with $y \neq 0$ if $x \leq 0$. Then D is a domain, and Log(z) is analytic in D. Further, for z in D,

$$\frac{d}{dz} \text{Log}(z) = \frac{1}{z}.$$

Conclusions (2) and (3) must be understood in the same sense as conclusions (3) and (4) of Theorem 19.1. For example, when we say that $\log(zw) = \log(z) + \log(w)$, we mean that any value of $\log(zw)$ can be written as a sum of values of $\log(z)$ and $\log(w)$.

Proof of (1)   First calculate

$$e^{\log(z)} = e^{\ln(|z|) + i\arg(z)} = e^{\ln(|z|)}e^{i\arg(z)}$$
$$= |z|e^{i\arg(z)} = z$$

because $|z|e^{i\arg(z)}$ is the polar form of z.

To show that $\log(e^z) = z + 2n\pi i$, write $z = x + iy$. Then

$$\log(e^z) = \ln(|e^z|) + i\arg(e^z) = \ln(|e^xe^{iy}|) + i\arg(e^xe^{iy})$$
$$= \ln(|e^x|) + i\arg(e^{iy}) = x + iy + 2n\pi i = z + 2n\pi i.$$

Here, we have used conclusion (6) of Theorem 19.6 to write $|e^{iy}| = 1$ and conclusion (5) of Theorem 19.1 to get $\arg(e^xe^{iy}) = \arg(e^{iy}) = y$, since $e^x > 0$.

Proof of (2)   Using properties of the magnitude and of the argument, we have

$$\log(zw) = \ln(|zw|) + i\arg(zw) = \ln(|z||w|) + i\arg(zw)$$
$$= \ln|z| + \ln|w| + i[\arg(z) + \arg(w)]$$
$$= \ln|z| + i\arg(z) + \ln|w| + i\arg(w)$$
$$= \log(z) + \log(w).$$

Conclusion (3) is proved by a similar argument.

Proof of (4)   We will use the facts that

$$|z^r| = |z|^r \quad \text{and} \quad \arg(z^r) = r\arg(z).$$

Thus,

$$\log(z^r) = \ln|z^r| + i \arg(z^r) = r \ln|z| + ir \arg(z)$$
$$= r[\ln|z| + i \arg(z)] = r \log(z).$$

Proof of (5)   We may think of D as the complex plane with the negative real axis and origin removed (Figure 19.36). Given any z_0 in D, there is a disk about z_0 containing only points of D, as illustrated in Figure 19.37. Further, as illustrated in Figure 19.38, any two points of D are connected by a piecewise-smooth curve in D. Thus, D is a domain.

Now let z be in D and let $w = \text{Log}(z)$. Then $z = e^w$. By the chain rule,

$$\frac{dz}{dz} = 1 = \frac{d}{dw}[e^w]\frac{dw}{dz} = e^w\frac{dw}{dz}.$$

Then

$$\frac{dw}{dz} = e^{-w} = e^{-\text{Log}(z)} = e^{\text{Log}(1/z)} = \frac{1}{z}. \quad \blacksquare$$

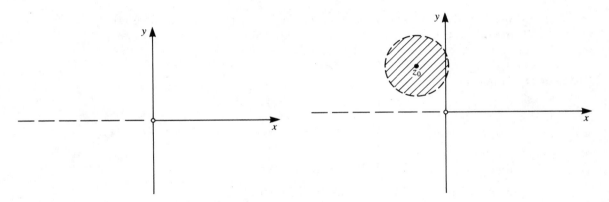

Figure 19.36   Figure 19.37

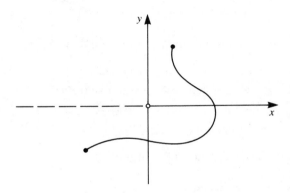

Figure 19.38

The formulas for differentiating e^z and Log(z) have the same form as derivative formulas for the real-valued exponential and logarithm functions. This is to be expected whenever we extend a real-valued function to complex values because the real and complex functions should agree if z is chosen to be real.

PROBLEMS FOR SECTION 19.7

In each of Problems 1 through 20, find all values of log(z) and determine Log(z).

1. $\log(2i)$ **2.** $\log(1 + i)$ **3.** $\log(-9)$ **4.** $\log(3 - 2i)$

5. $\log(1 - 2i)$ **6.** $\log[(1 + i)^{1/4}]$ **7.** $\log(i^{1/3})$ **8.** $\log[(2 - 2i)^{1/5}]$

9. $\log(-4 + 2i)$ **10.** $\log(-6 - 3i)$ **11.** $\log(2 + 4i)$ **12.** $\log[(1 - i)(2 + 3i)]$

13. $\log[(-1 - 2i)^2]$ **14.** $\log(-5i)$ **15.** $\log[(7 - 2i)^{1/8}]$ **16.** $\log(-8)$

17. $\log[(2i)^{1/2}]$ **18.** $\log[(6 - 18i)^{-3/4}]$ **19.** $\log[(2 + 2i)^{35}]$ **20.** $\log[(-4 + 3i)^{2/7}]$

21. Verify that

$$\log[(1 - i)(1 + i)] = \log(1 - i) + \log(1 + i)$$

by calculating each side of this equation independently.

22. Verify that

$$\log\left(\frac{1 - i}{1 + i}\right) = \log(1 - i) - \log(1 + i)$$

by calculating each side of this equation separately.

In each of Problems 23 through 27, show that $e^{\log(z)} = z$ and $\log(e^z) = z + 2n\pi i$ (n any integer) by direct computation of $e^{\log(z)}$ and $\log(e^z)$.

23. $z = -1 - i$ **24.** $z = i$ **25.** $z = -4$ **26.** $z = 3 + 3i$ **27.** $z = -5 + 5i$

28. Let $z = x + iy$, with x and y positive.

(a) Show that

$$\text{Log}(z) = \frac{1}{2}\ln(x^2 + y^2) + i\tan^{-1}\left(\frac{y}{x}\right).$$

(b) Verify that the real and imaginary parts of Log(z) satisfy the Cauchy-Riemann equations in the quarter-plane $x > 0$, $y > 0$.

19.8 *Powers of the Form z^w* _____

Using the complex exponential and logarithm functions, we can define z^w for any complex number $z \neq 0$ and any complex number w. Recall that, if $x > 0$ and y is any real number,

$$x^y = e^{y \ln(x)}.$$

Using this as a model, we define z^w by the equation

$$z^w = e^{w \log(z)} \qquad (z \neq 0).$$

Since log(z) has infinitely many values, so, in general, will z^w.

EXAMPLE 19.22

$$2^i = e^{i\log(2)} = e^{i[\ln(|2|) + i\arg(2)]} = e^{i[\ln(2) + i(0 + 2n\pi)]}$$
$$= e^{i[\ln(2) + 2n\pi i]} = e^{i\ln(2) - 2n\pi} = e^{-2n\pi}e^{i\ln(2)}$$
$$= e^{-2n\pi}[\cos(\ln(2)) + i\sin(\ln(2))]$$

for any integer n.   ∎

EXAMPLE 19.23

$$(1 - i)^{1+i} = e^{(1+i)\log(1-i)}$$
$$= e^{(1+i)[\ln(|1-i|) + i\arg(1-i)]}$$
$$= e^{(1+i)[\ln(\sqrt{2}) + i(-\pi/4 + 2n\pi)]}.$$

Now,

$$(1 + i)\left[\ln(\sqrt{2}) + i\left(-\frac{\pi}{4} + 2n\pi\right)\right] = \ln(\sqrt{2}) + \frac{\pi}{4} - 2n\pi + i\left[\ln(\sqrt{2}) - \frac{\pi}{4} + 2n\pi\right].$$

Therefore,

$$(1 - i)^{1+i} = e^{\ln(\sqrt{2}) + \pi/4 - 2n\pi + i[\ln(\sqrt{2}) - \pi/4 + 2n\pi]}$$
$$= e^{\ln(\sqrt{2}) + \pi/4 - 2n\pi}e^{i[\ln(\sqrt{2}) - \pi/4]} \qquad [\text{because } e^{2n\pi i} = 1]$$
$$= e^{\ln(\sqrt{2}) + \pi/4 - 2n\pi}\left\{\cos\left[\ln(\sqrt{2}) - \frac{\pi}{4}\right] + i\sin\left[\ln(\sqrt{2}) - \frac{\pi}{4}\right]\right\}. \quad ∎$$

In these examples, both 2^i and $(1 - i)^{1+i}$ have infinitely many values because n can be any integer.

Complex powers have many of the properties which hold for powers of real numbers.

THEOREM 19.8

Let z be any nonzero complex number, and let α and β be any numbers. Then

1. $z^\alpha z^\beta = z^{\alpha + \beta}$.
2. $z^\alpha / z^\beta = z^{\alpha - \beta}$.
3. $(z^\alpha)^\beta = z^{\alpha\beta}$.

In (1) through (3), an equality means that the quantity on the left yields the same values as the quantity on the right; usually, these powers will have infinitely many values.

Proof of (1)   Proceeding from the definition, we have

$$z^\alpha z^\beta = e^{\alpha\log(z)}e^{\beta\log(z)} = e^{(\alpha + \beta)\log(z)} = z^{\alpha + \beta}.$$

Parts (2) and (3) are proved similarly.   ∎

We can define a function called the *principal value of* z^w by

$$\Pr[z^w] = e^{w\operatorname{Log}(z)},$$

in which we use the principal logarithm function in place of the logarithm in the definition of powers.

EXAMPLE 19.24

From Examples 19.22 and 19.23 (with $n = 0$), we get

$$\Pr[2^i] = \cos[\ln(2)] + i \sin[\ln(\sqrt{2})]$$

and

$$\Pr[(1 - i)^{1+i}] = e^{\ln(\sqrt{2}) + \pi/4} \left\{ \cos\left[\ln(\sqrt{2}) - \frac{\pi}{4} \right] + i \sin\left[\ln(\sqrt{2}) - \frac{\pi}{4} \right] \right\}. \quad \blacksquare$$

THEOREM 19.9

Let D consist of all points in the complex plane except the origin and all points on the negative real axis. Then D is a domain. Further, if α is any complex number, $f(z) = \Pr[z^\alpha]$ is analytic on D, and

$$\frac{df}{dz} = \alpha \Pr[z^{\alpha-1}].$$

Proof   We have already observed that D is a domain [see conclusion (5) of Theorem 19.7]. For the derivative, compute

$$\frac{df}{dz} = \frac{d}{dz} [e^{\alpha \operatorname{Log}(z)}] = e^{\alpha \operatorname{Log}(z)} \frac{d}{dz} [\alpha \operatorname{Log}(z)]$$

$$= \Pr[z^\alpha] \alpha \frac{1}{z} = \alpha \Pr[z^{\alpha-1}]. \quad \blacksquare$$

PROBLEMS FOR SECTION 19.8

In each of Problems 1 through 15, determine all values of z^α and also determine $\Pr[z^\alpha]$.

1. i^{1+i} **2.** $(1 + i)^{2i}$ **3.** i^i **4.** $(1 + i)^{2-i}$ **5.** $(-1 + i)^{-3i}$

6. $(-4)^{2-i}$ **7.** 6^{-2-3i} **8.** $(7i)^{3i}$ **9.** $(1 - i)^{-2-2i}$ **10.** i^{2-4i}

11. 2^{3-i} **12.** 3^{5+i} **13.** $(3 - 2i)^i$ **14.** $(8 - 2i)^{1+2i}$ **15.** $(-3i)^{2i}$

16. Let $z = x + iy$, with $x > 0$ and $y > 0$. Determine u and v so that

$$\Pr[z^\alpha] = u(x, y) + iv(x, y),$$

with α an arbitrary but fixed complex number.

19.9 *Complex Trigonometric and*
Hyperbolic Functions _____

Euler's formula suggests a definition of the complex trigonometric functions. To motivate the definition, recall that, if y is real,

$$e^{iy} = \cos(y) + i \sin(y). \tag{19.10}$$

Replacing y with $-y$ yields

$$e^{-iy} = \cos(y) - i\sin(y) \tag{19.11}$$

because $\cos(-y) = \cos(y)$ and $\sin(-y) = \sin(y)$.

Solve equations (19.10) and (19.11) for $\cos(y)$ and $\sin(y)$ to get

$$\cos(y) = \tfrac{1}{2}(e^{iy} + e^{-iy})$$

and

$$\sin(y) = \frac{1}{2i}(e^{iy} - e^{-iy}).$$

These relationships suggest that we define, for any complex number z,

$$\cos(z) = \frac{1}{2}(e^{iz} + e^{-iz}) \quad \text{and} \quad \sin(z) = \frac{1}{2i}(e^{iz} - e^{-iz}).$$

The other trigonometric functions can be defined in obvious ways:

$$\tan(z) = \frac{\sin(z)}{\cos(z)}, \qquad \sec(z) = \frac{1}{\cos(z)}, \qquad \csc(z) = \frac{1}{\sin(z)}, \quad \text{and} \quad \cot(z) = \frac{1}{\tan(z)}$$

wherever the denominators are nonzero.

These complex trigonometric functions are extensions of the real trigonometric functions in the sense that each complex function agrees with its real counterpart if evaluated at a real value of z.

It is a simple matter to write these functions in the form $u(x, y) + iv(x, y)$ using Euler's formula. We will provide the details for $\sin(z)$. Let $z = x + iy$. Then

$$\sin(z) = \frac{1}{2i}(e^{iz} - e^{-iz})$$

$$= \frac{1}{2i}(e^{ix-y} - e^{-ix+y}) = \frac{1}{2i}(e^{-y}e^{ix} - e^{y}e^{-ix})$$

$$= \frac{1}{2i}e^{-y}[\cos(x) + i\sin(x)] - \frac{1}{2i}e^{y}[\cos(x) - i\sin(x)]$$

$$= \sin(x)\left(\frac{e^{y} + e^{-y}}{2}\right) + i\cos(x)\left(\frac{e^{y} - e^{-y}}{2}\right)$$

$$= \sin(x)\cosh(y) + i\cos(x)\sinh(y).$$

Therefore,

$$\sin(x + iy) = \sin(x)\cosh(y) + i\cos(x)\sinh(y). \tag{19.12}$$

A similar calculation yields

$$\cos(z) = \cos(x)\cosh(y) - i\sin(x)\sinh(y). \tag{19.13}$$

We use these equations to evaluate complex trigonometric functions at specific points in terms of real functions. For example, using equation (19.12) with $x = y = 1$, we have

$$\sin(1 + i) = \sin(1)\cosh(1) + i\cos(1)\sinh(1),$$

and, from equation (19.13), with $x = 0$ and $y = 2$,

$$\cos(2i) = \cos(0)\cosh(2) - i\sin(0)\sinh(2) = \cosh(2).$$

We can compute the other trigonometric functions as combinations of sines and cosines. For example,

$$\tan(2 + 6i) = \frac{\sin(2 + 6i)}{\cos(2 + 6i)} = \frac{\sin(2)\cosh(6) + i\cos(2)\sinh(6)}{\cos(2)\cosh(6) - i\sin(2)\sinh(6)},$$

a quotient of complex numbers we can manipulate into the form $a + bi$ if we wish.

By Theorem 19.4, it is routine to verify that $\sin(z)$ and $\cos(z)$ are analytic in the entire complex plane and that

$$\frac{d}{dz}\cos(z) = -\sin(z) \quad \text{and} \quad \frac{d}{dz}\sin(z) = \cos(z).$$

The formulas for the derivatives of the other trigonometric functions also mirror their real-valued counterparts. For example,

$$\frac{d}{dz}\tan(z) = \sec^2(z).$$

The complex trigonometric functions satisfy the identities we are familiar with for the real trigonometric functions. For example,

$$\sin^2(z) + \cos^2(z) = 1,$$
$$1 + \tan^2(z) = \sec^2(z),$$
$$\sin(z \pm w) = \sin(z)\cos(w) \pm \cos(z)\sin(w),$$
$$\cos(z \pm w) = \cos(z)\cos(w) \mp \sin(z)\sin(w),$$
$$\sin(2z) = 2\sin(z)\cos(z),$$
$$\cos(2z) = \cos^2(z) - \sin^2(z),$$

and so on. We will use such identities freely without proof, although the proofs are often easier for the complex case than for the real case because they can be done by routine manipulation of exponential functions. For example,

$$\sin(2z) = \frac{1}{2i}(e^{2iz} - e^{-2iz}) = \frac{1}{2i}(e^{iz} - e^{-iz})(e^{iz} + e^{-iz})$$

$$= 2\frac{1}{2i}(e^{iz} - e^{-iz})\frac{1}{2}(e^{iz} + e^{-iz}) = 2\sin(z)\cos(z).$$

The following theorem states that $\sin(z)$ and $\cos(z)$ have the same periods and zeros as their real-valued counterparts.

THEOREM 19.10

1. $\sin(z)$ and $\cos(z)$ are periodic of period 2π. That is,

$$\sin(z + 2n\pi) = \sin(z) \quad \text{and} \quad \cos(z + 2n\pi) = \cos(z)$$

for any z and any integer n.

2. $\sin(z) = 0$ if and only if $z = n\pi$ for some integer n.
3. $\cos(z) = 0$ if and only if $z = (2n + 1)\pi/2$ for some integer n.

Proof of (1)

$$\sin(z + 2n\pi) = \frac{1}{2i}(e^{iz + 2n\pi i} - e^{-iz - 2n\pi i})$$

$$= \frac{1}{2i}(e^{iz}e^{2n\pi i} - e^{-iz}e^{-2n\pi i}) = \frac{1}{2i}(e^{iz} - e^{-iz}) = \sin(z)$$

because $e^{2n\pi i} = e^{-2n\pi i} = 1$ for any integer n. A similar argument holds for $\cos(z)$.

Proof of (2)   First, if n is an integer,

$$\sin(n\pi) = \frac{1}{2i}(e^{in\pi} - e^{-in\pi}) = \frac{1}{2i} e^{-in\pi}(e^{2in\pi} - 1) = 0$$

because $e^{2in\pi} = 1$.

We must now show that, if $\sin(z) = 0$, z must be an integer multiple of π. To prove this, suppose that $z = x + iy$ and that $\sin(z) = 0$. By equation (19.12),

$$\sin(x)\cosh(y) + i \cos(x)\sinh(y) = 0.$$

Therefore,

$$\sin(x)\cosh(y) = 0 \quad \text{and} \quad \cos(x)\sinh(y) = 0.$$

But $\cosh(y) > 0$ for all real y; hence, $\sin(x) = 0$. Since the real-valued sine function is zero only if x is an integer multiple of π, $x = n\pi$ for some integer n. But then $\cos(x)\sinh(y) = 0$ implies that

$$\cos(n\pi)\sinh(y) = 0.$$

Since $\cos(n\pi) = (-1)^n \neq 0$, $\sinh(y) = 0$; hence, $y = 0$. Then $z = n\pi$, as was to be proved.
Part (3) of the theorem is proved similarly, using equation (19.13). ∎

The real-valued sine and cosine functions are bounded:

$$|\sin(x)| \leq 1 \quad \text{and} \quad |\cos(x)| \leq 1 \qquad \text{for real } x.$$

We will now show that $\sin(z)$ and $\cos(z)$ are *not* bounded for complex z. This is an important difference between the real and complex functions. To show that $\sin(z)$ is not bounded, put $z = ti$, with t real, into the definition

$$\sin(z) = \frac{1}{2i}(e^{iz} - e^{-iz}).$$

We get

$$\sin(ti) = \frac{1}{2i}(e^{-t} - e^{t}).$$

As $t \to \infty$ (so $z = ti$ moves up the imaginary axis),

$$|\sin(ti)| = \tfrac{1}{2}|e^{-t} - e^{t}| \to \infty$$

because $e^{-t} \to 0$ and $e^t \to \infty$. Similarly, if we let $t \to -\infty$, $e^{-t} \to \infty$ and $e^t \to 0$, so again $|\sin(ti)| \to \infty$. By a similar argument, $\cos(z)$ is also not bounded if z moves up or down the imaginary axis. We will see later that a complex function which is analytic for all z cannot be bounded unless it is a constant function.

The complex hyperbolic functions are defined by replacing real x with complex z in the usual definitions:

$$\cosh(z) = \frac{1}{2}(e^z + e^{-z}) \qquad \sinh(z) = \frac{1}{2}(e^z - e^{-z}),$$

$$\tanh(z) = \frac{\sinh(z)}{\cosh(z)}, \qquad \coth(z) = \frac{\cosh(z)}{\sinh(z)},$$

$$\operatorname{sech}(z) = \frac{1}{\cosh(z)}, \qquad \operatorname{csch}(z) = \frac{1}{\sinh(z)}.$$

Identities and derivative formulas carry over directly to these complex hyperbolic functions.

By extending trigonometric and hyperbolic functions to the complex plane, we obtain relationships which are not possible for the real-valued functions. For example,

$$\cos(iz) = \tfrac{1}{2}(e^{-z} + e^z) = \cosh(z)$$

and

$$-i\sin(iz) = -\tfrac{1}{2}(e^{-z} - e^z) = \sinh(z).$$

We will pursue other properties of these functions in the problems that follow.

PROBLEMS FOR SECTION 19.9

In each of Problems 1 through 12, write the function value as a complex (possibly real) number of the form $a + bi$. The numbers a and b may be written as real trigonometric or hyperbolic functions evaluated at a real number.

1. $\sin(i)$

2. $\cosh(1 - i)$

3. $\tan(2i)$

4. $\cos(-1 - i)$

5. $\sinh(4i)$

6. $\csc(2 + i)$

7. $\cos(-2 - 4i)$

8. $\sin(\pi + i)$

9. $\tanh(\pi i)$

10. $\cot\left[\left(\dfrac{\pi}{4}\right) + i\right]$

11. $\sin(e^i)$

12. $\cosh[\ln(i)]$

13. Prove that $\cos^2(z) + \sin^2(z) = 1$ for all complex z.

14. Prove that $\cos(z)$ and $\sin(z)$ are analytic in the entire complex plane. *Hint:* Use equations (19.12) and (19.13) and Theorem 19.4.

15. Write $\cosh(z)$ in the form $u(x, y) + iv(x, y)$. Use this result to show that $\cosh(z)$ is analytic in the entire complex plane.

16. Write $\sinh(z)$ in the form $u(x, y) + iv(x, y)$. Show that $\sinh(z)$ is analytic for all z.

17. We know that $\cosh(x) > 0$ for all real x. Is it possible for $\cosh(z)$ to be zero or negative for complex values of z?

18. Determine all complex numbers z such that $\sinh(z) = 0$.

19. Determine $u(x, y)$ and $v(x, y)$ such that $\tan(z) = u(x, y) + iv(x, y)$. Determine all z at which $\tan(z)$ is analytic.

20. Solve the equation $\sin(z) = \frac{1}{2}$. *Hint:* Write $(1/2i)(e^{iz} - e^{-iz}) = \frac{1}{2}$. Then $e^{iz} - e^{-iz} = i$. Multiply this equation by e^{iz} and treat it as a quadratic equation in e^{iz}. Solve this equation for e^{iz}, then solve the resulting equation for z.

21. Solve for all z such that $\sin(z) = i$.

22. Let $z = x + iy$. Show that

$$|\sinh(y)| \le |\sin(z)| \le \cosh(y).$$

23. Evaluate

$$\tan\left\{\frac{1}{i}\log\left[\left(\frac{1+iz}{1-iz}\right)^{1/2}\right]\right\}.$$

ADDITIONAL PROBLEMS

In each of Problems 1 through 25, carry out the indicated operations and write the final answer in the form $a + bi$. In each case, find all possible values.

1. e^{4-i}

2. $(3i)^{3/4}$

3. $\sinh(-5i)$

4. $\log[(3-2i)^2]$

5. $\cos[\log(1+i)]$

6. $3^{1/i}$

7. $\dfrac{2+4i}{e^{1-i}}$

8. $\sin(4i)$

9. $\cosh(-1+i)$

10. $2i - e^{2-i}$

11. $(2+i)^{4/7}$

12. $(i^i)^{2i}$

13. $\cosh(3^i)$

14. $\log(-4+7i)$

15. $(1+3i)^{21}$

16. $i^{\cos(i)}$

17. $(4+i)^{-3/7}$

18. $1^{1/9}$

19. $(2+i)^{3/8}$

20. $i^{1/5}$

21. $(2+i)^{2/3}$

22. $\log(2+8i)$

23. $\sinh[(2+5i)^2]$

24. $\log(e^{2-i})$

25. $\sin[\cos(2i)]$

26. Use de Moivre's theorem to write $\cos(5x)$ as a sum of constants times powers of $\cos(x)$ and $\sin(x)$, with x real.

27. Determine the locus of points satisfying

$$\left|\frac{z-i}{z+3-2i}\right| = 3.$$

28. Prove that $\cos(2z) = \cos^2(z) - \sin^2(z)$ for any complex number z.

29. Prove that $\cosh^2(z) - \sinh^2(z) = 1$ for any complex number z.

30. Write $\sec(z)$ in the form $u(x, y) + iv(x, y)$.

31. Find all solutions of the equation $z^5 = -4i$.

32. Prove that, for any nonzero complex number z,

$$\frac{1}{z} = \frac{\bar{z}}{|z|^2}.$$

33. Prove that, for any complex numbers z and w,

$$\cosh(z+w) = \cosh(z)\cosh(w) + \sinh(z)\sinh(w).$$

34. Let K be a positive constant. Show that the locus of points z such that

$$\left|\frac{z+1}{1-z}\right| = K$$

is a straight line if $K = 1$ and a circle if $K \neq 1$.

35. Let z and w be complex numbers such that

$$0 < \text{Arg}(z) - \text{Arg}(w) < \pi.$$

Prove that the triangle with vertices 0, z, and w has area $\frac{1}{2}\,\text{Im}(\bar{z}w)$.

36. Derive the identity

$$\cos(\theta) + \cos(3\theta) + \cdots + \cos[(2n-1)\theta] = \frac{\sin(2n\theta)}{2\sin(\theta)}$$

for any positive integer n and θ any real number such that $\sin(\theta) \neq 0$.

CHAPTER 20

Integration in the Complex Plane

20.1 Complex Line Integrals

Differentiation of the complex analogues of familiar real functions carries no surprises. For example, $(d/dz)\sin(z) = \cos(z)$, and $(d/dz)e^z = e^z$. We will now turn to integrals of complex functions, where the differences between the calculus of real functions and the calculus of complex functions are quite dramatic. We will begin with the antiderivative and then define the line integral of a complex function over a curve in the plane.

Given a complex function f, $\int f(z)\, dz$ is the *antiderivative*, or *indefinite integral*, of f and denotes any function whose derivative is f. For example,

$$\int e^z\, dz = e^z + C,$$

in which C is the constant of integration and can be any complex number. Similarly,

$$\int z^2\, dz = \tfrac{1}{3}z^3 + C.$$

The concept of an indefinite integral for complex functions is therefore entirely analogous to its real counterpart.

The definite (or Riemann) integral $\int_a^b f(x)\, dx$ of real calculus is replaced in complex analysis by the line integral of a complex function over a curve in the complex plane. We will now define this concept, assuming familiarity with line integrals of real functions of two variables over curves in the plane (see Section 16.1 and, for Green's theorem, Section 16.2).

Suppose we are given a curve C in the plane. Such a curve may be defined by parametric equations

$$x = x(t), \qquad y = y(t); \qquad a \le t \le b.$$

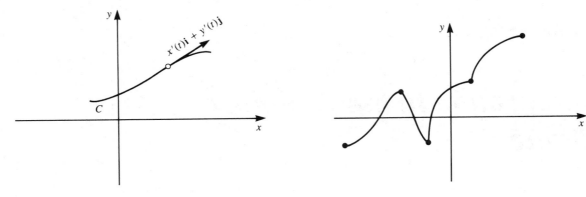

Figure 20.1   *Figure 20.2*. Typical piecewise-smooth curve.

C is *smooth* if $x'(t)$ and $y'(t)$ are continuous, and not simultaneously zero, on $[a, b]$. In this event, the vector $x'(t)\mathbf{i} + y'(t)\mathbf{j}$ is a continuous tangent to the curve at each point (Figure 20.1).

We say that C is *piecewise smooth* if x' and y' are continuous, except possibly at a finite number of points in $[a, b]$. A piecewise-smooth curve is therefore made up of a finite number of smooth pieces, but it may have points where there is no tangent, as in Figure 20.2.

Since we are identifying complex numbers with points in the plane, we will identify the point $(x(t), y(t))$ on the curve with the complex number $z(t) = x(t) + iy(t)$. For example, the unit circle $|z| = 1$ about the origin, oriented counterclockwise, is given parametrically by

$$z(t) = \cos(t) + i\sin(t); \qquad 0 \le t \le 2\pi.$$

As t increases from zero to 2π, the point $z(t)$ begins at 1 and moves counterclockwise around the circle to end at 1 when $t = 2\pi$.

We can also use the exponential function to write, more compactly,

$$z(t) = e^{it}; \qquad 0 \le t \le 2\pi.$$

These complex formulations are equivalent to giving the parametric equations

$$x(t) = \cos(t), \qquad y(t) = \sin(t); \qquad 0 \le t \le 2\pi.$$

If a curve is given parametrically by $z(t) = x(t) + iy(t)$ for $a \le t \le b$, $z(t)$ moves along the curve in a specific direction as t varies from a to b. This gives the curve an orientation, which we often indicate by placing an arrow on the graph, as in Figure 20.3.

Now suppose that f is a complex function, defined at least for points $z(t) = x(t) + iy(t)$ on the curve C. Subdivide $[a, b]$ on the t-axis by inserting points

$$a = t_0 < t_1 < t_2 < \cdots < t_{n-1} < t_n = b.$$

Let $z(t_j) = z_j$. We now have points

$$z_0 = x(a) + iy(a), \qquad z_1, z_2, \ldots, z_{n-1}, \qquad z_n = x(b) + iy(b)$$

on the curve, as shown in Figure 20.4. In each subinterval $[t_{j-1}, t_j]$, choose a point ξ_j.

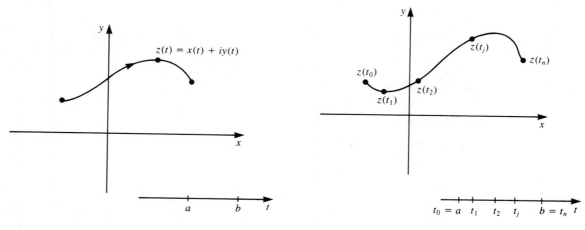

Figure 20.3   Figure 20.4

The point $w_j = z(\xi_j)$ lies on C between z_{j-1} and z_j, shown in Figure 20.5. Form a sum

$$\sum_{j=1}^{n} f(w_j)(z_j - z_{j-1}).$$ (20.1)

Take the limit as $n \to \infty$ and as each $|t_j - t_{j-1}| \to 0$. If the sums (20.1) approach a number L, we call L the *line integral of f over C* and denote it

$$\int_C f(z) \, dz.$$

Sometimes a curve C in the complex plane is called a *contour*, and the line integral is called a *contour integral*. We will sometimes refer to a line integral over C as just an integral over C.

We will state an existence theorem without proof.

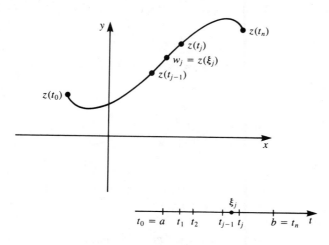

Figure 20.5

THEOREM 20.1

If C is piecewise smooth and f is continuous on C, $\int_C f(z)\, dz$ exists. ∎

Complex line integrals have some of the properties stated for line integrals of vector fields in Chapter 16.

THEOREM 20.2

Suppose that C is a piecewise-smooth curve and that $\int_C f(z)\, dz$ and $\int_C g(z)\, dz$ exist. Then

1. $\int_C [f(z) + g(z)]\, dz = \int_C f(z)\, dz + \int_C g(z)\, dz$.
2. For any complex number α, $\int_C \alpha f(z)\, dz = \alpha \int_C f(z)\, dz$.
3. If we denote by $-C$ the curve obtained by reversing orientation on C,

$$\int_{-C} f(z)\, dz = -\int_C f(z)\, dz. \quad ∎$$

We rarely use the limit definition to evaluate a line integral, although it can be done in simple cases. The following theorem provides an effective means of evaluating $\int_C f(z)\, dz$ in some instances.

THEOREM 20.3

If C is a smooth curve parametrized by $z = z(t)$ for $a \le t \le b$, and f is continuous on C,

$$\int_C f(z)\, dz = \int_a^b f(z(t))z'(t)\, dt. \quad ∎$$

EXAMPLE 20.1

Consider $\int_C \bar{z}\, dz$ if C is the straight line segment on the imaginary axis from 0 to $4i$.

Parametrize C by $z = it$, $0 \le t \le 4$. On C, f is continuous, and $f(z) = \bar{z} = \overline{it} = -it$. Further, C is smooth, and, on C, $z'(t) = i$. Then

$$f(z(t))z'(t) = (\overline{it})i = (-it)i = t;$$

hence,

$$\int_C \bar{z}\, dz = \int_0^4 t\, dt = 8. \quad ∎$$

EXAMPLE 20.2

Evaluate $\int_C (1/z)\, dz$, with C the unit circle about the origin, oriented counterclockwise.

We can parametrize C by $z = e^{it}$; $0 \le t \le 2\pi$. On C, $z'(t) = ie^{it}$, so

$$\int_C \frac{1}{z}\, dz = \int_0^{2\pi} \frac{1}{e^{it}} ie^{it}\, dt = \int_0^{2\pi} i\, dt = 2\pi i.$$

By the same token, for any integer $n \geq 2$,

$$
\begin{aligned}
\int_C \frac{1}{z^n} \, dz &= \int_0^{2\pi} \frac{1}{(e^{it})^n} \, ie^{it} \, dt = \int_0^{2\pi} e^{-int} ie^{it} \, dt \\
&= i \int_0^{2\pi} e^{i(1-n)t} \, dt = \frac{1}{1-n} e^{i(1-n)t} \Big]_0^{2\pi} \\
&= \frac{1}{1-n} [e^{2(1-n)\pi i} - 1] = 0
\end{aligned}
$$

because $e^{2(1-n)\pi i} = 1$ for any integer n. ∎

We will give an informal argument to suggest how $\int_C f(z) \, dz$ can be evaluated in terms of line integrals of the real and imaginary parts of f. Suppose that C is given by

$$ z(t) = x(t) + iy(t); \qquad a \leq t \leq b $$

and that $f(z)$ is written

$$ f(z) = u(x, y) + iv(x, y). $$

On C,

$$ f(z) = u(x(t), y(t)) + iv(x(t), y(t)) $$

and

$$ dz = dx + i \, dy. $$

Then

$$
\begin{aligned}
f(z) \, dz &= [u(x(t), y(t)) + iv(x(t), y(t))][dx + i \, dy] \\
&= [u(x(t), y(t)) \, dx - v(x(t), y(t)) \, dy] + i[v(x(t), y(t)) \, dx + u(x(t), y(t)) \, dy].
\end{aligned}
$$

This can be written more neatly by omitting the t-dependencies:

$$ f(z) \, dz = (u \, dx - v \, dy) + i(v \, dx + u \, dy). $$

Then

$$ \int_C f(z) \, dz = \int_C u \, dx - v \, dy + i \int_C v \, dx + u \, dy. $$

The right side of this equation gives $\int_C f(z) \, dz$ in terms of line integrals of real-valued functions. We will state this conclusion as a theorem.

THEOREM 20.4

Let $f = u + iv$ be continuous on the smooth curve C. Then

$$ \int_C f(z) \, dz = \int_C u \, dx - v \, dy + i \int_C v \, dx + u \, dy. \quad ∎ $$

In practice, Theorem 20.3 is often easier to use than Theorem 20.4, but we will use the latter formulation in certain discussions, including that of Cauchy's theorem in the next section.

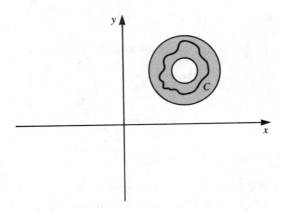

Figure 20.6. A nonsimply connected set.

To evaluate real integrals, we often can use the fundamental theorem of calculus, which states that $\int_a^b f(x)\,dx = F(b) - F(a)$ if f is continuous and $F'(x) = f(x)$ for $a \le x \le b$. To state a complex function version of this, we need a condition on the part of the plane in which the integration takes place. Recall from Section 16.3 that a set of points D in the plane is *simply connected* if every simple closed curve in D encloses only points of D, that is, if D has no "holes" in it. For example, a disk is simply connected. The annular region between two circles (Figure 20.6) is not simply connected because the closed curve C shown in Figure 20.6 encloses points not in D. The plane, with the origin removed, is not simply connected because a simple closed curve about the origin encloses a point (the origin) not in the set.

We can now state a complex version of the fundamental theorem of calculus.

THEOREM 20.5

Let D be a simply connected domain. Let F be analytic in D, and let $F'(z) = f(z)$ for z in D. Let C be a piecewise-smooth curve in D, with initial point z_1 and terminal point z_2. Then

$$\int_C f(z)\,dz = F(z_2) - F(z_1).$$

In particular, if C is a closed curve, $\int_C f(z)\,dz = 0$. ∎

The following argument shows intuitively why the conclusion of Theorem 20.5 is true. If C is smooth and is parametrized by $z = z(t)$ as t varies from a to b, we have

$$\int_C f(z)\,dz = \int_a^b f(z(t))z'(t)\,dt = \int_a^b F'(z(t))z'(t)\,dt$$

$$= \int_a^b \frac{d}{dt} F(z(t))\,dt = F(z(b)) - F(z(a))$$

$$= F(z_2) - F(z_1).$$

In particular, if C is closed, $z_2 = z_1$, and then $\int_C f(z)\,dz = 0$.

EXAMPLE 20.3

Evaluate $\int_C \cos(z)\,dz$, with C the line segment from 0 to $1 + i$.

Let $F(z) = \sin(z)$. Then F is analytic in the entire complex plane, which is a simply connected domain, and $F'(z) = \cos(z)$. Therefore,

$$\int_C \cos(z)\,dz = F(1 + i) - F(0) = \cos(1 + i) - \cos(0)$$

$$= \cos(1)\cosh(1) - 1 - i\sin(1)\sinh(1). \quad \blacksquare$$

EXAMPLE 20.4

Evaluate $\int_C (1/z)\,dz$ on any piecewise-smooth curve lying in the first quadrant and extending from $1 + i$ to $2 + 4i$.

The first quadrant consists of all (x, y) with $x > 0$ and $y > 0$ and is a simply connected domain. In the first quadrant, $\mathrm{Log}(z)$ is an analytic antiderivative of $1/z$. Therefore,

$$\int_C \frac{1}{z}\,dz = \mathrm{Log}(2 + 4i) - \mathrm{Log}(1 + i)$$

$$= \ln|2 + 4i| + i\,\mathrm{Arg}(2 + 4i) - \ln|1 + i| - i\,\mathrm{Arg}(1 + i)$$

$$= \ln(\sqrt{20}) - \ln(\sqrt{2}) + i\left[\tan^{-1}(2) - \frac{\pi}{4}\right]. \quad \blacksquare$$

EXAMPLE 20.5

$\int_C e^z\,dz = 0$ for any simple piecewise-smooth closed curve in the plane because $F(z) = e^z$ is analytic in the entire plane, and $F'(z) = e^z$. $\quad \blacksquare$

Theorem 20.5 has its limitations. For example, $f(z) = \bar{z}$ does not have an antiderivative. There is no F such that $F'(z) = \bar{z}$. Thus, we cannot use Theorem 20.5 to evaluate the integral in Example 20.1.

We conclude this section with a result which is used in making estimates.

THEOREM 20.6

Let f be continuous on the piecewise-smooth curve C parametrized by $z = z(t)$ as t varies from a to b. Then

1. $\left|\int_C f(z)\,dz\right| \le \int_C |f(z)||z'(t)|\,dt$.
2. If L is the length of C, and $|f(z)| \le M$ for z on C,

$$\left|\int_C f(z)\,dz\right| \le ML.$$

Proof of (1)   We may assume that C is smooth; if C is piecewise smooth, integrate over C by adding the integrals over each smooth piece of C. Write $\int_C f(z)\,dz = \int_a^b f(z(t))z'(t)\,dt$, by Theorem 20.3. We want to show that

$$\left|\int_a^b f(z(t))z'(t)\,dt\right| \le \int_a^b |f(z(t))||z'(t)|\,dt.$$

We know this result from calculus when f is a real-valued function, but here, f is complex valued, so we must prove the last inequality. Let

$$\int_a^b f(z(t))z'(t)\, dt = re^{i\theta}.$$

Then

$$r = e^{-i\theta}\int_a^b f(z(t))z'(t)\, dt = \int_a^b e^{-i\theta}f(z(t))z'(t)\, dt.$$

Then

$$r = \text{Re}(r) = \text{Re}\left[\int_a^b e^{-i\theta}f(z(t))z'(t)\, dt\right]$$

$$= \int_a^b \text{Re}[e^{-i\theta}f(z(t))z'(t)]\, dt.$$

Now, for any complex number w, $\text{Re}(w) \le |w|$, so, in particular,

$$\text{Re}(e^{-i\theta}f(z(t))z'(t)) \le |e^{-i\theta}f(z(t))z'(t)| = |f(z(t))z'(t)|,$$

since $|e^{-i\theta}| = 1$. Therefore,

$$\left|\int_C f(z)\, dz\right| = \left|\int_a^b f(z(t))z'(t)\, dt\right|$$

$$\le \int_a^b |f(z(t))z'(t)|\, dt = \int_a^b |f(z(t))||z'(t)|\, dt,$$

as we wanted to show.

Proof of (2)   Part (2) follows from part (1):

$$\left|\int_C f(z)\, dz\right| \le \int_a^b |f(z(t))||z'(t)|\, dt$$

$$\le \int_a^b M|z'(t)|\, dt = M\int_a^b |z'(t)|\, dt.$$

But $|z'(t)| = \sqrt{[x'(t)]^2 + [y'(t)]^2}$, so

$$\int_a^b |z'(t)|\, dt = \int_a^b \sqrt{[x'(t)]^2 + [y'(t)]^2}\, dt = \text{length of } C,$$

completing the proof of conclusion (2). ∎

EXAMPLE 20.6

Obtain a bound on $\left|\int_C e^{\text{Re}(z)}\, dz\right|$, with C the circle of radius 2 about the origin, oriented counterclockwise.

On C, $z = 2\cos(t) + 2i\sin(t)$ as t varies from 0 to 2π. Then $\text{Re}(z) = 2\cos(t)$ and $e^{\text{Re}(z)} = e^{2\cos(t)}$. This nonnegative quantity will have its maximum value when $2\cos(t)$ assumes its largest value, which is 2 at $t = 0$ and at $t = 2\pi$. Thus,

$$|e^{\text{Re}(z)}| \le e^2$$

on C. Since the length of C is 4π,

$$\left| \int_C e^{\text{Re}(z)} \, dz \right| \le 4\pi e^2. \quad \blacksquare$$

Note: This number is *not* an attempt at approximating the value of this integral; it is simply a bound on the magnitude of the integral.

PROBLEMS FOR SECTION 20.1

In each of Problems 1 through 25, evaluate $\int_C f(z) \, dz$.

1. $f(z) = 1$, $\quad z(t) = t^2 - it$; $\quad 1 \le t \le 3$

2. $f(z) = z^2 - iz$, $\quad C$ the quarter-circle counterclockwise about the origin from 2 to $2i$

3. $f(z) = \text{Re}(z)$, $\quad C$ the straight line segment from 1 to $2 + i$

4. $f(z) = 1/z$, $\quad C$ the part of the half-circle of radius 4 about the origin, counterclockwise from $4i$ to $-4i$

5. $f(z) = z - 1$, $\quad C$ any piecewise-smooth curve from $2i$ to $1 - 4i$

6. $f(z) = iz^2$, $\quad C$ the straight line segment from $1 + 2i$ to $3 + i$

7. $f(z) = \sin(2z)$, $\quad C$ the straight line segment from $-i$ to $-4i$

8. $f(z) = 1 + z^2$, $\quad C$ the part of the circle of radius 3 about the origin from -3 to $3i$, oriented counterclockwise

9. $f(z) = -i \cos(z)$, $\quad C$ any piecewise-smooth curve from 0 to $-2 + i$

10. $f(z) = |z|^2$, $\quad C$ the straight line segment from -4 to i

11. $f(z) = (z - i)^3$, $\quad C$ given by $z(t) = t - it^2$ as t varies from 1 to 2

12. $f(z) = e^{iz}$, $\quad C$ any piecewise-smooth curve from -2 to $-4 - i$

13. $f(z) = \sinh(z)$, $\quad C$ the ellipse $x^2 + \frac{1}{2}y^2 = 8$ from $-4i$ to $2\sqrt{2}$

14. $f(z) = i\bar{z}$, $\quad C$ the straight line segment from 0 to $-4 + 3i$

15. $f(z) = 1/(z - a)$, $\quad C$ the circle of radius r about a, oriented counterclockwise (a is any complex number)

16. $f(z) = \sinh(iz)$, $\quad C$ any piecewise-smooth curve from 0 to $3i$

17. $f(z) = \text{Im}(z)$, $\quad C$ the circle of radius 1 about i, oriented counterclockwise

18. $f(z) = |z|^2$, $\quad C$ the line segment from $-i$ to 1

19. $f(z) = z^2 - z + 8$, $\quad C$ any piecewise-smooth curve from 1 to $2 + 2i$

20. $f(z) = 1/z$, $\quad C$ any piecewise-smooth curve in the right quarter-plane $x > 0, y > 0$ and extending from $3 + i$ to $6 + 2i$

21. $f(z) = z\bar{z}$, $\quad z(t) = \sin(t) + i\cos(t)$ as t varies from 0 to π

22. $f(z) = z^2\cos(z^3)$, $\quad C$ any piecewise-smooth curve from 0 to i

23. $f(z) = (1 - i)z^2 + 2iz - 4$, $\quad C$ the straight line segment from 1 to $2i$

24. $f(z) = 1/\bar{z}$, $\quad C$ the straight line segment from 1 to 6

25. $f(z) = ze^{-z^2}$, $\quad C$ any piecewise-smooth curve from i to $1 + i$

26. Find a bound for $\left| \int_C \cos(z^2) \, dz \right|$ if C is the circle of radius 4 about the origin.

27. Find a bound for $\left| \int_C 1/(1 + z) \, dz \right|$ if C is the straight line between $2 + i$ and $4 + 2i$.

20.2 *The Cauchy Integral Theorem* _____

We will now discuss one of the fundamental theorems of complex function theory. The theorem is named after Augustin-Louis Cauchy, who first had the general idea of the theorem but was able to prove it only under unnecessarily restrictive assumptions. Later, Edouard Goursat proved the theorem under weaker hypotheses. For this

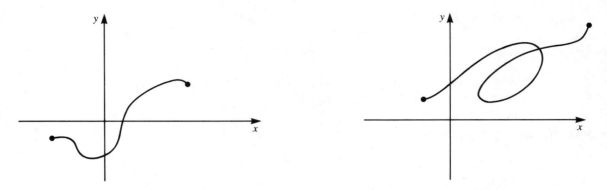

Figure 20.7. A simple curve. *Figure 20.8.* A nonsimple curve.

reason, the theorem is sometimes called the Cauchy-Goursat theorem. We will refer to it as simply Cauchy's theorem.

Recall that a curve in the plane is simple if it does not cross itself. If the curve is parametrized by $z(t) = x(t) + iy(t)$ for $a \le t \le b$, this means that there are not two distinct values of the parameter, t_1 and t_2, such that $z(t_1) = z(t_2)$. Figure 20.7 shows a typical simple curve, and Figure 20.8 shows a curve which is not simple.

A closed curve is not simple because the curve returns to the initial point as its terminal point. That is, $z(a) = z(b)$ even though $a \ne b$. We nevertheless call a closed curve simple if $z(t_1) \ne z(t_2)$ if t_1 and t_2 are distinct points in $(a, b]$. This means that the curve does not intersect itself except at the endpoints. Figure 20.9 shows a closed curve which is not simple.

As a convenience, we will save some writing by agreeing that *all curves considered are piecewise smooth.* As usual, we say that a curve is in a set D if the graph is in D. We will always orient a closed curve in the plane counterclockwise, unless specific exception is made.

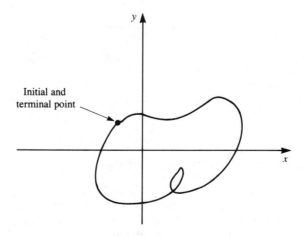

Initial and terminal point

Figure 20.9. A nonsimple closed curve.

We will also adopt the standard practice of denoting a line integral $\int_C f(z)\,dz$ as $\oint_C f(z)\,dz$ when C is a closed curve. The oval on the integral sign is nothing more than a reminder that the curve is closed.

THEOREM 20.7   The Cauchy Integral Theorem

Let f be analytic in a simply connected domain D. If C is any simple closed curve in D,

$$\oint_C f(z)\,dz = 0. \quad \blacksquare$$

A careful proof requires technical details, which are deferred to an appendix to this chapter. Here are two arguments which show why the conclusion is to be expected under more restrictive assumptions than those made in the statement of the theorem.

First, if f has an analytic antiderivative F for z in D, $\oint_C f(z)\,dz = 0$ by Theorem 20.5. This does not prove the general statement of Cauchy's theorem, however, because we do not know that f has an analytic antiderivative defined throughout D.

We can also "prove" Cauchy's theorem by using Green's theorem (Section 16.2). Write $f(z) = u(x, y) + iv(x, y)$, and apply Theorem 20.4 to write

$$\oint_C f(z)\,dz = \oint_C u\,dx - v\,dy + i\oint_C v\,dx + u\,dy.$$

Assume that u and v satisfy the hypotheses of Green's theorem, and apply this theorem to both line integrals on the right to get

$$\oint_C f(z)\,dz = \iint_M \left(\frac{\partial}{\partial x}(-v) - \frac{\partial u}{\partial y}\right) dA + i\iint_M \left(\frac{\partial u}{\partial x} - \frac{\partial v}{\partial y}\right) dA,$$

where M is the region enclosed by C (Figure 20.10). However, in both of these double integrals, the integrand is zero, because u and v satisfy the Cauchy-Riemann equations throughout D. Thus $\oint_C f(z)\,dz = 0$.

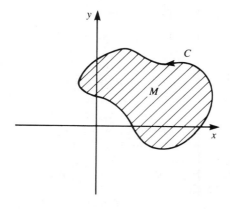

Figure 20.10

EXAMPLE 20.7

If n is a nonnegative integer, $\oint_C z^n \, dz = 0$ for any simple closed curve in the plane because z^n is analytic over the entire complex plane, which is a simply connected domain. ∎

EXAMPLE 20.8

We observed in Example 20.2 that

$$\oint_C \frac{1}{z} \, dz = 2\pi i$$

if C is the circle of radius 1 about the origin, oriented counterclockwise.

Notice that $1/z$ is analytic in the domain D formed by excluding the origin from the complex plane. However, D is not simply connected. In fact, C is a simple closed curve in D which encloses a point not in D (the origin). This shows that the conclusion of Cauchy's theorem may fail if D is not simply connected.

On the other hand, if we choose a simply connected domain D^* not containing the origin, $\oint_C (1/z) \, dz = 0$ for any simple closed curve in D^*. Figure 20.11 shows a typical simply connected domain D^* not containing the origin, with a simple closed curve C in D^*. ∎

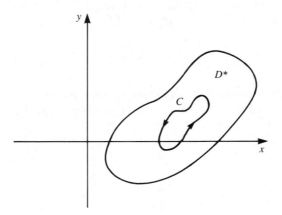

Figure 20.11. D^*—simply connected, not
containing the origin.

EXAMPLE 20.9

Cauchy's theorem gives sufficient conditions for a line integral around a simple closed curve to be zero. These conditions are not necessary—we may encounter a line integral which does not satisfy these hypotheses but which is zero anyway. From Example 20.2, $\oint_C (1/z^2) \, dz = 0$ if C is the circle of radius 1 about the origin. However, $1/z^2$ is not analytic at the origin, which is enclosed by C. In this example, C does not lie in a simply connected domain throughout which $1/z^2$ is analytic. ∎

In evaluating any line integral around a closed path, it is important to determine where the integrand is analytic. Even if the integrand fails to be analytic at certain points, we need be concerned only with points on and inside C in evaluating $\oint_C f(z)\, dz$.

EXAMPLE 20.10

Evaluate

$$\oint_C \frac{z}{\sin(z)(z - 2i)^3}\, dz,$$

with $C: |z - 8i| = 1$.

The integrand is analytic except at $2i$ and at points where $\sin(z) = 0$, which are the points $n\pi$ for n any integer.

In this example, C is the circle of radius 1 about $8i$ and does not pass through or enclose any of the points where f fails to be analytic. Thus, $\oint_C f(z)\, dz = 0$ by Cauchy's theorem. ∎

In Example 20.10, if C enclosed one or more of the points $2i$ or $n\pi$, with n an integer, we would not be able to use Cauchy's theorem and would have to evaluate this integral by some other means. For example, if C were the circle $|z| = 4$, C would enclose $0, \pi, -\pi$, and $2i$, where f is not analytic. In this case, we do not yet have a method for evaluating $\oint_C f(z)\, dz$. When we develop the residue theorem in Chapter 22, this integral will be easy to evaluate.

Cauchy's theorem can sometimes be used to help evaluate an integral even when f is not analytic within C.

EXAMPLE 20.11

Evaluate

$$\oint_C \frac{2z + 1}{z^3 - iz^2 + 6z}\, dz$$

if C is the circle of radius $\frac{1}{3}$ about $3i$.

C is shown in Figure 20.12. First, write

$$z^3 - iz^2 + 6z = z(z + 2i)(z - 3i).$$

Thus, the integrand f is not analytic at $0, -2i$, and $3i$. Use partial fractions (just as with quotients of real polynomials) to write

$$\frac{2z + 1}{z^3 - iz^2 + 6z} = \frac{1}{6}\frac{1}{z} + \frac{-1 + 4i}{10}\frac{1}{z + 2i} - \frac{1 + 6i}{15}\frac{1}{z - 3i}.$$

Then

$$\oint_C \frac{2z + 1}{z^3 - iz^2 + 6z}\, dz = \frac{1}{6}\oint_C \frac{1}{z}\, dz + \frac{-1 + 4i}{10}\oint_C \frac{1}{z + 2i}\, dz$$
$$- \frac{1 + 6i}{15}\oint_C \frac{1}{z - 3i}\, dz.$$

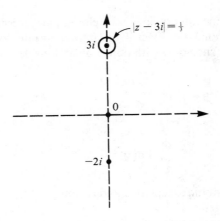

Figure 20.12

Now look at Figure 20.13, where we have drawn C and the points 0, $-2i$, and $3i$, where f is not analytic. Note that only $3i$ lies inside C and that $1/z$ and $1/(z + 2i)$ are analytic on and inside C. Thus,

$$\oint_C \frac{1}{z}\, dz = \oint_C \frac{1}{z + 2i}\, dz = 0$$

by Cauchy's theorem.

We now have only to evaluate $\oint_C 1/(z - 3i)\, dz$. Parametrize C by

$$z = 3i + \tfrac{1}{3}e^{it}; \quad 0 \le t \le 2\pi.$$

Then $z - 3i = \tfrac{1}{3}e^{it}$ and $z' = (i/3)e^{it}$ on C, and we have

$$\oint_C \frac{1}{z - 3i}\, dz = \int_0^{2\pi} \frac{1}{\tfrac{1}{3}e^{it}}\, \frac{i}{3}\, e^{it}\, dt = i \int_0^{2\pi} dt = 2\pi i.$$

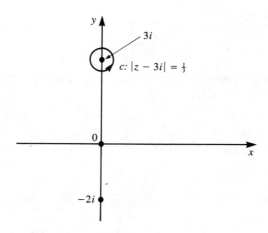

Figure 20.13

Therefore,

$$\oint_C \frac{2z+1}{z^3 - iz^2 + 6z}\, dz = -\frac{1+6i}{15}(2\pi i) = \frac{\pi}{15}(12-2i). \quad \blacksquare$$

EXAMPLE 20.12

Evaluate $\oint_C [z - \text{Re}(z)]$ on the circle $|z| = 2$.

First, $\oint_C z\, dz = 0$ because z is analytic on and inside the circle (in fact, in the entire plane). Cauchy's theorem does not apply to $\oint_C \text{Re}(z)\, dz$ because $\text{Re}(z)$ is not analytic at any point. Parametrize the circle by $z = 2\cos(t) + 2i\sin(t); 0 \le t \le 2\pi$. On C,

$$\text{Re}(z)z'(t) = 2\cos(t)[-2\sin(t) + 2i\cos(t)],$$

and we have

$$\oint_C \text{Re}(z)\, dz = \int_0^{2\pi} 2\cos(t)[-2\sin(t) + 2i\cos(t)]\, dt$$

$$= \int_0^{2\pi} -4\cos(t)\sin(t)\, dt + 4i\int_0^{2\pi} \cos^2(t)\, dt$$

$$= 2\cos^2(t)\Big]_0^{2\pi} + 2i[t + \tfrac{1}{2}\sin(2t)]\Big]_0^{2\pi} = 4\pi i.$$

Therefore,

$$\oint_C [z - \text{Re}(z)]\, dz = \oint_C z\, dz - \oint_C \text{Re}(z)\, dz = -4\pi i. \quad \blacksquare$$

In the next section, we will discuss some important ramifications of Cauchy's theorem.

PROBLEMS FOR SECTION 20.2

In each of Problems 1 through 15, evaluate $\oint_C f(z)\, dz$. All closed curves are oriented counterclockwise.

1. $\oint_C \sin(3z)\, dz, \quad C: |z| = 4$

2. $\oint_C \frac{2z}{z-i}\, dz, \quad C: |z-i| = 3$

3. $\oint_C \frac{1}{(z-2i)^3}\, dz, \quad C: |z-2i| = 2$

4. $\oint_C z^2\sin(z)\, dz, \quad C$ the square with vertices $0, 1, 1+i$, and i

5. $\oint_C \bar{z}\, dz, \quad C: |z| = 1$

6. $\oint_C \frac{1}{\bar{z}}\, dz, \quad C: |z| = 5$

7. $\oint_C ze^z\, dz, \quad C: |z-3i| = 8$

8. $\oint_C (z^2 - 4z + 8)\, dz, \quad C$ the rectangle with vertices $1, 8, 8+4i, 1+4i$

9. $\oint_C |z|^2\, dz, \quad C: |z| = 5$

10. $\oint_C \sin\left(\frac{1}{z}\right)\, dz, \quad C: |z-1+2i| = 1$

11. $\oint_C \text{Re}(z)\, dz, \quad C: |z| = 2$

12. $\oint_C [z^2 + \text{Im}(z)]\, dz$, C the square with vertices $0, -2i, 2-2i, 2$

13. $\oint_C (z^2 - |z|)\, dz$, $C: |z| = 4$

14. $\oint_C \dfrac{z^3 - 2z + 4}{z^8 - 2}\, dz$, $C: |z - 12| = 2$

15. $\oint_C \dfrac{z^2 - 8z}{\cos(z)}\, dz$, C the rectangle with vertices $\dfrac{-\pi i}{4}, \dfrac{\pi}{4} - \dfrac{\pi i}{4}, \dfrac{\pi}{4} + \dfrac{\pi i}{4}, \dfrac{\pi i}{4}$

16. Use partial fractions and Cauchy's theorem to evaluate

$$\oint_C \frac{1}{z^2 + 1}\, dz,$$

where C is the circle $|z - i| = \frac{1}{4}$.

17. Use partial fractions and Cauchy's theorem to evaluate

$$\oint_C \frac{1}{z^2 + 8z}\, dz,$$

where C is the circle $|z| = 1$.

20.3 *Some Consequences of Cauchy's Theorem*

Cauchy's theorem may be considered the fundamental theorem of complex integration in the sense that it has many important ramifications. In this section, we will develop six consequences of Cauchy's theorem.

I. INDEPENDENCE OF PATH

Suppose that f is analytic in a simply connected domain D and that z_0 and z_1 are points in D. Let C_1 and C_2 be curves from z_0 to z_1 in D, as shown in Figure 20.14. By reversing

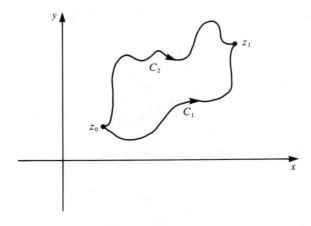

Figure 20.14

orientation on C_2, we obtain a closed curve K in D with initial and terminal points z_0. By Cauchy's theorem,

$$\oint_K f(z)\,dz = 0.$$

Now decompose K into C_1 and C_2, recalling that in forming K we reversed orientation on C_2. Then

$$\int_{C_1} f(z)\,dz - \int_{C_2} f(z)\,dz = 0,$$

implying that

$$\int_{C_1} f(z)\,dz = \int_{C_2} f(z)\,dz.$$

Thus, in D, $\int_C f(z)\,dz$ depends only on the endpoints of C and not on the path in D chosen between them. We say in this case that $\int_C f(z)\,dz$ is *independent of path in D*. We will summarize this discussion as a theorem.

THEOREM 20.8

If f is analytic in a simply connected domain D, $\int_C f(z)\,dz$ is independent of path in D. ∎

When $\int_C f(z)\,dz$ is independent of path in D, we usually write $\int_C f(z)\,dz$ as $\int_{z_0}^{z_1} f(z)\,dz$, where z_0 is the initial point of C and z_1 is the terminal point.

EXAMPLE 20.13

We know that $\sin(z)$ is analytic in the entire plane, which is simply connected. Thus, $\int_C \sin(z)\,dz$ is independent of path. For example, if C is any curve from $-i$ to $2 + i$,

$$\int_C \sin(z)\,dz = \int_{-i}^{2+i} \sin(z)\,dz = -\cos(z)\Big]_{-i}^{2+i}$$
$$= -\cos(2 + i) + \cos(-i). ∎$$

II.   EXISTENCE OF AN ANTIDERIVATIVE

By Theorem 20.5, we can evaluate $\int_C f(z)\,dz$ if we know an antiderivative of f. But not every function has an antiderivative—for example, $f(z) = \bar{z}$ does not. We will now give sufficient conditions for an antiderivative to exist.

THEOREM 20.9

Let f be analytic in a simply connected domain D. Then there is a function F which is analytic in D such that $F'(z) = f(z)$ for z in D.

Proof   Choose any z_0 in D. By Theorem 20.8, $\int_C f(z)\,dz$ is independent of path in D.

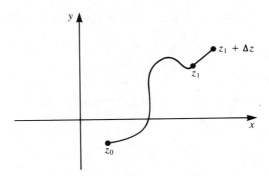

Figure 20.15

We can therefore define a function

$$F(z) = \int_{z_0}^{z} f(z) \, dz$$

for any z in D. We will prove that F is analytic in D and that $F'(z) = f(z)$.

Let z_1 be any point of D. Form a curve in D from z_0 to $z_1 + \Delta z$ as shown in Figure 20.15. This curve consists of any curve in D from z_0 to z_1 and then a straight line segment from z_1 to $z_1 + \Delta z$. This segment has length $|\Delta z|$. We have

$$F(z_1 + \Delta z) - F(z_1) = \int_{z_0}^{z_1 + \Delta z} f(z) \, dz - \int_{z_0}^{z_1} f(z) \, dz$$

$$= \int_{z_1}^{z_1 + \Delta z} f(z) \, dz.$$

Then

$$\frac{F(z_1 + \Delta z) - F(z_1)}{\Delta z} - f(z_1) = \frac{1}{\Delta z} \int_{z_1}^{z_1 + \Delta z} f(z) \, dz - f(z_1)$$

$$= \frac{1}{\Delta z} \int_{z_1}^{z_1 + \Delta z} [f(z) - f(z_1)] \, dz$$

because $f(z_1)$ is constant. Since f is analytic in D, f is certainly continuous at z_1. Given any $\epsilon > 0$, there is some $\delta > 0$ such that

$$|f(z) - f(z_1)| < \epsilon \quad \text{if} \quad |z - z_1| < \delta.$$

Then, for $|\Delta z| < \delta$, we have

$$\left| \frac{F(z_1 + \Delta z) - F(z_1)}{\Delta z} - f(z_1) \right| = \left| \frac{1}{\Delta z} \int_{z_1}^{z_1 + \Delta z} [f(z) - f(z_1)] \, dz \right|$$

$$= \frac{1}{|\Delta z|} \left| \int_{z_1}^{z_1 + \Delta z} [f(z) - f(z_1)] \, dz \right|$$

$$\leq \frac{1}{|\Delta z|} \epsilon |\Delta z| = \epsilon$$

by Theorem 20.6. This proves that

$$F'(z_1) = \lim_{\Delta z \to 0} \frac{F(z_1 + \Delta z) - F(z_1)}{\Delta z} = f(z_1).$$

Thus far, we have shown that F is differentiable at all points of D and that $F'(z) = f(z)$ for z in D. Now, D is a domain, so about any point in D there is an open disk containing only points of D. Since F is differentiable in this open disk, F is analytic at each point of D, completing the proof. ∎

EXAMPLE 20.14

Consider $\int_C (1/z)\, dz$ under two different circumstances.

Case 1   Suppose D is the right quarter-plane $x > 0$, $y > 0$. Then D is a simply connected domain, and $1/z$ is analytic in D, so $\int_C (1/z)\, dz$ is independent of path in D. By Theorem 20.9, $1/z$ has an analytic antiderivative in D; in fact Log(z) is such an antiderivative.

Case 2   Now let D be the entire plane minus the origin. Again, D is a domain, and $1/z$ is analytic in D, but D is not simply connected. Now, Log(z) is not an antiderivative of $1/z$ in D because Log(z) is analytic only for $-\pi < \text{Arg}(z) < \pi$, which excludes points on the negative real axis [recall conclusion (5) of Theorem 19.7]. In fact, $\oint_C (1/z)\, dz = 2\pi i \neq 0$ if C is the unit circle $|z| = 1$ (see Example 20.4). ∎

III.   THE DEFORMATION THEOREM

Consider the problem of evaluating $\oint_C f(z)\, dz$, where f is analytic in some domain D except at z_0 and C is a simple closed curve enclosing z_0, as in Figure 20.16. We will now demonstrate a remarkable fact: if we replace C by any other simple closed curve K in D enclosing z_0,

$$\oint_C f(z)\, dz = \oint_K f(z)\, dz.$$

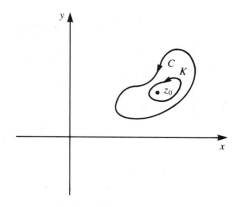

Figure 20.16. C encloses z_0, where f is not analytic.

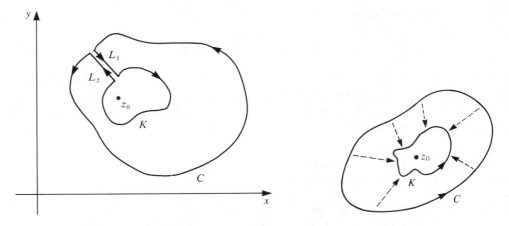

Figure 20.17   *Figure 20.18*

This means that, under some circumstances, we can replace a "difficult" curve with one that makes the integral easier to evaluate.

To see why this can be done, we will exploit the same device used in extending Green's theorem (see Appendix B to Part IV). With C and K as Figure 20.16, cut small arcs from C and K and form curves L_1 and L_2 connecting C and K, as shown in Figure 20.17. Then L_1, L_2, and the parts of C and K left after the cut form a simple closed curve C^* *which does not enclose* z_0. By Cauchy's theorem,

$$\oint_{C^*} f(z)\, dz = 0.$$

In integrating around C^* counterclockwise, we go around C (minus the excised arc) counterclockwise, around K (minus its excised arc) clockwise, and along L_1 and L_2. As L_1 and L_2 are chosen closer together, they merge to a single curve, and the line integral of f over L_1 cancels that over L_2 because we integrate over this curve once in each direction. In this limit as L_1 and L_2 merge, we obtain

$$\oint_C f(z)\, dz + \oint_{-K} f(z)\, dz = 0,$$

in which $\oint_{-K}$ denotes the line integral about K *clockwise*. If we integrate in the usual counterclockwise sense about K, we have

$$\oint_C f(z)\, dz - \oint_K f(z)\, dz = 0,$$

or

$$\oint_C f(z)\, dz = \oint_K f(z)\, dz.$$

This result is often called the deformation theorem. In effect, we are deforming C to K, as shown in Figure 20.18. Imagine a rubber band stretched into the shape of C and then continuously deformed into K. (In topology, C and K are said to be *homotopic curves*.) In this process, it is essential that as C is deformed into K, the intermediate curves do not pass over any points where the function is not analytic.

EXAMPLE 20.15

Evaluate $\oint_C 1/(z - z_0)\, dz$ over any simple closed curve enclosing z_0.

Note that $1/(z - z_0)$ is analytic everywhere except at z_0. Replace C by a circle K about z_0 of sufficiently small radius to be enclosed by C (Figure 20.19). We can explicitly evaluate $\oint_K 1/(z - z_0)\, dz$. Parametrize K by

$$z = z_0 + re^{it}; \qquad 0 \le t \le 2\pi.$$

Then

$$\oint_C \frac{1}{z - z_0}\, dz = \oint_K \frac{1}{z - z_0}\, dz = \int_0^{2\pi} \frac{1}{re^{it}} \, ire^{it}\, dt = \int_0^{2\pi} i\, dt = 2\pi i. \quad \blacksquare$$

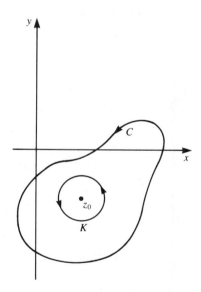

Figure 20.19

IV.   THE CAUCHY INTEGRAL FORMULA

We will now prove a remarkable fact which dramatizes one of the differences between complex functions and real functions. We will show that, under certain conditions, the value $f(z_0)$ of an analytic function at a point z_0 is completely determined by the values $f(z)$ for z on any simple closed curve enclosing z_0. There is no analogous result for real functions, since the value $g(x_0)$ of a real function at x_0 is not determined by values $g(x)$ in intervals away from x_0.

THEOREM 20.10 Cauchy Integral Formula

Let f be analytic in a simply connected domain D. Let z_0 be any point of D, and let C be any simple closed curve in D enclosing z_0. Then

$$f(z_0) = \frac{1}{2\pi i} \oint_C \frac{f(z)}{z - z_0} \, dz. \quad \blacksquare$$

This formula for $f(z_0)$ is called Cauchy's integral formula. Before proving it, we will observe that the formula can also be written

$$\oint_C \frac{f(z)}{z - z_0} \, dz = 2\pi i f(z_0). \tag{20.2}$$

In this form, the theorem provides a powerful tool for evaluating certain integrals. The value of the integral on the left side of equation (20.2) is just $2\pi i$ times the value of the function f at z_0.

EXAMPLE 20.16

Evaluate $\oint_C e^{z^2}/(z - i) \, dz$ if C is any simple closed curve in the plane and C does not pass through i.

First, $f(z) = e^{z^2}$ is analytic in the entire complex plane, which is simply connected. Now consider two cases.

Case 1 C does not enclose i.
Then $e^{z^2}/(z - i)$ is analytic on and inside C, so $\oint_C e^{z^2}/(z - i) \, dz = 0$ by Cauchy's theorem.

Case 2 C encloses i.
Now use Cauchy's integral formula [in the form of equation (20.2)] to write

$$\oint_C \frac{e^{z^2}}{z - i} \, dz = 2\pi i f(i) = 2\pi i e^{i^2} = \frac{2\pi i}{e}. \quad \blacksquare$$

EXAMPLE 20.17

Evaluate $\oint_C e^{2z} \sin(z^2)/(z - 2) \, dz$ for any simple closed curve C which does not pass through 2.

Observe that $f(z) = e^{2z} \sin(z^2)$ is analytic for all z, and consider two cases.

Case 1 C does not enclose 2.
Then

$$\oint_C \frac{e^{2z} \sin(z^2)}{z - 2} \, dz = 0$$

by Cauchy's theorem.

Case 2 C encloses 2.
Then

$$\oint_C \frac{e^{2z} \sin(z^2)}{z - 2} \, dz = 2\pi i f(2) = 2\pi i e^4 \sin(4). \quad \blacksquare$$

In applying Cauchy's integral formula, note that f is analytic in D; it is $f(z)/(z - z_0)$ which is not analytic at z_0 if C encloses z_0. We evaluate $\oint_C f(z)/(z - z_0)\,dz$ in this case by multiplying the analytic function f, evaluated at z_0, by $2\pi i$. We will now prove the theorem.

Proof of Theorem 20.10   Since $f(z_0)$ is constant, we can write

$$\oint_C \frac{f(z)}{z - z_0}\,dz = \oint_C \frac{f(z) - f(z_0) + f(z_0)}{z - z_0}\,dz$$

$$= f(z_0) \oint_C \frac{1}{z - z_0}\,dz + \oint_C \frac{f(z) - f(z_0)}{z - z_0}\,dz$$

$$= 2\pi i f(z_0) + \oint_C \frac{f(z) - f(z_0)}{z - z_0}\,dz,$$

in which we have used the result of Example 20.15.

Now consider

$$\oint_C \frac{f(z) - f(z_0)}{z - z_0}\,dz.$$

For Cauchy's integral formula to be true, this integral must be zero. We will prove this as follows. First, use the deformation theorem to replace C by a circle K about z_0, with the radius of K small enough that K is within C. Then

$$\oint_C \frac{f(z) - f(z_0)}{z - z_0}\,dz = \oint_K \frac{f(z) - f(z_0)}{z - z_0}\,dz.$$

Consider this line integral around K. Let $\epsilon > 0$. Since f is analytic at z_0, there is some $\delta > 0$ such that

$$|f(z) - f(z_0)| \le \frac{\epsilon}{2\pi} \quad \text{if} \quad |z - z_0| < \delta.$$

Choose K to have radius $\delta/2$. Then, for z on K,

$$\left| \frac{f(z) - f(z_0)}{z - z_0} \right| \le \frac{\epsilon}{2\pi} \left(\frac{2}{\delta} 2\pi \frac{\delta}{2} \right) = \epsilon,$$

in which we have used Theorem 20.6. Since ϵ is as small as we please, we conclude that

$$\oint_K \frac{f(z) - f(z_0)}{z - z_0}\,dz = 0,$$

proving the theorem.   ∎

V.   HIGHER DERIVATIVES OF ANALYTIC FUNCTIONS

If we know that a real-valued function g is differentiable at x_0, we can conclude nothing about existence of $g''(x_0)$. For complex functions, being analytic is much more powerful than simply being differentiable. We will now show that a function which is analytic at z_0 must have derivatives of all orders at z_0. Further, we can write an integral formula for these higher derivatives.

THEOREM 20.11 Cauchy's Integral Formula for Higher Derivatives

Let f be analytic in a simply connected domain D, and let z_0 be in D. Then f has derivatives of all orders at z_0. Further, the nth derivative of f at z_0 is

$$f^{(n)}(z_0) = \frac{n!}{2\pi i} \oint_C \frac{f(z)}{(z - z_0)^{n+1}} \, dz. \quad \blacksquare$$

We will sketch a proof at the end of this subsection. As with Cauchy's integral formula, this formula for higher derivatives can sometimes be used to evaluate an integral if we write the formula in the form

$$\oint_C \frac{f(z)}{(z - z_0)^{n+1}} \, dz = \frac{2\pi i}{n!} f^{(n)}(z_0). \tag{20.3}$$

EXAMPLE 20.18

Evaluate

$$\oint_C \frac{4e^{z^3}}{(z - i)^2} \, dz,$$

with C any simple closed curve not passing through i.

First, $f(z) = 4e^{z^3}$ is analytic for all z. Consider two cases.

Case 1 C does not enclose i.
Then $4e^{z^3}/(z - i)^2$ is analytic on and inside C, so

$$\oint_C \frac{4e^{z^3}}{(z - i)^2} \, dz = 0$$

by Cauchy's theorem.

Case 2 C encloses i.
Now match the integral we want to evaluate with equation (20.3). They match if we choose $f(z) = 4e^{z^3}$ and $n = 1$. Since $f'(z) = 12z^2 e^{z^3}$,

$$\oint_C \frac{4e^{z^3}}{(z - i)^2} \, dz = \frac{2\pi i}{1!} f'(i) = (2\pi i)(12)(i^2)e^{i^3} = -24\pi i e^{-i}. \quad \blacksquare$$

EXAMPLE 20.19

Evaluate

$$\oint_C \frac{2 \sin(z^2)}{(z - 1)^4} \, dz,$$

where C is any simple closed curve not passing through 1. Consider two cases.

Case 1 If C does not enclose 1, $2 \sin(z^2)/(z - 1)^4$ is analytic on and inside C, so

$$\oint_C \frac{2 \sin(z^2)}{(z - 1)^4} \, dz = 0.$$

Case 2   If C encloses 1, observe that $f(z) = 2 \sin(z^2)$ is analytic for all z, and use equation (20.3) with $n = 3$. We get

$$\oint_C \frac{2 \sin(z^2)}{(z - 1)^4} \, dz = \frac{2\pi i}{3!} f^{(3)}(1),$$

where $f^{(3)}(1)$ is the third derivative of f, evaluated at 1. Compute

$$f'(z) = 4z \cos(z^2), \qquad f''(z) = 4 \cos(z^2) - 8z^2 \sin(z^2),$$

and

$$f^{(3)}(z) = -24z \sin(z^2) - 16z^3 \cos(z^2).$$

Since $3! = 3 \cdot 2 = 6$, we have

$$\oint_C \frac{2 \sin(z^2)}{(z - 1)^4} \, dz = \frac{\pi i}{3} [-24 \sin(1) - 16 \cos(1)]. \quad \blacksquare$$

Cauchy's formula for higher derivatives also enables us to derive a bound on higher derivatives of an analytic function at a point.

COROLLARY TO THEOREM 20.11

Let f be analytic in a simply connected domain containing all points on and inside the circle C of radius r about z_0. Let $|f(z)| \le M$ for all z on C. Then

$$f^{(n)}(z_0) \le \frac{Mn!}{r^n}.$$

Proof   By Theorem 20.11 and Theorem 20.6,

$$|f^{(n)}(z_0)| = \frac{n!}{2\pi} \left| \oint_C \frac{f(z)}{(z - z_0)^{n+1}} \, dz \right| \le \frac{n!}{2\pi} \frac{M}{r^{n+1}} (2\pi r) = \frac{Mn!}{r^n}. \quad \blacksquare$$

We will now sketch a proof of Cauchy's integral formula for higher derivatives.

Sketch of a Proof of Theorem 20.11   Proceed by induction on n. We will first prove the formula for $n = 1$. Begin with

$$f'(z_0) = \lim_{\Delta z \to 0} \frac{f(z_0 + \Delta z) - f(z_0)}{\Delta z}$$

$$= \lim_{\Delta z \to 0} \frac{1}{2\pi i \, \Delta z} \left[\oint_C \frac{f(\xi)}{\xi - (z_0 + \Delta z)} \, d\xi - \oint_C \frac{f(\xi)}{\xi - z_0} \, d\xi \right],$$

by two applications of Cauchy's integral formula. Here, C is a simple closed curve in D enclosing z_0 and also $z_0 + \Delta z$ (Figure 20.20). Then

$$f'(z_0) = \frac{1}{2\pi i} \lim_{\Delta z \to 0} \oint_C \frac{f(\xi)}{\Delta z} \left[\frac{1}{\xi - (z_0 + \Delta z)} - \frac{1}{\xi - z_0} \right] d\xi.$$

Write

$$\frac{1}{\Delta z} \left[\frac{1}{\xi - (z_0 + \Delta z)} - \frac{1}{\xi - z_0} \right] = \frac{1}{(\xi - z_0)^2} + \frac{\Delta z}{(\xi - z_0 - \Delta z)(\xi - z_0)^2}.$$

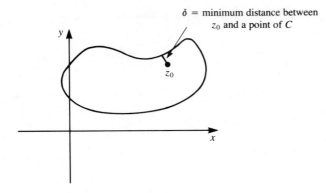

δ = minimum distance between z_0 and a point of C

Figure 20.20   Figure 20.21

Therefore,

$$f'(z_0) = \frac{1}{2\pi i} \oint_C \frac{f(\xi)}{(\xi - z_0)^2} \, d\xi + \frac{1}{2\pi i} \lim_{\Delta z \to 0} \Delta z \oint_C \frac{f(\xi)}{(\xi - z_0 - \Delta z)(\xi - z_0)^2} \, d\xi.$$

We must show that the last limit is zero. First, it can be shown that continuity of f implies that f is bounded on C, say, $|f(z)| \le M$. Now let δ be the smallest distance between z_0 and a point of C (see Figure 20.21). Then

$$|\xi - z_0| \ge \delta \qquad \text{for all } \xi \text{ on } C.$$

Then

$$\frac{1}{|\xi - z_0|^2} \le \frac{1}{\delta^2}.$$

Therefore, on C,

$$\frac{|f(\xi)|}{|\xi - z_0|^2} \le \frac{M}{\delta^2}.$$

Now choose $|\Delta z| \le \delta/2$. Then $z_0 + \Delta z$ is at least $\delta/2$ from any ξ on C:

$$|\xi - z_0 - \Delta z| \ge \frac{\delta}{2}.$$

Then

$$\frac{1}{|\xi - z_0 - \Delta z|} \le \frac{2}{\delta}$$

for z on C. Then

$$\left| \Delta z \oint_C \frac{f(\xi)}{(\xi - z_0 - \Delta z)(\xi - z_0)^2} \, d\xi \right| \le |\Delta z| \frac{M}{\delta^2} \frac{2}{\delta} (\text{length of } C)$$

for $|\Delta z| \le \delta/2$. As $\Delta z \to 0$, certainly $2|\Delta z| M/\delta^3 \to 0$; hence,

$$\lim_{\Delta z \to 0} \Delta z \oint_C \frac{f(\xi)}{(\xi - z_0 - \Delta z)(\xi - z_0)^2} \, d\xi = 0,$$

proving the theorem for $n = 1$. By a similar argument, we can prove Cauchy's formula for $f^{(n+1)}(z_0)$, assuming it true for $f^{(n)}(z_0)$. ∎

VI. AN EXTENSION OF CAUCHY'S INTEGRAL FORMULA

Cauchy's integral formula can be extended to a more general result in a way similar to the extension of Green's theorem in Appendix B to Part IV. We will use this extension when we discuss Laurent series in the next chapter.

Let w be a complex number, and let r and R be positive numbers, with $r < R$. Let D be the domain consisting of points lying between the concentric circles of radii r and R about w. D consists of all z satisfying

$$r < |z - w| < R.$$

Such a domain, bounded by two concentric circles, is called an *annulus*. Figure 20.22 shows a typical annulus.

Let C_r be the circle $|z - w| = r$, and let C_R be the circle $|z - w| = R$, oriented counterclockwise. Now suppose that f is analytic in D and on both of the circles. We claim that, for any z_0 in D,

$$f(z_0) = \frac{1}{2\pi i} \oint_{C_R} \frac{f(z)}{z - z_0}\, dz - \frac{1}{2\pi i} \oint_{C_r} \frac{f(z)}{z - z_0}\, dz. \tag{20.4}$$

Note that w does not appear in this formula. The number w is simply the center of the circles defining the annulus where the action occurs. The formula (20.4) gives the value of f at each point z_0 *in the annulus*. This formula is often used when w is a point at which f is not analytic (or perhaps is not even defined).

Here is a sketch of a proof of equation (20.4). Form curves L_1 and L_2 as shown in Figure 20.23(b) to form the new simple closed curve K of Figure 20.23(c). Note that z_0 is

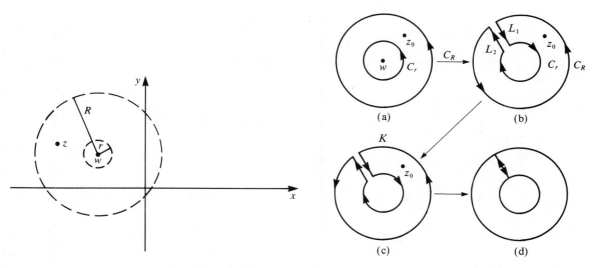

Figure 20.22. $r < |z - w| < R.$

Figure 20.23

interior to K. Since f is analytic on K and the simply connected domain it encloses, we can use Cauchy's integral formula to write

$$f(z_0) = \frac{1}{2\pi i} \oint_K \frac{f(z)}{z - z_0} \, dz.$$

As part of K, the integral about C_r is clockwise. Thus,

$$\oint_K = \int_{C_R} + \int_{-C_r} + \int_{L_1} + \int_{L_2}.$$

We write $-C_r$ here to indicate the clockwise orientation along C_r. Now take a limit as L_1 and L_2 merge to a curve between C_r and C_R [Figure 20.23(d)]. The integrals along the merged versions of L_1 and L_2 cancel, and we obtain

$$f(z_0) = \frac{1}{2\pi i} \oint_{C_R} \frac{f(z)}{z - z_0} \, dz + \frac{1}{2\pi i} \oint_{-C_r} \frac{f(z)}{z - z_0} \, dz.$$

With the usual counterclockwise orientation on C_r, we get

$$f(z_0) = \frac{1}{2\pi i} \oint_{C_R} \frac{f(z)}{z - z_0} \, dz - \frac{1}{2\pi i} \oint_{C_r} \frac{f(z)}{z - z_0} \, dz.$$

In the next chapter, we will develop the complex analogue of the Taylor series, as well as the Laurent series, which has no obvious analogue for real functions. The Laurent series will have important applications in evaluating both real and complex integrals and in a variety of other computations, such as the summation of series.

PROBLEMS FOR SECTION 20.3

1. Evaluate $\displaystyle\int_i^{1+2i} z^3 \, dz$.

2. Evaluate $\displaystyle\int_\pi^{\pi i} \cos(z) \, dz$.

3. Evaluate $\displaystyle\int_0^{4i} z e^{-z^2} \, dz$.

4. Evaluate $\displaystyle\int_{1-2i}^{4+i} iz \, dz$.

In each of Problems 5 through 16, evaluate the integral.

5. $\displaystyle\oint_c \frac{z^4}{z - 2i} \, dz$, C any simple closed curve enclosing $2i$

6. $\displaystyle\oint_c \frac{\sin(z^2)}{z - 5} \, dz$, C any simple closed curve enclosing 5

7. $\displaystyle\oint_c \frac{z^3 - 5z + i}{z - 1 + 2i} \, dz$, C the circle $|z| = 3$

8. $\displaystyle\oint_c \frac{2z^3}{(z - 2)^2} \, dz$, C the rectangle having vertices $4 - i, 4 + i, -4 - i, -4 + i$

9. $\displaystyle\oint_c \frac{ie^z}{(z - 2 + i)^4} \, dz$, C the circle $|z| = 2$

10. $\displaystyle\oint_c \frac{\cos(z - i)}{(z + 2i)^3} \, dz$, C any simple closed curve enclosing $-2i$

11. $\oint_C \dfrac{z \sin(3z)}{(z+4)^3} \, dz$, C the circle $|z - 2i| = 9$

12. $\oint_C \dfrac{z^2 - \cos(3z)}{(z + 4i)^4} \, dz$, C the triangle having vertices $0, -1 - 8i, 1 - 8i$

13. $\oint_C \dfrac{\sinh(z^3)}{(z+1)^2} \, dz$, C the circle $|z + 1 + 2i| = 1$

14. $\oint_C \dfrac{\cos(z) - \sin(z)}{(z+i)^4} \, dz$, C any simple closed curve enclosing $-i$

15. $\oint_C \dfrac{2z \cosh(z)}{(z - 2 + 4i)^2} \, dz$, C any simple closed curve not passing through $2 - 4i$

16. $\oint_C \dfrac{(z + i)^2}{(z + 3 - 2i)^3} \, dz$, C the circle $|z + 2| = 5$

17. Let f be analytic in a simply connected domain D. Let z_0 be in D, and suppose that the closed disk $|z - z_0| \le r$ is in D. Prove that

$$f(z_0) = \frac{1}{2\pi} \int_0^{2\pi} f(z_0 + re^{it}) \, dt.$$

[Thus, $f(z_0)$ is the average of the values of $f(z)$ on a circle centered at z_0.]

18. Under the conditions of Problems 17, show that

$$|f(z_0)| \le M,$$

where M is the maximum of the numbers $|f(z)|$ for z on the circle $|z - z_0| = r$.

19. Prove Morera's theorem, which is "almost" a converse of Cauchy's theorem. If f is continuous in a simply connected domain D, and $\oint_C f(z) \, dz = 0$ for every simple closed curve in D, f is analytic in D. *Hint:* Prove that f has an antiderivative in D.

20. Use the corollary to Theorem 20.11 to prove Liouville's theorem, which states: If f is analytic in the entire plane, and, for some M, $|f(z)| \le M$ for all z, $f(z)$ must be constant. (*Hint:* Show that $f'(z) = 0$ for all z.) Liouville's theorem enables us to conclude that a nonconstant function which is analytic for all z cannot be bounded. For example, e^z, $\sin(z)$, and $\cosh(z)$ are not bounded.

21. Use Liouville's theorem to prove the following: If f is analytic for all z, and $f(z)$ is not a constant function, then, given positive numbers M and r, there exists z with $|z| > r$ and $|f(z)| > M$. (That is, a function analytic in the entire plane can be made to take on values of arbitrarily large magnitude arbitrarily far from the origin.)

22. Use Liouville's theorem to prove the fundamental theorem of algebra: If $p(z) = a_0 + a_1 z + \cdots + a_n z^n$, with n a positive integer and $a_0, \ldots, a_n$ complex numbers and $a_n \ne 0$, there is some number z_0 such that $p(z_0) = 0$. (That is, every nonconstant polynomial with complex coefficients has a root.) *Hint:* Assume that $p(z) \ne 0$ for all z, and let $f(z) = 1/p(z)$. Show that f is analytic for all z but is also bounded, deriving a contradiction to Liouville's theorem.

ADDITIONAL PROBLEMS

In each of Problems 1 through 25, evaluate the integral.

1. $\oint_C \dfrac{e^{2z}}{(z - 1)^3} \, dz$, C any simple closed curve enclosing 1

2. $\displaystyle\int_{-2+3i}^{i} 4iz \sin(z^2) \, dz$

3. $\oint_C \dfrac{1}{(z-i)^3}\, dz$, $\quad C$ any simple closed curve not passing through i

4. $\displaystyle\int_C 2i\bar{z}|z|\, dz$, $\quad C$ the line segment from $-i$ to $4i$

5. $\displaystyle\int_C \operatorname{Im}(4z)\, dz$, $\quad C$ the line segment from 1 to $-i$

6. $\oint_C \dfrac{-(2+i)\sin(z^2)}{(z+4)^2}\, dz$, $\quad C$ any simple closed curve enclosing -4

7. $\displaystyle\int_{3+i}^{2-5i} (i-2z^2)\, dz$ $\qquad\qquad$ **8.** $\oint_C 3iz^2\, dz$, $\quad C$ the circle $|z+4i| = 2$

9. $\displaystyle\int_C [2z - i\operatorname{Im}(z)]\, dz$, $\quad C$ the line segment from $2 + 3i$ to $-1 + 2i$

10. $\displaystyle\int_C (z-i)^2\, dz$, $\quad C$ the semicircle of radius 1 about zero from i to $-i$

11. $\displaystyle\int_C \operatorname{Re}(z+4)\, dz$, $\quad C$ the line segment from $3 + i$ to $2 - 5i$

12. $\oint_C \dfrac{3z^2\cosh(z)}{(z+2i)^2}\, dz$, $\quad C$ any simple closed curve not enclosing $-2i$

13. $\displaystyle\int_C (z + i|z|)\, dz$, $\quad C$ the line segment from 0 to $-2i$

14. $\oint_C \dfrac{2iz^2}{(z+1-2i)^2}\, dz$, $\quad C$ any simple closed curve enclosing $-1 + 2i$

15. $\displaystyle\int_{2+i}^{-1} \cosh(3iz)\, dz$ $\qquad\qquad$ **16.** $\displaystyle\int_0^i -ze^{iz^2}\, dz$

17. $\displaystyle\int_C \operatorname{Re}(z)\, dz$, $\quad C$ the line segment from -2 to $3 + i$

18. $\oint_C (1-z^2)\, dz$, $\quad C$ any simple closed curve in the plane

19. $\displaystyle\int_2^4 (z^2 + 3z^4 - i)\, dz$

20. $\oint_C \dfrac{-2\cos(z^3)}{z+3i}\, dz$, $\quad C$ any simple closed curve not passing through $-3i$

21. $\oint_C e^{-z^2}\, dz$, $\quad C$ any simple closed curve about $2 - i$

22. $\displaystyle\int_C \operatorname{Im}(iz)\, dz$, $\quad C$ the semicircle $z = 3e^{it}$; $\quad 0 \le t \le \pi$

23. $\displaystyle\int_0^{4i} (3iz^2 - 1)\, dz$

24. $\oint_C \dfrac{\cosh[\sinh(z)]}{z+3}\, dz$, $\quad C$ any simple closed curve not passing through -3

25. $\displaystyle\int_{-5}^{2i} [z^3 + iz^2 - \sin(z)]\, dz$

26. Evaluate $\int_0^{2\pi} e^{\cos(\theta)}\cos[\sin(\theta)]\, d\theta$. *Hint:* Consider the integral $\oint_C (1/z)e^z\, dz$, with C the unit circle about the origin. Evaluate this integral once by using the Cauchy integral formula and then again by parametrizing the unit circle and substituting the parametric equations into the integral.

27. Let C be the circle $|z| = r$, oriented counterclockwise. Assume that $r > 1$. Prove that

$$\left| \oint_C \frac{1}{z^2} \operatorname{Log}(z)\, dz \right| \le 2\pi \left[\frac{\pi + \ln(r)}{r} \right].$$

Appendix: A Proof of Cauchy's Theorem

A rigorous proof of Cauchy's theorem requires a great deal of mathematics beyond what we have available in the prerequisites for this book. Indeed, even a very careful statement contains subtleties we have ignored or glossed over.

One subtlety is that of the set bounded by a simple closed continuous curve. It is intuitively apparent that such a curve separates the plane into two sets, one of which is bounded (that is, can be placed within a disk) while the other is not. If we took scissors and cut along the curve C of Figure 20.24, the plane would fall into two pieces, which are exactly these sets.

A proof of this intuitively appealing statement is quite difficult because of the generality of the notion of a simple closed continuous curve. It is possible for such a curve to have a tangent at none of its points, although certainly most curves we draw or encounter in everyday experience have a tangent at most points.

The modern definition of curve was devised by Camille Jordan (1838–1922), and the result we have just cited is known as the Jordan curve theorem, although the original proofs he offered contained errors. The first correct proof was due to Oswald Veblen in 1913. We have tacitly assumed the Jordan curve theorem every time we have spoken of the set of points enclosed by a curve (in connection with Green's theorem, Cauchy's theorem, and elsewhere).

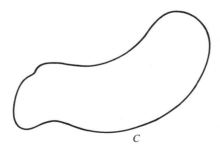

C

Figure 20.24

Cauchy's original version of his theorem assumed that $f'(z)$ is continuous in the set consisting of the curve and the points it enclosed. In fact, existence of the derivative is sufficient, and continuity need not be assumed. This was proved by Edouard Goursat (1858–1936).

Goursat's method is first to prove the theorem for the case in which the curve is a rectangle, using the method of bisection he devised about 1883. We will outline this argument. Suppose that C is a rectangular curve as shown in Figure 20.25, with sides parallel to the axes. We assume that f is analytic in a simply connected domain containing C and its interior. We want to show that $\oint_C f(z)\, dz = 0$.

Begin by dividing the rectangle enclosed by C into four congruent rectangles R_1, R_2, R_3, and R_4, as shown in Figure 20.26. Let C_j be the boundary curve of R_j. Then

$$\oint_C f(z)\, dz = \sum_{j=1}^{4} \oint_{C_j} f(z)\, dz.$$

Now observe that

$$\left| \oint_C f(z)\, dz \right| \le \sum_{j=1}^{4} \left| \oint_{C_j} f(z)\, dz \right|.$$

At least one term in the sum on the right must be at least as large as one-fourth of the term on the left. That is, for at least one of the rectangles $R_1, \ldots, R_4$, which we will relabel R^1, with boundary curve C^1, we must have

$$\left| \oint_{C^1} f(z)\, dz \right| \ge \tfrac{1}{4} \left| \oint_C f(z)\, dz \right|.$$

Figure 20.27 shows one of the rectangles shaded to represent R^1. Subdivide R^1 into four congruent rectangles, as shown in Figure 20.28, and repeat the above argument to conclude that for at least one of these rectangles, say, R^2, with boundary curve C^2, we have

$$\left| \oint_{C^2} f(z)\, dz \right| \ge \tfrac{1}{4} \left| \oint_{C^1} f(z)\, dz \right|.$$

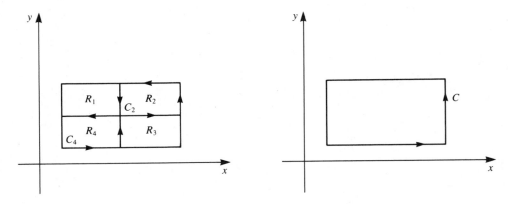

Figure 20.25   *Figure 20.26*

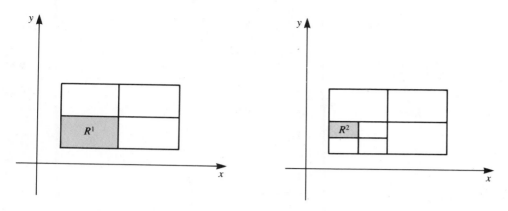

Figure 20.27   Figure 20.28

Then

$$\left| \oint_{C^2} f(z)\, dz \right| \geq \tfrac{1}{4} \left| \oint_{C^1} f(z)\, dz \right| \geq \frac{1}{4^2} \left| \oint_C f(z)\, dz \right|.$$

Continuing this process, we produce a sequence of rectangles $R^1, R^2, \ldots, R^n, \ldots$ such that each R^j is contained in R^{j-1} and

$$\left| \oint_{C^n} f(z)\, dz \right| \geq \frac{1}{4^n} \left| \oint_C f(z)\, dz \right|.$$

It is a routine geometric exercise to show that R^n has diameter $d/2^n$, where d is the diameter of the original rectangle, and perimeter $p/2^n$, where p is the perimeter of the original rectangle.

We now claim that there is exactly one point z_0 which is common to every rectangle R^n. To see this, note first that $d/2^n \to 0$ as $n \to \infty$. Thus, there cannot be more than one point common to every rectangle. If there were two such points, z_1 and z_2, at some stage the diameter of a rectangle would be smaller than $|z_1 - z_2|$, and both points could not be in the subsequent subdivided rectangle.

This shows that no more than one point can be common to every R^n. To show that there is a point common to them, choose any point z_j in R^j. Since the diameters of the rectangles go to zero as n increases, for any positive number ϵ, we can find some N such that the distance between z_k and z_j is less than ϵ if k and j are greater than N. If we write $z_j = x_j + iy_j$, we also have $|x_j - x_k| < \epsilon$ and $|y_j - y_k| < \epsilon$ if j and k are greater than N. Thus, the x_j's form a Cauchy sequence and converge to some number X, and the y_j's form a Cauchy sequence and converge to some number Y. Let $z_0 = X + iY$. It is routine to check that z_0 is indeed in every R^n.

Now, f is analytic at z_0, so

$$f'(z_0) = \lim_{z \to z_0} \frac{f(z) - f(z_0)}{z - z_0}.$$

Let $\epsilon > 0$. Then there is some positive number δ such that

$$\left| \frac{f(z) - f(z_0)}{z - z_0} - f'(z_0) \right| < \epsilon \quad \text{if} \quad |z - z_0| < \delta.$$

Then

$$|f(z) - f(z_0) - f'(z_0)(z - z_0)| < \epsilon |z - z_0| \quad \text{if} \quad |z - z_0| < \delta.$$

By a straightforward application of Theorem 20.4, we have

$$\oint_{C^j} dz = \oint_{C^j} z \, dz = 0$$

for each j. Since $f'(z_0)$ is constant,

$$\oint_{C^j} f(z) \, dz = \oint_{C^j} [f(z) - f(z_0) - (z - z_0)f'(z_0)] \, dz.$$

Choose N as a sufficiently large positive integer that $d/2^N < \delta$. Then, for z on C^N, $|z - z_0| < \delta$, and by Theorem 20.6, we have

$$\left| \oint_{C^N} f(z) \, dz \right| = \left| \oint_{C^N} [f(z) - f(z_0) - (z - z_0)f'(z_0)] \, dz \right|$$

$$\leq \oint_{C^N} |f(z) - f(z_0) - (z - z_0)f'(z_0)| \, dz \leq \epsilon \, \frac{d}{2^N} \frac{p}{2^N}.$$

We can therefore choose N large enough that

$$\frac{1}{4^N} \left| \oint_C f(z) \, dz \right| \leq \left| \oint_{C^N} f(z) \, dz \right| \leq \frac{\epsilon \, dp}{4^N};$$

hence,

$$\left| \oint_C f(z) \, dz \right| \leq dp\epsilon.$$

Since ϵ can be chosen as small as we like, we conclude that $\left| \oint_C f(z) \, dz \right| = 0$; hence, $\oint_C f(z) \, dz = 0$. This completes Goursat's proof of Cauchy's theorem for a rectangle.

From this argument, it is possible to prove Cauchy's theorem in general. We will not attempt the general case here. It is interesting to note, however, that for many applications of Cauchy's theorem, it is sufficient to know the theorem for the case in which C is a rectangle, along with the deformation theorem discussed in Section 20.3.

Complex Sequences and Series; Taylor and Laurent Expansions

21.1 Complex Sequences _____

The main objects of interest in this chapter are the Taylor and Laurent expansions of a complex function. The complex Taylor series is a direct generalization of the Taylor series of a real function. The Laurent expansion is different from any series seen in the calculus of real functions and has important ramifications in the evaluation of integrals, both real and complex, and in the summation of series.

Before considering infinite series of complex functions, we must consider infinite series of constants, which in turn requires that we look at infinite sequences.

A *complex sequence* is a function which assigns to each positive integer n (or nonnegative integer n) a complex number. The number assigned to n by this function is called the nth *term* of the sequence. If z_n is the nth term, we often denote the sequence as $\{z_n\}$. A sequence $\{z_n\}$ may be thought of as an ordered list

$$z_1, z_2, \ldots, z_n, \ldots.$$

For example, the sequence $\{i^n\}$ has i^n as its nth term. The terms of the sequence comprise the list

$$i, i^2, i^3, \ldots, i^n, \ldots,$$

or

$$i, -1, -i, \ldots, i^n, \ldots.$$

The terms $z_{n+1}, z_{n+2}, \ldots$ of a sequence $\{z_n\}$ are called the *terms beyond the nth term*. A sequence $\{z_n\}$ *converges* to a number L if, given any open disk D about L, there is some N such that the terms beyond the Nth term are in D (Figure 21.1). That is, given

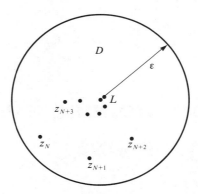

Figure 21.1. Convergence of
a complex sequence to L.

any positive number ϵ, there exists a positive number N such that

$$|z_n - L| < \epsilon$$

if n is an integer and $n \geq N$.

If $\{z_n\}$ converges to L, we call L the *limit* of the sequence and write

$$L = \lim_{n \to \infty} z_n,$$

or $z_n \to L$. If there is no such number L, we say that $\{z_n\}$ *diverges*.

The following theorem reduces the study of complex sequences to that of real sequences.

THEOREM 21.1

Let $z_n = x_n + iy_n$ for each positive integer n, and let $L = a + ib$. Then

$$\lim_{n \to \infty} z_n = L \quad \text{if and only if} \quad \lim_{n \to \infty} x_n = a \quad \text{and} \quad \lim_{n \to \infty} y_n = b. \quad \blacksquare$$

This means that a complex sequence $\{z_n\}$ converges if and only if the real sequences formed from the real and imaginary parts of z_n both converge. In theory, then, we can always consider the convergence of a complex sequence $\{z_n\}$ by looking at the real sequences $\{\mathrm{Re}(z_n)\}$ and $\{\mathrm{Im}(z_n)\}$. A proof of Theorem 21.1 follows Examples 21.1 and 21.2.

EXAMPLE 21.1

Let $z_n = 3/n + [(n+1)/(n+2)]i$. Since $\lim_{n \to \infty} 3/n = 0$ and $\lim_{n \to \infty} (n+1)/(n+2) = 1$, $\lim_{n \to \infty} z_n = i$. $\blacksquare$

EXAMPLE 21.2

Let $z_n = e^{in} = \cos(n) + i\sin(n)$. Since $\lim_{n \to \infty} \cos(n)$ does not exist, $\{z_n\}$ diverges. [In this example, $\lim_{n \to \infty} \sin(n)$ fails to exist also.] $\blacksquare$

Proof of Theorem 21.1   Suppose first that $z_n \to a + ib$. For some N,

$$|z_n - (a + ib)| < \epsilon \quad \text{if} \quad n \geq N.$$

Then $\sqrt{(x_n - a)^2 + (y_n - b)^2} < \epsilon$ if $n \geq N$. Then $|x_n - a| < \epsilon$ and $|y_n - b| < \epsilon$ if $n \geq N$; hence, $x_n \to a$ and $y_n \to b$.

Conversely, suppose that $x_n \to a$ and $y_n \to b$. Let $\epsilon > 0$. For some positive number N_1,

$$|x_n - a| < \frac{\epsilon}{2} \quad \text{if} \quad n \geq N_1.$$

For some positive number N_2, $|y_n - b| < \epsilon/2$ if $n \geq N_2$. Let $n \geq N_1 + N_2$. Then $n \geq N_1$ and $n \geq N_2$; hence,

$$|z_n - (a + ib)| = \sqrt{(x_n - a)^2 + (y_n - b)^2}$$
$$< \sqrt{\left(\frac{\epsilon}{2}\right)^2 + \left(\frac{\epsilon}{2}\right)^2} = \frac{1}{\sqrt{2}} \epsilon < \epsilon.$$

Hence, $z_n \to a + ib$. ∎

Complex sequences have many of the properties enjoyed by real sequences.

THEOREM 21.2

Let $z_n \to L$ and $w_n \to K$. Then

1. $z_n + w_n \to L + K$.

2. $\alpha z_n \to \alpha L$ for any number α.

3. $z_n w_n \to LK$.

4. $z_n/w_n \to L/K$, provided that each $w_n \neq 0$ and $K \neq 0$. ∎

We call $\{z_n\}$ a *Cauchy sequence* if, given $\epsilon > 0$, there exists a positive number N such that $|z_n - z_m| < \epsilon$ if $n \geq N$ and $m \geq N$. This means that, given any positive number ϵ, there is a cutoff term in the sequence such that any two terms beyond this term are within ϵ of each other. The Cauchy convergence criterion states that the convergent sequences are exactly the Cauchy sequences.

THEOREM 21.3

A complex sequence converges if and only if it is a Cauchy sequence.

Proof   We will assume familiarity with the Cauchy convergence criterion for real sequences. Suppose first that $z_n \to L$. For some N, $|z_n - L| < \epsilon/2$ if $n \geq N$. Then, for $n \geq N$ and $m \geq N$, we have

$$|z_n - z_m| = |(z_n - L) - (z_m - L)| \leq |z_n - L| + |z_m - L| < \frac{\epsilon}{2} + \frac{\epsilon}{2} = \epsilon,$$

proving that $\{z_n\}$ is a Cauchy sequence.

Conversely, suppose that $\{z_n\}$ is a Cauchy sequence. We must prove that $\{z_n\}$ converges. Let $z_n = x_n + iy_n$. For some N, $|z_n - z_m| < \epsilon$ if $n \geq N$ and $m \geq N$. Since

$|x_n - x_m| \leq |z_n - z_m|$, $\{x_n\}$ is a real Cauchy sequence and hence converges to some real number a. Similarly, $\{y_n\}$ is a real Cauchy sequence and so converges to some real number b. Then $z_n \to a + ib$. ∎

The following example illustrates how the Cauchy convergence criterion can be used to show that a sequence converges even when a candidate for the limit is not obvious.

EXAMPLE 21.3

Let $z_1 = 2i$, $z_2 = 1 - i$, and, for $n \geq 3$, $z_n = \frac{1}{2}(z_{n-2} + z_{n-1})$. Thus,

$$z_3 = \frac{1}{2}(z_1 + z_2) = \frac{1}{2}(1 + i), \qquad z_4 = \frac{1}{2}(z_2 + z_3) = \frac{1}{4}(3 - i),$$

and so on. We will show that this sequence converges even though it is not clear what the limit is.

First, observe that

$$|z_2 - z_1| = |1 + 3i| = \sqrt{10},$$
$$|z_3 - z_2| = |-\tfrac{1}{2} + \tfrac{3}{2}i| = \tfrac{1}{2}\sqrt{10},$$
$$|z_4 - z_3| = |\tfrac{1}{4} + \tfrac{3}{4}i| = \tfrac{1}{4}\sqrt{10},$$

and

$$|z_5 - z_4| = |-\tfrac{1}{8} + \tfrac{3}{8}i| = \tfrac{1}{8}\sqrt{10}.$$

A pattern suggests itself, and we conjecture that

$$|z_n - z_{n-1}| = \frac{1}{2^{n-2}}\sqrt{10}.$$

This fact can be verified by mathematical induction. Now let n and m be positive integers, with $m < n$. Write $n = m + t$. Then

$$
\begin{aligned}
|z_n - z_m| &= |(z_n - z_{n-1}) + (z_{n-1} - z_{n-2}) + \cdots + (z_{m+1} - z_m)| \\
&\leq |z_n - z_{n-1}| + |z_{n-1} - z_{n-2}| + \cdots + |z_{m+1} - z_m| \\
&= \frac{1}{2^{n-2}}\sqrt{10} + \frac{1}{2^{n-3}}\sqrt{10} + \cdots + \frac{1}{2^{m-1}}\sqrt{10} \\
&= \left(\frac{1}{2^{n-2}} + \frac{1}{2^{n-3}} + \cdots + \frac{1}{2^{m-1}}\right)\sqrt{10} \\
&= \left(\frac{1}{2^{m+t-2}} + \frac{1}{2^{m+t-3}} + \cdots + \frac{1}{2^{m-1}}\right)\sqrt{10} \\
&= \frac{1}{2^{m-1}}\left(\frac{1}{2^{t-1}} + \frac{1}{2^{t-2}} + \cdots + 1\right)\sqrt{10}.
\end{aligned}
$$

If $r \neq 1$,

$$\sum_{j=0}^{n} r^j = \frac{1 - r^{n+1}}{1 - r}.$$

(See Problem 21 at the end of this section.) Apply this formula to the last sum in parentheses, with $n = t - 1$, to get

$$1 + \frac{1}{2} + \cdots + \frac{1}{2^{t-2}} + \frac{1}{2^{t-1}} = \frac{1 - (\frac{1}{2})^t}{1 - \frac{1}{2}} = 2\left[1 - \left(\frac{1}{2}\right)^t\right].$$

Therefore,

$$|z_n - z_m| \le \frac{1}{2^{m-2}}\left[1 - \left(\frac{1}{2}\right)^t\right]\sqrt{10} \le \frac{1}{2^{m-2}}\sqrt{10}.$$

Certainly, $(1/2^{m-2})\sqrt{10} \to 0$ as $m \to \infty$. Given $\epsilon > 0$, there is some N such that $(1/2^{m-2})\sqrt{10} < \epsilon$ if $m \ge N$. Therefore, for $m \ge N$ and $n \ge N$, $|z_n - z_m| < \epsilon$. This proves that $\{z_n\}$ is a Cauchy sequence and hence that it converges. We will find in Problem 20 what the limit of this sequence is.   ∎

PROBLEMS FOR SECTION 21.1

In each of Problems 1 through 15, determine whether $\{z_n\}$ converges, with z_n the given complex number. In the case of convergence, find the limit.

1. $1 + \dfrac{2in}{1 + n}$

2. i^{2n}

3. $\dfrac{1 + 2n^2}{n^2} - \dfrac{n - 1}{n}i$

4. $e^{n\pi i/3}$

5. $(-1)^{in}$

6. $\left(1 + \dfrac{1}{n}\right)^n i$

7. $\dfrac{5n^2 i + 3}{n^2 + 1}$

8. $(1 - i)^{-n}$

9. $(-i)^{4n}$

10. $(-1)^n\dfrac{1}{n} + \dfrac{3n - 2}{n}i$

11. $\left(\dfrac{n^2}{n + i}\right)i$

12. $\tan(in)$

13. $e^{i/n}$

14. $\dfrac{1}{1 - i}\left(\dfrac{n^2 + 4}{n^2 - 3n}\right)$

15. $\left(-\dfrac{1}{n}\right)^n + \dfrac{3n^2}{n^2 + 2}i$

16. A sequence $\{z_n\}$ is bounded if there is some number M such that $|z_n| \le M$ for all n. Prove that a convergent sequence is bounded.

17. Give an example of a bounded sequence which does not converge.

18. Suppose that $\{z_n\}$ converges. Prove that $\{|z_n|\}$ converges. *Hint:* Use the inequality $||z| - |w|| \le |z - w|$.

19. Give an example in which $\{|z_n|\}$ converges but $\{z_n\}$ diverges.

20. Let z_1 and z_2 be given complex numbers. Let $z_n = \frac{1}{2}(z_{n-1} + z_{n-2})$. Prove that this sequence converges by showing that it is a Cauchy sequence. Then use the fact that it converges to show that the limit is $\frac{1}{3}(z_1 + 2z_2)$.

21. Verify that, for $r \ne 1$,

$$\sum_{j=0}^{n} r^j = \frac{1 - r^{n+1}}{1 - r}.$$

Hint: Let this sum be S_n and consider $rS_n - S_n$.

21.2 *Series of Complex Constants* _____

Let $\{z_n\}$ be a complex sequence. We want to assign a meaning to the symbol $\sum_{n=1}^{\infty} z_n$, which denotes the sum of the terms of the sequence. As in the real case, define the nth partial sum S_n of this series as the sum of the first n terms of the sequence $\{z_n\}$:

$$S_n = \sum_{j=1}^{n} z_j = z_1 + z_2 + \cdots + z_n.$$

$\{S_n\}$ is itself a complex sequence. If this sequence of partial sums converges, we say that the infinite series $\sum_{n=1}^{\infty} z_n$ *converges* and define its sum to be $\lim_{n\to\infty} S_n$. If $\{S_n\}$ diverges, we say that the infinite series $\sum_{n=1}^{\infty} z_n$ *diverges*.

As with sequences, the question of convergence of a complex series can always be reduced to a question about real series.

THEOREM 21.4

Let $z_n = x_n + iy_n$. Then

1. $\sum_{n=1}^{\infty} z_n$ converges if and only if $\sum_{n=1}^{\infty} x_n$ and $\sum_{n=1}^{\infty} y_n$ both converge.
2. If $\sum_{n=1}^{\infty} x_n$ converges to a and $\sum_{n=1}^{\infty} y_n$ converges to b, $\sum_{n=1}^{\infty} z_n$ converges to $a + ib$. ∎

We can therefore test a complex series $\sum_{n=1}^{\infty} z_n$ for convergence by applying tests for real series to both $\sum_{n=1}^{\infty} x_n$ and $\sum_{n=1}^{\infty} y_n$. We will, however, state some results for complex series which we will use in our study of Taylor and Laurent expansions.

THEOREM 21.5

If $\sum_{n=1}^{\infty} z_n$ converges, $z_n \to 0$. ∎

Often, this theorem is used to show that a series diverges. If $\{z_n\}$ has no limit or has a limit which is nonzero, $\sum_{n=1}^{\infty} z_n$ must diverge. If, however, $z_n \to 0$, all we can say is that $\sum_{n=1}^{\infty} z_n$ might converge. It is possible for $z_n \to 0$ and for $\sum_{n=1}^{\infty} z_n$ to diverge. For example, $\lim_{n\to\infty} i/n = 0$, but $\sum_{n=1}^{\infty} i/n$ diverges because this series is i times the divergent harmonic series.

The series $\sum_{n=1}^{\infty} z^n$, in which z is a given complex number, is called a *geometric series*. In the problems at the end of this section, we will show that this series converges to $z/(1 - z)$ if $|z| < 1$ and diverges if $|z| \geq 1$. The series $\sum_{n=0}^{\infty} z^n$ is also called a geometric series; it is just the series $\sum_{n=1}^{\infty} z^n$ with an additional term 1 (equal to z^0). This series converges to $[z/(1 - z)] + 1$, or $1/(1 - z)$ if $|z| < 1$, and diverges if $|z| \geq 1$.

If we apply the Cauchy convergence criterion for sequences to the sequence of partial sums of the complex series $\sum_{n=1}^{\infty} z_n$, we obtain the following criterion.

THEOREM 21.6   Cauchy Convergence Criterion for Series

$\sum_{n=1}^{\infty} z_n$ converges if and only if, given $\epsilon > 0$, there is some N such that, if $n \geq N$ and p is any positive integer, $|z_{n+1} + z_{n+2} + \cdots + z_{n+p}| < \epsilon$. ∎

If the real series $\sum_{n=1}^{\infty} |z_n|$ converges, the series $\sum_{n=1}^{\infty} z_n$ is said to *converge absolutely*. We will now show that a series which converges absolutely must itself converge.

THEOREM 21.7   Absolute Convergence Implies Convergence

If $\sum_{n=1}^{\infty} |z_n|$ converges, $\sum_{n=1}^{\infty} z_n$ converges also.

Proof   Assume that $\sum_{n=1}^{\infty} |z_n|$ converges. Let $\epsilon > 0$. For some N, if $n \geq N$ and p is any positive integer, we have

$$|z_{n+1}| + |z_{n+2}| + \cdots + |z_{n+p}| < \epsilon$$

by applying Theorem 21.6 to the series $\sum_{n=1}^{\infty} |z_n|$. Then

$$|z_{n+1} + z_{n+2} + \cdots + z_{n+p}| \leq |z_{n+1}| + |z_{n+2}| + \cdots + |z_{n+p}| < \epsilon;$$

hence, by Theorem 21.6, $\sum_{n=1}^{\infty} z_n$ converges also.   ∎

It is possible for $\sum_{n=1}^{\infty} z_n$ to converge and $\sum_{n=1}^{\infty} |z_n|$ to diverge, just as in the real case. When this occurs, we say that $\sum_{n=1}^{\infty} z_n$ *converges conditionally*. For example, $\sum_{n=1}^{\infty} (1/n)(i)^{2n}$ converges because $i^{2n} = (-1)^n$ and $\sum_{n=1}^{\infty} (-1)^n (1/n)$ is the alternating harmonic series. However, $|i^{2n}| = 1$, so for this series $\sum_{n=1}^{\infty} |z_n|$ is the divergent harmonic series.

One value of Theorem 21.7 is that tests for real series can be applied to $\sum_{n=1}^{\infty} |z_n|$. If they indicate that this series converges, we know that $\sum_{n=1}^{\infty} z_n$ converges also.

The following complex version of the ratio test will be used when power series are treated in the next section.

THEOREM 21.8

Let $z_n \neq 0$ for each n, and assume that $\lim_{n \to \infty} |z_{n+1}/z_n| = R$. Then

1. $\sum_{n=1}^{\infty} |z_n|$ converges, and hence also $\sum_{n=1}^{\infty} z_n$ converges, if $0 \leq R < 1$.
2. $\sum_{n=1}^{\infty} z_n$ diverges if $R > 1$.   ∎

PROBLEMS FOR SECTION 21.2

In each of Problems 1 through 6, test the series for convergence. Facts about real series may be assumed.

1. $\sum_{n=1}^{\infty} \left(\dfrac{1}{n^2} + \dfrac{1}{2^n} i \right)$

2. $\sum_{n=1}^{\infty} \left(\dfrac{2}{n} - \dfrac{1}{n^4} i \right)$

3. $\sum_{n=1}^{\infty} \left(-\dfrac{i}{n} \right)^n$

4. $\sum_{n=1}^{\infty} \tan(in)$

5. $\sum_{n=1}^{\infty} \left(\dfrac{\sin(n)}{n^2} + \dfrac{1}{3n} (i)^{3n} \right)$

6. $\sum_{n=1}^{\infty} \cosh(in)$

7. Prove Theorem 21.4. *Hint:* Apply Theorem 21.1 to the sequence of partial sums of the series.

8. Prove Theorem 21.5. *Hint:* Apply the Cauchy convergence criterion for sequences, and consider the terms S_n and S_{n-1} of the sequence of partial sums.

9. Prove Theorem 21.6. *Hint:* Apply the Cauchy convergence criterion for sequences to the sequence of partial sums.

10. Prove that $\sum_{n=1}^{\infty} z^n = z/(1-z)$ if $|z| < 1$ and that this series diverges if $|z| \geq 1$. *Hint:* Let $S_n = \sum_{j=1}^{\infty} z^n$. Derive a formula for S_n by using the reasoning suggested in Problem 21 of Section 21.1, then consider $\lim_{n \to \infty} S_n$. Consider separately the case in which $|z| = 1$.

21.3 *Complex Power Series*

Let $z_0, a_0, a_1, a_2, \ldots$ be given complex numbers. A series

$$\sum_{n=0}^{\infty} a_n(z - z_0)^n$$

is called a *power series* with *center z_0* and *coefficient sequence $\{a_n\}$*. The series begins at $n = 0$ in order to allow for a constant term:

$$\sum_{n=0}^{\infty} a_n(z - z_0)^n = a_0 + a_1(z - z_0) + a_2(z - z_0)^2 + \cdots.$$

Obviously, this series converges at z_0. We will now show that, if it converges at any other point z_1, it also converges absolutely at all points closer to z_0 than z_1 (Figure 21.2).

THEOREM 21.9

Suppose that $\sum_{n=0}^{\infty} a_n(z - z_0)^n$ converges for $z_1 \neq z_0$. Then the series converges absolutely for all z such that $|z - z_0| < |z_1 - z_0|$.

Proof   Let $|z - z_0| < |z_1 - z_0|$, with $z_1 \neq z_0$. Since $\sum_{n=0}^{\infty} a_n(z_1 - z_0)^n$ converges, we must have $\lim_{n \to \infty} a_n(z_1 - z_0)^n = 0$. Therefore, for some N,

$$|a_n(z_1 - z_0)|^n < 1 \quad \text{if} \quad n \geq N.$$

For $n \geq N$,

$$|a_n(z - z_0)|^n = |a_n(z - z_0)^n| \frac{|z_1 - z_0|^n}{|z_1 - z_0|^n}$$

$$= \left| \frac{z - z_0}{z_1 - z_0} \right|^n |a_n(z_1 - z_0)^n| < \left| \frac{z - z_0}{z_1 - z_0} \right|^n.$$

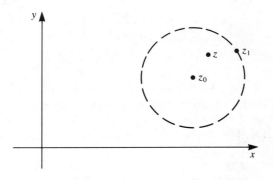

Figure 21.2. If $\sum_{n=0}^{\infty} a_n(z_1 - z_0)^n$ converges, $\sum_{n=0}^{\infty} |a_n(z - z_0)^n|$ converges for all z closer to z_0 than to z_1.

But $|(z - z_0)/(z_1 - z_0)| < 1$ because $|z - z_0| < |z_1 - z_0|$. Therefore, $\sum_{n=N}^{\infty} |(z - z_0)/(z_1 - z_0)|^n$ converges as part of the real, convergent geometric series $\sum_{n=0}^{\infty} r^n$ with $r = |(z - z_0)/(z_1 - z_0)| < 1$. By the comparison test for real series, $\sum_{n=N}^{\infty} |a_n(z - z_0)^n|$ converges also. But then $\sum_{n=0}^{\infty} |a_n(z - z_0)^n|$ converges; hence, $\sum_{n=0}^{\infty} a_n(z - z_0)^n$ converges absolutely. ∎

Theorem 21.9 can be used to show that there are exactly three possibilities for convergence of a power series $\sum_{n=0}^{\infty} a_n(z - z_0)^n$.

Case 1   The series may converge for all z. This occurs with $\sum_{n=0}^{\infty} (1/n!)z^n$ (by Theorem 21.8).

Case 2   The series may converge only for $z = z_0$. This occurs with $\sum_{n=0}^{\infty} n^n z^n$, which converges only at $z = 0$.

Case 3   If neither case 1 nor case 2 holds, the series converges for some $z_1 \neq z_0$ but does not converge for all z.

In this case use Theorem 21.9 to argue as follows. Consider the circle $|z - z_0| = r$ about z_0. At $r = 0$, this consists of just the point z_0. As r increases from zero, this circle must eventually encounter some point z_d at which the series diverges (or else it would converge for all z). Let R be the smallest value of r at which this occurs. The power series $\sum_{n=0}^{\infty} a_n(z - z_0)^n$ must converge for $|z - z_0| < R$ and diverge if $|z - z_0| > R$. Therefore, the power series converges in an open disk about z_0 and diverges outside of this disk. This disk is called the *disk of convergence* of the power series, and its radius R is called the *radius of convergence* of the power series.

We can combine all three cases by agreeing to call $R = \infty$ in case 1, in which the series converges for all z, and $R = 0$ in case 2, in which the series converges only for $z = z_0$. When $R = \infty$, the disk of convergence is the entire plane; when $R = 0$, the disk degenerates to the single point z_0. We will summarize this discussion in a theorem.

THEOREM 21.10

Given a power series $\sum_{n=0}^{\infty} a_n(z - z_0)^n$, which converges for some $z_1 \neq z_0$, there exists R (possibly $R = \infty$) such that the series converges absolutely if $|z - z_0| < R$ and diverges if $|z - z_0| > R$. ∎

Sometimes we can determine the radius of convergence of a power series by applying the ratio test (Theorem 21.8).

EXAMPLE 21.4

Consider the power series $\sum_{n=0}^{\infty} (n^n/n!)(z - i)^n$. Apply the ratio test to the real series $\sum_{n=0}^{\infty} (n^n/n!)|z - i|^n$. The ratio of the $(n + 1)$ term to the nth term is

$$\left| \frac{(n + 1)^{n+1}}{(n + 1)!} \frac{n!}{n^n} \frac{(z - i)^{n+1}}{(z - i)^n} \right| = \frac{(n + 1)^n}{n^n} |z - i| = \left(1 + \frac{1}{n} \right)^n |z - i|,$$

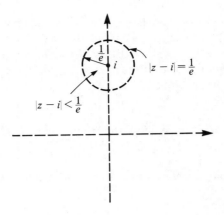

Figure 21.3

and this has limit $e|z - i|$ as $n \to \infty$. This limit is less than 1 if $e|z - i| < 1$, or $|z - i| < 1/e$. The power series therefore converges for $|z - i| < 1/e$ and diverges if $|z - i| > 1/e$. The radius of convergence is $1/e$, and the disk of convergence is the disk of radius $1/e$ about i (Figure 21.3). ■

EXAMPLE 21.5

Consider the power series

$$\sum_{n=0}^{\infty} \frac{(-1)^n}{n + 1} 2^n(z - 1 + 2i)^{2n}.$$

The center of this series is $1 - 2i$. Consider the absolute value of the ratio of successive terms of this series:

$$\left| \frac{\frac{(-1)^{n+1}}{(n + 1) + 1} 2^{n+1}(z - 1 + 2i)^{2n+2}}{\frac{(-1)^n}{n + 1} 2^n(z - 1 + 2i)^{2n}} \right| = 2\frac{n + 1}{n + 2}|z - 1 + 2i|^2$$

$$\longrightarrow 2|z - 1 + 2i|^2 \quad \text{as} \quad n \to \infty.$$

The limit of this ratio is less than 1 if $2|z - 1 + 2i|^2 < 1$, or

$$|z - 1 + 2i| < \frac{1}{\sqrt{2}}.$$

For all such z, the power series converges.

The limit of the ratio is greater than 1 if $|z - 1 + 2i| > 1/\sqrt{2}$; for all such z, the power series diverges.

The radius of convergence of this power series is $1/\sqrt{2}$. The disk of convergence is the disk $|z - 1 + 2i| < 1/\sqrt{2}$, which is the open disk of radius $1/\sqrt{2}$ and center $1 - 2i$ (Figure 21.4).

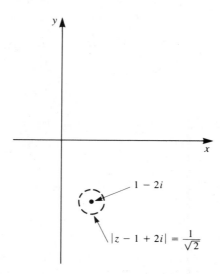

Figure 21.4   ∎

The radius of convergence of a power series depends only on the coefficients of the series. The center of the series is the center of the disk of convergence.

The next theorem states that, within the open disk of convergence of a power series, the series represents a function which is analytic and whose derivative may be computed by differentiating the series term by term. Further, we may integrate this function over any curve in the disk of convergence by integrating the series term by term.

THEOREM 21.11

Suppose that $\sum_{n=0}^{\infty} a_n(z - z_0)^n$ has radius of convergence R, with $R \neq 0$. For $|z - z_0| < R$, let $f(z) = \sum_{n=0}^{\infty} a_n(z - z_0)^n$. Then

1. f is analytic for $|z - z_0| < R$.

2. $f'(z) = \sum_{n=1}^{\infty} na_n(z - z_0)^{n-1}$ for $|z - z_0| < R$. This is exactly the result obtained by differentiating the power series for $f(z)$ term by term.

3. For any positive integer k and $|z - z_0| < R$,

$$f^{(k)}(z) = \sum_{n=k}^{\infty} n(n - 1) \cdots (n - k + 1)a_n(z - z_0)^{n-k},$$

in which $f^{(k)}(z)$ is the kth derivative of f evaluated at z.

4. If C is a piecewise-smooth curve whose graph lies within the disk of convergence of the power series,

$$\int_C f(z)\, dz = \sum_{n=0}^{\infty} a_n \int_C (z - z_0)^n\, dz. \quad \blacksquare$$

We will not prove this theorem, but we will use it in our treatment of Taylor series in the next section.

EXAMPLE 21.6

We know that, for $|z| < 1$, $\sum_{n=0}^{\infty} z^n = 1/(1-z)$ (geometric series). Let $f(z) = 1/(1-z)$ for $|z| < 1$. Then

$$f'(z) = \frac{1}{(1-z)^2} = \sum_{n=1}^{\infty} nz^{n-1},$$

$$f''(z) = \frac{2}{(1-z)^3} = \sum_{n=2}^{\infty} n(n-1)z^{n-2},$$

and, for any positive integer k,

$$f^{(k)}(z) = \frac{k!}{(1-z)^{k+1}} = \sum_{n=k}^{\infty} n(n-1)\cdots(n-k+1)z^{n-k}. \quad \blacksquare$$

PROBLEMS FOR SECTION 21.3

In each of Problems 1 through 10, find the radius of convergence and the disk of convergence of the power series. Denote the power series by $f(z)$ within the disk of convergence, and write power series for $f'(z)$ and $f''(z)$.

1. $\sum_{n=0}^{\infty} \frac{n+1}{2^n}(z+3i)^n$

2. $\sum_{n=0}^{\infty} \frac{(-1)^n}{(2n+1)^2}(z-i)^{2n}$

3. $\sum_{n=0}^{\infty} \frac{n^n}{(n+1)^n}(z-1+2i)^n$

4. $\sum_{n=0}^{\infty} \left(\frac{2}{3i}\right)^n (z+1+4i)^n$

5. $\sum_{n=0}^{\infty} \left(\frac{i^n}{2^{n+1}}\right)(z+4-i)^n$

6. $\sum_{n=0}^{\infty} \frac{(1-i)^n}{n+2}(z-3)^n$

7. $\sum_{n=0}^{\infty} \frac{n^2}{2n+1}(z+6+2i)^{2n}$

8. $\sum_{n=0}^{\infty} \left(\frac{n^3}{4^n}\right)(z-3)^{2n}$

9. $\sum_{n=0}^{\infty} \left(\frac{e^{in}}{2n+1}\right)(z+4)^n$

10. $\sum_{n=0}^{\infty} \left(\frac{1-i}{2+i}\right)^n (z-3)^{4n}$

11. Is it possible for $\sum_{n=0}^{\infty} a_n(z-2i)^n$ to converge at zero and diverge at i?

12. Is it possible for $\sum_{n=0}^{\infty} a_n(z-4+2i)^n$ to converge at i and diverge at $1+i$?

13. Suppose that $\sum_{n=0}^{\infty} a_n(z-z_0)^n$ has radius of convergence R_1 and $\sum_{n=0}^{\infty} b_n(z-z_0)^n$ has radius of convergence R_2. What can be said about the radius of convergence of $\sum_{n=0}^{\infty} (a_n + b_n)(z-z_0)^n$?

14. Suppose that $\lim_{n\to\infty} |a_n|^{1/n} = 1/R$. Prove that the radius of convergence of $\sum_{n=0}^{\infty} a_n(z-z_0)^n$ is R.

15. Suppose that $\sum_{n=0}^{\infty} a_n(z-z_0)^n = \sum_{n=0}^{\infty} b_n(z-z_0)^n$ for all z with $|z-z_0| < R$, with $R > 0$. Show that $a_n = b_n$ for $n = 0, 1, 2, \ldots$.

16. Consider the series $\sum_{n=0}^{\infty} a_n(z-z_0)^n$, where

$$a_n = \begin{cases} 2 & \text{if } n \text{ is even} \\ 1 & \text{if } n \text{ is odd}. \end{cases}$$

Show that the radius of convergence of this series is 1 but that this radius cannot be computed using the ratio test, as we did in Examples 21.4 and 21.5.

17. Let $f(z) = \sum_{n=0}^{\infty} a_n(z-z_0)^n$ and $g(z) = \sum_{n=0}^{\infty} b_n(z-z_0)^n$, and assume that both series converge for $|z-z_0| < R$, with $R > 0$. The Cauchy product of these series is defined to be $\sum_{n=0}^{\infty} c_n(z-z_0)^n$, where $c_n = \sum_{j=0}^{\infty} a_j b_{n-j}$. It can be proved that the Cauchy product series of these two power series converges to $f(z)g(z)$ if $|z-z_0| < R$. Assuming this fact, prove the following.

(a) $[1/(1-z)]^2 = \sum_{n=0}^{\infty} (n+1)z^n$ for $|z| < 1$. *Hint:* Use the fact that $\sum_{n=0}^{\infty} z^n = 1/(1-z)$ if $|z| < 1$.

(b) Let $f(z) = \sum_{n=0}^{\infty} (1/n!)z^n$. Prove that $[f(z)]^2 = f(2z)$. [We will see later that $f(z) = e^z$, so the relationship to be proved is that $(e^z)^2 = e^{2z}$.]

(c) Let

$$f(z) = \sum_{n=0}^{\infty} (-1)^n \frac{1}{(2n)!} z^{2n} \quad \text{and} \quad g(z) = \sum_{n=0}^{\infty} (-1)^n \frac{1}{(2n+1)!} z^{2n+1}.$$

Prove that $f(z)g(z) = \frac{1}{2}g(2z)$. [We will see later that $f(z) = \cos(z)$ and $g(z) = \sin(z)$, so the relationship to be proved is the identity $\sin(z)\cos(z) = \frac{1}{2}\sin(2z)$.]

21.4 *Complex Taylor Series*

We will now define the complex Taylor series and show that a function which is analytic at a point can be expanded in a Taylor series about the point. Suppose that $\sum_{n=0}^{\infty} a_n(z - z_0)^n$ converges in a disk $|z - z_0| < R$. The function f defined by $f(z) = \sum_{n=0}^{\infty} a_n(z - z_0)^n$ for z in this disk is analytic in this disk. We will determine the coefficients a_n in terms of f and its derivatives evaluated at z_0.

First,

$$f(z_0) = a_0.$$

Further, if k is any positive integer, we know from the preceding section that

$$f^{(k)}(z) = \sum_{n=k}^{\infty} n(n-1)\cdots(n-k+1)a_n(z - z_0)^{n-k} \quad \cdot$$

if $|z - z_0| < R$. In particular, letting $z = z_0$, all terms of this series are zero except possibly the first, constant term, which occurs when $n = k$. We have

$$f^{(k)}(z_0) = k(k-1)\cdots(1)a_k = k!a_k;$$

hence,

$$a_k = \frac{1}{k!} f^{(k)}(z_0)$$

for $k = 1, 2, 3, \ldots$. If we agree to let $f^{(0)}(z_0) = f(z_0)$ and recall that $0! = 1$ by definition, we have

$$a_k = \frac{1}{k!} f^{(k)}(z_0) \quad \text{for} \quad k = 0, 1, 2, \ldots. \tag{21.1}$$

These numbers are called the *Taylor coefficients* of f at z_0. The series for f is therefore

$$f(z) = \sum_{n=0}^{\infty} \frac{1}{n!} f^{(n)}(z_0)(z - z_0)^n, \tag{21.2}$$

which is called the *Taylor series for f about z_0*.

Thus far, we have said that any power series defining a function f in a disk about z_0 must be the Taylor series of f about z_0. Now consider the "reverse" question. Suppose we have a function f which is analytic at z_0. We claim that f is represented by its Taylor series (21.2) in some disk about z_0 [in the sense that the Taylor series (21.2) converges to $f(z)$ in this disk].

THEOREM 21.12

Let f be analytic at z_0. Then f has a Taylor series representation

$$f(z) = \sum_{n=0}^{\infty} \frac{1}{n!} f^{(n)}(z_0)(z - z_0)^n$$

for all z in some disk about z_0.

Proof   Since f is analytic at z_0, there is some disk $|z - z_0| < r$ in which f is differentiable. Let C be the circle $|z - z_0| = r/2$ (Figure 21.5). Then f is differentiable at all points on and enclosed by C. Let w be on C, and let z be any point enclosed by C. Write

$$\frac{1}{w - z} = \frac{1}{w - z_0} \frac{1}{1 - \dfrac{z - z_0}{w - z_0}}. \tag{21.3}$$

Since w is further from z_0 than z is, $|z - z_0| < |w - z_0|$; hence,

$$\left| \frac{z - z_0}{w - z_0} \right| < 1.$$

Use the geometric series to write

$$\frac{1}{1 - \dfrac{z - z_0}{w - z_0}} = \sum_{n=0}^{\infty} \left(\frac{z - z_0}{w - z_0} \right)^n.$$

Using equation (21.3), we can therefore write

$$\frac{1}{w - z} = \frac{1}{w - z_0} \sum_{n=0}^{\infty} \left(\frac{z - z_0}{w - z_0} \right)^n = \sum_{n=0}^{\infty} \frac{(z - z_0)^n}{(w - z_0)^{n+1}}.$$

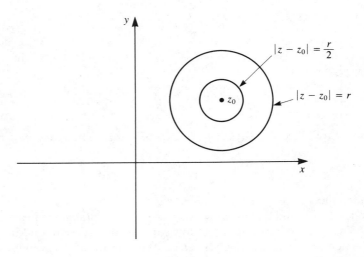

Figure 21.5

Now recall that, by Cauchy's integral formula,

$$f(z) = \frac{1}{2\pi i} \oint_C \frac{f(w)}{w - z} \, dw = \frac{1}{2\pi i} \oint_C \sum_{n=0}^{\infty} \frac{f(w)(z - z_0)^n}{(w - z_0)^{n+1}} \, dw.$$

We already know that a power series can be integrated term by term within its disk of convergence. It can be shown that a series obtained by multiplying a power series by an analytic function can still be integrated term by term within this disk. Therefore, we can interchange the sum and the integral in the last equation to get

$$f(z) = \sum_{n=0}^{\infty} \left[\frac{1}{2\pi i} \oint_C \frac{f(w)}{(w - z_0)^{n+1}} \, dw \right] (z - z_0)^n. \tag{21.4}$$

But, by Cauchy's integral formula for higher derivatives, recognize that

$$\frac{1}{2\pi i} \oint_C \frac{f(w)}{(w - z_0)^{n+1}} \, dw = \frac{1}{n!} f^{(n)}(z_0).$$

Therefore, equation (21.4) can be written

$$f(z) = \sum_{n=0}^{\infty} \frac{1}{n!} f^{(n)}(z_0)(z - z_0)^n,$$

which proves that f is represented by its Taylor series in a disk about z_0. ∎

When we write $f(z)$ as a Taylor series about z_0, we say that we have *expanded f in a Taylor series about* z_0. A Taylor series about zero is called a *Maclaurin series*.

When we expand $f(z)$ in a Taylor series about z_0, the Taylor series converges in a disk $|z - z_0| < R$, where R is the distance from z_0 to the nearest point at which f fails to be analytic. Indeed, if R were larger than this distance, the disk of convergence of the Taylor series would contain a point where f fails to be analytic, contradicting the fact that f must be analytic at every point of this disk. This means that we can often tell the radius of convergence of the Taylor series of f about z_0 just by looking at the points where f is not analytic and calculating the distance from z_0 to the nearest such point.

For example, let $f(z) = 1/(1 + z^2)$. Then f is analytic except at i and $-i$. The Taylor series of f about zero converges in $|z| < 1$ because 1 is the distance from zero (the center of the series) to i (or to $-i$). The disk of convergence cannot have radius larger than 1 because then it would contain points at which f is not analytic. In fact, we can use the geometric series $1/(1 - z) = \sum_{n=0}^{\infty} z^n$, and put $-z^2$ in place of z, to write the Taylor series of f about zero:

$$f(z) = \frac{1}{1 + z^2} = \sum_{n=0}^{\infty} (-z^2)^n = \sum_{n=0}^{\infty} (-1)^n z^{2n}.$$

This series converges if $|z^2| < 1$, which is the same as $|z| < 1$.

This discussion explains why the real Taylor series of $1/(1 + x^2)$ has radius of convergence 1 even though $1/(1 + x^2)$ is defined and differentiable for all real x. The real function $1/(1 + x^2)$ is simply $1/(1 + z^2)$ restricted to real values, and the Taylor series for this complex function $1/(1 + z^2)$ has radius of convergence 1.

In some instances, we compute a Taylor series by explicitly calculating the Taylor coefficients.

EXAMPLE 21.7

Let $f(z) = e^z$. Then $f^{(n)}(0) = 1$, so the Taylor series of f about zero (the Maclaurin series of f) is

$$\sum_{n=0}^{\infty} \frac{1}{n!} z^n,$$

which has infinite radius of convergence. ∎

In some instances, we can obtain a complex Taylor expansion from a real Taylor expansion simply by replacing x by z. For example, the real Maclaurin expansion (Taylor expansion about zero) of $\sin(x)$ is

$$\sin(x) = \sum_{n=0}^{\infty} \frac{1}{(2n+1)!} (-1)^n x^{2n+1}$$

for all real x. Since $\sin(x)$ must equal $\sin(z)$ for $z = x$ real, the Maclaurin expansion of $\sin(z)$ is

$$\sin(z) = \sum_{n=0}^{\infty} \frac{1}{(2n+1)!} (-1)^n z^{2n+1}.$$

Similarly, the Maclaurin series for $\cos(z)$ is

$$\cos(z) = \sum_{n=0}^{\infty} \frac{1}{(2n)!} (-1)^n z^{2n}.$$

When possible, we find the Taylor coefficients by some means other than differentiation.

EXAMPLE 21.8

Find the Taylor expansion of e^z about i.

We know that $e^z = \sum_{n=0}^{\infty} (1/n!) z^n$ is the Taylor expansion of e^z about zero. Replace z with $z - i$ in this expansion to get

$$e^{z-i} = \sum_{n=0}^{\infty} \frac{1}{n!} (z - i)^n.$$

But $e^{z-i} = e^z e^{-i}$, so

$$e^z = e^i e^{z-i} = \sum_{n=0}^{\infty} \frac{e^i}{n!} (z - i)^n.$$

This is a representation of e^z in a power series about i and thus is the Taylor series of e^z about i. This series converges for all z. ∎

EXAMPLE 21.9

Expand $\cos(z^3)$ in a Maclaurin series.

We know that $\cos(z) = \sum_{n=0}^{\infty} [1/(2n)!](-1)^n z^{2n}$. Replace z by z^3 to get

$$\cos(z^3) = \sum_{n=0}^{\infty} \frac{1}{(2n)!} (-1)^n z^{6n}.$$

This series converges and represents $\cos(z^3)$ for all z. ∎

EXAMPLE 21.10

Expand $1/(1 + z)$ in a Taylor series about $-2i$.

We will use the geometric series and algebraic manipulation. The Taylor series of $1/(1 + z)$ about $-2i$ must have the form $\sum_{n=0}^{\infty} a_n(z + 2i)^n$, so we will manipulate the geometric series to obtain powers of $z + 2i$. Write

$$\frac{1}{1 + z} = \frac{1}{1 + z + 2i - 2i} = \frac{1}{(1 - 2i) + (z + 2i)}$$

$$= \frac{1}{1 - 2i} \frac{1}{1 + \left(\dfrac{z + 2i}{1 - 2i}\right)} = \frac{1}{1 - 2i} \sum_{n=0}^{\infty} (-1)^n \left(\frac{z + 2i}{1 - 2i}\right)^n$$

if $|(z + 2i)/(1 - 2i)| < 1$. Here, we have used the fact that $1/(1 + t) = \sum_{n=0}^{\infty} (-1)^n t^n$ if $|t| < 1$ and have let $t = (z + 2i)/(1 - 2i)$. Therefore,

$$\frac{1}{1 + z} = \sum_{n=0}^{\infty} (-1)^n \frac{1}{(1 - 2i)^{n+1}} (z + 2i)^n$$

for z satisfying $|(z + 2i)/(1 - 2i)| < 1$, or $|z + 2i| < |1 - 2i| = \sqrt{5}$. This Taylor expansion represents $1/(1 + z)$ in the disk of radius $\sqrt{5}$ about $2i$. Notice that $\sqrt{5}$ is exactly the distance from $2i$ to -1 (see Figure 21.6), and -1 is the point nearest $2i$ at which $1/(1 + z)$ fails to be analytic (and, in fact, is the only point where this function is not analytic. ∎

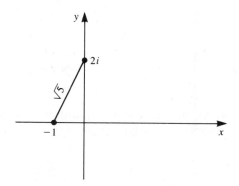

Figure 21.6

EXAMPLE 21.11

Find the Maclaurin series for $\tan(z)$.

We can compute some of the Taylor coefficients by evaluating derivatives of $\tan(z)$ at zero. Another way is to use the known Maclaurin series for $\sin(z)$ and $\cos(z)$, together with the fact that $\tan(z) = \sin(z)/\cos(z)$. Write

$$\tan(z) = \sum_{n=0}^{\infty} a_n z^n = \frac{\sin(z)}{\cos(z)} = \frac{\displaystyle\sum_{n=0}^{\infty} \frac{(-1)^n}{(2n + 1)!} z^{2n+1}}{\displaystyle\sum_{n=0}^{\infty} \frac{(-1)^n}{(2n)!} z^{2n}}.$$

Since it is usually easier to multiply than to divide, write

$$\sum_{n=0}^{\infty} a_n z^n \sum_{n=0}^{\infty} (-1)^n \frac{1}{(2n)!} z^{2n} = \sum_{n=0}^{\infty} (-1)^n \frac{1}{(2n+1)!} z^{2n+1},$$

or

$$(a_0 + a_1 z + a_2 z^2 + a_3 z^3 + a_4 z^4 + \cdots)(1 - \tfrac{1}{2} z^2 + \tfrac{1}{24} z^4 - \tfrac{1}{720} z^6 + \cdots)$$
$$= z - \tfrac{1}{6} z^3 + \tfrac{1}{120} z^5 - \cdots.$$

Collect some of the terms of this product to obtain

$$a_0 + a_1 z + (a_2 - \tfrac{1}{2} a_0) z^2 + (a_3 - \tfrac{1}{2} a_1) z^3 + (a_4 - \tfrac{1}{2} a_2 + \tfrac{1}{24} a_0) z^4$$
$$+ (a_5 - \tfrac{1}{2} a_3 + \tfrac{1}{24} a_1) z^5 + \cdots = z - \tfrac{1}{6} z^3 + \tfrac{1}{120} z^5 - \cdots.$$

Coefficients of like powers of z on both sides of this equation must be equal. Therefore,

$$\begin{aligned}
a_0 &= 0 & &\text{(constant term)}, \\
a_1 &= 1 & &\text{(coefficient of } z), \\
a_2 - \tfrac{1}{2} a_0 &= 0 & &\text{(coefficient of } z^2), \\
a_3 - \tfrac{1}{2} a_1 &= -\tfrac{1}{6} & &\text{(coefficient of } z^3), \\
a_4 - \tfrac{1}{2} a_2 + \tfrac{1}{24} a_0 &= 0 & &\text{(coefficient of } z^4), \\
a_5 - \tfrac{1}{2} a_3 + \tfrac{1}{24} a_1 &= \tfrac{1}{120} & &\text{(coefficient of } z^5),
\end{aligned}$$

and so on. Solve these equations to get $a_0 = a_2 = a_4 = 0, a_1 = 1, a_3 = \tfrac{1}{3}, a_5 = \tfrac{2}{15}$. Then

$$\tan(z) = z + \tfrac{1}{3} z^3 + \tfrac{2}{15} z^5 + \cdots.$$

In this way, we can compute as many terms of this Maclaurin series as we like.

This Maclaurin expansion of $\tan(z)$ has radius of convergence $\pi/2$, which is the distance from zero to the nearest point at which $\tan(z)$ fails to be analytic ($\pi/2$ or $-\pi/2$). ∎

In the next section, we will derive a type of expansion of a function in a series about a point at which the function is not analytic.

PROBLEMS FOR SECTION 21.4

In each of Problems 1 through 20, expand the function in a Taylor series about the point.

1. $\cos(2z)$; 0

2. e^z; $-3i$

3. $\sin(z^2)$; 0

4. $\dfrac{1}{1-z}$; $4i$

5. $\dfrac{1}{2+z}$; $1 - 8i$

6. $1 + \dfrac{1}{2+z^2}$; i

7. $\dfrac{1}{(1-z)^2}$; 0.   *Hint:* Differentiate $\dfrac{1}{1-z}$.

8. $e^z - \sin(z)$; 0

9. $\sinh(3z)$; 0

10. $z^2 - 3z + i$; $2 - i$

11. $\dfrac{3}{z - 4i}$; -5

12. $\cos(z^2) - \sin(z)$; 0

13. $\cos(z)$; i.   *Hint:* Use the definition of $\cos(z)$.

14. $(z - 9)^3$; $1 + i$

15. $\dfrac{1}{z - 2 - 4i}$; $-2i$

16. $\cosh(z + 1)$; -1

17. $e^{3-z}; \quad i$

18. $\sin(z + i); \quad -i$

19. $\operatorname{Log}(z); \quad 2i$

20. $e^z \cos(z); \quad 0$

21. Suppose that f is analytic at zero and satisfies $f''(z) = 2f(z) + 1$, $f(0) = 1$, and $f'(0) = i$. Find the Maclaurin expansion of f.

22. Let $f(z) = (2/\sqrt{\pi}) \int_0^z e^{-w^2} \, dw$. Find the Maclaurin expansion of f.

23. Find the first three nonzero terms of the Maclaurin expansion of $\sin^2(z)$
   (a) by computing the Taylor coefficients at zero;
   (b) by calculating enough terms of the product of the Maclaurin series of $\sin(z)$ with itself.

24. Use the method of Example 21.11 to find the first four nonzero terms of the Maclaurin expansion of $\sec(z)$.

25. Use the method of Example 21.11 to find the first four nonzero terms of the Maclaurin expansion of $\tanh(z)$.

26. Let the Maclaurin series for $1/(1 - z - z^2)$ be $\sum_{n=0}^{\infty} a_n z^n$. Prove that $a_0 = a_1 = 1$ and, for $n \geq 2$, $a_n = a_{n-1} + a_{n-2}$. The numbers $a_0, a_1, \ldots$ are called *Fibonacci numbers* and were discussed briefly in the context of difference equations.

21.5 *Laurent Series*

If f is analytic at z_0, we can expand f in a Taylor series about z_0 containing powers of $z - z_0$. If f is not analytic at z_0, we may still be able to represent f in a series about z_0 if we include powers of $1/(z - z_0)$. This is the idea behind the Laurent series.

THEOREM 21.13

Let f be analytic in the annulus $r_1 < |z - z_0| < r_2$. Then, for z in this annulus,

$$f(z) = \sum_{n=-\infty}^{\infty} a_n(z - z_0)^n,$$

where

$$a_n = \frac{1}{2\pi i} \oint_C \frac{f(w)}{(w - z_0)^{n+1}} \, dw$$

for $n = 0, \pm 1, \pm 2, \ldots$ and C is any circle $|z - z_0| = \rho$, with $r_1 < \rho < r_2$ (Figure 21.7).

The series $\sum_{n=-\infty}^{\infty} a_n(z - z_0)^n$ is called *Laurent expansion*, or *Laurent series*, of f about z_0 in the annulus $r_1 < |z - z_0| < r_2$. The numbers

$$a_n = \frac{1}{2\pi i} \oint_C \frac{f(w)}{(w - z_0)^{n+1}} \, dw$$

are called the *Laurent coefficients* of f at z_0.

Proof Let z be in the annulus. Choose numbers R_1 and R_2 so that $r_1 < R_1 < |z - z_0| < R_2 < r_2$, as in Figure 21.8. Let C_2 be the circle $|z - z_0| = R_2$, and let C_1 be the circle $|z - z_0| = R_1$. By the generalized form of Cauchy's integral theorem, we can write

$$f(z) = \frac{1}{2\pi i} \oint_{C_2} \frac{f(w)}{w - z} \, dw - \frac{1}{2\pi i} \oint_{C_1} \frac{f(w)}{w - z} \, dw,$$

with both integrations in a counterclockwise sense (this accounts for the minus sign

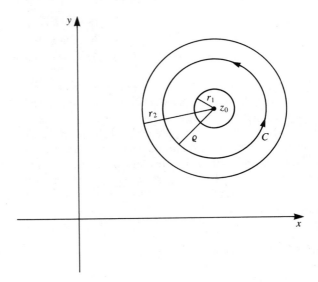

Figure 21.7

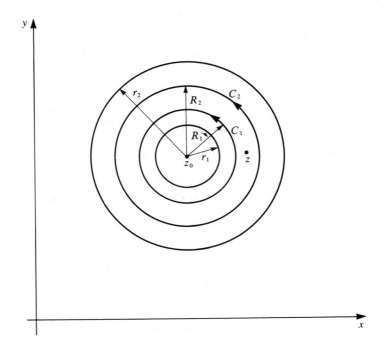

Figure 21.8

before the integral over C_1). The integration variable is w, while z is any given point in the annulus. Consider these line integrals individually.

For $\oint_{C_2} f(w)/(w - z) \, dw$, write (as in the proof of Theorem 21.12)

$$\frac{1}{w - z} = \sum_{n=0}^{\infty} \frac{(z - z_0)^n}{(w - z_0)^{n+1}} \, dw.$$

Then

$$\frac{1}{2\pi i} \oint_{C_2} \frac{f(w)}{w - z} \, dw = \sum_{n=0}^{\infty} \left\{ \frac{1}{2\pi i} \oint_{C_2} \left[\frac{f(w)}{(w - z_0)^{n+1}} \right] dw \right\} (z - z_0)^n$$

$$= \sum_{n=0}^{\infty} a_n (z - z_0)^n,$$

where

$$a_n = \frac{1}{2\pi i} \oint_{C_2} \frac{f(w)}{(w - z_0)^{n+1}} \, dw$$

for $n = 0, 1, 2, 3, \ldots$. For $\oint_{C_1} f(w)/(w - z) \, dw$, write

$$\frac{1}{w - z} = \frac{1}{w - z_0 - (z - z_0)} = -\frac{1}{z - z_0} \frac{1}{1 - \dfrac{w - z_0}{z - z_0}}$$

and note that, for w on C_1, $|(w - z_0)/(z - z_0)| < 1$. By the geometric series,

$$-\frac{1}{z - z_0} \frac{1}{1 - \dfrac{w - z_0}{z - z_0}} = -\frac{1}{z - z_0} \sum_{n=0}^{\infty} \left(\frac{w - z_0}{z - z_0} \right)^n$$

$$= -\sum_{n=1}^{\infty} \frac{(w - z_0)^{n-1}}{(z - z_0)^n}.$$

Therefore,

$$\frac{1}{2\pi i} \oint_{C_1} \frac{f(w)}{w - z} \, dw = \sum_{n=1}^{\infty} \left[\frac{1}{2\pi i} \oint_{C_1} f(w)(w - z_0)^{n-1} \, dw \right] \frac{1}{(z - z_0)^n}$$

$$= \sum_{n=1}^{\infty} a_{-n} (z - z_0)^{-n},$$

where

$$a_{-n} = \frac{1}{2\pi i} \oint_{C_1} f(w)(w - z_0)^{n-1} \, dw$$

for $n = 1, 2, 3, \ldots$.

Now use the deformation theorem (Theorem 20.10) to replace C_1 and C_2 with the circle C_ρ: $|z - z_0| = \rho$ in these line integrals. This enables us to consolidate the formulas for $a_0, a_1, a_2, \ldots$ and $a_{-1}, a_{-2}, a_{-3}, \ldots$ into the single formula

$$a_n = \frac{1}{2\pi i} \oint_{C_\rho} \frac{f(w)}{(w - z_0)^{n+1}} \, dw$$

for $n = 0, \pm 1, \pm 2, \ldots$.

With this choice of the coefficients, we have

$$f(z) = \frac{1}{2\pi i} \oint_{C_2} \frac{f(w)}{w-z} \, dw - \frac{1}{2\pi i} \int_{C_1} \frac{f(w)}{w-z} \, dw$$

$$= \sum_{n=0}^{\infty} a_n (z-z_0)^n - \sum_{n=1}^{\infty} a_{-n} \frac{-1}{(z-z_0)^n}$$

$$= \sum_{n=-\infty}^{\infty} a_n (z-z_0)^n,$$

as we wanted to show. ■

What does the Laurent expansion of f about z_0 tell us? For z in the annulus, it enables us to write $f(z)$ as

$$f(z) = g(z) + h(z),$$

where

$$g(z) = \sum_{n=0}^{\infty} a_n (z-z_0)^n$$

is analytic at z_0 (because it equals a Taylor series about z_0), and

$$h(z) = \sum_{n=1}^{\infty} a_{-n} \frac{1}{(z-z_0)^n}$$

is not analytic at z_0. Any "difficulty" f has at z_0, causing it to be nonanalytic there, is contained in h, which is expanded in a series of powers of $1/(z-z_0)$. In an intuitive sense, the higher the powers of $1/(z-z_0)$ appearing in this series for h, the "worse the difficulty" f has at z_0. We will make this idea precise when we classify singularities in the next chapter.

If f is analytic at z_0, $h(z) = 0$ and the Laurent expansion is simply the Taylor expansion of f about z_0.

In practice, often $r_1 = 0$ in the annulus used in the Laurent expansion. That is, we often make the expansion in the annulus $0 < |z - z_0| < R$, which is the annulus obtained by removing the single point z_0 from the open disk of radius R about z_0.

We almost never compute a Laurent expansion by computing the integrals in the formula for the coefficients. On the contrary, we usually find the Laurent expansion in some other way and then use certain of the coefficients to evaluate integrals appearing in this formula. This is the object of the residue theorem of the next chapter.

We will conclude this section with some examples of Laurent expansions.

EXAMPLE 21.12

$e^{1/z}$ is analytic in the annulus $0 < |z| < \infty$, which consists of the entire complex plane with the origin removed. Since $e^z = \sum_{n=0}^{\infty} (1/n!)z^n$ for all z,

$$e^{1/z} = \sum_{n=0}^{\infty} \frac{1}{n!} \frac{1}{z^n}$$

for all $z \neq 0$. ■

EXAMPLE 21.13

Since $\cos(z) = \sum_{n=0}^{\infty} (-1)^n [1/(2n)!] z^{2n}$ for all z,

$$\frac{1}{z^5} \cos(z) = \sum_{n=0}^{\infty} (-1)^n \frac{1}{(2n)!} z^{2n-5} = \frac{1}{z^5} - \frac{1}{2} \frac{1}{z^3} + \frac{1}{24} \frac{1}{z} - \frac{1}{720} z + \cdots.$$

This is the Laurent expansion of $(1/z^5)\cos(z)$ in the annulus $0 < |z| < \infty$. ∎

EXAMPLE 21.14

Expand $1/(1 + z^2)$ in a Laurent expansion about $-i$.

First, $1/(1 + z^2)$ is analytic in the annulus $0 < |z + i| < 2$ [the distance from $-i$ to the other point, i, at which $1/(1 + z^2)$ is not analytic]. The Laurent expansion must contain powers of $z + i$. Use partial fractions to write

$$\frac{1}{1 + z^2} = \frac{1}{(z + i)(z - i)} = \frac{i}{2} \frac{1}{z + i} - \frac{i}{2} \frac{1}{z - i}.$$

The term $(i/2)[1/(z + i)]$ already contains a power of $z + i$ and so is part of the Laurent expansion we want. Now look at the other term, $-(i/2)[1/(z - i)]$. Write

$$\frac{1}{z - i} = \frac{1}{z - i + i - i} = \frac{1}{-2i + (z + i)}$$

$$= \frac{1}{-2i\left(1 - \frac{z + i}{2i}\right)} = -\frac{1}{2i} \sum_{n=0}^{\infty} \left(\frac{z + i}{2i}\right)^n = \sum_{n=0}^{\infty} \frac{-1}{(2i)^{n+1}} (z + i)^n,$$

provided that $|(z + i)/2i| < 1$, or $|z + i| < 2$. Therefore, the Laurent expansion of $1/(1 + z^2)$ in $0 < |z + i| < 2$ is

$$\frac{1}{1 + z^2} = \frac{i}{2} \frac{1}{z + i} - \frac{i}{2} \frac{1}{z - i} = \frac{i}{2} \frac{1}{z + i} - \frac{i}{2} \sum_{n=0}^{\infty} \frac{-1}{(2i)^{n+1}} (z + i)^n$$

$$= \frac{i}{2} \frac{1}{z + i} + \frac{i}{2} \sum_{n=0}^{\infty} \frac{1}{(2i)^{n+1}} (z + i)^n.$$

This series is of the form $g(z) + h(z)$, where

$$g(z) = \frac{i}{2} \sum_{n=0}^{\infty} \frac{1}{(2i)^{n+1}} (z + i)^n$$

is analytic in $|z + i| < 2$ (because it equals a Taylor series) and

$$h(z) = \frac{i}{2} \frac{1}{z + i}$$

is analytic in $|z + i| > 0$ but not at $-i$. The points common to $|z + i| < 2$ and $|z + i| > 0$ form the annulus $0 < |z + i| < 2$. ∎

EXAMPLE 21.15

Expand $1/[(z + 1)(z - 3i)]$ in a Laurent series about -1.

First, write

$$\frac{1}{(z + 1)(z - 3i)} = \frac{-1 + 3i}{10}\frac{1}{z + 1} + \frac{1 - 3i}{10}\frac{1}{z - 3i}.$$

First, $1/(z - 3i)$ is analytic at -1 and so has a Taylor expansion about -1. To find this expansion, write

$$\frac{1}{z - 3i} = \frac{1}{z - 3i + 1 - 1} = \frac{1}{-1 - 3i + (z + 1)}$$

$$= \frac{1}{-1 - 3i}\frac{1}{1 - \left(\dfrac{z + 1}{1 + 3i}\right)} = \frac{-1}{1 + 3i}\sum_{n=0}^{\infty}\left(\frac{1}{1 + 3i}\right)^n (z + 1)^n$$

$$= -\sum_{n=0}^{\infty}\frac{1}{(1 + 3i)^{n+1}}(z + 1)^n$$

if $|(z + i)/(1 + 3i)| < 1$, or $|z + 1| < \sqrt{10}$. Since $1/(z + 1)$ is already a power of $z + 1$, the Laurent expansion of $1/[(z + 1)(z - 3i)]$ about -1 is

$$\frac{-1 + 3i}{10}\frac{1}{z + 1} - \frac{1 - 3i}{10}\sum_{n=0}^{\infty}\frac{1}{(1 + 3i)^{n+1}}(z + 1)^n.$$

This expansion is a sum of two terms, one valid in $|z + 1| < \sqrt{10}$ and the other valid in $|z + 1| > 0$. Their sum represents $f(z)$ in the common set $0 < |z + 1| < \sqrt{10}$. ∎

PROBLEMS FOR SECTION 21.5

In each of Problems 1 through 16, determine the Laurent expansion of the function about the given point z_0 and determine R so that this expansion is valid in the annulus $0 < |z - z_0| < R$.

1. $\dfrac{2z}{1 + z^2}$; $z_0 = i$

2. $\dfrac{1}{z^2}\sin(z)$; $z_0 = 0$

3. $\dfrac{1 - \cos(2z)}{z^2}$; $z_0 = 0$

4. $\cos\left(\dfrac{1}{z - i}\right)$; $z_0 = i$

5. $\dfrac{z^2}{1 - z}$; $z_0 = 1$

6. $\dfrac{1}{z^2}e^{1/z}$; $z_0 = 0$

7. $\dfrac{z + i}{z - i}$; $z_0 = i$

8. $\dfrac{z^2 + 1}{2z - 1}$; $z_0 = \dfrac{1}{2}$

9. $\dfrac{1}{z}\sin(4z)$; $z_0 = 0$

10. $z^2\cos\left(\dfrac{i}{z}\right)$; $z_0 = 0$

11. $\dfrac{1}{z^2 + 1}$; $z_0 = i$

12. $e^{1/(z + i)}$; $z_0 = -i$

13. $\dfrac{z + z_0}{z - z_0}$; z_0 arbitrary

14. $\sinh\left(\dfrac{1}{z^3}\right)$; $z_0 = 0$

15. $\dfrac{2i}{z - 1 + i}$; $z_0 = 1 - i$

16. $\dfrac{z}{z + 3 - 2i}$; $z_0 = -3 + 2i$

17. The Bessel function $J_n(z)$, for any integer n, can be defined by the equation

$$e^{z(w - 1/w)/2} = \sum_{n=-\infty}^{\infty} J_n(z)w^n.$$

(a) Use the integral formula for the Laurent coefficients to show that

$$J_n(z) = \frac{1}{\pi}\int_0^\pi \cos[n\theta - z\sin(\theta)]\, d\theta.$$

(b) Write $e^{z(w-1/w)/2} = e^{zw/2}e^{-z/2w}$ and multiply the expansions of $e^{zw/2}$ and $e^{-z/2w}$ about zero to obtain the expression

$$J_n(z) = \sum_{j=0}^{\infty} (-1)^j \frac{1}{j!(n+j)!} \left(\frac{z}{2}\right)^{n+2j}$$

ADDITIONAL PROBLEMS

In each of Problems 1 through 20, find the Taylor expansion of the function in some disk $|z - z_0| < R$ or the Laurent expansion in some annulus $0 < |z - z_0| < R$, whichever is appropriate, specifying a value of R for each problem.

1. $\dfrac{1}{z+4}$; $z_0 = 2 + i$

2. e^{2z}; $z_0 = -3i$

3. $\cos(z - 5i)$; $z_0 = 5i$

4. $\dfrac{1}{2z - 3 + i}$; $z_0 = -3$

5. $-iz^2 + (1 - i)z - 2$; $z_0 = 3 - i$

6. $\dfrac{1}{z^4}\cos(2z)$; $z_0 = 0$

7. $\dfrac{1}{z}e^{z^2}$; $z_0 = 0$

8. $\cosh(z - i)$; $z_0 = 0$

9. $\dfrac{1}{1 - iz}$; $z_0 = 4 + 5i$

10. $\dfrac{1}{i - z^2}$; $z_0 = 0$

11. $\dfrac{1}{z^2}\sin(iz^3)$; $z_0 = 0$

12. $\dfrac{i - \cos(iz)}{z^4}$; $z_0 = 0$

13. $\dfrac{1}{2z - 2 + i}$; $z_0 = 1 + i$

14. $\dfrac{2 + z^2}{3 + z}$; $z_0 = 0$

15. $\dfrac{\cosh(iz^2)}{3z}$; $z_0 = 0$

16. $\dfrac{1}{4 - 2z^2}$; $z_0 = i$

17. $\cos(3z^2) - ie^z$; $z_0 = 0$

18. $\dfrac{1}{1 - z}$; $z_0 = -2 + 3i$

19. e^{2z+1}; $z_0 = 0$

20. $\cosh(z) - i\sinh(z)$; $z_0 = 0$

In each of Problems 21 through 26, find the first four nonzero terms of the Taylor expansion of the function about the point.

21. $e^z\sin(iz)$; $z_0 = 0$

22. $\text{sech}(z)$; $z_0 = 0$

23. $\cos(z^2)$; $z_0 = i$

24. $z^2\sin(z)$; $z_0 = -3$

25. $\cosh\left(\dfrac{1}{z}\right)$; $z_0 = i$

26. e^{z^2}; $z_0 = 2i$

27. Derive a complex form of a Fourier series from the Taylor expansion as follows. Suppose that $f(z) = \sum_{n=0}^{\infty} a_n z^n$ in some disk $|z| < R$. Write z in polar form as $z = re^{i\theta}$, with $0 \le r < R$. Show that $f(z) = \sum_{n=0}^{\infty} a_n r^n e^{in\theta}$, where $a_n = (1/2\pi r^n)\int_0^{2\pi} f(re^{i\theta})e^{-in\theta}\, d\theta$.

28. Show that, in Problem 27,

$$\sum_{n=0}^{\infty} |a_n|^2 r^{2n} = \frac{1}{2\pi}\int_0^{2\pi} |f(re^{i\theta})|^2\, d\theta.$$

This relationship is called *Parseval's identity.*

29. Show that

$$\sum_{n=0}^{\infty} \frac{1}{(n!)^2} z^{2n} = \frac{1}{2\pi}\int_0^{2\pi} e^{2z\cos(\theta)}\, d\theta.$$

Hint: First show that

$$\left(\frac{z^n}{n!}\right)^2 = \frac{1}{2\pi i}\oint_C \frac{z^n}{n! w^{n+1}} e^{zw}\, dw$$

for $n = 0, 1, 2, 3, \ldots$, in which C is the unit circle about the origin, oriented counterclockwise.

CHAPTER 22

Singularities and the Residue Theorem

22.1 Classification of Singularities

If f is analytic in an annulus $0 < |z - z_0| < R$, but not at z_0, we say that f has an *isolated singularity* at z_0. Usually, we will shorten this and just say that f has a *singularity* at z_0. Intuitively, a singularity is a point where "something goes wrong with the function," in the sense that "right behavior" means analytic at z_0.

We will now distinguish several types of singularities at z_0 based on properties of the Laurent expansion of f about z_0. Suppose that f has a singularity at z_0. Expand f in a Laurent series

$$f(z) = \sum_{n=-\infty}^{\infty} a_n (z - z_0)^n, \qquad 0 < |z - z_0| < R. \tag{22.1}$$

We say that z_0 is

A *removable singularity* if no negative powers of $z - z_0$ appear in (22.1).

An *essential singularity* if infinitely many negative powers of $z - z_0$ appear.

A *pole of order m* if m is a positive integer and $(z - z_0)^{-m}$ appears in this series (so $a_{-m} \neq 0$) but no higher negative powers appear (so $a_{-m-1} = a_{-m-2} = \cdots = 0$).

Finally, z_0 is a *pole of f* if z_0 is a pole of order m for some m. A pole of order 1 is often called a *simple pole*, and a pole of order 2 is a *double pole*.

EXAMPLE 22.1

$\sin(z)/z$ has a removable singularity at zero. The Laurent expansion of $\sin(z)/z$ about zero is obtained by dividing the Maclaurin series for $\sin(z)$ by z to get

$$\frac{\sin(z)}{z} = \frac{1}{z} \sum_{n=0}^{\infty} (-1)^n \frac{1}{(2n+1)!} z^{2n+1}$$

$$= \sum_{n=0}^{\infty} (-1)^n \frac{1}{(2n+1)!} z^{2n} = 1 - \frac{1}{3!} z^2 + \frac{1}{5!} z^4 - \cdots,$$

with no negative powers of z in the expansion. Indeed, the series on the right in this equation is a Maclaurin series and so represents an analytic function, specifically, the analytic function g defined by

$$g(z) = \begin{cases} \sin(z)/z & \text{if } z \neq 0 \\ 1 & \text{if } z = 0. \end{cases}$$

The singularity of f at zero is removable in the sense that, by defining a new function which agrees with f if $z \neq 0$ and has the correct value at zero, we obtain a function g which is analytic at zero. ∎

EXAMPLE 22.2

Using the Taylor expansion of e^z about zero, we obtain

$$e^{1/(z-1)} = \sum_{n=0}^{\infty} \frac{1}{n!} \frac{1}{(z-1)^n}.$$

This is the Laurent expansion of $e^{1/(z-1)}$ about 1. Since infinitely many negative powers of $z - 1$ appear in this expansion, 1 is an essential singularity of $e^{1/(z-1)}$. ∎

EXAMPLE 22.3

$1/(z+i)^3$ has a pole of order 3 at $-i$. In fact, this function is its own Laurent expansion about $-i$ in the annulus $0 < |z + i| < \infty$. ∎

EXAMPLE 22.4

$\sin(z)/z^2$ has a pole of order 1 (a simple pole) at zero. To see this, divide the Maclaurin series for $\sin(z)$ by z^2 to get the Laurent expansion of $\sin(z)/z^2$ about zero,

$$\frac{\sin(z)}{z^2} = \sum_{n=0}^{\infty} (-1)^n \frac{1}{(2n+1)!} z^{2n-1} = \frac{1}{z} - \frac{1}{3!} z + \frac{1}{5!} z^3 - \frac{1}{7!} z^5 + \cdots$$

for $0 < |z| < \infty$. There is a $1/z$ term but no $1/z^2, 1/z^3, \ldots$; hence, $\sin(z)/z^2$ has a pole of order 1 at zero. ∎

The following theorem is sometimes useful in identifying the order of a pole.

THEOREM 22.1

Let $f(z) = g(z)h(z)$ and let g be analytic or have a removable singularity at z_0. Let $\lim_{z \to z_0} g(z) \neq 0$, while h has a pole of order m at z_0, then f has a pole of order m at z_0. ∎

For example, $\cos(z)/z^2$ has a double pole at zero because $\cos(z)$ is analytic at zero and $\cos(0) \neq 0$, while $1/z^2$ has a double pole there. (In fact, $1/z^2$ is its own Laurent expansion about zero in the annulus $0 < |z| < \infty$.)

It is often convenient to think of poles in terms of zeros of a function. A function f has a *zero* at z_0 if f is analytic at z_0 and $f(z_0) = 0$. We say that a zero z_0 is of order m if $f(z_0) = f'(z_0) = \cdots = f^{(m-1)}(z_0) = 0$ but $f^{(m)}(z_0) \neq 0$. For example, $f(z) = z^m$ has a zero of order m at zero, while $\sin^2(z)$ has a zero of order 2 at π. A zero of order 1 is called a *simple zero*.

We can tell the order of a zero of f at z_0 from the Taylor expansion of f about z_0. Write

$$f(z) = \sum_{n=0}^{\infty} a_n(z - z_0)^n.$$

We know that $a_n = (1/n!)f^{(n)}(z_0)$. Therefore, f has a zero of order m at z_0 exactly when the first nonzero coefficient in the Taylor expansion of f about z_0 is a_m. In this event, the Taylor series is

$$f(z) = \sum_{n=m}^{\infty} a_n(z - z_0)^n = a_m(z - z_0)^m + a_{m+1}(z - z_0)^{m+1} + \cdots,$$

with $a_m \neq 0$. These ideas relate to poles in the following way.

THOEREM 22.2

Let h have a zero of order m at z_0. Let g be analytic at z_0 or have a removable singularity at z_0, and let $\lim_{z \to z_0} g(z) \neq 0$. Then $f(z) = g(z)/h(z)$ has a pole of order m at z_0. ∎

A proof is left to the student. Intuitively, Theorem 22.2 states that a quotient $g(z)/h(z)$ has poles exactly where the denominator vanishes (assuming that the numerator does not also vanish there) and that the order of the pole equals the order of the zero of the denominator. For example, e^z/z^3 has a pole of order 3 at zero, and $\cos(z)/(z - i)^5$ has a pole of order 5 at i. Similarly, $\cot(z)$ has simple poles at $n\pi$, for any integer n, because $\cot(z) = \cos(z)/\sin(z)$ and $\cos(z)$ is analytic and nonzero at $n\pi$, while $\sin(z)$ is analytic at $n\pi$ with a simple zero there.

In the next section, we will exploit singularities to evaluate integrals.

PROBLEMS FOR SECTION 22.1

In each of Problems 1 through 20, determine all of the singularities of the function and classify each singularity.

1. $\dfrac{\cos(z)}{z^2}$

2. $\dfrac{1}{(z + i)^2(z - i)}$

3. $e^{1/z}(z - i)$

4. $\dfrac{\sin(z)}{z - \pi}$

5. $\dfrac{\cos(2z)}{(z - 1)^2(z^2 + 1)}$

6. $\dfrac{z}{(z + 1)^2}$

7. $\dfrac{z - i}{z^2 + 1}$

8. $\dfrac{\sin(z)}{\sinh(z)}$

9. $\dfrac{z}{z^4 - 1}$

10. $\tan(z)$

11. $\dfrac{1}{\cos(z)}$

12. $e^{1/z(z + 1)}$

13. $\dfrac{1}{z^2}e^{iz}$

14. $\dfrac{\sin(z)}{z(z - \pi)(z - i)^2}$

15. $\dfrac{e^{2z}}{(z + 1)^4}$

16. $\dfrac{1}{z^4}\sinh(z)$

17. $\cosh(z)$

18. $\dfrac{2i - 1}{(z^2 + 2z - 3)^2}$

19. $\dfrac{\sin(z)}{(z + \pi)^2}$

20. $\dfrac{1}{\sin^3(2z)}$

21. Let $f(z) = 1/\sin(1/z)$ for $z \neq 0$. Show that f is not analytic in any annulus $0 < |z| < r$. (This is an example of a singularity which is not isolated.)

22. Suppose that f has a pole at z_0. Prove that $|f(z)| \to \infty$ as $z \to z_0$ along any path.

23. Let f have a pole of order m at z_0. Let p be a polynomial of degree n. Prove that $p(f(z))$ has a pole of order nm at z_0.

24. Let f be analytic in the annulus $0 < |z - z_0| < R$. Prove that f has a removable singularity at z_0 if and only if $\lim_{z \to z_0}(z - z_0)f(z) = 0$.

25. Prove that, in every open disk about zero, $e^{1/z}$ takes on every nonzero value infinitely often. (The reason for this bizarre behavior is that $e^{1/z}$ has an essential singularity at zero.)

26. Prove that a rational function cannot have an essential singularity. (Recall that a rational function is a quotient of polynomials.)

22.2 *Residues and the Residue Theorem*_____

We will now develop a method of exploiting certain singularities of f to evaluate integrals $\oint_C f(z)\, dz$. Suppose that f has an isolated singularity at z_0. Then f is analytic in some annulus $0 < |z - z_0| < R$ and has a Laurent expansion

$$f(z) = \sum_{n=-\infty}^{\infty} a_n(z - z_0)^n$$

$$= \cdots + \frac{a_{-3}}{(z - z_0)^3} + \frac{a_{-2}}{(z - z_0)^2} + \frac{a_{-1}}{(z - z_0)^1} + a_0 + a_1(z - z_0) + \cdots$$

in this annulus. From the formula for the Laurent coefficients, we have

$$a_{-1} = \frac{1}{2\pi i} \oint_C f(z)\, dz,$$

in which C is any simple closed curve about z_0 in this annulus (Figure 22.1). Then

$$\oint_C f(z)\, dz = 2\pi i a_{-1}.$$

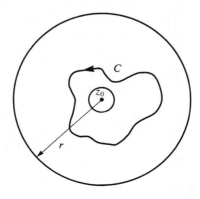

Figure 22.1

This means that we can evaluate $\oint_C f(z)\,dz$, when C encloses just one singularity z_0 of f, if we know just the coefficient of $1/(z - z_0)$ in the Laurent expansion of f about z_0. Since this number can often be found without performing any integration, this observation constitutes a potentially powerful method for evaluating an integral.

We will now see how to extend this idea to the case in which C encloses any finite number of singularities of f. First, define the *residue* of f at a singularity z_0 to be the coefficient of $1/(z - z_0)$ in the Laurent expansion of f about z_0. This residue is denoted $\operatorname*{Res}_{z_0} f$. From Example 22.2, $e^{1/(z-1)}$ has residue 1 at 1. Therefore,

$$\operatorname*{Res}_{1} e^{1/(z-1)} = 1.$$

From Example 22.3, $\operatorname*{Res}_{-i} 1/(z + i)^3 = 0$. For the residue of $i\cos(z)/3z$ at zero, look at the Laurent expansion about zero:

$$\frac{i\cos(3z)}{3z} = \frac{i}{3}\frac{1}{z}\sum_{n=0}^{\infty}(-1)^n\frac{1}{(2n)!}z^{2n} = \frac{i}{3}\frac{1}{z} - \cdots.$$

We need only the coefficient of the $1/z$ term, which is $i/3$. Therefore,

$$\operatorname*{Res}_{0}\frac{i\cos(3z)}{3z} = \frac{i}{3}.$$

THEOREM 22.3   The Residue Theorem

Let f be analytic in a domain D except at points $z_1, \ldots, z_n$, where f has singularities. Let C be a piecewise-smooth simple closed curve in D enclosing $z_1, \ldots, z_n$. Then

$$\oint_C f(z)\,dz = 2\pi i\sum_{j=1}^{n}\operatorname*{Res}_{z_j} f.$$

That is, the integral of f over a curve enclosing singularities $z_1, \ldots, z_n$ of f equals $2\pi i$ times the sum of the residues of f at these singularities. ∎

We will outline a proof based on an intuitive argument we have used before. Enclose each z_j in a circle C_j of radius sufficiently small that no two C_j's intersect and each C_j does not intersect C. Cut channels from C to C_1, from C_1 to $C_2, \ldots$, and from C_{n-1} to C_n, as shown in Figure 22.2. This forms a simple closed curve K. Note that f is analytic on K and the set enclosed by K (each z_j is outside K). Further, if we integrate f around K counterclockwise, we must integrate clockwise over the parts of the C_j's contained in K and over each channel cut in the directions indicated, as well as over C (minus the channel cut) counterclockwise (Figure 22.3). Since f is analytic on K and the set it encloses, by Cauchy's integral theorem,

$$\oint_K f(z)\,dz = 0.$$

In the limit as the channel cuts between the C_j's are merged together, we get the closed curve of Figure 22.4. The integral over each channel curve is zero because we integrate over it once in each direction. We therefore obtain

$$\oint_C f(z)\,dz + \sum_{j=1}^{n}\oint_{-C_j} f(z)\,dz = 0,$$

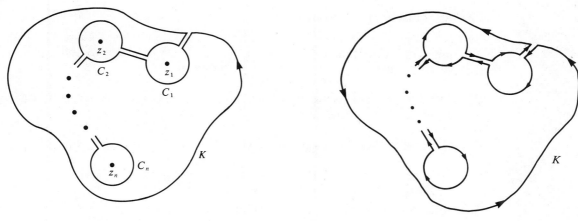

Figure 22.2

Figure 22.3

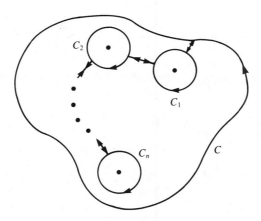

Figure 22.4

in which $-C_j$ denotes the curve C_j oriented clockwise. If we integrate counterclockwise on each C_j, then $\oint_{C_j} f(z)\, dz = -\oint_{C_j} f(z)\, dz$ and we obtain

$$\oint_C f(z)\, dz = \sum_{j=1}^{n} \oint_{C_j} f(z)\, dz.$$

But, for each j, $\oint_{C_j} f(z)\, dz = 2\pi i\ \text{Res}_{z_j} f$ because each C_j encloses only the singularity z_j of f. This yields the formula of the residue theorem.

In using the residue theorem to evaluate an integral, it is important to calculate residues only at those singularities enclosed by C. There may be other singularities of f elsewhere in the plane (though not on C itself), but these are not included in the calculation.

EXAMPLE 22.5

Evaluate $\oint_C \sin(z)/z^2 \, dz$ over any piecewise-smooth simple closed curve C which does not pass through the origin. $\sin(z)/z^2$ has a simple pole at zero and no other singularities.

Case 1   If C does not enclose the origin, $\sin(z)/z^2$ is analytic on and inside C; hence, $\oint_C \sin(z)/z^2 \, dz = 0$ by Cauchy's theorem.

Case 2   If C encloses the origin, $\oint_C \sin(z)/z^2 \, dz = 2\pi i \, \text{Res}_0 \sin(z)/z^2$. Now,

$$\frac{\sin(z)}{z^2} = \frac{1}{z^2} \sum_{n=0}^{\infty} (-1)^n \frac{1}{(2n+1)!} z^{2n+1} = \frac{1}{z} - \cdots.$$

Therefore, $\text{Res}_0 \sin(z)/z^2 = 1$, and $\oint_C \sin(z)/z^2 \, dz = 2\pi i$.   ∎

EXAMPLE 22.6

Evaluate $\oint_C 1/(1 + z^2) \, dz$ if C is any piecewise-smooth simple closed curve in the plane which does not pass through i or $-i$. Consider three cases.

Case 1 C does not enclose either i or $-i$ (Figure 22.5). Then $\oint_C 1/(1 + z^2) \, dz = 0$.

Case 2 C encloses one of the two singularities of $1/(1 + z^2)$ but not the other. Suppose that C encloses i but not $-i$, as in Figure 22.6. Write

$$\frac{1}{1 + z^2} = \frac{i}{2} \frac{1}{z + i} - \frac{i}{2} \frac{1}{z - i}. \tag{22.2}$$

Then

$$\oint_C \frac{1}{1 + z^2} \, dz = \frac{i}{2} \oint_C \frac{1}{z + i} \, dz - \frac{i}{2} \oint_C \frac{1}{z - i} \, dz.$$

Since C does not enclose $-i$, $1/(z + i)$ is analytic on and within C, so $\oint_C 1/(z + i) \, dz = 0$.

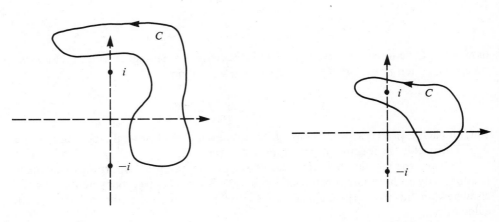

Figure 22.5   Figure 22.6

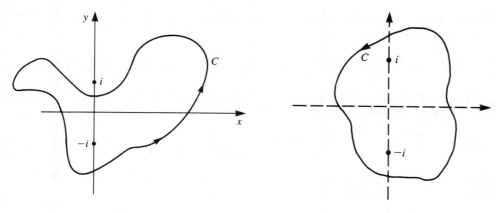

Figure 22.7

Figure 22.8

Further, $1/(z - i)$ is its own Laurent expansion about i, so $\operatorname{Res}_i 1/(z - i) = 1$. Then

$$\oint_C \frac{1}{z - i}\, dz = (2\pi i)(1) = 2\pi i.$$

Therefore,

$$\oint_C \frac{1}{1 + z^2}\, dz = -\frac{i}{2}(2\pi i) = \pi.$$

If C encloses $-i$ but not i, as in Figure 22.7, equation (22.2) yields $\operatorname{Res}_{-i} 1/(1 + z^2) = i/2$, and we obtain

$$\oint_C \frac{1}{1 + z^2}\, dz = 2\pi i\, \frac{i}{2} = -\pi.$$

Case 3 C encloses both i and $-i$ (Figure 22.8). In this event, $\oint_C 1/(1 + z^2)\, dz = 2\pi i[\operatorname{Res}_i f + \operatorname{Res}_{-i} f] = 0$ because $\operatorname{Res}_i f = -i/2 = -\operatorname{Res}_{-i} f$. ∎

The effectiveness of the residue theorem in evaluating an integral is directly related to how easy it is to compute the residue of f at the singularities enclosed by C. In the case in which z_0 is an essential singularity, we must often compute $\operatorname{Res}_{z_0} f$ by looking at the $1/(z - z_0)$ term in the Laurent expansion of f about z_0. If, however, z_0 is a pole of order m, there is a relatively simple formula for $\operatorname{Res}_{z_0} f$.

THEOREM 22.4

Let f have a pole of order m at z_0. Then

$$\operatorname*{Res}_{z_0} f = \frac{1}{(m - 1)!}\, \lim_{z \to z_0} \frac{d^{m-1}}{dz^{m-1}}[(z - z_0)^m f(z)].$$

In the case $m = 1$, d^{m-1}/dz^{m-1} is the "zero order" derivative, which is interpreted to be just 1. Further, $0! = 1$. Thus, if f has a simple pole at z_0, this formula reads

$$\operatorname*{Res}_{z_0} f = \lim_{z \to z_0} [(z - z_0)f(z)].$$

Proof Since f has a pole of order m at z_0, f has a Laurent expansion in some annulus $0 < |z - z_0| < R$:

$$f(z) = \frac{a_{-m}}{(z - z_0)^m} + \frac{a_{-m+1}}{(z - z_0)^{m-1}} + \cdots + \frac{a_{-1}}{z - z_0} + a_0 + a_1(z - z_0) + \cdots,$$

with $a_{-m} \neq 0$ because f has a pole of order m at z_0.

It is a_{-1}, the residue of f at z_0, that we want to evaluate. Write

$$(z - z_0)^m f(z) = a_{-m} + a_{-m+1}(z - z_0) + \cdots + a_{-1}(z - z_0)^{m-1}$$
$$+ a_0(z - z_0)^m + a_1(z - z_0)^{m+1} + \cdots.$$

Since the series on the right is a Taylor series about z_0, $(z - z_0)^m f(z)$ is analytic at z_0. Differentiate this equation $m - 1$ times to get

$$\frac{d^{m-1}}{dz^{m-1}}[(z - z_0)^m f(z)] = a_{-1}(m - 1)(m - 2) \cdots (1)$$

$$+ a_0 m(m - 1) \cdots (2)(z - z_0) + \cdots$$
$$= a_{-1}(m - 1)! + a_0 m!(z - z_0) + \cdots.$$

In the limit as z approaches z_0, all terms on the right vanish except possibly the first term, which is constant. Solving for a_{-1} yields

$$a_{-1} = \lim_{z \to z_0} \frac{1}{(m - 1)!} \frac{d^{m-1}}{dz^{m-1}}[(z - z_0)^m f(z)],$$

as we wanted to show. ∎

EXAMPLE 22.7

Evaluate $\oint_C \sin(2z)/(z - i)^3 \, dz$ if C is any piecewise-smooth simple closed curve enclosing i.

Since $\sin(2z)$ is analytic for all z and does not vanish at i, the quotient $\sin(2z)/(z - i)^3$ has a pole of order 3 at i. Let $m = 3$ in the formula of Theorem 22.4 to get

$$\operatorname*{Res}_{i} \frac{\sin(2z)}{(z - i)^3} = \lim_{z \to i} \frac{1}{2} \frac{d^2}{dz^2}\left[(z - i)^3 \frac{\sin(2z)}{(z - i)^3}\right]$$

$$= \lim_{z \to i} \frac{1}{2} \frac{d^2}{dz^2}[\sin(2z)] = \lim_{z \to i} \frac{1}{2}[-4\sin(2z)] = -2\sin(2i).$$

Therefore,

$$\oint_C \frac{\sin(2z)}{(z - i)^3} \, dz = 2\pi i[-2\sin(2i)] = -4\pi i \sin(2i).$$

This result can be simplified by writing

$$\sin(2i) = \frac{1}{2i}(e^{-2} - e^2) = i \sinh(2).$$

Then

$$\oint_C \frac{\sin(2z)}{(z - i)^3} \, dz = 4\pi \sinh(2). \quad \blacksquare$$

EXAMPLE 22.8

Evaluate $\oint_C \sin(z)/[z^2(z^2 + 4)]\, dz$ if C is any piecewise-smooth simple closed curve enclosing 0, $2i$, and $-2i$. A typical such curve is drawn in Figure 22.9.

It may at first appear that f has a pole of order 2 at zero because of the factor z^2 in the denominator. However, write

$$f(z) = \frac{\dfrac{\sin(z)}{z}}{z(z + 2i)(z - 2i)} = \frac{g(z)}{h(z)}$$

with $g(z) = \sin(z)/z$ and $h(z) = z(z + 2i)(z - 2i)$. Since g has a removable singularity at zero $[\lim_{z \to 0} \sin(z)/z = 1]$ and h has *simple* zeros at 0, $2i$, and $-2i$, the quotient $g(z)/h(z)$ has simple poles at 0, $-2i$, and $2i$ by Theorem 22.2.

We will calculate the residues of f at these singularities:

$$\operatorname*{Res}_{0} f = \lim_{z \to 0} z \frac{\sin(z)}{z^2(z^2 + 4)} = \lim_{z \to 0} \frac{\sin(z)}{z}\, \frac{1}{z^2 + 4} = \frac{1}{4};$$

$$\operatorname*{Res}_{2i} f = \lim_{z \to 2i} (z - 2i) \frac{\sin(z)}{z^2(z + 2i)(z - 2i)}$$

$$= \lim_{z \to 2i} \frac{\sin(z)}{z^2(z + 2i)} = \frac{i}{16} \sin(2i);$$

and

$$\operatorname*{Res}_{-2i} f = \lim_{z \to -2i} (z + 2i) \frac{\sin(z)}{z^2(z + 2i)(z - 2i)}$$

$$= \lim_{z \to -2i} \frac{\sin(z)}{z^2(z - 2i)} = \frac{i}{16} \sin(2i).$$

Therefore,

$$\oint_C \frac{\sin(z)}{z^2(z^2 + 4)}\, dz = 2\pi i \left[\frac{1}{4} + \frac{i\sin(2i)}{16} + \frac{i\sin(2i)}{16} \right] = \frac{\pi}{4}[2i - \sin(2i)]. \quad \blacksquare$$

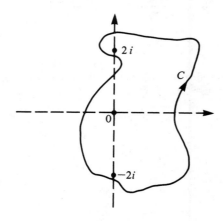

Figure 22.9

The following corollary gives a special case in which the residue formula of Theorem 22.4 has a particularly simple form.

COROLLARY TO THEOREM 22.4

Let $f(z) = g(z)/h(z)$, where g and h are analytic at z_0, $g(z_0) \neq 0$, and h has a simple zero at z_0. Then f has a simple pole at z_0 and

$$\mathop{\mathrm{Res}}_{z_0} f = \frac{g(z_0)}{h'(z_0)}.$$

Proof We know that f has a simple pole at z_0 by Theorem 22.2. Since h has a simple zero at z_0, $h(z_0) = 0$ but $h'(z_0) \neq 0$. Now apply Theorem 22.4 with $m = 1$ and $f(z) = g(z)/h(z)$ to get

$$\mathop{\mathrm{Res}}_{z_0} f = \lim_{z \to z_0} (z - z_0)\frac{g(z)}{h(z)} = \lim_{z \to z_0} g(z)\left[\frac{z - z_0}{h(z) - h(z_0)}\right]$$

$$= \lim_{z \to z_0} g(z)\frac{1}{\displaystyle\lim_{z \to z_0} \frac{h(z) - h(z_0)}{z - z_0}} = \frac{g(z_0)}{h'(z_0)}. \quad \blacksquare$$

EXAMPLE 22.9

Evaluate

$$\oint_C \frac{3z^2 + 1}{1 + z^2}\, dz$$

if C is any piecewise-smooth simple closed curve enclosing i but not $-i$.

Write $g(z) = 3z^2 + 1$ and $h(z) = 1 + z^2$. By the corollary,

$$\mathop{\mathrm{Res}}_{i} f = \frac{g(i)}{h'(i)} = \frac{3(i)^2 + 1}{2i} = i.$$

Therefore,

$$\oint_C \frac{3z^2 + 1}{1 + z^2}\, dz = 2\pi i(i) = -2\pi. \quad \blacksquare$$

PROBLEMS FOR SECTION 22.2

In each of Problems 1 through 20, find the residue of the function at each singularity. (The singularities were requested in Problems 1 through 20 of Section 22.1.)

1. $\dfrac{\cos(z)}{z^2}$

2. $\dfrac{1}{(z + i)^2(z - i)}$

3. $e^{1/z}(z - i)$

4. $\dfrac{\sin(z)}{z - \pi}$

5. $\dfrac{\cos(2z)}{(z - 1)^2(z^2 + 1)}$

6. $\dfrac{z}{(z + 1)^2}$

7. $\dfrac{z - i}{z^2 + 1}$

8. $\dfrac{\sin(z)}{\sinh(z)}$

9. $\dfrac{z}{z^4 - 1}$

10. $\tan(z)$

11. $\dfrac{1}{\cos(z)}$

12. $e^{1/z(z + 1)}$

13. $\dfrac{1}{z^2} e^{iz}$

14. $\dfrac{\sin(z)}{z(z - \pi)(z - i)^2}$

15. $\dfrac{e^{2z}}{(z + 1)^4}$

16. $\dfrac{1}{z^4} \sinh(z)$

17. $\cosh(z)$ 　　　　　　**18.** $\dfrac{2i-1}{(z^2+2z-3)^2}$ 　　　　**19.** $\dfrac{\sin(z)}{(z+\pi)^2}$ 　　　　**20.** $\dfrac{1}{\sin^3(2z)}$

In each of Problems 21 through 34, evaluate the integral with C any piecewise-smooth simple closed curve enclosing all of the singularities of the integrand.

21. $\displaystyle\oint_C \frac{2z}{(z-i)^2}\,dz$ 　　　　　**22.** $\displaystyle\oint_C \frac{1+z^2}{(z-1)^2(z+2i)}\,dz$ 　　　　**23.** $\displaystyle\oint_C \frac{\cos(z)}{4+z^2}\,dz$

24. $\displaystyle\oint_C \frac{1}{z}e^z\,dz$ 　　　　　**25.** $\displaystyle\oint_C \frac{z-i}{2z+1}\,dz$ 　　　　**26.** $\displaystyle\oint_C \frac{z+i}{z^2+6}\,dz$

27. $\displaystyle\oint_C \frac{\cos(z)}{ze^z}\,dz$ 　　　　　**28.** $\displaystyle\oint_C \frac{z}{\sinh^2(z)}\,dz$ 　　　　**29.** $\displaystyle\oint_C e^{2/z^3}\,dz$

30. $\displaystyle\oint_C \frac{iz}{(z^2+9)(z-i)^3}\,dz$ 　　　**31.** $\displaystyle\oint_C \frac{z^2}{z-1+2i}\,dz$ 　　　**32.** $\displaystyle\oint_C \frac{8z-4i+1}{z+4i}\,dz$

33. $\displaystyle\oint_C \frac{(1-z)^2}{z^3-8}\,dz$ 　　　　**34.** $\displaystyle\oint_C \coth(z)\,dz$

35. Evaluate $\oint_C \mathrm{Log}(z)/(z^2+1)\,dz$, with C the piecewise-smooth simple closed curve shown in Figure 22.8.

22.3 *Evaluation of Real Integrals* _____

We will now see how the residue theorem can be used to evaluate certain real integrals. There are several categories of such integrals we will consider.

INTEGRALS OF THE FORM $\int_0^{2\pi} K[\cos(\theta),\sin(\theta)]\,d\theta$

Suppose $K(x,y)$ is a rational function of x and y, such as

$$K(x,y)=\frac{x^2-4xy}{x+y}.$$

If we replace x with $\cos(\theta)$ and y with $\sin(\theta)$, we obtain a rational function of $\cos(\theta)$ and $\sin(\theta)$. With the example just given for K,

$$K[\cos(\theta),\sin(\theta)]=\frac{\cos^2(\theta)-4\cos(\theta)\sin(\theta)}{\cos(\theta)+\sin(\theta)}.$$

We will see how to evaluate an integral of the form

$$\int_0^{2\pi} K[\cos(\theta),\sin(\theta)]\,d\theta,$$

with K a rational function satisfying conditions we will state later.

Recall that $\cos(\theta)=\frac{1}{2}(e^{i\theta}+e^{-i\theta})$ and $\sin(\theta)=(1/2i)(e^{i\theta}-e^{-i\theta})$. Let $z=e^{i\theta}$. Since $|z|=1$, z is a complex number lying on the unit circle about the origin. Further,

$$\cos(\theta)=\frac{1}{2}\left(z+\frac{1}{z}\right)\quad\text{and}\quad\sin(\theta)=\frac{1}{2i}\left(z-\frac{1}{z}\right).$$

Parametrize C by $z = e^{i\theta}$, $0 \le \theta \le 2\pi$. Then $dz = ie^{i\theta}\, d\theta = iz\, d\theta$, so

$$d\theta = \frac{1}{iz}\, dz.$$

Therefore,

$$\int_0^{2\pi} K[\cos(\theta), \sin(\theta)]\, d\theta = \oint_C K\left[\frac{1}{2}\left(z + \frac{1}{z}\right), \frac{1}{2i}\left(z - \frac{1}{z}\right)\right]\frac{1}{iz}\, dz.$$

In order for the complex integral on the right to exist, we assume that

$$K\left[\frac{1}{2}\left(z + \frac{1}{z}\right), \frac{1}{2i}\left(z - \frac{1}{z}\right)\right]\frac{1}{iz}$$

has no singularities on C. If we can evaluate the complex integral

$$\oint_C K\left[\frac{1}{2}\left(z + \frac{1}{z}\right), \frac{1}{2i}\left(z - \frac{1}{z}\right)\right]\frac{1}{iz}\, dz$$

using the residue theorem, we also evaluate the real integral $\int_0^{2\pi} K[\cos(\theta), \sin(\theta)]\, d\theta$.

EXAMPLE 22.10

Evaluate

$$\int_0^{2\pi} \frac{\cos(\theta)}{1 + \frac{1}{4}\cos(\theta)}\, d\theta.$$

With C the circle $|z| = 1$, oriented counterclockwise, we have

$$\int_0^{2\pi} \frac{\cos(\theta)}{1 + \frac{1}{4}\cos(\theta)}\, d\theta = \oint_C \frac{\frac{1}{2}\left[z + \frac{1}{z}\right]}{1 + \frac{1}{8}\left[z + \frac{1}{z}\right]}\frac{1}{iz}\, dz = \oint_C \frac{4(z^2 + 1)}{iz(z^2 + 8z + 1)}\, dz.$$

The integrand has simple poles at 0, $-4 + \sqrt{15}$ (enclosed by C), and $-4 - \sqrt{15}$ (exterior to C). To compute the residues at the singularies inside C, we will use Theorem 22.4.

First,

$$\operatorname*{Res}_0 f = \lim_{z \to 0} zf(z) = \lim_{z \to 0} \frac{4(z^2 + 1)}{i(z^2 + 8z + 1)} = \frac{4}{i} = -4i.$$

For the residue at $-4 + \sqrt{15}$, write the integrand as

$$\frac{4(z^2 + 1)}{iz(z^2 + 8z + 1)} = \frac{4(z^2 + 1)}{iz[z - (-4 + \sqrt{15})][z - (-4 - \sqrt{15})]}.$$

Then

$$\operatorname*{Res}_{-4+\sqrt{15}} f = \lim_{z \to -4+\sqrt{15}} [z - (-4 + \sqrt{15})]\frac{4(z^2 + 1)}{iz[z - (-4 + \sqrt{15})][z - (-4 - \sqrt{15})]}$$

$$= \lim_{z \to -4+\sqrt{15}} \frac{4(z^2 + 1)}{iz[z + 4 + \sqrt{15}]} = \frac{-16}{\sqrt{15}i} = \frac{16i}{\sqrt{15}}.$$

By the residue theorem,

$$\oint_C \frac{4(z^2 + 1)}{iz(z^2 + 8z + 1)}\, dz = 2\pi i \left[-4i + \frac{16i}{\sqrt{15}} \right] = 2\pi \left[\frac{4\sqrt{15} - 16}{\sqrt{15}} \right].$$

Therefore,

$$\int_0^{2\pi} \frac{\cos(\theta)}{1 + \frac{1}{4}\cos(\theta)}\, d\theta = 2\pi \left[\frac{4\sqrt{15} - 16}{\sqrt{15}} \right],$$

which is approximately -0.8242. ∎

The answer must be a real number because it is the evaluation of a real integral (by complex methods).

INTEGRALS OF THE FORM $\int_{-\infty}^{\infty} p(x)/q(x)\, dx$

Consider a quotient $p(x)/q(x)$ of polynomials $p(x)$ and $q(x)$ having real coefficients. Assume that $q(x) \neq 0$ for all real x and that the degree of q exceeds that of p by at least 2. This ensures convergence of $\int_{-\infty}^{\infty} p(x)/q(x)\, dx$. We will derive a general formula for this integral.

Let $K(z) = p(z)/q(z)$. Assume that p and q have no common factors. The singularities of K are exactly the zeros of q, and these are poles of K. Since q has real coefficients, its zeros occur in conjugate pairs. Let the zeros of q be $z_1, \bar{z}_1, \ldots, z_n, \bar{z}_n$, with $z_1, \ldots, z_n$ in the upper half-plane and $\bar{z}_1, \ldots, \bar{z}_n$ their reflections across the real line (Figure 22.10).

Draw a circle $|z| = R$ of sufficiently large radius to enclose $z_1, \ldots, z_n$. Let C consist of the semicircle S of $|z| = R$ lying in the upper half-plane together with the segment L of the real line from $-R$ to R, as shown in Figure 22.11. C is oriented counterclockwise. By the residue theorem,

$$\oint_C K(z)\, dz = 2\pi i \sum_{j=1}^n \operatorname*{Res}_{z_j} K = \int_S K(z)\, dz + \int_L K(z)\, dz. \qquad (22.3)$$

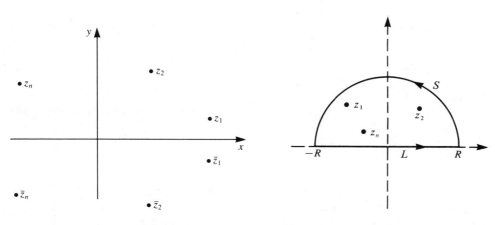

Figure 22.10. Zeros of q.   Figure 22.11

L can be parametrized $z = x$, $-R \leq x \leq R$. Then

$$\int_L K(z) \, dz = \int_{-R}^{R} K(x) \, dx,$$

a real integral over an integral on the real line.

Now consider $\int_S K(z) \, dz$. We have assumed that

$$\text{degree } q \geq 2 + \text{degree } p.$$

Then $z^2 p(z)/q(z)$ is bounded if we restrict z to lie on or outside the circle containing the zeros of q. That is, for some M,

$$\left| \frac{z^2 p(z)}{q(z)} \right| \leq M$$

for $|z| \geq R$. But then

$$\left| \frac{p(z)}{q(z)} \right| \leq \frac{1}{|z|^2} M$$

for $|z| \geq R$. On S, where $|z| = R$, we have

$$\left| \frac{p(z)}{q(z)} \right| \leq \frac{M}{R^2}.$$

By conclusion (2) of Theorem 20.6,

$$\left| \int_S \frac{p(z)}{q(z)} \, dz \right| \leq \frac{M}{R^2} (\pi R) = \pi \frac{M}{R}. \tag{22.4}$$

Now let $R \to \infty$ in equation (22.3). Then C expands to enclose the entire upper half-plane. Further, by equation (22.4), $\int_S K(z) \, dz \to 0$, and we also have $\int_{-R}^{R} K(x) \, dx \to \int_{-\infty}^{\infty} K(x) \, dx$. Thus, in the limit as $r \to \infty$, equation (22.3) yields

$$\int_{-\infty}^{\infty} K(x) \, dx = 2\pi i \sum_{j=1}^{n} \operatorname*{Res}_{z_j} K(z).$$

That is, under the conditions stated for K, $\int_{-\infty}^{\infty} K(x) \, dx$ equals $2\pi i$ times the sum of the residues of $K(z)$ at poles of K occurring in the upper half-plane.

EXAMPLE 22.11

Evaluate $\int_{-\infty}^{\infty} 1/(x^6 + 64) \, dx$.

Let $K(x) = 1/(x^6 + 64)$. Since K satisfies the conditions of the method just discussed,

$$\int_{-\infty}^{\infty} \frac{1}{x^6 + 64} \, dx = 2\pi i \sum_{z_j} \operatorname{Res} K,$$

with the summation taken over all poles of K in the upper half-plane. Poles of K are solutions of $z^6 + 64 = 0$, which are the sixth roots of -64. These sixth roots are $2e^{(\pi i + 2k\pi i)/6}$ for $k = 0, 1, 2, 3, 4, 5$. Three of these roots, namely, $2e^{\pi i/6}$, $2i$, and $2e^{5\pi i/6}$, occur in the upper half-plane. Therefore,

$$\int_{-\infty}^{\infty} \frac{1}{x^6 + 64}\, dx = 2\pi i \left[\operatorname*{Res}_{2e^{\pi i/6}} K + \operatorname*{Res}_{2i} K + \operatorname*{Res}_{2e^{5\pi i/6}} K \right].$$

It is convenient here to use the corollary to Theorem 22.4, with $g(z) = 1$ and $h(z) = z^6 + 64$, to compute these residues. We have

$$\operatorname*{Res}_{2e^{\pi i/6}} K = \frac{1}{6(2e^{\pi i/6})^5} = \frac{1}{192} e^{-5\pi i/6} = \frac{1}{192}\left[\cos\left(\frac{5\pi}{6}\right) - i \sin\left(\frac{5\pi}{6}\right) \right],$$

$$\operatorname*{Res}_{2i} K = \frac{1}{6(2i)^5} = -\frac{i}{192},$$

and

$$\operatorname*{Res}_{2e^{5\pi i/6}} K = \frac{1}{6(2e^{5\pi i/6})^5} = \frac{1}{192} e^{-\pi i/6} = \frac{1}{192}\left[\cos\left(\frac{\pi}{6}\right) - i \sin\left(\frac{\pi}{6}\right) \right].$$

Since $\cos(\pi/6) = -\cos(5\pi/6)$ and $\sin(\pi/6) = \sin(5\pi/6) = \frac{1}{2}$, we obtain, upon adding these residues, that

$$\int_{-\infty}^{\infty} \frac{1}{x^6 + 64}\, dx = \frac{2\pi i}{192}\left[\cos\left(\frac{5\pi}{6}\right) - i \sin\left(\frac{5\pi}{6}\right) - i + \cos\left(\frac{\pi}{6}\right) - i \sin\left(\frac{\pi}{6}\right) \right]$$

$$= \frac{\pi i}{96}[-2i] = \frac{\pi}{48}. \quad \blacksquare$$

INTEGRALS OF THE FORM $\int_{-\infty}^{\infty} [p(x)/q(x)]\cos(\alpha x)\, dx$ and $\int_{-\infty}^{\infty} [p(x)/q(x)]\sin(\alpha x)\, dx$

We will now consider integrals of these types, with α any positive constant. Let $K(x) = p(x)/q(x)$, a quotient of polynomials with real coefficients, as before. Assume that the degree of q exceeds that of p by at least 1 and that q has no real zeros.

We can evaluate the integrals by the following device. Consider $\oint_C K(z)e^{i\alpha z}\, dz$, where C is the closed curve used above, consisting of a semicircle of radius R and part of the real axis from $-R$ to R. Choose R large enough so that C encloses all poles of K in the upper half-plane. Poles of K are the only singularities of $K(z)e^{i\alpha z}$. By an argument like that used previously, we obtain in the limit as $R \to \infty$ that

$$\int_{-\infty}^{\infty} K(x)e^{i\alpha x}\, dx = 2\pi i \sum_{z_j} \operatorname{Res} K(z)e^{i\alpha z},$$

with the summation taken over all poles z_j of $K(z)$ occurring in the upper half-plane. For x real, $e^{i\alpha x} = \cos(\alpha x) + i \sin(\alpha x)$; hence,

$$\int_{-\infty}^{\infty} K(x)\cos(\alpha x)\, dx + i \int_{-\infty}^{\infty} K(x)\sin(\alpha x)\, dx = 2\pi i \sum_{z_j} \operatorname{Res} K(z)e^{i\alpha z}.$$

The real part of the quantity on the right is $\int_{-\infty}^{\infty} K(x)\cos(\alpha x)\, dx$, and the imaginary part is $\int_{-\infty}^{\infty} K(x)\sin(\alpha x)\, dx$.

EXAMPLE 22.12

Evaluate $\int_{-\infty}^{\infty} [x/(x^2 + 16)]\sin(\sqrt{3}x)\, dx$.

With $K(x) = x/(x^2 + 16)$, $K(z)$ satisfies the hypotheses we have stated, and

$$\int_{-\infty}^{\infty} \frac{x}{x^2 + 16} \cos(\sqrt{3}x)\, dx + i \int_{-\infty}^{\infty} \frac{x}{x^2 + 16} \sin(\sqrt{3}x)\, dx = 2\pi i \sum \text{Res} \frac{z}{z^2 + 16} e^{i\sqrt{3}z},$$

with the sum extended over all poles of $z/(z^2 + 16)$ in the upper half-plane. There is only one such pole, namely, $4i$. Use the corollary to Theorem 22.4 to compute

$$\text{Res}_{4i} \frac{z}{z^2 + 16} e^{i\sqrt{3}z} = \frac{4ie^{i\sqrt{3}(4i)}}{2 \cdot 4i} = \frac{1}{2} e^{-4\sqrt{3}}.$$

Therefore,

$$\int_{-\infty}^{\infty} \frac{x}{x^2 + 16} \cos(\sqrt{3}x)\, dx + i \int_{-\infty}^{\infty} \frac{x}{x^2 + 16} \sin(\sqrt{3}x)\, dx = 2\pi i \frac{1}{2} e^{-4\sqrt{3}} = \pi i e^{-4\sqrt{3}}.$$

The real part of the left side must equal the real part of the right side of this equation, and similarly for the imaginary parts. Therefore,

$$\int_{-\infty}^{\infty} \frac{x}{x^2 + 16} \cos(\sqrt{3}x)\, dx = 0 \quad \text{and} \quad \int_{-\infty}^{\infty} \frac{x}{x^2 + 16} \sin(\sqrt{3}x)\, dx = \pi e^{-4\sqrt{3}}. \quad \blacksquare$$

By choosing different curves in the plane, it is possible to convert many other kinds of improper real integrals into line integrals of complex functions and evaluate them using the residue theorem. We will pursue some of these possibilities in the problems that follow.

PROBLEMS FOR SECTION 22.3

In Problems 1 through 15, evaluate the integral by complex methods.

1. $\int_0^{2\pi} \frac{1}{2 - \cos(\theta)}\, d\theta$

2. $\int_{-\infty}^{\infty} \frac{1}{x^4 + 1}\, dx$

3. $\int_0^{\infty} \frac{1}{1 + x^6}\, dx$. *Hint:* Consider $\frac{1}{2} \int_{-\infty}^{\infty} \frac{1}{1 + x^6}\, dx$.

4. $\int_{-\infty}^{\infty} \frac{1}{x^2 - 2x + 6}\, dx$

5. $\int_{-\infty}^{\infty} \frac{x \sin(2x)}{x^4 + 16}\, dx$

6. $\int_{-\infty}^{\infty} \frac{x^2}{(x^2 + 1)(x^2 + 4)}\, dx$

7. $\int_{-\infty}^{\infty} \frac{\sin(x)}{x^2 - 4x + 5}\, dx$

8. $\int_0^{2\pi} \frac{2 \sin(\theta)}{2 + \sin^2(\theta)}\, d\theta$

9. $\int_{-\infty}^{\infty} \frac{\cos^2(x)}{(x^2 + 4)^2}\, dx$

10. $\int_0^{2\pi} \frac{\sin(\theta) + \cos(\theta)}{2 - \cos(\theta)}\, d\theta$

11. $\int_{-\infty}^{\infty} \frac{x^2}{(x^2 + 1)^2}\, dx$

12. $\int_0^{2\pi} \frac{1}{\cos^2(\theta) + 2 \sin^2(\theta)}\, d\theta$

13. $\int_{-\infty}^{\infty} \frac{1}{x^2 + x + 2}\, dx$

14. $\int_{-\infty}^{\infty} \frac{\cos(3x)}{x^2 + 9}\, dx$

15. $\int_0^{2\pi} \frac{\sin(\theta)}{4 + \sin(\theta)}\, d\theta$

16. Show that, for any positive constant α,

$$\int_{-\infty}^{\infty} \frac{\cos(\alpha x)}{x^2 + 1}\, dx = \pi e^{-\alpha}.$$

17. Let α and β be real numbers with $\alpha > \beta > 0$. Show that

$$\int_0^\pi \frac{1}{\alpha + \beta \cos(\theta)}\, d\theta = \frac{\pi}{\sqrt{\alpha^2 - \beta^2}}.$$

Hint: First show that the integral from zero to π is one-half the integral from zero to 2π.

18. Let α and β be distinct positive numbers. Show that

$$\int_{-\infty}^{\infty} \frac{1}{(x^2 + \alpha^2)(x^2 + \beta^2)}\, dx = \frac{\pi}{\alpha\beta(\alpha + \beta)}.$$

19. Let α and β be positive numbers. Show that

$$\int_{-\infty}^{\infty} \frac{\cos(\alpha x)}{(x^2 + \beta^2)^2}\, dx = \frac{\pi}{2\beta^3}(1 + \alpha\beta)e^{-\alpha\beta}.$$

20. Let α and β be distinct positive numbers. Show that

$$\int_{-\infty}^{\infty} \frac{\cos(x)}{(x^2 + \alpha^2)(x^2 + \beta^2)}\, dx = \frac{\pi}{\alpha^2 - \beta^2}\left(\frac{1}{\beta}e^{-\beta} - \frac{1}{\alpha}e^{-\alpha}\right).$$

21. Evaluate $\int_0^{2\pi} \cos^{2n}(\theta)\, d\theta$, in which n is any positive integer. *Hint:* Observe that, after converting the integral to a complex line integral, the integrand has a pole of order $2n + 1$ at the origin. The residue at this pole is the coefficient of z^{2n} in the binomial expansion of $(1 + z^2)^{2n}$.

22. Let α and β be distinct positive numbers. Show that

$$\int_0^{2\pi} \frac{1}{\alpha^2\cos^2(\theta) + \beta^2\sin^2(\theta)}\, d\theta = \frac{2\pi}{\alpha\beta}.$$

23. Let α be a positive number. Show that

$$\int_0^{\pi/2} \frac{1}{\alpha + \sin^2(\theta)}\, d\theta = \frac{\pi}{2\sqrt{\alpha^2 + \alpha}}.$$

24. Show that $\int_0^\infty e^{-x^2}\cos(2\beta x)\, dx = \frac{1}{2}\sqrt{\pi}e^{-\beta^2}$ for any positive number β. *Hint:* Integrate e^{-z^2} about the rectangular path having corners at $R, R + \beta i, -R + \beta i$, and $-R$, then let $R \to \infty$. In the course of the calculation, assume that $\int_0^\infty e^{-x^2}\, dx = \frac{1}{2}\sqrt{\pi}$.

25. Prove *Jordan's lemma:* $\int_0^{\pi/2} e^{-R\sin(\theta)}\, d\theta < (\pi/2)R$ for any positive number R. *Hint:* First show that $\sin(\theta) \geq 2\theta/\pi$ if $0 \leq \theta \leq \pi/2$ (consider the graph of the sine function). Then use the fact that $e^{-R\sin(\theta)} \leq e^{-2R\theta/\pi}$.

26. Derive *Fresnel's integrals:*

$$\int_0^\infty \cos(x^2)\, dx = \int_0^\infty \sin(x^2)\, dx = \frac{1}{2\sqrt{2}}\sqrt{\pi}.$$

Hint: Integrate e^{iz^2} about the curve bounding the sector $0 \leq x \leq R, 0 \leq \theta \leq \pi/4$, and write the integral as a sum of the integrals over the straight line segments and the circular arc of this curve. Use Jordan's lemma (Problem 25) to show that the integral along the circular arc tends to zero as $R \to \infty$, and parametrize the straight line segments to obtain Fresnel's integrals.

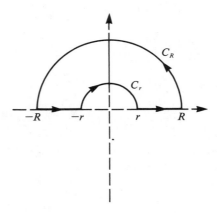

Figure 22.12

27. Use the residue theorem to show that

$$\int_{-\infty}^{\infty} \frac{1}{(1 + x^2)^n}\, dx = \frac{(2n - 2)!}{2^{2n-2}[(n-1)!]}$$

for any positive integer n.

28. Show that $\int_0^{\infty} \sin(x)/x\, dx = \pi/2$ by integrating $(1/z)e^{iz}$ over the contour shown in Figure 22.12 and then letting $r \to 0$ and $R \to \infty$. Show that, in these limits, the integrals over the semicircular arc of radius R tends to zero, while the integral over the semicircular arc of radius r tends to $-\pi i$.

22.4 *Summation of Series by the Residue Theorem*

We will now see how the residue theorem can sometimes be used to sum a real convergent series. Suppose that $P(z)$ is a polynomial of degree 2 or higher, with real coefficients and having no integer zeros (although it may possess other real zeros). Let $z_1, \ldots, z_m$ be the zeroes of $P(z)$. We claim that

$$\sum_{n=-\infty}^{\infty} \frac{1}{P(n)} = -\pi \sum_{j=1}^{m} \operatorname*{Res}_{z_j} \left(\frac{\cot(\pi z)}{P(z)} \right)$$

and

$$\sum_{n=-\infty}^{\infty} (-1)^n \frac{1}{P(n)} = -\pi \sum_{j=1}^{m} \operatorname*{Res}_{z_j} \left(\frac{\csc(\pi z)}{P(z)} \right).$$

We will look at two examples and then see how these formulas arise.

EXAMPLE 22.13

Let a be any positive constant. We will sum the series $\sum_{n=0}^{\infty} 1/(a^2 + n^2)$.

Let $P(z) = a^2 + z^2$. Then P is a polynomial of degree 2 with no integer zeros. The poles of P are ai and $-ai$. Therefore,

$$\sum_{n=-\infty}^{\infty} \frac{1}{P(n)} = \sum_{n=-\infty}^{\infty} \frac{1}{a^2 + n^2} = -\pi \left[\operatorname*{Res}_{ai} \frac{\cot(\pi z)}{a^2 + z^2} + \operatorname*{Res}_{-ai} \frac{\cot(\pi z)}{a^2 + z^2} \right].$$

Since ai and $-ai$ are simple poles of $\cot(\pi z)/(a^2 + z^2)$, compute

$$\operatorname*{Res}_{ai} \frac{\cot(\pi z)}{a^2 + z^2} = \frac{\cot(\pi ai)}{2ai} \quad \text{and} \quad \operatorname*{Res}_{-ai} \frac{\cot(\pi z)}{a^2 + z^2} = \frac{\cot(-\pi ai)}{-2ai} = \frac{\cot(\pi ai)}{2ai}.$$

Therefore,

$$\sum_{n=-\infty}^{\infty} \frac{1}{a^2 + n^2} = -\pi \left[\frac{\cot(\pi ai)}{ai} \right] = \frac{\pi}{a} i \cot(\pi ai).$$

The answer must be a real number. To see that it is, write

$$i \cot(\pi ai) = \frac{i \cos(\pi ai)}{\sin(\pi ai)} = i \frac{\frac{1}{2}(e^{-\pi a} + e^{\pi a})}{\frac{1}{2i}(e^{-\pi a} - e^{\pi a})} = \frac{\cosh(\pi a)}{\sinh(\pi a)} = \coth(\pi a).$$

Therefore,

$$\sum_{n=-\infty}^{\infty} \frac{1}{a^2 + n^2} = \frac{\pi}{a} \coth(\pi a).$$

Now use the fact that $(-n)^2 = n^2$ to write

$$\sum_{n=-\infty}^{\infty} \frac{1}{a^2 + n^2} = \sum_{n=-\infty}^{0} \frac{1}{a^2 + n^2} - \frac{1}{a^2} + \sum_{n=0}^{\infty} \frac{1}{a^2 + n^2}$$

$$= -\frac{1}{a^2} + 2 \sum_{n=0}^{\infty} \frac{1}{a^2 + n^2}$$

$$= \frac{\pi}{a} \coth(\pi a).$$

Therefore,

$$\sum_{n=0}^{\infty} \frac{1}{a^2 + n^2} = \frac{1}{2a^2} + \frac{\pi}{2a} \coth(\pi a).$$

For example, with $a = 1$, we have

$$\sum_{n=0}^{\infty} \frac{1}{1 + n^2} = 1 + \frac{1}{2} + \frac{1}{5} + \frac{1}{10} + \frac{1}{17} + \cdots = \frac{1}{2} + \frac{\pi}{2} \coth(\pi),$$

which is approximately 2.0076674. ∎

EXAMPLE 22.14

Sum the series $\sum_{n=0}^{\infty} (-1)^n / (2n + 1)^3$.

Let $P(z) = (2z + 1)^3$. Then P has degree 3 and no integer zeros. The only zero of $P(z)$ is $z = -\frac{1}{2}$, so $1/P(z)$ has a pole of order 3 at $-\frac{1}{2}$. Further, $\csc(-\pi/2) \neq 0$, so

$\csc(\pi z)/P(z)$ has a pole of order 3 at $-\frac{1}{2}$. Then

$$
\begin{aligned}
\sum_{n=-\infty}^{\infty} (-1)^n \frac{1}{(2n+1)^3} &= -\pi \operatorname*{Res}_{-1/2} \frac{\csc(\pi z)}{(2z+1)^3} \\
&= -\pi \lim_{z \to -1/2} \frac{1}{2} \frac{d^2}{dz^2} \left[\left(z+\frac{1}{2}\right)^3 \frac{\csc(\pi z)}{(2z+1)^3} \right] \\
&= -\pi \lim_{z \to -1/2} \frac{1}{2} \frac{d^2}{dz^2} \left[\left(z+\frac{1}{2}\right)^3 \frac{\csc(\pi z)}{[2(z+\frac{1}{2})]^3} \right] \\
&= -\frac{\pi}{16} \lim_{z \to -1/2} \frac{d^2}{dz^2} \csc(\pi z) \\
&= -\frac{\pi}{16} \lim_{z \to -1/2} [\pi^2 \csc^2(\pi z) \cot(\pi z) + \pi^2 \csc^3(\pi z)] = \frac{\pi^3}{16}.
\end{aligned}
$$

It is routine to verify that

$$
\sum_{n=-\infty}^{\infty} (-1)^n \frac{1}{(2n+1)^3} = 2 \sum_{n=0}^{\infty} (-1)^n \frac{1}{(2n+1)^3};
$$

therefore,

$$
\sum_{n=0}^{\infty} (-1)^n \frac{1}{(2n+1)^3} = \frac{\pi^3}{32}. \quad \blacksquare
$$

We will conclude this section with a sketch of the derivation of the formula

$$
\sum_{n=-\infty}^{\infty} \frac{1}{P(n)} = -\pi \sum_{j=1}^{m} \operatorname*{Res}_{z_j} \left[\frac{\cot(\pi z)}{P(z)} \right].
$$

Consider $\oint_{C_n} [\pi \cot(\pi z)]/P(z)\, dz$, in which C_n is the simple closed curve shown in Figure 22.13. C_n is a rectangle with sides parallel to the axes, passing through $n+\frac{1}{2}$ and $-n-\frac{1}{2}$ on the real axis and through the points in and $-in$ on the imaginary axis, with n a positive integer. Note that $[\pi \cot(\pi z)]/P(z)$ has simple poles at $0, \pm 1, \pm 2, \ldots, \pm n$

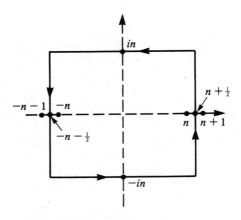

Figure 22.13

and also has poles at the zeros of $P(z)$. Therefore,

$$\oint_{C_n} \frac{\pi \cot(\pi z)}{P(z)} \, dz = 2\pi i \sum_{j=-n}^{n} \operatorname*{Res}_{j} \frac{\pi \cot(\pi z)}{P(z)} + 2\pi i \sum_{k=0}^{\infty} \operatorname*{Res}_{p} \frac{\pi \cot(\pi z)}{P(z)}, \qquad (22.5)$$

in which the second summation is taken over all poles of $\cot(\pi z)/P(z)$ occurring at zeros of $P(z)$. By making estimates of $|[\pi \cot(\pi z)]/P(z)|$ on the sides of C_n, it can be shown that the integral in (22.5) has limit zero as $n \to \infty$. By evaluating the residues on the right side of equation (22.5), we obtain the summation formula.

PROBLEMS FOR SECTION 22.4

Sum the series in each of Problems 1 through 6.

1. $\displaystyle\sum_{n=-\infty}^{\infty} \frac{1}{a^4 + n^4}$, with a any positive number

2. $\displaystyle\sum_{n=-\infty}^{\infty} (-1)^n \frac{1}{a^4 + n^4}$, with a any positive number

3. $\displaystyle\sum_{n=0}^{\infty} \frac{1}{(2n^2 + 1)^2}$

4. $\displaystyle\sum_{n=1}^{\infty} (-1)^n \frac{1}{n^4 - 2}$

5. Show that $\sum_{n=-\infty}^{\infty} (-1)^n 1/(a + n)^2 = \pi^2 \csc(\pi a)\cot(\pi a)$ for any positive number a which is not an integer.

22.5 *The Argument Principle*

In this section, we will prove the argument principle, which has a variety of uses, among them the evaluation of certain integrals and estimation of the number of zeros of a function which are enclosed by a given curve.

THEOREM 22.5 The Argument Principle

Let f be analytic in a domain D except at a finite number of poles. Let C be a piecewise-smooth simple closed curve in D not passing through any zeros or poles of f. Then

$$\frac{1}{2\pi i} \oint_C \frac{f'(z)}{f(z)} \, dz = Z - P,$$

where Z is the number of zeros of f enclosed by C and P is the number of poles of f enclosed by C, with each zero and pole counted according to its multiplicity. ■

We will prove the theorem after looking at a typical use.

EXAMPLE 22.15

Evaluate $\oint_C \cot(z) \, dz$, with C the simple closed curve shown in Figure 22.14.

Write $\cot(z) = \cos(z)/\sin(z) = f'(z)/f(z)$, where $f(z) = \sin(z)$. Since f is analytic for all z, f has no poles. The zeros of f are $n\pi$, with $n = 0, \pm 1, \pm 2, \ldots$, and these are all simple zeros. There are five zeros of f enclosed by C. Therefore,

$$\frac{1}{2\pi i} \oint_C \cot(z) \, dz = Z - P = 5 - 0 = 5;$$

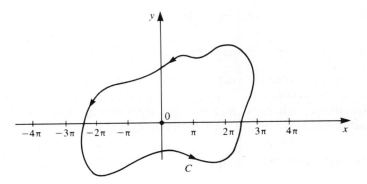

Figure 22.14

hence,

$$\oint_C \cot(z)\, dz = 10\pi i. \quad \blacksquare$$

Proof of Theorem 22.5   Suppose first that f has a pole of order m at w enclosed by C. The Laurent expansion of f about w has the form

$$f(z) = \frac{a_{-m}}{(z-w)^m} + \frac{a_{-m+1}}{(z-w)^{m-1}} + \cdots + \frac{a_{-1}}{z-w} + a_0 + a_1(z-w) + \cdots$$

in some annulus $0 < |z - w| < r$. For z in this annulus,

$$(z-w)^m f(z) = a_{-m} + a_{-m+1}(z-w) + \cdots + a_{-1}(z-w)^{m-1} + a_0(z-w)^m + \cdots.$$

Since this series for $(z-w)^m f(z)$ is a power series about w, $(z-w)^m f(z)$ is analytic at w. Further, a routine calculation yields

$$\frac{f'(z)}{f(z)} = \frac{-m}{z-w} + \frac{[(z-w)^m f(z)]'}{(z-w)^m f(z)}.$$

This is the Laurent expansion of $f'(z)/f(z)$ about w. The coefficient of the $1/(z-w)$ term shows that $\operatorname{Res}_w f'(z)/f(z) = -m$.

By a similar argument, if f has a zero of order n at w, then in some annulus about w we have $f'(z)/f(z) = n/(z-w) + g(z)$, where $g(z) = (z-w)^{-n}f(z)$. Then

$$\operatorname{Res}_{w} \frac{f'(z)}{f(z)} = n.$$

Now evaluate $\oint_C f'(z)/f(z)\, dz$ by the residue theorem. At each pole of order m of f, $f'(z)/f(z)$ has residue $-m$; and at each zero of order n of f, $f'(z)/f(z)$ has residue n. The sum of these residues is therefore $Z - P$, where Z is the number of zeros of f enclosed by C (each counted as many times as the multiplicity of the zero) and P is the number of poles of f within C (again, each counted according to its multiplicity). $\quad \blacksquare$

PROBLEMS FOR SECTION 22.5

1. Evaluate $\oint_C z/(1 + z^2)\, dz$, with C the circle $|z| = 2$, (a) by the residue theorem; (b) by the argument principle.

2. Evaluate $\oint_C \tan(z)\, dz$, with C the circle $|z| = \pi$, (a) by the residue theorem; (b) by the argument principle.

3. Evaluate

$$\oint_C \frac{z + 1}{z^2 + 2z + 4}\, dz,$$

with C the circle $|z| = 1$, (a) by the residue theorem; (b) by the argument principle.

4. Let $p(z) = (z - z_1)(z - z_2) \cdots (z - z_n)$, with $z_1, \ldots, z_n$ distinct complex numbers. Let C be a piecewise-smooth simple closed curve enclosing $z_1, \ldots, z_n$. Prove that

$$\oint_C \frac{p'(z)}{p(z)}\, dz = 2\pi i n$$

(a) by the residue theorem; (b) by the argument principle.

In Problems 5, 6, and 7, assume *Rouche's theorem*: Let f and g be analytic on and inside a simple closed curve C, and suppose that $|f(z) - g(z)| < |f(z)|$ for all z on C. Assume that C passes through no zero or pole of f or g. Then f and g have the same number of zeros (counting multiplicities) enclosed by C.

5. Use Rouche's theorem to find the number of zeros of $z^4 + 4z + 1$ enclosed by the unit circle $|z| = 1$. *Hint:* Let $f(z) = 4z$ and $g(z) = z^4 + 4z + 1$ in Rouche's theorem.

6. Let n be a positive integer. Use Rouche's theorem to show that the equation $e^z = 3z^n$ has n solutions enclosed by the unit circle $|z| = 1$. *Hint:* Let $f(z) = -3z^n$ and $g(z) = e^z - 3z^n$ in Rouche's theorem.

7. Use Rouche's theorem to prove that a polynomial of degree $n \geq 1$, with complex coefficients, has exactly n roots (counting multiplicities). *Hint:* Let $p(z) = a_0 + a_1 z + \cdots + a_n z^n$. Apply Rouche's theorem with $f(z) = z^n$ and $g(z) = p(z)$.

ADDITIONAL PROBLEMS

In each of Problems 1 through 30, evaluate the integral. C always denotes a piecewise-smooth simple closed curve, oriented counterclockwise.

1. $\oint_C \dfrac{z^2}{(z - 1)^4}\, dz$, $\quad C$ enclosing 1

2. $\displaystyle\int_0^{2\pi} \dfrac{1}{4 + \cos^2(\theta)}\, d\theta$

3. $\displaystyle\int_{-\infty}^{\infty} \dfrac{x^2}{(x^2 + 3)(x^2 + 2)}\, dx$

4. $\oint_C \dfrac{e^{2z}}{z(z - i)}\, dz$, $\quad C$ enclosing 0 and i

5. $\oint_C \dfrac{\sin(z)\cos(2z)}{1 + z^2}\, dz$, $\quad C$ enclosing i but not $-i$

6. $\displaystyle\int_0^{2\pi} \dfrac{\sin^2(\theta)}{2 + \sin(\theta)}\, d\theta$

7. $\oint_C \dfrac{\cosh(3z)}{z(z^2 + 9)}\, dz$, $\quad C$ the circle $|z - 2i| = 9$

8. $\displaystyle\int_0^{2\pi} \dfrac{\cos(\theta) - \sin(\theta)}{2 - \cos(\theta)}\, d\theta$

9. $\displaystyle\int_{-\infty}^{\infty} \dfrac{x^2\cos(x)}{1 + x^6}\, dx$

10. $\oint_C \dfrac{(z - i)^2}{\sin^2(z)}\, dz$, $\quad C$ the square with vertices $4 + 4i, 4 - 4i, -4 + 4i, -4 - 4i$

11. $\displaystyle\int_0^{2\pi} \dfrac{1}{1 + \cos^2(\theta)}\, d\theta$

12. $\oint_C \dfrac{e^z(z^2 - 4)^2}{(z + i)^2}\, dz$, $\quad C$ the circle $|z - 1 + 2i| = 4$

13. $\oint_C \dfrac{\cosh(2z)}{(z-i)(1+z^2)}\,dz,$ C the circle $|z-i| = 1$

14. $\oint_C \dfrac{1}{z^4}\,e^{2z}\,dz,$ C the circle $|z - \pi i| = 2\pi$

15. $\oint_C \dfrac{\sin(z)\sinh(z)}{z^3}\,dz,$ C the circle $|z+1| = \dfrac{3}{2}$

16. $\displaystyle\int_0^{2\pi} \dfrac{\sin^2(\theta)}{3+\cos^2(\theta)}\,d\theta$

17. $\oint_C \dfrac{ze^{-z}}{z^4+i}\,dz,$ C the circle $|z-5| = 1$

18. $\oint_C \dfrac{1}{1+z^5}\,dz,$ C the circle $|z+1| = 1$

19. $\displaystyle\int_{-\infty}^{\infty} \dfrac{x^4}{1+x^6}\,dx$

20. $\displaystyle\int_0^{2\pi} \dfrac{1}{4+2\sin^2(\theta)}\,d\theta$

21. $\oint_C \dfrac{(z-2i)^2}{z^2-2z+2}\,dz,$ C the circle $|z| = 8$

22. $\displaystyle\int_{-\infty}^{\infty} \dfrac{1}{x^2-x+5}\,dx$

23. $\oint_C \dfrac{1}{z^2}\,e^{\sin(z)}\,dz,$ C the circle $|z| = 2$

24. $\oint_C \dfrac{(z+1)\cosh(z)}{(z-i)(z+2i)}\,dz,$ C the circle $|z-3i| = \dfrac{3}{2}$

25. $\oint_C \dfrac{2z-3}{(z+i)^3}\,dz,$ C the circle $|z-2i| = 4$

26. $\displaystyle\int_0^{2\pi} \dfrac{1}{\cos^2(\theta)+\sin^2(\theta)}\,d\theta$

27. $\displaystyle\int_{-\infty}^{\infty} \dfrac{x^2-4}{(x^2+4)^2}\,dx$

28. $\displaystyle\int_{-\infty}^{\infty} \dfrac{\cos(4x)}{(x^2+1)^2}\,dx$

29. $\oint_C \dfrac{z^2-z^4}{(z-1)\cos(z)}\,dz,$ C the circle $|z| = 4$

30. $\oint_C \dfrac{z-\cos(4iz)}{(z^2+1)(z^2-1)}\,dz,$ C the circle $|z+i| = 1$

31. Suppose that g and h are analytic at z_0, $g(z_0) \neq 0$, $h''(z_0) \neq 0$, and $h(z_0) = h'(z_0) = 0$. Prove that $g(z)/h(z)$ has a pole of order 2 at z_0 and that

$$\operatorname*{Res}_{z_0} \frac{g(z)}{h(z)} = 2\,\frac{g'(z_0)}{h''(z_0)} - \frac{2}{3}\,\frac{g(z_0)h^{(3)}(z_0)}{[h''(z_0)]^2}.$$

32. Suppose that g and h are analytic at z_0, $g'(z_0) \neq 0$, and $h^{(3)}(z_0) \neq 0$, while $g(z_0) = h(z_0) = h'(z_0) = h''(z_0) = 0$. Prove that $g(z)/h(z)$ has a pole of order 2 at z_0 and that

$$\operatorname*{Res}_{z_0} \frac{g(z)}{h(z)} = 3\,\frac{g''(z_0)}{h^{(3)}(z_0)} - \frac{3}{2}\,\frac{g'(z_0)h^{(4)}(z_0)}{[h^{(3)}(z_0)]^2}.$$

33. Let α and β be positive numbers. Show that

$$\int_0^{\infty} \frac{x\sin(\alpha x)}{x^4+\beta^4}\,dx = \frac{\pi}{2\beta^2}\,e^{-\alpha\beta/\sqrt{2}}\sin\!\left(\frac{\alpha\beta}{\sqrt{2}}\right).$$

34. Let $\alpha > \beta > 0$. Show that

$$\int_0^{\pi} \frac{1}{[\alpha+\beta\cos(\theta)]^2}\,d\theta = \frac{\alpha\pi}{(\alpha^2-\beta^2)^{3/2}}.$$

CHAPTER 23

Conformal Mappings

23.0 Introduction

In the calculus of real functions, we may gain some insight into the behavior of a function by sketching its graph. By the same token, it is sometimes useful to think of a complex function f from a geometric point of view. Let $w = f(z)$ and make two copies of the complex plane, as in Figure 23.1. One copy is used for complex numbers denoted z; the other is used for complex numbers w corresponding to z through the function f. As z traces out certain curves or domains in the z-plane, we observe the curve or set traced out by the image point $w = f(z)$ in the w-plane.

When looked at in this way, a complex function f is called a *mapping*, or *transformation*. The point $w = f(z)$ in the w-plane is the *image* of z in the z-plane. Given

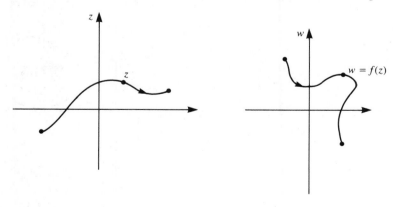

Figure 23.1

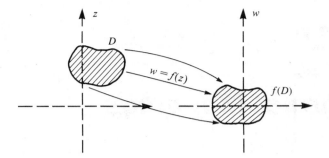

Figure 23.2

a domain D in the z-plane, the set of all images $f(z)$ of points z in D is denoted $f(D)$ and is called the image of D in the w-plane (Figure 23.2).

Thinking of a function as a mapping is important for more than simply analyzing functions. It can provide a powerful tool for solving problems. To illustrate, suppose we are able to solve some problem for a simple domain such as a disk D. If we want to solve the problem for another domain D^* and are able to find a suitable mapping of D onto D^*, we may be able to map the solution for D to a solution for D^*. This is an effective technique for solving certain boundary value problems in partial differentiation equations, as we will see in the next chapter.

In this chapter, we will discuss various types of mappings and their properties and see how we can sometimes construct mappings between given domains.

23.1 *Some Familiar Functions as Mappings*

Before studying mappings in general, we will look at some familiar functions viewed as mappings. First, we will develop some terminology.

Let f be a function. Corresponding to a complex number z, let $w = f(z)$. Think of z as a point in the z-plane and w as a point in the w-plane, as in Figure 23.1. Suppose that

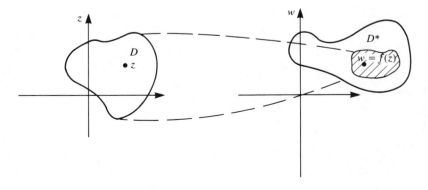

Figure 23.3. z in D maps to $w = f(z)$ in D^*.

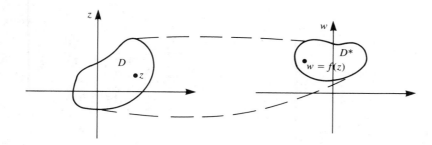

Figure 23.4. f is onto: every w in D^* is the image of some z in D.

D is a domain in the z-plane and D^* is a domain in the w-plane. We say that f *maps D into D^** if $f(z)$ is in D^* for each z in D. This is illustrated in Figure 23.3.

The points $f(z)$, for z in D, may not cover all of D^*. That is, there may be points of D^* which are not the image of any point in D. If, however, every w in D^* is the image of at least one z in D, we say that f is an *onto mapping* and *maps D onto D^**. This is illustrated in Figure 23.4.

We say that f is *one-to-one* if different numbers in D map to different images in D^*. This means that $f(z_1) \neq f(z_2)$ if $z_1 \neq z_2$.

We will now look at some familiar functions as mappings.

EXAMPLE 23.1

Let $f(z) = z^2$. Write $w = z^2$ and observe that

$$|w| = |z^2| = |z|^2$$

and

$$\arg(w) = \arg(z^2) = 2 \arg(z) \quad \text{if} \quad z \neq 0.$$

Thus, each nonzero z is mapped to a w having magnitude $|z|^2$ and twice the argument of z. Geometrically, we may think of the mapping as rotating points counterclockwise to double their argument, while squaring their magnitude. For example, $1 + i$ maps to $(1 + i)^2$, or $2i$. Notice that $1 + i$ has argument $\pi/4$, while $2i$ has argument $\pi/2$.

As a mapping of the entire complex z-plane to the entire complex w-plane, $w = z^2$ is onto. Given any w, there is some z with $w = z^2$. This mapping is not one-to-one because, for any $w \neq 0$, there are two values of z with $w = z^2$.

Suitably restricted, the mapping becomes one-to-one. For example, let D be the set of z defined by $0 \leq \text{Arg}(z) \leq \pi/4$, as shown in Figure 23.5. This set maps onto the right quarter-plane $0 \leq \text{Arg}(w) \leq \pi/2$ in the w-plane. This mapping is onto and is also one-to-one. Given any w in this right quarter-plane, there is exactly one z in D such that $w = z^2$.

It is often instructive to examine the effects of a mapping on particular curves in the z-plane. For example, the circle $|z| = r$ is mapped to the circle $|w| = r^2$ by the mapping $w = z^2$. However, as z traverses the circle $|z| = r$ once counterclockwise, w traverses $|w| = r^2$ twice counterclockwise because $\arg(w)$ increases by 4π if $\arg(z)$ increases by 2π.

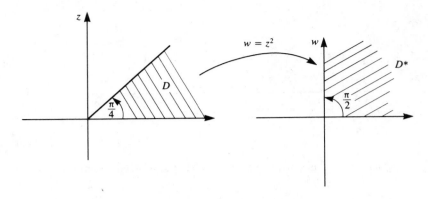

Figure 23.5. $w = z^2$ maps D one-to-one onto D^*.

A straight line parallel to the real or imaginary axis in the z-plane is mapped to a parabola in the w-plane. To see this, consider the vertical line $\text{Re}(z) = a$, with a constant. This line is shown in Figure 23.6. Write $z = x + iy$ and $w = u + iv$. On this line, $z = a + iy$, so the image of z is

$$w = u + iv = (a + iy)^2 = a^2 - y^2 + 2ayi.$$

Then

$$u = a^2 - y^2 \quad \text{and} \quad v = 2ay;$$

hence,

$$v^2 = 4a^2y^2 = 4a^2(a^2 - u),$$

which is the equation of a parabola in the uv-plane. This parabola is also shown in Figure 23.6.

Similarly, a horizontal line has equation $\text{Im}(z) = b$. With $z = x + ib$,

$$w = (x + ib)^2 = x^2 - b^2 + 2bxi,$$

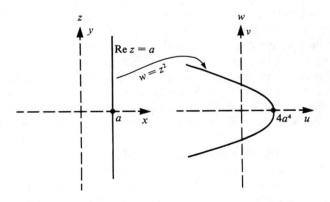

Figure 23.6. $w = z^2$ maps vertical lines to parabolas opening left.

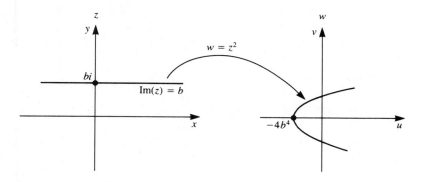

Figure 23.7

so

$$u = x^2 - b^2 \quad \text{and} \quad v = 2bx,$$

or

$$v^2 = 4b^2x^2 = 4b^2(b^2 + u),$$

again a parabola in the uv-plane. This parabola is shown in Figure 23.7. ∎

EXAMPLE 23.2

Consider the mapping $w = e^z$. Write

$$w = u + iv = e^{x+iy} = e^x\cos(y) + ie^x\sin(y).$$

Then

$$u = e^x\cos(y) \quad \text{and} \quad v = e^x\sin(y).$$

Then $|w| = e^x$ and $\arg(w) = y$. As a mapping of the z-plane to the w-plane, $w = e^z$ is not onto (no z maps to the origin) and is also not one-to-one (z and points $z + 2n\pi i$ have the same image for any integer n).

Consider a vertical line $\text{Re}(z) = a$. The image of this line is

$$w = u + iv = e^a\cos(y) + ie^a\sin(y).$$

Then

$$u^2 + v^2 = e^{2a},$$

so the vertical line $\text{Re}(z) = a$ maps to a circle of radius e^{2a} about the origin in the w-plane (Figure 23.8). As y varies from $-\infty$ to ∞, $z = a + iy$ varies once over the vertical line and $w = u + iv$ makes one complete circuit of the circle every time y varies by 2π. Thus, the image point traverses the circle infinitely many times as z moves over the line.

Points on a horizontal line $\text{Im}(z) = b$ map to

$$w = u + iv = e^x\cos(b) + ie^x\sin(b).$$

An argument of w is b, which is constant. As x varies from $-\infty$ to ∞, e^x varies over the interval $(0, \infty)$. Therefore, w varies over the half-line shown in Figure 23.9, emanating

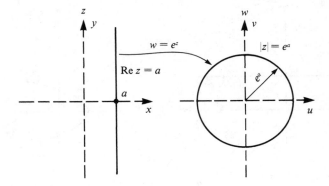

Figure 23.8. $w = e^z$ wraps a vertical line around a circle, covering the circle once for every interval of length 2π.

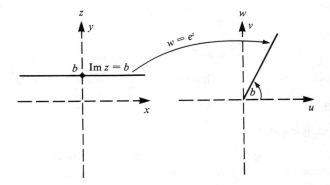

Figure 23.9. $w = e^z$ maps horizontal lines to half-rays from the origin.

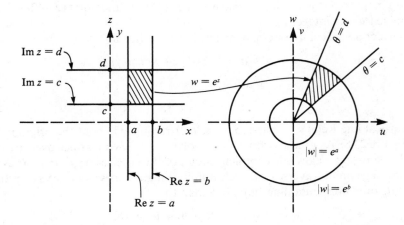

Figure 23.10. $w = e^z$ maps the rectangle shown to a wedge bounded by two half-rays and two circles.

from the origin and making an angle b with the positive real axis. Thus, vertical lines map to circles, and horizontal lines map to half-lines.

Using these results, we can find the image of any rectangle with sides parallel to the axes. Suppose R is the rectangle $a \leq x \leq b, c \leq y \leq d$. Think of R as bounded by the lines $x = a, x = b, y = c,$ and $y = d$ (Figure 23.10). The images of these lines are circles $x^2 + y^2 = e^{2a}, x^2 + y^2 = e^{2b}$ and half-lines making angles c and d with the positive real axis. These bound a sector R^* in the w-plane also shown in Figure 23.10. R^* is the image of R under the mapping $w = e^z$. ∎

EXAMPLE 23.3

Consider the mapping $w = \cos(z)$.

Certainly, this mapping is not one-to-one because z and $z + 2n\pi$ have the same image for any integer n. The mapping is onto. Given any w, we can find z such that $w = \cos(z)$. A vertical line $\operatorname{Re}(z) = a$ has image

$$w = u + iv = \cos(z) = \cos(a + iy)$$
$$= \cos(a)\cosh(y) - i\sin(a)\sinh(y).$$

Then

$$u = \cos(a)\cosh(y) \quad \text{and} \quad v = -\sin(a)\sinh(y).$$

Assuming that $\cos(a) \neq 0$ and $\sin(a) \neq 0$,

$$\frac{u^2}{\cos^2(a)} - \frac{v^2}{\sin^2(a)} = 1.$$

Since $\cosh(y) \geq 1, u + iv$ varies over one branch of this hyperbola in the w-plane as z varies over the line $\operatorname{Re}(z) = a$. The mapping is shown in Figures 23.11(a) and 23.11(b).

If $\cos(a) = 0, a = (2n + 1)\pi/2$ for some integer n. Then

$$u = 0 \quad \text{and} \quad v = (-1)^{n+1}\sinh(y).$$

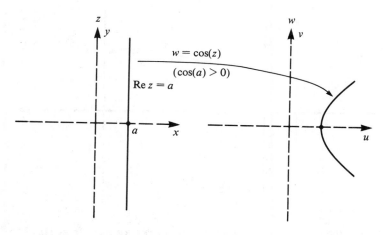

Figure 23.11(a). $w = \cos(z)$ maps a vertical line $\operatorname{Re} z = a$ to the right branch of a hyperbola if $\cos(a) > 0$ and $\sin(a) \neq 0$.

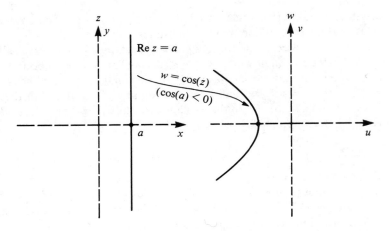

Figure 23.11(b). $w = \cos(z)$ maps the vertical line Re $z = a$ to the
left branch of a hyperbola if $\cos(a) < 0$ and $\sin(a) \neq 0$.

Now v varies between $-\infty$ and ∞ as y takes on all real values, and the point
$(u, v) = [0, (-1)^{n+1}\sinh(y)]$ varies over the imaginary axis in the w-plane, as shown
in Figure 23.12.

If $\sin(a) = 0$, $a = n\pi$ for some integer n. Now $v = 0$ and $u = (-1)^n\cosh(y)$. The
image point (u, v) now varies over the interval $u \geq 0$ (if n is even) or $u \leq -1$ (if n is odd)
as z varies over the line Re$(z) = a$. This is shown in Figures 23.13(a) and 23.13(b).

If Im$(z) = b$, $w = \cos(x)\cosh(b) - i \sin(x)\sinh(b)$, so

$$u = \cos(x)\cosh(b), \qquad v = -\sin(x)\sinh(b).$$

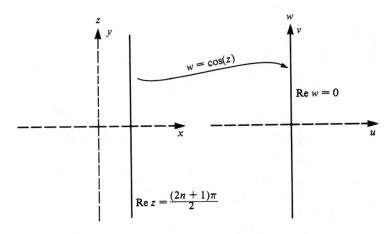

Figure 23.12. $w = \cos(z)$ maps vertical lines Re $z = (2n + 1)\pi/2$ to
the imaginary axis.

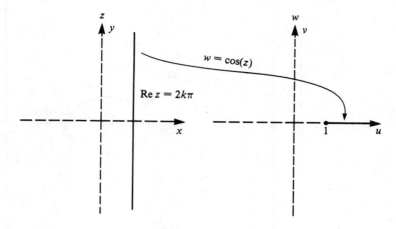

Figure 23.13(a). $w = \cos(z)$ maps vertical lines Re $z = 2k\pi$ to the
real axis from 1 to ∞.

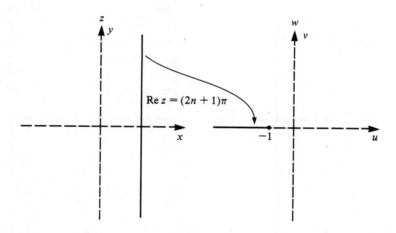

Figure 23.13(b). $w = \cos(z)$ maps vertical lines Re $z = (2k + 1)\pi$ to
the real axis from $-\infty$ to -1.

If $b \neq 0$, we can eliminate y and write

$$\frac{u^2}{\cosh^2(b)} + \frac{v^2}{\sinh^2(b)} = 1,$$

an ellipse with foci 1 and -1. This image is shown in Figure 23.14.

If $b = 0$, Im$(z) = 0$ is the real axis, and the image consists of points $w = \cos(x)$. The image point therefore varies over the interval $[-1, 1]$ on the real axis in the w-plane, as shown in Figure 23.15. ■

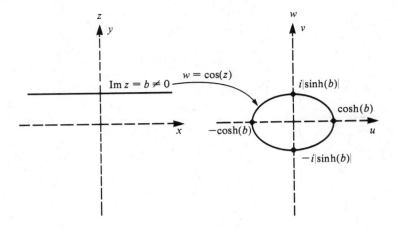

Figure 23.14. $w = \cos(z)$ maps horizontal lines Im $z = b \neq 0$
to ellipses.

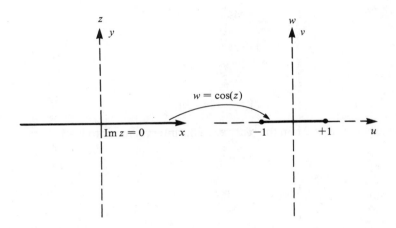

Figure 23.15. $w = \cos(z)$ maps the real axis Im $z = 0$ to the real
interval $[-1, 1]$.

EXAMPLE 23.4

Determine the image under the mapping $w = \sin(z)$ of the strip S consisting of all z
with $-\pi/2 < \mathrm{Re}(z) < \pi/2$ and $\mathrm{Im}(z) > 0$.

This strip is shown in Figure 23.16. The boundary of S consists of the vertical half-
lines $x = -\pi/2$ and $x = \pi/2$ for $y > 0$ and the segment $-\pi/2 \leq x \leq \pi/2$ on the real
axis. These boundary lines are not included in S and hence are shown as dashed lines in
Figure 23.16.

The strategy in determining the image of a set such as S is to find the image of each
part of the boundary of S. This determines the boundary curves of the image of S,
from which we can determine the image of S itself. The left boundary of S is the line

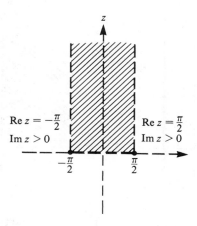

Figure 23.16. Strip bounded by vertical lines $x = -\pi/2$ and
$x = \pi/2$ and the x-axis.

$z = -\pi/2 + iy, 0 \leq y < \infty$. This line maps to

$$w = \sin\left(\frac{-\pi}{2} + iy\right) = \sin\left(\frac{-\pi}{2}\right)\cosh(y) + i\cos\left(\frac{-\pi}{2}\right)\sinh(y)$$

$$= -\cosh(y).$$

Since $0 \leq y < \infty$, $1 \leq \cosh(y) < \infty$, and $w = -\cosh(y)$ varies over $(-\infty, -1]$ on the real axis in the w-plane. Similarly, the right boundary line of S, $z = \pi/2 + iy$, maps to the interval $[1, \infty)$ on the real axis in the w-plane. The other part of the boundary of S is the interval $[-\pi/2, \pi/2]$ on the real axis. This interval maps to $[-1, 1]$ on the real axis in the w-plane.

In summary, as z moves over the boundary of S, the image point w covers the entire real axis in the w-plane. In fact, as we can follow from Figure 23.17, as z moves

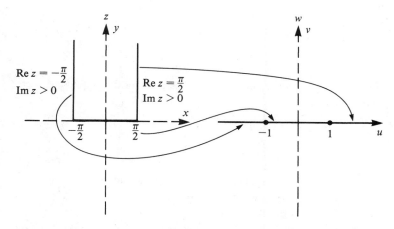

Figure 23.17. $w = \sin(z)$ maps $x = -\pi/2$, $y \geq 0$, to $u \leq -1$;
$-\pi/2 \leq x \leq \pi/2$ to $-1 \leq u \leq 1$; and $x = \pi/2$, $y \geq 0$, to $u \geq 1$.

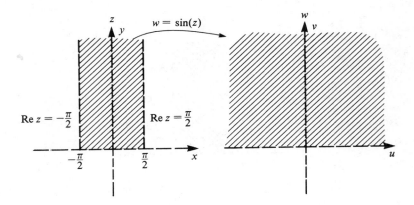

Figure 23.18. $w = \sin(z)$ maps the interior of the strip to the upper half-plane.

down the left boundary of S, w moves from $-\infty$ to -1 on the real axis in the w-plane; as z "turns the corner" and moves from $-\pi/2$ to $\pi/2$ on the real axis, w moves from -1 to 1; and as z "turns the other corner" and moves up the right boundary of S, w moves from 1 to ∞.

We now know that the real axis is the image of the boundary of S. The image of S must therefore be the upper half-plane or the lower half-plane, both of which have the real axis as boundary. Pick any point in S, say, i. Its image is $w = \sin(i) = i\sinh(1)$, which is in the upper half-plane. Thus, S maps to the upper half-plane $\text{Im}(w) > 0$ under the mapping $w = \sin(z)$. Figure 23.18 indicates the effect of this mapping on S. ∎

PROBLEMS FOR SECTION 23.1

1. In each of (a) through (e), find the image of the rectangle in the z-plane under the mapping $w = e^z$. In each case, sketch the rectangle and its image.

   (a) $0 < x < \pi, \quad 0 < y < \pi$

   (b) $-1 < x < 1, \quad \dfrac{-\pi}{2} < y < \dfrac{\pi}{2}$

   (c) $0 < x < 1, \quad 0 < y < \dfrac{\pi}{4}$

   (d) $1 < x < 2, \quad 0 < y < \pi$

   (e) $-1 < x < 2, \quad \dfrac{-\pi}{2} < y < \dfrac{\pi}{2}$

2. In each of (a) through (e), find the image of the rectangle in the z-plane under the mapping $w = \cos(z)$. In each case, sketch the rectangle and its image.

   (a) $0 < x < 1, \quad 1 < y < 2$

   (b) $\dfrac{\pi}{2} < x < \pi, \quad 1 < y < 3$

   (c) $0 < x < \pi, \quad \dfrac{\pi}{2} < y < \pi$

   (d) $\pi < x < 2\pi, \quad 1 < y < 2$

   (e) $0 < x < \dfrac{\pi}{2}, \quad 0 < y < 1$

3. Determine the images of lines parallel to the axes under the mapping $w = \sinh(z)$.

4. Determine the images of lines parallel to the axes under the mapping $w = \sin(z)$.

5. Determine the image of the sector $\pi/4 \le \arg(z) \le 5\pi/4$ under the mapping $w = z^2$.

6. Determine the image of the sector $\pi/6 < \arg(z) < \pi/3$ under the mapping $w = z^3$. Sketch the sector and its image under the mapping.

7. Show that the mapping $w = \frac{1}{2}(z + 1/z)$ maps the circle $|z| = r$ onto an ellipse with foci 1 and -1 in the w-plane.

8. Show that the mapping $w = \frac{1}{2}(z + 1/z)$ maps the half-line $\arg(z) = k$ (constant) onto a hyperbola with foci 1 and -1 in the w-plane.

9. Determine the image of $|z| > 1$ (exterior of the unit circle) under the mapping $w = 1/z$.

10. Determine the image of the right quarter-plane $\text{Re}(z) > 0$, $\text{Im}(z) > 0$ under the mapping $w = z^3$.

11. Consider the mapping $w = e^{\alpha z}$, where $\alpha = a + ib$ is a given number and a and b are nonzero.
    (a) Show that a horizontal line $\text{Im}(z) = c$ maps to the curve $r = e^d e^{a\theta/b}$ in polar coordinates, where $d = -(c/b)(a^2 + b^2)$.
    (b) Show that a vertical line $\text{Re}(z) = c$ maps to the curve $r = e^{-bd/a} e^{-b\theta/a}$ with d as in (a).
    (c) The image curves in (a) and (b) are logarithmic spirals. Sketch their graphs for $a = b = c = 1$.

12. Find the image of the upper half-plane $\text{Im}(z) > 0$ under the mapping $w = z^{1/2}$.

13. Show that the mapping $w = 1/z$ maps every straight line to a circle or straight line and every circle to a circle or straight line.

14. Determine the images of straight lines parallel to the axes under the mapping $w = \cosh(z)$.

15. Determine the image of the infinite strip $0 < \text{Im}(z) < 2\pi$ under the mapping $w = e^z$.

16. Determine the image under the mapping $w = \cos(z)$ of the set in the z-plane bounded by the rectangle with vertices αi, $-\alpha i$, $\pi + \alpha i$, and $\pi - \alpha i$. Here, α is a positive constant.

17. Determine the image under the mapping $w = z^{1/3}$ of the sector $\pi/9 \le \arg(z) \le \pi/3$.

23.2 Conformal Mappings and Linear Fractional Transformations _____

We will now concentrate on mappings which enjoy two important properties. We say that a mapping $w = f(z)$ *preserves angles* if curves intersecting at an angle θ in the z-plane map to curves intersecting at the same angle θ in the w-plane. As usual, we measure the angle between two curves at a point of intersection as the angle between their tangents at this point (shown in Figure 23.19).

A mapping *preserves orientation* if a counterclockwise rotation in the z-plane is mapped to a counterclockwise rotation in the w-plane. This means that, if a line L is rotated counterclockwise in the z-plane to form a new line L^*, the image $f(L)$ is rotated

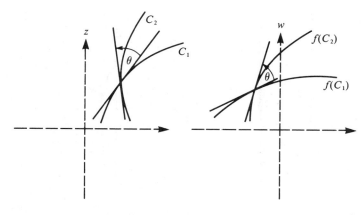

Figure 23.19. Angle-preserving mapping.

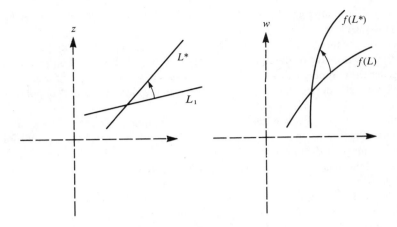

Figure 23.20. Orientation-preserving mapping.

counterclockwise in the w-plane to the image $f(L^*)$, as illustrated in Figure 23.20. A mapping which preserves orientation need not preserve angles but simply preserves the sense of rotation between lines and their images.

A *conformal mapping* is one which preserves both angles and orientation. The following theorem enables us to produce many examples of conformal mappings.

THEOREM 23.1

Suppose that f is analytic on a domain D in the z-plane and that $f'(z) \neq 0$ for z in D. Then $w = f(z)$ is a conformal mapping of D onto $f(D)$ in the w-plane.

Proof   Let z_0 be in D, and let C be a smooth curve in D through z_0. The image of C is a smooth curve $f(C)$ in the w-plane, and $f(C)$ passes through $w_0 = f(z_0)$, as shown in

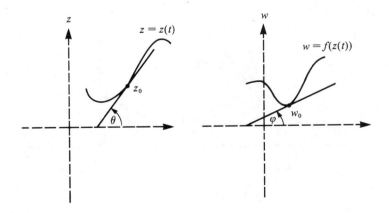

Figure 23.21

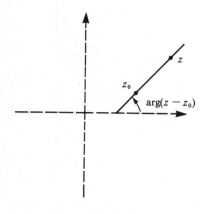

Figure 23.22

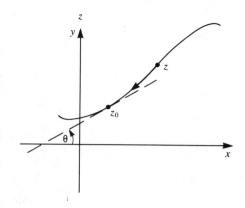

Figure 23.23. $\arg(z - z_0) \to \theta$ as $z \to z_0$ along C.

Figure 23.21. For any $w = f(z)$, calculate

$$w - w_0 = \frac{f(z) - f(z_0)}{z - z_0}(z - z_0).$$

Then

$$\arg(w - w_0) = \arg\left[\frac{f(z) - f(z_0)}{z - z_0}\right] + \arg(z - z_0). \qquad (23.1)$$

Now, $\arg(z - z_0)$ is the angle between the positive real axis and the line from z_0 to z (Figure 23.22), and $\arg(w - w_0)$ is the angle between the positive real axis in the w-plane and the line from w_0 to w. Now let $z \to z_0$. Then $\arg(z - z_0)$ approaches the angle θ between the positive real axis and the tangent to C at z_0 (Figure 23.23). At the same time, $\arg(w - w_0)$ approaches the angle φ between the positive real axis in the w-plane and the tangent to $f(C)$ at w_0. By equation (23.1), we have in the limit as $z \to z_0$ that

$$\varphi = f'(z_0) + \theta.$$

Therefore,

$$\varphi - \theta = f'(z_0).$$

This result is independent of C. If C^* is any other smooth curve through z_0, as in Figure 23.24, θ^* is the angle between the positive real axis and the tangent to C^* at z_0, and φ^* is the angle between the positive real axis in the w-plane and the tangent to $f(C^*)$ at w_0, we also have

$$\varphi^* - \theta^* = f'(z_0).$$

We conclude that

$$\varphi - \theta = \varphi^* - \theta^*,$$

or

$$\varphi - \varphi^* = \theta - \theta^*.$$

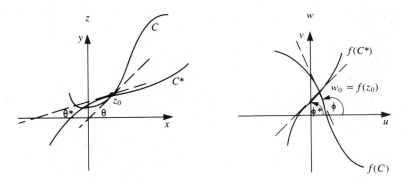

Figure 23.24. $\varphi - \theta = f'(z_0) = \varphi^* - \theta^*$.

But $\theta - \theta^*$ is the angle between the tangents to C and C^* at z_0, and $\varphi - \varphi^*$ is the angle between the tangents to $f(C)$ and $f(C^*)$ at w_0. Thus, f preserves angles. Further, f preserves orientation. If f reversed orientation, we would have obtained $\varphi - \varphi^* = \theta^* - \theta$. We conclude that f is a conformal mapping. ∎

Thus, for example, $w = \sin(z)$ is a conformal mapping of the strip $-\pi/2 <$ Re$(z) < \pi/2$ onto a set in the w-plane.

We will now consider an important class of mappings called *linear fractional transformations*, which are functions of the form

$$w = f(z) = \frac{az + b}{cz + d},$$

with a, b, c, and d constant and $ad - bc \neq 0$. Such functions are also called *Möbius transformations*, or *bilinear transformations*.

Compute

$$w' = \frac{ad - bc}{(cz + d)^2}.$$

Since $ad - bc \neq 0$ by assumption, $w'(z) \neq 0$ on any domain in which w' is defined (that is, $z \neq -d/c$). By Theorem 23.1, the transformation is conformal on any such domain. We will look at some examples of this type of mapping.

EXAMPLE 23.5

Let $w = z + b$, with b any complex number. Any mapping of this form is called a *translation*. Its effect is to shift z horizontally by Re(b) units and vertically by Im(b) units. For example, the mapping $w = z + 2 - i$ moves z to the right two units and downward one unit (Figure 23.25). This mapping sends

$$0 \rightarrow 2 - i, \quad 1 \rightarrow 3 - i, \quad i \rightarrow 2, \quad \text{and} \quad 4 + 3i \rightarrow 6 + 2i,$$

as shown in Figure 23.26. ∎

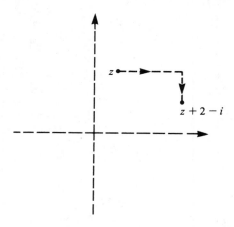

Figure 23.25. $w = z + 2 - i$ moves z two units right, one unit downward.

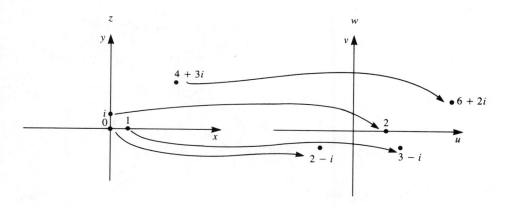

Figure 23.26

EXAMPLE 23.6

Let $w = az$, with $a \neq 0$. This mapping is called a *rotation-magnification*. First, observe that

$$|w| = |a||z|,$$

so the magnitude of the image point is the magnitude of z multiplied by the constant $|a|$. The vector from the origin to w in the w-plane has length $|a|$ times the length of the vector from the origin to z in the z-plane; hence the term magnification. Further,

$$\arg(w) = \arg(z) + \arg(a),$$

so the mapping rotates z counterclockwise by an angle $\arg(a)$, accounting for the name rotation. These effects are shown in Figure 23.27.

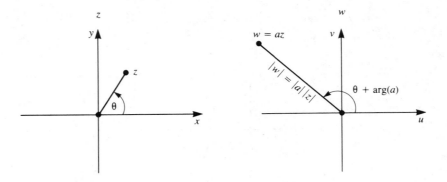

Figure 23.27

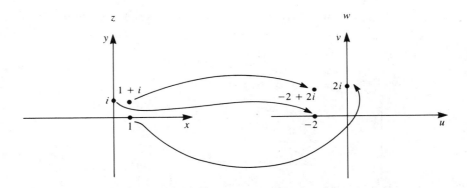

Figure 23.28

As a specific example, consider

$$w = 2iz.$$

Here, $a = 2i$. We have $|a| = 2$ and $\arg(a) = \pi/2$, so

$$|w| = 2|z| \quad \text{and} \quad \arg(w) = \arg(z) + \arg(2i) = \arg(z) + \frac{\pi}{2}.$$

The image point is twice as far from the origin as z is, and the line from the origin to w is the line from the origin to z rotated counterclockwise by $\pi/2$. For example,

$$1 \to 2i, \quad i \to -2, \quad \text{and} \quad 1 + i \to -2 + 2i,$$

as shown in Figure 23.28. ∎

If $|a| = 1$, the mapping $w = az$ is called a *pure rotation*. In this case, there is no magnification, and the mapping simply rotates points by an angle $\arg(a)$. If $\arg(a) = \theta$, we can write $a = e^{i\theta}$ (polar form), and a pure rotation by θ can be written

$$w = e^{i\theta}z.$$

EXAMPLE 23.7

Let $w = 1/z$ for $z \neq 0$. This mapping is called an *inversion*. We have

$$|w| = \frac{1}{|z|} \quad \text{and} \quad \arg(w) = \arg(1) - \arg(z) = -\arg(z).$$

If $z \neq 0$, $w = 1/z$ is obtained by moving $1/|z|$ units from the origin along the line from 0 to z, then reflecting across the real axis (Figure 23.29). The inversion map points inside the unit circle to points outside, and vice versa. Points on the unit circle are mapped to points on the unit circle, because $|w| = 1$ if $|z| = 1$. For example, $1 \rightarrow 1$ and $i \rightarrow 1/i = -i$ under this mapping. ∎

We will now show that any linear fractional transformation can be written as a sequence of mappings of these types: translation, rotation-magnification, and inversion. This means that these three transformations form the building blocks of all linear fractional transformations. Specifically, the linear fractional transformation $w = (az + b)/(cz + d)$ is the end result of the following sequence of individual mappings:

$$z \xrightarrow{\text{rotation-magnification}} cz \xrightarrow{\text{translation}} cz + d$$

$$\xrightarrow{\text{inversion}} \frac{1}{cz + d} \xrightarrow{\text{rotation-magnification}} \frac{bc - ad}{c} \frac{1}{cz + d}$$

$$\xrightarrow{\text{translation}} \frac{bc - ad}{c} \frac{1}{cz + d} + \frac{a}{c} = \frac{az + b}{cz + d}.$$

In solving certain types of problems, we will want to find a conformal mapping which maps a given domain or set of points onto another given set of points. The following three theorems will help us do this under certain circumstances.

THEOREM 23.2

Let z_1, z_2, and z_3 be distinct points in the z-plane, and let w_1, w_2, and w_3 be distinct points in the w-plane. Then there is a linear fractional transformation which maps

$$z_1 \rightarrow w_1, \qquad z_2 \rightarrow w_2, \quad \text{and} \quad z_3 \rightarrow w_3.$$

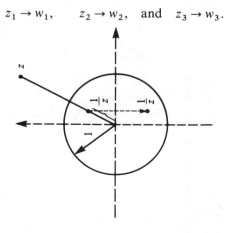

Figure 23.29. Inversion: $w = 1/z$.

Proof   We will show how to produce the linear fractional transformation which maps each z_j to w_j. This mapping is obtained by solving for w in terms of z in the equation

$$\frac{w_1 - w}{w_1 - w_2} \frac{w_3 - w_2}{w_3 - w} = \frac{z_1 - z}{z_1 - z_2} \frac{z_3 - z_2}{z_3 - z}. \tag{23.2}$$

Verification of this is left to the student.   ∎

EXAMPLE 23.8

Find a linear fractional transformation which maps

$$3 \to i, \qquad 1 - i \to 4, \quad \text{and} \quad 2 - i \to 6 + 2i.$$

Let $z_1 = 3$, $z_2 = 1 - i$, $z_3 = 2 - i$, $w_1 = i$, $w_2 = 4$, and $w_3 = 6 + 2i$ in equation (23.2). We get

$$\frac{i - w}{i - 4} \frac{2 + 2i}{6 + 2i - w} = \frac{3 - z}{2 + i} \frac{1}{2 - i - z}.$$

Solve for w in terms of z to get

$$w = \frac{(20 + 4i)z - (16i + 68)}{(6 + 5i)z - (22 + 7i)}.$$

It is routine to verify that this mapping sends each $z_j \to w_j$.   ∎

When dealing with mappings, it is often convenient to consider the extended complex plane formed by adjoining a point at infinity to the complex plane. This point is denoted ∞ and is thought of as the image of $z = -d/c$ under the linear fractional transformation $w = (az + b)/(cz + d)$. In this sense, zero maps to ∞ under the inversion $w = 1/z$.

The point at infinity can be visualized geometrically by looking at the stereographic projection of the complex plane to the sphere. In Figure 23.30, a sphere of

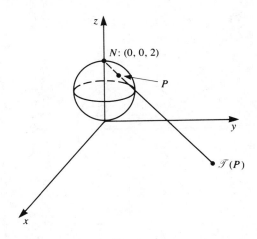

Figure 23.30

radius 1 is placed on the xy-plane with south pole at the origin and z-axis through the north pole N having coordinates $(0, 0, 2)$. Given a point P on the sphere with $P \neq N$, the line $\overline{NP}$ intersects the complex plane in exactly one point $\mathcal{T}(P)$. The mapping $\mathcal{T}$ associates with every point of the sphere (except N) a unique complex number, and conversely. $\mathcal{T}$ is called the *stereographic projection* of the sphere (minus the north pole) onto the plane. The point at infinity, which we have just adjoined to the complex plane, corresponds to N under this projection. As can be seen from Figure 23.30, as P moves along the sphere toward N, the line $\overline{NP}$ intersects the complex plane further from the origin; hence, it is natural to associate ∞ with N. In this way, we can envision the extended complex plane as the entire sphere, and the stereographic projection is a one-to-one mapping of the sphere onto the extended complex plane.

We will now show that we can map any three given complex numbers to two given complex numbers and ∞.

THEOREM 23.3

Let z_1, z_2, and z_3 be distinct complex numbers, and let w_1 and w_2 be distinct complex numbers. Then there is a linear fractional transformation which maps

$$z_1 \to w_1, \qquad z_2 \to w_2, \quad \text{and} \quad z_3 \to \infty. \quad \blacksquare$$

The required transformation is obtained by solving for w in the equation

$$\frac{w_1 - w}{w_1 - w_2} = \frac{z_1 - z}{z_1 - z_2} \frac{z_3 - z_2}{z_3 - z}, \tag{23.3}$$

obtained from equation (23.2) by deleting the factor containing w_3.

EXAMPLE 23.9

Find a linear fractional transformation mapping

$$i \to 4i, \qquad 1 \to 3 - i, \quad \text{and} \quad 2 + i \to \infty.$$

In equation (23.3), let $z_1 = i$, $z_2 = 1$, $z_3 = 2 + i$, $w_1 = 4i$, and $w_2 = 3 - i$. We get

$$\frac{4i - w}{-3 + 5i} = \frac{i - z}{i - 1} \frac{1 + i}{2 + i - z}.$$

Solve this equation for w to get

$$w = \frac{-1 + 3i + (5 - i)z}{2 + i - z}. \quad \blacksquare$$

Often, equation (23.3) is easier to work with than equation (23.2). Given a choice, we usually try to map one of the three given points to infinity and use (23.3) to construct the mapping, simply because equation (23.3) is easier to solve for w than equation (23.2) is.

THEOREM 23.4

A linear fractional transformation always maps a circle to a circle or straight line and a straight line to a circle or straight line.

Proof   It is geometrically apparent that any translation or rotation-magnification maps a circle to a circle and a straight line to a straight line. The image of a straight line or circle under an inversion is not so obvious, so we will investigate these.

First, observe that any circle or straight line in the plane has equation

$$A(x^2 + y^2) + Bx + Cy + R = 0, \tag{23.4}$$

in which A, B, C, and R are real numbers. This locus is a circle if $A \neq 0$ and a straight line if $A = 0$ but B and C are not both zero. Write $z = x + iy$. In terms of z, equation (23.4) can be written

$$A|z|^2 + B\frac{z + \bar{z}}{2} + C\frac{z - \bar{z}}{2i} + R = 0.$$

Let $w = 1/z$. Then $z = 1/w$, and the image of this locus in the w-plane has equation

$$A\left|\frac{1}{w}\right|^2 + \frac{1}{2}B\left(\frac{1}{w} + \frac{1}{\bar{w}}\right) + \frac{C}{2i}\left(\frac{1}{w} - \frac{1}{\bar{w}}\right) + R = 0.$$

Multiply this equation by $w\bar{w}$ and use the fact that $w\bar{w} = |w|^2$ to get

$$R|w|^2 + B\frac{w + \bar{w}}{2} - C\frac{w - \bar{w}}{2i} + A = 0.$$

The locus of this equation is a circle (if $R \neq 0$) or a straight line (if $R = 0$ and B and C are not both zero). Since any linear fractional transformation is a composition of translations, rotation-magnifications, and rotations, this completes the proof. ∎

Note that a translation or a rotation-magnification actually maps circles to circles and straight lines to straight lines. However, an inversion may map a circle to a circle or straight line and a straight line to a circle or straight line (see Example 23.11).

EXAMPLE 23.10

Consider the linear fractional transformation $w = iz + 3 - 2i = i(z - 2) + 3$. This is a translation (by 2) followed by a rotation ($z - 2$ multiplied by i) followed by another translation (by 3). Since this mapping does not involve an inversion, we expect it to map circles to circles and straight lines to straight lines.

As a specific example, let K be the circle $(x - 2)^2 + y^2 = 9$, with radius 3 and center $(2, 0)$. Write this equation as

$$x^2 + y^2 - 4x - 5 = 0,$$

or, in complex notation,

$$|z|^2 - 4\frac{z + \bar{z}}{2} - 5 = 0.$$

Since $w = iz + 3 - 2i$, $z = -i(w - 3 + 2i)$ and the last equation yields (after some algebra, which we omit)

$$|w|^2 - 6\frac{w + \bar{w}}{2} = 0.$$

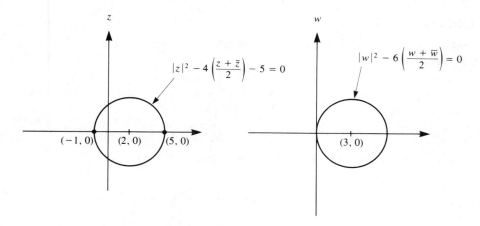

Figure 23.31

With $w = u + iv$, this equation can be written

$$(u - 3)^2 + v^2 = 9,$$

the equation of a circle of radius 3 and center (3, 0) in the uv-plane. From Figure 23.31, this is what we would expect if we translate z by two units to the left, rotate by $\pi/2$ radians, then translate the resulting point by three units to the right. ∎

EXAMPLE 23.11

We will examine the effects of the inversion $w = 1/z$ on the vertical line $\text{Re}(z) = a \neq 0$. This line may be written $z = a + iy$, and its image is

$$w = \frac{1}{a + iy} = \frac{a}{a^2 + y^2} - \frac{y}{a^2 + y^2}\, i.$$

If $w = u + iv$, $u = a/(a^2 + y^2)$ and $v = y/(a^2 + y^2)$. It is routine to check that

$$\left(u - \frac{1}{2a}\right)^2 + v^2 = \frac{1}{4a^2},$$

or

$$\left|w - \frac{1}{2a}\right| = \frac{1}{2a},$$

the equation of a circle of radius $1/2a$ about $1/2a$. Thus, the inversion $w = 1/z$ maps a vertical line $\text{Re}(z) = a \neq 0$ to a circle, not to another straight line. ∎

We are now prepared to consider the problem of constructing a mapping between given domains. This will occupy the next section.

PROBLEMS FOR SECTION 23.2

In each of Problems 1 through 10, construct a linear fractional transformation mapping the given points to the indicated images.

1. $1 \to 1$, $2 \to -i$, $3 \to 1 + i$

2. $i \to i$, $1 \to -i$, $2 \to 0$

3. $1 \to 1 + i$, $2i \to 3 - i$, $4 \to \infty$

4. $-5 + 2i \to 1$, $3i \to 0$, $-1 \to \infty$

5. $6 + i \to 2 - i$, $i \to 3i$, $4 \to -i$

6. $10 \to 8 - i$, $3 \to 2i$, $4 \to 3 - i$

7. $1 \to 6 - 4i$, $1 + i \to 2$, $3 + 4i \to \infty$

8. $8 - i \to 1 - i$, $i \to 1 + i$, $3i \to \infty$

9. $2i \to \frac{1}{2}$, $4i \to \frac{1}{3}$, $i \to 6i$

10. $-1 \to -i$, $i \to 2 - 3i$, $1 - 2i \to \infty$

In each of Problems 11 through 18, determine the image of the given circle or straight line under the given linear fractional transformation.

11. $w = \dfrac{2z + i}{z - i}$; $\text{Im}(z) = 2$

12. $w = 2i\dfrac{1}{z}$; $\text{Re}(z) = -4$

13. $w = 2iz - 4$; $\text{Re}(z) = 5$

14. $w = \dfrac{z - i}{iz}$; $\dfrac{z + \bar{z}}{2} + \dfrac{z - \bar{z}}{2i} = 4$

15. $w = \dfrac{z - 1 + i}{2z + 1}$; $|z| = 4$

16. $w = 3z - i$; $|z - 4| = 3$

17. $w = \dfrac{2z - 5}{z + i}$; $z + \bar{z} - 3\dfrac{z - \bar{z}}{2i} - 5 = 0$

18. $w = \dfrac{(3i + 1)z - 2}{z}$; $|z - i| = 1$

19. Determine the image of any line $\text{Im}(z) = a \neq 0$ under the inversion $w = 1/z$.

20. Let $w = f(z)$ be a conformal mapping of a domain D in the z-plane onto a domain D^* of the w-plane. Let $w^* = g(w)$ be a conformal mapping of D^* onto a domain D^{**} in the w^*-plane. Show that $w^* = g(f(z))$ is a conformal mapping of D onto D^{**}.

21. Prove that the mapping $w = \bar{z}$ is not conformal.

22. Prove that a composition of two linear fractional transformations is a linear fractional transformation.

23. Prove that the inverse of a linear fractional transformation is a linear fractional transformation.

24. Let

$$w = a\frac{z - b}{z - \bar{b}},$$

with $|a| = 1$ and $\text{Im}(b) > 0$. Show that this transformation maps the real line in the z-plane onto the unit circle $|w| = 1$. Show that points in the upper half of the z-plane map to points in the unit disk in the w-plane.

25. Consider the linear fractional transformation

$$w = a\frac{z - b}{\bar{b}z - 1},$$

with $|b| < 1$ and $|a| = 1$. Show that this transformation maps the unit circle $|z| = 1$ onto the unit circle $|w| = 1$. Show that points in the disk $|z| < 1$ map to points in the disk $|w| < 1$.

26. Show that there is no linear fractional transformation which maps the disk $|z| < 1$ onto the set of points bounded by the ellipse $\frac{1}{4}u^2 + v^2 = \frac{1}{16}$.

In Problems 27 and 28, the setting is the extended complex plane, with the point at infinity adjoined to the complex plane.

27. A point z_0 is called a *fixed point* of a mapping f if $f(z_0) = z_0$. Suppose that f is a linear fractional transformation which is not a translation or the identity $f(z) = z$. Prove that f must have either one or two fixed points. Why does this result fail if f is a translation?

28. Let f be a linear fractional transformation with three fixed points. Prove that f must be the identity transformation $f(z) = z$.

23.3 Construction of Mappings Between Given Domains

In solving certain types of problems, one strategy is to solve the problem for a relatively simple domain D, such as a disk or half-plane, and then conformally map this domain to the domain D^* on which the solution is sought. Such a mapping, applied to the solution for D, yields a solution for D^*. In order to implement this strategy, we must know how to construct a conformal mapping between two given domains. The following theorem suggests that this can be done, at least in theory, in a very general setting.

THEOREM 23.5   Riemann Mapping Theorem

Let $\mathscr{D}$ be a simply connected domain in the w-plane, and assume that $\mathscr{D}$ is not the entire plane. Then there exists a conformal mapping f which maps the unit disk $|z| < 1$ one-to-one onto $\mathscr{D}$. ∎

That is, we can map the unit disk conformally onto any simply connected domain which is not the entire w-plane.

If we have two simply connected domains D and D^* and want to map D onto D^*, we can attempt to map D onto the unit disk (using the inverse of the map of the unit disk onto D), then map the unit disk onto D^*. The composition of these mappings is shown in Figure 23.32 and is a mapping of D onto D^*.

In practice, constructing such mappings may be quite difficult. In this section, we will work with linear fractional transformations and other mappings we have constructed from familiar functions such as e^z and $\cos(z)$. We will also discuss the Schwarz-Christoffel transformation, which maps the upper half-plane onto the interior of a polygon.

In attempting to construct a conformal mapping between two given domains, the following observation is very useful. Any conformal mapping of a domain D onto a domain D^* will take the boundary C of D onto the boundary C^* of D^*. Thus, we first attempt to produce a conformal mapping sending C onto C^*. Now, C^* separates the

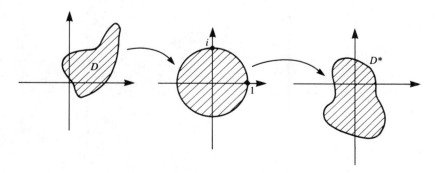

Figure 23.32. Mapping D onto D^* through the unit disk.

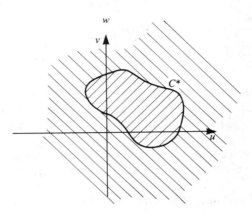

Figure 23.33. Complementary domains bounded by C^*.

w-plane into two domains, called *complementary domains* (see Figure 23.33), and C^* bounds both. One of these complementary domains is D^*. To determine which, choose any z_0 in D. If $f(z_0)$ is in D^*, f maps D onto D^* [Figure 23.34(a)]. If $f(z_0)$ is in the complement of D^*, f maps D onto the complement of D^*. In this event, $1/f$ maps D onto D^* [Figure 23.34(b)]. The following examples illustrate these ideas.

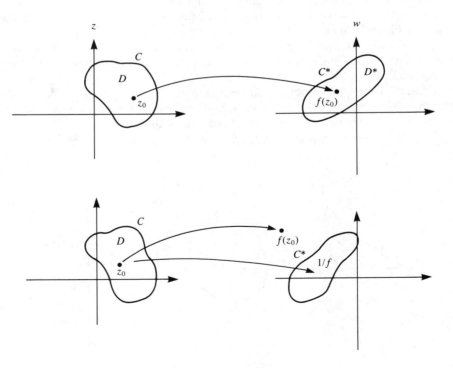

Figure 23.34

EXAMPLE 23.12

Map the unit disk onto the disk $|w| < 4$.

Obviously, the magnification $w = 4z$ will expand the unit disk to a disk of radius 4 (Figure 23.35). Notice that points on the boundary circle $|z| = 1$ of the unit disk map to points on the boundary $|w| = 4$ of the target disk in the w-plane. Further, points $|z| < 1$ are mapped to points $|w| < 4$. ∎

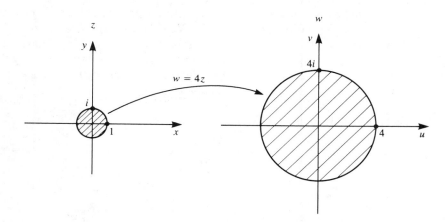

Figure 23.35. Mapping of $|z| < 1$ onto $|w| < 4$.

EXAMPLE 23.13

Map $|z| < 1$ to $|w| > 4$.

Now, we want to map the unit disk to the exterior of the disk $|w| = 4$, which is the complement of the disk $|w| < 4$ of Example 23.12. The domains are shown in Figure 23.36. We know that the inversion $w = 1/z$ maps the interior of the unit disk to the exterior of the unit disk. Thus, we need only multiply this mapping by 4 to expand

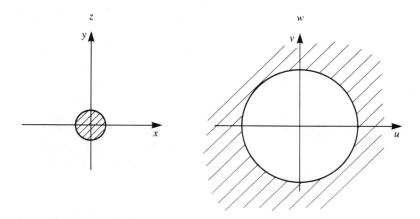

Figure 23.36. $w = 4/z$ maps $|z| < 1$ onto $|w| > 4$.

the boundary circle to $|w| = 4$ in the w-plane. A mapping which achieves this objective is $w = 4/z$. If $|z| = 1$, $|w| = 4$, so the unit circle maps to the circle $|w| = 4$ bounding the domain $|w| > 4$. Points inside the unit disk satisfy $|z| < 1$ and map to points satisfying $|w| = 4/|z| > 4$. ■

EXAMPLE 23.14

Map the unit disk $|z| < 1$ to the disk $|w - i| < 2$.

The domains are shown in Figure 23.37. As with the previous two examples, this transformation is easily visualized geometrically. We want to map the unit disk to the disk of radius 2 centered at i. Thus, we want to expand the unit disk to radius 2 and translate the resulting disk up one unit to have its center at i. We can do this by first magnifying the unit disk by 2, then translating the resulting disk by i:

$$z \xrightarrow{\text{magnification}} 2z \xrightarrow{\text{translation}} 2z + i,$$

with the effects shown in Figure 23.38. A suitable mapping is

$$w = 2z + i.$$

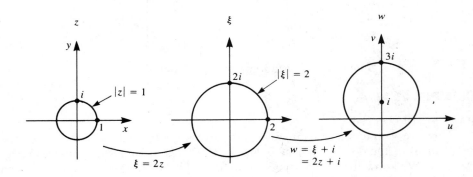

Figure 23.37

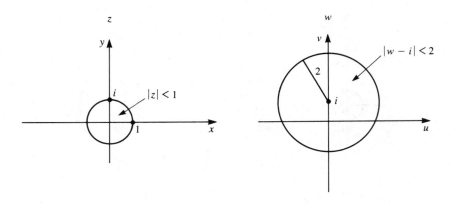

Figure 23.38

Notice that

$$|w - i| = |2z| = 2|z|;$$

therefore, the unit circle $|z| = 1$ maps to the circle $|w - i| = 2$. Further, since zero maps to i, the center of $|z| < 1$ maps to the center of $|w - i| < 2$, so points $|z| < 1$ map to points $|w - i| < 2$. ∎

Example 23.14 demonstrates the idea of building a mapping by stages. Whenever possible, we attempt to decompose a mapping into simple steps, then compose these individual steps to produce the final mapping.

EXAMPLE 23.15

Map the right half-plane $\text{Re}(z) > 0$ onto the unit disk $|w| < 1$.

The domains are shown in Figure 23.39. We will exploit the fact that, if one domain maps onto another, the boundary of one domain maps onto the boundary of the other.

The boundary of the right half-plane $\text{Re}(z) > 0$ in Figure 23.39 is the imaginary axis $\text{Re}(z) = 0$. We will map this boundary onto the boundary circle $|w| = 1$ of the domain $|w| < 1$. Choose three points on the imaginary axis in the z-plane and map these to three points on the boundary circle $|w| = 1$ in the w-plane. There are many ways to do this. To be specific, map

$$i \rightarrow 1, \qquad 0 \rightarrow i, \quad \text{and} \quad -i \rightarrow -i.$$

The linear fractional transformation mapping these points as indicated is

$$w = \frac{(-1 + 2i)z - i}{(-2 + i)z - 1}.$$

This mapping must take the right half-plane to the disk $|w| < 1$ or its complement $|w| > 1$. Choose any point in $\text{Re}(z) > 0$, say, $z = 1$. This point maps to $(-1 + i)/(3 + i)$, which lies in $|w| < 1$. Therefore, this linear fractional transformation maps $\text{Re}(z) > 0$

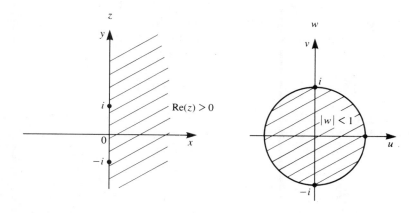

Figure 23.39

onto $|w| < 1$, as we wanted. If it had mapped $\text{Re}(z) > 0$ onto $|w| > 1$, we could have followed with an inversion to map the exterior of the unit disk onto its interior.   ∎

EXAMPLE 23.16

Map the right half-plane $\text{Re}(z) > 0$ onto the disk $|w - i| < 2$.

We can compose the mappings of Examples 23.14 and 23.15, as indicated in Figure 23.40:

$$w = 2\xi + i = 2\left[\frac{(-1 + 2i)z - i}{(-2 + i)z - 1}\right] + i.$$

A suitable mapping is

$$w = \frac{(-3 + 2i)z - 3i}{(-2 + i)z - 1}. \quad ■$$

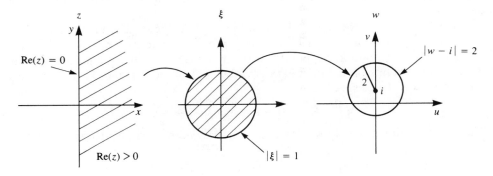

Figure 23.40. $\text{Re}(z) > 0 \rightarrow |\xi| < 1 \rightarrow |w - i| < 2$ yields a mapping of $\text{Re}(z) > 0$ onto $|w - i| < 2$.

EXAMPLE 23.17

Map the infinite strip S: $-\pi/2 < \text{Im}(z) < \pi/2$ onto the disk $|w| < 1$.

Recall from Example 23.2 that $w = e^z$ maps horizontal lines to half-lines. The boundary of the strip consists of the two horizontal lines $\text{Im}(z) = -\pi/2$ and $\text{Im}(z) = \pi/2$. If $\text{Im}(z) = \pi/2$, $z = x + i\pi/2$. This line maps to

$$w = e^{x + i\pi/2} = e^x\left[\cos\left(\frac{\pi}{2}\right) + i\sin\left(\frac{\pi}{2}\right)\right] = ie^x.$$

As z varies over $\text{Im}(z) = \pi/2$, $w = ie^x$ varies over the positive part of the imaginary axis in the w-plane. Similarly, if $\text{Im}(z) = -\pi/2$,

$$w = e^{x - i\pi/2} = e^x\left[\cos\left(\frac{\pi}{2}\right) - i\sin\left(\frac{\pi}{2}\right)\right] = -ie^x,$$

which varies over the negative imaginary axis in the w-plane. The origin 0 is in S and maps to $w = e^0 = 1$ in the right half-plane. Since the imaginary axis is the boundary of the right half-plane, $w = e^z$ maps S onto the right half-plane.

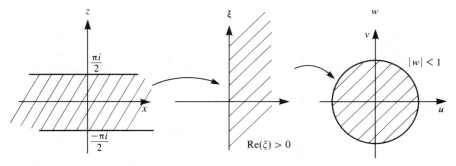

Figure 23.41. $|\text{Im}(z)| < \dfrac{\pi}{2} \to \text{Re}(\xi) > 0 \to |w| < 1$

yields a mapping of $|\text{Im}(z)| < \dfrac{\pi}{2}$ onto $|w| < 1$.

But we know how to map the right half-plane onto the unit disk, from Example 23.15. We will therefore map

$$S \to \text{right half-plane} \to \text{unit disk},$$

as illustrated in Figure 23.41. From Example 23.15, the composition is

$$w = \frac{(-1 + 2i)\xi - i}{(-2 + i)\xi - 1} = \frac{(-1 + 2i)e^z - i}{(-2 + i)e^z - 1}. \quad \blacksquare$$

EXAMPLE 23.18

Map the disk $|z| < 2$ onto the domain D^*: $u + v > 0$ in the w-plane.

The disk and the domain $u + v > 0$ are shown in Figure 23.42. The boundary of D^* is the straight line $u + v = 0$, or $v = -u$, in the w-plane. Notice that D^* is the half-plane $\text{Re}(w) > 0$, rotated counterclockwise by $\pi/4$ radians. We achieve this rotation by multiplying by $e^{i\pi/4}$.

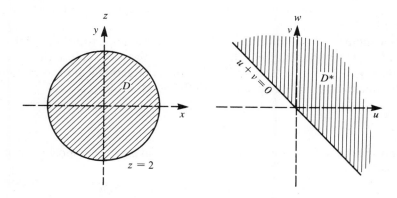

Figure·23.42

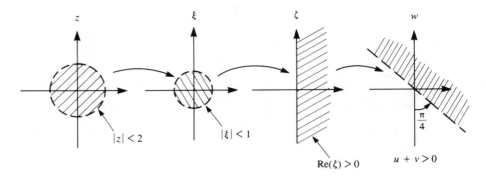

Figure 23.43

In order to exploit, as much as is possible, mappings we already know, we will envision the mapping in stages, as shown in Figure 23.43:

$$\text{disk } |z| < 2 \to \text{disk } |\xi| < 1 \to \text{half-plane } \text{Re}(\zeta) > 0 \to D^*.$$

Thus, map

$$w = \zeta e^{i\pi/4} = \left[\frac{(-1 + 2i)\xi - i}{(-2 + i)\xi - 1}\right] e^{i\pi/4} = \left[\frac{(-1 + 2i)(z/2) - i}{(-2 + i)(z/2) - 1}\right] e^{i\pi/4}.$$

One mapping of $|z| < 2$ onto D^* is

$$w = \left[\frac{(-1 + 2i)z - 2i}{(-2 + i)z - 2}\right] e^{i\pi/4}. \quad \blacksquare$$

EXAMPLE 23.19

Map $|z| < 1$ onto the domain D^* exterior to the ellipse

$$\frac{u^2}{\cosh^2(4)} + \frac{v^2}{\sinh^2(4)} = 1.$$

From Example 23.3, $w = \cos(\xi)$ maps the line $\text{Im}(\xi) = 4$ to this ellipse. Thus, one strategy is to find an intermediate mapping of $|z| < 1$ onto $\text{Im}(\xi) > 4$ [which has boundary $\text{Im}(\xi) = 4$].

We can map $|z| < 1$ onto $\text{Im}(\xi) > 4$ by a linear fractional transformation. Choose three points on the boundary of $|z| < 1$, say, 1, i, and $-i$. Choose three points on $\text{Im}(\xi) = 4$, say, $-1 + 4i$, $4i$, and $1 + 4i$. The linear fractional transformation mapping these points on $|z| = 1$, respectively, to the points on $\text{Im}(\xi) = 4$ is

$$\xi = \frac{(-4 - 7i)z + 9 + 4i}{(-2 + i)z + 1 - 2i}.$$

Now apply the cosine function to this mapping, achieving the composition depicted in Figure 23.44:

$$w = \cos(\xi) = \cos\left[\frac{(-4 - 7i)z + 9 + 4i}{(-2 + i)z + 1 - 2i}\right].$$

This maps the circle $|z| = 1$ onto the ellipse. Since zero (inside the unit disk) maps to

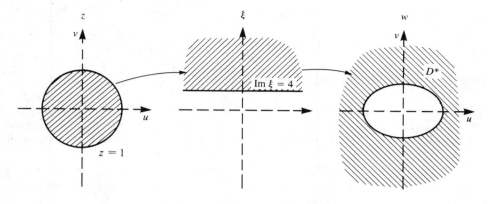

Figure 23.44. $|z| < 1 \rightarrow \text{Im}(\xi) > 4 \rightarrow D^*$
yields a mapping of $|z| < 1$ onto D^*.

$\cos[(-5 + i)/(1 + i)] = \cos(5i) = \cosh(5)$, which lies outside the ellipse, this mapping
sends the unit disk to the domain exterior to the ellipse, as we want. ∎

Thus far, we have constructed mappings using familiar analytic functions and
linear fractional transformations. We will now discuss the Schwarz-Christoffel transformation, which can be used to construct additional mappings. It gives an integral
formula for a conformal mapping of the upper half-plane onto the interior of any
domain bounded by a polygon.

THEOREM 23.6   Schwarz-Christoffel Transformation

Let P be a polygon in the w-plane with vertices $w_1, \ldots, w_n$. Let the exterior angles of P
be $\pi\alpha_1, \ldots, \pi\alpha_n$, as shown in Figure 23.45. Let $\text{Im}(z_0) \geq 0$. Then there are constants a

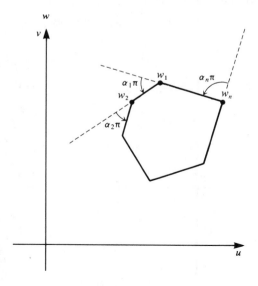

Figure 23.45

and b and real numbers $x_1, \ldots, x_n$ such that the function

$$f(z) = a \int_{z_0}^{z} (\xi - x_1)^{-\alpha_1}(\xi - x_2)^{-\alpha_2} \cdots (\xi - x_{n-1})^{-\alpha_{n-1}}(\xi - x_n)^{-\alpha_n}\, d\xi + b$$

maps the domain $\text{Im}(z) > 0$ onto the domain bounded by P. ∎

For $\text{Im}(z) > 0$, this integral is taken over any piecewise-smooth curve from z_0 to z in the upper half-plane. The powers appearing in the integrand are computed using the principal value of the logarithm function. Any function of the form of f in the theorem is called a *Schwarz-Christoffel transformation*.

To see the idea behind this function, suppose that $x_1, \ldots, x_{n-1}, x_n$ have been chosen on the x-axis with $x_j < x_{j+1}$. Figure 23.46 illustrates the following discussion.

Denote the upper half-plane $\text{Im}(z) > 0$ by D. For z in D, let

$$g(z) = a(z - x_1)^{-\alpha_1}(z - x_2)^{-\alpha_2} \cdots (z - x_{n-1})^{-\alpha_{n-1}}(z - x_n)^{-\alpha_n}.$$

Then $f'(z) = g(z)$ in D. Further,

$$\arg[f'(z)] = \arg(z) - \alpha_1 \arg(z - x_1) - \alpha_2 \arg(z - x_2) - \cdots - \alpha_n \arg(z - x_n).$$

As we saw in proving Theorem 23.1, $\arg[f'(z)]$ represents the number of radians by which the mapping f rotates tangent lines wherever $f'(z) \neq 0$.

Now envision z moving along the real axis from $-\infty$ toward ∞. On $(-\infty, x_1)$, $f(z)$ moves along a straight line (no change in angle). As z passes over x_1, $\arg[f'(z)]$ changes by an amount $\alpha_1 \pi$. This angle remains fixed as z moves from x_1 to x_2. As z passes over x_2, $\arg[f'(z)]$ changes by $\alpha_2 \pi$, then remains at this new value until z reaches x_3, where $\arg[f'(z)]$ changes by $\alpha_3 \pi$, and so on. Thus, $\arg[f'(z)]$ remains constant on intervals (x_{j-1}, x_j) and increases by $\alpha_j \pi$ as z passes over x_j. The net result is that the real axis is mapped to a polygon P^* with exterior angles $\alpha_1 \pi, \ldots, \alpha_n \pi$ determined by $\alpha_1, \ldots, \alpha_{n-1}$ because of the condition that the sum of the exterior angles of a closed polygon must equal 2π. This condition can be written

$$\alpha_1 \pi + \alpha_2 \pi + \cdots + \alpha_n \pi = 2\pi,$$

or

$$\alpha_1 + \alpha_2 + \cdots + \alpha_n = 2.$$

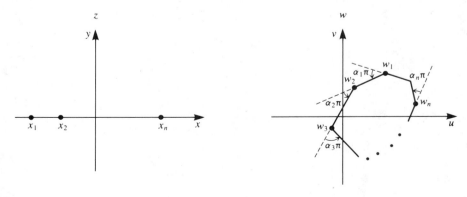

Figure 23.46

P^* has the same exterior angles as P but need not be the polygon P we began with because of its location in the plane and its "size." We may have to rotate, translate, and/or magnify P^* to obtain P. These effects are achieved by appropriately choosing $x_1, \ldots, x_n$ to make P^* similar to P, and the constants a (giving us a rotation-magnification factor) and b (a translation term) to make P^* the same as P.

If we choose $z_n = \infty$, so that $z_1, \ldots, z_{n-1}, \infty$ are mapped to the vertices of the polygon, the Schwarz-Christoffel transformation takes the form

$$f(z) = a \int_{z_0}^{z} (\xi - x_1)^{-\alpha_1}(\xi - x_2)^{-\alpha_2} \cdots (\xi - x_{n-1})^{-\alpha_{n-1}} \, d\xi + b,$$

with the factor $(\xi - x_n)^{-\alpha_n}$ omitted from the integrand.

We have not stated the Schwarz-Christoffel theorem in its complete generality. We can actually assert that *any* mapping of the upper half-plane onto a polygon must have the form of a Schwarz-Christoffel transformation.

In practice, a Schwarz-Christoffel transformation can be difficult or even impossible to find in explicit form because of the integral involved in its formulation. We will look at some examples involving this transformation.

EXAMPLE 23.20

Map the upper half-plane onto a triangle.

Of course, a triangle is a three-sided polygon, so Theorem 23.6 applies. Suppose that the triangle has vertices w_1, w_2, and w_3. The transformation has the general form

$$f(z) = a \int_{z_0}^{z} (\xi - x_1)^{-\alpha_1}(\xi - x_2)^{-\alpha_2}(\xi - x_3)^{-\alpha_3} \, d\xi + b,$$

where

$$\alpha_1 + \alpha_2 + \alpha_3 = 2.$$

If we let ∞ map to the third vertex w_3, using the extended complex plane, the mapping has the simpler form

$$f(z) = a \int_{z_0}^{z} (\xi - x_1)^{-\alpha_1}(\xi - x_2)^{-\alpha_2} \, d\xi + b. \tag{23.5}$$

We will illustrate this formula for a specific case. Choose $x_1 = -1$, $x_2 = 1$, and $x_3 = \infty$, so that the transformation has the form (23.5). Suppose that the triangle is equilateral, so $\alpha_1 = \alpha_2 = \alpha_3$. Since $\alpha_1 + \alpha_2 + \alpha_3 = 2$, each $\alpha_j = \frac{2}{3}$. Equation (23.5) becomes

$$f(z) = a \int_{z_0}^{z} (\xi + 1)^{-2/3}(\xi - 1)^{-2/3} \, d\xi + b.$$

If we choose $z_0 = 1$, $a = 1$, and $b = 0$, we get

$$f(z) = \int_{1}^{z} (\xi + 1)^{-2/3}(\xi - 1)^{-2/3} \, d\xi.$$

Note that $x_2 = 1$ maps to zero in the w-plane.

We chose $x_1 = -1$. When $z = -1$ in the integral, we have $\xi = x$ (real). When $-1 < x < 1$, $x + 1 > 0$ and $\arg(x + 1) = 0$. But $|x - 1| = 1 - x$, so $\arg(x - 1) = \pi$.

Therefore, the image of -1 is

$$
\begin{aligned}
f(-1) &= \int_{1}^{-1} (x+1)^{-2/3}(x-1)^{-2/3}\,dx \\
&= \int_{1}^{-1} (x+1)^{-2/3}[(-1)(1-x)]^{-2/3}\,dx \\
&= \int_{1}^{-1} (x+1)^{-2/3}[e^{\pi i}(1-x)]^{-2/3}\,dx \\
&= \int_{1}^{-1} (x+1)^{-2/3}(1-x)^{-2/3}e^{-2\pi i/3}\,dx \\
&= -e^{-2\pi i/3}\int_{-1}^{1} (1+x)^{-2/3}(1-x)^{-2/3}\,dx \\
&= e^{\pi i/3}\int_{-1}^{1} (1-x^2)^{-2/3}\,dx.
\end{aligned}
$$

The last integral is in the form of a beta function evaluated at $(\frac{1}{2}, \frac{1}{3})$; in terms of the gamma function, it is equal to $\Gamma(\frac{1}{2})\Gamma(\frac{1}{3})/\Gamma(\frac{5}{6})$. Then

$$
w_1 = f(-1) = \frac{\Gamma(\frac{1}{2})\Gamma(\frac{1}{3})}{\Gamma(\frac{5}{6})}\, e^{\pi i/3}.
$$

Finally, w_3, the image of ∞, is given by

$$
w_3 = \int_{1}^{\infty} (x+1)^{-2/3}(x-1)^{-2/3}\,dx = \int_{1}^{\infty} (x^2-1)^{-2/3}\,dx = \frac{\Gamma(\frac{1}{2})\Gamma(\frac{1}{3})}{\Gamma(\frac{5}{6})}.
$$

The image of the upper half-plane under this Schwarz-Christoffel transformation is therefore the equilateral triangle with side length $\Gamma(\frac{1}{2})\Gamma(\frac{1}{3})/\Gamma(\frac{5}{6})$ having one vertex at the origin, as shown in Figure 23.47.

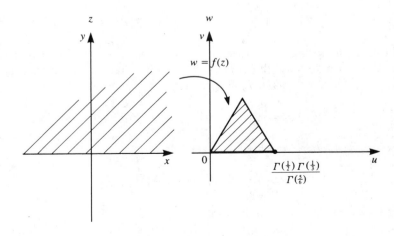

Figure 23.47

EXAMPLE 23.21

To map the upper half-plane onto a rectangle, we may choose $x_1 = 0, x_2 = 1, x_3$ as any real number greater than 1, and $x_4 = \infty$ to map to the corners of the rectangle. The Schwarz-Christoffel transformation has the form

$$f(z) = a \int_{z_0}^{z} \frac{1}{\sqrt{\xi(\xi - 1)(\xi - x_2)}} \, d\xi + b,$$

and a and b must be chosen to fit the size of the rectangle and its inclination with respect to the axes.

The radical appears in the integrand because the exterior angles of a rectangle are all equal to $\pi/2$; hence, $\alpha_1 + \alpha_2 + \alpha_3 + \alpha_4 = 4\alpha_j = 2$, and each $\alpha_j = \frac{1}{2}$.

The integral in this transformation is an elliptic integral and cannot be evaluated in closed form (that is, as a finite combination of elementary functions). ∎

Sometimes the Schwarz-Christoffel transformation can be used to map the upper half-plane to an infinite strip in the w-plane. We will discuss an example of this type of mapping.

EXAMPLE 23.22

Map the upper half-plane onto the strip S: $\text{Im}(w) > 0, -c < \text{Re}(w) < c$, with c a given positive constant.

S is shown in Figure 23.48. Our first impression might be that the Schwarz-Christoffel transformation does not apply because we are not mapping to a polygon. As we will see, however, we can still use the transformation if we envision S as a "polygon" with vertices $-c, c$, and ∞.

Choose $x_1 = -1$ to map to $-c$ and $x_2 = 1$ to map to c, and map ∞ to ∞. The interior angles of the strip are $\pi/2$ and $\pi/2$, so $\alpha_1 = \alpha_2 = \frac{1}{2}$. The Schwarz-Christoffel

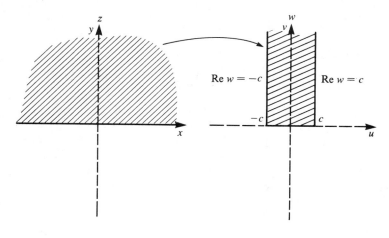

Figure 23.48

transformation for S has the form

$$w = f(z) = a \int_{z_0}^{z} (\xi + 1)^{-1/2} (\xi - 1)^{-1/2} \, d\xi + b.$$

Choose $z_0 = 0$ and $b = 0$. Now write

$$(\xi - 1)^{-1/2} = [-(1 - \xi)]^{-1/2} = -i(1 - \xi)^{-1/2}$$

and absorb the constant $-i$ into a (which we will still call a) to write

$$w = a \int_0^z (1 + \xi)^{-1/2} (1 - \xi)^{-1/2} \, d\xi = a \int_0^z \frac{1}{(1 - \xi^2)^{1/2}} \, d\xi.$$

This integral is reminiscent of the integral for the inverse sine function, and we can show that, in fact,

$$w = a \sin^{-1}(z).$$

We must choose a so that -1 maps to $-c$ and 1 maps to c. We need

$$a \sin^{-1}(1) = c$$

and hence

$$\sin\left(\frac{c}{a}\right) = 1.$$

We can choose $c/a = \pi/2$ or $a = 2c/\pi$. The mapping is therefore

$$w = \frac{2c}{\pi} \sin^{-1}(z).$$

If we choose $c = \pi/2$, this mapping is just $w = \sin^{-1}(z)$. This takes the upper half-plane onto the strip $\text{Im}(w) > 0$, $-\pi/2 < \text{Re}(w) < \pi/2$. This result is consistent with Example 23.4, with z and w interchanged from their roles here. ∎

In the next chapter, we will consider some uses of complex analysis and conformal mappings in a variety of settings.

PROBLEMS FOR SECTION 23.3

In each of Problems 1 through 10, find a linear fractional transformation mapping the first domain onto the second.

1. $|z| < 3$ onto $|w - i| < 4$
2. $|z| < 3$ onto $|w - i| > 4$
3. $|z + 2i| < 1$ onto $|w - 3| > 2$
4. $\text{Re}(z) > 1$ onto $\text{Im}(w) > 1$
5. $\text{Re}(z) < 0$ onto $|w| < 4$
6. $\text{Im}(z) > -4$ onto $|w - i| > 2$
7. $\text{Re}(z) > 0$ onto $\text{Im}(z) < 3$
8. $\text{Re}(z) < -4$ onto $|w + 1 - 2i| > 3$
9. $|z - 1| > 4$ onto $\text{Im}(w) < 2$
10. $|z - 1 + 3i| > 1$ onto $\text{Re}(w) < -5$
11. Find a conformal mapping of the upper half-plane $\text{Im}(z) > 0$ onto the wedge $0 < \arg(w) < \pi/3$.
12. Show that $w = \text{Log}(z)$ maps $\text{Im}(z) > 0$ onto the strip $0 < \text{Im}(w) < \pi$.
13. Let $w = \frac{1}{2}(z + 1/z)$.
    (a) Show that $|z| < 1$ and $|z| > 1$ are both mapped onto the complex w-plane with the interval $[-1, 1]$ of the real axis removed.

(b) Show that the circle $|z| = r$ maps to an ellipse if $0 < r < 1$.

14. Let

$$w = f(z) = \frac{z - z_0}{1 - \bar{z}_0 z} \, e^{i\theta},$$

in which z_0 is a given complex number with $|z_0| < 1$ and θ is a given real number in $[0, 2\pi)$. Show that f is a conformal mapping of the unit disk $|z| < 1$ onto itself. (In fact, it can be shown that any conformal mapping of the unit disk onto itself has this form for some z_0 and θ.)

15. Show that the Schwarz-Christoffel transformation

$$w = 2i \int_0^z (\xi + 1)^{-1/2} (\xi - 1)^{-1/2} \xi^{-1/2} \, d\xi$$

maps the upper half-plane onto the rectangle with vertices ci, 0, c, and $c + ic$, where $c = \Gamma(\frac{1}{2})\Gamma(\frac{1}{4})/\Gamma(\frac{3}{4})$.

16. Define the *cross ratio* of z_1, z_2, z_3, and z_4 to be the image of z_1 under the linear fractional transformation which maps $z_2 \to 1$, $z_3 \to 0$, $z_4 \to \infty$. Denote this cross ratio as $[z_1, z_2, z_3, z_4]$. Suppose that T is a linear fractional transformation. Prove that T preserves the cross ratio. That is, prove that

$$[z_1, z_2, z_3, z_4] = [T(z_1), T(z_2), T(z_3), T(z_4)].$$

17. Prove that $[z_1, z_2, z_3, z_4]$ is the image of z_1 under the linear fractional transformation defined by

$$w = 1 - \frac{z_3 - z_4}{z_3 - z_2} \cdot \frac{z - z_2}{z - z_4}.$$

18. Prove that $[z_1, z_2, z_3, z_4]$ is real if and only if the four points $z_1, z_2, z_3,$ and z_4 are collinear (lie on a straight line) or lie on a circle.

19. Two points z and z^* are *symmetric with respect to a circle* C passing through z_1, z_2, and z_3 if and only if $[z^*, z_1, z_2, z_3] = \overline{[z, z_1, z_2, z_3]}$. Using the result of Problem 18, prove that the points which are self-symmetric with respect to C (that is, $z = z^*$) are exactly the points on C itself.

20. Prove the *symmetry principle*: Suppose that a linear fractional transformation T maps a circle C onto a circle K. Let z and z^* be symmetric with respect to C. Then $T(z)$ and $T(z^*)$ are symmetric with respect to K. *Hint:* Use the result of Problem 19.

CHAPTER 24

Some Applications of Complex Analysis

24.1 Complex Analytic Methods in the Analysis of Fluid Flow

In this section, we will discuss how complex function theory can be used to model and analyze the flow of fluids. Consider an incompressible fluid (such as water, under normal conditions). Assume that we are given the velocity field $\mathbf{V}(x, y, z)$ of the fluid. $\mathbf{V}(x, y, z)$ is the velocity of the fluid at each point (x, y, z) in the region of flow. We will assume that $\mathbf{V}$ is independent of time. In this case, the fluid flow is called *stationary*.

In order to use complex function methods, we will also assume that $\mathbf{V}$ is independent of z. This means that the flow is the same in all planes parallel to the xy-plane. Such a flow is called *plane-parallel*.

A plane-parallel velocity field has the form $\mathbf{V}(x, y) = u(x, y)\mathbf{i} + v(x, y)\mathbf{j}$. We will identify vectors in the plane with complex numbers and, with a mild abuse of notation, write the velocity field in complex notation as

$$\mathbf{V}(z) = u(x, y) + iv(x, y).$$

Here, i is the imaginary unit, not the vector $\mathbf{i}$.

We will think of the complex plane as divided into two sets. First, we have a given domain D, where the velocity field is defined. On the complement of D (the set of points in the plane not in D), the flow is undefined. We think of the complement as channels confining the fluid to D, or perhaps as barriers or obstacles through which the fluid cannot flow. This enables us to model fluid flow through a variety of channels, with barriers of various shapes in the channels.

A flow $V(z) = u(x, y) + iv(x, y)$ is called *irrotational* in D if

$$\oint_C u\,dx + v\,dy = 0$$

for every simple closed piecewise-smooth curve in D. The flow is *solenoidal* if

$$\oint_C -v\,dx + u\,dy = 0$$

for every such C in D.

As we might expect, there are physical motivations for these terms. Suppose C is parametrized by $x = x(s)$, $y = y(s)$, with s arc length along the curve. Then the vector $x'(s)\mathbf{i} + y'(s)\mathbf{j}$ is a unit tangent to C, and

$$(u\mathbf{i} + v\mathbf{j}) \cdot \left(\frac{dx}{ds}\mathbf{i} + \frac{dy}{ds}\mathbf{j} \right)$$

is the tangential component of the velocity along the curve. Now observe that

$$(u\mathbf{i} + v\mathbf{j}) \cdot \left(\frac{dx}{ds}\mathbf{i} + \frac{dy}{ds}\mathbf{j} \right) ds = u\,dx + v\,dy.$$

Therefore,

$$\oint_C u\,dx + v\,dy$$

is a measure of the velocity of the fluid along C. The value of this integral is called the *circulation of the fluid* around C, and a fluid is irrotational when its circulation about an arbitrary closed curve is zero.

The vector $-(dy/ds)\mathbf{i} + (dx/ds)\mathbf{j}$ is a unit normal to C (it is the unit tangent vector rotated $\pi/2$ clockwise, as shown in Figure 24.1). Therefore,

$$-\oint_C (u\mathbf{i} + v\mathbf{j}) \cdot \left(-\frac{dy}{ds}\mathbf{i} + \frac{dx}{ds}\mathbf{j} \right) ds = \oint_C -v\,dx + u\,dy$$

is the negative of the integral of the normal component of the velocity field about C. When $\oint_C -v\,dx + u\,dy \neq 0$, the value of this integral is the *flux* of the velocity field across C.

A point $z_0 = (x_0, y_0)$ is called a *vortex* if the circulation $\oint_C u\,dx + v\,dy$ has a constant nonzero value for every piecewise-smooth simple closed curve in a disk about

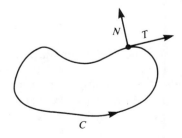

Figure 24.1

z_0 with z_0 removed (that is, in a deleted disk about z_0). The value of $\oint_C u\,dx + v\,dy$ is called the *strength of the vortex*.

If $\oint_C -v\,dx + u\,dy$ has the same positive value for all simple closed curves in a deleted disk about z_0, we call z_0 a *source of strength* $\oint_C -v\,dx + u\,dy$. Similarly, z_0 is a *sink of strength* $\left|\oint_C -v\,dx + u\,dy\right|$ if $\oint_C -v\,dx + u\,dy$ has the same negative value for all piecewise-smooth simple closed curves in some deleted disk about z_0.

The following theorem provides the connection between the velocity field of a fluid and analytic functions.

THEOREM 24.1

Let u and v have continuous first and second partial derivatives in a simply connected domain D. Suppose that the velocity field $u(x, y) + iv(x, y)$ is irrotational and solenoidal on D. Then u and $-v$ satisfy the Cauchy-Riemann equation on D.

Conversely, if u and v satisfy the Cauchy-Riemann equations on D, the analytic function $f(z) = u(x, y) - iv(x, y)$ defines an irrotational, solenoidal flow on D.   ∎

The proof is a routine calculation which is left to the student. To get further physical insight into a flow having a velocity field $u(x, y)\mathbf{i} + v(x, y)\mathbf{j}$, assume that this flow is irrotational. Now

$$\text{curl}(u\mathbf{i} + v\mathbf{j}) = \left(\frac{\partial v}{\partial x} - \frac{\partial u}{\partial y}\right)\mathbf{k},$$

a vector normal to the plane of the flow. From the argument at the end of Section 15.5, we can interpret this vector as twice the angular velocity of the particle of fluid at (x, y). However, since u and $-v$ satisfy the Cauchy-Riemann equations,

$$\frac{\partial u}{\partial y} = -\frac{\partial}{\partial x}[-v],$$

implying that $\text{curl}(u\mathbf{i} + v\mathbf{j}) = \mathbf{O}$. Thus, "irrotational" means that the particles of fluid experience translations (and possible distortions) in their motion but no rotation. Put another way, there is no swirling in the fluid.

Further, we can recognize

$$\text{div}(u\mathbf{i} + v\mathbf{j}) = \frac{\partial u}{\partial x} + \frac{\partial v}{\partial y}.$$

If the flow is solenoidal, u and $-v$ satisfy the Cauchy-Riemann equations and we can conclude that $\text{div}(u\mathbf{i} + v\mathbf{j}) = 0$.

THEOREM 24.2

Suppose that f is an analytic function on a domain D. Then $\overline{f'(z)}$ is an irrotational, solenoidal flow on D.

Conversely, suppose that $\mathbf{V}(x, y) = u\mathbf{i} + v\mathbf{j}$ is an irrotational, solenoidal vector field on a simply connected domain D. Then there is a unique analytic function f defined on D such that $\overline{f'(z)} = \mathbf{V}(x, y)$. Further, if $f(z) = \varphi(x, y) + i\psi(x, y)$,

$$\frac{\partial \varphi}{\partial x} = u, \qquad \frac{\partial \varphi}{\partial y} = v, \qquad \frac{\partial \psi}{\partial x} = -v, \quad \text{and} \quad \frac{\partial \psi}{\partial y} = u.   \blacksquare$$

Verification of these statements is left to the student. In view of the fact that the velocity of the flow is $\overline{f'(z)}$, we call f a *complex potential function*, or *complex potential*, for the flow.

Theorem 24.2 implies that any analytic function $f(z) = \varphi(x, y) + i\psi(x, y)$ defined on a simply connected domain D determines an irrotational, solenoidal flow

$$\overline{f'(z)} = \frac{\partial \varphi}{\partial x} + i\frac{\partial \psi}{\partial x} = \overline{u(x, y) - iv(x, y)} = u(x, y) + iv(x, y).$$

The function φ is called the *velocity potential* of the flow, and curves

$$\varphi(x, y) = \text{constant}$$

are called *equipotential lines*, or *equipotential curves*.

The function ψ is called the *stream function* of the flow, and curves

$$\psi(x, y) = \text{constant}$$

are called *streamlines*.

We may think of the mapping $w = f(z)$ as a conformal mapping wherever $f'(z) \neq 0$. We call a point at which $f'(z) = 0$ a *stagnation point of the flow*.

In the w-plane, write $w = f(z) = \varphi(x, y) + i\psi(x, y) = \alpha + i\beta$. Equipotential curves $\varphi(x, y) = k$ map under f to vertical lines $\alpha = k$, and streamlines $\psi(x, y) = c$ map to horizontal lines $\beta = k$. Since these lines form an orthogonal set of curves, the streamlines and equipotential curves in the z-plane are also orthogonal (because f is conformal). This means that, wherever a streamline and an equipotential line intersect, their tangents are orthogonal (provided that this point is not a stagnation point).

Along an equipotential curve, we have $\varphi(x, y) = k$ and

$$d\varphi = \frac{\partial \varphi}{\partial x}\, dx + \frac{\partial \varphi}{\partial y}\, dy = u\, dx + v\, dy = 0.$$

Now, $u\mathbf{i} + v\mathbf{j}$ is the velocity of the flow at (x, y), and $(dx/ds)\mathbf{i} + (dy/ds)\mathbf{j}$ is tangent to the equipotential curve at (x, y). Since the dot product of these two vectors is zero, we conclude that the velocity is orthogonal to the equipotential through (x, y), provided that (x, y) is not a stagnation point.

Similarly, along a streamline, $\psi(x, y) = c$, so

$$d\psi = \frac{\partial \psi}{\partial x}\, dx + \frac{\partial \psi}{\partial y}\, dy = -v\, dx + u\, dy = 0;$$

hence, the normal to the velocity is orthogonal to the streamline. This means that the velocity is tangent to the streamline and justifies the interpretation that the particle of fluid at (x, y) is moving in the direction of the streamline at that point. We may therefore think of the streamlines as the trajectories of particles in the fluid.

EXAMPLE 24.1

Consider the flow modeled by the complex potential $f(z) = -Ke^{i\theta}z$, in which K is a positive constant and $0 \leq \theta < 2\pi$. Write

$$f(z) = -K[\cos(\theta) + i\sin(\theta)][x + iy]$$
$$= -K[x\cos(\theta) - y\sin(\theta)] - iK[y\cos(\theta) + x\sin(\theta)].$$

If $f(z) = \varphi(x, y) + i\psi(x, y)$,

$$\varphi(x, y) = -K[x\cos(\theta) - y\sin(\theta)]$$

and

$$\psi(x, y) = -K[y\cos(\theta) + x\sin(\theta)].$$

Equipotential curves are curves having equation

$$\varphi(x, y) = -K[x\cos(\theta) - y\sin(\theta)] = \text{constant}.$$

Since K is constant, this equation has the form

$$x\cos(\theta) - y\sin(\theta) = k$$

or

$$y = \cot(\theta)x + b,$$

in which b is also constant $[b = k/\cos(\theta)]$. These are straight lines with slope $\cot(\theta)$.

Streamlines are graphs of

$$\psi(x, y) = -K[y\cos(\theta) + x\sin(\theta)] = \text{constant}.$$

This equation has the form

$$y = -\tan(\theta)x + d,$$

with d constant. These graphs are straight lines with slope $-\tan(\theta)$.

The streamlines are straight lines making an angle θ with the positive real axis (recall that θ is constant). Since the streamlines form trajectories of the flow, the particles of fluid may be thought of as moving along straight lines at an angle θ with the positive real axis, as in Figure 24.2. Notice also that the streamlines and equipotential lines are orthogonal, their slopes being negative reciprocals of each other.

Now compute

$$f'(z) = -Ke^{i\theta}.$$

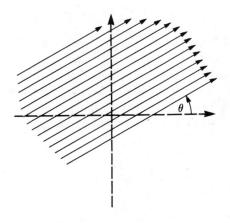

Figure 24.2. Streamlines of the flow with complex potential
$$f(z) = -Ke^{i\theta}z.$$

Then

$$\overline{f'(z)} = -Ke^{-i\theta}.$$

Therefore, the velocity is constant and of magnitude K. We may think of $f(z) = -Ke^{i\theta}z$ as modeling a uniform flow with velocity of constant magnitude K and making an angle θ with the positive real axis. ∎

EXAMPLE 24.2

Consider the complex potential $f(z) = z^2$. Then f is analytic in the entire complex plane, but $f'(0) = 0$, so the origin is a stagnation point. We will examine what effect this has on the flow and analyze the trajectories.

With $z = x + iy$, $f(z) = x^2 - y^2 + 2ixy$. Then

$$\varphi(x, y) = x^2 - y^2 \quad \text{and} \quad \psi(x, y) = 2xy.$$

The equipotential curves have equations of the form

$$x^2 - y^2 = k,$$

whose graphs are hyperbolas if $k \neq 0$. The streamlines have equations

$$2xy = c,$$

whose graphs are also hyperbolas (if $c \neq 0$). Figure 24.3 shows some equipotential curves and streamlines. The equipotential curves and streamlines form orthogonal families of curves in the plane.

If $k = 0$, we get the equation $x^2 - y^2 = 0$, whose graph consists of two straight lines, $y = x$ and $y = -x$. If $c = 0$, we get $xy = 0$ and hence $x = 0$ (the y-axis) or $y = 0$

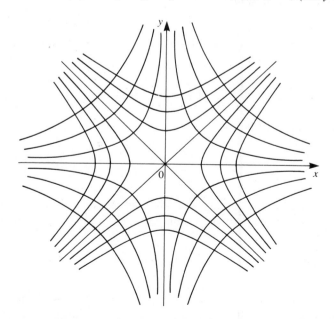

Figure 24.3. Equipotential curves and streamlines of the flow with complex potential $f(z) = z^2$.

(the x-axis). Thus, the axes are streamlines as well. The velocity of the flow is $\overline{f'(z)} = 2\overline{z}$. This is a nonuniform flow, having velocity at z of magnitude $2|z|$.

We can envision this flow as particles moving along the streamlines. In any quadrant, the particles of fluid move along the hyperbolas $xy = $ constant with the axes acting as barriers of the flow (or sides of a container holding the fluid). ∎

EXAMPLE 24.3

Consider the complex potential $f(z) = (iK/2\pi)\text{Log}(z)$, with K a positive constant. With $z = x + iy$, we have

$$f(z) = \frac{iK}{2\pi}[\ln(x^2 + y^2) + i\,\text{Arg}(z)] = \frac{-K}{2\pi}\,\text{Arg}(z) + \frac{iK}{2\pi}\ln(x^2 + y^2)$$

if $z \neq 0$. Therefore,

$$\varphi(x, y) = \frac{-K}{2\pi}\,\text{Arg}(z)$$

and

$$\psi(x, y) = \frac{K}{2\pi}\ln(x^2 + y^2).$$

Equipotential curves have equations

$$\text{Arg}(z) = k$$

which are half-lines emanating from the origin.

Streamlines have equations

$$\ln(x^2 + y^2) = c,$$

or $x^2 + y^2 = $ positive constant. These curves are circles about the origin. We have

$$f'(z) = \frac{iK}{2\pi}\frac{1}{z}$$

if $z \neq 0$. Thus, the velocity is

$$\overline{f'(z)} = -\frac{iK}{2\pi}\frac{1}{\overline{z}} = -\frac{iK}{2\pi}\frac{z}{|z|^2}$$

if $z \neq 0$.

Since streamlines are trajectories of the fluid, we may think of the particles of fluid as moving on circles about the origin. On a circle $|z| = r$, the magnitude of the velocity is

$$|\overline{f'(z)}| = \frac{K}{2\pi}\frac{1}{|z|} = \frac{K}{2\pi r}.$$

Therefore, the velocity is greater about circles nearer the origin. We may envision the particles of fluid as swirling about the origin, with faster motion as particles near the origin (Figure 24.4).

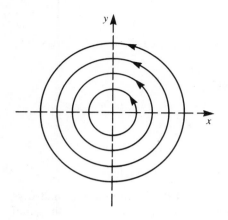

Figure 24.4. Streamlines of the flow having complex potential $f(z) = (iK/2\pi)\mathrm{Log}(z)$.

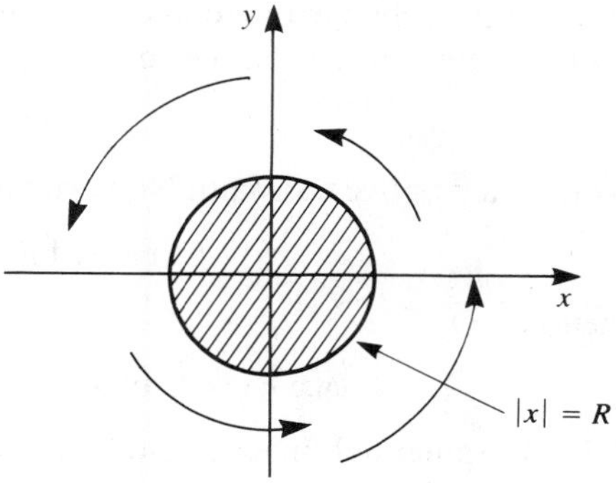

Figure 24.5. Flow around a cylindrical barrier of radius R.

We will calculate the circulation of the flow about the origin. Write

$$\overline{f'(z)} = -\frac{iK}{2\pi}\frac{z}{|z|^2} = \frac{K}{2\pi}\frac{y}{x^2+y^2} - \frac{iK}{2\pi}\frac{x}{x^2+y^2} = u + iv.$$

Compute $\oint_C u\,dx + v\,dy$ on the circle $|z| = r$, which we parametrize as

$$x = r\cos(\theta), \qquad y = r\sin(\theta)$$

for $0 \le \theta \le 2\pi$. We get

$$\oint_C u\,dx + v\,dy = \frac{K}{2\pi}\int_0^{2\pi}\frac{1}{r^2}r\sin(\theta)[-r\sin(\theta)]\,d\theta$$

$$-\frac{K}{2\pi}\int_0^{2\pi}\frac{1}{r^2}r\cos(\theta)[r\cos(\theta)]\,d\theta$$

$$= -\frac{K}{2\pi}\int_0^{2\pi}[\sin^2(\theta) + \cos^2(\theta)]\,d\theta = -K.$$

This is the value of the circulation of the flow about $|z| = r$. The origin is a vortex of the flow.

By a similar computation, we find that

$$\oint_C -v\,dx + u\,dy = 0;$$

hence, the origin is neither a source nor a sink.

We may also restrict $|z| > R$ for some positive constant R and think of the flow we have just described as fluid in a swirling motion about a solid cylindrical barrier of radius R about the origin (Figure 24.5). Imagine a cylindrical pipe about the z-axis, and think of fluid as flowing around this pipe in the xy-plane. ∎

EXAMPLE 24.4

We can interchange the roles of equipotential curves and streamlines in the preceding example by considering the complex potential

$$f(z) = K \, \mathrm{Log}(z),$$

with K a positive constant. Now we have

$$f(z) = K \ln(x^2 + y^2) + iK \, \mathrm{Arg}(z);$$

hence,

$$\varphi(x, y) = K \ln(x^2 + y^2) \quad \text{and} \quad \psi(x, y) = K \, \mathrm{Arg}(z).$$

The equipotential lines are circles about the origin, and the streamlines are half-rays emanating from the origin (Figure 24.6). As they must, these curves form orthogonal families in the plane. The velocity of this flow is

$$\overline{f'(z)} = \frac{K}{\overline{z}} = K \, \frac{x}{x^2 + y^2} + iK \, \frac{y}{x^2 + y^2} = u + iv.$$

Let C be the circle $|z| = r$. We find that

$$\oint_C u \, dx + v \, dy = 0$$

and

$$\oint_C -v \, dx + u \, dy = 2\pi K.$$

The origin is therefore a source of strength $2\pi K$.

We may think of this flow as having particles emanating from the source at the origin (with "infinite" velocity) and moving along straight line trajectories outward toward infinity along the streamlines, with decreasing velocity as their distance from the origin increases.

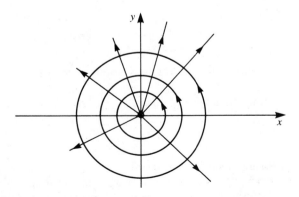

Figure 24.6. Streamlines and equipotential lines of the flow having complex potential $f(z) = K \, \mathrm{Log}(z)$.   ∎

EXAMPLE 24.5

We will model flow around an elliptic barrier. Recall from Example 24.3 that the complex potential $f(z) = (iK/2\pi)\text{Log}(z)$, for $|z| > R$, models flow with circulation $-K$ about a cylindrical barrier of radius R placed about the origin. We will model flow about an elliptic barrier by using a conformal mapping to send the circle $|z| = R$ to an ellipse. Consider the mapping

$$w = z + \frac{a^2}{z},$$

in which a is a positive constant. This mapping is called a *Joukowski transformation*, and it is used often in analyzing fluid flow around airplane wings.

Write $z = x + iy$ and $w = X + iY$. We find that the circle $x^2 + y^2 = R^2$ is mapped by the Joukowski transformation to the ellipse

$$\frac{X^2}{1 + \dfrac{a^2}{R^2}} + \frac{Y^2}{1 - \dfrac{a^2}{R^2}} = R^2$$

provided that $a \neq R$. If $a = R$, the circle $x^2 + y^2 = a^2$ maps to the interval $[-2a, 2a]$ on the real axis.

Solve for z in the Joukowski transformation. As a quadratic equation, it yields two solutions. Choose the solution

$$z = \frac{w + \sqrt{w^2 - 4a^2}}{2}.$$

Compose this mapping with the complex potential function for the circular barrier to get

$$F(w) = f(z(w)) = \frac{iK}{2\pi} \text{Log}\left(\frac{w + \sqrt{w^2 - 4a^2}}{2}\right).$$

This complex potential function models flow in the w-plane about an elliptic barrier if $R > a$ and about the flat plate $-2a \leq X \leq 2a$, $Y = 0$ if $R = a$. ∎

We will conclude this section with an application of complex integration to complex function modeling of fluid flow. Suppose that f is the complex potential for a flow about a barrier whose boundary is the piecewise-smooth simple closed curve C in the plane. Let the thrust of the fluid outside the barrier be the vector $A\mathbf{i} + B\mathbf{j}$. It can be shown that A and B are given by the equation

$$A - iB = \frac{1}{2} i\rho \oint_C [f'(z)]^2 \, dz,$$

in which ρ is the constant density of the fluid. Further, the moment of the thrust about the origin is

$$\text{Re}\left\{ -\frac{1}{2}\rho \oint_C z[f'(z)]^2 \, dz \right\}.$$

These results are part of a theorem due to Blasius. Often, these integrals are evaluated using the residue theorem.

PROBLEMS FOR SECTION 24.1

1. Analyze the flow modeled by the complex potential $f(z) = az$, in which a is a nonzero complex constant. Sketch some of the equipotential lines and streamlines, determine the velocity, and determine whether the flow has any sources or sinks.

2. Analyze the flow having complex potential $f(z) = z^3$. Sketch some of the equipotential lines and streamlines.

3. Sketch some of the streamlines and equipotential lines for the flow having complex potential $f(z) = \cos(z)$.

4. Sketch some of the streamlines and equipotential lines for the flow having complex potential $f(z) = z + iz^2$.

5. Analyze the flow given by the complex potential $f(z) = K \operatorname{Log}(z - z_0)$, where z_0 is a given complex number and K is a positive constant. Show that z_0 is a source of this flow if $K > 0$ and a sink if $K < 0$. Sketch some of the equipotential lines and streamlines.

6. Analyze the flow modeled by the complex potential

$$f(z) = K \operatorname{Log}\left(\frac{z - a}{z - b}\right),$$

with K a real, nonzero constant and a and b distinct complex constants. Sketch some of the equipotential lines and streamlines.

7. Let $f(z) = k(z + 1/z)$, with k a nonzero real constant. Sketch some streamlines and equipotential lines for this flow. Show that f is the complex potential for a flow around the upper half of the unit circle.

8. Consider the complex potential $f(z) = [(m - ik)/2\pi]\operatorname{Log}((z - a)/(z - b))$, in which m and k are nonzero real constants and a and b are distinct complex constants. Show that this flow has a source or sink of strength m and a vortex of strength k at both a and b. (A point combining properties of a source (or sink) and a vortex is called a *spiral vortex*. The behavior of the equipotential lines and streamlines near these points suggests this descriptive name.)

9. Analyze the flow having complex potential

$$f(z) = k\left(z + 1/z\right) + \frac{ib}{2\pi} \operatorname{Log}(z),$$

where k and b are nonzero real constants.

10. Consider the complex potential

$$f(z) = iKa\sqrt{3} \operatorname{Log}\left(\frac{2z - ia\sqrt{3}}{2z + ia\sqrt{3}}\right),$$

in which K and a are positive constants. Show that this potential models an irrotational flow around a cylinder $4x^2 + 4(y - a)^2 = a^2$ with a flat boundary along the y-axis.

11. Use the theorem of Blasius to show that the force per unit width on the cylinder in Problem 10 has y-component $2\sqrt{3}\pi\rho aK^2$, in which ρ is the constant density of the fluid.

24.2 *A Residue Formula for the Inverse Laplace Transform*

The Laplace transform of a complex function can be defined in a way analogous to that for the Laplace transform of a real function. If f is a complex function defined at least for all z on $[0, \infty)$, the Laplace transform of f is

$$\mathcal{L}[f](z) = \int_0^\infty e^{-z\xi} f(\xi)\, d\xi$$

for all z such that this integral exists.

Many identities and formulas for the Laplace transform of a real function extend to this complex version. In this section, we will discuss a residue formula for the inverse Laplace transform of a function. Not only is this formula important in carrying out specific computations, it also illustrates how we may gain tangible benefits by extending the domain of an operator to the complex plane.

If $\mathcal{L}[f] = F$, we call f an *inverse Laplace transform of* F and write $f = \mathcal{L}^{-1}[F]$. The following formula enables us to calculate $\mathcal{L}^{-1}[F]$ in terms of residues of F at certain singularities.

THEOREM 24.3

Let F be analytic over the entire complex plane except at a finite number of singularities $z_1, \ldots, z_n$, which are all poles. Assume that, for some real number σ, F is analytic on a domain consisting of all z with $\operatorname{Re}(z) > \sigma$. Assume also that, for some numbers M and R,

$$|zF(z)| \leq M \quad \text{if} \quad |z| \geq R.$$

For $t \geq 0$, let

$$f(t) = \sum_{j=1}^n \operatorname*{Res}_{z_j} e^{zt} F(z).$$

Then, for $\operatorname{Re}(z) > \sigma$, $\mathcal{L}[f] = F.$ ∎

The assumption that F is analytic for $\operatorname{Re}(z) > \sigma$ means that all the singularities of F lie to the left of the vertical line $\operatorname{Re}(z) = \sigma$, as shown in Figure 24.7. The assumption that $|zF(z)| \leq M$ if $|z| \geq R$ means that, on and outside a circle of sufficiently large

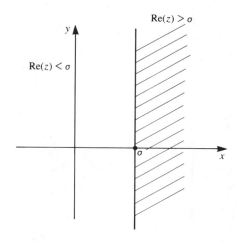

Figure 24.7

radius, the function $zF(z)$ is bounded. As an important special case, this condition will be satisfied by any rational function in which the degree of the denominator exceeds that of the numerator by at least 1. We will sketch a proof of Theorem 24.3 after illustrating the method.

EXAMPLE 24.6

Let a be any positive constant. Find the inverse Laplace transform of $1/(z^2 + a^2)$.

This result can be read from Table 4.1, but we will apply Theorem 24.3 as an illustration of the method. First, $F(z) = 1/(z^2 + a^2)$ has simple poles at ai and $-ai$ and satisfies the hypotheses of Theorem 24.3. Therefore, the inverse Laplace transform of F is

$$f(t) = \operatorname*{Res}_{ai} e^{zt} \frac{1}{z^2 + a^2} + \operatorname*{Res}_{-ai} e^{zt} \frac{1}{z^2 + a^2}.$$

$$= \frac{e^{ait}}{2ai} + \frac{e^{-ait}}{-2ai}$$

by the corollary to Theorem 22.4. Therefore,

$$f(t) = \frac{1}{2ai}(e^{iat} - e^{-iat}) = \frac{1}{a}\sin(at)$$

for $t \geq 0$. ■

EXAMPLE 24.7

Find the inverse Laplace transform of

$$F(z) = \frac{1}{(z^2 - 4)(z - 1)^2}.$$

This result can be obtained using partial fractions or Heaviside's formulas. We get the inverse Laplace transform with little effort using Theorem 24.3. F has simple poles at 2 and -2 and a pole of order 2 at 1. Compute

$$\operatorname*{Res}_{2} \frac{e^{zt}}{(z^2 - 4)(z - 1)^2} = \lim_{z \to 2} (z - 2)\frac{e^{zt}}{(z - 2)(z + 2)(z - 1)^2} = \frac{1}{4}e^{2t},$$

$$\operatorname*{Res}_{-2} \frac{e^{zt}}{(z^2 - 4)(z - 1)^2} = \lim_{z \to -2} (z + 2)\frac{e^{zt}}{(z - 2)(z + 2)(z - 1)^2} = \frac{-1}{36}e^{-2t},$$

and

$$\operatorname*{Res}_{1} \frac{e^{zt}}{(z^2 - 4)(z - 1)^2} = \lim_{z \to 1} \frac{d}{dz}\left(\frac{e^{zt}}{z^2 - 4}\right)$$

$$= \lim_{z \to 1} \frac{(z^2 - 4)te^{zt} - 2ze^{zt}}{(z^2 - 4)^2} = -\frac{1}{3}te^t - \frac{2}{9}e^t.$$

The inverse Laplace transform is

$$f(t) = \tfrac{1}{4}e^{2t} - \tfrac{1}{36}e^{-2t} - \tfrac{1}{3}te^t - \tfrac{2}{9}e^t$$

for $t > 2$ [since all poles occur on or to the left of the line $\operatorname{Re}(z) = 2$]. ■

We will conclude this section with a sketch of a proof of Theorem 24.3. Let $a > \sigma$ and consider $\oint_C e^{zt}F(z)\,dz$, with C the rectangle shown in Figure 24.8. C encloses all the poles of F. Suppose that C intersects the axes at $x = -c$, $x = a$, $y = b$, and $y = -b$, with b and c positive. By the residue theorem,

$$\oint_C e^{zt}F(z)\,dz = 2\pi i \sum_{j=1}^n \operatorname*{Res}_{z_j} e^{zt}F(z).$$

Use the fact that $|zF(z)| \le M$ for $|z|$ sufficiently large to argue that

$$\oint_C e^{zt}F(z)\,dz \to \int_{L_a} e^{zt}F(z)\,dz$$

as $c \to \infty$ and $b \to \infty$ [and the rectangle expands over the entire half-plane $\operatorname{Re}(z) \le a$]. Here, L_a is the vertical line $\operatorname{Re}(z) = a$. For $\operatorname{Re}(z) > a$, we have

$$2\pi i f(t) = \oint_C e^{zt}F(z)\,dz.$$

Now observe that

$$\int_0^r e^{-zt}\left[\oint_C e^{\xi t}F(\xi)\,d\xi\right]dt = \oint_C \int_0^r e^{-zt}e^{\xi t}F(\xi)\,dt\,d\xi$$

$$= \oint_C \int_0^r e^{(\xi-z)t}F(\xi)\,dt\,d\xi$$

$$= \oint_C \frac{1}{\xi-z}\left[e^{(\xi-z)t}\right]_0^r F(\xi)\,d\xi$$

$$= \oint_C \frac{1}{\xi-z}[e^{(\xi-z)r} - 1]F(\xi)\,d\xi.$$

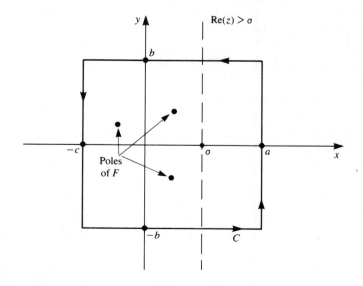

Figure 24.8

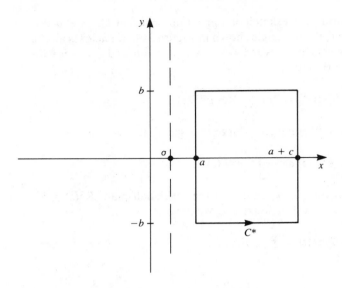

Figure 24.9. C^*—the reflection of C (Figure 24.8) about the
line $x = a$.

For ξ on C, $\text{Re}(\xi) < \text{Re}(z)$. Therefore, $e^{(\xi - z)r} \to 0$ as $r \to \infty$. Therefore,

$$2\pi i \mathcal{L}[f](z) = -\oint_C \frac{F(\xi)}{\xi - z} \, d\xi.$$

As $c \to \infty$ and $b \to \infty$, $-\oint_C [F(\xi)/(\xi - z)] \, d\xi \to -\int_{L_z} [F(\xi)/(\xi - z)] \, d\xi$.

Now reflect C across the line $x = a$ to obtain the curve C^* shown in Figure 24.9. By an argument similar to that just outlined, we obtain

$$\oint_{C^*} \frac{F(\xi)}{\xi - z} \, d\xi \to -\int_{L_a} \frac{F(\xi)}{\xi - z} \, d\xi.$$

But, by Cauchy's integral formula,

$$\oint_{C^*} \frac{F(\xi)}{\xi - z} \, d\xi = 2\pi i F(z).$$

Therefore, $2\pi i F(z) = 2\pi i \mathcal{L}[f](z)$.

PROBLEMS FOR SECTION 24.2

In each of Problems 1 through 10, use Theorem 24.3 to find the inverse Laplace transform of the function.

1. $\dfrac{z}{z^2 + 9}$

2. $\dfrac{1}{(z + 3)^2}$

3. $\dfrac{1}{(z - 2)^2(z + 4)}$

4. $\dfrac{1}{(z^2 + 9)(z - 2)^2}$

5. $\dfrac{1}{(z+5)^3}$ **6.** $\dfrac{\sin(z)}{z^2+16}$ **7.** $\dfrac{1}{z^4+1}$ **8.** $\dfrac{1}{e^z(z-1)}$

9. $\dfrac{\cos^2(z)}{(z-2)^3}$ **10.** $\dfrac{z+3}{(z^3-1)(z+2)}$

11. Use Theorem 24.3 to derive the Heaviside formulas given in Section 4.5. This should explain some of the similarities between the Heaviside formulas and formulas for residues.

24.3 *Harmonic Functions and the Dirichlet Problem* _____

Let D be a domain with boundary C. For example, D might consist of points $|z - z_0| < R$ in the disk of radius R about z_0. Then C is the circle $|z - z_0| = R$ bounding the disk. If D is the upper half-plane $\text{Im}(z) > 0$, C is the real axis $\text{Im}(z) = 0$.

A Dirichlet problem for D consists of finding a function $u(x, y)$ satisfying Laplace's equation in D and assuming prescribed values on the boundary of D. That is, we want u such that

$$\frac{\partial^2 u}{\partial x^2} + \frac{\partial^2 u}{\partial y^2} = 0$$

for (x, y) in D and

$$u(x, y) = g(x, y)$$

for (x, y) on C, with g a given function.

Any solution of Laplace's equation is called a *harmonic function*. Although ostensibly this Dirichlet problem is a problem in real analysis, we have already seen a connection between harmonic functions and analytic functions of a complex variable. By Theorem 19.5, the real and imaginary parts of an analytic function are harmonic.

This fact suggests a connection between Laplace's equation and complex functions but does not clarify how complex function theory might be of help in solving a Dirichlet problem. The following theorem is a first step in this direction.

THEOREM 24.4

Let u be harmonic in a simply connected domain D. Then there exists a function v such that $f(z) = u(x, y) + iv(x, y)$ is analytic in D. ∎

Such a function v is called a *harmonic conjugate* of u. The proof of Theorem 24.4 shows how to construct a harmonic conjugate.

Proof of Theorem 24.4 Let $g(z) = \partial u/\partial x - i(\partial u/\partial y)$ for (x, y) in D. We can check using the Cauchy-Riemann equations that g is analytic in D. Since D is simply connected, there is some analytic function G such that $G'(z) = g(z)$ for z in D. Write $G(z) = U(x, y) + iV(x, y)$.

From the derivative formulas at the end of Section 19.4, we can write

$$G'(z) = \frac{\partial U}{\partial x} - i\frac{\partial U}{\partial y} = g(z) = \frac{\partial u}{\partial x} - i\frac{\partial u}{\partial y}.$$

Therefore,

$$\frac{\partial U}{\partial x} = \frac{\partial u}{\partial x} \quad \text{and} \quad \frac{\partial U}{\partial y} = \frac{\partial u}{\partial y}.$$

For some constant K,

$$U(x, y) = u(x, y) + K.$$

Let $f(z) = G(z) - K$ for z in D. Then f is analytic in D. Further,

$$f(z) = G(z) - K = U(x, y) + iV(x, y) - K = u(x, y) + iV(x, y).$$

We may therefore choose $v(x, y) = V(x, y)$, proving the theorem. ■

We are rarely interested in actually computing a harmonic conjugate of a specific function. The value of knowing that a harmonic conjugate exists is that we can apply complex function theory to $f(z) = u(x, y) + iv(x, y)$. The harmonic function u, which is usually our objective, can be recovered from f as

$$u(x, y) = \text{Re}[f(z)].$$

We will now derive an integral representation for harmonic functions.

THEOREM 24.5

Let u be harmonic in a disk $|z - z_0| < R$. Let $z_0 = x_0 + iy_0$, and let r be any positive number with $0 < r < R$. Then

$$u(x_0, y_0) = \frac{1}{2\pi} \int_0^{2\pi} u(z_0 + re^{i\theta}) \, d\theta. \quad ■$$

This result is called the *mean value theorem for harmonic functions*. It states that the value of the harmonic function u at the center (x_0, y_0) of a disk is an average of values of u on a circle about (x_0, y_0). As θ varies from 0 to 2π, the point $z_0 + re^{i\theta}$ traverses the circle $|z - z_0| = r$ once counterclockwise.

Note that $z_0 + re^{i\theta} = x_0 + r\cos(\theta) + i[y_0 + r\sin(\theta)]$. Thus, in the theorem,

$$u(z_0 + re^{i\theta}) = u(x_0 + r\cos(\theta), y_0 + r\sin(\theta)),$$

and the conclusion of the theorem can be written

$$u(x_0, y_0) = \frac{1}{2\pi} \int_0^{2\pi} u(x_0 + r\cos(\theta), y_0 + r\sin(\theta)) \, d\theta.$$

Proof of Theorem 24.5   Let v be a harmonic conjugate of u. Then $f(z) = u(x, y) + iv(x, y)$ is analytic on $|z - z_0| < R$. By the Cauchy integral theorem,

$$f(z_0) = \frac{1}{2\pi i} \oint_{C_r} \frac{f(z)}{z - z_0} \, dz,$$

where C_r is the circle $|z - z_0| = r$. For z on C_r, $z = z_0 + re^{i\theta}$, with $0 \leq \theta \leq 2\pi$. Then

$$
f(z_0) = \frac{1}{2\pi i} \int_0^{2\pi} f(z_0 + re^{i\theta}) \frac{1}{re^{i\theta}} ire^{i\theta} \, d\theta
$$

$$
= \frac{1}{2\pi} \int_0^{2\pi} f(z_0 + re^{i\theta}) \, d\theta
$$

$$
= \frac{1}{2\pi} \int_0^{2\pi} [u(z_0 + re^{i\theta}) + iv(z_0 + re^{i\theta})] \, d\theta.
$$

We obtain the conclusion of the theorem upon taking the real part of this equation. ∎

We will use Theorem 24.5, in conjunction with conformal mappings, to write a formula for the solution of a Dirichlet problem as follows. Suppose that D is a simply connected domain with boundary C. Let T be a conformal mapping of D onto the unit disk $|w| < 1$ of the w-plane (Figure 24.10). Such a mapping exists by the Riemann mapping theorem if D is not the entire complex plane. Let z_0 be the point in D mapped to the origin by T, so that $T(z_0) = 0$. Then $z_0 = T^{-1}(0)$. If u is harmonic on D, by Theorem 24.5 we can write

$$
u(z_0) = u(T^{-1}(0)) = \frac{1}{2\pi} \int_0^{2\pi} u[T^{-1}(e^{i\theta})] \, d\theta.
$$

Let $w = e^{i\theta}$. Then

$$
u(z_0) = \frac{1}{2\pi i} \oint_{|w|=1} u[T^{-1}(w)] \frac{1}{w} \, dw.
$$

But $w = T(z)$, so

$$
\frac{1}{w} \, dw = \frac{T'(z)}{T(z)} \, dz
$$

and we can write this integral as an integral over the boundary C of D in the z-plane as

$$
u(z_0) = \frac{1}{2\pi i} \int_C u(\xi) \frac{T'(\xi)}{T(\xi)} \, d\xi.
$$

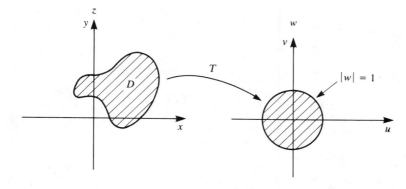

Figure 24.10

Finally, values of u are given on C. If $u(z) = g(z)$ on C,

$$u(z) = \frac{1}{2\pi i} \int_C g(\xi) \frac{T'(\xi)}{T(\xi)} \, d\xi. \tag{24.1}$$

Since g and T are presumed known, equation (24.1) is a formula for the solution of the Dirichlet problem on D.

We will illustrate the use of this formula by solving the Dirichlet problem for the disk $|z| < R$. The solution we will derive will be an integral formula called *Poisson's formula*. Let z_0 be any point of $|z| < R$. The mapping

$$T(z) = \frac{R(z - z_0)}{R^2 - z\bar{z}_0}$$

is a conformal mapping of $|z| < R$ onto $|w| < 1$ and maps z_0 to zero. Compute

$$\frac{T'(z)}{T(z)} = \frac{(R^2 - z\bar{z}_0)R - R(z - z_0)(-\bar{z}_0)}{(R^2 - z\bar{z}_0)^2} \frac{R^2 - z\bar{z}_0}{R(z - z_0)}$$

$$= \frac{R^2 - |z_0|^2}{(R^2 - z\bar{z}_0)(z - z_0)}.$$

In particular, if $|\xi| = R$, we get

$$\frac{T'(\xi)}{T(\xi)} = \frac{|\xi|^2 - |z_0|^2}{\xi|\xi - z_0|^2}.$$

Now equation (24.1) yields

$$u(z_0) = \frac{1}{2\pi i} \int_{|\xi| = R} g(\xi) \frac{|\xi|^2 - |z_0|^2}{\xi|\xi - z_0|^2} \, d\xi.$$

This is the solution of the Dirichlet problem for the disk $|z| < R$. To simplify this expression, write $\xi = Re^{i\varphi}$ on C ($0 \le \varphi \le 2\pi$) and write $z_0 = re^{i\theta}$. Then

$$u(re^{i\theta}) = \frac{1}{2\pi i} \int_0^{2\pi} g(Re^{i\varphi}) \frac{R^2 - r^2}{Re^{i\varphi}|Re^{i\varphi} - re^{i\theta}|^2} iRe^{i\varphi} \, d\varphi$$

$$= \frac{1}{2\pi} \int_0^{2\pi} g(Re^{i\varphi}) \frac{R^2 - r^2}{|Re^{i\varphi} - re^{i\theta}|^2} \, d\varphi.$$

Now compute

$$|Re^{i\varphi} - re^{i\theta}|^2 = (Re^{i\varphi} - re^{i\theta})(\overline{Re^{i\varphi} - re^{i\theta}})$$

$$= (Re^{i\varphi} - re^{i\theta})(Re^{-i\varphi} - re^{-i\theta})$$

$$= R^2 + r^2 - rR[e^{i(\varphi - \theta)} + e^{-i(\varphi - \theta)}]$$

$$= R^2 + r^2 - 2rR\cos(\varphi - \theta).$$

Therefore,

$$u(re^{i\theta}) = \frac{1}{2\pi} \int_0^{2\pi} g(Re^{i\varphi}) \frac{R^2 - r^2}{R^2 + r^2 - 2rR\cos(\varphi - \theta)} \, d\varphi. \tag{24.2}$$

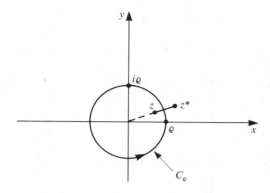

Figure 24.11

Equation (24.2) is *Poisson's integral formula* for the solution of the Dirichlet problem for the disk $|z| < R$.

Although we have used a conformal mapping to derive Poisson's integral formula, there are other ways to obtain the same result. We will discuss another technique, based on Cauchy's integral formula. To begin, note that u has a harmonic conjugate v. Consider the analytic function $f(z) = u(x, y) + iv(x, y)$ for $|z| < R$. Let C_ρ be the circle $|z| = \rho$, where $0 < \rho < R$. If $|z| < \rho$, use Cauchy's integral formula to write

$$f(z) = \frac{1}{2\pi i} \oint_C \frac{f(\xi)}{\xi - z}\, d\xi. \tag{24.3}$$

We will write this integral in a way which enables us to extract the real part. Let $z^* = \rho^2/\bar{z}$ for $|z| < \rho$. Then $|z^*| = \rho^2/|\bar{z}| > \rho$ because $|\bar{z}| = |z|$. (See Figure 24.11 for the geometric relationship between z and z^*.) Since $f(\xi)/(\xi - z^*)$ is analytic on $|z| < \rho$, we have, by Cauchy's theorem,

$$\frac{1}{2\pi i} \oint_{C_\rho} \frac{f(\xi)}{\xi - z^*}\, d\xi = 0. \tag{24.4}$$

Upon adding equations (24.3) and (24.4), we get

$$f(z) = \frac{1}{2\pi i} \oint_{C_\rho} f(\xi) \left[\frac{1}{\xi - z} - \frac{1}{\xi - z^*} \right] d\xi. \tag{24.5}$$

On C_ρ, we have $|\xi| = \rho$; hence,

$$\frac{1}{\xi - z} - \frac{1}{\xi - z^*} = \frac{1}{\xi - z} - \frac{1}{\xi - \dfrac{\rho^2}{\bar{z}}}$$

$$= \frac{1}{\xi - z} - \frac{1}{\xi - \dfrac{\xi\bar{\xi}}{\bar{z}}} = \frac{1}{\xi - z} - \frac{\bar{z}}{\xi(\bar{z} - \bar{\xi})}$$

$$= \frac{-\xi\bar{z} + |\xi|^2 + \xi\bar{z} - |z|^2}{\xi|\xi - z|^2} = \frac{|\xi|^2 - |z|^2}{\xi|\xi - z|^2}.$$

Substitute this result into equation (24.5) to get

$$f(z) = \frac{1}{2\pi i} \oint_{C_\rho} f(\xi) \frac{|\xi|^2 - |z|^2}{\xi|\xi - z|^2} \, d\xi.$$

This is similar to an expression derived previously using the conformal mapping T. Let $z = re^{i\theta}$ for some r in $(0, \rho)$ and θ in $[0, 2\pi)$. Let $\xi = \rho e^{i\varphi}$. We find that

$$f(z) = f(re^{i\theta}) = \frac{1}{2\pi} \int_0^{2\pi} f(\rho e^{i\varphi}) \frac{\rho^2 - r^2}{\rho^2 + r^2 - 2r\rho \cos(\varphi - \theta)} \, d\varphi.$$

Now take the real part of both sides of this equation. Since $\text{Re}[f(\rho e^{i\varphi})] = u(\rho e^{i\varphi})$, we get

$$u(re^{i\theta}) = \frac{1}{2\pi} \int_0^{2\pi} u(\rho e^{i\varphi}) \frac{\rho^2 - r^2}{\rho^2 + r^2 - 2r\rho \cos(\varphi - \theta)} \, d\varphi.$$

We must now invoke a fairly sensitive limit, which we will not rigorously justify. Recall that $\rho < R$. Let $\rho \to R$ to obtain the solution for u in the entire disk $|z| < R$. We obtain

$$u(re^{i\theta}) = \frac{1}{2\pi} \int_0^{2\pi} g(Re^{i\varphi}) \frac{R^2 - r^2}{R^2 + r^2 - 2rR \cos(\varphi - \theta)} \, d\varphi,$$

as before.

It is usually impossible to evaluate Poisson's integral in closed form. However, the integral formula for the solution does provide a means to obtain approximate numerical values of the solution. In some instances, we can actually evaluate Poisson's solution explicitly. As an example of this, imagine a cylinder aligned with the z-axis as in Figure 24.12. The boundary of the cylinder in the xy-plane is the circle $x^2 + y^2 = 1$. Suppose that the potential function is to be kept at zero on the upper half of this circle and 1 on the lower half. The solution for the potential function is

$$u(re^{i\theta}) = \frac{1}{2\pi} \int_\pi^{2\pi} \frac{1 - r^2}{1 + r^2 - 2r \cos(\varphi - \theta)} \, d\varphi$$

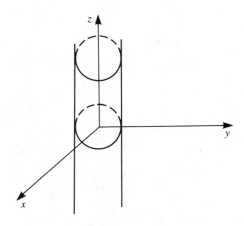

Figure 24.12

for $0 \le r < 1$ and $0 \le \theta < 2\pi$. With some effort, this integral can be evaluated as

$$u(re^{i\theta}) = \frac{1}{\pi} \tan^{-1}\left(\frac{1+r}{1-r} \tan\left[\frac{2\pi - \theta}{2}\right]\right)$$
$$- \frac{1}{\pi} \tan^{-1}\left(\frac{1+r}{1-r} \tan\left[\frac{\pi - \theta}{2}\right]\right).$$

As an illustration of the conformal mapping technique for solving a Dirichlet problem, we will consider the Dirichlet problem for the right half-plane $\text{Re}(z) > 0$. The boundary of the right half-plane is the imaginary axis $\text{Re}(z) = 0$. The Dirichlet problem is

$$\frac{\partial^2 u}{\partial x^2} + \frac{\partial^2 u}{\partial y^2} = 0 \qquad (x > 0, -\infty < y < \infty)$$

$$u(0, y) = g(y) \qquad (-\infty < y < \infty).$$

First, find a conformal mapping of the right half-plane onto the unit disk. Choose any z_0 in the right half-plane. The linear fractional transformation

$$w = T(z) = \frac{z - z_0}{z + z_0}$$

maps $\text{Re}(z) > 0$ onto $|w| < 1$, with z_0 mapped to the origin. We find by a routine computation that

$$\frac{T'(z)}{T(z)} = \frac{2z_0}{z^2 - z_0^2}.$$

Equation (24.1) yields the solution at z_0:

$$u(z_0) = \frac{1}{\pi i} z_0 \int_C u(\xi) \frac{1}{\xi^2 - z_0^2} \, d\xi,$$

in which C is the imaginary axis. Write $\xi = iy$. Then $u(\xi) = g(iy)$. With $z = x + iy$, the solution at z in the right half-plane is

$$u(x, y) = \frac{1}{\pi i}(x_0 + iy_0) \int_{-\infty}^{\infty} g(iy) \frac{1}{-y^2 - z^2} i \, dy$$

$$= \frac{1}{\pi}(-x_0 - iy_0) \int_{-\infty}^{\infty} g(iy) \frac{1}{y^2 + z^2} \, dy.$$

The solution $u(x, y)$ is the real part of the right side of this equation.

PROBLEMS FOR SECTION 24.3

1. Derive the solution

$$u(x, y) = \frac{y}{\pi} \int_{-\infty}^{\infty} \frac{g(t)}{(t - x)^2 + y^2} \, dt$$

for the Dirichlet problem for the upper half-plane: $\nabla^2 u = 0$ for $y > 0$, $u(x, 0) = g(x)$.

2. Derive the Schwarz integral formula for the solution of the Dirichlet problem for the unit disk $|z| < 1$:

$$u(x, y) = \text{Re}\left[\frac{1}{2\pi i} \oint_{|\xi|=1} u(\alpha, \beta)\left(\frac{\xi + z}{\xi - z}\right)\frac{1}{\xi} d\xi\right],$$

where $\xi = \alpha + i\beta$ is on the unit circle and $z = x + iy$ is in the disk $|z| < 1$. *Hint:* Suppose that u satisfies a Dirichlet problem for the disk $|z| < 1$. Write $f(z) = u(x, y) + iv(x, y)$. We may assume (by adding a constant if necessary) that $v(0, 0) = 0$. Expand f in a Maclaurin series $f(z) = \sum_{n=0}^{\infty} a_n z^n$. Show that

$$u(x, y) = \tfrac{1}{2}[f(z) + \overline{f(z)}] = u(0, 0) + \sum_{n=1}^{\infty} \tfrac{1}{2}(a_n z^n + \bar{a}_n \bar{z}^n).$$

Assume that the Maclaurin expansion of f is valid in some disk $|z| < 1 + \epsilon$, where $\epsilon > 0$. In this case, we can let $|z| = 1$ in the last series. Then $\bar{z} = 1/z$ and

$$u(x, y) = u(0, 0) + \sum_{n=1}^{\infty} \tfrac{1}{2}\left(a_n z^n + \bar{a}_n \frac{1}{z^n}\right)$$

Show that

$$a_n = \frac{1}{\pi i} \oint_{|z|=1} u(x, y) z^{-n-1} dz$$

for $n = 1, 2, 3, \ldots$. Now let $|z| < 1$ and $|\xi| = 1$. Write $z = x + iy$ and $\xi = \alpha + i\beta$. Let C denote the unit circle $|\xi| = 1$. Show that

$$f(z) = \frac{1}{2\pi i} \oint_C g(\alpha, \beta)\frac{1}{\xi} d\xi + \sum_{n=1}^{\infty}\left[\frac{1}{\pi i} \oint_C g(\alpha, \beta)\xi^{-n-1} d\xi\right] z^n$$

$$= \frac{1}{2\pi i} \oint_C g(\alpha, \beta)\left[1 + 2\sum_{n=1}^{\infty}\left(\frac{z}{\xi}\right)^n\right]\frac{1}{\xi} d\xi = \frac{1}{2\pi i} \oint_C g(\alpha, \beta)\frac{\xi + z}{\xi - z}\frac{1}{\xi} d\xi.$$

3. The Poisson formula for the solution of the Dirichlet problem for the disk $|z| < 1$ can be obtained from the Schwarz formula (Problem 2) by putting $\xi = e^{i\varphi} = \alpha + i\beta$ and $z = re^{i\theta}$. Carry out this calculation to derive the Poisson formula for the solution of the Dirichlet problem for the unit disk

$$u(r, \theta) = \frac{1 - r^2}{2\pi} \int_0^{2\pi} g(\cos(\varphi), \sin(\varphi))\frac{1}{1 + r^2 - 2r\cos(\varphi - \theta)} d\varphi.$$

4. Derive the solution

$$u(x, y) = \frac{y}{\pi} \int_0^{\infty} g(t)\left[\frac{1}{y^2 + (t - x)^2} - \frac{1}{y^2 + (t + x)^2}\right] dt$$

of the Dirichlet problem $\nabla^2 u = 0$ for $x > 0$ and $y > 0$, $u(0, y) = 0$, and $u(x, 0) = g(x)$.

5. Let $u(x, y) = (2/\pi)\tan^{-1}[(1 - x^2 - y^2)/2y]$. Show that $\nabla^2 u = 0$ in the upper half disk $y > 0$, $|z| < 1$, and $u(x, 0) = 1$ for $-1 < x < 1$, $u(e^{i\theta}) = 0$ for $0 < \theta < \pi$.

6. Write an integral formula for the solution of a Dirichlet problem for the domain $|z| > 1$ exterior to the unit circle.

*Notes on the History*_____
*of Complex Analysis*_____

The acceptance of complex numbers in algebra and analysis came only after a good deal of controversy among mathematicians. In 1770, the Swiss mathematician Leonhard Euler (1707–1783) wrote, "Because all conceivable numbers are either greater than zero, or less than zero, or equal to zero, then it is clear that the square roots of negative numbers cannot be included among the possible numbers." He went on to refer to complex numbers as impossible, or fancied, numbers.

However, complex numbers continued to occur in the solution of many problems, such as determining roots of polynomials. It would be only a matter of time before they were placed on a more solid foundation.

The major factors in the acceptance of complex numbers came in the nineteenth century as understanding of their geometric significance evolved. Caspar Wessel (1745–1818) was a Norwegian surveyor who wrote a paper in 1797 called "On the Analytic Representation of Direction; An Attempt." Wessel grasped the basic ideas of representing complex numbers as points in the plane and of adding them by the parallelogram law. His work went largely unnoticed until 1897, but it indicates how the germ of the idea was in the air and that its time was about to come.

In 1806, a Swiss bookkeeper, Jean-Robert Argand (1768–1822) wrote a short book on the geometric representation of complex numbers. Today, the complex plane is sometimes referred to as the Argand diagram in his honor. However, the major credit for the acceptance of complex numbers belongs to Carl Friedrich Gauss (1777–1855), who carried the prestige due the leading mathematician of his time. By 1815, it was clear that Gauss thoroughly understood the geometry of complex numbers, and by the 1830s, he included them in his writings without apology. As early as 1799, he had included many of his ideas about complex numbers in his proof of the fundamental theorem of algebra, which required the use of complex numbers as roots of polynomial equations.

Questions concerning complex functions and integrals began to be raised by Gauss and Siméon-Denis Poisson (1781–1840). Poisson originally worked on the recently developed Fourier series, but he was also the first to integrate along curves in the complex plane. It was left to Cauchy, however, to formulate and develop many of the properties of complex functions and integrals which bear his name today.

Augustin-Louis Cauchy (1789–1857) was born in Paris and become a professor at the École Polytechnique, the Sorbonne, and the Collège de France following a brief career as a military engineer. He made important contributions to the mechanics of waves in elastic media and to the theory of light, but his most important work was in mathematics, where he authored more than seven hundred papers, second only to Euler. His personal life was greatly influenced by his royalist sympathies, and he supported the Bourbons during a time of political upheaval in France. Napolean III excused Cauchy from the oath of allegiance required of all state employees of that time, possibly in recognition of Cauchy's stature as one of the leading mathematicians in Europe. Cauchy responded by donating his salary from the Sorbonne to the poor of Sceaux, the town in which he resided at the time.

In a sequence of papers from about 1814 to the early 1840s, Cauchy formulated his integral theorem and some of its consequences, including the concept of independence of path. He also grasped the ideas of poles and residues and the residue theorem, and he worked with power series and multiple-valued functions. (At this time, the only singularities that were understood were poles; the idea of an essential singularity was developed later.)

Pierre-Alphonse Laurent (1813–1854) developed his series about 1843. The result had also been known to the German mathematician Karl Weierstrass (1815–1897), who did not publish his ideas. Weierstrass studied the subtleties of power series representations of functions and helped establish a rigorous foundation for the theory of power series representations.

Georg Friedrich Riemann (1826–1866) may have more important theorems and concepts named for him than any other mathematician (the Riemann integral, Riemann sums, Riemann surfaces, the Riemann zeta function, . . .). Most of Riemann's work was in areas beyond an introductory treatment, but his name is found with Cauchy's on the fundamental Cauchy-Riemann equations. Riemann was also the first to realize that Δz must be allowed to go to zero along any path of approach to the origin in the limit of the difference quotient defining the complex derivative.

References and Further Reading

This section lists references for additional reading. They are listed by major area (differential equations, linear algebra, and so on). Of necessity, many books which could be read with profit are omitted from this list.

ORDINARY DIFFERENTIAL EQUATIONS

Birkhoff, G., and G. C. Rota. *Ordinary Differential Equations*. 3d ed. New York: Wiley, 1978.

Boyce, William E., and Richard C. DiPrima. *Elementary Differential Equations and Boundary Value Problems*. 4th ed. New York: Wiley, 1986.

Braun, Martin. *Differential Equations and Their Applications* (short version). New York: Springer-Verlag, 1978.

Campbell, Stephen L. *An Introduction to Differential Equations and Their Applications*. 2d ed. Belmont, CA: Wadsworth, 1990.

Goldberg, Jack L., and Arthur J. Schwartz. *Systems of Ordinary Differential Equations: An Introduction*. New York: Harper & Row, 1972.

Spiegel, Murray R. *Applied Differential Equations*. 3d ed. Englewood Cliffs, NJ: Prentice-Hall, 1981.

LINEAR ALGEBRA

Anton, Howard. *Elementary Linear Algebra*. 2d ed. New York: Wiley, 1977.

Kolman, Bernard. *Elementary Linear Algebra*. 2d ed. New York: Macmillan, 1977.

Noble, B., and J. W. Daniel. *Applied Linear Algebra*. 2d ed. Englewood Cliffs, NJ: Prentice-Hall, 1977.

Strang, Gilbert. *Linear Algebra and Its Applications*. New York: Academic Press, 1976.

VECTOR CALCULUS

Davis, Harry F., and Arthur David Snider. *Introduction to Vector Analysis.* 4th ed. Boston: Allyn & Bacon, 1979.

Schey, H. M. *Div, Grad, Curl, and All That.* New York: Norton, 1973.

Williamson, Richard E., Richard H. Crowell, and Hale F. Trotter. *Calculus of Vector Functions.* Englewood Cliffs, NJ: Prentice-Hall, 1968.

FOURIER ANALYSIS AND PARTIAL DIFFERENTIAL EQUATIONS

Berg, Paul W., and James L. McGregor. *Elementary Partial Differential Equations.* San Francisco: Holden-Day, 1966.

Broman, Arne. *An Introduction to Partial Differential Equations: From Fourier Series to Boundary Value Problems.* Reading, MA: Addison-Wesley, 1968.

Carrier, George F., and Carl E. Pearson. *Partial Differential Equations: Theory and Technique.* 2d ed. New York: Academic Press, 1988.

Colton, David. *Partial Differential Equations: An Introduction.* New York: Random House, 1988.

Davis, H. F. *Fourier Series and Orthogonal Functions.* Boston: Allyn & Bacon, 1963.

Franklin, Phillip. *An Introduction to Fourier Methods and the Laplace Transform.* New York: Dover, 1949.

Gustafson, Karl E. *Introduction to Partial Differential Equations and Hilbert Space Methods.* New York: Wiley, 1980.

Haberman, Richard. *Elementary Applied Partial Differential Equations, With Fourier Series and Boundary Value Problems.* Englewood Cliffs, NJ: Prentice-Hall, 1983.

Jackson, Dunham. *Fourier Series and Orthogonal Polynomials.* Carus Mathematical Monographs, no. 6. Washington, DC: Mathematical Association of America, 1941.

Powers, D. L. *Boundary Value Problems.* New York: Academic Press, 1972.

Seeley, R. T. *An Introduction to Fourier Series and Integrals.* New York: Benjamin, 1966.

Tolstov, Georgi P. *Fourier Series.* New York: Dover, 1962.

Weinberger, H. F. *A First Course in Partial Differential Equations.* Waltham, MA: Blaisdell, 1965.

Young, Eutiquio C. *Partial Differential Equations: An Introduction.* Boston: Allyn & Bacon, 1972.

Zachmanoglou, E. C., and Dale W. Thoe. *Introduction to Partial Differential Equations with Applications.* Baltimore: Williams & Wilkins, 1976.

COMPLEX ANALYSIS

Churchill, Ruel V., and James W. Brown. *Complex Variables and Applications.* 3d ed. New York: McGraw-Hill, 1984.

Derrick, William R. *Complex Analysis and Applications.* 2d ed. Belmont, CA: Wadsworth, 1984.

Dettman, J. W. *Applied Complex Variables.* New York: Macmillan, 1965.

Markushevich, A. I. *Theory of Functions of a Complex Variable*. 3 vols. Translated from the Russian by Richard A. Silverman. Englewood Cliffs, NJ: Prentice-Hall, 1965.

Marsden, Jerrold E. *Basic Complex Analysis*. San Francisco: W. H. Freeman, 1973.

Mathews, John H. *Basic Complex Variables for Mathematics and Engineering*. Boston: Allyn & Bacon, 1982.

Nehari, Z. *Introduction to Complex Analysis*. Boston: Allyn & Bacon, 1961.

Rothe, R. F., F. Ollendorf, and K. Pohlhausen. *Theory of Functions as Applied to Engineering Problems*. New York: Dover, 1961.

TRANSFORMS

Churchill, Ruel V. *Operational Mathematics*. 3d ed. New York: McGraw-Hill, 1972.

Jaeger, J. C. *Introduction to the Laplace Transform*. New York: Wiley, 1949.

Rainville, R. V. *The Laplace Transform*. New York: Macmillan, 1963.

Sneddon, Ian H. *The Use of Integral Transforms*. New York: McGraw-Hill, 1972.

CALCULUS OF VARIATIONS

Bliss, Gilbert Ames. *Calculus of Variations*. La Salle, IL: Open Court, 1925.

Sagan, Hans. *Introduction to the Calculus of Variations*. New York: McGraw-Hill, 1969.

Smith, Donald R. *Variational Methods in Optimization*. Englewood Cliffs, NJ: Prentice-Hall, 1974.

Young, L. C. *Lectures on the Calculus of Variations and Optimal Control Theory*. Philadelphia: W. B. Saunders, 1969.

NUMERICAL METHODS

Forsythe, G. E., and C. B. Moler. *Computer Solution of Linear Algebraic Systems*. Englewood Cliffs, NJ: Prentice-Hall, 1967.

Gear, C. W. *Numerical Initial Value Problems in Ordinary Differential Equations*. Englewood Cliffs, NJ: Prentice-Hall, 1971.

Henrici, Peter. *Elements of Numerical Analysis*. New York: Wiley, 1964.

Hildebrandt, F. B. *Introduction to Numerical Analysis*. New York: McGraw-Hill, 1964.

Maron, Melvin J. *Numerical Analysis: A Practical Approach*. New York: Macmillan, 1982.

Ortega, James M., and William G. Poole, Jr. *Numerical Methods for Differential Equations*. Belmont, CA: Pitman, 1981.

Ralston, A., and P. Rabinowitz. *A First Course in Numerical Analysis*. 2d ed. New York: McGraw-Hill, 1978.

Varga, R. S. *Matrix Iterative Analysis*. Englewood Cliffs, NJ: Prentice-Hall, 1962.

BESSEL FUNCTIONS

Bowman, F. *Introduction to Bessel Functions*. New York: Dover, 1958.

Gray, Andrew, and G. B. Mathews. *A Treatise on Bessel Functions and Their Applications to Physics.* New York: Dover, 1966.

Watson, G. N. *A Treatise on Bessel Functions.* 2d ed. London: Cambridge University Press, 1952.

CONFORMAL MAPPINGS

Bieberbach, L. *Conformal Mapping.* New York: Chelsea, 1953.

Kober, H. *Dictionary of Conformal Representations.* New York: Dover, 1957.

TABLES

Beyer, William H., ed. *CRC Standard Mathematical Tables.* 25th ed. Boca Raton, FL: CRC Press, 1981.

Answers and Solutions to Selected Odd-Numbered Problems

Chapter 0

1. $\dfrac{dT}{dt} = k(T - T_B)$, where T_B denotes the temperature of the surrounding medium.

3. $\dfrac{dP}{dt} = (b - d)P$, where b and d represent the birth and death rates, respectively.

5. From Newton's second law of motion, $my''(t) = -mg - \alpha y'(t)$, where m is the mass of the ball and α is its drag coefficient. Dividing by m and adding the initial conditions, the equation that describes the motion is $y'' = -g - \dfrac{\alpha}{m} y'$; $y(0) = 0$, $y'(0) = v_0$, where v_0 is the initial upward speed of the ball.

7. $\dfrac{dT}{dt} = k\sqrt{32 - t}$, where T is the thickness of the ice and t is the temperature.

9. $\dfrac{dP}{dt} = kP$. The constant of proportionality in this case is the interest rate, 6%. The solution of this differential equation is $P(t) = Ce^{kt} = Ce^{0.06t}$. Since the initial investment is $4000, we have $P(0) = Ce^0 = C = 4000$. Then the present value of the investment is $P(t) = 4000e^{0.06t}$.

11. $\dfrac{dS}{dt}$ = change in amount of salt in the tank with respect to time = input rate of salt − output rate of

salt $= \dfrac{2 \text{ pounds}}{10 \text{ gallons}} \cdot \dfrac{6 \text{ gallons}}{1 \text{ minute}} - \dfrac{S \text{ pounds}}{500 \text{ gallons}} \cdot \dfrac{6 \text{ gallons}}{1 \text{ minute}} = \dfrac{6}{5} - \dfrac{3S \text{ pounds}}{250 \text{ minutes}}$. The equation that describes the amount

of salt in the tank at time t is $\dfrac{dS}{dt} = \dfrac{6}{5} - \dfrac{3S}{250}$; $S(0) = S_0$.

13. Using Newton's laws of motion, we have $my'' = -mg + \rho \dfrac{4}{3}\pi R^3 + 6\pi\mu Ry'$, where m is the mass of the ball and ρ

represents the density of the grease. Letting $v = y'(t)$, we have $\dfrac{dv}{dt} = -g + \dfrac{4\rho\pi R^3}{3m} + \dfrac{6\pi\mu R}{m} v$. This assumes that the sphere is completely submersed in the grease. If the sphere is released into the grease from a position in which it is not completely submersed, the buoyant force will be proportional to the volume of the sphere that is below the surface of the grease. Of course, once the top of the ball reaches the grease surface, the given equation will apply.

15. Referring to Example 0.4 and Figure 0.4, we have the tangential displacement $s = \theta L$, where θ denotes the angular displacement from the vertical position. Here L is not constant, so $s' = \theta'L + \theta L'$ and $s'' = \theta''L + 2\theta'L' + \theta L''$. The differential equation that describes the motion as a function of time is $-mg\sin(\theta) = ms'' = m(\theta L)'' = m(\theta''L + 2\theta'L' + \theta L'')$. Dividing out the mass m and approximating $\sin(\theta)$ by θ, we have $\theta''L + 2\theta'L' + \theta L'' + g\theta = 0$. Since $L(t) = L_0 + Vt$, $L'(t) = V$ and $L''(t) = 0$. This reduces the equation to $(L_0 + Vt)\theta''(t) + 2V\theta'(t) + g\theta(t) = 0$, and, using $x = \dfrac{1}{V}(L_0 + Vt)$, we have $xV\theta''(t) + 2V\theta'(t) + g\theta(t) = 0$. To remove the time element from the equation, use chain rule differentiation to get $\dfrac{d\theta}{dt} = \dfrac{d\theta}{dx} \cdot \dfrac{dx}{dt}$ and $\dfrac{d^2\theta}{dt^2} = \dfrac{d}{dt}\left(\dfrac{d\theta}{dt}\right) = \dfrac{d}{dt}\left(\dfrac{d\theta}{dx} \cdot \dfrac{dx}{dt}\right) = \dfrac{d^2\theta}{dx^2}\left(\dfrac{dx}{dt}\right)^2 + \dfrac{d\theta}{dx} \cdot \dfrac{d^2x}{dt^2}$. Since $x = \dfrac{1}{V}(L_0 + Vt)$, $\dfrac{dx}{dt} = 1$ and $\dfrac{d^2x}{dt^2} = 0$. Put these values into the differential equation to get $xV\dfrac{d^2\theta}{dx^2} + 2V\dfrac{d\theta}{dx} + g\theta = 0$. Now divide by xV to obtain the requested form of the equation.

Chapter 1

Section 1.0

1. Yes; $\varphi'(x) = \dfrac{1}{2}\sqrt{x-1}$, so $2\varphi(x)\varphi'(x) = 2(\sqrt{x}-1)\left(\dfrac{1}{2\sqrt{x-1}}\right) = 1$ for all $x > 1$. **3.** Yes **5.** Yes

7. No. Since $\varphi(x) = (4 - x^2)/2x$, $\varphi'(x) = -(4 + x^2)/2x^2$. Then $x^2\varphi(x)\varphi'(x) = (x^4 - 16)/4x$; whereas $-1 - x\varphi^2(x) = -(x^4 - 8x^2 + 4x + 16)/4x$. Since these two functions are not equal on any interval, φ is not a solution of the equation $x^2yy' = -1 - xy^2$.

9. Yes **11.** $\dfrac{d}{dx}(y^2 + xy - 2x^2 - 3x - 2y) = 2yy' + y + xy' - 4x - 3 - 2y' = (y - 4x - 3) + (2y + x - 2)y' = 0 = \dfrac{d}{dx}($

13. $2xy' - 8x + ye^{xy} + xe^{xy}y' = 0$ **15.** $\dfrac{1}{1 + \left(\dfrac{y}{x}\right)^2}\left(\dfrac{xy^1 - y}{x^2}\right) + 2x = 0$ **17.** $y = C - e^{-x}; y = 3 - e^{-x}$

19. $y = 2\sin^2(x) + C; y = 2\sin^2(x) - 2$

Section 1.1

1. $\dfrac{1}{3}y^3 = \dfrac{1}{2}x^2 + \dfrac{1}{4}x^4 + C$ **3.** Not separable **5.** $2x^2 = y^3 + C$

7. $\ln|y| + \cos(x) = C$ or, equivalently, $y = Ke^{-\cos(x)}$ **9.** $y + 3\ln|y| = \dfrac{(x-1)^3}{3} + C$ or $y = 0$

11. $e^y = -2e^{-x}(x + 1) + C$ **13.** $\sin^2(y) = 2\ln(x) + C$ **15.** $\ln|y| - \dfrac{1}{2}\ln(2y^2 + 1) = \ln|x| - \ln|x + 1| + C$

17. $\ln|y + 2| = x^3 + C; \ln|y + 2| = x^3 + \ln(10) - 64$ **19.** $\dfrac{1}{2}y^2 = \dfrac{1}{3}x^3 + 2x + C; \dfrac{1}{2}y^2 = \dfrac{1}{3}x^3 + 2x + \dfrac{133}{6}$

21. $(y + 2)^2 = (x - 1)^2 + C; (y + 2)^2 = (x - 1)^2 + 60$ **23.** $-e^{-y} = \ln|x| + C; -e^{-y} = \ln|x| - e^{-4}, x > 0$

25. $\dfrac{1}{3}y^3 = \cos(x + 1) + C; \dfrac{1}{3}y^3 = 2\cos(x + 1) + \dfrac{64}{3} - 2\cos(\pi + 1)$

27. On the interval $[0, x]$, the y-coordinate of the centroid $= \dfrac{\int_0^x \dfrac{g^2(t)}{2}dt}{\int_0^x g(t)dt} = \dfrac{\int_0^x g(t)dt}{x - 0} = $ average value. $g(x) = kx^{1+\sqrt{2}}, k_1 > 0$.

Section 1.2

1. $i(t) = \dfrac{\alpha}{L\omega_1}\sin(\omega_1 t) - \dfrac{\beta}{L\omega_2}\cos(\omega_1 t) + \dfrac{\beta}{L\omega_2}$

3. (a) $\dfrac{dP}{dt} = k(P^2 - P)$; $\dfrac{P-1}{P} = Ke^{kt}$ or $P(t) = \dfrac{1}{1 - Ke^{kt}}$ **(b)** $K = \dfrac{999}{1000}$, so $\dfrac{P-1}{P} = \dfrac{999}{1000}e^{kt}$ or $P(t) = \dfrac{1000}{1000 - 999e^{kt}}$

(c) $k = \dfrac{1}{5}\ln\left(\dfrac{990}{999}\right)$, so $\dfrac{P-1}{P} = \dfrac{999}{1000}\left(\dfrac{990}{999}\right)^{t/5}$ or $P(t) = \dfrac{1000}{1000 - 999\left(\dfrac{990}{999}\right)^{t/5}}$ **(d)** $\lim\limits_{t \to \infty} P(t) = 1$

5. $\dfrac{1}{4}\ln(u) = \ln(T) + C$ or $\ln(u) = \ln(T^4) + C_1$; hence, $u(T) = kT^4$ for some constant k. This agrees with the Stefan-Boltzmann law if $k = \sigma A$.

7. $\dfrac{dV}{dh} = k \cdot (\text{cross-sectional area}) = k \cdot \pi r^2$. Since $V = \dfrac{4}{3}\pi r^3$, we have $r^2 = \left(\dfrac{3V}{4\pi}\right)^{2/3}$. Then $\dfrac{dV}{dh} = k\pi\left(\dfrac{3V}{4\pi}\right)^{2/3} = k\left(\dfrac{9\pi}{16}\right)^{1/3}V^{2/3}$.

$V(h) = \left[\left(\dfrac{\pi}{48}\right)^{1/3}kh + V_0\right]^3$, where h is the distance fallen and V_0 is the initial volume.

9. $\dfrac{dP}{dt} = kP$; $P(0) = 1600 \times 10^6$, where we start time in the year 1900. $P(t) = 1.6\left(\dfrac{251}{160}\right)^{t/50} \times 10^9$ people, with t in years after

1900. $P(100) = 1.6\left(\dfrac{251}{160}\right)^{100/50} \times 10^9 \approx 3940$ million people.

11. $\dfrac{dT}{dt} = k(T - 60)$; $T(0) = 90$. $T(t) = 60 + 30\left(\dfrac{2}{3}\right)^{t/10}$. $T(20) = \dfrac{220}{3} \approx 73.3$ degrees Fahrenheit. $T(t) = 65$ when

$t = 10\dfrac{\ln(\frac{1}{6})}{\ln(\frac{2}{3})} \approx 44.2$ minutes.

13. $T(t) = 70 + 24.6e^{t\ln(23.4/24.6)} = 70 + 24.6\left(\dfrac{23.4}{24.6}\right)^t$ degrees Fahrenheit, where t is in hours past noon. $T(t_m) = 98.6$ when

$t_m = \dfrac{\ln(28.6/24.6)}{\ln(23.4/24.6)} \approx -3$ hours. The crime was committed at approximately 9:00 A.M.

15. $P(t) = 10(\frac{1}{2})^{t/4.5} \times 10^9$ kilograms, with t in years. $P(10^9) = 10(\frac{1}{2})^{1/4.5} \approx 8.6$ kilograms.

17. $\dfrac{dP}{dh} = kP$, so $P(h) = Ke^{kh}$. $P(0) = 14.7 = K$; therefore, $P(h) = 14.7e^{kh}$. $P(18{,}000) = \dfrac{14.7}{2} = 14.7e^{18{,}000k}$, so $k = \dfrac{\ln(\frac{1}{2})}{18{,}000}$.

$P(h) = 14.7\left(\dfrac{1}{2}\right)^{h/18{,}000}$ pounds per square inch, with h in feet. The atmospheric pressure in Laramie is

$P(7200) = 14.7\left(\dfrac{1}{2}\right)^{7200/18{,}000} \approx 11.1$ pounds per square inch. An engine with a compression ratio of 7.9:1 should produce

$7.9 \times 11.1 \approx 88$ pounds per square inch at this altitude. This same engine would give a reading on a compression tester of $7.9 \times 14.7 \approx 116$ pounds per square inch at sea level.

Section 1.3

1. $\dfrac{1}{2}\dfrac{y^2}{x^2} = \ln|x| + C$ **3.** Not homogeneous **5.** $\ln|y| - \dfrac{x}{y} = C$ **7.** $\dfrac{1}{2}\dfrac{x^2}{y^2} = \ln|x| + C$

9. $\dfrac{2\sqrt{3}}{3}\arctan\left(\dfrac{2y-x}{\sqrt{3x}}\right) = \ln|x| + C$ **11.** $(y-3)\ln|y-3| = x + 2 + C(y-3)$ **13.** $(x - y + 3)^2 = 2x + C$

15. $14y - \ln|21x + 7y - 10| = 7x + C$ **17.** $3(x-2)^2 - 2(x-2)(y+3) - (y+3)^2 = C$

19. $(2x + y - 3)^2 = C(y - x + 3)$ **21.** $\dfrac{1}{4}x - y + \dfrac{1}{3}\ln|3(x-y)^2 - 2(x-y) - 1| - \dfrac{5}{6}\arctanh\left(\dfrac{3x - 3y - 1}{2}\right) = C$

23. $\sqrt{\dfrac{y}{x}} = 2\ln(x) + 1$ **25.** $\left(\dfrac{y}{x}\right)^3 = 3\ln(x) + 5$ **27.** $x^2 - 2xy + 2y^2 = 25$

29. $-\dfrac{2\sqrt{5}}{5}\arctanh\left(\dfrac{x + 2y}{\sqrt{5x}}\right) = \ln|x| - \dfrac{2\sqrt{5}}{5}\arctanh(9\sqrt{5}) - \ln(2)$

Section 1.4

1. $2xy^2 + e^{xy} + y^2 = C$ **3.** $xy + e^x = C$ **5.** Not exact **7.** $\ln(x^2 + y^2) = C$ or $x^2 + y^2 = K \neq 0$

9. $\cosh(x)\sinh(y) = C$ **11.** $y^3 + xy + \ln|x| = C$ **13.** $x\cos(4y^2) = C$ **15.** $-\arctan\left(\dfrac{s}{t}\right) + s^2 = C$

17. Not exact **19.** $t^2 = Cz^2$ **21.** $3xy^4 - x = C$; $3xy^4 - x = 47$ **23.** $xe^y - y = C$; $xe^y - y = 5$

25. $x^2 + y^3 + \cos(xy) = C$; $x^2 + y^3 + \cos(xy) = 9$ **27.** $-4x^2 + e^{xy} + y^2 = C$; $-4x^2 + e^{xy} + y^2 = 20 + e^{-12}$

29. $2xy - \tan(xy^2) = C$; $2xy - \tan(xy^2) = 4 - \tan(4)$ **31.** $\dfrac{\partial(F + C)}{\partial x} = \dfrac{\partial F}{\partial x} = M$ and $\dfrac{\partial(F + C)}{\partial y} = \dfrac{\partial F}{\partial y} = N$

Section 1.5

1. e^x; $(x + y - 1)e^x = C$ **3.** xy; $x^2y^3 - 3x^3y^2 = C$ **5.** e^{3y}; $xe^{3y} - e^y = C$ **7.** $\dfrac{1}{y + 1}$; $x^2y = C$ or $y = -1$

9. xe^{6x}; $e^{6x}(9x^2y + 12x - 2) = C$ **11.** e^{2y-x}; $(x^2 - 2xy)e^{2y-x} = C$ **13.** $\dfrac{1}{xy}$; $y = Cx^{-3/4}$, $x \neq 0$; $y = 6x^{-3/4}$, $x > 0$

15. ye^x; $xy^2e^x = C$; $xy^2e^x = 144e^4$ **17.** xe^{x^2}; $yx^2e^{x^2} = C$; $yx^2e^{x^2} = 12e^4$

19. $x^{2/3}y^{-1/3}$; $x^{2/3}y^{2/3}(5x + 12y) = C$; $x^{2/3}y^{2/3}(5x + 12y) = -29(2)^{2/3}$ **21.** $\dfrac{1}{y}$; $x^3 + xy^2 = C$ or $y = 0$; $x^3 + xy^2 = 8$ or $y = 0$

23. (a) $\dfrac{M_y - N_x}{N} = -\dfrac{2}{x}$ is a function of x alone. $\mu(x) = \dfrac{1}{x^2}$ (b) $\dfrac{N_x - M_y}{M} = -\dfrac{2}{y}$ is a function of y alone. $\mu(y) = \dfrac{1}{y^2}$

(c) $\mu(x, y) = x^ay^b$ is an integrating factor for any real numbers a and b such that $a + b = -2$. Note that in (a), $a = -2$
and $b = 0$, whereas in (b), $a = 0$ and $b = -2$.

25. $\dfrac{\partial}{\partial x}(\mu Ng(\varphi(x, y))) - \dfrac{\partial}{\partial y}(\mu Mg(\varphi(x, y))) = \left[\dfrac{\partial\mu}{\partial x}N + \mu\dfrac{\partial N}{\partial x} - \dfrac{\partial\mu}{\partial y}M - \mu\dfrac{\partial M}{\partial y}\right]g(\varphi(x, y)) + g'(\varphi(x, y))\left[\mu N\dfrac{\partial\varphi}{\partial x} - \mu M\dfrac{\partial\varphi}{\partial y}\right] = 0$

because $\mu M\,dx + \mu N\,dy = 0$ is exact and $\varphi(x, y) = C$ is a solution $\left(\text{hence } M\dfrac{\partial\varphi}{\partial y} - N\dfrac{\partial\varphi}{\partial x} = 0\right)$

29. $x^{-2/3}y^{4/3}$ is an integrating factor. $x^{7/3}y^{7/3} - 14x^{1/3} = C$ defines the general solution.

31. $x^{-7/4}y^{-3/2}$; $3x^{1/4}y^{-1/2} - x^{-3/4} = C$. **33.** y^{-3}; $y^2 = \dfrac{x^2}{c - 2x}$ **35.** $x^{-9/4}y^{3/4}$; $\dfrac{4}{7}x^{7/4}y^{7/4} + \dfrac{4}{5}x^{-5/4} = C$.

37. $x^{-9}y^{-4}$; $16x^{-6}y^{-3} + 15x^{-8} = C$.

Section 1.6

1. $y = \dfrac{1}{2}x^3 + 2x\ln|x| + Cx$ **3.** $y = \dfrac{1}{7}e^{2x} + Ce^{-3x/2}$ **5.** $y = \dfrac{1}{2}x - \dfrac{1}{4} + Ce^{-2x}$ **7.** $y = -\dfrac{3}{2}x^{-1}e^{-x^2} + Cx^{-1}$

9. $y = \dfrac{1}{4}\dfrac{(x - 2)^2}{x + 1} + C\dfrac{1}{(x + 1)(x - 2)^2}$ **11.** $y = -3 + Ce^{x^2/2}$ **13.** $y = \dfrac{2}{3}e^{4x} - \dfrac{11}{3}e^x$

15. $y = \dfrac{27}{41}x^4 + \dfrac{9}{23}x^2 - \dfrac{2782}{943}x^{-5/9}$ **17.** $y = \dfrac{2}{5}x - \dfrac{22}{5}x^{-4}$ **19.** $y = x^2 - x - 2$ **21.** $xe^{2y} - y = C$

23. Let $u = y^{1-\alpha}$. Then $u' = \dfrac{du}{dx} = (1 - \alpha)y^{-\alpha}\dfrac{dy}{dx}$. Multiply the differential equation by $(1 - \alpha)y^{-\alpha}$ to get

$(1 - \alpha)y^{-\alpha}y' + (1 - \alpha)P(x)y^{1-\alpha} = (1 - \alpha)Q(x)$. Now substitute u and u' into this equation to get
$u' + (1 - \alpha)P(x)u = (1 - \alpha)Q(x)$, a linear equation in $u(x)$.

25. $y = \dfrac{1}{Ce^{-2x} - 2x + 1}$ **27.** $y = \left(Ce^{-2x} + \dfrac{1}{4}x^3 - \dfrac{3}{8}x^2 + \dfrac{3}{8}x - \dfrac{3}{16}\right)^4$ **29.** $(xy)^{7/3} - 14x^{1/3} = C$

35. $R(x) = \dfrac{k}{c}x - \dfrac{Ik}{c^2} + Ke^{-cx/I}$; $R(x) = \dfrac{k}{c}x - \dfrac{Ik}{c^2}(1 - e^{-cx/I})$

37. Solving for i, we have $i(t) = \dfrac{E - E_c}{R} + Ke^{-Rt/L}$, where $K = i(0) - \dfrac{E - E_c}{R}$. Then

$$\dfrac{dN}{dt} + \dfrac{k}{M} = 2nrBsi = 2nrBs\left(\dfrac{E - E_c}{R} + Ke^{-Rt/L}\right).$$

Solving this yields

$$N(t) = \frac{2nrBs(E - E_c)}{kR} + \frac{2nrBsKL}{Lk - Mr}e^{-Rt/L} + Ce^{-kt/M}.$$

Section 1.7

1. $y = x + \left(\dfrac{C}{x} - \dfrac{1}{x}\ln(x)\right)^{-1}$; $y = x + \left(\dfrac{1}{2x} - \dfrac{1}{x}\ln(x)\right)^{-1}$ **3.** $S(x) = 2$ works; $y = 2 + \left(Cx^2 - \dfrac{1}{2}\right)^{-1}$; $y = 2 + \left(x^2 - \dfrac{1}{2}\right)^{-1}$

5. $y = -e^x + \left(Ce^{-3x} + \dfrac{1}{2}e^{-x}\right)^{-1}$; $y = -e^x + \left(-\dfrac{5}{14}e^{-3x} + \dfrac{1}{2}e^{-x}\right)^{-1}$

7. $S(x) = 8x^2$ works; $y = 8x^2 + \left(C + \dfrac{1}{16x}\right)^{-1}$; $y = 8x^2 + \left(-\dfrac{27}{176} + \dfrac{1}{16x}\right)^{-1}$

9. $S(x) = 4x$ works; $y = 4x + \left(\dfrac{C}{x} - \dfrac{1}{3}x^2\right)^{-1}$; $y = 4x + \left(\dfrac{5}{3x} - \dfrac{1}{3}x^2\right)^{-1}$

Section 1.8

1. The bag attains a maximum height of 342.25 feet. It strikes the ground at a speed of 148 feet/second, 4.75 seconds after being dropped.

3. The ship's velocity is $\frac{44}{3}(1 - e^{-t/1200})$ feet/second (t in seconds) $= 10(1 - e^{-3t})$ miles/hour (t in hours) with a terminal velocity of 10 miles/hour. A speed of 8.5 miles/hour is attained approximately 3.49 miles from the starting point.

5. $v(t) = 18 - 16e^{-2t/3}$ feet second; $s(t) = 18t + 24(e^{-2t/3} - 1)$ feet

7. $v(t) = 32 - 32e^{-t}$ feet/second for $0 \le t \le 4$ and $v(t) = \dfrac{8(1 + Ke^{-8t})}{1 - Ke^{-8t}}$ feet/second for $t \ge 4$, where $K = \dfrac{e^{32}(3e^4 - 4)}{5e^4 - 4}$.

$\lim\limits_{t \to \infty} v(t) = 8$ feet/second. $s(t) = 32(t + e^{-t} - 1)$ feet for $0 \le t \le 4$ and $s(t) = 8t + 2\ln\left(1 - \dfrac{(3e^4 - 4)e^{8(4-t)}}{5e^4 - 4}\right) + 64 + 32e^{-4} -$

$2\ln\left(\dfrac{2e^4}{5e^4 - 4}\right)$ feet for $t \ge 4$.

9. Denoting upward as positive and the bottom of the lake as zero with the surface at -100 feet, the box has a velocity of $v(t) = 46 - 46e^{t/22}$ feet/second with position $s(t) = 46(t - 22e^{t/22} + 22)$ feet. The box surfaces with an upward velocity of approximately 23.6 feet/second, about 9.1 seconds after leaving the bottom.

11. 64 pounds at 990 feet and 26.8 pounds at 3690 feet

13. $v_0 = \sqrt{2g \cdot \dfrac{R_e^2}{M_e} \cdot \dfrac{0.0123M_e}{0.2725R_e}} \approx 0.2125\sqrt{2gR_e} = 0.2125 \times$ the earth's escape velocity $\approx (0.2125)(6.95)$ miles/second $\approx$ 1.48 miles/second.

15. Letting θ represent the angle the chord makes with vertical, we have $m\dfrac{dv}{dt} = mg\cos(\theta)$; $v(0) = 0$. Then $s(t) = \dfrac{1}{2}gt^2\cos(\theta)$, so $t^2 = \dfrac{2s}{g}\cos(\theta)$ is the square of the time required to travel s units along the chord. By the law of cosines, the length of the chord is $s = 2R\cos(\theta)$, where R is the radius of the circle. Therefore, $t = 2\sqrt{\dfrac{R}{g}}$ independent of θ.

17. Assuming that upward is positive, an equation for velocity is $m\dfrac{dv}{dt} = -mg - kv|v|$; $v(0) = v_0$. The position $s(t)$ can be obtained through integrating v and using the initial condition, $s(0) = 0$. Consider two cases, $v > 0$ (upward flight) and $v < 0$ (downward flight). Find the rise time, t_{max}, by setting $v(t) = 0$ in the first equation and use this in the position function ($0 \le t \le t_{max}$) to find the maximum height, s_{max}. In the $v < 0$ case, use $v(t_{max}) = 0$ and $s(t_{max}) = s_{max}$ for initial conditions. The equations for the velocity and position are

$$v(t) = \begin{cases} \sqrt{\dfrac{mg}{k}}\tan\left(C - \sqrt{\dfrac{kg}{m}}\,t\right), & 0 \le t \le t_{max} \\[3mm] -\sqrt{\dfrac{mg}{k}}\tanh\left(\sqrt{\dfrac{kg}{m}}[t - t_{max}]\right), & t \ge t_{max} \end{cases}$$

and

$$
s(t) = \begin{cases}
\dfrac{m}{k} \ln\left[\dfrac{\sec(C)}{\sec\left(C - \sqrt{\dfrac{kg}{m}}\, t\right)} \right], & 0 \le t \le t_{max} \\[4ex]
\dfrac{m}{2k} \ln\left(\dfrac{mg + kv_0^2}{mg} \right) - \dfrac{m}{k} \ln\left[\cosh\left(\sqrt{\dfrac{kg}{m}}\, [t - t_{max}] \right) \right], & t \ge t_{max},
\end{cases}
$$

where $C = \arctan\left(\sqrt{\dfrac{k}{mg}}\, v_0 \right)$ and $t_{max} = \sqrt{\dfrac{m}{kg}} \arctan\left(\sqrt{\dfrac{k}{mg}}\, v_0 \right)$. To show that the rise time is less than the fall time, show that $s(2t_{max}) > 0$, which indicates that the ball is not yet back to ground level at this time. To show that $s(2t_{max}) > 0$, use the fact that $\sqrt{1 + x^2} \ge \cosh[\arctan(x)]$ for all x, with strict inequality whenever $x \ne 0$.

19. $V_c(t) = \dfrac{q_c(t)}{C} = 80(1 - e^{-2t})$ volts. $V_c(t) = 76$ when $t = \dfrac{1}{2} \ln(20) \approx 1.5$ seconds. $i(t) = q'(t) = \dfrac{1}{3125} e^{-2t}$.

$i\left(\dfrac{1}{2} \ln(20) \right) = 1.6 \times 10^{-5} = 16$ microamperes. Another approach is to note that when the capacitor has a potential of 76 volts, the resistor must be dropping 4 volts. Since the voltage drop across a resistor is the product of the current and its resistance, the current at this instant must be $\dfrac{4}{250,000} = 16$ microamperes.

21. $i_1(0+) = i_2(0+) = \dfrac{3}{20}$ ampere. $i_3(0+) = 0$

23. (a) If $E(t) = E$, then $q(t) = EC + Ke^{-t/RC}$. $q(0) = q_0 = EC + K$, so $q(t) = EC + (q_0 - EC)e^{-t/RC}$.
(b) $\lim\limits_{t \to \infty} q(t) = EC$ independent of q_0.
(c) $q'(t) = \dfrac{EC - q_0}{RC} e^{-t/Rc}$, which is never zero unless $EC = q_0$. In this case, $q(t)$ has a constant value of EC. The maximum value of $q(t)$ is q_0, occurring at time $t = 0$ if $q_0 > EC$; otherwise, if $Ec > q_0$, the charge never attains a maximum value but increases toward the value of EC.

25. Assuming that the initial current is zero, we have $i(t) = \dfrac{A}{23}(e^{-2t/25} - e^{-t})$ amperes.

27. $i(t) = \dfrac{AL}{R^2 + L^2\omega_1^2}\left[\dfrac{R}{L}\sin(\omega_1 t) - \omega_1\cos(\omega_1 t) \right] + \dfrac{BL}{R^2 + L^2\omega_2^2}\left[\dfrac{R}{L}\sin(\omega_2 t) - \omega_2\cos(\omega_2 t) \right]$
$\qquad + \left[\dfrac{AL\omega_1}{R^2 + L^2\omega_1^2} - \dfrac{BR}{R^2 + L^2\omega_2^2} \right]e^{-Rt/L}$

29. $i(t) = \dfrac{1}{R} + \dfrac{e^{-t}}{R - L} - \left(\dfrac{1}{R} + \dfrac{e^{-t}}{R - L} \right)e^{-Rt/L}$ if $R \ne L$ and $i(t) = \dfrac{1}{L} + \left(\dfrac{t - 1}{L} \right)e^{-t}$ if $R = L$.

Section 1.9

1. $x(t) = 4(t + 50) - 48,750\sqrt{2}(t + 50)^{-3/2}$ pounds. The tank overflows when $t = 200$ and $x(200) \approx 983$ pounds.

3. $x(t) = 50 - 30e^{-t/20}$ pounds (tank 1); $y(t) = 75 + 90e^{-t/20} - 75e^{-t/30}$ pounds (tank 2). $y(t)$ has a minimum value of $\dfrac{5450}{81}$ pounds at $60 \ln(\tfrac{9}{5})$ minutes.

5. (b) If $2x_0 = y_0$, then $z(t) = x_0 - \dfrac{1}{\sqrt{8kt + \dfrac{1}{x_0^2}}}$; so $x(t) = \dfrac{1}{\sqrt{8kt + \dfrac{1}{x_0^2}}}$ and $y(t) = 2x(t)$.

If $2x_0 \ne y_0$, $\dfrac{1}{y_0 - 2z} - \dfrac{1}{2x_0 - y_0} \ln\left(\dfrac{x_0 - z}{y_0 - 2z} \right) = (2x_0 - y_0)kt + \dfrac{1}{y_0} - \dfrac{1}{2x_0 - y_0}\ln\left(\dfrac{x_0}{y_0} \right)$.

(c) Assume that $x_0 < y_0 < 4x_0$. If $y_0 \le x_0$, chemical H vanishes before K is halved, and if $y_0 \ge 4x_0$, chemical K is gone before H is halved. If $2x_0 = y_0$, H is halved when $z = x = x_0/2$. This occurs when $t = \tfrac{3}{8}kx_0^2$. K is halved at the same time. If $2x_0 \ne y_0$, then H is halved when $z = x_0/2$, when

$$
t = \left. \frac{x_0}{y_0(y_0 - x_0)} + \frac{1}{2x_0 - y_0}\ln\left(\frac{2(x_0 - y_0)}{y_0} \right) \right/ k(2x_0 - y_0).
$$

K is halved when $z = \frac{1}{4}y_0$, when

$$t = \left.\left(\frac{1}{y_0} - \frac{1}{2x_0 - y_0}\ln\left(\frac{2x_0}{4x_0 - y_0}\right)\right)\right/ k(2x_0 - y_0).$$

7. $y = 2x + C$　　**9.** $y = Cx^2$　　**11.** $y = Cx^{-3}$　　**13.** $y = Ce^{4x}$　　**15.** $y^2[\ln(y^2) - 1] + 2x^2 = C$

17. $y - \frac{1}{3}y^3 = x^2 + C$　　**19.** $\frac{4}{3}y^{3/2} = C - x$　　**21.** $3y^2 + 2x^2 = C$　　**23.** $y^2 = \ln|x| - \frac{1}{2}x^2 + C$　　**27.** $y = C - 3x$

29. $x^2 + y^2 = C(y + \sqrt{3}x)$　　**33.** $r = Ce^{\theta}$　　**35.** $\frac{1}{2}\theta^2 = -2\ln|r| + C$

Section 1.10

1. Both $f(x, y) = 2y^2 + 3xe^y\sin(xy)$ and $\frac{\partial f}{\partial y}(x, y) = 4y + 3xe^y\sin(xy) + 3x^2e^y\cos(xy)$ are continuous everywhere.

3. Both $f(x, y) = (xy)^3 - \sin(y)$ and $\frac{\partial f}{\partial y}(x, y) = 3x^3y^2 - \cos(y)$ are continuous everywhere.

5. Both $f(x, y) = x^2ye^{-2x} + y^2$ and $\frac{\partial f}{\partial y}(x, y) = x^2e^{-2x} + 2y$ are continuous everywhere.

7. $y(x) = 0$ and $y(x) = \left(\frac{2x}{3}\right)^{3/2}$ are each solutions of the initial value problem. $f(x, y) = y^{1/3}$, so $\frac{\partial f}{\partial y}(x, y) = \frac{1}{3}y^{-2/3}$, which is not continuous in any rectangle containing the point $(0, 0)$.

9. (b) $y(x) = e^{x^2/2}$

(c) $y_1 = 1 + \frac{1}{2}x^2$, $y_2 = 1 + \frac{1}{2}x^2 + \frac{1}{8}x^4$, $y_3 = 1 + \frac{1}{2}x^2 + \frac{1}{8}x^4 + \frac{1}{48}x^6$, $y_4 = 1 + \frac{1}{2}x^2 + \frac{1}{8}x^4 + \frac{1}{48}x^6 + \frac{1}{348}x^8$

(d) $y_n = 1 + \frac{1}{2}x^2 + \frac{1}{8}x^4 + \frac{1}{48}x^6 + \frac{1}{348}x^8 + \cdots + \frac{x^{2n}}{2^n n!}$　　(e) $e^{x^2/2} = \sum_{n=0}^{\infty}\frac{x^{2n}}{2^n n!}$

11. (b) $y(x) = -4 + 2x - \frac{1}{2}x^2$　　(c) $y_1 = -4 + 2x - \frac{1}{2}x^2 = y_2 = y_3 = y_4$ (y_1 is the solution)　　(d) $y_n = y_1$

(e) $-4 + 2x - \frac{1}{2}x^2$

13. $y(x) = y_1 = \frac{7}{3} + \frac{2}{3}x^3 = y_2 = y_3 = y_4 = \cdots = y_n$ (y_1 is the solution)

Section 1.11

1. $y = 5e^{(x-2)} - x - 1$　　**3.** $y = (x + 4)e^{2x}$　　**5.** $y = 4e^{x^2/2}$　　**7.** $y = \frac{2}{3}x^2 - \frac{9}{x}$　　**9.** $y = \frac{1}{2}x^5 - x - \frac{233}{54}x^3$

17. Choose a point (x_0, z) on the line $x = x_0$. Then the tangent line to the solution curve through this point (the lineal element there) is of the form $y = m(x - x_0) + z$, where $m = y'(x_0, z) = q(x_0) - p(x_0)z$. That is, an equation of the tangent line to the solution curve through (x_0, z) is $y = [q(x_0) - p(x_0)z](x - x_0) + z$. We want to show that all of these tangent lines pass through the common point $\left(x_0 + \frac{1}{p(x_0)}, \frac{q(x_0)}{p(x_0)}\right)$. Inserting the value $x_0 + \frac{1}{p(x_0)}$ for x in the equation of the line yields $y = \frac{q(x_0)}{p(x_0)}$, the desired value.

Additional Problems for Chapter 1

1. $6xe^y - x^4y = C$　　**3.** $y = \frac{8}{27}[9x^3 - 9x^2 + 6x - 2] + Ce^{-3x}$　　**5.** $y^2 = x^2 - x + \frac{1}{2} + Ce^{-2x}$

7. $x^2 - 4xy + 8x + 4y^2 + 6y = C$　　**9.** $y = \frac{x^2}{C - x}$　　**11.** $(y - x)^2(x + y + 2)^3 = C$　　**13.** $y = x^{-2}(1 + Ce^{-x})$

15. $4x^3e^y + \cos(x - y) = C$　　**17.** $y^2 - 3x^2 = C$　　**19.** $y^5 + Cy^2 = x$　　**21.** $y = (x + 2)(x + 3)^2 + C(x + 2)$

23. $y = x + \frac{2}{C - x^2}$　　**25.** $x - 2y + \ln|1 + x - y| = C$　　**27.** $3(y + 3)^2 = 2\ln\left|\frac{x - 1}{x + 2}\right| + C$　　**29.** $x = (y + C)e^{y^2}$

31. $y = \dfrac{1}{4}x + Cx^{-3}$ **33.** $y = \dfrac{x}{C - \ln|x|}$ or $y = 0$ **35.** $y = 1 + \dfrac{1}{1 - x + Ce^{-x}}$ **37.** $x^2 = (C - 2e^x)y^2$ or $y = 0$

39. $\sec\left(\dfrac{y}{x}\right) + \tan\left(\dfrac{y}{x}\right) = Cx$ **41.** $3x^3y^2 + 4x^2 = C$ **43.** $x^4y = x\sin(x) + \cos(x) + C$ **45.** $8x^3y + e^x\sin(y) = C$

47. $y = \dfrac{1}{9}e^{-x} + Ce^{8x}$ **49.** $x^8y^4e^{-x} = C$ **51.** $(y + 1)e^{-y} = x^{-1} + C$ **53.** $x^2(x^2 + 2y^2) = C$

55. $2x + \sin(xy) = C$ **57.** $y^{-1/3} = Cx^{1/3} - \dfrac{1}{8}x^3$ or $y = 0$ **59.** $y = 3(x^2 + 2x + 2) + Ce^x$

61. $y = Ce^x - 2x^2 - 4x - 4$ **63.** $e^{x-y} + y = C$ **65.** $y = e^{5x}[C - \cos(x)]$ **67.** $\dfrac{2\sqrt{3}}{3}\arctan\left[\dfrac{2y + x}{\sqrt{3x}}\right] = \ln|x| + C$

69. $\dfrac{2\sqrt{7}}{7}\arctan\left[\dfrac{2y - x}{\sqrt{7x}}\right] = \ln|x| + C$ **71.** $y = x^{-3}[(4x^3 - 24x)\sin(x) - (x^4 - 12x^2 + 24)\cos(x) + C]$

73. $y^3 - 3xy = C$ **75.** $y = C[2x + \sin(x)]$

77. $\ln\left|\left(y - \dfrac{7}{3}\right)^2 + 3\left(y - \dfrac{7}{3}\right)\left(x + \dfrac{10}{3}\right) - \left(x + \dfrac{10}{3}\right)^2\right| + \dfrac{2\sqrt{13}}{13}\operatorname{arctanh}\left[\dfrac{9x + 6y + 16}{\sqrt{13}(3x + 10)}\right] = C$ **79.** $y = \tan(x + C) - x$

81. $C = -3, K = 2$ **83.** (a) $2\left(\dfrac{1.7}{2}\right)^{5/2} \approx 1.3322$ grams;   (b) $80\,\dfrac{\ln\left(\dfrac{1}{2}\right)}{\ln\left(\dfrac{1.7}{2}\right)} \approx 341.2$ years

Chapter 2

Section 2.1

1. $y = \dfrac{8}{3}e^x - \dfrac{5}{3}e^{4x}$ **3.** $y = 3e^{-4x} + 15xe^{-4x}$ **5.** $y = \dfrac{1}{4}e^{4x} + \dfrac{3}{4}e^{-4x}$ **7.** $y = -\dfrac{1}{4}\cos(2x) - \dfrac{\pi}{8}\sin(2x) + \dfrac{1}{2}x$

9. $y = 0$

Section 2.2

1. $W(x) = k \neq 0$; $y = c_1\cosh(kx) + c_2\sinh(kx)$; $y = \cosh(kx)$

3. $W = -8e^{-4x} \neq 0$; $y = c_1e^{2x} + c_2e^{-6x}$; $y = \dfrac{5}{8}e^{2x} + \dfrac{3}{8}e^{-6x}$

5. $W = 5e^x \neq 0$; $y = c_1e^{-2x} + c_2e^{3x}$; $y = \dfrac{12}{5}e^{3(x+1)} + \dfrac{3}{5}e^{-2(x+1)}$

7. $W = x^7 \neq 0$ for $x > 0$; $y = c_1x^4 + c_2x^4\ln(x)$; $y = 2x^4 - 4x^4\ln(x)$

9. x^2 and x^3 are not linearly independent *solutions* of any differential equation of the form $L[y] = y'' + p(x)y' + q(x)y = 0$ on an interval containing zero.

11. Let $p(x) = -3$ and $q(x) = 2$. $\varphi(x) = e^x$ and $\psi(x) = e^{2x}$ are each solutions of $y'' - 3y' + 2y = 0$; however, their product, e^{3x}, is not a solution of this equation.

13. $(a\varphi + b\psi)'' - 4(a\varphi + b\psi) = 8x(a + b) \neq 8x$ unless $a + b = 1$.

15. $0 \cdot \varphi(x) + a \cdot 0 = 0$ for all x in I for any nonzero number a.

17. If both φ and ψ have extrema at x_0, then $\varphi'(x_0) = \psi'(x_0) = 0$, so

$$W(\varphi, \psi)(x_0) = \begin{vmatrix} \varphi(x_0) & \psi(x_0) \\ \varphi'(x_0) & \psi'(x_0) \end{vmatrix} = \begin{vmatrix} \varphi(x_0) & \psi(x_0) \\ 0 & 0 \end{vmatrix} = 0.$$

19. If both φ and ψ vanish at x_0, then $W(\varphi, \psi)(x_0) = \begin{vmatrix} \varphi(x_0) & \psi(x_0) \\ \varphi'(x_0) & \psi'(x_0) \end{vmatrix} = \begin{vmatrix} 0 & 0 \\ \varphi'(x_0) & \psi'(x_0) \end{vmatrix} = 0.$

21. $W(x) = W[y_1, y_2](x_0)e^{-\int_{x_0}^x p(t)\,dt}$ is constant if and only if $F(x) = \int_{x_0}^x p(x)\,dt$ is constant if and only if $F'(x) = 0$ for all x in (a, b). Now use the fundamental theorem of calculus to get $F'(x) = p(x) = 0$ for all $x \in (a, b)$ if and only if $p(x) = 0$ for all $x \in (a, b)$.

Section 2.3

1. $y = c_1 + c_2 x^2 - 2x$ **3.** $y = c_1 + c_2 x^3 - \dfrac{1}{2}x$ **5.** $y = c_1 + c_2 e^{-2x/3} - 2x^2 + 5x$ **7.** $y = c_1 + c_2 x^{-1} + \dfrac{1}{6}x^2$

9. $x + c_2 = \sqrt{2}\,\ln|\sqrt{y^2 + 2y + c_1} + y + 1|$ **11.** $y = (c_1 x + c_2)^{1/2}$ **13.** $y = c_1 e^{2x} + c_2 e^{-2x}$

15. $y = \dfrac{c_1 e^{c_1 x}}{c_2 - e^{c_1 x}}$ or $y = \dfrac{1}{c_3 - x}$ **17.** $y = c_1 - \ln|c_2 - x|$

19. $y = c_1 \sin(kx + c_2) = c_1 \sin(kx)\cos(c_2) + c_1 \cos(kx)\sin(c_2) = c_3 \sin(kx) + c_4 \cos(kx)$ **21.** $y = \ln|\sec(x + c_1)| + c_2$

23. $e^{2y} + c_1 y = 4x + c_2$ **25.** $y = (c_1 + c_2 x)^2$ **27.** $y[\ln(y) + c_1] = c_2 - x$ **29.** $\psi(x) = x^{-2}; y = c_1 x^4 + c_2 x^{-2}$

31. $\psi(x) = 1 + x^2; y = c_1 x + c_2(1 + x^2)$ **33.** $\psi(x) = xe^{-x}; y = c_1 x + c_2 xe^{-x}$

35. $\psi(x) = x\tan(x); y = c_1 x + c_2 x\tan(x)$ **37.** $\psi(x) = \dfrac{1}{1 + x}; y = c_1 x + c_2\dfrac{1}{1 + x}$ **39.** $\psi(x) = xe^{-ax}$

41. $\dfrac{y''}{[1 + (y')^2]^{3/2}} = \cos(x); y = \ln[\sec(x)] + 2, -\dfrac{\pi}{2} < x < \dfrac{\pi}{2}$ **43.** $[2x^2 y']' + [xy]' = 0; y = c_1 x^{-1} + c_2 x^{-1/2}$

45. $[2x^2 y']' + [4xy]' = 0; y = c_1 x^{-1} + c_2 x^{-2}$

47. $[4xy']' + [(x^2 - 4)y]' = 0; y = c_1 xe^{-x^2/8}\displaystyle\int \dfrac{e^{x^2/8}}{x^2}\,dx + c_2 xe^{-x^2/8}$. The integral cannot be expressed in elementary terms.

49. $[e^{3x} y']' + [-2e^{3x} y]' = 0; y = c_1 e^{2x} + c_2 e^{-3x}$

Section 2.4

1. $y = c_1 e^{-2x} + c_2 e^{3x}$ **3.** $y = e^{8x}[c_1 + c_2 x]$ **5.** $y = c_1 + c_2 e^{3x}$ **7.** $y = c_1 e^{4x} + c_2 e^{5x}$

9. $y = e^{-x}[c_1 e^{\sqrt{17}x} + c_2 e^{-\sqrt{17}x}]$ **11.** $y = e^{7x}(c_1 + c_2 x)$ **13.** $y = e^{-6x}(c_1 + c_2 x)$

15. $y = e^{-7x/2}[c_1 e^{\sqrt{69}x/2} + c_2 e^{-\sqrt{69}x/2}]$ **17.** $y = e^{2x}[c_1 e^{\sqrt{3}x} + c_2 e^{-\sqrt{3}x}]$ **19.** $y = e^{5x}(c_1 + c_2 x)$ **21.** $y = 5 - 2e^{-3x}$

23. $y = e^{2x}(3 - x)$ **25.** $y = 0$ **27.** $y = \dfrac{\sqrt{6}}{4}e^x[e^{\sqrt{6}x} - e^{-\sqrt{6}x}]$

29. $y = \dfrac{\sqrt{41}}{82}e^{7(x-2)/2}[(\sqrt{41} - 7)e^{\sqrt{41}(x-2)/2} + (\sqrt{41} + 7)e^{-\sqrt{41}(x-2)/2}]$ **31.** $y = c_1\cosh(kx) + c_2\sinh(kx)$

33. (a) $mx''(t) = -g - kx'(t); x(0) = 0, x'(0) = v_0$

  (b) Solve $x''(t) + \dfrac{k}{m}x'(t) = -\dfrac{g}{m}$ as a first order linear equation in x' to get $x'(t) = Ce^{-kt/m} - \dfrac{g}{k}$. Now, using $x'(0) = v_0$

  yields $x'(t) = \left(v_0 + \dfrac{g}{k}\right)e^{-kt/m} - \dfrac{g}{k}$. Integration and the initial condition $x(0) = 0$ produce

$$x(t) = \dfrac{m}{k}\left(v_0 + \dfrac{g}{k}\right)(1 - e^{-kt/m}) - \dfrac{gt}{k}.$$

  (c) x is maximum when $x'(t) = 0$. This will occur when $t_{max} = -\dfrac{m}{k}\ln\left(\dfrac{g}{kv_0 + g}\right)$. The value of x at this time is

$$x_{max} = \dfrac{m}{k}\left[v_0 + \dfrac{g}{k}\ln\left(\dfrac{g}{kv_0 + g}\right)\right].$$

  (d) It is difficult to determine the time when $x = 0$. Either use an approximation technique such as Newton's method or write out the series for $e^{-kt/m}$ and use the first few terms. Since the problem is to show that it takes less time to rise than to fall back down, one attack is to show that x is still positive when $t = 2t_{max}$.

Section 2.5

1. $-e$ **3.** 1 **5.** -1 **7.** $e[\cos(4) + i\sin(4)]$ **9.** $\cos(1) + i\sin(1)$

Section 2.6

1. $y = e^{2x}[c_1\cos(2x) + c_2\sin(2x)]$ **3.** $y = e^{-11x}(c_1 + c_2x)$ **5.** $y = e^{-5x}[c_1\cos(x) + c_2\sin(x)]$

7. $y = e^{-5x}[c_1\cos(2x) + c_2\sin(2x)]$ **9.** $y = c_1 + c_2e^{4x}$ **11.** $y = e^{-x/2}\left[c_1\cos\left(\dfrac{\sqrt{3}x}{2}\right) + c_2\sin\left(\dfrac{\sqrt{3}x}{2}\right)\right]$

13. $y = e^{2x}(c_1e^{\sqrt{2}x} + c_2e^{-\sqrt{2}x})$ **15.** $y = e^{-5x}(c_1e^{\sqrt{26}x} + c_2e^{-\sqrt{26}x})$ **17.** $y = -\dfrac{\sqrt{3}}{3}e^{-x}(e^{\sqrt{3}x} - e^{-\sqrt{3}x})$

19. $y = -\cos(2x) - 4\sin(2x)$ **21.** $y = e^{2x}(3 - 4x)$ **23.** $y = e^{-x}\left[\cos(\sqrt{3}x) + \dfrac{1}{\sqrt{3}}\sin(\sqrt{3}x)\right]$

25. $y = 3e^{3(x-1)} + e^{-2(x-1)}$

27. $\theta(t) = c_1\cos\left(\sqrt{\dfrac{g}{L}}\,t\right) + c_2\sin\left(\sqrt{\dfrac{g}{L}}\,t\right) = A\cos\left(\sqrt{\dfrac{g}{L}}\,t + \delta\right)$. Since the period of $\cos(\omega t)$ is $\dfrac{2\pi}{\omega}$, the pendulum oscillates with

period $\dfrac{2\pi}{\sqrt{\dfrac{g}{L}}} = 2\pi\sqrt{\dfrac{L}{g}}$.

29. Let $\beta = \dfrac{\sqrt{A^2 - 4B}}{2}$. If $A^2 - 4B > 0$, the roots of the characteristic polynomial are $-\dfrac{A}{2} \pm \beta$, and the general solution is

expressible as $y = k_1e^{(-A/2 + \beta)x} + k_2e^{(-A/2 - \beta)x} = e^{-Ax/2}(k_1e^{\beta x} + k_2e^{-\beta x})$. Now let $c_1 = k_1 + k_2$ and $c_2 = k_1 - k_2$, so

$k_1 = \dfrac{c_1 + c_2}{2}$ and $k_2 = \dfrac{c_1 - c_2}{2}$. Making these substitutions yields $y = e^{-Ax/2}\left[\left(\dfrac{c_1 + c_2}{2}\right)e^{\beta x} + \left(\dfrac{c_1 - c_2}{2}\right)e^{-\beta x}\right] =$

$e^{-Ax/2}\left[c_1\left(\dfrac{e^{\beta x} + e^{-\beta x}}{2}\right) + c_2\left(\dfrac{e^{\beta x} - e^{-\beta x}}{2}\right)\right] = e^{-Ax/2}[c_1\cosh(\beta x) + c_2\sinh(\beta x)]$. If $A^2 - 4B < 0$, the roots of the

characteristic polynomial are still $-\dfrac{A}{2} \pm \beta$, but, since β is imaginary, we write the general solution as

$y = e^{-Ax/2}(k_1\cos(i\beta x) + k_2\sin(i\beta x))$. Now use the results of Problem 13 of Section 2.5 to get
$y = e^{-Ax/2}(k_1\cosh(\beta x) - k_2i\sinh(\beta x))$. Let $c_1 = k_1$ and $c_2 = -k_2i$ to obtain the desired result.

31. $8\sqrt{3}\cos\left(6x + \dfrac{4\pi}{3}\right)$ **33.** $8\sqrt{2}\cos\left(2x + \dfrac{\pi}{4}\right)$ **35.** $y'' - y' - 6y = 0$ **37.** $y'' + 8y' + 16y = 0$

Section 2.7

1. $y = c_1x^2 + c_2x^{-3}$ **3.** $y = x^{-1}[c_1 + c_2\ln(x)]$ **5.** $y = c_1\cos[2\ln(x)] + c_2\sin[2\ln(x)]$ **7.** $y = c_1x^2 + c_2x^{-2}$
9. $y = x^{-3}[c_1 + c_2\ln(x)]$ **11.** $y = x^{-1}\{c_1\cos[3\ln(x)] + c_2\sin[3\ln(x)]\}$ **13.** $y = c_1x^{-2} + c_2x^{-3}$
15. $y = x^3\{c_1\cos[7\ln(x)] + c_2\sin[7\ln(x)]\}$ **17.** $y = x^{-12}[c_1 + c_2\ln(x)]$ **19.** $y = c_1x^7 + c_2x^5$
21. $y = x^{-1}\{3\cos[\ln(x)] + 6\sin[\ln(x)]\}$ **23.** $y = \dfrac{7}{80}x^3 - \dfrac{192}{5}x^{-7}$ **25.** $y = -\dfrac{5}{4}x + 7x^{-1}$ **27.** $y = x^2[4 - 3\ln(x)]$

29. $y = x^{-3}\{\cos[2\ln(x)] + 3\sin[2\ln(x)]\}$ **33.** $y = x^2[5 - 3\ln(x)]$ **35.** $y = x^3 - 3x$

37. Let $t = \ln(3x - 4)$; $y = 3\cos[2\ln(3x - 4)] + 2\sin[2\ln(3x - 4)]$, $x > \dfrac{4}{3}$

39. $t = \ln(1 - x^2)$; $y = \sqrt{1 - x^2}[c_1 + c_2\ln(1 - x^2)]$ **41.** $t = e^x$; $y = e^{-e^x/2}(c_1 + c_2e^x)$
43. $t = \sin(x)$; $y = c_1\cos[\sin(x)] + c_2\sin[\sin(x)]$ **45.** $x^2y'' + 2xy' - 6y = 0$ **47.** $x^2y'' - 7xy' + 16y = 0$

Section 2.8

1. $\dfrac{1}{16}\left[\tan^{-1}\left(\dfrac{4}{3}\right) + \pi\right]$ second; 10 feet/second

3. (a) $\dfrac{1}{4}y'' + 16y = 0$; $y(0) = 0$, $y'(0) = 4$. $y(t) = \dfrac{1}{2}\sin(8t)$ feet **(b)** $\dfrac{\sqrt{63}}{2}$ feet/second **(c)** $\dfrac{\pi}{4}$ seconds/cycle

(d) $y = \dfrac{1}{2}\cos\left(8t - \dfrac{\pi}{2}\right)$

5. $y'' + 12y' + 36y = 0$; $y(0) = -\dfrac{1}{2}$, $y'(0) = 0$. $y(t) = -\dfrac{1}{2}e^{-6t}(1 + 6t)$ feet   **7.** $y(t) = \dfrac{19}{20}e^{-8t} - \dfrac{13}{15}e^{-3t}$ meter

13. If $c = 0$, $y = A\cos\left(\sqrt{\dfrac{k}{m}}\,t\right)$.

If $c^2 - 4mk > 0$, $y = \dfrac{A}{2\sqrt{c^2 - 4mk}}\,e^{-ct/2m}[(\sqrt{c^2 - 4mk} + c)e^{\sqrt{c^2 - 4mk}\,t/2m} - (\sqrt{c^2 - 4mk} - c)e^{-\sqrt{c^2 - 4mk}\,t/2m}]$.

If $c^2 - 4mk = 0$, $y = Ae^{-ct/2m}\left(1 + \dfrac{ct}{2m}\right)$.

If $c^2 - 4mk < 0$, $y = Ae^{-ct/2m}\left[\cos\left(\sqrt{4mk - c^2}\,\dfrac{t}{2m}\right) + \dfrac{c}{\sqrt{4mk - c^2}}\sin\left(\sqrt{4mk - c^2}\,\dfrac{t}{2m}\right)\right]$.

15. (a) $y = \dfrac{A}{2\sqrt{c^2 - 4mk}}\,e^{-ct/2m}[(\sqrt{c^2 - 4mk} + c)e^{\sqrt{c^2 - 4mk}\,t/2m} - (\sqrt{c^2 - 4mk} - c)e^{-\sqrt{c^2 - 4mk}\,t/2m}]$

(b) $y = \dfrac{mA}{\sqrt{c^2 - 4mk}}\,e^{-ct/2m}(e^{\sqrt{c^2 - 4mk}\,t/2m} - e^{-\sqrt{c^2 - 4mk}\,t/2m})$

17. (a) $y = Ae^{-ct/2m}\left[\cos\left(\sqrt{4mk - c^2}\,\dfrac{t}{2m}\right) + \dfrac{c}{\sqrt{4mk - c^2}}\sin\left(\sqrt{4mk - c^2}\,\dfrac{t}{2m}\right)\right]$

(b) $y = \dfrac{2mA}{\sqrt{4mk - c^2}}\,e^{-ct/2m}\sin\left(\sqrt{4mk - c^2}\,\dfrac{t}{2m}\right)$

19. Since the general solution is $y = Ae^{-ct/2m}\cos\left(\sqrt{4mk - c^2}\,\dfrac{t}{2m} + \delta\right)$, the position function is not periodic, but the

time between successive peaks (the pseudoperiod) is constant and has a value of $\dfrac{4m\pi}{\sqrt{4mk - c^2}}$. Inverting this gives us

$\dfrac{\sqrt{4mk - c^2}}{4m\pi}$ as a value whose physical units could be stated as peaks per unit time (pseudofrequency). As can be

seen, an increase in c will decrease this frequency value.

23. $d = \sqrt{c_1^2 + c_2^2}$ and $\delta = \tan^{-1}\left(\dfrac{-c_2}{c_1}\right)$

25. Since t does not appear explicitly in the equation $my'' + ky = 0$, let v (velocity) $= y' = \dfrac{dy}{dt}$. Then

$y'' = \dfrac{d^2y}{dt^2} = \dfrac{d}{dt}\left(\dfrac{dy}{dt}\right) = \dfrac{dv}{dt} = \dfrac{dv}{dy}\cdot\dfrac{dy}{dt} = \dfrac{dv}{dy}\cdot v$. Make these substitutions, separate variables, and integrate.

27. The period of the original system with mass m_1 was $p = 2\pi\sqrt{\dfrac{m_1}{k}}$. The new system with mass $m_1 + m_2$ has period

$2\pi\sqrt{\dfrac{m_1 + m_2}{k}} = 2\pi\sqrt{\dfrac{m_1}{k}}\sqrt{\dfrac{m_1 + m_2}{m_1}} = p\sqrt{1 + \dfrac{m_2}{m_1}}$.

29. $y = e^{0.4000t}[3.7200\cos(0.6460t) + 4.7800\sin(0.6460t)]$ **31.** $y = e^{-0.8847t}(3.2400 + 4.3864t)$

33. $y = e^{-0.5727t}[2.3600\cos(0.9896t) + 2.7805\sin(0.9896t)]$

Section 2.9

1. $y = c_1e^{3x} + c_2e^{-2x} + 4x - 5$; $y = 2e^{3x} + 3e^{-2x} + 4x - 5$

3. $y = c_1\cos(2x) + c_2\sin(2x) + \dfrac{1}{2}x^2 - 3x + 2$; $y = 3\cos(2x) + 4\sin(2x) + \dfrac{1}{2}x^2 - 3x + 2$

5. $y = c_1x^3 + c_2x^2 + \ln(x) + 2$; $y = x^3 + 2x^2 + \ln(x) + 2$

7. $y = x[c_1 + c_2\ln(x)] + 3x^2 + 3x\ln^2(x)$; $y = x[-1 + 14\ln(x)] + 3x^2 + 3x\ln^2(x)$

9. $y = c_1\cos(2x) + c_2\sin(2x) + 2\cos(x)$; $y = 4\cos(2x) + 3\sin(2x) + 2\cos(x)$

11. $y = e^{-x}(c_1 + c_2 x) + e^x$; $y = -e^{-x}(1 + 2x) + e^x$ **13.** $y = c_1 e^x + c_2 e^{2x} + \dfrac{1}{2}x + \dfrac{3}{4}$; $y = \dfrac{45}{4}e^{2x} - 17e^x + \dfrac{1}{2}x + \dfrac{3}{4}$

15. $y = c_1\cos(3x) + c_2\sin(3x) + \cos(2x) - 2\sin(2x)$; $y = \dfrac{2}{3}\cos(3x) - 3\sin(3x) + \cos(2x) - 2\sin(2x)$

17. $y = e^{2x}(c_1 e^{\sqrt{2}x} + c_2 e^{-\sqrt{2}x}) + \dfrac{3}{2}x + 4$; $y = \dfrac{1}{8}e^{2x}((9\sqrt{2} - 12)e^{\sqrt{2}x} - (9\sqrt{2} + 12)e^{-\sqrt{2}x}) + \dfrac{3}{2}x + 4$

19. $y = e^{-x}(c_1 e^{\sqrt{6}x} + c_2 e^{-\sqrt{6}x}) - 4$; $y = -4$

21. $y = c_1\cos(x) + c_2\sin(x) - \cos(x)\ln|\sec(x) + \tan(x)|$; $y = 2\cos(x) + \sin(x) - \cos(x)\ln|\sec(x) + \tan(x)|$

23. $y = c_1 x^2 + c_2 x^{-3} + \dfrac{1}{14}x^4 + \dfrac{2}{25}x^2 - \dfrac{2}{5}x^2\ln(x)$; $y = \dfrac{33}{10}x^2 + \dfrac{271}{175}x^{-3} + \dfrac{1}{14}x^4 - \dfrac{2}{5}x^2\ln(x)$

25. $y = c_1 e^x + c_2 e^{3x} - \dfrac{3}{5}\cos(x + 2) - \dfrac{3}{10}\sin(x + 2)$; $y = \dfrac{11}{4}e^{x+2} - \dfrac{3}{20}e^{3(x+2)} - \dfrac{3}{5}\cos(x + 2) - \dfrac{3}{10}\sin(x + 2)$

27. $b = \dfrac{a}{\alpha^2 + A\alpha + B}$ **29.** $y = \dfrac{2}{3}e^{3x}$ **31.** $y_p = -\dfrac{3}{2}e^x + \dfrac{5}{18}e^{-4x}$

Section 2.10

1. $y = c_1 e^{2x} + c_2 e^{-x} - x^2 + x - 4$ **3.** $y = c_1\cos(2x) + c_2\sin(2x) + 2x^3 - 5x^2 + x - 2$

5. $y = e^x[c_1\cos(3x) + c_2\sin(3x)] + 2x^2 + x - 1$ **7.** $y = c_1 e^{2x} + c_2 e^{4x} + e^x$ **9.** $y = c_1 e^x + c_2 e^{2x} + 3\cos(x) + \sin(x)$

11. $y = e^{2x}[c_1\cos(3x) + c_2\sin(3x)] + \dfrac{1}{3}e^{2x} - \dfrac{1}{2}e^{3x}$ **13.** $y = c_1 + c_2 e^{4x} - 3x^3 - 2x^2 - 7x$

15. $y = c_1 + c_2 e^{-4x} - \dfrac{1}{10}\cos(2x) - \dfrac{1}{20}\sin(2x) + \dfrac{3}{25}\cos(3x) - \dfrac{4}{25}\sin(3x)$ **17.** $y = e^{-x}(c_1 + c_2 x) - \dfrac{3}{2}x^2 e^{-x} + \dfrac{4}{3}x^3 e^{-x} + 1$

19. $y = c_1 e^{2x} + c_2 e^{-2x} + \dfrac{1}{250}e^x[(-50x^2 + 10x + 29)\cos(x) + (25x^2 + 70x + 28)\sin(x)]$

21. $y = c_1\cos(\sqrt{2}x) + c_2\sin(\sqrt{2}x) + 1 - \dfrac{1}{17}\cos(6x) = c_1\cos(\sqrt{2}x) + c_2\sin(\sqrt{2}x) + \dfrac{18}{17} - \dfrac{2}{17}\cos^2(3x)$

23. $y = c_1\cosh(2x) + c_2\sinh(2x) + \dfrac{5}{4}x\cosh(2x)$

25. $y = c_1\cos(x) + c_2\sin(x) + \dfrac{1}{5}e^{-x}[2\cos(x) + \sin(x)] + \dfrac{1}{75}e^{3x}[\cos(5x) - 2\sin(5x)]$

27. $y = 3 + 2e^{-4x} - 2\cos(x) + 8\sin(x) + 2x$ **29.** $y = 4\cos(x) + \dfrac{1}{3}\sin(x) - \dfrac{5}{3}\sin(2x) - 4$

31. $y = \dfrac{1}{5} + e^{3x} - \dfrac{1}{5}e^{2x}[\cos(x) + 3\sin(x)]$ **33.** $y = e^{3x} - xe^{3x} + 2x^2 e^{3x}$

35. For each $n = 1, 2, 3, \ldots$, try $y_n = a_n\cos(nx) + b_n\sin(nx)$ and set $y_n'' - 4y_n = \dfrac{1}{n}$. Then, for each n, $a_n = 0$ and $b_n = \dfrac{-1}{n(n^2 + 4)}$,

so $y_p = \sum y_n = \sum_{n=1}^{\infty}\dfrac{-\sin(nx)}{n(n^2 + 4)}$.

37. $y = \begin{cases} c_1 + c_2 e^{-4x} + \dfrac{3}{5}e^{3x} & \text{if } x \le 0 \\[2ex] \left(c_1 + \dfrac{67}{128}\right) + \left(c_2 + \dfrac{49}{640}\right)e^{-4x} + \dfrac{1}{12}x^3 - \dfrac{5}{16}x^2 + \dfrac{29}{32}x & \text{if } x \ge 0 \end{cases}$

39. $y = \begin{cases} e^{2x} + \dfrac{7}{8}e^{-2x} - \dfrac{1}{2}x & \text{if } x \le 2 \\[2ex] e^{2x} + \dfrac{7}{8}e^{-2x} - \dfrac{1}{8}e^{2(x-2)} + \dfrac{1}{8}e^{-2(x-2)} - 1 & \text{if } x \ge 2 \end{cases}$

41. $y = c_1 x^4 + c_2 x^{-3} - \dfrac{1}{3} x$ **43.** $y = c_1 x^2 + c_2 x^3 - \dfrac{1}{2} x - x^2 \ln(x)$ **45.** $y = c_1 x^{-4} + c_2 x^{-5} - \dfrac{1}{30} x + \dfrac{1}{5} \ln(x) - \dfrac{9}{100}$

47. The general solution is $\varphi(x) = (c_1 + c_2 x + x^2)e^x$. $\varphi(x) > 0$ for all x if and only if $c_1 + c_2 x + x^2 > 0$ for all x. This function's graph is a parabola that has a minimum value of $c_1 - (c_2/2)^2$ at $x = -c_2/2$; therefore, $\varphi(x) > 0$ for all x only if $c_1 - (c_2/2)^2 > 0$. $\varphi'(x) = [(c_1 + c_2) + (2 + c_2)x + x^2]e^x$. $\varphi'(x) > 0$ for all x if and only if $(c_1 + c_2) + (2 + c_2)x + x^2 > 0$ for all x. This function's graph is a parabola with minimum value of $c_1 - (c_2/2)^2 - 1$ at $x = -(2 + c_2)/2$. Thus, $\varphi'(x) > 0$ for all x if and only if $c_1 - (c_2/2)^2 > 1$. Part (a) is false. Let $c_1 = 1$ and $c_2 = 0$. $\varphi(x) = (1 + x^2)e^x$ is a solution and $\varphi(x) > 0$ for all x. However, $\varphi'(x) = (1 + x)^2 e^x = 0$ at $x = -1$. Part (b) is true. Suppose φ is a solution. Then $\varphi(x) = (c_1 + c_2 x + x^2)e^x$. Assume that $\varphi'(x) > 0$ for all x. Then $c_1 - (c_2/2)^2 > 1 > 0$. Thus, $c_1 + c_2 x + x^2 > 0$ for all x, so $\varphi(x) > 0$ for all x.

Section 2.11

1. $y = c_1 e^{2x} + c_2 e^{-x} + \dfrac{1}{3} x e^{2x}$ **3.** $y = c_1 e^{2x} + c_2 e^{-3x} + x$ **5.** $y = c_1 x^2 + c_2 x^4 + x$

7. $y = c_1 e^x + c_2 e^{3x} + \dfrac{1}{5} \cos(x + 3) - \dfrac{2}{5} \sin(x + 3)$ **9.** $y = c_1 \cos[2 \ln(x)] + c_2 \sin[2 \ln(x)] - \dfrac{1}{4} \cos[2 \ln(x)] \ln(x)$

11. $y = e^{2x}(c_1 + c_2 x) + 3x^2 e^{2x}$ **13.** $y = c_1 \cos(3x) + c_2 \sin(3x) + \dfrac{1}{3} \cos(3x) \ln|\cos(3x)| + x \sin(3x)$

15. $y = c_1 x + c_2 x \ln(x) + 3x \ln^2(x)$ **17.** $y = c_1 e^x + c_2 e^{2x} - e^{2x}\cos(e^{-x})$ **19.** $y = c_1 x^{-1} + c_2 x^{-1}\ln(x) + x^2 + 2x + 5$

21. $y = 4e^{-x} - \sin^2(x) - 2$ **23.** $y = 2x^3 + x^{-2} - 2x^2$ **25.** $y = x - x^2 + 3 \cos[\ln(x)] + \sin[\ln(x)]$

27. $y = c_1 x^2 + c_2(x - 1) + x^4 - 4x^3$

29. Using the fundamental theorem of calculus twice yields the initial value problem $x^2 y'' = \ln(x); y(1) = y'(1) = 0$, which has solution $y = x - 1 - \ln(x) - \frac{1}{2} \ln^2(x)$.

Section 2.12

1. Use l'Hôpital's rule (differentiating with respect to ω) to get

$$\lim_{\omega \to \omega_0} y(t) = \lim_{\omega \to \omega_0} \frac{A[\cos(\omega t) - \cos(\omega_0 t)]}{m(\omega_0^2 - \omega^2)} = \lim_{\omega \to \omega_0} \frac{-At \sin(\omega t)}{-2m\omega} = \frac{At \sin(\omega_0 t)}{2m\omega_0},$$

which agrees with the solution obtained in the case $\omega = \omega_0$.

3. $y = \dfrac{1}{5000} [1243e^{-8t} + 9800te^{-8t} + 7 \cos(6t) + 24 \sin(6t)]$ feet   **5.** $y = 3e^{-t} + 4te^{-t} - \cos(t)$

7. $y = \left(\dfrac{11 - 42A}{18}\right)e^{-t} - \left(\dfrac{5 - 15A}{9}\right)e^{-2t/5} + \dfrac{1}{2} t - \dfrac{1}{3} te^{-t}$

9. $i(t) = -0.0094e^{-0.6251t} + 0.0017e^{-3332.7082t} + 0.0076 \cos(20t) + 0.2998 \sin(20t)$

11. $i(t) = -0.0010e^{-0.3177t} - 0.0022e^{-473.3665t} + 0.0033e^{-t}$ **13.** $i_1(t) = \dfrac{2}{5} - 4te^{-5t}, i_2(t) = \dfrac{2}{5} e^{-5t} - 2te^{-5t}$

15. $i_1(t) = 2 - \dfrac{4}{5} e^{-4t/5}\left[2 \cos\left(\dfrac{2t}{5}\right) + \sin\left(\dfrac{2t}{5}\right)\right], i_2(t) = \dfrac{2}{5} e^{-4t/5}\left[\cos\left(\dfrac{2t}{5}\right) - \sin\left(\dfrac{2t}{5}\right)\right]$

17. The general solution is $y = c_1 e^{-ct/2m} + c_2 te^{-ct/2m} + k_1 \cos(\omega t) + k_2 \sin(\omega t)$, where

$$k_1 = \frac{A(k - m\omega^2)^2}{c^2\omega^2 + (k - m\omega^2)^2} \quad \text{and} \quad k_2 = \frac{Ac\omega}{c^2\omega^2 + (k - m\omega^2)^2}$$

(a) $c_1 = B - k_1$ and $c_2 = \dfrac{c}{2m}(B - k_1) - \omega k_2$ (b) $c_1 = -k_1$ and $c_2 = B - \omega k_2 - \dfrac{c}{2m} k_1$

19. If $c \neq m + k$, then, in all cases, $y_p = \dfrac{1}{m - c + k} e^{-t}$.

If $c^2 - 4mk > 0$, $y = e^{-ct/2m}(c_1 e^{\sqrt{c^2 - 4mk}\, t/2m} + c_2 e^{-\sqrt{c^2 - 4mk}\, t/2m}) + y_p$.
If $c^2 - 4mk = 0$, $y = e^{-ct/2m}(c_1 + c_2 t) + y_p$.

If $c^2 - 4mk < 0$, $y = e^{-ct/2m}\left[c_1\cos\left(\dfrac{\sqrt{4mk - c^2}\,t}{2m}\right) + c_2\sin\left(\dfrac{\sqrt{4mk - c^2}\,t}{2m}\right)\right] + y_p$.

If $c = 2m$ and $k = m$, we have critical damping and the forcing function, e^{-t}, is a solution of the associated homogeneous equation. The general solution in this case is $y = e^{-t}\left[c_1 + c_2t + \dfrac{1}{2m}t^2\right]$.

If $c = m + k$, but $c \neq 2m$, we have overdamping and e^{-t} is a solution of the associated homogeneous equation. Here, the general solution is $y = c_1e^{-t} + c_2^{-kt/m} + \dfrac{1}{k - m}\,te^{-t}$.

21. Unless $c = 2m$ and $\omega = \sqrt{\dfrac{m - k}{m}}$, in all cases

$$y_p = Ae^{-t}\frac{[\omega(2m - c)]\cos(\omega t) + [m(1 - \omega^2) - c + k]\sin(\omega t)}{[m(1 - \omega^2) - c + k]^2 + [\omega(2m - c)]^2}.$$

If $c^2 - 4mk > 0$, $y = e^{-ct/2m}(c_1e^{\sqrt{c^2 - 4mk}\,t/2m} + c_2e^{-\sqrt{c^2 - 4mk}\,t/2m}) + y_p$.
If $c^2 - 4mk = 0$, $y = e^{-ct/2m}(c_1 + c_2t) + y_p$.

If $c^2 - 4mk < 0$, $y = e^{-ct/2m}\left[c_1\cos\left(\dfrac{\sqrt{4mk - c^2}\,t}{2m}\right) + c_2\sin\left(\dfrac{\sqrt{4mk - c^2}\,t}{2m}\right)\right] + y_p$.

If $c = 2m$ and $\omega = \sqrt{\dfrac{m - k}{m}}$, the forcing function is a solution of the associated homogeneous equation and the general solution is

$$y = e^{-t}\left[c_1\cos\left(\sqrt{\frac{m - k}{m}}\,t\right) + c_2\sin\left(\sqrt{\frac{m - k}{m}}\,t\right)\right] - \frac{A}{2\sqrt{m(k - m)}}\,te^{-t}.$$

23. The transient solution will always have an $e^{-ct/2m}$ term that forces this part of the solution to decay to zero. This part is always finite even though it may be quite large with appropriate initial conditions. The steady-state solution is expressible as

$$\frac{A}{\sqrt{c^2\omega^2 + (k - m\omega^2)^2}}\,\cos(\omega t - \delta),$$

which is bounded as long as c and ω are not zero.

Additional Problems for Chapter 2

1. $y = e^{3x/2}\left[c_1\cos\left(\dfrac{\sqrt{23}x}{2}\right) + c_2\sin\left(\dfrac{\sqrt{23}x}{2}\right)\right]$ **3.** $y = x^5[c_1 + c_2\ln(x)]$ **5.** $y = c_1e^{c_2x}$ **7.** $y = e^{7x}(c_1 + c_2x)$

9. $y = c_1x^{2 + \sqrt{2}} + c_2x^{2 - \sqrt{2}}$ **11.** $y = c_1e^{c_2x} - \dfrac{2}{c_2}$ **13.** $y = c_1e^{2x} + c_2e^{-x} + \dfrac{1}{2}e^{4x} - 3x + \dfrac{3}{2}$

15. $y = c_1e^{2x} + c_2e^{-7x}$ **17.** $y = c_1x + c_2x^{-1} + e^{-x} + x^{-1}e^{-x}$ **19.** $y = c_1 + c_2e^{2x} + 3x^2 + 3x$

21. $y = x^{3/2}\left(c_1\cos\left[\dfrac{\sqrt{31}}{2}\ln(x)\right] + c_2\sin\left[\dfrac{\sqrt{31}}{2}\ln(x)\right]\right)$ **23.** $y = e^{3x}(c_1 + c_2x)$

25. $y = c_1\cos(x) + c_2\sin(x) - 2 + \sin(x)\ln|\sec(x) + \tan(x)|$ **27.** $y = c_1 + c_2e^{3x} - 2x^3 - 3x^2 + 3x$

29. $y = c_1x^{-2} + c_2x^4$ **31.** $y = x^5\{c_1\cos[4\ln(x)] + c_2\sin[4\ln(x)]\}$ **33.** $y = e^{2x}[c_1\cos(3x) + c_2\sin(3x)]$

35. $y = e^x(c_1e^{\sqrt{2}x} + c_2e^{-\sqrt{2}x})$ **37.** $y = e^{-2x}[c_1\cos(\sqrt{5}x) + c_2\sin(\sqrt{5}x)]$ **39.** $y = c_1e^x + c_2e^{-2x} - 4x^2 - 5x - 6$

41. $y = c_1 + c_2e^{-x} - x^3 + 3x^2 - 6x$ **43.** $y = e^{4x}(c_1 + c_2x) + \dfrac{1}{128}e^{-4x} + \dfrac{1}{4}x^2e^{4x}$

45. $y = c_1 + c_2e^{2x} + \dfrac{1}{6}x^3 + \dfrac{1}{4}x^2 + \dfrac{1}{4}x$ **47.** $y = x^2(c_1x^{\sqrt{2}} + c_2x^{-\sqrt{2}})$ **49.** $y = c_1x^4 + c_2x^2 + \dfrac{1}{15}x^{-1}$

51. $y = e^{-x}[c_1 + c_2x + x\ln(x)]$ **53.** $y = e^{-x}(c_1e^{\sqrt{5}x} + c_1e^{-\sqrt{5}x}) + e^x - \dfrac{13}{205}\cos(3x) + \dfrac{6}{205}\sin(3x)$

55. $y = e^{5x/2}\left[c_1\cos\left(\dfrac{\sqrt{11}x}{2}\right) + c_2\sin\left(\dfrac{\sqrt{11}x}{2}\right)\right] + \dfrac{1}{9}x^2 + \dfrac{10}{81}x + \dfrac{32}{729} - \dfrac{5}{89}\cos(x) - \dfrac{8}{89}\sin(x)$

57. $y = x^{3/2}\left(c_1\cos\left[\dfrac{\sqrt{47}\ln(x)}{2}\right] + c_2\sin\left[\dfrac{\sqrt{47}\ln(x)}{2}\right]\right)$ **59.** $y = c_1e^{-x} + c_2e^{-2x} + (e^{-x} + e^{-2x})\ln(1 + e^x)$

61. $y = -e^{-11x}\left(\dfrac{633}{121} + \dfrac{6721}{121}x\right) + \dfrac{28}{121}$

63. $y = \dfrac{1}{496,400}[-12,363e^{8x} - 1637e^{-12x} + 74,460xe^{8x} + 14,000\cos(3x) + 1600\sin(3x)]$

65. $y = e^x\left(\dfrac{47}{25} - \dfrac{2}{5}x\right) + \dfrac{3}{25}\cos(3x) - \dfrac{4}{25}\sin(3x)$ **67.** $y = \sin(x) - \cos(x)\ln|\sec(x) + \tan(x)|$

69. $y = 5x^{-1} - 3x^{-1}\ln(x) + x$ **71.** $y = -\dfrac{3}{20}e^{3(x+2)} + \dfrac{11}{4}e^{x+2} - \dfrac{3}{5}\cos(x + 2) - \dfrac{3}{10}\sin(x + 2)$

73. $y = 3x(x^{\sqrt{3}} + x^{-\sqrt{3}}) + x^3 - 2\ln(x) + 2$ **75.** $y = \dfrac{18}{3 - x}$ **77.** $y = \dfrac{1}{6}e^x[1 - 13\cos(\sqrt{6}x)]$

79. $y = 4\cos(x) + 2\sin(x) + \sin(x)\tan(x)$

81. (a) $mx''(t) = F[x(t)]$

(b) Integrate both sides of the equation in (a) to get $\dfrac{1}{2}m[x'(t)]^2 = \displaystyle\int_0^{x(t)} F(z)\,dz + C$, where

$C = \dfrac{1}{2}m[x'(0)]^2 + \displaystyle\int_0^{x(0)} F(z)\,dz = \text{constant}$. Then $\dfrac{1}{2}m[x'(t)]^2 - \displaystyle\int_0^{x(t)} F(z)\,dz = \text{kinetic energy} +$

potential energy $= $ constant.

83. Let $t = 0$ be the time that the ball is thrown. Then the initial value problem $mx''(t) = -mg; x(0) = h, x'(0) = -v_0$

describes the motion of the ball. Solve this to get $x(t) = \dfrac{1}{2}gt^2 - v_0t + h$. Then $x(t) = 0$ when $t = \dfrac{1}{g}[\pm\sqrt{v_0^2 + 2gh} - v_0]$.

Since t must be positive when the ball hits the ground, $t = \dfrac{1}{g}[\sqrt{v_0^2 + 2gh} - v_0]$.

85. The initial value problem is $\dfrac{d^2\theta}{dt^2} + \dfrac{g}{L}\sin(\theta) = 0; \theta(0) = -\alpha, \theta'(0) = 0$. Since no t occurs explicitly, let $u = \dfrac{d\theta}{dt}$ so

$\dfrac{d^2\theta}{dt^2} = u\dfrac{du}{d\theta}$. Make these substitutions to get $u\dfrac{du}{d\theta} + \dfrac{g}{L}\sin(\theta) = 0$. Separate variables and integrate to obtain the first

order equation $[\theta'(t)]^2 = \dfrac{2g}{L}\cos[\theta(t)] + C$. Since $\theta'(0) = 0$ and $\theta(0) = -\alpha$, $C = -\dfrac{2g}{L}\cos(-\alpha) = -\dfrac{2g}{L}\cos(\alpha)$. Thus far,

we have $[\theta']^2 = \dfrac{2g}{L}[\cos(\theta) - \cos(\alpha)]$. Since $\theta(0) = -\alpha < 0$, the pendulum was initially pulled off to the left and

consequently will have a positive angular velocity throughout the first half-swing. Use this fact to extract the positive

square root to get $\theta' = \sqrt{\dfrac{2g}{L}[\cos(\theta) - \cos(\alpha)]}$. Separate variables and integrate again to obtain the expression

$t = \sqrt{\dfrac{L}{2g}}\displaystyle\int_0^\theta \dfrac{1}{\sqrt{\cos(\varphi) - \cos(\alpha)}}\,d\varphi + C$. Let $t = 0$ and $\theta = -\alpha$ to evaluate $C = -\sqrt{\dfrac{L}{2g}}\displaystyle\int_0^{-\alpha} \dfrac{1}{\sqrt{\cos(\varphi) - \cos(\alpha)}}\,d\varphi$.

Now add the integrals to obtain the desired result.

Chapter 3

Section 3.1

1. If $L[y] = y^{(3)} - (\alpha + \beta + \gamma)y'' + (\alpha\beta + \alpha\gamma + \beta\gamma)y' - \alpha\beta\gamma$, then $e^{\alpha x}$, $e^{\beta x}$, and $e^{\gamma x}$ are each solutions of $L[y] = 0$ with
Wronskian $W(x) = [\alpha\beta(\beta - \alpha) + \alpha\gamma(\alpha - \gamma) + \beta\gamma(\gamma - \beta)]e^{(\alpha + \beta + \gamma)x}$. Since α, β, and γ are distinct, this function is never zero.
3. $W(x) = \pm 30e^{5x} \neq 0$, so the solutions are linearly independent. The general solution is $y = c_1e^{4x} + c_2e^{2x} + c_3e^{-x}$.
5. $W(0) = \pm 250$, $y = c_1e^{4x} + (c_2 + c_3x + c_4x^2)e^{-x}$ **7.** $W(0) = \pm 26$, $y = c_1e^{3x} + e^{-2x}[c_2\cos(x) + c_3\sin(x)]$
9. $W(0) = \pm 2e$, $y = c_1 + c_2x^2 + c_3(x - 1)e^x$ **17.** $y = \dfrac{11}{24}e^{2x} - \dfrac{31}{320}e^{4x} - \dfrac{8}{15}e^{-x} + \dfrac{1}{64}[8x^2 - 4x + 11]$

19. $y = \dfrac{89}{3200} e^{4x} + \left[\dfrac{362}{25} + \dfrac{32}{5} x + x^2\right]e^{-x} + \dfrac{1}{4} x^3 - \dfrac{33}{16} x^2 + \dfrac{255}{32} x - \dfrac{1729}{128}$

21. $y = -\dfrac{5}{78} e^{3x} + e^{-2x}\left[\dfrac{121}{130} \cos(x) + \dfrac{163}{130} \sin(x)\right] + \dfrac{2}{15} \cos(3x) - \dfrac{1}{15} \sin(3x)$ **23.** $y = 2 + 6(x-1)e^{x-1} - 2x^3$

Section 3.2

1. $y = e^{2x}(c_1 + c_2 x + c_3 x^2 + c_4 x^3 + c_5 x^4)$ **3.** $y = c_1 e^{-x} + c_2 e^x + c_3 e^{4x}$

5. $y = c_1\cos(2x) + c_2\sin(2x) + c_3\cos(x) + c_4\sin(x)$ **7.** $y = c_1 e^{-2x} + e^x[c_2\cos(x) + c_3\sin(x)]$

9. $y = c_1 e^{-6x} + c_2 e^x + c_3 x e^x$ **11.** $y = c_1 e^x + c_2 x e^x + c_3 e^{-6x} + c_4 e^{-x}$ **13.** $y = c_1 e^{2x} + c_2 e^{-2x} + c_3 e^x + c_4 e^{-x}$

15. $y = c_1 e^{2x} + e^{2x}[c_2\cos(3x) + c_3\sin(3x)]$ **17.** $y = e^x - 2\cos(\sqrt{2}x) + \dfrac{\sqrt{2}}{2} \sin(\sqrt{2}x)]$

19. $y = \dfrac{4}{183}(6e^{-4(x-\pi)} - e^{x-\pi}[6\cos(6x) - 5\sin(6x)])$ **21.** $y = \dfrac{1}{50}[101e^{-3x} + 99e^{7x} - 390xe^{7x}]$

23. $y = \dfrac{1}{15}[5e^{7x} - 12e^{3x} - 23e^{-2x}]$ **25.** $y = -\dfrac{1}{169}[121e^{3x} + 48e^{-10x} + 286xe^{-10x}]$

27. $y \simeq c_1 e^{-0.5468x} + e^{-2.2734x}[c_2\cos(0.5638x) + c_3\sin(0.5638x)]$

29. $y \simeq c_1 e^{1.4656x} + c_2 e^{2.4560x} + e^{-1.9608x}[c_3\cos(2.9891x) + c_4\sin(2.9891x)]$

31. $y^{(6)} + 18y^{(5)} + 131y^{(4)} + 496y^{(3)} + 1056y'' + 1280y' + 768y = 0$ **33.** $y^{(4)} + 13y'' + 36y = 0$

Section 3.3

1. $y = c_1 x^{-1} + c_2 x^3 + c_3 x$ **3.** $y = c_1 x^2 + x^4\{c_2\cos[2\ln(x)] + c_3\sin[2\ln(x)]\}$ **5.** $y = c_1 x^3 + c_2 x^{-2} + c_3 x^{-2}\ln(x)$

7. $y = c_1 x^{-3} + x^4\{c_2\cos[2\ln(x)] + c_3\sin[2\ln(x)]\}$ **9.** $y = x^{-3}[c_1 + c_2\ln(x) + c_3\ln^2(x)]$

11. $y = c_1 x^{-3} + x^2\{c_2\cos[5\ln(x)] + c_3\sin[5\ln(x)]\}$ **13.** $y = c_1 x^3 + x^4\{c_2\cos[\ln(x)] + c_3\sin[\ln(x)]\}$

15. $y = c_1 x^{-2} + x^6\{c_2\cos[4\ln(x)] + c_3\sin[4\ln(x)]\}$ **17.** $y = -\dfrac{174}{85} x^{-2} - x^4\left(\dfrac{81}{85} \cos[7\ln(x)] - \dfrac{279}{595} \sin[7\ln(x)]\right)$

19. $y = -\dfrac{3}{11}\left(\dfrac{x}{3}\right)^4 + \dfrac{x}{3}\left\{\dfrac{3}{11} \cos\left[\sqrt{2}\ln\left(\dfrac{x}{3}\right)\right] + \dfrac{21\sqrt{2}}{11} \sin\left[\sqrt{2}\ln\left(\dfrac{x}{3}\right)\right]\right\}$

21. $y = -\dfrac{33}{65} x^6 - x^{-2}\left(\dfrac{162}{65} \cos[\ln(x)] + \dfrac{321}{65} \sin[\ln(x)]\right)$ **23.** $y = x^{-2}\left[2 + 6\ln(x) + 8\ln^2(x) + \dfrac{17}{3} \ln^3(x)\right]$

25. $y = \dfrac{7}{67} x^{-6} - x^2\left(\dfrac{7}{67} \cos[\sqrt{3}\ln(x)] - \dfrac{56\sqrt{3}}{201} \sin[\sqrt{3}\ln(x)]\right)$ **27.** $x^5 y^{(5)} + 7x^4 y^{(4)} + 19x^3 y^{(3)} + 2x^2 y'' + 9xy' - 9y = 0$

Section 3.4

1. $y = c_1 e^x + c_2 e^{-x} + c_3\cos(x) + c_4\sin(x) - \dfrac{4}{15} \cosh(2x)$ **3.** $y = c_1 e^{2x} + e^{7x/2}[c_2 e^{\sqrt{41}x/2} + c_3 e^{-\sqrt{41}x/2}] - \dfrac{1}{4} x - \dfrac{3}{4}$

5. $y = c_1 + e^{2x}[c_2\cos(4x) + c_3\sin(4x)] + \dfrac{1}{60} x^3 + \dfrac{11}{100} x^2 - \dfrac{461}{1000} x$

7. $y = c_1 e^{2x} + c_2\cos(x) + c_3\sin(x) - \dfrac{1}{2} x^2 + \dfrac{1}{2} x + \dfrac{3}{4} + \dfrac{3}{10} x\cos(x) + \dfrac{3}{5} x\sin(x)$

9. $y = e^{-2x}[c_1 + c_2 x + c_3 x^2] + \dfrac{4}{3} x^3 e^{-2x}$

11. $y = c_1 e^{-2x} + c_2 e^{-3x} + c_3 e^{4x} - \dfrac{1}{24} x^3 + \dfrac{7}{96} x^2 - \dfrac{53}{576} x + \dfrac{355}{6912} + \dfrac{1}{6} xe^{4x} + \dfrac{1}{17} \cos(x) + \dfrac{3}{85} \sin(x)$

13. $y = c_1 + e^{-6x}[c_2 + c_3 x] - \dfrac{1}{400} \cos(2x) - \dfrac{7}{400} \sin(2x)$

15. $y = c_1 e^{3x} + c_2 e^{-x} + c_3 e^{-2x} + \dfrac{5}{6} + \dfrac{11}{260} \cos(2x) - \dfrac{3}{260} \sin(2x)$ **17.** $y = c_1\cos[2\ln(x)] + c_2\sin[2\ln(x)] + c_3 x - \dfrac{1}{8} x^{-2}$

19. $y = c_1 x^{\sqrt{2}} + c_2 x^{-\sqrt{2}} + c_3 x^{-2} - 2x \ln(x) - \dfrac{10}{3} x$ **21.** $y = \dfrac{29}{50} e^{-2x} - \dfrac{121}{675} e^{3x} + \dfrac{37}{45} x e^{3x} + \dfrac{1}{18} x + \dfrac{1}{108} - \dfrac{1}{4} e^{2x}$

23. $y = -\dfrac{25}{9} e^x - \dfrac{476}{5625} e^{-5x} - \dfrac{224}{375} x e^{-5x} - \dfrac{1}{25} x^2 - \dfrac{6}{125} x - \dfrac{86}{625}$

25. $y = e^{3x}\left[\dfrac{24}{1025} \cos(4x) + \dfrac{3}{4100} \sin(4x)\right] + \dfrac{43}{50} + \dfrac{3}{82} e^{-2x}$ **27.** $y = e^{-x}\left[\dfrac{1}{6} x^3 + \dfrac{1}{8} x^4\right]$

29. $y = \dfrac{5}{2} x - x^2\left(\dfrac{3}{2} \cos[\ln(x)] - \dfrac{1}{2} \sin[\ln(x)]\right) - 1$ **31.** $y = -\dfrac{1}{64} x^{-3} - x\left[\dfrac{127}{64} - \dfrac{45}{16} \ln(x)\right] + \dfrac{1}{8} x^{-3}\ln(x) + \dfrac{1}{8} x \ln^2(x)$

33. $y = \dfrac{5}{2} e^x - e^{2x}\left[\dfrac{3}{2} \cos(x) - \dfrac{1}{2} \sin(x)\right] - 1$

Additional Problems for Chapter 3

1. $y = \left(\dfrac{49}{48} + \dfrac{769}{108} x + \dfrac{1681}{72} x^2\right)e^{-6x} + \dfrac{1}{216} x - \dfrac{1}{48}$ **3.** $y = \left(\dfrac{25}{32} - \dfrac{1}{16} x\right)e^x + \left(\dfrac{39}{32} + \dfrac{31}{16} x - \dfrac{1}{2} x^2\right)e^{-3x}$

5. $y = -\dfrac{124}{65} e^{2x} - e^{3x}\left(\dfrac{6}{25} \cos(8x) - \dfrac{71}{520} \sin(8x)\right) - 1$ **7.** $y = 11x - \dfrac{41}{4} x^2 + \dfrac{32}{9} x^3 - \dfrac{1}{6} \ln(x) + \dfrac{11}{36}$

9. $y = c_1 x + c_2 x^{-1} + c_3 x^{-4} + \dfrac{1}{10} x \ln^2(x) - \dfrac{7}{50} x \ln(x)$ **11.** $y = -\dfrac{37}{15} e^{2(6x - \pi)/3} - e^{(6x - \pi)/6}\left[\dfrac{8}{15} \cos(6x) - \dfrac{26}{15} \sin(6x)\right]$

13. $y = -5x^2 + 10x^2\ln(x) - 10x^2\ln^2(x)$ **15.** $y = -\dfrac{35}{9} x + \dfrac{20}{3} x \ln(x) - x\left(\dfrac{1}{9} \cos[3 \ln(x)] + \dfrac{2}{9} \sin[\ln(x)]\right)$

17. $y = \dfrac{3}{400} e^{-4x} - e^{5x}\left[\dfrac{3}{400} \cos(2x) - \dfrac{2}{125} \sin(2x)\right] + \dfrac{3}{85} x e^{-4x}$ **19.** $y = \dfrac{5}{4} e^x + \dfrac{5}{8} e^{-x} + e^x\left[\dfrac{1}{8} \cos(2x) - \dfrac{3}{8} \sin(2x)\right] - 2$

21. $y = e^{5x}(c_1 + c_2 x) + e^{-6x}(c_3 + c_4 x)$ **23.** $y = x[c_1 + c_2\ln(x) + c_3\ln^2(x)]$

25. $y = c_1 e^{-2x} + e^{3x}[c_2\cos(4x) + c_3\sin(4x)] - \dfrac{11}{195} \cos(2x) - \dfrac{1}{65} \sin(2x)$

27. $y = c_1 e^{-3x} + e^x[c_2\cos(x) + c_3\sin(x)] + \dfrac{7}{30} e^{3x} - \dfrac{1}{3} x + \dfrac{4}{9}$ **29.** $y = c_1 e^{7x} + c_2 e^{-x} + c_3 e^{-2x}$

31. $y = c_1 e^{7x} + c_2 e^{-2x} + c_3 e^{-11x}$ **33.** $y = c_1 x^8 + c_2 x^{-1} + c_3 x^{-1}\ln(x)$

35. $y = c_1 e^x + c_2 e^{-x} + c_3 e^{2x} + \dfrac{1}{4} - \dfrac{1}{40} \cos(2x) + \dfrac{1}{40} \sin(2x)$ **37.** $y = \dfrac{5}{29} x^5 + \dfrac{24}{29} \cos[2 \ln(x)] - \dfrac{56}{29} \sin[2 \ln(x)]$

39. $y = c_1 e^{5x} + c_2 e^{-x} + c_3 x e^{-x} + c_4 x^2 e^{-x}$ **41.** $y = c_1 e^{-6x} + c_2\cos(\sqrt{6}x) + c_3\sin(\sqrt{6}x)$

43. $y = c_1 e^{-4x} + c_2 e^{3x} + c_3 e^x - \dfrac{1}{15} \cosh(2x) - \dfrac{1}{10} \sinh(2x)$

45. $y = c_1 e^{-4x} + c_2 e^{-2x} + c_3 e^{-3x} + \dfrac{1}{24} x^3 - \dfrac{13}{96} x^2 + \dfrac{115}{576} x - \dfrac{865}{6912}$ **47.** $y = c_1 e^x + c_2 e^{-x} + c_3 e^{2x} - \dfrac{1}{12} e^{-2x}$

49. $y = -\dfrac{1}{24} x^7 - \dfrac{1}{4} x^{-3} + \dfrac{7}{24} x^{-5}$ **51.** $y = c_1 x + c_2 x^2 + c_3 x^3 + x^4 + 6x \sin(x)$

53. $y = c_1 e^{-8x} + e^{7x}[c_2\cos(2x) + c_3\sin(2x)]$ **55.** $y = c_1 e^{-2.2455x} + c_2 e^{1.3686x} + c_3 e^{1.6270x}$

57. $y = 27e^{7x/3} - 18x e^{7x/3} - 12e^{\sqrt{3}x} - 12e^{-\sqrt{3}x}$

59. $y = c_1 x + c_2 x^{-1.7899} + x^{-2.6050}\{c_3\cos[10.6718 \ln(x)] + c_4\sin[10.6718 \ln(x)]\}$

Chapter 4

Section 4.1

1. $\dfrac{1}{s - 1} - \dfrac{1}{s + 1} - \dfrac{4}{s}$ **3.** $\dfrac{16s}{(s^2 + 4)^2}$ **5.** $\dfrac{1}{s^2} - \dfrac{s}{s^2 + 25}$ **7.** $\dfrac{2}{s^3} + \dfrac{8}{s^2} + \dfrac{16}{s}$ **9.** $\dfrac{6}{s^4} - \dfrac{3}{s^2} + \dfrac{s}{s^2 + 16}$

11. $\dfrac{2}{s} + \dfrac{2s}{s^2 + 36}$ **13.** $\dfrac{k}{s}(e^{-as} - e^{-bs})$ **15.** $-2e^{-16t}$ **17.** $2\cos(4t) - \dfrac{5}{4}\sin(4t)$ **19.** $3e^{7t} + t$ **21.** $e^{4t} - 6te^{4t}$

23. $\dfrac{\sqrt{5}}{10}\, t \sin(\sqrt{5}t) - \dfrac{2\sqrt{5}}{25}\sin(\sqrt{5}t) + \dfrac{2}{5}\, t\cos(\sqrt{5}t) + \cos(\sqrt{2}t)$ **33.** $M = 1; b = 0; t_0 = 0$ **35.** $M = 1; b = 5; t_0 = 0$

37. $M = 24; b = 1; t_0 = 0$

Section 4.2

1. $y = \dfrac{1}{4} - \dfrac{13}{4}e^{-4t}$ **3.** $y = -\dfrac{4}{17}e^{-4t} + \dfrac{4}{17}\cos(t) + \dfrac{1}{17}\sin(t)$ **5.** $y = -\dfrac{1}{4} + \dfrac{1}{2}t + \dfrac{5}{4}e^{2t}$

7. $y = \dfrac{1}{4} + \dfrac{3}{4}e^{2t} + \dfrac{5}{2}te^{2t}$ **9.** $-\dfrac{1}{16}\cos(4t) - \dfrac{1}{64}\sin(4t) + \dfrac{1}{16} + \dfrac{1}{16}t$

13. Since $\displaystyle\int_a^t f(r)\, dr = \int_0^t f(r)\, dr - \int_0^a f(r)\, dr$ and this last integral is constant, we have, by using Theorem 4.8,

$$\mathscr{L}\left\{\int_a^t f(r)\, dr\right\} = \mathscr{L}\left\{\int_0^t f(r)\, dr\right\} - \mathscr{L}\left\{\int_0^a f(r)\, dr\right\} = \frac{1}{s}F(s) - \frac{1}{s}\int_0^a f(r)\, dr.$$

15. $\dfrac{1}{2} - \dfrac{1}{2}\cos(2t)$

17. Let $u = t - nT$. Then $du = dt$, $u(nT) = 0$, $u[(n + 1)T] = T$, $e^{-st} = e^{-(u+nT)} = e^{-su}e^{-nT}$, and $f(t) = f(u + nT) = f(u)$, since f is periodic of period T. On making these substitutions, we have

$$\int_{nT}^{(n+1)T} e^{-st}f(t)\, dt = e^{-nT}\int_0^T e^{-su}f(u)\, du = e^{-nT}\int_0^T e^{-st}f(t)\, dt.$$

21. $\dfrac{E\omega(1 + e^{-s\pi/\omega})}{(s^2 + \omega^2)(1 - e^{-s\pi/\omega})}$ **23.** $\dfrac{E\omega}{s^2 + \omega^2}\dfrac{1}{1 - e^{-\pi s/\omega}}$ **25.** $\dfrac{3[1 - (2s + 1)]e^{-2s}}{2(1 - e^{-8s})s^2}$

27. $\dfrac{1}{1 - e^{-6s}}\left[-\dfrac{6s + 2}{s^2}e^{-3s} + \dfrac{1}{s^2} + \dfrac{6}{s}(e^{-3s} - e^{-6s}) + \dfrac{6s + 1}{s^2}e^{-6s}\right]$

Section 4.3

1. $\dfrac{1}{s^2}\Big|_{s\to s+2} = \dfrac{1}{(s + 2)^2}$ **3.** $\left[\dfrac{1}{s^2} - \dfrac{s}{s^2 + 1}\right]\Big|_{s\to s-4} = \dfrac{1}{(s - 4)^2} - \dfrac{s - 4}{s^2 - 8s + 17}$ **5.** $-\dfrac{1}{s^3 - 3s^2 + 2s}$

7. $\dfrac{\sqrt{2}}{4}e^{-2t}\sin(2\sqrt{2}t)$ **9.** $\dfrac{\sqrt{2}}{2}e^{-3t}\sinh(\sqrt{2}t)$ **11.** $e^{-3t}\cosh(2\sqrt{2}t) - \dfrac{\sqrt{2}}{4}e^{-3t}\sinh(2\sqrt{2}t)$ **13.** $2e^{2t} + 8te^{2t}$

15. $y = \dfrac{1}{2} + \dfrac{\sqrt{7}}{28}e^{-3t}[(\sqrt{7} - 9)e^{\sqrt{7}t} + (\sqrt{7} + 9)e^{-\sqrt{7}t}]$ **17.** $y = \dfrac{2}{15} + \dfrac{1}{6}e^{-3t} + \dfrac{7}{10}e^{-5t}$

19. $y = \dfrac{2}{13} + \dfrac{37}{13}e^{5t}\cos(t) + \dfrac{10}{13}e^{5t}\sin(t)$ **21.** $y = \dfrac{1}{3}e^t + e^{2t} + \dfrac{5}{3}e^{4t}$ **23.** $-\dfrac{4}{9} + \dfrac{1}{3}t + 5e^{-t} + \dfrac{40}{9}e^{-3t}$

25. $\dfrac{s}{(s + 2)(s^2 + 9)}$

Section 4.4

1. $6H(t - 3)$ **3.** $(1 - t)H(t - 5)$ **5.** $e^tH(t - 6)$ **7.** $tH(t - 1) + (2 - t)H(t - 5)$

9. $(t + t^2)H(t - 3) + (2 - t^2)H(t - 4)$ **11.** $4 - 4H(t - 6) = 4[1 - H(t - 6)]$ **13.** $e^{-t}[1 - H(t - 4)]$

15. $e^{-5s}\left[\dfrac{2}{s^3} + \dfrac{10}{s^2} + \dfrac{26}{s}\right]$ **17.** $e^{-s}\left[\dfrac{5}{s^2} + \dfrac{6}{s}\right]$ **19.** $-e^{3\pi s}\dfrac{1}{s^2 + 1}$ **21.** $\left[\dfrac{2}{s^3} + \dfrac{8}{s^2} + \dfrac{16}{s}\right](e^{-4s} - e^{-5s}) + e^{-5s}$

23. $\dfrac{1}{s}[3e^{-12s} - 5e^{-18s}]$ **25.** $\dfrac{1}{2}(t - 5)^2H(t - 5)$ **27.** $3e^{-2(t-5)}H(t - 5)$ **29.** $\dfrac{1}{16}\{1 - \cos[4(t - 4)]\}H(t - 4)$

31. $y = \cos(2t) + \dfrac{3}{4}\{1 - \cos[2(t - 4)]\}H(t - 4)$ **33.** $y = \left\{-\dfrac{1}{4} + \dfrac{1}{12}e^{2(t-4)} + \dfrac{1}{6}e^{-(t-4)}\cos[\sqrt{3}(t - 4)]\right\}H(t - 4)$

35. $y = -\dfrac{1}{4} + \dfrac{2}{5}e^{t} - \dfrac{3}{20}\cos(2t) - \dfrac{1}{5}\sin(2t) + \left\{ -\dfrac{1}{4} + \dfrac{2}{5}e^{t-5} - \dfrac{3}{20}\cos[2(t-5)] - \dfrac{1}{5}\sin[2(t-5)] \right\} H(t-5)$

37. $y = -\dfrac{1}{4}[(4-\sqrt{2})e^{-(1+2\sqrt{2})t} + (4+\sqrt{2})e^{-(1-2\sqrt{2})t}] - \dfrac{1}{28}[8 - (4-\sqrt{2})e^{-(1+2\sqrt{2})(t-5)} + (4+\sqrt{2})e^{-(1-2\sqrt{2})(t-5)}]H(t-5)$

39. $\dfrac{1}{4} + \dfrac{3}{4}e^{-2t} + \dfrac{7}{2}te^{-2t} + \left[-\dfrac{1}{4} + \dfrac{1}{4}e^{-2(t-2)} + \dfrac{1}{2}(t-2)e^{-2(t-2)} \right]H(t-2)$

41. $\varphi(t) = [1 - 2e^{t-3} + e^{2(t-3)}]H(t-3) = \begin{cases} 0 & \text{if } t \le 3 \\ 1 - 2e^{t-3} + e^{2(t-3)} & \text{if } t \ge 3 \end{cases}$

43. $E_{\text{out}} = 5e^{-4t} - 10\{[1 - e^{-4(t-2)}]H(t-2) - [1 - e^{-4(t-3)}]H(t-3)\} = \begin{cases} 5e^{-4t} & \text{if } 0 \le t \le 2 \\ 5e^{-4t} - 10[1 - e^{-4(t-2)}] & \text{if } 2 \le t \le 3 \\ 5e^{-4t} + 10[e^{-4(t-2)} - e^{-4(t-3)}] & \text{if } \quad t \ge 3 \end{cases}$

45. $i(t) = \dfrac{2}{R}[1 - e^{-R(t-5)/L}]H(t-5)$

47. $i(t) = \dfrac{A}{R-L}e^{-4}[e^{-(t-4)} - e^{-R(t-4)/L}]H(t-4)$ if $R \ne L$

and

$$i(t) = \dfrac{A}{R-L}(t-4)e^{-(t-4)}H(t-4) \quad \text{if} \quad R = L$$

49.
$$f(t) = \dfrac{k}{b-a}(t-a)[H(t-a) - H(t-b)] + \dfrac{k}{b-c}(t-c)[H(t-b) - H(t-c)]$$
$$= \dfrac{k}{b-a}(t-a)H(t-a) + \left[\dfrac{k}{b-c}(t-c) - \dfrac{k}{b-a}(t-a) \right]H(t-b) - \dfrac{k}{b-c}(t-c)H(t-c)$$
$$= \dfrac{k}{b-a}(t-a)H(t-a) + \left[\dfrac{k}{b-c} - \dfrac{k}{b-a} \right](t-b)H(t-b) - \dfrac{k}{b-c}(t-c)H(t-c);$$

thus,

$$\mathscr{L}\{f(t)\} = \dfrac{k}{b-a}\dfrac{e^{-as}}{s^2} + \left[\dfrac{k}{b-c} - \dfrac{k}{b-a} \right]\dfrac{e^{-bs}}{s^2} - \dfrac{k}{b-c}\dfrac{e^{-as}}{s^2}$$
$$= \dfrac{k}{b-a}\dfrac{e^{-as}}{s^2} + \dfrac{k(a-c)}{(c-b)(b-a)}\dfrac{e^{-bs}}{s^2} + \dfrac{k}{c-b}\dfrac{e^{-as}}{s^2}$$

51. $\mathscr{L}\left\{ \sin(t)H\left(t - \dfrac{\pi}{6} \right) \right\} = \mathscr{L}\left\{ \sin\left[\left(t - \dfrac{\pi}{6} \right) + \dfrac{\pi}{6} \right]H\left(t - \dfrac{\pi}{6} \right) \right\}$

$= \mathscr{L}\left\{ \left[\sin\left(t - \dfrac{\pi}{6} \right)\cos\left(\dfrac{\pi}{6} \right) + \cos\left(t - \dfrac{\pi}{6} \right)\sin\left(\dfrac{\pi}{6} \right) \right]H\left(t - \dfrac{\pi}{6} \right) \right\}$

$= e^{-\pi s/6}\left[\dfrac{\sqrt{3}}{2}\dfrac{1}{s^2+1} + \dfrac{1}{2}\dfrac{s}{s^2+1} \right]$

Section 4.5

1. $5e^{-4t} - 2e^{-2t}$ **3.** $4te^{5t} - 2te^{t}$ **5.** $\dfrac{1}{4} + \dfrac{11}{4}e^{-4t}$ **7.** $\dfrac{2}{3}e^{t} + \dfrac{1}{3}\cos(\sqrt{2}t) + \dfrac{\sqrt{2}}{6}\sin(\sqrt{2}t)$

9. $-\dfrac{7}{16}e^{-t} + \dfrac{9}{4}te^{-t} + \dfrac{7}{16}e^{3t}$ **11.** $\dfrac{7}{3}e^{-2t} - \dfrac{1}{3}e^{t}$ **13.** $\dfrac{32}{25}e^{t} + \dfrac{4}{5}te^{t} - \dfrac{32}{25}\cos(2t) + \dfrac{23}{50}\sin(2t)$

15. $\dfrac{3}{32}e^{-3t} + \dfrac{19}{16}te^{-3t} - \dfrac{3}{32}e^{t} + \dfrac{3}{16}te^{t}$

17. $\dfrac{13}{841}e^{5t} - \dfrac{13}{1682}[2\cos(2t) + 15\sin(2t)] - \dfrac{377}{3364}t\sin(2t) - \dfrac{203}{13456}[\sin(2t) - 2t\cos(2t)]$

19. $-\dfrac{7209}{882}e^{2t} - \dfrac{1890}{441}te^{2t} + \dfrac{983}{441}e^{-5t} + \dfrac{12299}{882}e^{4t}$ **21.** $\dfrac{10}{53}\cos(2t) - \dfrac{17}{106}\sin(2t) - \dfrac{10}{53}e^{-7t}$

23. $\frac{1}{9} e^{-2t} - \frac{1}{9} e^{t} + \frac{10}{3} te^{-2t}$ **25.** $10te^{-4t} - 3e^{-4t}$ **27.** $5e^{4t} + 4te^{4t} - 5e^{5t}$ **29.** $28e^{-2t} - 8te^{-2t} - 27e^{-3t} - 27te^{-3t}$

31. $y = \frac{1}{10} e^{t} - \frac{1}{6} e^{-t} + \frac{1}{15} e^{-4t}$ **33.** $y = \frac{1}{16} e^{t} - \frac{1}{16} e^{-3t} - \frac{1}{4} te^{-3t}$ **35.** $y = 2e^{-2t} - e^{-4t}$

37. $y = \frac{1}{21} e^{3t} - \frac{1}{21} e^{-4t} + \frac{1}{3} e^{-t} - \frac{1}{3}$ **39.** $\frac{3}{65} e^{3t} + \frac{1}{20} e^{-2t} - \frac{5}{52} \cos(2t) - \frac{1}{52} \sin(2t)$

Section 4.6

1. $\frac{1}{16}[\sinh(2t) - \sin(2t)]$ **3.** $\frac{\cos(at) - \cos(bt)}{(b-a)(b+a)}$ if $b^2 \neq a^2$ and $\frac{t \sin(at)}{2a}$ if $b^2 = a^2$

5. $\frac{9}{14} e^{3t} + \frac{5}{14} \cos(\sqrt{5}t) + \frac{3\sqrt{5}}{14} \sin(\sqrt{5}t)$ **7.** $\frac{1}{a^2 + b^2}[a \sin(at) + b \sinh(bt)]$ **9.** $\frac{1}{a^4}[1 - \cos(at)] - \frac{1}{2a^3} \sin(at)$

11. $\frac{1}{30} e^{3t} + \frac{1}{6} e^{-3t} - \frac{1}{5} e^{-2t}$ **13.** $\frac{1}{75} e^{-3t} - \frac{1}{3} e^{3t} + \frac{463}{900} e^{2t} + \frac{2}{5} te^{2t}$ **15.** $-\frac{2}{25} + \frac{1}{5} t + \frac{2}{25} \cos(\sqrt{5}t)$

17. $y = -\frac{1}{2} e^{-6t} * f(t) + \frac{1}{2} e^{-4t} * f(t) - 2e^{-6t} + 3e^{-4t}$

19. $y = \frac{1}{42} e^{3t} * f(t) - \frac{1}{42} e^{-3t} * f(t) - \frac{\sqrt{2}}{28} e^{\sqrt{2}t} * f(t) - \frac{\sqrt{2}}{28} e^{-\sqrt{2}t} * f(t)$

21. $y = \frac{1}{4} e^{6t} * f(t) - \frac{1}{4} e^{2t} * f(t) + 2e^{6t} - 5e^{2t}$ **23.** $y = \frac{1}{2} e^{-6t} * f(t) - \frac{1}{2} e^{6t} * f(t) + e^{-6t} - e^{-4t}$

25. $y = \frac{1}{15} e^{3t} * f(t) - \frac{1}{15} \cos(\sqrt{6}t) * f(t) - \frac{\sqrt{6}}{30} \sin(\sqrt{6}t) * f(t)$ **27.** $f(t) = -t - \frac{1}{6} t^3$ **29.** $f(t) = 2t^2 + \frac{2}{3} t^3$

31. $f(t) = -\frac{2}{3} \cos(\sqrt{3}t) + \frac{2\sqrt{3}}{9} \sin(\sqrt{3}t) - \frac{1}{3} + \frac{1}{3} t$ **33.** $f(t) = 1 + 3t + t^2$ **35.** $f(t) = 3 + \frac{2\sqrt{15}}{5} e^{t/2} \sin\left(\sqrt{15} \frac{t}{2}\right)$

45. $\frac{1}{2a} e^{at} * f(t) - \frac{1}{2a} e^{-at} * f(t)$, where $\mathcal{L}\{f(t)\} = F(s)$.

47. (a) $y = 29 - 8t + t^2 - \frac{1}{2}(29 + 50\sqrt{2})e^{(\sqrt{2}-2)t/2} - \frac{1}{2}(29 - 50\sqrt{2})e^{-(\sqrt{2}+2)t/2}$

(b) $y = 2 + \frac{1}{7}(3\sqrt{7} - 7)e^{(3+\sqrt{7})t} - \frac{1}{7}(3\sqrt{7} + 7)e^{(3-\sqrt{7})t}$ (c) $y = \frac{1}{5} e^{-t} + \frac{1}{50}(8\sqrt{5} - 5)e^{(3+\sqrt{5})t/2} - \frac{1}{50}(8\sqrt{5} + 5)e^{(3-\sqrt{5})t/2}$

Section 4.7

1. $y = -1 - Ct^2$ **3.** $y = 3t^2$ **5.** $y = \frac{3}{2} t^2$ **7.** $y = 10t$ **9.** $y = Ct^2 e^{-t}$

11. $Y(s) = \left(\frac{s-1}{s}\right)^n \frac{1}{s} = \left(1 - \frac{1}{s}\right)^n \frac{1}{s} = \frac{1}{s} \sum_{k=0}^{n} \binom{n}{k} 1^{n-k}\left(-\frac{1}{s}\right)^k = \sum_{k=0}^{n} \binom{n}{k}(-1)^k\left(\frac{1}{s}\right)^{k+1}$, where $\binom{n}{k} = \frac{n!}{n!(n-k)!}$.

Therefore, $y(t) = \sum_{k=0}^{n} \binom{n}{k} \frac{(-1)^k}{k!} t^k$.

15. The new equation is $s^2 Y'' + (4-A)sY' + (B - A + 2)Y = 0$, another Cauchy-Euler equation.

17. $y(x) = \frac{1}{6} F_0 x^3 + \frac{M}{6EI}(x-a)^3 H(x-a)$ **19.** $\frac{\pi}{2} - \arctan(s)$ **21.** $\frac{s^2 + 6s + 5}{(s^2 + 6s + 13)^2}$

23. $y = 3[e^{-2(t-2)} - e^{-3(t-2)}]H(t-2) - 4[e^{-2(t-5)} - e^{-3(t-5)}]H(t-5)$

25. $y = 6(e^{-2t} - e^{-t} + te^{-t})$ **29.** 0 **35.** $y(t) = \sqrt{\frac{m}{k}} v_0 \sin\left(\sqrt{\frac{k}{m}} t\right); mv_0$

37. $y(t) = \frac{1}{3} \sin(12t)$ feet; $y'(0) = 4$ feet/second; $\frac{6}{\pi}$ hertz; 4 inches

39. $y'(0) = 4$ meters/second downward; the maximum displacement first occurs at time $t = \frac{1}{5}\ln(\frac{8}{3})$ seconds; the maximum displacement is $\frac{1}{2}(\frac{3}{8})^{3/5}$ meter.

Section 4.8

1. $x(t) = -2 + 2e^{t/2} - t;\ y(t) = 1 + e^{t/2} - t$

3. $x(t) = \frac{4}{9} + \frac{1}{3}t - \frac{4}{9}e^{3t/4};\ y(t) = -\frac{2}{3} + \frac{2}{3}e^{3t/4}$

5. $x(t) = \frac{1}{2}e^{-t} + \frac{3}{2}e^{-3t};\ y(t) = e^{-t} - 1$

7. $x(t) = \frac{3}{4} - \frac{3}{4}e^{2t/3} + \frac{1}{2}t^2 + \frac{1}{2}t;\ y(t) = -\frac{3}{2}e^{2t/3} + t + \frac{3}{2}$

9. $x(t) = e^{-t}\cos(t) + t - 1;\ y(t) = e^{-t}\sin(t) + t^2 - t$

11. $x(t) = \frac{8}{27} + \frac{4}{9}t + \frac{1}{3}t^2 - \frac{8}{27}e^{3t/2};\ y(t) = -\frac{2}{9} - \frac{1}{3}t + \frac{2}{9}e^{3t/2}$

13. $x(t) = -\frac{1}{2} + \frac{1}{2}e^{-2t};\ y(t) = \frac{1}{4} + \frac{1}{2}t - \frac{1}{4}e^{-2t}$

15. $x(t) = -e^{-t}\cos(2t) - 32e^{-t}\sin(2t);\ y(t) = 7e^{-t}\cos(2t) + 19e^{-t}\sin(2t) - 6$

17. $i_1(t) = \frac{1}{5}\left\{2 - e^{-(t-4)/2}\left[\cos\left(\frac{t-4}{2}\right)\right] + \sin\left[\left(\frac{t-4}{2}\right)\right]\right\}H(t-4)$

$\qquad - \frac{1}{10}\left\{2 - e^{-(t-5)/2}\left[\cos\left(\frac{t-5}{2}\right)\right] + \sin\left[\left(\frac{t-5}{2}\right)\right]\right\}H(t-5);$

$\qquad i_2(t) = \frac{1}{5}e^{-(t-4)/2}\left[\cos\left(\frac{t-4}{2}\right)\right] + \sin\left[\left(\frac{t-4}{2}\right)\right]H(t-4)$

$\qquad - \frac{1}{10}e^{-(t-5)/2}\left[\cos\left(\frac{t-5}{2}\right)\right] + \sin\left[\left(\frac{t-5}{2}\right)\right]H(t-5)$

19. $i_1(t) = \frac{1}{1650}[143 - 135e^{-4(t-4)} - 8e^{-15(t-4)}]H(t-4);\ i_2(t) = \frac{1}{1650}[33 - 45e^{-4(t-4)} - 12e^{-15(t-4)}]H(t-4)$

21. $y_1 = (1.3578)(0.3482)\sin(0.7366t) - (0.6379)(0.4382)\sin(1.567t);$
$\quad y_2 = -(1.3578)(0.3159)\sin(0.7366t) - (0.6379)(0.2618)\sin(1.567t)$

23. (a) $My_1'' = -k_1y_1 - c_1y_1' + k_2(y_1 - y_2) + A\sin(\omega t),\ my_2'' = -k_2(y_2 - y_1);\ y_1(0) = y_1'(0) = y_2(0) = y_2'(0) = 0$

(b) $Y_1(s) = \dfrac{ms^2 + k_2}{k_2}Y_2(s)$ and $Y_2(s) = \dfrac{A\omega k_2}{(s^2 + \omega^2)[Mms^4 + mc_1s^3 + (mk_1 + mk_2 + Mk_2)s^2 + k_2c_1s + k_1k_2]}$

(c) If $\omega = \sqrt{\dfrac{k_2}{m}}$, then $Y_1(s) = \dfrac{m(s^2 + \omega^2)}{k_2}Y_2(s) = \dfrac{s^2 + \omega^2}{\omega^2}Y_2(s)$

$\quad = \dfrac{Ak_2}{\omega[Mms^4 + mc_1s^3 + (mk_1 + mk_2 + Mk_2)s^2 + k_2c_1s + k_1k_2]}$. The absence of an $s^2 + \omega^2$ term indicates that the $A\sin(\omega t)$ component has been absorbed.

25. $m_1y_1'' = k(y_2 - y_1),\ m_2y_2'' = -k(y_2 - y_1);\ y_1(0) = y_1'(0) = y_2'(0) = 0,\ y_2(0) = d.\ (m_1s^2 + k)Y_1 - kY_2 = 0,$
$(m_2s^2 + k)Y_2 - kY_1 = m_2ds.$ Replacing Y_2 with $\dfrac{m_1s^2 + k}{k}Y_1$ in the second equation, we have

$Y_1(s) = \dfrac{kd}{m_1s\left[s^2 + k\left(\dfrac{m_1 + m_2}{m_1m_2}\right)\right]}$. The quadratic term in the denominator indicates that the objects will oscillate with

period $2\pi\sqrt{\dfrac{m_1m_2}{k(m_1 + m_2)}}$.

27. $y_1(t) = \frac{3}{5}\cos(2t) + \frac{2}{5}\cos(3t);\ y_2(t) = \frac{6}{5}\cos(2t) - \frac{1}{5}\cos(3t)$

29. $y_1(t) = \frac{1}{18} - \frac{1}{10}\cos(2t) - \frac{2}{45}\cos(3t) - \left[\frac{1}{18} - \frac{1}{10}\cos[2(t-1)] - \frac{2}{45}\cos[3(t-1)]\right]H(t-1);$

$\quad y_2(t) = \frac{2}{9} + \frac{4}{45}\cos(2t) - \frac{14}{45}\cos(3t) - \left[\frac{2}{9} + \frac{4}{45}\cos[2(t-1)] - \frac{14}{45}\cos[3(t-1)]\right]H(t-1)$

31. $i_1(t) = \left[1 - \dfrac{1}{10}e^{-(t-2)} - \dfrac{9}{10}e^{-(t-2)/6}\right]H(t-2)$; $i_1(t) = \left[\dfrac{1}{2} + \dfrac{1}{10}e^{-(t-2)} - \dfrac{3}{5}e^{-(t-2)/6}\right]H(t-2)$

33. $10i_1' + 30i_1 - 20i_2 - 10i_2' = k$, $10i_1 + 10i_1 + 10\displaystyle\int i_2 = k$; $i_1(0+) = i_2(0+) = \dfrac{k}{20}$.

$i_1(t) = \dfrac{k}{120}[4 + (1 + \sqrt{3})e^{-(3-\sqrt{3})t/2} + (1 - \sqrt{3})e^{-(3+\sqrt{3})t/2}]$; $i_2(t)$

$= \dfrac{k}{120}[(3 + \sqrt{3})e^{-(3-\sqrt{3})t/2} + (3 - \sqrt{3})e^{-(3+\sqrt{3})t/2}]$.

35. $i_1(t) = \dfrac{A}{120}[4 + (1 + \sqrt{3})e^{-(3-\sqrt{3})(t-k)/2} + (1 - \sqrt{3})e^{-(3+\sqrt{3})(t-k)/2}]H(t-k)$;

$i_2(t) = \dfrac{A}{120}[(3 + \sqrt{3})e^{-(3-\sqrt{3})(t-k)/2} + (3 - \sqrt{3})e^{-(3+\sqrt{3})(t-k)/2}]H(t-k)$

37. $i_1 = 1.0000 - 0.4417e^{-3.5947t} - 0.5203e^{-2.6991t} - 0.0380e^{-0.2061t}$
$\qquad - [1.0000 - 0.4417e^{-3.5947(t-3)} - 0.5203e^{-2.6991(t-3)} - 0.0380e^{-0.2061(t-3)}]H(t-3)$,

$i_2 = -0.1960e^{-3.5947t} - 0.1348e^{-2.6991t} + 0.3307e^{-0.2061t}$
$\qquad - [-0.1960e^{-3.5947(t-3)} - 0.1348e^{-2.6991(t-3)} + 0.3307e^{-0.2061(t-3)}]H(t-3)$,

$i_3 = 0.3295e^{-3.5947t} - 0.4479e^{-2.6991t} + 0.1184e^{-0.2061t}$
$\qquad - [0.3295e^{-3.5947(t-3)} - 0.4479e^{-2.6991(t-3)} + 0.1184e^{-0.2061(t-3)}]H(t-3)$

39. $my_1'' = -ky_1 + k(y_2 - y_1) = -2ky_1 + ky_2$;
$my_2'' = -k(y_2 - y_1) + k(y_3 - y_2) = ky_1 - 2ky_2 + ky_3$;
$my_3'' = -k(y_3 - y_2) + k(y_4 - y_3) = ky_2 - 2ky_3 + ky_4$;
$my_4'' = -k(y_4 - y_3) + k(y_5 - y_4) = ky_3 - 2ky_4 + ky_5$;
$\vdots$
$my_{n-1}'' = -k(y_{n-1} - y_{n-2}) + k(y_n - y_{n-1}) = ky_{n-2} - 2ky_{n-1} + ky_n$;
$my_n'' = -k(y_n - y_{n-1}) - ky_n = ky_{n-1} - 2ky_n$;
$y_1(0) = 1, y_2(0) = y_3(0) = \cdots = y_n(0) = y_1'(0) = y_2'(0) = \cdots = y_n'(0) = 0$

41. $y_1 = 2.5714e^{-4t} + 0.6139e^{-0.2679t} - 3.3831e^{-3.7321t} + 0.1978e^{-1.5t}\cos(0.8660t) - 2.1698e^{-1.5t}\sin(0.8660t)$;
$y_2(t) = 0.2857e^{-4t} + 1.1456e^{-0.2679t} - 0.4533e^{-3.7321t} - 1.9780e^{-1.5t}\cos(0.8660t) + 0.9136e^{-1.5t}\sin(0.8660t)$

CHAPTER 5

Section 5.1

1. $R = 1; (3, 5]$ **3.** $R = 1; [-2, 0)$ **5.** $R = 1; (-1, 1)$ **7.** $R = \dfrac{2}{3}; \left(\dfrac{11}{6}, \dfrac{19}{6}\right)$ **9.** $R = 1; (-1, 1)$

11. $R = 1; [-1, 1)$ **13.** $R = 2; [-6, -2]$ **15.** $R = 3; [-1, 5]$

17. $\ln(2) + \dfrac{1}{2}x - \dfrac{1}{8}x^2 + \dfrac{1}{24}x^3 - \dfrac{1}{64}x^4 + \dfrac{1}{160}x^5 - \cdots$ **19.** $1 + x - \dfrac{1}{3}x^3 - \dfrac{1}{6}x^4 - \dfrac{1}{30}x^5 + \cdots$

21. $x - \dfrac{1}{3}x^3 + \dfrac{1}{5}x^5 - \cdots$ **23.** $\dfrac{\pi}{2} + 1 + \left(x - \dfrac{\pi}{2}\right) - 2\left(x - \dfrac{\pi}{2}\right)^2 - \dfrac{2}{3}\left(x - \dfrac{\pi}{2}\right)^4 - \cdots$

25. $\dfrac{\pi}{2} + \left(x - \dfrac{\pi}{2}\right) - \dfrac{\pi}{4}\left(x - \dfrac{\pi}{2}\right)^2 - \dfrac{1}{2}\left(x - \dfrac{\pi}{2}\right)^3 + \dfrac{\pi}{48}\left(x - \dfrac{\pi}{2}\right)^4 + \dfrac{1}{24}\left(x - \dfrac{\pi}{2}\right)^5 - \cdots$ **27.** $\dfrac{9}{4} + 3\left(x - \dfrac{1}{2}\right) + \left(x - \dfrac{1}{2}\right)^2$

29. $\displaystyle\sum_{n=2}^{\infty} \dfrac{(-1)^n x^n}{2n + 2}$ **31.** $\displaystyle\sum_{n=1}^{\infty} \dfrac{2n + 5}{n + 1} x^n$ **33.** $\displaystyle\sum_{n=2}^{\infty} \dfrac{2^{n-1}}{(n - 1)^3 - 3} x^n$ **35.** $1 + 2x + \displaystyle\sum_{n=2}^{\infty} (2^{n-1} + n + 1)x^n$

37. $1 + \dfrac{1}{2}x + \displaystyle\sum_{n=2}^{\infty} \left(\dfrac{n!}{n + 2} + 2^n\right)x^n$ **39.** $27 + \displaystyle\sum_{n=1}^{\infty} \left[\dfrac{1}{2n} - (n + 3)^{n+3}\right]x^n$ **41.** $\displaystyle\sum_{n=3}^{\infty} (n! + 1)x^{n+2}$

43. $\dfrac{1}{2} + \dfrac{1}{3} x + \displaystyle\sum_{n=2}^{\infty} \left[\dfrac{1}{n+2} + (-2)^{n-1} \right] x^n$ $\qquad$ **45.** $x + \dfrac{1}{3} x^3 + \dfrac{2}{15} x^5 + \dfrac{7}{315} x^7 + \cdots$ $\qquad$ **47.** $R = 7$ $\qquad$ **49.** $R = \sqrt{26}$

51. (a) $R = \sqrt{13}$ $\qquad$ (b) $R = 3$

Section 5.2

1. $a_{n+2} = -\dfrac{a_n}{(n+1)(n+2)}; n \geq 1$

$(a_1 = 0); y_1 = 1 - \dfrac{1}{6} x^3 + \dfrac{1}{180} x^6 - \dfrac{1}{12{,}960} x^9 + \dfrac{1}{1{,}610{,}720} x^{12} - \cdots$

$(a_0 = 0); y_2 = x - \dfrac{1}{12} x^4 + \dfrac{1}{504} x^7 - \dfrac{1}{45{,}360} x^{10} + \dfrac{1}{7{,}076{,}160} x^{13} - \cdots$

3. $a_{n+2} = \dfrac{2a_{n+1} - a_{n-1}}{(n+1)(n+2)}; n \geq 1$

$(a_1 = 0); y_1 = 1 - \dfrac{1}{6} x^3 - \dfrac{1}{12} x^4 - \dfrac{1}{30} x^5 - \dfrac{1}{180} x^6 - \cdots$

$(a_0 = 0); y_2 = x + x^2 + \dfrac{2}{3} x^3 + \dfrac{1}{4} x^4 + \dfrac{1}{20} x^5 + \cdots$

5. $a_{n+2} = \dfrac{a_{n-3}}{(n+1)(n+2)}; n \geq 3$

$(a_1 = 0); y_1 = 1 + \dfrac{1}{4 \cdot 5} x^5 + \dfrac{1}{4 \cdot 5 \cdot 9 \cdot 10} x^{10} + \dfrac{1}{4 \cdot 5 \cdot 9 \cdot 10 \cdot 14 \cdot 15} x^{15} + \dfrac{1}{4 \cdot 5 \cdot 9 \cdot 10 \cdot 14 \cdot 15 \cdot 19 \cdot 20} x^{20} + \cdots$

$(a_0 = 0); y_2 = x + \dfrac{1}{5 \cdot 6} x^6 + \dfrac{1}{5 \cdot 6 \cdot 10 \cdot 11} x^{11} + \dfrac{1}{5 \cdot 6 \cdot 10 \cdot 11 \cdot 15 \cdot 16} x^{16} + \dfrac{1}{5 \cdot 6 \cdot 10 \cdot 11 \cdot 15 \cdot 16 \cdot 20 \cdot 21} x^{21} + \cdots$

7. $a_{n+2} = -\dfrac{(n+1)a_{n+1} - na_n - 2a_{n-1}}{(n+1)(n+2)}; n \geq 1$

$(a_1 = 0); y_1 = 1 - \dfrac{1}{3} x^3 + \dfrac{1}{12} x^4 - \dfrac{1}{15} x^5 + \dfrac{1}{45} x^6 - \cdots$

$(a_0 = 0); y_2 = x - \dfrac{1}{2} x^2 + \dfrac{1}{3} x^3 - \dfrac{1}{3} x^4 + \dfrac{1}{6} x^5 - \cdots$

9. $a_{n+2} = -\dfrac{na_n + 2a_{n-1}}{(n+1)(n+2)}; n \geq 1$

$(a_1 = 0); y_1 = 1 - \dfrac{1}{3} x^3 + \dfrac{1}{20} x^5 + \dfrac{1}{45} x^6 - \dfrac{1}{168} x^7 + \cdots$

$(a_0 = 0); y_2 = x - \dfrac{1}{6} x^3 - \dfrac{1}{6} x^4 + \dfrac{1}{40} x^5 + \dfrac{1}{30} x^6 + \cdots$

11. $a_{n+2} = -\dfrac{(n+1)a_{n+1} - a_{n-2}}{(n+1)(n+2)}; n \geq 2$

$(a_1 = 0); y_1 = 1 + \dfrac{1}{12} x^4 - \dfrac{1}{60} x^5 + \dfrac{1}{360} x^6 - \dfrac{1}{2520} x^7 + \cdots$

$(a_0 = 0); y_2 = x - \dfrac{1}{2} x^2 - \dfrac{1}{6} x^3 - \dfrac{1}{24} x^4 + \dfrac{7}{120} x^5 - \cdots$

13. $a_{n+2} = \dfrac{8a_{n-1}}{(n+1)(n+2)}; n \geq 2$

$(a_1 = 0); y_1 = 1 + \dfrac{1}{2} x^2 + \dfrac{5}{3} x^3 + \dfrac{1}{5} x^5 + \dfrac{4}{9} x^6 + \cdots$

$(a_2 = 0); y_2 = x + \dfrac{1}{2} x^2 + \dfrac{1}{3} x^3 + \dfrac{2}{3} x^4 + \dfrac{1}{5} x^5 + \cdots$

15. $y = 1 + x^2 + \dfrac{1}{2} x^4 + \dfrac{1}{6} x^6 + \dfrac{1}{24} x^8 + \cdots$

17. $a_{n+2} = \dfrac{2na_n - 4a_{n-2}}{(n+1)(n+2)}; n \geq 2$ $y = a_0 + a_1 x + \dfrac{1}{3} a_1 x^3 - \dfrac{1}{3} a_0 x^4 - \dfrac{1}{10} a_1 x^5 + \cdots$

19. $a_{n+2} = -\dfrac{12(n+1)a_{n+1} + a_{n-2}}{(n+1)(n+2)}; n \geq 2$ $y = a_0 + a_1 x - 6a_1 x^2 + 24a_1 x^3 - \dfrac{1}{12}(864a_1 - a_0)x^4 + \cdots$

21. $y = a_0 + a_1 x - a_1 x^2 + \dfrac{1}{6}(1 + 4a_1)x^3 - \dfrac{1}{12}(1 + 3a_1)x^4 + \cdots$ **23.** $y = a_0 + a_1 x - \dfrac{1}{2} a_0 x^2 + \dfrac{1}{6} a_1 x^3 - \dfrac{1}{8} a_0 x^4 + \cdots$

25. $y = a_0 + a_1 x + \dfrac{1}{2} a_0 x^2 + \dfrac{1}{6}(a_1 - 3a_0)x^3 + \dfrac{1}{12}(2 + 5a_0 - 3a_1)x^4 + \cdots$

27. $y = -2 - \dfrac{1}{3} x^3 + \dfrac{1}{12} x^4 - \dfrac{1}{60} x^5 - \dfrac{1}{120} x^6 + \cdots$ **29.** $y = 1 + 2x + \dfrac{1}{2} x^2 - \dfrac{1}{2} x^3 - \dfrac{7}{24} x^4 + \cdots$

31. $y = 3 + \dfrac{5}{2}(x-1)^2 + \dfrac{5}{6}(x-1)^3 + \dfrac{5}{24}(x-1)^4 + \dfrac{1}{6}(x-1)^5 + \cdots$

33. $y = 1 - 4(x-2) + \dfrac{9}{2}(x-2)^2 - \dfrac{13}{6}(x-2)^3 + \dfrac{1}{3}(x-2)^4 + \cdots$

35. $y = 7 + 3(x-1) - 2(x-1)^2 - (x-1)^3 + (x-1)^4 + \cdots$ **37.** $y = -2 + 7x + \dfrac{1}{2} x^2 + \dfrac{1}{6} x^3 + \dfrac{5}{24} x^4 + \cdots$

39. $y = -3 + x + 4x^2 + \dfrac{7}{6} x^3 + \dfrac{1}{3} x^4 + \cdots$ **41.** $(-2, 2)$ **43.** $(0, 6)$ **45.** $(-1, 1)$ **47.** $(-4, 0)$

57. $(1 + x)^\alpha = y = 1 + \alpha x + \dfrac{\alpha(1 - \alpha)}{2} x^2 + \dfrac{\alpha(\alpha - 1)(\alpha - 2)}{3!} x^3 + \cdots + \dfrac{a(\alpha - 1)(\alpha - 2) \cdots (\alpha - n + 1)}{n!} x^n + \cdots$

Section 5.3

1. 0, regular; 3, regular   **3.** 0, regular; 2, regular   **5.** 0, irregular; 2, regular

7. $4r^2 - 2r = 0; c_n = \dfrac{c_{n-1}}{2(n+r)(2n+2r-1)}, n \geq 1$ $y_1 = \displaystyle\sum_{n=0}^{\infty} \dfrac{x^{n+1/2}}{(2n+1)!}$, $y_2 = \displaystyle\sum_{n=0}^{\infty} \dfrac{x^n}{(2n)!}$

9. $9r^2 - 9r + 2 = 0; c_n = \dfrac{-4c_{n-1}}{9(n+r)(n+r-1)+2}, n \geq 1$

$y_1 = x^{2/3}\left(1 - \dfrac{4}{3 \cdot 4} x + \dfrac{4^2}{3 \cdot 4 \cdot 6 \cdot 7} x^2 + \dfrac{4^3}{3 \cdot 4 \cdot 6 \cdot 7 \cdot 9 \cdot 10} x^3 + \dfrac{4^4}{3 \cdot 4 \cdot 6 \cdot 7 \cdot 9 \cdot 10 \cdot 12 \cdot 13} x^4\right.$

$\left. + \dfrac{4^5}{3 \cdot 4 \cdot 6 \cdot 7 \cdot 9 \cdot 10 \cdot 12 \cdot 13 \cdot 14 \cdot 15} x^5 + \cdots\right) = x^{2/3} = \displaystyle\sum_{n=1}^{\infty} \dfrac{2 \cdot 5 \cdot 8 \cdots (3n-1)(-1)^n 4^n x^{n+2/3}}{(3n+1)!}$;

$y_2 = x^{1/3} + \displaystyle\sum_{n=1}^{\infty} \dfrac{1 \cdot 4 \cdot 7 \cdots (3n-2)(-1)^n 4^n x^{n+1/3}}{(3n)!}$

11. $2r^2 - r = 0; c_n = \dfrac{-2(n+r-1)c_{n-1}}{(n+r)(2n+2r-1)}, n \geq 1$

$y_1 = x^{1/2} + \displaystyle\sum_{n=1}^{\infty} \dfrac{(-1)^n(2n+3)}{3n!} x^{n+1/2}$, $y_2 = 1 + \displaystyle\sum_{n=1}^{\infty} \dfrac{(-1)^n 2^n(n+1)}{1 \cdot 3 \cdot 5 \cdots (2n-1)} x^n$

13. $2r^2 - r - 1 = 0; c_1 = \dfrac{-2c_0}{2r+3}, c_n = \dfrac{2c_{n-2} - 2(n+r-1)c_{n-1}}{2(n+r)^2 - (n+r) - 1}, n \geq 2$

$y_1 = x - \dfrac{2}{1! \, 5} x^2 + \dfrac{18}{2! \, 5 \cdot 7} x^3 - \dfrac{164}{3! \, 5 \cdot 7 \cdot 9} x^4 + \dfrac{2284}{4! \, 5 \cdot 7 \cdot 9 \cdot 11} x^5 - \dfrac{37{,}272}{5! \, 5 \cdot 7 \cdot 9 \cdot 11 \cdot 13} x^6 + \cdots$

$y_2 = x^{-1/2} - x^{1/2} + \dfrac{3}{2} x^{3/2} - \dfrac{13}{3! \, 3} x^{5/2} + \dfrac{119}{4! \, 3 \cdot 5} x^{7/2} - \dfrac{1353}{5! \, 3 \cdot 5 \cdot 7} x^{9/2} + \cdots$

15. $9r^2 - 4 = 0; (9r^2 + 18r + 5)c_1 = 0, c_n = \dfrac{-9c_{n-2}}{9(n+r)^2 - 4}, n \geq 2$

$y_1 = x^{2/3} - \dfrac{3}{2^2 \, 1! \, 5} x^{8/3} + \dfrac{3^2}{2^4 \, 2! \, 5 \cdot 8} x^{14/3} - \dfrac{3^3}{2^6 \, 3! \, 5 \cdot 8 \cdot 11} x^{20/3} + \dfrac{3^4}{2^8 \, 4! \, 5 \cdot 8 \cdot 11 \cdot 14} x^{26/3}$

$- \dfrac{3^5}{2^{10} \, 5! \, 5 \cdot 8 \cdot 11 \cdot 14 \cdot 17} x^{32/3} \cdots$

$$y_2 = x^{-2/3} - \frac{3}{2^2\,1!\,1}\,x^{4/3} + \frac{3^2}{2^4\,2!\,1\cdot 4}\,x^{10/3} - \frac{3^3}{2^6\,3!\,1\cdot 4\cdot 7}\,x^{16/3} + \frac{3^4}{2^8\,4!\,1\cdot 4\cdot 7\cdot 10}\,x^{22/3}$$

$$- \frac{3^5}{2^{10}\,5!\,1\cdot 4\cdot 7\cdot 10\cdot 13}\,x^{28/3} + \cdots$$

17. $2r^2 - r = 0;\ (2r^2 + 3r + 1)c_1 = 0,\ c_n = \dfrac{-6c_{n-2}}{2(n+r)^2 - (n+r)},\ n \ge 2$

$$y_1 = x^{1/2} - \frac{3}{1!\,5}\,x^{3/2} + \frac{3^2}{2!\,5\cdot 9}\,x^{5/2} - \frac{3^3}{3!\,5\cdot 9\cdot 13}\,x^{7/2} + \frac{3^4}{4!\,5\cdot 9\cdot 13\cdot 17}\,x^{9/2} - \frac{3^5}{5!\,5\cdot 9\cdot 13\cdot 17\cdot 21}\,x^{11/2} + \cdots$$

$$y_2 = 1 - \frac{3}{1!\,3}\,x + \frac{3^2}{2!\,3\cdot 7}\,x^2 - \frac{3^3}{3!\,3\cdot 7\cdot 11}\,x^3 + \frac{3^4}{4!\,3\cdot 7\cdot 11\cdot 15}\,x^4 - \frac{3^5}{5!\,3\cdot 7\cdot 11\cdot 15\cdot 19}\,x^5 + \cdots$$

19. $2r^2 - 5r - 3 = 0;\ (2r^2 - r - 6)c_1 = 0,\ (2r^2 + 3r - 5)c_2 = 0,\ c_n = \dfrac{2c_{n-3}}{2(n+r)^2 - 5(n+r) - 3},\ n \ge 3$

$$y_1 = x^3 + \frac{2}{1!\,3^1\,13}\,x^6 + \frac{2^2}{2!\,3^2\,13\cdot 19}\,x^9 + \frac{2^3}{3!\,3^3\,13\cdot 19\cdot 25}\,x^{12} + \frac{2^4}{4!\,3^4\,13\cdot 19\cdot 25\cdot 31}\,x^{15}$$

$$+ \frac{2^5}{5!\,3^5\,13\cdot 19\cdot 25\cdot 31\cdot 37}\,x^{18} + \cdots$$

$$y_2 = x^{-1/2} - \frac{2}{1!\,3}\,x^{5/2} - \frac{2^2}{2!\,3^2\,5}\,x^{11/2} - \frac{2^3}{3!\,3^3\,5\cdot 11}\,x^{17/2} + \frac{2^4}{4!\,3^4\,5\cdot 11\cdot 17}\,x^{23/2}$$

$$- \frac{2^5}{5!\,3^5\,5\cdot 11\cdot 17\cdot 23}\,x^{29/2} + \cdots$$

21. $4r^2 - 7r - 2 = 0;\ c_n = \dfrac{(n+r-4)c_{n-1}}{4(n+r)^2 - 7(n+r) - 2},\ n \ge 1$

$$y_1 = x^2 - \frac{1}{13}\,x^3$$

$$y_2 = x^{-1/4} + \frac{13}{4\cdot 5}\,x^{3/4} + \frac{117}{4^2\,10}\,x^{7/4} - \frac{39}{4^3\,6}\,x^{11/4} + \frac{13}{4^4\,28}\,x^{15/4} + \frac{39}{4^5\,1540}\,x^{19/4} + \cdots$$

23. $2r^2 + 5r - 3 = 0;\ c_n = \dfrac{3[(n+r)^2 - 3(n+r) + 2]c_{n-1}}{2(n+r)^2 + 5(n+r) - 3},\ n \ge 1$

$$y_1 = x^{-3} - \frac{36}{5}\,x^{-2} + \frac{108}{5}\,x^{-1} - \frac{216}{5}$$

$$y_2 = x^{1/2} - \frac{3}{1!\,4^1\,9}\,x^{3/2} - \frac{3\cdot 3}{2!\,4^2\,9\cdot 11}\,x^{5/2} - \frac{3^2\cdot 3^2\cdot 5}{3!\,4^3\,9\cdot 11\cdot 13}\,x^{7/2} - \frac{3^3\cdot 3^2\cdot 5^2\cdot 7}{4!\,4^4\,9\cdot 11\cdot 13\cdot 15}\,x^{9/2}$$

$$- \frac{3^5\,3\cdot 5^2\cdot 7^2\cdot 9}{5!\,4^5\,9\cdot 11\cdot 13\cdot 15\cdot 17}\,x^{11/2} + \cdots$$

25. $r^2 - 6 = 0;\ (r^2 + 2r - 5)c_1 = 0,\ c_n = \dfrac{-6c_{n-2}}{(n+r)^2 - 6},\ n \ge 2$

$$y_1 = x^{\sqrt{6}}\Bigg[1 - \frac{1}{1!\,4^1(1+\sqrt{6})}\,x^2 + \frac{1}{2!\,4^2\,(1+\sqrt{6})(2+\sqrt{6})}\,x^4 - \frac{1}{3!\,4^3\,(1+\sqrt{6})(2+\sqrt{6})(3+\sqrt{6})}\,x^6$$

$$+ \frac{1}{4!\,4^4\,(1+\sqrt{6})(2+\sqrt{6})(3+\sqrt{6})(4+\sqrt{6})}\,x^6 - \frac{1}{5!\,4^5\,(1+\sqrt{6})(2+\sqrt{6})(3+\sqrt{6})(4+\sqrt{6})(5+\sqrt{6})}\,x^{10} + \cdots\Bigg]$$

$$y_1 = x^{-\sqrt{6}}\Bigg[1 - \frac{1}{1!\,4^1\,(1-\sqrt{6})}\,x^2 + \frac{1}{2!\,4^2\,(1-\sqrt{6})(2-\sqrt{6})}\,x^4 - \frac{1}{3!\,4^3\,(1-\sqrt{6})(2-\sqrt{6})(3-\sqrt{6})}\,x^6$$

$$+ \frac{1}{4!\,4^4\,(1-\sqrt{6})(2-\sqrt{6})(3-\sqrt{6})(4-\sqrt{6})}\,x^6 - \frac{1}{5!\,4^5\,(1-\sqrt{6})(2-\sqrt{6})(3-\sqrt{6})(4-\sqrt{6})(5-\sqrt{6})}\,x^{10} + \cdots\Bigg]$$

27. $r^2 + 1 = 0;\ c_n = \dfrac{2(n+r-1)c_{n-1}}{(n+r)^2 + 1},\ n \ge 1$

$$y_1 = x^i\Bigg[1 + \frac{2i}{1!(1+2i)}\,x + \frac{2^2\,i(1+i)}{2!(1+2i)(2+2i)}\,x^2 + \frac{2^3\,i(1+i)(2+i)}{3!(1+2i)(2+2i)(3+2i)}\,x^3$$

$$+ \frac{2^4\,i(1+i)(2+i)(3+i)(4+i)}{4!(1+2i)(2+2i)(3+2i)(4+2i)}\,x^4 + \frac{2^5\,i(1+i)(2+i)(3+i)(4+i)(5+i)}{5!(1+2i)(2+2i)(3+2i)(4+2i)(5+2i)}\,x^5 + \cdots\Bigg]$$

$$y_2 = x^{-i}\left[1 - \frac{2i}{1!(1-2i)}x - \frac{2^2\,i(1-i)}{2!(1-2i)(2-2i)}x^2 - \frac{2^3\,i(1-i)(2-i)}{3!(1-2i)(2-2i)(3-2i)}x^3\right.$$
$$\left. - \frac{2^4\,i(1-i)(2-i)(3-i)(4-i)}{4!(1-2i)(2-2i)(3-2i)(4-2i)}x^4 - \frac{2^5\,i(1-i)(2-i)(3-i)(4-i)(5-i)}{5!(1-2i)(2-2i)(3-2i)(4-2i)(5-2i)}x^5 + \cdots\right]$$

Section 5.4

1. $r_1 = \dfrac{1}{5}, r_2 = -\dfrac{4}{5}; y_1 = \displaystyle\sum_{n=0}^{\infty} c_n x^{n+1/5}, c_0 \neq 0; y_2 = ky_1(x)\ln(x) + \displaystyle\sum_{n=0}^{\infty} c_n^* x^{n-4/5}, c_0^* \neq 0$

3. $r_1 = \dfrac{2}{3}, r_2 = -\dfrac{1}{4}; y_1 = \displaystyle\sum_{n=0}^{\infty} c_n x^{n+2/3}, c_0 \neq 0; y_2 = \displaystyle\sum_{n=0}^{\infty} c_n^* x^{n-1/4}, c_0^* \neq 0$

5. $r_1 = r_2 = \dfrac{1-2\sqrt{2}}{2}; y_1 = \displaystyle\sum_{n=0}^{\infty} c_n x^{n+(1-2\sqrt{2})/2}, c_0 \neq 0; y_2 = y_1(x)\ln(x) + \displaystyle\sum_{n=1}^{\infty} c_n^* x^{n+(1-2\sqrt{2})/2}$

7. $y_1 = c_0(x-1); y_2 = c_0^*\left[(x-1)\ln(x) - 3x + \dfrac{1}{4}x^2 + \dfrac{1}{36}x^3 + \dfrac{1}{288}x^4 + \dfrac{1}{2400}x^5 + \cdots\right]$

9. $y_1 = c_0[x^4 + 2x^5 + 3x^6 + 4x^7 + 5x^8 + \cdots] = c_0\dfrac{x^4}{(x-1)^2}; y_2 = c_0^*\dfrac{3-4x}{(x-1)^2}$

11. $y_1 = c_0\left[x^{1/2} + \dfrac{1}{2^1\,1!\,3}x^{3/2} + \dfrac{1}{2^2\,2!\,3\cdot5}x^{5/2} + \dfrac{1}{2^3\,3!\,3\cdot5\cdot7}x^{7/2} + \dfrac{1}{2^4\,4!\,3\cdot5\cdot7\cdot9}x^{9/2} - \cdots\right]$

$\quad y_2 = c_0^*\left[1 - \dfrac{1}{2}x + \dfrac{1}{2^2\,2!\,3}x^2 + \dfrac{1}{2^3\,3!\,3\cdot5}x^3 + \dfrac{1}{2^4\,4!\,3\cdot5\cdot7}x^4 - \cdots\right]$

13. $y_1 = c_0\left[x^2 + \dfrac{1}{3!}x^4 + \dfrac{1}{5!}x^6 + \dfrac{1}{7!}x^8 + \dfrac{1}{9!}x^{10} + \cdots\right]$

$\quad y_2 = c_0^*\left[x - x^2 + \dfrac{1}{2!}x^3 - \dfrac{1}{3!}x^4 + \dfrac{1}{4!}x^5 - \cdots\right] = c_0^* x e^{-x}$

15. $y_1 = c_0(1-x); y_2 = c_0^*\left[1 + \dfrac{1}{2}(x-1)\ln\left(\dfrac{x-2}{x}\right)\right]$

17. $y_1 = c_0\left[x^2 - x^3 + \dfrac{1}{3}x^4 - \dfrac{1}{36}x^5 - \dfrac{7}{720}x^6 + \cdots\right]$

$\quad y_2 = c_0^*\left[y_1(x)\ln(x) - x + \dfrac{3}{2}x^3 - \dfrac{31}{36}x^4 + \dfrac{65}{432}x^5 + \dfrac{61}{4320}x^6 + \cdots\right]$

19. $y_1 = c_0 x^{2+2\sqrt{2}}\left[1 - \dfrac{3+2\sqrt{2}}{-5+2\sqrt{2}}x + \dfrac{(3+2\sqrt{2})(4+2\sqrt{2})}{(-5+2\sqrt{2})(-2+4\sqrt{2})}x^2\right.$

$\quad - \dfrac{(3+2\sqrt{2})(4+2\sqrt{2})(5+2\sqrt{2})}{(-5+2\sqrt{2})(-2+4\sqrt{2})(3+6\sqrt{2})}x^3 + \dfrac{(3+2\sqrt{2})(4+2\sqrt{2})(5+2\sqrt{2})(6+2\sqrt{2})}{(-5+2\sqrt{2})(-2+4\sqrt{2})(3+6\sqrt{2})(10+8\sqrt{2})}x^4$

$\quad \left. - \dfrac{(3+2\sqrt{2})(4+2\sqrt{2})(5+2\sqrt{2})(6+2\sqrt{2})(7+2\sqrt{2})}{(-5+2\sqrt{2})(-2+4\sqrt{2})(3+6\sqrt{2})(10+8\sqrt{2})(19+10\sqrt{2})}x^5 + \cdots\right]$

$\quad y_2 = c_0^* x^{2-2\sqrt{2}}\left[1 - \dfrac{3-2\sqrt{2}}{-5-2\sqrt{2}}x + \dfrac{(3-2\sqrt{2})(4-2\sqrt{2})}{(-5-2\sqrt{2})(-2-4\sqrt{2})}x^2\right.$

$\quad - \dfrac{(3-2\sqrt{2})(4-2\sqrt{2})(5-2\sqrt{2})}{(-5-2\sqrt{2})(-2-4\sqrt{2})(3-6\sqrt{2})}x^3 + \dfrac{(3-2\sqrt{2})(4-2\sqrt{2})(5-2\sqrt{2})(6-2\sqrt{2})}{(-5-2\sqrt{2})(-2-4\sqrt{2})(3-6\sqrt{2})(10-8\sqrt{2})}x^4$

$\quad \left. - \dfrac{(3-2\sqrt{2})(4-2\sqrt{2})(5-2\sqrt{2})(6-2\sqrt{2})(7-2\sqrt{2})}{(-5-2\sqrt{2})(-2-4\sqrt{2})(3-6\sqrt{2})(10-8\sqrt{2})(19-10\sqrt{2})}x^5 + \cdots\right]$

21. $y_1 = c_0\left[1 - \dfrac{1}{4}x^2 + \dfrac{1}{4^2(2!)^2}x^4 - \dfrac{1}{4^3(3!)^2}x^6 + \dfrac{1}{4^4(4!)^2}x^8 - \cdots\right]$

$$y_2 = c_0^*\left[y_1(x)\ln(x) + \frac{1}{4}x^2 + \frac{1+\frac{1}{2}}{4^2(2!)^2}x^4 + \frac{1+\frac{1}{2}+\frac{1}{3}}{4^3(3!)^2}x^6 - \frac{1+\frac{1}{2}+\frac{1}{3}+\frac{1}{4}}{4^4(4!)^2}x^8 + \frac{1+\frac{1}{2}+\frac{1}{3}+\frac{1}{4}+\frac{1}{5}}{4^5(5!)^2}x^{10} - \cdots\right]$$

23. $y_1(x) = c_0 x^{\sqrt{6}}\left[1 + \frac{1}{1!(1+2\sqrt{6})}x + \frac{1}{2!(1+2\sqrt{6})(2+2\sqrt{6})}x^2\right.$

$$+ \frac{1}{3!(1+2\sqrt{6})(2+2\sqrt{6})(3+2\sqrt{6})}x^3 + \frac{1}{4!(1+2\sqrt{6})(2+2\sqrt{6})(3+2\sqrt{6})(4+2\sqrt{6})}x^4$$

$$\left. + \frac{1}{5!(1+2\sqrt{6})(2+2\sqrt{6})(3+2\sqrt{6})(4+2\sqrt{6})(5+2\sqrt{6})}x^5 + \cdots\right]$$

$$y_2 = c_0^* x^{-\sqrt{6}}\left[1 + \frac{1}{1!(1-2\sqrt{6})}x + \frac{1}{2!(1-2\sqrt{6})(2-2\sqrt{6})}x^2 + \frac{1}{3!(1-2\sqrt{6})(2-2\sqrt{6})(3-2\sqrt{6})}x^3\right.$$

$$\left. + \frac{1}{4!(1-2\sqrt{6})(2-2\sqrt{6})(3-2\sqrt{6})(4-2\sqrt{6})}x^4 + \frac{1}{5!(1-2\sqrt{6})(2-2\sqrt{6})(3-2\sqrt{6})(4-2\sqrt{6})(5-2\sqrt{6})}x^5 + \cdots\right]$$

25. $y_1 = c_0(2 - 4x + x^2)$

$$y_2 = c_0^*\left[(2 - 4x + x^2)\ln(x) + 10x - \frac{9}{2}x^2 + \frac{1}{9}x^3 + \frac{1}{144}x^4 + \frac{1}{1800}x^5 + \cdots\right]$$

27. $y_1 = c_0 x^{1/4}\left[1 - \frac{1}{5}x^2 + \frac{1}{2!\,5\cdot9}x^4 - \frac{1}{3!\,5\cdot9\cdot13}x^6 + \frac{1}{4!\,5\cdot9\cdot13\cdot17}x^8 - \frac{1}{5!\,5\cdot9\cdot13\cdot17\cdot21}x^{10} + \cdots\right]$

$$y_2 = c_0^* x^{-1/4}\left[1 - \frac{1}{3}x^2 + \frac{1}{2!\,3\cdot7}x^4 - \frac{1}{3!\,3\cdot7\cdot11}x^6 + \frac{1}{4!\,3\cdot7\cdot11\cdot15}x^8 - \frac{1}{5!\,3\cdot7\cdot11\cdot15\cdot19}x^{10} + \cdots\right]$$

K	XK	YK
0	0	1
1	.314159	1
2	.628318	1.09708
3	.942477	1.299665
4	1.256636	1.629988
5	1.570795	2.117
6	1.884954	2.782075
7	2.199113	3.613312
8	2.513272	4.531673
9	2.827431	5.368485
10	3.14159	5.889663

K	XK	YK
0	0	4
1	.2	.8
2	.4	.712
3	.6	.6906111
4	.8	.7152224
5	1	.7729138
6	1.2	.8534346
7	1.4	.9477645
8	1.6	1.048113
9	1.8	1.148405
10	2	1.244638

K	XK	YK
0	0	2
1	.025	2.090672
2	.05	2.190881
3	.075	2.302522
4	.1	2.428139

5	.125	2.571258
6	.15	2.736956
7	.175	2.932896
8	.2	3.171379
9	.225	3.473859
10	.25	3.88236

7.

K	XK	YK
0	4	2
1	3.8	1.510102
2	3.6	1.049229
3	3.4	.6179875
4	3.2	.2170892
5	3	-.1526183
6	2.8	-.490102
7	2.6	-.794069
8	2.4	-1.062839
9	2.2	-1.29411
10	2	-1.484467

9.

K	XK	YK
0	0	1
1	.1	1.2
2	.2	1.39
3	.3	1.57
4	.4	1.74
5	.5	1.9
6	.6	2.05
7	.7	2.19
8	.8	2.32
9	.9	2.44
10	1	2.55

11. $y = e^{1 - \cos(x)}$ is the exact solution.

K	XK	YK	Y(XK)	% ERROR
0	0	1	1	0
1	.314159	1	1.050161	4.776485
2	.628318	1.09708	1.210439	9.365048
3	.942477	1.299665	1.510158	13.93846
4	1.256636	1.629988	1.995674	18.32396
5	1.570795	2.117	2.718278	22.11982
6	1.884954	2.782075	3.702527	24.8601
7	2.199113	3.613312	4.892894	26.15183
8	2.513272	4.531673	6.104436	25.76427
9	2.827431	5.368485	7.036112	23.70097
10	3.14159	5.889663	7.389056	20.29208

13. $y = 5e^{x^2}$ is the exact solution.

K	XK	YK	Y(XK)	% ERROR
0	0	5	5	0
1	.2	5	5.204054	3.921056
2	.4	5.4	5.867555	7.968474
3	.6	6.264	7.166647	12.59511
4	.8	7.76736	9.482404	18.0866
5	1	10.25291	13.59141	24.56327
6	1.2	14.35408	21.10348	31.98239
7	1.4	21.24404	35.49664	40.15197
8	1.6	33.1407	64.67911	48.76135
9	1.8	54.35076	127.6687	57.42828
10	2	93.4833	272.991	65.75591

15. $y = \dfrac{1}{2}[\sin(x-1) - \cos(x-1) - 5e^{x-1}]$ is the exact solution.

K	XK	YK	Y(XK)	% ERROR
0	1	-3	-3	0
1	1.2	-3.708061	-3.444206	7.660837
2	1.4	-4.522145	-3.995384	13.18424
3	1.6	-5.460567	-4.685644	16.53824
4	1.8	-6.54684	-5.553529	17.88613
5	2	-7.810767	-6.645122	17.54137
6	2.2	-9.289691	-8.015453	15.89726
7	2.4	-11.02993	-9.730261	13.35697
8	2.6	-13.08844	-11.8682	10.28158
9	2.8	-15.53475	-14.5236	6.962107
10	3.0	-18.45325	-17.80993	3.612154

17. $e \approx 2.704814$; this value is too small.   **19.** $\pi \approx 3.151577$; this value is too large.

Section 6.2

1.

K	XK	TAYLOR YK	MOD EULER YK	RK4 YK
0	0	2	2	2
1	.02	2.076019	2.076098	2.076222
2	.04	2.160259	2.160447	2.160747
3	.06	2.254413	2.254755	2.255306
4	.08	2.360755	2.361322	2.362243
5	.10	2.482449	2.483351	2.484839
6	.12	2.62407	2.625497	2.627895
7	.14	2.79258	2.794878	2.798837
8	.16	2.999317	3.003182	3.010064
9	.18	3.264538	3.271561	3.284718
10	.2	3.629717	3.644351	3.67433

3.

K	XK	TAYLOR YK	MOD EULER YK	RK4 YK
0	0	1	1	1
1	.1	1.14255	1.142519	1.142848
2	.2	1.264032	1.264029	1.264662
3	.3	1.366667	1.366721	1.367612
4	.4	1.452697	1.452815	1.453912
5	.5	1.524246	1.524428	1.525678
6	.6	1.583258	1.583495	1.584853
7	.7	1.631468	1.631751	1.633175
8	.8	1.670404	1.670723	1.672181
9	.9	1.701405	1.701751	1.703216
10	1	1.725636	1.726	1.727452

5.

K	XK	TAYLOR YK	MOD EULER YK	RK4 YK
0	0	4	4	4
1	.1	3.71	3.710123	3.709834
2	.2	3.438986	3.439208	3.43867
3	.3	3.185968	3.18627	3.185519
4	.4	2.949974	2.950339	2.949409
5	.5	2.730055	2.730468	2.729389
6	.6	2.525288	2.525736	2.524534
7	.7	2.334779	2.335251	2.333952
8	.8	2.157668	2.158156	2.15678
9	.9	1.993129	1.993626	1.992192
10	1	1.840373	1.840873	1.839398

		TAYLOR	MOD EULER	RK4
7. K	XK	YK	YK	YK
0	1	2	2	2
1	1.01	2.045243	2.045261	2.045297
2	1.02	2.093524	2.093565	2.093648
3	1.03	2.145252	2.145321	2.145464
4	1.04	2.200917	2.201023	2.201241
5	1.05	2.261117	2.26127	2.261587
6	1.06	2.326594	2.326809	2.327255
7	1.07	2.398276	2.398574	2.399192
8	1.08	2.477355	2.477764	2.478617
9	1.09	2.565385	2.565948	2.567126
10	1.1	2.66445	2.665234	2.666876

		TAYLOR	MOD EULER	RK4
9. K	XK	YK	YK	YK
0	0	1	1	1
1	.2	1.298227	1.298276	1.299406
2	.4	1.652986	1.653172	1.65589
3	.6	2.074123	2.074583	2.07957
4	.8	2.574838	2.57575	2.584005
5	1	3.172681	3.174249	3.187198
6	1.2	3.890515	3.892946	3.912565
7	1.4	4.757341	4.760816	4.789757
8	1.6	5.808976	5.813638	5.855365
9	1.8	7.088787	7.09476	7.153726
10	2	8.648814	8.656253	8.738156

11. The exact solution is $y = x^2 + 2x + 2 - 9e^{x-1}$.

Modified Euler

K	XK	YK	Y(XK)	% ERROR
0	1	-4	-4	0
1	1.2	-5.142	-5.152625	.2062034
2	1.4	-6.640041	-6.666424	.3957654
3	1.6	-8.59005	-8.639071	.5674311
4	1.8	-11.10906	-11.18987	.7221843
5	2	-14.33985	-14.46454	.8620197
6	2.2	-18.45662	-18.64106	.9894218
7	2.4	-23.67188	-23.93681	1.106792
8	2.6	-30.24489	-30.61731	1.21635
9	2.8	-38.49197	-39.00684	1.319957
10	3.0	-48.799	-49.50154	1.419224

Taylor

K	XK	YK	Y(XK)	% ERROR
0	1	-4	-4	0
1	1.2	-5.14	-5.152625	.245025
2	1.4	-6.6356	-6.666424	.4623796
3	1.6	-8.582631	-8.639071	.6533041
4	1.8	-11.09801	-11.18987	.8209363
5	2	-14.32437	-14.46454	.9690469
6	2.2	-18.43573	-18.64106	1.101472
7	2.4	-23.6444	-23.93681	1.221607
8	2.6	-30.20937	-30.61731	1.33239
9	2.8	-38.44663	-39.00684	1.436197
10	3.0	-48.74168	-49.50154	1.535017

RK4

K	XK	YK	Y(XK)	% ERROR
0	1	-4	-4	0
1	1.2	-5.152607	-5.152625	3.516618E-04
2	1.4	-6.666377	-6.666424	7.08129E-04

3	1.6	-8.638982	-8.639071	1.015596E-03
4	1.8	-11.18972	-11.18987	1.321012E-03
5	2	-14.46431	-14.46454	1.582365E-03
6	2.2	-18.64072	-18.64106	1.841756E-03
7	2.4	-23.93631	-23.93681	2.071749E-03
8	2.6	-30.61661	-30.61731	2.292508E-03
9	2.8	-39.00587	-39.00684	2.484008E-03
10	3.0	-49.5002	-49.50154	2.71259E-03

15. The exact solution is $y = \dfrac{1}{3}(x^2 + 2x^{-1})$.

K	XK	EULER YK	IMP EULER YK	MOD EULER YK	EXACT Y(XK)
0	1	1	1	1	1
1	1.1	1	1.009545	1.009762	1.009394
2	1.2	1.019091	1.035833	1.036165	1.035556
3	1.3	1.054167	1.076539	1.076926	1.076154
4	1.4	1.103077	1.13	1.130407	1.129524
5	1.5	1.164286	1.195	1.195404	1.194445
6	1.6	1.236667	1.270625	1.271014	1.27
7	1.7	1.319375	1.356177	1.356541	1.35549
8	1.8	1.411765	1.451111	1.451447	1.450371
9	1.9	1.513334	1.555	1.555305	1.554211
10	2	1.623684	1.6675	1.667773	1.666667

17. The exact solution is $y = \dfrac{4x}{2 - 3x^2}$.

K	XK	EULER YK	IMP EULER YK	MOD EULER YK	EXACT Y(XK)
0	1	-4	-4	-4	-4
1	1.1	-2	-2.743802	-2.897959	-2.699386
2	1.2	-1.487603	-2.10828	-2.245058	-2.068965
3	1.3	-1.209958	-1.72562	-1.836917	-1.693811
4	1.4	-1.030257	-1.469356	-1.561104	-1.443299
5	1.5	-.902513	-1.285036	-1.362513	-1.263158
6	1.6	-.8061442	-1.145556	-1.212447	-1.126761
7	1.7	-.7303747	-1.035952	-1.094776	-1.01949
8	1.8	-.6689531	-.9472891	-.9997991	-.9326423
9	1.9	-.6179773	-.8739021	-.9213536	-.860702
10	2	-.5748733	-.8120227	-.8553371	-.7999998

21. $e \approx 2.718282$ **23.** $\pi \approx 3.141593$

Section 6.3

1.

K	XK	YK
0	-2	-4
1	-1.9	-3.639341
2	-1.8	-3.354804
3	-1.7	-3.142715
4	-1.6	-2.999647
5	-1.5	-2.92244
6	-1.4	-2.908235
7	-1.3	-2.954503
8	-1.2	-3.05908
9	-1.1	-3.220205
10	-1	-3.436563

3.

K	XK	YK
0	2	1
1	2.1	1.611636
2	2.2	2.655995

3	2.3	4.511669
4	2.4	7.944402
5	2.5	14.58064
6	2.6	28.00753
7	2.7	56.48988
8	2.8	119.9536
9	2.9	268.7508
10	3.0	636.47

5.

K	XK	YK
0	1.570796	1
1	1.727876	.8409469
2	1.884956	.6286101
3	2.042036	.3565086
4	2.199116	3.091702E-02
5	2.356196	-.3225431
6	2.513276	-.6695841
7	2.670356	-.9912504
8	2.827436	-1.293384
9	2.984516	-1.59636
10	3.141595	-1.927221

7.

K	XK	YK
0	-1	0
1	-.9	9.531007E-02
2	-.8	.1823177
3	-.7	.2623378
4	-.6	.3363737
5	-.5	.4051919
6	-.4	.4693805
7	-.3	.5293821
8	-.2	.5855183
9	-.1	.6380028
10	0	.6869488

9.

K	XK	YK
0	0	2
1	.2	2.35643
2	.4	2.615313
3	.6	2.770547
4	.8	2.828727
5	1	2.807429
6	1.2	2.73163
7	1.4	2.629323
8	1.6	2.52748
9	1.8	2.449114
10	2	2.411831

11.

		TAYLOR	MOD EULER	RK4	PRE-COR
K	XK	YK	YK	YK	YK
0	3	0	0	0	0
1	3.1	-.305	-.296	-.293229	-.293229
2	3.2	-.5495096	-.5358391	-.5354929	-.5354929
3	3.3	-.709937	-.69781	-.702936	-.702936
4	3.4	-.8068716	-.7984615	-.806778	-.8045522
5	3.5	-.8657694	-.8605804	-.868927	-.8653003
6	3.6	-.9035699	-.9005458	-.9072745	-.9050242
7	3.7	-.9299139	-.9281996	-.9329938	-.9324062
8	3.8	-.950052	-.9490909	-.9522533	-.9521441
9	3.9	-.966807	-.9662674	-.9682499	-.9682393
10	4	-.9816891	-.9813816	-.9825829	-.9827527

13.

		TAYLOR	MOD EULER	RK4	PRE-COR
K	XK	YK	YK	YK	YK
0	0	-2	-2	-2	-2

1	.1	-2.96	-2.95975	-2.983097	-2.983097
2	.2	-4.3786	-4.37798	-4.447239	-4.447239
3	.3	-6.473528	-6.47236	-6.626514	-6.626514
4	.4	-9.567021	-9.565042	-9.870118	-9.869933
5	.5	-14.13599	-14.13281	-14.69896	-14.69838
6	.6	-20.88627	-20.88131	-21.89009	-21.8888
7	.7	-30.86248	-30.85489	-32.60267	-32.60009
8	.8	-45.61067	-45.59919	-48.56586	-48.56103
9	.9	-67.419	-67.40175	-72.35909	-72.35043
10	1	-99.67391	-99.64814	-107.8302	-107.8151

15.

K	XK	TAYLOR YK	MOD EULER YK	RK4 YK	PRE-COR YK
0	0	4	4	4	4
1	.1	3.912162	3.912411	3.912044	3.912044
2	.2	3.779433	3.780433	3.779998	3.779998
3	.3	3.607189	3.609488	3.609366	3.609366
4	.4	3.406391	3.410505	3.411048	3.411029
5	.5	3.193221	3.199508	3.20088	3.200805
6	.6	2.987425	2.995965	2.998028	2.997866
7	.7	2.80965	2.820236	2.822569	2.822326
8	.8	2.678885	2.691136	2.693191	2.69291
9	.9	2.611065	2.624587	2.625865	2.625603
10	1	2.619044	2.63356	2.633724	2.633523

17. A polynomial of degree $r = 2$ through the points $(x_k, f_k), (x_{k-1}, f_{k-1}), (x_{k-2}, f_{k-2})$ is

$$p(x) = \frac{1}{h^2}[(x - x_k)(x - x_{k-1})f_{k-2} - 2(x - x_k)(x - x_{k-2})f_{k-1} + (x - x_{k-1})(x - x_{k-2})f_k].$$ Now integrate to get

$$\int_{x_k}^{x_k + h} p(x)\, dx = \frac{h}{12}[23f_k - 16f_{k-1} + 5f_{k-2}],$$ and use 6.8 to obtain the desired result.

19. (a) $m = 1; \alpha_1 = 1; \varphi(x_k, y_k) = f(x_k, y_k)$ (b) $m = 1; \alpha_1 = 1; \varphi(x_k, y_k) = f\left(x_k + \frac{h}{2}, y_k + \frac{h}{2}\right)$

(c) $m = 1; \alpha_1 = 1; \varphi(x_k, y_k) = f(x_k, y_k) + \frac{h}{2}[f_x(x_k, y_k) + f(x_k, y_k)f_y(x_k, y_k)]$

(d) $m = 1; \alpha_1 = 1; \varphi(x_k, y_k) = \frac{1}{6}[W_1 + W_2 + W_3 + W_4]$, where $W_1 = f(x_k, y_k)$, $W_2 = f\left(x_k + \frac{h}{2}, y_k + \frac{h}{2}W_1\right)$,

$W_3 = f\left(x_k + \frac{h}{2}, y_k + \frac{h}{2}W_2\right)$, $W_4 = f(x_k + h, y_k + hW_3)$

(e) $m = 4; \alpha_1 = 1, \alpha_2 = \alpha_3 = \alpha_4 = 0; \varphi(x_{k-3}, x_{k-2}, x_{k-1}, x_k, y_{k-3}, y_{k-2}, y_{k-1}, y_k)$

$= \frac{1}{24}[55f(x_k, y_k) - 59f(x_{k-1}, y_{k-1}) + 37f(x_{k-2}, y_{k-2}) - 9f(x_{k-3}, y_{k-3})]$

Chapter 7

Section 7.1

1. If $y(x) = x^a J_v(bx^c)$, then $y'(x) = ax^{a-1}J_v(x) + x^a bcx^{c-1}J'_v(x)$ and $y''(x) = a(a - 1)x^{a-2}J_v(x) + 2ax^{a-1}bcx^{c-1}J'_v(x) + x^a J''_v(x)b^2c^2x^{2c-2} + x^a J'_v(x)bc(c - 1)x^{c-2}$. Substitute into the differential equation to get

$$y'' - \left(\frac{2a - 1}{x}\right)y' + \left[b^2c^2x^{2c-2} + \frac{a^2 - v^2c^2}{x^2}\right]y = a(a - 1)x^{a-2}J_v(x) + 2abcx^{a+c-2}J'_v(x) + bc(c - 1)x^{a+c-2}J'_v(x)$$

$$+ b^2c^2x^{a+2c-2}J''_v(x) - (2a - 1)ax^{a-2}J_v(x) - (2a - 1)bcx^{a+c-2}J'_v(x)$$

$$+ [b^2c^2x^{a+2c-2} + (a^2 - v^2c^2)x^{a-2}]J_v(x)$$

$$= c^2x^{a-2}(bx^c)^2J''_v(x) + [2abc + bc^2 - bc - 2abc + bc]x^{a+c-2}J'_v(x)$$

$$+ \{[a(a - 1) - (2a - 1)a + (a^2 - v^2c^2)]x^{a-2} + b^2c^2x^{a+2c-2}\}J_v(x)$$

$$= c^2x^{a-2}\{(bx^c)^2J''_v(x) + bx^cJ'_v(x) + [(bx^c)^2 - v^2]J_v(x)\} = 0,$$

because J_v is a solution of Bessel's equation.

3. $y = c_1 J_{1/3}(x^2) + c_2 J_{-1/3}(x^2)$ **5.** $y = c_1 x^{-1} J_{3/4}(2x^2) + c_2 x^{-1} J_{-3/4}(2x^2)$ **7.** $y = c_1 x^4 J_{3/4}(2x^3) + c_2 x^4 J_{-3/4}(2x^3)$

9. $y = c_1 x^{-2} J_{1/2}(3x^2) + c_2 x^{-2} J_{-1/2}(3x^2)$ **11.** $y = c_1 J_{3/5}(2x^3) + c_2 J_{-3/5}(2x^3)$

13. Write $x^{-v} J_v(x) = \sum_{n=0}^{\infty} \dfrac{(-1)^n}{2^{2n+v} n! \Gamma(n+v+1)} x^{2n}$. Differentiate to get

$$[x^{-v} J_v(x)]' = \sum_{n=1}^{\infty} \frac{(-1)^n}{2^{2n+v} n! \Gamma(n+v+1)} 2n x^{2n-1}. \text{ Let } n = m + 1 \text{ to get}$$

$$[x^{-v} J_v(x)]' = \sum_{m=0}^{\infty} \frac{(-1)^{m+1}}{2^{2m+v+1} m! \Gamma(m+v+2)} x^{2m+1} = -x^{-v} \sum_{m=0}^{\infty} \frac{(-1)^m}{2^{2m+(1+v)} m! \Gamma[m+(1+v)+1]} x^{2m+(1+v)} = -x^{-v} J_{v+1}(x).$$

Now calculate $[x^v J_v(x)]' = \sum_{n=0}^{\infty} \dfrac{(-1)^n}{2^{2n+v} n! \Gamma(n+v+1)} (2n+2v) x^{2n+2v-1}$. Since $\Gamma(n+v+1) = (n+v)\Gamma(n+v)$, we have

$$[x^v J_v(x)]' = \sum_{n=0}^{\infty} \frac{(-1)^n}{2^{2n+v-1} n! \Gamma(n+v)} x^{2n+2v-1} = x^v \sum_{n=0}^{\infty} \frac{(-1)^n}{2^{2n+(1+v)} n! \Gamma[n+(v-1)+1]} x^{2n+(v-1)} = x^v J_{v-1}(x).$$

15. Differentiate the second inequality in Problem 14 to get

$$x J_v''(x) + J_v'(x) = v J_v'(x) - x J_{v+1}'(x) - J_{v+1}(x).$$

Multiply this equation by x and rearrange terms to get

$$x^2 J_v''(x) = (v-1) x J_v'(x) - x^2 J_{v+1}'(x) - x J_{v+1}(x).$$

Into this equation, put $x J_v'(x) = v J_v(x) - x J_{v+1}(x)$ and $x J_{v+1}'(x) = -(v+1) J_{v+1}(x) + x J_v(x)$ to get
$x^2 J_v''(x) = (v-1)[v J_v(x) - x J_{v+1}(x)] - x[-(v+1) J_{v+1}(x) + x J_v(x)] - x J_{v+1}(x) = [(v-1)v - x^2] J_v(x) +$
$[(v+1) - (v-1) - 1] x J_{v+1}(x) = (v^2 - v - x^2) J_v(x) + x J_{v+1}(x).$

17. From the second formula in Problem 13, $x^{n+1} J_n(x) = [x^{n+1} J_{n+1}(x)]'$. Integrate to get $\int x^{n+1} J_n(x)\, dx = x^{n+1} J_{n+1}(x) + C.$

19. By the first equality in Problem 13,

$$x^{-3/2} J_{5/2}(x) = -x^{-3/2} J_{3/2}(x)]'$$

$$= -\left[x^{-2} \sqrt{\frac{2}{\pi}} \left(\frac{\sin(x)}{x} - \cos(x) \right) \right]'$$

$$= \sqrt{\frac{2}{\pi}} \left[2 x^{-3} \left(\frac{\sin(x)}{x} - \cos(x) \right) - x^{-2} \left(\frac{x \cos(x) - \sin(x)}{x^2} \right) + \cos(x) \right]$$

$$= \sqrt{\frac{2}{\pi}} x^{-2} \left[\left(\frac{3}{x^2} - 1 \right) \sin(x) - \frac{3}{x} \cos(x) \right],$$

and we get the final result by multiplying this equation by $x^{3/2}$.

21. From the first equality of Problem 13, with $v = 0$, we have $J_1(y) = -[J_0(y)]'$. Integrate from 0 to α to get

$$\int_0^{\alpha} J_1(y)\, dy = -J_0(y) \Big]_0^{\alpha} = J_0(0) - J_0(\alpha) = 1, \text{ because } J_0(0) = 1 \text{ and } J_0(\alpha) = 0. \text{ Now let } yn = \alpha x \text{ to get}$$

$$\int_0^{\alpha} J_1(y)\, dy = \alpha \int_0^1 J_1(\alpha x)\, dx = 1.$$

23. $J_v(x) = \sum_{n=0}^{\infty} \dfrac{(-1)^n}{2^{2n+v} n! \Gamma(n+v+1)} x^{2n+v}, \quad J_{-v}(x) = \sum_{n=0}^{\infty} \dfrac{(-1)^n}{2^{2n-v} n! \Gamma(n-v+1)} x^{2n-v},$

$$J_v'(x) = \sum_{n=0}^{\infty} \frac{(-1)^n (2n+v)}{2^{2n+v} n! \Gamma(n+v+1)} x^{2n+v-1}, \text{ and } J_{-v}'(x) = \sum_{n=0}^{\infty} \frac{(-1)^n (2n-v)}{2^{2n-v} n! \Gamma(n-v+1)} x^{2n-v-1}. \text{ Then}$$

$$xW[J_v, J_{-v}](x) = x[J_v(x) J_{-v}'(x) - J_{-v}(x) J_v'(x)] = x \left[\sum_{k=0}^{\infty} \sum_{m+n=k}^{\infty} \frac{(-1)^n}{2^{2n+v} n! \Gamma(n+v+1)} \frac{(-1)^m (2m-v)}{2^{2m-v} m! \Gamma(m-v+1)} x^{2(m+n)-1} \right.$$

$$\left. - \sum_{k=0}^{\infty} \sum_{m+n=k}^{\infty} \frac{(-1)^n (2n+v)}{2^{2n+v} n! \Gamma(n+v+1)} \frac{(-1)^m}{2^{2m-v} m! \Gamma(m-v+1)} x^{2(m+n)-1} \right]$$

$$= \sum_{k=0}^{\infty} \sum_{m+n=k}^{\infty} \frac{2(m-n) - 2v}{m! n! \Gamma(n+v+1) \Gamma(m-v+1)} \frac{(-1)^k}{2^{2k}} x^{2k}.$$

When $x = 0$, we get $xW[J_v, J_{-v}](x)\Big]_{x=0} = \dfrac{-2v}{\Gamma(v+1)\Gamma(-v+1)} = \dfrac{2}{\Gamma(v+1)\Gamma(-v)}.$

25. $e^{x(t-1/t)/2} = e^{xt/2}e^{-x/2t} = \displaystyle\sum_{n=0}^{\infty} \dfrac{1}{n!}\left(\dfrac{xt}{2}\right)^n \sum_{m=0}^{\infty} \dfrac{1}{m!}\left(\dfrac{-x}{2t}\right)^m = \sum_{n=-\infty}^{\infty} \sum_{\substack{k-m=n\\k,n\geq 0}}^{\infty} \dfrac{(-1)^m}{k!m!}\left(\dfrac{x}{2}\right)^{k+m} t^n.$

The coefficient of t^n is $\displaystyle\sum_{m=0}^{\infty} \dfrac{(-1)^m}{2^{2m+n}n!\,\Gamma(m+n+1)} x^{2m+n}$, which is $J_n(x)$.

27. (a) Use the result of Problem 1 to show that $u(x) = \sqrt{kx}J_n(kx)$ satisfies the given differential equation. Compute

$(uv' - u'v)' = uv'' - vu'' = \left(k^2 - 1 - \dfrac{n^2 - \frac14}{x^2}\right)uv.$ Then $\displaystyle\int_a^{a+\pi}\left(k^2 - 1 - \dfrac{n^2 - \frac14}{x^2}\right)uv\,dx = [uv' - vu']_a^{a+\pi} = -u(a+\pi) -$

$u(a)$, since $v'(a+\pi) = -1, v'(a) = 1, v(a) = v(a+\pi) = 0.$ For sufficiently large a, say, $a \geq A$, $k^2 - 1 - \dfrac{n^2 - \frac14}{x^2} > 0.$

Since $v(x) > 0$ in $(a, a+\pi), \left(k^2 - 1 - \dfrac{n^2 - \frac14}{x^2}\right)v(x) > 0$ in $(a, a+\pi)$ for $a \geq A$. By the mean value theorem

for integrals, for some α in $(a, a+\pi),$

$$\int_a^{a+\pi}\left(k^2 - 1 - \dfrac{n^2 - \frac14}{x^2}\right)u(x)v(x)\,dx = u(\alpha)\int_a^{a+\pi}\left(k^2 - 1 - \dfrac{n^2 - \frac14}{x^2}\right)v(x)\,dx.$$

This integral is positive for $a \geq A$, so $u(a), u(a+\pi)$, and $u(\alpha)$ cannot have the same sign. By the mean value theorem, there is at least one x_0 in $(a, a+\pi)$ such that $J_n(kx_0) = 0.$

(b) In (a), for any x in $(a + k\pi, a + k\pi + \pi), k^2 - 1 - \dfrac{n^2 - \frac14}{x^2} > 0.$ Thus, the argument in (a) is valid for this

interval, for $k = 0, 1, 2, \ldots.$ Then $J_n(x) = 0$ has infinitely many positive roots.

(c) If $J_n(\alpha) = J'_n(\alpha)$, then $J''_n(\alpha) = 0$ also from Bessel's equation. Now differentiate Bessel's equation and put $x = \alpha$ to observe that $J_n^{(3)}(\alpha) = 0.$ In this way, we find that $J_n^{(k)}(\alpha) = 0$ for $k = 0, 1, 2, 3, \ldots.$ The Maclaurin expansion of $J_n(x)$ is therefore zero for all x in some interval about zero, a contradiction.

(d) From Problem 14, $xJ'_n(x) = -nJ_n(x) + xJ_{n-1}(x)$ and $xJ'_n(x) = nJ_n(x) - xJ_{n+1}(x).$ If $J_n(x) = 0$ and $J_{n-1}(x) = 0$ have a common root α, then $J'_n(\alpha) = 0$ also, contradicting the result of (c).

(e) Let a and b be consecutive positive solutions of $J_n(x) = 0.$ Then a and b are also consecutive positive solutions of $f(x) = J_n(x)x^{-n} = 0.$ By Rolle's theorem, there is at least one number c between a and b such that $f'(c) = 0.$ By the first equality of Problem 13, $f'(x) = [x^{-n}J_n(x)]' = -x^{-n}J_{n+1}(x)$; hence, $J_{n+1}(c) = 0.$

(f) Suppose that c_1 and c_2 are positive solutions of $J_{n+1}(x) = 0$ and lie between a pair of consecutive positive solutions of $J_n(x) = 0.$ Let $g(x) = x^{n+1}J_{n+1}(x).$ Then $g(c_1) = g(c_2) = 0$ also. By Rolle's theorem, there is a number α between c_1 and c_2 such that $g'(\alpha) = 0.$ Then, by the second equality of Problem 13, $\alpha^{n+1}J_n(\alpha) = 0$, implying the contradiction that $J_n(\alpha) = 0.$

Section 7.2

1. $y = c_1 J_3(\sqrt{x}) + c_2 Y_3(\sqrt{x})$ **3.** $y = c_1 J_4(2x^{1/3}) + c_2 Y_4(2x^{1/3})$ **5.** $y = c_1 x^{2/3}J_{1/2}(x) + c_2 x^{2/3}Y_{1/2}(x)$

7. $y = c_1 x^{-2}J_3(3\sqrt{x}) + c_2 x^{-2}Y_3(3\sqrt{x})$

9. $y' = x^{1/2}J'_{1/3}\left(\dfrac{2kx^{3/2}}{3}\right)kx^{1/2} + \dfrac12 x^{-1/2}J_{1/3}\left(\dfrac{2kx^{3/2}}{3}\right); y'' = x^{1/2}J''_{1/3}\left(\dfrac{2kx^{3/2}}{3}\right)k^2 x + \dfrac32 kJ'_{1/3}\left(\dfrac{2kx^{3/2}}{3}\right) - \dfrac14 x^{-3/2}J_{1/3}\left(\dfrac{2kx^{3/2}}{3}\right).$

Then $y'' + k^2 xy = x^{1/2}J''_{1/3}\left(\dfrac{2kx^{3/2}}{3}\right)k^2 x + \dfrac32 kJ'_{1/3}\left(\dfrac{2kx^{3/2}}{3}\right) - \dfrac14 x^{-3/2}J_{1/3}\left(\dfrac{2kx^{3/2}}{3}\right) + k^2 x^{3/2}J_{1/3}\left(\dfrac{2kx^{3/2}}{3}\right) =$

$\dfrac94 x^{-3/2}\left\{\left(\dfrac{2kx^{3/2}}{3}\right)^2 J''_{1/3}\left(\dfrac{2kx^{3/2}}{3}\right) + \left(\dfrac{2kx^{3/2}}{3}\right)J'_{1/3}\left(\dfrac{2kx^{3/2}}{3}\right) + \left[\left(\dfrac{2kx^{3/2}}{3}\right)^2 - \dfrac19\right]J_{1/3}\left(\dfrac{2kx^{3/2}}{3}\right)\right\} = 0.$

11. $y = c_1 xJ_2(x) + c_2 xY_2(x)$ **13.** $y = c_1 x^2 J_2(\sqrt{x}) + c_2 x^2 Y_2(\sqrt{x})$ **15.** $y = c_1 x^4 J_1(x) + c_2 x^4 Y_1(x)$

17. $y = c_1 x^{-1}J_2\left(\dfrac{\sqrt{x}}{2}\right) + c_2 x^{-1}Y_2\left(\dfrac{\sqrt{x}}{2}\right)$ **19.** $y = c_1 x^2 J_4(x^2) + c_2 x^2 Y_4(x^2)$

21. $I_0(x) = J_0(ix) = \displaystyle\sum_{n=0}^{\infty} \dfrac{(-1)^n}{2^{2n}n!\,\Gamma(n+1)}(ix)^{2n} = \sum_{k=0}^{\infty} \dfrac{1}{2^{2n}(n!)^2} x^{2n}.$ Compute $I'_0(x) = \displaystyle\sum_{n=0}^{\infty} \dfrac{2n}{2^{2n}(n!)^2} x^{2n-1}$ and

$$I_0''(x) = \sum_{n=0}^{\infty} \frac{2n(2n-1)}{2^{2n}(n!)^2} x^{2n-2}, \text{ so } I_0''(x) + \frac{1}{x} I_0'(x) - I_0(x) = \sum_{n=0}^{\infty} \frac{2n(2n-1)}{2^{2n}(n!)^2} x^{2n-2} +$$

$$\sum_{n=0}^{\infty} \frac{2n}{2^{2n}(n!)^2} x^{2n-2} - \sum_{n=0}^{\infty} \frac{1}{2^{2n}(n!)^2} x^{2n}. \text{ Shift indices in the first two summations } (n = m + 1) \text{ to get}$$

$$\sum_{m=0}^{\infty} \frac{2(m+1)(2m+1)}{2^{2m+2}[(m+1)!]^2} x^{2m} + \sum_{m=0}^{\infty} \frac{2(m+1)}{2^{2m+2}[(m+1)!]^2} x^{2m} - \sum_{m=0}^{\infty} \frac{1}{2^{2m}(m!)^2} x^{2m}$$

$$= \sum_{m=0}^{\infty} \frac{1}{2^{2m}(m!)^2} \left[\frac{2(2m+1)}{2^2(m+1)} + \frac{2}{2^2(m+1)} - 1 \right] x^{2m} = 0.$$

23. From Problem 22, $[\alpha x I_0'(\alpha x)]' = \alpha x I_0(\alpha x)$. Then $\left[\left(\frac{x}{\alpha} \right) I_0'(\alpha x) \right]' = \frac{1}{\alpha^2} [\alpha x I_0'(\alpha x)]' = \frac{1}{\alpha^2} [\alpha x I_0'(\alpha x)]' \cdot \alpha = x I_0(\alpha x)$.

Section 7.3

1. $\frac{dy}{dx} = c_1 b^{-1/3} (bx^{3/2})^{1/3} J_{1/3}(bx^{3/2}) + c_2 b^{-1/3} (bx^{3/2})^{1/3} J_{-1/3}(bx^{3/2})$, where $b = \frac{2}{3} \sqrt{\frac{w}{EI}}$. By Problem 13 of Section 7.1,

$[(bx^{3/2})^{1/3} J_{1/3}(bx^{3/2})]' = (bx^{3/2})^{1/3} J_{-2/3}(bx^{3/2})$ and $[(bx^{3/2})^{1/3} J_{-1/3}(bx^{3/2})]' = -(bx^{3/2})^{1/3} J_{2/3}(bx^{3/2})$. Therefore, $\frac{d^2 y}{dx^2} =$

$c_1 b^{-1/3} [(bx^{3/2})^{1/3} J_{1/3}(bx^{3/2})]' \cdot b \frac{3}{2} x^{1/2} + c_2 b^{-1/3} [(bx^{3/2})^{1/3} J_{-1/3}(bx^{3/2})]' \cdot b \frac{3}{2} x^{1/2} = \frac{3}{2} c_1 bx J_{-2/3}(bx^{3/2}) -$

$\frac{3}{2} c_2 bx J_{2/3}(bx^{3/2})$. But $x J_{-2/3}(bx^{3/2}) \Big|_{x=0} \neq 0$ and $x J_{2/3}(bx^{3/2}) \Big|_{x=0} = 0$. Therefore, $\frac{d^2 y}{dx^2} \Big|_{x=0} = 0$ requires that $c_1 = 0$.

3. Since k^2 is imaginary, we need only calculate the imaginary part of $I_0^{-1}(kR) I_0)kr) e^{i\omega t}$. Let $I_0(kR) = a + bi$ and

$I_0(kr) = c + di$. Then $\text{Im}[I_0^{-1}(kR) I_0)kr) e^{i\omega t}] = \text{Im}\left[\frac{1}{a + bi} (c + di)(\cos(\omega t) + i \sin(\omega t)) \right] =$

$\frac{1}{a^2 + b^2} [(ac + bd)\sin(\omega t) + (ad - bc)\cos(\omega t)]$. Now manipulate the series to obtain

$$I_0(kr) = \sum_{n=0}^{\infty} \frac{1}{2^{2n}(n!)^2} \left[\sqrt{\frac{4\pi\mu\omega}{\rho}} \frac{1+i}{2} r \right]^{2n} = \sum_{n=0}^{\infty} \frac{(-1)^n}{2^{4n}[(2n)!]^2} \left(\frac{4\pi\mu\omega}{\rho} \right)^{2n} r^{4n} + i \sum_{n=0}^{\infty} \frac{(-1)^n}{2^{4n+2}[(2n+1)!]^2} \left(\frac{4\pi\mu\omega}{\rho} \right)^{2n+1} r^{4n+2}.$$

We also obtain

$$I_0'(kR) = \sqrt{\frac{\pi\mu\omega}{\rho}} R \left[\sum_{n=1}^{\infty} \frac{2n}{2^{2n}(n!)^2} \left(\frac{4\pi\mu\omega}{\rho} \right)^{n-1} (-1)^{n(n-1)/2} R^{2n-2} + i \sum_{n=1}^{\infty} \frac{2n}{2^{2n}(n!)^2} \left(\frac{4\pi\mu\omega}{\rho} \right)^{n-1} (-1)^{(n-1)(n-2)/2} R^{2n-2} \right].$$

Thus, $a = \sqrt{\frac{\pi\mu\omega}{\rho}} \sum_{n=1}^{\infty} \frac{2n}{2^{2n}(n!)^2} \left(\frac{4\pi\mu\omega}{\rho} \right)^{n-1} (-1)^{n(n-1)/2} R^{2n-1}$,

$b = \sqrt{\frac{\pi\mu\omega}{\rho}} \sum_{n=1}^{\infty} \frac{2n}{2^{2n}(n!)^2} \left(\frac{4\pi\mu\omega}{\rho} \right)^{n-1} (-1)^{(n-1)(n-2)/2} R^{2n-1}$,

$c = \sum_{n=0}^{\infty} \frac{(-1)^n}{2^{4n}[(2n)!]^2} \left(\frac{4\pi\mu\omega}{\rho} \right)^{2n} r^{4n}$,

and

$d = i \sum_{n=0}^{\infty} \frac{(-1)^n}{2^{4n+2}[(2n+1)!]^2} \left(\frac{4\pi\mu\omega}{\rho} \right)^{2n+1} r^{4n+2}$.

Section 7.4

1. A straightforward calculation **3.** Observe that $[(1 - x^2)y']' = (1 - x^2)y'' - 2xy'$.

5. From Problem 2, $P_n(x) = \sum_{k=0}^{[n/2]} \frac{(-1)^k (2n - 2k)!}{2^n k!(n - k)!(n - 2k)!} x^{n-2k}$

$$= \frac{1}{2^n n!} \sum_{k=0}^{[n/2]} \frac{(-1)^k}{k!} \frac{n!}{(n - k)!} \frac{(2n - 2k)!}{(n - 2k)!} x^{n-2k} = \frac{1}{2^n n!} \sum_{k=0}^{[n/2]} \frac{(-1)^k}{k!} \frac{n!}{(n - k)!} \frac{d^n}{dx^n} [x^{2n-2k}],$$

because $\dfrac{d^n}{dx^n}[x^{2n-2k}] = \dfrac{(2n-2k)!}{(n-2k)!}x^{n-2k}$. For $k > \left[\dfrac{n}{2}\right]$, $\dfrac{d^n}{dx^n}[x^{2n-2k}] = 0$; hence, replace $\left[\dfrac{n}{2}\right]$ by n in the summation to get

$$P_n(x) = \frac{1}{2^n n!}\sum_{k=0}^{n}\frac{(-1)^k}{k!}\frac{n!}{(n-k)!}\frac{d^n}{dx^n}[x^{2n-2k}]$$

$$= \frac{1}{2^n n!}\frac{d^n}{dx^n}\sum_{k=0}^{n}\frac{(-1)^k}{k!}\frac{n!}{(n-k)!}(x^2)^{n-k}.$$

But $(x^2-1)^n = \sum_{k=0}^{n}\dfrac{(-1)^k}{k!}\dfrac{n!}{(n-k)!}(x^2)^{n-k}$; hence, $P_n(x) = \dfrac{1}{2^n n!}\dfrac{d^n}{dx^n}[(x^2-1)^n]$.

7. A routine calculation. For example, with $n = 3$, we have $3P_3(x) + 2P_1(x) = \dfrac{3}{2}(5x^3 - 3x) + 2x = \dfrac{15}{2}x^3 - \dfrac{5}{2}x = 5xP_2(x)$.

9. $Q_0(x) = \displaystyle\int\frac{1}{1-x^2}\,dx = \frac{1}{2}\int\left[\frac{-1}{x-1} + \frac{1}{x+1}\right]dx = \frac{1}{2}\ln\left(\frac{1+x}{1-x}\right);$

$Q_1(x) = x\displaystyle\int\frac{1}{x^2(1-x^2)}\,dx = x\int\left[\frac{1}{x^2} + \frac{1}{2(x+1)} - \frac{1}{2(x-1)}\right]dx = x\left(-\frac{1}{x} + \frac{1}{2}\ln\left(\frac{1+x}{1-x}\right)\right) = -1 + \frac{x}{2}\ln\left(\frac{1+x}{1-x}\right);$

$Q_2(x) = 2(3x^2-1)\displaystyle\int\frac{1}{(3x^2-1)^2(1-x^2)}\,dx = \frac{1}{4}(3x^2-1)\int\left[\frac{1}{x+1} - \frac{1}{x-1} + \frac{1}{\left(x+\frac{1}{\sqrt{3}}\right)^2} + \frac{1}{\left(x-\frac{1}{\sqrt{3}}\right)^2}\right]dx$

$= \dfrac{1}{4}(3x^2-1)\ln\left(\dfrac{1+x}{1-x}\right) - \dfrac{3}{2}x$

Section 7.5

1. $P_6(x) = \dfrac{11}{6}xP_5(x) - \dfrac{5}{6}P_4(x) = \dfrac{1}{16}(231x^6 - 315x^4 + 105x^2 - 5)$; $P_7(x) = \dfrac{1}{16}(429x^7 - 693x^5 + 31x^3 - 35x)$;

$P_8(x) = \dfrac{1}{128}[6435x^8 - 12{,}012x^6 + 6930x^4 - 1260x^2 + 35]$

3. First, $P_0(-1) = 1 = (-1)^0$ and $P_1(-1) = -1 = (-1)^1$. Now let n be a positive integer, and suppose that $P_n(-1) = (-1)^n$ and $P_{n-1}(-1) = (-1)^{n-1}$. We will show that $P_{n+1}(-1) = (-1)^{n+1}$. By Theorem 7.2, $(n+1)P_{n+1}(-1) - (2n+1)(-1)P_n(-1) + nP_{n-1}(-1) = 0$. Therefore, $(n+1)P_{n+1}(-1) + (2n+1)(-1)^n + n(-1)^{n-1} = 0$. Then $P_{n+1}(-1) = \dfrac{1}{n+1}[(2n+1)(-1)^{n+1} - n(-1)^{n-1}] = (-1)^{n+1}$, since $(-1)^{n-1} = (-1)^{n+1}$. By complete induction, $P_n(-1) = (-1)^n$ for every nonnegative integer n.

5. Since P_n and P_0 are orthogonal on $[-1, 1]$, $\displaystyle\int_{-1}^{1}P_0(x)P_n(x)\,dx = \int_{-1}^{1}P_n(x)\,dx = 0$. Therefore, $P_n(x)$ changes sign at least once in $(-1, 1)$. By the intermediate value theorem, $P_n(\xi) = 0$ for some ξ in $(-1, 1)$. Let $x_1, \ldots, x_m$ be all the solutions of $P_n(x) = 0$ in $(-1, 1)$. Let $q(x) = (x - x_1)(x - x_2)\cdots(x - x_m)$, and write $P_n(x) = q(x)p(x)$. Since P_n has no repeated root in $(-1, 1)$, by Problem 4, p has no root in $(-1, 1)$. Therefore, $p(x)$ cannot change sign in $(-1, 1)$, so $P_n(x)q(x) = [q(x)]^2 p(x)$ also does not change sign in $(-1, 1)$. Therefore, $\displaystyle\int_{-1}^{1}P_n(x)q(x)\,dx$ is positive or negative. But if $m < n$, Theorem 7.8 implies that $\displaystyle\int_{-1}^{1}P_n(x)q(x)\,dx = 0$, a contradiction. Therefore, $m = n$, and $x_1, \ldots, x_n$ are all the zeros of P_n in $(-1, 1)$.

7. It is routine to check that $r\dfrac{\partial H}{\partial r} - (x - r)\dfrac{\partial H}{\partial x} = 0$. Substitute the partial derivatives into this equation to get

$$\sum_{n=0}^{\infty} nP_n(x)r^n - (x-r)\sum_{n=0}^{\infty}P_n'(x)r^n = 0, \text{ or } \sum_{n=0}^{\infty}nP_n(x)r^n - \sum_{n=0}^{\infty}xP_n'(x)r^n + \sum_{n=0}^{\infty}P_n'(x)r^{n+1} = 0. \text{ Shift indices in}$$

the third summation to get $\displaystyle\sum_{n=0}^{\infty}[nP_n(x) - xP_n'(x) + P_{n-1}'(x)]r^n = 0$. The coefficient of r^n must be zero for $n = 0, 1, 2, \ldots$, yielding the result.

9. If n is a positive integer, then $\int_{-1}^{1} P_n(x)\, dx = \int_{-1}^{1} P_0(x)P_n(x)\, dx = 0$, because P_0 and P_n are orthogonal on $[-1, 1]$.

11. Multiply the recurrence relation of Theorem 7.4 by $P_{n-1}(x)$ to get

$$(n + 1)P_{n+1}(x)P_{n-1}(x) - (2n + 1)xP_n(x)P_{n-1}(x) + nP_{n-1}^2(x) = 0.$$

Integrate from -1 to 1, noting that $\int_{-1}^{1} P_{n+1}(x)P_{n-1}(x)\, dx = 0$, to get

$$-(2n + 1) \int_{-1}^{1} xP_n(x)P_{n-1}(x)\, dx + n \int_{-1}^{1} P_{n-1}^2(x)\, dx = 0.$$

By Theorem 7.9, $\int_{-1}^{1} P_{n-1}^2(x)\, dx = \dfrac{2}{2n - 1}$. Therefore, $\int_{-1}^{1} xP_n(x)P_{n-1}(x)\, dx = \dfrac{2n}{2n - 1} \cdot \dfrac{1}{2n + 1} = \dfrac{2n}{4n^2 - 1}$.

13. Use the formula $P_m(x) = \sum_{k=0}^{[m/2]} \dfrac{1}{2^m k!} \dfrac{(2m - 2k)!}{(m - k)!(m - 2k)!} x^{m-2k}$. If $m = 2n + 1$, then $m - 2k > 0$ for $k = 0, 1, 2, \ldots$,

$\left[\dfrac{m}{2}\right]$, so $P_{2n+1}(0) = 0$. If $m = 2n$, then $m - 2k$ is positive for $k = 0, 1, \ldots, \left[\dfrac{m}{2}\right] - 1$, but $m - 2k = 0$ for

$m = \left[\dfrac{m}{2}\right] = n$; hence, $P_{2n}(0) = \dfrac{(-1)^n (4n - 2n)!}{2^{2n} n! (2n - n)!(2n - 2n)!} = (-1)^n \dfrac{(2n)!}{2^{2n}(n!)^2}$.

15. $2x + x^2 - 5x^3 = \dfrac{1}{3} P_0(x) - P_1(x) + \dfrac{2}{3} P_2(x) - 2P_3(x)$.

17. (a) If $k = 0$, $e^{ik\varphi} = 1$, so $\int_{-\pi}^{\pi} e^{ik\varphi}\, d\varphi = 2\pi$. If k is a positive integer, then

$$\int_{-\pi}^{\pi} e^{ik\varphi}\, d\varphi = \int_{-1}^{1} [\cos(k\varphi) + i\sin(k\varphi)]\, d\varphi = \left[\dfrac{1}{k}\sin(k\varphi) - \dfrac{i}{k}\cos(k\varphi)\right]_{-\pi}^{\pi} = 0.$$

(b) $\dfrac{1}{2\pi} \int_{-\pi}^{\pi} (a + be^{i\varphi})^k\, d\varphi = \dfrac{1}{2\pi} \int_{-\pi}^{\pi} \sum_{n=0}^{k} \dfrac{k(k - 1)\cdots(k - n + 1)}{n!} a^{k-n}b^n e^{in\varphi}\, d\varphi = \dfrac{1}{2\pi} a^k 2\pi = a^k$, by the result of (a).

(c) If $q(x) = a_0 + a_1 x + \cdots + a_k x^k$, then $\dfrac{1}{2\pi} \int_{-\pi}^{\pi} q(x + be^{i\varphi})\, d\varphi = \dfrac{1}{2\pi} \int_{-\pi}^{\pi} \sum_{k=0}^{n} a_k(x + be^{i\varphi})^k\, d\varphi = \dfrac{1}{2\pi} \sum_{k=0}^{n} a_k x^k 2\pi = q(x)$,

in which we used the result of (b).

(d) By (c), $p(x) = \dfrac{1}{2^n n!}(x^2 - 1)^n = \dfrac{1}{2\pi} \int_{-\pi}^{\pi} p(x + be^{i\varphi})\, d\varphi$. Then $P_n(x) = \dfrac{d^n}{dx^n}[p(x)] = \dfrac{1}{2\pi} \int_{-\pi}^{\pi} \left[\dfrac{d^n}{dx^n} p(x + be^{i\varphi})\right] d\varphi$.

(e) Note that $\dfrac{d}{d\varphi}\left[\dfrac{d^{n-k-1}}{dx^{n-k-1}} p(x + be^{i\varphi})\right] = \dfrac{d^{n-k}}{dx^{n-k}}[p(x + be^{i\varphi})]ibe^{i\varphi}$. Write

$$\int_{-\pi}^{\pi} e^{-ik\varphi} \dfrac{d^{n-k}}{dx^{n-k}}[p(x + be^{i\varphi})]\, d\varphi = \int_{-\pi}^{\pi} \dfrac{e^{-i(k+1)\varphi}}{bi} \dfrac{d}{d\varphi}\left[\dfrac{d^{n-k-1}}{dx^{n-k-1}} p(x + be^{i\varphi})\right] d\varphi$$

and integrate the right side by parts to get

$$\int_{-\pi}^{\pi} e^{-ik\varphi} \dfrac{d^{n-k}}{dx^{n-k}}[p(x + be^{i\varphi})]\, d\varphi = \dfrac{1}{ib}\left[e^{-i(k+1)\varphi} \dfrac{d^{n-k-1}}{dx^{n-k-1}}[p(x + be^{i\varphi})]\right]_{-\pi}^{\pi}$$
$$- \dfrac{1}{ib} \int_{-\pi}^{\pi} [-i(k + 1)]e^{-i(k+1)\varphi} \dfrac{d^{n-k-1}}{dx^{n-k-1}}[p(x + be^{i\varphi})]\, d\varphi$$
$$= \dfrac{k + 1}{b} \int_{-\pi}^{\pi} e^{-i(k+1)\varphi} \dfrac{d^{n-k-1}}{dx^{n-k-1}}[p(x + be^{i\varphi})]\, d\varphi,$$

in which k is any integer with $0 \le k \le n - 1$.

(f) Use the result of (e) and integrate by parts n times in the expression for $P_n(x)$ in (d) to get

$$P_n(x) = \dfrac{1}{2\pi} \int_{-\pi}^{\pi} \dfrac{d^n}{dx^n}[p(x + be^{i\varphi})]\, d\varphi = \dfrac{1}{2\pi b} \int_{-\pi}^{\pi} e^{-i\varphi} \dfrac{d^{n-1}}{dx^{n-1}}[p(x + be^{i\varphi})]\, d\varphi$$
$$= \dfrac{1 \cdot 2}{2\pi b^2} \int_{-\pi}^{\pi} e^{-2i\varphi} \dfrac{d^{n-2}}{dx^{n-2}}[p(x + be^{i\varphi})]\, d\varphi = \cdots = \dfrac{n!}{2\pi b^n} \int_{-\pi}^{\pi} e^{-in\varphi} p(x + be^{i\varphi})\, d\varphi.$$

(g) $\dfrac{n!}{b^n} e^{-in\varphi} p(x + be^{i\varphi}) = \dfrac{n!}{b^n} e^{-in\varphi} \dfrac{1}{2^n n!} [(x + be^{i\varphi})^2 - 1]^n = \dfrac{1}{(2b)^n} [e^{-i\varphi}(x^2 + 2bxe^{i\varphi} + b^2 e^{2i\varphi}) - e^{-i\varphi}]^n$

$= \left[\dfrac{1}{2b}(x^2 - 1)e^{-i\varphi} + \dfrac{b}{2} e^{i\varphi} + x \right]^n.$

Let $b = \sqrt{x^2 - 1}$ to get $\dfrac{n!}{b^n} e^{-in\varphi} p(x + be^{i\varphi}) = \left[\sqrt{x^2 - 1} \dfrac{e^{i\varphi} + e^{-i\varphi}}{2} + x \right]^n = [x + \sqrt{x^2 - 1} \, \cos(\varphi)]^n.$

(h) By (f) and (g), $P_n(x) = \dfrac{n!}{2\pi b^n} \displaystyle\int_{-\pi}^{\pi} e^{-in\varphi} p(x + be^{i\varphi}) \, d\varphi = \dfrac{1}{2\pi} \int_{-\pi}^{\pi} [x + \sqrt{x^2 - 1} \, \cos(\varphi)]^n \, d\varphi.$

(i) For $-1 < x < 1$, $P_n(x) = \dfrac{1}{2\pi} \displaystyle\int_{-\pi}^{\pi} [x + \sqrt{x^2 - 1} \, \cos(\varphi)]^n \, d\varphi$ by (h). If $x = 1$, then $\dfrac{1}{2\pi} \int_{-\pi}^{\pi} [x + \sqrt{x^2 - 1} \, \cos\varphi)]^n \, d\varphi =$

$\dfrac{1}{2\pi} \displaystyle\int_{-\pi}^{\pi} d\varphi = 1 = P_n(1).$ If $x = -1$, then $\dfrac{1}{2\pi} \int_{-\pi}^{\pi} [x + \sqrt{x^2 - 1} \, \cos(\varphi)]^n \, d\varphi = \dfrac{1}{2\pi} \int_{-\pi}^{\pi} (-1)^n \, d\varphi = (-1)^n = P_n(-1).$

19. $H_0(x) = 1,\ H_1(x) = x,\ H_2(x) = x^2 - 1,\ H_3(x) = x^3 - 3x,\ H_4(x) = x^4 - 6x^2 + 3,\ H_5(x) = x^5 - 10x^3 + 15x$

21. Write $\displaystyle\int_{-\infty}^{\infty} e^{-x^2/2} \left[e^{x^2/2} \dfrac{d^n}{dx^n} e^{-x^2/2} \right]^2 dx = \int_{-\infty}^{\infty} e^{x^2/2} \dfrac{d^n}{dx^n} e^{-x^2/2} \dfrac{d^n}{dx^n} e^{-x^2/2} \, dx$ and integrate by parts to get

$$e^{x^2/2} \dfrac{d^n}{dx^n} e^{-x^2/2} \dfrac{d^{n-1}}{dx^{n-1}} e^{-x^2/2} \Big|_{-\infty}^{\infty} - \int_{-\infty}^{\infty} \dfrac{d}{dx} \left[e^{x^2/2} \dfrac{d^n}{dx^n} e^{-x^2/2} \right] \cdot \dfrac{d^{n-1}}{dx^{n-1}} e^{-x^2/2} \, dx.$$

The term on the left is zero, being a sum of limits (as $x \to \pm\infty$) of a polynomial times $e^{-x^2/2}$. Continue to integrate by parts $n - 1$ times to get

$$\int_{-\infty}^{\infty} e^{-x^2/2} [H_n(x)]^2 \, dx = (-1)^n \int_{-\infty}^{\infty} \dfrac{d^n}{dx^n} \left[e^{x^2/2} \dfrac{d^n}{dx^n} e^{-x^2/2} \right] \cdot e^{-x^2/2} \, dx.$$

But $\dfrac{d^n}{dx^n} \left[(-1)^n e^{x^2/2} \dfrac{d^n}{dx^n} e^{-x^2/2} \right] = n!$, because $H_n(x)$ is a polynomial of degree n with the coefficient of x^n equal to 1.

Thus, $\displaystyle\int_{-\infty}^{\infty} e^{-x^2/2} [H_n(x)]^2 \, dx = n! \int_{-\infty}^{\infty} e^{-x^2/2} \, dx = \sqrt{2\pi} n!.$

23. Put $L(x) = \displaystyle\sum_{k=0}^{\infty} a_k x^k$ into Laguerre's differential equation and rearrange terms to get

$\displaystyle\sum_{k=0}^{\infty} \{[(k + 1)k + k + 1] a_{k+1} ((\lambda - k) a_k\} x^k = 0.$ Let the coefficient of x^k be zero to get $(k + 1)^2 a_{k+1} + \lambda - a_k = 0.$ When $\lambda = n$,

$a_{n+1} = a_{n+2} = \cdots = 0$, and we get a polynomial of degree n as a solution, with $a_n = -\dfrac{n - (n - 1)}{n^2} a_{n-1} = \cdots = \dfrac{(-1)^n}{n!} a_0.$

Choose $a_0 = (-1)^n n!$ to get $L_0(x) = 1,\ L_1(x) = x - 1,\ L_2(x) = x^2 - 4x + 2,\ L_3(x) = x^3 - 9x^2 + 18x - 6,$ $L_4(x) = x^4 - 16x^3 + 72x^2 - 96x + 24,$ and $L_5(x) = x^5 - 25x^4 + 200x^3 - 600x^2 + 600x - 120.$

25. Write $\displaystyle\int_0^{\infty} e^x (-1)^n e^x \dfrac{d^n}{dx^n} [x^n e^{-x}] L_n(x) \, dx = (-1)^n \int_0^{\infty} L_n(x) \dfrac{d^n}{dx^n} [x^n e^{-x}] \, dx$ and integrate by parts to get

$\displaystyle\int_0^{\infty} e^{-x} [L_n(x)]^2 \, dx = (-1)^n \left[L_n(x) \dfrac{d^{n-1}}{dx^{n-1}} [x^n e^{-x}] \right]_0^{\infty} - \int_0^{\infty} L_n'(x) \dfrac{d^{n-1}}{dx^{n-1}} [x^n e^{-x}] \, dx.$ The first term on the right is zero because

$\dfrac{d^k}{dx^k} [x^n e^{-x}]$ is a polynomial times xe^{-x} for $0 \le k \le n - 1$. Continue to integrate by parts, obtaining

$$\int_0^{\infty} e^{-x} [L_n(x)]^2 \, dx = (-1)^{2n} \int_0^{\infty} \dfrac{d^n}{dx^n} [L_n(x)] x^n e^{-x} \, dx.$$

But $\dfrac{d^n}{dx^n} L_n(x) = n!$ because $L_n(x)$ is a polynomial of degree n with leading coefficient 1. Thus,

$\displaystyle\int_0^{\infty} e^{-x} [L_n(x)]^2 \, dx = n! \int_0^{\infty} x^n e^{-x} \, dx = n! \Gamma(n + 1) = (n!)^2.$

27. (a) The total force exerted on q_1 is $2K \left[\dfrac{1}{x_1 + 1} - \dfrac{2}{x_1 - x_2} + \dfrac{1}{x_1 - 1} \right] = 0.$ Similarly, for q_2,

$2K\left[\dfrac{1}{x_2+1}-\dfrac{2}{x_2-x_1}+\dfrac{1}{x_2-1}\right]=0$. Add these two equations to get $\dfrac{2x_1}{x_1^2-1}+\dfrac{2x_2}{x_2^2-1}=0$. Multiply this equation by $(x_1^2-1)(x_2^2-1)$ to get $2(x_1x_2-1)(x_1+x_2)=0$; hence, $x_1=-x_2$. Substitute this into the first equation derived in this solution to get $\dfrac{1}{x_1+1}+\dfrac{2}{2x_1}+\dfrac{1}{x_1-1}=0$, or $3x_1^2-1=0$. Therefore, $x_1=\pm\dfrac{1}{\sqrt{3}}$ and $x_2=\mp\dfrac{1}{\sqrt{3}}$, the zeros of $P_2(x)$.

(b) The force on the bead located at x_k is the sum of $-\dfrac{4K}{x_r-x_k}$ (from the bead at x_r, $r>k$), $\dfrac{4K}{x_k-x_r}$ (from the bead at x_r, $r<k$), the $-\dfrac{2K}{1-x_k}$ from the charge at 1, and $\dfrac{2K}{x_k+1}$ from the charge at -1. The total force on the bead at x_k is

$$\sum_{\substack{i=0\\i\neq k}}^{N}\frac{4K}{x_k-x_i}+\frac{2K}{x_k-1}+\frac{2K}{x_k+1}=0.$$

Therefore, $\dfrac{2x_k}{1-x_k^2}=\displaystyle\sum_{\substack{i=0\\i\neq k}}^{N}\dfrac{2}{x_k-x_i}$.

(c) Since P_N satisfies Legendre's equation, we get, on dividing by $1-x^2$, that

$$P_N''(x)-\frac{2x}{1-x^2}P_N'(x)+\frac{N(N+1)}{1-x^2}P_N(x)=0.$$

If x_k is a zero of P_N, then $P_N(x_k)=0$ and $P_N'(x_k)\neq 0$ (the Legendre polynomials have no multiple roots). Divide the above equation by $P_N'(x)$ and let $x=x_k$ to get

$$\frac{P_N''(x_k)}{P_N'(x_k)}=\frac{2x_k}{1-x_k^2}-\frac{N(N+1)}{1-x^2}\frac{P_N(x_k)}{P_N'(x_k)}=\frac{2x_k}{1-x_k^2}.$$

Combining this result with the result of (b), we have that the positions of the beads are the zeros of P_N.

(d) $p'(x)=\displaystyle\sum_{i=0}^{N}(x-x_1)\cdots\overline{(x-x_i)}\cdots(x-x_N)$, in which the summation is over the product of terms indicated, with the term having the bar over it omitted for each i. Thus, $p'(x)=\displaystyle\sum_{i=1}^{N}\prod_{\substack{m=1\\m\neq i}}^{N}(x-x_m)$, in which $\displaystyle\prod_{\substack{m=1\\m\neq i}}^{N}$ denotes a product of terms, with m varying from 1 to N, but $m\neq i$ as the sum $\displaystyle\sum_{m=1}^{N}$ is executed. We now obtain the second derivative in similar fashion:

$$p''(x)=\sum_{i=1}^{N}\sum_{j=1}^{N}(x-x_1)\cdots\overline{(x-x_i)}\cdots\overline{(x-x_j)}\cdots(x-x_N).$$

Thus, $p''(x)=\displaystyle\sum_{i=1}^{N}\sum_{\substack{j=1\\j\neq i}}^{N}\prod_{\substack{m=1\\m\neq i\\m\neq j}}^{N}(x-x_m)$.

(e) If $x=x_k$ in the result of (d), the only term that does vanish is the only term not containing the $x-x_k$ factor. Thus, $p'(x_k)=\displaystyle\prod_{\substack{m=1\\m\neq k}}^{N}(x_k-x_m)$, and, similarly, $p''(x_k)=2\displaystyle\sum_{\substack{j=1\\j\neq k}}^{N}\prod_{\substack{m=1\\m\neq k\\m\neq j}}^{N}(x_k-x_m)$. Then

$$\frac{p''(x_k)}{p'(x_k)}=2\sum_{\substack{j=1\\j\neq k}}^{N}\frac{(x_k-x_1)\cdots\overline{(x_k-x_k)}\cdots\overline{(x_k-x_j)}\cdots(x_k-x_N)}{(x_k-x_1)\cdots\overline{(x_k-x_k)}\cdots(x_k-x_N)}=\sum_{\substack{j=1\\j\neq k}}^{N}\frac{2}{x_k-x_j}.$$

(f) Since P_N has N distinct roots $x_1, x_2, \ldots, x_N$ in $(-1,1)$, we can write $P_N(x)=C(x-x_1)\cdots(x-x_N)$ for some constant C. From (d) and (e),

$$\frac{P_N''(x_k)}{P_N'(x_k)}=\sum_{\substack{j=1\\j\neq k}}^{N}\frac{2}{x_k-x_j}.$$

Combined with (c), the zeros of P_N satisfy $\displaystyle\sum_{\substack{j=1\\j\neq k}}^{N}\frac{2}{x_k-x_j}=\frac{2x_k}{1-x_k^2}$. This implies that the beads are at the zeros of P_N.

Section 7.6

1. If m and n are positive integers and $m \neq n$,

$$\int_0^L \cos\left(\frac{n\pi x}{L}\right)\cos\left(\frac{m\pi x}{L}\right) dx = \frac{1}{2}\left[\frac{L}{(n+m)\pi}\sin\left(\frac{(n+m)\pi x}{L}\right) + \frac{L}{(n-m)\pi}\sin\left(\frac{(n-m)\pi x}{L}\right)\right]_0^L = 0.$$

If $n > 0$, then $\int_0^L 1 \cdot \cos\left(\frac{n\pi x}{L}\right) dx = \frac{L}{n\pi}\sin\left(\frac{n\pi x}{L}\right)\Big]_0^L = 0.$

3. First, $\int_{-L}^L 1 \cdot \cos\left(\frac{n\pi x}{L}\right) dx = \int_{-L}^L 1 \cdot \sin\left(\frac{n\pi x}{L}\right) dx = 0.$

If n and m are distinct positive integers,

$$\int_{-L}^L \sin\left(\frac{n\pi x}{L}\right)\cos\left(\frac{m\pi x}{L}\right) dx = -\frac{1}{2}\left[\frac{L}{(n+m)\pi}\cos\left(\frac{(n+m)\pi x}{L}\right) + \frac{L}{(n-m)\pi}\cos\left(\frac{(n-m)\pi x}{L}\right)\right]_{-L}^L = 0.$$

Next, $\int_{-L}^L \sin\left(\frac{n\pi x}{L}\right)\cos\left(\frac{n\pi x}{L}\right) dx = \frac{L}{2n\pi}\sin^2\left(\frac{n\pi x}{L}\right)\Big]_{-L}^L = 0.$

Finally, if n and m are distinct positive integers,

$$\int_{-L}^L \sin\left(\frac{n\pi x}{L}\right)\sin\left(\frac{m\pi x}{L}\right) dx = \frac{1}{2}\left[\frac{L}{(n-m)\pi}\sin\left(\frac{(n-m)\pi x}{L}\right) - \frac{L}{(n+m)\pi}\sin\left(\frac{(n+m)\pi x}{L}\right)\right]_{-L}^L = 0, \text{ and}$$

$$\int_{-L}^L \cos\left(\frac{n\pi x}{L}\right)\cos\left(\frac{m\pi x}{L}\right) dx = \frac{1}{2}\left[\frac{L}{(n+m)\pi}\sin\left(\frac{(n+m)\pi x}{L}\right) + \frac{L}{(n-m)\pi}\sin\left(\frac{(n-m)\pi x}{L}\right)\right]_{-L}^L = 0.$$

5. $b_n = \dfrac{\displaystyle\int_0^L e^x \sin\left(\frac{n\pi x}{L}\right) dx}{\displaystyle\int_0^L \sin^2\left(\frac{n\pi x}{L}\right) dx} = \dfrac{1}{1 - \dfrac{4}{n^2\pi^2}} \cdot \dfrac{2}{n\pi}[1 - (-1)^n]$ **7.** $\displaystyle\int_{-1}^1 x \cdot 1 \cdot x^2 \, dx = 0$

9. Choose $\alpha = \dfrac{1}{\sqrt{k_0}}\displaystyle\int_a^b p_0(x)p_1(x)\,dx$. Then choose $\beta = \dfrac{1}{\Delta}(I_3 I_5 - I_2 I_6)$ and $\gamma = \dfrac{1}{\Delta}(I_1 I_5 - I_2 I_4)$, where $\Delta = I_1 I_2 - I_4 I_5$,

$I_1 = \displaystyle\int_a^b q_0(x)q_1(x)\,dx, \; I_2 = \int_a^b q_0^2(x)\,dx, \; I_3 = \int_a^b q_0(x)p_2(x)\,dx, \; I_4 = \int_a^b [q_1(x)p_1(x) - \alpha q_0(x)q_1(x)]\,dx,$

$I_5 = \displaystyle\int_a^b [q_0(x)p_1(x) - \alpha q_0^2(x)]\,dx, \text{ and } I_6 = \int_a^b [p_1(x)p_2(x) - \alpha q_0(x)p_2(x)]\,dx.$

11. $q_1(x) = \dfrac{1}{2}, \; q_2(x) = \dfrac{3}{2}x, \; q_3(x) = \dfrac{45}{8}\left(x^2 - \dfrac{1}{3}\right), \; q_4(x) = \dfrac{135}{8}\left(x^3 - \dfrac{3}{5}x\right)$

Section 7.7

1. Regular on $[0, L]$; $n^2\pi^2/L^2$; $\sin\left(\dfrac{n\pi x}{L}\right)$ **3.** Regular on $[0, 4]$; $\left[\left(n + \dfrac{1}{2}\right)\dfrac{\pi}{4}\right]^2$; $\cos\left[\left(n + \dfrac{1}{2}\right)\dfrac{\pi x}{4}\right]$

5. Periodic on $[-3\pi, 3\pi]$; zero is an eigenvalue with eigenfunction 1; for $n = 1, 2, 3, \ldots$, $\dfrac{n^2}{9}$ is an eigenvalue with

eigenfunctions $\cos\left(\dfrac{nx}{3}\right)$ and $\sin\left(\dfrac{nx}{3}\right)$

7. Regular on $[0, 1]$; eigenvalues are positive solutions of $\tan(\sqrt{\lambda}) = \dfrac{1}{2}\sqrt{\lambda}$ (there are infinitely many such solutions);
if λ_n is the nth eigenvalue, a corresponding eigenfunction is $2\sqrt{\lambda_n}\cos(\sqrt{\lambda_n}x) + \sin(\sqrt{\lambda_n}x)$

9. Regular on $[0, 5]$; $2 + \dfrac{n^2\pi^2}{25}$; $e^{6x}\sin\left(\dfrac{n\pi x}{5}\right)$ **11.** Regular on $[1, e]$; $\left(n + \dfrac{1}{2}\right)^2\pi^2$; $\sin\left[\left(n + \dfrac{1}{2}\right)\pi \ln(x)\right]$

13. Regular on $[0, \pi]$; $1 + n^2$; $e^{-x}\sin(nx)$ **15.** Regular on $[1, e^3]$; $1 + \dfrac{n^2\pi^2}{9}$; $\dfrac{1}{x}\sin\left[\dfrac{n\pi \ln(x)}{3}\right]$

17. Regular on $[1, e^4]$; $-3 + \dfrac{n^2\pi^2}{16}$; $x\sin\left[\dfrac{n\pi \ln(x)}{4}\right]$

19. $(xe^{-x}y')' + \lambda ye^{-x} = 0$. Here $r(x) = xe^{-x}$ and $p(x) = e^{-x} > 0$ for all x. If $\lambda = n$, a nonnegative integer, an eigenfunction is $L_n(x)$, the nth Laguerre polynomial. By the Sturm-Liouville theorem, $\int_0^\infty p(x)L_n(x)L_m(x)\, dx = 0$ if $n \neq m$. Thus,

$$\int_0^\infty e^{-x}L_n(x)L_m(x)\, dx = 0 \text{ if } n \neq m.$$

Section 7.8

1. If n is an odd positive integer, then $P_n(x)$ contains only odd powers of x; hence, $P_n(-x) = -P_n(x)$. Further,

$\cos\left(\dfrac{-\pi x}{2}\right) = \cos\left(\dfrac{\pi x}{2}\right)$. Write

$$\int_{-1}^1 P_n(x)\cos\left(\frac{\pi x}{2}\right) dx = \int_{-1}^0 P_n(x)\cos\left(\frac{\pi x}{2}\right) dx + \int_0^1 P_n(x)\cos\left(\frac{\pi x}{2}\right) dx.$$

Let $t = -x$ in the first integral on the left to get $\displaystyle\int_{-1}^0 P_n(x)\cos\left(\frac{\pi x}{2}\right) dx = \int_1^0 P_n(-t)\cos\left(\frac{-\pi t}{2}\right)(-1)\, dt =$

$\displaystyle\int_1^0 -P_n(t)\cos\left(\frac{\pi t}{2}\right)(-1)\, dt = -\int_0^1 P_n(t)\cos\left(\frac{\pi t}{2}\right) dt = -\int_0^1 P_n(x)\cos\left(\frac{\pi x}{2}\right) dx$. Thus, $\displaystyle\int_{-1}^1 P_n(x)\cos\left(\frac{\pi x}{2}\right) dx = 0$.

3. $a_0 = 1$ and $a_n = 0$ for $n \geq 1$

5. $a_0 = 0$; $a_1 = \dfrac{3}{2}\displaystyle\int_{-1}^1 x \cdot x\, dx = 1$; and $a_n = \dfrac{2n+1}{2}\displaystyle\int_{-1}^1 g(x)P_n(x)\, dx = 0$ for $n = 2, 3, 4, \ldots$, because g is a polynomial of degree 1.

7. Assume that $4x - x^3 = \displaystyle\sum_{k=1}^\infty a_k J_1(\omega_k x)$ for $0 \leq x \leq 5$. Here, $R = 5$ and $\omega_k = \dfrac{1}{5}z_k$, with z_k the kth positive root of $J_1(x)$. Then

$$a_k = \frac{\dfrac{4}{25}\displaystyle\int_0^5 x(4x - x^3)J_1(\omega_k x)\, dx}{[J_2(5\omega_k)]^2} = \frac{2}{25[J_1(5\omega_k)]^2}\left[\int_0^5 4x^2 J_1(\omega_k x)\, dx - \int_0^5 x^4 J_1(\omega_k x)\, dx\right].$$

For the first integral,

$$\int_0^5 4x^2 J_1(\omega_k x)\, dx = \frac{4}{\omega_k^3}\int_0^{5\omega_k} t^2 J_1(t)\, dt \quad \text{(let } t = \omega_k x) = \frac{4}{\omega_k^3}\left[t^2 J_2(t)\right]_0^{5\omega_k} = \frac{100}{\omega_k}J_2(5\omega_k).$$

Next, $\displaystyle\int_0^5 x^4 J_1(\omega_k x)\, dx = \frac{1}{\omega_k^5}\int_0^{5\omega_k} t^4 J_1(t)\, dt$. Let $u = t^2$ and $dv = t^2 J_1(t)\, dt$. Then $v = t^2 J_2(t)$, so

$$\int_0^5 x^4 J_1(\omega_k x)\, dx = \frac{1}{\omega_k^5}\left[t^4 J_2(t)\right]_0^{5\omega_k} - \frac{2}{\omega_k^5}\int_0^{5\omega_k} t^3 J_2(t)\, dt = \frac{625}{\omega_k}J_2(5\omega_k) - \frac{2}{\omega_k^5}\left[t^3 J_3(t)\right]_0^{5\omega_k} = \frac{625}{\omega_k}J_2(5\omega_k) - \frac{250}{\omega_k^3}J_3(5\omega_k).$$

Now, $J_3(5\omega_k) = \dfrac{4}{5\omega_k}J_2(5\omega_k) - J_1(5\omega_k) = \dfrac{4}{5\omega_k}J_2(5\omega_k)$. Therefore, $\displaystyle\int_0^5 x^4 J_1(\omega_k x)\, dx = \frac{625}{\omega_k}J_2(5\omega_k) - \frac{200}{\omega_k^3}J_2(5\omega_k)$.

Putting these results together with the fact that $J_2(z_k) = -J_0(z_k)$, we get $a_k = \dfrac{10}{J_0(z_k)}\left[\dfrac{21}{z_k} - \dfrac{200}{z_k^3}\right]$. Then

$4x - x^3 = \displaystyle\sum_{k=1}^\infty \frac{10}{J_0(z_k)}\left[\frac{21}{z_k} - \frac{200}{z_k^3}\right]J_1\left(\frac{z_k x}{5}\right) \approx -47.8020J_1(0.7663x) + 80.4438J_1(1.4031x) - 75.0599J_1(2.0347x) + 68.2958J_1(2.6647) - 62.6075J_1(3.2914x)$.

9. Eigenvalues are positive solutions of $\tan(\sqrt{\lambda}) = -\dfrac{1}{3}\sqrt{\lambda}$. [f λ_n is the nth eigenvalue, a corresponding eigenfunction is $\sin(\sqrt{\lambda_n}x)$. Assume that $x^2 = \displaystyle\sum_{k=1}^\infty a_k \sin(\sqrt{\lambda_k}x)$. Then

$$a_k = \frac{\displaystyle\int_0^1 x^2 \sin(\sqrt{\lambda_k}x)\, dx}{\displaystyle\int_0^1 \sin^2(\sqrt{\lambda_k}x)\, dx} = \frac{-\dfrac{1}{\sqrt{\lambda_k}}\cos(\sqrt{\lambda_k}) + \dfrac{2}{\sqrt{\lambda_k}}\sin(\sqrt{\lambda_k}) + \dfrac{2}{\lambda_k\sqrt{\lambda_k}}\cos(\sqrt{\lambda_k}) - \dfrac{2}{\lambda_k\sqrt{\lambda_k}}}{\dfrac{1}{2} - \dfrac{1}{4\sqrt{\lambda_k}}\sin(2\sqrt{\lambda_k})}.$$

Now, $\sqrt{\lambda_1} \approx 2.4556$, $\sqrt{\lambda_2} \approx 5.2239$, $\sqrt{\lambda_3} \approx 8.2045$, and $\sqrt{\lambda_4} \approx 11.2560$, so $a_1 \approx 0.4762$, $a_2 \approx -0.3047$, $a_3 \approx 0.1249$ and $a_4 \approx -0.0767$. Then $x^2 \approx 0.4762 \sin(2.4556x) - 0.3047 \sin(5.2239x) + 0.1249 \sin(8.2045x) - 0.0767 \sin(11.2560x)$.

11. $2x^3 - 5x^2 + 3x - 4 = -9H_0(x) + 9H_1(x) - 5H_2(x) + 2H_3(x)$.

13. Assume that $e^{-2x} = \sum\limits_{k=0}^{\infty} a_k L_k(x)$. Then $a_k = \dfrac{\displaystyle\int_0^{\infty} e^{-3x} L_k(x)\, dx}{\displaystyle\int_0^{\infty} e^{-x} L_k^2(x)\, dx}$. We get $a_0 = \dfrac{1}{3}$, $a_1 = -\dfrac{2}{9}$, $a_2 = \dfrac{7}{27}$, and $a_3 = -\dfrac{7}{18 \cdot 27}$.

Then $e^{-2x} \approx 0.3333 - 0.2222L_1(x) + 0.2593L_2(x) - 0.0144L_3(x)$.

Chapter 8

Section 8.1

1. $X(t) = \dfrac{1 + Ce^{-2t}}{1 - Ce^{-2t}}$ or $X(t) = -1$; 1 is a stable equilibrium point and -1 is an unstable equilibrium point.

3. $X(t) = \dfrac{1}{1 - Ce^{-2t}}$ or $X(t) = 0$; 2 (stable) and 0 (unstable) are the critical points.

5. The general solution is $X(t) = 3 + Ce^t$; the initial value problem $X' = X - 3$; $X(0) = 3$ has solution $X(t) = 3$. The problem with $X(0) = 3.01$ has solution $X(t) = 3 + 0.01e^t \to \infty$ as $t \to \infty$, and the problem with $X(0) = 2.99$ has solution $X(t) = 3 - 0.01e^t \to -\infty$ as $t \to \infty$.

Section 8.2

1. $x = -c_1\cos(t) - c_2\sin(t)$; $y = (4c_1 - c_2)\cos(t) + (c_1 + 4c_2)\sin(t)$ **3.** $x = c_1e^{-3t} + 7c_2e^{2t}$; $y = c_1e^{-3t} + 2c_2e^{2t}$

5. $x = c_1e^{5t} + 2c_2e^{4t}$; $y = c_2e^{4t}$ **7.** $\dfrac{dy}{dx} = -\dfrac{4x}{9y}$; $9y^2 + 4x^2 = C$. The solution curves are ellipses.

9. $(x - 1)^2 - (y + 20)^2 = C$. The solution curves are hyperbolas.   **11.** $y = x \ln|Cx|$

13. The general solution is defined implicitly by the equation $73x^2 - 72xy + 52y^2 = C$. To determine the family of trajectories, calculate $2\alpha = \text{arccot}\left(\dfrac{73 - 52}{-72}\right) = \arctan\left(-\dfrac{24}{7}\right)$. Then $\cos(2\alpha) = \dfrac{7}{25}$ and $\sin(2\alpha) = -\dfrac{24}{25}$. Next, calculate $\sin(\alpha) = -\sqrt{\dfrac{1 - \cos(2\alpha)}{2}} = -\dfrac{3}{5}$ and $\cos(\alpha) = \sqrt{\dfrac{1 + \cos(2\alpha)}{2}} = \dfrac{4}{5}$. Putting these values into the equations for A^* and C^* produces the family of ellipses $4X^2 + Y^2 = K$. Rotating the graphs of the members of this family counterclockwise through the angle $\alpha = \arctan\left(-\dfrac{3}{4}\right)$, or, equivalently, clockwise by $\arctan\left(\dfrac{3}{4}\right)$, will generate a family of ellipses that are the trajectories of the system of differential equations.

15. $4x^2 - 3\sqrt{3}xy + y^2 = C$; rotate the family of hyperbolas $11X^2 - Y^2 = K$ $\dfrac{\pi}{6}$ clockwise.

17. $13x^2 - 6\sqrt{3}xy + 7y^2 = C$; rotate the family of ellipses $4X^2 + Y^2 = K$ $\dfrac{\pi}{6}$ counterclockwise.

19. They have the same curves, with the trajectories having arrows in opposite directions.

21. $x' = ax(K - x) - bxy$, $y' = cxy - ky$, where K is the carrying capacity of the prey's domain.

Section 8.3

1. $\lambda_1 = \lambda_2 = -2$; node; asymptotically stable   **3.** $\lambda_1 = 2i$, $\lambda_2 = -2i$; center; stable, but not asymptotically stable

5. $\lambda_1 = 4 + 5i$, $\lambda_2 = 4 - 5i$; spiral point; unstable   **7.** $\lambda_1 = \lambda_2 = 3$; node; unstable

9. $\lambda_1 = -2 + \sqrt{3}i$, $\lambda_2 = -2 - \sqrt{3}i$; spiral point; asymptotically stable

11. $\lambda_1 = \sqrt{5}$, $\lambda_2 = -\sqrt{5}$; saddle point; unstable   **13.** $\lambda_1 = -3 + \sqrt{7}$, $\lambda_2 = -3 - \sqrt{7}$; node; asymptotically stable

15. $\lambda_1 = 2 + \sqrt{3}$, $\lambda_2 = 2 - \sqrt{3}$; node; unstable

17. $\lambda_1 = \sqrt{\dfrac{k}{m}}\, i$, $\lambda_2 = -\sqrt{\dfrac{k}{m}}\, i$; the origin is a center; it is stable but not asymptotically stable.

Section 8.4

1. Unstable node   **3.** Asymptotically stable spiral point   **5.** No conclusion; $\det(A) = 0$

7. Asymptotically stable node   **9.** $(2, -3)$; $\lambda_1 = 2$, $\lambda_2 = -2$; unstable saddle point

11. $(-8, 3)$; $\lambda_1 = 2 + \sqrt{3}$, $\lambda_2 = 2 - \sqrt{3}$; unstable node

13. $(8, -20)$ is an unstable saddle point, $\lambda_1 = 2$, $\lambda_2 = -25$; $(-2, 50)$ is an asymptotically stable spiral point, $\lambda_1 = -1 + 7i$, $\lambda_2 = -1 - 7i$

15. $(2, 4)$ is an unstable node, $\lambda_1 = \lambda_2 = 1$; $(\frac{3}{2}, 3)$ is an unstable saddle point, $\lambda_1 = 2$, $\lambda_2 = -\frac{1}{2}$

17. $X' = 2X$, $Y' = 4X + 3Y$; $\lambda_1 = 2$, $\lambda_2 = 3$; unstable node

19. $X' = -4X + Y$, $Y' = -Y$; $\lambda_1 = -2 + \sqrt{3}$, $\lambda_2 = -2 - \sqrt{3}$; stable node

23. The equation of motion is $mz'' = mg - k(z + s) + \dfrac{K}{(L - z)^2}$, where $ks = mg + \dfrac{K}{L^2}$. Then

$$z'' = -\frac{k}{m}z + \frac{K}{m}\left[\frac{1}{(L - z)^2} - \frac{1}{L^2}\right] = -\frac{k}{m}z + \frac{K}{m}\frac{1}{L^2}\left[2\left(\frac{z}{L}\right) + 3\left(\frac{z}{L}\right)^2 + 4\left(\frac{z}{L}\right)^3 + \cdots\right].$$ Now let $x = z$ and $y = z'$ to

get the system $x' = y$, $y' = -\dfrac{k}{m}x + \dfrac{K}{m}\dfrac{1}{L^2}\left[2\left(\dfrac{x}{L}\right) + 3\left(\dfrac{x}{L}\right)^2 + 4\left(\dfrac{x}{L}\right)^3 + \cdots\right] \approx \left(\dfrac{2K}{mL^3} - \dfrac{k}{m}\right)x$. Then $\lambda^2 = \dfrac{2K}{mL^3} - \dfrac{k}{m}$,

and oscillations will occur only if this is negative. $\dfrac{kL^3}{2}$ is the minimum value of K that will prevent oscillations.

25. (a) $mz'' = -kz + cz' + \dfrac{Kz}{(r^2 + z^2)^{3/2}}$

   (b) Let $x = z$, and $y = z'$. Then we have $x' = y$ and $y' = -\dfrac{k}{m}x - \dfrac{c}{y} + \dfrac{Kx}{m(r^2 + x^2)^{3/2}} = -\dfrac{k}{m}x - \dfrac{c}{m}y +$

   $\dfrac{Kx}{mr^3}\left[1 - \dfrac{3}{2}\left(\dfrac{x}{r}\right)^2 + \dfrac{15}{8}\left(\dfrac{x}{r}\right)^4 + \cdots\right] \approx \left(\dfrac{2K}{mr^3} - \dfrac{k}{m}\right)x - \dfrac{c}{m}y.$

   (c) Here, $\lambda = \dfrac{1}{2}\left[-\dfrac{c}{m} \pm \sqrt{\dfrac{c^2}{m^2} - \dfrac{4k}{m} + \dfrac{4K}{mr^3}}\right].$

   If $K < \dfrac{r^3}{4m}(4mk - c^2)$, then the critical point is an asymptotically stable spiral point.

   If $K = \dfrac{r^3}{4m}(4mk - c^2)$, then the critical point is an asymptotically stable node.

   If $K > \dfrac{r^3}{4m}(4mk - c^2)$ and $K > kr^3$, then the critical point is an asymptotically stable node.

   If $K > \dfrac{r^3}{4m}(4mk - c^2)$ and $K < kr^3$, then the critical point is an unstable saddle point.

Section 8.5

1. (a) $f(x)$ is maximum when $x = \dfrac{k}{c}$ and has a value of $\left(\dfrac{k}{c}\right)^k e^{-k}$ at this point.

   (c) $g\left(\dfrac{a}{b}\right) = \left(\dfrac{a}{b}\right)^a e^{-a}$ is the maximum value of g.

3. $\bar{y} = \dfrac{a}{b}$

5. The system in this case is $x' = ax(K - x) - bxy$, $y' = -cy + kxy$. Here, $\bar{x}$ can be computed as in Problem 4 to get

   $\bar{x} = \dfrac{c}{k}$. Now evaluate $\bar{y}$ in a similar manner by noting that $0 = \displaystyle\int_{t=0}^{t=T} \frac{dx}{x} = \int_{t=0}^{t=T} [a(K - x) - by]\,dt = aKT - a\bar{x} - b\bar{y}.$

   Solve this for $\bar{y} = \dfrac{a}{b}\left(K - \dfrac{c}{k}\right)$. Unless the carrying capacity of the system K is larger than $\dfrac{c}{k}$, the predators will die off.

   If $K - \dfrac{c}{k} > 1$, the average predator population will be larger in the restricted-environment case than if the prey lives in an

   unrestricted environment. If $K < \dfrac{c}{k}$, there will be, on the average, fewer predators when the prey is restricted.

7. The system for this situation is $x' = ax - bxy - Hx$, $y' = -cy + kxy - Hy$. Using the results of Problems 3 and 4, we have that the average values of the populations are $\bar{x} = \dfrac{c + H}{k}$ and $\bar{y} = \dfrac{a - H}{b}$, so the prey benefits more than the predator with indiscriminate harvesting.

9. The system of equations is $x' = ax(K - x) - bxy$, $y' = cy(M - y) - kxy$.

Section 8.6

1. $\dfrac{\partial F}{\partial x} + \dfrac{\partial G}{\partial y} = -2 + 3x^2 + 5 + x^2 - 2\sin(x) = 3 + 4x^2 - 2\sin(x) \geq 1 + 4x^2 > 0$ **3.** $\dfrac{\partial F}{\partial x} + \dfrac{\partial G}{\partial y} \geq 15e^{3y} > 0$

7. $\dfrac{dr}{dt} = 0, \dfrac{d\theta}{dt} = r^2 - 1$; the trajectories consist of the origin (a critical point), all circles with center at the origin except the unit circle (each a closed trajectory), and each point on the unit circle (each a critical point). The only closed trajectories are the circles ($r \neq 1$), each of which is a neutrally stable limit cycle.

9. $\dfrac{dr}{dt} = r(1 - r^2)(4 - r^2)^2(9 - r^2), \dfrac{d\theta}{dt} = -1$; the trajectories consist of the origin (a critical point) and the three circles with centers at the origin and radii 1, 2, and 3. The only closed trajectories are these circles. The circle of radius 1 is a stable limit cycle, the circle of radius 2 is a semistable limit cycle, and the circle of radius 3 is an unstable limit cycle.

11. The ellipse defined by the equation $x^2 + 9y^2 = 4$ is a closed trajectory of the system. Now consider the annular region defined by $\dfrac{4}{9} \leq x^2 + y^2 \leq 4$. $\dfrac{dr}{dt} > 0$ on the inner circle $\left[\text{except at}\left(0, \pm\dfrac{3}{2}\right)\text{, where the limit cycle and circle meet}\right]$, and on the larger circle, $\dfrac{dr}{dt} < 0$ [except at $(\pm 2, 0)$, where the limit cycle and circle meet].

13. The ellipse defined by the equation $4x^2 + y^2 = 4$ is a limit cycle of the system. The annular region defined by $1 \leq x^2 + y^2 \leq 4$ has the property that all solutions within it remain in it.

15. No. $\dfrac{dr}{dt} = r(1 + r^2) > 0$ for all $r > 0$.

17. No. Since $y' > 0$ for all x and y, the system has no critical point, and every closed trajectory has to contain at least one critical point.

19. No. $F_x + F_y > 0$

21. Write $\dfrac{dy}{dx} = -\dfrac{x^3}{y}$, then solve the equation. The curve defined by the equation $2y^2 + x^4 = C$ is a closed trajectory for each value of $C > 0$.

25. Use Lienard's theorem.

27. Rewrite the equation as $z'' = -z^3$ and then let $x = z$ and $y = z'$. $\dfrac{dy}{dx} = -\dfrac{x^3}{y}$ has general solution $2y^2 + x^4 = C$, which is a closed trajectory for each positive value of C.

29. Rewrite the equation as $z'' = -\dfrac{z}{1 + z^2}$ and then let $x = z$ and $y = z'$. The general solution of $\dfrac{dy}{dx} = -\dfrac{x}{y(1 + x^2)}$ is $y^2 + \ln(1 + x^2) = C$, which is a closed trajectory for each positive value of C.

Chapter 9

Section 9.1

1. $y = -\dfrac{2}{x} + 2$ **3.** $y = -\dfrac{1}{5}x^4 + \dfrac{1}{5}$ **5.** $y = \dfrac{1}{2}(e^{x+k} - ce^{-(x+k)})$, where $k = \ln\left(\dfrac{2}{e^4 - e^{-4}}\right)$ and $c = \left(\dfrac{2}{e^4 - e^{-4}}\right)^2$

7. $y = \dfrac{1}{2}(3 - e^2)\ln(x) + x$ **9.** $y = \dfrac{1}{2}x^2 + \dfrac{7}{2}x$

11. With $f(x, y, y') = \alpha(x)(y')^2 + 2\beta(x)yy' + \gamma(x)y^2$, the Euler equation is

$$\alpha(x)y'' + \alpha'(x)y' - [\beta'(x) + \gamma(x)]y = 0,$$

a second order linear differential equation.

13. We can write equation (9.8) as $\dfrac{\partial f}{\partial y} - \dfrac{d}{dx}\left(\dfrac{\partial f}{\partial y'}\right) = 0$. It is routine to check that $\dfrac{d}{dx}\left(\dfrac{\partial f}{\partial y'}\right) = \dfrac{1}{y'}\left[\dfrac{d}{dx}\left(\dfrac{\partial f}{\partial y'}\,y'\right) - \dfrac{\partial f}{\partial y'}\,y''\right]$.

Substitute this result into the left side of the first equation to get

$$0 = \frac{\partial f}{\partial y} - \frac{1}{y'}\left[\frac{d}{dx}\left(\frac{\partial f}{\partial y'}\,y'\right) - \frac{\partial f}{\partial y'}\,y''\right] = \frac{1}{y'}\left[\frac{\partial f}{\partial y}\,y' + \frac{\partial f}{\partial y'}\,y'' - \frac{d}{dx}\left(\frac{\partial f}{\partial y'}\,y'\right)\right]$$

$$= \frac{1}{y'}\left[\frac{\partial f}{\partial x} - \frac{\partial f}{\partial x} - \frac{d}{dx}\left(\frac{\partial f}{\partial y'}\,y'\right)\right] = \frac{1}{y'}\left[\frac{d}{dx}\left(f - \frac{\partial f}{\partial y'}\,y'\right) - \frac{\partial f}{\partial x}\right].$$

Section 9.2

1. $y = \alpha x + \beta$ **3.** $y = A\ln(x) + B$ **5.** $y = -\dfrac{1}{24}x^4 + \dfrac{1}{6}x^3 + \dfrac{C}{2}x^2 + Dx + E$

7. Let u be a stationary function for this integral. Write $y(x) = u(x) + \epsilon\eta(x)$, where η has continuous first and second derivatives on $[a, b]$, and $\eta(x)$, $\eta'(x)$, and $\eta''(x)$ vanish at a and at b. Let

$$I(\epsilon) = \int_a^b f(x, u + \epsilon\eta, u' + \epsilon\eta', u'' + \epsilon\eta'', u^{(3)} + \epsilon\eta^{(3)})\, dx. \text{ Then}$$

$$I'(\epsilon) = \int_a^b\left[\frac{\partial f}{\partial x}\frac{\partial x}{\partial\epsilon} + \frac{\partial f}{\partial y}\frac{\partial y}{\partial\epsilon} + \frac{\partial f}{\partial y'}\frac{\partial y'}{\partial\epsilon} + \frac{\partial f}{\partial y''}\frac{\partial y''}{\partial e} + \frac{\partial f}{\partial y^{(3)}}\frac{\partial y^{(3)}}{\partial\epsilon}\right] dx. \text{ Let } I'(\epsilon)\bigg|_{\epsilon=0} = 0. \text{ Then}$$

$$\int_a^b\left[\frac{\partial f}{\partial x}\,\eta(x) + \frac{\partial f}{\partial y'}\,\eta'(x) + \frac{\partial f}{\partial y''}\,\eta''(x) + \frac{\partial f}{\partial y^{(3)}}\,\eta^{(3)}(x)\right] dx = 0. \text{ Integrate by parts, as we have done before, to get}$$

$$\int_a^b\frac{\partial f}{\partial y'}\,\eta'(x)\, dx = -\int_a^b\frac{d}{dx}\left(\frac{\partial f}{\partial y'}\right)\eta(x)\, dx, \quad \int_a^b\frac{\partial f}{\partial y''}\,\eta''(x)\, dx = -\int_a^b\frac{d^2}{dx^2}\left(\frac{\partial f}{\partial y''}\right)\eta(x)\, dx, \text{ and}$$

$$\int_a^b\frac{\partial f}{\partial y^{(3)}}\,\eta^{(3)}(x)\, dx = -\int_a^b\frac{d^3}{dx^3}\left(\frac{\partial f}{\partial y^{(3)}}\right)\eta(x)\, dx. \text{ Substitute these results into the equation } I'(0) = 0 \text{ to get}$$

$$\int_a^b\left[\frac{\partial f}{\partial y} - \frac{d}{dx}\left(\frac{\partial f}{\partial y'}\right) + \frac{d^2}{dx^2}\left(\frac{\partial f}{\partial y''}\right) - \frac{d^3}{dx^3}\left(\frac{\partial f}{\partial y^{(3)}}\right)\right]\eta(x)\, dx = 0.$$

One can show that because of the freedom of choice of η (subject to the conditions at a and b), vanishing of this integral implies that

$$\frac{\partial f}{\partial y} - \frac{d}{dx}\left(\frac{\partial f}{\partial y'}\right) + \frac{d^2}{dx^2}\left(\frac{\partial f}{\partial y''}\right) - \frac{d^3}{dx^3}\left(\frac{\partial f}{\partial y^{(3)}}\right) = 0.$$

Section 9.3

1. Let $f(x, y, u, u_x, u_y) = u_x^2 + u_y^2 + 2f(x, y)$. The Euler equation for $\displaystyle\iint_D f(x, y, u, u_x, u_y)\, dA$ is $2f(x, y) - 2u_{xx} - 2u_{yy} = 0$.

3. Let $w(x, y, z) = u(x, y, z) + \epsilon\eta(x, y, z)$, where u is a stationary function for the integral and η and its first partial derivatives are continuous and vanish at points on the surface bounding M. Let

$$I(\epsilon) = \iiint_M f(x, y, z, w, w_x, w_y, w_z, w_{xx}, w_{yy}, w_{zz}, w_{xy}, w_{xz}, w_{yz})\, dV.$$

Compute $I'(\epsilon)$ and let $I'(0) = 0$ to get

$$\iiint_M\left[\frac{\partial f}{\partial w}\,\eta + \frac{\partial f}{\partial w_x}\,\eta_x + \frac{\partial f}{\partial w_y}\,\eta_y + \frac{\partial f}{\partial w_z}\,\eta_z + \frac{\partial f}{\partial w_{xx}}\,\eta_{xx} + \frac{\partial f}{\partial w_{yy}}\,\eta_{yy} + \frac{\partial f}{\partial w_{zz}}\,\eta_{zz} + \frac{\partial f}{\partial w_{xy}}\,\eta_{xy} + \frac{\partial f}{\partial w_{xz}}\,\eta_{xz} + \frac{\partial f}{\partial w_{yz}}\,\eta_{yz}\right] dV = 0.$$

Using Gauss's divergence theorem, we get

$$\iiint_M\left[\frac{\partial f}{\partial w_x}\,\eta_x + \frac{\partial f}{\partial w_y}\,\eta_y + \frac{\partial f}{\partial w_z}\,\eta_z\right] dV = -\iiint_M\left[\frac{\partial}{\partial x}\left(\frac{\partial f}{\partial w_x}\right) + \frac{\partial}{\partial y}\left(\frac{\partial f}{\partial w_y}\right) + \frac{\partial}{\partial z}\left(\frac{\partial f}{\partial w_z}\right)\right]\eta(x, y, z)\, dV;$$

$$\iiint_M \left[\frac{\partial f}{\partial w_{xx}} \eta_{xx} + \frac{\partial f}{\partial w_{yy}} \eta_{yy} + \frac{\partial f}{\partial w_{zz}} \eta_{zz} \right] dV = \iiint_M \left[\frac{\partial^2}{\partial x^2} \left(\frac{\partial f}{\partial w_{xx}} \right) + \frac{\partial^2}{\partial y^2} \left(\frac{\partial f}{\partial w_{yy}} \right) + \frac{\partial^2}{\partial z^2} \left(\frac{\partial f}{\partial w_{zz}} \right) \right] \eta(x, y, z) \, dV;$$

and

$$\iiint_M \left[\frac{\partial f}{\partial w_{xy}} \eta_{xy} + \frac{\partial f}{\partial w_{xz}} \eta_{xz} + \frac{\partial f}{\partial w_{yz}} \eta_{yz} \right] dV = \iiint_M \left[\frac{\partial^2}{\partial x \, \partial y} \left(\frac{\partial f}{\partial w_{xy}} \right) + \frac{\partial^2}{\partial y \, \partial z} \left(\frac{\partial f}{\partial w_{yz}} \right) + \frac{\partial^2}{\partial z \, \partial x} \left(\frac{\partial f}{\partial w_{zx}} \right) \right] \eta(x, y, z) \, dV.$$

Substitute these results into $I'(\epsilon) = 0$ to get

$$\iiint_M \left[\frac{\partial f}{\partial w} + \frac{\partial}{\partial x} \left(\frac{\partial f}{\partial w_x} \right) + \frac{\partial}{\partial y} \left(\frac{\partial f}{\partial w_y} \right) + \frac{\partial}{\partial z} \left(\frac{\partial f}{\partial w_z} \right) + \frac{\partial^2}{\partial x^2} \left(\frac{\partial f}{\partial w_{xx}} \right) + \frac{\partial^2}{\partial y^2} \left(\frac{\partial f}{\partial w_{yy}} \right) \right.$$
$$\left. + \frac{\partial^2}{\partial z^2} \left(\frac{\partial f}{\partial w_{zz}} \right) + \frac{\partial^2}{\partial x \, \partial y} \left(\frac{\partial f}{\partial w_{xy}} \right) + \frac{\partial^2}{\partial y \, \partial z} \left(\frac{\partial f}{\partial w_{yz}} \right) + \frac{\partial^2}{\partial z \, \partial x} \left(\frac{\partial f}{\partial w_{zx}} \right) \right] \eta(x, y, z) \, dV = 0.$$

Because of the freedom of choice in selecting η, the quantity in brackets in this integral must vanish.

Section 9.4

1. Let $H(x, y, y', \lambda) = \int_a^b [p(x)(y')^2 - q(x)y^2 - \lambda r(x)y^2] \, dx$. With $f(x, y, y', \lambda) = p(x)(y')^2 - q(x)y^2 - \lambda r(x)y^2$, the Euler equation is $(py')' + (q + \lambda p)y = 0$.

3. The stationary function has the form $y(x) = c_1 e^{\sqrt{\lambda} x} + c_2 e^{-\sqrt{\lambda} x}$, in which c_1, c_2, and λ must be chosen so that $y(0) = y_0, y(1) = y_1$, and $\int_0^1 y^2 \, dx = 2$.

5. Let $H(x, y, y', \lambda_1, \ldots, \lambda_m) = \int_a^b \left[f(x, y, y') + \sum_{i=1}^m \lambda_i q_i(x, y, y') \right] dx$. Using the result obtained in this section, we get

$$\frac{\partial}{\partial y} \left[f + \sum_{i=1}^m \lambda_i q_i \right] - \frac{d}{dx} \left[\frac{\partial f}{\partial y'} + \sum_{i=1}^m \frac{\partial}{\partial y'} \lambda_i q_i \right] = 0.$$

7. Suppose we want a stationary function for $\int_a^b f(t, x(t), y(t), x'(t), y'(t)) \, dt$, subject to the constraints

$\int_a^b g_1(t, x(t), y(t), x'(t), y'(t)) \, dt = K_1$ and $\int_a^b g_2(t, x(t), y(t), x'(t), y'(t)) \, dt = K_2$. Show that a necessary condition for the functions x and y to be stationary functions for this problem is that x and y satisfy the system

$$\frac{\partial}{\partial x} [f + \lambda_1 x + \lambda_2 y] - \frac{d}{dt} \left(\frac{\partial}{\partial x'} [f + \lambda_1 x + \lambda_2 y] \right) = 0,$$

$$\frac{\partial}{\partial y} [f + \lambda_1 x + \lambda_2 y] - \frac{d}{dt} \left(\frac{\partial}{\partial y'} [f + \lambda_1 x + \lambda_2 y] \right) = 0.$$

9. Suppose that $x = u(t), y = v(t)$, and $z = w(t)$ form a solution of the problem. Write $x(t) = u(t) + \epsilon_1 \eta_1(t)$, $y(t) = v(t) + \epsilon_2 \eta_2(t)$, and $z(t) = w(t) + \epsilon_3 \eta_3(t)$, and let

$$I(\epsilon_1, \epsilon_2, \epsilon_3) = \int_a^b [f(t, u + \epsilon_1 \eta_1, v + \epsilon_2 \eta_2, w + \epsilon_3 \eta_3, u' + \epsilon_1 \eta_1', v' + \epsilon_2 \eta_2', w' + \epsilon_3 \eta_3')$$
$$+ \lambda g(u + \epsilon_1 \eta_1, v + \epsilon_2 \eta_2, w + \epsilon_3 \eta_3)] \, dt.$$

Calculate

$$\left. \frac{\partial I}{\partial \epsilon_1} \right|_{\epsilon_1 = 0} = \int_a^b \left[\frac{\partial f}{\partial x} \eta_1 + \frac{\partial f}{\partial x'} \eta_1' + \lambda \frac{\partial g}{\partial x} \eta_1 \right] dt = \int_a^b \left[\frac{\partial f}{\partial x} - \frac{d}{dt} \left(\frac{\partial f}{\partial x'} \right) + \lambda \frac{\partial g}{\partial x} \right] \eta_1(t) \, dt.$$

On letting $\left. \frac{\partial I}{\partial \epsilon_1} \right|_{\epsilon_1 = 0} = 0$, we get $\frac{\partial f}{\partial x} - \frac{d}{dt} \left(\frac{\partial f}{\partial x'} \right) + \lambda \frac{\partial g}{\partial x} = 0$. Similarly, by considering the equations $\left. \frac{\partial I}{\partial \epsilon_2} \right|_{\epsilon_2 = 0} = 0$ and

$\left.\dfrac{\partial I}{\partial \epsilon_3}\right|_{\epsilon_3=0} = 0$, we get

$$\frac{\partial f}{\partial y} - \frac{d}{dt}\left(\frac{\partial f}{\partial y'}\right) + \lambda \frac{\partial g}{\partial y} = 0 \text{ and } \frac{\partial f}{\partial z} - \frac{d}{dt}\left(\frac{\partial f}{\partial z'}\right) + \lambda \frac{\partial g}{\partial z} = 0.$$

11. Let $f(x, y, z) = \sqrt{(x')^2 + (y')^2 + (z')^2}$ and $g(x, y, z) = x^2 + y^2 + z^2 - R^2$. From the equation
$\dfrac{x''f - x'f'}{2xf^2} = \dfrac{y''f - y'f'}{2yf^2}$, we get $(x''f - x'f')y = x(y''f - y'f')$; hence, $\dfrac{yx'' - xy''}{yx' - xy'} = \dfrac{f'}{f}$. Similarly, $\dfrac{zy'' - yz''}{zy' - yz'} = \dfrac{f'}{f}$.

But $yx'' - xy'' = (yx' - y'x)'$. Thus, $\dfrac{(yx' - xy')'}{yx' - xy'} = \dfrac{(zy' - yz')'}{zy' - yz'}$. Integrate this equation to get

$$\ln|yx' - xy'| = \ln|zy' - yz'| + K,$$

or

$$yx' - xy' = C(zy' - yz').$$

Rearrange this equation to get

$$\frac{x' + Cz'}{x + Cz} = \frac{y'}{y},$$

with C constant. Integrate this equation to get $x + Cz = Ky$, the equation of a plane passing through the origin.

Chapter 10

Section 10.1

1. $y_k = A_0 + \dfrac{1}{3}(3^k - 1)$ **3.** $y_k = A_0(k - 1)!$ **5.** $y_k = A_1 + \displaystyle\sum_{j=1}^{k-1} \frac{j}{(j-1)!}$ for $k = 2, 2, 3, \ldots$

7. $y_k = \dfrac{1}{k} A_0 + \dfrac{1}{12}(k - 1)(k + 1)(3k - 2)$ **9.** $y_k = \dfrac{1}{3} k(k - 1)(k - 2)$

11. $D^3 y_k = y_{k+3} - 3y_{k+2} + 3y_{k+1} - y_k;\; D^4 y_k = y_{k+4} - 4y_{k+3} + 6y_{k+2} - 4y_{k+1} + y_k$

Section 10.2

1. $y_k = c_1(-6)^k + c_2 2^k$ **3.** $y_k = c_1(-1)^k 3^k + c_2(-1)^k$ **5.** $y_k = (c_1 + c_2 k)(-1)^k 4^k$

7. $y_k = (\sqrt{2})^k\left[c_1\cos\left(\dfrac{3\pi k}{4}\right) + c_2\sin\left(\dfrac{3\pi k}{4}\right)\right]$ **9.** $y_k = 2^k\left[c_1\cos\left(\dfrac{k\pi}{6}\right) + c_2\sin\left(\dfrac{k\pi}{6}\right)\right]$ **11.** $y_k = 2^{k+1} - 2^{2k}$

13. $y_k = \left(1 - \dfrac{12}{5} k\right)(-5)^k$ **15.** $y_k = \dfrac{1}{2}(-4)^k + \dfrac{3}{2}(2^k)$ **17.** $y_k = (\sqrt{2})^k\left[2\cos\left(\dfrac{3\pi k}{4}\right) + 4\sin\left(\dfrac{3\pi k}{4}\right)\right]$

19. $y_k = \dfrac{1}{2}(3^k) + \dfrac{3}{2}(-3)^k$ **21.** $y_k = \left(\dfrac{5 + \sqrt{5}}{10}\right)\left(\dfrac{1 + \sqrt{5}}{2}\right)^k + \left(\dfrac{5 - \sqrt{5}}{10}\right)\left(\dfrac{1 - \sqrt{5}}{2}\right)^k$

Section 10.3

1. $y_k = c_1(-1)^k + c_2(6)^k + \dfrac{1}{2} k + \dfrac{5}{4}$ **3.** $y_k = c_1(-6)^k + 1 + \dfrac{4}{7} k$ **5.** $y_k = c_1(-3)^k + c_2(-4)^k + \dfrac{1}{20}$

7. $y_k = c_1(-6)^k + c_2(-2)^k + \dfrac{1}{e^2 + 8e + 12} e^k$ **9.** $y_k = c_1 2^k + c_2 3^k - \dfrac{3}{2} - \dfrac{3}{2} k - \dfrac{1}{2} k^2$

11. $y_k = c_1 3^k + c_2(-3)^k + \left[\displaystyle\sum_{j=1}^{k-1} \frac{-\left(\dfrac{1}{j}\right)(-3)^{j+1}}{(-1)^j 2(3)^{2j+3}}\right] 3^k + \left[\displaystyle\sum_{j=1}^{k-1} \frac{\left(\dfrac{1}{j}\right)(3)^{j+1}}{(-1)^j 2(3)^{2j+3}}\right](-3)^k$

13. $y_k = c_1(-4)^k + c_2(3)^k + \left[\sum_{j=1}^{k-1} \dfrac{-\sin^2(j)3^{j+1}}{(-1)^j(12)^{j+1}(7)}\right](-4)^{k+1} + \left[\sum_{j=1}^{k-1} \dfrac{\sin^2(j)(-4)^{j+1}}{(-1)^j(12)^{j+1}(7)}\right](3)^{k+1}$

15. $y_k = c_1(-7)^k + c_2(-1)^k + \left[\sum_{j=1}^{k-1} \dfrac{(-je^j)(-1)^{j+1}}{(6)(7)^{j+1}}\right](-7)^{k+1} + \left[\sum_{j=1}^{k-1} \dfrac{(je^j)(-7)^{j+1}}{(6)(7)^{j+1}}\right](-1)^{k+1}$

Section 10.4

1. $y_k = c_1 \dfrac{\Gamma(k+3)}{\Gamma(k)} + c_2 \dfrac{\Gamma(k+4)}{\Gamma(k)} = c_1 k(k+1)(k+2) + c_2 k(k+1)(k+2)(k+3)$ **3.** $y_k = c[c_1 + c_2(k+1)(k+2)]$

5. $y_k = c_1 \dfrac{1}{k-\frac{1}{4}} \dfrac{\Gamma\left(k+\frac{3}{4}\right)}{\Gamma(k)} + c_2 k(k+1)(k+2)$ **7.** $y_k = k_1 k + c_2 \dfrac{1}{\left(k-\frac{4}{3}\right)\left(k-\frac{1}{3}\right)} \dfrac{\Gamma\left(k+\frac{2}{3}\right)}{\Gamma(k)}$

9. $y_k = c_1 \dfrac{1}{k-\frac{1}{2}} \dfrac{\Gamma\left(k+\frac{1}{2}\right)}{\Gamma(k)} + c_2 \dfrac{1}{\left(k-\frac{3}{2}\right)\left(k-\frac{1}{2}\right)} \dfrac{\Gamma\left(k+\frac{1}{2}\right)}{\Gamma(k)}$ **11.** $A_k = A_0 \prod_{j=0}^{k-1} \dfrac{b_j y_j^{(1)}}{y_{j+1}^{(1)}}$

13. $y_k^{(1)} = \dfrac{\Gamma\left(k+\frac{1}{2}\right)}{\Gamma(k)}$ and $y_k^{(2)} = A_1 \left[\prod_{j=1}^{k-1} \dfrac{\left(1+\frac{1}{4j(j+1)}\right)^j}{j+\frac{1}{2}}\right] \dfrac{\Gamma\left(k+\frac{1}{2}\right)}{\Gamma(k)}$

Section 10.5

1. $y_k = \dfrac{m^2\omega}{2Th} x_k(x_k - L)$ **3.** Since $m\dfrac{d^2(x_{k+1})}{dt^2} = k[x_{k+2} - 2x_{k+1} + x_k]$, then

$$mA_{k+1}[-\omega^2\cos(\omega t + \beta)] = k[A_{k+2}\cos(\omega t + \beta) - 2A_{k+1}\cos(\omega t + \beta) + A_k\cos(\omega t + \beta)].$$

Then $-\omega^2 mA_{k+1} = kA_{k+2} - 2kA_{k+1} + kA_k$; hence,

$$A_{k+2} + \left[\dfrac{\omega^2 m}{k} - 2\right]A_{k+1} + A_k = 0.$$

5. Let $A = 1$, $B = \cos(\theta)$, $C = 1$, and $F(t) = 1$ in the result of Problem 4. Since $F(t) = \sum_{k=0}^{\infty} F_k t^k = 1$, $F_0 = 1$ and $F_k = 0$ if $k = 1, 2, 3, \ldots$. Then f_k satisfies the difference equation $f_{k+2} - 2\cos(\theta)f_{k+1} + f_k = 0$, with initial conditions $f_0 = 1$ and $f_1 = 2\cos(\theta)$. Solve this difference equation to get $f_k = a\cos(k\theta) + b\sin(k\theta)$. The initial conditions imply that $a = 1$ and $b = \dfrac{\cos(\theta)}{\sin(\theta)}$, so $f_k = \dfrac{\sin(\theta)\cos(k\theta) + \cos(\theta)\sin(k\theta)}{\sin(\theta)} = \dfrac{\sin[(k+1)\theta]}{\sin(\theta)}$. Therefore,

$$\dfrac{1}{1 - 2\cos(\theta)t + t^2} = \sum_{k=0}^{\infty} f_k t^k = \sum_{k=0}^{\infty} \dfrac{\sin[(k+1)\theta]}{\sin(\theta)} t^k.$$

7. Using the result of Problem 5, we have

$$\dfrac{1}{1 + 2\cos(\theta)\dfrac{1}{r} + \dfrac{1}{r^2}} = \sum_{k=0}^{\infty} \dfrac{\sin[(k+1)\theta]}{\sin(\theta)}(-1)^k \dfrac{1}{r^k}$$

$$= \sum_{k=1}^{\infty} \dfrac{\sin(k\theta)}{\sin(\theta)}(-1)^{k+1} \dfrac{1}{r^{k-1}}.$$

Multiply this equation by $-r\sin(\theta)$ to get

$$\dfrac{-r\sin(\theta)}{r^2 + 2r\cos(\theta) + 1} = \sum_{k=1}^{\infty} \sin(k\theta)(-1)^k \dfrac{1}{r^k}.$$

Chapter 11

Section 11.1

1. $(2 + \sqrt{2})i + 3j; (2 - \sqrt{2})i - 9j + 10k; \sqrt{38}; \sqrt{63}; 4i - 6j + 10k; 3\sqrt{2}i + 18j - 15k$

3. $3i - k; i - 10j + k; \sqrt{29}; 3\sqrt{3}; 4i - 10j; 3i + 15j - 3k$

5. $3i - j + 3k; -i + 3j - k; \sqrt{3}; 2\sqrt{3}; 2i + 2j + 2k; 6i - 6j + 6k$

7. $F + G = 3i - 2j$ $F - G = i$

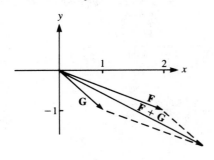

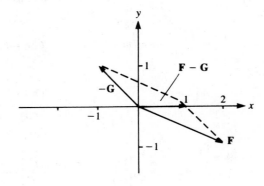

9. $F + G = 2i - 5j$ $F - G = j$

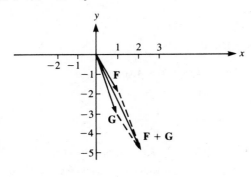

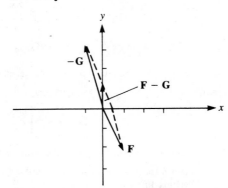

11. $\alpha F = -\dfrac{1}{2}i - \dfrac{1}{2}j$ 13. $\alpha F = 12j$

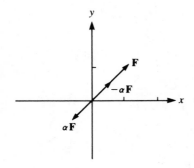

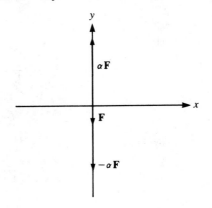

15. $\alpha\mathbf{F} = -9\mathbf{i} + 6\mathbf{j}$

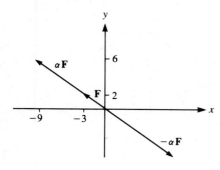

17. $x = 3 + 6t, y = -t, z = 0; -\infty < t < \infty$ **19.** $x = 0, y = 1 + t, z = 3 + 2t; -\infty < t < \infty$

21. $x = 2 + 3t, y = -3 - 9t, z = 6 + 2t; -\infty < t < \infty$ **23.** $x = 3 + t, y = 3 + 9t, z = -5 - 6t; -\infty < t < \infty$

25. $x = -1 + 5t, y = -8t, z = t; -\infty < t < \infty$

27. $\mathbf{F} = 3\mathbf{i} + 3\sqrt{3}\mathbf{j}$

29. $\mathbf{F} = \dfrac{1}{2}\sqrt{2}\mathbf{i} - \dfrac{1}{2}\sqrt{2}\mathbf{j}$

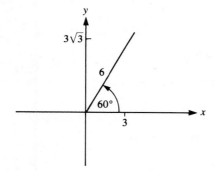

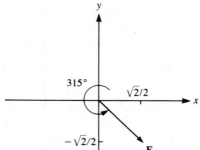

31. $\mathbf{F} = \dfrac{1}{2}\sqrt{6}\mathbf{i} + \dfrac{1}{2}\sqrt{2}\mathbf{j}$

33. $\mathbf{F} = -\sqrt{15}\cos(5°)\mathbf{i} + \sqrt{15}\sin(5°)\mathbf{j}$

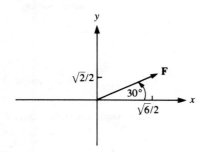

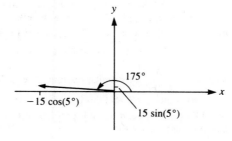

35. $\mathbf{F} = -25\mathbf{j}$

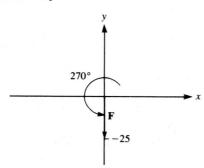

37. $t = \pm\dfrac{1}{\|\mathbf{A}\|}$ **39.** Since $\overline{RC} = \overline{CQ}$, $\overline{PC} = \dfrac{1}{2}(\mathbf{A} + \mathbf{B})$.

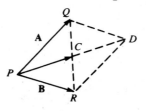

41. If $\alpha \neq 0$, then $\alpha\mathbf{F} + \beta\mathbf{G} = \mathbf{O}$ can be solved for $\mathbf{F}$ to yield $\mathbf{F} = -\dfrac{\beta}{\alpha}\mathbf{G}$. This implies that $\mathbf{F}$ and $\mathbf{G}$ are parallel,

a contradiction. If $\beta \neq 0$, then $\mathbf{G} = -\dfrac{\alpha}{\beta}\mathbf{F}$ again implies that $\mathbf{F}$ and $\mathbf{G}$ are parallel, a contradiction. Therefore, $\alpha = \beta = 0$.

Section 11.2

In solutions 1 through 9, $\mathbf{F}$ is the first vector listed in the problem and $\mathbf{G}$ is the second.

1. $\mathbf{F} \cdot \mathbf{G} = 2$, $\cos(\theta) = \dfrac{2}{\sqrt{14}}$; $\mathbf{F}$ and $\mathbf{G}$ are not orthogonal; $|\mathbf{F} \cdot \mathbf{G}| = 2 < \sqrt{14} = \|\mathbf{F}\|\|\mathbf{G}\|$

3. -23; $-\dfrac{23}{\sqrt{29}\sqrt{41}}$; not orthogonal; $23 < \sqrt{29}\sqrt{41}$ **5.** 0; 0; $\mathbf{F}$ and $\mathbf{G}$ are orthogonal; $0 < 6\sqrt{13}$

7. 5; 1; not orthogonal; $5 = \sqrt{5}\sqrt{5}$ **9.** -18; $-\dfrac{9}{10}$; not orthogonal; $18 < \sqrt{10}\sqrt{40}$

11. $3x - y + 4z = 4$ **13.** $4x - 3y + 2z = 25$ **15.** $8x - 3y + 4z = 19$ **17.** $2x - z = 0$

19. $3x - 7y - z = 5$ **21.** $\dfrac{112}{11\sqrt{3}\sqrt{35}}$ **23.** $\dfrac{-107}{\sqrt{108}\sqrt{101}}$ **25.** $\dfrac{44}{\sqrt{46}\sqrt{91}}$

27. Let $\mathbf{X} = a\mathbf{i} + b\mathbf{j} + c\mathbf{k}$. Then $\mathbf{F} \cdot \mathbf{X} = a\mathbf{F} \cdot \mathbf{i} + b\mathbf{F} \cdot \mathbf{j} + c\mathbf{F} \cdot \mathbf{k} = 0$. Therefore, $\mathbf{F} \cdot \mathbf{X} = 0$ for every vector $\mathbf{X}$. By the result of Problem 26, $\mathbf{F} = \mathbf{O}$.

29. Let $\mathbf{F} = a\mathbf{i} + b\mathbf{j} + c\mathbf{k}$. Then $\mathbf{F} \cdot \mathbf{i} = a\mathbf{i} \cdot \mathbf{i} + b\mathbf{i} \cdot \mathbf{j} + c\mathbf{i} \cdot \mathbf{k} = a$, because $\mathbf{i} \cdot \mathbf{i} = 1$ and $\mathbf{i} \cdot \mathbf{j} = \mathbf{i} \cdot \mathbf{k} = 0$. By similar reasoning, $\mathbf{F} \cdot \mathbf{j} = b$ and $\mathbf{F} \cdot \mathbf{k} = c$. Therefore, $\mathbf{F} = (\mathbf{F} \cdot \mathbf{i})\mathbf{i} + (\mathbf{F} \cdot \mathbf{j})\mathbf{j} + (\mathbf{F} \cdot \mathbf{k})\mathbf{k}$.

31. Write $\mathbf{A} = \alpha\mathbf{F} + \beta\mathbf{G} + \gamma\mathbf{H}$ and attempt to solve for α, β, and γ. First, because $\mathbf{F} \cdot \mathbf{G} = \mathbf{F} \cdot \mathbf{H} = 0$, we have

$$\mathbf{A} \cdot \mathbf{F} = \alpha\mathbf{F} \cdot \mathbf{F} + \beta\mathbf{G} \cdot \mathbf{F} + \gamma\mathbf{H} \cdot \mathbf{F} = \alpha\mathbf{F} \cdot \mathbf{F} = \alpha\|\mathbf{F}\|^2.$$

Therefore, we must choose $\alpha = \dfrac{\mathbf{A} \cdot \mathbf{F}}{\|\mathbf{F}\|^2}$. By similar reasoning, we must choose $\beta = \dfrac{\mathbf{A} \cdot \mathbf{G}}{\|\mathbf{G}\|^2}$ and $\gamma = \dfrac{\mathbf{A} \cdot \mathbf{H}}{\|\mathbf{H}\|^2}$. With this choice of constants, we have $\mathbf{A} = \alpha\mathbf{F} + \beta\mathbf{G} + \gamma\mathbf{H}$, and this is the only such choice for the constants.

33. We find that in general $\mathbf{F} \cdot \mathbf{F} \neq \|\mathbf{F}\|^2$, and that it is possible for $\mathbf{F} \cdot \mathbf{F} = 0$ without $\mathbf{F} = \mathbf{O}$. For example, with this definition, $(\mathbf{i} + \mathbf{j}) \cdot (\mathbf{i} + \mathbf{j}) = 0$, even though $\mathbf{i} + \mathbf{j} \neq \mathbf{O}$.

35. Let $\mathbf{F} = a_1\mathbf{i} + b_1\mathbf{j} + c_1\mathbf{k}$ and $\mathbf{G} = a_2\mathbf{i} + b_2\mathbf{j} + c_2\mathbf{k}$. Then

$$\alpha(\mathbf{F} \cdot \mathbf{G}) = \alpha(a_1 a_2 + b_1 b_2 + c_1 c_2) = (\alpha a_1)a_2 + (\alpha b_1)b_2 + (\alpha c_1)c_2$$
$$= (\alpha\mathbf{F}) \cdot \mathbf{G} = a_1(\alpha a_2) + b_1(\alpha b_2) + c_1(\alpha c_2) = \mathbf{F} \cdot (\alpha\mathbf{G}).$$

37. With the rhombus as shown, the diagonals are $\mathbf{A}_1 + \mathbf{A}_2$ and $\mathbf{A}_2 - \mathbf{A}_1$. Calculate $(\mathbf{A}_1 + \mathbf{A}_2) \cdot (\mathbf{A}_2 - \mathbf{A}_1) = \|\mathbf{A}_2\|^2 - \|\mathbf{A}_1\|^2 = 0$, because $\mathbf{A}_1$ and $\mathbf{A}_2$ have the same length. Therefore, vectors along the diagonals are perpendicular and hence the diagonals are perpendicular.

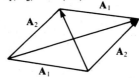

Section 11.3

1. $\mathbf{F} \times \mathbf{G} = 8\mathbf{i} + 2\mathbf{j} + 12\mathbf{k}$; $\cos(\theta) = -\dfrac{4}{\sqrt{23}\sqrt{3}}$; $\sin(\theta) = \dfrac{\sqrt{53}}{\sqrt{23}\sqrt{3}}$; $\|\mathbf{F}\|\|\mathbf{G}\|\sin(\theta) = 2\sqrt{53} = \|\mathbf{F} \times \mathbf{G}\|$

3. $\mathbf{F} \times \mathbf{G} = -8\mathbf{i} - 12\mathbf{j} - 5\mathbf{k}$; $\cos(\theta) = -\dfrac{12}{\sqrt{29}\sqrt{13}}$; $\sin(\theta) = \dfrac{\sqrt{233}}{\sqrt{29}\sqrt{13}}$; $\|\mathbf{F}\|\|\mathbf{G}\|\sin(\theta) = \sqrt{233} = \|\mathbf{F} \times \mathbf{G}\|$

5. $\mathbf{F} \times \mathbf{G} = 18\mathbf{i} + 50\mathbf{j} - 60\mathbf{k}$; $\cos(\theta) = \dfrac{62}{5\sqrt{2}\sqrt{109}}$; $\sin(\theta) = \dfrac{\sqrt{1606}}{5\sqrt{2}\sqrt{109}}$; $\|\mathbf{F}\|\|\mathbf{G}\|\sin(\theta) = 2\sqrt{1606} = \|\mathbf{F} \times \mathbf{G}\|$

7. $\mathbf{F} \times \mathbf{G} = -85\mathbf{i} + 18\mathbf{j} + 396\mathbf{k}$; $\cos(\theta) = -\dfrac{70}{\sqrt{349}\sqrt{485}}$; $\sin(\theta) = \dfrac{\sqrt{164{,}365}}{\sqrt{349}\sqrt{485}}$; $\|\mathbf{F}\|\|\mathbf{G}\|\sin(\theta) = \sqrt{164{,}365} = \|\mathbf{F} \times \mathbf{G}\|$

9. $\mathbf{F} \times \mathbf{G} = 12\mathbf{i} + 34\mathbf{j} + 8\mathbf{k}$; $\cos(\theta) = \dfrac{19}{3\sqrt{6}\sqrt{13}}$; $\sin(\theta) = \dfrac{\sqrt{341}}{3\sqrt{6}\sqrt{13}}$; $\|\mathbf{F}\|\|\mathbf{G}\|\sin(\theta) = 2\sqrt{341} = \|\mathbf{F} \times \mathbf{G}\|$

11. $\mathbf{F} \times \mathbf{G} = 3\mathbf{i} - \mathbf{j}$; $\cos(\theta) = \dfrac{21}{\sqrt{11}\sqrt{41}}$; $\sin(\theta) = \dfrac{10}{\sqrt{11}\sqrt{41}}$; $\|\mathbf{F}\|\|\mathbf{G}\|\sin(\theta) = 10 = \|\mathbf{F} \times \mathbf{G}\|$

13. $\mathbf{F} \times \mathbf{G} = -6\mathbf{i} + 21\mathbf{j} - 58\mathbf{k}$; $\cos(\theta) = \dfrac{9}{\sqrt{74}\sqrt{53}}$; $\sin(\theta) = \dfrac{\sqrt{3841}}{\sqrt{74}\sqrt{53}}$; $\|\mathbf{F}\|\|\mathbf{G}\|\sin(\theta) = \sqrt{3841} = \|\mathbf{F} \times \mathbf{G}\|$

15. $\mathbf{F} \times \mathbf{G} = -27\mathbf{i} + 12\mathbf{j} - 13\mathbf{k}$; $\cos(\theta) = \dfrac{6}{7\sqrt{22}}$; $\sin(\theta) = \dfrac{\sqrt{521}}{7\sqrt{11}}$; $\|\mathbf{F}\|\|\mathbf{G}\|\sin(\theta) = \sqrt{1042} = \|\mathbf{F} \times \mathbf{G}\|$

In solutions 17 and 19, any nonzero scalar multiple of the given vector is also perpendicular to the plane of the three given points.

17. (a) $2\mathbf{i} + 4\mathbf{j} + 12\mathbf{k}$ (b) $x + 2y + 6z = 12$ **19.** (a) $-5\mathbf{i} - 16\mathbf{j} + 2\mathbf{k}$ (b) $5x + 16y - 2z = -4$

21. $\sqrt{1013}$ **23.** 98   **25.** $38\sqrt{2}$ **27.** 14   **29.** 269

In solutions 31, 33, and 35, any nonzero scalar multiple of the vector is also normal to the given plane.

31. $8\mathbf{i} - \mathbf{j} + \mathbf{k}$ **33.** $\mathbf{i} - 3\mathbf{j} + 2\mathbf{k}$ **35.** $4\mathbf{i} + 6\mathbf{j} + 4\mathbf{k}$

37. Let $\mathbf{F} = a_1\mathbf{i} + b_1\mathbf{j} + c_1\mathbf{k}$ and $\mathbf{G} = a_2\mathbf{i} + b_2\mathbf{j} + c_2\mathbf{k}$. Then

$$(\alpha\mathbf{F}) \times \mathbf{G} = \begin{vmatrix} \mathbf{i} & \mathbf{j} & \mathbf{k} \\ \alpha a_1 & \alpha b_1 & \alpha c_1 \\ a_2 & b_2 & c_2 \end{vmatrix} = \alpha \begin{vmatrix} \mathbf{i} & \mathbf{j} & \mathbf{k} \\ a_1 & b_1 & c_1 \\ a_2 & b_2 & c_2 \end{vmatrix}$$

$$= \alpha(\mathbf{F} \times \mathbf{G}) = \begin{vmatrix} \mathbf{i} & \mathbf{j} & \mathbf{k} \\ a_1 & b_1 & c_1 \\ \alpha a_2 & \alpha b_2 & \alpha c_2 \end{vmatrix} = \mathbf{F} \times (\alpha\mathbf{G}).$$

39. Let $\mathbf{F} = a_1\mathbf{i} + b_1\mathbf{j} + c_1\mathbf{k}$, $\mathbf{G} = a_2\mathbf{i} + b_2\mathbf{j} + c_2\mathbf{k}$, and $\mathbf{H} = a_3\mathbf{i} + b_3\mathbf{j} + c_3\mathbf{k}$. Then

$$\mathbf{F} \cdot (\mathbf{G} \times \mathbf{H}) = (a_1\mathbf{i} + b_1\mathbf{j} + c_1\mathbf{k}) \cdot \begin{vmatrix} \mathbf{i} & \mathbf{j} & \mathbf{k} \\ a_2 & b_2 & c_2 \\ a_3 & b_3 & c_3 \end{vmatrix} = \begin{vmatrix} a_1 & b_1 & c_1 \\ a_2 & b_2 & c_2 \\ a_3 & b_3 & c_3 \end{vmatrix}$$

$$= \begin{vmatrix} a_2 & b_2 & c_2 \\ a_3 & b_3 & c_3 \\ a_1 & b_1 & c_1 \end{vmatrix} = \mathbf{G} \cdot (\mathbf{H} \times \mathbf{F}) = \begin{vmatrix} a_3 & b_3 & c_3 \\ a_1 & b_1 & c_1 \\ a_2 & b_2 & c_2 \end{vmatrix} = \mathbf{H} \cdot (\mathbf{F} \times \mathbf{G}).$$

41. $(\mathbf{F} \times \mathbf{G}) \times (\mathbf{H} \times \mathbf{K})$ must be orthogonal to the plane of $\mathbf{F} \times \mathbf{G}$ and $\mathbf{H} \times \mathbf{K}$, which is itself orthogonal to the plane of $\mathbf{F}$ and $\mathbf{G}$. Therefore, this vector must lie in the plane of $\mathbf{F}$ and $\mathbf{G}$.

43. By the conclusion of Problem 40,

$$(\mathbf{F} \times \mathbf{G}) \times (\mathbf{H} \times \mathbf{K}) = -(\mathbf{H} \times \mathbf{K}) \times (\mathbf{F} \times \mathbf{G}) = (-1)([(\mathbf{H} \times \mathbf{K}) \cdot \mathbf{G}]\mathbf{F} - [(\mathbf{H} \times \mathbf{K}) \cdot \mathbf{F}]\mathbf{G})$$
$$= [(\mathbf{H} \times \mathbf{K}) \cdot \mathbf{F}]\mathbf{G} - [(\mathbf{H} \times \mathbf{K}) \cdot \mathbf{G}]\mathbf{F} \quad \text{(by Problem 39)}$$
$$= [\mathbf{H} \cdot (\mathbf{K} \times \mathbf{F})]\mathbf{G} - [\mathbf{H} \cdot (\mathbf{K} \times \mathbf{G})\mathbf{F}.$$

45. If $\mathbf{F} = a_1\mathbf{i} + b_1\mathbf{j} + c_1\mathbf{k}$, $\mathbf{G} = a_2\mathbf{i} + b_2\mathbf{j} + c_2\mathbf{k}$, and $\mathbf{K} = a_3\mathbf{i} + b_3\mathbf{j} + c_3\mathbf{k}$, the area of the triangle is $\dfrac{1}{2}\mathbf{F} \cdot (\mathbf{G} \times \mathbf{K})$,

or $\dfrac{1}{2}\begin{vmatrix} a_1 & b_1 & c_1 \\ a_2 & b_2 & c_2 \\ a_3 & b_3 & c_3 \end{vmatrix}$. The points are collinear if this determinant is zero.

Section 11.4

1. 1   **3.** 38   **5.** 16   **7.** -304 **9.** -129

11. Since $\mathbf{X} - \mathbf{X}_0$ is parallel to the plane, and $\mathbf{F} \times \mathbf{G}$ is normal to the plane, then $\mathbf{X} - \mathbf{X}_0$ is orthogonal to $\mathbf{F} \times \mathbf{G}$. Hence, $(\mathbf{X} - \mathbf{X}_0) \cdot (\mathbf{F} \times \mathbf{G}) = 0$, or $[\mathbf{X} - \mathbf{X}_0, \mathbf{F}, \mathbf{G}] = 0$.

13. By Problem 39 of Section 11.3, $[\mathbf{F}, \mathbf{G}, \mathbf{H}] = \mathbf{F} \cdot (\mathbf{G} \times \mathbf{H}) = \mathbf{G} \cdot (\mathbf{H} \times \mathbf{F}) = [\mathbf{G}, \mathbf{H}, \mathbf{F}] = \mathbf{H} \cdot (\mathbf{F} \times \mathbf{G}) = [\mathbf{H}, \mathbf{F}, \mathbf{G}]$.

15. Let $\mathbf{F} = a_1\mathbf{i} + b_1\mathbf{j} + c_1\mathbf{k}$, $\mathbf{G} = a_2\mathbf{i} + b_2\mathbf{j} + c_2\mathbf{k}$ and $\mathbf{H} = a_3\mathbf{i} + b_3\mathbf{j} + c_3\mathbf{k}$. Then

$$[\mathbf{F}, \mathbf{G}, \mathbf{H}] = \mathbf{F} \cdot (\mathbf{G} \times \mathbf{H}) = \mathbf{F} \cdot [(b_2 c_3 - b_3 c_2)\mathbf{i} + (c_2 a_3 - c_3 a_2)\mathbf{j} + (a_2 b_3 - a_3 b_2)\mathbf{k}]$$
$$= a_1(b_2 c_3 - b_3 c_2) + b_1(c_2 a_3 - c_3 a_2) + c_1(a_2 b_3 - a_3 b_2)$$
$$= \begin{vmatrix} a_1 & b_1 & c_1 \\ a_2 & b_2 & c_2 \\ a_3 & b_3 & c_3 \end{vmatrix}.$$

17. We may assume that $\mathbf{F} = \alpha\mathbf{G} + \beta\mathbf{H}$. Then

$$[\mathbf{F}, \mathbf{G}, \mathbf{H}] = [\alpha\mathbf{G} + \beta\mathbf{H}, \mathbf{G}, \mathbf{H}] = \alpha[\mathbf{G}, \mathbf{G}, \mathbf{H}] + \beta[\mathbf{H}, \mathbf{G}, \mathbf{H}]$$
$$= \alpha\mathbf{G} \cdot (\mathbf{G} \times \mathbf{H}) + \beta\mathbf{H} \cdot (\mathbf{G} \times \mathbf{H}) = 0,$$

because $\mathbf{G}$ and $\mathbf{H}$ are perpendicular to $\mathbf{G} \times \mathbf{H}$.

Section 11.5

1. $5\mathbf{e}_1 + 5\mathbf{e}_2 + 6\mathbf{e}_3 + 5\mathbf{e}_4 + \mathbf{e}_5$; 0; $\cos(\theta) = 0$ **3.** $17\mathbf{e}_1 - 4\mathbf{e}_2 + 3\mathbf{e}_3 + 6\mathbf{e}_4$; 24; $\cos(\theta) = \dfrac{\sqrt{6}}{\sqrt{5}\sqrt{17}}$

5. $-5\mathbf{e}_1 + 3\mathbf{e}_2 + 5\mathbf{e}_3 + 2\mathbf{e}_4 + 4\mathbf{e}_5 + 15\mathbf{e}_7$; -93; $\cos(\theta) = -\dfrac{93}{8\sqrt{157}\sqrt{11}}$

7. $10\mathbf{e}_1 + 4\mathbf{e}_2 + 6\mathbf{e}_3$, where $\mathbf{e}_1 = \langle 1, 0, 0, 0, 0 \rangle$, $\mathbf{e}_2 = \langle 0, 1, 0, 0, 0 \rangle$, and $\mathbf{e}_3 = \langle 0, 0, 1, 0, 0 \rangle$; 27; $\cos(\theta) = \dfrac{27}{\sqrt{43}\sqrt{55}}$

9. $15\mathbf{e}_1 + 3\mathbf{e}_2 - 4\mathbf{e}_3 + 2\mathbf{e}_4 + 2\mathbf{e}_5$; -116; $\cos(\theta) = -\dfrac{116}{5\sqrt{11}\sqrt{807}}$ **11.** S is a subspace of R^5.

13. S is not a subspace.   **15.** S is a subspace.   **17.** S is a subspace.   **19.** S is a subspace.

21. $\|\mathbf{F} + \mathbf{G}\|^2 = \|\mathbf{F}\|^2 + 2\mathbf{F} \cdot \mathbf{G} + \|\mathbf{G}\|^2$ and $\|\mathbf{F} - \mathbf{G}\|^2 = \|\mathbf{F}\|^2 - 2\mathbf{F} \cdot \mathbf{G} + \|\mathbf{G}\|^2$. Therefore, $\|\mathbf{F} + \mathbf{G}\|^2 + \|\mathbf{F} - \mathbf{G}\|^2 = 2(\|\mathbf{F}\|^2 + \|\mathbf{G}\|^2)$.

23. Write $\|\mathbf{F} + \mathbf{G}\|^2 = \|\mathbf{F}\|^2 + 2\mathbf{F} \cdot \mathbf{G} + \|\mathbf{G}\|^2$. If also $\|\mathbf{F} + \mathbf{G}\|^2 = \|\mathbf{F}\|^2 + \|\mathbf{G}\|^2$, then we must have $\mathbf{F} \cdot \mathbf{G} = 0$; hence, $\mathbf{F}$ and $\mathbf{G}$ are orthogonal.

25. Let $\mathbf{F} = \langle x_1, x_2, \ldots, x_n \rangle$ and $\mathbf{G} = \langle y_1, y_2, \ldots, y_n \rangle$ in R^n. Since $|\mathbf{F} \cdot \mathbf{G}| \le \|\mathbf{F}\|\|\mathbf{G}\|$, then

$$\left| \sum_{j=1}^{n} x_j y_j \right| \le \sqrt{\sum_{j=1}^{n} x_j^2} \sqrt{\sum_{j=1}^{n} y_j^2}.$$ Now square both sides of this inequality.

Section 11.6

1. Independent   **3.** Independent   **5.** Independent   **7.** Dependent   **9.** Independent

11. If $\mathbf{F}$, $\mathbf{G}$, and $\mathbf{H}$ are linearly dependent, then there are numbers α, β, and γ, not all zero, such that $\alpha\mathbf{F} + \beta\mathbf{G} + \gamma\mathbf{H} = \mathbf{O}$.

Suppose, to be specific, that $\alpha \neq 0$. Then $\mathbf{F} = -\dfrac{\beta}{\alpha}\mathbf{G} - \dfrac{\gamma}{\alpha}\mathbf{H}$, so $[\mathbf{F}, \mathbf{G}, \mathbf{H}] = \left[-\dfrac{\beta}{\alpha}\mathbf{G} - \dfrac{\gamma}{\alpha}\mathbf{H}, \mathbf{G}, \mathbf{H}\right] = 0$ by conclusion (5) of Theorem 11.10. Conversely, suppose that $[\mathbf{F}, \mathbf{G}, \mathbf{H}] = 0$. Then $\mathbf{F} \cdot (\mathbf{G} \times \mathbf{H}) = 0$. Let $\mathbf{F} = a_1\mathbf{i} + b_1\mathbf{j} + c_1\mathbf{k}$,

$\mathbf{G} = a_2\mathbf{i} + b_2\mathbf{j} + c_2\mathbf{k}$, and $\mathbf{H} = a_3\mathbf{i} + b_3\mathbf{j} + c_3\mathbf{k}$. Then $\begin{vmatrix} a_1 & b_1 & c_1 \\ a_2 & b_2 & c_2 \\ a_3 & b_3 & c_3 \end{vmatrix} = 0$. Therefore, the system of equations

$a_1x + b_1y + c_1z = 0$, $a_2x + b_2y + c_2z = 0$, $a_3x + b_3y + c_3z = 0$ has a nontrivial solution $x = \alpha$, $y = \beta$, $z = \gamma$. Further, we then have $\alpha\mathbf{F} + \beta\mathbf{G} + \gamma\mathbf{H} = \mathbf{O}$, so $\mathbf{F}$, $\mathbf{G}$, and $\mathbf{H}$ are linearly dependent.

13. Linearly dependent   **15.** Linearly independent

17. A basis is $\langle 1, 0, 0, -1 \rangle$ and $\langle 0, 1, -1, 0 \rangle$; dimension = 2   **19.** A basis is $\langle 1, 0, -2 \rangle$ and $\langle 0, 1, 1 \rangle$; dimension = 2

21. A basis is $\langle 1, 0, 0, 0 \rangle$, $\langle 0, 0, 1, 0 \rangle$ and $\langle 0, 0, 0, 1 \rangle$; dimension = 3   **23.** A basis is $\langle 1, 4 \rangle$; dimension = 1

27. Let $\mathbf{F} = a_1\mathbf{i} + b_1\mathbf{j}$, $\mathbf{G} = a_2\mathbf{i} + b_2\mathbf{j}$, and $\mathbf{H} = a_3\mathbf{i} + b_3\mathbf{j}$. The system of equations $a_1x + a_2y + a_3z = 0$, $b_1x + b_2y + b_3z = 0$ has a nontrivial solution $x = \alpha$, $y = \beta$, $z = \gamma$ (because there are more unknowns than equations). Therefore, $\alpha\mathbf{F} + \beta\mathbf{G} + \gamma\mathbf{H} = \mathbf{O}$, with not all of α, β, and γ zero, so $\mathbf{F}$, $\mathbf{G}$, and $\mathbf{H}$ are linearly dependent.

29. Consider any k of these vectors, which we take for notational convenience to be $\mathbf{F}_1, \ldots, \mathbf{F}_k$ (by relabeling if necessary). If these k vectors are linearly dependent, then there are numbers $\alpha_1, \ldots, \alpha_k$, not all zero, such that $\alpha_1\mathbf{F}_1 + \cdots + \alpha_k\mathbf{F}_k = \mathbf{O}$. Then $\alpha_1\mathbf{F}_1 + \cdots + \alpha_k\mathbf{F}_k + 0\mathbf{F}_{k+1} + \cdots + 0\mathbf{F}_m = \mathbf{O}$, implying that $\mathbf{F}_1, \ldots, \mathbf{F}_m$ are linearly dependent, a contradiction. Therefore, $\mathbf{F}_1, \ldots, \mathbf{F}_k$ are linearly independent.

31. By Problem 28, $\mathbf{F}$, $\mathbf{G}$, $\mathbf{H}$, and $\mathbf{V}$ are linearly dependent. Thus, there are numbers α, β, γ, and δ, not all zero, such that $\alpha\mathbf{F} + \beta\mathbf{G} + \gamma\mathbf{H} + \delta\mathbf{V} = \mathbf{O}$. If $\delta = 0$, then this equation implies that $\mathbf{F}$, $\mathbf{G}$, and $\mathbf{H}$ are linearly dependent, a contradiction.

Thus, $\delta \neq 0$, so $\mathbf{V} = -\dfrac{\alpha}{\delta}\mathbf{F} - \dfrac{\beta}{\delta}\mathbf{G} - \dfrac{\gamma}{\delta}\mathbf{H} = a\mathbf{F} + b\mathbf{G} + c\mathbf{H}$. To solve for a, form a scalar triple product

$$[\mathbf{V}, \mathbf{G}, \mathbf{H}] = [a\mathbf{F} + b\mathbf{G} + c\mathbf{H}, \mathbf{G}, \mathbf{H}] = a[\mathbf{F}, \mathbf{G}, \mathbf{H}] + b[\mathbf{G}, \mathbf{G}, \mathbf{H}] + c[\mathbf{H}, \mathbf{G}, \mathbf{H}] = a[\mathbf{F}, \mathbf{G}, \mathbf{H}].$$

Therefore, $a = \dfrac{[\mathbf{V}, \mathbf{G}, \mathbf{H}]}{[\mathbf{F}, \mathbf{G}, \mathbf{H}]}$. In a similar way, solve for b and c.

Section 11.7

1. It is routine to verify that P_m satisfies the definition of a vector space. A basis consists of the polynomials $1, x, \ldots, x^m$.

3. If f and g are in V, then $(f + g)(1) = f(1) + g(1) = 2 \neq 1$, so $f + g$ is not in V. Therefore, V is not a vector space.

5. The difference of two polynomials of degree n may not be of degree n, so D_n is not a vector space. Conditions (1), (4), and (6) of the definition fail to hold.

7. A sum of solutions of this differential equation is not a solution.

9. No; in general $\langle x_1, y_1, z_1 \rangle + \langle x_2, y_2, z_2 \rangle \neq \langle x_2, y_2, z_2 \rangle + \langle x_1, y_1, z_1 \rangle$.

11. Yes; if f and g are in M, then $\alpha f + \beta g$ is in M for any numbers α and β.

13. Q is not a vector space because a product of a number in Q by an irrational number is not in Q [so condition (6) is violated].

15. P is not a vector space because conditions (1), (4), (5), and (6) are violated.

17. Conditions (7), (9), and (10) are violated.   **19.** Conditions (4) and (6)–(10) are violated.

21. Let θ and θ' be zero vectors in V. Then $\theta = \theta + \theta' = \theta'$.

Additional Problems for Chapter 11

1. -20; $6\mathbf{i} + 9\mathbf{j} + 15\mathbf{k}$; $\cos(\theta) = -\dfrac{20}{\sqrt{14}\sqrt{53}}$; $\theta \approx 2.40$ radians, or $137°30'$

3. -24; $15\mathbf{i} - 18\mathbf{j} - 8\mathbf{k}$; $-\dfrac{24}{\sqrt{41}\sqrt{29}}$; $\theta \approx 2.34$, or $134°7'$ **5.** 9; $12\mathbf{i} - 39\mathbf{j} - 18\mathbf{k}$; $\dfrac{3}{\sqrt{23}\sqrt{10}}$; $\theta \approx 1.37$, or $78°35'$

7. -7; $11\mathbf{i} - 32\mathbf{j} - 8\mathbf{k}$; $-\dfrac{7}{\sqrt{74}\sqrt{17}}$; $\theta \approx 1.37$, or $78°37'$ **9.** -46; $-10\mathbf{i} + 10\mathbf{j} + 40\mathbf{k}$; $-\dfrac{46}{\sqrt{89}\sqrt{44}}$; $\theta \approx 2.40$, or $137°19'$

11. 7; $20\mathbf{i} + 15\mathbf{j} + \mathbf{k}$; $\dfrac{7}{15\sqrt{3}}$; 1.30, or $74°22'$ **13.** 34; $42\mathbf{i} - 23\mathbf{j} + 18\mathbf{k}$; $\dfrac{34}{7\sqrt{77}}$; 0.98, or $56°23'$

15. $7; 56\mathbf{i} + 19\mathbf{j} + 24\mathbf{k}; \dfrac{7}{\sqrt{309}\,\sqrt{58}}; 1.52,$ or $87°$ **17.** $x = -2t, y = 2 + 2t, z = 3 - 2t$

19. $x = -2 + 8t, y = 1 + 6t, z = -5 + 7t$ **21.** $x = 7 + 5t, y = -14t, z = -t$ **23.** $4x - 5y - z = 0$

25. $4x - 2y - 3z = -13$ **27.** $x + 2z = -3$ **29.** $39x + 35y + 11z = 19$ **31.** $106x - 72y + 95z = 14$

33. $x + y - z = 10$ **35.** 124 **37.** 21 **39.** 15 **41.** 69 **43.** 164

45. Let $\mathbf{F} = \mathbf{i} - \mathbf{j}, \mathbf{G} = \mathbf{j} - \mathbf{k},$ and $\mathbf{H} = -\mathbf{i} - \mathbf{k}.$ Then $\mathbf{F} \times \mathbf{G} = \mathbf{i} + \mathbf{j} + \mathbf{k} = \mathbf{F} \times \mathbf{H},$ but $\mathbf{G} \neq \mathbf{H}.$

47. Area of $P = \|\mathbf{F}\|\|\mathbf{G}\|\sin(\theta) = \|\mathbf{F} \times \mathbf{G}\|.$ But

$$\|\mathbf{F} \times \mathbf{G}\|^2 = (\mathbf{F} \times \mathbf{G}) \cdot (\mathbf{F} \times \mathbf{G}) = \begin{vmatrix} \|\mathbf{F}\|^2 & \mathbf{F} \cdot \mathbf{G} \\ \mathbf{F} \cdot \mathbf{G} & \|\mathbf{G}\|^2 \end{vmatrix}.$$

49. This follows immediately from Lagrange's identity (Problem 44 of Section 11.3).

51. Without loss of generality, let $\mathbf{F} = \mathbf{i}, \mathbf{G} = \mathbf{j},$ and $\mathbf{H} = \mathbf{k}$ be vectors along the incident sides of a unit cube. Then $\mathbf{F} + \mathbf{G} + \mathbf{H} = \mathbf{i} + \mathbf{j} + \mathbf{k}$ is along a diagonal. The cosine of the angle between $\mathbf{i}$ and $\mathbf{i} + \mathbf{j} + \mathbf{k}$ is given by

$$\cos(\theta) = \frac{\mathbf{i} \cdot (\mathbf{i} + \mathbf{j} + \mathbf{k})}{\sqrt{3}} = \frac{1}{\sqrt{3}}; \text{ hence, } \theta \approx 54°44'.$$

53. Let $P_0 = (x_0, y_0, z_0),$ and let $P_1 = (x_1, y_1, z_1)$ be on the plane so that $\overline{P_0 P_1}$ is perpendicular to the plane. Let $N = a\mathbf{i} + b\mathbf{j} + c\mathbf{k}.$ Then any vector along $\overline{P_0 P_1}$ has cross product $\mathbf{O}$ with N; hence $c(y_1 - y_0) = b(z_1 - z_0),$ $c(x_1 - x_0) = a(z_1 - z_0),$ and $b(x_1 - x_0) = a(y_1 - y_0).$ Solve these equations to get

$$x_1 - x_0 = -\frac{a}{a^2 + b^2 + c^2}(ax_0 + by_0 + cz_0 + d),$$

$$y_1 - y_0 = -\frac{b}{a^2 + b^2 + c^2}(ax_0 + by_0 + cz_0 + d),$$

$$z_1 - z_0 = -\frac{c}{a^2 + b^2 + c^2}(ax_0 + by_0 + cz_0 + d).$$

These equations yield the formula to be derived.

55. From the diagram, we want to show that $(\overline{AC})^2 + (\overline{BD})^2 = 2[(\overline{EF})^2 + (\overline{GH})^2].$ But the quadrilateral $EGFH$ is a parallelogram, so $(\overline{EF})^2 + (\overline{GH})^2 = 2[(\overline{EG})^2 + (\overline{GF})^2],$ while $\overline{EG} = \dfrac{1}{2}\,\overline{BD}$ and $\overline{GF} = \dfrac{1}{2}\,\overline{AC}.$

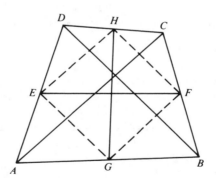

57. The general solution of $y'' - 4y = 0$ is $y = c_1 e^{2x} + c_2 e^{-2x},$ so every solution is a linear combination of e^{2x} and $e^{-2x}.$ Since e^{2x} and e^{-2x} are linearly independent, they form a basis for the solution space.

59. Let S consist of all vectors represented as arrows in the plane $x = 1.$ The sum of two vectors in S is in $S,$ but $0\mathbf{F}$ is not in S if $\mathbf{F}$ is in $S.$

61. Since S has dimension 3, there are linearly independent vectors $\mathbf{F}, \mathbf{G},$ and $\mathbf{H}$ in $S.$ By Problem 31 of Section 11.3, any vector in R^3 is a linear combination of $\mathbf{F}, \mathbf{G},$ and $\mathbf{H}$ and hence is in $S.$ Therefore, S must be all of $R^3.$

63. Suppose that $ae^x + be^{2x} + c \sin(x) + d \cos(x) = 0$ for all $x.$ Let $x = 0$ to get $a + b + d = 0,$ and let $x = \pi/2$ to get $ae^{\pi/2} + be^\pi + c = 0.$ Now let $x = \pi$ to get $ae^\pi + be^{2\pi} - d = 0,$ and let $x = 3\pi/2$ to get $ae^{3\pi/2} + be^{3\pi} - c = 0.$

The determinant of the coefficients of these equations is

$$\begin{vmatrix} 1 & 1 & 0 & 1 \\ e^{\pi/2} & e^{\pi} & 1 & 0 \\ e^{\pi} & e^{2\pi} & 0 & -1 \\ e^{3\pi/2} & e^{3\pi} & -1 & 0 \end{vmatrix} = e^{4\pi} + e^{2\pi} - e^{5\pi} - e^{3\pi} \neq 0.$$

Therefore, the equations for a, b, c, and d have only the trivial solution, $a = b = c = d = 0$, and the functions are linearly independent.

Chapter 12

Section 12.1

1. $\begin{bmatrix} 14 & 0 & 0 \\ 10 & -5 & -6 \\ -26 & -43 & -8 \end{bmatrix}$ **3.** $\begin{bmatrix} 44 & -8 & -30 & -14 & -94 \\ 54 & -32 & 32 & -28 & -110 \\ -162 & -38 & -134 & -10 & 0 \end{bmatrix}$ **5.** $[32 \ \ 58 \ \ 12]$ **7.** $\begin{bmatrix} 3 \\ 3 \\ -33 \end{bmatrix}$

9. $\begin{bmatrix} 4 & -3 \\ 5 & -1 \\ 8 & 2 \end{bmatrix}$ **11.** $AB = \begin{bmatrix} -10 & -34 & -16 & -30 & -14 \\ 10 & -2 & -11 & -8 & -45 \\ -5 & 1 & 15 & 61 & -63 \end{bmatrix}$; BA is not defined.

13. $AB = [115]$; $BA = \begin{bmatrix} 3 & -18 & -6 & -42 & 88 \\ -2 & 12 & 4 & 28 & -44 \\ -6 & 36 & 12 & 84 & -132 \\ 0 & 0 & 0 & 0 & 0 \\ -4 & 24 & 8 & 56 & -88 \end{bmatrix}$.   **15.** AB is not defined; $BA = \begin{bmatrix} -572 & 22 & 286 \\ -26 & 1 & 13 \\ -156 & 6 & 78 \\ -364 & 14 & 182 \\ -52 & 2 & 26 \end{bmatrix}$.

17. $AB = [-800 \ \ 28 \ \ -336]$; BA is not defined.   **19.** Neither BA nor AB is defined.

21. $AB = \begin{bmatrix} -22 & 30 & -10 & -4 \\ -42 & 45 & 30 & 6 \end{bmatrix}$; BA is not defined.   **23.** $AB = \begin{bmatrix} 9 & -6 & 21 \\ 0 & 0 & 0 \\ -3 & 2 & -7 \\ 12 & -8 & 28 \end{bmatrix}$; BA is not defined.

25. AB is not defined; $BA = [22 \ \ 34]$.   **27.** $AB = \begin{bmatrix} 2 & -6 \\ 0 & 0 \\ -6 & 18 \end{bmatrix}$; BA is not defined.

29. $AB = \begin{bmatrix} -1 & 5 & -10 & 9 & -22 & 14 \\ -2 & 2 & 28 & 18 & 28 & -4 \end{bmatrix}$; BA is not defined.

Section 12.2

1. $A^3 = \begin{bmatrix} 4 & 5 & 7 & 2 & 7 \\ 5 & 4 & 7 & 2 & 7 \\ 7 & 7 & 2 & 6 & 2 \\ 2 & 2 & 6 & 0 & 6 \\ 7 & 7 & 2 & 6 & 2 \end{bmatrix}$; $A^4 = \begin{bmatrix} 19 & 18 & 11 & 14 & 11 \\ 18 & 19 & 11 & 14 & 11 \\ 11 & 11 & 20 & 4 & 20 \\ 14 & 14 & 4 & 12 & 4 \\ 11 & 11 & 20 & 4 & 20 \end{bmatrix}$

The number of distinct v_1–v_4 walks of length 3 is $(A^3)_{14} = 2$; the number of v_2–v_3 walks of length 3 is $(A^3)_{23} = 7$; the number of v_1–v_4 walks of length 4 is $(A^4)_{14} = 14$; the number of v_2–v_4 walks of length 4 is $(A^4)_{24} = 14$.

3. $A^2 = \begin{bmatrix} 4 & 2 & 3 & 3 & 2 \\ 2 & 3 & 2 & 2 & 3 \\ 3 & 2 & 4 & 3 & 2 \\ 3 & 2 & 3 & 4 & 2 \\ 2 & 3 & 2 & 2 & 3 \end{bmatrix}$; $A^3 = \begin{bmatrix} 10 & 10 & 11 & 11 & 10 \\ 10 & 6 & 10 & 10 & 6 \\ 11 & 10 & 10 & 11 & 10 \\ 11 & 10 & 11 & 10 & 10 \\ 10 & 6 & 10 & 10 & 6 \end{bmatrix}$; $A^4 = \begin{bmatrix} 42 & 32 & 41 & 41 & 32 \\ 32 & 30 & 32 & 32 & 30 \\ 41 & 32 & 42 & 41 & 32 \\ 41 & 32 & 41 & 42 & 31 \\ 32 & 30 & 32 & 32 & 30 \end{bmatrix}$

There are ten v_2-v_3 walks of length 3; two v_4-v_5 walks of length 2; thirty-two v_1-v_2 walks of length 4; and thirty-one v_4-v_5 walks of length 4.

5. $(A^3)_{ii}$ is the number of v_i-v_i walks of length 3. Given a v_i-v_i walk of length 3, say, v_i, v_j, v_k, v_i, the vertices v_i, v_j, v_k form a triangle in the graph. However, this triangle is counted twice in the total number of v_i-v_i walks of length 3—namely as the walk v_i, v_j, v_k, v_i and also as the walk v_i, v_k, v_j, v_i (going around the triangle in the other direction). Therefore, $(A^3)_{ii}$ equals twice the number of triangles in G having v_i as a vertex.

Section 12.3

1. Let $D = [d_{ij}]$, a diagonal matrix with $d_{ii} \neq 0$. Let E be the $n \times n$ diagonal matrix having diagonal elements $E_{ii} = 1/d_{ii}$. It is easy to check that $DE = ED = I_n$.

3. Let $B = A + A^t$. Then $b_{ij} = a_{ij} + a_{ji}$; hence, $b_{ij} = b_{ji}$, and B is symmetric.

5. By choosing $B = I_n$, we have by assumption that $AI_n = I_n$. But $AI_n = A$; hence, $A = I_n$.

7. Choose $B = \begin{bmatrix} \dfrac{d}{\Delta} & -\dfrac{b}{\Delta} \\ -\dfrac{c}{\Delta} & \dfrac{a}{\Delta} \end{bmatrix}$, where $\Delta = ad - bc$.

9. For convenience, write $A = [a_{ij}]$ and $I_m = [\alpha_{ij}]$. To prove that $AI_m = A$, note first that A is $n \times m$ and I_m is $m \times m$; hence, AI_m is $n \times m$. Now write $(AI_m)_{ij} = \sum_{k=1}^{m} a_{ik}\alpha_{kj}$. Since $\alpha_{kj} = 0$ if $k \neq j$, this sum reduces to $a_{ij}\alpha_{jj}$. Since $\alpha_{jj} = 1$, we further have $(AI_m)_{ij} = a_{ij}$. Therefore, $AI_m = A$.
The proof that $I_n A = A$ is similar.

Section 12.4

1. $B = \begin{bmatrix} -2 & 1 & 4 & 2 \\ 0 & \sqrt{3} & 16\sqrt{3} & 3\sqrt{3} \\ 1 & -2 & 4 & 8 \end{bmatrix}$; $E = \begin{bmatrix} 1 & 0 & 0 \\ 0 & \sqrt{3} & 0 \\ 0 & 0 & 1 \end{bmatrix}$

3. $B = \begin{bmatrix} 13 & 1 & -4 & 77 & 23 \\ 3 & 0 & -1 & 14 & 2 \end{bmatrix}$; $E = \begin{bmatrix} 1 & 5 \\ 0 & 1 \end{bmatrix}$

5. $B = \begin{bmatrix} 2 & 2 & -6 \\ -2 & 4 & 3 \\ 1 & 0 & 3 \end{bmatrix}$; $E = \begin{bmatrix} 1 & 0 & 0 \\ 0 & 0 & 1 \\ 0 & 1 & 0 \end{bmatrix}$

7. $B = \begin{bmatrix} -45 & 0 & 20 \\ 3 & 5 & 2 \\ 14 & 4 & 4 \end{bmatrix}$; $E = \begin{bmatrix} 5 & 0 & 0 \\ 0 & 1 & 0 \\ 0 & 0 & 1 \end{bmatrix}$

9. $B = \begin{bmatrix} -6 & 8 & 0 \\ 3 & 5 & 12 \\ 22 & 13 & -9 \end{bmatrix}$; $E = \begin{bmatrix} 0 & 0 & 1 \\ 0 & 1 & 0 \\ 0 & 0 & 1 \end{bmatrix}$

11. $B = \begin{bmatrix} -10 + 10\sqrt{13} & 70 + 45\sqrt{13} & 30 + 25\sqrt{13} \\ 2 & 9 & 5 \\ 8 & 1 & -3 \end{bmatrix}$; $C = \begin{bmatrix} 5 & 0 & 5\sqrt{13} \\ 0 & 0 & 1 \\ 0 & 1 & 0 \end{bmatrix}$

13. $B = \begin{bmatrix} 30 & 120 \\ -3 + 2\sqrt{3} & 15 + 8\sqrt{3} \end{bmatrix}$; $C = \begin{bmatrix} 0 & 15 \\ 1 & \sqrt{3} \end{bmatrix}$

15. $B = \begin{bmatrix} -1 & 0 & 3 & 0 \\ -36 & 28 & -20 & 28 \\ -13 & 3 & 44 & 9 \end{bmatrix}$; $C = \begin{bmatrix} 1 & 0 & 0 \\ 0 & 0 & 4 \\ 14 & 1 & 0 \end{bmatrix}$

17. $B = \begin{bmatrix} -3 & 7 & 1 & 1 \\ 20 & -75 & 15 & -25 \\ 0 & 3 & 3 & -5 \end{bmatrix}$; $C = \begin{bmatrix} 1 & 0 & 0 \\ -10 & 0 & -5 \\ 0 & 1 & 0 \end{bmatrix}$

19. $B = \begin{bmatrix} -3 & 15 & 0 \\ 10 & -15 & 5 \\ 1 & -5 & 0 \end{bmatrix}$; $C = \begin{bmatrix} 0 & 1 & -3 \\ 5 & 0 & 0 \\ 0 & 0 & 1 \end{bmatrix}$

21. Suppose that A is $n \times m$ and that B is formed from A by interchanging rows r and s. Form E from I_n by interchanging rows r and s of I_n. We claim that $EA = B$. To prove this, write $E = [e_{ij}]$, $A = [a_{ij}]$ and $B = [b_{ij}]$.
Then $(EA)_{ij} = \sum_{k=1}^{n} e_{ik}a_{kj}$. If $i \neq r$ and $i \neq s$, then $e_{ik} = 1$ if $i = k$, and $i_{ik} = 0$ if $i \neq k$, so $(EA)_{ij} = e_{ii}a_{ij} = a_{ij} = b_{ij}$.
If $i = r$, then $(EA)_{rj} = \sum_{k=1}^{n} e_{rk}a_{kj}$. But $e_{rk} = 0$ if $k \neq s$, and $e_{rs} = 1$, so $(EA)_{rj} = a_{sj} = b_{rj}$. Similarly, if $i = s$, then $(EA)_{sj} = a_{rj} = b_{sj}$. Therefore, $EA = B$.

23. Suppose B is formed from A by adding α times row r to row s of A. Let $A = [a_{ij}]$ and $B = [b_{ij}]$. Then $b_{ij} = a_{ij}$ if $i \neq s$; and $b_{sj} = \alpha a_{rj} + a_{sj}$. Now, $(EA)_{ij} = \sum\limits_{k=1}^{n} e_{ik}a_{kj} = a_{ij} = b_{ij}$ if $i \neq s$, because then $e_{ik} = 1$ if $k = i$, and $e_{ik} = 0$ if $k \neq i$.

Finally, $(EA)_{sj} = \sum\limits_{k=1}^{n} e_{sk}a_{kj}$. But row s of E is formed as α times row r of I_n to row s of I_n. Therefore,

$$e_{sk} = \alpha(I_n)_{rk} + (I_n)_{sk} = \begin{cases} 0 & \text{if } k \neq s, \ k \neq r \\ \alpha & \text{if } k = r \\ 1 & \text{if } k = s. \end{cases}$$

Thus, $(EA)_{sj} = \alpha a_{rj} + a_{sj} = b_{sj}$; hence, $EA = B$.

Section 12.5

In these solutions, the matrix given in the problem is referred to as A.

1. A is not reduced, as condition (2) is violated. $A_R = \begin{bmatrix} 1 & 0 & 5 \\ 0 & 1 & 2 \\ 0 & 0 & 0 \end{bmatrix}$

3. A is not reduced; conditions (1), (2), and (3) are violated. $A_R = \begin{bmatrix} 1 & -4 & -1 & 0 \\ 0 & 0 & 0 & 1 \\ 0 & 0 & 0 & 0 \\ 0 & 0 & 0 & 0 \end{bmatrix}$

5. A is not reduced; conditions (1), (2), and (3) are violated. $A_R = \begin{bmatrix} 1 & 0 \\ 0 & 1 \\ 0 & 0 \\ 0 & 0 \end{bmatrix}$

7. A is not reduced; conditions (1) and (2) are violated. $A_R = I_3$

9. A is not reduced; conditions (1) and (2) are violated. $A_R = \begin{bmatrix} 1 & 0 & 0 & 0 \\ 0 & 1 & \frac{3}{2} & \frac{1}{2} \end{bmatrix}$

11. A is not reduced; conditions (1) and (2) are violated. $A_R = I_3$

13. A is not reduced; conditions (1) and (2) are violated. $A_R = \begin{bmatrix} 1 & 0 & 0 & 0 \\ 0 & 1 & 0 & 0 \\ 0 & 0 & 1 & -1 \end{bmatrix}$

15. A is not reduced; conditions (1), (2), (3), and (4) are violated.

$$A_R = \begin{bmatrix} 1 & -1 & 0 & -\frac{3}{2} & \frac{3}{2} \\ 0 & 0 & 1 & \frac{1}{2} & -\frac{1}{2} \end{bmatrix}$$

17. A is not reduced; conditions (1) and (2) are violated.

$$A_R = \begin{bmatrix} 1 & 0 & 0 & -\frac{6}{5} & \frac{2}{5} \\ 0 & 1 & 0 & \frac{21}{10} & \frac{14}{5} \\ 0 & 0 & 1 & -\frac{3}{10} & -\frac{7}{5} \end{bmatrix}$$

19. A is not reduced; conditions (1) and (2) are violated.

$$A_R = \begin{bmatrix} 1 & 0 & -\frac{44}{47} & -\frac{15}{47} & -\frac{54}{47} \\ 0 & 1 & -\frac{37}{47} & -\frac{3}{47} & -\frac{86}{47} \end{bmatrix}$$

21. A is not reduced; conditions (1) and (2) are violated.

$$A_R = \begin{bmatrix} 1 & 0 & 0 & \frac{9}{17} & \frac{2}{17} \\ 0 & 1 & \frac{1}{9} & \frac{74}{153} & \frac{92}{153} \end{bmatrix}$$

23. A is not reduced; conditions (1) and (2) are violated.

$$A_R = \begin{bmatrix} 1 & 0 & 0 \\ 0 & 1 & 0 \\ 0 & 0 & 1 \\ 0 & 0 & 0 \end{bmatrix}$$

25. A is not reduced; conditions (1) and (2) are violated. $A_R = I_4$

Section 12.6

1. $A_R = \begin{bmatrix} 1 & 0 & -\frac{3}{5} \\ 0 & 1 & \frac{3}{5} \end{bmatrix}$; rank$(A) = 2$; $\langle -4, 1, 3 \rangle$ and $\langle 2, 2, 0 \rangle$

3. $A_R = \begin{bmatrix} 1 & 0 \\ 0 & 1 \\ 0 & 0 \end{bmatrix}$; rank$(A) = 2$; $\langle -3, 1 \rangle$ and $\langle 2, 2 \rangle$

5. $A_R = \begin{bmatrix} 1 & 0 & \frac{15}{4} & \frac{1}{2} \\ 0 & 1 & \frac{11}{4} & \frac{1}{2} \end{bmatrix}$; rank$(A) = 2$; $\langle 8, -4, 3, 2 \rangle$ and $\langle 1, -1, 1, 0 \rangle$

7. $A_R = \begin{bmatrix} 1 & 0 & 0 \\ 0 & 1 & 0 \\ 0 & 0 & 1 \\ 0 & 0 & 0 \end{bmatrix}$; rank$(A) = 3$; $\langle 2, 2, 1 \rangle$, $\langle 1, -1, 3 \rangle$, $\langle 0, 0, 1 \rangle$

9. $A_R = I_3$; rank$(A) = 3$; $\langle 0, 4, 3 \rangle$, $\langle 6, 1, 0 \rangle$, $\langle 2, 2, 2 \rangle$ **11.** $A_R = I_3$; rank$(A) = 3$; $\langle -3, 2, 2 \rangle$, $\langle 1, 0, 5 \rangle$, $\langle 0, 0, 2 \rangle$

13. $A_R = \begin{bmatrix} 1 & 0 & -11 \\ 0 & 1 & -3 \\ 0 & 0 & 0 \end{bmatrix}$; rank$(A) = 2$; $\langle -2, 5, 7 \rangle$, $\langle 0, 1, -3 \rangle$

15. $A_R = \begin{bmatrix} 1 & 0 & 1 & \frac{1}{7} \\ 0 & 1 & 3 & \frac{3}{2} \\ 0 & 0 & 0 & 0 \\ 0 & 0 & 0 & 0 \end{bmatrix}$; rank$(A) = 2$; $\langle 7, -2, 1, -2 \rangle$, $\langle 0, 2, 6, 3 \rangle$

17. $A_R = \begin{bmatrix} 1 & 0 & 0 & -1 \\ 0 & 1 & 0 & 6 \\ 0 & 0 & 1 & -1 \end{bmatrix}$; rank$(A) = 3$; $\langle 4, 1, -3, 5 \rangle$, $\langle 2, 0, 0, -2 \rangle$, $\langle 13, 2, 0, -1 \rangle$

19. $A_R = \begin{bmatrix} 1 & 0 & -\frac{1}{2} & \frac{1}{2} & -\frac{3}{2} \\ 0 & 1 & -\frac{15}{4} & -\frac{7}{4} & -\frac{17}{4} \\ 0 & 0 & 0 & 0 & 0 \end{bmatrix}$; rank$(A) = 2$; $\langle 5, -2, 5, 6, 1 \rangle$, $\langle -2, 0, 1, -1, 3 \rangle$

21. Let the row vectors of A be $\mathbf{R}_1, \ldots, \mathbf{R}_n$. Suppose that the row space has dimension r. Then exactly r of these vectors are linearly independent. Suppose, for convenience, that $\mathbf{R}_1, \ldots, \mathbf{R}_r$ are linearly independent. Then

$$\mathbf{R}_{r+1} = \alpha_{r+1,1}\mathbf{R}_1 + \cdots + \alpha_{r+1,r}\mathbf{R}_r,$$
$$\mathbf{R}_{r+2} = \alpha_{r+2,1}\mathbf{R}_1 + \cdots + \alpha_{r+2,r}\mathbf{R}_r,$$
$$\vdots \qquad \vdots \qquad \qquad \vdots$$
$$\mathbf{R}_n = \alpha_{n,1}\mathbf{R}_1 + \cdots + \alpha_{n,r}\mathbf{R}_r.$$

Column j of A can be written

$$\begin{bmatrix} a_{1j} \\ a_{2j} \\ \vdots \\ a_{rj} \\ a_{r+1,j} \\ \vdots \\ a_{nj} \end{bmatrix} = a_{1j}\begin{bmatrix} 1 \\ 0 \\ \vdots \\ 0 \\ \alpha_{r+1,1} \\ \vdots \\ \alpha_{n1} \end{bmatrix} + a_{2j}\begin{bmatrix} 0 \\ 1 \\ \vdots \\ 0 \\ \alpha_{r+1,2} \\ \vdots \\ \alpha_{n2} \end{bmatrix} + \cdots + a_{rj}\begin{bmatrix} 0 \\ 0 \\ \vdots \\ 1 \\ \alpha_{r+1,r} \\ \vdots \\ \alpha_{nr} \end{bmatrix}.$$

Therefore, every column vector of A is a linear combination of r vectors in R^n; hence, the dimension of the column space cannot exceed r.

By reversing the roles of rows and columns in this argument, we find that the dimension of the row space cannot exceed the dimension of the column space. Therefore, the two spaces have equal dimensions.

23. The row vectors $\langle 1, 1, -4, 2 \rangle$ and $\langle 0, 1, 1, 3 \rangle$ are linearly independent and form a basis for the row space. The column space has basis $\begin{bmatrix} 1 \\ 0 \end{bmatrix}$ and $\begin{bmatrix} 1 \\ 1 \end{bmatrix}$.

25. The row vectors $\langle 8, 4 \rangle$ and $\langle 0, 3 \rangle$ are linearly independent and form a basis for the row space; the column space has basis $\begin{bmatrix} 8 \\ 2 \\ 0 \end{bmatrix}$ and $\begin{bmatrix} 4 \\ 1 \\ 3 \end{bmatrix}$.

27. A basis consists of all $n \times m$ matrices $E^{(r,s)}$, $1 \le r \le n$, $1 \le s \le m$, where the r, s element of $E^{(r,s)}$ is 1 and all other elements are zero. There are nm such matrices, so the dimension of this vector space is nm.

Section 12.7

1. $\alpha\begin{bmatrix} -1 \\ 1 \\ 1 \\ 0 \end{bmatrix} + \beta\begin{bmatrix} 1 \\ -1 \\ 0 \\ 1 \end{bmatrix}$; dimension 2   **3.** Only the trivial solution   **5.** $\alpha\begin{bmatrix} -\frac{9}{4} \\ -\frac{7}{4} \\ -\frac{5}{8} \\ \frac{13}{8} \\ 1 \end{bmatrix}$; dimension 1

7. $\alpha\begin{bmatrix} -\frac{5}{6} \\ -\frac{2}{3} \\ -\frac{8}{3} \\ -\frac{2}{3} \\ 1 \\ 0 \end{bmatrix} + \beta\begin{bmatrix} -\frac{5}{9} \\ -\frac{10}{9} \\ -\frac{13}{9} \\ -\frac{1}{9} \\ 0 \\ 1 \end{bmatrix}$; dimension 2   **9.** $\begin{bmatrix} x_1 \\ x_2 \\ x_4 \\ x_5 \end{bmatrix} = \alpha\begin{bmatrix} \frac{5}{14} \\ \frac{11}{7} \\ \frac{6}{7} \\ 1 \end{bmatrix}$; dimension 1

11. $\alpha\begin{bmatrix} 1 \\ 1 \\ 0 \\ 1 \\ 1 \\ 0 \\ 0 \end{bmatrix} + \beta\begin{bmatrix} -2 \\ \frac{3}{2} \\ \frac{2}{3} \\ -\frac{4}{3} \\ 0 \\ 1 \\ 0 \end{bmatrix} + \gamma\begin{bmatrix} 0 \\ \frac{1}{2} \\ -3 \\ 0 \\ 0 \\ 0 \\ 1 \end{bmatrix}$; dimension 3   **13.** $\alpha\begin{bmatrix} 8 \\ 5 \\ 2 \\ 1 \\ 0 \end{bmatrix} + \beta\begin{bmatrix} -1 \\ -6 \\ 0 \\ 0 \\ 1 \end{bmatrix}$; dimension 2

15. $\alpha\begin{bmatrix} \frac{2}{5} \\ 5 \\ 1 \\ 0 \\ 0 \end{bmatrix} + \beta\begin{bmatrix} \frac{1}{5} \\ 1 \\ 0 \\ 1 \\ 0 \end{bmatrix} + \gamma\begin{bmatrix} -\frac{3}{5} \\ -7 \\ 0 \\ 0 \\ 1 \end{bmatrix}$; dimension 3   **17.** $\alpha\begin{bmatrix} \frac{23}{9} \\ \frac{13}{9} \\ -8 \\ 1 \end{bmatrix}$; dimension 1

19. $\alpha\begin{bmatrix}\frac{21}{13}\\-1\\\frac{23}{13}\\1\\0\\0\end{bmatrix} + \beta\begin{bmatrix}\frac{6}{13}\\-6\\-\frac{12}{13}\\0\\1\\0\end{bmatrix} + \gamma\begin{bmatrix}-\frac{44}{13}\\1\\-\frac{55}{13}\\0\\0\\1\end{bmatrix}$; dimension 3

21. When the rank of the coefficient matrix is less than m, the system has a nontrivial solution.

Section 12.8

1. $\begin{bmatrix}1\\\frac{1}{2}\\4\end{bmatrix}$

3. $\alpha\begin{bmatrix}1\\1\\\frac{3}{2}\\1\\0\\0\end{bmatrix} + \beta\begin{bmatrix}0\\0\\\frac{1}{2}\\0\\1\\0\end{bmatrix} + \gamma\begin{bmatrix}-\frac{17}{2}\\-6\\-\frac{51}{4}\\0\\0\\1\end{bmatrix} + \begin{bmatrix}-\frac{9}{2}\\-3\\-\frac{29}{4}\\0\\0\\0\end{bmatrix}$

5. $\alpha\begin{bmatrix}2\\2\\7\\\frac{3}{2}\\1\\0\end{bmatrix} + \beta\begin{bmatrix}-2\\-1\\-\frac{9}{2}\\-\frac{3}{4}\\0\\1\end{bmatrix} + \begin{bmatrix}-4\\-4\\-38\\-\frac{11}{2}\\0\\0\end{bmatrix}$

7. $\alpha\begin{bmatrix}-\frac{1}{2}\\-1\\3\\1\\0\end{bmatrix} + \beta\begin{bmatrix}-\frac{3}{4}\\1\\-2\\0\\1\end{bmatrix} + \begin{bmatrix}\frac{9}{8}\\2\\0\\0\\0\end{bmatrix}$

9. $\alpha\begin{bmatrix}-1\\0\\1\\0\\0\\0\\0\end{bmatrix} + \beta\begin{bmatrix}1\\0\\0\\1\\0\\0\\0\end{bmatrix} + \gamma\begin{bmatrix}-\frac{3}{14}\\\frac{3}{14}\\0\\0\\1\\0\\0\end{bmatrix} + \delta\begin{bmatrix}-1\\0\\0\\0\\1\\0\\0\end{bmatrix} + \epsilon\begin{bmatrix}\frac{1}{14}\\-\frac{1}{14}\\0\\0\\0\\0\\1\end{bmatrix} + \begin{bmatrix}-\frac{29}{7}\\\frac{1}{7}\\0\\0\\0\\0\\0\end{bmatrix}$

11. $\alpha\begin{bmatrix}-\frac{19}{15}\\3\\\frac{67}{15}\\1\end{bmatrix} + \begin{bmatrix}\frac{22}{15}\\-5\\-\frac{121}{15}\\0\end{bmatrix}$

13. $\begin{bmatrix}-\frac{176}{63}\\\frac{61}{63}\\-\frac{23}{63}\end{bmatrix}$

15. No solution

17. $\alpha\begin{bmatrix}-1\\-1\\2\\1\end{bmatrix} + \begin{bmatrix}-\frac{21}{14}\\-4\\-\frac{63}{14}\\0\end{bmatrix}$

19. $\alpha\begin{bmatrix}\frac{2}{29}\\\frac{7}{29}\\\frac{18}{29}\\1\end{bmatrix} + \begin{bmatrix}\frac{67}{29}\\-\frac{7}{87}\\\frac{110}{29}\\0\end{bmatrix}$

21. The nonhomogeneous system $AX = B$ has a solution if and only if $\text{rank}(A) = \text{rank}([A \;\vdots\; B])$. Since A is reduced, this occurs only if the last $n - k$ rows of both A and $[A \;\vdots\; B]$ have only zero entries; hence, $b_{k+1} = \cdots = b_n = 0$.

23. $\begin{bmatrix}0\\\frac{1}{2}\\-\frac{3}{4}\\5\end{bmatrix}$

25. $\begin{bmatrix}-\frac{2}{19}\\-\frac{27}{19}\\\frac{9}{19}\end{bmatrix}$

27. $\begin{bmatrix}\frac{3}{2}\\\frac{3}{2}\\1\end{bmatrix}$

Section 12.9

1. $\begin{bmatrix}-\frac{1}{5} & \frac{2}{5}\\\frac{2}{5} & \frac{1}{5}\end{bmatrix}$

3. $\begin{bmatrix}-\frac{6}{31} & \frac{11}{31} & \frac{2}{31}\\\frac{3}{31} & \frac{10}{31} & -\frac{1}{31}\\\frac{1}{31} & -\frac{7}{31} & \frac{10}{31}\end{bmatrix}$

5. $\begin{bmatrix}-\frac{1}{2} & \frac{1}{2} & 0\\\frac{1}{4} & \frac{3}{4} & -\frac{1}{6}\\-\frac{1}{4} & \frac{1}{4} & \frac{1}{6}\end{bmatrix}$

7. A is singular.

9. $\begin{bmatrix}-1 & \frac{85}{2} & -\frac{55}{2} & 1\\0 & -\frac{9}{2} & \frac{7}{2} & 1\\0 & 3 & -2 & 0\\0 & -\frac{1}{2} & \frac{1}{2} & 0\end{bmatrix}$

11. $\begin{bmatrix}1 & -14\\0 & 1\end{bmatrix}$

13. A is singular.

15. $\begin{bmatrix}0 & \frac{1}{2} & 0\\\frac{5}{29} & \frac{4}{29} & -\frac{3}{29}\\\frac{3}{29} & -\frac{1}{58} & \frac{4}{29}\end{bmatrix}$

17. $\begin{bmatrix} 3 & -\frac{7}{5} & -\frac{1}{5} \\ 2 & -1 & 0 \\ 0 & \frac{2}{5} & \frac{1}{5} \end{bmatrix}$ **19.** $\begin{bmatrix} -\frac{7}{47} & \frac{6}{47} & \frac{24}{47} \\ -\frac{2}{47} & -\frac{5}{47} & \frac{27}{47} \\ \frac{4}{47} & \frac{10}{47} & -\frac{7}{47} \end{bmatrix}$ **21.** A is singular. **23.** $\begin{bmatrix} \frac{7}{2} & \frac{1}{2} & -\frac{3}{2} \\ 1 & 0 & 0 \\ -2 & 0 & 1 \end{bmatrix}$

25. $\begin{bmatrix} -\frac{2}{113} & -\frac{3}{226} & \frac{12}{113} \\ \frac{17}{113} & -\frac{31}{226} & \frac{11}{113} \\ \frac{18}{113} & \frac{27}{226} & \frac{5}{113} \end{bmatrix}$ **27.** $\begin{bmatrix} -\frac{7}{11} \\ -\frac{19}{11} \\ \frac{3}{11} \\ \frac{10}{11} \end{bmatrix}$ **29.** $\begin{bmatrix} 2 \\ -1 \\ -2 \end{bmatrix}$ **31.** $\begin{bmatrix} -\frac{21}{5} \\ \frac{14}{5} \\ 0 \end{bmatrix}$

33. By Theorem 12.22, A is nonsingular if and only if $\operatorname{rank}(A) = n$. By Theorem 12.12, the rank of A equals the dimension of the row space of A. Therefore, A is nonsingular if and only if the row vectors of A form a basis for R^n.

35. By conclusion (2) of Theorem 12.21, if A is nonsingular, so is AA, or A^2. By a simple induction argument, A^k is nonsingular for any positive integer k.

Section 12.10

1. -34 **3.** -58 **5.** -12 **7.** -35 **9.** 0 **11.** -25 **13.** -5 **15.** -11 **17.** -773

19. $52,883$ **21.** By repeated application of Theorem 12.25, $|B| = \alpha^n |A|$.

23. By Problem 21, $|A| = (-1)^n |A^t|$. Since $|A^t| = |A|$, $|A| = (-1)^n |A|$. If n is odd, then $|A| = -|A|$; hence, $|A| = 0$.

25. Since $A^{-1} = A^t$, $|A^{-1}| = |A^t| = |A|$. Then $|AA^t| = |I_n| = 1 = |A|^2$; hence, $|A| = \pm 1$.

27. -82 **29.** 23 **31.** 96 **33.** -572 **35.** 6882 **37.** 3372 **39.** 1033 **41.** $74,706$ **43.** 385

45. $4,730,516$ **47.** $27,246$ **49.** -4132 **51.** -314

53.
$$\begin{vmatrix} \alpha & \beta & \gamma & \delta \\ \beta & \gamma & \delta & \alpha \\ \gamma & \delta & \alpha & \beta \\ \delta & \alpha & \beta & \gamma \end{vmatrix} = \begin{vmatrix} \alpha+\beta+\gamma+\delta & \alpha+\beta+\gamma+\delta & \alpha+\beta+\gamma+\delta & \alpha+\beta+\gamma+\delta \\ \beta & \gamma & \delta & \alpha \\ \gamma & \delta & \alpha & \beta \\ \delta & \alpha & \beta & \gamma \end{vmatrix}$$

$$= (\alpha+\beta+\gamma+\delta) \begin{vmatrix} -1 & 1 & -1 & 1 \\ -\beta & \gamma & -\delta & \alpha \\ -\gamma & \delta & -\alpha & \beta \\ -\delta & \alpha & -\beta & \gamma \end{vmatrix}$$

$$= (\alpha+\beta+\gamma+\delta) \begin{vmatrix} 0 & 1 & -1 & 1 \\ \alpha-\beta+\gamma-\delta & \gamma & -\delta & \alpha \\ -(\alpha-\beta+\gamma-\delta) & \delta & -\alpha & \beta \\ \alpha-\beta+\gamma-\delta & \alpha & -\beta & \gamma \end{vmatrix}$$

$$= (\alpha+\beta+\gamma+\delta)(\beta-\alpha+\delta-\gamma) \begin{vmatrix} 0 & 1 & 1 & 1 \\ 1 & \gamma & \delta & \alpha \\ -1 & \delta & \alpha & \beta \\ 1 & \alpha & \beta & \gamma \end{vmatrix}$$

$$= (\alpha+\beta+\gamma+\delta)(\beta-\alpha+\delta-\gamma) \begin{vmatrix} 0 & 1 & -1 & 1 \\ 1 & \gamma & \delta & \alpha \\ 1 & \delta & \alpha & \beta \\ 1 & \alpha & \beta & \gamma \end{vmatrix}$$

55. It is easy to transform any square matrix B, by a sequence of elementary row and column operations, to a new matrix $\bar{B}$ that is upper triangular (and hence has determinant equal to the product of its main diagonal elements) and also $|B| = |\bar{B}|$. This process can be applied to each block D_j of A to produce a matrix $\bar{A}$ such that $|A| = |\bar{A}|$ and $|D_j| = |\bar{D}_j| = $ product of main diagonal elements of $\bar{D}_j$. Then $\bar{A}$ is upper triangular, and $|\bar{A}| = |\bar{D}_1| \cdots |\bar{D}_r| = |D_1| \cdots |D_r| = |A|$.

Section 12.11

1. 8 **3.** 40 **5.** 55 **7.** 64

9. The tree matrix is the $n \times n$ matrix

$$\begin{bmatrix} n-1 & -1 & -1 & \cdots & -1 \\ -1 & n-1 & -1 & \cdots & -1 \\ \vdots & \vdots & \vdots & & \vdots \\ -1 & -1 & -1 & \cdots & n-1 \end{bmatrix}.$$

Then

$$(-1)^{1+1}M_{11} = \begin{bmatrix} n-1 & -1 & -1 & \cdots & -1 \\ -1 & n-1 & -1 & \cdots & -1 \\ \vdots & \vdots & \vdots & & \vdots \\ -1 & -1 & -1 & \cdots & n-1 \end{bmatrix},$$

an $(n-1) \times (n-1)$ matrix. To expand the determinant of M_{11}, add the last $n-1$ rows to row 1 to get

$$|M_{11}| = \begin{vmatrix} 1 & 1 & 1 & \cdots & 1 \\ -1 & n-1 & -1 & \cdots & -1 \\ \vdots & \vdots & \vdots & & \vdots \\ -1 & -1 & -1 & \cdots & n-1 \end{vmatrix}.$$

Now subtract column 1 from each of the other columns to get

$$|M_{11}| = \begin{vmatrix} 1 & 0 & 0 & \cdots & 0 \\ -1 & n & 0 & \cdots & 0 \\ \vdots & \vdots & \vdots & & \vdots \\ -1 & 0 & 0 & \cdots & n \end{vmatrix} = n^{n-2}.$$

11. 29

Section 12.12

1. $\begin{bmatrix} \frac{6}{13} & \frac{1}{13} \\ -\frac{1}{13} & \frac{2}{13} \end{bmatrix}$ **3.** $\begin{bmatrix} -\frac{4}{5} & \frac{1}{5} \\ -\frac{1}{5} & \frac{1}{5} \end{bmatrix}$ **5.** $\begin{bmatrix} -\frac{6}{11} & \frac{5}{11} \\ \frac{1}{11} & \frac{1}{11} \end{bmatrix}$ **7.** $\begin{bmatrix} \frac{5}{32} & \frac{3}{32} & \frac{1}{3} \\ -\frac{8}{32} & -\frac{3}{4} & \frac{3}{4} \\ -\frac{1}{16} & -\frac{7}{16} & \frac{3}{16} \end{bmatrix}$

9. $\begin{bmatrix} -\frac{13}{83} & \frac{31}{83} & \frac{21}{83} \\ -\frac{20}{83} & \frac{3}{83} & -\frac{6}{83} \\ \frac{1}{83} & \frac{4}{83} & -\frac{8}{83} \end{bmatrix}$ **11.** $\begin{bmatrix} \frac{9}{43} & \frac{17}{43} & -\frac{7}{43} \\ -\frac{14}{43} & \frac{7}{43} & -\frac{13}{43} \\ -\frac{2}{43} & \frac{1}{43} & -\frac{8}{43} \end{bmatrix}$ **13.** $\dfrac{1}{733}\begin{bmatrix} 99 & -47 & -6 \\ 7 & -144 & 44 \\ -13 & 58 & 23 \end{bmatrix}$

15. $\dfrac{1}{784}\begin{bmatrix} -52 & 131 & -62 & 54 \\ 208 & -132 & 248 & -216 \\ -496 & 360 & -320 & 304 \\ -212 & 127 & -102 & 190 \end{bmatrix}$

Section 12.13

1. $x_1 = -\dfrac{1}{2}, x_2 = -\dfrac{19}{22}, x_3 = \dfrac{2}{11}$ **3.** $x_1 = -1, x_2 = 1$ **5.** $x_1 = \dfrac{5}{6}, x_2 = -\dfrac{10}{3}, x_3 = -\dfrac{5}{6}$

7. $x_1 = \dfrac{253}{128}, x_2 = -\dfrac{105}{128}, x_3 = -\dfrac{404}{128}, x_4 = -\dfrac{697}{128}$ **9.** $x_1 = -\dfrac{172}{2}, x_2 = -\dfrac{109}{2}, x_3 = -\dfrac{43}{2}, x_4 = \dfrac{37}{2}$

11. $x_1 = \dfrac{33}{93}, x_2 = -\dfrac{409}{93}, x_3 = -\dfrac{1}{93}, x_4 = \dfrac{116}{93}$ **13.** $x_1 = \dfrac{5}{16}, x_2 = -\dfrac{25}{8}, x_3 = -\dfrac{13}{8}, x_4 = -\dfrac{29}{16}$

15. Since each $|A(k; B)| = 0$, we obtain the solution $X = O$.

Additional Problems for Chapter 12

1. $\begin{bmatrix} 1 & 0 & 0 & \frac{271}{104} \\ 0 & 1 & 0 & \frac{11}{26} \\ 0 & 0 & 1 & \frac{121}{104} \end{bmatrix}; 3$
 3. $\begin{bmatrix} 1 & 0 \\ 0 & 1 \\ 0 & 0 \end{bmatrix}; 2$
 5. $\begin{bmatrix} 1 & -\frac{3}{8} & \frac{1}{4} & \frac{1}{8} & -\frac{5}{8} \\ 0 & 0 & 0 & 0 & 0 \end{bmatrix}; 1$

7. $\begin{bmatrix} 1 & 0 & -\frac{1}{3} \\ 0 & 1 & -1 \\ 0 & 0 & 0 \end{bmatrix}; 2$
 9. $\begin{bmatrix} 1 & 0 & -\frac{1}{3} & 0 \\ 0 & 1 & \frac{1}{2} & 0 \\ 0 & 0 & 0 & 1 \end{bmatrix}; 3$
 11. $\begin{bmatrix} \frac{1}{2} & \frac{5}{4} \\ \frac{1}{2} & \frac{3}{4} \end{bmatrix}$
 13. Singular

15. $\begin{bmatrix} -\frac{1}{3} & -\frac{1}{6} & -\frac{1}{6} \\ 0 & 1 & 0 \\ \frac{1}{3} & \frac{5}{3} & \frac{2}{3} \end{bmatrix}$
 17. $\begin{bmatrix} 0 & \frac{1}{2} & 1 \\ -\frac{2}{5} & \frac{6}{5} & \frac{11}{5} \\ -\frac{1}{5} & \frac{3}{5} & \frac{3}{5} \end{bmatrix}$
 19. $-\frac{1}{28}\begin{bmatrix} 10 & -10 & 10 & -4 \\ 4 & -4 & 4 & 4 \\ 15 & -43 & 29 & -6 \\ 7 & -7 & 21 & -14 \end{bmatrix}$

21. 0   **23.** 22   **25.** 70   **27.** 664   **29.** 48   **31.** $\alpha\begin{bmatrix} -\frac{15}{7} \\ -\frac{2}{7} \\ 1 \end{bmatrix}$ **33.** $\begin{bmatrix} 3 \\ \frac{13}{3} \\ \frac{34}{3} \\ 1 \end{bmatrix}$

35. $\alpha\begin{bmatrix} \frac{3}{4} \\ -\frac{1}{6} \\ \frac{5}{36} \\ 1 \end{bmatrix} + \begin{bmatrix} -\frac{1}{8} \\ \frac{7}{4} \\ \frac{7}{24} \\ 0 \end{bmatrix}$
 37. $\begin{bmatrix} -\frac{44}{17} \\ -\frac{7}{17} \\ \frac{16}{17} \end{bmatrix}$
 39. $\alpha\begin{bmatrix} \frac{29}{13} \\ -\frac{1}{13} \\ \frac{51}{13} \\ 1 \end{bmatrix} + \begin{bmatrix} -\frac{19}{13} \\ \frac{17}{26} \\ \frac{17}{26} \\ 0 \end{bmatrix}$

41. Since A is similar to B, $P^{-1}AP = B$ for P. Then $B = (P^{-1})^{-1}AP^{-1}$; hence, B is similar to A.

43. If $B = P^{-1}AP$ and $C = Q^{-1}BQ$, then $C = Q^{-1}(P^{-1}AP)Q = (PQ)^{-1}A(PQ)$; hence, A is similar to C.

45. If $B = P^{-1}AP$, then $B^{-1} = PA^{-1}P^{-1} = (P^{-1})^{-1}A^{-1}P^{-1}$, so A^{-1} is similar to B^{-1}.

47. If $B = P^{-1}AP$, then $\alpha B = P^{-1}(\alpha A)P$, so αA is similar to αB.

49. If $B = P^{-1}AP$, then $|B| = |P^{-1}AP| = |P^{-1}||A||P|$. But $|P^{-1}P| = |P^{-1}||P| = |I_n| = 1$, so $|B| = |A|$.

Chapter 13

Section 13.1

An eigenvector is given corresponding to each eigenvalue; any nonzero scalar multiple of an eigenvector is also an eigenvector. When there are more than one linearly independent eigenvectors corresponding to one eigenvalue, these are given.

1. $\lambda^2 - 2\lambda - 5; 1 + \sqrt{6}, \begin{bmatrix} \frac{\sqrt{6}}{2} \\ 1 \end{bmatrix}; 1 - \sqrt{6}, \begin{bmatrix} -\frac{\sqrt{6}}{2} \\ 1 \end{bmatrix}$
 3. $(\lambda + 5)(\lambda - 2); -5, \begin{bmatrix} 7 \\ 1 \end{bmatrix}; 2, \begin{bmatrix} 0 \\ 1 \end{bmatrix}$

5. $\lambda^2 - 3\lambda + 14; \dfrac{3 + \sqrt{47}i}{2}, \begin{bmatrix} \dfrac{-1 + \sqrt{47}i}{4} \\ 1 \end{bmatrix}; \dfrac{3 - \sqrt{47}i}{2}, \begin{bmatrix} \dfrac{-1 - \sqrt{47}i}{4} \\ 1 \end{bmatrix}$

7. $\lambda^2 + 9\lambda + 16; \dfrac{-9 + \sqrt{17}}{2}, \begin{bmatrix} \dfrac{-1 + \sqrt{17}}{4} \\ 1 \end{bmatrix}, \dfrac{-9 - \sqrt{17}}{2}, \begin{bmatrix} \dfrac{-1 - \sqrt{17}}{4} \\ 1 \end{bmatrix}$

9. $\lambda^2 + 15\lambda + 34; \dfrac{-15 + \sqrt{89}}{2}, \begin{bmatrix} \dfrac{5 + \sqrt{89}}{8} \\ 1 \end{bmatrix}; \dfrac{-15 - \sqrt{89}}{2}, \begin{bmatrix} \dfrac{5 - \sqrt{89}}{8} \\ 1 \end{bmatrix}$

11. $\lambda(\lambda - 2)(\lambda - 3)$; $0, \begin{bmatrix} 0 \\ 1 \\ 0 \end{bmatrix}$; $2, \begin{bmatrix} 2 \\ 1 \\ 0 \end{bmatrix}$; $3, \begin{bmatrix} 0 \\ 2 \\ 3 \end{bmatrix}$
  13. $\lambda^2(\lambda + 3)$; $0, \begin{bmatrix} 1 \\ 0 \\ 3 \end{bmatrix}$; $-3, \begin{bmatrix} 1 \\ 0 \\ 0 \end{bmatrix}$
  15. $(\lambda - 2)^2(\lambda + 14)$; $2, \begin{bmatrix} 0 \\ 0 \\ 1 \end{bmatrix}$; $-14, \begin{bmatrix} -16 \\ 0 \\ 1 \end{bmatrix}$

17. $\lambda(\lambda - 1)(\lambda - 7)$; $0, \begin{bmatrix} 14 \\ 7 \\ 10 \end{bmatrix}$; $1, \begin{bmatrix} 6 \\ 0 \\ 5 \end{bmatrix}$; $7, \begin{bmatrix} 0 \\ 0 \\ 1 \end{bmatrix}$
  19. $\lambda^2(\lambda - 1)(\lambda + 6)$; $0, \begin{bmatrix} 0 \\ 1 \\ 0 \\ 0 \end{bmatrix}$ and $\begin{bmatrix} 0 \\ 0 \\ 1 \\ 0 \end{bmatrix}$; $1, \begin{bmatrix} 1 \\ 1 \\ 0 \\ 7 \end{bmatrix}$; $-6, \begin{bmatrix} -6 \\ 1 \\ 0 \\ 0 \end{bmatrix}$

21. $(\lambda - 1)(\lambda - 2)(\lambda^2 + \lambda - 13)$; $1, \begin{bmatrix} -2 \\ -11 \\ 0 \\ 1 \end{bmatrix}$; $2, \begin{bmatrix} 0 \\ 0 \\ 1 \\ 0 \end{bmatrix}$; $\dfrac{-1 + \sqrt{53}}{2}, \begin{bmatrix} \dfrac{\sqrt{53} - 7}{2} \\ 0 \\ 0 \\ 1 \end{bmatrix}$; $\dfrac{-1 - \sqrt{53}}{2}, \begin{bmatrix} \dfrac{-\sqrt{53} - 7}{2} \\ 0 \\ 0 \\ 1 \end{bmatrix}$

23. $\lambda(\lambda - 1)(\lambda - 4)(\lambda^2 - 10\lambda + 23)$; $0, \begin{bmatrix} -5 \\ 0 \\ 23 \\ 0 \\ 1 \end{bmatrix}$; $1, \begin{bmatrix} 0 \\ 1 \\ 0 \\ 0 \\ 0 \end{bmatrix}$; $4, \begin{bmatrix} -1 \\ 0 \\ 1 \\ -2 \\ 1 \end{bmatrix}$; $5 + \sqrt{2}$; $\begin{bmatrix} \sqrt{2} \\ 0 \\ 0 \\ 0 \\ 1 \end{bmatrix}$; $5 - \sqrt{2}, \begin{bmatrix} -\sqrt{2} \\ 0 \\ 0 \\ 0 \\ 1 \end{bmatrix}$

25. $\lambda^3(\lambda - 1)$; $0, \begin{bmatrix} 0 \\ 1 \\ 0 \\ 0 \end{bmatrix}$; $1, \begin{bmatrix} 1 \\ 1 \\ 0 \\ 1 \end{bmatrix}$

27. $\lambda(\lambda - 3)(\lambda^2 - \lambda - 4)$; $0, \begin{bmatrix} -3 \\ 0 \\ 0 \\ 1 \end{bmatrix}$; $3, \begin{bmatrix} 0 \\ 0 \\ 0 \\ 1 \end{bmatrix}$; $\dfrac{1 + \sqrt{17}}{2}, \begin{bmatrix} 0 \\ \dfrac{\sqrt{17} - 1}{2} \\ 1 \\ \dfrac{3\sqrt{17} - 3}{\sqrt{17} - 5} \end{bmatrix}$; $\dfrac{1 - \sqrt{17}}{2}, \begin{bmatrix} 0 \\ \dfrac{-\sqrt{17} - 1}{2} \\ 1 \\ \dfrac{3\sqrt{17} + 3}{\sqrt{17} + 5} \end{bmatrix}$

29. $(\lambda - 1)^2(\lambda + 3)(\lambda + 6)$; $-6, \begin{bmatrix} -3 \\ \frac{10}{7} \\ -\frac{24}{49} \\ 1 \end{bmatrix}$; $-3, \begin{bmatrix} 0 \\ -\frac{1}{2} \\ -\frac{3}{8} \\ 1 \end{bmatrix}$; $1, \begin{bmatrix} 0 \\ 0 \\ 1 \\ 0 \end{bmatrix}$

31. $|\lambda I_2 - A| = \lambda^2 - (\alpha + \gamma)\lambda + \alpha\gamma - \beta^2$. Since $(\alpha + \gamma)^2 - 4(\alpha\gamma - \beta^2) = (\alpha - \gamma)^2 + \beta^2 \geq 0$, the characteristic equation has real roots.

33. Suppose that $AX = \lambda X$, with $X \neq 0$. Then $A^2 X = A(AX) = A(\lambda X) = \lambda(AX) = \lambda(\lambda X) = \lambda^2 X$. Therefore, λ^2 is an eigenvalue of A^2, with eigenvector X. By induction, λ^k is an eigenvalue of A^k, with eigenvector X, for any positive integer k.

35. If X_1 and X_2 are linearly dependent, there is a nonzero number α such that $X_1 = \alpha X_2$. But $AX_1 = \lambda_1 X_1$ and $AX_1 = A(\alpha X_2) = \alpha AX_2 = \alpha\lambda_2 X_2 = \lambda_2(\alpha X_2) = \lambda_2 X_1$. Therefore, $\lambda_1 X_1 = \lambda_2 X_1$, so $(\lambda_1 - \lambda_2)X_1 = 0$. Since $X_1 \neq 0$, $\lambda_1 = \lambda_2$, a contradiction. Therefore, X_1 and X_2 are linearly independent.

37. $A = \begin{bmatrix} 4 & 0 \\ c & -2 \end{bmatrix}$ or $A = \begin{bmatrix} -2 & 0 \\ c & 4 \end{bmatrix}$, with c any real number.

Section 13.2

In solutions 1 through 11, either a diagonalizing matrix is given or it is stated that the matrix is not diagonalizable.

1. $\begin{bmatrix} \dfrac{3 + \sqrt{7}i}{8} & \dfrac{3 - \sqrt{7}i}{8} \\ 1 & 1 \end{bmatrix}$
  3. Not diagonalizable
  5. $\begin{bmatrix} 0 & 5 & 0 \\ 1 & 1 & -\frac{3}{2} \\ 0 & 0 & 1 \end{bmatrix}$
  7. Not diagonalizable

9.
$$\begin{bmatrix} 1 & 0 & 0 & 0 \\ 0 & 1 & \dfrac{2-3\sqrt5}{41} & \dfrac{2+3\sqrt5}{41} \\ 0 & 0 & \dfrac{-1+\sqrt5}{2} & \dfrac{-1-\sqrt5}{2} \\ 0 & 0 & 1 & 1 \end{bmatrix}$$

11. Not diagonalizable

13. The characteristic polynomial of A is $\det(\lambda I - A) = 0$. The constant term in this polynomial is obtained by letting $\lambda = 0$, so this constant term is $\det(-A)$, which is $(-1)^n\det(A)$.

15. Since $P^{-1}AP = \begin{bmatrix} \lambda_1 & & & O \\ & \lambda_2 & & \\ & & \ddots & \\ O & & & \lambda_n \end{bmatrix}$, $A = P\begin{bmatrix} \lambda_1 & & & O \\ & \lambda_2 & & \\ & & \ddots & \\ O & & & \lambda_n \end{bmatrix}P^{-1}$, so

$$A^k = P\begin{bmatrix} \lambda_1 & & & O \\ & \lambda_2 & & \\ & & \ddots & \\ O & & & \lambda_n \end{bmatrix}P^{-1}P\begin{bmatrix} \lambda_1 & & & O \\ & \lambda_2 & & \\ & & \ddots & \\ O & & & \lambda_n \end{bmatrix}P^{-1}\cdots P\begin{bmatrix} \lambda_1 & & & O \\ & \lambda_2 & & \\ & & \ddots & \\ O & & & \lambda_n \end{bmatrix}P^{-1}$$

$$= P\begin{bmatrix} \lambda_1 & & & O \\ & \lambda_2 & & \\ & & \ddots & \\ O & & & \lambda_n \end{bmatrix}^k P^{-1} = P\begin{bmatrix} \lambda_1^k & & & O \\ & \lambda_2^k & & \\ & & \ddots & \\ O & & & \lambda_n^k \end{bmatrix}P^{-1}.$$

17. $\begin{bmatrix} 6(2^{16}) - 3^{16} & 3(2^{16}) - 3^{17} \\ -2^{17} + 2(3^{16}) & -2^{16} + 6(3^{16}) \end{bmatrix}$

19. $A^{31} = [k_{ij}]$, a 2×2 matrix, where

$$k_{11} = \frac{1}{2\sqrt{10}}[(1 + \sqrt{10})(-3 + \sqrt{10})^{31} + (\sqrt{10} - 1)(-3 - \sqrt{10})^{31}];$$

$$k_{12} = \frac{3}{2\sqrt{10}}[(-3 + \sqrt{10})^{31} + (3 + \sqrt{10})^{31}];$$

$$k_{22} = \frac{1}{2\sqrt{10}}[(-1 + \sqrt{10})(-3 + \sqrt{10})^{31} + (\sqrt{10} + 1)(3 + \sqrt{10})^{31}];$$

$$k_{21} = \frac{3}{2\sqrt{10}}[(-3 + \sqrt{10})^{31} + (3 + \sqrt{10})^{31}].$$

21. Suppose A is diagonalizable. For some P, $P^{-1}AP = D$, a diagonal matrix. Since $A = L^{-1}BL$, $P^{-1}L^{-1}BLP = (LP)^{-1}B(LP) = D$; hence, B is also diagonalizable.

Using this reasoning, it is easy to check that A and B are both diagonalizable or both nondiagonalizable. If A is diagonalizable and $B = L^{-1}AL$, then A and B have the same characteristic polynomials and hence the same eigenvalues.

Section 13.3

1. $0, \begin{bmatrix} 1 \\ 2 \end{bmatrix}$; $5, \begin{bmatrix} -2 \\ 1 \end{bmatrix}$ **3.** $5 + \sqrt{2}, \begin{bmatrix} 1 + \sqrt{2} \\ 1 \end{bmatrix}$; $5 - \sqrt{2}, \begin{bmatrix} 1 - \sqrt{2} \\ 1 \end{bmatrix}$

5. $0, \begin{bmatrix} 0 \\ 1 \\ 0 \end{bmatrix}$; $\dfrac{5 + \sqrt{41}}{2}, \begin{bmatrix} \dfrac{5 + \sqrt{41}}{2} \\ 0 \\ 1 \end{bmatrix}$; $\dfrac{5 - \sqrt{41}}{2}, \begin{bmatrix} \dfrac{5 - \sqrt{41}}{2} \\ 0 \\ 1 \end{bmatrix}$ **7.** $3, \begin{bmatrix} 0 \\ 0 \\ 1 \end{bmatrix}$; $-1 + \sqrt{2}, \begin{bmatrix} 1 + \sqrt{2} \\ 1 \\ 0 \end{bmatrix}$; $-1 - \sqrt{2}, \begin{bmatrix} 1 - \sqrt{2} \\ 1 \\ 0 \end{bmatrix}$

9. $-5, \begin{bmatrix} -\frac{3}{2} \\ 1 \end{bmatrix}$; $8, \begin{bmatrix} \frac{2}{3} \\ 1 \end{bmatrix}$ **11.** $\dfrac{1 + \sqrt{117}}{2}, \begin{bmatrix} -\dfrac{3 + \sqrt{13}}{2} \\ 1 \end{bmatrix}$; $\dfrac{1 - \sqrt{117}}{2}, \begin{bmatrix} -\dfrac{3 - \sqrt{13}}{2} \\ 1 \end{bmatrix}$ **13.** $3, \begin{bmatrix} 1 \\ 1 \end{bmatrix}$; $9, \begin{bmatrix} -1 \\ 1 \end{bmatrix}$

15. $0, \begin{bmatrix} 0 \\ 0 \\ 1 \end{bmatrix}; 1 + \sqrt{17}, \begin{bmatrix} \dfrac{-1 - \sqrt{17}}{4} \\ 1 \\ 0 \end{bmatrix}; 1 - \sqrt{17}, \begin{bmatrix} \dfrac{-1 + \sqrt{17}}{4} \\ 1 \\ 0 \end{bmatrix}$

17. Observe that $|\lambda I_n - A^t| = |(\lambda I_n - A)^t| = |\lambda I_n - A|$, so A and A^t have the same characteristic polynomial and hence the same eigenvalues.

Section 13.4

For Problems 1 and 3, verify that $AA^t = A^tA = I_n$ and hence A is orthogonal.

5. $\begin{bmatrix} \dfrac{\sqrt{29} - 5}{\sqrt{58 - 10\sqrt{29}}} & \dfrac{-\sqrt{29} - 5}{\sqrt{58 + 10\sqrt{29}}} \\ \dfrac{2}{\sqrt{58 - 10\sqrt{29}}} & \dfrac{2}{\sqrt{58 + 10\sqrt{29}}} \end{bmatrix}$

7. $\begin{bmatrix} \dfrac{1}{\sqrt{2}} & -\dfrac{1}{\sqrt{2}} \\ \dfrac{1}{\sqrt{2}} & \dfrac{1}{\sqrt{2}} \end{bmatrix}$

9. $\begin{bmatrix} 1 & 0 & 0 \\ 0 & -\dfrac{1 + \sqrt{17}}{\sqrt{34 + 2\sqrt{17}}} & -\dfrac{-1 + \sqrt{17}}{\sqrt{34 - 2\sqrt{17}}} \\ 0 & \dfrac{4}{\sqrt{34 + 2\sqrt{17}}} & \dfrac{4}{\sqrt{34 - 2\sqrt{17}}} \end{bmatrix}$

11. $\begin{bmatrix} 1 & 0 & 0 & 0 \\ 0 & 0 & -\dfrac{1}{\sqrt{2}} & \dfrac{1}{\sqrt{2}} \\ 0 & 0 & \dfrac{1}{\sqrt{2}} & \dfrac{1}{\sqrt{2}} \\ 0 & 1 & 0 & 0 \end{bmatrix}$

13. Suppose that A is positive definite. Let the eigenvalues of A be $\lambda_1, \ldots, \lambda_n$. These are all positive. There exists an orthogonal matrix R such that $R^{-1}AR$ is the diagonal matrix D having main diagonal elements $\lambda_1, \ldots, \lambda_n$. Let

$$D^* = \begin{bmatrix} \sqrt{\lambda_1} & & & \\ & \sqrt{\lambda_2} & & O \\ & & \ddots & \\ O & & & \sqrt{\lambda_n} \end{bmatrix}.$$

Then $A = RDR^{-1} = RD^*D^*R^{-1}$. Let $P = D^*R^{-1}$. Since $R^t = R^{-1}$ and $(D^*)^t = D^*$, $P^t = (D^*R^{-1})^t = (D^*R^t)^t = RD^*$; hence, $A = P^tP$.

Conversely, suppose that $A = P^tP$ for some nonsingular matrix P. Suppose that an eigenvalue λ of A is negative. Let X be an eigenvector corresponding to λ. Take the dot product of the two n-vectors AX and X^t to get $AX \cdot X^t = \lambda X \cdot X^t = \lambda \|X^t\|^2$. Since $\|X^t\|^2 > 0$, $\lambda\|X^t\|^2 < 0$, so $AX \cdot X^t < 0$. It is routine to check that $AX \cdot X^t = (P^tPX) \cdot X^t = PX \cdot (PX)^t = \|(PX)\|^2 \geq 0$, a contradiction.

Section 13.5

1. $\begin{bmatrix} 1 & 1 \\ 1 & 6 \end{bmatrix}$ **3.** $\begin{bmatrix} 1 & -2 \\ -2 & 1 \end{bmatrix}$ **5.** $\begin{bmatrix} -1 & 0 & -\frac{1}{2} & -1 \\ 0 & 0 & 2 & \frac{3}{2} \\ -\frac{1}{2} & 2 & 0 & 0 \\ -1 & \frac{3}{2} & 0 & 1 \end{bmatrix}$ **7.** $-2x_1^2 + 2x_1x_2 + 6x_2^2$

9. $6x_1^2 + 2x_1x_2 - 14x_1x_3 + 2x_2^2 + x_3^2$ **11.** $8x_1^2 + 2x_2^2 - 8x_2x_3 + 3x_3^2$ **13.** $(-1 + 2\sqrt{5})y_1^2 + (-1 - 2\sqrt{5})y_2^2$

15. $(2 + \sqrt{29})y_1^2 + (2 - \sqrt{29})y_2^2$ **17.** $(2 + \sqrt{13})y_1^2 + (2 - \sqrt{13})y_2^2$ **19.** $(1 + \sqrt{2})y_1^2 + (1 - \sqrt{2})y_2^2$

21. $\dfrac{\sqrt{61}}{2} y_1^2 - \dfrac{\sqrt{61}}{2} y_2^2 = 5$; hyperbola   **23.** $5y_1^2 - 5y_2^2 = 8$; hyperbola

25. Let the eigenvalues of A be $\lambda_1, \ldots, \lambda_n$ (all real). Let Q be an orthogonal matrix that diagonalizes A. Let $X = QZ$. Then $X^tAX = Z^tQ^tAQZ = Z^tDZ$, where D is the diagonal matrix having main diagonal elements $\lambda_1, \ldots, \lambda_n$. Since Q is nonsingular and $Q^tAQ = D$, $\text{rank}(A) = \text{rank}(D) = r = $ number of nonzero eigenvalues of A. Among these, suppose that $\lambda_1, \ldots, \lambda_p$ are positive, while $\lambda_{p+1}, \ldots, \lambda_r$ are negative. Then $\lambda_{r+1} = \cdots = \lambda_n = 0$. Let D^* be the $n \times n$ diagonal matrix whose main diagonal elements are $\sqrt{\lambda_1}, \ldots, \sqrt{\lambda_p}, \sqrt{-\lambda_{n+1}}, \ldots, \sqrt{-\lambda_r}, 1, 1, \ldots, 1$. Then $D = (D^*)^tMD^*$, where M is the $n \times n$ diagonal matrix having main diagonal elements $1, \ldots, 1, -1, \ldots, -1, 0, \ldots, 0$, with the first p elements equal to 1, the next $r - p$ equal to -1, and the remaining diagonal elements zero. Let $Z = D^*Y$. Then $X'AX = Y'(D^*)'Q'AQD^*Y = Y'MY = y_1^2 + \cdots + y_p^2 - y_{p+1}^2 - \cdots - y_r^2$. Now let $P = QD^*$. Then $X = PY$ transforms $X'AX$ into the form $y_1^2 + \cdots + y_p^2 - y_{p+1}^2 - \cdots - y_r^2$.

27. If A is positive definite, then all the eigenvalues $\lambda_1, \ldots, \lambda_n$ of A are positive. There exists an orthogonal matrix Q that diagonalizes A. Let $X = QY$. Then $X'AX = \lambda_1 y_1^2 + \cdots + \lambda_n y_n^2$. If $X \neq O$ (X is $n \times 1$), then the fact that rank$(Q) = n$ implies that $Y \neq O$, so $X'AX = \lambda_1 y_1^2 + \cdots + \lambda_n y_n^2 > 0$.

Conversely, suppose that $X'AX > 0$ for every nonzero $n \times 1$ matrix X. Let λ be any eigenvalue of A, and X a corresponding eigenvector. Then $X'AX > 0$. Therefore, $X'AX = X'\lambda X = \lambda X'X = \lambda\|X'\|^2 > 0$, where $\|X'\|$ is the length of X', thought of as a vector in R^n. Therefore, $\lambda > 0$; hence, A is positive definite.

Section 13.6

1. If A is hermitian, then $\bar{A} = A'$. Then $A = \overline{(\bar{A})} = (\bar{A})'$; hence, $A(\bar{A})' = A^2 = (\bar{A})'A$.
If A is skew-hermitian, then $\bar{A} = -A'$. Now, reasoning as in the first part, we get $A(\bar{A})' = -A^2 = (\bar{A})'A$.
If A is unitary, then $(\bar{A})^{-1} = A'$, so $[(\bar{A})^{-1}] = (\bar{A})'$; hence, $A^{-1} = (\bar{A})'$. Now $A(\bar{A})' = AA^{-1} = I_n = (\bar{A})'A$.

3. Unitary; $-1, -1$ **5.** Unitary; $1, \dfrac{1+\sqrt{3}}{2\sqrt{2}} + \dfrac{1-\sqrt{3}}{2\sqrt{2}}i, \dfrac{1-\sqrt{3}}{2\sqrt{2}} + \dfrac{1+\sqrt{3}}{2\sqrt{2}}i$ **7.** Hermitian; $1, \dfrac{-1 \pm \sqrt{41}}{2}$

9. None of these; $3i, \pm 1$

11. If A is skew-hermitian, then $\bar{A} = -A'$. If $A = [a_{ij}]$, then $\bar{a}_{kk} = -a_{kk}$. Let $a_{kk} = a_k + b_k i$. Then $a_k - b_k i = -a_k - b_k i$; therefore, $a_k = 0$ and a_{kk} is $b_k i$, pure imaginary. This is zero if $b_k = 0$.

13. Let U and V be $n \times n$ unitary matrices. Then $(\bar{U})^{-1} = U'$ and $(\bar{V})^{-1} = V'$. Then $(\overline{UV})^{-1} = (\bar{U} \cdot \bar{V})^{-1} = (\bar{V})^{-1}(\bar{U})^{-1} = V'U' = (UV)'$; hence, UV is unitary.

15. Let $A = [a_{kj}]$ be any $n \times n$ matrix. Suppose that $M = [m_{kj}]$ and $N = [n_{kj}]$ and N are $n \times n$ matrices, with M hermitian and N skew-hermitian. We ask what will be needed to have $A = M + N$. Of course, we need $a_{kj} = m_{kj} + n_{kj}$. We also need $a_{jk} = m_{jk} + n_{jk} = \bar{m}_{kj} - \bar{n}_{kj}$. For the moment, fix k and j and let $a_{kj} = a_1 + b_1 i, a_{jk} = a_2 + b_2 i, m_{kj} = m_1 + m_2 i$ and $n_{kj} = n_1 + n_2 i$. Then $a_1 + b_1 i = (m_1 + n_1) + (m_2 + n_2)i$ and $a_2 + b_2 i = (m_1 - n_1) + (-m_2 + n_2)i$. Therefore, $m_1 + n_1 = a_1$, $m_1 - n_1 = a_2, m_2 + n_2 = b_1$ and $-m_2 + n_2 = b_2$. This is a system of four equations in four unknowns (m_1, m_2, n_1, n_2), since a_1, a_2, b_1, and b_2 are presumed given with A. This system has a unique solution. If we solve this system for each k and j, we obtain a hermitian matrix M and a skew-hermitian matrix N such that $A = M + N$.

17. If S is skew-hermitian, then $\bar{S} = -S'$. If α is real, then $\overline{(\alpha S)} = \bar{\alpha} \cdot \bar{S} = \alpha(-S') = -(\alpha S)'$; hence, αS is skew-hermitian.

19. -43 **21.** $62i$

Additional Problems for Chapter 13

1. $\dfrac{5+3\sqrt{5}}{2}, \begin{bmatrix} \dfrac{1-\sqrt{5}}{2} \\ 1 \end{bmatrix}; \dfrac{5-3\sqrt{5}}{2}, \begin{bmatrix} \dfrac{1+\sqrt{5}}{2} \\ 1 \end{bmatrix}$ **3.** $2i, \begin{bmatrix} -i \\ 1 \end{bmatrix}$ **5.** $4, \begin{bmatrix} \dfrac{1+3i}{5} \\ 1 \end{bmatrix}; 1-i, \begin{bmatrix} 1 \\ 0 \end{bmatrix}$

7. $3 - 2i, \begin{bmatrix} \dfrac{3+2i}{1} \\ 1 \end{bmatrix}; -4i, \begin{bmatrix} 0 \\ 1 \end{bmatrix}$ **9.** $0, \begin{bmatrix} 0 \\ \dfrac{3}{5} \\ 1 \end{bmatrix}; \dfrac{1+\sqrt{23}i}{2}, \begin{bmatrix} \dfrac{-1-\sqrt{23}i}{4} \\ 0 \\ 1 \end{bmatrix}; \dfrac{1-\sqrt{23}i}{2}, \begin{bmatrix} \dfrac{-1+\sqrt{23}i}{4} \\ 0 \\ 1 \end{bmatrix}$

11. $\begin{bmatrix} \dfrac{1-\sqrt{5}}{\sqrt{10-2\sqrt{5}}} & \dfrac{1+\sqrt{5}}{\sqrt{10+2\sqrt{5}}} \\ \dfrac{2}{\sqrt{10-2\sqrt{5}}} & \dfrac{2}{\sqrt{10+2\sqrt{5}}} \end{bmatrix}$ **13.** $\begin{bmatrix} \dfrac{\sqrt{185}-13}{\sqrt{370-28\sqrt{185}}} & \dfrac{-\sqrt{185}-13}{\sqrt{370+28\sqrt{185}}} \\ \dfrac{4}{\sqrt{370-28\sqrt{185}}} & \dfrac{4}{\sqrt{370+28\sqrt{185}}} \end{bmatrix}$

15. $\begin{bmatrix} 0 & \dfrac{2-\sqrt{13}}{\sqrt{26-4\sqrt{13}}} & \dfrac{2+\sqrt{13}}{\sqrt{26+4\sqrt{13}}} \\ 1 & 0 & 0 \\ 0 & \dfrac{3}{\sqrt{26-4\sqrt{13}}} & \dfrac{3}{\sqrt{26+4\sqrt{13}}} \end{bmatrix}$

17. If A has real elements, then the characteristic polynomial of A has real coefficients, and any root $a + ib$ of this polynomial (hence any complex eigenvalue of A) is accompanied by $a - ib$ as another root (hence $a - ib$ is also an eigenvalue). This result is false in general if A has complex elements, as then the characteristic polynomial may have complex coefficients.

19. The characteristic equation of A is $|\lambda I_n - A| = 0$. The term in the expansion of this determinant yielding the higher power λ^n of λ is $(\lambda - a_{11})(\lambda - a_{22}) \cdots (\lambda - a_{nn})$, and the expansion of this product has λ^n with coefficient 1.

21. $\dfrac{3 + \sqrt{37}}{2} y_1^2 + \dfrac{3 - \sqrt{37}}{2} y_2^2$ **23.** $\dfrac{-3 + \sqrt{97}}{2} y_1^2 + \dfrac{-3 - \sqrt{97}}{2} y_2^2$ **25.** $-y_1^2 + \dfrac{-7 + \sqrt{53}}{2} y_2^2 + \dfrac{-7 - \sqrt{53}}{2} y_3^2$

27. Let rank$(\alpha) = k$. There are nonsingular matrices P and Q such that $PAQ = D$, where D is the diagonal matrix having the first k diagonal elements equal to 1 and the rest equal to zero. It is easy to prove that, for any matrix C, CD and CD have the same characteristic equation. Now, $(Q^{-1}BP^{-1})D = Q^{-1}BAQ$ and $D(Q^{-1}BP^{-1}) = PABP^{-1}$. Then $Q^{-1}BAQ$ and $PABP^{-1}$ have the same characteristic equation. But BA and $Q^{-1}BAQ$ have the same characteristic equation, as do AB and $PABP^{-1}$. Therefore, AB and BA have the same characteristic equation.

BA and AB may have different eigenvectors corresponding to the same eigenvalue.

29. (a) If $\lambda = 0$ is an eigenvalue of A with associated eigenvector X, then $AX = \lambda X = O$. Therefore, the homogeneous system $AX = O$ has nontrivial solution X. Then A must be singular. This contradicts the assumption (in the problem) that A is nonsingular.

(b) If λ is an eigenvalue of A with associated eigenvector X, then $AX = \lambda X$. Then $A^{-1}AX = X = X^{-1}(\lambda X) = \lambda A^{-1}X$. Since $\lambda \neq 0$, $A^{-1}X = (1/\lambda)X$; hence, $1/\lambda$ is an eigenvalue of A^{-1}, with eigenvector X.

31. Let λ be an eigenvalue of A, with associated eigenvector X. Then $AX = \lambda X$, so $A^2X = A(\lambda X) = \lambda(AX) = \lambda(\lambda X) = \lambda^2 X$. But $A^2 = I_n$, so $I_n X = X = \lambda^2 X$; hence, $\lambda^2 = 1$, and λ must equal 1 or -1.

33. Eigenvalues of A are $3 \pm \sqrt{2}$. The center of C_1 is $(2, 0)$, the radius is 1; the center of C_2 is $(4, 0)$, the radius is 1.

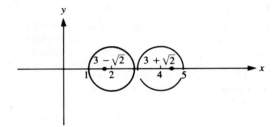

35. The eigenvalues are $3 \pm \sqrt{3}i$. The centers of C_1 and C_2 are $(0, 0)$; the radii are 12.

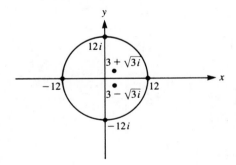

Chapter 14

Section 14.1

3. $x_0 = 1$; $y_0 = 0$; $x_1 = 1 + t$; $y_1 = 4t$; $x_2 = 1 + t + \dfrac{5}{2}t^2$; $y_2 = 4t + t^2$; $x_3 = 1 + t + \dfrac{5}{2}t^2 + \dfrac{7}{6}y^3$; $y_3 = 4t + t^2 + \dfrac{5}{3}t^3$

5. $x_0 = -1$; $y_0 = y_1 = -1$; $x_1 = x_2 = -3 - t + 2e^t$; $y_2 = y_3 = -3 - 2t - \dfrac{1}{2}t^2 + 2e^t$; $x_3 = -5 - 3t - t^2 - \dfrac{1}{6}t^3 + 4e^t$

7. $x(t) = 1 + t + \dfrac{5}{2}t^2 + \dfrac{7}{6}t^3 + \cdots$; $y(t) = 4t + t^2 + \dfrac{5}{3}t^3 + \cdots$

9. $x_0 = 4$; $y_0 = 4$; $z_0 = 3$; $x_1 = 4 + 5t$; $y_1 = 3 + 5t + e^{2t}$; $z_1 = 3 + 3t$;

$x_2 = \dfrac{9}{2} + 6t + \dfrac{7}{2}t^2 - \dfrac{1}{2}e^{2t}$; $y_2 = \dfrac{5}{2} + 4t + \dfrac{7}{2}t^2 + \dfrac{3}{2}e^{2t}$; $z_2 = \dfrac{7}{2} + 4t + \dfrac{3}{2}t^2 - \dfrac{1}{2}e^{2t}$;

$x_3 = \dfrac{21}{4} + \dfrac{15}{2}t + 5t^2 + \dfrac{11}{6}t^3 - \dfrac{5}{4}e^{2t}$; $y_3 = \dfrac{9}{4} + \dfrac{7}{2}t + 3t^2 + \dfrac{11}{6}t^3 + \dfrac{7}{4}e^{2t}$; $z_3 = \dfrac{17}{4} + \dfrac{11}{2}t + 3t^2 + \dfrac{1}{2}t^3 - \dfrac{5}{4}e^{2t}$

Section 14.2

1. $x_1 = 7c_1e^{3t}$; $x_2 = 5c_1e^{3t} + c_2e^{-4t}$ **3.** $x_1 = c_1 + c_2e^{2t}$; $x_2 = -c_1 + c_2e^{2t}$

5. $x_1 = c_1 + 2c_2e^{3t} - c_3e^{-4t}$; $x_2 = 6c_1 + 3c_2e^{3t} + 2c_3e^{-4t}$; $x_3 = -13c_1 - 2c_2e^{3t} + c_3e^{-4t}$

7. $x_1 = -c_1e^t + c_2e^{-2t} + c_3e^{3t}$; $x_2 = 4c_1e^t - c_2e^{-2t} + 2c_3e^{3t}$; $x_3 = c_1e^t - c_2e^{-2t} + c_3e^{3t}$

9. $x_1 = 6e^{4t} - 5e^{-3t}$; $x_2 = -9e^{4t} - 5e^{-3t}$ **11.** $x_1 = 4e^{2t} - 3e^{3t}$; $x_2 = 2 + 4e^{2t} - e^{3t}$; $x_3 = 2 - e^{3t}$

13. $x_1 = 5e^{2t} + 4e^{-t}$; $x_2 = 3e^{2t} - 4e^{-t}$; $x_3 = -3e^{2t}$ **15.** $x_1 = c_1t^{-4} + 3c_2t^{-2}$; $x_2 = c_2t^{-4} + c_2t^{-2}$

Section 14.3

1. $x_1 = e^{2t}[2c_1\cos(2t) + 2c_2\sin(2t)]$; $x_2 = e^{2t}[c_1\sin(2t) - c_2\cos(2t)]$

3. $x_1 = e^t[5c_1\cos(t) + 5c_2\sin(t)]$; $x_2 = e^t[(2c_1 + c_2)\cos(t) + (2c_2 - c_1)\sin(t)]$

5. $x_1 = e^{-t}[c_2\cos(2t) + c_3\sin(2t)]$; $x_2 = e^{-t}[(c_2 + 2c_3)\cos(2t) + (c_3 - 2c_2)\sin(2t)]$; $x_3 = c_1e^{-2t} + e^{-t}[3c_2\cos(2t) + 3c_3\sin(2t)]$

7. $x_1 = 2c_1\cos(t) + 2c_2\sin(t)$; $x_2 = (3c_1 - c_2)\cos(t) + (c_1 + 3c_2)\sin(t)$; $x_3 = 2c_3\cos(t) + 2c_4\sin(t)$; $x_4 = (3c_3 - c_4)\cos(t) + (c_3 + 3c_4)\sin(t)$

9. $x_1 = 2\cos(t) + 6\sin(t)$; $x_2 = 7\cos(t) + 10\sin(t)$

11. $x_1 = 2e^t + e^t[5\cos(t) + 5\sin(t)]$; $x_2 = 2e^t + e^t[2\cos(t) + 6\sin(t)]$; $x_3 = 2e^t + e^t[\cos(t) + 3\sin(t)]$

13. $x_1 = 6t^3\cos[2\ln(t)] - 4t^3\sin[2\ln(t)]$; $x_2 = 5t^3\cos[2\ln(t)] + t^3\sin[2\ln(t)]$

Section 14.4

1. $x_1 = c_1 + c_2e^{2t}$; $x_2 = -c_1 + c_2e^{2t}$ **3.** $x_1 = 2c_1e^t + c_2e^{4t}$; $x_2 = -c_1e^t + c_2e^{4t}$

5. $x_1 = c_1e^t + c_3e^{-3t}$; $x_2 = c_2e^t + 3c_3e^{-3t}$; $x_3 = (-c_1 + c_2)e^t + c_3e^{-3t}$

7. $x_1 = 2c_1e^t + c_2e^{6t}$; $x_2 = -3c_1e^t + c_2e^{6t}$; $x_3 = 2c_3e^t + c_4e^{-t}$; $x_4 = c_3e^t + c_4e^{-t}$

9. $x_1 = e^{3t}[c_1\cos(t) + c_2\sin(t)]$; $x_2 = e^{3t}[-c_2\cos(t) + c_1\sin(t)]$ **11.** $x_1 = 2e^{-4t} - 6e^{-2t}$; $x_2 = 2e^{-4t} - 2e^{-2t}$

13. $x_1 = 2e^t + 2e^{-2t} + 3e^{3t}$; $x_2 = -8e^t - 2e^{-2t} + 6e^{3t}$; $x_3 = -2e^t - 2e^{-2t} + 3e^{3t}$

Section 14.5

1. $x_1 = c_1e^{-t} + c_2e^{2t} - 3\cos(t) - \sin(t)$; $x_2 = c_1e^{-t} + 4c_2e^{2t} - 7\cos(t) + \sin(t)$

3. $x_1 = c_1 + c_2e^{2t} + 4e^{3t} - 1 - 2t$; $x_2 = -c_1 + c_2e^{2t} + 2e^{3t} - 1 + 2t$

5. $x_1 = 2c_1\cos(3t) + 2c_2\sin(3t) + e^{2t}$; $x_2 = (3c_1 - 3c_2)\cos(3t) + (3c_1 + 3c_2)\sin(3t) + 2e^{2t}$

7. $x_1 = c_2e^{2t} + 3c_3e^{3t} + 12e^{4t}$; $x_2 = c_1 + c_2e^{2t} + c_3e^{2t} + 3e^{4t} + 2\sin(2t)$; $x_3 = c_1 + 3e^{4t} + 2\sin(2t)$

9. $x_1 = c_1 + c_2e^{2t} + e^{2t}$; $x_2 = -c_1 + c_2e^{2t} - e^{2t}$; $x_3 = 2c_3e^t + 12c_4e^{6t} + 2e^{6t}$; $x_4 = -c_3e^t + 12c_4e^{6t} - 3e^{6t}$

11. $x_1 = 3e^{-t} + 10 - 2t + 2t^2$; $x_2 = 3e^{-t} + 9 - 7t + t^2$ **13.** $x_1 = \dfrac{1}{4}e^t[-\cos(t) + \sin(t)] + \dfrac{1}{4}$; $x_2 = \dfrac{1}{4}t - \dfrac{1}{4}e^t\sin(t)$

15. $x_1 = -\dfrac{1}{4}e^{2t} + (2 + 2t)e^t - \dfrac{3}{4} - \dfrac{1}{2}t$; $x_2 = e^{2t} + (2 + 2t)e^t - 1 - t$; $x_3 = -\dfrac{5}{4}e^{2t} + 2te^t - \dfrac{3}{4} - \dfrac{1}{2}t$

Section 14.6

1. $x_1 = c_1e^{3t} + 2c_2e^{3t}$; $x_2 = c_2te^{3t}$ **3.** $x_1 = 2c_1e^{4t} + c_2(1 - 2t)e^{4t}$; $x_2 = c_1e^{4t} + c_2te^{4t}$

5. $x_1 = c_1e^{2t} + 3c_2e^{5t} + c_3(1 - 9t)e^{5t}$; $x_2 = 3c_2e^{5t} - 9c_3te^{5t}$; $x_3 = -c_2e^{5t} - c_3(1 - 3t)e^{5t}$

7. $x_1 = (c_1 + 5c_2)e^{3t} + (c_3 - 6c_4t)e^t$; $x_2 = 2c_2e^{3t} + 4c_4e^t$; $x_3 = -c_1e^{3t} + 7c_4e^t$; $x_4 = -2c_4e^t$

9. $x_1 = (5 + 2t)e^{6t}$; $x_2 = (3 + 2t)e^{6t}$ **11.** $x_1 = -e^{-2t} + (1 + 22t)e^{-4t}$; $x_2 = -6e^{-2t} + 10e^{-4t}$; $x_3 = 12e^{-4t}$

13. $x_1 = 2\cos(t) + 6\sin(t)$; $x_2 = -2\cos(t) + 4\sin(t)$; $x_3 = (1 + 12t)e^{2t}$; $x_4 = (4 + 12t)e^{2t}$

15. $(A + B)^2 = (A + B)(A + B) = (A + B)A + (A + B)B = A^2 + BA + AB + B^2 = A^2 + AB + AB + B^2 = A^2 + 2AB + B^2$;

Let $A = \begin{pmatrix} 1 & 1 \\ 1 & 1 \end{pmatrix}$ and $B = \begin{pmatrix} 1 & 2 \\ 3 & 4 \end{pmatrix}$. Then $(A + B)^2 = \begin{pmatrix} 16 & 21 \\ 28 & 37 \end{pmatrix}$, whereas $A^2 + 2AB + B^2 = \begin{pmatrix} 17 & 24 \\ 25 & 36 \end{pmatrix}$.

Section 14.7

1. $x_1 = [c_1 + (1 + 2t)c_2 - (3t + 2t^2)]e^{3t}$; $x_2 = [-c_1 - 2c_2 t + (5t - 2t^2)]e^{3t}$

3. $x_1 = [c_1 + (1 + t)c_2 + (2t + t^2 - t^3)]e^{6t}$; $x_2 = [c_1 + c_2 t + (4t^2 - t^3)]e^{6t}$

5. $x_1 = c_2 e^t$; $x_2 = (1 - 2c_2)e^t + (c_3 - 9c_4)e^{3t}$; $x_3 = 2c_4 e^{3t}$; $x_4 = (c_1 - 5c_2 t + 1 + 3t)e^t + c_3 e^{3t}$

7. $x_1 = (-1 - 16t + t^2)e^t$; $x_2 = (3 - 16t + t^2)e^t$

9. $x_1 = (6 + 12t + t^2)e^{-2t}$; $x_2 = (2 + 12t + t^2)e^{-2t}$; $x_3 = (3 + 38t + 66t^2 + 11t^3)e^{-2t}$

13. $\begin{bmatrix} \dfrac{2}{5}e^t + \dfrac{3}{5}e^{6t} & -\dfrac{2}{5}e^t + \dfrac{2}{5}e^{6t} \\ -\dfrac{3}{5}e^t + \dfrac{3}{5}e^{6t} & \dfrac{3}{5}e^t + \dfrac{2}{5}e^{6t} \end{bmatrix}$

15. $\begin{bmatrix} 2e^t - e^{-3t} & e^{-3t} - e^t & e^t - e^{-3t} \\ 3e^t - 3e^{-3t} & 3e^{-3t} - 2e^t & 3e^t - 3e^{-3t} \\ e^t - e^{-3t} & e^{-3t} - e^t & 2e^t - e^{-3t} \end{bmatrix}$

Section 14.8

1. Let $x_1 = y$, $x_2 = y'$ and $x_3 = y''$. Then the new system of differential equations is $x_1' = x_2$, $x_2' = x_3$, $x_3 = 10x_1 - 8x_2 + 4x_3 + \cos(t)$.

3. Let $x_1 = y$, $x_2 = y'$, $x_3 = y''$, and $x_4 = y^{(3)}$. Then the new system is $x_1' = x_2$, $x_2' = x_3$, $x_3' = x_4$, $x_4' = -12x_1 - 8x_2 + 22x_3 + 2\cos(t) - e^t$.

5. Let $x_1 = y$, $x_2 = y'$, $x_3 = y''$, and $x_4 = y^{(3)}$. Then the new system is $x_1' = x_2$, $x_2' = x_3$, $x_3' = x_4$, $x_4' = 9x_1 - 4x_2 - 10x_3 + 6x_4 + t^3$.

7. Let $y_1 = x_1$, $y_2 = x_2$, and $y_3 = x_2'$. Then $A = \begin{bmatrix} 1 & -2 & 3 \\ 0 & 0 & 1 \\ 3 & -1 & 2 \end{bmatrix}$ and $G = \begin{bmatrix} e^{5t} + 6t \\ 0 \\ \cos(5t) \end{bmatrix}$

9. $A = \begin{pmatrix} 3 & 3 \\ 1 & 5 \end{pmatrix}$; $G = \begin{pmatrix} 8 \\ 4e^{3t} \end{pmatrix}$

11. Let $x_1 = y$, $x_2 = y'$, and $x_3 = y''$. Then $A = \begin{bmatrix} 0 & 1 & 0 \\ 0 & 0 & 1 \\ 5 & -3 & 2 \end{bmatrix}$ and $G = \begin{bmatrix} 0 \\ 0 \\ 6e^{2t} - 5t \end{bmatrix}$.

13. Let $y_1 = x_1$, $y_2 = x_1'$, $y_3 = x_2$, and $y_4 = x_2'$. Then $A = \begin{bmatrix} 0 & 1 & 0 & 0 \\ 2 & 3 & 1 & -4 \\ 0 & 0 & 0 & 1 \\ 5 & 3 & -7 & 4 \end{bmatrix}$ and $G = \begin{bmatrix} 0 \\ 8e^{4t} \\ 0 \\ -2e^{-t} \end{bmatrix}$.

15. $x_1 = c_1 e^{4t} + c_2(1 + t)e^{4t} - 2e^{2t}$, $x_2 = c_1 e^{4t} + c_2 t e^{4t} - 2e^{2t}$

17. $x_1 = c_1 e^t + c_2 e^{2t} + c_3 e^{2t} - 4e^t$, $x_2 = c_1 e^t + c_2 e^{2t} - 3e^t$, $x_3 = c_1 e^t + c_3 e^{2t} - e^t$

19. $x_1 = 2e^t + (1 - t)e^{-t} - 2e^{2t}$, $x_2 = (3 - 4t)e^{-t} - 2e^{2t}$

Section 14.9

1. $x_1 = 48e^{-t/10} + 22e^{-3t/5} + 80$ grams, $x_2 = 32e^{-t/10} - 22e^{-3t/5} + 40$ grams

3. $x_1 = 36e^{-t/50} + 44e^{-3t/25}$ pounds, $x_2 = 54e^{-t/50} - 44e^{-3t/25}$ pounds

5. $x_1 = \dfrac{80}{3}e^{-t/50} - \dfrac{130}{3}e^{-3t/25} + \dfrac{50}{3}$ pounds, $x_2 = 40e^{-t/50} + \dfrac{130}{3}e^{-3t/25} + \dfrac{50}{3}$ pounds

7. $i_1 = \dfrac{3}{10} - \dfrac{1}{10}e^{-40t/3}$ ampere, $i_2 = \dfrac{3}{10} - \dfrac{3}{10}e^{-40t/3}$ ampere

9. The steady-state currents are $i_1 = -0.2110\cos(3t) - 4.7535\sin(3t)$, $i_2 = 0.3251\cos(3t) + 1.6388\sin(3t)$, $i_3 = -0.1563\cos(3t) + 1.4044\sin(3t)$

11. $y_1 = -\dfrac{1}{5}\cos(2t) + \dfrac{6}{5}\cos(2\sqrt{6}t)$, $y_2 = -\dfrac{3}{5}\cos(2t) - \dfrac{2}{5}\cos(2\sqrt{6}t)$

13. Denoting right as positive, we have $y_1 = y_2 = -\cos(2t)$; $y_2 - y_1 = 0$

15. $y_1 = -\dfrac{1}{6}\sin(2t) - \dfrac{1}{24}\sin(3t) + \dfrac{11}{24}\sin(t)$, $y_2 = -\dfrac{1}{6}\sin(2t) + \dfrac{1}{24}\sin(3t) + \dfrac{5}{24}\sin(t)$

17. $i_1 = \dfrac{3}{1000}e^{-t} + \dfrac{1}{500}e^{-6t}$ ampere, $i_2 = \dfrac{1}{500}e^{-t} - \dfrac{1}{500}e^{-6t}$ ampere

19. $i_1 = \dfrac{1}{5} - \dfrac{1}{25}e^{-30t}[4\cos(10t) - 2\sin(10t)] + \left\{\dfrac{1}{5} - \dfrac{1}{25}e^{-30(t-3)}[4\cos\{10(t-3)\} - 2\sin\{10(t-3)\}]\right\}H(t-3)$ ampere,

$i_2 = \dfrac{1}{25}e^{-30t}[\cos(10t) - 3\sin(10t)] + \left\{\dfrac{1}{25}e^{-30(t-3)}[\cos\{10(t-3)\} - 3\sin\{10(t-3)\}]\right\}H(t-3)$ ampere

21. $y_1 = e^{-2t}[2\cos(t) + (8 + \sqrt{26})\sin(t)] - \dfrac{1}{2}e^{-t}[\cos(2t) + (5 + \sqrt{26})\sin(2t)]$, $y_2 = \dfrac{1}{13}e^{-2t}[(13 + 10\sqrt{26})\cos(t) +$

$(26 + 6\sqrt{26})\sin(t)] - \dfrac{1}{26}e^{-t}[(52 + 12\sqrt{26})\cos(2t) + (26 + \sqrt{26})\sin(2t)]$

Chapter 15

Section 15.1

In solutions 1, 3, and 5, $[f\mathbf{F}]'$ is given.

1. $-12\sin(3t)\mathbf{i} + 12[2\cos(3t) - 3t\sin(3t)]\mathbf{j} + 8[\cos(3t) - 3t\sin(3t)\mathbf{k}$

3. $4\sin(2t)\mathbf{i} + t^4[5\sin^2(t) + t\sin(2t)]\mathbf{k}$ [Note that $\sin(2t) = 2\sin(t)\cos(t)$.]　　**5.** $(-12t - 7)\mathbf{i} + 2t^3(5t + 6)\mathbf{j} - (4t + 3)\mathbf{k}$

In solutions 7 and 9, $(\mathbf{F} \cdot \mathbf{G})'$ is given.

7. $2e^t + 4t$　　**9.** $-16\cos(2t)$

In solutions 11, 13, and 15, $(\mathbf{F} \times \mathbf{G})'$ is given.

11. $[1 - 4\sin(t)]\mathbf{i} - 2t\mathbf{j} - [\cos(t) - t\sin(t)]\mathbf{k}$　　**13.** $te^t(2 + t)[\mathbf{j} - \mathbf{k}]$
15. $[t\sin(t) - \cos(t)]\mathbf{i} + t[2\cos(t) - t\sin(t)]\mathbf{j} - 8t\mathbf{k}$

In solutions 17 through 25, the position vector is given first, then the tangent vector.

17. $t\mathbf{i} + \sin(2\pi t)\mathbf{j} + \cos(2\pi t)\mathbf{k}$, $\mathbf{i} + 2\pi\cos(2\pi t)\mathbf{j} - 2\pi\sin(2\pi t)\mathbf{k}$

19. $e^t\cos(t)\mathbf{i} + e^t\sin(t)\mathbf{j} + e^t\mathbf{k}$, $e^t[\cos(t) - \sin(t)]\mathbf{i} + e^t[\sin(t) + \cos(t)]\mathbf{j} + e^t\mathbf{k}$　　**21.** $t^3(\mathbf{i} + \mathbf{j} + \mathbf{k})$; $3t^2(\mathbf{i} + \mathbf{j} + \mathbf{k})$

23. $4\ln(2t + 1)\mathbf{i} + 4\sinh(3t)\mathbf{j} + 2\mathbf{k}$, $\dfrac{8}{2t + 1}\mathbf{i} + 12\cosh(3t)\mathbf{j}$　　**25.** $(4t^2 - 5)\mathbf{i} + 3t\mathbf{j} + t\mathbf{k}$, $8t\mathbf{i} + 3\mathbf{j} + \mathbf{k}$

27. (a) $-4\sin(t)\mathbf{i} + 6\cos(3t)\mathbf{j}$　　(b) $\displaystyle\int_0^t \sqrt{16\sin^2(\xi) + 36\cos^2(3\xi)}\,d\xi$　　(c) $\sqrt{16\sin^2(t) + 36\cos^2(3t)}$

　　(d) $\dfrac{-4\sin(t(s))\mathbf{i} + 6\cos(3t(s))\mathbf{j}}{\sqrt{6\sin^2(t(s)) + 36\cos^2(t(s))}}$

29. (a) $e^t[\sin(t) + \cos(t)]\mathbf{i} + \mathbf{j} + e^t[\cos(t) - \sin(t)]\mathbf{k}$　　(b) $\displaystyle\int_0^t (2e^{2\xi} + 1)^{1/2}\,d\xi$　　(c) $(2e^{2t} + 1)^{1/2}$

　　(d) $\dfrac{1}{\sqrt{2e^{2t} + 1}}[e^{t(s)}[\sin(t(s)) + \cos(t(s))]\mathbf{i} + \mathbf{j} + e^{t(s)}[\cos(t(s)) - \sin(t(s))]\mathbf{k}]$

In solutions 31 and 33, $t(s)$ is given, then $\mathbf{F}(t(s))$.

31. $t = \sinh^{-1}(s)$, $\sinh^{-1}(s)\mathbf{i} + \cosh(\sinh^{-1}(s))\mathbf{j} + \mathbf{k}$　　**33.** $t = \left(\dfrac{s}{\sqrt{3}}\right)^{1/3}$, $\dfrac{s}{\sqrt{3}}(\mathbf{i} + \mathbf{j} + \mathbf{k})$

35. By equation (15.4), $(\mathbf{F} \times \mathbf{F}')' = \mathbf{F}' \times \mathbf{F}' + \mathbf{F} \times \mathbf{F}'' = \mathbf{O} + \mathbf{F} \times \mathbf{F}''$, because the cross product of any vector with itself is the zero vector.

37. (a) Referring to the diagram, the area of the triangle AOB is $\frac{1}{2}|OA||OB|\sin(\theta)$. Now, $\|\mathbf{F}(t) \times [\mathbf{F}(t + \Delta t) - \mathbf{F}(t)]\| = \|\mathbf{F}(t)\|\|\mathbf{F}(t + \Delta t) - \mathbf{F}(t)\|\sin(\theta)$, and this is twice the area of triangle AOB.

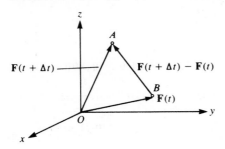

(b) Assume that $\mathbf{F} \times \mathbf{F}'' = \mathbf{O}$. By Problem 35, $(\mathbf{F} \times \mathbf{F}')' = \mathbf{O}$; hence, $\mathbf{F} \times \mathbf{F}' = \mathbf{a} = $ constant. Use this fact in conjunction with the conclusion of (a) to derive the result.

Section 15.2

1. $\mathbf{v} = 3\mathbf{i} + 2t\mathbf{k}$, $v = \sqrt{9 + 4t^2}$, $\mathbf{a} = 2\mathbf{k}$, $a_T = \dfrac{4t}{\sqrt{9 + 4t^2}}$, $a_N = \dfrac{36}{9 + 4t^2}$, $\kappa = \dfrac{6}{(9 + 4t^2)^{3/2}}$, $\mathbf{T} = \dfrac{1}{\sqrt{9 + 4t^2}}(3\mathbf{i} + 2t\mathbf{k})$,

$\mathbf{N} = \dfrac{1}{\sqrt{9 + 4t^2}}(-2t\mathbf{i} + 3\mathbf{k})$, $\mathbf{B} = -\mathbf{j}$

3. $\mathbf{v} = 2\mathbf{i} - 2\mathbf{j} + \mathbf{k}$, $v = 3$, $\mathbf{a} = \mathbf{O}$, $a_T = 0$, $a_N = 0$, $\kappa = 0$, $\mathbf{T} = \dfrac{1}{3}(2\mathbf{i} - 2\mathbf{j} + \mathbf{k})$, $\mathbf{N} = \dfrac{1}{\sqrt{2}}(\mathbf{i} - \mathbf{j})$ (or any unit vector

perpendicular to $\mathbf{T}$), $\mathbf{B} = \dfrac{\sqrt{2}}{6}(\mathbf{i} + \mathbf{j} - 4\mathbf{k})$

5. $\mathbf{v} = 2\cos(t)\mathbf{i} + \mathbf{j} - 2\sin(t)\mathbf{k}$, $v = \sqrt{5}$, $\mathbf{a} = -2\sin(t)\mathbf{i} - 2\cos(t)\mathbf{k}$, $a_T = 0$, $a_N = 2$, $\kappa = \dfrac{2}{5}$, $\mathbf{T} = \dfrac{1}{\sqrt{5}}[2\cos(t)\mathbf{i} + \mathbf{j} - 2\sin(t)\mathbf{k}]$,

$\mathbf{N} = \dfrac{1}{2}[-2\sin(t)\mathbf{i} - 2\cos(t)\mathbf{k}]$, $\mathbf{B} = \dfrac{1}{\sqrt{5}}[-\cos(t)\mathbf{i} + 2\mathbf{j} + \sin(t)\mathbf{k}]$

7. $\mathbf{v} = -e^{-t}(\mathbf{i} + \mathbf{j} - 2\mathbf{k})$, $v = e^{-t}\sqrt{6}$, $\mathbf{a} = e^{-t}(\mathbf{i} + \mathbf{j} - 2\mathbf{k})$, $a_T = -e^{-t}\sqrt{6}$, $a_N = 0$, $\kappa = 0$

9. $\mathbf{v} = 2\cosh(t)\mathbf{j} - 2\sinh(t)\mathbf{k}$, $v = 2\sqrt{\cosh(2t)}$, $\mathbf{a} = 2\sinh(t)\mathbf{j} - 2\cosh(t)\mathbf{k}$, $a_T = \dfrac{2\sinh(2t)}{\sqrt{\cosh(2t)}}$, $a_N = \dfrac{2}{\sqrt{\cosh(2t)}}$,

$\kappa = \dfrac{1}{2[\cosh(2t)]^{3/2}}$, $\mathbf{T} = \dfrac{1}{\sqrt{\cosh(2t)}}[\cosh(t)\mathbf{j} - \sinh(t)\mathbf{k}]$, $\mathbf{N} = \dfrac{1}{\sqrt{\cosh(2t)}}[-\sinh(t)\mathbf{j} - \cosh(t)\mathbf{k}]$, $\mathbf{B} = -\mathbf{i}$

11. $\mathbf{v} = 2\mathbf{i} + \sin(t)\mathbf{j} - \cos(t)\mathbf{k}$, $v = \sqrt{5}$, $\mathbf{a} = \cos(t)\mathbf{j} + \sin(t)\mathbf{k}$, $a_T = 0$, $a_N = 1$, $\kappa = \dfrac{1}{5}$, $\mathbf{T} = \dfrac{1}{\sqrt{5}}[2\mathbf{i} + \sin(t)\mathbf{j} - \cos(t)\mathbf{k}]$,

$\mathbf{N} = \cos(t)\mathbf{j} + \sin(t)\mathbf{k}$, $\mathbf{B} = \dfrac{1}{\sqrt{5}}[\mathbf{i} - 2\sin(t)\mathbf{j} + 2\cos(t)\mathbf{k}]$

13. $\mathbf{v} = 2t\mathbf{i} + 2t\mathbf{j} - 2\mathbf{k}$, $v = 2\sqrt{2t^2 + 1}$, $\mathbf{a} = 2\mathbf{i} + 2\mathbf{j}$, $a_T = \dfrac{4t}{\sqrt{2t^2 + 1}}$, $a_N = \dfrac{2\sqrt{2}}{\sqrt{2t^2 + 1}}$, $\kappa = \dfrac{\sqrt{2}}{2(2t^2 + 1)^{3/2}}$,

$\mathbf{T} = \dfrac{1}{\sqrt{2t^2 + 1}}(t\mathbf{i} + t\mathbf{j} - \mathbf{k})$, $\mathbf{N} = \dfrac{1}{\sqrt{2}(2t^2 + 1)^{3/2}}[\mathbf{i} + \mathbf{j} + 2t\mathbf{k}]$, $\mathbf{B} = \dfrac{1}{\sqrt{2}}[\mathbf{i} - \mathbf{j}]$

15. $\mathbf{v} = 2t(\alpha\mathbf{i} + \beta\mathbf{j} + \gamma\mathbf{k})$, $v = 2|t|\sqrt{\alpha^2 + \beta^2 + \gamma^2}$, $\mathbf{a} = 2(\alpha\mathbf{i} + \beta\mathbf{j} + \gamma\mathbf{k})$, $a_T = 2\sqrt{\alpha^2 + \beta^2 + \gamma^2}$, $a_N = 0$, $\kappa = 0$,

$\mathbf{T} = \dfrac{1}{\sqrt{\alpha^2 + \beta^2 + \gamma^2}}(\alpha\mathbf{i} + \beta\mathbf{j} + \gamma\mathbf{k})$, $\mathbf{N}$ can be any unit vector perpendicular to $\mathbf{T}$, and then $\mathbf{B} = \mathbf{T} \times \mathbf{N}$

17. Use equation (15.7) to get $\mathbf{B}'(s) \cdot \mathbf{N}(s) = -\tau(s)\mathbf{N}(s) \cdot \mathbf{N}(s) = -\tau(s)$, since $\|\mathbf{N}\|^2 = 1$.

19. $\tau(s) = -N(s) \cdot B'(s) = -N(s) \cdot [T(s) \times N(s)]' = -N(s) \cdot [T'(s) \times N(s) + T(s) \times N'(s)] = -[N(s), T'(s), N(s)] -$

$[N(s), T(s), N'(s)] = -[N(s), T(s), N'(s)] = [T(s), N(s), N'(s)]$. But $F'(s) = T(s)$, so $N(s) = \frac{1}{\kappa} T'(s) = \frac{1}{\kappa} F''(s)$. Then $N'(s) =$

$\left(\frac{1}{\kappa} F''(s)\right)' = \left(\frac{1}{\kappa}\right)' F''(s) + \frac{1}{\kappa} F^{(3)}(s)$. Therefore, $\tau(s) = \left[F'(s), \frac{1}{\kappa} F''(s), \left(\frac{1}{\kappa}\right)' F''(s) + \frac{1}{\kappa} F^{(3)}(s)\right] = \left[F'(s), \frac{1}{\kappa} F''(s), \left(\frac{1}{\kappa}\right)' F''(s)\right] +$

$\left[F'(s), \frac{1}{\kappa} F''(s), \frac{1}{\kappa} F^{(3)}(s)\right] = \frac{1}{\kappa}\left(\frac{1}{\kappa}\right)' [F'(s), F''(s), F''(s)] + \frac{1}{\kappa^2}[F'(s), F''(s), F^{(3)}(s)] = \frac{1}{\kappa^2}[F'(s), F''(s), F^{(3)}(s)]$.

Section 15.3

1. $\dfrac{\partial G}{\partial x} = 3i - 4yj, \dfrac{\partial G}{\partial y} = -4xj$

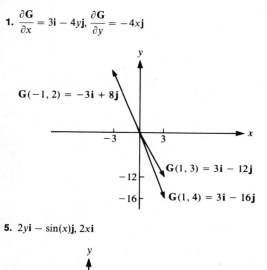

$G(-1, 2) = -3i + 8j$

$G(1, 3) = 3i - 12j$

$G(1, 4) = 3i - 16j$

3. $4yi, 4xi - j$

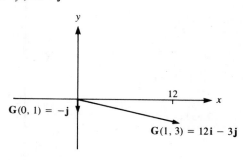

$G(0, 1) = -j$

$G(1, 3) = 12i - 3j$

5. $2yi - \sin(x)j, 2xi$

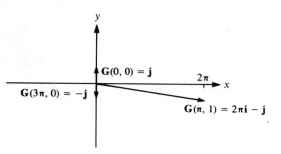

$G(0, 0) = j$

$G(3\pi, 0) = -j$

$G(\pi, 1) = 2\pi i - j$

7. $-e^{-x}yi + 8yj, e^{-x}i + 8xj$

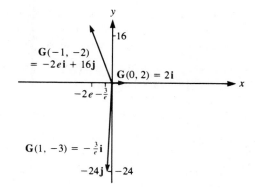

$G(-1, -2) = -2ei + 16j$

$G(0, 2) = 2i$

$G(1, -3) = -\frac{3}{e}i$

9. $\dfrac{2y^2}{x + 1}i + 8y^3j, 4\ln(x + 1)yi + 24xy^2j$

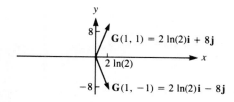

$G(1, 1) = 2\ln(2)i + 8j$

$G(1, -1) = 2\ln(2)i - 8j$

11. $ye^{xy}\mathbf{i} - 4xy\mathbf{j}, xe^{xy}\mathbf{i} - 2x^2\mathbf{j} + \sinh(z+y)\mathbf{k}, \sinh(z+y)\mathbf{k}$

13. $3y^3\mathbf{i} + \dfrac{1}{x+y+z}\mathbf{j} + yz\sinh(xyz)\mathbf{k}, 9xy^2\mathbf{i} + \dfrac{1}{x+y+z}\mathbf{j} + xz\sinh(xyz)\mathbf{k}, \dfrac{1}{x+y+z}\mathbf{j} + xy\sinh(xyz)\mathbf{k}$

15. $14\mathbf{i} + 2x\mathbf{j} + 5y\mathbf{k}, -2\mathbf{i} - 2y\mathbf{j} + 5x\mathbf{k}, -2z\mathbf{j}$ **17.** $x = z + c_1, y = -2z + c_2, z = z; x = z - 1, y = -2z + 3, z = z$

19. $x = x, y = \sin^{-1}[c_1 - \cos(x)], z = c_2; x = x, y = \sin^{-1}[-\cos(x)], z = -4$

21. $x = \dfrac{1}{3\ln(y) + c_1}, y = y, z = \dfrac{1}{\sqrt{2\ln(y) + c_2}}; x = \dfrac{1}{3\ln(y) + \frac{1}{2}}, y = y, z = \dfrac{1}{\sqrt{2\ln(y) + \frac{1}{36}}}$

23. $x = \dfrac{1}{c_1 - \frac{1}{z}}, y = \dfrac{1}{c_2 - \frac{1}{z}}, z = z; x = \dfrac{1}{2 - \frac{1}{z}}, y = \dfrac{1}{2 - \frac{1}{z}}, z = z$

25. $x = \dfrac{2}{c_1}\tan^{-1}\left[\dfrac{\tan\left(\frac{y}{2}\right) - \frac{1}{c_1}}{\sqrt{1 - \frac{1}{c_1^2}}}\right] + c_2, y = y, z = \ln\left(\dfrac{c_1 + \sin(y)}{2}\right);$ for the solution passing through $\left(2, \dfrac{\pi}{4}, 0\right)$, choose

$c_1 = 2 - \sqrt{2}$ and $c_2 = 1 - \dfrac{2}{2 - \sqrt{2}}\tan^{-1}\dfrac{\tan\left(\frac{\pi}{8}\right) - \frac{1}{2 - \sqrt{2}}}{\sqrt{1 - \frac{1}{(2 - \sqrt{2})^2}}}$

27. $y\mathbf{i} - x\mathbf{j}$

Section 15.4

1. $yz\mathbf{i} + xz\mathbf{j} + xy\mathbf{k}; \mathbf{i} + \mathbf{j} + \mathbf{k}$ **3.** $(2y + e^z)\mathbf{i} + 2x\mathbf{j} + xe^z\mathbf{k}; (2 + e^6)\mathbf{i} - 4\mathbf{j} - 2e^6\mathbf{k}$

5. $2y\sinh(2xy)\mathbf{i} + 2x\sinh(2xy)\mathbf{j} - \cosh(z)\mathbf{k}; -\cosh(1)\mathbf{k}$ **7.** $\dfrac{1}{x+y+z}[\mathbf{i} + \mathbf{j} + \mathbf{k}]; \dfrac{1}{2}[\mathbf{i} + \mathbf{j} + \mathbf{k}]$

9. $e^x\cos(y)\cos(z)\mathbf{i} - e^x\sin(y)\cos(z)\mathbf{j} - e^x\cos(y)\sin(z)\mathbf{k}; \dfrac{1}{2}[\mathbf{i} - \mathbf{j} - \mathbf{k}]$

11. $[2xy\cosh(xz) + x^2yz\sinh(xz)]\mathbf{i} + x^2\cosh(xz)\mathbf{j} + x^3y\sinh(xz)\mathbf{k}; \mathbf{O}$

13. $\mathbf{i} - 2\sinh(y+z)[\mathbf{j} + \mathbf{k}]; \mathbf{i} + 2\sinh(1)[\mathbf{j} + \mathbf{k}]$ **15.** $\sinh(x - y + 2z)\mathbf{i} - \sinh(x - y + 2z)\mathbf{j} + 2\sinh(x - y + 2z)\mathbf{k}; \mathbf{O}$

17. $-2x + y - z = 1; x = -1 - 2t, y = 1 + t, z = 2 - t$ **19.** $x + y + z = 0; x = y = z = t$

21. $x - y = 0; x = 1 + 2t, y = 1 - 2t, z = 0$ **23.** $x + y + 2z = 4; x = 1 + t, y = 1 + t, z = 1 + 2t$

25. $y + z = \pi; x = 0, y = \pi + t, z = t$ **27.** $\dfrac{\pi}{4}$ **29.** 0 **31.** The direction is that of $-\mathbf{i}$; the magnitude is e.

33. $\dfrac{-\frac{1}{4}\mathbf{i} - 3\mathbf{j} - 3\mathbf{k}}{\left\| -\frac{1}{4}\mathbf{i} - 3\mathbf{j} - 3\mathbf{k}\right\|}; \dfrac{17}{4}$ **35.** $\dfrac{16\mathbf{i} - 6\mathbf{j}}{\|16\mathbf{i} - 6\mathbf{j}\|}; 2\sqrt{73}$ **37.** $\dfrac{\sin(1)\mathbf{i} + \cos(1)\mathbf{j} - 36\mathbf{k}}{\|\sin(1)\mathbf{i} + \cos(1)\mathbf{j} - 36\mathbf{k}\|}; \sqrt{1297}$ **39.** $\dfrac{1}{\sqrt{2}}(\mathbf{i} + \mathbf{j}), \dfrac{1}{\sqrt{2}}$

41. $\varphi = $ constant $\neq 0$; no normal vector

Section 15.5

1. $4; \mathbf{O}$ **3.** $2y + e^y + 2; -2x\mathbf{k}$ **5.** $-2zy; -y^2\mathbf{i} - 2e^z\mathbf{j}$ **7.** $-1; \mathbf{O}$ **9.** $2(x + y + z); \mathbf{O}$

11. $\nabla\varphi = \mathbf{i} - \mathbf{j} + 4z\mathbf{k}$ **13.** $-6x^2yz^2\mathbf{i} - 2x^3z^2\mathbf{j} - 4x^3yz\mathbf{k}$ **15.** $3x^2y^2e^z\mathbf{i} + 2x^3ye^z\mathbf{j} + x^3y^2e^z$

17. $2e^x\ln(yz)\mathbf{i} + \dfrac{2e^x}{y}\mathbf{j} + \dfrac{2e^x}{z}\mathbf{k}$ **19.** $-\sin(x + y + z)(\mathbf{i} + \mathbf{j} + \mathbf{k})$ **21.** $3x\mathbf{j}$

23. $\nabla \cdot (\varphi\mathbf{F}) = \varphi(\nabla \cdot \mathbf{F}) + \mathbf{F} \cdot (\nabla\varphi); \varphi(\nabla \times \mathbf{F}) + (\nabla\varphi) \times \mathbf{F}$

25. Let $\mathbf{F} = f_1\mathbf{i} + f_2\mathbf{j} + f_3\mathbf{k}$ and $\mathbf{G} = g_1\mathbf{i} + g_2\mathbf{j} + g_3\mathbf{k}$. Then $\nabla \cdot (\mathbf{F} \times \mathbf{G}) = \nabla \cdot [(f_2g_3 - f_3g_2)\mathbf{i} + (f_3g_1 - f_1g_3)\mathbf{j} +$

$(f_1g_2 - f_2g_1)\mathbf{k}] = \dfrac{\partial}{\partial x}(f_2g_3 - f_3g_2) + \dfrac{\partial}{\partial y}(f_3g_1 - f_1g_3) + \dfrac{\partial}{\partial z}(f_1g_2 - f_2g_1) = f_2\dfrac{\partial g_3}{\partial x} - f_3\dfrac{\partial g_2}{\partial x} + g_3\dfrac{\partial f_2}{\partial x} - g_2\dfrac{\partial f_3}{\partial x} +$

$$f_3 \frac{\partial g_1}{\partial y} - f_1 \frac{\partial g_3}{\partial y} + g_1 \frac{\partial f_3}{\partial y} - g_3 \frac{\partial f_1}{\partial y} + f_1 \frac{\partial g_2}{\partial z} - f_2 \frac{\partial g_1}{\partial z} + g_2 \frac{\partial f_1}{\partial z} - g_1 \frac{\partial f_2}{\partial z} = g_1 \left[\frac{\partial f_3}{\partial y} - \frac{\partial f_2}{\partial z} \right] + g_2 \left[\frac{\partial f_1}{\partial z} - \frac{\partial f_3}{\partial x} \right] +$$

$$g_3 \left[\frac{\partial f_2}{\partial x} - \frac{\partial f_1}{\partial y} \right] + f_1 \left[\frac{\partial g_2}{\partial z} - \frac{\partial g_3}{\partial y} \right] + f_2 \left[\frac{\partial g_3}{\partial x} - \frac{\partial g_1}{\partial z} \right] + f_3 \left[\frac{\partial g_1}{\partial y} - \frac{\partial g_2}{\partial x} \right] = \mathbf{G} \cdot (\nabla \times \mathbf{F}) - \mathbf{F} \cdot (\nabla \times \mathbf{G}).$$

27. (a) $r^n = [x^2 + y^2 + z^2]^{n/2}$, so $\frac{\partial}{\partial x} r^n = nx[x^2 + y^2 + z^2]^{n/2-1}$. Then $\nabla r^n = [x^2 + y^2 + z^2]^{n/2-1} n(x\mathbf{i} + y\mathbf{j} + z\mathbf{k}) = nr^{n-2}\mathbf{R}$.

(b) $\nabla \times (\varphi(r)\mathbf{R}) = \left[\frac{\partial}{\partial y}[z\varphi(r)] - \frac{\partial}{\partial z}[y\varphi(r)] \right]\mathbf{i} + \left[\frac{\partial}{\partial z}[x\varphi(r)] - \frac{\partial}{\partial x}[z\varphi(r)] \right]\mathbf{j} + \left[\frac{\partial}{\partial x}[y\varphi(r)] - \frac{\partial}{\partial y}[x\varphi(r)] \right]\mathbf{k}$. The first

component of this vector is $z\frac{\partial \varphi}{\partial r}\frac{\partial r}{\partial y} - y\frac{\partial \varphi}{\partial r}\frac{\partial r}{\partial z} = z\varphi'(r)y(x^2 + y^2 + z^2)^{-1/2} - y\varphi'(r)z(x^2 + y^2 + z^2)^{-1/2} = 0$.

Similarly, the other components of $\nabla \times (\varphi(r)\mathbf{R})$ are zero; hence, $\nabla \times (\varphi(r)\mathbf{R}) = \mathbf{O}$.

29. (a) Let $\mathbf{A} = a_1\mathbf{i} + a_2\mathbf{j} + a_3\mathbf{k}$. Then $\mathbf{R} \cdot \mathbf{A} = a_1x + a_2y + a_3z$, so $\nabla(\mathbf{R} \cdot \mathbf{A}) = a_1\mathbf{i} + a_2\mathbf{j} + a_3\mathbf{k} = \mathbf{A}$.

(b) $\nabla \cdot (\mathbf{R} - \mathbf{A}) = \nabla \cdot \mathbf{R} - \nabla \cdot \mathbf{A} = \nabla \cdot \mathbf{R} = \frac{\partial x}{\partial x} + \frac{\partial y}{\partial y} + \frac{\partial z}{\partial z} = 3$

(c) $\nabla \times (\mathbf{R} - \mathbf{A}) = \left(\frac{\partial}{\partial y}(z - a_3) - \frac{\partial}{\partial z}(y - a_2) \right)\mathbf{i} + \left(\frac{\partial}{\partial z}(x - a_1) - \frac{\partial}{\partial x}(z - a_3) \right)\mathbf{j} + \left(\frac{\partial}{\partial x}(y - a_2) - \frac{\partial}{\partial y}(x - a_1) \right)\mathbf{k} = \mathbf{O}$

Additional Problems for Chapter 15

1. $(\mathbf{F} \cdot \mathbf{G})' = 4\cos(2t) - 8t\sin(2t) + 4$; $(\mathbf{F} \times \mathbf{G})' = -12t^2\mathbf{i} + [32t + 2\sin(2t)]\mathbf{j} + 2t[t\sin(2t) - \cos(2t)]\mathbf{k}$

3. $e^{-t}[t^{-1/2} - 2t^{1/2}] + \cosh(t) - 4t$; $[-2t\cosh(t) - 1 - 2\sinh(t)]\mathbf{i} +$
$[e^{-t} + 6t^{1/2} - te^{-t}]\mathbf{j} + [t^{-1/2} - e^{-t}\cosh(t) + e^{-t}\sinh(t)]\mathbf{k}$

5. $-\sinh(3t) - 3t\cosh(3t)$; $3\cosh(3t)\mathbf{i} - [2t\sinh(3t) + 3t^2\cosh(3t)]\mathbf{k}$

7. $2t - \sinh(t) - 4$; $[4\sinh(t) - 1]\mathbf{i} - 3t^2\mathbf{j} + t[2\cosh(t) - t\sinh(t)]\mathbf{k}$　　**9.** $3t^2 + \frac{1}{t} - 1$; $-4t^3\mathbf{i} - t^2[1 + 3\ln(t)]\mathbf{j} + \left(\frac{1}{t} + 1\right)\mathbf{k}$

11. $\mathbf{v} = 4t\mathbf{i} + 4t\mathbf{j} - \sin(t)\mathbf{k}$, $v = \sqrt{32t^2 + \cos^2(t)}$, $\mathbf{a} = 4\mathbf{i} + 4\mathbf{j} - \cos(t)\mathbf{k}$, $a_T = \frac{64t + \sin(2t)}{\sqrt{32t^2 + \sin^2(t)}}$, $a_N = \frac{4\sqrt{2}|\sin(t) - t\cos(t)|}{\sqrt{32t^2 + \sin^2(t)}}$,

$\kappa = \frac{4\sqrt{2}|\sin(t) - t\cos(t)|}{[32t^2 + \sin^2(t)]^{3/2}}$, $\rho = \frac{1}{\kappa}$, $\tau = 0$

13. $\mathbf{v} = 2t\mathbf{i} + \mathbf{j} + e^t\mathbf{k}$, $v = \sqrt{4t^2 + 1 + e^{2t}}$, $\mathbf{a} = 2\mathbf{i} + e^t\mathbf{k}$, $a_T = \frac{4t + e^{2t}}{\sqrt{4t^2 + 1 + e^{2t}}}$, $a_N = \frac{\sqrt{4 + e^{2t}(5 + 4t^2 - 8t)}}{\sqrt{4t^2 + 1 + e^{2t}}}$,

$\kappa = \frac{\sqrt{4 + e^{2t}(5 + 4t^2 - 8t)}}{(4t^2 + 1 + e^{2t})^{3/2}}$, $\tau = -2e^t\left[\frac{1 + e^{2t} + 4t^2}{4 + e^{2t}(5 + 4t^2 - 8t)}\right]^{3/2}$

15. $\mathbf{v} = 9t^2\mathbf{i} + 2\mathbf{j} + 2\mathbf{k}$, $v = \sqrt{81t^4 + 8}$, $\mathbf{a} = 18t\mathbf{i}$, $a_T = \frac{162t^3}{\sqrt{81t^4 + 8}}$, $a_N = \frac{36\sqrt{2}t}{\sqrt{81t^4 + 8}}$, $\kappa = \frac{36\sqrt{2}t}{(81t^4 + 8)^{3/2}}$, $\tau = 0$

17. $x = x$, $y = -\frac{1}{8}x^2 + c_1$, $z = c_2x^{1/4}$　　**19.** $x = -\frac{1}{z} + c_1$, $y = c_2e^{3/2z^2}$, $z = z$

21. The direction is that of $2\mathbf{i} + \mathbf{j}$; maximum rate of change is $\sqrt{5}$.　　**23.** $-\frac{1}{e}(\mathbf{i} + \mathbf{j} + \mathbf{k})$; $\frac{1}{e}\sqrt{3}$

25. $-\frac{\sqrt{2}}{2}\pi\mathbf{i} - \frac{\sqrt{2}}{2}\mathbf{k}$; $\frac{\sqrt{2}}{2}\sqrt{1 + \frac{\pi^2}{16}}$　　**27.** $x + y - \sqrt{2}z = 0$; $x = 1 + 2t = y$, $z = \sqrt{2}(1 - 2t)$

29. $2x - y - z = 0$; $x = 1 + 4t$, $y = 1 - 2t = z$　　**31.** $\nabla \cdot \mathbf{F} = y$; $\nabla \times \mathbf{F} = 3z^2\mathbf{i} - \mathbf{j} - x\mathbf{k}$　　**33.** 0; $(xe^{xy} + 3z^2)\mathbf{i} + -ye^{xy}\mathbf{j}$
35. 0; $-e^z\mathbf{i} + [-\cos(y - z) + 1]\mathbf{j} - \cos(y - z)\mathbf{k}$

Chapter 16

Section 16.1

1. 32　　**3.** $\sin(1) - \frac{1}{2}$　　**5.** -4　　**7.** 0　　**9.** $-\frac{422}{5}$　　**11.** $48\sqrt{2}$　　**13.** $-12e^4 + 4e^2$　　**15.** 2192　　**17.** 0

19. $\dfrac{58}{15}$ **21.** -129 **23.** 1980   **25.** 0   **27.** $\dfrac{1}{3} 26\sqrt{3}$ **29.** $27\sqrt{2}$ **31.** $\dfrac{1}{2} 27\sqrt{3}$ grams; $(2, 2, 2)$

33. $-\dfrac{27}{2}$ **35.** $64; \left(\dfrac{239}{64}, \dfrac{145}{64}, 3\right)$

37. Let C be the curve parametrized by $x = t$, $y = z = 0$, for $a \le t \le b$. Let $\mathbf{F} = f(x)\mathbf{i}$. Then $\displaystyle\int_C \mathbf{F} \cdot d\mathbf{R} = \int_a^b f(x)\,dx$.

Section 16.2

1. -48π **3.** 20   **5.** 125   **7.** 0   **9.** -2π **11.** $-\dfrac{4}{15}$ **13.** $\dfrac{15}{4} e^3 - \dfrac{63}{4} e$ **15.** -12π

17. $\displaystyle\oint_C x\,dy = \iint_D \left[\frac{\partial}{\partial x}[x] - \frac{\partial}{\partial y}[0] \right] dA = \iint_D dA = $ area of D.

19. The centroid of D has coordinates $\bar{x} = \dfrac{1}{A(D)} \displaystyle\iint_D x\,dA$ and $\bar{y} = \dfrac{1}{A(D)} \displaystyle\iint_D y\,dA$. By Green's theorem,

$$\bar{x} = \frac{1}{A(D)} \iint_D x\,dA = \frac{1}{2A(D)} \iint_D \frac{\partial}{\partial x}[x^2]\,dA = \frac{1}{2A(D)} \int_C x^2\,dy. \text{ Similarly,}$$

$$\bar{y} = \frac{1}{A(D)} \iint_D y\,dA = \frac{1}{2A(D)} \iint_D \frac{\partial}{\partial y}[y^2]\,dA = \frac{-1}{2A(D)} \int_C y\,dx.$$

21. By the result of Problem 18, $A(D) = \dfrac{1}{2} \displaystyle\oint_C -y\,dx + x\,dy$. Referring to the diagram, on $\overline{OA}$, $x = r\cos(\alpha)$, $y = r\sin(\alpha)$, $0 \le r \le f(\alpha)$.

On the curve C_{AB} from A to B, $x = f(\theta)\cos(\theta)$, $y = f(\theta)\sin(\theta)$, $\alpha \le \theta \le \beta$. On $\overline{BO}$, $x = r\cos(\beta)$, $y = r\sin(\beta)$, $0 \le r \le f(\beta)$.

Therefore, $\displaystyle\int_{\overline{OA}} -y\,dx + x\,dy = \int_0^{f(\alpha)} [-r\sin(\alpha)\cos(\alpha) + r\cos(\alpha)\sin(\alpha)]\,d\alpha = 0$, $\displaystyle\int_{\overline{BO}} -y\,dx + x\,dy =$

$\displaystyle\int_{f(\beta)}^0 [-r\sin(\beta)\cos(\beta) + r\cos(\beta)\sin(\beta)]\,d\beta = 0$, and $\displaystyle\int_{C_{AB}} -y\,dx + x\,dy = \int_\alpha^\beta [-f(\theta)\sin(\theta)[f'(\theta)\cos(\theta) - f(\theta)\sin(\theta)] +$

$f(\theta)\cos(\theta)[f'(\theta)\sin(\theta) + f(\theta)\cos(\theta)]]\,d\theta = \displaystyle\int_\alpha^\beta [f(\theta)]^2\,d\theta$. Thus, area of $D = \dfrac{1}{2} \displaystyle\int_\alpha^\beta [f(\theta)]^2\,d\theta$.

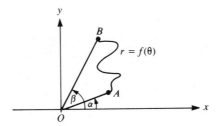

23. 72π

Section 16.3

1. Conservative   **3.** Conservative   **5.** Conservative   **7.** Conservative   **9.** Not conservative   **11.** -21

13. $5 + \ln\left(\dfrac{3}{2}\right)$ **15.** -5 **17.** $e^3 - 24e - 9e^{-1}$ **19.** $8\cosh(2) - 12$

21. (a) 2π (b) Both partial derivatives equal $-\dfrac{x^2 - y^2}{(x^2 + y^2)^2}$.

(c) $\dfrac{-y}{x^2 + y^2}$ and $\dfrac{x}{x^2 + y^2}$ are not continuous on the set enclosed by C (since the origin is in this set).

23. xyz^3 is one potential function; -24 **25.** e^{xyz}; $1 - e^{-2}$

27. If φ is a potential function for $\mathbf{F}$, then $\dfrac{\partial \varphi}{\partial x} = x^2$, so $\varphi(x, y, z) = \dfrac{1}{3} x^3 + c(y, z)$. Then $\dfrac{\partial \varphi}{\partial y} = \dfrac{\partial c}{\partial y} = -y$, so

$c(y, z) = -\dfrac{1}{2} y^2 + k(z)$. Thus far, $\varphi(x, y, z) = \dfrac{1}{3} x^3 - \dfrac{1}{2} y^2 + k(z)$. Then $\dfrac{\partial \varphi}{\partial z} = k'(z) = z \cos(x)$, impossible if k is a function of

z alone. Thus, no potential function exists.

Section 16.4

1. $\dfrac{\sqrt{2}}{8} (10)^3$ **3.** $\dfrac{\pi}{6} [29^{3/2} - 27]$ **5.** $\dfrac{28\pi}{3} \sqrt{2}$ **7.** $\dfrac{9}{8} \left[\sinh^{-1}(4) + \dfrac{1}{2} \sinh(2 \sinh^{-1}(4)) \right]$ **9.** $-10\sqrt{3}$ **11.** $\dfrac{7}{12}$

13. $9\pi\sqrt{2}$; $(0, 0, 2)$ **15.** 78π; $\left(0, 0, \dfrac{27}{13}\right)$ **17.** $\dfrac{128}{3}$ **19.** 0 **21.** 60 **23.** $\dfrac{64\sqrt{3}}{3}$ **25.** $\dfrac{1}{48} \left[\dfrac{391}{5} \sqrt{17} - 2\sqrt{5} \right]$

Section 16.5

1. By Green's theorem, $\displaystyle\oint_C -\varphi \dfrac{\partial \psi}{\partial y} \, dx + \varphi \dfrac{\partial \psi}{\partial x} \, dy = \iint_D \left(\dfrac{\partial}{\partial x} \left[\varphi \dfrac{\partial \psi}{\partial x} \right] + \dfrac{\partial}{\partial y} \left[\varphi \dfrac{\partial \psi}{\partial y} \right] \right) dA = \iint_D \left[\left(\dfrac{\partial \varphi}{\partial x} \dfrac{\partial \psi}{\partial x} + \varphi \dfrac{\partial^2 \psi}{\partial x^2} \right) + \right.$

$\left. \left(\dfrac{\partial \varphi}{\partial y} \dfrac{\partial \psi}{\partial y} + \varphi \dfrac{\partial^2 \psi}{\partial y^2} \right) \right] dA = \iint_D \left[\dfrac{\partial \varphi}{\partial x} \dfrac{\partial \psi}{\partial x} + \dfrac{\partial \varphi}{\partial y} \dfrac{\partial \psi}{\partial y} \right] dA + \iint_D \varphi \left[\dfrac{\partial^2 \psi}{\partial x^2} + \dfrac{\partial^2 \psi}{\partial y^2} \right] dA = \iint_D \nabla \varphi \cdot \nabla \psi \, dA + \iint_D \varphi \nabla^2 \psi \, dA.$

3. Since $D_{\mathbf{N}}\varphi(x, y) = \nabla\varphi(x, y) \cdot \mathbf{N}$, $\displaystyle\oint_C D_{\mathbf{N}}\varphi(x, y) \, ds = \oint_C \nabla\varphi \cdot \mathbf{N} \, ds = \oint_C -\dfrac{\partial \varphi}{\partial y} \, dx + \dfrac{\partial \varphi}{\partial x} \, dy =$

$\displaystyle\iint_D \left(\dfrac{\partial}{\partial x} \left[\dfrac{\partial \varphi}{\partial x} \right] - \dfrac{\partial}{\partial y} \left[-\dfrac{\partial \varphi}{\partial y} \right] \right) dA = \iint_D \left[\dfrac{\partial^2 \varphi}{\partial x^2} + \dfrac{\partial^2 \varphi}{\partial y^2} \right] dA = \iint_D \nabla^2\varphi \, dA.$

Section 16.6

1. $\dfrac{256}{3} \pi$ **3.** 0 **5.** $\dfrac{8}{3} \pi$ **7.** 8π **9.** 0

11. By Gauss's theorem, $\displaystyle\iint_\Sigma (\nabla \times \mathbf{F}) \cdot \mathbf{N} \, d\sigma = \iiint_M \nabla \cdot (\nabla \times \mathbf{F}) \, dV = 0$, because $\nabla \cdot (\nabla \times \mathbf{F}) = \mathbf{O}$.

13. By the result of Problem 12, applied once with φ and ψ interchanged, $\displaystyle\iint_\Sigma (\varphi \nabla \psi - \psi \nabla \varphi) \cdot \mathbf{N} \, d\sigma = \iiint_M (\varphi \nabla^2 \psi +$

$\nabla\varphi \cdot \nabla\psi) \, dV - \displaystyle\iiint_M (\psi \nabla^2 \varphi + \nabla\psi \cdot \nabla\varphi) \, dV = \iiint_M (\varphi \nabla^2 \psi - \psi \nabla^2 \varphi) \, dV.$

15. By Gauss's theorem, $\displaystyle\iint_\Sigma \mathbf{K} \cdot \mathbf{N} \, d\sigma = \iiint_M \nabla \cdot \mathbf{K} \, dV = 0$, since $\nabla \cdot \mathbf{K} = 0$.

17. On Σ_2, if $\mathbf{n}_2 = -\dfrac{\partial K}{\partial x} \mathbf{i} - \dfrac{\partial K}{\partial y} \mathbf{j} + \mathbf{k}$, $\mathbf{N}_2 = \dfrac{1}{\|\mathbf{n}_2\|} \mathbf{n}_2$ is a unit outer normal. On Σ_1, let $\mathbf{n}_1 = -\dfrac{\partial H}{\partial x} \mathbf{i} - \dfrac{\partial H}{\partial y} \mathbf{j} + \mathbf{k}$,

so $\mathbf{N}_1 = \dfrac{1}{\|\mathbf{n}_1\|} \mathbf{n}_1$ is a unit outer normal. On Σ_3, a unit outer normal is $\mathbf{N}_3 = \cos(\theta)\mathbf{i} + \cos(\theta)\mathbf{j}$. Now, $\displaystyle\iint_{\Sigma_1} h(x, y, z)\mathbf{k} \cdot \mathbf{N} \, d\sigma =$

$\displaystyle\iint_{\Sigma_1} h(x, y, H(x, y))\mathbf{k} \cdot \mathbf{N}_1 \, d\sigma = -\iint_D h(x, y, H(x, y)) \, dA.$ Similarly, $\displaystyle\iint_{\Sigma_2} h(x, y, z)\mathbf{k} \cdot \mathbf{N} \, d\sigma = \iint_D h(x, y, K(x, y)) \, dA.$

Further, $\displaystyle\iint_{\Sigma_3} h(x, y, z)\mathbf{k} \cdot \mathbf{N}\, d\sigma = 0$. Therefore, $\displaystyle\iint_{\Sigma} h(x, y, z)\mathbf{k} \cdot \mathbf{N}\, d\sigma = \iint_{\Sigma_1} h(x, y, z)\mathbf{k} \cdot \mathbf{N}_1\, d\sigma + \iint_{\Sigma_2} h(x, y, z)\mathbf{k} \cdot \mathbf{N}_2\, d\sigma +$

$\displaystyle\iint_{\Sigma_3} h(x, y, z)\mathbf{k} \cdot \mathbf{N}_3\, d\sigma = \iint_{D} [h(x, y, K(x, y)) - h(x, y, H(x, y))]\, dA$. But $\displaystyle\iiint_{M} \frac{\partial h}{\partial z}\, dV = \iint_{D} \int_{H(x,y)}^{K(x,y)} \frac{\partial h}{\partial z}\, dV =$

$\displaystyle\iint_{D} [h(x, y, K(x, y)) - h(x, y, H(x, y))]\, dA$. Therefore, $\displaystyle\iint_{\Sigma} h(x, y, z)\mathbf{k} \cdot \mathbf{N}\, d\sigma = \iiint_{M} \frac{\partial h}{\partial z}\, dV$. By a similar argument, we find that

$\displaystyle\iint_{\Sigma} f(x, y, z)\mathbf{i} \cdot \mathbf{N}\, d\sigma = \iiint_{M} \frac{\partial f}{\partial x}\, dV$ and $\displaystyle\iint_{\Sigma} g(x, y, z)\mathbf{j} \cdot \mathbf{N}\, d\sigma = \iiint_{M} \frac{\partial g}{\partial y}\, dV$. On adding these results, we obtain

the conclusion of Gauss's theorem.

Section 16.7

1. By Gauss's theorem, $\displaystyle\iint_{\Sigma} \nabla f \cdot \mathbf{N}\, d\sigma = \iiint_{M} \nabla^2 f\, dV$. By the mean value theorem for integrals, there is a point

(ξ, η, ζ) in M such that $\displaystyle\iiint_{M} \nabla^2 f\, dV = \nabla^2 f(\xi, \eta, \zeta)[\text{volume of } M] = \nabla^2 f(\xi, \eta, \zeta)\mathcal{V}$. Thus, $\displaystyle\frac{1}{\mathcal{V}}\iint_{\Sigma} \nabla f \cdot \mathbf{N}\, d\sigma = \nabla^2 f(\xi, \eta, \zeta)$,

and $\displaystyle\lim_{\mathcal{V} \to 0} \frac{1}{\mathcal{V}}\iint_{\Sigma} \nabla f \cdot \mathbf{N}\, d\sigma = \lim_{\mathcal{V} \to 0} \nabla^2 f(\xi, \eta, \zeta) = \nabla^2 f(P_0)$, by continuity of $\nabla^2 f$ at P_0.

For a physical interpretation of $\nabla^2 f$, let $f(x, y, z, t)$ be the temperature of the medium at time t and point (x, y, z). Let Σ be a smooth closed surface in the medium. By Faraday's law, the average amount of heat energy leaving M (the interior of Σ) across Σ in a time interval Δt is $\displaystyle\frac{1}{\mathcal{V}\Delta t}\iint_{\Sigma} \nabla f \cdot \mathbf{N}\, d\sigma\, \Delta t$. Thus, $\nabla^2 f(P_0)$ is the average amount

of heat energy leaving P_0.

3. Suppose that F and G are solutions, and let $w = F - G$. Then w is a solution of the problem $\dfrac{\partial u}{\partial t} = k\nabla^2 u + \varphi(x, y, z, t)$,

$u(x, y, z, t) = 0$ for (x, y, z) on Σ, and $w(x, y, z, 0) = 0$ for (x, y, z) in M. Let $\displaystyle I(t) = \frac{1}{2}\iiint_{M} w^2(x, y, z, t)\, dV$. Then

$\displaystyle I'(t) = \iiint_{M} w\frac{\partial w}{\partial t}\, dV = k\iiint_{M} w\nabla^2 w\, dV$. But $w\nabla^2 w = \nabla \cdot (w\nabla w) - \nabla w \cdot \nabla w$, and $\displaystyle\iiint_{M} \nabla \cdot (w\nabla w)\, dV = \iint_{\Sigma} (w\nabla w) \cdot \mathbf{N}\, d\sigma$

by Gauss's theorem. Since $w = 0$ on Σ, this surface integral is zero; hence, $\displaystyle\iiint_{M} \nabla \cdot (w\nabla w)\, dV = 0$. Therefore,

$\displaystyle I'(t) = -k\iiint_{M} \nabla w \cdot \nabla w\, dV = -k\iiint_{M} \left[\left(\frac{\partial w}{\partial x}\right)^2 + \left(\frac{\partial w}{\partial y}\right)^2 + \left(\frac{\partial w}{\partial y}\right)^2\right] dV \le 0$. Thus, $I(t)$ is nonincreasing for

$t > 0$. But $\displaystyle I(0) = \frac{1}{2}\iiint_{M} w^2(x, y, z, 0)\, dV = 0$. Thus, $I(t) \le 0$ for $t > 0$. Thus, $I(t) = 0$ for $t > 0$. Therefore,

$w = 0$ for $t > 0$ and (x, y, z) in M.

5. Let $w = f - g$. Then $\nabla^2 w = \nabla^2 f - \nabla^2 g = 0$ for (x, y, z) in M, and $\dfrac{\partial w}{\partial \eta} = 0$ for (x, y, z) on Σ. Now, $\nabla \cdot (w\nabla w) =$

$\nabla w \cdot \nabla w + w\nabla^2 w = \nabla w \cdot \nabla w$, since $\nabla^2 w = 0$. Therefore, $\displaystyle\iiint_{M} \nabla \cdot (w\nabla w)\, dV = \iiint_{M} \nabla w \cdot \nabla w\, dV = \iint_{\Sigma} w\nabla w \cdot \mathbf{N}\, d\sigma =$

$$\iint_{\Sigma} w \frac{\partial w}{\partial \eta} \, d\sigma = 0, \text{ using Gauss's theorem and the fact that } \frac{\partial w}{\partial \eta} = 0 \text{ on } \Sigma. \text{ Therefore, } \iiint_{M} \|\nabla w\|^2 \, dV = 0. \text{ Therefore,}$$

$\|\nabla w\|^2 = 0$; hence, $\|\nabla w\| = 0$ for (x, y, z) in M. Therefore, $\dfrac{\partial w}{\partial x} = \dfrac{\partial w}{\partial y} = \dfrac{\partial w}{\partial z} = 0$ in M, so $w = $ constant for (x, y, z) in M.

Section 16.8

1. -8π **3.** 16π **5.** $-\dfrac{32}{3}$ **7.** -3

9. A unit normal to Σ is $\mathbf{N} = \dfrac{1}{\sqrt{\left(\dfrac{\partial z}{\partial x}\right)^2 + \left(\dfrac{\partial z}{\partial y}\right)^2 + 1}} \left[-\dfrac{\partial z}{\partial x}\mathbf{i} - \dfrac{\partial z}{\partial y}\mathbf{j} + \mathbf{k}\right].$ Compute

$$(\nabla \times \mathbf{F}) \cdot \mathbf{N} = \frac{1}{\sqrt{\left(\dfrac{\partial z}{\partial x}\right)^2 + \left(\dfrac{\partial z}{\partial y}\right)^2 + 1}} \left[\left(\frac{\partial h}{\partial y} - \frac{\partial g}{\partial z}\right)\left(-\frac{\partial z}{\partial x}\right) + \left(\frac{\partial f}{\partial z} - \frac{\partial h}{\partial x}\right)\left(-\frac{\partial z}{\partial y}\right) + \left(\frac{\partial g}{\partial x} - \frac{\partial f}{\partial y}\right)\right].$$

Then

$$\iint_{\Sigma} (\nabla \times \mathbf{F}) \cdot \mathbf{N} \, d\sigma = \iint_{D} \left[\left(\frac{\partial h}{\partial y} - \frac{\partial g}{\partial z}\right)\left(-\frac{\partial z}{\partial x}\right) + \left(\frac{\partial f}{\partial z} - \frac{\partial h}{\partial x}\right)\left(-\frac{\partial z}{\partial y}\right) + \left(\frac{\partial g}{\partial x} - \frac{\partial f}{\partial y}\right)\right] dA.$$

If the boundary C^* of D is given by $x = x(t)$, $y = y(t)$, for $a \le t \le b$, the boundary C of Σ is given by $x = x(t)$, $y = y(t)$, $z = S(x(t), y(t))$, $a \le t \le b$. Then

$$\oint_{C} \mathbf{F} \cdot d\mathbf{R} = \int_{a}^{b} [f(x(t), y(t), S(x(t), y(t))x'(t) + g(x(t), y(t), S(x(t), y(t))y'(t)$$

$$+ h(x(t), y(t), S(x(t), y(t))x'(t)\left(\frac{\partial z}{\partial x}x'(t) + \frac{\partial z}{\partial y}y'(t)\right)\right] dt$$

$$= \int_{a}^{b} \left[\left(f + \frac{\partial z}{\partial x}h\right)x'(t) + \left(g + \frac{\partial z}{\partial y}h\right)y'(t)\right] dt$$

$$= \oint_{C^*} \left(f + \frac{\partial z}{\partial x}h\right) dx + \left(g + \frac{\partial z}{\partial y}h\right) dy$$

$$= \iint_{D} \left[\frac{\partial}{\partial x}\left(g + \frac{\partial z}{\partial y}h\right) - \frac{\partial}{\partial y}\left(f + \frac{\partial z}{\partial x}h\right)\right] dA \qquad \text{(by Green's theorem)}$$

$$= \iint_{D} \left[\left(\frac{\partial h}{\partial y} - \frac{\partial g}{\partial z}\right)\left(-\frac{\partial z}{\partial x}\right) + \left(\frac{\partial f}{\partial z} - \frac{\partial h}{\partial x}\right)\left(-\frac{\partial z}{\partial y}\right) + \left(\frac{\partial g}{\partial x} - \frac{\partial f}{\partial y}\right)\right] dA.$$

Therefore, $\iint_{\Sigma} (\nabla \times \mathbf{F}) \cdot \mathbf{N} \, d\sigma = \oint_{C} \mathbf{F} \cdot d\mathbf{R}.$

Section 16.9

1. Not conservative **3.** $x - 2y + z^2$ **5.** $x^2 - 2y + z$ **7.** Not conservative
9. $\sin(x) - \cos(xy) + z$ **11.** $x - 3y^3 + z$; -403 **13.** $2x^3 e^{yz}$; $2e^{-2}$ **15.** $-x + 2yz^2$; 71
17. $4y^3x - 4x^2 - 4z^2$; -92 **19.** $z \sinh(xy)$; $-7 \sinh(1)$

21. By Gauss's theorem, $\int_{C} f(\nabla g) \cdot d\mathbf{R} = \iint_{\Sigma} \nabla \times (f \nabla g) \cdot \mathbf{N} \, d\sigma.$ But a straightforward calculation shows that

$\nabla \times (f \nabla g) = \nabla f \times \nabla g.$

23. $\oint_C \mathbf{B} \cdot d\mathbf{R} = \iint_\Sigma (\nabla \times \mathbf{B}) \cdot \mathbf{N}\, d\sigma = \iint_\Sigma (\mu_0 \mathbf{J}) \cdot \mathbf{N}\, d\sigma = \mu_0 i$, since $i = \iint_\Sigma \mathbf{J} \cdot \mathbf{N}\, d\sigma.$

Additional Problems for Chapter 16

1. $\dfrac{23}{2}$ **3.** $-\dfrac{2}{3}$ **5.** $2\cos(1)$ **7.** $\dfrac{10}{3} - 2\sinh(1)$ **9.** 0   **11.** 0   **13.** $-\dfrac{1}{2}$ **15.** 0

17. 0   **19.** $\dfrac{16\pi}{3}$ **21.** -4π **23.** $\dfrac{40\pi}{3}$ **25.** 0   **27.** $x^2 - 2yz + \cos(z);\ -12 + \cos(4)$ **29.** $x - y^2 + z;\ -1$

31. (a) If $\mathbf{T} = x'(s)\mathbf{i} + y'(s)\mathbf{j}$, then $\mathbf{N} = y'(s)\mathbf{i} - x'(s)\mathbf{j}$, so $\oint_C \left(f\dfrac{\partial g}{\partial n} - g\dfrac{\partial f}{\partial n} \right) ds = \oint_C \left(f\left[\dfrac{\partial g}{\partial x}y' - \dfrac{\partial g}{\partial y}x' \right] - \right.$

$\left. g\left[\dfrac{\partial f}{\partial x}y' - \dfrac{\partial f}{\partial y}x' \right] \right) ds = \oint_C \left[g\dfrac{\partial f}{\partial y} - f\dfrac{\partial g}{\partial y} \right] dx + \left[f\dfrac{\partial g}{\partial x} - g\dfrac{\partial f}{\partial x} \right] dy = \iint_D \left[\dfrac{\partial}{\partial x}\left(f\dfrac{\partial g}{\partial x} - g\dfrac{\partial f}{\partial x} \right) - \right.$

$\left. \dfrac{\partial}{\partial y}\left(g\dfrac{\partial f}{\partial y} - f\dfrac{\partial g}{\partial y} \right) \right] dA$, by Green's theorem. On carrying out the indicated differentiations, the last double integral is

$\iint_D \left[f\left(\dfrac{\partial^2 g}{\partial x^2} + f\dfrac{\partial^2 g}{\partial y^2} \right) - g\left(\dfrac{\partial^2 f}{\partial x^2} + f\dfrac{\partial^2 f}{\partial y^2} \right) \right] dA$, and this completes the proof, because

$\nabla \cdot \nabla g = \dfrac{\partial^2 g}{\partial x^2} + f\dfrac{\partial^2 g}{\partial y^2}$ and $\nabla \cdot \nabla f = \dfrac{\partial^2 f}{\partial x^2} + f\dfrac{\partial^2 f}{\partial y^2}.$

(b) Assume that $\nabla \cdot \nabla g = 0$ for (x, y) in D. Let $f(x, y) = 1$ in the result of (a) to get $\oint_C \dfrac{\partial g}{\partial n}\, ds =$

$\iint_D f(\nabla \cdot \nabla g)\, dA = \iint_D (\nabla \cdot \nabla g)\, dA = 0.$

(c) By (a), if $\nabla \cdot \nabla f = \nabla \cdot \nabla g = 0$ in D, then $\oint_C \left(f\dfrac{\partial g}{\partial n} - g\dfrac{\partial f}{\partial n} \right) ds = 0.$

Chapter 16 Appendices

Appendix A

1. Refer to Section 18.0.

3. Curves $u = $ constant are ellipses; curves $v = $ constant are hyperbolas; scale factors are $h_u = \alpha\sqrt{\sinh^2(u) + \sin^2(v)} = h_v,\ h_z = 1.$

5. Curves $u = $ constant and $v = $ constant are parabolas; $h_u = h_v = \sqrt{u^2 + v^2},\ h_z = 1.$

Appendix B

1. 0   **3.** 0   **5.** 2π **7.** 0   **9.** 0

Appendix C

1. 25   **3.** $\dfrac{1}{3}$ **5.** $\dfrac{65}{3}$ **7.** $\dfrac{7}{60}$ **9.** $\dfrac{1}{4}(e^2 - 3)$ **11.** $\dfrac{-128\sqrt{2}}{7}$ **13.** $\dfrac{325}{8}$ **15.** $\dfrac{1}{24}$

17. $\dfrac{9}{2}(e - e^{-2})$ **19.** $\dfrac{32}{3}(\sqrt{3} - 1)$

Appendix D

1. 0   **3.** $9\sqrt{3}$ **5.** $\dfrac{40}{3}$ **7.** $\dfrac{1026}{5}$ **9.** $\dfrac{154}{3}$ **11.** 3   **13.** 60   **15.** 0   **17.** 392π **19.** $\dfrac{1}{2}$

21. $\dfrac{\sqrt{6}-\sqrt{2}}{\sqrt{2}}$ **23.** $\dfrac{18\ln(2)-9}{4}$

Appendix E

1. $\pi(e^4-1)$ **3.** $\dfrac{2\pi}{5}$ **5.** $14\sin(4)+8\cos(4)$ **7.** $32\sqrt{2}$ **9.** 8π **11.** 2π **13.** $3\pi[\sqrt{2}e^{\sqrt{2}}-e^{\sqrt{2}}+1]$

15. $\dfrac{81\pi}{2}$ **17.** $\dfrac{26\pi}{3}$ **19.** $\dfrac{32\pi\sqrt{2}}{15}(3\sqrt{3}-2\sqrt{2})$ **21.** $\dfrac{256\pi}{3}$ **23.** $\dfrac{\pi}{2}[2-\tan^{-1}(2)]$

Chapter 17

Section 17.1

1. 4   **3.** $\cos(\pi x)$ **5.** $(e^2-e^{-2})\left[\dfrac{1}{8}+\displaystyle\sum_{n=1}^{\infty}\dfrac{(-1)^n}{n^2\pi^2+4}\left[\cos(n\pi x)-\dfrac{1}{2}n\pi\sin(n\pi x)\right]\right]$ **7.** $3+\dfrac{36}{\pi^2}\displaystyle\sum_{n=1}^{\infty}\dfrac{(-1)^n}{n^2}\cos\left(\dfrac{n\pi x}{3}\right)$

9. $\dfrac{8}{\pi^2}\displaystyle\sum_{n=1}^{\infty}\dfrac{1}{(2n-1)^2}\cos\left[\dfrac{(2n-1)\pi x}{2}\right]$ **11.** $\dfrac{3}{4}-\displaystyle\sum_{n=1}^{\infty}\left[\left[\dfrac{1-(-1)^n}{n^2\pi^2}\right]\cos(n\pi x)+\left[\dfrac{1-2(-1)^n}{n\pi}\right]\sin(n\pi x)\right]$ **13.** $\sin(2x)$

15. $\dfrac{2}{\pi}-\sin(x)-\dfrac{4}{\pi}\displaystyle\sum_{n=1}^{\infty}\left[\dfrac{(-1)^n}{4n^2-1}\right]\cos(nx)$

17. Write $\displaystyle\int_{-L}^{L}f(x)\,dx=\int_{-L}^{0}f(x)\,dx+\int_{0}^{L}f(x)\,dx$. In the first integral on the right, let $t=-x$ to get $\displaystyle\int_{-L}^{0}f(x)\,dx=$
$\displaystyle\int_{L}^{0}f(-t)(-1)\,dt=\int_{0}^{L}f(-x)\,dx$, in which we have again used x as the dummy variable of integration in the last

integral. Thus, $\displaystyle\int_{-L}^{L}f(x)\,dx=\int_{0}^{L}f(-x)\,dx+\int_{0}^{L}f(x)\,dx$. Now, if f is odd, $f(-x)=-f(x)$, so $\displaystyle\int_{-L}^{L}f(x)\,dx=$
$\displaystyle\int_{0}^{L}-f(x)\,dx+\int_{0}^{L}f(x)\,dx=0$. If f is even, $f(-x)=f(x)$, so $\displaystyle\int_{-L}^{L}f(x)\,dx=\int_{0}^{L}f(x)\,dx+\int_{0}^{L}f(x)\,dx=2\int_{0}^{L}f(x)\,dx$.

Section 17.2

In solutions 1 through 15, $S(x)$ denotes the value of the Fourier series of the given function at x.

1. $f(-3+)=-6,\ f(3-)=9,\ f(-2-)=-4,\ f(-2+)=0,\ f(1-)=0,\ f(1+)=1;\ f'_{\mathscr{R}}(-3)=2,\ f'_{\mathscr{L}}(-2)=2,\ f'_{\mathscr{R}}(-2)=0,$
$f'_{\mathscr{L}}(1)=0,\ f'_{\mathscr{R}}(1)=2,\ f'_{\mathscr{L}}(3)=6;\ S(-3)=S(3)=\dfrac{3}{2},\ S(-2)=-2,\ S(1)=\dfrac{1}{2}$

3. $f(-3+)=9e^3,\ f(3-)=9e^{-3};\ f'_{\mathscr{R}}(-3)=-15e^3,\ f'_{\mathscr{L}}(3)=-3e^{-3};\ S(-3)=S(3)=\dfrac{9}{2}(e^3+e^{-3})=9\cosh(3)$

5. $f(-\pi+)=\pi^2,\ f(\pi-)=2,\ f(0-)=0,\ f(0+)=2;\ f'_{\mathscr{R}}(-\pi)=-2\pi,\ f'_{\mathscr{L}}(\pi)=0,\ f'_{\mathscr{R}}(0)=0,\ f'_{\mathscr{L}}(0)=0;$
$S(\pi)=S(-\pi)=\dfrac{1}{2}(\pi^2+2),\ S(0)=1$

7. $f(-\pi+)=-1,\ f(\pi-)=0,\ f(0-)=1,\ f(0+)=0;\ f'_{\mathscr{R}}(-\pi)=0,\ f'_{\mathscr{L}}(\pi)=-1,\ f'_{\mathscr{R}}(0)=1,\ f'_{\mathscr{L}}(0)=0;$
$S(-\pi)=S(\pi)=-\dfrac{1}{2},\ S(0)=\dfrac{1}{2}$

9. $f(-2+)=e^{-2},\ f(2-)=e^{-2};\ f'_{\mathscr{R}}(-2)=e^{-2},\ f'_{\mathscr{L}}(2)=-e^{-2};\ S(-2)=S(2)=e^{-2}$

11. $f(-2+)=-2,\ f(2-)=e^2,\ f(1-)=1,\ f(1+)=e;\ f'_{\mathscr{R}}(-2)=1,\ f'_{\mathscr{L}}(2)=e^2,\ f'_{\mathscr{R}}(1)=e,\ f'_{\mathscr{L}}(1)=1;$
$S(-2)=S(2)=\dfrac{1}{2}(-2+e^2),\ S(1)=\dfrac{1}{2}(1+e)$

13. $f(-5+)=1,\ f(5-)=25,\ f(-1-)=1,\ f(-1+)=0,\ f(3-)=0,\ f(3+)=9;\ f'_{\mathscr{R}}(-5)=0,\ f'_{\mathscr{L}}(5)=10,$
$f'_{\mathscr{R}}(-1)=f'_{\mathscr{L}}(-1)=0,\ f'_{\mathscr{R}}(3)=6,\ f'_{\mathscr{L}}(3)=0;\ S(-5)=S(5)=13,\ S(-1)=\dfrac{1}{2},\ S(3)=\dfrac{9}{2}$

15. $f(-1+) = 1, f(1-) = 4, f\left(\dfrac{1}{2}-\right) = \dfrac{1}{2}, f\left(\dfrac{1}{2}+\right) = 1, f\left(\dfrac{2}{3}-\right) = \dfrac{4}{3}, f\left(\dfrac{2}{3}+\right) = \dfrac{8}{3}; f'_{\mathcal{R}}(-1) = -1, f'_{\mathcal{L}}(1) = 4, f'_{\mathcal{R}}\left(\dfrac{1}{2}\right) = 2,$

$f'_{\mathcal{L}}\left(\dfrac{1}{2}\right) = 1, f'_{\mathcal{R}}\left(\dfrac{2}{3}\right) = 4, f'_{\mathcal{L}}\left(\dfrac{2}{3}\right) = 2; S(-1) = S(1) = \dfrac{5}{2}, S\left(\dfrac{1}{2}\right) = \dfrac{3}{4}, S\left(\dfrac{2}{3}\right) = 2$

17. $-\dfrac{\pi^2}{12}$

19. (a) $2 \displaystyle\sum_{n=1}^{\infty} \dfrac{(-1)^{n+1}}{n} \sin(nx)$

(b) Since f is continuous with a left and right derivative at each point of $(-\pi, \pi)$, the Fourier series for $f(x) = x$ converges to x on $(-\pi, \pi)$.

(c) The term by term derivative of the Fourier series is $2 \displaystyle\sum_{n=1}^{\infty} (-1)^{n+1} \cos(nx)$. This series diverges on $(-\pi, \pi)$ because, for such x, $(-1)^{n+1}\cos(nx)$ does not have limit zero as n approaches infinity.

21. (a) Certainly f is continuous on $[-1, 1]$, and $f(-1) = f(1) = 1$. Further,

$$f'(x) = \begin{cases} 1 & \text{if} \quad 0 < x < 1 \\ -1 & \text{if} \quad -1 < x < 0, \end{cases}$$

and $\displaystyle\lim_{x\to0-} f'(x) = -1$, $\displaystyle\lim_{x\to0+} f'(x) = 1$, $\displaystyle\lim_{x\to-1+} f'(x) = -1$, and $\displaystyle\lim_{x\to1-} f'(x) = 1$. Thus, f' is piecewise continuous on $[-1, 1]$.

(b) $\dfrac{1}{2} - \dfrac{4}{\pi^2} \displaystyle\sum_{n=1}^{\infty} \dfrac{1}{(2n-1)^2} \cos[(2n-1)\pi x]$

(c) The term by term derivative of this series is $\dfrac{4}{\pi} \displaystyle\sum_{n=1}^{\infty} \dfrac{1}{2n-1} \sin[(2n-1)\pi x]$.

23. (a) $\dfrac{\pi}{4} + \displaystyle\sum_{n=1}^{\infty} \left[\dfrac{(-1)^n - 1}{\pi n^2} \cos(nx) - \dfrac{(-1)^n}{n} \sin(nx) \right]$

(b) f is continuous (in fact, differentiable) on $(-\pi, 0)$ and on $(0, \pi)$. Thus, the series converges to $f(x)$ on $(-\pi, 0)$ and on $(0, \pi)$. The only point of discontinuity of f is 0, and $f'_{\mathcal{R}}(0) = 1$ and $f'_{\mathcal{L}}(0) = 0$, so the left and right derivatives of f exist there. Further, $f(0-) = f(0+) = 0$, so the Fourier series converges to zero at zero.

(c) Integrate the Fourier series in (a) from $-\pi$ to x to get

$$\dfrac{\pi}{4}x + \dfrac{\pi^2}{4} + \displaystyle\sum_{n=1}^{\infty} \dfrac{(-1)^n - 1}{n^3\pi} \sin(nx) + \dfrac{(-1)^n}{n^2} \cos(nx) - \dfrac{1}{n^2}.$$

Now, $\displaystyle\sum_{n=1}^{\infty} \dfrac{1}{n^2} = \dfrac{\pi^2}{6}$ from Problem 16, and the Fourier series for x on $[-\pi, \pi]$ is $\displaystyle\sum_{n=1}^{\infty} \dfrac{2(-1)^n}{n} \sin(nx)$. Thus, the

Fourier series for $\displaystyle\int_{-\pi}^{x} f(t)\, dt$ is

$$\dfrac{\pi^2}{12} + \displaystyle\sum_{n=1}^{\infty} \left[\dfrac{\pi(-1)^n}{2n} + \dfrac{(-1)^n - 1}{\pi n^3} \right] \sin(nx) + \dfrac{(-1)^n}{n^2} \cos(nx).$$

25. From the hint, we have $\displaystyle\int_{-L}^{L} [f(x)]^2\, dx - L\left[a_0^2 + \sum_{n=1}^{N} (a_n^2 + b_n^2) \right] \geq 0$. Therefore, $a_0^2 + \displaystyle\sum_{n=1}^{N} (a_n^2 + b_n^2) \leq \dfrac{1}{L} \int_{-L}^{L} [f(x)]^2\, dx$, for every positive integer N. We obtain Bessel's inequality by taking the limit as $N \to \infty$.

27. From Problem 26, $\displaystyle\sum_{n=1}^{\infty} a_n^2$ converges. Therefore, $\displaystyle\lim_{n\to\infty} a_n^2 = 0$; hence, $\displaystyle\lim_{n\to\infty} a_n = 0$ also. Therefore,

$$\lim_{n\to\infty} \int_{-L}^{L} f(x)\cos\left(\dfrac{n\pi x}{L}\right) dx = 0.$$ Since $\displaystyle\sum_{n=1}^{\infty} b_n^2$ converges, we can apply the same reasoning to b_n and conclude that

$$\lim_{n\to\infty} \int_{-L}^{L} f(x)\sin\left(\dfrac{n\pi x}{L}\right) dx = 0.$$

29. (a) $S_N(x) = a_0 + \sum\limits_{n=1}^{N} a_n\cos\left(\dfrac{n\pi x}{L}\right) + b_n\sin\left(\dfrac{n\pi x}{L}\right)$

$$= \frac{1}{2L}\int_{-L}^{L} f(t)\, dt + \sum_{n=1}^{N}\frac{1}{L}\left[\left[\int_{-L}^{L} f(t)\cos\left(\frac{n\pi t}{L}\right) dt\, \cos\left(\frac{n\pi x}{L}\right) + \int_{-L}^{L} f(t)\sin\left(\frac{n\pi t}{L}\right) dt\, \sin\left(\frac{n\pi x}{L}\right)\right]\right]$$

$$= \frac{1}{L}\int_{-L}^{L}\left\{\frac{1}{2}f(t) + \sum_{n=1}^{N} f(t)\left[\cos\left(\frac{n\pi t}{L}\right)\cos\left(\frac{n\pi x}{L}\right) + \sin\left(\frac{n\pi t}{L}\right)\sin\left(\frac{n\pi x}{L}\right)\right] dt\right\}$$

$$= \frac{1}{L}\int_{-L}^{L} f(t)\left\{\frac{1}{2} + \sum_{n=1}^{N}\cos\left(\frac{n\pi(t-x)}{L}\right)\right\} dt.$$

(b) From the hint, $\sum\limits_{n=1}^{N}\cos(n\xi) = \sum\limits_{n=1}^{N}\frac{1}{2}(e^{in\xi} + e^{in\xi}) = \frac{1}{2}\sum\limits_{n=1}^{N}(e^{i\xi})^n + \frac{1}{2}\sum\limits_{n=1}^{N}(e^{-i\xi})^n$

$$= \frac{1}{2}\frac{e^{i\xi}}{1 - e^{i\xi}}(1 - e^{iN\xi}) + \frac{1}{2}\frac{e^{-i\xi}}{1 - e^{-i\xi}}(1 - e^{-iN\xi}).$$

Using Euler's formula $[e^{iA} = \cos(A) + i\sin(A)]$, a long but routine manipulation reduces the last line to

$$\frac{-1 + \cos(\xi) + \cos(N\xi) - \cos(\xi)\cos(N\xi) + \sin(\xi)\sin(N\xi)}{2 - 2\cos(\xi)},$$

or

$$-\frac{1}{2} + \frac{\cos(N\xi) - \cos((N+1)\xi)}{2[1 - \cos(\xi)]}.$$

In the solution to Problem 21 in Section 19.1, it is shown that

$$\frac{\cos(N\xi) - \cos((N+1)\xi)}{1 - \cos(\xi)} = \frac{\sin\left[\left(N + \dfrac{1}{2}\right)\xi\right]}{\sin\left(\dfrac{\xi}{2}\right)}.$$

Put this into the preceding equation to complete the derivation of the identity.

(c) If $t \approx x$, $t - x \approx 0$ (where $\approx$ means "approximately equal to"). Then $\sin\left(\dfrac{\pi}{2L}(t-x)\right) \approx \dfrac{\pi}{2L}(t-x)$. From (a) and (b), with $\xi = \dfrac{\pi}{L}(t-x)$, we have

$$S_N(x) = \frac{1}{2L}\int_{-L}^{L} f(t)\frac{\sin\left[\dfrac{\pi\left(N + \dfrac{1}{2}\right)(t-x)}{L}\right]}{\sin\left[\dfrac{\pi(t-x)}{2L}\right]}\, dt$$

$$\approx \frac{1}{2L}\int_{-L}^{L} f(t)\frac{\sin\left[\dfrac{\pi\left(N + \dfrac{1}{2}\right)(t-x)}{L}\right]}{\dfrac{\pi}{2L}(t-x)}\, dt.$$

(d) With the given $f(x)$,

$$S_N(x) \approx \frac{1}{2L}\int_{x_0}^{L}\frac{\sin\left[\dfrac{\pi\left(N + \dfrac{1}{2}\right)(t-x)}{L}\right]}{\dfrac{\pi}{2L}(t-x)}\, dt.$$

Now let $s = \dfrac{\pi}{L}\left(N + \dfrac{1}{2}\right)(t - x)$. Then

$$S_N(x) \approx \frac{1}{2L} \int_{\pi(N+1/2)(x_0-x)/L}^{\pi(N+1/2)(L-x)/L} \frac{\sin(s)}{\dfrac{s}{N + \dfrac{1}{2}} \cdot \dfrac{1}{2}} \frac{1}{\dfrac{\pi}{L}\left(N + \dfrac{1}{2}\right)}\, ds$$

$$= \frac{1}{\pi} \int_{\pi(N+1/2)(x_0-x)/L}^{\pi(N+1/2)(L-x)/L} \frac{\sin(s)}{s}\, ds.$$

(e) If $x = x_0$, $\displaystyle \lim_{N\to\infty} S_N(x) = \lim_{N\to\infty} \frac{1}{\pi}\int_0^{\pi(N+1/2)(L-x_0)/L} \frac{\sin(s)}{s}\, ds = \frac{1}{\pi}\int_0^\infty \frac{\sin(s)}{s}\, ds = \frac{1}{\pi}\frac{\pi}{2} = \frac{1}{2}.$

[Here we used the fact that $\displaystyle\int_0^\infty \frac{\sin(s)}{s}\, ds = \frac{\pi}{2}$; see, for example, *CRC Standard Mathematical Tables*, 25th ed. (Boca Raton, FL: CRC Press, 1981), entry number 621.]

If $x < x_0$, $x_0 - x > 0$, so

$$\lim_{N\to\infty} S_N(x) = \lim_{N\to\infty} \frac{1}{\pi}\int_{\pi(N+1/2)(x_0-x)/L}^{\pi(N+1/2)(L-x)/L} \frac{\sin(s)}{s}\, ds$$

$$= \frac{1}{\pi}\int_\infty^\infty \frac{\sin(s)}{s}\, ds = 0.$$

If $x > x_0$, $x_0 - x < 0$, so $\displaystyle\lim_{N\to\infty} S_N(x) = \frac{1}{\pi}\int_{-\infty}^\infty \frac{\sin(s)}{s}\, ds = \frac{1}{\pi}(\pi) = 1.$

(f) For N large and x near x_0,

$$S_N(x) \approx \frac{1}{\pi}\int_{\pi(N+1/2)(x_0-x)/L}^{\pi(N+1/2)(L-x)/L} \frac{\sin(s)}{s}\, ds \approx \frac{1}{\pi}\int_{\pi(N+1/2)(x_0-x)/L}^\infty \frac{\sin(s)}{s}\, ds$$

$$= \frac{1}{\pi}\int_0^\infty \frac{\sin(s)}{s}\, ds - \frac{1}{\pi}\int_0^{\pi(N+1/2)(x_0-x)/L} \frac{\sin(s)}{s}\, ds$$

$$= \frac{1}{2} - \mathrm{Si}\!\left[\frac{\pi\left(N + \dfrac{1}{2}\right)(x_0 - x)}{L}\right],$$

where $\mathrm{Si}(x) = \displaystyle\int_0^x \frac{\sin(s)}{s}\, ds$. From the graphs, $\mathrm{Si}(x)$ has its maximum value at about $x = \pi$. If $\dfrac{\pi(N + \frac{1}{2})(x_0 - x)}{L} = \pi$,

then $x = x_0 - \dfrac{L}{N + \frac{1}{2}}$, which can be made arbitrarily close to x_0 by choosing N larger.

Section 17.3

In solutions 1 through 13, the cosine series is denoted $C(x)$ and the sine series $S(x)$.

1. $C(x) = 4$, converging to 4 for $0 \le x \le 3$; $S(x) = \dfrac{16}{\pi}\displaystyle\sum_{n=1}^\infty \frac{1}{2n-1}\sin\left[\frac{(2n-1)\pi x}{3}\right] = \begin{cases} 0 & \text{if } x = 0 \text{ or } x = 3 \\ 4 & \text{if } 0 < x < 3. \end{cases}$

3. $C(x) = \dfrac{1}{2}\cos(x) - \dfrac{2}{\pi}\displaystyle\sum_{n=1}^\infty \frac{(-1)^n(2n-1)}{(2n-3)(2n+1)}\cos\left[\frac{(2n-1)x}{2}\right] = \begin{cases} 0 & \text{if } 0 \le x < \pi \\ -\frac{1}{2} & \text{if } x = \pi \\ \cos(x) & \text{if } \pi < x \le 2\pi. \end{cases}$

$S(x) = \dfrac{2}{\pi}\displaystyle\sum_{n=1}^\infty \frac{(-1)^n(2n-1)}{(2n-3)(2n+1)}\sin\left[\frac{(2n-1)x}{2}\right] - \dfrac{2}{\pi}\displaystyle\sum_{n=2}^\infty \frac{n}{n^2-1}\sin(nx) = \begin{cases} 0 & \text{if } 0 \le x < \pi \\ -\frac{1}{2} & \text{if } x = \pi \\ \cos(x) & \text{if } \pi < x < 2\pi \\ 0 & \text{if } x = 2\pi. \end{cases}$

5. $C(x) = -\dfrac{3}{2} + \dfrac{20}{\pi^2} \displaystyle\sum_{n=1}^{\infty} \dfrac{1}{(2n-1)^2} \cos\left[\dfrac{(2n-1)\pi x}{5}\right] = 1 - x$ for $0 \le x \le 5$;

$S(x) = \dfrac{2}{\pi} \displaystyle\sum_{n=1}^{\infty} \dfrac{1 + 4(-1)^n}{n} \sin\left(\dfrac{n\pi x}{5}\right) = \begin{cases} 0 & \text{if} & x = 0 \\ 1 - x & \text{if} & 0 < x < 5 \\ 0 & \text{if} & x = 5. \end{cases}$

7. $C(x) = \dfrac{4}{3} + \dfrac{16}{\pi^2} \displaystyle\sum_{n=1}^{\infty} \dfrac{(-1)^n}{n^2} \cos\left(\dfrac{n\pi x}{2}\right) = x^2$ for $0 \le x \le 2$;

$S(x) = -\dfrac{8}{\pi} \displaystyle\sum_{n=1}^{\infty} \left[\dfrac{(-1)^n}{n} + \dfrac{2[1 - (-1)^n]}{n^3 \pi^3}\right] \sin\left(\dfrac{n\pi x}{3}\right) = \begin{cases} x^2 & \text{if} & 0 \le x < 2 \\ 0 & \text{if} & x = 2. \end{cases}$

9. $C(x) = \dfrac{1}{2} + \displaystyle\sum_{n=1}^{\infty} \left[\dfrac{2}{n\pi} \sin\left(\dfrac{2n\pi}{3}\right) + \dfrac{12}{n^2\pi^2} \cos\left(\dfrac{2n\pi}{3}\right) \dfrac{6}{n^2\pi^2}[1 + (-1)^n]\right] \cos\left(\dfrac{n\pi x}{3}\right) = \begin{cases} x & \text{if} & 0 \le x < 2 \\ 1 & \text{if} & x = 2 \\ 2 - x & \text{if} & 2 < x \le 3; \end{cases}$

$S(x) = \displaystyle\sum_{n=1}^{\infty} \left[\dfrac{18}{n^2\pi^2} \sin\left(\dfrac{2n\pi}{3}\right) - \dfrac{8}{n\pi} \cos\left(\dfrac{2n\pi}{3}\right) + \dfrac{5}{n\pi}(-1)^n\right] \sin\left(\dfrac{n\pi x}{3}\right) = \begin{cases} x & \text{if} & 0 \le x < 2 \\ 1 & \text{if} & x = 2 \\ 2 - x & \text{if} & 2 < x < 3 \\ 0 & \text{if} & x = 3. \end{cases}$

11. $C(x) = 1 + \dfrac{8}{\pi} \displaystyle\sum_{n=1}^{\infty} \dfrac{1}{4n^2 - 4n - 3} \cos\left[\dfrac{(2n-1)\pi x}{2}\right] = 1 - \sin(\pi x)$ for $0 \le x \le 2$;

$S(x) = -\sin(\pi x) + \dfrac{4}{\pi} \displaystyle\sum_{n=1}^{\infty} \dfrac{1}{2n-1} \sin\left[\dfrac{(2n-1)\pi x}{2}\right] = \begin{cases} 0 & \text{if} \quad x = 0 \quad \text{or} \quad x = 2 \\ 1 - \sin(\pi x) & \text{for} \quad 0 < x < 2. \end{cases}$

13. $C(x) = \dfrac{5}{6} + \dfrac{16}{\pi^2} \displaystyle\sum_{n=1}^{\infty} \left[\dfrac{1}{n^2} \cos\left(\dfrac{n\pi}{4}\right) - \dfrac{4}{n^3\pi} \sin\left(\dfrac{n\pi}{4}\right)\right] \cos\left(\dfrac{n\pi x}{4}\right) = \begin{cases} x^2 & \text{if} & 0 \le x \le 1 \\ 1 & \text{if} & 1 \le x \le 4; \end{cases}$

$S(x) = \displaystyle\sum_{n=1}^{\infty} \left[\dfrac{16}{n^2\pi^2} \sin\left(\dfrac{n\pi}{4}\right) + \dfrac{64}{n^3\pi^3}\left[\cos\left(\dfrac{n\pi}{4}\right) - 1\right] - \dfrac{2(-1)^n}{n\pi}\right] \sin\left(\dfrac{n\pi x}{4}\right) = \begin{cases} x^2 & \text{if} & 0 \le x \le 1 \\ 1 & \text{if} & 1 \le x < 4 \\ 0 & \text{if} & x = 4. \end{cases}$

15. $C(x) = \dfrac{2}{3\pi} - \dfrac{12}{\pi} \displaystyle\sum_{n=1}^{\infty} \dfrac{1}{4n^2 - 9} \cos(2nx) = \sin(3x)$ for $0 \le x \le \pi$; $S(x) = \sin(3x)$ for $0 \le x \le \pi$.

17. The only such function is the identically zero function, $f(x) = 0$ for $-L \le x \le L$.

Section 17.4

1. $\displaystyle\sum_{n=1}^{\infty} \sum_{m=1}^{\infty} \dfrac{4}{nm} \sin(nx)\sin(my)$

3. $\displaystyle\sum_{n=1}^{\infty} \sum_{m=1}^{\infty} \dfrac{4}{\pi^2} \left[3(-1)^n[(-1)^m - 1] - 4[1 - (-1)^n]\left((-1)^m + \dfrac{2}{m^2\pi^2}[(-1)^m - 1]\right)\right] \sin(n\pi x)\sin\left(\dfrac{m\pi y}{2}\right)$

5. $\displaystyle\sum_{n=1}^{\infty} \sum_{m=1}^{\infty} 4nm\pi^2 \dfrac{[e^4(-1)^n - 1][e^2(-1)^m - 1]}{(16 + n^2\pi^2)(4 + m^2\pi^2)} \sin\left(\dfrac{n\pi x}{4}\right) \sin\left(\dfrac{m\pi y}{3}\right)$

7. $\displaystyle\sum_{n=1}^{\infty} \sum_{m=1}^{\infty} \dfrac{16m(-1)^{n+m}}{n(4 + m^2\pi^2)} \sinh(2)\sin\left(\dfrac{n\pi x}{4}\right)\sin\left(\dfrac{m\pi y}{2}\right)$ **9.** $\dfrac{8}{\pi} \sin(2x) \displaystyle\sum_{m=1}^{\infty} (-1)^{m+1} \sin\left(\dfrac{m\pi y}{4}\right)$

11. Write $f(x, y) = \dfrac{1}{4} a_{00} + \displaystyle\sum_{m=1}^{\infty} \dfrac{1}{2} a_{0m} \cos\left(\dfrac{m\pi x}{L}\right) + \displaystyle\sum_{n=1}^{\infty} \dfrac{1}{2} a_{n0} \cos\left(\dfrac{n\pi x}{L}\right) + \displaystyle\sum_{n=1}^{\infty} \sum_{m=1}^{\infty} a_{nm} \cos\left(\dfrac{n\pi x}{L}\right) \cos\left(\dfrac{m\pi y}{K}\right)$. Choose positive

integers r and s, multiply the equation by $\cos\left(\dfrac{r\pi x}{L}\right) \cos\left(\dfrac{s\pi y}{K}\right)$, and integrate the resulting equation, interchanging the summations and the integrals, to get

$$\int_0^L \int_0^K f(x,y)\cos\left(\frac{r\pi x}{L}\right)\cos\left(\frac{s\pi y}{K}\right) dy\, dx = \frac{1}{4} a_{00} \int_0^L \int_0^K \cos\left(\frac{r\pi x}{L}\right)\cos\left(\frac{s\pi y}{K}\right) dy\, dx$$

$$+ \sum_{m=1}^{\infty} \frac{1}{2} a_{0m} \int_0^L \int_0^K \cos\left(\frac{m\pi y}{K}\right)\cos\left(\frac{r\pi x}{L}\right)\cos\left(\frac{s\pi y}{K}\right) dy\, dx$$

$$+ \sum_{n=1}^{\infty} \frac{1}{2} a_{n0} \int_0^L \int_0^K \cos\left(\frac{n\pi x}{L}\right)\cos\left(\frac{r\pi x}{L}\right)\cos\left(\frac{s\pi y}{K}\right) dy\, dx$$

$$+ \sum_{n=1}^{\infty} \sum_{m=1}^{\infty} a_{nm} \int_0^L \int_0^K \cos\left(\frac{n\pi x}{L}\right)\cos\left(\frac{m\pi y}{K}\right)\cos\left(\frac{r\pi x}{L}\right)\cos\left(\frac{s\pi y}{K}\right) dy\, dx.$$

Most of the integrals are zero by orthogonality of the cosine functions, and we obtain

$$\int_0^L \int_0^K f(x,y)\cos\left(\frac{r\pi x}{L}\right)\cos\left(\frac{s\pi y}{K}\right) dy\, dx = a_{rs} \int_0^L \cos^2\left(\frac{r\pi x}{L}\right) dx \int_0^K \cos^2\left(\frac{s\pi y}{K}\right) dy = \frac{LK}{4} a_{rs},$$ yielding the requested formula

if r and s are positive integers. The special cases of computing a_{rs} when r is positive and $s = 0$, when $r = 0$ and s is positive, and when $r = s = 0$, are handled similarly.

13. $\frac{1}{6} + \sum_{n=1}^{\infty} \frac{2(-1)^n}{n^2\pi^2}\cos(n\pi x) + \sum_{m=1}^{\infty} \frac{2}{3}\frac{(-1)^m - 1}{m^2\pi^2}\cos(m\pi y) + \sum_{n=1}^{\infty}\sum_{m=1}^{\infty} \frac{8(-1)^n[(-1)^m - 1]}{(nm\pi^2)^2}\cos(n\pi x)\cos(m\pi y)$

15. $\frac{7}{6} + \sum_{n=1}^{\infty} \frac{6[(-1)^n - 1]}{n^2\pi^2}\cos\left(\frac{n\pi x}{3}\right) + \sum_{m=1}^{\infty} \frac{4(-1)^{m+1}}{m^2\pi^2}\cos(m\pi y)$

Section 17.5

1. $\frac{K}{n}[1 - (-1)^n]$ **3.** $\pi(-1)^n\left[\frac{6}{n^3} - \frac{1}{n}\pi^2\right]$

5. If a is an integer, then

$$F_S(n) = \begin{cases} 0 & \text{if } n \neq |a| \\ \dfrac{\pi}{2} & \text{if } n = a \\ -\dfrac{\pi}{2} & \text{if } n = -a; \end{cases}$$

if a is not an integer, then $F_S(n) = \dfrac{(-1)^n n \sin(a\pi)}{a^2 - n^2}$

7. $\frac{n}{n^2 + 1}[1 - (-1)^n e^{-\pi}]$ **9.** $\frac{(-1)^n - 1}{n^2}$ **11.** $\frac{(-1)^n e - 1}{n^2 + 1}$ **13.** $\frac{3\pi^2}{n^2}(-1)^n + \frac{6}{n^2}[1 - (-1)^n]$

15. Integrate by parts to get

$$S_n\{f'(x)\} = \int_0^\pi f'(x)\sin(nx)\, dx = f(x)\sin(nx)\Big|_0^\pi - n\int_0^\pi f(x)\cos(nx)\, dx$$

$$= -nC_n\{f(x)\}.$$

17. $F_S(n + m) = \int_0^\pi f(x)\sin[(n + m)x]\, dx = \int_0^\pi [f(x)\cos(mx)]\sin(nx)\, dx + \int_0^\pi [f(x)\sin(mx)]\cos(nx)\, dx =$
$S_n\{f(x)\cos(mx)\} + C_n\{f(x)\sin(mx)\}.$

19. $F_C(n - m) + F_C(n + m) = \int_0^\pi f(x)[\cos(n - m)x + \cos(n + m)x]\, dx = \int_0^\pi 2f(x)\cos(nx)\cos(mx)\, dx =$
$\int_0^\pi [f(x)\cos(mx)]\cos(nx)\, dx = 2F_C\{f(x)\cos(mx)\}.$

Section 17.6

1. If $f(t + T) = f(t)$ and $g(t + T) = g(t)$ for any t, then $(\alpha f + \beta g)(t + T) = \alpha f(t + T) + \beta g(t + T) = \alpha f(t) + \beta g(t) = (\alpha f + \beta g)(t).$

3. We must show that $f'(t + T) = f'(t)$. We have $f'(t + T) = \lim\limits_{h \to 0} \dfrac{f(t + T + h) - f(t + T)}{h} = \lim\limits_{h \to 0} \dfrac{f(t + h) - f(t)}{h} = f'(t)$,

because $f(t + T + h) = f(t + h)$ and $f(t + T) = f(t)$ by periodicity of f.

5. $\dfrac{1}{2} K + \dfrac{2K}{\pi} \sum\limits_{n=1}^{\infty} \dfrac{1}{2n - 1} \sin[(2n - 1)x]$

7. Let the Fourier coefficients of kf be $a_0^*, a_1^*, \ldots, b_1^*, \ldots$. Then

$$a_0^* = \frac{1}{2L} \int_{-L}^{L} kf(x)\, dx = k\, \frac{1}{2L} \int_{-L}^{L} f(x)\, dx = ka_0,$$

and, for n any positive integer,

$$a_n^* = \frac{1}{L} \int_{-L}^{L} kf(x)\cos\left(\frac{n\pi x}{L}\right) dx = k\, \frac{1}{L} \int_{-L}^{L} f(x)\cos\left(\frac{n\pi x}{L}\right) dx = ka_n$$

and

$$b_n^* = \frac{1}{L} \int_{-L}^{L} kf(x)\sin\left(\frac{n\pi x}{L}\right) dx = k\, \frac{1}{L} \int_{-L}^{L} f(x)\sin\left(\frac{n\pi x}{L}\right) dx = kb_n.$$

9. Let f_5 be the function of Problem 5 and f_6 that of Problem 6. Then

$$f(t) = \frac{5}{2k} f_5(t) - \frac{5}{4} f_6(t)$$

$$= \frac{5}{4} + \sum_{n=1}^{\infty} \frac{5}{n\pi}\left[\sin\left(\frac{n\pi}{2}\right)\cos(nx) + \left[1 - \cos\left(\frac{n\pi}{2}\right)\right]\sin(nx)\right].$$

11. The Fourier series of f is $\dfrac{1}{2} + \sum\limits_{n=1}^{\infty} \dfrac{2}{(2n-1)\pi} \sin[(2n-1)\pi t]$, with phase angle form $\dfrac{1}{2} + \sum\limits_{n=1}^{\infty} \dfrac{2}{(2n-1)\pi} \cos\left[(2n-1)\pi t - \dfrac{\pi}{2}\right]$.

13. The Fourier series is $\dfrac{19}{8} + \sum\limits_{n=1}^{\infty} \dfrac{2}{n^2\pi^2}\left[n\pi \sin\left(\dfrac{3n\pi}{2}\right) + \cos\left(\dfrac{3n\pi}{2}\right) - 1\right]\cos\left(\dfrac{n\pi t}{2}\right) +$

$\left[\sin\left(\dfrac{3n\pi}{2}\right) - \dfrac{n\pi}{2} - n\pi \cos\left(\dfrac{3n\pi}{2}\right)\right]\sin\left(\dfrac{n\pi t}{2}\right)$, with phase angle form $\dfrac{19}{8} +$

$\sum\limits_{n=1}^{\infty} \dfrac{1}{n^2\pi^2} \sqrt{5n^2\pi^2 + 8 - 12n\pi \sin\left(\dfrac{3n\pi}{2}\right) + 4(n^2\pi^2 - 2)\cos\left(\dfrac{3n\pi}{2}\right)} \cos\left(\dfrac{n\pi t}{2} + \delta_n\right)$, where

$$\delta_n = \tan^{-1}\left[\frac{\dfrac{n\pi}{2} + n\pi \cos\left(\dfrac{3n\pi}{2}\right) - \sin\left(\dfrac{3n\pi}{2}\right)}{n\pi \sin\left(\dfrac{3n\pi}{2}\right) + \cos\left(\dfrac{3n\pi}{2}\right) - 1}\right].$$

15. The Fourier series of f is $\dfrac{k}{2} + \sum\limits_{n=1}^{\infty} \dfrac{2k}{(2n-1)\pi} \sin[(2n-1)\pi t]$, with phase angle form $\dfrac{k}{2} + \sum\limits_{n=1}^{\infty} \dfrac{2k}{(2n-1)\pi} \cos\left[(2n-1)\pi t - \dfrac{\pi}{2}\right]$.

17. The Fourier series is $\dfrac{2}{\pi} \sum\limits_{n=1}^{\infty} \dfrac{(-1)^{n+1}}{n} \sin(n\pi t)$, with phase angle form $\dfrac{2}{\pi} \sum\limits_{n=1}^{\infty} \dfrac{1}{n} \cos\left[n\pi t + \dfrac{(-1)^{n+1}\pi}{2}\right]$.

19. The Fourier series is $\dfrac{3}{2} + \sum\limits_{n=1}^{\infty} \dfrac{2(-1)^n}{(2n-1)\pi} \cos\left[\dfrac{(2n-1)\pi t}{2}\right]$, with phase angle form

$\dfrac{3}{2} + \sum\limits_{n=1}^{\infty} \dfrac{2}{(2n-1)\pi} \cos\left[\dfrac{(2n-1)\pi t}{2} + \dfrac{\pi(1 - (-1)^n)}{2}\right]$.

21. $\dfrac{4}{\pi} \sum\limits_{n=1}^{\infty} \dfrac{1}{(2n-1)^2} \cos[(2n-1)t]$

23. $\dfrac{\dfrac{C_3}{C}}{\dfrac{C_1}{C}} = \dfrac{C_3}{C_1} = \dfrac{\dfrac{9\pi\sqrt{\Delta_3}}{4k}}{\dfrac{\pi\sqrt{\Delta_1}}{}} = \sqrt{\dfrac{\Delta_1}{9\Delta_3}} = \left[\dfrac{(600)^2(800\pi)^2 + [6.25 \times 10^6 - (800\pi)^2]^2}{9[9(600)^2(800\pi)^2 + [6.25 \times 10^6 - 9(800\pi)^2]^2]}\right]^{1/2} \approx 0.00990.$

25. Under the given conditions, $f(t) = a_0 + \sum\limits_{n=1}^{\infty} a_n \cos\left(\dfrac{2n\pi t}{T}\right) + b_n \sin\left(\dfrac{2n\pi t}{T}\right)$ on $\left[\dfrac{-T}{2}, \dfrac{T}{2}\right]$. Write

$$[f(t)]^2 = a_0 f(t) + \sum_{n=1}^{\infty} a_n f(t)\cos\left(\frac{2n\pi t}{T}\right) + b_n f(t)\sin\left(\frac{2n\pi t}{T}\right)$$

and integrate this equation term by term to get

$$\frac{1}{T}\int_{a-T/2}^{a+T/2} [f(t)]^2\,dt = a_0 \frac{1}{T}\int_{a-T/2}^{a+T/2} f(t)\,dt + \sum_{n=1}^{\infty} a_n \frac{1}{T}\int_{a-T/2}^{a+T/2} f(t)\cos\left(\frac{2n\pi t}{T}\right) dt + b_n \frac{1}{T}\int_{a-T/2}^{a+T/2} f(t)\sin\left(\frac{2n\pi t}{T}\right) dt$$

$$= a_0^2 + \frac{1}{2}\sum_{n=1}^{\infty} (a_n^2 + b_n^2),$$

using the formulas for the Fourier coefficients, with $L = \dfrac{T}{2}$.

27. $y = \dfrac{k}{50} + \dfrac{2k}{\pi}\sum\limits_{n=1}^{\infty} \dfrac{(-0.04)\cos[(2n-1)\pi t] + [25 - (2n-1)^2\pi^2]\sin[(2n-1)\pi t]}{(2n-1)[(0.04)^2 + [25 - (2n-1)^2\pi^2]^2]} =$

$\dfrac{k}{50} + \dfrac{2k}{\pi}\sum\limits_{n=1}^{\infty} \dfrac{1}{(2n-1)\sqrt{\Delta_{2n-1}}}\cos[(2n-1)\pi t + \delta_{2n-1}]$, where $\Delta_n = (0.04)^2 + (25 - n^2\pi^2)^2$ and $\delta_n = \tan^{-1}\left[\dfrac{0.04}{25 - (n^2\pi^2)}\right]$

29. $y = \dfrac{3}{50} + \dfrac{8}{\pi}\sum\limits_{n=1}^{\infty} \dfrac{(-1)^n}{(2n-1)[100 - (2n-1)^2\pi^2]}\cos\left[\dfrac{(2n-1)\pi t}{2}\right]$

31. $f(t) = \sum\limits_{n=1}^{\infty} \dfrac{1200\pi(-1)^{n+1}}{n^3}\sin(nt)$, so $i'' + 10i' + 10i = \dfrac{1}{10}f'(t)$

and

$$i(t) = \sum_{n=1}^{\infty} \frac{120(-1)^{n+1}(10 - n^2)\pi}{n^2[(10n)^2 + (10 - n^2)^2]}\cos(nt) + \frac{1200(-1)^{n+1}\pi}{n[(10n)^2 + (10 - n^2)^2]}\sin(nt).$$

33. Power dissipation $= (i_{RMS})^2 R = \dfrac{100}{\pi}\sum\limits_{n=1}^{\infty} (a_n^2 + b_n^2) = \dfrac{100}{\pi}\sum\limits_{n=1}^{\infty}\left[\dfrac{(120)^2\pi^2(10 - n^2)^2}{n^4[(10n)^2 + (10 - n^2)^2]^2} + \dfrac{(1200)^2\pi^2}{n^2[(10n)^2 + (10 - n^2)^2]^2}\right] \approx$

$100(120)^2\pi\left[\dfrac{1}{181} + \dfrac{1}{6976}\right] \approx 25{,}642.4$ watts.

Section 17.7

1. $\overline{e^{i\theta}} = \overline{[\cos(\theta) + i\sin(\theta)]} = \cos(\theta) - i\sin(\theta) = \cos(-\theta) + i\sin(-\theta) = e^{-i\theta}$. This equation does not hold if θ is complex.

3. $\dfrac{4}{3} + \sum\limits_{\substack{n=-\infty \\ n\neq 0}}^{\infty}\left[\dfrac{2}{n^2\pi^2} - \dfrac{2i}{n\pi}\right]e^{n\pi it}$ **5.** $-\dfrac{1}{2} - \sum\limits_{\substack{n=-\infty \\ n\neq 0}}^{\infty} \dfrac{3i}{n\pi}e^{n\pi it/3}$ **7.** $\sum\limits_{n=-\infty}^{\infty} \dfrac{1 - e^{-5}}{5 + 2n\pi i}e^{2n\pi it/5}$

9. $\sum\limits_{n=-\infty}^{\infty} \dfrac{2n\pi i[\cos(1) - 1] + \sin(1)}{1 - 4n^2\pi^2}e^{2n\pi it}$ **11.** $4 + \sum\limits_{\substack{n=-\infty \\ n\neq 0}}^{\infty} \dfrac{16i}{n\pi}e^{n\pi it/2}$

13. $\varphi_0 = 0$, $\varphi_{2n-1} = -\dfrac{\pi}{2}$, $\varphi_{2n} = 0$

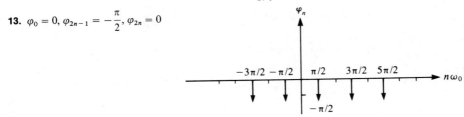

15. If n and m are distinct positive integers, then

$$\int_{-T/2}^{T/2} \varphi_n(t)\overline{\varphi_m(t)}\, dt = \int_{-T/2}^{T/2} e^{in\omega_0 t}e^{-im\omega_0 t}\, dt = \int_{-T/2}^{T/2} e^{2\pi i(n-m)t/T}$$

$$= \frac{T}{2\pi i(n-m)}e^{2\pi i(n-m)t/T}\Big]_{-T/2}^{T/2}$$

$$= \frac{T}{2\pi i(n-m)}\left[e^{\pi i(n-m)} - e^{-\pi i(n-m)}\right] = 0,$$

because $e^{\pi i} = \cos(\pi) + i\sin(\pi) = -1$.

17. Assume that $f(t) = \sum\limits_{n=-\infty}^{\infty} c_n e^{in\omega_0 t}$. Then $[f(t)]^2 = \sum\limits_{n=-\infty}^{\infty} c_n f(t) e^{in\omega_0 t}$. Assuming that we can interchange integration and summation in this equation, we have

$$\int_{-T/2}^{T/2} [f(t)]^2\, dt = \sum_{n=-\infty}^{\infty} c_n \int_{-T/2}^{T/2} f(t)e^{in\omega_0 t}\, dt.$$

But $\int_{-T/2}^{T/2} f(t)e^{in\omega_0 t}\, dt = \overline{\int_{-T/2}^{T/2} f(t)e^{-in\omega_0 t}\, dt} = T\overline{c_n}$. Therefore, $\dfrac{1}{T}\int_{-T/2}^{T/2} [f(t)]^2\, dt = \sum\limits_{n=-\infty}^{\infty} c_n\overline{c_n} = \sum\limits_{n=-\infty}^{\infty} |c_n|^2$.

Section 17.8

1. $\int_0^\infty \left[\dfrac{2\sin(\omega\pi)}{\pi\omega^2} - \dfrac{2\cos(\omega\pi)}{\omega}\right]\sin(\omega t)\, d\omega$, converging to t for $-\pi < t < \pi$, to zero for $|t| > \pi$, to $\dfrac{\pi}{2}$ for $t = \pi$,

and to $-\dfrac{\pi}{2}$ for $t = -\pi$

3. $\int_0^\infty \dfrac{2}{\pi\omega}[1 - \cos(\pi\omega)]\sin(\omega t)\, d\omega$, converging to -1 for $-\pi < t < 0$, to zero for $|t| > \pi$, to $\dfrac{1}{2}$ for $t = \pi$, and to $-\dfrac{1}{2}$ for $t = -\pi$.

5. $\int_0^\infty \left[\dfrac{400\cos(100\omega)}{\pi\omega^2} + \dfrac{20000\omega^2 - 4}{\pi\omega^3}\sin(100\omega)\right]\cos(\omega t)\, d\omega$, converging to t^2 for $|t| < 100$, to zero for $|t| > 100$, and to 5000 for $t = \pm 100$.

7. $\int_0^\infty \dfrac{2}{\pi}\dfrac{1}{1-\omega^2}\sin(\pi\omega)\sin(\omega t)\, d\omega$, converging to $\sin(t)$ for $|t| \le \pi$ and to zero for $|t| > \pi$.

9. Because f is an even function, $B(\omega) = 0$. From Example 17.17,

$$\frac{1}{\pi}\int_0^\infty \frac{2\sin(\omega)}{\omega}\cos(\omega t)\, d\omega = \begin{cases} 1 & \text{if } |t| < 1 \\ \frac{1}{2} & \text{if } t = 1 \\ 0 & \text{if } |t| > 1. \end{cases}$$

Thus, $A(\omega) = \int_{-\infty}^\infty \dfrac{\sin(\xi)}{\xi}\cos(\omega\xi)\, d\xi = 2\int_0^\infty \dfrac{\sin(\xi)}{\xi}\cos(\omega\xi)\, d\xi = \begin{cases} \pi & \text{if } 0 < \omega < 1 \\ \frac{1}{2} & \text{if } \omega = 1 \\ 0 & \text{if } \omega > 1. \end{cases}$

Thus, the Fourier integral of f is $\dfrac{1}{\pi}\int_0^\infty A(\omega)\cos(\omega t)\, d\omega = \int_0^1 \cos(\omega t)\, d\omega$, converging to $\dfrac{\sin(t)}{t}$ for $t \ne 0$ and to 1 for $t = 0$.

11. $f(t) = \dfrac{1}{\pi}\int_0^\infty [A(\omega)\cos(\omega t) + B(\omega)\sin(\omega t)]\, d\omega$

$$= \frac{1}{\pi}\int_0^\infty \left\{\left[\int_{-\infty}^\infty f(\xi)\cos(\omega\xi)\, d\xi\right]\cos(\omega t) + \left[\int_{-\infty}^\infty f(\xi)\sin(\omega\xi)\, d\xi\right]\sin(\omega t)\right\} d\omega$$

$$= \frac{1}{\pi}\int_0^\infty \int_{-\infty}^\infty f(\xi)[\cos(\omega\xi)\cos(\omega t) + \sin(\omega\xi)\sin(\omega t)]\, d\xi\, d\omega$$

$$= \frac{1}{\pi}\int_0^\infty \int_{-\infty}^\infty f(\xi)\cos[\omega(t - \xi)]\, d\xi\, d\omega.$$

13. From Problem 11, the Fourier integral of f is $\dfrac{1}{\pi}\displaystyle\int_0^\infty \int_{-\infty}^\infty f(\xi)\cos[\omega(t-\xi)]\,d\xi\,d\omega$, and this can be written

$\dfrac{1}{2\pi}\displaystyle\int_{-\infty}^\infty \int_{-\infty}^\infty f(\xi)\cos[\omega(t-\xi)]\,d\xi\,d\omega$ because $\cos[\omega(t-\xi)]$ is an even function.

In solutions 15 through 23, the Fourier cosine integral is denoted $C(t)$ and the sine integral $S(t)$.

15. $C(t) = \displaystyle\int_0^\infty \dfrac{4}{\pi\omega^3}[10\omega\cos(10\omega) + (50\omega^2 - 1)\sin(10\omega)]\cos(\omega t)\,d\omega$ and

$S(t) = \displaystyle\int_0^\infty \dfrac{4}{\pi\omega^3}[10\omega\sin(10\omega) - (50\omega^2 - 1)\cos(10\omega) - 1]\sin(\omega t)\,d\omega$, with both integrals converging to t^2 for $0 \le t < 10$, to zero for $t > 10$, and to 50 at $t = 10$.

17. $C(t) = \displaystyle\int_0^\infty \dfrac{2}{\pi\omega}[2\sin(4\omega) - \sin(\omega)]\cos(\omega t)\,d\omega$ and $S(t) = \displaystyle\int_0^\infty \dfrac{2}{\pi\omega}[1 + \cos(\omega) - 2\cos(4\omega)]\sin(\omega t)\,d\omega$, with both integrals

converging to 1 for $0 < t < 1$, to $\dfrac{3}{2}$ for $t = 1$, to 2 for $1 < t < 4$, to 1 for $t = 4$, and to zero for $t > 4$. The cosine integral converges to 1 for $t = 0$, while the sine integral converges to zero for $t = 0$.

19. $C(t) = \displaystyle\int_0^\infty \left\{\dfrac{2}{\pi\omega}[(2\pi - 1)\sin(\pi\omega) + \sin(3\pi\omega) + \sin(10\pi\omega)] + \dfrac{4}{\pi\omega^2}[\cos(\pi\omega) - 1]\right\}\cos(\omega t)\,d\omega$,

$S(t) = \displaystyle\int_0^\infty \left\{\dfrac{2}{\pi\omega}[1 - 2(\pi - 1)\cos(\pi\omega) - \cos(3\pi\omega) - \cos(10\pi\omega)] + \dfrac{4}{\pi\omega^2}\sin(\pi\omega)\right\}\sin(\omega t)\,d\omega$,

with both integrals converging to $2t + 1$ for $0 < t < \pi$, to $\dfrac{(2t+3)}{2}$ for $t = \pi$, to 2 for $\pi < t < 3\pi$, to $\dfrac{3}{2}$ for $t = 3\pi$, to 1 for

$3\pi < t < 10\pi$, to $\dfrac{1}{2}$ for $t = 10\pi$, and to zero for $t > 10\pi$. The cosine integral converges to 1 for $t = 0$, while the sine integral converges to zero for $t = 0$.

21. $C(t) = \displaystyle\int_0^\infty \dfrac{2}{\pi}\dfrac{2 + \omega^2}{4 + \omega^4}\cos(\omega t)\,d\omega$, $S(t) = \displaystyle\int_0^\infty \dfrac{2}{\pi}\dfrac{\omega^3}{4 + \omega^4}\sin(\omega t)\,d\omega$, with both integrals converging to $e^{-t}\cos(t)$

for $t > 0$. The cosine integral converges to 1 for $t = 0$, and the sine integral converges to zero for $t = 0$.

23. $C(t) = \displaystyle\int_0^\infty \dfrac{2}{\pi\omega}\sin(10\omega)\cos(\omega t)\,d\omega$, $S(t) = \displaystyle\int_0^\infty \dfrac{2}{\pi\omega}[1 - \cos(10\omega)]\sin(\omega t)\,d\omega$, with both integrals converging to 1 for

$0 < t < 10$, to $\dfrac{1}{2}$ for $t = 10$, and to zero for $t > 10$. The cosine integral converges to 1 at $t = 0$, and the sine integral converges to zero for $t = 0$.

25. (a) From Example 17.18, $2\displaystyle\int_0^\infty \dfrac{1}{k^2 + \omega^2}\cos(\omega t)\,d\omega = \dfrac{\pi}{k}e^{-k\omega}$. Let $k = 1$ and replace t with ω and ω with ξ to get

$A(\omega) = \pi e^{-\omega}$. Thus, the cosine integral is $\displaystyle\int_0^\infty e^{-\omega}\cos(\omega t)\,d\omega$, converging to $\dfrac{1}{1 + t^2}$ for all t.

(b) $B(\omega) = 2\displaystyle\int_0^\infty \dfrac{\xi}{1 + \xi^2}\sin(\omega\xi)\,d\xi = 2\dfrac{\pi}{2}e^{-\omega} = \pi e^{-\omega}$ if $\omega > 0$, from Example 17.18. Thus, the Fourier sine integral of

$\dfrac{t}{1 + t^2}$ is $\dfrac{1}{\pi}\displaystyle\int_0^\infty B(\omega)\sin(\omega t)\,d\omega = \displaystyle\int_0^\infty e^{-\omega}\sin(\omega t)\,d\omega$, converging to $\dfrac{t}{1 + t^2}$ for $t \ge 0$ [as can be verified

by explicitly evaluating $\displaystyle\int_0^\infty e^{-\omega}\sin(\omega t)\,d\omega$].

27. $A^*(\omega) = -\dfrac{d^2A}{d\omega^2} = -\dfrac{d^2}{d\omega^2}\displaystyle\int_{-\infty}^\infty f(\xi)\cos(\omega\xi)\,d\xi = \dfrac{d}{d\omega}\displaystyle\int_{-\infty}^\infty \xi f(\xi)\sin(\omega\xi)\,d\xi = \displaystyle\int_{-\infty}^\infty \xi^2 f(\xi)\cos(\omega\xi)\,d\xi$. This is the

coefficient of the cosine term in the Fourier integral for $t^2 f(t)$. We know that f is even because its Fourier integral has no sine component. Thus, $t^2 f(t)$ is also even and has no sine term in its representation.

29. $\displaystyle\int_0^\infty \dfrac{2}{\pi\omega}\sin(\omega)\cos(\omega t)\,d\omega$

Section 17.9

1. $\dfrac{2i}{\omega}[\cos(\omega) - 1]$ **3.** $\dfrac{10}{\omega}e^{-7i\omega}\sin(4\omega)$ **5.** $\dfrac{2}{1 + \omega^2}[\sinh(K)\cos(\omega K) + \cosh(K)\sin(\omega K)]$ **7.** $\pi e^{-|\omega|}$

9. $\dfrac{24}{16 + \omega^2}e^{2i\omega}$ **11.** $\dfrac{12}{16 + (\omega - 2)^2} + \dfrac{12}{16 + (\omega + 2)^2}$ **13.** $\dfrac{\pi i}{4}[e^{-2|\omega + 3|} - e^{-2|\omega - 3|}]$

15. $\dfrac{2k}{\omega}e^{-(a + b)i\omega/2}\sin\left[\dfrac{(b - a)\omega}{2}\right]$ **17.** $18\sqrt{2/\pi}e^{-4it}e^{-8t^2}$ **19.** $4H(t + 2)e^{-10 - (5 - 3i)t}$ **21.** $H(t)[2e^{-3t} - e^{-2t}]$

23. $\mathscr{F}\{e^{-a|t|}\} = \mathscr{F}\{H(t)e^{-at} + H(-t)e^{at}\} = \mathscr{F}\{H(t)e^{-at}\} + \mathscr{F}\{H(-t)e^{at}\}$. Now,

$$\mathscr{F}\{H(t)e^{-at}\} = \int_{-\infty}^{\infty} H(t)e^{-at}e^{-i\omega t}\,dt = \int_0^\infty e^{-(a + i\omega)t}\,dt = \dfrac{-1}{a + i\omega}\bigg]_0^\infty = \dfrac{1}{a + i\omega}.$$

Now apply time reversal (Theorem 17.15) with $f(t) = H(t)e^{-at}$ to get

$$\mathscr{F}\{H(-t)e^{at}\} = \mathscr{F}\{f(-t)\} = F(-\omega) = \dfrac{1}{a - i\omega}.$$

Therefore, $\mathscr{F}\{e^{-a|t|}\} = \dfrac{1}{a + i\omega} + \dfrac{1}{a - i\omega} = \dfrac{2a}{a^2 + \omega^2}$.

Section 17.10

1. $\pi i[H(-\omega)e^{3\omega} - H(\omega)e^{-3\omega}]$ **3.** $\dfrac{26}{(2 + i\omega)^2}$ **5.** $\dfrac{1}{(4 + i\omega)^2}e^{-3(4 + i\omega)}$ **7.** $\dfrac{i\omega}{3 + i\omega} - 1$

9. $\dfrac{2i}{\omega^2}[\omega \cos(\omega) - \sin(\omega)]$ **11.** $H(t)te^{-t}$ **13.** $\dfrac{1}{4}[1 - e^{-2(t + 3)}]H(t + 3) - \dfrac{1}{4}[1 - e^{-2(t - 3)}]H(t - 3)$

15. $\dfrac{3}{2\pi}e^{-4it}\left[\dfrac{1}{9 + (t + 2)^2} + \dfrac{1}{9 + (t - 2)^2}\right]$

17. By the frequency convolution theorem (the second conclusion of Theorem 17.22),

$$\mathscr{F}\{f(t)g(t)\} = \int_{-\infty}^{\infty} f(t)g(t)e^{-i\omega t}\,dt = \dfrac{1}{2\pi}[F * G](\omega) = \dfrac{1}{2\pi}\int_{-\infty}^{\infty} F(w)G(\omega - w)\,dw.$$

Now let $\omega = 0$ to get $\displaystyle\int_{-\infty}^{\infty} f(t)g(t)\,dt = \dfrac{1}{2\pi}\int_{-\infty}^{\infty} F(w)G(-w)\,dw.$

19. Let $g(t) = f(t)$ in Problem 18 to get $\displaystyle\int_{-\infty}^{\infty} [f(t)]^2\,dt = \dfrac{1}{2\pi}\int_{-\infty}^{\infty} F(\omega)\overline{F(\omega)}\,d\omega = \dfrac{1}{2\pi}\int_{-\infty}^{\infty} |F(\omega)|^2\,d\omega.$

21. 3π

25. Write

$$[F * G](\omega) = \int_{-\infty}^{\infty} F(\tau)G(\omega - \tau)\,d\tau = \int_{-\infty}^{\infty} F(\tau)\mathscr{F}\{e^{i\tau t}g(t)\}\,d\tau \qquad \text{(by frequency shifting)}$$

$$= \int_{-\infty}^{\infty} F(\tau)\left[\int_{-\infty}^{\infty} e^{i\tau t}g(t)e^{-i\omega t}\,dt\right]d\tau = \int_{-\infty}^{\infty}\left[\int_{-\infty}^{\infty} F(\tau)e^{i\tau t}\,d\tau\right]g(t)e^{-i\omega t}\,dt$$

$$= \int_{-\infty}^{\infty} 2\pi f(t)g(t)e^{-i\omega t}\,dt = 2\pi\mathscr{F}\{f(t)g(t)\}. \text{ Thus, } \mathscr{F}\{f(t)g(t)\} = \dfrac{1}{2\pi}[F * G](\omega).$$

27. 2   **29.** $4e^{-15}$

Section 17.11

1. $\mathscr{F}\{f(t)\} = \mathscr{F}\{m(t)\cos(\omega_0 t)\} = \dfrac{1}{2}M(\omega - \omega_0) + \dfrac{1}{2}M(\omega + \omega_0)$

5. $f(\theta) \approx 38.1333 + 36.5357 \sin(\theta) - 20.2333 \cos(2\theta) - 11.0333 \sin(3\theta) + 6.2667 \cos(4\theta) + 4.0309 \sin(5\theta) - 3.3333 \cos(6\theta)$

7. $g(x) \approx 30.0167 + 46.7372 \cos\left[\dfrac{\pi(x - 180)}{180}\right] + 17.9278 \sin\left[\dfrac{\pi(x - 180)}{180}\right]$, where x is in days after January 1.

9. $u(x, t) = 4e^{-3(x - 4t)^2} + 4e^{-3(x + 4t)^2}$ **11.** $u(x, t) = \tan^{-1}\left(\dfrac{x - 2t}{3}\right) - \tan^{-1}\left(\dfrac{x + 2t}{3}\right)$

13. Let $g(x, t) = \dfrac{1}{2}[f(x - 3t) + f(x + 3t)]$.

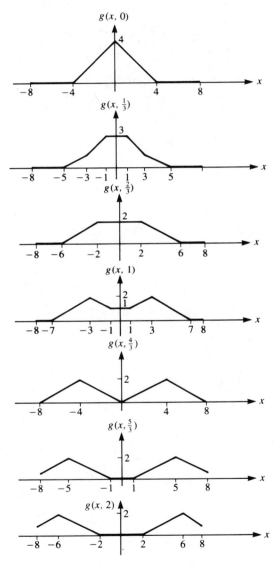

Section 17.12

1. $\dfrac{1}{1 + \omega^2}$ **3.** $\dfrac{1}{\omega^2}[2K \cos(\omega K) - (K^2\omega^2 + 2)\sin(\omega K)]$ **5.** $\dfrac{10}{\omega} \sin(5\omega) + \dfrac{2}{\omega^2}[\cos(5\omega) - 1]$ **7.** $\dfrac{a^2 - \omega^2}{(a^2 + \omega^2)^2}$

9. $\dfrac{2 + \omega^2}{4 + \omega^4}$ **11.** $\dfrac{1}{\omega}[2 - 3K \cos(\omega K) + \cos(2\omega K)]$ **13.** $\dfrac{2a\omega}{a^2 + \omega^2}$ **15.** $\dfrac{2\omega}{4 + \omega^4}$

17. $\dfrac{1}{2}\left[\dfrac{\sin(K-\omega)}{K-\omega} - \dfrac{\sin(K+\omega)}{K+\omega}\right]$ if $K \neq |\omega|$; $\dfrac{1}{4\omega}[2\omega - \sin(2\omega)]$ if $K = \omega$; $\dfrac{-1}{4\omega}[2\omega - \sin(2\omega)]$ if $K = -\omega$

19. Assume that f, f', f'', and $f^{(3)}$ are continuous on $[0, \infty)$, that $f(t) \to 0$, $f'(t) \to 0$, $f''(t) \to 0$, and $f^{(3)}(t) \to 0$ as $t \to \infty$, and that $f^{(4)}$ is piecewise continuous on $[0, k]$ for every $k > 0$. Integrate by parts and use Theorem 17.23 to get

$$\mathscr{F}_S\{f^{(4)}(t)\} = \int_0^\infty f^{(4)}(t)\sin(\omega t)\, dt = f^{(3)}(t)\sin(\omega t)\Big]_0^\infty - \omega \int_0^\infty f^{(3)}(t)\cos(\omega t)\, dt$$

$$= -\omega\left[f''(t)\cos(\omega t)\right]_0^\infty + \omega \int_0^\infty f''(t)[-\omega\sin(\omega t)]\, dt$$

$$= \omega f''(0) - \omega^2[-\omega^2 F_S(\omega) + \omega f(0)] = \omega^4 F_S(\omega) - \omega^3 f(0) + \omega f''(0).$$

Section 17.13

1. $G(0) = 6$, $G\left(\dfrac{\pi}{2}\right) = 2 - 2i$, $G(\pi) = 2$, $G\left(\dfrac{3\pi}{2}\right) = 2 + 2i$

3. $G(0) = 2$, $G(\pi) = 0$, $G(2\pi) = 1 + \sqrt{3}\,i$, $G(3\pi) = 0$, $G(4\pi) = 1 - \sqrt{3}\,i$, $G(5\pi) = 0$

5. $G(0) = 2$, $G\left(\dfrac{\pi}{2}\right) = -\dfrac{i}{2}(1 + \sqrt{2})$, $G(\pi) = 0$, $G\left(\dfrac{3\pi}{2}\right) = \dfrac{i}{2}(1 - \sqrt{2})$, $G(2\pi) = 0$, $G\left(\dfrac{5\pi}{2}\right) = -\dfrac{i}{2}(1 - \sqrt{2})$, $G(3\pi) = 0$,

$G\left(\dfrac{7\pi}{2}\right) = \dfrac{i}{2}(1 + \sqrt{2})$

7. The solution is the same as that of Problem 5.

9.

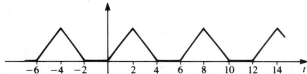

$g(t) * d(t)$

11. Write $f(t)$ in a complex Fourier series $f(t) = \displaystyle\sum_{n=-\infty}^{\infty} c_n e^{2n\pi it/T}$, where $c_n = \dfrac{1}{T}\displaystyle\int_0^T f(t)e^{-2n\pi it/T}\, dt$. Then

$$\mathscr{F}\{f(t)\} = \sum_{n=-\infty}^{\infty} c_n \mathscr{F}\{e^{2n\pi it/T}\} = \sum_{n=-\infty}^{\infty} c_n \cdot 2\pi\delta\left(\frac{\omega - 2n\pi}{T}\right).$$

Here we have used the frequency shifting theorem to write $\mathscr{F}\{e^{2n\pi it/T}\} = \mathscr{F}\{e^{2n\pi it/T} \cdot 1\} = 2\pi\delta\left(\omega - \dfrac{2n\pi}{T}\right)$, since $\mathscr{F}\{1\} = 2\pi\delta(\omega)$.

13. $G\left[\dfrac{2(N-n)\pi}{NT}\right] = T\displaystyle\sum_{k=0}^{N-1} f(kT)e^{-2(N-n)\pi ik/N} = T\displaystyle\sum_{k=0}^{N-1} f(kT)e^{-2\pi ik}e^{2n\pi ik/N}$. Now, $e^{-2\pi ik} = 1$ for any integer k, and $e^{2n\pi ik/N} = \overline{(e^{-2\pi ik/N})}$. Since f is real valued, $f(kT) = \overline{f(kT)}$. Hence, $G\left[\dfrac{2(N-n)\pi}{NT}\right] = \overline{G\left(\dfrac{2n\pi}{NT}\right)}$.

15. $\displaystyle\sum_{k=0}^{N-1} e^{2\pi ik(m-n)/N} = \displaystyle\sum_{k=0}^{N-1} [(e^{2\pi i/N})^{m-n}]^k$. Now use the familiar formula that $\displaystyle\sum_{k=0}^{N-1} r^k = \dfrac{1 - r^N}{1 - r}$ if $r \neq 1$. Assume first that $m \neq n$, so that $r = (e^{2\pi i/N})^{m-n} \neq 1$. Then $\displaystyle\sum_{k=0}^{N-1} e^{2\pi ik(m-n)/N} = 0$, because $r^N = (e^{2\pi i})^{m-n} = 1^{m-n} = 1$, so $1 - r^N = 0$. If $m = n$, then $r = 1$, and $\displaystyle\sum_{k=0}^{N-1} [(e^{2\pi i/N})^{m-n}]^k = 1 + 1 + \cdots + 1 = N$ (N terms).

Section 17.14

1. $W^{mN+k} = (e^{2\pi i/N})^{mN+k} = e^{2\pi im}e^{2\pi ik/N} = W^k$, because $e^{2\pi im} = \cos(2\pi m) + i\sin(2\pi m) = 1$ for any integer m.

3. Denoting the first row of the matrix as row 0, rows 1 and 4 have to be interchanged, as well as rows 3 and 6.

5.
$$
\begin{bmatrix} F(0) \\ F(1) \\ F(2) \\ F(3) \\ F(4) \\ F(5) \\ F(6) \\ F(7) \end{bmatrix} = \begin{bmatrix}
1 & 1 & 1 & 1 & 1 & 1 & 1 & 1 \\
1 & W & W^2 & W^3 & W^4 & W^5 & W^6 & W^7 \\
1 & W^2 & W^4 & W^6 & 1 & W^2 & W^4 & W^6 \\
1 & W^3 & W^6 & W & W^4 & W^7 & W^2 & W^5 \\
1 & W^4 & 1 & W^4 & 1 & W^4 & 1 & W^4 \\
1 & W^5 & W^2 & W^7 & W^4 & W & W^6 & W^3 \\
1 & W^6 & W^4 & W^2 & 1 & W^6 & W^4 & W^2 \\
1 & W^7 & W^6 & W^5 & W^4 & W^3 & W^2 & W
\end{bmatrix}
\begin{bmatrix} f(0) \\ f(1) \\ f(2) \\ f(3) \\ f(4) \\ f(5) \\ f(6) \\ f(7) \end{bmatrix}
$$

7.
$$
\begin{bmatrix} F(0) \\ F(1) \\ F(2) \\ F(3) \\ F(4) \\ F(5) \end{bmatrix} = \begin{bmatrix}
1 & 1 & 1 & 1 & 1 & 1 \\
1 & W & W^2 & W^3 & W^4 & W^5 \\
1 & W^2 & W^4 & 1 & W^2 & W^4 \\
1 & W^3 & 1 & W^3 & 1 & W^3 \\
1 & W^4 & W^2 & 1 & W^4 & W^2 \\
1 & W^5 & W^4 & W^3 & W^2 & W
\end{bmatrix}
\begin{bmatrix} f(1) \\ f(2) \\ f(3) \\ f(4) \\ f(5) \\ f(6) \end{bmatrix}
$$

Chapter 18

Section 18.0

For most of these problems, more than one answer is possible.

1. $u(x, y) = f(y)$ (any differentiable function of y) **3.** $u(x, y) = x + f(y)$ **5.** $u(x, y) = k(x - y)$, k constant

7. $u(x, y) = xy$ **9.** $u(x, y) = x$ **11.** Second order, nonlinear $\left(\text{because of the } \dfrac{\partial u}{\partial x} \dfrac{\partial u}{\partial y} \text{ term} \right)$

13. Second order, linear **15.** Second order, linear

17. $\dfrac{\partial^2 y}{\partial t^2} = -\dfrac{n^2 \pi^2 a^2}{L^2} \sin\left(\dfrac{n\pi x}{L}\right) \cos\left(\dfrac{n\pi at}{L}\right) = a^2 \left[\dfrac{n^2 \pi^2}{L^2} \sin\left(\dfrac{n\pi x}{L}\right) \cos\left(\dfrac{n\pi at}{L}\right) \right] = a^2 \dfrac{\partial^2 y}{\partial x^2}.$

19. $\dfrac{\partial y}{\partial t} = \dfrac{1}{2}[f'(x + at)a + f'(x - at)(-a)]$ and $\dfrac{\partial^2 y}{\partial t^2} = \dfrac{1}{2} a[f(x + at)(a) - f''(x - at)(-a)] =$

$a^2 \dfrac{1}{2}[f''(x + at) + f''(x - at)] = a^2 \dfrac{\partial^2 f}{\partial x^2}.$

21. (a) $\dfrac{\partial Y}{\partial \xi} = \dfrac{1}{2} \dfrac{\partial y}{\partial x} + \dfrac{1}{2a} \dfrac{\partial y}{\partial t}$, so

$$
\dfrac{\partial^2 Y}{\partial \xi \, \partial \eta} = \dfrac{1}{2}\left[\dfrac{1}{2} \dfrac{\partial^2 y}{\partial x^2} - \dfrac{1}{2a} \dfrac{\partial^2 y}{\partial x \, \partial t} \right] + \dfrac{1}{2a}\left[\dfrac{1}{2} \dfrac{\partial^2 y}{\partial t \, \partial x} - \dfrac{1}{2a} \dfrac{\partial^2 y}{\partial t^2} \right] = \dfrac{1}{4}\left[\dfrac{\partial^2 y}{\partial x^2} - \dfrac{1}{a^2} \dfrac{\partial^2 y}{\partial t^2} \right] = 0.
$$

(b) $\dfrac{\partial Y}{\partial \xi} = F'(\xi)$, so $\dfrac{\partial^2 Y}{\partial \xi \, \partial \eta} = \dfrac{d}{d\eta} F'(\xi) = 0$

(c) Since $x = \dfrac{1}{2}(\xi + \eta)$ and $t = \dfrac{1}{2a}(\xi - \eta)$, then $\xi = x + at$ and $\eta = x - at$, so $y(x, t) = F(x + at) + G(x - at)$ satisfies the wave equation if F and G are twice differentiable.

(d) From the initial conditions, $y(x, 0) = F(x) + G(x) = f(x)$ and $\dfrac{\partial y}{\partial t}(x, 0) = aF'(x) - aG'(x) = g(x)$. Then

$F(x) + G(x) = f(x)$ and $F'(x) - G'(x) = \dfrac{1}{a} g(x)$. Assuming that f is differentiable, we have two equations:

$$F'(x) + G'(x) = f'(x)$$
$$F'(x) - G'(x) = \dfrac{1}{a} g(x).$$

Solve these equations to get $F'(x) = \frac{1}{2}\left[f'(x) + \frac{1}{a}g(x)\right]$ and $G'(x) = \frac{1}{2}\left[f'(x) - \frac{1}{a}g(x)\right]$. Then

$$F(\xi) = \int_0^\xi F'(\alpha)\,d\alpha + F(0) = \frac{1}{2}f(\xi) - \frac{1}{2}f(0) + \frac{1}{2a}\int_0^\xi g(\alpha)\,d\alpha, \text{ and}$$

$$G(\eta) = \int_0^\eta G(\alpha)\,d\alpha + G(0) = \frac{1}{2}f(\eta) - \frac{1}{2}f(0) - \frac{1}{2a}\int_0^\eta g(\alpha)\,d\alpha.$$

Then

$$y(x, t) = F(x + at) + G(x - at) = \frac{1}{2}f(x + at) - \frac{1}{2}f(0) + \frac{1}{2a}\int_0^{x+at} g(\alpha)\,d\alpha + \frac{1}{2}f(x - at) - \frac{1}{2}f(0) - \frac{1}{2a}\int_0^{x-at} g(\alpha)\,d\alpha$$

$$= \frac{1}{2}[f(x + at) + f(x - at)] + \frac{1}{2a}\int_{x-at}^{x+at} g(\alpha)\,d\alpha,$$

since we must have $f(0) = 0$ to be consistent with $y(x, 0) = f(x)$ and $y(0, t) = 0$.

23. $\dfrac{\partial^2 u}{\partial x^2} + \dfrac{\partial^2 u}{\partial x^2} = 0 \qquad (0 < x < \alpha, 0 < y < \beta)$

$u(x, 0) = u(x, \beta) = 0 \qquad (0 < x < \alpha)$

$u(0, y) = u(\alpha, y) = T \qquad (0 < y < \beta)$

25. $\dfrac{\partial^2 u}{\partial t^2} = a^2\dfrac{\partial^2 u}{\partial x^2} - k\left(\dfrac{\partial y}{\partial t}\right)^2 \qquad (0 < x < L, t > 0)$

$y(0, t) = y(L, t) = 0 \qquad (t > 0)$

$y(x, 0) = f(x), \qquad \dfrac{\partial y}{\partial t}(x, 0) = 0 \qquad (0 < x < L)$

Section 18.1

1. $y(x, t) = \sum\limits_{n=1}^\infty \dfrac{16}{n^2\pi^2 a}(-1)^{n+1}\sin\left(\dfrac{n\pi x}{2}\right)\sin\left(\dfrac{n\pi a t}{2}\right)$

3. $y(x, t) = \sum\limits_{n=1}^\infty \dfrac{9}{n^2\pi^2}(-1)^{n+1}\sin\left(\dfrac{n\pi x}{3}\right)\sin\left(\dfrac{2n\pi t}{3}\right)$

5. $y(x, t) = \sum\limits_{n=1}^\infty \dfrac{8}{(2n-1)^2\pi}(-1)^{n+1}\sin\left[\dfrac{(2n-1)\pi x}{2}\right]\cos[(2n-1)\sqrt{2}t]$

7. $y(x, t) = \sum\limits_{n=1}^\infty \dfrac{32}{(2n-1)^3\pi^3}\sin\left[\dfrac{(2n-1)\pi x}{2}\right]\cos\left[\dfrac{3(2n-1)\pi t}{2}\right] + \sum\limits_{n=1}^\infty \dfrac{32}{3\pi(2n-1)^2}\sin\left[\dfrac{(2n-1)\pi x}{2}\right]\sin\left[\dfrac{3(2n-1)\pi t}{2}\right]$

9. $y(x, t) = \sum\limits_{n=1}^\infty \dfrac{256}{n^3\pi^3}[2(-1)^n + 1]\sin\left(\dfrac{n\pi x}{4}\right)\cos\left(\dfrac{n\pi t}{\sqrt{2}}\right) + \sum\limits_{n=1}^\infty \dfrac{8}{\sqrt{2}(2n-1)^2\pi^2}\sin\left[\dfrac{(2n-1)\pi x}{4}\right]\sin\left[\dfrac{(2n-1)\pi t}{\sqrt{2}}\right]$

11. Let $Y(x, t) = y(x, t) + h(x)$ and choose $h(x) = \dfrac{1}{9}x^3 - \dfrac{4}{9}x$. Then $Y_{tt} = 3Y_{xx}$; $Y(0, t) = Y(2, t)$; $Y(x, 0) = \dfrac{1}{9}x^3 - \dfrac{4}{9}x$, and $Y_t(x, 0) = 0$. We find that

$$Y(x, t) = \sum_{n=1}^\infty \dfrac{16}{9n^3\pi^3}(n^2\pi^2 + 6)(-1)^n\sin\left(\dfrac{n\pi x}{2}\right)\cos\left(\dfrac{n\pi\sqrt{3}t}{2}\right);$$

hence, $y(x, t) = Y(x, t) - \dfrac{1}{9}x^3 + \dfrac{4}{9}x$.

13. Let $Y(x, t) = y(x, t) + h(x)$ and choose $h(x) = \cos(x) - 1$. Then $Y_{tt} = Y_{xx}$; $Y(0, t) = Y(2\pi, t) = 0$; $y(x, 0) = \cos(x) - 1$; and $Y_t(x, 0) = 0$. Solve for $Y(x, t)$ to get

$$Y(x, t) = \sum_{n=1}^\infty \dfrac{8}{\pi}\dfrac{1}{4n^2 - 4n - 3}\sin\left[\dfrac{(2n-1)x}{2}\right]\sin\left[\dfrac{(2n-1)t}{2}\right].$$

Then $y(x, t) = Y(x, t) - \cos(x) + 1$.

15. $y(x, t) = \dfrac{1}{2}(A + 1)L - \sum\limits_{n=1}^\infty \dfrac{4(A + 1)L}{(2n-1)^2\pi^2}\cos\left[\dfrac{(2n-1)\pi x}{L}\right]\cos\left[\dfrac{(2n-1)\pi a t}{L}\right]$

17. $u(x, t) = e^{-At/2}\sum\limits_{n=1}^\infty C_n\sin\left(\dfrac{n\pi x}{L}\right)\left[\dfrac{1}{AL}r_n\cos\left(\dfrac{r_n t}{2L}\right) + \sin\left(\dfrac{r_n t}{2L}\right)\right]$, where $r_n = [4(B^2L^2 - n^2\pi^2a^2) - A^2L^2]^{1/2}$ and

$C_n = \dfrac{2A}{\sqrt{r_n}}\int_0^L f(x)\sin\left(\dfrac{n\pi x}{L}\right)dx.$

19. With $\theta(x, t) = X(x)T(t)$, we get $X'' + \lambda X = 0$, $X'(0) - \alpha X(0) = 0$, $X'(L) + \alpha X(L) = 0$, and $T'' + a^2\lambda T = 0$. The problem in X is a regular Sturm-Liouville problem, and we get $X(x) = A\cos(\sqrt{\lambda}x) + B\sin(\sqrt{\lambda}x)$. The boundary value problem has no negative eigenvalue, nor is zero an eigenvalue. Put $\lambda = \omega^2$ to get $0 = X'(0) - \alpha X(0) = \omega B - \alpha A$ and $0 = X'(L) + \alpha X(L) = -A\omega\sin(\omega L) + B\omega\cos(\omega L) + \alpha A\cos(\omega L) + \alpha B\sin(\omega L)$. For nontrivial solutions, we need to have $A \neq 0$ and ω satisfy $\tan(\omega L) = \dfrac{2\alpha\omega}{\omega^2 - \alpha^2}$. A sketch of the graphs of $\tan(\omega L)$ and $\dfrac{2\alpha\omega}{\omega^2 - \alpha^2}$ indicates the existence of an infinite number of positive solutions, $0 < \omega_1 < \omega_2 < \cdots$. Then $\lambda_n = \omega_n^2$ are the eigenvalues, with eigenfunctions $\varphi_n(x) = \omega_n\cos(\omega_n x) + \alpha\sin(\omega_n x)$. These eigenfunctions are orthogonal $\left[\displaystyle\int_0^L \varphi_n(x)\varphi_m(x)\, dx = 0 \text{ if } n \neq m\right]$. Write

$$\theta(x, t) = \sum_{n=1}^{\infty} A_n\varphi_n(x)\cos(a\omega_n t), \text{ where } A_n = \frac{\displaystyle\int_0^L f(x)\varphi_n(x)\, dx}{\displaystyle\int_0^L \varphi_n^2(x)\, dx}.$$

Section 18.2

1. $u(x, t) = \displaystyle\sum_{n=1}^{\infty} \frac{8L^2}{(2n-1)^3\pi^2} \sin\left[\frac{(2n-1)\pi x}{L}\right]\exp\left[\frac{-(2n-1)^2\pi^2a^2t}{L^2}\right]$, where $\exp[A] = e^A$.

3. $u(x, t) = \displaystyle\sum_{n=1}^{\infty} \frac{-16L}{(2n-1)\pi[(2n-1)^2 - 4]} \sin\left[\frac{(2n-1)\pi x}{L}\right]\exp\left[\frac{-3(2n-1)^2\pi^2 t}{L^2}\right]$

5. $u(x, t) = \dfrac{2\pi^2}{3} - \displaystyle\sum_{n=1}^{\infty} \frac{4}{n^2} \cos(nx)e^{-4n^2 t}$ **7.** $u(x, t) = \displaystyle\sum_{n=1}^{\infty} \frac{48}{(2n-1)\pi} \sin\left[\frac{(2n-1)\pi x}{2L}\right]\exp\left[\frac{-(2n-1)^2a^2\pi^2 t}{4L^2}\right]$

9. The eigenvalues are $\lambda_n = z_n^2$, where $\tan(z_n) = -\dfrac{z_n}{3}$. Associated eigenfunctions are $\sin\left(\dfrac{z_n x}{6}\right)$, so

$$u(x, t) = \sum_{n=1}^{\infty} A_n\sin\left(\frac{z_n x}{6}\right)e^{-z_n^2 t/9} \text{ and } A_n = \frac{\displaystyle\int_0^6 x(6-x)\sin\left(\frac{z_n x}{6}\right)dx}{\displaystyle\int_0^6 \sin^2\left(\frac{z_n x}{6}\right)dx}.$$

$u(x, t) \approx 8.0749\sin(0.4093x)e^{-2.6801t} + 2.5739\sin(0.8732x)e^{-12.2005t} + 0.6295\sin(1.3674x)e^{-29.9175t} - 0.6106\sin(1.8760x)e^{-56.3104t} + 0.2799\sin(2.3906x)e^{-91.4363t}$.

11. Let $\alpha = \dfrac{A}{2}$ and $\beta = k\left(B + \dfrac{3A^2}{4}\right)$.

Section 18.3

1. $u(x, y) = \displaystyle\sum_{n=1}^{\infty} \frac{4K}{(2n-1)\pi\, \sinh\left[\dfrac{(2n-1)\pi b}{a}\right]} \sin\left[\frac{(2n-1)\pi x}{a}\right]\sinh\left[\frac{(2n-1)\pi y}{a}\right]$

3. $u(x, y) = \displaystyle\sum_{n=1}^{\infty} \frac{4T}{(2n-1)\pi} \sin[(2n-1)\pi y]e^{-(2n-1)\pi x}$ **5.** $u(x, y) = T - \displaystyle\sum_{n=1}^{\infty} \frac{4T}{(2n-1)\pi} \sin\left[\frac{(2n-1)\pi x}{4}\right]e^{-(2n-1)\pi y/4}$

7. $u(r, \theta) = \dfrac{1}{2} + \dfrac{1}{2}\left(\dfrac{r}{R}\right)^2\cos(2\theta)$ **9.** $u(r, \theta) = T$

11. (a) Write $u(r, \theta) = a_0 + \displaystyle\sum_{n=1}^{\infty} r^n[a_n\cos(n\theta) + b_n\sin(n\theta)]$

$$= \frac{1}{2\pi}\int_0^{2\pi} f(\xi)\, d\xi + \sum_{n=1}^{\infty} r^n\left[\frac{1}{\pi}\int_0^{2\pi} f(\xi)\cos(n\xi)\, d\xi\,\cos(n\theta) + \frac{1}{\pi}\int_0^{2\pi} f(\xi)\sin(n\xi)\, d\xi\,\sin(n\theta)\right]$$

$$= \frac{1}{2\pi}\int_0^{2\pi} f(\xi)\left[1 + 2\sum_{n=1}^{\infty} r^n[\cos(n\xi)\cos(n\theta) + \sin(n\xi)\sin(n\theta)]\right]d\xi$$

$$= \frac{1}{2\pi}\int_0^{2\pi} f(\xi)[1 + 2r^n\cos[n(\theta - \xi)]]\, d\xi.$$

(b) Let $z = re^{i\xi}$ to get $z^n = r^n e^{in\xi} = r^n[\cos(n\xi) + i\sin(n\xi)]$. Then $r^n\cos(n\xi) = \text{Re}(z^n)$. Then

$$1 + 2\sum_{n=1}^{\infty} r^n\cos(n\xi) = \text{Re}\left(1 + 2\sum_{n=1}^{\infty} z^n\right).$$

(c) If $0 \le r < 1$, then $|re^{i\xi}| = r < 1$. Let $z = re^{i\xi}$ to get

$$1 + 2\sum_{n=1}^{\infty} r^n\cos(n\xi) = \text{Re}\left(1 + 2\sum_{n=1}^{\infty} z^n\right) = \text{Re}\left(1 + 2\frac{re^{i\xi}}{1 - re^{i\xi}}\right) = \text{Re}\left(\frac{1 + re^{i\xi}}{1 - re^{i\xi}}\right).$$

(d) $\dfrac{1 + re^{i\xi}}{1 - re^{i\xi}} = \dfrac{1 + re^{i\xi}}{1 - re^{i\xi}}\dfrac{1 - re^{-i\xi}}{1 - re^{-i\xi}} = \dfrac{1 - r^2 + re^{i\xi} - re^{-i\xi}}{1 + r^2 - re^{-i\xi} - re^{i\xi}} = \dfrac{1 - r^2 + 2ir\sin(\xi)}{1 + r^2 - 2r\cos(\xi)}.$

Then $1 + 2\sum_{n=1}^{\infty} r^n\cos(n\xi) = \text{Re}\left(\dfrac{1 + re^{i\xi}}{1 - re^{i\xi}}\right) = \dfrac{1 - r^2}{1 + r^2 - 2r\cos(\xi)}.$

(e) Replacing ξ with $\theta - \xi$ in the result of (d), we have

$$1 + 2\sum_{n=1}^{\infty} r^n\cos[n(\theta - \xi)] = \frac{1 - r^2}{1 + r^2 - 2r\cos(\theta - \xi)}.$$

From (a), we now have

$$u(r, \theta) = \frac{1}{2\pi}\int_0^{2\pi} f(\xi)\{1 + 2r^n\cos[n(\theta - \xi)]\}\, d\xi = \frac{1}{2\pi}\int_0^{2\pi} f(\xi)\frac{1 - r^2}{1 + r^2 - 2r\cos(\theta - \xi)}\, d\xi.$$

13. Since $u(r, \theta) = a_0 + \sum_{n=1}^{\infty} \dfrac{r^n}{R^n}[A_n\cos(n\theta) + B_n\sin(n\theta)]$, the temperature at the center of the disk is

$$u(0, \theta) = a_0 = \frac{1}{2\pi}\int_{-\pi}^{\pi} f(\theta)\, d\theta, \text{ where } f(\theta) \text{ is the boundary temperature.}$$

Section 18.4

1. Let $V(x, t) = u(x, t) + f(x)$ and choose $f(x) = \dfrac{2}{27}(x^3 - x)$. Then $V_{tt} = 3V_{xx}$, $V(0, t) = V(1, t) = 0$, $V(x, 0) = \dfrac{2}{27}(x^3 - x)$, $V_t(x, 0) = 1$. Solve for V to get

$$V(x, t) = \sum_{n=1}^{\infty} \frac{2}{3n^2\pi^2}\sin(n\pi x)\{12(-1)^n\cos(3n\pi t) + [1 - (-1)^n]\sin(3n\pi t)\},$$

and $u(x, t) = V(x, t) - \dfrac{2}{27}(x^3 - x)$.

3. Let $Y(x, t) = y(x, t) + f(x)$ and choose $f(x) = -\dfrac{1}{9}x$. Then $Y_{tt} = 4Y_{xx}$, $Y(0, t) = Y(9, t) = 0$, $Y(x, 0) = -\dfrac{x}{9}$, $Y_t(x, 0) = x$.
Solve for Y to get

$$Y(x, t) = \sum_{n=1}^{\infty} \frac{9(-1)^{n+1}}{n\pi}\sin\left(\frac{n\pi x}{9}\right)\left[9\cos\left(\frac{2n\pi x}{9}\right) - \sin\left(\frac{2n\pi x}{9}\right)\right].$$

Finally, $y(x, t) = Y(x, t) + \dfrac{x}{9}$.

5. Let $V(x, t) = u(x, t) + f(x)$ and choose $f(x) = -\dfrac{3x}{5}$. Then $V_t = 9V_{xx}$, $V(0, t) = V(5, t) = 0$, $V(x, 0) = -\dfrac{3x}{5}$.
Solve for V to get

$$V(x, t) = \sum_{n=1}^{\infty} \frac{6(-1)^n}{n\pi}\sin\left(\frac{n\pi x}{5}\right)e^{-9n^2\pi^2 t/25},$$

and $u(x, t) = V(x, t) + \dfrac{3x}{5}$.

7. Let $w(x, t) = e^{\alpha t}u(x, t)$ and choose $\alpha = A$. Then $w_t = 4w_{xx}$, $w(0, t) = w(9, t) = 0$, and $w(x, 0) = 3x$. We get

$$w(x, t) = \sum_{n=1}^{\infty} \frac{54(-1)^{n+1}}{n\pi}\sin\left(\frac{n\pi x}{9}\right)e^{-4n^2\pi^2 t/81} \text{ and } u(x, t) = e^{-At}w(x, t).$$

9. $u(x, t) = 50x + 100 + \sum_{n=1}^{\infty} \frac{4}{n\pi}[99(-1)^n - 50]\sin\left(\frac{n\pi x}{2}\right)e^{-9n^2\pi^2 t/4}$; $\lim_{t \to \infty} u(x, t) = 50x + 100$.

Section 18.5

1. The problem to be solved is $u_{rr} + \frac{1}{r}u_r + u_{zz} = 0$ $(0 < r < 2, z > 0)$; $u(R, z) = 0$ $(z > 0)$; $u(r, 0) = K$ $(0 < r < 2)$. We get

$u(r, z) = \sum_{n=1}^{\infty} A_n J_0\left(\frac{z_n r}{R}\right)e^{-z_n z/R}$, where z_n is the nth positive zero of J_0 and $A_n = \frac{2}{R^2 J_1^2(z_n)}$, with $R = 2$. To evaluate this

integral, first let $x = \frac{z_n r}{2}$ to get $A_n = \frac{2K}{z_n^2 J_1^2(z_n)}\int_0^{z_n} xJ_0(x)\, dx$, then use the identity $\int xJ_0(x)\, dx = xJ_1(x)$. The

steady-state temperature is

$$u(r, z) = \sum_{n=1}^{\infty} \frac{2K}{z_n J_1(z_n)} J_0\left(\frac{z_n r}{2}\right)e^{-z_n z/2}.$$

3. The problem is $u_t = a^2\left(u_{rr} + \frac{1}{r}u_r\right) - Au$ $(0 < r < R, t > 0)$; $u(R, t) = 0$ $(t > 0)$; $u(r, 0) = K$ $(0 < r < R)$. Let

$v(r, t) = e^{At}u(r, t)$ to obtain the separable problem $v_t = a^2\left(v_{rr} + \frac{1}{r}v_r\right)$, $v(R, t) = 0$, $v(r, 0) = K$. We get

$$v(r, t) = \sum_{n=1}^{\infty} \frac{2K}{z_n J_1(z_n)} J_1\left(\frac{z_n r}{R}\right)e^{-a^2 z_n^2 t/R^2},$$

where z_n is the nth positive zero of J_0. Then $u(r, t) = e^{-At}v(r, t)$.

Section 18.4

5. The problem is $u_{rr} + \frac{1}{r}u_r + u_{zz} = 0$ $(0 < r < R, 0 < z < L)$; $u_r(R, 0) = 0$ $(0 < z < L)$; $u(r, 0) = 0$, $u(r, L) = 2$ $(0 < r < R)$.
Zero is an eigenvalue with eigenfunction $A_0 z$. The requirement that $u_r(R, 0) = 0$ forces $J_1(kR) = 0$; hence, the remaining

eigenvalues are $\lambda_n = \frac{z_n^2}{R^2}$, where z_n is the nth positive zero of J_1. Associated eigenfunctions are $J_0\left(\frac{z_n r}{R}\right)$. Then

$u(r, z) = A_0 z + \sum_{n=1}^{\infty} J_0\left(\frac{z_n r}{R}\right)[A_n e^{z_n z/R} + B_n e^{-z_n z/R}]$. Since $u(r, 0) = 0$, $B_n = -A_n$, so

$u(r, z) = A_0 z + \sum_{n=1}^{\infty} C_n J_0\left(\frac{z_n r}{R}\right)\sinh\left(\frac{z_n z}{R}\right)$. Since $u(r, L) = 2$, $A_0 = \dfrac{\int_0^R 2r\, dr}{\int_0^R Lr\, dr} = \frac{2}{L}$ and

$C_n = \dfrac{1}{\sinh\left(\frac{z_n L}{R}\right)} \cdot \dfrac{\int_0^R 2rJ_0\left(\frac{z_n r}{R}\right)dr}{\int_0^R rJ_0^2\left(\frac{z_n r}{R}\right)dr}$. Since $\int_0^R rJ_0\left(\frac{z_n r}{R}\right)dr = \frac{4}{z_n^2}\int_0^{z_n} xJ_0(x)\, dx = \frac{4x}{z_n^2}J_1(x)\Big]_0^{z_n} = 0$ (z_n being a zero of J_1),

we have $C_n = 0$ for $n = 1, 2, 3, \ldots$; hence, $u(r, z) = \frac{2}{L} z$.

7. Let $x = \lambda r$. Then $F(r) = F\left(\frac{\lambda}{x}\right) = Y(x)$; $F'(x) = \frac{dF}{dr} = \frac{dF}{dx}\frac{dx}{dr}$; and $F''(x) = \frac{d^2 F}{dx^2} = \frac{d}{dr}\left[\frac{dY}{dx}\frac{dx}{dr}\right] = \frac{d^2 Y}{dx^2}\left[\frac{dx}{dr}\right]^2 + \frac{dY}{dx}\frac{d^2 x}{dr^2}$.

Since $\frac{dx}{dr} = \lambda$ and $\frac{d^2 x}{dr^2} = 0$, we get $F'(r) = \lambda Y'(x)$ and $F''(r) = \lambda^2 Y''(x)$. Then

$$0 = r^2 F''(r) + rF'(r) + \lambda^2 r^2 F(r) = \left(\frac{x^2}{\lambda^2}\right)\lambda^2 Y''(x) + \left(\frac{x}{\lambda}\right)\lambda Y'(x) + \lambda^2\left(\frac{x^2}{\lambda^2}\right)Y(x)$$

$$= x^2 Y''(x) + xY'(x) + x^2 Y(x),$$

a Bessel equation of order zero with general solution $Y(x) = aJ_0(x) + bY_0(x)$. Therefore, $F(r) = Y(x) = Y(\lambda r) = aJ_0(\lambda r) + bY_0(\lambda r)$.

Section 18.6

1. $u(\rho, \varphi) = \sum_{n=1}^{\infty} A_{2n-1}\rho^{2n-1}P_{2n-1}[\cos(\varphi)]$, where $A_{2n-1} = \dfrac{\displaystyle\int_0^1 AP_{2n-1}(x)\,dx}{R^{2n-1}\displaystyle\int_0^1 [P_{2n-1}(x)]^2\,dx} = \dfrac{(4n-1)A}{R^{2n-1}}\displaystyle\int_0^1 P_{2n-1}(x)\,dx$

3. $u(\rho, \varphi) = \sum_{n=1}^{\infty} A_{2n-1}\rho^{2n-1}P_{2n-1}[\cos(\varphi)]$, with $A_{2n-1} = \dfrac{4n-1}{R^{2n-1}}\displaystyle\int_0^1 f(x)P_{2n-1}(x)\,dx$.   **5.** $u(\rho, \varphi) = T\left[\dfrac{R_2 - \rho}{R_2 - R_1}\right]$

Section 18.7

1. $u(x, y, z) = \sum_{n=1}^{\infty}\sum_{m=1}^{\infty} \dfrac{4(-1)^{n+m}}{nm\pi^2\sinh(\pi\sqrt{n^2 + m^2})}\sin(n\pi x)\sin(m\pi y)\sinh(\pi\sqrt{n^2 + m^2}\,z)$

3. $u(x, y, t) = \sum_{n=1}^{\infty}\dfrac{1}{\pi}\left\{\dfrac{8(-1)^{n+1}\pi^2}{n} + \dfrac{16}{n^3}[(-1)^n - 1]\right\}\sin\left(\dfrac{nx}{2}\right)\sin(y)\cos\left(\dfrac{1}{2}\sqrt{n^2 + 4}\,t\right)$

5. $u(x, y, t) = \sum_{n=1}^{\infty}\sum_{m=1}^{\infty}\left[\dfrac{16}{\pi^2(2n-1)(2m-1)\sqrt{(2n-1)^2 + (2m-1)^2}}\right]$

$\times \sin\left[\dfrac{(2n-1)x}{2}\right]\sin\left[\dfrac{(2m-1)y}{2}\right]\sin(\sqrt{(2n-1)^2 + (2m-1)^2}\,t)$

7. $u(x, y, z) = \sum_{n=1}^{\infty}\sum_{m=1}^{\infty}\left\{\dfrac{32}{(2n-1)(2m-1)\pi^2\sinh(2\pi\sqrt{(2m-1)^2 + \pi^2(2n-1)^2})}\right\}$

$\times \{\sin[(2n-1)\pi x]\sin[(2m-1)z]\sinh(\sqrt{(2m-1)^2 + \pi^2(2n-1)^2}\,y)\}$

$+ \sum_{n=1}^{\infty}\sum_{m=1}^{\infty}\dfrac{16}{(2n-1)(2m-1)\pi^2}\dfrac{1}{\sinh\left(\pi\sqrt{\dfrac{(2m-1)^2}{4} + \pi^2(2n-1)^2}\right)}$

$\times \left\{\sin[(2n-1)\pi x]\sin\left[\dfrac{(2m-1)y}{2}\right]\sinh\left(\sqrt{\dfrac{(2m-1)^2}{4} + \pi^2(2n-1)^2}\,z\right)\right\}$

9. $u(x, y, z, t) = \dfrac{1}{8}a_{000} + \dfrac{1}{4}\sum_{n=1}^{\infty}a_{n00}\cos\left(\dfrac{n\pi z}{L}\right)\exp\left[-\dfrac{n^2\pi^2 a^2 t}{L^2}\right]$

$+ \dfrac{1}{4}\sum_{m=1}^{\infty}a_{0m0}\cos\left(\dfrac{m\pi y}{L}\right)\exp\left[-\dfrac{m^2\pi^2 a^2 t}{L^2}\right] + \dfrac{1}{4}\sum_{r=1}^{\infty}a_{00r}\cos\left(\dfrac{r\pi z}{L}\right)\exp\left[-\dfrac{r^2\pi^2 a^2 t}{L^2}\right]$

$+ \dfrac{1}{2}\sum_{n=1}^{\infty}\sum_{m=1}^{\infty}a_{nm0}\cos\left(\dfrac{n\pi x}{L}\right)\cos\left(\dfrac{m\pi y}{L}\right)\exp\left[-\dfrac{(n^2 + m^2)\pi^2 a^2 t}{L^2}\right]$

$+ \dfrac{1}{2}\sum_{n=1}^{\infty}\sum_{r=1}^{\infty}a_{n0r}\cos\left(\dfrac{n\pi x}{L}\right)\cos\left(\dfrac{r\pi z}{L}\right)\exp\left[-\dfrac{(n^2 + r^2)\pi^2 a^2 t}{L^2}\right]$

$+ \dfrac{1}{2}\sum_{m=1}^{\infty}\sum_{r=1}^{\infty}a_{0mr}\cos\left(\dfrac{m\pi y}{L}\right)\cos\left(\dfrac{r\pi z}{L}\right)\exp\left[-\dfrac{(m^2 + r^2)\pi^2 a^2 t}{L^2}\right]$

$+ \sum_{n=1}^{\infty}\sum_{m=1}^{\infty}\sum_{r=1}^{\infty}a_{nmr}\cos\left(\dfrac{n\pi x}{L}\right)\cos\left(\dfrac{m\pi y}{L}\right)\cos\left(\dfrac{r\pi z}{L}\right)\exp\left[\dfrac{(-n^2 - m^2 - r^2)\pi^2 a^2 t}{L^2}\right],$

with $a_{nmr} = \dfrac{8}{L^3}\displaystyle\int_0^L\int_0^L\int_0^L f(x, y, z)\cos\left(\dfrac{n\pi x}{L}\right)\cos\left(\dfrac{m\pi y}{L}\right)\cos\left(\dfrac{r\pi z}{L}\right)dx\,dy\,dz$

11. Suppose the room occupies $0 \le x \le L, 0 \le y \le L, 0 \le z \le L$. We find that

$$u(x, y, z, t) = \sum_{n=1}^{\infty} \sum_{m=1}^{\infty} \sum_{r=1}^{\infty} b_{nmr} \sin\left(\frac{n\pi x}{L}\right) \sin\left(\frac{m\pi y}{L}\right) \sin\left(\frac{r\pi z}{L}\right) \cos\left(\frac{\sqrt{n^2 + m^2 + r^2}\pi at}{L}\right)$$

$$+ \sum_{n=1}^{\infty} \sum_{m=1}^{\infty} \sum_{r=1}^{\infty} c_{nmr} \sin\left(\frac{n\pi x}{L}\right) \sin\left(\frac{m\pi y}{L}\right) \sin\left(\frac{r\pi z}{L}\right) \sin\left(\frac{\sqrt{n^2 + m^2 + r^2}\pi at}{L}\right),$$

with

$$b_{nmr} = \frac{8}{L^3} \int_0^L \int_0^L \int_0^L f(x, y, z) \sin\left(\frac{n\pi x}{L}\right) \sin\left(\frac{m\pi y}{L}\right) \sin\left(\frac{r\pi z}{L}\right) dx \, dy \, dz$$

and

$$c_{nmr} = \frac{8}{L^3} \int_0^L \int_0^L \int_0^L g(x, y, z) \sin\left(\frac{n\pi x}{L}\right) \sin\left(\frac{m\pi y}{L}\right) \sin\left(\frac{r\pi z}{L}\right) dx \, dy \, dz.$$

13. $u(x, y, z) = \sum_{n=1}^{\infty} \sum_{m=1}^{\infty} \frac{4}{\sinh(2\pi\sqrt{n^2 + m^2})} \left(\frac{(-1)^{n+1}}{n\pi}\right)\left(\frac{(-1)^{m+1}}{m\pi} + \frac{2}{m^3\pi^3}[(-1)^m - 1]\right) \sin(n\pi x)\sin(m\pi y)\sinh(\pi\sqrt{n^2 + m^2}z)$

15. The problem is $u_t = k^2[u_{xx} + u_{yy}] - Au \ (0 < x < a, 0 < y < b, t > 0); u_x(0, y, t) = u_x(a, y, t) = u_y(x, 0, t) = u_y(x, b, t) = 0,$
$u(x, y, 0) = x(a - x)$. Let $v(x, y, t) = e^{At}u(x, y, t)$. Then v satisfies $v_t = k^2[v_{xx} + v_{yy}]$,
$v_x(0, y, t) = v_x(a, y, t) = v_y(x, 0, t) = v_y(x, b, t) = 0, v(x, y, 0) = x(a - x)$. Solve for $v(x, y, t)$ to get

$$v(x, y, t) = \sum_{n=1}^{\infty} \sum_{m=1}^{\infty} \frac{32a^2}{\pi^4(2n - 1)^3(2m - 1)} \sin\left[\frac{(2n - 1)\pi x}{a}\right] \sin\left[\frac{(2m - 1)\pi y}{b}\right]$$

$$\times \exp\left\{-\left[\frac{(2n - 1)^2\pi^2}{a^2} + \frac{k^2(2m - 1)^2\pi^2}{b^2}\right]t\right\}$$

and $u(x, y, t) = e^{At}v(x, y, t)$.

Section 18.8

1. $u(r, \theta, t) = \sum_{n=1}^{\infty} \sum_{k=1}^{\infty} [a_{nk}\cos(n\theta) + b_{nk}\sin(n\theta)]J_n(\omega_{nk}r)\sin(\omega_{nk}at)$, where

$$a_{nk} = \frac{2}{a\omega_{nk}\pi R^2[J_1(\omega_{nk}R)]^2} \int_0^R \int_{-\pi}^{\pi} rJ_n(\omega_{nk}r)g(r, \theta)\cos(n\theta) \, d\theta \, dr$$

and

$$b_{nk} = \frac{2}{a\omega_{nk}\pi R^2[J_1(\omega_{nk}R)]^2} \int_0^R \int_{-\pi}^{\pi} rJ_n(\omega_{nk}r)g(r, \theta)\sin(n\theta) \, d\theta \, dr.$$

3. Since $J_0(0) = 1$ and $J_n(0) = 0$ if $n = 1, 2, 3, \ldots$, the deflection at the center of the membrane is given by

$$u(0, \theta, t) = \sum_{k=1}^{\infty} a_{0k}\cos(\omega_{0k}at) = \sum_{k=1}^{\infty} a_{0k}\cos\left(\frac{z_k at}{2}\right), \text{ where } z_k \text{ is the } k\text{th positive zero of } J_0 \text{ and}$$

$$a_{0k} = \frac{1}{4\pi[J_1(z_k)]^2} \int_0^2 \int_{-\pi}^{\pi} rJ_0\left(\frac{z_k r}{2}\right)(4 - r^2)\sin^2(\theta) \, d\theta \, dr$$

$$= \frac{1}{4[J_1(z_k)]^2} \int_0^2 (4r - r^3)J_0\left(\frac{z_k r}{2}\right) dr.$$

Let $x = \frac{z_k r}{2}$ to get $a_{0k} = \frac{4}{z_k^4[J_1(z_k)]^2} \int_0^{z_k} (z_k^2 x - x^3)J_0(x) \, dx$. Now use the two identities $\int xJ_0(x) \, dx = xJ_1(x)$

and $\int x^3J_0(x) \, dx = x(x^2 - 4)J_1(x) + 2x^2J_0(x)$ to get $a_{0k} = \frac{16}{z_k^3J_1(z_k)}$. Therefore,

$$u(0, \theta, t) = \sum_{k=1}^{\infty} \frac{16}{z_k^3 J_1(z_k)} \cos\left(\frac{z_k at}{2}\right) \approx 2.2163 \cos(1.2024at) + 0.2795 \cos(2.700at) + 0.0909 \cos(4.3269at).$$

Section 18.9

1. $y(x, t) = \dfrac{2}{\pi} \displaystyle\int_0^\infty \left\{ -\dfrac{\sin(k)}{k^2} + \dfrac{2[1 - \cos(k)]}{k^3} \right\} \sin(kx)\cos(akt)\, dk$ **3.** $y(x, t) = \dfrac{2}{\pi} \displaystyle\int_0^\infty \dfrac{1}{1 - k^2} \sin(k\pi)\sin(kx)\cos(akt)\, dk$

5. $y(x, t) = \dfrac{2}{\pi a} \displaystyle\int_0^\infty \dfrac{1}{k^3} [\sin(k) - k \cos(k)]\sin(kx)\sin(akt)\, dk$

7. The problem is $y_{tt} = a^2 y_{xx}$ $(-\infty < x < \infty, t > 0)$; $y(x, 0) = f(x)$, $y_t(x, 0) = 0$, $y(x, t)$ bounded for $-\infty < x < \infty$ and $t > 0$. Then $y(x, t) = \dfrac{2}{\pi} \displaystyle\int_0^\infty \dfrac{1 - \cos(k)}{k^2} \cos(kx)\cos(akt)\, dk$.

9. The problem is $y_{tt} = a^2 y_{xx}$ $(-\infty < x < \infty, t > 0)$; $y(x, 0) = 0$, $y_t(x, 0) = g(x)$. We get
$$y(x, t) = \dfrac{2}{\pi a} \int_0^\infty \dfrac{\sin(2k)}{k^2} \cos(kx)\sin(akt)\, dk.$$

11. For the semi-infinite string, the solution is
$$y(x, t) = \int_0^\infty A_k \sin(kx)\cos(kat)\, dk + \int_0^\infty B_k \sin(kx)\sin(kat)\, dk,$$

where $A_k = \dfrac{2}{\pi} \displaystyle\int_0^\infty f(\xi)\sin(k\xi)\, d\xi$ and $B_k = \dfrac{2}{\pi a k} \displaystyle\int_0^\infty g(\xi)\sin(k\xi)\, d\xi$. For the infinite string, the solution is $Y(x, t) = $

$\displaystyle\int_0^\infty [\mathscr{A}_k\cos(kx) + \mathscr{B}_k\sin(kx)][\mathscr{C}_k\cos(kat) + \mathscr{D}_k\sin(kat)]\, dk$, where

$$\mathscr{A}_k\mathscr{C}_k = \dfrac{1}{\pi} \int_{-\infty}^\infty F(\xi)\cos(k\xi)\, d\xi, \quad \mathscr{B}_k\mathscr{C}_k = \dfrac{1}{\pi} \int_{-\infty}^\infty F(\xi)\sin(k\xi)\, d\xi,$$

$$ka\, \mathscr{A}_k\mathscr{D}_k = \dfrac{1}{\pi} \int_{-\infty}^\infty G(\xi)\cos(k\xi)\, d\xi, \text{ and } ka\, \mathscr{B}_k\mathscr{D}_k = \dfrac{1}{\pi} \int_{-\infty}^\infty G(\xi)\sin(k\xi)\, d\xi.$$

We can obtain $y(x, t)$ from $Y(x, t)$ by choosing F to be an odd extension of f to the entire real line and G an odd extension of g.

13. $u(x, t) = \displaystyle\int_0^\infty \dfrac{4k}{\pi(k^4 + 4)} \sin(kx)e^{-4k^2 t}\, dk$ **15.** $u(x, y) = \dfrac{4}{\pi} \displaystyle\int_0^\infty \dfrac{1}{k} [\cos(7k) - \cos(5k)]e^{-kx}\sin(ky)\, dk$

17. $u(x, t) = \dfrac{2}{\pi} \displaystyle\int_0^\infty \left[\dfrac{1}{2}\sin(4k) - \dfrac{4}{k}\cos(4k) \right] \sin(kx)e^{-a^2 k^2 t}\, dk$ **19.** $u(x, t) = \dfrac{2}{\pi} \displaystyle\int_0^\infty \dfrac{1}{1 + k^2} \cos(kx)e^{-a^2 k^2 t}\, dk$

21. $u(x, y) = \dfrac{2}{\pi} \displaystyle\int_0^\infty \dfrac{1}{k} \sin(k)\cos(ky)e^{-kx}\, dk$ **23.** $u(x, y) = \displaystyle\int_0^\infty B_k\sin(kx)\sinh(ky)\, dk$, where $B_k = \dfrac{2}{\pi} \dfrac{1}{\sinh(k)} \displaystyle\int_0^\infty f(x)\sin(kx)\, dx$

25. $u(x, t) = \displaystyle\int_0^\infty A_k\cos(kx)e^{-a^2 k^2 t}\, dk$, where $A_k = \dfrac{2}{\pi} \displaystyle\int_0^\infty f(x)\cos(kx)\, dx$ **27.** $u(x, y) = \dfrac{2}{\pi} \displaystyle\int_0^\infty \dfrac{1}{1 + k^2} e^{-kx}\cos(ky)\, dk$

Section 18.10

1. $y(x, t) = At + (1 - A)\left(t - \dfrac{x}{a} \right) H\left(t - \dfrac{x}{a} \right) = \begin{cases} At & \text{if } t < \dfrac{x}{a} \\[2mm] t + (A - 1)\dfrac{x}{a} & \text{if } t > \dfrac{x}{a} \end{cases}$

3. $u(x, t) = (4t - 2x^2)H\left(t - \dfrac{x^2}{2} \right) = \begin{cases} 0 & \text{if } t < \dfrac{x^2}{2} \\[2mm] 4t - 2x^2 & \text{if } t > \dfrac{x^2}{2} \end{cases}$

5. $u(x, y) = \dfrac{y^2}{2b} + \dfrac{(ay - bx)^2}{2a^2 b} H(ay - bx) = \begin{cases} \dfrac{y^2}{2b} & \text{if } ay < bx \\[2mm] \dfrac{xy}{a} - \dfrac{bx^2}{2a^2} & \text{if } ay > bx \end{cases}$

7. $y(x, t) = \begin{cases} \dfrac{1}{2} e^{-x}\sinh(2t) & \text{if } t < \dfrac{x}{2} \\[2mm] \dfrac{1}{2} e^{-x}\sinh(2t) + \sin\left(t - \dfrac{x}{2}\right) - \dfrac{1}{2}\sinh\left[2\left(t - \dfrac{x}{2}\right)\right] & \text{if } t > \dfrac{x}{2} \end{cases}$

9. $u(x, t) = \begin{cases} 1 - at & \text{if } t < \dfrac{x}{\sqrt{ab}} \\[2mm] 1 - at + \sqrt{\dfrac{a}{b}}\left(t - \dfrac{x}{\sqrt{ab}}\right) & \text{if } t > \dfrac{x}{\sqrt{ab}}, \end{cases}$

$v(x, t) = \begin{cases} x & \text{if } t < \dfrac{x}{\sqrt{ab}} \\[2mm] x\left(1 - \dfrac{1}{\sqrt{ab}}\right) + t & \text{if } t > \dfrac{x}{\sqrt{ab}} \end{cases}$

Section 18.11

1. $u(x, y) = \dfrac{1}{2\pi} \displaystyle\int_{-\infty}^{\infty} F(\omega)e^{-\omega y}e^{i\omega x}\, d\omega$ **3.** $u(x, y) = Be^{-y}\sin(x)$

5. Using a finite Fourier sine transform in x, we get

$$u(x, y) = \sum_{n=1}^{\infty} \frac{2}{\pi}\left\{\left[-\frac{4}{n} + \frac{6(-1)^n}{n}\right]e^{-ny} - \frac{2(-1)^n}{n}\right\}\sin(nx).$$

Using a Fourier sine transform in y, we get

$$u(x, y) = \frac{2}{\pi}\int_0^{\infty}\left[\left(\frac{\frac{4}{\omega}e^{-\omega\pi} - 2 - \frac{4}{\omega}}{e^{-\omega\pi} - e^{\omega\pi}}\right)e^{\omega x} + \left(\frac{2 + \frac{4}{\omega} - \frac{4}{\omega}e^{\omega\pi}}{e^{-\omega\pi} - e^{\omega\pi}}\right)e^{-\omega x} - \frac{4}{\omega}\right]\sin(\omega x)\, d\omega.$$

7. Using a finite Fourier sine transform in x, we get

$$u(x, y) = \sum_{n=1}^{\infty} \frac{8}{\pi}\left[\frac{(-1)^n}{n}\frac{\cosh(ny)}{\cosh(2y)}\right]\sin(nx).$$

9. Using a Fourier sine transform in x, we get

$$u(x, y) = \frac{2}{\pi}\int_0^{\infty}\left[\frac{2}{\omega^3}\left(\frac{2 + e^{-\omega}}{e^{\omega} - e^{-\omega}}\right)e^{\omega y} - \frac{2}{\omega^3}\left(\frac{2 + e^{\omega}}{e^{\omega} - e^{-\omega}}\right)e^{-\omega y} - \frac{y^3}{\omega} + \frac{y^2}{\omega} - \frac{6y}{\omega^3} + \frac{2}{\omega^3}\right]\sin(\omega x)\, d\omega.$$

11. $u(x, t) = -\dfrac{4}{\pi}\displaystyle\int_0^{\infty}\frac{\omega}{(1 + \omega^2)^2}\sin(\omega x)e^{-\omega^2 t - t^2/2}\, d\omega$ **13.** Using the Fourier transform in x, we get

$$u(x, y) = \frac{1}{2\sqrt{\pi}} = \int_{-\infty}^{\infty}\frac{\cosh(\omega y)}{\cosh(\omega)}e^{-\omega^2/4}e^{i\omega x}\, d\omega$$

Section 18.12

1. Suppose that the boundary value problem with initial data $y(x, 0) = f_1(x)$, $\dfrac{\partial y}{\partial t}(x, 0) = g_1(x)$ has solution $Y_1(x, t)$.

Suppose that the problem with initial data $y(x, 0) = f_2(x)$, $\dfrac{\partial y}{\partial t}(x, 0) = g_2(x)$ has solution $Y_2(x, t)$. Then the problem depends continuously on the initial data if, given $\epsilon > 0$ and any $T > 0$, there is some $\delta > 0$ such that

$$|Y_1(x, t) - Y_2(x, t)| < \epsilon$$

if $|f_1(x) - f_2(x)| < \delta$ and $|g_1(t) - g_2(t)| < \delta$, for $0 \leq x \leq L$ and $0 \leq t \leq T$.

To prove that the given problem depends continuously on the data, use d'Alembert's formula to write

$$Y_1(x, t) = \frac{1}{2}[f_1(x + at) + f_1(x - at)] + \frac{1}{2a}\int_{x-at}^{x+at} g_1(\alpha)\, d\alpha, \text{ in which } f_1 \text{ and } g_1 \text{ have been extended periodically over the}$$

entire real line, as discussed in Section 18.1. Similarly, $Y_2(x, t) = \frac{1}{2}[f_2(x + at) + f_2(x - at)] + \frac{1}{2a}\int_{x-at}^{x+at} g_2(\alpha)\, d\alpha$. Now write

$$|Y_1(x, t) - Y_2(x, t)| \le \frac{1}{2}|f_1(x + at) - f_2(x + at)| + \frac{1}{2}|f_1(x - at) - f_2(x - at)| + \frac{1}{2a}\int_{x-at}^{x+at} |g_1(\alpha) - g_2(\alpha)|\, d\alpha$$

$$\le \frac{1}{2}\delta + \frac{1}{2}\delta + \frac{1}{2a}\delta(2at) \le \delta + \delta T = (1 + T)\delta < \epsilon$$

if $\delta < \dfrac{\epsilon}{1 + T}$.

3. (a) It is routine to check that $\dfrac{\partial^2 u}{\partial x^2} = -\sinh(ny)\sin(nx)$ and $\dfrac{\partial^2 u}{\partial y^2} = \sinh(ny)\sin(nx)$.

(b) Observe that $B(x, y) - A(x, y)$ is the unique solution of $\dfrac{\partial^2 u}{\partial x^2} + \dfrac{\partial^2 u}{\partial y^2} = 0$, $u(x, 0) = 0$, $\dfrac{\partial u}{\partial y}(x, 0) = \dfrac{1}{n}\sin(nx)$.

By (a), $B(x, y) - A(x, y) = \dfrac{1}{n^2}\sinh(ny)\sin(nx)$.

(c) Consider the two problems in (b). In both problems, $u(x, 0) = f(x)$. Further, the difference between the data functions for $\dfrac{\partial u}{\partial y}(x, 0)$ is $\dfrac{1}{n}\sin(nx)$, which can be made as small as we like by choosing n sufficiently large.

Nevertheless, the difference in the solutions is $\dfrac{1}{n^2}\sinh(ny)\sin(nx)$, and this can be made arbitrarily large for any choice of n by choosing y sufficiently large. Thus, the solution of this problem does not depend continuously on the data.

5. Suppose that there are two solutions, y_1 and y_2, and let $u(x, t) = y_1(x, t) - y_2(x, t)$. It is routine to verify that $\dfrac{\partial^2 u}{\partial t^2} = a^2\dfrac{\partial^2 u}{\partial x^2}$, $\dfrac{\partial u}{\partial t}(0, t) = 0$, $\dfrac{\partial u}{\partial t}(L, t) = 0$, $u(x, 0) = 0$, and $\dfrac{\partial u}{\partial t}(x, 0) = 0$. Let $I(t) = \dfrac{1}{2}\int_0^L \left[\left(\dfrac{\partial u}{\partial x}\right)^2 + \dfrac{1}{a^2}\left(\dfrac{\partial u}{\partial t}\right)^2\right] dx$. Then

$$I'(t) = \int_0^L \left[\dfrac{\partial u}{\partial x}\dfrac{\partial^2 u}{\partial x\,\partial t} + \dfrac{1}{a^2}\dfrac{\partial u}{\partial t}\dfrac{\partial^2 u}{\partial t^2}\right] dx$$

$$= \int_0^L \left[\dfrac{1}{a^2}\dfrac{\partial^2 u}{\partial t^2} - \dfrac{\partial^2 u}{\partial x^2}\right]\dfrac{\partial u}{\partial t}\, dx + \int_0^L \left[\dfrac{\partial u}{\partial x}\dfrac{\partial^2 u}{\partial x\,\partial t} + \dfrac{\partial^2 u}{\partial x^2}\right] dx$$

$$= 0 + \int_0^L \dfrac{\partial}{\partial x}\left(\dfrac{\partial u}{\partial t}\dfrac{\partial u}{\partial x}\right) dx = \dfrac{\partial u}{\partial t}\dfrac{\partial u}{\partial x}\bigg]_0^L = 0.$$

Thus, $I(t) = $ constant. But $I(0) = 0$; hence, $I(t) = 0$ for $t \ge 0$. Therefore, $\dfrac{\partial u}{\partial x} = 0$ and $\dfrac{\partial u}{\partial t} = 0$, so $u(x, t) = $ constant also. But $u(x, 0) = 0$; hence, $u(x, t) = 0$ for $0 \le x \le L$ and $t \ge 0$, proving that $y_1(x, t) = y_2(x, t)$.

7. Suppose that u_1 and u_2 are continuous solutions with continuous first and second partial derivatives throughout M. Let $w(x, y, z) = u_1(x, y, z) - u_2(x, y, z)$. Then $\nabla^2 w = 0$ in M, and $a\dfrac{\partial w}{\partial \eta}(x, y, z) + bw(x, y, z) = 0$ for (x, y, z) on Σ.

By Gauss's theorem, write

$$\iint_\Sigma w\dfrac{\partial w}{\partial \eta}\, d\sigma = \iiint_M [w_x^2 + w_y^2 + w_z^2]\, dx\, dy\, dz + \iiint_M w\nabla^2 w\, dx\, dy\, dz.$$

Now, $\nabla^2 w = 0$ throughout M; hence, $\iiint_M w\nabla^2 w\, dx\, dy\, dz = 0$. Further, $w\dfrac{\partial w}{\partial \eta} = -\dfrac{b}{a}w^2$, so $\iint_\Sigma w\dfrac{\partial w}{\partial \eta}\, d\sigma = -\dfrac{b}{a}\iint_\Sigma w^2\, d\sigma \le 0$. But $\iiint_M [w_x^2 + w_y^2 + w_z^2]\, dx\, dy\, dz \ge 0$. Therefore, $\iiint_M [w_x^2 + w_y^2 + w_z^2]\, dx\, dy\, dz = 0$; hence,

$w_x = w_y = w_z = 0$ for (x, y, z) in M. Then $w(x, y, z) = $ constant for (x, y, z) in M. Therefore, $\dfrac{\partial w}{\partial \eta} = 0$ on Σ; hence, $bw(x, y, z) = 0$ on Σ, so $w(x, y, z) = 0$ on Σ. Since $w(x, y, z) = $ constant for (x, y, z) in M, this constant must be zero; hence, $u_1 = u_2$.

Chapter 19

Section 19.1

1. $26 - 18i$ **3.** $2 + 6i$ **5.** $\dfrac{1}{17}(7 - 6i)$ **7.** $-i$ **9.** $30i$ **11.** $\dfrac{1}{37}(-8 + 48i)$ **13.** $8 + 6i$

15. $\dfrac{1}{629}(368 - 826i)$ **17.** $-14 + 8i$ **19.** $\dfrac{1}{2}(-12 + 5i)$ **21.** $\text{Re}(z^2) = a^2 - b^2,\ \text{Im}(z^2) = 2ab$ **23.** $4 - a$

25. $\dfrac{-2b}{\sqrt{a^2 + b^2}}$ **27.** The line $y = \dfrac{1}{2}$ **29.** The circle of radius 9 about $(8, -4)$ **31.** The line $y = x + 2$

33. Let z be pure imaginary, say, $z = ti$ for some real t. Then $z^2 = -t^2$ and $(\bar{z})^2 = (-ti)^2 = -t^2$, so $z^2 = (\bar{z})^2$. If z is real, then $\bar{z} = z$, so $z^2 = (\bar{z})^2$ in this case also.
Conversely, suppose that $z^2 = (\bar{z})^2$. Let $z = a + ib$. Then $a^2 - b^2 + 2iab = a^2 - b^2 - 2iab$. Therefore, $2ab = -2ab$. Then $a = 0$ or $b = 0$. If $a = 0$, z is pure imaginary; if $b = 0$, z is real.

35. If $z = a + ib$, then $\text{Re}(iz) = \text{Re}(ia - b) = -b = -\text{Im}(z)$, and $\text{Im}(iz) = \text{Im}(ia - b) = a = \text{Re}(z)$.

37. Circle of radius 2 about $(0, 3)$

39. $|z| = |z + w - w| = |w + (z - w)| \le |w| + |z - w|$; hence, $|z| - |w| \le |z - w|$. Similarly, $|w| - |z| \le |w - z| = |z - w|$. Therefore, $-(|z| - |w|) \le |z - w|$. Then $||z| - |w|| \le |z - w|$.

41. If $|z| = 1$, then $z\bar{z} = 1$, so $\left|\dfrac{az + b}{\bar{b}z + \bar{a}}\right| = \left|\dfrac{az + b}{(\bar{b} + \overline{az})z}\right| = \left|\dfrac{az + b}{\overline{az + b}}\right|\dfrac{1}{|z|} = 1$.

Section 19.2

1. $\sqrt{17}$; $\tan^{-1}(4) + 2n\pi$; $\tan^{-1}(4)$; $\sqrt{17}[\cos(\tan^{-1}(4)) + i\sin(\tan^{-1}(4))]$

3. $2\sqrt{17}$; $-\tan^{-1}\left(\dfrac{1}{4}\right) + 2n\pi$; $-\tan^{-1}\left(\dfrac{1}{4}\right)$; $2\sqrt{10}\left[\cos\left(\tan^{-1}\left(\dfrac{1}{4}\right)\right) - i\sin\left(\tan^{-1}\left(\dfrac{1}{4}\right)\right)\right]$

5. 14; $\dfrac{-\pi}{2} + 2n\pi$; $\dfrac{-\pi}{2}$; $14\left[\cos\left(\dfrac{\pi}{2}\right) - i\sin\left(\dfrac{\pi}{2}\right)\right]$

7. $3\sqrt{10}$; $\tan^{-1}(3) + 2n\pi$; $\tan^{-1}(3)$; $3\sqrt{10}[\cos(\tan^{-1}(3)) + i\sin(\tan^{-1}(3))]$

9. $\sqrt{73}$; $\dfrac{-\pi}{2} - \tan^{-1}\left(\dfrac{8}{3}\right) + 2n\pi$; $\dfrac{-\pi}{2} - \tan^{-1}\left(\dfrac{8}{3}\right)$; $\sqrt{73}\left[\cos\left(\dfrac{\pi}{2} + \tan^{-1}\left(\dfrac{8}{3}\right)\right) - i\sin\left(\dfrac{\pi}{2} + \tan^{-1}\left(\dfrac{8}{3}\right)\right)\right]$

11. $\dfrac{3\sqrt{2}}{2}(1 + i)$ **13.** $-4 + 4\sqrt{3}i$ **15.** $2\sqrt{2}(-1 + i)$ **17.** $\dfrac{5\sqrt{2}}{2}(-1 - i)$ **19.** $-\dfrac{7}{2} + \dfrac{7\sqrt{3}}{2}i$

21. Let $z = \cos(\theta) + i\sin(\theta)$. Then $\displaystyle\sum_{j=0}^{n} z^j = 1 + \sum_{j=1}^{n} [\cos(\theta) + i\sin(\theta)]^j = 1 + \sum_{j=1}^{n} [\cos(j\theta) + i\sin(j\theta)] = 1 + \sum_{j=1}^{n} \cos(j\theta) +$

$i\displaystyle\sum_{j=1}^{n} \sin(j\theta) = \dfrac{1 - \cos((n + 1)\theta) - i\sin((n + 1)\theta)}{1 - \cos(\theta) - i\sin(\theta)}$. Take the real part of this equation to get

$$1 + \sum_{j=1}^{n} \cos(j\theta) = \dfrac{1 - \cos(\theta) - \cos((n + 1)\theta) + \cos(\theta)\cos((n + 1)\theta) + \sin(\theta)\sin((n + 1)\theta)}{(1 - \cos(\theta))^2 + \sin^2(\theta)}.$$

After some manipulation, the last quantity can be written

$$\dfrac{1}{2} + \dfrac{\cos(n\theta) - \cos((n + 1)\theta)}{2 - 2\cos(\theta)}.$$

To complete the proof, it is enough to show that

$$\dfrac{\cos(n\theta) - \cos((n + 1)\theta)}{1 - \cos(\theta)} = \dfrac{\sin\left(\left(n + \dfrac{1}{2}\right)\theta\right)}{\sin\left(\dfrac{\theta}{2}\right)}.$$

But

$$
\frac{\sin\left(\left(n+\frac{1}{2}\right)\theta\right)}{\sin\left(\frac{\theta}{2}\right)} = \frac{\sin(n\theta)\cos\left(\frac{\theta}{2}\right) + \cos(n\theta)\sin\left(\frac{\theta}{2}\right)}{\sin\left(\frac{\theta}{2}\right)}
$$

$$
= \frac{\sin(n\theta)\cos\left(\frac{\theta}{2}\right)\sin\left(\frac{\theta}{2}\right) + \cos(n\theta)\sin^2\left(\frac{\theta}{2}\right)}{\sin^2\left(\frac{\theta}{2}\right)}
$$

$$
= \frac{\sin(n\theta)\sin(\theta) + \cos(n\theta)(1 - \cos(\theta))}{1 - \cos(\theta)}
$$

$$
= \frac{\sin(n\theta)\sin(\theta) - \cos(n\theta)\cos(\theta) + \cos(n\theta)}{1 - \cos(\theta)}
$$

$$
= \frac{\cos(n\theta) - \cos((n+1)\theta)}{1 - \cos(\theta)}.
$$

Section 19.3

1. $2(1 + i)$ **3.** $\frac{1}{9}$ **5.** Does not exist   **7.** 1   **9.** Does not exist

11. $f(z) = \mathrm{Im}(z)$ is one of infinitely many possible examples.

Section 19.4

1. $u = x$, $v = y - 1$; analytic   **3.** $u = 2 + \dfrac{x}{x^2 + y^2}$, $v = \dfrac{-y}{x^2 + y^2}$; analytic   **5.** $u = 1$, $v = \dfrac{y}{x}$; not analytic

7. $u = (x + 2)^2 - y^2$, $v = 2(x + 2)y$; analytic   **9.** $u = -y + \sqrt{x^2 + y^2}$, $v = x$; not analytic

11. $u = 1 - \dfrac{2y}{x^2 + y^2}$, $v = \dfrac{-2x}{x^2 + y^2}$; analytic

13. $u = \dfrac{x^5 - 10x^3y^2 + 5xy^4}{(x^2 + y^2)^2}$, $v = \dfrac{5x^4y - 10x^2y^3 + y^5}{(x^2 + y^2)^2}$; $\dfrac{\partial u}{\partial x}(0, 0) = 1 = \dfrac{\partial v}{\partial y}(0, 0)$; $\dfrac{\partial u}{\partial y}(0, 0) = 0 = -\dfrac{\partial v}{\partial x}(0, 0)$. There is no

contradiction because u and its first partial derivatives are not continuous within a disk about the origin.

15. Let z_0 and z_1 be in D, and let C be a smooth path from z_0 to z_1 in D. Let C be parametrized by

$z = z(t) = x(t) + iy(t)$, for $a \le t \le b$. On C, $f(z) = u(x(t), y(t)) + iv(x(t), y(t))$. Then $f'(z(t)) = \dfrac{d}{dt} u(x(t), y(t)) +$

$i\dfrac{d}{dt} v(x(t), y(t)) = 0$, so $u(x(t), y(t)) = $ constant and $v(x(t), y(t)) = $ constant for $a \le t \le b$ (that is, for z on C). Thus,

$f(z_0) = f(z_1)$. If C is piecewise smooth, we can apply this argument to each smooth component of C.

Section 19.5

1. $2^{1/4}\left[\cos\left(\dfrac{\pi}{8}\right) - i\sin\left(\dfrac{\pi}{8}\right)\right]$ and $2^{1/4}\left[\cos\left(\dfrac{7\pi}{8}\right) + i\sin\left(\dfrac{7\pi}{8}\right)\right]$; $1.0987 - 0.4551i$ and $-1.0987 + 0.4551i$

3. $2, 2i, -2, -2i$

5. $2\left[\cos\left(\dfrac{(2k+1)\pi}{4}\right) + i\sin\left(\dfrac{(2k+1)\pi}{4}\right)\right]$, $k = 0, 1, 2, 3$; $1.4142 + 1.4142i$, $-1.4142 + 1.4142i$,

$-1.4142 - 1.4142i$, $1.4142 - 1.4142i$

7. $\cos\left(\dfrac{k\pi}{3}\right) + i\sin\left(\dfrac{k\pi}{3}\right)$, $k = 0, 1, 2, 3, 4, 5$; $1, 0.5 + 0.8660i$, $-0.5 + 0.8660i$, $-1, -0.5 - 0.8660i$, $0.5 - 0.8660i$

9. $8^{1/8}\left[\cos\left(\dfrac{\dfrac{-3\pi}{4}+2k\pi}{4}\right)+i\sin\left(\dfrac{\dfrac{-3\pi}{4}+2k\pi}{4}\right)\right]$, $k=0,1,2,3$; $1.0783-0.7205i$, $0.7205+1.0783i$, $-1.0783+0.7205i$, $-0.7205-1.0783i$

11. $\cos\left(\dfrac{\dfrac{-\pi}{2}+2k\pi}{5}\right)+i\sin\left(\dfrac{\dfrac{-\pi}{2}+2k\pi}{5}\right)$, $k=0,1,2,3$; $-i$, $0.9511-0.3090i$, $0.5878+0.8090i$, $-0.9511-0.3090i$, $-0.5878+0.8090i$

13. $\dfrac{1}{2}\left[\cos\left(\dfrac{\pi}{4}\right)-i\sin\left(\dfrac{\pi}{4}\right)\right]$, $\dfrac{1}{2}\left[\cos\left(\dfrac{3\pi}{4}\right)+i\sin\left(\dfrac{3\pi}{4}\right)\right]$; $0.3536-0.3536i$, $-0.3536+0.3536i$

15. $17^{1/6}\left[\cos\left(\dfrac{\tan^{-1}(4)+2k\pi}{3}\right)+i\sin\left(\dfrac{\tan^{-1}(4)+2k\pi}{3}\right)\right]$, $k=0,1,2$; $1.4495+0.6858i$, $-1.3187+0.9124i$, $-0.1308-1.5982i$

17. Write $z^2+\dfrac{b}{a}z+\dfrac{c}{a}=0$, or $\left(z+\dfrac{b}{2a}\right)^2+\dfrac{c}{a}-\dfrac{b^2}{4a}=0$. Then $\left(z+\dfrac{b}{2a}\right)^2=\dfrac{b^2-4ac}{4a^2}$. Then $z+\dfrac{b}{2a}=\pm\dfrac{1}{2a}\sqrt{b^2-4ac}$; hence, $z=-\dfrac{b}{2a}\pm\dfrac{1}{2a}\sqrt{b^2-4ac}$.

19. $\dfrac{\sqrt{3}-1+(1-\sqrt{3})i}{2}$ and $\dfrac{-\sqrt{3}-1+(1+\sqrt{3})i}{2}$ **21.** The coefficients are not real numbers.

23. Write $z=r[\cos(\theta+2k\pi)+i\sin(\theta+2k\pi)]$. Then $z^m=r^m[\cos(m(\theta+2k\pi))+i\sin(m(\theta+2k\pi))]$. Then $(z^m)^{1/n}=r^{m/n}\left[\cos\left(\dfrac{m(\theta+2k\pi)}{n}\right)+i\sin\left(\dfrac{m(\theta+2k\pi)}{n}\right)\right]$, for $k=0,1,\ldots,n-1$. Next, $z^{1/n}=r^{1/n}\left[\cos\left(\dfrac{\theta+2k\pi}{n}\right)+i\sin\left(\dfrac{\theta+2k\pi}{n}\right)\right]$, so $(z^{1/n})^m=r^{m/n}\left[\cos\left(\dfrac{m(\theta+2k\pi)}{n}\right)+i\sin\left(\dfrac{m(\theta+2k\pi)}{n}\right)\right]$ for $k=0,1,\ldots,n-1$.

Section 19.6

1. $\cos(1)+i\sin(1)$ **3.** $e^\pi[\cos(1)-i\sin(1)]$ **5.** $e^2[\cos(2)-i\sin(2)]$ **7.** $\dfrac{\sqrt{2}}{2}e^{\pi/4}(1+i)$

9. $e^{-5}[\cos(7)+i\sin(7)]$ **11.** $3e^{i\pi/2}$ **13.** $\sqrt{2}e^{-i\pi/4}$ **15.** $\sqrt{10}e^{i\tan^{-1}(1/3)}$

17. Let $z=a+ib$. Then
$$\dfrac{1}{e^z}=\dfrac{1}{e^a}\dfrac{1}{\cos(b)+i\sin(b)}=\dfrac{1}{e^a}\dfrac{1}{\cos(b)+i\sin(b)}\dfrac{\cos(b)-i\sin(b)}{\cos(b)-i\sin(b)}$$
$$=e^{-a}[\cos(b)-i\sin(b)]=e^{-a}e^{-ib}=e^{-(a+ib)}=e^{-z}.$$

19. Write $1=e^{2k\pi i}$. The nth roots of 1 are therefore $e^{2k\pi i/n}$, for $k=0,1,2,\ldots,n-1$.

21. $e^{1/z}=e^{x/(x^2+y^2)}e^{-yi/(x^2+y^2)}=e^{x/(x^2+y^2)}\left[\cos\left(\dfrac{y}{x^2+y^2}\right)-i\sin\left(\dfrac{y}{x^2+y^2}\right)\right]$. Let $u(x,y)=e^{x/(x^2+y^2)}\cos\left(\dfrac{y}{x^2+y^2}\right)$ and $v(x,y)=-e^{x/(x^2+y^2)}\sin\left(\dfrac{y}{x^2+y^2}\right)$. Then $\dfrac{\partial u}{\partial x}=e^{x/(x^2+y^2)}\left[\dfrac{y^2-x^2}{(x^2+y^2)^2}\right]\cos\left(\dfrac{y}{x^2+y^2}\right)-e^{x/(x^2+y^2)}\left[\dfrac{-2xy}{(x^2+y^2)^2}\right]\sin\left(\dfrac{y}{x^2+y^2}\right)=\dfrac{\partial v}{\partial y}$ and $\dfrac{\partial u}{\partial y}=e^{x/(x^2+y^2)}\left[\dfrac{-2xy}{(x^2+y^2)^2}\right]\cos\left(\dfrac{y}{x^2+y^2}\right)-e^{x/(x^2+y^2)}\left[\dfrac{y^2-x^2}{(x^2+y^2)^2}\right]\sin\left(\dfrac{y}{x^2+y^2}\right)=-\dfrac{\partial v}{\partial x}$.

Section 19.7

1. $\ln(2)+i\left(\dfrac{\pi}{2}+2k\pi\right)$; $\ln(2)+\dfrac{i\pi}{2}$ **3.** $\ln(9)+(2k+1)\pi i$; $\ln(9)+\pi i$

5. $\frac{1}{2}\ln(5) + i[\tan^{-1}(2) + 2k\pi]; \frac{1}{2}\ln(5) + i\tan^{-1}(2)$ **7.** $i\left(\frac{\pi}{6} + \frac{2k\pi}{3}\right); \frac{i\pi}{6}$

9. $\frac{1}{2}\ln(20) + i\left[\tan^{-1}(2) + \frac{\pi}{2} + 2k\pi\right]; \frac{1}{2}\ln(20) + i\left[\tan^{-1}(2) + \frac{\pi}{2}\right]$ **11.** $\frac{1}{2}\ln(20) + i[\tan^{-1}(2) + 2k\pi]; \frac{1}{2}\ln(20) + i\tan^{-1}(2)$

13. $\ln(5) + i\left[\frac{\pi}{2} + \tan^{-1}\left(\frac{3}{4}\right) + 2k\pi\right]; \ln(5) + i\left[\frac{\pi}{2} + \tan^{-1}\left(\frac{3}{4}\right)\right]$

15. $\frac{1}{16}\ln(53) + \frac{i}{8}\left[-\tan^{-1}\left(\frac{2}{7}\right) + 2k\pi\right]; \frac{1}{16}\ln(53) - \frac{1}{8}\tan^{-1}\left(\frac{2}{7}\right)i$

17. $\frac{1}{2}\ln(2) + \frac{i}{2}\left(\frac{\pi}{2} + 2k\pi\right); \frac{1}{2}\ln(2) + \frac{\pi}{4}i$ **19.** $\frac{35}{2}\ln(8) + 35i\left(\frac{\pi}{4} + 2k\pi\right); \frac{35}{2}\ln(8) + \frac{35\pi}{4}i$

21. $\log[(1-i)(1+i)] = \log(2) = \ln(2) + 2k\pi i$. Next, $\log(1-i) = \frac{1}{2}\ln(2) + i\left[\frac{-\pi}{4} + 2m\pi\right]$ and $\log(1+i) = $

$\frac{1}{2}\ln(2) + i\left[\frac{\pi}{4} + 2n\pi\right]$. Thus, $\log(1-i) + \log(1+i) = \ln(2) + i(2m+2n)\pi = \ln(2) + 2k\pi i$.

23. $e^{-1-i} = e^{-1}[\cos(1) - i\sin(1)]$, so $\log(e^{-1-i}) = \ln(e^{-1}) + i(-1 + 2n\pi) = -1 - i + 2n\pi i$.

$\log(-1-i) = \frac{1}{2}\ln(2) + i\left(\frac{-3\pi}{4} + 2k\pi\right)$, so $e^{\log(-1-i)} = e^{\ln(\sqrt{2})}\left[\cos\left(\frac{-3\pi}{4} + 2k\pi\right) + i\sin\left(\frac{-3\pi}{4} + 2k\pi\right)\right] = -1 - i$.

25. $\log(e^{-4}) = -4 + i(0 + 2k\pi) = -4 + 2k\pi i$; $e^{\log(-4)} = e^{\ln(4) + i(\pi + 2k\pi)} = 4[\cos(\pi) + i\sin(\pi)] = -4$

27. $\log(e^{-5+5i}) = \ln(e^{-5}) + i(5 + 2k\pi) = -5 + 5i + 2k\pi i$; $e^{\log(-5+5i)} = \sqrt{50}\left[\cos\left(\frac{3\pi}{4}\right) + i\sin\left(\frac{3\pi}{4}\right)\right] = $

$\sqrt{50}\left[-\frac{\sqrt{2}}{2} + \frac{\sqrt{2}}{2}i\right] = -5 + 5i$

Section 19.8

1. $ie^{-2k\pi - \pi/2}; ie^{-\pi/2}$ **3.** $e^{-\pi/2 - 2k\pi}; e^{-\pi/2}$

5. $e^{-3(\pi/4 + 2k\pi)}\left[\cos\left(\frac{3\ln(2)}{2}\right) + i\sin\left(\frac{3\ln(2)}{2}\right)\right]; e^{-3\pi/4}e^{3i\ln(2)/2}\left[\cos\left(\frac{3\ln(2)}{2}\right) + i\sin\left(\frac{3\ln(2)}{2}\right)\right]$

7. $e^{-2\ln(6) + 6k\pi}[\cos(3\ln(6)) - i\sin(3\ln(6))]; e^{-2\ln(6)}[\cos(3\ln(6)) - i\sin(3\ln(6))]$

9. $e^{-\ln(2) - \pi/2 + 4k\pi}\left[\cos\left(\frac{\pi}{2} - \ln(2)\right) + i\sin\left(\frac{\pi}{2} - \ln(2)\right)\right]; e^{-\ln(2) - \pi/2}\left[\cos\left(\frac{\pi}{2} - \ln(2)\right) + i\sin\left(\frac{\pi}{2} - \ln(2)\right)\right]$

11. $8e^{2k\pi}[\cos(\ln(2)) - i\sin(\ln(2))]; 8[\cos(\ln(2)) - i\sin(\ln(2))]$

13. $e^{-\tan^{-1}(2/3) - 2k\pi}\left[\cos\left(\frac{\ln(13)}{2}\right) + i\sin\left(\frac{\ln(13)}{2}\right)\right]; e^{-\tan^{-1}(2/3)}\left[\cos\left(\frac{\ln(13)}{2}\right) + i\sin\left(\frac{\ln(13)}{2}\right)\right]$

15. $e^{\pi - 4k\pi}[\cos(2\ln(3)) + i\sin(2\ln(3))]; e^{\pi}[\cos(2\ln(3)) + i\sin(2\ln(3))]$

Section 19.9

1. $\frac{i}{2}\sinh(1)$ **3.** $i\tanh(2)$ **5.** $i\sin(4)$ **7.** $\cos(2)\cosh(4) - i\sin(2)\sinh(4)$ **9.** 0

11. $\sin(\cos(1))\cosh(\sin(1)) + i\cos(\cos(1))\sinh(\sin(1))$

13. $\cos^2(z) + \sin^2(z) = \frac{1}{4}(e^{2iz} + e^{-2iz} + 2) - \frac{1}{4}(e^{2iz} + e^{-2iz} - 2) = 1$

15. $\cosh(z) = \cosh(z)\cos(y) + i\sinh(x)\sin(y)$. With $u(x, y) = \cosh(x)\cos(y)$ and $v(x, y) = \sinh(x)\sin(y)$, u and v are continuous with continuous partial derivatives for all (x, y), and $\frac{\partial u}{\partial x} = \sinh(x)\cos(y) = \frac{\partial v}{\partial y}$ and $\frac{\partial u}{\partial y} = -\cosh(x)\sin(y) = -\frac{\partial v}{\partial x}$.

17. If $z = iy$, and $\cos(y) < 0$, then $\cosh(z) < 0$. Further, $\cosh(x + in\pi) < 0$ if n is an odd integer. Finally, $\cosh(z) = 0$ if and only if $z = \frac{2n+1}{2}\pi i$.

19. $u(x, y) = \dfrac{\sin(x)\cos(x)[\cosh^2(y) + \sinh^2(y)]}{\cos^2(x)\cosh^2(y) + \sin^2(x)\sinh^2(y)}$, $v(x, y) = \dfrac{\sinh(y)\cosh(y)[\cos^2(x) - \sin^2(x)]}{\cos^2(x)\cosh^2(y) + \sin^2(x)\sinh^2(y)}$; $\tan(z)$ is analytic for all z such that $\cos(z) \neq 0$.

21. $(2k + 1)\pi - i \ln(1 + \sqrt{2})$ or $2k\pi - i \ln(\sqrt{2} - 1)$, k any integer   **23.** $\tan^{-1}(z)$

Additional Problems for Chapter 19

1. $e^4[\cos(1) - i \sin(1)]$ **3.** $-i \sin(5)$ **5.** $\cos\left(\dfrac{\ln(2)}{2}\right)\cosh\left(\dfrac{\pi}{4} + 2k\pi\right) - i \sin\left(\dfrac{\ln(2)}{2}\right)\sinh\left(\dfrac{\pi}{4} + 2k\pi\right)$

7. $e^{-1}[2\cos(1) - 4\sin(1) + i(2\sin(1) + 4\cos(1))]$ **9.** $\cosh(1)\cos(1) - i \sinh(1)\sin(1)$

11. $(25)^{1/7}\left[\cos\left(\dfrac{\dfrac{\pi}{2} + \tan^{-1}\left(\dfrac{7}{24}\right) + 2k\pi}{7}\right) + i \sin\left(\dfrac{\dfrac{\pi}{2} + \tan^{-1}\left(\dfrac{7}{24}\right) + 2k\pi}{7}\right)\right]$, $k = 0, 1, \ldots, 6$

13. $\cosh[e^{-2k\pi}\cos(\ln(3))]\cos[e^{-2k\pi}\sin(\ln(3))] + i \sinh[e^{-2k\pi}\cos(\ln(3))]\sin[e^{-2k\pi}\sin(\ln(3))]$

15. $e^{21 \ln(1)/2}[\cos(21 \tan^{-1}(3)) + i \sin(21 \tan^{-1}(3))]$

17. $e^{-3 \ln(17)/14}\left[\cos\left(\dfrac{-3 \tan^{-1}\left(\dfrac{1}{4}\right) - 6k\pi}{7}\right) + i \sin\left(\dfrac{-3 \tan^{-1}\left(\dfrac{1}{4}\right) - 6k\pi}{7}\right)\right]$

19. $e^{3 \ln(5)/16}\left[\cos\left(\dfrac{3}{8}\tan^{-1}\left(\dfrac{1}{2}\right) + \dfrac{3k\pi}{4}\right) + i \sin\left(\dfrac{3}{8}\tan^{-1}\left(\dfrac{1}{2}\right) + \dfrac{3k\pi}{4}\right)\right]$

21. $e^{\ln(5)/3}\left[\cos\left(\dfrac{2}{3}\tan^{-1}\left(\dfrac{1}{2}\right) + \dfrac{4k\pi}{3}\right) + i \sin\left(\dfrac{2}{3}\tan^{-1}\left(\dfrac{1}{2}\right) + \dfrac{4k\pi}{3}\right)\right]$

23. $-\sinh(21)\cos(20) + i \cosh(21)\sin(20)$ **25.** $\sin(\cosh(2))$ **27.** Ellipse: $8(x^2 + y^2) + 54x - 34y + 117 = 0$

29. $\cosh^2(z) - \sinh^2(z) = \dfrac{1}{4}[e^{2z} + e^{-2z} + 2 - e^{2z} - e^{-2z} + 2] = 1$

31. $e^{\ln(4)/5}\left[\cos\left(-\dfrac{\pi}{10} + \dfrac{2k\pi}{5}\right) + i \sin\left(-\dfrac{\pi}{10} + \dfrac{2k\pi}{5}\right)\right]$, $k = 0, 1, 2, 3, 4$

33. $\cosh(z)\cosh(w) + \sinh(z)\sinh(w) = \dfrac{1}{4}(e^z + e^{-z})(e^w + e^{-w}) + \dfrac{1}{4}(e^z - e^{-z})(e^w - e^{-w}) = \dfrac{1}{4}[2e^{z+w} + 2e^{-z-w}] = \cosh(z + w)$

35. Let $z = a + ib$, $w = \alpha + i\beta$. Referring to the diagram, the area of the triangle Ozw is equal to one-half the area of the rectangle, or $\dfrac{1}{2}|(a\mathbf{i} + b\mathbf{j}) \times (\alpha\mathbf{i} + \beta\mathbf{j})|$. This equals $\dfrac{1}{2}|(a\beta - \alpha b)\mathbf{k}|$, or $\dfrac{1}{2}|a\beta - \alpha b|$. Since $\text{Im}(\hat{z}w) = \text{Im}[(a - ib)(\alpha + i\beta)] = a\beta - \alpha b$, we are done.

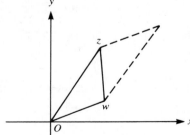

Chapter 20

Section 20.1

1. $8 - 2i$ **3.** $\dfrac{3}{2}(1 + i)$ **5.** $\dfrac{1}{2}(-13 + 4i)$ **7.** $-\dfrac{1}{2}[\cosh(8) - \cosh(2)]$

9. $-\dfrac{1}{2}\{e^{-1}[\cos(2) - i\sin(2)] - e[\cos(2) + i\sin(2)]\}$ **11.** $12 + 204i$ **13.** $\cosh(2\sqrt{2}) - \cos(4)$ **15.** $2\pi i$

17. $-\pi$ **19.** $\dfrac{17}{6} + \dfrac{52}{3}i$ **21.** $-2i$ **23.** $1 - \dfrac{46}{3}i$ **25.** $\dfrac{e}{2} - \dfrac{1}{2}\cos(2) + \dfrac{i}{2}\sin(2)$

27. $\sqrt{5}$ (there are infinitely many other bounds)

Section 20.2

1. 0 **3.** 0 **5.** $2\pi i$ **7.** 0 **9.** 0 **11.** $4\pi i$ **13.** 0 **15.** 0 **17.** $\dfrac{\pi i}{4}$

Section 20.3

1. $-2 - 3i$ **3.** $-\dfrac{1}{2}(e^{16} - 1)$ **5.** $32\pi i$ **7.** $2\pi i[-16 + 13i]$ **9.** 0 **11.** $\pi i[6\cos(12) - 36\sin(12)]$ **13.** 0

15. 0, if C does not enclose $2 - 4i$; $4\pi i[\cosh(2 - 4i) + (2 - 4i)\sinh(2 - 4i)]$ if C does enclose $2 - 4i$

17. On C, $z = z_0 + re^{it}$, $0 \le t \le 2\pi$, so $f(z_0) = \dfrac{1}{2\pi i}\oint_C \dfrac{f(z)}{z - z_0}\,dz = \dfrac{1}{2\pi i}\oint_C \dfrac{f(z_0 + re^{it})}{re^{it}}ire^{it}\,dt = \dfrac{1}{2\pi}\int_0^{2\pi} f(z_0 + re^{it})\,dt.$

19. Choose any z_0 in D. If z is in D, let $F(z) = \displaystyle\int_{z_0}^{z} f(z)\,dz$. A straightforward modification of the proof of Theorem 20.9 shows that $F'(z) = f(z)$ for z in D. Since F' is analytic in D, f is analytic in D.

21. Suppose that f is analytic for all z. Let $M > 0$ and $r > 0$ be given. We know that f is bounded on $|z| \le r$ [because $f(z) = u(x, y) + iv(x, y)$, and u and v are bounded on $|z| \le R$ [because u and v are continuous]. If $|f(z)| \le M$ for all $|z| > r$, f is bounded for all z; hence, f is a constant function by Liouville's theorem (Problem 20). Thus, $|f(z)| > M$ for some z with $|z| > r$.

Additional Problems for Chapter 20

1. $4\pi i e^2$ **3.** 0 **5.** $2(1 + i)$ **7.** $\dfrac{338}{3} - 27i$ **9.** $-\dfrac{1}{2}[1 + 17i]$ **11.** $-\dfrac{13}{2} - 39i$ **13.** 0

15. $\dfrac{1}{3i}[\sinh(-3i) - \sinh(-3 + 6i)]$ **17.** $\dfrac{1}{2}(5 + i)$ **19.** $\dfrac{9208}{15} - 2i$ **21.** 0 **23.** $64 - 4i$

25. $-\dfrac{1795}{12} - \cos(5) + i\left[\dfrac{125}{3} + \cosh(2)\right]$

27. On C, $\text{Log}(z) = \ln|z| + i\,\text{Arg}(z) = \ln(r) + i\,\text{Arg}(z)$, so

$$\left|\oint_C \frac{1}{z^2}\text{Log}(z)\,dz\right| \le 2\pi r\frac{1}{r^2}\left[\underset{z \text{ on } C}{\text{Max}}|\ln(r) + i\,\text{Arg}(z)|\right]$$

$$\le \frac{2\pi}{r}[\ln(r) + \pi].$$

Chapter 21

Section 21.1

1. $1 + 2i$ **3.** $2 - i$ **5.** 0 **7.** $5i$ **9.** 1 **11.** Diverges **13.** 1 **15.** $3i$
17. $\{(-1)^n\}$ (there are infinitely many different examples) **19.** $\{e^{in\theta}\}$, with θ any number in $(0, 2\pi)$

21. $S_n - rS_n = 1 - r^{n+1}$, so $S_n = \dfrac{1 - r^{n+1}}{1 - r}$ if $r \ne 1$.

Section 21.2

1. Converges **3.** Converges **5.** Diverges

7. Let $S_n = \sum_{j=1}^{n} z_j$, $X_n = \sum_{j=1}^{n} x_j$ and $Y_n = \sum_{j=1}^{n} y_j$. By Theorem 21, $\{S_n\}$ converges if and only if both $\{X_n\}$ and $\{Y_n\}$ converge, and, in this event, $\lim_{n\to\infty} S_n = \lim_{n\to\infty} X_n + i \lim_{n\to\infty} Y_n$. Therefore, when $\{X_n\}$ and $\{Y_n\}$ converge,

$$\sum_{n=1}^{\infty} z_n = \sum_{n=1}^{\infty} x_n + i \sum_{n=1}^{\infty} y_n.$$

9. Suppose first that $\sum_{n=1}^{\infty} z_n$ converges. Then the sequence $\{S_n\}$ of partial sums converges. Then $\{S_n\}$ is a Cauchy sequence. Let $\epsilon > 0$. For some, N, $|S_n - S_m| < \epsilon$ if $m \le N$ and $n \le N$. Let $m = n + p$ to get $|z_{n+1} + z_{n+2} + \cdots + z_{n+p}| < \epsilon$ if $n \le N$. Conversely, suppose, given $\epsilon > 0$, there is some N such that $|z_{n+1} + z_{n+2} + \cdots + z_{n+p}| < \epsilon$. Then, if m and n exceed N, we have $|S_n - S_m| < \epsilon$, so $\{S_n\}$ is a Cauchy sequence and therefore converges.

Section 21.3

1. 2; $|z + 3i| < 2$; $f'(z) = \sum_{n=1}^{\infty} \dfrac{n(n+1)}{2^n}(z + 3i)^{n-1}$, $f''(z) = \sum_{n=2}^{\infty} \dfrac{n(n+1)(n-1)}{2^n}(z + 3i)^{n-2}$

3. 1; $|z - 1 + 2i| < 1$; $f'(z) = \sum_{n=1}^{\infty} \dfrac{n^{n+1}}{(n+1)^n}(z - 1 + 2i)^{n-1}$, $f''(z) = \sum_{n=2}^{\infty} \dfrac{n^{n+1}(n-1)}{(n+1)^n}(z - 1 + 2i)^{n-2}$

5. 2; $|z + 4 - i| < 2$; $f'(z) = \sum_{n=1}^{\infty} \left(\dfrac{i^n}{2^{n+1}}\right)n(z + 4 - i)^{n-1}$; $f''(z) = \sum_{n=2}^{\infty} \left(\dfrac{i^n}{2^{n+1}}\right)n(n-1)(z + 4 - i)^{n-2}$

7. 1; $|z + 6 + 2i| < 1$; $f'(z) = \sum_{n=1}^{\infty} \dfrac{2n^3}{2n+1}(z + 6 + 2i)^{2n-1}$; $f''(z) = \sum_{n=2}^{\infty} \dfrac{2n^3}{2n+1}(2n-1)(z + 6 + 2i)^{2n-2}$

9. 1; $|z + 4| < 1$; $f'(z) = \sum_{n=1}^{\infty} \left(\dfrac{ne^{in}}{2n+1}\right)(z + 4)^{n-1}$; $f''(z) = \sum_{n=2}^{\infty} \left(\dfrac{ne^{in}}{2n+1}\right)(n-1)(z + 4)^{n-2}$

11. The series converges at zero and diverges at i if $0 < \dfrac{a_n}{a_{n+1}} < 1$.

13. The radius of the sum cannot exceed the smaller of R_1 and R_2.

15. First, $f(z_0) = a_0 = h(z_0) = b_0$. Now let k be any positive integer. Then

$$f^{(k)}(z) = \sum_{n=k}^{\infty} n(n-1) \cdots (n - k + 1)a_n(z - z_0)^{n-k}$$

$$= \sum_{n=k}^{\infty} n(n-1) \cdots (n - k + 1)b_n(z - z_0)^{n-k} = h^{(k)}(z).$$

Then $f^{(k)}(z_0) = k(k-1) \cdots (k - k + 1)a_k = k(k-1) \cdots (k - k + 1)b_k = h^{(k)}(z_0)$; therefore, $a_k = b_k$.

17. (a) $\left(\dfrac{1}{1-z}\right)^2 = \left(\sum_{n=0}^{\infty} z^n\right)\left(\sum_{n=0}^{\infty} z^n\right)$ if $|z| < 1$. Here each $a_n = b_n = 1$, so $c_n = \sum_{j=0}^{\infty} a_j b_{n-j} = n + 1$.

(b) $[f(z)]^2 = \left(\sum_{n=0}^{\infty} \dfrac{1}{n!} z^n\right)\left(\sum_{n=0}^{\infty} \dfrac{1}{n!} z^n\right)$. With $a_n = b_n = \dfrac{1}{n!}$, we have $c_n = \sum_{j=0}^{\infty} a_j b_{n-j} = \sum_{j=0}^{\infty} \dfrac{1}{j!(n-j)!}$. By the binomial theorem,

$(1 + 1)^n = \sum_{j=0}^{n} \dfrac{n!}{j!(n-j)!} = 2^n$. Therefore, $\sum_{j=0}^{n} \dfrac{1}{j!(n-j)!} = \dfrac{2^n}{n!}$, and $[f(z)]^2 = \sum_{n=0}^{\infty} \dfrac{2^n}{n!} z^n = \sum_{n=0}^{\infty} \dfrac{1}{n!}(2z)^n = f(2z)$.

(c) Write $\left(\sum_{n=0}^{\infty} (-1)^n \dfrac{1}{(2n)!} z^{2n}\right)\left(\sum_{n=0}^{\infty} (-1)^n \dfrac{1}{(2n+1)!} z^{2n+1}\right) = \sum_{n=0}^{\infty} c_n z^n$. Write $a_{2n} = (-1)^n \dfrac{1}{(2n)!}$ and $a_{2n+1} = 0$,

$b_{2n+1} = (-1)^n \dfrac{1}{(2n+1)!}$ and $b_{2n} = 0$, and $c_n = \sum_{j=0}^{n} a_j b_{n-j}$. If n is even, $c_n = 0$ (because then either

a_j or b_{n-j} is zero for $j = 0, 1, \ldots, n$). Now write $c_{2n+1} = \sum_{j=0}^{n} a_{2j} b_{2n+1-j} = \sum_{j=0}^{n} \dfrac{(-1)^j}{(2j)!} \dfrac{(-1)^{n-j}}{(2n+1-2j)!} =$

$(-1)^n \sum_{j=0}^{n} \dfrac{1}{(2j)!(2n+1-2j)!}$. By the binomial theorem, $\sum_{j=0}^{2n+1} \dfrac{(2n+1)!}{j!(2n+1-j)!} = (1 + 1)^{2n+1} = 2^{2n+1}$, and

$\sum_{j=0}^{2n+1} \dfrac{(-1)^j(2n+1)!}{j!(2n+1-j)!} = (1 - 1)^{2n+1} = 0$. Add the last two summations to get $2 \sum_{j=0}^{n} \dfrac{(2n+1)!}{(2j)!(2n+1-2j)!} = 2^{2n+1}$, the terms with j odd canceling. Therefore, $c_n = \sum_{j=0}^{n} \dfrac{(-1)^n}{(2j)!(2n+1-2j)!} = \dfrac{(-1)^n 2^{2n+1}}{2(2n+1)!}$; hence, $f(z)g(z) =$

$$\sum_{n=0}^{\infty} \frac{(-1)^n 2^{2n+1}}{2(2n+1)!} z^{2n+1} = \frac{1}{2} \sum_{n=0}^{\infty} \frac{(-1)^n}{(2n+1)!} (2z)^{2n+1} = \frac{1}{2} g(2z).$$

Section 21.4

1. $\displaystyle\sum_{n=0}^{\infty} \frac{(-1)^n}{(2n)!} (2z)^{2n}$ **3.** $\displaystyle\sum_{n=0}^{\infty} \frac{(-1)^n}{(2n+1)!} z^{4n+2}$ **5.** $\dfrac{1}{3-8i} \displaystyle\sum_{n=0}^{\infty} (-1)^n \dfrac{1}{(3-8i)^n} (z-1+8i)^n$

7. $\displaystyle\sum_{n=0}^{\infty} (n+1)z^n$ **9.** $\displaystyle\sum_{n=0}^{\infty} \frac{3^{2n+1}}{(2n+1)!} z^{2n+1}$ **11.** $\dfrac{-3}{5+4i} \displaystyle\sum_{n=0}^{\infty} \dfrac{1}{(5+4i)^n} (z+5)^n$

13. $\dfrac{1}{2} \displaystyle\sum_{n=0}^{\infty} \left[\dfrac{i^n}{en!} + \dfrac{e(-i)^n}{n!} \right] (z-i)^n$ **15.** $\dfrac{-1}{2+6i} \displaystyle\sum_{n=0}^{\infty} \dfrac{1}{(2+6i)^n} (z+2i)^n$ **17.** $e^{3-i} \displaystyle\sum_{n=0}^{\infty} (-1)^n \dfrac{1}{n!} (z-i)^n$

19. $\mathrm{Log}(2i) + \displaystyle\sum_{n=1}^{\infty} \frac{(-1)^{n+1}}{n} \frac{1}{(2i)^n} (z-2i)^n$

21. Find that $f^{(2n)}(z) = 2^n f(z) + 2^{n-1}$ for $n = 1, 2, 3, \ldots$; the Maclaurin series is $f(z) = 1 + iz +$

$$\sum_{n=1}^{\infty} \left[\frac{2^n + 2^{n-1}}{(2n)!} z^{2n} + \frac{2^n i}{(2n+1)!} z^{2n+1} \right].$$

23. $z^2 - \dfrac{1}{3} z^4 + \dfrac{2}{45} z^6 - \cdots$ **25.** $z - \dfrac{1}{3} z^3 + \dfrac{2}{15} z^5 - \dfrac{17}{315} z^7 + \cdots$

Section 21.5

1. $\dfrac{1}{z-i} + \dfrac{1}{2i} \displaystyle\sum_{n=0}^{\infty} \left(\frac{1}{2i} \right)^n (z-i)^n$ **3.** $\displaystyle\sum_{n=1}^{\infty} (-1)^n \frac{1}{(2n)!} (2z)^{2n}$ **5.** $\dfrac{-1}{z-1} - 2 - (z-1)$ **7.** $1 + \dfrac{2i}{z-i}$

9. $\displaystyle\sum_{n=0}^{\infty} \frac{(-1)^n 4^{2n+1}}{(2n+1)!} z^{2n}$ **11.** $-\dfrac{i}{2} \dfrac{1}{z-i} - \displaystyle\sum_{n=0}^{\infty} (-1)^n \left(\frac{z-i}{2i} \right)^n$ **13.** $1 + \dfrac{2z_0}{z-z_0}$ **15.** $\dfrac{2i}{z-1+i}$

17. (a) $J_n(z) = \dfrac{1}{2\pi i} \displaystyle\oint_C w^{-n-1} e^{z(w-1/w)/2} \, dw$, where C is a circle about the origin. Parametrize C as $z = e^{i\theta}$, $-\pi \le \theta \le \pi$. Then

$$J_n(z) = \frac{1}{2\pi i} \int_{-\pi}^{\pi} e^{-in\theta} e^{-i\theta} e^{z(e^{i\theta} - e^{-i\theta})/2} i e^{i\theta} \, d\theta = \frac{1}{2\pi} \int_{-\pi}^{\pi} e^{-in\theta} e^{iz\sin(\theta)} \, d\theta$$

$$= \frac{1}{2\pi} \int_{-\pi}^{\pi} \cos(z \sin(\theta) - n\theta) \, d\theta + \frac{i}{2\pi} \int_{-\pi}^{\pi} \sin(z \sin(\theta) - n\theta) \, d\theta.$$

Since $\cos(z \sin(\theta) - n\theta)$ is even and $\sin(z \sin(\theta) - n\theta)$ is odd, the second integral is zero, and we obtain

$$J_n(z) = \frac{1}{\pi} \int_0^{\pi} \cos(z \sin(\theta) - n\theta) \, d\theta.$$

(b) $e^{zw/2} = \displaystyle\sum_{n=0}^{\infty} \frac{1}{n!} \left(\frac{z}{2} \right)^n w^n$ and $e^{-z/2} = \displaystyle\sum_{n=0}^{\infty} \frac{1}{n!} \left(-\frac{z}{2} \right)^n w^n$. In the product of these series, we find the coefficient

of w^n by finding the coefficient of $w^{n+j} w^{-j}$ for each j and summing. The coefficient of $w^{n+j} w^{-j}$ is

$\left(\dfrac{z}{2} \right)^{n+j} \dfrac{1}{(n+j)!} \left(-\dfrac{z}{2} \right)^j \dfrac{1}{j!}$, or $(-1)^j \left(\dfrac{z}{2} \right)^{n+2j} \dfrac{1}{j!(n+j)!}$. Thus, $J_n(z) = \displaystyle\sum_{n=0}^{\infty} (-1)^j \dfrac{1}{j!(n+j)!} \left(\dfrac{z}{2} \right)^{n+2j}$.

Additional Problems for Chapter 21

1. $\dfrac{1}{6+i} \displaystyle\sum_{n=0}^{\infty} (-1)^n \left(\frac{1}{6+i} \right)^n (z-2-i)^n$; $R = \sqrt{37}$ **3.** $\displaystyle\sum_{n=0}^{\infty} (-1)^n \frac{1}{(2n)!} (z-5i)^{2n}$; $R = \infty$

5. $-6 + 4i + (-1 - 7i)(z - 3 + 4i) - i(z - 3 + 4i)^2$; $R = \infty$ **7.** $\displaystyle\sum_{n=0}^{\infty} \frac{1}{n!} z^{2n-1}$; $R = \infty$

9. $\dfrac{1}{5-4i}\displaystyle\sum_{n=0}^{\infty}\left(\dfrac{i}{5-4i}\right)^n(z-4-5i)^n; R=\sqrt{41}$ **11.** $i\displaystyle\sum_{n=0}^{\infty}\dfrac{1}{(2n+1)!}z^{6n+1}; R=\infty$

13. $\dfrac{1}{3i}\displaystyle\sum_{n=0}^{\infty}\left(\dfrac{2i}{3}\right)^n(z-1-i)^n; R=\dfrac{3}{2}$ **15.** $\dfrac{1}{3}\displaystyle\sum_{n=0}^{\infty}\dfrac{(-1)^n}{(2n)!}z^{4n-1}; R=\infty$ **17.** $\displaystyle\sum_{n=0}^{\infty}\dfrac{(-9)^n}{(2n)!}z^{4n}-i\displaystyle\sum_{n=0}^{\infty}\dfrac{1}{n!}z^n; R=\infty$

19. $e\displaystyle\sum_{n=0}^{\infty}\dfrac{2^n}{n!}z^n; R=\infty$ **21.** $iz+iz^2+\dfrac{2}{3}iz^3+\dfrac{2}{3}iz^4+\cdots$

23. $\cos(1)+2i\sin(1)(z-i)+[2\cos(1)+\sin(1)](z-i)^2+\left(\dfrac{4}{3}i\sin(1)-2i\cos(1)\right)(z-i)^3+\cdots$

25. $\cos(1)-i\sin(1)(z-i)+[\sin(1)+\dfrac{1}{2}\cos(1)](z-i)^2+\left(\dfrac{5}{6}i\sin(1)+i\cos(1)\right)(z-i)^3+\cdots$

27. By Theorem 21.13, $a_n=\dfrac{1}{2\pi i}\displaystyle\oint_C\dfrac{1}{w^{n+1}}f(w)\,dw$. Let $w=re^{i\theta}, 0\le\theta\le 2\pi$, to get

$$a_n=\dfrac{1}{2\pi i}\int_0^{2\pi}\dfrac{1}{r^{n+1}e^{i(n+1)\theta}}f(re^{i\theta})ire^{i\theta}\,d\theta=\dfrac{1}{2\pi r^n}\int_0^{2\pi}f(re^{i\theta})e^{-in\theta}\,d\theta.$$

29. $\dfrac{1}{2\pi i}\displaystyle\oint_C\dfrac{1}{n!w^{n+1}}z^ne^{wz}\,dw=\dfrac{1}{2\pi i}\displaystyle\oint_C\sum_{n=0}^{\infty}\dfrac{1}{(n!)(j!)}z^{n+j}w^{j-n-1}\,dw$. Let $w=e^{i\theta}, 0\le\theta\le 2\pi$, in the last integral to get

$\dfrac{1}{2\pi i}\displaystyle\int_0^{2\pi}\sum_{n=0}^{\infty}\dfrac{z^{n+j}e^{i(j-n+1)\theta}}{(n!)(j)}ie^{i\theta}\,d\theta=\dfrac{1}{2\pi}\displaystyle\sum_{n=0}^{\infty}\int_0^{2\pi}\dfrac{z^{n+j}e^{i\theta(j-n)}}{(n!)(j!)}\,d\theta$. Every integral is zero except the term having $n=j$; hence,

we get $\dfrac{1}{2\pi}\displaystyle\int_0^{2\pi}\dfrac{1}{(n!)^2}z^{2n}\,d\theta$, which is $\dfrac{1}{(n!)^2}z^{2n}$ for $n\ge 0$. If $n<0$, we never have $n=j$, and all the integrals in the above

summation are zero. Therefore, we have $\displaystyle\sum_{n=0}^{\infty}\dfrac{1}{(n!)^2}z^{2n}=\dfrac{1}{2\pi i}\displaystyle\sum_{n=0}^{\infty}\oint_C\dfrac{z^ne^{zw}}{n!w^{n+1}}\,dw=\dfrac{1}{2\pi}\displaystyle\sum_{n=0}^{\infty}\int_0^{2\pi}\dfrac{z^ne^{ze^{i\theta}}}{n!e^{in\theta}}\,d\theta=$

$\dfrac{1}{2\pi}\displaystyle\int_0^{2\pi}\sum_{n=0}^{\infty}\dfrac{(ze^{-i\theta})^n}{n!}e^{ze^{i\theta}}\,d\theta=\dfrac{1}{2\pi}\displaystyle\int_0^{2\pi}e^{ze^{-i\theta}}e^{ze^{i\theta}}\,d\theta=\dfrac{1}{2\pi}\displaystyle\int_0^{2\pi}e^{z(e^{i\theta}+e^{-i\theta})}\,d\theta=\dfrac{1}{2\pi}\displaystyle\int_0^{2\pi}e^{2z\cos(\theta)}\,d\theta.$

Chapter 22

Section 22.1

1. Pole of order 2 at zero   **3.** Essential singularity at zero   **5.** Simple poles at $i, -i$; pole of order 2 at 1
7. Simple pole at $-i$, removable singularity at i **9.** Simple poles at $1, -1, i, -i$

11. Simple poles at $\dfrac{\pi}{2}(2n+1)$, n any integer   **13.** Pole of order 2 at zero   **15.** Pole of order 4 at -1

17. No singularities   **19.** Simple pole at $-\pi$

21. f has isolated singularities at $\dfrac{1}{n\pi}$, and these numbers can be made as close as one likes to the origin by choosing n larger.
Thus, f cannot be analytic in any annulus about the origin.

23. Write $f(z)=a_{-m}(z-z_0)^{-m}+\cdots+a_{-1}(z-z_0)^{-1}+\displaystyle\sum_{n=0}^{\infty}a_k(z-z_0)^k=a_{-m}(z-z_0)^{-m}+g(z)$, with $g(z)$ containing
only powers of $z-z_0$ greater than $-m$. Let $p(z)=\alpha_nz^n+\alpha_{n-1}a^{n-1}+\cdots+\alpha_1z+\alpha_0$, with $\alpha_n\ne 0$. Then
$p(f(z))=\alpha_na^n_{-m}(z-z_0)^{-nm}+$ [powers of $(z-z_0)$ greater than $-m$]; hence, $p(f(z))$ has a pole of order nm at z_0.

25. Let $\alpha=re^{i\theta}\ne 0$. Let $e^{1/z}=\alpha$ to obtain $z=\dfrac{1}{\ln(r)+i(\theta+2k\pi)}$, with k any integer and θ any argument of z. As

$k\to\infty, \left|\dfrac{1}{\ln(r)+i(\theta+2k\pi)}\right|\to 0$. Thus, any open disk about zero contains infinitely many numbers of the form

$\dfrac{1}{\ln(r)+i(\theta+2k\pi)}$, and, at each of these, $e^{1/z}=\alpha$.

Section 22.2

1. Residue 0 at zero   **3.** Residue $\dfrac{1}{2} - i$ at zero   **5.** Residue $\dfrac{\cosh(2)}{4}$ at i and at $-i$; $-\sin(2) - \dfrac{1}{2}\cos(2)$ at 1

7. Residue 1 at $-i$ **9.** Residue $\dfrac{1}{4}$ at 1 and at -1; residue $-\dfrac{1}{4}$ at i and at $-i$ **11.** Residue $(-1)^{n+1}$ at $\dfrac{\pi}{2}(2n+1)$

13. Residue i at zero   **15.** Residue $\dfrac{4}{3}e^{-2}$ at -1 **17.** No singularities   **19.** Residue -1 at $-\pi$ **21.** $4\pi i$

23. 0   **25.** $\pi - \dfrac{\pi}{2}i$ **27.** $2\pi i$ **29.** 0   **31.** $8\pi - 6\pi i$ **33.** $2\pi i$ **35.** $\pi^2 i$

Section 22.3

1. $\dfrac{2\pi}{\sqrt{3}}$ **3.** $\dfrac{\pi}{3}$ **5.** $\dfrac{1}{4}\pi e^{-2\sqrt{2}}\sin(2\sqrt{2})$ **7.** $\dfrac{\pi}{e}\sin(2)$ **9.** $\dfrac{\pi}{32}(1 + 5e^{-4})$ **11.** $\dfrac{\pi}{2}$

13. $\dfrac{2\pi}{\sqrt{7}}$ **15.** $2\pi\left(1 - \dfrac{4}{15}\sqrt{15}\right)$

17. First, $\displaystyle\int_0^\pi \frac{1}{\alpha + \beta\cos(\theta)}\,d\theta = \int_0^{2\pi} \frac{1}{\alpha + \beta\cos(\theta)}\,d\theta - \int_\pi^{2\pi} \frac{1}{\alpha + \beta\cos(\theta)}\,d\theta$. Let $\theta = 2\pi - u$ in the last integral

to get $\displaystyle\int_0^\pi \frac{1}{\alpha + \beta\cos(\theta)}\,d\theta = \int_0^{2\pi} \frac{1}{\alpha + \beta\cos(\theta)}\,d\theta - \int_0^\pi \frac{1}{\alpha + \beta\cos(u)}\,du$; hence,

$$\int_0^\pi \frac{1}{\alpha + \beta\cos(\theta)}\,d\theta = \frac{1}{2}\int_0^{2\pi} \frac{1}{\alpha + \beta\cos(\theta)}\,d\theta = \frac{1}{2}\oint_C \frac{1}{\alpha + \dfrac{\beta\left(z + \dfrac{1}{z}\right)}{2}}\frac{1}{iz}\,dz = -i\oint_C \frac{1}{2\alpha z + \beta z^2 + \beta}\,dz.$$

The only singularity within C is $\dfrac{-\alpha + \sqrt{\alpha^2 - \beta^2}}{\beta}$, and the residue there is $\dfrac{1}{2\sqrt{\alpha^2 - \beta^2}}$. Therefore,

the integral is $-i(2\pi i)\dfrac{1}{2\sqrt{\alpha^2 - \beta^2}}$, or $\dfrac{\pi}{\sqrt{\alpha^2 - \beta^2}}$.

19. The residue at βi is $\dfrac{\alpha\beta e^{-\alpha\beta} + e^{-\alpha\beta}}{4i\beta^3}$, so the integral is Re $2\pi i\,\dfrac{(\alpha\beta + 1)e^{-\alpha\beta}}{4i\beta^3}$, or $\dfrac{\pi}{2\beta^3}(1 + \alpha\beta)e^{-\alpha\beta}$.   **21.** $\dfrac{\pi}{2^{2n-1}}\dfrac{(2n)!}{(n!)^2}$

23. $\displaystyle\int_0^{2\pi} \frac{1}{\alpha + \sin^2(\theta)}\,d\theta = \int_0^{\pi/2} \frac{1}{\alpha + \sin^2(\theta)}\,d\theta + \int_{\pi/2}^\pi \frac{1}{\alpha + \sin^2(\theta)}\,d\theta + \int_\pi^{3\pi/2} \frac{1}{\alpha + \sin^2(\theta)}\,d\theta + \int_{3\pi/2}^{2\pi} \frac{1}{\alpha + \sin^2(\theta\Omega}\,d\theta$. Let

$\theta = \pi - u$ in the second integral, $\theta = \pi + u$ in the third, and $\theta = 2\pi - u$ in the fourth to show that $\displaystyle\int_0^{\pi/2} \frac{1}{\alpha + \sin^2(\theta)}\,d\theta =$

$\dfrac{1}{4}\displaystyle\int_0^{2\pi} \frac{1}{\alpha + \sin^2(\theta)}\,d\theta = \frac{1}{4}\oint_C \frac{-i}{\alpha - \dfrac{1}{4}\left(z - \dfrac{1}{z}\right)^2}\frac{1}{z}\,dz = \oint_C \frac{1}{z^4 - (2 + 4\alpha)z^2 + 1}\,dz$. Singularities are at $\pm\sqrt{1 + 2\alpha - 2\sqrt{\alpha + \alpha^2}}$,

and the residue at each singularity is $\dfrac{-i}{-8\sqrt{\alpha + \alpha^2}}$. Thus, $\displaystyle\int_0^{\pi/2} \frac{1}{\alpha + \sin^2(\theta)}\,d\theta = 2\pi i\,\frac{-2i}{-8\sqrt{\alpha + \alpha^2}} = \frac{\pi}{2\sqrt{\alpha + \alpha^2}}$.

25. First, $\sin(\theta) \geq \dfrac{2\theta}{\pi}$ for $0 \leq \theta \leq \dfrac{\pi}{2}$. $\left[\text{To prove this, let } f(\theta) = \sin(\theta) - \dfrac{2\pi}{n} \text{ and observe that } f \text{ is concave}\right.$

down on $\left(0, \dfrac{\pi}{2}\right)$ and that $f(0) = f\left(\dfrac{\pi}{2}\right) = 0.\Big]$ Then $-R\sin(\theta) \leq -\dfrac{2R}{\pi}$, so $e^{-R\sin(\theta)} \leq e^{-2R\theta/\pi}$. Thus,

$$\int_0^{\pi/2} e^{-R\sin(\theta)\,d\theta} \leq \int_0^{\pi/2} e^{-2R\theta/\pi}\,d\theta = \frac{\pi}{2R}(1 - e^{-\alpha}) < \frac{\pi}{2R}.$$

27. $\dfrac{1}{(1+z^2)^n}$ has a pole of order n at i. We find that $\dfrac{d^{n-1}}{dz^{n-1}}\dfrac{1}{(z+i)^n} = \dfrac{-n(-n-1)\cdots(-n-(n-2))}{(z+i)^{n+n-1}} = \dfrac{(-1)^{n-1}(2n-2)!}{(n-1)!(z+i)^{2n-1}}$,

so the residue at i is $\dfrac{-i(2n-2)!}{2^{2n-1}[(n-1)!]^2}$. Thus, $\displaystyle\int_{-\infty}^{\infty}\dfrac{1}{(1+x^2)^n}\,dx = \dfrac{\pi(2n-2)!}{2^{2n-2}[(n-1)!]^2}$.

Section 22.4

1. $-\dfrac{\pi}{a^3}\operatorname{Re}\{e^{-3\pi i/4}\cot(a\pi e^{\pi i/4})\}$ **3.** $\dfrac{1}{8}\left[\pi^2\operatorname{csch}^2\left(\dfrac{\pi}{\sqrt{2}}\right) + \pi\sqrt{2}\,\coth\left(\dfrac{\pi}{\sqrt{2}}\right) + 4\right]$

5. $P(n) = (a+n)^2$, $\dfrac{1}{P(n)}$ has a double zero at $-a$, and the residue at $-a$ of $\dfrac{\csc(\pi z)}{P(n)}$ is $-\pi\csc(\pi a)\cot(\pi a)$.

Section 22.5

1. $2\pi i$ **3.** 0

5. Let $f(z) = 4z$ and $g(z) = z^4 + 4z + 1$ in Rouche's theorem. We have $|f(z) - g(z)| = |z^4 + 1| \le 2$ for z with $|z| = 1$, and $|f(z)| = |4z| = 4$. Thus, $|f(z) - g(z)| \le |f(z)|$. Further, f and g are analytic with no poles on the unit circle. If $|z| = 1$ and $z^4 + 4z + 1 = 0$, then $|z^3 + 4| = 1$, impossible if $|z^3| = 1$. Thus, g has no zeros on the unit circle. Since f has no zeros on the unit circle and one zero inside, the unit circle encloses exactly one zero of g.

7. Let $f(z) = a_n z^n$ and $g(z) = p(z)$, and let C be the circle $|z| = R$. On C, $\left|\dfrac{f(z) - g(z)}{f(z)}\right| = \left|\dfrac{a_0}{a_n z^n} + \dfrac{a_1}{a_n z^{n-1}} + \cdots + \dfrac{a_{n-1}}{a_n z}\right| =$

$\dfrac{1}{|z|}\left|\dfrac{a_0}{a_n z^{n-1}} + \cdots + \dfrac{a_{n-1}}{a_n}\right| = \dfrac{1}{R}\left|\dfrac{a_0}{a_n z^{n-1}} + \cdots + \dfrac{a_{n-1}}{a_n}\right|$.

As $R \to \infty$, the last expression has limit zero. We may therefore choose R sufficiently large that $\left|\dfrac{f(z) - g(z)}{f(z)}\right| \le 1$ if $|z| = R$, and also that C encloses all zeros of g. Since f has at least one zero of order n, g must have n zeros, counting multiplicities.

Additional Problems for Chapter 22

1. 0 **3.** $\pi\left(\dfrac{3\sqrt{2} - 2\sqrt{3}}{\sqrt{6}}\right)$ **5.** $\pi i\cosh(2)\sinh(1)$ **7.** $\dfrac{2\pi i}{9}[1 - \cos(9)]$ **9.** $-\dfrac{\pi}{3}\cosh(1) + \dfrac{2\pi}{3}\cos\left(\dfrac{\sqrt{3}}{2}\right)\cosh\left(\dfrac{1}{2}\right)$

11. $\pi\sqrt{2}$ **13.** $-2\pi\sin(2) + \dfrac{\pi}{2}i\cos(2)$ **15.** $2\pi i$ **17.** 0 **19.** $\dfrac{2\pi}{3}$ **21.** $4\pi(2 + i)$ **23.** $2\pi i\cos(2)e^{\sin(2)}$

25. 0 **27.** 0 **29.** $4\pi i\left[\left(\dfrac{\pi}{2}\right)^2 - \left(\dfrac{\pi}{2}\right)^4\right]\dfrac{1}{1 - \dfrac{\pi^2}{4}}$

31. Write $h(z) = a_2(z - z_0)^2 + a_3(z - z_0)^3 + \cdots$, since $h(z_0) = h'(z_0) = 0$. Since $h''(z_0) \ne 0$, $a_2 \ne 0$, so h has a zero of order 2 at z_0. Since $g(z_0) \ne 0$, $\dfrac{g}{h}$ has a pole of order 2 at z_0. Write $g(z) = b_0 + b_1(z - z_0) + \cdots$ and

$\dfrac{g(z)}{h(z)} = \dfrac{c_{-2}}{(z - z_0)^2} + \dfrac{c_{-1}}{z - z_0} + c_0 + c_1(z - z_0) + \cdots$. Then

$b_0 + b_1(z - z_0) + \cdots = \left[\dfrac{c_{-2}}{(z - z_0)^2} + \dfrac{c_{-1}}{z - z_0} + c_0 + c_1(z - z_0) + \cdots\right][a_2(z - z_0)^2 + \cdots]$.

Therefore, $b_0 = c_2 a_2$, $b_1 = a_2 c_1 + a_3 c_2$, and $c_{-1} = \dfrac{b_1 a_3 c_{-2}}{a_2} = \dfrac{b_1}{a_2} - \dfrac{a_3 b_0}{a_2^2}$. Further, $f(z_0) = b_0$, $g'(z_0) = b_1$,

$h''(z_0) = 2a_2$ and $h^{(3)}(z_0) = 6a_3$. Then the residue of $\dfrac{g}{h}$ at z_0 is $\dfrac{2g'(z_0)}{h''(z_0)} - \dfrac{4(\frac{1}{6})h^{(3)}(z_0)g(z_0)}{[h''(z_0)]^2}$.

33. First, $\dfrac{ze^{iaz}}{z^4 + \beta^4}$ has simple poles in the upper half-plane at $\dfrac{\beta}{\sqrt{2}}(1 + i)$ and at $\dfrac{\beta}{\sqrt{2}}(-1 + i)$. The residue at $\dfrac{\beta}{\sqrt{2}}(1 + i)$ is

$$\dfrac{\left(\dfrac{\beta}{\sqrt{2}}\right)e^{i\alpha\beta/\sqrt{2}}e^{-\alpha\beta/\sqrt{2}}}{4\beta^3\left(\dfrac{1}{\sqrt{2}}\right)^3(1+i)^2},\ \text{ while the residue at } \dfrac{\beta}{\sqrt{2}}(-1+i)\text{ is } \dfrac{\left(\dfrac{\beta}{\sqrt{2}}\right)e^{-i\alpha\beta/\sqrt{2}}e^{-\alpha\beta/\sqrt{2}}}{4\beta^3\left(\dfrac{1}{\sqrt{2}}\right)^3(-1+i)^2}.\ \text{ After squaring }(1+i)\text{ and }(-1+i)$$

in the denominators, as indicated, we get

$$\int_0^\infty \frac{x\sin(\alpha x)}{x^4+\beta^4}\,dx = \frac{1}{2}\int_{-\infty}^\infty \frac{x\sin(\alpha x)}{x^4+\beta^4}\,dx$$

$$= \operatorname{Im}\left\{\pi i\left[\frac{\left(\dfrac{\beta}{\sqrt{2}}\right)e^{-\alpha\beta/\sqrt{2}}}{4\beta^3\left(\dfrac{1}{\sqrt{2}}\right)^3 2i}\right][e^{i\alpha\beta/\sqrt{2}}-e^{-i\alpha\beta/\sqrt{2}}]\right\}$$

$$= \operatorname{Im}\left\{\pi i\,\frac{e^{-\alpha\beta/\sqrt{2}}}{2\beta^2}\sin\!\left(\frac{\alpha\beta}{\sqrt{2}}\right)\right\} = \frac{\pi e^{-\alpha\beta/\sqrt{2}}}{2\beta^2}\sin\!\left(\frac{\alpha\beta}{\sqrt{2}}\right).$$

Chapter 23

Section 23.1

1. (a)

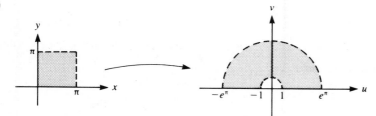

(b)

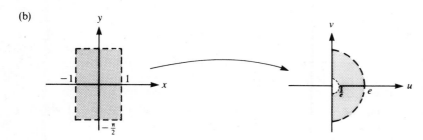

(c)

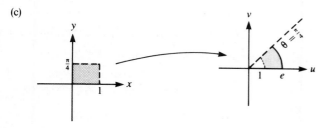

(d)

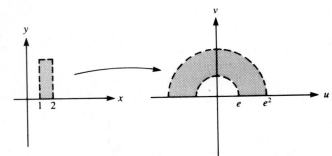

(e)

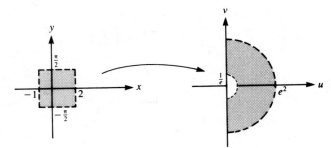

3. A line $x = a \neq 0$ maps to an ellipse $\dfrac{u^2}{\sinh^2(a)} + \dfrac{v^2}{\cosh^2(a)} = 1$. The line $x = 0$ maps to the interval from i to $-i$ on the imaginary axis, covering it infinitely many times.

5. The image is the entire w-plane.

7. Let $z = re^{i\theta}$ in $w = \dfrac{1}{2}\left(z + \dfrac{1}{z}\right) = u + iv$ to get $u = \dfrac{1}{2}\left(r + \dfrac{1}{r}\right)\cos(\theta)$ and $v = \dfrac{1}{2}\left(r - \dfrac{1}{r}\right)\sin(\theta)$. Then

$$\frac{u^2}{\frac{1}{4}(r + 1/r)^2} = \cos^2(\theta) \text{ and } \frac{v^2}{\frac{1}{4}(r - 1/r)^2} = \sin^2(\theta). \text{ Thus,}$$

$$\frac{u^2}{\frac{1}{4}\left(r + \dfrac{1}{r}\right)^2} + \frac{v^2}{\frac{1}{4}\left(r - \dfrac{1}{r}\right)^2} = 1,$$

the equation of an ellipse. The foci in the w-plane are at $(\pm c, 0)$, where $c^2 = a^2 - b^2$ and $a^2 = \left(\dfrac{1}{2}\left(r + \dfrac{1}{r}\right)\right)^2$ and $b^2 = \left(\dfrac{1}{2}\left(r - \dfrac{1}{r}\right)\right)^2$. We find that $a^2 - b^2 = 1$; hence, $c = 1$.

9. $|z| > 1$ maps to $|w| < 1$.

11. (a) $w = e^{(a + ib)z}$. If $\text{Im}(z) = c$, then $z = t + ic$, so $w = e^{(a + ib)(t + ic)} = e^{(at - bc) + i(bt + ac)} = u + iv$. Then $u = e^{at - bc}\cos(bt + ac)$ and $v = e^{at - bc}\sin(bt + ac)$. If $w = re^{i\theta}$, then $r = e^{at - bc}$ and $\theta = bt + ac$. Thus, $t = \dfrac{\theta - ac}{b}$, and $r = e^{a[(\theta - ac)/b] - bc} = e^{a\theta/b}e^d$, where $d = bc - a^2c/b$.

(b) If $\text{Re}(z) = c$, then $z = c + it$ and $w = e^{(a + ib)(c + it)} = e^{ac - bt}e^{i(at + bc)} = e^{ac - bt}\cos(at + bc) + ie^{ac - bt}\sin(at + bc) = u + iv = re^{i\theta}$. Then $r = e^{ac - bt}$ and $t = \dfrac{\theta - bc}{a}$, so $r = e^{ac - b[(\theta - bc)/a]} = e^{-b\theta/a}e^{-db/a}$.

13. The equation of any circle or straight line in the xy-plane can be written

$$A|z|^2 + B\frac{z + \bar{z}}{2} + C\frac{z - \bar{z}}{2} + R = 0.$$

If $w = \dfrac{1}{z}$, then $A\left|\dfrac{1}{w}\right|^2 + \dfrac{B}{2}\left(\dfrac{1}{w} + \dfrac{1}{\bar{w}}\right) + \dfrac{C}{2}\left(\dfrac{1}{w} - \dfrac{1}{\bar{w}}\right) + R = 0$. Multiply this equation by $w\bar{w}$ and recall that $|w|^2 = w\bar{w}$ to get

$R|w|^2 + B\dfrac{w + \bar{w}}{2} + C\dfrac{w - \bar{w}}{2} + A = 0$, the equation of a circle or straight line in the w-plane.

15. The image is the entire w-plane, except for the nonnegative real axis $u \geq 0$.

17. The image is the sector $\dfrac{\pi}{27} \leq \arg(w) \leq \dfrac{\pi}{9}$ in the w-plane.

Section 23.2

1. $w = \dfrac{3 + 8i - (1 + 4i)z}{4 + 7i - (2 + 3i)z}$ **3.** $w = \dfrac{-48 - 16i + (33 + i)z}{5(z - 4)}$ **5.** $w = \dfrac{4 - 75i + (3 + 22i)z}{-21 + 4i + (2 + 3i)z}$

7. $w = \dfrac{8 + 38i - (24 + 10i)z}{-4 + 3i + iz}$ **9.** $w = \dfrac{\dfrac{1}{3} + \left(3 + \dfrac{1}{3}i\right)z}{\dfrac{5}{3} - 12i + \left(12 + \dfrac{7}{6}i\right)z}$ **11.** $\left(u - \dfrac{7}{2}\right)^2 + v^2 = \left(\dfrac{3}{2}\right)^2$ **13.** $v = 10$

15. $\left(u - \dfrac{11}{21}\right)^2 + \left(v + \dfrac{1}{63}\right)^2 = \dfrac{208}{3969}$ **17.** $(u - 1)^2 + \left(v + \dfrac{19}{4}\right)^2 = \dfrac{377}{16}$ **19.** $u^2 + \left(v + \dfrac{1}{2a}\right)^2 = \left(\dfrac{1}{2a}\right)^2$

21. The map $w = \bar{z}$ reverses orientation. For example, the half-line $x \geq 0$ maps to the half-line $u \geq 0$, while the half-line $y \geq 0$

$(z = ti, t \geq 0)$ maps to the half-line $v \leq 0$ $(v = -ti, t \geq 0)$. Thus, an orientation by $\dfrac{\pi}{2}$ in the z-plane maps to an

orientation by $\dfrac{-\pi}{2}$ in the w-plane.

23. If $w = \dfrac{az + b}{cz + d}$ with $ad - bc \neq 0$, then $z = \dfrac{-dw + b}{cw - a}$, also a linear fractional transformation.

25. Let $w = a\dfrac{z - b}{bz - 1}$. If $|z| = |a| = 1$, then $\bar{z} = \dfrac{1}{z}$, so $|w| = \left|\dfrac{a(z - b)}{z\left(\bar{b} - \dfrac{1}{z}\right)}\right| = \dfrac{|a|}{|z|}\left|\dfrac{z - b}{\bar{b} - \dfrac{1}{z}}\right| = \left|\dfrac{z - b}{\bar{b} - \bar{z}}\right| = 1$. The origin maps to $w = b$,

and $|b| < 1$ by assumption, so $|z| < 1$ maps to $|w| < 1$.

27. z is a fixed point if and only if $\dfrac{az + b}{cz + d} = z$, or $cz^2 + (a - d)z - b = 0$. If $c \neq 0$, so that the transformation is not a translation, this quadratic equation has one or two roots; hence, the mapping has one or two fixed points.

Section 23.3

In Problems 1 through 10, many mappings exist having the requested property.

1. $w = \dfrac{4}{3}z + i$ **3.** $w = \dfrac{3z + 2 + 6i}{z + 2i}$ **5.** $w = \dfrac{-4i + (4 - 8i)z}{-1 + (2 - 8i)z}$ **7.** $w = 3i - iz$

9. $w = \dfrac{-11 - 22i + (3 + 2i)z}{-9 + 6i + (1 - 2i)z}$ **11.** $w = z^{1/3}$

13. If $|z| = 1$, then $w = \dfrac{1}{2}\left(z + \dfrac{1}{z}\right) = \dfrac{1}{2}\left(x + iy + \dfrac{1}{x + iy}\right) = \dfrac{1}{2}\left(x + iy + \dfrac{x + iy}{x^2 + y^2}\right) = \dfrac{1}{2}(x + iy + x - iy) = x$, so w is real and

$-1 \leq w \leq 1$.

    Conversely, if w is real and $-1 \leq w \leq 1$, then $\sqrt{w^2 - 1}$ is on the imaginary axis. Solve for z in

$w = \dfrac{1}{2}\left(z + \dfrac{1}{z}\right)$ to get $z = w \pm \sqrt{w^2 - 1}$. Further, the conjugate of $w + \sqrt{w^2 - 1}$ is $w - \sqrt{w^2 - 1}$. Then

$|z|^2 = (w + \sqrt{w^2 - 1})(w - \sqrt{w^2 - 1}) = 1$. Thus, $-1 \leq w \leq 1$ if and only if $|z| = 1$.

    For (b), let $|z| = 1$, with $0 < r < 1$. Then $w = \dfrac{1}{2}\left(z + \dfrac{1}{z}\right) = \dfrac{x}{r}\left(\dfrac{r^2 + 1}{2r}\right) + i\dfrac{y}{r}\left(\dfrac{r^2 - 1}{2r}\right) = u + iv$. Then

$\dfrac{u^2}{\left(\dfrac{r^2 + 1}{2r}\right)^2} + \dfrac{v^2}{\left(\dfrac{r^2 - 1}{2r}\right)^2} = 1$, the equation of an ellipse.

15. $w(1) = 2i \int_0^1 (\xi + 1)^{-1/2}(\xi - 1)^{-1/2}\xi^{-1/2} \, d\xi = 2i \int_0^1 (\xi^2 - 1)^{-1/2}\xi^{-1/2} \, d\xi = 2i \int_0^1 (-1)^{-1/2}(1 - \xi^2)^{-1/2}\xi^{-1/2} \, d\xi =$

$2 \int_0^1 (1 - \xi^2)^{-1/2}\xi^{-1/2} \, d\xi$. Let $\xi = u^{1/2}$ to get $w(1) = \int_0^1 (1 - u)^{-1/2}u^{-3/4} \, du = \dfrac{\Gamma\left(\frac{1}{2}\right)\Gamma\left(\frac{1}{4}\right)}{\Gamma\left(\frac{3}{4}\right)}$.

Next, $w(-1) = 2i \int_0^{-1} (\xi + 1)^{-1/2}(\xi - 1)^{-1/2}\xi^{-1/2} \, d\xi$. Let $\xi = -u$ to get $w(-1) =$

$2i \int_0^1 (1 - u)^{-1/2}(-1)^{-1/2}(1 + u)^{-1/2}(-1)^{-1/2}u(-1) \, du = 2i \int_0^1 (1 - u^2)^{-1/2}u^{-1/2} \, du = i\,\dfrac{\Gamma\left(\frac{1}{2}\right)\Gamma\left(\frac{1}{4}\right)}{\Gamma\left(\frac{3}{4}\right)}$.

Clearly, $w(0) = 0$. Finally, $w(\infty) = 2i \int_0^\infty (\xi + 1)^{-1/2}(\xi - 1)^{-1/2}\xi^{-1/2} \, d\xi = 2i \int_0^1 (\xi + 1)^{-1/2}(\xi - 1)^{-1/2}\xi^{-1/2} \, d\xi +$

$2i \int_1^\infty (\xi + 1)^{-1/2}(\xi - 1)^{-1/2}\xi^{-1/2} \, d\xi = \dfrac{\Gamma\left(\frac{1}{2}\right)\Gamma\left(\frac{1}{4}\right)}{\Gamma\left(\frac{3}{4}\right)} + 2i \int_1^\infty (\xi + 1)^{-1/2}(\xi - 1)^{-1/2}\xi^{-1/2} \, d\xi$. Let $\xi = u^{-1}$ in this integral

to get $-2i \int_1^0 \left(\dfrac{1 + u}{u}\right)^{-1/2}\left(\dfrac{1 - u}{u}\right)^{-1/2} u^{1/2}u^{-2} \, du = 2i \int_0^1 (1 - u^2)^{-1/2}u^{-1/2} \, du = i\,\dfrac{\Gamma\left(\frac{1}{2}\right)\Gamma\left(\frac{1}{4}\right)}{\Gamma\left(\frac{3}{4}\right)}$. Thus,

$w(\infty) = (1 + i)\,\dfrac{\Gamma\left(\frac{1}{2}\right)\Gamma\left(\frac{1}{4}\right)}{\Gamma\left(\frac{3}{4}\right)}$, and the upper half-plane is mapped to the square with vertices $0, c, ic, (1 + i)c$,

with $c = \dfrac{\Gamma\left(\frac{1}{2}\right)\Gamma\left(\frac{1}{4}\right)}{\Gamma\left(\frac{3}{4}\right)}$.

17. From Theorem 23.3, the requested mapping is defined by $\dfrac{-1}{w - 1} = \dfrac{z_3 - z_4}{z_3 - z_2}\,\dfrac{z - z_2}{z - z_4}$.

19. If $z = z^*$, then $[z, z_1, z_2, z_3] = \overline{[z, z_1, z_2, z_3]}$, implying that $[z, z_1, z_2, z_3]$ is real; hence, z must lie on the circle passing through z_1, z_2, and z_3 by the result of Problem 18. Conversely, if z is a point on the circle containing z_1, z_2, and z_3, then $[z, z_1, z_2, z_3]$ is real by Problem 18. Then $[z, z_1, z_2, z_3] = \overline{[z, z_1, z_2, z_3]}$; hence, z is self-symmetric.

Chapter 24

Section 24.1

1. Let $a = Ke^{i\theta}$. Equipotential lines are graphs of $\varphi(x, y) = K[\cos(\theta)x - \sin(\theta)y] = $ constant [or $y = \cot(\theta)x + b$, straight lines with slope $\cot(\theta)$]; streamlines are graphs of $\psi(x, y) = K[\cos(\theta)y + \sin(\theta)x] = $ constant [or $y = \tan(\theta)x + b$, straight lines with slope $\tan(\theta)$].

3. $\varphi(x, y) = \cos(x)\cosh(y)$ and $\psi(x, y) = -\sin(x)\sinh(y)$. Equipotential lines are graphs of $y = \cosh^{-1}\left[\dfrac{K}{\cos(x)}\right]$, and streamlines are graphs of $y = \sinh^{-1}\left[\dfrac{C}{\sin(x)}\right]$.

5. $\varphi(x, y) = K \ln|z - z_0|$ and $\psi(x, y) = K \arg(z - z_0)$. Streamlines are circles about z_0; equipotential lines are half-lines emanating from z_0.

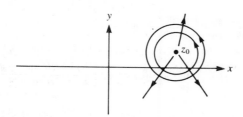

7. $f(z) = k\left[x + \dfrac{x}{x^2 + y^2} + i\left(y - \dfrac{y}{x^2 + y^2}\right)\right]$. Equipotential lines are graphs of $x + \dfrac{x}{x^2 + y^2} = $ constant, and streamlines are graphs of $y - \dfrac{y}{x^2 + y^2} = $ constant.

9. Equipotential lines are graphs of $k\left[x + \dfrac{x}{x^2 + y^2}\right] - \dfrac{b}{2\pi}$ Arg$(z) = $ constant, and streamlines are graphs of $k\left[y - \dfrac{y}{x^2 + y^2}\right] + \dfrac{b}{2\pi}$ In$|z| = $ constant. Stagnation points occur where $f'(z) = 0$, or $z = \dfrac{-ib}{4\pi k} \pm \sqrt{1 - \dfrac{b^2}{16\pi^2 K}}$.

11. Find $f'(z) = \dfrac{9a^4 K^2}{\left(z - \dfrac{ia\sqrt{3}}{2}\right)^2 \left(z + \dfrac{ia\sqrt{3}}{2}\right)^2}$. The net fluid force is $\dfrac{1}{2} i\sigma \oint_C \dfrac{9a^4 K^2}{\left(z - \dfrac{ia\sqrt{3}}{2}\right)^2 \left(z + \dfrac{ia\sqrt{3}}{2}\right)^2} dz = $

$\dfrac{1}{2} i\sigma[9a^4 K^2]\left[\dfrac{4\pi}{a^3 3\sqrt{3}}\right]$, using the residue theorem. The y-component of the force is $2\sqrt{3}\pi\sigma a K^2$.

Section 24.2

1. $\dfrac{1}{2}(e^{3it} + e^{-3it})$ **3.** $\dfrac{1}{36}[6te^{2t} - e^{2t} + e^{-4t}]$ **5.** $\dfrac{1}{2}t^2 e^{-5t}$

7. $\dfrac{1}{2\sqrt{2}(-1 + i)} e^{(1+i)t/\sqrt{2}} + \dfrac{1}{2\sqrt{2}(1 + i)} e^{(-1+i)t/\sqrt{2}} + \dfrac{1}{2\sqrt{2}(1 - i)} e^{(-1-i)t/\sqrt{2}} + \dfrac{1}{2\sqrt{2}(-1 - i)} e^{(1-i)t/\sqrt{2}}$

9. $\dfrac{1}{4} e^{2t}[t^2 + t^2\cos(4) - 4t \sin(4) - 4 \cos(4)]$

11. To illustrate, we prove conclusion (3) of Theorem 4.13. Write $\dfrac{P(s)}{Q(s)} = \dfrac{P(s)}{[(s - a)^2 + b^2]R(s)}$. Then Res$_{a + ib} \dfrac{P(z)e^{tz}}{Q(z)} = $

$\lim\limits_{z \to a + ib} \dfrac{[z - (a + ib)]P(z)e^{tz}}{[(z - a)^2 + b^2]R(z)} = \dfrac{e^{(a + ib)t}(\alpha + i\beta)}{a + ib - a + ib} = \dfrac{(\alpha + i\beta)e^{at}[\cos(bt) + i \sin(bt)]}{2ib}$. Similarly, Res$_{a - ib} \dfrac{R(z)e^{tz}}{Q(z)} = $

$\dfrac{(\alpha - i\beta)e^{at}[\cos(bt) - i \sin(bt)]}{-2ib}$. The sum of these residues is $\dfrac{1}{b} e^{at}[\alpha \sin(bt) + \beta \cos(bt)]$.

Section 24.3

1. With $T(z) = \dfrac{z - z_0}{z - \bar{z}_0}$, we get $\dfrac{T'(z)}{T(z)} = \dfrac{z_0 - \bar{z}_0}{(z - \bar{z}_0)(z - z_0)}$, so $u(z_0) = \dfrac{1}{2\pi i} \int_C g(z) \dfrac{T'(z)}{T(z)} dz$, with C the real axis, Thus,

$$u(z_0) = \dfrac{1}{\pi} y_0 \int_{-\infty}^{\infty} \dfrac{g(t)}{(t - \bar{z}_0)(t - z_0)} dt,$$

from which we get $u(x, y) = \dfrac{y}{\pi} \int_{-\infty}^{\infty} \dfrac{g(t)}{(t - x)^2 + y^2} dt$.

3. Let $\xi = e^{i\varphi} = \alpha + i\beta$ and $z = re^{i\theta}$ in Schwarz's formula (Problem 2) to get $f(z) = \dfrac{1}{2\pi i} \oint_C g(\alpha, \beta) \dfrac{\xi + z}{\xi - z} \dfrac{1}{\xi} d\xi = $

$\dfrac{1}{2\pi i} \int_0^{2\pi} g(\cos(\varphi), \sin(\varphi)) \dfrac{e^{i\varphi} + e^{i\theta}}{e^{i\varphi} - e^{i\theta}} \dfrac{1}{e^{i\varphi}} ie^{i\varphi} d\varphi = \dfrac{1}{2\pi} \int_0^{2\pi} g(\cos(\varphi), \sin(\varphi)) \dfrac{1 + e^{i(\theta - \varphi)}}{1 - e^{i(\theta - \varphi)}} d\varphi = $

$$\frac{1}{2\pi} \int_0^{2\pi} g(\cos(\varphi), \sin(\varphi)) \frac{1 + r\cos(\theta - \varphi) + ir\sin(\theta - \varphi)}{1 - r\cos(\theta - \varphi) - ir\sin(\theta - \varphi)} \, d\varphi.$$

On multiplying numerator and denominator of the integrand by $1 - r\cos(\theta - \varphi) + ir\sin(\theta - \varphi)$, we get

$$f(z) = \frac{1}{2\pi} \int_0^{2\pi} g(\cos(\varphi), \sin(\varphi)) \frac{1 - r^2 + 2ir\sin(\theta - \varphi)}{1 + r^2 - 2r\cos(\theta - \varphi)} \, d\varphi.$$

Finally, $u(r, \theta) = \text{Re}[f(z)]$.

5. It is routine to show that $\nabla^2 u = 0$. If (x, y) is on the upper half of the unit circle, then $1 - x^2 - y^2 = 0$, so $u(x, y) = \frac{2}{\pi} \tan^{-1}(0) = 0$. If $-1 < x < 1$ and $y \to 0$, then $u(x, y) \to \frac{2}{\pi} \frac{\pi}{2} = 1$.

Index